Y0-AAX-566

PLEASE STAMP DATE DUE, BOTH BELOW AND ON CARD

| DATE

```
SFL
TA464 .W67 2000
Worldwide guide to equivale
irons and steels
```

Worldwide Guide to Equivalent Irons and Steels

Fourth Edition

ASM Materials Data Series

Prepared under the direction of the
ASM International Materials Properties Database Committee

William C. Mack, Coordinating Editor

Charles Moosbrugger, Technical Editor
Fran Cverna, Manager, Center for Materials Data
Sandy Dunigan, Associate Editor
Jill Schneider, Associate Editor
Theresa Nelson, Editorial Assistant
Bonnie Sanders, Manager of Production
William W. Scott, Jr., Director of Technical Publications

Editorial Assistance
Hugh Baker
Corrina Eastman
Jan Horesh
Heather Lampman
Lillian McLeandon
Tom Moosbrugger
Annette Schell
Judy Snyder

The Materials Information Society

ASM International®
Materials Park, Ohio 44073-0002
http://www.asm-intl.org

Library of Congress Cataloging-in-Publication Data

ASM International

Worldwide guide to equivalent irons and steels.—4th ed./contributors to the 4th ed.,
William C. Mack, coordinating editor... [et al.].
p. cm.
"Prepared under the direction of the ASM International Materials Properties Database Committee."
Includes bibliographical references and index.
1. Iron—Specifications. 2. Steel—Specifications.
I. Mack, William C. II. ASM International. Materials Properties Database Committee.
TA466.W67 1999   620.1'7—dc21   99-047360
ISBN: 0-87170-635-0
SAN: 204-7586

ASM International®
Materials Park, OH 44073-0002
http://www.asm-intl.org

Printed in the United States of America

Preface

This Fourth Edition of the *Worldwide Guide to Equivalent Irons and Steels* carries on a tradition begun twenty years ago. At its inception, the Worldwide Guide was the first such book to present so thorough and comprehensive a selection of ferrous alloys. With this edition, over 25,000 alloy designations are presented. The listing of each designation has been expanded to include more mechanical properties, text, and specification issue dates when available.

Because specifications are dynamic documents that change with time, we have included obsolete or inactive specifications and former alloy designations to aid those who need to locate alloys cited in older documents.

As done in the previous edition, similar alloys from around the world are listed together. The alloys are grouped on the basis of chemical composition. This provides the reader with a starting point for the further investigation of equivalency. Equivalency could be judged by any or all of the methods of classifying irons and steels: composition, manufacturing method, finishing method, product shape, microstructure, deoxidization method, mechanical equivalence, corrosion and heat resistance, quality, and cost. It is important to recognize that a definitive judgment on the equivalency of two irons or steels requires the study of the specifications by a qualified individual. Most specifications are complex documents that cannot be condensed into a single line of text. A listing of Standards Organizations with contact information is provided in the Appendix to assist the reader.

Each entry in the book contains the following information as applicable and available.

Specification is usually an acronym of the issuing standards organization with a specific document number, revision indication, and date. The format of the document number in the book will differ slightly from that on the document itself. Spaces and revision letters have sometimes been eliminated, and the date is always two digits in parentheses. Some organizations use Roman and Arabic numerals interchangeably for the same specification. We use the Arabic.

Designation is a numeric or alphanumeric symbol for a particular alloy, strength, or form of the metal. It is how the alloy is identified in the specification. In some cases a specification defines a unique alloy. In this case the specification number is the designation. Designations such as the Unified Numbering System (UNS) and American Iron and Steel Institute (AISI) numbers are given without a specification because they do not by themselves specify an alloy. An asterisk (*) follows the designation if it or the specification is known to be some way inactive or obsolete.

Notes are properties, uses, available forms, processing information, size, and status. Abbreviations are used extensively to pack condense the information. The key to the abbreviations is found in the Appendix.

Chemical compositions are given as the specification lists them. They are given in conventional weight percent. Records revised or new to this fourth edition list all the significant figures that the parent document lists. The elements common to a material family are listed in columnar fashion. Others are listed in a group under "other elements." A maximum, minimum, range, or nominal value is given. Incidental elements are not given. Because of space limitations, not all specified functional relationships between chemicals can be presented. Inequality expressions have been used to define the chemical limits: 10xC<=Nb+Ta<=1.10, means the sum of niobium and tantalum is greater than or equal to 10 times the weight per cent of carbon and less than or equal to 1.10%. Note that Nb is used for niobium (columbium) everywhere except where Cb is a part of the alloy designation such as TP309Cb. It is beyond the scope of this book to differentiate between mandatory compositions and those offered in a specification as typical "for information only." The latter is true in the case of specifications based on mechanical properties rather than composition. Specifications often allow compositions as agreed upon contractually between the buyer and seller. When a choice was offered in a specification, the ladle analysis composition was preferred.

Properties are ultimate tensile strength (UTS) and yield strength (YS) given in SI units (MPa) (1 MPa = 1 N/mm^2 = 0.1450377 ksi). Elongation is expressed in percent (of gage length). It is the longitudinal elongation unless a T (transverse) follows the number. The hardness units are listed in the individual records. These mechanical values are often functions of product thickness or diameter. A representative size is often listed in the notes. This size is in no way a limit on the size available for the products defined by the specification. Room temperature values are given unless noted otherwise. Since these values are typical, they must not be used for design purposes.

This guide is arranged with alloys of similar chemical composition grouped together, headed by the common United States name. These similar alloys are grouped into Material Families, based on common alloying elements being in a proscribed range. The classification follows the SAE-AISI system for Carbon and Alloy Steels. These Material Families are found in the Table of Contents. Collections of Material Families comprise Material Groups, which make up the first eight sections of the book. Although the underlying structure is chemical equivalency, historically steels have been classified by application as well. For this reason, Tool Steels and High Strength Steels are in their own Material Groups.

Various organizations have struggled with the equivalency of designations. Reaching a consensus on the definition of terms such as similarity, equivalency, and interchangeability is itself difficult. The disclaimer of all previous editions of this book must be repeated. The groupings of designations serve as a useful guide to materials having similar compositions. The grouping is therefore not intended to be used for design interchangeability, nor is functional equivalency implied. Substitution of one material for another is solely the responsibility of the reader.

Charles Moosbrugger

Contributors to the 4th Edition

Graham MacIntosh
Standards Australia

Lin Huiguo
Chinese Society for Metals

Kaare A. Johansson
STATOIL, Norway

Nihei Masatoshi
National Research Institute for Metals, Japan

Jose Angel Ugarriza
Instituto Argentino de Siderugia

Ingrid Rossouw
Bushi Molathwa
South African Bureau of Standards

Turkish Standards Institute

Members of the ASM Materials Properties Database Committee (1999–2000)

Bruce E. Boardman, *Chair*
Deere & Company Technical Ctr.

J. Gilbert Kaufman, *Past Chair*
Kaufman Associates Ltd.

Min Berbon
Rockwell International Science Center

David R. Bourell
University of Texas at Austin

Horst H. Cerjak
Technical University Graz

Bryan R. Cope
America Spring Wire Corporation

Stephen R. Crosby
Stanley

Barry Hindin
Battelle Memorial Institute

David Kaaret
Copper Development Association Inc.

Gordon Lippa
North Star Casteel

Guiri Liu Nash
General Motors Corporation

Charles A. Parker
AlliedSignal Aircraft Landing Systems

Paul J. Sikorsky
The Trane Company

Howard W. Sizek
Special Metals Corporation

Jason M. Smith
Lexmark International

John E. Smugeresky
Sandia National Laboratories

Gary A. Volk
Illinois Central College

Jack H. Westbrook
Brookline Technologies

Contents

Specification	Designation	Notes	C	Cr	Mg	Mn	Ni	P	S	Si	Other	UTS	YS	El	Hard
Cast Iron, Ductile, 100-70-03															
Australia															
AS 1831(85)	700-2											700		2	
AS 1831(85)	800-2											800		2	
Belgium															
NBN 830-02	FNG70-2														
NBN 830-02	FNG80-2														
China															
GB 1348(88)	QT 700-2												700	420	2	225-305 HB
Denmark															
DS 11303	707														
DS 11303	708														
France															
AFNOR NFA32201	FGS700-2														
AFNOR NFA32201	FGS800-2														
Germany															
DIN	WNr 0.7070														
DIN 1693	GGG-70														
India															
IS 1865	SG700/2														
IS 1865	SG800/2														
International															
ISO 1083(87)	700-2	Spheroidal graphite, SCS										700	420	2	225-305 HB
ISO 1083(87)	800-2	Spheroidal graphite, SCS										800	480	2	245-335 HB
Italy															
UNI 4544	GS700-2														
UNI 4544	GS800-2														
Japan															
JIS G5502(95)	FCD70 Class 5														
JIS G5502(95)	FCD80 Class 6														
Netherlands															
NEN 2733	GN700-2														
NEN 2733	GN800-2														
Norway															
NS 11301	NS11370														225-280 HB
Russia															
GOST 7293	VCh70														
GOST 7293	VCh80														
South Africa															
SABS 936	SG80												785	490	2	260-330 HB
UK															
BS 2789	700/2												700	420	2	
BS 2789	800/2												800	480	2	
USA															
	UNS F34800												689 Min	483 Min		241-302 HB
ASTM A536(93)	100-70-03												689	483	3.0	
SAE J434	D7003	Prl	3.20-4.10			0.10-1.00		0.015-0.10	0.005-0.035	1.80-3.00	bal Fe	689	483	3	241-302 HB
Cast Iron, Ductile, 120-90-02 (Continued from previous page)															
China															
GB 1348(88)	QT 800-2												800	480	2	245-335 HB
GB 1348(88)	QT 900-2												900	600	2	280-360 HB

UNS numbers and US grades are provided as a means of cross referencing chemically similar alloys. Exchangability is only possible after independent examination of specifications. Tensile properties are minimum or typical as specified. UTS and YS as MPa. El as %. See Appendix for list of abbreviations used in Notes. * indicates obsolete material.

Specification	Designation	Notes	C	Cr	Mg	Mn	Ni	P	S	Si	Other	UTS	YS	El	Hard
Cast Iron, Ductile, 120-90-02 (Continued from previous page)															
France															
AFNOR NFA32201	FGS900-2														
Germany															
DIN	WNr 0.7080														
DIN 1693	GGG-80														
International															
ISO 1083(87)	900-2	Spheroidal graphite, SCS										900	600	2	280-360 HB
Norway															
NS 11301	NS11380														265-300 HB
Russia															
GOST 7293	VCh100														
UK															
BS 2789	900/2												900	600	2	
USA															
	UNS F36200												827 Min	621 Min		
ASTM A536(93)	120-90-02												827	621	2.0	
Cast Iron, Ductile, 5315															
China															
GB 1348(88)	QT 400-15												400	250	15	130-180 HB
GB 1348(88)	QT 450-10												450	310	10	160-210 BH
USA															
	UNS F33101			3.0 min							2.50 max	bal Fe	415 Min	310 Min		190 Max HB
AMS 5315D(94)*		Noncurrent;	3.2-4.0			0.8 max		0.08 max		1.7-2.5	bal Fe	415	310	10	190 HB max
MIL-I-24137	- - -*	Obs													
Cast Iron, Ductile, 5394															
USA															
	UNS F47006	Corr res; Heat res	2.4-3.0	1.7-2.4		0.8-1.6	18.0-22.0	0.25 max		2.0-3.2	Cu <=0.50; Pb <=0.003; bal Fe	379 Min	227 Min		
Cast Iron, Ductile, 5395															
USA															
	UNS F43030		2.5-3.0	0.50 max		1.9-2.5	20.0-24.0	0.15 max	0.05 max	2.0-3.0	Mo <=0.30; bal Fe	345 Min	173 Min		
Cast Iron, Ductile, 60-40-18															
Australia															
AS 1831(85)	370-17												370		17	
Belgium															
NBN 830-02	FNG38-17	Charpy Impact 13J + 20C Avg														
China															
GB 1348(88)	QT 400-15												400	250	15	130-180 HB
GB 1348(88)	QT 400-18												400	250	18	130-180 HB
Denmark															
DS 11303	715	Charpy Impact 13J +20C Avg														
DS 11303	716															
France															
AFNOR NFA32201	FGS350-22	Charpy Impact 17J+23C Avg														
AFNOR NFA32201	FGS350-22L	Charpy Impact 12J-40C Avg														
AFNOR NFA32201	FGS400-15															

UNS numbers and US grades are provided as a means of cross referencing chemically similar alloys. Exchangability is only possible after independent examination of specifications. Tensile properties are minimum or typical as specified. UTS and YS as MPa. El as %. See Appendix for list of abbreviations used in Notes. * indicates obsolete material.

Specification	Designation	Notes	C	Cr	Mg	Mn	Ni	P	S	Si	Other	UTS	YS	El	Hard

Cast Iron, Ductile, 60-40-18 (Continued from previous page)

France

Specification	Designation	Notes	C	Cr	Mg	Mn	Ni	P	S	Si	Other	UTS	YS	El	Hard
AFNOR NFA32201	FGS400-18	Charpy Impact 14J+23C Avg													
AFNOR NFA32201	FGS400-18L	Charpy Impact 12J-20C Avg													

Germany

Specification	Designation	Notes	UTS	YS	El	Hard
DIN	WNr 0.7040					
DIN	WNr 0.7043					
DIN 1693	GGG-35.3	Charpy Impact 19J + 20C, 14J - 20				
DIN 1693	GGG-40					
DIN 1693	GGG-40.3	Charpy Impact 16J + 20C, 14J - 20				

India

Specification	Designation	Notes	UTS	YS	El	Hard
IS 1865	SG370/17	Charpy Impact 13J+20C Avg				

International

Specification	Designation	Notes	UTS	YS	El	Hard
ISO 1083(87)	350-22	Spheroidal graphite, SCS	350	220	22	150 HB max
ISO 1083(87)	350-22L	Spheroidal graphite, SCS	350	220	22	150 HB max
ISO 1083(87)	400-15	Spheroidal graphite, SCS	400	250	15	130-180 HB
ISO 1083(87)	400-18	Spheroidal graphite, SCS	400	250	18	130-180 HB
ISO 1083(87)	400-18L	Spheroidal graphite, SCS	400	250	18	130-180 HB

Italy

Specification	Designation	Notes	UTS	YS	El	Hard
UNI 4544	GS370-17					

Japan

Specification	Designation	Notes	UTS	YS	El	Hard
JIS G5502(95)	FCD37 Class 0	Charpy Impact 12.7J + 20C Avg				

Netherlands

Specification	Designation	Notes	UTS	YS	El	Hard
NEN 2733	GN350-22L	Charpy Impact 14J - 40C Avg				
NEN 2733	GN400-15					
NEN 2733	GN400-18L	Charpy Impact 14J - 20C Avg				

Norway

Specification	Designation	Notes	UTS	YS	El	Hard
NS 11301	NS11335	Charpy Impact 14J - 40C Avg				180 HB max
NS 11301	NS11338	Charpy Impact 14J - 20C Avg				130-170 HB
NS 11301	NS11342					135-180 HB

Russia

Specification	Designation	Notes	UTS	YS	El	Hard
GOST 7293	VCh35					
GOST 7293	VCh40					

South Africa

Specification	Designation	Notes	UTS	YS	El	Hard
SABS 936	SG38		375	245	17	180 HB max

Sweden

Specification	Designation	Notes	UTS	YS	El	Hard
SIS 140717	0717-00		400	250	15	
SIS 140717	0717-02		400	250	18	
SIS 140717	0717-15		350	220	22	

UK

Specification	Designation	Notes	UTS	YS	El	Hard
BS 2789	350/22		350	220	22	
BS 2789	350/22L40	Charpy Impact 17J + 20C Avg	350	220	22	
BS 2789	400/18		400	250	18	
BS 2789	400/18L20	Charpy Impact 14J + 20C Avg	400	250	18	

Specification	Designation	Notes	C	Cr	Mg	Mn	Ni	P	S	Si	Other	UTS	YS	El	Hard

Cast Iron, Ductile, 60-40-18 (Continued from previous page)

USA

Specification	Designation	Notes	C	Cr	Mg	Mn	Ni	P	S	Si	Other	UTS	YS	El	Hard
	UNS F32800											414 Min	276 Min		170 HB
ASME SA395	60-40-18	Refer to ASTM A395(93)													
ASTM A356(84)	60-40-18*														
ASTM A395/A395(98)	60-40-18											414	276		170 HB
ASTM A536(93)	60-40-18											414	276	18	
MIL-C-24707/5(89)	60-45-15	As spec in ASTM A395										414	276	18	143-187 Brinell
SAE J434	D4018	Ferr	3.20-4.10			0.10-1.00		0.015-0.10	0.005-0.035	1.80-3.00	bal Fe	414	276	18	170 HB max

Cast Iron, Ductile, 65-45-12

Australia

Specification	Designation	Notes	C	Cr	Mg	Mn	Ni	P	S	Si	Other	UTS	YS	El	Hard
AS 1831(85)	400-12											400		12	

Belgium

NBN 830-02	FNG42-12														

China

GB 1348(88)	QT 400-15											400	250	15	130-180 HB
GB 1348(88)	QT 450-10											450	310	10	160-210 BH

India

IS 1865	SG400/12														

Italy

UNI 4544	GS400-12														

Japan

JIS G5502(95)	FCD40 Class 1														

South Africa

SABS 936	SG42											410	275	12	200 HB max

UK

BS 2789	420/12											420	270	12	

USA

	UNS F33100											448 Min	310 Min		156-217 HB
ASTM A536(93)	65-45-12											448	310	12	
SAE J434	D4512	Ferr, Prl	3.20-4.10			0.10-1.00		0.015-0.10	0.005-0.035	1.80-3.00	bal Fe	448	310	12	156-217 HB

Cast Iron, Ductile, 70-50-05

China

Specification	Designation	Notes	C	Cr	Mg	Mn	Ni	P	S	Si	Other	UTS	YS	El	Hard
GB 1348(88)	QT 600-3											600	370	3	190-270 HB

South Africa

SABS 936	SG60											590	390	4	210-280 HB

Sweden

SIS 140732	0732-03											600	380	5	

USA

ASTM A536(93)	70-50-05											485	345	5	

Cast Iron, Ductile, 80-55-06

Australia

Specification	Designation	Notes	C	Cr	Mg	Mn	Ni	P	S	Si	Other	UTS	YS	El	Hard
AS 1831(85)	500-7											500		7	

Belgium

NBN 830-02	FNG50-7														

China

GB 1348(88)	QT 500-7											500	320	7	170-230 HB

UNS numbers and US grades are provided as a means of cross referencing chemically similar alloys. Exchangability is only possible after independent examination of specifications. Tensile properties are minimum or typical as specified. UTS and YS as MPa. El as %. See Appendix for list of abbreviations used in Notes. * indicates obsolete material.

Specification	Designation	Notes	C	Cr	Mg	Mn	Ni	P	S	Si	Other	UTS	YS	El	Hard

Cast Iron, Ductile, 80-55-06 (Continued from previous page)

Denmark

| DS 11303 | 727 | | | | | | | | | | | | | | |

France

| AFNOR NFA32201 | FGS500-7 | | | | | | | | | | | | | | |

Germany

| DIN | WNr 0.7050 | | | | | | | | | | | | | | |
| DIN 1693 | GGG-50 | | | | | | | | | | | | | | |

India

| IS 1865 | SG500/7 | | | | | | | | | | | | | | |

International

| ISO 1083(87) | 500-7 | Spheroidal graphite, SCS | | | | | | | | | | 500 | 320 | 7 | 170-230 HB |

Italy

| UNI 4544 | GS500-7 | | | | | | | | | | | | | | |

Japan

| JIS G5502(95) | FCD50 Class 3 | | | | | | | | | | | | | | |

Netherlands

| NEN 2733 | GN500-7 | | | | | | | | | | | | | | |

Norway

| NS 11301 | NS11350 | | | | | | | | | | | | | | 170-230 HB |

Russia

| GOST 7293 | VCh50 | | | | | | | | | | | | | | |

South Africa

| SABS 936 | SG50 | | | | | | | | | | | 490 | 345 | 7 | 170-240 HB |

Sweden

| SIS 140727 | 0727-02 | | | | | | | | | | | 500 | 320 | 7 | |

UK

| BS 2789 | 500/7 | | | | | | | | | | | 500 | 320 | 7 | |

USA

	UNS F33800											552 Min	379 Min		187-255 HB
ASTM A536(93)	80-55-06											552	379	6.0	
SAE J434	D5506	Ferr, Prl	3.20-4.10			0.10-1.00		0.015-0.10	0.005-0.035	1.80-3.00	bal Fe	552	379	6	187-255 HB

Cast Iron, Ductile, A476

Australia

| AS 1831(85) | 600-3 | | | | | | | | | | | 600 | | 3 | |

Belgium

| NBN 830-02 | FNG60-2 | | | | | | | | | | | | | | |

China

| GB 1348(88) | QT 600-3 | | | | | | | | | | | 600 | 370 | 3 | 190-270 HB |

France

| AFNOR NFA32201 | FSG600-3 | | | | | | | | | | | | | | |

Germany

| DIN | WNr 0.7060 | | | | | | | | | | | | | | |
| DIN 1693 | GGG-60 | | | | | | | | | | | | | | |

India

| IS 1865 | SG600/3 | | | | | | | | | | | | | | |

International

| ISO 1083(87) | 600-3 | Spheroidal graphite, SCS | | | | | | | | | | 600 | 370 | 3 | 190-270 HB |

Italy

| UNI 4544 | GS600-3 | | | | | | | | | | | | | | |

Japan

| JIS G5502(95) | FCD60 Class 4 | | | | | | | | | | | | | | |

Netherlands

| NEN 2733 | GN600-3 | | | | | | | | | | | | | | |

Specification	Designation	Notes	C	Cr	Mg	Mn	Ni	P	S	Si	Other	UTS	YS	El	Hard
Cast Iron, Ductile, A476 (Continued from previous page)															
Norway															
NS 11301	NS11360														
Russia															
GOST 7293	VCh60														
South Africa															
SABS 936	SG70												685	440	3	230-300 HB
UK															
BS 2789	600/3												600	370	3	
USA															
	UNS F34100		3 min					0.08 max	0.05 max	3.0 max	bal Fe	552 Min	414 Min		
ASTM A536(93)	80-60-03											555	415	3	
Cast Iron, Ductile, A716															
China															
GB 1348(88)	QT 450-10												450	310	10	160-210 HB
France															
AFNOR NFA32201	FGS450-10															
International															
ISO 1083(87)	450-10	Spheroidal graphite, SCS										450	310	10	160-210 HB
Japan															
JIS G5502(95)	FCD45 Class 2															
Russia															
GOST 7293	VCh45															
Sweden															
SIS 140722	0722-00												450	310	10	
UK															
BS 2789	450/10												450	320	10	
USA															
	UNS F32900												414 Min	290 Min		
ASTM A536(93)	60-42-10												415	290	10	
ASTM A716(95)	A716	Culvert pipe											413.7	289.6	10	
Cast Iron, Ductile, DQ&T															
USA															
	UNS F30000	Martensitic														as specified (Q and T)
SAE J434	DQ&T	Mart; hard by agree, mech prop vary	3.20-4.10			0.10-1.00		0.015-0.10	0.005-0.035	1.80-3.00	bal Fe				
Cast Iron, Ductile, No equivalents identified															
Japan															
JIS G5502(95)	FCD350-22	SCS	2.5 min		0.09 max	0.4 max		0.08 max	0.02 max	2.7 max	bal Fe	350	220	22	150 HB
JIS G5502(95)	FCD350-22L	SCS	2.5 min		0.09 max	0.4 max		0.08 max	0.02 max	2.7 max	bal Fe	350	220	22	150 HB
JIS G5502(95)	FCD400-15	SCS	2.5 min		0.09 max				0.02 max		bal Fe	400	250	15	130-180 HB
JIS G5502(95)	FCD400-15A	Cast-on test sample; 30<t<=60mm	2.5 min		0.09 max				0.02 max		bal Fe	390	250	15	120-180 HB
JIS G5502(95)	FCD400-15A	Cast-on test sample; 60<t<=200mm	2.5 min		0.09 max				0.02 max		bal Fe	370	240	12	120-180 HB
JIS G5502(95)	FCD400-18	SCS	2.5 min		0.09 max	0.4 max		0.08 max	0.02 max	2.7 max	bal Fe	400	250	18	130-180 HB
JIS G5502(95)	FCD400-18A	Cast-on test sample; 30<t<=60mm	2.5 min		0.09 max	0.4 max		0.08 max	0.02 max	2.7 max	bal Fe	390	250	15	120-180 HB
JIS G5502(95)	FCD400-18A	Cast-on test sample; 60<t<=200mm	2.5 min		0.09 max	0.4 max		0.08 max	0.02 max	2.7 max	bal Fe	370	240	12	120-180 HB

UNS numbers and US grades are provided as a means of cross referencing chemically similar alloys. Exchangability is only possible after independent examination of specifications. Tensile properties are minimum or typical as specified. UTS and YS as MPa. El as %. See Appendix for list of abbreviations used in Notes. * indicates obsolete material.

Specification	Designation	Notes	C	Cr	Mg	Mn	Ni	P	S	Si	Other	UTS	YS	El	Hard

Cast Iron, Ductile, No equivalents identified

Japan

Specification	Designation	Notes	C	Cr	Mg	Mn	Ni	P	S	Si	Other	UTS	YS	El	Hard
JIS G5502(95)	FCD400-18AL	Cast-on test sample; 30<t<=60mm	2.5 min		0.09 max	0.4 max		0.08 max	0.02 max	2.7 max	bal Fe	390	250	15	120-180 HB
JIS G5502(95)	FCD400-18AL	Cast-on test sample; 60<t<=200mm	2.5 min		0.09 max	0.4 max		0.08 max	0.02 max	2.7 max	bal Fe	370	240	12	120-180 HB
JIS G5502(95)	FCD400-18L	SCS	2.5 min		0.09 max	0.4 max		0.08 max	0.02 max	2.7 max	bal Fe	400	250	18	130-180 HB
JIS G5502(95)	FCD450-10	SCS	2.5 min		0.09 max				0.02 max		bal Fe	450	280	10	140-210 HB
JIS G5502(95)	FCD500-7	SCS	2.5 min		0.09 max				0.02 max		bal Fe	500	320	7	150-230 HB
JIS G5502(95)	FCD500-7A	Cast-on test sample; 30<t<=60mm	2.5 min		0.09 max				0.02 max		bal Fe	450	300	7	130-230 HB
JIS G5502(95)	FCD500-7A	Cast-on test sample; 60<t<=200mm	2.5 min		0.09 max				0.02 max		bal Fe	420	290	5	130-230 HB
JIS G5502(95)	FCD600-3	SCS	2.5 min		0.09 max				0.02 max		bal Fe	600	370	3	170-270 HB
JIS G5502(95)	FCD600-3A	Cast-on test sample; 30<t<=60mm	2.5 min		0.09 max				0.02 max		bal Fe	600	360	2	160-270 HB
JIS G5502(95)	FCD600-3A	Cast-on test sample; 60<t<=200mm	2.5 min		0.09 max				0.02 max		bal Fe	550	340	1	160-270 HB
JIS G5502(95)	FCD700-2	SCS	2.5 min		0.09 max				0.02 max		bal Fe	700	420	2	180-300 HB
JIS G5502(95)	FCD800-2	SCS	2.5 min		0.09 max				0.02 max		bal Fe	800	480	2	200-330 HB
JIS G5527(89)	DF	Fittings										420		10	230 HB

USA

Specification	Designation	Notes	C	Cr	Mg	Mn	Ni	P	S	Si	Other	UTS	YS	El	Hard
ASTM A874/A874M(93)	A874	Ferr; Low-Temp service	3.0-3.7	0.07 max	0.07 max	0.25 max	1.0 max	0.03 max		1.2-2.3	Cu <=0.1; C + 1/3Si<=4.5; bal Fe	300	200	12	
ASTM A897/A897M(97)	1050/700/7	Austempered										1050	700	7	302-363 HB
ASTM A897/A897M(97)	1200/850/4	Austempered										1200	850	4	341-444 HB
ASTM A897/A897M(97)	1400/1100/1	Austempered										1400	1100	1	388-477 HB
ASTM A897/A897M(97)	1600/1300/-	Austempered										1600	1300	not specified	444-555 HB
ASTM A897/A897M(97)	850/550/10	Austempered										850	550	10	269-321 HB

Specification	Designation	Notes	C	Cr	Mn	Mo	Ni	P	S	Si	Other	UTS	YS	El	Hard

Cast Iron, Gray, 20 (Continued from previous page)

Australia

Specification	Designation	Notes	C	Cr	Mn	Mo	Ni	P	S	Si	Other	UTS	YS	El	Hard
AS 1830(86)	T150	as cast; test bar 30-32mm diam										150			

Belgium

| NBN 83001 | FGG10 | as cast; test bar 30-32mm diam | | | | | | | | | | | | | |
| NBN 83001 | FGG15 | as cast; test bar 30-32mm diam | | | | | | | | | | | | | |

China

GB 9439(88)	HT 100	as cast; test bar 30mm diam; 2.5<t<=10mm										130			
GB 9439(88)	HT 100	as cast; test bar 30mm diam; 10<t<=20mm										100			
GB 9439(88)	HT 100	as cast; test bar 30mm diam; 20<t<=30mm										90			
GB 9439(88)	HT 100	as cast; test bar 30mm diam; 30<t<=40mm										80			

Denmark

| DS 11301 | GG10 | as cast; test bar 30-32mm diam | | | | | | | | | | | | | |
| DS 11301 | GG15 | as cast; test bar 30-32mm diam | | | | | | | | | | | | | |

France

AFNOR NFA32101	FGL150	as cast; test bar 30mm diam													
AFNOR NFA32101	FGL150A	as cast; test bar 30mm diam; 150<t<=300mm													
AFNOR NFA32101	FGL150A	as cast; test bar 30mm diam; 80<t<=150mm													
AFNOR NFA32101	FGL150A	as cast; test bar 20mm diam; 20<t<=40mm													
AFNOR NFA32101	FGL150A	as cast; test bar 20mm diam; 40<t<=80mm													

Germany

| DIN | WNr 0.6010 | | | | | | | | | | | | | | |
| DIN 1691 | GG10 | as cast; test bar 30mm diam | | | | | | | | | | | | | |

India

| IS 210 | FG150 | | | | | | | | | | | | | | 130-180 HB |

International

| ISO 185(88) | 100 | SCS | | | | | | | | | | 100 | | | |
| ISO 185(88) | 150 | SCS | | | | | | | | | | 150 | | | |

Italy

| UNI 5007 | G10 | 7.5<t<=15mm | | | | | | | | | | | | | 175 HB |
| UNI 5007 | G10 | 3.5<t<=7.5mm | | | | | | | | | | | | | 215 HB |

Japan

| JIS G5501(95) | FC100 | SCS | | | | | | | | | | 100 | | | 201 HB |
| JIS G5501(95) | FC150 | SCS | | | | | | | | | | 150 | | | 212 HB |

Netherlands

| NEN 6002A | GG15 | as cast; test bar 30-32mm diam | | | | | | | | | | | | | |

Norway

| NS 722 | SjG15 | as cast; test bar 30-32mm diam | | | | | | | | | | | | | |
| NS 722 | SjG150 | as cast; test bar 30mm diam; Cast-on; 40<t<=80mm | | | | | | | | | | | | | |

UNS numbers and US grades are provided as a means of cross referencing chemically similar alloys. Exchangability is only possible after independent examination of specifications. Tensile properties are minimum or typical as specified. UTS and YS as MPa. El as %. See Appendix for list of abbreviations used in Notes. * indicates obsolete material.

Specification	Designation	Notes	C	Cr	Mn	Mo	Ni	P	S	Si	Other	UTS	YS	El	Hard

Cast Iron, Gray, 20 (Continued from previous page)

Norway

Specification	Designation	Notes	C	Cr	Mn	Mo	Ni	P	S	Si	Other	UTS	YS	El	Hard
NS 722	SjG150	as cast; test bar 50mm diam; Cast-on; 80<t<=150mm													
NS 722	SjG150	as cast; test bar 30mm diam; Cast-on; 20<t<=40mm													
NS 722	SjG150	as cast; test bar 50mm diam; Cast-on; 150<t<=300mm													

Russia

Specification	Designation	Notes	C	Cr	Mn	Mo	Ni	P	S	Si	Other	UTS	YS	El	Hard
GOST 1412	SCh10	as cast; test bar 30mm diam														143-229 HB
GOST 1412	SCh15	as cast; test bar 30mm diam														163-229 HB

UK

Specification	Designation	Notes	C	Cr	Mn	Mo	Ni	P	S	Si	Other	UTS	YS	El	Hard
BS 1452	100	as cast; test bar 30-32mm diam											100			
BS 1452	150	as cast; test bar 30-32mm diam											150			

USA

Specification	Designation	Notes	C	Cr	Mn	Mo	Ni	P	S	Si	Other	UTS	YS	El	Hard
	UNS F11401											138 Min			
ASME SA278	20	Refer to ASTM A278(93)													
ASTM A278(93)	20												150			
ASTM A278M(93)	150												150			
ASTM A48(94)	20A	SCS; 22.4mm diam											138			
ASTM A48(94)	20B	SCS; 30.5mm diam											138			
ASTM A48(94)	20C	SCS; 50.8mm diam											138			
ASTM A48(94)	20S	SCS; as agreed											138			

Cast Iron, Gray, 25

Australia

Specification	Designation	Notes	C	Cr	Mn	Mo	Ni	P	S	Si	Other	UTS	YS	El	Hard
AS 1830(86)	T180	as cast; test bar 30-32mm diam											180			

China

Specification	Designation	Notes	C	Cr	Mn	Mo	Ni	P	S	Si	Other	UTS	YS	El	Hard
GB 9439(88)	HT 150	as cast; test bar 30mm diam; 20<t<=30mm											130			
GB 9439(88)	HT 150	as cast; test bar 30mm diam; 10<t<=20mm											145			
GB 9439(88)	HT 150	as cast; test bar 30mm diam; 2.5<t<=10mm											175			
GB 9439(88)	HT 150	as cast; test bar 30mm diam; 30<t<=40mm											120			

France

Specification	Designation	Notes	C	Cr	Mn	Mo	Ni	P	S	Si	Other	UTS	YS	El	Hard
AFNOR NFA32101	FGL200A	as cast; test bar 20mm diam; 40<t<=80mm														
AFNOR NFA32101	FGL200A	as cast; test bar 20mm diam; 20<t<=40mm														
AFNOR NFA32101	FGL200A	as cast; test bar 30mm diam; 80<t<=150mm														
AFNOR NFA32101	FGL200A	as cast; test bar 30mm diam; 150<t<=300mm														
AFNOR NFA32101	FGL250A	as cast; test bar 20mm diam; 40<t<=80mm														
AFNOR NFA32101	FGL250A	as cast; test bar 20mm diam; 20<t<=40mm														

Specification	Designation	Notes	C	Cr	Mn	Mo	Ni	P	S	Si	Other	UTS	YS	El	Hard
Cast Iron, Gray, 25 (Continued from previous page)															
France															
AFNOR NFA32101	FGL250A	as cast; test bar 30mm diam; 80<t<=150mm													
AFNOR NFA32101	FGL250A	as cast; test bar 30mm diam; 150<t<=300mm													
Germany															
DIN	WNr 0.6015														
DIN 1691	GG15	as cast; test bar 30mm diam													
Italy															
UNI 5007	G15	15<t<=30mm													165 HB
UNI 5007	G15	7.5<t<=15mm													185 HB
UNI 5007	G15	3.5<t<=7.5mm													225 HB
Norway															
NS 722	SjG200	as cast; test bar 50mm diam; Cast-on; 150<t<=300mm													
NS 722	SjG200	as cast; test bar 30mm diam; Cast-on; 40<t<=80mm													
NS 722	SjG200	as cast; test bar 30mm diam; Cast-on; 20<t<=40mm													
NS 722	SjG200	as cast; test bar 50mm diam; Cast-on; 80<t<=150mm													
Russia															
GOST 1412	SCh18	as cast; test bar 30mm diam													170-229 HB
Sweden															
SS 140115	115	as cast; test bar 30-32mm diam											200			
UK															
BS 1452	180	as cast; test bar 30-32mm diam											180			
USA															
	UNS F11701												172 Min			
ASME SA278	25	Refer to ASTM A278(93)														
ASTM A278(93)	25												175			
ASTM A278M(93)	175												175			
ASTM A48(94)	25A	SCS; 22.4mm diam											172			
ASTM A48(94)	25B	SCS; 30.5mm diam											172			
ASTM A48(94)	25C	SCS; 50.8mm diam											172			
ASTM A48(94)	25S	SCS; as agreed											172			
Cast Iron, Gray, 30															
Belgium															
NBN 83001	FGG20	as cast; test bar 30-32mm diam														
China															
GB 9439(88)	HT 200	as cast; test bar 30mm diam; 30<t<=40mm											160			
GB 9439(88)	HT 200	as cast; test bar 30mm diam; 10<t<=20mm											195			
GB 9439(88)	HT 200	as cast; test bar 30mm diam; 2.5<t<=10mm											220			
GB 9439(88)	HT 200	as cast; test bar 30mm diam; 20<t<=30mm											170			

Specification	Designation	Notes	C	Cr	Mn	Mo	Ni	P	S	Si	Other	UTS	YS	El	Hard

Cast Iron, Gray, 30 (Continued from previous page)

Denmark

Specification	Designation	Notes	C	Cr	Mn	Mo	Ni	P	S	Si	Other	UTS	YS	El	Hard
DS 11301	GG20	as cast; test bar 30-32mm diam													

France

| AFNOR NFA32101 | FGL200 | as cast; test bar 30mm diam | | | | | | | | | | | | | |

Germany

| DIN 1691 | GG15 | as cast; test bar 30mm diam | | | | | | | | | | | | | |

India

| IS 210 | FG200 | | | | | | | | | | | | | | 160-220 HB |

International

| ISO 185(88) | 200 | SCS | | | | | | | | | | 200 | | | |

Italy

UNI 5007	G20	7.5<t<=15mm													205 HB
UNI 5007	G20	3.5<t<=7.5mm													235 HB
UNI 5007	G20	15<t<=30mm													175 HB
UNI 5007	G20	30<t<=50mm													155 HB

Japan

| JIS G5501(95) | FC200 | SCS | | | | | | | | | | 200 | | | 223 HB |

Netherlands

| NEN 6002A | GG20 | as cast; test bar 30-32mm diam | | | | | | | | | | | | | |

Norway

NS 722	SjG20	as cast; test bar 30-32mm diam													
NS 722	SjG250	as cast; test bar 50mm diam; Cast-on; 150<t<=300mm													
NS 722	SjG250	as cast; test bar 30mm diam; Cast-on; 40<t<=80mm													
NS 722	SjG250	as cast; test bar 50mm diam; Cast-on; 80<t<=150mm													
NS 722	SjG250	as cast; test bar 30mm diam; Cast-on; 20<t<=40mm													

Russia

| GOST 1412 | SCh20 | as cast; test bar 30mm diam | | | | | | | | | | | | | 170-241 HB |

South Africa

SABS 1034	200	as cast; test bar 55mm diam; 40<t<=55mm										185			156-217 HB
SABS 1034	200	as cast; test bar 15mm diam; t<10mm										225			156-217 HB
SABS 1034	200	as cast; test bar 40mm diam; 30<t<=40mm										195			156-217 HB
SABS 1034	200	as cast; test bar 22mm diam; 10<t<=20mm										215			156-217 HB
SABS 1034	200	as cast; test bar 30mm diam; 20<t<=30mm										200			156-217 HB

UK

| BS 1452 | 200 | as cast; test bar 30-32mm diam | | | | | | | | | | 200 | | | |

USA

	UNS F12101											207 Min			
ASME SA278	30	Refer to ASTM A278(93)													
ASTM A126(95)	B														

UNS numbers and US grades are provided as a means of cross referencing chemically similar alloys. Exchangability is only possible after independent examination of specifications. Tensile properties are minimum or typical as specified. UTS and YS as MPa. El as %. See Appendix for list of abbreviations used in Notes. * indicates obsolete material.

Specification	Designation	Notes	C	Cr	Mn	Mo	Ni	P	S	Si	Other	UTS	YS	El	Hard
Cast Iron, Gray, 30 (Continued from previous page)															
USA															
ASTM A278(93)	30											200			
ASTM A278M(93)	200											200			
ASTM A48(94)	30A	SCS; 22.4mm diam										207			
ASTM A48(94)	30B	SCS; 30.5mm diam										207			
ASTM A48(94)	30C	SCS; 50.8mm diam										207			
ASTM A48(94)	30S	SCS; as agreed										207			
Cast Iron, Gray, 319(I)															
USA															
	UNS F10001		3.50 min								bal Fe				
Cast Iron, Gray, 319(II)															
USA															
	UNS F10002		3.20 min								bal Fe				
Cast Iron, Gray, 319(III)															
USA															
	UNS F10003		2.80 min								bal Fe				
Cast Iron, Gray, 35															
Australia															
AS 1830(86)	T260	as cast; test bar 30-32mm diam										260			
Belgium															
NBN 83001	FGG25	as cast; test bar 30-32mm diam													
China															
GB 9439(88)	HT 250	as cast; test bar 30mm diam; 20<t<=30mm										220			
GB 9439(88)	HT 250	as cast; test bar 30mm diam; 30<t<=40mm										200			
GB 9439(88)	HT 250	as cast; test bar 30mm diam; 4.0<t<=10mm										270			
GB 9439(88)	HT 250	as cast; test bar 30mm diam; 10<t<=20mm										240			
Denmark															
DS 11301	GG25	as cast; test bar 30-32mm diam													
France															
AFNOR NFA32101	FGL250	as cast; test bar 30mm diam													
AFNOR NFA32101	FGL300A	as cast; test bar 20mm diam; 20<t<=40mm													
AFNOR NFA32101	FGL300A	as cast; test bar 30mm diam; 80<t<=150mm													
AFNOR NFA32101	FGL300A	as cast; test bar 20mm diam; 40<t<=80mm													
AFNOR NFA32101	FGL300A	as cast; test bar 30mm diam; 150<t<=300mm													
Germany															
DIN	WNr 0.6020														
DIN 1691	GG20	as cast; test bar 30mm diam													

Specification	Designation	Notes	C	Cr	Mn	Mo	Ni	P	S	Si	Other	UTS	YS	El	Hard

Cast Iron, Gray, 35 (Continued from previous page)

India

Specification	Designation	Notes	C	Cr	Mn	Mo	Ni	P	S	Si	Other	UTS	YS	El	Hard
IS 210	FG220														180-220 HB
IS 210	FG260														180-230 HB

International

Specification	Designation	Notes	UTS	Hard
ISO 185(88)	250	SCS	250	

Italy

Specification	Designation	Notes	UTS	Hard
UNI 5007	G25	30<t<=50mm		170 HB
UNI 5007	G25	3.5<t<=7.5mm		250 HB
UNI 5007	G25	15<t<=30mm		195 HB
UNI 5007	G25	7.5<t<=15mm		220 HB

Japan

Specification	Designation	Notes	UTS	Hard
JIS G5501(95)	FC250	SCS	200	241 HB

Netherlands

Specification	Designation	Notes	UTS	Hard
NEN 6002A	GG25	as cast; test bar 30-32mm diam		

Norway

Specification	Designation	Notes	UTS	Hard
NS 722	SjG25	as cast; test bar 30-32mm diam		
NS 722	SjG300	as cast; test bar 30mm diam; Cast-on; 20<t<=40mm		
NS 722	SjG300	as cast; test bar 50mm diam; Cast-on; 150<t<=300mm		
NS 722	SjG300	as cast; test bar 30mm diam; Cast-on; 40<t<=80mm		
NS 722	SjG300	as cast; test bar 50mm diam; Cast-on; 80<t<=150mm		

Russia

Specification	Designation	Notes	UTS	Hard
GOST 1412	SCh25	as cast; test bar 30mm diam		180-250 HB

South Africa

Specification	Designation	Notes	UTS	Hard
SABS 1034	250	as cast; test bar 15mm diam; t<10mm	280	187-241 HB
SABS 1034	250	as cast; test bar 55mm diam; t>40mm	220	187-241 HB
SABS 1034	250	as cast; test bar 30mm diam; 20<t<=30mm	250	187-241 HB
SABS 1034	250	as cast; test bar 22mm diam; 10<t<=20mm	265	187-241 HB
SABS 1034	250	as cast; test bar 40mm diam; 30<t<=40mm	235	187-241 HB

Sweden

Specification	Designation	Notes	UTS	Hard
SS 140120	120	as cast; test bar 30-32mm diam	250	

UK

Specification	Designation	Notes	UTS	Hard
BS 1452	220	as cast; test bar 30-32mm diam	220	
BS 1452	250	as cast; test bar 30-32mm diam	250	
BS 1452	260	as cast; test bar 30-32mm diam	260	

USA

Specification	Designation	Notes	UTS	Hard
	UNS F12401		241 Min	
ASME SA278	35	Refer to ASTM A278(93)		
ASTM A278(93)	35		225	
ASTM A278M(93)	225		225	

Specification	Designation	Notes	C	Cr	Mn	Mo	Ni	P	S	Si	Other	UTS	YS	El	Hard

Cast Iron, Gray, 35 (Continued from previous page)

USA

Specification	Designation	Notes	UTS
ASTM A278M(93)	250	If for temp >230 C, CEV<=3.8	250
ASTM A278M(93)	275	If for temp >230 C, CEV<=3.8	275
ASTM A48(94)	35A	SCS; 22.4mm diam	241
ASTM A48(94)	35B	SCS; 30.5mm diam	241
ASTM A48(94)	35C	SCS; 50.8mm diam	241
ASTM A48(94)	35S	SCS; as agreed	241

Cast Iron, Gray, 40

USA

Specification	Designation	Notes	UTS
	UNS F12801		276 Min
	UNS F12803		276 Min
ASME SA278	40	Refer to ASTM A278(93)	
ASTM A126(95)	C		
ASTM A278(93)	40	If for temp >230 C, CEV<=3.8	300
ASTM A48(94)	40A	SCS; 22.4mm diam	276
ASTM A48(94)	40B	SCS; 30.5mm diam	276
ASTM A48(94)	40C	SCS; 50.8mm diam	276
ASTM A48(94)	40S	SCS; as agreed	276

Cast Iron, Gray, 45

Australia

Specification	Designation	Notes	UTS
AS 1830(86)	T300	as cast; test bar 30-32mm diam	300

Belgium

Specification	Designation	Notes	UTS
NBN 83001	FGG30	as cast; test bar 30-32mm diam	

China

Specification	Designation	Notes	UTS
GB 9439(88)	HT 300	as cast; test bar 30mm diam; 10<t<=20mm	290
GB 9439(88)	HT 300	as cast; test bar 30mm diam; 20<t<=30mm	250
GB 9439(88)	HT 300	as cast; test bar 30mm diam; 30<t<=40mm	230
GB 9439(88)	HT 300	as cast; test bar 30mm diam; 30<t<=40mm	230
GB 9439(88)	HT 300	as cast; test bar 30mm diam	300
GB 9439(88)	HT 300	as cast; test bar 30mm diam; 20<t<=30mm	250
GB 9439(88)	HT 300	as cast; test bar 30mm diam	300
GB 9439(88)	HT 300	as cast; test bar 30mm diam; 10<t<=20mm	290

Denmark

Specification	Designation	Notes	UTS
DS 11301	GG30	as cast; test bar 30-32mm diam	

France

Specification	Designation	Notes	UTS
AFNOR NFA32101	FGL300	as cast; test bar 30mm diam	
AFNOR NFA32101	FGL350A	as cast; test bar 20mm diam; 40<t<=80mm	

Specification	Designation	Notes	C	Cr	Mn	Mo	Ni	P	S	Si	Other	UTS	YS	El	Hard

Cast Iron, Gray, 45 (Continued from previous page)

France

Specification	Designation	Notes	UTS	Hard
AFNOR NFA32101	FGL350A	as cast; test bar 20mm diam; 20<t<=40mm		
AFNOR NFA32101	FGL350A	as cast; test bar 30mm diam; 80<t<=150mm		
AFNOR NFA32101	FGL350A	as cast; test bar 30mm diam; 150<t<=300mm		
AFNOR NFA32101	FGL400A	as cast; test bar 20mm diam; 20<t<=40mm		
AFNOR NFA32101	FGL400A	as cast; test bar 30mm diam; 80<t<=150mm		
AFNOR NFA32101	FGL400A	as cast; test bar 20mm diam; 40<t<=80mm		

Germany

| DIN | WNr 0.6025 | | | |
| DIN 1691 | GG25 | as cast; test bar 30mm diam | | |

India

| IS 210 | FG300 | | | 180-230 HB |

International

| ISO 185(88) | 300 | SCS | 300 | |

Italy

UNI 5007	G30	15<t<=30mm		215 HB
UNI 5007	G30	7.5<t<=15mm		235 HB
UNI 5007	G30	30<t<=50mm		190 HB

Japan

| JIS G5501(95) | FC300 | SCS | 250 | 262 HB |

Netherlands

| NEN 6002A | GG30 | as cast; test bar 30-32mm diam | | |

Norway

NS 722	SjG30	as cast; test bar 30-32mm diam		
NS 722	SjG350	as cast; test bar 30mm diam; Cast-on; 40<t<=80mm		
NS 722	SjG350	as cast; test bar 50mm diam; Cast-on; 150<t<=300mm		
NS 722	SjG350	as cast; test bar 50mm diam; Cast-on; 80<t<=150mm		
NS 722	SjG350	as cast; test bar 30mm diam; Cast-on; 20<t<=40mm		

Russia

| GOST 1412 | SCh30 | as cast; test bar 30mm diam | | 181-255 HB |

South Africa

SABS 1034	300	as cast; test bar 40mm diam; 30<t<=40mm	285	207-269 HB
SABS 1034	300	as cast; test bar 15mm diam; t<10mm	330	207-269 HB
SABS 1034	300	as cast; test bar 30mm diam; 20<t<=30mm	300	207-269 HB
SABS 1034	300	as cast; test bar 55mm diam; t>40mm	270	207-269 HB

UNS numbers and US grades are provided as a means of cross referencing chemically similar alloys. Exchangability is only possible after independent examination of specifications. Tensile properties are minimum or typical as specified. UTS and YS as MPa. El as %. See Appendix for list of abbreviations used in Notes. * indicates obsolete material.

Specification	Designation	Notes	C	Cr	Mn	Mo	Ni	P	S	Si	Other	UTS	YS	El	Hard

Cast Iron, Gray, 45 (Continued from previous page)

South Africa

Specification	Designation	Notes	C	Cr	Mn	Mo	Ni	P	S	Si	Other	UTS	YS	El	Hard
SABS 1034	300	as cast; test bar 22mm diam; 10<t<=20mm										315			207-269 HB

UK

BS 1452	300	as cast; test bar 30-32mm diam										300			

USA

	UNS F13101											310 Min			
	UNS F13102											310 Min			
ASME SA278	45	Refer to ASTM A278(93)													
ASTM A278(93)	45	If for temp >230 C, CEV<=3.8										325			
ASTM A278M(93)	300	If for temp >230 C, CEV<=3.8										300			
ASTM A278M(93)	325	If for temp >230 C, CEV<=3.8										325			
ASTM A48(94)	45A	SCS; 22.4mm diam										310			
ASTM A48(94)	45B	SCS; 30.5mm diam										310			
ASTM A48(94)	45C	SCS; 50.8mm diam										310			
ASTM A48(94)	45S	SCS; as agreed										310			

Cast Iron, Gray, 50

Australia

Specification	Designation	Notes	C	Cr	Mn	Mo	Ni	P	S	Si	Other	UTS	YS	El	Hard
AS 1830(86)	T350	as cast; test bar 30-32mm diam										350			

Belgium

NBN 83001	FGG35	as cast; test bar 30-32mm diam													

China

GB 9439(88)	HT 350	as cast; test bar 30mm diam; 10<t<=20mm										340			
GB 9439(88)	HT 350	as cast; test bar 30mm diam; 20<t<=30mm										290			
GB 9439(88)	HT 350	as cast; test bar 30mm diam										350			
GB 9439(88)	HT 350	as cast; test bar 30mm diam; 30<t<=40mm										260			
GB 9439(88)	HT 350	as cast; test bar 30mm diam; 20<t<=30mm										290			
GB 9439(88)	HT 350	as cast; test bar 30mm diam; 10<t<=20mm										340			
GB 9439(88)	HT 350	as cast; test bar 30mm diam										350			
GB 9439(88)	HT 350	as cast; test bar 30mm diam; 30<t<=40mm										260			

Denmark

DS 11301	GG35	as cast; test bar 30-32mm diam													

France

AFNOR NFA32101	FGL350	as cast; test bar 30mm diam													

Germany

DIN	WNr 0.6030														
DIN 1691	GG30	as cast; test bar 30mm diam													

Specification	Designation	Notes	C	Cr	Mn	Mo	Ni	P	S	Si	Other	UTS	YS	El	Hard

Cast Iron, Gray, 50 (Continued from previous page)

India

| IS 210 | FG350 | | | | | | | | | | | | | | 207-241 HB |

International

| ISO 185(88) | 350 | SCS | | | | | | | | | | 350 | | | |

Italy

UNI 5007	G35	30<t<=50mm													210 HB
UNI 5007	G35	7.5<t<=15mm													
UNI 5007	G35	15<t<=30mm													235 HB

Japan

| JIS G5501(95) | FC350 | SCS | | | | | | | | | | 350 | | | 277 HB |

Netherlands

| NEN 6002A | GG35 | as cast; test bar 30-32mm diam | | | | | | | | | | | | | |

Norway

| NS 722 | SjG35 | as cast; test bar 30-32mm diam | | | | | | | | | | | | | |

Russia

| GOST 1412 | SCh35 | as cast; test bar 30mm diam | | | | | | | | | | | | | 197-269 HB |

South Africa

SABS 1034	350	as cast; test bar 30mm diam; 20<t<=30mm										350			217-285 HB
SABS 1034	350	as cast; test bar 53mm diam; t>40mm										325			217-285 HB
SABS 1034	350	as cast; test bar 40mm diam; 30<t<=40mm										340			217-285 HB
SABS 1034	350	as cast; test bar 15mm diam; t<10mm										385			217-285 HB
SABS 1034	350	as cast; test bar 22mm diam; 10<t<=20mm										370			217-285 HB

Sweden

| SS 140130 | 130 | as cast; test bar 30-32mm diam | | | | | | | | | | 350 | | | |

UK

| BS 1452 | 350 | as cast; test bar 30-32mm diam | | | | | | | | | | 350 | | | |

USA

	UNS F13501											345 Min			
	UNS F13502											345 Min			
ASME SA278	50	Refer to ASTM A278(93)													
ASTM A278(93)	50	If for temp >230 C, CEV<=3.8										350			
ASTM A278M(93)	350	If for temp >230 C, CEV<=3.8										350			
ASTM A48(94)	50A	SCS; 22.4mm diam										345			
ASTM A48(94)	50B	SCS; 30.5mm diam										345			
ASTM A48(94)	50C	SCS; 50.8mm diam										345			
ASTM A48(94)	50S	SCS; as agreed										345			

Cast Iron, Gray, 55

Belgium

| NBN 83001 | FGG40 | as cast; test bar 30-32mm diam | | | | | | | | | | | | | |

Denmark

| DS 11301 | GG40 | as cast; test bar 30-32mm diam | | | | | | | | | | | | | |

Specification	Designation	Notes	C	Cr	Mn	Mo	Ni	P	S	Si	Other	UTS	YS	El	Hard

Cast Iron, Gray, 55 (Continued from previous page)

Norway

Specification	Designation	Notes	C	Cr	Mn	Mo	Ni	P	S	Si	Other	UTS	YS	El	Hard
NS 722	SjG40	as cast; test bar 30-32mm diam													

Russia

| GOST 1412 | SCh40 | as cast; test bar 30mm diam | | | | | | | | | | | | | 207-285 HB |

USA

	UNS F13801											379 Min			
	UNS F13802											379 Min			
ASME SA278	55	Refer to ASTM A278(93)													
ASTM A278(93)	55	If for temp >230 C, CEV<=3.8											380			
ASTM A278M(93)	380	If for temp >230 C, CEV<=3.8											380			
ASTM A48(94)	55A	SCS; 22.4mm diam											379			
ASTM A48(94)	55B	SCS; 30.5mm diam											379			
ASTM A48(94)	55C	SCS; 50.8mm diam											379			
ASTM A48(94)	55S	SCS; as agreed											379			

Cast Iron, Gray, 60

Australia

Specification	Designation	Notes	C	Cr	Mn	Mo	Ni	P	S	Si	Other	UTS	YS	El	Hard
AS 1830(86)	T400	as cast; test bar 30-32mm diam											400			

France

| AFNOR NFA32101 | FGL400 | as cast; test bar 30mm diam | | | | | | | | | | | | | | |

Germany

| DIN | WNr 0.6035 | | | | | | | | | | | | | | | |
| DIN 1691 | GG35 | as cast; test bar 30mm diam | | | | | | | | | | | | | | |

India

| IS 210 | FG400 | | | | | | | | | | | | | | | 207-270 HB |

Russia

| GOST 1412 | SCh45 | as cast; test bar 30mm diam | | | | | | | | | | | | | | 229-289 HB |

Sweden

| SS 140135 | 135 | as cast; test bar 30-32mm diam | | | | | | | | | | | | | | |

UK

| BS 1452 | 400 | as cast; test bar 30-32mm diam | | | | | | | | | | | 400 | | | |

USA

	UNS F14101												414 Min			
	UNS F14102												414 Min			
ASME SA278	60	Refer to ASTM A278(93)														
ASTM A278(93)	60	If for temp >230 C, CEV<=3.8											415			
ASTM A278M(93)	415	If for temp >230 C, CEV<=3.8											415			
ASTM A48(94)	60A	SCS; 22.4mm diam											414			
ASTM A48(94)	60B	SCS; 30.5mm diam											414			
ASTM A48(94)	60C	SCS; 50.8mm diam											414			
ASTM A48(94)	60S	SCS; as agreed											414			

Cast Iron, Gray, 70

USA

Specification	Designation	Notes	C	Cr	Mn	Mo	Ni	P	S	Si	Other	UTS	YS	El	Hard
	UNS F14801												483 Min			

UNS numbers and US grades are provided as a means of cross referencing chemically similar alloys. Exchangability is only possible after independent examination of specifications. Tensile properties are minimum or typical as specified. UTS and YS as MPa. El as %. See Appendix for list of abbreviations used in Notes. * indicates obsolete material.

Specification	Designation	Notes	C	Cr	Mn	Mo	Ni	P	S	Si	Other	UTS	YS	El	Hard

Cast Iron, Gray, 80

USA

| | UNS F15501 | | | | | | | | | | | 552 Min | | | |

Cast Iron, Gray, A

South Africa

SABS 1034	150	as cast; test bar 15mm diam; t<10mm										170			131-187 HB
SABS 1034	150	as cast; test bar 30mm diam; 20<t<=30mm										150			131-187 HB
SABS 1034	150	as cast; test bar 40mm diam; 30<t<=40mm										145			131-187 HB
SABS 1034	150	as cast; test bar 22mm diam; 10<t<=20mm										160			131-187 HB
SABS 1034	150	as cast; test bar 55mm diam; t>40mm										135			131-187 HB

USA

| | UNS F11501 | | | | | | | | | | | 145 Min | | | |
| ASTM A126(95) | A | | | | | | | | | | | | | | |

Cast Iron, Gray, B

Australia

| AS 1830(86) | T220 | as cast; test bar 30-32mm diam | | | | | | | | | | 220 | | | |

China

GB 9439(88)	HT 200	as cast; test bar 30mm diam; 2.5<t<=10mm										220			
GB 9439(88)	HT 200	as cast; test bar 30mm diam; 30<t<=40mm										160			
GB 9439(88)	HT 200	as cast; test bar 30mm diam; 20<t<=30mm										170			
GB 9439(88)	HT 200	as cast; test bar 30mm diam; 10<t<=20mm										195			

USA

| | UNS F12102 | | | | | | | | | | | 214 Min | | | |

Cast Iron, Gray, C

USA

| | UNS F12802 | | | | | | | | | | | 283 Min | | | |

Cast Iron, Gray, G1800

Australia

| AS 1830(86) | H187 | | | | | | | | | | | | | | 143-187 HB |

Germany

DIN 1691	GG150HB	20<t<=40mm													160 max HB
DIN 1691	GG150HB	10<t<=20mm													170 max HB
DIN 1691	GG150HB	5<t<=10mm													185 max HB
DIN 1691	GG150HB	40<t<=80mm													150 max HB
DIN 1691	GG150HB	2.5<t<=5mm													210 max HB

Italy

| UNI 3330 | Gh130 | sand | | | | | | | | | | | | | 130-180 HB |

Specification	Designation	Notes	C	Cr	Mn	Mo	Ni	P	S	Si	Other	UTS	YS	El	Hard

Cast Iron, Gray, G1800 (Continued from previous page)

Japan

| JIS G5501(95) | FC100 | SCS | | | | | | | | | | | 100 | | 201 HB |

Norway

| NS 722 | H145 | | | | | | | | | | | | | | 170 HB max |
| NS 722 | SjG10 | | | | | | | | | | | | | | 180 HB max |

Sweden

| SS 140212 | 212 | | | | | | | | | | | | | | 170 B max |

USA

	UNS F10004		3.40-3.70		0.50-0.80			0.25 max	0.15 max	2.30-2.80	Low C with high Si & vice versa				187 max HB
ASTM A159(93)	G1800	Auto	3.40-3.70		0.50-0.80			0.25 max	0.15 max	2.30-2.80	bal Fe		137		187 HB max
SAE J431(96)	G1800	sand mold auto, Ferr-Prl	3.30-3.60		0.50-0.80			0.25 max	0.15 max	2.10-2.50	All chem comp typical	123			120-187 HB
SAE J431(96)	G1800h	sand mold auto, Ferr-Prl	3.40 max		0.50-0.80			0.20 max	0.14 max	2.30-2.80	All chem comp typical	121			163-223 HB

Cast Iron, Gray, G2500

Australia

| AS 1830(86) | H229 | | | | | | | | | | | | | | 170-229 HB |

Germany

DIN 1691	GG170HB	40<t<=80mm													100-170 HB
DIN 1691	GG170HB	5<t<=10mm													140-225 HB
DIN 1691	GG170HB	20<t<=40mm													110-185 HB
DIN 1691	GG170HB	2.5<t<=5mm													170-260 HB
DIN 1691	GG170HB	10<t<=20mm													125-205 HB

Italy

| UNI 5330 | Gh170 | sand | | | | | | | | | | | | | 170-220 HB |

Japan

| JIS G5501(95) | FC150 | SCS | | | | | | | | | | | 150 | | 212 HB |

Norway

| NS 722 | H175 | | | | | | | | | | | | | | 150-200 HB |
| NS 722 | H195 | | | | | | | | | | | | | | 170-220 HB |

Sweden

| SS 140215 | 215 | | | | | | | | | | | | | | 150-200 HB |
| SS 140217 | 217 | | | | | | | | | | | | | | 170-220 HB |

USA

	UNS F10005		3.20-3.50		0.60-0.90			0.20 max	0.15 max	2.00-2.40	Low C with high Si & vice versa				170-229 HB
ASTM A159(93)	G2500	Auto	3.20-3.50		0.60-0.90			0.20 max	0.15 max	2.00-2.40	bal Fe		172		170-229 HB
ASTM A159(93)	G2500a	Auto brake/clutch	3.40 min		0.60-0.90			0.15 max	0.15 max	1.60-2.10	bal Fe		172		170-229 HB
SAE J431(96)	G2500	sand mold auto, Prl-Ferr	3.30-3.60		0.60-0.90			0.20 max	0.15 max	2.10-2.50	All chem comp typical	169			170-229 HB
SAE J431(96)	G2500a	sand mold auto, Prl-Ferr	3.40 max		0.60-0.90			0.15 max	0.12 max	2.30-2.80	All chem comp typical	169			170-229 HB

Cast Iron, Gray, G2500a

USA

| | UNS F10009 | | 3.40 min | | 0.60-0.90 | | | 0.15 max | 0.12 max | 1.60-2.10 | bal Fe | | | | 170-229 HB |

UNS numbers and US grades are provided as a means of cross referencing chemically similar alloys. Exchangability is only possible after independent examination of specifications. Tensile properties are minimum or typical as specified. UTS and YS as MPa. El as %. See Appendix for list of abbreviations used in Notes. * indicates obsolete material.

Specification	Designation	Notes	C	Cr	Mn	Mo	Ni	P	S	Si	Other	UTS	YS	El	Hard

Cast Iron, Gray, G3000

Australia

Specification	Designation	Notes	C	Cr	Mn	Mo	Ni	P	S	Si	Other	UTS	YS	El	Hard
AS 1830(86)	H241														187-241 HB

Germany

Specification	Designation	Notes	C	Cr	Mn	Mo	Ni	P	S	Si	Other	UTS	YS	El	Hard
DIN 1691	GG190HB	10<t<=20mm													150-230 HB
DIN 1691	GG190HB	20<t<=40mm													135-210 HB
DIN 1691	GG190HB	5<t<=10mm													170-260 HB
DIN 1691	GG190HB	40<t<=80mm													120-190 HB
DIN 1691	GG190HB	4<t<=5mm													190-275 HB
DIN 1691	GG220HB	5<t<=10mm													200-275 HB
DIN 1691	GG220HB	20<t<=40mm													160-235 HB
DIN 1691	GG220HB	10<t<=20mm													180-250 HB
DIN 1691	GG220HB	40<t<=80mm													145-220 HB

Italy

Specification	Designation	Notes	C	Cr	Mn	Mo	Ni	P	S	Si	Other	UTS	YS	El	Hard
UNI 5330	Gh190	sand													190-240 HB
UNI 5330	Gh210	sand													220-260 HB

Japan

Specification	Designation	Notes	C	Cr	Mn	Mo	Ni	P	S	Si	Other	UTS	YS	El	Hard
JIS G5501(95)	FC250	SCS											250			241 HB
JIS G5501(95)	FC300	SCS											300			262 HB

Norway

Specification	Designation	Notes	C	Cr	Mn	Mo	Ni	P	S	Si	Other	UTS	YS	El	Hard
NS 722	H215	Cast-on														190-240 HB
NS 722	H235															210-260 HB

Sweden

Specification	Designation	Notes	C	Cr	Mn	Mo	Ni	P	S	Si	Other	UTS	YS	El	Hard
SS 140219	219															190-240 HB
SS 140221	221															210-260 HB

USA

Specification	Designation	Notes	C	Cr	Mn	Mo	Ni	P	S	Si	Other	UTS	YS	El	Hard
	UNS F10006		3.10-3.40		0.60-0.90			0.15 max	0.15 max	1.90-2.30	Low C with high Si & vice versa				187-241 HB
ASTM A159(93)	G3000	Auto	3.10-3.40		0.60-0.90			0.15 max	0.15 max	1.90-2.30	bal Fe	206			187-241 HB
ASTM A159(93)	G3500	Auto	3.00-3.30		0.60-0.90			0.12 max	0.15 max	1.80-3.30	bal Fe	240			207-255 HB
ASTM A159(93)	G3500b	Auto brake/clutch	3.40 min		0.60-0.90			0.15 max	0.15 max	1.30-1.80	bal Fe	240			207-255 HB
ASTM A159(93)	G3500c	Auto brake/clutch	3.50 min		0.60-0.90			0.15 max	0.15 max	1.30-1.80	bal Fe	240			207-255 HB
SAE J431(96)	G3000	sand mold auto, Prl	3.20-3.50		0.60-0.90			0.15 max	0.15 max	1.90-2.30	All chem comp typical 204				187-241 HB

Cast Iron, Gray, G3500

USA

Specification	Designation	Notes	C	Cr	Mn	Mo	Ni	P	S	Si	Other	UTS	YS	El	Hard
	UNS F10007		3.00-3.30		0.60-0.90			0.12 max	0.15 max	1.80-2.20	Low C with high Si & vice versa				207-255 HB
SAE J431(96)	G3500	sand mold auto, Prl	3.10-3.45		0.60-0.90			0.12 max	0.15 max	1.80-2.20	All chem comp typical 244				207-255 HB
SAE J431(96)	G3500b	sand mold auto, lammellar Prl	3.10-3.45		0.60-0.90			0.15 max	0.12 max	1.30-1.80	All chem comp typical 223				207-255 HB
SAE J431(96)	G3500c	sand mold auto, lammellar Prl	3.10-3.45		0.60-0.90			0.15 max	0.12 max	1.30-1.80	All chem comp typical 212				207-255 HB

Specification	Designation	Notes	C	Cr	Mn	Mo	Ni	P	S	Si	Other	UTS	YS	El	Hard

Cast Iron, Gray, G3500b

USA

Specification	Designation	Notes	C	Cr	Mn	Mo	Ni	P	S	Si	Other	UTS	YS	El	Hard
	UNS F10010		3.40 min		0.60-0.90			0.15 max	0.12 max	1.30-1.80	bal Fe				207-255 HB

Cast Iron, Gray, G3500c

USA

Specification	Designation	Notes	C	Cr	Mn	Mo	Ni	P	S	Si	Other	UTS	YS	El	Hard
	UNS F10011		3.50 min		0.60-0.90			0.15 max	0.12 max	1.30-1.80	bal Fe				207-255 HB

Cast Iron, Gray, G4000

Australia

Specification	Designation	Notes	C	Cr	Mn	Mo	Ni	P	S	Si	Other	UTS	YS	El	Hard
AS 1830(86)	H269														217-269 HB

Germany

Specification	Designation	Notes	C	Cr	Mn	Mo	Ni	P	S	Si	Other	UTS	YS	El	Hard
DIN 1691	GG220HB	5<t<=10mm													200-275 HB
DIN 1691	GG220HB	20<t<=40mm													160-235 HB
DIN 1691	GG220HB	40<t<=80mm													145-220 HB
DIN 1691	GG220HB	10<t<=20mm													180-250 HB
DIN 1691	GG240HB	20<t<=40mm													180-255 HB
DIN 1691	GG240HB	10<t<=20mm													200-275 HB
DIN 1691	GG240HB	40<t<=80mm													165-240 HB

Italy

Specification	Designation	Notes	C	Cr	Mn	Mo	Ni	P	S	Si	Other	UTS	YS	El	Hard
UNI 5330	Gh230	sand													230-280 HB

Japan

Specification	Designation	Notes	C	Cr	Mn	Mo	Ni	P	S	Si	Other	UTS	YS	El	Hard
JIS G5501(95)	FC350	SCS											350			277 HB

Norway

Specification	Designation	Notes	C	Cr	Mn	Mo	Ni	P	S	Si	Other	UTS	YS	El	Hard
NS 722	H255														230-260 HB

Sweden

Specification	Designation	Notes	C	Cr	Mn	Mo	Ni	P	S	Si	Other	UTS	YS	El	Hard
SS 140223	223														230-280 HB

USA

Specification	Designation	Notes	C	Cr	Mn	Mo	Ni	P	S	Si	Other	UTS	YS	El	Hard
	UNS F10008		3.00-3.30		0.70-1.00			0.10 max	0.15 max	1.80-2.10	Low C with high Si & vice versa				217-269 HB
ASTM A159(93)	G4000	Auto	3.00-3.30		0.70-1.00			0.10 max	0.15 max	3.00-3.30	bal Fe	275			217-269 HB
SAE J125(88)	G4000	Currently under revision	3.27	0.08 max	0.72 max	0.07 max	0.15 max	0.26 max	0.156 max	1.74 max	bal Fe				
SAE J431(96)	G4000	sand mold auto, Prl	3.10-3.45		0.70-1.00			0.08 max	0.15 max	1.80-2.10	All chem comp typical 273				217-269 HB

Cast Iron, Gray, G4000d

USA

Specification	Designation	Notes	C	Cr	Mn	Mo	Ni	P	S	Si	Other	UTS	YS	El	Hard
	UNS F10012		3.10-3.60	0.85-1.25	0.60-0.90	0.40-0.60	0.20-0.45	0.10 max	0.15 max	1.95-2.40	Ni optional; bal Fe				241-321 HB
SAE J431(96)	G4000D	sand mold auto, lammellar Prl, camshaft	3.10-3.60	0.85-1.50	0.60-0.90		0.20-0.45	0.10 max	0.15 max	1.95-2.40	All chem comp typical 290				241-321 HB

Cast Iron, Gray, No equivalents identified

USA

Specification	Designation	Notes	C	Cr	Mn	Mo	Ni	P	S	Si	Other	UTS	YS	El	Hard
ASTM A823(95)	A-CA	Statically cast permanent mold, Cored, Ann										207			143-207 HB
ASTM A823(95)	A-CB	Statically cast permanent mold, Cored, Ann										172			143-207 HB

UNS numbers and US grades are provided as a means of cross referencing chemically similar alloys. Exchangability is only possible after independent examination of specifications. Tensile properties are minimum or typical as specified. UTS and YS as MPa. El as %. See Appendix for list of abbreviations used in Notes. * indicates obsolete material.

Specification	Designation	Notes	C	Cr	Mn	Mo	Ni	P	S	Si	Other	UTS	YS	El	Hard

Cast Iron, Gray, No equivalents identified

USA

Specification	Designation	Notes	C	Cr	Mn	Mo	Ni	P	S	Si	Other	UTS	YS	El	Hard
ASTM A823(95)	A-CC	Statically cast permanent mold, Cored, Ann										138			143-207 HB
ASTM A823(95)	A-SA	Statically cast permanent mold, Uncored, Ann										207			163-207 HB
ASTM A823(95)	A-SB	Statically cast permanent mold, Uncored, Ann										172			163-207 HB
ASTM A823(95)	A-SC	Statically cast permanent mold, Uncored, Ann										138			163-207 HB
ASTM A823(95)	A-SS	Statically cast permanent mold, Uncored, Ann										124			143-207 HB
ASTM A823(95)	N-CA	Statically cast permanent mold, Cored, Norm										207			170-229 HB
ASTM A823(95)	N-CB	Statically cast permanent mold, Cored, Norm										172			170-229 HB
ASTM A823(95)	N-CC	Statically cast permanent mold, Cored, Norm										138			170-229 HB
ASTM A823(95)	N-SA	Statically cast permanent mold, Uncored, Norm										207			170-229 HB
ASTM A823(95)	N-SB	Statically cast permanent mold, Uncored, Norm										172			170-229 HB
ASTM A823(95)	N-SC	Statically cast permanent mold, Uncored, Norm										138			170-229 HB
ASTM A823(95)	N-SS	Statically cast permanent mold, Uncored, Norm										124			149-229 HB
ASTM A888(98)	A888	Hubless soil pipe/fittings										145			
FED QQ-I-652C(70)	- - -*	Obs; see ASTM A48													
FED QQ-I-653A(69)	- - -*	Obs; Automotive													
SAE J1236(95)		Sealing rings (Metric)	3.50-3.95		0.40-0.80			0.30-0.80	0.13 max	2.20-3.10	bal Fe				95-107 HRB
SAE J125(88)	G4500	Currently under revision	2.84	0.31 max	1.05 max	0.20 max		0.07 max	0.124 max	1.52 max	bal Fe				

Specification	Designation	Notes	C	Cr	Mn	Mo	Ni	P	S	Si	Other	UTS	YS	El	Hard

Cast Iron, Malleable, 32510

China

Specification	Designation	Notes	UTS	YS	El	Hard
GB 9440(88)	HT 350-10	Ferr 12 or 15mm test bar diam	350	200	10	150 max HB
GB 9440(88)	HT 370-12	Ferr 12 or 15mm test bar diam	370		12	150 max HB

France

Specification	Designation	Notes	UTS	YS	El	Hard
AFNOR NFA32702	MN350-10	Ferritic Prl, 15mm test bar diam				150 HB max
AFNOR NFA32702	MN380-18	Ferritic Prl, 15mm test bar diam				156 HB max

Russia

Specification	Designation	Notes	UTS	YS	El	Hard
GOST 1215	KCh30-6	8, 12 or 16 mm test bar diam				100-163 HB
GOST 1215	KCh33-8	8, 12 or 16 mm test bar diam				100-163 HB
GOST 1215	KCh35-10	8, 12 or 16 mm test bar diam				100-163 HB

USA

Specification	Designation	Notes	UTS	YS	El	Hard
	UNS F22200		345 Min	224 Min		
ASTM A47(95)	32510	Ferr, 5/8 in. (5.9mm) est diam, metric units A47M	345	224	10	

Cast Iron, Malleable, 35018

China

Specification	Designation	Notes	UTS	YS	El	Hard
GB 9440(88)	HT 350-10	Ferr 12 or 15mm test bar diam	350	200	10	150 max HB
GB 9440(88)	HT 370-12	Ferr 12 or 15mm test bar diam	370		12	150 max HB

Russia

Specification	Designation	Notes	UTS	YS	El	Hard
GOST 1215	KCh37-12	8, 12 or 16 mm test bar diam				110-163 HB

USA

Specification	Designation	Notes	UTS	YS	El	Hard
	UNS F22400		365 Min	241 Min		
ASTM A47(95)	35018	Ferr, 5/8 in. (5.9mm) est diam, metric units A47M	365	241	18	

Cast Iron, Malleable, 40010

Sweden

Specification	Designation	Notes	UTS	YS	El	Hard
SIS 140852	0852(400-7)	Prl	400	240	7	137-179 HB

USA

Specification	Designation	Notes	UTS	YS	El	Hard
	UNS F22830	Prl	414 Min	276 Min		
ASTM A220(93)	40010	Prl, 16mm test bar diam	400	280	10	149-197 HB
ASTM A220M(93)	280M10	Prl, 16mm test bar diam	400	280	10	149-197 HB

Cast Iron, Malleable, 45006/45008 (Continued from previous page)

China

Specification	Designation	Notes	UTS	YS	El	Hard
GB 9440(88)	KTZ 450-06	Prl, 12 or 15mm test bar diam	450	270	6	
GB 9440(88)	KTZ 450-06	Prl, 12 or 15mm test bar diam	450	270	6	

France

Specification	Designation	Notes	UTS	YS	El	Hard
AFNOR NFA32702	MN450-6	Ferritic Prl, 15mm test bar diam				150-210 HB

India

Specification	Designation	Notes	UTS	YS	El	Hard
IS 2640	PM440	Prl, 15mm test bar diam				149-201 HB

International

Specification	Designation	Notes	UTS	YS	El	Hard
ISO 5922(81)	P45-06	Prl, 12 or 15mm test bar diam				150-200 HB

Specification	Designation	Notes	C	Cr	Mn	Mo	Ni	P	S	Si	Other	UTS	YS	El	Hard

Cast Iron, Malleable, 45006/45008 (Continued from previous page)

International

| ISO 5922(81) | P45-06 | Prl, 12 or 15mm test bar diam | | | | | | | | | | | | | 150-200 HB |

Japan

JIS G5704(88)	FCMP440	Prl										440	265	6	149-207 HB
JIS G5704(88)	FCMP440	Prl										440	265	6	149-207 HB
JIS G5704(88)	FCMP45*	Obs; See JIS FCMP 440; Prl Class 1													149-207 HB
JIS G5704(88)	FCMP45*	Obs; See JIS FCMP 440; Prl Class 1													149-207 HB

Netherlands

| NEN 6002-C | GSMp45 | Prl | | | | | | | | | | | | | 150-200 HB |
| NEN 6002-C | GSMp45 | Prl | | | | | | | | | | | | | 150-200 HB |

Russia

| GOST 1215 | KCh45-7 | 8, 12 or 16 mm test bar diam | | | | | | | | | | | | | 150-207 HB |

UK

| BS 6681 | P45-06 | Prl, 12 or 15mm test bar diam | | | | | | | | | | 450 | 270 | 6 | 150-200 HB |

USA

	UNS F23130	Prl										448 Min	310 Min	8 Min	
	UNS F23131	Prl										448 Min	310 Min	6 Min	
ASTM A220(93)	45006	Prl, 16mm test bar diam										450	310	6	156-207 HB
ASTM A220(93)	45008	Prl, 16mm test bar diam										450	310	8	156-197 HB
ASTM A220M(93)	310M6	Prl, 16mm test bar diam										450	310	6	156-207 HB
ASTM A220M(93)	310M8	Prl, 16mm test bar diam										450	310	8	156-197 HB

Cast Iron, Malleable, 50005

India

| IS 2640 | PM490 | Prl, 15mm test bar diam | | | | | | | | | | | | | 170-229 HB |

Japan

| JIS G5704(88) | FCMP490 | Prl | | | | | | | | | | 490 | 305 | 4 | 167-229 HB |
| JIS G5704(88) | FCMP50* | Obs; See JIS FCMP 490; Prl Class 2 | | | | | | | | | | | | | 167-229 HB |

Netherlands

| NEN 6002-C | GSmP50 | Prl | | | | | | | | | | | | | 170-230 HB |

Russia

| GOST 1215 | KCh50-5 | 8, 12 or 16 mm test bar diam | | | | | | | | | | | | | 170-230 HB |

UK

| BS 6681 | P50-05 | Prl, 12 or 15mm test bar diam | | | | | | | | | | 500 | 300 | 5 | 160-220 HB |

USA

	UNS F23530	Prl										483 Min	345 Min		
ASTM A220(93)	50005	Prl, 16mm test bar diam										480	340	5	179-229 HB
ASTM A220M(93)	340M5	Prl, 16mm test bar diam										480	340	5	179-229 HB

Cast Iron, Malleable, 5310

International

| ISO 5922(81) | P50-05 | Prl, 12 or 15mm test bar diam | | | | | | | | | | | | | 100-270 HB |

UNS numbers and US grades are provided as a means of cross referencing chemically similar alloys. Exchangability is only possible after independent examination of specifications. Tensile properties are minimum or typical as specified. UTS and YS as MPa. El as %. See Appendix for list of abbreviations used in Notes. * indicates obsolete material.

Specification	Designation	Notes	C	Cr	Mn	Mo	Ni	P	S	Si	Other	UTS	YS	El	Hard

Cast Iron, Malleable, 5310 (Continued from previous page)

Sweden

| SIS 140854 | 0854(500-05) | Prl | | | | | | | | | | 500 | 300 | 5 | 170-217 HB |

USA

| | UNS F23330 | Prl | | | | | | | | | | 483 Min | 331 Min | | |
| AMS 5310F(94)* | | Noncurrent; Prl, no chem compound specified | | | | | | | | | | 360 | 250 | 4 | 163-229 HB |

Cast Iron, Malleable, 60004

China

| GB 9440(88) | KTZ 550-04 | Prl, 12 or 15mm test bar diam | | | | | | | | | | 550 | 340 | 4 | 160-230 HB |

France

| AFNOR NFA32702 | MN550-4 | Ferritic Prl, 15mm test bar diam | | | | | | | | | | | | | 180-240 HB |

India

| IS 2640 | PM540 | Prl, 15mm test bar diam | | | | | | | | | | | | | 180-230 HB |

International

| ISO 5922(81) | P55-04 | Prl, 12 or 15mm test bar diam | | | | | | | | | | | | | 180-230 HB |

Japan

| JIS G5704(88) | FCMP540 | Prl | | | | | | | | | | 540 | 345 | 3 | 183-241 HB |
| JIS G5704(88) | FCMP55* | Obs; See JIS FCMP 540; Prl Class 3 | | | | | | | | | | | | | 183-241 HB |

Netherlands

| NEN 6002-C | GSmP55 | Prl | | | | | | | | | | | | | 180-240 HB |

Russia

| GOST 1215 | KCh55-4 | 8, 12 or 16 mm test bar diam | | | | | | | | | | | | | 192-241 HB |

UK

| BS 6681 | P55-04 | Prl, 12 or 15mm test bar diam | | | | | | | | | | 550 | 340 | 4 | 180-230 HB |

USA

	UNS F24130	Prl										552 Min	414 Min		
ASTM A220(93)	60004	Prl, 16mm test bar diam										550	410	4	197-241 HB
ASTM A220M(93)	410M4	Prl, 16mm test bar diam										550	410	4	197-241 HB

Cast Iron, Malleable, 70003

India

| IS 2640 | PM590 | Prl, 15mm test bar diam | | | | | | | | | | | | | 200-248 HB |

International

| ISO 5922(81) | P60-03 | Prl, 12 or 15mm test bar diam | | | | | | | | | | | | | 200-250 HB |

Japan

| JIS G5704(88) | FCMP590 | Prl | | | | | | | | | | 590 | 390 | 3 | 207-269 HB |
| JIS G5704(88) | FCMP60* | Obs; See JIS FCMP 590; Prl Class 4 | | | | | | | | | | | | | 207-269 HB |

Russia

| GOST 1215 | KCh60-3 | 8, 12 or 16 mm test bar diam | | | | | | | | | | | | | 200-269 HB |

Sweden

| SIS 140856 | 0856(600-04) | Prl | | | | | | | | | | 600 | 380 | 4 | 201-248 HB |

UK

| BS 6681 | P60-03 | Prl, 12 or 15mm test bar diam | | | | | | | | | | 600 | 390 | 3 | 200-250 HB |

UNS numbers and US grades are provided as a means of cross referencing chemically similar alloys. Exchangability is only possible after independent examination of specifications. Tensile properties are minimum or typical as specified. UTS and YS as MPa. El as %. See Appendix for list of abbreviations used in Notes. * indicates obsolete material.

Specification	Designation	Notes	C	Cr	Mn	Mo	Ni	P	S	Si	Other	UTS	YS	El	Hard

Cast Iron, Malleable, 70003 (Continued from previous page)

USA

Specification	Designation	Notes	UTS	YS	El	Hard
	UNS F24830	Prl	586 Min	483 Min		
ASTM A220(93)	70003	Prl, 16mm test bar diam	590	480	3	217-269 HB
ASTM A220M(93)	480M3	Prl, 16mm test bar diam	590	480	3	217-269 HB

Cast Iron, Malleable, 80002

China

Specification	Designation	Notes	UTS	YS	El	Hard
GB 9440(88)	KTZ 650-02	Prl, 12 or 15mm test bar diam	650	430	2	210-260 HB
GB 9440(88)	KTZ 700-02	Prl, 12 or 15mm test bar diam	700	530	2	240-290 HB

France

Specification	Designation	Notes	UTS	YS	El	Hard
AFNOR NFA32702	MN650-3	Ferritic Prl, 15mm test bar diam				210-270 HB

India

Specification	Designation	Notes	UTS	YS	El	Hard
IS 2640	PM690	Prl, 15mm test bar diam				241-265 HB

International

Specification	Designation	Notes	UTS	YS	El	Hard
ISO 5922(81)	P65-02	Prl, 12 or 15mm test bar diam				210-260 HB
ISO 5922(81)	P65-02	Prl, 12 or 15mm test bar diam				210-260 HB

UK

Specification	Designation	Notes	UTS	YS	El	Hard
BS 6681	P65-02	Prl, 12 or 15mm test bar diam	650	430	2	210-260 HB

USA

Specification	Designation	Notes	UTS	YS	El	Hard
	UNS F25530	Prl	655 Min	552 Min		
ASTM A220(93)	80002	Prl, 16mm test bar diam	650	550	2	241-285 HB
ASTM A220M(93)	550M2	Prl, 16mm test bar diam	650	550	2	241-285 HB

Cast Iron, Malleable, 90001

China

Specification	Designation	Notes	UTS	YS	El	Hard
GB 9440(88)	KTZ 650-02	Prl, 12 or 15mm test bar diam	650	430	2	210-260 HB
GB 9440(88)	KTZ 700-02	Prl, 12 or 15mm test bar diam	700	530	2	240-290 HB

France

Specification	Designation	Notes	UTS	YS	El	Hard
AFNOR NFA32702	MN700-2	Ferritic Prl, 15mm test bar diam				240-290 HB

International

Specification	Designation	Notes	UTS	YS	El	Hard
ISO 5922(81)	P70-02	Prl, 12 or 15mm test bar diam				240-290 HB
ISO 5922(81)	P70-02	Prl, 12 or 15mm test bar diam				240-290 HB
ISO 5922(81)	P80-01	Prl, 12 or 15mm test bar diam				270-310 HB
ISO 5922(81)	P80-01	Prl, 12 or 15mm test bar diam				270-310 HB

Japan

Specification	Designation	Notes	UTS	YS	El	Hard
JIS G5704(88)	FCMP690	Prl	690	510	2	229-285 HB
JIS G5704(88)	FCMP690	Prl	690	510	2	229-285 HB
JIS G5704(88)	FCMP70*	Obs; See JIS FCMP 690; Prl Class 5				229-285 HB
JIS G5704(88)	FCMP70*	Obs; See JIS FCMP 690; Prl Class 5				229-285 HB

Netherlands

Specification	Designation	Notes	UTS	YS	El	Hard
NEN 6002-C	GSmP60	Prl				210-250 HB

UNS numbers and US grades are provided as a means of cross referencing chemically similar alloys. Exchangability is only possible after independent examination of specifications. Tensile properties are minimum or typical as specified. UTS and YS as MPa. El as %. See Appendix for list of abbreviations used in Notes. * indicates obsolete material.

Specification	Designation	Notes	C	Cr	Mn	Mo	Ni	P	S	Si	Other	UTS	YS	El	Hard

Cast Iron, Malleable, 90001 (Continued from previous page)

Specification	Designation	Notes	UTS	YS	El	Hard
Netherlands						
NEN 6002-C	GSmP60	Prl				210-250 HB
NEN 6002-C	GSmP70	Prl				240-285 HB
NEN 6002-C	GSmP70	Prl				240-285 HB
Russia						
GOST 1215	KCh65-3	8, 12 or 16 mm test bar diam				212-269 HB
GOST 1215	KCh70-2	8, 12 or 16 mm test bar diam				241-285 HB
GOST 1215	KCh80-1.5	8, 12 or 16 mm test bar diam				270-320 HB
Sweden						
SIS 140862	0862(700-03)	Prl	700	530	3	241-285 HB
SIS 140864	0864(800-02)	Prl	800	600	2	269-311 HB
UK						
BS 6681	P70-02	Prl, 12 or 15mm test bar diam	700	530	2	240-290 HB
USA						
	UNS F26230	Prl	724 Min	621 Min		
ASTM A220(93)	90001	Prl, 16mm test bar diam	720	620	1	269-321 HB
ASTM A220M(93)	620M1	Prl, 16mm test bar diam	720	620	1	269-321 HB

Cast Iron, Malleable, A197

Specification	Designation	Notes	UTS	YS	El	Hard
USA						
	UNS F22000	Cupola	276 Min	207 Min		
ASTM A197(92)		Cupola; 15.9mm test bar diam				

Cast Iron, Malleable, Blackheart

Specification	Designation	Notes	UTS	YS	El	Hard
Belgium						
NBN A22-204	MAS35-12	15mm test bar diam				150 HB
NBN A22-204	MAS45-7	15mm test bar diam				150-200 HB
NBN A22-204	MAS55-4	15mm test bar diam				190-240 HB
NBN A22-204	MAS65-3	15mm test bar diam				210-250 HB
NBN A22-204	MAS70-2	15mm test bar diam				240-285 HB
China						
GB 9440(88)	KTH 330-06	Ferr; 12 or 15mm test bar diam	300		6	150 max HB
GB 9440(88)	KTH 330-08	Ferr; 12 or 15mm test bar diam	330		8	150 max HB
Denmark						
DS 11302	814	15mm test bar diam				160 HB
DS 11302	815	15mm test bar diam				160 HB
DS 11302	816	15mm test bar diam				160 HB
Germany						
DIN 1692	GTS-35-10	12 or 15mm test bar diam				150 HB
DIN 1692	GTS-45-06	12 or 15mm test bar diam				150-200 HB
DIN 1692	GTS-55-04	12 or 15mm test bar diam				180-230 HB
DIN 1692	GTS-65-02	12 or 15mm test bar diam				210-260 HB

Specification	Designation	Notes	C	Cr	Mn	Mo	Ni	P	S	Si	Other	UTS	YS	El	Hard

Cast Iron, Malleable, Blackheart (Continued from previous page)

Germany

Specification	Designation	Notes	C	Cr	Mn	Mo	Ni	P	S	Si	Other	UTS	YS	El	Hard
DIN 1692	GTS-70-02	12 or 15mm test bar diam													240-290 HB

India

Specification	Designation	Notes	UTS	YS	El	Hard
IS 2108	BM290	15mm test bar diam				160 HB max
IS 2108	BM310	15mm test bar diam				150 HB max
IS 2108	BM340	15mm test bar diam				150 HB max

International

Specification	Designation	Notes	UTS	YS	El	Hard
ISO 5922(81)	B30-06	12 or 15mm test bar diam				150 HB max
ISO 5922(81)	B32-12	12 or 15mm test bar diam				180 HB max
ISO 5922(81)	B35-10	12 or 15mm test bar diam				180 HB max

Japan

Specification	Designation	Notes	UTS	YS	El	Hard
JIS G5702(88)	FCMB270	Class 1	270	165	5	163 HB max
JIS G5702(88)	FCMB310	Class 2	310	185	8	163 HB max
JIS G5702(88)	FCMB340	Class 3	340	205	10	163 HB max
JIS G5702(88)	FCMB360	Class 4	360	215	14	163 HB max

Netherlands

Specification	Designation	Notes	UTS	YS	El	Hard
NEN 6002-C	GSmT32	15mm test bar diam				160 HB max
NEN 6002-C	GSmT35	15mm test bar diam				150 HB max

Norway

Specification	Designation	Notes	UTS	YS	El	Hard
NS 11501	NS11533					150 HB max
NS 11501	NS11550					160-210 HB
NS 11501	NS11560					190-250 HB

Sweden

Specification	Designation	Notes	UTS	YS	El	Hard
SIS 140814	0814(300-06)		300		6	149 HB max
SIS 140815	0815(300-12)		320	190	12	107-149 HB

UK

Specification	Designation	Notes	UTS	YS	El	Hard
BS 6681	B30-06	12 or 15mm test bar diam	300		6	150 HB max
BS 6681	B32-10	12 or 15mm test bar diam	320	190	10	150 HB max
BS 6681	B35-12	12 or 15mm test bar diam	350	200	12	150 HB max

Cast Iron, Malleable, M3210

USA

Specification	Designation	Notes	C	Cr	Mn	Mo	Ni	P	S	Si	Other	UTS	YS	El	Hard
	UNS F20000		2.20-2.90		0.15-1.25			0.02-0.15	0.02-0.20	0.90-1.90	bal Fe	345 Min	220.5 Min		
ASTM A602(94)	M3210	Auto, Ann	2.20-2.90		0.15-1.25			0.02-0.15	0.02-0.20	0.90-1.90	bal Fe	345	221	10	156 HB max
SAE J158(86)	M3210	Auto, Ann	2.20-2.90		0.15-1.25			0.02-0.10	0.02-0.20	0.90-1.90	bal Fe	345	224	25	156 HB max

Cast Iron, Malleable, M4504

USA

Specification	Designation	Notes	C	Cr	Mn	Mo	Ni	P	S	Si	Other	UTS	YS	El	Hard
	UNS F20001		2.20-2.90		0.15-1.25			0.02-0.15	0.02-0.20	0.90-1.90	bal Fe	447.9 Min	309.7 Min		
ASTM A602(94)	M4504	Auto, air or liquid Q/A	2.20-2.90		0.15-1.25			0.02-0.15	0.02-0.20	0.90-1.90	bal Fe	448	310	4	163-217 HB

UNS numbers and US grades are provided as a means of cross referencing chemically similar alloys. Exchangability is only possible after independent examination of specifications. Tensile properties are minimum or typical as specified. UTS and YS as MPa. El as %. See Appendix for list of abbreviations used in Notes. * indicates obsolete material.

Specification	Designation	Notes	C	Cr	Mn	Mo	Ni	P	S	Si	Other	UTS	YS	El	Hard

Cast Iron, Malleable, M4504 (Continued from previous page)

USA

| SAE J158(86) | M4504 | Auto, air Q/T | 2.20-2.90 | | 0.15-1.25 | | | 0.02-0.10 | 0.02-0.20 | 0.90-1.90 | bal Fe | 448 | 310 | 26 | 163-217 HB |

Cast Iron, Malleable, M5003

USA

	UNS F20002		2.20-2.90		0.15-1.25			0.02-0.15	0.02-0.20	0.90-1.90	bal Fe	516.5 Min	345 Min		
ASTM A602(94)	M5003	Auto, air or liquid Q/A	2.20-2.90		0.15-1.25			0.02-0.15	0.02-0.20	0.90-1.90	bal Fe	517	345	3	187-241 HB
SAE J158(86)	M5003	Auto, air Q/T	2.20-2.90		0.15-1.25			0.02-0.10	0.02-0.20	0.90-1.90	bal Fe	517	345	26	187-241 HB

Cast Iron, Malleable, M5503

USA

	UNS F20003		2.20-2.90		0.15-1.25			0.02-0.15	0.02-0.20	0.90-1.90	bal Fe	516.5 Min	379.3 Min		
ASTM A602(94)	M5503	Auto, liquid Q/A	2.20-2.90		0.15-1.25			0.02-0.15	0.02-0.20	0.90-1.90	bal Fe	517	379	3	187-241 HB
SAE J158(86)	M5503	Auto, liquid Q/T	2.20-2.90		0.15-1.25			0.02-0.10	0.02-0.20	0.90-1.90	bal Fe	517	379	26	187-241 HB

Cast Iron, Malleable, M7002

USA

	UNS F20004		2.20-2.90		0.15-1.25			0.02-0.15	0.02-0.20	0.90-1.90	bal Fe	620.3 Min	482.2 Min		
ASTM A602(94)	M7002	Auto, liquid Q/T	2.20-2.90		0.15-1.25			0.02-0.15	0.02-0.20	0.90-1.90	bal Fe	621	483	2	229-269 HB
SAE J158(86)	M7002	Auto, liquid Q/T	2.20-2.90		0.15-1.25			0.02-0.10	0.02-0.20	0.90-1.90	bal Fe	621	483	26	229-269 HB

Cast Iron, Malleable, M8501

USA

	UNS F20005		2.20-2.90		0.15-1.25			0.02-0.15	0.02-0.20	0.90-1.90	bal Fe	723.2 Min	586.0 Min		
ASTM A602(94)	M8501	Auto, liquid Q/T	2.20-2.90		0.15-1.25			0.02-0.15	0.02-0.20	0.90-1.90	bal Fe	724	586	1	269-302 HB
SAE J158(86)	M8501	Auto, liquid Q/T	2.20-2.90		0.15-1.25			0.02-0.10	0.02-0.20	0.90-1.90	bal Fe	724	586	26	269-302 HB

Cast Iron, Malleable, No equivalents identified

USA

| FED QQ-I-666D(69) | - - -* | Obs; see ASTM A47; Ferritic | | | | | | | | | | | | | |

Cast Iron, Malleable, Whiteheart

Belgium

NBN A22-204	MES35-10	9mm test bar diam													150 HB
NBN A22-204	MES45-7	9mm test bar diam													150-200 HB
NBN A22-204	MES55-4	9mm test bar diam													190-240 HB
NBN A22-204	MES65-3	9mm test bar diam													210-250 HB
NBN A22-204	MES70-2	9mm test bar diam													240-285 HB

China

GB 9440(88)	KTB 350-04	15mm test bar diam										360		3	
GB 9440(88)	KTB 350-04	12mm test bar diam										350		4	230 max HB
GB 9440(88)	KTB 350-04	9mm test bar diam										340		5	
GB 9440(88)	KTB 380-12	15mm test bar diam										400	210	8	
GB 9440(88)	KTB 380-12	12mm test bar diam										380	200	12	200 max HB
GB 9440(88)	KTB 380-12	9mm test bar diam										320	170	15	

UNS numbers and US grades are provided as a means of cross referencing chemically similar alloys. Exchangability is only possible after independent examination of specifications. Tensile properties are minimum or typical as specified. UTS and YS as MPa. El as %. See Appendix for list of abbreviations used in Notes. * indicates obsolete material.

Specification	Designation	Notes	C	Cr	Mn	Mo	Ni	P	S	Si	Other	UTS	YS	El	Hard

Cast Iron, Malleable, Whiteheart (Continued from previous page)

China

Specification	Designation	Notes	C	Cr	Mn	Mo	Ni	P	S	Si	Other	UTS	YS	El	Hard
GB 9440(88)	KTB 400-05	12mm test bar diam										400	220	5	220 max HB
GB 9440(88)	KTB 400-05	15mm test bar diam										420	230	4	
GB 9440(88)	KTB 400-05	9mm test bar diam										360	200	8	
GB 9440(88)	KTB 450-07	9mm test bar diam										400	230	10	
GB 9440(88)	KTB 450-07	15mm test bar diam										480	280	4	
GB 9440(88)	KTB 450-07	12mm test bar diam										450	260	7	220 max HB

France

Specification	Designation	Notes	C	Cr	Mn	Mo	Ni	P	S	Si	Other	UTS	YS	El	Hard
AFNOR NFA32701	MB380-12	12mm test bar diam													200 HB max
AFNOR NFA32701	MB380-12	15mm test bar diam													200 HB max
AFNOR NFA32701	MB380-12	9mm test bar diam													200 HB max
AFNOR NFA32701	MB400-5	9mm test bar diam													200 HB max
AFNOR NFA32701	MB400-5	15mm test bar diam													200 HB max
AFNOR NFA32701	MB400-5	12mm test bar diam													200 HB max
AFNOR NFA32701	MB450-7	9mm test bar diam													200 HB max
AFNOR NFA32701	MB450-7	12mm test bar diam													200 HB max
AFNOR NFA32701	MB450-7	15mm test bar diam													200 HB max

Germany

Specification	Designation	Notes	C	Cr	Mn	Mo	Ni	P	S	Si	Other	UTS	YS	El	Hard
DIN 1692	GTW-35-04	12mm test bar diam													230 HB
DIN 1692	GTW-35-04	9mm test bar diam													230 HB
DIN 1692	GTW-35-04	15mm test bar diam													230 HB
DIN 1692	GTW-40-5	12mm test bar diam													220 HB
DIN 1692	GTW-40-5	15mm test bar diam													220 HB
DIN 1692	GTW-40-5	9mm test bar diam													220 HB
DIN 1692	GTW-45-07	12mm test bar diam													220 HB
DIN 1692	GTW-45-07	9mm test bar diam													220 HB
DIN 1692	GTW-45-07	15mm test bar diam													220 HB
DIN 1692	GTW-538-12	Weld; 9mm test bar diam													200 HB
DIN 1692	GTW-538-12	Weld; 12mm test bar diam													200 HB
DIN 1692	GTW-538-12	Weld; 15mm test bar diam													200 HB

India

Specification	Designation	Notes	C	Cr	Mn	Mo	Ni	P	S	Si	Other	UTS	YS	El	Hard
IS 2107	WM340	15mm test bar diam													220 HB max
IS 2107	WM340	12mm test bar diam													220 HB max
IS 2107	WM340	9mm test bar diam													220 HB max
IS 2107	WM410	12mm test bar diam													220 HB max
IS 2107	WM410	15mm test bar diam													220 HB max
IS 2107	WM410	9mm test bar diam													220 HB max

International

Specification	Designation	Notes	C	Cr	Mn	Mo	Ni	P	S	Si	Other	UTS	YS	El	Hard
ISO 5922(81)	W35-04	9mm test bar diam													
ISO 5922(81)	W35-04	15mm test bar diam													
ISO 5922(81)	W35-04	12mm test bar diam													230 HB max
ISO 5922(81)	W38-12	15mm test bar diam													
ISO 5922(81)	W38-12	12mm test bar diam													200 HB max

UNS numbers and US grades are provided as a means of cross referencing chemically similar alloys. Exchangability is only possible after independent examination of specifications. Tensile properties are minimum or typical as specified. UTS and YS as MPa. El as %. See Appendix for list of abbreviations used in Notes. * indicates obsolete material.

Specification	Designation	Notes	C	Cr	Mn	Mo	Ni	P	S	Si	Other	UTS	YS	El	Hard

Cast Iron, Malleable, Whiteheart (Continued from previous page)

International

Specification	Designation	Notes	C	Cr	Mn	Mo	Ni	P	S	Si	Other	UTS	YS	El	Hard
ISO 5922(81)	W38-12	9mm test bar diam													
ISO 5922(81)	W40-05	15mm test bar diam													
ISO 5922(81)	W40-05	9mm test bar diam													
ISO 5922(81)	W40-05	12mm test bar diam													200 HB max
ISO 5922(81)	W45-07	9mm test bar diam													
ISO 5922(81)	W45-07	12mm test bar diam													150 HB max
ISO 5922(81)	W45-07	15mm test bar diam													

Japan

Specification	Designation	Notes	C	Cr	Mn	Mo	Ni	P	S	Si	Other	UTS	YS	El	Hard
JIS G5702(88)	FCMB28*	Obs; See JIS FCMB 270; Class 1													163 max HB
JIS G5702(88)	FCMB32*	Obs; See JIS FCMB 310; Class 2													163 max HB
JIS G5702(88)	FCMB35*	Obs; See JIS FCMB 340; Class 3													163 max HB
JIS G5702(88)	FCMB37*	Obs; See JIS FCMB 360; Class 4													163 max HB
JIS G5702(88)	FCMW34	6mm test bar diam													207 max HB
JIS G5702(88)	FCMW34	10mm test bar diam													207 max HB
JIS G5702(88)	FCMW34	14mm test bar diam													207 max HB
JIS G5702(88)	FCMW38	6mm test bar diam													192 max HB
JIS G5702(88)	FCMW38	14mm test bar diam													192 max HB
JIS G5702(88)	FCMW38	10mm test bar diam													192 max HB
JIS G5702(88)	FCMWP45	14mm test bar diam													149-207 HB
JIS G5702(88)	FCMWP50	14mm test bar diam													167-229 HB
JIS G5702(88)	FCMWP55	14mm test bar diam													183-241 HB

Netherlands

Specification	Designation	Notes	C	Cr	Mn	Mo	Ni	P	S	Si	Other	UTS	YS	El	Hard
NEN 6002-C	GSmF	12mm test bar diam													220 HB max
NEN 6002-C	GSmF	15mm test bar diam													220 HB max
NEN 6002-C	GSmF	9mm test bar diam													220 HB max

UK

Specification	Designation	Notes	C	Cr	Mn	Mo	Ni	P	S	Si	Other	UTS	YS	El	Hard
BS 6681	W35-04	9mm test bar diam											340		5	230 HB max
BS 6681	W35-04	12mm test bar diam											350		4	230 HB max
BS 6681	W35-04	15mm test bar diam											360		3	230 HB max
BS 6681	W38-12	Weld; 12mm test bar diam											380	200	12	200 HB max
BS 6681	W38-12	Weld; 9mm test bar diam											320	170	15	200 HB max
BS 6681	W38-12	Weld; 15mm test bar diam											320	170	15	200 HB max
BS 6681	W40-05	9mm test bar diam											360	200	8	220 HB max
BS 6681	W40-05	12mm test bar diam											400	220	5	220 HB max
BS 6681	W40-05	15mm test bar diam											420	230	4	220 HB max
BS 6681	W45-07	12mm test bar diam											450	260	7	220 HB max
BS 6681	W45-07	15mm test bar diam											480	280	4	220 HB max

UNS numbers and US grades are provided as a means of cross referencing chemically similar alloys. Exchangability is only possible after independent examination of specifications. Tensile properties are minimum or typical as specified. UTS and YS as MPa. El as %. See Appendix for list of abbreviations used in Notes. * indicates obsolete material.

Specification	Designation	Notes	C	Cr	Mn	Mo	Ni	P	S	Si	Other	UTS	YS	El	Hard

Cast Iron, Malleable, Whiteheart (Continued from previous page)

Specification	Designation	Notes	C	Cr	Mn	Mo	Ni	P	S	Si	Other	UTS	YS	El	Hard
UK															
BS 6681	W45-07	9mm test bar diam										400	230	10	220 HB max

Specification	Designation	Notes	C	Cr	Cu	Mn	Ni	P	S	Si	Other	UTS	YS	El	Hard

Cast Iron, Austenitic with Graphite, 5

France

Specification	Designation	Notes	C	Cr	Cu	Mn	Ni	P	S	Si	Other	UTS	YS	El	Hard
AFNOR NFA32301	LN35	As ISO 2892													

International

| ISO 2892(73) | L-Ni35 | Flake graphite | 2.4 max | 0.2 max | 0.5 max | 0.2-1.5 | 34.0-36.0 | | | 1.0-2.0 | bal Fe | 120 | | | |

UK

| BS 3468 | L-Ni35 | As ISO 2892 | 2.4 | 0.2 max | 0.5 max | 0.2-1.5 | 34-36 | | | 1-2 | bal Fe | | | | |

USA

| | UNS F41006 | | 2.40 max | 0.10 max | 0.50 max | 0.5-1.5 | 34.00-36.00 | | 0.12 max | 1.00-2.00 | bal Fe | 138 Min | | | |
| ASTM A436(97) | 5 | Gray with flake graphite | 2.40 max | 0.10 max | 0.50 max | 0.5-1.5 | 34.00-36.00 | | 0.12 max | 1.00-2.00 | bal Fe | 138 | | | 99-124 HB |

Cast Iron, Austenitic with Graphite, 6

USA

Specification	Designation	Notes	C	Cr	Cu	Mn	Ni	P	S	Si	Other	UTS	YS	El	Hard
	UNS F41007		3.00 max	1.00-2.00	3.50-5.50	0.5-1.5	18.00-22.00		0.12 max	1.50-2.50	Mo <=1.00; bal Fe	172 Min			
ASTM A436(97)	6	Gray with flake graphite	3.00 max	1.00-2.00	3.50-5.50	0.5-1.5	18.00-22.00		0.12 max	1.50-2.50	Mo <=1.00; bal Fe	172			124-174 HB

Cast Iron, Austenitic with Graphite, D2

France

Specification	Designation	Notes	C	Cr	Cu	Mn	Ni	P	S	Si	Other	UTS	YS	El	Hard
AFNOR NFA32301	SNC202	As ISO 2892													

Germany

DIN	WNr 0.7660														
DIN 1694	GGG-NiCr202														
DIN 1694	GGG-NiCrNb202	Spheroidal graphite	3 max	2.5-3.5		0.5-1.5	18-22	0.08 max		1.5-2.4	Mg <=0.8; Nb 0.1-0.2; bal Fe				

International

| ISO 2892(73) | S-NiCr202 | Spheroidal graphite | 3.0 max | 1.0-2.5 | 0.5 max | 0.5-1.5 | 18.0-22.0 | 0.080 max | | 1.5-3.0 | bal Fe | 370 | 210 | 7 | |

Sweden

| SIS 140776 | 0776-03 | Spheroidal graphite | 3 max | 1-2.5 | 0.5 max | 0.7-1.25 | 18-22 | 0.08 max | 0.02-0.08 | 1.5-3 | Mg 0.02; bal Fe | 380 | 210 | 7 | 140-200 HB |

UK

BS 3468	S-2	Spheroidal	3 max	1.5-2.5	0.5 max	0.5-1.5	18-22	0.08 max		1.5-2.8	bal Fe				140-230 HB
BS 3468	S-2W	Spheroidal	3 max	1.5-2.2	0.5 max	0.5-1.5	18-22	0.05 max		1.5-2.2	Mg <=0.06; Nb 0.12-0.2; bal Fe				140-200 HB
BS 3468	S-NiCr202	As ISO 2892													

USA

	UNS F43000		3.00 max	1.75-2.75		0.70-1.25	18.00-22.00	0.08 max		1.50-3.00	bal Fe	400 Min	207 Min		
	UNS F43020		2.40-3.00	1.70-2.40		0.80-1.50	18.00-22.00	0.20 max		1.80-3.20	bal Fe	379 Min	207 Min		
	UNS F43021		2.70-3.10	0.50 max		1.90-2.50	20.00-23.00	0.15		2.00-3.00	bal Fe	345 Min	172 Min		
ASTM A439(94)	D-2	Ductile nodular spheroidal	3.00 max	1.75-2.75		0.70-1.25	18.00-22.00	0.08 max		1.50-3.00	bal Fe	400	207	8.0	139-202 HB
MIL-C-24707/5(89)	D-2	As spec in ASTM A439; Type I no magnetic restriction, Type II low rel magnetic perm													
MIL-I-24137	B*	Obs	2.4-3	1.7-2.4		0.8-1.5	18-22		0.02 max	1.8-3.2	bal Fe				

Cast Iron, Austenitic with Graphite, D2B (Continued from previous page)

France

Specification	Designation	Notes	C	Cr	Cu	Mn	Ni	P	S	Si	Other	UTS	YS	El	Hard
AFNOR NFA32301	SNC203	As ISO 2892													

Germany

| DIN | WNr 0.7661 | | | | | | | | | | | | | | |
| DIN 1694 | GGG-NiCr203 | | | | | | | | | | | | | | |

UNS numbers and US grades are provided as a means of cross referencing chemically similar alloys. Exchangability is only possible after independent examination of specifications. Tensile properties are minimum or typical as specified. UTS and YS as MPa. El as %. See Appendix for list of abbreviations used in Notes. * indicates obsolete material.

Specification	Designation	Notes	C	Cr	Cu	Mn	Ni	P	S	Si	Other	UTS	YS	El	Hard

Cast Iron, Austenitic with Graphite, D2B (Continued from previous page)

International

| ISO 2892(73) | S-NiCr203 | Spheroidal graphite | 3.0 max | 2.5-3.5 | 0.5 max | 0.5-1.5 | 18.0-22.0 | .0.080 max | | 1.5-3.0 | bal Fe | 390 | 210 | 7 | |

UK

| BS 3468 | S-2B | Spheroidal | 3 max | 2.5-3.5 | 0.5 max | 0.5-1.5 | 18-22 | 0.08 max | | 1.5-2.8 | bal Fe | | | | 140-230 HB |
| BS 3468 | S-NiCr203 | As ISO 2892 | | | | | | | | | | | | | |

USA

| | UNS F43001 | | 3.00 max | 2.75-4.00 | | 0.70-1.25 | 18.00-22.00 | 0.08 max | | 1.50-3.00 | bal Fe | 400 Min | 207 Min | | |
| ASTM A439(94) | D-2B | Ductile nodular spheroidal | 3.00 max | 2.75-4.00 | | 0.70-1.25 | 18.00-22.00 | 0.08 max | | 1.50-3.00 | bal Fe | 400 | 207 | 7.0 | 148-211 HB |

Cast Iron, Austenitic with Graphite, D2C

France

| AFNOR NFA32301 | SN22 | As ISO 2892 | | | | | | | | | | | | | |

Germany

| DIN | WNr 0.7670 | | | | | | | | | | | | | | |
| DIN 1694 | GGG-Ni22 | | | | | | | | | | | | | | |

International

| ISO 2892(73) | S-Ni22 | Spheroidal graphite | 3.0 max | 0.5 max | 0.5 max | 1.5-2.5 | 21.0-24.0 | 0.080 max | | 1.0-3.0 | bal Fe | 370 | 170 | 20 | |

UK

| BS 3468 | S-2C | Spheroidal | 3 max | 0.5 max | 0.5 max | 1.5-2.5 | 21-24 | 0.08 max | | 1.5-2.8 | bal Fe | | | | 130-170 HB |
| BS 3468 | S-Ni22 | As ISO 2892 | | | | | | | | | | | | | |

USA

	UNS F43002		2.90 max	0.50 max		1.80-2.40	21.00-24.00	0.08 max		1.00-3.00	bal Fe	400 Min	193 Min		
ASTM A439(94)	D-2C	Ductile nodular spheroidal	2.90 max	0.50 max		1.80-2.40	21.00-24.00	0.08 max		1.00-3.00	bal Fe	400	193	20.0	121-171 HB
MIL-C-24707/5(89)	D-2C	As spec in ASTM A439; Type I no magnetic restriction, Type II low rel magnetic perm													
MIL-I-24137	C*	Obs	2.7-3.1	0.5 max		1.9-2.5	20-23	0.15		2-3	bal Fe				

Cast Iron, Austenitic with Graphite, D2M

France

| AFNOR NFA32301 | SNM234 | As ISO 2892 | | | | | | | | | | | | | |

Germany

| DIN | WNr 0.7673 | | | | | | | | | | | | | | |
| DIN 1694 | GGG-NiMn234 | | | | | | | | | | | | | | |

International

| ISO 2892(73) | S-NiMn234 | Spheroidal graphite | 3.0 max | 0.2 max | 0.5 max | 4.0-4.5 | 22.0-24.0 | 0.080 max | | 1.5-2.5 | bal Fe | 440 | 210 | 25 | |

UK

| BS 3468 | S-2M | Spheroidal | 3 max | 0.2 max | 0.5 max | 4-4.5 | 21-24 | 0.08 max | | 1.5-2.5 | bal Fe | | | | 150-180 HB |
| BS 3468 | S-NiMn234 | As ISO 2892 | | | | | | | | | | | | | |

USA

	UNS F43010		2.20-2.70	0.20 max		3.75-4.50	21.0-24.0	0.08 max		1.50-2.50	bal Fe	448 Min	207 Min		
ASTM A571(97)	A571(1)	Press containing parts at low-temp	2.2-2.7	0.20 max			21.0-24.0	0.08 max		1.5-2.50	Mg 3.75-4.5; bal Fe	450	205	30	121-171 HB
ASTM A571(97)	A571(2)	Press containing parts at low-temp	2.2-2.7	0.20 max			21.0-24.0	0.08 max		1.5-2.50	Mg 3.75-4.5; bal Fe	415	170	25	111-171 HB

Cast Iron, Austenitic with Graphite, D3

France

| AFNOR NFA32301 | SNC303 | As ISO 2892 | | | | | | | | | | | | | |

Germany

| DIN | WNr 0.7676 | | | | | | | | | | | | | | |

UNS numbers and US grades are provided as a means of cross referencing chemically similar alloys. Exchangability is only possible after independent examination of specifications. Tensile properties are minimum or typical as specified. UTS and YS as MPa. El as %. See Appendix for list of abbreviations used in Notes. * indicates obsolete material.

Specification	Designation	Notes	C	Cr	Cu	Mn	Ni	P	S	Si	Other	UTS	YS	El	Hard

Cast Iron, Austenitic with Graphite, D3 (Continued from previous page)

Germany

Specification	Designation	Notes	C	Cr	Cu	Mn	Ni	P	S	Si	Other	UTS	YS	El	Hard
DIN 1694	GGG-NiCr303														
DIN 1694	GGG-NiSiCr3052	Spheroidal graphite	2.6 max	1.5-2.5		0.5-1.5	29-32	0.08 max		4-6	bal Fe				

International

| ISO 2892(73) | S-NiCr303 | Spheroidal graphite | 2.6 max | 2.5-3.5 | 0.5 max | 0.5-1.5 | 28.0-32.0 | 0.080 max | | 1.5-3.0 | bal Fe | 370 | 210 | 7 | |

UK

| BS 3468 | S-NiCr303 | As ISO 2892 | | | | | | | | | | | | | |

USA

| | UNS F43003 | | 2.60 max | 2.50-3.50 | | 1.00 max | 28.00-32.00 | 0.80 max | | 1.00-2.80 | bal Fe | 379 Min | 207 Min | | |
| ASTM A439(94) | D-3 | Ductile nodular spheroidal | 2.60 max | 2.50-3.50 | | 1.00 max | 28.00-32.00 | 0.08 max | | 1.00-2.80 | bal Fe | 379 | 207 | 6.0 | 139-202 HB |

Cast Iron, Austenitic with Graphite, D3A

France

| AFNOR NFA32301 | SNC301 | As ISO 2892 | | | | | | | | | | | | | |

Germany

| DIN | WNr 0.7677 | | | | | | | | | | | | | | |
| DIN 1694 | GGG-NiCr301 | | | | | | | | | | | | | | |

International

| ISO 2892(73) | S-NiCr301 | Spheroidal graphite | 2.6 max | 1.0-1.5 | 0.5 max | 0.5-1.5 | 28.0-32.0 | 0.080 max | | 1.5-3.0 | bal Fe | 370 | 210 | 13 | |

UK

| BS 3468 | S-3 | Spheroidal | 2.5 max | 2.5-3.5 | 0.05 max | 0.5-1.5 | 28-32 | 0.08 max | | 1.5-2.8 | bal Fe | | | | 130-200 HB |
| BS 3468 | S-NiCr301 | As ISO 2892 | | | | | | | | | | | | | |

USA

| | UNS F43004 | | 2.60 max | 1.00-1.50 | | 1.00 max | 28.00-32.00 | 0.08 max | | 1.00-2.80 | bal Fe | 379 Min | 207 Min | | |
| ASTM A439(94) | D-3A | Ductile nodular spheroidal | 2.60 max | 1.00-1.50 | | 1.00 max | 28.00-32.00 | 0.08 max | | 1.00-2.80 | bal Fe | 379 | 207 | 10.0 | 131-193 HB |

Cast Iron, Austenitic with Graphite, D4

France

| AFNOR NFA32301 | SNSC3055 | As ISO 2892 | | | | | | | | | | | | | |

Germany

| DIN | WNr 0.7680 | | | | | | | | | | | | | | |
| DIN 1694 | GGG-NiSiCr3055 | | | | | | | | | | | | | | |

International

| ISO 2892(73) | S-NiSiCr3055 | Spheroidal graphite | 2.6 max | 4.5-5.5 | 0.5 max | 0.5-1.5 | 28.0-32.0 | 0.080 max | | 5.0-6.0 | bal Fe | 390 | 240 | | |

UK

| BS 3468 | S-NiSiCr3055 | As ISO 2892 | | | | | | | | | | | | | |

USA

| | UNS F43005 | | 2.60 max | 4.50-5.50 | | 1.00 max | 28.00-32.00 | 0.08 max | | 5.00-6.00 | bal Fe | 414 Min | | | 202-273 HB |
| ASTM A439(94) | D-4 | Ductile nodular spheroidal | 2.60 max | 4.50-5.50 | | 1.00 max | 28.00-32.00 | 0.08 max | | 5.00-6.00 | bal Fe | 414 | | | 202-273 HB |

Cast Iron, Austenitic with Graphite, D5

France

| AFNOR NFA32301 | SN35 | As ISO 2892 | | | | | | | | | | | | | |

Germany

| DIN | WNr 0.7683 | | | | | | | | | | | | | | |
| DIN 1694 | GGG-Ni35 | | | | | | | | | | | | | | |

International

| ISO 2892(73) | S-Ni35 | Spheroidal graphite | 2.4 max | 0.2 max | 0.5 max | 0.5-1.5 | 34.0-36.0 | 0.080 max | | 1.5-3.0 | bal Fe | 370 | 210 | 20 | |

UK

| BS 3468 | S-Ni35 | As ISO 2892 | | | | | | | | | | | | | |

UNS numbers and US grades are provided as a means of cross referencing chemically similar alloys. Exchangability is only possible after independent examination of specifications. Tensile properties are minimum or typical as specified. UTS and YS as MPa. El as %. See Appendix for list of abbreviations used in Notes. * indicates obsolete material.

Specification	Designation	Notes	C	Cr	Cu	Mn	Ni	P	S	Si	Other	UTS	YS	El	Hard

Cast Iron, Austenitic with Graphite, D5 (Continued from previous page)

USA

| | UNS F43006 | | 2.40 max | 0.10 max | | 1.00 max | 34.00- 36.00 | 0.08 max | | 1.00- 2.80 | bal Fe | 379 Min | 207 Min | | |
| ASTM A439(94) | D-5 | Ductile nodular spheroidal | 2.40 max | 0.10 max | | 1.00 max | 34.00- 36.00 | 0.08 max | | 1.00- 2.80 | bal Fe | 379 | 207 | 20.0 | 131-185 HB |

Cast Iron, Austenitic with Graphite, D5B

France

| AFNOR NFA32301 | SNC353 | As ISO 2892 | | | | | | | | | | | | | |

Germany

DIN	WNr 0.7685														
DIN 1694	GGG-NiCr353														
DIN 1694	GGG-NiSiCr3552	Spheroidal graphite	2 max	1.5-2.5		0.5-1.5	34-36	0.08 max		4-6	bal Fe				

International

| ISO 2892(73) | S-NiCr353 | Spheroidal graphite | 2.4 max | 2.0-3.0 | 0.5 max | 0.5-1.5 | 34.0- 36.0 | 0.080 max | | 1.5-3.0 | bal Fe | 370 | 210 | 7 | |

UK

| BS 3468 | S-5S | Spheroidal | 2.2 max | 1.5-2.5 | 0.5 max | 1 max | 34-36 | 0.08 max | | 4.8-5.4 | bal Fe | | | | 130-180 HB |
| BS 3468 | S-NiCr353 | As ISO 2892 | | | | | | | | | | | | | |

USA

	UNS F43007		2.40 max	2.00- 3.00		1 max	34.00- 36.00	0.08 max		1.00- 2.80	bal Fe	379 Min	207 Min		
ASTM A439(94)	D-5B	Ductile nodular spheroidal	2.40 max	2.00- 3.00		1.00 max	34.00- 36.00	0.08 max		1.00- 2.80	bal Fe	379	207	6.0	139-193 HB
ASTM A439(94)	D-5S	Ductile nodular spheroidal	2.30 max	1.75- 2.25		1.00 max	34.00- 37.00	0.08 max		4.90- 5.50	bal Fe	449	207	10	131-193 HB

Cast Iron, Austenitic with Graphite, Ni-Resist 1

France

| AFNOR NFA32301 | LNUC1562 | As ISO 2892 | | | | | | | | | | | | | |

Germany

| DIN | WNr 0.6655 | | | | | | | | | | | | | | |
| DIN 1694 | GGL-NiCuCr1562 | | | | | | | | | | | | | | |

International

| ISO 2892(73) | L-NiCuCr1562 | Flake graphite | 3.0 max | 1.0-2.5 | 5.5-7.5 | 0.5-1.5 | 13.5- 17.5 | | | 1.0-2.8 | bal Fe | 170 | | | |

UK

| BS 3468 | F1 | Flake graphite | 3 max | 1.5-2.5 | 5.5-7.5 | 0.5-1.5 | 13.5- 17.5 | 0.2 max | | 1.5-2.8 | bal Fe | | | | 140-220 HB |

USA

| | UNS F41000 | | 3.00 max | 1.5-2.5 | 5.50- 7.50 | 0.5-1.5 | 13.50- 17.50 | | 0.12 max | 1.00- 2.80 | bal Fe | 172 Min | | | |
| ASTM A436(97) | 1 | Gray with flake graphite | 3.00 max | 1.5-2.5 | 5.50- 7.50 | 0.5-1.5 | 13.50- 17.50 | | 0.12 max | 1.00- 2.80 | bal Fe | 172 | | | 131-183 HB |

Cast Iron, Austenitic with Graphite, Ni-Resist 1b

France

| AFNOR NFA32301 | LNUC1563 | As ISO 2892 | | | | | | | | | | | | | |

Germany

| DIN | WNr 0.6656 | | | | | | | | | | | | | | |
| DIN 1694 | GGL-NiCuCr1563 | | | | | | | | | | | | | | |

International

| ISO 2892(73) | L-NiCuCr1563 | Flake graphite | 3.0 max | 2.5-3.5 | 5.5-7.5 | 0.5-1.5 | 13.5- 17.5 | | | 1.0-2.8 | bal Fe | 190 | | | |

UK

| BS 3468 | L-NiCuCr1563 | As ISO 2892 | | | | | | | | | | | | | |

USA

| | UNS F41001 | | 3.00 max | 2.50- 3.50 | 5.50- 7.50 | 0.5-1.5 | 13.50- 17.50 | | 0.12 max | 1.00- 2.80 | bal Fe | 207 Min | | | |
| ASTM A436(97) | 1b | Gray with flake graphite | 3.00 max | 2.50- 3.50 | 5.50- 7.50 | 0.5-1.5 | 13.50- 17.50 | | 0.12 max | 1.00- 2.80 | bal Fe | 207 | | | 149-212 HB |

UNS numbers and US grades are provided as a means of cross referencing chemically similar alloys. Exchangability is only possible after independent examination of specifications. Tensile properties are minimum or typical as specified. UTS and YS as MPa. El as %. See Appendix for list of abbreviations used in Notes. * indicates obsolete material.

Specification	Designation	Notes	C	Cr	Cu	Mn	Ni	P	S	Si	Other	UTS	YS	El	Hard

Cast Iron, Austenitic with Graphite, Ni-Resist 2

France

| AFNOR NFA32301 | LNC202 | As ISO 2892 | | | | | | | | | | | | | |

Germany

| DIN | WNr 0.6660 | | | | | | | | | | | | | | |
| DIN 1694 | GGL-NiCr202 | | | | | | | | | | | | | | |

International

| ISO 2892(73) | L-NiCr202 | Flake graphite | 3.0 max | 1.0-2.5 | 0.5 max | 0.5-1.5 | 18.0-22.0 | | | 1.0-2.8 | bal Fe | 170 | | | |

Sweden

| SIS 140523 | 0523-00 | Flake graphite | 3 max | 1-2.5 | 0.5 max | 1-1.5 | 18-22 | | | 1-2.8 | bal Fe | | | | 150-215 HB |

UK

| BS 3468 | F2 | Flake graphite | 3 | 1.5-2.5 | 0.5 max | 0.5-1.5 | 18-22 | 0.2 max | | 1.5-2.8 | bal Fe | | | | 140-220 HB |

USA

| | UNS F41002 | | 3.00 max | 1.5-2.5 | 0.50 max | 0.50-1.50 | 18.00-22.00 | | 0.12 max | 1.00-2.80 | bal Fe | 172 Min | | | |
| ASTM A436(97) | 2 | Gray with flake graphite | 3.00 max | 1.5-2.5 | 0.50 max | 0.5-1.5 | 18.00-22.00 | | 0.12 max | 1.00-2.80 | bal Fe | 172 | | | 118-174 HB |

Cast Iron, Austenitic with Graphite, Ni-Resist 2b

France

| AFNOR NFA32301 | LNC203 | As ISO 2892 | | | | | | | | | | | | | |

Germany

| DIN | WNr 0.6661 | | | | | | | | | | | | | | |
| DIN 1694 | GGL-NiCr203 | | | | | | | | | | | | | | |

International

| ISO 2892(73) | L-NiCr203 | Flake graphite | 3.0 max | 2.5-3.5 | 0.5 max | 0.5-1.5 | 18.0-22.0 | | | 1.0-2.8 | bal Fe | 190 | | | |

UK

| BS 3468 | L-NiCr203 | As ISO 2892 | | | | | | | | | | | | | |

USA

| | UNS F41003 | | 3.00 max | 3.00-6.00 | 0.50 max | 0.5-1.5 | 18.00-22.00 | | 0.12 max | 1.00-2.80 | bal Fe | 207 Min | | | |
| ASTM A436(97) | 2b | Gray with flake graphite | 3.00 max | 3.00-6.00 | 0.50 max | 0.5-1.5 | 18.00-22.00 | | 0.12 max | 1.00-2.80 | bal Fe | 207 | | | 171-248 HB |

Cast Iron, Austenitic with Graphite, Ni-Resist 3

France

| AFNOR NFA32301 | LNC303 | As ISO 2892 | | | | | | | | | | | | | |

Germany

| DIN | WNr 0.6676 | | | | | | | | | | | | | | |
| DIN 1694 | GGL-NiCr303 | | | | | | | | | | | | | | |

International

| ISO 2892(73) | L-NiCr303 | Flake graphite | 2.5 max | 2.5-3.5 | 0.5 max | 0.5-1.5 | 28.0-32.0 | | | 1.0-2.0 | bal Fe | 190 | | | |

UK

| BS 3468 | F3 | Flake graphite | 2.5 max | 2.5-3.5 | 0.5 max | 0.5-1.5 | 28-32 | 0.2 max | | 1.5-2.8 | bal Fe | | | | 120-215 HB |

USA

| | UNS F41004 | | 2.60 max | 2.50-3.50 | 0.50 max | 0.5-1.5 | 28.00-32.00 | | 0.12 max | 1.00-2.00 | bal Fe | 172 Min | | | |
| ASTM A436(97) | 3 | Gray with flake graphite | 2.60 max | 2.50-3.50 | 0.50 max | 0.5-1.5 | 28.00-32.00 | | 0.12 max | 1.00-2.00 | bal Fe | 172 | | | 118-159 HB |

Cast Iron, Austenitic with Graphite, Ni-Resist 4

France

| AFNOR NFA32301 | LNSC3055 | As ISO 2892 | | | | | | | | | | | | | |

Germany

| DIN | WNr 0.6680 | | | | | | | | | | | | | | |
| DIN 1694 | GGL-NiSiCr3055 | | | | | | | | | | | | | | |

Specification	Designation	Notes	C	Cr	Cu	Mn	Ni	P	S	Si	Other	UTS	YS	El	Hard

Cast Iron, Austenitic with Graphite, Ni-Resist 4 (Continued from previous page)

International

Specification	Designation	Notes	C	Cr	Cu	Mn	Ni	P	S	Si	Other	UTS	YS	El	Hard
ISO 2892(73)	L-NiSiCr3055	Flake graphite	2.5 max	4.5-5.5	0.5 max	0.5-1.5	29.0-32.0			5.0-6.0	bal Fe	170			

UK

BS 3468	L-NiSiCr3055	As ISO 2892													

USA

Specification	Designation	Notes	C	Cr	Cu	Mn	Ni	P	S	Si	Other	UTS	YS	El	Hard
	UNS F41005		2.60 max	4.50-5.50	0.50 max	0.5-1.5	29.00-32.00		0.12 max	5.00-6.00	bal Fe	172 Min			
ASTM A436(97)	4	Gray with flake graphite	2.60 max	4.50-5.50	0.50 max	0.5-1.5	29.00-32.00		0.12 max	5.00-6.00	bal Fe	172			149-212 HB

Cast Iron, Austenitic with Graphite, No equivalents identified

France

Specification	Designation	Notes
AFNOR NFA32301	LNM137	As ISO 2892
AFNOR NFA32301	LNSC2053	As ISO 2892; Nicrosilal
AFNOR NFA32301	SNM137	As ISO 2892; SNiMn137
AFNOR NFA32301	SNSC2052	As ISO 2892; Nicrosilal Spheronic

Germany

Specification	Designation
DIN 1694	GGG-NiMn137
DIN 1694	GGG-NiSiCr2052
DIN 1694	GGL-NiMn137
DIN 1694	GGL-NiSiCr2053

Hungary

Specification	Designation	Notes	C	Cr	Cu	Mn	Ni	P	S	Si	Other	UTS	YS	El	Hard
MSZ ISO 2892(92)	L-Ni35	as cast	2.4 max	0.2 max	0.5 max	0.5-1.5	34-36	0.5 max	0.5 max	1-2	bal Fe	120-180		1L	120-140 HB
MSZ ISO 2892(92)	L-NiCr202	as cast	3.0 max	1-2.5	0.5 max	0.5-1.5	18-22	0.5 max	0.5 max	1-2.8	bal Fe	170-210		2L	120-215 HB
MSZ ISO 2892(92)	L-NiCr203	as cast	3.0 max	2.5-3.5	0.5 max	0.5-1.5	18-22	0.5 max	0.5 max	1-2.8	bal Fe	190-240		1L	160-250 HB
MSZ ISO 2892(92)	L-NiCr303	as cast	2.5 max	2.5-3.5	0.5 max	0.5-1.5	28-32	0.5 max	0.5 max	1-2	bal Fe	190-240		1L	120-215 HB
MSZ ISO 2892(92)	L-NiCuCr1562	as cast	3.0 max	1-2.5	5.5-7.5	0.5-1.5	13.5-17.5	0.5 max	0.5 max	1-2.8	bal Fe	170-210		2L	140-200 HB
MSZ ISO 2892(92)	L-NiCuCr1563	as cast	3.0 max	2.5-3.5	5.5-7.5	0.5-1.5	13.5-17.5	0.5 max	0.5 max	1-2.8	bal Fe	190-240		1L	150-250 HB
MSZ ISO 2892(92)	L-NiMn137	as cast	3 max	0.2 max	0.5 max	6-7	12-14	0.5 max	0.5 max	1.5-3.0	bal Fe	140-200			120-150 HB
MSZ ISO 2892(92)	L-NiSiCr2053	as cast	2.5 max	1.5-4.5	0.5 max	0.5-1.5	18-22	0.5 max	0.5 max	4.5-5.5	bal Fe	190-280		2L	140-250 HB
MSZ ISO 2892(92)	L-NiSiCr3055	as cast	2.5 max	4.5-5.5	05 max	0.5-1.5	29-32	0.5 max	0.5 max	5-6	bal Fe	170-240			150-210 HB
MSZ ISO 2892(92)	S-Ni22	as cast	3.0 max	0.5 max	0.5 max	1.5-2.5	21-24	0.08 max	0.5 max	1-3.0	bal Fe	370-440	170	20L	130-170 HB
MSZ ISO 2892(92)	S-Ni35	as cast	2.4 max	0.2 max	0.5 max	0.5-1.5	34-36	0.08 max	0.5 max	1.5-3.0	bal Fe	370-410	210	20L	130-180 HB
MSZ ISO 2892(92)	S-NiCr202	as cast	3.0 max	1-2.5	0.5 max	0.5-1.5	18-22	0.08 max	0.5 max	1.5-3.0	bal Fe	370-440	210	7L	140-200 HB
MSZ ISO 2892(92)	S-NiCr203	as cast	3.0 max	2.5-3.5	0.5 max	0.5-1.5	18-22	0.08 max	0.5 max	1.5-3.0	bal Fe	390-490	210	7L	150-255 HB
MSZ ISO 2892(92)	S-NiCr301	as cast	2.6 max	1-1.5	0.5 max	0.5-1.5	28-32	0.08 max	0.5 max	1.5-3.0	bal Fe	370-440	210	13L	130-190 HB
MSZ ISO 2892(92)	S-NiCr303	as cast	2.6 max	2.5-3.5	0.5 max	0.5-1.5	28-32	0.08 max	0.5 max	1.5-3.0	bal Fe	370-470	210	7L	140-200 HB
MSZ ISO 2892(92)	S-NiCr353	as cast	2.4 max	2-3.0	0.5 max	0.5-1.5	34-36	0.08 max	0.5 max	1.5-3.0	bal Fe	370-440	210	7L	140-170 HB
MSZ ISO 2892(92)	S-NiMn137	as cast	3 max	0.2 max	0.5 max	6-7	12-14	0.08 max	0.5 max	2-3.0	bal Fe	390-460	210	15L	130-170 HB
MSZ ISO 2892(92)	S-NiMn234	as cast	2.6 max	0.2 max	0.5 max	3.5-4.5	22-24	0.08 max	0.05 max	1.5-2.5	bal Fe	440-470	210	25L	150-180 HB
MSZ ISO 2892(92)	S-NiSiCr2052	as cast	3.0 max	1-2.5	0.5 max	0.5-1.5	18-22	0.08 max	0.05 max	4.5-5.5	bal Fe	370-430	210	10L	180-230 HB
MSZ ISO 2892(92)	S-NiSiCr2055	as cast	2.6 max	4.5-5.5	0.5 max	0.5-1.5	28-32	0.08 max	0.5 max	5-6	bal Fe	370-490	240	1L	170-250 HB

UNS numbers and US grades are provided as a means of cross referencing chemically similar alloys. Exchangability is only possible after independent examination of specifications. Tensile properties are minimum or typical as specified. UTS and YS as MPa. El as %. See Appendix for list of abbreviations used in Notes. * indicates obsolete material.

Specification	Designation	Notes	C	Cr	Cu	Mn	Ni	P	S	Si	Other	UTS	YS	El	Hard

Cast Iron, Austenitic with Graphite, No equivalents identified

Specification	Designation	Notes	C	Cr	Cu	Mn	Ni	P	S	Si	Other	UTS	YS	El	Hard
International															
ISO 2892(73)	L-NiMn137	Flake graphite	3.0 max	0.2 max	0.5 max	6.0-7.0	12.0-14.0			1.5-3.0	bal Fe	140			
ISO 2892(73)	L-NiSiCr2053	Flake graphite	2.5 max	1.5-4.5	0.5 max	0.5-1.5	18.0-22.0			4.5-5.5	bal Fe	190			
ISO 2892(73)	S-NiMn137	Spheroidal graphite	3.0 max	0.2 max	0.5 max	6.0-7.0	12.0-14.0	0.080 max		2.0-3.0	bal Fe	390	210	15	
ISO 2892(73)	S-NiSiCr2052	Spheroidal graphite	3.0 max	1.0-2.5	0.5 max	0.5-1.5	18.0-22.0	0.080 max		4.5-5.5	bal Fe	370	210	10	
Japan															
JIS G5503(95)	FCAD1000-5	SCS; Austempered Ductile										1000	700	5	
JIS G5503(95)	FCAD1200-2	SCS; Austempered Ductile										1200	900	2	341 HB
JIS G5503(95)	FCAD1400-1	SCS; Austempered Ductile										1400	1100	1	401 HB
JIS G5503(95)	FCAD900-4	SCS; Austempered Ductile										900	600	4	
JIS G5503(95)	FCAD900-8	SCS; Austempered Ductile										900	600	8	
Sweden															
SIS 140772	0772-00	Spheroidal graphite	3 max		0.5 max	6-7	12-14	0.08 max	0.02 max	2-3	Mg <=0.02; bal Fe	400	210	10	150 HB max
UK															
BS 3468	L-NiMn137	As ISO 2892	3 max	0.2 max	0.5 max	6-7	12-14			1.5-3	bal Fe				
BS 3468	L-NiSiCr2053	As ISO 2892; Nicrosilal 2.5		1.5-4.5	0.5 max	0.5-1.5	18-22			4.5-5.5	bal Fe				
BS 3468	S-6	Spheroidal; S-NiMn137	3 max	0.5 max	0.2 max	6-7	12-14	0.08 max		1.5-2.8	bal Fe				130-170 HB
BS 3468	S-NiMn137	As ISO 2892; S-NiMn137													
BS 3468	S-NiSiCr2052	As ISO 2892; Nicrosilal Spheronic													

Specification	Designation	Notes	C	Cr	Mn	Mo	Ni	P	S	Si	Other	UTS	YS	El	Hard

Cast Iron, Unclassified, 5392

USA

| | UNS F47004 | Corr res | 2.4-2.8 | 1.8-2.4 | 1.0-1.5 | | 14.0-16.0 | 0.30 max | 0.12 max | 1.5-2.5 | Cu 6.0-7.0; Pb <=0.003; bal Fe | 207 Min | | | |

Cast Iron, Unclassified, 5393

USA

| | UNS F47005 | Corr res; Heat res | 2.4-2.8 | 1.7-2.4 | 0.8-1.6 | | 18.0-22.0 | 0.30 max | | 1.5-2.8 | Cu 0.5; Pb <=0.003; bal Fe | 207 Min | | | |

Cast Iron, Unclassified, F47001

USA

| | UNS F47001 | Corr res | 2.60-3.00 | 1.80-3.50 | 1.00-1.50 | | 13.5-17.5 | 0.20 max | 0.10 max | 1.25-2.20 | Cu 5.50-7.50; bal Fe | 172 Min | | | |

Cast Iron, Unclassified, F47002

USA

| | UNS F47002 | Corr res | 2.60-3.00 | 1.75-3.50 | 0.80-1.30 | | 18.0-22.0 | 0.20 max | 0.10 max | 1.25-2.20 | Cu <=0.50; bal Fe | 172 Min | | | |

Cast Iron, Unclassified, F47003

USA

| | UNS F47003 | Corr res | 0.70-1.10 | 0.50 max | 1.50 max | 0.50 max | | | | 14.20-14.75 | Cu <=0.50; bal Fe | | | | |

Cast Iron, Unclassified, IA

China

| GB 8491(87) | KmTBNi4Cr2-GT | White; Abr res; Hard | 3.2-3.6 | 2.0-3.0 | 0.3-0.8 | 1.0 max | 3.0-5.0 | 0.15 max | 0.10 max | 0.3-0.8 | bal Fe | | | | 55 HRC |

France

| AFNOR NFA32401 | FBNi4Cr2HC | White; Abr res | 3.2-3.6 | 1.5-2.5 | 0.3-0.7 | 1 max | 3-5.5 | | | 0.2-0.8 | Cu may part replace Ni; bal Fe | | | | 500-700 HB |

Germany

| DIN | G-X330NiCr42 | | | | | | | | | | | | | | |
| DIN | WNr 0.9625 | | | | | | | | | | | | | | |

India

| IS 4771-1A | NiLCr34/550 | White; Abr res; as cast | 3.2-3.6 | 1.5-2.5 | 0.3-0.6 | 0.5 max | 3-5.5 | 0.03 max | 0.15 max | 0.3-0.6 | bal Fe | | | | 550 HB |

Sweden

| SIS 140513 | 34102 | White; Abr res | 3.2-3.6 | 1.5-2.5 | 0.3-0.6 | | 3-5.5 | 0.03 max | 0.15 max | 0.3-0.6 | bal Fe | | | | 550 HB as Cast |

UK

| BS 4844 | 2B | White; Abr res; HT | 3.2-3.6 | 1.5-3.5 | 0.2-0.8 | 0.5 max | 3-5.5 | 0.15 max | 0.15 max | 0.3-0.8 | bal Fe | | | | 550 HB |

USA

	UNS F45000	White; chill cast	3.0-3.6	1.4-4.0	1.3 max	1.0 max	3.3-5.0	0.30 max	0.15 max	0.8 max	bal Fe				600 Min HB
	UNS F45000	White; sand cast	3.0-3.6	1.4-4.0	1.3 max	1.0 max	3.3-5.0	0.30 max	0.15 max	0.8 max	bal Fe				550 Min HB
ASTM A532/A532M(93a)	Ni-Cr-Hc	White; sand/chill cast	2.8-3.6	1.4-4.0	2.0 max	1.0 max	3.3-5.0	0.3 max	0.15 max	0.8 max	bal Fe				550 HB

Cast Iron, Unclassified, IB (Continued from previous page)

China

| GB 8491(87) | KmTBNi4Cr2-DT | White; Abr res; Hard | 2.7-3.2 | 2.0-3.0 | 0.3-0.8 | 1.0 max | 3.0-5.0 | 0.15 max | 0.10 max | 0.3-0.8 | bal Fe | | | | 53 HRC |

France

| AFNOR NFA32401 | FBNi4Cr2BC | White; Abr res | 2.7-3.2 | 1.5-2.5 | 0.3-0.7 | 1 max | 3-5.5 | | | 0.2-0.8 | Cu may part replace Ni; bal Fe | | | | 450-650 HB |

Germany

| DIN | G-X260NiCr42 | | | | | | | | | | | | | | |
| DIN | WNr 0.9620 | | | | | | | | | | | | | | |

India

| IS 4771-1A | NiLCr30/500 | White; Abr res; as cast | 2.7-3.3 | 1.5-2.5 | 0.3-0.6 | 0.5 max | 3-5.5 | 0.3 max | 0.15 max | 0.3-0.6 | bal Fe | | | | 500 HB |

Specification	Designation	Notes	C	Cr	Mn	Mo	Ni	P	S	Si	Other	UTS	YS	El	Hard
Cast Iron, Unclassified, IB (Continued from previous page)															
Sweden															
SIS 140512	512	White; Abr res	2.7-3.3	1.5-2.5	0.3-0.6		3-5.5	0.3 max	0.15 max	0.3-0.6	bal Fe				500 HB as Cast
UK															
BS 4844	2A	White; Abr res; HT	2.7-3.2	1.5-3.5	0.2-0.8	0.5 max	3-5.5	0.15 max	0.15 max	0.3-0.8	bal Fe				500 HB
USA															
	UNS F45001	White; chill cast	2.5-3.0	1.4-4.0	1.3 max	1.0 max	3.3-5.0	0.30 max	0.15 max	0.8 max	bal Fe				600 Min HB
	UNS F45001	White; sand cast	2.5-3.0	1.4-4.0	1.3 max	1.0 max	3.3-5.0	0.30 max	0.15 max	0.8 max	bal Fe				550 Min HB
ASTM A532/A532M(93a)	Ni-Cr-Lc	White; sand/chill cast	2.4-3.0	1.4-4.0	2.0 max	1.0 max	3.3-5.0	0.3 max	0.15 max	0.8 max	bal Fe				550 HB
Cast Iron, Unclassified, IC															
China															
GB 1504(91)	KmTGNi-Cr-Mo		2.9-3.7	1.0-1.5	0.6-1.2	0.2-0.6	3.0-4.5	0.30 max	0.12 max	0.6-1.2	bal Fe				
USA															
	UNS F45002	White; chill cast	2.9-3.7	1.1-1.5	1.3 max	1.0 max	2.7-4.0	0.30 max	0.15 max	0.8 max	bal Fe				600 Min HB
	UNS F45002	White; sand cast	2.9-3.7	1.1-1.5	1.3 max	1.0 max	2.7-4.0	0.30 max	0.15 max	0.8 max	bal Fe				550 Min HB
ASTM A532/A532M(93a)	Ni-Cr-BG	White; sand/chill cast	2.5-3.7	1.0-2.5	2.0 max	1.0 max	4.0 max	0.3 max	0.15 max	0.8 max	bal Fe				550 HB
Cast Iron, Unclassified, ID															
China															
GB 8491(87)	KmTBCr9Ni5Si2	White; Abr res; Hard	2.5-3.6	8.0-10.0	0.3-0.8	1.0 max	4.5-6.5	0.15 max	0.10 max	1.5-2.2	bal Fe				55 HRC
France															
AFNOR NFA32401	FBCr9Ni5	White; Abr res	2.5-3.6	5-11	0.3-0.7	0.5 max	4-6			1.5-2.2	bal Fe				550-750 HB
Germany															
DIN	G-X300CrNiSi952														
DIN	WNr 0.9630														
India															
IS 4771-1B	NiHCr27/500	White; Abr res; as cast	2.5-2.9	8-10	0.3-0.6	0.5 max	4-6	0.3 max	0.15 max	1.5-2.2	bal Fe				500 HB
IS 4771-1B	NiHCr30/550	White; Abr res; as cast	2.8-3.3	8-10	0.3-0.6	0.5 max	4-6	0.3 max	0.15 max	1.5-2.2	bal Fe				550 HB
IS 4771-1B	NiHCr34/600	White; Abr res; as cast	3.2-3.6	7.5-9.5	0.3-0.6	0.5 max	4-6	0.3 max	0.15 max	1.5-2.2	bal Fe				600 HB
Sweden															
SIS 140457	457	White; Abr res	2.8-3.3	8-10	0.3-0.6		4-6	0.3 max	0.15 max	1.5-2.2	bal Fe				550 HB as Cast
UK															
BS 4844	2C	White; Abr res; HT	2.4-2.8	8-10	0.2-0.8	0.5 max	4-6	0.1 max	0.15 max	1.5-2.2	bal Fe				500 HB
BS 4844	2D	White; Abr res; HT	2.8-3.2	8-10	0.2-0.8	0.5 max	4-6	0.1 max	0.15 max	1.5-2.2	bal Fe				550 HB
BS 4844	2E	White; Abr res; HT	3.2-3.6	8-10	0.2-0.8	0.5 max	4-6	0.1 max	0.15 max	1.5-2.2	bal Fe				600 HB
USA															
	UNS F45003	White	2.5-3.6	7.0-11.0	1.3 max	1.0 max	4.5-7.0	0.10 max	0.15 max	1.0-2.2	bal Fe				400-600 HB
ASTM A532/A532M(93a)	Ni-HiCr	White	2.5-3.6	7.0-11.0	2.0 max	1.5 max	4.5-7.0	0.10 max	0.15 max	2.0 max	bal Fe				500 HB
Cast Iron, Unclassified, IIA															
France															
AFNOR NFA32401	FBCr12MoNi	White; Abr res	2-3.6	11-14	0.5-1	0.5-3	0.2 max			0.2-0.8	bal Fe				500-800 HB
UK															
BS 4844	3F	White; Abr res; HT	2-2.7	11-13	0.5-1.5	2.5 max	2 max	0.1 max	0.1 max	1 max	Cu <=2; bal Fe				600 HB
BS 4844	3G	White; Abr res; HT	2.7-3.4	11-13	0.5-1.5	3 max	2 max	0.1 max	0.1 max	1 max	Cu <=2; bal Fe				650 HB

UNS numbers and US grades are provided as a means of cross referencing chemically similar alloys. Exchangability is only possible after independent examination of specifications. Tensile properties are minimum or typical as specified. UTS and YS as MPa. El as %. See Appendix for list of abbreviations used in Notes. * indicates obsolete material.

Specification	Designation	Notes	C	Cr	Mn	Mo	Ni	P	S	Si	Other	UTS	YS	El	Hard

Cast Iron, Unclassified, IIA (Continued from previous page)

USA

Specification	Designation	Notes	C	Cr	Mn	Mo	Ni	P	S	Si	Other	UTS	YS	El	Hard
	UNS F45004	White	2.4-2.8	11.0-14.0	0.5-1.5	0.5-1.0	0.5 max	0.10 max	0.06 max	1.0 max	Cu <=1.2; bal Fe				400-600 HB
ASTM A532/A532M(93a)	12% Cr	White	2.0-3.3	11.0-14.0	2.0 max	3.0 max	2.5 max	0.10 max	0.06 max	1.5 max	Cu <=1.2; bal Fe				550HB

Cast Iron, Unclassified, IIB

China

Specification	Designation	Notes	C	Cr	Mn	Mo	Ni	P	S	Si	Other	UTS	YS	El	Hard
GB 8491(87)	KmTBCr15Mo2-DT	White; Abr res; as cast	2.0-2.8	13.0-18.0	0.5-1.0	0.5-2.5	1.0 max	0.10 max	0.06 max	1.0 max	Cu <=1.2; bal Fe				40-56 HRC
GB 8491(87)	KmTBCr15Mo2-GT	White; Abr res; as cast	2.8-3.5	13.0-18.0	0.5-1.0	0.5-3.0	1.0 max	0.10 max	0.06 max	1.0 max	Cu <=1.2; bal Fe				50-58 HRC

India

Specification	Designation	Notes	C	Cr	Mn	Mo	Ni	P	S	Si	Other	UTS	YS	El	Hard
IS 4771-2	CrMoLC28/500	White; Abr res; Ann	2.4-3.1	14-18	0.4-0.9	2.5-3.5	0.5 max	0.3 max	0.15 max	0.3-0.8	bal Fe				380 HB
IS 4771-2	CrMoLC28/500	White; Abr res; as cast	2.4-3.1	14-18	0.4-0.9	2.5-3.5	0.5 max	0.3 max	0.15 max	0.3-0.8	Cu <=1.2; bal Fe				500 HB
IS 4771-2	CrMoLC28/500	White; Abr res; Hard	2.4-3.1	14-18	0.4-0.9	2.5-3.5	0.5 max	0.3 max	0.15 max	0.3-0.8	bal Fe				550 HB

UK

Specification	Designation	Notes	C	Cr	Mn	Mo	Ni	P	S	Si	Other	UTS	YS	El	Hard
BS 4844	3A	White; Abr res; HT	1.8-3	14-17	0.5-1.5	2.5 max	2 max	0.1 max	0.1 max	1 max	Cu <=2; bal Fe				600 HB

USA

Specification	Designation	Notes	C	Cr	Mn	Mo	Ni	P	S	Si	Other	UTS	YS	El	Hard
	UNS F45005	White	2.4-2.8	14.0-18.0	0.5-1.5	1.0-3.0	0.5 max	0.10 max	0.06 max	1.0 max	Cu <=1.2; bal Fe				400-600 HB
ASTM A532/A532M(93a)	15% Cr-Mo	White	2.0-3.3	14.0-18.0	2.0 max	3.0 max	2.5 max	0.10 max	0.06 max	1.5 max	Cu <=1.2; bal Fe				450 HB

Cast Iron, Unclassified, IIC

China

Specification	Designation	Notes	C	Cr	Mn	Mo	Ni	P	S	Si	Other	UTS	YS	El	Hard
GB 8491(87)	KmTBCr15Mo2-DT	White; Abr res; as cast	2.0-2.8	13.0-18.0	0.5-1.0	0.5-2.5	1.0 max	0.10 max	0.06 max	1.0 max	Cu <=1.2; bal Fe				40-56 HRC
GB 8491(87)	KmTBCr15Mo2-GT	White; Abr res; as cast	2.8-3.5	13.0-18.0	0.5-1.0	0.5-3.0	1.0 max	0.10 max	0.06 max	1.0 max	Cu <=1.2; bal Fe				50-58 HRC

France

Specification	Designation	Notes	C	Cr	Mn	Mo	Ni	P	S	Si	Other	UTS	YS	El	Hard
AFNOR NFA32401	FBCr15MoNi	White; Abr res	2-3.6	14-17	0.5-1	0.5-3	2.5 max			0.2-0.8	bal Fe				500-800 HB

Germany

Specification	Designation	Notes	C	Cr	Mn	Mo	Ni	P	S	Si	Other	UTS	YS	El	Hard
DIN	G-X300CrMo153														
DIN	G-X300CrMoNi1521														
DIN	WNr 0.9635														
DIN	WNr 0.9640														

India

Specification	Designation	Notes	C	Cr	Mn	Mo	Ni	P	S	Si	Other	UTS	YS	El	Hard
IS 4771-2	CrMoHC34/500	White; Abr res; Hard	3.1-3.6	14-18	0.4-0.9	2.5-3.5	0.5 max	0.3 max	0.15 max	0.3-0.8	bal Fe				600 HB
IS 4771-2	CrMoHC34/500	White; Abr res; as cast	3.1-3.6	14-18	0.4-0.9	2.5-3.5	0.5 max	0.3 max	0.15 max	0.3-0.8	bal Fe				500 HB
IS 4771-2	CrMoHC34/500	White; Abr res; Ann	3.1-3.6	14-18	0.4-0.9	2.5-3.5	0.5 max	0.3 max	0.15 max	0.3-0.8	bal Fe				380 HB

UK

Specification	Designation	Notes	C	Cr	Mn	Mo	Ni	P	S	Si	Other	UTS	YS	El	Hard
BS 4844	3B	White; Abr res; HT	3-3.6	14-17	0.5-1.5	3 max	2 max	0.1 max	0.1 max	0.3-0.8	Cu <=2; bal Fe				650 HB

USA

Specification	Designation	Notes	C	Cr	Mn	Mo	Ni	P	S	Si	Other	UTS	YS	El	Hard
	UNS F45006	White	2.8-3.6	14.0-18.0	0.5-1.5	2.3-3.5	0.5 max	0.10 max	0.06 max	1.0 max	Cu <=1.2; bal Fe				400-600 HB

Cast Iron, Unclassified, IID

China

Specification	Designation	Notes	C	Cr	Mn	Mo	Ni	P	S	Si	Other	UTS	YS	El	Hard
GB 8491(87)	KmTBCr20Mo2Cu1	White; Abr res; as cast	2.0-3.0	18.0-22.0	0.5-1.0	1.5-2.5	1.5 max	0.10 max	0.06 max	1.0 max	Cu 0.8-1.2; bal Fe				50-58 HRC

USA

Specification	Designation	Notes	C	Cr	Mn	Mo	Ni	P	S	Si	Other	UTS	YS	El	Hard
	UNS F45007	White	2.0-2.6	18.0-23.00	0.5-1.5	1.5 max	1.5 max	0.10 max	0.06 max	1.0 max	Cu <=1.2; bal Fe				400-600 HB
ASTM A532/A532M(93a)	20% Cr-Mo	White	2.0-3.3	18.0-23.0	2.0 max	3.0 max	2.5 max	0.10 max	0.06 max	1.0-2.2	Cu <=1.2; bal Fe				450 HB

Specification	Designation	Notes	C	Cr	Mn	Mo	Ni	P	S	Si	Other	UTS	YS	El	Hard

Cast Iron, Unclassified, IIE

China

| GB 8491(87) | KmTBCr20Mo2Cu1 | White; Abr res; as cast | 2.0-3.0 | 18.0-22.0 | 0.5-1.0 | 1.5-2.5 | 1.5 max | 0.10 max | 0.06 max | 1.0 max | Cu 0.8-1.2; bal Fe | | | | 50-58 HRC |

France

| AFNOR NFA32401 | FBCr20MoNi | White; Abr res | 2-3.6 | 17-22 | 0.5-1.5 | 0.5-3 | 2.5 max | | | 0.2-1.2 | Cu <=1.5; bal Fe | | | | 500-800 HB |

Germany

| DIN | G-X260CrMoNi2021 | | | | | | | | | | | | | | |
| DIN | WNr 0.9645 | | | | | | | | | | | | | | |

UK

| BS 4844 | 3C | White; Abr res; HT | 1.8-3 | 17-22 | 0.5-1.5 | 3 max | 2 max | 0.1 max | 0.1 max | 1 max | Cu <=2; bal Fe | | | | 600 HB |

USA

| | UNS F45008 | White | 2.6-3.2 | 18.0-23.0 | 0.5-1.5 | 1.0-2.0 | 1.5 max | 0.10 max | 0.06 max | 1.0 max | Cu <=1.2; bal Fe | | | | 400-600 HB |

Cast Iron, Unclassified, IIIA

China

| GB 8491(87) | KmTBCr26 | White; Abr res; as cast | 2.3-3.0 | 23.0-28.0 | 0.5-1.0 | 1.0 max | 1.5 max | 0.10 max | 0.06 max | 1.0 max | Cu <=2.0; bal Fe | | | | 50-58 HRC |

France

| AFNOR NFA32401 | FBCr26MoNi | White; Abr res | 1.5-3.5 | 22-28 | 0.5-1.5 | 0.5-3 | 2.5 max | | | 0.2-1.2 | Cu <=1.5; Mn 4 max for hard aust; bal Fe | | | | 450-650 HB |

Germany

DIN	G-X260Cr27														
DIN	G-X300CrMo271														
DIN	WNr 0.9650														
DIN	WNr 0.9655														

India

IS 4771-3	HCr27/400	White; Abr res; Hard	2.3-3	24-28	1.5 max	0.6 max	0.5 max	0.3 max	0.15 max	0.2-1.5	bal Fe				550 HB
IS 4771-3	HCr27/400	White; Abr res; as cast	2.3-3	24-28	1.5 max	0.6 max	0.5 max	0.3 max	0.15 max	0.2-1.5	bal Fe				400 HB
IS 4771-3	HCrNi27/400	White; Abr res; Hard	2.3-3	24-28	1.5 max	0.6 max	1.2 max	0.3 max	0.15 max	0.2-1.5	bal Fe				550 HB
IS 4771-3	HCrNi27/400	White; Abr res; as cast	2.3-3	24-28	1.5 max	0.6 max	1.2 max	0.3 max	0.15 max	0.2-1.5	bal Fe				400 HB

Sweden

| SIS 140466 | 466 | White; Abr res | 2.5 min | 24-30 | 0.5 | | | 0.3 max | | 1 | bal Fe | | | | 550 HB as Cast |

UK

| BS 4844 | 3D | White; Abr res; HT | 2-2.8 | 22-28 | 0.5-1.5 | 1.5 max | 2 max | 0.1 max | 0.1 max | 1 max | Cu <=2; bal Fe | | | | 600 HB |
| BS 4844 | 3E | White; Abr res; HT | 2.8-3.5 | 22-28 | 0.5-1.5 | 1.5 max | 2 max | 0.1 max | 0.1 max | 1 max | Cu <=2; bal Fe | | | | 600 HB |

USA

| | UNS F45009 | White | 2.3-3.0 | 23.0-28.0 | 0.5-1.5 | 1.5 max | 1.5 max | 0.10 max | 0.06 max | 1.0 max | Cu <=1.2; bal Fe | | | | 400-600 HB |

Cast Iron, Unclassified, No equivalents identified

China

GB 8491(87)	KmTBCr2Mo1Cu1	White; Abr res; as cast	2.4-3.6	2.0-3.0	1.0-2.0	0.5-1.0		0.15 max	0.10 max	1.0 max	Cu 0.8-1.2; bal Fe				50-56 HRC
GB 8491(87)	KmTBMn5W3	White; Abr res; as cast	3.0-3.5		4.0-6.0			0.15 max	0.10 max	0.8-1.3	W 2.5-3.5; bal Fe				50-60 HRC
GB 8491(87)	KmTBW5Cr4	White; Abr res; as cast	2.5-3.5	3.5-4.5	0.5-1.0			0.15 max	0.10 max	0.5-1.0	W 4.5-5.5; bal Fe				50-65 HRC

France

| AFNOR NFA32401 | FBA | White; Abr res | 2.7-3.9 | | 0.2-0.8 | | | | | 0.4-1.5 | Ni Cr Mo Cu optional; bal Fe | | | | 400-600 HB |
| AFNOR NFA32401 | FBO | White; Abr res | 2.7-3.9 | | 0.2-0.8 | | | | | 0.4-1.5 | bal Fe | | | | 350-500 HB |

Germany

DIN	WNr 0.6652														
DIN	WNr 0.6667														
DIN	WNr 0.7652														
DIN	WNr 0.7665														

UNS numbers and US grades are provided as a means of cross referencing chemically similar alloys. Exchangability is only possible after independent examination of specifications. Tensile properties are minimum or typical as specified. UTS and YS as MPa. El as %. See Appendix for list of abbreviations used in Notes. * indicates obsolete material.

Specification	Designation	Notes	C	Cr	Mn	Mo	Ni	P	S	Si	Other	UTS	YS	El	Hard

Cast Iron, Unclassified, No equivalents identified

Hungary

Specification	Designation	Notes	C	Cr	Mn	Mo	Ni	P	S	Si	Other	UTS	YS	El	Hard
MSZ 8273(91)	FOX290NiCr42	Abr res; alloyed; 30-30mm; HT	2.7-3.2	1.5-2.5	0.3-0.7	0.2-0.5	3.3-5	0.3 max	0.15 max	0.3-0.5	150mm Cr=1.5-3.5; 0-15mm Ni=2.5-5; bal Fe	270-340			513 HB; Rmh=490-610
MSZ 8273(91)	FOX290NiCr42	Abr res; alloyed; 30-30mm; HT	2.7-3.2	1.5-2.5	0.3-0.7	0.2-0.5	3.3-5	0.3 max	0.15 max	0.3-0.5	150mm Cr=1.5-3.5; 0-15mm Ni=2.5-5; bal Fe	270-340			52 HRC; Rmh=490-610
MSZ 8273(91)	FOX300CrNi95	Abr res; alloyed; 30-30mm; HT	2.8-3.2	8-10	0.3-0.7	0.2-0.5	4-6	0.3 max	0.15 max	1.5-2.2	bal Fe	490-590			53 HRC; Rmh=610-740
MSZ 8273(91)	FOX300CrNi95	Abr res; alloyed; 30-30mm; HT	2.8-3.2	8-10	0.3-0.7	0.2-0.5	4-6	0.3 max	0.15 max	1.5-2.2	bal Fe	490-590			532 HB; Rmh=610-740
MSZ 8273(91)	FOX340CrNi95	Abr res; alloyed; 30-30mm; HT	3.2-3.6	8-10	0.3-0.7	0.2-0.5	4-6	0.3 max	0.15 max	1.5-2	bal Fe	490-590			532 HB; Rmh=610-740
MSZ 8273(91)	FOX340CrNi95	Abr res; alloyed; 30-30mm; HT	3.2-3.6	8-10	0.3-0.7	0.2-0.5	4-6	0.3 max	0.15 max	1.5-2	bal Fe	490-590			53 HRC; Rmh=610-740
MSZ 8273(91)	FOX340NiCr42	Abr res; alloyed; 30-30mm; HT	3.2-3.6	1.5-2.5	0.3-0.7	0.2-0.5	3.3-5	0.3 max	0.15 max	0.3-0.5	150mm Cr=1.5-3.5; 0-15mm Ni=2.5-5; bal Fe	270-340			53 HRC; Rmh=610-740
MSZ 8273(91)	FOX340NiCr42	Abr res; alloyed; 30-30mm; HT	3.2-3.6	1.5-2.5	0.3-0.7	0.2-0.5	3.3-5	0.3 max	0.15 max	0.3-0.5	150mm Cr=1.5-3.5; 0-15mm Ni=2.5-5; bal Fe	270-340			532 HB; Rmh=610-740
MSZ 8273(91)	OX100CrMo6	Abr res; alloyed; HT	0.8-1.1	5-6	0.4-0.8	0.8-1.2		0.3 max	0.3 max	0.5-1	bal Fe				53-64 HRC
MSZ 8273(91)	OX190CrW21	Abr res; alloyed; HT	1.6-2.2	19-23	0.4-0.8			0.3 max	0.3 max	0.5-1	W 1.2-1.6; bal Fe				40-58 HRC
MSZ 8273(91)	OX215CrMoW21	Abr res; alloyed; HT	2.0-2.3	20-23	0.5-0.8	0.15-0.25		0.3 max	0.3 max	0.5-1	W 1.5-2; bal Fe				40-58 HRC
MSZ 8273(91)	OX260Cr27	Abr res; alloyed; HT	2.3-2.9	24-28	0.5-1.5	1 max	1.2 max	0.3 max	0.3 max	0.5-1.5	bal Fe	560-960			39-62 HRC
MSZ 8273(91)	OX260CrMoNi2021	Abr res; alloyed; HT	2.3-2.9	18-22	0.5-1	1.4-2	0.8-1.2	03 max	03 max	02-0.8	bal Fe	450-1000			39-62 HRC
MSZ 8273(91)	OX260NiCr42	Abr res; alloyed; HT	2.6-2.9	1.4-2.4	0.3-0.7	0.5 max	3.3-5	0.3 max	0.3 max	0.2-0.8	bal Fe	320-390			45-65 HRC
MSZ 8273(91)	OX270CrMo12	Abr res; alloyed; HT	2.5-2.9	11-13	0.4-0.8	0.5-0.8		0.3 max	0.3 max	0.5-1	Pb <=015; bal Fe				58-64 HRC
MSZ 8273(91)	OX270CrV25	Abr res; alloyed; HT	2.4-3.0	22-27	0.4-0.8			0.3 max	0.3 max	0.8-1.2	V 0.2-0.4; bal Fe				52-64 HRC
MSZ 8273(91)	OX300CrMo153	Abr res; alloyed; HT	2.3-3.6	14-17	0.5-1	1-3	0.7 max	0.3 max	0.3 max	0.2-0.8	bal Fe	450-1000			39-62 HRC
MSZ 8273(91)	OX300CrMo271	Abr res; alloyed; HT	3.0-3.5	23-28	0.5-1	1-2	1.2 max	0.3 max	0.3 max	0.2-1	bal Fe	450-1000			39-62 HRC
MSZ 8273(91)	OX300CrMoNi1521	Abr res; alloyed; HT	2.3-3.6	14-17	0.5-1.2	1.8-2.2	0.8-1.2	0.3 max	0.3 max	0.2-0.8	bal Fe	450-1000			39-62 HRC
MSZ 8273(91)	OX300CrNiSi952	Abr res; alloyed; HT	2.5-3.5	8-10	0.3-0.7	0.5 max	4.5-6.5	0.3 max	0.3 max	1.5-2.2	bal Fe	500-600			45-62 HRC
MSZ 8273(91)	OX300NiMo3Mg	Abr res; alloyed; HT	2.8-3.5		0.2-0.5	0.5-0.8	1.5-4.5	0.3 max	0.3 max	2-2.6	bal Fe	700-1300	600	1L	30-58 HRC
MSZ 8273(91)	OX330CrV25	Abr res; alloyed; HT	3.0-3.6	22-27	0.6-1			0.3 max	0.3 max	1-1.5	V 0.2-0.4; bal Fe				52-64 HRC
MSZ 8273(91)	OX330NiCr42	Abr res; alloyed; HT	3.0-3.6	1.4-2.4	0.3-0.7	0.5 max	3.3-5	0.3 max	0.3 max	0.2-0.8	bal Fe	280-350			45-62 HRC
MSZ 8274(81)	OV340NiCr	Corr res; as cast	3.4 max	0.2-0.5	1-1.6		0.5-1	0.3 max	0.1 max	1.4-2.5	bal Fe	240			290 HB max
MSZ 8274(81)	OVX300NiCr202	Corr res; as cast	3.0 max	1-2.5	0.5-1.5		18-22	0.1 max	0.1 max	1-2.8	bal Fe	170			230 HB max
MSZ 8274(81)	OVX300NiCuCr1562	Corr res; as cast	3.0 max	1-2.5	0.5-1.5		14.5-17.5	0.1 max	0.1 max	1-2.8	Cu 5.5-7.5; bal Fe	170			210 HB max
MSZ 8274(81)	OVX65Si15	Corr res; as cast	0.4-0.9		0.3-0.8			0.1 max	0.07 max	14-16	bal Fe				400 HB max
MSZ 8274(81)	OVX65SiMo153	Corr res; as cast	0.4-0.9		0.3-0.8	2-4		0.1 max	0.07 max	14-16	bal Fe				400 HB max
MSZ 8278(81)	OV340Cr6	Heat res	3.0-3.8	1-2	1 max			0.3 max	0.12 max	2-3.0	bal Fe				
MSZ 8278(81)	OV340Cr9	Heat res	3.0-3.8	2-2.7	1 max			0.3 max	0.12 max	2.5-3.5	bal Fe				

Specification	Designation	Notes	C	Cr	Mn	Mo	Ni	P	S	Si	Other	UTS	YS	El	Hard

Cast Iron, Unclassified, No equivalents identified

Hungary

Specification	Designation	Notes	C	Cr	Mn	Mo	Ni	P	S	Si	Other	UTS	YS	El	Hard
MSZ 8278(91)	GOVX200A122	Heat res; as cast/HT	1.6-2.5		0.8 max			0.2 max	0.03 max	1-2	Al 19.0-25.0; bal Fe	250		5L	180-360 HB
MSZ 8278(91)	GOVX300Si5	Heat res; as cast/HT	2.7-3.3		0.5 max			0.1 max	0.02 max	4.5-6	bal Fe	500		2L	260-320 HB
MSZ 8278(91)	OV340Cr	Heat res; as cast/HT	3.0-3.8	0.5-1	1 max			0.3 max	0.12 max	1.5-2.5	bal Fe	175			210-290 HB
MSZ 8278(91)	OV340Cr1.5	Heat res; as cast/HT	3.0-3.8	1-2	1 max			0.3 max	0.12 max	2-3.0	bal Fe	150			210-290 HB
MSZ 8278(91)	OV340Cr2.5	Heat res; as cast/HT	3.0-3.8	2-2.7	1 max			0.3 max	0.12 max	2.5-3.5	bal Fe	150			230-370 HB
MSZ 8278(91)	OVX110A130	Heat res; as cast/HT	1.0-1.2		0.7 max			0.04 max	0.1 max	0.5 max	Al 29.0-31.0; bal Fe	200			360-540 HB
MSZ 8278(91)	OVX200A122	Heat res; as cast/HT	1.6-2.5		0.8 max			0.2 max	0.08 max	1-2	Al 19.0-25.0; bal Fe	90			290 HB max
MSZ 8278(91)	OVX200Cr16	Heat res; as cast/HT	1.6-2.4	15-18	1 max			0.1 max	0.05 max	1.5-2.2	bal Fe	350			400-450 HB
MSZ 8278(91)	OVX210A1Si65	Heat res; as cast/HT	1.8-2.4		0.8 max			0.3 max	0.12 max	4.5-6	Al 5.5-7; bal Fe	120			235-300 HB
MSZ 8278(91)	OVX230Cr30	Heat res; as cast/HT	1.6-3.0	28-32	0.7 max			0.1 max	0.06 max	1.5-2	bal Fe	300			360-550 HB
MSZ 8278(91)	OVX285A1Cr72	Heat res; as cast/HT	2.5-3.2	1.5-3.0	1 max			0.3 max	0.12 max	1.5-3.0	Al 5-9; bal Fe	120			240-290 HB
MSZ 8278(91)	OVX285SiCr5	Heat res; as cast/HT	2.5-3.2	0.5-1	0.8 max			0.3 max	0.12 max	4.5-6	bal Fe	150			140-300 HB

India

Specification	Designation	Notes	C	Cr	Mn	Mo	Ni	P	S	Si	Other	UTS	YS	El	Hard
IS 224/I(79)	PG10Mn5P38		0.3 max	0.3 max	2 max	0.15 max	0.4 max	0.05 max	0.04 max	2.25-2.75	Co <=0.1; Cu <=0.3; Pb <=0.15; W <=0.1; bal Fe				
IS 224/I(79)	PG12Mn5P38		0.3 max	0.3 max	2 max	0.15 max	0.4 max	0.05 max	0.04 max	2.75-3.25	Co <=0.1; Cu <=0.3; Pb <=0.15; W <=0.1; bal Fe				
IS 224/I(79)	PG6Mn5P38		0.3 max	0.3 max	2 max	0.15 max	0.4 max	0.05 max	0.04 max	1.25-1.75	Co <=0.1; Cu <=0.3; Pb <=0.15; W <=0.1; bal Fe				
IS 224/I(79)	PG8Mn5P38		0.3 max	0.3 max	2 max	0.15 max	0.4 max	0.05 max	0.04 max	1.75-2.25	Co <=0.1; Cu <=0.3; Pb <=0.15; W <=0.1; bal Fe				
IS 2842/II(80)	PG6Mn7P38		0.3 max	0.3 max	1.5-9.99	0.15 max	0.4 max	0.06 max	0.04 max	1.25-1.75	Co <=0.1; Cu <=0.3; Pb <=0.15; W <=0.1; bal Fe				
IS 7925	1	White; Abr res; as cast	2.4-3.4	2 max	0.2-0.8			0.15 max		0.5-1.5	bal Fe				400 HB
IS 7925	2	White; Abr res; as cast	2.4-3.4	2 max	0.2-0.8			0.15 max		0.5-1.5	bal Fe				400 HB
IS 7925	3	White; Abr res; as cast	2.4-3	2 max	0.2-0.8			0.15 max		0.5-1.5	bal Fe				250 HB

Japan

Specification	Designation	Notes	C	Cr	Mn	Mo	Ni	P	S	Si	Other	UTS	YS	El	Hard
JIS G5503(95)	FCD1000A	Unnotched Charpy 80 J; AustmpDuctile													
JIS G5503(95)	FCD1200A	Unnotched Charpy 80 J; AustmpDuctile													
JIS G5503(95)	FCD900A	Unnotched Charpy 80 J; AustmpDuctile													

Sweden

Specification	Designation	Notes	C	Cr	Mn	Mo	Ni	P	S	Si	Other	UTS	YS	El	Hard
SS 140747	0747-03	Austempered Ductile											900	600	8	280-320 HB
SS 140747	0747-04	Austempered Ductile											1400	1200	1	

UK

Specification	Designation	Notes	C	Cr	Mn	Mo	Ni	P	S	Si	Other	UTS	YS	El	Hard
BS 4844	1A	White; Abr res; as cast	2.4-3.4	2 max	0.2-0.8			0.15 max		0.5-1.5	bal Fe				400 HB
BS 4844	1B	White; Abr res; as cast	2.4-3.4	2 max	0.2-0.8			0.5 max		0.5-1.5	bal Fe				400 HB
BS 4844	1C	White; Abr res; HT	2.4-3	2 max	0.2-0.8			0.15 max		0.5-1.5	bal Fe				250 HB

USA

Specification	Designation	Notes	C	Cr	Mn	Mo	Ni	P	S	Si	Other	UTS	YS	El	Hard
ASTM A518/A518M(97)	A518(1)	Corr res; High-Silicon	0.65-1.10	0.50 max	1.50 max	0.50 max				14.20-14.75	Cu <=0.50; bal Fe				

UNS numbers and US grades are provided as a means of cross referencing chemically similar alloys. Exchangability is only possible after independent examination of specifications. Tensile properties are minimum or typical as specified. UTS and YS as MPa. El as %. See Appendix for list of abbreviations used in Notes. * indicates obsolete material.

Specification	Designation	Notes	C	Cr	Mn	Mo	Ni	P	S	Si	Other	UTS	YS	El	Hard
Cast Iron, Unclassified, No equivalents identified															
USA															
ASTM A518/A518M(97)	A518(2)	Corr res; High-Silicon	0.75-1.15	3.25-5.00	1.50 max	0.40-0.60			14.20-14.75		Cu <=0.50; bal Fe				
ASTM A518/A518M(97)	A518(3)	Corr res; High-Silicon	0.70-1.10	3.25-5.00	1.50 max	0.20 max			14.20-14.75		Cu <=0.50; bal Fe				
ASTM A532/A532M(93a)	25% Cr	White	2.0-3.3	23.0-30.0	2.0 max	3.0 max	2.5 max	0.10 max	0.06 max	1.5 max	Cu <=1.2; bal Fe				450 HB
ASTM A842(97)	250	Compacted graphite; Ferr										250	175	3.0	
ASTM A842(97)	300	Compacted graphite										300	210	1.5	
ASTM A842(97)	350	Compacted graphite										350	245	1.0	
ASTM A842(97)	400	Compacted graphite										400	280	1.0	
ASTM A842(97)	450	Compacted graphite; Prl										450	315	1.0	
ASTM A861(94)	A861(1)	High-Silicon iron pipe and fittings	0.65-1.10	0.50 max	1.50 max	0.50 max			14.20-14.75		Cu <=0.50; bal Fe				
ASTM A861(94)	A861-2	High-Silicon iron pipe and fittings	0.75-1.15	3.25-5.00	1.50 max	0.40-0.60			14.20-14.75		Cu <=0.50; bal Fe				

Specification	Designation	Notes	C	Cr	Mn	Mo	Ni	P	S	Si	Other	UTS	YS	El	Hard
Cast Stainless Steel, Austenitic, 302															
Romania															
STAS 10718(88)	T15NiCr180	Corr res	0.15 max	17-20	2 max		8-11	0.04 max	0.03 max	0.5-2	Pb <=0.15; Ti>=5C; bal Fe				
USA															
	UNS J92501		0.15 max	17.0-19.0	2.00 max		8.0-10.0	0.04 max	0.03 max	1.00 max	bal Fe				
MIL-S-81591(93)	IC-302*	Inv, C/Corr res; Obs for new design see AMS specs									bal Fe				
Cast Stainless Steel, Austenitic, 303															
USA															
	UNS J92511		0.12 max	17.0-20.0	2.00 max	0.60 max	8.0-10.0	0.17 max	0.15-0.35	1.00 max	Cu <=0.50; bal Fe				
	UNS J92711	60303	0.16 max	18.00-21.00	2.00 max	0.80 max	9.00-21.00	0.04 max	0.15-0.35	2.00 max	Cu <=0.50; bal Fe				
AMS 5341C(95)*		Noncurrent; Inv cast; Corr res	0.16 max	18.00-21.00	2.00 max	0.75 max	9.00-12.00	0.04 max	0.15-0.35	2.00 max	Cu <=0.75; bal Fe				180 HB max
MIL-S-81591(93)	IC-303*	Inv, C/Corr res; Obs for new design see AMS specs									bal Fe				
Cast Stainless Steel, Austenitic, 304															
UK															
BS 1504(76)	364C11*	Press ves	0.07 max	20.0-24.0	2.0 max	3.0-6.0	20.0-26.0	0.03 max	0.03 max	2.5 max	Cu <=2.0; Nb <=0.5; Ti <=0.05; W <=0.1; bal Fe				
USA															
	UNS J92610		0.08 max	18.0-20.0	2.00 max		8.0-12.0	0.04 max	0.03 max	1.00 max	bal Fe				
MIL-S-81591(93)	IC-304*	Inv, C/Corr res; Obs for new design see AMS specs									bal Fe				
Cast Stainless Steel, Austenitic, 304L															
Bulgaria															
BDS 6738(72)	000CH18N11	Corr res	0.03 max	17-19	2 max	0.3 max	10-12.5	0.035 max	0.02 max	0.8 max	Ti <=0.05; V <=0.1; W <=0.1; bal Fe				
Italy															
UNI 3161(83)	GX2CrNi1910	Sand; Heat res	0.03 max	17-21	1.5 max		8-12	0.04 max	0.035 max	2 max	bal Fe				
UK															
BS 1504(76)	304S12*		0.03 max	17-21	2 max		8	0.04 max	0.04 max	1.5 max	bal Fe				
USA															
	UNS J92620		0.05 max	18.0-21.0	1.00-2.00	0.50 max	8.0-11.0	0.04 max	0.03 max	1.00 max	Cu <=0.50; bal Fe				
MIL-S-81591(93)	IC-304L*	Inv, C/Corr res; Obs for new design see AMS specs									bal Fe				
Cast Stainless Steel, Austenitic, 310															
UK															
BS 1504(76)	310C40*	Press ves	0.3-0.5	24.0-27.0	2.0 max	0.15 max	19.0-22.0	0.04 max	0.04 max	1.5 max	Cu <=0.3; Ti <=0.05; W <=0.1; bal Fe				
USA															
	UNS J94302	310	0.25 max	24.0-26.0	2.00 max		19.0-22.0	0.04 max	0.03 max	1.00 max	bal Fe				
MIL-S-81591(93)	IC-310*	Inv, C/Corr res; Obs for new design see AMS specs									bal Fe				
Cast Stainless Steel, Austenitic, 316															
Germany															
DIN SEW 390(91)	G-X2CrNiMoN18 14	Non-magnetic	0.03 max	16.5-18.5	2.00 max	2.50-3.00	13.0-15.0	0.045 max	0.015 max	1.00 max	N 0.15-0.25; bal Fe	490-690	295	30	

UNS numbers and US grades are provided as a means of cross referencing chemically similar alloys. Exchangability is only possible after independent examination of specifications. Tensile properties are minimum or typical as specified. UTS and YS as MPa. El as %. See Appendix for list of abbreviations used in Notes. * indicates obsolete material.

Specification	Designation	Notes	C	Cr	Mn	Mo	Ni	P	S	Si	Other	UTS	YS	El	Hard

Cast Stainless Steel, Austenitic, 316 (Continued from previous page)

Germany

Specification	Designation	Notes	C	Cr	Mn	Mo	Ni	P	S	Si	Other	UTS	YS	El	Hard
DIN SEW 390(91)	WNr 1.3952	Non-magnetic	0.03 max	16.5-18.5	2.00 max	2.50-3.00	13.0-15.0	0.045 max	0.015 max	1.00 max	N 0.15-0.25; bal Fe	490-690	295	30	

UK

| BS 1504(76) | 316C71* | Press ves | 0.08 max | 17.0-21.0 | 2.0 max | 2.0-3.0 | 8.0-14.00 | 0.04 max | 0.04 max | 1.5 max | Cu <=0.3; Ti <=0.05; W <=0.1; bal Fe | | | | |

USA

| | UNS J92810 | 316 | 0.08 max | 16.0-18.0 | 2.00 max | 2.0-3.0 | 10.0-14.0 | 0.04 max | 0.03 max | 1.00 max | bal Fe | | | | |

Cast Stainless Steel, Austenitic, 316H

Romania

Specification	Designation	Notes	C	Cr	Mn	Mo	Ni	P	S	Si	Other	UTS	YS	El	Hard
STAS 9277(84)	OTA10NbMoNiCr170	Corr res	0.1 max	16-18	2 max	2-2.5	10-13	0.035 max	0.03 max	1 max	Pb <=0.15; Nb>=8C; bal Fe				
STAS 9277(84)	OTA10TiMoNiCr170	Corr res	0.1 max	16-18	2 max	2-2.5	10-13	0.035 max	0.03 max	1 max	Pb <=0.15; Ti <=1.30; Ti>=5C; bal Fe				

USA

| | UNS J92920 | 316H | 0.04-0.10 | 16.0-18.0 | 2.00 max | 2.00-3.00 | 11.0-14.0 | 0.040 max | 0.030 max | 0.75 max | bal Fe | | | | |

Cast Stainless Steel, Austenitic, 316N

Bulgaria

Specification	Designation	Notes	C	Cr	Mn	Mo	Ni	P	S	Si	Other	UTS	YS	El	Hard
BDS 6738(72)	000CH17N14M2	Corr res	0.03 max	16-18	2 max	2-2.5	12-15	0.035 max	0.02 max	0.8 max	Cu <=0.30; Ti <=0.05; V <=0.1; W <=0.1; bal Fe				

USA

| | UNS J92804 | CF3MN | 0.03 max | 17.0-22.0 | 1.50 max | 2.0-3.0 | 9.0-13.0 | 0.040 max | 0.040 max | 1.50 max | N 0.10-0.20; bal Fe | | | | |
| ACI | CF3MN | Corr res | 0.03 max | 17.0-22.0 | 1.50 max | 2.0-3.0 | 9.0-13.0 | 0.040 max | 0.040 max | 1.50 max | N 0.10-0.20; bal Fe | | | | |

Cast Stainless Steel, Austenitic, 321

Hungary

Specification	Designation	Notes	C	Cr	Mn	Mo	Ni	P	S	Si	Other	UTS	YS	El	Hard
MSZ 21053(82)	AoX12CrNiMoNb1810F	Corr res; 0-30.0mm; SA	0.15 max	17-19	2 max	2-3	9-11	0.04 max	0.04 max	1.5 max	Nb=8C-1.3; bal Fe	460	200	20L	140-210HB
MSZ 21053(82)	AoX12CrNiMoTi1810	Corr res; 0-30.0mm; as cast	0.15 max	17-19	2 max	2-3	9-11	0.04 max	0.04 max	1.5 max	Ti <=0.9; Ti=5C-0.9; bal Fe				160-230 HB
MSZ 21053(82)	AoX12CrNiMoTi1810F	Corr res; 0-30.0mm; SA	0.15 max	17-19	2 max	2-3	9-11	0.04 max	0.04 max	1.5 max	Ti <=0.9; Ti=5C-0.9; bal Fe	460	200	20L	140-210 HB
MSZ 21053(82)	AoX12CrNiTi189	Corr res; 0-30.0mm; as cast	0.15 max	17-19	2 max		8-11	0.04 max	0.04 max	1.5 max	Ti <=0.9; Ti+5C-0.9; bal Fe				160-230 HB
MSZ 21053(82)	AoX12CrNiTi189F	Corr res; 0-30.0mm; SA	0.15 max	17-19	2 max		8-11	0.04 max	0.04 max	1.5 max	Ti <=0.9; Ti+5C-0.9; bal Fe	460	200	20L	140-210 HB

USA

| | UNS J92630 | | 0.08 max | 17.0-19.0 | 2.00 max | | 9.0-12.0 | 0.04 max | 0.03 max | 1.00 max | Ti 5xC min; bal Fe | | | | |
| MIL-S-81591(93) | IC-321* | Inv, C/Corr res; Obs for new design see AMS specs | | | | | | | | | bal Fe | | | | |

Cast Stainless Steel, Austenitic, 347

USA

Specification	Designation	Notes	C	Cr	Mn	Mo	Ni	P	S	Si	Other	UTS	YS	El	Hard
	UNS J92640		0.08 max	17.0-19.5	2.00 max		9.0-13.0	0.04 max	0.03 max	1.00 max	Nb+Ta 10xC min, 1.5 max; bal Fe				
MIL-S-81591(93)	IC-347*	Inv, C/Corr res; Obs for new design see AMS specs									bal Fe				

Cast Stainless Steel, Austenitic, 347H

USA

Specification	Designation	Notes	C	Cr	Mn	Mo	Ni	P	S	Si	Other	UTS	YS	El	Hard
	UNS J92660		0.04-0.10	17.0-20.0	2.00 max		9.00-13.0	0.040 max	0.030 max	0.75 max	(Nb/Ta) 8xC min-1.0 max; bal Fe				

Specification	Designation	Notes	C	Cr	Mn	Mo	Ni	P	S	Si	Other	UTS	YS	El	Hard

Cast Stainless Steel, Austenitic, CF10M

USA

| | UNS J92901 | Cr-Ni-Mo CF-10M | 0.04-0.10 | 18.0-21.0 | 1.50 max | 2.0-3.0 | 9.0-12.0 | 0.040 max | 0.040 max | 1.50 max | bal Fe | | | | |
| ACI | CF-10M | Cr-Ni-Mo | 0.04-0.10 | 18.0-21.0 | 1.50 max | 2.0-3.0 | 9.0-12.0 | 0.040 max | 0.040 max | 1.50 max | bal Fe | | | | |

Cast Stainless Steel, Austenitic, CF16F

Bulgaria

| BDS 9631 | Ch18N10SL | Corr res | 0.15 max | 17-19 | 2 max | | 8-12 | 0.04 | 0.04 | 2 max | bal Fe | | | | |
| BDS 9631 | Ch19N9L | Corr res | 0.15 max | 17-19 | 2 max | | 8-11 | | | 1.5 max | bal Fe | | | | |

China

| GB 2100(80) | ZG1Cr18Ni9 | Corr res; Quen | 0.12 max | 17.0-20.0 | 0.8-2.0 | | 8.0-11.0 | 0.045 max | 0.030 max | 1.5 max | bal Fe | 441 | 196 | 25 | |

France

| AFNOR NFA35586 | Z10CN18.9M | Corr res | 0.12 max | 17-19.5 | 1.5 max | | 8-10 | | | 2 max | bal Fe | | | | |

Germany

DIN	GX10CrNi18-8		0.12 max	17.0-19.5	1.50 max		8.00-10.0	0.045 max	0.030 max	2.00 max	bal Fe	440-640	175	20	
DIN	WNr 1.4312		0.12 max	17.0-19.5	1.50 max		8.00-10.0	0.045 max	0.030 max	2.00 max	bal Fe	440-640	175	20	
DIN SEW 395(87)	G-X12CrNi18 11	Non-magnetic	0.15 max	16.5-18.5	2.00 max	0.75 max	10.0-12.0	0.045 max	0.030 max	1.00 max	bal Fe				
DIN SEW 395(87)	WNr 1.3955	Non-magnetic	0.15 max	16.5-18.5	2.00 max	0.75 max	10.0-12.0	0.045 max	0.030 max	1.00 max					
TGL 10414	GS-X15CrNiSi19.10*	Corr res	0.1-0.2	18-20	1.5 max		9-11			1.5-2.5	bal Fe				
TGL 14415	GSF-X12CrNi189*		0.16 max	17-19	1 max		8-10			1.5 max	bal Fe				

Mexico

| DGN B-354 | CF-16F* | Obs; Corr res | 0.16 | 18-21 | 1.5 | | 9-12 | 0.04 | 0.04 | 2 | bal Fe | | | | |

Poland

| PNH83158(86) | LH18N9 | Corr res | 0.15 max | 17-19 | 2 max | | 8-11 | 0.035 max | 0.035 max | 2 max | bal Fe | | | | |

Romania

STAS 6855	T15NiCr180X	Corr res	0.15 max	17-19	2 max		8-10			0.5-2	bal Fe				
STAS 6855(86)	T15NiCr180	Corr res; Heat res	0.15 max	17-20	2 max		8-10	0.04 max	0.03 max	0.5-2	Pb <=0.15; bal Fe				
STAS 6855(86)	T15TiNiCr180	Corr res; Heat res	0.15 max	17-19	2 max		8-12	0.035 max	0.03 max	2 max	Pb <=0.15; Ti <=0.8; 5(C-0.03)<=Ti<=0.8; bal Fe				

Russia

| GOST | 10Ch18N9L | Corr res | 0.07-0.14 | 17-20 | 1-2 | | 8-11 | 0.035 | 0.03 | 0.2-1 | bal Fe | | | | |
| GOST | 10Ch18N9MLS | Corr res | 0.14-0.7 | 17-20 | 1-1.8 | 0.1-0.2 | 8-11 | 0.02 | 0.02 | 0.2-1 | Cu 0.3; bal Fe | | | | |

UK

| BS 3100(91) | 302C25 | Corr res; old En 1631D | 0.12 max | 17.0-21.0 | 2.00 max | | 8.00 min | 0.040 max | 0.040 max | 1.50 max | bal Fe | 480 | | 26 | |
| BS 3146/2(75) | ANC3(A) | Corr res | 0.12 max | 17.0-20.0 | 0.20-2.00 | | 8.00-12.0 | 0.035 max | 0.035 max | 0.20-2.00 | | | | | |

USA

	UNS J92701		0.16 max	18.0-21.0	1.50 max		9.0-12.0	0.04 max	0.04 max	2.00 max	bal Fe				
ACI	CF-16F	Corr res	0.16 max	18.0-21.0	1.50 max	3.0 max	9.0-12.0	0.04 max	0.04 max	2.00 max	bal Fe				
AMS 5341C(95)*		Noncurrent; Corr res	0.16 max	18.00-21.00	2.00 max	0.75 max	9.00-12.00	0.04 max	0.15-0.35	2.00 max	Cu <=0.75; bal Fe				
ASTM A743/A743M(98)	CF-16F	Corr res; Solution Q/A	0.16 max	18.0-21.0	1.5 max	1.50 max	9.0-12.0	0.17 max	0.04 max	2.00 max	Se 0.2-0.35; bal Fe	485	205	25	
ASTM A743/A743M(98)	CF-16Fa	Corr res; Solution Q/A	0.16 max	18.0-21.0	1.5 max	0.40-0.80	9.0-12.0	0.04 max	0.20-0.40	2.00 max	bal Fe	485	205	25	

UNS numbers and US grades are provided as a means of cross referencing chemically similar alloys. Exchangability is only possible after independent examination of specifications. Tensile properties are minimum or typical as specified. UTS and YS as MPa. El as %. See Appendix for list of abbreviations used in Notes. * indicates obsolete material.

Specification	Designation	Notes	C	Cr	Mn	Mo	Ni	P	S	Si	Other	UTS	YS	El	Hard

Cast Stainless Steel, Austenitic, CF20

Bulgaria

Specification	Designation	Notes	C	Cr	Mn	Mo	Ni	P	S	Si	Other	UTS	YS	El	Hard
BDS 9631	Ch18N10SL	Corr res	0.15 max	17-19	2 max		8-12	0.04	0.04	2 max	bal Fe				
China															
GB 2100(80)	ZG1Cr18Ni9	Corr res; Quen	0.12 max	17.0-20.0	0.8-2.0		8.0-11.0	0.045 max	0.030 max	1.5 max	bal Fe	441	196	25	
Czech Republic															
CSN 422931	422931	Corr res	0.15 max	18-21	1.5 max		8-11	0.045 max	0.04 max	1.5 max	bal Fe				
CSN 422932	422932	Corr res	0.15-0.35	17-20	1.5 max		8-11			1-2	bal Fe				
CSN 422933	422933	Corr res	0.12 max	17.0-19.0	1.5 max		9.0-11.0	0.045 max	0.04 max	2.0 max	Ti <=0.8; Ti=5C-0.8; bal Fe				
France															
AFNOR NFA35586	Z10CN189M	Corr res	0.12 max	17-19.5	1.5 max		8-10			2 max	bal Fe				
Germany															
TGL 10414	GS-X15CrNiSi1910*	Corr res	0.1-0.2	18-20	1.5 max		9-11			1.5-2.5	bal Fe				
TGL 14394	GS-X10CrNiN197*	Obs	0.12 max	18-20	2 max		6-8			1.5 max	N 0.1-0.2; bal Fe				
TGL 14415	GSF-X12CrNi189*	Corr res	0.16 max	17-19	1 max		8-10			1.5 max	bal Fe				
Hungary															
MSZ 21053(82)	AoX12CrNi189	Corr res; 0-30.0mm; as cast	0.15 max	17-19	2 max		8-11	0.04 max	0.04 max	1.5 max	bal Fe				160-230 HB
MSZ 21053(82)	AoX12CrNi189F	Corr res; 0-30.0mm; SA	0.15 max	17-19	2 max		8-11	0.04 max	0.04 max	1.5 max	bal Fe	450	200	20L	140-210 HB
Italy															
UNI 3161(83)	GX16CrNi2010	Sand; Heat res	0.2 max	18-21.	1.5 max		8-11	0.04 max	0.035 max	2 max	bal Fe				
Japan															
JIS G5121(91)	SCS12	Corr res; Pipe SA	0.20 max	18.00-21.00	2.00 max		8.00-11.00	0.040 max	0.040 max	2.00 max	bal Fe	480	205	28	183 HB max
Mexico															
DGN B-354	CF-20*	Obs; Corr res	0.2	18-21	1.5		8-11	0.04	0.04	2	bal Fe				
Poland															
PNH83158(86)	LH18N9	Corr res	0.15 max	17-19	2 max		8-11			1.5 max	bal Fe				
Russia															
GOST	10Ch18N9MLS	Corr res	0.14-0.7	17-20	1-1.8	0.1-0.2	8-11	0.02	0.02	0.2-1	Cu 0.3; bal Fe				
Spain															
UNE 36257(74)	AM-X7CrNi20-10	Corr res	0.8 max	18-21	1.5 max		8-11	0.040	0.040	2 max	bal Fe				
UNE 36257(74)	F.8411*	see AM-X7CrNi20-10	0.8 max	18-21	1.5 max		8-11	0.040	0.040	2 max	bal Fe				
UK															
BS 3146/2(75)	ANC3(B)	Corr res	0.12 max	17.0-20.0	0.20-2.00		8.50-12.0	0.035 max	0.035 max	0.20-2.00	bal Fe				
USA															
	UNS J92602		0.20 max	18.0-21.0	1.50 max		8.0-11.0	0.04 max	0.04 max	2.00 max	bal Fe				
ACI	CF-20	Corr res	0.20 max	18.0-21.0	1.50 max		8.0-11.0	0.04 max	0.04 max	2.00 max	bal Fe				
AMS 5358C(99)*		Noncurrent; Inv cast; Corr res	0.25 max	17.00-19.00	1.50 max	0.75 max	8.00-10.00	0.04 max	0.03 max	2.00 max	Cu <=0.75; bal Fe				
ASTM A743/A743M(98)	CF-20	Corr res; Solution Q/A	0.20 max	18.0-21.0	1.5 max		8.0-11.0	0.04 max	0.04 max	2.00 max	bal Fe	485	205	30	

Cast Stainless Steel, Austenitic, CH20

Specification	Designation	Notes	C	Cr	Mn	Mo	Ni	P	S	Si	Other	UTS	YS	El	Hard
Australia															
AS 2074(82)	H8A	Corr res	0.2 max	22-27	2 max		10-14	0.04 max	0.04 max	2 max	bal Fe				
Italy															
UNI 3161(83)	GX16CrNi2414	Sand; Heat res	0.2 max	22-26	1.5 max		12-15	0.04 max	0.035 max	2 max	bal Fe				

UNS numbers and US grades are provided as a means of cross referencing chemically similar alloys. Exchangability is only possible after independent examination of specifications. Tensile properties are minimum or typical as specified. UTS and YS as MPa. El as %. See Appendix for list of abbreviations used in Notes. * indicates obsolete material.

Specification	Designation	Notes	C	Cr	Mn	Mo	Ni	P	S	Si	Other	UTS	YS	El	Hard

Cast Stainless Steel, Austenitic, CH20 (Continued from previous page)

Japan

Specification	Designation	Notes	C	Cr	Mn	Mo	Ni	P	S	Si	Other	UTS	YS	El	Hard
JIS G5121(91)	SCS17	Corr res; Tmp	0.20 max	22.00-26.00	2.00 max		12.00-15.00	0.040 max	0.040 max	2.00 max	bal Fe	480	205	28	183 HB max

Mexico

Specification	Designation	Notes	C	Cr	Mn	Mo	Ni	P	S	Si	Other	UTS	YS	El	Hard
DGN B-354	CH-20*	Obs; Corr res	0.2	22-26	1.5		12-15	0.04	0.04	2	bal Fe				

USA

Specification	Designation	Notes	C	Cr	Mn	Mo	Ni	P	S	Si	Other	UTS	YS	El	Hard
	UNS J93402		0.20 max	22.0-26.0	1.50 max		12.0-15.0	0.04 max	0.04 max	2.00 max	bal Fe				
ACI	CH-20	Corr res	0.20 max	22.0-26.0	00-1.50		12.0-15.0	0.04 max	0.04 max	2.00 max	bal Fe				
ASTM A351(94)	CH-20	Press ves	0.04-0.20	22.0-26.0	1.50 max	0.50 max	12.0-15.0	0.040 max	0.040 max	2.00 max	bal Fe	485	205	30.0	
ASTM A451(97)	CPH20	Pipe; SA Quen; High-temp	0.20 max	22.0-26.0	1.50 max		12.0-15.0	0.040 max	0.040 max	2.00 max	bal Fe	485	205	30.0	
ASTM A743/A743M(98)	CH-20	Corr res; Solution Q/A	0.20 max	22.0-26.0	1.5 max		12.0-15.0	0.04 max	0.04 max	2.00 max	bal Fe	485	205	30	

Cast Stainless Steel, Austenitic, CK20

Germany

Specification	Designation	Notes	C	Cr	Mn	Mo	Ni	P	S	Si	Other	UTS	YS	El	Hard
DIN 17470(84)	CrNi2520	Heat conducting	0.20 max	22.0-25.0	2.00 max		19.0-22.0	0.045 max	0.030 max	1.50-2.50	bal Fe				
DIN 17470(84)	WNr 1.4843	Heating conductor	0.20 max	22.0-25.0	2.00 max		19.0-22.0	0.045 max	0.030 max	1.50-2.50	bal Fe				
DIN SEW 595(76)	GX15CrNi25-20	Petrochemical use	0.10-0.20	24.0-26.0	0.50-1.50		19.0-21.0	0.045 max	0.030 max	0.50-1.50	bal Fe				
DIN SEW 595(96)	WNr 1.4840	Petrochemical use	0.10-0.20	24.0-26.0	0.50-1.50		19.0-21.0	0.045 max	0.030 max	0.50-1.50	bal Fe				
TGL 10414	GS-X15CrNiSi2419*	Corr res	0.1-0.2	23-25	1.5 max		18-20			1-2.5	bal Fe				
TGL 14415	GSF-X15CrNiSi2419*	Corr res	0.1-0.2	23-25	1.5 max		18-20			1.51-2	bal Fe				

International

Specification	Designation	Notes	C	Cr	Mn	Mo	Ni	P	S	Si	Other	UTS	YS	El	Hard
ISO DP4991	C68	Corr res													

Italy

Specification	Designation	Notes	C	Cr	Mn	Mo	Ni	P	S	Si	Other	UTS	YS	El	Hard
UNI 3161(83)	GX16CrNi2521	Sand; Heat res	0.2 max	23.-27	1.5 max		19-22	0.04 max	0.035 max	2 max	bal Fe				

Japan

Specification	Designation	Notes	C	Cr	Mn	Mo	Ni	P	S	Si	Other	UTS	YS	El	Hard
JIS G5121(91)	SCS18	Corr res; Tmp	0.20 max	23.00-27.00	2.00 max		19.00-22.00	0.040 max	0.040 max	2.00 max	bal Fe	450	195	28	183 HB max

Mexico

Specification	Designation	Notes	C	Cr	Mn	Mo	Ni	P	S	Si	Other	UTS	YS	El	Hard
DGN B-354	CK-20*	Obs; Corr res	0.2	23-27	2		19-22	0.04	0.04	2	bal Fe				

Poland

Specification	Designation	Notes	C	Cr	Mn	Mo	Ni	P	S	Si	Other	UTS	YS	El	Hard
PNH83159(90)	LH25N19S2	Heat res	0.25 max	24-26	0.6 max		17.5-19.5	0.04 max	0.035 max	2-3	bal Fe				

Romania

Specification	Designation	Notes	C	Cr	Mn	Mo	Ni	P	S	Si	Other	UTS	YS	El	Hard
STAS 6855	T25NiCr250X	Corr res	0.25 max	23-27	1.5 max		18-21			1-2.5	bal Fe				
STAS 6855(86)	T25NiCr250	Corr res; Heat res	0.25 max	23-27	1.5 max		18-21	0.035 max	0.03 max	1-2.5	Pb <=0.15; bal Fe				

USA

Specification	Designation	Notes	C	Cr	Mn	Mo	Ni	P	S	Si	Other	UTS	YS	El	Hard
	UNS J94202		0.20 max	23.0-27.0	2.00 max		19.0-22.0	0.04 max	0.04 max	2.00 max	bal Fe				
ACI	CK-20	Corr res	0.20 max	23.0-27.0	2.00 max		19.0-22.0	0.04 max	0.04 max	2.00 max	bal Fe				
ASTM A351(94)	CK-20	Press ves	0.04-0.20	23.0-27.0	1.50 max	0.50 max	19.0-22.0	0.040 max	0.040 max	1.75 max	bal Fe	450	195	30.0	
ASTM A451(97)	CPK20	Pipe; SA Quen; High-temp	0.20 max	23.0-27.0	1.50 max		19.0-22.0	0.040 max	0.040 max	1.75 max	bal Fe	448	195	30.0	
ASTM A743/A743M(98)	CK-20	Corr res; Solution Q/A	0.20 max	23.0-27.0	2.00 max		19.0-22.0	0.04 max	0.04 max	2.00 max	bal Fe	450	195	30	

Cast Stainless Steel, Austenitic, CN3M

USA

Specification	Designation	Notes	C	Cr	Mn	Mo	Ni	P	S	Si	Other	UTS	YS	El	Hard
	UNS J94652 CN-3M	Austenitic Cr-Ni-Mo	0.03 max	20.0-22.0	2.0 max	4.5-5.5	23.0-27.0	0.03 max	0.03 max	1.0 max	bal Fe				
ACI	CN-3M	Corr res; Cr-Ni-Mo	0.03 max	20.0-22.0	2.0 max	4.5-5.5	23.0-27.0	0.03 max	0.03 max	1.0 max	bal Fe				

UNS numbers and US grades are provided as a means of cross referencing chemically similar alloys. Exchangability is only possible after independent examination of specifications. Tensile properties are minimum or typical as specified. UTS and YS as MPa. El as %. See Appendix for list of abbreviations used in Notes. * indicates obsolete material.

Specification	Designation	Notes	C	Cr	Mn	Mo	Ni	P	S	Si	Other	UTS	YS	El	Hard

Cast Stainless Steel, Austenitic, CN3M (Continued from previous page)

USA

Specification	Designation	Notes	C	Cr	Mn	Mo	Ni	P	S	Si	Other	UTS	YS	El	Hard
ASTM A743/A743M(98)	CN-3M	Corr res	0.03 max	20.0-22.0	2.0 max	4.5-5.5	23.0-27.0	0.03 max	0.03 max	1.0 max	bal Fe	435	170	30	

Cast Stainless Steel, Austenitic, CN3MN

USA

Specification	Designation	Notes	C	Cr	Mn	Mo	Ni	P	S	Si	Other	UTS	YS	El	Hard
	UNS J94651	Austenitic Cr-Ni-Mo-N CN-3MN, AL-6XN	0.03 max	20.0-22.0	2.00 max	6.0-7.0	23.5-25.5	0.040 max	0.010 max	1.00 max	Cu <=0.75; N 0.18-0.26; bal Fe				
ACI	CN-3MN	Corr res; Cr-Ni-Mo-N	0.03 max	20.0-22.0	2.00 max	6.0-7.0	23.5-25.5	0.040 max	0.010 max	1.00 max	Cu <=0.75; N 0.18-0.26; bal Fe				
ACI	CN3MN	Corr res	0.03 max	20.0-22.0	2.00 max	6.0-7.0	23.5-25.5	0.040 max	0.10 max	1.00 max	Cu <=0.75; N 0.18-0.26; bal Fe				
ASTM A743/A743M(98)	CN-3MN	Corr res	0.03 max	20.0-22.0	2.00 max	6.0-7.0	23.5-25.5	0.040 max	0.010 max	1.00 max	Cu <=0.75; N 0.18-0.26; bal Fe	550	260	35	
ASTM A744/A744M(98)	CN-3MN	Corr res; severe service	0.03 max	20.0-22.0	2.00 max	6.00-7.00	23.5-25.5	0.040 max	0.010 max	1.00 max	Cu <=0.75; N 0.18-0.26; bal Fe	550	260	35	

Cast Stainless Steel, Austenitic, CN7M

Finland

Specification	Designation	Notes	C	Cr	Mn	Mo	Ni	P	S	Si	Other	UTS	YS	El	Hard
SFS 390(79)	G-X8NiCrMoCuNb252032	Corr res	0.08 max	19.0-21.0	2.0 max	2.5-3.5	24.0-26.0	0.045 max	0.03 max	1.5 max	Cu 1.5-2.5; Nb>=8C; bal Fe				

France

Specification	Designation	Notes	C	Cr	Mn	Mo	Ni	P	S	Si	Other	UTS	YS	El	Hard
AFNOR NFA32056	Z6NCDU252004M	Corr res; SA	0.08	18-22	1.5	2.5-6	23-27	0.04	0.03	1.2	Cu 1.5-3.5; bal Fe				

Germany

Specification	Designation	Notes	C	Cr	Mn	Mo	Ni	P	S	Si	Other	UTS	YS	El	Hard
DIN	G-XNiCrMoCuNb25-20		0.08 max	19.0-21.0	2.00 max	2.50-3.50	24.0-26.0	0.045 max	0.030 max	1.50 max	Cu 1.50-2.50; Nb>=8xC; bal Fe				
DIN	WNr 1.4500		0.08 max	19.0-21.0	2.00 max	2.50-3.50	24.0-26.0	0.045 max	0.030 max	1.50 max	Cu 1.50-2.50; Nb>=8xC; bal Fe				

Italy

Specification	Designation	Notes	C	Cr	Mn	Mo	Ni	P	S	Si	Other	UTS	YS	El	Hard
UNI 3161(83)	GX5NiCrCuMo2921	Sand; Heat res	0.07 max	19-21.	1.5 max	2-3	27-31	0.04 max	0.035 max	1.5 max	Cu 3-4; bal Fe				
UNI 3161(83)	GX5NiCrSiMoCu2419	Sand; Heat res	0.07 max	18-20	1.5 max	2.5-3	22-25	0.04 max	0.035 max	2.5-3.5	Cu 1.5-2; bal Fe				

Japan

Specification	Designation	Notes	C	Cr	Mn	Mo	Ni	P	S	Si	Other	UTS	YS	El	Hard
JIS G5121(91)	SCS23	Corr res; Pipe SA	0.07 max	19.00-22.00	2.00 max	2.00-3.00	27.50-30.00	0.040 max	0.040 max	2.00 max	Cu 3.00-4.00; bal Fe	165	390	30	183 HB max

Mexico

Specification	Designation	Notes	C	Cr	Mn	Mo	Ni	P	S	Si	Other	UTS	YS	El	Hard
DGN B-354	CN-7M*	Obs; Corr res; Q/A	0.07	19-22	1.5	2-3	27.5-30.5	0.04	0.04	1.5	bal Fe				

Russia

Specification	Designation	Notes	C	Cr	Mn	Mo	Ni	P	S	Si	Other	UTS	YS	El	Hard
GOST	O3ChN28MDT	Corr res	0.03 max	22-25	0.8	2.5-3	26-29			0.8	Cu 2.5-3.5; Ti 0.5-0.9; bal Fe				
GOST	O6ChN28MDT	Corr res	0.06 max	22-25	0.8	2.5-3	26-29			0.8	Cu 2.5-3.5; Ti 0.5-0.9; bal Fe				
GOST 2246	Sw-01Ch23N28M3D3T	Corr res	0.03 max	22-25	0.55	2.5-3	26-29	0.03	0.018	0.55	Cu 2.5-3.5; Ti 0.5-0.9; bal Fe				

Spain

Specification	Designation	Notes	C	Cr	Mn	Mo	Ni	P	S	Si	Other	UTS	YS	El	Hard
UNE 36257(74)	AM-X6NiCrMoCu29-20	Corr res	0.07 max	19.0-22.0	1.5 max	2.0-3.0	27.0-31.0	0.040 max	0.040 max	1.5 max	Cu 3.0-4.0; bal Fe				
UNE 36257(74)	F.8417*	see AM-X6NiCrMoCu29-20	0.07 max	19.0-22.0	1.5 max	2.0-3.0	27.0-31.0	0.040 max	0.040 max	1.5 max	Cu 3.0-4.0; bal Fe				

USA

Specification	Designation	Notes	C	Cr	Mn	Mo	Ni	P	S	Si	Other	UTS	YS	El	Hard
	UNS N08007		0.07 max	19-22	1.5 max	2-3	27.5-30.5			1.5 max	Cu 3-4; bal Fe				
ACI	CN-7M	Corr res	0.07 max	19-22	1.5 max	2-3	27.5-30.5	0.04	0.04	1.5 max	Cu 3-4; bal Fe				
ASTM A351(94)	CN-7M	Press ves	0.07 max	19.0-22.0	1.50 max	2.0-3.0	27.5-30.5	0.040 max	0.040 max	1.50 max	Cu 3.0-4.0; bal Fe	425	170	35.0	
ASTM A743/A743M(98)	CN-7M	Corr res; Solution Q/A	0.07 max	19.0-22.0	1.50 max	2.0-3.0	27.5-30.5	0.04 max	0.04 max	1.50 max	Cu 3.0-4.0; bal Fe	425	170	35	
ASTM A744/A744M(98)	CN-7M	Corr res; severe service	0.07 max	19.0-22.0	1.50 max	2.0-3.0	27.5-30.5	0.04 max	0.04 max	1.50 max	Cu 3.0-4.0; bal Fe	425	170	35	

Specification	Designation	Notes	C	Cr	Mn	Mo	Ni	P	S	Si	Other	UTS	YS	El	Hard

Cast Stainless Steel, Austenitic, CN7MS

France

Specification	Designation	Notes	C	Cr	Mn	Mo	Ni	P	S	Si	Other	UTS	YS	El	Hard
AFNOR	Z6NCDU252004M	Corr res	0.08 max	18-22	1.5 max	2.5-6	23-27			1.2 max	Cu 1.5-3.5; bal Fe				

Germany

Specification	Designation	Notes	C	Cr	Mn	Mo	Ni	P	S	Si	Other	UTS	YS	El	Hard
DIN SEW 410(88)	GX2NiCrMoCuN25-20		0.03 max	19.0-21.0	1.00 max	2.50-3.50	24.0-26.0	0.035 max	0.020 max	1.00 max	Cu 1.50-2.00; N 0.10-0.20; bal Fe	440-640	200	20	
DIN SEW 410(88)	WNr 1.4536		0.03 max	19.0-21.0	1.00 max	2.50-3.50	24.0-26.0	0.035 max	0.020 max	1.00 max	Cu 1.50-2.00; N 0.10-0.20; bal Fe	440-640	200	20	

Romania

Specification	Designation	Notes	C	Cr	Mn	Mo	Ni	P	S	Si	Other	UTS	YS	El	Hard
STAS 10718(88)	T6CuMoNiCr200	Corr res	0.06 max	18-22	2 max	2-3	24-26	0.04 max	0.03 max	0.5-1.5	Cu 1.5-3; Pb <=0.15; Ti <=0.5; bal Fe				

USA

Specification	Designation	Notes	C	Cr	Mn	Mo	Ni	P	S	Si	Other	UTS	YS	El	Hard
	UNS J94650		0.07 max	18.0-20.0	1.00 max	2.5-3.0	22.0-25.0	0.04 max	0.03 max	2.50-3.50	Cu 1.5-2.0; bal Fe				
ASTM A743/A743M(98)	CN-7MS	Corr res; Solution Q/A	0.07 max	18.0-20.0	1.00 max	2.5-3.0	22.0-25.0	0.04 max	0.03 max	2.50-3.50	Cu 1.5-2.0; bal Fe	485	205	35	
ASTM A744/A744M(98)	CN-7MS	Corr res; severe service	0.07 max	18.0-20.0	1.0 max	2.5-3.0	22.0-25.0	0.04 max	0.03 max	2.50-3.50	Cu 1.5-2.0; bal Fe	485	205	35	

Cast Stainless Steel, Austenitic, CT15C

USA

Specification	Designation	Notes	C	Cr	Mn	Mo	Ni	P	S	Si	Other	UTS	YS	El	Hard
ACI	CT-15C	Corr res	0.05-0.15	19.0-21.0	0.15-1.50		31.0-34.0	0.03 max	0.03 max	0.50-1.50	Nb 0.50-1.50; bal Fe				

Cast Stainless Steel, Austenitic, F745

USA

Specification	Designation	Notes	C	Cr	Mn	Mo	Ni	P	S	Si	Other	UTS	YS	El	Hard
	UNS J31670	Cr-Ni-Mo Surgical Implant 316	0.06 max	17.00-19.00	2.0 max	2.00-3.00	11.00-14.00	0.045 max	0.030 max	1.0 max	bal Fe				

Cast Stainless Steel, Austenitic, HF

Czech Republic

Specification	Designation	Notes	C	Cr	Mn	Mo	Ni	P	S	Si	Other	UTS	YS	El	Hard
CSN 422934	422934	Heat res	0.25-0.45	20.0-23.0	1.5 max		9.0-11.0	0.045 max	0.04 max	1.0-2.0	bal Fe				

Germany

Specification	Designation	Notes	C	Cr	Mn	Mo	Ni	P	S	Si	Other	UTS	YS	El	Hard
DIN 17465(93)	GX25CrNiSi18-9	Heat res	0.15-0.30	17.0-19.0	1.50 max		8.00-10.0	0.035 max	0.030 max	1.00-2.50	bal Fe				
DIN 17465(93)	GX40CrNiSi22-9		0.30-0.50	21.0-23.0	1.50 max		9.00-11.0	0.035 max	0.030 max	1.00-2.50	bal Fe				
DIN 17465(93)	WNr 1.4825		0.15-0.30	17.0-19.0	1.50 max		8.00-10.0	0.035 max	0.030 max	1.00-2.50	bal Fe				
DIN 17465(93)	WNr 1.4826		0.30-0.50	21.0-23.0	1.50 max		9.00-11.0	0.035 max	0.030 max	1.00-2.50	bal Fe				

Italy

Specification	Designation	Notes	C	Cr	Mn	Mo	Ni	P	S	Si	Other	UTS	YS	El	Hard
UNI 3159	GX30CrNi2010	Heat res	0.2-0.4	18-22	2 max	0.5 max	8-12			2.5 max	bal Fe				

Japan

Specification	Designation	Notes	C	Cr	Mn	Mo	Ni	P	S	Si	Other	UTS	YS	El	Hard
JIS G5122(91)	SCH12	Heat res	0.20-0.40	18.00-23.00	2.00 max	0.50 max	8.00-12.00	0.040 max	0.040 max	2.00 max	bal Fe	490	235	23	

Mexico

Specification	Designation	Notes	C	Cr	Mn	Mo	Ni	P	S	Si	Other	UTS	YS	El	Hard
DGN B-355	HF*	Obs; Heat res; Q/A, Tmp	0.2-0.4	18-23	2	0.5	8-12	0.04	0.04	2	bal Fe				

Spain

Specification	Designation	Notes	C	Cr	Mn	Mo	Ni	P	S	Si	Other	UTS	YS	El	Hard
UNE 36258(74)	AM-X30CrNi20-10	Heat res	0.2-0.4	18-23	2 max	0.5 max	8-12			2 max	bal Fe				
UNE 36258(74)	F.8450*	see AM-X30CrNi20-10	0.2-0.4	18-23	2 max	0.5 max	8-12			2 max	bal Fe				

UK

Specification	Designation	Notes	C	Cr	Mn	Mo	Ni	P	S	Si	Other	UTS	YS	El	Hard
BS 3100(91)	302C35	Heat res; old En1648D	0.20-0.40	17.0-22.0	2.00 max	1.50 max	6.00-10.0	0.050 max	0.050 max	2.00 max	bal Fe				

USA

Specification	Designation	Notes	C	Cr	Mn	Mo	Ni	P	S	Si	Other	UTS	YS	El	Hard
	UNS J92603		0.20-0.40	18.0-23.0	2.00 max	0.50 max	8.0-12.0	0.04 max	0.04 max	2.00 max	bal Fe				
ACI	HF	Heat res	0.20-0.40	18.0-23.0	2.00 max	0.50 max	8.0-12.0	0.04 max	0.04 max	2.00 max	bal Fe				
ASTM A297/A297M(98)	HF	Heat res; as cast	0.2-0.4	18-23	2 max	0.5 max	8-12	0.04 max	0.04 max	2 max	bal Fe				

Specification	Designation	Notes	C	Cr	Mn	Mo	Ni	P	S	Si	Other	UTS	YS	El	Hard

Cast Stainless Steel, Austenitic, HF30

USA

| | UNS J92803 | | 0.25-0.35 | 19.0-23.0 | 1.50 max | 0.50 max | 9.0-12.0 | 0.04 max | 0.04 max | 0.50-2.00 | bal Fe | | | | |
| ASTM A608(98) | HF30 | Heat res; Cent Tub | 0.25-0.35 | 19-23 | 1.50 max | 0.50 max | 9-12 | 0.04 max | 0.04 max | 0.50-2.00 | bal Fe | | | | |

Cast Stainless Steel, Austenitic, HI

China

| GB 8492(87) | ZG40Cr28Ni16 | Heat res; Ann | 0.20-0.50 | 26.0-30.0 | 2.00 max | 0.50 max | 14.0-18.0 | 0.040 max | 0.040 max | 2.00 max | bal Fe | 490 | 235 | 8 | |

Czech Republic

| CSN 422936 | 422936 | Heat res | 0.25-0.5 | 24-27 | 1.5 max | 0.5 max | 12-14 | 0.045 max | 0.04 max | 2 max | bal Fe | | | | |

Italy

| UNI 3159 | GX35CrNi2816 | Heat res | 0.2-0.5 | 26-30 | 2 max | 0.5 max | 14-18 | | | 2.5 max | bal Fe | | | | |

Japan

| JIS G5122(91) | SCH18 | Heat res | 0.20-0.50 | 26.00-30.00 | 2.00 max | 0.50 max | 14.00-18.00 | 0.040 max | 0.040 max | 2.00 max | bal Fe | 490 | 235 | 5 | |

Mexico

| DGN B-355 | HI* | Obs; Heat res | 0.2-0.5 | 26-30 | 2 | 0.5 | 14-18 | 0.04 | 0.04 | 2 | bal Fe | | | | |

USA

	UNS J94003		0.20-0.50	26.0-30.0	2.00 max	0.50 max	14.0-18.0	0.04 max	0.04 max	2.00 max	bal Fe				
ACI	HI	Heat res	0.20-0.50	26.0-30.0	2.00 max	0.50 max	14.0-18.0	0.04 max	0.04 max	2.00 max	bal Fe				
ASTM A297/A297M(98)	HI	Heat res; as cast	0.2-0.5	26-30	2 max	0.5 max	14-18	0.04 max	0.04 max	2 max	bal Fe				

Cast Stainless Steel, Austenitic, HI35

USA

| | UNS J94013 | | 0.30-0.40 | 26.0-30.0 | 1.50 max | 0.50 max | 14.0-18.0 | 0.04 max | 0.04 max | 0.50-2.00 | bal Fe | | | | |
| ASTM A608(98) | HI35 | Heat res; Cent Tub | 0.30-0.40 | 26-30 | 1.50 max | 0.50 max | 14-18 | 0.04 max | 0.04 max | 0.50-2.00 | bal Fe | | | | |

Cast Stainless Steel, Austenitic, HK

Czech Republic

| CSN 422952 | 422952 | Heat res | 0.3-0.45 | 24.0-27.0 | 1.5 max | 0.5 max | 20.0-22.0 | 0.045 max | 0.04 max | 0.75-1.75 | bal Fe | | | | |

Finland

| SFS 393(79) | G-X40CrNi2520 | Heat res | 0.2-0.6 | 24.0-28.0 | 2.0 max | 0.5 max | 18.0-22.0 | 0.04 max | 0.04 max | 2.0 max | bal Fe | | | | |

Germany

| DIN 17465(93) | GX40CrNiSi25-20 | | 0.30-0.50 | 24.0-26.0 | 1.50 max | | 19.0-21.0 | 0.035 max | 0.030 max | 1.00-2.50 | bal Fe | | | | |
| DIN 17465(93) | WNr 1.4848 | | 0.30-0.50 | 24.0-26.0 | 1.50 max | | 19.0-21.0 | 0.035 max | 0.030 max | 1.00-2.50 | bal Fe | 440 | 230 | 6 | |

Hungary

| MSZ 4357 | 40CrNiSi2520 | Heat res | 0.2-0.5 | 24-26 | 0.5-1.5 | | 19-21 | 0.04 | 0.03 | 1-2.5 | bal Fe | | | | |

Italy

| UNI 3159 | GX40CrNi2620 | Heat res | 0.2-0.6 | 24-28 | 2 max | 0.5 max | 18-22 | 0.04 | 0.035 | 2.5 max | bal Fe | | | | |

Mexico

| DGN B-355 | HK* | Obs; Heat res; as cast | 0.2-0.6 | 24-28 | 2 | 0.5 | 18-22 | 0.04 | 0.04 | 2 | bal Fe | | | | |

UK

BS 310	310C40	Heat res; En4238FC	0.3-0.5	24-27	2 max	1.5 max	19-22	0.04	0.04	1.5 max	bal Fe				
BS 310	310C45	Heat res; En1648F	0.5 max	22-27	2 max	1.5 max	17-22	0.06	0.06	3 max	bal Fe				
BS 3100(91)	310C40	Heat res; old En4238FC	0.30-0.50	24.0-27.0	2.00 max	1.50 max	19.0-22.0	0.040 max	0.040 max	1.50 max	bal Fe	450		7	
BS 3100(91)	310C45	Heat res; En1648F	0.50 max	22.0-27.0	2.00 max	1.50 max	17.0-22.0	0.050 max	0.050 max	3.00 max	bal Fe				

USA

| | UNS J94224 | | 0.20-0.60 | 24.0-28.0 | 2.00 max | 0.50 max | 18.0-22.0 | 0.04 max | 0.04 max | 2.00 max | bal Fe | | | | |
| ACI | HK | Heat res | 0.20-0.60 | 24.0-28.0 | 2.00 max | 0.50 max | 18.0-22.0 | 0.04 max | 0.04 max | 2.00 max | bal Fe | | | | |

UNS numbers and US grades are provided as a means of cross referencing chemically similar alloys. Exchangability is only possible after independent examination of specifications. Tensile properties are minimum or typical as specified. UTS and YS as MPa. El as %. See Appendix for list of abbreviations used in Notes. * indicates obsolete material.

Specification	Designation	Notes	C	Cr	Mn	Mo	Ni	P	S	Si	Other	UTS	YS	El	Hard

Cast Stainless Steel, Austenitic, HK (Continued from previous page)

USA

Specification	Designation	Notes	C	Cr	Mn	Mo	Ni	P	S	Si	Other	UTS	YS	El	Hard
ASTM A297/A297M(98)	HK	Heat res	0.2-0.6	24-28	2 max	0.5 max	18-22	0.04 max	0.04 max	2 max	bal Fe				

Cast Stainless Steel, Austenitic, HK30

China

Specification	Designation	Notes	C	Cr	Mn	Mo	Ni	P	S	Si	Other	UTS	YS	El	Hard
GB 8492(87)	ZG40Cr25Ni20	Heat res; Ann	0.35-0.45	23.0-27.0	1.50 max	0.50 max	19.0-22.0	0.040 max	0.040 max	1.75 max	bal Fe	440	235	8	
JB 4298(86)	ZG3Cr25Ni20Si2	Heat res; Ann	0.20-0.35	24.0-28.0	2.00 max	0.50 max	18.0-22.0	0.040 max	0.040 max	2.00 max	bal Fe				

Germany

Specification	Designation	Notes	C	Cr	Mn	Mo	Ni	P	S	Si	Other	UTS	YS	El	Hard
DIN 17465(93)	GX30CrNiSiNb24-24		0.25-0.40	23.0-25.0	1.50 max		23.0-25.0	0.035 max	0.030 max	0.50-2.00	Nb 1.20-1.80; bal Fe	440	230	5	
DIN 17465(93)	WNr 1.4855		0.25-0.40	23.0-25.0	1.50 max		23.0-25.0	0.035 max	0.030 max	0.50-2.00	Nb 1.20-1.80; bal Fe	440	230	5	

Japan

Specification	Designation	Notes	C	Cr	Mn	Mo	Ni	P	S	Si	Other	UTS	YS	El	Hard
JIS G5122(91)	SCH21	Heat res	0.25-0.35	23.00-27.00	1.50 max	0.50 max	19.00-22.00	0.040 max	0.040 max	1.75 max	bal Fe	440	235	4	

USA

Specification	Designation	Notes	C	Cr	Mn	Mo	Ni	P	S	Si	Other	UTS	YS	El	Hard
	UNS J94203		0.25-0.35	23.0-27.0	1.50 max		19.0-22.0	0.040 max	0.040 max	1.75 max	bal Fe				
ACI	HK-30	Heat res	0.25-0.35	23.0-27.0	1.50 max		19.0-22.0	0.040 max	0.040 max	1.75 max	bal Fe				
ASTM A351(94)	HK30	Press ves	0.25-0.35	23.0-27.0	1.50 max	0.50 max	19.0-22.0	0.040 max	0.040 max	1.75 max	bal Fe	450	240	10.0	
ASTM A608(98)	HK30	Heat res; Cent Tub	0.25-0.35	23-27	1.50 max	0.50 max	19-22	0.04 max	0.04 max	0.50-2.00	bal Fe				

Cast Stainless Steel, Austenitic, HK40

China

Specification	Designation	Notes	C	Cr	Mn	Mo	Ni	P	S	Si	Other	UTS	YS	El	Hard
GB 8492(87)	ZG40Cr25Ni20	Heat res; Ann	0.35-0.45	23.0-27.0	1.50 max	0.50 max	19.0-22.0	0.040 max	0.040 max	1.75 max	bal Fe	440	235	8	
JB 4298(86)	ZG3Cr25Ni20Si2	Heat res; Ann	0.20-0.35	24.0-28.0	2.00 max	0.50 max	18.0-22.0	0.040 max	0.040 max	2.00 max	bal Fe				

Hungary

Specification	Designation	Notes	C	Cr	Mn	Mo	Ni	P	S	Si	Other	UTS	YS	El	Hard
MSZ 4357(81)	AoX40CrNiSi2520	Heat res; as cast	0.3-0.5	24-26	0.5-1.5		19-21	0.04 max	0.03 max	1-2.5	bal Fe	500		5L	180-230 HB
MSZ 4357(81)	AoX40CrNiSi2520F	Heat res; soft ann	0.3-0.5	24-26	0.5-1.5		19-21	0.04 max	0.03 max	1-2.5	bal Fe	450	250	10L	150-200 HB

Japan

Specification	Designation	Notes	C	Cr	Mn	Mo	Ni	P	S	Si	Other	UTS	YS	El	Hard
JIS G5122(91)	SCH22	Heat res	0.35-0.45	23.00-27.00	1.50 max	0.50 max	19.00-22.00	0.040 max	0.040 max	1.75 max	bal Fe	440	235	8	

Spain

Specification	Designation	Notes	C	Cr	Mn	Mo	Ni	P	S	Si	Other	UTS	YS	El	Hard
UNE 36258(74)	AM-X40CrNi25-20	Heat res	0.2-0.6	24-28	2 max	0.5 max	18-22	0.040 max	0.040 max	2 max	Nb 1.5; bal Fe				
UNE 36258(74)	F.8452*	see AM-X40CrNi25-20	0.2-0.6	24-28	2 max	0.5 max	18-22	0.040 max	0.040 max	2 max	Nb 1.5; bal Fe				

USA

Specification	Designation	Notes	C	Cr	Mn	Mo	Ni	P	S	Si	Other	UTS	YS	El	Hard
	UNS J94204		0.35-0.45	23.0-27.0	1.50 max		19.0-22.0	0.040 max	0.040 max	1.75 max	bal Fe				
ACI	HK-40	Heat res	0.35-0.45	23.0-27.0	1.50 max		19.0-22.0	0.040 max	0.040 max	1.75 max	bal Fe				
ASTM A351(94)	HK40	Press ves	0.35-0.45	23.0-27.0	1.50 max	0.50 max	19.0-22.0	0.040 max	0.040 max	1.75 max	bal Fe	425	240	10.0	
ASTM A608(98)	HK40	Heat res; Cent Tub	0.35-0.45	23-27	1.50 max	0.50 max	19-22	0.04 max	0.04 max	0.50-2.00	bal Fe				

Cast Stainless Steel, Austenitic, HL

China

Specification	Designation	Notes	C	Cr	Mn	Mo	Ni	P	S	Si	Other	UTS	YS	El	Hard
GB 8492(87)	ZG40Cr30Ni20	Heat res; Ann	0.20-0.60	28.0-32.0	2.00 max	0.50 max	18.0-22.0	0.040 max	0.040 max	2.00 max	bal Fe	450	245	8	

India

Specification	Designation	Notes	C	Cr	Mn	Mo	Ni	P	S	Si	Other	UTS	YS	El	Hard
IS 4522	10	Heat res; as cast	0.2-0.6	28-32	2		18-22	0.05	0.05	2.5	bal Fe				

Japan

Specification	Designation	Notes	C	Cr	Mn	Mo	Ni	P	S	Si	Other	UTS	YS	El	Hard
JIS G5122(91)	SCH23	Heat res; Ann	0.20-0.60	28.00-32.00	2.00 max		18.00-22.00	0.040 max	0.040 max	2.00 max	bal Fe	450	245	8	

UNS numbers and US grades are provided as a means of cross referencing chemically similar alloys. Exchangability is only possible after independent examination of specifications. Tensile properties are minimum or typical as specified. UTS and YS as MPa. El as %. See Appendix for list of abbreviations used in Notes. * indicates obsolete material.

Specification	Designation	Notes	C	Cr	Mn	Mo	Ni	P	S	Si	Other	UTS	YS	El	Hard

Cast Stainless Steel, Austenitic, HL (Continued from previous page)

Mexico

Specification	Designation	Notes	C	Cr	Mn	Mo	Ni	P	S	Si	Other	UTS	YS	El	Hard
DGN B-355	HL*	Obs; Heat res; as cast	0.2-0.6	28-32	2	0.5	18-22	0.04	0.04	2	bal Fe				

USA

Specification	Designation	Notes	C	Cr	Mn	Mo	Ni	P	S	Si	Other	UTS	YS	El	Hard
	UNS N08604		0.2 0.6	28-32	2 max	0.5 max	18-22	0.04 max	0.04 max	2 max	bal Fe				
	UNS N08613		0.25-0.35	28-32	1.5 max	0.5 max	18-22	0.04 max	0.04 max	0.5-2	bal Fe				
	UNS N08614		0.35-0.45	28-32	1.5 max	0.5 max	18-22	0.04 max	0.04 max	0.5-2	bal Fe				
ACI	HL	Heat res	0.2-0.6	28-32	2 max	0.5 max	18-22	0.04 max	0.04 max	2 max	bal Fe				
ACI	HL-30	Heat res	0.25-0.35	28.0-32.0	1.50 max	0.50 max	18.0-22.0	0.04 max	0.04 max	0.50-2.00	bal Fe				
ACI	HL-40	Heat res	0.35-0.45	28.0-32.0	1.50 max	0.50 max	18.0-22.0	0.04 max	0.04 max	0.50-2.00	bal Fe				
ASTM A297/A297M(98)	HL	Heat res	0.2-0.6	28-32	2	0.5	18-22	0.04	0.04	2	bal Fe				
ASTM A608(98)	HL30	Heat res; Cent Tub	0.25-0.35	28-32	1.50 max	0.50 max	18-22	0.04 max	0.04 max	0.50-2.00	bal Fe				
ASTM A608(98)	HL40	Heat res; Cent Tub	0.35-0.45	28-32	1.50 max	0.50 max	18-22	0.04 max	0.04 max	0.50-2.00	bal Fe				

Cast Stainless Steel, Austenitic, HN

Australia

Specification	Designation	Notes	C	Cr	Mn	Mo	Ni	P	S	Si	Other	UTS	YS	El	Hard
AS 2074(82)	H8F	Heat res	0.5 max	17-23	2 max	1 max	23-28	0.06 max	0.06 max	3 max	bal Fe				

China

Specification	Designation	Notes	C	Cr	Mn	Mo	Ni	P	S	Si	Other	UTS	YS	El	Hard
GB 8492(87)	ZG35Ni24Cr18Si2	Heat res; Ann	0.30-0.40	17.0-20.0	1.50 max		23.0-26.0	0.035 max	0.030 max	1.50-2.50	bal Fe	390	195	5	

India

Specification	Designation	Notes	C	Cr	Mn	Mo	Ni	P	S	Si	Other	UTS	YS	El	Hard
IS 4522	11	Heat res; as cast	0.2-0.5	19-23	2		23-27	0.05	0.05	2	bal Fe				

Japan

Specification	Designation	Notes	C	Cr	Mn	Mo	Ni	P	S	Si	Other	UTS	YS	El	Hard
JIS G5122(91)	SCH19	Heat res; as cast; Pipe	0.20-0.50	19.00-23.00	2.00 max	0.50 max	23.00-27.00	0.040 max	0.040 max	2.00 max	bal Fe	390		8	

Mexico

Specification	Designation	Notes	C	Cr	Mn	Mo	Ni	P	S	Si	Other	UTS	YS	El	Hard
DGN B-355	HN*	Obs; Heat res; as cast	0.2-0.5	19-23	2	0.5	23-27	0.04	0.04	2	bal Fe				

USA

Specification	Designation	Notes	C	Cr	Mn	Mo	Ni	P	S	Si	Other	UTS	YS	El	Hard
	UNS J94213		0.20-0.50	19.0-23.0	2.00 max	0.50 max	23.0-27.0	0.04 max	0.04 max	2.00 max	bal Fe				
ACI	HN	Heat res	0.20-0.50	19.0-23.0	2.00 max	0.50 max	23.0-27.0	0.04 max	0.04 max	2.00 max	bal Fe				
ASTM A297/A297M(98)	HN	Heat res	0.2-0.5	19-23	2 max	0.5 max	23-27	0.04 max	0.04 max	2 max	bal Fe				

Cast Stainless Steel, Austenitic, HN40

USA

Specification	Designation	Notes	C	Cr	Mn	Mo	Ni	P	S	Si	Other	UTS	YS	El	Hard
	UNS J94214		0.35-0.45	19.0-23.0	1.50 max	0.50 max	23.0-27.0	0.04 max	0.04 max	0.50-2.00	bal Fe				
ASTM A608(98)	HN40	Heat res; Cent Tub	0.35-0.45	19-23	1.50 max	0.50 max	23-27	0.04 max	0.04 max	0.50-2.00	bal Fe				

Cast Stainless Steel, Austenitic, HP

China

Specification	Designation	Notes	C	Cr	Mn	Mo	Ni	P	S	Si	Other	UTS	YS	El	Hard
GB 8492(87)	ZG45Ni35Cr26	Heat res; Ann	0.35-0.75	24.0-28.0	2.00 max		33.0-37.0	0.040 max	0.040 max	2.00 max	bal Fe	440	235	5	

Germany

Specification	Designation	Notes	C	Cr	Mn	Mo	Ni	P	S	Si	Other	UTS	YS	El	Hard
DIN 17465(93)	GX40NiCrSi35-25		0.30-0.50	24.0-26.0	1.50 max		34.0-36.0	0.035 max	0.030 max	1.00-2.50	bal Fe	440	230	5	
DIN 17465(93)	WNr 1.4857		0.30-0.50	24.0-26.0	1.50 max		34.0-36.0	0.035 max	0.030 max	1.00-2.50	bal Fe	440	230	5	

Hungary

Specification	Designation	Notes	C	Cr	Mn	Mo	Ni	P	S	Si	Other	UTS	YS	El	Hard
MSZ 4357(81)	AoX40NiCrSi3525	Heat res; as cast	0.3-0.5	24-26	0.5-1.5		33-36	0.04 max	0.03 max	1-2.5	bal Fe	500		5L	180-230 HB
MSZ 4357(81)	AoX40NiCrSi3525F	Heat res; SA	0.3-0.5	24-26	0.5-1.5		33-36	0.04 max	0.03 max	1-2.5	bal Fe	450	250	10L	150-200 HB

Cast Stainless Steel, Austenitic, HP (Continued from previous page)

Specification	Designation	Notes	C	Cr	Mn	Mo	Ni	P	S	Si	Other	UTS	YS	El	Hard
Italy															
UNI 3159	GX50NiCr3525	Heat res; Sand	0.3-0.7	23-27	2 max	0.5 max	33-37	0.04	0.035	2.5 max	bal Fe				
Japan															
JIS G5122(91)	SCS24	Heat res; Ann	0.35-0.75	24.00-28.00	2.00 max		33.00-37.00	0.040 max	0.040 max	2.00 max	bal Fe	440	235	5	
Mexico															
DGN B-355	HP*	Obs; Heat res; as cast	0.35-0.75	24-28	2 max	0.5 max	33-37	0.04	0.04	2	bal Fe				
USA															
	UNS N08705		0.35-0.75	24-28	2 max	0.5 max	33-37	0.04 max	0.04 max	2.5 max	bal Fe				
ASTM A297/A297M(98)	HP	Heat res; Sand	0.35-0.75	24-28	2 max	0.5 max	33-37	0.04 max	0.04 max	2.5 max	bal Fe				

Cast Stainless Steel, Austenitic, HT

Specification	Designation	Notes	C	Cr	Mn	Mo	Ni	P	S	Si	Other	UTS	YS	El	Hard
China															
GB 8492(87)	ZG30Ni35Cr15	Heat res; Ann	0.20-0.35	13.0-17.0	2.00 max		33.0-37.0	0.040 max	0.040 max	2.50 max	bal Fe	440	195	13	
Germany															
DIN 17465(93)	GX40NiCrSi38-18		0.30-0.50	17.0-19.0	1.50 max		36.0-39.0	0.035 max	0.030 max	1.00-2.50	bal Fe	400	230	5	
DIN 17465(93)	WNr 1.4865		0.30-0.50	17.0-19.0	1.50 max		36.0-39.0	0.035 max	0.030 max	1.00-2.50	bal Fe	440	230	5	
India															
IS 4522	120	Heat res; as cast	0.35-0.75	13-17	2		33-37	0.05	0.04	2.5	bal Fe				
IS 7806	12	Heat res; as cast	0.25-0.35	13-17	2	0.5	33-37	0.04	0.04	2.5	bal Fe				
Italy															
UNI 3159	GX50NiCr3919	Heat res	0.3-0.7	17-21	2 max	0.5 max	37-41	0.04	0.035	2.5 max	bal Fe				
Japan															
JIS G5122(91)	SCH15	Heat res; as cast; Pipe	0.35-0.70	15.00-19.00	2.00 max	0.50 max	33.00-37.00	0.040 max	0.040 max	2.50 max	bal Fe	440		4	
JIS G5122(91)	SCH16	Heat res; as cast; Pipe	0.20-0.35	13.00-17.00	2.00 max	0.50 max	33.00-37.00	0.040 max	0.040 max	2.50 max	bal Fe	440	195	13	
Mexico															
DGN B-355	HT*	Obs; Heat res; Q/A, Tmp	0.35-0.75	13-17	2	0.5	33-37	0.04	0.04	2.5	bal Fe				
Spain															
UNE 36258(74)	AM-X55NiCr35--15	Heat res	0.35-0.75	12-17	2 max	0.5 max	33-37	0.040 max	0.040 max	2.5 max	Nb 1.5; bal Fe				
UNE 36258(74)	F.8453*	see AM-X55NiCr35--15	0.35-0.75	12-17	2 max	0.5 max	33-37	0.040 max	0.040 max	2.5 max	Nb 1.5; bal Fe				
UK															
BS 3100(91)	330C11	old En4238(H1C)	0.35-0.55	13.0-17.0	2.00 max	1.50 max	33.0-37.0	0.040 max	0.040 max	1.50 max	bal Fe	450		3	
USA															
	UNS N08002		0.35-0.75	13-17	2 max		33-37	0.04 max	0.04 max	2.5 max	bal Fe				
	UNS N08030		0.25-0.35	13-17	2 max	0.5 max	33-37	0.04 max	0.04 max	2.5 max	bal Fe				
ACI	HT	Heat res	0.35-0.75	15.0-19.0	2.00 max		33.0-37.0	0.04 max	0.04 max	2.50 max	bal Fe				
ACI	HT	Heat res	0.35-0.75	15.0-19.0	2.00 max	0.5 max	33.0-37.0	0.04 max	0.04 max	2.50 max	bal Fe				
ACI	HT-30	Heat res	0.25-0.35	13.0-17.0	2.00 max	0.50 max	33.0-37.0	0.040 max	0.040 max	2.50 max	bal Fe				
ASTM A297/A297M(98)	HT	Heat res; Sand	0.35-0.75	15-19	2 max	0.5 max	33-37	0.04 max	0.04 max	2.5 max	bal Fe				
ASTM A351(94)	HT30	Press ves	0.25-0.35	13.0-17.0	2.00 max	0.50 max	33.0-37.0	0.040 max	0.040 max	2.50 max	bal Fe	450	195	15.0	
ASTM A608(98)	HT50	Heat res; Cent Tub	0.40-0.60	15-19	1.50 max	0.50 max	33-37	0.04 max	0.04 max	0.50-2.00	bal Fe				

UNS numbers and US grades are provided as a means of cross referencing chemically similar alloys. Exchangability is only possible after independent examination of specifications. Tensile properties are minimum or typical as specified. UTS and YS as MPa. El as %. See Appendix for list of abbreviations used in Notes. * indicates obsolete material.

Specification	Designation	Notes	C	Cr	Mn	Mo	Ni	P	S	Si	Other	UTS	YS	El	Hard
Cast Stainless Steel, Austenitic, HU															
Australia															
AS 2074(82)	H8H	Heat res	0.75 max	15-25	2 max	1 max	36-46	0.06 max	0.06 max	3 max	bal Fe				
Czech Republic															
CSN 422955	422955	Heat res	0.2-0.6	20.0-22.0	1.5 max		37.0-40.0	0.045 max	0.04 max	1.0-2.0	bal Fe				
Hungary															
MSZ 4357(81)	AoX40NiCrSi3717	Heat res; as cast	0.3-0.5	16-18	0.5-1.5		36-39	0.04 max	0.03 max	1-2.5	bal Fe	450		6L	180-230 HB
MSZ 4357(81)	AoX40NiCrSi3717F	Heat res; SA	0.3-0.5	16-18	0.5-1.5		36-39	0.04 max	0.03 max	1-2.5	bal Fe	450	250	10L	150-200 HB
India															
IS 4522	IS4522	Heat res; as cast	0.35-0.75	17-21	2		37-41	0.05	0.05	2.5	bal Fe				
Italy															
UNI 3159(83)	GX50NiCr3919	Heat res; Sand	0.3-0.7	17-21	2 max	0.5 max	37-41	0.04	0.035	2.5 max	bal Fe				
Japan															
JIS G5122(91)	SCH20	Heat res; as cast; Pipe	0.35-0.75	17.00-21.00	2.00 max	0.50 max	37.00-41.00	0.040 max	0.040 max	2.50 max	bal Fe	390			
Mexico															
DGN B-355	HU*	Obs; Heat res; Q/A, Tmp	0.35-0.75	17-21	2	0.5	37-41	0.04	0.04	2.5	bal Fe				
Romania															
STAS 6855(86)	T35CrNi370	Corr res; Heat res	0.2-0.5	16-20	1.5 max		35-39	0.035 max	0.03 max	1-2.5	Pb <=0.15; bal Fe				
UK															
BS 3100(91)	331C40	old En4238(H2C)	0.35-0.55	17.0-21.0	2.00 max	1.50 max	37.0-41.0	0.040 max	0.040 max	1.50 max	bal Fe	450		3	
BS 3100(91)	331C60	old En1648H2	0.75 max	15.0-25.0	2.00 max	1.50 max	36.0-46.0	0.050 max	0.050 max	3.00 max	bal Fe				
BS 3146/2(75)	ANC5(B)	Corr res	0.50 max	15.0-25.0	0.20-2.00		36.0-46.0			0.20-3.00	bal Fe				
USA															
	UNS N08004		0.35-0.75	17-21	2 max	0.5 max	37-41	0.04 max	0.04 max	2.5 max	bal Fe				
	UNS N08005		0.4-0.6	17-21	1.5 max	0.5 max	37-41	0.04 max	0.04 max	0.5-2	bal Fe				
ACI	HU	Heat res	0.35-0.75	17-21	2 max	0.5 max	37-41	0.04 max	0.04 max	2.5 max	bal Fe				
ACI	HU-50	Heat res	0.40-0.60	17.0-21.0	1.50 max	0.50 max	37.0-41.0	0.04 max	0.04 max	0.50-2.00	bal Fe				
ASTM A297/A297M(98)	HU	Heat res; Sand	0.35-0.75	17-21	2 max	0.5 max	37-41	0.04 max	0.04 max	2.5 max	bal Fe				
ASTM A608(98)	HU50	Heat res; Cent Tub	0.40-0.60	17-21	1.50 max	0.50 max	37-41	0.04 max	0.04 max	0.50-2.00	bal Fe				
Cast Stainless Steel, Austenitic, HW															
Mexico															
DGN B-355	HW*	Obs; Heat res; as cast	0.35-0.75	10-14	2	0.5	58-62	0.04	0.04	2.5	bal Fe				
USA															
	UNS N08001		0.35-0.75	10-14	2 max	0.5 max	58-62	0.04 max	0.04 max	2.5 max	bal Fe				
	UNS N08006		0.4-0.6	10-14	1.5 max	0.5 max	58-62	0.04 max	0.04 max	0.5-2	bal Fe				
ACI	HW	Heat res	0.35-0.75	10-14	2 max	0.5 max	58-62	0.04 max	0.04 max	2.5 max	bal Fe				
ASTM A297/A297M(98)	HW	Heat res; as cast	0.35-0.75	10-14	2 max	0.5 max	58-62	0.04 max	0.04 max	2.5 max	bal Fe				
ASTM A608(98)	HW50	Heat res; Cent Tub	0.40-0.60	10-14	1.50 max	0.50 max	58-62	0.04 max	0.04 max	0.50-2.00	bal Fe				
Cast Stainless Steel, Austenitic, HX															
Mexico															
DGN B-355	HX*	Obs; Heat res; as cast	0.35-0.75	15-19	2	0.5	64-66	0.04	0.04	2.5	bal Fe				

UNS numbers and US grades are provided as a means of cross referencing chemically similar alloys. Exchangability is only possible after independent examination of specifications. Tensile properties are minimum or typical as specified. UTS and YS as MPa. El as %. See Appendix for list of abbreviations used in Notes. * indicates obsolete material.

Specification	Designation	Notes	C	Cr	Mn	Mo	Ni	P	S	Si	Other	UTS	YS	El	Hard

Cast Stainless Steel, Austenitic, HX (Continued from previous page)

USA

	UNS N06006		0.35-0.75	15-19	2 max	0.5 max	64-68	0.04 max	0.04 max	2.5 max	bal Fe				
ACI	HX	Heat res	0.35-0.75	15-19	2 max	0.5 max	64-68	0.04 max	0.04 max	2.5 max	bal Fe				
ASTM A297/A297M(98)	HX	Heat res; Sand	0.35-0.75	15-19	2 max	0.5 max	64-68	0.04 max	0.04 max	2.5 max	bal Fe				
ASTM A608(98)	HX50	Heat res; Cent Tub	0.40-0.60	15-19	1.50 max	0.50 max	64-68	0.04 max	0.04 max	0.50-2.00	bal Fe				

Cast Stainless Steel, Austenitic, J92502

USA

| | UNS J92502 | | 0.15-0.30 | 17.00-20.00 | 1.50 max | 0.50 max | 8.00-11.00 | 0.05 max | 0.05 max | 1.50 max | bal Fe | | | | |
| MIL-S-17509 | I* | Obs; Heat res; as cast | 0.15-0.3 | 17-20 | 1.5 max | 0.5 max | 8-11 | 0.05 max | 0.05 max | 1.5 max | bal Fe | | | | |

Cast Stainless Steel, Austenitic, No equivalents identified

Czech Republic

CSN 422931	422931	Corr res	0.15 max	18.0-21.0	1.5 max		8.0-11.0	0.045 max	0.04 max	1.5 max	bal Fe				
CSN 422938	422938	Corr res	0.12 max	20.0-22.0	2.0 max		4.5-6.0	0.045 max	0.035 max	1.5 max	Ti <=0.7; Ti=4C-0.7; P+S<=0.07; bal Fe				
CSN 422939	422939	Corr res	0.07-0.12	15.0-17.0	0.8-1.5	0.4-0.8	11.0-13.0	0.03 max	0.025 max	0.6 max	Nb=8C-1.2; P+S<=0.05; bal Fe				
CSN 422941	422941	Corr res	0.05-0.15	17.0-19.0	1.5 max	2.0-2.5	9.0-11.0	0.045 max	0.04 max	2.0 max	Ti <=0.8; Ti=5C-0.8; bal Fe				
CSN 422942	422942	Corr res	0.2 max	18.0-20.0	1.5 max	2.0-2.5	9.0-11.0	0.045 max	0.04 max	2.0 max	bal Fe				
CSN 422943	422943	Corr res	0.12 max	20.0-22.0	2.0 max	1.8-2.2	4.5-6.0	0.045 max	0.035 max	1.2 max	P+S<=0.07; bal Fe				
CSN 422944	422944	Corr res	0.35-0.6	26.0-28.5	1.0 max		8.0-10.0	0.045 max	0.04 max	2.0 max	bal Fe				
CSN 422950	422950	Corr res	0.3-0.4	23.0-25.0	0.5 max		20.0-22.0	0.04 max	0.04 max	1.0-1.5	Al 1.0-1.5; Ti 0.05-0.2; bal Fe				
CSN 422951	422951	Corr res	0.2-0.3	20.0-22.0	0.5 max		37.0-40.0	0.04 max	0.04 max	1.5-2.0	Al 1.0-1.5; Ti 0.05-0.2; bal Fe				
CSN 422953	422953	Corr res	0.12 max	20.0-22.0	2.0 max		4.5-6.0	0.045 max	0.035 max	1.2 max	bal Fe				
CSN 422956	422956	Corr res	0.2 max	20.0-22.0	1.0 max	4.8-5.5	37.0-40.0	0.045 max	0.04 max	1.5 max	Cu 2.8-3.3; bal Fe				
CSN 422957	422957	Corr res	0.07-0.12	15.0-17.0	0.8-1.5	0.4-0.8	11.0-13.0	0.03 max	0.025 max	0.6 max	Ti <=1.20; W 3.0-5.0; P+S<=0.05; Ti=8C-1.2; bal Fe				
CSN 422958	422958	Corr res	0.1 max	20.0-22.5	2.0 max	2.5-3.5	17.0-19.0	0.04 max	0.035 max	1.5 max	Cu 2.5-3.5; Ti 0.2-0.5; V 0.2-0.5; bal Fe				

Poland

PNH83158(86)	L0H18N9M	Corr res	0.07 max	17-19	2 max	0.5 max	8-11	0.035 max	0.035 max	2 max	bal Fe				
PNH83158(86)	LH16N5G6	Corr res	0.2 max	15-17	6-7		4.5-5.5	0.05 max	0.035 max	1-2	bal Fe				
PNH83158(86)	LH17N37S2G	Corr res	0.2-0.5	15-19	1.5 max		37-39	0.035 max	0.035 max	1-2.5	bal Fe				
PNH83158(86)	LH17N8G	Corr res	0.25 max	16.5-18.5	1.5 max		7.5-8.5	0.04 max	0.035 max	1.5 max	bal Fe				
PNH83158(86)	LH18N10M2	Corr res	0.15 max	17-19	2 max	2-2.5	9-11	0.035 max	0.035 max	2 max	bal Fe				
PNH83158(86)	LH18N10M2T	Corr res	0.15 max	17-19	2 max	2-2.5	9-11	0.035 max	0.035 max	2 max	Ti <=0.8; Ti=5(C-0.03)-0.8; bal Fe				
PNH83158(86)	LH18N9T	Corr res	0.15 max	17-19	2 max		8-11	0.035 max	0.035 max	2 max	Ti <=0.8; Ti=5(C-0.03)-0.8; bal Fe				
PNH83159(90)	LH19N14	Heat res	0.25 max	18-20	1 max		13-15	0.04 max	0.035 max	1.5 max	bal Fe				
PNH83159(90)	LH23N18G	Heat res	0.25 max	22-24	1.5 max		16.5-18.5	0.04 max	0.035 max	1.8 max	bal Fe				

Russia

| GOST | 10Ch18N10T | Corr res | 0.1 max | 17.0-20.0 | 1.0-2.0 | 0.15 max | 10.0-11.0 | 0.035 max | 0.02 max | 0.8 max | Ti <=0.6; Ti=(5C-0.02)-0.6; bal Fe | | | | |

UNS numbers and US grades are provided as a means of cross referencing chemically similar alloys. Exchangability is only possible after independent examination of specifications. Tensile properties are minimum or typical as specified. UTS and YS as MPa. El as %. See Appendix for list of abbreviations used in Notes. * indicates obsolete material.

Cast Stainless Steel, Austenitic, No equivalents identified

Specification	Designation	Notes	C	Cr	Mn	Mo	Ni	P	S	Si	Other	UTS	YS	El	Hard
Russia															
GOST 977	10Ch18N11BL	Corr res	0.1 max	17.0-20.0	1.0-2.0	0.15 max	8.0-12.0	0.035 max	0.03 max	0.2-1.0	Nb 0.45-0.9; bal Fe				
GOST 977	10Ch18N3G3D2L	Corr res	0.1 max	17.0-19.0	2.3-3.0	0.15 max	3.0-3.5	0.03 max	0.03 max	0.6 max	Cu 1.8-2.2; bal Fe				
UK															
BS 3100(91)	311C11	old En1648G	0.50 max	17.0-23.0	2.00 max	1.50 max	23.0-28.0	0.050 max	0.050 max	3.00 max	bal Fe				
BS 3100(91)	330C12		0.75 max	13.0-20.0	2.00 max	1.50 max	30.0-40.0	0.050 max	0.050 max	3.00 max	bal Fe				
BS 3100(91)	332C11		0.07 max	19.0-22.0	1.50 max	2.00-3.00	27.5-30.5	0.040 max	0.040 max	1.50 max	Cu 3.00-4.00; bal Fe	425		34	
BS 3100(91)	332C13		0.04 max	24.5-26.5	1.00 max	1.75-2.25	4.75-6.00	0.040 max	0.040 max	1.00 max	Cu 2.75-3.25; bal Fe	690		16	
BS 3100(91)	332C15		0.08 max	21.0-27.0	1.50 max	1.75-3.00	4.00-7.00	0.040 max	0.040 max	1.50 max	N 0.10-0.25; bal Fe	640		30	
BS 3100(91)	334C11	old En1648(K)	0.75 max	10.0-20.0	2.00 max	1.50 max	55.0-65.0	0.050 max	0.050 max	3.00 max	bal Fe				
USA															
ASTM A351(94)	CT-15C	Press ves	0.05-0.15	19.0-21.0	0.15-1.50		31.0-34.0	0.03 max	0.03 max	1.50 max	Nb 0.50-1.50; bal Fe	435	170	20.0	
ASTM A451(97)	CPE20N	Pipe; SA Quen; High-temp	0.20 max	23.0-26.0	1.50 max		8.0-11.0	0.040 max	0.040 max	1.50 max	bal Fe	550	275	30.0	
ASTM A743/A743M(98)	CK-35MN	Corr res	0.035 max	22.0-24.0	2.00 max	6.0-6.8	20.0-22.0	0.035 max	0.020 max	1.00 max	Cu <=0.40; N 0.21-0.32; bal Fe	570	280	35	

Cast Stainless Steel, Ferritic, CC50

Specification	Designation	Notes	C	Cr	Mn	Mo	Ni	P	S	Si	Other	UTS	YS	El	Hard
Australia															
AS 2074(82)	H4A	Corr res	0.5 max	26-30	1 max	0.50 max	4 max	0.04 max	0.04 max	2 max	bal Fe				
China															
GB 2100(80)	ZGCr28	Corr res; Ann	0.50-1.00	26.0-30.0	0.5-0.8			0.10 max	0.035 max	0.5-1.3	bal Fe	343			
Czech Republic															
CSN 422913	422913	Corr res	0.4-0.7	24-26.5	0.9 max		1.0 max	0.045 max	0.04 max	1-2	Cu <=0.5; bal Fe				
Germany															
DIN SEW 410(88)	GX40CrNi27-4		0.30-0.50	26.0-28.0	1.50 max		3.50-5.50	0.045 max	0.030 max	2.00 max	bal Fe	440-640			230-300 HB
DIN SEW 410(88)	WNr 1.4340		0.30-0.50	26.0-28.0	1.50 max		3.50-5.50	0.045 max	0.030 max	2.00 max	bal Fe	440-640			230-300 HB
Italy															
UNI 3161(83)	GX40Cr28	Sand; Heat res	0.5 max	26-30	1 max		4 max	0.04 max	0.035 max	1.5 max	bal Fe				
Poland															
PNH83159(90)	LH26	Heat res	0.4-0.6	25-27	0.8 max		0.4 max	0.04 max	0.035 max	1 max	bal Fe				
Spain															
UNE 36257(74)	AM-X40Cr28	Heat res	0.5 max	25-30	1 max		4 max	0.040	0.040	1.5 max	bal Fe				
UNE 36257(74)	F.8404*	see AM-X40Cr28	0.5 max	25-30	1 max		4 max	0.040	0.040	1.5 max	bal Fe				
UNE 36258(74)	AM-X35CrNi28-6	Heat res	0.2-0.5	26-30	1 max	0.5 max	4-7	0.040	0.040	2 max	bal Fe				
UNE 36258(74)	AM-X45Cr22	Heat res	0.3-0.6	20-24	1 max	0.5 max	4 max	0.040 max	0.040 max	2 max	bal Fe				
UNE 36258(74)	AM-X45Cr28	Heat res	0.3-0.6	26-30	1.5 max	0.5 max	4 max	0.040	0.040	2 max	bal Fe				
UNE 36258(74)	F.8454*	see AM-X45Cr22	0.3-0.6	20-24	1 max	0.5 max	4 max	0.040 max	0.040 max	2 max	bal Fe				
UNE 36258(74)	F.8455*	see AM-X45Cr28	0.3-0.6	26-30	1.5 max	0.5 max	4 max	0.040	0.040	2 max	bal Fe				
UNE 36258(74)	F.8456*	see AM-X35CrNi28-6	0.2-0.5	26-30	1 max	0.5 max	4-7	0.040	0.040	2 max	bal Fe				
USA															
	UNS J92615		0.50 max	26.0-30.0	1.00 max		4.00 max	0.04 max	0.04 max	1.50 max	bal Fe				
ACI	CC-50	Corr res	0.50 max	26.0-30.0	1.00 max		4.00 max	0.04 max	0.04 max	1.50 max	bal Fe				
ASTM A743/A743M(98)	CC-50	Corr res; Norm, Ann	0.50 max	26.0-30.0	1.00 max		4.00 max	0.04 max	0.04 max	1.50 max	bal Fe	380			

Cast Stainless Steel, Ferritic, HC (Continued from previous page)

Specification	Designation	Notes	C	Cr	Mn	Mo	Ni	P	S	Si	Other	UTS	YS	El	Hard
Bulgaria															
BDS 9631	4Ch26SL	Heat res	0.3-0.6	25-27	0.5-1		1 max			1-1.7	bal Fe				
Czech Republic															
CSN 422913	422913	Heat res	0.4-0.7	24.0-26.5	0.9 max		1.0 max	0.045 max	0.04 max	1.0-2.0	Cu <=0.5; bal Fe				
Finland															
SFS 392	G-X30CrNi275	Heat res	0.5 max	26-30	1.5 max	0.5 max	4-7	0.04	0.04	2 max	bal Fe				
SFS 392(79)	X30CrNi275	Heat res	0.5 max	26.0-30.0	1.5 max	0.5 max	4.0-7.0	0.04 max	0.04 max	2.0 max	bal Fe				
Germany															
DIN	GX40CrNi24-5		0.30-0.50	23.0-25.0	1.50 max		3.50-5.50	0.045 max	0.030 max	1.00-2.00	bal Fe				
DIN	WNr 1.4822		0.30-0.50	23.0-25.0	1.50 max		3.50-5.50	0.045 max	0.030 max	1.00-2.00	bal Fe				
DIN 17465(93)	GX40CrNiSi27-4		0.35-0.50	25.0-28.0	1.50 max		3.50-5.50	0.035 max	0.030 max	1.00-2.50	bal Fe				
DIN 17465(93)	GX40CrSi29		0.30-0.45	27.0-30.0	1.00 max			0.035 max	0.030 max	1.00-2.50	bal Fe				
DIN 17465(93)	WNr 1.4776		0.30-0.45	27.0-30.0	0.50-1.00			0.035 max	0.030 max	1.00-2.50	bal Fe				
DIN 17465(93)	WNr 1.4823		0.35-0.50	25.0-28.0	1.50 max		3.50-5.50	0.035 max	0.030 max	1.00-2.50	bal Fe				
DIN SEW 410(88)	G-X120Cr29		0.90-1.30	27.0-30.0	1.00 max			0.045 max	0.030 max	2.00 max	bal Fe	880-1080			260-330 HB

UNS numbers and US grades are provided as a means of cross referencing chemically similar alloys. Exchangability is only possible after independent examination of specifications. Tensile properties are minimum or typical as specified. UTS and YS as MPa. El as %. See Appendix for list of abbreviations used in Notes. * indicates obsolete material.

Specification	Designation	Notes	C	Cr	Mn	Mo	Ni	P	S	Si	Other	UTS	YS	El	Hard

Cast Stainless Steel, Ferritic, HC (Continued from previous page)

Germany

Specification	Designation	Notes	C	Cr	Mn	Mo	Ni	P	S	Si	Other	UTS	YS	El	Hard
DIN SEW 410(88)	WNr 1.4086		0.90-1.30	27.0-30.0	1.00 max			0.045 max	0.030 max	2.00 max	bal Fe	880-1080			260-330 HB
TGL 10414	GS-X45CrNi24.4*	Heat res	0.4-0.5	23-25	1.5 max		3-5	0.045	0.035	1.5 max	bal Fe				

Hungary

Specification	Designation	Notes	C	Cr	Mn	Mo	Ni	P	S	Si	Other	UTS	YS	El	Hard
MSZ 4357	40CrSi29	Heat res	0.2-0.5	27-30	0.4-1		1 max	0.04	0.04	1-2.5	bal Fe				

Italy

Specification	Designation	Notes	C	Cr	Mn	Mo	Ni	P	S	Si	Other	UTS	YS	El	Hard
UNI 3159(83)	GX35Cr28	Heat res	0.2-0.5	26-30	1 max	0.5 max	4 max	0.04	0.035	2.5 max	bal Fe				
UNI 3159(83)	GX35CrNi2805	Heat res; Sand	0.2-0.5	26-30	1.5 max	0.5 max	4-7	0.04	0.035	2.5 max	bal Fe				

Japan

Specification	Designation	Notes	C	Cr	Mn	Mo	Ni	P	S	Si	Other	UTS	YS	El	Hard
JIS G5122(91)	SCH2	Heat res	0.40 max	25.00-28.00	1.00 max	0.50 max	1.00 max	0.040 max	0.040 max	2.00 max	bal Fe	340		5	

Poland

Specification	Designation	Notes	C	Cr	Mn	Mo	Ni	P	S	Si	Other	UTS	YS	El	Hard
PNH83158(86)	LH26	Heat res	0.4-0.6	25-27	0.8 max					1 max	bal Fe				

Romania

Specification	Designation	Notes	C	Cr	Mn	Mo	Ni	P	S	Si	Other	UTS	YS	El	Hard
STAS 6855(86)	T75Cr280	Corr res; Heat res	0.5-1	26-30	1 max		0.4 max	0.035 max	0.03 max	0.5-1.5	Pb <=0.15; bal Fe				

Spain

Specification	Designation	Notes	C	Cr	Mn	Mo	Ni	P	S	Si	Other	UTS	YS	El	Hard
UNE 36016/1(89)	F.3553*	Obs; EN10088-3(96); X20CrNi25-4	0.15-0.25	24.0-27.0	2.0 max	0.15 max	3.0-6.0	0.045 max	0.03 max	0.8-1.5	N <=0.1; bal Fe				
UNE 36022(91)	F.3214*	see X55CrMnNiN20-08	0.5-0.6	19.5-21.5	7-9.5	0.15 max	2-2.8	0.04 max	0.035 max	0.25 max	N 0.3-0.4; bal Fe				
UNE 36022(91)	X55CrMnNiN20-08	Valve	0.5-0.6	19.5-21.5	7-9.5	0.15 max	2-2.8	0.04 max	0.035 max	0.25 max	N 0.3-0.4; bal Fe				
UNE 36257(74)	AM-X40Cr28	Heat res; Corr res	0.5 max	25-35	1 max		4 max	0.040	0.040	1.5	bal Fe				
UNE 36257(74)	F.8404*	see AM-X40Cr28	0.5 max	25-35	1 max		4 max	0.040	0.040	1.5	bal Fe				
UNE 36258(74)	AM-X35CrNi28-6	Heat res	0.2-0.5	26-30	1 max	0.5 max	4-7	0.040	0.040	2 max	bal Fe				
UNE 36258(74)	AM-X45Cr28	Heat res	0.3-0.6	26-30	1.5 max	0.5 max	4 max	0.040 max	0.040 max	2 max	bal Fe				
UNE 36258(74)	F.8455*	see AM-X45Cr28	0.3-0.6	26-30	1.5 max	0.5 max	4 max	0.040 max	0.040 max	2 max	bal Fe				
UNE 36258(74)	F.8456*	see AM-X35CrNi28-6	0.2-0.5	26-30	1 max	0.5 max	4-7	0.040	0.040	2 max	bal Fe				

UK

Specification	Designation	Notes	C	Cr	Mn	Mo	Ni	P	S	Si	Other	UTS	YS	El	Hard
BS 3100(91)	452C11	old En1648(B1)	1.00 max	25.0-30.0	1.00 max	1.50 max	4.00 max	0.050 max	0.050 max	2.00 max	bal Fe				

USA

Specification	Designation	Notes	C	Cr	Mn	Mo	Ni	P	S	Si	Other	UTS	YS	El	Hard
	UNS J92605		0.50 max	26.0-30.0	1.00 max	0.50 max	4.00 max	0.04 max	0.04 max	2.00 max	bal Fe				
ACI	HC	Heat res	0.50 max	26.0-30.0	1.00 max	0.50 max	4.00 max	0.04 max	0.04 max	2.00 max	bal Fe				
ASTM A297/A297M(98)	HC	Heat res	0.5 max	26-30	1 max	0.5 max	4 max	0.04 max	0.04 max	2 max	bal Fe				
ASTM A608(98)	HC30	Heat res; Cent Tub	0.25-0.35	26-30	0.5-1.0	0.50 max	4.0 max	0.04 max	0.04 max	0.50-2.00	bal Fe				

Cast Stainless Steel, Ferritic, HC30

USA

Specification	Designation	Notes	C	Cr	Mn	Mo	Ni	P	S	Si	Other	UTS	YS	El	Hard
	UNS J92613		0.25-0.35	26.0-30.0	0.5-1.0	0.50 max	4.0 max	0.04 max	0.04 max	0.50-2.00	bal Fe				
ACI	HC-30	Heat res	0.25-0.35	26.0-30.0	0.5-1.0	0.50 max	4.0 max	0.04 max	0.04 max	0.50-2.00	bal Fe				

Cast Stainless Steel, Ferritic, No equivalents identified

Poland

Specification	Designation	Notes	C	Cr	Mn	Mo	Ni	P	S	Si	Other	UTS	YS	El	Hard
PNH83158(86)	L0H13	Corr res	0.08 max	12-14	0.4-0.8		1 max	0.035 max	0.035 max	0.7 max	bal Fe				
PNH83158(86)	LH2GN4S2	Corr res	0.2 max	24-28	0.8 max		3-5	0.035 max	0.03	2.5 max	bal Fe				
PNH83159(90)	LH18S2	Heat res	1.35-1.5	17-19	0.4 max		0.4 max	0.04 max	0.035 max	1.9-2.5	bal Fe				
PNH83159(90)	LH29S2	Heat res	1.45-1.6	28-30	1 max		0.4 max	0.04 max	0.035 max	1.5-2.1	bal Fe				

Spain

Specification	Designation	Notes	C	Cr	Mn	Mo	Ni	P	S	Si	Other	UTS	YS	El	Hard
UNE 36257(74)	AM-X12Cr13	Corr res	0.15 max	12.0-14.0	1.0 max	0.5 max	1.0 max	0.040 max	0.040 max	1.5 max	bal Fe				

UNS numbers and US grades are provided as a means of cross referencing chemically similar alloys. Exchangability is only possible after independent examination of specifications. Tensile properties are minimum or typical as specified. UTS and YS as MPa. El as %. See Appendix for list of abbreviations used in Notes. * indicates obsolete material.

Specification	Designation	Notes	C	Cr	Mn	Mo	Ni	P	S	Si	Other	UTS	YS	El	Hard

Cast Stainless Steel, Ferritic, No equivalents identified

UK

Specification	Designation	Notes	C	Cr	Mn	Mo	Ni	P	S	Si	Other	UTS	YS	El	Hard
BS 3100(91)	452C12	old En1648(C)	1.00-2.00	25.0-30.0	1.00 max	1.50 max	4.00 max	0.050 max	0.050 max	2.00 max	bal Fe				

Specification	Designation	Notes	C	Cr	Mn	Mo	Ni	P	S	Si	Other	UTS	YS	El	Hard

Cast Stainless Steel, Martensitic, 420

Germany

Specification	Designation	Notes	C	Cr	Mn	Mo	Ni	P	S	Si	Other	UTS	YS	El	Hard
DIN 17441(97)	X15Cr13	CR Strp Plt Sh	0.12-0.17	12.0-14.0	1.00 max			0.045 max	0.030 max	1.00 max	bal Fe				

Poland

| PNH86020 | 3H13 | Corr res | 0.26-0.35 | 12-14 | 0.8 max | | 0.6 max | 0.04 max | 0.03 max | 0.8 max | bal Fe | | | | |

USA

| | UNS J91201 | | 0.15 max | 12.0-14.0 | 1.00 max | | | 0.04 max | 0.03 max | 1.00 max | bal Fe | | | | |
| MIL-S-81591(93) | IC-420* | Inv, C/Corr res; Obs for new design see AMS specs | | | | | | | | | bal Fe | | | | |

Cast Stainless Steel, Martensitic, 440C

Germany

Specification	Designation	Notes	C	Cr	Mn	Mo	Ni	P	S	Si	Other	UTS	YS	El	Hard
DIN 17230(80)	WNr 1.3543	Bearing Q/T 200C	0.95-1.10	16.0-18.0	1.00 max	0.35-0.75	0.50 max	0.040 max	0.030 max	1.00 max	Cu <=0.30; bal Fe				59 HRC
DIN 17230(80)	X102CrMo17	Q/T to 200C	0.95-1.10	16.0-18.0	1.00 max	0.35-0.75	0.50 max	0.040 max	0.030 max	1.00 max	Cu <=0.30; bal Fe				59 HRC

USA

	UNS J91639		0.95-1.20	16.0-18.0	1.00 max	0.35-0.75	0.75 max	0.04 max	0.03 max	1.00 max	bal Fe				
AMS 5352C(82)		Inv cast, as Ann	0.95-1.20	16.00-18.00	1.00 max	0.35-0.75	0.75 max	0.04 max	0.03 max	1.00 max	Cu <=0.75; bal Fe				30 HRC max
MIL-S-81591(93)	IC-440C*	Inv, C/Corr res; Obs for new design see AMS specs	0.95-1.2	16-18	1	0.35-0.75	0.75	0.04	0.03	1	bal Fe				

Cast Stainless Steel, Martensitic, 5349

USA

Specification	Designation	Notes	C	Cr	Mn	Mo	Ni	P	S	Si	Other	UTS	YS	El	Hard
	UNS J91161		0.15 max	11.5-14.0	1.25 max	0.5 max	0.5 max	0.05 max	0.15-0.35	1.5 max	Cu <=0.5; Zr 0.5 max; bal Fe				

Cast Stainless Steel, Martensitic, 5350

USA

| | UNS J91152 | | 0.05-0.15 | 11.5-13.5 | 1.00 max | 0.50 max | 0.50 max | 0.04 max | 0.03 max | 1.00 max | Cu <=0.50; bal Fe | | | | |

Cast Stainless Steel, Martensitic, CA15

Australia

Specification	Designation	Notes	C	Cr	Mn	Mo	Ni	P	S	Si	Other	UTS	YS	El	Hard
AS 2074(82)	H3A	Corr res	0.15 max	11.5-13.5	1 max		1 max	0.04 max	0.04 max	1 max	bal Fe				

Bulgaria

BDS 6738(72)	1Ch13	Corr res	0.09-0.15	12.0-14.0	0.80 max	0.30 max	0.60 max	0.035 max	0.025 max	0.80 max	bal Fe				
BDS 9631	1Ch13L	Corr res	0.15 max	12-14	0.6 max		1 max			0.7 max	bal Fe				
BDS 9631	Ch13L	Corr res	0.1 max	12-14	0.6 max		1 max			0.7 max	bal Fe				

China

| GB 2100(80) | ZG1Cr13 | Corr res; Quen | 0.08-0.15 | 12.0-14.0 | 0.6 max | | | 0.040 max | 0.040 max | 1.0 max | bal Fe | 549 | 392 | 20 | |
| JB/ZQ 4299(88) | ZG1Cr12Mo | Corr res; Q/T | 0.15 max | 11.5-14.0 | 1.00 max | 0.50 max | 1.00 max | 0.040 max | 0.040 max | 1.50 max | bal Fe | 620 | 450 | 18 | |

Czech Republic

CSN 422904	422904	Corr res	0.15 max	11.5-14.0	0.5-0.9		0.7-1.2	0.035 max	0.035 max	0.6 max	Cu <=0.5; bal Fe				
CSN 422905	422905	Corr res	0.15 max	12.0-14.0	0.7 max		1.0 max	0.04 max	0.04 max	0.7 max	Cu <=0.5; bal Fe				
CSN 422906	422906	Corr res	0.15-0.3	12.0-14.0	0.7 max		1.0 max	0.04 max	0.04 max	0.7 max	Cu <=0.5; bal Fe				

Finland

| SFS 385(79) | G-X12Cr13 | Corr res | 0.15 max | 11.5-14.0 | 1.0 max | 0.5 max | 1.0 max | 0.04 max | 0.04 max | 1.0 max | bal Fe | | | | |

France

| AFNOR | Z12CN13M | Corr res | 0.08-0.15 | 12-14 | 1 max | | 0.5-1.5 | | | 1 max | bal Fe | | | | |

UNS numbers and US grades are provided as a means of cross referencing chemically similar alloys. Exchangability is only possible after independent examination of specifications. Tensile properties are minimum or typical as specified. UTS and YS as MPa. El as %. See Appendix for list of abbreviations used in Notes. * indicates obsolete material.

Cast Stainless Steel, Martensitic, CA15 (Continued from previous page)

Specification	Designation	Notes	C	Cr	Mn	Mo	Ni	P	S	Si	Other	UTS	YS	El	Hard
France															
AFNOR NFA32056	Z12C13M	Corr res; Q/A Tmp	0.15 max	11.5-13.5	1.5 max					1.2 max	bal Fe				
Germany															
DIN	WNr 1.4106		0.03 max	17.0-19.0	1.00 max	1.50-2.50			0.25-0.35	1.00-1.70	bal Fe				
DIN 17445(84)	WNr 1.4008		0.06-0.12	12.0-13.5	1.00 max	0.50 max	1.00-2.00	0.045 max	0.030 max	1.00 max	bal Fe	590-780	440	15	
DIN EN 10213(96)	GX20Cr14	HT	0.16-0.23	12.5-14.5	1.00 max		1.00 max	0.045 max	0.030 max	1.00 max	bal Fe	590-790	440	12	
DIN EN 10213(96)	GX7CrNiMo12-1		0.06-0.12	12.0-13.5	1.00 max	0.50 max	1.00-2.00	0.045 max	0.030 max	1.00 max	bal Fe	590-780	440	15	
DIN EN 10213(96)	WNr 1.4027		0.16-0.23	12.5-14.5	1.00 max		1.00 max	0.045 max	0.030 max	1.00 max	bal Fe	590-790	440	12	
Hungary															
MSZ 21053(82)	AoX12Cr13	Corr res; 0-30.0mm; soft ann	0.15 max	12-14	1 max		1 max	0.04 max	0.04 max	1.5 max	bal Fe				140-180 HB
Italy															
UNI 3161(83)	GX12Cr13	Sand; Heat res	0.15 max	11.5-13.5	1 max	0.5 max	1 max	0.04 max	0.035 max	1.5 max	bal Fe				
UNI 6901(71)	X12Cr13	Corr res	0.09-0.15	11.5-14	1 max		1 max	0.04 max	0.03 max	1 max	bal Fe				
UNI 6942(71)	X10Cr13	Corr res	0.08-0.14	11.5-13.5	1 max		1 max	0.03 max	0.025 max	0.8 max	bal Fe				
Japan															
JIS G5121(91)	SCS1	Corr res; Solution HT	0.15 max	11.50-14.00	1.00 max	0.50 max	1.00 max	0.040 max	0.040 max	1.50 max	bal Fe				
Mexico															
DGN B-354	CA-15*	Obs; Corr res	0.15 max	11.5-14	1	0.5	1	0.04	0.04	1.5	bal Fe				
Romania															
STAS 3583(87)	12C130	Corr res; Heat res	0.09-0.15	12-14	0.6 max		0.6 max	0.035 max	0.03 max	0.6 max	Pb <=0.15; bal Fe				
STAS 6855(86)	T20Cr130	Corr res; Heat res	0.15-0.25	12-14	1 max		1 max	0.035 max	0.03 max	1 max	Pb <=0.15; bal Fe				
STAS6855(86)	T15Cr130	Corr res; Heat res	0.15 max	12-14	1 max		1 max	0.035 max	0.03 max	1 max	Pb <=0.15; bal Fe				
Russia															
GOST 5632	12Ch13	Corr res	0.09-0.15	12-14	0.8 max		0.6 max	0.03	0.025	0.8 max	Cu 0.3; Ti <=0.2; bal Fe				
GOST 977	15Ch13L	Corr res	0.15 max	12-14	0.3-0.8		0.5 max	0.03	0.025	0.2-0.8	Cu 0.3; bal Fe				
Spain															
UNE 36257(74)	AM-X12Cr13	Corr res	0.15 max	12-14	1 max	0.5 max	1 max			1.5 max	bal Fe				
UNE 36259(79)	AM-X20Cr13*		0.25 max	12.0-14.0	1.0 max	0.5 max		0.040 max	0.040 max	2.0 max	bal Fe				
UNE 36259(79)	F.8387*	see AM-X20Cr13*	0.25 max	12.0-14.0	1.0 max	0.5 max		0.040 max	0.040 max	2.0 max	bal Fe				
Sweden															
SS 142302	2302	Corr res	0.09-0.15	12-14	1 max		1 max	0.04 max	0.03 max	1 max	bal Fe				
UK															
BS 1504(76)	420C29*	En713 Corr res	0.2 max	11.5-13.5	1 max		1 max			1 max	bal Fe				
BS 3100(91)	410C21	old En1630A; Corr res	0.15 max	11.5-13.5	1.00 max		1.00 max	0.040 max	0.040 max	1.00 max	Cu <=0.30; bal Fe	540	370	15	152-207 HB
BS 3100(91)	420C29	old En1630(B)	0.20 max	11.5-13.5	1.00 max		1.00 max	0.040 max	0.040 max	1.00 max	Cu <=0.30; bal Fe	690	465	11	201-255 HB
BS 3146/2(75)	ANC1(A)	Corr res	0.15 max	11.5-13.5	0.20-1.00		1.00 max			0.20-1.20	bal Fe				
USA															
	UNS J91150		0.15 max	11.5-14.0	1.00 max	0.50 max	1.00 max	0.040 max	0.040 max	1.50 max	bal Fe				
	UNS J91171		0.15 max	11.5-14.0	1.00 max	0.50 max		0.040 max	0.040 max	1.50 max	Cu <=0.50; N <=1.00; V <=0.03; W <=0.10; bal Fe				

Specification	Designation	Notes	C	Cr	Mn	Mo	Ni	P	S	Si	Other	UTS	YS	El	Hard

Cast Stainless Steel, Martensitic, CA15 (Continued from previous page)

USA

Specification	Designation	Notes	C	Cr	Mn	Mo	Ni	P	S	Si	Other	UTS	YS	El	Hard
ACI	CA-15	Corr res	0.15 max	11.5-14.0	1.00 max	0.50 max	1.00 max	0.04 max	0.040 max	1.50 max	bal Fe				
AMS 5351F(95)*		Corr res; N/T	0.15 max	11.50-14.00	1.00 max	0.50 max	1.00 max	0.04 max	0.03 max	1.50 max	Al <=0.050; Cu <=0.50, Sn 0.05 max, bal Fe				217-248 HB
ASTM A217/A217M(95)	CA-15	Corr res; N/T	0.15 max	11.5-14	1 max	0.5 max	1 max	0.04 max	0.04 max	1.5 max	bal Fe				
ASTM A426(97)	CPCA15	Pipe; Heat res	0.15 max	11.5-14.0	1 max	0.5 max	1.00 max	0.040 max	0.04 max	1.50 max	Cu <=0.50; W <=0.10; Cu+Ni+W<=1.50; bal Fe	620	450	18	225 HB max
ASTM A487	CA15*	N/T, Q/T	0.15 max	11.5-14.0	1.00 max	0.50 max	1.00 max	0.040 max	0.040 max	1.50 max	Cu <=0.50; V <=0.05; W <=0.10; Cu+W+V<=0.50; bal Fe	620-795	450	18	
ASTM A487	CA15A	N/T, Q/T	0.15 max	11.5-14.0	1.00 max	0.50 max	1.00 max	0.040 max	0.040 max	1.50 max	Cu <=0.50; V <=0.05; W <=0.10; Cu+W+V<=0.50; bal Fe	965-1170	766-895	10	
ASTM A487	CA15B	N/T, Q/T	0.15 max	11.5-14.0	1.00 max	0.50 max	1.00 max	0.040 max	0.040 max	1.50 max	Cu <=0.50; V <=0.05; W <=0.10; Cu+W+V<=0.50; bal Fe	620-795	450	18	
ASTM A487	CA15C	N/T, Q/T	0.15 max	11.5-14.0	1.00 max	0.50 max	1.00 max	0.040 max	0.040 max	1.50 max	Cu <=0.50; V <=0.05; W <=0.10; Cu+W+V<=0.50; bal Fe	620	415	18	22 HRC
ASTM A487	CA15D	N/T, Q/T	0.15 max	11.5-14.0	1.00 max	0.50 max	1.00 max	0.040 max	0.040 max	1.50 max	Cu <=0.50; V <=0.05; W <=0.10; Cu+W+V<=0.50; bal Fe	690	515	17	22 HRC
ASTM A487	CA6NMB	N/T, Q/T	0.06 max	11.5-14.0	1.00 max	0.4-1.0	3.5-4.5	0.04 max	0.03 max	1.00 max	Cu <=0.50; V <=0.05; W <=0.10; Cu+W+V<=0.50; bal Fe	690	520	17	23 HRC
ASTM A743/A743M(98)	CA-15	Corr res; N/T, Ann	0.15 max	11.5-14.0	1.00 max	0.50 max	1.00 max	0.04 max	0.04 max	1.50 max	bal Fe	620	450	18	
MIL-C-24707/6(89)	CA-15	As spec in ASTM A217; pressure containers, high-temp service									bal Fe				
MIL-S-16993A(89)	Class 1*	Corr res; Obs; see MIL-C-24707	0.15 max	11.5-14	1 max	0.5 max	1 max	0.04 max	0.04 max	1.5 max	bal Fe				
MIL-S-81591(93)	IC-410*	Inv; C/Corr res; Obs for new design see AMS specs	0.05-0.15	11.5-13.5	1 max	0.5 max	0.5 max			1 max	Cu 0.5; bal Fe				

Yugoslavia

Specification	Designation	Notes	C	Cr	Mn	Mo	Ni	P	S	Si	Other	UTS	YS	El	Hard
	CL.4575	Corr res	0.08-0.15	11.5-13.5	0.35-0.5	0.15 max	0.8-1.3	0.035 max	0.035 max	0.25-0.5	bal Fe				

Cast Stainless Steel, Martensitic, CA15M

Japan

Specification	Designation	Notes	C	Cr	Mn	Mo	Ni	P	S	Si	Other	UTS	YS	El	Hard
JIS G5121(91)	SCS3	Corr res; Tmp	0.15 max	11.50-14.00	1.00 max		1.50-2.50	0.040 max	0.040 max	1.50 max	bal Fe	640	490	13	192-255 HB
JIS G5121(91)	SCS4	Corr res; Tmp	0.15 max	11.50-14.00	1.00 max		1.50-2.50	0.040 max	0.040 max	1.50 max	bal Fe	640	490	13	192-255 HB

USA

Specification	Designation	Notes	C	Cr	Mn	Mo	Ni	P	S	Si	Other	UTS	YS	El	Hard
	UNS J91151		0.15 max	11.5-14.0	1.00 max	0.15-1.0	1.0 max	0.040 max	0.040 max	0.65 max	bal Fe				
ACI	CA-15M	Corr res	0.15 max	11.5-14.0	1.00 max	0.15-1.00	1.00 max	0.040 max	0.040 max	0.65 max	bal Fe				
ASTM A487	CA15M*	N/T, Q/T	0.15 max	11.5-14.0	1.00 max	0.15-1.0	1.0 max	0.040 max	0.040 max	0.65 max	Cu <=0.50; V <=0.05; W <=0.10; Cu+W+V<=0.50; bal Fe	620-795	450	18	
ASTM A487	CA15MA	N/T, Q/T	0.15 max	11.5-14.0	1.00 max	0.15-1.0	1.0 max	0.040 max	0.040 max	0.65 max	Cu <=0.50; V <=0.05; W <=0.10; Cu+W+V<=0.50; bal Fe	620-795	450	18	

UNS numbers and US grades are provided as a means of cross referencing chemically similar alloys. Exchangability is only possible after independent examination of specifications. Tensile properties are minimum or typical as specified. UTS and YS as MPa. El as %. See Appendix for list of abbreviations used in Notes. * indicates obsolete material.

Specification	Designation	Notes	C	Cr	Mn	Mo	Ni	P	S	Si	Other	UTS	YS	El	Hard

Cast Stainless Steel, Martensitic, CA15M (Continued from previous page)

USA

ASTM A487	CA6NMA	N/T, Q/T	0.06 max	11.5-14.0	1.00 max	0.4-1.0	3.5-4.5	0.04 max	0.03 max	1.00 max	Cu <=0.50; V <=0.05; W <=0.10; Cu+W+V<=0.50; bal Fe	760-930	515	15	
ASTM A743/A743M(98)	CA-15M	Corr res	0.15 max	11.5-14.0	1.00 max	0.15-1.0	1.00 max	0.040 max	0.040 max	0.65 max	bal Fe	620	450	18	
MIL-C-24707/6(89)	CA-15M	As spec in ASTM A487; pressure containers, high-temp service									bal Fe				

Cast Stainless Steel, Martensitic, CA28MWV

Romania

| STAS 9277(84) | OTA23VMoCr120 | | 0.2-0.26 | 11.5-12.5 | 0.5-0.8 | 1-1.2 | 0.7-1 | 0.045 max | 0.03 max | 0.2-0.4 | Pb <=0.15; V 0.2-0.4; bal Fe | | | | |
| STAS 9277(84) | OTA23VWMoCr120 | | 0.2-0.26 | 11.5-12.5 | 0.5-0.8 | 1-1.2 | 0.7-1 | 0.045 max | 0.03 max | 0.2-0.4 | Pb <=0.15; V 0.2-0.4; W 0.4-0.6; bal Fe | | | | |

UK

| BS 1503(89) | 762-690 | Frg Press ves; 0-250mm; N/T Q/T | 0.17-0.23 | 11.0-12.5 | 0.3-1.0 | 0.7-1.2 | 0.3-0.8 | 0.04 max | 0.025 max | 0.15-0.4 | Al <=0.02; Co <=0.1; Cu <=0.3; Pb <=0.15; Ti <=0.05; V 0.2-0.35; W <=0.1; bal Fe | 690-840 | 490 | 14 | |
| BS 3059/2(90) | 762 | Boiler, Superheater; High-temp; Smls Tub; 2-12.5mm | 0.17-0.23 | 10.0-12.5 | 1.00 max | 0.8-1.20 | 0.30-0.80 | 0.030 max | 0.030 max | 0.50 max | V 0.25-0.35; bal Fe | 720-870 | 470 | 15 | |

USA

	UNS J91422		0.20-0.28	11.0-12.5	0.50-1.00	0.90-1.25	0.50-1.00	0.030 max	0.030 max	1.0 max	V 0.20-0.30; W 0.90-1.25; bal Fe				
ACI	CA28MWV	Corr res	0.20-0.28	11.0-12.5	0.50-1.00	0.90-1.25	0.50-1.00	0.030 max	0.030 max	1.00 max	V 0.20-0.30; W 0.90-1.25; bal Fe				
ASTM A743/A743M(98)	CA-28MWV	Corr res; Q/A, Tmp	0.20-0.28	11.0-12.5	0.50-1.00	0.90-1.25	0.5-1.00	0.030 max	0.030 max	1.0 max	V 0.2-0.3; W 0.90-1.25; bal Fe	965	760	10	
SAE J467(68)	422M (Cast)		0.26	13.0	1.00	2.50				0.40	V 0.50; W 1.50; bal Fe				

Cast Stainless Steel, Martensitic, CA40

Bulgaria

| BDS 6738(72) | 3Ch13 | Corr res | 0.25-0.35 | 12.0-14.0 | 0.80 max | 0.30 max | 0.60 max | 0.035 max | 0.025 max | 0.80 max | bal Fe | | | | |
| BDS 9631 | 3Ch14L | Corr res | 0.26-0.35 | 13-15 | 0.3-0.7 | | 1 max | | | 0.3-0.7 | bal Fe | | | | |

China

| GB 2100(80) | ZG2Cr13 | Corr res; Quen | 0.16-0.24 | 12.0-14.0 | 0.6 max | | | 0.040 max | 0.040 max | 1.0 max | bal Fe | 618 | 441 | 16 | |
| JB/ZQ 4299(88) | ZG3Cr12Mo | Corr res; Q/T | 0.20-0.40 | 11.5-14.0 | 1.00 max | 0.50 max | 1.00 max | 0.040 max | 0.040 max | 1.50 max | bal Fe | 690 | 480 | 15 | |

France

| AFNOR | Z30C13M | Corr res | 0.25-0.35 | 13-15 | 1 max | | | | | 1 max | bal Fe | | | | |
| AFNOR NFA32056 | Z28C13M | Corr res | 0.3 | 11.5-13.5 | 1.5 | 0.5 max | 1 max | 0.04 | 0.03 | 1.2 | bal Fe | | | | |

Germany

| DIN 17465(93) | GX40CrSi13 | | 0.30-0.45 | 12.0-14.0 | 1.00 max | | | 0.035 max | 0.030 max | 1.00-2.50 | bal Fe | | | | 300 HB |
| DIN 17465(93) | WNr 1.4729 | | 0.30-0.45 | 12.0-14.0 | 0.50-1.00 | | | 0.035 max | 0.030 max | 1.00-2.50 | bal Fe | | | | 300 HB |

Hungary

| MSZ 4357(81) | AoX40CrSi13 | Heat res; soft ann | 0.3-0.5 | 12-14 | 0.4-1 | | 1 max | 0.04 max | 0.04 max | 1-2.5 | bal Fe | | | | 210-300 HB |
| MSZ 4357(81) | AoX40CrSi13F | Heat res; soft ann | 0.3-0.5 | 12-14 | 0.4-1 | | 1 max | 0.04 max | 0.04 max | 1-2.5 | bal Fe | 500 | | 5L | 210-300 HB |

Italy

| UNI 3159 | GX35Cr13 | Corr res | 0.2-0.5 | 12-14 | 1 max | 0.5 max | | | | 2.5 max | bal Fe | | | | |
| UNI 3161(83) | GX30Cr13 | Sand; Heat res | 0.2-0.4 | 11.5-14 | 1 max | 0.5 max | 1 max | 0.04 max | 0.035 max | 1.5 max | bal Fe | | | | |

UNS numbers and US grades are provided as a means of cross referencing chemically similar alloys. Exchangability is only possible after independent examination of specifications. Tensile properties are minimum or typical as specified. UTS and YS as MPa. El as %. See Appendix for list of abbreviations used in Notes. * indicates obsolete material.

Specification	Designation	Notes	C	Cr	Mn	Mo	Ni	P	S	Si	Other	UTS	YS	El	Hard

Cast Stainless Steel, Martensitic, CA40 (Continued from previous page)

Japan

Specification	Designation	Notes	C	Cr	Mn	Mo	Ni	P	S	Si	Other	UTS	YS	El	Hard
JIS G5121(91)	SCS2	Corr res; Tmp	0.16-0.24	11.50-14.00	1.00 max	0.50 max	1.00 max	0.040 max	0.040 max	1.50 max	bal Fe	590	390	16	170-235 HB
JIS G5121(91)	SCS2A	Corr res; Tmp	0.25-0.40	11.50-14.00	1.00 max	0.50 max	1.00 max	0.040 max	0.040 max	1.50 max	bal Fe	690	485	15	269 HB max
JIS G5122(91)	SCH3	Corr res; Heat Res; Ann	0.40 max	12.00-15.00	1.00 max	0.50 max	1.00 max	0.040 max	0.040 max	2.00 max	bal Fe	490			

Mexico

Specification	Designation	Notes	C	Cr	Mn	Mo	Ni	P	S	Si	Other	UTS	YS	El	Hard
DGN B-354	CA-40*	Obs; Corr res	0.2-0.4	11.5-14	1	0.5	1	0.04	0.04	1.5	bal Fe				

Poland

Specification	Designation	Notes	C	Cr	Mn	Mo	Ni	P	S	Si	Other	UTS	YS	El	Hard
PNH86020	3H13	Corr res	0.26-0.35	12-14	0.8 max		0.6 max	0.04	0.03	0.8 max	bal Fe				

Romania

Specification	Designation	Notes	C	Cr	Mn	Mo	Ni	P	S	Si	Other	UTS	YS	El	Hard
STAS 6855(86)	T40SiCr130	Corr res; Heat res	0.3-0.5	12-14	1 max		0.4 max	0.035 max	0.03 max	2-3	Pb <=0.15; bal Fe				

Russia

Specification	Designation	Notes	C	Cr	Mn	Mo	Ni	P	S	Si	Other	UTS	YS	El	Hard
GOST 5632	30Ch13	Corr res	0.26-0.35	12-14	0.8 max		0.6 max	0.03	0.025	0.8 max	bal Fe				

Spain

Specification	Designation	Notes	C	Cr	Mn	Mo	Ni	P	S	Si	Other	UTS	YS	El	Hard
UNE 36257(74)	AM-X30Cr13	Corr res	0.2-0.4	12-14	1 max	0.5 max	1 max			1.5 max	bal Fe				
UNE 36257(74)	F.8402*	see AM-X30Cr13	0.2-0.4	12-14	1 max	0.5 max	1 max			1.5 max	bal Fe				

UK

Specification	Designation	Notes	C	Cr	Mn	Mo	Ni	P	S	Si	Other	UTS	YS	El	Hard
BS 3146/2(75)	ANC1(C)	Corr res	0.20-0.30	11.5-13.5	0.20-1.00		1.00 max			0.20-1.20	bal Fe				

USA

Specification	Designation	Notes	C	Cr	Mn	Mo	Ni	P	S	Si	Other	UTS	YS	El	Hard
	UNS J91153		0.2-0.4	11.5-14.0	1.00 max	0.5 max	1.0 max	0.04 max	0.04 max	1.50 max	bal Fe				
ACI	CA-40	Corr res	0.20-0.40	11.5-14.0	1.00 max	0.50 max	1.0 max	0.04 max	0.04 max	1.50 max	bal Fe				
ASTM A743/A743M(98)	CA-40	Corr res; N/T	0.20-0.40	11.5-14.0	1.00 max	0.5 max	1.0 max	0.04 max	0.04 max	1.50 max	bal Fe	690	485	15	

Yugoslavia

Specification	Designation	Notes	C	Cr	Mn	Mo	Ni	P	S	Si	Other	UTS	YS	El	Hard
	CL.4275	Heat res	0.3-0.6	12.0-14.0	0.3-1.0	0.15 max		0.045 max	0.03 max	1.0-2.5	bal Fe				

Cast Stainless Steel, Martensitic, CA40F

Japan

Specification	Designation	Notes	C	Cr	Mn	Mo	Ni	P	S	Si	Other	UTS	YS	El	Hard
JIS G5122(91)	SCH1	Corr res	0.20-0.40	12.00-15.00	1.00 max	0.50 max	1.00 max	0.040 max	0.040 max	1.50-3.00	bal Fe	490			

Romania

Specification	Designation	Notes	C	Cr	Mn	Mo	Ni	P	S	Si	Other	UTS	YS	El	Hard
STAS 6855(86)	T22Cr135	Corr res; Heat res	0.15-0.35	12-15	0.4-0.8		1 max	0.035 max	0.03 max	0.7 max	Pb <=0.15; bal Fe				
STAS 6855(86)	T40NiCr130	Corr res; Heat res	0.3-0.5	12-14	1 max		1 max	0.035 max	0.03 max	0.15-1.2	Pb <=0.15; bal Fe				

USA

Specification	Designation	Notes	C	Cr	Mn	Mo	Ni	P	S	Si	Other	UTS	YS	El	Hard
	UNS J91154		0.20-0.40	11.5-14.0	1.00 max	0.5 max	1.00 max	0.04 max	0.20-0.40	1.50 max	bal Fe				
ACI	CA40F	Corr res	0.20-0.40	11.5-14.0	1.00 max	0.50 max	1.00 max	0.04 max	0.20-0.40	1.50 max	bal Fe				
ASTM A743/A743M(98)	CA-40F	Corr res; N/T	0.20-0.40	11.5-14.0	1.00 max	0.5 max	1.0 max	0.40 max	0.20-0.40	1.50 max	bal Fe	690	485	12	

Cast Stainless Steel, Martensitic, CA6N

USA

Specification	Designation	Notes	C	Cr	Mn	Mo	Ni	P	S	Si	Other	UTS	YS	El	Hard
	UNS J91650		0.06 max	10.5-12.5	0.50 max		6.0-8.0	0.02 max	0.02 max	1.00 max	bal Fe				
ACI	CA-6N	Corr res	0.06 max	10.5-12.5	0.50 max		6.0-8.0	0.02 max	0.02 max	1.00 max	bal Fe				
ASTM A743/A743M(98)	CA-6N	Corr res; Solution Age	0.06 max	10.5-12.5	0.5 max		6.0-8.0	0.02 max	0.02 max	1.00 max	bal Fe	965	930	15	

Cast Stainless Steel, Martensitic, CA6NM

Bulgaria

Specification	Designation	Notes	C	Cr	Mn	Mo	Ni	P	S	Si	Other	UTS	YS	El	Hard
BDS 9631	0Ch12N4ML	Corr res	0.08 max	11.5-13.5	1.5 max	1 max	3.5-5	0.035	0.035	1 max	bal Fe				

UNS numbers and US grades are provided as a means of cross referencing chemically similar alloys. Exchangability is only possible after independent examination of specifications. Tensile properties are minimum or typical as specified. UTS and YS as MPa. El as %. See Appendix for list of abbreviations used in Notes. * indicates obsolete material.

Cast Stainless Steel, Martensitic, CA6NM (Continued from previous page)

Specification	Designation	Notes	C	Cr	Mn	Mo	Ni	P	S	Si	Other	UTS	YS	El	Hard
China															
GB 6967(86)	ZG06Cr13Ni4Mo	Corr res; Quen	0.07 max	11.5-13.5	1.00 max	0.40-1.00	3.5-5.0	0.035 max	0.030 max	1.00 max	bal Fe	750	550	15	
JB/ZQ 4299(88)	ZG0Cr13Ni4Mo	Corr res; Q/T	0.06 max	11.5-14.0	1.00 max	0.40-1.00	3.50-4.50	0.030 max	0.030 max	1.00 max	bal Fe	760	550	15	
Finland															
SFS 386(79)	G-X6CrNi134	Corr res	0.06 max	11.5-14.0	1.0 max	0.4-1.0	3.5-4.5	0.04 max	0.03 max	1.0 max	bal Fe				
France															
AFNOR	Z4CND134M	Corr res	0.06 max	12-13.5	1 max	0.4-0.7	3.5-5			1 max	bal Fe				
AFNOR NFA32055	Z6CND1304M	Corr res; Norm, Q/A	0.08	11.5-13.5	1.5	0.5-1.5	2-4	0.04	0.03	1.2	bal Fe				
AFNOR NFA32056	Z6CND1304M	Corr res; Q/A Tmp	0.08 max	11.5-13.5	1.5	0.4-1.5	3-5	0.04	0.03	1.2	bal Fe				
Germany															
DIN SEW 520(96)	G-X5CrNi13-4		0.05 max	12.0-14.0	1.50 max	0.30-0.70	3.50-4.50	0.040 max	0.015 max	0.70 max	bal Fe				
DIN SEW 520(96)	GX5CrNi13-4		0.05 max	12.0-14.0	1.50 max	0.30-0.70	3.50-4.50	0.040 max	0.015 max	0.70 max	bal Fe				
DIN SEW 520(96)	WNr 1.4313		0.05 max	12.0-14.0	1.50 max	0.30-0.70	3.50-4.50	0.040 max	0.015 max	0.70 max					
DIN(Military Hdbk)	GX4CrNiMo13-4		0.03-0.07	12.0-13.0	0.30-0.70	0.40-0.80	3.80-4.50	0.040 max	0.035 max	0.30-0.70	bal Fe				
DIN(Military Hdbk)	WNr 1.4414		0.03-0.07	12.0-13.0	0.30-0.70	0.40-0.80	3.80-4.50	0.040 max	0.035 max	0.30-0.70	bal Fe				
Italy															
UNI 3161(83)	GX6CrNi1304	Sand; Heat res	0.08 max	11.5-13.5	1.5 max	0.4-0.7	3.5-5	0.04 max	0.035 max	1 max	bal Fe				
Japan															
JIS G5121(91)	SCS5	Corr res; Solution HT	0.06 max	11.50-14.00	1.00 max		3.50-4.50	0.040 max	0.040 max	1.00 max	bal Fe	740	540	13	217-277 HB
JIS G5121(91)	SCS6	Corr res; Tmp	0.06 max	11.50-14.00	1.00 max	0.40-1.00	3.50-4.50	0.040 max	0.030 max	1.00 max	bal Fe	750	550	15	285 HB max
Romania															
STAS 6855(86)	T8MoNiCr125	Corr res; Heat res	0.08 max	11.5-13.5	1.5 max	1 max	3.5-5	0.035 max	0.03 max	1 max	Pb <=0.15; bal Fe				
Sweden															
SS 142385	2385	Corr res	0.1 max	12-14	1 max		5-6	0.045 max	0.03 max	1 max	bal Fe				
UK															
BS 1504(76)	425C11*	Corr res	0.1 max	11.5-13.5	1 max		0.6 max	3.4-4.2			1 max	bal Fe				
BS 3100(91)	425C11	old En1630(B); Corr res	0.10 max	11.5-13.5	1.00 max	0.60 max	3.40-4.20	0.040 max	0.030 max	1.00 max	bal Fe	770	620	12	321 HB
USA															
	UNS J91540		0.06 max	11.5-14.0	1.00 max	0.40-1.0	3.5-4.5	0.04 max	0.03 max		bal Fe				
ACI	CA-6NM	Corr res	0.06 max	11.5-14.0	1.00 max	0.40-1.0	3.5-4.5	0.04 max	0.04 max	1.00 max	bal Fe				
ASTM A352/A352M(98)	UNS J91540	Press; Low-temp	0.06 max	11.5-14.0	1.00 max	0.4-1.0	3.5-4.5	0.04 max	0.03 max	1.00 max	bal Fe	760-930	550	15	
ASTM A356/A356M(98)	CA6NM	Heavy walled for steam turbines	0.06 max	11.5-14.0	1.00 max	0.4-1.0		0.035 max	0.030 max	1.00 max	bal Fe	760	550	15.0	
ASTM A743/A743M(98)	CA-6NM	Corr res; Q/A, N/T	0.06 max	11.5-14.0	1.00 max	0.40-1.0	3.5-4.5	0.04 max	0.03 max	1.00 max	bal Fe	755	550	15	
MIL-C-24707/6(89)	E3N	As spec in ASTM A757; pressure containers, high-temp service									bal Fe				
Yugoslavia															
	CL.4579	Corr res	0.08 max	11.5-13.5	0.35-0.65	0.4-0.6	3.5-4.1	0.035 max	0.035 max	0.25-0.5	bal Fe				

Specification	Designation	Notes	C	Cr	Mn	Mo	Ni	P	S	Si	Other	UTS	YS	El	Hard
Cast Stainless Steel, Martensitic, CB7Cu-1															
China															
GB 2100(80)	ZG0Cr17Ni4Cu4Nb	Corr res; Sol Aged	0.07 max	15.5-17.5	1.0 max		3.0-5.0	0.035 max	0.030 max	1.0 max	Cu 2.6-4.6; Nb 0.15-0.45; bal Fe	981	785	5	
France															
AFNOR NFA35586	Z4CNU17.4M	Corr res	0.06 max	15.5-17.5	1.5 max		3-5	0.04	0.03	1.2 max	Cu 3-5; Nb 0.15-0.35; bal Fe				
Japan															
JIS G5121(91)	SCS24	Corr res; Case	0.07 max	15.50-17.50	1.00 max		3.00-5.00	0.040 max	0.040 max	1.00 max	Cu 2.50-4.00; Nb 0.15-0.45; bal Fe				
USA															
	UNS J92180		0.07 max	15.50-17.70	0.70 max		3.60-4.60	0.035 max	0.03 max	1.00 max	Cu 2.50-3.20; N <=0.05; Nb 0.15-0.35; bal Fe				
AMS 5398B		Corr res; Solution Age	0.07 max	15.5-17.7	0.7 max		3.6-4.6	0.040-0.04	0.03	0.5-1	Cu 2.5-3.20; N <=0.050; Nb+Ta 0.15-0.35; bal Fe				
ASTM A747/A747M(98)	CB7Cu-1	PH	0.07 max	15.50-17.70	0.70 max		3.60-4.60	0.035 max	0.03 max	1.00 max	Cu 2.50-3.20; N <=0.05; Nb 0.15-0.35; bal Fe	860-1170	670-1000	5-10	269-375 HB
Cast Stainless Steel, Martensitic, CB7Cu-2															
Germany															
DIN	GX4CrNiCuNb16-4		0.06 max	15.0-17.0	1.00 max		3.50-5.00			1.00 max	Cu 2.50-4.00; Nb 0.15-0.40; bal Fe	1035	965	9	
DIN	WNr 1.4540		0.06 max	15.0-17.0	1.00 max		3.50-5.00			1.00 max	Cu 2.50-4.00; N <=0.05; Nb 0.15-0.40; bal Fe	1035	965	9	
USA															
	UNS J92110	Precipitation Hardening (15-5 PH)	0.07 max	14.0-15.50	0.70 max		4.20-5.50	0.035 max	0.03 max	1.00 max	Cu 2.50-3.20; N <=0.05; Nb 0.15-0.35; bal Fe				
	UNS J92130*	Obs; see J92110; 15-5 PH									bal Fe				
AMS 5346B(96)		Corr res; Solution Age, SCS	0.050 max	14.00-15.50	0.60 max		4.20-5.00	0.025 max	0.025 max	0.50-1.00	Cu 2.50-3.20; N <=0.050; Nb 0.15-0.30; Nb+Ta 0.15-0.3; Ta<=0.05; bal Fe	1241	1103	6	40-47 HRC
AMS 5347A(95)		Corr res; Solution Age, SCS	0.050 max	14.00-15.50	0.60 max		4.20-5.00	0.025 max	0.025 max	0.50-1.00	Cu 2.50-3.20; N <=0.050; Nb 0.15-0.30; Nb+Ta 0.15-0.3; Ta<=0.05; bal Fe	1034	896	8	35-42 HRC
AMS 5348B(95)*		Corr res; Solution Age	0.050 max	14.00-15.50	0.60 max		4.20-5.00	0.025 max	0.025 max	0.50-1.00	Cu 2.50-3.20; N <=0.050; Nb+Ta 0.15-0.30; bal Fe	1240	1035	6	40-45 HRC
AMS 5356A(95)		Corr res; Solution Age	0.050 max	14.00-15.50	0.60 max		4.20-5.00	0.025 max	0.025 max	0.50-1.00	Cu 2.50-3.20; N <=0.050; Nb 0.15-0.30; bal Fe	896	827	8	33-40 HRC
AMS 5357A		Corr res; Solution Age	0.050 max	14.00-15.50	0.60 max		4.2-5	0.025	0.025	0.5-1	Cu 2.5-3.20; N <=0.050; Nb+Ta 0.15-0.3; bal Fe				
AMS 5357A(95)		Corr res; Inv cast; Physicals depend on HT	0.05 max	14.00-15.50	0.60 max		4.20-5.00	0.025 max	0.025 max	0.50-1.00	Cu 2.50-3.20; N <=0.05; Nb 0.15-0.30; Ta 0.05max; bal Fe	860-1240	760-1100	6-12	
AMS 5400		Corr res; Solution Age	0.050 max	14-15.5	0.60 max		4.2-5	0.025	0.025	0.5-1	Cu 2.5-3.20; N 0.050; Nb+Ta 0.15-0.3; bal Fe				
ASTM A747/A747M(98)	CB7Cu-2	PH	0.07 max	14.0-15.50	0.70 max		4.50-5.50	0.035 max	0.03 max	1.00 max	Cu 2.50-3.20; N <=0.05; Nb 0.15-0.35; bal Fe	860-1170	670-1000	5-10	269-375 HB
Cast Stainless Steel, Martensitic, E3N															
Germany															
DIN	GX5CrNiMo13-4		0.08 max	11.5-13.5	1.50 max	0.50-2.00	4.00-5.00	0.045 max	0.030 max	1.50 max	bal Fe				
DIN	WNr 1.4407		0.08 max	11.5-13.5	1.50 max	0.50-2.00	4.00-5.00	0.045 max	0.030 max	1.50 max	bal Fe				

UNS numbers and US grades are provided as a means of cross referencing chemically similar alloys. Exchangability is only possible after independent examination of specifications. Tensile properties are minimum or typical as specified. UTS and YS as MPa. El as %. See Appendix for list of abbreviations used in Notes. * indicates obsolete material.

Specification	Designation	Notes	C	Cr	Mn	Mo	Ni	P	S	Si	Other	UTS	YS	El	Hard

Cast Stainless Steel, Martensitic, E3N (Continued from previous page)

USA

Specification	Designation	Notes	C	Cr	Mn	Mo	Ni	P	S	Si	Other	UTS	YS	El	Hard
	UNS J91550		0.06 max	11.5-14.0	1.00 max	0.40-1.0	3.5-4.5	0.030 max	0.030 max	1.00 max	Cu <=0.50; W <=0.10; bal Fe				
ASTM A757/A757M(96)	E3N	Corr res; N/T	0.06 max	11.5-14.0	1.00 max	0.40-1.0	3.5-4.5	0.030 max	0.030 max	1.00 max	Cu+W<=0.50; bal Fe	760	550	15	

Cast Stainless Steel, Martensitic, No equivalents identified

France

Specification	Designation	Notes	C	Cr	Mn	Mo	Ni	P	S	Si	Other	UTS	YS	El	Hard
AFNOR	Z12CN13.O2M	Bar Bil	0.14 max	11.5-14	1.2 max		1.5-2.5	0.04 max	0.03 max	0.6 max	bal Fe				

Poland

Specification	Designation	Notes	C	Cr	Mn	Mo	Ni	P	S	Si	Other	UTS	YS	El	Hard
PNH83158(86)	LH14	Corr res	0.15-0.3	12-15	0.4-0.8		1 max	0.035 max	0.035 max	0.7 max	bal Fe				
PNH83158(86)	LH14N	Corr res	0.15 max	12-15	0.4-0.9		0.7-1.2	0.035 max	0.035 max	0.6 max	bal Fe				

Russia

Specification	Designation	Notes	C	Cr	Mn	Mo	Ni	P	S	Si	Other	UTS	YS	El	Hard
GOST 977	15Ch13L	Corr res	0.15 max	12.0-14.0	0.3-0.8	0.15 max	0.5 max	0.03 max	0.025 max	0.2-0.8	bal Fe				

Spain

Specification	Designation	Notes	C	Cr	Mn	Mo	Ni	P	S	Si	Other	UTS	YS	El	Hard
UNE 36257(74)	AM-X12Cr13	Corr res	0.15 max	12.0-14.0	1.0 max	0.5 max	1.0 max	0.040 max	0.040 max	1.5 max	bal Fe				

UK

Specification	Designation	Notes	C	Cr	Mn	Mo	Ni	P	S	Si	Other	UTS	YS	El	Hard
BS 3100(91)	420C24	old En1648(A)	0.25 max	12.0-16.0	1.00 max			0.050 max	0.050 max	2.00 max	bal Fe				
BS 3100(91)	420C28		0.20 max	11.5-13.5	1.00 max		1.00 max	0.040 max	0.040 max	1.00 max	Cu <=0.30; bal Fe	620	450	13	179-235 HB
BS 3100(91)	425C12		0.06 max	11.5-14.0	1.00 max	0.40-1.00	3.50-4.50	0.040 max	0.030 max	1.00 max	bal Fe	755	550	15	262 HB

USA

Specification	Designation	Notes	C	Cr	Mn	Mo	Ni	P	S	Si	Other	UTS	YS	El	Hard
MIL-S-81591(93)	IC-416*	Inv, C/Corr res; Obs for new design see AMS specs									bal Fe				

Specification	Designation	Notes	C	Cr	Mn	Mo	Ni	P	S	Si	Other	UTS	YS	El	Hard

Cast Stainless Steel, Unclassified, 17-4

USA

Specification	Designation	Notes	C	Cr	Mn	Mo	Ni	P	S	Si	Other	UTS	YS	El	Hard
	UNS J92150		0.08 max	15.5-17.5	1.00 max		3.0-5.0	0.04 max	0.04 max	1.00 max	Cu 3.0-5.0; bal Fe				
MIL-S-81591(93)	IC-17-4*	Inv, C/Corr res; Obs for new design see AMS specs	0.07 max	15.5-17.5	1 max		3-5	0.04 max	0.03 max	1 max	Cu 3-5; Nb 0.15-0.45; bal Fe				

Cast Stainless Steel, Unclassified, 2A

USA

Specification	Designation	Notes	C	Cr	Mn	Mo	Ni	P	S	Si	Other	UTS	YS	El	Hard
	UNS J93345	Duplex Austenitic-Ferritic Cr-Ni-Mo-N ESCOLOY 45 D	0.08 max	22.5-25.5	1.00 max	3.0-4.5	8.0-11.0	0.04 max	0.025 max	1.50 max	N 0.10-0.30; bal Fe				
ASTM A890/A890M(97)	2A	Corr res; duplex	0.08 max	22.5-25.5	1.00 max	3.00-4.50	8.00-11.00	0.04 max	0.04 max	1.50 max	N 0.10-0.30; bal Fe	655	450	25	

Cast Stainless Steel, Unclassified, 3A

USA

Specification	Designation	Notes	C	Cr	Mn	Mo	Ni	P	S	Si	Other	UTS	YS	El	Hard
	UNS J93371	Duplex Alloy 3A	0.06 max	24.0-27.0	1.00 max	1.75-2.50	4.00-6.00	0.040 max	0.040 max	1.00 max	N 0.15-0.25; bal Fe				
ASTM A890/A890M(97)	3A	Corr res; duplex	0.06 max	24.0-27.0	1.00 max	1.75-2.50	4.00-6.00	0.040 max	0.040 max	1.00 max	N 0.15-0.25; bal Fe	620	415	25	

Cast Stainless Steel, Unclassified, 440A

USA

Specification	Designation	Notes	C	Cr	Mn	Mo	Ni	P	S	Si	Other	UTS	YS	El	Hard
	UNS J91606		0.60-0.75	16.0-18.0	1.00 max	0.75 max		0.04 max	0.03 max	1.00 max	bal Fe				
MIL-S-81591(93)	IC-440A*	Inv, C/Corr res; Obs for new design see AMS specs	0.6-0.75	16.0-18.0	1	0.75		0.04	0.03	1	bal Fe				

Cast Stainless Steel, Unclassified, 5340

USA

Specification	Designation	Notes	C	Cr	Mn	Mo	Ni	P	S	Si	Other	UTS	YS	El	Hard
	UNS J92240		0.06 max	13.5-14.7	0.7 max	2.0-2.75	3.75-4.75	0.02 max	0.025 max	0.5-1.0	Cu 3.0-3.5; N <=0.05; Nb+Ta 0.15-0.35; bal Fe				

Cast Stainless Steel, Unclassified, 5342

USA

Specification	Designation	Notes	C	Cr	Mn	Mo	Ni	P	S	Si	Other	UTS	YS	El	Hard
	UNS J92200		0.06 max	15.5-16.7	0.70 max		3.6-4.6	0.04 max	0.03 max	0.5-1.0	Cu 2.8-3.5; N <=0.05; Nb+Ta 0.15-0.40; bal Fe				
AMS 5342D(94)		Corr res, Inv cast, SCS	0.06 max	15.50-16.70	0.70 max		3.60-4.60	0.025 max	0.025 max	0.50-1.00	Al <=0.05; Cu 2.80-3.50; N <=0.05; Nb 0.15-0.40; bal Fe	896	827	10	30-38 HRC

Cast Stainless Steel, Unclassified, 5354

USA

Specification	Designation	Notes	C	Cr	Mn	Mo	Ni	P	S	Si	Other	UTS	YS	El	Hard
	UNS J91631		0.12-0.20	12.0-14.0	1.0 max	0.50 max	1.8-2.2	0.04 max	0.03 max	1.0 max	Cu <=0.50; W 2.5-3.5; bal Fe				

Cast Stainless Steel, Unclassified, 5359

USA

Specification	Designation	Notes	C	Cr	Mn	Mo	Ni	P	S	Si	Other	UTS	YS	El	Hard
	UNS J92001		0.08-0.15	14.5-15.5	0.40-1.10	2.0-2.6	3.5-4.5	0.04 max	0.03 max	0.75 max	N 0.05-0.13; bal Fe				

Cast Stainless Steel, Unclassified, 5360

USA

Specification	Designation	Notes	C	Cr	Mn	Mo	Ni	P	S	Si	Other	UTS	YS	El	Hard
	UNS J92951		0.15 max	16.0-18.0	2.0 max	1.5-2.25	12.0-14.0	0.04 max	0.03 max	0.75 max	Cu <=0.50; bal Fe				

Cast Stainless Steel, Unclassified, 5361

USA

Specification	Designation	Notes	C	Cr	Mn	Mo	Ni	P	S	Si	Other	UTS	YS	El	Hard
	UNS J93072		0.15-0.25	17.0-20.0	2.0 max	1.75-2.5	12.0-15.0	0.04 max	0.04 max	1.0 max	bal Fe				

Specification	Designation	Notes	C	Cr	Mn	Mo	Ni	P	S	Si	Other	UTS	YS	El	Hard

Cast Stainless Steel, Unclassified, 5362

USA

| | UNS J92811 | | 0.12 max | 18.0-19.5 | 2.0 max | 0.5 max | 10.0-14.0 | 0.04 max | 0.03 max | 1.5 max | Cu <=0.5; Ti 0.15-0.50; Nb+Ta 10xC-1.5; bal Fe | | | | |

Cast Stainless Steel, Unclassified, 5363

USA

| | UNS J92641 | | 0.10 max | 17.00-20.00 | 2.00 max | 0.50 max | 9.00-12.00 | 0.04 max | 0.04 max | 1.50 max | Cu <=0.50; Nb+Ta 10xC min, 1.35 max; bal Fe | | | | |

Cast Stainless Steel, Unclassified, 5369

USA

| | UNS J92843 | | 0.28-0.35 | 18.00-21.00 | 0.75-1.50 | 1.00-1.75 | 8.00-11.00 | 0.04 max | 0.04 max | 1.00 max | Cu <=0.50; Ti 0.15-0.50; W 1.00-1.75; Nb+Ta 0.30-0.70; bal Fe | | | | |

Cast Stainless Steel, Unclassified, 5372

USA

| | UNS J91601 | | 0.12-0.20 | 14.5-17.0 | 1.0 max | 0.50 max | 1.5-2.25 | 0.04 max | 0.04 max | 1.0 max | Cu <=0.50; bal Fe | | | | |

Cast Stainless Steel, Unclassified, 5A

USA

| | UNS J93404 | Ferritic-Austenitic Alloy 958 | 0.03 max | 24.0-26.0 | 1.50 max | 4.0-5.0 | 6.0-8.0 | | | 1.00 max | N 0.10-0.30; bal Fe | | | | |
| ASTM A890/A890M(97) | 5A | Corr res; duplex | 0.03 max | 24.0-26.0 | 1.50 max | 4.0-5.0 | 6.0-8.0 | 0.04 max | 0.04 max | 1.00 max | N 0.10-0.30; bal Fe | 690 | 485 | 25 | |

Cast Stainless Steel, Unclassified, A351

USA

| | UNS J93380 | Ferritic-Austenitic Cr-Ni-Mo-W-N Zeron 100 | 0.03 max | 24.0-26.0 | 1.0 max | 3.0-4.0 | 6.5-8.5 | 0.030 max | 0.025 max | 1.0 max | Cu 0.5-1.0; N 0.2-0.3; bal Fe | | | | |
| ASTM A890/A890M(97) | 6A | Corr res; duplex | 0.03 max | 24.0-26.0 | 1.00 max | 3.0-4.0 | 6.5-8.5 | 0.030 max | 0.025 max | 1.00 max | Cu 0.5-1.0; N 0.20-0.30; W 0.5-1.0; bal Fe | 690 | 485 | 16 | |

Cast Stainless Steel, Unclassified, A447

USA

| | UNS J93303 | | 0.20-0.45 | 23.00-28.00 | 2.50 max | | 10.0-14.0 | 0.05 max | 0.05 max | 1.75 max | N <=0.20; bal Fe | | | | |

Cast Stainless Steel, Unclassified, A872

USA

	UNS J93183	Ferritic-Austenitic KCR-D183	0.03 max	20.0-23.0	2.0 max	2.0-4.0	4.0-6.0	0.040 max	0.03 max	2.0 max	Co 0.5-1.5; Cu <=1.0; N 0.08-0.25; bal Fe				
	UNS J93550	Ferritic-Austenitic KCR-D283	0.03 max	23.0-26.0	2.0 max	5.0-8.0		0.040 max	0.03 max	2.0 max	Co 0.5-1.5; Cu <=1.0; N 0.08-0.25; bal Fe				
ASTM A872(97)	A872	Centrifugally cast Ferr-Aust Pipe; Corr res; KCR-D183	0.030 max	23.0-26.0	2.0 max	2.00-4.00	5.00-8.00	0.040 max	0.030 max	2.0 max	Co 0.50-1.50; Cu <=1.00; N 0.08-0.25; bal Fe	620	450	20	297 HB max

Cast Stainless Steel, Unclassified, CB30

Bulgaria

| BDS 9631 | 2Ch17N2L | Corr res | 0.15-0.25 | 16-18 | 0.5-0.9 | | 1.5-2.5 | | | 0.2-0.6 | bal Fe | | | | |

Czech Republic

| CSN 422911 | 422911 | Heat res | 0.25 max | 17.0-19.0 | 0.9 max | | 1.0 max | 0.04 max | 0.04 max | 1.5 max | Cu <=0.5; bal Fe | | | | |
| CSN 422912 | 422912 | Heat res | 0.5 max | 19.0-22.0 | 0.9 max | | 1.0 max | 0.045 max | 0.04 max | 1.5 max | Cu <=0.5; bal Fe | | | | |

France

| AFNOR | Z20CN17-2M | | 0.15-0.25 | 15-18 | 1 | | 1.5-3 | | | 1 | bal Fe | | | | |

UNS numbers and US grades are provided as a means of cross referencing chemically similar alloys. Exchangability is only possible after independent examination of specifications. Tensile properties are minimum or typical as specified. UTS and YS as MPa. El as %. See Appendix for list of abbreviations used in Notes. * indicates obsolete material.

Specification	Designation	Notes	C	Cr	Mn	Mo	Ni	P	S	Si	Other	UTS	YS	El	Hard

Cast Stainless Steel, Unclassified, CB30 (Continued from previous page)

Germany

Specification	Designation	Notes	C	Cr	Mn	Mo	Ni	P	S	Si	Other	UTS	YS	El	Hard
DIN 17445(84)	G-X22CrNi17		0.20-0.27	16.0-18.0	1.00 max		1.00-2.00	0.045 max	0.030 max	1.00 max	bal Fe	780-980	590	4	
DIN 17445(84)	WNr 1.4059		0.20-0.27	16.0-18.0	1.00 max		1.00-2.00	0.045 max	0.030 max	1.00 max	bal Fe	780-980	590	4	

Hungary

Specification	Designation	Notes	C	Cr	Mn	Mo	Ni	P	S	Si	Other	UTS	YS	El	Hard
MSZ 21053(82)	AoX20CrNi172	Corr res; 0-30.0mm; soft ann	0.15-0.25	16-18	1 max		1.5-2.5	0.04 max	0.04 max	1.5 max	bal Fe				150-220 HB
MSZ 21053(82)	AoX20CrNi172F	Corr res; 0-30.0mm; Q/T	0.15-0.25	16-18	1 max		1.5-2.5	0.04 max	0.04 max	1.5 max	bal Fe	590	390	6L	200-280 HB

Italy

Specification	Designation	Notes	C	Cr	Mn	Mo	Ni	P	S	Si	Other	UTS	YS	El	Hard
UNI 3161(83)	GX25Cr19	Sand; Heat res	0.3 max	18-21.	1 max		2 max	0.04 max	0.035 max	1.5 max	bal Fe				

Poland

Specification	Designation	Notes	C	Cr	Mn	Mo	Ni	P	S	Si	Other	UTS	YS	El	Hard
PNH83158(86)	LH17N		0.15-0.25	16-18	0.5-0.8		1.5-2.5			0.2-0.6	bal Fe				

Spain

Specification	Designation	Notes	C	Cr	Mn	Mo	Ni	P	S	Si	Other	UTS	YS	El	Hard
UNE 36257(74)	AM-X15CrNi17	Corr res	0.25 max	16.0-18.0	1.0 max	0.5 max	1.5-2.5	0.040 max	0.040 max	1.5 max	bal Fe				
UNE 36257(74)	F.8403*	see AM-X15CrNi17	0.25 max	16.0-18.0	1.0 max	0.5 max	1.5-2.5	0.040 max	0.040 max	1.5 max	bal Fe				

UK

Specification	Designation	Notes	C	Cr	Mn	Mo	Ni	P	S	Si	Other	UTS	YS	El	Hard
BS 3146/2(75)	ANC2	Corr res	0.12-0.25	15.5-20.0	0.20-1.00		1.50-3.00			0.20-1.00	bal Fe				

USA

Specification	Designation	Notes	C	Cr	Mn	Mo	Ni	P	S	Si	Other	UTS	YS	El	Hard
	UNS J91803		0.30 max	18.0-21.0	1.00 max		2.00 max	0.04 max	0.04 max	1.50 max	bal Fe				
ACI	CB-30	Corr res	0.30 max	18.0-21.0	1.00 max		2.00 max	0.04 max	0.04 max	1.50 max	bal Fe				
ASTM A296	CB30	Obs; Sand; SA; see A743 A744	0.3	18-21	1		2	0.04	0.04	1.5	bal Fe				
ASTM A743/A743M(98)	CB-30	Corr res; Norm, Ann	0.30 max	18.0-21.0	1.00 max		2.00 max	0.04 max	0.04 max	1.50 max	Cu optional 0.90-1.20; bal Fe	450	205		
MIL-S-81591(93)	IC-431*	Inv, C/Corr res; Obs for new design see AMS specs									bal Fe				

Cast Stainless Steel, Unclassified, CB6

Italy

Specification	Designation	Notes	C	Cr	Mn	Mo	Ni	P	S	Si	Other	UTS	YS	El	Hard
UNI 3161(83)	GX5CrNi1704	Sand; Heat res	0.06 max	16-18	1.5 max	0.4-0.7	3.5-5	0.04 max	0.035 max	1 max	bal Fe				

USA

Specification	Designation	Notes	C	Cr	Mn	Mo	Ni	P	S	Si	Other	UTS	YS	El	Hard
	UNS J91804	Cr-Ni Nb-6	0.06 max	15.5-17.5	1.00 max	0.5 max	3.5-5.5	0.04 max	0.03 max	1.00 max	bal Fe				
ACI	CB-6	Corr res	0.06 max	15.5-17.5	1.00 max	0.5 max	3.5-5.5	0.04 max	0.03 max	1.00 max	bal Fe				
ASTM A743/A743M(98)	CB-6	Corr res	0.06 max	15.5-17.5	1.00 max	0.5 max	3.5-5.5	0.04 max	0.03 max	1.00 max	bal Fe	790	580	16	

Cast Stainless Steel, Unclassified, CD3MN

USA

Specification	Designation	Notes	C	Cr	Mn	Mo	Ni	P	S	Si	Other	UTS	YS	El	Hard
	UNS J92205	Duplex Alloy 2205	0.03 max	21.0-23.5	1.50 max	2.5-3.5	4.5-6.5	0.04 max	0.020 max	1.00 max	Cu <=1.00; N 0.10-0.30; bal Fe				
ACI	CD-3MN	Duplex Alloy	0.03 max	21.0-23.5	1.50 max	2.5-3.5	4.5-6.5	0.04 max	0.020 max	1.00 max	Cu <=1.00; N 0.10-0.30; bal Fe				
ASTM A890/A890M(97)	4A	Corr res; duplex	0.03 max	21.0-23.5	1.50 max	2.5-3.5	4.5-6.5	0.04 max	0.020 max	1.00 max	Cu <=1.00; N 0.10-0.30; bal Fe	690	515	18	

Cast Stainless Steel, Unclassified, CD4MCu

France

Specification	Designation	Notes	C	Cr	Mn	Mo	Ni	P	S	Si	Other	UTS	YS	El	Hard
AFNOR	Z3CNUD26.5M		0.05	25-27	2	1.5-2.5	4.5-6			1.5	Cu 2.5-3.5; bal Fe				

UK

Specification	Designation	Notes	C	Cr	Mn	Mo	Ni	P	S	Si	Other	UTS	YS	El	Hard
BS 3146/2(75)	ANC21	Corr res	0.05 max	25.0-27.0	0.75 max	1.75-2.75	4.75-6.00	0.050 max	0.050 max	0.75 max	Cu 2.75-3.25; N <=0.10; bal Fe				

Specification	Designation	Notes	C	Cr	Mn	Mo	Ni	P	S	Si	Other	UTS	YS	El	Hard

Cast Stainless Steel, Unclassified, CD4MCu (Continued from previous page)

USA

Specification	Designation	Notes	C	Cr	Mn	Mo	Ni	P	S	Si	Other	UTS	YS	El	Hard
	UNS J93370		0.04 max	24.5-26.5	1.00 max	1.75-2.25	4.75-6.00	0.04 max	0.04 max	1.00 max	Cu 2.75-3.25; bal Fe				
ASTM A351(94)	CD4MCU	Press ves	0.04 max	24.5-26.5	1.00 max	1.75-2.25	4.75-6.0	0.04 max	0.04 max	1.00 max	Cu 2.75-3.25; bal Fe	690	485	16.0	
ASTM A744/A744M(98)	CD4MCU	Corr res; severe service	0.04 max	24.5-26.5	1.0 max	1.75-2.25	4.75-6.00	0.04 max	0.04 max	1.0 max	Cu 2.75-3.25; bal Fe	690	485	16	
ASTM A890/A890M(97)	1A	Corr res; duplex	0.04 max	24.5-26.5	1.00 max	1.75-2.25	4.75-6.00	0.040 max	0.040 max	1.00 max	Cu 2.75-3.25; bal Fe	690	485	16	

Cast Stainless Steel, Unclassified, CD4MCuN

USA

Specification	Designation	Notes	C	Cr	Mn	Mo	Ni	P	S	Si	Other	UTS	YS	El	Hard
	UNS J93372	Duplex Austenitic-Ferritic Cr-Ni-Cu-Mo-N CD-4MCuN	0.04 max	24.5-26.5	1.0 max	1.7-2.3	4.7-6.0	0.04 max	0.04 max	1.0 max	Cu 2.7-3.3; N 0.10-0.25; bal Fe				
ACI	CD-4MCuN	Duplex Austenitic-Ferritic Cr-Ni-Cu-Mo-N	0.04 max	24.5-26.5	1.0 max	1.7-2.3	4.7-6.0	0.04 max	0.04 max	1.0 max	Cu 2.7-3.3; N 0.10-0.25; bal Fe				
ASTM A890/A890M(97)	1B	Corr res; duplex	0.04 max	24.5-26.5	1.0 max	1.7-2.3	4.7-6.0	0.04 max	0.04 max	1.0 max	Cu 2.75-3.3; N 0.10-0.25; bal Fe	655	450	25	

Cast Stainless Steel, Unclassified, CE20N

USA

Specification	Designation	Notes	C	Cr	Mn	Mo	Ni	P	S	Si	Other	UTS	YS	El	Hard
	UNS J92802	CE20N	0.20 max	23.0-36.0	1.50 max		8.0-11.0	0.040 max	0.040 max	1.50 max	N 0.08-0.20; bal Fe				
ACI	CE20N	Corr res	0.20 max	23.0-36.0	1.50 max		8.0-11.0	0.040 max	0.040 max	1.50 max	N 0.08-0.20; bal Fe				

Cast Stainless Steel, Unclassified, CE30

USA

Specification	Designation	Notes	C	Cr	Mn	Mo	Ni	P	S	Si	Other	UTS	YS	El	Hard
	UNS J93423		0.30 max	26.0-30.0	1.50 max		8.0-11.0	0.04 max	0.04 max	2.00 max	bal Fe				
ACI	CE-30	Corr res	0.30 max	26.0-30.0	2.00 max		8.0-11.0	0.04 max	0.04 max	2.00 max	bal Fe				
ASTM A296	CE30	Obs; Sand; SA; see A743 A744	0.3	26-30	1.5		8-11	0.04	0.04	2	bal Fe				
ASTM A743/A743M(98)	CE-30	Corr res; Solution Q/A	0.30 max	26.0-30.0	1.5 max		8.0-11.0	0.04 max	0.04 max	2.00 max	bal Fe	550	275	10	

Cast Stainless Steel, Unclassified, CE8N

Japan

Specification	Designation	Notes	C	Cr	Mn	Mo	Ni	P	S	Si	Other	UTS	YS	El	Hard
JIS G5121(91)	SCS10	Corr res; Tmp; Solution HT	0.03 max	21.00-26.00	1.50 max	2.50-4.00	4.50-8.50	0.040 max	0.030 max	1.50 max	N 0.08-0.30; bal Fe	620	390	15	302 HB max

USA

Specification	Designation	Notes	C	Cr	Mn	Mo	Ni	P	S	Si	Other	UTS	YS	El	Hard
	UNS J92805	CE-8N	0.08 max	23.0-26.0	1.50 max	0.50 max	8.0-11.0	0.040 max	0.040 max	1.50 max	N 0.20-0.30; bal Fe				
ACI	CE-8N	Corr res	0.08 max	23.0-26.0	1.50 max	0.50 max	8.0-11.0	0.040 max	0.040 max	1.50 max	N 0.20-0.30; bal Fe				

Cast Stainless Steel, Unclassified, CF10

USA

Specification	Designation	Notes	C	Cr	Mn	Mo	Ni	P	S	Si	Other	UTS	YS	El	Hard
	UNS J92590		0.04-0.10	18.0-20.0	2.00 max		8.00-11.0	0.040 max	0.030 max	0.75 max	bal Fe				

Cast Stainless Steel, Unclassified, CF10MC

China

Specification	Designation	Notes	C	Cr	Mn	Mo	Ni	P	S	Si	Other	UTS	YS	El	Hard
GB 2100(80)	ZG0Cr18Ni12Mo2Ti	Corr res; Quen	0.08 max	16.0-19.0	0.8-2.0	2.0-3.0	11.0-13.0	0.040 max	0.030 max	1.5 max	5xC<=Ti<=0.7; bal Fe	490	216	30	

France

Specification	Designation	Notes	C	Cr	Mn	Mo	Ni	P	S	Si	Other	UTS	YS	El	Hard
AFNOR	Z4CNDNb18.12M		0.08	17-19.5	1.5	2-2.5	10.5-12.5			1.5	8xC-1.00 Nb+Ta; bal Fe				
AFNOR	Z6CNDNb18.12M		0.08	17-20	1.5	2-3	9-13			1.2	10xC-1.2 Nb+Ta; bal Fe				

Germany

Specification	Designation	Notes	C	Cr	Mn	Mo	Ni	P	S	Si	Other	UTS	YS	El	Hard
DIN EN 10213(96)	G-X5CrNiMoNb1810	Press purposes	0.07 max	18.0-20.0	1.50 max	2.00-2.50	9.0-12.0	0.040 max	0.030 max	1.50 max	8xC<=Nb<=1.00; bal Fe	440-640	185	20	
DIN EN 10213(96)	WNr 1.4581		0.07 max	18.0-20.0	1.50 max	2.00-2.50	9.0-12.0	0.040 max	0.030 max	1.50 max	8xC<=Nb<=1.00; bal Fe	440-640	185	20	

UNS numbers and US grades are provided as a means of cross referencing chemically similar alloys. Exchangability is only possible after independent examination of specifications. Tensile properties are minimum or typical as specified. UTS and YS as MPa. El as %. See Appendix for list of abbreviations used in Notes. * indicates obsolete material.

Specification	Designation	Notes	C	Cr	Mn	Mo	Ni	P	S	Si	Other	UTS	YS	El	Hard

Cast Stainless Steel, Unclassified, CF10MC (Continued from previous page)

Hungary

Specification	Designation	Notes	C	Cr	Mn	Mo	Ni	P	S	Si	Other	UTS	YS	El	Hard
MSZ 21053(82)	AoX10CrNiMoNb1812	Corr res; 0-30.0mm; as cast	0.12 max	17-19	2 max	3-4	11-14	0.04 max	0.03 max	1.5 max	Nb=8C-1.1; bal Fe				160-230 HB
MSZ 21053(82)	AoX12CrNiMoNb1810	Corr res; 0-30.0mm; as cast	0.15 max	17-19	2 max	2-3	9-11	0.04 max	0.04 max	1.5 max	Nb=8C-1.3; bal Fe				160-230 HB

Italy

Specification	Designation	Notes	C	Cr	Mn	Mo	Ni	P	S	Si	Other	UTS	YS	El	Hard
UNI 3161(83)	GX6CrNiMoNb2011	Sand; Heat res	0.08 max	18-21.	1.5 max	2-3	9-13	0.04 max	0.035 max	2 max	Nb+Ta=8C-1; bal Fe				

Romania

Specification	Designation	Notes	C	Cr	Mn	Mo	Ni	P	S	Si	Other	UTS	YS	El	Hard
STAS 9277	OTA10N6MoNiCr170		0.1	16-18	2	2-2.5	10-13			1	Cu 0.3; 8xC Nb+Ta; bal Fe				

UK

Specification	Designation	Notes	C	Cr	Mn	Mo	Ni	P	S	Si	Other	UTS	YS	El	Hard
BS 1504(76)	318C17 (En845 N6)*		0.08	17-21	2	2-3	10 min			1.5	8xC-1 Nb + Ta or 5xC-0.7 Ti; bal Fe				
BS 3100(91)	318C17	old En1632(C)	0.080 max	17.0-21.0	2.00	2.00-3.00	9.00 min	0.040 max	0.040 max	1.50 max	8xC<=Nb<=1.00; bal Fe	480		18	
BS 3146/2(75)	ANC4(C)	Corr res	0.12 max	17.0-20.0	0.20-2.00	2.00-3.00	10.0 min	0.035 max	0.035 max	0.20-1.50	8xC<=Nb<=1.10; bal Fe				

USA

Specification	Designation	Notes	C	Cr	Mn	Mo	Ni	P	S	Si	Other	UTS	YS	El	Hard
	UNS J92971		0.10 max	15.0-18.0	1.50 max	1.75-2.25	13.0-16.0	0.040 max	0.040 max	1.5 max	Nb <=1.35; Nb>=10xC; bal Fe				
ACI	CF-10MC	Corr res	0.10 max	15.0-18.0	1.50 max	1.75-2.25	13.0-16.0	0.040 max	0.040 max	1.50 max	Nb 10xC-1.20; bal Fe				
ASTM A351(94)	CF10MC	Press ves	0.10 max	15.0-18.0	1.50 max	1.75-2.25	13.0-16.0	0.040 max	0.040 max	1.50 max	10xC<=Nb+Ta<=1.20; bal Fe	485	205	20.0	
ASTM A451(97)	CPF10MC	Pipe; SA Quen; High-temp	0.10 max	15.0-18.0	1.50 max	1.75-2.25	13.0-16.0	0.040 max	0.040 max	1.00 max	10xC<=Nb<=1.2; bal Fe	485	205	20.0	

Cast Stainless Steel, Unclassified, CF10SMnN

USA

Specification	Designation	Notes	C	Cr	Mn	Mo	Ni	P	S	Si	Other	UTS	YS	El	Hard
	UNS J92972		0.10 max	16.0-18.0	7.00-9.00		8.0-9.0	0.060 max	0.030 max	3.50-4.50	N 0.08-0.18; bal Fe				
ACI	CF10SMnN	Corr res	0.10 max	16.0-18.0	7.00-9.00		8.0-9.0	0.060 max	0.030 max	3.50-4.50	N 0.08-0.18; bal Fe				
ASTM A743/A743M(98)	CF10SMnN	Corr res; HT, Quen	0.10 max	16.0-18.0	7.00-9.00		8.0-9.0	0.060 max	0.030 max	3.50-4.50	N 0.08-0.18; bal Fe	585	290	30	

Cast Stainless Steel, Unclassified, CF3

Bulgaria

Specification	Designation	Notes	C	Cr	Mn	Mo	Ni	P	S	Si	Other	UTS	YS	El	Hard
BDS 6738(72)	000CH18N11	Corr res	0.03 max	17.0-19.0	2.00 max	0.30 max	10.0-12.5	0.035	0.02	0.80 max	bal Fe				

China

Specification	Designation	Notes	C	Cr	Mn	Mo	Ni	P	S	Si	Other	UTS	YS	El	Hard
GB 2100(80)	ZG00Cr18Ni10	Corr res; Quen	0.03 max	17.0-20.0	0.8-2.0		8.0-12.0	0.040 max	0.030 max	1.5 max	bal Fe	392	177	25	

France

Specification	Designation	Notes	C	Cr	Mn	Mo	Ni	P	S	Si	Other	UTS	YS	El	Hard
AFNOR	Z2CN18.10M		0.03	17-20	1.5		8-12			1.2	bal Fe				
AFNOR	Z2CND18.12M		0.03	17-20	1.5	2-3	9-13			1.2	bal Fe				
AFNOR	Z3CN19.10M		0.04	18.5-21	1.5		9-10			1.5	Co 0.2; Cu 0.5; N 0.08; Ta 0.15; bal Fe				
AFNOR	Z3CN19.9M		0.03	18-21	1.5		8-12			2	bal Fe				
AFNOR	Z3CND20.10M		0.03	17-21	1.5	2-3	9-13			1.5	bal Fe				
AFNOR NFA32055	Z6CNNb1810M	SA	0.08	17-20	1.5		8-11	0.04	0.03	1.2	bal Fe				
AFNOR NFA32056	Z2CND1812M	SA	0.03	17-20	1.5	2-3	9-13	0.04	0.03	1.2	bal Fe				

Japan

Specification	Designation	Notes	C	Cr	Mn	Mo	Ni	P	S	Si	Other	UTS	YS	El	Hard
JIS G5121(91)	SCS19	Corr res; Tmp	0.03 max	17.00-21.00	2.00 max		8.00-12.00	0.040 max	0.040 max	2.00 max	bal Fe	390	185	33	183 HB max
JIS G5121(91)	SCS19A	Corr res; Tmp	0.03 max	17.00-21.00	1.50 max		8.00-12.00	0.040 max	0.040 max	2.00 max	bal Fe	480	205	33	183 HB max

Romania

Specification	Designation	Notes	C	Cr	Mn	Mo	Ni	P	S	Si	Other	UTS	YS	El	Hard
STAS 6855(86)	T15MoNiCr180	Corr res; Heat res	0.15 max	17-20	2 max	2-3	9-12	0.04 max	0.03 max	0.5-2	Pb <=0.15; bal Fe				
STAS 6855(86)	T15TiMoNiCr180	Corr res; Heat res	0.15 max	17-19	2 max	2-2.5	9-12	0.035 max	0.03 max	2 max	Pb <=0.15; Ti <=0.8; 5(C-0.03)<=Ti<=0.8; bal Fe				

Spain

Specification	Designation	Notes	C	Cr	Mn	Mo	Ni	P	S	Si	Other	UTS	YS	El	Hard
UNE 36257(74)	AM-X2CrNi1910	Corr res	0.03	17-21	1.5		8-12			2	bal Fe				

UNS numbers and US grades are provided as a means of cross referencing chemically similar alloys. Exchangability is only possible after independent examination of specifications. Tensile properties are minimum or typical as specified. UTS and YS as MPa. El as %. See Appendix for list of abbreviations used in Notes. * indicates obsolete material.

Specification	Designation	Notes	C	Cr	Mn	Mo	Ni	P	S	Si	Other	UTS	YS	El	Hard

Cast Stainless Steel, Unclassified, CF3 (Continued from previous page)

Spain

Specification	Designation	Notes	C	Cr	Mn	Mo	Ni	P	S	Si	Other	UTS	YS	El	Hard
UNE 36257(74)	AM-X2CrNiMo19-11	Corr res	0.03	17-21	1.5	2-3	9-13			2	bal Fe				
UNE 36257(74)	F.8412*	see AM-X2CrNi1910	0.03	17-21	1.5		8-12			2	bal Fe				
UNE 36257(74)	F.8415*	see AM-X2CrNiMo19-11	0.03	17-21	1.5	2-3	9-13			2	bal Fe				

UK

Specification	Designation	Notes	C	Cr	Mn	Mo	Ni	P	S	Si	Other	UTS	YS	El	Hard
BS 1504(76)	304C12*	Press ves	0.03 max	17.0-21.0	2.0 max		8.0-14.00	0.040 max	0.04 max	1.5 max	Cu <=0.3; Ti <=0.05; W <=0.1; bal Fe				
BS 1504(76)	316C12*		0.03	17-21	2	2-3	10 min			1.5					
BS 3100(91)	304C12	old En162C	0.03 max	17.0-21.0	2.00 max		8.00-12.0	0.040 max	0.040 max	1.50 max	bal Fe	430		26	
BS 3100(91)	304C12LT196	Corr res	0.03 max	17.0-21.0	2.00 max		8.00-12.0	0.040 max	0.040 max	1.50 max	bal Fe	430		26	

USA

Specification	Designation	Notes	C	Cr	Mn	Mo	Ni	P	S	Si	Other	UTS	YS	El	Hard
	UNS J92500		0.03 max	17.0-21.0	1.50 max		8.0-12.0	0.040 max	0.040 max	2.00 max	bal Fe				
	UNS J92700		0.03 max	17.0-21.0	1.50 max	2.0-3.0	8.0-12.0	0.04 max	0.04 max	1.50 max	bal Fe				
ACI	CF3	Corr res	0.03 max	17.0-21.0	1.50 max		8.0-12.0	0.040 max	0.040 max	2.00 max	bal Fe				
ASTM A296	CF3*	Obs; Sand; SA; see A743 A744	0.03	17-21	1.5		8-12	0.04	0.04	2	bal Fe				
ASTM A351(94)	CF3	Press ves	0.03 max	17.0-21.0	1.50 max	0.50 max	8.0-12.0	0.040 max	0.040 max	2.00 max	bal Fe	485	205	35.0	
ASTM A351(94)	CF3A	Press ves	0.03 max	17.0-21.0	1.50 max	0.50 max	8.0-12.0	0.040 max	0.040 max	2.00 max	bal Fe	530	240	35.0	
ASTM A451(97)	CPF3	Pipe; SA Quen; High-temp	0.03 max	17.0-21.0	1.50 max		8.0-12.0	0.040 max	0.040 max	2.00 max	bal Fe	485	205	35	
ASTM A451(97)	CPF3A	Pipe; SA Quen; High-temp	0.03 max	17.0-21.0	1.50 max		8.0-12.0	0.040 max	0.040 max	2.00 max	bal Fe	535	240	35	
ASTM A743/A743M(98)	CF-3	Corr res; as cast; Solution Q/A	0.03 max	17.0-21.0	1.50 max		8.0-12.0	0.04 max	0.04 max	2.00 max	bal Fe	485	205	35	
ASTM A744/A744M(98)	CF3	Corr res; severe service	0.03 max	17.0-21.0	1.50 max		8.0-12.0	0.04 max	0.04 max	2.0 max	bal Fe	485	205	35	
ASTM A744/A744M(98)	CF3M	Corr res; severe service	0.03 max	17.0-21.0	1.50 max	2.0-3.0	9.0-13.0	0.04 max	0.04 max	1.50 max	bal Fe	485	205	30	

Cast Stainless Steel, Unclassified, CF3M

France

Specification	Designation	Notes	C	Cr	Mn	Mo	Ni	P	S	Si	Other	UTS	YS	El	Hard
AFNOR	Z2CND18.12M		0.03	17-20	1.5	2-3	9-13			1.2	bal Fe				
AFNOR	Z3CND19.10M		0.045	17-21	1.5	2.3-2.8	10-11.5			1.5	Co 0.2; Cu 0.5; N 0.08; bal Fe				
AFNOR	Z3CND20.10M		0.03	17-21	1.5	2-3	9-13			1.5	bal Fe				
AFNOR NFA32055	Z2CND1812M	SA	0.03	17-20	1.5	2-3	9-13	0.04	0.03	1.2	bal Fe				
AFNOR NFA32056	Z6CND1812M	SA	0.08	17-20	1.5	2-3	9-13	0.04	0.03	1.2	bal Fe				

Italy

Specification	Designation	Notes	C	Cr	Mn	Mo	Ni	P	S	Si	Other	UTS	YS	El	Hard
UNI 3161(83)	GX2CrNiMo1911	Sand; Heat res	0.03 max	17-21	1.5 max	2-3	9-13	0.04 max	0.035 max	2 max	bal Fe				

Japan

Specification	Designation	Notes	C	Cr	Mn	Mo	Ni	P	S	Si	Other	UTS	YS	El	Hard
JIS G5121(91)	SCS16	Corr res; Tmp	0.03 max	17.00-20.00	2.00 max	2.00-3.00	12.00-16.00	0.040 max	0.040 max	1.50 max	bal Fe	380	175	33	183 HB max
JIS G5121(91)	SCS16A	Corr res; Tmp	0.03 max	17.0-21.00	1.50 max	2.00-3.00	9.00-13.00	0.040 max	0.040 max	1.50 max	bal Fe	480	205	33	183 HB max

Spain

Specification	Designation	Notes	C	Cr	Mn	Mo	Ni	P	S	Si	Other	UTS	YS	El	Hard
UNE 36257(74)	AM-X2CrNiMo19-11	Corr res	0.03	17-21	1.5	2-3	9-13			2	bal Fe				

UK

Specification	Designation	Notes	C	Cr	Mn	Mo	Ni	P	S	Si	Other	UTS	YS	El	Hard
BS 3100(91)	316C12	old En1632F	0.030 max	17.0-21.0	2.00 max	2.00-3.00	9.00 min	0.040 max	0.040 max	1.50 max	bal Fe	430		26	
BS 3100(91)	316C12LT196		0.030 max	17.0-21.0	2.00 max	2.00-3.00	9.00 min	0.040 max	0.040 max	1.50 max	bal Fe	430		26	

USA

Specification	Designation	Notes	C	Cr	Mn	Mo	Ni	P	S	Si	Other	UTS	YS	El	Hard
	UNS J92800		0.03 max	17.0-21.0	1.50 max	2.0-3.0	9.0-13.0	0.04 max	0.04 max	1.50 max	bal Fe				
ACI	CF3M	Corr res	0.03 max	17.0-21.0	1.50 max	2.0-3.0	9.0-13.0	0.04 max	0.04 max	1.50 max	bal Fe				

UNS numbers and US grades are provided as a means of cross referencing chemically similar alloys. Exchangability is only possible after independent examination of specifications. Tensile properties are minimum or typical as specified. UTS and YS as MPa. El as %. See Appendix for list of abbreviations used in Notes. * indicates obsolete material.

Specification	Designation	Notes	C	Cr	Mn	Mo	Ni	P	S	Si	Other	UTS	YS	El	Hard
Cast Stainless Steel, Unclassified, CF3M (Continued from previous page)															
USA															
ASTM A351(94)	CF3M	Press ves	0.03 max	17.0-21.0	1.50 max	2.0-3.0	9.0-13.0	0.040 max	0.040 max	1.50 max	bal Fe	485	205	30.0	
ASTM A351(94)	CF3MA	Press ves	0.03 max	17.0-21.0	1.50 max	2.0-3.0	9.0-13.0	0.040 max	0.040 max	1.50 max	bal Fe	550	255	30.0	
ASTM A451(97)	CPF3M	Pipe; SA Quen; High-temp	0.03 max	17.0-21.0	1.50 max	2.0-3.0	9.0-13.0	0.040 max	0.040 max	1.50 max	bal Fe	485	205	30	
ASTM A743/A743M(98)	CF-3M	Corr res; as cast; Solution Q/A	0.03 max	17.0-21.0	1.50 max	2.0-3.0	9.0-13.0	0.04 max	0.04 max	1.50 max	bal Fe	485	205	30	
ASTM A743/A743M(98)	CF-3MN	Corr res; as cast; Solution Q/A	0.03 max	17.0-22.0	1.50 max	2.0-3.0	9.0-13.0	0.040 max	0.040 max	1.50 max	N 0.10-0.20; bal Fe	515	255	35	
Cast Stainless Steel, Unclassified, CF8															
Australia															
AS 2074(82)	H5A		0.08 max	17-21	2 max		8 min	0.04 max	0.04 max	1.5 max	bal Fe				
Bulgaria															
BDS 9631	0Ch18N10SL	Corr res	0.07 max	17-19	2 max	0.5 max	8-12	0.04 max	0.04 max	2 max	Al <=0.1; Cu <=0.30; bal Fe				
BDS 9631	0Ch18N9L	Corr res	0.07	17-20	2		8-11			1.5	bal Fe				
China															
GB 2100(80)	ZG0Cr18Ni9	Corr res; Quen	0.08 max	17.0-20.0	0.8-2.0		8.0-11.0	0.040 max	0.030 max	1.5 max	bal Fe	441	196	25	
Finland															
SFS 387(79)	G-X8CrNi199	Corr res	0.08 max	18.0-21.0	1.5 max		8.0-11.0	0.04 max	0.04 max	2.0 max	bal Fe				
France															
AFNOR	Z6CN18.10M		0.08	17-20	1.5		8-12			1.2	bal Fe				
AFNOR	Z6CN19.9M		0.08	18-21	1.5		8-11			2	bal Fe				
Germany															
DIN EN 10213(96)	GX5CrNi19-10	Q	0.07 max	18.0-20.0	1.50 max		8.0-11.0	0.040 max	0.030 max	1.50 max	bal Fe	440-640	175	20	
DIN EN 10213(96)	WNr 1.4308	Press ves parts	0.07 max	18.0-20.0	1.50 max		8.0-11.0	0.040 max	0.030 max	1.50 max	bal Fe	440-640	175	20	
DIN SEW 595(76)	GX8CrNi19-10	Petrochemical use	0.08 max	18.0-20.0	0.50-1.50		9.00-11.0	0.045 max	0.030 max	0.50-1.50	bal Fe				
DIN SEW 595(76)	WNr 1.4815	Petrochemical use	0.08 max	18.0-20.0	0.50-1.50		9.00-11.0	0.045 max	0.030 max	0.50-1.50	bal Fe				
Hungary															
MSZ 21053(82)	AoX7CrNi189	Corr res; 0-30.0mm; as cast	0.08 max	17-19	2 max		8-11	0.04 max	0.04 max	1.5 max	bal Fe				160-220 HB
MSZ 21053(82)	AoX7CrNi189F	Corr res; 0-30.0mm; SA	0.08 max	17-19	2 max		8-11	0.04 max	0.04 max	1.5 max	bal Fe	440	190	25L	130-200 HB
Japan															
JIS G5121(91)	SCS13	Corr res; Tmp	0.08 max	18.00-21.00	2.00 max		8.00-11.00	0.040 max	0.040 max	2.00 max	bal Fe	440	185	30	183 HB max
JIS G5121(91)	SCS13A	Corr res; Tmp	0.08 max	18.00-21.00	1.50 max		8.00-11.00	0.040 max	0.040 max	2.00 max	bal Fe	480	205	33	183 HB max
Romania															
STAS 10718(88)	T6NiCr180	Corr res	0.06 max	17-20	2 max		8-11	0.04 max	0.03 max	0.5-2	Pb <=0.15; bal Fe				
STAS 6855(86)	T7NiCr180	Corr res; Heat res	0.07 max	17-19	2 max	0.5 max	8-12	0.035 max	0.03 max	2 max	Pb <=0.15; bal Fe				
Russia															
GOST 997	07Ch18N9L	Corr res	0.07 max	17.0-20.0	1.0-2.0	0.15 max	8.0-11.0	0.035 max	0.03 max	0.2-1.0	bal Fe				
Spain															
UNE 36257(74)	AM-X7CrNi20-10	Corr res	0.08	18-21	1.5		8-11			2	bal Fe				
Sweden															
SS 142333	2333	Corr res	0.05 max	17-19	2 max		8-11	0.045 max	0.03 max	1 max	bal Fe				
UK															
BS 3100(91)	304C15	old En1631A	0.08 max	18.0-21.0	2.00 max		8.00-11.0	0.040 max	0.040 max	1.50 max	bal Fe	480		26	
BS 3100(91)	304C15LT196		0.08 max	18.0-21.0	2.00 max		8.00-11.0	0.040 max	0.040 max	1.50 max	bal Fe	480		26	

UNS numbers and US grades are provided as a means of cross referencing chemically similar alloys. Exchangability is only possible after independent examination of specifications. Tensile properties are minimum or typical as specified. UTS and YS as MPa. El as %. See Appendix for list of abbreviations used in Notes. * indicates obsolete material.

Specification	Designation	Notes	C	Cr	Mn	Mo	Ni	P	S	Si	Other	UTS	YS	El	Hard

Cast Stainless Steel, Unclassified, CF8 (Continued from previous page)

USA

Specification	Designation	Notes	C	Cr	Mn	Mo	Ni	P	S	Si	Other	UTS	YS	El	Hard
	UNS J92600		0.08 max	18.0-21.0	1.50 max		8.0-11.0	0.04 max	0.04 max	2.00 max	bal Fe				
ACI	CF8	Corr res	0.08 max	18.0-21.0	1.50 max		8.0-11.0	0.04 max	0.04 max	2.00 max	bal Fe				
ASTM A351(94)	CF8	Press ves	0.08 max	18.0-21.0	1.50 max	0.50 max	8.0-11.0	0.040 max	0.040 max	2.00 max	bal Fe	485	205	35.0	
ASTM A351(94)	CF8A	Press ves	0.08 max	18.0-21.0	1.50 max	0.50 max	8.0-11.0	0.040 max	0.040 max	2.00 max	bal Fe	530	240	35.0	
ASTM A451(97)	CPF8	Pipe; SA Quen; High-temp	0.08 max	18.0-21.0	1.50 max		8.0-11.0	0.040 max	0.040 max	1.00 max	bal Fe	485	205	35	
ASTM A451(97)	CPF8A	Pipe; SA Quen; High-temp	0.08 max	18.0-21.0	1.50 max		8.0-11.0	0.040 max	0.040 max	2.00 max	bal Fe	535	240	35	
ASTM A743/A743M(98)	CF-8	Corr res; Solution Q/A at 1040 C	0.08 max	18.0-21.0	1.50 max		8.0-11.0	0.04 max	0.04 max	2.00 max	bal Fe	485	205	35	
ASTM A744/A744M(98)	CF8	Corr res; severe service	0.08 max	18.0-21.0	1.50 max		8.0-11.0	0.04 max	0.04 max	2.0 max	bal Fe	485	205	35	
MIL-C-24707/3(89)	CF-8	As spec in ASTM A744; Corr res; low magnetic perm apps									bal Fe				

Cast Stainless Steel, Unclassified, CF8C

Bulgaria

Specification	Designation	Notes	C	Cr	Mn	Mo	Ni	P	S	Si	Other	UTS	YS	El	Hard
BDS 9631	0Ch18N9L	Corr res	0.07	17-20	2		8-11			1.5	bal Fe				

China

| GB 2100(80) | ZG0Cr18Ni9Ti | Corr res; Quen | 0.08 max | 17.0-20.0 | 0.8-2.0 | | 8.0-11.0 | 0.040 max | 0.030 max | 1.5 max | 5x(C-0.02)<=Ti<=0.7; bal Fe | 441 | 196 | 25 | |

France

| AFNOR | Z6CN18.10M | | 0.08 | 17-20 | 1.5 | | 8-12 | | | 1.2 | bal Fe | | | | |
| AFNOR | Z6CN19.9M | | 0.08 | 18-21 | 1.5 | | 8-11 | | | 2 | bal Fe | | | | |

Hungary

| MSZ 21053(82) | AoX12CrNiNb189 | Corr res; 0-30.0mm; as cast | 0.15 max | 17-19 | 2 max | 2 max | 8-11 | 0.04 max | 0.04 max | 1.5 max | Nb=8C-1.3; bal Fe | | | | 160-230 HB |
| MSZ 21053(82) | AoX12CrNiNb189F | Corr res; 0-30.0mm; SA | 0.15 max | 17-19 | 2 max | | 8-11 | 0.04 max | 0.04 max | 1.5 max | Nb=8C-1.3; bal Fe | 460 | 200 | 20L | 140-210 HB |

Italy

| UNI 3161(83) | GX6CrNiNb2011 | Sand; Heat res | 0.08 max | 18-21. | 1.5 max | | 9-12 | 0.04 max | 0.035 max | 2 max | Nb+Ta=8C-1; bal Fe | | | | |

Japan

| JIS G5121(91) | SCS21 | | 0.08 max | 18.00-21.00 | 2.00 max | | 9.00-12.00 | 0.040 max | 0.040 max | 2.00 max | Nb <=1.35; Nb>=10xC; bal Fe | 480 | 205 | 28 | 183 HB max |

Spain

| UNE 36257(74) | AM-X7CrNi20-10 | Corr res | 0.08 | 18-21 | 1.5 | | 8-11 | | | 2 | bal Fe | | | | |

UK

BS 1504(76)	304C15 (En801)*		0.08	17-21	2		8.0 min			1.5	bal Fe				
BS 1504(76)	347C17;En821grade Nb*	Press ves	0.08 max	17.0-21.0	2.0 max	0.15 max	8.5-14.00	0.04 max	0.04 max	1.5 max	Cu <=0.3; Ti <=0.7; W <=0.1; Nb=8C-1.0/Ti=5C-0.7; bal Fe				
BS 3100(91)	347C17	old En1631(B)	0.08 max	18.0-21.0	2.00 max		9.00-12.0	0.040 max	0.040 max	1.50 max	8xC<=Nb<=1.00; bal Fe	480		22	

USA

	UNS J92710		0.08 max	18.0-21.0	1.50 max		9.0-12.0	0.04 max	0.04 max	2.00 max	Nb 8xC-1.00 or Nb+Ta 9xC-1.1; bal Fe				
ACI	CF8C	Corr res	0.08 max	18.0-21.0	1.50 max		9.0-12.0	0.04 max	0.04 max	2.00 max	bal Fe				
ASTM A351(94)	CF8C	Press ves	0.08 max	18.0-21.0	1.50 max	0.50 max	9.0-12.0			2.00 max	8xC<=Nb+Ta<=1.00; bal Fe	485	205	30.0	
ASTM A451(97)	CPF8C	Pipe; SA Quen; High-temp	0.08 max	18.0-21.0	1.50 max		9.0-12.0	0.040 max	0.040 max	1.00 max	8xC<=Nb<=1; bal Fe	485	205	30.0	
ASTM A451(97)	CPF8C (Ta max)	Pipe; SA Quen; High-temp	0.08 max	18.0-21.0	1.50 max		9.0-12.0	0.040 max	0.040 max	1.00 max	Ta 0.1; 8xC<=Nb<=1; bal Fe	485	205	30.0	
ASTM A743/A743M(98)	CF-8C	Corr res; Solution Q/A at 1040 C	0.08 max	18.0-21.0	1.50 max		9.0-12.0	0.04 max	0.04 max	2.00 max	Nb>=8xC<1.0; Nb+Ta>=9xC<1.1; bal Fe	485	205	30	

Specification	Designation	Notes	C	Cr	Mn	Mo	Ni	P	S	Si	Other	UTS	YS	El	Hard

Cast Stainless Steel, Unclassified, CF8C (Continued from previous page)

USA

Specification	Designation	Notes	C	Cr	Mn	Mo	Ni	P	S	Si	Other	UTS	YS	El	Hard
ASTM A744/A744M(98)	CF8C	Corr res; severe service	0.08 max	18.0-21.0	1.50 max		9.0-12.0	0.04 max	0.04 max	2.0 max	Nb>=8xC<1.0; Nb+Ta=9xC<1.1; bal Fe	485	205	30	
MIL-C-24707/3(89)	CF-8C	As spec in ASTM A744; Corr res; low magnetic perm apps									bal Fe				

Cast Stainless Steel, Unclassified, CF8M

Australia

Specification	Designation	Notes	C	Cr	Mn	Mo	Ni	P	S	Si	Other	UTS	YS	El	Hard
AS 2074(82)	H6C		0.08 max	17-20	2 max	2-3	10 min	0.04 max	0.04 max	1.5 max	Nb=8xC - 0.1; bal Fe				

Bulgaria

Specification	Designation	Notes	C	Cr	Mn	Mo	Ni	P	S	Si	Other	UTS	YS	El	Hard
BDS 9631	0Ch18N10M2SL	Corr res	0.07 max	17-19	2 max	2-2.5	9-12	0.04 max	0.04 max	2 max	Al <=0.1; Cu <=0.30; bal Fe				
BDS 9631	0Ch18NTM2L	Corr res	0.07	17-19	2	2-2.5	9-11			1.5	bal Fe				

China

Specification	Designation	Notes	C	Cr	Mn	Mo	Ni	P	S	Si	Other	UTS	YS	El	Hard
GB 2100(80)	ZG0Cr18Ni12Mo2Ti	Corr res; Quen	0.08 max	16.0-19.0	0.8-2.0	2.0-3.0	11.0-13.0	0.040 max	0.030 max	1.5 max	5x(C-0.02)<=Ti<=0.7; bal Fe	490	216	30	

Finland

Specification	Designation	Notes	C	Cr	Mn	Mo	Ni	P	S	Si	Other	UTS	YS	El	Hard
SFS 388(79)	G-X8CrNiMo19102	Corr res	0.08 max	18.0-21.0	1.5 max	2.0-3.0	9.0-12.0	0.04 max	0.04 max	2.0 max	bal Fe				

France

Specification	Designation	Notes	C	Cr	Mn	Mo	Ni	P	S	Si	Other	UTS	YS	El	Hard
AFNOR	Z5CND20.10M		0.08	18-21	1.5	2-3	9-12			1.5	bal Fe				
AFNOR	Z5CND20.8M		0.07	20-22	2	2.2-2.8	7-9			1.5	Cu 0.5; bal Fe				
AFNOR	Z6CND18.12M		0.08	17-20	1.5	2-3	9-13			1.2	bal Fe				

Germany

Specification	Designation	Notes	C	Cr	Mn	Mo	Ni	P	S	Si	Other	UTS	YS	El	Hard
DIN EN 10213(96)	GX5CrNiMo19-11-2	Q	0.07 max	18.0-20.0	1.50 max	2.00-2.50	9.0-12.0	0.040 max	0.030 max	1.50 max	bal Fe	440-640	185	20	
DIN EN 10213(96)	WNr 1.4408	Press ves; Quen	0.07 max	18.0-20.0	1.50 max	2.00-2.50	9.0-12.0	0.040 max	0.030 max	1.50 max	bal Fe	440-640	185	20	

Hungary

Specification	Designation	Notes	C	Cr	Mn	Mo	Ni	P	S	Si	Other	UTS	YS	El	Hard
MSZ 21053(82)	AoX7CrNiMo1810	Corr res; 0-30.0mm; as cast	0.08 max	17-19	2 max	2-3	9-11	0.04 max	0.04 max	1.5 max	bal Fe				160-220 HB
MSZ 21053(82)	AoX7CrNiMo1810F	Corr res; 0-30.0mm; SA	0.08 max	17-19	2 max	2-3	9-11	0.04 max	0.04 max	1.5 max	bal Fe	440	190	25L	130-200 HB

Italy

Specification	Designation	Notes	C	Cr	Mn	Mo	Ni	P	S	Si	Other	UTS	YS	El	Hard
UNI 3161(83)	GX6CrNiMo2011	Sand; Heat res	0.08 max	18-21.	1.5 max	2-3	9-12	0.04 max	0.035 max	2 max	bal Fe				

Japan

Specification	Designation	Notes	C	Cr	Mn	Mo	Ni	P	S	Si	Other	UTS	YS	El	Hard
JIS G5121(91)	SCS14	Corr res; Tmp	0.08 max	17.00-20.00	2.00 max	2.00-3.00	10.00-14.00	0.040 max	0.040 max	2.00 max	bal Fe	440	185	28	183 HB max
JIS G5121(91)	SCS14A		0.08 max	18.00-23.00	1.50 max	2.00-3.00	9.00-12.00	0.040 max	0.040 max	1.50 max	bal Fe	480	205	33	183 HB max
JIS G5121(91)	SCS22	Pipe; SA	0.08 max	17.00-20.00	2.00 max	2.00-3.00	10.00-14.00	0.040 max	0.040 max	2.00 max	Nb <=1.35; Nb>=10xC; bal Fe	440	205	28	183 HB max

Poland

Specification	Designation	Notes	C	Cr	Mn	Mo	Ni	P	S	Si	Other	UTS	YS	El	Hard
PNH83158(86)	L0H18N10M2	Corr res	0.07 max	17-19	2 max	0.5 max	8-11	0.035 max	0.035 max	2 max	bal Fe				
PNH83158(86)	LH18N10M		0.15	17-19	2	2-2.5	9-11			1.5	bal Fe				
PNH83158(86)	LOH18N10M		0.07	17-19	2	2-2.5	9-11			1.5	bal Fe				

Romania

Specification	Designation	Notes	C	Cr	Mn	Mo	Ni	P	S	Si	Other	UTS	YS	El	Hard
STAS 10718(88)	T6MoNiCr180	Corr res	0.06 max	17-20	2 max	2-3	9-12	0.04 max	0.03 max	0.5-2	Pb <=0.15; bal Fe				
STAS 6855(86)	T7MoNiCr180	Corr res; Heat res	0.07 max	17-19	2 max	2-2.5	9-12	0.035 max	0.03 max	2 max	Pb <=0.15; bal Fe				

Russia

Specification	Designation	Notes	C	Cr	Mn	Mo	Ni	P	S	Si	Other	UTS	YS	El	Hard
GOST 5632	07Ch18N10G2S2M2L	Corr res	0.07 max	17.0-19.0	2.0 max	2.1-2.5	9.0-12.0	0.04 max	0.04 max	2.0 max	Ti <=0.2; W <=0.2; bal Fe				

Spain

Specification	Designation	Notes	C	Cr	Mn	Mo	Ni	P	S	Si	Other	UTS	YS	El	Hard
UNE 36257(74)	AM-X7CrNiMo20-10	Corr res	0.08 max	18.0-21.0	1.5 max	2.0-3.0	9.0-12.0	0.040 max	0.040 max	2.0 max	bal Fe				
UNE 36257(74)	AM-X7CrNiNb20-10	Corr res	0.08 max	18.0-21.0	1.5 max	0.15 max	8.0-12.0	0.040 max	0.040 max	2.0 max	8xC<=Nb<=1.0; bal Fe				

Specification	Designation	Notes	C	Cr	Mn	Mo	Ni	P	S	Si	Other	UTS	YS	El	Hard

Cast Stainless Steel, Unclassified, CF8M (Continued from previous page)

Spain

Specification	Designation	Notes	C	Cr	Mn	Mo	Ni	P	S	Si	Other	UTS	YS	El	Hard
UNE 36257(74)	F.8413*	see AM-X7CrNiNb20-10	0.08 max	18.0-21.0	1.5 max	0.15 max	8.0-12.0	0.040 max	0.040 max	2.0 max	8xC<=Nb<=1.0; bal Fe				
UNE 36257(74)	F.8414*	see AM-X7CrNiMo20-10	0.08 max	18.0-21.0	1.5 max	2.0-3.0	9.0-12.0	0.040 max	0.040 max	2.0 max	bal Fe				

Sweden

Specification	Designation	Notes	C	Cr	Mn	Mo	Ni	P	S	Si	Other	UTS	YS	El	Hard
SIS 142343	2343-12	SA	0.06	17-20	2	2.5-3.2	10-13.5	0.05	0.03	1.5	bal Fe	440	200	35	
SS 142343	2343	Corr res	0.05 max	16-18.5	2 max	2.5-3	10.5-14	0.045 max	0.03 max	1 max	bal Fe				

UK

Specification	Designation	Notes	C	Cr	Mn	Mo	Ni	P	S	Si	Other	UTS	YS	El	Hard
BS 1504(76)	315C16*	Press ves	0.08 max	17.0-21.0	2.0 max	1.0-1.75	8.0-14.00	0.04 max	0.04 max	1.5 max	Cu <=0.3; Ti <=0.05; W <=0.1; bal Fe				
BS 1504(76)	316C16 (En845gradeB)*	Press ves	0.08 max	17.0-21.0	2.0 max	2.0-3.0	10.0-16.00	0.04 max	0.04 max	1.5 max	Cu <=0.3; Ti <=0.05; W <=0.1; bal Fe				
BS 3100	315C71*	old En1632D	0.08	17-21	2	2-3	8 max			1.5	bal Fe				
BS 3100(91)	316C16	En1632(B)	0.080 max	17.0-21.0	2.00 max	2.00-3.00	9.00 min	0.040 max	0.040 max	1.50 max	bal Fe	480		26	
BS 3100(91)	316C16LT196		0.080 max	17.0-21.0	2.00 max	2.00-3.00	9.00 min	0.040 max	0.040 max	1.50 max	bal Fe	480		26	
BS 3146/2(75)	ANC4(B)	Corr res	0.08 max	17.0-20.0	0.20-2.00	2.00-3.00	10.0 min	0.035 max	0.035 max	0.20-1.50	bal Fe				

USA

Specification	Designation	Notes	C	Cr	Mn	Mo	Ni	P	S	Si	Other	UTS	YS	El	Hard
	UNS J92900		0.08 max	18.0-21.0	1.50 max	2.0-3.0	9.0-12.0	0.04 max	0.04 max	2.00 max	bal Fe				
ACI	CF8M	Corr res	0.08 max	18.0-21.0	1.50 max	2.0-3.0	9.0-12.0	0.04 max	0.04 max	2.00 max	bal Fe				
ASTM A296	CF-8M	Obs; Sand; SA; see A743 A744	0.08	18-21	1.5		9-12		0.04	2	bal Fe				
ASTM A351(94)	CF-8M	Press ves	0.08 max	18.0-21.0	1.50 max	2.0-3.0	9.0-12.0	0.040 max	0.040 max	1.50 max	bal Fe	485	205	30.0	
ASTM A451(97)	CPF-8M	Pipe; SA Quen; High-temp	0.08 max	18.0-21.0	1.50 max	2.0-3.0	9.0-12.0	0.040 max	0.040 max	1.00 max	bal Fe	485	205	30.0	
ASTM A743/A743M(98)	CF-8M	Corr res; Solution Q/A at 1040 C	0.08 max	18.0-21.0	1.50 max	2.0-3.0	9.0-12.0	0.04 max	0.04 max	2.00 max	bal Fe	485	205	30	
ASTM A744/A744M(98)	CF-8M	Corr res; severe service	0.08 max	18.0-21.0	1.50 max	2.0-3.0	9.0-12.0	0.04 max	0.04 max	2.0 max	bal Fe	485	205	30	
MIL-C-24707/3(89)	CF-8M	As spec in ASTM A744; Corr res; low magnetic perm apps									bal Fe				
MIL-S-81591(93)	IC-316*	Inv, C/Corr res; Obs for new design see AMS specs									bal Fe				

Cast Stainless Steel, Unclassified, CG12

Germany

Specification	Designation	Notes	C	Cr	Mn	Mo	Ni	P	S	Si	Other	UTS	YS	El	Hard
DIN	G-X5CrNi2210		0.07 max	22.0-23.0	1.50-2.00	0.75 max	9.50-10.5	0.035 max	0.025 max	0.50-1.20	Cu <=0.30; bal Fe				
DIN	WNr 1.4947		0.07 max	22.0-23.0	1.50-2.00	0.75 max	9.50-10.5	0.035 max	0.025 max	0.50-1.20	Cu <=0.30; bal Fe				

USA

Specification	Designation	Notes	C	Cr	Mn	Mo	Ni	P	S	Si	Other	UTS	YS	El	Hard
	UNS J93001		0.12 max	20.0-23.0	1.50 max		10.0-13.0	0.04 max	0.04 max	2.00 max	bal Fe				
ACI	CG12	Corr res	0.12 max	20.0-23.0	1.50 max		10.0-13.0	0.04 max	0.040 max	2.00 max	bal Fe				
ASTM A743/A743M(98)	CG-12	Corr res; Solution Q/A at 1040 C	0.12 max	20.0-23.0	1.50 max		10.0-13.0	0.04 max	0.04 max	2.00 max	bal Fe	485	195	35	

Cast Stainless Steel, Unclassified, CG3M

USA

Specification	Designation	Notes	C	Cr	Mn	Mo	Ni	P	S	Si	Other	UTS	YS	El	Hard
	UNS J92999	Cr-Ni-Mo Stainless Steel CG-3M	0.03 max	18.0-21.0	1.50 max	3.0-4.0	9.0-13.0	0.04 max	0.04 max	1.50 max	bal Fe				
ACI	CG-3M	Corr res, Cr-Ni-Mo	0.03 max	18.0-21.0	1.50 max	3.0-4.0	9.0-13.0	0.04 max	0.04 max	1.50 max	bal Fe				
ASTM A744/A744M(98)	CG-3M	Corr res; severe service	0.03 max	18.0-21.0	1.50 max	3.0-4.0	9.0-13.0	0.04 max	0.04 max	1.50 max	bal Fe	515	240	25	

Specification	Designation	Notes	C	Cr	Mn	Mo	Ni	P	S	Si	Other	UTS	YS	El	Hard
Cast Stainless Steel, Unclassified, CG6MMN															
USA															
	UNS J93790	CG6MMN	0.06 max	20.50-23.50	4.00-6.00	1.50-3.00	11.50-13.50	0.040 max	0.030 max	1.00 max	N 0.20-0.40; Nb 0.10-0.30; V 0.10-0.30; bal Fe				
ASTM A743/A743M(98)	CG-3M	Corr res; HT, Quen	0.03 max	18.0-21.0	1.50 max	3.0-4.0	9.0-13.0	0.04 max	0.04 max	1.50 max	bal Fe	515	240	25	
ASTM A743/A743M(98)	CG6MMN	Corr res; HT, Quen	0.06 max	20.5-23.5	4.00-6.00	1.50-3.00	11.5-13.5	0.04 max	0.03 max	1.00 max	N 0.20-0.40; Nb 0.10-0.30; V 0.10-0.30; bal Fe	585	290	30	
Cast Stainless Steel, Unclassified, CG8M															
Finland															
SFS 389(79)	G-X8CrNiMo19114	Corr res	0.08 max	18.0-21.0	1.5 max	3.0-4.0	9.0-13.0	0.04 max	0.04 max	1.5 max	bal Fe				
France															
AFNOR	Z8CND18.10.3M		0.1	17-21	1.5	3-3.5	9-11			1.5	bal Fe				
Germany															
DIN	GX6CrNiMo17-13		0.07 max	16.0-18.0	2.00 max	4.00-5.00	12.5-14.5	0.045 max	0.030 max	1.00 max	bal Fe	440-640	185	20	
DIN	WNr 1.4448		0.07 max	16.0-18.0	2.00 max	4.00-5.00	12.5-14.5	0.045 max	0.030 max	1.00 max	bal Fe	440-640	185	20	
Italy															
UNI 3161(83)	GX6CrNiMo201103	Sand; Heat res	0.08 max	18-21.	1.5 max	3-4	9-13	0.04 max	0.035 max	1.5 max	bal Fe				
Mexico															
DGN B-354	CG8M*	Obs; as Agreed	0.08	18-21	1.5	3-4	9-13	0.04	0.04	1.5	bal Fe				
Spain															
UNE 36257(74)	AM-X7CrNiMo20-11	Corr res	0.08	18-21	1.5	3-4	9-13			1.5	bal Fe				
UNE 36257(74)	F.8416*	see AM-X7CrNiMo20-11	0.08	18-21	1.5	3-4	9-13			1.5	bal Fe				
UK															
BS 1504(76)	317C16 (En846)*	Press ves	0.08 max	17.0-21.0	2.0 max	3.0-4.0	10.0-16.0	0.04 max	0.04 max	1.5 max	Cu <=0.3; Ti <=0.05; W <=0.1; bal Fe				
BS 3100(91)	317C16	old En1632(A)	0.080 max	17.0-21.0	2.00 max	3.00-4.00	9.00 min	0.040 max	0.040 max	1.50 max	bal Fe	480		22	
BS 3146/2(75)	ANC4(A)	Corr res	0.08 max	18.0-20.0	0.20-2.00	3.00-4.00	11.0-14.0	0.035 max	0.035 max	0.20-1.50	bal Fe				
USA															
	UNS J93000		0.08 max	18.0-21.0	1.50 max	3.0-4.0	9.0-13.0	0.04 max	0.04 max	1.50 max	bal Fe				
ASTM A743/A743M(98)	CG-8M	Corr res; Solution Q/A	0.08 max	18.0-21.0	1.50 max	3.0-4.0	9.0-13.0	0.04 max	0.04 max	1.50 max	bal Fe	520	240	25	
ASTM A744/A744M(98)	CG-8M	Corr res; severe service	0.08 max	18.0-21.0	1.50 max	3.0-4.0	9.0-13.0	0.04 max	0.04 max	1.50 max	bal Fe	520	240	25	
Cast Stainless Steel, Unclassified, CH10															
USA															
	UNS J93401		0.10 max	22.0-26.0	1.50 max		12.0-15.0	0.040 max	0.040 max	2.00 max	bal Fe				
ACI	CH10	Corr res	0.10 max	22.0-26.0	1.50 max		12.0-15.0	0.040 max	0.040 max	2.00 max	bal Fe				
ASTM A351(94)	CH-10	Press ves	0.10 max	22.0-26.0	1.50 max	0.50 max	12.0-15.0	0.040 max	0.040 max	2.00 max	bal Fe	485	205	30.0	
ASTM A451(97)	CPH10	Pipe; SA Quen; High-temp	0.10 max	22.0-26.0	1.50 max		12.0-15.0	0.040 max	0.040 max	1.00 max	bal Fe	485	205	30.0	
ASTM A743/A743M(98)	CH-10	Corr res; HT, Quen	0.10 max	22.0-26.0	1.5 max		12.0-15.0	0.04 max	0.04 max	2.00 max	bal Fe	485	205	30	
Cast Stainless Steel, Unclassified, CH8															
USA															
	UNS J93400		0.08 max	22.0-26.0	1.50 max		12.0-15.0	0.040 max	0.040 max	1.50 max	bal Fe				
ACI	CH8	Corr res	0.08 max	22.0-26.0	1.50 max		12.0-15.0	0.040 max	0.04 max	1.50 max	bal Fe				
ASTM A351(94)	CH8	Press ves	0.08 max	22.0-26.0	1.50 max	0.50 max	12.0-15.0	0.040 max	0.040 max	1.50 max	bal Fe	450	195	30.0	

UNS numbers and US grades are provided as a means of cross referencing chemically similar alloys. Exchangability is only possible after independent examination of specifications. Tensile properties are minimum or typical as specified. UTS and YS as MPa. El as %. See Appendix for list of abbreviations used in Notes. * indicates obsolete material.

Specification	Designation	Notes	C	Cr	Mn	Mo	Ni	P	S	Si	Other	UTS	YS	El	Hard

Cast Stainless Steel, Unclassified, CH8 (Continued from previous page)

USA

Specification	Designation	Notes	C	Cr	Mn	Mo	Ni	P	S	Si	Other	UTS	YS	El	Hard
ASTM A451(97)	CPH8	Pipe; SA Quen; High-temp	0.08 max	22.0-26.0	1.50 max		12.0-15.0	0.040 max	0.040 max	1.50 max	bal Fe	448	195	30.0	

Cast Stainless Steel, Unclassified, CK3MCuN

USA

Specification	Designation	Notes	C	Cr	Mn	Mo	Ni	P	S	Si	Other	UTS	YS	El	Hard
	UNS J93254	Austenitic Cr-Ni-Mo-Cu-N CK-3MCuN, 254 SMO	0.025 max	19.5-20.5	1.20 max	6.0-7.0	17.5-19.7	0.045 max	0.010 max	1.0 max	Cu 0.50-1.0; N 0.180-0.240; bal Fe				
ACI	CK-3MCuN	Ferrite in Austenite, Cr-Ni-Mo-Cu-N	0.025 max	19.5-20.5	1.20 max	6.0-7.0	17.5-19.7	0.045 max	0.010 max	1.0 max	Cu 0.50-1.0; N 0.180-0.240; bal Fe				
ASTM A743/A743M(98)	CK-3MCuN	Corr res	0.025 max	19.5-20.5	1.20 max	6.0-7.0	17.5-19.5	0.045 max	0.010 max	1.00 max	Cu 0.50-1.00; N 0.180-0.240; bal Fe	550	260	35	
ASTM A744/A744M(98)	CK-3MCuN	Corr res; severe service	0.025 max	19.5-20.5	1.20 max	6.0-7.0	17.5-19.5	0.045 max	0.010 max	1.00 max	Cu 0.50-1.00; N 0.180-0.240; bal Fe	550	260	35	

Cast Stainless Steel, Unclassified, CK3MN

USA

Specification	Designation	Notes	C	Cr	Mn	Mo	Ni	P	S	Si	Other	UTS	YS	El	Hard
	UNS J94653	Austenitic Cr-Ni-Mo-N CK-3MN, cast SR50A	0.035 max	22.0-24.0	2.00 max	6.0-6.8	20.0-22.0	0.035 max	0.020 max	1.00 max	Cu <=0.40; N 0.21-0.32; bal Fe				
ACI	CK-3MN	Corr res	0.035 max	22.0-24.0	2.00 max	6.0-6.8	20.0-22.0	0.035 max	0.020 max	1.00 max	Cu <=0.40; N 0.21-0.32; bal Fe				

Cast Stainless Steel, Unclassified, Ducomet 5

USA

Specification	Designation	Notes	C	Cr	Mn	Mo	Ni	P	S	Si	Other	UTS	YS	El	Hard
	UNS J93900	Durcomet 5	0.025 max	20.0-22.0	1.50 max		15.0-17.0	0.040 max	0.040 max	4.00-6.00	N 0.08-0.20; bal Fe				

Cast Stainless Steel, Unclassified, HD

China

Specification	Designation	Notes	C	Cr	Mn	Mo	Ni	P	S	Si	Other	UTS	YS	El	Hard
GB 8492(87)	ZG30Cr26Ni5	Heat res; Ann	0.20-0.40	24.0-28.0	1.00 max	0.50 max	4.00-6.00	0.040 max	0.040 max	2.00 max	bal Fe	590			

Japan

| JIS G5122(91) | SCHII | Pipe; as cast | 0.40 max | 24.00-28.00 | 1.00 max | 0.50 max | 4.00-6.00 | 0.040 max | 0.040 max | 2.00 max | bal Fe | 590 | | | |

Mexico

| DGN B-354 | CC50* | Obs; Norm | 0.5 | 26-30 | 1 | | 4 | 0.04 | 0.04 | 1.5 | bal Fe | | | | |
| DGN B-355 | HC* | Obs; as cast | 0.5 | 26-30 | 1 | 0.5 | 4 | 0.04 | 0.04 | 2 | bal Fe | | | | |

USA

	UNS J93005		0.50 max	26.0-30.0	1.50 max	0.50 max	4.0-7.0	0.04 max	0.04 max	2.00 max	bal Fe				
ACI	HD	Heat res	0.50 max	26.0-30.0	1.50 max	0.50 max	4.0-7.0	0.04 max	0.04 max	2.00 max	bal Fe				
ASTM A297/A297M(98)	HD	as agreed	0.5	26-30	1.5	0.5	4-7	0.04	0.04	2	bal Fe				

Cast Stainless Steel, Unclassified, HD50

USA

Specification	Designation	Notes	C	Cr	Mn	Mo	Ni	P	S	Si	Other	UTS	YS	El	Hard
	UNS J93015	HD-50	0.45-0.55	26.0-30.0	1.50 max	0.50 max	4.0-7.0	0.04 max	0.04 max	0.50-2.00	bal Fe				
ASTM A608(98)	HD50	Heat res; Cent Tub	0.45-0.55	26-30	1.50 max	0.50 max	4-7	0.04 max	0.04 max	0.50-2.00	bal Fe				

Cast Stainless Steel, Unclassified, HE

Japan

Specification	Designation	Notes	C	Cr	Mn	Mo	Ni	P	S	Si	Other	UTS	YS	El	Hard
JIS G5122(91)	SCH13A		0.25-0.50	23.00-26.00	2.50 max	0.50 max	12.00-14.00	0.040 max	0.040 max	1.75 max	bal Fe	490	235	8	
JIS G5122(91)	SCH17		0.20-0.50	26.00-30.00	2.00 max	0.50 max	8.00-11.00	0.040 max	0.040 max	2.00 max	bal Fe	540	275		

Mexico

| DGN B-355 | HE* | Obs; as cast | 0.2-0.5 | 24-28 | 2 | 0.5 | 8-11 | 0.04 | 0.04 | 2 | bal Fe | | | | |

Spain

| UNE 36258(74) | AM-X35CrNi25-12 | Heat res | 0.2-0.5 | 24-28 | 2 | 0.5 | 11-14 | | | 2 | bal Fe | | | | |
| UNE 36258(74) | F.8451* | see AM-X35CrNi25-12 | 0.2-0.5 | 24-28 | 2 | 0.5 | 11-14 | | | 2 | bal Fe | | | | |

Specification	Designation	Notes	C	Cr	Mn	Mo	Ni	P	S	Si	Other	UTS	YS	El	Hard

Cast Stainless Steel, Unclassified, HE (Continued from previous page)

UK

Specification	Designation	Notes	C	Cr	Mn	Mo	Ni	P	S	Si	Other	UTS	YS	El	Hard
BS 3100(91)	309C40	old En1648 B2	0.50 max	25.0-30.0	2.00 max	1.50 max	8.00-12.0	0.050 max	0.050 max	2.00 max	bal Fe	450		7	

USA

Specification	Designation	Notes	C	Cr	Mn	Mo	Ni	P	S	Si	Other	UTS	YS	El	Hard
	UNS J93403		0.20-0.50	26.0-30.0	2.00 max	0.50 max	8.0-11.0	0.04 max	0.04 max	2.00 max	bal Fe				
ACI	HE	Heat res	0.20-0.50	26.0-30.0	2.00 max	0.50 max	8.0-11.0	0.04 max	0.04 max	2.00 max	bal Fe				
ASTM A297/A297M(98)	HE	Sand	0.2-0.5	26-30	2	0.5	8-11	0.04	0.04	2	bal Fe				

Cast Stainless Steel, Unclassified, HE35

USA

Specification	Designation	Notes	C	Cr	Mn	Mo	Ni	P	S	Si	Other	UTS	YS	El	Hard
	UNS J93413	HE-35	0.30-0.40	26.0-30.0	1.50 max	0.50 max	8.0-11.0	0.04 max	0.04 max	0.50-2.00	bal Fe				
ACI	HE-35	Heat res	0.30-0.40	26.0-30.0	1.50 max	0.50 max	8.0-11.0	0.04 max	0.04 max	0.50-2.00	bal Fe				
ASTM A608(98)	HE35	Heat res; Cent Tub	0.30-0.40	26-30	1.50 max	0.50 max	8-11	0.04 max	0.04 max	0.50-2.00	bal Fe				

Cast Stainless Steel, Unclassified, HH

Australia

Specification	Designation	Notes	C	Cr	Mn	Mo	Ni	P	S	Si	Other	UTS	YS	El	Hard
AS 2074(82)	H8B		0.2-0.45	24-28	2.5 max	0.5 max	11-14	0.04 max	0.04 max	1.5 max	N <=0.2; bal Fe				

China

Specification	Designation	Notes	C	Cr	Mn	Mo	Ni	P	S	Si	Other	UTS	YS	El	Hard
GB 8492(87)	ZG35Cr26Ni12	Heat res; Ann	0.20-0.50	24.0-28.0	2.00 max		11.00-14.00	0.040 max	0.030 max	2.00 max	bal Fe	490	235	8	

Czech Republic

Specification	Designation	Notes	C	Cr	Mn	Mo	Ni	P	S	Si	Other	UTS	YS	El	Hard
CSN 422936	422936		0.25-0.5	24.0-27.0	1.5 max	0.5 max	12.0-14.0	0.045 max	0.04 max	2.0 max	bal Fe				

Germany

Specification	Designation	Notes	C	Cr	Mn	Mo	Ni	P	S	Si	Other	UTS	YS	El	Hard
DIN 17465(93)	GX40CrNiSi25-12		0.30-0.50	24.0-26.0	1.50 max		11.0-14.0	0.035 max	0.030 max	1.00-2.50	bal Fe				
DIN 17465(93)	WNr 1.4837		0.30-0.50	24.0-26.0	1.50 max		11.0-14.0	0.035 max	0.030 max	1.00-2.50	bal Fe				

India

Specification	Designation	Notes	C	Cr	Mn	Mo	Ni	P	S	Si	Other	UTS	YS	El	Hard
IS 4522	7 Type 1	as cast	0.2-0.5	23-27	2		11-14	0.05	0.05	2	bal Fe				
IS 4522	7 Type 2	as cast	0.2-0.5	23-27	2		11-14	0.05	0.05	2	bal Fe				

Italy

Specification	Designation	Notes	C	Cr	Mn	Mo	Ni	P	S	Si	Other	UTS	YS	El	Hard
UNI 3159	GX35CrNi2512		0.25-0.5	24-27	2	0.5	11-14			2.5	bal Fe				

Japan

Specification	Designation	Notes	C	Cr	Mn	Mo	Ni	P	S	Si	Other	UTS	YS	El	Hard
JIS G5122(91)	SCH13		0.20-0.50	24.00-28.00	2.00 max	0.50 max	11.00-14.00	0.040 max	0.040 max	2.00 max	bal Fe	490	235	8	

Mexico

Specification	Designation	Notes	C	Cr	Mn	Mo	Ni	P	S	Si	Other	UTS	YS	El	Hard
DGN B-355	HH*	Obs; Q/A, Tmp	0.2-0.5	24-28	2	0.5	11-14	0.04	0.04	2	bal Fe				

Romania

Specification	Designation	Notes	C	Cr	Mn	Mo	Ni	P	S	Si	Other	UTS	YS	El	Hard
STAS 6855	T35NiCr260X		0.3-0.4	24-28	1.5		13-15			1-2.5	bal Fe				
STAS 6855(86)	T35NiCr260	Corr res; Heat res	0.3-0.4	24-28	1.5 max		13-15	0.035 max	0.03 max	1-2.5	Pb <=0.15; bal Fe				

Spain

Specification	Designation	Notes	C	Cr	Mn	Mo	Ni	P	S	Si	Other	UTS	YS	El	Hard
UNE 36258(74)	AM-X35CrNi25-12	Heat res	0.2-0.5	24-28	2	0.5	11-14			2	bal Fe				
UNE 36258(74)	F.8451*	see AM-X35CrNi25-12	0.2-0.5	24-28	2	0.5	11-14			2	bal Fe				

Sweden

Specification	Designation	Notes	C	Cr	Mn	Mo	Ni	P	S	Si	Other	UTS	YS	El	Hard
SS 14	(USA HH)		0.2	23	1.5		11.5			1.5	bal Fe				

UK

Specification	Designation	Notes	C	Cr	Mn	Mo	Ni	P	S	Si	Other	UTS	YS	El	Hard
BS 3100(91)	309C30	old En1648E	0.50 max	22.0-27.0	2.00 max	1.50 max	10.0-14.0	0.050 max	0.050 max	2.50 max	bal Fe				
BS 3100(91)	309C32	old En4238 EC1	0.20-0.45	24.0-28.0	2.50 max	1.50 max	11.0-14.0	0.040 max	0.040 max	1.50 max	bal Fe	550		3	
BS 3100(91)	309C35	old En4238EC2	0.20-0.50	24.0-28.0	2.00 max	1.50 max	11.0-14.0	0.040 max	0.040 max	2.00 max	bal Fe	510		7	

USA

Specification	Designation	Notes	C	Cr	Mn	Mo	Ni	P	S	Si	Other	UTS	YS	El	Hard
	UNS J93503		0.20-0.50	24.0-28.0	2.00 max	0.50 max	11.0-14.0	0.04 max	0.04 max	2.00 max	bal Fe				

UNS numbers and US grades are provided as a means of cross referencing chemically similar alloys. Exchangability is only possible after independent examination of specifications. Tensile properties are minimum or typical as specified. UTS and YS as MPa. El as %. See Appendix for list of abbreviations used in Notes. * indicates obsolete material.

Specification	Designation	Notes	C	Cr	Mn	Mo	Ni	P	S	Si	Other	UTS	YS	El	Hard

Cast Stainless Steel, Unclassified, HH (Continued from previous page)

USA

Specification	Designation	Notes	C	Cr	Mn	Mo	Ni	P	S	Si	Other	UTS	YS	El	Hard
ACI	HH	Heat res	0.20-0.50	24.0-28.0	2.00 max	0.50 max	11.0-14.0	0.04 max	0.04 max	2.00 max	bal Fe				
ASTM A297/A297M(98)	HH		0.2-0.5	24-28	2	0.5	11-14			2	bal Fe				
ASTM A567	HH90		0.8-1	24-28	2	0.5	11-14	0.04	0.04	2	bal Fe				
ASTM A567	HH90		0.8-1	24-28	2	0.5	11-14	0.04	0.04	2	bal Fe				

Cast Stainless Steel, Unclassified, HH30

USA

Specification	Designation	Notes	C	Cr	Mn	Mo	Ni	P	S	Si	Other	UTS	YS	El	Hard
	UNS J93513	HH-30	0.25-0.35	24.0-28.0	1.50 max	0.50 max	11.0-14.0	0.04 max	0.04 max	0.50-2.00	bal Fe				
ACI	HH-30	Heat res	0.25-0.35	24.0-28.0	1.50 max	0.50 max	11.0-14.0	0.04 max	0.04 max	0.50-2.00	bal Fe				
ASTM A608(98)	HH30	Heat res; Cent Tub	0.25-0.35	24-28	1.50 max	0.50 max	11-14	0.04 max	0.04 max	0.50-2.00	bal Fe				

Cast Stainless Steel, Unclassified, HH33

Hungary

Specification	Designation	Notes	C	Cr	Mn	Mo	Ni	P	S	Si	Other	UTS	YS	El	Hard
MSZ 4357(81)	AoX40CrNiSi2512	Heat res; as cast	0.3-0.5	24-26	0.5-1.5		11.5-13.5	0.04 max	0.03 max	1-2.5	bal Fe	550		4L	180-230 HB
MSZ 4357(81)	AoX40CrNiSi2512F	Heat res; soft ann	0.3-0.5	24-26	0.5-1.5		11.5-13.5	0.04 max	0.03 max	1-2.5	bal Fe	500	250	10L	150-200 HB

USA

Specification	Designation	Notes	C	Cr	Mn	Mo	Ni	P	S	Si	Other	UTS	YS	El	Hard
	UNS J93633	HH-33	0.28-0.38	24.0-26.0	1.50 max	0.50 max	12.0-14.0	0.04 max	0.04 max	0.50-2.00	bal Fe				
ASTM A608(98)	HH33	Heat res; Cent Tub	0.28-0.38	24-26	1.50 max	0.50 max	12-14	0.04 max	0.04 max	0.50-2.00	bal Fe				

Cast Stainless Steel, Unclassified, J91261

USA

Specification	Designation	Notes	C	Cr	Mn	Mo	Ni	P	S	Si	Other	UTS	YS	El	Hard
	UNS J91261		0.15 max	11.50-14.00	1.00 max	0.50-0.70	0.65-1.00	0.05 max	0.05 max	0.50 max	bal Fe				

Cast Stainless Steel, Unclassified, J92170

USA

Specification	Designation	Notes	C	Cr	Mn	Mo	Ni	P	S	Si	Other	UTS	YS	El	Hard
	UNS J92170	Precipitation Hardening	0.06 max	15.5-16.7	0.70 max		3.6-4.6	0.04 max	0.03 max	0.50-1.0	Cu 2.5-3.2; bal Fe				

Cast Stainless Steel, Unclassified, J92512

USA

Specification	Designation	Notes	C	Cr	Mn	Mo	Ni	P	S	Si	Other	UTS	YS	El	Hard
	UNS J92512		0.25 max	17.0-19.0	2.0 max	0.5 max	8.0-10.0	0.04 max	0.03 max	1.0 max	Cu <=0.5; bal Fe				

Cast Stainless Steel, Unclassified, J92650

Italy

Specification	Designation	Notes	C	Cr	Mn	Mo	Ni	P	S	Si	Other	UTS	YS	El	Hard
UNI 3161(83)	GX6CrNi2010	Sand; Heat res	0.08 max	18-21.	1.5 max		8-11	0.04 max	0.035 max	2 max	bal Fe				

Romania

Specification	Designation	Notes	C	Cr	Mn	Mo	Ni	P	S	Si	Other	UTS	YS	El	Hard
STAS 10718(88)	T10NiCr180	Corr res	0.1 max	17-20	2 max		8-11	0.04 max	0.03 max	0.5-2	Pb <=0.15; Ti <=1.2; 5C<=Ti<=1.2; bal Fe				

USA

Specification	Designation	Notes	C	Cr	Mn	Mo	Ni	P	S	Si	Other	UTS	YS	El	Hard
	UNS J92650		0.08 max	18.00-21.00	1.50 max		8.00-11.00	0.05 max	0.05 max	2.00 max	bal Fe				

Cast Stainless Steel, Unclassified, J92720

USA

Specification	Designation	Notes	C	Cr	Mn	Mo	Ni	P	S	Si	Other	UTS	YS	El	Hard
	UNS J92720		0.08 max	17.00-20.00	1.50 max	0.50 max	10.00 min	0.05 max	0.05 max	1.50 max	bal Fe				

Cast Stainless Steel, Unclassified, J92730

USA

Specification	Designation	Notes	C	Cr	Mn	Mo	Ni	P	S	Si	Other	UTS	YS	El	Hard
	UNS J92730		0.08 max	18.00-21.00	1.50 max		9.00-12.00	0.05 max	0.05 max	2.00 max	bal Fe				

UNS numbers and US grades are provided as a means of cross referencing chemically similar alloys. Exchangability is only possible after independent examination of specifications. Tensile properties are minimum or typical as specified. UTS and YS as MPa. El as %. See Appendix for list of abbreviations used in Notes. * indicates obsolete material.

Specification	Designation	Notes	C	Cr	Mn	Mo	Ni	P	S	Si	Other	UTS	YS	El	Hard
Cast Stainless Steel, Unclassified, J92740															
USA															
	UNS J92740	Durcomet 101	0.07 max	22.00-24.00	0.50 max	0.85-1.15	4.50-5.50	0.04 max	0.04 max	1.70 max	Cu <=0.50; bal Fe				
Cast Stainless Steel, Unclassified, J92801															
USA															
	UNS J92801		0.10 max	17.00-20.00	1.50 max	0.50 max	11.00 min	0.05 max	0.05 max	1.50 max	bal Fe				
Cast Stainless Steel, Unclassified, J92910															
Romania															
STAS 10718(88)	T10MoNiCr180	Corr res	0.1 max	17-20	2 max	2-3	9-12	0.04 max	0.03 max	0.5-2	Pb <=0.15; Nb=8C-1.2; bal Fe				
USA															
	UNS J92910		0.08 max	18.00-21.00	1.50 max	2.00-3.00	9.00-12.00	0.05 max	0.05 max	2.00 max	bal Fe				
Cast Stainless Steel, Unclassified, No equivalents identified															
Czech Republic															
CSN 422914	422914	Heat res	0.5-0.8	27.5-30.0	0.9 max		2.0 max	0.045 max	0.04 max	1.0-2.0	Cu <=0.5; bal Fe				
CSN 422916	422916	Corr res	0.16-0.22	10.2-11.8	0.4-0.7	0.9-1.2	0.2-0.6	0.035 max	0.03 max	0.1-0.4	Cu <=0.5; V 0.2-0.35; bal Fe				
CSN 422917	422917	Corr res	0.1-0.18	11.5-13.2	0.6-1.0		0.7-1.1	0.045 max	0.04 max	0.2-0.6	Cu <=0.5; V 0.1-0.2; W 0.5-0.8; bal Fe				
CSN 422992	422992		0.75-0.9	3.8-4.5	0.5 max	0.7-1.0	0.25 max	0.04 max	0.04 max	0.7 max	V 2.0-2.7; W 9.5-11.0; P+S<=0.07; bal Fe				
France															
AFNOR NFA32057(81)	KC30Fe20M	28mm	0.3-0.6	25-30	1 max	0.15 max	3 max	0.02 max	0.02 max	1 max	Co 48-52; Nb <=2; Fe<=20; Class II; bal Ni	540	350	3	200 HB max
Hungary															
MSZ 21053(82)	AoX10CrNiMoCuNb1818	Corr res; 0-30.0mm; as cast	0.12 max	17-19	2 max	2.5-3.5	17-19	0.04 max	0.03 max	1.5 max	Cu 1.8-2.3; Nb=8C-1.1; bal Fe				160-230 HB
MSZ 21053(82)	AoX10CrNiMoCuNb1818F	Corr res; 0-30.0mm; SA	0.12 max	17-19	2 max	2.5-3.5	17-19	0.04 max	0.03 max	1.5 max	Cu 1.8-2.3; Nb=8C-1.1; bal Fe	470	200	20L	140-200 HB
MSZ 21053(82)	AoX10CrNiMoCuTi1818	Corr res; 0-30.0mm; as cast	0.12 max	17-19	2 max	2.5-3.5	17-19	0.04 max	0.03 max	1.5 max	Cu 1.8-2.3; Ti <=0.8; Ti=5C-0.8; bal Fe				160-230 HB
MSZ 21053(82)	AoX10CrNiMoCuTi1818F	Corr res; 0-30.0mm; SA	0.12 max	17-19	2 max	2.5-3.5	17-19	0.04 max	0.03 max	1.5 max	Cu 1.8-2.3; Ti <=0.8; Ti=5C-0.8; bal Fe	470	200	20L	160-230 HB
MSZ 21053(82)	AoX10CrNiMoNb1812F	Corr res; 0-30.0mm; SA	0.12 max	17-19	2 max	3-4	11-14	0.04 max	0.03 max	1.5 max	Nb=8C-1.1; bal Fe	470	220	20L	140-200 HB
MSZ 21053(82)	AoX10CrNiMoTi1812	Corr res; 0-30.0mm; as cast	0.12 max	17-19	2 max	3-4	11-14	0.04 max	0.03 max	1.5 max	Ti <=0.8; Ti=8C-0.8; bal Fe				160-230 HB
MSZ 21053(82)	AoX10CrNiMoTi1812F	Corr res; 0-30.0mm; SA	0.12 max	17-19	2 max	3-4	11-14	0.04 max	0.03 max	1.5 max	Ti <=0.8; Ti=8C-0.8; bal Fe	470	220	20L	140-200 HB
MSZ 21053(82)	AoX120CrMo292	Corr res; 0-30.0mm; as cast	1.1-1.3	28-32	0.3-0.6	1.8-2.2	1 max	0.04 max	0.04 max	1.5 max	bal Fe				280-350 HB
MSZ 21053(82)	AoX120CrMo292F	Corr res; 0-30.0mm; soft ann	1.1-1.3	28-32	0.3-0.6	1.8-2.2	1 max	0.04 max	0.04 max	1.5 max	bal Fe	390	290		250-320 HB
MSZ 21053(82)	AoX12Cr13F	Corr res; 0-30.0mm; Q/T	0.15 max	12-14	1 max		1 max	0.04 max	0.04 max	1.5 max	bal Fe	530	350	16L	140-180 HB
MSZ 21053(82)	AoX20CrNi14	Corr res; 0-30.0mm; soft ann	0.15-0.25	13-15	1 max		1-1.5	0.04 max	0.04 max	1.5 max	bal Fe				180-230 HB
MSZ 21053(82)	AoX20CrNi14F	Corr res; 0-30.0mm; Q/T	0.15-0.25	13-15	1 max		1-1.5	0.04 max	0.04 max	1.5 max	bal Fe	540	350	12L	200-250 HB
MSZ 4357(81)	AoX30CrSi7	Heat res; soft ann	0.2-0.4	6-8	0.4-1		1 max	0.04 max	0.04 max	1-2.5	bal Fe				210-290 HB
MSZ 4357(81)	AoX30CrSi7F	Heat res; soft ann	0.2-0.4	6-8	0.4-1		1 max	0.04 max	0.04 max	1-2.5	bal Fe	500		5L	210-290 HB
MSZ 4357(81)	AoX40CrNiSi254	Heat res; as cast	0.3-0.5	24-26	0.4-1		3.5-5	0.04 max	0.04 max	1-2.5	bal Fe				220-300 HB
MSZ 4357(81)	AoX40CrNiSi254F	Heat res; soft ann	0.3-0.5	24-26	0.4-1		3.5-5	0.04 max	0.04 max	1-2.5	bal Fe	600		4L	200-280 HB
MSZ 4357(81)	AoX40CrSi18	Heat res; soft ann	0.3-0.5	17-19	0.4-1		1 max	0.04 max	0.04 max	1-2.5	bal Fe				210-300 HB

Specification	Designation	Notes	C	Cr	Mn	Mo	Ni	P	S	Si	Other	UTS	YS	El	Hard

Cast Stainless Steel, Unclassified, No equivalents identified

Hungary

Specification	Designation	Notes	C	Cr	Mn	Mo	Ni	P	S	Si	Other	UTS	YS	El	Hard
MSZ 4357(81)	AoX40CrSi18F	Heat res; soft ann	0.3-0.5	17-19	0.4-1		1 max	0.04 max	0.04 max	1-2.5	bal Fe	450		2L	210-300 HB
MSZ 4357(81)	AoX40CrSi23	Heat res; as cast	0.3-0.5	22-24	0.4-1		1 max	0.04 max	0.04 max	1-2.5	bal Fe				220-300 HB
MSZ 4357(81)	AoX40CrSi23F	Heat res; soft ann	0.3-0.5	22-24	0.4-1		1 max	0.04 max	0.04 max	1-2.5	bal Fe	400		2L	220-280 HB
MSZ 4357(81)	AoX40CrSi29	Heat res; as cast	0.3-0.5	27-30	0.4-1		1 max	0.04 max	0.04 max	1-2.5	bal Fe				220-300 HB
MSZ 4357(81)	AoX40CrSi29F	Heat res; soft ann	0.3-0.5	27-30	0.4-1		1 max	0.04 max	0.04 max	1-2.5	bal Fe	300			220-280 HB

Italy

Specification	Designation	Notes	C	Cr	Mn	Mo	Ni	P	S	Si	Other	UTS	YS	El	Hard
UNI 3160(83)	GX200Cr13	Wear res	2-2.2	12-14	0.6 max			0.035 max	0.035 max	0.5 max	bal Fe				
UNI 3161(83)	GX6CrNi1301	Sand; Heat res	0.08 max	11.5-13.5	1 max	0.5 max	1.4-1.8	0.04 max	0.035 max	1 max	bal Fe				

Japan

Specification	Designation	Notes	C	Cr	Mn	Mo	Ni	P	S	Si	Other	UTS	YS	El	Hard
JIS G5121(91)	SCS11	Corr res; Duplex	0.08 max	23.00-27.00	1.00 max	1.50-2.50	4.00-7.00	0.040 max	0.030 max	1.50 max	bal Fe	590	345	13	241 HB max
JIS G5121(91)	SCS15	Corr res; Solution HT	0.08 max	17.00-20.00	2.00 max	1.75-2.75	10.00-14.00	0.040 max	0.040 max	2.00 max	Cu 1.00-2.50; bal Fe	440	185	28	183 HB max
JIS G5121(91)	SCS20	Corr res; Solution HT	0.03 max	17.00-20.00	2.00 max	1.75-2.75	12.00-16.00	0.040 max	0.040 max	2.00 max	Cu 1.00-2.50; bal Fe	390	175	33	183 HB max

Romania

Specification	Designation	Notes	C	Cr	Mn	Mo	Ni	P	S	Si	Other	UTS	YS	El	Hard
STAS 10718(88)	T15Cr170	Corr res	0.1-0.2	16-18	0.6 max		1 max	0.04 max	0.03 max	1.5-2	Pb <=0.15; bal Fe				
STAS 10718(88)	T15Cr280	Corr res	0.1-0.2	27-29	1 max		1 max	0.04 max	0.03 max	1 max	Pb <=0.15; bal Fe				
STAS 6855(86)	T12MoNiCr210	Corr res; Heat res	0.12 max	20-22	2 max	1.8-2.2	4.5-6	0.035 max	0.03 max	1.5 max	Pb <=0.15; bal Fe				
STAS 6855(86)	T12NiCr210	Corr res; Heat res	0.12 max	20-22	2 max		4.5-6	0.035 max	0.03 max	1.5 max	Pb <=0.15; bal Fe				
STAS 6855(86)	T12NNiCr190	Corr res; Heat res	0.12 max	18-20	2 max		6-8	0.035 max	0.03 max	1.5 max	N 0.1-0.2; Pb <=0.15; bal Fe				
STAS 6855(86)	T12NNiCr210	Corr res; Heat res	0.12 max	20-22	2 max		4-6	0.035 max	0.03 max	1.5 max	N 0.1-0.2; Pb <=0.15; bal Fe				
STAS 6855(86)	T12TiMoNiCr175	Corr res; Heat res	0.12 max	16-19	1-2	3-4	11-13	0.035 max	0.03 max	1 max	Pb <=0.15; Ti <=0.80; Ti>=5C; bal Fe				
STAS 6855(86)	T12TiNiCr210	Corr res; Heat res	0.12 max	20-22	2 max		4.5-6	0.035 max	0.03 max	1.5 max	Pb <=0.15; Ti <=0.7; 4C<=Ti<=0.7; bal Fe				
STAS 6855(86)	T15NiCr135	Corr res; Heat res	0.15 max	12-15	0.4-0.9		0.7-1.2	0.035 max	0.03 max	0.6 max	Pb <=0.15; bal Fe				
STAS 6855(86)	T20CrNi370	Corr res; Heat res	0.2 max	16-20	1-2		35-39	0.035 max	0.03 max	1-2.5	Pb <=0.15; bal Fe				
STAS 6855(86)	T70MoCr280	Corr res; Heat res	0.5-0.9	27-30	1 max	2-2.5	0.4 max	0.035 max	0.03 max	1-2	Pb <=0.15; bal Fe				

UK

Specification	Designation	Notes	C	Cr	Mn	Mo	Ni	P	S	Si	Other	UTS	YS	El	Hard
BS 1504(76)	317C12*	Press ves	0.03 max	17.0-21.0	2.0 max	3.0-4.0	10.0-16.00	0.04 max	0.04 max	1.5 max	Cu <=0.3; Ti <=0.05; W <=0.1; bal Fe				

USA

Specification	Designation	Notes	C	Cr	Mn	Mo	Ni	P	S	Si	Other	UTS	YS	El	Hard
	UNS J94603*	Obs; see N08030									bal Fe				
	UNS J94604*	Obs; see N08604									bal Fe				
	UNS J94605*	Obs; see N08002									bal Fe				
	UNS J94613*	Obs; see N08613									bal Fe				
	UNS J94614*	Obs; see N08614									bal Fe				
	UNS J94805*	Obs; see N08050									bal Fe				
	UNS J95150*	Obs; see N08007									bal Fe				
	UNS J95151*	Obs; see N08151									bal Fe				
	UNS J95404*	Obs; see N08005									bal Fe				
	UNS J95405*	Obs; see N08004									bal Fe				
	UNS J95705*	Obs; see N08705									bal Fe				

UNS numbers and US grades are provided as a means of cross referencing chemically similar alloys. Exchangability is only possible after independent examination of specifications. Tensile properties are minimum or typical as specified. UTS and YS as MPa. El as %. See Appendix for list of abbreviations used in Notes. * indicates obsolete material.

Specification	Designation	Notes	C	Cr	Mn	Mo	Ni	P	S	Si	Other	UTS	YS	El	Hard

Steel Casting, Low-Carbon, 0022

Bulgaria

Specification	Designation	Notes	C	Cr	Mn	Mo	Ni	P	S	Si	Other	UTS	YS	El	Hard
BDS 3492	15LI		0.10-0.20	0.30 max	0.40-0.80			0.060 max	0.060 max	0.20-0.60	bal Fe				
BDS 3492	15LII		0.10-0.20	0.30 max	0.40-0.80		0.40 max	0.060 max	0.060 max	0.20-0.60	Cu 0.3-0.30; bal Fe				
BDS 3492	15LIII		0.10-0.20	0.30 max	0.40-0.80			0.050 max	0.050 max	0.25-0.50	bal Fe				
BDS 6550(86)	15ChL		0.12-0.18	0.50-0.80	0.40-0.60	0.15 max	0.40 max	0.040 max	0.040 max	0.20-0.50	Cu <=0.30; bal Fe				
BDS 6550(86)	15GL		0.12-0.18	0.30 max	0.70-1.00	0.15 max	0.30 max	0.040 max	0.040 max	0.30-0.60	Cu <=0.30; bal Fe				

Canada

Specification	Designation	Notes	C	Cr	Mn	Mo	Ni	P	S	Si	Other	UTS	YS	El	Hard
CSA 422630	22630		0.1-0.2		0.4-0.8			0.05 max	0.05 max	0.2-0.5	P+S=0.09 max; bal Fe				

China

Specification	Designation	Notes	C	Cr	Mn	Mo	Ni	P	S	Si	Other	UTS	YS	El	Hard
GB 11352(89)	ZG200-400(ZG15)*	Obs	0.20 max	0.35 max	0.80 max	0.20 max	0.30 max	0.040 max	0.040 max	0.50 max	Cu <=0.30; V <=0.05; bal Fe	400	200	25	

Finland

Specification	Designation	Notes	C	Cr	Mn	Mo	Ni	P	S	Si	Other	UTS	YS	El	Hard
SFS 355(79)	G-20-40	Struct, Const	0.18 max		0.7 max			0.04 max	0.04 max	0.5 max	bal Fe				

France

Specification	Designation	Notes	C	Cr	Mn	Mo	Ni	P	S	Si	Other	UTS	YS	El	Hard
AFNOR NFA32051	E20-40M	Ann-Norm, or Q/A Tmp	0.18					0.04	0.04	0.5	bal Fe				

Germany

Specification	Designation	Notes	C	Cr	Mn	Mo	Ni	P	S	Si	Other	UTS	YS	El	Hard
DIN 1681(85)	GS-38										bal Fe	360	200	25	
DIN 1681(85)	WNr 1.0420										bal Fe	360	200	25	

Japan

Specification	Designation	Notes	C	Cr	Mn	Mo	Ni	P	S	Si	Other	UTS	YS	El	Hard
JIS G5101(91)	SC360		0.2					0.04	0.04		bal Fe				

Poland

Specification	Designation	Notes	C	Cr	Mn	Mo	Ni	P	S	Si	Other	UTS	YS	El	Hard
PNH83152(80)	L400		0.1-0.2	0.4 max	0.35-0.8		0.4 max	0.05 max	0.05 max	0.2-0.5	bal Fe				

Russia

Specification	Designation	Notes	C	Cr	Mn	Mo	Ni	P	S	Si	Other	UTS	YS	El	Hard
GOST 977(88)	15L		0.12-0.2	0.3 max	0.45-0.9		0.4 max	0.04 max	0.04 max	0.2-0.52	bal Fe				
GOST 977(88)	15L-2		0.12-0.2	0.3 max	0.45-0.9		0.4 max	0.035 max	0.035 max	0.2-0.52	bal Fe				
GOST 977(88)	15L-3		0.12-0.2	0.3 max	0.45-0.9		0.4 max	0.03 max	0.03 max	0.2-0.52	bal Fe				

Spain

Specification	Designation	Notes	C	Cr	Mn	Mo	Ni	P	S	Si	Other	UTS	YS	El	Hard
UNE 36256(73)	AMC15K*	Q/T	0.2		1.3			0.035	0.035	0.3-0.6	bal Fe				
UNE 36256(73)	F.8371*	see AMC15K*	0.2		1.3			0.035	0.035	0.3-0.6	bal Fe				

UK

Specification	Designation	Notes	C	Cr	Mn	Mo	Ni	P	S	Si	Other	UTS	YS	El	Hard
BS 3146/1(74)	CLA9	Inv casting	0.10-0.18	0.25 max	0.60-1.00	0.15 max	0.40 max	0.035 max	0.035 max	0.20-0.60	Cu <=0.3; bal Fe				

USA

Specification	Designation	Notes	C	Cr	Mn	Mo	Ni	P	S	Si	Other	UTS	YS	El	Hard
	UNS J01700		0.12-0.22		0.50-0.90			0.040 max	0.045 max	0.60 max	bal Fe				
SAE J435(74)	0022	Auto	0.12-0.22		0.50-0.90			0.040 max	0.045 max	0.60 max	bal Fe				187 HB max

Steel Casting, Low-Carbon, 0025

China

Specification	Designation	Notes	C	Cr	Mn	Mo	Ni	P	S	Si	Other	UTS	YS	El	Hard
GB 11352(89)	ZG200-400(ZG15)*	Obs	0.20 max	0.35 max	0.80 max	0.20 max	0.30 max	0.040 max	0.040 max	0.50 max	Cu <=0.30; V <=0.05; bal Fe	400	200	25	

Czech Republic

Specification	Designation	Notes	C	Cr	Mn	Mo	Ni	P	S	Si	Other	UTS	YS	El	Hard
CSN 422714	422714		0.15-0.22	0.3 max	1.0-1.5		0.5 max	0.03 max	0.02 max	0.45 max	bal Fe				

Italy

Specification	Designation	Notes	C	Cr	Mn	Mo	Ni	P	S	Si	Other	UTS	YS	El	Hard
UNI 3608	CG20	As Agreed	0.25		0.8			0.035	0.035	0.6	bal Fe				

Japan

Specification	Designation	Notes	C	Cr	Mn	Mo	Ni	P	S	Si	Other	UTS	YS	El	Hard
JIS G5102(91)	SCW42*	Obs; see SCW410	0.22					0.04	0.04		Others each 0.4 max; bal Fe				
JIS G5102(91)	SCW49*	Obs; see SCW480	0.22					0.04	0.04		Others each 0.43 max; bal Fe				
JIS G5102(91)	SCW56*	Obs; see SCW550	0.2					0.04	0.04		Others each 0.44 max; bal Fe				

UNS numbers and US grades are provided as a means of cross referencing chemically similar alloys. Exchangability is only possible after independent examination of specifications. Tensile properties are minimum or typical as specified. UTS and YS as MPa. El as %. See Appendix for list of abbreviations used in Notes. * indicates obsolete material.

Specification	Designation	Notes	C	Cr	Mn	Mo	Ni	P	S	Si	Other	UTS	YS	El	Hard

Steel Casting, Low-Carbon, 0025 (Continued from previous page)

Japan

Specification	Designation	Notes	C	Cr	Mn	Mo	Ni	P	S	Si	Other	UTS	YS	El	Hard
JIS G5102(91)	SCW63*	Obs; see SCW620	0.2					0.04	0.04		Others each 0.44 max; bal Fe				
JIS G5201(91)	SCW42-CF		0.22					0.04	0.04		Others each 0.4 max; bal Fe				
JIS G5201(91)	SCW49-CF		0.22					0.04	0.04		Others each 0.43 max; bal Fe				
JIS G5201(91)	SCW50-CF		0.2					0.04	0.04		Others each 0.44 max; bal Fe				
JIS G5201(91)	SCW53-CF		0.2					0.04	0.04		Others each 0.44 max; bal Fe				

USA

Specification	Designation	Notes	C	Cr	Mn	Mo	Ni	P	S	Si	Other	UTS	YS	El	Hard
	UNS J02507		0.25 max		0.75 max			0.040 max	0.045 max	0.80 max	bal Fe				
SAE J435(74)	0025	Auto	0.25 max		0.75 max			0.040 max	0.045 max	0.80 max	bal Fe	413.7	206.8	22	187 HB max

Steel Casting, Low-Carbon, LCC

Hungary

Specification	Designation	Notes	C	Cr	Mn	Mo	Ni	P	S	Si	Other	UTS	YS	El	Hard
MSZ 8267(85)	Ao20Mn5	Tough at subzero; 0-30mm; HT	0.23 max		1-1.3			0.03 max	0.03 max	0.65 max	bal Fe	450	280	20L	

Italy

Specification	Designation	Notes	C	Cr	Mn	Mo	Ni	P	S	Si	Other	UTS	YS	El	Hard
UNI 4010(75)	FeG52	HS	0.25 max		1-1.5			0.035 max	0.035 max	0.5 max	bal Fe				
UNI 7317(74)	G22Mn3	Tough at subzero	0.25 max		1 max			0.035 max	0.035 max	0.6 max	bal Fe				

Japan

Specification	Designation	Notes	C	Cr	Mn	Mo	Ni	P	S	Si	Other	UTS	YS	El	Hard
JIS G5201(91)	SCW410-CF	Cent cast weld	0.22 max		1.50 max			0.040 max	0.040 max	0.80 max	bal Fe	410	235	21	
JIS G5201(91)	SCW42-CF*	Obs; see SCW410-CF	0.22					0.04	0.04		Others each 0.4 max; bal Fe				
JIS G5201(91)	SCW480-CF	Cent cast weld	0.22 max		1.50 max			0.040 max	0.040 max	0.80 max	bal Fe	480	275	20	
JIS G5201(91)	SCW49-CF*	Obs; see SCW480-CF	0.22					0.04	0.04		Others each 0.43 max; bal Fe				
JIS G5201(91)	SCW490-CF	Cent cast weld	0.20 max		1.50 max			0.040 max	0.040 max	0.80 max	bal Fe	490	315	20	
JIS G5201(91)	SCW50-CF*	Obs; see SCW490-CF	0.2					0.04	0.04		Others each 0.44 max; bal Fe				
JIS G5201(91)	SCW520-CF	Cent cast weld	0.20 max	0.50 max	1.50 max		0.50 max	0.040 max	0.040 max	0.80 max	bal Fe	520	355	18	
JIS G5201(91)	SCW53-CF*	Obs; see SCW520-CF	0.2					0.04	0.04		Others each 0.44 max; bal Fe				
JIS G5201(91)	SCW570-CF	Cent cast weld	0.20 max	0.50 max	1.50 max	0.50 max	2.50 max	0.040 max	0.040 max	1.00 max	V <=0.20; bal Fe	570	430	17	
JIS G5201(91)	SCW58-CF*	Obs; see SCW570-CF	0.2					0.04	0.04		Others each 0.44 max; bal Fe				
JIS G5202(91)	SCPH1-CF	Cont cast; High-temp	0.22 max	0.25 max	1.10 max	0.25 max	0.50 max	0.040 max	0.040 max	0.60 max	Cu <=0.50; bal Fe	410	245	21	
JIS G5202(91)	SCPH2-CF	Cont cast; High-temp	0.30 max	0.25 max	1.10 max	0.25 max	0.50 max	0.040 max	0.040 max	0.60 max	Cu <=0.50; Others 1 max; bal Fe	480	275	19	

Romania

Specification	Designation	Notes	C	Cr	Mn	Mo	Ni	P	S	Si	Other	UTS	YS	El	Hard
STAS 1773(82)	T20Mn14		0.15-0.25	0.3 max	1.2-1.6			0.04 max	0.04 max	0.2-0.45	Pb <=0.15; bal Fe				

Sweden

Specification	Designation	Notes	C	Cr	Mn	Mo	Ni	P	S	Si	Other	UTS	YS	El	Hard
SS 141305	1305	Gen Struct	0.25 max		2 max			0.04 max	0.04 max	0.6 max	Si=0.5; Mn:0.7				

USA

Specification	Designation	Notes	C	Cr	Mn	Mo	Ni	P	S	Si	Other	UTS	YS	El	Hard
	UNS J02505		0.25 max		1.20 max			0.04 max	0.045 max	0.60 max	bal Fe				
ASTM A352/A352M(98)	LCC	Press; low-temp	0.25 max	0.50 max	1.20 max	0.20 max	0.50 max	0.04 max	0.045 max	0.60 max	Cu <=0.30; V <=0.03; bal Fe	485-655	275	22	
ASTM A660(96)	WCC	Centrifugally cast pipe; high-temp	0.25 max		1.20 max			0.035 max	0.035 max	0.60 max	bal Fe	483	276	22	

Yugoslavia

Specification	Designation	Notes	C	Cr	Mn	Mo	Ni	P	S	Si	Other	UTS	YS	El	Hard
	CL.3132		0.17-0.23		0.3-0.6	0.15 max		0.035 max	0.035 max	1.0-1.3	bal Fe				

UNS numbers and US grades are provided as a means of cross referencing chemically similar alloys. Exchangability is only possible after independent examination of specifications. Tensile properties are minimum or typical as specified. UTS and YS as MPa. El as %. See Appendix for list of abbreviations used in Notes. * indicates obsolete material.

Specification	Designation	Notes	C	Cr	Mn	Mo	Ni	P	S	Si	Other	UTS	YS	El	Hard

Steel Casting, Low-Carbon, No equivalents identified

Germany

Specification	Designation	Notes	C	Cr	Mn	Mo	Ni	P	S	Si	Other	UTS	YS	El	Hard
DIN EN 10213(96)	GS-C25	see GC25E	0.18-0.23		0.50-1.20			0.030 max	0.020 max	0.60 max	bal Fe				

Poland

Specification	Designation	Notes	C	Cr	Mn	Mo	Ni	P	S	Si	Other	UTS	YS	El	Hard
PNH83152(80)	LII400		0.1-0.2	0.4 max	0.4-0.9	0.1 max	0.35 max	0.04 max	0.04 max	0.2-0.5	bal Fe				
PNH83152(80)	LII450		0.15-0.30	0.4 max	0.4-0.9	0.1 max	0.35 max	0.04 max	0.04 max	0.2-0.5	bal Fe				
PNH83156(87)	L20G		0.15-0.25	0.3 max	1.2-1.6		0.3 max	0.04 max	0.04 max	0.2-0.5	bal Fe				
PNH83156(87)	L35G		0.3-0.4	0.3 max	1.2-1.6		0.3 max	0.04 max	0.04 max	0.2-0.5	bal Fe				
PNH83157(89)	L20	High-temp const	0.18-0.23	0.3 max	0.5-0.8		0.3 max	0.03 max	0.03 max	0.3-0.6	bal Fe				
PNH83157(89)	L20M	High-temp const	0.18-0.23	0.3 max	0.5-0.8	0.35-0.45	0.4 max	0.03 max	0.03 max	0.3-0.6	bal Fe				
PNH83160(88)	L30GS		0.25-0.35	0.3 max	1.1-1.4		0.3 max	0.04 max	0.04 max	0.6-0.8	bal Fe				

Romania

Specification	Designation	Notes	C	Cr	Mn	Mo	Ni	P	S	Si	Other	UTS	YS	El	Hard
STAS 1773(82)	T20TiMn12		0.16-0.24	0.3 max	1-1.3			0.03 max	0.03 max	0.6-0.8	Pb <=0.15; Ti 0.04-0.08; bal Fe				

Russia

Specification	Designation	Notes	C	Cr	Mn	Mo	Ni	P	S	Si	Other	UTS	YS	El	Hard
GOST 21357	12ChNDMFLS	Tough at subzero; Wear res	0.10-0.15	1.2-1.7	0.3-0.55	0.2-0.3	1.4-1.8	0.02 max	0.02 max	0.2-0.4	Cu 0.4-0.65; Pb <=0.15; V 0.08-0.15; bal Fe				
GOST 21357	13ChNDTFLS	Tough at subzero; Wear res	0.16 max	0.8-1.1	0.7-1.0	0.15 max	1.2-1.6	0.035 max	0.035 max	0.2-0.4	Cu 0.65-0.9; Pb <=0.15; Ti 0.04-0.1; V 0.06-0.12; bal Fe				
GOST 21357(87)	08G2DNFL	Tough at subzero; Wear res	0.05-0.10	0.30 max	1.30-1.70		1.15-1.55	0.020 max	0.020 max	0.15-0.40	Cu 0.80-1.10; V 0.02-0.08; bal Fe				
GOST 21357(87)	12ChGFL	Tough at subzero; Wear res	0.10-0.16	0.20-0.60	0.90-1.40		0.30 max	0.020 max	0.020 max	0.30-0.50	Cu <=0.30; Mg <=1; V 0.05-0.10; bal Fe				
GOST 21357(87)	14Ch2GMRL	Tough at subzero; Wear res	0.10-0.17	1.40-1.70	0.90-1.20	0.45-0.55	0.30 max	0.020 max	0.020 max	0.20-0.42	B <=0.004; Cu <=0.30; bal Fe				
GOST 977(88)	03N12Ch5M3TJuL		0.01-0.04	4.5-5	0.2 max	2.5-3	12-12.5	0.015 max	0.015 max	0.2 max	Al 0.25-0.45; Ti 0.7-0.9; bal Fe				
GOST 977(88)	03N12Ch5M3TL		0.01-0.04	4.5-5	0.2 max	2.5-3	12-12.5	0.015 max	0.015 max	0.2 max	Ti 0.7-0.9; bal Fe				
GOST 977(88)	13ChNDFTL		0.16 max	0.15-0.4	0.4-0.9		1.2-1.6	0.03 max	0.03 max	0.2-0.4	Cu 0.65-0.9; Ti 0.04-0.1; V 0.06-0.12; bal Fe				
GOST 977(88)	20GSL		0.16-0.22	0.3 max	1-1.3		0.4 max	0.03 max	0.03 max	0.6-0.8	bal Fe				

UK

Specification	Designation	Notes	C	Cr	Mn	Mo	Ni	P	S	Si	Other	UTS	YS	El	Hard
BS 1504(76)	625*	Press ves	0.2 max	4.0-6.0	0.4-0.7	0.45-0.65	0.4 max	0.04 max	0.04 max	0.75 max	Cu <=0.3; Pb <=0.15; V <=0.1; W <=0.1; bal Fe				
BS 1504(76)	660*	Press ves	0.10-0.15	0.3-0.5	0.4-0.7	0.4-0.6	0.3 max	0.03 max	0.03 max	0.45 max	Cu <=0.3; Pb <=0.15; V 0.22-0.3; W <=0.1; Sn<=0.05; bal Fe				

USA

Specification	Designation	Notes	C	Cr	Mn	Mo	Ni	P	S	Si	Other	UTS	YS	El	Hard
AMS 2431/5(88)		CH Peening Balls									bal Fe				57-62 HRC
ASTM A587(96)	A587	ERW pipe	0.15 max		0.27-0.63			0.035 max	0.035 max		Al 0.02-0.100; bal Fe	331	207	40	
SAE J2175(91)		Shot peen	0.10-0.15		1.20-1.50			0.035 max	0.035 max	0.10-0.25	Al 0.05-0.15; bal Fe				40-50 HRC

Steel Casting, Low-Carbon, WCC

Hungary

Specification	Designation	Notes	C	Cr	Mn	Mo	Ni	P	S	Si	Other	UTS	YS	El	Hard
MSZ 1749(89)	Ao21Mn	High-temp const; 0-30.00mm; HT	0.17-0.25		0.9-1.4			0.03 max	0.03 max	0.3-0.6	bal Fe	500-650	280	20L	
MSZ 8272(89)	Ao20Mn6	0-30.00mm; norm	0.15-0.25		1.2-1.6		0.3 max	0.05 max	0.05 max	0.2-0.4	bal Fe	500	300	18L	
MSZ 8272(89)	Ao20Mn6ne	0-30.00mm; Q/T	0.15-0.25		1.2-1.6		0.3 max	0.05 max	0.05 max	0.2-0.4	bal Fe	550	400	14L	

UNS numbers and US grades are provided as a means of cross referencing chemically similar alloys. Exchangability is only possible after independent examination of specifications. Tensile properties are minimum or typical as specified. UTS and YS are MPa. El as %. See Appendix for list of abbreviations used in Notes. * indicates obsolete material.

Specification	Designation	Notes	C	Cr	Mn	Mo	Ni	P	S	Si	Other	UTS	YS	El	Hard

Steel Casting, Low-Carbon, WCC (Continued from previous page)

Russia

Specification	Designation	Notes	C	Cr	Mn	Mo	Ni	P	S	Si	Other	UTS	YS	El	Hard
GOST 21357(87)	20GL	Tough at subzero; Wear res	0.17-0.25	0.30 max	1.10-1.40		0.30 max	0.020 max	0.020 max	0.20-0.50	Cu <=0.30; bal Fe				

Specification	Designation	Notes	C	Cr	Mn	Mo	Ni	P	S	Si	Other	UTS	YS	El	Hard

Steel Casting, Medium-Carbon, 0030

Belgium

Specification	Designation	Notes	C	Cr	Mn	Mo	Ni	P	S	Si	Other	UTS	YS	El	Hard
NBN 253-02	C25-1		0.22-0.29		0.4-0.7			0.045	0.045	0.15-0.4	bal Fe				

Canada

Specification	Designation	Notes	C	Cr	Mn	Mo	Ni	P	S	Si	Other	UTS	YS	El	Hard
CSA G28	60-30	Full Ann-Norm, N/T, Q/A	0.3		0.75			0.04	0.04	0.8	bal Fe				
CSA G28	65-35	Full Ann-Norm, N/T, Q/A	0.3		0.75			0.04	0.04	0.8	bal Fe				

China

Specification	Designation	Notes	C	Cr	Mn	Mo	Ni	P	S	Si	Other	UTS	YS	El	Hard
GB 11352(89)	ZG230-450(ZG25)	Ann or N/T	0.30 max	0.35 max	0.90 max	0.20 max	0.30 max	0.040 max	0.040 max	0.50 max	Cu <=0.30; V <=0.05; bal Fe	450	230	22	

Czech Republic

Specification	Designation	Notes	C	Cr	Mn	Mo	Ni	P	S	Si	Other	UTS	YS	El	Hard
CSN 422640	422640		0.20-0.28	0.3 max	0.4-0.8			0.05 max	0.05 max	0.2-0.5	P+S<=0.09; bal Fe				
CSN 422643	422643		0.17-0.25	0.3 max	0.5-0.9			0.04 max	0.04 max	0.2-0.5	P+S<=0.07; Ni+Cu+Cr<=0.9; bal Fe				

Russia

Specification	Designation	Notes	C	Cr	Mn	Mo	Ni	P	S	Si	Other	UTS	YS	El	Hard
GOST 977(88)	20L-3		0.17-0.25	0.3 max	0.45-0.9		0.4 max	0.03 max	0.03 max	0.2-0.52	bal Fe				
GOST 977(88)	23ChGS2MFL		0.18-0.24	0.6-0.9	0.5-0.8	0.25-0.3	0.4 max	0.025 max	0.025 max	1.8-2	V 0.1-0.15; bal Fe				
GOST 977(88)	25L		0.22-0.3	0.3 max	0.45-0.9		0.4 max	0.04 max	0.04 max	0.2-0.52	bal Fe				
GOST 977(88)	25L-2		0.22-0.3	0.3 max	0.45-0.9		0.4 max	0.035 max	0.035 max	0.2-0.52	bal Fe				
GOST 977(88)	25L-3		0.22-0.3	0.3 max	0.45-0.9		0.4 max	0.03 max	0.03 max	0.2-0.52	bal Fe				

USA

Specification	Designation	Notes	C	Cr	Mn	Mo	Ni	P	S	Si	Other	UTS	YS	El	Hard
	UNS J03010		0.30 max		0.70 max			0.040 max	0.045 max	0.80 max	bal Fe				
SAE J435(74)	0030	Auto	0.30 max		0.70 max			0.040 max	0.045 max	0.80 max	bal Fe	448.2	241.3	24	131-187 HB

Steel Casting, Medium-Carbon, 0050A

Belgium

Specification	Designation	Notes	C	Cr	Mn	Mo	Ni	P	S	Si	Other	UTS	YS	El	Hard
NBN 253-02	C45-2		0.42-0.5		0.5-0.8			0.035	0.035	0.15-0.4	bal Fe				

Bulgaria

Specification	Designation	Notes	C	Cr	Mn	Mo	Ni	P	S	Si	Other	UTS	YS	El	Hard
BDS 3492	45LI		0.40-0.50		0.40-0.80			0.060 max	0.060 max	0.20-0.60	bal Fe				
BDS 3492	45LII		0.40-0.50		0.40-0.80			0.060 max	0.060 max	0.20-0.60	bal Fe				
BDS 3492	45LIII		0.40-0.50		0.40-0.80			0.050 max	0.050 max	0.25-0.50	bal Fe				

China

Specification	Designation	Notes	C	Cr	Mn	Mo	Ni	P	S	Si	Other	UTS	YS	El	Hard
GB 11352(89)	ZG310-570	Ann or N/T	0.50 max	0.35 max	0.90 max	0.20 max	0.30 max	0.040 max	0.040 max	0.60 max	Cu <=0.30; V <=0.05; bal Fe	570	310	15	

Czech Republic

Specification	Designation	Notes	C	Cr	Mn	Mo	Ni	P	S	Si	Other	UTS	YS	El	Hard
CSN 422660	422660		0.4-0.5	0.3 max	0.4-0.8			0.05 max	0.05 max	0.2-0.5	P+S<=0.09; bal Fe				
CSN 422670	422670		0.5-0.6	0.3 max	0.4-0.8			0.05 max	0.05 max	0.2-0.5	P+S<=0.09; bal Fe				

Germany

Specification	Designation	Notes	C	Cr	Mn	Mo	Ni	P	S	Si	Other	UTS	YS	El	Hard
DIN	GE300										bal Fe	600	300	15	
DIN 1681(85)	WNr 1.0558										bal Fe	600	300	15	
DIN SEW 835(95)	GS-46Mn4		0.42-0.50		0.90-1.20			0.035 max	0.035 max	0.25-0.50	bal Fe				
DIN SEW 835(95)	WNr 1.1159	F/IH	0.42-0.50		0.90-1.20			0.035 max	0.035 max	0.25-0.50	bal Fe				

Japan

Specification	Designation	Notes	C	Cr	Mn	Mo	Ni	P	S	Si	Other	UTS	YS	El	Hard
JIS G5111(91)	SCC5	Pipe; Q/A Tmp	0.40-0.50		0.50-0.80			0.040 max	0.040 max	0.30-0.60	bal Fe	690	440	9	201 HB
JIS G5111(91)	SCC5	Pipe; N/T	0.40-0.50		0.50-0.80			0.040 max	0.040 max	0.30-0.60	bal Fe	620	-295	9	163 HB
JIS G5111(91)	SCMn5	Pipe; Q/A Tmp	0.40-0.50		1.00-1.60			0.040 max	0.040 max	0.30-0.60	bal Fe	740	540	9	212 HB

UNS numbers and US grades are provided as a means of cross referencing chemically similar alloys. Exchangability is only possible after independent examination of specifications. Tensile properties are minimum or typical as specified. UTS and YS as MPa. El as %. See Appendix for list of abbreviations used in Notes. * indicates obsolete material.

Specification	Designation	Notes	C	Cr	Mn	Mo	Ni	P	S	Si	Other	UTS	YS	El	Hard

Steel Casting, Medium-Carbon, 0050A (Continued from previous page)

Romania

Specification	Designation	Notes	C	Cr	Mn	Mo	Ni	P	S	Si	Other	UTS	YS	El	Hard
STAS 600	OT600-1		0.4-0.5	0.3	0.4-0.8		0.3	0.06	0.06	0.25-0.5	Cu 0.3; bal Fe				
STAS 600	OT700-2		0.5-0.6	0.3	0.4-0.8		0.3	0.04	0.04	0.25-0.5	Cu 0.3; bal Fe				
STAS 600(82)	OT600-1	Unalloyed	0.4-0.5	0.3 max	0.4-0.8			0.06 max	0.06 max	0.25-0.5	Pb <=0.15; bal Fe				
STAS 600(82)	OT600-2	Unalloyed	0.4-0.5	0.3 max	0.4-0.8			0.04 max	0.04 max	0.25-0.5	Pb <=0.15; bal Fe				
STAS 600(82)	OT600-3	Unalloyed	0.4-0.5	0.3 max	0.4-0.8			0.04 max	0.04 max	0.25-0.5	Pb <=0.15; bal Fe				

Russia

Specification	Designation	Notes	C	Cr	Mn	Mo	Ni	P	S	Si	Other	UTS	YS	El	Hard
GOST 977(88)	45FL		0.42-0.5	0.3 max	0.4-0.9		0.4 max	0.04 max	0.04 max	0.2-0.52	bal Fe				
GOST 977(88)	45FL-2		0.42-0.5	0.3 max	0.4-0.9		0.4 max	0.035 max	0.035 max	0.2-0.52	bal Fe				
GOST 977(88)	45FL-3		0.42-0.5	0.3 max	0.4-0.9		0.4 max	0.03 max	0.03 max	0.2-0.52	bal Fe				
GOST 977(88)	45L		0.42-0.5	0.3 max	0.45-0.9		0.4 max	0.04 max	0.04 max	0.2-0.52	bal Fe				
GOST 977(88)	45L-2		0.42-0.5	0.3 max	0.45-0.9		0.4 max	0.035 max	0.035 max	0.2-0.52	bal Fe				

Spain

Specification	Designation	Notes	C	Cr	Mn	Mo	Ni	P	S	Si	Other	UTS	YS	El	Hard
UNE 36254(79)	46Mn4	Q/T	0.42-0.5		0.9-1.2			0.04	0.04	0.3-0.5	bal Fe				
UNE 36254(79)	F.8213*	see 46Mn4	0.42-0.50		0.9-1.2			0.04	0.04	0.3-0.5	bal Fe				

UK

Specification	Designation	Notes	C	Cr	Mn	Mo	Ni	P	S	Si	Other	UTS	YS	El	Hard
BS 3146/1(74)	CLA1(C)	Inv casting	0.35-0.45	0.30 max	0.40-1.00	0.10 max	0.40 max	0.035 max	0.035 max	0.20-0.60	Cu <=0.30; Cr+Cu+Mo+Ni<=0.8; bal Fe				

USA

Specification	Designation	Notes	C	Cr	Mn	Mo	Ni	P	S	Si	Other	UTS	YS	El	Hard
	UNS J04501		0.40-0.50		0.50-0.90			0.040 max	0.045 max	0.80 max	bal Fe				
SAE J435(74)	0050A	Auto	0.40-0.50		0.50-0.90			0.040 max	0.045 max	0.80 max	bal Fe	586.0	310.3	16	170-229 HB
SAE J435(74)	0050B	Auto	0.40-0.50		0.50-0.90			0.040 max	0.045 max	0.80 max	bal Fe	689.5	482.6	10	207-255 HB

Steel Casting, Medium-Carbon, 1

Canada

Specification	Designation	Notes	C	Cr	Mn	Mo	Ni	P	S	Si	Other	UTS	YS	El	Hard
CSA G28	70-36	Full Ann-Norm, N/T, Q/A	0.35		0.75			0.04	0.04	0.8	bal Fe				

USA

Specification	Designation	Notes	C	Cr	Mn	Mo	Ni	P	S	Si	Other	UTS	YS	El	Hard
	UNS J03502		0.35 max		0.70 max			0.035 max	0.030 max	0.60 max	bal Fe				
ASTM A356/A356M(98)	1	Heavy wall; steam turbines	0.35 max		0.70 max			0.035 max	0.030 max	0.60 max	bal Fe	485	250	20.0	

Steel Casting, Medium-Carbon, 1020

Bulgaria

Specification	Designation	Notes	C	Cr	Mn	Mo	Ni	P	S	Si	Other	UTS	YS	El	Hard
BDS 7478	20L	High-temp const	0.17-0.25	0.30 max	0.80 max		0.40	0.035	0.035	0.30-0.50	Cu <=0.30; bal Fe				

Germany

Specification	Designation	Notes	C	Cr	Mn	Mo	Ni	P	S	Si	Other	UTS	YS	El	Hard
DIN EN 10213(96)	WNr 1.0619		0.18-0.23		0.50-1.20			0.030 max	0.020 max	0.60 max	bal Fe				

Hungary

Specification	Designation	Notes	C	Cr	Mn	Mo	Ni	P	S	Si	Other	UTS	YS	El	Hard
MSZ 1749(89)	Ao21C	High-temp,const; 0-30.00mm; HT	0.17-0.25		0.5-0.8			0.03 max	0.03 max	0.3-0.6	bal Fe	450-600	250	22L	

Japan

Specification	Designation	Notes	C	Cr	Mn	Mo	Ni	P	S	Si	Other	UTS	YS	El	Hard
JIS G5151(91)	SCPH1	High-temp; High-press	0.25 max	0.25 max	0.70 max	0.25 max	0.50 max	0.040 max	0.040 max	0.60 max	Cu <=0.50; Cr+Cu+Mo+Ni<=1.00; bal Fe	410	205	21	

Romania

Specification	Designation	Notes	C	Cr	Mn	Mo	Ni	P	S	Si	Other	UTS	YS	El	Hard
STAS 9277(84)	OTA20		0.17-0.23	0.3 max	0.5-0.8			0.05 max	0.05 max	0.3-0.5	Pb <=0.15; bal Fe				

UNS numbers and US grades are provided as a means of cross referencing chemically similar alloys. Exchangability is only possible after independent examination of specifications. Tensile properties are minimum or typical as specified. UTS and YS as MPa. El as %. See Appendix for list of abbreviations used in Notes. * indicates obsolete material.

Specification	Designation	Notes	C	Cr	Mn	Mo	Ni	P	S	Si	Other	UTS	YS	El	Hard
Steel Casting, Medium-Carbon, 1020 (Continued from previous page)															
Russia															
GOST 977(88)	20L		0.17-0.25	0.3 max	0.45-0.9		0.4 max	0.04 max	0.04 max	0.2-0.52	bal Fe				
GOST 977(88)	20L-2		0.17-0.25	0.3 max	0.45-0.9		0.4 max	0.035 max	0.035 max	0.2-0.52	bal Fe				
Spain															
UNE 36259(79)	AM-X18CrMo5*		0.2 max	4.0-6.5	0.4-0.7	0.45-0.65		0.040 max	0.040 max	0.75 max	bal Fe				
UNE 36259(79)	AM18Mo5*		0.25 max		0.5-0.8	0.45-0.65		0.040 max	0.040 max	0.6 max	bal Fe				
UNE 36259(79)	F.8381*	see AM18Mo5*	0.25 max		0.5-0.8	0.45-0.65		0.040 max	0.040 max	0.6 max	bal Fe				
UNE 36259(79)	F.8385*	see AM-X18CrMo5*	0.2 max	4.0-6.5	0.4-0.7	0.45-0.65		0.040 max	0.040 max	0.75 max	bal Fe				
USA															
	UNS J02000		0.15-0.25		0.30-0.60			0.04	0.04	0.20-1.00	bal Fe				
	UNS J02001		0.15-0.25		0.30-0.60			0.025 max	0.025 max	0.20-1.00	bal Fe				
	UNS J02003		0.18-0.23		0.40-0.80			0.040 max	0.040 max	0.30-0.60	bal Fe				
ASTM A915/A915M(98)	SC1020		0.18-0.23		0.40-0.80			0.040 max	0.040 max	0.30-0.60	bal Fe				
ASTM A958(96)	SC1020		0.18-0.23		0.40-0.80			0.040 max	0.040 max	0.30-0.60	bal Fe	450-485	240-250	22-24	
MIL-S-22141B(84)	IC-1020*	Inv Ann; Obs for new design	0.15-0.25		0.30-0.60			0.04 max	0.04 max	0.20-1.00	bal Fe	414	276	35	
MIL-S-22141B(84)	IC-1020*	Inv cast; Obs for new design	0.15-0.25		0.30-0.60			0.04 max	0.04 max	0.20-1.00	bal Fe				80 HRB
MIL-S-81591(93)	IC-1020*	Inv, C/Corr res; Obs for new design see AMS specs	0.15-0.25		0.3-0.6			0.01	0.01	0.2-1	bal Fe				
Yugoslavia															
	CL.1330		0.18-0.23		0.5-0.8	0.15 max		0.03 max	0.03 max	0.3-0.5	bal Fe				
Steel Casting, Medium-Carbon, 1025															
USA															
	UNS J02508		0.22-0.28		0.40-0.80			0.040 max	0.040 max	0.30-0.60	bal Fe				
ASTM A915/A915M(98)	SC1025		0.22-0.28		0.40-0.80			0.040 max	0.040 max	0.30-0.60	bal Fe				
ASTM A958(96)	SC1025		0.22-0.28		0.40-0.80			0.040 max	0.040 max	0.30-0.60	bal Fe	450-485	240-250	22-24	
Steel Casting, Medium-Carbon, 1030															
China															
JB/T 6402(92)	ZG30Mn	N/T	0.27-0.34		1.20-1.50			0.035 max	0.035 max	0.30-0.50	bal Fe	560	300	18	
Hungary															
MSZ 8272(89)	Ao30MnSi5	0-30.00mm; norm	0.25-0.35		1.1-1.4		0.3 max	0.04 max	0.04 max	0.6-0.8	bal Fe	590	340	14L	
MSZ 8272(89)	Ao30MnSi5ne	0-30.00mm; Q/T	0.25-0.35		1.1-1.4		0.3 max	0.04 max	0.04 max	0.6-0.8	bal Fe	640	390	14L	
Japan															
JIS G5111(91)	SCMnCr2	N/T	0.25-0.35	0.40-0.80	1.20-1.60			0.040 max	0.040 max	0.30-0.60	bal Fe	590	370	13	170 HB
JIS G5111(91)	SCMnCr2	Q/A Tmp	0.25-0.35	0.40-0.80	1.20-1.60			0.040 max	0.040 max	0.30-0.60	bal Fe	640	440	17	183 HB
JIS G5111(91)	SCSiMn2	Q/A Tmp	0.25-0.35		0.90-1.20			0.040 max	0.040 max	0.50-0.80	bal Fe	640	440	17	183 HB
JIS G5111(91)	SCSiMn2	N/T	0.25-0.35		0.90-1.20			0.040 max	0.040 max	0.50-0.80	bal Fe	590	295	13	163 HB
Romania															
STAS 1773(82)	T30SiMn12		0.25-0.35	0.3 max	1.1-1.4			0.04 max	0.04 max	0.5-0.8	Pb <=0.15; bal Fe				

Specification	Designation	Notes	C	Cr	Mn	Mo	Ni	P	S	Si	Other	UTS	YS	El	Hard

Steel Casting, Medium-Carbon, 1030 (Continued from previous page)

Romania

Specification	Designation	Notes	C	Cr	Mn	Mo	Ni	P	S	Si	Other	UTS	YS	El	Hard
STAS 600(82)	OT500-1	Unalloyed	0.3-0.4	0.3 max	0.4-0.8			0.06 max	0.06 max	0.25-0.5	Pb <=0.15; bal Fe				
STAS 600(82)	OT500-2	Unalloyed	0.3-0.4	0.3 max	0.4-0.8			0.04 max	0.04 max	0.25-0.5	Pb <=0.15; bal Fe				
STAS 600(82)	OT500-3	Unalloyed	0.3-0.4	0.3 max	0.4-0.8			0.04 max	0.04 max	0.25-0.5	Pb <=0.15; bal Fe				

Russia

Specification	Designation	Notes	C	Cr	Mn	Mo	Ni	P	S	Si	Other	UTS	YS	El	Hard
GOST 977(88)	30L		0.27-0.35	0.3	0.4-0.9		0.3	0.05	0.05	0.2-0.52	bal Fe				
GOST 977(88)	30L		0.27-0.35	0.3 max	0.45-0.9		0.4 max	0.04 max	0.04 max	0.2-0.52	bal Fe				
GOST 977(88)	30L-2		0.27-0.35	0.3 max	0.45-0.9		0.4 max	0.035 max	0.035 max	0.2-0.52	bal Fe				
GOST 977(88)	30L-3		0.27-0.35	0.3 max	0.45-0.9		0.4 max	0.03 max	0.03 max	0.2-0.52	bal Fe				

UK

Specification	Designation	Notes	C	Cr	Mn	Mo	Ni	P	S	Si	Other	UTS	YS	El	Hard
BS 3100(91)	A5	0-999mm	0.25-0.33		1.20-1.60			0.050 max	0.050 max	0.60 max	bal Fe	620-770	370	13	179-229 HB
BS 3100(91)	A6	0-63mm	0.25-0.33		1.20-1.60			0.050 max	0.050 max	0.60 max	bal Fe	690-850	495	13	201-255 HB
BS 3146/1(74)	CLA1(B)	Inv casting	0.25-0.35	0.30 max	0.40-1.00	0.10 max	0.40 max	0.035 max	0.035 max	0.20-0.60	Cu <=0.30; Cr+Cu+Mo+Ni<=0.8; bal Fe				

USA

Specification	Designation	Notes	C	Cr	Mn	Mo	Ni	P	S	Si	Other	UTS	YS	El	Hard
	UNS J03005		0.25-0.35		0.70-1.00			0.025 max	0.025 max	0.20-1.00	bal Fe				
	UNS J03006		0.25-0.35		0.70-1.00			0.04 max	0.04 max	0.20-1.00	bal Fe				
	UNS J03012		0.28-0.34		0.50-0.90			0.040 max	0.040 max	0.30-0.60	bal Fe				
ASTM A915/A915M(98)	SC1030		0.28-0.34		0.50-0.90			0.040 max	0.040 max	0.30-0.60	bal Fe				
ASTM A958(96)	SC1030		0.28-0.34		0.50-0.90			0.040 max	0.040 max	0.30-0.60	bal Fe	450-550	240-345	22-24	
MIL-S-22141B(84)	IC-1030*	Inv cast; Obs for new design	0.25-0.35		0.70-1.00			0.04 max	0.04 max	0.20-1.00	bal Fe				85 HRB
MIL-S-22141B(84)	IC-1030*	Inv HT; Obs for new design	0.25-0.35		0.70-1.00			0.04 max	0.04 max	0.20-1.00	bal Fe	586	414	10	
MIL-S-22141B(84)	IC-1030*	Inv Ann; Obs for new design	0.25-0.35		0.70-1.00			0.04 max	0.04 max	0.20-1.00	bal Fe	448	310	25	75 HRB
MIL-S-81591(93)	IC-1030*	Inv, C/Corr res; Obs for new design see AMS specs									bal Fe				

Yugoslavia

Specification	Designation	Notes	C	Cr	Mn	Mo	Ni	P	S	Si	Other	UTS	YS	El	Hard
	CL.3130		0.25-0.35		1.1-1.4	0.15 max		0.04 max	0.04 max	0.6-0.8	bal Fe				

Steel Casting, Medium-Carbon, 1040

Finland

Specification	Designation	Notes	C	Cr	Mn	Mo	Ni	P	S	Si	Other	UTS	YS	El	Hard
SFS 358(79)	G-30-57	Struct, Const	0.4 max		0.7 max			0.04 max	0.04 max	0.5 max	bal Fe				
SFS 366(77)	G-41Mn5		0.38-0.45		1.1-1.4			0.035 max	0.035 max	0.6 max	bal Fe				

Italy

Specification	Designation	Notes	C	Cr	Mn	Mo	Ni	P	S	Si	Other	UTS	YS	El	Hard
UNI 3160(83)	GC37	Wear res	0.34-0.4		0.9-1.2			0.035 max	0.035 max	0.5 max	bal Fe				

Japan

Specification	Designation	Notes	C	Cr	Mn	Mo	Ni	P	S	Si	Other	UTS	YS	El	Hard
JIS G5111(91)	SCC3	Q/A Tmp	0.30-0.40		0.50-0.80			0.040 max	0.040 max	0.30-0.60	bal Fe	620	370	13	183 HB

Romania

Specification	Designation	Notes	C	Cr	Mn	Mo	Ni	P	S	Si	Other	UTS	YS	El	Hard
STAS 1773(82)	T40Mn11		0.35-0.45	0.3 max	1-1.3			0.04 max	0.04 max	0.2-0.45	Pb <=0.15; bal Fe				
STAS 600(82)	OT550-1	Unalloyed	0.35-0.45	0.3 max	0.4-0.8			0.06 max	0.06 max	0.25-0.5	Pb <=0.15; bal Fe				
STAS 600(82)	OT550-2	Unalloyed	0.35-0.45	0.3 max	0.4-0.8			0.04 max	0.04 max	0.25-0.5	Pb <=0.15; bal Fe				

UNS numbers and US grades are provided as a means of cross referencing chemically similar alloys. Exchangability is only possible after independent examination of specifications. Tensile properties are minimum or typical as specified. UTS and YS as MPa. El as %. See Appendix for list of abbreviations used in Notes. * indicates obsolete material.

Specification	Designation	Notes	C	Cr	Mn	Mo	Ni	P	S	Si	Other	UTS	YS	El	Hard
Steel Casting, Medium-Carbon, 1040 (Continued from previous page)															
Romania															
STAS 600(82)	OT550-3	Unalloyed	0.35-0.45	0.3 max	0.4-0.8			0.04 max	0.04 max	0.25-0.5	Pb <=0.15; bal Fe				
Russia															
GOST 977(88)	35L		0.32-0.4	0.3 max	0.45-0.9		0.4 max	0.04 max	0.04 max	0.2-0.52	bal Fe				
GOST 977(88)	35L-2		0.32-0.4	0.3 max	0.45-0.9		0.4 max	0.035 max	0.035 max	0.2-0.52	bal Fe				
GOST 977(88)	35L-3		0.32-0.4	0.3 max	0.45-0.9		0.4 max	0.03 max	0.03 max	0.2-0.52	bal Fe				
GOST 977(88)	40L		0.37-0.45	0.3 max	0.45-0.9		0.4 max	0.04 max	0.04 max	0.2-0.52	bal Fe				
GOST 977(88)	40L-2		0.37-0.45	0.3 max	0.45-0.9		0.4 max	0.035 max	0.035 max	0.2-0.52	bal Fe				
GOST 977(88)	40L-3		0.37-0.45	0.3 max	0.45-0.9		0.4 max	0.03 max	0.03 max	0.2-0.52	bal Fe				
UK															
BS 3100(91)	A3	0-999mm	0.45 max		1.00 max			0.050 max	0.050 max	0.60 max	bal Fe	540	295	14	
BS 3146/1(74)	CLA1(B)	Inv casting	0.35-0.45	0.30 max	0.40-1.00	0.10 max	0.40 max	0.035 max	0.035 max	0.20-0.60	Cu <=0.30; Cr+Cu+Mo+Ni<=0.8; bal Fe				
BS 3146/1(74)	CLA1(C)	Inv casting	0.35-0.45	0.30 max	0.40-1.00	0.10 max	0.40 max	0.035 max	0.035 max	0.20-0.60	Cu <=0.30; Cr+Cu+Mo+Ni<=0.8; bal Fe				
BS 3146/1(74)	CLA8	Inv casting	0.35-0.45	0.30 max	0.50-0.80	0.10 max	0.40 max	0.035 max	0.035 max	0.20-0.60	Cu <=0.30; Cr+Cu+Mo+Ni<=0.8; bal Fe				
USA															
	UNS J04000		0.35-0.45		0.70-1.00			0.04 max	0.04 max	0.20-1.00	bal Fe				
	UNS J04001		0.35-0.45		0.70-1.00			0.025 max	0.025 max	0.20-1.00	bal Fe				
	UNS J04003		0.37-0.44		0.50-0.90			0.040 max	0.040 max	0.30-0.60	bal Fe				
ASTM A915/A915M(98)	SC1040		0.37-0.44		0.50-0.90			0.040 max	0.040 max	0.30-0.60	bal Fe				
ASTM A958(96)	SC1040		0.37-0.44		0.50-0.90			0.040 max	0.040 max	0.30-0.60	bal Fe	450-620	240-415	18-24	
MIL-S-22141B(84)	IC-1040*	Inv Ann; Obs for new design	0.35-0.45		0.70-1.00			0.04 max	0.04 max	0.20-1.00	bal Fe	517	331	25	85 HRB
MIL-S-22141B(84)	IC-1040*	Inv cast; Obs for new design	0.35-0.45		0.70-1.00			0.04 max	0.04 max	0.20-1.00	bal Fe				95 HRB
MIL-S-22141B(84)	IC-1040*	Inv HT; Obs for new design	0.35-0.45		0.70-1.00			0.04 max	0.04 max	0.20-1.00	bal Fe	690	621	10	
MIL-S-22146	CS1040*	Obs; see MIL-S-22121	0.37-0.44		0.6-0.9			0.04 max	0.05 max		bal Fe				
MIL-S-81591(93)	IC-1040*	Inv, C/Corr res; Obs for new design see AMS specs	0.35-0.45		0.7-1			0.04	0.04	0.2-1	bal Fe				
Yugoslavia															
	CL.3131		0.3-0.4		1.2-1.6	0.15 max		0.04 max	0.04 max	0.17-0.37	bal Fe				
Steel Casting, Medium-Carbon, 1A															
USA															
	UNS J02002		0.15-0.25		0.20-0.60			0.04 max	0.045 max	0.20-1.00	Cu <=0.50; Mo+W<=0.25 max				
ASTM A732/A732M(98)	1A	Ann	0.15-0.25		0.20-0.60			0.04 max	0.045 max	0.20-1.00	Cu+Ni+Cr+Mo+W<=1.00; bal Fe	414	276	24	
Steel Casting, Medium-Carbon, 2A, 2Q															
Japan															
JIS G5111(91)	SCMnM3	Q/A Tmp	0.30-0.40	0.20 max	1.20-1.60	0.15-0.35		0.040 max	0.040 max	0.30-0.60	bal Fe	740	490	13	212 HB
USA															
	UNS J03011		0.25-0.35	0.35 max	0.70-1.00		0.50 max	0.04 max	0.045 max	0.20-1.00	Cu <=0.50; W <=0.10; bal Fe				

UNS numbers and US grades are provided as a means of cross referencing chemically similar alloys. Exchangability is only possible after independent examination of specifications. Tensile properties are minimum or typical as specified. UTS and YS as MPa. El as %. See Appendix for list of abbreviations used in Notes. * indicates obsolete material.

Specification	Designation	Notes	C	Cr	Mn	Mo	Ni	P	S	Si	Other	UTS	YS	El	Hard
Steel Casting, Medium-Carbon, 2A, 2Q (Continued from previous page)															
USA															
ASTM A732/A732M(98)	2A/2Q	Ann, Q/A	0.25-0.35		0.70-1.00			0.04 max	0.045 max	0.20-1.00	Cu+Ni+Cr+W<=1.00; bal Fe	448-586	310-414	10-25	
Steel Casting, Medium-Carbon, 3A															
Japan															
JIS G5111(91)	SCC3	N/T	0.30-0.40		0.50-0.80			0.040 max	0.040 max	0.30-0.60	bal Fe	520	265	13	143 HB
UK															
BS 3146/1(74)	CLA1(C)	Inv casting	0.35-0.45	0.30 max	0.40-1.00	0.10 max	0.40 max	0.035 max	0.035 max	0.20-0.60	Cu <=0.30; Cr+Cu+Mo+Ni<=0.8; bal Fe				
BS 3146/1(74)	CLA8	Inv casting	0.37-0.45	0.30 max	0.50-0.80	0.10 max	0.40 max	0.035 max	0.035 max	0.20-0.60	Cu 0.3; Cr+Cu+Mo+Ni<=0.8; bal Fe				
USA															
	UNS J04002		0.35-0.45	0.30 max	0.70-1.00		0.50 max	0.04 max	0.045 max	0.20-1.00	Cu <=0.50; W <=0.10; bal Fe				
ASTM A732/A732M(98)	3A/3Q	Ann, Q/A	0.35-0.45		0.70-1.00			0.04 max	0.045 max	0.20-1.00	Cu+Ni+Cr+W<=1.00; bal Fe	517-689	331-621	10-25	
Steel Casting, Medium-Carbon, 60-30															
Belgium															
NBN 253-02	C25-1		0.22-0.29		0.4-0.7			0.045	0.045	0.15-0.40	bal Fe				
China															
GB 11352(89)	ZG230-450(ZG25)	Ann or N/T	0.30 max	0.35 max	0.90 max	0.20 max	0.30 max	0.040 max	0.040 max	0.50 max	Cu <=0.30; V <=0.05; bal Fe	450	230	22	
Japan															
JIS G5101(91)	SC410		0.30					0.040	0.040		bal Fe				
USA															
	UNS J03000		0.30 max		0.60 max			0.050 max	0.060 max	0.80 max	bal Fe				
ASTM A27/A27M(95)	60-30	Q/T, norm, ann	0.30 max		0.60 max			0.050 max	0.060 max	0.80 max	bal Fe	415	205	22	
Steel Casting, Medium-Carbon, 65-35															
Belgium															
NBN 253-02	C25-1		0.22-0.29		0.4-0.7			0.045	0.045	0.15-0.4	bal Fe				
Bulgaria															
BDS 3492	25LI		0.20-0.30		0.40-0.80			0.060 max	0.060 max	0.20-0.60	Cu 0.3-0.30; bal Fe				
BDS 3492	25LII		0.20-0.30		0.40-0.80		0.40 max	0.060 max	0.060 max	0.20-0.60	bal Fe				
BDS 3492	25LIII		0.20-0.30		0.40-0.80			0.050 max	0.050 max	0.25-0.50	bal Fe				
China															
GB 11352(89)	ZG230-450(ZG25)	Ann or N/T	0.30 max	0.35 max	0.90 max	0.20 max	0.30 max	0.040 max	0.040 max	0.50 max	Cu <=0.30; V <=0.05; bal Fe	450	230	22	
Hungary															
MSZ 8276(89)	Ao450	0-30.0mm; HT	0.21-0.3		0.4-0.9		0.3 max	0.06 max	0.06 max	0.2-0.52	P<=0.06/P<=0.05; S<=0.06/S<=0.05; bal Fe	450		18L	125-175 HB
MSZ 8276(89)	Ao450F	0-30.0mm; HT	0.21-0.3		0.4-0.9		0.3 max	0.06 max	0.06 max	0.2-0.52	P<=0.06/P<=0.04; S<=0.06/S<=0.045; bal Fe	450	240	22L	125-175 HB
MSZ 8276(89)	Ao450FK	0-30.0mm; Q/T	0.21-0.3		0.4-0.9		0.3 max	0.05 max	0.06 max	0.2-0.52	P<=0.05/P<=0.04; S<=0.06/S<=0.045; bal Fe	500	300	22L	
MSZ 8276(89)	Ao450FK	0-30.0mm; HT	0.21-0.3		0.4-0.9		0.3 max	0.05 max	0.06 max	0.2-0.52	P<=0.05/P<=0.04; S<=0.06/S<=0.045; bal Fe	450	240	22L	125-175 HB
Japan															
JIS G5101(91)	SC450	Gen Struct	0.35 max					0.040 max	0.040 max		bal Fe	450	225	19	

Specification	Designation	Notes	C	Cr	Mn	Mo	Ni	P	S	Si	Other	UTS	YS	El	Hard
Steel Casting, Medium-Carbon, 65-35 (Continued from previous page)															
Japan															
JIS G5151(91)	SCPH2	High-temp; High-press	0.30 max	0.25 max	1.00 max	0.25 max	0.50 max	0.040 max	0.040 max	0.60 max	Cu <=0.50; Cr+Cu+Mo+Ni<=1.00; bal Fe	480	245	19	
Romania															
STAS 600(82)	OT450-1	Unalloyed	0.2-0.3	0.3 max	0.4-0.8			0.06 max	0.06 max	0.25-0.5	Pb <=0.15; bal Fe				
STAS 600(82)	OT450-2	Unalloyed	0.2-0.3	0.3 max	0.4-0.8			0.04 max	0.04 max	0.25-0.5	Pb <=0.15; bal Fe				
STAS 600(82)	OT450-3	Unalloyed	0.2-0.3	0.3 max	0.4-0.8			0.06 max	0.04 max	0.25-0.5	Pb <=0.15; bal Fe				
UK															
BS 1504(76)	480*	Austenitic	0.3	0.25	0.9	0.15	0.4	0.05	0.05	0.6	Cu 0.3; bal Fe				
USA															
	UNS J03001		0.30 max		0.70 max			0.05 max	0.06 max	0.80 max	bal Fe				
	UNS J03009		0.30 max	0.35 max	0.70 max	0.20 max	0.50 max	0.05 max	0.06 max	0.80 max	Cu <=0.30; Cu+Mo+Ni<=1.00; bal Fe				
ASTM A27/A27M(95)	65-35	Q/T, norm, ann	0.30 max		0.70 max			0.05 max	0.06 max	0.8 max	bal Fe	450	240	22	
MIL-S-15083	65-35*	Obs; see MIL-C-24707/1, ASTM A757 A216									bal Fe				
Steel Casting, Medium-Carbon, 70															
UK															
BS 3100(91)	A2	0-999mm	0.35 max		1.00 max			0.050 max	0.050 max	0.60 max	bal Fe	490	260	18	
USA															
	UNS J03503		0.35 max		0.90 max			0.05 max	0.06 max	0.80 max	bal Fe				
Steel Casting, Medium-Carbon, 70-36															
Bulgaria															
BDS 3492	35LI		0.30-0.40		0.40-0.80			0.060 max	0.060 max	0.20-0.60	Cu 0.3-0.30; bal Fe				
BDS 3492	35LII		0.30-0.40		0.40-0.80			0.060 max	0.060 max	0.20-0.60	Cu 0.3-0.30; bal Fe				
BDS 3492	35LIII		0.30-0.40		0.40-0.80			0.050 max	0.050 max	0.25-0.40	bal Fe				
China															
GB 11352(89)	ZG270-500(ZG35)	Ann or N/T	0.40 max	0.35 max	0.90 max	0.20 max	0.30 max	0.040 max	0.040 max	0.50 max	Cu <=0.30; V <=0.05; bal Fe	500	270	18	
Japan															
JIS G5101(91)	SC450		0.35					0.04	0.04		bal Fe				
JIS G5101(91)	SC480	Weldable	0.40 max					0.040 max	0.040 max		bal Fe	480	245	17	
USA															
	UNS J03501		0.35 max		0.70 max			0.05 max	0.06 max	0.80 max	bal Fe				
ASTM A27/A27M(95)	70-36	Q/T, norm, ann	0.35 max		0.70 max			0.05 max	0.06 max	0.8 max	bal Fe	485	250	24	
MIL-S-15083	70-36*	Obs; see MIL-C-24707/1, ASTM A757 A216	0.35					0.05	0.05		bal Fe				
Steel Casting, Medium-Carbon, 70-40															
China															
GB 7659(87)	ZG230-450H		0.20 max	0.30 max	1.20 max	0.15 max	0.30 max	0.040 max	0.040 max	0.50 max	Cu <=0.30; V <=0.05; Cr+Cu+Mo+Ni+V<=0.8; bal Fe	450	230	22	
Germany															
DIN 17243(90)	GS-20Mn5		0.17-0.23		1.00-1.30			0.035 max	0.035 max	0.30-0.60	bal Fe				
DIN EN 10213(96)	WNr 1.0625		0.18-0.25		0.80-1.20			0.030 max	0.020 max	0.60 max	bal Fe				

UNS numbers and US grades are provided as a means of cross referencing chemically similar alloys. Exchangability is only possible after independent examination of specifications. Tensile properties are minimum or typical as specified. UTS and YS as MPa. El as %. See Appendix for list of abbreviations used in Notes. * indicates obsolete material.

Specification	Designation	Notes	C	Cr	Mn	Mo	Ni	P	S	Si	Other	UTS	YS	El	Hard

Steel Casting, Medium-Carbon, 70-40 (Continued from previous page)

Germany

| DIN EN 10213(96) | WNr 1.1138 | Low-temp; Press ves | 0.17-0.23 | | 1.10-1.60 | | | 0.020 max | 0.020 max | 0.60 max | bal Fe | | | | |

UK

BS 3100(91)	A4	0-999mm	0.18-0.25		1.20-1.60			0.050 max	0.050 max	0.60 max	bal Fe	540-690	320	16	152-207 HB
BS 3100(91)	AL2	0-999mm	0.25 max	0.30 max	1.20 max	0.15 max	0.40 max	0.040 max	0.040 max	0.60 max	Cu <=0.30; Cr+Mo+Ni+Cu<=0.8; bal Fe	485-655	275	22	
BS 3100(91)	AL3	0-999mm	0.25 max	0.30 max	1.20 max	0.15 max	0.40 max	0.040 max	0.040 max	0.60 max	Cu <=0.30; Cr+Mo+Ni+Cu<=0.8; bal Fe	485-655	275	22	

USA

| | UNS J02501 | | 0.25 max | | 1.20 max | | | 0.05 max | 0.06 max | 0.80 max | bal Fe | | | | |
| ASTM A27/A27M(95) | 70-40 | Q/T, norm, ann | 0.25 max | | 1.20 max | | | 0.05 max | 0.06 max | 0.8 max | bal Fe | 485 | 275 | 24 | |

Steel Casting, Medium-Carbon, 80-40

Belgium

| NBN 253-02 | C45-2 | | 0.42-0.5 | | 0.5-0.8 | | | 0.035 | 0.035 | 0.15-0.4 | bal Fe | | | | |

USA

| | UNS J05002 | | 0.50 max | | 0.90 max | | | 0.05 max | 0.06 max | 0.80 max | bal Fe | | | | |
| MIL-S-15083 | 80-40* | Obs; see MIL-C-24707/1, ASTM A757 A216 | 0.5 | | 0.9 | | | 0.05 | 0.06 | 0.8 | bal Fe | | | | |

Steel Casting, Medium-Carbon, DN

Australia

| AS 2074(82) | C6 | | 0.4-0.5 | 0.25 max | 1 max | 0.15 max | 0.4 max | 0.05 max | 0.05 max | 0.6 max | Cu <=0.3; bal Fe | | | | |

Belgium

| NBN 253-02 | C45-2 | | 0.42-0.5 | | 0.5-0.8 | | | 0.035 | 0.035 | 0.15-0.4 | bal Fe | | | | |

China

| GB 11352(89) | ZG310-570 | Ann or N/T | 0.50 max | 0.35 max | 0.90 max | 0.20 max | 0.30 max | 0.040 max | 0.040 max | 0.60 max | Cu <=0.30; V <=0.05; bal Fe | 570 | 310 | 15 | |

Poland

| PNH83152(80) | L600 | | 0.4-0.5 | 0.4 max | 0.4-0.9 | | 0.4 max | 0.05 max | 0.05 max | 0.2-0.5 | bal Fe | | | | |

UK

| BS 3100(91) | AW2 | 0-999mm | 0.40-0.50 | 0.30 max | 1.00 max | 0.15 max | 0.40 max | 0.050 max | 0.050 max | 0.60 max | Cu <=0.30; Cr+Mo+Ni+Cu<=0.8; bal Fe | 620 | 325 | 12 | |

USA

| | UNS J04500 | | 0.40-0.50 | 0.35 max | 0.50-0.90 | | 0.50 max | 0.04 max | 0.045 max | 0.80 max | Cu <=0.35; V <=0.03; W <=0.10; Mo+W<=0.10; bal Fe | | | | |
| ASTM A487 | DN* | N/T, Q/T | 0.4-0.5 | 0.35 | 0.5-0.9 | | 0.5 | 0.04 | 0.05 | 0.8 | Cu >=0.35; V 0.03; W 0.1; bal Fe | | | | |

Steel Casting, Medium-Carbon, J03007

USA

| | UNS J03007 | | 0.30 max | | 0.70 max | | | 0.07 max | 0.06 max | | bal Fe | | | | |

Steel Casting, Medium-Carbon, J03008

USA

| | UNS J03008 | | 0.30 max | 0.20 max | 0.60 max | 0.20 max | 0.50 max | 0.05 max | 0.05 max | 0.20-0.60 | Cu <=0.30; bal Fe | | | | |

Steel Casting, Medium-Carbon, J03504

USA

| | UNS J03504 | | 0.35 max | | | | | 0.05 max | 0.05 max | | bal Fe | | | | |

UNS numbers and US grades are provided as a means of cross referencing chemically similar alloys. Exchangability is only possible after independent examination of specifications. Tensile properties are minimum or typical as specified. UTS and YS as MPa. El as %. See Appendix for list of abbreviations used in Notes. * indicates obsolete material.

Specification	Designation	Notes	C	Cr	Mn	Mo	Ni	P	S	Si	Other	UTS	YS	El	Hard

Steel Casting, Medium-Carbon, J05003

USA

Specification	Designation	Notes	C	Cr	Mn	Mo	Ni	P	S	Si	Other	UTS	YS	El	Hard
	UNS J05003		0.30 max		0.70-1.00			0.04 max	0.045 max	0.80 max	Cu <=0.50; W <=0.10; bal Fe				

Steel Casting, Medium-Carbon, LCA

Czech Republic

Specification	Designation	Notes	C	Cr	Mn	Mo	Ni	P	S	Si	Other	UTS	YS	El	Hard
CSN 422630	422630		0.1-0.2	0.3 max	0.4-0.8			0.05 max	0.05 max	0.2-0.5	P+S<=0.09; bal Fe				
CSN 422633	422633		0.10-0.18	0.3 max	0.5-0.9			0.04 max	0.04 max	0.2-0.5	P+S<=0.07; Ni+Cr+Cu<=0.9; bal Fe				

France

Specification	Designation	Notes	C	Cr	Mn	Mo	Ni	P	S	Si	Other	UTS	YS	El	Hard
AFNOR NFA32055(85)	A420APM	t<1000mm; Norm	0.23 max	0.4 max	1 max	0.2 max	0.3 max	0.03 max	0.03 max	0.6 max	Cu <=0.3; Pb <=0.15; V <=0.04; W <=0.1; Cr+Mo+Ni+V+Cu<=1; bal Fe	420-530	240	25	130-165 HB
AFNOR NFA32055(85)	A420CPM	t<1000mm; Norm	0.23 max	0.4 max	1 max	0.2 max	0.3 max	0.03 max	0.03 max	0.6 max	Cu <=0.3; Pb <=0.15; V <=0.04; W <=0.1; Cu+6Sn<=0.33; Cr+Mo+Ni+V+Cu<=1; bal Fe	420-530	240	25	130-165 HB

Hungary

Specification	Designation	Notes	C	Cr	Mn	Mo	Ni	P	S	Si	Other	UTS	YS	El	Hard
MSZ 8276(89)	Ao400	0-30.0mm; HT	0.11-0.2		0.4-0.9		0.3 max	0.06 max	0.06 max	0.2-0.52	P<=0.06/P<=0.05; S<=0.05; bal Fe	400		20L	110-160 HB
MSZ 8276(89)	Ao400F	0-30.0mm; HT	0.11-0.2		0.4-0.9			0.06 max	0.06 max	0.2-0.52	P<=0.06/P<=0.05; S<=0.05; bal Fe	400	200	25L	110-160 HB
MSZ 8276(89)	Ao400FK	0-30.0mm; HT	0.11-0.2		0.4-0.9		0.3 max	0.05 max	0.06 max	0.2-0.52	P<=0.06/P<=0.04; S<=0.045; bal Fe	400	200	25L	110-160 HB

International

Specification	Designation	Notes	C	Cr	Mn	Mo	Ni	P	S	Si	Other	UTS	YS	El	Hard
ISO 3755	23-45*	Sand, As Cast Ann, Norm, Q/A, Tmp	0.25					0.04	0.04		bal Fe				

Italy

Specification	Designation	Notes	C	Cr	Mn	Mo	Ni	P	S	Si	Other	UTS	YS	El	Hard
UNI 3608	CG20	As Agreed	0.25		0.8			0.04	0.04	0.5	bal Fe				

Japan

Specification	Designation	Notes	C	Cr	Mn	Mo	Ni	P	S	Si	Other	UTS	YS	El	Hard
JIS G5201(91)	SCW42-CF		0.22					0.04	0.04		Others each 0.4 max; bal Fe				
JIS G5201(91)	SCW49-CF		0.22					0.04	0.04		Others each 0.43 max; bal Fe				
JIS G5201(91)	SCW50-CF		0.2					0.04	0.04		Others each 0.44 max; bal Fe				
JIS G5201(91)	SCW53-CF		0.2					0.04	0.04		Others each 0.44 max; bal Fe				

Mexico

Specification	Designation	Notes	C	Cr	Mn	Mo	Ni	P	S	Si	Other	UTS	YS	El	Hard
DGN B-356	WCA*	Ann-Norm, Norm, Q/A; Obs	0.25		0.7			0.04	0.05	0.6	bal Fe				

Romania

Specification	Designation	Notes	C	Cr	Mn	Mo	Ni	P	S	Si	Other	UTS	YS	El	Hard
STAS 600(82)	OT400-1	Unalloyed	0.1-0.2	0.3 max	0.4-0.8			0.06 max	0.06 max	0.25-0.5	Pb <=0.15; bal Fe				
STAS 600(82)	OT400-2	Unalloyed	0.1-0.2	0.3 max	0.4-0.8			0.04 max	0.04 max	0.25-0.5	Pb <=0.15; bal Fe				
STAS 600(82)	OT400-3	Unalloyed	0.1-0.2	0.3 max	0.4-0.8			0.04 max	0.04 max	0.25-0.5	Pb <=0.15; bal Fe				

Sweden

Specification	Designation	Notes	C	Cr	Mn	Mo	Ni	P	S	Si	Other	UTS	YS	El	Hard
SIS 141305	1305-02	Ann	0.25		0.7			0.04	0.04	0.5	bal Fe				

USA

Specification	Designation	Notes	C	Cr	Mn	Mo	Ni	P	S	Si	Other	UTS	YS	El	Hard
	UNS J02504		0.25 max		0.70 max			0.04 *max	0.045 max	0.60 max	bal Fe				
ASTM A352/A352M(98)	LCA	Press; low-temp	0.25 max	0.50 max	0.70 max	0.20 max	0.50 max	0.04 max	0.045 max	0.60 max	Cu <=0.30; V <=0.03; Cr+Cu+Ni+Mo+V<=1.00; bal Fe	415-485	205	24	
ASTM A660(96)	WCA	Centrifugally cast pipe; high-temp	0.25 max		0.70 max			0.035 max	0.035 max	0.60 max	Cr+Cu+Ni+Mo+V = 1 max; bal Fe	414	207	24	

Steel Casting, Medium-Carbon, LCB

Australia

Specification	Designation	Notes	C	Cr	Mn	Mo	Ni	P	S	Si	Other	UTS	YS	El	Hard
AS 2074(82)	C4-1		0.3 max		1 max			0.06 max	0.06 max	0.6 max	bal Fe				

UNS numbers and US grades are provided as a means of cross referencing chemically similar alloys. Exchangability is only possible after independent examination of specifications. Tensile properties are minimum or typical as specified. UTS and YS as MPa. El as %. See Appendix for list of abbreviations used in Notes. * indicates obsolete material.

Specification	Designation	Notes	C	Cr	Mn	Mo	Ni	P	S	Si	Other	UTS	YS	El	Hard
Steel Casting, Medium-Carbon, LCB (Continued from previous page)															
Belgium															
NBN 253-02	C25-1		0.22-0.29		0.4-0.7			0.045	0.045	0.15-0.4	bal Fe				
France															
AFNOR NFA32052(75)	M10M	Magnetic, HT	0.3 max	0.3 max	2 max	0.15 max	0.4 max	0.04 max	0.04 max	0.6 max	Cu <=0.3; W <=0.1; bal Fe	320	160	25	
AFNOR NFA32052(75)	M20M	Magnetic, HT	0.3 max	0.3 max	2 max	0.15 max	0.4 max	0.04 max	0.04 max	0.6 max	Cu <=0.3; W <=0.1; bal Fe	440	220	22	
AFNOR NFA32052(75)	M20SM	Magnetic, HT	0.3 max	0.3 max	2 max	0.15 max	0.4 max	0.04 max	0.04 max	0.6 max	Cu <=0.3; W <=0.1; bal Fe	500	290	18	
AFNOR NFA32052(75)	M30M	Magnetic, HT	0.3 max	0.3 max	2 max	0.15 max	0.4 max	0.04 max	0.04 max	0.6 max	Cu <=0.3; W <=0.1; bal Fe	500	250	18	
AFNOR NFA32052(75)	M40M	Magnetic, HT	0.3 max	0.3 max	2 max	0.15 max	0.4 max	0.04 max	0.04 max	0.6 max	Cu <=0.3; W <=0.1; bal Fe	560	280	15	
AFNOR NFA32055(85)	A420FPM	t<1000mm; Norm	0.23 max	0.4 max	1.2 max	0.2 max	1 max	0.03 max	0.03 max	0.6 max	Cu <=0.3; Pb <=0.15; V <=0.04; W <=0.1; Cr+Mo+V+Cu<=1; bal Fe	420-530	240	25	130-165 HB
Germany															
DIN	GS-24Mn4		0.20-0.28		0.90-1.20			0.035 max	0.035 max	0.30-0.60	bal Fe				
DIN	GS-Ck24		0.20-0.28	0.30 max	0.50-0.80			0.030 max	0.030 max	0.30-0.50	N <=0.007; N depends on P; bal Fe				
DIN	WNr 1.1136		0.20-0.28		0.90-1.20			0.035 max	0.035 max	0.30-0.60	bal Fe				
DIN	WNr 1.1156		0.20-0.28	0.30 max	0.50-0.80			0.030 max	0.030 max	0.30-0.50	N <=0.007; N depends on P; bal Fe				
DIN 17205(92)	GS-30Mn5	Q/T 101-160mm	0.27-0.34	0.30 max	1.20-1.50			0.035 max	0.035 max	0.15-0.40	bal Fe	640-780	440	16	
DIN 17205(92)	GS-30Mn5	Q/T 17-40mm	0.27-0.34	0.30 max	1.20-1.50			0.035 max	0.035 max	0.15-0.40	bal Fe	690-830	440	15	
DIN 17205(92)	WNr 1.1165	Q/T 101-160mm	0.27-0.34	0.30 max	1.20-1.50			0.035 max	0.035 max	0.15-0.40	bal Fe	640-780	440	16	
DIN 17205(92)	WNr 1.1165	Q/T 17-40mm	0.27-0.34	0.30 max	1.20-1.50			0.035 max	0.035 max	0.15-0.40	bal Fe	690-830	440	15	
DIN SEW 835(95)	GS-36Mn5	Q/T 17-40mm	0.32-0.40		1.20-1.50			0.035 max	0.035 max	0.40 max	bal Fe	830-980	590	10	
Japan															
JIS G5101	SC42		0.3					0.04	0.04		bal Fe				
JIS G5152(91)	SCPL1	Low-temp; High-press	0.30 max	0.25 max	1.00 max		0.50 max	0.040 max	0.040 max	0.60 max	Cu <=0.50; Cu+Cr+Ni<=1.00; bal Fe	450	245	21	
Mexico															
DGN B-356	WCB*	Ann-Norm, Norm, Q/A; Obs	0.3		1			0.04	0.05	0.6	bal Fe				
Russia															
GOST 21357(87)	27ChGSNMDTL	Tough at subzero; Wear res	0.22-0.31	0.70-1.30	0.90-1.50	0.10-0.30	0.70-1.20	0.020 max	0.020 max	0.70-1.30	Cu 0.30-0.50; Ti 0.03-0.07; bal Fe				
GOST 977(88)	30GSL		0.25-0.35	0.3 max	1.1-1.4		0.4 max	0.04 max	0.04 max	0.6-0.8	bal Fe				
Spain															
UNE 36255(79)	AM30Mn5*	Q/T	0.25-0.34		1.2-1.6	0.15 max		0.040 max	0.040 max	0.3-0.5	bal Fe				
UNE 36255(79)	F.8311*	see AM30Mn5*	0.25-0.34		1.2-1.6	0.15 max		0.040 max	0.040 max	0.3-0.5	bal Fe				
USA															
	UNS J03003		0.30 max	*	1 max			0.04 max	0.045 max	0.60 max	bal Fe				
ASTM A352/A352M(98)	LCB	Press; low-temp	0.30 max	0.50 max	1.00 max	0.20 max	0.50 max	0.04 max	0.045 max	0.60 max	Cu <=0.30; V <=0.03; Cr+Cu+Ni+Mo+V<=1.00; bal Fe	450-620	240	24	
ASTM A660(96)	WCB	Centrifugally cast pipe; high-temp	0.30 max		1.00 max			0.035 max	0.035 max	0.60 max	Cr+Cu+Mo+Ni+V = 1 max;bal Fe	483	248	22	
Steel Casting, Medium-Carbon, N1															
China															
GB 11352(89)	ZG200-400(ZG15)	Ann or norm	0.20 max	0.35 max	0.80 max	0.20 max	0.30 max	0.040 max	0.040 max	0.50 max	Cu <=0.30; V <=0.05; bal Fe	400	200	25	

UNS numbers and US grades are provided as a means of cross referencing chemically similar alloys. Exchangability is only possible after independent examination of specifications. Tensile properties are minimum or typical as specified. UTS and YS as MPa. El as %. See Appendix for list of abbreviations used in Notes. * indicates obsolete material.

Specification	Designation	Notes	C	Cr	Mn	Mo	Ni	P	S	Si	Other	UTS	YS	El	Hard

Steel Casting, Medium-Carbon, N1 (Continued from previous page)

France

Specification	Designation	Notes	C	Cr	Mn	Mo	Ni	P	S	Si	Other	UTS	YS	El	Hard
AFNOR NFA32051	E23-45M	Ann-Norm, or Q/A Tmp	0.25					0.04	0.04	0.5	bal Fe				

Italy

Specification	Designation	Notes	C	Cr	Mn	Mo	Ni	P	S	Si	Other	UTS	YS	El	Hard
UNI 3608	CG20	As Agreed	0.25		0.8			0.035	0.035	0.6	bal Fe				

Japan

Specification	Designation	Notes	C	Cr	Mn	Mo	Ni	P	S	Si	Other	UTS	YS	El	Hard
JIS G5101(91)	SC360	Gen Struct	0.20 max					0.040 max	0.040 max		bal Fe	360	175	23	
JIS G5101(91)	SC410	Gen Struct	0.30 max					0.040 max	0.040 max		bal Fe	410	205	21	
JIS G5102(91)	SCW42		0.22					0.04	0.04		Others each 0.4 max; bal Fe				
JIS G5102(91)	SCW49		0.22					0.04	0.04		Others each 0.43 max; bal Fe				
JIS G5102(91)	SCW56		0.2					0.04	0.04		Others each 0.44 max; bal Fe				
JIS G5102(91)	SCW63		0.2					0.04	0.04		Others each 0.44 max; bal Fe				

Russia

Specification	Designation	Notes	C	Cr	Mn	Mo	Ni	P	S	Si	Other	UTS	YS	El	Hard
GOST 977(88)	20ChMFL		0.18-0.25	0.9-1.2	0.6-0.9	0.5-0.7	0.4 max	0.025 max	0.025 max	0.2-0.4	V 0.2-0.3; bal Fe				

UK

Specification	Designation	Notes	C	Cr	Mn	Mo	Ni	P	S	Si	Other	UTS	YS	El	Hard
BS 3100(91)	A1	0-999mm	0.25 max	0.30 max	0.90 max	0.15 max	0.40 max	0.050 max	0.050 max	0.60 max	Cu <=0.30; Cr+Mo+Ni+Cu<=0.8; bal Fe	430	230	22	
BS 3146/1(74)	CLA1(A)	Inv casting	0.15-0.25	0.30 max	0.40-0.60	0.10 max	0.40 max	0.035 max	0.035 max	0.20-0.60	Cu <=0.30; Cr+Cu+Mo+Ni<=0.8; bal Fe				

USA

Specification	Designation	Notes	C	Cr	Mn	Mo	Ni	P	S	Si	Other	UTS	YS	El	Hard
	UNS J02500		0.25 max		0.75 max			0.05 max	0.06 max	0.80 max	bal Fe				
ASTM A27/A27M(95)	N1	Q/T, norm, ann	0.25 max		0.75 max			0.05 max	0.06 max	0.8 max	bal Fe				
ASTM A27/A27M(95)	U-60-30	Q/T, norm, ann	0.25 max		0.75 max			0.05 max	0.06 max	0.8 max	bal Fe	415	205	22	

Steel Casting, Medium-Carbon, N2

China

Specification	Designation	Notes	C	Cr	Mn	Mo	Ni	P	S	Si	Other	UTS	YS	El	Hard
GB 11352(89)	ZG270-500(ZG35)	Ann or N/T	0.40 max	0.35 max	0.90 max	0.20 max	0.30 max	0.040 max	0.040 max	0.50 max	Cu <=0.30; V <=0.05; bal Fe	500	270	18	

Finland

Specification	Designation	Notes	C	Cr	Mn	Mo	Ni	P	S	Si	Other	UTS	YS	El	Hard
SFS 356(79)	G-23-45	Struct, Const	0.25 max		0.7 max			0.04 max	0.04 max	0.5 max	bal Fe				
SFS 357(79)	G-26-52	Struct, Const	0.25 max		0.7 max			0.04 max	0.04 max	0.5 max	bal Fe				

Germany

Specification	Designation	Notes	C	Cr	Mn	Mo	Ni	P	S	Si	Other	UTS	YS	El	Hard
DIN	GS-45		0.25 max		0.20-0.50			0.040 max	0.040 max	0.60 max	bal Fe				
DIN	WNr 1.0443		0.25 max		0.20-0.50			0.040 max	0.040 max	0.60 max	bal Fe				

UK

Specification	Designation	Notes	C	Cr	Mn	Mo	Ni	P	S	Si	Other	UTS	YS	El	Hard
BS 1504(76)	161 540*	Press ves	0.35	0.25	1.6	0.15	0.4	0.05	0.05	0.6	Cu 0.3; Cr+Cu+Mo+Ni<=0.8; bal Fe				

USA

Specification	Designation	Notes	C	Cr	Mn	Mo	Ni	P	S	Si	Other	UTS	YS	El	Hard
	UNS J03500		0.35 max		0.60 max			0.05 max	0.06 max	0.80 max	bal Fe				
ASTM A27/A27M(95)	N2	Q/T, norm, ann	0.35 max		0.60 max			0.05 max	0.06 max	0.8 max	bal Fe				

Yugoslavia

Specification	Designation	Notes	C	Cr	Mn	Mo	Ni	P	S	Si	Other	UTS	YS	El	Hard
	CL.0400	Reinforcing	0.30 max		1.60 max	0.15 max		0.070 max	0.060 max	0.60 max	bal Fe				

Steel Casting, Medium-Carbon, No equivalents identified

Czech Republic

Specification	Designation	Notes	C	Cr	Mn	Mo	Ni	P	S	Si	Other	UTS	YS	El	Hard
CSN 422650	422650		0.28-0.38	0.3 max	0.4-0.8			0.05 max	0.05 max	0.2-0.5	P+S<=0.09; bal Fe				

UNS numbers and US grades are provided as a means of cross referencing chemically similar alloys. Exchangability is only possible after independent examination of specifications. Tensile properties are minimum or typical as specified. UTS and YS as MPa. El as %. See Appendix for list of abbreviations used in Notes. * indicates obsolete material.

Specification	Designation	Notes	C	Cr	Mn	Mo	Ni	P	S	Si	Other	UTS	YS	El	Hard

Steel Casting, Medium-Carbon, No equivalents identified

Czech Republic

Specification	Designation	Notes	C	Cr	Mn	Mo	Ni	P	S	Si	Other	UTS	YS	El	Hard
CSN 422711	422711		0.65-0.80	0.3 max	1.1-1.6			0.045 max	0.045 max	0.2-0.5	P+S<=0.08; bal Fe				
CSN 422712	422712		0.17-0.25	0.3 max	0.9-1.4			0.04 max	0.04 max	0.2-0.5	P+S<=0.07; bal Fe				
CSN 422713	422713		0.10-0.18	0.3 max	0.9-1.4			0.04 max	0.04 max	0.2-0.5	P+S<=0.07; bal Fe				
CSN 422715	422715		0.34-0.42	0.3 max	1.2-1.6		0.5 max	0.045 max	0.045 max	0.2-0.5	P+S<=0.08; bal Fe				
CSN 422719	422719		0.52-0.60	0.3 max	0.5-0.8			0.045 max	0.045 max	0.5-.7	Ti 0.02-0.1; P+S<=0.08; bal Fe				

Hungary

Specification	Designation	Notes	C	Cr	Mn	Mo	Ni	P	S	Si	Other	UTS	YS	El	Hard
MSZ 8276(89)	Ao500	0-30.0mm; HT	0.31-0.4		0.4-0.9		0.3 max	0.06 max	0.06 max	0.2-0.52	P<=0.06/P<=0.05; S<=0.06/S<=0.05; bal Fe	500		15L	140-190 HB
MSZ 8276(89)	Ao500F	0-30.0mm; HT	0.31-0.4		0.4-0.9		0.3 max	0.06 max	0.06 max	0.2-0.52	P<=0.06/P<=0.04; S<=0.06/S<=0.045; bal Fe	500	280	20L	140-190 HB
MSZ 8276(89)	Ao500FK	0-30.0mm; Q/T	0.31-0.4		0.4-0.9		0.3 max	0.05 max	0.06 max	0.2-0.52	P<=0.05/P<=0.04; S<=0.045; bal Fe	550	350	16L	
MSZ 8276(89)	Ao500FK	0-30.0mm; HT	0.31-0.4		0.4-0.9		0.3 max	0.05 max	0.06 max	0.2-0.52	P<=0.05/P<=0.04; S<=0.045; bal Fe	500	280	20L	140-190 HB

International

Specification	Designation	Notes	C	Cr	Mn	Mo	Ni	P	S	Si	Other	UTS	YS	El	Hard
ISO 3755(91)	200-400	HT						0.035 max	0.035 max		bal Fe	400-550	200	25	
ISO 3755(91)	200-400W	HT, weld	0.25 max	0.35 max	1.00 max	0.15 max	0.40 max	0.035 max	0.035 max	0.60 max	Cu <=0.40; V <=0.05; bal Fe	400-550	200	25	
ISO 3755(91)	230-450	HT						0.035 max	0.035 max		bal Fe	450-600	230	22	
ISO 3755(91)	230-450W	HT, weld	0.25 max	0.35 max	1.20 max	0.15 max	0.40 max	0.035 max	0.035 max	0.60 max	Cu <=0.40; V <=0.05; bal Fe	450-600	230	22	
ISO 3755(91)	270-480	HT						0.035 max	0.035 max		bal Fe	450-600	270	18	
ISO 3755(91)	270-480W	HT, weld	0.25 max	0.35 max	1.20 max	0.15 max	0.40 max	0.035 max	0.035 max	0.60 max	Cu <=0.40; V <=0.05; bal Fe	480-630	270	18	
ISO 3755(91)	340-550	HT						0.035 max	0.035 max		bal Fe	550-700	340	15	
ISO 3755(91)	340-550W	HT, weld	0.25 max	0.35 max	1.50 max	0.15 max	0.40 max	0.035 max	0.035 max	0.60 max	Cu <=0.40; V <=0.05; bal Fe	550-700	340	15	

Italy

Specification	Designation	Notes	C	Cr	Mn	Mo	Ni	P	S	Si	Other	UTS	YS	El	Hard
UNI 4010(75)	FeG60	HS	0.35 max		1-1.5			0.035 max	0.035 max	0.5 max	bal Fe				

Japan

Specification	Designation	Notes	C	Cr	Mn	Mo	Ni	P	S	Si	Other	UTS	YS	El	Hard
JIS G5102(91)	SCW410	Weldable	0.22 max		1.50 max			0.040 max	0.040 max	0.80 max	bal Fe	410	235	21	
JIS G5102(91)	SCW450	Weldable	0.22 max		1.50 max			0.040 max	0.040 max	0.80 max	bal Fe	450	255	20	
JIS G5102(91)	SCW480	Weldable	0.22 max	0.50 max	1.50 max		0.50 max	0.040 max	0.040 max	0.80 max	bal Fe	480	275	20	
JIS G5102(91)	SCW550	Weldable	0.22 max	0.50 max	1.50 max	0.30 max	2.50 max	0.040 max	0.040 max	0.80 max	V <=0.20; bal Fe	550	355	18	
JIS G5102(91)	SCW620	Weldable	0.22 max	0.50 max	1.50 max	0.30 max	2.50 max	0.040 max	0.040 max	0.80 max	V <=0.20; bal Fe	620	430	17	

Poland

Specification	Designation	Notes	C	Cr	Mn	Mo	Ni	P	S	Si	Other	UTS	YS	El	Hard
PNH83152(80)	L500		0.3-0.4	0.4 max	0.4-0.9		0.4 max	0.05 max	0.05 max	0.2-0.5	bal Fe				
PNH83152(80)	LII500		0.3-0.4	0.4 max	0.4-0.9	0.1 max	0.35	0.04 max	0.04 max	0.2-0.5	bal Fe				
PNH83156(87)	L35GM		0.3-0.4	0.3 max	1-1.4	0.2-0.3	0.3 max	0.04 max	0.04 max	0.2-0.6	bal Fe				
PNH83156(87)	L35HGS		0.3-0.4	0.6-0.9	1.1-1.5		0.3 max	0.04 max	0.04 max	0.6-0.8	bal Fe				
PNH83156(87)	L35HM	Q/T	0.3-0.4	0.8-1.1	0.5-0.8	0.2-0.3	0.3 max	0.04 max	0.04 max	0.2-0.5	bal Fe				
PNH83160(88)	L35GSM	Wear res	0.32-0.4	0.3 max	1.2-1.4	0.3-0.4	0.3 max	0.03 max	0.03 max	0.6-0.8	bal Fe				

Specification	Designation	Notes	C	Cr	Mn	Mo	Ni	P	S	Si	Other	UTS	YS	El	Hard

Steel Casting, Medium-Carbon, No equivalents identified

Romania

Specification	Designation	Notes	C	Cr	Mn	Mo	Ni	P	S	Si	Other	UTS	YS	El	Hard
STAS 1773(82)	T35Mn14	Q/T	0.3-0.4	0.3 max	1.2-1.6			0.04 max	0.04 max	0.2-0.45	Pb <=0.15; bal Fe				

Russia

Specification	Designation	Notes	C	Cr	Mn	Mo	Ni	P	S	Si	Other	UTS	YS	El	Hard
GOST 21357(87)	20ChGSFL	Tough at subzero; Wear res	0.14-0.22	0.30-0.60	0.90-1.30		0.40 max	0.020 max	0.020 max	0.50-0.70	Cu <=0.30; V 0.07-0.13; bal Fe				
GOST 21357(87)	20FTL	Tough at subzero; Wear res	0.17-0.25	0.30 max	0.80-1.20		0.30 max	0.020 max	0.020 max	0.30-0.50	Cu <=0.30; Ti 0.010-0.025; V 0.01-0.06; bal Fe				
GOST 21357(87)	25Ch2NML	Tough at subzero; Wear res	0.22-0.30	1.60-1.90	0.50-0.80	0.20-0.30	0.60-0.90	0.020 max	0.020 max	0.20-0.40	Cu <=0.30; bal Fe				
GOST 21357(87)	27ChN2MFL	Tough at subzero; Wear res	0.23-0.30	0.80-1.20	0.60-0.90	0.30-0.50	1.65-2.00	0.020 max	0.020 max	0.20-0.42	Cu <=0.30; V 0.08-0.15; bal Fe				
GOST 21357(87)	30ChG2STL	Tough at subzero; Wear res	0.25-0.35	0.60-1.00	1.50-1.80		0.30 max	0.020 max	0.020 max	0.40-0.80	Cu <=0.30; Ti 0.01-0.04; bal Fe				
GOST 21357(87)	30ChL	Tough at subzero; Wear res	0.25-0.35	0.30 max	0.50-0.90		0.30 max	0.020 max	0.020 max	0.20-0.50	Cu <=0.30; bal Fe				
GOST 21357(87)	30GL	Tough at subzero; Wear res	0.25-0.35	0.30 max	1.20-1.60		0.30 max	0.020 max	0.020 max	0.20-0.50	Cu <=0.30; bal Fe				
GOST 21357(87)	35ChMFL	Tough at subzero; Wear res	0.30-0.40	0.80-1.10	0.40-0.90	0.08-0.15	0.30 max	0.020 max	0.020 max	0.20-0.40	Cu <=0.30; V 0.06-0.12; bal Fe				
GOST 21357(87)	35ChML	Tough at subzero; Wear res	0.30-0.40	0.90-1.10	0.40-0.90	0.20-0.30	0.30 max	0.020 max	0.020 max	0.20-0.40	Cu <=0.30; bal Fe				
GOST 977	20Ch5ML	Corr res	0.15-0.25	4.0-6.5	0.4-0.6	0.4-0.65	0.5 max	0.04 max	0.04 max	0.35-0.7	Cu <=0.3; Pb <=0.15; V <=0.1; W <=0.1; bal Fe				
GOST 977	20Ch8WL	Corr res	0.15-0.25	7.5-9.0	0.3-0.5	0.15 max	0.5 max	0.04 max	0.035 max	0.3-0.6	Cu <=0.3; Pb <=0.15; V <=0.1; W 1.25-1.75; bal Fe				
GOST 977	20ChGSNDML		0.18-0.24	0.6-0.9	0.9-1.3	0.1-0.15	1.1-1.5	0.05 max	0.045 max	0.9-1.2	Cu 0.4-0.6; Pb <=0.1; Ti 0.03-0.07; V <=0.1; W <=0.1; bal Fe				
GOST 977	25Ch2G2FL		0.2-0.25	1.7-2	1.6-1.8	0.15 max	0.2 max	0.02 max	0.02 max	0.6-0.8	Cu <=0.3; Pb <=0.15; V 0.15-0.2; W <=0.1; bal Fe				
GOST 977	40Ch9S2L	Heat res	0.35-0.5	8.0-10.0	0.3-0.7	0.15 max	0.4 max	0.035 max	0.03 max	2.0-3.0	Cu <=0.3; Pb <=0.15; V <=0.1; W <=0.1; bal Fe				
GOST 977(88)	25Ch2GNMFL		0.22-0.3	1.4-2	0.7-1.1	0.2-0.5	0.04-0.2	0.025 max	0.025 max	0.3-0.7	bal Fe				
GOST 977(88)	27Ch5GSML		0.24-0.28	5-5.5	0.9-1.2	0.55-0.6	0.4 max	0.02 max	0.02 max	0.9-1.2	bal Fe				
GOST 977(88)	30Ch3S3GML		0.29-0.33	2.8-3.2	0.7-1.2	0.5-0.6	0.4 max	0.02 max	0.02 max	2.8-3.2	bal Fe				

UK

Specification	Designation	Notes	C	Cr	Mn	Mo	Ni	P	S	Si	Other	UTS	YS	El	Hard
BS 1504(76)	623*	Press ves	0.25 max	2.5-3.5	0.3-0.7	0.35-0.6	0.4 max	0.04 max	0.04 max	0.75 max	Cu <=0.3; Pb <=0.15; V <=0.1; W <=0.1; bal Fe				

USA

Specification	Designation	Notes	C	Cr	Mn	Mo	Ni	P	S	Si	Other	UTS	YS	El	Hard
ASTM A583(93)	Class L	HT, wheels for railway service	0.47 max		0.60-0.85			0.05 max	0.05 max	0.15 min	bal Fe				197-277 HB
ASTM A915/A915M(98)	SC1045		0.43-0.50		0.50-0.90			0.040 max	0.040 max	0.30-0.60	bal Fe				
ASTM A958(96)	SC1045		0.43-0.50		0.50-0.90			0.040 max	0.040 max	0.30-0.60	bal Fe	450-795	240-655	14-24	

Steel Casting, Medium-Carbon, WCA

Australia

Specification	Designation	Notes	C	Cr	Mn	Mo	Ni	P	S	Si	Other	UTS	YS	El	Hard
AS 2074(82)	C3		0.25 max	0.25 max	0.9 max	0.15 max	0.4 max	0.06 max	0.06 max	0.6 max	Cu <=0.3; bal Fe				

China

Specification	Designation	Notes	C	Cr	Mn	Mo	Ni	P	S	Si	Other	UTS	YS	El	Hard
GB 7659(87)	ZG200-400H(ZG15)		0.20 max	0.30 max	0.80 max	0.15 max	0.30 max	0.040 max	0.040 max	0.50 max	Cu <=0.30; V <=0.05; Cr+Cu+Mo+Ni+V<=0.8; bal Fe	400	200	25	

Czech Republic

Specification	Designation	Notes	C	Cr	Mn	Mo	Ni	P	S	Si	Other	UTS	YS	El	Hard
CSN 422709	422709		0.20-0.28	0.3 max	1.2-1.6			0.05 max	0.05 max	0.2-0.5	P+S<=0.09; bal Fe				

UNS numbers and US grades are provided as a means of cross referencing chemically similar alloys. Exchangability is only possible after independent examination of specifications. Tensile properties are minimum or typical as specified. UTS and YS as MPa. El as %. See Appendix for list of abbreviations used in Notes. * indicates obsolete material.

Specification	Designation	Notes	C	Cr	Mn	Mo	Ni	P	S	Si	Other	UTS	YS	El	Hard

Steel Casting, Medium-Carbon, WCA (Continued from previous page)

Finland
| SFS 365(77) | G-20Mn5 | Norm, Ann | 0.2 max | | 1.5 max | | | 0.035 max | 0.035 max | 0.6 max | Cu <=0.4; bal Fe | | | | |

Italy
| UNI 4010 | .FeGS52 | | 0.25 | | 1-1.5 | | | 0.035 | 0.035 | 0.5 | bal Fe | | | | |
| UNI 7316 | FeG42 | | 0.23 | 0.4 | 1 | | 0.5 | 0.04 | 0.045 | 0.5 | Cu 0.5; Mo+W =0.25; Others 1 max; bal Fe | | | | |

Japan
JIS G5201(91)	SCW42-CF		0.22					0.04	0.04		Others each 0.4 max; bal Fe				
JIS G5201(91)	SCW49-CF		0.22					0.04	0.04		Others each 0.43 max; bal Fe				
JIS G5201(91)	SCW50-CF		0.2					0.04	0.04		Others each 0.44 max; bal Fe				
JIS G5201(91)	SCW53-CF		0.2					0.04	0.04		Others each 0.44 max; bal Fe				

Russia
| GOST 977(88) | 20GSL | | 0.16-0.22 | 0.3 | 1-1.3 | | 0.3 | 0.03 | 0.03 | 0.6-0.8 | bal Fe | | | | |

UK
| BS 1504(76) | 161 430* | Press ves | 0.25 | 0.25 | 1.1 | 0.15 | 0.4 | 0.05 | 0.05 | 0.6 | Cu 0.3; Cr+Cu+Mo+Ni<=0.8; bal Fe | | | | |
| BS 3100(91) | AM2 | 0-999mm | 0.25 max | 0.30 max | 0.50 max | 0.15 max | 0.40 max | 0.050 max | 0.050 max | 0.60 max | Cu <=0.30; Cr+Mo+Ni+Cu<=0.8; bal Fe | 400-490 | 215 | 22 | |

USA
	UNS J02502		0.25 max	0.40 max	0.70 max	0.20 max	0.50 max	0.04 max	0.045 max	0.60 max	Cu <=0.50; V <=0.03; Cr+Cu+Mo+Ni+V<=1.00; bal Fe				
ASTM A216/A216M(98)	WCA		0.25	0.5	0.7	0.2	0.5	0.04	0.045	0.6	Cu 0.3; V 0.03; Cr+Cu+Mo+Ni+V<=1; bal Fe				
MIL-C-24707/1(89)	WCA	As spec in ASTM A216; Mach, Struct									bal Fe				

Steel Casting, Medium-Carbon, WCB

Belgium
| NBN 253-02 | C25-1 | | 0.22-0.29 | | 0.4-0.7 | | | 0.045 | 0.045 | 0.15-0.4 | bal Fe | | | | |

China
| GB 7659(87) | ZG275-485H | | 0.25 max | 0.30 max | 1.20 max | 0.15 max | 0.30 max | 0.040 max | 0.040 max | 0.50 max | Cu <=0.30; V <=0.05; Cr+Cu+Mo+Ni+V<=0.80; bal Fe | 485 | 275 | 20 | |

Italy
| UNI 7316 | FeG49-1 | | 0.3 | 0.4 | 1 | | 0.5 | 0.04 | 0.045 | 0.5 | Cu 0.5; Mo+W =0.25; others 1.00 max; bal Fe | | | | |

Japan
| JIS G5101 | SC42 | | 0.3 | | | | | 0.04 | 0.04 | | bal Fe | | | | |

Poland
| PNH83152(80) | L450 | | 0.2-0.3 | 0.4 max | 0.35-0.8 | | 0.4 max | 0.05 max | 0.05 max | 0.2-0.5 | bal Fe | | | | |

UK
| BS 1504(76) | 161 480* | Press ves | 0.3 | 0.25 | 1.1 | 0.15 | 0.4 | 0.05 | 0.05 | 0.6 | Cu 0.3; Cr+Cu+Mo+Ni<=0.8; bal Fe | | | | |

USA
	UNS J03002		0.30 max	0.50 max	1.00 max	0.20 max	0.50 max	0.04 max	0.045 max	0.60 max	Cu <=0.30; V <=0.03; Cr+Cu+Mo+Ni+V<=1.00 max; bal Fe				
ASTM A216/A216M(98)	WCB		0.3	0.5	1	0.2	0.5	0.04	0.045	0.6	Cu 0.3; V 0.03; Cu+Ni+Cr+Mo+V<=1; bal Fe				
ASTM A216/A216M(98)	WCV		0.3	0.5	1	0.2	0.5	0.04	0.045	0.6	Cu 0.3; V 0.03; Cu+Ni+Cr+Mo+V=1; bal Fe				

UNS numbers and US grades are provided as a means of cross referencing chemically similar alloys. Exchangability is only possible after independent examination of specifications. Tensile properties are minimum or typical as specified. UTS and YS as MPa. El as %. See Appendix for list of abbreviations used in Notes. * indicates obsolete material.

Specification	Designation	Notes	C	Cr	Mn	Mo	Ni	P	S	Si	Other	UTS	YS	El	Hard

Steel Casting, Medium-Carbon, WCB (Continued from previous page)

USA

Specification	Designation	Notes	C	Cr	Mn	Mo	Ni	P	S	Si	Other	UTS	YS	El	Hard
ASTM A757/A757M(96)	A1Q	Corr res, Q/T	0.30 max		1.00 max			0.025 max	0.025 max	0.60 max	V+Cu+Ni+Cr+Mo<=1.00; bal Fe	450	240	24	
MIL-C-24707/1(89)	A1Q	As spec in ASTM A757; Mach, Struct									bal Fe				
MIL-C-24707/1(89)	A2Q	As spec in ASTM A757; Mach, Struct									bal Fe				
MIL-C-24707/1(89)	WCB	As spec in ASTM A216; Mach, Struct									bal Fe				

Steel Casting, Medium-Carbon, WCC

China

Specification	Designation	Notes	C	Cr	Mn	Mo	Ni	P	S	Si	Other	UTS	YS	El	Hard
GB 7659(87)	ZG230-450H		0.20 max	0.30 max	1.20 max	0.15 max	0.30 max	0.040 max	0.040 max	0.50 max	Cu <=0.30; V <=0.05; Cr+Cu+Mo+Ni+V<=0.8; bal Fe	450	230	22	

Germany

Specification	Designation	Notes	C	Cr	Mn	Mo	Ni	P	S	Si	Other	UTS	YS	El	Hard
DIN EN 10213(96)	GS-20Mn5		0.17-0.23	0.30 max	1.00-1.50	0.15 max	0.40 max	0.020 max	0.015 max	0.60 max	bal Fe				
DIN EN 10213(96)	WNr 1.1120		0.17-0.23	0.30 max	1.00-1.50	0.15 max	0.40 max	0.020 max	0.015 max	0.60 max	bal Fe				

Hungary

Specification	Designation	Notes	C	Cr	Mn	Mo	Ni	P	S	Si	Other	UTS	YS	El	Hard
MSZ 8267	Ao21Mn		0.2		0.8-1.1			0.03	0.03	0.6	Cu 0.3; bal Fe				

Italy

Specification	Designation	Notes	C	Cr	Mn	Mo	Ni	P	S	Si	Other	UTS	YS	El	Hard
UNI 4010	FeGS52		0.25		1-1.5			0.035	0.035	0.5	bal Fe				

Switzerland

Specification	Designation	Notes	C	Cr	Mn	Mo	Ni	P	S	Si	Other	UTS	YS	El	Hard
VSM 10698	GS-20Mn5V48	Norm	0.18-0.23	0.4	1-1.4	0.25	0.5	0.03	0.03	0.3-0.5	bal Fe				

UK

Specification	Designation	Notes	C	Cr	Mn	Mo	Ni	P	S	Si	Other	UTS	YS	El	Hard
BS 1504(76)	430*	Austenitic	0.25	0.25	0.9	0.15	0.4	0.05	0.05	0.6	Cu 0.3; bal Fe				

USA

Specification	Designation	Notes	C	Cr	Mn	Mo	Ni	P	S	Si	Other	UTS	YS	El	Hard
	UNS J02503		0.25 max	0.50 max	1.20 max	0.20 max	0.50 max	0.04 max	0.045 max	0.60 max	Cu <=0.30; V <=0.03; Cr+Cu+Mo+Ni+V<=1.00; bal Fe				
ASTM A216/A216M(98)	WCC		0.25	0.5	1.2	0.2-0.5	0.5	0.04	0.045	0.6	Cu 0.3; V 0.03; Cr+Cu+Mo+Ni+V=1; bal Fe				
ASTM A757/A757M(96)	A2Q	Corr res, Q/T	0.25 max		1.20 max			0.025 max	0.025 max	0.60 max	V+Cu+Ni+Cr+Mo<=1.00; bal Fe	485	275	22	
MIL-C-24707/1(89)	WCC	As spec in ASTM A216; Mach, Struct									bal Fe				

Specification	Designation	Notes	C	Cr	Mn	Mo	Ni	P	S	Si	Other	UTS	YS	El	Hard

Steel Casting, High-Carbon, 1050

China

Specification	Designation	Notes	C	Cr	Mn	Mo	Ni	P	S	Si	Other	UTS	YS	El	Hard
GB 11352(89)	ZG340-640	Ann or N/T	0.60 max	0.35 max	0.90 max	0.20 max	0.30 max	0.040 max	0.040 max	0.60 max	Cu <=0.30; V <=0.05; bal Fe	640	340	10	

Germany

Specification	Designation	Notes	C	Cr	Mn	Mo	Ni	P	S	Si	Other	UTS	YS	El	Hard
DIN 17140	D53-2*	Obs; see C52D	0.5-0.55		0.3-0.7			0.04 max	0.04 max	0.1-0.3	bal Fe				

Hungary

Specification	Designation	Notes	C	Cr	Mn	Mo	Ni	P	S	Si	Other	UTS	YS	El	Hard
MSZ 8276(89)	Ao600	0-30.0mm; HT	0.51-0.6		0.4-0.9		0.3 max	0.06 max	0.06 max	0.2-0.9	P<=0.06/P<=0.05; S<=0.06/S<=0.05; bal Fe	600		10L	170-220 HB
MSZ 8276(89)	Ao600F	0-30.0mm; HT	0.51-0.6		0.4-0.9		0.3 max	0.06 max	0.06 max	0.2-0.9	P<=0.06/P<=0.04; S<=0.06/S<=0.045; bal Fe	600	350	15L	170-220 HB
MSZ 8276(89)	Ao600FK	0-30.0mm; Q/T	0.51-0.6		0.4-0.9		0.3 max	0.05 max	0.06 max	0.2-0.9	P<=0.05/P<=0.04; S<=0.06/S<=0.045; bal Fe	860	470	12L	
MSZ 8276(89)	Ao600FK	0-30.0mm; HT	0.51-0.6		0.4-0.9		0.3 max	0.05 max	0.06 max	0.2-0.9	P<=0.05/P<=0.04; S<=0.06/S<=0.045; bal Fe	600	350	15L	170-220 HB

Romania

Specification	Designation	Notes	C	Cr	Mn	Mo	Ni	P	S	Si	Other	UTS	YS	El	Hard
STAS 600(82)	OT700-1	Unalloyed	0.5-0.6	0.3 max	0.4-0.8			0.06 max	0.06 max	0.25-0.5	Pb <=0.15; bal Fe				
STAS 600(82)	OT700-2	Unalloyed	0.5-0.6	0.3 max	0.4-0.8			0.04 max	0.04 max	0.25-0.5	Pb <=0.15; bal Fe				

Russia

Specification	Designation	Notes	C	Cr	Mn	Mo	Ni	P	S	Si	Other	UTS	YS	El	Hard
GOST 977(88)	45L-3		0.42-0.5	0.3 max	0.45-0.9		0.4 max	0.03 max	0.03 max	0.2-0.52	bal Fe				
GOST 977(88)	50L		0.47-0.55	0.3 max	0.45-0.9		0.4 max	0.04 max	0.04 max	0.2-0.52	bal Fe				
GOST 977(88)	50L-2		0.47-0.55	0.3 max	0.45-0.9		0.4 max	0.035 max	0.035 max	0.2-0.52	bal Fe				
GOST 977(88)	50L-3		0.47-0.55	0.3 max	0.45-0.9		0.4 max	0.03 max	0.03 max	0.2-0.52	bal Fe				

Spain

Specification	Designation	Notes	C	Cr	Mn	Mo	Ni	P	S	Si	Other	UTS	YS	El	Hard
UNE 36254(79)	46Mn4	Q/T	0.42-0.5		0.9-1.2			0.04	0.04	0.3-0.5	bal Fe				
UNE 36254(79)	F.8213*	see 46Mn4	0.42-0.5		0.9-1.2			0.04	0.04	0.3-0.5	bal Fe				

Sweden

Specification	Designation	Notes	C	Cr	Mn	Mo	Ni	P	S	Si	Other	UTS	YS	El	Hard
SIS 141606	1606-02	Ann	0.5		0.7			0.04	0.04	0.5	bal Fe				

UK

Specification	Designation	Notes	C	Cr	Mn	Mo	Ni	P	S	Si	Other	UTS	YS	El	Hard
BS 3100(91)	AW3	0-999mm	0.50-0.60	0.30 max	1.00 max	0.15 max	0.40 max	0.050 max	0.050 max	0.60 max	Cu <=0.30; Cr+Mo+Ni+Cu<=0.8; bal Fe	690	370	8	

USA

Specification	Designation	Notes	C	Cr	Mn	Mo	Ni	P	S	Si	Other	UTS	YS	El	Hard
	UNS J05000		0.45-0.55		0.70-1.00			0.04 max	0.04 max	0.20-1.00	bal Fe				
	UNS J05001		0.45-0.55		0.70-1.00			0.025 max	0.025 max	0.20-1.00	bal Fe				
ASTM A732/A732M(98)	4A/4Q	Inv, Ann, Q/A	0.45-0.55		0.70-1.00			0.04 max	0.045 max	0.20-1.00	Cu+W<=0.60; bal Fe	621-862	345-689	5-20	
MIL-S-22141B(84)	IC-1050*	Inv Ann; Obs for new design	0.45-0.55		0.70-1.00			0.04 max	0.04 max	0.20-1.00	bal Fe	621	345	20	95 HRB
MIL-S-22141B(84)	IC-1050*	Inv cast; Obs for new design	0.45-0.55		0.70-1.00			0.04 max	0.04 max	0.20-1.00	bal Fe				20 HRC
MIL-S-22141B(84)	IC-1050*	Inv HT; Obs for new design	0.45-0.55		0.70-1.00			0.04 max	0.04 max	0.20-1.00	bal Fe	862	690	5	
MIL-S-81591(93)	IC-1050*	Inv, C/Corr res; Obs for new design see AMS specs	0.45-0.55		0.7-1			0.04	0.04	0.2-1	bal Fe				

Steel Casting, High-Carbon, No equivalents identified

Hungary

Specification	Designation	Notes	C	Cr	Mn	Mo	Ni	P	S	Si	Other	UTS	YS	El	Hard
MSZ 8276(89)	Ao550	0-30.0mm; HT	0.41-0.5		0.4-0.9		0.3 max	0.06 max	0.06 max	0.2-0.52	P<=0.06/P<=0.05; S<=0.06/S<=0.05; bal Fe	550		12L	155-205 HB

UNS numbers and US grades are provided as a means of cross referencing chemically similar alloys. Exchangability is only possible after independent examination of specifications. Tensile properties are minimum or typical as specified. UTS and YS as MPa. El as %. See Appendix for list of abbreviations used in Notes. * indicates obsolete material.

Specification	Designation	Notes	C	Cr	Mn	Mo	Ni	P	S	Si	Other	UTS	YS	El	Hard

Steel Casting, High-Carbon, No equivalents identified

Hungary

Specification	Designation	Notes	C	Cr	Mn	Mo	Ni	P	S	Si	Other	UTS	YS	El	Hard
MSZ 8276(89)	Ao550F	0-30.0mm; HT	0.41-0.5		0.4-0.9		0.3 max	0.06 max	0.06 max	0.2-0.52	P<=0.06/P<=0.04; S<=0.06/S<=0.045; bal Fe	550	320	17L	155-205 HB
MSZ 8276(89)	Ao550FK	0-30.0mm; HT	0.41-0.5		0.4-0.9		0.3 max	0.05 max	0.06 max	0.2-0.52	P<=0.05/P<=0.04; S<=0.06/S<=0.045; bal Fe	550	320	17L	155-205 HB
MSZ 8276(89)	Ao550FK	0-30.0mm; Q/T	0.41-0.5		0.4-0.9		0.3 max	0.05 max	0.06 max	0.2-0.52	P<=0.05/P<=0.04; S<=0.06/S<=0.045; bal Fe	600	400	14L	

Poland

Specification	Designation	Notes	C	Cr	Mn	Mo	Ni	P	S	Si	Other	UTS	YS	El	Hard
PNH83152(80)	L650		0.5-0.6	0.4 max	0.4-0.9		0.4 max	0.05 max	0.05 max	0.2-0.5	bal Fe				
PNH83152(80)	LII600		0.4-0.5	0.4 max	0.4-0.9	0.1 max	0.35 max	0.04 max	0.04 max	0.2-0.5	bal Fe				
PNH83152(80)	LII650		0.5-0.6	0.4 max	0.4-0.9	0.1 max	0.35 max	0.04 max	0.04 max	0.2-0.5	bal Fe				
PNH83160(88)	L40GM	Wear res	0.35-0.45	0.35 max	1.4-1.8	0.15-0.3	0.35 max	0.04 max	0.035 max	0.2-0.4	bal Fe				

Russia

Specification	Designation	Notes	C	Cr	Mn	Mo	Ni	P	S	Si	Other	UTS	YS	El	Hard
GOST 977	110G13Ch2BRL		0.9-1.5	1-2	11.5-14.5	0.15 max	0.5 max	0.12 max	0.05 max	0.3-1	B 0.001-0.006; Cu <=0.3; Nb 0.08-0.12; Pb <=0.15; V <=0.1; W <=0.1; bal Fe				
GOST 977	110G13FTL		0.9-1.3	0.3 max	11.5-14.5	0.15 max	0.4 max	0.12 max	0.05 max	0.4-0.9	Cu <=0.3; Pb <=0.15; Ti 0.01-0.05; V 0.1-0.3; W <=0.1; bal Fe				
GOST 977	120G10FL		0.9-1.4	1 max	8.5-12	0.15 max	1 max	0.12 max	0.05 max	0.2-0.9	Cu <=0.7; N <=0.03; Nb <=0.01; Pb <=0.15; Ti <=0.15; V 0.03-0.12; W <=0.1; bal Fe				
GOST 977	84Ch4M5F2W6L	Tool	0.82-0.9	3.8-4.4	0.5 max	4.8-5.3	0.4 max	0.03 max	0.02 max	0.5 max	Cu <=0.3; Pb <=0.15; V 1.7-2.1; W 5.5-6.5; bal Fe				
GOST 977	90Ch4M4F2W6L	Tool	0.85-0.95	3-4	0.4-0.7	3.1-4	0.4 max	0.04 max	0.04 max	0.2-0.4	Cu <=0.3; Pb <=0.15; V 2-2.6; W 5-7; bal Fe				

USA

Specification	Designation	Notes	C	Cr	Mn	Mo	Ni	P	S	Si	Other	UTS	YS	El	Hard
AMS 2431/1B(97)		Shot peening media	0.85-1.20		0.35-1.20			0.05 max	0.05 max	0.40-1.50	bal Fe				45-52 HRC
AMS 2431/2B(97)		Shot peening media	0.85-1.20		0.35-1.20			0.05 max	0.05 max	0.40-1.50	bal Fe				52-62 HRC
ASTM A583(93)	Class A	HT, wheels for railway service	0.47-0.57		0.60-0.85			0.05 max	0.05 max	0.15 min	bal Fe				255-321 HB
ASTM A583(93)	Class B	HT, wheels for railway service	0.57-0.67		0.60-0.85			0.05 max	0.05 max	0.15 min	bal Fe				277-341 HB
ASTM A583(93)	Class C	HT, wheels for railway service	0.67-0.77		0.60-0.85			0.05 max	0.05 max	0.15 min	bal Fe				321-363 HB
ASTM A583(93)	Class U	Wheels for railway service	0.65-0.77		0.60-0.85			0.05 max	0.05 max	0.15 min	bal Fe				
SAE J1993(96)	HCS G(X)H	Grit for blast cleaning/etching; x=sieve designation	0.80-1.20		0.60-1.20			0.05 max	0.05 max	0.40 max	bal Fe				60 HRC
SAE J1993(96)	HCS G(X)L	Grit for blast cleaning/etching; x=sieve designation	0.80-1.20		0.60-1.20			0.05 max	0.05 max	0.40 max	bal Fe				54-61 HRC
SAE J1993(96)	HCS G(X)M	Grit for blast cleaning/etching; x=sieve designation	0.80-1.20		0.60-1.20			0.05 max	0.05 max	0.40 max	bal Fe				47-56 HRC
SAE J1993(96)	HCS G(X)S	Grit for blast cleaning/etching; x=sieve designation	0.80-1.20		0.60-1.20			0.05 max	0.05 max	0.40 max	bal Fe				40-51 HRC
SAE J827(96)	HCS S170	Peening, blast cleaning shot	0.80-1.2		0.5-1.2			0.050 max	0.050 max	0.4 min	bal Fe				40-51 HRC
SAE J827(96)	HCS S230 and up	Peening, blast cleaning shot	0.80-1.2		0.6-1.2			0.050 max	0.050 max	0.4 min	bal Fe				40-51 HRC
SAE J827(96)	HCS S70 to S110	Peening, blast cleaning shot	0.80-1.2		0.35-1.2			0.050 max	0.050 max	0.4 min	bal Fe				40-51 HRC

UNS numbers and US grades are provided as a means of cross referencing chemically similar alloys. Exchangability is only possible after independent examination of specifications. Tensile properties are minimum or typical as specified. UTS and YS as MPa. El as %. See Appendix for list of abbreviations used in Notes. * indicates obsolete material.

Specification	Designation	Notes	C	Cr	Mn	Mo	Ni	P	S	Si	Other	UTS	YS	El	Hard
Steel Casting, Low-Alloy, 10Q															
USA															
	UNS J24054		0.35-0.45	0.70-0.90	0.70-1.00	0.20-0.30	1.65-2.00	0.04 max	0.045 max	0.20-0.80	Cu <=0.50; W <=0.10; bal Fe				
ASTM A732/A732M(98)	10Q	Q/A, Cr Ni Mo	0.35-0.45	0.70-0.90	0.70-1.00	0.20-0.30	1.65-2.00	0.04 max	0.045 max	0.20-0.80	Cu+W<=1.00; bal Fe	1241	1000	5	
Steel Casting, Low-Alloy, 11Q															
USA															
	UNS J12094		0.15-0.25	0.35 max	0.40-0.70	0.20-0.30	1.65-2.00	0.04 max	0.045 max	0.20-0.80	Cu <=0.50; W <=0.10; bal Fe				
ASTM A732/A732M(98)	11Q	Q/A, Ni Mo	0.15-0.25		0.40-0.70	0.20-0.30	1.65-2.00	0.04 max	0.045 max	0.20-0.80	Cu+Cr+W<=1.00; bal Fe	827	689	10	
Steel Casting, Low-Alloy, 12Q															
UK															
BS 3100(91)	BW2		0.45-0.60	0.80-1.50	0.50-1.00	0.40 max		0.040 max	0.040 max	0.75 max	bal Fe				201-255 HB
BS 3100(91)	BW3		0.45-0.60	0.80-1.50	0.50-1.00	0.40 max		0.040 max	0.040 max	0.75 max	bal Fe				293 HB
BS 3100(91)	BW4		0.45-0.60	0.80-1.50	0.50-1.00	0.40 max		0.040 max	0.040 max	0.75 max	bal Fe				341 HB
USA															
	UNS J15048		0.45-0.55	0.80-1.10	0.65-0.95		0.50 max	0.04 max	0.045 max	0.20-0.80	Cu <=0.50; V >=0.15; W <=0.10; Mo+W 0.10 max				
ASTM A732/A732M(98)	12Q	Q/A, Cr V	0.45-0.55	0.80-1.10	0.65-0.95			0.04 max	0.045 max	0.20-0.80	V >=0.15; Cu+Ni+Mo+W<=1.00; bal Fe	1310	1172	4	
Steel Casting, Low-Alloy, 13Q															
USA															
	UNS J12048		0.15-0.25	0.40-0.70	0.65-0.95	0.15-0.25	0.40-0.70	0.04 max	0.045 max	0.20-0.80	Cu <=0.50; W <=0.10; bal Fe				
ASTM A732/A732M(98)	13Q	Q/A, Cr Ni Mo	0.15-0.25	0.40-0.70	0.65-0.95	0.15-0.25	0.40-0.70	0.04 max	0.045 max	0.20-0.80	Cu+W<=1.00; bal Fe	724	586	10	
Steel Casting, Low-Alloy, 14Q															
Russia															
GOST 977(88)	32Ch06L		0.25-0.35	0.5-0.8	0.4-0.9		0.4 max	0.05 max	0.05 max	0.2-0.4	bal Fe				
USA															
	UNS J13051		0.25-0.35	0.40-0.70	0.65-0.95	0.15-0.25	0.40-0.70	0.04 max	0.045 max	0.20-0.80	Cu <=0.50; W <=0.10; bal Fe				
ASTM A732/A732M(98)	14Q	Q/A, Cr Ni Mo, Inv	0.25-0.35	0.40-0.70	0.65-0.95	0.15-0.25	0.40-0.70	0.04 max	0.045 max	0.20-0.80	Cu+W<=1.00; bal Fe	1030	793	7	
Steel Casting, Low-Alloy, 15A															
Bulgaria															
BDS 12731(75)	SchCh15	Bearing; smls tub	0.95-1.05	1.30-1.65	0.20-0.45		0.30 max	0.027 max	0.020 max	0.17-0.37	Cu <=0.25; bal Fe				
USA															
	UNS J19966		0.95-1.10	1.30-1.60	0.25-0.55		0.50 max	0.04 max	0.045 max	0.20-0.80	Cu <=0.50; W <=0.10; bal Fe				
ASTM A732/A732M(98)	15A	Cr, Inv, Q/A, Tmp	0.95-1.10	1.30-1.60	0.25-0.55			0.04 max	0.045 max	0.20-0.80	Cu+Ni+W<=0.60; bal Fe				100 HRB max
Steel Casting, Low-Alloy, 16															
Bulgaria															
BDS 6550(86)	08GDNFL		0.10 max	0.3 max	0.6-1.1	0.15 max	1.15-1.55	0.035 max	0.035 max	0.2-0.5	Cu 0.8-1.2; V 0.06-0.15; W <=0.1; bal Fe				
Russia															
GOST 977(88)	08GDNFL		0.1 max	0.3 max	0.6-1		1.15-1.55	0.035 max	0.035 max	0.15-0.4	Cu 0.8-1.2; bal Fe				
USA															
	UNS J31200		0.12 max	0.20 max	2.10 max	0.10 max	1.00-1.40	0.02 max	0.02 max	0.50 max	Cu <=0.20; V <=0.02; W <=0.10; bal Fe				

UNS numbers and US grades are provided as a means of cross referencing chemically similar alloys. Exchangability is only possible after independent examination of specifications. Tensile properties are minimum or typical as specified. UTS and YS as MPa. El as %. See Appendix for list of abbreviations used in Notes. * indicates obsolete material.

Specification	Designation	Notes	C	Cr	Mn	Mo	Ni	P	S	Si	Other	UTS	YS	El	Hard

Steel Casting, Low-Alloy, 2

France

Specification	Designation	Notes	C	Cr	Mn	Mo	Ni	P	S	Si	Other	UTS	YS	El	Hard
AFNOR NFA32055(85)	20D5M	t<1000mm	0.23 max	0.4 max	1 max	0.4-0.7	0.3 max	0.03 max	0.03 max	0.6 max	Cu <=0.3; Pb <=0.15; V <=0.04; W <=0.1; Cu+Ni+Cr+V<=1; bal Fe	450-600	250	21	135-180 HB

USA

Specification	Designation	Notes	C	Cr	Mn	Mo	Ni	P	S	Si	Other	UTS	YS	El	Hard
	UNS J12523		0.25 max		0.70 max	0.45-0.65		0.035 max	0.030 max	0.60 max	bal Fe				
ASTM A356/A356M(98)	2	Heavy wall; steam turbines	0.25 max		0.70 max	0.45-0.65		0.035 max	0.030 max	0.60 max	bal Fe	450	240	22.0	

Steel Casting, Low-Alloy, 4130

Bulgaria

Specification	Designation	Notes	C	Cr	Mn	Mo	Ni	P	S	Si	Other	UTS	YS	El	Hard
BDS 6550(86)	25ChML		0.22-0.29	0.80-1.20	0.50-0.80	0.20-0.30	0.40 max	0.040 max	0.040 max	0.20-0.50	Cu <=0.30; bal Fe				

China

Specification	Designation	Notes	C	Cr	Mn	Mo	Ni	P	S	Si	Other	UTS	YS	El	Hard
JB/T 6402(92)	ZG35Cr1Mo	Q/T	0.30-0.37	0.80-1.20	0.50-0.80	0.20-0.30		0.035 max	0.035 max	0.30-0.50	bal Fe	740-880	510	12	

Finland

Specification	Designation	Notes	C	Cr	Mn	Mo	Ni	P	S	Si	Other	UTS	YS	El	Hard
SFS 367(77)	G-25CrMo4		0.22-0.29	0.9-1.2	0.5-0.8	0.15-0.25	0.3 max	0.035 max	0.035 max	0.6 max	bal Fe				

Japan

Specification	Designation	Notes	C	Cr	Mn	Mo	Ni	P	S	Si	Other	UTS	YS	El	Hard
JIS G5111(91)	SCCrM1	N/T	0.20-0.30	0.80-1.20	0.50-0.80	0.15-0.35		0.040 max	0.040 max	0.30-0.60	bal Fe	590	390	13	170 HB
JIS G5111(91)	SCCrM1	Q/A Tmp	0.20-0.30	0.80-1.20	0.50-0.80	0.15-0.35		0.040 max	0.040 max	0.30-0.60	bal Fe	690	490	13	201 HB
JIS G5111(91)	SCCrMI	Q/A Tmp	0.20-0.30	0.80-1.20	0.50-0.80	0.15-0.35		0.040 max	0.040 max	0.30-0.60	bal Fe	690	490	13	201 HB

Romania

Specification	Designation	Notes	C	Cr	Mn	Mo	Ni	P	S	Si	Other	UTS	YS	El	Hard
STAS 11534(82)	T30CrMo3	Q/T	0.25-0.35	0.8-1.2	0.5-0.8	0.2-0.3		0.04 max	0.04 max	0.2-0.4	Pb <=0.15; V 0.08-0.16; bal Fe				

Russia

Specification	Designation	Notes	C	Cr	Mn	Mo	Ni	P	S	Si	Other	UTS	YS	El	Hard
GOST 977(88)	30ChNML		0.25-0.35	1.3-1.6	0.4-0.9	0.2-0.3	1.3-1.6	0.04 max	0.04 max	0.2-0.4	bal Fe				

Spain

Specification	Designation	Notes	C	Cr	Mn	Mo	Ni	P	S	Si	Other	UTS	YS	El	Hard
UNE 36254(79)	34CrMo4		0.3-0.38	0.8-1.2	0.5-0.8	0.15-0.3		0.04 max	0.04 max	0.3-0.5	bal Fe				
UNE 36255(79)	AM25CrMo4*	Q/T	0.22-0.3	0.8-1.2	0.5-0.8	0.15-0.3		0.040 max	0.040 max	0.3-0.5	bal Fe				
UNE 36255(79)	AM34CrMo4*		0.3-0.38	0.8-1.2	0.5-0.8	0.15-0.3		0.040 max	0.040 max	0.3-0.5	bal Fe				
UNE 36255(79)	F.8330*	see AM25CrMo4*	0.22-0.3	0.8-1.2	0.5-0.8	0.15-0.3		0.040 max	0.040 max	0.3-0.5	bal Fe				
UNE 36256(73)	AM26CrMo4*	Q/T	0.22-0.29	0.8-1.2	0.5-0.8	0.2-0.3		0.035 max	0.035 max	0.3-0.6	bal Fe				
UNE 36256(73)	F.8372*	see AM26CrMo4*	0.22-0.29	0.8-1.2	0.5-0.8	0.2-0.3		0.035 max	0.035 max	0.3-0.6	bal Fe				

USA

Specification	Designation	Notes	C	Cr	Mn	Mo	Ni	P	S	Si	Other	UTS	YS	El	Hard
	UNS J13048		0.25-0.35	0.80-1.10	0.40-0.70	0.15-0.25		0.025 max	0.025 max	0.20-0.80	bal Fe				
	UNS J13502		0.28-0.33	0.80-1.10	0.40-0.80	0.15-0.25		0.035 max	0.040 max	0.20-0.60	bal Fe				
ASTM A915/A915M(98)	SC4130		0.28-0.33	0.80-1.10	0.40-0.80	0.15-0.25		0.035 max	0.040 max	0.30-0.60	bal Fe				
ASTM A958(96)	SC4130		0.28-0.33	0.80-1.10	0.40-0.80	0.15-0.25		0.035 max	0.040 max	0.30-0.60	bal Fe	450-1035	240-930	7-24	
MIL-S-22141B(84)	IC-4130*	Inv Ann; Obs for new design	0.25-0.35	0.80-1.10	0.40-0.70	0.15-0.25		0.04 max	0.04 max	0.20-0.80	bal Fe				90 HRB
MIL-S-22141B(84)	IC-4130*	Inv HT; Obs for new design	0.25-0.35	0.80-1.10	0.40-0.70	0.15-0.25		0.04 max	0.04 max	0.20-0.80	bal Fe		792	7	

Steel Casting, Low-Alloy, 4140

China

Specification	Designation	Notes	C	Cr	Mn	Mo	Ni	P	S	Si	Other	UTS	YS	El	Hard
JB/T 6402(92)	ZG42Cr1Mo	Q/T	0.38-0.45	0.80-1.20	0.60-1.00	0.20-0.30		0.035 max	0.035 max	0.30-0.60	bal Fe	690-830	490	11	

UNS numbers and US grades are provided as a means of cross referencing chemically similar alloys. Exchangability is only possible after independent examination of specifications. Tensile properties are minimum or typical as specified. UTS and YS as MPa. El as %. See Appendix for list of abbreviations used in Notes. * indicates obsolete material.

Specification	Designation	Notes	C	Cr	Mn	Mo	Ni	P	S	Si	Other	UTS	YS	El	Hard

Steel Casting, Low-Alloy, 4140 (Continued from previous page)

Germany

Specification	Designation	Notes	C	Cr	Mn	Mo	Ni	P	S	Si	Other	UTS	YS	El	Hard
DIN SEW 835(95)	WNr 1.7228	Cast for hardening	0.46-0.54	0.90-1.20	0.50-0.80	0.15-0.30		0.035 max	0.035 max	0.40 max	bal Fe				

Italy

Specification	Designation	Notes	C	Cr	Mn	Mo	Ni	P	S	Si	Other	UTS	YS	El	Hard
UNI 3160(83)	G40CrMo4	Wear res	0.37-0.42	0.8-1.2	0.75-1	0.15-0.25		0.035 max	0.035 max	0.2-0.35	bal Fe				

Japan

Specification	Designation	Notes	C	Cr	Mn	Mo	Ni	P	S	Si	Other	UTS	YS	El	Hard
JIS G5111(91)	SCMnCr4	Q/A Tmp	0.35-0.45	0.40-0.80	1.20-1.60			0.040 max	0.040 max	0.30-0.60	bal Fe	740	540	13	223 HB

USA

Specification	Designation	Notes	C	Cr	Mn	Mo	Ni	P	S	Si	Other	UTS	YS	El	Hard
	UNS J14047		0.35-0.45	0.80-1.10	0.70-1.35	0.15-0.25		0.025 max	0.025 max	0.20-0.80	bal Fe				
ASTM A915/A915M(98)	SC4140		0.38-0.43	0.80-1.10	0.70-1.10	0.15-0.25		0.035 max	0.040 max	0.30-0.60	bal Fe				
ASTM A958(96)	SC4140		0.38-0.43	0.80-1.10	0.70-1.10	0.15-0.25		0.035 max	0.040 max	0.30-0.60	bal Fe	450-1140	240-1035	5-24	
MIL-S-22141B(84)	IC-4140*	Inv HT; Obs for new design	0.35-0.45	0.80-1.10	0.70-1.05	0.15-0.25		0.04 max	0.04 max	0.20-0.80	bal Fe	1241	1000	5	
MIL-S-22141B(84)	IC-4140*	Inv Ann; Obs for new design	0.35-0.45	0.80-1.10	0.70-1.05	0.15-0.25		0.04 max	0.04 max	0.20-0.80	bal Fe				20 HRC

Steel Casting, Low-Alloy, 4330

Germany

Specification	Designation	Notes	C	Cr	Mn	Mo	Ni	P	S	Si	Other	UTS	YS	El	Hard
DIN 17205(92)	G33NiCrMo7-4-4	HT	0.30-0.36	0.90-1.20	0.50-0.80	0.35-0.50	1.50-1.80	0.015 max	0.007 max	0.60 max	bal Fe				
DIN 17205(92)	WNr 1.6740	HT	0.30-0.36	0.90-1.20	0.50-0.80	0.35-0.50	1.50-1.80	0.015 max	0.007 max	0.60 max	bal Fe				

Japan

Specification	Designation	Notes	C	Cr	Mn	Mo	Ni	P	S	Si	Other	UTS	YS	El	Hard
JIS G5111(91)	SCNCrM2	N/T	0.25-0.35	0.30-0.90	0.90-1.50	0.15-0.35	1.60-2.00	0.040 max	0.040 max	0.30-0.60	bal Fe	780	590	9	223 HB
JIS G5111(91)	SCNCrM2	Q/A Tmp	0.25-0.35	0.30-0.90	0.90-1.50	0.15-0.35	1.60-2.00	0.040 max	0.040 max	0.30-0.60	bal Fe	880	685	9	269 HB

USA

Specification	Designation	Notes	C	Cr	Mn	Mo	Ni	P	S	Si	Other	UTS	YS	El	Hard
	UNS J23259		0.28-0.33	0.70-0.90	0.60-0.90	0.20-0.30	1.65-2.00	0.035 max	0.040 max	0.30-0.60	bal Fe				
ASTM A915/A915M(98)	SC4330		0.28-0.33	0.70-0.90	0.60-0.90	0.20-0.30	1.65-2.00	0.035 max	0.040 max	0.30-0.60	bal Fe				
ASTM A958(96)	SC4330		0.28-0.33	0.70-0.90	0.60-0.90	0.20-0.30	1.65-2.00	0.035 max	0.040 max	0.30-0.60	bal Fe	450-1450	240-1240	4-24	

Steel Casting, Low-Alloy, 4335M

Russia

Specification	Designation	Notes	C	Cr	Mn	Mo	Ni	P	S	Si	Other	UTS	YS	El	Hard
GOST 977(88)	35NGML		0.32-0.42	0.3 max	0.8-1.2	0.15-0.25	0.8-1.2	0.04 max	0.04 max	0.2-0.4	bal Fe				

USA

Specification	Designation	Notes	C	Cr	Mn	Mo	Ni	P	S	Si	Other	UTS	YS	El	Hard
	UNS J13432		0.30-0.38		0.60-1.00	0.65-1.00		0.04 max	0.04 max	0.50-1.00	Al <=0.05; V <=0.14; bal Fe				
MIL-S-22141B(84)	IC-4335M*	Inv Ann; Obs for new design	0.30-0.38		0.60-1.00	0.65-1.00		0.025 max	0.025 max	0.50-1.00	V <=0.14; bal Fe				20 HRC
MIL-S-22141B(84)	IC-4335M*	Inv HT; Obs for new design	0.30-0.38		0.60-1.00	0.65-1.00		0.025 max	0.025 max	0.50-1.00	V <=0.14; bal Fe	1379	1241	5	

Steel Casting, Low-Alloy, 4340

USA

Specification	Designation	Notes	C	Cr	Mn	Mo	Ni	P	S	Si	Other	UTS	YS	El	Hard
	UNS J24053		0.38-0.43	0.70-0.90	0.60-0.90	0.20-0.30	1.65-2.00	0.035 max	0.040 max	0.30-0.60	bal Fe				
	UNS J24055		0.36-0.44	0.70-0.90	0.60-0.90	0.20-0.30	1.65-2.00	0.04 max	0.04 max	0.20-0.80	bal Fe				
ASTM A915/A915M(98)	SC4340		0.38-0.43	0.70-0.90	0.60-0.90	0.20-0.30	1.65-2.00	0.035 max	0.040 max	0.30-0.60	bal Fe				
ASTM A958(96)	SC4340		0.38-0.43	0.70-0.90	0.60-0.90	0.20-0.30	1.65-2.00	0.035 max	0.040 max	0.30-0.60	bal Fe	450-1450	240-1240	4-24	
MIL-S-22141B(84)	IC-4340*	Inv Ann; Obs for new design	0.36-0.44	0.70-0.90	0.60-0.90	0.20-0.30	1.65-2.00	0.025 max	0.025 max	0.20-0.80	bal Fe				20 HRC
MIL-S-22141B(84)	IC-4340*	Inv HT; Obs for new design	0.36-0.44	0.70-0.90	0.60-0.90	0.20-0.30	1.65-2.00	0.025 max	0.025 max	0.20-0.80	bal Fe	1379	1241		

UNS numbers and US grades are provided as a means of cross referencing chemically similar alloys. Exchangability is only possible after independent examination of specifications. Tensile properties are minimum or typical as specified. UTS and YS as MPa. El as %. See Appendix for list of abbreviations used in Notes. * indicates obsolete material.

Specification	Designation	Notes	C	Cr	Mn	Mo	Ni	P	S	Si	Other	UTS	YS	El	Hard

Steel Casting, Low-Alloy, 4620

France

Specification	Designation	Notes	C	Cr	Mn	Mo	Ni	P	S	Si	Other	UTS	YS	El	Hard
AFNOR NFA32053(74)	FBIM	28mm; HT	0.22 max	0.3 max	1.5 max	0.15 max	0.5-2	0.04 max	0.035 max	0.5 max	W <=0.1; bal Fe	450-600	230	18	

USA

Specification	Designation	Notes	C	Cr	Mn	Mo	Ni	P	S	Si	Other	UTS	YS	El	Hard
	UNS J12093		0.15-0.25		0.40-0.70	0.20-0.30	1.65-2.00	0.025 max	0.025 max	0.20-0.80	bal Fe				
MIL-S-22141B(84)	IC-4620*	Inv Ann; Obs for new design	0.15-0.25		0.40-0.70	0.20-0.30	1.65-2.00	0.04 max	0.04 max	0.20-0.80	bal Fe				95 HRB
MIL-S-22141B(84)	IC-4620*	Inv HT; Obs for new design	0.15-0.25		0.40-0.70	0.20-0.30	1.65-2.00	0.04 max	0.04 max	0.20-0.80	bal Fe	827	690	10	

Steel Casting, Low-Alloy, 5

Russia

Specification	Designation	Notes	C	Cr	Mn	Mo	Ni	P	S	Si	Other	UTS	YS	El	Hard
GOST 977(88)	20ChML		0.15-0.25	0.4-0.7	0.4-0.9		0.4 max	0.04 max	0.04 max	0.2-0.4	bal Fe				
GOST 977(88)	20DChL		0.15-0.25	0.8-1.1	0.5-0.8		0.4 max	0.04 max	0.04 max	0.2-0.4	V 1.4-1.6; bal Fe				

USA

Specification	Designation	Notes	C	Cr	Mn	Mo	Ni	P	S	Si	Other	UTS	YS	El	Hard
	UNS J12540		0.25 max	0.40-0.70	0.70 max	0.40-0.60		0.035 max	0.030 max	0.60 max	bal Fe				
ASTM A356/A356M(98)	5	Heavy wall; steam turbines	0.25 max	0.40-0.70	0.70 max	0.40-0.60		0.035 max	0.030 max	0.60 max	bal Fe	485	275	22.0	

Steel Casting, Low-Alloy, 52100

Russia

Specification	Designation	Notes	C	Cr	Mn	Mo	Ni	P	S	Si	Other	UTS	YS	El	Hard
GOST 801(60)	SHKH15	Ball & roller bearing	0.95-1.05	1.3-1.65	0.2-0.4	0.15 max	0.3 max	0.027 max	0.02 max	0.17-0.37	Cu <=0.3; Pb <=0.15; V <=0.1; W <=0.1; bal Fe				
GOST 801(78)	SChCh15	Ball & roller bearing	0.95-1.05	1.3-1.65	0.2-0.4	0.15 max	0.3 max	0.027 max	0.02 max	0.17-0.37	Cu <=0.25; Pb <=0.15; V <=0.1; W <=0.1; Ni+Cu<=0.5; bal Fe				

USA

Specification	Designation	Notes	C	Cr	Mn	Mo	Ni	P	S	Si	Other	UTS	YS	El	Hard
	UNS J19965		0.95-1.10	1.30-1.60	0.25-0.55			0.025 max	0.025 max	0.20-0.80	bal Fe				
MIL-S-22141B(84)	IC-52100*	Inv Ann; Obs for new design	0.95-1.10	1.30-1.60	0.25-0.55			0.04 max	0.04 max	0.20-0.80	bal Fe				100 HRB

Steel Casting, Low-Alloy, 5328

Russia

Specification	Designation	Notes	C	Cr	Mn	Mo	Ni	P	S	Si	Other	UTS	YS	El	Hard
GOST 977(88)	35ChML	Q/T	0.3-0.4	0.8-1.1	0.4-0.9	0.2-0.3	0.4 max	0.04 max	0.04 max	0.2-0.4	bal Fe				

USA

Specification	Designation	Notes	C	Cr	Mn	Mo	Ni	P	S	Si	Other	UTS	YS	El	Hard
	UNS J23260	4330 Modified	0.28-0.36	0.65-1.00	0.60-0.90	0.30-0.45	1.65-2.00	0.025 max	0.025 max	0.50-1.00	bal Fe				

Steel Casting, Low-Alloy, 5330

USA

Specification	Designation	Notes	C	Cr	Mn	Mo	Ni	P	S	Si	Other	UTS	YS	El	Hard
	UNS J24060*	Obs; see J23260									bal Fe				

Steel Casting, Low-Alloy, 5333

Russia

Specification	Designation	Notes	C	Cr	Mn	Mo	Ni	P	S	Si	Other	UTS	YS	El	Hard
GOST 977(88)	12DN2FL		0.08-0.16	0.3 max	0.4-0.9		1.8-2.2	0.035 max	0.035 max	0.2-0.4	Cu 1.2-1.5; V 0.08-0.15; bal Fe				

UK

Specification	Designation	Notes	C	Cr	Mn	Mo	Ni	P	S	Si	Other	UTS	YS	El	Hard
BS 1502(82)	271	Bar Shp Press ves	0.17 max	0.50-1.00	1.00-1.50	0.20-0.35	0.30-0.70	0.040 max	0.040 max	0.15-0.40	Al <=0.020; W <=0.1; bal Fe				
BS 3100(91)	AW1	0-999mm	0.10-0.18	0.30 max	0.60-1.00	0.15 max	0.40 max	0.050 max	0.050 max	0.60 max	Cu <=0.30; Cr+Mo+Ni+Cu<=0.8; bal Fe	460		12	

USA

Specification	Designation	Notes	C	Cr	Mn	Mo	Ni	P	S	Si	Other	UTS	YS	El	Hard
	UNS J11442		0.11-0.17	0.35-0.65	0.65-1.00	0.15-0.35	0.35-0.75	0.04 max	0.04 max	0.50-1.00	Cu <=0.35; bal Fe				

UNS numbers and US grades are provided as a means of cross referencing chemically similar alloys. Exchangability is only possible after independent examination of specifications. Tensile properties are minimum or typical as specified. UTS and YS as MPa. El as %. See Appendix for list of abbreviations used in Notes. * indicates obsolete material.

Specification	Designation	Notes	C	Cr	Mn	Mo	Ni	P	S	Si	Other	UTS	YS	El	Hard

Steel Casting, Low-Alloy, 5334

USA

| | UNS J13042 | | 0.25-0.35 | 0.35-0.65 | 0.60-0.95 | 0.15-0.30 | 0.35-0.75 | 0.04 max | 0.04 max | 1.00 max | Cu <=0.35; bal Fe | | | | |

Steel Casting, Low-Alloy, 5335

USA

| | UNS J13050 | | 0.25-0.33 | 0.40-0.90 | 0.60-0.95 | 0.15-0.25 | 0.40-1.10 | 0.025 max | 0.025 max | 0.50-0.90 | Cu <=0.35; bal Fe | | | | |

Steel Casting, Low-Alloy, 5336

USA

| | UNS J13046 | | 0.25-0.35 | 0.80-1.10 | 0.40-0.80 | 0.15-0.25 | 0.25 max | 0.04 max | 0.04 max | 1.00 max | Cu <=0.35; bal Fe | | | | |

Steel Casting, Low-Alloy, 5338

USA

| | UNS J14046 | | 0.35-0.45 | 0.80-1.10 | 0.75-1.00 | 0.15-0.25 | 0.25 max | 0.04 max | 0.04 max | 1.00 max | Cu <=0.35; bal Fe | | | | |

Steel Casting, Low-Alloy, 5N

USA

| | UNS J13052 | | 0.30 max | 0.35 max | 0.70-1.00 | 0.15-0.25 | | 0.04 max | 0.045 max | 0.20-0.80 | Cu <=0.50; W <=0.10; bal Fe | | | | |
| ASTM A732/A732M(98) | 5N | N/T V | 0.30 max | | 0.70-1.00 | | | 0.04 max | 0.045 max | 0.20-0.80 | V 0.05-0.15; Cu+Ni+Cr+Mo+W<=1.00; bal Fe | 586 | 379 | 22 | |

Steel Casting, Low-Alloy, 6150

China

| JB/T 6402(92) | ZG40Cr1 | N/T | 0.35-0.45 | 0.80-1.10 | 0.50-0.80 | 0.05 max | 0.30 max | 0.035 max | 0.035 max | 0.20-0.40 | V <=0.03; bal Fe | 630 | 345 | 18 | |

Poland

| PNH83156(87) | L40HF | | 0.35-0.45 | 1-1.4 | 0.5-0.8 | | 0.3 max | 0.04 max | 0.04 max | 0.3-0.5 | V 0.15-0.3; bal Fe | | | | |
| PNH83160(88) | L40HF | | 0.35-0.45 | 1-1.4 | 0.5-0.8 | | 0.3 | | | 0.3-0.5 | Cu 0.3; V 0.15-0.3; bal Fe | | | | |

UK

BS 1956A	1956B		0.45-0.55	0.8-1.2	0.5-1			0.06	0.06	0.75	bal Fe				
BS 3146/1(74)	CLA12(A)	Inv casting	0.45-0.55	0.80-1.20	0.50-1.00	0.10 max	0.40 max	0.035 max	0.035 max	0.30-0.80	Cu <=0.30; bal Fe				
BS 3146/1(74)	CLA12(B)	Inv casting	0.45-0.55	0.80-1.20	0.50-1.00	0.10 max	0.40 max	0.035 max	0.035 max	0.30-0.80	Cu <=0.30; bal Fe				

USA

	UNS J15047		0.45-0.55	0.80-1.10	0.65-0.95			0.025 max	0.025 max	0.20-0.80	V >=0.15; bal Fe				
MIL-S-22141B(84)	IC-6150*	Inv HT; Obs for new design	0.45-0.55	0.80-1.10	0.65-0.95			0.04 max	0.04 max	0.20-0.80	V >=0.15; bal Fe	1310	1172	4	
MIL-S-22141B(84)	IC-6150*	Inv Ann; Obs for new design	0.45-0.55	0.80-1.10	0.65-0.95			0.04 max	0.04 max	0.20-0.80	V >=0.15; bal Fe				20 HRC

Steel Casting, Low-Alloy, 8620

France

AFNOR NFA32053(74)	FA-M	28mm; HT	0.23 max	0.3 max	1 max	0.15 max	0.4 max	0.04 max	0.035 max	0.5 max	W <=0.1; bal Fe	380-530	200	18	
AFNOR NFA32053(74)	FB-M	28mm; HT	0.23 max	0.3 max	1.2 max	0.15 max	1 max	0.04 max	0.035 max	0.5 max	W <=0.1; bal Fe	450-600	230	16	
AFNOR NFA32053(74)	FCIM	28mm; HT	0.23 max	0.3 max	0.8 max	0.45-0.65	0.4 max	0.04 max	0.035 max	0.5 max	W <=0.1; bal Fe	450-600	230	18	
AFNOR NFA32053(74)	FCM	28mm; HT	0.23 max	0.3 max	1.5 max	0.15 max	1 max	0.04 max	0.035 max	0.5 max	W <=0.1; bal Fe	520-670	260	16	

USA

| | UNS J12047 | | 0.15-0.25 | 0.40-0.60 | 0.65-0.95 | 0.15-0.25 | 0.40-0.70 | 0.025 max | 0.025 max | 0.20-0.80 | Cu <=0.50; bal Fe | | | | |
| | UNS J12095 | | 0.18-0.23 | 0.40-0.60 | 0.60-1.00 | 0.15-0.25 | 0.40-0.70 | 0.035 max | 0.040 max | 0.30-0.60 | bal Fe | | | | |

UNS numbers and US grades are provided as a means of cross referencing chemically similar alloys. Exchangability is only possible after independent examination of specifications. Tensile properties are minimum or typical as specified. UTS and YS as MPa. El as %. See Appendix for list of abbreviations used in Notes. * indicates obsolete material.

Specification	Designation	Notes	C	Cr	Mn	Mo	Ni	P	S	Si	Other	UTS	YS	El	Hard

Steel Casting, Low-Alloy, 8620 (Continued from previous page)

USA

Specification	Designation	Notes	C	Cr	Mn	Mo	Ni	P	S	Si	Other	UTS	YS	El	Hard
ASTM A915/A915M(98)	SC8620		0.18-0.23	0.40-0.60	0.60-1.00	0.15-0.25	0.40-0.70	0.035 max	0.040 max	0.30-0.60	bal Fe				
ASTM A958(96)	SC8620		0.18-0.23	0.40-0.60	0.60-1.00	0.15-0.25	0.40-0.70	0.035 max	0.040 max	0.30-0.60	bal Fe	450-795	240-655	14-24	
MIL-S-22141B(84)	IC-8620*	Inv HT; Obs for new design	0.15-0.25	0.40-0.60	0.65-0.95	0.15-0.25	0.40-0.70	0.04 max	0.04 max	0.20-0.80	bal Fe	724	586	10	
MIL-S-22141B(84)	IC-8620*	Inv Ann; Obs for new design	0.15-0.25	0.40-0.60	0.65-0.95	0.15-0.25	0.40-0.70	0.04 max	0.04 max	0.20-0.80	bal Fe				90 HRB

Steel Casting, Low-Alloy, 8625

USA

Specification	Designation	Notes	C	Cr	Mn	Mo	Ni	P	S	Si	Other	UTS	YS	El	Hard
	UNS J12595		0.23-0.28	0.40-0.60	0.60-1.00	0.15-0.25	0.40-0.70	0.035 max	0.040 max	0.30-0.60	bal Fe				
ASTM A915/A915M(98)	SC8625		0.23-0.28	0.40-0.60	0.60-1.00	0.15-0.25	0.40-0.70	0.035 max	0.040 max	0.30-0.60	bal Fe				
ASTM A958(96)	SC8625		0.23-0.28	0.40-0.60	0.60-1.00	0.15-0.25	0.40-0.70	0.035 max	0.040 max	0.30-0.60	bal Fe	450-930	240-860	9-24	

Steel Casting, Low-Alloy, 8630

Italy

Specification	Designation	Notes	C	Cr	Mn	Mo	Ni	P	S	Si	Other	UTS	YS	El	Hard
UNI 7316	FeG63-2		0.35	0.4-0.8	1		0.15-0.3	0.4-0.8		0.8	bal Fe				
UNI 7316	FeG74-2		0.35	0.4-0.8	1		0.15-0.3	0.4-0.8		0.8	bal Fe				

Japan

Specification	Designation	Notes	C	Cr	Mn	Mo	Ni	P	S	Si	Other	UTS	YS	El	Hard
JIS G5111(91)	SCMnCrM2	N/T	0.25-0.35	0.30-0.70	1.20-1.60	0.15-0.35		0.040 max	0.040 max	0.30-0.60	bal Fe	690	440	13	201 HB
JIS G5111(91)	SCMnCrM2	Q/A Tmp	0.25-0.35	0.30-0.70	1.20-1.60	0.15-0.35		0.040 max	0.040 max	0.30-0.60	bal Fe	740	540	13	212 HB

Poland

Specification	Designation	Notes	C	Cr	Mn	Mo	Ni	P	S	Si	Other	UTS	YS	El	Hard
PNH83156(87)	L30HNM		0.25-0.35	0.3-0.7	0.7-1	0.35-0.45	0.4-0.6			0.3-0.5	Cu 0.3; bal Fe				

USA

Specification	Designation	Notes	C	Cr	Mn	Mo	Ni	P	S	Si	Other	UTS	YS	El	Hard
	UNS J13049		0.25-0.35	0.40-0.60	0.65-0.95	0.15-0.25	0.40-0.70	0.025 max	0.025 max	0.20-0.80	bal Fe				
	UNS J13095		0.28-0.33	0.40-0.60	0.60-1.00	0.15-0.25	0.40-0.70	0.035 max	0.040 max	0.30-0.60	bal Fe				
AMS 5335E(94)*		Noncurrent; SCS	0.25-0.33	0.40-0.90	0.60-0.95	0.15-0.25	0.40-1.10	0.025 max	0.025 max	0.50-0.90	Cu <=0.35; bal Fe	1140	1035	8	331 HB
ASTM A915/A915M(98)	SC8630		0.28-0.33	0.40-0.60	0.60-1.00	0.15-0.25	0.40-0.70	0.035 max	0.040 max	0.30-0.60	bal Fe				
ASTM A958(96)	SC8630		0.28-0.33	0.40-0.60	0.60-1.00	0.15-0.25	0.40-0.70	0.035 max	0.040 max	0.30-0.60	bal Fe	450-1035	240-930	7-24	
MIL-S-22141B(84)	IC-8630*	Inv Ann; Obs for new design	0.25-0.35	0.40-0.60	0.65-0.95	0.15-0.25	0.40-0.70	0.04 max	0.04 max	0.20-0.80	bal Fe				90 HRB
MIL-S-22141B(84)	IC-8630*	Inv HT; Obs for new design	0.25-0.35	0.40-0.60	0.65-0.95	0.15-0.25	0.40-0.70	0.04 max	0.04 max	0.20-0.80	bal Fe	1034	792	7	

Steel Casting, Low-Alloy, 8640

USA

Specification	Designation	Notes	C	Cr	Mn	Mo	Ni	P	S	Si	Other	UTS	YS	El	Hard
	UNS J14048		0.35-0.45	0.40-0.60	0.70-1.05	0.15-0.25	0.40-0.70	0.025 max	0.025 max	0.20-0.80	bal Fe				
MIL-S-22141B(84)	IC-8640*	Inv Ann; Obs for new design	0.35-0.45	0.40-0.60	0.70-1.05	0.15-0.25	0.40-0.70	0.04 max	0.04 max	0.20-0.80	bal Fe				20 HRC
MIL-S-22141B(84)	IC-8640*	Inv HT; Obs for new design	0.35-0.45	0.40-0.60	0.70-1.05	0.15-0.25	0.40-0.70	0.04 max	0.04 max	0.20-0.80	bal Fe	1241	1000	5	

Steel Casting, Low-Alloy, 8735

China

Specification	Designation	Notes	C	Cr	Mn	Mo	Ni	P	S	Si	Other	UTS	YS	El	Hard
JB/T 6402(92)	ZG35NiCrMo	HT	0.30-0.37	0.40-0.90	0.70-1.00	0.40-0.50	0.60-0.90	0.035 max	0.035 max	0.60-0.90	bal Fe	830	660	14	

Finland

Specification	Designation	Notes	C	Cr	Mn	Mo	Ni	P	S	Si	Other	UTS	YS	El	Hard
SFS 368(77)	G-34CrMo4		0.3-0.37	0.9-1.2	0.5-0.8	0.15-0.25	0.3 max	0.035 max	0.035 max	0.6 max	bal Fe				

Specification	Designation	Notes	C	Cr	Mn	Mo	Ni	P	S	Si	Other	UTS	YS	El	Hard

Steel Casting, Low-Alloy, 8735 (Continued from previous page)

Japan

Specification	Designation	Notes	C	Cr	Mn	Mo	Ni	P	S	Si	Other	UTS	YS	El	Hard
JIS G5111(91)	SCCrM3	Q/A Tmp	0.30-0.40	0.80-1.20	0.50-0.80	0.15-0.35		0.040 max	0.040 max	0.30-0.60	bal Fe	740	540	9	217 HB
JIS G5111(91)	SCCrM3	N/T	0.30-0.40	0.80-1.20	0.50-0.80	0.15-0.35		0.040 max	0.040 max	0.30-0.60	bal Fe	690	440	9	201 HB

Poland

Specification	Designation	Notes	C	Cr	Mn	Mo	Ni	P	S	Si	Other	UTS	YS	El	Hard
PNH83156(87)	L35HNM		0.3-0.4	0.3-0.7	0.5-0.8	0.35-0.45	0.4-0.6	0.035 max	0.035 max	0.2-0.5	bal Fe				

Romania

Specification	Designation	Notes	C	Cr	Mn	Mo	Ni	P	S	Si	Other	UTS	YS	El	Hard
STAS 1773(82)	T35MoCrNi08	Q/T	0.3-0.4	0.6-1	0.5-0.8	0.15-0.25	0.6-1	0.03 max	0.03 max	0.45 max	Pb <=0.15; bal Fe				

Russia

Specification	Designation	Notes	C	Cr	Mn	Mo	Ni	P	S	Si	Other	UTS	YS	El	Hard
GOST 977(88)	35ChGSL		0.3-0.4	0.6-0.9	1-1.3		0.4 max	0.04 max	0.04 max	0.6-0.8	bal Fe				

USA

Specification	Designation	Notes	C	Cr	Mn	Mo	Ni	P	S	Si	Other	UTS	YS	El	Hard
	UNS J13442		0.30-0.38	0.35-0.90	0.30-0.70	0.15-0.40	0.35-0.75	0.04 max	0.04 max	0.20-1.00	bal Fe				
AMS 6282G(89)		Mech tub, HF	0.33-0.38	0.40-0.60	0.75-1.00	0.20-0.30	0.40-0.70	0.025 max	0.025 max	0.15-0.35	Cu <=0.35; bal Fe				40-99 HRC
AMS 6282G(89)		Mech tub, CF	0.33-0.38	0.40-0.60	0.75-1.00	0.20-0.30	0.40-0.70	0.025 max	0.025 max	0.15-0.35	Cu <=0.35; bal Fe				25 HRC max
MIL-S-22141B(84)	IC-8735*	Inv Ann; Obs for new design	0.30-0.38	0.35-0.90	0.30-0.70	0.15-0.40	0.35-0.75	0.025 max	0.025 max	0.20-1.00	bal Fe				90 HRB
MIL-S-22141B(84)	IC-8735*	Inv HT; Obs for new design	0.30-0.38	0.35-0.90	0.30-0.70	0.15-0.40	0.35-0.75	0.025 max	0.025 max	0.20-1.00	bal Fe	1379	1241	5	

Steel Casting, Low-Alloy, 8Q

USA

Specification	Designation	Notes	C	Cr	Mn	Mo	Ni	P	S	Si	Other	UTS	YS	El	Hard
	UNS J14049		0.35-0.45	0.80-1.10	0.70-1.00	0.15-0.25	0.50 max	0.04 max	0.045 max	0.20-0.80	Cu <=0.50; W <=0.10; bal Fe				
ASTM A732/A732M(98)	8Q	Q/A, Cr Mo	0.35-0.45	0.80-1.10	0.70-1.00	0.15-0.25		0.04 max	0.045 max	0.20-0.80	Cu+Ni+W<=1.00; bal Fe	1241	1000	5	

Steel Casting, Low-Alloy, 9Q

USA

Specification	Designation	Notes	C	Cr	Mn	Mo	Ni	P	S	Si	Other	UTS	YS	El	Hard
	UNS J23055		0.25-0.35	0.70-0.90	0.40-0.70	0.20-0.30	1.65-2.00	0.04 max	0.045 max	0.20-0.80	Cu <=0.50; W <=0.10; bal Fe				
ASTM A732/A732M(98)	9Q	Q/A, Cr Ni Mo	0.25-0.35	0.70-0.90	0.40-0.70	0.20-0.30	1.65-2.00	0.04 max	0.045 max	0.20-0.80	Cu+W<=0.60; bal Fe	1030	793	7	

Steel Casting, Low-Alloy, A356(6)

France

Specification	Designation	Notes	C	Cr	Mn	Mo	Ni	P	S	Si	Other	UTS	YS	El	Hard
AFNOR NFA32055(85)	15CD5.05M	t<1000mm	0.12-0.2	1-1.5	1 max	0.45-0.65	0.3 max	0.03 max	0.03 max	0.6 max	Cu <=0.3; Pb <=0.15; V <=0.04; W <=0.1; bal Fe	500-650	300	18	155-200 HB

Hungary

Specification	Designation	Notes	C	Cr	Mn	Mo	Ni	P	S	Si	Other	UTS	YS	El	Hard
MSZ 1749(89)	Ao16CrMo	High-temp const; 0-30.00mm; HT	0.12-0.2	0.9-1.5	0.5-0.8	0.4-0.6		0.03 max	0.03 max	0.3-0.6	bal Fe	600-650	300	20L	

Poland

Specification	Designation	Notes	C	Cr	Mn	Mo	Ni	P	S	Si	Other	UTS	YS	El	Hard
PNH83157(89)	L18HM	High-temp const	0.15-0.2	1-1.5	0.5-0.8	0.45-0.55	0.4 max	0.03 max	0.03 max	0.3-0.6	bal Fe				

USA

Specification	Designation	Notes	C	Cr	Mn	Mo	Ni	P	S	Si	Other	UTS	YS	El	Hard
	UNS J12073		0.20 max	1.00-1.50	0.50-0.80	0.45-0.65	0.50 max	0.035 max	0.030 max	0.60 max	Cu <=0.50; W <=0.10; bal Fe				
ASTM A356/A356M(98)	6	Heavy wall; steam turbines	0.20 max	1.00-1.50	0.50-0.80	0.45-0.65		0.035 max	0.030 max	0.60 max	bal Fe	485	310	22.0	

Steel Casting, Low-Alloy, A356(8)

USA

Specification	Designation	Notes	C	Cr	Mn	Mo	Ni	P	S	Si	Other	UTS	YS	El	Hard
	UNS J11697		0.13-0.20	1.00-1.50	0.50-0.90	0.90-1.20		0.035 max	0.030 max	0.20-0.60	V 0.05-0.15; bal Fe				
ASTM A356/A356M(98)	8	Heavy wall; steam turbines	0.20 max	1.00-1.50	0.50-0.90	0.90-1.20		0.035 max	0.030 max	0.20-0.60	bal Fe	550	345	18.0	

UNS numbers and US grades are provided as a means of cross referencing chemically similar alloys. Exchangability is only possible after independent examination of specifications. Tensile properties are minimum or typical as specified. UTS and YS as MPa. El as %. See Appendix for list of abbreviations used in Notes. * indicates obsolete material.

Specification	Designation	Notes	C	Cr	Mn	Mo	Ni	P	S	Si	Other	UTS	YS	El	Hard
Steel Casting, Low-Alloy, A356(9)															
Hungary															
MSZ 1749(89)	Ao17CrMoV	High-temp const; 0-30.00mm; HT	0.14-0.2	1.2-1.7	0.6-0.9	0.9-1.2		0.03 max	0.03 max	0.3-0.6	V 0.2-0.4; bal Fe	640-750	440	15L	
Japan															
JIS G5151(91)	SCPH23	High-temp; High-press	0.20 max	1.00-1.50	0.50-0.80	0.90-1.20	0.50 max	0.040 max	0.040 max	0.60 max	Cu <=0.50; V 0.15-0.25; W <=0.10; Cr+Cu+W<=1.00; bal Fe	550	345	13	
Poland															
PNH83157(89)	L17HMF	High-temp const	0.15-0.2	1.2-1.5	0.5-0.8	0.9-1.1	0.4 max	0.03 max	0.03 max	0.3-0.6	V 0.2-0.3; bal Fe				
USA															
	UNS J21610		0.13-0.20	1.00-1.50	0.50-0.90	0.90-1.20		0.035 max	0.030 max	0.20-0.60	V 0.20-0.35; bal Fe				
ASTM A356/A356M(98)	9	Heavy wall; steam turbines	0.20 max	1.00-1.50	0.50-0.90	0.90-1.20		0.035 max	0.030 max	0.20-0.60	bal Fe	585	415	15.0	
Yugoslavia															
	CL.7434		0.15-0.2	1.2-1.5	0.5-0.8	0.9-1.1		0.04 max	0.04 max	0.3-0.5	V 0.2-0.3; bal Fe				
Steel Casting, Low-Alloy, A487(1)															
Czech Republic															
CSN 422723	422723		0.15-0.25	0.3 max	1.1-1.5			0.04 max	0.04 max	0.2-0.5	V 0.1-0.25; P+S<= 0.07; bal Fe				
CSN 422724	422724		0.28-0.38	0.2-0.6	1.1-1.4		0.3 max	0.04 max	0.04 max	0.2-0.5	V 0.1-0.25; P+S<=0.07; bal Fe				
India															
IS 7899	1N	Q/A, Tmp	0.3		1			0.04	0.05	0.8	V 0.07-0.15; bal Fe	620	450	20	
IS 7899	1N	N/T	0.3		1			0.04	0.05	0.8	V 0.07-0.15; bal Fe	590	380	20	
IS 7899	1Q		0.3		1			0.04	0.05	0.8	V 0.07-0.15; bal Fe				
Romania															
STAS 11534	OTA18Mn14		0.14-0.22	0.3	1.2-1.6		0.3			0.2-0.5	Cu 0.3; V 0.08-0.15; bal Fe				
STAS 1773(82)	T26VSiMn14		0.22-0.3	0.3 max	1.2-1.6			0.04 max	0.04 max	0.6-0.9	Pb <=0.15; V 0.07-0.15; bal Fe				
USA															
	UNS J02506*	Obs; see J22500									bal Fe				
	UNS J13002		0.3 max	0.35 max	1.00 max		0.50 max	0.04 max	0.045 max	0.80 max	Cu <=0.50; V 0.04-0.12; Mo+W 0.25 max; bal Fe				
ASTM A487	1A	N/T, Q/T	0.30 max		1.00 max			0.04 max	0.045 max	0.80 max	Cu <=0.500; Mo+W<=0.25; Cu+Ni+Cr+Mo+W<=1.00; bal Fe	585-760	380	22	
ASTM A487	1B	N/T, Q/T	0.30 max		1.00 max			0.04 max	0.045 max	0.80 max	Cu <=0.50; Cu+Ni+Cr+Mo+W<=1.00; bal Fe	620-795	450	22	
ASTM A487	1C	N/T, Q/T	0.30 max		1.00 max			0.04 max	0.045 max	0.80 max	Cu <=0.50; Cu+Ni+Cr+Mo+W<=1.00; bal Fe	620	450	22	22 HRC
ASTM A487	1N*	N/T, Q/T	0.30 max		1.00 max			0.04 max	0.045 max	0.80 max	Cu <=0.50; Cu+Ni+Cr+Mo+W<=1.00; bal Fe	585-760	380	22	
ASTM A487	1Q*	N/T, Q/T	0.30 max		1.00 max			0.04 max	0.045 max	0.80 max	Cu <=0.50; Cu+Ni+Cr+Mo+W<=1.00; bal Fe	620-795	450	22	
Steel Casting, Low-Alloy, A487(10)															
Hungary															
MSZ 8272(89)	Ao30CrNiMo66	0-30.00mm; norm	0.25-0.35	1.3-1.6	0.4-0.9	0.2-0.3	1.3-1.6	0.04 max	0.04 max	0.2-0.4	bal Fe	690	540	12L	
MSZ 8272(89)	Ao30CrNiMo66ne	0-30.00mm; Q/T	0.25-0.35	1.3-1.6	0.4-0.9	0.2-0.3	1.3-1.6	0.04 max	0.04 max	0.2-0.4	bal Fe	780	640	10L	
India															
IS 7899	10N	N/T	0.3	0.55-0.9		0.2-0.4	1.4-2	0.04	0.05	0.8	bal Fe	690	480	16	

UNS numbers and US grades are provided as a means of cross referencing chemically similar alloys. Exchangability is only possible after independent examination of specifications. Tensile properties are minimum or typical as specified. UTS and YS as MPa. El as %. See Appendix for list of abbreviations used in Notes. * indicates obsolete material.

Specification	Designation	Notes	C	Cr	Mn	Mo	Ni	P	S	Si	Other	UTS	YS	El	Hard

Steel Casting, Low-Alloy, A487(10) (Continued from previous page)

India

Specification	Designation	Notes	C	Cr	Mn	Mo	Ni	P	S	Si	Other	UTS	YS	El	Hard
IS 7899	10N	Q/A, Tmp	0.3	0.55-0.9		0.2-0.4	1.4-2	0.04	0.05	0.8	bal Fe	860	690	14	
IS 7899	10Q		0.3	0.55-0.9		0.2-0.4	1.4-2	0.04	0.05	0.8	bal Fe				

Poland

Specification	Designation	Notes	C	Cr	Mn	Mo	Ni	P	S	Si	Other	UTS	YS	El	Hard
PNH83160(88)	L25SHNM	Wear res	0.2-0.33	0.6-0.9	0.5-0.9	0.2-0.5	1.5-2	0.035 max	0.035 max	1.1-1.4	bal Fe				

Spain

Specification	Designation	Notes	C	Cr	Mn	Mo	Ni	P	S	Si	Other	UTS	YS	El	Hard
UNE 36255(79)	AM30NiCrMo7*		0.27-0.34	0.6-0.9	0.5-0.8	0.2-0.4	1.4-2			0.3-0.5	bal Fe				
UNE 36255(79)	F.8351*	see AM30NiCrMo7*	0.27-0.34	0.6-0.9	0.5-0.8	0.2-0.4	1.4-2			0.3-0.5	bal Fe				

USA

Specification	Designation	Notes	C	Cr	Mn	Mo	Ni	P	S	Si	Other	UTS	YS	El	Hard
	UNS J22090		0.20 max	2.00-2.75	0.50-0.80	0.90-1.20		0.035 max	0.030 max	0.60 max	bal Fe				
	UNS J23015		0.30 max	0.55-0.90		0.20-0.40	1.40-2.00	0.04 max	0.045 max	0.80 max	Cu <=0.50; V <=0.03; W <=0.10; bal Fe				
ASTM A356/A356M(98)	10	Heavy wall; steam turbines	0.20 max	2.00-2.75	0.50-0.80	0.90-1.20		0.035 max	0.030 max	0.20-0.60	bal Fe	585	380	20.0	
ASTM A487	10A	N/T, Q/T	0.30 max	0.55-0.90	0.60-1.00	0.20-0.40	1.40-2.00	0.04 max	0.045 max	0.80 max	Cu <=0.50; V <=0.03; W <=0.10; Cu+W+V<=0.60; bal Fe	690	485	18	
ASTM A487	10B	N/T, Q/T	0.30 max	0.55-0.90	0.60-1.00	0.20-0.40	1.40-2.00	0.04 max	0.045 max	0.80 max	Cu <=0.50; V <=0.03; W <=0.10; Cu+W+V<=0.60; bal Fe	860	690	15	
ASTM A487	10N*	N/T, Q/T	0.30 max	0.55-0.90	0.60-1.00	0.20-0.40	1.40-2.00	0.04 max	0.045 max	0.80 max	Cu <=0.50; V <=0.03; W <=0.10; Cu+W+V<=0.60; bal Fe	690	485	18	
ASTM A487	10Q*	N/T, Q/T	0.30 max	0.55-0.90	0.60-1.00	0.20-0.40	1.40-2.00	0.04 max	0.045 max	0.80 max	Cu <=0.50; V <=0.03; W <=0.10; Cu+W+V<=0.60; bal Fe	860	690	15	

Steel Casting, Low-Alloy, A487(12)

Mexico

Specification	Designation	Notes	C	Cr	Mn	Mo	Ni	P	S	Si	Other	UTS	YS	El	Hard
DGN B-141	WC5*	Norm, Ann; Obs	0.2	0.5-0.9	0.4-0.7	0.9-1.2	0.6-1	0.04	0.05	0.6	bal Fe				

USA

Specification	Designation	Notes	C	Cr	Mn	Mo	Ni	P	S	Si	Other	UTS	YS	El	Hard
	UNS J22000		0.20 max	0.50-0.90	0.40-0.70	0.90-1.20	0.60-1.00	0.04 max	0.045 max	0.60 max	Cu <=0.50; V <=0.03; W <=0.10; bal Fe				
ASTM A217/A217M(95)	WC5	N/T	0.05-0.2	0.5-0.9	0.4-0.7	0.9-1.2	0.6-1	0.04	0.045	0.6	Cu 0.5; W 0.1; bal Fe				
ASTM A487	12A	N/T, Q/T	0.05-0.20	0.50-0.90	0.40-0.70	0.90-1.20	0.60-1.00	0.04 max	0.045 max	0.60 max	Cu <=0.50; V <=0.03; W <=0.10; Cu+W+V<=0.50; bal Fe	485-655	275	20	
ASTM A487	12B	N/T, Q/T	0.05-0.20	0.50-0.90	0.40-0.70	0.90-1.20	0.60-1.00	0.04 max	0.045 max	0.60 max	Cu <=0.50; V <=0.03; W <=0.10; Cu+W+V<=0.50; bal Fe	725-895	585	17	
ASTM A487	12N*	N/T, Q/T	0.05-0.20	0.50-0.90	0.40-0.70	0.90-1.20	0.60-1.00	0.04 max	0.045 max	0.60 max	Cu <=0.50; V <=0.03; W <=0.10; Cu+W+V<=0.50; bal Fe	485-655	275	20	
ASTM A487	12Q*	N/T, Q/T	0.05-0.20	0.50-0.90	0.40-0.70	0.90-1.20	0.60-1.00	0.04 max	0.045 max	0.60 max	Cu <=0.50; V <=0.03; W <=0.10; Cu+W+V<=0.50; bal Fe	725-895	585	17	

Steel Casting, Low-Alloy, A487(13)

USA

Specification	Designation	Notes	C	Cr	Mn	Mo	Ni	P	S	Si	Other	UTS	YS	El	Hard
	UNS J13080		0.30 max	0.40 max	0.80-1.10	0.2-0.3	1.40-1.75	0.04 max	0.045 max	0.60 max	Cu <=0.50; V <=0.03; W <=0.1; bal Fe				

UNS numbers and US grades are provided as a means of cross referencing chemically similar alloys. Exchangability is only possible after independent examination of specifications. Tensile properties are minimum or typical as specified. UTS and YS as MPa. El as %. See Appendix for list of abbreviations used in Notes. * indicates obsolete material.

Specification	Designation	Notes	C	Cr	Mn	Mo	Ni	P	S	Si	Other	UTS	YS	El	Hard

Steel Casting, Low-Alloy, A487(13) (Continued from previous page)

USA

Specification	Designation	Notes	C	Cr	Mn	Mo	Ni	P	S	Si	Other	UTS	YS	El	Hard
ASTM A487	13A	N/T, Q/T	0.30 max		0.80-1.10	0.20-0.30	1.40-1.75	0.04 max	0.045 max	0.60 max	Cu <=0.50; V <=0.03; W <=0.10; Cu+Cr+W+V<=0.75; bal Fe	620-795	415	18	
ASTM A487	13B	N/T, Q/T	0.30 max		0.80-1.10	0.20-0.30	1.40-1.75	0.04 max	0.045 max	0.60 max	Cu <=0.50; V <=0.03; W <=0.10; Cu+Cr+W+V<=0.75; bal Fe	725-895	585	17	
ASTM A487	13N*	N/T, Q/T	0.30 max		0.80-1.10	0.20-0.30	1.40-1.75	0.04 max	0.045 max	0.60 max	Cu <=0.50; V <=0.03; W <=0.10; Cu+Cr+W+V<=0.75; bal Fe	620-795	415	18	
ASTM A487	13Q*	N/T, Q/T	0.30 max		0.80-1.10	0.20-0.30	1.40-1.75	0.04 max	0.045 max	0.60 max	Cu <=0.50; V <=0.03; W <=0.10; Cu+Cr+W+V<=0.75; bal Fe	725-895	585	17	

Steel Casting, Low-Alloy, A487(14)

USA

Specification	Designation	Notes	C	Cr	Mn	Mo	Ni	P	S	Si	Other	UTS	YS	El	Hard
	UNS J15580		0.55 max	0.40 max	0.80-1.10	0.20-0.30	1.40-1.75	0.04 max	0.045 max	0.60 max	Cu <=0.50; V <=0.03; W <=0.10; bal Fe				
ASTM A487	14A	N/T, Q/T	0.55 max		0.80-1.10	0.20-0.30	1.40-1.75	0.04 max	0.045 max	0.60 max	Cu <=0.50; V <=0.03; W <=0.10; Cu+Cr+W+V<=0.75; bal Fe	825-1000	655	14	
ASTM A487	14Q*	N/T, Q/T	0.55 max		0.80-1.10	0.20-0.30	1.40-1.75	0.04 max	0.045 max	0.60 max	Cu <=0.50; V <=0.03; W <=0.10; Cu+Cr+W+V<=0.75; bal Fe	825-1000	655	14	
MIL-C-24707/1(89)	14A	As spec in ASTM A487; Mach, Struct									bal Fe				

Steel Casting, Low-Alloy, A487(2)

China

Specification	Designation	Notes	C	Cr	Mn	Mo	Ni	P	S	Si	Other	UTS	YS	El	Hard
JB/ZQ 4297(83)	ZG20SiMn	N/T	0.23 max	0.30 max	1.00-1.50	0.15 max	0.40 max	0.025 max	0.025 max	0.60 max	Cu <=0.25; V <=0.05; bal Fe	510	295	14	
JB/ZQ 4297(83)	ZG35CrMnSi	N/T	0.30-0.40	0.50-0.80	0.90-1.20	0.15 max	0.30 max	0.030 max	0.030 max	0.50-0.75	Cu <=0.25; V <=0.05; bal Fe	690	345	14	

India

Specification	Designation	Notes	C	Cr	Mn	Mo	Ni	P	S	Si	Other	UTS	YS	El	Hard
IS 7899	2N	Q/A, Tmp	0.3		1-1.4	0.1-0.3		0.04	0.05	0.8	bal Fe	620	450	20	
IS 7899	2N	N/T	0.3		1-1.4	0.1-0.3		0.04	0.05	0.8	bal Fe	590	360	20	
IS 7899	2Q		0.3		1-1.4	0.1-0.3		0.04	0.05	0.8	bal Fe				

USA

Specification	Designation	Notes	C	Cr	Mn	Mo	Ni	P	S	Si	Other	UTS	YS	El	Hard
	UNS J13005		0.3 max	0.35 max	1.00-1.40	0.10-0.30	0.50 max	0.04 max	0.045 max	0.80 max	Cu <=0.50; V <=0.03; W <=0.10; bal Fe				
ASTM A487	2A	N/T, Q/T	0.30 max		1.00-1.40	0.10-0.30		0.04 max	0.045 max	0.80 max	Cu <=0.50; V <=0.03; W <=0.10; Cu+Ni+Cr+Mo+W<=1.00; bal Fe	585-760	365	22	
ASTM A487	2B	N/T, Q/T	0.30 max		1.00-1.40	0.10-0.30		0.04 max	0.045 max	0.80 max	Cu <=0.50; V <=0.03; W <=0.10; Cu+Ni+Cr+Mo+W<=1.00; bal Fe	620-795	450	22	
ASTM A487	2C	N/T, Q/T	0.30 max		1.00-1.40	0.10-0.30		0.04 max	0.045 max	0.80 max	Cu <=0.50; V <=0.03; W <=0.10; Cu+Ni+Cr+Mo+W<=1.00; bal Fe	620	450	22	22 HRC
ASTM A487	2N*	N/T, Q/T	0.30 max		1.00-1.40			0.04 max	0.045 max	0.80 max	Cu <=0.50; V <=0.03; W <=0.10; Cu+Ni+Cr+Mo+W<=1.00; bal Fe	585-760	365	22	
ASTM A487	2Q*	N/T, Q/T	0.30 max		1.00-1.40	0.10-0.30		0.04 max	0.045 max	0.80 max	Cu <=0.50; V <=0.03; W <=0.10; Cu+Ni+Cr+Mo+W<=1.00; bal Fe	620-795	450	22	

Steel Casting, Low-Alloy, A487(4)

Specification	Designation	Notes	C	Cr	Mn	Mo	Ni	P	S	Si	Other	UTS	YS	El	Hard
India															
IS 7899	4N	Q/A, Tmp	0.3	0.4-0.8	1	0.15-0.3	0.4-0.8	0.04	0.05	0.8	bal Fe	720	590	15	
IS 7899	4N	N/T	0.3	0.4-0.8	1	0.15-0.3	0.4-0.8	0.04	0.05	0.8	bal Fe	620	410	18	
IS 7899	4Q		0.3	0.4-0.8	1	0.15-0.3	0.4-0.8	0.04	0.05	0.8	bal Fe				
Italy															
UNI 4010	FeGb5-2		0.25	0.4-0.8	0.8	0.15-0.35	0.4-0.8			0.5	bal Fe				
UNI 4010(75)	FeG70	HS	0.3 max	0.4-0.8	1-1.5	0.15-0.35	0.4-0.8	0.035 max	0.035 max	0.5 max	bal Fe				
UNI 4010(75)	FeG80	HS	0.35 max	0.4-0.8	1-1.5	0.15-0.35	1.5-2	0.035 max	0.035 max	0.5 max	bal Fe				
USA															
	UNS J13047		0.30 max	0.40-0.80	1.00 max	0.15-0.30	0.40-0.80	0.04 max	0.045 max	0.80 max	Cu <=0.50; V <=0.03; W <=0.10; bal Fe				
ASTM A487	4A	N/T, Q/T	0.30 max	0.40-0.80	1.00 max	0.15-0.30	0.40-0.80	0.04 max	0.045 max	0.80 max	Cu <=0.50; V <=0.03; W <=0.10; Cu+W+V<=0.60; bal Fe	620-795	415	18	
ASTM A487	4B	N/T, Q/T	0.30 max	0.40-0.80	1.00 max	0.15-0.30	0.40-0.80	0.04 max	0.045 max	0.80 max	Cu <=0.50; V <=0.03; W <=0.10; Cu+W+V<=0.60; bal Fe	725-895	585	17	
ASTM A487	4C	N/T, Q/T	0.30 max	0.40-0.80	1.00 max	0.15-0.30	0.40-0.80	0.04 max	0.045 max	0.80 max	Cu <=0.50; V <=0.03; W <=0.10; Cu+W+V<=0.60; bal Fe	620	415	18	22 HRC
ASTM A487	4D	N/T, Q/T	0.30 max	0.40-0.80	1.00 max	0.15-0.30	0.40-0.80	0.04 max	0.045 max	0.80 max	Cu <=0.50; V <=0.03; W <=0.10; Cu+W+V<=0.60; bal Fe	690	515	17	22 HRC
ASTM A487	4E	N/T, Q/T	0.30 max	0.40-0.80	1.00 max	0.15-0.30	0.40-0.80	0.04 max	0.045 max	0.80 max	Cu <=0.50; V <=0.03; W <=0.10; Cu+W+V<=0.60; bal Fe	795	655	15	
ASTM A487	4N*	N/T, Q/T	0.30 max	0.40-0.80	1.00 max	0.15-0.30	0.40-0.80	0.04 max	0.045 max	0.80 max	Cu <=0.50; V <=0.03; W <=0.10; Cu+W+V<=0.60; bal Fe	620-795	415	18	
ASTM A487	4Q*	N/T, Q/T	0.30 max	0.40-0.80	1.00 max	0.15-0.30	0.40-0.80	0.04 max	0.045 max	0.80 max	Cu <=0.50; V <=0.03; W <=0.10; Cu+W+V<=0.60; bal Fe	725-895	585	17	
ASTM A487	4QA*	N/T, Q/T	0.30 max	0.40-0.80	1.00 max	0.15-0.30	0.40-0.80	0.04 max	0.045 max	0.80 max	Cu <=0.50; V <=0.03; W <=0.10; Cu+W+V<=0.60; bal Fe	795	655	15	
MIL-C-24707/1(89)	4B	As spec in ASTM A487; Mach, Struct									bal Fe				

Steel Casting, Low-Alloy, A487(7)

Specification	Designation	Notes	C	Cr	Mn	Mo	Ni	P	S	Si	Other	UTS	YS	El	Hard
France															
AFNOR NFA32053(92)	16M5M	t<=30mm	0.20 max	0.4 max	1.1-1.5	0.2 max	1 max	0.025 max	0.02 max	0.6 max	V <=0.04; W <=0.1; Ni+Cr+Cu+Mo+V<=1; bal Fe	450	240	24	125 HB
India															
IS 7899	7N	Q/A, Tmp, 63.5mm diam	0.2	0.4-0.8	0.6-1	0.4-0.6	0.7-1	0.04	0.05	0.8	B 0.002-0.006; Cu 0.15-0.5; V 0.03-0.10; bal Fe	790	690	14	
IS 7899	7Q	Q/A, Tmp, 63.5mm diam	0.2	0.4-0.8	0.6-1	0.4-0.6	0.7-1	0.04	0.05	0.8	B 0.002-0.006; Cu 0.15-0.5; V 0.03-0.10; bal Fe	790	690	14	
UK															
BS 3100(91)	AL1	0-999mm	0.20 max	0.30 max	1.10 max	0.15 max	0.40 max	0.040 max	0.040 max	0.60 max	Cu <=0.30; Cr+Mo+Ni+Cu<=0.8; bal Fe	430	230	22	

Specification	Designation	Notes	C	Cr	Mn	Mo	Ni	P	S	Si	Other	UTS	YS	El	Hard

Steel Casting, Low-Alloy, A487(7) (Continued from previous page)

UK

Specification	Designation	Notes	C	Cr	Mn	Mo	Ni	P	S	Si	Other	UTS	YS	El	Hard
BS 3100(91)	B1	0-999mm	0.20 max	0.30 max	0.40-1.00	0.45-0.65	0.40 max	0.040 max	0.040 max	0.20-0.60	Cu <=0.30; Cr+Mo+Ni+Cu<=0.8; bal Fe	460	260	18	

USA

Specification	Designation	Notes	C	Cr	Mn	Mo	Ni	P	S	Si	Other	UTS	YS	El	Hard
	UNS J12084		0.20 max	0.40-0.80	0.60-1.00	0.40-0.60	0.70-1.00	0.04 max	0.045 max	0	B 0.002-0.006; Cu 0.15-0.50; V 0.03-0.10; W <=0.10; bal Fe				
ASTM A487	7A	N/T, Q/T	0.05-0.20	0.40-0.80	0.60-1.00	0.40-0.60	0.70-1.00	0.04 max	0.045 max	0.80 max	B 0.002-0.006; Cu <=0.50; W <=0.10; Cu+W<=0.60; bal Fe	795	690	15	
ASTM A487	7Q*	N/T, Q/T	0.05-0.20	0.40-0.80	0.60-1.00	0.40-0.60	0.70-1.00	0.04 max	0.045 max	0.80 max	B 0.002-0.006; Cu <=0.50; W <=0.10; Cu+W<=0.60; bal Fe	795	690	15	

Steel Casting, Low-Alloy, A487(8)

Finland

Specification	Designation	Notes	C	Cr	Mn	Mo	Ni	P	S	Si	Other	UTS	YS	El	Hard
SFS 370(77)	G-16CrMo99		0.18 max	2.0-2.5	0.5-0.8	0.9-1.1		0.04 max	0.04 max	0.6 max	bal Fe				

Germany

Specification	Designation	Notes	C	Cr	Mn	Mo	Ni	P	S	Si	Other	UTS	YS	El	Hard
DIN 10213(96)	WNr 1.7379	Press purposes	0.13-0.20	2.00-2.50	0.50-0.90	0.90-1.10		0.020 max	0.020 max	0.60 max	bal Fe				
DIN EN 10213(96)	G17CrMo9-10	Press ves	0.13-0.20	2.00-2.50	0.50-0.90	0.90-1.10		0.020 max	0.020 max	0.60 max	bal Fe				
DIN SEW 520(96)	G19CrMo9-10		0.15-0.22	2.00-2.50	0.60-1.00	0.90-1.10		0.015 max	0.010 max	0.60 max	bal Fe				
DIN SEW 520(96)	WNr 1.7382		0.15-0.22	2.00-2.50	0.60-1.00	0.90-1.10		0.015 max	0.010 max	0.60 max	bal Fe				
DIN SEW 685(89)	G15CrMo9-10		0.12-0.19	2.00-2.50	0.60-1.00	0.90-1.10		0.015 max	0.015 max	0.60 max	bal Fe				
DIN SEW 685(89)	WNr 1.7377		0.12-0.19	2.00-2.50	0.60-1.00	0.90-1.10		0.015 max	0.015 max	0.60 max	bal Fe				

Japan

Specification	Designation	Notes	C	Cr	Mn	Mo	Ni	P	S	Si	Other	UTS	YS	El	Hard
JIS G5151(91)	SCPH32	High-temp; High-press	0.20 max	2.00-2.75	0.50-0.80	0.90-1.20	0.50 max	0.040 max	0.040 max	0.60 max	Cu <=0.50; W <=0.10; Cr+Cu+W<=1.00; bal Fe	480	275	17	

Spain

Specification	Designation	Notes	C	Cr	Mn	Mo	Ni	P	S	Si	Other	UTS	YS	El	Hard
UNE 36259(79)	AM16CrMo10-10*		0.18 max	2.0-2.75	0.4-0.7	0.9-1.2		0.040 max	0.040 max	0.6 max	bal Fe				
UNE 36259(79)	F.8384*	see AM16CrMo10-10*	0.18 max	2.0-2.75	0.4-0.7	0.9-1.2		0.040 max	0.040 max	0.6 max	bal Fe				

UK

Specification	Designation	Notes	C	Cr	Mn	Mo	Ni	P	S	Si	Other	UTS	YS	El	Hard
BS 1504(76)	622 (BS 1398C)*		0.18	2-2.75	0.4-0.7	0.9-1.2	0.4			0.6	Cu 0.3; bal Fe				

USA

Specification	Designation	Notes	C	Cr	Mn	Mo	Ni	P	S	Si	Other	UTS	YS	El	Hard
	UNS J22091		0.20 max	2.00-2.75	0.50-0.90	0.90-1.10		0.04 max	0.045 max	0.80 max	Cu <=0.50; V <=0.03; W <=0.10; bal Fe				
ASTM A487	8A	N/T, Q/T	0.05-0.20	2.00-2.75	0.50-0.90	0.90-1.10		0.04 max	0.045 max	0.80 max	Cu <=0.50; V <=0.03; W <=0.10; Cu+W+V<=0.60; bal Fe	585-760	380	20	
ASTM A487	8B	N/T, Q/T	0.05-0.20	2.00-2.75	0.50-0.90	0.90-1.10		0.04 max	0.045 max	0.80 max	Cu <=0.50; V <=0.03; W <=0.10; Cu+W+V<=0.60; bal Fe	725	585	17	
ASTM A487	8C	N/T, Q/T	0.05-0.20	2.00-2.75	0.50-0.90	0.90-1.10		0.04 max	0.045 max	0.80 max	Cu <=0.50; V <=0.03; W <=0.10; Cu+W+V<=0.60; bal Fe	690	515	17	22 HRC
ASTM A487	8N*	N/T, Q/T	0.05-0.20	2.00-2.75	0.50-0.90	0.90-1.10		0.04 max	0.045 max	0.80 max	Cu <=0.50; V <=0.03; W <=0.10; Cu+W+V<=0.60; bal Fe	585-760	380	20	
ASTM A487	8Q*	N/T, Q/T	0.05-0.20	2.00-2.75	0.50-0.90	0.90-1.10		0.04 max	0.045 max	0.80 max	Cu <=0.50; V <=0.03; W <=0.10; Cu+W+V<=0.60; bal Fe	725	585	17	

UNS numbers and US grades are provided as a means of cross referencing chemically similar alloys. Exchangability is only possible after independent examination of specifications. Tensile properties are minimum or typical as specified. UTS and YS as MPa. El as %. See Appendix for list of abbreviations used in Notes. * indicates obsolete material.

Specification	Designation	Notes	C	Cr	Mn	Mo	Ni	P	S	Si	Other	UTS	YS	El	Hard
Steel Casting, Low-Alloy, A487(9)															
USA															
	UNS J13345		0.33 max	0.75-1.10	0.60-0.90	0.15-0.30	0.50 max	0.04 max	0.045 max	0.80 max	Cu <=0.50; V <=0.03; W <=0.1; bal Fo				
ASTM A487	9A	N/T, Q/T	0.05-0.33	0.75-1.10	0.60-1.00	0.15-0.30		0.04 max	0.045 max	0.80 max	Cu <=0.50; V <=0.03; W <=0.10; Cu+Ni+W+V=1.00; bal Fe	620	415	18	
ASTM A487	9B	N/T, Q/T	0.05-0.33	0.75-1.10	0.60-1.00	0.15-0.30		0.04 max	0.045 max	0.80 max	Cu <=0.50; V <=0.03; W <=0.10; Cu+Ni+W+V=1.00; bal Fe	725	585	16	
ASTM A487	9C	N/T, Q/T	0.05-0.33	0.75-1.10	0.60-1.00	0.15-0.30		0.04 max	0.045 max	0.80 max	Cu <=0.50; V <=0.03; W <=0.10; Cu+Ni+W+V<=1.00; bal Fe	620	415	18	22 HRC
ASTM A487	9D	N/T, Q/T	0.05-0.33	0.75-1.10	0.60-1.00	0.15-0.30		0.04 max	0.045 max	0.80 max	Cu <=0.50; V <=0.03; W <=0.10; Cu+Ni+W+V<=1.00; bal Fe	690	515	17	22 HRC
ASTM A487	9E	N/T, Q/T	0.05-0.33	0.75-1.10	0.60-1.00	0.15-0.30		0.04 max	0.045 max	0.80 max	Cu <=0.50; V <=0.03; W <=0.10; Cu+Ni+W+V=1.00; bal Fe	795	655	15	
ASTM A487	9N*	N/T, Q/T	0.05-0.33	0.75-1.10	0.60-1.00	0.15-0.30		0.04 max	0.045 max	0.80 max	Cu <=0.50; V <=0.03; W <=0.10; Cu+Ni+W+V=1.00; bal Fe	620	415	18	
ASTM A487	9Q*	N/T, Q/T	0.05-0.33	0.75-1.10	0.60-1.00	0.15-0.30		0.04 max	0.045 max	0.80 max	Cu <=0.50; V <=0.03; W <=0.10; Cu+Ni+W+V=1.00; bal Fe	725	585	16	
Steel Casting, Low-Alloy, A732(7Q)															
USA															
	UNS J13045		0.25-0.35	0.80-1.10	0.40-0.70	0.15-0.25		0.04 max	0.045 max	0.20-0.80	Cu <=0.50; W <=0.10; bal Fe				
ASTM A732/A732M(98)	7Q	Q/A, Cr Mo	0.25-0.35	0.80-1.10	0.40-0.70	0.15-0.25		0.04 max	0.045 max	0.20-0.80	Cu+W<=0.60; bal Fe	1030	793	7	
Steel Casting, Low-Alloy, C															
USA															
	UNS J22092		0.20 max	2.00-2.75	0.40-0.80	0.90-1.20	0.50 max	0.035 max	0.035 max	0.60 max	Cu <=0.50; V <=0.03; W <=0.10; bal Fe				
ASTM A757/A757M(96)	D1N1	Corr res, N/T	0.20 max	2.0-2.75	0.40-0.80	0.90-1.20		0.025 max	0.025 max	0.60 max	V+Cu+Ni+W<=1.00; bal Fe	585	380	20	
ASTM A757/A757M(96)	D1N2	Corr res, N/T	0.20 max	2.0-2.75	0.40-0.80	0.90-1.20		0.025 max	0.025 max	0.60 max	V+Cu+Ni+W<=1.00; bal Fe	655	515	18	
ASTM A757/A757M(96)	D1N3	Corr res, N/T	0.20 max	2.0-2.75	0.40-0.80	0.90-1.20		0.025 max	0.025 max	0.60 max	V+Cu+Ni+W<=1.00; bal Fe	725	585	15	
ASTM A757/A757M(96)	D1Q1	Corr res, Q/T	0.20 max	2.0-2.75	0.40-0.80	0.90-1.20		0.025 max	0.025 max	0.60 max	V+Cu+Ni+W<=1.00; bal Fe	795	380	20	
ASTM A757/A757M(96)	D1Q2	Corr res, Q/T	0.20 max	2.0-2.75	0.40-0.80	0.90-1.20		0.025 max	0.025 max	0.60 max	V+Cu+Ni+W<=1.00; bal Fe	860	515	18	
ASTM A757/A757M(96)	D1Q3	Corr res, Q/T	0.20 max	2.0-2.75	0.40-0.80	0.90-1.20		0.025 max	0.025 max	0.60 max	V+Cu+Ni+W<=1.00; bal Fe	930	585	15	
Steel Casting, Low-Alloy, C1Q															
Russia															
GOST 977(88)	20FL		0.14-0.25	0.3 max	0.7-1.2		0.4 max	0.05 max	0.05 max	0.2-0.52	V 0.06-0.12; bal Fe				
GOST 977(88)	20G1FL		0.16-0.25	0.3 max	0.9-1.4		0.4 max	0.05 max	0.05 max	0.2-0.5	V 0.06-0.12; bal Fe				
GOST 977(88)	20GL		0.15-0.25	0.3 max	1.2-1.6		0.4 max	0.04 max	0.04 max	0.2-0.4	bal Fe				
GOST 977(88)	20GNMFL		0.14-0.22	0.3 max	0.7-1.2	0.15-0.25	0.7-1	0.03 max	0.03 max	0.2-0.4	V 0.06-0.12; bal Fe				
USA															
	UNS J12582		0.25 max	0.40 max	1.20 max		1.5-2.0	0.025 max	0.025 max	0.60 max	Cu <=0.50; V <=0.03; bal Fe				

Specification	Designation	Notes	C	Cr	Mn	Mo	Ni	P	S	Si	Other	UTS	YS	El	Hard
Steel Casting, Low-Alloy, C1Q (Continued from previous page)															
USA															
ASTM A757/A757M(96)	C1Q	Corr res, Q/T	0.25 max		1.20 max	0.15-0.30	1.5-2.0	0.025 max	0.025 max	0.60 max	V+Cu+Cr<=1.00; bal Fe	515	380	22	
MIL-C-24707/1(89)	C1Q	As spec in ASTM A757; Mach, Struct									bal Fe				
Steel Casting, Low-Alloy, C23															
Bulgaria															
BDS 7478	16ChML	High-temp const	0.12-0.20	0.90-1.30	0.80 max	0.40-0.60	0.30 max	0.030 max	0.030 max	0.35-0.50	Cu <=0.30; bal Fe				
Czech Republic															
CSN 422740	422740		0.11-0.19	1.0-1.5	0.6-1.0	0.4-0.6	0.3 max	0.035 max	0.035 max	0.2-0.5	V 0.5-0.7; W 0.4-0.7; bal Fe				
Poland															
PNH83157(89)	L21HMF	High-temp const	0.18-0.25	0.9-1.2	0.4-0.7	0.5-0.7	0.3 max	0.03 max	0.03 max	0.2-0.5	V 0.2-0.35; bal Fe				
UK															
BS 3100(91)	B2	0-999mm	0.20 max	1.00-1.50	0.50-0.80	0.45-0.65	0.40 max	0.040 max	0.040 max	0.60 max	Cu <=0.30; bal Fe	480	280	17	140-212 HB
USA															
	UNS J12080		0.20 max	1.00-1.50	0.30-0.80	0.45-0.65		0.04 max	0.045 max	0.60 max	V 0.15-0.25; bal Fe				
ASTM A389/A389M(98)	C23	Sand, N/T	0.2	1-1.5	0.3-0.8	0.45-0.65		0.04	0.045	0.6	V 0.15-0.25; bal Fe				
MIL-C-24707/2(89)	C23	As spec in ASTM A389; Press, High-temp									bal Fe				
Steel Casting, Low-Alloy, C24															
USA															
	UNS J12092		0.20 max	0.80-1.25	0.30-0.80	0.90-1.20		0.04 max	0.045 max	0.60 max	V 0.15-0.25; bal Fe				
ASTM A389/A389M(98)	C24		0.20 max	0.80-1.25	0.30-0.80	0.90-1.20		0.04 max	0.045 max	0.60 max	V 0.15-0.25; bal Fe				
Steel Casting, Low-Alloy, C5															
Italy															
UNI 3608(83)	GX15CrMo5		0.2 max	4-6.5	0.8 max	0.45-0.65		0.035 max	0.035 max	0.75 max	bal Fe				
Japan															
JIS G5151(91)	SCPH61	High-temp; High-press	0.20 max	4.00-6.50	0.50-0.80	0.45-0.65	0.50 max	0.040 max	0.040 max	0.75 max	Cu <=0.50; W <=0.10; Cr+Cu+W<=1.00; bal Fe	620	410	17	
UK															
BS 3100(91)	B5	0-999mm	0.20 max	4.00-6.00	0.40-0.70	0.45-0.65	0.40 max	0.040 max	0.040 max	0.75 max	Cu <=0.30; bal Fe	620	420	13	179-255 HB
USA															
	UNS J42045		0.20 max	4.00-6.50	0.40-0.70	0.45-0.65	0.50 max	0.04 max	0.045 max	0.75 max	Cu <=0.50; W <=0.10; bal Fe				
ASTM A426(97)	CP5	Pipe; Heat res	0.20 max	4.00-6.50	0.30-0.70	0.45-0.65	0.50 max	0.040 max	0.045 max	0.75 max	Cu <=0.50; W <=0.10; Cu+Ni+W<=1.00; bal Fe	620	415	18	225 HB max
Steel Casting, Low-Alloy, CP1															
Japan															
JIS G5202(91)	SCPH11-CF	Cont cast; High-temp	0.20 max	0.35 max	0.30-0.60	0.45-0.65	0.50 max	0.035 max	0.035 max	0.60 max	Cu <=0.50; W <=0.10; bal Fe	380	205	19	
USA															
	UNS J12521		0.25 max		0.30-0.80	0.44-0.65		0.040 max	0.045 max	0.10-0.50	bal Fe				
ASTM A426(97)	CP1	Pipe; Heat res	0.25 max		0.30-0.80	0.44-0.65	0.50 max	0.040 max	0.045 max	0.10-0.50	Cu <=0.50; W <=0.10; Cu+Ni+W<=1.00; bal Fe	450	240	24	201 HB max

UNS numbers and US grades are provided as a means of cross referencing chemically similar alloys. Exchangability is only possible after independent examination of specifications. Tensile properties are minimum or typical as specified. UTS and YS as MPa. El as %. See Appendix for list of abbreviations used in Notes. * indicates obsolete material.

Specification	Designation	Notes	C	Cr	Mn	Mo	Ni	P	S	Si	Other	UTS	YS	El	Hard

Steel Casting, Low-Alloy, CP12

Bulgaria

Specification	Designation	Notes	C	Cr	Mn	Mo	Ni	P	S	Si	Other	UTS	YS	El	Hard
BDS 6550(86)	12DChNMFL		0.10-0.15	1.20-1.70	0.30-0.55	0.20-0.30	1.30-1.80	0.030 max	0.030 max	0.20-0.50	Cu 0.40-0.65; V 0.08-0.15; bal Fe				

Sweden

Specification	Designation	Notes	C	Cr	Mn	Mo	Ni	P	S	Si	Other	UTS	YS	El	Hard
SS 142223	2223		0.18 max	0.7-1.3	0.8 max	0.5-0.7		0.04 max	0.04 max	0.6 max	bal Fe				

USA

Specification	Designation	Notes	C	Cr	Mn	Mo	Ni	P	S	Si	Other	UTS	YS	El	Hard
	UNS J11562		0.15 max	0.80-1.25	0.30-0.61	0.44-0.65	0.50 max	0.040 max	0.045 max	0.50 max	Cu <=0.5; W <=0.10; bal Fe				
ASTM A426(97)	CP12	Pipe; Heat res	0.15 max	0.80-1.25	0.30-0.61	0.44-0.65	0.50 max	0.040 max	0.045 max	0.50 max	Cu <=0.50; W <=0.10; Cu+Ni+W<=1.00; bal Fe	415	205	22	201 HB max

Steel Casting, Low-Alloy, CP15

Bulgaria

Specification	Designation	Notes	C	Cr	Mn	Mo	Ni	P	S	Si	Other	UTS	YS	El	Hard
BDS 5930	09G2S	Boiler	0.12 max	0.30 max	1.30-1.70	0.15 max	0.30 max	0.035 max	0.040 max	0.50-0.80	Cu <=0.30; bal Fe				

China

Specification	Designation	Notes	C	Cr	Mn	Mo	Ni	P	S	Si	Other	UTS	YS	El	Hard
JB/T 6402(92)	ZG20CrMo	Q/T	0.17-0.25	0.50-0.80	0.50-0.80	0.40-0.60	0.30 max	0.035 max	0.035 max	0.20-0.45	Cu <=0.25; V <=0.05; bal Fe	460	245	18	

Hungary

Specification	Designation	Notes	C	Cr	Mn	Mo	Ni	P	S	Si	Other	UTS	YS	El	Hard
MSZ 8267	Ao21Mo5		0.2		0.6-0.9	0.45-0.65				0.6	Cu 0.3; bal Fe				

Spain

Specification	Designation	Notes	C	Cr	Mn	Mo	Ni	P	S	Si	Other	UTS	YS	El	Hard
UNE 36259(79)	AM18Mo5*		0.25		0.5-0.9	0.45-0.65				0.6	bal Fe				

UK

Specification	Designation	Notes	C	Cr	Mn	Mo	Ni	P	S	Si	Other	UTS	YS	El	Hard
BS 1504(76)	245*	Press ves	0.2 max	0.25	0.5-1	0.45-0.65	0.4 max	0.04 max	0.04 max	0.2-0.6	Cu <=0.3; Pb <=0.15; V <=0.1; W <=0.1; Cr+Ni+Cu<=0.8; bal Fe				
BS 3100(91)	AM1	0-999mm	0.15 max	0.30 max	0.50 max	0.15 max	0.40 max	0.050 max	0.050 max	0.60 max	Cu <=0.30; Cr+Mo+Ni+Cu<=0.8; bal Fe	340-430	185	22	
BS 3100(91)	B7	0-999mm	0.10-0.15	0.30-0.50	0.40-0.70	0.40-0.60	0.30 max	0.030 max	0.030 max	0.45 max	Cu <=0.30; V 0.22-0.30; Sn<=0.025; bal Fe	510	295	17	
BS 3604/1(90)	660	Pipe; Tub; High-press/temp smls weld t<=200mm	0.10-0.15	0.30-0.60	0.40-0.70	0.50-0.70	0.30 max	0.030 max	0.030 max	0.10-0.35	Al <=0.02; Cu <=0.25; V 0.22-0.28; Sn<=0.025; bal Fe	460-660	300	20	

USA

Specification	Designation	Notes	C	Cr	Mn	Mo	Ni	P	S	Si	Other	UTS	YS	El	Hard
	UNS J11522		0.15 max	0.35 max	0.30-0.60	0.44-0.65	0.50 max	0.040 max	0.045 max	0.15-1.65	Cu <=0.5; W <=0.10; bal Fe				
ASTM A426(97)	CP15	Pipe; Heat res	0.15 max	0.35 max	0.30-0.60	0.44-0.65	0.50 max	0.040 max	0.045 max	0.15-1.65	Cu <=0.50; W <=0.10; Cu+Ni+W<=1.00; bal Fe	415	205	22	201 HB max

Steel Casting, Low-Alloy, CP2

Italy

Specification	Designation	Notes	C	Cr	Mn	Mo	Ni	P	S	Si	Other	UTS	YS	El	Hard
UNI 3608(83)	G20CrMo52		0.25 max	0.4-0.7	0.8 max	0.4-0.6		0.035 max	0.035 max	0.6 max	bal Fe				

Spain

Specification	Designation	Notes	C	Cr	Mn	Mo	Ni	P	S	Si	Other	UTS	YS	El	Hard
UNE 36256(73)	AMC15K*	Q/T	0.2 max		1.3 max	0.15 max		0.035 max	0.035 max	0.3-0.6	bal Fe				
UNE 36256(73)	F.8371*	see AMC15K*	0.2 max		1.3 max	0.15 max		0.035 max	0.035 max	0.3-0.6	bal Fe				
UNE 36259(79)	AM17CrMo02-05*		0.2	0.4-0.7	0.5-0.8	0.45-0.65				0.6	bal Fe				
UNE 36259(79)	F.8382*	see AM17CrMo02-05*	0.2	0.4-0.7	0.5-0.8	0.45-0.65				0.6	bal Fe				

USA

Specification	Designation	Notes	C	Cr	Mn	Mo	Ni	P	S	Si	Other	UTS	YS	El	Hard
	UNS J11547		0.10-0.20	0.50-0.81	0.30-0.61	0.44-0.65	0.50 max	0.040 max	0.045 max	0.10-0.50	Cu <=0.5; W <=0.10; bal Fe				

Specification	Designation	Notes	C	Cr	Mn	Mo	Ni	P	S	Si	Other	UTS	YS	El	Hard

Steel Casting, Low-Alloy, CP2 (Continued from previous page)

USA

Specification	Designation	Notes	C	Cr	Mn	Mo	Ni	P	S	Si	Other	UTS	YS	El	Hard
ASTM A426(97)	CP2	Pipe; Heat res	0.10-0.20	0.50-0.81	0.30-0.61	0.44-0.65	0.50 max	0.040 max	0.045 max	0.10-0.50	Cu <=0.50; W <=0.10; Cu+Ni+W<=1.00; bal Fe	415	205	22	201 HB max

Steel Casting, Low-Alloy, CP21

USA

Specification	Designation	Notes	C	Cr	Mn	Mo	Ni	P	S	Si	Other	UTS	YS	El	Hard
	UNS J31545		0.15 max	2.65-3.35	0.30-0.60	0.80-1.06	0.50 max	0.040 max	0.045 max	0.50 max	Cu <=0.50; W <=0.10; bal Fe				
ASTM A426(97)	CP21	Pipe; Heat res	0.15 max	2.65-3.35	0.30-0.60	0.80-1.06	0.50 max	0.040 max	0.045 max	0.50 max	Cu <=0.50; W <=0.10; Cu+Ni+W<=1.00; bal Fe	415	205	22	201 HB max

Steel Casting, Low-Alloy, CP7

Russia

Specification	Designation	Notes	C	Cr	Mn	Mo	Ni	P	S	Si	Other	UTS	YS	El	Hard
GOST 977(88)	12Ch7G3SL		0.10-0.15	7-7.5	3-3.5		0.4 max	0.02 max	0.02 max	0.8-1.2	bal Fe				

USA

Specification	Designation	Notes	C	Cr	Mn	Mo	Ni	P	S	Si	Other	UTS	YS	El	Hard
	UNS J61594		0.15 max	6.00-8.00	0.30-0.60	0.44-0.65	0.50 max	0.040 max	0.045 max	0.50-1.00	Cu <=0.50; W <=0.10; bal Fe				
ASTM A426(97)	CP9	Pipe; Heat res	0.2 max	8-10	0.3-0.65	0.9-1.2	0.50 max	0.040 max	0.045	0.25-1.00	Cu <=0.50; W <=0.10; Cu+Ni+W<=1.00; bal Fe	620	415	18	225 HB max

Steel Casting, Low-Alloy, J11875

Russia

Specification	Designation	Notes	C	Cr	Mn	Mo	Ni	P	S	Si	Other	UTS	YS	El	Hard
GOST 977(88)	12DChN1MFL		0.10-0.18	1.2-1.7	0.3-0.55	0.2-0.3	1.4-1.8	0.03 max	0.03 max	0.2-0.4	Cu 0.4-0.65; V 0.08-0.15; bal Fe				

USA

Specification	Designation	Notes	C	Cr	Mn	Mo	Ni	P	S	Si	Other	UTS	YS	El	Hard
	UNS J11875		0.18 max	1.00-1.60	0.40-0.70	0.40-0.60	0.60 max	0.05 max	0.05 max	0.60 max	Cu <=0.50; V 0.10-0.20; bal Fe				

Steel Casting, Low-Alloy, J12070

USA

Specification	Designation	Notes	C	Cr	Mn	Mo	Ni	P	S	Si	Other	UTS	YS	El	Hard
	UNS J12070		0.20 max	1.00-1.60	0.50-0.80	0.40-0.60	0.60 max	0.05 max	0.05 max	0.20-0.60	Cu <=0.50; bal Fe				

Steel Casting, Low-Alloy, J12545

USA

Specification	Designation	Notes	C	Cr	Mn	Mo	Ni	P	S	Si	Other	UTS	YS	El	Hard
	UNS J12545		0.25 max	0.40 max	1.15-1.50	0.45-0.60	0.45-1.00	0.035 max	0.035 max	0.60 max	Cu <=0.50; V <=0.03; bal Fe				

Steel Casting, Low-Alloy, J21880

USA

Specification	Designation	Notes	C	Cr	Mn	Mo	Ni	P	S	Si	Other	UTS	YS	El	Hard
	UNS J21880		0.18 max	2.00-2.75	0.40-0.70	0.80-1.10	0.60 max	0.05 max	0.05 max	0.20-0.60	Cu <=0.50; bal Fe				

Steel Casting, Low-Alloy, LC1

Germany

Specification	Designation	Notes	C	Cr	Mn	Mo	Ni	P	S	Si	Other	UTS	YS	El	Hard
DIN EN 10213(96)	22Mo4	Quen <=60mm	0.18-0.25	0.30 max	0.40-0.70	0.30-0.40		0.030 max	0.030 max	0.20-0.40	bal Fe	490-590	295	20	
DIN EN 10213(96)	WNr 1.5419	Bar Q<=60mm	0.18-0.25	0.30 max	0.40-0.70	0.30-0.40		0.030 max	0.030 max	0.20-0.40	bal Fe	490-590	295	20	

Hungary

Specification	Designation	Notes	C	Cr	Mn	Mo	Ni	P	S	Si	Other	UTS	YS	El	Hard
MSZ 1749(89)	Ao18Mo	High-temp const; 0-30.00mm; HT	0.14-0.22		0.5-0.8	0.4-0.6		0.03 max	0.03 max	0.3-0.6	bal Fe	450-600	250	22L	

Italy

Specification	Designation	Notes	C	Cr	Mn	Mo	Ni	P	S	Si	Other	UTS	YS	El	Hard
UNI 3608(83)	G20Mo5		0.25 max		0.8 max	0.4-0.6		0.035 max	0.035 max	0.6 max	bal Fe				
UNI 7317(74)	G22Mo5	Tough at subzero	0.25 max		0.8 max	0.45-0.65		0.035 max	0.035 max	0.6 max	bal Fe				

Japan

Specification	Designation	Notes	C	Cr	Mn	Mo	Ni	P	S	Si	Other	UTS	YS	El	Hard
JIS G5152(91)	SCPL11	Low-temp; High-press	0.25 max	0.35 max	0.50-0.80	0.45-0.65		0.040 max	0.040 max	0.60 max	Cu <=0.50; bal Fe	450	245	21	

UNS numbers and US grades are provided as a means of cross referencing chemically similar alloys. Exchangability is only possible after independent examination of specifications. Tensile properties are minimum or typical as specified. UTS and YS as MPa. El as %. See Appendix for list of abbreviations used in Notes. * indicates obsolete material.

Specification	Designation	Notes	C	Cr	Mn	Mo	Ni	P	S	Si	Other	UTS	YS	El	Hard

Steel Casting, Low-Alloy, LC1 (Continued from previous page)

Spain

Specification	Designation	Notes	C	Cr	Mn	Mo	Ni	P	S	Si	Other	UTS	YS	El	Hard
UNE 36259(79)	AMC20*		0.18-0.23		0.5-0.8	0.15 max		0.040 max	0.040 max	0.6 max	bal Fe				
UNE 36259(79)	F.8380*	see AMC20*	0.18-0.23		0.5-0.0	0.15 max		0.040 max	0.040 max	0.6 max	bal Fe				

USA

Specification	Designation	Notes	C	Cr	Mn	Mo	Ni	P	S	Si	Other	UTS	YS	El	Hard
	UNS J12522		0.25 max		0.50-0.80	0.45-0.65		0.04 max	0.045 max	0.60 max	bal Fe				
ASTM A352/A352M(98)	LC1	Press; low-temp	0.25 max		0.50-0.80	0.45-0.65		0.04 max	0.045 max	0.60 max	bal Fe	450-620	240	24	

Steel Casting, Low-Alloy, LC2

France

Specification	Designation	Notes	C	Cr	Mn	Mo	Ni	P	S	Si	Other	UTS	YS	El	Hard
AFNOR NFA32055(85)	20N12M	t<1000mm	0.23 max	0.3 max	0.5-0.8	3.1-3.7	2.9-3.5	0.025 max	0.025 max	0.1-0.4	Cu <=0.3; Pb <=0.15; V <=0.1; W <=0.1; bal Fe	450-600	230	18	135-180 HB
AFNOR NFA32055(85)	20N12M	t<1000mm	0.23 max	0.4 max	1.5 max	0.3 max	2.4-4	0.03 max	0.03 max	0.6 max	Cu <=0.3; Pb <=0.15; V <=0.04; W <=0.1; Cr+Mo+V+Cu<=1; bal Fe	450-600	230	18	135-180 HB

USA

Specification	Designation	Notes	C	Cr	Mn	Mo	Ni	P	S	Si	Other	UTS	YS	El	Hard
	UNS J22500		0.25 max		0.50-0.80		2.0-3.0	0.04 max	0.045 max	0.60 max	bal Fe				
ASTM A352/A352M(98)	UNS J22500	Press; low-temp	0.25 max		0.50-0.80		2.00-3.00	0.04 max	0.045 max	0.60 max	bal Fe	485-655	275	24	

Steel Casting, Low-Alloy, MIL-A-11356

USA

Specification	Designation	Notes	C	Cr	Mn	Mo	Ni	P	S	Si	Other	UTS	YS	El	Hard
	UNS J13025		0.30 max	0.50-0.70	0.30-0.50	0.20 max	0.70 max			0.50 max	V <=0.10; P+S 0.07 max				

Steel Casting, Low-Alloy, MIL-S-870

USA

Specification	Designation	Notes	C	Cr	Mn	Mo	Ni	P	S	Si	Other	UTS	YS	El	Hard
	UNS J12520	*	0.25 max	0.35 max	0.50-0.80	0.45-0.65	0.50 max	0.04 max	0.45 max	0.30-0.60	Cu <=0.50; bal Fe				

Steel Casting, Low-Alloy, Nitralloy

USA

Specification	Designation	Notes	C	Cr	Mn	Mo	Ni	P	S	Si	Other	UTS	YS	El	Hard
	UNS J24056		0.35-0.45	1.40-1.80	0.40-0.70	0.30-0.45		0.025 max	0.025 max	0.20-0.80	Al 0.85-1.20; bal Fe				
MIL-S-22141B(84)	IC-Nitralloy 135M*	Inv HT; Obs for new design	0.35-0.45	1.40-1.80	0.40-0.70	0.30-0.45		0.04 max	0.04 max	0.20-0.80	bal Fe	931	690	8	

Steel Casting, Low-Alloy, No equivalents identified

Bulgaria

Specification	Designation	Notes	C	Cr	Mn	Mo	Ni	P	S	Si	Other	UTS	YS	El	Hard
BDS 5930	14ChM	Boiler	0.10-0.18	0.70-1.10	0.40-0.70	0.15 max	0.30 max	0.040 max	0.040 max	0.15-0.37	Cu <=0.30; V 0.20-0.35; bal Fe				

Czech Republic

Specification	Designation	Notes	C	Cr	Mn	Mo	Ni	P	S	Si	Other	UTS	YS	El	Hard
CSN 422726	422726		0.3-0.4	0.6-0.9	1.1-1.3			0.04 max	0.04 max	0.6-0.8	P+S<=0.08; bal Fe				
CSN 422731	422731		0.11-0.19	0.9-1.3	0.6-1.0			0.035 max	0.035 max	0.2-0.5	V 0.2-0.35; P+S<=0.06; bal Fe				
CSN 422733	422733		0.17-0.25	0.9-1.3	1.1-1.5			0.04 max	0.04 max	0.2-0.5	V 0.15-0.25; P+S<=0.07; bal Fe				
CSN 422735	422735		0.5-0.65	0.9-1.3	0.9-1.3			0.05 max	0.05 max	0.2-0.5	P+S<=0.09; bal Fe				
CSN 422736	422736		0.6-0.75	1.8-2.2	0.8-1.2			0.05 max	0.05 max	0.2-0.5	P+S<=0.09; bal Fe				
CSN 422739	422739		0.45-0.6	0.6-0.9	0.5-0.9			0.045 max	0.045 max	0.2-0.5	P+S<=0.09; bal Fe				
CSN 422743	422743		0.11-0.19	0.5-0.7	0.4-0.7	0.2-0.35		0.035 max	0.035 max	0.2-0.5	V 0.2-0.35; bal Fe				
CSN 422744	422744		0.11-0.19	0.5-0.7	0.4-0.7	0.4-0.6		0.035 max	0.035 max	0.2-0.5	V 0.2-0.35; bal Fe				
CSN 422745	422745		0.11-0.19	0.4-0.6	0.4-0.8	0.9-1.1		0.035 max	0.035 max	0.2-0.5	V 0.2-0.35; bal Fe				
CSN 422750	422750		0.35-0.45	0.4-0.7	0.6-0.9		1.6-2.0	0.045 max	0.045 max	0.2-0.5	P+S<=0.09; bal Fe				

UNS numbers and US grades are provided as a means of cross referencing chemically similar alloys. Exchangability is only possible after independent examination of specifications. Tensile properties are minimum or typical as specified. UTS and YS as MPa. El as %. See Appendix for list of abbreviations used in Notes. * indicates obsolete material.

Specification	Designation	Notes	C	Cr	Mn	Mo	Ni	P	S	Si	Other	UTS	YS	El	Hard

Steel Casting, Low-Alloy, No equivalents identified

Czech Republic

Specification	Designation	Notes	C	Cr	Mn	Mo	Ni	P	S	Si	Other	UTS	YS	El	Hard
CSN 422753	422753		0.7-0.85	1.3-1.7	0.6-0.9	0.4-0.6	0.5-0.8	0.05 max	0.05 max	0.2-0.5	V 0.1-0.25; P+S<=0.09; bal Fe				
CSN 422771	422771		0.15-0.22	4.0-6.0	0.4-0.7	0.45-0.65	0.5 max	0.035 max	0.035 max	0.2-0.5	bal Fe				

France

Specification	Designation	Notes	C	Cr	Mn	Mo	Ni	P	S	Si	Other	UTS	YS	El	Hard
AFNOR NFA32054(78)	12MDV6M	HT; t<1000mm; Air Hard & Tmp	0.15 max	0.5 max	1.2-1.5	0.2-0.4	0.5 max	0.04 max	0.035 max	0.6 max	Cu <=0.5; Pb <=0.15; V 0.05-0.1; W <=0.1; Cu+Ni+Cr<=1; bal Fe	500	400	18	
AFNOR NFA32054(78)	20M6M	HT Const; t<1000mm	0.23 max	0.5 max	1.7 max	0.025 max	0.5 max	0.04 max	0.035 max	0.6 max	Cu <=0.5; Pb <=0.15; V <=0.04; W <=0.1; Cr+Mo+Ni+V+Cu<=1; bal Fe	510	310	16	
AFNOR NFA32054(78)	22MD5M	t<1000mm	0.25 max	0.5 max	1.4 max	0.2-0.4	0.5 max	0.04 max	0.035 max	0.6 max	Cu <=0.5; Pb <=0.15; V <=0.04; W <=0.1; Cu+Ni+Cr+V<=1; bal Fe	500	350	18	
AFNOR NFA32054(78)	32NCD14M	t<1000mm	0.35 max	0.8-1.2	0.8 max	0.3-0.6	3-4	0.04 max	0.035 max	0.6 max	Cu <=0.5; Pb <=0.15; V <=0.120; W <=0.1; V<=0.04/0.05-0.12; bal Fe	1100	1000	7	
AFNOR NFA32054(78)	35CD4M	t<1000mm	0.38 max	0.8-1.2	1 max	0.15-0.35	0.5 max	0.04 max	0.035 max	0.6 max	Cu <=0.5; Pb <=0.15; V <=0.04; W <=0.1; Cu+Ni+V<=1; bal Fe	750	520	12	
AFNOR NFA32055(85)	20M5M	Weld, Boiler & Press ves; Norm; Air Hard Temp; Q/T; t<1000mm	0.22 max		0.4 max	1.2 max	0.2 max	0.3 max	0.03 max	0.03 max	0.5 max	Cu <=0.3; Pb <=0.15; V <=0.04; W <=0.1; Cr+Mo+Ni+V+Cu<=1; bal Fe	470-590	235	20	140-135 HB
AFNOR NFA32055(85)	20MN5M	Boiler, Pres ves; t<1000mm	0.22 max		0.4 max	1.2 max	0.3 max	0.5 max	0.03 max	0.03 max	0.6 max	Cu <=0.3; Pb <=0.15; V <=0.04; W <=0.1; Cr+V+Cu<=1; bal Fe	485-610	280	22	145-190 HB

Hungary

Specification	Designation	Notes	C	Cr	Mn	Mo	Ni	P	S	Si	Other	UTS	YS	El	Hard
MSZ 8272(89)	Ao35MnCrSi5	0-30.00mm; norm	0.3-0.4	0.6-0.9	1-1.3		0.3 max	0.04 max	0.04 max	0.6-0.8	bal Fe	590	340	14L	
MSZ 8272(89)	Ao35MnCrSi5ne	0-30.00mm; Q/T	0.3-0.4	0.6-0.9	1-1.3		0.3 max	0.04 max	0.04 max	0.6-0.8	bal Fe	780	590	10L	
MSZ 8272(89)	Ao40Cr4ne	0-30.00mm; Q/T	0.35-0.45	0.8-1.1	0.4-0.9		0.3 max	0.05 max	0.05 max	0.2-0.4	bal Fe	640	490	12L	
MSZ 8272(89)	Ao40CrV5	0-30.00mm; norm	0.35-0.45	1-1.4	0.5-0.8		03 max	0.04 max	0.04 max	0.2-0.5	V 0.15-0.3; bal Fe	785	590	10L	
MSZ 8272(89)	Ao40CrV5ne	0-30.00mm; Q/T	0.35-0.45	1-1.4	0.5-0.8		0.3 max	0.04 max	0.04 max	0.2-0.5	V 0.15-0.3; bal Fe	1200	900	4L	

Italy

Specification	Designation	Notes	C	Cr	Mn	Mo	Ni	P	S	Si	Other	UTS	YS	El	Hard
UNI 4010(75)	FeG65-1	HS	0.35 max	0.7-1.2	1-1.5			0.035 max	0.035 max	0.5 max	bal Fe				
UNI 4010(75)	FeG65-2	HS	0.25 max	0.4-0.8	0.8 max	0.15-0.35	0.4-0.8	0.035 max	0.035 max	0.5 max	bal Fe				

Japan

Specification	Designation	Notes	C	Cr	Mn	Mo	Ni	P	S	Si	Other	UTS	YS	El	Hard
JIS G5151(91)	SCPH11	High-temp; High-press	0.25 max	0.35 max	0.50-0.80	0.45-0.65	0.50 max	0.040 max	0.040 max	0.60 max	Cu <=0.50; W <=0.10; Cr+Cu+Ni+W<=1.00; bal Fe	450	245	22	

Poland

Specification	Designation	Notes	C	Cr	Mn	Mo	Ni	P	S	Si	Other	UTS	YS	El	Hard
PNH83156(87)	L17HM		0.13-0.22	0.5-0.7	0.5-0.8	0.4-0.6	0.3 max	0.035 max	0.035 max	0.3-0.5	bal Fe				
PNH83156(87)	L25HM		0.22-0.29	0.8-1.2	0.5-0.8	0.2-0.3	0.4 max	0.035 max	0.35 max	0.2-0.5	bal Fe				
PNH83156(87)	L25HN		0.2-0.3	0.4-0.6	0.4-0.9		0.6-0.8	0.04 max	0.04 max	0.2-0.5	bal Fe				
PNH83156(87)	L30H		0.25-0.35	0.5-0.8	0.5-0.9		0.3 max	0.04 max	0.04 max	0.2-0.5	bal Fe				
PNH83156(87)	L30NM		0.25-0.3	0.3-0.7	0.7-1.1	0.4-0.55	0.4-0.6	0.04 max	0.04 max	0.3-0.5	bal Fe				
PNH83156(87)	L40H		0.35-0.45	0.8-1.1	0.5-0.8		0.3	0.04	0.04	0.1-0.4	Cu 0.3; bal Fe				
PNH83156(87)	L40H		0.35-0.45	0.8-1.1	0.4-0.9		0.3 max	0.04 max	0.04 max	0.2-0.5	bal Fe				
PNH83157(73)	L17M	High-temp const	0.14-0.2	0.3 max	0.3-0.5	0.4-0.6	0.3 max	0.035 max	0.035 max	0.17-0.37	bal Fe				

3-40/Steel Casting

Steel Casting, Low-Alloy, No equivalents identified

Specification	Designation	Notes	C	Cr	Mn	Mo	Ni	P	S	Si	Other	UTS	YS	El	Hard
Poland															
PNH83157(89)	L15HMF	High-temp const	0.11-0.19	0.4-0.6	0.4-0.8	0.2-0.35	0.4 max	0.03 max	0.03 max	0.2-0.5	V 0.2-0.35; bal Fe				
PNH83157(89)	L16M	High-temp const	0.11-0.19	0.4-0.6	0.4-0.8	0.2-0.35	0.4 max	0.03 max	0.03 max	0.2-0.5	V 0.2-0.35; bal Fe				
PNH83157(89)	L18H2M	High-temp const	0.15-0.2	2-2.5	0.5-0.8	0.9-1.1	0.4 max	0.03 max	0.03 max	0.3-0.6	bal Fe				
PNH83157(89)	L20HM	High-temp const	0.15-0.25	0.4-0.7	0.5-0.8	0.4-0.6	0.3 max	0.03 max	0.03 max	0.2-0.5	bal Fe				
PNH83158(86)	L0H12N4M	Corr res	0.6 max	0.3 max	2 max	1 max	3.5-5	0.035 max	0.035 max	0.6 max	bal Fe				
PNH83160(88)	L100H6M	Wear res	0.9-1.1	5.5-6.5	0.2-0.5	0.8-1.1	0.4 max	0.03 max	0.03 max	0.2-0.7	bal Fe				
PNH83160(88)	L20HGSNM	Wear res	0.18-0.25	0.6-0.9	0.8-1.1	0.1-0.2	0.9-1.2	0.035 max	0.035 max	0.7-1	bal Fe				
PNH83160(88)	L40HM	Wear res	0.32-0.4	1-1.3	0.6-0.9	0.4-0.6	0.6 max	0.035 max	0.035 max	0.3-0.4	bal Fe				
Romania															
STAS 1773(82)	T14VMoCuCrNi16		0.1-0.18	1.2-1.7	0.3-0.55	0.2-0.3	1.4-1.8	0.03 max	0.03 max	0.2-0.45	Cu 0.4-0.65; Pb <=0.15; V 0.08-0.15; bal Fe				
STAS 1773(82)	T30MoCrNi14		0.25-0.35	1.3-1.6	0.4-0.9	0.2-0.3	1.3-1.6	0.04 max	0.04 max	0.2-0.45	Pb <=0.15; bal Fe				
STAS 1773(82)	T34MoCr09	Corr res; Heat res	0.3-0.38	0.8-1.1	0.5-0.8	0.2-0.3		0.04 max	0.04 max	0.2-0.45	Pb <=0.15; bal Fe				
STAS 1773(82)	T35SiCrMn11	Q/T	0.3-0.4	0.6-0.9	1-1.3			0.04 max	0.04 max	0.5-0.8	Pb <=0.15; bal Fe				
STAS 1773(82)	T40MnNi07		0.35-0.45	0.3 max	1-1.2		0.6-0.9	0.04 max	0.04 max	0.2-0.45	Pb <=0.15; bal Fe				
STAS 1773(82)	T40TiCrNi17		0.35-0.45	0.7-1	0.4-0.9		1.5-1.9	0.04 max	0.04 max	0.2-0.45	Pb <=0.15; Ti 0.02-0.1; bal Fe				
STAS 1773(82)	T40VMn17	Corr res	0.35-0.45	0.3 max	2 max	2-3	9-12	0.04 max	0.03 max	0.5-2	Pb <=0.15; bal Fe				
STAS 9277(84)	OTA30VMn13		0.25-0.35	0.3 max	1.2-1.4		0.7-1	0.04 max	0.035 max	0.3-0.5	Pb <=0.15; V 0.2-0.4; bal Fe				
STAS 9277(84)	OTA40Mn13		0.35-0.45	0.3 max	1.2-1.4			0.04 max	0.035 max	0.3-0.5	Pb <=0.15; bal Fe				
Spain															
UNE 36027(80)	100Cr6	Ball & roller bearing	0.95-1.1	1.35-1.65	0.25-0.45	0.15 max		0.03 max	0.025 max	0.15-0.35	bal Fe				
UNE 36027(80)	F.1310*	see 100Cr6	0.95-1.1	1.35-1.65	0.25-0.45	0.15 max		0.03 max	0.025 max	0.15-0.35	bal Fe				
UNE 36254(79)	34CrMo4	Cast	0.3-0.38	0.8-1.2	0.5-0.8	0.15-0.3		0.04 max	0.04 max	0.3-0.5	bal Fe				
UNE 36255(79)	AM34CrMo4*	Q/T	0.3-0.38	0.8-1.2	0.5-0.8	0.15-0.3		0.040 max	0.040 max	0.3-0.5	bal Fe				
UNE 36255(79)	AM42CrMo4*	Q/T	0.38-0.45	0.8-1.2	0.5-0.8	0.15-0.3		0.040 max	0.040 max	0.3-0.5	bal Fe				
UNE 36255(79)	F.8331*	see AM34CrMo4*	0.3-0.38	0.8-1.2	0.5-0.8	0.15-0.3		0.040 max	0.040 max	0.3-0.5	bal Fe				
UNE 36255(79)	F.8332*	see AM42CrMo4*	0.38-0.45	0.8-1.2	0.5-0.8	0.15-0.3		0.040 max	0.040 max	0.3-0.5	bal Fe				
USA															
ASTM A487	16A	N/T, Q/T	0.12 max	0.20 max	2.10 max	0.10 max	1.00-1.40	0.02 max	0.02 max	0.50 max	Cu <=0.20; V <=0.02; W <=0.10; Cu+Cr+Mo+W+V<=1.00; bal Fe	485-655	275	22	
ASTM A487	16N*	N/T, Q/T	0.12 max	0.20 max	2.10 max	0.10 max	1.00-1.40	0.02 max	0.02 max	0.50 max	Cu <=0.20; V <=0.02; W <=0.10; Cu+Cr+Mo+W+V<=1.00; bal Fe	485-655	275	22	
Yugoslavia															
	CL.4100		0.25-0.35	0.5-0.8	0.5-0.9	0.15 max		0.05 max	0.05 max	0.17-0.37	bal Fe				

Steel Casting, Low-Alloy, WC1

Specification	Designation	Notes	C	Cr	Mn	Mo	Ni	P	S	Si	Other	UTS	YS	El	Hard
Spain															
UNE 36259(79)	AM18Mo5*		0.25		0.5-0.8	0.45-0.65				0.6	bal Fe				

UNS numbers and US grades are provided as a means of cross referencing chemically similar alloys. Exchangability is only possible after independent examination of specifications. Tensile properties are minimum or typical as specified. UTS and YS as MPa. El as %. See Appendix for list of abbreviations used in Notes. * indicates obsolete material.

Specification	Designation	Notes	C	Cr	Mn	Mo	Ni	P	S	Si	Other	UTS	YS	El	Hard

Steel Casting, Low-Alloy, WC1 (Continued from previous page)

Spain

Specification	Designation	Notes	C	Cr	Mn	Mo	Ni	P	S	Si	Other	UTS	YS	El	Hard
UNE 36259(79)	F.8381*	see AM18Mo5*	0.25		0.5-0.8	0.45-0.65				0.6	bal Fe				

UK

Specification	Designation	Notes	C	Cr	Mn	Mo	Ni	P	S	Si	Other	UTS	YS	El	Hard
BS 1504(76)	240*		0.15-0.25	0.25	0.5-1	0.4-0.7	0.4			0.2-0.5	Cu 0.3; bal Fe				
BS 3100(91)	BL1	old En4242B	0.2		1	0.45-0.65				0.6	bal Fe				

USA

Specification	Designation	Notes	C	Cr	Mn	Mo	Ni	P	S	Si	Other	UTS	YS	El	Hard
	UNS J12524		0.25 max		0.50-0.80	0.45-0.65		0.04 max	0.045 max	0.60 max	Cu <=0.50; W <=0.10; bal Fe				
ASTM A217/A217M(95)	WC1		0.25	0.35	0.5-0.8	0.45-0.65	0.5	0.04	0.045	0.6	Cu 0.5; W 0.1; Cu+Ni+Cr+W<=1; bal Fe				
MIL-C-24707/2(89)	WC1	As spec in ASTM A217; Press, High-temp									bal Fe				

Steel Casting, Low-Alloy, WC11

Bulgaria

Specification	Designation	Notes	C	Cr	Mn	Mo	Ni	P	S	Si	Other	UTS	YS	El	Hard
BDS 7478	15ChGFL	High-temp const	0.10-0.20	0.95-1.30	0.60-1.00		0.30 max	0.030 max	0.030 max	0.20-0.50	Cu <=0.30; bal Fe				
BDS 7478	16ChML	High-temp const	0.12-0.20	0.90-1.30	0.80 max	0.40-0.60	0.30 max	0.030 max	0.030 max	0.35-0.50	Cu <=0.30; bal Fe				

China

Specification	Designation	Notes	C	Cr	Mn	Mo	Ni	P	S	Si	Other	UTS	YS	El	Hard
GB/T 14408(93)	ZGD270-480	N/T	0.20 max	1.00-1.50	0.30-0.80	0.45-0.65		0.040 max	0.040 max	0.60 max	V 0.15-0.25; W <=0.10; bal Fe	480	270	18	

Japan

Specification	Designation	Notes	C	Cr	Mn	Mo	Ni	P	S	Si	Other	UTS	YS	El	Hard
JIS G5151(91)	SCPH22	High-temp; High-press	0.25 max	1.00-1.50	0.50-0.80	0.90-1.20	0.50 max	0.040 max	0.040 max	0.60 max	Cu <=0.50; W <=0.10; Cr+Cu+W<=1.00; bal Fe	550	345	16	

Romania

Specification	Designation	Notes	C	Cr	Mn	Mo	Ni	P	S	Si	Other	UTS	YS	El	Hard
STAS 9277(84)	OTA17MoCr13		0.15-0.2	1-1.5	0.5-0.8	0.4-0.6		0.04 max	0.045 max	0.3-0.5	Pb <=0.15; bal Fe				

Spain

Specification	Designation	Notes	C	Cr	Mn	Mo	Ni	P	S	Si	Other	UTS	YS	El	Hard
UNE 36259(79)	AM18CrMo05-05*		0.2	1-1.5	0.5-0.8	0.45-0.65				0.6	bal Fe				

UK

Specification	Designation	Notes	C	Cr	Mn	Mo	Ni	P	S	Si	Other	UTS	YS	El	Hard
BS 3100(91)	B4	0-999mm	0.25 max	2.50-3.50	0.30-0.70	0.35-0.60	0.40 max	0.040 max	0.040 max	0.75 max	Cu <=0.30; bal Fe	620	370	13	179-255 HB

USA

Specification	Designation	Notes	C	Cr	Mn	Mo	Ni	P	S	Si	Other	UTS	YS	El	Hard
	UNS J11872		0.15-0.21	1.00-1.50	0.50-0.80	0.45-0.65	0.50 max	0.020 max	0.015 max	0.30-0.60	Al <=0.01; Cu <=0.35; V <=0.03; bal Fe				
ASTM A217/A217M(95)	WC11		0.15-0.21	1-1.5	0.5-0.8	0.45-0.65	0.5	0.02	0.015	0.3-0.6	Al 0.01; Cu 0.35; V 0.03; Cu+Ni+V<=1; bal Fe				

Steel Casting, Low-Alloy, WC4

Germany

Specification	Designation	Notes	C	Cr	Mn	Mo	Ni	P	S	Si	Other	UTS	YS	El	Hard
DIN	G20NiMoCr3-7	CH 30mm	0.17-0.23	0.30-0.50	0.70-1.10	0.40-0.80	0.60-1.10	0.015 max	0.015 max	0.60 max	bal Fe	980-1270	685	8	
DIN SEW 685(89)	WNr 1.6750		0.17-0.23	0.30-0.50	0.70-1.10	0.40-0.80	0.60-1.10	0.015 max	0.015 max	0.60 max	bal Fe				

Spain

Specification	Designation	Notes	C	Cr	Mn	Mo	Ni	P	S	Si	Other	UTS	YS	El	Hard
UNE 36256(73)	AM10Ni10*	Q/T	0.2 max		0.5-0.8	0.15 max	2.0-3.0	0.035 max	0.035 max	0.3-0.6	bal Fe				
UNE 36256(73)	F.8373*	see AM10Ni10*	0.2 max		0.5-0.8	0.15 max	2.0-3.0	0.035 max	0.035 max	0.3-0.6	bal Fe				

USA

Specification	Designation	Notes	C	Cr	Mn	Mo	Ni	P	S	Si	Other	UTS	YS	El	Hard
	UNS J12082		0.20 max	0.50-0.80	0.50-0.80	0.45-0.65	0.70-1.10	0.04 max	0.045 max	0.60 max	Cu <=0.50; V <=0.03; W <=0.10; bal Fe				
ASTM A217/A217M(95)	WC4	N/T	0.2	0.5-0.8	0.5-0.8	0.45-0.65	0.7-1.1	0.04	0.045	0.6	Cu 0.5; W <=0.1; Cu+W<=0.60; bal Fe				

UNS numbers and US grades are provided as a means of cross referencing chemically similar alloys. Exchangability is only possible after independent examination of specifications. Tensile properties are minimum or typical as specified. UTS and YS as MPa. El as %. See Appendix for list of abbreviations used in Notes. * indicates obsolete material.

Specification	Designation	Notes	C	Cr	Mn	Mo	Ni	P	S	Si	Other	UTS	YS	El	Hard

Steel Casting, Low-Alloy, WC4 (Continued from previous page)

USA

Specification	Designation	Notes	C	Cr	Mn	Mo	Ni	P	S	Si	Other	UTS	YS	El	Hard
ASTM A487	11A	N/T, Q/T	0.05-0.20	0.50-0.80	0.50-0.80	0.45-0.65	0.70-1.10	0.04 max	0.045 max	0.60 max	Cu <=0.50; V <=0.03; W <=0.10; Cu+W+V<=0.50; bal Fe	484-655	275	20	
ASTM A487	11B	N/T, Q/T	0.05-0.20	0.50-0.80	0.50-0.80	0.45-0.65	0.70-1.10	0.04 max	0.045 max	0.60 max	Cu <=0.50; V <=0.03; W <=0.10; Cu+W+V<=0.50; bal Fe	725-895	585	17	
ASTM A487	11N*	N/T, Q/T	0.05-0.20	0.50-0.80	0.50-0.80	0.45-0.65	0.70-1.10	0.04 max	0.045 max	0.60 max	Cu <=0.50; V <=0.03; W <=0.10; Cu+W+V<=0.50; bal Fe	484-655	275	20	
ASTM A487	11Q*	N/T, Q/T	0.05-0.20	0.50-0.80	0.50-0.80	0.45-0.65	0.70-1.10	0.04 max	0.045 max	0.60 max	Cu <=0.50; V <=0.03; W <=0.10; Cu+W+V<=0.50; bal Fe	725-895	585	17	

Steel Casting, Low-Alloy, WC6

Australia

Specification	Designation	Notes	C	Cr	Mn	Mo	Ni	P	S	Si	Other	UTS	YS	El	Hard
AS 2074(82)	L5B		0.2 max	1-1.5	0.5-0.8	0.45-0.65	0.4 max	0.05 max	0.05 max	0.6 max	Cu <=0.3; bal Fe				

Bulgaria

Specification	Designation	Notes	C	Cr	Mn	Mo	Ni	P	S	Si	Other	UTS	YS	El	Hard
BDS 7478	16ChML	High-temp const	0.12-0.20	0.90-1.30	0.80 max	0.40-0.60	0.30 max	0.030 max	0.030 max	0.35-0.50	Cu <=0.30; bal Fe				

Finland

Specification	Designation	Notes	C	Cr	Mn	Mo	Ni	P	S	Si	Other	UTS	YS	El	Hard
SFS 369(77)	G-17CrMo55		0.2 max	1.0-1.5	0.5-0.8	0.45-0.55		0.04 max	0.04 max	0.6 max	Cu <=0.4; bal Fe				

Germany

Specification	Designation	Notes	C	Cr	Mn	Mo	Ni	P	S	Si	Other	UTS	YS	El	Hard
DIN EN 10213(96)	G17CrMo5-5	Press purposes	0.15-0.20	1.00-1.50	0.50-1.00	0.45-0.65		0.020 max	0.020 max	0.60 max	bal Fe				
DIN EN 10213(96)	WNr 1.7357	Press purposes	0.15-0.20	1.00-1.50	0.50-1.00	0.45-0.65		0.020 max	0.020 max	0.60 max	bal Fe				

Italy

Specification	Designation	Notes	C	Cr	Mn	Mo	Ni	P	S	Si	Other	UTS	YS	El	Hard
UNI 3608(83)	G15CrMo55		0.2 max	1-1.5	0.8 max	0.45-0.65		0.035 max	0.035 max	0.6 max	bal Fe				

Japan

Specification	Designation	Notes	C	Cr	Mn	Mo	Ni	P	S	Si	Other	UTS	YS	El	Hard
JIS G5151(91)	SCPH21	High-temp; High-press	0.20 max	1.00-1.50	0.50-0.80	0.45-0.65	0.50 max	0.040 max	0.040 max	0.6 max	Cu <=0.50; W <=0.10; Cr+Cu+W<=1.00; bal Fe	480	275	17	
JIS G5202(91)	SCPH21-CF	Cont cast; High-temp	0.15 max	1.00-1.50	0.30-0.60	0.45-0.65	0.50 max	0.030 max	0.030 max	0.60 max	Cu <=0.50; W <=0.10; bal Fe	410	205	19	

Spain

Specification	Designation	Notes	C	Cr	Mn	Mo	Ni	P	S	Si	Other	UTS	YS	El	Hard
UNE 36259(79)	AM18CrMo05-05*		0.2	1-1.5	0.5-0.8	0.45-0.65				0.6	bal Fe				
UNE 36259(79)	F.8383*	see AM18CrMo05-05*	0.2	1-1.5	0.5-0.8	0.45-0.65				0.6	bal Fe				

UK

Specification	Designation	Notes	C	Cr	Mn	Mo	Ni	P	S	Si	Other	UTS	YS	El	Hard
BS 1504(76)	621 (BS1398B)*		0.2	1-1.5	0.5-0.8	0.45-0.65	0.4			0.6	Cu 0.3; bal Fe				

USA

Specification	Designation	Notes	C	Cr	Mn	Mo	Ni	P	S	Si	Other	UTS	YS	El	Hard
	UNS J12072		0.20 max	1.00-1.50	0.50-0.80	0.45-0.65	0.50 max	0.04 max	0.045 max	0.60 max	Cu <=0.50; W <=0.10; bal Fe				
ASTM A217/A217M(95)	WC6	N/T	0.2	1-1.5	0.5-0.8	0.45-0.65	0.5	0.04	0.045	0.6	Cu 0.5; W 0.1; Cu+Ni+W<=1; bal Fe				
ASTM A426(97)	CP11	Pipe; Heat res	0.05-0.2	1-1.5	0.3-0.8	0.44-0.65	0.50 max	0.040 max	0.045	0.60 max	Cu <=0.50; W <=0.10; Cu+Ni+W<=1.00; bal Fe	485	275	20	201 HB max
MIL-C-24707/2(89)	WC6	As spec in ASTM A217; Press, High-temp									bal Fe				

Steel Casting, Low-Alloy, WC9

Bulgaria

Specification	Designation	Notes	C	Cr	Mn	Mo	Ni	P	S	Si	Other	UTS	YS	El	Hard
BDS 6609(73)	12Ch2M		0.08-0.15	2.00-2.5	0.40-0.70	0.90-1.10	0.30 max	0.040 max	0.040 max	0.15-0.50	Cu <=0.30; bal Fe				

UNS numbers and US grades are provided as a means of cross referencing chemically similar alloys. Exchangability is only possible after independent examination of specifications. Tensile properties are minimum or typical as specified. UTS and YS as MPa. El as %. See Appendix for list of abbreviations used in Notes. * indicates obsolete material.

Specification	Designation	Notes	C	Cr	Mn	Mo	Ni	P	S	Si	Other	UTS	YS	El	Hard

Steel Casting, Low-Alloy, WC9 (Continued from previous page)

Germany

Specification	Designation	Notes	C	Cr	Mn	Mo	Ni	P	S	Si	Other	UTS	YS	El	Hard
DIN	GS-12CrMo910*	Obs; see 10CrMo9-10	0.08-0.15	2-2.5	0.4-0.7	0.9-1.1		0.04 max	0.04 max	0.3-0.5	bal Fe				

Italy

UNI 3608(83)	G14CrMo910	High-temp const	0.18 max	2-2.75	0.8 max	0.9-1.2		0.035 max	0.035 max	0.6 max	bal Fe				
UNI 3608(83)	G14CrMoV9102		0.18 max	2-2.75	0.8 max	0.9-1.2		0.035 max	0.035 max	0.6 max	V 0.15-0.25; bal Fe				

Japan

JIS G5202(91)	SCPH32-CF	Cont cast; High-temp	0.15 max	1.90-2.60	0.30-0.60	0.90-1.20	0.50 max	0.030 max	0.030 max	0.60 max	Cu <=0.50; W <=0.10; bal Fe	410	205	19	

Mexico

DGN B-141	WC9*	Norm, Ann; Obs	0.18	2-2.75	0.4-0.7	0.9-1.2		0.04	0.05		bal Fe				

Spain

UNE 36259(79)	AM16CrMo10-10*		0.18	2-2.75	0.4-0.7	0.9-1.2				0.6	bal Fe				

Sweden

SS 142224	2224		0.18 max	2-2.5	0.7 max	0.9-1.2		0.04 max	004 max	0.6 max	bal Fe				

UK

BS 1503(89)	622-490	Frg Press ves; 0-999mm; N/T Q/T	0.15 max	2.0-2.5	0.4-0.7	0.9-1.2	0.4 max	0.03 max	0.025 max	0.15-0.4	Al <=0.02; Cu <=0.3; Pb <=0.15; V <=0.1; W <=0.1; bal Fe	490-640	275	18L	
BS 1503(89)	622-560	Frg Press ves; 0-999mm; N/T Q/T	0.15 max	2.0-2.5	0.4-0.7	0.9-1.2	0.4 max	0.03 max	0.025 max	0.15-0.4	Al <=0.02; Cu <=0.3; Pb <=0.15; V <=0.1; W <=0.1; bal Fe	560-710	370	17	
BS 1503(89)	622-650	Frg Press ves; 0-999mm; Q/T	0.15 max	2.0-2.5	0.4-0.7	0.9-1.2	0.4 max	0.03 max	0.025 max	0.15-0.4	Al <=0.02; Cu <=0.3; Pb <=0.15; V <=0.1; W <=0.1; bal Fe	650-800	475	16	
BS 1504(76)	622 (BS1398C)*		0.18	2-2.75	0.4-0.7	0.9-1.2	0.4			0.6	Cu 0.3; bal Fe				
BS 3100(91)	B3	0-999mm	0.18 max	2.00-2.75	0.40-0.70	0.90-1.20	0.40 max	0.040 max	0.040 max	0.60 max	Cu <=0.30; bal Fe	540	325	17	156-235 HB

USA

	UNS J21890		0.18 max	2.00-2.75	0.40-0.70	0.90-1.20	0.50 max	0.04 max	0.045 max	0.60 max	Cu <=0.50; W <=0.10; bal Fe				
ASTM A217/A217M(95)	WC9	N/T	0.18	2-2.75	0.4-0.7	0.9-1.2	0.5	0.04	0.045	0.6	Cu 0.5; W 0.1; Cu+Ni+W<=1; bal Fe				
ASTM A426(97)	CP22	Pipe; Heat res	0.18	2-2.75	0.3-0.7	0.9-1.2	0.50 max	0.040 max	0.03	0.60 max	Cu <=0.50; W <=0.10; Cu+Ni+W<=1.00; bal Fe	485	275	20	201 HB max
MIL-C-24707/2(89)	WC9	As spec in ASTM A217; Press, High-temp									bal Fe				

Specification	Designation	Notes	C	Cr	Mn	Mo	Ni	P	S	Si	Other	UTS	YS	El	Hard

Steel Casting, Austenitic Manganese, 6N

France

Specification	Designation	Notes	C	Cr	Mn	Mo	Ni	P	S	Si	Other	UTS	YS	El	Hard
AFNOR NFA32051(81)	230-400-M2	28mm	0.20 max		1.20 max			0.04 max	0.04 max	0.6 max	Cu <=0.4; V <=0.05; Ni+Cr+Cu+Mo+V<=1; bal Fe	400	230	25	120-165 IIB
AFNOR NFA32051(81)	230-400-M3	28mm	0.20 max		1.20 max			0.04 max	0.04 max	0.6 max	Cu <=0.4; V <=0.05; Ni+Cr+Cu+Mo+V<=1; bal Fe	400	230	25	120-165 HB
AFNOR NFA32051(81)	280-480-M2	28mm	0.25 max		1.20 max			0.04 max	0.04 max	0.6 max	Cu <=0.4; V <=0.05; Ni+Cr+Cu+Mo+V<=1; bal Fe	480	280	20	145-190 HB
AFNOR NFA32051(81)	280-480-M3	28mm	0.25 max		1.20 max			0.04 max	0.04 max	0.6 max	Cu <=0.4; V <=0.05; Ni+Cr+Cu+Mo+V<=1; bal Fe	480	280	20	145-190 HB
AFNOR NFA32051(81)	320-560-M2	28mm	0.32 max		1.20 max			0.04 max	0.04 max	0.6 max	Cu <=0.4; V <=0.05; Ni+Cr+Cu+Mo+V<=1; bal Fe	560	320	15	165-215 HB
AFNOR NFA32051(81)	320-560-M3	28mm	0.32 max		1.20 max			0.04 max	0.04 max	0.6 max	Cu <=0.4; V <=0.05; Ni+Cr+Cu+Mo+V<=1; bal Fe	560	320	15	165-215 HB

Russia

Specification	Designation	Notes	C	Cr	Mn	Mo	Ni	P	S	Si	Other	UTS	YS	El	Hard
GOST 977(88)	30ChGSFL		0.25-0.35	0.3-0.5	1-1.5		0.4 max	0.05 max	0.05 max	0.4-0.6	V 0.06-0.12; bal Fe				

USA

Specification	Designation	Notes	C	Cr	Mn	Mo	Ni	P	S	Si	Other	UTS	YS	El	Hard
	UNS J13512		0.35 max		1.35-1.75	0.25-0.55		0.04 max	0.045 max	0.20-0.80	bal Fe				
ASTM A732/A732M(98)	6N	N/T Mn Mo	0.35 max		1.35-1.75	0.25-0.55		0.04 max	0.045 max	0.20-0.80	Cu+Ni+Cr+W<=1.00; bal Fe	621	414	20	

Steel Casting, Austenitic Manganese, A

Australia

Specification	Designation	Notes	C	Cr	Mn	Mo	Ni	P	S	Si	Other	UTS	YS	El	Hard
AS 2074(82)	H1A		1-1.35		11 min			0.1 max		1 max	bal Fe				

China

Specification	Designation	Notes	C	Cr	Mn	Mo	Ni	P	S	Si	Other	UTS	YS	El	Hard
GB/T 5680(98)	ZGMn13-2	Water Toughening	0.90-1.35		11.0-14.0			0.070 max	0.040 max	0.30-1.00	bal Fe	685		20	229 HB max
JB/T 6404(92)	ZGMn13-2	Water Toughening	1.00-1.40		11.0-14.0			0.090 max	0.050 max	0.30-1.00	bal Fe	637		20	229 HB max
YB/T 036.4(92)	ZGMn13-2	Water Toughening	1.00-1.40		11.0-14.0			0.090 max	0.040 max	0.30-0.80	bal Fe	635		20	230 HB max

Czech Republic

Specification	Designation	Notes	C	Cr	Mn	Mo	Ni	P	S	Si	Other	UTS	YS	El	Hard
CSN 422920	422920	C-Mn	1.1-1.5		12-14			0.1 max		0.7	bal Fe				

Hungary

Specification	Designation	Notes	C	Cr	Mn	Mo	Ni	P	S	Si	Other	UTS	YS	El	Hard
MSZ 520(79)	Mn13	Wear res; SA	1.0-1.3		12-14			0.1 max	0.04 max	1 max	bal Fe	780	390	39L	250 HB max

Italy

Specification	Designation	Notes	C	Cr	Mn	Mo	Ni	P	S	Si	Other	UTS	YS	El	Hard
UNI M13	120Mn13		1.2	0.5	12			0.1		0.5	bal Fe				

Japan

Specification	Designation	Notes	C	Cr	Mn	Mo	Ni	P	S	Si	Other	UTS	YS	El	Hard
JIS G5131(91)	SCMnH1	Gen	0.9-1.3		11-14			0.1			bal Fe				
JIS G5131(91)	SCMnH2	Gen	0.9-1.2		11-14			0.070 max	0.040 max	0.80 max	bal Fe	740		35	
JIS G5131(91)	SCMnH3	Rail Crossings	0.9-1.2		11-14			0.050 max	0.035 max	0.30-0.80	bal Fe	740		35	

Romania

Specification	Designation	Notes	C	Cr	Mn	Mo	Ni	P	S	Si	Other	UTS	YS	El	Hard
STAS 3718(76)	T110Mn110	C-Mn	1-1.4	0.3 max	11-21.0		0.4 max	0.1 max	0.06 max	1 max	Pb <=0.15; Mn>=11; bal Fe				

Russia

Specification	Designation	Notes	C	Cr	Mn	Mo	Ni	P	S	Si	Other	UTS	YS	El	Hard
GOST 977 (88)	110G13L		0.9-1.5	1.0 max	11.5-15.0		1.0 max	0.12 max	0.05 max	0.3-1.0	bal Fe				

Spain

Specification	Designation	Notes	C	Cr	Mn	Mo	Ni	P	S	Si	Other	UTS	YS	El	Hard
UNE 36253(71)	AM-X120Mn12		1.00-1.40		11.5-14.0			0.10 max	0.06 max	1.00 max	P+S<=0.14; bal Fe				
UNE 36253(71)	AM-X120MnCr12-2		1.00-1.40	1.50-2.50	11.5-14.0			0.10 max	0.06 max	1.00 max	P+S<=0.14; bal Fe				

Sweden

Specification	Designation	Notes	C	Cr	Mn	Mo	Ni	P	S	Si	Other	UTS	YS	El	Hard
SS 142183	2183		1-1.35		11-14			0.08 max	0.06 max	1 max	bal Fe				

Specification	Designation	Notes	C	Cr	Mn	Mo	Ni	P	S	Si	Other	UTS	YS	El	Hard
Steel Casting, Austenitic Manganese, A (Continued from previous page)															
UK															
BS 3100(91)	BW10		1.00-1.35		11.00 min			0.050 max	0.050 max	1.00 max	bal Fe				
USA															
	UNS J91109		1.05-1.35		11.0 min			0.07 max		1.00 max	bal Fe				
ASTM A128/A128M(98)	A		1.05-1.35		11 min			0.07		1	bal Fe				
Steel Casting, Austenitic Manganese, A487(6)															
France															
AFNOR NFA32051(81)	230-400-M1	28mm	0.50 max		2.00 max			0.07 max	0.06 max	0.6 max	Cu <=0.3; V <=0.1; bal Fe	400	230	25	120-165 HB
AFNOR NFA32051(81)	280-480-M1	28mm	0.50 max		2.00 max			0.07 max	0.06 max	0.6 max	Cu <=0.3; V <=0.1; bal Fe	480	280	20	145-190 HB
AFNOR NFA32051(81)	320-560-M1	28mm	0.50 max		2.00 max			0.07 max	0.06 max	0.6 max	Cu <=0.3; V <=0.1; bal Fe	560	320	15	165-215 HB
AFNOR NFA32051(81)	370-650-M1	28mm; N/T	0.50 max		2.00 max			0.07 max	0.06 max	0.6 max	Cu <=0.3; V <=0.1; bal Fe	650	370	10	195-240 HB
AFNOR NFA32051(81)	370-650-M2	28mm; N/T	0.50 max		1.20 max			0.04 max	0.04 max	0.6 max	Cu <=0.4; V <=0.05; Ni+Cr+Cu+Mo+V<=1; bal Fe	650	370	10	195-240 HB
AFNOR NFA32051(81)	370-650-M3	28mm; N/T	0.50 max		1.20 max			0.04 max	0.04 max	0.6 max	Cu <=0.4; V <=0.05; Ni+Cr+Cu+Mo+V<=1; bal Fe	650	370	10	195-240 HB
Japan															
JIS G5111(91)	3SCMnCr3	Pipe; N/T	0.30-0.40	0.40-0.80	1.20-1.60			0.04 max	0.04 max	0.30-0.60	bal Fe	640	390	9	183 HB
JIS G5111(91)	3SCMnCr3	Pipe; Q/A Tmp	0.30-0.40	0.40-0.80	1.20-1.60			0.04 max	0.04 max	0.30-0.60	bal Fe	690	490	13	207 HB
JIS G5111(91)	3SCMnCrM3	Pipe; Q/A Tmp	0.30-0.40	0.30-0.70	1.20-1.60	0.15-0.35		0.04 max	0.04 max	0.30-0.60	bal Fe	830	635	9	223 HB
JIS G5111(91)	3SCMnCrM3	Pipe; N/T	0.30-0.40	0.30-0.70	1.20-1.60	0.15-0.35		0.04 max	0.04 max	0.30-0.60	bal Fe	740	540	9	212 HB
Poland															
PNH83156(87)	L35HNM		0.3-0.4	0.3-0.7	0.5-0.8	0.35-0.45	0.4-0.6			0.2-0.5	Cu 0.3; bal Fe				
USA															
	UNS J13855		0.38 max	0.40-0.80	1.30-1.70	0.30-0.40	0.40-0.80	0.04 max	0.045 max	0.80 max	Cu <=0.50; V <=0.03; W <=0.1; bal Fe				
ASTM A487	6A	N/T, Q/T	0.05-0.38	0.40-0.80	1.30-1.70	0.30-0.40	0.40-0.80	0.04 max	0.045 max	0.80 max	Cu <=0.50; V <=0.03; W <=0.10; Cu+W+V<=0.60; bal Fe	795	550	18	
ASTM A487	6B	N/T, Q/T	0.05-0.38	0.40-0.80	1.30-1.70	0.30-0.40	0.40-0.80	0.04 max	0.045 max	0.80 max	Cu <=0.50; V <=0.03; W <=0.10; Cu+W+V<=0.60; bal Fe	825	655	12	
ASTM A487	6N*	N/T, Q/T	0.05-0.38	0.40-0.80	1.30-1.70	0.30-0.40	0.40-0.80	0.04 max	0.045 max	0.80 max	Cu <=0.50; V <=0.03; W <=0.10; Cu+W+V<=0.60; bal Fe	795	550	18	
ASTM A487	6Q*	N/T, Q/T	0.05-0.38	0.40-0.80	1.30-1.70	0.30-0.40	0.40-0.80	0.04 max	0.045 max	0.80 max	Cu <=0.50; V <=0.03; W <=0.10; Cu+W+V<=0.60; bal Fe	825	655	12	
Steel Casting, Austenitic Manganese, B1															
China															
JB/T 6404(92)	ZGMn13-4	Water Toughening	0.90-1.20		11.0-14.0			0.070 max	0.050 max	0.30-0.80	bal Fe	735		35	229 HB max
YB/T 036.4(92)	ZGMn13-4	Water Toughening	0.90-1.20		11.0-14.0			0.070 max	0.040 max	0.30-0.60	bal Fe	735		35	230 HB max
Czech Republic															
CSN 422920	422920	C-Mn	1.1-1.5	0.3 max	12.0-14.0			0.1 max	0.05 max	0.7 max	bal Fe				

UNS numbers and US grades are provided as a means of cross referencing chemically similar alloys. Exchangability is only possible after independent examination of specifications. Tensile properties are minimum or typical as specified. UTS and YS as MPa. El as %. See Appendix for list of abbreviations used in Notes. * indicates obsolete material.

Specification	Designation	Notes	C	Cr	Mn	Mo	Ni	P	S	Si	Other	UTS	YS	El	Hard
Steel Casting, Austenitic Manganese, B1 (Continued from previous page)															
Germany															
DIN 17145(80)	WNr 1.3402	Rod	1.00-1.25		13.5-14.5			0.080 max	0.020 max	0.35-0.70	bal Fe				
DIN 17145(80)	X110Mn14	Rod	1.00-1.25		13.5-14.5			0.080 max	0.020 max	0.35-0.70	bal Fe				
Japan															
JIS G5131(91)	SCMnH2	Gen	0.9-1.2		11-14			0.070 max	0.040 max	0.80 max	bal Fe	740		35	
JIS G5131(91)	SCMnH3	Rail Crossings	0.9-1.2		11-14			0.050 max	0.035 max	0.30-0.80	bal Fe	740		35	
Romania															
STAS 3718	T100Mn120		0.9-1.05		11.5-14			0.1		1	bal Fe				
STAS 3718(76)	T120Mn110	C-Mn	1-1.4	0.3 max	11-21.0		0.4 max	0.1 max	0.6 max	1 max	Pb <=0.15; Mn>=11; bal Fe				
Spain															
UNE 36253(71)	AM-X120MnCr12-2		1.00-1.40	1.50-2.50	11.5-14.0			0.10 max	0.06 max	1.00 max	P+S<=0.14; bal Fe				
UNE 36253(71)	F.8252*	see AM-X120MnCr12-2	1.00-1.40	1.50-2.50	11.5-14.0			0.10 max	0.06 max	1.00 max	P+S<=0.14; bal Fe				
USA															
	UNS J91119		0.9-1.05		11.5-14.0			0.07 max		1.00 max	bal Fe				
ASTM A128/A128M(98)	B1		0.9-1.05		11.5-14			0.07		1	bal Fe				
Steel Casting, Austenitic Manganese, B2															
China															
GB/T 5680(98)	ZGMn13-3	Water Toughening	0.90-1.35		11.0-14.0			0.070 max	0.035 max	0.30-0.80	bal Fe	735		25	229 HB max
JB/T 6404(92)	ZGMn13-3	Water Toughening	0.90-1.30		11.0-14.0			0.080 max	0.050 max	0.30-0.80	bal Fe	686		25	229 HB max
YB/T 036.4(92)	ZGMn13-3	Water Toughening	0.90-1.30		11.0-14.0			0.080 max	0.040 max	0.30-0.80	bal Fe	685		25	230 HB max
Czech Republic															
CSN 422920	422920	C-Mn	1.1-1.5		12-14			0.1 max		0.7	bal Fe				
Japan															
JIS G5131(91)	SCMnH1	Gen	0.9-1.3		11-14			0.1			bal Fe				
JIS G5131(91)	SCMnH2	Gen	0.90-1.20		11.00-14.00			0.070 max	0.040 max	0.80 max	bal Fe	740		35	
JIS G5131(91)	SCMnH3	Rail Crossings	0.9-1.2		11-14			0.050 max	0.035 max	0.30-0.80	bal Fe	740		35	
Poland															
PNH83160(88)	L120G13	Wear res	1-1.4	1	12-14		1	0.1		0.3-1	bal Fe				
PNH83160(88)	L120G13T	Wear res	1-1.4	1 max	12-14		1 max	0.1 max	0.03 max	0.3-1	Ti 0.1-0.3; bal Fe				
Romania															
STAS 3718(88)	T105Mn120	C-Mn	0.9-1.2	0.3 max	11.5-13.5		0.8 max	0.11 max	0.05 max	1 max	Pb <=0.15; bal Fe				
Spain															
UNE 36253(71)	AM-X120Mn12		1.00-1.40		11.5-14.0			0.10 max	0.06 max	1.00 max	P+S<=0.14; bal Fe				
UNE 36253(71)	AM-X120MnCr12-2		1.00-1.40	1.50-2.50	11.5-14.0			0.10 max	0.06 max	1.00 max	P+S<=0.14; bal Fe				
USA															
	UNS J91129		1.05-1.2		11.5-14.0			0.07 max		1.00 max	bal Fe				
ASTM A128/A128M(98)	B2		1.05-1.2		11.5-14			0.07		1	bal Fe				
Steel Casting, Austenitic Manganese, B3															
China															
GB/T 5680(98)	ZGMn13-2	Water Toughening	0.90-1.35		11.0-14.0			0.070 max	0.040 max	0.30-1.00	bal Fe	685		20	229 HB max
JB/T 6404(92)	ZGMn13-2	Water Toughening	1.00-1.40		11.0-14.0			0.090 max	0.050 max	0.30-1.00	bal Fe	637		20	229 HB max

UNS numbers and US grades are provided as a means of cross referencing chemically similar alloys. Exchangability is only possible after independent examination of specifications. Tensile properties are minimum or typical as specified. UTS and YS as MPa. El as %. See Appendix for list of abbreviations used in Notes. * indicates obsolete material.

Steel Casting, Austenitic Manganese, B3 (Continued from previous page)

Specification	Designation	Notes	C	Cr	Mn	Mo	Ni	P	S	Si	Other	UTS	YS	El	Hard
China															
YB/T 036.4(92)	ZGMn13-2	Water Toughening	1.00-1.40		11.0-14.0			0.090 max	0.040 max	0.30-0.80	bal Fe	635		20	230 HB max
Czech Republic															
CSN 422920	422920	C-Mn	1.1-1.5		12-14			0.1 max		0.7	bal Fe				
Finland															
SFS 394(79)	G-X120Mn13	Q/T	1.05-1.35		11.0-15.00			0.07 max	0.060 max	1.0 max	bal Fe				
Germany															
DIN	GX120Mn12	Q/T 17-40mm	1.10-1.30	1.50 max	12.0-13.0			0.100 max	0.040 max	0.30-0.50	bal Fe	830-1080	390	42	
DIN	WNr 1.3401	Q/T 17-40mm	1.10-1.30	1.50 max	12.0-13.0			0.100 max	0.040 max	0.30-0.50	bal Fe	830-1080	390	42	
Italy															
UNI 3160(83)	GX120Mn12	Wear res	1-1.4		10-14			0.1 max	0.05 max	1 max	bal Fe				
UNI M13	120Mn13		1.2	0.5	12			0.1		0.5	bal Fe				
Japan															
JIS G5131(91)	SCMnH1	Gen	0.9-1.3		11-14			0.1			bal Fe				
JIS G5131(91)	SCMnH2	Gen	0.9-1.2		11-14			0.070 max	0.040 max	0.80 max	bal Fe	740		35	
JIS G5131(91)	SCMnH3	Rail Crossings	0.90-1.20		11.00-14.00			0.050 max	0.035 max	0.30-0.80	bal Fe	740		35	
Poland															
PNH83160(88)	L120G13	Wear res	1-1.4	1	12-14		1	0.1		0.3-1	bal Fe				
PNH83160(88)	L120G13T	Wear res	1-1.4	1	12-14		1	0.1		0.3-1	Ti 0.1-0.3; bal Fe				
Spain															
UNE 36253(71)	AM-X120Mn12		1.00-1.40		11.5-14.0			0.10 max	0.06 max	1.00 max	P+S<=0.14; bal Fe				
UNE 36253(71)	F.8251*	see AM-X120Mn12	1.00-1.40		11.5-14.0			0.10 max	0.06 max	1.00 max	P+S<=0.14; bal Fe				
USA															
	UNS J91139		1.12-1.28		11.5-14.0			0.07 max		1.00 max	bal Fe				
ASTM A128/A128M(98)	B3		1.12-1.28		11.5-14			0.07		1	bal Fe				

Steel Casting, Austenitic Manganese, B4

Specification	Designation	Notes	C	Cr	Mn	Mo	Ni	P	S	Si	Other	UTS	YS	El	Hard
China															
GB/T 5680(98)	ZGMn13-1	Water Toughening	1.00-1.45		11.0-14.0			0.090 max	0.040 max	0.30-1.00	bal Fe	635		20	229 HB max
JB/T 6404(92)	ZGMn13-1	Water Toughening	1.10-1.50		11.0-14.0			0.090 max	0.050 max	0.30-1.00	bal Fe	637		20	229 HB max
YB/T 036.4(92)	ZGMn13-1	Water Toughening	1.10-1.50		11.0-14.0			0.090 max	0.040 max	0.30-0.80	bal Fe	635		20	230 HB max
Czech Republic															
CSN 422920	422920	C-Mn	1.1-1.5		12-14			0.1 max		0.7	bal Fe				
Germany															
DIN SEW 395(87)	GX120Mn12	Non-magnetic	1.10-1.30	0.50 max	11.5-13.5			0.060 max	0.030 max	0.50 max	bal Fe	780-1080	345	40	
DIN SEW 395(87)	WNr 1.3802	Non-magnetic	1.10-1.30	0.50 max	11.5-13.5			0.060 max	0.030 max	0.50 max	bal Fe	780-1080	345	40	
Italy															
UNI M13	120Mn13		1.2	0.5	12			0.1		0.5	bal Fe				
Japan															
JIS G5131(91)	SCMnH1	Gen	0.90-1.30		11.00-14.00			0.100 max	0.050 max		bal Fe				
JIS G5131(91)	SCMnH2	Gen	0.9-1.2		11.00-14.00			0.070 max	0.040 max	0.80 max	bal Fe	740		35	
JIS G5131(91)	SCMnH3	Rail Crossings	0.9-1.2		11-14			0.050 max	0.035 max	0.30-0.80	bal Fe	740		35	
Norway															
NS 1699	1699		1.00-1.35		11.0-14.0			0.08 max	0.04 max	1.00 max	bal Fe				

UNS numbers and US grades are provided as a means of cross referencing chemically similar alloys. Exchangability is only possible after independent examination of specifications. Tensile properties are minimum or typical as specified. UTS and YS as MPa. El as %. See Appendix for list of abbreviations used in Notes. * indicates obsolete material.

Specification	Designation	Notes	C	Cr	Mn	Mo	Ni	P	S	Si	Other	UTS	YS	El	Hard

Steel Casting, Austenitic Manganese, B4 (Continued from previous page)

Poland

Specification	Designation	Notes	C	Cr	Mn	Mo	Ni	P	S	Si	Other	UTS	YS	El	Hard
PNH83160(88)	L120G13	Wear res	1-1.4	1	12-14		1	0.1		0.3-1	bal Fe				
PNH83160(88)	L120G13T	Wear res	1-1.4	1	12-14		1	0.1		0.3-1	Ti 0.1-0.3; bal Fe				

Romania

| STAS 3718(88) | T130Mn135 | C-Mn | 1.25-1.4 | 0.3 max | 12.5-14.5 | | 0.8 max | 0.11 max | 0.05 max | 1 max | Pb <=0.15; bal Fe | | | | |

Spain

| UNE 36253(71) | AM-X120Mn12 | | 1.00-1.40 | | 11.5-14.0 | | | 0.10 max | 0.06 max | 1.00 max | P+S<=0.14; bal Fe | | | | |

USA

| | UNS J91149 | | 1.2-1.35 | | 11.5-14.0 | | | 0.07 max | | 1.00 max | bal Fe | | | | |
| ASTM A128/A128M(98) | B4 | | 1.2-1.35 | | 11.5-14 | | | 0.07 | | 1 | bal Fe | | | | |

Steel Casting, Austenitic Manganese, C

Australia

| AS 2074(82) | H1B | | 1-1.35 | 1.5-2.5 | 11 min | | | 0.1 max | | 1 max | bal Fe | | | | |

China

GB/T 5680(98)	ZGMn13-4	Water Toughening	0.90-1.35	1.50-2.50	11.0-14.0			0.070 max	0.040 max	0.30-0.80	bal Fe	735	390	30	300 HB max
JB/T 6404(92)	ZGMn13Cr2	Water Toughening	1.05-1.35	1.50-2.50	11.0-14.0			0.070 max	0.050 max	0.30-1.00	bal Fe	655-1000		27-63	220 HB max
YB/T 036.4(92)	ZGMn13-5	Water Toughening	0.90-1.30	1.50-2.50	11.0-14.0			0.070 max	0.040 max	0.30-0.60	bal Fe	735		15	

Hungary

| MSZ 17742 | AoX120Mn12 | | 0.95-1.45 | 1.5 | 10-14 | | | 0.12 | 0.07 | 1.5 | bal Fe | | | | |
| MSZ 17742 | AoX120Mn13 | | 1.00-1.35 | 1 | 12-15 | | | 0.1 | | 1 | bal Fe | | | | |

Italy

| UNI 3160(83) | GX120Mn1202 | Wear res | 1-1.4 | 1.5-2.5 | 11-14 | | | 0.1 max | 0.05 max | 1 max | bal Fe | | | | |

Japan

| JIS G5131(91) | SCMnH11 | Abr res | 0.90-1.30 | 1.50-2.50 | 11.00-14.00 | | | 0.070 max | 0.040 max | 0.80 max | bal Fe | 740 | 390 | 20 | |
| JIS G5131(91) | SCMnH21 | Caterpillar slides | 1.00-1.35 | 2.00-3.00 | 11.00-14.00 | | | 0.070 max | 0.040 max | 0.80 max | V 0.40-0.70; bal Fe | 740 | 440 | 10 | |

Romania

| STAS 3718(88) | T120CrMn130 | C-Mn | 1.05-1.35 | 1.5-2.5 | 11.5-14 | | 0.4 max | 0.07 max | 0.05 max | 1 max | Pb <=0.15; bal Fe | | | | |

Spain

| UNE 36253(71) | AM-X120MnCr12-2 | | 1.00-1.40 | 1.50-2.50 | 11.5-14.0 | | | 0.10 max | 0.06 max | 1.00 max | P+S<=0.14; bal Fe | | | | |

USA

| | UNS J91309 | | 1.05-1.35 | 1.5-2.5 | 11.5-14.0 | | | 0.07 max | | 1.00 max | bal Fe | | | | |
| ASTM A128/A128M(98) | C | | 1.05-1.35 | 1.5-2.5 | 11.5-14 | | | 0.07 | | 1 | bal Fe | | | | |

Steel Casting, Austenitic Manganese, D

Romania

| STAS 3718(88) | T100NiMn130 | C-Mn | 0.7-1.3 | 0.3 max | 11.5-14 | | 3-4 | 0.07 max | 0.05 max | 1 max | Pb <=0.15; bal Fe | | | | |

USA

| | UNS J91459 | | 0.7-1.3 | | 11.5-14.0 | | 3.0-4.0 | 0.07 max | | 1.00 max | bal Fe | | | | |
| ASTM A128/A128M(98) | D | | 0.7-1.3 | | 11.5-14 | | 3-4 | 0.07 | | 1 | bal Fe | | | | |

Steel Casting, Austenitic Manganese, E1

Romania

| STAS 3718(88) | T100MoMn130 | C-Mn | 0.7-1.3 | 0.3 max | 11.5-14 | 0.9-1.2 | 0.4 max | 0.07 max | 0.05 max | 1 max | Pb <=0.15; bal Fe | | | | |

Specification	Designation	Notes	C	Cr	Mn	Mo	Ni	P	S	Si	Other	UTS	YS	El	Hard

Steel Casting, Austenitic Manganese, E1 (Continued from previous page)

USA

| USA | UNS J91249 | | 0.7-1.3 | | 11.5-14.0 | 0.9-1.2 | | 0.07 | | 1.00 max | bal Fe | | | | |
| ASTM A128/A128M(98) | E1 | | 0.7-1.3 | | 11.5-14 | 0.9-1.2 | | 0.07 | | 1 | bal Fe | | | | |

Steel Casting, Austenitic Manganese, E2

Romania

| STAS 3718 | T130MoMn135 | | 1.25-1.4 | | 12.5-14.5 | 0.9-1.1 | 0.8 | 0.11 | | 0.5-1 | bal Fe | | | | |

USA

| | UNS J91339 | | 1.05-1.45 | | 11.5-14.0 | 1.8-2.1 | | 0.07 max | | 1.00 max | bal Fe | | | | |
| ASTM A128/A128M(98) | E2 | | 1.05-1.45 | | 11.5-14 | 1.8-2.1 | | 0.07 | | 1 | bal Fe | | | | |

Steel Casting, Austenitic Manganese, F

USA

| | UNS J91340 | | 1.05-1.35 | | 6.0-8.0 | 0.9-1.2 | | 0.07 max | | 1.00 max | bal Fe | | | | |

Steel Casting, Austenitic Manganese, J91209

Romania

| STAS 3718(88) | T70Mn140 | C-Mn | 0.6-0.8 | 0.6-1 | 12.5-15.5 | | 0.4 max | 0.1 max | 0.05 max | 1 max | Pb <=0.15; bal Fe | | | | |

USA

| | UNS J91209 | | 1.00-1.35 | 0.75 max | 12.00-14.00 | 0.50 max | 1.00 max | 0.060 max | | 0.40-1.00 | bal Fe | | | | |

Steel Casting, Austenitic Manganese, No equivalents identified

Japan

JIS G5111(91)	SCC5							0.05	0.05	0.30-0.60	bal Fe				
JIS G5111(91)	SCMn1	Q/A Tmp	0.20-0.30		1.00-1.60			0.040 max	0.040 max	0.30-0.60	bal Fe	590	390	17	170 HB
JIS G5111(91)	SCMn1	N/T	0.20-0.30		1.00-1.60			0.040 max	0.040 max	0.30-0.60	bal Fe	540	275	17	143 HB
JIS G5111(91)	SCMn2	N/T	0.25-0.35		1.00-1.60			0.040 max	0.040 max	0.30-0.60	bal Fe	590	345	16	163 HB
JIS G5111(91)	SCMn2	Q/A Tmp	0.25-0.35		1.00-1.60			0.040 max	0.040 max	0.30-0.60	bal Fe	640	440	16	183 HB
JIS G5111(91)	SCMn3	N/T	0.30-0.40		1.00-1.60			0.040 max	0.040 max	0.30-0.60	bal Fe	640	370	13	170 HB
JIS G5111(91)	SCMn3	Q/A Tmp	0.30-0.40		1.00-1.60			0.040 max	0.040 max	0.30-0.60	bal Fe	690	490	13	197 HB
JIS G5111(91)	SCMnCr4	N/T	0.35-0.45	0.4-0.8	1.20-1.60			0.040 max	0.040 max	0.30-0.60	bal Fe	690	410	9	201 HB
JIS G5111(91)	SCMnM3	N/T	0.30-0.40	0.20 max	1.20-1.60	0.15-0.35		0.040 max	0.040 max	0.30-0.60	bal Fe	690	390	13	183 HB

Poland

| PNH83160(88) | L120G13 | Wear res | 1-1.4 | 1 max | 12-14 | | 1 max | 0.1 max | 0.03 max | 0.3-1 | bal Fe | | | | |
| PNH83160(88) | L120G13H | Wear res | 1-1.4 | 0.6-1.3 | 12-14 | | 0.5 max | 0.1 max | 0.03 max | 0.3-1 | bal Fe | | | | |

Russia

| GOST 2176 | 110G13L | | 0.9-1.4 | 1.0 max | 11.5-15.0 | | 1.0 max | 0.12 max | 0.05 max | 0.3-1.0 | bal Fe | | | | |

Spain

UNE 36254(79)	30Mn5	Q/T	0.25-0.34		1.2-1.6	0.15 max		0.04 max	0.04 max	0.3-0.5	bal Fe				
UNE 36254(79)	36Mn5	Q/T	0.32-0.4		1.2-1.6	0.15 max		0.04 max	0.04 max	0.3-0.5	bal Fe				
UNE 36254(79)	C70n	Q/T	0.6-0.8		0.5-0.9	0.15 max		0.04 max	0.04 max	0.3-0.5	bal Fe				
UNE 36254(79)	F.8201*	see C70n	0.6-0.8		0.5-0.9	0.15 max		0.04 max	0.04 max	0.3-0.5	bal Fe				

UNS numbers and US grades are provided as a means of cross referencing chemically similar alloys. Exchangability is only possible after independent examination of specifications. Tensile properties are minimum or typical as specified. UTS and YS as MPa. El as %. See Appendix for list of abbreviations used in Notes. * indicates obsolete material.

Steel Casting, Austenitic Manganese, No equivalents identified

Specification	Designation	Notes	C	Cr	Mn	Mo	Ni	P	S	Si	Other	UTS	YS	El	Hard
Spain															
UNE 36254(79)	F.8211*	see 30Mn5	0.25-0.34		1.2-1.6			0.04 max	0.04 max	0.3-0.5	bal Fe				
UNE 36254(79)	F.8212*	see 36Mn5	0.32-0.4		1.2-1.6	0.15 max		0.04 max	0.04 max	0.3-0.5	bal Fe				
USA															
MIL-C-24707/4(89)	A(128)	As spec in ASTM A128; ship non-magnetic, Wear res									bal Fe				

Specification	Designation	Notes	C	Cr	Mn	Mo	Ni	P	S	Si	Other	UTS	YS	El	Hard

Steel Casting, Nickel Based, 4892

USA

Specification	Designation	Notes	C	Cr	Mn	Mo	Ni	P	S	Si	Other	UTS	YS	El	Hard
FED QQ-N-288(91)	A*	Obs; see ASTM A494, MIL-C-24733	0.35		1.5		62-68			2	Al 0.5; Cu 26-33; Fe 2.5; bal Ni				
FED QQ-N-288(91)	B*	Obs; see ASTM A494, MIL-C-24733	0.3		1.5		61-68			2.7-3.7	Al 0.5; Cu 27-33; Fe 2.5; bal Ni				
FED QQ-N-288(91)	C*	Obs; see ASTM A494, MIL-C-24733	0.2		1.5		60 min			3.3-4.3	Al 0.5; Cu 27-31; Fe 2.5; bal Ni				
FED QQ-N-288(91)	D*	Obs; see ASTM A494, MIL-C-24733	0.25		1.5		60 min			3.5-4.5	Al 0.5; Cu 27-31; Fe 2.5; bal Ni				

Steel Casting, Nickel Based, 5337

USA

Specification	Designation	Notes	C	Cr	Mn	Mo	Ni	P	S	Si	Other	UTS	YS	El	Hard
	UNS J93150	Maraging	0.03 max	0.10 max		4.50-5.50	18.0-19.0	0.010 max	0.010 max	0.10 max	Al 0.05-0.20; Co 8.50-9.50; Ti 0.55-0.85; bal Fe				

Steel Casting, Nickel Based, 5339

USA

Specification	Designation	Notes	C	Cr	Mn	Mo	Ni	P	S	Si	Other	UTS	YS	El	Hard
	UNS J93010	Maraging	0.03 max	0.10 max		4.40-4.80	16.00-17.50	0.01 max	0.01 max	0.10 max	Al 0.02-0.10; Co 9.50-11.00; Ti 0.15-0.45; bal Fe				

Steel Casting, Nickel Based, B2N

France

Specification	Designation	Notes	C	Cr	Mn	Mo	Ni	P	S	Si	Other	UTS	YS	El	Hard
AFNOR NFA32053(74)	C2M	28mm; HT	0.23 max	0.3 max	0.8 max	0.15 max	2.5-4	0.04 max	0.035 max	0.5 max	W <=0.1; bal Fe	450-600	230	18	

Japan

| JIS G5152(91) | SCPL21 | Low-temp; High-press | 0.25 max | 0.35 max | 0.50-0.80 | | 2.00-3.00 | 0.040 max | 0.040 max | 0.60 max | Cu <=0.50; bal Fe | 480 | 275 | 21 | |

USA

| | UNS J22501 | | 0.25 max | 0.40 max | 0.50-0.80 | 0.25 max | 2.0-3.0 | 0.025 max | 0.025 max | 0.60 max | Cu <=0.50; V <=0.03; bal Fe | | | | |
| ASTM A757/A757M(96) | B2N B2Q | Corr res, N/T, Q/T | 0.25 max | | 0.50-0.80 | | 2.0-3.0 | 0.025 max | 0.025 max | 0.60 max | V+Cu+Cr+Mo<=1.00; bal Fe | 485 | 275 | 24 | |

Steel Casting, Nickel Based, B3N

Japan

Specification	Designation	Notes	C	Cr	Mn	Mo	Ni	P	S	Si	Other	UTS	YS	El	Hard
JIS G5152(91)	SCPL31	Low-temp; High-press	0.15 max	0.35 max	0.50-0.80		3.00-4.00	0.040 max	0.040 max	0.60 max	Cu <=0.50; bal Fe	480	275	21	

USA

| | UNS J31500 | | 0.15 max | 0.40 max | 0.50-0.80 | 0.25 max | 3.0-4.0 | 0.025 max | 0.025 max | 0.60 max | Cu <=0.50; V <=0.03; bal Fe | | | | |
| ASTM A757/A757M(96) | B3N B3Q | Corr res, N/T, Q/T | 0.15 max | | 0.50-0.80 | | 3.0-4.0 | 0.025 max | 0.025 max | 0.60 max | V+Cu+Cr+Mo<=1.00; bal Fe | 485 | 275 | 24 | |

Steel Casting, Nickel Based, B4N B4Q

USA

Specification	Designation	Notes	C	Cr	Mn	Mo	Ni	P	S	Si	Other	UTS	YS	El	Hard
	UNS J41501		0.15 max	0.40 max	0.50-0.80	0.25 max	4.0-5.0	0.025 max	0.025 max	0.60 max	Cu <=0.50; V <=0.03; bal Fe				
ASTM A757/A757M(96)	B4N B4Q	Corr res, N/T, Q/T	0.15 max		0.50-0.80		4.0-5.0	0.025 max	0.025 max	0.60 max	V+Cu+Cr+Mo<=1.00; bal Fe	485	275	24	

Steel Casting, Nickel Based, E1Q

USA

Specification	Designation	Notes	C	Cr	Mn	Mo	Ni	P	S	Si	Other	UTS	YS	El	Hard
	UNS J42220		0.22 max	1.35-1.85	0.50-0.80	0.35-0.60	2.5-3.5	0.025 max	0.025 max	0.60 max	Cu <=0.50; V <=0.03; bal Fe				
ASTM A757/A757M(96)	E1Q	Corr res, Q/T	0.22 max	1.35-1.85	0.50-0.80	0.35-0.60	2.5-3.5	0.025 max	0.025 max	0.60 max	V+Cu<=0.70; bal Fe	620	450	22	
MIL-C-24707/1(89)	E1Q	As spec in ASTM A757; Mach, Struct									bal Fe				

Steel Casting, Nickel Based, E2N

USA

Specification	Designation	Notes	C	Cr	Mn	Mo	Ni	P	S	Si	Other	UTS	YS	El	Hard
	UNS J42065		0.20 max	1.50-2.00	0.40-0.70	0.40-0.60	2.75-3.90	0.020 max	0.020 max	0.60 max	Cu <=0.50; V <=0.03; W <=0.10; bal Fe				
ASTM A757/A757M(96)	E2N	Corr res, N/T	0.20 max	1.50-2.0	0.40-0.70	0.40-0.60	2.75-3.90	0.020 max	0.020 max	0.60 max	V+Cu+W<=0.70; bal Fe	620-795	485-690	13-18	

UNS numbers and US grades are provided as a means of cross referencing chemically similar alloys. Exchangability is only possible after independent examination of specifications. Tensile properties are minimum or typical as specified. UTS and YS as MPa. El as %. See Appendix for list of abbreviations used in Notes. * indicates obsolete material.

3-52/Steel Casting

Specification	Designation	Notes	C	Cr	Mn	Mo	Ni	P	S	Si	Other	UTS	YS	El	Hard
Steel Casting, Nickel Based, E2N (Continued from previous page)															
USA															
ASTM A757/A757M(96)	E2Q	Corr res, Q/T	0.20 max	1.50-2.0	0.40-0.70	0.40-0.60	2.75-3.90	0.020 max	0.020 max	0.60 max	V+Cu+W<=0.70; bal Fe	825-1000	485-690	13-18	
MIL-C-24707/1(89)	E2N	As spec in ASTM A757; Mach, Struct									bal Fe				
MIL-C-24707/1(89)	E2Q	As spec in ASTM A757; Mach, Struct									bal Fe				
Steel Casting, Nickel Based, HY100															
Germany															
DIN	G19NiCrMo12-6		0.22 max	1.35-1.85	0.50-0.80	0.35-0.60	2.50-3.50	0.025 max	0.020 max	0.60 max	bal Fe				
DIN	WNr 1.6783		0.22 max	1.35-1.85	0.50-0.80	0.35-0.60	2.50-3.50	0.025 max	0.020 max	0.60 max	bal Fe				
USA															
	UNS J42240		0.22 max	1.35-1.85	0.55-0.75	0.30-0.60	2.75-3.50	0.02 max	0.015 max	0.5 max	bal Fe				
MIL-S-23008D(93)	HY-100	Struct	0.22 max	1.35-1.65	0.55-0.75	0.30-0.60	3.00-3.50	0.014 max	0.008 max	0.50 max	Cu <=0.25; Ti <=0.02; V <=0.03; Sb 0.025 max; As 0.025 max; Sn 0.030 max; bal Fe	690	690-793	18	
Steel Casting, Nickel Based, HY80															
France															
AFNOR NFA32053(74)	FC2IM	28mm; HT	0.20 max	1-2	0.8 max	0.3-0.6	3-4	0.04 max	0.035 max	0.5 max	W <=0.1; bal Fe	700-850	500	12	
AFNOR NFA32053(92)	18NCD12.6M	t<=200mm	0.15-0.20	1.3-1.8	0.6-0.8	0.45-0.6	3-3.5	0.025 max	0.02 max	0.6 max	V <=0.04; W <=0.1; Ni+Cr+Cu+Mo+V<=1; bal Fe	720	550	15	230 HB
USA															
	UNS J42015		0.20 max	1.35-1.65	0.55-0.75	0.30-0.60	2.50-3.25	0.02 max	0.015 max	0.50 max	bal Fe				
MIL-S-23008D(93)	HY-80	Struct	0.20 max	1.35-1.65	0.55-0.75	0.30-0.60	2.75-3.25	0.014 max	0.008 max	0.50 max	Cu <=0.25; Ti <=0.02; V <=0.03; Sb 0.025 max; As 0.025 max; Sn 0.030 max; bal Fe	552	552-686	20	
Steel Casting, Nickel Based, J32075															
Germany															
DIN EN 10213(96)	G17NiCrMo13-6	Press	0.15-0.19	1.30-1.80	0.55-0.80	0.45-0.60	3.00-3.50	0.015 max	0.015 max	0.50 max	bal Fe				
DIN EN 10213(96)	WNr 1.6781	Press purposes	0.15-0.19	1.30-1.80	0.55-0.80	0.45-0.60	3.00-3.50	0.015 max	0.015 max	0.50 max	bal Fe				
Italy															
UNI 3160(83)	G35NiCrMo16	Wear res	0.4 max	1.2-1.5	0.6 max	0.2-0.4	3.8-4.2	0.035 max	0.035 max	0.35 max	bal Fe				
USA															
	UNS J32075		0.20 max	1.15-1.65	0.55-0.75	0.10-0.60	2.50-3.25	0.02 max	0.015 max	0.50 max	Cu <=0.20; bal Fe				
Yugoslavia															
	CL.5430		0.34-0.4	1.4-1.7	0.5-0.8	0.15-0.25	1.4-1.7	0.035 max	0.035 max	0.3-0.5	bal Fe				
Steel Casting, Nickel Based, LC2															
USA															
	UNS J03004*	Obs; see J13002									bal Fe				
Steel Casting, Nickel Based, LC2-1															
UK															
BS 1504(76)	503LT60*	Press ves	0.12 max	0.3 max	0.8 max	0.15 max	3.0-4.0	0.03 max	0.03 max	0.6 max	Cu <=0.3; Pb <=0.15; V <=0.1; W <=0.1; bal Fe				
USA															
	UNS J42215		0.22 max	1.35-1.85	0.55-0.75	0.30-0.60	2.50-3.50	0.04 max	0.045 max	0.50 max	bal Fe				

UNS numbers and US grades are provided as a means of cross referencing chemically similar alloys. Exchangability is only possible after independent examination of specifications. Tensile properties are minimum or typical as specified. UTS and YS as MPa. El as %. See Appendix for list of abbreviations used in Notes. * indicates obsolete material.

Specification	Designation	Notes	C	Cr	Mn	Mo	Ni	P	S	Si	Other	UTS	YS	El	Hard

Steel Casting, Nickel Based, LC2-1 (Continued from previous page)

USA

Specification	Designation	Notes	C	Cr	Mn	Mo	Ni	P	S	Si	Other	UTS	YS	El	Hard
ASTM A352/A352M(98)	UNS J42215	Press; low-temp	0.22 max	1.35-1.85	0.55-0.75	0.30-0.60	2.50-3.50	0.04 max	0.045 max	0.50 max	bal Fe	725-895	550	18	

Steel Casting, Nickel Based, LC3

Bulgaria

Specification	Designation	Notes	C	Cr	Mn	Mo	Ni	P	S	Si	Other	UTS	YS	El	Hard
BDS 6550(86)	12DChN2FL		0.08-0.16	0.30 max	0.40-0.90	0.15 max	1.80-2.20	0.035 max	0.035 max	0.20-0.50	Cu 1.20-1.50; V 0.08-0.15; bal Fe				
BDS 6550(86)	13ChNDFTL		0.16 max	0.15-0.40	0.40-0.90	0.15 max	1.20-1.60	0.030 max	0.030 max	0.20-0.50	Cu 0.65-0.90; Ti 0.04-0.10; V 0.06-0.12; bal Fe				

France

Specification	Designation	Notes	C	Cr	Mn	Mo	Ni	P	S	Si	Other	UTS	YS	El	Hard
AFNOR NFA32053(74)	FC3M	28mm; HT	0.15 max	0.3 max	0.8 max	0.15 max	3.5-4.5	0.04 max	0.035 max	0.5 max	W <=0.1; bal Fe	450-600	230	18	

Hungary

Specification	Designation	Notes	C	Cr	Mn	Mo	Ni	P	S	Si	Other	UTS	YS	El	Hard
MSZ 8267(85)	Ao10Ni14	Tough at subzero; 0-30mm; HT	0.12 max		0.5-0.8		3.3-3.8	0.025 max	0.025 max	0.6 max	bal Fe	500	360	20L	

Italy

Specification	Designation	Notes	C	Cr	Mn	Mo	Ni	P	S	Si	Other	UTS	YS	El	Hard
UNI 7317(74)	G12Ni14	Tough at subzero	0.15 max		0.8 max		3-4	0.035 max	0.035 max	0.6 max	bal Fe				

Romania

Specification	Designation	Notes	C	Cr	Mn	Mo	Ni	P	S	Si	Other	UTS	YS	El	Hard
STAS 9277(84)	OTA15Ni35		0.15 max	0.3 max	0.5 max		3-4	0.04 max	0.03 max	0.6 max	Pb <=0.15; bal Fe				

Spain

Specification	Designation	Notes	C	Cr	Mn	Mo	Ni	P	S	Si	Other	UTS	YS	El	Hard
UNE 36256(73)	AM8Ni14*	Q/T	0.15 max		0.5-0.8	0.15 max	3.0-4.0	0.035 max	0.035 max	0.3-0.6	bal Fe				
UNE 36256(73)	F.8374*	see AM8Ni14*	0.15 max		0.5-0.8	0.15 max	3.0-4.0	0.035 max	0.035 max	0.3-0.6	bal Fe				

UK

Specification	Designation	Notes	C	Cr	Mn	Mo	Ni	P	S	Si	Other	UTS	YS	El	Hard
BS 3100(91)	BL2	0-999mm	0.12 max		0.80 max		3.00-4.00	0.030 max	0.030 max	0.60 max	bal Fe	460	280	20	

USA

Specification	Designation	Notes	C	Cr	Mn	Mo	Ni	P	S	Si	Other	UTS	YS	El	Hard
	UNS J31550		0.15 max		0.50-0.80		3.00-4.00	0.04 max	0.045 max	0.60 max	bal Fe				
ASTM A352/A352M(98)	UNS J31550	Press; low-temp	0.15 max		0.50-0.80		3.00-4.00	0.04 max	0.045 max	0.60 max	bal Fe	485-655	275	24	

Steel Casting, Nickel Based, LC4

Hungary

Specification	Designation	Notes	C	Cr	Mn	Mo	Ni	P	S	Si	Other	UTS	YS	El	Hard
MSZ 8267(85)	Ao10Ni20	Tough at subzero; 0-30mm; HT	0.12 max		0.5-0.8		4.5-5.5	0.025 max	0.025 max	0.6 max	bal Fe	550	400	18L	

Spain

Specification	Designation	Notes	C	Cr	Mn	Mo	Ni	P	S	Si	Other	UTS	YS	El	Hard
UNE 36256(73)	AM8Ni18*	Q/T	0.15 max		0.5-0.8	0.15 max	4.0-5.0	0.035 max	0.035 max	0.3-0.6	bal Fe				
UNE 36256(73)	F.8375*	see AM8Ni18*	0.15 max		0.5-0.8	0.15 max	4.0-5.0	0.035 max	0.035 max	0.3-0.6	bal Fe				

USA

Specification	Designation	Notes	C	Cr	Mn	Mo	Ni	P	S	Si	Other	UTS	YS	El	Hard
	UNS J41500		0.15 max		0.50-0.80		4.00-5.00	0.04 max	0.045 max	0.60 max	bal Fe				
ASTM A352/A352M(98)	UNS J41500	Press; low-temp	0.15 max		0.50-0.80		4.00-5.00	0.04 max	0.045 max	0.60 max	bal Fe	485-655	275	24	

Steel Casting, Nickel Based, LC9

France

Specification	Designation	Notes	C	Cr	Mn	Mo	Ni	P	S	Si	Other	UTS	YS	El	Hard
AFNOR NFA32053(92)	10N14M	Low-temp; Q/T; t<=30mm	0.06-0.12	0.4 max	0.5-0.8	0.2 max	3-4	0.025 max	0.02 max	0.6 max	V <=0.04; W <=0.1; Ni+Cr+Cu+Mo+V<=1; bal Fe	500	360	18	
AFNOR NFA32053(92)	10N19M	Low-temp; Q/T; t<=30mm	0.06-0.12	0.4 max	0.5-0.8	0.2 max	4.5-5.5	0.025 max	0.02 max	0.6 max	V <=0.04; W <=0.1; Ni+Cr+Cu+Mo+V<=1; bal Fe	550	380	18	
AFNOR NFA32053(92)	10N6M	Low-temp; Q/T; t<=30mm	0.06-0.12	0.4 max	0.5-0.8	0.2 max	1.3-1.8	0.025 max	0.02 max	0.6 max	V <=0.04; W <=0.1; Ni+Cr+Cu+Mo+V<=1; bal Fe	400	250	18	

USA

Specification	Designation	Notes	C	Cr	Mn	Mo	Ni	P	S	Si	Other	UTS	YS	El	Hard
	UNS J31300		0.13 max	0.50 max	0.90 max	0.20 max	8.50-10.0	0.04 max	0.045 max	0.45 max	V <=0.03; bal Fe				

UNS numbers and US grades are provided as a means of cross referencing chemically similar alloys. Exchangability is only possible after independent examination of specifications. Tensile properties are minimum or typical as specified. UTS and YS as MPa. El as %. See Appendix for list of abbreviations used in Notes. * indicates obsolete material.

3-54/Steel Casting

Specification	Designation	Notes	C	Cr	Mn	Mo	Ni	P	S	Si	Other	UTS	YS	El	Hard
Steel Casting, Nickel Based, LC9 (Continued from previous page)															
USA															
ASTM A352/A352M(98)	LC9	Press; low-temp	0.13 max	0.50 max	0.90 max	0.20 max	8.50-10.0	0.04 max	0.045 max	0.45 max	Cu <=0.30; V <=0.03; bal Fe	585	515	20	
Steel Casting, Nickel Based, No equivalents identified															
Russia															
GOST 977	08Ch12N4GSML	Corr res	0.08 max	11.5-13.5	1.5 max	1.0 max	3.5-5.5	0.035 max	0.035 max	1.0 max	Cu <=0.3; Pb <=0.15; V <=0.1; W <=0.1; bal Fe				
GOST 977	08Ch14N7ML	Corr res	0.08 max	13.0-15.0	0.3-0.9	0.5-1.0	6.0-8.5	0.03 max	0.03 max	0.2-0.75	Cu <=0.3; Pb <=0.15; V <=0.1; W <=0.1; bal Fe				
GOST 977	08Ch15N4DML	Corr res	0.08 max	14.0-16.0	1.0-1.5	0.3-0.45	3.5-3.9	0.025 max	0.025 max	0.4 max	Cu 1.0-1.4; Pb <=0.15; V <=0.1; W <=0.1; bal Fe				
GOST 977	08Ch17N34W5T3Ju2L	Corr res	0.08 max	15.0-18.0	0.3-0.6	0.15 max	32.0-35.0	0.01 max	0.01 max	0.2-0.5	Al 1.7-2.1; B <=0.05; Cu <=0.3; Pb <=0.15; Ti 2.6-3.2; V <=0.1; W 4.5-5.5; Ce<=0.1; bal Fe				
GOST 977	09Ch16N4BL	Corr res	0.05-0.13	15.0-17.0	0.3-0.6	0.3 max	3.5-4.5	0.03 max	0.025 max	0.2-0.6	Cu <=0.3; Nb 0.05-0.2; Pb <=0.15; Ti <=0.2; V <=0.1; W <=0.2; bal Fe				
GOST 977	09Ch17N3SL	Corr res	0.05-0.12	15.0-18.0	0.3-0.8	0.3 max	2.8-3.8	0.035 max	0.03 max	0.8-1.5	Cu <=0.3; Pb <=0.15; Ti <=0.2; V <=0.1; W <=0.2; bal Fe				
GOST 977	10Ch18N9L	Corr res	0.14 max	17.0-20.0	1.0-2.0	0.15 max	8.0-11.0	0.035 max	0.03 max	0.2-1.0	Cu <=0.3; Pb <=0.15; V <=0.1; W <=0.1; bal Fe				
GOST 977	12Ch18N9TL	Corr res	0.12 max	17.0-20.0	1.0-2.0	0.3 max	8.0-11.0	0.035 max	0.03 max	0.2-1.0	Cu <=0.3; Pb <=0.15; Ti <=0.7; V <=0.1; W <=0.2; Ti=5C-0.7; bal Fe				
GOST 977	12Ch19N7G2SAL	Corr res	0.12 max	20.0-22.0	2.0 max	0.15 max	4.5-6.0	0.045 max	0.035 max	1.5 max	Cu <=0.3; Pb <=0.15; Ti <=0.7; V <=0.1; W <=0.1; Ti=4C-0.7; bal Fe				
GOST 977	12Ch21N5G2SAL	Corr res	0.12 max	20.2-22.0	2.0 max	0.15 max	4.0-6.0	0.04 max	0.04 max	1.5 max	Cu <=0.3; N 0.1-0.2; Pb <=0.15; V <=0.1; W <=0.1; bal Fe				
GOST 977	12Ch21N5G2SM2L	Corr res	0.12 max	20.0-22.0	2.0 max	1.8-2.2	4.5-6.0	0.045 max	0.035 max	1.5 max	Cu <=0.3; Pb <=0.15; V <=0.1; W <=0.1; bal Fe				
GOST 977	12Ch21N5G2STL	Corr res	0.12 max	20.0-22.0	2.0 max	0.15 max	4.5-6.0	0.045 max	0.035 max	1.5 max	Cu <=0.3; Pb <=0.15; Ti <=0.7; V <=0.1; W <=0.1; Ti=4C-0.7; bal Fe				
GOST 977	14Ch18N4G4L	Corr res	0.14 max	16.0-20.0	4.0-5.0	0.06-0.12	4.0-5.0	0.035 max	0.03 max	0.2-1.0	Cu <=0.3; Pb <=0.15; V <=0.1; W <=0.1; bal Fe				
GOST 977	15Ch18N10G2S2M2L	Corr res	0.15 max	17.0-19.0	2.0 max	2.0-2.5	9.0-12.0	0.04 max	0.04 max	2.0 max	Cu <=0.3; Pb <=0.15; V <=0.1; W <=0.1; bal Fe				
GOST 977	15Ch18N10G2S2M2TL	Corr res	0.12 max	17.0-19.0	2.0 max	2.0-2.5	9.0-12.0	0.04 max	0.04 max	2.0 max	Cu <=0.3; Pb <=0.15; Ti <=0.8; V <=0.1; W <=0.1; Ti=5(C-0.03)-0.8; bal Fe				*
GOST 977	20Ch21N46W8RL	Corr res	0.1-0.25	19.0-22.0	0.3-0.8	0.15 max	43.0-48.0	0.04 max	0.035 max	0.2-0.8	B <=0.06; Cu <=0.3; Pb <=0.15; V <=0.1; W 7.0-9.0; bal Fe				
GOST 977	31Ch19N9MWBTL	Corr res	0.26-0.35	18.0-20.0	0.8-1.5	1.0-1.5	8.0-10.0	0.035 max	0.02 max	0.8 max	Cu <=0.3; Nb 0.2-0.5; Pb <=0.15; Ti 0.2-0.5; V <=0.1; W 1.0-1.5; bal Fe				
GOST 977	35Ch23N7SL	Corr res	0.35 max	21.0-25.0	0.5-0.85	0.15 max	6.0-8.0	0.035 max	0.035 max	0.5-1.2	Cu <=0.3; Pb <=0.15; V <=0.1; W <=0.1; bal Fe				
GOST 977	45Ch17G13N3JuL	Corr res	0.4-0.5	16.0-18.0	12.0-15.0	0.15 max	2.5-3.5	0.035 max	0.03 max	0.8-1.5	Al 0.6-1.0; Cu <=0.3; Pb <=0.15; V <=0.1; W <=0.1; bal Fe				

Specification	Designation	Notes	C	Cr	Mn	Mo	Ni	P	S	Si	Other	UTS	YS	El	Hard

Steel Casting, Nickel Based, No equivalents identified

Russia

Specification	Designation	Notes	C	Cr	Mn	Mo	Ni	P	S	Si	Other	UTS	YS	El	Hard
GOST977	12Ch25N5TMFL	Corr res	0.12 max	23.5-26.0	0.3-0.8	0.06-0.12	5.0-6.5	0.03 max	0.03 max	0.2-1.0	Cu <=0.3; N 0.8-0.2*; Pb <=0.15; Ti 0.08-0.2; V 0.07-0.15; W <=0.1; bal Fe				

Specification	Designation	Notes	C	Cr	Mn	Mo	Ni	P	S	Si	Other	UTS	YS	El	Hard

Steel Casting, Unclassified, 105-85

Russia

Specification	Designation	Notes	C	Cr	Mn	Mo	Ni	P	S	Si	Other	UTS	YS	El	Hard
GOST 977(88)	40ChL		0.35-0.45	0.8-1.1	0.4-0.9		0.4 max	0.04 max	0.04 max	0.2-0.4	bal Fe				

USA

Specification	Designation	Notes	C	Cr	Mn	Mo	Ni	P	S	Si	Other	UTS	YS	El	Hard
	UNS J31575		0.43 max	1.10	1.00	0.25 max		0.04 max	0.05 max	0.50 max	total others 1.70 max; bal Fe				
SAE J435(74)	0105	Auto						0.040 max	0.045 max		bal Fe	723.9	586.0	17	217-248 HB
SAE J435(74)	0120	Auto						0.040 max	0.045 max		bal Fe	827.4	655.0	14	248-311 HB
SAE J435(74)	0150	Auto						0.040 max	0.045 max		bal Fe	1034.2	861.8	9	311-363 HB
SAE J435(74)	0175	Auto						0.040 max	0.045 max		bal Fe	1206.6	999.7	6	363-415 HB
SAE J435(74)	080	Auto						0.040 max	0.045 max		bal Fe	551.6	344.7	22	163-207 HB
SAE J435(74)	090	Auto						0.040 max	0.045 max		bal Fe	620.5	413.7	20	187-241 HB

Steel Casting, Unclassified, 12

USA

Specification	Designation	Notes	C	Cr	Mn	Mo	Ni	P	S	Si	Other	UTS	YS	El	Hard
	UNS J80490		0.08-0.12	8.0-9.5	0.30-0.60	0.85-1.05	0.40 max	0.02 max	0.010 max	0.20-0.50	Al <=0.04; N 0.03-0.07; V 0.18-0.25; Cb 0.06-0.10; bal Fe				

Steel Casting, Unclassified, 5365

UK

Specification	Designation	Notes	C	Cr	Mn	Mo	Ni	P	S	Si	Other	UTS	YS	El	Hard
BS 1502(82)	625-590	Bar Shp Press ves	0.10-0.18	4.00-6.00	0.30-0.60	0.45-0.65		0.030 max	0.030 max	0.15-0.50	Al <=0.020; W <=0.1; bal Fe				
BS 1502(82)	625-640	Bar Shp Press ves	0.10-0.18	4.00-6.00	0.30-0.60	0.45-0.65		0.030 max	0.030 max	0.15-0.50	Al <=0.020; W <=0.1; bal Fe				
BS 1503(89)	625-590	Frg Press ves; 0-999mm; N/T Q/T	0.18 max	4.0-6.0	0.3-0.8	0.45-0.65	0.4 max	0.03 max	0.025 max	0.15-0.4	Al <=0.02; Cu <=0.3; Pb <=0.15; V <=0.1; W <=0.1; bal Fe	590-740	450	18L	

USA

Specification	Designation	Notes	C	Cr	Mn	Mo	Ni	P	S	Si	Other	UTS	YS	El	Hard
	UNS J94211		0.10-0.18	23.0-26.0	2.0 max	0.50 max	19.0-22.0	0.04 max	0.04 max	0.50-1.50	Cu <=0.50; bal Fe				

Steel Casting, Unclassified, C12

Australia

Specification	Designation	Notes	C	Cr	Mn	Mo	Ni	P	S	Si	Other	UTS	YS	El	Hard
AS 2074(82)	H2A	Heat res	0.2 max	8-10	0.3-0.7	0.9-1.2	0.4 max	0.04 max	0.04 max	1 max	Cu <=0.3; bal Fe				

International

Specification	Designation	Notes	C	Cr	Mn	Mo	Ni	P	S	Si	Other	UTS	YS	El	Hard
ISO DIS4991	C38H	Heat res	0.1-0.17	8-10	0.5-0.8	1-1.3			0.035	0.8 max	P 0.035; bal Fe				

Italy

Specification	Designation	Notes	C	Cr	Mn	Mo	Ni	P	S	Si	Other	UTS	YS	El	Hard
UNI 3608(83)	GX15CrMo9		0.2 max	8-10	0.8 max	0.9-1.2		0.035 max	0.035 max	1 max	bal Fe				

Mexico

Specification	Designation	Notes	C	Cr	Mn	Mo	Ni	P	S	Si	Other	UTS	YS	El	Hard
DGN B-141	C12*	Obs; Heat res; Norm, Ann	0.2	8-10	0.35-0.65	0.9-1.2		0.04	0.05	1	bal Fe				

Romania

Specification	Designation	Notes	C	Cr	Mn	Mo	Ni	P	S	Si	Other	UTS	YS	El	Hard
STAS 6855(86)	T20MoCr90	Corr res; Heat res	0.2 max	8-10	0.35-0.65	0.9-1.2	0.4 max	0.035 max	0.03 max	1 max	Pb <=0.15; bal Fe				

Spain

Specification	Designation	Notes	C	Cr	Mn	Mo	Ni	P	S	Si	Other	UTS	YS	El	Hard
UNE 36259(79)	AM-X18CrMo09-01*		0.2 max	8.0-10.0	0.35-0.65	0.9-1.2		0.040 max	0.040 max	1.0 max	bal Fe				
UNE 36259(79)	F.8386*	see AM-X18CrMo09-01*	0.2 max	8.0-10.0	0.35-0.65	0.9-1.2		0.040 max	0.040 max	1.0 max	bal Fe				

UK

Specification	Designation	Notes	C	Cr	Mn	Mo	Ni	P	S	Si	Other	UTS	YS	El	Hard
BS 1504(76)	629*	Heat res; Press ves	0.2 max	8-10	0.3-0.7	0.9-1.2	0.4 max	0.04	0.04	1 max	Cu <=0.3; bal Fe				
BS 3100(91)	B6	0-999mm	0.20 max	8.00-10.00	0.30-0.70	0.90-1.20	0.40 max	0.040 max	0.040 max	1.00 max	Cu <=0.30; bal Fe	620	420	13	179-255 HB

USA

Specification	Designation	Notes	C	Cr	Mn	Mo	Ni	P	S	Si	Other	UTS	YS	El	Hard
	UNS J82090		0.20 max	8.00-10.00	0.35-0.65	0.90-1.20	0.50 max	0.04 max	0.045 max	1.00 max	Cu <=0.50; W <=0.10; bal Fe				

UNS numbers and US grades are provided as a means of cross referencing chemically similar alloys. Exchangability is only possible after independent examination of specifications. Tensile properties are minimum or typical as specified. UTS and YS as MPa. El as %. See Appendix for list of abbreviations used in Notes. * indicates obsolete material.

Specification	Designation	Notes	C	Cr	Mn	Mo	Ni	P	S	Si	Other	UTS	YS	El	Hard

Steel Casting, Unclassified, C12 (Continued from previous page)

USA

Specification	Designation	Notes	C	Cr	Mn	Mo	Ni	P	S	Si	Other	UTS	YS	El	Hard
ACI	HA*	Heat res	0.2 max	8-10	0.35-0.65	0.9-1.2	0.5 max	0.04 max	0.04 max	1	bal Fe				
ASTM A217/A217M(95)	C12	Heat res	0.2 max	8-10	0.35-0.65	0.9-1.2	0.5 max	0.04	0.045	1	Cu <=0.5; W 0.1; bal Fe				

Steel Casting, Unclassified, C12A

USA

Specification	Designation	Notes	C	Cr	Mn	Mo	Ni	P	S	Si	Other	UTS	YS	El	Hard
	UNS J84090		0.12 max	8.0-9.5	0.30-0.60	0.85-1.05	0.40 max	0.020 max	0.018 max	0.20-0.50	Al <=0.040; N 0.030-0.070; V 0.18-0.25; Cb 0.060-0.10; bal Fe				

Steel Casting, Unclassified, CP5b

USA

Specification	Designation	Notes	C	Cr	Mn	Mo	Ni	P	S	Si	Other	UTS	YS	El	Hard
	UNS J51545		0.15 max	4.00-6.00	0.30-0.60	0.45-0.65	0.50 max	0.040 max	0.045 max	1.00-2.00	Cu <=0.50; W <=0.10; bal Fe				
ASTM A426(97)	CP5b	Pipe; Heat res	0.15 max	4.00-6.00	0.30-0.60	0.45-0.65	0.50 max	0.040 max	0.045 max	1.00-2.00	Cu <=0.50; W <=0.10; Cu+Ni+W<=1.00; bal Fe	415	205	22	225 HB max

Steel Casting, Unclassified, LF3

France

Specification	Designation	Notes	C	Cr	Mn	Mo	Ni	P	S	Si	Other	UTS	YS	El	Hard
AFNOR NFA36208(82)	3.5Ni285	t<=30mm	0.15 max	0.25 max	0.3-0.8	0.1 max	3.25-3.75	0.025 max	0.02 max	0.35 max	Al >=0.015; V <=0.04; W <=0.1; Cr+Cu+Mo<=0.5; bal Fe	460-570	285	23	
AFNOR NFA36208(82)	3.5Ni355	t<=30mm	0.15 max	0.25 max	0.3-0.8	0.1 max	3.25-3.75	0.25 max	0.02 max	0.35 max	Al >=0.015; V <=0.04; W <=0.1; Cr+Cu+Mo<=0.5; bal Fe	490-610	355	22	

Steel Casting, Unclassified, No equivalents identified

Czech Republic

Specification	Designation	Notes	C	Cr	Mn	Mo	Ni	P	S	Si	Other	UTS	YS	El	Hard
CSN 422880	422880		0.1 max	0.1 max	0.35 max		22.0-25.5	0.070 max	0.05 max	0.3 max	Al 11.0-14.0; Cu 3.0-4.0; bal Fe				
CSN 422881	422881		0.1 max	0.1 max	0.35 max		25.5-28.5	0.070 max	0.05 max	0.3 max	Al 11.0-14.0; Cu 5.0-7.0; bal Fe				
CSN 422882	422882		0.1 max	0.1 max	0.35 max		26.0-29.0	0.070 max	0.05 max	0.3 max	Al 10.0-13.0; bal Fe				
CSN 422891	422891		0.1 max	0.06 max	0.3 max		12.0-15.0	0.500 max		0.5 max	0.3 max Al 7.0-9.0; Co 22.5-25.0; Cu 2.0-4.0; Nb <=1.2; Ti <=0.6; bal Fe				
CSN 422895	422895		0.1 max	0.06 max	0.3 max		13.0-16.0	0.500 max		0.5 max	0.3 max Al 7.5-8.5; Co 23.0-26.0; Cu 2.5-4.0; Nb <=1.2; Ti <=0.5; bal Fe				
CSN 422921	422921		1.1-1.5	0.7-1.2	12.0-14.0			0.1 max	0.05 max	0.7 max	bal Fe				

France

Specification	Designation	Notes	C	Cr	Mn	Mo	Ni	P	S	Si	Other	UTS	YS	El	Hard
AFNOR NFA32053(92)	20M5M	Q/T; Low temp; t<=30mm	0.17-0.23	0.4 max	1.1-1.5	0.2 max	0.3 max	0.025 max	0.02 max	0.6 max	V <=0.04; W <=0.1; Ni+Cr+Cu+Mo+V<=1; bal Fe	500	300	24	150 HB
AFNOR NFA32053(92)	20NCD4M	t<=150mm	0.17-0.23	0.3-0.5	0.8-1.2	0.4-0.8	0.8-1.2	0.025 max	0.02 max	0.6 max	V <=0.04; W <=0.1; Ni+Cr+Cu+Mo+V<=1; bal Fe	570	410	16	170 HB
AFNOR NFA32054(78)	20NCD12M	t<1000mm	0.23 max	1.35-1.85	0.8 max	0.3-0.6	2.5-3.5	0.04 max	0.035 max	0.6 max	Cu <=0.5; Pb <=0.15; V <=0.04; W <=0.1; bal Fe	750	650	14	
AFNOR NFA32054(78)	25CD4M	t<1000mm	0.28 max	0.8-1.2	1 max	0.15-0.35	0.5 max	0.04 max	0.035 max	0.6 max	Cu <=0.5; Pb <=0.15; V <=0.04; W <=0.1; bal Fe	600	400	15	
AFNOR NFA32054(78)	25NCD2M	t<1000mm	0.28 max	0.4-0.8	1 max	0.3-0.6	0.4-0.8	0.04 max	0.035 max	0.6 max	Cu <=0.5; Pb <=0.15; V <=0.04; W <=0.1; bal Fe	620	380	17	
AFNOR NFA32054(78)	30M6M	t<1000mm	0.33 max	0.5 max	1.7 max	0.25 max	0.5 max	0.04 max	0.035 max	0.6 max	Cu <=0.5; Pb <=0.15; V <=0.04; W <=0.1; bal Fe	580	350	16	

UNS numbers and US grades are provided as a means of cross referencing chemically similar alloys. Exchangability is only possible after independent examination of specifications. Tensile properties are minimum or typical as specified. UTS and YS as MPa. El as %. See Appendix for list of abbreviations used in Notes. * indicates obsolete material.

Specification	Designation	Notes	C	Cr	Mn	Mo	Ni	P	S	Si	Other	UTS	YS	El	Hard

Steel Casting, Unclassified, No equivalents identified

France

Specification	Designation	Notes	C	Cr	Mn	Mo	Ni	P	S	Si	Other	UTS	YS	El	Hard
AFNOR NFA32054(78)	30MV6M	t<1000mm	0.33 max	0.5 max	1.7 max	0.25 max	0.5 max	0.04 max	0.035 max	0.6 max	Cu <=0.5; Pb <=0.15; V 0.08-0.12; W <=0.1; Cu+Ni+Cr+Mo<=1; bal Fe	620	450	14	
AFNOR NFA32054(78)	30NCD8M	t<1000mm	0.33 max	0.8-1.2	1 max	0.3-0.6	1.7-2.3	0.04 max	0.035 max	0.6 max	Cu <=0.5; Pb <=0.15; V <=0.04; W <=0.1; bal Fe	720	500	15	
AFNOR NFA32054(78)	33MD6M	t<1000mm	0.36 max	0.5 max	1.7 max	0.2-0.4	0.5 max	0.04 max	0.035 max	0.6 max	Cu <=0.5; Pb <=0.15; V <=0.04; W <=0.1; Cu+Ni+Cr+V<=1; bal Fe	600	400	16	
AFNOR NFA32054(94)	G10MnMoV6	28-50mm; Norm	0.12 max	0.3 max	1.8 max	0.2-0.4	0.4 max	0.03 max	0.02 max	0.6 max	Cu <=0.3; Pb <=0.15; V 0.05-0.1; W <=0.1; bal Fe	500	380	22	
AFNOR NFA32054(94)	G10MnMoV6	150<t<=250mm; Norm	0.12 max	0.3 max	1.8 max	0.2-0.4	0.4 max	0.03 max	0.02 max	0.6 max	Cu <=0.3; Pb <=0.15; V 0.05-0.1; W <=0.1; bal Fe	460	350	18	
AFNOR NFA32054(94)	G15CrMoV6	28-50mm; Q/T	0.12-0.18	1.3-1.8	1 max	0.8-1	0.4 max	0.03 max	0.02 max	0.6 max	Cu <=0.3; Pb <=0.15; V 0.15-0.25; W <=0.1; bal Fe	980	930	4	
AFNOR NFA32054(94)	G16Mn5	50<t<=100mm; Norm	0.13-0.2	0.3 max	1.6 max	0.15 max	0.4 max	0.03 max	0.025 max	0.6 max	Cu <=0.3; Pb <=0.15; V <=0.1; W <=0.1; bal Fe	430	230	24	
AFNOR NFA32054(94)	G16Mn5	28-50mm; Norm	0.13-0.2	0.3 max	1.6 max	0.15 max	0.4 max	0.03 max	0.025 max	0.6 max	Cu <=0.3; Pb <=0.15; V <=0.1; W <=0.1; bal Fe	430	250	24T	
AFNOR NFA32054(94)	G20Mn6	28-50mm; Norm	0.17-0.23	0.3 max	1.8 max	0.15 max	0.4 max	0.03 max	0.025 max	0.6 max	Cu <=0.3; Pb <=0.15; V <=0.1; W <=0.1; bal Fe	500	300	22	
AFNOR NFA32054(94)	G20Mn6	100<t<=150mm; Norm	0.17-0.23	0.3 max	1.8 max	0.15 max	0.4 max	0.03 max	0.025 max	0.6 max	Cu <=0.3; Pb <=0.15; V <=0.1; W <=0.1; bal Fe	480	260	20	
AFNOR NFA32054(94)	G20NiCrMo12	28-100mm; Q/T	0.22 max	1.3-1.8	1 max	0.45-0.6	3-3.5	0.03 max	0.02 max	0.6 max	Cu <=0.3; Pb <=0.15; V <=0.1; W <=0.1; bal Fe	750	650	16	
AFNOR NFA32054(94)	G20NiCrMo12	150<t<=250mm; Q/T	0.22 max	1.3-1.8	1 max	0.45-0.6	3-3.5	0.03 max	0.02 max	0.6 max	Cu <=0.3; Pb <=0.15; V <=0.1; W <=0.1; bal Fe	700	600	14	
AFNOR NFA32054(94)	G25CrMo4	28-50mm; Norm	0.22-0.28	0.8-1.2	1 max	0.15-0.35	0.4 max	0.03 max	0.02 max	0.6 max	Cu <=0.3; Pb <=0.15; V <=0.1; W <=0.1; bal Fe	580	380	18	
AFNOR NFA32054(94)	G25CrMo4	100<t<=250mm; Norm	0.22-0.28	0.8-1.2	1 max	0.15-0.35	0.4 max	0.03 max	0.02 max	0.6 max	Cu <=0.3; Pb <=0.15; V <=0.1; W <=0.1; bal Fe	550	250	14	
AFNOR NFA32054(94)	G30Mn6	28-50mm; Norm	0.25-0.32	0.3 max	1.8 max	0.15 max	0.4 max	0.03 max	0.025 max	0.6 max	Cu <=0.3; Pb <=0.15; V <=0.1; W <=0.1; bal Fe	580	350	16	
AFNOR NFA32054(94)	G30Mn6	150<t<=400mm; Norm	0.25-0.32	0.3 max	1.8 max	0.15 max	0.4 max	0.03 max	0.025 max	0.6 max	Cu <=0.3; Pb <=0.15; V <=0.1; W <=0.1; bal Fe	520	250	14	
AFNOR NFA32054(94)	G30NiCrMo14	28-50mm; Q/T	0.33 max	0.8-1.2	1 max	0.3-0.6	3-4	0.03 max	0.02 max	0.6 max	Cu <=0.3; Pb <=0.15; V <=0.1; W <=0.1; bal Fe	1100	1000	7	
AFNOR NFA32054(94)	G30NiCrMo14	150<t<=250mm; Q/T	0.33 max	0.8-1.2	1 max	0.3-0.6	3-4	0.03 max	0.02 max	0.6 max	Cu <=0.3; Pb <=0.15; V <=0.1; W <=0.1; bal Fe	800	600	7	
AFNOR NFA32054(94)	G30NiCrMo8	150<t<=400mm; Norm	0.33 max	0.8-1.2	1 max	0.3-0.6	1.7-2.3	0.03 max	0.02 max	0.6 max	Cu <=0.3; Pb <=0.15; V <=0.1; W <=0.1; bal Fe	700	500	12	
AFNOR NFA32054(94)	G30NiCrMo8	28-50mm; Norm	0.33 max	0.8-1.2	1 max	0.3-0.6	1.7-2.3	0.03 max	0.02 max	0.6 max	Cu <=0.3; Pb <=0.15; V <=0.1; W <=0.1; bal Fe	750	550	15	
AFNOR NFA32054(94)	G35CrMo4	28-50mm; Norm	0.3-0.38	0.8-1.2	1 max	0.15-0.35	0.4 max	0.03 max	0.02 max	0.6 max	Cu <=0.3; Pb <=0.15; V <=0.1; W <=0.1; bal Fe	750	520	12	
AFNOR NFA32054(94)	G35CrMo4	150<t<=400mm; Norm	0.3-0.38	0.8-1.2	1 max	0.15-0.35	0.4 max	0.03 max	0.02 max	0.6 max	Cu <=0.3; Pb <=0.15; V <=0.1; W <=0.1; bal Fe	620	330	10	

Specification	Designation	Notes	C	Cr	Mn	Mo	Ni	P	S	Si	Other	UTS	YS	El	Hard

Steel Casting, Unclassified, No equivalents identified

France

Specification	Designation	Notes	C	Cr	Mn	Mo	Ni	P	S	Si	Other	UTS	YS	El	Hard
AFNOR NFA32054(94)	G35NiCrMo6	28-150mm; Norm	0.38 max	1.4-1.7	1 max	0.15-0.35	1.4-1.7	0.03 max	0.02 max	0.6 max	Cu <=0.3; Pb <=0.15; V <=0.1; W <=0.1; bal Fe	800	550	12	
AFNOR NFA32054(94)	G35NiCrMo6	15t<=400mm; Norm	0.38 max	1.4-1.7	1 max	0.15-0.35	1.4-1.7	0.03 max	0.02 max	0.6 max	Cu <=0.3; Pb <=0.15; V <=0.1; W <=0.1; bal Fe	750	500	12	
AFNOR NFA32054(94)	G42CrMo4	28-50mm; Norm	0.39-0.45	0.8-1.2	1 max	0.15-0.35	0.4 max	0.03 max	0.02 max	0.6 max	Cu <=0.3; Pb <=0.15; V <=0.1; W <=0.1; bal Fe	780	580	10	
AFNOR NFA32054(94)	G42CrMo4	150<t<=400mm; Norm	0.39-0.45	0.8-1.2	1 max	0.15-0.35	0.4 max	0.03 max	0.02 max	0.6 max	Cu <=0.3; Pb <=0.15; V <=0.1; W <=0.1; bal Fe	650	350	10	
AFNOR NFA32054(94)	GE230	28-50mm; Norm	0.2 max	0.3 max	1.2 max	0.15 max	0.4 max	0.035 max	0.03 max	0.6 max	Cu <=0.3; Pb <=0.15; V <=0.05; W <=0.1; Cr+Mo+Ni+V<=1; bal Fe	400	230	25	
AFNOR NFA32054(94)	GE230	50<t<=100mm; Norm	0.2 max	0.3 max	1.2 max	0.15 max	0.4 max	0.035 max	0.03 max	0.6 max	Cu <=0.3; Pb <=0.15; V <=0.05; W <=0.1; Cr+Mo+Ni+V<=1; bal Fe	400	210	23	
AFNOR NFA32054(94)	GE280	500<t<=100mm; Norm	0.25 max	0.3 max	1.2 max	0.15 max	0.4 max	0.035 max	0.03 max	0.6 max	Cu <=0.3; Pb <=0.15; V <=0.05; W <=0.1; Cr+Mo+Ni+V<=1; bal Fe	480	260	18	
AFNOR NFA32054(94)	GE280	28-50mm; Norm	0.25 max	0.3 max	1.2 max	0.15 max	0.4 max	0.035 max	0.03 max	0.6 max	Cu <=0.3; Pb <=0.15; V <=0.05; W <=0.1; Cr+Mo+Ni+V<=1; bal Fe	480	280	20	
AFNOR NFA32054(94)	GE320	28-50mm; Norm	0.32 max	0.3 max	1.2 max	0.15 max	0.4 max	0.035 max	0.03 max	0.6 max	Cu <=0.3; Pb <=0.15; V <=0.05; W <=0.1; Cr+Mo+Ni+V<=1; bal Fe	560	320	16	
AFNOR NFA32054(94)	GE320	50<t<=100mm; Norm	0.32 max	0.3 max	1.2 max	0.15 max	0.4 max	0.035 max	0.03 max	0.6 max	Cu <=0.3; Pb <=0.15; V <=0.05; W <=0.1; Cr+Mo+Ni+V<=1; bal Fe	560	300	14	
AFNOR NFA32054(94)	GE370	28-50mm; Norm	0.45 max	0.3 max	1.2 max	0.15 max	0.4 max	0.035 max	0.03 max	0.6 max	Cu <=0.3; Pb <=0.15; V <=0.05; W <=0.1; Cr+Mo+Ni+V<=1; bal Fe	650	370	12	
AFNOR NFA32054(94)	GE370	50<t<=100mm; Norm	0.45 max	0.3 max	1.2 max	0.15 max	0.4 max	0.035 max	0.03 max	0.6 max	Cu <=0.3; Pb <=0.15; V <=0.05; W <=0.1; Cr+Mo+Ni+V<=1; bal Fe	650	320	10	
AFNOR NFA32054(94)	GX4CrNi13-4	100<t<=400mm; Q/T	0.06 max	12-13.5	1 max	0.15 max	3.5-4.5	0.035 max	0.02 max	0.8 max	Cu <=0.3; Pb <=0.15; V <=0.1; W <=0.1; bal Fe	700	500	16	
AFNOR NFA32054(94)	GX4CrNi13-4	28<t<=400mm; Temp	0.06 max	12-13.5	1 max	0.15 max	3.5-4.5	0.035 max	0.02 max	0.8 max	bal Fe	900	800	12	
AFNOR NFA32054(94)	GX4CrNi16-4	28<t<=250mm	0.06 max	15.5-17	1 max	0.15 max	4-5.5	0.035 max	0.02 max	0.8 max	bal Fe	1000	830	10	
AFNOR NFA32054(94)	GX4CrNi16-4	28<t<=250mm	0.06 max	15.5-17	1 max	0.15 max	4-5.5	0.035 max	0.02 max	0.8 max	bal Fe	780	540	15	
AFNOR NFA32055(85)	15CD9.10M	t<1000mm	0.1-0.18	2-2.5	1.1 max	0.9-1.1	0.3 max	0.03 max	0.03 max	0.6 max	Cu <=0.3; Pb <=0.15; V <=0.04; W <=0.1; bal Fe	550-700	325	17	160-200 HB
AFNOR NFA32055(85)	15CDV4.10M	t<1000mm	0.12-0.2	1-1.5	1 max	0.85-1.15	0.3 max	0.03 max	0.03 max	0.6 max	Cu <=0.3; Pb <=0.15; V 0.15-0.3; W <=0.1; bal Fe	600-750	350	15	
AFNOR NFA32055(85)	15CDV9.10M	t<1000mm	0.1-0.18	2-2.75	1 max	0.9-1.2	0.3 max	0.03 max	0.03 max	0.6 max	Cu <=0.3; Pb <=0.15; V 0.15-0.3; W <=0.1; bal Fe	600-750	350	15	180-220 HB
AFNOR NFA32055(85)	18CD2.05M	t<1000mm	0.14-0.22	0.4-0.65	1 max	0.45-0.7	0.3 max	0.03 max	0.03 max	0.6 max	Cu <=0.3; Pb <=0.15; V <=0.04; W <=0.1; bal Fe	500-650	300	18	155-200 HB

Hungary

Specification	Designation	Notes	C	Cr	Mn	Mo	Ni	P	S	Si	Other	UTS	YS	El	Hard
MSZ 1749(89)	Ao15CrMoVW	High-temp const; 0-30.00mm; HT	0.12-0.18	1-1.5	0.6-1	0.4-0.6		0.03 max	0.03 max	0.3-0.6	V 0.5-0.7; W 0.4-0.7; bal Fe	600-750	400	18L	
MSZ 8272(89)	Ao34Mn6	0-30.00mm; Q/T	0.3-0.4		1.2-1.6		0.3 max	0.05 max	0.05 max	0.2-0.4	bal Fe	620	400	14L	

UNS numbers and US grades are provided as a means of cross referencing chemically similar alloys. Exchangability is only possible after independent examination of specifications. Tensile properties are minimum or typical as specified. UTS and YS as MPa. El as %. See Appendix for list of abbreviations used in Notes. * indicates obsolete material.

Steel Casting, Unclassified, No equivalents identified

Specification	Designation	Notes	C	Cr	Mn	Mo	Ni	P	S	Si	Other	UTS	YS	El	Hard
Hungary															
MSZ 8272(89)	Ao34Mn6	0-30.00mm; norm	0.3-0.4		1.2-1.6		0.3 max	0.05 max	0.05 max	0.2-0.4	bal Fe	540	290	12L	
MSZ 8272(89)	Ao35Mn6ne		0.3-0.4		1.2-1.6		0.3 max	0.05 max	0.05 max	0.2-0.4	bal Fe				
MSZ 8272(89)	Ao40MnVne	0-30.00mm; Q/T	0.35-0.45		1.6-1.9		0.3 max	0.04 max	0.04 max	0.2-0.42	V 0.1-0.2; bal Fe	850	540	8L	
Italy															
UNI 3160(83)	G90Cr4	Wear res	0.8-1	1-1.2	0.8 max			0.035 max	0.035 max	0.5 max	bal Fe				
UNI 3160(83)	GC20		0.25 max		0.8 max			0.035 max	0.035 max	0.6 max	bal Fe				
UNI 3608(83)	GC20		0.25 max		0.8 max			0.035 max	0.035 max	0.6 max	bal Fe				
UNI 7317(74)	G22Ni10	Tough at subzero	0.25 max		0.8 max		2-3	0.035 max	0.035 max	0.6 max	bal Fe				
Spain															
UNE 36254(79)	25MnCrMo5	Q/T	0.22-0.3	0.7-0.9	1.2-1.6	0.45-0.65		0.04 max	0.04 max	0.3-0.5	bal Fe				
UNE 36254(79)	80CrMo8	Q/T	0.7-0.9	1.75-2.25	0.6-0.9	0.35-0.45		0.04 max	0.04 max	0.3-0.5	bal Fe				
UNE 36254(79)	90Cr4	Q/T	0.8-1.0	0.8-1.2	0.5-0.8	0.15 max		0.04 max	0.04 max	0.3-0.5	bal Fe				
UNE 36254(79)	F.8233*	see 80CrMo8	0.7-0.9	1.75-2.25	0.6-0.9	0.35-0.45		0.04 max	0.04 max	0.3-0.5	bal Fe				
UNE 36254(79)	F.8241*	see 25MnCrMo5	0.22-0.3	0.7-0.9	1.2-1.6	0.45-0.65		0.04 max	0.04 max	0.3-0.5	bal Fe				
UK															
BS 3100(91)	BT1	0-999mm						0.040 max	0.040 max		bal Fe	690	495	11	201-279 HB
BS 3100(91)	BT2	0-999mm						0.040 max	0.040 max		bal Fe	850	585	8	248-327 HB
BS 3100(91)	BT3	0-999mm						0.030 max	0.030 max		bal Fe	1000	695	6	293-362 HB
USA															
ASTM A426(97)	CP7	Pipe; Heat res	0.15 max	6.00-8.00	0.30-0.60	0.44-0.65	0.50 max	0.040 max	0.045 max	0.50-1.00	Cu <=0.50; W <=0.10; Cu+Ni+W<=1.00; bal Fe	415	205	22	201 HB max
FED QQ-S-681F(85) - - -*		Obs; see ASTM A24, A148									bal Fe				

Specification	Designation	Notes	C	Cr	Cu	Mn	Ni	P	S	Si	Other	UTS	YS	El	Hard

Carbon Steel, Nonresulfurized, 1005 (Continued from previous page)

Specification	Designation	Notes	C	Cr	Cu	Mn	Ni	P	S	Si	Other	UTS	YS	El	Hard
Australia															
AS 1443(94)	1004	CF bar (may treat w/microalloying elements)	0.06 max			0.25-0.50		0.040 max	0.040 max	0.10-0.35	Si<=0.10 for Al-killed; bal Fe				
Bulgaria															
BDS 5785(83)	05kp	Struct	0.06 max	0.10 max	0.25 max	0.40 max	0.25 max	0.035 max	0.040 max	0.03 max	bal Fe				
Czech Republic															
CSN 411304	11304		0.06 max	0.05 max	0.05 max	0.35 max		0.02 max	0.02 max	0.01 max	Al <=0.005; N <=0.006; bal Fe				
Europe															
EN 10016/3(94)	1.1185	Rod	0.03 max	0.10 max	0.10 max	0.20-0.35	0.10 max	0.020 max	0.020 max	0.05 max	Al <=0.01; Mo <=0.03; N <=0.007; Cr+Ni+Cu<=0.25; bal Fe				
EN 10016/3(94)	1.1187	Rod	0.05 max	0.10 max	0.15 max	0.20-0.4	0.10 max	0.025 max	0.025 max	0.05 max	Al <=0.05; Mo <=0.03; Cr+Ni+Cu<=0.30; bal Fe				
EN 10016/3(94)	C2D1	Rod	0.03 max	0.10 max	0.10 max	0.20-0.35	0.10 max	0.020 max	0.020 max	0.05 max	Al <=0.01; Mo <=0.03; N <=0.007; Cr+Ni+Cu<=0.25; bal Fe				
EN 10016/3(94)	C3D1	Rod	0.05 max	0.10 max	0.15 max	0.20-0.4	0.10 max	0.025 max	0.025 max	0.05 max	Al <=0.05; Mo <=0.03; Cr+Ni+Cu<=0.30; bal Fe				
EN 10016/4(94)	1.1110	Rod	0.05 max	0.10 max	0.15 max	0.30-0.50	0.10 max	0.020 max	0.025 max	0.30 max	Al <=0.01; Mo <=0.05; N <=0.007; Cr+Ni+Cu<=0.30, Cu+Sn<=0.15; bal Fe				
EN 10016/4(94)	C3D2	Rod	0.05 max	0.10 max	0.15 max	0.30-0.50	0.10 max	0.020 max	0.025 max	0.30 max	Al <=0.01; Mo <=0.05; N <=0.007; Cr+Ni+Cu<=0.30, Cu+Sn<=0.15; bal Fe				
EN 10130(91)	1.0873	Sh Strp, CR t<=3mm	0.02 max			0.25 max		0.020 max	0.020 max		Ti <=0.30; bal Fe	270-350	120	38	
EN 10130(91)	FeP06	Sh Strp, CR t<=3mm	0.02 max			0.25 max		0.020 max	0.020 max		Ti <=0.30; bal Fe	270-350	120	38	
India															
IS 1570/2(79)	2C2	Sh Plt Sect Shp Bar Bil Frg Tub Pip	0.05 max	0.3 max	0.3 max	0.4 max	0.4 max	0.055 max	0.055 max	0.6 max	Co <=0.1; Mo <=0.15; Pb <=0.15; W <=0.1; P:S Varies; bal Fe				
Italy															
UNI 5598(71)	3CD5	Wir rod	0.06 max			0.25-0.5		0.035 max	0.035 max	0.6 max	N <=0.012; bal Fe				
UNI 7356(74)	CB4FU	Q/T	0.08 max			0.2-0.4		0.04 max	0.04 max	0.6 max	N <=0.007; trace Si/Spuren/nyomok; bal Fe				
Mexico															
NMX-B-301(86)	1005	Bar	0.06 max			0.35 max		0.040 max	0.050 max		bal Fe				
USA															
	AISI 1005	Wir rod	0.06 max			0.35 max		0.040 max	0.050 max		bal Fe				
	AISI 1005	Bar	0.06 max			0.35 max		0.040 max	0.050 max	0.050 max	bal Fe				
	UNS G10050		0.06 max			0.35 max		0.040 max	0.050 max		bal Fe				
ASTM A29/A29M(93)	1005	Bar	0.06 max			0.35 max		0.040 max	0.050 max	0.050 max	bal Fe				
ASTM A510(96)	1005	Wir rod	0.06 max			0.35 max		0.040 max	0.050 max		bal Fe				
SAE J403(95)	1005	Bar Wir rod Smls Tub HR CF	0.06 max			0.35 max		0.030 max	0.050 max		bal Fe				

Specification	Designation	Notes	C	Cr	Cu	Mn	Ni	P	S	Si	Other	UTS	YS	El	Hard

Carbon Steel, Nonresulfurized, 1006

Australia

Specification	Designation	Notes	C	Cr	Cu	Mn	Ni	P	S	Si	Other	UTS	YS	El	Hard
AS 1442(92)	1006	HR bar, Semifinished (may treat w/microalloying elements)	0.08 max			0.25-0.50		0.040 max	0.040 max	0.10-0.35	Si<=0.10 for Al-killed; bal Fe				
AS/NZS 1594(97)	HA1006	HR flat products	0.08 max	0.30 max	0.35 max	0.40 max	0.35 max	0.040 max	0.030 max	0.03 max	Al <=0.100; Mo <=0.10; Ti <=0.040; (Cu+Ni+Cr+Mo) 1 max; bal Fe				

Bulgaria

Specification	Designation	Notes	C	Cr	Cu	Mn	Ni	P	S	Si	Other	UTS	YS	El	Hard
BDS 14351	08kp	Plt HR	0.10 max	0.10 max	0.20 max	0.25-0.45	0.15 max	0.025 max	0.03 max	0.03 max	Mo <=0.15; bal Fe				
BDS 9609(72)	08Fkp	Strp, CR	0.08 max	0.03 max	0.15 max	0.3-0.45	0.10 max	0.020 max	0.03 max	0.01 max	Mo <=0.15; Ti <=0.05; V 0.02-0.04; W <=0.1; bal Fe				

Czech Republic

Specification	Designation	Notes	C	Cr	Cu	Mn	Ni	P	S	Si	Other	UTS	YS	El	Hard
CSN 411301	11301	Unalloyed mild	0.08 max	0.3 max		0.4 max		0.025 max	0.025 max	0.60 max	bal Fe				

Europe

Specification	Designation	Notes	C	Cr	Cu	Mn	Ni	P	S	Si	Other	UTS	YS	El	Hard
EN 10016/2(94)	1.0300	Rod	0.06 max	0.20 max	0.30 max	0.30-0.60	0.25 max	0.035 max	0.035 max	0.30 max	Al <=0.01; Mo <=0.05; bal Fe				
EN 10016/2(94)	C4D	Rod	0.06 max	0.20 max	0.30 max	0.30-0.60	0.25 max	0.035 max	0.035 max	0.30 max	Al <=0.01; Mo <=0.05; bal Fe				
EN 10016/3(94)	1.1188	Rod	0.06 max	0.15 max	0.15 max	0.20-0.45	0.15 max	0.025 max	0.025 max	0.10 max	Al <=0.05; Mo <=0.03; Cr+Ni+Cu<=0.35; bal Fe				
EN 10016/3(94)	C4D1	Rod	0.06 max	0.15 max	0.15 max	0.20-0.45	0.15 max	0.025 max	0.025 max	0.10 max	Al <=0.05; Mo <=0.03; Cr+Ni+Cu<=0.35; bal Fe				
EN 10016/4(94)	1.1111	Rod	0.07 max	0.10 max	0.15 max	0.30-0.50	0.10 max	0.020 max	0.025 max	0.30 max	Al <=0.01; Mo <=0.05; N <=0.007; Cr+Ni+Cu<=0.30, Cu+Sn<=0.15; bal Fe				
EN 10016/4(94)	C5D2	Rod	0.07 max	0.10 max	0.15 max	0.30-0.50	0.10 max	0.020 max	0.025 max	0.30 max	Al <=0.01; Mo <=0.05; N <=0.007; Cr+Ni+Cu<=0.30, Cu+Sn<=0.15; bal Fe				

Italy

Specification	Designation	Notes	C	Cr	Cu	Mn	Ni	P	S	Si	Other	UTS	YS	El	Hard
UNI 5598(71)	3CD6	Wir rod	0.08 max			0.25-0.5		0.035 max	0.035 max	0.6 max	N <=0.012; bal Fe				
UNI 5771(66)	C8	Chain	0.08 max			0.35-0.45		0.035 max	0.035 max	0.6 max	trace Si; bal Fe				

Japan

Specification	Designation	Notes	C	Cr	Cu	Mn	Ni	P	S	Si	Other	UTS	YS	El	Hard
JIS G3507(91)	SWRCH6R	Wir rod FU	0.08 max			0.60 max		0.040 max	0.040 max		bal Fe				

Mexico

Specification	Designation	Notes	C	Cr	Cu	Mn	Ni	P	S	Si	Other	UTS	YS	El	Hard
NMX-B-301(86)	1006	Bar	0.08 max			0.25-0.40		0.040 max	0.050 max		bal Fe				

Russia

Specification	Designation	Notes	C	Cr	Cu	Mn	Ni	P	S	Si	Other	UTS	YS	El	Hard
GOST 9045	08Fkp	Body panel Sh	0.08 max	0.04 max	0.15 max	0.2-0.4	0.1 max	0.02 max	0.025 max	0.03 max	Al <=0.1; V 0.02-0.04; bal Fe				

Spain

Specification	Designation	Notes	C	Cr	Cu	Mn	Ni	P	S	Si	Other	UTS	YS	El	Hard
UNE 36086(75)	AP13	Unalloyed mild; Sh CR	0.08 max			0.45 max		0.03 max	0.03 max	0.60 max	Mo <=0.15; bal Fe				

UK

Specification	Designation	Notes	C	Cr	Cu	Mn	Ni	P	S	Si	Other	UTS	YS	El	Hard
BS 970/1(96)	040A04	Blm Bil Slab Bar Rod Frg	0.08 max			0.30-0.50		0.050 max	0.050 max	0.10-0.40	bal Fe				

USA

Specification	Designation	Notes	C	Cr	Cu	Mn	Ni	P	S	Si	Other	UTS	YS	El	Hard
	AISI 1006	Sh Strp Plt	0.08 max			0.45 max		0.030 max	0.035 max		bal Fe				
	AISI 1006	Wir rod	0.08 max			0.25-0.40		0.040 max	0.050 max		bal Fe				
	AISI 1006	Struct shp	0.08 max			0.45 max		0.030 max	0.035 max		bal Fe				
	AISI 1006	Bar	0.08 max			0.25-0.40		0.040 max	0.050 max	0.050 max	bal Fe				

UNS numbers and US grades are provided as a means of cross referencing chemically similar alloys. Exchangability is only possible after independent examination of specifications. Tensile properties are minimum or typical as specified. UTS and YS as MPa. El as %. See Appendix for list of abbreviations used in Notes. * indicates obsolete material.

Specification	Designation	Notes	C	Cr	Cu	Mn	Ni	P	S	Si	Other	UTS	YS	El	Hard

Carbon Steel, Nonresulfurized, 1006 (Continued from previous page)

USA

Specification	Designation	Notes	C	Cr	Cu	Mn	Ni	P	S	Si	Other	UTS	YS	El	Hard
	UNS G10060		0.08 max			0.25-0.40		0.040 max	0.050 max		Sheets and Plates, Mn 0.25-0.45; bal Fe				
ASTM A29/A29M(93)	1006	Bar	0.08 max			0.25-0.40		0.040 max	0.050 max	0.050 max	bal Fe				
ASTM A510(96)	1006	Wir rod	0.08 max			0.25-0.40		0.040 max	0.050 max		bal Fe				
ASTM A635/A635M(98)	1006	Sh Strp, Coil, HR	0.08 max	0.20 min		0.45 max		0.030 max	0.035 max		bal Fe				
ASTM A830/A830M(98)	1006	Plt	0.08 max			0.45 max		0.035 max	0.04 max		bal Fe				
SAE J1397(92)	1006	Bar CD, est mech prop									bal Fe	330	280	20	95 HB
SAE J1397(92)	1006	Bar HR, est mech prop									bal Fe	300	170	30	86 HB
SAE J403(95)	1006	Struct Shps Plt Strp Sh Weld Tub	0.08 max			0.45 max		0.030 max	0.035 max		bal Fe				
SAE J403(95)	1006	Bar Wir rod Smls Tub HR CF	0.08 max			0.25-0.40		0.030 max	0.050 max		bal Fe				

Carbon Steel, Nonresulfurized, 1008

Argentina

Specification	Designation	Notes	C	Cr	Cu	Mn	Ni	P	S	Si	Other	UTS	YS	El	Hard
IAS	IRAM 1008		0.10 max			0.30-0.70		0.040 max	0.050 max	0.10 max	bal Fe	300-380	190-250	30-38	85-110 HB

Australia

Specification	Designation	Notes	C	Cr	Cu	Mn	Ni	P	S	Si	Other	UTS	YS	El	Hard
AS 1442	K1008*	Obs; Bar Bil	0.1			0.25-0.5		0.05	0.05	0.1-0.35	bal Fe				
AS 1442	R1008*	Obs; Bar Bil	0.1			0.25-0.5		0.04	0.05		bal Fe				
AS 1442	S1008*	Obs; Bar Bil	0.1			0.25-0.5		0.05	0.05	0.35	bal Fe				
AS 1442(92)	1008	HR bar, Semifinished (may treat w/microalloying elements)	0.10 max			0.25-0.50		0.040 max	0.040 max	0.10-0.35	Si<=0.10 for Al-killed; bal Fe				
AS 1443	K1008*	Obs; Bar	0.1			0.25-0.5		0.05	0.05	0.1-0.35	bal Fe				
AS 1443	R1008*	Obs	0.1			0.25-0.5		0.04	0.05		bal Fe				
AS 1443	S1008*	Obs; Bar	0.1			0.25-0.5		0.05	0.05	0.35	bal Fe				
AS 1446	C1008*	Withdrawn, see AS/NZS 1594(97)	0.1			0.25-0.5		0.04	0.04		bal Fe				
AS 1446	R1008*	Withdrawn, see AS/NZS 1594(97)	0.1			0.25-0.5		0.04	0.05		bal Fe				
AS 1446	S1008*	Withdrawn, see AS/NZS 1594(97)	0.1			0.25-0.5		0.05	0.05	0.35	bal Fe				
AS 1585 Part 2	Temper 4*	Withdrawn	0.1			0.5		0.04	0.04		bal Fe				
AS 1585 Part 2	Temper 5*	Withdrawn, Sh Strp	0.1			0.5		0.04	0.04		bal Fe				
AS 1585 Part 2	Temper 6*	Withdrawn, Sh Strp	0.1			0.5		0.04	0.04		bal Fe				
AS 1594	HRC*	Withdrawn, Sh Strp, HR	0.1			0.5		0.03	0.035		bal Fe				
AS 1595	ICRC*	Obs	0.1			0.5		0.03	0.035		bal Fe				

Austria

Specification	Designation	Notes	C	Cr	Cu	Mn	Ni	P	S	Si	Other	UTS	YS	El	Hard
ONORM M3110	RC10	Wir	0.06-0.10			0.5 max		0.040 max	0.040 max	0.3 max	bal Fe				
ONORM M3110	UC10	Wir	0.06-0.10			0.5 max		0.040 max	0.040 max	0.15	bal Fe				
ONORM M3124	St12F	Sh Strp	0.1 max			0.20-0.45		0.040 max	0.040 max	0.15 max	bal Fe				
ONORM M3124	St13F	Sh Strp	0.1 max			0.20-0.45		0.030 max	0.030 max	0.15 max	bal Fe				
ONORM M3124	St22F	Sh Strp	0.1 max			0.20-0.45		0.040 max	0.040 max	0.15 max	bal Fe				
ONORM M3124	St23F	Sh Strp	0.1 max			0.20-0.45		0.030 max	0.030 max	0.15 max	bal Fe				
ONORM M3124	StO2F	Sh Strp	0.1 max			0.20-0.45		0.040 max	0.040 max	0.15 max	bal Fe				

Carbon Steel, Nonresulfurized, 1008 (Continued from previous page)

Specification	Designation	Notes	C	Cr	Cu	Mn	Ni	P	S	Si	Other	UTS	YS	El	Hard
Austria															
ONORM M3124	StO2FK32	Sh Strp	0.1 max			0.20-0.45		0.040 max	0.040 max	0.15 max	bal Fe				
ONORM M3124	StO2FK40	Sh Strp	0.1 max			0.20-0.45		0.040 max	0.040 max	0.15 max	bal Fe				
ONORM M3124	StO2FK50	Sh Strp	0.1 max			0.20-0.45		0.040 max	0.040 max	0.15 max	bal Fe				
ONORM M3124	StO2FK60	Sh Strp	0.1 max			0.20-0.45		0.040 max	0.040 max	0.15 max	bal Fe				
ONORM M3124	StO2FK70	Sh Strp	0.1 max			0.20-0.45		0.040 max	0.040 max	0.15 max	bal Fe				
ONORM M3124	StO3F	Sh Strp	0.1 max			0.20-0.45		0.030 max	0.030 max	0.15 max	bal Fe				
ONORM M3124	StO3FK32	Sh Strp	0.1 max			0.20-0.45		0.030 max	0.030 max	0.15 max	bal Fe				
ONORM M3124	StO3FK40	Sh Strp	0.1 max			0.20-0.45		0.030 max	0.030 max	0.15 max	bal Fe				
ONORM M3124	StO3FK50	Sh Strp	0.1 max			0.20-0.45		0.030 max	0.030 max	0.15 max	bal Fe				
ONORM M3124	StO3FK60	Sh Strp.	0.1 max			0.20-0.45		0.030 max	0.030 max	0.15 max	bal Fe				
ONORM M3124	StO3FK70	Sh Strp	0.1 max			0.20-0.45		0.030 max	0.030 max	0.15 max	bal Fe				
Bulgaria															
BDS 11488(83)	08JuA	Sh, CR	0.07 max	0.30 max	0.06 max	0.2-0.4	0.06 max	0.020 max	0.025 max	0.01 max	Al 0.02-0.07; Mo <=0.15; bal Fe				
BDS 11488(83)	08kpA	Sh, CR	0.10 max	0.10 max	0.15 max	0.2-0.4	0.10 max	0.025 max	0.03 max	0.03 max	Mo <=0.15; bal Fe				
BDS 11488(83)	08psA	Sh, CR	0.09 max	0.10 max	0.15 max	0.2-0.45	0.10 max	0.025 max	0.03 max	0.04 max	Mo <=0.15; bal Fe				
BDS 14351	08Ju	Plt HR	0.07 max	0.30 max	0.06 max	0.2-0.4	0.06 max	0.025 max	0.025 max	0.01 max	Al 0.02-0.07; Mo <=0.15; bal Fe				
BDS 14351	08ps	Plt HR	0.09 max	0.10 max	0.15 max	0.25-0.45	0.10 max	0.025 max	0.03 max	0.04 max	Mo <=0.15; bal Fe				
BDS 5785(83)	08kp	Struct	0.05-0.11	0.10 max	0.25 max	0.25-0.50	0.25 max	0.035 max	0.040 max	0.03 max	bal Fe				
BDS 5785(83)	08ps	Struct	0.05-0.11	0.10 max	0.25 max	0.35-0.65	0.25 max	0.035 max	0.040 max	0.05-0.17	bal Fe				
BDS 9609	08kpselekt	Strp, CR	0.08 max	0.03 max	0.15 max	0.3-0.45	0.10 max	0.020 max	0.03 max	0.01 max	Mo <=0.15; Ti <=0.05; V <=0.1; W <=0.1; bal Fe				
BDS 9609(72)	08Ju	Strp, CR	0.08 max	0.03 max	0.15 max	0.3-0.45	0.10 max	0.020 max	0.03 max	0.01 max	Al 0.02-0.07; Mo <=0.15; Ti <=0.05; V <=0.1; W <=0.1; bal Fe				
China															
GB 13237(91)	08	Sh Strp CR HT	0.05-0.12	0.10 max	0.25 max	0.35-0.65	0.25 max	0.035 max	0.035 max	0.17-0.37	bal Fe	275-410		28	
GB 13237(91)	08F	Sh Strp CR HT	0.05-0.11	0.10 max	0.25 max	0.25-0.50	0.25 max	0.035 max	0.035 max	0.03 max	bal Fe	275-380		30	
GB 3275(91)	08	Plt Strp HR HT	0.05-0.12	0.10 max	0.25 max	0.35-0.65	0.25 max	0.035 max	0.035 max	0.17-0.37	bal Fe	275-410		27	
GB 3275(91)	08F	Plt Strp HR HT	0.05-0.11	0.10 max	0.25 max	0.25-0.50	0.25 max	0.035 max	0.035 max	0.03 max	bal Fe	275-370		30	
GB 6478(86)	ML 08	Bar HR Norm	0.05-0.12	0.20 max	0.20 max	0.35-0.65	0.25 max	0.035 max	0.035 max	0.03 max	bal Fe	325	195	33	
GB 699(88)	08	Bar HR Norm 25mm diam	0.05-0.12	0.10 max	0.25 max	0.35-0.65	0.25 max	0.035 max	0.035 max	0.17-0.37	bal Fe	325	195	33	
GB 699(88)	08F	Bar HR Norm 25mm diam	0.05-0.11	0.10 max	0.25 max	0.25-0.50	0.25 max	0.035 max	0.035 max	0.03 max	bal Fe	295		35	
GB 710(91)	08	Sh Strp HR HT	0.05-0.12	0.10 max	0.25 max	0.35-0.65	0.25 max	0.035 max	0.035 max	0.17-0.37	bal Fe	275-410		25	
GB 710(91)	08F	Sh Strp HR HT	0.05-0.11	0.10 max	0.25 max	0.25-0.50	0.25 max	0.035 max	0.035 max	0.03 max	bal Fe	275-380	195	27	
GB/T 13796(92)	08	Bar Wir CD Ann	0.05-0.12	0.10 max	0.25 max	0.35-0.65	0.25 max	0.035 max	0.035 max	0.17-0.37	bal Fe	295-440			
GB/T 8164(93)	08	Strp HR/CR	0.05-0.12	0.10 max	0.25 max	0.20-0.50	0.20 max	0.035 max	0.035 max	0.17-0.37	bal Fe	325		33	

Specification	Designation	Notes	C	Cr	Cu	Mn	Ni	P	S	Si	Other	UTS	YS	El	Hard

Carbon Steel, Nonresulfurized, 1008 (Continued from previous page)

Czech Republic

Specification	Designation	Notes	C	Cr	Cu	Mn	Ni	P	S	Si	Other	UTS	YS	El	Hard
CSN 411320	11320	Unalloyed mild	0.11 max	0.3 max	0.03 max	1.60 max		0.045 max	0.045 max	0.60 max	bal Fe				
CSN 411321	11321	Unalloyed mild	0.1 max	0.3 max		0.45 max		0.03 max	0.03 max	0.60 max	bal Fe				
CSN 411325	11325	Unalloyed mild	0.1 max	0.3 max		0.45 max		0.035 max	0.035 max	0.60 max	Al 0.02-0.120; bal Fe				
CSN 411330	11330	Unalloyed mild; DDS sheet	0.09 max	0.3 max		2 max		0.04 max	0.04 max	0.6 max	bal Fe				
CSN 411331	11331	Unalloyed mild	0.11 max	0.3 max		0.45 max		0.035 max	0.035 max	0.60 max	bal Fe				
CSN 411342	11342		0.1			0.45		0.03 max	0.025 max	0.15	Ti 0.06-0.12; bal Fe				

Europe

Specification	Designation	Notes	C	Cr	Cu	Mn	Ni	P	S	Si	Other	UTS	YS	El	Hard
EN 10016/2(94)	1.0304	Rod	0.10 max	0.25 max	0.30 max	0.60 max	0.25 max	0.035 max	0.035 max	0.30 max	Al <=0.01; Mo <=0.08; bal Fe				
EN 10016/2(94)	1.0313	Rod	0.05-0.09	0.20 max	0.30 max	0.30-0.60	0.25 max	0.035 max	0.035 max	0.30 max	Al <=0.01; Mo <=0.05; bal Fe				
EN 10016/2(94)	C7D	Rod	0.05-0.09	0.20 max	0.30 max	0.30-0.60	0.25 max	0.035 max	0.035 max	0.30 max	Al <=0.01; Mo <=0.05; bal Fe				
EN 10016/2(94)	C9D	Rod	0.10 max	0.25 max	0.30 max	0.60 max	0.25 max	0.035 max	0.035 max	0.30 max	Al <=0.01; Mo <=0.08; bal Fe				
EN 10016/4(94)	1.1113	Rod	0.06-0.10	0.10 max	0.15 max	0.30-0.50	0.10 max	0.020 max	0.025 max	0.10-0.30	Al <=0.01; Mo <=0.05; N <=0.007; Cr+Ni+Cu<=0.30, Cu+Sn<=0.15; bal Fe				
EN 10016/4(94)	C8D2	Rod	0.06-0.10	0.10 max	0.15 max	0.30-0.50	0.10 max	0.020 max	0.025 max	0.10-0.30	Al <=0.01; Mo <=0.05; N <=0.007; Cr+Ni+Cu<=0.30, Cu+Sn<=0.15; bal Fe				
EN 10130(91)	1.0330	Sh Strp, CR t<=3mm	0.12 max			0.60 max		0.045 max	0.045 max		bal Fe	270-410	140	28	
EN 10130(91)	1.0347	Sh Strp, CR t<=3mm	0.10 max			0.45 max		0.035 max	0.035 max		bal Fe	270-370	140	34	
EN 10130(91)	FeP01	Sh Strp, CR t<=3mm	0.12 max			0.60 max		0.045 max	0.045 max		bal Fe	270-410	140	28	
EN 10130(91)	FeP03	Sh Strp, CR t<=3mm	0.10 max			0.45 max		0.035 max	0.035 max		bal Fe	270-370	140	34	

Finland

Specification	Designation	Notes	C	Cr	Cu	Mn	Ni	P	S	Si	Other	UTS	YS	El	Hard
SFS 600(81)	CR2	Unalloyed mild	0.1 max			0.45 max		0.07 max	0.06 max	0.6 max	bal Fe				
SFS 600(81)	CR3	Unalloyed mild	0.1 max			0.45 max		0.07 max	0.06 max	0.6 max	bal Fe				
SFS 600(81)	CR4	Unalloyed mild	0.08 max			0.45 max		0.070 max	0.060 max	0.60 max	Al 0.02-0.120; bal Fe				
SFS 650	Z02	Hot dip zinc coated	0.12 max			0.6 max		0.04 max	0.04 max	0.60 max	bal Fe				
SFS 679(86)	CR280	Gen struct	0.1 max			0.8 max		0.03 max	0.03 max	0.5 max	bal Fe				
SFS 679(86)	CR320	Gen struct	0.1 max			1.0 max		0.03 max	0.03 max	0.5 max	bal Fe				

France

Specification	Designation	Notes	C	Cr	Cu	Mn	Ni	P	S	Si	Other	UTS	YS	El	Hard
AFNOR	XC6		0.04-0.09			0.25-0.45		0.03	0.03	0.1	bal Fe				
AFNOR	XC6FF		0.04-0.09			0.25-0.45		0.03	0.03	0.1	bal Fe				
AFNOR NFA35551	XC10		0.06-0.12			0.3-0.6		0.035	0.035	0.15-0.35	bal Fe				
AFNOR NFA36102(83)	C01RR		0.01 max	0.08 max		0.4 max	0.08 max	0.025 max	0.01 max	0.03 max	Al 0.02-0.06; Mo <=0.040; bal Fe				
AFNOR NFA36102(83)	XC01		0.01 max	0.08 max		0.4 max	0.08 max	0.025 max	0.01 max	0.03 max	Al 0.02-0.06; Mo <=0.040; bal Fe				
AFNOR NFA36102(87)	Fd2*		0.1 max	0.3 max		0.2-0.45	0.40 max	0.035 max	0.03 max	0.6 max	N <=0.007; bal Fe				
AFNOR NFA36102(87)	Fd3*		0.08 max	0.3 max		0.02-0.4	0.40 max	0.025 max	0.025 max	0.6 max	N <=0.006; bal Fe				
AFNOR NFA36102(87)	Fd4*		0.07 max	0.3 max		0.4 max	0.40 max	0.025 max	0.025 max	0.6 max	Al >=0.02; bal Fe				

UNS numbers and US grades are provided as a means of cross referencing chemically similar alloys. Exchangability is only possible after independent examination of specifications. Tensile properties are minimum or typical as specified. UTS and YS as MPa. El as %. See Appendix for list of abbreviations used in Notes. * indicates obsolete material.

Specification	Designation	Notes	C	Cr	Cu	Mn	Ni	P	S	Si	Other	UTS	YS	El	Hard

Carbon Steel, Nonresulfurized, 1008 (Continued from previous page)

France

Specification	Designation	Notes	C	Cr	Cu	Mn	Ni	P	S	Si	Other	UTS	YS	El	Hard
AFNOR NFA36102(87)	FdTu3*		0.1 max	0.3 max		0.25-0.5	0.40 max	0.04 max	0.04 max	0.6 max	N <=0.007; bal Fe				
AFNOR NFA36102(87)	FdTu4*		0.09 max	0.3 max		0.25-0.5	0.40 max	0.03 max	0.03 max	0.6 max	N <=0.006; bal Fe				
AFNOR NFA36102(87)	FdTu5*		0.09 max	0.3 max		0.25-0.5	0.40 max	0.025 max	0.025 max	0.6 max	N <=0.006; bal Fe				
AFNOR NFA36102(93)	C02RR	HR; Strp; CR	0.02 max	0.08 max		0.25 max	0.08 max	0.02 max	0.02 max	0.03 max	Al 0.02-0.06; Mo <=0.040; bal Fe				
AFNOR NFA36102(93)	C05RR	HR; Strp; CR	0.06 max	0.08 max		0.35 max	0.08 max	0.025 max	0.02 max	0.03 max	Al 0.02-0.06; Mo <=0.040; bal Fe				
AFNOR NFA36102(93)	C08RR	HR; Strp; CR	0.1 max	0.08 max		0.15-0.45	0.08 max	0.025 max	0.02 max	0.04 max	Al 0.02-0.06; Mo <=0.040; bal Fe				
AFNOR NFA36102(93)	Fd2*	HR; Strp; CR	0.1 max	0.08 max		0.15-0.45	0.08 max	0.025 max	0.02 max	0.04 max	Al 0.02-0.06; Mo <=0.040; bal Fe				
AFNOR NFA36102(93)	Fd4*	HR; Strp; CR	0.06 max	0.08 max		0.35 max	0.08 max	0.025 max	0.02 max	0.03 max	Al 0.02-0.06; Mo <=0.040; bal Fe				
AFNOR NFA36102(93)	XC02*	HR; Strp; CR	0.02 max	0.08 max		0.25 max	0.08 max	0.02 max	0.02 max	0.03 max	Al 0.02-0.06; Mo <=0.040; bal Fe				
AFNOR NFA36232(92)	E240C	t<3mm; CR	0.10 max	0.3 max		0.5 max	0.4 max	0.025 max	0.025 max	0.4 max	Al >=0.02; Nb <=0.03; bal Fe	345-415	240	29	
AFNOR NFA36232(92)	E260C	t<3mm; CR	0.10 max	0.3 max		0.5 max	0.4 max	0.025 max	0.025 max	0.4 max	Al >=0.02; Nb <=0.03; bal Fe	365-435	260	27	
AFNOR NFA36232(92)	E280C	t<3mm; CR	0.10 max	0.3 max		0.5 max	0.4 max	0.025 max	0.025 max	0.4 max	Al >=0.02; Nb <=0.05; bal Fe	375-455	280	24	
AFNOR NFA36232(92)	H240M	t<3mm; CR Plt	0.10 max	0.3 max		0.5 max	0.4 max	0.025 max	0.025 max	0.4 max	Al >=0.02; Nb <=0.03; bal Fe	345-415	240	29	
AFNOR NFA36232(92)	H260M	t<3mm; CR Plt	0.10 max	0.3 max		0.5 max	0.4 max	0.025 max	0.025 max	0.4 max	Al >=0.02; Nb <=0.03; bal Fe	365-435	260	27	
AFNOR NFA36232(92)	H280M	t<3mm; CR Plt	0.10 max	0.3 max		0.5 max	0.4 max	0.025 max	0.025 max	0.4 max	Al >=0.02; Nb <=0.05; bal Fe	375-455	280	24	
AFNOR NFA36232(92)	H315M	t<3mm; CR Plt	0.10 max	0.3 max		0.6 max	0.4 max	0.025 max	0.025 max	0.4 max	Al >=0.02; N <=0.009; bal Fe	400-490	3115	22	
AFNOR NFA36301(92)	3C	t<3mm; HR	0.08 max	0.3 max		0.4 max	0.4 max	0.025 max	0.025 max	0.4 max	Al >=0.02; bal Fe	290-370	200	30T	59 HRB
AFNOR NFA36301(92)	3CT	t<3mm; HR	0.08 max	0.3 max		0.4 max	0.4 max	0.025 max	0.025 max	0.04 max	Al >=0.02; B <=0.005; Ti <=0.03; bal Fe	290-370	200	30T	59 HRB

Germany

Specification	Designation	Notes	C	Cr	Cu	Mn	Ni	P	S	Si	Other	UTS	YS	El	Hard
DIN	RoSt2		0.10 max			0.30-0.60		0.045 max	0.045 max		N <=0.007; trace Si; bal Fe				
DIN	RoSt4		0.10 max			0.30-0.60		0.030 max	0.035 max	0.10 max	N <=0.007; bal Fe				
DIN	USD8		0.06-0.10	0.12 max	0.17 max	0.45-0.65	0.12 max	0.030 max	0.030 max		N <=0.007; trace Si; N depends on P; bal Fe				
DIN	USt14		0.09 max			0.25-0.50		0.030 max	0.030 max		N <=0.007; trace Si; bal Fe				
DIN	USt4		0.09 max			0.25-0.50		0.030 max	0.030 max		N <=0.007; trace Si; bal Fe				
DIN	WNr 1.0010		0.10 max			0.50 max		0.070 max	0.060 max	0.30 max	bal Fe				
DIN	WNr 1.0322		0.06-0.10	0.12 max	0.17 max	0.45-0.65	0.12 max	0.030 max	0.030 max		N <=0.007; trace Si; N depends on P; bal Fe				
DIN	WNr 1.0331		0.10 max			0.30-0.60		0.045 max	0.045 max		N <=0.007; trace Si; N depends on P; bal Fe				
DIN	WNr 1.0336		0.09 max			0.25-0.50		0.030 max	0.030 max		N <=0.007; trace Si; bal Fe				
DIN	WNr 1.0337		0.10 max			0.30-0.60		0.030 max	0.035 max	0.10 max	N <=0.007; bal Fe				
DIN	WNr 1.0744*	Obs	0.09			0.2-0.45		0.08-0.15	0.05		bal Fe				
DIN	WNr 1.0746*	Obs	0.09 max			0.2-0.45		0.15-0.25	0.05 max		bal Fe				
DIN 1614(86)	RRSt23	Sht Strp HR	0.10 max			0.45 max		0.025 max	0.025 max	0.03-0.10	Al >=0.020; N <=0.007; bal Fe				

Specification	Designation	Notes	C	Cr	Cu	Mn	Ni	P	S	Si	Other	UTS	YS	El	Hard

Carbon Steel, Nonresulfurized, 1008 (Continued from previous page)

Germany

Specification	Designation	Notes	C	Cr	Cu	Mn	Ni	P	S	Si	Other	UTS	YS	El	Hard
DIN 1614(86)	St22	Sht Strp HR	0.10 max			0.45 max		0.035 max	0.035 max		N <=0.007; trace Si; bal Fe				
DIN 1614(86)	StW22	Sht Strp HR	0.10 max			0.20-0.45		0.035 max	0.035 max		N <=0.007; trace Si; bal Fe				
DIN 1614(86)	WNr 1.0320	Tub Sh Strp HR	0.10 max			0.45 max		0.035 max	0.035 max		N <=0.007; trace Si; bal Fe				
DIN 1614(86)	WNr 1.0332	Sh Strp HR	0.10 max			0.20-0.45		0.035 max	0.035 max		N <=0.007; trace Si; bal Fe				
DIN 1614(86)	WNr 1.0359	Sh Strp HR	0.10 max			0.45 max		0.025 max	0.025 max	0.03-0.10	Al >=0.020; N <=0.007; bal Fe				
DIN 1624(87)	USt3	Strp<=650mm	0.08 max								N <=0.007; bal Fe				
DIN 1624(87)	WNr 1.0333	Strp<=650mm	0.08 max								N <=0.007; bal Fe				
DIN 17111(80)	UQSt36		0.14 max			0.25-0.50		0.040 max	0.040 max		trace Si; C <=0.18 if thk>=22mm; bal Fe				
DIN 17111(80)	WNr 1.0204		0.14 max			0.25-0.50		0.040 max	0.040 max		trace Si; C<=0.18 if >=22mm Thk; bal Fe				
DIN 17140	C8D		0.10 max			0.50 max		0.07 max	0.06 max	0.3 max	bal Fe				
DIN 17145(80)	USD6	Wir rod for weld filler	0.06-0.10	0.12 max	0.17 max	0.45-0.65	0.12 max	0.020 max	0.020 max		trace Si; bal Fe				
DIN 2393(94)	RSt28	Weld precision tub	0.13 max			0.45 max		0.050 max	0.050 max	0.05 max	Al >=0.025; bal Fe				
DIN 2393(94)	St28	Tube	0.13 max					0.050 max	0.050 max		bal Fe				
DIN 2393(94)	USt28	Weld precision tub	0.13 max			0.45 max		0.050 max	0.050 max		trace Si; bal Fe				
DIN 2393(94)	WNr 1.0318		0.13 max					0.050 max	0.050 max		bal Fe				
DIN 2393(94)	WNr 1.0326	Weld precision tub	0.13 max			0.45 max		0.050 max	0.050 max	0.05 max	Al >=0.0025; bal Fe				
DIN 2393(94)	WNr 1.0357	Weld precision tub	0.13 max			0.45 max		0.050 max	0.050 max		trace Si; bal Fe				
DIN 2394(94)	RSt28	As-weld, sized precision tube	0.13 max			0.45 max		0.050 max	0.050 max	0.05 max	Al >=0.025; bal Fe				
DIN 2394(94)	St28	Tube	0.13 max					0.050 max	0.050 max		bal Fe				
DIN 2394(94)	WNr 1.0318	Tube	0.13 max					0.050 max	0.050 max		bal Fe				
DIN 2394(94)	WNr 1.0326	As-weld, sized precision tube	0.13 max			0.45 max		0.050 max	0.050 max	0.05 max	Al >=0.025; bal Fe				
DIN 5512(97)	St12Cu3	Flat product <=3mm for rail vehicles	0.10 max		0.25-0.35	0.20-0.45		0.050 max	0.050 max		N <=0.008; trace Si; bal Fe				
DIN 5512(97)	WNr 1.0344	Flat product <=3mm for rail vehicles	0.10 max		0.25-0.35	0.20-0.45		0.050 max	0.050 max		N <=0.008; trace Si; bal Fe				
DIN E EN 10139(92)	St12	CR Strp	0.12 max			0.60 max		0.045 max	0.045 max		bal Fe				
DIN E EN 10139(92)	St2	CR Strp	0.12 max			0.60 max		0.045 max	0.045 max		bal Fe				
DIN E EN 10139(92)	WNr 1.0330	CR Strp	0.12 max			0.60 max		0.045 max	0.045 max		bal Fe				
DIN EN 10130(91)	RRSt13	Strp	0.10 max			0.45 max		0.035 max	0.035 max		bal Fe				
DIN EN 10130(91)	WNr 1.0347	Strp	0.10 max			0.45 max		0.035 max	0.035 max		bal Fe				

Hungary

Specification	Designation	Notes	C	Cr	Cu	Mn	Ni	P	S	Si	Other	UTS	YS	El	Hard
MSZ 23(83)	K1H	Sh Strp; mild; 0.5-0.7mm; CR	0.08 max			0.45 max		0.03 max	0.035 max	0.1 max	Al <=0.120; bal Fe	270-370		30L	0-55 HR30T<= 55
MSZ 23(83)	K1H	Sh Strp; mild; 0.71-3mm; CR	0.08 max			0.45 max		0.03 max	0.035 max	0.1 max	Al <=0.120; bal Fe	270-370		32L	0-57 HR30T<= 55
MSZ 23(83)	K2H	Sh Strp; mild; 0.71-3mm; CR	0.08 max			0.45 max		0.03 max	0.035 max	0.1 max	Al <=0.120; bal Fe	270-370		32L	0-57 HR30T<= 55

UNS numbers and US grades are provided as a means of cross referencing chemically similar alloys. Exchangability is only possible after independent examination of specifications. Tensile properties are minimum or typical as specified. UTS and YS as MPa. El as %. See Appendix for list of abbreviations used in Notes. * indicates obsolete material.

Carbon Steel, Nonresulfurized, 1008 (Continued from previous page)

Specification	Designation	Notes	C	Cr	Cu	Mn	Ni	P	S	Si	Other	UTS	YS	El	Hard
Hungary															
MSZ 23(83)	K2H	Sh Strp; mild; 0.5-0.7mm; CR	0.08 max			0.45 max		0.03 max	0.035 max	0.1 max	Al <=0.120; bal Fe	270-370		30L	0-55 HR30T<= 55
MSZ 23(83)	KO1H	Sh Strp; mild; 0.5-0.7mm; CR	0.08 max			0.45 max		0.03 max	0.035 max	0.1 max	Al 0.025-0.120; bal Fe	270-370		32L	0-53 HR30T
MSZ 23(83)	KO1H	Sh Strp; mild; 0.71-3mm; CR	0.08 max			0.45 max		0.03 max	0.035 max	0.1 max	Al 0.025-0.120; bal Fe	270-370		34L	0-55 HR30T<= 60
MSZ 23(83)	KO2H	Sh Strp; mild; 0.5-0.7mm; CR	0.08 max			0.45 max		0.03 max	0.035 max	0.1 max	Al 0.025-0.120; bal Fe	270-370		32L	0-53 HR30T
MSZ 23(83)	KO2H	Sh Strp; mild; 0.71-3mm; CR	0.08 max			0.45 max		0.03 max	0.035 max	0.1 max	Al 0.025-0.120; bal Fe	270-370		34L	0-55 HR30T<= 60
MSZ 4213(72)	ASZ1	CR Strp	0.12 max			0.2-0.45		0.05 max	0.05 max	0.2 max	bal Fe				
MSZ 4213(72)	ASZ2	CR, Strp	0.1 max			0.2-0.45		0.04 max	0.05 max	0.15 max	bal Fe				
MSZ 4213(72)	ASZ3	CR Strp	0.1 max			0.2-0.45		0.03 max	0.04 max	0.6 max	bal Fe				
MSZ 5736(88)	D08	Wir rod; for CD	0.09 max		0.25 max	0.5 max	0.3 max	0.04 max	0.04 max	0.04 max	bal Fe				
MSZ 5736(88)	D08K	Wir rod; for CD	0.09 max		0.25 max	0.5 max	0.3 max	0.03 max	0.03 max	0.04 max	bal Fe				
MSZ 6251(87)	D08Z	CF; 0-36mm; HR; HR/soft ann; drawn/soft ann; drawn/bright ann; soft ann/ground	0.1 max			0.5 max	0.3 max	0.04 max	0.04 max	0.04 max	bal Fe			0-450	
MSZ 6251(87)	D08Z	CF; 0-36mm; drawn, half-hard	0.1 max			0.5 max	0.3 max	0.04 max	0.04 max	0.04 max	bal Fe			0-520	
MSZ 6447	F1		0.1	0.2	0.25	0.35-0.6	0.2	0.03	0.03	0.07	Mo 0.05; bal Fe				
India															
IS 1570	C05		0.1			0.5					bal Fe				
IS 1570	C07		0.1			0.5					bal Fe				
IS 1570/2(79)	5C4		0.1 max	0.3 max	0.3 max	0.5 max	0.4 max	0.055 max	0.055 max	0.6 max	Co <=0.1; Mo <=0.15; Pb <=0.15; W <=0.1; P:S Varies; bal Fe				
IS 1570/2(79)	C05		0.1 max	0.3 max	0.3 max	0.5 max	0.4 max	0.055 max	0.055 max	0.6 max	Co <=0.1; Mo <=0.15; Pb <=0.15; W <=0.1; P:S Varies; bal Fe				
International															
ISO 3573(99)	HR3	Sh, HR, t=3mm, DD	0.08 max			0.45 max		0.030 max	0.030 max		bal Fe	400		29	
ISO 3574(99)	CR1	Sh, comm	0.15 max			0.60 max		0.05 max	0.05 max		Co <=0.1; Pb <=0.15; W <=0.1; bal Fe	410 max	280 max	28 min	
ISO 3574(99)	CR2	Sh, DS	0.12 max			0.50 max		0.04 max	0.04 max		Co <=0.1; Pb <=0.15; W <=0.1; bal Fe	371 max	240 max	31 min	
ISO 3574(99)	CR3	Sh, DDS	0.10 max			0.45 max		0.03 max	0.03 max		bal Fe	350 max	220 max	35 min	
Italy															
UNI 5598(71)	3CD8	Wir rod	0.1 max			0.25-0.6		0.035 max	0.035 max	0.6 max	N <=0.012; bal Fe				
UNI 5771(66)	C12	Chain	0.12 max			0.35-0.6		0.035 max	0.035 max	0.3 max	bal Fe				
Japan															
JIS G3141(96)	CR1	CR Sh Coil	0.15 max			0.60 max		0.05 max	0.05 max		bal Fe				
JIS G3141(96)	CR2	CR Sh Coil	0.12 max			0.50 max		0.04 max	0.04 max		bal Fe	370 max		31	57 HRB max
JIS G3141(96)	SPCC	CR Sh Coil Comm	0.12 max			0.45 max		0.040 max	0.045 max		bal Fe	270		32-39	95-170 HV
JIS G3141(96)	SPCD	CR Sh Coil Drawing	0.10 max			0.45 max		0.035 max	0.035 max		bal Fe	270		34-41	95-170
JIS G3445(88)	STKM11A	Tube	0.12 max			0.60 max		0.040 max	0.040 max	0.35 max	bal Fe	294		35L	

UNS numbers and US grades are provided as a means of cross referencing chemically similar alloys. Exchangability is only possible after independent examination of specifications. Tensile properties are minimum or typical as specified. UTS and YS as MPa. El as %. See Appendix for list of abbreviations used in Notes. * indicates obsolete material.

Specification	Designation	Notes	C	Cr	Cu	Mn	Ni	P	S	Si	Other	UTS	YS	El	Hard

Carbon Steel, Nonresulfurized, 1008 (Continued from previous page)

Japan

Specification	Designation	Notes	C	Cr	Cu	Mn	Ni	P	S	Si	Other	UTS	YS	El	Hard
JIS G3507(91)	SWRCH8R	Wir rod FU	0.10 max			0.60 max		0.040 max	0.040 max		bal Fe				

Mexico

NMX-B-265(89)	LEEP	Sh for enamelling	0.08 max			0.40 max		0.030 max	0.030 max		bal Fe				
NMX-B-301(86)	1008	Bar	0.10 max			0.30-0.50		0.040 max	0.050 max		bal Fe				

Pan America

COPANT 330	1008	Bar	0.1			0.3-0.5		0.04	0.05		bal Fe				
COPANT 331	1008	Bar	0.1			0.3-0.5		0.04	0.05		bal Fe				
COPANT 333	1008	Wir rod	0.1			0.35-0.5		0.04	0.05		bal Fe				
COPANT 38	EP	Sh, HR, 1.5 3mm diam	0.1			0.2-0.6		0.035	0.04		bal Fe				

Poland

PNH84019	8		0.05-0.11	0.1	0.25	0.35-0.65	0.25	0.035	0.04	0.17-0.37	bal Fe				
PNH84022	St2N		0.1			0.3-0.5		0.05	0.05		bal Fe				
PNH84023	08XA		0.05-0.1	0.15	0.25	0.25-0.45	0.3	0.03	0.03	0.05	bal Fe				
PNH84023	09XA		0.1	0.06	0.06	0.25-0.42		0.02	0.025	0.02	bal Fe				
PNH84023	MSt1		0.06-0.12	0.3	0.3	0.1		0.05	0.05	0.05	Mo 0.3; bal Fe				
PNH84023	St1S		0.1	0.3	0.3	0.5	0.3	0.045	0.045	0.05	Mo 0.1; bal Fe				
PNH84023	St2N		0.1			0.3-0.5		0.05	0.05		bal Fe				
PNH84023/03	08XA		0.08 max	0.1 max	0.1 max	0.25-0.45	0.1 max	0.25 max	0.03 max	0.03 max	bal Fe				

Romania

STAS 10318	A4		0.1	0.03	0.1	0.5	0.1	0.035	0.035	0.05	bal Fe				
STAS 11501	A21		0.1	0.03	0.1	0.4-1	0.1	0.035	0.035	0.3	bal Fe				
STAS 9485	A1		0.1	0.03	0.1	0.2-0.45	0.1	0.05	0.05		bal Fe				
STAS 9485(80)	A1k		0.1 max	0.3 max		0.2-0.45		0.05 max	0.05 max	0.03-0.05	Al 0.015-0.08; Pb <=0.15; As<=0.08; bal Fe				
STAS 9485(80)	A1n		0.1 max	0.03 max	0.1 max	0.2-0.45	0.1 max	0.05 max	0.05 max	0.6 max	Pb <=0.15; trace Si; bal Fe				
STAS 9485(80)	A2k		0.1 max	0.03 max	0.1 max	0.2-0.45	0.1 max	0.04 max	0.04 max	0.03-0.08	Al 0.015-0.07; Pb <=0.15; bal Fe				
STAS 9485(80)	A2n		0.1 max	0.03 max	0.1 max	0.2-0.45	0.1 max	0.04 max	0.04 max	0.6 max	Pb <=0.15; trace Si; bal Fe				
STAS 9485(80)	A3k		0.08 max	0.03 max	0.1 max	0.2-0.45	0.1 max	0.03 max	0.03 max	0.03-0.08	Al 0.02-0.07; Pb <=0.15; bal Fe				
STAS 9485(80)	A3n		0.08 max	0.03 max	0.1 max	0.2-0.45	0.1 max	0.035 max	0.035 max	0.6 max	Pb <=0.15; trace Si; bal Fe				

Russia

GOST	08KPVG		0.05-0.08	0.15 max	0.15 max	0.3-0.42	0.3 max	0.03 max	0.03 max	0.08 max	bal Fe				
GOST	M39kp		0.06-0.1			0.25-0.5		0.045	0.05	0.05	bal Fe				
GOST 1050	8		0.05-0.12	0.1 max	0.25 max	0.25-0.5	0.25 max	0.04 max	0.04 max	0.17-0.37	bal Fe				
GOST 1050(88)	08kp	High-grade struct; Unkilles; Rimming; R; FU	0.05-0.12			0.25-0.5	0.3 max	0.035 max	0.04 max	0.03 max	As<=0.08; bal Fe				
GOST 1050(88)	08ps	High-grade struct; Semi-killed	0.05-0.11			0.35-0.65	0.3 max	0.035 max	0.04 max	0.05-0.17	As<=0.08; bal Fe				
GOST 1050(88)	10kp	High-grade struct; Unkilled; Rimming; R; FU	0.07-0.14			0.25-0.5	0.3 max	0.035 max	0.04 max	0.07 max	As<=0.08; bal Fe				
GOST 24244(80)	10kp		0.1	0.15	0.25	0.25-0.5	0.25	0.035	0.04	0.03	Al 0.01; As 0.08; bal Fe				
GOST 24244(80)	St1kp	Gen struct; Semi-Finished; Unkilled; R; FU	0.1			0.7		0.04	0.04	0.03	Al 0.01; bal Fe				

UNS numbers and US grades are provided as a means of cross referencing chemically similar alloys. Exchangability is only possible after independent examination of specifications. Tensile properties are minimum or typical as specified. UTS and YS as MPa. El as %. See Appendix for list of abbreviations used in Notes. * indicates obsolete material.

Specification	Designation	Notes	C	Cr	Cu	Mn	Ni	P	S	Si	Other	UTS	YS	El	Hard

Carbon Steel, Nonresulfurized, 1008 (Continued from previous page)

Russia

Specification	Designation	Notes	C	Cr	Cu	Mn	Ni	P	S	Si	Other	UTS	YS	El	Hard
GOST 380	BSt1kp		0.06-0.12			0.25-0.5		0.04 max	0.05 max	0.05 max	bal Fe				
GOST 380	St1kp	Gen struct; Semi-Finished; Unkilled; Rimming; R; FU	0.09			0.5		0.03	0.04	0.04	bal Fe				
GOST 380(71)	BSt1kp2	Gen struct; Unkilled; Rimming; R; FU	0.06-0.12	0.3 max	0.3 max	0.25-0.5	0.3 max	0.04 max	0.05 max	0.05 max	Al <=0.1; N <=0.008; As<=0.08; bal Fe				
GOST 380(71)	BSt1ps	Gen struct; Semi-killed	0.06-0.12	0.3 max	0.3 max	0.25-0.5	0.3 max	0.04 max	0.05 max	0.05-0.17	Al <=0.1; N <=0.008; As<=0.08; bal Fe				
GOST 380(71)	BSt1ps2	Gen struct; Semi-killed	0.06-0.12	0.3 max	0.3 max	0.25-0.5	0.3 max	0.04 max	0.05 max	0.05-0.17	Al <=0.1; N <=0.008; As<=0.08; bal Fe				
GOST 380(71)	BSt1sp	Gen struct; Non-rimming; FF	0.06-0.12	0.3 max	0.3 max	0.25-0.5	0.3 max	0.04 max	0.05 max	0.12-0.3	Al <=0.1; N <=0.008; As<=0.08; bal Fe				
GOST 380(71)	BSt1sp2	Gen struct; Non-rimming; FF	0.06-0.12	0.3 max	0.3 max	0.25-0.5	0.3 max	0.04 max	0.05 max	0.12-0.3	Al <=0.1; N <=0.008; As<=0.08; bal Fe				
GOST 380(71)	St1kp	Gen struct; Semi-Finished; Unkilled; Rimming; R; FU	0.3 max	0.3 max	0.3 max	2 max	0.4 max	0.07 max	0.06 max	0.6 max	Al <=0.1; bal Fe				
GOST 380(71)	St1kp	Gen struct; Semi-Finished; Unkilled; Rimming; R; FU; 0-20mm	0.3 max	0.3 max	0.3 max	2 max	0.4 max	0.07 max	0.06 max	0.6 max	Al <=0.1; bal Fe	300-390		35L	
GOST 380(71)	St1kp	Gen struct; Semi-Finished; Unkilled; Rimming; R; FU; >40mm	0.3 max	0.3 max	0.3 max	2 max	0.4 max	0.07 max	0.06 max	0.6 max	Al <=0.1; bal Fe	300-390		32L	
GOST 380(88)	St1kp	Gen struct; Semi-Finished; Unkilled; R; FU	0.06-0.12	0.3 max	0.3 max	0.25-0.5	0.3 max	0.04 max	0.05 max	0.05 max	Al <=0.1; N <=0.008; Ti <=0.1; As<=0.08; bal Fe				
GOST 4041	08kp		0.10 max	0.1 max	0.2 max	0.25-0.45	0.15 max	0.025 max	0.030	0.03 max	bal Fe				
GOST 4041	08ps		0.09 max	0.1 max	0.2 max	0.25-0.45	0.15 max	0.025 max	0.030	0.04 max	bal Fe				
GOST 4041	08Yu		0.10 max	0.1 max	0.2 max	0.25-0.45	0.15 max	0.025 max	0.030	0.03 max	Al 0.02-0.08; bal Fe				
GOST 4041	08YuA		0.10 max	0.1 max	0.2 max	0.20-0.40	0.15 max	0.020 max	0.025	0.03 max	Al 0.02-0.08; bal Fe				
GOST 4041(71)	08Ju	High-grade Plts for cold pressing	0.10 max	0.1 max	0.2 max	0.25-0.45	0.15 max	0.025 max	0.030 max	0.03 max	Al 0.02-0.08; As<=0.08; bal Fe				
GOST 4041(71)	08kp-3	High-grade for cold pressing	0.10 max	0.1 max	0.2 max	0.25-0.50	0.15 max	0.025 max	0.030 max	0.03 max	As<=0.08; bal Fe				
GOST 4041(71)	08ps-3	High-grade Plts for cold pressing	0.09 max	0.1 max	0.2 max	0.25-0.45	0.15 max	0.025 max	0.030 max	0.04 max	As<=0.08; bal Fe				
GOST 9045	08Ju	Body panel Sh	0.07 max	0.04 max	0.15 max	0.2-0.35	0.1 max	0.02 max	0.025 max	0.03 max	Al 0.02-0.07; bal Fe				
GOST 9045	08kp	Body panel Sh	0.1 max	0.1 max	0.15 max	0.25-0.45	0.1 max	0.025 max	0.03 max	0.03 max	Al <=0.1; bal Fe				
GOST 9045	08ps	Body panel Sh	0.09 max	0.1 max	0.15 max	0.2-0.4	0.1 max	0.025 max	0.03 max	0.04 max	Al <=0.1; bal Fe				

Spain

Specification	Designation	Notes	C	Cr	Cu	Mn	Ni	P	S	Si	Other	UTS	YS	El	Hard
UNE 36086(75)	AP00	Unalloyed mild; Sh CR	0.15 max			0.6 max		0.05 max	0.05 max	0.60 max	Mo <=0.15; bal Fe				
UNE 36086(75)	AP00	Unalloyed mild; Sh CR	0.15 max			0.6 max		0.05 max	0.05 max	0.60 max	Mo <=0.15; bal Fe				
UNE 36086(75)	AP01	Unalloyed mild; Sh CR	0.12 max			0.5 max		0.04 max	0.04 max	0.60 max	Mo <=0.15; bal Fe				
UNE 36086(75)	AP03	Unalloyed mild; Sh CR	0.1 max			0.45 max		0.03 max	0.03 max	0.60 max	Mo <=0.15; bal Fe				
UNE 36086(75)	AP04	Unalloyed mild; Sh CR	0.08 max			0.45 max		0.03 max	0.03 max	0.60 max	Mo <=0.15; bal Fe				
UNE 36086(75)	AP10	Unalloyed mild; Sh CR	0.15 max			0.6 max		0.05 max	0.05 max	0.60 max	Mo <=0.15; bal Fe				
UNE 36086(75)	AP12	Unalloyed mild; Sh CR	0.1 max			0.45 max		0.03 max	0.03 max	0.60 max	Mo <=0.15; bal Fe				

Sweden

Specification	Designation	Notes	C	Cr	Cu	Mn	Ni	P	S	Si	Other	UTS	YS	El	Hard
SIS 141142	1142-42	Plt, ST, CR, 3mm diam	0.1			0.5		0.04	0.04		bal Fe	270		26	
SIS 141146	1146-32	Sh, ST, CR, 3mm diam	0.1			0.5		0.04	0.04		bal Fe	270		33	
SIS 141146	1146-42	Sh, ST, CR, 3mm diam	0.1			0.5		0.04	0.04		bal Fe	270		33	

UNS numbers and US grades are provided as a means of cross referencing chemically similar alloys. Exchangability is only possible after independent examination of specifications. Tensile properties are minimum or typical as specified. UTS and YS as MPa. El as %. See Appendix for list of abbreviations used in Notes. * indicates obsolete material.

Specification	Designation	Notes	C	Cr	Cu	Mn	Ni	P	S	Si	Other	UTS	YS	El	Hard

Carbon Steel, Nonresulfurized, 1008 (Continued from previous page)

Sweden

Specification	Designation	Notes	C	Cr	Cu	Mn	Ni	P	S	Si	Other	UTS	YS	El	Hard
SIS 141147	1147-32	Sh, ST, CR, 3mm diam	0.08			0.45		0.04	0.04		bal Fe	270		37	
SIS 141147	1147-42	Sh, ST, CR, 3mm diam	0.08			0.45		0.04	0.04		bal Fe	270		37	
SIS 141225	1225-00	Frg Rod Wir	0.08	0.1	0.2	0.4-0.6		0.03	0.03		N 0.01; bal Fe				
SIS 141225	1225-01	Frg Rod Wir, Norm 100mm diam	0.08	0.1	0.2	0.4-0.6		0.03	0.03		N 0.01; bal Fe	310		30	
SIS 141232	1232-03	Tub, as Drawn	0.13					0.05 max			N 0.01; bal Fe	310	180	25	
SIS 141232	1232-04	Tub, as weld	0.13					0.05 max			N 0.01; bal Fe	310	180	25	
SIS 141232	1232-08	Tube, Cold drawn	0.13					0.05 max			N 0.01; bal Fe	400	360	12	
SIS 141311E	1311-00	Bar, Plt, as rolled 40mm diam	0.12			0.4-0.7		0.08	0.06	0.05	bal Fe	360	220	24	
SIS 141311E	1311-00	Bar, Plt, as rolled 40mm diam	0.15			0.3-0.6		0.08	0.06	0.02	bal Fe	360	220	24	
SIS 141311E	1311-10	Bar, as rolled 6/32mm diam	0.12			0.4-0.7		0.08	0.06	0.25	bal Fe		220	20	
SIS 141311E	1311-10	Bar, as rolled 6/32mm diam	0.15			0.3-0.6		0.08	0.06	0.02	bal Fe		220	20	
SS 141142	1142		0.1			0.5		0.04	0.04		bal Fe				
SS 141146	1146		0.1			0.5		0.04	0.04		bal Fe				
SS 141232E	1232-03	Tub, as rolled or Q/A	0.13	0.25	0.3	0.3-0.7		0.05	0.05		N 0.01; bal Fe	320	200	25	
SS 141232E	1232-04	Tub, as rolled or Q/A	0.15	0.25	0.3	0.3-0.7		0.05	0.05		N 0.01; bal Fe	320	200	25	
SS 141232E	1232-04	Tub, as rolled or Q/A	0.13	0.25	0.3	0.3-0.7		0.05	0.05		N 0.01; bal Fe	320	200	25	
SS 141232E	1232-05	Tub, as Q/A, or Rolled	0.13	0.25	0.3	0.3-0.7		0.05	0.05		N 0.01; bal Fe	320	200	25	
SS 141232E	1232-06	Tub, as rolled or Q/A	0.15	0.25	0.3	0.3-0.7		0.05	0.05		N 0.01; bal Fe	320	200	25	
SS 141232E	1232-06	Tube, Q/A	0.13	0.25	0.3	0.3-0.7		0.05	0.05		N 0.01; bal Fe	320	200	25	
SS 141232E	1232-08	Tub, as rolled or Q/A	0.13	0.25	0.3	0.3-0.7		0.05	0.05		N 0.01; bal Fe	400	360	12	

Turkey

Specification	Designation	Notes	C	Cr	Cu	Mn	Ni	P	S	Si	Other	UTS	YS	El	Hard
TS 2348(76)	C-8-2/1.0313	Wir rod	0.08						0.04		N 0.01; bal Fe				
TS 2348(76)	C12-1/1.0012	Wir rod	0.12			0.5		0.05	0.05		N 0.007; bal Fe				
TS 2348(76)	C9-1/1.0010	Wir rod	0.1			0.5		0.07	0.06		bal Fe				
TS 2348(76)	CF-1/1.0311	Wir rod	0.08			0.45		0.06	0.05		bal Fe				
TS 924(71)	UDK669.14.418/Fe-0	CR strip	0.12			0.2-0.45		0.08	0.06		bal Fe				
TS 924(71)	UDK669.14.418/Fe-1	CR strip	0.12			0.2-0.45		0.07	0.06	0.03-0.2	bal Fe				
TS 924(71)	UDK669.14.418/Fe-2	CR strip	0.1			0.2-0.45		0.06	0.06	0.03-0.2	bal Fe				
TS 924(71)	UDK669.14.418/Fe-4	CR strip	0.1			0.2-0.45		0.03	0.035	0.05-0.1	bal Fe				

UK

Specification	Designation	Notes	C	Cr	Cu	Mn	Ni	P	S	Si	Other	UTS	YS	El	Hard
BS 1449/1(91)	1CR	Plt sht strp; CR wide for vitreous enamelling; t<=16mm	0.08 max			0.45 max		0.025 max	0.03 max	0.6 max	bal Fe	280	140	29-38	
BS 1449/1(91)	1CS	Plt Sh Strp; t<=3mm	0.08 max			0.45 max		0.025 max	0.03 max	0.6 max	bal Fe	270	140	34-38	
BS 1449/1(91)	1HR	Plt Sh Strp HR wide; t<=16mm	0.08 max			0.45 max		0.025 max	0.03 max	0.6 max	bal Fe	290	170	25-34	
BS 1449/1(91)	1HS	Plt Sh Strp HR; t<=16mm	0.08 max			0.45 max		0.025 max	0.03 max	0.6 max	bal Fe	290	170	32-34	
BS 1449/1(91)	2CR	Plt sht strp; CR wide for vitreous enamelling; t<=16mm	0.08 max			0.45 max		0.030 max	0.035 max		bal Fe	280	140	27-36	
BS 1449/1(91)	2CS	CR; t<=3mm	0.08 max			0.45 max		0.030 max	0.035 max		bal Fe	270	140	34-36	
BS 1449/1(91)	2HR	Plt Sh Strp HR wide; t<=16mm	0.08 max			0.45 max		0.030 max	0.035 max		bal Fe	290	170	25-34	
BS 1449/1(91)	2HS	Plt Sh Strp HR; t<=16mm	0.08 max			0.45 max		0.030 max	0.035 max		bal Fe	290	170	32-34	
BS 1449/1(91)	3CR	Sh Strp	0.10 max			0.50 max		0.040 max	0.040 max		bal Fe	280	140	25-34	
BS 1449/1(91)	3CS	CR; t<=3mm	0.10 max			0.50 max		0.040 max	0.040 max		bal Fe	280	140	32-34	

UNS numbers and US grades are provided as a means of cross referencing chemically similar alloys. Exchangability is only possible after independent examination of specifications. Tensile properties are minimum or typical as specified. UTS and YS as MPa. El as %. See Appendix for list of abbreviations used in Notes. * indicates obsolete material.

Specification	Designation	Notes	C	Cr	Cu	Mn	Ni	P	S	Si	Other	UTS	YS	El	Hard

Carbon Steel, Nonresulfurized, 1008 (Continued from previous page)

UK

Specification	Designation	Notes	C	Cr	Cu	Mn	Ni	P	S	Si	Other	UTS	YS	El	Hard
BS 1449/1(91)	3HR	Plt Sh Strp HR wide; t<=16mm	0.10 max			0.50 max		0.040 max	0.040 max		bal Fe	290	170	21-28	
BS 1449/1(91)	3HS	Plt Sh Strp HR; t<=16mm	0.10 max			0.50 max		0.040 max	0.040 max		bal Fe	290	170	26-28	
BS 1717(83)	ERW101	Tub	0.10 max			0.60		0.060 max	0.060 max		bal Fe				
BS 3606(92)	261	Heat exch Tub; t<=3.2mm	0.06-0.10	0.20 max		0.60-0.80		0.020 max	0.020 max	0.10-0.35	Al <=0.06; B 0.002-0.006; Mo 0.40-0.60; Ti <=0.06; Sn<=0.03; bal Fe	540-690	400	17	

USA

Specification	Designation	Notes	C	Cr	Cu	Mn	Ni	P	S	Si	Other	UTS	YS	El	Hard
	AISI 1008	Sh Strp Plt	0.10 max			0.50 max		0.030 max	0.035 max		bal Fe				
	AISI 1008	Wir rod	0.10 max			0.30-0.50		0.040 max	0.050 max		bal Fe				
	AISI 1008	Struct shp	0.10 max			0.50 max		0.030 max	0.035 max		bal Fe				
	AISI M1008	Bar, merchant qual	0.10 max			0.25-0.60		0.04 max	0.05 max		bal Fe				
	UNS G10080		0.10 max			0.30-0.50		0.040 max	0.050 max		Sheets, Mn 0.25-0.50; ERW Tubing, Mn 0.25-0.50; bal Fe				
ASTM A108(95)	1008	Bar, CF	0.10 max			0.30-0.50		0.040 max	0.050 max		bal Fe				
ASTM A29/A29M(93)	1008	Bar	0.10 max			0.30-0.50		0.040 max	0.050 max		bal Fe				
ASTM A29/A29M(93)	M1008	Bar, merchant qual	0.10 max			0.25-0.60		0.04 max	0.05 max		bal Fe				
ASTM A510(96)	1008	Wir rod	0.10 max			0.30-0.50		0.040 max	0.050 max		bal Fe				
ASTM A512(96)	1008	Buttweld mech tub, CD	0.10 max			0.50 max		0.040 max	0.045 max		bal Fe				
ASTM A513(97a)	1008	ERW Mech tub	0.10 max			0.50 max		0.035 max	0.035 max		bal Fe				
ASTM A519(96)	1008	Smls mech tub	0.10 max			0.30-0.50		0.040 max	0.050 max		bal Fe				
ASTM A575(96)	M1008	Merchant qual bar	0.10 max			0.25-0.60		0.04 max	0.05 max		Mn can vary with C; bal Fe				
ASTM A576(95)	1008	Special qual HW bar	0.10 max			0.30-0.50		0.040 max	0.050 max		Si Cu Pb B Bi Ca Se Te if spec'd; bal Fe				
ASTM A619	A619*	Obs, 1997; Sh, CR,	0.1 max			0.5 max		0.025 max	0.03 max		bal Fe				
ASTM A621	A621*	Sh Strp, HR	0.1 max			0.5 max		0.025 max	0.03 max		bal Fe				
ASTM A635/A635M(98)	1008	Sh Strp, Coil, HR	0.10 max		0.20 min	0.50 max		0.030 max	0.035 max		bal Fe				
ASTM A787(96)	1008	ERW metallic-coated mech tub	0.10 max			0.50 max		0.035 max	0.035 max		bal Fe				
ASTM A830/A830M(98)	1008	Plt	0.10 max			0.50 max		0.035 max	0.04 max		bal Fe				
FED QQ-S-637A(70)	C1008*	Obs; see ASTM A108; CF Bar std qual, free mach									bal Fe				
FED QQ-W-461H(88)	1008*	Obs; Wir	0.1 max			0.3-0.5		0.04	0.05		bal Fe				
MIL-R-8814	1008*	Obs; Rvt	0.1 max			0.3-0.5		0.04	0.05		bal Fe				
MIL-S-11310E(76)	1008*	Bar HR cold shape cold ext; Obs for new design	0.1 max	0.07	0.2	0.3-0.5	0.15	0.04 max	0.05 max	0.2	Mo 0.05; bal Fe				
MIL-S-11310E(76)	CS1008*	Bar HR cold shape cold ext; Obs for new design									bal Fe				
MIL-T-3520	1008*	Obs; Tube	0.1 max			0.25-0.5		0.04	0.05		bal Fe				
SAE J1397(92)	1008	Bar HR, est mech prop	0.1 max			0.3-0.5		0.04 max	0.05 max		bal Fe	303	170	30	86 HB
SAE J1397(92)	1008	Bar CD, est mech prop	0.1 max			0.3-0.5		0.04 max	0.05 max		bal Fe	340	290	20	95 HB

UNS numbers and US grades are provided as a means of cross referencing chemically similar alloys. Exchangability is only possible after independent examination of specifications. Tensile properties are minimum or typical as specified. UTS and YS as MPa. El as %. See Appendix for list of abbreviations used in Notes. * indicates obsolete material.

Specification	Designation	Notes	C	Cr	Cu	Mn	Ni	P	S	Si	Other	UTS	YS	El	Hard

Carbon Steel, Nonresulfurized, 1008 (Continued from previous page)

USA

Specification	Designation	Notes	C	Cr	Cu	Mn	Ni	P	S	Si	Other	UTS	YS	El	Hard
SAE J403(95)	1008	Bar Wir rod Smls Tub HR CF	0.10 max			0.30-0.50		0.030 max	0.050 max		bal Fe				
SAE J403(95)	1008	Struct Shps Plt Strp Sh Weld Tub	0.10 max			0.50 max		0.030 max	0.035 max		bal Fe				
SAE J403(95)	M1008	Merchant qual	0.10 max			0.25-0.60		0.04 max	0.05 max		bal Fe				
SAE J526(96)		Weld Tub	0.10 max			0.25-0.50		0.040 max	0.050 max		bal Fe	290	170	14	65 HR30T
SAE J527(96)		Brazed dbl wall tubing	0.10 max			0.30-0.50		0.040 max	0.050 max		bal Fe	290	170	14	65 HR30T

Yugoslavia

Specification	Designation	Notes	C	Cr	Cu	Mn	Ni	P	S	Si	Other	UTS	YS	El	Hard
	C.0146	Unalloyed mild	0.12 max	0.12 max		0.5 max		0.04 max	0.04 max	0.60 max	bal Fe				
	C.0147	Unalloyed mild	0.08 max	0.12 max		0.45 max		0.03 max	0.03 max	0.60 max	Al 0.02-0.120; bal Fe				
	C.0148	Unalloyed mild	0.08 max	0.12 max		0.45 max		0.03 max	0.03 max	0.60 max	Al 0.02-0.120; bal Fe				
	C.0246	Unalloyed; rivets	0.09 max			0.2-0.45		0.15 max	0.05 max	0.60 max	bal Fe				
	C.0265	Unalloyed; rivets	0.13 max			0.2-0.45		0.04 max	0.04 max	0.60 max	N <=0.007; bal Fe				
	C.0275	Unalloyed; rivets	0.13 max			0.25-0.5		0.05 max	0.05 max	0.4 max	N <=0.007; bal Fe				
	C.0446	Unalloyed; rivets	0.09 max			0.2-0.45		0.25 max	0.05 max	0.60 max	bal Fe				
	Z10N		0.1			0.5		0.04	0.04		bal Fe				

Carbon Steel, Nonresulfurized, 1009

Czech Republic

Specification	Designation	Notes	C	Cr	Cu	Mn	Ni	P	S	Si	Other	UTS	YS	El	Hard
CSN 411343	11343		0.17 max	0.3 max		1.60 max		0.05 max	0.05 max	0.60 max	bal Fe				
CSN 411369	11369	Low-temp	0.14 max	0.3 max		0.8 max	0.3 max	0.04 max	0.04 max	0.35 max	Al 0.02-0.120; bal Fe				
CSN 411448	11448	Gen struct	0.16 max	0.3 max		1.3 max	0.2 max	0.04 max	0.04 max	0.4 max	Al <=0.10; P+S<=0.07; bal Fe				

Finland

Specification	Designation	Notes	C	Cr	Cu	Mn	Ni	P	S	Si	Other	UTS	YS	El	Hard
SFS 650	Z01	Hot dip zinc coated	0.15 max			0.6 max		0.05 max	0.05 max	0.60 max	bal Fe				
SFS 679(86)	CR700	Gen struct	0.15 max			0.9 max		0.05 max	0.05-0.03	0.5 max	bal Fe				

Mexico

Specification	Designation	Notes	C	Cr	Cu	Mn	Ni	P	S	Si	Other	UTS	YS	El	Hard
NMX-B-028-SCFI(98)		Sheet, CR, Comm quality	0.15 max			0.60 max		0.030 max	0.035 max		Cu if spec'd>=0.20; bal Fe				

Romania

Specification	Designation	Notes	C	Cr	Cu	Mn	Ni	P	S	Si	Other	UTS	YS	El	Hard
STAS 880(88)	OLC8	Q/T	0.05-0.12	0.3 max		0.35-0.65		0.04 max	0.045 max	0.17-0.37	Pb <=0.15; As<=0.05; bal Fe				
STAS 880(88)	OLC8S	Q/T	0.05-0.12	0.3 max		0.35-0.65		0.04 max	0.02-0.045	0.17-0.37	Pb <=0.15; As<=0.05; bal Fe				
STAS 880(88)	OLC8X	Q/T	0.05-0.12	0.3 max		0.35-0.65		0.035 max	0.02-0.035	0.17-0.37	Pb <=0.15; As<=0.05; bal Fe				
STAS 880(88)	OLC8XS	Q/T	0.05-0.12	0.3 max		0.35-0.6		0.035 max	0.02-0.04	0.17-0.37	Pb <=0.15; As<=0.05; bal Fe				

UK

Specification	Designation	Notes	C	Cr	Cu	Mn	Ni	P	S	Si	Other	UTS	YS	El	Hard
BS 1449/1(91)	14CS	Plt Sh Strp	0.15 max			0.60 max		0.050 max	0.050 max		bal Fe				
BS 1449/1(91)	14HR	Plt Sh Strp HR wide; t<=16mm	0.15 max			0.60 max		0.050 max	0.050 max		bal Fe	280	170	18-25	
BS 1449/1(91)	14HS	Plt Sh Strp HR; t<=16mm	0.15 max			0.60 max		0.050 max	0.050 max		bal Fe	280	170	23-25	

USA

Specification	Designation	Notes	C	Cr	Cu	Mn	Ni	P	S	Si	Other	UTS	YS	El	Hard
	AISI 1009	Sh Strp Plt	0.15 max			0.60 max		0.030 max	0.035 max		bal Fe				
	AISI 1009	Struct shp	0.15 max			0.60 max		0.030 max	0.035 max		bal Fe				
	UNS G10090		0.15 max			0.60 max		0.040 max	0.050 max		bal Fe				

UNS numbers and US grades are provided as a means of cross referencing chemically similar alloys. Exchangability is only possible after independent examination of specifications. Tensile properties are minimum or typical as specified. UTS and YS as MPa. El as %. See Appendix for list of abbreviations used in Notes. * indicates obsolete material.

Specification	Designation	Notes	C	Cr	Cu	Mn	Ni	P	S	Si	Other	UTS	YS	El	Hard

Carbon Steel, Nonresulfurized, 1009 (Continued from previous page)

USA

ASTM A635/A635M(98)	1009	Sh Strp, Coil, HR	0.15 max		0.20 min	0.60 max		0.030 max	0.035 max		bal Fe				
ASTM A827(98)	1009	Plt, frg	0.15 max			0.60 max		0.035 max	0.040 max	0.15-0.40	bal Fe				
ASTM A830/A830M(98)	1009	Plt	0.15 max			0.60 max		0.035 max	0.04 max		bal Fe				
SAE J403(95)	1009	Struct Shps Plt Strp Sh Weld Tub	0.15 max			0.60 max		0.030 max	0.035 max		bal Fe				

Carbon Steel, Nonresulfurized, 1010

Argentina

| IAS | IRAM 1010 | | 0.08-0.13 | | | 0.30-0.60 | | 0.040 max | 0.050 max | 0.10 max | bal Fe | 330-430 | 210-280 | 28-38 | 95-124 HB |

Australia

AS 1442(92)	1010	HR bar, Semifinished (may treat w/microalloying elements)	0.08-0.13			0.30-0.60		0.040 max	0.040 max	0.10-0.35	Si<=0.10 for Al-killed; bal Fe				
AS 1443(94)	1010	CF bar (may treat w/microalloying elements)	0.08-0.13			0.30-0.60		0.040 max	0.040 max	0.10-0.35	Si<=0.10 for Al-killed; bal Fe				
AS/NZS 1594(97)	HA1010	HR flat products	0.08-0.13	0.30 max	0.35 max	0.30-0.60	0.35 max	0.040 max	0.030 max	0.03 max	Al <=0.100; Mo <=0.10; Ti <=0.040; (Cu+Ni+Cr+Mo) 1 max; bal Fe				

Bulgaria

| BDS 5785(83) | 08 | Struct | 0.05-0.12 | 0.10 max | 0.25 max | 0.35-0.65 | 0.25 max | 0.035 max | 0.040 max | 0.17-0.37 | bal Fe | | | | |
| BDS 5785(83) | 10 | Struct | 0.07-0.14 | 0.15 max | 0.25 max | 0.35-0.65 | 0.25 max | 0.035 max | 0.040 max | 0.17-0.37 | bal Fe | | | | |

China

GB 13237(91)	10	Sh Strp CD HT	0.07-0.14	0.15 max	0.25 max	0.35-0.65	0.25 max	0.035 max	0.035 max	0.17-0.37	bal Fe	295-430		28	
GB 13237(91)	10F	Sh Strp CR HT	0.07-0.14	0.15 max	0.25 max	0.25-0.50	0.25 max	0.035 max	0.035 max	0.07 max	bal Fe	275-410		28	
GB 3275(91)	10	Plt Strp HR HT	0.07-0.14	0.15 max	0.25 max	0.35-0.65	0.25 max	0.035 max	0.035 max	0.17-0.37	bal Fe	275-410		27	
GB 5953(86)	ML 10	Wir CD	0.07-0.14	0.20 max	0.20 max	0.20-0.50	0.20 max	0.035 max	0.035 max	0.03 max	bal Fe	335	205	31	
GB 6479(86)	10	Smls Tub HR/CD Norm	0.07-0.14		0.25 max	0.35-0.65		0.035 max	0.040 max	0.17-0.37	bal Fe	335-490	205	24	
GB 699(88)	10	Bar HR Norm 25mm diam	0.07-0.14	0.15 max	0.25 max	0.35-0.65	0.25 max	0.035 max	0.035 max	0.17-0.37	bal Fe	335	205	31	
GB 710(91)	10	Sh Strp HR HT	0.07-0.14	0.15 max	0.25 max	0.35-0.65	0.25 max	0.035 max	0.035 max	0.17-0.37	bal Fe	295-430		24	
GB 710(91)	10F	Sh Strp HR HT	0.07-0.14	0.15 max	0.25 max	0.25-0.50	0.25 max	0.035 max	0.035 max	0.07 max	bal Fe	275-410		25	
GB 8162(87)	10	Smls Tub HR/CD Q/T	0.07-0.14	0.15 max	0.25 max	0.35-0.65	0.25 max	0.035 max	0.035 max	0.17-0.37	bal Fe	335	205	24	
GB 8163(87)	10	Smls Pip HR/CD	0.07-0.14	0.15 max	0.25 max	0.35-0.65	0.25 max	0.035 max	0.035 max	0.17-0.37	bal Fe	335	205	24	
GB 9948(88)	10	Smls Tub HR/CD Norm	0.07-0.14	0.15 max	0.25 max	0.35-0.65	0.25 max	0.035 max	0.035 max	0.17-0.37	bal Fe	330-490	205	24	
GB/T 13795(92)	10	Strp CR Ann	0.07-0.14	0.15 max	0.25 max	0.35-0.65	0.25 max	0.035 max	0.035 max	0.17-0.37	bal Fe	400-700		15	
GB/T 13796(92)	10	Bar Wir CD Ann	0.07-0.14	0.15 max	0.25 max	0.35-0.65	0.25 max	0.035 max	0.035 max	0.17-0.37	bal Fe	295-440			
GB/T 3078(94)	10	Bar CD Ann	0.07-0.14	0.15 max	0.25 max	0.35-0.65	0.25 max	0.035 max	0.035 max	0.17-0.37	bal Fe	295		26	
GB/T 3078(94)	10F	Bar CD Norm 25mm diam	0.07-0.13			0.50 max		0.035 max	0.035 max	0.03 max	bal Fe	315	185	33	
GB/T 3275(91)	10F	Plt Strp HR HT	0.07-0.14	0.15 max	0.25 max	0.25-0.50	0.25 max	0.035 max	0.035 max	0.07 max	bal Fe	275-410		27	
GB/T 699(88)	10F	Bar HR Norm 25mm diam	0.07-0.14	0.15 max	0.25 max	0.25-0.50	0.25 max	0.035 max	0.035 max	0.07 max	bal Fe	315	185	33	
GB/T 8164(93)	10	Strp HR/CR	0.07-0.14	0.15 max	0.25 max	0.35-0.65	0.25 max	0.035 max	0.035 max	0.17-0.37	bal Fe	335		31	

Carbon Steel, Nonresulfurized, 1010 (Continued from previous page)

Specification	Designation	Notes	C	Cr	Cu	Mn	Ni	P	S	Si	Other	UTS	YS	El	Hard
Czech Republic															
CSN 412010	12010	CH	0.07-0.14	0.15 max		0.35-0.65	0.3 max	0.04 max	0.04 max	0.15-0.4	bal Fe				
Europe															
EN 10016/2(94)	1.0310	Rod	0.08-0.13	0.20 max	0.30 max	0.30-0.60	0.25 max	0.035 max	0.035 max	0.30 max	Al <=0.01; Mo <=0.05; bal Fe				
EN 10016/2(94)	C10D	Rod	0.08-0.13	0.20 max	0.30 max	0.30-0.60	0.25 max	0.035 max	0.035 max	0.30 max	Al <=0.01; Mo <=0.05; bal Fe				
EN 10084(98)	1.1121	CH, Ann	0.07-0.13			0.30-0.60		0.035 max	0.035 max	0.40 max	bal Fe				131 HB
EN 10084(98)	C10E	CH, Ann	0.07-0.13			0.30-0.60		0.035 max	0.035 max	0.40 max	bal Fe				131 HB
France															
AFNOR NFA33101(82)	AF34C10	t<=450mm; Norm	0.12 max	0.3 max	0.3 max	0.3-0.6	0.4 max	0.04 max	0.04 max	0.3 max	bal Fe	330-420	195	31	
AFNOR NFA35551(75)	CC10		0.05-0.15	0.3 max	0.3 max	0.3-0.5	0.4 max	0.04 max	0.04 max	0.3 max	bal Fe				
AFNOR NFA36102(87)	FdTu1*		0.12 max	0.3 max		0.2-0.5	0.40 max	0.05 max	0.04 max	0.6 max	N <=0.008; bal Fe				
AFNOR NFA36102(93)	C10RR	HR; Strp; CR	0.06-0.12	0.08 max		0.3-0.6	0.08 max	0.025 max	0.02 max	0.04 max	Al 0.015-0.06; Mo <=0.040; bal Fe				
AFNOR NFA36102(93)	XC10*	HR; Strp; CR	0.06-0.12	0.08 max		0.3-0.6	0.08 max	0.025 max	0.02 max	0.04 max	Al 0.015-0.06; Mo <=0.040; bal Fe				
AFNOR NFA36232(92)	E315C	t<3mm; CR	0.10 max	0.3 max		0.6 max	0.4 max	0.025 max	0.025 max	0.4 max	Al >=0.02; N <=0.009; bal Fe	400-490	3115	22	
Hungary															
MSZ 23(83)	M1H	Sh Strp; mild; 0.71-3mm; rolled	0.1 max			0.45 max		0.03 max	0.04 max	0.12 max	Al <=0.120; bal Fe	270-410		28L	0-65 1.3mm HRB; 0.71-1.3mm HR30T<=60
MSZ 23(83)	M1H	Sh Strp; mild; 0.5-0.7mm; rolled	0.1 max			0.45 max		0.03 max	0.04 max	0.12 max	Al <=0.120; bal Fe	270-410		26L	0-60 HR30T
MSZ 23(83)	M2H	Sh Strp; mild; 0.71-3mm; rolled	0.1 max			0.45 max		0.03 max	0.04 max	0.12 max	Al <=0.120; bal Fe	270-410		28L	0-65 1.3mm HRB; 0.71-1.3mm HR30T<=60
MSZ 23(83)	M2H	Sh Strp; mild; 0.5-0.7mm; rolled	0.1 max			0.45 max		0.03 max	0.04 max	0.12 max	Al <=0.120; bal Fe	270-410		26L	0-60 HR30T
MSZ 23(83)	M2P	Sh Strp; mild; 0.5-0.7mm; rolled	0.1 max			0.45 max		0.03 max	0.04 max	0.12 max	Al <=0.120; bal Fe	270-410		26L	0-60 HR30T
MSZ 23(83)	M2P	Sh Strp; mild; 0.71-3mm; rolled	0.1 max			0.45 max		0.03 max	0.04 max	0.12 max	Al <=0.120; bal Fe	270-410		28L	0-65 1.3mm HRB; 0.71-1.3mm HR30T<=60
MSZ 23(83)	M3H	Sh Strp; mild; 0.5-0.7mm; rolled	0.1 max			0.45 max		0.03 max	0.04 max	0.12 max	Al <=0.120; bal Fe	270-410		26L	0-60 HR30T
MSZ 23(83)	M3H	Sh Strp; mild; 0.71-3mm; rolled	0.1 max			0.45 max		0.03 max	0.04 max	0.12 max	Al <=0.120; bal Fe	270-410		28L	0-65 1.3mm HRB; 0.71-1.3mm HR30T<=60
MSZ 23(83)	M3P	Sh Strp; mild; 0.5-0.7mm; rolled	0.1 max			0.45 max		0.03 max	0.04 max	0.12 max	Al <=0.120; bal Fe	270-410		26L	0-60 HR30T
MSZ 23(83)	M3P	Sh Strp; mild; 0.71-3mm; rolled	0.1 max			0.45 max		0.03 max	0.04 max	0.12 max	Al <=0.120; bal Fe	270-410		28L	0-65 1.3mm HRB; 0.71-1.3mm HR30T<=60

UNS numbers and US grades are provided as a means of cross referencing chemically similar alloys. Exchangability is only possible after independent examination of specifications. Tensile properties are minimum or typical as specified. UTS and YS as MPa. El as %. See Appendix for list of abbreviations used in Notes. * indicates obsolete material.

Carbon Steel, Nonresulfurized, 1010 (Continued from previous page)

Specification	Designation	Notes	C	Cr	Cu	Mn	Ni	P	S	Si	Other	UTS	YS	El	Hard
Hungary															
MSZ 2978/2(82)	A34	Longitudinal weld/CF tub; gen struct; 0.5-3mm; bright-drawn, hard	0.15 max			2 max	0.3 max	0.045 max	0.05 max	0.6 max	bal Fe	410	330	6L	
MSZ 2978/2(82)	A34	Longitudinal weld/CF tub; gen struct; 0.5-3mm; bright-drawn, soft	0.15 max			2 max	0.3 max	0.045 max	0.05 max	0.6 max	bal Fe	350	245	10L	
MSZ 31(85)	C10	CH; 30-30mm; blank carburized	0.07-0.13			0.3-0.6		0.035 max	0.035 max	0.4 max	bal Fe	450-650	300	16L	
MSZ 31(85)	C10	CH; 0-11mm; blank carburized	0.07-0.13			0.3-0.6		0.035 max	0.035 max	0.4 max	bal Fe	500-700	330	14L	
MSZ 500(81)	A34	Gen struct; 0-16mm; HR; Frg	0.15 max			2 max	0.3 max	0.045 max	0.05 max	0.6 max	bal Fe	330-430	215	32L	
MSZ 500(81)	A34	Gen struct; 40.1-100mm; HR; Frg	0.15 max			2 max	0.3 max	0.045 max	0.05 max	0.6 max	bal Fe	330-430	195	32L	
MSZ 500(81)	A34B	Gen struct; Non-rimming; FF; 40.1-100mm; HR/Frg	0.15 max			2 max		0.045 max	0.05 max	0.12-0.60	bal Fe	330-430	195	32L	
MSZ 500(81)	A34B	Gen struct; Non-rimming; FF; 0-16mm; HR/Frg	0.15 max			2 max		0.045 max	0.05 max	0.12-0.60	bal Fe	330-430	215	32L	
MSZ 500(81)	A34X	Gen struct; Unkilled; R; FU; 0-16mm; HR; Frg	0.15 max			2 max		0.045 max	0.05 max	0.07 max	bal Fe	330-430	215	32L	
MSZ 500(81)	A34X	Gen struct; Unkilled; R; FU; 40.1-100mm; HR; Frg	0.15 max			2 max		0.045 max	0.05 max	0.07 max	bal Fe	330-430	195	32L	
MSZ 5736(88)	D10	Wir rod; for CD	0.08-0.13		0.25 max	0.35-0.6	0.3 max	0.04 max	0.04 max	0.1-0.35	bal Fe				
MSZ 5736(88)	D10K	Wir rod; for CD	0.08-0.13		0.25 max	0.35-0.6	0.3 max	0.03 max	0.03 max	0.1-0.35	bal Fe				
MSZ 5736(88)	D10X	Wir rod; for CD	0.08-0.13		0.25 max	0.5 max	0.3 max	0.04 max	0.04 max	0.05 max	bal Fe				
MSZ 6251(87)	C10Z	CF; CH; 0-36mm; drawn, half-hard	0.07-0.13			0.3-0.6		0.035 max	0.035 max	0.4 max	bal Fe	0-470			
MSZ 6251(87)	C10Z	CF; CH; 0-36mm; HR; HR/soft ann; drawn/soft ann; drawn/bright ann; soft ann/ground	0.07-0.13			0.3-0.6		0.035 max	0.035 max	0.4 max	bal Fe	0-440			
MSZ 6251(87)	D10Z	CF; 0-36mm; HR; HR/soft ann; drawn/soft ann; drawn/bright ann; soft ann/ground	0.12 max			0.25-0.5	0.3 max	0.04 max	0.04 max	0.1 max	bal Fe	0-460			
MSZ 6251(87)	D10Z	CF; 0-36mm; drawn, half-hard	0.12 max			0.25-0.5	0.3 max	0.04 max	0.04 max	0.1 max	bal Fe	0-520			
India															
IS 1570/2(79)	7C4	Sh Plt Strp Sect Shp Bar Bil Frg	0.12 max	0.3 max	0.3 max	0.5 max	0.4 max	0.055 max	0.055 max	0.6 max	Co <=0.1; Mo <=0.15; Pb <=0.15; W <=0.1; P:S Varies; bal Fe	320-400	176	27	
IS 1570/2(79)	C07	Sh Plt Strp Sect Shp Bar Bil Frg	0.12 max	0.3 max	0.3 max	0.5 max	0.4 max *	0.055 max	0.055 max	0.6 max	Co <=0.1; Mo <=0.15; Pb <=0.15; W <=0.1; P:S Varies; bal Fe	320-400	176	27	
International															
ISO 3305(85)	R33	Weld Tub, CF	0.16 max			0.70 max		0.050 max	0.050 max		Co <=0.1; Pb <=0.15; W <=0.1; bal Fe	420		6	
ISO 3305(85)	R37	Weld Tub, CF	0.17 max			0.8 max		0.050 max	0.050 max	0.35 max	Co <=0.1; Pb <=0.15; W <=0.1; bal Fe	450		6	
ISO 683-11(87)	C10	CH, 16mm test	0.07-0.13		0.3 max	0.30-0.60		0.035 max	0.035 max	0.15-0.40	Al <=0.1; Co <=0.1; bal Fe	450-800	270		
Italy															
UNI 5598(71)	1CD10	Wir rod	0.12 max			0.25-0.6		0.05 max	0.05 max	0.6 max	bal Fe				
UNI 5598(71)	3CD12	Wir rod	0.08-0.13			0.3-0.6		0.035 max	0.035 max	0.6 max	N <=0.012; bal Fe				

UNS numbers and US grades are provided as a means of cross referencing chemically similar alloys. Exchangability is only possible after independent examination of specifications. Tensile properties are minimum or typical as specified. UTS and YS as MPa. El as %. See Appendix for list of abbreviations used in Notes. * indicates obsolete material.

Specification	Designation	Notes	C	Cr	Cu	Mn	Ni	P	S	Si	Other	UTS	YS	El	Hard

Carbon Steel, Nonresulfurized, 1010 (Continued from previous page)

Italy

Specification	Designation	Notes	C	Cr	Cu	Mn	Ni	P	S	Si	Other	UTS	YS	El	Hard
UNI 6403(86)	C10	Q/T	0.07-0.13			0.3-0.6		0.035 max	0.035 max	0.15-0.35	bal Fe				
UNI 7065(72)	C10	Strp	0.07-0.12			0.3-0.7		0.035 max	0.035 max	0.35 max	bal Fe				
UNI 7356(74)	CB10FF	Q/T	0.08-0.13			0.3-0.6		0.04 max	0.04 max	0.1 max	bal Fe				
UNI 7356(74)	CB10FU	Q/T	0.08-0.13			0.3-0.6		0.04 max	0.04 max	0.6 max	N <=0.007; trace Si; bal Fe				
UNI 7846(78)	C10	CH	0.07-0.13			0.3-0.6		0.035 max	0.035 max	0.15-0.35	bal Fe				
UNI 8550(84)	C10	CH	0.07-0.13			0.3-0.6		0.035 max	0.035 max	0.15-0.35	bal Fe				
UNI 8788(85)	1C10	CH	0.06-0.14			0.3-0.6		0.045 max	0.045 max	0.15-0.35	bal Fe				
UNI 8788(85)	2C10	CH	0.07-0.13			0.3-0.6		0.035 max	0.035 max	0.15-0.35	bal Fe				
UNI 8913(87)	C10	Q/T	0.07-0.13			0.3-0.6		0.035 max	0.035 max	0.35 max	bal Fe				

Japan

Specification	Designation	Notes	C	Cr	Cu	Mn	Ni	P	S	Si	Other	UTS	YS	El	Hard
JIS G3507(91)	SWRCH10A	Wir rod Al killed	0.08-0.13			0.30-0.60		0.030 max	0.035 max	0.10 max	Al >=0.02; bal Fe				
JIS G3507(91)	SWRCH10K	Wir rod FF	0.08-0.13			0.30-0.60		0.030 max	0.035 max	0.10-0.35	bal Fe				
JIS G3507(91)	SWRCH10R	Wir rod FU	0.08-0.13			0.30-0.60		0.040 max	0.040 max		bal Fe				
JIS G4051(79)	S09CK	CH	0.07-0.12	0.20 max	0.25 max	0.30-0.60	0.20 max	0.025 max	0.025 max	0.10-0.35	Ni+Cr=0.30 max; bal Fe				
JIS G4051(79)	S10C	Bar Wir rod	0.08-0.13	0.20 max	0.30 max	0.30-0.60	0.20 max	0.030 max	0.035 max	0.15-0.35	Ni+Cr<=0.35; bal Fe				

Mexico

Specification	Designation	Notes	C	Cr	Cu	Mn	Ni	P	S	Si	Other	UTS	YS	El	Hard
NMX-B-201(68)	MT1010	CD buttweld mech tub	0.05-0.15			0.30-0.60		0.04 max	0.05 max		bal Fe				
NMX-B-265(89)	LEP	Sh for enamelling	0.10 max			0.50 max		0.030 max	0.030 max		bal Fe				
NMX-B-267-SCFI(98)		CR Sh, drawing quality	0.10 max	0.15 max	0.20 max	0.50 max		0.020 max	0.030 max		Mo <=0.06; N <=0.20; Nb <=0.008; V <=0.008; bal Fe				
NMX-B-301(86)	1010	Bar	0.08-0.13			0.30-0.60		0.040 max	0.050 max		bal Fe				

Norway

Specification	Designation	Notes	C	Cr	Cu	Mn	Ni	P	S	Si	Other	UTS	YS	El	Hard
NS 12111	12111	Struct const	0.18 max					0.05 max	0.05 max		bal Fe				

Poland

Specification	Designation	Notes	C	Cr	Cu	Mn	Ni	P	S	Si	Other	UTS	YS	El	Hard
PNH84019	10	CH; gear, gear wheel; Plt, drill jig	0.07-0.14	0.25 max		0.35-0.65	0.3 max	0.04 max	0.04 max	0.17-0.37	bal Fe				

Romania

Specification	Designation	Notes	C	Cr	Cu	Mn	Ni	P	S	Si	Other	UTS	YS	El	Hard
STAS 500/2(88)	OL32.1	Gen struct	0.12 max	0.3 max		0.55 max	0.4 max	0.05 max	0.05 max	0.6 max	Pb <=0.15; bal Fe				
STAS 500/2(88)	OL32.1a	Gen struct	0.12 max	0.3 max		0.55 max	0.4 max	0.05 max	0.05 max	0.6 max	Pb <=0.15; bal Fe				
STAS 500/2(88)	OL32.1b	Gen struct	0.12 max	0.3 max		0.55 max	0.4 max	0.05 max	0.05 max	0.6 max	Pb <=0.15; bal Fe				
STAS 500/2(88)	OL34.1	Gen struct	0.15 max	0.3 max		0.55 max	0.4 max	0.05 max	0.05 max	0.6 max	Pb <=0.15; bal Fe				
STAS 500/2(88)	OL34.1a	Gen struct	0.15 max	0.3 max		0.55 max	0.4 max	0.05 max	0.05 max	0.6 max	Pb <=0.15; bal Fe				
STAS 500/2(88)	OL34.1b	Gen struct	0.15 max	0.3 max		0.55 max	0.4 max	0.05 max	0.05 max	0.6 max	Pb <=0.15; bal Fe				
STAS 880(88)	OLC10	CH	0.07-0.13	0.3 max		0.35-0.65		0.04 max	0.045 max	0.17-0.37	Pb <=0.15; As<=0.05; bal Fe				
STAS 880(88)	OLC10S	CH	0.07-0.14	0.3 max		0.35-0.65		0.04 max	0.02-0.045	0.17-0.37	Pb <=0.15; As<=0.05; bal Fe				
STAS 880(88)	OLC10X	CH	0.07-0.14	0.3 max		0.35-0.6		0.035 max	0.02-0.035	0.17-0.37	Pb <=0.15; As<=0.05; bal Fe				
STAS 880(88)	OLC10XS	Q/T	0.07-0.14	0.3 max		0.35-0.6		0.035 max	0.02-0.04	0.17-0.37	Pb <=0.15; As<=0.05; bal Fe				

UNS numbers and US grades are provided as a means of cross referencing chemically similar alloys. Exchangability is only possible after independent examination of specifications. Tensile properties are minimum or typical as specified. UTS and YS as MPa. El as %. See Appendix for list of abbreviations used in Notes. * indicates obsolete material.

Carbon Steel, Nonresulfurized, 1010 (Continued from previous page)

Specification	Designation	Notes	C	Cr	Cu	Mn	Ni	P	S	Si	Other	UTS	YS	El	Hard
Russia															
GOST 1050(88)	08	High-grade struct; CH; Norm; 0-100mm	0.05-0.12			0.35-0.65	0.3 max	0.035 max	0.04 max	0.17-0.37	As<=0.08; bal Fe	320	196	33L	
GOST 1050(88)	08	High-grade struct; CH; Full hard; 0-100mm	0.05-0.12			0.35-0.65	0.3 max	0.035 max	0.04 max	0.17-0.37	As<=0.08; bal Fe				179 HB max
GOST 1050(88)	10	High-grade struct; Norm; 0-100mm	0.07-0.14			0.35-0.65	0.3 max	0.035 max	0.04 max	0.17-0.37	As<=0.08; bal Fe	330	205	34 L	
GOST 1050(88)	10	High-grade struct; Full hard; 0-100mm	0.07-0.14			0.35-0.65	0.3 max	0.035 max	0.04 max	0.17-0.37	As<=0.08; bal Fe	410		8L	0-187 HB
GOST 1050(88)	11kp	High-grade struct; Unkilled; Rimming; R; FU	0.05-0.12		0.2 max	0.3-0.5	0.3 max	0.035 max	0.04 max	0.06 max	As<=0.08; bal Fe				
GOST 4041(71)	10JuA	High-grade Plts for cold pressing	0.07-0.14	0.1 max	0.2 max	0.20-0.40	0.15 max	0.020 max	0.025 max	0.07 max	Al 0.02-0.08; As<=0.08; bal Fe				
Spain															
UNE	F.151	CH	0.08-0.12			0.3-0.4		0.04 max	0.04 max	0.15-0.35	Mo <=0.15; bal Fe				
UNE 36013(76)	C10k	CH	0.07-0.13			0.3-0.6		0.035 max	0.035 max	0.15-0.4	Mo <=0.15; bal Fe				
UNE 36013(76)	C10k-1	CH	0.07-0.13			0.3-0.6		0.035 max	0.02-0.035	0.15-0.4	Mo <=0.15; bal Fe				
UNE 36013(76)	F.1510*	see C10k	0.07-0.13			0.3-0.6		0.035 max	0.035 max	0.15-0.4	Mo <=0.15; bal Fe				
UNE 36013(76)	F.1512*	see C10k-1	0.07-0.13			0.3-0.6		0.035 max	0.02-0.035	0.15-0.4	Mo <=0.15; bal Fe				
UNE 36032(85)	10E-DF	Screw	0.08-0.13			0.30-0.60		0.040 max	0.040 max	0.60 max	N <=0.008; bal Fe				
UNE 36032(85)	10KA-DF	Screw	0.08-0.13			0.30-0.60		0.040 max	0.040 max	0.10 max	Al >=0.020; bal Fe				
UNE 36032(85)	F.7503*	see 10E-DF	0.08-0.13			0.30-0.60		0.040 max	0.040 max	0.60 max	N <=0.008; bal Fe				
UNE 36032(85)	F.7513*	see 10KA-DF	0.08-0.13			0.30-0.60		0.040 max	0.040 max	0.10 max	Al >=0.020; bal Fe				
UNE 36082(84)	AE235-W-B	Struct; Corr res	0.13 max	0.40-0.80	0.25-0.55	0.20-0.60	0.65 max	0.05 max	0.035 max	0.4 max	Mo <=0.15; N <=0.009; bal Fe				
UNE 36082(84)	AE235-W-C	Struct; Corr res	0.13 max	0.40-0.80	0.25-0.55	0.20-0.60	0.65 max	0.05 max	0.035 max	0.4 max	Mo <=0.15; N <=0.009; bal Fe				
UNE 36082(84)	AE235-W-D	Struct; Corr res	0.13 max	0.40-0.80	0.25-0.55	0.20-0.60	0.65 max	0.04 max	0.035 max	0.4 max	Mo <=0.15; bal Fe				
UNE 36082(84)	F.6431*	see AE235-W-B	0.13 max	0.40-0.80	0.25-0.55	0.20-0.60	0.65 max	0.05 max	0.035 max	0.4 max	Mo <=0.15; N <=0.009; bal Fe				
UNE 36082(84)	F.6432*	see AE235-W-C	0.13 max	0.40-0.80	0.25-0.55	0.20-0.60	0.65 max	0.05 max	0.035 max	0.4 max	Mo <=0.15; N <=0.009; bal Fe				
UNE 36082(84)	F.6433*	see AE235-W-D	0.13 max	0.40-0.80	0.25-0.55	0.20-0.60	0.65 max	0.04 max	0.035 max	0.4 max	Mo <=0.15; bal Fe				
Sweden															
SS 141265	1265	CH; FF; Non-rimming	0.07-0.13			0.25-0.45		0.03 max	0.04 max	0.3 max	bal Fe				
UK															
BS 1449/1(83)	10CS	Plt Sh Strp; CH Quen; t<=16mm	0.08-0.15			0.60-0.90		0.045 max	0.045 max	0.10-0.35	bal Fe				190-285 HV
BS 1449/1(91)	10CS	Plt Sh Strp; CR Ann; t<=16mm	0.08-0.15			0.60-0.90		0.045 max	0.045 max	0.10-0.35	bal Fe				114 HV
BS 1449/1(91)	10HS	Plt sht strp; CR Ann; t<=16mm	0.08-0.15			0.60-0.90		0.045 max	0.045 max	0.10-0.35	bal Fe				114 HV
BS 1449/1(91)	4CR	Plt Sh Strp; CR wide for vitreous enamelling; 0-16mm	0.12 max			0.60 max		0.050 max	0.050 max		bal Fe	280	140		
BS 1449/1(91)	4CS	Plt Sh Strp; CR; t<=16mm	0.12 max			0.60 max		0.050 max	0.050 max		bal Fe	280-350	140	28-30	
BS 1449/1(91)	4CS	CR; t<=3mm	0.12 max			0.60 max		0.050 max	0.050 max		bal Fe	280	140		
BS 1449/1(91)	4HR	Plt Sh Strp HR wide; t<=16mm	0.12 max			0.60 max		0.050 max	0.050 max		bal Fe	280	170	18-25	
BS 1449/1(91)	4HS	Plt Sh Strp HR; t<=16mm	0.12 max			0.60 max		0.050 max	0.050 max		bal Fe	280	170	23-25	
BS 1717(83)	CEWC1	Tub	0.13 max			0.60 max		0.050 max	0.050 max		W <=0.1; bal Fe				

UNS numbers and US grades are provided as a means of cross referencing chemically similar alloys. Exchangability is only possible after independent examination of specifications. Tensile properties are minimum or typical as specified. UTS and YS as MPa. El as %. See Appendix for list of abbreviations used in Notes. * indicates obsolete material.

Specification	Designation	Notes	C	Cr	Cu	Mn	Ni	P	S	Si	Other	UTS	YS	El	Hard

Carbon Steel, Nonresulfurized, 1010 (Continued from previous page)

UK

Specification	Designation	Notes	C	Cr	Cu	Mn	Ni	P	S	Si	Other	UTS	YS	El	Hard
BS 1717(83)	ERWC1	Tub	0.13 max			0.60 max		0.050 max	0.050 max		W <=0.1; bal Fe				
BS 970/1(83)	045M10	Blm Bil Slab Bar Rod Frg CH	0.07-0.13			0.30-0.60					bal Fe	430		18	
BS 970/1(96)	040A10	Blm Bil Slab Bar Rod Frg	0.08-0.13			0.30-0.50		0.050 max	0.050 max	0.10-0.40	bal Fe				
BS 970/1(96)	040A12	Blm Bil Slab Bar Rod Frg	0.10-0.15			0.30-0.50		0.050 max	0.050 max	0.10-0.40	bal Fe				
BS 970/1(96)	045A10	Blm Bil Slab Bar Rod Frg	0.08-0.13			0.30-0.60		0.050 max	0.050 max	0.10-0.40	bal Fe				
BS 970/1(96)	045M10	Blm Bil Slab Bar Rod Frg	0.07-0.13			0.30-0.60		0.050 max	0.050 max	0.10-0.40	bal Fe				
BS 970/1(96)	055M15	Blm Bil Slab Bar Rod Frg HR	0.20 max			0.80 max		0.050 max	0.050 max	0.10-0.40	bal Fe				121 HB

USA

Specification	Designation	Notes	C	Cr	Cu	Mn	Ni	P	S	Si	Other	UTS	YS	El	Hard
	AISI 1010	Sh Strp Plt	0.08-0.13			0.30-0.60		0.030 max	0.035 max		bal Fe				
	AISI 1010	Wir rod	0.08-0.13			0.30-0.60		0.040 max	0.050 max		bal Fe				
	AISI 1010	Struct shp	0.08-0.13			0.30-0.60		0.030 max	0.035 max		bal Fe				
	AISI 1010	Bar	0.08-0.13			0.30-0.60		0.040 max	0.050 max	0.050 max	bal Fe				
	AISI M1010	Bar, merchant qual	0.07-0.14			0.25-0.60		0.04 max	0.05 max		bal Fe				
	UNS G10100		0.08-0.13			0.30-0.60		0.040 max	0.050 max		bal Fe				
AMS 5050J(89)		Smls tub, Ann; 12.7<t<=139.7	0.15 max			0.30-0.60		0.040 max	0.050 max		bal Fe			35	
AMS 5053G(90)		Weld Tub, 12.7<t<=139.7	0.13 max			0.30-0.60		0.04 max	0.05 max		bal Fe			35	
ASTM A108(95)	1010	Bar, CF	0.08-0.13			0.30-0.60		0.040 max	0.050 max		bal Fe				
ASTM A29/A29M(93)	1010	Bar	0.08-0.13			0.30-0.60		0.040 max	0.050 max	0.050 max	bal Fe				
ASTM A29/A29M(93)	M1010	Bar, merchant qual	0.07-0.14			0.25-0.60		0.04 max	0.05 max		bal Fe				
ASTM A510(96)	1010	Wir rod	0.08-0.13			0.30-0.60		0.040 max	0.050 max		bal Fe				
ASTM A512(96)	1010	Buttweld mech tub, CD	0.08-0.13			0.30-0.60		0.040 max	0.045 max		bal Fe				
ASTM A512(96)	MT1010	Buttweld mech tub, CD	0.05-0.15			0.30-0.60		0.040 max	0.045 max		bal Fe				
ASTM A513(97a)	1010	ERW Mech tub	0.08-0.13			0.30-0.60		0.035 max	0.035 max		bal Fe				
ASTM A513(97a)	MT1010	ERW Mech tub	0.05-0.15			0.30-0.60		0.035 max	0.035 max		bal Fe				
ASTM A519(96)	1010	Smls mech tub	0.08-0.13			0.30-0.60		0.040 max	0.050 max		bal Fe				
ASTM A519(96)	MT1010	Smls Low-C Mech tub	0.05-0.15			0.30-0.60		0.040 max	0.050 max		bal Fe				
ASTM A575(96)	M1010	Merchant qual bar	0.07-0.14			0.25-0.60		0.04 max	0.05 max		Mn can vary with C; bal Fe				
ASTM A576(95)	1010	Special qual HW bar	0.08-0.13			0.30-0.60		0.040 max	0.050 max		Si Cu Pb B Bi Ca Se Te if spec'd; bal Fe				
ASTM A635/A635M(98)	1010	Sh Strp, Coil, HR	0.08-0.13	0.20 min		0.30-0.60		0.030 max	0.035 max		bal Fe				
ASTM A787(96)	1010	ERW metallic-coated mech tub	0.08-0.13			0.30-0.60		0.035 max	0.035 max		bal Fe				
ASTM A787(96)	MT1010	ERW metallic-coated mech tub, Low-C	0.05-0.15			0.30-0.60		0.035 max	0.035 max		bal Fe				
ASTM A830/A830M(98)	1010	Plt	0.80-0.13			0.30-0.60		0.035 max	0.04 max		bal Fe				
MIL-S-16788A(86)	C1*	HR, frg, bloom bil slab for refrg; Obs for new design	0.08-0.13			0.30-0.60		0.040 max	0.050 max		bal Fe				
SAE J1397(92)	1010	Bar HR, est mech prop									bal Fe	320	180	28	95 HB
SAE J1397(92)	1010	Bar CD, est mech prop									bal Fe	370	300	20	105 HB

UNS numbers and US grades are provided as a means of cross referencing chemically similar alloys. Exchangability is only possible after independent examination of specifications. Tensile properties are minimum or typical as specified. UTS and YS as MPa. El as %. See Appendix for list of abbreviations used in Notes. * indicates obsolete material.

Specification	Designation	Notes	C	Cr	Cu	Mn	Ni	P	S	Si	Other	UTS	YS	El	Hard
Carbon Steel, Nonresulfurized, 1010 (Continued from previous page)															
USA															
SAE J403(95)	1010	Struct Shps Plt Strp Sh Weld Tub	0.08-0.13			0.30-0.60		0.030 max	0.035 max		bal Fe				
SAE J403(95)	1010	Bar Wir rod Smls Tub HR CF	0.08-0.13			0.30-0.60		0.030 max	0.050 max		bal Fe				
SAE J403(95)	M1010	Merchant qual	0.08-0.13			0.30-0.60		0.04 max	0.05 max		bal Fe				
Yugoslavia															
	C.0210	Chain	0.06-0.12			0.4-0.6		0.045 max	0.045 max	0.60 max	N <=0.007; bal Fe				
	C.0211	Chain	0.06-0.12			0.4-0.6		0.045 max	0.045 max	0.12-0.3	N <=0.007; bal Fe				
	C.1100	Struct	0.08-0.12			0.3-0.6		0.06 max	0.06 max	0.60 max	bal Fe				
	C.1101	Struct	0.08-0.14			0.3-0.6		0.06 max	0.06 max	0.35 max	bal Fe				
	C.1120	CH	0.07-0.13			0.3-0.6		0.045 max	0.045 max	0.15-0.35	Mo <=0.15; bal Fe				
	C.1121	CH	0.07-0.13			0.3-0.6		0.035 max	0.035 max	0.15-0.35	Mo <=0.15; bal Fe				
Carbon Steel, Nonresulfurized, 1011															
Argentina															
IAS	IRAM 1011		0.08-0.13			0.30-0.60		0.040 max	0.050 max	0.10 max	bal Fe	360-450	230-290	28-38	110-140 HB
Europe															
EN 10016/4(94)	1.1114	Rod	0.08-0.12	0.10 max	0.15 max	0.30-0.50	0.10 max	0.020 max	0.025 max	0.30 max	Al <=0.01; Mo <=0.05; N <=0.007; Cr+Ni+Cu<=0.30, Cu+Sn<=0.15; bal Fe				
EN 10016/4(94)	C10D2	Rod	0.08-0.12	0.10 max	0.15 max	0.30-0.50	0.10 max	0.020 max	0.025 max	0.30 max	Al <=0.01; Mo <=0.05; N <=0.007; Cr+Ni+Cu<=0.30, Cu+Sn<=0.15; bal Fe				
EN 10084(98)	1.1207	CH, Ann	0.07-0.13			0.30-0.60		0.035 max	0.020-0.040	0.40 max	bal Fe				131 HB
EN 10084(98)	C10R	CH, Ann	0.07-0.13			0.30-0.60		0.035 max	0.020-0.040	0.40 max	bal Fe				131 HB
France															
AFNOR NFA36232(92)	E355C	t<3mm; CR	0.10 max	0.3 max		0.9 max	0.4 max	0.025 max	0.025 max	0.35 max	Al >=0.02; Nb <=0.08; bal Fe	430-530	355	20	
AFNOR NFA36232(92)	H355M	t<3mm; CR Plt	0.10 max	0.3 max		0.9 max	0.4 max	0.025 max	0.025 max	0.35 max	Al >=0.02; Nb <=0.08; bal Fe	430-530	355	20	
Mexico															
NMX-B-301(86)	1011	Bar	0.08-0.13			0.60-0.90		0.040 max	0.050 max		bal Fe				
USA															
	AISI 1011	Bar	0.08-0.13			0.60-0.90		0.040 max	0.050 max	0.050 max	bal Fe				
	UNS G10110		0.08-0.13			0.60-0.90		0.040 max	0.050 max		bal Fe				
ASTM A29/A29M(93)	1011	Bar	0.08-0.13			0.60-0.90		0.040 max	0.050 max	0.050 max	bal Fe				
ASTM A510(96)	1011	Wir rod	0.08-0.13			0.60-0.90		0.040 max	0.050 max		bal Fe				
Carbon Steel, Nonresulfurized, 1012															
Australia															
AS 1442	K1012*	Obs	0.1-0.15			0.3-0.6		0.05	0.05	0.1-0.35	bal Fe				
AS 1442	S1012*	Obs; Bar Bil	0.1-0.15			0.3-0.6		0.05	0.05	0.35	bal Fe				
AS 1443	K1012*	Obs	0.1-0.15			0.3-0.6		0.05	0.05	0.1-0.35	bal Fe				
AS 1443	S1012*	Obs; Bar	0.1-0.15			0.3-0.6		0.05	0.05	0.35	bal Fe				
AS 1446	S1012*	Withdrawn, see AS/NZS 1594(97)	0.1-0.15			0.3-0.6		0.05	0.05	0.35	bal Fe				

UNS numbers and US grades are provided as a means of cross referencing chemically similar alloys. Exchangability is only possible after independent examination of specifications. Tensile properties are minimum or typical as specified. UTS and YS as MPa. El as %. See Appendix for list of abbreviations used in Notes. * indicates obsolete material.

Specification	Designation	Notes	C	Cr	Cu	Mn	Ni	P	S	Si	Other	UTS	YS	El	Hard

Carbon Steel, Nonresulfurized, 1012 (Continued from previous page)

Austria

Specification	Designation	Notes	C	Cr	Cu	Mn	Ni	P	S	Si	Other	UTS	YS	El	Hard
ONORM M3110	RC14	Wir	0.10-0.14			0.5 max		0.040 max	0.040 max	0.3 max	bal Fe				
ONORM M3110	UC14		0.10-0.14			0.5		0.4	0.040-0.400		bal Fe				

Bulgaria

Specification	Designation	Notes	C	Cr	Cu	Mn	Ni	P	S	Si	Other	UTS	YS	El	Hard
BDS 14351	10JuA	Plt HR	0.07-0.14	0.10 max	0.20 max	0.2-0.4	0.15 max	0.020 max	0.025 max	0.07 max	Al 0.02-0.07; Mo <=0.15; bal Fe				
BDS 2592(71)	BSt2kp	Struct	0.09-0.15	0.30 max	0.30 max	0.25-0.50	0.30 max	0.045 max	0.055 max	0.07	bal Fe				
BDS 2592(71)	BSt2ps	Struct	0.09-0.15	0.30 max	0.30 max	0.25-0.50	0.30 max	0.045 max	0.055 max	0.05-0.17	bal Fe				
BDS 2592(71)	BSt2sp	Struct	0.09-0.15	0.30 max	0.30 max	0.25-0.50	0.30 max	0.045 max	0.055 max	0.12-0.35	bal Fe				
BDS 5785(83)	10kp	Struct	0.07-0.14	0.15 max	0.25 max	0.25-0.50	0.25 max	0.035 max	0.040 max	0.07 max	bal Fe				
BDS 5785(83)	10ps	Struct	0.07-0.14	0.15 max	0.25 max	0.35-0.65	0.25 max	0.035 max	0.040 max	0.05-0.17	bal Fe				

Canada

Specification	Designation	Notes	C	Cr	Cu	Mn	Ni	P	S	Si	Other	UTS	YS	El	Hard
CSA 95-1/1	C1012		0.1-0.15			0.3-0.6		0.04	0.05	0.1	bal Fe				
CSA STAN95-1-1	C1012	Bar	0.1-0.15			0.3-0.6		0.04	0.05	0.1	bal Fe				

China

Specification	Designation	Notes	C	Cr	Cu	Mn	Ni	P	S	Si	Other	UTS	YS	El	Hard
GB 700(88)	Q215A	Bar Plt Sh HR 16-40mm diam	0.09-0.15	0.30 max	0.30 max	0.25-0.55	0.30 max	0.045 max	0.050 max	0.30 max	bal Fe	335-450	205	30	
GB 700(88)	Q215B	Bar Plt Sh HR 16-40mm diam	0.09-0.15	0.30 max	0.30 max	0.25-0.55	0.30 max	0.045 max	0.045 max	0.30 max	bal Fe	335-450	205	30	
GB 715(89)	BL2	Bar HR	0.09-0.15		0.25 max	0.25-0.55		0.040 max	0.040 max	0.07 max	bal Fe	335-410	215	33	
GB/T 701(97)	Q215A-b	Wir rod HR; semikilled	0.09-0.15	0.30 max	0.30 max	0.25-0.55	0.30 max	0.045 max	0.050 max	0.17 max	As<=0.080; bal Fe	375	215	27	
GB/T 701(97)	Q215A-F	Wir rod HR; rimmed	0.09-0.15	0.30 max	0.30 max	0.25-0.55	0.30 max	0.045 max	0.050 max	0.07 max	As<=0.080; bal Fe	375	215	27	
GB/T 701(97)	Q215A-Z	Wir rod HR; killed	0.09-0.15	0.30 max	0.30 max	0.25-0.55	0.30 max	0.045 max	0.050 max	0.12 max	As<=0.080; bal Fe	375	215	27	
GB/T 701(97)	Q215B-b	Wir rod HR; semikilled	0.09-0.15	0.30 max	0.30 max	0.25-0.55	0.30 max	0.045 max	0.045 max	0.17 max	As<=0.080; bal Fe	375	215	27	
GB/T 701(97)	Q215B-F	Wir rod HR; rimmed	0.09-0.15	0.30 max	0.30 max	0.25-0.55	0.30 max	0.045 max	0.045 max	0.07 max	As<=0.080; bal Fe	375	215	27	
GB/T 701(97)	Q215B-Z	Wir rod HR; killed	0.09-0.15	0.30 max	0.30 max	0.25-0.55	0.30 max	0.045 max	0.045 max	0.12 max	As<=0.080; bal Fe	375	215	27	
GB/T 8164(93)	Q215A	Strp HR/CR	0.09-0.15	0.30 max	0.30 max	0.25-0.55	0.30 max	0.045 max	0.050 max	0.30 max	bal Fe	335-410	205	30	
GB/T 8164(93)	Q215B	Strp HR	0.09-0.15	0.30 max	0.30 max	0.25-0.55	0.30 max	0.045 max	0.045 max	0.30 max	bal Fe	335-410	205	30	

Europe

Specification	Designation	Notes	C	Cr	Cu	Mn	Ni	P	S	Si	Other	UTS	YS	El	Hard
EN 10016/2(94)	1.0311	Rod	0.10-0.15	0.20 max	0.30 max	0.30-0.60	0.25 max	0.035 max	0.035 max	0.30 max	Al <=0.01; Mo <=0.05; bal Fe				
EN 10016/2(94)	C12D	Rod	0.10-0.15	0.20 max	0.30 max	0.30-0.60	0.25 max	0.035 max	0.035 max	0.30 max	Al <=0.01; Mo <=0.05; bal Fe				
EN 10016/4(94)	1.1124	Rod	0.10-0.14	0.10 max	0.15 max	0.30-0.50	0.10 max	0.020 max	0.025 max	0.30 max	Al <=0.01; Mo <=0.05; N <=0.007; Cr+Ni+Cu<=0.30, Cu+Sn<=0.15; bal Fe				
EN 10016/4(94)	C12D2	Rod	0.10-0.14	0.10 max	0.15 max	0.30-0.50	0.10 max	0.020 max	0.025 max	0.30 max	Al <=0.01; Mo <=0.05; N <=0.007; Cr+Ni+Cu<=0.30, Cu+Sn<=0.15; bal Fe				

Finland

Specification	Designation	Notes	C	Cr	Cu	Mn	Ni	P	S	Si	Other	UTS	YS	El	Hard
SFS 679(86)	CR220	Gen struct	0.15 max			0.9 max		0.05 max	0.05 max	0.5 max	bal Fe				

France

Specification	Designation	Notes	C	Cr	Cu	Mn	Ni	P	S	Si	Other	UTS	YS	El	Hard
AFNOR NFA35551	XC12		0.1-0.16			0.3-0.6		0.035	0.035	0.15-0.35	bal Fe				
AFNOR NFA36102(93)	XC12*	HR; Strp; CR	0.1-0.14	0.08 max		0.3-0.6	0.08 max	0.025 max	0.02 max	0.15-0.35	Al <=0.03; Mo <=0.040; bal Fe				

UNS numbers and US grades are provided as a means of cross referencing chemically similar alloys. Exchangability is only possible after independent examination of specifications. Tensile properties are minimum or typical as specified. UTS and YS as MPa. El as %. See Appendix for list of abbreviations used in Notes. * indicates obsolete material.

Carbon Steel, Nonresulfurized, 1012 (Continued from previous page)

Specification	Designation	Notes	C	Cr	Cu	Mn	Ni	P	S	Si	Other	UTS	YS	El	Hard
Germany															
DIN 2393(94)	S250G1T	Tube	0.15 max			0.20-0.50		0.050 max	0.050 max		N <=0.007; N depends on P; trace Si; bal Fe				
DIN 2393(94)	WNr 1.0028	Tube	0.15 max			0.20-0.50		0.050 max	0.050 max		N <=0.007; N depends on P; trace Si; bal Fe				
DIN 2394(94)	S250G1T	Tube	0.15 max			0.20-0.50		0.050 max	0.050 max		N <=0.007; N depends on P; trace Si; bal Fe				
DIN 2394(94)	WNr 1.0028	Tube	0.15 max			0.20-0.50		0.050 max	0.050 max		N <=0.007; N depends on P; trace Si; bal Fe				
DIN V ENV 10080(95)	B500H	Bar coil welded fabric, for conc reinforce	0.22 max					0.050 max	0.050 max		N <=0.012; bal Fe				
DIN V ENV 10080(95)	WNr 1.0439	Bar, coil weld for conc reinforcement	0.22 max					0.050 max	0.050 max		N <=0.012; bal Fe				
Hungary															
MSZ 113(88)	A34Sz/SzZ	Rivets; CHd; 0-36mm; HR/CD	0.09-0.15			0.3-0.5		0.04 max	0.04 max	0.15-0.35	bal Fe	340-440	220	50L	
MSZ 23(83)	S1H	Sh Strp; mild; 0.71-3mm; rolled	0.12 max			0.5 max		0.04 max	0.05 max	0.2 max	Al <=0.120; bal Fe	270-410		25L	
MSZ 23(83)	S1H	Sh Strp; mild; 0.5-0.7mm; rolled	0.12 max			0.5 max		0.04 max	0.05 max	0.2 max	Al <=0.120; bal Fe	270-410		23L	
MSZ 23(83)	S2H	Sh Strp; mild; 0.5-0.7mm; rolled	0.12 max			0.5 max		0.04 max	0.05 max	0.2 max	Al <=0.120; bal Fe	270-410		23L	
MSZ 23(83)	S2H	Sh Strp; mild; 0.71-3mm; rolled	0.12 max			0.5 max		0.04 max	0.05 max	0.2 max	Al <=0.120; bal Fe	270-410		25L	
MSZ 23(83)	S2P	Sh Strp; mild; 0.5-0.7mm; rolled	0.12 max			0.5 max		0.04 max	0.05 max	0.2 max	Al <=0.120; bal Fe	270-410		23L	
MSZ 23(83)	S2P	Sh Strp; mild; 0.71-3mm; rolled	0.12 max			0.5 max		0.04 max	0.05 max	0.2 max	Al <=0.120; bal Fe	270-410		25L	
MSZ 23(83)	S3F	Sh Strp; mild; 0.71-3mm; rolled	0.12 max			0.5 max		0.04 max	0.05 max	0.2 max	Al <=0.120; bal Fe	270-410		25L	
MSZ 23(83)	S3F	Sh Strp; mild; 0.5-0.7mm; rolled	0.12 max			0.5 max		0.04 max	0.05 max	0.2 max	Al <=0.120; bal Fe	270-410		23L	
MSZ 23(83)	S3H	Sh Strp; mild; 0.5-0.7mm; rolled	0.12 max			0.5 max		0.04 max	0.05 max	0.2 max	Al <=0.120; bal Fe	270-410		23L	
MSZ 23(83)	S3H	Sh Strp; mild; 0.71-3mm; rolled	0.12 max			0.5 max		0.04 max	0.05 max	0.2 max	Al <=0.120; bal Fe	270-410		25L	
MSZ 23(83)	S3P	Sh Strp; mild; 0.71-3mm; rolled	0.12 max			0.5 max		0.04 max	0.05 max	0.2 max	Al <=0.120; bal Fe	270-410		25L	
MSZ 23(83)	S3P	Sh Strp; mild; 0.5-0.7mm; rolled	0.12 max			0.5 max		0.04 max	0.05 max	0.2 max	Al <=0.120; bal Fe	270-410		23L	
MSZ 500(81)	B34B	Gen struct; Non-rimming; FF	0.09-0.15			0.25-0.5		0.045 max	0.05 max	0.12-0.3	bal Fe				
MSZ 500(81)	B34X	Gen struct	0.09-0.15			0.25-0.5		0.045 max	0.05 max	0.07 max	bal Fe				
India															
IS 1570	C14		0.1-0.18			0.4-0.7					bal Fe				
IS 1570/2(79)	10C4	Sh Plt Strp Sect Shp Bar Bil Frg	0.15 max	0.3 max	0.3 max	0.3-0.6	0.4 max	0.055 max	0.055 max	0.6 max	Co <=0.1; Mo <=0.15; Pb <=0.15; W <=0.1; P:S Varies; bal Fe	340-420	187	26	
IS 1570/2(79)	C10	Sh Plt Strp Sect Shp Bar Bil Frg	0.15 max	0.3 max	0.3 max	0.3-0.6	0.4 max	0.055 max	0.055 max	0.6 max	Co <=0.1; Mo <=0.15; Pb <=0.15; W <=0.1; P:S Varies; bal Fe	340-420	187	26	
Japan															
JIS G3507(91)	SWRCH12A	Wir rod Al killed	0.10-0.15			0.30-0.60		0.030 max	0.035 max	0.10 max	Al >=0.02; bal Fe				
JIS G3507(91)	SWRCH12K	Wir rod FF	0.10-0.15			0.30-0.60		0.030 max	0.035 max	0.10-0.35	bal Fe				
JIS G3507(91)	SWRCH12R	Wir rod FU	0.10-0.15			0.30-0.60		0.040 max	0.040 max		bal Fe				
JIS G4051(79)	S12C	Bar Wir rod	0.10-0.15	0.20 max	0.30 max	0.30-0.60	0.20 max	0.030 max	0.035 max	0.15-0.35	Ni+Cr<=0.35; bal Fe				

Specification	Designation	Notes	C	Cr	Cu	Mn	Ni	P	S	Si	Other	UTS	YS	El	Hard
Carbon Steel, Nonresulfurized, 1012 (Continued from previous page)															
Mexico															
NMX-B-301(86)	1012	Bar	0.10-0.15			0.30-0.60		0.040 max	0.050 max		bal Fe				
Pan America															
COPANT 333	1012	Wir rod	0.1-0.15			0.3-0.6		0.04	0.05		bal Fe				
Poland															
PNH74244(79)	G205	Struct; Weld tube	0.09-0.15	0.3 max		0.35-0.6	0.4 max	0.04 max	0.04 max	0.3 max	bal Fe				
PNH84023	14G		0.11-0.15			0.8-1		0.035	0.035	0.1-0.2	bal Fe				
PNH84023	St36K	High-temp const	0.08-0.16	0.3	0.3	0.4	0.3	0.045	0.045	0.15-0.35	bal Fe				
PNH84023/05	12X		0.09-0.15	0.3 max		0.35-0.6	0.4 max	0.04 max	0.04 max	0.05 max	Al <=0.07; bal Fe				
Romania															
STAS 10382	OLT35R		0.16	0.3	0.3	0.4	0.3	0.035	0.035	0.15-0.35	Al 0.02-0.06; bal Fe				
STAS 9382/2	OL87q		0.1-0.17	0.2	0.2	0.25-0.45	0.2	0.04	0.04	0.07	N <=0.007; bal Fe				
Russia															
GOST	M12KP		0.09-0.16			0.3-0.5		0.04 max	0.05 max	0.05 max	bal Fe				
GOST	MSt2		0.09-0.15			0.35-0.5		0.04 max	0.05 max		bal Fe				
GOST	St2		0.09-0.15			0.35-0.5		0.05 max	0.05 max		bal Fe				
GOST	St2S		0.09-0.15	0.3	0.25	0.35-0.5	0.3	0.05	0.05		bal Fe				
GOST 1050(88)	10ps	High-grade struct; Semi-killed	0.07-0.14			0.35-0.65	0.3 max	0.035 max	0.04 max	0.05-0.17	As<=0.08; bal Fe				
GOST 14637	VStTkp		0.1-0.21			0.3-0.6		0.04	0.05	0.05	N 0.008; As 0.08; bal Fe				
GOST 14637	VStTps		0.1-0.21			0.4-0.65		0.04	0.05	0.05-0.15	N 0.008; As 0.08; bal Fe				
GOST 14637	VStTsp		0.1-0.21			0.4-0.65		0.04	0.05	0.15-0.3	N 0.008; As 0.08; bal Fe				
GOST 14637	WStTkp		0.1-0.21			0.3-0.6		0.04	0.05	0.05	N 0.008; As 0.08; bal Fe				
GOST 14637	WStTps		0.1-0.21			0.4-0.65		0.04	0.05	0.05-0.15	N 0.008; As 0.08; bal Fe				
GOST 14637	WStTsp		0.1-0.21			0.4-0.65		0.04	0.05	0.15-0.3	N 0.008; As 0.08; bal Fe				
GOST 380	BSt2Gps		0.09-0.15			0.7-1.1		0.04	0.05	0.15	N 0.008; As 0.08; bal Fe				
GOST 380	BSt2Gps2		0.09-0.15	0.3	0.3	0.7-1.1	0.3	0.04	0.05	0.15-0.3	N 0.008; As 0.08; bal Fe				
GOST 380	BSt2kp		0.09-0.15			0.25-0.5		0.04 max	0.05 max	0.07 max	N <=0.01; As<=0.08; bal Fe				
GOST 380	BSt2ps		0.09-0.15			0.25-0.5		0.04	0.05	0.05-0.17	N 0.008; As 0.08; bal Fe				
GOST 380	BSt2sp		0.09-0.15			0.25-0.5		0.04 max	0.05 max	0.12-0.3	N <=0.008; As 0.08; bal Fe				
GOST 380	KSt2kp		0.09-0.15			0.25-0.5		0.04	0.05	0.07	N 0.008; As 0.08; bal Fe				
GOST 380	KSt2ps		0.09-0.15			0.25-0.5		0.04	0.05	0.05-0.17	N 0.008; As 0.08; bal Fe				
GOST 380	KSt2sp		0.09-0.15			0.25-0.5		0.04 max	0.05 max	0.12-0.3	N <=0.008; As 0.08; bal Fe				
GOST 380	MSt2kp		0.09-0.15			0.25-0.5		0.04	0.05	0.07	N 0.008; As 0.08; bal Fe				
GOST 380	MSt2ps		0.09-0.15			0.25-0.5		0.04	0.05	0.05-0.17	N 0.008; As 0.08; bal Fe				
GOST 380	MSt2sp		0.09-0.15			0.25-0.5		0.04 max	0.05 max	0.12-0.3	N <=0.008; As 0.08; bal Fe				
GOST 380	VSt2kp		0.09-0.15			0.25-0.5		0.04	0.05	0.07	N 0.008; As 0.08; bal Fe				

Specification	Designation	Notes	C	Cr	Cu	Mn	Ni	P	S	Si	Other	UTS	YS	El	Hard

Carbon Steel, Nonresulfurized, 1012 (Continued from previous page)

Russia

Specification	Designation	Notes	C	Cr	Cu	Mn	Ni	P	S	Si	Other	UTS	YS	El	Hard
GOST 380	VSt2ps		0.09-0.15			0.25-0.5		0.04	0.05	0.05-0.17	N 0.008; As 0.08; bal Fe				
GOST 380	VSt2sp		0.09-0.15			0.25-0.5		0.04 max	0.05 max	0.12-0.3	N <=0.008; As 0.08; bal Fe				
GOST 380	WSt2kp		0.09-0.15			0.25-0.5		0.04	0.05	0.07	N 0.008; As 0.08; bal Fe				
GOST 380	WSt2ps		0.09-0.15			0.25-0.5		0.04	0.05	0.05-0.17	N 0.008; As 0.08; bal Fe				
GOST 380	WSt2sp		0.09-0.15			0.25-0.5		0.04 max	0.05 max	0.12-0.3	N <=0.008; As 0.08; bal Fe				
GOST 380(71)	BSt2kp2	Gen struct; Unkilled; Rimming; R; FU	0.09-0.15	0.3 max	0.3 max	0.25-0.5	0.3 max	0.04 max	0.05 max	0.07 max	Al <=0.1; N <=0.008; As<=0.08; bal Fe				
GOST 380(71)	BSt2sp2	Gen struct; Non-rimming; FF	0.09-0.15	0.3 max	0.3 max	0.25-0.5	0.3 max	0.04 max	0.05 max	0.12-0.3	Al <=0.1; N <=0.008; As<=0.08; bal Fe				
GOST 380(71)	WSt2sp2	Gen struct; Non-rimming;K; FF; >100mm	0.15 max	0.3 max	0.3 max	0.25-0.7	0.3 max	0.04 max	0.05 max	0.12-0.3	Al <=0.1; N <=0.008; As<=0.08; bal Fe	330-430	195	29L	
GOST 380(71)	WSt2sp2	Gen struct; Non-rimming;K; FF; 0-20mm	0.15 max	0.3 max	0.3 max	0.25-0.7	0.3 max	0.04 max	0.05 max	0.12-0.3	Al <=0.1; N <=0.008; As<=0.08; bal Fe	330-430	225	32L	

Spain

Specification	Designation	Notes	C	Cr	Cu	Mn	Ni	P	S	Si	Other	UTS	YS	El	Hard
UNE 36032(85)	12KA-DF	Screw	0.10-0.15			0.30-0.60		0.040 max	0.040 max	0.10 max	Al >=0.020; bal Fe				
UNE 36032(85)	F.7514*	see 12KA-DF	0.10-0.15			0.30-0.60		0.040 max	0.040 max	0.10 max	Al >=0.020; bal Fe				
UNE 36081(76)	AE285	Fine grain Struct	0.16	0.25	0.35	0.5-1.4	0.3	0.03	0.03	0.4	Al 0.015; Mo 0.1; Nb 0.015-0.05; V 0.02-0.1; bal Fe				
UNE 36086(75)	AP02	Unalloyed mild; Sh CR	0.12 max			0.5 max		0.04 max	0.04 max	0.60 max	Mo <=0.15; bal Fe				
UNE 36086(75)	AP11	Unalloyed mild; Sh CR	0.12 max			0.5 max		0.04 max	0.04 max	0.60 max	Mo <=0.15; bal Fe				

Sweden

Specification	Designation	Notes	C	Cr	Cu	Mn	Ni	P	S	Si	Other	UTS	YS	El	Hard
SS 141332	1332		0.1-0.14	0.25	0.4	0.5-0.8		0.04	0.04	0.15-0.4	N 0.009-0.015; bal Fe				

Turkey

Specification	Designation	Notes	C	Cr	Cu	Mn	Ni	P	S	Si	Other	UTS	YS	El	Hard
TS 2348(76)	C12-1 1.0012	Wir rod	0.12						0.05		N 0.01; bal Fe				

UK

Specification	Designation	Notes	C	Cr	Cu	Mn	Ni	P	S	Si	Other	UTS	YS	El	Hard
BS 1449/1(91)	12CS	Plt Sh Strp; SP; t <=16mm	0.10-0.15			0.40-0.60		0.050 max	0.050 max		bal Fe	310-410	170	26-28	
BS 1449/1(91)	12HS	Sh Strp	0.10-0.15			0.40-0.60		0.050 max	0.050 max		bal Fe	310	170	23-25	
BS 1501(86)	141-360	Plate, rimmed, for enamelling; 3<t<=16mm	0.16 max	0.25 max	0.30 max	0.50 max	0.30 max	0.050 max	0.050 max		Mo <=0.10; bal Fe	360-480	205	25	
BS 970/1(83)	050A12	Blm Bil Slab Bar Rod Frg	0.10-0.15			0.40-0.60		0.050 max	0.050 max	0.4	bal Fe				
BS 970/1(83)	060A12	Blm Bil Slab Bar Rod Frg	0.10-0.15			0.50-0.70		0.050 max	0.050 max	0.4	bal Fe				

USA

Specification	Designation	Notes	C	Cr	Cu	Mn	Ni	P	S	Si	Other	UTS	YS	El	Hard
	AISI 1012	Wir rod	0.10-0.15			0.30-0.60		0.040 max	0.050 max		bal Fe				
	AISI 1012	Bar	0.10-0.15			0.30-0.60		0.040 max	0.050 max		bal Fe				
	AISI 1012	Sh Strp Plt	0.10-0.15			0.30-0.60		0.030 max	0.035 max		bal Fe				
	AISI 1012	Struct shp	0.10-0.15			0.30-0.60		0.030 max	0.035 max		bal Fe				
	AISI M1012	Bar, merchant qual	0.09-0.16			0.25-0.60		0.04 max	0.05 max		bal Fe				
	UNS G10120		0.10-0.15			0.30-0.60		0.040 max	0.050 max		bal Fe				
ASTM A29/A29M(93)	1012	Bar	0.10-0.15			0.30-0.60		0.040 max	0.050 max		bal Fe				
ASTM A29/A29M(93)	M1012	Bar, merchant qual	0.09-0.16			0.25-0.6		0.040 max	0.050 max		bal Fe				

Specification	Designation	Notes	C	Cr	Cu	Mn	Ni	P	S	Si	Other	UTS	YS	El	Hard

Carbon Steel, Nonresulfurized, 1012 (Continued from previous page)

USA

Specification	Designation	Notes	C	Cr	Cu	Mn	Ni	P	S	Si	Other	UTS	YS	El	Hard
ASTM A510(96)	1012	Wir rod	0.10-0.15			0.30-0.60		0.040 max	0.050 max		bal Fe				
ASTM A512(96)	1012	Buttweld mech tub, CD	0.10-0.15			0.30-0.60		0.040 max	0.045 max		bal Fe				
ASTM A513(97a)	1012	ERW Mech tub	0.10-0.15			0.30-0.60		0.035 max	0.035 max		bal Fe				
ASTM A519(96)	1012	Smls mech tub	0.10-0.15			0.30-0.60		0.040 max	0.050 max		bal Fe				
ASTM A575(96)	M1012	Merchant qual bar	0.09-0.16			0.25-0.60		0.04 max	0.05 max		Mn can vary with C; bal Fe				
ASTM A576(95)	1012	Special qual HW bar	0.10-0.15			0.30-0.60		0.040 max	0.050 max		Si Cu Pb B Bi Ca Se Te if spec'd; bal Fe				
ASTM A611(97)	A611(C,D,2)	SS Sh, CR	0.15 max		0.20 min	0.60 max		0.020 max	0.035 max		Cu if spec'd; bal Fe	330-360	230-275	20-22	
ASTM A635/A635M(98)	1012	Sh Strp, Coil, HR	0.10-0.15		0.20 min	0.30-0.60		0.030 max	0.035 max		bal Fe				
ASTM A830/A830M(98)	1012	Plt	0.10-0.15			0.30-0.60		0.035 max	0.04 max		bal Fe				
MIL-R-8814	1012*	Obs; Rvt	0.1-0.15			0.3-0.6		0.04	0.05		bal Fe				
MIL-S-11310E(76)	1012*	Bar HR cold shape cold ext; Obs for new design	0.1-0.15	0.07	0.2	0.3-0.6	0.15	0.04 max	0.05 max	0.2	Mo 0.05; bal Fe				
MIL-S-11310E(76)	CS1012*	Bar HR cold shape cold ext; Obs for new design									bal Fe				
SAE J1397(92)	1012	Bar CD, est mech prop	0.1-0.15			0.3-0.6		0.04 max	0.05 max		bal Fe	370	310	19	105 HB
SAE J1397(92)	1012	Bar HR, est mech prop	0.1-0.15			0.3-0.6		0.04 max	0.05 max		bal Fe	330	180	28	95 HB
SAE J403(95)	1012	Struct Shps Plt Strp Sh Weld Tub	0.10-0.15			0.30-0.60		0.030 max	0.035 max		bal Fe				
SAE J403(95)	1012	Bar Wir rod Smls Tub HR CF	0.10-0.15			0.30-0.60		0.030 max	0.050 max		bal Fe				
SAE J403(95)	M1012	Merchant qual	0.10-0.15			0.30-0.60		0.04 max	0.05 max		bal Fe				

Yugoslavia

Specification	Designation	Notes	C	Cr	Cu	Mn	Ni	P	S	Si	Other	UTS	YS	El	Hard
	C.0271	Struct	0.15 max			1.60 max		0.06 max	0.05 max	0.60 max	N <=0.07; bal Fe				
	C.1209	Struct	0.1-0.15			0.3-0.6		0.06 max	0.06 max	0.60 max	bal Fe				

Carbon Steel, Nonresulfurized, 1013

Australia

Specification	Designation	Notes	C	Cr	Cu	Mn	Ni	P	S	Si	Other	UTS	YS	El	Hard
AS/NZS 1594(97)	HA1016	HR flat products	0.12-0.18	0.30 max	0.35 max	0.60-0.90	0.35 max	0.040 max	0.035 max	0.03 max	Al <=0.100; Mo <=0.10; Ti <=0.040; (Cu+Ni+Cr+Mo) 1 max; bal Fe				

Mexico

Specification	Designation	Notes	C	Cr	Cu	Mn	Ni	P	S	Si	Other	UTS	YS	El	Hard
NMX-B-301(86)	1013	Bar	0.11-0.16			0.50-0.80		0.040 max	0.050 max		bal Fe				

UK

Specification	Designation	Notes	C	Cr	Cu	Mn	Ni	P	S	Si	Other	UTS	YS	El	Hard
BS 1717(83)	CEWC2	Tub	0.16 max			0.70 max		0.050 max	0.050 max		W <=0.1; bal Fe				
BS 1717(83)	ERWC2	Tub	0.16 max			0.70 max		0.050 max	0.050 max		W <=0.1; bal Fe				
BS 3606(92)	320	Heat exch Tub; t<=3.2mm	0.16 max			0.30-0.70	0.40 max	0.040 max	0.040 max		Mo <=0.15; bal Fe	320-460	195	25	

USA

Specification	Designation	Notes	C	Cr	Cu	Mn	Ni	P	S	Si	Other	UTS	YS	El	Hard
	AISI 1013	Wir rod	0.11-0.16			0.50-0.80		0.040 max	0.050 max		bal Fe				
	AISI 1013	Bar	0.11-0.16			0.50-0.80		0.040 max	0.050 max	0.050 max	bal Fe				
	UNS G10130		0.11-0.16			0.50-0.80		0.040 max	0.050 max		bal Fe				
ASTM A29/A29M(93)	1013	Bar	0.11-0.16			0.50-0.80		0.040 max	0.050 max	0.050 max	bal Fe				

UNS numbers and US grades are provided as a means of cross referencing chemically similar alloys. Exchangability is only possible after independent examination of specifications. Tensile properties are minimum or typical as specified. UTS and YS as MPa. El as %. See Appendix for list of abbreviations used in Notes. * indicates obsolete material.

Specification	Designation	Notes	C	Cr	Cu	Mn	Ni	P	S	Si	Other	UTS	YS	El	Hard

Carbon Steel, Nonresulfurized, 1013 (Continued from previous page)

USA

Specification	Designation	Notes	C	Cr	Cu	Mn	Ni	P	S	Si	Other	UTS	YS	El	Hard
ASTM A510(96)	1013	Wir rod	0.11-0.16			0.50-0.80		0.040 max	0.050 max		bal Fe				
SAE J1249(95)	1013	Former SAE std valid for Wir rod	0.11-0.16			0.50-0.80		0.040 max	0.050 max		bal Fe				

Carbon Steel, Nonresulfurized, 1015

Argentina

Specification	Designation	Notes	C	Cr	Cu	Mn	Ni	P	S	Si	Other	UTS	YS	El	Hard
IAS	IRAM 1015		0.13-0.18			0.30-0.60		0.040 max	0.050 max	0.15-0.30	bal Fe	400-500	260-320	24-34	120-150 HB

Australia

Specification	Designation	Notes	C	Cr	Cu	Mn	Ni	P	S	Si	Other	UTS	YS	El	Hard
AS 1442	S1015*	Obs; Bar Bil	0.13-0.18			0.3-0.6		0.05	0.05	0.35	bal Fe				
AS 1443	S1015*	Obs; Bar	0.13-0.18			0.3-0.6		0.05	0.05	0.35	bal Fe				
AS 1446	S1015*	Withdrawn, see AS/NZS 1594(97)	0.13-0.18			0.3-0.6		0.05	0.05	0.35	bal Fe				

Austria

Specification	Designation	Notes	C	Cr	Cu	Mn	Ni	P	S	Si	Other	UTS	YS	El	Hard
ONORM M3161	C16		0.13-0.18			0.3-0.5		0.040 max	0.040 max	0.15-0.30	bal Fe				

Belgium

Specification	Designation	Notes	C	Cr	Cu	Mn	Ni	P	S	Si	Other	UTS	YS	El	Hard
NBN 253-03	C16-2		0.12-0.18			0.3-0.6		0.035	0.035	0.1-0.4	bal Fe				
NBN 253-03	C16-3		0.12-0.18			0.3-0.6		0.035	0.02-0.035	0.1-0.4	bal Fe				

Bulgaria

Specification	Designation	Notes	C	Cr	Cu	Mn	Ni	P	S	Si	Other	UTS	YS	El	Hard
BDS 2592(71)	ASt3	Struct									bal Fe				
BDS 5785(83)	15kp	Struct	0.12-0.19	0.25 max	0.25 max	0.25-0.50	0.25 max	0.035 max	0.040 max	0.07	bal Fe				

China

Specification	Designation	Notes	C	Cr	Cu	Mn	Ni	P	S	Si	Other	UTS	YS	El	Hard
GB 13237(91)	15	Sh Strp CR HT	0.12-0.19	0.25 max	0.25 max	0.35-0.65	0.25 max	0.035 max	0.035 max	0.17-0.37	bal Fe	335-470		25	
GB 13237(91)	15F	Sh Strp CR HT	0.12-0.19	0.25 max	0.25 max	0.25-0.50	0.25 max	0.035 max	0.035 max	0.07 max	bal Fe	315-450		27	
GB 13796(92)	15	Bar Wir CD Ann	0.12-0.19	0.25 max	0.25 max	0.35-0.65	0.25 max	0.035 max	0.035 max	0.17-0.37	bal Fe	340-470			
GB 3078(94)	15F	Bar CD Norm 25mm diam	0.12-0.18			0.50 max		0.035 max	0.035 max	0.07 max	bal Fe	355	205	29	
GB 3275(91)	15	Plt Strp HR HT	0.12-0.19	0.25 max	0.25 max	0.35-0.65	0.25 max	0.035 max	0.035 max	0.17-0.37	bal Fe	315-440		26	
GB 3275(91)	15F	Plt Strp HR HT	0.12-0.19	0.25 max	0.25 max	0.25-0.50	0.25 max	0.035 max	0.035 max	0.07 max	bal Fe	315-440		26	
GB 3522(83)	15	Strp CR Ann	0.12-0.19	0.25 max	0.25 max	0.35-0.65	0.25 max	0.035 max	0.035 max	0.17-0.37	bal Fe	315-490		22	
GB 5953(86)	ML 15	Wir CD	0.12-0.19	0.20 max	0.20 max	0.60 max		0.035 max	0.035 max	0.20 max	bal Fe	440-635			
GB 6478(86)	ML 15	Bar HR Norm	0.12-0.19	0.20 max	0.20 max	0.20-0.50	0.20 max	0.035 max	0.035 max	0.07 max	bal Fe	375	225	27	
GB 699(88)	15	Bar HR Norm 25mm diam	0.12-0.19	0.25 max	0.25 max	0.35-0.65	0.25 max	0.035 max	0.035 max	0.17-0.37	bal Fe	375	225	27	
GB 699(88)	15F	Bar HR Norm 25mm diam	0.12-0.19	0.25 max	0.25 max	0.25-0.50	0.25 max	0.035 max	0.035 max	0.07 max	bal Fe	355	205	29	
GB 700	BJ3*	Obs	0.1-0.2			0.3-0.6		0.045	0.045	0.07-0.3	bal Fe				
GB 700	BJ3F*	Obs	0.1-0.2			0.3-0.6		0.045	0.045	0.07-0.3	bal Fe				
GB 700	BS4*	Obs	0.12-0.22			0.35-0.55		0.085	0.065	0.07-0.3	bal Fe				
GB 700	BS4F*	Obs	0.12-0.22			0.35-0.55		0.085	0.065	0.07-0.3	bal Fe				
GB 700	CJ3*	Obs	0.1-0.2			0.3-0.6		0.045	0.045	0.07-0.3	bal Fe				
GB 700	CJ3F*	Obs	0.1-0.2			0.3-0.6		0.045	0.045	0.07-0.3	bal Fe				
GB 710(91)	15	Sh Strp HR HT	0.12-0.19	0.25 max	0.25 max	0.35-0.65	0.25 max	0.035 max	0.035 max	0.17-0.37	bal Fe	335-470		24	
GB 710(91)	15F	Sh Strp HR HT	0.12-0.19	0.25 max	0.25 max	0.25-0.50	0.25 max	0.035 max	0.035 max	0.07 max	bal Fe	315-450		24	

UNS numbers and US grades are provided as a means of cross referencing chemically similar alloys. Exchangability is only possible after independent examination of specifications. Tensile properties are minimum or typical as specified. UTS and YS as MPa. El as %. See Appendix for list of abbreviations used in Notes. * indicates obsolete material.

Specification	Designation	Notes	C	Cr	Cu	Mn	Ni	P	S	Si	Other	UTS	YS	El	Hard

Carbon Steel, Nonresulfurized, 1015 (Continued from previous page)

China

Specification	Designation	Notes	C	Cr	Cu	Mn	Ni	P	S	Si	Other	UTS	YS	El	Hard
GB/T 13795(92)	15	Strp CR Ann	0.12-0.19	0.25 max	0.25 max	0.35-0.65	0.25 max	0.035 max	0.035 max	0.17-0.37	bal Fe	400-700		15	
GB/T 3078(94)	15	Bar CD Ann	0.12-0.19	0.25 max	0.25 max	0.35-0.65	0.25 max	0.035 max	0.035 max	0.17-0.37	bal Fe	345		23	
GB/T 8164(93)	15	Strp HR/CR	0.12-0.19	0.25 max	0.30 max	0.35-0.65	0.30 max	0.035 max	0.035 max	0.17-0.37	bal Fe	370		30	

Europe

Specification	Designation	Notes	C	Cr	Cu	Mn	Ni	P	S	Si	Other	UTS	YS	El	Hard
EN 10016/2(94)	1.0413	Rod	0.12-0.17	0.20 max	0.30 max	0.30-0.60	0.25 max	0.035 max	0.035 max	0.30 max	Al <=0.01; Mo <=0.05; bal Fe				
EN 10016/2(94)	C15D	Rod	0.12-0.17	0.20 max	0.30 max	0.30-0.60	0.25 max	0.035 max	0.035 max	0.30 max	Al <=0.01; Mo <=0.05; bal Fe				
EN 10016/4(94)	1.1126	Rod	0.13-0.17	0.10 max	0.15 max	0.30-0.50	0.10 max	0.020 max	0.025 max	0.30 max	Al <=0.01; Mo <=0.05; N <=0.007; Cr+Ni+Cu<=0.30, Cu+Sn<=0.15; bal Fe				
EN 10016/4(94)	C15D2	Rod	0.13-0.17	0.10 max	0.15 max	0.30-0.50	0.10 max	0.020 max	0.025 max	0.30 max	Al <=0.01; Mo <=0.05; N <=0.007; Cr+Ni+Cu<=0.30, Cu+Sn<=0.15; bal Fe				
EN 10025(90)	Fe360B	Obs EU desig; struct HR; BS t<=16mm	0.17 max					0.045 max	0.045 max		N <=0.009; N max na if Al>=0.020; bal Fe	340-510	235	17-26	
EN 10025(90)	Fe360B	Obs EU desig; struct HR; BS 16-25mm	0.20 max					0.045 max	0.045 max		N <=0.009; N max na if Al>=0.020; bal Fe	340-470	225	26	
EN 10084(98)	1.1141	CH, Ann	0.12-0.18			0.30-0.60		0.035 max	0.035 max	0.40 max	bal Fe				143 HB
EN 10084(98)	C15E	CH, Ann	0.12-0.18			0.30-0.60		0.035 max	0.035 max	0.40 max	bal Fe				143 HB

Finland

Specification	Designation	Notes	C	Cr	Cu	Mn	Ni	P	S	Si	Other	UTS	YS	El	Hard
SFS 505	C15	Bar Frg Q/A, 890C 11mm diam	0.12-0.18			0.6-0.9		0.035	0.02-0.035	0.1-0.4	bal Fe	880	590	7	
SFS 505	C15	Bar Frg Q/A, 780C 11mm diam	0.12-0.18			0.6-0.9		0.035	0.02-0.035	0.1-0.4	bal Fe	690	390	9	
SFS 506(76)	C15	CH	0.12-0.18			0.6-0.9		0.035 max	0.02-0.035	0.1-0.4	bal Fe				
SFS 600(81)	CR1	Gen struct	0.15 max			0.5 max		0.07 max	0.06 max	0.6 max	bal Fe				

France

Specification	Designation	Notes	C	Cr	Cu	Mn	Ni	P	S	Si	Other	UTS	YS	El	Hard
AFNOR	XC15		0.12-0.18			0.3-0.7		0.04	0.035	0.35	bal Fe				
AFNOR NFA33101(82)	AF37C12	t<=450mm; Norm	0.08-0.15	0.3 max	0.3 max	0.3-0.6	0.4 max	0.04 max	0.04 max	0.3 max	bal Fe	360-450	235	30	
AFNOR NFA35552	XC18	Bar, Norm, 16mm diam									bal Fe				
AFNOR NFA35590	4312	Bar Frg	0.17 max			0.3-0.6		0.04 max		0.15-0.4	bal Fe				
AFNOR NFA36102(93)	C12RR	HR; Strp; CR	0.1-0.14	0.08 max		0.3-0.6	0.08 max	0.025 max	0.02 max	0.15-0.35	Al <=0.03; Mo <=0.040; bal Fe				
AFNOR NFA36102(93)	C18RR	HR; Strp; CR	0.17-0.24	0.08 max		0.4-0.7	0.08 max	0.025 max	0.02 max	0.15-0.35	Al <=0.03; Mo <=0.040; bal Fe				
AFNOR NFA36102(93)	XC18*	HR; Strp; CR	0.17-0.24	0.08 max		0.4-0.7	0.08 max	0.025 max	0.02 max	0.15-0.35	Al <=0.03; Mo <=0.040; bal Fe				

Germany

Specification	Designation	Notes	C	Cr	Cu	Mn	Ni	P	S	Si	Other	UTS	YS	El	Hard
DIN	Ck16Al		0.13-0.18		0.15 max	0.40-0.60	0.10 max	0.025 max	0.025 max	0.15 max	Al 0.03-0.08; N <=0.007; N depends on P; bal Fe				
DIN	WNr 1.1135		0.13-0.18	0.15 max		0.40-0.60	0.10 max	0.025 max	0.025 max	0.15 max	Al 0.03-0.08; N <=0.007; N depends on P; bal Fe				
DIN 1626(84)	P235T1	Weld tube	0.17 max					0.040 max	0.040 max		N <=0.009; N depends on P; bal Fe				
DIN 1626(84)	WNr 1.0254	Weld tube	0.17 max					0.040 max	0.040 max		N <=0.009; N depends on P; bal Fe				
DIN 1629(84)	P235T1	Smls tube	0.17 max					0.040 max	0.040 max		N <=0.009; N depends on P; bal Fe				
DIN 1629(84)	WNr 1.0254	Smls tube	0.17 max					0.040 max	0.040 max		N <=0.009; N depends on P; bal Fe				
DIN 1652(90)	C15	Bright	0.12-0.18			0.30-0.60		0.045 max	0.045 max	0.40 max	bal Fe	590	355	14	30

UNS numbers and US grades are provided as a means of cross referencing chemically similar alloys. Exchangability is only possible after independent examination of specifications. Tensile properties are minimum or typical as specified. UTS and YS as MPa. El as %. See Appendix for list of abbreviations used in Notes. * indicates obsolete material.

Specification	Designation	Notes	C	Cr	Cu	Mn	Ni	P	S	Si	Other	UTS	YS	El	Hard

Carbon Steel, Nonresulfurized, 1015 (Continued from previous page)

Germany

Specification	Designation	Notes	C	Cr	Cu	Mn	Ni	P	S	Si	Other	UTS	YS	El	Hard
DIN 1652(90)	Ck15	CH, 30mm diam	0.12-0.18			0.30-0.60		0.035 max	0.035 max	0.40 max	bal Fe	590-780	355	14	
DIN 1652(90)	Cm15		0.12-0.18			0.30-0.60		0.035 max	0.02-0.040	0.40 max	bal Fe				
DIN 1652(90)	WNr 1.0401	CH, 30mm	0.12-0.18			0.30-0.60		0.045 max	0.045 max	0.40 max	bal Fe	590	355	14	30
DIN 1652(90)	WNr 1.1140	CH, 30mm diam	0.12-0.18			0.30-0.60		0.035 max	0.020-0.040	0.40 max	bal Fe	590-780	355	14	30
DIN 1652(90)	WNr 1.1141	CH, 30mm diam	0.12-0.18			0.30-0.60		0.035 max	0.035 max	0.40 max	bal Fe	590-780	355	14	
DIN 1654(89)	Cq15	CHd, cold ext, CH	0.12-0.18			0.25-0.50		0.035 max	0.035 max	0.40 max	bal Fe				
DIN 1654(89)	WNr 1.1132	CHd, cold ext, CH	0.12-0.18			0.25-0.50		0.035 max	0.035 max	0.40 max	bal Fe				
DIN 17119(84)	UkSt37-2	Weld CF sq rect tub	0.17 max			0.20-0.50		0.050 max	0.050 max		N <=0.007; N depends on P;C 0.20 max if >=16 mm thk; trace Si; bal Fe				
DIN 17119(84)	WNr 1.0124	Weld CF sq rect tub	0.17 max			0.20-0.50		0.050 max	0.050 max		N <=0.007; N depends on P; C<=0.20 if >=16mm thk; trace Si; bal Fe				
DIN 17210(86)	C16E	CH	0.12-0.18			0.60-0.90		0.035 max	0.035 max	0.40 max	bal Fe				
DIN 17210(86)	WNr 1.1148	CH	0.12-0.18			0.60-0.90		0.035 max	0.035 max	0.40 max	bal Fe				
DIN 2393(94)	RSt34-2	Tube	0.15 max			0.20-0.50		0.050 max	0.050 max	0.03-0.30	N <=0.007; N depends on P; bal Fe				
DIN 2393(94)	S250GT	Tube	0.15 max			0.20-0.50		0.050 max	0.050 max	0.30 max	N <=0.007; N depends on P; bal Fe				
DIN 2393(94)	WNr 1.0032	Tube	0.15 max			0.20-0.50		0.050 max	0.050 max	0.30 max	N <=0.007; N depends on P; bal Fe				
DIN 2393(94)	WNr 1.0034	Tube	0.15 max			0.20-0.50		0.050 max	0.050 max	0.03-0.30	N <=0.007; N depends on P; bal Fe				
DIN 2394(94)	RSt34-2	Tube	0.15 max			0.20-0.50		0.050 max	0.050 max	0.03-0.30	N <=0.007; N depends on P; bal Fe				
DIN 2394(94)	S250GT	Tube	0.15 max			0.20-0.50		0.050 max	0.050 max	0.30 max	N <=0.007; N depends on P; bal Fe				
DIN 2394(94)	WNr 1.0032	Tube	0.15 max			0.20-0.50		0.050 max	0.050 max	0.30 max	N <=0.007; N depends on P; bal Fe				
DIN 2394(94)	WNr 1.0034	Tube	0.15 max			0.20-0.50		0.050 max	0.050 max	0.03-0.30	N <=0.007; N depends on P; bal Fe				
DIN E EN 10216(95)	P235T1	Smls tube for Press	0.17 max					0.040 max	0.040 max		N <=0.009; N depends on P; bal Fe				
DIN E EN 10216(95)	WNr 1.0254	Smls tube for Press	0.17 max					0.040 max	0.040 max		N <=0.009; N depends on P; bal Fe				
DIN E EN 10217(95)	P235T1	Weld tube for Press	0.17 max					0.040 max	0.040 max		N <=0.009; N depends on P; bal Fe				
DIN E EN 10217(95)	WNr 1.0254	Weld tube for Press	0.17 max					0.040 max	0.040 max		N <=0.009; N depends on P; bal Fe				
DIN(Aviation Hdbk)	WNr 1.1144		0.12-0.18			0.30-0.60		0.035 max	0.035 max	0.15-0.35	bal Fe				

Hungary

Specification	Designation	Notes	C	Cr	Cu	Mn	Ni	P	S	Si	Other	UTS	YS	El	Hard
MSZ 186/2(82)	A34	Longitudinal weld tub; gen struct; 1.4-2.9mm; weld	0.15 max			2 max	0.3 max	0.045 max	0.05 max	0.6 max	bal Fe	330-490	220	26L	
MSZ 186/2(82)	A34	Longitudinal weld tub; gen struct; 3-12.5mm; weld	0.15 max			2 max	0.3 max	0.045 max	0.05 max	0.6 max	bal Fe	330-460	215	26L	
MSZ 186/2(82)	A38	Longitudinal weld tub; gen struct; 1.4-2.9mm; weld	0.22 max			2 max	0.3 max	0.045 max	0.05 max	0.6 max	0-3mm MSZ23; 3mm -MSZ500; bal Fe	370-540	240	23L	
MSZ 186/2(82)	A38	Longitudinal weld tub; gen struct; 3-12.5mm; weld	0.22 max			2 max	0.3 max	0.045 max	0.05 max	0.6 max	0-3mm MSZ23; 3mm -MSZ500; bal Fe	370-510	235	23L	
MSZ 23(83)	H2H	Sh Strp; mild; 0.5-3mm; rolled	0.15 max			0.6 max		0.05 max	0.05 max	0.6 max	Al <=0.120; bal Fe	0-490			
MSZ 23(83)	H3F	Sh Strp; mild; 0.5-3mm; rolled	0.15 max			0.6 max		0.05 max	0.05 max	0.6 max	Al <=0.120; bal Fe	0-490			

UNS numbers and US grades are provided as a means of cross referencing chemically similar alloys. Exchangability is only possible after independent examination of specifications. Tensile properties are minimum or typical as specified. UTS and YS as MPa. El as %. See Appendix for list of abbreviations used in Notes. * indicates obsolete material.

Specification	Designation	Notes	C	Cr	Cu	Mn	Ni	P	S	Si	Other	UTS	YS	El	Hard

Carbon Steel, Nonresulfurized, 1015 (Continued from previous page)

Hungary

Specification	Designation	Notes	C	Cr	Cu	Mn	Ni	P	S	Si	Other	UTS	YS	El	Hard
MSZ 23(83)	H3H	Sh Strp; mild; 0.5-3mm; rolled	0.15 max			0.6 max		0.05 max	0.05 max	0.6 max	Al <=0.120; bal Fe	0-490			
MSZ 2978/2(82)	A38	Longitudinal weld/CF tub; gen struct; 0.5-3mm; bright-drawn, hard	0.22 max			2 max	0.3 max	0.045 max	0.05 max	0.6 max	bal Fe	440	350	6L	
MSZ 2978/2(82)	A38	Longitudinal weld/CF tub; gen struct; 0.5-3mm; bright-drawn, soft	0.22 max			2 max	0.3 max	0.045 max	0.05 max	0.6 max	bal Fe	370	260	10L	
MSZ 31	C15EK		0.12-0.18	0.25	0.3	0.3-0.6	0.3	0.035	0.05	0.17-0.37	bal Fe				
MSZ 31(85)	C15	CH; 0-11mm; blank carburized	0.12-0.18			0.3-0.6		0.035 max	0.035 max	0.4 max	bal Fe	530-730	400	13L	
MSZ 31(85)	C15	CH; 30mm; blank carburized	0.12-0.18			0.3-0.6		0.035 max	0.035 max	0.4 max	bal Fe	500-700	330	14	
MSZ 31(85)	C15E	CH; 0-11mm; blank carburized	0.12-0.18			0.3-0.6		0.035 max	0.02-0.035	0.4 max	bal Fe	530-730	400	13L	
MSZ 31(85)	C15E	CH; 30-30mm; blank carburized	0.12-0.18			0.3-0.6		0.035 max	0.02-0.035	0.4 max	bal Fe	500-700	330	14L	
MSZ 500(81)	A38	Gen struct; 40.1-100mm; HR/Frg	0.22 max			2 max	0.3 max	0.045 max	0.05 max	0.6 max	bal Fe	370-480	215	26L	
MSZ 500(81)	A38	Gen struct; 0-16mm; HR/Frg	0.22 max			2 max	0.3 max	0.045 max	0.05 max	0.6 max	bal Fe	370-480	235	26L	
MSZ 500(89)	Fe235B	Gen struct; Base; BS; Size; 16.1-25mm; HR/Frg	0.2 max			2 max		0.045 max	0.045 max	0.6 max	Al 0.02-0.1; N <=0.009; 0-16mm C<=0.17; 16-25mm C<=0.2; bal Fe	340-470	225	26L; 24T	
MSZ 500(89)	Fe235B	Gen struct; Base; BS; Size; 0-1mm; HR/Frg	0.2 max			2 max		0.045 max	0.045 max	0.6 max	Al 0.02-0.1; N <=0.009; 0-16mm C<=0.17; 16-25mm C<=0.2; bal Fe	360-510	235	17L; 15T	
MSZ 5736(88)	D15	Wir rod; for CD	0.12-0.18		0.25 max	0.3-0.6	0.3 max	0.04 max	0.04 max	0.1-0.35	bal Fe				
MSZ 5736(88)	D15K	Wir rod; for CD	0.12-0.18		0.25 max	0.3-0.6	0.3 max	0.03 max	0.03 max	0.1-0.35	bal Fe				
MSZ 6251(87)	C15Z	CF; CH; 0-36mm; HR; HR/soft ann; drawn/soft ann; drawn/bright ann; soft ann/ground	0.12-0.18			0.3-0.6		0.035 max	0.035 max	0.4 max	bal Fe	0-480			
MSZ 6251(87)	C15Z	CF; CH; 0-36mm; drawn, half-hard	0.12-0.18			0.3-0.6		0.035 max	0.035 max	0.4 max	bal Fe	0-510			

India

Specification	Designation	Notes	C	Cr	Cu	Mn	Ni	P	S	Si	Other	UTS	YS	El	Hard
IS 1570	C14		0.1-0.18			0.4-0.7					bal Fe				
IS 2100	Grade 2		0.1-0.2			0.6-0.9		0.05	0.05		bal Fe				

International

Specification	Designation	Notes	C	Cr	Cu	Mn	Ni	P	S	Si	Other	UTS	YS	El	Hard
ISO 683-11(87)	C15E4	CH, 16mm test	0.12-0.18		0.3 max	0.30-0.60		0.035 max	0.035 max	0.15-0.40		500-850	300		
ISO 683-11(87)	C15M2	CH, 16mm test	0.12-0.18		0.3 max	0.30-0.60		0.035 max	0.020-0.040	0.15-0.40	Al <=0.1; Co <=0.1; bal Fe	500-850	300		
ISO 683-18(96)	C10	Bar Wir, CD, t<=5mm	0.07-0.13			0.30-0.60		0.035 max	0.035 max	0.15-0.40	bal Fe	500	400	7	131 HB
ISO 683-18(96)	C15E4	Bar Wir, CD, t<=5mm	0.12-0.18			0.30-0.60		0.035 max	0.035 max	0.15-0.40	bal Fe	540	430	6	143 HB
ISO 683-18(96)	C15M2	Bar Wir, CD, t<=5mm	0.12-0.18			0.30-0.60		0.035 max	0.020-0.040	0.15-0.40	bal Fe	540	430	6	143 HB
ISO R683-11(70)	2*	Bar, Carburized Hardness	0.12-0.18			0.3-0.6		0.035	0.035	0.15-0.4	bal Fe				
ISO R683-11(70)	2a*	Bar, Carburized Hardness	0.12-0.18			0.3-0.6		0.035	0.02-0.035	0.15-0.4	bal Fe				

Italy

Specification	Designation	Notes	C	Cr	Cu	Mn	Ni	P	S	Si	Other	UTS	YS	El	Hard
UNI 5331(64)	C16	CH	0.12-0.18			0.3-0.7		0.035 max	0.035 max	0.35 max	bal Fe				
UNI 5598(71)	1CD15	Wir rod	0.12-0.19			0.3-0.6		0.05 max	0.05 max	0.35 max	bal Fe				

Carbon Steel, Nonresulfurized, 1015 (Continued from previous page)

Specification	Designation	Notes	C	Cr	Cu	Mn	Ni	P	S	Si	Other	UTS	YS	El	Hard
Italy															
UNI 5598(71)	3CD15	Wir rod	0.13-0.18			0.3-0.6		0.035 max	0.035 max	0.35 max	N <=0.012; bal Fe				
UNI 5949(67)	C15	Smls tube w/low-temp impact test	0.15 max			1 max		0.035 max	0.035 max	0.15-0.35	bal Fe				
UNI 7065(72)	C16	CH	0.12-0.18			0.3-0.7		0.035 max	0.035 max	0.35 max	bal Fe				
UNI 7356(74)	CB15	Q/T	0.12-0.18			0.3-0.6		0.035 max	0.035 max	0.1-0.4	bal Fe				
UNI 7746(77)	Fe360B	Wrought	0.17 max			0.35-0.7		0.035 max	0.035 max	0.35 max	N <=0.009; P+S<=0.04; bal Fe				
UNI 7846(78)	C15	CH	0.12-0.18			0.3-0.6		0.035 max	0.035 max	0.15-0.35	bal Fe				
UNI 8550(84)	C15	CH	0.12-0.18			0.3-0.6		0.035 max	0.035 max	0.15-0.35	bal Fe				
UNI 8788(85)	1C15	CH	0.11-0.19			0.3-0.6		0.045 max	0.045 max	0.15-0.35	bal Fe				
UNI 8788(85)	2C15	CH	0.12-0.18			0.3-0.6		0.035 max	0.035 max	0.15-0.35	bal Fe				
Japan															
JIS G3445(88)	STKM12A	Tube	0.20 max			0.60 max		0.040 max	0.040 max	0.35 max	bal Fe	343	177	35L	
JIS G3445(88)	STKM12B	Tube	0.20 max			0.60 max		0.040 max	0.040 max	0.35 max	bal Fe	392	275	25L	
JIS G3445(88)	STKM12C	Tube	0.20 max			0.60 max		0.040 max	0.040 max	0.35 max	bal Fe	471	353	20L	
JIS G3507(91)	SWRCH15A	Wir rod Al killed	0.13-0.18			0.30-0.60		0.030 max	0.035 max	0.10 max	Al >=0.02; bal Fe				
JIS G3507(91)	SWRCH15K	Wir rod FF	0.13-0.18			0.30-0.60		0.030 max	0.035 max	0.10-0.35	bal Fe				
JIS G3507(91)	SWRCH15R	Wir rod FU	0.13-0.18			0.30-0.60		0.040 max	0.040 max		bal Fe				
JIS G4051(79)	S15C	Bar Wir rod	0.13-0.18	0.20 max	0.30 max	0.30-0.60	0.20 max	0.030 max	0.035 max	0.15-0.35	Ni+Cr<=0.35; bal Fe				
JIS G4051(79)	S15CK	CH	0.13-0.18	0.20 max	0.25 max	0.30-0.60	0.20 max	0.025 max	0.025 max	0.15-0.35	Ni+Cr=0.30 max; bal Fe				
Mexico															
DGN B-13	C Grade C1*	Wir, Obs	0.13-0.18			0.33-0.43		0.04	0.06		bal Fe				
DGN B-13	C Grade C2*	Wir, Obs	0.13-0.18			0.33-0.43		0.04	0.06		bal Fe				
DGN B-13	C Grade C3*	Wir, Obs	0.13-0.18			0.33-0.43		0.04	0.06		bal Fe				
DGN B-13	C Grade C4*	Wir, Obs	0.13-0.18			0.33-0.43		0.04	0.06		bal Fe				
DGN B-203	MT1015*	Tube, Obs	0.1-0.2			0.3-0.6		0.04	0.05		bal Fe				
DGN B-203	MTX1015*	Tube, Obs	0.1-0.2			0.6-0.9		0.04	0.05		bal Fe				
NMX-B-142-SCFI(94)		Smls tub for refinery service	0.10-0.20			0.30-0.80		0.035 max	0.035 max	0.25 max	bal Fe	324	179	35	
NMX-B-201(68)	MT1015	CD buttweld mech tub	0.10-0.20			0.30-0.60		0.04 max	0.05 max		bal Fe				
NMX-B-201(68)	MTX1015	CD buttweld mech tub	0.10-0.20			0.60-0.90		0.04 max	0.05 max		bal Fe				
NMX-B-248-SCFI(98)		HR Sh strip comm quality	0.15 max	0.15 max	0.20 max	0.60 max	0.20 max	0.035 max	0.035 max		Mo <=0.06; Nb <=0.008; V <=0.008; Cu if spec'd>=0.20; bal Fe				
NMX-B-301(86)	1015	Bar	0.13-0.18			0.30-0.60		0.040 max	0.050 max		bal Fe				
NOM-060-SCFI(94)	Type 2b	HR for portable gasoline containers	0.15 max			0.40-1.00		0.04 max	0.04 max	0.30 max	bal Fe	420	283	19.0	
Norway															
NS 13101	13115	Struct const	0.12-0.18			0.30-0.60		0.035 max	0.035 max	0.15-0.35	bal Fe				
Pan America															
COPANT 330	1015	Bar	0.13-0.18			0.3-0.6		0.04	0.05		bal Fe				

UNS numbers and US grades are provided as a means of cross referencing chemically similar alloys. Exchangability is only possible after independent examination of specifications. Tensile properties are minimum or typical as specified. UTS and YS as MPa. El as %. See Appendix for list of abbreviations used in Notes. * indicates obsolete material.

Specification	Designation	Notes	C	Cr	Cu	Mn	Ni	P	S	Si	Other	UTS	YS	El	Hard

Carbon Steel, Nonresulfurized, 1015 (Continued from previous page)

Pan America

Specification	Designation	Notes	C	Cr	Cu	Mn	Ni	P	S	Si	Other	UTS	YS	El	Hard
COPANT 331	1015	Bar	0.13-0.18			0.3-0.6		0.04	0.05		bal Fe				
COPANT 333	1015	Wir rod	0.13-0.18			0.3-0.6		0.04	0.05		bal Fe				
COPANT 514	TMX1015	Ann	0.1-0.2			0.6-0.9		0.04	0.05	0.2-0.35	bal Fe				
COPANT 514	TMX1015	Norm	0.1-0.2			0.6-0.9		0.04	0.05	0.2-0.35	bal Fe				
COPANT 514	TMX1015	HF, 323.9mm diam	0.1-0.2			0.6-0.9		0.04	0.05	0.2-0.35	bal Fe				
COPANT 514	TMX1015	CF	0.1-0.2			0.6-0.9		0.04	0.05	0.2-0.35	bal Fe				
COPANT R193	BC	Tube, HF, CF, 5.6mm diam	0.1-0.2			0.3-0.8		0.048	0.058	0.25	bal Fe				

Poland

Specification	Designation	Notes	C	Cr	Cu	Mn	Ni	P	S	Si	Other	UTS	YS	El	Hard
PNH84019	15	CH	0.12-0.19	0.3 max		0.35-0.65	0.3 max	0.04 max	0.04 max	0.17-0.37	bal Fe				
PNH84019	15X	CH	0.12-0.19	0.3 max		0.25-0.5	0.3 max	0.04 max	0.04 max	0.07 max	Nb 0.02-0.06; bal Fe				
PNH84019	15Y	CH	0.12-0.19	0.3 max		0.35-0.65	0.3 max	0.04 max	0.04 max	0.17 max	bal Fe				
PNH84020	St3S	Struct	0.22 max	0.3 max		1.1 max	0.3 max	0.05 max	0.05 max	0.1-0.35	Al 0.02-0.120; Al>=0.02; bal Fe				

Romania

Specification	Designation	Notes	C	Cr	Cu	Mn	Ni	P	S	Si	Other	UTS	YS	El	Hard
STAS 880(88)	OLC15	CH	0.12-0.18	0.3 max		0.35-0.65		0.04 max	0.045 max	0.17-0.37	Pb <=0.15; As<=0.05; bal Fe				
STAS 880(88)	OLC15S	CH	0.12-0.18	0.3 max		0.35-0.65		0.045 max	0.04 max	0.17-0.37	Pb <=0.15; As<=0.05; bal Fe				
STAS 880(88)	OLC15X	CH	0.12-0.18	0.3 max		0.35-0.6		0.035 max	0.02-0.035	0.17-0.37	Pb <=0.15; As<=0.05; bal Fe				
STAS 880(88)	OLC15XS	CH	0.12-0.18	0.3 max		0.35-0.6		0.035 max	0.02-0.04	0.17-0.37	Pb <=0.15; As<=0.05; bal Fe				
STAS 9282/3	OLC15q		0.12-0.18	0.3	0.3	0.35-0.65	0.3	0.045	0.04	0.17-0.37	As 0.05; bal Fe				

Russia

Specification	Designation	Notes	C	Cr	Cu	Mn	Ni	P	S	Si	Other	UTS	YS	El	Hard
GOST	3TS		0.12-0.19			0.25-0.65		0.04	0.04	0.17-0.37	bal Fe				
GOST	KSt3		0.12-0.22			0.35-0.5		0.05 max	0.05 max	0.12-0.3	bal Fe				
GOST	M16		0.12-0.19			0.4-0.65		0.04 max	0.05 max	0.12-0.3	bal Fe				
GOST	M16S		0.12-0.2	0.3	0.3	0.4-0.7	0.3	0.04	0.045	0.12-0.15	bal Fe				
GOST 1050	14kp		0.17 max			0.2 max		0.03 max	0.03 max	0.2 max	bal Fe				
GOST 1050	15		0.12-0.19	0.25	0.25	0.25-0.65	0.25	0.04	0.04	0.17-0.37	bal Fe				
GOST 1050	15K		0.12-0.2	0.3	0.30 max	0.35-0.5	0.3	0.045	0.045	0.15-0.3	bal Fe				
GOST 1050	15kp		0.12-0.2	0.25	0.25 max	0.25-0.5	0.25	0.04	0.04	0.07	bal Fe				
GOST 1050	15LB		0.17			0.38		0.042	0.021	0.26	bal Fe				
GOST 1050(88)	15	High-grade struct; CH; Full hard; 0-100mm	0.12-0.19	0.25 max		0.35-0.65	0.3 max	0.035 max	0.04 max	0.17-0.37	As<=0.08; bal Fe	440		8L	0-197 HB
GOST 1050(88)	15	High-grade struct; CH; Norm; 0-100mm	0.12-0.19	0.25 max		0.35-0.65	0.3 max	0.035 max	0.04 max	0.17-0.37	As<=0.08; bal Fe	370	225	27L	
GOST 1050(88)	15kp	High-grade struct; Unkilled; Rimming; R; FU	0.12-0.19	0.25 max		0.25-0.5	0.3 max	0.035 max	0.04 max	0.07 max	As<=0.08; bal Fe				
GOST 1050(88)	15ps	High-grade struct; Semi-killed	0.12-0.19	0.25 max	0.25 max	0.35-0.65	0.3 max	0.035 max	0.04 max	0.05-0.17	As<=0.08; bal Fe				
GOST 14637	VStTkp		0.1-0.21			0.3-0.6		0.04	0.05	0.05	N 0.008; As 0.08; bal Fe				
GOST 14637	VStTps		0.1-0.21			0.4-0.65		0.04	0.05	0.05-0.15	N 0.008; As 0.08; bal Fe				
GOST 14637	VStTsp		0.1-0.21			0.4-0.65		0.04	0.05	0.15-0.3	N 0.008; As 0.08; bal Fe				

Carbon Steel, Nonresulfurized, 1015 (Continued from previous page)

Specification	Designation	Notes	C	Cr	Cu	Mn	Ni	P	S	Si	Other	UTS	YS	El	Hard
Russia															
GOST 14637	WStTkp		0.1-0.21			0.3-0.6		0.04	0.05	0.05	N 0.008; As 0.08; bal Fe				
GOST 14637	WStTps		0.1-0.21			0.4-0.65		0.04	0.05	0.05-0.15	N 0.008; As 0.08; bal Fe				
GOST 14637	WStTsp		0.1-0.21			0.4-0.65		0.04	0.05	0.15-0.3	N 0.008; As 0.08; bal Fe				
GOST 4041	15YuA		0.12-0.18	0.3 max	0.2 max	0.25-0.45	0.15 max	0.020 max	0.025	0.07 max	Al 0.02-0.08; bal Fe				
GOST 4041(71)	15JuA	High-grade Plts for cold pressing	0.12-0.18	0.3 max	0.2 max	0.25-0.45	0.15 max	0.020 max	0.025 max	0.07 max	Al 0.02-0.08; As<=0.08; bal Fe				
GOST 4041(71)	15JuA	High-grade Plts for cold pressing	0.12-0.18	0.3 max	0.2 max	0.25-0.45	0.15 max	0.020 max	0.025 max	0.07 max	Al 0.02-0.08; bal Fe				
Spain															
UNE	F.111		0.10-0.20			0.30-0.50		0.040 max	0.040 max	0.15-0.30	bal Fe				
UNE 36011(75)	C15k	CH	0.1-0.2			0.4-0.7		0.035 max	0.035 max	0.15-0.4	Mn <=0.15; bal Fe				
UNE 36011(75)	F.1110*	see C15k	0.1-0.2			0.4-0.7		0.035 max	0.035 max	0.15-0.4	Mo <=0.15; bal Fe				
UNE 36013(60)	C16k		0.12-0.18			0.3-0.6		0.035	0.035	0.15-0.4	bal Fe				
UNE 36013(60)	C16k-1		0.12-0.18			0.13-0.6		0.035	0.02-0.035	0.15-0.4	bal Fe				
UNE 36013(60)	F.1511*	see C16k	0.12-0.18			0.3-0.6		0.035	0.035	0.15-0.4	bal Fe				
UNE 36013(60)	F.1513*	see C16k-1	0.12-0.18			0.13-0.6		0.035	0.02-0.035	0.15-0.4	bal Fe				
UNE 36013(76)	C16k	CH	0.12-0.18			0.3-0.6		0.035 max	0.035 max	0.15-0.4	Mo <=0.15; bal Fe				
UNE 36013(76)	F.1511*	see C16k	0.12-0.18			0.3-0.6		0.035 max	0.035 max	0.15-0.4	Mo <=0.15; bal Fe				
UNE 36080(85)	AE235-B	Gen struct	0.2 max			1.60 max		0.045 max	0.045 max	0.60 max	Mo <=0.15; N <=0.009; bal Fe				
UNE 36080(85)	F.6201*	see AE235-B	0.2 max			1.60 max		0.045 max	0.045 max	0.60 max	Mo <=0.15; N <=0.009; bal Fe				
Sweden															
SIS*141370	1370-00	Multiple Forms as roll or Frg	0.12-0.18			0.5-0.9		0.04	0.02-0.04	0.1-0.4	bal Fe	360	210	22	
SIS 141370	1370-03	Bar Frg CH Q/A, Ann	0.12-0.18			0.5-0.9		0.04	0.02-0.04	0.1-0.4	bal Fe	690	390	9	
SIS 141370	1370-04	Bar Frg CH Q/A, Ann	0.12-0.18			0.5-0.9		0.04	0.02-0.04	0.1-0.4	bal Fe	880	590	7	
SS 141311	1311	Gen struct	0.3 max			2 max		0.06 max	0.05 max	0.6 max	bal Fe				
SS 141350	1350		0.2 max			0.4-0.8		0.05 max	0.05 max	0.1-0.4	C:0.15				
SS 141370	1370	CH	0.12-0.18			0.6-0.9		0.035 max	0.02-0.035	0.15-0.4	bal Fe				
Switzerland															
VSM 10648	C15	Bar Sh Plt, CD, 5mm diam	0.12-0.18			0.3-0.6		0.045	0.045	0.15-0.35	bal Fe				
UK															
BS 1449/1(91)	37/23CR	Plt Sh Strp; CR wide based on min strength; 0-16mm	0.2 max			1.20 max		0.05 max	0.05 max	0.6 max	bal Fe	370	230	20-28	
BS 1449/1(91)	37/23CS	Plt Sh Strp; CR t<=16mm	0.2 max			1.20 max		0.05 max	0.05 max	0.6 max	bal Fe	370	230	20-28	
BS 1449/1(91)	37/23HR	Plt Sh Strp HR wide; t<=16mm	0.2 max			1.20 max		0.05 max	0.05 max	0.6 max	bal Fe	370	230	20-28	
BS 1449/1(91)	37/23HS	Plt Sh Strp HR t<=8mm	0.2 max			1.20 max		0.05 max	0.05 max	0.6 max	bal Fe	370	230	20-28	
BS 970/3(91)	080M15	Bright bar; 63<t<=150mm; norm	0.12-0.18			0.60-1.00			0.050 max	0.10-0.40	Mo <=0.15; bal Fe	330	165	22	101-152 HB
BS 970/3(91)	080M15	Bright bar; 6<t<=13mm; HR CD	0.12-0.18			0.60-1.00			0.050 max	0.10-0.40	Mo <=0.15; bal Fe	450	330	10	

Specification	Designation	Notes	C	Cr	Cu	Mn	Ni	P	S	Si	Other	UTS	YS	El	Hard

Carbon Steel, Nonresulfurized, 1015 (Continued from previous page)

USA

Specification	Designation	Notes	C	Cr	Cu	Mn	Ni	P	S	Si	Other	UTS	YS	El	Hard
	AISI 1015	Sh Strp Plt	0.12-0.18			0.30-0.60		0.030 max	0.035 max		bal Fe				
	AISI 1015	Wir rod	0.13-0.18			0.30-0.60		0.040 max	0.050 max		bal Fe				
	AISI 1015	Bar	0.13-0.18			0.30-0.60		0.040 max	0.050 max		bal Fe				
	AISI 1015	Struct shp	0.13-0.18			0.30-0.60		0.030 max	0.035 max		bal Fe				
	AISI M1015	Bar, merchant qual	0.12-0.19			0.25-0.60		0.04 max	0.05 max		bal Fe				
	UNS G10150		0.13-0.18			0.30-0.60		0.040 max	0.050 max		Sheets, C 0.12-0.18; bal Fe				
AMS 5060G(97)		Bar Frg, Tub, HB varies with form	0.13-0.18			0.30-0.60		0.040 max	0.050 max	0.10-0.35	bal Fe				240 HB max
ASTM A108(95)	1015	Bar, CF	0.13-0.18			0.3-0.6		0.04	0.050 max		bal Fe				
ASTM A29/A29M(93)	1015	Bar	0.13-0.18			0.30-0.60		0.040 max	0.050 max		bal Fe				
ASTM A29/A29M(93)	M1015	Bar, merchant qual	0.12-0.19			0.25-0.6		0.040 max	0.050 max		bal Fe				
ASTM A510(96)	1015	Wir rod	0.13-0.18			0.30-0.60		0.040 max	0.050 max		bal Fe				
ASTM A512(96)	1015	Buttweld mech tub, CD	0.12-0.18			0.30-0.60		0.040 max	0.045 max		bal Fe				
ASTM A512(96)	MT1015	Buttweld mech tub, CD	0.10-0.20			0.30-0.60		0.040 max	0.045 max		bal Fe				
ASTM A512(96)	MTX1015	Buttweld mech tub, CD	0.10-0.20			0.60-0.90		0.040 max	0.045 max		bal Fe				
ASTM A513(97a)	1015	ERW Mech tub	0.12-0.18			0.30-0.60		0.035 max	0.035 max		bal Fe				
ASTM A513(97a)	MT1015	ERW Mech tub	0.10-0.20			0.30-0.60		0.035 max	0.035 max		bal Fe				
ASTM A513(97a)	MTX1015	ERW Mech tub	0.10-0.20			0.60-0.90		0.035 max	0.035 max		bal Fe				
ASTM A519(96)	1015	Smls mech tub	0.13-0.18			0.30-0.60		0.040 max	0.050 max		bal Fe				
ASTM A519(96)	MT1015	Smls Low-C Mech tub	0.10-0.20			0.30-0.60		0.040 max	0.050 max		bal Fe				
ASTM A519(96)	MTX1015	Smls Low-C Mech tub	0.10-0.20			0.60-0.90		0.040 max	0.050 max		bal Fe				
ASTM A575(96)	M1015	Merchant qual bar	0.12-0.19			0.25-0.60		0.04 max	0.05 max		Mn can vary with C; bal Fe				
ASTM A576(95)	1015	Special qual HW bar	0.13-0.18			0.30-0.60		0.040 max	0.050 max		Si Cu Pb B Bi Ca Se Te if spec'd; bal Fe				
ASTM A635/A635M(98)	1015	Sh Strp, Coil, HR	0.12-0.18		0.20 min	0.30-0.60		0.030 max	0.035 max		bal Fe				
ASTM A659/A659M(97)	1015	Sh Strp, HR	0.12-0.18			0.30-0.60		0.030 max	0.035 max		bal Fe				
ASTM A787(96)	1015	ERW metallic-coated mech tub	0.12-0.18			0.30-0.60		0.035 max	0.035 max		bal Fe				
ASTM A787(96)	MT1015	ERW metallic-coated mech tub, Low-C	0.10-0.20			0.30-0.60		0.035 max	0.035 max		bal Fe				
ASTM A794(97)	1015	Sh, CS, CR	0.12-0.18			0.30-0.60		0.030 max	0.035 max		Cu >=0.20 if spec; bal Fe				
ASTM A830/A830M(98)	1015	Plt	0.12-0.18			0.30-0.60		0.035 max	0.04 max		bal Fe				
DoD-F-24669/1(86)	1015	Bar Bil; Supersedes MIL-S-866 & MIL-S-16974	0.12-0.18			0.3-0.6		0.04	0.05		bal Fe				
FED QQ-S-698(88)	1008	Sh Strp	0.10 max			0.25-0.50		0.040 max	0.050 max		bal Fe				
FED QQ-S-698(88)	1009	Sh Strp	0.15			0.50 max		0.040 max	0.050 max		bal Fe				
FED QQ-S-698(88)	1015	Sh Strp	0.12-0.19			0.30-0.60		0.040 max	0.050 max		bal Fe				
FED QQ-S-698(88)	1018 mod	Sh Strp	0.13-0.20			0.60-0.90		0.040 max	0.050 max		bal Fe				

UNS numbers and US grades are provided as a means of cross referencing chemically similar alloys. Exchangability is only possible after independent examination of specifications. Tensile properties are minimum or typical as specified. UTS and YS as MPa. El as %. See Appendix for list of abbreviations used in Notes. * indicates obsolete material.

Specification	Designation	Notes	C	Cr	Cu	Mn	Ni	P	S	Si	Other	UTS	YS	El	Hard

Carbon Steel, Nonresulfurized, 1015 (Continued from previous page)

USA

Specification	Designation	Notes	C	Cr	Cu	Mn	Ni	P	S	Si	Other	UTS	YS	El	Hard
FED QQ-W-461H(88)	1015*	Obs; Wir	0.13-0.18			0.3-0.6		0.04	0.05		bal Fe				
MIL-S-16974E(86)	1015*	Obs; Bar bil boom slab for refrg									bal Fe				
MIL-S-645C(87)		CR Strp, drawing qual, for ammunition components	0.13-0.19			0.30-0.60		0.025 max	0.035 max	0.10 max	Al 0.02-0.07; bal Fe				
MIL-T-3520	1015*	Obs; Tube	0.13-0.18			0.3-0.6		0.04	0.05		bal Fe				
SAE J1397(92)	1015	Bar HR, est mech prop	0.13-0.18			0.3-0.6		0.04 max	0.05 max		bal Fe	340	190	28	101 HB
SAE J1397(92)	1015	Bar CD, est mech prop	0.13-0.18			0.3-0.6		0.04 max	0.05 max		bal Fe	390	320	18	111 HB
SAE J403(95)	1015	Bar Wir rod Smls Tub HR CF	0.13-0.18			0.30-0.60		0.030 max	0.050 max		bal Fe				
SAE J403(95)	1015	Struct Shps Plt Strp Sh Weld Tub	0.12-0.10			0.30-0.60		0.030 max	0.035 max		bal Fe				
SAE J403(95)	M1015	Merchant qual	0.13-0.18			0.30-0.60		0.04 max	0.05 max		bal Fe				

Yugoslavia

Specification	Designation	Notes	C	Cr	Cu	Mn	Ni	P	S	Si	Other	UTS	YS	El	Hard
	C.0245	Unalloyed; rivets	0.14 max			0.25-0.45		0.08 max	0.05 max	0.60 max	bal Fe				
	C.0255	Unalloyed; rivets	0.14 max			0.25-0.5		0.05 max	0.05 max	0.60 max	N <=0.007; bal Fe				
	C.0257	Unalloyed; rivets	0.15 max			0.3-0.6		0.08 max	0.08 max	0.60 max	bal Fe				
	C.0261	Struct	0.15 max			1.60 max		0.06 max	0.05 max	0.60 max	N <=0.07; bal Fe				
	C.0345	Unalloyed; rivets	0.19 max			0.5 max		0.08 max	0.05 max	0.60 max	bal Fe				
	C.0355	Unalloyed; rivets	0.19 max			0.5 max		0.05 max	0.05 max	0.60 max	N <=0.007; bal Fe				
	C.0365	Unalloyed; rivets	0.19 max			0.25-0.45		0.04 max	0.04 max	0.60 max	N <=0.007; bal Fe				
	C.1210	Struct	0.12-0.18			0.3-0.6		0.06 max	0.06 max	0.35 max	Mo <=0.15; bal Fe				
	C.1220	CH	0.12-0.18			0.3-0.6		0.045 max	0.045 max	0.15-0.35	Mo <=0.15; bal Fe				
	C.1221	CH	0.12-0.18			0.3-0.6		0.035 max	0.035 max	0.15-0.35	Mo <=0.15; bal Fe				
	C.1281	CH	0.12-0.18			0.3-0.6		0.035 max	0.035 max	0.15-0.35	Mo <=0.15; bal Fe				
	PZ		0.12-0.17	0.15	0.025	0.3-0.6	0.015	0.04	0.04	0.1-0.3	bal Fe				

Carbon Steel, Nonresulfurized, 1016

Argentina

Specification	Designation	Notes	C	Cr	Cu	Mn	Ni	P	S	Si	Other	UTS	YS	El	Hard
IAS	IRAM 1016		0.13-0.18			0.60-0.90		0.040 max	0.050 max	0.15-0.30	bal Fe	450-520	270-330	25-39	131-156 HB

Australia

Specification	Designation	Notes	C	Cr	Cu	Mn	Ni	P	S	Si	Other	UTS	YS	El	Hard
AS 1442	K1016*	Obs; Bar Bil	0.13-0.18			0.6-0.9		0.05	0.05	0.1-0.35	bal Fe				
AS 1442	S1016*	Obs; Bar Bil	0.13-0.18			0.6-0.9		0.05	0.05	0.35	bal Fe				
AS 1442(92)	1016	HR bar, Semifinished (may treat w/microalloying elements)	0.13-0.18			0.60-0.90		0.040 max	0.040 max	0.10-0.35	Si<=0.10 for Al-killed; bal Fe				
AS 1443	K1016*	Obs; Bar	0.13-0.18			0.6-0.9		0.05	0.05	0.1-0.35	bal Fe				
AS 1443	S1016*	Obs; Bar	0.13-0.18			0.6-0.9		0.05	0.05	0.35	bal Fe				
AS 1443(94)	1016	CF bar (may treat w/microalloying elements)	0.13-0.18			0.60-0.90		0.040 max	0.040 max	0.10-0.35	Si<=0.10 for Al-killed; bal Fe				

UNS numbers and US grades are provided as a means of cross referencing chemically similar alloys. Exchangability is only possible after independent examination of specifications. Tensile properties are minimum or typical as specified. UTS and YS as MPa. El as %. See Appendix for list of abbreviations used in Notes. * indicates obsolete material.

Specification	Designation	Notes	C	Cr	Cu	Mn	Ni	P	S	Si	Other	UTS	YS	El	Hard

Carbon Steel, Nonresulfurized, 1016 (Continued from previous page)

Australia

Specification	Designation	Notes	C	Cr	Cu	Mn	Ni	P	S	Si	Other	UTS	YS	El	Hard
AS/NZS 1594(97)	HXA1016	HR flat products	0.12-0.18	0.30 max	0.35 max	0.80-1.20	0.35 max	0.040 max	0.035 max	0.03 max	Al <=0.100; Mo <=0.10; Ti <=0.040; (Cu+Ni+Cr+Mo) 1 max; bal Fe				

Belgium

Specification	Designation	Notes	C	Cr	Cu	Mn	Ni	P	S	Si	Other	UTS	YS	El	Hard
NBN A25-102	D35	Tube, HF, CF Norm	0.17			0.36-0.84		0.05	0.05	0.12-0.38	bal Fe				

Bulgaria

Specification	Designation	Notes	C	Cr	Cu	Mn	Ni	P	S	Si	Other	UTS	YS	El	Hard
BDS 5785(83)	15	Struct	0.12-0.19	0.25 max	0.25 max	0.35-0.55	0.25 max	0.035 max	0.040 max	0.17-0.37	bal Fe				
BDS 5785(83)	15ps	Struct	0.12-0.19	0.25 max	0.25 max	0.35-0.65	0.25 max	0.035 max	0.040 max	0.05-0.17	bal Fe				
BDS 6354	15CLGM	Struct	0.12-0.19		0.3	0.8-1.1	0.3	0.035	0.035	0.17-0.37	bal Fe				
BDS 9801	14G	Struct, weld ships	0.12-0.18	0.3	0.3	0.7-1	0.3	0.045	0.04	0.15-0.37	bal Fe				

Canada

Specification	Designation	Notes	C	Cr	Cu	Mn	Ni	P	S	Si	Other	UTS	YS	El	Hard
CSA 95-1/1	C1016		0.11-0.2			0.57-0.93		0.048	0.058	0.18-0.37	bal Fe				
CSA STAN95-1-1	1016	Bar	0.11-0.2			0.57-0.93		0.05	0.06	0.18-0.37	bal Fe				

China

Specification	Designation	Notes	C	Cr	Cu	Mn	Ni	P	S	Si	Other	UTS	YS	El	Hard
GB 699(88)	15Mn	Bar HR Norm 25mm diam	0.12-0.19	0.25 max	0.25 max	0.70-1.00	0.25 max	0.035 max	0.035 max	0.17-0.37	bal Fe	410	245	26	
GB/T 3078(94)	15Mn	Bar CD Ann	0.12-0.19	0.25 max	0.25 max	0.70-1.00	0.25 max	0.035 max	0.035 max	0.17-0.37	bal Fe	390		21	

Czech Republic

Specification	Designation	Notes	C	Cr	Cu	Mn	Ni	P	S	Si	Other	UTS	YS	El	Hard
CSN 412023	12023	CH	0.12-0.19	0.25 max		0.35-0.65	0.3 max	0.04 max	0.04 max	0.15-0.4	bal Fe				

Europe

Specification	Designation	Notes	C	Cr	Cu	Mn	Ni	P	S	Si	Other	UTS	YS	El	Hard
EN 10084(98)	1.1140	CH, Ann	0.12-0.18			0.30-0.60		0.035 max	0.020-0.040	0.40 max	bal Fe				143 HB
EN 10084(98)	1.1148	CH, Ann	0.12-0.18			0.60-0.90		0.035 max	0.035 max	0.40 max	bal Fe				156 HB
EN 10084(98)	1.1208	CH, Ann	0.12-0.18			0.60-0.90		0.035 max	0.020-0.040	0.40 max	bal Fe				156 HB
EN 10084(98)	C15R	CH, Ann	0.12-0.18			0.30-0.60		0.035 max	0.020-0.040	0.40 max	bal Fe				143 HB
EN 10084(98)	C16E	CH, Ann	0.12-0.18			0.60-0.90		0.035 max	0.035 max	0.40 max	bal Fe				156 HB
EN 10084(98)	C16R	CH, Ann	0.12-0.18			0.60-0.90		0.035 max	0.020-0.040	0.40 max	bal Fe				156 HB

Finland

Specification	Designation	Notes	C	Cr	Cu	Mn	Ni	P	S	Si	Other	UTS	YS	El	Hard
SFS 505	C15	HR	0.12-0.18			0.6-0.9		0.04	0.02-0.04	0.1-0.4	bal Fe	370	205	22	

Germany

Specification	Designation	Notes	C	Cr	Cu	Mn	Ni	P	S	Si	Other	UTS	YS	El	Hard
DIN	15Mn3		0.12-0.18			0.70-0.90		0.040 max	0.040 max	0.10-0.20	C 0.20 max for diam >=26 mm; bal Fe				
DIN	GS-Ck16		0.12-0.19	0.30 max		0.50-0.80		0.030 max	0.030 max	0.30-0.50	N <=0.007; N depends on P; Cu+Ni+Cr<=0.30; Cu+Sn<=0.15; bal Fe				
DIN	RSt44-2		0.18 max			0.80 max		0.050 max	0.050 max	0.45 max	N <=0.007; bal Fe				
DIN	WNr 1.0419		0.18 max			0.80 max		0.050 max	0.050 max	0.45 max	N <=0.007; bal Fe				
DIN	WNr 1.0467		0.12-0.18			0.70-0.90		0.040 max	0.040 max	0.10-0.20	C <=0.20 if diam >=26mm; bal Fe				
DIN	WNr 1.1142		0.12-0.19	0.30 max		0.50-0.80		0.030 max	0.030 max	0.30-0.50	N <=0.007; N depends on P; Cu+Sn<=0.15; Cu+Ni+Cr<=0.30; bal Fe				
DIN 17115(87)	15Mn3Al	Weld round link chain	0.12-0.18	0.25		0.70-0.90		0.035 max	0.035 max	0.20 max	Al 0.020-0.050; N <=0.012; C 0.20 max for diam >=26 mm; bal Fe				

UNS numbers and US grades are provided as a means of cross referencing chemically similar alloys. Exchangability is only possible after independent examination of specifications. Tensile properties are minimum or typical as specified. UTS and YS as MPa. El as %. See Appendix for list of abbreviations used in Notes. * indicates obsolete material.

Carbon Steel, Nonresulfurized, 1016 (Continued from previous page)

Specification	Designation	Notes	C	Cr	Cu	Mn	Ni	P	S	Si	Other	UTS	YS	El	Hard
Germany															
DIN 17115(87)	WNr 1.0468	Weld round link chains	0.12-0.18	0.25		0.70-0.90		0.035 max	0.035 max	0.20 max	Al 0.020-0.050; N <=0.012; C <=0.20 if diam >=26mm; bal Fe				
DIN EN 10025(94)	S235J0	HR 100-150mm thk	0.17 max			1.40 max		0.040 max	0.040 max		N <=0.009; N depends on P; bal Fe	340 470	195	22	
DIN EN 10025(94)	WNr 1.0114	HR 100-150mm thk	0.17 max			1.40 max		0.040 max	0.040 max		N <=0.009; N depends on P; bal Fe	340 470	195	22	
Hungary															
MSZ 113(88)	A44Sz/SzK	Rivets; 0-36mm; HR/CD	0.1-0.18			0.5-0.8		0.04 max	0.04 max	0.25-0.5	bal Fe	440 540	260	21L	
India															
IS 1570	C15Mn75		0.1-0.2			0.6-0.9					bal Fe				
IS 1570/2(79)	14C6		0.1-0.18	0.3 max	0.3 max	0.4-0.6	0.4 max	0.055 max	0.055 max	0.6 max	Co <=0.1; Mo <=0.15; Pb <=0.15; W <=0.1; P:S Varies; bal Fe	370 450	203	26	
IS 1570/2(79)	C14		0.1-0.18	0.3 max	0.3 max	0.4-0.6	0.4 max	0.055 max	0.055 max	0.6 max	Co <=0.1; Mo <=0.15; Pb <=0.15; W <=0.1; P:S Varies; bal Fe	370 450	203	20	
IS 2100	Grade 2		0.1-0.2			0.6-0.9		0.05	0.05		bal Fe				
International															
ISO 683-11(87)	C15E4	CH, 16mm test	0.12-0.18		0.3 max	0.30-0.60		0.035 max	0.035 max	0.15-0.40	Al <=0.1; Co <=0.1; bal Fe	500 850	300		
ISO 683-11(87)	C15M2	CH, 16mm test	0.12-0.18		0.3 max	0.30-0.60		0.035 max	0.020-0.040	0.15-0.40		500 850	300		
ISO R683-11(70)	3*	Bar, Carburized Hardness	0.12-0.18			0.6-0.9		0.04	0.03-0.05	0.15-0.4	bal Fe				
ISO R683-11(70)	3a*	Bar, Carburized Hardness	0.12-0.18			0.6-0.9		0.04	0.02-0.04	0.15-0.4	bal Fe				
Japan															
JIS G3507(91)	SWRCH16A	Wir rod Al killed	0.13-0.18			0.60-0.90		0.030 max	0.035 max	0.10 max	Al >=0.02; bal Fe				
JIS G3507(91)	SWRCH16K	Wir rod FF	0.13-0.18			0.60-0.90		0.030 max	0.035 max	0.10-0.35	bal Fe				
Mexico															
NMX-B-301(86)	1016	Bar	0.13-0.18			0.60-0.90		0.040 max	0.050 max		bal Fe				
Pan America															
COPANT 330	1016	Bar	0.13-0.18			0.6-0.9		0.04	0.05		bal Fe				
COPANT 331	1016	Bar	0.13-0.18			0.6-0.9		0.04	0.05		bal Fe				
COPANT 333	1016	Wir rod	0.13-0.18			0.6-0.9		0.04	0.05		bal Fe				
COPANT 514	TMX1015	Tube, Ann	0.1-0.2			0.6-0.9		0.04	0.05	0.2-0.35	bal Fe				
COPANT 514	TMX1015	Tube, CF	0.1-0.2			0.6-0.9		0.04	0.05	0.2-0.35	bal Fe				
COPANT 514	TMX1015	Tube, Norm	0.1-0.2			0.6-0.9		0.04	0.05	0.2-0.35	bal Fe				
COPANT 514	TMX1015	Tube, HF, 323.9mm diam	0.1-0.2			0.6-0.9		0.04	0.05	0.2-0.35	bal Fe				
Poland															
PNH84018(86)	15GA	HS weld const	0.18 max	0.3 max		0.7-1.3	0.3 max	0.04 max	0.04 max	0.15-0.5	Al 0.02-0.120; As<=0.08; CEV<=0.045; bal Fe				
PNH84019	15G	CH	0.12-0.19	0.3 max		0.7-1	0.3 max	0.04 max	0.04 max	0.17-0.37	bal Fe				
PNH84023/08	15GJ	Chain	0.13-0.18	0.3 max		0.9-1.2	0.3 max	0.04 max	0.04 max	0.1-0.2	Al <=0.02; bal Fe				
PNH93027	15GA	Struct; Wir Bar	0.12-0.19	0.2 max		0.7-1	0.3 max	0.04 max	0.04 max	0.17-0.37	Al 0.02-0.120; bal Fe				
Russia															
GOST	M16S		0.12-0.2	0.3	0.3	0.4-0.7	0.3	0.04	0.045	0.12-0.15	bal Fe				

UNS numbers and US grades are provided as a means of cross referencing chemically similar alloys. Exchangability is only possible after independent examination of specifications. Tensile properties are minimum or typical as specified. UTS and YS as MPa. El as %. See Appendix for list of abbreviations used in Notes. * indicates obsolete material.

Specification	Designation	Notes	C	Cr	Cu	Mn	Ni	P	S	Si	Other	UTS	YS	El	Hard

Carbon Steel, Nonresulfurized, 1016 (Continued from previous page)

Specification	Designation	Notes	C	Cr	Cu	Mn	Ni	P	S	Si	Other	UTS	YS	El	Hard
Russia															
GOST 4543	15G		0.12-0.19	0.3	0.3	0.7-1	0.3	0.035	0.035	0.17-0.37	bal Fe				
GOST 4543(71)	15G	CH	0.12-0.19	0.3 max	0.3 max	0.7-1.0	0.3 max	0.035 max	0.035 max	0.17-0.37	Al <=0.1; bal Fe				
GOST 5520(79)	16K	Boiler, Press ves	0.12-0.20	0.30 max	0.30 max	0.45-0.75	0.30 max	0.040 max	0.040 max	0.17-0.37	N <=0.012; As<=0.08; bal Fe				
Spain															
UNE 36011(75)	C15k	CH	0.1-0.2			0.4-0.7		0.035 max	0.035 max	0.15-0.4	Mo <=0.15; bal Fe				
UNE 36013(76)	C16k	CH	0.12-0.18			0.3-0.6		0.035 max	0.035 max	0.15-0.4	Mo <=0.15; bal Fe				
UK															
BS 3059	440	Tube, Norm	0.12-0.18			0.9-1.2		0.04	0.035	0.1-0.35	bal Fe				
BS 970/1(83)	080M15	Blm Bil Slab Bar Rod Frg CH	0.12-0.18			0.60-1.00					bal Fe	460		16	
BS 970/1(96)	080A15	Blm Bil Slab Bar Rod Frg CH	0.13-0.18			0.70-0.90		0.050 max	0.050 max	0.10-0.40	bal Fe				
BS 970/1(96)	080A20	Blm Bil Slab Bar Rod Frg CH	0.18-0.23			0.70-0.90		0.050 max	0.050 max	0.10-0.40	bal Fe				
USA															
	AISI 1016	Struct shp	0.13-0.18			0.60-0.90		0.030 max	0.035 max		bal Fe				
	AISI 1016	Bar HW CF	0.13-0.18			0.60-0.90		0.040 max	0.050 max		bal Fe				
	AISI 1016	Sh Strp Plt	0.12-0.18			0.60-0.90		0.030 max	0.035 max		bal Fe				
	AISI 1016	Wir rod	0.13-0.18			0.60-0.90		0.040 max	0.050 max		bal Fe				
	UNS G10160		0.13-0.18			0.60-0.90		0.040 max	0.050 max		Sheets, C 0.12-0.18; ERW Tubing, C 0.12-0.19; bal Fe				
ASTM A108(95)	1016	Bar, CF	0.13-0.18			0.60-0.90		0.040 max	0.050 max		bal Fe				
ASTM A29/A29M(93)	1016	Bar, HW, CF	0.13-0.18			0.60-0.90		0.040 max	0.050 max		bal Fe				
ASTM A510(96)	1016	Wir rod	0.13-0.18			0.60-0.90		0.040 max	0.050 max		bal Fe				
ASTM A512(96)	1016	Buttweld mech tub, CD	0.12-0.18			0.60-0.90		0.040 max	0.045 max		bal Fe				
ASTM A513(97a)	1016	ERW Mech tub	0.12-0.18			0.60-0.90		0.035 max	0.035 max		bal Fe				
ASTM A519(96)	1016	Smls mech tub	0.13-0.18			0.60-0.90		0.040 max	0.050 max		bal Fe				
ASTM A576(95)	1016	Special qual HW bar	0.13-0.18			0.60-0.90		0.040 max	0.050 max		Si Cu Pb B Bi Ca Se Te if spec'd; bal Fe				
ASTM A635/A635M(98)	1016	Sh Strp, Coil, HR	0.12-0.18		0.20 min	0.60-0.90		0.030 max	0.035 max		bal Fe				
ASTM A659/A659M(97)	1016	Sh Strp, HR	0.12-0.18			0.60-0.90		0.030 max	0.035 max		bal Fe				
ASTM A787(96)	1016	ERW metallic-coated mech tub	0.12-0.19			0.60-0.90		0.035 max	0.035 max		bal Fe				
ASTM A794(97)	1016	Sh, CS, CR	0.12-0.18			0.60-0.90		0.030 max	0.035 max		Cu >=0.20 if spec; bal Fe				
ASTM A830/A830M(98)	1016	Plt	0.12-0.18			0.60-0.90		0.035 max	0.04 max		bal Fe				
DoD-F-24669/1(86)	1016	Bar Bil; Supersedes MIL-S-866 & MIL-S-16974	0.13-0.18			0.6-0.9		0.04	0.05		bal Fe				
MIL-S-866C	- - -*	Obs	0.13-0.18			0.6-0.9		0.04	0.05		bal Fe				
SAE J1397(92)	1016	Bar CD, est mech prop	0.13-0.18			0.6-0.9		0.04	0.05		bal Fe	420	350	18	121 HB
SAE J1397(92)	1016	Bar HR, est mech prop	0.13-0.18			0.6-0.9		0.04	0.05		bal Fe	380	210	25	111 HB
SAE J403(95)	1016	Struct Shps Plt Strp Sh Weld Tub	0.12-0.18			0.60-0.90		0.030 max	0.035 max		bal Fe				

UNS numbers and US grades are provided as a means of cross referencing chemically similar alloys. Exchangability is only possible after independent examination of specifications. Tensile properties are minimum or typical as specified. UTS and YS as MPa. El as %. See Appendix for list of abbreviations used in Notes. * indicates obsolete material.

Specification	Designation	Notes	C	Cr	Cu	Mn	Ni	P	S	Si	Other	UTS	YS	El	Hard

Carbon Steel, Nonresulfurized, 1016 (Continued from previous page)

USA

Specification	Designation	Notes	C	Cr	Cu	Mn	Ni	P	S	Si	Other	UTS	YS	El	Hard
SAE J403(95)	1016	Bar Wir rod Smls Tub HR CF	0.13-0.18			0.60-0.90		0.030 max	0.050 max		bal Fe				

Yugoslavia

| | C.0411 | Chain | 0.1-0.18 | | | 0.4-0.6 | | 0.04 max | 0.04 max | 0.12-0.3 | Mo <=0.15; N <=0.007; bal Fe | | | | |

Carbon Steel, Nonresulfurized, 1017

Austria

Specification	Designation	Notes	C	Cr	Cu	Mn	Ni	P	S	Si	Other	UTS	YS	El	Hard
ONORM M3110	RC15	Wir	0.15-0.20			0.4-0.7		0.040 max	0.040 max	0.15-0.4	bal Fe				

Belgium

| NBN 253-03 | C16-2 | | 0.12-0.18 | | | 0.3-0.6 | | 0.04 max | 0.04 max | 0.1-0.4 | bal Fe | | | | |

Bulgaria

BDS 2592(71)	BSt3kp	Struct	0.14-0.22	0.30 max	0.30 max	0.30-0.60	0.30 max	0.045 max	0.055 max	0.07	bal Fe				
BDS 2592(71)	BSt3ps	Struct	0.14-0.22	0.30 max	0.30 max	0.40-0.65	0.30 max	0.045 max	0.055 max	0.05-0.17	bal Fe				
BDS 2592(71)	BSt3sp	Struct	0.14-0.22	0.30 max	0.30 max	0.40-0.65	0.30 max	0.045 max	0.055 max	0.12-0.35	bal Fe				
BDS 9801	VSt3kp	Struct, weld ships	0.14-0.22	0.3	0.3	0.3-0.6	0.3	0.045	0.055	0.07	bal Fe				
BDS 9801	VSt3ps	Struct, weld ships	0.14-0.22	0.3	0.3	0.4-0.65	0.3	0.045	0.055	0.05-0.17	bal Fe				
BDS 9801	VSt3sp	Struct, weld ships	0.14-0.22	0.3	0.3	0.4-0.65	0.3	0.045	0.055	0.12-0.35	bal Fe				
BDS 9801	WSt3kp	Struct, weld ships	0.14-0.22	0.3	0.3	0.3-0.6	0.3	0.045	0.055	0.07	bal Fe				
BDS 9801	WSt3ps	Struct weld ships	0.14-0.22	0.3	0.3	0.4-0.65	0.3	0.045	0.055	0.05-0.17	bal Fe				
BDS 9801	WSt3sp	Struct, weld ships	0.14-0.22	0.3	0.3	0.4-0.65	0.3	0.045	0.055	0.12-0.35	bal Fe				

China

GB 5953(86)	ML 18	Wir CD	0.15-0.20	0.20 max	0.20 max	0.60 max		0.035 max	0.035 max	0.20 max	bal Fe	440-635			
GB 700(88)	Q235A	Bar Plt Sh HR 16-40mm diam	0.14-0.22	0.30 max	0.30 max	0.30-0.65	0.30 max	0.045 max	0.050 max	0.30 max	bal Fe	375-500	225	25	
GB 700(88)	Q235B	Bar Plt Sh HR 16-40mm diam	0.12-0.20	0.30 max	0.30 max	0.30-0.70	0.30 max	0.045 max	0.045 max	0.30 max	bal Fe	375-500	225	25	
GB 700(88)	Q235C	Bar Plt Sh HR 16-40mm diam	0.18 max	0.30 max	0.30 max	0.35-0.80	0.30 max	0.040 max	0.040 max	0.30 max	bal Fe	375-500	225	25	
GB 700(88)	Q235D	Bar Plt Sh HR 16-40mm diam	0.17 max	0.30 max	0.30 max	0.35-0.80	0.30 max	0.035 max	0.035 max	0.30 max	bal Fe	375-500	225	25	
GB 715(89)	BL3	Bar HR	0.14-0.22		0.25 max	0.30-0.60		0.040 max	0.040 max	0.07 max	bal Fe	370-460	235	28	
GB 900	B3*	Obs	0.14-0.22			0.3-0.65		0.045	0.055	0.07-0.3	bal Fe				
GB 900	B3F*	Obs	0.14-0.22			0.3-0.65		0.045	0.055	0.07-0.3	bal Fe				
GB 900	C3*	Obs	0.14-0.22			0.3-0.65		0.045	0.055	0.07-0.3	bal Fe				
GB 900	C3F*	Obs	0.14-0.22			0.3-0.65		0.045	0.055	0.07-0.3	bal Fe				
GB/T 13013(91)	Q235	Bar HR	0.14-0.22	0.30 max	0.30 max	0.30-0.65	0.30 max	0.045 max	0.050 max	0.12-0.30	bal Fe	370	235	25	
GB/T 701(97)	Q235A-b	Wir rod HR	0.14-0.22	0.30 max	0.30 max	0.30-0.65	0.30 max	0.045 max	0.050 max	0.17 max	As<=0.080; bal Fe	410	235	23	
GB/T 701(97)	Q235A-F	Wir rod HR	0.14-0.22	0.30 max	0.30 max	0.30-0.65	0.30 max	0.045 max	0.050 max	0.07 max	As<=0.080; bal Fe	410	235	23	
GB/T 701(97)	Q235A-Z	Wir rod HR	0.14-0.22	0.30 max	0.30 max	0.30-0.65	0.30 max	0.045 max	0.050 max	0.12 max	As<=0.080; bal Fe	410	235	23	
GB/T 701(97)	Q235B-b	Wir rod HR	0.12-0.20	0.30 max	0.30 max	0.30-0.70	0.30 max	0.045 max	0.045 max	0.17 max	As<=0.080; bal Fe	410	235	23	
GB/T 701(97)	Q235B-F	Wir rod HR	0.12-0.20	0.30 max	0.30 max	0.30-0.70	0.30 max	0.045 max	0.045 max	0.07 max	As<=0.080; bal Fe	410	235	23	
GB/T 701(97)	Q235B-Z	Wir rod HR	0.12-0.20	0.30 max	0.30 max	0.30-0.70	0.30 max	0.045 max	0.045 max	0.12 max	As<=0.080; bal Fe	410	235	23	

UNS numbers and US grades are provided as a means of cross referencing chemically similar alloys. Exchangability is only possible after independent examination of specifications. Tensile properties are minimum or typical as specified. UTS and YS as MPa. El as %. See Appendix for list of abbreviations used in Notes. * indicates obsolete material.

Specification	Designation	Notes	C	Cr	Cu	Mn	Ni	P	S	Si	Other	UTS	YS	El	Hard

Carbon Steel, Nonresulfurized, 1017 (Continued from previous page)

China

Specification	Designation	Notes	C	Cr	Cu	Mn	Ni	P	S	Si	Other	UTS	YS	El	Hard
GB/T 8164(93)	Q235A	Strp HR/CR	0.14-0.22	0.30 max	0.30 max	0.30-0.65	0.30 max	0.045 max	0.050 max	0.30 max	bal Fe	375-405	225	25	
GB/T 8164(93)	Q235B	Strp HR/CR	0.12-0.20	0.30 max	0.30 max	0.30-0.70	0.30 max	0.045 max	0.045 max	0.30 max	bal Fe	375-405	225	25	
GB/T 8164(93)	Q235C	Strp HR/CR	0.18 max	0.30 max	0.30 max	0.35-0.80	0.30 max	0.040 max	0.040 max	0.30 max	bal Fe	375-405	225	25	
GB/T 8164(93)	Q235D	Strp HR/CR	0.17 max	0.30 max	0.30 max	0.35-0.80	0.30 max	0.035 max	0.035 max	0.30 max	bal Fe	375-405	225	25	

Czech Republic

Specification	Designation	Notes	C	Cr	Cu	Mn	Ni	P	S	Si	Other	UTS	YS	El	Hard
CSN 412025	12025		0.14-0.2	0.25 max	0.25 max	0.6-1.0	0.25 max	0.04 max	0.04 max	0.17-0.37	V 0.05-0.09; P+S<=0.07; bal Fe				

Finland

Specification	Designation	Notes	C	Cr	Cu	Mn	Ni	P	S	Si	Other	UTS	YS	El	Hard
SFS 1100	Fe37BP	Gen struct	0.17 max		0.4 max	0.4-1.2		0.05 max	0.05 max	0.15-0.55	N <=0.009; As 0-0.08; bal Fe				
SFS 1100	Fe37DP	Gen struct	0.17 max		0.4 max	0.4-1.2		0.05 max	0.05 max	0.15-0.55	N <=0.009; As 0-0.08; bal Fe				
SFS 200	Fe37B	Gen struct	0.17 max			1.60 max		0.05 max	0.05 max	0.60 max	N <=0.009; bal Fe				
SFS 200	Fe37C	Gen struct	0.17 max			1.60 max		0.045 max	0.045 max	0.60 max	Al <=0.10; N <=0.009; bal Fe				
SFS 200	Fe37D	Gen struct	0.17 max		0.4 max	1.60 max		0.04 max	0.04 max	0.60 max	Al 0.02; N <=0.015; bal Fe				

France

Specification	Designation	Notes	C	Cr	Cu	Mn	Ni	P	S	Si	Other	UTS	YS	El	Hard
AFNOR NFA35551	XC18		0.16-0.22			0.4-0.7		0.035	0.035	0.15-0.35	Al 0.02; bal Fe				
AFNOR NFA35552	XC18	Bar, as roll, Norm; 16mm diam	0.16-0.22			0.4-0.7		0.035	0.035	0.15-0.35	Al 0.02; bal Fe				
AFNOR NFA35553	XC18S	Strp, Q/A Tmp	0.15-0.22			0.4-0.65		0.035	0.035	0.25	bal Fe				
AFNOR NFA35553	XC18S	Strp, Norm	0.15-0.22			0.4-0.65		0.035	0.035	0.25	bal Fe				
AFNOR NFA35554	XC18S	Plt, HR, Norm, 16mm diam	0.15-0.22			0.4-0.65		0.035	0.035	0.25	bal Fe				
AFNOR NFA35566	XC18		0.16-0.22			0.4-0.7		0.035	0.035	0.15-0.35	Al 0.02; bal Fe				
AIR 9160C21	9160C011	Bar Frg Q/A Tmp, 16mm diam	0.15-0.22			0.4-0.65		0.04 max	0.04 max	0.25	bal Fe				

Hungary

Specification	Designation	Notes	C	Cr	Cu	Mn	Ni	P	S	Si	Other	UTS	YS	El	Hard
MSZ 500(81)	B38X	Gen struct; unkilled; R; FU	0.14-0.22			0.3-0.6		0.045 max	0.05 max	0.07 max	bal Fe				

Japan

Specification	Designation	Notes	C	Cr	Cu	Mn	Ni	P	S	Si	Other	UTS	YS	El	Hard
JIS G3507(91)	SWRCH17K	Wir rod FF	0.15-0.20			0.30-0.60		0.030 max	0.035 max	0.10-0.35	bal Fe				
JIS G3507(91)	SWRCH17R	Wir rod FU	0.15-0.20			0.30-0.60		0.040 max	0.040 max		bal Fe				
JIS G4051(79)	S17C	Bar Wir rod	0.15-0.20	0.20 max	0.30 max	0.30-0.60	0.20 max	0.030 max	0.035 max	0.15-0.35	Ni+Cr<=0.35; bal Fe				

Mexico

Specification	Designation	Notes	C	Cr	Cu	Mn	Ni	P	S	Si	Other	UTS	YS	El	Hard
NMX-B-301(86)	1017	Bar	0.15-0.20			0.30-0.60		0.040 max	0.050 max		bal Fe				

Pan America

Specification	Designation	Notes	C	Cr	Cu	Mn	Ni	P	S	Si	Other	UTS	YS	El	Hard
COPANT 333	1017		0.15-0.2			0.3-0.6		0.04	0.05		bal Fe				

Poland

Specification	Designation	Notes	C	Cr	Cu	Mn	Ni	P	S	Si	Other	UTS	YS	El	Hard
PNH84023	St3NO		0.14-0.2	0.3 max		0.35-0.6	0.3 max	0.05 max	0.05 max	0.12-0.35	bal Fe				
PNH84023/08	18A		0.15-0.2	0.2	0.3	0.35-0.65	0.3	0.035	0.035	0.17-0.37	bal Fe				
PNH84024	R45	High-temp const	0.14-0.2			0.45-0.7		0.05	0.05	0.15-0.35	bal Fe				
PNH84024	R45A	High-temp const	0.14-0.2	0.3 max		0.35-0.6	0.4 max	0.05 max	0.05 max	0.15-0.3	Al 0.02-0.120; bal Fe				
PNH84024	St44K	High-temp const	0.114-0.22	0.3 max		0.55-1.6	0.3 max	0.045 max	0.045 max	0.15-0.35	bal Fe				
PNH84030/05	18A	Const	0.15-0.2	0.2 max		0.35-0.65	0.3 max	0.035 max	0.035 max	0.17-0.37	bal Fe				

4-40/Carbon Steel

Specification	Designation	Notes	C	Cr	Cu	Mn	Ni	P	S	Si	Other	UTS	YS	El	Hard

Carbon Steel, Nonresulfurized, 1017 (Continued from previous page)

Romania

Specification	Designation	Notes	C	Cr	Cu	Mn	Ni	P	S	Si	Other	UTS	YS	El	Hard
STAS 10607	16		0.12-0.2	0.3	0.3	0.4-0.7	0.3	0.04	0.04	0.12-0.4	As 0.08; bal Fe				

Russia

Specification	Designation	Notes	C	Cr	Cu	Mn	Ni	P	S	Si	Other	UTS	YS	El	Hard
GOST	KSt3		0.12-0.22			0.35-0.5		0.05	0.055	0.12-0.3	bal Fe				
GOST	KSt3kp		0.12-0.22			0.35-0.5		0.05	0.055	0.07	bal Fe				
GOST	KSt3ps		0.14-0.22			0.4-0.65		0.045	0.055	0.17	bal Fe				
GOST	MSt3		0.14-0.22			0.4-0.65		0.04 max	0.05 max	0.12-0.22	bal Fe				
GOST	MSt3PS		0.14-0.22			0.4-0.65		0.05 max	0.05 max	0.17 max	bal Fe				
GOST	MU		0.17 max			0.44 max		0.1 max	0.04 max	0.19 max	bal Fe				
GOST 1050	10kp		0.12-0.2	0.15	0.25 max	0.3-0.5	0.25	0.035	0.04	0.06	As 0.08; bal Fe				
GOST 1050(88)	18kp	High-grade struct; Unkilled; Rimming; R; FU	0.12-0.2	0.15 max	0.25 max	0.3-0.5	0.3 max	0.035 max	0.04 max	0.06 max	As<=0.08; bal Fe				
GOST 23570	18kp		0.14-0.22	0.3	0.3	0.3-0.6	0.3	0.04	0.05	0.05	bal Fe				
GOST 380	BSt3kp		0.14-0.22			0.3-0.6		0.04 max	0.05 max	0.07 max	N <=0.008; As<=0.08; bal Fe				
GOST 380	BSt3kp2		0.14-0.22	0.3	0.3	0.3-0.6	0.3	0.04	0.05	0.07	N 0.008; As 0.08; bal Fe				
GOST 380	BSt3ps		0.14-0.22			0.4-0.65		0.04	0.05	0.05-0.17	N 0.008; As 0.08; bal Fe				
GOST 380	BSt3ps2		0.14-0.22	0.3	0.3	0.4-0.65	0.3	0.04	0.05	0.05-0.17	N 0.008; As 0.08; bal Fe				
GOST 380	BSt3sp		0.14-0.22			0.4-0.65		0.04	0.05	0.12-0.3	N 0.008; As 0.08; bal Fe				
GOST 380	BSt3sp2		0.14-0.22	0.3	0.3	0.4-0.65	0.3	0.04	0.05	0.12-0.3	N 0.008; As 0.08; bal Fe				
GOST 380(71)	WSt3Gps	Gen struct; Semi-killed; 0-20mm	0.22 max	0.3 max	0.3 max	0.8-1.1	0.3 max	0.04 max	0.05 max	0.15 max	Al <=0.1; N <=0.008; As<=0.08; bal Fe	370-490		26L	
GOST 380(71)	WSt3Gps	Gen struct; Semi-killed; >40mm	0.22 max	0.3 max	0.3 max	0.8-1.1	0.3 max	0.04 max	0.05 max	0.15 max	Al <=0.1; N <=0.008; As<=0.08; bal Fe	370-490		23L	
GOST 380(71)	WSt3Gps2	Gen struct; Semi-killed; 0-20mm	0.22 max	0.3 max	0.3 max	0.8-1.1	0.3 max	0.04 max	0.05 max	0.15 max	Al <=0.1; N <=0.008; As<=0.08; bal Fe	370-490	245	26L	
GOST 380(71)	WSt3Gps2	Gen struct; Semi-killed; >100mm	0.22 max	0.3 max	0.3 max	0.8-1.1	0.3 max	0.04 max	0.05 max	0.15 max	Al <=0.1; N <=0.008; As<=0.08; bal Fe	370-490	205	23L	
GOST 380(71)	WSt3Gps3	Gen struct; Semi-killed; 0-20mm	0.22 max	0.3 max	0.3 max	0.8-1.1	0.3 max	0.04 max	0.05 max	0.15 max	Al <=0.1; N <=0.008; As<=0.08; bal Fe	370-490	245	26L	
GOST 380(71)	WSt3Gps3	Gen struct; Semi-killed; >100mm	0.22 max	0.3 max	0.3 max	0.8-1.1	0.3 max	0.04 max	0.05 max	0.15 max	Al <=0.1; N <=0.008; As<=0.08; bal Fe	370-490	205	23L	
GOST 380(71)	WSt3Gps4	Gen struct; Semi-killed; >100mm	0.22 max	0.3 max	0.3 max	0.8-1.1	0.3 max	0.04 max	0.05 max	0.15 max	Al <=0.1; N <=0.008; As<=0.08; bal Fe	370-490	205	23L	
GOST 380(71)	WSt3Gps4	Gen struct; Semi-killed; 0-20mm	0.22 max	0.3 max	0.3 max	0.8-1.1	0.3 max	0.04 max	0.05 max	0.15 max	Al <=0.1; N <=0.008; As<=0.08; bal Fe	370-490	245	26L	
GOST 380(71)	WSt3Gps5	Gen struct; Semi-killed; 0-20mm	0.22 max	0.3 max	0.3 max	0.8-1.1	0.3 max	0.04 max	0.05 max	0.15 max	Al <=0.1; N <=0.008; As<=0.08; bal Fe	370-490	245	26L	
GOST 380(71)	WSt3Gps5	Gen struct; Semi-killed; >100mm	0.22 max	0.3 max	0.3 max	0.8-1.1	0.3 max	0.04 max	0.05 max	0.15 max	Al <=0.1; N <=0.008; As<=0.08; bal Fe	370-490	205	23L	
GOST 380(71)	WSt3Gps6	Gen struct; Semi-killed; 0-20mm	0.22 max	0.3 max	0.3 max	0.8-1.1	0.3 max	0.04 max	0.05 max	0.15 max	Al <=0.1; N <=0.008; As<=0.08; bal Fe	370-490	245	26L	
GOST 380(71)	WSt3Gps6	Gen struct; Semi-killed; >100mm	0.22 max	0.3 max	0.3 max	0.8-1.1	0.3 max	0.04 max	0.05 max	0.15 max	Al <=0.1; N <=0.008; As<=0.08; bal Fe	370-490	205	23L	
GOST 380(71)	WSt3kp	Gen struct; Unkilled; Rimming; R; FU; 0-20mm	0.22 max	0.3 max	0.3 max	0.3-0.8	0.3 max	0.04 max	0.05 max	0.07 max	Al <=0.1; N <=0.008; As<=0.08; bal Fe	360-460		27L	
GOST 380(71)	WSt3kp	Gen struct; Unkilled; Rimming; R; FU; >40mm	0.22 max	0.3 max	0.3 max	0.3-0.8	0.3 max	0.04 max	0.05 max	0.07 max	Al <=0.1; N <=0.008; As<=0.08; bal Fe	360-460		24L	
GOST 380(71)	WSt3kp2	Gen struct; Unkilled; Rimming; R; FU; 0-20mm	0.22 max	0.3 max	0.3 max	0.3-0.8	0.3 max	0.04 max	0.05 max	0.07 max	Al <=0.1; N <=0.008; As<=0.08; bal Fe	360-460	235	27L	

Specification	Designation	Notes	C	Cr	Cu	Mn	Ni	P	S	Si	Other	UTS	YS	El	Hard

Carbon Steel, Nonresulfurized, 1017 (Continued from previous page)

Russia

Specification	Designation	Notes	C	Cr	Cu	Mn	Ni	P	S	Si	Other	UTS	YS	El	Hard
GOST 380(71)	WSt3kp2	Gen struct; Unkilled; Rimming; R; FU; >100mm	0.22 max	0.3 max	0.3 max	0.3-0.8	0.3 max	0.04 max	0.05 max	0.07 max	Al <=0.1; N <=0.008; As<=0.08; bal Fe	360-460	195	24L	
GOST 380(71)	WSt3ps	Gen struct; Semi-killed; 0-20mm	0.22 max	0.3 max	0.3 max	0.4-0.85	0.3 max	0.04 max	0.05 max	0.05-0.17	Al <=0.1; N <=0.008; As<=0.08; bal Fe	370-480		26L	
GOST 380(71)	WSt3ps	Gen struct; Semi-killed; >40mm	0.22 max	0.3 max	0.3 max	0.4-0.85	0.3 max	0.04 max	0.05 max	0.05-0.17	Al <=0.1; N <=0.008; As<=0.08; bal Fe	370-480		23L	
GOST 380(71)	WSt3ps2	Gen struct; Semi-killed; >100mm	0.22 max	0.3 max	0.3 max	0.4-0.85	0.3 max	0.04 max	0.05 max	0.05-0.17	Al <=0.1; N <=0.008; As<=0.08; bal Fe	370-480	205	23L	
GOST 380(71)	WSt3ps2	Gen struct; Semi-killed; 0-20mm	0.22 max	0.3 max	0.3 max	0.4-0.85	0.3 max	0.04 max	0.05 max	0.05-0.17	Al <=0.1; N <=0.008; As<=0.08; bal Fe	370-480	245	26L	
GOST 380(71)	WSt3ps3	Gen struct; Semi-killed; 0-20mm	0.22 max	0.3 max	0.3 max	0.4-0.85	0.3 max	0.04 max	0.05 max	0.05-0.17	Al <=0.1; N <=0.008; As<=0.08; bal Fe	370-480	235	25L	
GOST 380(71)	WSt3ps3	Gen struct; Semi-killed; >100mm	0.22 max	0.3 max	0.3 max	0.4-0.85	0.3 max	0.04 max	0.05 max	0.05-0.17	Al <=0.1; N <=0.008; As<=0.08; bal Fe	370-480	205	23L	
GOST 380(71)	WSt3ps4	Gen struct; Semi-killed; >100mm	0.22 max	0.3 max	0.3 max	0.4-0.85	0.3 max	0.04 max	0.05 max	0.05-0.17	Al <=0.1; N <=0.008; As<=0.08; bal Fe	370-480	205	23L	
GOST 380(71)	WSt3ps4	Gen struct; Semi-killed; 0-20mm	0.22 max	0.3 max	0.3 max	0.4-0.85	0.3 max	0.04 max	0.05 max	0.05-0.17	Al <=0.1; N <=0.008; As<=0.08; bal Fe	370-480	245	26L	
GOST 380(71)	WSt3ps5	Gen struct; Semi-killed; >100mm	0.22 max	0.3 max	0.3 max	0.4-0.85	0.3 max	0.04 max	0.05 max	0.05-0.17	Al <=0.1; N <=0.008; As<=0.08; bal Fe	370-480	205	23L	
GOST 380(71)	WSt3ps5	Gen struct; Semi-killed; 0-20mm	0.22 max	0.3 max	0.3 max	0.4-0.85	0.3 max	0.04 max	0.05 max	0.05-0.17	Al <=0.1; N <=0.008; As<=0.08; bal Fe	370-480	245	26L	
GOST 380(71)	WSt3ps6	Gen struct; Semi-killed; >100mm	0.22 max	0.3 max	0.3 max	0.4-0.85	0.3 max	0.04 max	0.05 max	0.05-0.17	Al <=0.1; N <=0.008; As<=0.08; bal Fe	370-480	205	23L	
GOST 380(71)	WSt3ps6	Gen struct; Semi-killed; 0-20mm	0.22 max	0.3 max	0.3 max	0.4-0.85	0.3 max	0.04 max	0.05 max	0.05-0.17	Al <=0.1; N <=0.008; As<=0.08; bal Fe	370-480	245	26L	
GOST 380(71)	WSt3sp	Gen struct; Non-rimming; FF; 0-20mm	0.22 max	0.3 max	0.3 max	0.4-0.85	0.3 max	0.04 max	0.05 max	0.12-0.3	Al <=0.1; N <=0.008; As<=0.08; bal Fe	370-480		26L	
GOST 380(71)	WSt3sp	Gen struct; Non-rimming; FF; >40mm	0.22 max	0.3 max	0.3 max	0.4-0.85	0.3 max	0.04 max	0.05 max	0.12-0.3	Al <=0.1; N <=0.008; As<=0.08; bal Fe	370-480		23L	
GOST 380(71)	WSt3sp4	Gen struct; Non-rimming; FF; 0-20mm	0.22 max	0.3 max	0.3 max	0.4-0.85	0.3 max	0.04 max	0.05 max	0.12-0.3	Al <=0.1; N <=0.008; As<=0.08; bal Fe	370-480	245	26L	
GOST 380(71)	WSt3sp4	Gen struct; Non-rimming; FF; >100mm	0.22 max	0.3 max	0.3 max	0.4-0.85	0.3 max	0.04 max	0.05 max	0.12-0.3	Al <=0.1; N <=0.008; As<=0.08; bal Fe	370-480	205	23L	
GOST 380(71)	WSt3sp5	Gen struct; Non-rimming; FF; >100mm	0.22 max	0.3 max	0.3 max	0.4-0.85	0.3 max	0.04 max	0.05 max	0.12-0.3	Al <=0.1; N <=0.008; As<=0.08; bal Fe	370-480	205	23L	
GOST 380(71)	WSt3sp5	Gen struct; Non-rimming; FF; 0-20mm	0.22 max	0.3 max	0.3 max	0.4-0.85	0.3 max	0.04 max	0.05 max	0.12-0.3	Al <=0.1; N <=0.008; As<=0.08; bal Fe	370-480	245	26L	
GOST 380(71)	WSt3sp6	Gen struct; Non-rimming; FF; >100mm	0.22 max	0.3 max	0.3 max	0.4-0.85	0.3 max	0.04 max	0.05 max	0.12-0.3	Al <=0.1; N <=0.008; As<=0.08; bal Fe	370-480	205	23L	
GOST 380(71)	WSt3sp6	Gen struct; Non-rimming; FF; 0-2-mm	0.22 max	0.3 max	0.3 max	0.4-0.85	0.3 max	0.04 max	0.05 max	0.12-0.3	Al <=0.1; N <=0.008; As<=0.08; bal Fe	370-480	245	26L	
GOST 5781(82)	WSt3Gps2; A-I	HR conc-reinf	0.22 max	0.3 max	0.3 max	0.8-1.1	0.3 max	0.04 max	0.05 max	0.15 max	Al <=0.1; N <=0.008; As<=0.08; bal Fe				
GOST 5781(82)	WSt3kp2; A-I	HR conc-reinf	0.22 max	0.3 max	0.3 max	0.3-0.8	0.3 max	0.04 max	0.05 max	0.07 max	Al <=0.1; N <=0.008; As<=0.08; bal Fe				
GOST 5781(82)	WSt3ps2; A-I	HR conc-reinf	0.22 max	0.3 max	0.3 max	0.4-0.85	0.3 max	0.04 max	0.05 max	0.05-0.17	Al <=0.1; N <=0.008; As<=0.08; bal Fe				
GOST 924	VMSt3		0.14-0.22	0.3	0.3	0.4-0.65	0.3	0.045	0.05	0.12-0.22	N 0.008; As 0.08; bal Fe				
GOST 924	WMSt3		0.14-0.22	0.3	0.3	0.4-0.65	0.3	0.045	0.05	0.12-0.22	N 0.008; As 0.08; bal Fe				

Spain

Specification	Designation	Notes	C	Cr	Cu	Mn	Ni	P	S	Si	Other	UTS	YS	El	Hard
UNE 36080(80)	A360C	Gen struct	0.17 max			1.60 max		0.045 max	0.045 max	0.60 max	Mo <=0.15; N <=0.009; bal Fe				
UNE 36080(80)	A360D	Gen struct	0.17 max			1.60 max		0.04 max	0.04 max	0.60 max	Mo <=0.15; +N; bal Fe				
UNE 36080(85)	AE235-C	Gen struct	0.17 max			1.60 max		0.04 max	0.04 max	0.60 max	Mo <=0.15; N <=0.009; bal Fe				
UNE 36080(85)	AE235-D	Gen struct	0.17 max			1.60 max		0.035 max	0.035 max	0.60 max	Mo <=0.15; bal Fe				

Sweden

Specification	Designation	Notes	C	Cr	Cu	Mn	Ni	P	S	Si	Other	UTS	YS	El	Hard
SS 141312	1312	Semi-killed; Gen struct	0.2			0.4-0.7		0.05	0.05	0.05	N 0.009; bal Fe				

Turkey

Specification	Designation	Notes	C	Cr	Cu	Mn	Ni	P	S	Si	Other	UTS	YS	El	Hard
TS 381	UDK621.643.2/Fe35.8*	Pipe as drawn	0.17			0.4		0.05	0.05	0.35	bal Fe				

Specification	Designation	Notes	C	Cr	Cu	Mn	Ni	P	S	Si	Other	UTS	YS	El	Hard

Carbon Steel, Nonresulfurized, 1017 (Continued from previous page)

Turkey

Specification	Designation	Notes	C	Cr	Cu	Mn	Ni	P	S	Si	Other	UTS	YS	El	Hard
TS 416	Fe34.2*	Weld Pipe	0.17					0.05	0.05		N 0.01; bal Fe				
TS 911(86)	UDK669.14.423/Fe34	I IП, T-bar	0.17					0.06	0.06		bal Fe				

UK

Specification	Designation	Notes	C	Cr	Cu	Mn	Ni	P	S	Si	Other	UTS	YS	El	Hard
BS 1449/1(83)	17CS	Plt Sh Strp	0.15-0.2			0.40-0.60		0.05 max	0.05 max	0.60 max	bal Fe				
BS 1449/1(91)	17CS	Plt Sh Strp	0.15-0.2			0.40-0.60		0.05	0.05		bal Fe	340	190	24-26	
BS 1449/1(91)	17HS	Sh Strp	0.15-0.2			0.40-0.60		0.05	0.05		bal Fe	350	200	23-25	
BS 970/1(83)	040A17	Blm Bil Slab Bar Rod Frg	0.15-0.20			0.30-0.50		0.050 max	0.050 max		bal Fe				
BS 970/1(83)	050A17	Blm Bil Slab Bar Rod Frg	0.15-0.20			0.40-0.60		0.050 max	0.050 max	0.4	bal Fe				
BS 970/1(83)	060A17	Blm Bil Slab Bar Rod Frg	0.15-0.20			0.50-0.70		0.050 max	0.050 max	0.4	bal Fe				

USA

Specification	Designation	Notes	C	Cr	Cu	Mn	Ni	P	S	Si	Other	UTS	YS	El	Hard
	AISI 1017	Sh Strp Plt	0.14-0.20			0.30-0.60		0.030 max	0.035 max		bal Fe				
	AISI 1017	Wir rod	0.15-0.20			0.30-0.60		0.040 max	0.050 max		bal Fe				
	AISI 1017	Bar	0.14-0.20			0.30-0.60		0.040 max	0.050 max		bal Fe				
	AISI 1017	Struct shp	0.15-0.20			0.30-0.60		0.030 max	0.035 max		bal Fe				
	AISI M1017	Bar, merchant qual	0.14-0.21			0.25-0.60		0.04 max	0.05 max		bal Fe				
	UNS G10170		0.15-0.20			0.30-0.60		0.040 max	0.050 max		Sheets and Plates, C 0.14-0.20; ERW Tubing, C 0.14-0.21; bal Fe				
ASTM A29/A29M(93)	1017	Bar	0.14-0.20			0.30-0.60		0.040 max	0.050 max		bal Fe				
ASTM A29/A29M(93)	M1017	Bar, merchant qual	0.14-0.21			0.25-0.6		0.040 max	0.050 max		bal Fe				
ASTM A510(96)	1017	Wir rod	0.15-0.20			0.30-0.60		0.040 max	0.050 max		bal Fe				
ASTM A513(97a)	1017	ERW Mech tub	0.14-0.20			0.30-0.60		0.035 max	0.035 max		bal Fe				
ASTM A519(96)	1017	Smls mech tub	0.15-0.20			0.30-0.60		0.040 max	0.050 max		bal Fe				
ASTM A575(96)	M1017	Merchant qual bar	0.14-0.21			0.25-0.60		0.04 max	0.05 max		Mn can vary with C; bal Fe				
ASTM A576(95)	1017	Special qual HW bar	0.15-0.20			0.30-0.60		0.040 max	0.050 max		Si Cu Pb B Bi Ca Se Te if spec'd; bal Fe				
ASTM A611(97)	A611(A,B,C,1,E)	SS Sh, CR	0.20 max		0.20 min	0.60 max		0.035 max	0.035 max		Cu if spec'd; bal Fe	290-565	170-550	20-26	
ASTM A635/A635M(98)	1017	Sh Strp, Coil, HR	0.14-0.20		0.20 min	0.30-0.60		0.030 max	0.03 max		bal Fe				
ASTM A659/A659M(97)	1017	Sh Strp, HR	0.14-0.20			0.30-0.60		0.030 max	0.035 max		bal Fe				
ASTM A787(96)	1017	ERW metallic-coated mech tub	0.14-0.21			0.30-0.60		0.035 max	0.035 max		bal Fe				
ASTM A794(97)	1017	Sh, CS, CR	0.14-0.20			0.30-0.60		0.030 max	0.035 max		Cu >=0.20 if spec; bal Fe				
ASTM A830/A830M(98)	1017	Plt	0.14-0.20			0.30-0.60		0.035 max	0.04 max		bal Fe				
SAE J1397(92)	1017	Bar CD, est mech prop									bal Fe	410	340	18	116 HB
SAE J1397(92)	1017	Bar HR, est mech prop									bal Fe	370	200	26	105 HB
SAE J403(95)	1017	Bar Wir rod Smls Tub HR CF	0.15-0.20			0.30-0.60		0.030 max	0.050 max		bal Fe				
SAE J403(95)	1017	Struct Shps Plt Strp Sh Weld Tub	0.14-0.20			0.30-0.60		0.030 max	0.035 max		bal Fe				

Yugoslavia

Specification	Designation	Notes	C	Cr	Cu	Mn	Ni	P	S	Si	Other	UTS	YS	El	Hard
	C.0270	Struct	0.17 max			1.60 max		0.06 max	0.06 max	0.60 max	bal Fe				
	C.0361	Struct	0.17 max			1.60 max		0.05 max	0.05 max	0.60 max	N <=0.007; bal Fe				

UNS numbers and US grades are provided as a means of cross referencing chemically similar alloys. Exchangability is only possible after independent examination of specifications. Tensile properties are minimum or typical as specified. UTS and YS as MPa. El as %. See Appendix for list of abbreviations used in Notes. * indicates obsolete material.

Specification	Designation	Notes	C	Cr	Cu	Mn	Ni	P	S	Si	Other	UTS	YS	El	Hard

Carbon Steel, Nonresulfurized, 1017 (Continued from previous page)

Yugoslavia

Specification	Designation	Notes	C	Cr	Cu	Mn	Ni	P	S	Si	Other	UTS	YS	El	Hard
	C.0362	Struct	0.17 max			1.60 max		0.045 max	0.045 max	0.60 max	N <=0.009; bal Fe				
	C.0363	Struct	0.17 max			1.60 max		0.045 max	0.045 max	0.60 max	N <=0.009; bal Fe				
	C.1211	Struct	0.14-0.22			0.3-0.6		0.06 max	0.06 max	0.60 max	bal Fe				

Carbon Steel, Nonresulfurized, 1018

Argentina

Specification	Designation	Notes	C	Cr	Cu	Mn	Ni	P	S	Si	Other	UTS	YS	El	Hard
IAS	IRAM 1018		0.15-0.20			0.60-0.90		0.040 max	0.050 max		bal Fe	470	345	36	143 HB

Australia

Specification	Designation	Notes	C	Cr	Cu	Mn	Ni	P	S	Si	Other	UTS	YS	El	Hard
AS 1442	K1018*	Obs; Bar Bil	0.15-0.2			0.6-0.9		0.05	0.05	0.1-0.35	bal Fe				
AS 1442(92)	U1	Rnd<= 50mm, HR as rolled or norm	0.20 max			0.40-1.20		0.040 max	0.040 max	0.40 max	bal Fe	400	220	26	
AS 1443	K1018*	Obs; Bar	0.15-0.2			0.6-0.9		0.05	0.05	0.1-0.35	bal Fe				

Belgium

Specification	Designation	Notes	C	Cr	Cu	Mn	Ni	P	S	Si	Other	UTS	YS	El	Hard
NBN 629	D37-2	Strip As roll 63mm diam	0.17			0.4-1.2		0.04	0.04	0.1-0.35	bal Fe				
NBN 630	E37-1	Strip As roll 3/63mm diam	0.17			0.4-1.2		0.05	0.05	0.35	bal Fe				
NBN 630	E37-2	Strip As roll 3/40mm diam	0.17			0.4-1.2		0.04	0.04	0.1-0.35	bal Fe				
NBN A21-221	C17KD	Wir	0.14-0.2			0.6-0.9		0.04	0.05	0.1-0.4	bal Fe				

Bulgaria

Specification	Designation	Notes	C	Cr	Cu	Mn	Ni	P	S	Si	Other	UTS	YS	El	Hard
BDS 9801	S	Struct, weld ships	0.14-0.2	0.3	0.3	0.5-0.8	0.3	0.045	0.04	0.12-0.35	bal Fe				

Canada

Specification	Designation	Notes	C	Cr	Cu	Mn	Ni	P	S	Si	Other	UTS	YS	El	Hard
CSA 3TAN95-1-I	C1018	Bar	0.15-0.2			0.6-0.9		0.04	0.05	0.1-0.3	bal Fe				

Czech Republic

Specification	Designation	Notes	C	Cr	Cu	Mn	Ni	P	S	Si	Other	UTS	YS	El	Hard
CSN 412020	12020	CH	0.13-0.2	0.25 max		0.6-0.9	0.3 max	0.04 max	0.04 max	0.15-0.4	bal Fe				
CSN 412022	12022	High-temp const; Boiler tube	0.15-0.22	0.25 max	0.25 max	0.5-0.8	0.25 max	0.04 max	0.04 max	0.17-0.37	bal Fe				

Europe

Specification	Designation	Notes	C	Cr	Cu	Mn	Ni	P	S	Si	Other	UTS	YS	El	Hard
EN 10016/2(94)	1.0416	Rod	0.15-0.20	0.20 max	0.30 max	0.30-0.60	0.25 max	0.035 max	0.035 max	0.30 max	Al <=0.01; Mo <=0.05; bal Fe				
EN 10016/2(94)	C18D	Rod	0.15-0.20	0.20 max	0.30 max	0.30-0.60	0.25 max	0.035 max	0.035 max	0.30 max	Al <=0.01; Mo <=0.05; bal Fe				
EN 10016/4(94)	1.1129	Rod	0.16-0.20	0.10 max	0.15 max	0.30-0.50	0.10 max	0.020 max	0.025 max	0.30 max	Al <=0.01; Mo <=0.05; N <=0.007; Cr+Ni+Cu<=0.30, Cu+Sn<=0.15; bal Fe				
EN 10016/4(94)	C18D2	Rod	0.16-0.20	0.10 max	0.15 max	0.30-0.50	0.10 max	0.020 max	0.025 max	0.30 max	Al <=0.01; Mo <=0.05; N <=0.007; Cr+Ni+Cu<=0.30, Cu+Sn<=0.15; bal Fe				

Finland

Specification	Designation	Notes	C	Cr	Cu	Mn	Ni	P	S	Si	Other	UTS	YS	El	Hard
SFS 1100	Fe44DP	Gen struct; press ves	0.18 max		0.4 max	0.4-1.2		0.045 max	0.045 max	0.15-0.55	N <=0.009; As 0-0.08; bal Fe				
SFS 1255(77)	SFS1255	Reinforcing	0.18 max			0.5 max		0.06 max	0.05 max	0.6 max	N <=0.01; As 0-0.08; S/Mn=0.1; bal Fe				
SFS 1256(80)	SFS1256	Reinforcing	0.18 max			0.5 max		0.06 max	0.05 max	0.6 max	N <=0.01; As 0-0.08; S/Mn=0.1; bal Fe				

Germany

Specification	Designation	Notes	C	Cr	Cu	Mn	Ni	P	S	Si	Other	UTS	YS	El	Hard
DIN	C16.8		0.14-0.19	0.30 max		0.40-0.80		0.040 max	0.040 max	0.15-0.35	bal Fe				
DIN	WNr 1.0453		0.14-0.19	0.30 max		0.40-0.80		0.040 max	0.040 max	0.15-0.35	bal Fe				

Japan

Specification	Designation	Notes	C	Cr	Cu	Mn	Ni	P	S	Si	Other	UTS	YS	El	Hard
JIS G3507(91)	SWRCH18A	Wir rod Al killed	0.15-0.20			0.60-0.90		0.030 max	0.035 max	0.10 max	Al >=0.02; bal Fe				

UNS numbers and US grades are provided as a means of cross referencing chemically similar alloys. Exchangability is only possible after independent examination of specifications. Tensile properties are minimum or typical as specified. UTS and YS as MPa. El as %. See Appendix for list of abbreviations used in Notes. * indicates obsolete material.

Specification	Designation	Notes	C	Cr	Cu	Mn	Ni	P	S	Si	Other	UTS	YS	El	Hard

Carbon Steel, Nonresulfurized, 1018 (Continued from previous page)

Japan

Specification	Designation	Notes	C	Cr	Cu	Mn	Ni	P	S	Si	Other	UTS	YS	El	Hard
JIS G3507(91)	SWRCH18K	Wir rod FF	0.15-0.20			0.60-0.90		0.030 max	0.035 max	0.10-0.35	bal Fe				

Mexico

NMX-B-301(86)	1018	Bar	0.15-0.20			0.60-0.90		0.040 max	0.050 max		bal Fe				

Pan America

COPANT 331	1018		0.15-0.2			0.6-0.9		0.04	0.05		bal Fe				
COPANT 333	1018		0.15-0.2			0.6-0.9		0.04	0.05		bal Fe				

Russia

GOST	M18S		0.14-0.22	0.3	0.4	0.4-0.8	0.3	0.05	0.05	0.12-0.35	bal Fe				
GOST	MStT		0.09-0.22			0.3-0.5		0.05	0.05	0.12	bal Fe				
GOST	St20A		0.11-0.22	0.1-0.3							bal Fe				
GOST	St3		0.14-0.22			0.3-0.5		0.05 max	0.05 max		bal Fe				
GOST	St3T		0.12-0.22					0.04 max	0.04 max		bal Fe				
GOST 23570	18ps	Gen struct	0.14-0.22	0.3 max	0.3 max	0.5-0.8	0.3 max	0.04 max	0.045 max	0.05-0.15	Al <=0.1; bal Fe				
GOST 23570	18ps		0.14-0.22	0.3	0.3	0.5-0.8	0.3	0.04	0.045	0.05-0.15	bal Fe				
GOST 23570	18sp		0.14-0.22	0.3	0.3	0.5-0.8	0.3	0.04	0.045	0.15-0.3	bal Fe				
GOST 23570	18sp	Gen struct	0.14-0.22	0.3 max	0.3 max	0.5-0.8	0.3 max	0.04 max	0.045 max	0.15-0.3	Al <=0.1; bal Fe				
GOST 5521	S		0.14-0.2	0.30	0.3-0.35	0.5-0.9	0.3	0.04	0.04	0.12-0.35	N 0.008; bal Fe				

Spain

UNE 36080(80)	A430C	Gen struct	0.2 max			1.60 max		0.045 max	0.045 max	0.60 max	Mo <=0.15; N <=0.009; bal Fe				
UNE 36080(80)	A430D	Gen struct	0.2 max			1.60 max		0.04 max	0.04 max	0.60 max	Mo <=0.15; +N; bal Fe				
UNE 36080(85)	AE275-C	Gen struct	0.2 max			1.60 max		0.04 max	0.04 max	0.60 max	Mo <=0.15; N <=0.009; bal Fe				
UNE 36080(85)	AE275-D	Gen struct	0.2 max			1.60 max		0.035 max	0.035 max	0.60 max	Mo <=0.15; bal Fe				
UNE 36087(74)	A37RBII	Sh Plt	0.2 max			0.3-1.3		0.045 max	0.04 max	0.4 max	Al 0.015-0.10; Mo <=0.15; 0-40mm Mn<=1.3; 40mm Mn<=1.1; bal Fe				

Turkey

TS 302	Fe35.2*	Pipe as drawn or weld	0.18					0.05	0.05		bal Fe				
TS 346	Fe35*	Pipe as drawn	0.18			0.4		0.05 max		0.35	bal Fe				

UK

BS 1717(83)	CEWC3	Tub	0.20 max			0.90 max		0.050 max	0.050 max	0.35 max	W <=0.1; bal Fe				
BS 1717(83)	CFSC3	Tub	0.20 max			0.90 max		0.050 max	0.050 max	0.35 max	W <=0.1; bal Fe				
BS 1717(83)	ERWC3	Tub	0.20 max			0.90 max		0.050 max	0.050 max	0.35 max	W <=0.1; bal Fe				
BS 970/1(83)	080A17	Blm Bil Slab Bar Rod Frg	0.15-0.20			0.70-0.90		0.050 max	0.050 max	0.4	bal Fe				

USA

	AISI 1018	Bar	0.15-0.20			0.60-0.90		0.040 max	0.050 max		bal Fe				
	AISI 1018	Struct shp	0.15-0.20			0.60-0.90		0.030 max	0.035 max		bal Fe				
	AISI 1018	Sh Strp Plt	0.14-0.20			0.60-0.90		0.030 max	0.035 max		bal Fe				
	AISI 1018	Wir rod	0.15-0.20			0.60-0.90		0.040 max	0.050 max		bal Fe				

UNS numbers and US grades are provided as a means of cross referencing chemically similar alloys. Exchangability is only possible after independent examination of specifications. Tensile properties are minimum or typical as specified. UTS and YS as MPa. El as %. See Appendix for list of abbreviations used in Notes. * indicates obsolete material.

Specification	Designation	Notes	C	Cr	Cu	Mn	Ni	P	S	Si	Other	UTS	YS	El	Hard

Carbon Steel, Nonresulfurized, 1018 (Continued from previous page)

USA

Specification	Designation	Notes	C	Cr	Cu	Mn	Ni	P	S	Si	Other	UTS	YS	El	Hard
	UNS G10180		0.15-0.20			0.60-0.90		0.040 max	0.050 max		Sheets and Plates, C 0.14-0.20; ERW Tubing, C 0.14-0.21; bal Fe				
AMS 5069E		Bar Frg, Tub, HB varies with form	0.15-0.20			0.60-0.90		0.040 max	0.050 max	0.15-0.35	bal Fe				240 HB max
ASTM A108(95)	1018	Bar, CF	0.15-0.20			0.60-0.90		0.040 max	0.050 max		bal Fe				
ASTM A29/A29M(93)	1018	Bar	0.15-0.20			0.60-0.90		0.040 max	0.050 max		bal Fe				
ASTM A311/A311M(95)	1018 Class A	Bar CD SR ann, t<=20mm	0.15-0.20			0.60-0.90		0.040 max	0.050 max		bal Fe	485	415	18,R A=40	
ASTM A311/A311M(95)	1018 Class A	Bar CD SR ann, 20<t<=30mm	0.15-0.20			0.60-0.90		0.040 max	0.050 max		bal Fe	450	380	16,R A=40	
ASTM A311/A311M(95)	1018 Class A	Bar CD SR ann, 50<t<=75mm	0.15-0.20			0.60-0.90		0.040 max	0.050 max		bal Fe	380	310	15,R A=35	
ASTM A311/A311M(95)	1018 Class A	Bar CD SR ann, 30<t<=50mm	0.15-0.20			0.60-0.90		0.040 max	0.050 max		bal Fe	415	345	15,R A=35	
ASTM A510(96)	1018	Wir rod	0.15-0.20			0.60-0.90		0.040 max	0.050 max		bal Fe				
ASTM A512(96)	1018	Buttweld mech tub, CD	0.14-0.20			0.60-0.90		0.040 max	0.045 max		bal Fe				
ASTM A513(97a)	1018	ERW Mech tub	0.14-0.20			0.60-0.90		0.035 max	0.035 max		bal Fe				
ASTM A519(96)	1018	Smls mech tub	0.15-0.20			0.60-0.90		0.040 max	0.050 max		bal Fe				
ASTM A576(95)	1018	Special qual HW bar	0.15-0.20			0.60-0.90		0.040 max	0.050 max		Si Cu Pb B Bi Ca Se Te if spec'd; bal Fe				
ASTM A611(97)	A611(D,1)	SS Sh, CR	0.20 max		0.20 min	0.90 max		0.035 max	0.035 max		Cu if spec'd; bal Fe	360	275	20	
ASTM A635/A635M(98)	1018	Sh Strp, Coil, HR	0.14-0.20		0.20 min	0.60-0.90		0.030 max	0.035 max		bal Fe				
ASTM A659/A659M(97)	1018	Sh Strp, HR	0.14-0.20			0.60-0.90		0.030 max	0.05-0.035		bal Fe				
ASTM A787(96)	1018	ERW metallic-coated mech tub	0.14-0.21			0.60-0.90		0.035 max	0.035 max		bal Fe				
ASTM A794(97)	1018	Sh, CS, CR	0.14-0.20			0.60-0.90		0.030 max	0.035 max		Cu >=0.20 if spec; bal Fe				
ASTM A830/A830M(98)	1018	Plt	0.14-0.20			0.60-0.90		0.035 max	0.04 max		bal Fe				
FED QQ-S-698(88)	1018	Sh Strp	0.14-0.21			0.60-0.90		0.040 max	0.050 max		bal Fe				
FED QQ-W-461H(88)	1018*	Obs; Wir	0.15-0.2			0.6-0.9		0.04	0.05		bal Fe				
MIL-R-8814	1018*	Obs; Rvt	0.15-0.2			0.6-0.9		0.04	0.05		bal Fe				
MIL-S-10520D(AR)(88)	1*	Frg; Obs; see ASTM A711, A575	0.2	0.2	0.5	0.9	0.25	0.04	0.05	0.2	Mo 0.06; bal Fe				
MIL-S-11310E(76)	1018*	Bar HR cold shape cold ext; Obs for new design	0.15-0.2	0.07	0.2	0.6-0.9	0.15	0.04	0.05	0.2	Mo 0.05; bal Fe				
MIL-S-11310E(76)	CS1018*	Bar HR cold shape cold ext; Obs for new design									bal Fe				
SAE J1397(92)	1018	Bar HR, est mech prop									bal Fe	400	220	25	116 HB
SAE J1397(92)	1018	Bar CD, est mech prop									bal Fe	440	370	15	126 HB
SAE J403(95)	1018	Bar Wir rod Smls Tub HR CF	0.15-0.20			0.60-0.90		0.030 max	0.050 max		bal Fe				
SAE J403(95)	1018	Struct Shps Plt Strp Sh Weld Tub	0.14-0.20			0.60-0.90		0.030 max	0.035 max		bal Fe				

Yugoslavia

Specification	Designation	Notes	C	Cr	Cu	Mn	Ni	P	S	Si	Other	UTS	YS	El	Hard
	C.0375	Unalloyed; rivets	0.18 max			0.25-0.5		0.05 max	0.05 max	0.4 max	N <=0.007; bal Fe				
	C.0445	Unalloyed; rivets	0.18 max			0.8 max		0.05 max	0.05 max	0.45 max	N <=0.007; bal Fe				

UNS numbers and US grades are provided as a means of cross referencing chemically similar alloys. Exchangability is only possible after independent examination of specifications. Tensile properties are minimum or typical as specified. UTS and YS as MPa. El as %. See Appendix for list of abbreviations used in Notes. * indicates obsolete material.

Specification	Designation	Notes	C	Cr	Cu	Mn	Ni	P	S	Si	Other	UTS	YS	El	Hard

Carbon Steel, Nonresulfurized, 1019

Austria

Specification	Designation	Notes	C	Cr	Cu	Mn	Ni	P	S	Si	Other	UTS	YS	El	Hard
ONORM M3121	17Mn4KK		0.14-0.20	0.3		0.9-1.2		0.040	0.040	0.2-0.4	bal Fe				
ONORM M3121	17Mn4KKW		0.14-0.20	0.3		0.9-1.2		0.040	0.040	0.2-0.4	bal Fe				
ONORM M3121	17Mn4KW		0.14-0.20	0.3		0.9-1.2		0.040	0.040	0.2-0.4	bal Fe				

Finland

Specification	Designation	Notes	C	Cr	Cu	Mn	Ni	P	S	Si	Other	UTS	YS	El	Hard
SFS 200	Fe44C	Gen struct	0.18 max			1.60 max		0.045 max	0.045 max	0.55 max	Al <=0.10; N <=0.009; bal Fe				
SFS 200	Fe44D	Gen struct	0.20 max		0.4 max	1.60 max		0.04 max	0.04 max	0.55 max	Al 0.02-0.10; N <=0.015; bal Fe				

France

Specification	Designation	Notes	C	Cr	Cu	Mn	Ni	P	S	Si	Other	UTS	YS	El	Hard
AFNOR NFA35553	XC18S	Strp, Norm	0.15-0.22			0.4-0.65		0.04	0.04	0.25	bal Fe				
AFNOR NFA35554	XC18S	Plt, HR, Norm, 16mm diam	0.15-0.22			0.4-0.65		0.04	0.04	0.25	bal Fe				
AIR 9113A	XC18S	Sh Tub Bar Wir, Norm, Ann 1.5mm	0.15-0.22			0.45-0.65		0.04	0.04	0.25	bal Fe				

Germany

Specification	Designation	Notes	C	Cr	Cu	Mn	Ni	P	S	Si	Other	UTS	YS	El	Hard
DIN 1623(86)	S275J2G3	CR Sht Strp 100-150mm	0.18 max			1.50 max		0.035 max	0.035 max		C 0.20 max if >=150 mm thk; bal Fe	400-540	225	18	
DIN 1623(86)	WNr 1.0036	Flat product CR Sh Strp	0.17 max			1.40 max		0.045 max	0.045 max		N <=0.007; N depends on P; C<=0.20 if >=16mm; bal Fe				
DIN 1623(86)	WNr 1.0144	CR Sh Strp 100-150mm	0.18 max			1.50 max		0.035 max	0.035 max		C<=0.20 if >=150mm thk; bal Fe	400-540	225	18	
DIN 1652(90)	WNr 1.0036	Bright struct	0.17 max			1.40 max		0.045 max	0.045 max		N <=0.007; N depends on P; C<=0.20 if >=16mm; bal Fe				
DIN 17119(84)	S275J2G3	Weld CF sq rect tub 100-150mm	0.18 max			1.50 max		0.035 max	0.035 max		C 0.20 max if >=150 mm thk; bal Fe	400-540	225	18	
DIN 17119(84)	WNr 1.0036	Tube	0.17 max			1.40 max		0.045 max	0.045 max		N <=0.007; N depends on P; C<=0.20 if >=16mm; bal Fe				
DIN 17119(84)	WNr 1.0144	Weld CF sq rect tub 100-150mm	0.18 max			1.50 max		0.035 max	0.035 max		C<=0.20 if >=150mm thk; bal Fe	400-540	225	18	
DIN 2393(94)	WNr 1.0036	Tube	0.17 max			1.40 max		0.045 max	0.045 max		N <=0.007; N depends on P; C<=0.20 if >=16mm; bal Fe				
DIN 2394(94)	WNr 1.0036	Tube	0.17 max			1.40 max		0.045 max	0.045 max		N <=0.007; N depends on P; C<=0.20 if >=16mm; bal Fe				
DIN EN 10025(94)	S235JRG1	HR	0.17 max			1.40 max		0.045 max	0.045 max		N <=0.007; N depends on P; C<=0.20 if >=16 mm; bal Fe				
DIN EN 10025(94)	S275J2G3	HR 100-150mm thk	0.18 max			1.50 max		0.035 max	0.035 max		C 0.20 max if >=150 mm thk; bal Fe	400-540	225	18	
DIN EN 10025(94)	WNr 1.0144	HR 100-150mm thk	0.18 max			1.50 max		0.035 max	0.035 max		C<=0.20 if >=150mm thk; bal Fe	400-540	225	18	
DIN EN 10210(94)	S275J2G3	Smls tube 100-150mm	0.18 max			1.50 max		0.035 max	0.035 max		C 0.20 max if >=150 mm thk; bal Fe	400-540	225	18	
DIN EN 10210(94)	WNr 1.0144	Smls tube 100-150mm	0.18 max			1.50 max		0.035 max	0.035 max		C<=0.20 if >=150mm thk; bal Fe	400-540	225	18	

Hungary

Specification	Designation	Notes	C	Cr	Cu	Mn	Ni	P	S	Si	Other	UTS	YS	El	Hard
MSZ 500(81)	B44B	Gen struct; FF	0.14-0.21			0.7-1		0.045 max	0.05 max	0.12-0.35	bal Fe				

India

Specification	Designation	Notes	C	Cr	Cu	Mn	Ni	P	S	Si	Other	UTS	YS	El	Hard
IS 1570/2(79)	C15Mn75	Sh Plt Sect Shp Bar Bil Frg	0.1-0.2	0.3 max	0.3 max	0.6-0.9	0.4 max	0.055 max	0.055 max	0.6 max	Co <=0.1; Mo <=0.15; Pb <=0.15; W <=0.1; P:S Varies; bal Fe	420-500	231	25	

UNS numbers and US grades are provided as a means of cross referencing chemically similar alloys. Exchangability is only possible after independent examination of specifications. Tensile properties are minimum or typical as specified. UTS and YS as MPa. El as %. See Appendix for list of abbreviations used in Notes. * indicates obsolete material.

Specification	Designation	Notes	C	Cr	Cu	Mn	Ni	P	S	Si	Other	UTS	YS	El	Hard
Carbon Steel, Nonresulfurized, 1019 (Continued from previous page)															
India															
IS 1570/2(79)	C15Mn75	Sh Plt Sect Shp Bar Bil Frg	0.1-0.2	0.3 max	0.3 max	0.6-0.9	0.4 max	0.055 max	0.055 max	0.6 max	Co <=0.1; Mo <=0.15; Pb <=0.15; W <=0.1; P:S Varies; bal Fe	420-500	231	25	
Mexico															
NMX-B-301(86)	1019	Bar	0.15-0.20			0.70-1.00		0.040 max	0.050 max		bal Fe				
Pan America															
COPANT 333	1019		0.15-0.2			0.7-1		0.04	0.05		bal Fe				
Poland															
PNH84018(86)	18G2	HS weld const	0.22 max	0.3 max		1-1.6	0.3 max	0.05 max	0.05 max	0.2-0.55	bal Fe				
PNH84023/08	18G2		0.15-0.22	0.3	0.35	1-1.5	0.3	0.05	0.05	0.2-0.55	Mo 0.1; Ce 0.48; bal Fe				
Russia															
GOST 19282	17GS		0.14-0.22	0.3	0.3	1-1.4	0.3	0.035	0.04	0.4-0.6	bal Fe				
GOST 23570	18Gps		0.14-0.22	0.3	0.3	0.8-1.1	0.03	0.04	0.045	0.15	bal Fe				
GOST 23570	18Gsp		0.14-0.2	0.3	0.3	0.8-1.1	0.3	0.04	0.045	0.15-0.3	bal Fe				
GOST 380	BSt3Gps		0.14-0.22			0.8-1.1		0.04	0.05	0.15	N 0.008; As 0.08; bal Fe				
GOST 380	BSt3Gps2		0.14-0.22	0.3	0.3	0.8-1.1	0.3	0.04	0.05	0.15	N 0.008; As 0.08; bal Fe				
Spain															
UNE 36032(85)	18KA-DF	Screw	0.16-0.21			0.30-0.60		0.040 max	0.040 max	0.10 max	Al >=0.020; bal Fe				
UNE 36032(85)	F.7516*	see 18KA-DF	0.16-0.21			0.30-0.60		0.040 max	0.040 max	0.10 max	Al >=0.020; bal Fe				
UNE 36080(73)	A430B		0.22					0.05	0.05		N 0.009; bal Fe				
UNE 36080(73)	F.6210*	see A430B	0.22					0.05	0.05		N 0.009; bal Fe				
USA															
	AISI 1019	Wir rod	0.15-0.20			0.70-1.00		0.040 max	0.050 max		bal Fe				
	AISI 1019	Sh Strp Plt	0.14-0.20			0.70-1.00		0.030 max	0.035 max		bal Fe				
	AISI 1019	Struct shp	0.15-0.20			0.70-1.00		0.030 max	0.035 max		bal Fe				
	UNS G10190		0.15-0.20			0.70-1.00		0.040 max	0.050 max		Sheets and Plates, C 0.14-0.20; ERW Tubing, C 0.14-0.21; bal Fe				
ASTM A29/A29M(93)	1019	Bar	0.15-0.20			0.70-1.00		0.040 max	0.050 max		bal Fe				
ASTM A510(96)	1019	Wir rod	0.15-0.20			0.70-1.00		0.040 max	0.050 max		bal Fe				
ASTM A512(96)	1019	Buttweld mech tub, CD	0.14-0.20			0.70-1.00		0.040 max	0.045 max		bal Fe				
ASTM A513(97a)	1019	ERW Mech tub	0.14-0.20			0.70-1.00		0.035 max	0.035 max		bal Fe				
ASTM A519(96)	1019	Smls mech tub	0.15-0.20			0.70-1.00		0.040 max	0.050 max		bal Fe				
ASTM A576(95)	1019	Special qual HW bar	0.15-0.20			0.70-1.00		0.040 max	0.050 max		Si Cu Pb B Bi Ca Se Te if spec'd; bal Fe				
ASTM A635/A635M(98)	1019	Sh Strp, Coil, HR	0.14-0.20		0.20 min	0.70-1.00		0.030 max	0.035 max		bal Fe				
ASTM A787(96)	1019	ERW metallic-coated mech tub	0.14-0.21			0.70-1.00		0.035 max	0.035 max		bal Fe				
ASTM A830/A830M(98)	1019	Plt	0.14-0.20			0.70-1.00		0.035 max	0.04 max		bal Fe				
FED QQ-S-633A(63)	1019*	Obs; CF HR Bar	0.15-0.2			0.7-1		0.04 max	0.05 max		bal Fe				
MIL-R-908	- - -*	Obs; Weld rod	0.15-0.2			0.7-1		0.04 max	0.05 max		bal Fe				

UNS numbers and US grades are provided as a means of cross referencing chemically similar alloys. Exchangability is only possible after independent examination of specifications. Tensile properties are minimum or typical as specified. UTS and YS as MPa. El as %. See Appendix for list of abbreviations used in Notes. * indicates obsolete material.

Specification	Designation	Notes	C	Cr	Cu	Mn	Ni	P	S	Si	Other	UTS	YS	El	Hard

Carbon Steel, Nonresulfurized, 1019 (Continued from previous page)

USA

Specification	Designation	Notes	C	Cr	Cu	Mn	Ni	P	S	Si	Other	UTS	YS	El	Hard
MIL-S-16113	- - -*	Obs; Plt	0.15-0.2			0.7-1		0.04 max	0.05 max		bal Fe				
MIL-S-20166B	- - -*	Obs; Bar shape	0.15-0.2			0.7-1		0.04 max	0.05 max		bal Fe				
MIL-S-7809	- - -*	Obs	0.15-0.2			0.7-1		0.04 max	0.05 max		bal Fe				
SAE J1249(95)	1019	Former SAE std valid for Wir rod	0.15-0.20			0.70-1.00		0.040 max	0.050 max		bal Fe				
SAE J1397(92)	1019	Bar HR, est mech prop	0.15-0.2			0.7-1		0.04 max	0.05 max		bal Fe	410	220	25	116 HB
SAE J1397(92)	1019	Bar CD, est mech prop	0.15-0.2			0.7-1		0.04 max	0.05 max		bal Fe	460	380	15	131 HB
SAE J403(95)	1019	Struct Shps Plt Strp Sh Weld Tub	0.14-0.20			0.70-1.00		0.030 max	0.035 max		bal Fe				

Yugoslavia

Specification	Designation	Notes	C	Cr	Cu	Mn	Ni	P	S	Si	Other	UTS	YS	El	Hard
	C.0345V	Struct	0.30 max			1.60 max		0.05 max	0.05 max	0.60 max	P+S<=0.09; bal Fe				
	C.0371	Struct	0.17 max			1.60 max		0.05 max	0.05 max	0.60 max	N <=0.007; bal Fe				
	C.0461	Struct	0.2 max			1.60 max		0.05 max	0.05 max	0.60 max	Mo <=0.15; N <=0.007; bal Fe				
	C.0482	Struct	0.2 max			1.60 max		0.045 max	0.045 max	0.60 max	N <=0.009; bal Fe				
	C.0483	Struct	0.2 max			1.60 max		0.045 max	0.045 max	0.60 max	N <=0.009; bal Fe				

Carbon Steel, Nonresulfurized, 1020

Argentina

Specification	Designation	Notes	C	Cr	Cu	Mn	Ni	P	S	Si	Other	UTS	YS	El	Hard
IAS	IRAM 1020		0.18-0.23			0.30-0.60		0.040 max	0.050 max	0.15-0.30	bal Fe	450-550	280-350	22-36	131-163 HB

Australia

Specification	Designation	Notes	C	Cr	Cu	Mn	Ni	P	S	Si	Other	UTS	YS	El	Hard
AS 1442	CS1020*	Obs; Bar Bil	0.15-0.25			0.3-0.9		0.06	0.06	0.35	bal Fe				
AS 1442	K1020*	Obs; Bar Bil	0.18-0.23			0.3-0.6		0.05	0.05	0.1-0.35	bal Fe				
AS 1442	S1020*	Obs; Bar Bil	0.18-0.23			0.3-0.6		0.05	0.05		bal Fe				
AS 1442(92)	1020	HR bar, Semifinished (may treat w/microalloying elements)	0.18-0.23			0.30-0.60		0.040 max	0.040 max	0.10-0.35	Si<=0.10 for Al-killed; bal Fe				
AS 1442(92)	M1020	Merchant qual, HR bar, Semifinished	0.15-0.25			0.30-0.90		0.050 max	0.050 max	0.35 max	bal Fe				
AS 1443	CD1*	Obs; Bar CD or CR	0.15-0.25			0.3-0.9		0.05	0.05	0.35	bal Fe				
AS 1443	CS1020*	Obs	0.15-0.25			0.3-0.9		0.06	0.06	0.35	bal Fe				
AS 1443	K1020*	Obs; Bar	0.18-0.23			0.3-0.6		0.05	0.05	0.1-0.35	bal Fe				
AS 1443	S1020*	Obs; Bar	0.18-0.23			0.3-0.6		0.05	0.05		bal Fe				
AS 1443(94)	1020	CF bar (may treat w/microalloying elements)	0.18-0.23			0.30-0.60		0.040 max	0.040 max	0.10-0.35	Si<=0.10 for Al-killed; bal Fe				
AS 1443(94)	D3	CR or CR bar, 16-38mm	0.15-0.25			0.30-0.90		0.050 max	0.050 max	0.35 max	bal Fe	460	370	12	
AS 1443(94)	M1020	CF bar, Merchant quality	0.15-0.25			0.30-0.90		0.050 max	0.050 max	0.35 max	bal Fe				
AS 1443(94)	T3	CF bar, not CD or CR; 50-250mm	0.15-0.25			0.30-0.90		0.050 max	0.050 max	0.35 max	bal Fe	410	230	22	
AS 1446	CS1020*	Withdrawn, see AS/NZS 1594(97)	0.15-0.25			0.3-0.9		0.06	0.06	0.35	bal Fe				
AS 1446	S1020*	Withdrawn, see AS/NZS 1594(97)	0.18-0.23			0.3-0.6		0.05	0.05		bal Fe				

Austria

Specification	Designation	Notes	C	Cr	Cu	Mn	Ni	P	S	Si	Other	UTS	YS	El	Hard
ONORM M3167	C22SP	Bar	0.18-0.25			0.4-0.6		0.040 max	0.040 max	0.30 max	bal Fe				

Specification	Designation	Notes	C	Cr	Cu	Mn	Ni	P	S	Si	Other	UTS	YS	El	Hard

Carbon Steel, Nonresulfurized, 1020 (Continued from previous page)

Bulgaria

Specification	Designation	Notes	C	Cr	Cu	Mn	Ni	P	S	Si	Other	UTS	YS	El	Hard
BDS 5785(83)	20kp	Struct	0.17-0.24	0.25 max	0.25 max	0.25-0.50	0.25 max	0.035 max	0.040 max	0.07	bal Fe				

Canada

| CSA STAN95-1-1 | C1020 | Bar | 0.18-0.23 | | | 0.3-0.6 | | 0.04 | 0.05 | 0.1-0.3 | bal Fe | | | | |

China

Specification	Designation	Notes	C	Cr	Cu	Mn	Ni	P	S	Si	Other	UTS	YS	El	Hard
GB 13237(91)	20	Sh Strp CR HT	0.17-0.24	0.25 max	0.25 max	0.35-0.65	0.25 max	0.035 max	0.035 max	0.17-0.37	bal Fe	355-500		24	
GB 3275(91)	20	Plt Strp HR HT	0.17-0.24	0.25 max	0.25 max	0.35-0.65	0.25 max	0.035 max	0.035 max	0.17-0.37	bal Fe	345-490		24	
GB 3522(83)	20	Strp CR Ann	0.17-0.24	0.25 max	0.25 max	0.35-0.65	0.25 max	0.035 max	0.035 max	0.17-0.37	bal Fe	315-540		20	
GB 5953(86)	ML 20	Wir CD	0.17-0.24	0.20 max	0.20 max	0.60 max		0.035 max	0.035 max	0.20 max	bal Fe	440-635			
GB 6478(86)	ML 20	Bar HR Norm	0.17-0.24	0.20 max	0.20 max	0.20-0.50	0.20 max	0.035 max	0.035 max	0.07 max	bal Fe	410	245	25	
GB 6479(89)	20g	Smls Tub HR/CD Norm	0.17-0.24		0.25 max	0.35-0.60		0.035 max	0.035 max	0.17-0.37	bal Fe	410-550	245	24	
GB 699(88)	20	Bar HR Norm 25mm diam	0.17-0.24	0.25 max	0.25 max	0.35-0.65	0.25 max	0.035 max	0.035 max	0.17-0.37	bal Fe	410	245	25	
GB 700	BJ4*	Obs	0.16-0.26			0.3-0.7		0.045	0.055	0.07-0.35	bal Fe				
GB 700	BJ4F*	Obs	0.16-0.26			0.3-0.7		0.045	0.055	0.07-0.35	bal Fe				
GB 700	CJ4*	Obs	0.16-0.26			0.3-0.7		0.045	0.055	0.07-0.35	bal Fe				
GB 700	CJ4F*	Obs	0.16-0.26			0.3-0.7		0.045	0.055	0.07-0.35	bal Fe				
GB 710(91)	20	Sh Strp HR HT	0.17-0.24	0.25 max	0.25 max	0.35-0.65	0.25 max	0.035 max	0.035 max	0.17-0.37	bal Fe	355-500		24	
GB 8162(87)	20	Smls Tub HR/CD Q/T	0.17-0.24	0.25 max	0.25 max	0.35-0.65	0.25 max	0.035 max	0.035 max	0.17-0.37	bal Fe	390	245	20	
GB 8163(87)	20	Smls Pip HR/CD	0.17-0.24	0.25 max	0.25 max	0.35-0.65	0.25 max	0.035 max	0.035 max	0.17-0.37	bal Fe	390	245	20	
GB 9948(88)	20	Smls Tub HR/CD Norm	0.17-0.24	0.25 max	0.25 max	0.35-0.65	0.25 max	0.035 max	0.035 max	0.17-0.37	bal Fe	410-550	245	21	
GB/T 13795(92)	20	Strp CR Ann	0.17-0.24	0.25 max	0.25 max	0.35-0.65	0.25 max	0.035 max	0.035 max	0.17-0.37	bal Fe	400-700		15	
GB/T 3078(94)	20	Bar CR Ann	0.17-0.24	0.25 max	0.25 max	0.35-0.65	0.25 max	0.035 max	0.035 max	0.17-0.37	bal Fe	390		21	
GB/T 8164(93)	20	Strp HR/CR	0.17-0.24	0.25 max	0.25 max	0.35-0.65	0.25 max	0.035 max	0.035 max	0.17-0.37	bal Fe	410		28	

Czech Republic

CSN 411353	11353	Gen struct	0.18 max	0.3 max		1.60 max		0.05 max	0.05 max	0.60 max	bal Fe				
CSN 411423	11423	Gen struct	0.24 max	0.3 max		1.60 max		0.05 max	0.05 max	0.60 max	bal Fe				
CSN 412024	12024	CH	0.17-0.24	0.25 max		0.35-0.65	0.3 max	0.04 max	0.04 max	0.15-0.4	bal Fe				

Denmark

| DS 12011 | St42A | Rod Bar, HR 16mm diam | 0.22 | 0.3 | 0.4 | | | 0.08 | 0.05 | 0.5 | bal Fe | | | | |
| DS 12011 | St42B | | 0.22 | 0.3 | 0.4 | | | 0.06 | 0.05 | 0.5 | N 0.009; bal Fe | | | | |

Europe

EN 10016/2(94)	1.0414	Rod	0.18-0.23	0.20 max	0.30 max	0.30-0.60	0.25 max	0.035 max	0.035 max	0.30 max	Al <=0.01; Mo <=0.05; bal Fe				
EN 10016/2(94)	C20D	Rod	0.18-0.23	0.20 max	0.30 max	0.30-0.60	0.25 max	0.035 max	0.035 max	0.30 max	Al <=0.01; Mo <=0.05; bal Fe				
EN 10016/4(94)	1.1137	Rod	0.18-0.23	0.10 max	0.15 max	0.30-0.50	0.10 max	0.020 max	0.025 max	0.30 max	Al <=0.01; Mo <=0.05; N <=0.007; Cr+Ni+Cu<=0.30, Cu+Sn<=0.15; bal Fe				
EN 10016/4(94)	C20D2	Rod	0.18-0.23	0.10 max	0.15 max	0.30-0.50	0.10 max	0.020 max	0.025 max	0.30 max	Al <=0.01; Mo <=0.05; N <=0.007; Cr+Ni+Cu<=0.30, Cu+Sn<=0.15; bal Fe				

UNS numbers and US grades are provided as a means of cross referencing chemically similar alloys. Exchangability is only possible after independent examination of specifications. Tensile properties are minimum or typical as specified. UTS and YS as MPa. El as %. See Appendix for list of abbreviations used in Notes. * indicates obsolete material.

Specification	Designation	Notes	C	Cr	Cu	Mn	Ni	P	S	Si	Other	UTS	YS	El	Hard

Carbon Steel, Nonresulfurized, 1020 (Continued from previous page)

Europe

Specification	Designation	Notes	C	Cr	Cu	Mn	Ni	P	S	Si	Other	UTS	YS	El	Hard
EN 10025(90)	Fe430B	Obs EU desig; struct HR FN; BS 16<t<=40mm	0.21 max					0.045 max	0.045 max		N <=0.009; N max na if Al>=0.020; bal Fe	410-560	265	22	
EN 10025(90)	Fe430B	Obs EU desig; struct HR FN; BS t<=16mm	0.21 max					0.045 max	0.045 max		N <=0.009; N max na if Al>=0.020; bal Fe	410-580	275	14-22	
EN 10025(90)	Fe430B	Obs EU desig; struct HR FN; BS 4t<=100mm	0.22 max					0.045 max	0.045 max		N <=0.009; N max na if Al>=0.020; bal Fe	410-560	235-255	20-21	
EN 10025(90)	Fe430C	Obs EU desig; struct HR FN; QS t<=16mm	0.18 max					0.040 max	0.040 max		N <=0.009; N max na if Al>=0.020; bal Fe	410-580	275	14-22	
EN 10025(90)	Fe430C	Obs EU desig; struct HR FN; QS150<t<=250mm	0.20 max					0.040 max	0.040 max		N <=0.009; N max na if Al>=0.020; bal Fe	380-540	215	17	
EN 10025(90)	Fe430C	Obs EU desig; struct HR FN; QS 16<t<=40mm	0.18 max					0.040 max	0.040 max		N <=0.009; N max na if Al>=0.020; bal Fe	410-560	265	22	
EN 10025(90)	Fe430C	Obs EU desig; struct HR FN; QS 4t<=100mm	0.18 max					0.040 max	0.040 max		N <=0.009; N max na if Al>=0.020; bal Fe	410-560	235-255	20-21	
EN 10025(90)	Fe430D1	Obs EU desig; struct HR FF; QS 150<t<=250mm	0.20 max					0.035 max	0.035 max		bal Fe	380-540	215	17	
EN 10025(90)	Fe430D1	Obs EU desig; struct HR FF; QS 40<t<=100mm	0.18 max					0.035 max	0.035 max		bal Fe	410-560	235-255	20-21	
EN 10025(90)	Fe430D1	Obs EU desig; struct HR FF; QS 16<t<=40mm	0.18 max					0.035 max	0.035 max		bal Fe	410-560	265	22	
EN 10025(90)	Fe430D1	Obs EU desig; struct HR FF; QS t<=16mm	0.18 max					0.035 max	0.035 max		bal Fe	410-580	275	14-22	
EN 10025(90)	Fe430D1	Obs EU desig; struct HR FF; QS 16<t<=40mm	0.18 max					0.035 max	0.035 max		bal Fe	410-560	265	22	
EN 10025(90)	Fe430D1	Obs EU desig; struct HR FF; QS t<=16mm	0.18 max					0.035 max	0.035 max		bal Fe	410-580	275	14-22	
EN 10025(90)	Fe430D1	Obs EU desig; struct HR FF; QS t<=16mm	0.18 max					0.035 max	0.035 max		bal Fe	410-580	275	14-22	
EN 10025(90)	Fe430D1	Obs EU desig; struct HR FF; QS 16<t<=40mm	0.18 max					0.035 max	0.035 max		bal Fe	410-560	265	22	
EN 10025(90)	Fe430D2	Obs EU desig; struct HR FF; QS t<=16mm	0.18 max					0.035 max	0.035 max		bal Fe	410-580	275	12-20	
EN 10025(90)	Fe430D2	Obs EU desig; struct HR FF; QS 16<t<=40mm	0.18 max					0.035 max	0.035 max		bal Fe	410-560	265	20	
EN 10025(90)	Fe430D2	Obs EU desig; struct HR FF; QS 150<t<=250mm	0.20 max					0.035 max	0.035 max		bal Fe	380-540	205-215	17	
EN 10025(90)	Fe430D2	Obs EU desig; struct HR FF; QS 40<t<=100mm	0.18 max					0.035 max	0.035 max		bal Fe	410-560	235-255	18-19	
EN 10083/1(91)A1(96)	1.1149	Q/T t<=16mm	0.17-0.24	0.40 max		0.40-0.70	0.40 max	0.035 max	0.020-0.040	0.40 max	Mo <=0.10; Cr+Mo+Ni<=0.63; bal Fe	500-650	340	20	
EN 10083/1(91)A1(96)	1.1151	Q/T t<=16mm	0.17-0.24	0.40 max		0.40-0.70	0.40 max	0.035 max	0.035 max	0.40 max	Mo <=0.10; Cr+Mo+Ni<=0.63; bal Fe	500-650	340	20	
EN 10083/1(91)A1(96)	C22E	Q/T t<=16mm	0.17-0.24	0.40 max		0.40-0.70	0.40 max	0.035 max	0.035 max	0.40 max	Mo <=0.10; Cr+Mo+Ni<=0.63; bal Fe	500-650	340	20	
EN 10083/1(91)A1(96)	C22R	Q/T t<=16mm	0.17-0.24	0.40 max		0.40-0.70	0.40 max	0.035 max	0.020-0.040	0.40 max	Mo <=0.10; Cr+Mo+Ni<=0.63; bal Fe	500-650	340	20	
EN 10083/2(91)A1(96)	1.0402	Q/T t<=16mm	0.17-0.24	0.40 max		0.40-0.70	0.40 max	0.045 max	0.045 max	0.40 max	Mo <=0.10; Cr+Mo+Ni<=0.63; bal Fe	500-650	340	20	
EN 10083/2(91)A1(96)	C22	Q/T t<=16mm	0.17-0.24	0.40 max		0.40-0.70	0.40 max	0.045 max	0.045 max	0.40 max	Mo <=0.10; Cr+Mo+Ni<=0.63; bal Fe	500-650	340	20	

UNS numbers and US grades are provided as a means of cross referencing chemically similar alloys. Exchangability is only possible after independent examination of specifications. Tensile properties are minimum or typical as specified. UTS and YS as MPa. El as %. See Appendix for list of abbreviations used in Notes. * indicates obsolete material.

Specification	Designation	Notes	C	Cr	Cu	Mn	Ni	P	S	Si	Other	UTS	YS	El	Hard
Carbon Steel, Nonresulfurized, 1020 (Continued from previous page)															
Finland															
SFS 1100	Fe44BP	Gen struct; press ves	0.2 max			0.4 max 0.4-1.2		0.05 max	0.05 max	0.15-0.55	N <=0.009; As 0-0.08; bal Fe				
SFS 1100	Fe52BP	Gen struct; press ves	0.2 max			0.4 max 0.8-1.5		0.045 max	0.045 max	0.15-0.55	N <=0.009; As 0-0.08; bal Fe				
SFS 1100	Fe52DP	Gen struct	0.20 max			0.4 max 0.8-1.5		0.045 max	0.045 max	0.15-0.55	N <=0.009; As 0-0.08; bal Fe				
SFS 1213(80)	SFS1213	Reinforcing	0.2 max			1.6 max		0.06 max	0.05 max	0.15-0.55	N <=0.01; Ce C+Mn/6+(Cr+Mo+V)/5+(Ni+Cu)/15; bal Fe				
SFS 1215(89)	SFS1215	Reinforcing	0.2 max			1.6 max		0.06 max	0.05 max	0.15-0.55	N <=0.01; Ce C+Mn/6+(Cr+Mo+V)/5+(Ni+Cu)/15; bal Fe				
SFS 1257(89)	SFS1257	Reinforcing	0.2 max			0.5 max		0.06 max	0.05 max	0.6 max	N <=0.01; S/Mn=0.1; bal Fe				
SFS 1258(90)	SFS1258		0.2 max			0.5 max		0.06 max	0.05 max	0.6 max	N <=0.01; S/Mn=0.1; bal Fe				
SFS 200	Fe44B	Gen struct	0.21 max		0.4 max	1.60 max		0.05 max	0.05 max	0.55 max	N <=0.009; bal Fe				
France															
AFNOR NFA33101(82)	AF42C20	t<=450mm; Norm	0.14-0.21	0.3 max	0.3 max	0.5-0.8	0.4 max	0.04 max	0.04 max	0.1-0.4	bal Fe	410-510	255	26	
AFNOR NFA35551	XC18		0.16-0.22			0.4-0.7		0.035	0.035	0.15-0.35	Al 0.02; bal Fe				
AFNOR NFA35551(75)	CC20		0.15-0.25	0.3 max	0.3 max	0.4-0.7	0.4 max	0.04 max	0.04 max	0.1-0.4	bal Fe				
AFNOR NFA35552	XC18		0.16-0.22			0.4-0.7		0.035	0.035	0.15-0.35	Al 0.02; bal Fe				
AFNOR NFA35553	C20		0.15-0.25			0.4-0.7		0.04	0.04	0.1-0.4	bal Fe				
AFNOR NFA35566	XC18		0.16-0.22			0.4-0.7		0.035	0.035	0.15-0.35	Al 0.02; bal Fe				
AFNOR NFA36211(90)	BS3*	t<3mm; Norm	0.20 max			0.7 max		0.025 max	0.02 max	0.3 max	Al >=0.015; N <=0.009; Nb <=0.050; bal Fe	460-540	310	21	
AFNOR NFA36211(90)	BS4*	t<3mm; Norm	0.20 max			0.7 max		0.025 max	0.02 max	0.45 max	Al >=0.015; N <=0.009; Nb <=0.050; bal Fe	510-610	355	19	
Germany															
DIN 1652(90)	S275JR	100-150mm	0.21 max			1.50 max		0.045 max	0.045 max		N <=0.009; N depends on P; C<= 0.22 if >=40 mm thk; bal Fe	400-540	225	18	
DIN 1652(90)	WNr 1.0044	100-150mm	0.21 max			1.50 max		0.045 max	0.045 max		N <=0.009; N depends on P; C<=0.22 if >=40mm thk; bal Fe	400-540	225	18	
DIN 1652(90)	WNr 1.0129		0.21 max					0.050 max	0.050 max		N <=0.009; N depends on P; C<=0.22 if >=40mm thk; bal Fe				
DIN 1652(90)	ZSt44-2		0.21 max					0.050 max	0.050 max		N <=0.009; N depends on P; C<= 0.22 if >=40 mm thk; bal Fe				
DIN 17119(84)	S275JR	Weld CF tub 100-150mm	0.21 max			1.50 max		0.045 max	0.045 max		N <=0.009; N depends on P; C<= 0.22 if >=40 mm thk; bal Fe	400-540	225	18	
DIN 17119(84)	WNr 1.0044	Weld CF tub 100-150mm	0.21 max			1.50 max		0.045 max	0.045 max		N <=0.009; N depends on P; C<=0.22 if >=40mm thk; bal Fe	400-540	225	18	
DIN 17140	D20-2*	Obs; see C20D	0.18-0.23			0.3-0.6		0.04 max	0.04 max	0.1-0.3	bal Fe				
DIN 17204(90)	WNr 1.0402	Smls tube; Q/T 17-40mm	0.17-0.24	0.40 max		0.40-0.70	0.40 max	0.045 max	0.045 max	0.40 max	Mo <=0.10; Cr+Mo+Ni<=0.63; bal Fe	470	290	22	
DIN 17243(87)	C22.8	Frg HR bar for elev temp	0.18-0.23	0.30 max		0.40-0.90		0.035 max	0.030 max	0.40 max	Al 0.015-0.050; bal Fe				

UNS numbers and US grades are provided as a means of cross referencing chemically similar alloys. Exchangability is only possible after independent examination of specifications. Tensile properties are minimum or typical as specified. UTS and YS as MPa. El as %. See Appendix for list of abbreviations used in Notes. * indicates obsolete material.

Specification	Designation	Notes	C	Cr	Cu	Mn	Ni	P	S	Si	Other	UTS	YS	El	Hard

Carbon Steel, Nonresulfurized, 1020 (Continued from previous page)

Germany

Specification	Designation	Notes	C	Cr	Cu	Mn	Ni	P	S	Si	Other	UTS	YS	El	Hard
DIN 17243(87)	WNr 1.0460	Frg HR bar for elev temp	0.18-0.23	0.30 max		0.40-0.90		0.035 max	0.030 max	0.40 max	Al 0.015-0.050; bal Fe				
DIN 2393(94)	S275JR	Weld tube 100-150mm	0.21 max			1.50 max		0.045 max	0.045 max		N <=0.009; N depends on P; C<= 0.22 if >=40 mm thk; bal Fe	400-540	225	18	
DIN 2393(94)	WNr 1.0044	Weld tube 100-150mm	0.21 max			1.50 max		0.045 max	0.045 max		N <=0.009; N depends on P; C<=0.22 if >=40mm thk; bal Fe	400-540	225	18	
DIN 2394(94)	S275JR	As-weld, sized tube 100-150mm	0.21 max			1.50 max		0.045 max	0.045 max		N <=0.009; N depends on P; C<= 0.22 if >=40 mm thk; bal Fe	400-540	225	18	
DIN 2394(94)	WNr 1.0044	As-weld, sized tube 100-150mm	0.21 max			1.50 max		0.045 max	0.045 max		N <=0.009; N depends on P; C<=0.22 if >=40mm thk; bal Fe	400-540	225	18	
DIN 2528(91)	C22.3	Flanges	0.18-0.23	0.30 max		0.30-0.60		0.045 max	0.045 max	0.15-0.35	bal Fe				
DIN 2528(91)	WNr 1.0427	Flanges	0.18-0.23	0.30 max		0.30-0.60		0.045 max	0.045 max	0.15-0.35	bal Fe				
DIN EN 10025(94)	S275JR	HR 100-150mm thk	0.21 max			1.50 max		0.045 max	0.045 max		N <=0.009; N depends on P; C<= 0.22 if >=40 mm thk; bal Fe	400-540	225	18	
DIN EN 10025(94)	WNr 1.0044	HR 100-150mm thk	0.21 max			1.50 max		0.045 max	0.045 max		N <=0.009; N depends on P; C<=0.22 if >=40mm thk; bal Fe	400-540	225	18	
DIN EN 10025(94)	WNr 1.0519	HR	0.20 max		0.25-0.40	1.60 max		0.035 max	0.035 max	0.55 max	bal Fe				
DIN EN 10025(94)	WNr 1.0553	HR 100-150mm	0.20 max			1.60 max		0.040 max	0.040 max	0.55 max	N <=0.009; N depends on P; Cu<=0.22 thkness >30mm; bal Fe	470-630	295	18	
DIN EN 10025(94)	WNr 1.0554	HR CW 100-150mm	0.20 max			1.60 max		0.040 max	0.040 max	0.55 max	N depends on P; Cu<=0.22 thkness >30mm; bal Fe	470-630	295	18	
DIN EN 10083(91)	Cm22	Q/T	0.17-0.24	0.40 max		0.40-0.70	0.40 max	0.035 max	0.020-0.035	0.40 max	Mo <=0.10; Cr+Mo+Ni<=0.63; bal Fe				
DIN EN 10083(91)	WNr 1.1149	Q/T	0.17-0.24	0.40 max		0.40-0.70	0.40 max	0.035 max	0.020-0.040	0.40 max	Mo <=0.10; Cr+Mo+Ni<=0.63; bal Fe				
DIN EN 10210(94)	S275JR	Tub 100-150mm	0.21 max			1.50 max		0.045 max	0.045 max		N <=0.009; N depends on P; C<= 0.22 if >=40 mm thk; bal Fe	400-540	225	18	
DIN EN 10210(94)	WNr 1.0044	Tub 100-150mm	0.21 max			1.50 max		0.045 max	0.045 max		N <=0.009; N depends on P; C<=0.22 if >=40mm thk; bal Fe	400-540	225	18	
DIN EN 10248(95)	WNr 1.0021	HR Sh piling	0.20 max					0.045 max	0.045 max		N <=0.009; bal Fe				

Hungary

Specification	Designation	Notes	C	Cr	Cu	Mn	Ni	P	S	Si	Other	UTS	YS	El	Hard
MSZ 120/1(82)	A37	Tub; screwing; Thk wall tube; 2.9-5.4mm; HF/CF	0.17 max			1 max		0.04 max	0.04 max	0.1-0.35	MSZ29; bal Fe	350-500	235	25L	
MSZ 120/1(82)	A37X	Tub; screwing; Thk wall tube; 2.9-5.4mm; HF/CF	0.17 max			1 max		0.04 max	0.04 max	0.1 max	MSZ29; bal Fe	350-500	235	25L	
MSZ 120/2(82)	A37	Tub; screwing; med series tube; 2.3-5mm; HF/CF	0.17 max			1 max		0.04 max	0.04 max	0.1-0.35	MSZ29; bal Fe	350-500	235	25L	
MSZ 120/2(82)	A37X	Tub; screwing; med series tube; 2.3-5mm; HF/CF	0.17 max			1 max		0.04 max	0.04 max	0.1 max	MSZ29; bal Fe	350-500	235	25L	
MSZ 29(86)	A37	Smls tube; Non-rimming; FF; 0-16mm	0.17 max			1 max		0.04 max	0.04 max	0.1-0.35	bal Fe	350-500	235	25L	

Specification	Designation	Notes	C	Cr	Cu	Mn	Ni	P	S	Si	Other	UTS	YS	El	Hard

Carbon Steel, Nonresulfurized, 1020 (Continued from previous page)

Hungary

Specification	Designation	Notes	C	Cr	Cu	Mn	Ni	P	S	Si	Other	UTS	YS	El	Hard
MSZ 29(86)	A37	Smls tube; Non-rimming; FF; 40.1-100.00mm	0.17 max			1 max		0.04 max	0.04 max	0.1-0.35	bal Fe	350-500	215	25L	
MSZ 29(86)	A37X	Smls tube; Unkilled; R; FU; 40.1-100.00mm	0.17 max			1 max		0.04 max	0.04 max	0.1 max	bal Fe	350-500	215	25L	
MSZ 29(86)	A37X	Smls tube; Unkilled; R; FU; 0-16mm	0.17 max			1 max		0.04 max	0.04 max	0.1 max	bal Fe	350-500	235	25L	
MSZ 500(81)	A0	Gen struct; 0-100mm; Frg	0.3 max			2 max		0.07 max	0.06 max	0.6 max	bal Fe	310-490		22L	
MSZ 5736(88)	D20	Wir rod; for CD	0.17-0.24		0.25 max	0.3-0.6	0.3 max	0.04 max	0.04 max	0.1-0.35	bal Fe				
MSZ 5736(88)	D20K	Wir rod; for CD	0.17-0.24		0.25 max	0.3-0.6	0.3 max	0.03 max	0.03 max	0.1-0.35	bal Fe				

International

Specification	Designation	Notes	C	Cr	Cu	Mn	Ni	P	S	Si	Other	UTS	YS	El	Hard
ISO 3304(85)	R37	Smls Tub, CF	0.17 max			0.8 max		0.050 max	0.050 max	0.35 max	Pb <=0.15;	450		6	
ISO 3304(85)	R37	Smls Tub, CF	0.17 max			0.8 max		0.050 max	0.050 max	0.35 max	Co <=0.1; Pb <=0.15; W <=0.1; bal Fe	450		6	
ISO 3304(85)	R37	Smls Tub, CF	0.17 max			0.8 max		0.050 max	0.050 max	0.35 max	Pb <=0.15;	450		6	
ISO 3304(85)	R37	Smls Tub, CF	0.17 max			0.8 max		0.050 max	0.050 max	0.35 max	Pb <=0.15;	450		6	
ISO 3304(85)	R37	Smls Tub, CF	0.17 max			0.8 max		0.050 max	0.050 max	0.35 max	Pb <=0.15;	450		6	
ISO 683-1(87)	C20*		0.17-0.23			0.3-0.6		0.05	0.05	0.15-0.4	bal Fe				
ISO 683-18(96)	C20	Bar Wir, CD, t<=5mm	0.17-0.23			0.30-0.60		0.045 max	0.045 max	0.10-0.40	bal Fe	580	460	5	156 HB
ISO 683-18(96)	C20E4	Bar Wir, CD, t<=5mm	0.17-0.23			0.30-0.60		0.035 max	0.035 max	0.10-0.40	bal Fe	580	460	5	156 HB
ISO 683-18(96)	C20M2	Bar Wir, CD, t<=5mm	0.17-0.23			0.30-0.60		0.035 max	0.020-0.040	0.10-0.40	bal Fe	580	460	5	156 HB

Italy

Specification	Designation	Notes	C	Cr	Cu	Mn	Ni	P	S	Si	Other	UTS	YS	El	Hard
UNI 5332(64)	C20	Q/T	0.18-0.24			0.4-0.8		0.035 max	0.035 max	0.4 max	bal Fe				
UNI 5598	ICD20		0.17-0.24			0.4-0.7		0.05	0.05	0.35	bal Fe				
UNI 5598(71)	1CD20	Wir rod	0.17-0.24			0.4-0.7		0.05 max	0.05 max	0.35 max	bal Fe				
UNI 5598(71)	3CD20	Wir rod	0.18-0.23			0.4-0.7		0.035 max	0.035 max	0.35 max	N <=0.012; bal Fe				
UNI 5771(66)	20Mn4	Chain	0.16-0.24			1.5 max		0.035 max	0.035 max	0.35 max	26mm+ Si<=0.35 Mn<=1.5; bal Fe				
UNI 5771(66)	20Mn4	Chain	0.16-0.24			1.5 max		0.035 max	0.035 max	0.35 max	0-26mm Si<=0.3 Mn 0.8-1; bal Fe				
UNI 5949(67)	C20	Smls tube w/low-temp test	0.2 max			1 max		0.035 max	0.035 max	0.15-0.35	bal Fe				
UNI 6403(86)	C20	Q/T	0.18-0.24		*	0.4-0.8		0.035 max	0.035 max	0.15-0.35	bal Fe				
UNI 6922(71)	C21	Q/T; Aircraft material	0.18-0.24			0.3-0.6		0.035 max	0.035 max	0.1 max	bal Fe				
UNI 7065(72)	C20	Strp	0.18-0.24			0.4-0.8		0.035 max	0.035 max	0.4 max	bal Fe				
UNI 7356(74)	CB20FF	Q/T	0.18-0.23			0.3-0.6		0.04 max	0.04 max	0.1 max	bal Fe				
UNI 7874(79)	C20	Q/T	0.18-0.24			0.4-0.8		0.035 max	0.035 max	0.15-0.4	bal Fe				
UNI 8913(87)	C20	Q/T	0.18-0.24			0.4-0.8		0.035 max	0.035 max	0.35 max	bal Fe				

Japan

Specification	Designation	Notes	C	Cr	Cu	Mn	Ni	P	S	Si	Other	UTS	YS	El	Hard
JIS G3507(91)	SWRCH20A	Wir rod Al killed	0.18-0.23			0.30-0.60		0.030 max	0.035 max	0.10 max	Al >=0.02; bal Fe				
JIS G3507(91)	SWRCH20K	Wir rod FF	0.18-0.23			0.30-0.60		0.030 max	0.035 max	0.10-0.35	bal Fe				
JIS G4051(79)	S20C	Bar Wir rod	0.18-0.23	0.20 max	0.30 max	0.30-0.60	0.20 max	0.030 max	0.035 max	0.15-0.35	Ni+Cr<=0.35; bal Fe				
JIS G4051(79)	S20CK	CH	0.18-0.23	0.20 max	0.25 max	0.30-0.60	0.20 max	0.025 max	0.025 max	0.15-0.35	Ni+Cr=0.30 max; bal Fe				

UNS numbers and US grades are provided as a means of cross referencing chemically similar alloys. Exchangability is only possible after independent examination of specifications. Tensile properties are minimum or typical as specified. UTS and YS as MPa. El as %. See Appendix for list of abbreviations used in Notes. * indicates obsolete material.

Specification	Designation	Notes	C	Cr	Cu	Mn	Ni	P	S	Si	Other	UTS	YS	El	Hard
Carbon Steel, Nonresulfurized, 1020 (Continued from previous page)															
Mexico															
DGN B-203	MT1020*	Tube, Obs	0.15-0.25			0.3-0.6		0.04	0.05		bal Fe				
DGN B-203	MTX1020*	Tube, Obs	0.15-0.25			0.7-1		0.04	0.05		bal Fe				
NMX-B-201(68)	MT1020	CD buttweld mech tub	0.15-0.25			0.30-0.60		0.04 max	0.05 max		bal Fe				
NMX-B-201(68)	MTX1020	CD buttweld mech tub	0.15-0.25			0.70-1.00		0.04 max	0.05 max		bal Fe				
NMX-B-301(86)	1020	Bar	0.18-0.23			0.30-0.60		0.040 max	0.050 max		bal Fe				
NOM-060-SCFI(94)	Type 1	HR for portable gasoline containers	0.20 max			0.40-0.80		0.04 max	0.04 max	0.40 max	bal Fe	360	182	20.0	
Pan America															
COPANT 330	1020	Bar	0.13-0.23			0.3-0.6		0.04	0.05		bal Fe				
COPANT 331	1020		0.18-0.23			0.3-0.6		0.04	0.05		bal Fe				
COPANT 333	1020		0.18-0.23			0.3-0.6		0.04	0.05		bal Fe				
COPANT 514	TM1020	Tube, Ann	0.15-0.25			0.3-0.6		0.04	0.05	0.1-0.2	bal Fe				
COPANT 514	TM1020	Tube, Norm	0.15-0.25			0.3-0.6		0.04	0.05	0.1-0.2	bal Fe				
COPANT 514	TM1020	Tube, HF, 323.9mm diam	0.15-0.25			0.3-0.6		0.04	0.05	0.1-0.2	bal Fe				
COPANT 514	TM1020	Tube, CF	0.15-0.25			0.3-0.6		0.04	0.05	0.1-0.2	bal Fe				
COPANT 514	TMX1020	Tube, CF	0.15-0.25	0.05		0.6-1		0.04	0.05	0.2-0.35	bal Fe				
COPANT 514	TMX1020	Tube, Norm	0.15-0.25	0.05		0.6-1		0.04	0.05	0.2-0.35	bal Fe				
COPANT 514	TMX1020	Tube, Ann	0.15-0.25	0.05		0.6-1		0.04	0.05	0.2-0.35	bal Fe				
COPANT 514	TMX1020	Tube, HF, 323.9mm diam	0.15-0.25	0.05		0.6-1		0.04	0.05	0.2-0.35	bal Fe				
Poland															
PNH84019	20	CH	0.17-0.24	0.3 max		0.35-0.65	0.3 max	0.04 max	0.04 max	0.17-0.37	bal Fe				
PNH84019	20Y	CH; High-temp const	0.17-0.24	0.3 max		0.35-0.65	0.3 max	0.04 max	0.04 max	0.65 max	bal Fe				
PNH84023/07	R35	High-temp const	0.07-0.16	0.3 max	0.25 max	0.4-0.75	0.4 max	0.04 max	0.04 max	0.12-0.35	bal Fe				
PNH84024	K18	High-temp const	0.16-0.22	0.2	0.25	0.6	0.35	0.045	0.045	0.1-0.35	bal Fe				
PNH84024	R35	High-temp const	0.07-0.13	0.3 max		0.35-0.5	0.4 max	0.05 max	0.05 max	0.15-0.35	bal Fe				
Romania															
STAS 438/1(89)	OB37	Reinf; HR	0.23 max	0.3 max		0.75 max		0.045 max	0.045 max	0.07 max	Pb <=0.15; bal Fe				
STAS 500/2(88)	OL42.1	Gen struct	0.25 max	0.3 max		0.8 max	0.4 max	0.06 max	0.06 max	0.4 max	Pb <=0.15; bal Fe				
STAS 500/2(88)	OL42.1a	Gen struct	0.25 max	0.3 max		0.8 max	0.4 max	0.06 max	0.06 max	0.4 max	Pb <=0.15; bal Fe				
STAS 500/2(88)	OL42.1b	Gen struct	0.25 max	0.3 max		0.8 max	0.4 max	0.06 max	0.06 max	0.4 max	Pb <=0.15; bal Fe				
STAS 500/2(88)	OL42.2	Gen struct	0.25 max	0.3 max		0.8 max	0.4 max	0.05 max	0.05 max	0.4 max	Pb <=0.15; bal Fe				
STAS 500/2(88)	OL42.3k	Gen struct; FF; Non-rimming	0.23 max	0.3 max		0.8 max		0.045 max	0.045 max	0.4 max	Pb <=0.15; As<=0.08; bal Fe				
STAS 500/2(88)	OL42.3kf	Gen struct; FF; Non-rimming	0.23 max	0.3 max		0.8 max		0.045 max	0.045 max	0.4 max	Pb <=0.15; As<=0.08; bal Fe				
STAS 500/2(88)	OL42.3kf	Gen struct; Specially killed	0.23 max	0.3 max		0.8 max		0.045 max	0.045 max	0.4 max	Pb <=0.15; As<=0.08; bal Fe				
STAS 880(88)	OLC20	Q/T	0.17-0.24	0.3 max		0.3-0.6		0.04 max	0.045 max	0.17-0.37	Pb <=0.15; As<=0.05; bal Fe				
STAS 880(88)	OLC20S	Q/T	0.17-0.24	0.3 max		0.4-0.7		0.04 max	0.02-0.045	0.17-0.37	Pb <=0.15; As<=0.05; bal Fe				

UNS numbers and US grades are provided as a means of cross referencing chemically similar alloys. Exchangability is only possible after independent examination of specifications. Tensile properties are minimum or typical as specified. UTS and YS as MPa. El as %. See Appendix for list of abbreviations used in Notes. * indicates obsolete material.

Carbon Steel, Nonresulfurized, 1020 (Continued from previous page)

Specification	Designation	Notes	C	Cr	Cu	Mn	Ni	P	S	Si	Other	UTS	YS	El	Hard
Romania															
STAS 880(88)	OLC20X	Q/T	0.17-0.24	0.3 max		0.3-0.6		0.035 max	0.035 max	0.17-0.37	Pb <=0.15; As<=0.05; bal Fe				
STAS 880(88)	OLC20XS	Q/T	0.17-0.24	0.3 max		0.3-0.6		0.035 max	0.02-0.04	0.17-0.37	Pb <=0.15; As<=0.05; bal Fe				
Russia															
GOST	MSt0		0.23 max					0.07 max	0.06 max		bal Fe				
GOST 1050	20		0.17-0.24	0.25 max	0.25 max	0.35-0.65	0.25 max	0.04 max	0.04 max	0.17-0.37	bal Fe				
GOST 1050	20kp		0.17-0.24	0.25	0.25 max	0.25-0.5	0.3	0.035	0.04	0.07	bal Fe				
GOST 1050	20ps		0.17-0.24	0.25	0.25 max	0.35-0.65	0.25	0.035	0.04	0.05-0.17	As 0.08; bal Fe				
GOST 1050(88)	20	High-grade struct; Q/T; Full hard; 0-100mm	0.17-0.24	0.25 max		0.35-0.65	0.3 max	0.035 max	0.04 max	0.17-0.37	As<=0.08; bal Fe	490		7L	207 HB max
GOST 1050(88)	20	High-grade struct; Q/T; Norm; 0-100mm	0.17-0.24	0.25 max		0.35-0.65	0.3 max	0.035 max	0.04 max	0.17-0.37	As<=0.08; bal Fe	410	245	25L	
GOST 1050(88)	20kp	High-grade struct; Unkilled; Rimming; R; FU	0.17-0.24	0.25 max		0.25-0.5	0.3 max	0.035 max	0.04 max	0.07 max	As<=0.08; bal Fe				
GOST 1050(88)	20ps	High-grade struct; Semi-killed	0.17-0.24	0.25 max		0.35-0.65	0.3 max	0.035 max	0.04 max	0.05-0.17	As<=0.08; bal Fe				
GOST 19277	20A		0.17-0.24	0.25	0.2	0.35-0.65	0.25	0.04	0.035	0.17-0.37	bal Fe				
GOST 4041	20YuA		0.16-0.22	0.1 max	0.2 max	0.25-0.45	0.15 max	0.020 max	0.025	0.07 max	Al 0.02-0.08; bal Fe				
GOST 4041(71)	20JuA	High-grade Plts for cold pressing	0.16-0.22	0.1 max	0.2 max	0.25-0.45	0.15 max	0.020 max	0.025 max	0.07 max	Al 0.02-0.08; bal Fe				
GOST 4041(71)	20JuA	High-grade Plts for cold pressing	0.16-0.22	0.1 max	0.2 max	0.25-0.45	0.15 max	0.020 max	0.025 max	0.07 max	Al 0.02-0.08; As<=0.08; bal Fe				
GOST 803	18JuA	High-grade Plts for cold pressing	0.16-0.22	0.15 max	0.25 max	0.2-0.4	0.25 max	0.025 max	0.03 max	0.13 max	Al 0.02-0.07; bal Fe				
GOST 803	18YuA		0.16-0.22	0.15	0.25	0.2-0.4	0.25	0.025	0.03	0.13	Al 0.02-0.07; bal Fe				
GOST 8731(74)	10		0.30 max	0.3 max	0.3 max	1.60 max	0.4 max	0.070 max	0.060 max	0.60 max	Al <=0.1; bal Fe				
GOST MRTU	20		0.17-0.24	0.25	0.3	0.35-0.65	0.25	0.03	0.025	0.17-0.37	bal Fe				
Spain															
UNE 36011(75)	C25k	Unalloyed; Q/T	0.2-0.3			0.5-0.8		0.035 max	0.035 max	0.15-0.4	Mo <=0.15; bal Fe				
UNE 36032(85)	20KA-DF	Screw	0.18-0.23			0.30-0.60		0.040 max	0.040 max	0.10 max	Al >=0.020; bal Fe				
UNE 36032(85)	20KX-DF	Screw	0.18-0.23			0.30-0.60		0.040 max	0.040 max	0.15-0.35	bal Fe				
UNE 36032(85)	F.7505*	see 20KX-DF	0.18-0.23			0.30-0.60		0.040 max	0.040 max	0.15-0.35	bal Fe				
UNE 36032(85)	F.7517*	see 20KA-DF	0.18-0.23			0.30-0.60		0.040 max	0.040 max	0.10 max	Al >=0.020; bal Fe				
Sweden															
SS 141357	1357		0.17-0.23			0.25-0.45		0.03 max	0.03 max	0.1-0.3	bal Fe				
SS 141450	1450	FF; Non-rimming	0.16-0.28			0.4-0.9		0.05	0.05		bal Fe	430	250	24	
Turkey															
TS 416	Fe37.2*	Weld Pipe	0.2					0.06	0.05		N 0.01; bal Fe				
TS 911(86)	UDK669.14.423/Fe37	HR, T-bar	0.2					0.06	0.06		bal Fe				
UK															
BS 3601(87)	S360	Pip Tub, press, t<=16mm	0.17 max			0.40-0.80		0.040 max	0.040 max	0.35 max	bal Fe	360-500	235	25	
BS 970/1(83)	040A20	Blm Bil Slab Bar Rod Frg	0.18-0.23			0.30-0.50		0.050 max	0.050 max		bal Fe				
BS 970/1(83)	050A20	Blm Bil Slab Bar Rod Frg	0.18-0.23			0.40-0.60		0.050 max	0.050 max	0.4	bal Fe				
BS 970/1(83)	055M15	Blm Bil Slab Bar Rod Frg	0.20 max			0.80 max		0.050 max	0.050 max	0.10-0.40	bal Fe	310		25	121 HB

UNS numbers and US grades are provided as a means of cross referencing chemically similar alloys. Exchangability is only possible after independent examination of specifications. Tensile properties are minimum or typical as specified. UTS and YS as MPa. El as %. See Appendix for list of abbreviations used in Notes. * indicates obsolete material.

Specification	Designation	Notes	C	Cr	Cu	Mn	Ni	P	S	Si	Other	UTS	YS	El	Hard

Carbon Steel, Nonresulfurized, 1020 (Continued from previous page)

UK

Specification	Designation	Notes	C	Cr	Cu	Mn	Ni	P	S	Si	Other	UTS	YS	El	Hard
BS 970/1(83)	060A20	Blm Bil Slab Bar Rod Frg	0.18-0.23			0.50-0.70		0.050 max	0.050 max	0.4	bal Fe				
BS 970/3(91)	070M20	Bright bar; 6<t<=13mm; HR CD	0.16-0.24			0.50-0.90			0.050 max	0.10-0.40	Mo <=0.15; bal Fe	560	440	10	
BS 970/3(91)	070M20	Bright bar	0.16-0.24			0.50-0.90			0.050 max	0.10-0.40	Mo <=0.15; bal Fe	400	200	21	126-250 HB

USA

Specification	Designation	Notes	C	Cr	Cu	Mn	Ni	P	S	Si	Other	UTS	YS	El	Hard
	AISI 1020	Bar	0.18-0.23			0.30-0.60		0.040 max	0.050 max		bal Fe				
	AISI 1020	Sh Strp Plt	0.17-0.23			0.30-0.60		0.030 max	0.035 max		bal Fe				
	AISI 1020	Struct shp	0.18-0.23			0.30-0.60		0.030 max	0.035 max		bal Fe				
	AISI 1020	Wir rod	0.18-0.23			0.30-0.60		0.040 max	0.050 max		bal Fe				
	AISI M1020	Bar, merchant qual	0.17-0.24			0.25-0.60		0.04 max	0.05 max		bal Fe				
	UNS G10200		0.18-0.23			0.30-0.60		0.040 max	0.050 max		Sheets and Plates, C 0.17-0.23; bal Fe				
AMS 5032E(90)	1020	Wir	0.18-0.23		0.15 max	0.30-0.60		0.040 max	0.050 max		bal Fe	345-517		15	
AMS 5045G(94)	1020	Sh, Strp, Hard tmp, t<=1.75	0.25 max			0.30-0.60		0.035 max	0.040 max		bal Fe				84-96 HRB
ASTM A108(95)	1020	Bar, CF	0.18-0.23			0.30-0.60		0.040 max	0.050 max		bal Fe				
ASTM A29/A29M(93)	1020	Bar	0.18-0.23			0.30-0.60		0.040 max	0.050 max		bal Fe				
ASTM A29/A29M(93)	M1020	Bar, merchant qual	0.17-0.24			0.25-0.6		0.040 max	0.050 max		bal Fe				
ASTM A510(96)	1020	Wir rod	0.18-0.23			0.30-0.60		0.040 max	0.050 max		bal Fe				
ASTM A512(96)	1020	Buttweld mech tub, CD	0.17-0.23			0.30-0.60		0.040 max	0.045 max		bal Fe				
ASTM A512(96)	M1020	Buttweld mech tub, CD	0.15-0.25			0.30-0.60		0.040 max	0.045 max		bal Fe				
ASTM A512(96)	MTX1020	Buttweld mech tub, CD	0.15-0.25			0.70-1.00		0.040 max	0.045 max		bal Fe				
ASTM A513(97a)	1020	ERW Mech tub	0.17-0.23			0.30-0.60		0.035 max	0.035 max		bal Fe				
ASTM A513(97a)	MT1020	ERW Mech tub	0.15-0.25			0.30-0.60		0.035 max	0.035 max		bal Fe				
ASTM A513(97a)	MTX1020	ERW Mech tub	0.15-0.25			0.70-1.00		0.035 max	0.035 max		bal Fe				
ASTM A519(96)	1020	Smls mech tub, HR	0.18-0.23			0.30-0.60		0.040 max	0.050 max		bal Fe	345	221	25	55 HRB
ASTM A519(96)	1020	Smls mech tub, CW	0.18-0.23			0.30-0.60		0.040 max	0.050 max		bal Fe	483	414	5	75 HRB
ASTM A519(96)	1020	Smls mech tub, SR	0.18-0.23			0.30-0.60		0.040 max	0.050 max		bal Fe	448	345	10	72 HRB
ASTM A519(96)	1020	Smls mech tub, Ann	0.18-0.23			0.30-0.60		0.040 max	0.050 max		bal Fe	331	193	30	50 HRB
ASTM A519(96)	1020	Smls mech tub, Norm	0.18-0.23			0.30-0.60		0.040 max	0.050 max		bal Fe	379	234	22	60 HRB
ASTM A519(96)	MT1020	Smls Low-C Mech tub	0.15-0.25			0.30-0.60		0.040 max	0.050 max		bal Fe				
ASTM A519(96)	MTX1020	Smls Low-C Mech tub	0.15-0.25			0.70-1.00		0.040 max	0.050 max		bal Fe				
ASTM A575(96)	M1020	Merchant qual bar	0.17-0.24			0.25-0.60		0.04 max	0.05 max		Mn can vary with C; bal Fe				
ASTM A576(95)	1020	Special qual HW bar	0.18-0.23			0.30-0.60		0.040 max	0.050 max		Si Cu Pb B Bi Ca Se Te if spec'd; bal Fe				
ASTM A635/A635M(98)	1020	Sh Strp, Coil, HR	0.17-0.23		0.20 min	0.30-0.60		0.030 max	0.035 max		bal Fe				
ASTM A659/A659M(97)	1020	Sh Strp, HR	0.17-0.23			0.30-0.60		0.030 max	0.035 max		bal Fe				
ASTM A787(96)	MT1020	ERW metallic-coated mech tub, Low-C	0.15-0.25			0.30-0.60		0.035 max	0.035 max		bal Fe				

Specification	Designation	Notes	C	Cr	Cu	Mn	Ni	P	S	Si	Other	UTS	YS	El	Hard

Carbon Steel, Nonresulfurized, 1020 (Continued from previous page)

USA

Specification	Designation	Notes	C	Cr	Cu	Mn	Ni	P	S	Si	Other	UTS	YS	El	Hard
ASTM A787(96)	MTX1015	ERW metallic-coated mech tub, Low-C	0.10-0.20			0.60-0.90		0.035 max	0.035 max		bal Fe				
ASTM A787(96)	MTX1020	ERW metallic-coated mech tub, Low-C	0.15-0.25			0.70-1.00		0.035 max	0.035 max		bal Fe				
ASTM A794(97)	1020	Sh, CS, CR	0.17-0.23			0.30-0.60		0.030 max	0.035 max		Cu >=0.20 if spec; bal Fe				
ASTM A827(98)	1020	Plt, frg	0.17-0.23			0.30-0.60		0.035 max	0.040 max	0.15-0.40	bal Fe				
ASTM A830/A830M(98)	1020	Plt	0.17-0.23			0.30-0.60		0.035 max	0.04 max		bal Fe				
DoD-F-24669/1(86)	1020	Bar Bil; Supersedes MIL-S-866 & MIL-S-16974	0.18-0.23			0.3-0.6		0.04	0.05		bal Fe				
FED QQ-S-635B(88)	1020*	Obs; see ASTM A827; Plt	0.17-0.23			0.3-0.6		0.04	0.05		bal Fe				
FED QQ-S-698(88)	1020	Sh Strp	0.17-0.21			0.30-0.60		0.040 max	0.050 max		bal Fe				
FED QQ-W-461H(88)	1020*	Obs; Wir	0.18-0.23			0.3-0.6		0.04	0.05		bal Fe				
MIL-S-11310E(76)	1020*	Bar HR cold shape cold ext; Obs for new design	0.18-0.23	0.07-0.2	0.2	0.3-0.6		0.04	0.05	0.2	bal Fe				
MIL-S-11310E(76)	CS1020*	Bar HR cold shape cold ext; Obs for new design	0.18-0.23			0.3-0.6		0.04 max	0.05 max		bal Fe				
MIL-S-16788A(86)	C2*	HR, frg, bloom bil slab for refrg; Obs for new design	0.18-0.23			0.30-0.60		0.040 max	0.050 max		bal Fe				
MIL-S-46059	G10200*	Obs; Shape	0.18-0.23			0.3-0.6		0.04	0.05		bal Fe				
MIL-S-7952A(63)	1020*	Uncoated Sh Strp, aircraft qual; Obs for new design see AMS 5046	0.18-0.25			0.3-0.6		0.04	0.05		bal Fe				
MIL-T-3520	1020*	Obs; Tube	0.18-0.23			0.3-0.6		0.04	0.05		bal Fe				
SAE J1397(92)	1020	Bar CD, est mech prop	0.18-0.23			0.3-0.6		0.04 max	0.05 max		bal Fe	420	350	15	121 HB
SAE J1397(92)	1020	Bar HR, est mech prop	0.18-0.23			0.3-0.6		0.04 max	0.05 max		bal Fe	380	210	25	111 HB
SAE J403(95)	1020	Struct Shps Plt Strp Sh Weld Tub	0.17-0.23			0.30-0.60		0.030 max	0.035 max		bal Fe				
SAE J403(95)	1020	Bar Wir rod Smls Tub HR CF	0.18-0.23			0.30-0.60		0.030 max	0.050 max		bal Fe				
SAE J403(95)	M1020	Merchant qual	0.18-0.23			0.30-0.60		0.04 max	0.05 max		bal Fe				
Yugoslavia															
	C.0445V	Struct	0.30 max			1.60 max		0.05 max	0.05 max	0.60 max	P+S<=0.09; bal Fe				
	C.0471	Struct	0.2 max			1.60 max		0.05 max	0.05 max	0.60 max	Mo <=0.15; N <=0.007; bal Fe				
	C.0481	Struct	0.2 max			1.60 max		0.05 max	0.05 max	0.60 max	N <=0.007; bal Fe				
	C.0561	Struct	0.2 max			1.5 max		0.05 max	0.05 max	0.55 max	Mo <=0.15; N <=0.007; bal Fe				
	C.1330	Q/T	0.17-0.24			0.3-0.6		0.045 max	0.045 max	0.4 max	Mo <=0.15; bal Fe				
	C.1331	Q/T	0.17-0.24			0.3-0.6		0.035 max	0.03 max	0.4 max	Mo <=0.15; bal Fe				
	CL.0600	Reinforcing	0.30 max			1.60 max		0.070 max	0.060 max	0.60 max	Mo <=0.15; bal Fe				
	CL.0601	Reinforcing	0.30 max			1.60 max		0.070 max	0.060 max	0.60 max	Mo <=0.15; bal Fe				
	CL.0700	Reinforcing	0.30 max			1.60 max		0.070 max	0.060 max	0.60 max	Mo <=0.15; bal Fe				
	PZ18		0.18-0.23	0.15	0.025	0.3-0.6	0.015	0.04	0.04	0.1-0.3	bal Fe				

UNS numbers and US grades are provided as a means of cross referencing chemically similar alloys. Exchangability is only possible after independent examination of specifications. Tensile properties are minimum or typical as specified. UTS and YS as MPa. El as %. See Appendix for list of abbreviations used in Notes. * indicates obsolete material.

Carbon Steel, Nonresulfurized, 1021

Specification	Designation	Notes	C	Cr	Cu	Mn	Ni	P	S	Si	Other	UTS	YS	El	Hard
Australia															
AS 1442	S1021*	Obs; Bar Bil	0.18-0.23			0.6-0.9		0.05	0.05	0.35	bal Fe				
AS 1442(92)	1021	HR bar, Semifinished (may treat w/microalloying elements)	0.18-0.23			0.60-0.90		0.040 max	0.040 max	0.10-0.35	Si<=0.10 for Al-killed; bal Fe				
AS 1443	S1021*	Obs	0.18-0.23			0.6-0.9		0.05	0.05	0.35	bal Fe				
Bulgaria															
BDS 7478	20L	High-temp const	0.17-0.25	0.30 max	0.30 max	0.80 max	0.40	0.035	0.035	0.30-0.50	bal Fe				
Canada															
CSA B193		Pipe, Mill Elect Weld G 6.35mm	0.21			0.9		0.04	0.05		bal Fe				
China															
GB 6654-96	20R	Plt HR Norm 16-25mm Thk	0.20 max	0.30 max	0.30 max	0.40-0.90	0.30 max	0.035 max	0.035 max	0.15-0.30	bal Fe	400-520	235	25	
GB 713(97)	20g	Plt HR Norm 16-25mm Thk	0.20 max	0.30 max	0.30 max	0.50-0.90	0.30 max	0.035 max	0.035 max	0.15-0.30	bal Fe	400-520	235	25	
France															
AFNOR NFA35551	21B3		0.18-0.24			0.6-0.9		0.035	0.035	0.1-0.4	B 0.0008-0.005; bal Fe				
AFNOR NFA35552	21B3		0.18-0.24			0.6-0.9		0.035	0.035	0.1-0.4	B 0.0008-0.005; bal Fe				
AFNOR NFA35553	21B3		0.18-0.24			0.6-0.9		0.035	0.035	0.1-0.4	B 0.0008-0.005; bal Fe				
AFNOR NFA35557	21B3		0.18-0.24			0.6-0.9		0.035	0.035	0.1-0.4	B 0.0008-0.005; bal Fe				
AFNOR NFA35566	21B3		0.18-0.24			0.6-0.9		0.035	0.035	0.1-0.4	B 0.0008-0.005; bal Fe				
Hungary															
MSZ 500(81)	B44	Gen struct; Non-rimming; FF	0.17-0.25			0.5-0.8		0.045 max	0.05 max	0.12-0.35	bal Fe				
Japan															
JIS G3445(88)	STKM13A	Tube	0.25 max			0.30-0.90		0.040 max	0.040 max	0.35 max	bal Fe	373	216	30L	
JIS G3445(88)	STKM13B	Tube	0.25 max			0.30-0.90		0.040 max	0.040 max	0.35 max	bal Fe	441	304	20L	
JIS G3445(88)	STKM13C	Tube	0.25 max			0.30-0.90		0.040 max	0.040 max	0.35 max	bal Fe	510	382	15L	
JIS G4051(79)	S22C	Bar Wir rod	0.20-0.25	0.20 max	0.30 max	0.30-0.60	0.20 max	0.030 max	0.035 max	0.15-0.35	Ni+Cr<=0.35; bal Fe				
Mexico															
NMX-B-301(86)	1021	Bar	0.18-0.23			0.60-0.90		0.040 max	0.050 max		bal Fe				
Pan America															
COPANT 331	1021		0.18-0.23			0.6-0.9		0.04	0.05		bal Fe				
COPANT 333	1021		0.18-0.23			0.6-0.9		0.04	0.05		bal Fe				
COPANT 514	TM1021	Tube, HF, 323.9mm diam	0.18-0.23			0.6-0.9		0.04	0.05	0.2-0.35	bal Fe				
COPANT 514	TM1021	Tube, CF	0.18-0.23			0.6-0.9		0.04	0.05	0.2-0.35	bal Fe				
COPANT 514	TM1021	Tube, Norm	0.18-0.23			0.6-0.9		0.04	0.05	0.2-0.35	bal Fe				
COPANT 514	TM1021	Tube, Ann	0.18-0.23			0.6-0.9		0.04	0.05	0.2-0.35	bal Fe				
Romania															
STAS 8183	OLT45		0.17-0.24	0.3		0.4-0.8	0.3	0.04	0.045	0.15-0.35	Mo 0.06; bal Fe				
Russia															
GOST	M21		0.17-0.25	0.3 max	0.25 max	0.4-0.7	0.3 max	0.05 max	0.05 max	0.12-0.3	Cr+Cu+Ni<=0.3; bal Fe				
GOST 977	20L-II		0.17-0.25	0.3	0.3	0.35-0.9	0.3	0.035	0.045	0.2-0.52	bal Fe				

UNS numbers and US grades are provided as a means of cross referencing chemically similar alloys. Exchangability is only possible after independent examination of specifications. Tensile properties are minimum or typical as specified. UTS and YS as MPa. El as %. See Appendix for list of abbreviations used in Notes. * indicates obsolete material.

Specification	Designation	Notes	C	Cr	Cu	Mn	Ni	P	S	Si	Other	UTS	YS	El	Hard
Carbon Steel, Nonresulfurized, 1021 (Continued from previous page)															
Russia															
GOST 977	20L-III		0.17-0.25	0.3	0.3	0.35-0.9	0.3	0.035	0.045	0.2-0.52	bal Fe				
GOST 977(88)	20L		0.17-0.25	0.3		0.35-0.9	0.3	0.035	0.045	0.2-0.52	bal Fe				
Turkey															
TS 908(86)	UDK669.14.423/Fe34	HR, equal angles	0.21					0.06	0.06		bal Fe				
TS 909(86)	UDK669.14.423/Fe31	HR, unequal angles	0.21					0.06	0.06		bal Fe				
TS 912(86)	UDK669.14.423/Fe34	HR, channel	0.21					0.06	0.06		bal Fe				
TS 913(86)	UDK669.14.423/Fe34	HR, Z-bar	0.21					0.06	0.06		bal Fe				
UK															
BS 970/1(83)	080M20	Blm Bil Slab Bar Rod Frg	0.18-0.23			0.70-0.90		0.050 max	0.050 max	0.1-0.4	bal Fe				
BS 970/1(96)	080M15	Blm Bil Slab Bar Rod Frg	0.12-0.18			0.60-1.00		0.050 max	0.050 max	0.10-0.40	bal Fe				
USA															
	AISI 1021	Wir rod	0.18-0.23			0.60-0.90		0.040 max	0.050 max		bal Fe				
	AISI 1021	Sh Strp Plt	0.17-0.23			0.60-0.90		0.030 max	0.035 max		bal Fe				
	AISI 1021	Struct shp	0.18-0.23			0.60-0.90		0.030 max	0.035 max		bal Fe				
	AISI 1021	Bar	0.18-0.23			0.60-0.90		0.040 max	0.050 max		bal Fe				
	UNS G10210		0.18-0.23			0.60-0.90		0.040 max	0.050 max		Sheets, C 0.17-0.23; ERW Tubing, C 0.17-0.24; bal Fe				
ASTM A29/A29M(93)	1021	Bar	0.18-0.23			0.60-0.90		0.040 max	0.050 max		bal Fe				
ASTM A510(96)	1021	Wir rod	0.18-0.23			0.60-0.90		0.040 max	0.050 max		bal Fe				
ASTM A512(96)	1021	Buttweld mech tub, CD	0.17-0.23			0.60-0.90		0.040 max	0.045 max		bal Fe				
ASTM A513(97a)	1021	ERW Mech tub	0.17-0.23			0.60-0.90		0.035 max	0.035 max		bal Fe				
ASTM A519(96)	1021	Smls mech tub	0.18-0.23			0.60-0.90		0.040 max	0.050 max		bal Fe				
ASTM A576(95)	1021	Special qual HW bar	0.18-0.23			0.60-0.90		0.040 max	0.050 max		Si Cu Pb B Bi Ca Se Te if spec'd; bal Fe				
ASTM A635/A635M(98)	1021	Sh Strp, Coil, HR	0.17-0.23		0.20 min	0.60-0.90		0.030 max	0.035 max		bal Fe				
ASTM A659/A659M(97)	1021	Sh Strp, HR	0.17-0.23			0.60-0.90		0.030 max	0.035 max		bal Fe				
ASTM A787(96)	1021	ERW metallic-coated mech tub	0.17-0.24			0.60-0.90		0.035 max	0.035 max		bal Fe				
ASTM A794(97)	1021	Sh, CS, CR	0.17-0.23			0.60-0.90		0.030 max	0.035 max		Cu >=0.20 if spec; bal Fe				
ASTM A830/A830M(98)	1021	Plt	0.17-0.23			0.60-0.90		0.035 max	0.04 max		bal Fe				
SAE J1397(92)	1021	Bar HR, est mech prop									bal Fe	420	230	24	116 HB
SAE J1397(92)	1021	Bar CD, est mech prop									bal Fe	470	390	15	131 HB
SAE J403(95)	1021	Frg Bar, HR, CR Rod, Smls Tub	0.18-0.23			0.6-0.9		0.04 max	0.05 max		bal Fe				
SAE J403(95)	1021	Shp Plt Strp Sh Weld Tub	0.17-0.23			0.6-0.9		0.03 max	0.035 max		bal Fe				
SAE J403(95)	1021	Bar Wir rod Smls Tub HR CF	0.18-0.23			0.60-0.90		0.030 max	0.050 max		bal Fe				
SAE J403(95)	1021	Struct Shps Plt Strp Sh Weld Tub	0.17-0.23			0.60-0.90		0.030 max	0.035 max		bal Fe				
Carbon Steel, Nonresulfurized, 1022															
Australia															
AS 1442	K1022*	Obs; Bar Bil	0.18-0.23			0.7-1		0.05	0.05	0.1-0.35	bal Fe				

Specification	Designation	Notes	C	Cr	Cu	Mn	Ni	P	S	Si	Other	UTS	YS	El	Hard
Carbon Steel, Nonresulfurized, 1022 (Continued from previous page)															
Australia															
AS 1442(92)	1022	HR bar, Semifinished (may treat w/microalloying elements)	0.18-0.23			0.70-1.00		0.040 max	0.040 max	0.10-0.35	Si<=0.10 for Al-killed; bal Fe				
AS 1442(92)	3	Rnd/sq/hex 50-215mm, Flat bar 40-60mm, Bloom/bil/slab<=250mm, HR as rolled or norm	0.25 max			1.40 max		0.040 max	0.040 max	0.10-0.40	bal Fe	410	230	22	
AS 1442(92)	U3	Rnd/sq/hex 50-170mm, Flat bar 40-60mm, Bil<=120mm, HR as rolled or norm	0.25 max			1.40 max		0.040 max	0.040 max	0.40 max	bal Fe	410	230	22	
AS 1443	K1022*	Obs; Bar	0.18-0.23			0.7-1		0.05	0.05	0.1-0.35	bal Fe				
AS 1443(94)	1022	CF bar (may treat w/microalloying elements)	0.18-0.23			0.70-1.00		0.040 max	0.040 max	0.10-0.35	Si<=0.10 for Al-killed; bal Fe				
AS 1446	K1022*	Withdrawn, see AS/NZS 1594(97)	0.18-0.23			0.7-1		0.05	0.05	0.1-0.35	bal Fe				
Austria															
ONORM N3121	19Mn5KK		0.17-0.23	0.3		1.0-1.3		0.040	0.040	0.4-0.6	bal Fe				
ONORM N3121	19Mn5KKW		0.17-0.23	0.3		1.0-1.3		0.040	0.040	0.4-0.6	bal Fe				
ONORM N3121	19Mn5KW		0.17-0.23	0.3		1.0-1.3		0.040	0.040-0.04	0.4-0.6	bal Fe				
Belgium															
NBN 629	D42-1	Strip	0.2			0.5-1.3		0.05	0.05	0.35	bal Fe				
NBN 629	D42-2	Strip	0.2			0.5-1.3		0.04	0.04	0.1-0.35	bal Fe				
NBN 629	D47-1	Strip	0.2			0.6-1.4		0.05	0.05	0.35	bal Fe				
NBN 630	E42-1	Strip	0.2			0.5-1.3		0.05	0.05	0.35	bal Fe				
NBN 630	E42-2	Strip	0.2			0.5-1.3		0.04	0.04	0.1-0.35	bal Fe				
NBN 630	E47-1	Strip	0.2			0.6-1.4		0.05	0.05	0.35	bal Fe				
NBN 630	E47-2	Strip	0.2			0.6-1.4		0.04	0.04	0.1-0.35	bal Fe				
NBN 837	D45	Tube	0.22			0.45-1		0.04	0.04	0.15-0.35	bal Fe				
NBN A21-221	C20KD	Wir	0.17-0.23			0.7-1		0.04	0.05	0.1-0.4	bal Fe				
Bulgaria															
BDS 5785(83)	20ps	Struct	0.17-0.24	0.25 max	0.25 max	0.35-0.65	0.25 max	0.035 max	0.040 max	0.05-0.17	bal Fe				
Canada															
CSA B193	A	Pipe	0.22			0.9		0.04	0.05		bal Fe				
China															
GB 699(88)	20Mn	Bar HR Norm 25mm diam	0.17-0.24	0.25 max	0.25 max	0.70-1.00	0.25 max	0.035 max	0.035 max	0.17-0.37	bal Fe	450	275	24	
GB/T 3078(94)	20Mn	Bar CR Norm 25mm diam	0.17-0.24	0.25 max	0.25 max	0.70-1.00	0.25 max	0.035 max	0.035 max	0.17-0.37	bal Fe	450	275	24	
Finland															
SFS 1206(74)	SFS1206	Reinforcing	0.22 max			1.8 max		0.06 max	0.05 max	0.55 max	N <=0.01; As 0-0.08; Ce C+Mn/6+(Cr+Mo+V)/5+(Ni+Cu)/15; bal Fe				
SFS 200	Fe33	Gen struct	0.23 max			1.60 max		0.06 max	0.05 max	0.60 max	bal Fe				
SFS 200	Fe52C	Gen struct	0.20 max		0.4 max	1.6 max		0.045 max	0.045 max	0.55 max	Al 0.02-0.10; N <=0.015; bal Fe				
SFS 200	Fe52D	Gen struct	0.20 max		0.4 max	1.6 max		0.04 max	0.04 max	0.55 max	Al 0.02-0.10; N <=0.015; bal Fe				
SFS 2146	Fe45		0.25		0.3			0.05	0.05		bal Fe				

Specification	Designation	Notes	C	Cr	Cu	Mn	Ni	P	S	Si	Other	UTS	YS	El	Hard

Carbon Steel, Nonresulfurized, 1022 (Continued from previous page)

France

Specification	Designation	Notes	C	Cr	Cu	Mn	Ni	P	S	Si	Other	UTS	YS	El	Hard
AFNOR	20MB4		0.17-0.23			0.9-1.2		0.035	0.035	0.1-0.35	Al 0.02; B 0.0008; bal Fe				
AFNOR NFA35551	20MB5	CH; Bar Rod Wir	0.16-0.22			1.1-1.4		0.035	0.035	0.1-0.4	B 0.0008-0.005; bal Fe				
AFNOR NFA35552	20MB5	Bar Rod Wir	0.16-0.22			1.1-1.4		0.035	0.035	0.1-0.4	B 0.0008-0.005; bal Fe				
AFNOR NFA35553	20MB5	Bar Rod Wir	0.16-0.22			1.1-1.4		0.035	0.035	0.1-0.4	B 0.0008-0.005; bal Fe				
AFNOR NFA35556(84)	20MB5	CH	0.16-0.22			1.1-1.4		0.035 max	0.035 max	0.1-0.4	B 0.0008-0.005; bal Fe				
AFNOR NFA35557	20MB5	Q/T	0.16-0.22			1.1-1.4		0.035	0.035	0.1-0.4	B 0.0008-0.005; bal Fe				
AFNOR NFA35557(83)	20MB5	Q/T; t<=16mm	0.16-0.22	0.3 max	0.3 max	1.1-1.4	0.4 max	0.035 max	0.035 max	0.1-0.4	B 0.0008-0.005; bal Fe	820-1020	660	12	
AFNOR NFA35566	20MB5	Q/T; Chains	0.16-0.22			1.1-1.4		0.035	0.035	0.1-0.35	Al 0.02; B 0.0008-0.005; bal Fe				

Germany

Specification	Designation	Notes	C	Cr	Cu	Mn	Ni	P	S	Si	Other	UTS	YS	El	Hard
DIN	21Mn4		0.16-0.24			0.80-1.10		0.035 max	0.035 max	0.10-0.25	Si 0.35 max; Mn 1.30 max if diam >=26mm; bal Fe				
DIN	WNr 1.0469		0.16-0.24			0.80-1.10		0.035 max	0.03 max	0.10-0.25	Si<=0.35, Mn<=1.30 if diam>=26mm; bal Fe				
DIN 1623(86)	S355J2G3	Flat product HR Sht Strp 100-150mm	0.20 max			1.60 max		0.035 max	0.035 max	0.55 max	C depends on thkness; bal Fe	470-630	295	18	
DIN 1623(86)	WNr 1.0570	Flat product HR Sh Strp 100-150mm	0.20 max			1.60 max		0.035 max	0.035 max	0.55 max	C depends on thkness; bal Fe	470-630	295	18	
DIN 17119(84)	S355J2G3	Weld CF sq rect tub	0.20 max			1.60 max		0.035 max	0.035 max	0.55 max	C depends on thkness; bal Fe				
DIN 17119(84)	WNr 1.0570	Weld CF sq rect tub	0.20 max			1.60 max		0.035 max	0.035 max	0.55 max	C depends on thkness; bal Fe				
DIN 17175(79)	19Mn5	Smls tube, long sample bar <=60mm HW norm Q/T	0.17-0.22	0.30 max		1.00-1.30		0.045 max	0.045 max	0.30-0.60	bal Fe	510	315	20	150 HB
DIN 17175(79)	WNr 1.0482	Smls tube; Long <=60mm, HW norm Q/T	0.17-0.22	0.30 max		1.00-1.30		0.045 max	0.045 max	0.30-0.60	bal Fe	510	315	20	150 HB
DIN 17243(90)	20Mn5	HR bar Q/T 41-100mm	0.17-0.23	0.30 max		1.00-1.50		0.035 max	0.030 max	0.60 max	Al <=0.015; bal Fe	490-590	295	18	
DIN 17243(90)	WNr 1.1133	HR bar Q/T 41-100mm	0.17-0.23	0.30 max		1.00-1.50		0.035 max	0.030 max	0.60 max	Al 0.015-0.050; bal Fe	490-590	295	18	
DIN 17243(90)	WNr 1.1133	HR bar Q/T <=16mm	0.17-0.23	0.30 max		1.00-1.50		0.035 max	0.030 max	0.60 max	Al 0.015-0.050; bal Fe	540-690	390	22	
DIN EN 10025(94)	S355J2G3	HR 100-150mm	0.20 max			1.60 max		0.035 max	0.035 max	0.55 max	C depends on thkness; bal Fe	470-630	295	18	
DIN EN 10025(94)	WNr 1.0570	HR 100-150mm	0.20 max			1.60 max		0.035 max	0.035 max	0.55 max	C depends on thkness; bal Fe	470-630	295	18	
DIN(Military Hdbk)	Ck19		0.15-0.23	0.05 max	0.15 max	0.40-0.60	0.10 max	0.030 max	0.025 max	0.15 max	Al 0.03-0.08; bal Fe				
DIN(Military Hdbk)	WNr 1.1134		0.15-0.23	0.05 max	0.15 max	0.40-0.60	0.10 max	0.030 max	0.025 max	0.15 max	Al 0.03-0.08; bal Fe				

Hungary

Specification	Designation	Notes	C	Cr	Cu	Mn	Ni	P	S	Si	Other	UTS	YS	El	Hard
MSZ 61(85)	C22	Q/T; t<=16mm	0.17-0.24			0.3-0.6		0.035 max	0.035 max	0.4 max	bal Fe	550-700	350	20L	
MSZ 61(85)	C22	Q/T; 16<t<=40mm	0.17-0.24			0.3-0.6		0.035 max	0.035 max	0.4 max	bal Fe	500-650	300	22L	
MSZ 61(85)	C22E	Q/T; 16<t<=40mm	0.17-0.24			0.3-0.6		0.035 max	0.02-0.035	0.4 max	bal Fe	500-650	300	22L	
MSZ 61(85)	C22E	Q/T; t<=16mm	0.17-0.24			0.3-0.6		0.035 max	0.02-0.035	0.4 max	bal Fe	550-700	350	20L	
MSZ 6251(87)	C22Z	CF; Q/T; 0-36mm; drawn, half-hard	0.17-0.24			0.3-0.6		0.035 max	0.035 max	0.4 max	bal Fe	0-530			
MSZ 6251(87)	C22Z	CF; Q/T; 0-36mm; HR; HR/soft ann; drawn/soft ann; drawn/bright ann; soft ann/ground	0.17-0.24			0.3-0.6		0.035 max	0.035 max	0.4 max	bal Fe	0-500			

Specification	Designation	Notes	C	Cr	Cu	Mn	Ni	P	S	Si	Other	UTS	YS	El	Hard

Carbon Steel, Nonresulfurized, 1022 (Continued from previous page)

Specification	Designation	Notes	C	Cr	Cu	Mn	Ni	P	S	Si	Other	UTS	YS	El	Hard
India															
IS 1914/IV			0.25					0.05	0.05		bal Fe				
Italy															
UNI 6363(84)	Fe510	Weld/Smls tub; water main	0.26 max			1.3 max		0.04 max	0.04 max	0.5 max	bal Fe				
UNI 6403(86)	Fe510	Q/T	0.2 max			1-1.5		0.04 max	0.04 max	0.15-0.55	bal Fe				
UNI 8913(87)	Fe510	Q/T	0.22 max			0.8-1.5		0.04 max	0.04 max	0.35 max	bal Fe				
Japan															
JIS G3444(94)	STK500	Tub Gen struct	0.24 max			0.30-1.30		0.040 max	0.040 max	0.35 max	bal Fe	500	355	15	
JIS G3507(91)	SWRCH22K	Wir rod FF	0.18-0.23			0.70-1.00		0.030 max	0.035 max	0.10-0.35	bal Fe				
Mexico															
NMX-B-301(86)	1022	Bar	0.18-0.23			0.70-1.00		0.040 max	0.050 max		bal Fe				
Pan America															
COPANT 330	1022		0.18-0.23			0.7-1		0.04	0.05		bal Fe				
COPANT 331	1022		0.18-0.23			0.7-1		0.04	0.05		bal Fe				
COPANT 333	1022		0.18-0.23			0.7-1		0.04	0.05		bal Fe				
COPANT 514	TMX1020		0.15-0.25	0.05		0.6-1		0.04	0.05	0.2-0.35	bal Fe				
Poland															
PNH84019	20G	CH; Q/T	0.17-0.24	0.3 max		0.7-1	0.3 max	0.04 max	0.04 max	0.17-0.37	bal Fe				
PNH84023	R45	High-temp const	0.16-0.22		0.3	0.6-1.2		0.04	0.04	0.12-0.35	bal Fe				
PNH84023	R45A	High-temp const	0.16-0.22			0.6-1.2		0.04	0.04	0.12-0.35	Al 0.02-0.06; bal Fe				
PNH84024	19G2		0.16-0.22	0.3	0.25	1-1.4	0.25	0.045	0.045	0.4-0.6	bal Fe				
Romania															
STAS 11511	20MnB5		0.17-0.23	0.3	0.3	1.1-1.4	0.3	0.035	0.035	0.1-0.4	Al 0.02; B 0.003-0.005; Ti 0.01-0.04; bal Fe				
STAS 15513	20Mn10		0.17-0.24	0.3	0.3	0.7-1	0.3	0.035	0.035	0.17-0.37	bal Fe				
Russia															
GOST 380	BS+4Gps		0.18-0.27			0.8-1.2		0.04	0.05	0.15	N 0.008; As 0.08; bal Fe				
GOST 380	Bst4Gps2		0.18-0.27	0.3	0.3	0.8-1.2	0.3	0.04	0.05	0.15	N 0.008; As 0.08; bal Fe				
GOST 4543	20G		0.17-0.24	0.3	0.3	0.7-1	0.3	0.035	0.035	0.17-0.37	bal Fe				
GOST 4543(71)	20G	CH	0.17-0.24	0.3 max	0.3 max	0.7-1.0	0.3 max	0.035 max	0.035 max	0.17-0.37	Al <=0.1; bal Fe				
GOST 977(88)	20GSL		0.16-0.22	0.3		1-1.3	0.3	0.03	0.03	0.6-0.8	bal Fe				
Spain															
UNE	F.220.A		0.17-0.23			1-1.4		0.035	0.035	0.13-0.38	bal Fe				
UNE 36013(76)	20Mn6	CH	0.18-0.25			1.3-1.6		0.035 max	0.035 max	0.15-0.4	Mo <=0.15; bal Fe				
UNE 36013(76)	F.1515*	see 20Mn6	0.18-0.25			1.3-1.6		0.035 max	0.035 max	0.15-0.4	Mo <=0.15; bal Fe				
UNE 36034(85)	20MnB4DF	CHd; Cold ext	0.17-0.23			0.8-1.1		0.035	0.035	0.15-0.35	B 0.0008-0.005; bal Fe				
Turkey															
TS 302	Fe45.2*	Pipe as drawn or weld	0.25					0.05	0.05		bal Fe				
TS 346	Fe45*	Pipe as drawn	0.25			0.9		0.05	0.05	0.35	bal Fe				
UK															
BS 3111/1(87)	9/1	Q/T Wir	0.17-0.23			0.75-0.95		0.035 max	0.035 max	0.40 max	Al 0.02-0.120; B 0.0008-0.005; bal Fe				

UNS numbers and US grades are provided as a means of cross referencing chemically similar alloys. Exchangability is only possible after independent examination of specifications. Tensile properties are minimum or typical as specified. UTS and YS as MPa. El as %. See Appendix for list of abbreviations used in Notes. * indicates obsolete material.

Specification	Designation	Notes	C	Cr	Cu	Mn	Ni	P	S	Si	Other	UTS	YS	El	Hard

Carbon Steel, Nonresulfurized, 1022 (Continued from previous page)

UK

Specification	Designation	Notes	C	Cr	Cu	Mn	Ni	P	S	Si	Other	UTS	YS	El	Hard
BS 970/1(83)	120M19	Blm Bil Slab Bar Rod Frg Hard Tmp; diam<100mm	0.15-0.23			1.00-1.40		0.050 max	0.050 max		bal Fe				

USA

Specification	Designation	Notes	C	Cr	Cu	Mn	Ni	P	S	Si	Other	UTS	YS	El	Hard
	AISI 1022	Struct shp	0.18-0.23			0.70-1.00		0.030 max	0.035 max		bal Fe				
	AISI 1022	Bar	0.18-0.23			0.70-1.00		0.040 max	0.050 max		bal Fe				
	AISI 1022	Sh Strp Plt	0.17-0.23			0.70-1.00		0.030 max	0.035 max		bal Fe				
	AISI 1022	Wir rod	0.18-0.23			0.70-1.00		0.040 max	0.050 max		bal Fe				
	UNS G10220		0.18-0.23			0.70-1.00		0.040 max	0.050 max		Sheets, C 0.17-0.23; ERW Tubing, C 0.17-0.24; bal Fe				
AMS 5070G(90)		Bar, CF t<=22.22	0.18-0.23			0.70-1.00		0.040 max	0.050 max	0.15-0.35	bal Fe	482	414	18	
AMS 5070G(90)		Bar, 50.8< t<=76.2	0.18-0.23			0.70-1.00		0.040 max	0.050 max	0.15-0.35	bal Fe	379	310	15	
AMS 5070G(90)		Frg, Norm	0.18-0.23			0.70-1.00		0.040 max	0.050 max	0.15-0.35	bal Fe	379	248	22	
ASTM A108(95)	1022	Bar, CF	0.18-0.23			0.70-1.00		0.040 max	0.050 max		bal Fe				
ASTM A29/A29M(93)	1022	Bar	0.18-0.23			0.70-1.00		0.040 max	0.050 max		bal Fe				
ASTM A510(96)	1022	Wir rod	0.18-0.23			0.70-1.00		0.040 max	0.050 max		bal Fe				
ASTM A513(97a)	1022	ERW Mech tub	0.17-0.23			0.70-1.00		0.035 max	0.035 max		bal Fe				
ASTM A519(96)	1022	Smls mech tub	0.18-0.23			0.70-1.00		0.040 max	0.050 max		bal Fe				
ASTM A576(95)	1022	Special qual HW bar	0.18-0.23			0.70-1.00		0.040 max	0.050 max		Si Cu Pb B Bi Ca Se Te if spec'd; bal Fe				
ASTM A635/A635M(98)	1022	Sh Strp, Coil, HR	0.17-0.23		0.20 min	0.70-1.00		0.030 max	0.035 max		bal Fe				
ASTM A830/A830M(98)	1022	Plt	0.17-0.23			0.70-1.00		0.035 max	0.04 max		bal Fe				
DoD-F-24669/1(86)	1022	Bar Bil; Supersedes MIL-S-866 & MIL-S-16974	0.18-0.23			0.7-1		0.04	0.05		bal Fe				
MIL-S-11310E(76)	1022*	Bar HR cold shape cold ext; Obs for new design	0.18-0.23	0.2	0.35	0.7-1	0.25	0.04	0.05	0.2	Mo 0.06; bal Fe				
MIL-S-11310E(76)	CS1022*	Bar HR cold shape cold ext; Obs for new design									bal Fe				
SAE J1397(92)	1022	Bar CD, est mech prop	0.18-0.23			0.7-1		0.04 max	0.05 max		bal Fe	480	400	15	137 HB
SAE J1397(92)	1022	Bar HR, est mech prop	0.18-0.23			0.7-1		0.04 max	0.05 max		bal Fe	430	230	23	121 HB
SAE J403(95)	1022	Struct Shps Plt Strp Sh Weld Tub	0.17-0.23			0.70-1.00		0.030 max	0.035 max		bal Fe				
SAE J403(95)	1022	Bar Wir rod Smls Tub HR CF	0.18-0.23			0.70-1.00		0.030 max	0.050 max		bal Fe				

Yugoslavia

Specification	Designation	Notes	C	Cr	Cu	Mn	Ni	P	S	Si	Other	UTS	YS	El	Hard
	C.0462	Struct	0.22 max			1.60 max		0.045 max	0.045 max	0.60 max	Mo <=0.15; N <=0.007; bal Fe				
	C.0463	Struct	0.22 max			1.60 max		0.045 max	0.045 max	0.60 max	Mo <=0.15; N <=0.007; bal Fe				
	C.0562	Struct	0.2 max			1.5 max		0.045 max	0.045 max	0.55 max	N <=0.009; bal Fe				
	C.0563	Struct	0.2 max			1.5 max		0.045 max	0.045 max	0.55 max	N <=0.009; bal Fe				
	C.3111	Chain	0.15-0.23			0.8-1.3		0.04 max	0.04 max	0.15-0.35	N <=0.007; bal Fe				

UNS numbers and US grades are provided as a means of cross referencing chemically similar alloys. Exchangability is only possible after independent examination of specifications. Tensile properties are minimum or typical as specified. UTS and YS as MPa. El as %. See Appendix for list of abbreviations used in Notes. * indicates obsolete material.

Specification	Designation	Notes	C	Cr	Cu	Mn	Ni	P	S	Si	Other	UTS	YS	El	Hard
Carbon Steel, Nonresulfurized, 1023															
Australia															
AS 1302(91)	400Y	Deformed reinf bar for conc	0.22 max	0.25 max	0.50 max		0.35 max	0.040 max	0.040 max	0.40 max	Mo <=0.10; bal Fe; CE 0.39	440	400	16	
Austria															
ONORM M3110	RC20	Wir	0.20-0.25			0.4-0.7		0.040 max	0.050 max	0.15-0.4	P+S<=0.06; bal Fe				
ONORM M3167	C22SP	Bar	0.18-0.25			0.4-0.6		0.040 max	0.050 max	0.30 max	bal Fe				
Bulgaria															
BDS 2592(71)	BST4kp	Struct	0.18-0.27	0.30 max	0.30 max	0.40-0.70	0.30 max	0.045 max	0.055 max	0.07	bal Fe				
BDS 2592(71)	BST4ps	Struct	0.18-0.27	0.30 max	0.30 max	0.40-0.70	0.30 max	0.045 max	0.055 max	0.05-0.17	bal Fe				
BDS 2592(71)	BST4sp	Struct	0.18-0.27	0.30 max	0.30 max	0.40-0.70	0.30 max	0.045 max	0.055 max	0.12-0.35	bal Fe				
BDS 5785(83)	20	Struct	0.18-0.24	0.25 max	0.25 max	0.35-0.65	0.25 max	0.035 max	0.040 max	0.17-0.37	bal Fe				
China															
GB 700	B4*	Obs	0.18-0.24			0.4-0.7		0.045	0.05	0.07-0.3	bal Fe				
GB 700	B4F*	Obs	0.18-0.24			0.4-0.7		0.045	0.05	0.07-0.3	bal Fe				
GB 700	BY4*	Obs	0.18-0.24			0.4-0.7		0.045	0.05	0.07-0.3	bal Fe				
GB 700	BY4F*	Obs	0.18-0.24			0.4-0.7		0.045	0.05	0.07-0.3	bal Fe				
GB 700	C4*	Obs	0.18-0.24			0.4-0.7		0.045	0.05	0.07-0.3	bal Fe				
GB 700	C4F*	Obs	0.18-0.24			0.4-0.7		0.045	0.05	0.07-0.3	bal Fe				
GB 700	CY4*	Obs	0.18-0.24			0.4-0.7		0.045	0.05	0.07-0.3	bal Fe				
GB 700	CY4F*	Obs	0.18-0.24			0.4-0.7		0.045	0.05	0.07-0.3	bal Fe				
GB 700(88)	Q255A	Bar Plt Sh HR 16-40mm diam	0.18-0.28	0.30 max	0.30 max	0.40-0.70	0.30 max	0.045 max	0.050 max	0.30 max	bal Fe	410-510	245	23	
GB 700(88)	Q255B	Bar Plt Sh HR 16-40mm diam	0.18-0.28	0.30 max	0.30 max	0.40-0.70	0.30 max	0.045 max	0.045 max	0.30 max	bal Fe	410-510	245	23	
France															
AFNOR NFA35553	XC18S	Strp, Q/A Tmp	0.15-0.22			0.4-0.65		0.04	0.04	0.25	bal Fe				
AFNOR NFA35553	XC18S	Strp, Norm	0.15-0.22			0.4-0.65		0.04	0.04	0.25	bal Fe				
Germany															
DIN 1654(89)	C22C	CD soft ann	0.18-0.24			0.30-0.60		0.035 max	0.035 max	0.40 max	bal Fe	480			
DIN 1654(89)	Cq22	CD soft ann	0.18-0.24			0.30-0.60		0.035 max	0.035 max	0.40 max	bal Fe	480			
DIN 1654(89)	WNr 1.1152	CD soft ann	0.18-0.24			0.30-0.60		0.035 max	0.035 max	0.40 max	bal Fe	480			
DIN EN 10083(91)	C22E	Q/T 17-40mm	0.17-0.24	0.40 max		0.40-0.70	0.40 max	0.035 max	0.035 max	0.40 max	Mo <=0.10; Cr+Mo+Ni<=0.63; bal Fe	470-620	290	22	
DIN EN 10083(91)	Ck22	Q/T 17-40mm	0.17-0.24	0.40 max		0.40-0.70	0.40 max	0.035 max	0.035 max	0.40 max	Mo <=0.10; Cr+Mo+Ni<=0.63; bal Fe	470-620	290	22	
DIN EN 10083(91)	WNr 1.1151	Q/T 17-40mm	0.17-0.24	0.40 max		0.40-0.70	0.40 max	0.035 max	0.035 max	0.40 max	Mo <=0.10; Cr+Mo+Ni<=0.63; bal Fe	470-620	290	22	
DIN EN 10248(95)	WNr 1.0023	HR Sh piling	0.25 max					0.045 max	0.045 max		N <=0.009; bal Fe				
Mexico															
NMX-B-301(86)	1023	Bar	0.20-0.25			0.30-0.60		0.040 max	0.050 max		bal Fe				
Pan America															
COPANT 333	1023		0.2-0.25			0.3-0.6		0.04	0.05		bal Fe				

UNS numbers and US grades are provided as a means of cross referencing chemically similar alloys. Exchangability is only possible after independent examination of specifications. Tensile properties are minimum or typical as specified. UTS and YS as MPa. El as %. See Appendix for list of abbreviations used in Notes. * indicates obsolete material.

Specification	Designation	Notes	C	Cr	Cu	Mn	Ni	P	S	Si	Other	UTS	YS	El	Hard

Carbon Steel, Nonresulfurized, 1023 (Continued from previous page)

Pan America

Specification	Designation	Notes	C	Cr	Cu	Mn	Ni	P	S	Si	Other	UTS	YS	El	Hard
COPANT 37 II	AT-23	Plt, As roll 25mm diam	0.24					0.04	0.05	0.15-0.35	bal Fe				

Poland

| PNH84023 | ST4A | | 0.2-0.28 | | | 0.5-0.75 | | 0.04 | 0.05 | 0.07-0.15 | bal Fe | | | | |

Russia

GOST	KSt4		0.18-0.27			0.4-0.7		0.05	0.055	0.12-0.3	bal Fe				
GOST	KSt4kp		0.18-0.27			0.4-0.7		0.05	0.055	0.07	bal Fe				
GOST	KSt4ps		0.18-0.27			0.4-0.7		0.05	0.05	0.17	bal Fe				
GOST	KStO		0.23 max					0.07 max	0.06 max		bal Fe				
GOST	MSt4		0.18-0.27			0.4-0.7		0.04	0.05	0.12-0.3					
GOST	MSt4kp		0.18-0.27			0.4-0.7		0.04	0.055	0.07	bal Fe				
GOST	MSt4ps		0.18-0.27			0.4-0.7		0.04	0.055	0.17	bal Fe				
GOST	St4		0.18-0.27			0.4-0.7		0.05	0.055	0.12-0.3	bal Fe				
GOST	St4F		0.18-0.27			0.4-0.7		0.05	0.05		bal Fe				
GOST	St4L		0.18-0.27	0.3	0.25	0.4-0.7	0.3 *	0.05	0.05	0.12-0.35	bal Fe				
GOST	St4S		0.18-0.27	0.3	0.25	0.4-0.7	0.3	0.05	0.05	0.12-0.35	bal Fe				
GOST 380	Bst4Gps		0.18-0.27			0.8-1.2		0.04	0.05	0.15	N 0.008; As 0.08; bal Fe				
GOST 380	Bst4Gps2		0.18-0.27	0.3	0.3	0.8-1.2	0.3	0.04	0.05	0.15	N 0.008; As 0.08; bal Fe				
GOST 380	Bst4kp		0.18-0.27			0.4-0.7		0.04 max	0.05 max	0.07 max	N <=0.01; As<=0.08; bal Fe				
GOST 380	Bst4kp2		0.18-0.27	0.3	0.3	0.4-0.7	0.3	0.04	0.05	0.07	N 0.008; As 0.08; bal Fe				
GOST 380	Bst4ps2		0.18-0.27	0.3	0.3	0.4-0.7	0.3	0.04	0.05	0.05-0.17	N 0.008; As 0.08; bal Fe				
GOST 380	Bst4sp2		0.17-0.27	0.3	0.3	0.4-0.7	0.3	0.04	0.05	0.12-0.3	N 0.008; As 0.08; bal Fe				

Spain

UNE 36011(75)	C25k	Unalloyed; Q/T	0.2-0.3			0.5-0.8		0.035 max	0.035 max	0.15-0.4	Mo <=0.15; bal Fe				
UNE 36034(85)	21B3DF	CHd; Cold ext	0.19-0.25			0.50-0.80		0.035 max	0.035 max	0.15-0.35	B 0.0008-0.005; bal Fe				
UNE 36034(85)	23MnCrB4DF	CHd; Cold ext	0.19-0.25	0.2-0.4		0.8-1.1		0.035 max	0.035 max	0.15-0.35	B 0.0008-0.005; Mo <=0.15; Al 0.02; bal Fe				
UNE 36034(85)	F.1291*	see 21B3DF	0.19-0.25			0.50-0.80		0.035 max	0.035 max	0.15-0.35	B 0.0008-0.005; bal Fe				
UNE 36034(85)	F.1294*	see 23MnCrB4DF	0.19-0.25	0.2-0.4		0.8-1.1		0.035 max	0.035 max	0.15-0.35	B 0.0008-0.005; Mo <=0.15; Al 0.02; bal Fe				

Turkey

| TS 381 | UDK621.643/Fe45.8* | Pipe as drawn | 0.22 | | | 0.45 | | 0.05 | 0.05 | 0.1-0.35 | bal Fe | | | | |

UK

BS 1449/1(91)	22CS	Plt Sh Strp; CR Ann	0.20-0.25			0.40-0.60		0.050 max	0.050 max		bal Fe	370	200	23-25	
BS 1449/1(91)	22HS	Plt Sh Strp HR; t<=16mm	0.20-0.25			0.40-0.60		0.050 max	0.050 max		bal Fe	400	230	22-24	
BS 970/1(83)	040A22	Blm Bil Slab Bar Rod Frg	0.20-0.25			0.30-0.50		0.050 max	0.050 max	0.4	bal Fe				
BS 970/1(83)	050A22	Blm Bil Slab Bar Rod Frg	0.22-0.25			0.40-0.60		0.050 max	0.050 max	0.4	bal Fe				
BS 970/1(83)	060A22	Blm Bil Slab Bar Rod Frg	0.20-0.25			0.50-0.70		0.050 max	0.050 max	0.04	bal Fe				

UNS numbers and US grades are provided as a means of cross referencing chemically similar alloys. Exchangability is only possible after independent examination of specifications. Tensile properties are minimum or typical as specified. UTS and YS as MPa. El as %. See Appendix for list of abbreviations used in Notes. * indicates obsolete material.

Specification	Designation	Notes	C	Cr	Cu	Mn	Ni	P	S	Si	Other	UTS	YS	El	Hard

Carbon Steel, Nonresulfurized, 1023 (Continued from previous page)

Specification	Designation	Notes	C	Cr	Cu	Mn	Ni	P	S	Si	Other	UTS	YS	El	Hard
UK															
BS 970/1(83)	080A22	Blm Bil Slab Bar Rod Frg	0.20-0.25			0.70-0.90		0.050 max	0.050 max	0.1-0.4	bal Fe				
USA															
	AISI 1023	Bar	0.20-0.25			0.30-0.60		0.040 max	0.050 max		bal Fe				
	AISI 1023	Struct shp	0.20-0.25			0.30-0.60		0.030 max	0.035 max		bal Fe				
	AISI 1023	Wir rod	0.20-0.25			0.30-0.60		0.040 max	0.050 max		bal Fe				
	AISI 1023	Sh Strp Plt	0.19-0.25			0.30-0.60		0.030 max	0.035 max		bal Fe				
	AISI M1023	Bar, merchant qual	0.19-0.27			0.25-0.60		0.04 max	0.05 max		bal Fe				
	UNS G10230		0.20-0.25			0.30-0.60		0.04 max	0.05 max		Sheets, C 0.19-0.25; ERW Tubing, C 0.19-0.26; bal Fe				
ASTM A29/A29M(93)	1023	Bar	0.20-0.25			0.30-0.60		0.040 max	0.050 max		bal Fe				
ASTM A29/A29M(93)	M1023	Bar, merchant qual	0.19-0.27			0.25-0.60		0.04 max	0.05 max		bal Fe				
ASTM A510(96)	1023	Wir rod	0.20-0.25			0.30-0.60		0.040 max	0.050 max		bal Fe				
ASTM A513(97a)	1023	ERW Mech tub	0.19-0.25			0.30-0.60		0.035 max	0.035 max		bal Fe				
ASTM A575(96)	M1023	Merchant qual bar	0.19-0.27			0.25-0.60		0.04 max	0.05 max		Mn can vary with C; bal Fe				
ASTM A576(95)	1023	Special qual HW bar	0.20-0.25			0.30-0.60		0.040 max	0.050 max		Si Cu Pb B Bi Ca Se Te if spec'd; bal Fe				
ASTM A635/A635M(98)	1023	Sh Strp, Coil, HR	0.19-0.25		0.20 min	0.30-0.60		0.030 max	0.035 max		bal Fe				
ASTM A659/A659M(97)	1023	Sh Strp, HR	0.19-0.25			0.30-0.60		0.030 max	0.035 max		bal Fe				
ASTM A794(97)	1023	Sh, CS, CR	0.19-0.25			0.30-0.60		0.030 max	0.035 max		Cu >=0.20 if spec; bal Fe				
ASTM A830/A830M(98)	1023	Plt	0.19-0.25			0.30-0.60		0.035 max	0.04 max		bal Fe				
SAE J1397(92)	1023	Bar CD, est mech prop	0.2-0.25			0.3-0.6		0.04 max	0.05 max		bal Fe	430	360	15	121 HB
SAE J1397(92)	1023	Bar HR, est mech prop	0.2-0.25			0.3-0.6		0.04 max	0.05 max		bal Fe	370	210	25	111 HB
SAE J403(95)	1023	Bar Wir rod Smls Tub HR CF	0.20-0.25			0.30-0.60		0.030 max	0.050 max		bal Fe				
SAE J403(95)	1023	Struct Shps Plt Strp Sh Weld Tub	0.19-0.25			0.30-0.60		0.030 max	0.035 max		bal Fe				
SAE J403(95)	M1023	Merchant qual	0.20-0.25			0.30-0.60		0.04 max	0.05 max		bal Fe				
Yugoslavia															
	C.1300	Struct	0.18-0.25			0.3-0.6		0.06 max	0.06 max	0.60 max	bal Fe				
	C.1301	Struct	0.18-0.25			0.35-0.6		0.06 max	0.06 max	0.35 max	bal Fe				

Carbon Steel, Nonresulfurized, 1025

Specification	Designation	Notes	C	Cr	Cu	Mn	Ni	P	S	Si	Other	UTS	YS	El	Hard
Argentina															
IAS	IRAM 1025		0.22-0.28			0.30-0.60		0.040 max	0.050 max	0.15-0.30	bal Fe	470-570	300-360	20-34	137-170 HB
Australia															
AS 1302(91)	250R	Plain round reinf bar for conc	0.25 max	0.25 max	0.50 max		0.35 max	0.040 max	0.040 max	0.40 max	Mo <=0.10; bal Fe; CE 0.43	275	250	22	
AS 1302(91)	250S	Deformed reinf bar for conc	0.25 max	0.25 max	0.50 max		0.35 max	0.040 max	0.040 max	0.40 max	Mo <=0.10; bal Fe; CE 0.43	275	250	22	
AS 1442	S1025*	Obs; Bar Bil	0.22-0.28			0.3-0.6		0.05	0.05	0.35	bal Fe				
AS 1443	S1025*	Obs; Bar	0.22-0.28			0.3-0.6		0.05	0.05	0.35	bal Fe				
AS 1446	CS1025*	Withdrawn, see AS/NZS 1594(97)	0.2-0.3			0.4-0.9		0.06	0.06	0.35	bal Fe				

UNS numbers and US grades are provided as a means of cross referencing chemically similar alloys. Exchangability is only possible after independent examination of specifications. Tensile properties are minimum or typical as specified. UTS and YS as MPa. El as %. See Appendix for list of abbreviations used in Notes. * indicates obsolete material.

Specification	Designation	Notes	C	Cr	Cu	Mn	Ni	P	S	Si	Other	UTS	YS	El	Hard

Carbon Steel, Nonresulfurized, 1025 (Continued from previous page)

Austria

Specification	Designation	Notes	C	Cr	Cu	Mn	Ni	P	S	Si	Other	UTS	YS	El	Hard
ONORM M3110	RC25	Wir	0.25-0.30			0.4-0.7		0.035	0.035	0.15	bal Fe				
ONORM M3161	C25		0.21-0.30			0.3-0.5		0.040	0.040	0.2-0.4	bal Fe				

Belgium

Specification	Designation	Notes	C	Cr	Cu	Mn	Ni	P	S	Si	Other	UTS	YS	El	Hard
NBN 251-02	C25-1		0.22-0.29			0.4-0.7		0.045	0.045	0.15-0.4	bal Fe				
NBN 253-02	C25-2		0.22-0.29			0.4-0.7		0.035	0.035	0.15-0.4	bal Fe				
NBN 253-02	C25-3		0.22-0.29			0.4-0.7		0.035	0.02-0.035	0.15-0.4	bal Fe				

Bulgaria

Specification	Designation	Notes	C	Cr	Cu	Mn	Ni	P	S	Si	Other	UTS	YS	El	Hard
BDS 14351	25ps	Plt HR	0.22-0.27	0.25	0.3	0.25-0.5	0.25	0.04	0.04	0.03	bal Fe				
BDS 5785(83)	25	Struct	0.22-0.30	0.25 max	0.25 max	0.50-0.80	0.25 max	0.035 max	0.040 max	0.17-0.37	bal Fe				

Canada

Specification	Designation	Notes	C	Cr	Cu	Mn	Ni	P	S	Si	Other	UTS	YS	El	Hard
CSA 94-1/1	C1025		0.22-0.28			0.3-0.6		0.04	0.05	0.1-0.3	bal Fe				
CSA STAN95-1-1	C1025	Bar	0.22-0.28			0.3-0.6		0.04	0.05	0.1-0.3	bal Fe				

China

Specification	Designation	Notes	C	Cr	Cu	Mn	Ni	P	S	Si	Other	UTS	YS	El	Hard
GB 13237(91)	25	Sh Strp CR HT	0.22-0.30	0.25 max	0.25 max	0.50-0.80	0.25 max	0.035 max	0.035 max	0.17-0.37	bal Fe	390-540		23	
GB 3275(91)	25	Plt Strp HR HT	0.22-0.30	0.25 max	0.25 max	0.50-0.80	0.25 max	0.035 max	0.035 max	0.17-0.37	bal Fe	390-540		23	
GB 3522(83)	25	Strp CR Ann	0.22-0.30	0.25 max	0.25 max	0.50-0.80	0.25 max	0.035 max	0.035 max	0.17-0.37	bal Fe	345-590		18	
GB 5953(86)	ML 25	Wir CD	0.22-0.30	0.20 max	0.20 max	0.60 max		0.035 max	0.035 max	0.20 max	bal Fe	490-685			
GB 6478(86)	ML 25	Bar HR Norm	0.22-0.30	0.20 max	0.20 max	0.30-0.60	0.20 max	0.035 max	0.035 max	0.20 max	bal Fe	450	275	23	
GB 699(88)	25	Bar HR Norm Q/T 25mm diam	0.22-0.30	0.25 max	0.25 max	0.50-0.80	0.25 max	0.035 max	0.035 max	0.17-0.37	bal Fe	450	275	23	
GB 710(91)	25	Sh Strp HR HT	0.22-0.30	0.25 max	0.25 max	0.50-0.80	0.25 max	0.035 max	0.035 max	0.17-0.37	bal Fe	390-540		23	
GB/T 3078(94)	25	Bar CR Ann	0.22-0.30	0.25 max	0.25 max	0.50-0.80	0.25 max	0.035 max	0.035 max	0.17-0.37	bal Fe	410		19	

Czech Republic

Specification	Designation	Notes	C	Cr	Cu	Mn	Ni	P	S	Si	Other	UTS	YS	El	Hard
CSN 412030	12030	Q/T	0.22-0.3	0.25 max		0.5-0.8	0.3 max	0.04 max	0.04 max	0.15-0.4	bal Fe				

Europe

Specification	Designation	Notes	C	Cr	Cu	Mn	Ni	P	S	Si	Other	UTS	YS	El	Hard
EN 10083/1(91)A1(96)	1.1158	Q/T t<=16mm	0.22-0.29	0.40 max		0.40-0.70	0.40 max	0.035 max	0.035 max	0.40 max	Mo <=0.10; Cr+Mo+Ni<=0.63; bal Fe	550-700	320	19	
EN 10083/1(91)A1(96)	1.1163	Q/T t<=16mm	0.22-0.29	0.40 max		0.40-0.70	0.40 max	0.035 max	0.020-0.040	0.40 max	Mo <=0.10; Cr+Mo+Ni<=0.63; bal Fe	550-700	320	19	
EN 10083/1(91)A1(96)	C25E	Q/T t<=16mm	0.22-0.29	0.40 max		0.40-0.70	0.40 max	0.035 max	0.035 max	0.40 max	Mo <=0.10; Cr+Mo+Ni<=0.63; bal Fe	550-700	320	19	
EN 10083/1(91)A1(96)	C25R	Q/T t<=16mm	0.22-0.29	0.40 max		0.40-0.70	0.40 max	0.035 max	0.020-0.040	0.40 max	Mo <=0.10; Cr+Mo+Ni<=0.63; bal Fe	550-700	320	19	
EN 10083/2(91)A1(96)	1.0406	Q/T t<=16mm	0.22-0.29	0.40 max		0.40-0.70	0.40 max	0.045 max	0.045 max	0.40 max	Mo <=0.10; Cr+Mo+Ni<=0.63; bal Fe	550-700	370	19	
EN 10083/2(91)A1(96)	C25	Q/T t<=16mm	0.22-0.29	0.40 max		0.40-0.70	0.40 max	0.045 max	0.045 max	0.40 max	Mo <=0.10; Cr+Mo+Ni<=0.63; bal Fe	550-700	370	19	

France

Specification	Designation	Notes	C	Cr	Cu	Mn	Ni	P	S	Si	Other	UTS	YS	El	Hard
AFNOR NFA35552	XC25	Bar, Q/A Tmp, 16mm diam	0.23-0.29			0.4-0.7		0.035	0.035	0.1-0.35	Al 0.02; bal Fe				
AFNOR NFA35566	XC25		0.23-0.29			0.4-0.7		0.035	0.035	0.1-0.35	Al 0.02; bal Fe				

UNS numbers and US grades are provided as a means of cross referencing chemically similar alloys. Exchangability is only possible after independent examination of specifications. Tensile properties are minimum or typical as specified. UTS and YS as MPa. El as %. See Appendix for list of abbreviations used in Notes. * indicates obsolete material.

Specification	Designation	Notes	C	Cr	Cu	Mn	Ni	P	S	Si	Other	UTS	YS	El	Hard

Carbon Steel, Nonresulfurized, 1025 (Continued from previous page)

Germany

Specification	Designation	Notes	C	Cr	Cu	Mn	Ni	P	S	Si	Other	UTS	YS	El	Hard
DIN 17140	D25-2*	Obs; see C26D	0.23-0.28			0.3-0.6		0.04 max	0.04 max	0.1-0.3	bal Fe				
DIN 2391(94)	St45	Smls tube	0.21 max			0.40 min		0.040 max	0.040 max	0.35 max	bal Fe				
DIN 2391(94)	WNr 1.0408	Smls tube	0.21 max			0.40 min		0.040 max	0.040 max	0.35 max	bal Fe				
DIN EN 10083(91)	C25E	Q/T 17-40mm	0.22-0.29	0.40 max		0.40-0.70	0.40 max	0.035 max	0.035 max	0.40 max	Mo <=0.10; Cr+Mo+Ni<=0.63; bal Fe	500-650	320	21	
DIN EN 10083(91)	Ck25	Q/T 17-40mm	0.22-0.29	0.40 max		0.40-0.70	0.40 max	0.035 max	0.035 max	0.40 max	Mo <=0.10; Cr+Mo+Ni<=0.63; bal Fe	500-650	320	21	
DIN EN 10083(91)	WNr 1.0406	Sh Plt Strp; Q/T 17-40mm	0.22-0.29	0.40 max		0.40-0.70	0.40 max	0.045 max	0.045 max	0.40 max	Mo <=0.10; Cr+Mo+Ni<=0.63; bal Fe	500	320	21	
DIN EN 10083(91)	WNr 1.1158	Q/T 17-40mm	0.22-0.29	0.40 max		0.40-0.70	0.40 max	0.035 max	0.035 max	0.40 max	Mo <=0.10; Cr+Mo+Ni<=0.63; bal Fe	500-650	320	21	

Hungary

Specification	Designation	Notes	C	Cr	Cu	Mn	Ni	P	S	Si	Other	UTS	YS	El	Hard
MSZ 5736(88)	D26	Wir rod; for CD	0.23-0.3		0.25 max	0.3-0.6	0.3 max	0.04 max	0.04 max	0.1-0.35	bal Fe				
MSZ 5736(88)	D26K	Wir rod; for CD	0.23-0.3		0.25 max	0.3-0.6	0.3 max	0.03 max	0.03 max	0.1-0.35	bal Fe				
MSZ 61(85)	C25	Q/T; 16<t<=40mm	0.22-0.29			0.4-0.7		0.035 max	0.035 max	0.4 max	bal Fe	500-650	320	21L	
MSZ 61(85)	C25E	Q/T; 16<t<=40mm	0.22-0.29			0.4-0.7		0.035 max	0.02-0.035	0.4 max	bal Fe	500-650	320	21L	
MSZ 61(85)	C25E	Q/T; t<=16mm	0.22-0.29			0.4-0.7		0.035 max	0.02-0.035	0.4 max	bal Fe	550-700	370	19L	
MSZ 6251	C25EK		0.22-0.29			0.4-0.7		0.035	0.035	0.17-0.37	bal Fe				
MSZ 6251(87)	C25Z	CF	0.22-0.29			0.4-0.7		0.035 max	0.035 max	0.17-0.37	bal Fe				

India

Specification	Designation	Notes	C	Cr	Cu	Mn	Ni	P	S	Si	Other	UTS	YS	El	Hard
IS 1570	C25		0.2-0.3			0.3-0.6					bal Fe				

International

Specification	Designation	Notes	C	Cr	Cu	Mn	Ni	P	S	Si	Other	UTS	YS	El	Hard
ISO 683-1(87)	C25	Bar Frg Plt Wir rod, Blts, Slbs; Q/T, t<16mm	0.22-0.29			0.40-0.70		0.045 max	0.045 max	0.10-0.40	bal Fe	550-700	370		
ISO 683-1(87)	C25E4	Bar Frg Plt Wir rod, Blts, Slbs; Q/T, t<16mm	0.22-0.29			0.40-0.70		0.035 max	0.035 max	0.10-0.40	bal Fe	550-700	370		
ISO 683-1(87)	C25M2	Bar Frg Plt Wir rod, Blts, Slbs; Q/T, t<16mm	0.22-0.29			0.40-0.70		0.035 max	0.020-0.040	0.10-0.40	bal Fe	550-700	370		
ISO 683-18(96)	C25	Bar Wir, CD, t<=5mm	0.22-0.29			0.40-0.70		0.045 max	0.045 max	0.10-0.40	bal Fe	610	490	5	156 HB
ISO 683-18(96)	C25E4	Bar Wir, CD, t<=5mm	0.22-0.29			0.40-0.70		0.035 max	0.035 max	0.10-0.40	bal Fe	610	490	5	156 HB
ISO 683-18(96)	C25M2	Bar Wir, CD, t<=5mm	0.22-0.29			0.40-0.70		0.035 max	0.020-0.040	0.10-0.40	bal Fe	610	490	5	156 HB
ISO R683-3(70)	C25ea	Bar Rod, Q/A, Tmp, 16mm diam	0.22-0.29	0.3 max	0.3 max	0.4-0.7	0.4 max	0.035 max	0.02-0.035	0.15-0.4	Al <=0.1; Co <=0.1; Mo <=0.15; Pb <=0.15; Ti <=0.05; V <=0.1; W <=0.1; bal Fe				
ISO R683-3(70)	C25eb	Bar Rod, Q/A, Tmp, 16mm diam	0.22-0.29			0.4-0.7		0.04	0.03-0.05	0.15-0.4	bal Fe				

Italy

Specification	Designation	Notes	C	Cr	Cu	Mn	Ni	P	S	Si	Other	UTS	YS	El	Hard
UNI 5598(71)	1CD25	Wir rod	0.22-0.29			0.4-0.7		0.05 max	0.05 max	0.15-0.35	bal Fe				
UNI 5598(71)	3CD25	Wir rod	0.23-0.28			0.4-0.7		0.035 max	0.035 max	0.15-0.35	N <=0.012; bal Fe				
UNI 7845(78)	C25	Q/T	0.22-0.29			0.4-0.8		0.035 max	0.035 max	0.15-0.4	bal Fe				
UNI 7874(79)	C25	Q/T	0.22-0.29			0.5-0.8		0.035 max	0.035 max	0.15-0.4	bal Fe				

UNS numbers and US grades are provided as a means of cross referencing chemically similar alloys. Exchangability is only possible after independent examination of specifications. Tensile properties are minimum or typical as specified. UTS and YS as MPa. El as %. See Appendix for list of abbreviations used in Notes. * indicates obsolete material.

Specification	Designation	Notes	C	Cr	Cu	Mn	Ni	P	S	Si	Other	UTS	YS	El	Hard

Carbon Steel, Nonresulfurized, 1025 (Continued from previous page)

Japan

Specification	Designation	Notes	C	Cr	Cu	Mn	Ni	P	S	Si	Other	UTS	YS	El	Hard
JIS G3507(91)	SWRCH25K	Wir rod FF	0.22-0.28			0.30-0.60		0.030 max	0.035 max	0.10-0.35	bal Fe				
JIS G4051(79)	S25C	Bar Wir rod	0.22-0.28	0.20 max	0.30 max	0.30-0.60	0.20 max	0.030 max	0.035 max	0.15-0.35	Ni+Cr<=0.35; bal Fe				

Mexico

Specification	Designation	Notes	C	Cr	Cu	Mn	Ni	P	S	Si	Other	UTS	YS	El	Hard
DGN B-203	1025*	Tube, Obs	0.22-0.28			0.3-0.6		0.04	0.05		bal Fe				
NMX-B-178(90)	Grade A	Pipe for high-temp service	0.25 max	0.40 max	0.40 max	0.27-0.93	0.40 max	0.025 max	0.025 max	0.10 min	Mo <=0.15; V <=0.08; Mn increases with a decrease in C up to 1.35; (Cr+Co+Mo+Ni+V) 1 max; bal Fe				
NMX-B-201(68)	1025	CD buttweld mech tub	0.22-0.28			0.30-0.60		0.04 max	0.05 max		bal Fe				
NMX-B-301(86)	1025	Bar	0.22-0.28			0.30-0.60		0.040 max	0.050 max		bal Fe				
NOM-060-SCFI(94)	Type 2a	HR for portable gasoline containers	0.24 max			0.50-1.00		0.04 max	0.04 max	0.40 max	Nb 0.01-0.04; V 0.02-0.09; (Nb+V) 0.01-0.04; Mn up to 1.40 if C<=0.15; bal Fe	420	283	19.0	

Pan America

Specification	Designation	Notes	C	Cr	Cu	Mn	Ni	P	S	Si	Other	UTS	YS	El	Hard
COPANT 330	1025		0.22-0.28			0.3-0.6		0.04	0.05		bal Fe				
COPANT 331	1025		0.22-0.28			0.3-0.6		0.04	0.05		bal Fe				
COPANT 333	1025		0.22-0.28			0.3-0.6		0.04	0.05		bal Fe				
COPANT 514	1025	Tube, Norm	0.22-0.28			0.3-0.6		0.04	0.05	0.15-0.3	bal Fe				
COPANT 514	1025	Tube, Ann	0.22-0.28			0.3-0.6		0.04	0.05	0.15-0.3	bal Fe				
COPANT 514	1025	Tube, HF, 323.9mm diam	0.22-0.28			0.3-0.6		0.04	0.05	0.15-0.3	bal Fe				
COPANT 514	1025	Tube, CF	0.22-0.28			0.3-0.6		0.04	0.05	0.15-0.3	bal Fe				

Poland

Specification	Designation	Notes	C	Cr	Cu	Mn	Ni	P	S	Si	Other	UTS	YS	El	Hard
PNH84023	ST4A		0.2-0.28			0.5-0.75		0.04	0.05	0.07-0.15	bal Fe				

Romania

Specification	Designation	Notes	C	Cr	Cu	Mn	Ni	P	S	Si	Other	UTS	YS	El	Hard
STAS 880(88)	OLC25	Q/T	0.22-0.29	0.3 max		0.5-0.8		0.04 max	0.045 max	0.17-0.37	Pb <=0.15; As<=0.05; bal Fe				
STAS 880(88)	OLC25S	Q/T	0.22-0.29	0.3 max		0.4-0.7		0.04 max	0.02-0.045	0.17-0.37	Pb <=0.15; As<=0.05; bal Fe				
STAS 880(88)	OLC25X	Q/T	0.22-0.29	0.3 max		0.4-0.7		0.035 max	0.035 max	0.17-0.37	Pb <=0.15; As<=0.05; bal Fe				
STAS 880(88)	OLC25XS	Q/T	0.22-0.29	0.3 max		0.4		0.035 max	0.02-0.04	0.17-0.37	Pb <=0.15; As<=0.05; bal Fe				

Russia

Specification	Designation	Notes	C	Cr	Cu	Mn	Ni	P	S	Si	Other	UTS	YS	El	Hard
GOST	25										bal Fe				
GOST	KSt4		0.18-0.27			0.4-0.7		0.05	0.055	0.12-0.3	bal Fe				
GOST	KSt4kp		0.18-0.27			0.4-0.7		0.05	0.055	0.07	bal Fe				
GOST	KSt4ps		0.18-0.27			0.4-0.7		0.05	0.05	0.17	bal Fe				
GOST	M26		0.22-0.3			0.4-0.7		0.05 max	0.05 max	0.12-0.3	bal Fe				
GOST	M26A		0.22-0.3			0.4-0.7		0.04	0.04	0.12-0.3	bal Fe				
GOST	MSt4		0.18-0.27			0.4-0.7		0.04 max	0.05 max	0.12-0.3	bal Fe				
GOST	MSt4kp		0.18-0.27			0.4-0.7		0.04	0.055	0.07	bal Fe				
GOST	MSt4ps		0.18-0.27			0.4-0.7		0.04	0.055	0.17	bal Fe				

UNS numbers and US grades are provided as a means of cross referencing chemically similar alloys. Exchangability is only possible after independent examination of specifications. Tensile properties are minimum or typical as specified. UTS and YS as MPa. El as %. See Appendix for list of abbreviations used in Notes. * indicates obsolete material.

Specification	Designation	Notes	C	Cr	Cu	Mn	Ni	P	S	Si	Other	UTS	YS	El	Hard

Carbon Steel, Nonresulfurized, 1025 (Continued from previous page)

Russia

Specification	Designation	Notes	C	Cr	Cu	Mn	Ni	P	S	Si	Other	UTS	YS	El	Hard
GOST	St4		0.18-0.27			0.4-0.7		0.05	0.055	0.12-0.3	bal Fe				
GOST	St4F		0.18-0.27			0.4-0.7		0.05	0.05		bal Fe				
GOST	St4L		0.18-0.27			0.4-0.7		0.05	0.05	0.12-0.35	bal Fe				
GOST	St4S		0.18-0.27			0.4-0.7		0.05	0.05	0.12-0.35	bal Fe				
GOST 1050	25		0.22-0.3	0.25 max	0.25 max	0.35-0.65	0.25 max	0.04 max	0.04 max	0.17-0.37	bal Fe				
GOST 1050(88)	25	High-grade struct; Q/T; Norm; 0-100mm	0.22-0.3	0.25 max		0.5-0.8	0.3 max	0.035 max	0.04 max	0.17-0.37	As<=0.08; bal Fe	450	275	23L	
GOST 1050(88)	25	High-grade struct; Q/T; Full hard; 0-100mm	0.22-0.3	0.25 max		0.5-0.8	0.3 max	0.035 max	0.04 max	0.17-0.37	As<=0.08; bal Fe	450		7L	0-217 HB
GOST 4041(71)	25ps	High-grade Plts for cold pressing	0.22-0.27	0.25 max	0.3 max	0.25-0.50	0.25 max	0.040 max	0.040	0.03 max	bal Fe				
GOST 4041(71)	25ps	High-grade Plts for cold pressing	0.22-0.27	0.25 max	0.3 max	0.25-0.50	0.30 max	0.040 max	0.040 max	0.03 max	As<=0.08; bal Fe				
GOST 977	25L-II		0.22-0.3	0.3	0.3	0.35-0.9	0.3	0.04	0.05	0.2-0.52	bal Fe				
GOST 977	25L-III		0.22-0.3	0.3	0.3	0.35-0.9	0.3	0.04	0.05	0.2-0.52	bal Fe				
GOST 977(88)	25L		0.22-0.3	0.3		0.35-0.9	0.3	0.05	0.05	0.2-0.52	bal Fe				

Spain

Specification	Designation	Notes	C	Cr	Cu	Mn	Ni	P	S	Si	Other	UTS	YS	El	Hard
UNE	F.112		0.2-0.3			0.4-0.7		0.04	0.04	0.15-0.3	bal Fe				
UNE 36011(75)	C25k	Unalloyed; Q/T	0.2-0.3			0.5-0.8		0.035 max	0.035 max	0.15-0.4	Mo <=0.15; bal Fe				

Turkey

Specification	Designation	Notes	C	Cr	Cu	Mn	Ni	P	S	Si	Other	UTS	YS	El	Hard
TS 302	Fe45.2*	Pipe as drawn or weld	0.25					0.05	0.05		bal Fe				
TS 416	Fe42.2*	Weld Pipe	0.25					0.06	0.05		bal Fe				
TS 908(86)	UDK669.14.423/Fe37	HR, equal angles	0.25					0.06	0.06		bal Fe				
TS 911(86)	UDK669.14.423/Fe42	HR, T-bar	0.25					0.06	0.06		bal Fe				
TS 912(86)	UDK669.14.423/Fe37	HR, channel	0.25					0.06	0.06		bal Fe				
TS 913(86)	UDK669.14.423/Fe37	HR, Z-bar	0.25					0.06	0.06		bal Fe				

UK

Specification	Designation	Notes	C	Cr	Cu	Mn	Ni	P	S	Si	Other	UTS	YS	El	Hard
BS 1717(83)	CEW-103	Tub	0.30 max			0.60 max		0.060 max	0.060 max		bal Fe				
BS 1717(83)	CEW-104	Tub	0.30 max			0.60 max		0.060 max	0.060 max		bal Fe				

USA

Specification	Designation	Notes	C	Cr	Cu	Mn	Ni	P	S	Si	Other	UTS	YS	El	Hard
	AISI 1025	Wir rod	0.22-0.28			0.30-0.60		0.040 max	0.050 max		bal Fe				
	AISI 1025	Sh Strp Plt	0.22-0.28			0.30-0.60		0.030 max	0.035 max		bal Fe				
	AISI 1025	Struct shp	0.22-0.28			0.30-0.60		0.030 max	0.035 max		bal Fe				
	AISI 1025	Bar	0.22-0.28			0.30-0.60		0.040 max	0.050 max		bal Fe				
	AISI M1025	Bar, merchant qual	0.20-0.30			0.25-0.60		0.04 max	0.05 max		bal Fe				
	UNS G10250		0.22-0.60			0.30-0.60		0.04 max	0.05 max		ERW Tubing, C 0.21-0.28; bal Fe				
AMS 5075E(88)		Norm or Stress relieved Weld tub	0.22-0.28			0.30-0.60		0.040 max	0.050 max	0.10-0.30	bal Fe	379	248	22	
ASTM A108(95)	1025	Bar, CF	0.22-0.28			0.30-0.60		0.040 max	0.050 max		bal Fe				
ASTM A29/A29M(93)	1025	Bar	0.22-0.28			0.30-0.60		0.040 max	0.050 max		bal Fe				
ASTM A29/A29M(93)	M1025	Bar, merchant qual	0.20-0.30			0.25-0.60		0.04 max	0.05 max		bal Fe				
ASTM A510(96)	1025	Wir rod	0.22-0.28			0.30-0.60		0.040 max	0.050 max		bal Fe				

UNS numbers and US grades are provided as a means of cross referencing chemically similar alloys. Exchangability is only possible after independent examination of specifications. Tensile properties are minimum or typical as specified. UTS and YS as MPa. El as %. See Appendix for list of abbreviations used in Notes. * indicates obsolete material.

Specification	Designation	Notes	C	Cr	Cu	Mn	Ni	P	S	Si	Other	UTS	YS	El	Hard

Carbon Steel, Nonresulfurized, 1025 (Continued from previous page)

USA

Specification	Designation	Notes	C	Cr	Cu	Mn	Ni	P	S	Si	Other	UTS	YS	El	Hard
ASTM A512(96)	1025	Buttweld mech tub, CD	0.22-0.28			0.30-0.60		0.040 max	0.045 max		bal Fe				
ASTM A513(97a)	1025	ERW Mech tub	0.22-0.28			0.30-0.60		0.035 max	0.035 max		bal Fe				
ASTM A519(96)	1025	Smls mech tub, Norm	0.22-0.28			0.30-0.60		0.040 max	0.050 max		bal Fe	379	248	22	60 HRB
ASTM A519(96)	1025	Smls mech tub, Ann	0.22-0.28			0.30-0.60		0.040 max	0.050 max		bal Fe	365	207	25	57 HRB
ASTM A519(96)	1025	Smls mech tub, SR	0.22-0.28			0.30-0.60		0.040 max	0.050 max		bal Fe	483	379	8	75 HRB
ASTM A519(96)	1025	Smls mech tub, CW	0.22-0.28			0.30-0.60		0.040 max	0.050 max		bal Fe	517	448	5	80 HRB
ASTM A519(96)	1025	Smls mech tub, HR	0.22-0.28			0.30-0.60		0.040 max	0.050 max		bal Fe	379	241	25	60 HRB
ASTM A575(96)	M1025	Merchant qual bar	0.20-0.30			0.25-0.60		0.04 max	0.05 max		Mn can vary with C; bal Fe				
ASTM A576(95)	1025	Special qual HW bar	0.22-0.28			0.30-0.60		0.040 max	0.050 max		Si Cu Pb B Bi Ca Se Te if spec'd; bal Fe				
ASTM A830/A830M(98)	1025	Plt	0.22-0.28			0.30-0.60		0.035 max	0.04 max		bal Fe				
FED QQ-S-700D(91)	G10250*	Obs; Sh Strp	0.22-0.28			0.3-0.6		0.04 max	0.05 max		bal Fe				
MIL-S-11310E(76)	1025*	Bar HR cold shape cold ext; Obs for new design	0.22-0.28			0.3-0.6		0.04 max	0.05 max		bal Fe				
MIL-S-11310E(76)	CS1025*	Bar HR cold shape cold ext; Obs for new design									bal Fe				
MIL-S-7952A(63)	1025*	Uncoated Sh Strp, aircraft qual; Obs for new design see AMS 5046	0.22-0.3			0.3-0.6		0.04	0.05		bal Fe				
MIL-T-3520	1025*	Obs; Tube	0.21-0.28			0.3-0.6		0.04	0.05		bal Fe				
MIL-T-5066B(67)	1025*	Obs; Smls weld tube aircraft qual	0.22-0.28			0.3-0.6		0.25	0.25		bal Fe				
SAE J1397(92)	1025	Bar CD, est mech prop	0.22-0.28			0.3-0.6		0.04 max	0.05 max		bal Fe	440	370	15	126 HB
SAE J1397(92)	1025	Bar HR, est mech prop	0.22-0.28			0.3-0.6		0.04 max	0.05 max		bal Fe	400	220	25	116 HB
SAE J403(95)	1025	Bar Wir rod Smls Tub HR CF	0.22-0.28			0.30-0.60		0.030 max	0.050 max		bal Fe				
SAE J403(95)	1025	Struct Shps Plt Strp Sh Weld Tub	0.22-0.28			0.30-0.60		0.030 max	0.035 max		bal Fe				
SAE J403(95)	M1025	Merchant qual	0.22-0.28			0.30-0.60		0.04 max	0.05 max		bal Fe				

Yugoslavia

Specification	Designation	Notes	C	Cr	Cu	Mn	Ni	P	S	Si	Other	UTS	YS	El	Hard
	C.0460	Struct	0.25 max			1.60 max		0.06 max	0.06 max	0.60 max	Mo <=0.15; bal Fe				
	C.1332	Q/T	0.22-0.29			0.4-0.7		0.035 max	0.03 max	0.4 max	Mo <=0.15; bal Fe				
	PZ24		0.24-0.29	0.15	0.25	0.3-0.6	0.15	0.04	0.04	0.1-0.3	bal Fe				

Carbon Steel, Nonresulfurized, 1026

Australia

Specification	Designation	Notes	C	Cr	Cu	Mn	Ni	P	S	Si	Other	UTS	YS	El	Hard
AS 1442	K1026*	Obs; Bar Bil	0.22-0.28			0.6-0.9		0.05	0.05	0.1-0.35	bal Fe				
AS 1443	K1026*	Obs; Bar	0.22-0.28			0.6-0.9		0.05	0.05	0.1-0.35	bal Fe				

Belgium

Specification	Designation	Notes	C	Cr	Cu	Mn	Ni	P	S	Si	Other	UTS	YS	El	Hard
NBN A25-102	D45	Tube, HF, CF Norm	0.24			0.4-1.05		0.05	0.05	0.12-0.38	bal Fe				

Bulgaria

Specification	Designation	Notes	C	Cr	Cu	Mn	Ni	P	S	Si	Other	UTS	YS	El	Hard
BDS 3492(86)	25LI		0.20-0.30		0.3	0.4-0.8		0.06	0.06	0.25-0.5	bal Fe				
BDS 3492(86)	25LII		0.20-0.30	0.3-0.30	0.3	0.4-0.8	0.40 max	0.06	0.06	0.25-0.5	bal Fe				

UNS numbers and US grades are provided as a means of cross referencing chemically similar alloys. Exchangability is only possible after independent examination of specifications. Tensile properties are minimum or typical as specified. UTS and YS as MPa. El as %. See Appendix for list of abbreviations used in Notes. * indicates obsolete material.

Specification	Designation	Notes	C	Cr	Cu	Mn	Ni	P	S	Si	Other	UTS	YS	El	Hard

Carbon Steel, Nonresulfurized, 1026 (Continued from previous page)

Specification	Designation	Notes	C	Cr	Cu	Mn	Ni	P	S	Si	Other	UTS	YS	El	Hard
Bulgaria															
BDS 3492(86)	25LIII		0.20-0.30		0.3	0.4-0.8		0.05	0.05	0.25-0.5	bal Fe				
China															
GB 6478(86)	ML 25Mn	Bar HR Norm	0.22-0.30	0.20 max	0.20 max	0.50-0.80	0.20 max	0.035 max	0.035 max	0.25 max	bal Fe	450	275	23	
GB 699(88)	25 Mn	Bar HR Q/T 25mm diam	0.22-0.30	0.25 max	0.25 max	0.70-1.00	0.25 max	0.035 max	0.035 max	0.17-0.37	bal Fe	490	295	22	
GB/T 3078(94)	25 Mn	Bar CR Q/T 25mm diam	0.22-0.30	0.25 max	0.25 max	0.70-1.00	0.25 max	0.035 max	0.035 max	0.17-0.37	bal Fe	490	295	22	
Europe															
EN 10016/2(94)	1.0415	Rod	0.24-0.29	0.20 max	0.30 max	0.50-0.80	0.25 max	0.035 max	0.035 max	0.10-0.30	Al <=0.01; Mo <=0.05; bal Fe				
EN 10016/2(94)	C26D	Rod	0.24-0.29	0.20 max	0.30 max	0.50-0.80	0.25 max	0.035 max	0.035 max	0.10-0.30	Al <=0.01; Mo <=0.05; bal Fe				
EN 10016/4(94)	1.1139	Rod	0.24-0.29	0.10 max	0.15 max	0.50-0.70	0.10 max	0.020 max	0.025 max	0.10-0.30	Al <=0.01; Mo <=0.03; N <=0.007; Cr+Ni+Cu<=0.30, Cu+Sn<=0.15; bal Fe				
EN 10016/4(94)	C26D2	Rod	0.24-0.29	0.10 max	0.15 max	0.50-0.70	0.10 max	0.020 max	0.025 max	0.10-0.30	Al <=0.01; Mo <=0.03; N <=0.007; Cr+Ni+Cu<=0.30, Cu+Sn<=0.15; bal Fe				
Finland															
SFS 1211(74)	SFS1211	Reinforcing	0.28 max			1.6 max		0.06 max	0.05 max	0.15-0.55	N <=0.01; As 0-0.08; Ce C+Mn/6+(Cr+Mo+V)/5+(Ni+Cu)/15; bal Fe				
Germany															
DIN	GS-Ck25		0.20-0.28			0.50-0.80		0.035 max	0.035 max	0.30-0.50	bal Fe				
DIN	WNr 1.1155		0.20-0.28			0.50-0.80		0.035 max	0.035 max	0.30-0.50	bal Fe				
India															
IS 1570	C25Mn75		0.2-0.3			0.6-0.9					bal Fe				
IS 1570/2(79)	25C8	Sh Plt Sect Shp Bar Bil Frg Tub Pip	0.2-0.3	0.3 max	0.3 max	0.6-0.9	0.4 max	0.055 max	0.055 max	0.6 max	Co <=0.1; Mo <=0.15; Pb <=0.15; W <=0.1; P:S Varies; bal Fe	470-570	258	22	
IS 1570/2(79)	C25Mn75	Sh Plt Sect Shp Bar Bil Frg Tub Pip	0.2-0.3	0.3 max	0.3 max	0.6-0.9	0.4 max	0.055 max	0.055 max	0.6 max	Co <=0.1; Mo <=0.15; Pb <=0.15; W <=0.1; P:S Varies; bal Fe	470-570	258	22	
Japan															
JIS G3445(88)	STKM14A	Tube	0.30 max			0.30-1.00		0.040 max	0.040 max	0.35 max	bal Fe	412	245	25L	
JIS G3445(88)	STKM14B	Tube	0.30 max			0.30-1.00		0.040 max	0.040 max	0.35 max	bal Fe	500	353	15L	
JIS G3445(88)	STKM14C	Tube	0.30 max			0.30-1.00		0.040 max	0.040 max	0.35 max	bal Fe	549	412	15L	
Mexico															
DGN B-203	1026*	Tube, Obs	0.22-0.28			0.6-0.9		0.04	0.05		bal Fe				
NMX-B-301(86)	1026	Bar	0.22-0.28			0.60-0.90		0.040 max	0.050 max		bal Fe				
Pan America															
COPANT 331	1026		0.22-0.28			0.6-0.9		0.04	0.05		bal Fe				
COPANT 333	1026		0.22-0.28			0.6-0.9		0.04	0.05	0.2-0.35	bal Fe				
COPANT 514	1026		0.22-0.28			0.6-0.9		0.04	0.05		bal Fe				
Poland															
PNH84019	25	CH; Q/T									bal Fe				
Romania															
STAS 600	OT45-1		0.2-0.3	0.3	0.3	0.4-0.5	0.3	0.05	0.05	0.25-0.5	bal Fe				

UNS numbers and US grades are provided as a means of cross referencing chemically similar alloys. Exchangability is only possible after independent examination of specifications. Tensile properties are minimum or typical as specified. UTS and YS as MPa. El as %. See Appendix for list of abbreviations used in Notes. * indicates obsolete material.

Specification	Designation	Notes	C	Cr	Cu	Mn	Ni	P	S	Si	Other	UTS	YS	El	Hard

Carbon Steel, Nonresulfurized, 1026 (Continued from previous page)

Romania

Specification	Designation	Notes	C	Cr	Cu	Mn	Ni	P	S	Si	Other	UTS	YS	El	Hard
STAS 600	OT45-2		0.2-0.3	0.3	0.3	0.4-0.5	0.3	0.04	0.045	0.25-0.5	bal Fe				
STAS 600	OT45-3		0.2-0.3	0.3	0.3	0.4-0.5	0.3	0.04	0.045	0.25-0.5	bal Fe				
STAS 9382/4	OLC25q		0.22-0.29	0.3	0.3	0.5-0.8	0.3	0.035	0.035	0.17-0.37	As 0.05; bal Fe				

Russia

Specification	Designation	Notes	C	Cr	Cu	Mn	Ni	P	S	Si	Other	UTS	YS	El	Hard
GOST	22K		0.21-0.28	0.3	0.3	0.65-1.08	0.3	0.045	0.045	0.15-0.3	bal Fe				
GOST	25K		0.21-0.28	0.3	0.3	0.8	0.3	0.045	0.045	0.15-0.3	bal Fe				
GOST 1050	25		0.22-0.3	0.25	0.25 max	0.5-0.8	0.25	0.035	0.04	0.17-0.3	As 0.08; bal Fe				
GOST 380	Bst5Gps		0.22-0.3			0.8-1.2		0.04	0.05	0.15	N 0.008; As 0.08; bal Fe				
GOST 380	Bst5Gps2		0.22-0.3	0.3	0.3	0.8-1.2	0.3	0.04	0.05	0.15	N 0.008; As 0.08; bal Fe				
GOST 4543	25G		0.22-0.3	0.3	0.3	0.7-1	0.3	0.035	0.035	0.17-0.37	bal Fe				
GOST 4543(71)	25G	Q/T	0.22-0.3	0.3 max	0.3 max	0.7-1.0	0.3 max	0.035 max	0.035 max	0.17-0.37	Al <=0.1; bal Fe				

Spain

Specification	Designation	Notes	C	Cr	Cu	Mn	Ni	P	S	Si	Other	UTS	YS	El	Hard
UNE 36011(75)	C25k		0.2-0.3			0.5-0.8		0.035	0.035	0.15-0.4	bal Fe				
UNE 36011(75)	C25k-1		0.2-0.3			0.5-0.8		0.035	0.02-0.035	0.15-0.4	bal Fe				
UNE 36011(75)	F.1120*	see C25k	0.2-0.3			0.5-0.8		0.035	0.035	0.15-0.4	bal Fe				
UNE 36011(75)	F.1125*	see C25k-1	0.2-0.3			0.5-0.8		0.035	0.02-0.035	0.15-0.4	bal Fe				

Turkey

Specification	Designation	Notes	C	Cr	Cu	Mn	Ni	P	S	Si	Other	UTS	YS	El	Hard
TS 346	Fe45*	Pipe as drawn	0.25			0.9		0.05 max		0.35	bal Fe				

UK

Specification	Designation	Notes	C	Cr	Cu	Mn	Ni	P	S	Si	Other	UTS	YS	El	Hard
BS 1717(83)	CDS-103	Tub	0.30 max			0.30-0.90		0.050 max	0.050 max	0.35 max	bal Fe				
BS 1717(83)	CDS-104	Tub	0.30 max			0.30-0.90		0.050 max	0.050 max	0.35 max	bal Fe				
BS 970/1(83)	070M26	Blm Bil Slab Bar Rod Frg HR Norm t<=63mm	0.22-0.30			0.50-0.90		0.050 max	0.050 max		bal Fe	490	245	20	143-192 HB
BS 970/1(83)	080A25	Blm Bil Slab Bar Rod Frg	0.23-0.28			0.70-0.90		0.050 max	0.050 max	0.1-0.4	bal Fe				
BS 970/1(83)	080A27	Blm Bil Slab Bar Rod Frg	0.25-0.30			0.70-0.90		0.050 max	0.050 max	0.1-0.4	bal Fe				

USA

Specification	Designation	Notes	C	Cr	Cu	Mn	Ni	P	S	Si	Other	UTS	YS	El	Hard
	AISI 1026	Wir rod	0.22-0.28			0.60-0.90		0.040 max	0.050 max		bal Fe				
	AISI 1026	Struct shp	0.22-0.28			0.60-0.90		0.030 max	0.035 max		bal Fe				
	AISI 1026	Sh Strp Plt	0.22-0.28			0.60-0.90		0.030 max	0.035 max		bal Fe				
	AISI 1026	Bar	0.22-0.28			0.60-0.90		0.040 max	0.050 max		bal Fe				
	UNS G10260		0.22-0.28			0.60-0.90		0.04 max	0.05 max		ERW Tubing, C 0.21-0.28; bal Fe				
ASTM A273	1026*	Obs, 1975; see A711	0.22-0.28			0.6-0.9		0.04 max	0.05 max		bal Fe				
ASTM A29/A29M(93)	1026	Bar	0.22-0.28			0.60-0.90		0.040 max	0.050 max		bal Fe				
ASTM A510(96)	1026	Wir rod	0.22-0.28			0.60-0.90		0.040 max	0.050 max		bal Fe				
ASTM A512(96)	1026	Buttweld mech tub, CD	0.22-0.28			0.60-0.90		0.040 max	0.045 max		bal Fe				
ASTM A513(97a)	1026	ERW Mech tub	0.22-0.28			0.60-0.90		0.035 max	0.035 max		bal Fe				

UNS numbers and US grades are provided as a means of cross referencing chemically similar alloys. Exchangability is only possible after independent examination of specifications. Tensile properties are minimum or typical as specified. UTS and YS as MPa. El as %. See Appendix for list of abbreviations used in Notes. * indicates obsolete material.

Specification	Designation	Notes	C	Cr	Cu	Mn	Ni	P	S	Si	Other	UTS	YS	El	Hard

Carbon Steel, Nonresulfurized, 1026 (Continued from previous page)

USA

Specification	Designation	Notes	C	Cr	Cu	Mn	Ni	P	S	Si	Other	UTS	YS	El	Hard
ASTM A519(96)	1026	Smls mech tub	0.22-0.28			0.60-0.90		0.040 max	0.050 max		bal Fe				
ASTM A576(95)	1026	Special qual HW bar	0.22-0.28			0.60-0.90		0.040 max	0.050 max		Si Cu Pb B Bi Ca Se Te if spec'd; bal Fe				
ASTM A830/A830M(98)	1026	Plt	0.22-0.28			0.60-0.90		0.035 max	0.04 max		bal Fe				
MIL-S-22698	1026*	Obs									bal Fe				
MIL-S-24093A(SH)(91)	Type V Class H	Frg for shipboard apps	0.30 max		0.25 max	0.90 max	0.25 max	0.04 max	0.04 max	0.10-0.30	bal Fe	621 max	207	30	
MIL-T-20157	A*	Obs; Tube	0.25			0.27-0.93		0.048	0.045		bal Fe				
MIL-T-20157	B*	Obs; Tube	0.25			0.27-0.93		0.048	0.045		bal Fe				
MIL-T-20157	C*	Obs; Tube	0.25			0.27-0.93		0.048	0.045		bal Fe				
MIL-T-20157	D*	Obs; Tube	0.25			0.27-0.93		0.048	0.045		bal Fe				
MIL-T-20157	E*	Obs; Tube	0.3			0.29-1.06		0.048	0.045		bal Fe				
SAE J1397(92)	1026	Bar CD, est mech prop	0.22-0.28			0.6-0.9		0.04 max	0.05 max		bal Fe	490	410	15	143 HB
SAE J1397(92)	1026	Bar HR, est mech prop	0.22-0.28			0.6-0.9		0.04 max	0.05 max		bal Fe	440	240	24	126 HB
SAE J403(95)	1026	Bar Wir rod Smls Tub HR CF	0.22-0.28			0.60-0.90		0.030 max	0.050 max		bal Fe				
SAE J403(95)	1026	Struct Shps Plt Strp Sh Weld Tub	0.22-0.28			0.60-0.90		0.030 max	0.035 max		bal Fe				

Carbon Steel, Nonresulfurized, 1029

Australia

Specification	Designation	Notes	C	Cr	Cu	Mn	Ni	P	S	Si	Other	UTS	YS	El	Hard
AS 1302	230R*	Obs; Bar As rolled	0.3					0.05	0.05		bal Fe				
AS 1302	230S*	Obs; Bar As rolled	0.3					0.05	0.05		bal Fe				
AS 1302	410C*	Obs; Bar As rolled	0.3					0.05	0.05		bal Fe				

Austria

ONORM M3110	RC25	Wir	0.25-0.30			0.4-0.7		0.035	0.035	0.15-0.40	bal Fe				

Canada

CSA B193	X46		0.32			1.35		0.04	0.05		bal Fe				
CSA B193	X52		0.32			1.35		0.04	0.05		bal Fe				

China

GB 700	BJ5*	Obs	0.24-0.37			0.5-0.8		0.045	0.055	0.12-0.35	bal Fe				
GB 700	CJ5*	Obs	0.24-0.37			0.5-0.8		0.045	0.055	0.12-0.35	bal Fe				

Finland

SFS 200	Fe50	Gen struct	0.3 max			1.60 max		0.06 max	0.05 max	0.60 max	N <=0.009; bal Fe				
SFS 200:E	Fe50	Bar Plt Strp Wir Frg HW	0.3					0.05	0.05		N 0.01; bal Fe				

France

AFNOR	CC28		0.25-0.3			0.4-0.7		0.05	0.05	0.1-0.4	bal Fe				
AFNOR NFA33101(82)	AF50C30	t<=450mm; Norm	0.25-0.33	0.3 max	0.3 max	0.5-0.8	0.4 max	0.04 max	0.04 max	0.1-0.4	bal Fe	490-590	290	23	

India

IS 6286	1		0.3			0.4-1.06		0.05	0.05		bal Fe				
IS 6286	2		0.3			0.29-1.06		0.05	0.05	0.1	bal Fe				

Japan

JIS G4051(79)	S28C	Bar Wir rod	0.25-0.31	0.20 max	0.30 max	0.60-0.90	0.20 max	0.030 max	0.035 max	0.15-0.35	Ni+Cr<=0.35; bal Fe				

Mexico

NMX-B-301(86)	1029	Bar	0.25-0.31			0.60-0.90		0.040 max	0.050 max		bal Fe				

UNS numbers and US grades are provided as a means of cross referencing chemically similar alloys. Exchangability is only possible after independent examination of specifications. Tensile properties are minimum or typical as specified. UTS and YS as MPa. El as %. See Appendix for list of abbreviations used in Notes. * indicates obsolete material.

Specification	Designation	Notes	C	Cr	Cu	Mn	Ni	P	S	Si	Other	UTS	YS	El	Hard

Carbon Steel, Nonresulfurized, 1029 (Continued from previous page)

Mexico

Specification	Designation	Notes	C	Cr	Cu	Mn	Ni	P	S	Si	Other	UTS	YS	El	Hard
NOM-061-SCFI(94)	Type I	Plate for non-portable gasoline containers	0.28 max			0.90 max		0.035 max	0.04 max	0.4 max	bal Fe	382-520	206	27	
NOM-061-SCFI(94)	Type III-1	Plate for non-portable gasoline containers	0.28 max			0.79-1.40		0.035 max	0.04 max	0.4 max	bal Fe	520-657	284	22	

Norway

Specification	Designation	Notes	C	Cr	Cu	Mn	Ni	P	S	Si	Other	UTS	YS	El	Hard
NS 13234	St50-2		0.3					0.05	0.05		N 0.007; bal Fe				

Pan America

Specification	Designation	Notes	C	Cr	Cu	Mn	Ni	P	S	Si	Other	UTS	YS	El	Hard
COPANT 333	1029		0.25-0.31			0.6-0.9		0.04	0.05		bal Fe				
COPANT 37 II	AT-27	Plt, As roll 25mm diam	0.31					0.04	0.05	0.15-0.35	bal Fe				

Poland

Specification	Designation	Notes	C	Cr	Cu	Mn	Ni	P	S	Si	Other	UTS	YS	El	Hard
PNH84023	30GS		0.27-0.32	0.3	0.3	1-1.25	0.3	0.05	0.05	0.4-0.6	Mo 0.1; bal Fe				
PNH84023	R50		0.23-0.32			0.5-0.8		0.045	0.045	0.15-0.35	bal Fe				

Switzerland

Specification	Designation	Notes	C	Cr	Cu	Mn	Ni	P	S	Si	Other	UTS	YS	El	Hard
VSM 10648	StAc20-2		0.3					0.055	0.055		bal Fe				

Turkey

Specification	Designation	Notes	C	Cr	Cu	Mn	Ni	P	S	Si	Other	UTS	YS	El	Hard
TS 909(86)	UDK669.14.423/Fe42	HR, unequal angles	0.31					0.063	0.063		bal Fe				
TS 911(86)	UDK669.14.423/Fe50	HR, T-bar	0.3					0.055	0.055		bal Fe				
TS 912(86)	UDK669.14.423/Fe42	HR, channel	0.31					0.063	0.063		bal Fe				
TS 913(86)	UDK669.14.423/Fe42	HR, Z-bar	0.31					0.063	0.063		bal Fe				
TS 980	UDK669.14.423/Fe42*	HR	0.31					0.063	0.063		bal Fe				

UK

Specification	Designation	Notes	C	Cr	Cu	Mn	Ni	P	S	Si	Other	UTS	YS	El	Hard
BS 970/1(83)	060A27	Blm Bil Slab Bar Rod Frg	0.25-0.30			0.50-0.70		0.050 max	0.050 max	0.1-0.4	bal Fe				

USA

Specification	Designation	Notes	C	Cr	Cu	Mn	Ni	P	S	Si	Other	UTS	YS	El	Hard
	AISI 1029	Wir rod	0.25-0.31			0.60-0.90		0.040 max	0.050 max		bal Fe				
	AISI 1029	Bar	0.25-0.31			0.60-0.90		0.040 max	0.050 max		bal Fe				
	UNS G10290		0.25-0.31			0.60-0.90		0.040 max	0.050 max		bal Fe				
ASTM A29/A29M(93)	1029	Bar	0.25-0.31			0.60-0.90		0.040 max	0.050 max		bal Fe				
ASTM A510(96)	1029	Wir rod	0.25-0.31			0.60-0.90		0.040 max	0.050 max		bal Fe				
ASTM A576(95)	1029	Special qual HW bar	0.25-0.31			0.60-0.90		0.040 max	0.050 max		Si Cu Pb B Bi Ca Se Te if spec'd; bal Fe				
SAE J403(95)	1029	Bar Wir rod Smls Tub HR CF	0.25-0.31			0.60-0.90		0.030 max	0.050 max		bal Fe				
SAE J403(95)	1029	Bar Rod, Smls Tub	0.25-0.31			0.6-0.9		0.04 max	0.05 max		bal Fe				

Yugoslavia

Specification	Designation	Notes	C	Cr	Cu	Mn	Ni	P	S	Si	Other	UTS	YS	El	Hard
	C.1302	Struct	0.26-0.32			0.4-0.6		0.06 max	0.06 max	0.35 max	bal Fe				

Carbon Steel, Nonresulfurized, 1030

Argentina

Specification	Designation	Notes	C	Cr	Cu	Mn	Ni	P	S	Si	Other	UTS	YS	El	Hard
IAS	IRAM 1030		0.28-0.34			0.60-0.90		0.040 max	0.050 max	0.15-0.30	bal Fe	550-650	340-400	20-32	163-195 HB

Australia

Specification	Designation	Notes	C	Cr	Cu	Mn	Ni	P	S	Si	Other	UTS	YS	El	Hard
AS 1442	CS1030*	Obs; Bar Bil	0.25-0.35			0.3-0.9		0.06	0.06	0.35	bal Fe				
AS 1442	K1030*	Obs; Bar Bil	0.28-0.34			0.6-0.9		0.05	0.05	0.1-0.35	bal Fe				
AS 1442	S1030*	Obs; Bar Bil	0.28-0.34			0.6-0.9		0.05	0.05	0.35	bal Fe				
AS 1442(92)	1030	HR bar, Semifinished (may treat w/microalloying elements)	0.28-0.34			0.60-0.90		0.040 max	0.040 max	0.10-0.35	Si<=0.10 for Al-killed; bal Fe				

Specification	Designation	Notes	C	Cr	Cu	Mn	Ni	P	S	Si	Other	UTS	YS	El	Hard

Carbon Steel, Nonresulfurized, 1030 (Continued from previous page)

Australia

Specification	Designation	Notes	C	Cr	Cu	Mn	Ni	P	S	Si	Other	UTS	YS	El	Hard
AS 1442(92)	M1030	Merchant qual, HR bar, Semifinished	0.25-0.35			0.30-0.90		0.050 max	0.050 max	0.35 max	bal Fe				
AS 1443	CS1030*	Obs; Bar	0.25-0.35			0.3-0.9		0.06	0.06	0.35	bal Fe				
AS 1443	K1030*	Obs; Bar	0.28-0.34			0.6-0.9		0.05	0.05	0.1-0.35	bal Fe				
AS 1443	S1030*	Obs; Bar	0.28-0.34			0.6-0.9		0.05	0.05	0.35	bal Fe				
AS 1443(94)	1030	CF bar (may treat w/microalloying elements)	0.28-0.34			0.60-0.90		0.040 max	0.040 max	0.10-0.35	Si<=0.10 for Al-killed; bal Fe				
AS 1443(94)	D4	CR or CR bar, 16-38mm	0.25-0.35			0.30-0.55		0.050 max	0.050 max	0.35 max	bal Fe	540	430	11	
AS 1443(94)	M1030	CF bar, Merchant quality	0.25-0.35			0.30-0.90		0.050 max	0.050 max	0.35 max	bal Fe				
AS 1443(94)	T4	CF bar, not CD or CR;<=260mm	0.25-0.35			0.30-0.90		0.050 max	0.050 max	0.35 max	bal Fe	500	250	20	

Austria

Specification	Designation	Notes	C	Cr	Cu	Mn	Ni	P	S	Si	Other	UTS	YS	El	Hard
ONORM M3110	RC30	Wir	0.30-0.35			0.4-0.7		0.035	0.035	0.15-0.40	bal Fe				
ONORM M3167	C355P		0.28-0.34			0.6-0.9		0.040	0.050	0.15-0.30	bal Fe				

Bulgaria

Specification	Designation	Notes	C	Cr	Cu	Mn	Ni	P	S	Si	Other	UTS	YS	El	Hard
BDS 5785(83)	30	Struct	0.27-0.35	0.25 max	0.25 max	0.50-0.80	0.25 max	0.035 max	0.040 max	0.17-0.37	bal Fe				
BDS 5785(83)	30G	Struct	0.27-0.35	0.25 max	0.25 max	0.70-1.00	0.25 max	0.035 max	0.040 max	0.17-0.37	bal Fe				

China

Specification	Designation	Notes	C	Cr	Cu	Mn	Ni	P	S	Si	Other	UTS	YS	El	Hard
GB 13237(91)	30	Sh Strp CR HT	0.27-0.35	0.25 max	0.25 max	0.50-0.80	0.25 max	0.035 max	0.035 max	0.17-0.37	bal Fe	440-590		21	
GB 3275(91)	30	Plt Strp HR HT	0.27-0.35	0.25 max	0.25 max	0.50-0.80	0.25 max	0.035 max	0.035 max	0.17-0.37	bal Fe	440-590		21	
GB 3522(83)	30	Strp CR Ann	0.27-0.35	0.25 max	0.25 max	0.50-0.80	0.25 max	0.035 max	0.035 max	0.17-0.37	bal Fe	395-590		16	
GB 5953(86)	ML 30	Wir CD	0.27-0.35	0.20 max	0.20 max	0.60 max		0.035 max	0.035 max	0.20 max	bal Fe	490-685			
GB 6478(86)	ML 30	Bar HR Norm	0.27-0.35	0.20 max	0.20 max	0.30-0.60	0.20 max	0.035 max	0.035 max	0.20 max	bal Fe	490	295	21	
GB 699(88)	30	Bar HR Norm Q/T 25mm diam	0.27-0.35	0.25 max	0.25 max	0.50-0.80	0.25 max	0.035 max	0.035 max	0.17-0.37	bal Fe	490	295	21	
GB 710(91)	30	Sh Strp HR HT	0.27-0.35	0.25 max	0.25 max	0.50-0.80	0.25 max	0.035 max	0.035 max	0.17-0.37	bal Fe	440-590		21	
GB/T 3078(94)	30	Bar CR Ann	0.27-0.34	0.25 max	0.25 max	0.60 max	0.25 max	0.035 max	0.035 max	0.20 max	bal Fe	440		17	

Czech Republic

Specification	Designation	Notes	C	Cr	Cu	Mn	Ni	P	S	Si	Other	UTS	YS	El	Hard
CSN 412031	12031	Q/T	0.27-0.35	0.25 max		0.5-0.8	0.3 max	0.04 max	0.04 max	0.15-0.4	bal Fe				

Europe

Specification	Designation	Notes	C	Cr	Cu	Mn	Ni	P	S	Si	Other	UTS	YS	El	Hard
EN 10083/1(91)A1(96)	1.1178	Q/T t<=16mm	0.27-0.34	0.40 max		0.50-0.80	0.40 max	0.035 max	0.035 max	0.40 max	Mo <=0.10; Cr+Mo+Ni<=0.63; bal Fe	600-750	400	18	
EN 10083/1(91)A1(96)	1.1179	Q/T t<=16mm	0.27-0.34	0.40 max		0.50-0.80	0.40 max	0.035 max	0.020-0.040	0.40 max	Mo <=0.10; Cr+Mo+Ni<=0.63; bal Fe	600-750	400	18	
EN 10083/1(91)A1(96)	C30E	Q/T t<=16mm	0.27-0.34	0.40 max		0.50-0.80	0.40 max	0.035 max	0.035 max	0.40 max	Mo <=0.10; Cr+Mo+Ni<=0.63; bal Fe	600-750	400	18	
EN 10083/1(91)A1(96)	C30R	Q/T t<=16mm	0.27-0.34	0.40 max		0.50-0.80	0.40 max	0.035 max	0.020-0.040	0.40 max	Mo <=0.10; Cr+Mo+Ni<=0.63; bal Fe	600-750	400	18	
EN 10083/2(91)A1(96)	1.0528	Q/T t<=16mm	0.27-0.34	0.40 max		0.50-0.80	0.40 max	0.045 max	0.045 max	0.40 max	Mo <=0.10; Cr+Mo+Ni<=0.63; bal Fe	600-750	400	18	
EN 10083/2(91)A1(96)	C30	Q/T t<=16mm	0.27-0.34	0.40 max		0.50-0.80	0.40 max	0.045 max	0.045 max	0.40 max	Mo <=0.10; Cr+Mo+Ni<=0.63; bal Fe	600-750	400	18	

Specification	Designation	Notes	C	Cr	Cu	Mn	Ni	P	S	Si	Other	UTS	YS	El	Hard

Carbon Steel, Nonresulfurized, 1030 (Continued from previous page)

Finland

Specification	Designation	Notes	C	Cr	Cu	Mn	Ni	P	S	Si	Other	UTS	YS	El	Hard
SFS 1205(74)	SFS1205	Reinforcing	0.3 max			0.5 max		0.06 max	0.06 max	0.4 max	N <=0.01; As 0-0.08; bal Fe				
SFS 1212(80)	SFS1210	Reinforcing	0.3 max			0.5 max		0.06 max	0.06 max	0.4 max	N <=0.01; bal Fe				
SFS 1212(80)	SFS1212	Reinforcing	0.3 max			0.5 max		0.06 max	0.06 max	0.4 max	N <=0.01; As 0-0.08; bal Fe				
SFS 1214(83)	SFS1214	Reinforcing	0.3 max			0.5 max		0.06 max	0.06 max	0.4 max	N <=0.01; bal Fe				
SFS 2148	Fe55	Tub, as Drawn 16mm diam	0.36					0.05	0.05		bal Fe				

France

Specification	Designation	Notes	C	Cr	Cu	Mn	Ni	P	S	Si	Other	UTS	YS	El	Hard
AFNOR NFA35552	XC32		0.3-0.35			0.5-0.8		0.035	0.035	0.1-0.35	bal Fe				
AFNOR NFA35553	XC32	Strp, Norm	0.3-0.35			0.5-0.8		0.035	0.035	0.1-0.35	bal Fe				
AFNOR NFA35553	XC32	Strp, Q/A Tmp	0.3-0.35			0.5-0.8		0.035	0.035	0.1-0.35	bal Fe				
AFNOR NFA36102(93)	25B3*	HR; Strp; CR	0.22-0.28	0.15-0.25		0.6-0.9	0.20 max	0.025 max	0.02 max	0.1-0.4	B 0.0008-0.005; Mo <=0.10; bal Fe				
AFNOR NFA36102(93)	25B3RR	HR; Strp; CR	0.22-0.28	0.15-0.25		0.6-0.9	0.20 max	0.025 max	0.02 max	0.1-0.4	B 0.0008-0.005; Mo <=0.10; bal Fe				

Germany

Specification	Designation	Notes	C	Cr	Cu	Mn	Ni	P	S	Si	Other	UTS	YS	El	Hard
DIN 1654(89)	28B2	CHd, cold extrude for Q/T	0.25-0.32			0.50-0.80		0.035 max	0.035 max	0.40 max	B 0.0008-0.005; bal Fe				
DIN 1654(89)	WNr 1.5510	CHd, cold ext for Q/T	0.25-0.32			0.50-0.80		0.035 max	0.035 max	0.40 max	B 0.0008-0.005; bal Fe				
DIN 17140	D30-2*	Obs; see C32D	0.28-0.33			0.3-0.6		0.04 max	0.04 max	0.1-0.3	bal Fe				
DIN EN 10083(91)	C30E	Q/T 17-40mm	0.27-0.34	0.40 max		0.50-0.80	0.40 max	0.035 max	0.035 max	0.40 max	Mo <=0.10; Cr+Mo+Ni<=0.63; bal Fe	550-700	350	20	
DIN EN 10083(91)	Ck30	Q/T 17-40mm	0.27-0.34	0.40 max		0.50-0.80	0.40 max	0.035 max	0.35 max	0.40 max	Mo <=0.10; Cr+Mo+Ni<=0.63; bal Fe	550-700	350	20	
DIN EN 10083(91)	Cm30	Q/T	0.27-0.34	0.40 max		0.50-0.80	0.40 max	0.035 max	0.020-0.040	0.40 max	Mo <=0.10; Cr+Mo+Ni<=0.63; bal Fe				
DIN EN 10083(91)	WNr 1.0528	Q/T 17-40mm	0.27-0.34	0.40 max		0.50-0.80	0.40 max	0.045 max	0.045 max	0.40 max	Mo <=0.10; Cr+Mo+Ni<=0.63; bal Fe	550-700	350	20	
DIN EN 10083(91)	WNr 1.1178	Q/T 17-40mm	0.27-0.34	0.40 max		0.50-0.80	0.40 max	0.035 max	0.035 max	0.40 max	Mo <=0.10; Cr+Mo+Ni<=0.63; bal Fe	550-700	350	20	
DIN EN 10083(91)	WNr 1.1179	Q/T	0.27-0.34	0.40 max		0.50-0.80	0.40 max	0.035 max	0.020-0.040	0.40 max	Mo <=0.10; Cr+Mo+Ni<=0.63; bal Fe				

Hungary

Specification	Designation	Notes	C	Cr	Cu	Mn	Ni	P	S	Si	Other	UTS	YS	El	Hard
MSZ 3156(84)	D	Drill pipe; 6.45-12.7mm	0.3 max			2 max		0.04 max	0.06 max	0.6 max	bal Fe	580	380	0-270 HB; A>=6.33S Lo=50.8	
MSZ 3156(84)	E	Drill pipe; 6.45-12.7mm; HT	0.3 max			2 max		0.04 max	0.06 max	0.6 max	bal Fe	690	515	0-270 HB; A>=5.42S Lo=50.8	
MSZ 61(78)	C30	Q/T; t<=16mm	0.27-0.34			0.5-0.8		0.035 max	0.035 max	0.4 max	bal Fe	600-750	400	18L	
MSZ 61(85)	C30	Q/T; 40<t<=100mm	0.27-0.34			0.5-0.8		0.035 max	0.035 max	0.4 max	bal Fe	500-650	300	21L	
MSZ 61(85)	C30E	Q/T; 40<t<=100mm	0.27-0.34			0.5-0.8		0.035 max	0.02-0.035	0.4 max	bal Fe	500-650	300	21L	
MSZ 61(85)	C30E	Q/T; t<=16mm	0.27-0.34			0.5-0.8		0.035 max	0.02-0.035	0.4 max	bal Fe	600-750	400	18L	

India

Specification	Designation	Notes	C	Cr	Cu	Mn	Ni	P	S	Si	Other	UTS	YS	El	Hard
IS 1570/2(79)	30C8	Sh Plt Sect Shp Bar Bil Frg Tub Pip	0.25-0.35	0.3 max	0.3 max	0.6-0.9	0.4 max	0.055 max	0.055 max	0.6 max	Co <=0.1; Mo <=0.15; Pb <=0.15; W <=0.1; P:S Varies; bal Fe	500-600	275	21	

Specification	Designation	Notes	C	Cr	Cu	Mn	Ni	P	S	Si	Other	UTS	YS	El	Hard
Carbon Steel, Nonresulfurized, 1030 (Continued from previous page)															
India															
IS 1570/2(79)	C30	Sh Plt Sect Shp Bar Bil Frg Tub Pip	0.25-0.35	0.3 max	0.3 max	0.6-0.9	0.4 max	0.055 max	0.055 max	0.6 max	Co <=0.1; Mo <=0.15; Pb <=0.15; W <=0.1; P:S Varies; bal Fe	500-600	275	21	
IS 5517	C30	Bar, Norm, Ann or Hard Tmp	0.25-0.35			0.6-0.9				0.1-0.35	bal Fe				
International															
ISO 683-1(87)	C30	Bar Frg Plt Wir rod, Blts, Slbs; Q/T, t<16mm	0.27-0.34			0.50-0.80		0.045 max	0.045 max	0.10-0.40	bal Fe	600-750	400		
ISO 683-1(87)	C30E4	Bar Frg Plt Wir rod, Blts, Slbs; Q/T, t<16mm	0.27-0.34			0.50-0.80		0.035 max	0.035 max	0.10-0.40	bal Fe	600-750	400		
ISO 683-1(87)	C30M2	Bar Frg Plt Wir rod, Blts, Slbs; Q/T, t<16mm	0.27-0.34			0.50-0.80		0.035 max	0.020-0.040	0.10-0.40	bal Fe	600-750	400		
ISO 683-18(96)	C30	Bar Wir, CD, t<=5mm	0.27-0.34			0.50-0.80		0.045 max	0.045 max	0.10-0.40	bal Fe	650	520	5	170 HB
ISO 683-18(96)	C30E4	Bar Wir, CD, t<=5mm	0.27-0.34			0.50-0.80		0.035 max	0.035 max	0.10-0.40	bal Fe	650	520	5	170 HB
ISO 683-18(96)	C30M2	Bar Wir, CD, t<=5mm	0.27-0.34			0.50-0.80		0.035 max	0.020-0.040	0.10-0.40	bal Fe	650	520	5	170 HB
ISO R683-1	C30e	Bar Frg Q/A, Tmp	0.27-0.34			0.5-0.8		0.04	0.04	0.15-0.4	bal Fe				
ISO R683-3(70)	C30ea	Bar Rod, Q/A, Tmp	0.27-0.34	0.3 max	0.3 max	0.5-0.8	0.4 max	0.035 max	0.02-0.035	0.15-0.4	Al <=0.1; Co <=0.1; Mo <=0.15; Pb <=0.15; Ti <=0.05; V <=0.1; W <=0.1; bal Fe				
ISO R683-3(70)	C30eb	Bar Rod, Q/A, Tmp, 16mm diam	0.27-0.34			0.5-0.8		0.04	0.03-0.05	0.15-0.4	bal Fe				
Italy															
UNI 5332	C30		0.27-0.34			0.5-0.8		0.035	0.035	0.4	bal Fe				
UNI 5598(71)	1CD30	Wir rod	0.27-0.34			0.4-0.7		0.05 max	0.05 max	0.15-0.35	bal Fe				
UNI 5598(71)	3CD30	Wir rod	0.28-0.33			0.4-0.7		0.035 max	0.035 max	0.15-0.35	N <=0.012; bal Fe				
UNI 6403(86)	C30	Q/T	0.27-0.34			0.5-0.8		0.035 max	0.035 max	0.15-0.35	bal Fe				
UNI 6783	Fe50-3		0.25-0.35			0.4-0.8		0.04	0.04	0.4	bal Fe				
UNI 7065(72)	C30	Strp	0.28-0.33			0.4-0.65		0.02 max	0.02 max	0.35 max	bal Fe				
UNI 7065(72)	C31	Strp	0.27-0.34			0.5-0.8		0.035 max	0.035 max	0.4 max	bal Fe				
UNI 7845(78)	C30	Q/T	0.27-0.34			0.5-0.8		0.035 max	0.035 max	0.15-0.4	bal Fe				
UNI 7874(79)	C30	Q/T	0.27-0.34			0.5-0.8		0.035 max	0.035 max	0.15-0.4	bal Fe				
Japan															
JIS G3311(88)	S30CM	CR Strp	0.27-0.33	0.20 max	0.30 max	0.60-0.90	0.20 max	0.030 max	0.035 max	0.15-0.35	Ni+Cr=0.35 max, bal Fe				160-230 HV
JIS G3445(88)	STKM15A	Tube	0.25-0.35			0.30-1.00		0.040 max	0.040 max	0.35 max	bal Fe	471	275	22L	
JIS G3445(88)	STKM15C	Tube	0.25-0.35			0.30-1.00		0.040 max	0.040 max	0.35 max	bal Fe	579	431	12L	
JIS G3507(91)	SWRCH30K	Wir rod FF	0.27-0.33			0.60-0.90		0.030 max	0.035 max	0.10-0.35	bal Fe				
JIS G4051(79)	S30C	Bar Wir rod	0.27-0.33	0.20 max	0.30 max	0.60-0.90	0.20 max	0.030 max	0.035 max	0.15-0.35	Ni+Cr<=0.35; bal Fe				
JIS G4051(79)	S33C	Bar Wir rod	0.30-0.36	0.20 max	0.30 max	0.60-0.90	0.20 max	0.030 max	0.035 max	0.15-0.35	Ni+Cr<=0.35; bal Fe				
Mexico															
DGN B-203	1030*	Tube, Obs	0.28-0.34			0.6-0.9		0.04	0.05		bal Fe				

UNS numbers and US grades are provided as a means of cross referencing chemically similar alloys. Exchangability is only possible after independent examination of specifications. Tensile properties are minimum or typical as specified. UTS and YS as MPa. El as %. See Appendix for list of abbreviations used in Notes. * indicates obsolete material.

Specification	Designation	Notes	C	Cr	Cu	Mn	Ni	P	S	Si	Other	UTS	YS	El	Hard

Carbon Steel, Nonresulfurized, 1030 (Continued from previous page)

Mexico

Specification	Designation	Notes	C	Cr	Cu	Mn	Ni	P	S	Si	Other	UTS	YS	El	Hard
NMX-B-178(90)	Grade B	Pipe for high-temp service	0.30 max	0.40 max	0.40 max	0.29-1.06	0.40 max	0.025 max	0.025 max	0.10 min	Mo <=0.15; V <=0.08; Mn increases with a decrease in C up to 1.35; (Cr+Co+Mo+Ni+V) 1 max; bal Fe				
NMX-B-197(85)	Grade 1	Smls weld tub for low-temp	0.30 max			0.40-1.06		0.05 max	0.06 max		Mn increases to 1.35 with a decrease in C; bal Fe	379	207	35	
NMX-B-197(85)	Grade 6	Smls weld tub for low-temp	0.30 max			0.29-1.06		0.048 max	0.058 max	0.10 min	Mn increases to 1.35 with a decrease in C; bal Fe	414	241	30	
NMX-B-201(68)	1030	CD buttweld mech tub	0.28-0.34			0.60-0.90		0.04 max	0.05 max		bal Fe				
NMX-B-301(86)	1030	Bar	0.28-0.34			0.60-0.90		0.040 max	0.050 max		bal Fe				

Norway

Specification	Designation	Notes	C	Cr	Cu	Mn	Ni	P	S	Si	Other	UTS	YS	El	Hard
NS	SE3		0.35			0.65		0.03	0.03	0.25	bal Fe				

Pan America

Specification	Designation	Notes	C	Cr	Cu	Mn	Ni	P	S	Si	Other	UTS	YS	El	Hard
COPANT 330	1030		0.28-0.34			0.6-0.9		0.04	0.05		bal Fe				
COPANT 331	1030		0.28-0.34			0.6-0.9		0.04	0.05		bal Fe				
COPANT 333	1030		0.28-0.34			0.6-0.9		0.04	0.05		bal Fe				
COPANT 514	1030		0.28-0.34			0.6-0.9		0.04	0.05	0.2-0.35	bal Fe				

Poland

Specification	Designation	Notes	C	Cr	Cu	Mn	Ni	P	S	Si	Other	UTS	YS	El	Hard
PNH84019	30	Q/T	0.27-0.35	0.3 max		0.5-0.8	0.3 max	0.04 max	0.04 max	0.17-0.37	bal Fe				
PNH84023/06	34GS	Reinforcing	0.3-0.36	0.3 max	0.35 max	0.8-1.2	0.3 max	0.05 max	0.05 max	0.4-0.7	Mo <=0.1; CEV<=0.59; bal Fe				
PNH84023/07	R34GS		0.3-0.35	0.3 max		0.9-1.2	0.3 max	0.035 max	0.035 max	0.6-0.9	bal Fe				

Romania

Specification	Designation	Notes	C	Cr	Cu	Mn	Ni	P	S	Si	Other	UTS	YS	El	Hard
STAS 880(88)	OLC30	Q/T	0.27-0.34	0.3 max		0.5-0.8		0.04 max	0.045 max	0.17-0.37	Pb <=0.15; As<=0.05; bal Fe				
STAS 880(88)	OLC30S	Q/T	0.27-0.34	0.3 max		0.5-0.8		0.04 max	0.02-0.045	0.17-0.37	Pb <=0.15; As<=0.05; bal Fe				
STAS 880(88)	OLC30X	Q/T	0.27-0.34	0.3 max		0.5-0.8		0.035 max	0.035 max	0.17-0.37	Pb <=0.15; As<=0.05; bal Fe				
STAS 880(88)	OLC30XS	Q/T	0.27-0.34	0.3 max		0.5-0.8		0.035 max	0.02-0.04	0.17-0.37	Pb <=0.15; As<=0.05; bal Fe				

Russia

Specification	Designation	Notes	C	Cr	Cu	Mn	Ni	P	S	Si	Other	UTS	YS	El	Hard
GOST	2Kh		0.31	0.14		0.66		0.013	0.028	0.17	bal Fe				
GOST	5K		0.3	0.3		0.6-0.8	0.3	0.05	0.05	0.4	bal Fe				
GOST	5sp		0.33			0.6		0.017	0.035	0.27	bal Fe				
GOST	KSt5		0.28-0.37			0.5-0.8		0.05 max	0.05 max	0.17-0.35	bal Fe				
GOST	KSt5kp		0.28-0.37			0.5-0.8		0.05	0.055	0.07	bal Fe				
GOST	KSt5ps		0.28-0.37			0.5-0.8		0.05	0.055	0.17	bal Fe				
GOST	M31		0.27-0.35			0.5-0.8		0.045	0.05	0.15-0.32	bal Fe				
GOST	M31A		0.27-0.35			0.5-0.8		0.04	0.04	0.15-0.32	bal Fe				
GOST	MSt5		0.28-0.37			0.5-0.8		0.05	0.055	0.17-0.35	bal Fe				
GOST	St5		0.28-0.37			0.5-0.8		0.05	0.055	0.17-0.35	bal Fe				
GOST	St5K		0.3					0.05	0.05		bal Fe				
GOST	VK		0.2-0.4			0.5					bal Fe				
GOST 1050	30		0.27-0.35	0.25	0.25 max	0.5-0.8	0.25	0.04	0.035	0.17-0.37	As 0.08; bal Fe				

UNS numbers and US grades are provided as a means of cross referencing chemically similar alloys. Exchangability is only possible after independent examination of specifications. Tensile properties are minimum or typical as specified. UTS and YS as MPa. El as %. See Appendix for list of abbreviations used in Notes. * indicates obsolete material.

Specification	Designation	Notes	C	Cr	Cu	Mn	Ni	P	S	Si	Other	UTS	YS	El	Hard

Carbon Steel, Nonresulfurized, 1030 (Continued from previous page)

Russia

Specification	Designation	Notes	C	Cr	Cu	Mn	Ni	P	S	Si	Other	UTS	YS	El	Hard
GOST 1050(88)	30	High-grade struct; Norm; 0-100mm	0.27-0.35	0.25 max		0.5-0.8	0.3 max	0.035 max	0.04 max	0.17-0.37	As<=0.08; bal Fe	490	295	21L	
GOST 1050(88)	30	High-grade struct; Full hard; 0-100mm	0.27-0.35	0.25 max		0.5-0.8	0.3 max	0.035 max	0.04 max	0.17-0.37	As<=0.08; bal Fe	560		7L	0-229 HB
GOST 380	BSt5sp		0.28-0.38			0.5-0.8		0.04	0.05	0.15-0.35	N 0.008; As 0.08; bal Fe				
GOST 380	St5S		0.28-0.32	0.3	0.25	0.8	0.3	0.05	0.05	0.17-0.35	bal Fe				
GOST 380	St5S		0.28-0.32	0.3	0.4	0.8	0.3	0.05	0.05	0.17-0.35	bal Fe				
GOST 4543	30G		0.27-0.35	0.3	0.3	0.7-1	0.3	0.035	0.035	0.17-0.37	bal Fe				
GOST 977	30L-II		0.27-0.35	0.3	0.3	0.4-0.9	0.3	0.04	0.045	0.2-0.52	bal Fe				
GOST 977	30L-III		0.27-0.35	0.3	0.3	0.4-0.9	0.3	0.04	0.045	0.2-0.52	bal Fe				

Sweden

Specification	Designation	Notes	C	Cr	Cu	Mn	Ni	P	S	Si	Other	UTS	YS	El	Hard
SS 141505	1505	Gen struct	0.3			0.7		0.04	0.04	0.5	bal Fe				

Turkey

Specification	Designation	Notes	C	Cr	Cu	Mn	Ni	P	S	Si	Other	UTS	YS	El	Hard
TS 302	Fe55.2*	Pipe as drawn or weld	0.36					0.05	0.05		bal Fe				
TS 346	Fe55*	Pipe as drawn	0.36			0.9		0.05	0.05	0.35	bal Fe				
TS 909(86)	UDK629.14.423/Fe42	HR, unequal angles	0.31					0.06	0.06		bal Fe				
TS 911(86)	UDK669.14.423/Fe50	HR, T-bar	0.3					0.05	0.05		bal Fe				
TS 912(86)	UDK669.14.423/Fe42	HR, channel	0.31					0.06	0.06		bal Fe				
TS 913(86)	UDK669.14.423/Fe42	HR, Z-bar	0.31					0.06	0.06		bal Fe				

UK

Specification	Designation	Notes	C	Cr	Cu	Mn	Ni	P	S	Si	Other	UTS	YS	El	Hard
BS 1449/1(91)	30CS	t<=16mm	0.25-0.35			0.50-0.90		0.045 max	0.045 max	0.05-0.35	bal Fe	380	230	18-20	
BS 1449/1(91)	30HS	Plt Sh Strp HR; t<=16mm	0.25-0.35			0.50-0.90		0.045 max	0.045 max	0.05-0.35	bal Fe	500	280	16-18	
BS 970/1(83)	060A30	Blm Bil Slab Bar Rod Frg	0.28-0.33			0.50-0.70		0.050 max	0.050 max	0.4	bal Fe				
BS 970/1(83)	080A30	Blm Bil Slab Bar Rod Frg	0.28-0.33			0.70-0.90		0.050 max	0.050 max	0.4	bal Fe				
BS 970/1(83)	080M30	Blm Bil Slab Bar Rod Frg Hard Tmp	0.26-0.34			0.60-1.00		0.050 max	0.050 max		bal Fe				
BS 970/3(91)	080M30	Bright bar; 150<t<=250mm; norm	0.26-0.34			0.60-1.00			0.050 max	0.10-0.40	Mo <=0.15; bal Fe	460	230	19	134-183 HB
BS 970/3(91)	080M30	Bright bar; 6<t<=13mm; HR CD	0.26-0.34			0.60-1.00			0.050 max	0.10-0.40	Mo <=0.15; bal Fe	620	480	9	

USA

Specification	Designation	Notes	C	Cr	Cu	Mn	Ni	P	S	Si	Other	UTS	YS	El	Hard
	AISI 1030	Wir rod	0.28-0.34			0.60-0.90		0.040 max	0.050 max		bal Fe				
	AISI 1030	Sh Strp Plt	0.27-0.34			0.60-0.90		0.030 max	0.035 max		bal Fe				
	AISI 1030	Struct shp	0.28-0.34			0.60-0.90		0.030 max	0.035 max		bal Fe				
	AISI 1030	Bar	0.28-0.34			0.60-0.90		0.040 max	0.050 max		bal Fe				
	UNS G10300		0.28-0.34			0.60-0.90		0.040 max	0.050 max		Sheets, C 0.27-0.34; ERW Tubing, C 0.27-0.35; bal Fe				
ASTM A108(95)	1030	Bar, CF	0.28-0.34			0.60-0.90		0.040 max	0.050 max		bal Fe				
ASTM A29/A29M(93)	1030	Bar	0.28-0.34			0.60-0.90		0.040 max	0.050 max		bal Fe				
ASTM A510(96)	1030	Wir rod	0.28-0.34			0.60-0.90		0.040 max	0.050 max		bal Fe				
ASTM A512(96)	1030	Buttweld mech tub, CD	0.27-0.34			0.60-0.90		0.040 max	0.045 max		bal Fe				
ASTM A513(97a)	1030	ERW Mech tub	0.27-0.34			0.60-0.90		0.035 max	0.035 max		bal Fe				
ASTM A519(96)	1030	Smls mech tub	0.28-0.34			0.60-0.90		0.040 max	0.050 max		bal Fe				
ASTM A575(96)	M1031	Merchant qual bar	0.26-0.36			0.25-0.60		0.04 max	0.05 max		Mn can vary with C; bal Fe				

UNS numbers and US grades are provided as a means of cross referencing chemically similar alloys. Exchangability is only possible after independent examination of specifications. Tensile properties are minimum or typical as specified. UTS and YS as MPa. El as %. See Appendix for list of abbreviations used in Notes. * indicates obsolete material.

Specification	Designation	Notes	C	Cr	Cu	Mn	Ni	P	S	Si	Other	UTS	YS	El	Hard

Carbon Steel, Nonresulfurized, 1030 (Continued from previous page)

USA

Specification	Designation	Notes	C	Cr	Cu	Mn	Ni	P	S	Si	Other	UTS	YS	El	Hard
ASTM A576(95)	1030	Special qual HW bar	0.28-0.34			0.60-0.90		0.040 max	0.050 max		Si Cu Pb B Bi Ca Se Te if spec'd; bal Fe				
ASTM A682/A682M(98)	1030	Strp, high-C, CR, spring	0.27-0.34			0.60-0.90		0.035 max	0.040 max	0.15-0.30	bal Fe				
ASTM A684A684M(86)	1030	Strp, high-C, CR, Ann	0.27-0.34			0.60-0.90		0.035 max	0.040 max	0.15-0.30	bal Fe				80 HRB
ASTM A830/A830M(98)	1030	Plt	0.27-0.34			0.60-0.90		0.035 max	0.04		bal Fe				
ASTM A866(94)	1030	Anti-friction bearings	0.28-0.34			0.60-0.90		0.025 max	0.025 max	0.15-0.35	bal Fe				
FED QQ-S-635B(88)	C1030*	Obs; see ASTM A827; Plt	0.28-0.34			0.6-0.9		0.04 max	0.05 max		bal Fe				
FED QQ-S-700D(91)	C1030*	Obs; Sh Strp	0.28-0.34			0.6-0.9		0.04 max	0.05 max		bal Fe				
MIL-F-20670B(61)		Flange, pipe for naval shipboard use; Obs for new design	0.35 max	0.25 max	0.25 max	0.91 max	0.50 max	0.05 max	0.05 max	0.30 max	Mo <=0.25; bal Fe	414	207	25	
MIL-S-10520D(AR)(88)	2*	Frg; Obs; see ASTM A711, A575	0.28-0.34	0.2	0.5	0.6-0.9	0.25	0.04	0.05	0.15-0.3	Mo 0.06; bal Fe				
MIL-S-11310E(76)	CS1030*	Bar HR cold shape cold ext; Obs for new design	0.28-0.34			0.6-0.9		0.04 max	0.05 max		bal Fe				
MIL-S-16788A(86)	C3*	HR, frg, bloom bil slab for refrg; Obs for new design	0.28-0.34			0.60-0.90		0.040 max	0.050 max		bal Fe				
MIL-S-3289B(84)		Frg qual Plt disk for cartridge cases	0.26-0.33			0.60-0.90		0.035 max	0.040 max	0.10 max	bal Fe				
MIL-S-46070	1030*	Obs for new design	0.28-0.34			0.6-0.9		0.04	0.05	0.3	bal Fe				
SAE J1397(92)	1030	Bar HR, est mech prop	0.28-0.34			0.6-0.9		0.04 max	0.05 max		bal Fe	470	260	20	137 HB
SAE J1397(92)	1030	Bar CD, est mech prop	0.28-0.34			0.6-0.9		0.04 max	0.05 max		bal Fe	520	440	12	149 HB
SAE J403(95)	1030	Struct Shps Plt Strp Sh Weld Tub	0.27-0.34			0.60-0.90		0.030 max	0.035 max		bal Fe				
SAE J403(95)	1030	Bar Wir rod Smls Tub HR CF	0.28-0.34			0.60-0.90		0.030 max	0.050 max		bal Fe				

Yugoslavia

Specification	Designation	Notes	C	Cr	Cu	Mn	Ni	P	S	Si	Other	UTS	YS	El	Hard
	C.0545	Struct	0.3			1.60 max		0.05 max	0.05 max	0.60 max	bal Fe				

Carbon Steel, Nonresulfurized, 1031

USA

Specification	Designation	Notes	C	Cr	Cu	Mn	Ni	P	S	Si	Other	UTS	YS	El	Hard
	AISI M1031	Bar, merchant qual	0.26-0.36			0.25-0.60		0.04 max	0.05 max		bal Fe				
ASTM A29/A29M(93)	M1031	Bar, merchant qual	0.26-0.36			0.25-0.60		0.04 max	0.05 max		bal Fe				
SAE J403(95)	M1031	Merchant qual	0.26-0.36			0.25-0.60		0.04 max	0.05 max		bal Fe				

Carbon Steel, Nonresulfurized, 1033

China

Specification	Designation	Notes	C	Cr	Cu	Mn	Ni	P	S	Si	Other	UTS	YS	El	Hard
GB 6478(86)	ML 30 Mn	Bar HR Norm	0.27-0.35	0.20 max	0.20 max	0.50-0.80	0.20 max	0.035 max	0.035 max	0.25 max	bal Fe	490	295	21	
GB 699(88)	30 Mn	Bar HR Q/T 25mm diam	0.27-0.35	0.25 max	0.25 max	0.70-1.00	0.25 max	0.035 max	0.035 max	0.17-0.37	bal Fe	540	315	20	

Europe

Specification	Designation	Notes	C	Cr	Cu	Mn	Ni	P	S	Si	Other	UTS	YS	El	Hard
EN 10016/2(94)	1.0530	Rod	0.30-0.35	0.20 max	0.30 max	0.50-0.80	0.25 max	0.035 max	0.035 max	0.10-0.30	Al <=0.01; Mo <=0.05; bal Fe				
EN 10016/2(94)	C32D	Rod	0.30-0.35	0.20 max	0.30 max	0.50-0.80	0.25 max	0.035 max	0.035 max	0.10-0.30	Al <=0.01; Mo <=0.05; bal Fe				
EN 10016/4(94)	1.1143	Rod	0.30-0.34	0.10 max	0.15 max	0.50-0.70	0.10 max	0.020 max	0.025 max	0.10-0.30	Al <=0.01; Mo <=0.03; N <=0.007; Cr+Ni+Cu<=0.30, Cu+Sn<=0.15; bal Fe				

Specification	Designation	Notes	C	Cr	Cu	Mn	Ni	P	S	Si	Other	UTS	YS	El	Hard

Carbon Steel, Nonresulfurized, 1033 (Continued from previous page)

Europe

Specification	Designation	Notes	C	Cr	Cu	Mn	Ni	P	S	Si	Other	UTS	YS	El	Hard
EN 10016/4(94)	C32D2	Rod	0.30-0.34	0.10 max	0.15 max	0.50-0.70	0.10 max	0.020 max	0.025 max	0.10-0.30	Al <=0.01; Mo <=0.03; N <=0.007; Cr+Ni+Cu<=0.30, Cu+Sn<=0.15; bal Fe				

Japan

Specification	Designation	Notes	C	Cr	Cu	Mn	Ni	P	S	Si	Other	UTS	YS	El	Hard
JIS G3507(91)	SWRCH33K	Wir rod FF	0.30-0.36			0.60-0.90		0.030 max	0.035 max	0.10-0.35	bal Fe				

USA

Specification	Designation	Notes	C	Cr	Cu	Mn	Ni	P	S	Si	Other	UTS	YS	El	Hard
	AISI 1033	Sh Strp Plt	0.29-0.36			0.70-1.00		0.030 max	0.035 max		bal Fe				
	AISI 1033	Struct shp	0.29-0.36			0.70-1.00		0.030 max	0.035 max		bal Fe				
	UNS G10330		0.29-0.36			0.70-1.00		0.040 max	0.050 max		bal Fe				
ASTM A513(97a)	1033	ERW Mech tub	0.29-0.36			0.70-1.00		0.035 max	0.035 max		bal Fe				
ASTM A830/A830M(98)	1033	Plt	0.29-0.36			0.70-1.00		0.035 max	0.04 max		bal Fe				
SAE J403(95)	1033	Struct Shps Plt Strp Sh Weld Tub	0.29-0.36			0.70-1.00		0.030 max	0.035 max		bal Fe				

Carbon Steel, Nonresulfurized, 1034

India

Specification	Designation	Notes	C	Cr	Cu	Mn	Ni	P	S	Si	Other	UTS	YS	El	Hard
IS 1570/2(79)	35C4	Bar Frg Tub Pipe	0.3-0.4	0.3 max	0.3 max	0.3-0.6	0.4 max	0.055 max	0.055 max	0.6 max	Co <=0.1; Mo <=0.15; Pb <=0.15; W <=0.1; P:S Varies; bal Fe	520-620	286	20	
IS 1570/2(79)	C35	Bar Frg Tub Pipe	0.3-0.4	0.3 max	0.3 max	0.3-0.6	0.4 max	0.055 max	0.055 max	0.6 max	Co <=0.1; Mo <=0.15; Pb <=0.15; W <=0.1; P:S Varies; bal Fe	520-620	286	20	

USA

Specification	Designation	Notes	C	Cr	Cu	Mn	Ni	P	S	Si	Other	UTS	YS	El	Hard
	UNS G10340		0.32-0.38			0.50-0.80		0.040 max	0.050 max		bal Fe				
ASTM A29/A29M(93)	1034	Bar	0.32-0.38			0.50-0.80		0.040 max	0.050 max	0.050 max	bal Fe				
ASTM A510(96)	1034	Wir rod	0.32			0.50-0.80		0.040 max	0.050 max		bal Fe				

Carbon Steel, Nonresulfurized, 1035

Argentina

Specification	Designation	Notes	C	Cr	Cu	Mn	Ni	P	S	Si	Other	UTS	YS	El	Hard
IAS	IRAM 1035		0.32-0.38			0.60-0.90		0.040 max	0.050 max	0.10-0.30	bal Fe	590-690	360-420	18-30	174-201 HB

Australia

Specification	Designation	Notes	C	Cr	Cu	Mn	Ni	P	S	Si	Other	UTS	YS	El	Hard
AS 1442	K1035*	Obs; Bar Bil	0.32-0.38			0.6-0.9		0.05	0.05	0.1-0.35	bal Fe				
AS 1442	S1035*	Obs; Bar Bil	0.32-0.38			0.6-0.9		0.05	0.05	0.35	bal Fe				
AS 1442(92)	1035	HR bar, Semifinished (may treat w/microalloying elements)	0.32-0.38			0.60-0.90		0.040 max	0.040 max	0.10-0.35	Si<=0.10 for Al-killed; bal Fe				
AS 1442(92)	4	Bar<=215mm, Bloom/bil/slab<=250mm, HR as rolled or norm	0.25-0.38			0.40-1.00		0.040 max	0.040 max	0.10-0.40	bal Fe	500	250	20	
AS 1442(92)	U4	Bar<=100mm, HR as rolled or norm	0.25-0.38			0.40-1.00		0.040 max	0.040 max	0.40 max	bal Fe	500	250	20	
AS 1443	K1035*	Obs; Bar	0.32-0.38			0.6-0.9		0.05	0.05	0.1-0.35	bal Fe				
AS 1443	S1035*	Obs; Bar	0.32-0.38			0.6-0.9		0.05	0.05	0.35	bal Fe				
AS 1443(94)	1035	CF bar (may treat w/microalloying elements)	0.32-0.38			0.60-0.90		0.040 max	0.040 max	0.10-0.35	Si<=0.10 for Al-killed; bal Fe				
AS 1446	S1035*	Withdrawn, see AS/NZS 1594(97)	0.32-0.38			0.6-0.9		0.05	0.05	0.35	bal Fe				

UNS numbers and US grades are provided as a means of cross referencing chemically similar alloys. Exchangability is only possible after independent examination of specifications. Tensile properties are minimum or typical as specified. UTS and YS as MPa. El as %. See Appendix for list of abbreviations used in Notes. * indicates obsolete material.

Specification	Designation	Notes	C	Cr	Cu	Mn	Ni	P	S	Si	Other	UTS	YS	El	Hard

Carbon Steel, Nonresulfurized, 1035 (Continued from previous page)

Austria

Specification	Designation	Notes	C	Cr	Cu	Mn	Ni	P	S	Si	Other	UTS	YS	El	Hard
ONORM M3108	C35SW		0.32-0.40			0.5-0.7		0.035	0.035	0.15-0.35	bal Fe				
ONORM M3110	RC35		0.35-0.40			0.4-0.7		0.035	0.035	0.15-0.40	bal Fe				
ONORM M3161	C35		0.31-0.40			0.5-0.7		0.040	0.040	0.2-0.4	bal Fe				
ONORM M3167	C35SP		0.3-0.4			0.5-0.9		0.040	0.040	0.3	bal Fe				

Belgium

Specification	Designation	Notes	C	Cr	Cu	Mn	Ni	P	S	Si	Other	UTS	YS	El	Hard
NBN 253-02	C35-1		0.32-0.39			0.5-0.8		0.05	0.05	0.15-0.4	bal Fe				
NBN 253-02	C35-2		0.32-0.39			0.5-0.8		0.035	0.035	0.15-0.4	bal Fe				
NBN 253-02	C35-3		0.32-0.39			0.5-0.8		0.035	0.02-0.035	0.15-0.4	bal Fe				
NBN 253-06	C36		0.33-0.39			0.5-0.8		0.025	0.035	0.15-0.4	bal Fe				

Bulgaria

Specification	Designation	Notes	C	Cr	Cu	Mn	Ni	P	S	Si	Other	UTS	YS	El	Hard
BDS 2592(71)	BSt5ps	Struct	0.28-0.37	0.30 max	0.30 max	0.50-0.80	0.30 max	0.045 max	0.055 max	0.05-0.17	bal Fe				
BDS 2592(71)	BSt5sp	Struct	0.28-0.37	0.30 max	0.30 max	0.50-0.80	0.30 max	0.045 max	0.055 max	0.15-0.35	bal Fe				
BDS 5785(83)	35	Struct	0.32-0.40	0.25 max	0.25 max	0.50-0.80	0.25 max	0.035 max	0.040 max	0.17-0.37	bal Fe				

China

Specification	Designation	Notes	C	Cr	Cu	Mn	Ni	P	S	Si	Other	UTS	YS	El	Hard
GB 13237(91)	35	Sh Strp CR HT	0.32-0.40	0.25 max	0.25 max	0.50-0.80	0.25 max	0.035 max	0.035 max	0.17-0.37	bal Fe	490-635		19	
GB 3275(91)	35	Plt Strp HR HT	0.32-0.40	0.25 max	0.25 max	0.50-0.80	0.25 max	0.035 max	0.035 max	0.17-0.37	bal Fe	490-635		18	
GB 3522(83)	35	Strp CR Ann	0.32-0.40	0.25 max	0.25 max	0.50-0.80	0.25 max	0.035 max	0.035 max	0.17-0.37	bal Fe	395-640		16	
GB 5953(86)	ML 35	Wir CD	0.32-0.40	0.20 max	0.20 max	0.60 max		0.035 max	0.035 max	0.20 max	bal Fe	590-735			
GB 6478(86)	ML 35	Bar HR Norm	0.32-0.40	0.20 max	0.20 max	0.30-0.60	0.20 max	0.035 max	0.035 max	0.20 max	bal Fe	530	315	20	
GB 699(88)	35	Bar HR Norm Q/T 25mm diam	0.32-0.40	0.25 max	0.25 max	0.50-0.80	0.25 max	0.035 max	0.035 max	0.17-0.37	bal Fe	530	315	20	
GB 710(91)	35	Sh Strp HR HT	0.32-0.40	0.25 max	0.25 max	0.50-0.80	0.25 max	0.035 max	0.035 max	0.17-0.37	bal Fe	490-635		19	
GB 8162(87)	35	Smls Tub HR/CD Q/T	0.32-0.40	0.25 max	0.25 max	0.50-0.80	0.25 max	0.035 max	0.035 max	0.17-0.37	bal Fe	510	305	17	
GB/T 3078(94)	35	Bar CR Ann	0.32-0.39	0.25 max	0.25 max	0.60 max	0.25 max	0.035 max	0.035 max	0.20 max	bal Fe	470		15	

Czech Republic

Specification	Designation	Notes	C	Cr	Cu	Mn	Ni	P	S	Si	Other	UTS	YS	El	Hard
CSN 411500	11500	Gen struct	0.38 max	0.3 max		1.60 max		0.05 max	0.05 max	0.60 max	bal Fe				
CSN 412040	12040	Q/T	0.32-0.4	0.25 max		0.5-0.8	0.3 max	0.04 max	0.04 max	0.15-0.4	bal Fe				
CSN 412042	12042	Q/T; Unalloyed	0.32-0.4	0.3 max		0.5-0.8	0.4 max	0.04 max	0.04 max	0.35 max	B 0.001-0.005; bal Fe				
CSN 412140	12140	Q/T; Unalloyed	0.3-0.38	0.25 max		0.6-0.85	0.3 max	0.035 max	0.035 max	0.17-0.37	V 0.08-0.15; bal Fe				

Europe

Specification	Designation	Notes	C	Cr	Cu	Mn	Ni	P	S	Si	Other	UTS	YS	El	Hard
EN 10083/1(91)A1(96)	1.1180	Q/T t<=16mm	0.32-0.39	0.40 max		0.50-0.80	0.40 max	0.035 max	0.020-0.040	0.40 max	Mo <=0.10; Cr+Mo+Ni<=0.63; bal Fe	630-780	430	17	
EN 10083/1(91)A1(96)	1.1181	Q/T t<=16mm	0.32-0.39	0.40 max		0.50-0.80	0.40 max	0.035 max	0.035 max	0.40 max	Mo <=0.10; Cr+Mo+Ni<=0.63; bal Fe	630-780	430	17	
EN 10083/1(91)A1(96)	C35E	Q/T t<=16mm	0.32-0.39	0.40 max		0.50-0.80	0.40 max	0.035 max	0.035 max	0.40 max	Mo <=0.10; Cr+Mo+Ni<=0.63; bal Fe	630-780	430	17	
EN 10083/1(91)A1(96)	C35R	Q/T t<=16mm	0.32-0.39	0.40 max		0.50-0.80	0.40 max	0.035 max	0.020-0.040	0.40 max	Mo <=0.10; Cr+Mo+Ni<=0.63; bal Fe	630-780	430	17	
EN 10083/2(91)A1(96)	1.0501	Q/T t<=16mm	0.32-0.39	0.40 max		0.50-0.80	0.40 max	0.045 max	0.045 max	0.40 max	Mo <=0.10; Cr+Mo+Ni<=0.63; bal Fe	630-780	430	17	

Carbon Steel, Nonresulfurized, 1035 (Continued from previous page)

Specification	Designation	Notes	C	Cr	Cu	Mn	Ni	P	S	Si	Other	UTS	YS	El	Hard
Europe															
EN 10083/2(91)A1(96)	C35	Q/T t<=16mm	0.32-0.39	0.40 max		0.50-0.80	0.40 max	0.045 max	0.045 max	0.40 max	Mo <=0.10; Cr+Mo+Ni<=0.63; bal Fe	630-780	430	17	
Finland															
SFS 200:E	Fe60	Bar Plt Strp Wir Frg HW	0.4					0.05	0.05		N 0.009; bal Fe				
SFS 455	CO 35	Bar Frg Tube Q/A Tmp, 30mm	0.32-0.39			0.5-0.8		0.04	0.04	0.15-0.4	bal Fe				
SFS 455(73)	SFS455	Q/T	0.32-0.39			0.5-0.8		0.035 max	0.035 max	0.15-0.4	bal Fe				
France															
AFNOR	XC38TS		0.35-0.4			0.5-0.8		0.025	0.03	0.1-0.4	bal Fe				
AFNOR NFA33101(82)	AF55C35	t<=450mm; Norm	0.31-0.39	0.3 max	0.3 max	0.5-0.8	0.4 max	0.04 max	0.04 max	0.1-0.4	bal Fe	540-660	325	21	
AFNOR NFA35551(75)	CC35		0.3-0.4	0.3 max	0.3 max	0.5-0.8	0.4 max	0.04 max	0.04 max	0.1-0.4	bal Fe				
AFNOR NFA35553	C35	Strp	0.3-0.4			0.5-0.8		0.04	0.04	0.1-0.4	bal Fe				
AFNOR NFA35553	XC38	Strp, Norm	0.35-0.4			0.5-0.8		0.04	0.04	0.1-0.35	bal Fe				
AFNOR NFA35553	XC38	Strp, Q/A Tmp	0.35-0.4			0.5-0.8		0.04	0.04	0.1-0.35	bal Fe				
AFNOR NFA35554	XC38	Plt, HR, Norm, 16mm diam	0.35-0.4			0.5-0.8		0.04	0.04	0.1-0.35	bal Fe				
AFNOR NFA36102(93)	C35RR	HR; Strp; CR	0.32-0.39	0.35 max		0.5-0.8	0.40 max	0.025 max	0.02 max	0.15-0.35	Al <=0.03; Mo <=0.10; bal Fe				
AFNOR NFA36102(93)	XC35*	HR; Strp; CR	0.32-0.39	0.35 max		0.5-0.8	0.40 max	0.025 max	0.02 max	0.15-0.35	Al <=0.03; Mo <=0.10; bal Fe				
Germany															
DIN	Ck34		0.31-0.38			0.45-0.55		0.025 max	0.025 max	0.20 max	N <=0.007; N depends on P; bal Fe				
DIN	WNr 1.1173		0.31-0.38			0.45-0.55		0.025 max	0.025 max	0.20 max	N <=0.007; N depends on P; bal Fe				
DIN 1654(89)	Cq35	CHd, ext, soft ann	0.32-0.39			0.50-0.80		0.035 max	0.035 max	0.40 max	bal Fe	650-700			
DIN 1654(89)	WNr 1.1172	CHd ext, soft ann	0.32-0.39			0.50-0.80		0.035 max	0.035 max	0.40 max	bal Fe	650-700			
DIN 17140	D35-2*	Obs; see C38D	0.33-0.38			0.3-0.6		0.04 max	0.04 max	0.1-0.3	bal Fe				
DIN EN 10083(91)	Ck35	Q/T	0.32-0.39	0.40 max		0.50-0.80	0.40 max	0.035 max	0.03 max	0.40 max	Mo <=0.10; Cr+Mo+Ni<=0.63; bal Fe	600-750	380	19	
DIN EN 10083(91)	Cm35	Q/T	0.32-0.39	0.40 max		0.50-0.80	0.40 max	0.035 max	0.020-0.040	0.40 max	Mo <=0.10; Cr+Mo+Ni<=0.63; bal Fe				
DIN EN 10083(91)	WNr 1.0501	Q/T 17-40mm	0.32-0.39	0.40 max		0.50-0.80	0.40 max	0.045 max	0.045 max	0.40 max	Mo <=0.10; Cr+Mo+Ni<=0.63; bal Fe	600	380	19	
DIN EN 10083(91)	WNr 1.1180		0.32-0.39	0.40 max		0.50-0.80	0.40 max	0.035 max	0.020-0.040	0.40 max	Mo <=0.10; Cr+Mo+Ni<=0.63; bal Fe				
DIN EN 10083(91)	WNr 1.1181	Q/T 17-40mm	0.32-0.39	0.40 max		0.50-0.80	0.40 max	0.035 max	0.035 max	0.40 max	Mo <=0.10; Cr+Mo+Ni<=0.63; bal Fe	600-750	380	19	
Hungary															
MSZ 1745(79)	MC	High-temp const; 0-80mm; Q/T	0.45-0.4	0.5 max		0.5-0.8		0.035 max	0.035 max	0.17-0.37	bal Fe	490-640	270	20L	
MSZ 1745(79)	MC	High-temp const; 80.1-160mm; Q/T	0.32-0.4	0.5 max		0.5-0.8		0.035 max	0.035 max	0.17-0.37	bal Fe	490-640	270	20L	
MSZ 2898/2(80)	A45	Smls CF tube; Smls tubes w/special delivery conds; 0.5-10mm; bright-drawn, soft	0.25 max			2 max		0.05 max	0.05 max	0.6 max	bal Fe	470	330	8L	

Carbon Steel, Nonresulfurized, 1035 (Continued from previous page)

Specification	Designation	Notes	C	Cr	Cu	Mn	Ni	P	S	Si	Other	UTS	YS	El	Hard
Hungary															
MSZ 2898/2(80)	A45	Smls CF tube; Smls tubes w/special delivery conds; 0.5-10mm; bright-drawn, hard	0.25 max			2 max		0.05 max	0.05 max	0.6 max	bal Fe	540	430	5L	
MSZ 29(86)	A44	Smls tube; Non-rimming; FF; 40.1-100.00mm	0.21 max			1.3 max		0.04 max	0.04 max	0.1-0.35	bal Fe	420-580	255	21L	
MSZ 29(86)	A44	Smls tube; Non-rimming; FF; 0-16mm	0.21 max			1.3 max		0.04 max	0.04 max	0.1-0.35	bal Fe	420-580	275	21L	
MSZ 29(86)	A44X	Smls tube; Unkilled; R; FU; 40.1-100.00mm	0.21 max			1.3 max		0.04 max	0.04 max	0.1 max	bal Fe	420-580	255	21L	
MSZ 29(86)	A44X	Smls tube; Unkilled; R; FU; 0-16mm	0.21 max			1.3 max		0.04 max	0.04 max	0.1 max	bal Fe	420-580	275	21L	
MSZ 29/2(64)	A45		0.25 max			2 max		0.05 max	0.05 max	0.6 max	bal Fe				
MSZ 4217(85)	C35	CR, Q/T or spring Strp, Q/T; 0-2mm	0.32-0.39			0.5-0.8		0.035 max	0.035 max	0.4 max	bal Fe	980-1280		5L	
MSZ 4217(85)	C35	CR, Q/T or spring Strp, Q/T; 0-2mm; rolled, hard	0.32-0.39			0.5-0.8		0.035 max	0.035 max	0.4 max	bal Fe	630-940			
MSZ 500(81)	A44	Gen struct; Non-rimming; FF; 0-16mm; HR/Frg	0.25 max			2 max	0.3 max	0.045 max	0.05 max	0.12 min	bal Fe	430-550	275	24L	
MSZ 500(81)	A44	Gen struct; Non-rimming; FF; 40.1-100mm; HR/Frg	0.25 max			2 max	0.3 max	0.045 max	0.05 max	0.12 min	bal Fe	430-550	255	24L	
MSZ 5736(88)	D35	Wir rod; for CD	0.31-0.38	0.2 max	0.25 max	0.3-0.7	0.2 max	0.04 max	0.04 max	0.1-0.35	bal Fe				
MSZ 5736(88)	D35K	Wir rod; for CD	0.31-0.38	0.2 max	0.25 max	0.3-0.7	0.2 max	0.03 max	0.03 max	0.1-0.35	bal Fe				
MSZ 61(85)	C35	Q/T; 40<t<=100mm	0.32-0.39			0.5-0.8		0.035 max	0.035 max	0.4 max	bal Fe	550-700	320	20L	
MSZ 61(85)	C35	Q/T; t<=16mm	0.32-0.39			0.5-0.8		0.035 max	0.035 max	0.4 max	bal Fe	630-780	430	17L	
MSZ 61(85)	C35E	Q/T; t<=16mm	0.32-0.39			0.5-0.8		0.035 max	0.02-0.035	0.4 max	bal Fe	630-780	430	17L	
MSZ 61(85)	C35E	Q/T; 40<t<=100mm	0.32-0.39			0.5-0.8		0.035 max	0.02-0.035	0.4 max	bal Fe	550-700	320	20L	
MSZ 6251	C35E		0.32-0.39			0.5-0.8		0.035	0.035	0.17-0.37	bal Fe				
MSZ 6251	C35EK		0.32-0.39			0.5-0.8		0.035	0.035	0.17-0.37	bal Fe				
MSZ 6251(87)	C35Z	CF; Q/T; 0-36mm; drawn, half-hard	0.32-0.39			0.5-0.8		0.035 max	0.035 max	0.4 max	bal Fe	0-600			
MSZ 6251(87)	C35Z	CF; Q/T; 0-36mm; HR/HR/ann; drawn/soft ann; drawn/bright ann; soft ann/ground	0.32-0.39			0.5-0.8		0.035 max	0.035 max	0.4 max	bal Fe	0-570			
India															
IS 1570	C35Mn75		0.3-0.4			0.6-0.9				0.1-0.35	bal Fe				
IS 1570/2(79)	35C8	Bar Frg Tub Pipe	0.3-0.4	0.3 max	0.3 max	0.6-0.9	0.4 max	0.055 max	0.055 max	0.6 max	Co <=0.1; Mo <=0.15; Pb <=0.15; W <=0.1; P:S Varies; bal Fe	550-650	302	20	
IS 1570/2(79)	C35Mn75	Bar Frg Tub Pipe	0.3-0.4	0.3 max	0.3 max	0.6-0.9	0.4 max	0.055 max	0.055 max	0.6 max	Co <=0.1; Mo <=0.15; Pb <=0.15; W <=0.1; P:S Varies; bal Fe	550-650	302	20	
IS 5517	C35Mn75	Bar, Norm, Ann or Hard Tmp	0.3-0.4			0.6-0.9				0.1-0.35	bal Fe				
International															
ISO 2938(74)	2	Hollow bar, HF, t<=16mm	0.32-0.39			0.50-0.80		0.035 max	0.035 max	0.15-0.40	bal Fe	490-640	275	21	
ISO 3304(85)	R44	Smls Tub, CF	0.21 max			1.2 max		0.050 max	0.050 max	0.35 max	Pb <=0.15;	520		5	

UNS numbers and US grades are provided as a means of cross referencing chemically similar alloys. Exchangability is only possible after independent examination of specifications. Tensile properties are minimum or typical as specified. UTS and YS as MPa. El as %. See Appendix for list of abbreviations used in Notes. * indicates obsolete material.

Carbon Steel, Nonresulfurized, 1035 (Continued from previous page)

Specification	Designation	Notes	C	Cr	Cu	Mn	Ni	P	S	Si	Other	UTS	YS	El	Hard
International															
ISO 3304(85)	R44	Smls Tub, CF	0.21 max			1.2 max		0.050 max	0.050 max	0.35 max	Co <=0.1; Pb <=0.15; W <=0.1; bal Fe	520		5	
ISO 3304(85)	R44	Smls Tub, CF	0.21 max			1.2 max		0.050 max	0.050 max	0.35 max	Pb <=0.15;	520		5	
ISO 3304(85)	R44	Smls Tub, CF	0.21 max			1.2 max		0.050 max	0.050 max	0.35 max	Pb <=0.15;	520		5	
ISO 683-1(87)	C35	Bar Frg Plt Wir rod, Blts, Slbs; Q/T, t<16mm	0.32-0.39			0.50-0.80		0.045 max	0.045 max	0.10-0.40	bal Fe	630-780	430		
ISO 683-1(87)	C35E4	Bar Frg Plt Wir rod, Blts, Slbs; Q/T, t<16mm	0.32-0.39			0.50-0.80		0.035 max	0.035 max	0.10-0.40	bal Fe	630-780	430		
ISO 683-1(87)	C35M2	Bar Frg Plt Wir rod, Blts, Slbs; Q/T, t<16mm	0.32-0.39			0.50-0.80		0.035 max	0.020-0.040	0.10-0.40	bal Fe	630-780	430		
ISO 683-12	C35EB*		0.33-0.39			0.5-0.8		0.035	0.035	0.15-0.4	bal Fe				
ISO 683-12(72)	1*	F/IH	0.33-0.39	0.3 max	0.3 max	0.5-0.8	0.4 max	0.035 max	0.035 max	0.15-0.4	Al <=0.1; Co <=0.1; Mo <=0.15; Pb <=0.15; Ti <=0.05; V <=0.1; W <=0.1; bal Fe				
ISO 683-18(96)	C35	Bar Wir, CD, t<=5mm	0.32-0.39			0.50-0.80		0.045 max	0.045 max	0.10-0.40	bal Fe	680	550	5	183 HB
ISO 683-18(96)	C35E4	Bar Wir, CD, t<=5mm	0.32-0.39			0.50-0.80		0.035 max	0.035 max	0.10-0.40	bal Fe	680	550	5	183 HB
ISO 683-18(96)	C35M2	Bar Wir, CD, t<=5mm	0.32-0.39			0.50-0.80		0.020-0.040	0.035 max	0.10-0.40	bal Fe	680	550	5	183 HB
ISO 683-3(70)	C35eb*		0.32-0.39			0.5-0.8		0.035	0.03-0.05	0.15-0.4	bal Fe				
ISO R683-3(70)	C35ea	Q/T	0.32-0.39	0.3 max	0.3 max	0.5-0.8	0.4 max	0.035 max	0.02-0.035	0.15-0.4	Al <=0.1; Co <=0.1; Mo <=0.15; Pb <=0.15; Ti <=0.05; V <=0.1; W <=0.1; bal Fe				
Italy															
UNI 5333	C33		0.3-0.36	0.25	0.25	0.6-0.9		0.035	0.035	0.4	bal Fe				
UNI 5333(64)	C38	F/IH	0.35-0.41	0.25 max	0.25 max	0.6-0.9		0.03 max	0.035 max	0.4 max	bal Fe				
UNI 5598(71)	1CD35	Wir rod	0.32-0.39			0.4-0.7		0.05 max	0.05 max	0.15-0.35	bal Fe				
UNI 5598(71)	3CD35	Wir rod	0.33-0.38			0.4-0.7		0.035 max	0.035 max	0.15-0.35	N <=0.012; bal Fe				
UNI 6403(86)	C35	Q/T	0.32-0.39			0.5-0.8		0.035 max	0.035 max	0.15-0.35	bal Fe				
UNI 7065(72)	C35	Strp	0.33-0.38			0.4-0.65		0.02 max	0.02 max	0.35 max	bal Fe				
UNI 7065(72)	C36	Strp	0.32-0.39			0.5-0.9		0.035 max	0.035 max	0.4 max	bal Fe				
UNI 7356(74)	CB35	Q/T	0.34-0.39			0.5-0.8		0.035 max	0.035 max	0.15-0.4	bal Fe				
UNI 7845(78)	C35	Q/T	0.32-0.39			0.5-0.8		0.035 max	0.035 max	0.15-0.4	bal Fe				
UNI 7847(79)	C36	Surf Hard	0.33-0.39			0.5-0.8		0.03 max	0.03 max	0.15-0.4	bal Fe				
UNI 7874(79)	C35	Wrought	0.32-0.39			0.5-0.8		0.035 max	0.035 max	0.15-0.4	bal Fe				
UNI 8551(84)	C36	Surf Hard	0.33-0.39			0.5-0.8		0.03 max	0.025 max	0.15-0.4	bal Fe				
Japan															
JIS G3311(88)	S35CM	CR Strp	0.32-0.38	0.20 max	0.30 max	0.60-0.90	0.20 max	0.030 max		0.15-0.35	Ni+Cr=0.35 max, bal Fe				170-250HV
JIS G3507(91)	SWRCH35K	Wir rod FF	0.32-0.38			0.60-0.90		0.030 max	0.035 max	0.10-0.35	bal Fe				
Mexico															
DGN B-203	1035*	Tube, Obs	0.32-0.38			0.6-0.9		0.04	0.05		bal Fe				

UNS numbers and US grades are provided as a means of cross referencing chemically similar alloys. Exchangability is only possible after independent examination of specifications. Tensile properties are minimum or typical as specified. UTS and YS as MPa. El as %. See Appendix for list of abbreviations used in Notes. * indicates obsolete material.

Specification	Designation	Notes	C	Cr	Cu	Mn	Ni	P	S	Si	Other	UTS	YS	El	Hard

Carbon Steel, Nonresulfurized, 1035 (Continued from previous page)

Specification	Designation	Notes	C	Cr	Cu	Mn	Ni	P	S	Si	Other	UTS	YS	El	Hard
Mexico															
NMX-B-178(90)	Grade C	Pipe for high-temp service	0.35 max	0.40 max	0.40 max	0.29-1.06	0.40 max	0.025 max	0.025 max	0.10 min	Mo <=0.15; V <=0.08; Mn increases with a decrease in C up to 1.35; (Cr+Co+Mo+Ni+V) 1 max; bal Fe				
NMX-B-301(86)	1034	Bar	0.32-0.38			0.50-0.80		0.040 max	0.050 max		bal Fe				
NMX-B-301(86)	1035	Bar	0.32-0.38			0.60-0.90		0.040 max	0.050 max		bal Fe				
Norway															
NS 13205	St60-2		0.4					0.05	0.05		N 0.007; bal Fe				
Pan America															
COPANT 330	1035		0.32-0.35			0.6-0.9		0.04	0.05		bal Fe				
COPANT 331	1035		0.32-0.38			0.6-0.9		0.04	0.05		bal Fe				
COPANT 333	1034		0.32-0.38			0.5-0.8		0.04	0.05		bal Fe				
COPANT 333	1035		0.32-0.38			0.6-0.9		0.04	0.05		bal Fe				
COPANT 514	1035	Tube, CF	0.32-0.38			0.6-0.9		0.04	0.05	0.1-0.2	bal Fe				
COPANT 514	1035	Tube, Ann	0.32-0.38			0.6-0.9		0.04	0.05	0.1-0.2	bal Fe				
COPANT 514	1035	Tube, Norm	0.32-0.38			0.6-0.9		0.04	0.05	0.1-0.2	bal Fe				
COPANT 514	1035	Tube, HF, 323.9mm diam	0.32-0.38			0.6-0.9		0.04	0.05	0.1-0.2	bal Fe				
Poland															
PNH84019	35	Q/T	0.32-0.4	0.3 max		0.5-0.8	0.3 max	0.04 max	0.04 max	0.17-0.37	bal Fe				
PNH84023	R55		0.32-0.4			0.6-0.8		0.045	0.045	0.2-0.35	bal Fe				
PNH84023/07	R45	High-temp const	0.16-0.22	0.3 max		0.6-1.2	0.4 max	0.04 max	0.04 max	0.12-0.35	bal Fe				
PNH84024	K10	High-temp const	0.17 max	0.2 max	0.25 max	0.4-1.60	0.35 max	0.045 max	0.045 max	0.35 max	bal Fe				
PNH84024	K18	High-temp const	0.16-0.22	0.2 max	0.25 max	0.6-1.6	0.35 max	0.045 max	0.045 max	0.15-0.35	Pb <=0.1; bal Fe				
PNH84024	R45	High-temp const	0.14-0.2	0.3 max		0.45-0.7	0.4 max	0.05 max	0.05 max	0.15-0.35	bal Fe				
PNH84028	D35	Wir rod	0.33-0.38	0.2 max	0.2 max	0.3-0.6	0.2 max	0.35 max	0.035 max	0.1-0.3	Mo <=0.08; P+S<=0.06; bal Fe				
Romania															
STAS 10677	OLC35CS		0.35-0.38	0.2	0.3	0.5-0.8	0.2	0.035	0.035	0.17-0.37	Ti 0.02; As 0.05; bal Fe				
STAS 500/2(88)	OL50.1ak	Gen struct; FF; Non-rimming	0.3-0.40	0.3 max		0.8 max	0.4 max	0.05 max	0.05 max	0.4 max	Pb <=0.15; bal Fe				
STAS 500/2(88)	OL50.1bk	Gen struct; FF; Non-rimming	0.3-0.40	0.3 max		0.8 max	0.4 max	0.05 max	0.05 max	0.4 max	Pb <=0.15; bal Fe				
STAS 500/2(88)	OL50.1k	Gen struct; FF; Non-rimming	0.3-0.40	0.3 max		0.8 max	0.4 max	0.05 max	0.05 max	0.4 max	Pb <=0.15; bal Fe				
STAS 600	OT50-1		0.3-0.4	0.3	0.3	0.4-0.8	0.3	0.04	0.045	0.35-0.5	bal Fe				
STAS 600	OT50-2		0.3-0.4	0.3	0.3	0.4-0.8	0.3	0.04	0.045	0.35-0.5	bal Fe				
STAS 600	OT50-3		0.3-0.4	0.3	0.3	0.4-0.8	0.3	0.04	0.045	0.35-0.5	bal Fe				
STAS 880(88)	OLC35	Q/T	0.32-0.39	0.3 max		0.5-0.8		0.04 max	0.045 max	0.17-0.37	Pb <=0.15; As<=0.05; bal Fe				
STAS 880(88)	OLC35S	Q/T	0.32-0.39	0.3 max		0.5-0.8		0.04 max	0.02-0.045	0.17-0.37	Pb <=0.15; As<=0.05; bal Fe				
STAS 880(88)	OLC35X	Q/T	0.32-0.39	0.3 max		0.5-0.8		0.035 max	0.035 max	0.17-0.37	Pb <=0.15; As<=0.05; bal Fe				
STAS 880(88)	OLC35XS	Q/T	0.32-0.39	0.3 max		0.5-0.8		0.035 max	0.02-0.04	0.17-0.37	Pb <=0.15; As<=0.05; bal Fe				

UNS numbers and US grades are provided as a means of cross referencing chemically similar alloys. Exchangability is only possible after independent examination of specifications. Tensile properties are minimum or typical as specified. UTS and YS as MPa. El as %. See Appendix for list of abbreviations used in Notes. * indicates obsolete material.

Specification	Designation	Notes	C	Cr	Cu	Mn	Ni	P	S	Si	Other	UTS	YS	El	Hard

Carbon Steel, Nonresulfurized, 1035 (Continued from previous page)

Specification	Designation	Notes	C	Cr	Cu	Mn	Ni	P	S	Si	Other	UTS	YS	El	Hard
Romania															
STAS 9382/4	OLC35Q		0.32-0.39	0.3	0.3	0.5-0.8	0.3	0.035	0.035	0.17-0.37	As 0.05; bal Fe				
Russia															
GOST	35LK		0.3-0.4	0.25		0.5-0.8	0.5	0.04	0.04	0.25-0.45	bal Fe				
GOST	M34		0.29-0.39			0.5-0.8		0.04 max	0.05 max	0.15-0.32	bal Fe				
GOST	MSt5ps		0.28-0.39			0.5-0.8		0.045	0.055	0.17	bal Fe				
GOST 1050	35		0.32-0.4	0.25	0.25 max	0.5-0.8	0.25	0.035	0.04	0.17-0.37	bal Fe				
GOST 1050(88)	35	High-grade struct; Q/T; Full hard; 0-100mm	0.32-0.4	0.25 max		0.5-0.8	0.3 max	0.035 max	0.04 max	0.17-0.37	As<=0.08; bal Fe	590		6L	0-229 HB
GOST 1050(88)	35	High-grade struct; Q/T; Norm; 0-100mm	0.32-0.4	0.25 max		0.5-0.8	0.3 max	0.035 max	0.04 max	0.17-0.37	As<=0.08; bal Fe	530	315	20L	
GOST 380	St5ps	Gen struct; Semi-Finished; Semi-killed	0.3-0.39			0.5-0.8		0.045	0.055	0.05-0.12	bal Fe				
GOST 380	St5sp	Gen struct; Semi-Finished; Non-rimming; FF	0.33			0.79		0.03	0.04	0.25	bal Fe				
GOST 8731	35		0.32-0.39	0.25	0.25	0.5-0.8	0.25	0.04	0.04	0.17-0.37	bal Fe				
GOST 8731(74)	20		0.30 max	0.3 max	0.3 max	1.60 max	0.4 max	0.070 max	0.060 max	0.60 max	Al <=0.1; bal Fe				
GOST 977	35L-II		0.32-0.4	0.3	0.3	0.4-0.9	0.3	0.04	0.045	0.2-0.52	bal Fe				
GOST 977	35L-III		0.32-0.4	0.3	0.3	0.4-0.9	0.3	0.04	0.045	0.2-0.52	bal Fe				
GOST 977(88)	35L		0.32-0.4	0.3		0.4-0.9	0.3	0.05	0.05	0.2-0.52	bal Fe				
Spain															
UNE	F.113		0.3-0.4			0.4-0.7		0.04	0.04	0.15-0.3	bal Fe				
UNE 36011(75)	C32k		0.3-0.35			0.5-0.8		0.035	0.035	0.15-0.4	bal Fe				
UNE 36011(75)	C32k-1	Unalloyed; Q/T	0.3-0.35			0.5-0.8		0.035 max	0.02-0.035	0.15-0.4	Mo <=0.15; bal Fe				
UNE 36011(75)	C35k	Unalloyed; Q/T	0.3-0.4			0.5-0.8		0.035 max	0.035 max	0.15-0.4	Mo <=0.15; bal Fe				
UNE 36011(75)	C35k-1	Unalloyed; Q/T	0.3-0.4			0.5-0.8		0.035 max	0.02-0.035	0.15-0.4	Mo <=0.15; bal Fe				
UNE 36011(75)	C38k		0.35-0.40			0.5-0.8		0.035	0.035	0.15-0.4	bal Fe				
UNE 36011(75)	F.1130*	see C35k	0.3-0.4			0.5-0.8		0.035 max	0.035 max	0.15-0.4	Mo <=0.15; bal Fe				
UNE 36011(75)	F.1131*	see C32k	0.3-0.35			0.5-0.8		0.035	0.035	0.15-0.4	bal Fe				
UNE 36011(75)	F.1132*	see C38k	0.35-0.40			0.5-0.8		0.035	0.035	0.15-0.4	bal Fe				
UNE 36011(75)	F.1135*	see C35k-1	0.3-0.4			0.5-0.8		0.035 max	0.02-0.035	0.15-0.4	Mo <=0.15; bal Fe				
UNE 36011(75)	F.1136*	see C32k-1	0.3-0.35			0.5-0.8		0.035 max	0.02-0.035	0.15-0.4	Mo <=0.15; bal Fe				
UNE 36034(85)	35B3DF	CHd; Cold ext	0.32-0.39			0.5-0.8		0.035	0.035	0.15-0.35	B 0.0008-0.005; bal Fe				
UNE 36034(85)	C35DF	CHd; Cold ext	0.32-0.39			0.5-0.8		0.035 max	0.035 max	0.15-0.4	Mo <=0.15; bal Fe				
UNE 36034(85)	F.1133*	see C35DF	0.32-0.39			0.5-0.8		0.035 max	0.035 max	0.15-0.4	Mo <=0.15; bal Fe				
UNE 36034(85)	F.1295*	see 35B3DF	0.32-0.39			0.5-0.8		0.035	0.035	0.15-0.35	B 0.0008-0.005; bal Fe				
Sweden															
SIS 141572	1572-03	Bar Frg Plt Sh, Q/A, Tmp	0.32-0.39			0.5-0.8		0.04	0.04	0.15-0.4	bal Fe	540	320	30	
SIS 141572	1572-04	Bar Frg Plt Sh, Q/A, Tmp	0.32-0.39			0.5-0.8		0.04	0.04	0.15-0.4	bal Fe	580	360	19	

UNS numbers and US grades are provided as a means of cross referencing chemically similar alloys. Exchangability is only possible after independent examination of specifications. Tensile properties are minimum or typical as specified. UTS and YS as MPa. El as %. See Appendix for list of abbreviations used in Notes. * indicates obsolete material.

Specification	Designation	Notes	C	Cr	Cu	Mn	Ni	P	S	Si	Other	UTS	YS	El	Hard
Carbon Steel, Nonresulfurized, 1035 (Continued from previous page)														
Sweden														
SIS 141572	1572-05	Bar Frg Plt Sh, Q/A, Tmp	0.32-0.39			0.5-0.8		0.04	0.04	0.15-0.4	bal Fe	620	420	17	
SIS 141572	1572-08	Bar Frg Plt Sh, as rolled	0.32-0.39			0.5-0.8		0.04	0.04	0.15-0.4	bal Fe	490	270	21	
SS 141550	1550	Non-rimming; Q/T	0.28-0.4			0.4-0.9		0.05	0.05		bal Fe				
SS 141572	1572		0.32-0.39			0.5-0.8		0.035	0.035	0.15-0.4	bal Fe				
Switzerland														
VSM 10648	C35	Bar Sh Plt, as roll 5/16mm	0.32-0.39			0.5-0.8		0.045	0.045	0.15-0.35	bal Fe				
VSM 10648	C35	Bar Sh Plt, CD, 5mm diam	0.32-0.39			0.5-0.8		0.045	0.045	0.15-0.35	bal Fe				
VSM 10648	C35	Bar Sh Plt, CD, Norm 16/100mm	0.32-0.39			0.5-0.8		0.045	0.045	0.15-0.35	bal Fe				
VSM 10648	Ck35	Bar Sh Plt, CD, Norm 16/100mm	0.32-0.39			0.5-0.8		0.035	0.015-0.035	0.15-0.35	bal Fe				
VSM 10648	Ck35	Bar Sh Plt, CD, 5mm diam	0.32-0.39			0.5-0.8		0.035	0.015-0.035	0.15-0.35	bal Fe				
VSM 10648	Ck35	Bar Sh Plt, as roll 5/16mm	0.32-0.39			0.5-0.8		0.035	0.015-0.035	0.15-0.35	bal Fe				
VSM 10648	StAc60-2	Bar Sh Plt, CD, Ann 25mm diam	0.4					0.05	0.05		bal Fe				
VSM 10648	StAc60-2	Bar Sh Plt, CF and Norm	0.4					0.05	0.05		bal Fe				
VSM 10648	StAc60-2	Bar Sh Plt, CD, 5mm diam	0.4					0.05	0.05		bal Fe				
VSM 10648	StAc60-2	Bar Sh Plt, CF 10/100mm diam	0.4					0.05	0.05		bal Fe				
Turkey														
TS 302	Fe55.2*	Pipe as drawn or weld	0.36					0.05	0.05		bal Fe				
TS 911(86)	UDK669.14.423/Fe60	HR, T-bar	0.4					0.055	0.055		bal Fe				
UK														
BS 1449/1(83)	40CS	Plt sht strp; CR Ann; t<=16mm	0.35-0.45			0.50-0.90		0.045 max	0.045 max	0.05-0.35	bal Fe	420	250	16-18	0-147 HB
BS 1449/1(91)	40HS	Plt sht strp; CR Ann; t<=16mm	0.35-0.45			0.50-0.90		0.045 max	0.045 max	0.05-0.35	bal Fe	540	300	14-16	0-219 HB
BS 1717(83)	CDS105/106	Tub	0.40 max			0.30-0.90		0.050 max	0.050 max	0.35 max	bal Fe				
BS 3601(87)	S430	Pip Tub, press, t<=16mm	0.21 max			0.40-1.20		0.040 max	0.040 max	0.35 max	bal Fe	430-570	255	22	
BS 3601(87)	SAW430	Pip Tub, press, t<=16mm	0.25 max			1.20 max		0.040 max	0.040 max	0.50 max	bal Fe	430-570	255	22	
BS 970/1(83)	060A35	Blm Bil Slab Bar Rod Frg	0.33-0.38			0.50-0.70		0.050 max	0.050 max	0.1-0.4	bal Fe				
BS 970/1(83)	080A32	Blm Bil Slab Bar Rod Frg	0.30-0.35			0.70-0.90		0.050 max	0.050 max	0.1-0.4	bal Fe				
BS 970/1(83)	080A35	Blm Bil Slab Bar Rod Frg	0.33-0.38			0.70-0.90		0.050 max	0.050 max	0.1-0.4	bal Fe				
BS 970/1(83)	080A37	Blm Bil Slab Bar Rod Frg	0.35-0.40			0.70 min		0.050 max	0.050 max		bal Fe				
BS 980	CFS6		0.3-0.4			0.5-0.8		0.05	0.05	0.35	bal Fe				
USA														
	AISI 1035	Bar	0.32-0.38			0.60-0.90		0.040 max	0.050 max		bal Fe				
	AISI 1035	Struct shp	0.32-0.38			0.60-0.90		0.030 max	0.035 max		bal Fe				
	AISI 1035	Wir rod	0.32-0.38			0.60-0.90		0.040 max	0.050 max		bal Fe				
	AISI 1035	Sh Strp Plt	0.31-0.38			0.60-0.90		0.030 max	0.035 max		bal Fe				
	UNS G10350		0.32-0.38			0.60-0.90		0.040 max	0.050 max		Sheets, C 0.31-0.38; ERW Tubing, C 0.31-0.39; bal Fe				
AMS 5080H(90)		Frg as ordered; CF bars and Mech tub	0.31-0.38			0.60-0.90		0.040 max	0.050 max	0.15-0.30	bal Fe				241 HB max

Specification	Designation	Notes	C	Cr	Cu	Mn	Ni	P	S	Si	Other	UTS	YS	El	Hard
Carbon Steel, Nonresulfurized, 1035 (Continued from previous page)															
USA															
AMS 5080H(90)		HF and Ann bars, Mech tub	0.31-0.38			0.60-0.90		0.040 max	0.050 max	0.15-0.30	bal Fe				229 HB max
AMS 5082E(89)		Smls, stress relieved tub, can be ordered stronger	0.31-0.38			0.60-0.90		0.040 max	0.050 max	0.10-0.35	bal Fe	621	483	8-10	
ASTM A108(95)	1035	Bar, CF	0.32-0.38			0.60-0.90		0.040 max	0.050 max		bal Fe				
ASTM A29/A29M(93)	1035	Bar	0.32-0.38			0.60-0.90		0.040 max	0.050 max		bal Fe				
ASTM A311/A311M(95)	1035 Class A	Bar CD SR ann, 30<t<=50mm	0.32-0.38			0.60-0.90		0.040 max	0.050 max		bal Fe	520	450	12,RA=35	
ASTM A311/A311M(95)	1035 Class A	Bar CD SR ann, 20<t<=30mm	0.32-0.38			0.60-0.90		0.040 max	0.050 max		bal Fe	550	485	12,RA=35	
ASTM A311/A311M(95)	1035 Class A	Bar CD SR ann, t<=20mm	0.32-0.38			0.60-0.90		0.040 max	0.050 max		bal Fe	585	520	13,RA=35	
ASTM A311/A311M(95)	1035 Class A	Bar CD SR ann, 50<t<=75mm	0.32-0.38			0.60-0.90		0.040 max	0.050 max		bal Fe	485	415	10,RA=30	
ASTM A510(96)	1035	Wir rod	0.32-0.38			0.60-0.90		0.040 max	0.050 max		bal Fe				
ASTM A512(96)	1035	Buttweld mech tub, CD	0.31-0.38			0.60-0.90		0.040 max	0.045 max		bal Fe				
ASTM A513(97a)	1035	ERW Mech tub	0.31-0.38			0.60-0.90		0.035 max	0.035 max		bal Fe				
ASTM A519(96)	1035	Smls mech tub, SR	0.32-0.38			0.60-0.90		0.040 max	0.050 max		bal Fe	517	448	8	80 HRB
ASTM A519(96)	1035	Smls mech tub, HR	0.32-0.38			0.60-0.90		0.040 max	0.050 max		bal Fe	448	276	20	72 HRB
ASTM A519(96)	1035	Smls mech tub, CW	0.32-0.38			0.60-0.90		0.040 max	0.050 max		bal Fe	586	517	5	88 HRB
ASTM A519(96)	1035	Smls mech tub, Ann	0.32-0.38			0.60-0.90		0.040 max	0.050 max		bal Fe	414	228	25	67 HRB
ASTM A519(96)	1035	Smls mech tub, Norm	0.32-0.38			0.60-0.90		0.040 max	0.050 max		bal Fe	448	276	20	72 HRB
ASTM A576(95)	1035	Special qual HW bar	0.32-0.38			0.60-0.90		0.040 max	0.050 max		Si Cu Pb B Bi Ca Se Te if spec'd; bal Fe				
ASTM A682/A682M(98)	1035	Strp, high-C, CR, spring	0.31-0.38			0.60-0.90		0.035 max	0.040 max	0.15-0.30	bal Fe				
ASTM A684A684M(86)	1035	Strp, high-C, CR, Ann	0.31-0.38			0.60-0.90		0.035 max	0.040 max	0.15-0.30	bal Fe				82 HRB
ASTM A827(98)	1035	Plt, frg	0.31-0.38			0.60-0.90		0.035 max	0.040 max	0.15-0.40	bal Fe				
ASTM A830/A830M(98)	1035	Plt	0.31-0.38			0.60-0.90		0.035 max	0.04 max		bal Fe				
DoD-F-24669/1(86)	1035	Bar Bil; Supersedes MIL-S-866 & MIL-S-16974	0.32-0.38			0.6-0.9		0.04	0.05		bal Fe				
FED QQ-S-635B(88)	C1035*	Obs; see ASTM A827; Plt	0.32-0.38			0.6-0.9		0.04 max	0.05 max		bal Fe				
FED QQ-S-700D(91)	C1035*	Obs; Sh Strp	0.32-0.38			0.6-0.9		0.04 max	0.05 max		bal Fe				
FED QQ-W-461H(88)	1035*	Obs; Wir	0.32-0.38			0.6-0.9		0.04	0.05		bal Fe				
MIL-S-19434B(SH)(90)	Class 1	Gear pinion frg HT; shipboard propulsion and turbine	0.45 max			0.55-0.90		0.040 max	0.040 max	0.15-0.35	V <=0.10; Si 0.10 max if VCD used; bal Fe				163-193 Brinell
MIL-S-46070	1035*	Obs for new design	0.32-0.38			0.6-0.9		0.04	0.05	0.3	bal Fe				
SAE J1397(92)	1035	Bar HR, est mech prop	0.32-0.38			0.6-0.9		0.04 max	0.05 max		bal Fe	500	270	18	143 HB
SAE J1397(92)	1035	Bar CD, est mech prop	0.32-0.38			0.6-0.9		0.04 max	0.05 max		bal Fe	550	460	12	163 HB
SAE J403(95)	1035	Struct Shps Plt Strp Sh Weld Tub	0.31-0.38			0.60-0.90		0.030 max	0.035 max		bal Fe				
SAE J403(95)	1035	Bar Wir rod Smls Tub HR CF	0.32-0.38			0.60-0.90		0.030 max	0.050 max		bal Fe				
Yugoslavia															
	C.1213	Tub w/special req	0.22 max			0.4-1.60		0.05 max	0.05 max	0.1-0.35	Mo <=0.15; bal Fe				

Specification	Designation	Notes	C	Cr	Cu	Mn	Ni	P	S	Si	Other	UTS	YS	El	Hard

Carbon Steel, Nonresulfurized, 1035 (Continued from previous page)

Yugoslavia

Specification	Designation	Notes	C	Cr	Cu	Mn	Ni	P	S	Si	Other	UTS	YS	El	Hard
	C.1400	Struct	0.32-0.38			0.5-0.7		0.06 max	0.06 max	0.35 max	bal Fe				
	C.1430	Q/T	0.32-0.39			0.5-0.8		0.045 max	0.045 max	0.4 max	Mo <=0.15; bal Fe				
	C.1431	Q/T	0.32-0.39			0.5-0.8		0.035 max	0.03 max	0.4 max	Mo <=0.15; bal Fe				
	C.1436	F/IH	0.33-0.39	0.2 max		0.5-0.8	0.25 max	0.025 max	0.035 max	0.15-0.35	Cr+Mo+Ni<=0.45; bal Fe				
	C.1480	Q/T	0.32-0.39			0.5-0.8		0.035 max	0.02-0.035	0.4 max	Mo <=0.15; bal Fe				
	PZ30		0.3-0.39	0.15	0.25	0.3-0.6	0.15	0.04	0.04	0.1-0.3	bal Fe				

Carbon Steel, Nonresulfurized, 1037

Australia

Specification	Designation	Notes	C	Cr	Cu	Mn	Ni	P	S	Si	Other	UTS	YS	El	Hard
AS 1442	K1037*	Obs; Bar Bil	0.32-0.38			0.7-1		0.05	0.05	0.1-0.35	bal Fe				
AS 1443	K1037*	Obs; Bar	0.32-0.38			0.7-1		0.05	0.05	0.1-0.35	bal Fe				

Canada

Specification	Designation	Notes	C	Cr	Cu	Mn	Ni	P	S	Si	Other	UTS	YS	El	Hard
CSA B193	X46		0.32			1.35		0.04	0.05		bal Fe				
CSA B193	X52		0.32			1.35		0.04	0.05		bal Fe				

China

Specification	Designation	Notes	C	Cr	Cu	Mn	Ni	P	S	Si	Other	UTS	YS	El	Hard
GB 6478(86)	ML 35 Mn	Bar HR Norm	0.32-0.40	0.20 max	0.20 max	0.50-0.80	0.20 max	0.035 max	0.035 max	0.25 max	bal Fe	530	315	20	
GB 699(88)	35 Mn	Bar HR Q/T 25mm diam	0.32-0.40	0.25 max	0.25 max	0.70-1.00	0.25 max	0.035 max	0.035 max	0.17-0.37	bal Fe	560	335	18	
GB/T 3078(94)	35 Mn	Bar CR Q/T 25mm diam	0.32-0.40	0.25 max	0.25 max	0.70-1.00	0.25 max	0.035 max	0.035 max	0.17-0.37	bal Fe	560	335	18	

Europe

Specification	Designation	Notes	C	Cr	Cu	Mn	Ni	P	S	Si	Other	UTS	YS	El	Hard
EN 10016/4(94)	1.1145	Rod	0.34-0.38	0.10 max	0.15 max	0.50-0.70	0.10 max	0.020 max	0.025 max	0.10-0.30	Al <=0.01; Mo <=0.03; N <=0.007; Cr+Ni+Cu<=0.30, Cu+Sn<=0.15; bal Fe				
EN 10016/4(94)	C36D2	Rod	0.34-0.38	0.10 max	0.15 max	0.50-0.70	0.10 max	0.020 max	0.025 max	0.10-0.30	Al <=0.01; Mo <=0.03; N <=0.007; Cr+Ni+Cu<=0.30, Cu+Sn<=0.15; bal Fe				

Germany

Specification	Designation	Notes	C	Cr	Cu	Mn	Ni	P	S	Si	Other	UTS	YS	El	Hard
DIN 17204(90)	36Mn4	Smls tube	0.32-0.40			0.90-1.20		0.040 max	0.040 max	0.40 max	bal Fe				
DIN 17204(90)	WNr 1.0561	Smls tube	0.32-0.40			0.90-1.20		0.040 max	0.040 max	0.40 max	bal Fe				
DIN 21544(90)	31Mn4	Mine support	0.28-0.36			0.80-1.10		0.045 max	0.045 max	0.20-0.50	Al >=0.020; bal Fe				
DIN 21544(90)	WNr 1.0520	Mine supports	0.28-0.36			0.80-1.10		0.045 max	0.045 max	0.20-0.50	Al >=0.020; bal Fe				

Hungary

Specification	Designation	Notes	C	Cr	Cu	Mn	Ni	P	S	Si	Other	UTS	YS	El	Hard
MSZ 2751	TNTV		0.3-0.4	0.3	0.3	1.1	0.3	0.05	0.05	0.5	Mo 0.05; V 0.05; bal Fe				

Japan

Specification	Designation	Notes	C	Cr	Cu	Mn	Ni	P	S	Si	Other	UTS	YS	El	Hard
JIS G4051(79)	S35C	Bar Wir rod	0.32-0.38	0.20 max	0.30 max	0.60-0.90	0.20 max	0.030 max	0.035 max	0.15-0.35	Ni+Cr<=0.35; bal Fe				

Mexico

Specification	Designation	Notes	C	Cr	Cu	Mn	Ni	P	S	Si	Other	UTS	YS	El	Hard
NMX-B-301(86)	1037	Bar	0.32-0.38			0.70-1.00		0.040 max	0.050 max		bal Fe				

Poland

Specification	Designation	Notes	C	Cr	Cu	Mn	Ni	P	S	Si	Other	UTS	YS	El	Hard
PNH93215(82)	34GS	Reinforcing	0.3-0.37			0.9-1.2		0.05	0.05	0.6-0.8	bal Fe				

Russia

Specification	Designation	Notes	C	Cr	Cu	Mn	Ni	P	S	Si	Other	UTS	YS	El	Hard
GOST 4543	35G		0.32-0.4	0.3	0.3	0.7-1	0.3	0.035	0.035	0.17-0.37	bal Fe				
GOST 4543(71)	35G	Q/T	0.32-0.4	0.3 max	0.3 max	0.7-1.0	0.3 max	0.035 max	0.035 max	0.17-0.37	Al <=0.1; bal Fe				
GOST 5781	K-35GS		0.3-0.37	0.3	0.3	0.8-1.2	0.3	0.04	0.04	0.6-0.9	bal Fe				

UNS numbers and US grades are provided as a means of cross referencing chemically similar alloys. Exchangability is only possible after independent examination of specifications. Tensile properties are minimum or typical as specified. UTS and YS as MPa. El as %. See Appendix for list of abbreviations used in Notes. * indicates obsolete material.

Specification	Designation	Notes	C	Cr	Cu	Mn	Ni	P	S	Si	Other	UTS	YS	El	Hard
Carbon Steel, Nonresulfurized, 1037 (Continued from previous page)															
Russia															
GOST 5781(82)	35GS		0.3-0.37	0.3	0.3	0.8-1.2	0.3	0.04	0.04	0.6-0.9	bal Fe				
Turkey															
TS 346	Fe55*	Pipe as drawn	0.36			0.9		0.05 max		0.35	bal Fe				
UK															
BS 3111/1(87)	10/1	Q/T Wir	0.32-0.39			0.70-1.00		0.035 max	0.035 max	0.40 max	Al 0.02-0.120; B 0.0008-0.005; bal Fe				
BS 3111/1(87)	10/2	Q/T Wir	0.32-0.39	0.15-0.30		0.80-1.10		0.035 max	0.035 max	0.40 max	Al 0.02-0.120; B 0.0008-0.005; bal Fe				
BS 970/1(83)	080M36	Blm Bil Slab Bar Rod Frg CD HR	0.32-0.40			0.60-1.00		0.050 max	0.050 max		bal Fe				
USA															
	AISI 1037	Wir rod	0.32-0.38			0.70-1.00		0.040 max	0.050 max		bal Fe				
	AISI 1037	Struct shp	0.32-0.38			0.70-1.00		0.030 max	0.035 max		bal Fe				
	AISI 1037	Sh Strp Plt	0.31-0.38			0.70-1.00		0.030 max	0.035 max		bal Fe				
	UNS G10370		0.32-0.38			0.70-1.00		0.040 max	0.050 max		Sheets, C 0.31-0.38; bal Fe				
ASTM A29/A29M(93)	1037	Bar	0.32-0.38			0.70-1.00		0.040 max	0.050 max		bal Fe				
ASTM A510(96)	1037	Wir rod	0.32-0.38			0.70-1.00		0.040 max	0.050 max		bal Fe				
ASTM A576(95)	1037	Special qual HW bar	0.32-0.38			0.70-1.00		0.040 max	0.050 max		Si Cu Pb B Bi Ca Se Te if spec'd; bal Fe				
ASTM A830/A830M(98)	1037	Plt	0.31-0.38			0.70-1.00		0.035 max	0.04 max		bal Fe				
SAE J1249(95)	1037	Former SAE std valid for Wir rod	0.32-0.38			0.70-1.00		0.040 max	0.050 max		bal Fe				
SAE J1397(92)	1037	Bar CD, est mech prop	0.32-0.38			0.7-1		0.04 max	0.05 max		bal Fe	570	480	12	167 HB
SAE J1397(92)	1037	Bar HR, est mech prop	0.32-0.38			0.7-1		0.04 max	0.05 max		bal Fe	510	280	18	143 HB
SAE J403(95)	1037	Struct Shps Plt Strp Sh Weld Tub	0.31-0.38			0.70-1.00		0.030 max	0.035 max		bal Fe				
Carbon Steel, Nonresulfurized, 1038/1038H															
Argentina															
IAS	IRAM 1038		0.35-0.42			0.60-0.90		0.040 max	0.050 max	0.10-0.30	bal Fe	610-720	370-440	18-30	179-217 HB
Australia															
AS 1442	K1038*	Obs; Bar Bil	0.35-0.42			0.6-0.9		0.05	0.05	0.1-0.35	bal Fe				
AS 1442	XK1038*	Obs; Bar Bil	0.35-0.42			0.85-1.1		0.05	0.05	0.1-0.35	bal Fe				
AS 1442(92)	5	Bar<=215mm, Bloom/bil/slab<=250mm, HR as rolled or norm	0.35-0.45			0.50-1.00		0.040 max	0.040 max	0.10-0.40	bal Fe	540	270	16	
AS 1442(92)	U5	Bar<=100mm, HR as rolled or norm	0.35-0.45			0.50-1.00		0.040 max	0.040 max	0.40 max	bal Fe	540	270	16	
AS 1442(92)	X1038	HR bar, Semifinished (may treat w/microalloying elements)	0.35-0.42			0.70-1.00		0.040 max	0.040 max	0.10-0.35	Si<=0.10 for Al-killed; bal Fe				
AS 1443	K1038*	Obs; Bar	0.35-0.42			0.6-0.9		0.05	0.05	0.1-0.35	bal Fe				
AS 1443	XK1038*	Obs; Bar	0.35-0.42			0.85-1.1		0.05	0.05	0.1-0.35	bal Fe				
AS 1443(94)	X1038	CF bar (may treat w/microalloying elements)	0.35-0.42			0.70-1.00		0.040 max	0.040 max	0.10-0.35	Si<=0.10 for Al-killed; bal Fe				
Europe															
EN 10016/2(94)	1.0516	Rod	0.35-0.40	0.20 max	0.30 max	0.50-0.80	0.25 max	0.035 max	0.035 max	0.10-0.30	Al <=0.01; Mo <=0.05; bal Fe				

UNS numbers and US grades are provided as a means of cross referencing chemically similar alloys. Exchangability is only possible after independent examination of specifications. Tensile properties are minimum or typical as specified. UTS and YS as MPa. El as %. See Appendix for list of abbreviations used in Notes. * indicates obsolete material.

Carbon Steel, Nonresulfurized, 1038/1038H (Continued from previous page)

Specification	Designation	Notes	C	Cr	Cu	Mn	Ni	P	S	Si	Other	UTS	YS	El	Hard
Europe															
EN 10016/2(94)	C38D	Rod	0.35-0.40	0.20 max	0.30 max	0.50-0.80	0.25 max	0.035 max	0.035 max	0.10-0.30	Al <=0.01; Mo <=0.05; bal Fe				
EN 10016/4(94)	1.1150	Rod	0.36-0.40	0.10 max	0.15 max	0.50-0.70	0.10 max	0.020 max	0.025 max	0.10-0.30	Al <=0.01; Mo <=0.03; N <=0.007; Cr+Ni+Cu<=0.30, Cu+Sn<=0.15; bal Fe				
EN 10016/4(94)	C38D2	Rod	0.36-0.40	0.10 max	0.15 max	0.50-0.70	0.10 max	0.020 max	0.025 max	0.10-0.30	Al <=0.01; Mo <=0.03; N <=0.007; Cr+Ni+Cu<=0.30, Cu+Sn<=0.15; bal Fe				
France															
AFNOR	XC38TS		0.35-0.4			0.5-0.8		0.025	0.03	0.1-0.4	bal Fe				
AFNOR NFA35552	XC38Hi	Bar Rod Wir	0.35-0.4			0.5-0.8		0.03 max	0.04 max	0.15-0.35	bal Fe				
Germany															
DIN	Ck38		0.35-0.40			0.50-0.70		0.035 max	0.035 max	0.35-0.50	N depends on P; bal Fe				
DIN	WNr 1.1176		0.35-0.40			0.50-0.70		0.035 max	0.035 max	0.35-0.50	N <=0.007; N depends on P; bal Fe				
Japan															
JIS G3507(91)	SWRCH38K	Wir rod FF	0.35-0.41			0.60-0.90		0.030 max	0.035 max	0.10-0.35	bal Fe				
JIS G4051(79)	S38C	Bar Wir rod	0.35-0.41	0.20 max	0.30 max	0.60-0.90	0.20 max	0.030 max	0.035 max	0.15-0.35	Ni+Cr<=0.35; bal Fe				
Mexico															
NMX-B-301(86)	1038	Bar	0.35-0.42			0.60-0.90		0.040 max	0.050 max		bal Fe				
NMX-B-301(86)	1038H	Bar	0.34-0.43			0.50-1.00		0.040 max	0.050 max	0.15-0.30	bal Fe				
Pan America															
COPANT 331	1038	Bar	0.35-0.42			0.6-0.9		0.04	0.05		bal Fe				
COPANT 333	1038	Wir rod	0.35-0.42			0.6-0.9		0.04	0.05		bal Fe				
Spain															
UNE 36011(75)	C38k-1	Unalloyed; Q/T	0.35-0.4			0.5-0.8		0.035 max	0.02-0.035	0.15-0.4	Mo <=0.15; bal Fe				
UNE 36011(75)	F.1137*	see C38k-1	0.35-0.4			0.5-0.8		0.035 max	0.02-0.035	0.15-0.4	Mo <=0.15; bal Fe				
USA															
	AISI 1038	Sh Strp Plt	0.34-0.42			0.60-0.90		0.030 max	0.035 max		bal Fe				
	AISI 1038	Wir rod	0.35-0.42			0.60-0.90		0.040 max	0.050 max		bal Fe				
	AISI 1038	Struct shp	0.35-0.42			0.60-0.90		0.030 max	0.035 max		bal Fe				
	AISI 1038	Bar	0.35-0.42			0.60-0.90		0.040 max	0.050 max		bal Fe				
	AISI 1038H	Wir rod, Hard	0.34-0.43			0.50-1.00		0.040 max	0.050 max	0.15-0.35	bal Fe				
	UNS G10380		0.35-0.42			0.60-0.90		0.040 max	0.050 max		Sheets, C 0.34-0.42; bal Fe				
	UNS H10380		0.34-0.43			0.50-1.00		0.040 max	0.050 max	0.15-0.35	bal Fe				
ASTM A29/A29M(93)	1038	Bar	0.35-0.42			0.60-0.90		0.040 max	0.050 max		bal Fe				
ASTM A304(96)	1038H	Bar, hard bands spec	0.34-0.43			0.50-1.00		0.040 max	0.050 max	0.15-0.30	Cu Ni Cr Mo trace allowed; bal Fe				
ASTM A510(96)	1038	Wir rod	0.35-0.42			0.60-0.90		0.040 max	0.050 max		bal Fe				
ASTM A576(95)	1038	Special qual HW bar	0.35-0.42			0.60-0.90		0.040 max	0.050 max		Si Cu Pb B Bi Ca Se Te if spec'd; bal Fe				
ASTM A830/A830M(98)	1038	Plt	0.34-0.42			0.60-0.90		0.035 max	0.04		bal Fe				
SAE J1268(95)	1038H	Bar Rod, Frg; Hard see std	0.34-0.43			0.50-1.00		0.030 max	0.050 max	0.15-0.35	bal Fe				

UNS numbers and US grades are provided as a means of cross referencing chemically similar alloys. Exchangability is only possible after independent examination of specifications. Tensile properties are minimum or typical as specified. UTS and YS as MPa. El as %. See Appendix for list of abbreviations used in Notes. * indicates obsolete material.

Carbon Steel, Nonresulfurized, 1038/1038H (Continued from previous page)

Specification	Designation	Notes	C	Cr	Cu	Mn	Ni	P	S	Si	Other	UTS	YS	El	Hard
USA															
SAE J1397(92)	1038	Bar HR, est mech prop									bal Fe	520	280	18	149 HB
SAE J1397(92)	1038	Bar CD, est mech prop									bal Fe	570	480	12	163 HB
SAE J403(95)	1038	Bar Wir rod Smls Tub HR CF	0.35-0.42			0.60-0.90		0.030 max	0.050 max		bal Fe				
SAE J403(95)	1038	Shp Plt Strp Sh Weld Tub	0.34-0.42			0.6-0.9		0.03 max	0.035 max		bal Fe				
SAE J403(95)	1038	Struct Shps Plt Strp Sh Weld Tub	0.34-0.42			0.60-0.90		0.030 max	0.035 max		bal Fe				
SAE J403(95)	1038	Bar Rod, Smls Tub	0.35-0.42			0.6-0.9		0.04 max	0.05 max		bal Fe				
SAE J414	1038*	Obs see J1397	0.34-0.43			0.5-1.1		0.04	0.05	0.15-0.3	bal Fe				

Carbon Steel, Nonresulfurized, 1039

Specification	Designation	Notes	C	Cr	Cu	Mn	Ni	P	S	Si	Other	UTS	YS	El	Hard
Australia															
AS 1442	K1039*	Obs; Bar Bil	0.37-0.44			0.7-1		0.05	0.05	0.1-0.35	bal Fe				
AS 1443	K1039*	Obs; Bar	0.37-0.44			0.7-1		0.05	0.05	0.1-0.35	bal Fe				
Bulgaria															
BDS 5785(83)	40G	Struct	0.37-0.45	0.25 max	0.25 max	0.70-1.00	0.25 max	0.035 max	0.040 max	0.17-0.37	bal Fe				
China															
GB 13795(92)	40 Mn	Strp CR Ann	0.37-0.45	0.25 max	0.25 max	0.70-1.00	0.25 max	0.035 max	0.035 max	0.17-0.37	bal Fe	400-700		15	
GB 6478(86)	ML 40 Mn	Bar HR Norm	0.37-0.45	0.20 max	0.20 max	0.50-0.80	0.20 max	0.035 max	0.035 max	0.25 max	bal Fe	570	335	19	
GB 699(88)	40 Mn	Bar HR Q/T 25mm diam	0.37-0.45	0.25 max	0.25 max	0.70-1.00	0.25 max	0.035 max	0.035 max	0.17-0.37	bal Fe	590	355	17	
GB/T 3078(94)	40 Mn	Bar CR Q/T 25mm diam	0.37-0.44	0.25 max	0.25 max	0.70-1.00	0.25 max	0.035 max	0.035 max	0.17-0.37	bal Fe	590	355	17	
Finland															
SFS 457	CO.40Mn1.25	Bar Frg Tube Q/A Tmp, 25mm	0.38-0.45			1.1-1.4		0.035	0.035	0.15-0.4	bal Fe				
France															
AFNOR	40M5		0.36-0.44			1-1.35		0.04	0.035	0.1-0.4	bal Fe				
AFNOR NFA35552	XC38H2		0.35-0.4			1.2		0.035	0.035	0.15-0.35	bal Fe				
AFNOR NFA35552(84)	38C2u	t<16mm; Q/T	0.35-0.4	0.4-0.6	0.3 max	0.6-0.9	0.4 max	0.035 max	0.02-0.04	0.1-0.4	bal Fe	750-900	560	14	
AFNOR NFA35552(84)	38C2u	40<t<=100mm; Q/T	0.35-0.4	0.4-0.6	0.3 max	0.6-0.9	0.4 max	0.035 max	0.02-0.04	0.1-0.4	bal Fe	690-840	510	15	
AFNOR NFA35553	38MB5		0.34-0.4			1.1-1.4		0.035	0.035	0.1-0.4	B 0.0008-0.005; bal Fe				
AFNOR NFA35553(82)	38CB1	t<4.5mm; Q	0.34-0.4	0.2-0.4	0.3 max	0.6-0.9	0.4 max	0.035 max	0.035 max	0.1-0.4	B 0.0008-0.005; bal Fe				49 HRC
AFNOR NFA35556(84)	38CB1	t<16mm; Q/T	0.34-0.4	0.2-0.4	0.3 max	0.6-0.9		0.035 max	0.035 max	0.1-0.4	B 0.0008-0.005; bal Fe	880-1080	670	11	
AFNOR NFA35556(84)	38CB1	16<t<=100mm; Q/T	0.34-0.4	0.2-0.4	0.3 max	0.6-0.9		0.035 max	0.035 max	0.1-0.4	B 0.0008-0.005; bal Fe	840-990	650	11	
AFNOR NFA35556(84)	38MB5		0.34-0.4			1.1-1.4		0.035 max	0.035 max	0.1-0.4	B 0.0008-0.005; bal Fe				
AFNOR NFA35557	35M5										bal Fe				
AFNOR NFA35557	38MB5		0.34-0.4			1.1-1.4		0.035	0.035	0.1-0.4	B 0.0008-0.005; bal Fe				
AFNOR NFA35557	XC38H2		0.35-0.4	0.4		0.5-1.2	0.4	0.035	0.4	0.1-0.35	Mo 0.1; bal Fe				
AFNOR NFA35557	XC42		0.4-0.45			0.5-0.8		0.025	0.035	0.1-0.4	bal Fe				
AFNOR NFA35557	XC42TS		0.4-0.45			0.5-0.8		0.025	0.035	0.1-0.4	bal Fe				
AFNOR NFA35557(83)	38CB1	t<16mm; Q/T	0.34-0.4	0.2-0.4	0.3 max	0.6-0.9	0.4 max	0.035 max	0.035 max	0.1-0.4	B 0.0008-0.005; bal Fe	880-1080	670	11	
AFNOR NFA35557(83)	38CB1	16<t<=22mm; Q/T	0.34-0.4	0.2-0.4	0.3 max	0.6-0.9	0.4 max	0.035 max	0.035 max	0.1-0.4	B 0.0008-0.005; bal Fe	840-990	650	11	

Carbon Steel, Nonresulfurized, 1039 (Continued from previous page)

Specification	Designation	Notes	C	Cr	Cu	Mn	Ni	P	S	Si	Other	UTS	YS	El	Hard
France															
AFNOR NFA35563(83)	38C2TS	HT	0.35-0.4	0.4-0.6	0.3 max	0.6-0.9	0.4 max	0.025 max	0.035 max	0.1-0.4	bal Fe				55 HRC
AFNOR NFA35563(83)	38CB1TS	HT	0.34-0.4	0.2-0.4	0.3 max	0.6-0.9	0.4 max	0.025 max	0.035 max	0.1-0.4	B 0.0008-0.005; bal Fe				55 HRC
AFNOR NFA35564(83)	38C2FF	Soft ann	0.35-0.4	0.4-0.6		0.6-0.9	0.4 max	0.03 max	0.03 max	0.35 max	Al >=0.02; bal Fe	0-570			207 HB max
AFNOR NFA35564(83)	38CB1FF	Soft ann	0.34-0.4	0.2-0.4		0.6-0.9	0.4 max	0.03 max	0.03 max	0.35 max	Al >=0.02; B 0.0008-0.005; bal Fe	0-550			207 HB max
Germany															
DIN	40Mn4	Q/T 17-40mm	0.36-0.44			0.80-1.10		0.035 max	0.035 max	0.25-0.50	bal Fe	780-930	540	14	
DIN	C42EAl		0.39-0.44			0.75-0.90		0.035 max	0.035 max	0.25-0.40	N <=0.007; N depends on P; bal Fe				
DIN	WNr 1.1157	Q/T 17-40mm	0.36-0.44			0.80-1.10		0.035 max	0.035 max	0.25-0.50	bal Fe	780-930	540	14	
DIN	WNr 1.1190		0.39-0.44			0.75-0.90		0.035 max	0.035 max	0.25-0.40	N <=0.007; N depends on P; bal Fe				
Mexico															
NMX-B-301(86)	1039	Bar	0.37-0.44			0.70-1.00		0.040 max	0.050 max		bal Fe				
Pan America															
COPANT 333	1039	Wir rod	0.37-0.44			0.7-1		0.04	0.05		bal Fe				
Romania															
STAS 11513	40Mn10		0.37-0.45	0.3	0.3	0.7-1	0.3	0.035	0.035	0.17-0.37	bal Fe				
Russia															
GOST 4543(71)	40G	Q/T	0.37-0.45	0.3 max	0.3 max	0.7-1.0	0.3 max	0.035 max	0.035 max	0.17-0.37	Al <=0.1; bal Fe				
Spain															
UNE 36011(75)	C42k	Unalloyed; Q/T	0.4-0.45			0.5-0.8		0.035 max	0.035 max	0.15-0.4	Mo <=0.15; bal Fe				
UK															
BS 970/1(83)	080M40	Blm Bil Slab Bar Rod Frg	0.36-0.44			0.60-1.00		0.050 max	0.050 max		bal Fe				
BS 970/1(83)	120M36	Rod, Hard, Tmp, diam<150mm	0.32-0.40			1.00-1.40		0.050 max	0.050 max		bal Fe				
BS 970/1(83)	120M36	Bil Bar; Norm; diam<150mm	0.32-0.40			1.00-1.40		0.050 max	0.050 max		bal Fe				
USA															
	AISI 1039	Bar	0.37-0.44			0.70-1.00		0.040 max	0.050 max		bal Fe				
	AISI 1039	Wir rod	0.37-0.44			0.70-1.00		0.040 max	0.050 max		bal Fe				
	AISI 1039	Sh Strp Plt	0.36-0.44			0.70-1.00		0.030 max	0.035 max		bal Fe				
	AISI 1039	Struct shp	0.37-0.44			0.70-1.00		0.030 max	0.035 max		bal Fe				
	UNS G10390		0.37-0.44			0.70-1.00		0.040 max	0.050 max		Sheets, C 0.36-0.44; bal Fe				
ASTM A29/A29M(93)	1039	Bar	0.37-0.44			0.70-1.00		0.040 max	0.050 max		bal Fe				
ASTM A510(96)	1039	Wir rod	0.37-0.44			0.70-1.00		0.040 max	0.050 max		bal Fe				
ASTM A576(95)	1039	Special qual HW bar	0.37-0.44			0.70-1.00		0.040 max	0.050 max		Si Cu Pb B Bi Ca Se Te if spec'd; bal Fe				
ASTM A830/A830M(98)	1039	Plt	0.36-0.44			0.70-1.00		0.035 max	0.04 max	0.15-0.40	bal Fe				
SAE J1397(92)	1039	Bar CD, est mech prop	0.37-0.44			0.7-1		0.04 max	0.05 max		bal Fe	610	510	12	179 HB
SAE J1397(92)	1039	Bar HR, est mech prop	0.37-0.44			0.7-1		0.04 max	0.05 max		bal Fe	540	300	16	156 HB
SAE J403(95)	1039	Bar Wir rod Smls Tub HR CF	0.37-0.44			0.70-1.00		0.030 max	0.050 max		bal Fe				
SAE J403(95)	1039	Struct Shps Plt Strp Sh Weld Tub	0.36-0.44			0.70-1.00		0.030 max	0.035 max		bal Fe				

UNS numbers and US grades are provided as a means of cross referencing chemically similar alloys. Exchangability is only possible after independent examination of specifications. Tensile properties are minimum or typical as specified. UTS and YS as MPa. El as %. See Appendix for list of abbreviations used in Notes. * indicates obsolete material.

Specification	Designation	Notes	C	Cr	Cu	Mn	Ni	P	S	Si	Other	UTS	YS	El	Hard
Carbon Steel, Nonresulfurized, 1039 (Continued from previous page)															
Yugoslavia															
	C.3130	Q/T	0.36-0.44			0.8-1.1		0.035 max	0.035 max	0.25-0.5	Mo <=0.15; bal Fe				
Carbon Steel, Nonresulfurized, 1040															
Australia															
AS 1442	CS1040*	Obs; Bar Bil	0.35-0.45			0.4-0.9		0.06	0.06	0.35	bal Fe				
AS 1442	K1040*	Obs; Bar Bil	0.37-0.44			0.6-0.9		0.05	0.05	0.1-0.35	bal Fe				
AS 1442	S1040*	Obs; Bar Bil	0.37-0.44			0.6-0.9		0.05	0.05	0.1-0.35	bal Fe				
AS 1442(92)	1040	HR bar, Semifinished (may treat w/microalloying elements)	0.37-0.44			0.60-0.90		0.040 max	0.040 max	0.10-0.35	Si<=0.10 for Al-killed; bal Fe				
AS 1442(92)	M1040	Merchant qual, HR bar, Semifinished	0.35-0.45			0.40-0.90		0.050 max	0.050 max	0.35 max	bal Fe				
AS 1443	CS1040*	Obs; Bar	0.35-0.45			0.4-0.9		0.06	0.06	0.35	bal Fe				
AS 1443	K1040*	Obs; Bar	0.37-0.44			0.6-0.9		0.05	0.05	0.1-0.35	bal Fe				
AS 1443	S1040*	Obs; Bar	0.37-0.44			0.6-0.9		0.05	0.05	0.1-0.35	bal Fe				
AS 1443(94)	1040	CF bar (may treat w/microalloying elements)	0.37-0.44			0.60-0.90		0.040 max	0.040 max	0.10-0.35	Si<=0.10 for Al-killed; bal Fe				
AS 1443(94)	D5	CR or CR bar, 16-38mm	0.35-0.45			0.40-0.90		0.050 max	0.050 max	0.35 max	bal Fe	610	480	9	
AS 1443(94)	M1040	CF bar, Merchant quality	0.35-0.45			0.40-0.90		0.050 max	0.050 max	0.35 max	bal Fe				
AS 1443(94)	T5	CF bar, not CD or CR;<=260mm	0.35-0.45			0.40-0.90		0.050 max	0.050 max	0.35 max	bal Fe	540	270	16	
AS 1446	K1040*	Withdrawn, see AS/NZS 1594(97)	0.37-0.44			0.6-0.9		0.05	0.05	0.1-0.35	bal Fe				
AS 1446	S1040*	Withdrawn, see AS/NZS 1594(97)	0.37-0.44			0.6-0.9		0.05	0.05	0.1-0.35	bal Fe				
Austria															
ONORM M3110	RC40	Wir	0.40-0.45			0.4-0.7		0.035 max	0.035 max	0.15-0.40	P+S=0.06; bal Fe				
Belgium															
NBN 253-02	C40-1		0.37-0.44			0.5-0.8		0.045	0.045	0.15-0.4	bal Fe				
NBN 253-02	C40-2		0.37-0.44			0.6-0.9		0.04 max	0.05 max	0.15-0.3	bal Fe				
NBN 253-02	C40-3		0.37-0.44			0.5-0.8		0.035	0.02-0.035	0.15-0.4	bal Fe				
Bulgaria															
BDS 5785(83)	40	Struct	0.37-0.45	0.25 max	0.25 max	0.50-0.80	0.25 max	0.035 max	0.040 max	0.17-0.37	bal Fe				
China															
GB 13237(91)	40	Sh Strp CR HT	0.37-0.45	0.25 max	0.25 max	0.50-0.80	0.25 max	0.035 max	0.035 max	0.17-0.37	bal Fe	510-650		18	
GB 3275(91)	40	Plt Strp HR HT	0.37-0.45	0.25 max	0.25 max	0.50-0.80	0.25 max	0.035 max	0.035 max	0.17-0.37	bal Fe	510-655		17	
GB 3522(83)	40	Strp CR Ann	0.37-0.45	0.25 max	0.25 max	0.50-0.80	0.25 max	0.035 max	0.035 max	0.17-0.37	bal Fe	440-685		15	
GB 5953(83)	ML 40	Wir CD	0.37-0.45	0.20 max	0.20 max	0.60 max		0.035 max	0.035 max	0.20 max	bal Fe	590-735			
GB 6478(86)	ML 40	Bar HR Norm	0.37-0.45	0.20 max	0.20 max	0.30-0.60	0.20 max	0.035 max	0.035 max	0.20 max	bal Fe	570	335	19	
GB 699(88)	40	Bar HR Norm Q/T 25mm diam	0.37-0.45	0.25 max	0.25 max	0.50-0.80	0.25 max	0.035 max	0.035 max	0.17-0.37	bal Fe	570	335	19	
GB 700	B6*	Obs	0.38-0.5			0.5-0.8		0.045	0.055	0.15-0.35	bal Fe				
GB 700	BJ6*	Obs	0.38-0.5			0.5-0.8		0.045	0.055	0.15-0.35	bal Fe				

Specification	Designation	Notes	C	Cr	Cu	Mn	Ni	P	S	Si	Other	UTS	YS	El	Hard

Carbon Steel, Nonresulfurized, 1040 (Continued from previous page)

Specification	Designation	Notes	C	Cr	Cu	Mn	Ni	P	S	Si	Other	UTS	YS	El	Hard
China															
GB 700	BY6*	Obs	0.38-0.5			0.5-0.8		0.045	0.055	0.15-0.35	bal Fe				
GB 710(91)	40	Sh Strp HR HT	0.37-0.45	0.25 max	0.25 max	0.50-0.80	0.25 max	0.035 max	0.035 max	0.17-0.37	bal Fe	510-650		17	
GB/T 3078(94)	40	Bar CR Ann	0.37-0.45	0.25 max	0.25 max	0.60 max	0.25 max	0.035 max	0.035 max	0.20 max	bal Fe	510		14	
Czech Republic															
CSN 411550	11550	Gen struct; Tub Pipe	0.4 max	0.3 max		1.60 max		0.05 max	0.05 max	0.60 max	bal Fe				
CSN 412041	12041	Q/T	0.37-0.45	0.25 max		0.5-0.8	0.3 max	0.04 max	0.04 max	0.15-0.4	bal Fe				
Europe															
EN 10016/4(94)	1.1153	Rod	0.38-0.42	0.10 max	0.15 max	0.50-0.70	0.10 max	0.020 max	0.025 max	0.10-0.30	Al <=0.01; Mo <=0.03; N <=0.007; Cr+Ni+Cu<=0.30, Cu+Sn<=0.15; bal Fe				
EN 10016/4(94)	C40D2	Rod	0.38-0.42	0.10 max	0.15 max	0.50-0.70	0.10 max	0.020 max	0.025 max	0.10-0.30	Al <=0.01; Mo <=0.03; N <=0.007; Cr+Ni+Cu<=0.30, Cu+Sn<=0.15; bal Fe				
EN 10083/1(91)A1(96)	1.1186	Q/T t<=16mm	0.37-0.44	0.40 max		0.50-0.80	0.40 max	0.035 max	0.035 max	0.40 max	Mo <=0.10; Cr+Mo+Ni<=0.63; bal Fe	650-800	460	16	
EN 10083/1(91)A1(96)	1.1189	Q/T t<=16mm	0.37-0.44	0.40 max		0.50-0.80	0.40 max	0.035 max	0.020-0.040	0.40 max	Mo <=0.10; Cr+Mo+Ni<=0.63; bal Fe	650-800	460	16	
EN 10083/1(91)A1(96)	C40E	Q/T t<=16mm	0.37-0.44	0.40 max		0.50-0.80	0.40 max	0.035 max	0.035 max	0.40 max	Mo <=0.10; Cr+Mo+Ni<=0.63; bal Fe	650-800	460	16	
EN 10083/1(91)A1(96)	C40R	Q/T t<=16mm	0.37-0.44	0.40 max		0.50-0.80	0.40 max	0.035 max	0.020-0.040	0.40 max	Mo <=0.10; Cr+Mo+Ni<=0.63; bal Fe	650-800	460	16	
EN 10083/2(91)A1(96)	1.0511	Q/T t<=16mm	0.37-0.44	0.40 max		0.50-0.80	0.40 max	0.045 max	0.045 max	0.40 max	Mo <=0.10; Cr+Mo+Ni<=0.63; bal Fe	650-800	460	16	
EN 10083/2(91)A1(96)	C40	Q/T t<=16mm	0.37-0.44	0.40 max		0.50-0.80	0.40 max	0.045 max	0.045 max	0.40 max	Mo <=0.10; Cr+Mo+Ni<=0.63; bal Fe	650-800	460	16	
France															
AFNOR NFA33101(82)	AF60C40	t<=450mm; Norm	0.37-0.45	0.3 max	0.3 max	0.5-0.8	0.4 max	0.04 max	0.04 max	0.1-0.4	bal Fe	590-710	345	19	
AFNOR NFA36102(93)	C40RR	HR; Strp; CR	0.37-0.42	0.35 max		0.5-0.8	0.40 max	0.025 max	0.02 max	0.15-0.35	Al <=0.03; Mo <=0.10; bal Fe				
AFNOR NFA36102(93)	XC40*	HR; Strp; CR	0.37-0.42	0.35 max		0.5-0.8	0.40 max	0.025 max	0.02 max	0.15-0.35	Al <=0.03; Mo <=0.10; bal Fe				
AFNOR NFA36612(82)	F60	t<=300mm; HT	0.37-0.46	0.3 max	0.3 max	0.5-0.9	0.4 max	0.04 max	0.04 max	0.1-0.4	bal Fe	590-710	310	17	
Germany															
DIN	38MnSi4	Q/T 101-160mm	0.34-0.42			0.90-1.20		0.035 max	0.035 max	0.70-0.90	bal Fe	640-740	440	14	
DIN	38MnSi4	Q/T 17-40mm	0.34-0.42			0.90-1.20		0.035 max	0.035 max	0.70-0.90	bal Fe	830-1030	635	12	
DIN	WNr 1.5120	Q/T 101-160mm	0.34-0.42			0.90-1.20		0.035 max	0.035 max	0.70-0.90	bal Fe	640-740	440	14	
DIN	WNr 1.5120	Q/T 17-40mm	0.34-0.42			0.90-1.20		0.035 max	0.035 max	0.70-0.90	bal Fe	830-1030	635	12	
DIN 17140	D40-2*	Obs; see C42D	0.38-0.43			0.3-0.6		0.04 max	0.04 max	0.1-0.3	bal Fe				
DIN EN 10083(91)	Ck40	Q/T 17-40mm	0.37-0.44			0.50-0.80	0.40 max	0.035 max	0.035 max	0.40 max	Mo <=0.10; Cr+Mo+Ni<=0.63; bal Fe	630-780	400	18	
DIN EN 10083(91)	Cm40	Q/T	0.37-0.44	0.40 max		0.50-0.80	0.40 max	0.035 max	0.020-0.040	0.40 max	Mo <=0.10; Cr+Mo+Ni<=0.63; bal Fe				
DIN EN 10083(91)	WNr 1.0511	Q/T 17-40mm	0.37-0.44	0.40 max		0.50-0.80	0.40 max	0.045 max	0.045 max	0.40 max	Mo <=0.10; Cr+Mo+Ni<=0.63; bal Fe	630	400	18	

Specification	Designation	Notes	C	Cr	Cu	Mn	Ni	P	S	Si	Other	UTS	YS	El	Hard

Carbon Steel, Nonresulfurized, 1040 (Continued from previous page)

Specification	Designation	Notes	C	Cr	Cu	Mn	Ni	P	S	Si	Other	UTS	YS	El	Hard
Germany															
DIN EN 10083(91)	WNr 1.1186	Q/T 17-40mm	0.37-0.44	0.40 max		0.50-0.80	0.40 max	0.035 max	0.035 max	0.40 max	Mo <=0.10; Cr+Mo+Ni<=0.63; bal Fe	630-780	400	18	
DIN EN 10083(91)	WNr 1.1189	Q/T	0.37-0.44	0.40 max		0.50-0.80	0.40 max	0.035 max	0.020-0.040	0.40 max	Mo <=0.10; Cr+Mo+Ni<=0.63; bal Fe				
Hungary															
MSZ 61(85)	C40	Q/T; 40<t<=100mm	0.37-0.44			0.5-0.8		0.035 max	0.035 max	0.4 max	bal Fe	600-750	350	19L	
MSZ 61(85)	C40	Q/T; t<=16mm	0.37-0.44			0.5-0.8		0.035 max	0.035 max	0.4 max	bal Fe	650-800	460	16L	
MSZ 61(85)	C40E	Q/T; t<=16mm	0.37-0.44			0.5-0.8		0.035 max	0.02-0.035	0.4 max	bal Fe	650-800	460	16L	
MSZ 61(85)	C40E	Q/T; 40<t<=100mm	0.37-0.44			0.5-0.8		0.035 max	0.02-0.035	0.4 max	bal Fe	600-750	350	19L	
India															
IS 1570/2(79)	40C8	Sh Plt Smls tub	0.35-0.45	0.3 max	0.3 max	0.6-0.9	0.4 max	0.055 max	0.055 max	0.6 max	Co <=0.1; Mo <=0.15; Pb <=0.15; W <=0.1; P:S Varies; bal Fe	580-680	319	18	
IS 1570/2(79)	C40	Sh Plt Smls tub	0.35-0.45	0.3 max	0.3 max	0.6-0.9	0.4 max	0.055 max	0.055 max	0.6 max	Co <=0.1; Mo <=0.15; Pb <=0.15; W <=0.1; P:S Varies; bal Fe	580-680	319	18	
IS 5517	C40	Bar, Norm, Ann or Hard, Tmp, 30mm	0.35-0.45			0.6-0.9				0.1-0.35	bal Fe				
International															
ISO 683-1(87)	C40	Bar Frg Plt Wir rod, Blts, Slbs; Q/T, t<16mm	0.37-0.44			0.50-0.80		0.045 max	0.045 max	0.10-0.40	bal Fe	650-800	460		
ISO 683-1(87)	C40E4	Bar Frg Plt Wir rod, Blts, Slbs; Q/T, t<16mm	0.37-0.44			0.50-0.80		0.035 max	0.035 max	0.10-0.40	bal Fe	650-800	460		
ISO 683-1(87)	C40M2	Bar Frg Plt Wir rod, Blts, Slbs; Q/T, t<16mm	0.37-0.44			0.50-0.80		0.035 max	0.020-0.040	0.10-0.40	bal Fe	650-800	460		
ISO 683-18(96)	C40	Bar Wir, CD, t<=5mm	0.37-0.44			0.50-0.80		0.045 max	0.045 max	0.10-0.40	bal Fe	720	580	4	197 HB
ISO 683-18(96)	C40E4	Bar Wir, CD, t<=5mm	0.37-0.44			0.50-0.80		0.035 max	0.035 max	0.10-0.40	bal Fe	720	580	4	197 HB
ISO 683-18(96)	C40M2	Bar Wir, CD, t<=5mm	0.37-0.44			0.50-0.80		0.035 max	0.020-0.040	0.10-0.40	bal Fe	720	580	4	197 HB
ISO R683-3(70)	C40ea	Bar, Q/A, Tmp, 16mm diam	0.37-0.44	0.3 max	0.3 max	0.5-0.8	0.4 max	0.035 max	0.02-0.035	0.15-0.4	Al <=0.1; Co <=0.1; Mo <=0.15; Pb <=0.15; Ti <=0.05; V <=0.1; W <=0.1; bal Fe				
ISO R683-3(70)	C40eb	Bar Rod, Q/A, Tmp, 16mm diam	0.37-0.44			0.5-0.8		0.04	0.03-0.05	0.15-0.4	bal Fe				
Italy															
UNI 5598	ICD40		0.37-0.44			0.4-0.7		0.05	0.05	0.15-0.35	bal Fe				
UNI 5598(71)	1CD40	Wir rod	0.37-0.44			0.4-0.7		0.05 max	0.05 max	0.15-0.35	bal Fe				
UNI 5598(71)	3CD40	Wir rod	0.38-0.43			0.4-0.7		0.035 max	0.035 max	0.15-0.35	N <=0.012; bal Fe				
UNI 6403(86)	C40	Q/T	0.37-0.44			0.6-0.9		0.035 max	0.035 max	0.15-0.35	bal Fe				
UNI 6783	Fe60-3		0.35-0.45			0.4-0.8		0.04	0.04	0.04	bal Fe				
UNI 6923(71)	C40		0.37-0.44	0.3		0.5-0.9		0.035	0.035	0.15-0.4	bal Fe				
UNI 7065(72)	C40	Strp	0.38-0.43			0.4-0.65		0.02 max	0.02 max	0.35 max	bal Fe				
UNI 7065(72)	C41	Strp	0.37-0.44			0.5-0.9		0.035 max	0.035 max	0.4 max	bal Fe				
UNI 7845(78)	C40	Q/T	0.37-0.44			0.5-0.8		0.035 max	0.035 max	0.15-0.4	Pb 0.15-0.3; bal Fe				

Specification	Designation	Notes	C	Cr	Cu	Mn	Ni	P	S	Si	Other	UTS	YS	El	Hard

Carbon Steel, Nonresulfurized, 1040 (Continued from previous page)

Specification	Designation	Notes	C	Cr	Cu	Mn	Ni	P	S	Si	Other	UTS	YS	El	Hard
Italy															
UNI 7874(79)	C40	Q/T	0.37-0.44			0.5-0.8		0.035 max	0.035 max	0.15-0.4	Pb 0.15-0.3; bal Fe				
Japan															
JIS G3445(88)	STKM16A	Tube	0.35-0.40			0.40-1.00		0.040 max	0.040 max	0.40 max	bal Fe	510	324	20L	
JIS G3445(88)	STKM16C	Tube	0.35-0.45			0.40-1.00		0.040 max	0.040 max	0.40 max	bal Fe	618	461	12L	
JIS G3507(91)	SWRCH40K	Wir rod FF	0.37-0.43			0.60-0.90		0.030 max	0.035 max	0.10-0.35	bal Fe				
JIS G4051(79)	S40C	Bar Wir rod	0.37-0.43	0.20 max	0.30 max	0.60-0.90	0.20 max	0.030 max	0.035 max	0.15-0.35	Ni+Cr<=0.35; bal Fe				
Mexico															
DGN B-203	1040*	Tube, Obs	0.37-0.44			0.6-0.9		0.04	0.05		bal Fe				
NMX-B-301(86)	1040	Bar	0.37-0.44			0.60-0.90		0.040 max	0.050 max		bal Fe				
Pan America															
COPANT 330	1040	Bar	0.37-0.4			0.6-0.9		0.04	0.05		bal Fe				
COPANT 331	1040	Bar	0.37-0.44			0.6-0.9		0.04	0.05		bal Fe				
COPANT 333	1040	Wir rod	0.37-0.44			0.6-0.9		0.04	0.05		bal Fe				
COPANT 514	1040	Tube	0.37-0.44			0.6-0.9		0.04	0.05	0.2-0.35	bal Fe				
Poland															
PNH84019	40	Q/T	0.37-0.45	0.3 max		0.5-0.8	0.3 max	0.04 max	0.04 max	0.17-0.37	bal Fe				
PNH91046	P40	Q/T	0.37-0.45	0.3 max	0.25 max	0.5-0.8	0.3 max	0.04 max	0.04 max	0.15-0.35	bal Fe				
Romania															
STAS 880(88)	OLC40	Q/T	0.37-0.44	0.3 max		0.5-0.8		0.04 max	0.045 max	0.17-0.37	Pb <=0.15; As<=0.05; bal Fe				
STAS 880(88)	OLC40S	Q/T	0.37-0.44	0.3 max		0.5-0.8		0.04 max	0.02-0.045	0.17-0.37	Pb <=0.15; As<=0.05; bal Fe				
STAS 880(88)	OLC40X	Q/T	0.37-0.44	0.3 max		0.5-0.8		0.035 max	0.035 max	0.17-0.37	Pb <=0.15; As<=0.05; bal Fe				
STAS 880(88)	OLC40XS	Q/T	0.37-0.44	0.3 max		0.5-0.8		0.035 max	0.02-0.04	0.17-0.37	Pb <=0.15; As<=0.05; bal Fe				
Russia															
GOST	40R		0.35-0.45	0.3		0.5-0.8	0.3	0.04	0.045	0.17-0.37	B 0.002-0.005; bal Fe				
GOST	45R		0.35-0.45	0.3		0.5-0.8	0.3	0.04	0.045	0.17-0.37	B 0.002-0.005; bal Fe				
GOST	45TR		0.44			0.69		0.016	0.029	0.28	bal Fe				
GOST 1050	40		0.37-0.45	0.25	0.25	0.5-0.8	0.25	0.035	0.04	0.17-0.37	bal Fe				
GOST 1050(88)	40	High-grade struct; Q/T; Full hard; 0-100mm	0.37-0.45	0.25 max		0.5-0.8	0.3 max	0.035 max	0.04 max	0.17-0.37	As<=0.08; bal Fe	610		6L	0-241 HB
GOST 1050(88)	40	High-grade struct; Q/T; Norm; 0-100mm	0.37-0.45	0.25 max		0.5-0.8	0.3 max	0.035 max	0.04 max	0.17-0.37	As<=0.08; bal Fe	570	335	19L	
GOST 977	40L-II		0.37-0.45	0.3	0.3	0.4-0.9	0.3	0.04	0.045	0.2-0.52	bal Fe				
GOST 977	40L-III		0.37-0.45	0.3	0.3	0.4-0.9	0.3	0.04	0.045	0.2-0.52	bal Fe				
GOST 977(88)	40L		0.37-0.45			0.4-0.9		0.05	0.05	0.2-0.52	bal Fe				
Spain															
UNE 36011(75)	C42k		0.4-0.45			0.5-0.8		0.035	0.035	0.15-0.4	bal Fe				
UNE 36011(75)	F.1141*	see C42k	0.4-0.45			0.5-0.8		0.035 max	0.035 max	0.15-0.4	bal Fe				
Turkey															
TS 911(86)	UDK669.14.423/Fe60	HR, T-bar	0.4					0.05	0.05		bal Fe				

UNS numbers and US grades are provided as a means of cross referencing chemically similar alloys. Exchangability is only possible after independent examination of specifications. Tensile properties are minimum or typical as specified. UTS and YS as MPa. El as %. See Appendix for list of abbreviations used in Notes. * indicates obsolete material.

Specification	Designation	Notes	C	Cr	Cu	Mn	Ni	P	S	Si	Other	UTS	YS	El	Hard

Carbon Steel, Nonresulfurized, 1040 (Continued from previous page)

UK

Specification	Designation	Notes	C	Cr	Cu	Mn	Ni	P	S	Si	Other	UTS	YS	El	Hard
BS 970/1(83)	060A40	Blm Bil Slab Bar Rod Frg	0.38-0.43			0.50-0.70		0.050 max	0.050 max	0.4	bal Fe				
BS 970/1(83)	080A40	Blm Bil Slab Bar Rod Frg	0.38-0.43			0.70-0.90		0.050 max	0.050 max	0.4	bal Fe				
BS 970/3(91)	080M40	Bright bar; 6-13mm; HR CD	0.36-0.44			0.60-1.00			0.050 max	0.10-0.40	Mo <=0.15; bal Fe	660	530	7	
BS 970/3(91)	080M40	Bright bar; 150<t<=250mm; norm	0.36-0.44			0.60-1.00			0.050 max	0.10-0.40	Mo <=0.15; bal Fe	510	245	17	146-197 HB

USA

Specification	Designation	Notes	C	Cr	Cu	Mn	Ni	P	S	Si	Other	UTS	YS	El	Hard
	AISI 1040	Bar	0.37-0.44			0.60-0.90		0.040 max	0.050 max		bal Fe				
	AISI 1040	Wir rod	0.37-0.44			0.60-0.90		0.040 max	0.050 max		bal Fe				
	AISI 1040	Struct shp	0.37-0.44			0.60-0.90		0.030 max	0.035 max		bal Fe				
	AISI 1040	Sh Strp Plt	0.36-0.44			0.60-0.90		0.030 max	0.035 max		bal Fe				
	UNS G10400		0.37-0.44			0.60-0.90		0.040 max	0.050 max		Sheets, C 0.36-0.44; bal Fe				
ASTM A108(95)	1040	Bar, CF	0.37-0.44			0.60-0.90		0.040 max	0.050 max		bal Fe				
ASTM A29/A29M(93)	1040	Bar	0.37-0.44			0.60-0.90		0.040 max	0.050 max		bal Fe				
ASTM A449	1040		0.37-0.44			0.6-0.9		0.04 max	0.05 max		bal Fe				
ASTM A510(96)	1040	Wir rod	0.37-0.44			0.60-0.90		0.040 max	0.050 max		bal Fe				
ASTM A513(97a)	1040	ERW Mech tub	0.36-0.44			0.60-0.90		0.040 max	0.050 max		bal Fe				
ASTM A519(96)	1040	Smls mech tub	0.37-0.44			0.60-0.90		0.040 max	0.050 max		bal Fe				
ASTM A576(95)	1040	Special qual HW bar	0.37-0.44			0.60-0.90		0.040 max	0.050 max		Si Cu Pb B Bi Ca Se Te if spec'd; bal Fe				
ASTM A682/A682M(98)	1040	Strp, high-C, CR, spring	0.36-0.44			0.60-0.90		0.035 max	0.040 max	0.15-0.30	bal Fe				
ASTM A684A684M(86)	1040	Strp, high-C, CR, Ann	0.36-0.44			0.60-0.90		0.035 max	0.040 max	0.15-0.30	bal Fe				83 HRB
ASTM A827(98)	1040	Plt, frg	0.36-0.44			0.60-0.90		0.035 max	0.040 max	0.15-0.40	bal Fe				
ASTM A830/A830M(98)	1040	Plt	0.36-0.44			0.60-0.90		0.035 max	0.04 max	0.15-0.40	bal Fe				
ASTM A866(94)	1040	Anti-friction bearings	0.37-0.44			0.60-0.90		0.025 max	0.025 max	0.15-0.35	bal Fe				
DoD-F-24669/1(86)	1040	Bar Bil; Supersedes MIL-S-866 & MIL-S-16974	0.37-0.44			0.6-0.9		0.04	0.05		bal Fe				
FED QQ-S-635B(88)	1040*	Obs; see ASTM A827; Plt	0.37-0.44			0.6-0.9		0.04	0.05		bal Fe				
MIL-S-11310E(76)	CS1040*	Bar HR cold shape cold ext; Obs for new design	0.37-0.44			0.6-0.9		0.04 max	0.05 max		bal Fe				
MIL-S-16788A(86)	C4*	HR, frg, bloom bil slab for refrg; Obs for new design	0.37-0.44			0.60-0.90		0.040 max	0.050 max		bal Fe				
MIL-S-46070	1040*	Obs for new design	0.37-0.44			0.6-0.9		0.04	0.05	0.3	bal Fe				
SAE J1397(92)	1040	Bar HR, est mech prop									bal Fe	520	290	18	149 HB
SAE J1397(92)	1040	Bar CD, est mech prop									bal Fe	590	490	12	170 HB
SAE J403(95)	1040	Struct Shps Plt Strp Sh Weld Tub	0.38-0.44			0.60-0.90		0.030 max	0.035 max		bal Fe				
SAE J403(95)	1040	Bar Wir rod Smls Tub HR CF	0.37-0.44			0.60-0.90		0.030 max	0.050 max		bal Fe				

Carbon Steel, Nonresulfurized, 1042

Australia

Specification	Designation	Notes	C	Cr	Cu	Mn	Ni	P	S	Si	Other	UTS	YS	El	Hard
AS 1442	K1042*	Obs; Bar Bil	0.4-0.47			0.6-0.9		0.05	0.05	0.1-0.35	bal Fe				

UNS numbers and US grades are provided as a means of cross referencing chemically similar alloys. Exchangability is only possible after independent examination of specifications. Tensile properties are minimum or typical as specified. UTS and YS as MPa. El as %. See Appendix for list of abbreviations used in Notes. * indicates obsolete material.

Specification	Designation	Notes	C	Cr	Cu	Mn	Ni	P	S	Si	Other	UTS	YS	El	Hard

Carbon Steel, Nonresulfurized, 1042 (Continued from previous page)

Australia

Specification	Designation	Notes	C	Cr	Cu	Mn	Ni	P	S	Si	Other	UTS	YS	El	Hard
AS 1442	K6*	Obs	0.4-0.5			0.5-1		0.05	0.05	0.1-0.35	bal Fe				
AS 1442	S6*	Obs	0.4-0.5			0.5-0.8		0.05	0.05	0.35	bal Fe				
AS 1443	K1042*	Obs	0.4-0.47			0.6-0.9		0.05	0.05	0.1-0.35	bal Fe				
AS 1443	K6*	Obs	0.4-0.5			0.5-1		0.05	0.05	0.1-0.35	bal Fe				
AS 1443	S6*	Obs	0.4-0.5			0.5-0.8		0.05	0.05	0.35	bal Fe				
AS 1446	K1042*	Withdrawn, see AS/NZS 1594(97)	0.4-0.47			0.6-0.9		0.05	0.05	0.1-0.35	bal Fe				
AS 1446	S1042*	Withdrawn, see AS/NZS 1594(97)	0.4-0.47			0.6-0.9		0.05	0.05	0.35	bal Fe				
AS/NZS 1594(97)	HK1042	HR flat products	0.39-0.47	0.30 max	0.35 max	0.60-0.90	0.35 max	0.040 max	0.035 max	0.50 max	Al <=0.100; Mo <=0.10; Ti <=0.040; (Cu+Ni+Cr+Mo) 1 max; bal Fe				

Austria

Specification	Designation	Notes	C	Cr	Cu	Mn	Ni	P	S	Si	Other	UTS	YS	El	Hard
ONORM M3110	RC40		0.40-0.45			0.4-0.7		0.035 max	0.035 max	0.15-0.40	P+S=0.06; bal Fe				

Bulgaria

Specification	Designation	Notes	C	Cr	Cu	Mn	Ni	P	S	Si	Other	UTS	YS	El	Hard
BDS 2592(71)	BSt6sp	Struct	0.38-0.49	0.30 max	0.30 max	0.50-0.80	0.30 max	0.045 max	0.055 max	0.15-0.35	bal Fe				

China

Specification	Designation	Notes	C	Cr	Cu	Mn	Ni	P	S	Si	Other	UTS	YS	El	Hard
GB 700	B6*	Obs	0.38-0.5			0.5-0.8		0.045	0.055	0.15-0.35	bal Fe				
GB 700	BJ6*	Obs	0.38-0.5			0.5-0.8		0.045	0.055	0.15-0.35	bal Fe				
GB 700	BY6*	Obs	0.38-0.5			0.5-0.8		0.045	0.055	0.15-0.35	bal Fe				

Europe

Specification	Designation	Notes	C	Cr	Cu	Mn	Ni	P	S	Si	Other	UTS	YS	El	Hard
EN 10016/2(94)	1.0541	Rod	0.40-0.45	0.20 max	0.30 max	0.50-0.80	0.25 max	0.035 max	0.035 max	0.10-0.30	Al <=0.01; Mo <-0.05; bal Fe				
EN 10016/2(94)	C42D	Rod	0.40-0.45	0.20 max	0.30 max	0.50-0.80	0.25 max	0.035 max	0.035 max	0.10-0.30	Al <=0.01; Mo <=0.05; bal Fe				
EN 10016/4(94)	1.1154	Rod	0.40-0.44	0.10 max	0.15 max	0.50-0.70	0.10 max	0.020 max	0.025 max	0.10-0.30	Al <=0.01; Mo <=0.03; N <=0.007; Cr+Ni+Cu<=0.30, Cu+Sn<=0.15; bal Fe				
EN 10016/4(94)	C42D2	Rod	0.40-0.44	0.10 max	0.15 max	0.50-0.70	0.10 max	0.020 max	0.025 max	0.10-0.30	Al <=0.01; Mo <=0.03; N <=0.007; Cr+Ni+Cu<=0.30, Cu+Sn<=0.15; bal Fe				

France

Specification	Designation	Notes	C	Cr	Cu	Mn	Ni	P	S	Si	Other	UTS	YS	El	Hard
AFNOR	XC42		0.4-0.45			0.5-0.8		0.025	0.035	0.1-0.4	bal Fe				
AFNOR	XC42TS		0.4-0.45			0.5-0.8		0.025	0.035	0.1-0.4	bal Fe				
AFNOR NFA35552	XC42H1		0.4-0.45			0.5-0.8		0.03 max	0.04 max	0.15-0.35	Cr+Ni+Mo=0.6; bal Fe				
AFNOR NFA35553	C40		0.4-0.5			0.5-0.8		0.04	0.04	0.1-0.4	bal Fe				
AFNOR NFA36102(93)	XC45*	HR; Strp; CR	0.42-0.48	0.3 max		0.5-0.8	0.40 max	0.025 max	0.02 max	0.15-0.35	Al <=0.03; Mo <=0.10; bal Fe				

Germany

Specification	Designation	Notes	C	Cr	Cu	Mn	Ni	P	S	Si	Other	UTS	YS	El	Hard
DIN 17140	D45-2*	Obs; see C48D	0.43-0.48			0.3-0.7		0.04 max	0.04 max	0.1-0.3	bal Fe				

Hungary

Specification	Designation	Notes	C	Cr	Cu	Mn	Ni	P	S	Si	Other	UTS	YS	El	Hard
MSZ 339(87)	B60.40	HR conc-reinf; 8-40mm; ageing treated	0.3 max			2 max		0.06 max	0.06 max	0.6 max	bal Fe	590	390	14L	
MSZ 339(87)	B60.50	HR; weld conc-reinf; 8-28mm; ageing treated	0.21 max			1.5 max		0.035 max	0.035 max	0.6 max	22-28mm C<=0.22; bal Fe	590	490	18L	
MSZ 339(87)	B60.50S	HR; weld conc-reinf; 6-12mm; ageing treated	0.21 max			1.5 max		0.035 max	0.035 max	0.6 max	bal Fe	590	490	14L	
MSZ 500(81)	B60	Gen struct; Non-rimming; FF	0.38-0.49			0.5-0.8		0.045 max	0.05 max	0.12-0.35	bal Fe				
MSZ 5736(88)	D43	Wir rod; for CD	0.39-0.46	0.2 max	0.25 max	0.3-0.7	0.2 max	0.04 max	0.04 max	0.1-0.35	bal Fe				

Specification	Designation	Notes	C	Cr	Cu	Mn	Ni	P	S	Si	Other	UTS	YS	El	Hard
Carbon Steel, Nonresulfurized, 1042 (Continued from previous page)															
Hungary															
MSZ 5736(88)	D43K	Wir rod; for CD	0.39-0.46	0.2 max	0.25 max	0.3-0.7	0.2 max	0.03 max	0.03 max	0.1-0.35	bal Fe				
India															
IS 1570	C45		0.4-0.5			0.6-0.9				0.1-0.35	bal Fe				
IS 5517	C45	Bar, Norm, Ann or Hard Tmp, 30mm	0.4-0.5			0.6-0.9				0.1-0.35	bal Fe				
Japan															
JIS G3507(91)	SWRCH43K	Wir rod FF	0.40-0.46			0.60-0.90		0.030 max	0.035 max	0.10-0.35	bal Fe				
Mexico															
NMX-B-301(86)	1042	Bar	0.40-0.47			0.60-0.90		0.040 max	0.050 max		bal Fe				
Poland															
PNH84023	R60		0.4-0.46			0.6-0.85		0.045	0.045	0.2-0.35	bal Fe				
Romania															
STAS 5750	OLC42		0.39-0.47			0.5-0.8		0.03	0.02-0.04	0.15-0.4	bal Fe				
Russia															
GOST	M44		0.38-0.49			0.5-0.8		0.045	0.05	0.15-0.32	bal Fe				
GOST	M44A		0.38-0.44			0.5-0.8		0.045	0.04	0.15-0.32	bal Fe				
GOST	OSM		0.37-0.47			0.5-0.8		0.04	0.05	0.15-0.35	bal Fe				
GOST 380	BSt6ps		0.38-0.49			0.5-0.8		0.04	0.05	0.15-0.3	N 0.008; As 0.08; bal Fe				
GOST 380	BSt6ps2		0.38-0.49	0.3	0.3	0.5-0.8	0.3	0.04	0.05	0.15-0.3	N 0.008; As 0.08; bal Fe				
GOST 380	BSt6sp		0.38-0.49	*		0.5-0.8		0.04	0.05	0.15-0.3	N 0.008; As 0.08; bal Fe				
GOST 380	BSt6sp2		0.38-0.49	0.3	0.3	0.5-0.8	0.3	0.04	0.05	0.15-0.3	N 0.008; As 0.08; bal Fe				
GOST 4728	OsV		0.4-0.48	0.3	0.25	0.55-0.85	0.3	0.04	0.045	0.15-0.35	bal Fe				
GOST 4728	OsW		0.4-0.48	0.3	0.25	0.55-0.85	0.3	0.04	0.045	0.15-0.35	bal Fe				
Spain															
UNE	F.114.A		0.4-0.46	0.05		0.8		0.035	0.035	0.35	bal Fe				
UNE 36011(75)	C42k1		0.4-0.45			0.5-0.8		0.035	0.02-0.035	0.15-0.4	bal Fe				
UNE 36011(75)	C45K		0.4-0.5			0.5-0.8		0.035	0.035	0.15-0.4	bal Fe				
UNE 36011(75)	F.1140*	see C45k	0.4-0.5			0.5-0.8		0.035 max	0.035 max	0.15-0.4	bal Fe				
UNE 36011(75)	F.1146*	see C42k1	0.4-0.45			0.5-0.8		0.035 max	0.02-0.035	0.15-0.4	bal Fe				
UK															
BS 970/1(83)	060A42	Blm Bil Slab Bar Rod Frg	0.40-0.45			0.50-0.70		0.050 max	0.050 max	0.4	bal Fe				
USA															
	AISI 1042	Wir rod	0.40-0.47			0.60-0.90		0.040 max	0.050 max		bal Fe				
	AISI 1042	Bar	0.40-0.47			0.60-0.90		0.040 max	0.050 max		bal Fe				
	AISI 1042	Sh Strp Plt	0.39-0.47			0.60-0.90		0.030 max	0.035 max		bal Fe				
	AISI 1042	Struct shp	0.40-0.47			0.60-0.90		0.030 max	0.035 max		bal Fe				
	UNS G10420		0.40-0.47			0.60-0.90		0.040 max	0.050 max		Sheets, C 0.39-0.47; bal Fe				
ASTM A183(98)	Grade 2 Nut	Rail Nuts	0.40-0.55					0.05 max	0.06 max		bal Fe	760	550	12	

UNS numbers and US grades are provided as a means of cross referencing chemically similar alloys. Exchangability is only possible after independent examination of specifications. Tensile properties are minimum or typical as specified. UTS and YS as MPa. El as %. See Appendix for list of abbreviations used in Notes. * indicates obsolete material.

Specification	Designation	Notes	C	Cr	Cu	Mn	Ni	P	S	Si	Other	UTS	YS	El	Hard

Carbon Steel, Nonresulfurized, 1042 (Continued from previous page)

USA

Specification	Designation	Notes	C	Cr	Cu	Mn	Ni	P	S	Si	Other	UTS	YS	El	Hard
ASTM A29/A29M(93)	1042	Bar	0.40-0.47			0.60-0.90		0.040 max	0.050 max		bal Fe				
ASTM A510(96)	1042	Wir rod	0.40-0.47			0.60-0.90		0.040 max	0.050 max		bal Fe				
ASTM A576(95)	1042	Special qual HW bar	0.40-0.47			0.60-0.90		0.040 max	0.050 max		Si Cu Pb B Bi Ca Se Te if spec'd; bal Fe				
ASTM A830/A830M(98)	1042	Plt	0.38-0.47			0.60-0.90		0.035 max	0.04 max	0.15-0.40	bal Fe				
FED QQ-S-635B(88)	C1042*	Obs; see ASTM A827; Plt	0.4-0.47			0.6-0.9		0.04 max	0.05 max		bal Fe				
SAE J1397(92)	1042	Bar CD, est mech prop	0.4-0.47			0.6-0.9		0.04 max	0.05 max		bal Fe	610	520	12	179 HB
SAE J1397(92)	1042	Bar norm D, est mech prop	0.4-0.47			0.6-0.9		0.04 max	0.05 max		bal Fe	590	500	12	179 HB
SAE J1397(92)	1042	Bar HR, est mech prop	0.4-0.47			0.6-0.9		0.04 max	0.05 max		bal Fe	550	300	16	163 HB
SAE J403(95)	1042	Struct Shps Plt Strp Sh Weld Tub	0.39-0.47			0.60-0.90		0.030 max	0.035 max		bal Fe				
SAE J403(95)	1042	Bar Wir rod Smls Tub HR CF	0.40-0.47			0.60-0.90		0.030 max	0.050 max		bal Fe				

Yugoslavia

Specification	Designation	Notes	C	Cr	Cu	Mn	Ni	P	S	Si	Other	UTS	YS	El	Hard
	C.1500	Struct	0.38-0.45			0.5-0.8		0.06 max	0.06 max	0.35 max	bal Fe				
	PZ40		0.4-0.49	0.15	0.025	0.3-0.7	0.015	0.04	0.04	0.1-0.3	bal Fe				

Carbon Steel, Nonresulfurized, 1043

Australia

Specification	Designation	Notes	C	Cr	Cu	Mn	Ni	P	S	Si	Other	UTS	YS	El	Hard
AS 1442(92)	6	Bar<=215mm, Bloom/bil/slab<=250mm, HR as rolled or norm	0.40-0.50			0.50-1.00		0.040 max	0.040 max	0.10-0.40	bal Fe	600	300	14	
AS 1442(92)	U6	Bar<=100mm, HR as rolled or norm	0.40-0.50			0.50-1.00		0.040 max	0.040 max	0.40 max	bal Fe	600	300	14	

Austria

Specification	Designation	Notes	C	Cr	Cu	Mn	Ni	P	S	Si	Other	UTS	YS	El	Hard
ONORM M3167	C45SP	Bar	0.4-0.5			0.6-1.0		0.040 max	0.040 max	0.3 max	bal Fe				

Bulgaria

Specification	Designation	Notes	C	Cr	Cu	Mn	Ni	P	S	Si	Other	UTS	YS	El	Hard
BDS 5785(83)	45	Struct	0.42-0.50	0.25 max	0.25 max	0.50-0.80	0.25 max	0.040 max	0.040 max	0.17-0.37	bal Fe				

China

Specification	Designation	Notes	C	Cr	Cu	Mn	Ni	P	S	Si	Other	UTS	YS	El	Hard
GB 13795(92)	40 Mn	Strp CR Ann	0.37-0.45	0.25 max	0.25 max	0.70-1.00	0.25 max	0.035 max	0.035 max	0.17-0.37	bal Fe	400-700		15	
GB 6478(86)	ML 40 Mn	Bar HR Norm	0.37-0.45	0.20 max	0.20 max	0.50-0.80	0.20 max	0.035 max	0.035 max	0.25 max	bal Fe	570	335	19	
GB 699(88)	40 Mn	Bar CR Q/T 25mm diam	0.37-0.45	0.25 max	0.25 max	0.70-1.00	0.25 max	0.035 max	0.035 max	0.17-0.37	bal Fe	590	355	17	
GB/T 3078(94)	40 Mn	Bar CR Q/T 25mm diam	0.37-0.45	0.25 max	0.25 max	0.70-1.00	0.25 max	0.035 max	0.035 max	0.17-0.37	bal Fe	590	355	17	

France

Specification	Designation	Notes	C	Cr	Cu	Mn	Ni	P	S	Si	Other	UTS	YS	El	Hard
AFNOR	CC45		0.4-0.5			0.5-0.8		0.04	0.04	0.1-0.4	bal Fe				
AFNOR NFA35552	XC42H2		0.4-0.45			1.2		0.035	0.035	0.15-0.35	bal Fe				

Italy

Specification	Designation	Notes	C	Cr	Cu	Mn	Ni	P	S	Si	Other	UTS	YS	El	Hard
UNI 3545	C45		0.42-0.5			0.4-0.9		0.035	0.035	0.15-0.4	bal Fe				
UNI 7847(79)	C43	Surf Hard	0.4-0.46			0.5-0.8		0.03 max	0.03 max	0.15-0.4	bal Fe				
UNI 8551(84)	C43	Surf Hard	0.4-0.46			0.5-0.8		0.03 max	0.025 max	0.15-0.4	Pb 0.15-0.25; bal Fe				

Japan

Specification	Designation	Notes	C	Cr	Cu	Mn	Ni	P	S	Si	Other	UTS	YS	El	Hard
JIS G4051(79)	S43C	Bar Wir rod	0.40-0.46	0.20 max	0.30 max	0.60-0.90	0.20 max	0.030 max	0.035 max	0.15-0.35	Ni+Cr<=0.35; bal Fe				

Mexico

Specification	Designation	Notes	C	Cr	Cu	Mn	Ni	P	S	Si	Other	UTS	YS	El	Hard
NMX-B-301(86)	1043	Bar	0.40-0.47			0.70-1.00		0.040 max	0.050 max		bal Fe				

UNS numbers and US grades are provided as a means of cross referencing chemically similar alloys. Exchangability is only possible after independent examination of specifications. Tensile properties are minimum or typical as specified. UTS and YS as MPa. El as %. See Appendix for list of abbreviations used in Notes. * indicates obsolete material.

Specification	Designation	Notes	C	Cr	Cu	Mn	Ni	P	S	Si	Other	UTS	YS	El	Hard

Carbon Steel, Nonresulfurized, 1043 (Continued from previous page)

Romania

Specification	Designation	Notes	C	Cr	Cu	Mn	Ni	P	S	Si	Other	UTS	YS	El	Hard
STAS 8185	44Mn11		0.4-0.48	0.3	0.3	0.9-1.2	0.3	0.04	0.04	0.17-0.37	Mo 0.06; bal Fe *				

UK

Specification	Designation	Notes	C	Cr	Cu	Mn	Ni	P	S	Si	Other	UTS	YS	El	Hard
BS 970/1(83)	080A42	Blm Bil Slab Bar Rod Frg	0.40-0.45			0.70-0.90		0.050 max	0.050 max	0.4	bal Fe				
BS 970/1(83)	080H46	Blm Bil Slab Bar Rod Frg	0.43-0.50			0.60-1.00		0.050 max	0.050 max		bal Fe				

USA

Specification	Designation	Notes	C	Cr	Cu	Mn	Ni	P	S	Si	Other	UTS	YS	El	Hard
	AISI 1043	Wir rod	0.40-0.47			0.70-1.00		0.040 max	0.050 max		bal Fe				
	AISI 1043	Bar	0.40-0.47			0.70-1.00		0.040 max	0.050 max		bal Fe				
	AISI 1043	Struct shp	0.40-0.47			0.70-1.00		0.030 max	0.035 max		bal Fe				
	AISI 1043	Sh Strp Plt	0.39-0.47			0.70-1.00		0.030 max	0.035 max		bal Fe				
	UNS G10430		0.40-0.47			0.70-1.00		0.040 max	0.050 max		Sheets, C 0.39-0.47; bal Fe				
ASTM A29/A29M(93)	1043	Bar	0.40-0.47			0.70-1.00		0.040 max	0.050 max		bal Fe				
ASTM A510(96)	1043	Wir rod	0.40-0.47			0.70-1.00		0.040 max	0.050 max		bal Fe				
ASTM A576(95)	1043	Special qual HW bar	0.40-0.47			0.70-1.00		0.040 max	0.050 max		Si Cu Pb B Bi Ca Se Te if spec'd; bal Fe				
ASTM A830/A830M(98)	1043	Plt	0.39-0.47			0.70-1.00		0.035 max	0.04 max	0.15-0.40	bal Fe				
SAE J1397(92)	1043	Bar norm D, est mech prop	0.4-0.47			0.7-1		0.04 max	0.05 max		bal Fe	600	520	12	179 HB
SAE J1397(92)	1043	Bar CD, est mech prop	0.4-0.47			0.7-1		0.04 max	0.05 max		bal Fe	630	530	12	179 HB
SAE J1397(92)	1043	Bar HR, est mech prop	0.4-0.47			0.7-1		0.04 max	0.05 max		bal Fe	570	310	16	163 HB
SAE J403(95)	1043	Bar Wir rod Smls Tub HR CF	0.40-0.47			0.70-1.00		0.030 max	0.050 max		bal Fe				
SAE J403(95)	1043	Struct Shps Plt Strp Sh Weld Tub	0.40-0.47			0.70-1.00		0.030 max	0.035 max		bal Fe				

Carbon Steel, Nonresulfurized, 1044

Austria

Specification	Designation	Notes	C	Cr	Cu	Mn	Ni	P	S	Si	Other	UTS	YS	El	Hard
ONORM M3108	C45SW		0.42-0.50			0.5-0.7		0.040 max	0.040 max	0.15-0.35	bal Fe				
ONORM M3110	RC45	Wir	0.45-0.50			0.4-0.7		0.040 max	0.040 max	0.15-0.40	P+S<=0.06; bal Fe				

Mexico

Specification	Designation	Notes	C	Cr	Cu	Mn	Ni	P	S	Si	Other	UTS	YS	El	Hard
NMX-B-301(86)	1044	Bar	0.43-0.50			0.30-0.60		0.040 max	0.050 max		bal Fe				

Poland

Specification	Designation	Notes	C	Cr	Cu	Mn	Ni	P	S	Si	Other	UTS	YS	El	Hard
PNH84023	45YA		0.42-0.5	0.3	0.3	0.4	0.3	0.03	0.04	0.2	bal Fe				
PNH84028	D45A	Wir rod	0.43-0.48	0.1 max	0.2 max	0.3-0.6	0.15 max	0.03 max	0.03 max	0.1-0.25	Mo <=0.05; P+S<=0.05; bal Fe				
PNH84028	DW45		0.44-0.49			0.4-0.6		0.03	0.03	0.25	bal Fe				

Romania

Specification	Designation	Notes	C	Cr	Cu	Mn	Ni	P	S	Si	Other	UTS	YS	El	Hard
STAS 500/2(88)	OL60.1ak	Gen struct; FF; Non-rimming	0.4-0.50	0.3 max		0.8 max	0.4 max	0.05 max	0.05 max	0.4 max	Pb <=0.15; bal Fe				
STAS 500/2(88)	OL60.1bk	Gen struct; FF; Non-rimming	0.4-0.50	0.3 max		0.8 max	0.4 max	0.05 max	0.05 max	0.4 max	Pb <=0.15; bal Fe				
STAS 500/2(88)	OL60.1k	Gen struct; FF; Non-rimming	0.4-0.50	0.3 max		0.8 max	0.4 max	0.05 max	0.05 max	0.4 max	Pb <=0.15; bal Fe				

USA

Specification	Designation	Notes	C	Cr	Cu	Mn	Ni	P	S	Si	Other	UTS	YS	El	Hard
	AISI 1044	Wir rod	0.43-0.50			0.30-0.60		0.040 max	0.050 max		bal Fe				
	AISI 1044	Bar	0.43-0.50			0.30-0.60		0.040 max	0.050 max		bal Fe				
	AISI M1044	Bar, merchant qual	0.40-0.50			0.25-0.60		0.04 max	0.05 max		bal Fe				

UNS numbers and US grades are provided as a means of cross referencing chemically similar alloys. Exchangability is only possible after independent examination of specifications. Tensile properties are minimum or typical as specified. UTS and YS as MPa. El as %. See Appendix for list of abbreviations used in Notes. * indicates obsolete material.

Specification	Designation	Notes	C	Cr	Cu	Mn	Ni	P	S	Si	Other	UTS	YS	El	Hard

Carbon Steel, Nonresulfurized, 1044 (Continued from previous page)

USA

Specification	Designation	Notes	C	Cr	Cu	Mn	Ni	P	S	Si	Other	UTS	YS	El	Hard
	UNS G10440		0.43-0.50			0.30-0.60		0.040 max	0.050 max		bal Fe				
ASTM A29/A29M(93)	1044	Bar	0.43-0.50			0.30-0.60		0.040 max	0.050 max		bal Fe				
ASTM A29/A29M(93)	M1044	Bar, merchant qual	0.40-0.50			0.25-0.60		0.04 max	0.05 max		bal Fe				
ASTM A510(96)	1044	Wir rod	0.43-0.50			0.30-0.60		0.040 max	0.050 max		bal Fe				
ASTM A575(96)	M1044	Merchant qual bar	0.40-0.50			0.25-0.60		0.04 max	0.05 max		Mn can vary with C; bal Fe				
ASTM A576(95)	1044	Special qual HW bar	0.43-0.50			0.30-0.60		0.040 max	0.050 max		Si Cu Pb B Bi Ca Se Te if spec'd; bal Fe				
SAE J1397(92)	1044	Bar HR, est mech prop	0.43-0.5			0.3-0.6		0.04 max	0.05 max		bal Fe	550	300	16	163 HB
SAE J403(95)	1044	Bar Wir rod Smls Tub HR CF	0.43-0.50			0.30-0.60		0.030 max	0.050 max		bal Fe				
SAE J403(95)	M1044	Merchant qual	0.43-0.50			0.30-0.60		0.04 max	0.05 max		bal Fe				

Carbon Steel, Nonresulfurized, 1045/1045H

Argentina

Specification	Designation	Notes	C	Cr	Cu	Mn	Ni	P	S	Si	Other	UTS	YS	El	Hard
IAS	IRAM 1045		0.43-0.50			0.60-0.90		0.040 max	0.050 max	0.10-0.30	bal Fe	650-770	390-460	16-24	197-229 HB

Australia

Specification	Designation	Notes	C	Cr	Cu	Mn	Ni	P	S	Si	Other	UTS	YS	El	Hard
AS 1442	K1045*	Obs; Bar Bil	0.43-0.5			0.6-0.9		0.05	0.05	0.1-0.35	bal Fe				
AS 1442	S1045*	Obs; Bar Bil	0.43-0.5			0.6-0.9		0.05	0.05	0.35	bal Fe				
AS 1442(92)	1045	HR bar, Semifinished (may treat w/microalloying elements)	0.43-0.50			0.60-0.90		0.040 max	0.040 max	0.10-0.35	Si<=0.10 for Al-killed; bal Fe				
AS 1443	K1045*	Obs; Bar	0.43-0.5			0.6-0.9		0.05	0.05	0.1-0.35	bal Fe				
AS 1443	S1045*	Obs; Bar	0.43-0.5			0.6-0.9		0.05	0.05	0.35	bal Fe				
AS 1443(94)	1045	CF bar (may treat w/microalloying elements)	0.43-0.50			0.60-0.90		0.040 max	0.040 max	0.10-0.35	Si<=0.10 for Al-killed; bal Fe				
AS 1443(94)	D6	CR or CR bar, 16-38mm	0.43-0.50			0.60-0.90		0.040 max	0.040 max	0.10-0.35	Si<=0.10 for Al-killed; bal Fe	650	510	8	
AS 1443(94)	T6	CF bar, not CD or CR;<=260mm	0.43-0.50			0.60-0.90		0.040 max	0.040 max	0.10-0.35	Si<=0.10 for Al-killed; bal Fe	600	300	14	
AS 1446	K1045*	Withdrawn, see AS/NZS 1594(97)	0.43-0.5			0.6-0.9		0.05	0.05	0.1-0.35	bal Fe				
AS 1446	S1045*	Withdrawn, see AS/NZS 1594(97)	0.43-0.5			0.6-0.9		0.05	0.05	0.35	bal Fe				

Austria

Specification	Designation	Notes	C	Cr	Cu	Mn	Ni	P	S	Si	Other	UTS	YS	El	Hard
ONORM M3108	C45SW	Bar	0.42-0.50			0.5-0.8		0.040 max	0.040 max	0.15-0.35	bal Fe				
ONORM M3110	RC45	Wir	0.45-0.50			0.4-0.7		0.035 max	0.035 max	0.15-0.40	P+S=0.06; bal Fe				
ONORM M3161	C45		0.41-0.50			0.5-0.7		0.040 max	0.040 max	0.2-0.4	P+S=0.07; bal Fe				

Belgium

Specification	Designation	Notes	C	Cr	Cu	Mn	Ni	P	S	Si	Other	UTS	YS	El	Hard
NBN 253-02	C45-1		0.42-0.5			0.5-0.8		0.045	0.045	0.15-0.4	bal Fe				
NBN 253-02	C45-2		0.42-0.5			0.5-0.8		0.035	0.02-0.035	0.15-0.4	bal Fe				
NBN 253-02	C45-3		0.42-0.5			0.5-0.8		0.035	0.02-0.035	0.15-0.4	bal Fe				
NBN 253-06	C46		0.43-0.49			0.5-0.8		0.025	0.035	0.15-0.4	bal Fe				

Bulgaria

Specification	Designation	Notes	C	Cr	Cu	Mn	Ni	P	S	Si	Other	UTS	YS	El	Hard
BDS 3492(86)	45LI		0.4-0.50	0.3		0.5-0.8		0.05	0.05	0.25-0.5	bal Fe				

UNS numbers and US grades are provided as a means of cross referencing chemically similar alloys. Exchangability is only possible after independent examination of specifications. Tensile properties are minimum or typical as specified. UTS and YS as MPa. El as %. See Appendix for list of abbreviations used in Notes. * indicates obsolete material.

Carbon Steel, Nonresulfurized, 1045/1045H (Continued from previous page)

Specification	Designation	Notes	C	Cr	Cu	Mn	Ni	P	S	Si	Other	UTS	YS	El	Hard
Bulgaria															
BDS 3492(86)	45LII		0.40-0.50		0.3	0.5-0.8		0.05	0.05	0.25-0.5	bal Fe				
BDS 3492(86)	45I III		0.40-0.50		0.3	0.5-0.8		0.05	0.05	0.25-0.5	bal Fe				
BDS 6354	45G2		0.41-0.49			0.5-0.8		0.025	0.025	0.17-0.37	bal Fe				
BDS 6354	45G2A		0.41-0.49			0.5-0.8		0.025	0.025	0.17-0.37	bal Fe				
BDS 6354	45G2K2	Struct	0.41-0.49			0.5-0.8		0.025	0.025	0.17-0.37	bal Fe				
BDS 6354	45G2K3	Struct	0.41-0.49			0.5-0.8		0.025	0.025	0.17-0.37	bal Fe				
China															
GB 13237(91)	45	Sh Strp CR HT	0.42-0.50	0.25 max	0.25 max	0.50-0.80	0.25 max	0.035 max	0.035 max	0.17-0.37	bal Fe	530-685		16	
GB 13795(92)	45	Strp CR Ann	0.42-0.50	0.25 max	0.25 max	0.50-0.80	0.25 max	0.035 max	0.035 max	0.17-0.37	bal Fe	400-700		15	
GB 3275(91)	45	Strp CR Ann	0.42-0.50	0.25 max	0.25 max	0.50-0.80	0.25 max	0.035 max	0.035 max	0.17-0.37	bal Fe	440-685		15	
GB 3275(91)	45	Plt Strp HR HT	0.42-0.50	0.25 max	0.25 max	0.50-0.80	0.25 max	0.035 max	0.035 max	0.17-0.37	bal Fe	540-685		15	
GB 5216(85)	45H	Bar Rod HR/Frg Ann	0.42-0.50	0.25 max	0.25 max	0.50-0.85	0.25 max	0.035 max	0.035 max	0.17-0.37	bal Fe				197 max HB
GB 5953(86)	ML 45	Wir CD	0.42-0.50	0.20 max	0.20 max	0.60 max		0.035 max	0.035 max	0.20 max	bal Fe	590-735			
GB 6478(86)	ML 45	Bar HR Norm	0.42-0.50	0.20 max	0.20 max	0.30-0.60	0.20 max	0.035 max	0.035 max	0.20 max	bal Fe	600	355	16	
GB 699(88)	45	Bar HR Norm Q/T 25mm diam	0.42-0.50	0.25 max	0.25 max	0.50-0.80	0.25 max	0.035 max	0.035 max	0.17-0.37	bal Fe	600	355	16	
GB 710(91)	45	Sh Strp HR HT	0.42-0.50	0.25 max	0.25 max	0.50-0.80	0.25 max	0.035 max	0.035 max	0.17-0.37	bal Fe	530-685		15	
GB 8162(87)	45	Smls Tub HR/CD Q/T	0.42-0.50	0.25 max	0.25 max	0.50-0.80	0.25 max	0.035 max	0.035 max	0.17-0.37	bal Fe	590	335	14	
GB/T 3078(94)	45	Bar CR Ann	0.42-0.49	0.25 max	0.25 max	0.60 max	0.25 max	0.035 max	0.035 max	0.20 max	bal Fe	540		13	
Czech Republic															
CSN 412050	12050	Q/T	0.42-0.5	0.25 max		0.5-0.8	0.3 max	0.04 max	0.04 max	0.17-0.37	bal Fe				
CSN 412052	12052	Q/T; Unalloyed	0.42-0.5	0.3 max		0.5-0.8	0.4 max	0.04 max	0.04 max	0.35 max	B 0.001-0.005; bal Fe				
CSN 413151	13151	Q/T	0.42-0.5	0.3 max		0.5-0.8		0.04 max	0.04 max	1.3-1.7	bal Fe				
Europe															
EN 10016/4(94)	1.1162	Rod	0.44-0.48	0.10 max	0.15 max	0.50-0.70	0.10 max	0.020 max	0.025 max	0.10-0.30	Al <=0.01; Mo <=0.03; N <=0.007; Cr+Ni+Cu<=0.30, Cu+Sn<=0.15; bal Fe				
EN 10016/4(94)	C46D2	Rod	0.44-0.48	0.10 max	0.15 max	0.50-0.70	0.10 max	0.020 max	0.025 max	0.10-0.30	Al <=0.01; Mo <=0.03; N <=0.007; Cr+Ni+Cu<=0.30, Cu+Sn<=0.15; bal Fe				
EN 10083/1(91)A1(96)	1.1191	Q/T t<=16mm	0.42-0.50	0.40 max		0.50-0.80	0.40 max	0.035 max	0.035 max	0.40 max	Mo <=0.10; Cr+Mo+Ni<=0.63; bal Fe	700-850	490	14	
EN 10083/1(91)A1(96)	1.1201	Q/T t<=16mm	0.42-0.50	0.40 max		0.50-0.80	0.40 max	0.035 max	0.020-0.040	0.40 max	Mo <=0.10; Cr+Mo+Ni<=0.63; bal Fe	700-850	490	14	
EN 10083/1(91)A1(96)	C45E	Q/T t<=16mm	0.42-0.50	0.40 max		0.50-0.80	0.40 max	0.035 max	0.035 max	0.40 max	Mo <=0.10; Cr+Mo+Ni<=0.63; bal Fe	700-850	490	14	
EN 10083/1(91)A1(96)	C45R	Q/T t<=16mm	0.42-0.50	0.40 max		0.50-0.80	0.40 max	0.035 max	0.020-0.040	0.40 max	Mo <=0.10; Cr+Mo+Ni<=0.63; bal Fe	700-850	490	14	
EN 10083/2(91)A1(96)	1.0503	Q/T t<=16mm	0.42-0.50	0.40 max		0.50-0.80	0.40 max	0.045 max	0.045 max	0.40 max	Mo <=0.10; Cr+Mo+Ni<=0.63; bal Fe	700-850	490	14	

Carbon Steel, Nonresulfurized, 1045/1045H (Continued from previous page)

Specification	Designation	Notes	C	Cr	Cu	Mn	Ni	P	S	Si	Other	UTS	YS	El	Hard
Europe															
EN 10083/2(91)A1(96)	C45	Q/T t<=16mm	0.42-0.50	0.40 max		0.50-0.80	0.40 max	0.045 max	0.045 max	0.40 max	Mo <=0.10; Cr+Mo+Ni<=0.63; bal Fe	700-850	490	14	
Finland															
SFS 456(73)	SFS456	Q/T	0.43-0.5			0.5-0.8		0.035 max	0.035 max	0.15-0.4	bal Fe				
SFS M56	CO 45	Bar Frg Tube Q/A Tmp, 60mm	0.43-0.5			0.5-0.8		0.04	0.04	0.15-0.4	bal Fe				
France															
AFNOR	XC42TS		0.4-0.45			0.5-0.8		0.025	0.035	0.1-0.4	bal Fe				
AFNOR NFA33101(82)	AF65C45	t<=450mm; Norm	0.43-0.51	0.3 max	0.3 max	0.5-0.8	0.4 max	0.04 max	0.04 max	0.1-0.4	bal Fe	640-760	355	17	
AFNOR NFA35551(75)	CC45		0.4-0.5	0.3 max	0.3 max	0.5-0.8	0.4 max	0.04 max	0.04 max	0.1-0.4	bal Fe				
AFNOR NFA35552	XC48H1		0.45-0.51			0.5-0.8		0.03	0.035	0.15-0.35	bal Fe				
AFNOR NFA35553	XC45	Strp, Norm	0.42-0.48			0.5-0.8		0.04	0.04	0.1-0.35	bal Fe				
AFNOR NFA35554	XC48	Plt, HR, Norm, 16mm diam	0.45-0.51			0.5-0.8		0.035	0.035	0.1-0.4	bal Fe				
AFNOR NFA36102(93)	C45RR	HR; Strp; CR	0.42-0.48	0.3 max		0.5-0.8	0.40 max	0.025 max	0.02 max	0.15-0.35	Al <=0.03; Mo <=0.10; bal Fe				
Germany															
DIN 1654(89)	Cq45	CHd, ext, soft ann	0.42-0.50			0.50-0.80		0.035 max	0.035 max	0.40 max	bal Fe	590			
DIN 1654(89)	WNr 1.1192	CHd ext, soft ann	0.42-0.50			0.50-0.80		0.035 max	0.035 max	0.40 max	bal Fe	590			
DIN 17212(72)	C45G	HT 17-40mm	0.43-0.49			0.50-0.80		0.025 max	0.035 max	0.15-0.35	bal Fe	660-800	410	16	
DIN 17212(72)	Cf45	HT 17-40mm	0.43-0.49			0.50-0.80		0.025 max	0.035 max	0.15-0.35	bal Fe	660-800	410	16	
DIN 17212(72)	WNr 1.1193	HT 17-40mm	0.43-0.49			0.50-0.80		0.025 max	0.035 max	0.15-0.35	bal Fe	660-800	410	16	
DIN EN 10083(91)	CK45	Q/T 17-40mm	0.42-0.50	0.40 max		0.50-0.80	0.40 max	0.035 max	0.035 max	0.40 max	Mo <=0.10; Cr+Mo+Ni<=0.63; bal Fe	650-800	430	16	
DIN EN 10083(91)	Cm45	Q/T	0.42-0.50			0.50-0.80	0.40 max	0.035 max	0.020-0.040	0.40 max	Mo <=0.10; Cr+Mo+Ni<=0.63; bal Fe				
DIN EN 10083(91)	GS-Ck45	Q/T 17-40mm	0.42-0.50	0.40 max		0.50-0.80	0.40 max	0.035 max	0.035 max	0.40 max	Mo <=0.10; Cr+Mo+Ni<=0.63; bal Fe	650-800	430	16	
DIN EN 10083(91)	WNr 1.0503	Q/T 17-40mm	0.42-0.50	0.40 max		0.50-0.80	0.40 max	0.045 max	0.045 max	0.40 max	Mo <=0.10; Cr+Mo+Ni<=0.63; bal Fe	650	430	16	
DIN EN 10083(91)	WNr 1.1191	Q/T 17-40mm	0.42-0.50	0.40 max		0.50-0.80	0.40 max	0.035 max	0.035 max	0.40 max	Mo <=0.10; Cr+Mo+Ni<=0.63; bal Fe	650-800	430	16	
DIN EN 10083(91)	WNr 1.1201	Q/T	0.42-0.50	0.40 max		0.50-0.80	0.40 max	0.035 max	0.020-0.040	0.40 max	Mo <=0.10; Cr+Mo+Ni<=0.63; bal Fe				
DIN(Aviation Hdbk)	WNr 1.1194		0.42-0.50			0.50-0.80		0.035 max	0.035 max	0.15-0.35	bal Fe				
DIN(Military Hdbk)	CK46		0.42-0.50			0.50-0.80		0.025 max	0.025 max	0.15-0.35	bal Fe				
DIN(Military Hdbk)	WNr 1.1184		0.42-0.50			0.50-0.80		0.025 max	0.025 max	0.15-0.35	bal Fe				
Hungary															
MSZ 2752(89)	B1	Semifinished railway tire; Non-rimming; FF	0.42-0.48			0.7-1.2	0.3 max	0.04 max	0.04 max	0.2-0.5	Mo <=0.08; V <=0.05; P<=0.04/P<=0.035; S<=0.04/S<=0.035; bal Fe	600-720		12L	
MSZ 2752(89)	B1	Semifinished railway tire; Non-rimming; FF	0.42-0.48			0.7-1.2	0.3 max	0.04 max	0.04 max	0.2-0.5	Mo <=0.08; V <=0.05; P<=0.04/P<=0.035; S<=0.04/S<=0.035; bal Fe	750-850		14L	

UNS numbers and US grades are provided as a means of cross referencing chemically similar alloys. Exchangability is only possible after independent examination of specifications. Tensile properties are minimum or typical as specified. UTS and YS as MPa. El as %. See Appendix for list of abbreviations used in Notes. * indicates obsolete material.

Specification	Designation	Notes	C	Cr	Cu	Mn	Ni	P	S	Si	Other	UTS	YS	El	Hard

Carbon Steel, Nonresulfurized, 1045/1045H (Continued from previous page)

Hungary

Specification	Designation	Notes	C	Cr	Cu	Mn	Ni	P	S	Si	Other	UTS	YS	El	Hard
MSZ 4217(85)	C45	CR, Q/T or spring Strp; Q/T; 0-2mm; rolled, hard	0.42-0.5			0.5-0.8		0.035 max	0.035 max	0.4 max	bal Fe	680-1030			
MSZ 4217(85)	C45	CR, Q/T or spring Strp; Q/T; 0-2mm	0.42-0.5			0.5-0.8		0.035 max	0.035 max	0.4 max	bal Fe	980-1280		5L	
MSZ 61	C45EK		0.42-0.5			0.5-0.8		0.035	0.035	0.17-0.37	bal Fe				
MSZ 61(85)	C45	Q/T; 40<t<=100mm	0.42-0.5			0.5-0.8		0.035 max	0.035 max	0.4 max	bal Fe	630-780	370	17L	
MSZ 61(85)	C45	Q/T; t<=16mm	0.42-0.5			0.5-0.8		0.035 max	0.035 max	0.4 max	bal Fe	700-850	500	14L	
MSZ 61(85)	C45E	Q/T; 40<t<=100mm	0.42-0.5			0.5-0.8		0.035 max	0.02-0.035	0.4 max	bal Fe	630-780	370	17L	
MSZ 61(85)	C45E	Q/T; t<=16mm	0.42-0.5			0.5-0.8		0.035 max	0.02-0.035	0.4 max	bal Fe	700-850	500	14L	
MSZ 6251(87)	C45Z	CF; Q/T; 0-36mm; HR; HR/soft ann; drawn/soft ann; drawn/bright ann; soft ann/ground	0.42-0.5			0.5-0.8		0.035 max	0.035 max	0.4 max	bal Fe	0-600			
MSZ 6251(87)	C45Z	CF; Q/T; 0-36mm; drawn, half-hard	0.42-0.5			0.5-0.8		0.035 max	0.035 max	0.4 max	bal Fe	0-630			
MSZ 8270	Ao.55		0.41-0.5			0.4-0.9		0.05	0.05	0.2-0.42	bal Fe				
MSZ 8270	Ao.55F		0.41-0.5			0.4-0.9		0.05	0.05	0.2-0.42	bal Fe				
MSZ 8270	Ao.55FK		0.41-0.5			0.4-0.9		0.05	0.05	0.2-0.42	bal Fe				

India

Specification	Designation	Notes	C	Cr	Cu	Mn	Ni	P	S	Si	Other	UTS	YS	El	Hard
IS 1570/2(79)	45C8	Sh Plt Sect Shp Bar Bil Frg Tub Pip	0.4-0.5	0.3 max	0.3 max	0.6-0.9	0.4 max	0.055 max	0.055 max	0.6 max	Co <=0.1; Mo <=0.15; Pb <=0.15; W <=0.1; P:S Varies; bal Fe	630-710	346	15	
IS 1570/2(79)	C45	Sh Plt Sect Shp Bar Bil Frg Tub Pip	0.4-0.5	0.3 max	0.3 max	0.6-0.9	0.4 max	0.055 max	0.055 max	0.6 max	Co <=0.1; Mo <=0.15; Pb <=0.15; W <=0.1; P:S Varies; bal Fe	630-710	346	15	
IS 5517	C45	Bar, Norm, Ann or Hard, Tmp	0.4-0.5			0.6-0.9				0.1-0.35	bal Fe				

International

Specification	Designation	Notes	C	Cr	Cu	Mn	Ni	P	S	Si	Other	UTS	YS	El	Hard
ISO 683-1(87)	C45	Bar Frg Plt Wir rod, Blts, Slbs; Q/T, t<16mm	0.42-0.50			0.50-0.80		0.045 max	0.045 max	0.10-0.40	bal Fe	700-850	490		
ISO 683-1(87)	C45E4	Bar Frg Plt Wir rod, Blts, Slbs; Q/T, t<16mm	0.42-0.50			0.50-0.80		0.035 max	0.035 max	0.10-0.40	bal Fe	700-850	490		
ISO 683-1(87)	C45M2	Bar Frg Plt Wir rod, Blts, Slbs; Q/T, t<16mm	0.42-0.50			0.50-0.80		0.035 max	0.020-0.040	0.10-0.40	bal Fe	700-850	490		
ISO 683-12	3*		0.43-0.49			0.5-0.8		0.035	0.035	0.15-0.4	bal Fe				
ISO 683-12(72)	3*	F/IH	0.43-0.49	0.3 max	0.3 max	0.5-0.8	0.4 max	0.035 max	0.035 max	0.15-0.4	Al <=0.1; Co <=0.1; Mo <=0.15; Pb <=0.15; Ti <=0.05; V <=0.1; W <=0.1; bal Fe				
ISO 683-18(96)	C45	Bar Wir, CD, t<=5mm	0.42-0.50			0.50-0.80		0.045 max	0.045 max	0.10-0.40	bal Fe	750	610	4	207 HB
ISO 683-18(96)	C45E4	Bar Wir, CD, t<=5mm	0.42-0.50			0.50-0.80		0.035 max	0.035 max	0.10-0.40	bal Fe	750	610	4	207 HB
ISO 683-18(96)	C45M2	Bar Wir, CD, t<=5mm	0.42-0.50			0.50-0.80		0.035 max	0.020-0.040	0.10-0.40	bal Fe	750	610	4	207 HB
ISO R683-3(70)	C45ea	Bar Rod, Q/A, Tmp, 16mm diam	0.42-0.5	0.3 max	0.3 max	0.5-0.8	0.4 max	0.035 max	0.02-0.035	0.15-0.4	Al <=0.1; Co <=0.1; Mo <=0.15; Pb <=0.15; Ti <=0.05; V <=0.1; W <=0.1; bal Fe				
ISO R683-3(70)	C45eb	Bar Rod, Q/A, Tmp, 16mm diam	0.42-0.5			0.5-0.8		0.04	0.03-0.05	0.15-0.4	bal Fe				

UNS numbers and US grades are provided as a means of cross referencing chemically similar alloys. Exchangability is only possible after independent examination of specifications. Tensile properties are minimum or typical as specified. UTS and YS as MPa. El as %. See Appendix for list of abbreviations used in Notes. * indicates obsolete material.

Specification	Designation	Notes	C	Cr	Cu	Mn	Ni	P	S	Si	Other	UTS	YS	El	Hard

Carbon Steel, Nonresulfurized, 1045/1045H (Continued from previous page)

Italy

Specification	Designation	Notes	C	Cr	Cu	Mn	Ni	P	S	Si	Other	UTS	YS	El	Hard
UNI 5332	C45		0.42-0.5			0.4-0.9		0.035	0.035	0.15-0.4	bal Fe				
UNI 5598	ICD45		0.42-0.49			0.4-0.7		0.05	0.05	0.15-0.35	bal Fe				
UNI 5598(71)	1CD45	Wir rod	0.42-0.49			0.4-0.7		0.05 max	0.05 max	0.15-0.35	bal Fe				
UNI 5598(71)	3CD45	Wir rod	0.43-0.48			0.4-0.7		0.035 max	0.035 max	0.15-0.35	N <=0.012; bal Fe				
UNI 6403(86)	C45	Q/T	0.42-0.5			0.5-0.8		0.035 max	0.035 max	0.15-0.35	bal Fe				
UNI 7065(72)	C45	Strp	0.43-0.48			0.4-0.65		0.02 max	0.02 max	0.35 max	bal Fe				
UNI 7845(78)	C45	Q/T	0.42-0.5			0.5-0.8		0.035 max	0.035 max	0.15-0.4	Pb 0.15-0.3; bal Fe				
UNI 7874(79)	C45	Q/T	0.42-0.5			0.5-0.8		0.035 max	0.035 max	0.15-0.4	Pb 0.15-0.3; bal Fe				
UNI 8893(86)	C45	Spring	0.42-0.5			0.5-0.8		0.035 max	0.035 max	0.15-0.4	bal Fe				

Japan

Specification	Designation	Notes	C	Cr	Cu	Mn	Ni	P	S	Si	Other	UTS	YS	El	Hard
JIS G3311(88)	S45CM	CR Strp	0.42-0.48	0.20 max	0.30 max	0.60-0.90	0.20 max	0.030 max	0.035 max	0.15-0.35	Ni+Cr=0.35 max, bal Fe				170-260 HV
JIS G3507(91)	SWRCH45K	Wir rod FF	0.42-0.48			0.60-0.90		0.030 max	0.035 max	0.10-0.35	bal Fe				
JIS G4051(79)	S45C	Bar Rod Wir	0.42-0.48	0.20 max	0.30 max	0.60-0.90	0.20 max	0.030 max	0.035 max	0.15-0.35	Ni+Cr<=0.35; bal Fe				
JIS G4051(79)	S48C	Bar Rod Wir	0.45-0.51	0.20 max	0.30 max	0.60-0.90	0.20 max	0.030 max	0.035 max	0.15-0.35	Ni+Cr<=0.35; bal Fe				

Mexico

Specification	Designation	Notes	C	Cr	Cu	Mn	Ni	P	S	Si	Other	UTS	YS	El	Hard
DGN B-203	1045*	Tube, Obs	0.43-0.5			0.6-0.9		0.04	0.05		bal Fe				
NMX-B-301(86)	1045	Bar	0.43-0.50			0.60-0.90		0.040 max	0.050 max		bal Fe				
NMX-B-301(86)	1045H	Bar	0.42-0.51			0.50-1.00		0.040 max	0.050 max	0.15-0.30	bal Fe				

Pan America

Specification	Designation	Notes	C	Cr	Cu	Mn	Ni	P	S	Si	Other	UTS	YS	El	Hard
COPANT 330	1045	Bar	0.43-0.5			0.6-0.9		0.04	0.05		bal Fe				
COPANT 331	1045	Bar	0.43-0.5			0.6-0.9		0.04	0.04		bal Fe				
COPANT 333	1045	Wir rod	0.43-0.5			0.6-0.9		0.04	0.05		bal Fe				
COPANT 514	1045	Tube, Ann	0.43-0.5			0.6-0.9		0.04	0.05	0.15-0.3	bal Fe				
COPANT 514	1045	Tube, CF	0.43-0.5			0.6-0.9		0.04	0.05	0.15-0.3	bal Fe				
COPANT 514	1045	Tube, HF, 324mm diam	0.43-0.5			0.6-0.9		0.04	0.05	0.15-0.3	bal Fe				
COPANT 514	1045	Tube, Norm	0.43-0.5			0.6-0.9		0.04	0.05	0.15-0.3	bal Fe				

Poland

Specification	Designation	Notes	C	Cr	Cu	Mn	Ni	P	S	Si	Other	UTS	YS	El	Hard
PNH84019	45	Q/T; Hand tools	0.42-0.5	0.3 max		0.5-0.8	0.3 max	0.04 max	0.04 max	0.17-0.37	bal Fe				

Romania

Specification	Designation	Notes	C	Cr	Cu	Mn	Ni	P	S	Si	Other	UTS	YS	El	Hard
STAS 10677	OLC45CS		0.43-0.48	0.2	0.3	0.5-0.8	0.3	0.035	0.035	0.17-0.37	Ti 0.02; As 0.05; bal Fe				
STAS 2470	OLC45q		0.42-0.5	0.3	0.3	0.5-0.8	0.3	0.035	0.035	0.17-0.37	As 0.05; bal Fe				
STAS 600	OT60-1		0.4-0.5	0.3	0.3	0.4-0.8	0.3	0.04	0.045	0.25-0.5	bal Fe				
STAS 600	OT60-2		0.4-0.5	0.3	0.3	0.4-0.8	0.3	0.04	0.045	0.25-0.5	bal Fe				
STAS 600	OT60-3		0.4-0.5	0.3	0.3	0.4-0.8	0.3	0.04	0.045	0.25-0.5	bal Fe				
STAS 880(88)	OLC45	Q/T	0.42-0.5	0.3 max		0.5-0.8		0.04 max	0.045 max	0.17-0.37	Pb <=0.15; As<=0.05; bal Fe				
STAS 880(88)	OLC45S	Q/T	0.42-0.5	0.3 max		0.5-0.8		0.04 max	0.02-0.045	0.17-0.37	Pb <=0.15; As<=0.05; bal Fe				

UNS numbers and US grades are provided as a means of cross referencing chemically similar alloys. Exchangability is only possible after independent examination of specifications. Tensile properties are minimum or typical as specified. UTS and YS as MPa. El as %. See Appendix for list of abbreviations used in Notes. * indicates obsolete material.

Carbon Steel, Nonresulfurized, 1045/1045H (Continued from previous page)

Specification	Designation	Notes	C	Cr	Cu	Mn	Ni	P	S	Si	Other	UTS	YS	El	Hard
Romania															
STAS 880(88)	OLC45X	Q/T	0.42-0.5	0.3 max		0.5-0.8		0.035 max	0.035 max	0.17-0.37	Pb <=0.15; As<=0.05; bal Fe				
STAS 880(88)	OLC45XS	Q/T	0.42-0.5	0.3 max		0.5-0.8		0.035 max	0.02-0.04	0.17-0.37	Pb <=0.15; As<=0.05; bal Fe				
Russia															
GOST	KSt6		0.38-0.5			0.5-0.8		0.05	0.055	0.17-0.35	bal Fe				
GOST	KSt6ps		0.38-0.5			0.5-0.8		0.05	0.055	0.17	bal Fe				
GOST	OSL		0.42-0.5			0.6-0.9		0.04	0.05	0.15-0.35	bal Fe				
GOST	St6		0.38-0.5			0.5-0.8		0.05	0.05	0.17-0.4	bal Fe				
GOST 1050	45		0.42-0.5	0.25	0.25	0.5-0.8	0.25	0.035	0.04	0.17-0.37	bal Fe				
GOST 1050(88)	45	High-grade struct; Q/T; Full hard; 0-100mm	0.42-0.5	0.25 max		0.5-0.8	0.3 max	0.035 max	0.04 max	0.17-0.37	As<=0.08; bal Fe	640		6L	0-241 HB
GOST 1050(88)	45	High-grade struct; Q/T; Norm; 0-100mm	0.42-0.5	0.25 max		0.5-0.8	0.3 max	0.035 max	0.04 max	0.17-0.37	As<=0.08; bal Fe	600	355	16L	
GOST 977	45L-II		0.42-0.5	0.3	0.3	0.4-0.9	0.3	0.04	0.045	0.2-0.52	bal Fe				
GOST 977	45L-III		0.42-0.5	0.3	0.3	0.4-0.9	0.3	0.04	0.045	0.2-0.52	bal Fe				
GOST 977(88)	45L		0.42-0.5	0.3		0.4-0.9	0.3	0.05	0.05	0.2-0.52	V 0.06-0.15; bal Fe				
Spain															
UNE 36011(75)	C15k-1	CH	0.1-0.2			0.4-0.7		0.035 max	0.02-0.035	0.15-0.4	Mo <=0.15; bal Fe				
UNE 36011(75)	C42k-1	Unalloyed; Q/T	0.4-0.45			0.5-0.8		0.035 max	0.02-0.035	0.15-0.4	Mo <=0.15; bal Fe				
UNE 36011(75)	C45k	Unalloyed; Q/T	0.4-0.5			0.5-0.8		0.035 max	0.035 max	0.15-0.4	Mo <=0.15; bal Fe				
UNE 36011(75)	C45k-1	Unalloyed; Q/T	0.4-0.5			0.5-0.8		0.035 max	0.02-0.035	0.15-0.4	Mo <=0.15; bal Fe				
UNE 36011(75)	C48k	Unalloyed; Q/T	0.45-0.5			0.5-0.8		0.035 max	0.035 max	0.15-0.4	Mo <=0.15; bal Fe				
UNE 36011(75)	C48k-1	Unalloyed; Q/T	0.45-0.5			0.5-0.8		0.035 max	0.02-0.035	0.15-0.4	Mo <=0.15; bal Fe				
UNE 36011(75)	F.1115*	see C15k-1	0.1-0.2			0.4-0.7		0.035 max	0.02-0.035	0.15-0.4	Mo <=0.15; bal Fe				
UNE 36011(75)	F.1145*	see C45k-1	0.4-0.5			0.5-0.8		0.035 max	0.02-0.035	0.15-0.4	Mo <=0.15; bal Fe				
UNE 36011(75)	F.1147*	see C48k-1	0.45-0.5			0.5-0.8		0.035 max	0.02-0.035	0.15-0.4	bal Fe				
Sweden															
SIS 141672	1672-01	Bar Frg Plt Sh, as rolled	0.43-0.5			0.5-0.8		0.04	0.04	0.15-0.4	bal Fe	590	326	16	
SIS 141672	1672-03	Bar Frg Plt Sh, Q/A, Tmp	0.43-0.5			0.5-0.8		0.04	0.04	0.15-0.4	bal Fe	620	370	17	
SIS 141672	1672-04	Bar Frg Plt Sh, Q/A, Tmp	0.43-0.5			0.5-0.8		0.04	0.04	0.15-0.4	bal Fe	660	370	16	
SIS 141672	1672-05	Bar Frg Plt Sh, Q/A, Tmp	0.43-0.5			0.5-0.8		0.04	0.04	0.15-0.4	bal Fe	700	480	14	
SS 141650	1650	Non-rimming; Q/T	0.38-0.5			0.4-0.9		0.05 max	0.05 max	0.6 max	bal Fe				
SS 141660	1660	Q/T	0.42-0.49			0.3-0.5		0.03 max	0.03 max	0.15-0.35	bal Fe				
SS 141672	1672	Q/T	0.42-0.5			0.8 max		0.035 max	0.035 max	0.1-0.4	bal Fe				
Switzerland															
VSM 10648	C45	Bar Sh Plt, as roll 5/16mm	0.42-0.5			0.5-0.8		0.05	0.05	0.15-0.35	bal Fe				
VSM 10648	C45	Bar Sh Plt, CD, 5mm diam	0.42-0.5			0.5-0.8		0.05	0.05	0.15-0.35	bal Fe				
VSM 10648	C45	Bar Sh Plt, CD, Norm 16/100mm	0.42-0.5			0.5-0.8		0.05	0.05	0.15-0.35	bal Fe				

UNS numbers and US grades are provided as a means of cross referencing chemically similar alloys. Exchangability is only possible after independent examination of specifications. Tensile properties are minimum or typical as specified. UTS and YS as MPa. El as %. See Appendix for list of abbreviations used in Notes. * indicates obsolete material.

Specification	Designation	Notes	C	Cr	Cu	Mn	Ni	P	S	Si	Other	UTS	YS	El	Hard

Carbon Steel, Nonresulfurized, 1045/1045H (Continued from previous page)

UK

Specification	Designation	Notes	C	Cr	Cu	Mn	Ni	P	S	Si	Other	UTS	YS	El	Hard
BS 1449/1(91)	50CS	Plt Sht Strp; CD norm t<=16mm	0.45-0.55			0.50-0.90		0.045 max	0.045 max	0.05-0.35	bal Fe				0-219 HB
BS 1449/1(91)	50HS	Plt Sh Strp HR; t<=16mm	0.45-0.55			0.50-0.90		0.045 max	0.045 max	0.05-0.35	bal Fe				0-219 HB
BS 970/1(83)	080A47	Blm Bil Slab Bar Rod Frg	0.45-0.50			0.70-0.90		0.050 max	0.050 max	0.1-0.4	bal Fe				
BS 970/1(83)	080M46	Blm Bil Slab Bar Rod Frg Hard Tmp; 4 in.	0.42-0.50			0.60-1.00		0.050 max	0.050 max		bal Fe				

USA

Specification	Designation	Notes	C	Cr	Cu	Mn	Ni	P	S	Si	Other	UTS	YS	El	Hard
	1045	CD	0.43-0.5			0.6-0.9		0.04 max	0.05 max		bal Fe	825	690	10.0	241-321 HB
	AISI 1045	Struct shp	0.43-0.5			0.60-0.90		0.030 max	0.035 max		bal Fe				
	AISI 1045	Bar	0.43-0.50			0.60-0.90		0.040 max	0.050 max		bal Fe				
	AISI 1045	Wir rod	0.43-0.50			0.60-0.90		0.040 max	0.050 max		bal Fe				
	AISI 1045	Sh Strp Plt	0.42-0.50			0.60-0.90		0.030 max	0.035 max		bal Fe				
	AISI 1045H	Wir rod, Hard	0.42-0.51			0.50-1.00		0.040 max	0.050 max	0.15-0.35	bal Fe				
	UNS G10450		0.43-0.50			0.60-0.90		0.040 max	0.050 max		Sheets, C 0.42-0.50; bal Fe				
	UNS H10450		0.42-0.51			0.50-1.00		0.040 max	0.050 max	0.15-0.35	bal Fe				
ASTM A108(95)	1045	Bar, CF	0.43-0.50			0.60-0.90		0.040 max	0.050 max		bal Fe				
ASTM A183(98)	Grade 2 Nut	Rail Nuts	0.40-0.55					0.05 max	0.06 max		bal Fe	760	550	12	
ASTM A29/A29M(93)	1045	Bar	0.43-0.50			0.60-0.90		0.040 max	0.050 max		bal Fe				
ASTM A304(96)	1045H	Bar, hard bands spec	0.42-0.51			0.50-1.00		0.040 max	0.050 max	0.15-0.30	Cu Ni Cr Mo trace allowed; bal Fe				
ASTM A311/A311M(95)	1045 Class A	Bar CD SR ann, 50<t<=75mm	0.43-0.50			0.60-0.90		0.040 max	0.050 max		bal Fe	550	485	10,RA=30	
ASTM A311/A311M(95)	1045 Class A	Bar CD SR ann, 20<t<=30mm	0.43-0.50			0.60-0.90		0.040 max	0.050 max		bal Fe	620	550	11,RA=30	
ASTM A311/A311M(95)	1045 Class A	Bar CD SR ann, 30<t<=50mm	0.43-0.50			0.60-0.90		0.040 max	0.050 max		bal Fe	585	520	10,RA=30	
ASTM A311/A311M(95)	1045 Class A	Bar CD SR ann, t<=20mm	0.43-0.50			0.60-0.90		0.040 max	0.050 max		bal Fe	655	585	12,RA=35	
ASTM A311/A311M(95)	1045 Class B	Bar CD SR ann, 30<t<=50mm	0.43-0.50			0.60-0.90		0.040 max	0.050 max		bal Fe	795	690	9,RA=25	
ASTM A311/A311M(95)	1045 Class B	Bar CD SR ann, t<=57mm	0.43-0.50			0.60-0.90		0.040 max	0.050 max		bal Fe	795	690	10,RA=25	
ASTM A311/A311M(95)	1045 Class B	Bar CD SR ann, 75<t<=102mm	0.43-0.50			0.60-0.90		0.040 max	0.050 max		bal Fe	725	620	7,RA=20	
ASTM A510(96)	1045	Wir rod	0.43-0.5			0.60-0.90		0.040 max	0.050 max		bal Fe				
ASTM A519(96)	1045	Smls mech tub, HR	0.43-0.50			0.60-0.90		0.040 max	0.050 max		bal Fe	517	310	15	80 HRB
ASTM A519(96)	1045	Smls mech tub, Ann	0.43-0.50			0.60-0.90		0.040 max	0.050 max		bal Fe	448	241	20	72 HRB
ASTM A519(96)	1045	Smls mech tub, SR	0.43-0.50			0.60-0.90		0.040 max	0.050 max		bal Fe	552	483	8	85 HRB
ASTM A519(96)	1045	Smls mech tub, CW	0.43-0.50			0.60-0.90		0.040 max	0.050 max		bal Fe	621	552	5	90 HRB
ASTM A519(96)	1045	Smls mech tub, Norm	0.43-0.50			0.60-0.90		0.040 max	0.050 max		bal Fe	517	331	15	80 HRB
ASTM A576(95)	1045	Special qual HW bar	0.43-0.50			0.60-0.90		0.040 max	0.050 max		Si Cu Pb B Bi Ca Se Te if spec'd; bal Fe				
ASTM A682/A682M(98)	1045	Strp, high-C, CR, spring	0.42-0.5			0.60-0.90		0.035 max	0.040 max	0.15-0.30	bal Fe				
ASTM A684A684M(86)	1045	Strp, high-C, CR, Ann	0.42-0.5			0.60-0.90		0.035 max	0.040 max	0.15-0.30	bal Fe				85 HRB
ASTM A827(98)	1045	Plt, frg	0.42-0.50			0.60-0.90		0.035 max	0.040 max	0.15-0.40	bal Fe				

Carbon Steel, Nonresulfurized, 1045/1045H (Continued from previous page)

Specification	Designation	Notes	C	Cr	Cu	Mn	Ni	P	S	Si	Other	UTS	YS	El	Hard
USA															
ASTM A830/A830M(98)	1045	Plt	0.42-0.50			0.60-0.90		0.035 max	0.04 max	0.15-0.40	bal Fe				
FED QQ-S-635B(88)	C1045*	Obs; see ASTM A827; Plt	0.43-0.5			0.6-0.9		0.04 max	0.05 max	*	bal Fe				
FED QQ-S-700D(91)	1045*	Obs; Sh Strp	0.43-0.5			0.6-0.9		0.04 max	0.05 max		bal Fe				
FED QQ-W-461H(88)	1045*	Obs; Wir	0.43-0.5			0.6-0.9		0.04 max	0.05 max		bal Fe				
MIL-S-24093A(SH)(91)	Type IV Class F	Frg for shipboard apps	0.44 max			0.90 max		0.04 max	0.04 max	0.10-0.30	bal Fe	621 max	310	22	
MIL-S-24093A(SH)(91)	Type IV Class G	Frg for shipboard apps	0.44 max			0.90 max		0.04 max	0.04 max	0.10-0.30	bal Fe	621 max	276	22	
MIL-S-3039C(MR)(88)		Spheroid ann Strp for ammunition cartridge clips	0.43-0.50			0.45-0.75		0.030 max	0.035 max	0.15-0.35	bal Fe				
MIL-S-46070	1045*	Obs for new design	0.43-0.5			0.6-0.9		0.04	0.05	0.3	bal Fe				
SAE J1268(95)	1045H	Hard see std	0.42-0.51			0.50-1.00		0.030 max	0.050 max	0.15-0.35	bal Fe				
SAE J1397(92)	1045	Bar CD, est mech prop	0.43-0.5			0.6-0.9		0.04 max	0.05 max		bal Fe	630	530	12	179 HB
SAE J1397(92)	1045	Bar HR, est mech prop	0.43-0.5			0.6-0.9		0.04 max	0.05 max		bal Fe	570	310	16	163 HB
SAE J1397(92)	1045	Bar Ann CD, est mech prop	0.43-0.5			0.6-0.9		0.04 max	0.05 max		bal Fe	590	500	12	170 HB
SAE J403(95)	1045	Bar Wir rod Smls Tub HR CF	0.43-0.50			0.60-0.90		0.030 max	0.050 max		bal Fe				
SAE J403(95)	1045	Struct Shps Plt Strp Sh Weld Tub	0.42-0.50			0.60-0.90		0.030 max	0.035 max		bal Fe				
Yugoslavia															
	C.0645	Struct	0.4			1.60 max		0.05 max	0.05 max	0.60 max	bal Fe				
	C.1501	Struct	0.42-0.48			0.5-0.8		0.06 max	0.06 max	0.35	bal Fe				
	C.1530	Q/T	0.42-0.5			0.5-0.8		0.045 max	0.045 max	0.4 max	Mo <=0.15; bal Fe				
	C.1531	Q/T	0.42-0.5			0.5-0.8		0.035 max	0.03 max	0.4 max	Mo <=0.15; bal Fe				
	C.1534	F/IH	0.43-0.49	0.2 max		0.5-0.8	0.25 max	0.025 max	0.035 max	0.15-0.35	Cr+Mo+Ni<=0.45; bal Fe				
	C.1580	Q/T	0.42-0.5			0.5-0.8		0.035 max	0.02-0.035	0.4 max	Mo <=0.15; bal Fe				
	C.2131	Spring	0.45-0.5	0.4 max		0.5-0.8		0.05 max	0.05 max	1.5-1.8	Mo <=0.15; bal Fe				

Carbon Steel, Nonresulfurized, 1046

Specification	Designation	Notes	C	Cr	Cu	Mn	Ni	P	S	Si	Other	UTS	YS	El	Hard
Australia															
AS 1442	K1046*	Obs; Bar Bil	0.43-0.5			0.7-1		0.05	0.05	0.1-0.35	bal Fe				
AS 1443	K1046*	Obs; Bar	0.43-0.5			0.7-1		0.05	0.05	0.1-0.35	bal Fe				
Bulgaria															
BDS 5785(83)	45G	Struct	0.42-0.50	0.25 max	0.25 max	0.70-1.00	0.25 max	0.040 max	0.040 max	0.17-0.37	bal Fe				
China															
GB 13795(92)	45 Mn	Bar HR Ann	0.42-0.50	0.25 max	0.25 max	0.70-1.00	0.25 max	0.035 max	0.035 max	0.17-0.37	bal Fe	400-700		15	
GB 6478(86)	ML 45 Mn	Bar HR Norm	0.42-0.50	0.20 max	0.20 max	0.50-0.80	0.20 max	0.035 max	0.035 max	0.25 max	bal Fe	600	355	16	
GB 699(88)	45 Mn	Bar HR Q/T 25mm diam	0.42-0.50	0.25 max	0.25 max	0.70-1.00	0.25 max	0.035 max	0.035 max	0.17-0.37	bal Fe	620	375	15	
GB/T 3078(94)	45 Mn	Bar CR Q/T 25mm diam	0.42-0.50	0.25 max	0.25 max	0.70-1.00	0.25 max	0.035 max	0.035 max	0.17-0.37	bal Fe	620	375	15	
Europe															
EN 10016/2(94)	1.0517	Rod	0.45-0.50	0.15 max	0.25 max	0.50-0.80	0.20 max	0.035 max	0.035 max	0.10-0.30	Al <=0.01; Mo <=0.05; Cu+Sn<=0.25: bal Fe				

UNS numbers and US grades are provided as a means of cross referencing chemically similar alloys. Exchangability is only possible after independent examination of specifications. Tensile properties are minimum or typical as specified. UTS and YS as MPa. El as %. See Appendix for list of abbreviations used in Notes. * indicates obsolete material.

Specification	Designation	Notes	C	Cr	Cu	Mn	Ni	P	S	Si	Other	UTS	YS	El	Hard

Carbon Steel, Nonresulfurized, 1046 (Continued from previous page)

Europe

Specification	Designation	Notes	C	Cr	Cu	Mn	Ni	P	S	Si	Other	UTS	YS	El	Hard
EN 10016/2(94)	C48D	Rod	0.45-0.50	0.15 max	0.25 max	0.50-0.80	0.20 max	0.035 max	0.035 max	0.10-0.30	Al <=0.01; Mo <=0.05; Cu+Sn<=0.25; bal Fe				
EN 10016/4(94)	1.1164	Rod	0.46-0.50	0.10 max	0.15 max	0.50-0.70	0.10 max	0.020 max	0.025 max	0.10-0.30	Al <=0.01; Mo <=0.03; N <=0.007; Cr+Ni+Cu<=0.30, Cu+Sn<=0.15; bal Fe				
EN 10016/4(94)	C48D2	Rod	0.46-0.50	0.10 max	0.15 max	0.50-0.70	0.10 max	0.020 max	0.025 max	0.10-0.30	Al <=0.01; Mo <=0.03; N <=0.007; Cr+Ni+Cu<=0.30, Cu+Sn<=0.15; bal Fe				

Finland

Specification	Designation	Notes	C	Cr	Cu	Mn	Ni	P	S	Si	Other	UTS	YS	El	Hard
SFS 200:E	Fe70		0.5					0.05	0.05		N 0.009; bal Fe				

France

Specification	Designation	Notes	C	Cr	Cu	Mn	Ni	P	S	Si	Other	UTS	YS	El	Hard
AFNOR	45MATS		0.43-0.49			0.8-1.1		0.035	0.035	0.1-0.4	bal Fe				
AFNOR NFA35552	XC48H2		0.45-0.51			1.2		0.035	0.035	0.15-0.35	bal Fe				
AFNOR NFA35552	XC48HI		0.45-0.51			0.5-0.8		0.03	0.035	0.15-0.35	bal Fe				
AFNOR NFA35552	XC48TS		0.45-0.51			0.5-0.8		0.025	0.03	0.1-0.4	bal Fe				

Germany

Specification	Designation	Notes	C	Cr	Cu	Mn	Ni	P	S	Si	Other	UTS	YS	El	Hard
DIN	46MnSi4	Q/T 17-40mm	0.42-0.50			0.90-1.20		0.035 max	0.035 max	0.70-0.90	bal Fe	930-1130	735	12	
DIN	WNr 1.5121	Q/T 101-160mm	0.42-0.50			0.90-1.20		0.035 max	0.035 max	0.70-0.90	bal Fe	640-780	490	15	
DIN	WNr 1.5121	Q/T 17-40mm	0.42-0.50			0.90-1.20		0.035 max	0.035 max	0.70-0.90	bal Fe	930-1130	735	12	

Italy

Specification	Designation	Notes	C	Cr	Cu	Mn	Ni	P	S	Si	Other	UTS	YS	El	Hard
UNI 7065(72)	C46	Strp	0.42-0.5			0.5-0.9		0.035 max	0.035 max	0.4 max	bal Fe				
UNI 7847(79)	C46		0.42-0.5			0.5-0.9		0.035	0.035	0.4	bal Fe				
UNI 8551(84)	C46	Surf Hard	0.43-0.49			0.5-0.8		0.03 max	0.025 max	0.15-0.4	Pb 0.15-0.25; bal Fe				

Mexico

Specification	Designation	Notes	C	Cr	Cu	Mn	Ni	P	S	Si	Other	UTS	YS	El	Hard
NMX-B-301(86)	1046	Bar	0.43-0.50			0.70-1.00		0.040 max	0.050 max		bal Fe				

Norway

Specification	Designation	Notes	C	Cr	Cu	Mn	Ni	P	S	Si	Other	UTS	YS	El	Hard
NS 13205	St70-2		0.5					0.05	0.05		N 0.007; bal Fe				

Pan America

Specification	Designation	Notes	C	Cr	Cu	Mn	Ni	P	S	Si	Other	UTS	YS	El	Hard
COPANT 333	1046	Wir rod	0.43-0.5			0.7-1		0.04	0.05	0.1-0.2	bal Fe				

Poland

Specification	Designation	Notes	C	Cr	Cu	Mn	Ni	P	S	Si	Other	UTS	YS	El	Hard
PNH84019	45G	Q/T	0.42-0.5	0.3 max		0.5-0.8	0.3 max	0.04 max	0.04 max	0.17-0.37	bal Fe				

Romania

Specification	Designation	Notes	C	Cr	Cu	Mn	Ni	P	S	Si	Other	UTS	YS	El	Hard
STAS 8183	OLT65		0.4-0.5	0.3	0.3	0.7-1	0.3	0.04	0.045	0.17-0.37	Mo 0.06; bal Fe				
STAS 8185	OLT65		0.4-0.5	0.3	0.3	0.7-1	0.3	0.04	0.045	0.17-0.37	Mo 0.06; bal Fe				

Russia

Specification	Designation	Notes	C	Cr	Cu	Mn	Ni	P	S	Si	Other	UTS	YS	El	Hard
GOST 4543	45G		0.42-0.5	0.3	0.3	0.7-1	0.3	0.035	0.035	0.17-0.37	bal Fe				
GOST 4543(71)	45G	Q/T	0.42-0.5	0.3 max	0.3 max	0.7-1.0	0.3 max	0.035 max	0.035 max	0.17-0.37	Al <=0.1; bal Fe				

Turkey

Specification	Designation	Notes	C	Cr	Cu	Mn	Ni	P	S	Si	Other	UTS	YS	El	Hard
TS 911(86)	UDK699.14.423/Fe70	HR, T-bar	0.5					0.055	0.055		N 0.008; bal Fe				

UK

Specification	Designation	Notes	C	Cr	Cu	Mn	Ni	P	S	Si	Other	UTS	YS	El	Hard
BS 970/1(83)	060A47	Blm Bil Slab Bar Rod Frg	0.45-0.50			0.50-0.70		0.050 max	0.050 max	0.1-0.4	bal Fe				

USA

Specification	Designation	Notes	C	Cr	Cu	Mn	Ni	P	S	Si	Other	UTS	YS	El	Hard
	AISI 1046	Struct shp	0.43-0.5			0.70-1.00		0.030 max	0.035 max		bal Fe				

UNS numbers and US grades are provided as a means of cross referencing chemically similar alloys. Exchangability is only possible after independent examination of specifications. Tensile properties are minimum or typical as specified. UTS and YS as MPa. El as %. See Appendix for list of abbreviations used in Notes. * indicates obsolete material.

Specification	Designation	Notes	C	Cr	Cu	Mn	Ni	P	S	Si	Other	UTS	YS	El	Hard

Carbon Steel, Nonresulfurized, 1046 (Continued from previous page)

USA

Specification	Designation	Notes	C	Cr	Cu	Mn	Ni	P	S	Si	Other	UTS	YS	El	Hard
	AISI 1046	Bar	0.43-0.50			0.70-1.00		0.040 max	0.050 max		bal Fe				
	AISI 1046	Sh Strp Plt	0.42-0.50			0.70-1.00		0.030 max	0.035 max		bal Fe				
	AISI 1046	Wir rod	0.43-0.50			0.70-1.00		0.040 max	0.050 max		bal Fe				
	UNS G10460		0.43-0.50			0.70-1.00		0.040 max	0.050 max		bal Fe				
ASTM A29/A29M(93)	1046	Bar	0.43-0.50			0.70-1.00		0.040 max	0.050 max		bal Fe				
ASTM A510(96)	1046	Wir rod	0.43-0.5			0.70-1.00		0.040 max	0.050 max		bal Fe				
ASTM A576(95)	1046	Special qual HW bar	0.43-0.50			0.70-1.00		0.040 max	0.050 max		Si Cu Pb B Bi Ca Se Te if spec'd; bal Fe				
ASTM A830/A830M(98)	1046	Plt	0.42-0.50			0.70-1.00		0.035 max	0.04 max	0.15-0.40	bal Fe				
ASTM A983/983M(98)	Grade 1	Continuous grain flow frg for crankshafts	043-0.53			0.60-1.10		0.025 max	0.025 max	0.15-0.40	V <=0.10; bal Fe				
SAE J1397(92)	1046	Bar CD, est mech prop	0.43-0.5			0.7-1		0.04 max	0.05 max		bal Fe	650	540	12	187 HB
SAE J1397(92)	1046	Bar Ann CD, est mech prop	0.43-0.5			0.7-1		0.04 max	0.05 max		bal Fe	620	520	12	179 HB
SAE J1397(92)	1046	Bar HR, est mech prop	0.43-0.5			0.7-1		0.04 max	0.05 max		bal Fe	590	320	15	170 HB
SAE J403(95)	1046	Bar Wir rod Smls Tub HR CF	0.43-0.50			0.70-1.00		0.030 max	0.050 max		bal Fe				
SAE J403(95)	1046	Struct Shps Plt Strp Sh Weld Tub	0.42-0.50			0.70-1.00		0.030 max	0.035 max		bal Fe				

Carbon Steel, Nonresulfurized, 1049

France

Specification	Designation	Notes	C	Cr	Cu	Mn	Ni	P	S	Si	Other	UTS	YS	El	Hard
AFNOR NFA35552	XC48HI		0.45-0.51			0.5-0.8		0.03	0.035	0.15-0.35	bal Fe				
AFNOR NFA35554	XC48	Plt, HR, Norm, 16mm diam	0.45-0.51			0.5-0.8		0.04	0.04	0.1-0.4	bal Fe				
AFNOR NFA35565(94)	C48E3	Ball & roller bearing	0.45-0.52	0.25 max	0.3 max	0.5-0.9	0.25 max	0.025 max	0.015 max	0.15-0.35	Al <=0.05; Mo <=0.1; bal Fe				
AFNOR NFA35565(94)	XC48*	see C48E3	0.45-0.52	0.25 max	0.3 max	0.5-0.9	0.25 max	0.025 max	0.015 max	0.15-0.35	Al <=0.05; Mo <=0.1; bal Fe				

India

Specification	Designation	Notes	C	Cr	Cu	Mn	Ni	P	S	Si	Other	UTS	YS	El	Hard
IS 1570	C50		0.45-0.55			0.6-0.9				0.1-0.35	bal Fe				
IS 5517	C50	Bar, Norm, Ann or Hard, Tmp, 30mm	0.45-0.55			0.6-0.9				0.1-0.35	bal Fe				

Italy

Specification	Designation	Notes	C	Cr	Cu	Mn	Ni	P	S	Si	Other	UTS	YS	El	Hard
UNI 6403(86)	C48	Bar	0.45-0.52	0.25	0.25	0.6-0.9	0.25			0.4	bal Fe				
UNI 7847(79)	C48	Surf Hard	0.45-0.52			0.5-0.8		0.03 max	0.03 max	0.15-0.4	bal Fe				
UNI 8551(84)	C48	Surf Hard	0.45-0.52			0.5-0.8		0.03 max	0.025 max	0.15-0.4	Pb 0.15-0.25; bal Fe				

Japan

Specification	Designation	Notes	C	Cr	Cu	Mn	Ni	P	S	Si	Other	UTS	YS	El	Hard
JIS G3311(88)	S50CM	CR Strp	0.47-0.53	0.20 max	0.30 max	0.60-0.90	0.20 max	0.030 max	0.035 max	0.15-0.35	Ni+Cr=0.35 max; bal Fe				180-270 HV
JIS G3445(88)	STKM17A	Tube	0.45-0.55			0.40-1.00		0.040 max	0.040 max	0.040 max	bal Fe	549	343	20L	
JIS G3445(88)	STKM17C	Tube	0.45-0.55			0.40-1.00		0.040 max	0.040 max	0.040	bal Fe	647	481	10L	
JIS G3507(91)	SWRCH48K	Wir rod FF	0.45-0.51			0.60-0.90		0.030 max	0.035 max	0.10-0.35	bal Fe				
JIS G3507(91)	SWRCH50K	Wir rod FF	0.47-0.53			0.60-0.90		0.030 max	0.035 max	0.10-0.35	bal Fe				

Mexico

Specification	Designation	Notes	C	Cr	Cu	Mn	Ni	P	S	Si	Other	UTS	YS	El	Hard
NMX-B-301(86)	1049	Bar	0.46-0.53			0.60-0.90		0.040 max	0.050 max		bal Fe				

Specification	Designation	Notes	C	Cr	Cu	Mn	Ni	P	S	Si	Other	UTS	YS	El	Hard

Carbon Steel, Nonresulfurized, 1049 (Continued from previous page)

Poland

Specification	Designation	Notes	C	Cr	Cu	Mn	Ni	P	S	Si	Other	UTS	YS	El	Hard
PNH84023/07	R65		0.45-0.52	0.3 max		0.6-0.85	0.4 max	0.045 max	0.045 max	0.2-0.35	bal Fe				
PNH84028	D50	Wir rod	0.46-0.53	0.15	0.2	0.3-0.6	0.2	0.035	0.035	0.35	P+S+.06;bal Fe				

Russia

Specification	Designation	Notes	C	Cr	Cu	Mn	Ni	P	S	Si	Other	UTS	YS	El	Hard
GOST 10543	Np-50G		0.45-0.55	0.25		0.7-1	0.25	0.04	0.04	0.17-0.37	bal Fe				

Spain

Specification	Designation	Notes	C	Cr	Cu	Mn	Ni	P	S	Si	Other	UTS	YS	El	Hard
UNE 36011(75)	C48k		0.45-0.5			0.5-0.8		0.035	0.035	0.15-0.4	bal Fe				
UNE 36011(75)	C48k-1		0.45-0.5			0.5-0.8		0.035	0.02-0.035	0.15-0.4	bal Fe				
UNE 36011(75)	F.1142*	see C48k	0.45-0.5			0.5-0.8		0.035	0.035	0.15-0.4	bal Fe				

USA

Specification	Designation	Notes	C	Cr	Cu	Mn	Ni	P	S	Si	Other	UTS	YS	El	Hard
	AISI 1049	Struct shp	0.46-0.53			0.60-0.90		0.030 max	0.035 max		bal Fe				
	AISI 1049	Bar	0.46-0.53			0.60-0.90		0.040 max	0.050 max		bal Fe				
	AISI 1049	Sh Strp Plt	0.45-0.53			0.60-0.90		0.030 max	0.035 max		bal Fe				
	AISI 1049	Wir rod	0.46-0.53			0.60-0.90		0.040 max	0.050 max		bal Fe				
	UNS G10490		0.46-0.53			0.60-0.90		0.040 max	0.050 max		bal Fe				
ASTM A29/A29M(93)	1049	Bar	0.46-0.53			0.60-0.90		0.040 max	0.050 max		bal Fe				
ASTM A510(96)	1049	Wir rod	0.46-0.53			0.60-0.90		0.040 max	0.050 max		bal Fe				
ASTM A576(95)	1049	Special qual HW bar	0.46-0.53			0.60-0.90		0.040 max	0.050 max		Si Cu Pb B Bi Ca Se Te if spec'd; bal Fe				
ASTM A830/A830M(98)	1049	Plt	0.45-0.53			0.60-0.90		0.035 max	0.04 max	0.15-0.40	bal Fe				
SAE J1397(92)	1049	Bar HR, est mech prop	0.46-0.53			0.6-0.9		0.04 max	0.05 max		bal Fe	600	330	15	179 HB
SAE J1397(92)	1049	Bar CD, est mech prop	0.46-0.53			0.6-0.9		0.04 max	0.05 max		bal Fe	670	560	10	197 HB
SAE J1397(92)	1049	Bar Ann CD, est mech prop	0.46-0.53			0.6-0.9		0.04 max	0.05 max		bal Fe	630	530	10	187 HB
SAE J403(95)	1049	Struct Shps Plt Strp Sh Weld Tub	0.45-0.53			0.60-0.90		0.030 max	0.035 max		bal Fe				
SAE J403(95)	1049	Bar Wir rod Smls Tub HR CF	0.46-0.53			0.60-0.90		0.030 max	0.050 max		bal Fe				

Carbon Steel, Nonresulfurized, 1050

Argentina

Specification	Designation	Notes	C	Cr	Cu	Mn	Ni	P	S	Si	Other	UTS	YS	El	Hard
IAS	IRAM 1050		0.48-0.55			0.60-0.90		0.040 max	0.050 max	0.10-0.30	bal Fe	700-820	420-490	15-22	212-248 HB

Australia

Specification	Designation	Notes	C	Cr	Cu	Mn	Ni	P	S	Si	Other	UTS	YS	El	Hard
AS 1442	K1050*	Obs; Bar Bil	0.48-0.55			0.6-0.9		0.05	0.05	0.1-0.35	bal Fe				
AS 1442	S1050*	Obs; Bar Bil	0.48-0.55			0.6-0.9		0.05	0.05	0.35	bal Fe				
AS 1442(92)	1050	HR bar, Semifinished (may treat w/microalloying elements)	0.48-0.55			0.60-0.90		0.040 max	0.040 max	0.10-0.35	Si<=0.10 for Al-killed; bal Fe				
AS 1443	K1050*	Obs; Bar	0.48-0.55			0.6-0.9		0.05	0.05	0.1-0.35	bal Fe				
AS 1443	S1050*	Obs; Bar	0.48-0.55			0.6-0.9		0.05	0.05	0.35	bal Fe				
AS 1443(94)	1050	CF bar (may treat w/microalloying elements)	0.48-0.55			0.60-0.90		0.040 max	0.040 max	0.10-0.35	Si<=0.10 for Al-killed; bal Fe				

Austria

Specification	Designation	Notes	C	Cr	Cu	Mn	Ni	P	S	Si	Other	UTS	YS	El	Hard
ONORM M3110	RC50	Wir	0.50-0.55			0.4-0.7		0.035 max	0.035 max	0.15-0.40	P+S=0.06; bal Fe				

UNS numbers and US grades are provided as a means of cross referencing chemically similar alloys. Exchangability is only possible after independent examination of specifications. Tensile properties are minimum or typical as specified. UTS and YS as MPa. El as %. See Appendix for list of abbreviations used in Notes. * indicates obsolete material.

Specification	Designation	Notes	C	Cr	Cu	Mn	Ni	P	S	Si	Other	UTS	YS	El	Hard
Carbon Steel, Nonresulfurized, 1050 (Continued from previous page)															
Belgium															
NBN 253-06	C53		0.5-0.57			0.4-0.7		0.025	0.035	0.15-0.4	bal Fe				
Bulgaria															
BDS 3492(86)	55LI		0.50-0.60		0.3	0.4-0.8		0.06	0.06	0.25-0.5	bal Fe				
BDS 3492(86)	55LII		0.50-0.60		0.3	0.4-0.8		0.06	0.06	0.25-0.5	bal Fe				
BDS 3492(86)	55LIII		0.50-0.60		0.3	0.4-0.8		0.06	0.06	0.25-0.5	bal Fe				
BDS 5785(83)	50	Struct	0.47-0.55	0.25 max	0.25 max	0.50-0.80	0.25 max	0.040 max	0.040 max	0.17-0.37	bal Fe				
China															
GB 13237(91)	50	Sh Strp CR HT	0.47-0.55	0.25 max	0.25 max	0.50-0.80	0.25 max	0.035 max	0.035 max	0.17-0.37	bal Fe	540-715		14	
GB 3275(91)	50	Sh Strp HR HT	0.47-0.55	0.25 max	0.25 max	0.50-0.80	0.25 max	0.035 max	0.035 max	0.17-0.37	bal Fe	540-735		13	
GB 3522(83)	50	Strp CR Ann	0.47-0.55	0.25 max	0.25 max	0.50-0.80	0.25 max	0.035 max	0.035 max	0.17-0.37	bal Fe	540-735		13	
GB 699(88)	50	Bar HR Norm Q/T 25mm diam	0.47-0.55	0.25 max	0.25 max	0.50-0.80	0.25 max	0.035 max	0.035 max	0.17-0.37	bal Fe	630	375	14	
GB 710(91)	50	Sh Strp HR HT	0.47-0.55	0.25 max	0.25 max	0.50-0.80	0.25 max	0.035 max	0.035 max	0.17-0.37	bal Fe	540-715		13	
GB/T 3078(94)	50	Bar CR Ann	0.47-0.55	0.25 max	0.25 max	0.50-0.80	0.25 max	0.035 max	0.035 max	0.17-0.37	bal Fe	560		12	
Czech Republic															
CSN 411600	11600	Gen struct	0.5 max	0.3 max		1.60 max		0.055 max	0.05 max	0.60 max	bal Fe				
CSN 411650	11650	Gen struct; Tub Pipe	0.55 max	0.3 max		1.60 max		0.05 max	0.05 max	0.60 max	bal Fe				
CSN 412051	12051	Q/T; Unalloyed	0.47-0.55	0.25 max		0.5-0.8	0.3 max	0.04 max	0.04 max	0.15-0.4	bal Fe				
Europe															
EN 10016/2(94)	1.0586	Rod	0.48-0.53	0.15 max	0.25 max	0.50-0.80	0.20 max	0.035 max	0.035 max	0.10-0.30	Al <=0.01; Mo <=0.05; Cu+Sn<=0.25; bal Fe				
EN 10016/2(94)	C50D	Rod	0.48-0.53	0.15 max	0.25 max	0.50-0.80	0.20 max	0.035 max	0.035 max	0.10-0.30	Al <=0.01; Mo <=0.05; Cu+Sn<=0.25; bal Fe				
EN 10016/4(94)	1.1171	Rod	0.48-0.52	0.10 max	0.15 max	0.50-0.70	0.10 max	0.020 max	0.025 max	0.10-0.30	Al <=0.01; Mo <=0.03; N <=0.007; Cr+Ni+Cu<=0.30, Cu+Sn<=0.15; bal Fe				
EN 10016/4(94)	C50D2	Rod	0.48-0.52	0.10 max	0.15 max	0.50-0.70	0.10 max	0.020 max	0.025 max	0.10-0.30	Al <=0.01; Mo <=0.03; N <=0.007; Cr+Ni+Cu<=0.30, Cu+Sn<=0.15; bal Fe				
EN 10083/1(91)A1(96)	1.1206	Q/T t<=16mm	0.47-0.55	0.40 max		0.60-0.90	0.40 max	0.035 max	0.035 max	0.40 max	Mo <=0.10; Cr+Mo+Ni<=0.63; bal Fe	750-900	520	13	
EN 10083/1(91)A1(96)	1.1241	Q/T t<=16mm	0.47-0.55	0.40 max		0.60-0.90	0.40 max	0.035 max	0.035 max	0.40 max	Mo <=0.10; Cr+Mo+Ni<=0.63; bal Fe	750-900	520	13	
EN 10083/1(91)A1(96)	C50E	Q/T t<=16mm	0.47-0.55	0.40 max		0.60-0.90	0.40 max	0.035 max	0.035 max	0.40 max	Mo <=0.10; Cr+Mo+Ni<=0.63; bal Fe	750-900	520	13	
EN 10083/1(91)A1(96)	C50R	Q/T t<=16mm	0.47-0.55	0.40 max		0.60-0.90	0.40 max	0.035 max	0.035 max	0.40 max	Mo <=0.10; Cr+Mo+Ni<=0.63; bal Fe	750-900	520	13	
EN 10083/2(91)A1(96)	1.0540	Q/T t<=16mm	0.47-0.55	0.40 max		0.60-0.90	0.40 max	0.045 max	0.045 max	0.40 max	Mo <=0.10; Cr+Mo+Ni<=0.63; bal Fe	750-900	520	13	
EN 10083/2(91)A1(96)	C50	Q/T t<=16mm	0.47-0.55	0.40 max		0.60-0.90	0.40 max	0.045 max	0.045 max	0.40 max	Mo <=0.10; Cr+Mo+Ni<=0.63; bal Fe	750-900	520	13	
France															
AFNOR NFA35553	XC50	Strp	0.46-0.52			0.5-0.8		0.04	0.04	0.15-0.35	bal Fe				

Specification	Designation	Notes	C	Cr	Cu	Mn	Ni	P	S	Si	Other	UTS	YS	El	Hard

Carbon Steel, Nonresulfurized, 1050 (Continued from previous page)

France

Specification	Designation	Notes	C	Cr	Cu	Mn	Ni	P	S	Si	Other	UTS	YS	El	Hard
AFNOR NFA36102(93)	C50RR	HR; Strp; CR	0.47-0.52	0.3 max		0.5-0.8	0.40 max	0.025 max	0.02 max	0.15-0.35	Al <=0.03; Mo <=0.10; bal Fe				
AFNOR NFA36102(93)	XC50*	HR; Strp; CR	0.47-0.52	0.3 max		0.5-0.8	0.40 max	0.025 max	0.02 max	0.15-0.35	Al <=0.03; Mo <=0.10; bal Fe				
AFNOR NFA36612(82)	F70	t<=300mm; HT	0.47-0.58	0.3 max	0.3 max	0.5-0.9	0.4 max	0.04 max	0.04 max	0.1-0.4	bal Fe	690-830	360	15	

Germany

Specification	Designation	Notes	C	Cr	Cu	Mn	Ni	P	S	Si	Other	UTS	YS	El	Hard
DIN	D53-3*	Obs; see C52D2	0.5-0.55			0.3-0.7		0.03 max	0.03 max	0.1-0.3	bal Fe				
DIN 17230(80)	Cf54	Ball & roller bearing	0.50-0.57		0.30 max	0.40-0.70		0.025 max	0.035 max	0.40 max	bal Fe				
DIN EN 10016(95)	WNr 1.0588		0.50-0.55	0.15 max	0.25 max	0.50-0.80	0.20 max	0.035 max	0.035 max	0.10-0.30	Al <=0.010; Mo <=0.05; bal Fe				
DIN EN 10083(91)	Ck50	Q/T 17-40mm	0.47-0.55	0.40 max		0.60-0.90	0.40 max	0.035 max	0.035 max	0.40 max	Mo <=0.10; Cr+Mo+Ni<=0.63; bal Fe	700-850	460	15	
DIN EN 10083(91)	Cm50	Q/T	0.47-0.55			0.60-0.90	0.40 max	0.035 max	0.020-0.040	0.40 max	Mo <=0.10; Cr+Mo+Ni<=0.63; bal Fe				
DIN EN 10083(91)	WNr 1.0540	Q/T 17-40mm	0.47-0.55	0.40 max		0.60-0.90	0.40 max	0.045 max	0.045 max	0.40 max	Mo <=0.10; Cr+Mo+Ni<=0.63; bal Fe	700-850	460	15	
DIN EN 10083(91)	WNr 1.1206	Q/T 17-40mm	0.47-0.55	0.40 max		0.60-0.90	0.40 max	0.035 max	0.035 max	0.40 max	Mo <=0.10; Cr+Mo+Ni<=0.63; bal Fe	700-850	460	15	
DIN EN 10083(91)	WNr 1.1241	Q/T	0.47-0.55	0.40 max		0.60-0.90	0.40 max	0.035 max	0.020-0.040	0.40 max	Mo <=0.10; Cr+Mo+Ni<=0.63; bal Fe				
DIN SEW 085(88)	WNr 1.0555	Sect, bar	0.16 max	0.20 max	0.20 max	1.60 max	0.20 max	0.025 max	0.025 max	0.50 max	Al >=0.020; Mo <=0.08; N >=0.015; Nb <=0.040; Ti <=0.05; V <=0.60; bal Fe				

Hungary

Specification	Designation	Notes	C	Cr	Cu	Mn	Ni	P	S	Si	Other	UTS	YS	El	Hard
MSZ 2752(89)	B1V	Semifinished railway tire; Non-rimming; FF	0.47-0.53			0.6-0.9	0.3 max	0.04 max	0.04 max	0.2-0.5	Mo <=0.08; V <=0.05; P<=0.04/P<=0.035; S<=0.04/S<=0.035; bal Fe				
MSZ 2898/2(80)	A55	Smls CF tube; Smls tubes w/special delivery conds; 0.5-10mm; norm free from scale	0.36 max			2 max		0.05 max	0.05 max	0.6 max	bal Fe	540-640	290	17L	
MSZ 2898/2(80)	A55	Smls CF tube; Smls tubes w/special delivery conds; 0.5-10mm; bright-drawn, hard	0.36 max			2 max		0.05 max	0.05 max	0.6 max	bal Fe	640	510	4L	
MSZ 29(86)	A52	Smls tube; FF; 0-16mm	0.22 max			1.6 max		0.04 max	0.035 max	0.1-0.55	bal Fe	500-680	355	21L	
MSZ 29(86)	A52	Smls tube; FF; 40.1-100.00mm	0.22 max			1.6 max		0.04 max	0.035 max	0.1-0.55	bal Fe	500-650	355	21L	
MSZ 29(86)	A55	Smls tube; Non-rimming; FF; 0-16mm	0.27 max			1.6 max		0.04 max	0.04 max	0.1-0.55	bal Fe	540-720	355	19L	
MSZ 29(86)	A55	Smls tube; Non-rimming; FF; 40.1-100.00mm	0.27 max			1.6 max		0.04 max	0.04 max	0.1-0.55	bal Fe	540-720	335	19L	
MSZ 29/2(64)	A55		0.36 max			2 max		0.05 max	0.05 max	0.6 max	bal Fe				
MSZ 3160(87)	A55	casing tub w/Whitworth thread; gen struct; non-rimming; FF; 4-8mm HF	0.27 max			1.6 max		0.04 max	0.04 max	0.1-0.55	MSZ29; bal Fe	540-720	355	19L	
MSZ 5736(88)	D51	Wir rod; for CD	0.47-0.54	0.2 max	0.25 max	0.3-0.7	0.2 max	0.04 max	0.04 max	0.1-0.35	bal Fe				
MSZ 5736(88)	D51K	Wir rod; for CD	0.47-0.54	0.2 max	0.15 max	0.3-0.7	0.2 max	0.025 max	0.025 max	0.1-0.35	bal Fe				

Specification	Designation	Notes	C	Cr	Cu	Mn	Ni	P	S	Si	Other	UTS	YS	El	Hard

Carbon Steel, Nonresulfurized, 1050 (Continued from previous page)

Hungary

Specification	Designation	Notes	C	Cr	Cu	Mn	Ni	P	S	Si	Other	UTS	YS	El	Hard
MSZ 61(85)	C50	Q/T; t<=16mm	0.47-0.55			0.6-0.9		0.035 max	0.035 max	0.4 max	bal Fe	750-900	520	13L	
MSZ 61(85)	C50	Q/T; 40<t<=100mm	0.47-0.55			06-0.9		0.035 max	0.035 max	0.4 max	bal Fe	650-800	400	16L	
MSZ 61(85)	C50E	Q/T; t<=16mm	0.47-0.55			0.6-0.9		0.035 max	0.02-0.035	0.4 max	bal Fe	750-900	520	13L	
MSZ 61(85)	C50E	Q/T; 40<t<=100mm	0.47-0.55			0.6-0.9		0.035 max	0.02-0.035	0.4 max	Ti <=0.0.05; bal Fe	650-800	400	16L	

India

Specification	Designation	Notes	C	Cr	Cu	Mn	Ni	P	S	Si	Other	UTS	YS	El	Hard
IS 1570/2(79)	50C4	Sh Plt Strp	0.45-0.55	0.3 max	0.3 max	0.3-0.6	0.4 max	0.055 max	0.055 max	0.6 max	Co <=0.1; Mo <=0.15; Pb <=0.15; W <=0.1; P:S Varies; bal Fe	660-780	363	13	
IS 1570/2(79)	50C8	Sh Plt Sect Shp Bar Bil Frg	0.45-0.55	0.3 max	0.3 max	0.6-0.9	0.4 max	0.055 max	0.055 max	0.6 max	Co <=0.1; Mo <=0.15; Pb <=0.15; W <=0.1; P:S Varies; bal Fe				
IS 1570/2(79)	C50	Sh Plt Strp	0.45-0.55	0.3 max	0.3 max	0.3-0.6	0.4 max	0.055 max	0.055 max	0.6 max	Co <=0.1; Mo <=0.15; Pb <=0.15; W <=0.1; P:S Varies; bal Fe	660-780	363	13	
IS 3749	T50	Bar	0.45-0.55			0.6-0.9		0.04	0.04		bal Fe				
IS 5517	C50	Bar	0.45-0.55			0.6-0.9				0.1-0.35	bal Fe				

International

Specification	Designation	Notes	C	Cr	Cu	Mn	Ni	P	S	Si	Other	UTS	YS	El	Hard
ISO 3304(85)	R50	Smls Tub, CF	0.23 max			1.6 max		0.050 max	0.050 max	0.55 max	Co <=0.1; Pb <=0.15; W <=0.1; bal Fe	600		4	
ISO 683-1(87)	C50	Bar Frg Plt Wir rod, Blts, Slbs; Q/T, t<16mm	0.47-0.55			0.60-0.90		0.045 max	0.045 max	0.10-0.40	bal Fe	750-900	520		
ISO 683-1(87)	C50E4	Bar Frg Plt Wir rod, Blts, Slbs; Q/T, t<16mm	0.47-0.55			0.60-0.90		0.035 max	0.035 max	0.10-0.40	bal Fe	750-900	520		
ISO 683-1(87)	C50M2	Bar Frg Plt Wir rod, Blts, Slbs; Q/T, t<16mm	0.47-0.55			0.60-0.90		0.035 max	0.020-0.040	0.10-0.40	bal Fe	750-900	520		
ISO 683-12	4*		0.48-0.55			0.6-0.9		0.035	0.035	0.15-0.4	bal Fe				
ISO 683-12(72)	5*	F/IH	0.5-0.57	0.3 max	0.3 max	0.4-0.7	0.4 max	0.035 max	0.035 max	0.15-0.4	Al <=0.1; Co <=0.1; Mo <=0.15; Pb <=0.15; Ti <=0.05; V <=0.1; W <=0.1; bal Fe				
ISO 683-18(96)	C50	Bar Wir, CD, t<=5mm	0.47-0.55			0.60-0.90		0.045 max	0.045 max	0.10-0.40	bal Fe	790	640	4	217 HB
ISO 683-18(96)	C50E4	Bar Wir, CD, t<=5mm	0.47-0.55			0.60-0.90		0.035 max	0.035 max	0.10-0.40	bal Fe	790	640	4	217 HB
ISO 683-18(96)	C50M2	Bar Wir, CD, t<=5mm	0.47-0.55			0.60-0.90		0.035 max	0.020-0.040	0.10-0.40	bal Fe	790	640	4	217 HB
ISO R683-3(70)	C50ea	Bar Rod, Q/A, Tmp, 16mm diam	0.47-0.55	0.3 max	0.3 max	0.6-0.9	0.4 max	0.035 max	0.02-0.035	0.15-0.4	Al <=0.1; Co <=0.1; Mo <=0.15; Pb <=0.15; Ti <=0.05; V <=0.1; W <=0.1; bal Fe				
ISO R683-3(70)	C50eb	Bar Rod, Q/A, Tmp, 16mm diam	0.47-0.55			0.6-0.9		0.04	0.03-0.05	0.15-0.4	bal Fe				

Italy

Specification	Designation	Notes	C	Cr	Cu	Mn	Ni	P	S	Si	Other	UTS	YS	El	Hard
UNI 5332	C50		0.47-0.55			0.6-0.9		0.035	0.035	0.4	bal Fe				
UNI 5598	ICD50		0.47-0.54			0.4-0.7		0.05	0.05	0.15-0.35	bal Fe				
UNI 5598(71)	1CD50	Wir rod	0.47-0.54			0.4-0.7		0.05 max	0.05 max	0.15-0.35	bal Fe				
UNI 5598(71)	3CD50	Wir rod	0.48-0.53			0.4-0.7		0.035 max	0.035 max	0.15-0.35	N <=0.012; bal Fe				
UNI 6783	Fe70-3		0.45-0.55			0.4-0.8		0.04	0.04	0.4	bal Fe				

Specification	Designation	Notes	C	Cr	Cu	Mn	Ni	P	S	Si	Other	UTS	YS	El	Hard
Carbon Steel, Nonresulfurized, 1050 (Continued from previous page)															
Italy															
UNI 7065(72)	C50	Strp	0.48-0.53			0.4-0.65		0.02 max	0.02 max	0.35 max	bal Fe				
UNI 7065(72)	C51	Strp	0.47-0.55			0.6-0.9		0.035 max	0.035 max	0.4 max	bal Fe				
UNI 7845(78)	C50	Q/T	0.47-0.55			0.6-0.9		0.035 max	0.035 max	0.15-0.4	Pb 0.15-0.3; bal Fe				
UNI 7874(79)	C50	Q/T	0.47-0.55			0.6-0.9		0.035 max	0.035 max	0.15-0.4	Pb 0.15-0.3; bal Fe				
UNI 8893(86)	C50	Spring	0.47-0.55			0.6-0.9		0.035 max	0.035 max	0.15-0.4	bal Fe				
Japan															
JIS G4051(79)	S50C	Bar Wir rod	0.47-0.53	0.20 max	0.30 max	0.60-0.90	0.20 max	0.030 max	0.035 max	0.15-0.35	Ni+Cr<=0.35; bal Fe				
JIS G4051(79)	S55C	Bar Wir rod	0.52-0.58	0.20 max	0.30 max	0.60-0.90	0.20 max	0.030 max	0.035 max	0.15-0.35	Ni+Cr<=0.35; bal Fe				
Mexico															
NMX-B-096-SCFI(97)		Smls tub for high-pressure	0.06-0.18			0.27-0.63		0.035 max	0.035 max	0.25 max	bal Fe	324	179	35	
NMX-B-138(86)		Weld Tub for high-press	0.06-0.18			0.27-0.63		0.050 max	0.060 max	0.25 max	bal Fe				
NMX-B-301(86)	1050	Bar	0.48-0.55			0.60-0.90		0.040 max	0.050 max		bal Fe				
Pan America															
COPANT 330	1050	Bar	0.46-0.55			0.6-0.9		0.04	0.05		bal Fe				
COPANT 331	1050	Bar	0.48-0.55			0.6-0.9		0.04	0.05		bal Fe				
COPANT 333	1050	Wir, HF, 324mm diam	0.48-0.55			0.6-0.9		0.04	0.05		bal Fe				
COPANT 514	1050	Tube, Norm	0.48-0.55			0.6-0.9		0.04	0.05	0.15-0.3	bal Fe				
COPANT 514	1050	Tube	0.48-0.55			0.6-0.9		0.04	0.05	0.15-0.3	bal Fe				
COPANT 514	1050	Tube, Ann	0.48-0.55			0.6-0.9		0.04	0.05	0.15-0.3	bal Fe				
Poland															
PNH84023/07	R55		0.32-0.4	0.3 max		0.6-0.85	0.4 max	0.045 max	0.045 max	0.2-0.35	bal Fe				
PNH84028	D50	Wir rod	0.46-0.53	0.2 max	0.2 max	0.3-0.6	0.2 max	0.035 max	0.035 max	0.1-0.3	Mo <=0.08; P+S<=0.06; bal Fe				
PNH84028	D50A	Wir rod	0.5-0.53	0.1 max	0.2 max	0.3-0.6	0.15 max	0.03 max	0.03 max	0.1-0.25	Mo <=0.05; P+S<=0.05; bal Fe				
PNH84028	DW52		0.5-0.55	0.15	0.2	0.4-0.6	0.25	0.03 max	0.03 max	0.25	P+S=0.05; bal Fe				
PNH84032	50S	Spring	0.45-0.55	0.3 max	0.25 max	0.5-0.8	0.4 max	0.05 max	0.05 max	0.3-0.6	bal Fe				
Romania															
STAS 880(88)	OLC50	Q/T	0.47-0.55	0.3 max		0.6-0.9		0.04 max	0.045 max	0.17-0.37	Pb <=0.15; As<=0.05; bal Fe				
STAS 880(88)	OLC50S	Q/T	0.47-0.55	0.3 max		0.6-0.9	*	0.04 max	0.02-0.045	0.17-0.37	Pb <=0.15; As<=0.05; bal Fe				
STAS 880(88)	OLC50X	Q/T	0.47-0.55	0.3 max		0.6-0.9		0.035 max	0.035 max	0.17-0.37	Pb <=0.15; As<=0.05; bal Fe				
STAS 880(88)	OLC50XS	Q/T	0.47-0.55	0.3 max		0.6-0.9		0.035 max	0.02-0.04	0.17-0.37	Pb <=0.15; As<=0.05; bal Fe				
Russia															
GOST	KSt7		0.5-0.52			0.5-0.8		0.045	0.055	0.15-0.35	bal Fe				
GOST	St7		0.48-0.6					0.05	0.05		bal Fe				
GOST 1050	50		0.47-0.55	0.25	0.25	0.5-0.8	0.25	0.04	0.045	0.17-0.37	bal Fe				
GOST 1050(88)	05kp	High-grade struct; Unkilled; Rimming; R; FU	0.06 max			0.4 max	0.3 max	0.035 max	0.04 max	0.03 max	As<=0.08; bal Fe				
GOST 1050(88)	50	High-grade struct; Q/T; Norm; 0-100mm	0.47-0.55	0.25 max		0.5-0.8	0.3 max	0.035 max	0.04 max	0.17-0.37	As<=0.08; bal Fe	630	375	14L	

Carbon Steel, Nonresulfurized, 1050 (Continued from previous page)

Specification	Designation	Notes	C	Cr	Cu	Mn	Ni	P	S	Si	Other	UTS	YS	El	Hard
Russia															
GOST 1050(88)	50	High-grade struct; Q/T; Full hard; 0-100mm	0.47-0.55	0.25 max		0.5-0.8	0.3 max	0.035 max	0.04 max	0.17-0.37	As<=0.08; bal Fe	660		6L	0-255 HB
GOST 10543	Np-50G		0.45-0.55	0.25		0.7-1	0.25	0.04	0.04	0.17-0.37	bal Fc				
GOST 7521	B56		0.48-0.65			0.6-1		0.075	0.06	0.15-0.3	bal Fe				
GOST 8531	L53		0.47-0.57			0.5-0.8		0.045	0.05	0.15-0.4	bal Fe				
GOST 8731(74)	35		0.30 max	0.3 max	0.3 max	1.60 max	0.4 max	0.070 max	0.060 max	0.60 max	Al <=0.1; bal Fe				
GOST 977(88)	50L		0.47-0.55	0.3		0.4-0.9	0.3	0.05	0.05	0.2-0.52	bal Fe				
Spain															
UNE 36011(75)	C55k		0.5-0.6			0.6-0.9		0.035	0.02-0.035	0.15-0.4	bal Fe				
Sweden															
SIS 141674	1674-01	Bar Frg Plt Sh, Norm	0.48-0.55			0.6-0.9		0.04	0.04	0.15-0.4	bal Fe	650	355	12	
SIS 141674	1674-03	Bar Frg Plt Sh, Q/A, Tmp	0.48-0.55			0.6-0.9		0.04	0.04	0.15-0.4	bal Fe	660	400	16	
SIS 141674	1674-04	Bar Frg Plt Sh, Q/A, Tmp	0.48-0.55			0.6-0.9		0.04	0.04	0.15-0.4	bal Fe	700	440	15	
SIS 141674	1674-05	Bar Frg Plt Sh, Q/A, Tmp	0.48-0.55			0.6-0.9		0.04	0.04	0.15-0.4	bal Fe	740	510	13	
SIS 141674	1674-08	Bar Frg Plt Sh, as rolled, As Frg	0.48-0.55			0.6-0.9		0.04	0.04	0.15-0.4	bal Fe	650	355	12	
SS 141606	1606	Gen struct	0.6 max			2 max		0.04 max	0.04 max	0.6 max	C:0.5; Si:0.5; Mn:0.7				
Turkey															
TS 911(86)	UDK669.14.423	HR, T-bar	0.5					0.05	0.05		bal Fe				
UK															
BS 1549	50CS		0.45-0.55			0.5-0.9		0.045	0.045	0.05-0.35	bal Fe				
BS 1549	50HS		0.45-0.55			0.5-0.9		0.045	0.045	0.05-0.35	bal Fe				
BS 970/1(83)	060A52	Blm Bil Slab Bar Rod Frg	0.45-0.50			0.50-0.70		0.050 max	0.050 max		bal Fe				
BS 970/1(83)	080M50	Blm Bil Slab Bar Rod Frg; 6-150mm; Hard Tmp	0.45-0.55			0.60-1.00		0.050 max	0.050 max	0.1-0.4	Mo <=0.15; bal Fe	625-775	390	15	179-229 HB
BS 970/3(91)	080M50	Bright bar; 150<t<=250mm; norm	0.45-0.55			0.60-1.00		0.050 max	0.050 max	0.10-0.40	Mo <=0.15; bal Fe	570	295	14	163-217 HB
USA															
	AISI 1050	Struct shp	0.48-0.55			0.60-0.90		0.030 max	0.035 max		bal Fe				
	AISI 1050	Wir rod	0.48-0.55			0.60-0.90		0.040 max	0.050 max		bal Fe				
	AISI 1050	Sh Strp Plt	0.47-0.55			0.60-0.90		0.030 max	0.035 max		bal Fe				
	AISI 1050	Bar	0.48-0.55			0.60-0.90		0.040 max	0.050 max		bal Fe				
	UNS G10500		0.48-0.55			0.60-0.90		0.040 max	0.050 max		bal Fe				
AMS 5085E(97)		Ann Sh, Strp, Plt	0.47-0.55			0.60-0.90		0.040 max	0.050 max	0.10-0.30	bal Fe				
ASTM A108(95)	1050	Bar, CF	0.48-0.55			0.60-0.90		0.040 max	0.050 max		bal Fe				
ASTM A29/A29M(93)	1050	Bar	0.48-0.55			0.60-0.90		0.040 max	0.050 max		bal Fe				
ASTM A311/A311M(95)	1050 Class A	Bar CD SR ann, t<=20mm	0.48-0.55			0.60-0.90		0.040 max	0.050 max		bal Fe	690	620	11,RA=35	
ASTM A311/A311M(95)	1050 Class A	Bar CD SR ann, 20<t<=30mm	0.48-0.55			0.60-0.90		0.040 max	0.050 max		bal Fe	655	585	11,RA=30	
ASTM A311/A311M(95)	1050 Class A	Bar CD SR ann, 30<t<=50mm	0.48-0.55			0.60-0.90		0.040 max	0.050 max		bal Fe	620	550	10,RA=30	

UNS numbers and US grades are provided as a means of cross referencing chemically similar alloys. Exchangability is only possible after independent examination of specifications. Tensile properties are minimum or typical as specified. UTS and YS as MPa. El as %. See Appendix for list of abbreviations used in Notes. * indicates obsolete material.

Specification	Designation	Notes	C	Cr	Cu	Mn	Ni	P	S	Si	Other	UTS	YS	El	Hard

Carbon Steel, Nonresulfurized, 1050 (Continued from previous page)

USA

Specification	Designation	Notes	C	Cr	Cu	Mn	Ni	P	S	Si	Other	UTS	YS	El	Hard
ASTM A311/A311M(95)	1050 Class A	Bar CD SR ann, 50<t<=75mm	0.48-0.55			0.60-0.90		0.040 max	0.050 max		bal Fe	585	520	10,RA=30	
ASTM A311/A311M(95)	1050 Class B	Bar CD SR ann, 75<t<=115mm	0.48-0.55			0.60-0.90		0.040 max	0.050 max		bal Fe	725	620	7,RA=20	
ASTM A311/A311M(95)	1050 Class B	Bar CD SR ann, t<=50mm	0.48-0.55			0.60-0.90		0.040 max	0.050 max		bal Fe	795	690	10,RA=25	
ASTM A311/A311M(95)	1050 Class B	Bar CD SR ann, 50<t<=75mm	0.48-0.55			0.60-0.90		0.040 max	0.050 max		bal Fe	795	690	9,RA=25	
ASTM A510(96)	1050	Wir rod	0.48-0.55			0.60-0.90		0.040 max	0.050 max		bal Fe				
ASTM A513(97a)	1050	ERW Mech tub	0.47-0.55			0.60-0.90		0.040 max	0.050 max		bal Fe				
ASTM A519(96)	1050	Smls mech tub, Norm	0.48-0.55			0.60-0.90		0.040 max	0.050 max		bal Fe	538	345	12	82 HRB
ASTM A519(96)	1050	Smls mech tub, Ann	0.48-0.55			0.60-0.90		0.040 max	0.050 max		bal Fe	469	262	18	74 HRB
ASTM A519(96)	1050	Smls mech tub, SR	0.48-0.55			0.60-0.90		0.040 max	0.050 max		bal Fe	565	483	6	86 HRB
ASTM A519(96)	1050	Smls mech tub, HR	0.48-0.55			0.60-0.90		0.040 max	0.050 max		bal Fe	552	345	10	85 HRB
ASTM A576(95)	1050	Special qual HW bar	0.48-0.55			0.60-0.90		0.040 max	0.050 max		Si Cu Pb B Bi Ca Se Te if spec'd; bal Fe				
ASTM A682/A682M(98)	1050	Strp, high-C, CR, spring	0.47-0.55			0.60-0.90		0.035 max	0.040 max	0.15-0.30	bal Fe				
ASTM A684A684M(86)	1050	Strp, high-C, CR, Ann	0.47-0.55			0.60-0.90		0.035 max	0.040 max	0.15-0.30	bal Fe				87 HRB
ASTM A827(98)	1050	Plt, frg	0.47-0.55			0.60-0.90		0.035 max	0.040 max	0.15-0.40	bal Fe				
ASTM A830/A830M(98)	1050	Plt	0.47-0.55			0.60-0.90		0.035 max	0.04 max	0.15-0.40	bal Fe				
ASTM A866(94)	1050	Anti-friction bearings	0.48-0.55			0.60-0.90		0.025 max	0.025 max	0.15-0.35	bal Fe				
DoD-F-24669/1(86)	1050	Bar Bil; Supersedes MIL-S-866 & MIL-S-16974	0.48-0.55			0.6-0.9		0.04	0.05		bal Fe				
FED QQ-S-635B(88)	C1050*	Obs; see ASTM A827; Plt	0.48-0.55			0.6-0.9		0.04 max	0.05 max		bal Fe				
FED QQ-S-700D(91)	C1050*	Obs; Sh Strp	0.48-0.55			0.6-0.9		0.04 max	0.05 max		bal Fe				
MIL-S-10520D(AR)(88)	6*	Frg; Obs; see ASTM A711, A575	0.55			1		0.04	0.05	0.15-0.3	bal Fe				
MIL-S-16788A(86)	C5*	HR, frg, bloom bil slab for refrg; Obs for new design	0.48-0.55			0.60-0.90		0.040 max	0.050 max		bal Fe				
MIL-S-16974E(86)	1050*	Obs; Bar bil boom slab for refrg	0.48-0.55			0.6-0.9		0.04 max	0.05 max		bal Fe				
MIL-S-46059	G10500*	Obs; Shape	0.48-0.55			0.6-0.9		0.04	0.05		bal Fe				
SAE J1397(92)	1050	Bar HR, est mech prop	0.48-0.55			0.6-0.9		0.04 max	0.05 max		bal Fe	620	340	15	179 HB
SAE J1397(92)	1050	Bar CD, est mech prop	0.48-0.55			0.6-0.9		0.04 max	0.05 max		bal Fe	690	580	10	197 HB
SAE J1397(92)	1050	Bar Ann CD, est mech prop	0.48-0.55			0.6-0.9		0.04 max	0.05 max		bal Fe	660	550	10	189 HB
SAE J403(95)	1050	Struct Shps Plt Strp Sh Weld Tub	0.47-0.55			0.60-0.90		0.030 max	0.035 max		bal Fe				
SAE J403(95)	1050	Bar Wir rod Smls Tub HR CF	0.48-0.55			0.60-0.90		0.030 max	0.050 max		bal Fe				

Yugoslavia

Specification	Designation	Notes	C	Cr	Cu	Mn	Ni	P	S	Si	Other	UTS	YS	El	Hard
	C.0745	Struct	0.5	0.3 max		1.60 max		0.05 max	0.05 max	0.60 max	bal Fe				
	C.1600	Struct	0.48-0.55			0.5-0.8		0.06 max	0.06 max	0.35 max	bal Fe				
	PZ50		0.5-0.59	0.15	0.025	0.3-0.7	0.015	0.04	0.04	0.1-0.3	bal Fe				

UNS numbers and US grades are provided as a means of cross referencing chemically similar alloys. Exchangability is only possible after independent examination of specifications. Tensile properties are minimum or typical as specified. UTS and YS as MPa. El as %. See Appendix for list of abbreviations used in Notes. * indicates obsolete material.

Carbon Steel, Nonresulfurized, 1053

Specification	Designation	Notes	C	Cr	Cu	Mn	Ni	P	S	Si	Other	UTS	YS	El	Hard
Belgium															
NBN 253-06	C53		0.5-0.57			0.4-0.7		0.025	0.035	0.15-0.4	bal Fe				
Bulgaria															
BDS 5785(83)	50G	Struct	0.48-0.56	0.25 max	0.25 max	0.70-1.00	0.25 max	0.040 max	0.040 max	0.17-0.37	bal Fe				
China															
GB 699(88)	50 Mn	Bar HR Norm Q/T 25mm diam	0.48-0.56	0.25 max	0.25 max	0.70-1.00	0.25 max	0.035 max	0.035 max	0.17-0.37	bal Fe	645	390	13	
GB/T 3078(94)	50 Mn	Bar CR Ann	0.48-0.56	0.25 max	0.25 max	0.70-1.00	0.25 max	0.035 max	0.035 max	0.17-0.37	bal Fe	590		10	
YB/T 5052(93)	50 Mn	Tub HR Norm	0.48-0.56			0.70-1.00		0.040 max	0.040 max	0.17-0.37	bal Fe	640	395	14	
Europe															
EN 10016/2(94)	1.0586	Rod	0.50-0.55	0.15 max	0.25 max	0.50-0.80	0.20 max	0.035 max	0.035 max	0.10-0.30	Al <=0.01; Mo <=0.05; Cu+Sn<=0.25; bal Fe				
EN 10016/2(94)	C52D	Rod	0.50-0.55	0.15 max	0.25 max	0.50-0.80	0.20 max	0.035 max	0.035 max	0.10-0.30	Al <=0.01; Mo <=0.05; Cu+Sn<=0.25; bal Fe				
EN 10016/4(94)	1.1202	Rod	0.50-0.54	0.10 max	0.15 max	0.50-0.70	0.10 max	0.020 max	0.025 max	0.10-0.30	Al <=0.01; Mo <=0.03; N <=0.007; Cr+Ni+Cu<=0.30, Cu+Sn<=0.15; bal Fe				
EN 10016/4(94)	C52D2	Rod	0.50-0.54	0.10 max	0.15 max	0.50-0.70	0.10 max	0.020 max	0.025 max	0.10-0.30	Al <=0.01; Mo <=0.03; N <=0.007; Cr+Ni+Cu<=0.30, Cu+Sn<=0.15; bal Fe				
France															
AFNOR	52M4TS		0.49-0.55			0.8-1.1		0.025	0.035	0.1-0.4	bal Fe				
AFNOR NFA35553	XC54	Strp	0.5-0.57			0.4-0.7		0.035	0.035	0.15-0.35	bal Fe				
AFNOR NFA35563(76)	52M4TS		0.49-0.55	0.3 max	0.3 max	0.8-1.1	0.4 max	0.025 max	0.035 max	0.1-0.4	bal Fe				
AFNOR NFA35565(94)	C54ER	Ball & roller bearing	0.5-0.57	0.3 max	0.3 max	0.4-0.7	0.4 max	0.025 max	0.035 max	0.4 max	bal Fe				
AFNOR NFA35565(94)	XC54*	see C54ER	0.5-0.57	0.3 max	0.3 max	0.4-0.7	0.4 max	0.025 max	0.035 max	0.4 max	bal Fe				
Germany															
DIN 17201(89)	Ck53	Frg	0.50-0.57			0.60-0.90		0.035 max	0.035 max	0.40 max	bal Fe				
DIN 17212(72)	Cf53	HT 17-40mm	0.50-0.57			0.40-0.70		0.025 max	0.035 max	0.15-0.35	bal Fe	690-830	430	14	
DIN 17212(72)	WNr 1.1213	HT 17-40mm	0.50-0.57			0.40-0.70		0.025 max	0.035 max	0.15-0.35	bal Fe	690-830	430	14	
DIN 17230(80)	C54G	Ball & roller bearing	0.50-0.57	0.30 max		0.40-0.70		0.025 max	0.035 max	0.40 max	bal Fe				
DIN 17230(80)	WNr 1.1219	Ball & roller bearing	0.50-0.57	0.30 max		0.40-0.70		0.025 max	0.035 max	0.40 max	bal Fe				
DIN E 17201(89)	C53E	Frg	0.50-0.57			0.60-0.90		0.035 max	0.035 max	0.40 max	bal Fe				
DIN E 17201(89)	WNr 1.1210	Frg	0.50-0.57			0.60-0.90		0.035 max	0.035 max	0.40 max	bal Fe				
Italy															
UNI 7847(79)	C53	Surf Hard	0.5-0.57			0.5-0.8		0.03 max	0.03 max	0.15-0.4	bal Fe				
UNI 8551(84)	C53	Surf Hard	0.5-0.57			0.5-0.8		0.03 max	0.025 max	0.15-0.4	bal Fe				
Mexico															
DGN B-203	1050*	Tube, Obs	0.48-0.55			0.6-0.9		0.04	0.05		bal Fe				
NMX-B-301(86)	1053	Bar	0.48-0.55			0.70-1.00		0.040 max	0.050 max		bal Fe				
Pan America															
COPANT 330	1050		0.48-0.55			0.6-0.9		0.04	0.05		bal Fe				

Specification	Designation	Notes	C	Cr	Cu	Mn	Ni	P	S	Si	Other	UTS	YS	El	Hard

Carbon Steel, Nonresulfurized, 1053 (Continued from previous page)

Pan America

Specification	Designation	Notes	C	Cr	Cu	Mn	Ni	P	S	Si	Other	UTS	YS	El	Hard
COPANT 331	1050		0.48-0.55			0.6-0.9		0.04	0.05		bal Fe				
COPANT 333	1050		0.48-0.55			0.6-0.9		0.04	0.05		bal Fe				
COPANT 333	1053		0.48-0.55		0.7-1	0.6-0.9		0.04	0.05		bal Fe				
COPANT 514	1050		0.48-0.55		0.7-1	0.6-0.9		0.04	0.05	0.15-0.3	bal Fe				

Russia

Specification	Designation	Notes	C	Cr	Cu	Mn	Ni	P	S	Si	Other	UTS	YS	El	Hard
GOST 4543	50G		0.48-0.56	0.3	0.3	0.7-1	0.3	0.035	0.035	0.17-0.37	bal Fe				
GOST 4543(71)	50G	Q/T	0.48-0.56	0.3 max	0.3 max	0.7-1.0	0.3 max	0.035 max	0.035 max	0.17-0.37	Al <=0.1; bal Fe				
GOST 8531	L53		0.47-0.57			0.5-0.8		0.045	0.05	0.15-0.4	bal Fe				
GOST 977	50L-II		0.47-0.55	0.3	0.3	0.4-0.9	0.3	0.04	0.045	0.2-0.52	bal Fe				
GOST 977	50L-III		0.47-0.55	0.3	0.3	0.4-0.9	0.3	0.04	0.045	0.2-0.52	bal Fe				

UK

Specification	Designation	Notes	C	Cr	Cu	Mn	Ni	P	S	Si	Other	UTS	YS	El	Hard
BS 970/1(83)	080A52	Blm Bil Slab Bar Rod Frg	0.50-0.55			0.70-0.90		0.050 max	0.050 max		bal Fe				
BS 970/1(83)	080A52*	Blm Bil Slab Bar Rod Frg	0.50-0.55			0.70-0.90		0.050 max	0.050 max		bal Fe				

USA

Specification	Designation	Notes	C	Cr	Cu	Mn	Ni	P	S	Si	Other	UTS	YS	El	Hard
	AISI 1053	Wir rod	0.48-0.55			0.70-1.00		0.040 max	0.050 max		bal Fe				
	AISI 1053	Bar	0.48-0.55			0.10-1.00		0.040 max	0.050 max		bal Fe				
	UNS G10530		0.48-0.55			0.70-1.00		0.040 max	0.050 max		bal Fe				
ASTM A29/A29M(93)	1053	Bar	0.48-0.55			0.10-1.00		0.040 max	0.050 max		bal Fe				
ASTM A510(96)	1053	Wir rod	0.48-0.55			0.70-1.00		0.040 max	0.050 max		bal Fe				
ASTM A576(95)	1053	Special qual HW bar	0.48-0.55			0.70-1.00		0.040 max	0.050 max		Si Cu Pb B Bi Ca Se Te if spec'd; bal Fe				
SAE J403(95)	1053	Frg Bar, HR, CF, Rod, Smls Tub	0.48-0.55			0.7-1		0.04 max	0.05 max		bal Fe				
SAE J403(95)	1053	Bar Wir rod Smls Tub HR CF	0.48-0.55			0.70-1.00		0.030 max	0.050 max		bal Fe				

Carbon Steel, Nonresulfurized, 1055

Argentina

Specification	Designation	Notes	C	Cr	Cu	Mn	Ni	P	S	Si	Other	UTS	YS	El	Hard
IAS	IRAM 1055		0.50-0.60			0.60-0.90		0.040 max	0.050 max	0.10-0.30	bal Fe	740-880	440-520	12-20	223-262 HB

Australia

Specification	Designation	Notes	C	Cr	Cu	Mn	Ni	P	S	Si	Other	UTS	YS	El	Hard
AS 1442	K1055*	Obs; Bar Bil	0.5-0.6			0.6-0.9		0.05	0.05	0.1-0.35	bal Fe				
AS 1442(92)	1055	HR bar, Semifinished (may treat w/microalloying elements)	0.50-0.60			0.60-0.90		0.040 max	0.040 max	0.10-0.35	Si<=0.10 for Al-killed; bal Fe				
AS 1443	K1055*	Obs; Bar	0.5-0.6			0.6-0.9		0.05	0.05	0.1-0.35	bal Fe				
AS 1443(94)	1055	CF bar (may treat w/microalloying elements)	0.50-0.60			0.60-0.90		0.040 max	0.040 max	0.10-0.35	Si<=0.10 for Al-killed; bal Fe				
AS/NZS 1594(97)	HK10B55	HR flat products	0.50-0.60	0.30 max	0.35 max	0.60-0.90	0.35 max	0.040 max	0.035 max	0.50 max	Al <=0.100; B <=0.0008; Mo <=0.10; Ti <=0.060; (Cu+Ni+Cr+Mo) 1 max; bal Fe				

Austria

Specification	Designation	Notes	C	Cr	Cu	Mn	Ni	P	S	Si	Other	UTS	YS	El	Hard
ONORM M3110	RC50	Wir	0.50-0.55			0.4-0.7		0.035 max	0.035 max	0.15-0.40	P+S<=0.06; bal Fe				

Carbon Steel, Nonresulfurized, 1055 (Continued from previous page)

Specification	Designation	Notes	C	Cr	Cu	Mn	Ni	P	S	Si	Other	UTS	YS	El	Hard
Austria															
ONORM M3110	RC55	Wir	0.50-0.55			0.4-0.7		0.035 max	0.035 max	0.15-0.40	P+S<=0.06; bal Fe				
Belgium															
NBN 253-02	C55-1		0.52-0.6			0.6-0.9		0.045	0.045	0.15-0.4	bal Fe				
NBN 253-02	C55-2		0.52-0.6			0.6-0.9		0.04	0.05		bal Fe				
NBN 253-02	C55-3		0.52-0.6			0.6-0.9		0.04	0.05		bal Fe				
Bulgaria															
BDS 3492(86)	55LI		0.50-0.60		0.3	0.4-0.8		0.06	0.06	0.25-0.5	bal Fe				
BDS 5785(83)	55	Struct	0.52-0.60	0.25 max	0.25 max	0.50-0.80	0.25 max	0.040 max	0.040 max	0.17-0.37	bal Fe				
China															
GB 3522(83)	55	Strp CR Ann	0.52-0.60	0.25 max	0.25 max	0.50-0.80	0.25 max	0.035 max	0.035 max	0.17-0.37	bal Fe	440-735		12	
GB 699(88)	55	Bar HR Norm Q/T 25mm diam	0.52-0.60	0.25 max	0.25 max	0.50-0.80	0.25 max	0.035 max	0.035 max	0.17-0.37	bal Fe	645	380	13	
GB 700	B7*	Obs	0.5-0.62			0.5-0.8		0.045	0.055	0.15-0.35	bal Fe				
GB 700	BJ7*	Obs	0.5-0.62			0.5-0.8		0.045	0.055	0.15-0.35	bal Fe				
GB 700	BY7*	Obs	0.5-0.62			0.5-0.8		0.045	0.055	0.15-0.35	bal Fe				
GB/T 3078(94)	55	Bar CR Norm/Q/T 25mm diam	0.52-0.60	0.25 max	0.25 max	0.50-0.80	0.25 max	0.035 max	0.035 max	0.17-0.37	bal Fe	645	380	13	
YB/T 5103(93)	55	Wir CD Q/T 2.0-4.0mm diam	0.52-0.60	0.25 max	0.25 max	0.50-0.80	0.25 max	0.035 max	0.035 max	0.17-0.37	bal Fe	1425-1765			
Czech Republic															
CSN 412060	12060	Q/T	0.52-0.6	0.25 max		0.5-0.8	0.3 max	0.04 max	0.4 max	0.15-0.4	bal Fe				
Europe															
EN 10016/2(94)	1.0518	Rod	0.53-0.58	0.15 max	0.25 max	0.50-0.80	0.20 max	0.035 max	0.035 max	0.10-0.30	Al <=0.01; Mo <=0.05; Cu+Sn<=0.25; bal Fe				
EN 10016/2(94)	C56D	Rod	0.53-0.58	0.15 max	0.25 max	0.50-0.80	0.20 max	0.035 max	0.035 max	0.10-0.30	Al <=0.01; Mo <=0.05; Cu+Sn<=0.25; bal Fe				
EN 10016/4(94)	1.1220	Rod	0.54-0.58	0.10 max	0.15 max	0.50-0.70	0.10 max	0.020 max	0.025 max	0.10-0.30	Al <=0.01; Mo <=0.03; N <=0.007; Cr+Ni+Cu<=0.30, Cu+Sn<=0.15; bal Fe				
EN 10016/4(94)	C56D2	Rod	0.54-0.58	0.10 max	0.15 max	0.50-0.70	0.10 max	0.020 max	0.025 max	0.10-0.30	Al <=0.01; Mo <=0.03; N <=0.007; Cr+Ni+Cu<=0.30, Cu+Sn<=0.15; bal Fe				
EN 10083/1(91)A1(96)	1.1203	Q/T t<=16mm	0.52-0.60	0.40 max		0.60-0.90	0.40 max	0.035 max	0.035 max	0.40 max	Mo <=0.10; Cr+Mo+Ni<=0.63; bal Fe	800-950	550	12	
EN 10083/1(91)A1(96)	1.1209	Q/T t<=16mm	0.52-0.60	0.40 max		0.60-0.90	0.40 max	0.035 max	0.020-0.040	0.40 max	Mo <=0.10; Cr+Mo+Ni<=0.63; bal Fe	800-950	550	12	
EN 10083/1(91)A1(96)	C55E	Q/T t<=16mm	0.52-0.60	0.40 max		0.60-0.90	0.40 max	0.035 max	0.035 max	0.40 max	Mo <=0.10; Cr+Mo+Ni<=0.63; bal Fe	800-950	550	12	
EN 10083/1(91)A1(96)	C55R	Q/T t<=16mm	0.52-0.60	0.40 max		0.60-0.90	0.40 max	0.035 max	0.020-0.040	0.40 max	Mo <=0.10; Cr+Mo+Ni<=0.63; bal Fe	800-950	550	12	
EN 10083/2(91)A1(96)	1.0535	Q/T t<=16mm	0.52-0.60	0.40 max		0.60-0.90	0.40 max	0.045 max	0.045 max	0.40 max	Mo <=0.10; Cr+Mo+Ni<=0.63; bal Fe	800-950	550	12	
EN 10083/2(91)A1(96)	C55	Q/T t<=16mm	0.52-0.60	0.40 max		0.60-0.90	0.40 max	0.045 max	0.045 max	0.40 max	Mo <=0.10; Cr+Mo+Ni<=0.63; bal Fe	800-950	550	12	
France															
AFNOR NFA33101(82)	AF70C55	t<=450mm; Norm	0.5-0.58	0.3 max	0.3 max	0.5-0.8	0.4 max	0.04 max	0.04 max	0.1-0.4	bal Fe	710-860	410	15	

UNS numbers and US grades are provided as a means of cross referencing chemically similar alloys. Exchangability is only possible after independent examination of specifications. Tensile properties are minimum or typical as specified. UTS and YS as MPa. El as %. See Appendix for list of abbreviations used in Notes. * indicates obsolete material.

Specification	Designation	Notes	C	Cr	Cu	Mn	Ni	P	S	Si	Other	UTS	YS	El	Hard

Carbon Steel, Nonresulfurized, 1055 (Continued from previous page)

France

Specification	Designation	Notes	C	Cr	Cu	Mn	Ni	P	S	Si	Other	UTS	YS	El	Hard
AFNOR NFA35552	XC55H1	Bar Rod Wir	0.52-0.6			0.5-0.8		0.03	0.035	0.15-0.35	Cr+Ni+Mo=0.6 max; bal Fe				
AFNOR NFA35552	XC55H2	Bar Rod Wir	0.52-0.6			1.2		0.035	0.035	0.15-0.35	Cr+Ni+Mo=0.6 max; bal Fe				
AFNOR NFA35553	XC54	Strp	0.5-0.57			0.4-0.7		0.035	0.035	0.15-0.35	bal Fe				
AFNOR NFA35565(94)	C55E3	Ball & roller bearing	0.55-0.6	0.25 max	0.3 max	0.6-0.8	0.25 max	0.025 max	0.015 max	0.1-0.25	Al <=0.05; Mo <=0.1; bal Fe				
AFNOR NFA35565(94)	XC55*	see C55E3	0.55-0.6	0.25 max	0.3 max	0.6-0.8	0.25 max	0.025 max	0.015 max	0.1-0.25	Al <=0.05; Mo <=0.1; bal Fe				
AFNOR NFA36102(93)	C55RR	HR; Strp; CR	0.52-0.58	0.3 max		0.5-0.8	0.40 max	0.025 max	0.02 max	0.15-0.35	Al <=0.03; Mo <=0.10; bal Fe				
AFNOR NFA36102(93)	XC55*	HR; Strp; CR	0.52-0.58	0.3 max		0.5-0.8	0.40 max	0.025 max	0.02 max	0.15-0.35	Al <=0.03; Mo <=0.10; bal Fe				

Germany

Specification	Designation	Notes	C	Cr	Cu	Mn	Ni	P	S	Si	Other	UTS	YS	El	Hard
DIN	D55-3*	Obs; see C56D2	0.53-0.58			0.3-0.7		0.03 max	0.03 max	0.1-0.3	bal Fe				
DIN 17140	D55-2*	Obs; see C56D	0.53-0.58			0.3-0.7		0.04 max	0.04 max	0.1-0.3	bal Fe				
DIN EN 10083(91)	C55E	Q/T 17-40mm	0.52-0.60	0.40 max		0.60	0.40 max	0.035 max	0.035 max	0.40 max	Mo <=0.10; Cr+Mo+Ni<=0.63; bal Fe	750-900	490	14	
DIN EN 10083(91)	Ck55	Q/T 17-40mm	0.52-0.60	0.40 max		0.60-0.90	0.40 max	0.035 max	0.035 max	0.40 max	Mo <=0.10; Cr+Mo+Ni<=0.63; bal Fe	750-900	490	14	
DIN EN 10083(91)	Cm55	Q/T	0.52-0.60	0.40 max		0.60-0.90	0.40 max	0.035 max	0.020-0.040	0.40 max	Mo <=0.10; Cr+Mo+Ni<=0.63; bal Fe				
DIN EN 10083(91)	WNr 1.0535	Q/T 17-40mm	0.52-0.60	0.40 max		0.60-0.90	0.40 max	0.045 max	0.045 max	0.40 max	Mo <=0.10; Cr+Mo+Ni<=0.63; bal Fe	750-900	490	14	
DIN EN 10083(91)	WNr 1.1203	Q/T 17-40mm	0.52-0.60	0.40 max		0.60-0.90	0.40 max	0.035 max	0.035 max	0.40 max	Mo <=0.10; Cr+Mo+Ni<=0.63; bal Fe	750-900	490	14	
DIN EN 10083(91)	WNr 1.1209	Q/T	0.52-0.60	0.40 max		0.60-0.90	0.40 max	0.035 max	0.020-0.040	0.40 max	Mo <=0.10; Cr+Mo+Ni<=0.63; bal Fe				

Hungary

Specification	Designation	Notes	C	Cr	Cu	Mn	Ni	P	S	Si	Other	UTS	YS	El	Hard
MSZ 12043	VhV		0.52-0.6	0.3	0.3	0.6-0.9	0.3	0.045	0.045	0.15-0.4	Mo 0.1; V 0.05; bal Fe				
MSZ 2752(89)	B2	Semifinished railway tire; Non-rimming; FF	0.52-0.58			0.6-0.9	0.3 max	0.04 max	0.04 max	0.2-0.5	Mo <=0.08; V <=0.05; P<=0.04/P<=0.035; S<=0.04/S<=0.035; bal Fe	700-820		9L	
MSZ 2752(89)	B2	Semifinished railway tire; Non-rimming; FF	0.52-0.58			0.6-0.9	0.3 max	0.04 max	0.04 max	0.2-0.5	Mo <=0.08; V <=0.05; P<=0.04/P<=0.035; S<=0.04/S<=0.035; bal Fe	700-820		14L	
MSZ 2752(89)	B3	Semifinished railway tire; Non-rimming; FF	0.54-0.6			0.8-1.1	0.3 max	0.04 max	0.04 max	0.2-0.5	Mo <=0.08; V <=0.05; P<=0.04/P<=0.035; S<=0.04/S<=0.035; bal Fe	750-880		12L	
MSZ 2752(89)	B5	Semifinished railway tire; Non-rimming; FF	0.54-0.6			0.5-0.8	0.3 max	0.04 max	0.04 max	0.2-0.5	Mo <=0.08; V <=0.05; P<=0.04/P<=0.035; S<=0.04/S<=0.035; bal Fe	800-920		14L	
MSZ 4217(85)	C55	CR, Q/T or spring Strp; 0-2mm; rolled, hard	0.52-0.6			0.6-0.9		0.035 max	0.035 max	0.4 max	bal Fe	730-1080			
MSZ 4217(85)	C55	CR, Q/T or spring Strp; 0-2mm	0.52-0.6			0.6-0.9		0.035 max	0.035 max	0.4 max	bal Fe	980-1280		5L	
MSZ 500(81)	B70	Gen struct; Non-rimming; FF	0.5-0.62			0.5-0.8		0.045 max	0.05 max	0.12-0.35	bal Fe				
MSZ 61	C55EK		0.52-0.6			0.6-0.9		0.035	0.035	0.17-0.37	bal Fe				

UNS numbers and US grades are provided as a means of cross referencing chemically similar alloys. Exchangability is only possible after independent examination of specifications. Tensile properties are minimum or typical as specified. UTS and YS as MPa. El as %. See Appendix for list of abbreviations used in Notes. * indicates obsolete material.

Specification	Designation	Notes	C	Cr	Cu	Mn	Ni	P	S	Si	Other	UTS	YS	El	Hard

Carbon Steel, Nonresulfurized, 1055 (Continued from previous page)

Specification	Designation	Notes	C	Cr	Cu	Mn	Ni	P	S	Si	Other	UTS	YS	El	Hard
Hungary															
MSZ 61(85)	C55	Q/T; 40<t<=100mm	0.52-0.6			0.6-0.9		0.035 max	0.035 max	0.4 max	bal Fe	700-850		15L	
MSZ 61(85)	C55	Q/T; t<=16mm	0.52-0.6			0.6-0.9		0.035 max	0.035 max	0.4 max	bal Fe	800-950	550	12L	
MSZ 61(85)	C55E	Q/T; t<=16mm	0.52-0.6			0.6-0.9		0.035 max	0.02-0.035	0.4 max	bal Fe	800-950	550	12L	
MSZ 61(85)	C55E	Q/T; 40<t<=100mm	0.52-0.6			0.6-0.9		0.035 max	0.02-0.035	0.4 max	bal Fe	700-850	430	15L	
MSZ 8270	Ao.60		0.51-0.6			0.4-0.9		0.06	0.06	0.2-0.4	bal Fe				
MSZ 8270	Ao.60F		0.51-0.6			0.4-0.9		0.06	0.06	0.2-0.4	bal Fe				
MSZ 8270	Ao.60FK		0.51-0.6			0.4-0.9		0.05	0.05	0.2-0.42	bal Fe				
India															
IS 1570	C55		0.5-0.6			0.5-0.65					bal Fe				
IS 1570	C55Mn75		0.5-0.6			0.6-0.9				0.1-0.35	bal Fe				
IS 1570/2(79)	55C4		0.5-0.6	0.3 max	0.3 max	0.3-0.6	0.4 max	0.055 max	0.055 max	0.6 max	Co <=0.1; Mo <=0.15; Pb <=0.15; W <=0.1; P:S Varies; bal Fe				
IS 1570/2(79)	55C8	Sh Plt Sect Shp Bar Bil Frg	0.5-0.6	0.3 max	0.3 max	0.6-0.9	0.4 max	0.055 max	0.055 max	0.6 max	Co <=0.1; Mo <=0.15; Pb <=0.15; W <=0.1; P:S Varies; bal Fe	720	396	13	
IS 1570/2(79)	C55		0.5-0.6	0.3 max	0.3 max	0.3-0.6	0.4 max	0.055 max	0.055 max	0.6 max	Co <=0.1; Mo <=0.15; Pb <=0.15; W <=0.1; P:S Varies; bal Fe				
IS 1570/2(79)	C55Mn75	Sh Plt Sect Shp Bar Bil Frg	0.5-0.6	0.3 max	0.3 max	0.6-0.9	0.4 max	0.055 max	0.055 max	0.6 max	Co <=0.1; Mo <=0.15; Pb <=0.15; W <=0.1; P:S Varies; bal Fe	720	396	13	
IS 5517	C55Mn75	Bar, Norm, Ann or Hard, Tmp, 30mm	0.5-0.6			0.6-0.9				0.1-0.35	bal Fe				
International															
ISO 683-1(87)	C55	Bar Frg Plt Wir rod, Blts, Slbs; Q/T, t<16mm	0.52-0.60			0.60-0.90		0.045 max	0.045 max	0.10-0.40	bal Fe	800-950	550		
ISO 683-1(87)	C55E4	Bar Frg Plt Wir rod, Blts, Slbs; Q/T, t<16mm	0.52-0.60			0.60-0.90		0.035 max	0.035 max	0.10-0.40	bal Fe	800-950	550		
ISO 683-1(87)	C55M2	Bar Frg Plt Wir rod, Blts, Slbs; Q/T, t<16mm	0.52-0.60			0.60-0.90		0.035 max	0.020-0.040	0.10-0.40	bal Fe	800-950	550		
ISO 683-18(96)	C55	Bar Wir, CD, t<=5mm	0.52-0.60			0.60-0.90		0.045 max	0.045 max	0.10-0.40	bal Fe	820	670	4	229 HB
ISO 683-18(96)	C55E4	Bar Wir, CD, t<=5mm	0.52-0.60			0.60-0.90		0.035 max	0.035 max	0.10-0.40	bal Fe	820	670	4	229 HB
ISO 683-18(96)	C55M2	Bar Wir, CD, t<=5mm	0.52-0.60			0.60-0.90		0.035 max	0.020-0.040	0.10-0.40	bal Fe	820	670	4	229 HB
ISO R683-3(70)	C55ea	Bar Rod, Q/A, Tmp, 16mm diam	0.52-0.6	0.3 max	0.3 max	0.6-0.9	0.4 max	0.035 max	0.02-0.035	0.15-0.4	Al <=0.1; Co <=0.1; Mo <=0.15; Pb <=0.15; Ti <=0.05; V <=0.1; W <=0.1; bal Fe				
ISO R683-3(70)	C55eb	Bar Rod, Q/A, Tmp, 16mm diam	0.52-0.6			0.6-0.9		0.04	0.03-0.05	0.15-0.4	bal Fe				
Italy															
UNI 5598(71)	3CD55	Wir rod	0.53-0.58			0.4-0.7		0.035 max	0.035 max	0.15-0.35	N <=0.012; bal Fe				
UNI 7064(82)	C55	Spring	0.52-0.6			0.6-0.9		0.035 max	0.035 max	0.15-0.4	bal Fe				
UNI 7065(72)	C55	Strp	0.53-0.58			0.4-0.65		0.02 max	0.02 max	0.35 max	bal Fe				
UNI 7065(72)	C56	Strp	0.52-0.6			0.6-0.9		0.035 max	0.035 max	0.4 max	bal Fe				

UNS numbers and US grades are provided as a means of cross referencing chemically similar alloys. Exchangability is only possible after independent examination of specifications. Tensile properties are minimum or typical as specified. UTS and YS as MPa. El as %. See Appendix for list of abbreviations used in Notes. * indicates obsolete material.

Carbon Steel, Nonresulfurized, 1055 (Continued from previous page)

Specification	Designation	Notes	C	Cr	Cu	Mn	Ni	P	S	Si	Other	UTS	YS	El	Hard
Italy															
UNI 7845(78)	C55	Q/T	0.52-0.6			0.6-0.9		0.035 max	0.035 max	0.15-0.4	bal Fe				
UNI 7874(79)	C55	Q/T	0.52-0.6			0.6-0.9		0.035 max	0.035 max	0.15-0.4	bal Fe				
UNI 8893(86)	C55	Spring	0.52-0.6			0.6-0.9		0.035 max	0.035 max	0.15-0.4	bal Fe				
Japan															
JIS G3311(88)	S55CM	CR Strp	0.52-0.58	0.20 max	0.30 max	0.60-0.90	0.20 max	0.030 max	0.035 max	0.15-0.35	Ni+Cr=0.35 max, bal Fe				180-270 HV
JIS G4051(79)	S53C	Bar Wir rod	0.50-0.56	0.20 max	0.30 max	0.60-0.90	0.20 max	0.030 max	0.035 max	0.15-0.35	Ni+Cr<=0.35; bal Fe				
Mexico															
NMX-B-301(86)	1055	Bar	0.50-0.60			0.60-0.90		0.040 max	0.050 max		bal Fe				
Pan America															
COPANT 330	1055	Bar	0.5-0.6			0.6-0.9		0.04	0.05		bal Fe				
COPANT 331	1055	Bar	0.5-0.6			0.6-0.9		0.04	0.05		bal Fe				
COPANT 333	1055	Wir rod	0.5-0.6			0.6-0.9		0.04	0.05		bal Fe				
Poland															
PNH84019	55	Q/T	0.52-0.6	0.3 max		0.5-0.8	0.3 max	0.04 max	0.04 max	0.17-0.37	bal Fe				
PNH84027	P55		0.5-0.6			0.5-0.8		0.04	0.04	0.17-0.37	bal Fe				
PNH84027	P55G		0.5-0.6			0.6-1		0.04	0.04	0.17-0.37	bal Fe				
PNH84028	D55A	Const; Wir rod	0.53-0.58	0.1 max	0.2 max	0.3-0.6	0.15 max	0.03 max	0.03 max	0.1-0.25	Mo <=0.05; P+S<=0.05; bal Fe				
Romania															
STAS 10677	OLC55CS		0.52-0.57	0.2		0.5-0.8		0.035	0.035	0.17-0.37	Ti 0.02; As 0.05; bal Fe				
STAS 500/2(88)	OL70.1ak	Gen struct; FF; Non-rimming	0.4-0.60	0.3 max		0.8 max	0.4 max	0.05 max	0.05 max	0.4 max	Pb <=0.15; bal Fe				
STAS 500/2(88)	OL70.1bk	Gen struct; FF; Non-rimming	0.4-0.60	0.3 max		0.8 max	0.4 max	0.05 max	0.05 max	0.4 max	Pb <=0.15; bal Fe				
STAS 500/2(88)	OL70.1k	Gen struct; FF; Non-rimming	0.4-0.60	0.3 max		0.8 max	0.4 max	0.05 max	0.05 max	0.4 max	Pb <=0.15; bal Fe				
STAS 60	OT70-1		0.5-0.6	0.3	0.3	0.4-0.8	0.3	0.04	0.045	0.25-0.5	bal Fe				
STAS 60	OT70-2		0.5-0.6	0.3	0.3	0.4-0.8	0.3	0.04	0.045	0.25-0.5	bal Fe				
STAS 795(92)	OLC55A	Spring; HF	0.5-0.6	0.3 max	0.25 max	0.5-0.8		0.04 max	0.04 max	0.17-0.37	Pb <=0.15; bal Fe				
STAS 880(88)	OLC55	Q/T	0.52-0.6	0.3 max		0.5-0.8		0.04 max	0.045 max	0.17-0.37	Pb <=0.15; As<=0.05; bal Fe				
STAS 880(88)	OLC55S	Q/T	0.52-0.6	0.3 max		0.6-0.9		0.04 max	0.02-0.045	0.17-0.37	Pb <=0.15; As<=0.05; bal Fe				
STAS 880(88)	OLC55X	Q/T	0.52-0.6	0.3 max		0.6-0.9		0.035 max	0.035 max	0.17-0.37	Pb <=0.15; As<=0.05; bal Fe				
STAS 880(88)	OLC55XS	Q/T	0.52-0.6	0.3 max		0.6-0.9		0.035 max	0.02-0.04	0.17-0.37	Pb <=0.15; As<=0.05; bal Fe				
Russia															
GOST	L53		0.52			0.7				0.16	bal Fe				
GOST	M56		0.5-0.62			0.5-0.8		0.045	0.05	0.15-0.32	bal Fe				
GOST	M56A		0.5-0.62			0.5-0.8		0.04	0.04	0.15-0.32	bal Fe				
GOST 1050	55		0.52-0.6	0.25	0.25	0.5-0.8	0.25	0.04	0.04	0.17-0.3	As 0.08; bal Fe				
GOST 1050(88)	55	High-grade struct; Q/T; Norm; 0-100mm	0.52-0.6	0.25 max		0.5-0.8	0.3 max	0.035 max	0.04 max	0.17-0.37	As<=0.08; bal Fe	650	380	13L	
GOST 1050(88)	55	High-grade struct; Q/T; Full hard; 0-100mm	0.52-0.6	0.25 max		0.5-0.8	0.3 max	0.035 max	0.04 max	0.17-0.37	As<=0.08; bal Fe				0-269 HB
GOST 398	I		0.5-0.6	0.25	0.3	0.6-0.9	0.25	0.035	0.04	0.2-0.42	V 0.1; bal Fe				

UNS numbers and US grades are provided as a means of cross referencing chemically similar alloys. Exchangability is only possible after independent examination of specifications. Tensile properties are minimum or typical as specified. UTS and YS as MPa. El as %. See Appendix for list of abbreviations used in Notes. * indicates obsolete material.

Specification	Designation	Notes	C	Cr	Cu	Mn	Ni	P	S	Si	Other	UTS	YS	El	Hard

Carbon Steel, Nonresulfurized, 1055 (Continued from previous page)

Russia

Specification	Designation	Notes	C	Cr	Cu	Mn	Ni	P	S	Si	Other	UTS	YS	El	Hard
GOST 7521	B56		0.48-0.65			0.6-1		0.075	0.06	0.15-0.3	bal Fe				
GOST 977	55L		0.52-0.62	0.3	0.3	0.4-0.9	0.3	0.05	0.05	0.2-0.52	bal Fe				
GOST 977	55L-II		0.52-0.6	0.3	0.3	0.4-0.9	0.3	0.04	0.045	0.2-0.52	bal Fe				
GOST 977	55L-III		0.52-0.6	0.3	0.3	0.4-0.9	0.3	0.04	0.045	0.2-0.52	bal Fe				

Spain

Specification	Designation	Notes	C	Cr	Cu	Mn	Ni	P	S	Si	Other	UTS	YS	El	Hard
UNE	F.115		0.5-0.6			0.4-0.7		0.04	0.04	0.15-0.3	bal Fe				
UNE 36011(75)	C55k		0.5-0.6			0.6-0.9		0.035	0.02-0.035	0.15-0.4	bal Fe				
UNE 36011(75)	C55k-1		0.5-0.6			0.6-0.9		0.035	0.02-0.035	0.15-0.4	bal Fe				
UNE 36011(75)	F.1150*	see C55k	0.5-0.6			0.6-0.9		0.035	0.02-0.035	0.15-0.4	bal Fe				
UNE 36011(75)	F.1155*	see C55k-1	0.5-0.6			0.6-0.9		0.035	0.02-0.035	0.15-0.4	bal Fe				

Sweden

Specification	Designation	Notes	C	Cr	Cu	Mn	Ni	P	S	Si	Other	UTS	YS	El	Hard
SS 141655	1655	Non-rimming; Q/T	0.48-0.6			0.4-0.9		0.05 max	0.05 max	0.6 max	bal Fe				
SS 141674	1674	Q/T	0.47-0.55			0.8-0.9		0.035 max	0.035 max	0.1-0.4	bal Fe				

UK

Specification	Designation	Notes	C	Cr	Cu	Mn	Ni	P	S	Si	Other	UTS	YS	El	Hard
BS 970/1(83)	070M55	Blm Bil Slab Bar Rod Frg Norm, 6<t<=63mm	0.50-0.60			0.50-0.90		0.050 max	0.050 max		bal Fe	700	355	12	201-255 HB
BS 970/3(91)	070M55	Bright bar; 29<t<=100mm; HR CD	0.50-0.60			0.50-0.90		0.050 max	0.050 max	0.10-0.40	Mo <=0.15; bal Fe	700-850	415	14	201-255 HB
BS 970/3(91)	070M55	Bright bar; 63<t<=76mm; norm CD	0.50-0.60			0.50-0.90		0.050 max	0.050 max	0.10-0.40	Mo <=0.15; bal Fe	670	530	9	

USA

Specification	Designation	Notes	C	Cr	Cu	Mn	Ni	P	S	Si	Other	UTS	YS	El	Hard
	AISI 1055	Wir rod	0.50-0.60			0.60-0.90		0.040 max	0.050 max		bal Fe				
	AISI 1055	Struct shp	0.50-0.60			0.60-0.90		0.030 max	0.035 max		bal Fe				
	AISI 1055	Plt	0.52-0.60			0.60-0.90		0.030 max	0.035 max		bal Fe				
	AISI 1055	Bar	0.50-0.60			0.60-0.90		0.040 max	0.050 max		bal Fe				
	AISI 1055	Sh Strp	0.50-0.60			0.60-0.90		0.030 max	0.035 max		bal Fe				
	UNS G10550		0.50-0.60			0.60-0.90		0.040 max	0.050 max		bal Fe				
ASTM A29/A29M(93)	1055	Bar	0.50-0.60			0.60-0.90		0.040 max	0.050 max		bal Fe				
ASTM A510(96)	1055	Wir rod	0.50-0.60			0.60-0.90		0.040 max	0.050 max		bal Fe				
ASTM A576(95)	1055	Special qual HW bar	0.50-0.60			0.60-0.90		0.040 max	0.050 max		Si Cu Pb B Bi Ca Se Te if spec'd; bal Fe				
ASTM A682/A682M(98)	1055	Strp, high-C, CR, spring	0.52-0.6			0.60-0.90		0.035 max	0.040 max	0.15-0.30	bal Fe				
ASTM A684A684M(86)	1055	Strp, high-C, CR, Ann	0.52-0.6			0.60-0.90		0.035 max	0.040 max	0.15-0.30	bal Fe				88 HRB
ASTM A713(93)	1055	Wir, High-C spring, HT	0.50-0.60			0.60-0.90		0.040 max	0.050 max		bal Fe				
ASTM A830/A830M(98)	1055	Plt	0.52-0.60			0.60-0.90		0.035 max	0.04 max	0.15-0.40	bal Fe				
FED QQ-S-700D(91)	C1055*	Obs; Sh Strp	0.5-0.6			0.6-0.9		0.04 max	0.05 max		bal Fe				
MIL-S-10520D(AR)(88)	3*	Frg; Obs; see ASTM A711, A575	0.6			1		0.04	0.05	0.15-0.3	bal Fe				
SAE J1397(92)	1055	Bar Ann CD, est mech prop	0.5-0.6			0.6-0.9		0.04 max	0.05 max		bal Fe	660	560	10	197 HB
SAE J1397(92)	1055	Bar HR, est mech prop	0.5-0.6			0.6-0.9		0.04 max	0.05 max		bal Fe	650	360	12	192 HB

UNS numbers and US grades are provided as a means of cross referencing chemically similar alloys. Exchangability is only possible after independent examination of specifications. Tensile properties are minimum or typical as specified. UTS and YS as MPa. El as %. See Appendix for list of abbreviations used in Notes. * indicates obsolete material.

Specification	Designation	Notes	C	Cr	Cu	Mn	Ni	P	S	Si	Other	UTS	YS	El	Hard
Carbon Steel, Nonresulfurized, 1055 (Continued from previous page)															
USA															
SAE J403(95)	1055	Struct Shps Plt Strp Sh Weld Tub	0.52-0.60			0.60-0.90		0.030 max	0.035 max		bal Fe				
SAE J403(95)	1055	Bar Wir rod Smls Tub HR CF	0.50-0.60			0.60-0.90		0.030 max	0.050 max		bal Fe				
Yugoslavia															
	C.1601	Struct	0.52-0.58			0.5-0.8		0.06 max	0.06 max	0.35 max	bal Fe				
	C.1630	Q/T	0.52-0.6			0.6-0.9		0.045 max	0.045 max	0.4 max	Mo <=0.15; bal Fe				
	C.1631	Q/T	0.52-0.6			0.6-0.9		0.035 max	0.03 max	0.4 max	Mo <=0.15; bal Fe				
	C.1633	F/IH	0.5-0.57	0.2 max		0.4-0.7	0.25 max	0.025 max	0.035 max	0.15-0.35	Cr+Mo+Ni<=0.45; bal Fe				
	C.1680	Q/T	0.52-0.6			0.6-0.9		0.035 max	0.02-0.035	0.4 max	Mo <=0.15; bal Fe				
	PZ50		0.5-0.59	0.15 min	0.025	0.3-0.37	0.015	0.04	0.04	0.1-0.3	bal Fe				
Carbon Steel, Nonresulfurized, 1058															
Australia															
AS 1442(92)	1058	HR bar, Semifinished (may treat w/microalloying elements)	0.56-0.63			0.30-0.55		0.040 max	0.040 max	0.10-0.35	Si<=0.10 for Al-killed; bal Fe				
AS 1443(94)	1058	CF bar (may treat w/microalloying elements)	0.56-0.63			0.30-0.55		0.040 max	0.040 max	0.10-0.35	Si<=0.10 for Al-killed; bal Fe				
Carbon Steel, Nonresulfurized, 1059															
Europe															
EN 10016/2(94)	1.0609	Rod	0.55-0.60	0.15 max	0.25 max	0.50-0.80	0.20 max	0.035 max	0.035 max	0.10-0.30	Al <=0.01; Mo <=0.05; Cu+Sn<=0.25; bal Fe				
EN 10016/2(94)	C58D	Rod	0.55-0.60	0.15 max	0.25 max	0.50-0.80	0.20 max	0.035 max	0.035 max	0.10-0.30	Al <=0.01; Mo <=0.05; Cu+Sn<=0.25; bal Fe				
EN 10016/4(94)	1.1212	Rod	0.56-0.60	0.10 max	0.15 max	0.50-0.70	0.10 max	0.020 max	0.025 max	0.10-0.30	Al <=0.01; Mo <=0.03; N <=0.007; Cr+Ni+Cu<=0.30, Cu+Sn<=0.15; bal Fe				
EN 10016/4(94)	C58D2	Rod	0.56-0.60	0.10 max	0.15 max	0.50-0.70	0.10 max	0.020 max	0.025 max	0.10-0.30	Al <=0.01; Mo <=0.03; N <=0.007; Cr+Ni+Cu<=0.30, Cu+Sn<=0.15; bal Fe				
Hungary															
MSZ 5736(88)	D59	Wir rod; for CD	0.55-0.62	0.2 max	0.2 max	0.3-0.7	0.2 max	0.04 max	0.04 max	0.1-0.35	bal Fe				
MSZ 5736(88)	D59K	Wir rod; for CD	0.55-0.62	0.2 max	0.15 max	0.3-0.7	0.2 max	0.025 max	0.025 max	0.1-0.35	bal Fe				
India															
IS 1570/2(79)	60C4		0.55-0.65	0.3 max	0.3 max	0.3-0.6	0.4 max	0.055 max	0.055 max	0.6 max	Co <=0.1; Mo <=0.15; Pb <=0.15; W <=0.1; P:S Varies; bal Fe	750	412	11	
IS 1570/2(79)	60C6	Sh Plt Sect Shp Bar Bil Frg	0.55-0.65	0.3 max	0.3 max	0.5-0.8	0.4 max	0.055 max	0.055 max	0.6 max	Co <=0.1; Mo <=0.15; Pb <=0.15; W <=0.1; P:S Varies; bal Fe				
IS 1570/2(79)	C60		0.55-0.65	0.3 max	0.3 max	0.3-0.6	0.4 max	0.055 max	0.055 max	0.6 max	Co <=0.1; Mo <=0.15; Pb <=0.15; W <=0.1; P:S Varies; bal Fe	750	412	11	
Mexico															
NMX-B-301(86)	1059	Bar	0.55-0.65			0.50-0.80		0.040 max	0.050 max		bal Fe				
Russia															
GOST 1050(88)	58	High-grade struct; Untreated; 0-250mm	0.55-0.63	0.15 max		0.2 max	0.3 max	0.035 max	0.04 max	0.1-0.3	As<=0.08; bal Fe				0-255 HB

UNS numbers and US grades are provided as a means of cross referencing chemically similar alloys. Exchangability is only possible after independent examination of specifications. Tensile properties are minimum or typical as specified. UTS and YS as MPa. El as %. See Appendix for list of abbreviations used in Notes. * indicates obsolete material.

Specification	Designation	Notes	C	Cr	Cu	Mn	Ni	P	S	Si	Other	UTS	YS	El	Hard

Carbon Steel, Nonresulfurized, 1059 (Continued from previous page)

Russia

Specification	Designation	Notes	C	Cr	Cu	Mn	Ni	P	S	Si	Other	UTS	YS	El	Hard
GOST 1050(88)	58	High-grade struct; Norm; 0-100mm	0.55-0.63	0.15 max		0.2 max	0.3 max	0.035 max	0.04 max	0.1-0.3	As<=0.08; bal Fe	600	315	12L	

USA

Specification	Designation	Notes	C	Cr	Cu	Mn	Ni	P	S	Si	Other	UTS	YS	El	Hard
	AISI 1059	Wir rod	0.55-0.65			0.50-0.80		0.040 max	0.050 max		bal Fe				
	AISI 1059	Bar	0.55-0.65			0.50-0.80		0.040 max	0.050 max		bal Fe				
	UNS G10590		0.55-0.65			0.50-0.80		0.040 max	0.050 max		bal Fe				
ASTM A29/A29M(93)	1059	Bar	0.55-0.65			0.50-0.80		0.040 max	0.050 max		bal Fe				
ASTM A510(96)	1059	Wir rod	0.55-0.65			0.50-0.80		0.040 max	0.050 max		bal Fe				
ASTM A713(93)	1059	Wir, High-C spring, HT	0.55-0.65			0.50-0.80		0.040 max	0.050 max		bal Fe				
SAE J1249(95)	1059	Former SAE std valid for Wir rod	0.55-0.65			0.50-0.80		0.040 max	0.050 max		bal Fe				

Carbon Steel, Nonresulfurized, 1060

Argentina

Specification	Designation	Notes	C	Cr	Cu	Mn	Ni	P	S	Si	Other	UTS	YS	El	Hard
IAS	IRAM 1060		0.55-0.66			0.60-0.90		0.040 max	0.050 max	0.10-0.30	bal Fe	780-920	450-540	12-17	233-275 HB

Australia

Specification	Designation	Notes	C	Cr	Cu	Mn	Ni	P	S	Si	Other	UTS	YS	El	Hard
AS 1442	K1060*	Obs; Bar Bil	0.55-0.65			0.6-0.9		0.05	0.05	0.1-0.35	bal Fe				
AS 1442(92)	1060	HR bar, Semifinished (may treat w/microalloying elements)	0.55-0.65			0.60-0.90		0.040 max	0.040 max	0.10-0.35	Si<=0.10 for Al-killed; bal Fe				

Austria

Specification	Designation	Notes	C	Cr	Cu	Mn	Ni	P	S	Si	Other	UTS	YS	El	Hard
ONORM M3161	C60		0.56-0.65			0.5-0.7		0.040	0.040	0.2-0.4	bal Fe				

Belgium

Specification	Designation	Notes	C	Cr	Cu	Mn	Ni	P	S	Si	Other	UTS	YS	El	Hard
NBN 253-02	C60-1		0.57-0.65			0.6-0.9		0.045	0.045	0.15-0.4	bal Fe				
NBN 253-02	C60-2		0.57-0.65			0.6-0.9		0.035	0.02-0.035	0.15-0.4	bal Fe				
NBN 253-02	C60-3		0.57-0.65			0.6-0.9		0.035	0.02-0.035	0.15-0.4	bal Fe				

Bulgaria

Specification	Designation	Notes	C	Cr	Cu	Mn	Ni	P	S	Si	Other	UTS	YS	El	Hard
BDS 5785(83)	60	Struct	0.57-0.65	0.25 max	0.25 max	0.50-0.80	0.25 max	0.040 max	0.040 max	0.17-0.37	bal Fe				
BDS 6742(82)	60G	Struct	0.57-0.65	0.25 max	0.25 max	0.70-1.00	0.25	0.04	0.04	0.17-0.37	bal Fe				

China

Specification	Designation	Notes	C	Cr	Cu	Mn	Ni	P	S	Si	Other	UTS	YS	El	Hard
GB 3522(83)	60	Strp CR Ann	0.57-0.65	0.25 max	0.25 max	0.50-0.80	0.25 max	0.035 max	0.035 max	0.17-0.37	bal Fe	440-735		12	
GB 699(88)	60	Bar HR Norm 25mm diam	0.57-0.65	0.25 max	0.25 max	0.50-0.80	0.25 max	0.035 max	0.035 max	0.17-0.37	bal Fe	675	400	12	
GB/T 3078(94)	60	Bar HR Norm 25mm diam	0.57-0.65	0.25 max	0.25 max	0.50-0.80	0.25 max	0.035 max	0.035 max	0.17-0.37	bal Fe	675	400	12	
YB/T 5101(93)	60	Wir CD Q/T 0.10-2.00mm diam	0.58-0.64	0.10 max	0.20 max	0.50-0.80	0.15 max	0.025 max	0.020 max	0.17-0.37	bal Fe	1815-3090			
YB/T 5103(93)	60	Wir CD Q/T 2.0-4.0mm diam	0.57-0.65	0.25 max	0.25 max	0.50-0.80	0.25 max	0.035 max	0.035 max	0.17-0.37	bal Fe	1425-1765			

Czech Republic

Specification	Designation	Notes	C	Cr	Cu	Mn	Ni	P	S	Si	Other	UTS	YS	El	Hard
CSN 411700	11700	Gen struct	0.65 max	0.3 max		1.60 max		0.055 max	0.05 max	0.60 max	bal Fe				
CSN 412061	12061	Q/T	0.57-0.65	0.25 max		0.5-0.8	0.3 max	0.4 max	0.4 max	0.15-0.4	bal Fe				

Europe

Specification	Designation	Notes	C	Cr	Cu	Mn	Ni	P	S	Si	Other	UTS	YS	El	Hard
EN 10016/2(94)	1.0610	Rod	0.58-0.63	0.15 max	0.25 max	0.50-0.80	0.20 max	0.035 max	0.035 max	0.10-0.30	Al <=0.01; Mo <=0.05; Cu+Sn<=0.25; bal Fe				
EN 10016/2(94)	C60D	Rod	0.58-0.63	0.15 max	0.25 max	0.50-0.80	0.20 max	0.035 max	0.035 max	0.10-0.30	Al <=0.01; Mo <=0.05; Cu+Sn<=0.25; bal Fe				

UNS numbers and US grades are provided as a means of cross referencing chemically similar alloys. Exchangability is only possible after independent examination of specifications. Tensile properties are minimum or typical as specified. UTS and YS as MPa. El as %. See Appendix for list of abbreviations used in Notes. * indicates obsolete material.

Carbon Steel, Nonresulfurized, 1060 (Continued from previous page)

Specification	Designation	Notes	C	Cr	Cu	Mn	Ni	P	S	Si	Other	UTS	YS	El	Hard
Europe															
EN 10016/4(94)	1.1226	Rod	0.58-0.62	0.10 max	0.15 max	0.50-0.70	0.10 max	0.020 max	0.025 max	0.10-0.30	Al <=0.01; Mo <=0.03; N <=0.007; Cr+Ni+Cu<=0.30, Cu+Sn<=0.15; bal Fe				
EN 10016/4(94)	C60D2	Rod	0.58-0.62	0.10 max	0.15 max	0.50-0.70	0.10 max	0.020 max	0.025 max	0.10-0.30	Al <=0.01; Mo <=0.03; N <=0.007; Cr+Ni+Cu<=0.30, Cu+Sn<=0.15; bal Fe				
EN 10083/1(91)A1(96)	1.1221	Q/T t<=16mm	0.57-0.65	0.40 max		0.60-0.90	0.40 max	0.035 max	0.020-0.040	0.40 max	Mo <=0.10; Cr+Mo+Ni<=0.63; bal Fe	850-1000	580	11	
EN 10083/1(91)A1(96)	1.1223	Q/T t<=16mm	0.57-0.65	0.40 max		0.60-0.90	0.40 max	0.035 max	0.020-0.040	0.40 max	Mo <=0.10; Cr+Mo+Ni<=0.63; bal Fe	850-1000	580	11	
EN 10083/1(91)A1(96)	C60E	Q/T t<=16mm	0.57-0.65	0.40 max		0.60-0.90	0.40 max	0.035 max	0.020-0.040	0.40 max	Mo <=0.10; Cr+Mo+Ni<=0.63; bal Fe	850-1000	580	11	
EN 10083/1(91)A1(96)	C60R	Q/T t<=16mm	0.57-0.65	0.40 max		0.60-0.90	0.40 max	0.035 max	0.020-0.040	0.40 max	Mo <=0.10; Cr+Mo+Ni<=0.63; bal Fe	850-1000	580	11	
EN 10083/2(91)A1(96)	1.0601	Q/T t<=16mm	0.57-0.65	0.40 max		0.60-0.90	0.40 max	0.045 max	0.045 max	0.40 max	Mo <=0.10; Cr+Mo+Ni<=0.63; bal Fe	850-1000	580	11	
EN 10083/2(91)A1(96)	C60	Q/T t<=16mm	0.57-0.65	0.40 max		0.60-0.90	0.40 max	0.045 max	0.045 max	0.40 max	Mo <=0.10; Cr+Mo+Ni<=0.63; bal Fe	850-1000	580	11	
France															
AFNOR	CC55										bal Fe				
AFNOR NFA35553	XC60	Strp	0.57-0.65			0.4-0.7		0.04	0.04	0.15-0.35	bal Fe				
Germany															
DIN	60Mn3		0.57-0.65			0.70-0.90		0.050 max	0.050 max	0.20-0.40	N <=0.007; bal Fe				
DIN	WNr 1.0642		0.57-0.65			0.70-0.90		0.050 max	0.050 max	0.20-0.40	N <=0.007; bal Fe				
DIN EN 10083(91)	CK60	Q/T 17-40mm	0.57-0.65	0.40 max		0.60-0.90	0.40 max	0.035 max	0.035 max	0.40 max	Mo <=0.10; Cr+Mo+Ni<=0.63; bal Fe	800-950	520	13	
DIN EN 10083(91)	Cm60	Q/T	0.57-0.65	0.40 max		0.60-0.90	0.40 max	0.035 max	0.020-0.040	0.40 max	Mo <=0.10; Cr+Mo+Ni<=0.63; bal Fe				
DIN EN 10083(91)	WNr 1.0601	Q/T 17-40mm	0.57-0.65	0.40 max		0.60-0.90	0.40 max	0.045 max	0.045 max	0.40 max	Mo <=0.10; Cr+Mo+Ni<=0.63; bal Fe	800-950	520	13	
DIN EN 10083(91)	WNr 1.1223	Q/T	0.57-0.65	0.40 max		0.60-0.90	0.40 max	0.035 max	0.020-0.040	0.40 max	Mo <=0.10; Cr+Mo+Ni<=0.63; bal Fe				
Hungary															
MSZ 4217(85)	C60	CR, Q/T or spring Strp; Q/T; 0-2mm; rolled, hard	0.57-0.65			0.6-0.9		0.035 max	0.035 max	0.4 max	bal Fe	730-1130			
MSZ 4217(85)	C60	CR, Q/T or spring Strp; Q/T; 0-2mm	0.57-0.65			0.6-0.9		0.035 max	0.035 max	0.4 max	bal Fe	980-1280		5L	
MSZ 61	C60Ek		0.57-0.65			0.6-0.9		0.035	0.035	0.17-0.37	bal Fe				
MSZ 61(85)	C60	Q/T; t<=16mm	0.57-0.65			0.6-0.9		0.035 max	0.035 max	0.4 max	bal Fe	850-1000	580	11L	
MSZ 61(85)	C60	Q/T; 40<t<=100mm	0.57-0.65			0.6-0.9		0.035 max	0.035 max	0.4 max	bal Fe	750-900	450	14L	
MSZ 61(85)	C60E	Q/T; 40<t<=100mm	0.57-0.65			0.6-0.9		0.035 max	0.02-0.035	0.4 max	bal Fe	750-900	450	14L	
MSZ 61(85)	C60E	Q/T; t<=16mm	0.57-0.65			0.6-0.9		0.035 max	0.02-0.035	0.4 max	bal Fe	850-1000	580	11L	
India															
IS 1570	C60		0.55-0.65			0.5-0.8					bal Fe				
IS 4454	SW		0.55-0.75		0.15	0.6-0.9		0.04	0.04	0.1-0.35	bal Fe				

UNS numbers and US grades are provided as a means of cross referencing chemically similar alloys. Exchangability is only possible after independent examination of specifications. Tensile properties are minimum or typical as specified. UTS and YS as MPa. El as %. See Appendix for list of abbreviations used in Notes. * indicates obsolete material.

Specification	Designation	Notes	C	Cr	Cu	Mn	Ni	P	S	Si	Other	UTS	YS	El	Hard
Carbon Steel, Nonresulfurized, 1060 (Continued from previous page)															
International															
ISO 683-1(87)	C60	Bar Frg Plt Wir rod, Blts, Slbs; Q/T, t<16mm	0.57-0.65			0.60-0.90		0.045 max	0.045 max	0.10-0.40	bal Fe	850-1000	580		
ISO 683-1(87)	C60E4	Bar Frg Plt Wir rod, Blts, Slbs; Q/T, t<16mm	0.57-0.65			0.60-0.90		0.035 max	0.035 max	0.10-0.40	bal Fe	850-1000	580		
ISO 683-1(87)	C60M2	Bar Frg Plt Wir rod, Blts, Slbs; Q/T, t<16mm	0.57-0.65			0.60-0.90		0.035 max	0.020-0.040	0.10-0.40	bal Fe	850-1000	580		
ISO 683-18(96)	C60	Bar Wir, CD, t<=5mm	0.57-0.65			0.60-0.90		0.045 max	0.045 max	0.10-0.40	bal Fe	860	700	4	241 HB
ISO 683-18(96)	C60E4	Bar Wir, CD, t<=5mm	0.57-0.65			0.60-0.90		0.035 max	0.035 max	0.10-0.40	bal Fe	860	700	4	241 HB
ISO 683-18(96)	C60M2	Bar Wir, CD, t<=5mm	0.57-0.65			0.60-0.90		0.035 max	0.020-0.040	0.10-0.40	bal Fe	860	700	4	241 HB
ISO 683-3(70)	C60ea*	Bar Rod, Q/A, Tmp, 16mm diam	0.57-0.65	0.3 max	0.3 max	0.6-0.9	0.4 max	0.035 max	0.02-0.035	0.15-0.4	Al <=0.1; Co <=0.1; Mo <=0.15; Pb <=0.15; Ti <=0.05; V <=0.1; W <=0.1; bal Fe				
ISO R683-3(70)	C60eb	Bar Rod, Q/A, Tmp, 16mm diam	0.57-0.65			0.6-0.9		0.04	0.03-0.05	0.15-0.4	bal Fe				
Italy															
UNI 3545	C60		0.57-0.65			0.5-0.9		0.035	0.035	0.15-0.4	bal Fe				
UNI 5598(71)	3CD60	Wir rod	0.58-0.63			0.4-0.7		0.035 max	0.035 max	0.15-0.35	N <=0.012; bal Fe				
UNI 7064(82)	C60	Spring	0.57-0.65			0.6-0.9		0.035 max	0.035 max	0.15-0.4	bal Fe				
UNI 7065(72)	C60	Strp	0.58-0.63			0.4-0.65		0.02 max	0.02 max	0.35 max	bal Fe				
UNI 7065(72)	C61	Strp	0.57-0.65			0.6-0.9		0.035 max	0.035 max	0.4 max	bal Fe				
UNI 7845(78)	C60	Q/T	0.57-0.65			0.6-0.9		0.035 max	0.035 max	0.15-0.4	bal Fe				
UNI 7874(79)	C60	Q/T	0.57-0.65			0.6-0.9		0.035 max	0.035 max	0.15-0.4	bal Fe				
UNI 8893(86)	C60	Spring	0.57-0.63			0.6-0.9		0.035 max	0.035 max	0.15-0.4	bal Fe				
Japan															
JIS G3311(88)	S60CM	CR Strp	0.55-0.65	0.20 max	0.30 max	0.60-0.90	0.20 max	0.030 max	0.035 max	0.15-0.35	bal Fe				190-280 HV
JIS G3311(88)	S65CM	CR Strp	0.60-0.70	0.20 max	0.30 max	0.60-0.90	0.20 max	0.030 max	0.035 max	0.15-0.35	bal Fe				190-280 HV
JIS G4051(79)	S58C	Bar Wir rod	0.55-0.61	0.20 max	0.30 max	0.60-0.90	0.20 max	0.030 max	0.035 max	0.15-0.35	Ni+Cr<=0.35; bal Fe				
Mexico															
NMX-B-301(86)	1060	Bar	0.55-0.65			0.60-0.90		0.040 max	0.050 max		bal Fe				
Pan America															
COPANT 330	1060	Bar	0.55-0.65			0.6-0.9		0.04	0.05		bal Fe				
COPANT 331	1060	Bar	0.55-0.65			0.6-0.9		0.04	0.05		bal Fe				
COPANT 333	1060	Wir rod	0.55-0.65			0.6-0.9		0.04	0.05		bal Fe				
Poland															
PNH84019	60	Q/T; Hand tools	0.57-0.65	0.3 max		0.5-0.8	0.3 max	0.04 max	0.04 max	0.17-0.37	bal Fe				
PNH84019	60G	Q/T	0.47-0.65	0.3 max		0.7-1	0.3 max	0.04 max	0.04 max	0.17-0.37	bal Fe				
PNH84028	D55	Wir rod	0.53-0.58	0.2 max	0.2 max	0.3-0.6	0.2 max	0.035 max	0.035	0.1-0.3	Mo <=0.08; P+S<=0.06; bal Fe				
Romania															
STAS 880(88)	OLC60	Q/T	0.57-0.65	0.3 max		0.6-0.9		0.04 max	0.045 max	0.17-0.37	Pb <=0.15; As<=0.05; bal Fe				

UNS numbers and US grades are provided as a means of cross referencing chemically similar alloys. Exchangability is only possible after independent examination of specifications. Tensile properties are minimum or typical as specified. UTS and YS as MPa. El as %. See Appendix for list of abbreviations used in Notes. * indicates obsolete material.

Specification	Designation	Notes	C	Cr	Cu	Mn	Ni	P	S	Si	Other	UTS	YS	El	Hard

Carbon Steel, Nonresulfurized, 1060 (Continued from previous page)

Romania

Specification	Designation	Notes	C	Cr	Cu	Mn	Ni	P	S	Si	Other	UTS	YS	El	Hard
STAS 880(88)	OLC60S	Q/T	0.57-0.65	0.3 max		0.6-0.9		0.04 max	0.02-0.045	0.17-0.37	Pb <=0.15; As<=0.05; bal Fe				
STAS 880(88)	OLC60X	Q/T	0.57-0.65	0.3 max		0.6-0.9		0.035 max	0.035 max	0.17-0.37	Pb <=0.15; As<=0.05; bal Fe				
STAS 880(88)	OLC60XS	Q/T	0.57-0.65	0.3 max		0.6-0.9		0.035 max	0.02-0.04	0.17-0.37	Pb <=0.15; As<=0.05; bal Fe				

Russia

Specification	Designation	Notes	C	Cr	Cu	Mn	Ni	P	S	Si	Other	UTS	YS	El	Hard
GOST 1050(74)	60G	High-grade struct; Untreated; 0-250mm	0.57-0.65	0.25 max	0.25 max	0.7-1	0.25 max	0.035 max	0.04 max	0.17-0.37	As<=0.08; bal Fe				0-269 HB
GOST 1050(74)	60G	High-grade struct; Norm; 0-100mm	0.57-0.65	0.25 max	0.25 max	0.7-1	0.25 max	0.035 max	0.04 max	0.17-0.37	As<=0.08; bal Fe	700	710	11L	
GOST 1050(88)	60	High-grade struct; Q/T; Full hard; 0-100mm	0.57-0.65	0.25 max		0.5-0.8	0.3 max	0.035 max	0.04 max	0.17-0.37	As<=0.08; bal Fe				269 HB max
GOST 1050(88)	60	High-grade struct; Q/T; Norm; 0-100mm	0.57-0.65	0.25 max		0.5-0.8	0.3 max	0.035 max	0.04 max	0.17-0.37	As<=0.08; bal Fe	680	400	12L	
GOST 10791	2		0.55-0.65	0.25	0.25	0.5-0.9	0.25	0.035	0.04	0.2-0.42	bal Fe				
GOST 14959	60G	Spring; Q/T	0.57-0.65	0.25 max	0.2 max	0.7-1	0.25 max	0.035 max	0.035 max	0.17-0.37	bal Fe				
GOST 14959	60G		0.57-0.65	0.25	0.2	0.7-1	0.25	0.04	0.035	0.17-0.37	bal Fe				
GOST 398	II		0.57-0.65	0.25	0.3	0.6-0.9	0.25	0.04	0.04	0.25-0.42	V 0.1; bal Fe				
GOST 398	III		0.57-0.65	0.25	0.3	0.6-0.9	0.25	0.04	0.04	0.25-0.42	V 0.1; bal Fe				
GOST 4121	K62		0.5-0.73			0.6-1		0.055	0.05	0.15-0.3	bal Fe				
GOST 4121	K63	Crane rail	0.53-0.73	0.3 max	0.3 max	0.6-1	0.3 max	0.05 max	0.05 max	0.15-0.35	Al <=0.1; As<=0.08; bal Fe				

Sweden

Specification	Designation	Notes	C	Cr	Cu	Mn	Ni	P	S	Si	Other	UTS	YS	El	Hard
SS 141665	1665	Q/T	0.5-0.6			0.3-0.6		0.03 max	0.03 max	0.15-0.35	bal Fe				
SS 141678	1678	Q/T	0.57-0.65			0.6-0.9		0.035 max	0.035 max	0.15-0.4	bal Fe				

Switzerland

Specification	Designation	Notes	C	Cr	Cu	Mn	Ni	P	S	Si	Other	UTS	YS	El	Hard
VSM 10648	C60	Bar Sh Plt, as roll 5/16mm	0.57-0.65			0.6-0.9		0.05	0.05	0.15-0.35	bal Fe				
VSM 10648	C60	Bar Sh Plt, CD, 5mm diam	0.57-0.65			0.6-0.9		0.05	0.05	0.15-0.35	bal Fe				
VSM 10648	C60	Bar Sh Plt, CD, Norm 16/100mm	0.57-0.65			0.6-0.9		0.05	0.05	0.15-0.35	bal Fe				
VSM 10648	CK60	Bar Sh Plt, as roll 5/16mm	0.57-0.65			0.6-0.9		0.04	0.04	0.15-0.35	bal Fe				
VSM 10648	CK60	Bar Sh Plt, CD, Norm 16/100mm	0.57-0.65			0.6-0.9		0.04	0.04	0.15-0.35	bal Fe				
VSM 10648	CK60	Bar Sh Plt, CD, 5mm diam	0.57-0.65			0.6-0.9		0.04	0.04	0.15-0.35	bal Fe				

UK

Specification	Designation	Notes	C	Cr	Cu	Mn	Ni	P	S	Si	Other	UTS	YS	El	Hard
BS 1449/1(91)	60CS	Plt Sht Strp; HR Q/T t<=16mm	0.55-0.65			0.50-0.90		0.045 max	0.045 max	0.05-0.35	bal Fe				0-166 HB
BS 1449/1(91)	60CS	Plt Sht Strp; CD norm t<=16mm	0.55-0.65			0.50-0.90		0.045 max	0.045 max	0.05-0.35	bal Fe				0-166 HB
BS 1449/1(91)	60HS	Plt Sh Strp HR hot; t<=16mm	0.55-0.65			0.50-0.90		0.045 max	0.045 max	0.05-0.35	bal Fe				0-257 HB
BS 1449/1(91)	60HS	Plt Sht Strp; CD norm t<16mm	0.55-0.65			0.50-0.90		0.045 max	0.045 max	0.05-0.35	bal Fe				
BS 970/1(83)	060A57	Blm Bil Slab Bar Rod Frg	0.55-0.60			0.50-0.70		0.050 max	0.050 max	0.4	bal Fe				
BS 970/1(83)	080A57	Blm Bil Slab Bar Rod Frg	0.55-0.60			0.70-0.90		0.050 max	0.050 max	0.4	bal Fe				
BS 970/1(96)	060A62	Blm Bil Slab Bar Rod Frg	0.60-0.65			0.50-0.70		0.050 max	0.050 max	0.10-0.40	bal Fe				207 HB max

USA

Specification	Designation	Notes	C	Cr	Cu	Mn	Ni	P	S	Si	Other	UTS	YS	El	Hard
	AISI 1060	Struct shp	0.55-0.65			0.60-0.90		0.030 max	0.035 max		bal Fe				

UNS numbers and US grades are provided as a means of cross referencing chemically similar alloys. Exchangability is only possible after independent examination of specifications. Tensile properties are minimum or typical as specified. UTS and YS as MPa. El as %. See Appendix for list of abbreviations used in Notes. * indicates obsolete material.

Specification	Designation	Notes	C	Cr	Cu	Mn	Ni	P	S	Si	Other	UTS	YS	El	Hard

Carbon Steel, Nonresulfurized, 1060 (Continued from previous page)

USA

Specification	Designation	Notes	C	Cr	Cu	Mn	Ni	P	S	Si	Other	UTS	YS	El	Hard
	AISI 1060	Bar	0.55-0.65			0.60-0.90		0.040 max	0.050 max		bal Fe				
	AISI 1060	Wir rod	0.55-0.65			0.60-0.90		0.040 max	0.050 max		bal Fe				
	AISI 1060	Sh Strp Plt	0.55-0.66			0.60-0.90		0.030 max	0.035 max		bal Fe				
	UNS G10600		0.55-0.65			0.60-0.90		0.040 max	0.050 max		bal Fe				
AMS 7240		Bar Rod, Strp, Bil, Frg	0.55-0.65			0.60-0.90		0.040 max	0.050 max		bal Fe				
ASTM A29/A29M(93)	1060	Bar	0.55-0.65			0.60-0.90		0.040 max	0.050 max		bal Fe				
ASTM A510(96)	1060	Wir rod	0.55-0.65			0.60-0.90		0.040 max	0.050 max		bal Fe				
ASTM A513(97a)	1060	ERW Mech tub	0.55-0.66			0.60-0.90		0.040 max	0.050 max		bal Fe				
ASTM A576(95)	1060	Special qual HW bar	0.55-0.65			0.60-0.90		0.040 max	0.050 max		Si Cu Pb B Bi Ca Se Te if spec'd; bal Fe				
ASTM A682/A682M(98)	1060	Strp, high-C, CR, spring	0.55-0.65			0.60-0.90		0.035 max	0.040 max	0.15-0.30	bal Fe				
ASTM A684A684M(86)	1060	Strp, high-C, CR, Ann	0.55-0.65			0.60-0.90		0.035 max	0.040 max	0.15-0.30	bal Fe				90 HRB
ASTM A713(93)	1060	Wir, High-C spring, HT	0.55-0.65			0.60-0.90		0.040 max	0.050 max		bal Fe				
ASTM A830/A830M(98)	1060	Plt	0.55-0.66			0.60-0.90		0.035 max	0.04 max	0.15-0.40	bal Fe				
DoD-F-24669/1(86)	1060	Bar Bil; Supersedes MIL-S-866 & MIL-S-16974	0.55-0.65			0.6-0.9		0.04	0.05		bal Fe				
FED QQ-S-00640(63)	1060*	Obs; Sh Strp	0.55-0.65			0.6-0.9		0.04 max	0.05 max		bal Fe				
FED QQ-S-633A(63)	1060*	Obs; CF HR Bar	0.55-0.65			0.6-0.9		0.04 max	0.05 max		bal Fe				
MIL-S-10520D(AR)(88)	5*	Frg; Obs; see ASTM A711, A575	0.65			1		0.04	0.05	0.15-0.3	bal Fe				
MIL-S-10520D(AR)(88)	7*	Frg; Obs; see ASTM A711, A575	0.65			1.3		0.04	0.05	0.15-0.3	bal Fe				
MIL-S-16788A(86)	C6*	HR, frg, bloom bil slab for refrg; Obs for new design	0.55-0.65			0.60-0.90		0.040 max	0.050 max		bal Fe				
MIL-S-16974E(86)	1060*	Obs; Bar bil boom slab for refrg	0.55-0.65			0.6-0.9		0.04 max	0.05 max		bal Fe				
SAE J1397(92)	1060	Bar HR, est mech prop	0.55-0.65			0.6-0.9		0.04 max	0.05 max		bal Fe	680	370	12	201 HB
SAE J1397(92)	1060	Bar Ann CD, est mech prop	0.55-0.65			0.6-0.9		0.04 max	0.05 max		bal Fe	620	480	10	183 HB
SAE J403(95)	1060	Struct Shps Plt Strp Sh Weld Tub	0.55-0.66			0.60-0.90		0.030 max	0.035 max		bal Fe				
SAE J403(95)	1060	Bar Wir rod Smls Tub HR CF	0.55-0.65			0.60-0.90		0.030 max	0.050 max		bal Fe				

Yugoslavia

Specification	Designation	Notes	C	Cr	Cu	Mn	Ni	P	S	Si	Other	UTS	YS	El	Hard
	C.1700	Struct	0.58-0.65			0.5-0.8		0.06 max	0.06 max	0.35 max	bal Fe				
	C.1730	Q/T	0.57-0.65			0.6-0.9		0.045 max	0.045 max	0.4 max	Mo <=0.15; bal Fe				
	C.1731	Q/T	0.57-0.65			0.6-0.9		0.035 max	0.03 max	0.4 max	Mo <=0.15; bal Fe				
	C.1780	Q/T	0.57-0.65			0.6-0.9		0.035 max	0.02-0.035	0.4 max	Mo <=0.15; bal Fe				

Carbon Steel, Nonresulfurized, 1062

China

Specification	Designation	Notes	C	Cr	Cu	Mn	Ni	P	S	Si	Other	UTS	YS	El	Hard
GB 699(88)	60 Mn	Bar HR Norm 25mm diam	0.57-0.65	0.25 max	0.25 max	0.70-1.00	0.25 max	0.035 max	0.035 max	0.17-0.37	bal Fe	695	410	11	
GB/T 3078(94)	60 Mn	Bar CR Norm 25mm diam	0.57-0.65	0.25 max	0.25 max	0.70-1.00	0.25 max	0.035 max	0.035 max	0.17-0.37	bal Fe	695	410	11	
YB/T 5101(93)	60 Mn	Wir CD 0.10-2.00mm diam	0.58-0.64	0.10 max	0.20 max	0.70-1.00	0.15 max	0.025 max	0.020 max	0.17-0.37	bal Fe	1815-3090			

UNS numbers and US grades are provided as a means of cross referencing chemically similar alloys. Exchangability is only possible after independent examination of specifications. Tensile properties are minimum or typical as specified. UTS and YS as MPa. El as %. See Appendix for list of abbreviations used in Notes. * indicates obsolete material.

Specification	Designation	Notes	C	Cr	Cu	Mn	Ni	P	S	Si	Other	UTS	YS	El	Hard

Carbon Steel, Nonresulfurized, 1062 (Continued from previous page)

China

Specification	Designation	Notes	C	Cr	Cu	Mn	Ni	P	S	Si	Other	UTS	YS	El	Hard
YB/T 5103(93)	60 Mn	Wir CD Q/T 2.0-4.0mm diam	0.57-0.65	0.25 max	0.25 max	0.70-1.00	0.25 max	0.035 max	0.035 max	0.17-0.37	bal Fe	1425-1765			

Carbon Steel, Nonresulfurized, 1064

Europe

Specification	Designation	Notes	C	Cr	Cu	Mn	Ni	P	S	Si	Other	UTS	YS	El	Hard
EN 10016/2(94)	1.0611	Rod	0.60-0.65	0.15 max	0.25 max	0.50-0.80	0.20 max	0.035 max	0.035 max	0.10-0.30	Al <=0.01; Mo <=0.05; Cu+Sn<=0.25; bal Fe				
EN 10016/2(94)	C62D	Rod	0.60-0.65	0.15 max	0.25 max	0.50-0.80	0.20 max	0.035 max	0.035 max	0.10-0.30	Al <=0.01; Mo <=0.05; Cu+Sn<=0.25; bal Fe				
EN 10016/4(94)	1.1222	Rod	0.60-0.64	0.10 max	0.15 max	0.50-0.70	0.10 max	0.020 max	0.025 max	0.10-0.30	Al <=0.01; Mo <=0.03; N <=0.007; Cr+Ni+Cu<=0.30, Cu+Sn<=0.15; bal Fe				
EN 10016/4(94)	C62D2	Rod	0.60-0.64	0.10 max	0.15 max	0.50-0.70	0.10 max	0.020 max	0.025 max	0.10-0.30	Al <=0.01; Mo <=0.03; N <=0.007; Cr+Ni+Cu<=0.30, Cu+Sn<=0.15; bal Fe				

France

Specification	Designation	Notes	C	Cr	Cu	Mn	Ni	P	S	Si	Other	UTS	YS	El	Hard
AFNOR NFA36102(93)	C60RR	HR; Strp; CR	0.57-0.65	0.3 max		0.5-0.8	0.40 max	0.025 max	0.02 max	0.15-0.35	Al <=0.03; Mo <=0.10; bal Fe				
AFNOR NFA36102(93)	XC60*	HR; Strp; CR	0.57-0.65	0.3 max		0.5-0.8	0.40 max	0.025 max	0.02 max	0.15-0.35	Al <=0.03; Mo <=0.10; bal Fe				

Germany

Specification	Designation	Notes	C	Cr	Cu	Mn	Ni	P	S	Si	Other	UTS	YS	El	Hard
DIN EN 10083(91)	WNr 1.1221	Q/T 17-40mm	0.57-0.65	0.40 max		0.60-0.90	0.40 max	0.035 max	0.035 max	0.40 max	Mo <=0.10; Cr+Mo+Ni<=0.63; bal Fe	800-950	520	13	

Hungary

Specification	Designation	Notes	C	Cr	Cu	Mn	Ni	P	S	Si	Other	UTS	YS	El	Hard
MSZ 2752(89)	B6	Semifinished railway tire; Non-rimming; FF	0.58-0.65			0.6-0.9	0.3 max	0.04 max	0.04 max	0.2-0.5	Mo <=0.08; V <=0.05; bal Fe	920-1050		12L	

India

Specification	Designation	Notes	C	Cr	Cu	Mn	Ni	P	S	Si	Other	UTS	YS	El	Hard
IS 1570/2(79)	65C6		0.6-0.7	0.3 max	0.3 max	0.5-0.8	0.4 max	0.055 max	0.055 max	0.6 max	Co <=0.1; Mo <=0.15; Pb <=0.15; W <=0.1; P:S Varies; bal Fe	750	412	10	
IS 1570/2(79)	C65		0.6-0.7	0.3 max	0.3 max	0.5-0.8	0.4 max	0.055 max	0.055 max	0.6 max	Co <=0.1; Mo <=0.15; Pb <=0.15; W <=0.1; P:S Varies; bal Fe	750	412	10	

Mexico

Specification	Designation	Notes	C	Cr	Cu	Mn	Ni	P	S	Si	Other	UTS	YS	El	Hard
NMX-B-301(86)	1064	Bar	0.60-0.70			0.50-0.80		0.040 max	0.050 max		bal Fe				

Russia

Specification	Designation	Notes	C	Cr	Cu	Mn	Ni	P	S	Si	Other	UTS	YS	El	Hard
GOST 1050(74)	65	High-grade struct; Norm; 0-100mm	0.62-0.7	0.25 max	0.25 max	0.5-0.8	0.25 max	0.035 max	0.04 max	0.17-0.37	As<=0.08; bal Fe	700	410	10L	
GOST 1050(74)	65	High-grade struct; Untreated; 0-250mm	0.62-0.7	0.25 max	0.25 max	0.5-0.8	0.25 max	0.035 max	0.04 max	0.17-0.37	As<=0.08; bal Fe				0-255 HB
GOST 14959	60GA	Spring	0.57-0.65	0.25 max	0.2 max	0.7-1.0	0.25 max	0.025 max	0.025 max	0.17-0.37	Al <=0.1; bal Fe				

USA

Specification	Designation	Notes	C	Cr	Cu	Mn	Ni	P	S	Si	Other	UTS	YS	El	Hard
	AISI 1064	Sh Strp Plt	0.59-0.70			0.50-0.80		0.030 max	0.035 max		bal Fe				
	AISI 1064	Struct shp	0.60-0.70			0.50-0.80		0.030 max	0.035 max		bal Fe				
	AISI 1064	Wir rod	0.60-0.70			0.50-0.80		0.040 max	0.050 max		bal Fe				
	UNS G10640		0.60-0.70			0.50-0.80		0.040 max	0.050 max		Sheets and Plates, C 0.59-0.70; bal Fe				
ASTM A29/A29M(93)	1064	Bar	0.60-0.70			0.50-0.80		0.040 max	0.050 max		bal Fe				
ASTM A510(96)	1064	Wir rod	0.60-0.70			0.50-0.80		0.040 max	0.050 max		bal Fe				
ASTM A682/A682M(98)	1064	Strp, high-C, CR, spring	0.59-0.70			0.50-0.80		0.035 max	0.040 max	0.15-0.30	bal Fe				
ASTM A684A684M(86)	1064	Strp, high-C, CR, Ann	0.59-0.70			0.50-0.80		0.035 max	0.040 max	0.15-0.30	bal Fe				92 HRB

UNS numbers and US grades are provided as a means of cross referencing chemically similar alloys. Exchangability is only possible after independent examination of specifications. Tensile properties are minimum or typical as specified. UTS and YS as MPa. El as %. See Appendix for list of abbreviations used in Notes. * indicates obsolete material.

Specification	Designation	Notes	C	Cr	Cu	Mn	Ni	P	S	Si	Other	UTS	YS	El	Hard

Carbon Steel, Nonresulfurized, 1064 (Continued from previous page)

USA

Specification	Designation	Notes	C	Cr	Cu	Mn	Ni	P	S	Si	Other	UTS	YS	El	Hard
ASTM A713(93)	1064	Wir, High-C spring, HT	0.60-0.70			0.50-0.80		0.040 max	0.050 max		bal Fe				
ASTM A830/A830M(98)	1064	Plt	0.59-0.70			0.50-0.80		0.035 max	0.04 max	0.15-0.40	bal Fe				
SAE J1249(95)	1064	Former SAE std valid for Wir rod	0.60-0.70			0.50-0.80		0.040 max	0.050 max		bal Fe				
SAE J1249(95)	1069	Former SAE std valid for Wir rod	0.65-0.75			0.40-0.70		0.040 max	0.050 max		bal Fe				
SAE J1397(92)	1064	Bar CD, est mech prop									bal Fe	610	480	10	183 HB
SAE J1397(92)	1064	Bar HR, est mech prop									bal Fe	670	370	12	201 HB

Carbon Steel, Nonresulfurized, 1065

Australia

Specification	Designation	Notes	C	Cr	Cu	Mn	Ni	P	S	Si	Other	UTS	YS	El	Hard
AS 1442(92)	1065	HR bar, Semifinished (may treat w/microalloying elements)	0.60-0.70			0.60-0.90		0.040 max	0.040 max	0.10-0.35	Si<=0.10 for Al-killed; bal Fe				

China

Specification	Designation	Notes	C	Cr	Cu	Mn	Ni	P	S	Si	Other	UTS	YS	El	Hard
GB 1222(84)	65	Bar Flat HR Q/T	0.62-0.70	0.25 max	0.25 max	0.50-0.80	0.25 max	0.035 max	0.035 max	0.17-0.37	bal Fe	980	785	9	
GB 3522(83)	65	Strp CR Ann	0.62-0.70	0.25 max	0.25 max	0.50-0.80	0.25 max	0.035 max	0.035 max	0.17-0.37	bal Fe	440-735		10	
GB 699(88)	65	Bar HR Norm 25mm diam	0.62-0.70	0.25 max	0.25 max	0.50-0.80	0.25 max	0.035 max	0.035 max	0.17-0.37	bal Fe	695	410	10	
GB/T 3078(94)	65	Bar CR Norm 25mm diam	0.62-0.70	0.25 max	0.25 max	0.50-0.80	0.25 max	0.035 max	0.035 max	0.17-0.37	bal Fe	695	410	10	
YB/T 5101(93)	65	Wir CD 0.10-2.00mm diam	0.63-0.69	0.10 max	0.20 max	0.50-0.80	0.15 max	0.025 max	0.020 max	0.17-0.37	bal Fe	1815-3090			
YB/T 5103(93)	65	Wir CD Q/T 2.0-4.0mm diam	0.62-0.70	0.25 max	0.25 max	0.50-0.80	0.25 max	0.035 max	0.035 max	0.17-0.37	bal Fe	1425-1765			

Czech Republic

Specification	Designation	Notes	C	Cr	Cu	Mn	Ni	P	S	Si	Other	UTS	YS	El	Hard
CSN 412071	12071	Q/T; Unalloyed	0.6-0.7	0.3 max		0.6-0.8		0.035 max	0.035 max	0.35 max	bal Fe				

Europe

Specification	Designation	Notes	C	Cr	Cu	Mn	Ni	P	S	Si	Other	UTS	YS	El	Hard
EN 10016/2(94)	1.0612	Rod	0.63-0.68	0.15 max	0.25 max	0.50-0.80	0.20 max	0.035 max	0.035 max	0.10-0.30	Al <=0.01; Mo <=0.05; Cu+Sn<=0.25; bal Fe				
EN 10016/2(94)	C66D	Rod	0.63-0.68	0.15 max	0.25 max	0.50-0.80	0.20 max	0.035 max	0.035 max	0.10-0.30	Al <=0.01; Mo <=0.05; Cu+Sn<=0.25; bal Fe				
EN 10016/4(94)	1.1236	Rod	0.64-0.68	0.10 max	0.15 max	0.50-0.70	0.10 max	0.020 max	0.025 max	0.10-0.30	Al <=0.01; Mo <=0.03; N <=0.007; Cr+Ni+Cu<=0.30, Cu+Sn<=0.15; bal Fe				
EN 10016/4(94)	C66D2	Rod	0.64-0.68	0.10 max	0.15 max	0.50-0.70	0.10 max	0.020 max	0.025 max	0.10-0.30	Al <=0.01; Mo <=0.03; N <=0.007; Cr+Ni+Cu<=0.30, Cu+Sn<=0.15; bal Fe				

France

Specification	Designation	Notes	C	Cr	Cu	Mn	Ni	P	S	Si	Other	UTS	YS	El	Hard
AFNOR NFA36102(93)	C68RR	HR; Strp; CR	0.65-0.73	0.3 max		0.5-0.8	0.40 max	0.025 max	0.02 max	0.15-0.35	Al <=0.03; Mo <=0.10; bal Fe				
AFNOR NFA36102(93)	XC68*	HR; Strp; CR	0.65-0.73	0.3 max		0.5-0.8	0.40 max	0.025 max	0.02 max	0.15-0.35	Al <=0.03; Mo <=0.10; bal Fe				

Hungary

Specification	Designation	Notes	C	Cr	Cu	Mn	Ni	P	S	Si	Other	UTS	YS	El	Hard
MSZ 2752(89)	B4	Semifinished railway tire; Non-rimming; FF	0.6-0.7			0.6-0.9	0.3 max	0.04 max	0.04 max	0.2-0.5	Mo <=0.08; V <=0.05; P<=0.04/P<=0.035; S<=0.04/S<=0.035; bal Fe	800-940		10L	
MSZ 5736(88)	D67	Wir rod; for CD	0.63-0.7	0.2 max	0.2 max	0.3-0.7	0.2 max	0.035 max	0.04 max	0.1-0.35	D67V:V<=0.15; bal Fe				
MSZ 5736(88)	D67K	Wir rod; for CD	0.63-0.7	0.2 max	0.15 max	0.3-0.7	0.2 max	0.025 max	0.025 max	0.1-0.35	bal Fe				

Italy

Specification	Designation	Notes	C	Cr	Cu	Mn	Ni	P	S	Si	Other	UTS	YS	El	Hard
UNI 5598(71)	3CD65	Wir rod	0.63-0.68			0.4-0.7		0.035 max	0.035 max	0.15-0.35	N <=0.012; bal Fe				

UNS numbers and US grades are provided as a means of cross referencing chemically similar alloys. Exchangability is only possible after independent examination of specifications. Tensile properties are minimum or typical as specified. UTS and YS as MPa. El as %. See Appendix for list of abbreviations used in Notes. * indicates obsolete material.

Specification	Designation	Notes	C	Cr	Cu	Mn	Ni	P	S	Si	Other	UTS	YS	El	Hard

Carbon Steel, Nonresulfurized, 1065 (Continued from previous page)

Mexico

Specification	Designation	Notes	C	Cr	Cu	Mn	Ni	P	S	Si	Other	UTS	YS	El	Hard
NMX-B-151(90)		CD wire for mech spring	0.45-0.85			0.30-1.30		0.040 max	0.050 max	0.15-0.35	bal Fe				
NMX-B-301(86)	1065	Bar	0.60-0.70			0.60-0.90		0.040 max	0.050 max		bal Fe				

Romania

Specification	Designation	Notes	C	Cr	Cu	Mn	Ni	P	S	Si	Other	UTS	YS	El	Hard
STAS 795(92)	OLC65A	Spring; HF	0.62-0.7	0.3 max	0.25 max	0.5-0.8		0.04 max	0.04 max	0.17-0.37	Pb <=0.15; bal Fe				

Russia

Specification	Designation	Notes	C	Cr	Cu	Mn	Ni	P	S	Si	Other	UTS	YS	El	Hard
GOST 1071(81)	65GA	Spring	0.65-0.7	0.15 max	0.2 max	0.7-1	0.2 max	0.025 max	0.025 max	0.15-0.3	Al <=0.08; bal Fe				
GOST 1071(81)	68GA	Spring	0.65-0.7	0.12 max	0.15 max	0.7-1	0.2 max	0.025 max	0.025 max	0.15-0.25	Al <=0.05; bal Fe				

UK

Specification	Designation	Notes	C	Cr	Cu	Mn	Ni	P	S	Si	Other	UTS	YS	El	Hard
BS 970/1(96)	060A67	Blm Bil Slab Bar Rod Frg	0.65-0.70			0.50-0.70		0.050 max	0.050 max	0.10-0.40	bal Fe				217 HB max
BS 970/1(96)	080A67	Blm Bil Slab Bar Rod Frg Q/T	0.65-0.70			0.70-0.90			0.050 max	0.10-0.40	bal Fe				229 HB max

USA

Specification	Designation	Notes	C	Cr	Cu	Mn	Ni	P	S	Si	Other	UTS	YS	El	Hard
	AISI 1065	Wir rod	0.60-0.70			0.60-0.90		0.040 max	0.050 max		bal Fe				
	AISI 1065	Bar	0.60-0.70			0.60-0.90		0.040 max	0.050 max		bal Fe				
	AISI 1065	Struct shp	0.60-0.70			0.60-0.90		0.030 max	0.035 max		bal Fe				
	AISI 1065	Sh Strp Plt	0.59-0.70			0.60-0.90		0.030 max	0.035 max		bal Fe				
	UNS G10650		0.60-0.70			0.60-0.90		0.040 max	0.050 max		Sheets and Plates, C 0.54-0.70; bal Fe				
ASTM A29/A29M(93)	1065	Bar	0.60-0.70			0.60-0.90		0.040 max	0.050 max		bal Fe				
ASTM A510(96)	1065	Wir rod	0.60-0.70			0.60-0.90		0.040 max	0.050 max		bal Fe				
ASTM A682/A682M(98)	1065	Strp, high-C, CR, spring	0.59-0.70			0.60-0.90		0.035 max	0.040 max	0.15-0.30	bal Fe				
ASTM A684A684M(86)	1065	Strp, high-C, CR, Ann	0.59-0.70			0.60-0.90		0.035 max	0.040 max	0.15-0.30	bal Fe				92 HRB
ASTM A713(93)	1065	Wir, High-C spring, HT	0.60-0.70			0.60-0.90		0.040 max	0.050 max		bal Fe				
ASTM A830/A830M(98)	1065	Plt	0.59-0.70			0.60-0.90		0.035 max	0.04 max	0.15-0.40	bal Fe				
MIL-S-46049C(98)	1065	Strp CR hard temp, spring	0.59-0.70			0.60-0.90		0.040 max	0.050 max	0.15-0.30	bal Fe				
SAE J1397(92)	1065	Bar CD, est mech prop									bal Fe	630	490	10	187 HB
SAE J1397(92)	1065	Bar HR, est mech prop									bal Fe	690	380	12	207 HB
SAE J403(95)	1065	Bar Wir rod Smls Tub HR CF	0.60-0.70			0.60-0.90		0.030 max	0.050 max		bal Fe				
SAE J403(95)	1065	Struct Shps Plt Strp Sh Weld Tub	0.59-0.70			0.60-0.90		0.030 max	0.035 max		bal Fe				

Yugoslavia

Specification	Designation	Notes	C	Cr	Cu	Mn	Ni	P	S	Si	Other	UTS	YS	El	Hard
	C.1701	Struct	0.62-0.7			0.5-0.8		0.06 max	0.06 max	0.35 max	bal Fe				

Carbon Steel, Nonresulfurized, 1069

Europe

Specification	Designation	Notes	C	Cr	Cu	Mn	Ni	P	S	Si	Other	UTS	YS	El	Hard
EN 10016/2(94)	1.0613	Rod	0.65-0.70	0.15 max	0.25 max	0.50-0.80	0.20 max	0.035 max	0.035 max	0.10-0.30	Al <=0.01; Mo <=0.05; Cu+Sn<=0.25; bal Fe				
EN 10016/2(94)	C68D	Rod	0.65-0.70	0.15 max	0.25 max	0.50-0.80	0.20 max	0.035 max	0.035 max	0.10-0.30	Al <=0.01; Mo <=0.05; Cu+Sn<=0.25; bal Fe				
EN 10016/4(94)	1.1232	Rod	0.66-0.70	0.10 max	0.15 max	0.50-0.70	0.10 max	0.020 max	0.025 max	0.10-0.30	Al <=0.01; Mo <=0.03; N <=0.007; Cr+Ni+Cu<=0.30, Cu+Sn<=0.15; bal Fe				

Specification	Designation	Notes	C	Cr	Cu	Mn	Ni	P	S	Si	Other	UTS	YS	El	Hard

Carbon Steel, Nonresulfurized, 1069 (Continued from previous page)

Europe

Specification	Designation	Notes	C	Cr	Cu	Mn	Ni	P	S	Si	Other	UTS	YS	El	Hard
EN 10016/4(94)	C68D2	Rod	0.66-0.70	0.10 max	0.15 max	0.50-0.70	0.10 max	0.020 max	0.025 max	0.10-0.30	Al <=0.01; Mo <=0.03; N <=0.007; Cr+Ni+Cu<=0.30, Cu+Sn<=0.15; bal Fe				

Italy

Specification	Designation	Notes	C	Cr	Cu	Mn	Ni	P	S	Si	Other	UTS	YS	El	Hard
UNI 7064(82)	C67	Strp	0.65-0.72			0.6-0.9		0.035 max	0.035 max	0.15-0.4	bal Fe				
UNI 8893(86)	C67	Spring	0.65-0.72			0.6-0.9		0.035 max	0.035 max	0.15-0.4	bal Fe				

Mexico

Specification	Designation	Notes	C	Cr	Cu	Mn	Ni	P	S	Si	Other	UTS	YS	El	Hard
NMX-B-301(86)	1069	Bar	0.65-0.75			0.40-0.70		0.040 max	0.050 max		bal Fe				

USA

Specification	Designation	Notes	C	Cr	Cu	Mn	Ni	P	S	Si	Other	UTS	YS	El	Hard
	AISI 1069	Wir rod	0.65-0.75			0.40-0.70		0.040 max	0.050 max		bal Fe				
	UNS G10690		0.65-0.75			0.40-0.70		0.040 max	0.050 max		bal Fe				
ASTM A29/A29M(93)	1069	Bar	0.65-0.75			0.40-0.70		0.040 max	0.050 max		bal Fe				
ASTM A510(96)	1069	Wir rod	0.65-0.75			0.40-0.70		0.040 max	0.050 max		bal Fe				
ASTM A713(93)	1069	Wir, High-C spring, HT	0.65-0.75			0.40-0.70		0.040 max	0.050 max		bal Fe				

Carbon Steel, Nonresulfurized, 1070

Argentina

Specification	Designation	Notes	C	Cr	Cu	Mn	Ni	P	S	Si	Other	UTS	YS	El	Hard
IAS	IRAM 1070		0.65-0.75			0.60-0.90		0.040 max	0.050 max	0.10-0.30	bal Fe	870-1030	500-590	8-16	261-313 HB

Australia

Specification	Designation	Notes	C	Cr	Cu	Mn	Ni	P	S	Si	Other	UTS	YS	El	Hard
AS 1442	K1070*	Obs; Bar Bil	0.65-0.75			0.6-0.9		0.05	0.05	0.1-0.35	bal Fe				
AS 1442	S1070*	Obs; Bar Bil	0.65-0.75			0.6-0.9		0.05	0.05	0.35	bal Fe				
AS 1442(92)	1070	HR bar, Semifinished (may treat w/microalloying elements)	0.65-0.75			0.60-0.90		0.040 max	0.040 max	0.10-0.35	Si<=0.10 for Al-killed; bal Fe				
AS 1446	S1070*	Withdrawn, see AS/NZS 1594(97)	0.65-0.75			0.6-0.9		0.05	0.05	0.35	bal Fe				

Bulgaria

Specification	Designation	Notes	C	Cr	Cu	Mn	Ni	P	S	Si	Other	UTS	YS	El	Hard
BDS 5785(83)	65	Struct	0.62-0.70	0.25 max	0.25	0.50-0.80	0.25	0.035	0.035	0.17-0.37	bal Fe				
BDS 5785(83)	65G	Struct	0.62-0.70	0.25 max	0.25	0.90-1.20	0.25	0.035	0.035	0.17-0.37	bal Fe				
BDS 6742	65	Spring	0.62-0.70	0.25 max	0.30 max	0.50-0.80	0.30 max	0.035 max	0.035 max	0.17-0.37	bal Fe				
BDS 6742	65G	Spring	0.62-0.70	0.25 max	0.30 max	0.90-1.20	0.30 max	0.035 max	0.035 max	0.17-0.37	bal Fe				

China

Specification	Designation	Notes	C	Cr	Cu	Mn	Ni	P	S	Si	Other	UTS	YS	El	Hard
GB 1222(84)	70	Bar Flat HR Q/T	0.67-0.75	0.25 max	0.25 max	0.50-0.80	0.25 max	0.035 max	0.035 max	0.17-0.37	bal Fe	1030	835	8	
GB 3522(83)	70	Strp CR Ann	0.67-0.75	0.25 max	0.25 max	0.50-0.80	0.25 max	0.035 max	0.035 max	0.17-0.37	bal Fe	440-735		10	
GB 699(88)	70	Bar HR Norm 25mm diam	0.67-0.75	0.25 max	0.25 max	0.50-0.80	0.25 max	0.035 max	0.035 max	0.17-0.37	bal Fe	715	420	9	
GB/T 4358(95)	70	Wir CD 0.10-2.00mm diam	0.67-0.74	0.10 max	0.20 max	0.30-0.60	0.15 max	0.025 max	0.020 max	0.17-0.37	bal Fe	1760-2690			
YB/T 5063(93)	70	Strp HR HT	0.65-0.74	0.25 max	0.30 max	0.40 max	0.20 max	0.030 max	0.020 max	0.35 max	bal Fe	1275-1570			
YB/T 5101(93)	70	Wir CD 0.10-2.00mm diam	0.68-0.74	0.10 max	0.20 max	0.50-0.80	0.15 max	0.025 max	0.020 max	0.17-0.37	bal Fe	1815-3090			
YB/T 5103(93)	70	Wir CD Q/T 2.0-4.0mm diam	0.67-0.75	0.25 max	0.25 max	0.50-0.80	0.25 max	0.035 max	0.035 max	0.17-0.37	bal Fe	1425-1765			

Czech Republic

Specification	Designation	Notes	C	Cr	Cu	Mn	Ni	P	S	Si	Other	UTS	YS	El	Hard
CSN 411800	11800	Gen struct	0.75 max	0.3 max		1.60 max		0.055 max	0.05 max	0.60 max	bal Fe				

UNS numbers and US grades are provided as a means of cross referencing chemically similar alloys. Exchangability is only possible after independent examination of specifications. Tensile properties are minimum or typical as specified. UTS and YS as MPa. El as %. See Appendix for list of abbreviations used in Notes. * indicates obsolete material.

Specification	Designation	Notes	C	Cr	Cu	Mn	Ni	P	S	Si	Other	UTS	YS	El	Hard

Carbon Steel, Nonresulfurized, 1070 (Continued from previous page)

Europe

Specification	Designation	Notes	C	Cr	Cu	Mn	Ni	P	S	Si	Other	UTS	YS	El	Hard
EN 10016/2(94)	1.0615	Rod	0.68-0.73	0.15 max	0.25 max	0.50-0.80	0.20 max	0.035 max	0.035 max	0.10-0.30	Al <=0.01; Mo <=0.05; Cu+Sn<=0.25; bal Fe				
EN 10016/2(94)	1.0617	Rod	0.70-0.75	0.15 max	0.25 max	0.50-0.80	0.20 max	0.035 max	0.035 max	0.10-0.30	Al <=0.01; Mo <=0.05; Cu+Sn<=0.25; bal Fe				
EN 10016/2(94)	C70D	Rod	0.68-0.73	0.15 max	0.25 max	0.50-0.80	0.20 max	0.035 max	0.035 max	0.10-0.30	Al <=0.01; Mo <=0.05; Cu+Sn<=0.25; bal Fe				
EN 10016/2(94)	C72D	Rod	0.70-0.75	0.15 max	0.25 max	0.50-0.80	0.20 max	0.035 max	0.035 max	0.10-0.30	Al <=0.01; Mo <=0.05; Cu+Sn<=0.25; bal Fe				
EN 10016/4(94)	1.1242	Rod	0.70-0.74	0.10 max	0.15 max	0.50-0.70	0.10 max	0.020 max	0.025 max	0.10-0.30	Al <=0.01; Mo <=0.02; N <=0.007; Cr+Ni+Cu<=0.30, Cu+Sn<=0.15; bal Fe				
EN 10016/4(94)	1.1251	Rod	0.68-0.72	0.10 max	0.15 max	0.50-0.70	0.10 max	0.020 max	0.025 max	0.10-0.30	Al <=0.01; Mo <=0.03; N <=0.007; Cr+Ni+Cu<=0.30, Cu+Sn<=0.15; bal Fe				
EN 10016/4(94)	C70D2	Rod	0.68-0.72	0.10 max	0.15 max	0.50-0.70	0.10 max	0.020 max	0.025 max	0.10-0.30	Al <=0.01; Mo <=0.03; N <=0.007; Cr+Ni+Cu<=0.30, Cu+Sn<=0.15; bal Fe				
EN 10016/4(94)	C72D2	Rod	0.70-0.74	0.10 max	0.15 max	0.50-0.70	0.10 max	0.020 max	0.025 max	0.10-0.30	Al <=0.01; Mo <=0.02; N <=0.007; Cr+Ni+Cu<=0.30, Cu+Sn<=0.15; bal Fe				

France

Specification	Designation	Notes	C	Cr	Cu	Mn	Ni	P	S	Si	Other	UTS	YS	El	Hard
AFNOR NFA35553	XC68		0.65-0.73			0.4-0.7		0.035	0.035	0.15-0.35	bal Fe				
AFNOR NFA35565(94)	C70E3	Ball & roller bearing	0.65-0.75	0.25 max	0.35 max	0.8-1.1	0.25 max	0.025 max	0.015 max	0.15-0.35	Al <=0.05; Mo <=0.1; bal Fe				
AFNOR NFA35565(94)	XC70*	see C70E3	0.65-0.75	0.25 max	0.35 max	0.8-1.1	0.25 max	0.025 max	0.015 max	0.15-0.35	Al <=0.05; Mo <=0.1; bal Fe				

Germany

Specification	Designation	Notes	C	Cr	Cu	Mn	Ni	P	S	Si	Other	UTS	YS	El	Hard
DIN	70Mn3		0.65-0.75			0.60-0.90		0.050 max	0.050 max	0.20-0.40	N <=0.007; bal Fe				
DIN	WNr 1.0643		0.65-0.75			0.60-0.90		0.050 max	0.050 max	0.20-0.40	N <=0.007; bal Fe				
DIN 17222(79)	C67	CR Strp for spring	0.65-0.72			0.60-0.90		0.045 max	0.045 max	0.15-0.35	bal Fe				
DIN 17222(79)	C67E	CR Strp spring hard <=2.5mm	0.65-0.72			0.60-0.90		0.035 max	0.035 max	0.15-0.35	bal Fe	1230-1770	1275	6	
DIN 17222(79)	Ck67	CR Strp spring hard <=2.5mm	0.65-0.72			0.60-0.90		0.035 max	0.035 max	0.15-0.35	bal Fe	1230-1770	1275	6	
DIN 17222(79)	WNr 1.0603	CR Strp for springs	0.65-0.72			0.60-0.90		0.045 max	0.045 max	0.15-0.35	bal Fe				
DIN 17222(79)	WNr 1.1231	CR Strp spring hard <=2.5mm	0.65-0.72			0.60-0.90		0.035 max	0.035 max	0.15-0.35	bal Fe	1230-1770	1275	6	

India

Specification	Designation	Notes	C	Cr	Cu	Mn	Ni	P	S	Si	Other	UTS	YS	El	Hard
IS 1570	C70		0.65-0.75			0.5-0.8					bal Fe				
IS 1570/2(79)	70C6		0.65-0.75	0.3 max	0.3 max	0.5-0.8	0.4 max	0.055 max	0.055 max	0.6 max	Co <=0.1; Mo <=0.15; Pb <=0.15; W <=0.1; P:S Varies; bal Fe				
IS 1570/2(79)	C70		0.65-0.75	0.3 max	0.3 max	0.5-0.8	0.4 max	0.055 max	0.055 max	0.6 max	Co <=0.1; Mo <=0.15; Pb <=0.15; W <=0.1; P:S Varies; bal Fe				

Italy

Specification	Designation	Notes	C	Cr	Cu	Mn	Ni	P	S	Si	Other	UTS	YS	El	Hard
UNI 3545	C70		0.65-0.72			0.5-0.9		0.035	0.035	0.15-0.4	bal Fe				
UNI 5598(71)	3CD70	Wir rod	0.68-0.73			0.4-0.7		0.035 max	0.035 max	0.15-0.35	N <=0.012; bal Fe				

Japan

Specification	Designation	Notes	C	Cr	Cu	Mn	Ni	P	S	Si	Other	UTS	YS	El	Hard
JIS G3311(88)	S70CM	CR Strp	0.65-0.75	0.20 max	0.30 max	0.60-0.90	0.20 max	0.030 max	0.035 max	0.15-0.35	bal Fe				190-280 HV

UNS numbers and US grades are provided as a means of cross referencing chemically similar alloys. Exchangability is only possible after independent examination of specifications. Tensile properties are minimum or typical as specified. UTS and YS as MPa. El as %. See Appendix for list of abbreviations used in Notes. * indicates obsolete material.

4-140/Carbon Steel

Specification	Designation	Notes	C	Cr	Cu	Mn	Ni	P	S	Si	Other	UTS	YS	El	Hard
Carbon Steel, Nonresulfurized, 1070 (Continued from previous page)															
Mexico															
NMX-B-301(86)	1070	Bar	0.65-0.75			0.60-0.90		0.040 max	0.050 max		bal Fe				
Pan America															
COPANT 330	1070	Bar	0.65-0.75			0.6-0.9		0.04	0.05		bal Fe				
COPANT 331	1070	Bar	0.65-0.75			0.6-0.9		0.04	0.05		bal Fe				
COPANT 333	1070	Wir rod	0.65-0.75			0.6-0.9		0.04	0.05		bal Fe				
COPANT 333	1071	Wir rod	0.65-0.75			0.75-1.05		0.04	0.05		bal Fe				
Romania															
STAS 795(92)	OLC70A	Spring; HF	0.65-0.75	0.3 max	0.25 max	0.5-0.8		0.04 max	0.04 max	0.17-0.37	Pb <=0.15; bal Fe				
Russia															
GOST	M71		0.64-0.74			0.6-0.9		0.04	0.05	0.13-0.28	bal Fe				
GOST	M73		0.67-0.8			0.7-1		0.04	0.05	0.13-0.28	bal Fe				
GOST	St65A		0.65-0.7	0.2	0.15	0.4-0.55	0.12	0.025	0.025	0.15-0.25	Al 0.05; bal Fe				
GOST 1050	70		0.67-0.75	0.25	0.25	0.5-0.8	0.25	0.04	0.04	0.17-0.37	bal Fe				
GOST 1050(74)	70	High-grade struct; Untreated; 0-250mm	0.67-0.75	0.25 max	0.25 max	0.5-0.8	0.25 max	0.035 max	0.04 max	0.17-0.37	As<=0.08; bal Fe				0-269 HB
GOST 1050(74)	70	High-grade struct; Norm; 0-100mm	0.67-0.75	0.25 max	0.25 max	0.5-0.8	0.25 max	0.035 max	0.04 max	0.17-0.37	As<=0.08; bal Fe	720	420	9L	
GOST 1050(74)	70G	High-grade struct; Norm; 0-100mm	0.67-0.75	0.25 max	0.25 max	0.9-1.2	0.25 max	0.035 max	0.04 max	0.17-0.37	As<=0.08; bal Fe	780	450	8L	
GOST 1050(74)	70G	High-grade struct; Untreated; 0-250mm	0.67-0.75	0.25 max	0.25 max	0.9-1.2	0.25 max	0.035 max	0.04 max	0.17-0.37	As<=0.08; bal Fe				0-285 HB
GOST 14959	65GA	Spring	0.62-0.7	0.25 max	0.2 max	0.9-1.2	0.25 max	0.025 max	0.025 max	0.17-0.37	Al <=0.1; bal Fe				
GOST 14959	70		0.67-0.75	0.25	0.2	0.5-0.8	0.25	0.035	0.035	0.17-0.37	bal Fe				
GOST 14959	70	Spring	0.67-0.75	0.25 max	0.2 max	0.5-0.8	0.25 max	0.035 max	0.035 max	0.17-0.37	bal Fe				
GOST 14959	70G		0.67-0.75	0.25	0.2	0.9-1.2	0.25	0.035	0.035	0.17-0.37	bal Fe				
GOST 14959(79)	70G	Spring	0.67-0.75	0.25 max	0.2 max	0.9-1.2	0.25 max	0.035 max	0.035 max	0.17-0.37	Al <=0.1; bal Fe				
GOST 7521	M67		0.6-0.75			0.6-0.9		0.04	0.05	0.13-0.28	bal Fe				
GOST 9960	M73		0.67-0.78			0.75-1.05		0.035	0.04	0.18-0.45	V 0.03; bal Fe				
Sweden															
SS 141770	1770	Unalloyed high qual	0.55-0.8			0.5-0.9		0.035	0.035	0.15-0.8	bal Fe				
UK															
BS 1429(80)	070A72	Spring	0.70-0.75			0.60-0.80				0.10-0.35	bal Fe				
BS 1449/1(91)	70CS	Plt Sht Strp; HR Q/T t<=16mm	0.65-0.75			0.50-0.90		0.045 max	0.045 max	0.05-0.35	bal Fe				352-518 HB
BS 1449/1(91)	70HS	Plt Sht Strp; CD norm t<16mm	0.65-0.75			0.50-0.90		0.045 max	0.045 max	0.05-0.35	bal Fe				238-285 HB
BS 970/1(83)	080A72	Blm Bil Slab Bar Rod Frg HR	0.70-0.75			0.70-0.90		0.050 max	0.050 max	0.4	bal Fe				241 HB max
BS 970/1(96)	060A72	Blm Bil Slab Bar Rod Frg	0.70-0.75			0.50-0.70		0.050 max	0.050 max	0.10-0.40	bal Fe				241HB max
BS 970/5(72)	070A72*	Obs; Blm Bil Slab Bar Rod Frg	0.70-0.75			0.60-0.80			0.05	0.1-0.35	bal Fe				
USA															
	AISI 1070	Wir rod	0.65-0.75			0.60-0.90		0.040 max	0.050 max		bal Fe				
	AISI 1070	Plt	0.65-0.76			0.60-0.90		0.030 max	0.035 max		bal Fe				

UNS numbers and US grades are provided as a means of cross referencing chemically similar alloys. Exchangability is only possible after independent examination of specifications. Tensile properties are minimum or typical as specified. UTS and YS as MPa. El as %. See Appendix for list of abbreviations used in Notes. * indicates obsolete material.

Specification	Designation	Notes	C	Cr	Cu	Mn	Ni	P	S	Si	Other	UTS	YS	El	Hard

Carbon Steel, Nonresulfurized, 1070 (Continued from previous page)

USA

Specification	Designation	Notes	C	Cr	Cu	Mn	Ni	P	S	Si	Other	UTS	YS	El	Hard
	AISI 1070	Bar	0.65-0.75			0.60-0.90		0.040 max	0.050 max		bal Fe				
	AISI 1070	Struct shp	0.65-0.75			0.60-0.90		0.030 max	0.035 max		bal Fe				
	AISI 1070	Sh Strp	0.65-0.75			0.60-0.90		0.030 max	0.035 max		bal Fe				
	UNS G10700		0.65-0.75			0.60-0.90		0.040 max	0.050 max		Plates, C 0.65-0.76; bal Fe				
AMS 5115H(98)*		Noncurrent; Wir, 2.36<D<=3.05	0.60-0.75			0.50-0.90		0.025 max	0.030 max	0.10-0.30	bal Fe	1448-1586			
ASTM A29/A29M(93)	1070	Bar	0.65-0.75			0.60-0.90		0.040 max	0.050 max		bal Fe				
ASTM A295(94)	1070M	Bearing, coil bar tub, CD ann	0.65-0.75	0.20 max	0.35 max	0.80-1.10	0.25 max	0.025 max	0.025 max	0.15-0.35	Mo <=0.10; bal Fe				250 HB max
ASTM A504(93)	Class U	Wheels	0.65-0.77			0.60-0.85		0.05 max	0.05 max	0.15 min	bal Fe				
ASTM A510(96)	1070	Wir rod	0.65-0.75			0.60-0.90		0.040 max	0.050 max		bal Fe				
ASTM A576(95)	1070	Special qual HW bar	0.65-0.75			0.60-0.90		0.040 max	0.050 max		Si Cu Pb B Bi Ca Se Te if spec'd; bal Fe				
ASTM A682/A682M(98)	1070	Strp, high-C, CR, spring	0.65-0.75			0.60-0.90		0.035 max	0.040 max	0.15-0.30	bal Fe				
ASTM A684A684M(86)	1070	Strp, high-C, CR, Ann	0.65-0.75			0.60-0.90		0.035 max	0.040 max	0.15-0.30	bal Fe				92 HRB
ASTM A713(93)	1070	Wir, High-C spring, HT	0.65-0.75			0.60-0.90		0.040 max	0.050 max		bal Fe				
ASTM A830/A830M(98)	1070	Plt	0.65-0.76			0.60-0.90		0.035 max	0.04 max	0.15-0.40	bal Fe				
MIL-B-12504E(90)	G10700	CD Bar wire for bullets	0.65-0.75		0.35 max	0.60-0.90		0.04 max	0.05 max	0.15-0.30	bal Fe				24-26 HRC
MIL-S-11713C(88)	1	Spring Strp	0.65-0.75			0.30-0.60		0.025 max	0.025 max	0.15-0.25	bal Fe				
MIL-S-11713C(88)	2	Spring Strp	0.65-0.75			0.60-0.90		0.040 max	0.050 max	0.15-0.25	bal Fe				
MIL-S-12504E(87)	G10700	Bar	0.65-0.75		0.35	0.6-0.9		0.04	0.05	0.15-0.3	bal Fe				
SAE J1397(92)	1070	Bar HR, est mech prop	0.65-0.75			0.6-0.9		0.04 max	0.05 max		bal Fe	700	390	12	212 HB
SAE J1397(92)	1070	Bar Ann CD, est mech prop	0.65-0.75			0.6-0.9		0.04 max	0.05 max		bal Fe	640	500	10	192 HB
SAE J403(95)	1070	Struct Shps Plt Strp Sh Weld Tub	0.65-0.76			0.60-0.90		0.030 max	0.035 max		bal Fe				
SAE J403(95)	1070	Bar Wir rod Smls Tub HR CF	0.65-0.75			0.60-0.90		0.030 max	0.050 max		bal Fe				

Yugoslavia

Specification	Designation	Notes	C	Cr	Cu	Mn	Ni	P	S	Si	Other	UTS	YS	El	Hard
	C.1833	F/IH	0.68-0.75	0.2 max		0.2-0.35	0.25 max	0.025 max	0.035 max	0.15-0.35	Cr+Mo+Ni<=0.45; bal Fe				

Carbon Steel, Nonresulfurized, 1074

Australia

Specification	Designation	Notes	C	Cr	Cu	Mn	Ni	P	S	Si	Other	UTS	YS	El	Hard
AS/NZS 1594(97)	HK1073	HR flat products	0.68-0.78	0.30 max	0.35 max	0.70-1.00	0.35 max	0.040 max	0.035 max	0.50 max	Al <=0.100; Mo <=0.10; Ti <=0.040; (Cu+Ni+Cr+Mo) 1 max; bal Fe				

China

Specification	Designation	Notes	C	Cr	Cu	Mn	Ni	P	S	Si	Other	UTS	YS	El	Hard
GB 699(88)	75	Bar HR Q/T	0.72-0.80	0.25 max	0.25 max	0.50-0.80	0.25 max	0.035 max	0.035 max	0.17-0.37	bal Fe	1080	880	7	
YB/T 5101(93)	75	Wir CD 0.10-2.00mm diam	0.73-0.79	0.25 max	0.20 max	0.50-0.80	0.25 max	0.025 max	0.020 max	0.17-0.37	bal Fe	1815-3090			
YB/T 5103(93)	75	Wir CD Q/T 2.0-4.0mm diam	0.72-0.80	0.25 max	0.25 max	0.50-0.80	0.25 max	0.035 max	0.035 max	0.17-0.37	bal Fe	1425-1765			

Japan

Specification	Designation	Notes	C	Cr	Cu	Mn	Ni	P	S	Si	Other	UTS	YS	El	Hard
JIS G3311(88)	S75CM	CR Strp	0.70-0.80	0.20 max	0.30 max	0.60-0.90	0.20 max	0.030 max	0.035 max	0.15-0.35	bal Fe				200-290 HV

Mexico

Specification	Designation	Notes	C	Cr	Cu	Mn	Ni	P	S	Si	Other	UTS	YS	El	Hard
NMX-B-301(86)	1074	Bar	0.70-0.80			0.50-0.80		0.040 max	0.050 max		bal Fe				

UNS numbers and US grades are provided as a means of cross referencing chemically similar alloys. Exchangability is only possible after independent examination of specifications. Tensile properties are minimum or typical as specified. UTS and YS as MPa. El as %. See Appendix for list of abbreviations used in Notes. * indicates obsolete material.

Specification	Designation	Notes	C	Cr	Cu	Mn	Ni	P	S	Si	Other	UTS	YS	El	Hard

Carbon Steel, Nonresulfurized, 1074 (Continued from previous page)

USA

Specification	Designation	Notes	C	Cr	Cu	Mn	Ni	P	S	Si	Other	UTS	YS	El	Hard
	AISI 1074	Struct shp	0.70-0.80			0.50-0.80		0.030 max	0.035 max		bal Fe				
	AISI 1074	Wir rod	0.70-0.80			0.50-0.80		0.040 max	0.050 max		bal Fe				
	AISI 1074	Sh Strp Plt	0.69-0.80			0.50-0.80		0.030 max	0.035 max		bal Fe				
	AISI 1074	Bar	0.70-0.80			0.50-0.80		0.040 max	0.050 max		bal Fe				
	UNS G10740		0.70-0.80			0.50-0.80		0.040 max	0.050 max		Sheets and Plates, C 0.69-0.80; bal Fe				
AMS 5120J(88)		Strp, CR and Ann	0.68-0.80			0.50-0.80		0.040 max	0.050 max	0.10-0.30	bal Fe				85 HRB max
ASTM A29/A29M(93)	1074	Bar	0.70-0.80			0.50-0.80		0.040 max	0.050 max		bal Fe				
ASTM A510(96)	1074	Wir rod	0.70-0.80			0.50-0.80		0.040 max	0.050 max		bal Fe				
ASTM A682/A682M(98)	1074	Strp, high-C, CR, spring	0.69-0.80			0.50-0.80		0.035 max	0.040 max	0.15-0.30	bal Fe				
ASTM A684A684M(86)	1074	Strp, high-C, CR, Ann	0.69-0.80			0.50-0.80		0.035 max	0.040 max	0.15-0.30	bal Fe				93 HRB
ASTM A713(93)	1074	Wir, High-C spring, HT	0.70-0.80			0.50-0.80		0.040 max	0.050 max		bal Fe				
ASTM A830/A830M(98)	1074	Plt	0.69-0.80			0.50-0.80		0.035 max	0.04 max	0.15-0.40	bal Fe				
MIL-S-46049C(98)	1074	Strp CR hard temp, spring	0.69-0.80			0.50-0.80		0.040 max	0.050 max	0.15-0.30	bal Fe				
SAE J1397(92)	1074	Bar CD, est mech prop									bal Fe	650	500	10	192 HB
SAE J1397(92)	1074	Bar HR, est mech prop									bal Fe	720	400	12	217 HB
SAE J403(95)	1074	Struct Shps Plt Strp Sh Weld Tub	0.69-0.80			0.50-0.80		0.030 max	0.035 max		bal Fe				

Carbon Steel, Nonresulfurized, 1075

Czech Republic

Specification	Designation	Notes	C	Cr	Cu	Mn	Ni	P	S	Si	Other	UTS	YS	El	Hard
CSN 412081	12081	Q/T	0.7-0.8	0.3 max		0.4-0.65		0.035 max	0.035 max	0.35 max	bal Fe				

Europe

Specification	Designation	Notes	C	Cr	Cu	Mn	Ni	P	S	Si	Other	UTS	YS	El	Hard
EN 10016/2(94)	1.0614	Rod	0.73-0.78	0.15 max	0.25 max	0.50-0.80	0.20 max	0.035 max	0.035 max	0.10-0.30	Al <=0.01; Mo <=0.05; Cu+Sn<=0.25; bal Fe				
EN 10016/2(94)	C76D	Rod	0.73-0.78	0.15 max	0.25 max	0.50-0.80	0.20 max	0.035 max	0.035 max	0.10-0.30	Al <=0.01; Mo <=0.05; Cu+Sn<=0.25; bal Fe				
EN 10016/4(94)	1.1253	Rod	0.74-0.78	0.10 max	0.15 max	0.50-0.70	0.10 max	0.020 max	0.025 max	0.10-0.30	Al <=0.01; Mo <=0.02; N <=0.007; Cr+Ni+Cu<=0.30, Cu+Sn<=0.15; bal Fe				
EN 10016/4(94)	C76D2	Rod	0.74-0.78	0.10 max	0.15 max	0.50-0.70	0.10 max	0.020 max	0.025 max	0.10-0.30	Al <=0.01; Mo <=0.02; N <=0.007; Cr+Ni+Cu<=0.30, Cu+Sn<=0.15; bal Fe				

Hungary

Specification	Designation	Notes	C	Cr	Cu	Mn	Ni	P	S	Si	Other	UTS	YS	El	Hard
MSZ 4217(85)	75	CR, Q/T or spring Strp; 0-2mm	0.72-0.8			0.5-0.8		0.04 max	0.04 max	0.17-0.37	bal Fe	1280-1580		4L	
MSZ 4217(85)	75	CR, Q/T or spring Strp; 0-2mm; rolled, hard	0.72-0.8			0.5-0.8		0.04 max	0.04 max	0.17-0.37	bal Fe	780-1180			
MSZ 5736(88)	D75	Wir rod; for CD	0.71-0.78	0.2 max	0.2 max	0.3-0.7	0.2 max	0.035 max	0.04 max	0.1-0.35	D75V:V<=0.15; bal Fe				
MSZ 5736(88)	D75K	Wir rod; for CD	0.71-0.78	0.2 max	0.15 max	0.3-0.7	0.2 max	0.025 max	0.025 max	0.1-0.35	bal Fe				

India

Specification	Designation	Notes	C	Cr	Cu	Mn	Ni	P	S	Si	Other	UTS	YS	El	Hard
IS 1570/2(79)	75C6	Sh Plt Sect Shp Bar Bil Frg	0.7-0.8	0.3 max	0.3 max	0.5-0.8	0.4 max	0.055 max	0.055 max	0.6 max	Co <=0.1; Mo <=0.15; Pb <=0.15; W <=0.1; P:S Varies; bal Fe				

Specification	Designation	Notes	C	Cr	Cu	Mn	Ni	P	S	Si	Other	UTS	YS	El	Hard

Carbon Steel, Nonresulfurized, 1075 (Continued from previous page)

Specification	Designation	Notes	C	Cr	Cu	Mn	Ni	P	S	Si	Other	UTS	YS	El	Hard
India															
IS 1570/2(79)	C75	Sh Plt Sect Shp Bar Bil Frg	0.7-0.8	0.3 max	0.3 max	0.5-0.8	0.4 max	0.055 max	0.055 max	0.6 max	Co <=0.1; Mo <=0.15; Pb <=0.15; W <=0.1; P:S Varies; bal Fe				
Italy															
UNI 5598(71)	3CD75	Wir rod	0.73-0.78			0.4-0.7		0.035 max	0.035 max	0.15-0.35	N <=0.012; bal Fe				
UNI 7064(82)	C75	Spring	0.7-0.8			0.6-0.8		0.035 max	0.035 max	0.15-0.4	bal Fe				
UNI 8893(86)	C75	Spring	0.7-0.8			0.6-0.8		0.035 max	0.035 max	0.15-0.4	bal Fe				
Mexico															
NMX-B-301(86)	1075	Bar	0.70-0.80			0.40-0.70		0.040 max	0.050 max		bal Fe				
Romania															
STAS 795(92)	OLC75A	Spring; HF	0.7-0.8	0.3 max	0.25 max	0.5-0.8		0.04 max	0.04 max	0.17-0.37	Pb <=0.15; bal Fe				
Russia															
GOST 1050(74)	75	High-grade struct; Q/T; 0-100mm	0.72-0.8	0.25 max	0.25 max	0.5-0.8	0.25 max	0.035 max	0.04 max	0.17-0.37	As<=0.08; bal Fe	1080	885	7L	
GOST 1050(74)	75	High-grade struct; Untreated; 0-250mm	0.72-0.8	0.25 max	0.25 max	0.5-0.8	0.25 max	0.035 max	0.04 max	0.17-0.37	As<=0.08; bal Fe				0-285 HB
GOST 14959	75A	Spring	0.72-0.8	0.25 max	0.2 max	0.5-0.8	0.25 max	0.025 max	0.025 max	0.17-0.37	Al <=0.1; bal Fe				
GOST 14959(79)	75	Spring	0.72-0.8	0.25 max	0.2 max	0.5-0.8	0.25 max	0.035 max	0.035 max	0.17-0.37	Al <=0.1; bal Fe				
UK															
BS 1449/1(91)	80CS	Plt Sht Strp; CD norm t<=16mm	0.75-0.85			0.50-0.90		0.045 max	0.045 max	0.05-0.35	bal Fe				257-304 HB
BS 1449/1(91)	80HS	Plt Sht Strp; CD norm t<16mm	0.75-0.85			0.50-0.90		0.045 max	0.045 max	0.05-0.35	bal Fe				257-304 HB
USA															
	AISI 1075	Wir rod	0.70-0.80			0.40-0.70		0.040 max	0.050 max		bal Fe				
	AISI 1075	Plt	0.69-0.80			0.40-0.70		0.030 max	0.035 max		bal Fe				
	AISI 1075	Struct shp	0.70-0.80			0.40-0.70		0.030 max	0.035 max		bal Fe				
	UNS G10750		0.70-0.80			0.40-0.70		0.040 max	0.050 max		bal Fe				
ASTM A29/A29M(93)	1075	Bar	0.70-0.80			0.40-0.70		0.040 max	0.050 max		bal Fe				
ASTM A510(96)	1075	Wir rod	0.70-0.80			0.40-0.70		0.040 max	0.050 max		bal Fe				
ASTM A713(93)	1075	Wir, High-C spring, HT	0.70-0.80			0.40-0.70		0.040 max	0.050 max		bal Fe				
SAE J1249(95)	1075	Former SAE std valid for Wir rod	0.70-0.80			0.40-0.70		0.040 max	0.050 max		bal Fe				
SAE J403(95)	1075	Struct Shps Plt Strp Sh Weld Tub	0.69-0.80			0.40-0.70		0.030 max	0.035 max		bal Fe				

Carbon Steel, Nonresulfurized, 1078

Specification	Designation	Notes	C	Cr	Cu	Mn	Ni	P	S	Si	Other	UTS	YS	El	Hard
Austria															
ONORM M3110	RC75	Wir	0.75-0.80			0.4-0.7		0.035 max	0.035 max	0.15-0.40	P+S=0.06; bal Fe				
ONORM M3110	RC80	Wir	0.80-0.85			0.4-0.7		0.035 max	0.035 max	0.15-0.40	P+S=0.06; bal Fe				
Bulgaria															
BDS 5785(83)	75	Struct	0.72-0.80	0.25 max	0.3	0.50-0.80	0.3	0.035 max	0.035 max	0.17-0.37	bal Fe				
BDS 6742	75	Spring	0.72-0.80	0.25 max	0.30 max	0.50-0.80	0.30 max	0.035 max	0.035 max	0.17-0.37	bal Fe				
China															
GB/T 4358(95)	T8MnA	Wir CD 0.10-2.00mm diam	0.80-0.89	0.10 max	0.20 max	0.40-0.60	0.12 max	0.025 max	0.020 max	0.35 max	bal Fe	1760-3040			

UNS numbers and US grades are provided as a means of cross referencing chemically similar alloys. Exchangability is only possible after independent examination of specifications. Tensile properties are minimum or typical as specified. UTS and YS as MPa. El as %. See Appendix for list of abbreviations used in Notes. * indicates obsolete material.

Carbon Steel, Nonresulfurized, 1078 (Continued from previous page)

Specification	Designation	Notes	C	Cr	Cu	Mn	Ni	P	S	Si	Other	UTS	YS	El	Hard
China															
YB/T 5101(93)	T8MnA	Wir CD 0.10-2.00mm diam	0.81-0.89	0.10 max	0.20 max	0.40-0.60	0.12 max	0.025 max	0.020 max	0.35 max	bal Fe	1760-3040			
Europe															
EN 10016/2(94)	1.0620	Rod	0.75-0.80	0.15 max	0.25 max	0.50-0.80	0.20 max	0.035 max	0.035 max	0.10-0.30	Al <=0.01; Mo <=0.05; Cu+Sn<=0.25; bal Fe				
EN 10016/2(94)	C78D	Rod	0.75-0.80	0.15 max	0.25 max	0.50-0.80	0.20 max	0.035 max	0.035 max	0.10-0.30	Al <=0.01; Mo <=0.05; Cu+Sn<=0.25; bal Fe				
EN 10016/4(94)	1.1252	Rod	0.76-0.80	0.10 max	0.15 max	0.50-0.70	0.10 max	0.020 max	0.025 max	0.10-0.30	Al <=0.01; Mo <=0.02; N <=0.007; Cr+Ni+Cu<=0.30, Cu+Sn<=0.15; bal Fe				
EN 10016/4(94)	C78D2	Rod	0.76-0.80	0.10 max	0.15 max	0.50-0.70	0.10 max	0.020 max	0.025 max	0.10-0.30	Al <=0.01; Mo <=0.02; N <=0.007; Cr+Ni+Cu<=0.30, Cu+Sn<=0.15; bal Fe				
France															
AFNOR NFA35553	XC75	Strp	0.7-0.8			0.4-0.7		0.035	0.035	0.15-0.3	bal Fe				
AFNOR NFA36102(93)	C75RR	HR; Strp; CR	0.7-0.8	0.3 max		0.5-0.8	0.40 max	0.025 max	0.02 max	0.15-0.35	Al <=0.03; Mo <=0.10; bal Fe				
AFNOR NFA36102(93)	XC75*	HR; Strp; CR	0.7-0.8	0.3 max		0.5-0.8	0.40 max	0.025 max	0.02 max	0.15-0.35	Al <=0.03; Mo <=0.10; bal Fe				
Germany															
DIN	D75-3*	Obs; see C76D2	0.73-0.78			0.3-0.7		0.03 max	0.03 max	0.1-0.3	bal Fe				
DIN	D78-3*	Obs; see C78D2	0.75-0.8			0.3-0.7		0.03 max	0.03 max	0.1-0.3	bal Fe				
DIN	D80-3*	Obs; see C80D2	0.78-0.83			0.3-0.7		0.03 max	0.03 max	0.1-0.3	bal Fe				
DIN	D83-3*	Obs; see C82D2	0.8-0.85			0.3-0.7		0.03 max	0.03 max	0.1-0.3	bal Fe				
DIN 17140	D78-2*	Obs; see C78D	0.75-0.8			0.3-0.7		0.04 max	0.04 max	0.1-0.3	bal Fe				
DIN 17140	D80-2*	Obs; see C80D	0.78-0.83			0.3-0.7		0.04 max	0.04 max	0.1-0.3	bal Fe				
DIN 17140	D83-2*	Obs; see C82D	0.8-0.85			0.3-0.7		0.04 max	0.04 max	0.1-0.3	bal Fe				
DIN 17222(79)	Ck75	CR Strp spring hard <=2.5mm	0.70-0.80			0.60-0.80		0.035 max	0.035 max	0.15-0.35	bal Fe	1320-1870	1275	6	
DIN 17222(79)	WNr 1.1248	CR Strp spring hard <=2.5mm	0.70-0.80			0.60-0.80		0.035 max	0.035 max	0.15-0.35	bal Fe	1320-1870	1275	6	
India															
IS 1570/2(79)	80C6		0.75-0.85	0.3 max	0.3 max	0.5-0.8	0.4 max	0.055 max	0.055 max	0.6 max	Co <=0.1; Mo <=0.15; Pb <=0.15; W <=0.1; P:S Varies; bal Fe				
IS 1570/2(79)	C80		0.75-0.85	0.3 max	0.3 max	0.5-0.8	0.4 max	0.055 max	0.055 max	0.6 max	Co <=0.1; Mo <=0.15; Pb <=0.15; W <=0.1; P:S Varies; bal Fe				
Japan															
JIS G4801(84)	SUP 3	Bar Spring Q/T	0.75-0.90		0.30 max	0.30-0.60		0.035 max	0.035 max	0.15-0.35	bal Fe	1079	834	8	341-401 HB
Mexico															
NMX-B-301(86)	1078	Bar	0.72-0.85			0.30-0.60		0.040 max	0.050 max		bal Fe				
Poland															
PNH84028	D80	Wir rod	0.78-0.83	0.2 max	0.2 max	0.3-0.6	0.2 max	0.035 max	0.035 max	0.1-0.3	Mo <=0.08; P+S<=0.06; bal Fe				
PNH84028	D80A	Wir rod	0.78-0.83	0.1 max	0.2 max	0.3-0.6	0.15 max	0.03 max	0.03 max	0.1-0.25	Mo <=0.05; P+S<=0.05; bal Fe				
PNH84028	DB80		0.75-0.82	0.1	0.2	0.4-0.8	0.2	0.04 max	0.04 max	0.25	P+S=0.07; bal Fe				
PNH84028	DS82		0.79-0.85	0.1	0.15	0.3-0.6	0.15	0.02	0.02	0.25	bal Fe				

UNS numbers and US grades are provided as a means of cross referencing chemically similar alloys. Exchangability is only possible after independent examination of specifications. Tensile properties are minimum or typical as specified. UTS and YS as MPa. El as %. See Appendix for list of abbreviations used in Notes. * indicates obsolete material.

Specification	Designation	Notes	C	Cr	Cu	Mn	Ni	P	S	Si	Other	UTS	YS	El	Hard

Carbon Steel, Nonresulfurized, 1078 (Continued from previous page)

Poland

Specification	Designation	Notes	C	Cr	Cu	Mn	Ni	P	S	Si	Other	UTS	YS	El	Hard
PNH84028	DW82	Wir rod	0.8-0.85	0.12 max	0.2 max	0.3-0.5	0.15 max	0.02 max	0.02 max	0.25 max	P+S<=0.035; bal Fe				
PNH84032	75	Spring	0.72-0.8	0.3 max	0.25 max	0.5-0.8	0.3 max	0.04 max	0.04 max	0.17-0.37	bal Fe				
PNH92602	DS82		0.79-0.85	0.1	0.15	0.3-0.6	0.15	0.02	0.02	0.25	bal Fe				

Russia

Specification	Designation	Notes	C	Cr	Cu	Mn	Ni	P	S	Si	Other	UTS	YS	El	Hard
GOST 14959	75		0.72-0.8	0.25	0.25	0.5-0.8	0.25	0.04	0.04	0.17-0.37	bal Fe				
GOST 24182(80)	M76		0.71-0.82			0.75-1.05		0.035	0.045	0.18-0.40	bal Fe				

USA

Specification	Designation	Notes	C	Cr	Cu	Mn	Ni	P	S	Si	Other	UTS	YS	El	Hard
	AISI 1078	Struct shp	0.72-0.85			0.30-0.60		0.030 max	0.035 max		bal Fe				
	AISI 1078	Sh Strp Plt	0.72-0.86			0.30-0.60		0.030 max	0.035 max		bal Fe				
	AISI 1078	Wir rod	0.72-0.85			0.30-0.60		0.040 max	0.050 max		bal Fe				
	AISI 1078	Bar	0.72-0.85			0.30-0.60		0.040 max	0.050 max		bal Fe				
	UNS G10780		0.72-0.85			0.30-0.60		0.040 max	0.050 max		Plates, C 0.72-0.86; bal Fe				
ASTM A29/A29M(93)	1078	Bar	0.72-0.85			0.30-0.60		0.040 max	0.050 max		bal Fe				
ASTM A510(96)	1078	Wir rod	0.72-0.85			0.30-0.60		0.040 max	0.050 max		bal Fe				
ASTM A576(95)	1078	Special qual HW bar	0.72-0.85			0.30-0.60		0.040 max	0.050 max		Si Cu Pb B Bi Ca Se Te if spec'd; bal Fe				
ASTM A713(93)	1078	Wir, High-C spring, HT	0.72-0.85			0.30-0.60		0.040 max	0.050 max		bal Fe				
ASTM A830/A830M(98)	1078	Plt	0.72-0.86			0.30-0.60		0.035 max	0.04 max	0.15-0.40	bal Fe				
SAE J1397(92)	1078	Bar Ann CD, est mech prop	0.72-0.85			0.3-0.6		0.04 max	0.05 max		bal Fe	650	500	10	192 HB
SAE J1397(92)	1078	Bar HR, est mech prop	0.72-0.85			0.3-0.6		0.04 max	0.05 max		bal Fe	690	380	12	207 HB
SAE J403(95)	1078	Bar Wir rod Smls Tub HR CF	0.72-0.85			0.30-0.60		0.030 max	0.050 max		bal Fe				
SAE J403(95)	1078	Struct Shps Plt Strp Sh Weld Tub	0.72-0.86			0.30-0.60		0.030 max	0.035 max		bal Fe				

Yugoslavia

Specification	Designation	Notes	C	Cr	Cu	Mn	Ni	P	S	Si	Other	UTS	YS	El	Hard
	PZ80		0.8-0.9	0.15	0.025	0.3-0.7	0.05	0.04	0.04	0.1-0.3	bal Fe				

Carbon Steel, Nonresulfurized, 1080

Argentina

Specification	Designation	Notes	C	Cr	Cu	Mn	Ni	P	S	Si	Other	UTS	YS	El	Hard
IAS	IRAM 1080		0.75-0.88			0.60-0.90		0.040 max	0.050 max	0.10-0.30	bal Fe	950-1100	530-610	8-15	284-331 HB

Australia

Specification	Designation	Notes	C	Cr	Cu	Mn	Ni	P	S	Si	Other	UTS	YS	El	Hard
AS 1442	K1082*	Obs; Bar Bil	0.78-0.9			0.6-0.9		0.05	0.05	0.1-0.35	bal Fe				
AS 1442(92)	1080	HR bar, Semifinished (may treat w/microalloying elements)	0.75-0.88			0.60-0.90		0.040 max	0.040 max	0.10-0.35	Si<=0.10 for Al-killed; bal Fe				

Belgium

Specification	Designation	Notes	C	Cr	Cu	Mn	Ni	P	S	Si	Other	UTS	YS	El	Hard
NBN 253-05	C79	Wir, Ann or Q/A Tmp 12mm	0.72-0.85			0.5-0.8		0.04	0.04	0.15-0.4	bal Fe				

Bulgaria

Specification	Designation	Notes	C	Cr	Cu	Mn	Ni	P	S	Si	Other	UTS	YS	El	Hard
BDS 6742(83)	80	Struct	0.77-0.85	0.25 max	0.25	0.50-0.80	0.25	0.04	0.04	0.17-0.37	bal Fe				

China

Specification	Designation	Notes	C	Cr	Cu	Mn	Ni	P	S	Si	Other	UTS	YS	El	Hard
GB 699(88)	80	Bar HR Q/T	0.77-0.85	0.25 max	0.25 max	0.50-0.80	0.25 max	0.035 max	0.035 max	0.17-0.37	bal Fe	1080	930	6	
YB/T 5063(93)	T8A	Strp HR HT	0.75-0.84	0.25 max	0.20 max	0.40 max	0.30 max	0.025 max	0.020 max	0.35 max	bal Fe	1275-1570			

UNS numbers and US grades are provided as a means of cross referencing chemically similar alloys. Exchangability is only possible after independent examination of specifications. Tensile properties are minimum or typical as specified. UTS and YS as MPa. El as %. See Appendix for list of abbreviations used in Notes. * indicates obsolete material.

Specification	Designation	Notes	C	Cr	Cu	Mn	Ni	P	S	Si	Other	UTS	YS	El	Hard
Carbon Steel, Nonresulfurized, 1080 (Continued from previous page)															
China															
YB/T 5101(93)	80	Wir CD 0.10-2.00mm diam	0.78-0.84	0.25 max	0.20 max	0.50-0.80	0.25 max	0.025 max	0.020 max	0.17-0.37	bal Fe	1815-3090			
YB/T 5103(93)	80	Wir CD Q/T 2.0-4.0mm diam	0.77-0.85	0.25 max	0.25 max	0.50-0.80	0.25 max	0.035 max	0.035 max	0.17-0.37	bal Fe	1425-1765			
Europe															
EN 10016/2(94)	1.0622	Rod	0.78-0.83	0.15 max	0.25 max	0.50-0.80	0.20 max	0.035 max	0.035 max	0.10-0.30	Al <=0.01; Mo <=0.05; Cu+Sn<=0.25; bal Fe				
EN 10016/2(94)	1.0626	Rod	0.80-0.85	0.15 max	0.25 max	0.50-0.80	0.20 max	0.035 max	0.035 max	0.10-0.30	Al <=0.01; Mo <=0.05; bal Fe				
EN 10016/2(94)	C80D	Rod	0.78-0.83	0.15 max	0.25 max	0.50-0.80	0.20 max	0.035 max	0.035 max	0.10-0.30	Al <=0.01; Mo <=0.05; Cu+Sn<=0.25; bal Fe				
EN 10016/2(94)	C82D	Rod	0.80-0.85	0.15 max	0.25 max	0.50-0.80	0.20 max	0.035 max	0.035 max	0.10-0.30	Al <=0.01; Mo <=0.05; bal Fe				
EN 10016/4(94)	1.1255	Rod	0.78-0.82	0.10 max	0.15 max	0.50-0.70	0.10 max	0.020 max	0.025 max	0.10-0.30	Al <=0.01; Mo <=0.02; N <=0.007; Cr+Ni+Cu<=0.30, Cu+Sn<=0.15; bal Fe				
EN 10016/4(94)	1.1262	Rod	0.80-0.84	0.10 max	0.15 max	0.50-0.70	0.10 max	0.020 max	0.025 max	0.10-0.30	Al <=0.01; Mo <=0.02; N <=0.007; Cr+Ni+Cu<=0.30, Cu+Sn<=0.15; bal Fe				
EN 10016/4(94)	C80D2	Rod	0.78-0.82	0.10 max	0.15 max	0.50-0.70	0.10 max	0.020 max	0.025 max	0.10-0.30	Al <=0.01; Mo <=0.02; N <=0.007; Cr+Ni+Cu<=0.30, Cu+Sn<=0.15; bal Fe				
EN 10016/4(94)	C82D2	Rod	0.80-0.84	0.10 max	0.15 max	0.50-0.70	0.10 max	0.020 max	0.025 max	0.10-0.30	Al <=0.01; Mo <=0.02; N <=0.007; Cr+Ni+Cu<=0.30, Cu+Sn<=0.15; bal Fe				
Finland															
SFS 1265(87)	SFS1265	Pressed conc wire	0.6-0.9			0.5-0.9		0.04 max	0.04 max	0.1-0.35	bal Fe				
France															
AFNOR	XC80		0.75-0.85	0.12		0.5-0.8		0.035	0.035	0.1-0.4	bal Fe				
Germany															
DIN	80Mn4		0.75-0.85			0.90-1.20		0.035 max	0.035 max	0.25-0.50	bal Fe				
DIN	D85-3*	Obs; see C86D2	0.83-0.88			0.3-0.7		0.03 max	0.03 max	0.1-0.3	bal Fe				
DIN	WNr 1.1259		0.75-0.85			0.90-1.20		0.035 max	0.035 max	0.25-0.50	bal Fe				
Hungary															
MSZ 5736(88)	D80	Wir rod; for CD	0.77-0.84	0.2 max	0.2 max	0.3-0.7	0.2 max	0.035 max	0.04 max	0.1-0.35	D80V:V<=0.15; bal Fe				
MSZ 5736(88)	D80K	Wir rod; for CD	0.77-0.84	0.2 max	0.15 max	0.3-0.7	0.2 max	0.025 max	0.025 max	0.1-0.35	bal Fe				
MSZ 5736(88)	D83	Wir rod; for CD	0.79-0.86	0.2 max	0.2 max	0.3-0.7	0.2 max	0.035 max	0.04 max	0.1-0.35	D83V:V<=0.15; bal Fe				
MSZ 5736(88)	D83K	Wir rod; for CD	0.79-0.86	0.2 max	0.15 max	0.3-0.7	0.2 max	0.025 max	0.025 max	0.1-0.35	bal Fe				
India															
IS 1570	C80		0.75-0.85			0.5-0.8					bal Fe				
IS 3749	T80Mn65	Bar	0.75-0.85			0.5-0.8		0.04	0.04	0.1-0.35	bal Fe				
IS 4454	3		0.75-1			0.8		0.03	0.03	0.1-0.35	bal Fe				
IS 4454	4		0.75-1			0.8		0.03	0.03	0.1-0.35	bal Fe				
Italy															
UNI 5598(71)	3CD80	Wir rod	0.78-0.83			0.4-0.7		0.035 max	0.035 max	0.15-0.35	N <=0.012; bal Fe				
Mexico															
NMX-B-301(86)	1080	Bar	0.75-0.88			0.60-0.90		0.040 max	0.050 max		bal Fe				

UNS numbers and US grades are provided as a means of cross referencing chemically similar alloys. Exchangability is only possible after independent examination of specifications. Tensile properties are minimum or typical as specified. UTS and YS as MPa. El as %. See Appendix for list of abbreviations used in Notes. * indicates obsolete material.

Specification	Designation	Notes	C	Cr	Cu	Mn	Ni	P	S	Si	Other	UTS	YS	El	Hard

Carbon Steel, Nonresulfurized, 1080 (Continued from previous page)

Pan America

Specification	Designation	Notes	C	Cr	Cu	Mn	Ni	P	S	Si	Other	UTS	YS	El	Hard
COPANT 330	1080	Bar	0.75-0.88			0.6-0.9		0.04	0.05		bal Fe				
COPANT 331	1080	Bar	0.75-0.88			0.6-0.9		0.04	0.05		bal Fe				
Poland															
PNH84028	DB80		0.75-0.82	0.1	0.2	0.4-0.8	0.2	0.04 max	0.04 max	0.25	P+S=0.07; bal Fe				
PNH84028	DB85		0.82-0.88	0.1	0.2	0.4-0.8	0.2	0.04 max	0.04 max	0.25	P+S=0.07; bal Fe				
PNH84032	75	Spring	0.72-0.82	0.3	0.25	0.5-0.8	0.3	0.04 max	0.04 max	0.17-0.37	bal Fe				
Russia															
GOST 1050	80		0.77-0.85	0.25	0.25	0.5-0.8	0.25	0.04	0.04	0.17-0.37	As 0.05; bal Fe				
GOST 1050(74)	80	High-grade struct; Q/T; 0-100mpa	0.77-0.85	0.25 max	0.25 max	0.5-0.8	0.25 max	0.035 max	0.04 max	0.17-0.37	As<=0.08; bal Fe	1080	930	6L	
GOST 1050(74)	80	High-grade struct; Untreated; 0-250mm	0.77-0.85	0.25 max	0.25 max	0.5-0.8	0.25 max	0.035 max	0.04 max	0.17-0.37	As<=0.08; bal Fe				0-285 HB
GOST 14959	80		0.77-0.85	0.25	0.2	0.5-0.8	0.25	0.03	0.03	0.17-0.37	bal Fe				
GOST 14959(79)	80	Spring	0.77-0.85	0.25 max	0.2 max	0.5-0.8	0.25 max	0.035 max	0.035 max	0.17-0.37	Al <=0.1; bal Fe				
Spain															
UNE 36015(60)	C79	Spring	0.72-0.85			0.5-0.8		0.035	0.035	0.15-0.4	bal Fe				
UNE 36015(60)	F.1410*	see C79	0.72-0.85			0.5-0.8		0.035	0.035	0.15-0.4	bal Fe				
Sweden															
SIS 141778	1778-02	Strp, Ann, 2mm diam	0.66-0.8			0.4-0.9		0.03	0.03	0.15-0.4	bal Fe		370	26	
SIS 141778	1778-04	Strp, Q/A, 0.125mm diam	0.66-0.8			0.4-0.9		0.03	0.03	0.15-0.4	bal Fe	860			
UK															
BS 970/1(83)	060A83	Blm Bil Slab Bar Rod Frg	0.80-0.87			0.50-0.70		0.050 max	0.050 max	0.4	bal Fe				
BS 970/1(83)	070A78	Blm Bil Slab Bar Rod Frg	0.75-0.82			0.60-0.80		0.050 max	0.050 max	0.1-0.4	bal Fe				
BS 970/1(83)	080A78	Blm Bil Slab Bar Rod Frg	0.75-0.82			0.70-0.90		0.050 max	0.050 max	0.4	bal Fe				
BS 970/1(83)	080A83	Blm Bil Slab Bar Rod Frg	0.80-0.87			0.70-0.90		0.050 max	0.050 max	0.4	bal Fe				
BS 970/1(96)	060A78	Blm Bil Slab Bar Rod Frg	0.75-0.82			0.50-0.70		0.050 max	0.050 max	0.10-0.40	bal Fe				
BS 970/1(96)	060A81	Blm Bil Slab Bar Rod Frg	0.78-0.85			0.50-0.70		0.050 max	0.050 max	0.10-0.40	bal Fe				269 HB max
USA															
	AISI 1080	Wir rod	0.75-0.88			0.60-0.90		0.040 max	0.050 max		bal Fe				
	AISI 1080	Struct shp	0.75-0.88			0.60-0.90		0.030 max	0.035 max		bal Fe				
	AISI 1080	Sh Strp Plt	0.74-0.88			0.60-0.90		0.030 max	0.035 max		bal Fe				
	AISI 1080	Bar	0.75-0.88			0.60-0.90		0.040 max	0.050 max		bal Fe				
	UNS G10800		0.75-0.88			0.60-0.90		0.040 max	0.050 max		Plates, C 0.74-0.88; bal Fe				
AMS 5110G(95)		Wir Spring tmp, CD, D<=1.57	0.75-0.88			0.60-0.90		0.040 max	0.050 max	0.10-0.30	bal Fe	2068 min			
AMS 5110G(95)		Wir Spring tmp, CD, 4.85<D<6.35	0.75-0.88			0.60-0.90		0.040 max	0.050 max	0.10-0.30	bal Fe	1379-1724			
ASTM A29/A29M(93)	1080	Bar	0.75-0.88			0.60-0.90		0.040 max	0.050 max		bal Fe				
ASTM A510(96)	1080	Wir rod	0.75-0.88			0.60-0.90		0.040 max	0.050 max		bal Fe				
ASTM A576(95)	1080	Special qual HW bar	0.75-0.88			0.60-0.90		0.040 max	0.050 max		Si Cu Pb B Bi Ca Se Te if spec'd; bal Fe				

Specification	Designation	Notes	C	Cr	Cu	Mn	Ni	P	S	Si	Other	UTS	YS	El	Hard

Carbon Steel, Nonresulfurized, 1080 (Continued from previous page)

USA

Specification	Designation	Notes	C	Cr	Cu	Mn	Ni	P	S	Si	Other	UTS	YS	El	Hard
ASTM A682/A682M(98)	1080	Strp, high-C, CR, spring	0.74-0.88			0.60-0.90		0.035 max	0.040 max	0.15-0.30	bal Fe				
ASTM A684A684M(86)	1080	Strp, high-C, CR, Ann	0.74-0.88			0.00-0.90		0.035 max	0.040 max	0.15-0.30	bal Fe				94 HRB
ASTM A713(93)	1080	Wir, High-C spring, HT	0.75-0.88			0.60-0.90		0.040 max	0.050 max		bal Fe				
ASTM A830/A830M(98)	1080	Plt	0.74-0.88			0.60-0.90		0.035 max	0.04 max	0.15-0.40	bal Fe				
ASTM A911/A911M(92)		SR bar prestress con	0.70-0.90	0.15 max	0.30 max	0.50-0.90		0.030 max	0.035 max	0.10-0.35	bal Fe	1570	1375	6.0	
DoD-F-24669/1(86)	1080	Bar Bil; Supersedes MIL-S-866 & MIL-S-16974	0.75-0.88			0.6-0.9		0.04	0.05		bal Fe				
FED QQ-S-700D(91)	C1080*	Obs; Sh Strp	0.75-0.88			0.6-0.9		0.04 max	0.05 max		bal Fe				
FED QQ-W-470B(85)	1080*	Obs; Wir	0.75-0.88			0.6-0.9		0.04 max	0.05 max		bal Fe				
MIL-S-16788A(86)	C8*	HR, frg, bloom bil slab for refrg; Obs for new design	0.75-0.88			0.60-0.90		0.040 max	0.050 max		bal Fe				
MIL-S-16974E(86)	1080*	Obs; Bar bil boom slab for refrg	0.75-0.88			0.6-0.9		0.04 max	0.05 max		bal Fe				
SAE J1397(92)	1080	Bar HR, est mech prop	0.75-0.88			0.6-0.9		0.04 max	0.05 max		bal Fe	770	420	10	229 HB
SAE J1397(92)	1080	Bar Ann CD, est mech prop	0.75-0.88			0.6-0.9		0.04 max	0.05 max		bal Fe	680	520	10	192 HB
SAE J403(95)	1080	Struct Shps Plt Strp Sh Weld Tub	0.74-0.88			0.60-0.90		0.030 max	0.035 max		bal Fe				
SAE J403(95)	1080	Bar Wir rod Smls Tub HR CF	0.75-0.88			0.60-0.90		0.030 max	0.050 max		bal Fe				

Carbon Steel, Nonresulfurized, 1084

Australia

Specification	Designation	Notes	C	Cr	Cu	Mn	Ni	P	S	Si	Other	UTS	YS	El	Hard
AS 1442(92)	1084	HR bar, Semifinished (may treat w/microalloying elements)	0.80-0.93			0.60-0.90		0.040 max	0.040 max	0.10-0.35	Si<=0.10 for Al-killed; bal Fe				
AS 1446	K1084*	Withdrawn, see AS/NZS 1594(97)	0.8-0.93			0.6-0.9		0.05	0.05	0.1-0.35	bal Fe				

Bulgaria

Specification	Designation	Notes	C	Cr	Cu	Mn	Ni	P	S	Si	Other	UTS	YS	El	Hard
BDS 5785(83)	85	Struct	0.82-0.90	0.30 max	0.3	0.50-0.80	0.3	0.035	0.035	0.17-0.37	bal Fe				
BDS 6742	85	Spring	0.82-0.90	0.25 max	0.25 max	0.50-0.80	0.25 max	0.035 max	0.035 max	0.17-0.37	bal Fe				

China

Specification	Designation	Notes	C	Cr	Cu	Mn	Ni	P	S	Si	Other	UTS	YS	El	Hard
GB 1222(84)	85	Bar Flat HR Q/T	0.82-0.90	0.25 max	0.25 max	0.50-0.80	0.25 max	0.035 max	0.035 max	0.17-0.37	bal Fe	1130	980	6	
GB 3279(89)	85	Sh HR Ann	0.82-0.90	0.25 max	0.25 max	0.50-0.80	0.25 max	0.035 max	0.035 max	0.17-0.37	bal Fe	800		10	
GB 699(88)	85	Bar HR Q/T	0.82-0.90	0.25 max	0.25 max	0.50-0.80	0.25 max	0.035 max	0.035 max	0.17-0.37	bal Fe	1130	980	6	

Europe

Specification	Designation	Notes	C	Cr	Cu	Mn	Ni	P	S	Si	Other	UTS	YS	El	Hard
EN 10016/2(94)	1.0616	Rod	0.83-0.88	0.15 max	0.25 max	0.50-0.80	0.20 max	0.035 max	0.035 max	0.10-0.30	Al <=0.01; Mo <=0.05; Cu+Sn<=0.25; bal Fe				
EN 10016/2(94)	C86D	Rod	0.83-0.88	0.15 max	0.25 max	0.50-0.80	0.20 max	0.035 max	0.035 max	0.10-0.30	Al <=0.01; Mo <=0.05; Cu+Sn<=0.25; bal Fe				
EN 10016/4(94)	1.1265	Rod	0.84-0.88	0.10 max	0.15 max	0.50-0.70	0.10 max	0.020 max	0.025 max	0.10-0.30	Al <=0.01; Mo <=0.02; N <=0.007; Cr+Ni+Cu<=0.30, Cu+Sn<=0.15; bal Fe				
EN 10016/4(94)	C86D2	Rod	0.84-0.88	0.10 max	0.15 max	0.50-0.70	0.10 max	0.020 max	0.025 max	0.10-0.30	Al <=0.01; Mo <=0.02; N <=0.007; Cr+Ni+Cu<=0.30, Cu+Sn<=0.15; bal Fe				

Specification	Designation	Notes	C	Cr	Cu	Mn	Ni	P	S	Si	Other	UTS	YS	El	Hard

Carbon Steel, Nonresulfurized, 1084 (Continued from previous page)

France

Specification	Designation	Notes	C	Cr	Cu	Mn	Ni	P	S	Si	Other	UTS	YS	El	Hard
AFNOR	XC85		0.8-0.98			0.4-0.7		0.04	0.04	0.2-0.4	bal Fe				

Germany

Specification	Designation	Notes	C	Cr	Cu	Mn	Ni	P	S	Si	Other	UTS	YS	El	Hard
DIN	85Mn3		0.80-0.90			0.70-0.90		0.050 max	0.050 max	0.15-0.35	N <=0.007; bal Fe				
DIN	WNr 1.0647		0.80-0.90			0.70-0.90		0.050 max	0.050 max	0.15-0.35	N <=0.007; bal Fe				

India

Specification	Designation	Notes	C	Cr	Cu	Mn	Ni	P	S	Si	Other	UTS	YS	El	Hard
IS 1570	C85		0.8-0.9			0.5-0.8					bal Fe				
IS 1570/2(79)	85C6	Sh Plt Sect Shp Bar Bil Frg	0.8-0.9	0.3 max	0.3 max	0.5-0.8	0.4 max	0.055 max	0.055 max	0.6 max	Co <=0.1; Mo <=0.15; Pb <=0.15; W <=0.1; P:S Varies; bal Fe				
IS 1570/2(79)	C85	Sh Plt Sect Shp Bar Bil Frg	0.8-0.9	0.3 max	0.3 max	0.5-0.8	0.4 max	0.055 max	0.055 max	0.6 max	Co <=0.1; Mo <=0.15; Pb <=0.15; W <=0.1; P:S Varies; bal Fe				
IS 3749	T85	Bar	0.8-0.9			0.5-0.8		0.04	0.04	0.1-0.35	bal Fe				
IS 4367	T85	Frg	0.8-0.9			0.5-0.8				0.1-0.35	bal Fe				

Mexico

Specification	Designation	Notes	C	Cr	Cu	Mn	Ni	P	S	Si	Other	UTS	YS	El	Hard
NMX-B-301(86)	1084	Bar	0.80-0.93			0.60-0.90		0.040 max	0.050 max		bal Fe				

Poland

Specification	Designation	Notes	C	Cr	Cu	Mn	Ni	P	S	Si	Other	UTS	YS	El	Hard
PNH84028	DB85		0.82-0.88	0.1	0.2	0.4-0.8	0.2	0.04 max	0.04 max	0.25	P+S=0.07; bal Fe				
PNH84032	85	Spring	0.82-0.9	0.3	0.25	0.5-0.8	0.3	0.04 max	0.04 max	0.17-0.37	bal Fe				

Romania

Specification	Designation	Notes	C	Cr	Cu	Mn	Ni	P	S	Si	Other	UTS	YS	El	Hard
STAS 795(92)	OLC85A	Spring; HF	0.82-0.9	0.3 max	0.25 max	0.5-0.8		0.04 max	0.04 max	0.17-0.37	Pb <=0.15; bal Fe				

Russia

Specification	Designation	Notes	C	Cr	Cu	Mn	Ni	P	S	Si	Other	UTS	YS	El	Hard
GOST 14959	85		0.82-0.9	0.25	0.2	0.5-0.8	0.25	0.035	0.035	0.17-0.35	bal Fe				

UK

Specification	Designation	Notes	C	Cr	Cu	Mn	Ni	P	S	Si	Other	UTS	YS	El	Hard
BS 970/1(83)	060A86	Blm Bil Slab Bar Rod Frg	0.83-0.90			0.50-0.70		0.050 max	0.050 max	0.4	bal Fe				
BS 970/1(83)	080A86	Blm Bil Slab Bar Rod Frg	0.83-0.90			0.70-0.90		0.050 max	0.050 max	0.4	bal Fe				

USA

Specification	Designation	Notes	C	Cr	Cu	Mn	Ni	P	S	Si	Other	UTS	YS	El	Hard
	AISI 1084	Wir rod	0.80-0.93			0.60-0.90		0.040 max	0.050 max		bal Fe				
	AISI 1084	Sh Strp Plt	0.80-0.94			0.60-0.90		0.030 max	0.035 max		bal Fe				
	AISI 1084	Struct shp	0.80-0.93			0.60-0.90		0.030 max	0.035 max		bal Fe				
	UNS G10840		0.80-0.93			0.60-0.90		0.040 max	0.050 max		Plates, C 0.80-0.94; bal Fe				
ASTM A29/A29M(93)	1084	Bar	0.80-0.93			0.60-0.90		0.040 max	0.050 max		bal Fe				
ASTM A510(96)	1084	Wir rod	0.80-0.93			0.60-0.90		0.040 max	0.050 max		bal Fe				
ASTM A576(95)	1084	Special qual HW bar	0.80-0.93			0.60-0.90		0.040 max	0.050 max		Si Cu Pb B Bi Ca Se Te if spec'd; bal Fe				
ASTM A713(93)	1084	Wir, High-C spring, HT	0.80-0.93			0.60-0.90		0.040 max	0.050 max		bal Fe				
ASTM A830/A830M(98)	1084	Plt	0.80-0.94			0.60-0.90		0.035 max	0.04 max	0.15-0.40	bal Fe				
FED QQ-S-700D(91)	C1084*	Obs; Sh Strp	0.8-0.93			0.6-0.9		0.04 max	0.05 max		bal Fe				
SAE J1249(95)	1084	Former SAE std valid for Wir rod	0.80-0.93			0.60-0.90		0.040 max	0.050 max		bal Fe				
SAE J1397(92)	1084	Bar HR, est mech prop	0.8-0.93			0.6-0.9		0.04 max	0.05 max		bal Fe	820	450	10	241 HB
SAE J1397(92)	1084	Bar Ann CD, est mech prop	0.8-0.93			0.6-0.9		0.04 max	0.05 max		bal Fe	690	530	10	192 HB

UNS numbers and US grades are provided as a means of cross referencing chemically similar alloys. Exchangability is only possible after independent examination of specifications. Tensile properties are minimum or typical as specified. UTS and YS as MPa. El as %. See Appendix for list of abbreviations used in Notes. * indicates obsolete material.

Specification	Designation	Notes	C	Cr	Cu	Mn	Ni	P	S	Si	Other	UTS	YS	El	Hard

Carbon Steel, Nonresulfurized, 1084 (Continued from previous page)

USA

Specification	Designation	Notes	C	Cr	Cu	Mn	Ni	P	S	Si	Other	UTS	YS	El	Hard
SAE J403(95)	1084	Struct Shps Plt Strp Sh Weld Tub	0.80-0.94			0.60-0.90		0.030 max	0.035 max		bal Fe				

Carbon Steel, Nonresulfurized, 1085

Argentina

Specification	Designation	Notes	C	Cr	Cu	Mn	Ni	P	S	Si	Other	UTS	YS	El	Hard
IAS	IRAM 1085		0.80-0.93			0.70-1.00		0.040 max	0.050 max	0.10-0.30	bal Fe	1050-1200	580-660	8-15	321-363 HB

Czech Republic

Specification	Designation	Notes	C	Cr	Cu	Mn	Ni	P	S	Si	Other	UTS	YS	El	Hard
CSN 412090	12090	Q/T	0.8-0.9	0.3 max		0.2-0.6		0.03 max	0.035 max	0.35 max	P+S<=0.06; bal Fe				

Europe

Specification	Designation	Notes	C	Cr	Cu	Mn	Ni	P	S	Si	Other	UTS	YS	El	Hard
EN 10016/2(94)	1.0628	Rod	0.85-0.90	0.15 max	0.25 max	0.50-0.80	0.20 max	0.035 max	0.035 max	0.10-0.30	Al <=0.01; Mo <=0.05; Cu+Sn<=0.25; bal Fe				
EN 10016/2(94)	C88D	Rod	0.85-0.90	0.15 max	0.25 max	0.50-0.80	0.20 max	0.035 max	0.035 max	0.10-0.30	Al <=0.01; Mo <=0.05; Cu+Sn<=0.25; bal Fe				
EN 10016/4(94)	1.1272	Rod	0.86-0.90	0.10 max	0.15 max	0.50-0.70	0.10 max	0.020 max	0.025 max	0.10-0.30	Al <=0.01; Mo <=0.02; N <=0.007; Cr+Ni+Cu<=0.30, Cu+Sn<=0.15; bal Fe				
EN 10016/4(94)	C88D2	Rod	0.86-0.90	0.10 max	0.15 max	0.50-0.70	0.10 max	0.020 max	0.025 max	0.10-0.30	Al <=0.01; Mo <=0.02; N <=0.007; Cr+Ni+Cu<=0.30, Cu+Sn<=0.15; bal Fe				

Italy

Specification	Designation	Notes	C	Cr	Cu	Mn	Ni	P	S	Si	Other	UTS	YS	El	Hard
UNI 5598(71)	3CD85	Wir rod	0.83-0.88			0.4-0.7		0.035 max	0.035 max	0.15-0.35	N <=0.012; bal Fe				
UNI 7064(82)	C85	Spring	0.8-0.9			0.45-0.65		0.035 max	0.035 max	0.15-0.4	bal Fe				
UNI 8893(86)	C85	Spring	0.8-0.9			0.45-0.65		0.035 max	0.035 max	0.15-0.4	bal Fe				

Mexico

Specification	Designation	Notes	C	Cr	Cu	Mn	Ni	P	S	Si	Other	UTS	YS	El	Hard
NMX-B-301(86)	1085	Bar	0.80-0.93			0.70-1.00		0.040 max	0.050 max		bal Fe				

USA

Specification	Designation	Notes	C	Cr	Cu	Mn	Ni	P	S	Si	Other	UTS	YS	El	Hard
	AISI 1085	Struct shp	0.80-0.93			0.70-1.00		0.030 max	0.035 max		bal Fe				
	AISI 1085	Sh Strp Plt	0.80-0.94			0.70-1.00		0.030 max	0.035 max		bal Fe				
	UNS G10850		0.80-0.94			0.70-1.00		0.040 max	0.050 max		bal Fe				
ASTM A510(96)	1085	Wir rod	0.80-0.93			0.70-1.00		0.040 max	0.050 max		bal Fe				
ASTM A682/A682M(98)	1085	Strp, high-C, CR, spring	0.80-0.94			0.70-1.00		0.035 max	0.040 max	0.15-0.30	bal Fe				
ASTM A684A684M(86)	1085	Strp, high-C, CR, Ann	0.80-0.94			0.70-1.00		0.035 max	0.040 max	0.15-0.30	bal Fe				95 HRB
ASTM A830/A830M(98)	1085	Plt	0.80-0.94			0.70-1.00		0.035 max	0.04 max	0.15-0.40	bal Fe				
MIL-S-46049C(98)	1085	Strp CR hard temp, spring	0.80-0.94			0.70-1.00		0.040 max	0.050 max	0.15-0.30	bal Fe				
SAE J1397(92)	1085	Bar CD, est mech prop									bal Fe	690	540	10	192 HB
SAE J1397(92)	1085	Bar HR, est mech prop									bal Fe	830	460	10	248 HB
SAE J403(95)	1085	Struct Shps Plt Strp Sh Weld Tub	0.80-0.94			0.70-1.00		0.030 max	0.035 max		bal Fe				

Carbon Steel, Nonresulfurized, 1086

Mexico

Specification	Designation	Notes	C	Cr	Cu	Mn	Ni	P	S	Si	Other	UTS	YS	El	Hard
NMX-B-301(86)	1086	Bar	0.80-0.93			0.30-0.50		0.040 max	0.050 max		bal Fe				

Poland

Specification	Designation	Notes	C	Cr	Cu	Mn	Ni	P	S	Si	Other	UTS	YS	El	Hard
PNH84032	85	Spring	0.82-0.9	0.3 max	0.25 max	0.5-0.8	0.3 max	0.04 max	0.04 max	0.17-0.37	bal Fe				

Specification	Designation	Notes	C	Cr	Cu	Mn	Ni	P	S	Si	Other	UTS	YS	El	Hard

Carbon Steel, Nonresulfurized, 1086 (Continued from previous page)

Russia

Specification	Designation	Notes	C	Cr	Cu	Mn	Ni	P	S	Si	Other	UTS	YS	El	Hard
GOST 1050(74)	85	High-grade struct; Q/T; 0-100mm	0.82-0.9	0.25 max	0.25 max	0.5-0.8	0.25 max	0.035 max	0.04 max	0.17-0.37	As<=0.08; bal Fe	1130	980	6L	
GOST 1050(74)	85	High-grade struct; Untreated; 0-250mm	0.82-0.9	0.25 max	0.25 max	0.5-0.8	0.25 max	0.035 max	0.04 max	0.17-0.37	As<=0.08; bal Fe				0-302 HB
GOST 14959	85A	Spring	0.82-0.9	0.25 max	0.2 max	0.5-0.8	0.25 max	0.035 max	0.025 max	0.17-0.37	Al <=0.1; bal Fe				
GOST 14959(79)	85	Spring	0.82-0.9	0.25 max	0.2 max	0.5-0.8	0.25 max	0.035 max	0.035 max	0.17-0.37	Al <=0.1; bal Fe				

Sweden

Specification	Designation	Notes	C	Cr	Cu	Mn	Ni	P	S	Si	Other	UTS	YS	El	Hard
SS 141774	1774	Spring; Unalloyed high qual	0.6-0.95			0.3-0.8		0.035 max	0.035 max	0.15-0.4	bal Fe				

USA

Specification	Designation	Notes	C	Cr	Cu	Mn	Ni	P	S	Si	Other	UTS	YS	El	Hard
	AISI 1086	Struct shp	0.80-0.93			0.30-0.50		0.030 max	0.035 max		bal Fe				
	AISI 1086	Wir rod	0.80-0.93			0.30-0.50		0.040 max	0.050 max		bal Fe				
	AISI 1086	Sh Strp Plt	0.80-0.94			0.30-0.50		0.030 max	0.035 max		bal Fe				
	AISI 1086	Bar	0.80-0.93			0.30-0.50		0.040 max	0.050 max		bal Fe				
	UNS G10860		0.80-0.93			0.30-0.50		0.040 max	0.050 max		Sheets and Plates, C 0.80-0.94; bal Fe				
ASTM A29/A29M(93)	1086	Bar	0.80-0.93			0.30-0.50		0.040 max	0.050 max		bal Fe				
ASTM A510(96)	1086	Wir rod	0.80-0.93			0.30-0.50		0.040 max	0.050 max		bal Fe				
ASTM A682/A682M(98)	1086	Strp, high-C, CR, spring	0.80-0.94			0.30-0.50		0.035 max	0.040 max	0.15-0.30	bal Fe				
ASTM A684A684M(86)	1086	Strp, high-C, CR, Ann	0.80-0.94			0.30-0.50		0.035 max	0.040 max	0.15-0.30	bal Fe				95 HRB
ASTM A713(93)	1086	Wir, High-C spring, HT	0.80-0.93			0.30-0.50		0.040 max	0.050 max		bal Fe				
ASTM A830/A830M(98)	1086	Plt	0.80-0.94			0.30-0.50		0.035 max	0.04 max	0.15-0.40	bal Fe				
SAE J1249(95)	1086	Former SAE std valid for Wir rod	0.80-0.94			0.30-0.50		0.040 max	0.050 max		bal Fe				
SAE J1397(92)	1086	Bar HR, est mech prop									bal Fe	770	420	10	229 HB
SAE J1397(92)	1086	Bar CD, est mech prop									bal Fe	670	510	10	192 HB
SAE J403(95)	1086	Struct Shps Plt Strp Sh Weld Tub	0.80-0.94			0.30-0.50		0.030 max	0.035 max		bal Fe				
SAE J403(95)	1086	Bar Wir rod Smls Tub HR CF	0.80-0.93			0.30-0.50		0.030 max	0.050 max		bal Fe				

Carbon Steel, Nonresulfurized, 1090

Austria

Specification	Designation	Notes	C	Cr	Cu	Mn	Ni	P	S	Si	Other	UTS	YS	El	Hard
ONORM M3110	RC90		0.90-0.95			0.4-0.7		0.035 max	0.035 max	0.15-0.40	P+S=0.06; bal Fe				

China

Specification	Designation	Notes	C	Cr	Cu	Mn	Ni	P	S	Si	Other	UTS	YS	El	Hard
GB/T 4358(95)	T9A	Wir CD 0.10-2.00mm diam	0.85-0.93	0.10 max	0.20 max	0.40 max	0.12 max	0.025 max	0.020 max	0.35 max	bal Fe	1760-3040			
YB/T 5063(93)	T9A	Strp HR HT	0.85-0.94	0.25 max	0.30 max	0.40 max	0.20 max	0.030 max	0.020 max	0.35 max	bal Fe	1275-1570			
YB/T 5101(93)	T9A	Wir CD 0.10-2.00mm diam	0.86-0.93	0.10 max	0.20 max	0.40 max	0.12 max	0.025 max	0.020 max	0.35 max	bal Fe	1815-3090			

Europe

Specification	Designation	Notes	C	Cr	Cu	Mn	Ni	P	S	Si	Other	UTS	YS	El	Hard
EN 10016/2(94)	1.0618	Rod	0.90-0.95	0.15 max	0.25 max	0.50-0.80	0.20 max	0.035 max	0.035 max	0.10-0.30	Al <=0.01; Mo <=0.05; Cu+Sn<=0.25; bal Fe				
EN 10016/2(94)	C92D	Rod	0.90-0.95	0.15 max	0.25 max	0.50-0.80	0.20 max	0.035 max	0.035 max	0.10-0.30	Al <=0.01; Mo <=0.05; Cu+Sn<=0.25; bal Fe				
EN 10016/4(94)	1.1282	Rod	0.90-0.95	0.10 max	0.15 max	0.50-0.70	0.10 max	0.020 max	0.025 max	0.10-0.30	Al <=0.01; Mo <=0.02; N <=0.007; Cr+Ni+Cu<=0.30, Cu+Sn<=0.15; bal Fe				

Specification	Designation	Notes	C	Cr	Cu	Mn	Ni	P	S	Si	Other	UTS	YS	El	Hard
Carbon Steel, Nonresulfurized, 1090 (Continued from previous page)															
Europe															
EN 10016/4(94)	C92D2	Rod	0.90-0.95	0.10 max	0.15 max	0.50-0.70	0.10 max	0.020 max	0.025 max	0.10-0.30	Al <=0.01; Mo <=0.02; N <=0.007; Cr+Ni+Cu<=0.30, Cu+Sn<=0.15; bal Fe				
France															
AFNOR NFA35553	XC90	Strp	0.85-0.95			0.3-0.5		0.03	0.03	0.15-0.3	bal Fe				
AFNOR NFA36102(93)	C90RR	HR; Strp; CR	0.85-0.95	0.3 max		0.4-0.7	0.40 max	0.025 max	0.02 max	0.15-0.3	Al <=0.03; Mo <=0.10; bal Fe				
AFNOR NFA36102(93)	XC90*	HR; Strp; CR	0.85-0.95	0.3 max		0.4-0.7	0.40 max	0.025 max	0.02 max	0.15-0.3	Al <=0.03; Mo <=0.10; bal Fe				
Germany															
DIN	90Mn4	Q/T 17-40mm	0.85-0.95			0.90-1.10		0.035 max	0.035 max	0.25-0.50	bal Fe	1670	1325	5	
DIN	WNr 1.1273	Q/T 17-40mm	0.85-0.95			0.90-1.10		0.035 max	0.035 max	0.25-0.50	bal Fe	1670	1325	5	
Hungary															
MSZ 5736(88)	D91	Wir rod; for CD	0.87-0.94	0.2 max	0.2 max	0.3-0.7	0.2 max	0.035 max	0.04 max	0.1-0.35	D91V:V<=0.15; bal Fe				
MSZ 5736(88)	D91K	Wir rod; for CD	0.87-0.94	0.2 max	0.15 max	0.3-0.7	0.2 max	0.025 max	0.025 max	0.1-0.35	bal Fe				
Italy															
UNI 3545	C90		0.85-0.95			0.4-0.7		0.035	0.035	0.15-0.3	bal Fe				
UNI 5598(71)	3CD90	Wir rod	0.88-0.93			0.4-0.7		0.035 max	0.035 max	0.15-0.35	N <=0.012; bal Fe				
UNI 7064	C90		0.85-0.95			0.4-0.7		0.035	0.035	0.15-0.3	bal Fe				
UNI 8893(86)	C90	Spring	0.85-0.95			0.4-0.6		0.035 max	0.035 max	0.15-0.4	bal Fe				
Mexico															
NMX-B-301(86)	1090	Bar	0.85-0.98			0.60-0.90		0.040 max	0.050 max		bal Fe				
Pan America															
COPANT 331	1090	Bar	0.85-0.98			0.6-0.9		0.04	0.05		bal Fe				
Poland															
PNH84028	DB90	Wir rod	0.88-0.94	0.1 max	0.2 max	0.4-0.8	0.2 max	0.04 max	0.04 max	0.25 max	P+S<=0.07; bal Fe				
Romania															
STAS 795(92)	OLC90A	Spring; HF	0.85-0.95	0.3 max	0.25 max	0.2-0.5		0.04 max	0.04 max	0.17-0.37	Pb <=0.15; bal Fe				
UK															
BS 1429(80)	060A96	Spring	0.93-1.00			0.50-0.70				0.10-0.35	bal Fe				
BS 1449/1(91)	95CS	Plt Sht Strp; CD norm t<=16mm	0.90-1.00			0.30-0.90		0.040 max	0.040 max	0.05-0.35	bal Fe				314-333 HB
BS 1449/1(91)	95HS	Plt Sht Strp; CD norm t<16mm	0.90-1.00			0.30-0.90		0.040 max	0.040 max	0.05-0.35	bal Fe				314-333 HB
USA															
	AISI 1090	Sh Strp Plt	0.84-0.98			0.60-0.90		0.030 max	0.035 max		bal Fe				
	AISI 1090	Wir rod	0.85-0.98			0.60-0.90		0.040 max	0.050 max		bal Fe				
	AISI 1090	Struct shp	0.85-0.98			0.60-0.90		0.030 max	0.035 max		bal Fe				
	AISI 1090	Bar	0.85-0.98			0.60-0.90		0.040 max	0.050 max		bal Fe				
	UNS G10900		0.85-0.98			0.60-0.90		0.040 max	0.050 max		Plates, C 0.84-0.98; bal Fe				
AMS 5112K(96)		Bar, Spring quality music wire CD, D=0.10	0.70-1.00			0.20-0.60		0.025 max	0.030 max	0.10-0.30	bal Fe	3027-3344			
AMS 5112K(96)		Spring quality music wire CD, D=6.35	0.70-1.00			0.20-0.60		0.025 max	0.030 max	0.10-0.30	bal Fe	1586-1751			
ASTM A29/A29M(93)	1090	Bar	0.85-0.98			0.60-0.90		0.040 max	0.050 max		bal Fe				

UNS numbers and US grades are provided as a means of cross referencing chemically similar alloys. Exchangability is only possible after independent examination of specifications. Tensile properties are minimum or typical as specified. UTS and YS as MPa. El as %. See Appendix for list of abbreviations used in Notes. * indicates obsolete material.

Specification	Designation	Notes	C	Cr	Cu	Mn	Ni	P	S	Si	Other	UTS	YS	El	Hard

Carbon Steel, Nonresulfurized, 1090 (Continued from previous page)

USA

Specification	Designation	Notes	C	Cr	Cu	Mn	Ni	P	S	Si	Other	UTS	YS	El	Hard
ASTM A510(96)	1090	Wir rod	0.85-0.98			0.60-0.90		0.040 max	0.050 max		bal Fe				
ASTM A576(95)	1090	Special qual HW bar	0.85-0.98			0.60-0.90		0.040 max	0.050 max		Si Cu Pb B Bi Ca Se Te if spec'd; bal Fe				
ASTM A713(93)	1090	Wir, High-C spring, HT	0.85-0.98			0.60-0.90		0.040 max	0.050 max		bal Fe				
ASTM A830/A830M(98)	1090	Plt	0.84-0.98			0.60-0.90		0.035 max	0.04 max	0.15-0.40	bal Fe				
SAE J1397(92)	1090	Bar Ann CD, est mech prop	0.85-0.98			0.6-0.9		0.04 max	0.05 max		bal Fe	700	540	10	197 HB
SAE J1397(92)	1090	Bar HR, est mech prop	0.85-0.98			0.6-0.9		0.04 max	0.05 max		bal Fe	840	460	10	248 HB
SAE J403(95)	1090	Bar Wir rod Smls Tub HR CF	0.85-0.98			0.60-0.90		0.030 max	0.050 max		bal Fe				
SAE J403(95)	1090	Struct Shps Plt Strp Sh Weld Tub	0.84-0.98			0.60-0.90		0.030 max	0.035 max		bal Fe				

Carbon Steel, Nonresulfurized, 1095

Australia

Specification	Designation	Notes	C	Cr	Cu	Mn	Ni	P	S	Si	Other	UTS	YS	El	Hard
AS 1442	K1095*	Obs; Bar Bil	0.9-1.03			0.4-0.7		0.05	0.05	0.1-0.35	bal Fe				
AS 1442(92)	1095	HR bar, Semifinished (may treat w/microalloying elements)	0.90-1.03			0.40-0.70		0.040 max	0.040 max	0.10-0.35	Si<=0.10 for Al-killed; bal Fe				

Austria

Specification	Designation	Notes	C	Cr	Cu	Mn	Ni	P	S	Si	Other	UTS	YS	El	Hard
ONORM M3110	RC95	Wir	0.95-1.00			0.4-0.7		0.040 max	0.040 max	0.15-0.40	P+S=0.06; bal Fe				
ONORM M3161	C60		0.56-0.65			0.5-0.7		0.040 max	0.040 max	0.2-0.4	bal Fe				

Canada

Specification	Designation	Notes	C	Cr	Cu	Mn	Ni	P	S	Si	Other	UTS	YS	El	Hard
CSA 95-1/1	C1095		0.9-1.03			0.3-0.5		0.04	0.05	0.15-0.3	bal Fe				

China

Specification	Designation	Notes	C	Cr	Cu	Mn	Ni	P	S	Si	Other	UTS	YS	El	Hard
YB/T 5063(93)	T10A	Strp HR HT	0.95-1.04	0.25 max	0.30 max	0.40 max	0.20 max	0.030 max	0.020 max	0.35 max	bal Fe	1275-1570			

Europe

Specification	Designation	Notes	C	Cr	Cu	Mn	Ni	P	S	Si	Other	UTS	YS	El	Hard
EN 10016/4(94)	1.1283	Rod	0.96-1.00	0.10 max	0.15 max	0.50-0.70	0.10 max	0.020 max	0.025 max	0.10-0.30	Al <=0.01; Mo <=0.02; N <=0.007; Cr+Ni+Cu<=0.30, Cu+Sn<=0.15; bal Fe				
EN 10016/4(94)	C98D2	Rod	0.96-1.00	0.10 max	0.15 max	0.50-0.70	0.10 max	0.020 max	0.025 max	0.10-0.30	Al <=0.01; Mo <=0.02; N <=0.007; Cr+Ni+Cu<=0.30, Cu+Sn<=0.15; bal Fe				

Finland

Specification	Designation	Notes	C	Cr	Cu	Mn	Ni	P	S	Si	Other	UTS	YS	El	Hard
SFS 906(73)	SFS906		0.95-1.1			0.2-0.4		0.03 max	0.02 max	0.1-0.3	bal Fe				

France

Specification	Designation	Notes	C	Cr	Cu	Mn	Ni	P	S	Si	Other	UTS	YS	El	Hard
AFNOR NFA35553	XC100	Strp	0.95-1.05			0.25-0.45		0.03	0.025	0.15-0.3	bal Fe				
AFNOR NFA36102(93)	C100RR	HR; Strp; CR	0.95-1.05	0.3 max		0.3-0.6	0.40 max	0.025 max	0.02 max	0.15-0.3	Al <=0.03; Mo <=0.10; bal Fe				
AFNOR NFA36102(93)	C125RR	HR; Strp; CR	1.2-1.3	0.3 max		0.3-0.6	0.40 max	0.025 max	0.02 max	0.15-0.3	Al <=0.03; Mo <=0.10; bal Fe				
AFNOR NFA36102(93)	XC100*	HR; Strp; CR	0.95-1.05	0.3 max		0.3-0.6	0.40 max	0.025 max	0.02 max	0.15-0.3	Al <=0.03; Mo <=0.10; bal Fe				
AFNOR NFA36102(93)	XC125*	HR; Strp; CR	1.2-1.3	0.3 max		0.3-0.6	0.40 max	0.025 max	0.02 max	0.15-0.3	Al <=0.03; Mo <=0.10; bal Fe				

Germany

Specification	Designation	Notes	C	Cr	Cu	Mn	Ni	P	S	Si	Other	UTS	YS	El	Hard
DIN	Ck100		0.98-1.05			0.25 max		0.025 max	0.025 max	0.15-0.25	N <=0.007; N depends on P; bal Fe				
DIN	WNr 1.1275		0.98-1.05			0.25 max		0.025 max	0.025 max	0.15-0.25	N <=0.007; N depends on P; bal Fe				
DIN 17140	D95-2*	Obs; see C92D	0.9-0.99			0.3-0.7		0.04 max	0.04 max	0.1-0.3	bal Fe				

Carbon Steel, Nonresulfurized, 1095 (Continued from previous page)

Specification	Designation	Notes	C	Cr	Cu	Mn	Ni	P	S	Si	Other	UTS	YS	El	Hard
Germany															
DIN 17222(79)	C101E	CR Strp spring hard <=2.5mm	0.95-1.05			0.40-0.60		0.035 max	0.035 max	0.15-0.35	bal Fe	1470-1670	1275	6	
DIN 17222(79)	Ck101	CR Strp spring hard <=2.5mm	0.95-1.05			0.40-0.60		0.035 max	0.035 max	0.15-0.35	bal Fe	1470-1670	1275	6	
DIN 17222(79)	WNr 1.1274	CR Strp spring hard <=2.5mm	0.95-1.05			0.40-0.60		0.035 max	0.035 max	0.15-0.35	bal Fe	1470-1670	1275	6	
India															
IS 1570	C98		0.9-1.05			0.5-0.8					bal Fe				
IS 1570/2(79)	98C6	Sh Plt Sect Shp Bar Bil Frg	0.9-1.05	0.3 max	0.3 max	0.5-0.8	0.4 max	0.055 max	0.055 max	0.6 max	Co <=0.1; Mo <=0.15; Pb <=0.15; W <=0.1; P:S Varies; bal Fe				
IS 1570/2(79)	C98	Sh Plt Sect Shp Bar Bil Frg	0.9-1.05	0.3 max	0.3 max	0.5-0.8	0.4 max	0.055 max	0.055 max	0.6 max	Co <=0.1; Mo <=0.15; Pb <=0.15; W <=0.1; P:S Varies; bal Fe				
IS 3749	T103	Bar	0.95-1.1			0.2-0.35		0.035	0.04	0.1-0.35	bal Fe				
Italy															
UNI 3545	C100		0.98-1.05			0.35-0.6		0.035	0.035	0.15-0.3	bal Fe				
UNI 5598(71)	3CD95	Wir rod	0.93-0.98			0.4-0.7		0.035 max	0.035 max	0.15-0.35	N <=0.012; bal Fe				
UNI 7064(82)	C100	Spring	0.95-1.05			0.4-0.6		0.035 max	0.035 max	0.15-0.4	bal Fe				
UNI 8893(86)	C100	Spring	0.95-1.05			0.4-0.6		0.035 max	0.035 max	0.15-0.4	bal Fe				
Mexico															
NMX-B-301(86)	1095	Bar	0.90-1.03			0.30-0.50		0.040 max	0.050 max		bal Fe				
Pan America															
COPANT 330	1095	Bar	0.9-1.03			0.3-0.5		0.04	0.05		bal Fe				
COPANT 331	1095	Bar	0.9-1.03			0.3-0.5		0.04	0.05		bal Fe				
Poland															
PNH84028	DS105		1.01-1.1	0.1	0.15	0.3-0.6	0.15	0.02 max	0.02 max	0.25	P+S=0.035; bal Fe				
PNH84028	DS95	Unalloyed; Wir rod	0.93-0.98	0.1 max	0.15 max	0.3-0.6	0.15 max	0.02 max	0.02 max	0.1-0.25	Mo <=0.05; P+S<=0.035; bal Fe				
PNH84028	DS96		0.93-1	0.1	0.15	0.3-0.6	0.15	0.02 max	0.02 max	0.25	P+S=0.035; bal Fe				
PNH84028	DW100		0.98-1.03	0.12	0.2	0.3-0.5	0.15	0.02 max	0.02 max	0.25	P+S=0.035; bal Fe				
PNH84028	DW95		0.92-0.97	0.12	0.2	0.3-0.5	0.15	0.02 max	0.02 max	0.25	P+S=0.035; bal Fe				
PNH92602	DS105		1.01-1.1	0.1	0.15	0.3-0.6	0.15	0.02 max	0.02 max	0.25	P+S=0.035; bal Fe				
PNH92602	DS95		0.93-1	0.1	0.15	0.3-0.6	0.15	0.02 max	0.02 max	0.25	P+S=0.035; bal Fe				
Sweden															
SIS 141870	1870-02	Strp, Ann, 2mm diam	0.94-1.06			0.3-0.6		0.03	0.03	0.1-0.35	bal Fe		410	23	
SS 141870	1870	Spring	0.95-1.06			0.4-0.6		0.03 max	0.03 max	0.15-0.35	bal Fe				
UK															
BS 1449/1(91)	95CS	Plt Sht Strp; CD norm t<=16mm	0.90-1.00			0.30-0.90		0.040 max	0.040 max	0.05-0.35	bal Fe				314-333 HB
BS 1449/1(91)	95HS	Plt Sht Strp; CD norm t<16mm	0.90-1.00			0.30-0.90		0.040 max	0.040 max	0.05-0.35	bal Fe				314-333 HB
BS 970/1(83)	060A96	Blm Bil Slab Bar Rod Frg	0.93-1.00			0.50-0.70		0.050 max	0.050 max		bal Fe				
BS 970/1(83)	060A99	Blm Bil Slab Bar Rod Frg	0.95-1.05			0.50-0.70		0.050 max	0.050 max	0.4	bal Fe				
BS 970/5(72)	060A96*	Obs; Blm Bil Slab Bar Rod Frg	0.93-1.00			0.50-0.70				0.1-0.35	bal Fe				

UNS numbers and US grades are provided as a means of cross referencing chemically similar alloys. Exchangability is only possible after independent examination of specifications. Tensile properties are minimum or typical as specified. UTS and YS as MPa. El as %. See Appendix for list of abbreviations used in Notes. * indicates obsolete material.

Specification	Designation	Notes	C	Cr	Cu	Mn	Ni	P	S	Si	Other	UTS	YS	El	Hard

Carbon Steel, Nonresulfurized, 1095 (Continued from previous page)

UK

Specification	Designation	Notes	C	Cr	Cu	Mn	Ni	P	S	Si	Other	UTS	YS	El	Hard
BS DEF STAN 95-1-1	C1095	Bar	0.9-1.03			0.3-0.5		0.04	0.05	0.15-0.3	bal Fe				

USA

Specification	Designation	Notes	C	Cr	Cu	Mn	Ni	P	S	Si	Other	UTS	YS	El	Hard
	AISI 1095	Bar	0.90-1.03			0.30-0.50		0.040 max	0.050 max		bal Fe				
	AISI 1095	Struct shp	0.90-1.03			0.30-0.50		0.030 max	0.035 max		bal Fe				
	AISI 1095	Wir rod	0.90-1.03			0.30-0.50		0.040 max	0.050 max		bal Fe				
	AISI 1095	Sh Strp Plt	0.90-1.04			0.30-0.50		0.030 max	0.035 max		bal Fe				
	UNS G10950		0.90-1.03			0.30-0.50		0.040 max	0.050 max		Plates, C 0.90-1.04; bal Fe				
AMS 5121G(90)		Sh, Strp	0.90-1.04			0.30-0.50		0.040 max	0.050 max	0.15-0.35	bal Fe				85 HRB max
AMS 5122G(90)		Strp CF hard tmp	0.90-1.04			0.30-0.50		0.040 max	0.050 max	0.15-0.35	bal Fe				47-52 HRC
AMS 5132H(93)		Bar, t<=3.18	0.90-1.30			0.30-0.50		0.040 max	0.050 max	0.15-0.35	bal Fe				302 HB max
AMS 5132H(93)		Bar, t>12.70	0.90-1.30			0.30-0.50		0.040 max	0.050 max	0.15-0.35	bal Fe				207 HB max
AMS 7304		Bar Rod, Wir, Plt, Strp, Frg	0.90-1.03			0.30-0.50		0.040 max	0.050 max		bal Fe				
AMS 8559(98)		Bar, Aircraft Quality	0.90-1.05			0.30-0.50		0.040 max	0.050 max		bal Fe				207 or 229 HB max
ASTM A108(95)	1095	Bar, CF	0.90-1.03			0.30-0.50		0.040 max	0.050 max		bal Fe				
ASTM A29/A29M(93)	1095	Bar	0.90-1.03			0.30-0.50		0.040 max	0.050 max		bal Fe				
ASTM A510(96)	1095	Wir rod	0.90-1.03			0.30-0.50		0.040 max	0.050 max		bal Fe				
ASTM A576(95)	1095	Special qual HW bar	0.90-1.03			0.30-0.50		0.040 max	0.050 max		Si Cu Pb B Bi Ca Se Te if spec'd; bal Fe				
ASTM A682/A682M(98)	1095	Strp, high-C, CR, spring	0.9-1.04			0.30-0.50		0.035 max	0.040 max	0.15-0.30	bal Fe				
ASTM A684A684M(86)	1095	Strp, high-C, CR, Ann	0.9-1.04			0.30-0.50		0.035 max	0.040 max	0.15-0.30	bal Fe				96 HRB
ASTM A713(93)	1095	Wir, High-C spring, HT	0.90-1.03			0.30-0.50		0.040 max	0.050 max		bal Fe				
ASTM A830/A830M(98)	1095	Plt	0.90-1.04			0.30-0.50		0.035 max	0.04 max	0.15-0.40	bal Fe				
FED QQ-S-700D(91)	C1095*	Obs; Sh Strp	0.9-1.03			0.3-0.5		0.04	0.05		bal Fe				
MIL-S-11713C(88)	3	Spring Strp	0.90-1.03			0.30-0.50		0.025 max	0.025 max	0.15-0.25	bal Fe				
MIL-S-16788A(86)	C10*	HR, frg, bloom bil slab for refrg; Obs for new design	0.90-1.03			0.30-0.50		0.040 max	0.050 max		bal Fe				
MIL-S-46049C(98)	1095	Strp CR hard temp, spring	0.91-1.04			0.30-0.50		0.040 max	0.050 max	0.15-0.30	bal Fe				
MIL-S-7947A(63)	1095*	Sh Strp, aircraft qual; Obs for new design see AMS 5121 5122	0.9-1.03			0.3-0.5		0.04 max	0.05 max		bal Fe				
MIL-S-8559A1(81)	1095	Bar aircraft qual	0.9-1.05			0.30-0.50		0.040 max	0.050 max	0.15-0.35	bal Fe				
SAE J1397(92)	1095	Bar HR, est mech prop	0.9-1.03			0.3-0.5		0.04	0.05		bal Fe	830	480	10	248 HB
SAE J1397(92)	1095	Bar Ann CD, est mech prop	0.9-1.03			0.3-0.5		0.04	0.05		bal Fe	680	520	10	197 HB
SAE J403(95)	1095	Struct Shps Plt Strp Sh Weld Tub	0.90-1.04			0.30-0.50		0.030 max	0.035 max		bal Fe				
SAE J403(95)	1095	Bar Wir rod Smls Tub HR CF	0.90-1.03			0.30-0.50		0.030 max	0.050 max		bal Fe				

Yugoslavia

Specification	Designation	Notes	C	Cr	Cu	Mn	Ni	P	S	Si	Other	UTS	YS	El	Hard
	PZ90		0.9-0.99	0.15	0.025	0.3-0.7	0.15	0.04	0.04	0.1-0.3	bal Fe				

UNS numbers and US grades are provided as a means of cross referencing chemically similar alloys. Exchangability is only possible after independent examination of specifications. Tensile properties are minimum or typical as specified. UTS and YS as MPa. El as %. See Appendix for list of abbreviations used in Notes. * indicates obsolete material.

Specification	Designation	Notes	C	Cr	Cu	Mn	Ni	P	S	Si	Other	UTS	YS	El	Hard
Carbon Steel, Nonresulfurized, 5061															
USA															
	UNS K00802	Low C	0.08-0.20			0.40-0.80		0.040 max	0.050 max		bal Fe				
AMS 5061E(94)		Bar and Wir, CD	0.08-0.20			0.40-0.80		0.040 max	0.050 max		bal Fe	483		10	80-100 HRB
Carbon Steel, Nonresulfurized, 5062															
UK															
BS 1717(83)	CFSC4	Tub	0.25 max			0.60-1.00		0.050 max	0.050 max	0.35 max	W <=0.1; bal Fe				
USA															
	UNS K02508		0.25 max			1.00 max		0.400 max	0.050 max		bal Fe				
Carbon Steel, Nonresulfurized, 5112															
Germany															
DIN	Mk97		0.95-0.99			0.25-0.45		0.025 max	0.025 max	0.10-0.25	N <=0.007; N depends on P; bal Fe				
DIN	WNr 1.1291		0.95-0.99			0.25-0.45		0.025 max	0.025 max	0.10-0.25	N <=0.007; N depends on P; bal Fe				
USA															
	UNS K08500		0.70-1.00			0.20-0.60		0.025 max	0.03 max	0.12-0.30	bal Fe				
Carbon Steel, Nonresulfurized, A1(61-80)															
USA															
	UNS K06100		0.55-0.68			0.60-0.90		0.04 max	0.05 max	0.10-0.25	bal Fe				
ASTM A1(92)	60-84#/yd	Rails	0.55-0.68			0.60-0.90		0.040 max	0.050 max	0.10-0.50	bal Fe				201 HB
ASTM A616/A616M(96)	Grade 50	Std section T-rails, bar for conc reinf	0.55-0.82			0.60-1.10		0.040 max	0.050 max	0.10-0.50	bal Fe	550	350	5-7	
ASTM A616/A616M(96)	Grade 60	Std section T-rails, bar for conc reinf	0.55-0.82			0.60-1.10		0.040 max	0.050 max	0.10-0.50	bal Fe	620	420	4.5-6	
Carbon Steel, Nonresulfurized, A1(81-89)															
USA															
	UNS K07000		0.64-0.77			0.60-0.90		0.04 max	0.05 max	0.10-0.25	bal Fe				
ASTM A1(92)	85-114#/yd	Rails	0.67-0.80			0.70-0.100		0.035 max	0.040 max	0.10-0.50					248 HB
Carbon Steel, Nonresulfurized, A1(90-114)															
USA															
	UNS K07301		0.67-0.80			0.70-1.00		0.04 max	0.05 max	0.10-0.25	bal Fe				
ASTM A1(92)	>=115#/yd	Rails	0.72-0.82			0.80-0.110		0.035 max	0.040 max	0.10-0.50					285 HB
Carbon Steel, Nonresulfurized, A109(1,2,3)															
USA															
	UNS K02500		0.25 max			0.60 max		0.035 max	0.04 max		Cu 0.20 min (when spec'd); bal Fe				
Carbon Steel, Nonresulfurized, A109(4,5)															
USA															
	UNS K01507		0.15 max			0.60 max		0.035 max	0.04 max		Cu 0.20 min (when spec'd); bal Fe				
Carbon Steel, Nonresulfurized, A131(A)															
USA															
	UNS K02300		0.23 max					0.05 max	0.05 max		bal Fe				
ASTM A131/A131M(94)	A	Ship struct	0.23 max					0.035 max	0.04 max		bal Fe	400-490	235	24	
ASTM A131/A131M(94)	A	Ship rivet/cold flanging stl	0.23 max					0.035 max	0.04 max		bal Fe	380-450	205	26	

UNS numbers and US grades are provided as a means of cross referencing chemically similar alloys. Exchangability is only possible after independent examination of specifications. Tensile properties are minimum or typical as specified. UTS and YS as MPa. El as %. See Appendix for list of abbreviations used in Notes. * indicates obsolete material.

Specification	Designation	Notes	C	Cr	Cu	Mn	Ni	P	S	Si	Other	UTS	YS	El	Hard
Carbon Steel, Nonresulfurized, A139(B)															
USA															
	UNS K03003		0.30 max			1.00 max		0.040 max	0.050 max		bal Fe				
ASTM A139	B	Weld Pipe	0.30 max			1.00 max		0.035 max	0.035 max		bal Fe	415	240	35L	
Carbon Steel, Nonresulfurized, A194(1)															
USA															
	UNS K01503		0.15 max								bal Fe				
ASTM A194/A194M(98)	1	Nuts, high-temp press	0.15 min			1.00 max		0.040 max	0.050 max	0.40 max	bal Fe				121 HB
Carbon Steel, Nonresulfurized, A194(2)															
USA															
	UNS K04002		0.40 max					0.040 max	0.050 max		bal Fe				
ASTM A194/A194M(98)	2	Nuts, high-temp press	0.40 min			1.00 max		0.040 max	0.050 max	0.40 max	bal Fe				159-352 HB
ASTM A194/A194M(98)	2H	Nuts, high-temp press	0.40 min			1.00 max		0.040 max	0.050 max	0.40 max	bal Fe				248-352 HB
ASTM A194/A194M(98)	2HM	Nuts, high-temp press	0.40 min			1.00 max		0.040 max	0.050 max	0.40 max	bal Fe				159-237 HB
Carbon Steel, Nonresulfurized, A2(A)															
USA															
	UNS K06703		0.60-0.75			0.60-0.90		0.04 max		0.10-0.40	bal Fe				
ASTM A2(97)	Class A	Girder rails	0.60-0.75			0.60-0.90		0.04 max		0.10-0.40	bal Fe				
Carbon Steel, Nonresulfurized, A2(B)															
USA															
	UNS K07700		0.70-0.85			0.60-0.90		0.04 max		0.10-0.40	bal Fe				
ASTM A2(97)	Class B	Girder rails	0.70-0.85			0.60-0.90		0.04 max		0.10-0.40	bal Fe				
Carbon Steel, Nonresulfurized, A2(C)															
USA															
	UNS K08201		0.75-0.90			0.60-0.90		0.04 max		0.10-0.40	bal Fe				
ASTM A2(97)	Class C	Girder rails	0.75-0.90			0.60-0.90		0.04 max		0.10-0.40	bal Fe				
Carbon Steel, Nonresulfurized, A228															
USA															
	UNS K08501	Wir, Spring Quality Music Wir	0.70-1.00			0.20-0.60		0.025 max	0.030 max	0.10-0.30	bal Fe				
Carbon Steel, Nonresulfurized, A230															
USA															
	UNS K06701		0.60-0.75			0.60-0.90		0.025 max	0.030 max	0.15-0.35	bal Fe				
Carbon Steel, Nonresulfurized, A235(A)															
Hungary															
MSZ 4747(85)	A35.47	High-temp, Tub; Non-rimming; FF; 0-16mm; norm	0.17 max			0.4-0.8		0.04 max	0.04 max	0.35 max	bal Fe	360-480	235	25L	
MSZ 4747(85)	A35.47	High-temp, Tub; Non-rimming; FF; 40.1-60mm; norm	0.17 max			0.4-0.8		0.04 max	0.04 max	0.35 max	bal Fe	360-480	215	25L	
UK															
BS 3059/1(87)	320	Boiler, Superheater; Smls Tub; 2-12.5mm	0.16 max			0.30-0.70		0.040 max	0.040 max	0.10-0.35	bal Fe	320-480	195	25	

UNS numbers and US grades are provided as a means of cross referencing chemically similar alloys. Exchangability is only possible after independent examination of specifications. Tensile properties are minimum or typical as specified. UTS and YS as MPa. El as %. See Appendix for list of abbreviations used in Notes. * indicates obsolete material.

Specification	Designation	Notes	C	Cr	Cu	Mn	Ni	P	S	Si	Other	UTS	YS	El	Hard

Carbon Steel, Nonresulfurized, A235(A) (Continued from previous page)

UK

Specification	Designation	Notes	C	Cr	Cu	Mn	Ni	P	S	Si	Other	UTS	YS	El	Hard
BS 3059/2(90)	360	Boiler, Superheater; High-temp; Smls Tub; Tub, Weld; 2-12.5mm	0.17 max			0.40-0.80		0.035 max	0.035 max	0.10-0.35	bal Fe	360-500	235	24	

Carbon Steel, Nonresulfurized, A25(A)

USA

Specification	Designation	Notes	C	Cr	Cu	Mn	Ni	P	S	Si	Other	UTS	YS	El	Hard
	UNS K05700		0.57 max			0.60-0.85		0.05 max	0.05 max	0.15 max	bal Fe				
ASTM A504(93)	Class L	Wheels	0.47 max			0.60-0.85		0.05 max	0.05 max	0.15 min	bal Fe				197-277 HB
ASTM A504(93)	Grade A		0.47-0.57			0.60-0.85		0.05 max	0.05 max	0.15 min	bal Fe				255-321 HB

Carbon Steel, Nonresulfurized, A254

USA

Specification	Designation	Notes	C	Cr	Cu	Mn	Ni	P	S	Si	Other	UTS	YS	El	Hard
	UNS K01001		0.05-0.15			0.27-0.63		0.050 max	0.060 max		bal Fe				

Carbon Steel, Nonresulfurized, A266(3)

UK

Specification	Designation	Notes	C	Cr	Cu	Mn	Ni	P	S	Si	Other	UTS	YS	El	Hard
BS 3111/1(87)	v	Q/T Wir	0.30-0.35			0.70-1.00		0.035 max	0.035 max	0.10-0.35	bal Fe				

USA

Specification	Designation	Notes	C	Cr	Cu	Mn	Ni	P	S	Si	Other	UTS	YS	El	Hard
	UNS K05001		0.50 max			0.50-0.90		0.040 max	0.040 max	0.35 max	bal Fe				
ASTM A266/A266M(96)	3	Frg	0.45 max			0.50-0.90		0.025 max	0.025 max	0.35 max	bal Fe	75-100	37.5	19, RA=30	156-207 HB

Carbon Steel, Nonresulfurized, A283(C)

Bulgaria

Specification	Designation	Notes	C	Cr	Cu	Mn	Ni	P	S	Si	Other	UTS	YS	El	Hard
BDS 2592(71)	ASt0	Struct									bal Fe				
BDS 2592(71)	ASt3	Struct									bal Fe				

France

Specification	Designation	Notes	C	Cr	Cu	Mn	Ni	P	S	Si	Other	UTS	YS	El	Hard
AFNOR NFA35501(83)	A33*	30<t<=50mm; roll	0.3 max	0.3 max	0.3 max	2 max	0.4 max	0.07 max	0.06 max	0.6 max	bal Fe	300-540	175	17	
AFNOR NFA35501(83)	A33*	t<=30mm; roll	0.3 max	0.3 max	0.3 max	2 max	0.4 max	0.07 max	0.06 max	0.6 max	bal Fe	300-540	175	18	
AFNOR NFA35501(83)	E24-2*	t<=30mm; roll	0.17 max	0.3 max	0.3 max	2 max	0.4 max	0.045 max	0.045 max	0.6 max	N <=0.008; bal Fe	340-440	235	28	
AFNOR NFA35501(83)	E24-2*	10t<=350mm; roll	0.17 max	0.3 max	0.3 max	2 max	0.4 max	0.045 max	0.045 max	0.6 max	N <=0.008; bal Fe	320	195	26	
AFNOR NFA35501(83)	E24-2*	50<t<=100mm; roll	0.17 max	0.3 max	0.3 max	2 max	0.4 max	0.045 max	0.045 max	0.6 max	N <=0.008; bal Fe	340-440	215	27	
AFNOR NFA35501(83)	E24-2*	50<t<=100mm; roll	0.17 max	0.3 max	0.3 max	2 max	0.4 max	0.045 max	0.045 max	0.6 max	N <=0.008; bal Fe	340-440	215	27	
AFNOR NFA35501(83)	E24-2*	50<t<=100mm; roll	0.17 max	0.3 max	0.3 max	2 max	0.4 max	0.045 max	0.045 max	0.6 max	N <=0.007; bal Fe	340-440	215	27	
AFNOR NFA35501(83)	E24-2*	t<=30mm; roll	0.17 max	0.3 max	0.3 max	2 max	0.4 max	0.045 max	0.045 max	0.6 max	N <=0.007; bal Fe	340-440	235	28	
AFNOR NFA35501(83)	E24-2*	110<t<=150mm; roll	0.17 max	0.3 max	0.3 max	2 max	0.4 max	0.045 max	0.045 max	0.6 max	N <=0.008; bal Fe	340-460	185	20	
AFNOR NFA35501(83)	E24-2*	t<=2.99mm; roll	0.17 max	0.3 max	0.3 max	2 max	0.4 max	0.045 max	0.045 max	0.6 max	N <=0.008; bal Fe	360-480	215	22	
AFNOR NFA35501(83)	E24-2*	100<t<=150mm; roll	0.17 max	0.3 max	0.3 max	2 max	0.4 max	0.045 max	0.045 max	0.6 max	N <=0.008; bal Fe	340-460	185	20	
AFNOR NFA35501(83)	E24-2*	t<=2.99mm; roll	0.17 max	0.3 max	0.3 max	2 max	0.4 max	0.045 max	0.045 max	0.6 max	N <=0.008; bal Fe	360-480	215	22	
AFNOR NFA35501(83)	E24-2*	110<t<=150mm; roll	0.17 max	0.3 max	0.3 max	1.6 max	0.4 max	0.045 max	0.045 max	0.6 max	N <=0.007; bal Fe	340-460	185	20	
AFNOR NFA35501(83)	E24-2*	t<=2.99mm; roll	0.17 max	0.3 max	0.3 max	1.6 max	0.4 max	0.045 max	0.045 max	0.6 max	N <=0.007; bal Fe	360-480	215	22	
AFNOR NFA35501(83)	E24-2*	t<=30mm; roll	0.17 max	0.3 max	0.3 max	2 max	0.4 max	0.045 max	0.045 max	0.6 max	N <=0.008; bal Fe	340-440	235	28	

UNS numbers and US grades are provided as a means of cross referencing chemically similar alloys. Exchangability is only possible after independent examination of specifications. Tensile properties are minimum or typical as specified. UTS and YS as MPa. El as %. See Appendix for list of abbreviations used in Notes. * indicates obsolete material.

Specification	Designation	Notes	C	Cr	Cu	Mn	Ni	P	S	Si	Other	UTS	YS	El	Hard

Carbon Steel, Nonresulfurized, A283(C) (Continued from previous page)

Russia

Specification	Designation	Notes	C	Cr	Cu	Mn	Ni	P	S	Si	Other	UTS	YS	El	Hard
GOST 380(71)	St0	Gen struct; Semi-Finished; >40mm	0.3 max	0.3 max	0.3 max	2 max	0.4 max	0.07 max	0.06 max	0.6 max	Al <=0.1; bal Fe	300		20L	
GOST 380(71)	St0	Gen struct; Semi-Finished; 0-20mm	0.3 max	0.3 max	0.3 max	2 max	0.4 max	0.07 max	0.06 max	0.6 max	Al <=0.1; bal Fe	300		23L	
GOST 380(71)	WSt2kp2	Gen struct; Unkilled; Rimming; R; FU; 0-20mm	0.15 max	0.3 max	0.3 max	0.25-0.7	0.3 max	0.04 max	0.05 max	0.07 max	Al <=0.1; N <=0.008; As<=0.08; bal Fe	320-410	215	33L	
GOST 380(71)	WSt2kp2	Gen struct; Unkilled; Rimming; R; FU; >100mm	0.15 max	0.3 max	0.3 max	0.25-0.7	0.3 max	0.04 max	0.05 max	0.07 max	Al <=0.1; N <=0.008; As<=0.08; bal Fe	320-410	185	30L	
GOST 380(88)	St0	Gen struct; Semi-Finished	0.23 max	0.3 max	0.3 max	2 max	0.3 max	0.07 max	0.06 max	0.6 max	Al <=0.1; N <=0.008; As<=0.08; bal Fe				
GOST 6713	16D	Gen struct; Bridges	0.1-0.18	0.3 max	0.2-0.35	0.4-0.7	0.3 max	0.035 max	0.04 max	0.12-0.25	Al <=0.1; bal Fe				

Spain

Specification	Designation	Notes	C	Cr	Cu	Mn	Ni	P	S	Si	Other	UTS	YS	El	Hard
UNE 36080(80)	A310	Gen struct	0.3 max			1.6 max		0.07 max	0.06 max	0.6 max	Mo <=0.15; bal Fe				
UNE 36080(80)	F.6200*	see A310	0.3 max			1.6 max		0.07 max	0.06 max	0.6 max	Mo <=0.15; bal Fe				
UNE 36080(85)	AE235-B	Gen struct	0.2 max			1.60 max		0.045 max	0.045 max	0.60 max	Mo <=0.15; N <=0.009; bal Fe				

USA

Specification	Designation	Notes	C	Cr	Cu	Mn	Ni	P	S	Si	Other	UTS	YS	El	Hard
	UNS K02401		0.24 max			0.90 max		0.04 max	0.05 max	0.15-0.30	bal Fe				
ASTM A283/A283M(97)	C	Struct plt, t>40mm	0.24 max		0.20	0.90 max		0.035 max	0.04 max	0.15-0.40	bal Fe	380-515	205	25	
ASTM A283/A283M(97)	C	Struct plt, t<=40mm	0.24 max		0.20	0.90 max		0.035 max	0.04 max	0.40 max	bal Fe	380-515	205	25	
ASTM A515(92)	Grade 60	Press ves plt, int/High-temp, t<=25mm	0.24 max			0.90 max		0.035 max	0.035 max	0.15-0.40	bal Fe	415-550	220	25	
ASTM A515(92)	Grade 60	Press ves plt, int/High-temp, t>25mm-50mm	0.27 max			0.90 max		0.035 max	0.035 max	0.15-0.40	bal Fe	415-550	220	25	
ASTM A515(92)	Grade 60	Press ves plt, int/High-temp, t>50mm-100mm	0.29 max			0.90 max		0.035 max	0.035 max	0.15-0.40	bal Fe	415-550	220	25	
ASTM A515(92)	Grade 60	Press ves plt, int/High-temp, t>100mm-200mm	0.31 max			0.90 max		0.035 max	0.035 max	0.15-0.40	bal Fe	415-550	220	25	
ASTM A515(92)	Grade 60	Press ves plt, int/High-temp, t>200mm	0.31 max			0.90 max		0.035 max	0.035 max	0.15-0.40	bal Fe	415-550	220	25	

Carbon Steel, Nonresulfurized, A283(D)

Czech Republic

Specification	Designation	Notes	C	Cr	Cu	Mn	Ni	P	S	Si	Other	UTS	YS	El	Hard
CSN 411425	11425	Gen struct	0.22 max	0.3 max		1.60 max		0.05 max	0.05 max	0.60 max	bal Fe				

France

Specification	Designation	Notes	C	Cr	Cu	Mn	Ni	P	S	Si	Other	UTS	YS	El	Hard
AFNOR NFA35501(79)	E26-1		0.2 max	0.3 max	0.3 max	2 max	0.4 max	0.06 max	0.05 max	0.6 max	bal Fe				
AFNOR NFA35501(79)	E26-2		0.2 max	0.3 max	0.3 max	2 max	0.4 max	0.05 max	0.05 max	0.6 max	N <=0.007; bal Fe				
AFNOR NFA35501(83)	E26-2	t<=30mm; roll	0.24 max	0.3 max	0.3 max	1.6 max	0.4 max	0.045 max	0.045 max	0.55 max	bal Fe	490-630	355	22	
AFNOR NFA35501(83)	E26-2	50<t<=100mm; roll	0.24 max	0.3 max	0.3 max	1.6 max	0.4 max	0.045 max	0.045 max	0.55 max	bal Fe	490-630	335	21	
AFNOR NFA35501(83)	E28-2	t<=2.99mm; roll	0.2 max	0.3 max	0.3 max	1.3 max	0.4 max	0.045 max	0.045 max	0.4 max	N <=0.008; bal Fe	420-560	255	19	
AFNOR NFA35501(83)	E28-2	t<=30mm; roll	0.2 max	0.3 max	0.3 max	1.3 max	0.4 max	0.045 max	0.045 max	0.4 max	N <=0.008; bal Fe	400-540	275	24	
AFNOR NFA35501(83)	E28-2	110<t<=150mm; roll	0.2 max	0.3 max	0.3 max	1.3 max	0.4 max	0.045 max	0.045 max	0.4 max	N <=0.008; bal Fe	400-540	225	16	
AFNOR NFA35501(83)	E28-2	50<t<=100mm; roll	0.2 max	0.3 max	0.3 max	1.3 max	0.4 max	0.045 max	0.045 max	0.4 max	N <=0.008; bal Fe	400-540	255	22	
AFNOR NFA35501(83)	E28-2	10t<=350mm; roll	0.2 max	0.3 max	0.3 max	1.3 max	0.4 max	0.045 max	0.045 max	0.4 max	N <=0.008; bal Fe	380	235	20	
AFNOR NFA35501(83)	E36-2	t<=30mm; roll	0.24 max	0.3 max	0.3 max	1.6 max	0.4 max	0.045 max	0.045 max	0.55 max	bal Fe	490-630	355	22	
AFNOR NFA35501(83)	E36-2	50<t<=100mm; roll	0.24 max	0.3 max	0.3 max	1.6 max	0.4 max	0.045 max	0.045 max	0.55 max	bal Fe	490-630	335	21	

Carbon Steel, Nonresulfurized, A283(D) (Continued from previous page)

Specification	Designation	Notes	C	Cr	Cu	Mn	Ni	P	S	Si	Other	UTS	YS	El	Hard
Hungary															
MSZ 17(86)	A44B	Tub; w/special req; specially killed; 0-16mm; HF/CF	0.2 max			1.3 max		0.04 max	0.04 max	0.1-0.35	Al 0.015-0.10; Nb 0.01-0.06; Ti 0.02-0.06; V 0.02-0.1; Zr=0.015-0.06; Al+Nb+V+Ti+Zr<=0.15; bal Fe	420-550	275	21L	
MSZ 17(86)	A44B	Tub; w/special req; specially killed; >40mm; HF/CF	0.2 max			1.3 max		0.04 max	0.04 max	0.1-0.35	Al 0.015-0.10; Nb 0.01-0.06; Ti 0.02-0.06; V 0.02-0.1; Zr=0.015-0.06; Al+Nb+V+Ti+Zr<=0.15; bal Fe	420-550	255	21L	
MSZ 500(81)	A44B	Gen struct; Non-rimming; FF; 0-16mm; HR/Frg	0.21 max			2 max	0.3 max	0.045 max	0.05 max	0.12 min	bal Fe	430-590	275	24L	
MSZ 500(81)	A44B	Gen struct; Non-rimming; FF; 40.1-100mm; HR/Frg	0.21 max			2 max	0.3 max	0.045 max	0.05 max	0.12 min	bal Fe	430-550	255	24L	
MSZ 500(89)	Fe275B	Gen struct; Non-rimming; FF; Base; BS; 200.1-250mm; HR/Frg	0.22 max			2 max		0.045 max	0.045 max	0.6 max	0-40mm C<=0.21; 40-100mm C<=0.22; N<=0.009/Al>=0.02; bal Fe	380-540	205	17L; 17T	
MSZ 500(89)	Fe275B	Gen struct; Non-rimming; FF; Base; BS; 0-1mm; HR/Frg	0.22 max			2 max		0.045 max	0.045 max	0.6 max	0-40mm C<=0.21; 40-100mm C<=0.22; N<=0.009/Al>=0.02; bal Fe	430-580	275	14L; 12T	
MSZ 6280(82)	45B	Weld const; 40<t<=60mm; rolled/norm	0.2 max	0.25 max		1.3 max	0.3 max	0.045 max	0.045 max	0.15-0.5	Mo <=0.1; bal Fe	440-550	255	24L	
MSZ 6280(82)	45B	Weld const; 3-16mm; rolled/norm	0.2 max	0.25 max		1.3 max	0.3 max	0.045 max	0.045 max	0.15-0.5	Mo <=0.1; bal Fe	440-550	295	24L	
International															
ISO 630(80)	Fe360C	Struct; FN; FF	0.27 max	0.3 max	0.3 max	2 max	0.4 max	0.045 max	0.05 max	0.4 max	Mo <=0.15; N <=0.009; Pb <=0.15; Ti <=0.05; V <=0.1;				
ISO 630(80)	Fe360C	Struct; FN; FF	0.27 max	0.3 max	0.3 max	2 max	0.4 max	0.045 max	0.05 max	0.4 max	Mo <=0.15; N <=0.009; Pb <=0.15; Ti <=0.05; V <=0.1;				
ISO 630(80)	Fe360C	Struct; FN; FF	0.27 max	0.3 max	0.3 max	2 max	0.4 max	0.045 max	0.05 max	0.4 max	Mo <=0.15; N <=0.009; Pb <=0.15; Ti <=0.05; V <=0.1;				
ISO 630(80)	Fe360D	Struct; FF	0.17 max	0.3 max	0.3 max	2 max	0.4 max	0.04 max	0.04 max	0.6 max	Mo <=0.15; Pb <=0.15; Ti <=0.05; V <=0.1;				
ISO 630(80)	Fe360D	Struct; FF	0.17 max	0.3 max	0.3 max	2 max	0.4 max	0.04 max	0.04 max	0.6 max	Mo <=0.15; Pb <=0.15; Ti <=0.05; V <=0.1;				
ISO R630(67)	Fe42B	Struct	0.25 max	0.3 max	0.3 max	1.60 max	0.4 max	0.065 max	0.055 max	0.60 max	Al <=0.1; Co <=0.1; Mo <=0.15; Pb <=0.15; Ti <=0.05; V <=0.1; W <=0.1; bal Fe				
Italy															
UNI 7070(82)	Fe430BFN	Gen struct	0.21 max			2 max		0.045 max	0.045 max	0.6 max	bal Fe				
Japan															
JIS G3101(95)	SS330	HR Sh Plt Bar						0.050 max	0.050 max		bal Fe	330-430	175-205	21-30	
JIS G3101(95)	SS400	HR Sh Plt Bar						0.050 max	0.050 max		bal Fe	400-510	215-245	17-24	
Russia															
GOST 380(71)	WSt4ps3	Gen struct; Semi-killed; 0-20mm	0.27 max	0.3 max	0.3 max	0.4-0.9	0.3 max	0.04 max	0.05 max	0.05-0.17	Al <=0.1; N <=0.008; As<=0.08; bal Fe	410-530	265	24L	
GOST 380(71)	WSt4ps3	Gen struct; Semi-killed; >100mm	0.27 max	0.3 max	0.3 max	0.4-0.9	0.3 max	0.04 max	0.05 max	0.05-0.17	Al <=0.1; N <=0.008; As<=0.08; bal Fe	410-530	235	21L	
GOST 380(71)	WSt4sp3	Gen struct; Non-rimming; FF; 0-20mm	0.27 max	0.3 max	0.3 max	0.4-0.9	0.3 max	0.04 max	0.05 max	0.12-0.3	Al <=0.1; N <=0.008; As<=0.08; bal Fe	410-530	265	24L	
GOST 380(71)	WSt4sp3	Gen struct; Non-rimming; FF; >100mm	0.27 max	0.3 max	0.3 max	0.4-0.9	0.3 max	0.04 max	0.05 max	0.12-0.3	Al <=0.1; N <=0.008; As<=0.08; bal Fe	410-530	235	21L	

Specification	Designation	Notes	C	Cr	Cu	Mn	Ni	P	S	Si	Other	UTS	YS	El	Hard

Carbon Steel, Nonresulfurized, A283(D) (Continued from previous page)

Spain

Specification	Designation	Notes	C	Cr	Cu	Mn	Ni	P	S	Si	Other	UTS	YS	El	Hard
UNE 36080(80)	A410B	Gen struct	0.24 max			1.60 max		0.05 max	0.05 max	0.60 max	Mo <=0.15; N <=0.009; bal Fe				
UNE 36080(80)	A430B	Gen struct	0.22 max			1.60 max		0.05 max	0.05 max	0.60 max	Mo <=0.15; N <=0.009; bal Fe				
UNE 36080(80)	F.6210*	see A430B	0.22 max			1.60 max		0.05 max	0.05 max	0.60 max	Mo <=0.15; N <=0.009; bal Fe				
UNE 36080(85)	AE275-B	Gen struct	0.22 max			1.60 max		0.045 max	0.045 max	0.60 max	Mo <=0.15; N <=0.009; bal Fe				

USA

Specification	Designation	Notes	C	Cr	Cu	Mn	Ni	P	S	Si	Other	UTS	YS	El	Hard
	UNS K02702		0.27 max			0.90 max		0.04 max	0.05 max	0.15-0.40	bal Fe				
ASTM A283/A283M(97)	D	Struct plt, t<=40mm	0.27 max	0.20		0.90 max		0.035 max	0.04 max	0.40	bal Fe	415-550	230	23	
ASTM A283/A283M(97)	D	Struct plt, t>40mm	0.27 max	0.20		0.90 max		0.035 max	0.04 max	0.15-0.40	bal Fe	415-550	230	23	

Carbon Steel, Nonresulfurized, A284

International

Specification	Designation	Notes	C	Cr	Cu	Mn	Ni	P	S	Si	Other	UTS	YS	El	Hard
ISO 630(80)	Fe360C	Struct; FN; FF	0.27 max	0.3 max	0.3 max	2 max	0.4 max	0.045 max	0.05 max	0.4 max	Al <=0.1; Co <=0.1; Mo <=0.15; N <=0.009; Pb <=0.15; Ti <=0.05; V <=0.1; W <=0.1; bal Fe				
ISO 630(80)	Fe360D	Struct; FF	0.17 max	0.3 max	0.3 max	2 max	0.4 max	0.04 max	0.04 max	0.6 max	Al <=0.1; Co <=0.1; Mo <=0.15; Pb <=0.15; Ti <=0.05; V <=0.1; W <=0.1; bal Fe				
ISO 630(80)	Fe430B	Struct; FF, FN; Thk>40mm	0.22 max	0.3 max	0.3 max	2 max	0.4 max	0.05 max	0.05 max	0.6 max	Al <=0.1; Co <=0.1; Mo <=0.15; N <=0.009; Pb <=0.15; Ti <=0.05; V <=0.1; W <=0.1; bal Fe				
ISO 630(80)	Fe430B	Struct; FF, FN; 0-40mm	0.21 max	0.3 max	0.3 max	2 max	0.4 max	0.05 max	0.05 max	0.6 max	Mo <=0.15; Pb <=0.15; Ti <=0.05; V <=0.1;				

Italy

Specification	Designation	Notes	C	Cr	Cu	Mn	Ni	P	S	Si	Other	UTS	YS	El	Hard
UNI 7070(72)	Fe37CFN	Gen struct	0.2 max			2 max		0.04 max	0.045 max	0.6 max	N <=0.009; bal Fe				
UNI 7070(72)	Fe42BFN	Gen struct	0.2 max			2 max		0.045 max	0.045 max	0.6 max	N <=0.009; bal Fe				
UNI 7070(82)	Fe360CFN	Gen struct	0.17 max			2 max		0.04 max	0.045 max	0.6 max	bal Fe				
UNI 7070(82)	Fe360DFF	Gen struct	0.17 max			2 max		0.04 max	0.04 max	0.6 max	Al <=0.015; bal Fe				
UNI 7070(82)	Fe430BFN	Gen struct	0.21 max			2 max		0.045 max	0.045 max	0.6 max	bal Fe				

USA

Specification	Designation	Notes	C	Cr	Cu	Mn	Ni	P	S	Si	Other	UTS	YS	El	Hard
	UNS K01804		0.18 max			0.90 max		0.04 max	0.05 max	0.10-0.30	bal Fe				

Carbon Steel, Nonresulfurized, A284(D)

Europe

Specification	Designation	Notes	C	Cr	Cu	Mn	Ni	P	S	Si	Other	UTS	YS	El	Hard
EN 10025(90)	Fe360C	Obs EU desig; struct HR FN; QS, 40 <t<=100mm	0.17 max					0.040 max	0.040 max		N <=0.009; N max na if Al>=0.020; bal Fe	340-470	215	24-25	
EN 10025(90)	Fe360C	Obs EU desig; struct HR FN; QS, 16<t<=40mm	0.17 max					0.040 max	0.040 max		N <=0.009; N max na if Al>=0.020; bal Fe	340-470	225	26	
EN 10025(90)	Fe360D1	Obs EU desig; struct HR FF; QS, t<=16mm	0.17 max					0.035 max	0.035 max		bal Fe	340-510	235	17-26	
EN 10025(90)	Fe360D1	Obs EU desig; struct HR FF; QS, 16<t<=40mm	0.17 max					0.035 max	0.035 max		bal Fe	340-470	225	26	
EN 10025(90)	Fe360D1	Obs EU desig; struct HR FF; QS, 16<t<=40mm	0.17 max					0.035 max	0.035 max		bal Fe	340-470	215	24-25	

Specification	Designation	Notes	C	Cr	Cu	Mn	Ni	P	S	Si	Other	UTS	YS	El	Hard

Carbon Steel, Nonresulfurized, A284(D) (Continued from previous page)

Europe

Specification	Designation	Notes	C	Cr	Cu	Mn	Ni	P	S	Si	Other	UTS	YS	El	Hard
EN 10025(90)	Fe360D2	Obs EU desig; struct HR FF; QS, 40t<=100mm	0.17 max					0.035 max	0.035 max		bal Fe	347-470	215	18-19	
EN 10025(90)	Fe360D2	Obs EU desig; struct HR FF, QS, t<=16mm	0.17 max					0.035 max	0.035 max		bal Fe	340-510	235	12-20	
EN 10025(90)	Fe360D2	Obs EU desig; struct HR FF, QS, 16<t<=40mm	0,17 max					0.035 max	0.035 max		bal Fe	340-470	225	20	

Spain

Specification	Designation	Notes	C	Cr	Cu	Mn	Ni	P	S	Si	Other	UTS	YS	El	Hard
UNE 36080(80)	A360C	Gen struct	0.17 max			1.60 max		0.045 max	0.045 max	0.60 max	Mo <=0.15; N <=0.009; bal Fe				
UNE 36080(80)	A360D	Gen struct	0.17 max			1.60 max		0.04 max	0.04 max	0.60 max	Mo <=0.15; +N; bal Fe				
UNE 36080(80)	A410B	Gen struct	0.24 max			1.60 max		0.05 max	0.05 max	0.60 max	Mo <=0.15; N <=0.009; bal Fe				
UNE 36080(85)	AE235-C	Gen struct	0.17 max			1.60 max		0.04 max	0.04 max	0.60 max	Mo <=0.15; N <=0.009; bal Fe				
UNE 36080(85)	AE235-D	Gen struct	0.17 max			1.60 max		0.035 max	0.035 max	0.60 max	Mo <=0.15; bal Fe				

Carbon Steel, Nonresulfurized, A285(A)

UK

Specification	Designation	Notes	C	Cr	Cu	Mn	Ni	P	S	Si	Other	UTS	YS	El	Hard
BS 3601(87)	BW320	Pip Tub, press, t<=16mm	0.16 max			0.30-0.70		0.040 max	0.040 max		bal Fe	320-460	195	25	
BS 3601(87)	ERW320	Pip Tub, press, t<=16mm	0.16 max			0.30-0.70		0.040 max	0.040 max		bal Fe	320-460	195	25	
BS 3601(87)	FRW360	Pip Tub, press, t<=16mm	0.17 max			0.40-0.80		0.040 max	0.040 max	0.35 max	bal Fe	360-500	235	25	
BS 3602/1(87)	ERW360; CEW360	ERW IW smls Tub; press High-temp; t<=16mm	0.17 max			0.30-0.80		0.035 max	0.035 max	0.35 max	Al <=0.06; bal Fe	360-500	235	25	
BS 3602/1(87)	HFS360; CFS360	ERW IW smls Tub; press High-temp; t<=16mm	0.17 max			0.30-0.80		0.035 max	0.035 max	0.35 max	Al <=0.06; bal Fe	360-500	235	25	

USA

Specification	Designation	Notes	C	Cr	Cu	Mn	Ni	P	S	Si	Other	UTS	YS	El	Hard
	UNS K01700		0.17 max			0.90 max		0.035 max	0.045 max		bal Fe				
ASTM A285/A285M(96)	A	Press ves plt	0.17 max			0.90 max		0.035 max	0.035 max		bal Fe	310-450	165	30	

Carbon Steel, Nonresulfurized, A285(B)

USA

Specification	Designation	Notes	C	Cr	Cu	Mn	Ni	P	S	Si	Other	UTS	YS	El	Hard
	UNS K02200		0.22 max			0.90 max		0.035 max	0.045 max		bal Fe				
ASTM A285/A285M(96)	B	Press ves plt	0.22 max			0.90 max		0.035 max	0.035 max		bal Fe	345-485	185	28	
ASTM A984(98)	35	CEV Req'd; ERW	0.22 max					0.025 max	0.015 max		B <=0.0005; bal Fe	415	245-450		
ASTM A984(98)	45	CEV Req'd; ERW	0.22 max					0.025 max	0.015 max		B <=0.0005; bal Fe	450	315-500		
ASTM A984(98)	65	CEV Req'd; ERW	0.22 max					0.025 max	0.015 max		B <=0.0005; bal Fe	520	450-570		
ASTM A984(98)	80	CEV Req'd; ERW	0.22 max					0.025 max	0.015 max		B <=0.0005; bal Fe	625	550-670		

Carbon Steel, Nonresulfurized, A285(C)

Europe

Specification	Designation	Notes	C	Cr	Cu	Mn	Ni	P	S	Si	Other	UTS	YS	El	Hard
EN 10028/2(92)	1.0345	High-temp, Press; 16<=t<=40mm	0.16 max	0.30 max	0.30 max	0.40-1.20	0.30 max	0.030 max	0.025 max	0.35 max	Al >=0.020; Mo <=0.08; Nb <=0.010; Ti <=0.03; V <=0.02; Cr+Cu+Mo+Ni<=0.70; bal Fe	360-480	225	25T	
EN 10028/2(92)	1.0345	High-temp, Press; 3<=t<=16mm	0.16 max	0.30 max	0.30 max	0.40-1.20	0.30 max	0.030 max	0.025 max	0.35 max	Al >=0.020; Mo <=0.08; Nb <=0.010; Ti <=0.03; V <=0.02; Cr+Cu+Mo+Ni <=0.70: bal Fe	360-480	235	25T	

Specification	Designation	Notes	C	Cr	Cu	Mn	Ni	P	S	Si	Other	UTS	YS	El	Hard

Carbon Steel, Nonresulfurized, A285(C) (Continued from previous page)

Europe

Specification	Designation	Notes	C	Cr	Cu	Mn	Ni	P	S	Si	Other	UTS	YS	El	Hard
EN 10028/2(92)	1.0345	High-temp, Press; 100<=t<=150mm	0.16 max	0.30 max	0.30 max	0.40-1.20	0.30 max	0.030 max	0.025 max	0.35 max	Al >=0.020; Mo <=0.08; Nb <=0.010; Ti <=0.03; V <=0.02; Cr+Cu+Mo+Ni<=0.70 ; bal Fe	360-480	185	24T	
EN 10028/2(92)	1.0345	High-temp, Press; 60<=t<=100mm	0.16 max	0.30 max	0.30 max	0.40-1.20	0.30 max	0.030 max	0.025 max	0.35 max	Al >=0.020; Mo <=0.08; Nb <=0.010; Ti <=0.03; V <=0.02; Cr+Cu+Mo+Ni<=0.70 ; bal Fe	360-480	200	24T	
EN 10028/2(92)	1.0345	High-temp, Press; 40<=t<=60mm	0.16 max	0.30 max	0.30 max	0.40-1.20	0.30 max	0.030 max	0.025 max	0.35 max	Al >=0.020; Mo <=0.08; Nb <=0.010; Ti <=0.03; V <=0.02; Cr+Cu+Mo+Ni<=0.70 : bal Fe	360-480	215	25T	
EN 10028/2(92)	P235GH	High-temp, Press; 16<=t<=40mm	0.16 max	0.30 max	0.30 max	0.40-1.20	0.30 max	0.030 max	0.025 max	0.35 max	Al >=0.020; Mo <=0.08; Nb <=0.010; Ti <=0.03; V <=0.02; Cr+Cu+Mo+Ni<=0.70 ; bal Fe	360-480	225	25T	
EN 10028/2(92)	P235GH	High-temp, Press; 40<=t<=60mm	0.16 max	0.30 max	0.30 max	0.40-1.20	0.30 max	0.030 max	0.025 max	0.35 max	Al >=0.020; Mo <=0.08; Nb <=0.010; Ti <=0.03; V <=0.02; Cr+Cu+Mo+Ni<=0.70 : bal Fe	360-480	215	25T	
EN 10028/2(92)	P235GH	High-temp, Press; 60<=t<=100mm	0.16 max	0.30 max	0.30 max	0.40-1.20	0.30 max	0.030 max	0.025 max	0.35 max	Al >=0.020; Mo <=0.08; Nb <=0.010; Ti <=0.03; V <=0.02; Cr+Cu+Mo+Ni<=0.70 ; bal Fe	360-480	200	24T	
EN 10028/2(92)	P235GH	High-temp, Press; 3<=t<=16mm	0.16 max	0.30 max	0.30 max	0.40-1.20	0.30 max	0.030 max	0.025 max	0.35 max	Al >=0.020; Mo <=0.08; Nb <=0.010; Ti <=0.03; V <=0.02; Cr+Cu+Mo+Ni <=0.70: bal Fe	360-480	235	25T	
EN 10028/2(92)	P235GH	High-temp, Press; 100<=t<=150mm	0.16 max	0.30 max	0.30 max	0.40-1.20	0.30 max	0.030 max	0.025 max	0.35 max	Al >=0.020; Mo <=0.08; Nb <=0.010; Ti <=0.03; V <=0.02; Cr+Cu+Mo+Ni<=0.70 ; bal Fe	360-480	185	24T	

USA

Specification	Designation	Notes	C	Cr	Cu	Mn	Ni	P	S	Si	Other	UTS	YS	El	Hard
	UNS K02801		0.28 max			0.90 max		0.035 max	0.040 max		bal Fe				
ASTM A285/A285M(96)	C	Press ves plt	0.28 max			0.90 max		0.035 max	0.035 max		bal Fe	380-515	205	27	

Carbon Steel, Nonresulfurized, A288(1)

USA

Specification	Designation	Notes	C	Cr	Cu	Mn	Ni	P	S	Si	Other	UTS	YS	El	Hard
	UNS K05002		0.50 max			0.60-1.00		0.025 max	0.025 max	0.15-0.30	bal Fe				
ASTM A288(98)	Class 1	Frg	0.50 max			0.80-1.00		0.025 max	0.025 max	0.15-0.30	bal Fe	485	310	18	

Carbon Steel, Nonresulfurized, A291(1)

USA

Specification	Designation	Notes	C	Cr	Cu	Mn	Ni	P	S	Si	Other	UTS	YS	El	Hard
	UNS K05500		0.55 max	0.25-0.35	0.35	0.60-0.90	0.30 max	0.040 max	0.040 max	0.15-0.35	Mo <=0.10; V 0.06; bal Fe				
ASTM A291(95)	1	Gear pinion frg, t<=250mm	0.55 max	0.25 max	0.35 max	0.60-0.90	0.30 max	0.040 max	0.040 max	0.35 max	Mo <=0.10; V <=0.06; bal Fe	585	345	22 L 16T	170-223 HB
ASTM A291(95)	1	Gear pinion frg, t>250mm	0.55 max	0.25 max	0.35 max	0.60-0.90	0.30 max	0.040 max	0.040 max	0.35 max	Mo <=0.10; V <=0.06; bal Fe	550	310	20 L 16T	170-223 HB

Carbon Steel, Nonresulfurized, A291(2)

USA

Specification	Designation	Notes	C	Cr	Cu	Mn	Ni	P	S	Si	Other	UTS	YS	El	Hard
	UNS K05000		0.50 max		0.35	0.40-0.90		0.040 max	0.040 max	0.15-0.35	V <=0.10; bal Fe				
ASTM A291(95)	2	Gear pinion frg, t>510mm	0.50 max		0.35 max	0.40-0.90		0.040 max	0.040 max	0.35 max	V <=0.10; bal Fe	655	485	18 L 16T	201-241 HB
ASTM A291(95)	2	Gear pinion frg, 250< t<=510mm	0.50 max		0.35 max	0.40-0.90		0.040 max	0.040 max	0.35 max	V <=0.10; bal Fe	655	485	20 L 18T	201-241 HB

UNS numbers and US grades are provided as a means of cross referencing chemically similar alloys. Exchangability is only possible after independent examination of specifications. Tensile properties are minimum or typical as specified. UTS and YS as MPa. El as %. See Appendix for list of abbreviations used in Notes. * indicates obsolete material.

Specification	Designation	Notes	C	Cr	Cu	Mn	Ni	P	S	Si	Other	UTS	YS	El	Hard

Carbon Steel, Nonresulfurized, A291(2) (Continued from previous page)

USA

Specification	Designation	Notes	C	Cr	Cu	Mn	Ni	P	S	Si	Other	UTS	YS	El	Hard
ASTM A291(95)	2	Gear pinion frg, t<=250mm	0.50 max		0.35 max	0.40-0.90		0.040 max	0.040 max	0.35 max	V <=0.10; bal Fe	655	485	20 L	201-241 HB
ASTM A649/649M(98)	Class 2	Frg roll for paperboard machinery	0.55 max			0.50-0.90		0.025 max	0.025 max	0.15-0.35	Si<=0.10 with VCD; bal Fe	515	260	20	

Carbon Steel, Nonresulfurized, A321

USA

Specification	Designation	Notes	C	Cr	Cu	Mn	Ni	P	S	Si	Other	UTS	YS	El	Hard
	UNS K05501		0.55 max			0.60-0.90		0.040 max	0.050 max	0.15-0.35	Pb 0.15-0.35 (when spec'd); bal Fe				
ASTM A321(95)		Bars, Q/T, 101.6<t<=152.4mm	0.55 max			0.60-0.90		0.040 max	0.050 max	0.15-0.35	Pb may be spec; bal Fe	620	415	18	
ASTM A321(95)		Bars, Q/T, 152.4<t<=241.3mm	0.55 max			0.60-0.90		0.040 max	0.050 max	0.15-0.35	Pb may be spec; bal Fe	590	345	18	
ASTM A321(95)		Bars, Q/T, 63.5<t<=101.6mm	0.55 max			0.60-0.90		0.040 max	0.050 max	0.15-0.35	Pb may be spec; bal Fe	660	450	18	
ASTM A321(95)		Bars, Q/T, 25.4<t<=63.5mm	0.55 max			0.60-0.90		0.040 max	0.050 max	0.15-0.35	Pb may be spec; bal Fe	7205	485	18	
ASTM A321(95)		Bars, Q/T, t<=25.4mm	0.55 max			0.60-0.90		0.040 max	0.050 max	0.15-0.35	Pb may be spec; bal Fe	760	520	18	

Carbon Steel, Nonresulfurized, A354

USA

Specification	Designation	Notes	C	Cr	Cu	Mn	Ni	P	S	Si	Other	UTS	YS	El	Hard
	UNS K04100		0.28-0.55					0.040 max	0.045 max		bal Fe				

Carbon Steel, Nonresulfurized, A36

USA

Specification	Designation	Notes	C	Cr	Cu	Mn	Ni	P	S	Si	Other	UTS	YS	El	Hard
	UNS K02599	Plate < 7.5 in.	0.25 max					0.04 max	0.05 max		bal Fe				
ASTM A36/A36M(97)	A36	Struct bar, t<=20mm	0.26 max		0.20 min			0.04 max	0.05 max	0.40 max	bal Fe	400-550	250	21	
ASTM A36/A36M(97)	A36	Struct bar, 20<t<=40mm	0.27 max		0.20 min	0.60-0.90		0.04 max	0.05 max	0.40 max	bal Fe	400-550	250	21	
ASTM A36/A36M(97)	A36	Struct bar, 40<t<=100mm	0.28 max		0.20 min	0.60-0.90		0.04 max	0.05 max	0.40 max	bal Fe	400-550	250	21	
ASTM A36/A36M(97)	A36	Struct bar, t>100mm	0.29 max		0.20 min	0.60-0.90		0.04 max	0.05 max	0.40 max	bal Fe	400-550	250	21	
ASTM A36/A36M(97)	A36	Struct plt, t<=20mm	0.25 max		0.20 min			0.04 max	0.05 max	0.40 max	bal Fe	400-550	250	21	

Carbon Steel, Nonresulfurized, A36(Shapes)

Mexico

Specification	Designation	Notes	C	Cr	Cu	Mn	Ni	P	S	Si	Other	UTS	YS	El	Hard
NMX-B-199(86)	Grade A	CF weld Smls struct tub	0.26 max					0.04 max	0.05 max		Cu if spec'd>=0.20; bal Fe	310	228	25	
NMX-B-199(86)	Grade B	CF weld Smls struct tub	0.26 max					0.04 max	0.05 max		Cu if spec'd>=0.20; bal Fe	400	290	23	
NMX-B-200(90)		HF weld Smls struct tub	0.26 max					0.04 max	0.05 max		Cu if spec>=0.20; bal Fe	402	245	23	

USA

Specification	Designation	Notes	C	Cr	Cu	Mn	Ni	P	S	Si	Other	UTS	YS	El	Hard
	UNS K02600		0.26 max					0.04 max	0.05 max		Cu 0.20 min (when spec'd); bal Fe				
ASTM A36/A36M(97)	A36	Struct Shp	0.26 max		0.20 min			0.04 max	0.05 max	0.40 max	bal Fe	400-550	250	23	

Carbon Steel, Nonresulfurized, A372(I)

USA

Specification	Designation	Notes	C	Cr	Cu	Mn	Ni	P	S	Si	Other	UTS	YS	El	Hard
	UNS K03002		0.30 max			1.00 max		0.04 max	0.05 max	0.15-0.30	bal Fe				
ASTM A372/372M(95)	Grade A	Thin wall frg press ves	0.30 max			1.00 max		0.025 max	0.025 max	0.15-0.35	bal Fe	415-485	240	20	121 HB
ASTM A372/372M(95)	Type I*	Thin wall frg press ves	0.30 max			1.00 max		0.025 max	0.025 max	0.15-0.35	bal Fe	415-485	240	20	121 HB

Specification	Designation	Notes	C	Cr	Cu	Mn	Ni	P	S	Si	Other	UTS	YS	El	Hard
Carbon Steel, Nonresulfurized, A414(A)															
USA															
	UNS K01501		0.15 max			0.90 max		0.035 max	0.040 max		Cu 0.20 min (when spec'd); bal Fe				
Carbon Steel, Nonresulfurized, A414(B)															
USA															
	UNS K02201		0.22 max			0.90 max		0.035 max	0.040 max		Cu 0.20 min (when spec'd); bal Fe				
Carbon Steel, Nonresulfurized, A414(C)															
Czech Republic															
CSN 411366	11366	High-temp const	0.15 max	0.3 max		0.65 max	0.3 max	0.045 max	0.04 max	0.35 max	bal Fe				
Italy															
UNI 5869(75)	Fe360-1KG	Press ves	0.17 max			0.4-1.60		0.035 max	0.04 max	0.35 max	bal Fe				
UNI 5869(75)	Fe360-1KW	Press ves	0.17 max			0.4-1.60		0.04 max	0.04 max	0.35 max	bal Fe				
UNI 5869(75)	Fe360-2KG	Press ves	0.17 max			0.4-1.60		0.03 max	0.035 max	0.35 max	bal Fe				
UNI 5869(75)	Fe360-2KW	Press ves	0.17 max			0.4-1.60		0.035 max	0.035 max	0.35 max	bal Fe				
Spain															
UNE 36087(74)	A37RAII	Sh Plt	0.2 max			0.3-1.3		0.045 max	0.04 max	0.4 max	Al 0.015-0.120; Mo <=0.15; 0-40mm Mn<=1.3; 40mm Mn<=1.1; bal Fe				
UNE 36087(74)	A37RCI	Sh Plt	0.2 max			0.3-1.3		0.055 max	0.05 max	0.4 max	Mo <=0.15; N <=0.009; 0-40mm Mn<=1.3; 40mm Mn<=1.1; bal Fe				
USA															
	UNS K02503		0.25 max			0.90 max		0.035 max	0.040 max		Cu 0.20 min (when spec'd); bal Fe				
Carbon Steel, Nonresulfurized, A422(60)															
Spain															
UNE 36087(74)	A42RCI	Sh Plt	0.23 max			0.4-1.4		0.055 max	0.05 max	0.4 max	Al <=0.01; Mo <=0.15; N <=0.009; bal Fe				
UNE 36087(74)	A42RCII	Sh Plt	0.23 max			0.4-1.4		0.045 max	0.04 max	0.4 max	Al 0.015-0.120; Mo <=0.15; bal Fe				
Carbon Steel, Nonresulfurized, A442(60)															
Europe															
EN 10028/2(92)	1.0425	High-temp, Press; 100<=t<=150mm	0.20 max	0.30 max	0.30 max	0.50-1.40	0.30 max	0.030 max	0.025 max	0.40 max	Al >=0.020; Mo <=0.08; Nb <=0.010; Ti <=0.03; V <=0.02; Cr+Cu+Mo+Ni<=0.70; bal Fe	400-530	200	22T	
EN 10028/2(92)	1.0425	High-temp, Press; 40<=t<=60mm	0.20 max	0.30 max	0.30 max	0.50-1.40	0.30 max	0.030 max	0.025 max	0.40 max	Al >=0.020; Mo <=0.08; Nb <=0.010; Ti <=0.03; V <=0.02; Cr+Cu+Mo+Ni<=0.70; bal Fe	410-530	245	23T	
EN 10028/2(92)	1.0425	High-temp, Press; 16<=t<=40mm	0.20 max	0.30 max	0.30 max	0.50-1.40	0.30 max	0.030 max	0.025 max	0.40 max	Al >=0.020; Mo <=0.08; Nb <=0.010; Ti <=0.03; V <=0.02; Cr+Cu+Mo+Ni<=0.70; bal Fe	410-530	255	23T	
EN 10028/2(92)	1.0425	High-temp, Press; 3<=t<=16mm	0.20 max	0.30 max	0.30 max	0.50-1.40	0.30 max	0.030 max	0.025 max	0.40 max	Al >=0.020; Mo <=0.08; Nb <=0.010; Ti <=0.03; V <=0.02; Cr+Cu+Mo+Ni<=0.70; bal Fe	410-530	265	23T	
EN 10028/2(92)	1.0425	High-temp, Press; 60<=t<=100mm	0.20 max	0.30 max	0.30 max	0.50-1.40	0.30 max	0.030 max	0.025 max	0.40 max	Al >=0.020; Mo <=0.08; Nb <=0.010; Ti <=0.03; V <=0.02; Cr+Cu+Mo+Ni<=0.70; bal Fe	410-530	215	22T	

Specification	Designation	Notes	C	Cr	Cu	Mn	Ni	P	S	Si	Other	UTS	YS	El	Hard
Carbon Steel, Nonresulfurized, A442(60) (Continued from previous page)															
Europe															
EN 10028/2(92)	P265GH	High-temp, Press; 100<=t<=150mm	0.20 max	0.30 max	0.30 max	0.50-1.40	0.30 max	0.030 max	0.025 max	0.40 max	Al >=0.020; Mo <=0.08; Nb <=0.010; Ti <=0.03; V <=0.02; Cr+Cu+Mo+Ni<=0.70; bal Fe	400-530	200	22T	
EN 10028/2(92)	P265GH	High-temp, Press; 3<=t<=16mm	0.20 max	0.30 max	0.30 max	0.50-1.40	0.30 max	0.030 max	0.025 max	0.40 max	Al >=0.020; Mo <=0.08; Nb <=0.010; Ti <=0.03; V <=0.02; Cr+Cu+Mo+Ni<=0.70; bal Fe	410-530	265	23T	
EN 10028/2(92)	P265GH	High-temp, Press; 16<=t<=40mm	0.20 max	0.30 max	0.30 max	0.50-1.40	0.30 max	0.030 max	0.025 max	0.40 max	Al >=0.020; Mo <=0.08; Nb <=0.010; Ti <=0.03; V <=0.02; Cr+Cu+Mo+Ni<=0.70; bal Fe	410-530	255	23T	
EN 10028/2(92)	P265GH	High-temp, Press; 60<=t<=100mm	0.20 max	0.30 max	0.30 max	0.50-1.40	0.30 max	0.030 max	0.025 max	0.40 max	Al >=0.020; Mo <=0.08; Nb <=0.010; Ti <=0.03; V <=0.02; Cr+Cu+Mo+Ni<=0.70; bal Fe	410-530	215	22T	
EN 10028/2(92)	P265GH	High-temp, Press; 40<=t<=60mm	0.20 max	0.30 max	0.30 max	0.50-1.40	0.30 max	0.030 max	0.025 max	0.40 max	Al >=0.020; Mo <=0.08; Nb <=0.010; Ti <=0.03; V <=0.02; Cr+Cu+Mo+Ni<=0.70; bal Fe	410-530	245	23T	
Carbon Steel, Nonresulfurized, A489															
USA															
	UNS K04800		0.48 max			1.00 max		0.040 max	0.050 max	0.15-0.35	bal Fe				
Carbon Steel, Nonresulfurized, A49															
USA															
	UNS K04701		0.35-0.60			1.00 max		0.04 max			bal Fe				
ASTM A49(95)	A49	Rail joint bars	0.35-0.60			1.20 max		0.04 max	0.050 max		bal Fe	690	485	12	
Carbon Steel, Nonresulfurized, A504(C)															
USA															
	UNS K07201		0.67-0.77			0.60-0.85		0.05 max	0.05 max	0.15 min	bal Fe				
ASTM A504(93)	Class C		0.67-0.77			0.60-0.85		0.05 max	0.05 max	0.15 min	bal Fe				321-363 HB
Carbon Steel, Nonresulfurized, A515(55)															
India															
IS 1570/2(79)	15C4	Sh Plt Strp Sect Shp Bar Bil Frg Tub Pip	0.2 max	0.3 max	0.3 max	0.3-0.6	0.4 max	0.055 max	0.055 max	0.6 max	Co <=0.1; Mo <=0.15; Pb <=0.15; W <=0.1; P:S Varies; bal Fe	370-490	203	25	
IS 1570/2(79)	C15	Sh Plt Strp Sect Shp Bar Bil Frg Tub Pip	0.2 max	0.3 max	0.3 max	0.3-0.6	0.4 max	0.055 max	0.055 max	0.6 max	Co <=0.1; Mo <=0.15; Pb <=0.15; W <=0.1; P:S Varies; bal Fe	370-490	203	25	
Spain															
UNE 36087(74)	A37RAII	Sh Plt	0.2 max			0.3-1.3		0.045 max	0.04 max	0.4 max	Al 0.015-0.120; Mo <=0.15; 0-40mm Mn<=1.3; 40mm Mn<=1.1; bal Fe				
UNE 36087(74)	A37RCI	Sh Plt	0.2 max			0.3-1.3		0.055 max	0.05 max	0.4 max	Mo <=0.15; N <=0.009; 0-40mm Mn<=1.3; 40mm Mn<=1.1; bal Fe				
UK															
BS 1449/1(91) *	15CS	Plt Sh Strp	0.20 max			0.90 max		0.060 max	0.050 max		bal Fe				
BS 1449/1(91)	15HR	Plt Sh Strp HR wide; t<=16mm	0.20 max			0.90 max		0.060 max	0.050 max		bal Fe	280	170		

UNS numbers and US grades are provided as a means of cross referencing chemically similar alloys. Exchangability is only possible after independent examination of specifications. Tensile properties are minimum or typical as specified. UTS and YS as MPa. El as %. See Appendix for list of abbreviations used in Notes. * indicates obsolete material.

Specification	Designation	Notes	C	Cr	Cu	Mn	Ni	P	S	Si	Other	UTS	YS	El	Hard
Carbon Steel, Nonresulfurized, A515(55) (Continued from previous page)															
UK															
BS 1449/1(91)	15HS	Plt Sh Strp HR; t<=16mm	0.20 max			0.90 max		0.060 max	0.050 max		bal Fe	280	170		
USA															
	UNS K02001		0.20 max			0.90 max		0.04 max	0.05 max	0.15-0.30	bal Fe				
ASTM A515(90)	Grade 55	Press ves plt, mod/low-temp, t>50-100mm	0.22 max			0.60-1.20		0.035 max	0.035 max	0.15-0.40	bal Fe	380-515	205	27	
ASTM A984(98)	55	CEV Req'd; ERW	0.22 max					0.025 max	0.015 max		B <=0.0005; bal Fe	485	380-520		
Carbon Steel, Nonresulfurized, A515(65)															
Czech Republic															
CSN 411366	11366	High-temp const	0.15 max	0.3 max		0.65 max	0.3 max	0.045 max	0.04 max	0.35 max	bal Fe				
CSN 411444	11444	High-temp const	0.22 max	0.3 max		0.55 max	0.3 max	0.04 max	0.04 max	0.35 max	bal Fe				
France															
AFNOR NFA36205(82)	A37AP*	Press ves; 80<t<=110mm; Norm	0.16 max	0.25 max	0.30 max	0.40-1.60	0.3 max	0.035 max	0.03 max	0.3 max	Mo 0.-0.07; Nb <=0.015; V <=0.02; bal Fe	360-430	185	26	
AFNOR NFA36205(82)	A37AP*	Press ves; t<=30mm; Norm	0.16 max	0.25 max	0.30 max	0.40-1.60	0.3 max	0.035 max	0.03 max	0.3 max	Mo 0.-0.07; Nb <=0.015; V <=0.02; bal Fe	360-430	225	28	
AFNOR NFA36205(82)	A37CP*	Press ves; 80<t<=110mm; Norm	0.16 max	0.25 max	0.30 max	0.40-1.60	0.3 max	0.035 max	0.03 max	0.3 max	Mo <=0.07; Nb <=0.015; V <=0.02; bal Fe	360-430	185	26	
AFNOR NFA36205(82)	A37CP*	Press ves; t<=30mm; Norm	0.16 max	0.25 max	0.30 max	0.40-1.60	0.3 max	0.035 max	0.03 max	0.3 max	Mo <=0.07; Nb <=0.015; V <=0.02; bal Fe	360-430	225	28	
AFNOR NFA36601(80)	A37AP	t<=30mm; Norm	0.16 max	0.3 max	0.18 max	0.4-1.60	0.5 max	0.035 max	0.03 max	0.3 max	Cu+6Sn<=0.33; Mn>=0.4; bal Fe	360-430	225	28T	
AFNOR NFA36601(80)	A37CP	t<=30mm; Norm	0.16 max	0.15 max	0.1 max	0.4-1.60	0.1 max	0.035 max	0.03 max	0.3 max	Al <=0.18; Mo <=0.5; W <=0.005; Cu+6Sn<=0.33; Mn>=0.4; bal Fe	360-430	225	30	
Hungary															
MSZ 1741(89)	KL1	Broiler, Press ves; 40.1-60mm; HT	0.16 max		0.35 max	0.4-1	0.35 max	0.035 max	0.03 max	0.15-0.4	Al 0.02-0.06; Nb 0.01-0.06; Ti 0.02-0.06; V 0.02-0.1; Zr=0.015-0.06; Al+Nb+V+Ti+Zr <=0.15; N<=Al/2+V/4 +Nb/7+Ti/3.5<=0.015; bal Fe	350-430	215	24L	
MSZ 1741(89)	KL1	Broiler, Press ves; 0-40mm; HT	0.16 max		0.35 max	0.4-1	0.35 max	0.035 max	0.03 max	0.15-0.4	Al 0.02-0.06; Nb 0.01-0.06; Ti 0.02-0.06; V 0.02-0.1; Zr=0.015-0.06; Al+Nb+V+Ti+Zr <=0.15; N<=Al/2+V/4 +Nb/7+Ti/3.5<=0.015; bal Fe	350-430	235	24L	
MSZ 1741(89)	KL1C	Broiler, Press ves; 40.1-60mm; HT	0.16 max		0.35 max	0.4-1	0.35 max	0.035 max	0.03 max	0.15-0.4	Al 0.02-0.06; Nb 0.01-0.06; Ti 0.02-0.06; V 0.02-0.1; Zr=0.015-0.06; Al+Nb+V+Ti+Zr <=0.15; N<=Al/2+V/4 +Nb/7+Ti/3.5<=0.015; bal Fe	350-430	215	24L	
MSZ 1741(89)	KL1C	Broiler, Press ves; 0-40mm; HT	0.16 max		0.35 max	0.4-1	0.35 max	0.035 max	0.03 max	0.15-0.4	Al 0.02-0.06; Nb 0.01-0.06; Ti 0.02-0.06; V 0.02-0.1; Zr=0.015-0.06; Al+Nb+V+Ti+Zr <=0.15; N<=Al/2+V/4 +Nb/7+Ti/3.5<=0.015; bal Fe	350-430	235	24L	

Carbon Steel, Nonresulfurized, A515(65) (Continued from previous page)

Specification	Designation	Notes	C	Cr	Cu	Mn	Ni	P	S	Si	Other	UTS	YS	El	Hard
Hungary															
MSZ 1741(89)	KL1D	Broiler, Press ves; 40.1-60mm; HT	0.16 max		0.35 max	0.4-1	0.35 max	0.035 max	0.03 max	0.15-0.4	Al 0.02-0.06; Nb 0.01-0.06; Ti 0.02-0.06; V 0.02-0.1; Zr=0.015-0.06; Al+Nb+V+Ti+Zr <=0.15; N<=Al/2+V/4 +Nb/7+Ti/3.5<=0.015; bal Fe	350-430	215	24L	
MSZ 1741(89)	KL1D	Broiler, Press ves; 0-40mm; HT	0.16 max		0.35 max	0.4-1	0.35 max	0.035 max	0.03 max	0.15-0.4	Al 0.02-0.06; Nb 0.01-0.06; Ti 0.02-0.06; V 0.02-0.1; Zr=0.015-0.06; Al+Nb+V+Ti+Zr <=0.15; N<=Al/2+V/4 +Nb/7+Ti/3.5<=0.015; bal Fe	350-430	235	24L	
MSZ 1741(89)	KL3	Boiler, Press ves; 0-20mm; HT	0.22 max		0.35 max	0.5-1.3	0.35 max	0.035 max	0.03 max	0.15-0.4	Al 0.02-0.06; Nb 0.01-0.06; Ti 0.02-0.06; V 0.02-0.1; Zr=0.015-0.06; Al+Nb+V+Ti+Zr <=0.15; N<=Al/2+V/4 +Nb/7+Ti/3.5<=0.015; bal Fe	440-550	295	20L	
MSZ 1741(89)	KL3	Boiler, Press ves; 40.1-60mm; HT	0.22 max		0.35 max	0.5-1.3	0.35 max	0.035 max	0.03 max	0.15-0.4	Al 0.02-0.06; Nb 0.01-0.06; Ti 0.02-0.06; V 0.02-0.1; Zr=0.015-0.06; Al+Nb+V+Ti+Zr <=0.15; N<=Al/2+V/4 +Nb/7+Ti/3.5<=0.015; bal Fe	440-550	255	20L	
MSZ 1741(89)	KL3C	Boiler, Press ves; 40.1-60mm; HT	0.22 max		0.35 max	0.5-1.3	0.35 max	0.035 max	0.03 max	0.15-0.4	Al 0.02-0.06; Nb 0.01-0.06; Ti 0.02-0.06; V 0.02-0.1; Zr=0.015-0.06; Al+Nb+V+Ti+Zr <=0.15; N<=Al/2+V/4 +Nb/7+Ti/3.5<=0.015; bal Fe	440-550	255	20L	
MSZ 1741(89)	KL3D	Boiler, Press ves; 0-20mm; HT	0.22 max		0.35 max	0.5-1.3	0.35 max	0.035 max	0.03 max	0.15-0.4	Al <=0.06; Nb 0.01-0.06; Ti <=0.06; Zr=0.015-0.06; Al+Ti+Nb+V+Zr<= 0.15; N<=Al/2+V/4+V/4+ Nb/7+Ti/3.5<=0.015; bal Fe	440-550	295	20L	
Italy															
UNI 5869(75)	Fe360-1KG	Press ves	0.17 max			0.4-1.60		0.035 max	0.04 max	0.35 max	bal Fe				
UNI 5869(75)	Fe360-1KW	Press ves	0.17 max			0.4-1.60		0.04 max	0.04 max	0.35 max	bal Fe				
UNI 5869(75)	Fe360-2KG	Press ves	0.17 max			0.4-1.60		0.03 max	0.035 max	0.35 max	bal Fe				
UNI 5869(75)	Fe360-2KW	Press ves	0.17 max			0.4-1.60		0.035 max	0.035 max	0.35 max	bal Fe				
Romania															
STAS 2883/2(80)	R37	Heat res; Press ves	0.17 max	0.3 max		0.3-0.8	0.4 max	0.04 max	0.04 max	0.17-0.4	Al 0.025-0.120; N <=0.009; Pb <=0.15; As<=0.06; bal Fe				
STAS 2883/2(80)	R44	Heat res; Press ves	0.2 max	0.3 max		0.8-1.1	0.4 max	0.04 max	0.04 max	0.17-0.45	Al 0.025-0.120; N <=0.009; Pb <=0.15; As<=0.06; bal Fe				
Russia															
GOST 5520(79)	18K	Boiler, Press ves	0.14-0.22	0.30 max	0.30 max	0.55-0.85	0.30 max	0.040 max	0.040 max	0.17-0.37	N <=0.012; As<=0.08; bal Fe				
Spain															
UNE 36087(74)	A37RAII	Sh Plt	0.2 max			0.3-1.3		0.045 max	0.04 max	0.4 max	Al 0.015-0.120; Mo <=0.15; 0-40mm Mn<=1.3; 40mm Mn<=1.1; bal Fe				
UNE 36087(74)	A37RCI	Sh Plt	0.2 max			0.3-1.3		0.055 max	0.05 max	0.4 max	Mo <=0.15; N <=0.009; 0-40mm Mn<=1.3; 40mm Mn<=1.1; bal Fe				

Specification	Designation	Notes	C	Cr	Cu	Mn	Ni	P	S	Si	Other	UTS	YS	El	Hard

Carbon Steel, Nonresulfurized, A515(65) (Continued from previous page)

USA

Specification	Designation	Notes	C	Cr	Cu	Mn	Ni	P	S	Si	Other	UTS	YS	El	Hard
	UNS K02800		0.28 max			0.90 max		0.035 max	0.04 max	0.13-0.45	bal Fe				
ASTM A515(92)	Grade 65	Press ves plt, int/High-temp, t>50mm	0.31 max			0.90 max		0.035 max	0.035 max	0.15-0.40	bal Fe	450-585	240	23	
ASTM A515(92)	Grade 65	Press ves plt, int/High-temp, t>25-50mm	0.31 max			0.90 max		0.035 max	0.035 max	0.15-0.40	bal Fe	450-585	240	23	
ASTM A515(92)	Grade 65	Press ves plt, int/High-temp, t<=25mm	0.25 max			0.90 max		0.035 max	0.035 max	0.15-0.40	bal Fe	450-585	240	23	

Yugoslavia

Specification	Designation	Notes	C	Cr	Cu	Mn	Ni	P	S	Si	Other	UTS	YS	El	Hard
	C.1206	High-temp const	0.22 max			0.55 max		0.05 max	0.05 max	0.35 max	Mo <=0.15; bal Fe				

Carbon Steel, Nonresulfurized, A515(70)

Europe

Specification	Designation	Notes	C	Cr	Cu	Mn	Ni	P	S	Si	Other	UTS	YS	El	Hard
EN 10028/3(92)	1.0566	High-temp, Press; 35<t<=50mm	0.18 max	0.30 max	0.30 max	0.90-1.70	0.50 max	0.030 max	0.020 max	0.50 max	Al >=0.020; Mo <=0.08; N <=0.020; Nb <=0.05; Ti <=0.03; V <=0.10; Nb+Ti+V<=0.12, Cr+Cu+Mo<=0.45; bal Fe	490-630	345	22	
EN 10028/3(92)	1.0566	High-temp, Press; 0<=t<=35mm	0.18 max	0.30 max	0.30 max	0.90-1.70	0.50 max	0.030 max	0.020 max	0.50 max	Al >=0.020; Mo <=0.08; N <=0.020; Nb <=0.05; Ti <=0.03; V <=0.10; Nb+Ti+V<=0.12, Cr+Cu+Mo<=0.45; bal Fe	490-630	355	22	
EN 10028/3(92)	1.0566	High-temp, Press; 50<t<=70mm	0.18 max	0.30 max	0.30 max	0.90-1.70	0.50 max	0.030 max	0.020 max	0.50 max	Al >=0.020; Mo <=0.08; N <=0.020; Nb <=0.05; Ti <=0.03; V <=0.10; Nb+Ti+V<=0.12, Cr+Cu+Mo<=0.45; bal Fe	490-630	325	22	
EN 10028/3(92)	1.0566	High-temp, Press; 70<t<=100mm	0.18 max	0.30 max	0.30 max	0.90-1.70	0.50 max	0.030 max	0.020 max	0.50 max	Al >=0.020; Mo <=0.08; N <=0.020; Nb <=0.05; Ti <=0.03; V <=0.10; Nb+Ti+V<=0.12, Cr+Cu+Mo<=0.45; bal Fe	470-610	315	21	
EN 10028/3(92)	1.0566	High-temp, Press; 100<t<=150mm	0.18 max	0.30 max	0.30 max	0.90-1.70	0.50 max	0.030 max	0.020 max	0.50 max	Al >=0.020; Mo <=0.08; N <=0.020; Nb <=0.05; Ti <=0.03; V <=0.10; Nb+Ti+V<=0.12, Cr+Cu+Mo<=0.45; bal Fe	450-590	295	21	
EN 10028/3(92)	P355NL1	High-temp, Press; 70<t<=100mm	0.18 max	0.30 max	0.30 max	0.90-1.70	0.50 max	0.030 max	0.020 max	0.50 max	Al >=0.020; Mo <=0.08; N <=0.020; Nb <=0.05; Ti <=0.03; V <=0.10; Nb+Ti+V<=0.12, Cr+Cu+Mo<=0.45; bal Fe	470-610	315	21	
EN 10028/3(92)	P355NL1	High-temp, Press; 35<t<=50mm	0.18 max	0.30 max	0.30 max	0.90-1.70	0.50 max	0.030 max	0.020 max	0.50 max	Al >=0.020; Mo <=0.08; N <=0.020; Nb <=0.05; Ti <=0.03; V <=0.10; Nb+Ti+V<=0.12, Cr+Cu+Mo<=0.45; bal Fe	490-630	345	22	
EN 10028/3(92)	P355NL1	High-temp, Press; 0<=t<=35mm	0.18 max	0.30 max	0.30 max	0.90-1.70	0.50 max	0.030 max	0.020 max	0.50 max	Al >=0.020; Mo <=0.08; N <=0.020; Nb <=0.05; Ti <=0.03; V <=0.10; Nb+Ti+V<=0.12, Cr+Cu+Mo<=0.45; bal Fe	490-630	355	22	

Specification	Designation	Notes	C	Cr	Cu	Mn	Ni	P	S	Si	Other	UTS	YS	El	Hard

Carbon Steel, Nonresulfurized, A515(70) (Continued from previous page)

Europe

Specification	Designation	Notes	C	Cr	Cu	Mn	Ni	P	S	Si	Other	UTS	YS	El	Hard
EN 10028/3(92)	P355NL1	High-temp, Press; 100<t<=150mm	0.18 max	0.30 max	0.30 max	0.90-1.70	0.50 max	0.030 max	0.020 max	0.50 max	Al >=0.020; Mo <=0.08; N <=0.020; Nb <=0.05; Ti <=0.03; V <=0.10; Nb+Ti+V<=0.12, Cr+Cu+Mo<=0.45; bal Fe	450-590	295	21	
EN 10028/3(92)	P355NL1	High-temp, Press; 50<t<=70mm	0.18 max	0.30 max	0.30 max	0.90-1.70	0.50 max	0.030 max	0.020 max	0.50 max	Al >=0.020; Mo <=0.08; N <=0.020; Nb <=0.05; Ti <=0.03; V <=0.10; Nb+Ti+V<=0.12, Cr+Cu+Mo<=0.45; bal Fe	490-630	325	22	

France

Specification	Designation	Notes	C	Cr	Cu	Mn	Ni	P	S	Si	Other	UTS	YS	El	Hard
AFNOR NFA36205(82)	A48AP*	Press ves; 3<=t<=30mm; Norm	0.20 max	025 max	0.30 max	0.80-1.50	0.3 max	0.035 max	0.03 max	0.35 max	Mo <=0.10; Nb <=0.010; V <=0.05; Cu+6Sn<=0.33; bal Fe	470-560	285	23	
AFNOR NFA36205(82)	A48AP*	Press ves; 80<t<=300mm; Q/T	0.20 max	025 max	0.30 max	0.80-1.50	0.3 max	0.035 max	0.03 max	0.35 max	Mo <=0.10; Nb <=0.010; V <=0.05; Cu+6Sn<=0.33; bal Fe	470-560	255	23L; 21T	
AFNOR NFA36205(82)	A48APR*	Press ves; 80<t<=110mm; Norm	0.20 max	0.25 max	0.30 max	1.00-1.60	0.3 max	0.035 max	0.03 max	0.5 max	Nb <=0.010; bal Fe	490-620	285	20	
AFNOR NFA36205(82)	A48APR*	Press ves; 3<=t<=30mm; Norm	0.20 max	0.25 max	0.30 max	1.00-1.60	0.3 max	0.035 max	0.03 max	0.5 max	Nb <=0.010; bal Fe	490-620	315	22	
AFNOR NFA36205(82)	A48CP*	Press ves; 80<t<=110mm, Norm	0.20 max	0.25 max	0.30 max	0.80-1.50	0.3 max	0.035 max	0.03 max	0.35 max	Nb <=0.010; bal Fe	470-560	255	21	
AFNOR NFA36205(82)	A48CP*	Press ves; 3<=t<=30mm; Norm	0.20 max	0.25 max	0.30 max	0.80-1.50	0.3 max	0.035 max	0.03 max	0.35 max	Nb <=0.010; bal Fe	470-560	285	23	
AFNOR NFA36205(82)	A48CPR*	Press ves; 3<=t<=30mm; Norm	0.20 max	0.25 max	0.30 max	1.00-1.60	0.3 max	0.035 max	0.03 max	0.5 max	Nb <=0.010; bal Fe	490-620	315	22	
AFNOR NFA36205(82)	A48CPR*	Press ves; 80<t<=110mm; Norm	0.20 max	0.25 max	0.30 max	1.00-1.60	0.3 max	0.035 max	0.03 max	0.5 max	Nb <=0.010; bal Fe	490-620	285	20	
AFNOR NFA36205(82)	A48FP*	Press ves; 3<=t<=30mm; Norm	0.20 max	0.25 max	0.30 max	0.80-1.50	0.4 max	0.03 max	0.02 max	0.35 max	Nb <=0.010; bal Fe	470-560	285	23	
AFNOR NFA36205(82)	A48FP*	Press ves; 80<t<=300mm; Norm	0.20 max	0.25 max	0.30 max	0.80-1.50	0.4 max	0.03 max	0.02 max	0.35 max	Nb <=0.010; bal Fe	510-620	305	22L; 20T	
AFNOR NFA36205(82)	A48FP*	Press ves; 0<t<=30mm; Norm	0.20 max	0.25 max	0.30 max	0.80-1.50	0.4 max	0.03 max	0.02 max	0.35 max	Nb <=0.010; bal Fe	470-560	285	23	
AFNOR NFA36205(82)	A48FP*	Press ves; 80<t<=110mm; Norm	0.20 max	0.25 max	0.30 max	0.80-1.50	0.4 max	0.03 max	0.02 max	0.35 max	Nb <=0.010; bal Fe	470-560	255	21	
AFNOR NFA36205(82)	A48FPR*	Press ves; 3<=t<=30mm; Norm	0.20 max	0.25 max	0.30 max	1.00-1.60	0.4 max	0.03 max	0.02 max	0.5 max	Nb <=0.010; bal Fe	490-620	315	22	
AFNOR NFA36205(82)	A48FPR*	Press ves; 50<t<=80mm; Norm	0.20 max	0.25 max	0.30 max	1.00-1.60	0.4 max	0.03 max	0.02 max	0.5 max	Nb <=0.010; bal Fe	490-620	295	20	
AFNOR NFA36601(80)	A48CP	80<t<=300mm; Q/T	0.2 max	0.25 max	0.18 max	0.8-1.5	0.5 max	0.035 max	0.03 max	0.35 max	Cu+6Sn<=0.33; bal Fe	470-560	255	23L; 21T	
AFNOR NFA36601(80)	A48CP	0<t<=30mm; Norm	0.2 max	0.25 max	0.18 max	0.8-1.5	0.5 max	0.035 max	0.03 max	0.35 max	Cu+6Sn<=0.33; bal Fe	470-560	285	25L; 23T	

Italy

Specification	Designation	Notes	C	Cr	Cu	Mn	Ni	P	S	Si	Other	UTS	YS	El	Hard
UNI 5869(75)	Fe510-1KG	Press ves	0.22 max			1.6 max		0.035 max	0.04 max	0.4 max	0-50mm C<=0.2; bal Fe				
UNI 5869(75)	Fe510-1KW	Press ves	0.22 max			1.6 max		0.04 max	0.04 max	0.4 max	0-50mm C<=0.2; bal Fe				
UNI 5869(75)	Fe510-2KG	Press ves	0.22 max			1.6 max		0.03 max	0.035 max	0.4 max	0-50mm C<=0.2; bal Fe				
UNI 5869(75)	Fe510-2KW	Press ves	0.22 max			1.6 max		0.035 max	0.035 max	0.4 max	0-50mm C<=0.2; bal Fe				

Spain

Specification	Designation	Notes	C	Cr	Cu	Mn	Ni	P	S	Si	Other	UTS	YS	El	Hard
UNE 36087(74)	A47RAII	Sh Plt	0.23 max			0.5-1.5		0.045 max	0.04 max	0.4 max	Al 0.015-0.120; Mo <=0.15; bal Fe				
UNE 36087(74)	A47RCI	Sh Plt	0.23 max			0.7-1.5		0.055 max	0.05 max	0.4 max	Mo <=0.15; N <=0.009; bal Fe				

USA

Specification	Designation	Notes	C	Cr	Cu	Mn	Ni	P	S	Si	Other	UTS	YS	El	Hard
	UNS K03101		0.31 max			0.90 max		0.035 max	0.04 max	0.13-0.33	bal Fe				

UNS numbers and US grades are provided as a means of cross referencing chemically similar alloys. Exchangability is only possible after independent examination of specifications. Tensile properties are minimum or typical as specified. UTS and YS as MPa. El as %. See Appendix for list of abbreviations used in Notes. * indicates obsolete material.

Specification	Designation	Notes	C	Cr	Cu	Mn	Ni	P	S	Si	Other	UTS	YS	El	Hard

Carbon Steel, Nonresulfurized, A515(70) (Continued from previous page)

USA

Specification	Designation	Notes	C	Cr	Cu	Mn	Ni	P	S	Si	Other	UTS	YS	El	Hard
ASTM A515(92)	Grade 70	Press ves plt, int/High-temp, t>25-50mm	0.33 max			1.20 max		0.035 max	0.035 max	0.15-0.40	bal Fe	485-620	260	21	
ASTM A515(92)	Grade 70	Press ves plt, int/High-temp, t<=25mm	0.31 max			1.20 max		0.035 max	0.035 max	0.15-0.40	bal Fe	485-620	260	21	
ASTM A515(92)	Grade 70	Press ves plt, int/High-temp, t>50mm	0.35 max			1.20 max		0.035 max	0.035 max	0.15-0.40	bal Fe	485-620	260	21	

Carbon Steel, Nonresulfurized, A516(55)

Bulgaria

Specification	Designation	Notes	C	Cr	Cu	Mn	Ni	P	S	Si	Other	UTS	YS	El	Hard
BDS 6609(73)	12K	Heat res; boiler tube	0.16 max	0.30 max	0.30 max	0.30-0.65	0.30 max	0.045 max	0.045 max	0.15-0.35	Mo <=0.15; bal Fe				

Spain

Specification	Designation	Notes	C	Cr	Cu	Mn	Ni	P	S	Si	Other	UTS	YS	El	Hard
UNE 36032(85)	16KX-DF	Screw, bolt	0.12-0.19			0.30-0.60		0.040 max	0.040 max	0.15-0.35	bal Fe				
UNE 36032(85)	F.7504*	see 16KX-DF	0.12-0.19			0.30-0.60		0.040 max	0.040 max	0.15-0.35	bal Fe				

USA

Specification	Designation	Notes	C	Cr	Cu	Mn	Ni	P	S	Si	Other	UTS	YS	El	Hard
	UNS K01800		0.18 max			0.55-0.98		0.035 max	0.04 max	0.13-0.45	bal Fe				
ASTM A516(90)	Grade 55	Press ves plt, mod/low-temp, t<=12.5mm	0.18 max			0.60-0.90		0.035 max	0.035 max	0.15-0.40	bal Fe	380-515	205	27	
ASTM A516(90)	Grade 55	Press ves plt, mod/low-temp, t>12.5-50mm	0.20 max			0.60-1.20		0.035 max	0.035 max	0.15-0.40	bal Fe	380-515	205	27	
ASTM A516(90)	Grade 55	Press ves plt, mod/low-temp, t>200mm	0.26 max			0.60-1.20		0.035 max	0.035 max	0.15-0.40	bal Fe	380-515	205	27	
ASTM A516(90)	Grade 55	Press ves plt, mod/low-temp, t>100-200	0.24 max			0.60-1.20		0.035 max	0.035 max	0.15-0.40	bal Fe	380-515	205	27	

Carbon Steel, Nonresulfurized, A516(60)

Europe

Specification	Designation	Notes	C	Cr	Cu	Mn	Ni	P	S	Si	Other	UTS	YS	El	Hard
EN 10028/3(92)	1.0486	High-temp, Press; 100<t<=150mm	0.18 max	0.30 max	0.30 max	0.50-1.40	0.50 max	0.030 max	0.025 max	0.40 max	Al >=0.020; Mo <=0.08; N <=0.020; Nb <=0.05; Ti <=0.03; V <=0.05; Nb+Ti+V<=0.05, Cr+Cu+Mo<=0.45; bal Fe	390-510	225	23	
EN 10028/3(92)	1.0486	High-temp, Press; 70<t<=100mm	0.18 max	0.30 max	0.30 max	0.50-1.40	0.50 max	0.030 max	0.025 max	0.40 max	Al >=0.020; Mo <=0.08; N <=0.020; Nb <=0.05; Ti <=0.03; V <=0.05; Nb+Ti+V<=0.05, Cr+Cu+Mo<=0.45; bal Fe	390-510	235	23	
EN 10028/3(92)	1.0486	High-temp, Press; 50<t<=70mm	0.18 max	0.30 max	0.30 max	0.50-1.40	0.50 max	0.030 max	0.025 max	0.40 max	Al >=0.020; Mo <=0.08; N <=0.020; Nb <=0.05; Ti <=0.03; V <=0.05; Nb+Ti+V<=0.05, Cr+Cu+Mo<=0.45; bal Fe	390-510	255	24	
EN 10028/3(92)	1.0486	High-temp, Press; 35<t<=50mm	0.18 max	0.30 max	0.30 max	0.50-1.40	0.50 max	0.030 max	0.025 max	0.40 max	Al >=0.020; Mo <=0.08; N <=0.020; Nb <=0.05; Ti <=0.03; V <=0.05; Nb+Ti+V<=0.05, Cr+Cu+Mo<=0.45; bal Fe	390-510	265	24	
EN 10028/3(92)	1.0486	High-temp, Press; 0<=t<=35mm	0.18 max	0.30 max	0.30 max	0.50-1.40	0.50 max	0.030 max	0.025 max	0.40 max	Al >=0.020; Mo <=0.08; N <=0.020; Nb <=0.05; Ti <=0.03; V <=0.05; Nb+Ti+V<=0.05, Cr+Cu+Mo<=0.45; bal Fe	390-510	275	24	

Specification	Designation	Notes	C	Cr	Cu	Mn	Ni	P	S	Si	Other	UTS	YS	El	Hard

Carbon Steel, Nonresulfurized, A516(60) (Continued from previous page)

Europe

Specification	Designation	Notes	C	Cr	Cu	Mn	Ni	P	S	Si	Other	UTS	YS	El	Hard
EN 10028/3(92)	1.0487	High-temp, Press; 70<t<=100mm	0.18 max	0.30 max	0.30 max	0.50-1.50	0.50 max	0.030 max	0.025 max	0.40 max	Al >=0.020; Mo <=0.08; N <=0.020; Nb <=0.05; Ti <=0.03; V <=0.05; Nb+Ti+V<=0.05, Cr+Cu+Mo<=0.45; bal Fe	390-510	235	23	
EN 10028/3(92)	1.0487	High-temp, Press; 50<t<=70mm	0.18 max	0.30 max	0.30 max	0.50-1.50	0.50 max	0.030 max	0.025 max	0.40 max	Al >=0.020; Mo <=0.08; N <=0.020; Nb <=0.05; Ti <=0.03; V <=0.05; Nb+Ti+V<=0.05, Cr+Cu+Mo<=0.45; bal Fe	390-510	255	24	
EN 10028/3(92)	1.0487	High-temp, Press; 100<t<=150mm	0.18 max	0.30 max	0.30 max	0.50-1.50	0.50 max	0.030 max	0.025 max	0.40 max	Al >=0.020; Mo <=0.08; N <=0.020; Nb <=0.05; Ti <=0.03; V <=0.05; Nb+Ti+V<=0.05, Cr+Cu+Mo<=0.45; bal Fe	390-510	225	23	
EN 10028/3(92)	1.0487	High-temp, Press; 35<t<=50mm	0.18 max	0.30 max	0.30 max	0.50-1.50	0.50 max	0.030 max	0.025 max	0.40 max	Al >=0.020; Mo <=0.08; N <=0.020; Nb <=0.05; Ti <=0.03; V <=0.05; Nb+Ti+V<=0.05, Cr+Cu Mo<=0.45; bal Fe	390-510	265	24	
EN 10028/3(92)	1.0487	High-temp, Press; 0<=t<=35mm	0.18 max	0.30 max	0.30 max	0.50-1.50	0.50 max	0.030 max	0.025 max	0.40 max	Al >=0.020; Mo <=0.08, N <=0.020; Nb <=0.05; Ti <=0.03; V <=0.05; Nb+Ti+V<=0.05, Cr+Cu+Mo<=0.45; bal Fe	390-510	275	24	
EN 10028/3(92)	1.0488	High-temp, Press; 35<t<=50mm	0.16 max	0.30 max	0.30 max	0.50-1.50	0.50 max	0.030 max	0.020 max	0.40 max	Al >=0.020; Mo <=0.08; N <=0.020; Nb <=0.05; Ti <=0.03; V <=0.05; Nb+Ti+V<=0.05, Cr+Cu+Mo<=0.45; bal Fe	390-510	265	24	
EN 10028/3(92)	1.0488	High-temp, Press; 0<=t<=35mm	0.16 max	0.30 max	0.30 max	0.50-1.50	0.50 max	0.030 max	0.020 max	0.40 max	Al >=0.020; Mo <=0.08; N <=0.020; Nb <=0.05; Ti <=0.03; V <=0.05; Nb+Ti+V<=0.05, Cr+Cu+Mo<=0.45; bal Fe	390-510	275	24	
EN 10028/3(92)	1.0488	High-temp, Press; 50<t<=70mm	0.16 max	0.30 max	0.30 max	0.50-1.50	0.50 max	0.030 max	0.020 max	0.40 max	Al >=0.020; Mo <=0.08; N <=0.020; Nb <=0.05; Ti <=0.03; V <=0.05; Nb+Ti+V<=0.05, Cr+Cu+Mo<=0.45; bal Fe	390-510	255	24	
EN 10028/3(92)	1.0488	High-temp, Press; 70<t<=100mm	0.16 max	0.30 max	0.30 max	0.50-1.50	0.50 max	0.030 max	0.020 max	0.40 max	Al >=0.020; Mo <=0.08; N <=0.020; Nb <=0.05; Ti <=0.03; V <=0.05; Nb+Ti+V<=0.05, Cr+Cu+Mo<=0.45; bal Fe	390-510	235	23	
EN 10028/3(92)	1.0488	High-temp, Press; 100<t<=150mm	0.16 max	0.30 max	0.30 max	0.50-1.50	0.50 max	0.030 max	0.020 max	0.40 max	Al >=0.020; Mo <=0.08; N <=0.020; Nb <=0.05; Ti <=0.03; V <=0.05; Nb+Ti+V<=0.05, Cr+Cu+Mo<=0.45; bal Fe	390-510	225	23	

Carbon Steel, Nonresulfurized, A516(60) (Continued from previous page)

Specification	Designation	Notes	C	Cr	Cu	Mn	Ni	P	S	Si	Other	UTS	YS	El	Hard
Europe															
EN 10028/3(92)	P275N	High-temp, Press; 0<=t<=35mm	0.18 max	0.30 max	0.30 max	0.50-1.40	0.50 max	0.030 max	0.025 max	0.40 max	Al >=0.020; Mo <=0.08; N <=0.020; Nb <=0.05; Ti <=0.03; V <=0.05; Nb+Ti+V<=0.05, Cr+Cu+Mo<=0.45; bal Fe	390-510	275	24	
EN 10028/3(92)	P275N	High-temp, Press; 35<t<=50mm	0.18 max	0.30 max	0.30 max	0.50-1.40	0.50 max	0.030 max	0.025 max	0.40 max	Al >=0.020; Mo <=0.08; N <=0.020; Nb <=0.05; Ti <=0.03; V <=0.05; Nb+Ti+V<=0.05, Cr+Cu+Mo<=0.45; bal Fe	390-510	265	24	
EN 10028/3(92)	P275N	High-temp, Press; 50<t<=70mm	0.18 max	0.30 max	0.30 max	0.50-1.40	0.50 max	0.030 max	0.025 max	0.40 max	Al >=0.020; Mo <=0.08; N <=0.020; Nb <=0.05; Ti <=0.03; V <=0.05; Nb+Ti+V<=0.05, Cr+Cu+Mo<=0.45; bal Fe	390-510	255	24	
EN 10028/3(92)	P275N	High-temp, Press; 70<t<=100mm	0.18 max	0.30 max	0.30 max	0.50-1.40	0.50 max	0.030 max	0.025 max	0.40 max	Al >=0.020; Mo <=0.08; N <=0.020; Nb <=0.05; Ti <=0.03; V <=0.05; Nb+Ti+V<=0.05, Cr+Cu+Mo<=0.45; bal Fe	390-510	235	23	
EN 10028/3(92)	P275N	High-temp, Press; 100<t<=150mm	0.18 max	0.30 max	0.30 max	0.50-1.40	0.50 max	0.030 max	0.025 max	0.40 max	Al >=0.020; Mo <=0.08; N <=0.020; Nb <=0.05; Ti <=0.03; V <=0.05; Nb+Ti+V<=0.05, Cr+Cu+Mo<=0.45; bal Fe	390-510	225	23	
EN 10028/3(92)	P275NH	High-temp, Press; 35<t<=50mm	0.18 max	0.30 max	0.30 max	0.50-1.50	0.50 max	0.030 max	0.025 max	0.40 max	Al >=0.020; Mo <=0.08; N <=0.020; Nb <=0.05; Ti <=0.03; V <=0.05; Nb+Ti+V<=0.05, Cr+Cu+Mo<=0.45; bal Fe	390-510	265	24	
EN 10028/3(92)	P275NH	High-temp, Press; 70<t<=100mm	0.18 max	0.30 max	0.30 max	0.50-1.50	0.50 max	0.030 max	0.025 max	0.40 max	Al >=0.020; Mo <=0.08; N <=0.020; Nb <=0.05; Ti <=0.03; V <=0.05; Nb+Ti+V<=0.05, Cr+Cu+Mo<=0.45; bal Fe	390-510	235	23	
EN 10028/3(92)	P275NH	High-temp, Press; 50<t<=70mm	0.18 max	0.30 max	0.30 max	0.50-1.50	0.50 max	0.030 max	0.025 max	0.40 max	Al >=0.020; Mo <=0.08; N <=0.020; Nb <=0.05; Ti <=0.03; V <=0.05; Nb+Ti+V<=0.05, Cr+Cu+Mo<=0.45; bal Fe	390-510	255	24	
EN 10028/3(92)	P275NH	High-temp, Press; 0<=t<=35mm	0.18 max	0.30 max	0.30 max	0.50-1.50	0.50 max	0.030 max	0.025 max	0.40 max	Al >=0.020; Mo <=0.08; N <=0.020; Nb <=0.05; Ti <=0.03; V <=0.05; Nb+Ti+V<=0.05, Cr+Cu+Mo<=0.45; bal Fe	390-510	275	24	
EN 10028/3(92)	P275NH	High-temp, Press; 100<t<=150mm	0.18 max	0.30 max	0.30 max	0.50-1.50	0.50 max	0.030 max	0.025 max	0.40 max	Al >=0.020; Mo <=0.08; N <=0.020; Nb <=0.05; Ti <=0.03; V <=0.05; Nb+Ti+V<=0.05, Cr+Cu+Mo<=0.45; bal Fe	390-510	225	23	

Carbon Steel, Nonresulfurized, A516(60) (Continued from previous page)

Specification	Designation	Notes	C	Cr	Cu	Mn	Ni	P	S	Si	Other	UTS	YS	El	Hard
Europe															
EN 10028/3(92)	P275NL1	High-temp, Press; 35<t<=50mm	0.16 max	0.30 max	0.30 max	0.50-1.50	0.50 max	0.030 max	0.020 max	0.40 max	Al >=0.020; Mo <=0.08; N <=0.020; Nb <=0.05; Ti <=0.03; V <=0.05; Nb+Ti+V<=0.05, Cr+Cu+Mo<=0.45; bal Fe	390-510	265	24	
EN 10028/3(92)	P275NL1	High-temp, Press; 100<t<=150mm	0.16 max	0.30 max	0.30 max	0.50-1.50	0.50 max	0.030 max	0.020 max	0.40 max	Al >=0.020; Mo <=0.08; N <=0.020; Nb <=0.05; Ti <=0.03; V <=0.05; Nb+Ti+V<=0.05, Cr+Cu+Mo<=0.45; bal Fe	390-510	225	23	
EN 10028/3(92)	P275NL1	High-temp, Press; 70<t<=100mm	0.16 max	0.30 max	0.30 max	0.50-1.50	0.50 max	0.030 max	0.020 max	0.40 max	Al >=0.020; Mo <=0.08; N <=0.020; Nb <=0.05; Ti <=0.03; V <=0.05; Nb+Ti+V<=0.05, Cr+Cu+Mo<=0.45; bal Fe	390-510	235	23	
EN 10028/3(92)	P275NL1.	High-temp, Press; 50<t<=70mm	0.16 max	0.30 max	0.30 max	0.50-1.50	0.50 max	0.030 max	0.020 max	0.40 max	Al >=0.020; Mo <=0.08; N <=0.020; Nb <=0.05; Ti <=0.03; V <=0.05; Nb+Ti+V<=0.05, Cr+Cu+Mo<=0.45; bal Fe	390-510	255	24	
EN 10028/3(92)	P275NL1	High-temp, Press; 0<=t<=35mm	0.16 max	0.30 max	0.30 max	0.50-1.50	0.50 max	0.030 max	0.020 max	0.40 max	Al >=0.020; Mo <=0.08; N <=0.020; Nb <=0.05; Ti <=0.03; V <=0.05; Nb+Ti+V<=0.05, Cr+Cu+Mo<=0.45; bal Fe	390-510	275	24	
EN 10113/2(93)	1.0490	HR weld t<=16mm	0.18 max	0.30 max	0.35 max	0.50-1.40	0.30 max	0.035 max	0.030 max	0.40 max	Al >=0.02; Mo <=0.10; N <=0.015; Nb <=0.05; Ti <=0.03; V <=0.05; CEV; bal Fe	370-510	275	24	
EN 10113/2(93)	1.0491	HR weld t<=16mm	0.16 max	0.30 max	0.35 max	0.50-1.40	0.30 max	0.030 max	0.025 max	0.40 max	Al >=0.02; Mo <=0.10; N <=0.015; Nb <=0.05; Ti <=0.03; V <=0.05; CEV; bal Fe	370-510	275	24	
EN 10113/2(93)	S275N	HR weld t<=16mm	0.18 max	0.30 max	0.35 max	0.50-1.40	0.30 max	0.035 max	0.030 max	0.40 max	Al >=0.02; Mo <=0.10; N <=0.015; Nb <=0.05; Ti <=0.03; V <=0.05; CEV; bal Fe	370-510	275	24	
EN 10113/2(93)	S275NL	HR weld t<=16mm	0.16 max	0.30 max	0.35 max	0.50-1.40	0.30 max	0.030 max	0.025 max	0.40 max	Al >=0.02; Mo <=0.10; N <=0.015; Nb <=0.05; Ti <=0.03; V <=0.05; CEV; bal Fe	370-510	275	24	
Romania															
STAS 9021/1(89)	OCS285.5a	Weld const; Fine	0.2 max	0.3 max		0.8-1.1		0.04 max	0.04 max	0.5 max	Al 0.015-0.120; Nb >=0.02; Pb <=0.15; V 0.03-0.130; bal Fe				
STAS 9021/1(89)	OCS285.5b	Weld const; Fine	0.2 max	0.3 max		0.8-1.1		0.04 max	0.04 max	0.5 max	Al 0.015-0.120; Nb >=0.02; Pb <=0.15; V 0.03-0.130; bal Fe				
STAS 9021/1(89)	OCS285.6a	Weld const; Fine	0.2 max	0.3 max		0.8-1.1		0.04 max	0.04 max	0.5 max	Al 0.015-0.120; Nb >=0.02; Pb <=0.15; V 0.03-0.130; bal Fe				
STAS 9021/1(89)	OCS285.6b	Weld const; Fine	0.2 max	0.3 max		0.8-1.1		0.04 max	0.04 max	0.5 max	Al 0.015-0.120; Nb >=0.02; Pb <=0.15; V 0.03-0.130; bal Fe				

Specification	Designation	Notes	C	Cr	Cu	Mn	Ni	P	S	Si	Other	UTS	YS	El	Hard

Carbon Steel, Nonresulfurized, A516(60) (Continued from previous page)

Spain

Specification	Designation	Notes	C	Cr	Cu	Mn	Ni	P	S	Si	Other	UTS	YS	El	Hard
UNE 36081(76)	AE285KG	Fine grain Struct	0.18 max	0.25 max	0.35 max	0.5-1.4	0.3 max	0.035 max	0.035 max	0.4 max	Al <=0.015; Mo <=0.1; Nb 0.015-0.05; Ti <=0.5; V 0.02-0.1; bal Fe				
UNE 36081(76)	AE285KT	Fine grain Struct	0.16 max	0.25 max		0.5-1.4	0.3 max	0.03 max	0.03 max	0.4 max	Mo <=0.1; +Al+V+Nb; bal Fe				
UNE 36081(76)	AE285KW	Fine grain Struct	0.18 max	0.25 max	0.35 max	0.5-1.4	0.3 max	0.035 max	0.035 max	0.4 max	Al <=0.015; Mo <=0.1; Nb 0.015-0.05; V 0.02-0.1; bal Fe				
UNE 36081(76)	F.6404*	see AE285KG	0.18 max	0.25 max	0.35 max	0.5-1.4	0.3 max	0.035 max	0.035 max	0.4 max	Al <=0.015; Mo <=0.1; Nb 0.015-0.05; Ti <=0.5; V 0.02-0.1; bal Fe				
UNE 36081(76)	F.6405*	see AE285KW	0.18 max	0.25 max	0.35 max	0.5-1.4	0.3 max	0.035 max	0.035 max	0.4 max	Al <=0.015; Mo <=0.1; Nb 0.015-0.05; V 0.02-0.1; bal Fe				
UNE 36081(76)	F.6406*	see AE285KT	0.16 max	0.25 max		0.5-1.4	0.3 max	0.03 max	0.03 max	0.4 max	Mo <=0.1; +Al+V+Nb; bal Fe				

USA

Specification	Designation	Notes	C	Cr	Cu	Mn	Ni	P	S	Si	Other	UTS	YS	El	Hard
	UNS K02100		0.21 max			0.55-0.98		0.035 max	0.04 max	0.13-0.45	bal Fe				
ASTM A516(90)	Grade 60	Press ves plt, mod/low-temp, t>50-100mm	0.25 max			0.85-1.20		0.035 max	0.035 max	0.15-0.40	bal Fe	415-550	220	21	
ASTM A516(90)	Grade 60	Press ves plt, mod/low-temp, t>100mm	0.27 max			0.85-1.20		0.035 max	0.035 max	0.15-0.40	bal Fe	415-550	220	21	
ASTM A516(90)	Grade 60	Press ves plt, mod/low-temp, t<=12.5mm	0.21 max			0.60-0.90		0.035 max	0.035 max	0.15-0.40	bal Fe	415-550	220	21	
ASTM A516(90)	Grade 60	Press ves plt, mod/low-temp, t>12.5-50mm	0.23 max			0.85-1.20		0.035 max	0.035 max	0.15-0.40	bal Fe	415-550	220	21	

Carbon Steel, Nonresulfurized, A551(A)

USA

Specification	Designation	Notes	C	Cr	Cu	Mn	Ni	P	S	Si	Other	UTS	YS	El	Hard
	UNS K05701		0.50-0.65			0.60-0.90		0.050 max	0.050 max	0.15-0.35	bal Fe				
ASTM A551(94)	Class A	Tires	0.50-0.65			0.60-0.90		0.050 max	0.050 max	0.15-0.35	bal Fe				
ASTM A551(94)	Class AHT	HT, tires	0.50-0.65			0.60-0.90		0.050 max	0.050 max	0.15-0.35	bal Fe	760		16.0	223-277HB

Carbon Steel, Nonresulfurized, A551(B)

USA

Specification	Designation	Notes	C	Cr	Cu	Mn	Ni	P	S	Si	Other	UTS	YS	El	Hard
	UNS K06702		0.60-0.75			0.60-0.90		0.050 max	0.050 max	0.15-0.35	bal Fe				
ASTM A551(94)	Class B	Tires	0.60-0.75			0.60-0.90		0.050 max	0.050 max	0.15-0.35	bal Fe				
ASTM A551(94)	Class BHT	HT, tires	0.60-0.75			0.60-0.90		0.050 max	0.050 max	0.15-0.35	bal Fe	860		14.0	255-302HB

Carbon Steel, Nonresulfurized, A551(C)

USA

Specification	Designation	Notes	C	Cr	Cu	Mn	Ni	P	S	Si	Other	UTS	YS	El	Hard
	UNS K07701		0.70-0.85			0.60-0.90		0.050 max	0.050 max	0.15-0.35	bal Fe				
ASTM A551(94)	Class C	Tires	0.70-0.85			0.60-0.90		0.050 max	0.050 max	0.15-0.35	bal Fe				
ASTM A551(94)	Class CHT	HT, tires	0.70-0.85			0.60-0.90		0.050 max	0.050 max	0.15-0.35	bal Fe	965		12.0	285-331 HB
ASTM A551(94)	Class DHT	HT, tires	0.70-0.85			0.60-0.90		0.050 max	0.050 max	0.15-0.35	bal Fe	1070		10.0	321-363 HB

Carbon Steel, Nonresulfurized, A563(O)

USA

Specification	Designation	Notes	C	Cr	Cu	Mn	Ni	P	S	Si	Other	UTS	YS	El	Hard
	UNS K05802		0.58 max					0.13 max	0.15 max		bal Fe				

UNS numbers and US grades are provided as a means of cross referencing chemically similar alloys. Exchangability is only possible after independent examination of specifications. Tensile properties are minimum or typical as specified. UTS and YS as MPa. El as %. See Appendix for list of abbreviations used in Notes. * indicates obsolete material.

Specification	Designation	Notes	C	Cr	Cu	Mn	Ni	P	S	Si	Other	UTS	YS	El	Hard

Carbon Steel, Nonresulfurized, A570(30)

Czech Republic

Specification	Designation	Notes	C	Cr	Cu	Mn	Ni	P	S	Si	Other	UTS	YS	El	Hard
CSN 411373	11373	Gen struct	0.22 max	0.3 max		1.60 max		0.05 max	0.05 max	0.60 max	bal Fe				
CSN 411375	11375	Gen struct	0.2 max	0.3 max		1.60 max		0.05 max	0.05 max	0.60 max	bal Fe				

Europe

Specification	Designation	Notes	C	Cr	Cu	Mn	Ni	P	S	Si	Other	UTS	YS	El	Hard
EN 10025(90)	Fe360BFN	Obs EU desig; struct HR. FN; BS >40mm	0.20 max					0.045 max	0.045 max		N <=0.009; N max na if Al>=0.020; bal Fe				
EN 10025(90)	Fe360BFN	Obs EU desig; struct HR. FN; BS t<=40mm	0.17 max					0.045 max	0.045 max		N <=0.009; N max na if Al>=0.020; bal Fe				
EN 10025(90)	Fe360BFU	Obs EU desig; struct HR R; FU; BS t<=16mm	0.17 max					0.045 max	0.045 max		N <=0.007; N max na if Al>=0.020; bal Fe	340-510	235	17-26	
EN 10025(90)	Fe360BFU	Obs EU desig; struct HR R; FU; BS 16-25mm	0.20 max					0.045 max	0.045 max		N <=0.007; N max na if Al>=0.020; bal Fe	340-470	225	26	

France

Specification	Designation	Notes	C	Cr	Cu	Mn	Ni	P	S	Si	Other	UTS	YS	El	Hard
AFNOR NFA35501(83)	E24-3*	110<t<=150mm; roll	0.16 max	0.3 max	0.3 max	2 max	0.4 max	0.04 max	0.04 max	0.6 max	bal Fe	340-460	185	21	
AFNOR NFA35501(83)	E24-3*	t<=30mm; roll	0.16 max	0.3 max	0.3 max	2 max	0.4 max	0.04 max	0.04 max	0.6 max	bal Fe	340-440	235	28	
AFNOR NFA35501(83)	E24-3*	10t<=350mm; roll	0.16 max	0.3 max	0.3 max	2 max	0.4 max	0.04 max	0.04 max			320	195	26	
AFNOR NFA35501(83)	E24-3*	t<=2.99mm; roll	0.16 max	0.3 max	0.3 max	2 max	0.4 max	0.04 max	0.04 max	0.6 max	bal Fe	360-480	215	22	
AFNOR NFA35501(83)	E24-3*	30<t<=100mm; roll	0.16 max	0.3 max	0.3 max	2 max	0.4 max	0.04 max	0.04 max	0.6 max	bal Fe	340-440	215	27	
AFNOR NFA35501(83)	E24-4*	10t<=350mm; roll	0.16 max	0.3 max		2 max	0.4 max	0.035 max	0.035 max	0.6 max	Al >=0.02; bal Fe	320	195	27	
AFNOR NFA35501(83)	E24-4*	30<t<=100mm; roll	0.16 max	0.3 max		2 max	0.4 max	0.035 max	0.035 max	0.6 max	Al >=0.02; bal Fe	340-440	215	28	
AFNOR NFA35501(83)	E24-4*	t<=30mm; roll	0.16 max	0.3 max		2 max	0.4 max	0.035 max	0.035 max	0.6 max	Al >=0.02; bal Fe	340-440	235	28	
AFNOR NFA35501(83)	E24-4*	t<3mm; roll	0.16 max	0.3 max		2 max	0.4 max	0.035 max	0.035 max	0.6 max	Al >=0.02; bal Fe	360-480	215	24	
AFNOR NFA35501(83)	E24-4*	110<t<=150mm; roll	0.16 max	0.3 max		2 max	0.4 max	0.035 max	0.035 max	0.6 max	Al >=0.02; bal Fe	340-460	185	23	

Hungary

Specification	Designation	Notes	C	Cr	Cu	Mn	Ni	P	S	Si	Other	UTS	YS	El	Hard
MSZ 500(81)	A38B	Gen struct; Non-rimming; FF; 40.1-100mm; HR/Frg	0.2 max			2 max		0.045 max	0.05 max	0.12 min	bal Fe	370-480	215	26L	
MSZ 500(81)	A38B	Gen struct; Non-rimming; FF; 0-16mm; HR/Frg	0.2 max			2 max		0.045 max	0.05 max	0.12 min	bal Fe	370-480	235	26L	
MSZ 500(81)	A38X	Gen struct; FF; R; FU; 0-16mm; HR/Frg	0.22 max			2 max		0.045 max	0.05 max	0.07 max	bal Fe	370-480	235	26L	
MSZ 500(81)	A38X	Gen struct; FF; R; FU; 40.1-100mm; HR/Frg	0.22 max			2 max		0.045 max	0.05 max	0.07 max	bal Fe	370-480	215	26L	
MSZ 500(89)	Fe235BFN	Gen struct; Non-rimming; FF; Base; BS; 0-1mm; HR/Frg	0.2 max			2 max		0.045 max	0.045 max	0.6 max	0-40mm C<=0.17; 40-100mm C<=0.2; N<=0.009/Al>=0.02; bal Fe	360-510	235	17L; 15T	
MSZ 500(89)	Fe235BFN	Gen struct; Non-rimming; FF; Base; BS; 200.1-250mm; HR/Frg	0.2 max			2 max		0.045 max	0.045 max	0.6 max	0-40mm C<=0.17; 40-100mm C<=0.2; N<=0.009/Al>=0.02; bal Fe	320-470	175	21L; 21T	
MSZ 500(89)	Fe235BFU	Gen struct; unkilled; R; FU; 16.1-25mm; HR/Frg	0.2 max			2 max		0.045 max	0.045 max	0.6 max	Al 0.02-0.1; N <=0.07; 0-16mm C<=0.17; 16-25mm C<=0.2; bal Fe	340-470	225	26L; 24T	
MSZ 500(89)	Fe235BFU	Gen struct; unkilled; R; FU; 0-1mm; HR/Frg	0.2 max			2 max		0.045 max	0.045 max	0.6 max	Al 0.02-0.1; N <=0.07; 0-16mm C<=0.17; 16-25mm C<=0.2; bal Fe	360-510	235	17L; 15T	
MSZ 6280(82)	37B	Weld const; 3-40mm; rolled/norm	0.18 max	0.25 max		1 max	0.3 max	0.045 max	0.045 max	0.15-0.5	Mo <=0.1; bal Fe	360-460	235	26L	
MSZ 6280(82)	37B	Weld const; 40<t<=60mm; rolled/norm	0.18 max	0.25 max		1 max	0.3 max	0.045 max	0.045 max	0.15-0.5	Mo <=0.1; bal Fe	360-460	215	26L	

Specification	Designation	Notes	C	Cr	Cu	Mn	Ni	P	S	Si	Other	UTS	YS	El	Hard

Carbon Steel, Nonresulfurized, A570(30) (Continued from previous page)

International

Specification	Designation	Notes	C	Cr	Cu	Mn	Ni	P	S	Si	Other	UTS	YS	El	Hard
ISO 630(80)	Fe360B	Struct, Thk<=16mm	0.18 max	0.3 max	0.3 max	2 max	0.4 max	0.05 max	0.05 max	0.6 max	Mo <=0.15; Pb <=0.15; Ti <=0.05; V <=0.1;				
ISO 630(80)	Fe360B	Struct, Thk>16mm	0.25 max	0.3 max	0.3 max	2 max	0.4 max	0.05 max	0.05 max	0.6 max	Al <=0.1; Co <=0.1; Mo <=0.15; N <=0.009; Pb <=0.15; Ti <=0.05; V <=0.1; W <=0.1; bal Fe				
ISO 630(80)	Fe360B	Struct, Thk>16mm	0.25 max	0.3 max	0.3 max	2 max	0.4 max	0.05 max	0.05 max	0.6 max	Mo <=0.15; N <=0.009; Pb <=0.15; Ti <=0.05; V <=0.1;				
ISO 630(80)	Fe360B	Struct, Thk<=16mm	0.18 max	0.3 max	0.3 max	2 max	0.4 max	0.05 max	0.05 max	0.6 max	Al <=0.1; Co <=0.1; Mo <=0.15; Pb <=0.15; Ti <=0.05; V <=0.1; W <=0.1; bal Fe				

Italy

Specification	Designation	Notes	C	Cr	Cu	Mn	Ni	P	S	Si	Other	UTS	YS	El	Hard
UNI 7070(72)	Fe37BFN	Gen struct	0.2 max			2 max		0.045 max	0.045 max	0.6 max	N <=0.009; bal Fe				
UNI 7070(72)	Fe37BFU	Gen struct	0.2 max			2 max		0.045 max	0.045 max	0.6 max	N <=0.007; bal Fe				
UNI 7070(82)	Fe360BFN	Gen struct	0.19 max			2 max		0.045 max	0.045 max	0.6 max	bal Fe				
UNI 7070(82)	Fe360BFU	Gen struct	0.19 max			2 max		0.045 max	0.045 max	0.6 max	bal Fe				

Romania

Specification	Designation	Notes	C	Cr	Cu	Mn	Ni	P	S	Si	Other	UTS	YS	El	Hard
STAS 500/2(88)	OL37.1	Gen struct	0.2 max	0.3 max		0.8 max	0.4 max	0.06 max	0.06 max	0.6 max	Pb <=0.15; bal Fe				
STAS 500/2(88)	OL37.1a	Gen struct	0.2 max	0.3 max		0.8 max	0.4 max	0.06 max	0.06 max	0.6 max	Pb <=0.15; bal Fe				
STAS 500/2(88)	OL37.1b	Gen struct	0.2 max	0.3 max		0.8 max	0.4 max	0.06 max	0.06 max	0.6 max	Pb <=0.15; bal Fe				
STAS 500/2(88)	OL37.2	Gen struct	0.18 max	0.3 max		0.8 max	0.4 max	0.05 max	0.05 max	0.6 max	Pb <=0.15; bal Fe				

Russia

Specification	Designation	Notes	C	Cr	Cu	Mn	Ni	P	S	Si	Other	UTS	YS	El	Hard
GOST 19281	16GS	Weld const	0.12-0.18	0.3 max	0.3 max	0.9-1.2	0.3 max	0.035 max	0.04 max	0.4-0.7	Al <=0.1; N <=0.012; As<=0.08; bal Fe				
GOST 19282(73)	17GS	Weld const	0.14-0.2	0.3 max	0.3 max	1-1.4	0.3 max	0.035 max	0.04 max	0.4-0.6	Al <=0.1; bal Fe				
GOST 23570	18kp	Gen struct	0.14-0.22	0.3 max	0.3 max	0.3-0.6	0.3 max	0.04 max	0.05 max	0.05 max	Al <=0.1; bal Fe				
GOST 380(71)	BSt5ps	Gen struct; Semi-killed	0.28-0.37	0.3 max	0.3 max	0.5-0.8	0.3 max	0.04 max	0.05 max	0.05-0.17	Al <=0.1; N <=0.008; As<=0.08; bal Fe				
GOST 380(71)	BSt5ps2	Gen struct; Semi-killed	0.28-0.37	0.3 max	0.3 max	0.5-0.8	0.3 max	0.04 max	0.05 max	0.05-0.17	Al <=0.1; N <=0.008; As<=0.08; bal Fe				
GOST 380(71)	BSt5sp2	Gen struct; Non-rimming; FF	0.28-0.37	0.3 max	0.3 max	0.5-0.8	0.3 max	0.04 max	0.05 max	0.15-0.35	Al <=0.1; N <=0.008; As<=0.08; bal Fe				
GOST 380(71)	St3kp	Gen struct; Semi-Finished; Unkilled; Rimming; R; FU; 0-20mm	0.3 max	0.3 max	0.3 max	2 max	0.4 max	0.07 max	0.06 max	0.6 max	Al <=0.1; bal Fe	360-460		27L	
GOST 380(71)	St3kp	Gen struct; Semi-Finished; Unkilled; Rimming; R; FU; >40mm	0.3 max	0.3 max	0.3 max	2 max	0.4 max	0.07 max	0.06 max	0.6 max	Al <=0.1; bal Fe	360-460		24L	
GOST 380(71)	St3ps	Gen struct; Semi-Finished; Semi-killed; >40mm	0.3 max	0.3 max	0.3 max	2 max	0.4 max	0.07 max	0.06 max	0.6 max	Al <=0.1; bal Fe	370-480		23L	
GOST 380(71)	St3ps	Gen struct; Semi-Finished; Semi-killed; 0-20mm	0.3 max	0.3 max	0.3 max	2 max	0.4 max	0.07 max	0.06 max	0.6 max	Al <=0.1; bal Fe	370-480		26L	
GOST 380(71)	St5ps	Gen struct; Semi-Finished; Semi-killed; 0-20mm	0.3 max	0.3 max	0.3 max	2 max	0.4 max	0.07 max	0.06 max	0.6 max	Al <=0.1; bal Fe	490-630		20L	
GOST 380(71)	St5ps	Gen struct; Semi-Finished; Semi-killed; >40mm	0.3 max	0.3 max	0.3 max	2 max	0.4 max	0.07 max	0.06 max	0.6 max	Al <=0.1; bal Fe	490-630		17L	

Specification	Designation	Notes	C	Cr	Cu	Mn	Ni	P	S	Si	Other	UTS	YS	El	Hard

Carbon Steel, Nonresulfurized, A570(30) (Continued from previous page)

Russia

Specification	Designation	Notes	C	Cr	Cu	Mn	Ni	P	S	Si	Other	UTS	YS	El	Hard
GOST 380(71)	WSt2kp	Gen struct; Unkilled; Rimming; R; FU; 0-20mm	0.15 max	0.3 max	0.3 max	0.25-0.7	0.3 max	0.04 max	0.05 max	0.07 max	Al <=0.1; N <=0.008; As<=0.08; bal Fe	320-410		33L	
GOST 380(71)	WSt2kp	Gen struct; Unkilled; Rimming; R; FU; >40mm	0.15 max	0.3 max	0.3 max	0.25-0.7	0.3 max	0.04 max	0.05 max	0.07 max	Al <=0.1; N <=0.008; As<=0.08; bal Fe	320-410		30L	
GOST 380(71)	WSt2ps	Gen struct; Semi-killed; >40mm	0.15 max	0.3 max	0.3 max	0.25-0.7	0.3 max	0.04 max	0.05 max	0.05-0.17	Al <=0.1; N <=0.008; As<=0.08; bal Fe	330-430		29L	
GOST 380(71)	WSt2ps	Gen struct; Semi-killed stee; l0-20mm	0.15 max	0.3 max	0.3 max	0.25-0.7	0.3 max	0.04 max	0.05 max	0.05-0.17	Al <=0.1; N <=0.008; As<=0.08; bal Fe	330-430		32L	
GOST 380(71)	WSt2ps2	Gen struct; Semi-killed; >100mm	0.15 max	0.3 max	0.3 max	0.25-0.7	0.3 max	0.04 max	0.05 max	0.05-0.17	Al <=0.1; N <=0.008; As<=0.08; bal Fe	330-430	195	29L	
GOST 380(71)	WSt2sp	Gen struct; Non-rimming;K; FF; >40mm	0.15 max			0.25-0.7	0.3 max	0.04 max	0.05 max	0.12-0.3	N <=0.008; As<=0.08; bal Fe	330-430		-	
GOST 380(71)	WSt2sp	Gen struct; Non-rimming;K; FF; 0-20mm	0.15 max			0.25-0.7	0.3 max	0.04 max	0.05 max	0.12-0.3	N <=0.008; As<=0.08; bal Fe	330-430		32L	
GOST 380(71)	WSt3sp2	Gen struct; Non-rimming; FF; >100mm	0.22 max	0.3 max	0.3 max	0.4-0.85	0.3 max	0.04 max	0.05 max	0.12-0.3	Al <=0.1; N <=0.008; As<=0.08; bal Fe	370-480	205	23L	
GOST 380(71)	WSt3sp2	Gen struct; Non-rimming; FF; 0-20mm	0.22 max	0.3 max	0.3 max	0.4-0.85	0.3 max	0.04 max	0.05 max	0.12-0.3	Al <=0.1; N <=0.008; As<=0.08; bal Fe	370-480	245	26L	
GOST 380(71)	WSt3sp3	Gen struct; Non-rimming; FF; 0-20mm	0.22 max	0.3 max	0.3 max	0.4-0.85	0.3 max	0.04 max	0.05 max	0.12-0.3	Al <=0.1; N <=0.008; As<=0.08; bal Fe	370-480	245	26L	
GOST 380(71)	WSt3sp3	Gen struct; Non-rimming; FF; >100mm	0.22 max	0.3 max	0.3 max	0.4-0.85	0.3 max	0.04 max	0.05 max	0.12-0.3	Al <=0.1; N <=0.008; As<=0.08; bal Fe	370-480	205	23L	
GOST 380(88)	St3kp	Gen struct; Semi-Finished; Unkilled; R; FU	0.14-0.22	0.3 max	0.3 max	0.3-0.6	0.3 max	0.04 max	0.05 max	0.05 max	Al <=0.1; N <=0.008; As<=0.08; bal Fe				
GOST 380(88)	St3ps	Gen struct; Semi-Finished; Semi-killed	0.14-0.22	0.3 max	0.3 max	0.4-0.65	0.3 max	0.04 max	0.05 max	0.05-0.15	Al <=0.1; N <=0.008; As<=0.08; bal Fe				
GOST 380(88)	St5ps	Gen struct; Semi-Finished; Semi-killed	0.28-0.37	0.3 max	0.3 max	0.5-0.8	0.4 max	0.04 max	0.05 max	0.05-0.15	Al <=0.1; N <=0.008; As<=0.08; bal Fe				
GOST 380(88)	St5sp	Gen struct; Semi-Finished; Non-rimming; FF	0.28-0.37	0.3 max	0.3 max	0.5-0.8	0.4 max	0.04 max	0.05 max	0.15-0.3	Al <=0.1; N <=0.008; As<=0.08; bal Fe				
GOST 5781(82)	WSt3sp2; A-I	HR conc-reinf	0.22 max	0.3 max	0.3 max	0.4-0.85	0.3 max	0.04 max	0.05 max	0.12-0.3	Al <=0.1; N <=0.008; As<=0.08; bal Fe				

Spain

Specification	Designation	Notes	C	Cr	Cu	Mn	Ni	P	S	Si	Other	UTS	YS	El	Hard
UNE 36080(80)	A360B	Gen struct	0.2 max			1.60 max		0.05 max	0.05 max	0.60 max	Mo <=0.15; N <=0.009; bal Fe				
UNE 36080(80)	A360C	Gen struct	0.17 max			1.60 max		0.045 max	0.045 max	0.60 max	Mo <=0.15; N <=0.009; bal Fe				
UNE 36080(80)	A360D	Gen struct	0.17 max			1.60 max		0.04 max	0.04 max	0.60 max	Mo <=0.15; +N; bal Fe				
UNE 36080(85)	AE235-B	Gen struct	0.2 max			1.60 max		0.045 max	0.045 max	0.60 max	Mo <=0.15; N <=0.009; bal Fe				
UNE 36080(85)	AE235-C	Gen struct	0.17 max			1.60 max		0.04 max	0.04 max	0.60 max	Mo <=0.15; N <=0.009; bal Fe				
UNE 36080(85)	AE235-D	Gen struct	0.17 max			1.60 max		0.035 max	0.035 max	0.60 max	Mo <=0.15; bal Fe				
UNE 36080(85)	F.6202*	see AE235-B	0.2 max			1.60 max		0.045 max	0.045 max	0.60 max	Mo <=0.15; N <=0.009; bal Fe				

USA

Specification	Designation	Notes	C	Cr	Cu	Mn	Ni	P	S	Si	Other	UTS	YS	El	Hard
	UNS K02502		0.25 max			0.90 max		0.04 max	0.05 max		Cu 0.20 min (when spec'd); bal Fe				
ASTM A570/A570M(98)	30	HR Sh Strp	0.25 max			0.90 max		0.035 max	0.04 max		If Cu bearing reqd, Cu >=0.20; bal Fe	340	205	21.0-25.0	
ASTM A570/A570M(98)	33	HR Sh Strp	0.25 max			0.90 max		0.035 max	0.04 max		If Cu bearing reqd, Cu >=0.20; bal Fe	360	230	18.0-23.0	
ASTM A570/A570M(98)	36 Type 1	HR Sh Strp	0.25 max			0.90 max		0.035 max	0.04 max		If Cu bearing reqd, Cu >=0.20; bal Fe	365	250	17.0-22.0	
ASTM A570/A570M(98)	40	HR Sh Strp	0.25 max			0.90 max		0.035 max	0.04 max		If Cu bearing reqd, Cu >=0.20; bal Fe	380	275	16.0-21.0	
ASTM A709(97)	Grade 36	t<=100mm	0.26 max		0.20 min			0.04 max	0.05 max	0.40 max	bal Fe	400-550	250	23	

Yugoslavia

Specification	Designation	Notes	C	Cr	Cu	Mn	Ni	P	S	Si	Other	UTS	YS	El	Hard
	C.0370	Struct	0.2 max			1.60 max		0.06 max	0.06 max	0.60 max	bal Fe				

UNS numbers and US grades are provided as a means of cross referencing chemically similar alloys. Exchangability is only possible after independent examination of specifications. Tensile properties are minimum or typical as specified. UTS and YS as MPa. El as %. See Appendix for list of abbreviations used in Notes. * indicates obsolete material.

Specification	Designation	Notes	C	Cr	Cu	Mn	Ni	P	S	Si	Other	UTS	YS	El	Hard

Carbon Steel, Nonresulfurized, A573(58)

Czech Republic

| CSN 411378 | 11378 | Weld Const | 0.16 max | 0.3 max | | 1.60 max | | 0.045 max | 0.045 max | 0.60 max | bal Fe | | | | |

Hungary

MSZ 500(89)	Fe235C	Gen struct; QS; 0-1mm; HR/Frg	0.17 max			2 max		0.04 max	0.04 max	0.6 max	Al >=0.02; N <=0.009; bal Fe	360-510	235	17L; 15T	
MSZ 500(89)	Fe235C	Gen struct; QS; 150.1-200mm; HR/Frg	0.17 max			2 max		0.04 max	0.04 max	0.6 max	Al >=0.02; N <=0.009; bal Fe	320-470	185	21L; 21T	
MSZ 500(89)	Fe235D	Gen struct; specially killed; QS; 200.1-250mm; HR/Frg	0.17 max			2 max		0.035 max	0.035 max	0.6 max	bal Fe	320-470	175	21L; 21T	
MSZ 500(89)	Fe235D	Gen struct; specially killed; QS; 0-1mm; HR/Frg	0.17 max			2 max		0.035 max	0.035 max	0.6 max	bal Fe	360-510	235	17L; 15T	
MSZ 6280(82)	37C	Weld const; 40<t<=60mm; rolled/norm	0.16 max	0.25 max		1 max	0.3 max	0.04 max	0.04 max	0.15-0.5	Al 0.015-0.120; Mo <=0.1; Nb 0.01-0.06; Ti <=0.06; V 0.02-0.06; Zr=0.01-0.06; Al+V>=0.02; Al+Nb+V+Zr<=0.15; N<=Al/2+V/4+Nb/7+Ti/3.5<=0.015; bal Fe	360-460	215	26L	
MSZ 6280(82)	37C	Weld const; 3-40mm; rolled/norm	0.16 max	0.25 max		1 max	0.3 max	0.04 max	0.04 max	0.15-0.5	Al 0.015-0.120; Mo <=0.1; Nb 0.01-0.06; Ti <=0.06; V 0.02-0.06; Zr=0.01-0.06; Al+V>=0.02; Al+Nb+V+Zr<=0.15; N<=Al/2+V/4+Nb/7+Ti/3.5<=0.015; bal Fe	360-460	235	26L	

Italy

UNI 7070(72)	Fe37CFN	Gen struct	0.2 max			2 max		0.04 max	0.045 max	0.6 max	N <=0.009; bal Fe				
UNI 7070(82)	Fe360CFN	Gen struct	0.17 max			2 max		0.04 max	0.045 max	0.6 max	bal Fe				
UNI 7070(82)	Fe360DFF	Gen struct	0.17 max			2 max		0.04 max	0.04 max	0.6 max	Al <=0.015; bal Fe				

Romania

| STAS 500/2(88) | OL37.3k | Gen struct; FF; Non-rimming | 0.17 max | 0.3 max | | 0.8 max | | 0.045 max | 0.045 max | 0.4 max | Pb <=0.15; As<=0.08; bal Fe | | | | |

Spain

UNE 36080(80)	A360C	Gen struct	0.17 max			1.60 max		0.045 max	0.045 max	0.60 max	Mo <=0.15; N <=0.009; bal Fe				
UNE 36080(80)	A360D	Gen struct	0.17 max			1.60 max		0.04 max	0.04 max	0.60 max	Mo <=0.15; +N; bal Fe				
UNE 36080(85)	AE235-C	Gen struct	0.17 max			1.60 max		0.04 max	0.04 max	0.60 max	Mo <=0.15; N <=0.009; bal Fe				
UNE 36080(85)	AE235-D	Gen struct	0.17 max			1.60 max		0.035 max	0.035 max	0.60 max	Mo <=0.15; bal Fe				
UNE 36080(85)	F.6203*	see AE235-C	0.17 max			1.60 max		0.04 max	0.04 max	0.60 max	Mo <=0.15; N <=0.009; bal Fe				
UNE 36080(85)	F.6204*	see AE235-D	0.17 max			1.60 max		0.035 max	0.035 max	0.60 max	Mo <=0.15; bal Fe				

Sweden

| SS 141313 | 1313 | Gen struct; FF; Non-rimming | 0.2 max | | | 2 max | | 0.06 max | 0.05 max | 0.6 max | N<=0.009/N<=0.012; Si 0.25; Mn 0.5-0.8 | | | | |

USA

| | UNS K02301 | | 0.23 max | | | 0.60-0.90 | | 0.04 max | 0.05 max | 0.10-0.35 | bal Fe | | | | |
| ASTM A573/A753M(93) | 58 | Struct plt | 0.23 max | | | 0.60-0.90 | | 0.035 max | 0.04 max | 0.10-0.35 | bal Fe | 400-490 | 220 | 24 | |

Carbon Steel, Nonresulfurized, A67

USA

| | UNS K01505 | | 0.15 max | | | | | 0.05 max | | | Cu 0.20 min (when spec'd); bal Fe | | | | |

Specification	Designation	Notes	C	Cr	Cu	Mn	Ni	P	S	Si	Other	UTS	YS	El	Hard
Carbon Steel, Nonresulfurized, A67(2)															
USA															
	UNS K06002		0.35-0.85					0.05 max			Cu 0.20 min (when spec'd); bal Fe				
Carbon Steel, Nonresulfurized, A730(A)															
USA															
	UNS K01502		0.15 max			0.30-0.60		0.045 max	0.050 max		bal Fe				
ASTM A730(93)	Grade A	Frg for railway	0.15 max			0.30-0.60		0.045 max	0.050 max		bal Fe				
Carbon Steel, Nonresulfurized, A730(B)															
UK															
BS 3111/1(87)	0	Q/T Wir	0.25 max			0.25-1.00		0.050 max	0.050 max	0.40 max	bal Fe				
USA															
*	UNS K02000		0.15-0.25			0.30-0.60		0.045 max	0.050 max		bal Fe				
ASTM A730(93)	Grade B	Frg for railway	0.15-0.25			0.30-0.60		0.045 max	0.050 max		bal Fe				
Carbon Steel, Nonresulfurized, A759															
USA															
	UNS K07500		0.69-0.82			0.70-1.00		0.04 max	0.05 max	0.10-0.25	bal Fe				
ASTM A759(92)		Crane rails	0.67 max			0.70-1.00		0.04 max	0.05 max	0.10-0.50	bal Fe				
Carbon Steel, Nonresulfurized, B															
USA															
	UNS K06200		0.57-0.67			0.60-0.85		0.05 max	0.05 max	0.15 min	bal Fe				
ASTM A504(93)	Class B	Wheels	0.57-0.67			0.60-0.85		0.05 max	0.05 max	0.15 min	bal Fe				277-341 HB
Carbon Steel, Nonresulfurized, I															
Europe															
EN 10130(91)	1.0312	Sh Strp, CR t<=3mm	0.06 max			0.35 max		0.025 max	0.025 max		bal Fe	270-330	140	40	
EN 10130(91)	FeP05	Sh Strp, CR t<=3mm	0.06 max			0.35 max		0.025 max	0.025 max		bal Fe	270-330	140	40	
USA															
	UNS K00100	Enameling	0.008 max			0.60 max			0.040 max		bal Fe				
Carbon Steel, Nonresulfurized, IIA															
USA															
	UNS K00400	Enameling	0.04 max			0.12 max		0.015 max	0.040 max		bal Fe				
Carbon Steel, Nonresulfurized, IIB															
USA															
	UNS K00801	Enameling	0.08 max			0.20 max		0.015 max	0.040 max		bal Fe				
Carbon Steel, Nonresulfurized, K00600															
USA															
	UNS K00600 *	Special Magnetic Properties	0.06 max			0.40 max		0.015 max	0.015 max	0.20 max	Al <=0.015; bal Fe				
Carbon Steel, Nonresulfurized, K00800															
India															
IS 1570/2(79)	4C2	Sh Plt Sect Shp Bar Bil Frg	0.08 max	0.3 max	0.3 max	0.4 max	0.4 max	0.055 max	0.055 max	0.6 max	Co <=0.1; Mo <=0.15; Pb <=0.15; W <=0.1; P:S Varies; bal Fe				

Specification	Designation	Notes	C	Cr	Cu	Mn	Ni	P	S	Si	Other	UTS	YS	El	Hard

Carbon Steel, Nonresulfurized, K00800 (Continued from previous page)

India

| IS 1570/2(79) | C04 | Sh Plt Sect Shp Bar Bil Frg | 0.08 max | 0.3 max | 0.3 max | 0.4 max | 0.4 max | 0.055 max | 0.055 max | 0.6 max | Co <=0.1; Mo <=0.15; Pb <=0.15; W <=0.1; P:S Varies; bal Fe | | | | |

USA

| | UNS K00800 | Special Magnetic Properties | 0.08 max | | | 0.40 max | | 0.025 max | 0.025 max | 0.20 max | Al <=0.02; bal Fe | | | | |

Carbon Steel, Nonresulfurized, K01000

Hungary

MSZ 4213(85)	ASZ1K32	CR Strp; 0-3mm; CRR (K32)	0.12 max			0.6 max		0.05 max	0.05 max	0.60 max	Al <=0.120; bal Fe	310-450			
MSZ 4213(85)	ASZ1K40	CR Strp; mild; 0-3mm; CRR (K40)	0.12 max			0.6 max		0.05 max	0.05 max	0.60 max	Al <=0.120; bal Fe	390-540			
MSZ 4213(85)	ASZ1K50	CR Strp; mild; 0-2mm; CRR (K50)	0.12 max			0.6 max		0.05 max	0.05 max	0.60 max	Al <=0.120; bal Fe	490-640			
MSZ 4213(85)	ASZ1K60	CR Strp; mild; 0-1.5mm; CRR (K60)	0.12 max			0.6 max		0.05 max	0.05 max	0.60 max	Al <=0.120; bal Fe	590			
MSZ 4213(85)	ASZ1L	CR Strp; mild; 0-3mm; soft ann	0.12 max			0.6 max		0.05 max	0.05 max	0.60 max	Al <=0.120; bal Fe	0-410		28L	
MSZ 4213(85)	ASZ1LS	CR Strp; mild; 0-3mm; soft ann; skin rolled	0.12 max			0.6 max		0.05 max	0.05 max	0.60 max	Al <=0.120; bal Fe	0-440		26L	
MSZ 4213(85)	ASZ2K32	CR Strp; mild; 0-3mm; CRR (K32)	0.12 max			0.5 max		0.04 max	0.05 max	0.2 max	Al <=0.120; bal Fe	290-430		18L	
MSZ 4213(85)	ASZ2K40	CR Strp; mild; 0-3mm; CRR (K40)	0.12 max			0.5 max		0.04 max	0.05 max	0.2 max	Al <=0.120; bal Fe	390-540		4L	
MSZ 4213(85)	ASZ2K50	CR Strp; mild; 0-2mm; CRR (K50)	0.12 max			0.5 max		0.04 max	0.05 max	0.2 max	Al <=0.120; bal Fe	490-640			
MSZ 4213(85)	ASZ2K60	CR Strp; mild; 0-1.5mm; CRR (K60)	0.12 max			0.5 max		0.04 max	0.05 max	0.2 max	Al <=0.120; bal Fe	590-740			
MSZ 4213(85)	ASZ2K70	CR Strp; mild; 0-1mm; CRR (K70)	0.12 max			0.5 max		0.04 max	0.05 max	0.2 max	Al <=0.120; bal Fe	690			
MSZ 4213(85)	ASZ2L	CR Strp; mild; 0-3mm; soft ann	0.12 max			0.5 max		0.04 max	0.05 max	0.2 max	Al <=0.120; bal Fe	270-390		28L	
MSZ 4213(85)	ASZ2LS	CR Strp; mild; 0-3mm; soft ann, skin rolled	0.12 max			0.5 max		0.04 max	0.05 max	0.2 max	Al <=0.120; bal Fe	270-410		28L	
MSZ 4213(85)	ASZ3K32	CR Strp; mild; 0-3mm; CRR (K32)	0.1 max			0.45 max		0.03 max	0.04 max	0.12 max	Al <=0.120; bal Fe	310-410		22L	
MSZ 4213(85)	ASZ3K40	CR Strp; mild; 0-3mm; CRR (K40)	0.1 max			0.45 max		0.03 max	0.04 max	0.12 max	Al <=0.120; bal Fe	390-490		5L	
MSZ 4213(85)	ASZ3K50	CR Strp; mild; 0-2mm; CRR (K50)	0.1 max			0.45 max		0.03 max	0.04 max	0.12 max	Al <=0.120; bal Fe	490-590			
MSZ 4213(85)	ASZ3K60	CR Strp; mild; 0-1.5mm; CRR (K60)	0.1 max			0.45 max		0.03 max	0.04 max	0.12 max	Al <=0.120; bal Fe	590-690			
MSZ 4213(85)	ASZ3L	CR Strp; mild; 0-3mm; soft ann	0.1 max			0.45 max		0.03 max	0.04 max	0.12 max	Al <=0.120; bal Fe	270-370		32L	
MSZ 4213(85)	ASZ3LS	CR Strp; mild; 0-3mm; soft ann, skin rolled	0.1 max			0.45 max		0.03 max	0.04 max	0.12 max	Al <=0.120; bal Fe	290-390		32L	
MSZ 4213(85)	ASZ4K32	CR Strp; mild; ageing res; 0-3mm; CRR (K32)	0.8 max			0.45 max		0.03 max	0.035 max	0.1 max	Al 0.025-0.120; bal Fe	290-390		24L	
MSZ 4213(85)	ASZ4K40	CR Strp; mild; ageing res; 0-3mm; CRR (K40)	0.8 max			0.45 max		0.03 max	0.035 max	0.1 max	Al 0.025-0.120; bal Fe	390-490		6L	
MSZ 4213(85)	ASZ4K50	CR Strp; mild; ageing res; 0-2mm; CRR (K50)	0.8 max			0.45 max		0.03 max	0.035 max	0.1 max	Al 0.025-0.120; bal Fe	490-590			
MSZ 4213(85)	ASZ4K60	CR Strp; mild; ageing res; 0-1.5mm; CRR (K60)	0.8 max			0.45 max		0.03 max	0.035 max	0.1 max	Al 0.025-0.120; bal Fe	590-690			
MSZ 4213(85)	ASZ4L	CR Strp; mild; ageing res; 0-3mm; soft ann	0.8 max			0.45 max		0.03 max	0.035 max	0.1 max	Al 0.025-0.120; bal Fe	270-350		36L	
MSZ 4213(85)	ASZ4LS	CR Strp; mild; ageing res; 0-3mm; soft ann, skin rolled	0.8 max			0.45 max		0.03 max	0.035 max	0.1 max	Al 0.025-0.120; bal Fe	270-350		36L	

Spain

| UNE 36032(85) | 8E-DF | Screw | 0.05-0.10 | | | 0.30-0.60 | | 0.040 max | 0.040 max | 0.60 max | N <=0.008; bal Fe | | | | |

UNS numbers and US grades are provided as a means of cross referencing chemically similar alloys. Exchangability is only possible after independent examination of specifications. Tensile properties are minimum or typical as specified. UTS and YS as MPa. El as %. See Appendix for list of abbreviations used in Notes. * indicates obsolete material.

Specification	Designation	Notes	C	Cr	Cu	Mn	Ni	P	S	Si	Other	UTS	YS	El	Hard

Carbon Steel, Nonresulfurized, K01000 (Continued from previous page)

Spain

Specification	Designation	Notes	C	Cr	Cu	Mn	Ni	P	S	Si	Other	UTS	YS	El	Hard
UNE 36032(85)	8KA-DF	Screw	0.05-0.10			0.30-0.60		0.040 max	0.040 max	0.10 max	Al >=0.020; bal Fe				
UNE 36032(85)	F.7502*	see 8E-DF	0.05-0.10			0.30-0.60		0.040 max	0.040 max	0.60 max	N <=0.008; bal Fe				
UNE 36032(85)	F.7512*	see 8KA-DF	0.05-0.10			0.30-0.60		0.040 max	0.040 max	0.10 max	Al >=0.020; bal Fe				

UK

Specification	Designation	Notes	C	Cr	Cu	Mn	Ni	P	S	Si	Other	UTS	YS	El	Hard
BS 1502(82)	509-650	Bar Shp Press ves	0.10 max	0.25 max		0.30-0.80	8.50-10.0	0.025 max	0.020 max	0.15-0.35	Al >=0.015; Co <=0.1; Mo <=0.10; W <=0.1; bal Fe				

USA

Specification	Designation	Notes	C	Cr	Cu	Mn	Ni	P	S	Si	Other	UTS	YS	El	Hard
	UNS K01000	Special Magnetic Properties	0.10 max			0.60 max		0.04 max	0.04 max	0.20 max	Al <=0.02; bal Fe				

Carbon Steel, Nonresulfurized, K01500

Europe

Specification	Designation	Notes	C	Cr	Cu	Mn	Ni	P	S	Si	Other	UTS	YS	El	Hard
EN 10120(96)	1.0111	Weld gas cylinder; norm; 3<t<=5mm	0.16 max			0.30 min		0.025 max	0.015 max	0.25 max	Al >=0.020; N <=0.009; Nb <=0.050; Ti <=0.030; bal Fe	360-450	245	34	
EN 10120(96)	1.0111	Weld gas cylinder; norm; t<=3mm	0.16 max			0.30 min		0.025 max	0.015 max	0.25 max	Al >=0.020; N <=0.009; Nb <=0.050; Ti <=0.030; bal Fe	360-450	245	26	
EN 10120(96)	1.0423	Weld gas cylinder; norm; 3<t<=5mm	0.19 max			0.40 min		0.025 max	0.015 max	0.25 max	Al >=0.020; N <=0.009; Nb <=0.050; Ti <=0.030; bal Fe	410-500	265	32	
EN 10120(96)	1.0423	Weld gas cylinder; norm; t<=3mm	0.19 max			0.40 min		0.025 max	0.015 max	0.25 max	Al >=0.020; N <=0.009; Nb <=0.050; Ti <=0.030; bal Fe	410-500	265	24	
EN 10120(96)	P245NB	Weld gas cylinder; norm; t<=3mm	0.16 max			0.30 min		0.025 max	0.015 max	0.25 max	Al >=0.020; N <=0.009; Nb <=0.050; Ti <=0.030; bal Fe	360-450	245	26	
EN 10120(96)	P245NB	Weld gas cylinder; norm; 3<t<=5mm	0.16 max			0.30 min		0.025 max	0.015 max	0.25 max	Al >=0.020; N <=0.009; Nb <=0.050; Ti <=0.030; bal Fe	360-450	245	34	
EN 10120(96)	P265NB	Weld gas cylinder; norm; t<=3mm	0.19 max			0.40 min		0.025 max	0.015 max	0.25 max	Al >=0.020; N <=0.009; Nb <=0.050; Ti <=0.030; bal Fe	410-500	265	24	
EN 10120(96)	P265NB	Weld gas cylinder; norm; 3<t<=5mm	0.19 max			0.40 min		0.025 max	0.015 max	0.25 max	Al >=0.020; N <=0.009; Nb <=0.050; Ti <=0.030; bal Fe	410-500	265	32	

France

Specification	Designation	Notes	C	Cr	Cu	Mn	Ni	P	S	Si	Other	UTS	YS	El	Hard
AFNOR NFA36211(90)	BS1*	t<3mm; Norm	0.16 max			0.3 max		0.025 max	0.02 max	0.2 max	Al >=0.015; N <=0.009; Nb <=0.050; bal Fe	360-430	245	26	
AFNOR NFA36211(90)	BS2*	t<3mm; Norm	0.19 max			0.4 max		0.025 max	0.02 max	0.2 max	Al >=0.015; N <=0.009; Nb <=0.050; bal Fe	410-480	265	24	

USA

Specification	Designation	Notes	C	Cr	Cu	Mn	Ni	P	S	Si	Other	UTS	YS	El	Hard
	UNS K01500	Special Magnetic Properties	0.15 max			0.50 max		0.04 max	0.04 max	0.20 max	Al <=0.02; bal Fe				

Carbon Steel, Nonresulfurized, K02603

USA

Specification	Designation	Notes	C	Cr	Cu	Mn	Ni	P	S	Si	Other	UTS	YS	El	Hard
	UNS K02603		0.26 max			0.60-0.90		0.035 max	0.040 max		bal Fe				

Carbon Steel, Nonresulfurized, K03016

USA

Specification	Designation	Notes	C	Cr	Cu	Mn	Ni	P	S	Si	Other	UTS	YS	El	Hard
	UNS K03016		0.30 max			0.60-0.90		0.04 max	0.05 max	0.15-0.30	bal Fe				

UNS numbers and US grades are provided as a means of cross referencing chemically similar alloys. Exchangability is only possible after independent examination of specifications. Tensile properties are minimum or typical as specified. UTS and YS as MPa. El as %. See Appendix for list of abbreviations used in Notes. * indicates obsolete material.

Specification	Designation	Notes	C	Cr	Cu	Mn	Ni	P	S	Si	Other	UTS	YS	El	Hard
Carbon Steel, Nonresulfurized, MIL-S-22698															
USA															
	UNS K02708		0.24-0.30			0.60-0.90		0.04 max	0.05 max	0.15-0.30	bal Fe				
Carbon Steel, Nonresulfurized, MIL-S-3039															
USA															
	UNS K04600		0.43-0.50			0.45-0.75		0.030 max	0.035 max	0.15-0.35	bal Fe				
Carbon Steel, Nonresulfurized, MIL-S-3289															
USA															
	UNS K02901		0.26-0.33			0.60-0.90		0.040 max	0.040 max	0.10 max	bal Fe				
Carbon Steel, Nonresulfurized, MIL-S-645															
Spain															
UNE 36032(85)	16KA-DF	Screw, bolt	0.12-0.19			0.30-0.60		0.040 max	0.040 max	0.10 max	Al >=0.020; bal Fe				
UNE 36032(85)	F.7515*	see 16KA-DF	0.12-0.19			0.30-0.60		0.040 max	0.040 max	0.10 max	Al >=0.020; bal Fe				
USA															
	UNS K01602		0.13-0.19			0.30-0.60		0.04 max	0.05 max	0.10 max	bal Fe				
Carbon Steel, Nonresulfurized, MIL-W-8957															
USA															
	UNS K08700		0.85-0.90	0.04 max	0.08 max	0.25-0.40	0.03 max	0.025 max	0.020 max	0.15-0.25	Mo <=0.04; bal Fe				
Carbon Steel, Nonresulfurized, No equivalents identified															
Czech Republic															
CSN 411364	11364	High-temp const	0.2 max	0.3 max		0.6 max	0.3 max	0.045 max	0.045 max	0.60 max	bal Fe				
CSN 411368	11368	High-temp const	0.15 max	0.3 max		0.65 max	0.3 max	0.04 max	0.04 max	0.35 max	bal Fe				
CSN 411416	11416	High-temp const	0.21 max	0.3 max		0.65 max	0.3 max	0.045 max	0.04 max	0.35 max	bal Fe				
CSN 411418	11418	High-temp const	0.2 max	0.3 max		0.7 max	0.3 max	0.04 max	0.04 max	0.35 max	bal Fe				
CSN 411443	11443		0.2 max	0.3 max		1.60 max		0.05 max	0.05 max	0.60 max	P+S<=0.085; bal Fe				
France															
AFNOR NFA35556(84)	38B3	t<16mm; Q/T	0.34-0.4			0.3 max 0.6-0.9		0.035 max	0.035 max	0.1-0.4	B 0.0008-0.005; bal Fe	860-1070	660	11	
Hungary															
MSZ 120/3(82)	A34	Tub; screwing; 2-4mm; weld	0.15 max			2 max	0.3 max	0.045 max	0.05 max	0.6 max	bal Fe	330-430	215	32L	
MSZ 17(86)	A37B	Tub; w/special req; specially killed; 0-16mm; HF/CF	0.17 max			1 max		0.04 max	0.04 max	0.1-0.35	Al 0.015-0.10; Nb 0.01-0.06; Ti 0.02-0.06; V 0.02-0.1; Zr=0.015-0.06; Al+V>=0.02; Al+Nb+V+Ti+Zr<=0.15; bal Fe	350-480	235	25L	
MSZ 17(86)	A37B	Tub; w/special req; specially killed; >40mm; HF/CF	0.17 max			1 max		0.04 max	0.04 max	0.1-0.35	Al 0.015-0.10; Nb 0.01-0.06; Ti 0.02-0.06; V 0.02-0.1; Zr=0.015-0.06; Al+V>=0.02; Al+Nb+V+Ti+Zr<=0.15; bal Fe	350-480	215	25L	
MSZ 17(86)	A37C	Tub; w/special req; specially killed; 0-16mm; HF/CF	0.17 max			1 max		0.04 max	0.04 max	0.1-0.35	Al 0.015-0.10; Nb 0.01-0.06; Ti 0.02-0.06; V 0.02-0.1; Zr=0.015-0.06; Al+V>=0.02; Al+Nb+V+Ti+Zr<=0.15; bal Fe	350-480	235	25L	

UNS numbers and US grades are provided as a means of cross referencing chemically similar alloys. Exchangability is only possible after independent examination of specifications. Tensile properties are minimum or typical as specified. UTS and YS as MPa. El as %. See Appendix for list of abbreviations used in Notes. * indicates obsolete material.

Specification	Designation	Notes	C	Cr	Cu	Mn	Ni	P	S	Si	Other	UTS	YS	El	Hard

Carbon Steel, Nonresulfurized, No equivalents identified

Hungary

Specification	Designation	Notes	C	Cr	Cu	Mn	Ni	P	S	Si	Other	UTS	YS	El	Hard
MSZ 17(86)	A37C	Tub; w/special req; specially killed; >40mm; HF/CF	0.17 max			1 max		0.04 max	0.04 max	0.1-0.35	Al 0.015-0.10; Nb 0.01-0.06; Ti 0.02-0.06; V 0.02-0.1; Zr=0.015-0.06; Al+V>=0.02; Al+Nb+V+Ti+Zr<=0.15; bal Fe	350-480	215	25L	
MSZ 17(86)	A44C	Tub; w/special req; specially killed; 0-16mm; HF/CF	0.2 max			1.3 max		0.04 max	0.04 max	0.1-0.35	Al 0.015-0.10; Nb 0.01-0.06; Ti 0.02-0.06; V 0.02-0.1; Zr=0.015-0.06; Al+Nb+V+Ti+Zr<=0.15; bal Fe	420-550	275	21L	
MSZ 17(86)	A44C	Tub; w/special req; specially killed; >40mm; HF/CF	0.2 max			1.3 max		0.04 max	0.04 max	0.1-0.35	Al 0.015-0.10; Nb 0.01-0.06; Ti 0.02-0.06; V 0.02-0.1; Zr=0.015-0.06; Al+Nb+V+Ti+Zr<=0.15; bal Fe	420-550	255	21L	
MSZ 17(86)	A52B	Tub; w/special req; specially killed; >40mm; HF/CF	0.22 max			1.6 max		0.04 max	0.035 max	0.1-0.55	Al 0.015-0.10; Nb 0.01-0.06; Ti 0.02-0.06; V 0.02-0.1; Zr=0.015-0.06; Al+Nb+V+Ti+Zr<=0.15; bal Fe	500-650	335	21L	
MSZ 17(86)	A52B	Tub; w/special req; specially killed; 0-16mm; HF/CF	0.22 max			1.6 max		0.04 max	0.035 max	0.1-0.55	Al 0.015-0.10; Nb 0.01-0.06; Ti 0.02-0.06; V 0.02-0.1; Zr=0.015-0.06; Al+Nb+V+Ti+Zr<=0.15; bal Fe	500-650	355	21L	
MSZ 17(86)	A52C	Tub; w/special req; specially killed; >40mm; HF/CF	0.2 max			1.6 max		0.04 max	0.035 max	0.1-0.55	Al 0.015-0.10; Nb 0.01-0.06; Ti 0.02-0.06; V 0.02-0.1; Zr=0.015-0.06; Al+Nb+V+Ti+Zr<=0.15; bal Fe	500-650	335	21L	
MSZ 17(86)	A52C	Tub; w/special req; specially killed; 0-16mm; HF/CF	0.2 max			1.6 max		0.04 max	0.035 max	0.1-0.55	Al 0.015-0.10; Nb 0.01-0.06; Ti 0.02-0.06; V 0.02-0.1; Zr=0.015-0.06; Al+Nb+V+Ti+Zr<=0.15; bal Fe	500-650	355	21L	
MSZ 186/2(82)	52C	Longitudinal weld tub; weld const; 3-12.5mm	0.2 max	0.25 max		1.5 max	0.3 max	0.04 max	0.04 max	0.15-0.5	Al 0.015-0.10; Mo <=0.1; Nb 0.01-0.06; Ti 0.02-0.06; V 0.02-0.1; Zr=0.015-0.06; Al+Nb+V+Ti+Zr <=0.15; N<=Al/2+V/4 +Nb/7+Ti/3.5<=0.015; bal Fe	490-640	355	22L	
MSZ 382/2(83)	FK	Longitudinal weld tub; weld precision tub w/special accuracy; 0.8-2.5mm; untreated, CR (bright, hard)	0.15 max			0.2-0.5		0.05 max	0.05 max	0.12 max	bal Fe	440		10L	
MSZ 382/2(83)	FL	Mild; weld precision tubes w/special accuracy; 0.8-2.5mm; untreated (bright, soft)	0.15 max			0.2-0.5		0.05 max	0.05 max	0.12 max	bal Fe	310-490		14L	
MSZ 500(81)	A50	Gen struct; Non-rimming; FF; 0-16mm; HR/Frg	0.3 max			2 max		0.045 max	0.05 max	0.12 min	bal Fe	490-640	295	21L	
MSZ 500(81)	A50	Gen struct; Non-rimming; FF; 40.1-100mm; HR/Frg	0.3 max			2 max		0.045 max	0.05 max	0.12 min	bal Fe	490-640	275	21L	
MSZ 500(81)	A60	Gen struct; Non-rimming; FF; 40.1-100mm; HR/Frg	0.3 max			2 max		0.045 max	0.05 max	0.12 min	bal Fe	590-740	315	16L	
MSZ 500(81)	A60	Gen struct; Non-rimming; FF; 0-16mm; HR/Frg	0.3 max			2 max		0.045 max	0.05 max	0.12 min	bal Fe	590-740	335	16L	

Specification	Designation	Notes	C	Cr	Cu	Mn	Ni	P	S	Si	Other	UTS	YS	El	Hard

Carbon Steel, Nonresulfurized, No equivalents identified

Hungary

Specification	Designation	Notes	C	Cr	Cu	Mn	Ni	P	S	Si	Other	UTS	YS	El	Hard
MSZ 500(81)	A70	Gen struct; Non-rimming; FF; 0-100mm; HR/Frg	0.3 max			2 max		0.045 max	0.05 max	0.12 min	bal Fe	670		10L	
MSZ 500(81)	B0	Gen struct	0.23 max			2 max		0.07 max	0.06 max	0.6 max	bal Fe				
MSZ 500(81)	B38B	Gen struct; Non-rimming; FF	0.13-0.2			0.4-0.65		0.045 max	0.05 max	0.12-0.3	bal Fe				
MSZ 500(81)	B50	Gen struct; Non-rimming; FF	0.28-0.37			0.5-0.8		0.045 max	0.05 max	0.12-0.35	bal Fe				
MSZ 500(89)	A0	Gen struct; Base; BS; 16.1-25mm; HR/Frg	0.3 max			2 max		0.07 max	0.06 max	0.6 max	bal Fe	290-510	175	18L; 16T	
MSZ 500(89)	A0	Gen struct; Base; BS; 0-1mm; HR/Frg	0.3 max			2 max		0.07 max	0.06 max	0.6 max	bal Fe	310-540	185	10L; 8T	
MSZ 5750(86)	A60	Crane rail; 40.1-100mm; HR/Frg	0.3 max			2 max		0.045 max	0.05 max	0.12-0.60	bal Fe	590-740	315	16L	
MSZ 5750(86)	A60	Crane rail; 0-16mm; HR/Frg	0.3 max			2 max		0.045 max	0.05 max	0.12-0.60	bal Fe	590-740	335	16L	
MSZ 5750(86)	DS	Crane rail; HR	0.53-0.73			0.6-1		0.045 max	0.045 max	0.15-0.35	bal Fe	730	370	10L	210 HB
MSZ 6280(82)	37D	Weld const; 40<t<=60mm; rolled/norm	0.16 max	0.25 max		1 max	0.3 max	0.035 max	0.035 max	0.15-0.5	Al 0.015-0.120; Mo <=0.1; Nb 0.01-0.06; Ti 0.02-0.06; V 0.02-0.1; Zr=0.01-0.06; Al+V>=0.02; Al+Nb+V+Zr<=0.15; N<=Al/2+V/4+Nb/7+Ti/3.5<=0.015; bal Fe	360-460	215	26L	
MSZ 6280(82)	37D	Weld const; 3-40mm; rolled/norm	0.16 max	0.25 max		1 max	0.3 max	0.035 max	0.035 max	0.15-0.5	Al 0.015-0.120; Mo <=0.1; Nb 0.01-0.06; Ti 0.02-0.06; V 0.02-0.1; Zr=0.01-0.06; Al+V>=0.02; Al+Nb+V+Zr<=0.15; N<=Al/2+V/4+Nb/7+Ti/3.5<=0.015; bal Fe	360-460	235	26L	
MSZ 982(87)	BHB55.50	CF conc wire; 6-12mm; CF	0.18 max			0.6 max		0.05 max	0.05 max	0.35 max	bal Fe	560	500	10L	
MSZ 982(87)	BHS55.50	CF conc wire; 4-12mm; CF	0.18 max			0.6 max		0.05 max	0.05 max	0.35 max	bal Fe	560	500	10L	

International

Specification	Designation	Notes	C	Cr	Cu	Mn	Ni	P	S	Si	Other	UTS	YS	El	Hard
ISO 2604-2(75)	TS26	Smls Tub	0.12-0.2	0.3 max	0.3 max	0.4-0.8	0.4 max	0.04 max	0.04 max	0.1-0.35	Al <=0.012; Co <=0.1; Mo 0.25-0.35; Pb <=0.15; Ti <=0.05; V <=0.1; W <=0.1; bal Fe				
ISO 2604-3(75)	TW1	Weld Tub for press	0.16 max			0.30-0.70		0.050 max	0.050 max		Bal Fe	320-440	195	25	
ISO 2604-3(75)	TW10	Weld Tub for press	0.19 max			0.60-1.20		0.045 max	0.045 max	0.35 max	Al <=0.015; Bal Fe	410-530	235	22	
ISO 2604-3(75)	TW13	Weld Tub for press	0.22 max			0.60-1.40		0.045 max	0.045 max	0.35 max	Bal Fe	460-580	265	21	
ISO 2604-3(75)	TW14	Weld Tub for press	0.22 max			0.80-1.40		0.045 max	0.045 max	0.35 max	Bal Fe	460-580	265	21	
ISO 2604-3(75)	TW15	Weld Tub for press	0.20 max			0.80-1.40		0.045 max	0.045 max	0.35 max	Al <=0.015; Bal Fe	460-580	265	21	
ISO 2604-3(75)	TW2	Weld Tub for press	0.16 max			0.30-0.70		0.050 max	0.050 max		Bal Fe	320-440	195	25	
ISO 2604-3(75)	TW26	Weld Tub for press	0.12-0.20			0.40-0.80		0.040 max	0.040 max	0.10-0.35	Al <=0.012; Mo 0.25-0.35; Bal Fe	450-800	250	22	
ISO 2604-3(75)	TW32	Weld Tub for press	0.10-0.18	0.70-1.10		0.40-0.70		0.040 max	0.040 max	0.10-0.35	Al <=0.020; Mo 0.45-0.65; Bal Fe	440-590	275	22	
ISO 2604-3(75)	TW4	Weld Tub for press	0.17 max			0.40-0.80		0.045 max	0.045 max	0.35 max	Bal Fe	360-480	215	24	
ISO 2604-3(75)	TW5	Weld Tub for press	0.17 max			0.40-0.80		0.045 max	0.045 max	0.35 max	Bal Fe	360-480	215	24	
ISO 2604-3(75)	TW6	Weld Tub for press	0.17 max			0.40-1.00		0.045 max	0.045 max	0.35 max	Al <=0.015; Bal Fe	360-480	215	24	
ISO 2604-3(75)	TW9	Weld Tub for press	0.21 max			0.40-1.20		0.045 max	0.045 max	0.35 max	Bal Fe	410-530	235	22	

Specification	Designation	Notes	C	Cr	Cu	Mn	Ni	P	S	Si	Other	UTS	YS	El	Hard

Carbon Steel, Nonresulfurized, No equivalents identified

International

Specification	Designation	Notes	C	Cr	Cu	Mn	Ni	P	S	Si	Other	UTS	YS	El	Hard
ISO 2604-3(75)	TW9H	Weld Tub for press	0.21 max			0.40-1.20		0.045 max	0.045 max	0.35 max	Bal Fe	410-530	235	22	
ISO 2938(74)	1	Hollow bar, HF, t<=16mm	0.20 max			1.6 max		0.045 max	0.045 max	0.50 max	bal Fe	490-610	335	21	
ISO 3304(85)	R28	Smls Tub, CF	0.10 max			0.30 max		0.040 max	0.040 max		bal Fe	400		8	
ISO 3304(85)	R33	Smls Tub, CF	0.16 max			0.70 max		0.050 max	0.040 max		bal Fe	420		6	
ISO 3305(85)	R28	Weld Tub, CF	0.13 max			0.60 max		0.050 max	0.050 max		Co <=0.1; Pb <=0.15; W <=0.1; bal Fe	400		8	
ISO 3305(85)	R44	Weld Tub, CF	0.21 max			1.2 max		0.050 max	0.050 max	0.35 max	Co <=0.1; Pb <=0.15; W <=0.1; bal Fe	520		5	
ISO 3305(85)	R50	Weld Tub, CF	0.23 max			1.6 max		0.050 max	0.050 max	0.55 max	Co <=0.1; Pb <=0.15; W <=0.1; bal Fe	600		4	
ISO 3573(99)	HR1	Sh, HR, t=3mm, comm	0.12 max			0.60 max		0.045 max	0.045 max		bal Fe	440		24	
ISO 3573(99)	HR2	Sh, HR, t=3mm, DS	0.10 max			0.45 max		0.035 max	0.035 max		bal Fe	420		26	
ISO 3573(99)	HR4	Sh, HR, t=3mm, DD, FF	0.08 max			0.35 max		0.025 max	0.025 max		bal Fe	380		32	
ISO 3574(99)	CR1	Sh, comm	0.15 max			0.60 max		0.05 max	0.05 max		Pb <=0.15;	410 max	280 max	28	
ISO 3574(99)	CR4	Sh, DDS, FF	0.08 max			0.45 max		0.03 max	0.03 max		Co <=0.1; Pb <=0.15; W <=0.1; bal Fe	350 max	210 max	37 min	
ISO 3574(99)	CR5	Sh, extra DDS, stabalized	0.02 max			0.25 max		0.02 max	0.02 max		Ti <=0.3; bal Fe	350 max	190 max	38 min	
ISO 4954(93)	35MnB5E	Wir rod Bar, CH ext, boron	0.32-0.39			1.10-1.40		0.035 max	0.035 max	0.40 max	Al >=0.020; B 0.0008-0.005; bal Fe	600			
ISO 4954(93)	CC11A	Wir rod Bar, CH ext	0.08-0.13			0.30-0.60		0.040 max	0.040 max	0.10 max	Al >=0.020; bal Fe				
ISO 4954(93)	CC11X	Wir rod Bar, CH ext	0.08-0.13			0.30-0.60		0.040 max	0.040 max	0.10 max	Al <=0.020; bal Fe				
ISO 4954(93)	CC15A	Wir rod Bar, CH ext	0.12-0.19			0.30-0.60		0.040 max	0.040 max	0.10 max	Al >=0.020; bal Fe				
ISO 4954(93)	CC15K	Wir rod Bar, CH ext	0.12-0.19			0.30-0.60		0.040 max	0.040 max	0.15-0.35	bal Fe				
ISO 4954(93)	CC15X	Wir rod Bar, CH ext	0.12-0.19			0.30-0.60		0.040 max	0.040 max	0.10 max	Al <=0.020; bal Fe				
ISO 4954(93)	CC21A	Wir rod Bar, CH ext	0.18-0.23			0.30-0.60		0.040 max	0.040 max	0.10 max	Al >=0.020; bal Fe				
ISO 4954(93)	CC21K	Wir rod Bar, CH ext	0.18-0.23			0.30-0.60		0.040 max	0.040 max	0.15-0.35	bal Fe				
ISO 4954(93)	CC4A	Wir rod Bar, CH ext	0.06 max			0.20-0.40		0.040 max	0.040 max	0.10 max	Al >=0.020; bal Fe				
ISO 4954(93)	CC4X	Wir rod Bar, CH ext	0.06 max			0.20-0.40		0.040 max	0.040 max	0.10 max	Al <=0.020; bal Fe				
ISO 4954(93)	CC8A	Wir rod Bar, CH ext	0.05-0.10			0.30-0.60		0.040 max	0.040 max	0.10 max	Al >=0.020; bal Fe				
ISO 4954(93)	CC8X	Wir rod Bar, CH ext	0.05-0.10			0.30-0.60		0.040 max	0.040 max	0.10 max	Al <=0.020; bal Fe				
ISO 4954(93)	CE10	Wir rod Bar, CH ext	0.07-0.13			0.30-0.60		0.035 max	0.035 max	0.40 max	bal Fe	450			
ISO 4954(93)	CE15E4	Wir rod Bar, CH ext	0.12-0.18			0.30-0.60		0.035 max	0.035 max	0.40 max	bal Fe	470			
ISO 4954(93)	CE16E4	Wir rod Bar, CH ext	0.12-0.18			0.60-0.90		0.035 max	0.035 max	0.40 max	bal Fe	490			
ISO 4954(93)	CE20BG1	Wir rod Bar, CH ext, boron	0.17-0.24			0.50-0.80		0.035 max	0.035 max	0.40 max	Al >=0.020; B 0.0008-0.005; bal Fe	500			
ISO 4954(93)	CE20BG2	Wir rod Bar, CH ext, boron	0.17-0.24			0.80-1.20		0.035 max	0.035 max	0.40 max	Al >=0.020; B 0.0008-0.005; bal Fe	520			
ISO 4954(93)	CE20E4	Wir rod Bar, CH ext	0.17-0.23			0.30-0.60		0.035 max	0.035 max	0.40 max	bal Fe	490			
ISO 4954(93)	CE28B	Wir rod Bar, CH ext, boron	0.25-0.32			0.60-0.90		0.035 max	0.035 max	0.40 max	Al >=0.020; B 0.0008-0.005; bal Fe	530			
ISO 4954(93)	CE28E4	Wir rod Bar, CH ext	0.25-0.32			0.60-0.90		0.035 max	0.035 max	0.40 max	bal Fe	540			
ISO 4954(93)	CE35B	Wir rod Bar, CH ext, boron	0.32-0.39			0.50-0.80		0.035 max	0.035 max	0.40 max	Al >=0.020; B 0.0008-0.005; bal Fe	570			

UNS numbers and US grades are provided as a means of cross referencing chemically similar alloys. Exchangability is only possible after independent examination of specifications. Tensile properties are minimum or typical as specified. UTS and YS as MPa. El as %. See Appendix for list of abbreviations used in Notes. * indicates obsolete material.

Specification	Designation	Notes	C	Cr	Cu	Mn	Ni	P	S	Si	Other	UTS	YS	El	Hard

Carbon Steel, Nonresulfurized, No equivalents identified

International

Specification	Designation	Notes	C	Cr	Cu	Mn	Ni	P	S	Si	Other	UTS	YS	El	Hard
ISO 4954(93)	CE35E4	Wir rod Bar, CH ext	0.32-0.39			0.50-0.80		0.035 max	0.035 max	0.40 max	bal Fe	560			
ISO 4954(93)	CE40E4	Wir rod Bar, CH ext	0.37-0.44			0.50-0.80		0.035 max	0.035 max	0.40 max	bal Fe	580			
ISO 4954(93)	CE45E4	Wir rod Bar, CH ext	0.42-0.50			0.50-0.80		0.035 max	0.035 max	0.40 max	bal Fe	600			
ISO 4960(99)	CS30	CD strp, Ann	0.27-0.34			0.60-0.90		0.035 max	0.03 max	0.10-0.35	bal Fe	585		18	160 HV
ISO 4960(99)	CS35	CD strp, Ann	0.31-0.38			0.60-0.90		0.035 max	0.03 max	0.10-0.35	bal Fe	590		17	170 HV
ISO 4960(99)	CS40	CD strp, Ann	0.36-0.44			0.60-0.90		0.035 max	0.03 max	0.10-0.35	bal Fe	595		16	170 HV
ISO 4960(99)	CS45	CD strp, Ann	0.42-0.50			0.60-0.90		0.035 max	0.03 max	0.10-0.35	bal Fe	600		16	175 HV
ISO 4960(99)	CS50	CD strp, Ann	0.47-0.55			0.60-0.90		0.035 max	0.03 max	0.10-0.35	bal Fe	605		15	180 HV
ISO 4960(99)	CS55	CD strp, Ann	0.52-0.60			0.60-0.90		0.035 max	0.03 max	0.10-0.35	bal Fe	610		15	180 HV
ISO 4960(99)	CS60	CD strp, Ann	0.55-0.66			0.60-0.90		0.035 max	0.03 max	0.10-0.35	bal Fe	620		14	185 HV
ISO 4960(99)	CS65	CD strp, Ann	0.59-0.70			0.60-0.90		0.035 max	0.03 max	0.10-0.35	bal Fe	630		13	185 HV
ISO 4960(99)	CS70	CD strp, Ann	0.65-0.76			0.60-0.90		0.035 max	0.03 max	0.10-0.35	bal Fe	640		12	190 HV
ISO 4960(99)	CS75	CD strp, Ann	0.69-0.80			0.40-0.70		0.035 max	0.03 max	0.10-0.35	bal Fe	640		12	190 HV
ISO 4960(99)	CS85	CD strp, Ann	0.80-0.91			0.70-1.00		0.035 max	0.03 max	0.10-0.35	bal Fe	670		12	205 HV
ISO 4960(99)	CS95	CD strp, Ann	0.90-1.04			0.30-0.60		0.035 max	0.03 max	0.10-0.35	bal Fe	680		12	210 HV
ISO 630(80)	Fe360A	Struct	0.2 max	0.3 max	0.3 max	2 max	0.4 max	0.06 max	0.05 max	0.6 max	Al <=0.1; Co <=0.1; Mo <=0.15; Pb <=0.15; Ti <=0.05; V <=0.1; W <=0.1; bal Fe				
ISO 630(80)	Fe430B	Struct; FF, FN; 0-40mm	0.21 max	0.3 max	0.3 max	2 max	0.4 max	0.05 max	0.05 max	0.6 max	Mo <=0.15; Pb <=0.15; Ti <=0.05; V <=0.1;				
ISO 630(80)	Fe430B	Struct; FF, FN; Thk>40mm	0.22 max	0.3 max	0.3 max	2 max	0.4 max	0.05 max	0.05 max	0.6 max	Mo <=0.15; N <=0.009; Pb <=0.15; Ti <=0.05; V <=0.1;				
ISO 683-18(96)	C18	Bar Wir, CD, t<=5mm	0.15-0.20			0.60-0.90		0.045 max	0.045 max	0.10-0.40	bal Fe	500	430	7	156 HB

Italy

Specification	Designation	Notes	C	Cr	Cu	Mn	Ni	P	S	Si	Other	UTS	YS	El	Hard
UNI 7070(72)	Fe33	Gen struct	0.3 max			2 max		0.06 max	0.06 max	0.6 max	bal Fe				
UNI 7070(72)	Fe37A	Gen struct	0.2 max			2 max		0.05 max	0.05 max	0.6 max	bal Fe				
UNI 7070(82)	Fe320	Gen struct	0.3 max			2 max		0.055 max	0.055 max	0.6 max	bal Fe				

Japan

Specification	Designation	Notes	C	Cr	Cu	Mn	Ni	P	S	Si	Other	UTS	YS	El	Hard
JIS G3101(95)	SS490	HR Sh Plt Bar									bal Fe	490-610	255-285	15-21	
JIS G3131(96)	SPHC	HR Sh comm	0.15 max			0.60 max		0.050 max	0.050 max		bal Fe	270		27-31	
JIS G3131(96)	SPHD	HR Sh Drawing	0.10 max			0.50 max		0.040 max	0.040 max		bal Fe	270		30-39	
JIS G3131(96)	SPHE	HR Sh DDS	0.10 max			0.50 max		0.030 max	0.035 max		bal Fe	270		31-41	

Mexico

Specification	Designation	Notes	C	Cr	Cu	Mn	Ni	P	S	Si	Other	UTS	YS	El	Hard
NMX-B-301(86)	1071	Bar	0.65-0.70			0.75-1.05		0.040 max	0.050 max		bal Fe				

Poland

Specification	Designation	Notes	C	Cr	Cu	Mn	Ni	P	S	Si	Other	UTS	YS	El	Hard
PNH74244(79)	G235	Struct; Weld tube	0.22 max	0.3 max		1.1 max	0.4 max	0.05 max	0.05 max	0.1-0.35	bal Fe				
PNH74244(79)	G390	Struct; Weld tube; Fine grain	0.15-0.19	0.3 max		1.2-1.5	0.4 max	0.04 max	0.04 max	0.3-0.5	Nb 0.015-0.025; bal Fe				

UNS numbers and US grades are provided as a means of cross referencing chemically similar alloys. Exchangability is only possible after independent examination of specifications. Tensile properties are minimum or typical as specified. UTS and YS as MPa. El as %. See Appendix for list of abbreviations used in Notes. * indicates obsolete material.

Specification	Designation	Notes	C	Cr	Cu	Mn	Ni	P	S	Si	Other	UTS	YS	El	Hard

Carbon Steel, Nonresulfurized, No equivalents identified

Poland

Specification	Designation	Notes	C	Cr	Cu	Mn	Ni	P	S	Si	Other	UTS	YS	El	Hard
PNH84018(86)	09G2	HS weld const	0.12 max	0.3 max		1.2-1.8	0.3 max	0.04 max	0.04 max	0.15-0.4	Al 0.02-0.120; Mo <=0.08; V <=0.05; As<=0.08; CEV<=0.44; bal Fe				
PNH84019	08X	CH	0.05-0.11	0.15 max	0.25 max	0.25-0.5	0.25 max	0.04 max	0.04 max	0.04 max	bal Fe				
PNH84019	08Y	CH	0.05-0.11	0.15 max	0.25 max	0.35-0.65	0.25 max	0.04 max	0.04 max	0.12 max	bal Fe				
PNH84019	10X	CH	0.07-0.14	0.2 max		0.25-0.5	0.3 max	0.04 max	0.04 max	0.04 max	bal Fe				
PNH84019	10Y	CH	0.07-0.14	0.2 max		0.35-0.65	0.3 max	0.04 max	0.04 max	0.17 max	bal Fe				
PNH84019	50G	Q/T	0.48-0.56	0.3 max		0.7-1	0.3 max	0.04 max	0.04 max	0.17-0.37	bal Fe				
PNH84019	65	High-grade struct	0.62-0.7	0.3 max		0.5-0.8	0.3 max	0.04 max	0.04 max	0.17-0.37	bal Fe				
PNH84019	A620M		0.05-0.11	0.15 max	0.25 max	0.25-0.5	0.25 max	0.4 max	0.04 max	0.04 max	bal Fe				
PNH84020	MSt5	Struct	0.26-0.37	0.3 max		0.8 max	0.4 max	0.05 max	0.05 max	0.35 max	bal Fe				
PNH84020	MSt6	Struct	0.38-0.49	0.3 max		0.8 max	0.4 max	0.05 max	0.05 max	0.35 max	bal Fe				
PNH84020	MSt7	Struct	0.5-0.62	0.3 max		0.8 max	0.4 max	0.05 max	0.05 max	0.35 max	bal Fe				
PNH84020	St0S	Struct	0.23 max	0.3 max		1.3 max	0.4 max	0.07 max	0.065 max	0.4 max	bal Fe				
PNH84020	St2S	Struct	0.15 max	0.3 max		2 max	0.3 max	0.05 max	0.05 max	0.07 max	bal Fe				
PNH84020	St3SCu	Struct	0.22 max	0.3 max		1.1 max	0.3 max	0.05 max	0.05 max	0.1-0.35	bal Fe				
PNH84020	St3SCuX	Struct	0.22 max	0.3 max		1.1 max	0.3 max	0.0.-05	0.05 max	0.07 max	bal Fe				
PNH84020	St3SCuY	Struct	0.22 max	0.3 max		1.1 max	0.3 max	0.05 max	0.5 max	0.15 max	bal Fe				
PNH84020	St3SX	Struct	0.22 max	0.3 max		1.1 max	0.3 max	0.05 max	0.05 max	0.07 max	bal Fe				
PNH84020	St3SY	Struct	0.22 max	0.3 max		1.1 max	0.3 max	0.05 max	0.05 max	0.15 max	bal Fe				
PNH84020	St3V	Struct	0.2 max	0.3 max		1.2 max	0.3 max	0.045 max	0.045 max	0.1-0.35	bal Fe				
PNH84020	St3VX	Struct	0.2 max	0.3 max		1.2 max	0.3 max	0.045 max	0.045 max	0.07 max	bal Fe				
PNH84020	St3VY	Struct	0.2 max	0.3 max		1.2 max	0.3 max	0.045 max	0.045 max	0.15 max	bal Fe				
PNH84020	St3W	Struct	0.17 max	0.3 max		1.3 max	0.3 max	0.04 max	0.04 max	0.1-0.35	Al 0.02-0.120; bal Fe				
PNH84020	St4S	Struct	0.24 max	0.3 max		1.1 max	0.3 max	0.05 max	0.05 max	0.1-0.35	bal Fe				
PNH84020	St4SCu	Struct	0.24 max	0.3 max		1.1 max	0.3 max	0.05 max	0.05 max	0.1-0.35	bal Fe				
PNH84020	St4SCuX	Struct	0.24 max	0.3 max		1.1 max	0.3 max	0.05 max	0.05 max	0.07 max	bal Fe				
PNH84020	St4SCuY	Struct	0.24 max	0.3 max		1.1 max	0.3 max	0.05 max	0.05 max	0.15 max	bal Fe				
PNH84020	St4SX	Struct	0.24 max	0.3 max		1.1 max	0.3 max	0.05 max	0.05 max	0.07 max	bal Fe				
PNH84020	St4SY	Struct	0.24 max	0.3 max		1.1 max	0.3 max	0.05 max	0.05 max	0.15 max	bal Fe				
PNH84020	St4V	Struct	0.22 max	0.3 max		1.2 max	0.3 max	0.045 max	0.045 max	0.1-0.35	bal Fe				
PNH84020	St4VX	Struct	0.22 max	0.3 max		1.2 max	0.3 max	0.045 max	0.045 max	0.07 max	bal Fe				
PNH84020	St4VY	Struct	0.22 max	0.3 max		1.2 max	0.3 max	0.045 max	0.045 max	0.15 max	bal Fe				
PNH84020	St4W	Struct	0.2 max	0.3 max		1.3 max	0.3 max	0.04 max	0.04 max	0.1-0.35	Al 0.02-0.120; bal Fe				
PNH84020	St5	Struct	0.3 max	0.3 max		2 max	0.4 max	0.05 max	0.05 max	0.6 max	bal Fe				

Specification	Designation	Notes	C	Cr	Cu	Mn	Ni	P	S	Si	Other	UTS	YS	El	Hard

Carbon Steel, Nonresulfurized, No equivalents identified

Poland

Specification	Designation	Notes	C	Cr	Cu	Mn	Ni	P	S	Si	Other	UTS	YS	El	Hard
PNH84020	St6	Struct	0.3 max	0.3 max		2 max	0.4 max	0.05 max	0.05 max	0.6 max	bal Fe				
PNH84020	St7	Struct	0.3 max	0.3 max		2 max	0.4 max	0.05 max	0.05 max	0.6 max	bal Fe				
PNH84023/02	03J		0.03 max	0.05 max	0.08 max	0.15 max	0.08 max	0.015 max	0.02 max	0.03 max	Al 0.02-0.07; N <=0.012; O<=0.03; As<=0.03; Sn<=0.01; bal Fe				
PNH84023/02	03JA		0.03 max	0.05 max	0.08 max	0.1 max	0.08 max	0.01 max	0.015 max	0.03 max	Al 0.02-0.07; N <=0.012; O<=0.03; As<=0.03; Sn<=0.01; bal Fe				
PNH84023/02	04J		0.035 max	0.1 max	0.1 max	0.25 max	0.1 max	0.025 max	0.03 max	0.02 max	Al <=0.02; N <=0.012; O<=0.03; bal Fe				
PNH84023/02	04JA		0.035 max	0.1 max	0.1 max	0.2 max	0.1 max	0.02 max	0.025 max	0.02 max	Al 0.02-0.07; N <=0.012; O<=0.03; bal Fe				
PNH84023/02	06JA		0.06 max	0.1 max	0.1 max	0.3 max	0.1 max	0.05 max	0.025 max	0.02 max	Al 0.025-0.130; bal Fe				
PNH84023/02	07X		0.08 max	0.05 max	0.15 max	0.35 max	0.1 max	0.04 max	0.04 max	0.02 max	Al 0.02-0.07; P+S<=0.07; bal Fe				
PNH84023/02	07XA		0.08 max	0.1 max	0.1 max	0.25-0.45	0.1 max	0.02 max	0.025 max	0.03 max	Mo <=0.05; bal Fe				
PNH84023/02	08J		0.08 max	0.1 max	0.1 max	0.2-0.45	0.1 max	0.025 max	.03 max	0.03 max	Al 0.02-0.07; N <=0.01; O<=0.03; bal Fe				
PNH84023/02	08JA		0.08 max	0.1 max	0.1 max	0.2 max	0.1 max	0.02 max	0.025 max	0.05 max	Al 0.02-0.07; N <=0.01; O<=0.03; bal Fe				
PNH84023/03	04		0.04 max	0.1 max	0.1 max	0.2 max	0.1 max	0.25 max	0.03 max	0.02 max	Pb <=0.1; 5N<=0.012; O<=0.03; bal Fe				
PNH84023/03	04A		0.04 max	0.1 max	0.1 max	0.2 max	0.1 max	0.02 max	0.25 max	0.02 max	N <=0.012; Ti <=0.5; O<=0.03; bal Fe				
PNH84023/03	05XA		0.07 max	0.06 max	0.06 max	0.35 max	0.06 max	0.3 max	0.03 max	0.01 max	Al <=0.02; N <=0.007; bal Fe				
PNH84023/03	08F		0.08 max	0.1 max	0.1 max	0.25-0.45	0.1 max	0.025 max	0.03 max	0.03 max	bal Fe				
PNH84023/03	08YA		0.08 max	0.1 max	0.1 max	0.25-0.45	0.1 max	0.03 max	0.03 max	0.1 max	bal Fe				
PNH84023/03	10J		0.08-0.15	0.3 max		0.3-0.6	0.4 max	0.035 max	0.045 max	0.05 max	Al 0.02-0.120; bal Fe				
PNH84023/04	St0		0.25 max	0.3 max		2 max	0.4 max	0.07 max	0.06 max	0.6 max	bal Fe				
PNH84023/04	St1X		0.12 max	0.3 max		0.25-0.5	0.4 max	0.05 max	0.05 max	0.05 max	bal Fe				
PNH84023/04	St2NY		0.07-0.13	0.3 max		0.25-0.45	0.4 max	0.05 max	0.05 max	0.15 max	bal Fe				
PNH84023/04	St2SX		0.15 max	0.3 max		1 max	0.3 max	0.05 max	0.05 max	0.07 max	Mo <=0.3; bal Fe				
PNH84023/04	St3M		0.2 max	0.3 max		0.4 max	0.3 max	0.05 max	0.05 max	0.12-0.3	Al <=0.02; Cr+Ni<=0.5; [C+Mn]/6<=0.4				
PNH84023/04	St3NY		0.14-0.2	0.3 max		0.25-0.45	0.4 max	0.05 max	0.05 max	0.15 max	bal Fe				
PNH84023/04	St44N		0.18 max	0.3 max		1.2 max	0.4 max	0.05 max	0.05 max	0.2 max	bal Fe				
PNH84023/05	06X		0.08 max	0.3 max	0.25 max	0.2-0.4	0.4 max	0.04 max	0.04 max	0.3 max	bal Fe				
PNH84023/05	06XA		0.08 max	0.3 max		0.35 max	0.4 max	0.035 max	0.035 max	0.35 max	bal Fe				
PNH84023/05	08Z		0.1 max	0.05 max	0.25 max	0.4-0.55	0.4 max	0.04 max	0.04 max	0.03 max	bal Fe				
PNH84023/05	16G2		0.15-0.19	0.3 max	0.25 max	1.2-1.45	0.3 max	0.04 max	0.04 max	0.3-0.5	Al <=0.02; N <=0.012; C+Mn/6<=0.43; bal Fe				

Specification	Designation	Notes	C	Cr	Cu	Mn	Ni	P	S	Si	Other	UTS	YS	El	Hard
Carbon Steel, Nonresulfurized, No equivalents identified															
Poland															
PNH84023/05	34GJ		0.3-0.38	0.3 max	0.25 max	0.7-0.95	0.4 max	0.045 max	0.045 max	0.15-0.3	Al 0.02-0.120; bal Fe				
PNH84023/06	St0S-b	Reinforcing	0.23 max	0.3 max		1 max	0.3 max	0.07 max	0.06 max	0.4 max	Mo <=0.1; bal Fe				
PNH84023/06	St3S-b	Reinforcing	0.22 max	0.3 max	0.35 max	1 max	0.3 max	0.05 max	0.05 max	0.1-0.35	Al 0.02-0.120; Mo <=0.1; bal Fe				
PNH84023/06	St3SX-b	Reinforcing	0.22 max	0.3 max	0.35 max	1 max	0.3 max	0.05 max	0.05 max	0.07 max	Mo <=0.1; bal Fe				
PNH84023/06	St3SY-b	Reinforcing	0.22 max	0.3 max	0.35 max	1 max	0.3 max	0.05 max	0.05 max	0.03-0.15	Mo <=0.1; bal Fe				
PNH84023/06	St50B	Reinforcing	0.4 max	0.3 max		1.6 max	0.4 max	0.06 max	0.06 max	0.9 max	bal Fe				
PNH84023/07	40G2M	Const	0.37-0.45	0.8-1.2		1.3-1.7	0.3 max	0.04 max	0.035 max	0.2-0.4	Mo 0.15-0.25; bal Fe				
PNH84023/07	R		0.25 max	0.3 max		0.25-1	0.4 max	0.06 max	0.06 max	0.12-0.35	bal Fe				
PNH84023/07	R35Y	Tube, Pip	0.07-0.16	0.3 max	0.25 max	0.4-0.75	0.4 max	0.04 max	0.04 max	0.15 max	bal Fe				
PNH84023/07	R45J		0.16-0.22	0.3 max		0.6-1	0.4 max	0.04 max	0.04 max	0.12-0.35	Al 0.02-0.06; bal Fe				
PNH84023/07	R45Y	Tube, Pip	0.16-0.22	0.3 max		0.6-1.2	0.4 max	0.04 max	0.04 max	0.15 max	bal Fe				
PNH84023/08	St1Z	Chain	0.07-0.12	0.1 max		0.35-0.5	0.3 max	0.05 max	0.05 max	0.05 max	P+S<=0.09				
PNH84024	R55W	High-temp const	0.36-0.44	0.3 max		0.6-0.8	0.4 max	0.045 max	0.045 max	0.2-0.35	bal Fe				
PNH84024	St36K	High-temp const	0.08-0.16	0.3 max		0.4-1.6	0.3 max	0.045 max	0.045 max	0.15-0.35	bal Fe				
PNH84028	D40	Wir rod	0.38-0.43	0.2 max	0.2 max	0.3-0.6	0.2 max	0.035 max	0.035 max	0.1-0.3	Mo <=0.08; P+S<=0.06; bal Fe				
PNH84028	D40A	Wir rod	0.38-0.43	0.1 max	0.2 max	0.3-0.6	0.15 max	0.03 max	0.03 max	0.1-0.25	Mo <=0.05; P+S<=0.05; bal Fe				
PNH84028	D45	Wir rod	0.43-0.48	0.2 max	0.2 max	0.3-0.6	0.2 max	0.035 max	0.035 max	0.1-0.3	Mo <=0.08; P+S<=0.06; bal Fe				
PNH84028	D53	Const; Wir rod	0.5-0.55	0.2 max	0.2 max	0.3-0.6	0.2 max	0.035 max	0.035 max	0.1-0.3	Mo <=0.08; P+S<=0.06; bal Fe				
PNH84028	D53A	Wir rod	0.5-0.55	0.1 max	0.2 max	0.3-0.6	0.15 max	0.03 max	0.03 max	0.1-0.25	Mo <=0.05; P+S<=0.05; bal Fe				
PNH84028	D58	Const; Wir rod	0.55-0.6	0.2 max	0.2 max	0.3-0.6	0.2 max	0.035 max	0.035 max	0.1-0.3	Mo <=0.08; P+S<=0.065; bal Fe				
PNH84028	D58A	Const; Wir rod	0.55-0.6	0.1 max	0.2 max	0.3-0.6	0.15 max	0.03 max	0.03 max	0.1-0.25	Mo <=0.05; P+S<=0.05; bal Fe				
PNH84028	D60	Wir rod	0.58-0.63	0.2 max	0.2 max	0.3-0.6	0.2 max	0.035 max	0.035 max	0.1-0.3	Mo <=0.08; P+S<=0.06; bal Fe				
PNH84028	D60A	Wir rod	0.58-0.63	0.1 max	0.2 max	0.3-0.6	0.15 max	0.03 max	0.03 max	0.1-0.25	Mo <=0.05; P+S<=0.05; bal Fe				
PNH84028	D63A	Wir rod	0.6-0.65	0.1 max	0.2 max	0.3-0.6	0.15 max	0.03 max	0.03 max	0.1-0.25	Mo <=0.05; P+S<=0.05; bal Fe				
PNH84028	D65	Wir rod	0.63-0.68	0.2 max	0.2 max	0.3-0.6	0.2 max	0.035 max	0.35 max	0.1-0.3	Mo <=0.08; P+S<=0.05; bal Fe				
PNH84028	D68A	Wir rod	0.65-0.7	0.1 max	0.2 max	0.3-0.6	0.15 max	0.03 max	0.03 max	0.1-0.25	Mo <=0.05; P+S<=0.05; bal Fe				
PNH84028	D70	Wir rod	0.68-0.73	0.2 max	0.2 max	0.3-0.6	0.2 max	0.035 max	0.035 max	0.1-0.3	Mo <=0.08; P+S<=0.06; bal Fe				
PNH84028	D70A	Wir rod	0.68-0.73	0.1 max	0.2 max	0.3-0.6	0.15 max	0.03 max	0.03 max	0.1-0.25	Mo <=0.05; P+S<=0.06; bal Fe				
PNH84028	D73	Wir rod	0.7-0.75	0.2 max	0.2 max	0.3-0.6	0.2 max	0.035 max	0.035 max	0.1-0.3	Mo <=0.08; P+S<=0.06; bal Fe				
PNH84028	D73A	Wir rod	0.7-0.75	0.1 max	0.2 max	0.3-0.6	0.15 max	0.03 max	0.03 max	0.1-0.25	Mo <=0.05; P+S<=0.05; bal Fe				
PNH84028	D75	Wir rod	0.73-0.78	0.2 max	0.2 max	0.3-0.6	0.2 max	0.035 max	0.035 max	0.1-0.3	Mo <=0.08; P+S<=0.06; bal Fe				
PNH84028	D75A	Wir rod	0.73-0.78	0.1 max	0.2 max	0.3-0.6	0.15 max	0.03 max	0.03 max	0.1-0.25	Mo <=0.05; P+S<=0.05; bal Fe				
PNH84028	D78	Wir rod	0.75-0.8	0.2 max	0.2 max	0.3-0.6	0.2 max	0.035 max	0.035 max	0.1-0.3	Mo <=0.08; P+S<=0.06; bal Fe				
PNH84028	D78A	Wir rod	0.75-0.8	0.1 max	0.2 max	0.3-0.6	0.15 max	0.03 max	0.03 max	0.1-0.25	Mo <=0.05; P+S<=0.05; bal Fe				

UNS numbers and US grades are provided as a means of cross referencing chemically similar alloys. Exchangability is only possible after independent examination of specifications. Tensile properties are minimum or typical as specified. UTS and YS as MPa. El as %. See Appendix for list of abbreviations used in Notes. * indicates obsolete material.

Specification	Designation	Notes	C	Cr	Cu	Mn	Ni	P	S	Si	Other	UTS	YS	El	Hard

Carbon Steel, Nonresulfurized, No equivalents identified

Poland

Specification	Designation	Notes	C	Cr	Cu	Mn	Ni	P	S	Si	Other	UTS	YS	El	Hard
PNH84028	D83A	Unalloyed; Wir rod	0.8-0.85	0.1 max	0.2 max	0.3-0.6	0.15 max	0.03 max	0.03 max	0.1-0.3	Mo <=0.05; P+S<=0.05; bal Fe				
PNH84028	D85	Unalloyed; Wir rod	0.83-0.88	0.2 max	0.2 max	0.3-0.6	0.2 max	0.035 max	0.035 max	0.1-0.25	Mo <=0.08; P+S<=0.06; bal Fe				
PNH84028	D85A	Unalloyed; Wir rod	0.83-0.88	0.1 max	0.2 max	0.3-0.6	0.15 max	0.03 max	0.03 max	0.1-0.25	Mo <=0.05; P+S<=0.05; bal Fe				
PNH84028	D88A	Unalloyed; Wir rod	0.85-0.9	0.1 max	0.15 max	0.3-0.6	0.15 max	0.03 max	0.03 max	0.1-0.25	Mo <=0.05; P+S<=0.05; bal Fe				
PNH84028	D94A	Unalloyed; Wir rod	0.93-0.98	0.1 max	0.2 max	0.3-0.6	0.15 max	0.03 max	0.03 max	0.1-0.25	Mo <=0.05; P+S<=0.05; bal Fe				
PNH84028	DS65	Unalloyed; Wir rod	0.63-0.68	0.1 max	0.15 max	0.3-0.6	0.15 max	0.02 max	0.02 max	0.1-0.25	Mo <=0.05; P+S<=0.035; bal Fe				
PNH84028	DS70	Unalloyed; Wir rod	0.68-0.73	0.1 max	0.15 max	0.3-0.6	0.15 max	0.02 max	0.02 max	0.1-0.25	Mo <=0.05; P+S<=0.035; bal Fe				
PNH84028	DS75	Unalloyed; Wir rod	0.73-0.78	0.1 max	0.15 max	0.3-0.6	0.15 max	0.02 max	0.2 max	0.1-0.25	Mo <=0.05; P+S<=0.035; bal Fe				
PNH84028	DS80	Unalloyed; Wir rod	0.78-0.83	0.1 max	0.15 max	0.3-0.6	0.15 max	0.2 max	0.02 max	0.1-0.3	Mo <=0.05; P+S<=0.035; bal Fe				
PNH84028	DS85	Unalloyed; Wir rod	0.83-0.88	0.1 max	0.15 max	0.3-0.6	0.15 max	0.02 max	0.02 max	0.1-0.25	Mo <=0.05; P+S<=0.035; bal Fe				
PNH84028	DT06	Unalloyed; Wir rod	0.08 max	0.3 max	0.2 max	0.2-0.4	0.4 max	0.04 max	0.025 max	0.25 max	bal Fe				
PNH84028	DW60	Unalloyed; Wir rod	0.56-0.61	0.15 max	0.2 max	0.4-0.6	0.25 max	0.025 max	0.025 max	0.25 max	P+S<=0.045; bal Fe				
PNH84028	DW65	Unalloyed; Wir rod	0.62-0.67	0.15 max	0.2 max	0.4-0.6	0.25 max	0.025 max	0.025 max	0.25 max	P+S<=0.045; bal Fe				
PNH84028	DW90	Unalloyed; Wir rod	0.86-0.91	0.12 max	0.2 max	0.3-0.5	0.15 max	0.02 max	0.02 max	0.25 max	P+S<=0.035; bal Fe				
PNH84030/04	45HN	Q/T	0.41-0.49	1.3-1.6	0.2 max	0.5-0.8	1.6-2.1	0.025 max	0.025 max	0.17-0.37	bal Fe				
PNH84030/04	45HN2A	Q/T	0.43-0.49	1.3-1.6	0.2 max	0.5-0.8	1.6-2.1	0.025 max	0.025 max	0.17-0.37	Al 0.015-0.05; bal Fe				
PNH84030/04	45HNMF	Q/T	0.42-0.5	0.8-1.1		0.5-0.8	1.3-1.8	0.035 max	0.035 max	0.17-0.37	Mo 0.2-0.3; V 0.1-0.2; bal Fe				
PNH84032	40S2	Spring	0.35-0.42	0.3 max	0.25 max	0.6-0.8	0.4 max	0.04 max	0.04 max	1.5-1.9	V 0.15-0.25; bal Fe				
PNH84032	45S	Spring	0.4-0.5	0.3 max	0.25 max	0.6-0.9	0.4 max	0.05 max	0.05 max	1-1.3	bal Fe				
PNH84032	60S2	Spring	0.57-0.65	0.3 max	0.25 max	0.6-0.9	0.4 max	0.04 max	0.04 max	1.5-1.8	bal Fe				
PNH84035	30HN2MFA	Q/T	0.26-0.33	0.6 max	0.2 max	0.3-0.6	2 max	0.03 max	0.03 max	0.17-0.37	Mo <=0.2; V 0.15-0.3; bal Fe				
PNH84035	65S2WA	Q/T	0.61-0.69	0.3 max	0.2 max	0.7-1	0.3 max	0.03 max	0.03 max	1.5-2	bal Fe				
PNH92147	A	Struct/welded ships; Sh, Plt	0.23 max	0.3 max		0.60 max	0.3 max	0.04 max	0.04 max	0.35 max	Mn<=5.2C; CEV<=0.4; bal Fe				
PNH92147	B	Struct/welded ships; Sh, Plt	0.21 max	0.3 max		0.8 max	0.3 max	0.04 max	0.04 max	0.35 max	CEV<=0.4; bal Fe				
PNH92147	DH36	Struct/welded ships; Sh, Plt	0.18 max	0.2 max	0.35 max	0.9-1.6	0.4 max	0.04 max	0.04 max	0.1-0.5	Al <=0.15; Mo <=0.08; V 0.05-0.1; CEV<=0.45; bal Fe				
PNH92147	DH40	Struct/welded ships; Sh, Plt	0.18 max	0.2 max	0.35 max	0.8-1.6	0.4 max	0.04 max	0.04 max	0.1-0.5	Al <=0.15; Mo <=0.08; V 0.05-0.1; CEV<=0.45; bal Fe				
PNH93027	15E	Struct; Wir Bar	0.18 max	0.2 max		0.35-0.6	0.3 max	0.04 max	0.04 max	0.10 max	bal Fe; Si:Traces				
PNH93027	18G2AA	Struct; Wir Bar	0.2 max	0.2 max	0.2 max	1-1.5	0.2 max	0.04 max	0.04 max	0.55	Al 0.2-0.40; bal Fe				
PNH93215(82)	34GS A-III	HR reinf; Ribbed	0.3-0.36	0.3 max	0.035 max	0.8-1.2	0.3 max	0.05 max	0.05 max	0.4-0.7	Mo <=0.1; bal Fe				
PNH93215(82)	St3S; A-1	HR reinf; Round bar	0.22 max	0.3 max	0.35 max	1 max	0.3 max	0.05 max	0.05 max	0.1-0.35	Mo <=0.1; bal Fe				
PNH93215(82)	St3SX; A-1	HR reinf; Round bar	0.22 max	0.3 max	0.35 max	1 max	0.3 max	0.05 max	0.05 max	0.07 max	Mo <=0.1; bal Fe				
PNH93215(82)	St3SY; A-1	HR reinf; Round bar	0.22 max	0.3 max	0.35 max	1 max	0.3 max	0.05 max	0.05 max	0.03-0.17	Mo <=0.1; bal Fe				

Specification	Designation	Notes	C	Cr	Cu	Mn	Ni	P	S	Si	Other	UTS	YS	El	Hard

Carbon Steel, Nonresulfurized, No equivalents identified

Romania

Specification	Designation	Notes	C	Cr	Cu	Mn	Ni	P	S	Si	Other	UTS	YS	El	Hard
STAS 500/2(88)	OL37.3kf	Gen struct; Specially killed	0.17 max	0.3 max		0.8 max		0.045 max	0.045 max	0.4 max	Pb <=0.15; As<=0.08; bal Fe				
STAS 500/2(88)	OL37.4kf	Gen struct; Specially killed	0.17 max	0.3 max		0.8 max		0.04 max	0.04 max	0.4 max	Al 0.02-0.120; Pb <=0.15; As<=0.08; bal Fe				
STAS 8184(87)	OLT35K	High-temp const	0.17 max	0.3 max		0.4-2.0		0.04 max	0.045 max	0.15-0.35	Pb <=0.15; As<=0.08; bal Fe				
STAS 8184(87)	OLT45K	High-temp const	0.23 max	0.3 max		0.4-2.0		0.04 max	0.045 max	0.15-0.35	Pb <=0.15; As<=0.08; bal Fe				

Russia

Specification	Designation	Notes	C	Cr	Cu	Mn	Ni	P	S	Si	Other	UTS	YS	El	Hard
GOST 1071(81)	68A	Spring	0.65-0.7	0.12 max	0.15 max	0.4-0.55	0.2 max	0.025 max	0.025 max	0.15-0.25	Al <=0.05; bal Fe				
GOST 1071(81)	70ChGFA	Spring	0.65-0.72	0.3-0.5	0.15 max	0.5-0.8	0.2 max	0.025 max	0.025 max	0.15-0.3	Al <=0.05; V 0.1-0.2; bal Fe				
GOST 18232	M68	Rail	0.62-0.73	0.3 max	0.3 max	0.7-1.0	0.4 max	0.035 max	0.045 max	0.13-0.28	Al <=0.1; bal Fe				
GOST 19281	14G2	Weld const	0.12-0.18	0.3 max	0.3 max	1.2-1.6	0.3 max	0.035 max	0.04 max	0.17-0.37	Al <=0.1; N <=0.012; As<=0.08; bal Fe				
GOST 380(71)	BSt0	Gen struct	0.25 max	0.3 max	0.3 max	2 max	0.4 max	0.07 max	0.06 max	0.6 max	Al <=0.1; bal Fe				
GOST 380(71)	BSt4sp	Gen struct; Non-rimming; FF	0.18-0.27	0.3 max	0.3 max	0.4-0.7	0.3 max	0.04 max	0.05 max	0.12-0.3	Al <=0.1; N <=0.008; As<=0.08; bal Fe				
GOST 380(71)	BSt5Gps2	Gen struct; Semi-killed	0.22-0.3	0.3 max	0.3 max	0.8-1.2	0.3 max	0.04 max	0.05 max	0.15 max	Al <=0.1; N <=0.008; As<=0.08; bal Fe				
GOST 380(71)	St0-2	Gen struct; 0-20mm	0.3 max	0.3 max	0.3 max	2 max	0.4 max	0.07 max	0.06 max	0.6 max	Al <=0.1; bal Fe	300		23L	
GOST 300(71)	St0 2	Gen struct; >40mm	0.3 max	0.3 max	0.3 max	2 max	0.4 max	0.07 max	0.06 max	0.6 max	Al <=0.1; bal Fe	300		20L	
GOST 380(71)	St1kp2	Gen struct; Semi-Finished; Unkilled; Rimming; R; FU; 0-20mm	0.3 max	0.3 max	0.3 max	2 max	0.4 max	0.07 max	0.06 max	0.6 max	Al <=0.1; bal Fe	30390		35L	
GOST 380(71)	St1kp2	Gen struct; Semi-Finished; Unkilled; Rimming; R; FU; >40mm	0.3 max	0.3 max	0.3 max	2 max	0.4 max	0.07 max	0.06 max	0.6 max	Al <=0.1; bal Fe	300-390		32L	
GOST 380(71)	St1ps	Gen struct; Semi-Finished; Semi-killed; 0-20mm	0.3 max	0.3 max	0.3 max	2 max	0.4 max	0.07 max	0.06 max	0.6 max	Al <=0.1; bal Fe	310-410		34L	
GOST 380(71)	St1ps	Gen struct; Semi-Finished; Semi-killed; >40mm	0.3 max	0.3 max	0.3 max	2 max	0.4 max	0.07 max	0.06 max	0.6 max	Al <=0.1; bal Fe	310-410		31L	
GOST 380(71)	St1ps2	Gen struct; Semi-killed; 0-20mm	0.3 max	0.3 max	0.3 max	2 max	0.4 max	0.07 max	0.06 max	0.6 max	Al <=0.1; bal Fe	310-410		34L	
GOST 380(71)	St1ps2	Gen struct; Semi-killed; >40mm	0.3 max	0.3 max	0.3 max	2 max	0.4 max	0.07 max	0.06 max	0.6 max	Al <=0.1; bal Fe	310-410		31L	
GOST 380(71)	St1sp	Gen struct; Semi-Finished; Non-rimming; FF; >40mm	0.3 max	0.3 max	0.3 max	2 max	0.4 max	0.07 max	0.06 max	0.6 max	Al <=0.1; bal Fe	310-410		31L	
GOST 380(71)	St1sp	Gen struct; Semi-Finished; Non-rimming; FF; 0-20mm	0.3 max	0.3 max	0.3 max	2 max	0.4 max	0.07 max	0.06 max	0.6 max	Al <=0.1; bal Fe	310-410		34L	
GOST 380(71)	St1sp2	Gen struct; Non-rimming; FF; >40mm	0.3 max	0.3 max	0.3 max	2 max	0.4 max	0.07 max	0.06 max	0.6 max	Al <=0.1; bal Fe	310-410		31L	
GOST 380(71)	St1sp2	Gen struct; Non-rimming; FF; 0-20mm	0.3 max	0.3 max	0.3 max	2 max	0.4 max	0.07 max	0.06 max	0.6 max	Al <=0.1; bal Fe	310-410		34L	
GOST 380(71)	St5sp	Gen struct; Non-rimming; FF; 0-20mm	0.3 max	0.3 max	0.3 max	2 max	0.4 max	0.07 max	0.06 max	0.6 max	Al <=0.1; bal Fe	490-630		20L	
GOST 380(71)	St5sp	Gen struct; Non-rimming; FF; >40mm	0.3 max	0.3 max	0.3 max	2 max	0.4 max	0.07 max	0.06 max	0.6 max	Al <=0.1; bal Fe	490-630		17L	
GOST 380(71)	WSt2ps2	Gen struct; Semi-killed; 0-20mm	0.15 max	0.3 max	0.3 max	0.25-0.7	0.3 max	0.04 max	0.05 max	0.05-0.17	Al <=0.1; N <=0.008; As<=0.08; bal Fe	330-430	225	32L	
GOST 380(71)	WSt4kp	Gen struct; Unkilled; Rimming; R; FU; >40mm	0.27 max	0.3 max	0.3 max	0.4-0.9	0.3 max	0.04 max	0.05 max	0.07 max	Al <=0.1; N <=0.008; As<=0.08; bal Fe	400-510		22L	
GOST 380(71)	WSt4kp	Gen struct; Unkilled; Rimming; R; FU; 0-20mm	0.27 max	0.3 max	0.3 max	0.4-0.9	0.3 max	0.04 max	0.05 max	0.07 max	Al <=0.1; N <=0.008; As<=0.08; bal Fe	400-510		25L	

UNS numbers and US grades are provided as a means of cross referencing chemically similar alloys. Exchangability is only possible after independent examination of specifications. Tensile properties are minimum or typical as specified. UTS and YS as MPa. El as %. See Appendix for list of abbreviations used in Notes. * indicates obsolete material.

Specification	Designation	Notes	C	Cr	Cu	Mn	Ni	P	S	Si	Other	UTS	YS	El	Hard

Carbon Steel, Nonresulfurized, No equivalents identified

Russia

Specification	Designation	Notes	C	Cr	Cu	Mn	Ni	P	S	Si	Other	UTS	YS	El	Hard
GOST 380(71)	WSt4kp2	Gen struct; Unkilled; Rimming; R; FU; 0-20mm	0.27 max	0.3 max	0.3 max	0.4-0.9	0.3 max	0.04 max	0.05 max	0.07 max	Al <=0.1; N <=0.008; As<=0.08; bal Fe	400-510	255	25L	
GOST 380(71)	WSt4kp2	Gen struct; Unkilled; Rimming; R; FU; >100mm	0.27 max	0.3 max	0.3 max	0.4-0.9	0.3 max	0.04 max	0.05 max	0.07 max	Al <=0.1; N <=0.008; As<=0.08; bal Fe	400-510	225	22L	
GOST 380(71)	WSt4ps	Gen struct; Semi-killed >40mm	0.27 max	0.3 max	0.3 max	0.4-0.9	0.3 max	0.04 max	0.05 max	0.05-0.17	Al <=0.1; N <=0.008; As<=0.08; bal Fe	410-530		21L	
GOST 380(71)	WSt4ps	Gen struct; Semi-killed 0-20mm	0.27 max	0.3 max	0.3 max	0.4-0.9	0.3 max	0.04 max	0.05 max	0.05-0.17	Al <=0.1; N <=0.008; As<=0.08; bal Fe	410-530		24L	
GOST 380(71)	WSt4ps2	Gen struct; Semi-killed; 0-20mm	0.27 max	0.3 max	0.3 max	0.4-0.9	0.3 max	0.04 max	0.05 max	0.05-0.17	Al <=0.1; N <=0.008; As<=0.08; bal Fe	410-530	265	24L	
GOST 380(71)	WSt4ps2	Gen struct; Semi-killed; >100mm	0.27 max	0.3 max	0.3 max	0.4-0.9	0.3 max	0.04 max	0.05 max	0.05-0.17	Al <=0.1; N <=0.008; As<=0.08; bal Fe	410-530	235	21L	
GOST 380(71)	WSt4sp	Gen struct; Non-rimming; FF; 0-20mm	0.27 max	0.3 max	0.3 max	0.4-0.9	0.3 max	0.04 max	0.05 max	0.12-0.3	Al <=0.1; N <=0.008; Pb <=0.1; As<=0.08; bal Fe	410-530		24L	
GOST 380(71)	WSt4sp	Gen struct; Non-rimming; FF; >40mm	0.27 max	0.3 max	0.3 max	0.4-0.9	0.3 max	0.04 max	0.05 max	0.12-0.3	Al <=0.1; N <=0.008; Pb <=0.1; As<=0.08; bal Fe	410-530		21L	
GOST 380(71)	WSt4sp2	Gen struct; Non-rimming; FF; >100mm	0.27 max	0.3 max	0.3 max	0.4-0.9	0.3 max	0.04 max	0.05 max	0.12-0.3	Al <=0.1; N <=0.008; Pb <=0.1; As<=0.08; bal Fe	410-530	235	21L	
GOST 380(71)	WSt4sp2	Gen struct; Non-rimming; FF; 0-20mm	0.27 max	0.3 max	0.3 max	0.4-0.9	0.3 max	0.04 max	0.05 max	0.12-0.3	Al <=0.1; N <=0.008; Pb <=0.1; As<=0.08; bal Fe	410-530	265	24L	
GOST 380(71)	WSt5Gps	Gen struct; Semi-killed; 0-20mm	0.3 max	0.3 max	0.3 max	0.8-1.2	0.3 max	0.04 max	0.05 max	0.15 max	Al <=0.1; N <=0.008; As<=0.08; bal Fe	450-590		20L	
GOST 380(71)	WSt5Gps	Gen struct; Semi-killed; >40mm	0.3 max	0.3 max	0.3 max	0.8-1.2	0.3 max	0.04 max	0.05 max	0.15 max	Al <=0.1; N <=0.008; As<=0.08; bal Fe	450-590		17L	
GOST 380(71)	WSt5Gps2	Gen struct; Semi-killed; 0-20mm	0.3 max	0.3 max	0.3 max	0.8-1.2	0.3 max	0.04 max	0.05 max	0.15 max	Al <=0.1; N <=0.008; As<=0.08; bal Fe	450-590	285	20L	
GOST 380(71)	WSt5Gps2	Gen struct; Semi-killed; >100mm	0.3 max	0.3 max	0.3 max	0.8-1.2	0.3 max	0.04 max	0.05 max	0.15 max	Al <=0.1; N <=0.008; As<=0.08; bal Fe	450-590	255	17L	
GOST 380(71)	WSt5ps	Gen struct; Semi-killed; >40mm	0.37 max	0.3 max	0.3 max	0.5-1	0.3 max	0.04 max	0.05 max	0.05-0.17	Al <=0.1; N <=0.008; As<=0.08; Betonst: 28mm C 0.28-0.37 Mn 0.8-1.1; bal Fe	490-630		17L	
GOST 380(71)	WSt5ps	Gen struct; Semi-killed; >40mm	0.37 max	0.3 max	0.3 max	0.5-1	0.3 max	0.04 max	0.05 max	0.05-0.17	Al <=0.1; N <=0.008; As<=0.08; Betonst: 1<=28mm C 0.3-0.39 Mn 0.6-0.9; bal Fe	490-630		17L	
GOST 380(71)	WSt5ps	Gen struct; Semi-killed; 0-20mm	0.37 max	0.3 max	0.3 max	0.5-1	0.3 max	0.04 max	0.05 max	0.05-0.17	Al <=0.1; N <=0.008; As<=0.08; Betonst: 1<=28mm C 0.3-0.39 Mn 0.6-0.9; bal Fe	490-630		20L	
GOST 380(71)	WSt5ps	Gen struct; Semi-killed; 0-20mm	0.37 max	0.3 max	0.3 max	0.5-1	0.3 max	0.04 max	0.05 max	0.05-0.17	Al <=0.1; N <=0.008; As<=0.08; Betonst: 28mm C 0.28-0.37 Mn 0.8-1.1; bal Fe	490-630		20L	
GOST 380(71)	WSt5ps2	Gen struct; Semi-killed; >100mm	0.37 max	0.3 max	0.3 max	0.5-1	0.3 max	0.04 max	0.05 max	0.05-0.17	Al <=0.1; N <=0.008; As<=0.08; Betonst: 1<=28mm C=0.3-0.39 Mn=006-0.9; bal Fe	490-630	255	17L	
GOST 380(71)	WSt5ps2	Gen struct; Semi-killed; 0-20mm	0.37 max	0.3 max	0.3 max	0.5-1	0.3 max	0.04 max	0.05 max	0.05-0.17	Al <=0.1; N <=0.008; As<=0.08; Betonst: 1<=28mm C=0.3-0.39 Mn=006-0.9; bal Fe	490-630	285	20L	
GOST 380(71)	WSt5sp	Gen struct; Non-rimming; FF; >40mm	0.37 max	0.3 max	0.3 max	0.5-1	0.3 max	0.04 max	0.05 max	0.15-0.35	Al <=0.1; N <=0.008; As<=0.08; bal Fe	490-630		17L	
GOST 380(71)	WSt5sp	Gen struct; Non-rimming; FF; 0-20mm	0.37 max	0.3 max	0.3 max	0.5-1	0.3 max	0.04 max	0.05 max	0.15-0.35	Al <=0.1; N <=0.008; As<=0.08; bal Fe	490-630		20L	
GOST 380(71)	WSt5sp2	Gen struct; Non-rimming; FF; >100mm	0.37 max	0.3 max	0.3 max	0.5-1	0.3 max	0.04 max	0.05 max	0.15-0.35	Al <=0.1; N <=0.008; As<=0.08; bal Fe	490-630	255	17L	
GOST 380(71)	WSt5sp2	Gen struct; Non-rimming; FF; 0-20mm	0.37 max	0.3 max	0.3 max	0.5-1	0.3 max	0.04 max	0.05 max	0.15-0.35	Al <=0.1; N <=0.008; As<=0.08; bal Fe	490-630	285	20L	

Specification	Designation	Notes	C	Cr	Cu	Mn	Ni	P	S	Si	Other	UTS	YS	El	Hard

Carbon Steel, Nonresulfurized, No equivalents identified

Russia

Specification	Designation	Notes	C	Cr	Cu	Mn	Ni	P	S	Si	Other	UTS	YS	El	Hard
GOST 380(88)	St1ps	Gen struct; Semi-Finished; Semi-killed	0.06-0.12	0.3 max	0.3 max	0.25-0.5	0.3 max	0.04 max	0.05 max	0.05-0.15	Al <=0.1; N <=0.008; Ti <=0.1; As<=0.08; bal Fe				
GOST 380(88)	St1sp	Gen struct; Semi-Finished; Non-rimming; FF	0.06-0.12	0.3 max	0.3 max	0.25-0.5	0.3 max	0.04 max	0.05 max	0.15-0.3	Al <=0.1; N <=0.008; Ti <=0.1; As<=0.08; bal Fe				
GOST 5058	14G	struct, const; weld	0.12-0.18	0.3 max	0.3 max	0.7-1	0.3 max	0.035 max	0.04 max	0.17-0.37	Al <=0.1; Ti 0.01-0.03; As<=0.08; bal Fe				
GOST 5520(79)	12K	Boiler, Press ves	0.08-0.16	0.30 max	0.30 max	0.40-0.70	0.30 max	0.030 max	0.035 max	0.17-0.37	N <=0.008; As<=0.08; bal Fe				
GOST 5520(79)	15K	Boiler, Press ves	0.12-0.20	0.30 max	0.30 max	0.35-0.65	0.30 max	0.040 max	0.040 max	0.15-0.30	N <=0.012; As<=0.08; bal Fe				
GOST 5520(79)	20K	Boiler, Press ves	0.16-0.24	0.30 max	0.30 max	0.35-0.65	0.30 max	0.040 max	0.040 max	0.15-0.30	N <=0.012; As<=0.08; bal Fe				
GOST 5520(79)	22K	Boiler, Press ves	0.19-0.26	0.30 max	0.30 max	0.70-1.00	0.30 max	0.040 max	0.035 max	0.17-0.40	N <=0.012; As<=0.08; bal Fe				
GOST 5781(82)	WSt5sp2	HR conc-reinf	0.37 max	0.3 max	0.3 max	0.5-1	0.3 max	0.04 max	0.05 max	0.15-0.35	Al <=0.1; N <=0.008; As<=0.08; bal Fe				
GOST 9960	M73T	Point rail (types: OP65,0P50,0P43	0.67-0.78	0.3 max	0.3 max	0.75-1.05	0.4 max	0.035 max	0.04 max	0.18-0.45	Al <=0.1; Ti 0.007-0.015; bal Fe				
GOST 9960	M73W	Point rail (types: OP65,0P50,0P43	0.67-0.78	0.3 max	0.3 max	0.75-1.05	0.4 max	0.035 max	0.04 max	0.18-0.45	Al <=0.1; V <=0.03; bal Fe				
GOST 9960	M73Z	Point rail (types: OP65,0P50,0P43	0.67-0.78	0.3 max	0.3 max	0.75-1.05	0.4 max	0.035 max	0.04 max	0.18-0.45	Al <=0.1; Zr 0.001-0.05; bal Fe				

Spain

Specification	Designation	Notes	C	Cr	Cu	Mn	Ni	P	S	Si	Other	UTS	YS	El	Hard
UNE 36123(90)	AE275HC	HR; Strp	0.09 max			0.9 max		0.03 max	0.03 max	0.5 max	Al 0.02-0.08; Mo <=0.15; Nb 0.01-0.06; Ti 0.01-0.12; V 0.01-0.08; bal Fe				
UNE 36254(79)	46M4	Q/T	0.42-0.5			0.9-1.2		0.04 max	0.04 max	0.3-0.5	bal Fe				

Sweden

Specification	Designation	Notes	C	Cr	Cu	Mn	Ni	P	S	Si	Other	UTS	YS	El	Hard
MNC810E(83)	1300-00	Gen struct	0.3 max			2 max		0.07 max	0.06 max	0.6 max	bal Fe				

UK

Specification	Designation	Notes	C	Cr	Cu	Mn	Ni	P	S	Si	Other	UTS	YS	El	Hard
BS 3602/1(87)	ERW430; CEW430	ERW IW smls Tub; press High-temp; t<=16mm	0.17 max			0.30-0.80		0.035 max	0.035 max	0.35 max	Al <=0.06; bal Fe	430-570	275	22	
BS 3603(91)	HFS509LT; CFS509LT	Pip Tub press low temp t<=200mm	0.10 max			0.30-0.80	8.50-9.50	0.025 max	0.020 max	0.10-0.30	Al >=0.020; bal Fe	690-840	510	15	

USA

Specification	Designation	Notes	C	Cr	Cu	Mn	Ni	P	S	Si	Other	UTS	YS	El	Hard
AMS 7732A(87)		Cu Clad Wir	0.08 max			0.25-0.40		0.04 max	0.05 max		bal Fe	345-690			
AMS 7733B(87)		Cu Clad Wir	0.13 max			0.30-0.60		0.04 max	0.05 max		bal Fe		250	15-20min	
ASTM A135	A	Weld Pipe	0.25 max			0.95 max		0.035 max	0.035 max		bal Fe	331	207	35L	
ASTM A161(94)	Low C	Smls Low C Still tubs	0.10-0.20			0.30-0.80		0.035 max	0.035 max	0.25 max	bal Fe	324	179	35	125 or 137 HB max
ASTM A183(98)	Grade 2 Bolt	Rail Bolts	0.30 min					0.05 max	0.33 max		bal Fe	760	550	12	
ASTM A192/A192M(96)		Smls Boiler tubs	0.06-0.18			0.27-0.63		0.035 max	0.035 max	0.25 max	bal Fe	325	180	35	77 HRB max
ASTM A210/A210M(96)	C	Tub	0.35 max			0.29-1.06		0.035 max	0.035 max	0.10 min	bal Fe	485	275	30	89 HRB max
ASTM A283/A283M(97)	A	Struct plt, t>40mm	0.14 max		0.20	0.90 max		0.035 max	0.04 max	0.15-0.40	bal Fe	310-415	165	30	
ASTM A283/A283M(97)	A	Struct plt, t<=40mm	0.14 max		0.20	0.90 max		0.035 max	0.04 max	0.40 max	bal Fe	310-415	165	30	
ASTM A283/A283M(97)	B	Struct plt, t>40mm	0.17 max		0.20	0.90 max		0.035 max	0.04 max	0.15-0.40	bal Fe	345-450	185	28	
ASTM A283/A283M(97)	B	Struct plt, t<=40mm	0.17 max		0.20	0.90 max		0.035 max	0.04 max	0.40 max	bal Fe	345-450	185	28	
ASTM A29/A29M(93)	1071	Bar	0.65-0.75			0.75-1.05		0.040 max	0.050 max		bal Fe				

UNS numbers and US grades are provided as a means of cross referencing chemically similar alloys. Exchangability is only possible after independent examination of specifications. Tensile properties are minimum or typical as specified. UTS and YS as MPa. El as %. See Appendix for list of abbreviations used in Notes. * indicates obsolete material.

Specification	Designation	Notes	C	Cr	Cu	Mn	Ni	P	S	Si	Other	UTS	YS	El	Hard

Carbon Steel, Nonresulfurized, No equivalents identified

USA

Specification	Designation	Notes	C	Cr	Cu	Mn	Ni	P	S	Si	Other	UTS	YS	El	Hard
ASTM A3(95)	Grade 1	Rail joint bars, no HT						0.05 max			bal Fe	380		22	
ASTM A3(95)	Grade 2	Rail joint bars, no HT	0.30 min					0.05 max			bal Fe	470		20	
ASTM A3(95)	Grade 3	Rail joint bars, no HT	0.45 min					0.04 max			bal Fe	585		15	
ASTM A328/A328M(96)		Sht Piling			0.20 min			0.035 max	0.04 max		bal Fe	485	270	20	
ASTM A421/A421M(98)	BA 0.196"dia	Wir, conc reinf, YS at 1% exp						0.040 max	0.050 max		bal Fe	1655	1407		
ASTM A421/A421M(98)	BA 0.250"dia	Wir, conc reinf, YS at 1% exp						0.040 max	0.050 max		bal Fe	1655	1407		
ASTM A421/A421M(98)	BA 0.276"dia	Wir, conc reinf, YS at 1% exp						0.040 max	0.050 max		bal Fe	1620	1377		
ASTM A421/A421M(98)	WA 0.192"dia	Wir, conc reinf, YS at 1% exp						0.040 max	0.050 max		bal Fe	1725	1465		
ASTM A421/A421M(98)	WA 0.196"dia	Wir, conc reinf, YS at 1% exp						0.040 max	0.050 max		bal Fe	1725	1465		
ASTM A421/A421M(98)	WA 0.250"dia	Wir, conc reinf, YS at 1% exp						0.040 max	0.050 max		bal Fe	1655	1407		
ASTM A421/A421M(98)	WA 0.276"dia	Wir, conc reinf, YS at 1% exp						0.040 max	0.050 max		bal Fe	1690	1377		
ASTM A569/A569M(98)	A	HR Sh Strp, when Cu not spec'd	0.10 max	0.15 max	0.20 max	0.60 max	0.20 max	0.030 max	0.035		Mo <=0.06; Nb <=0.008; Ti <=0.008; V <=0.008; Cu+Ni+Cr+Mo<=0.5; bal Fe		205-345	>=25	
ASTM A569/A569M(98)	A(Cu)	HR Sh Strp, when Cu spec'd	0.10 max	0.15 max	0.20 min	0.60 max	0.20 max	0.030 max	0.035 max		Mo <=0.06; Nb <=0.008; Ti <=0.008; V <=0.008; bal Fe		205-345	>=25	
ASTM A569/A569M(98)	B	HR Sh Strp, when Cu not spec'd	0.02-0.15	0.15 max	0.20 max	0.60 max	0.20 max	0.030 max	0.035 max		Mo <=0.06; Nb <=0.008; Ti <=0.008; V <=0.008; Cu+Ni+Cr+Mo<=0.5; bal Fe		205-345	>=25	
ASTM A569/A569M(98)	B(Cu)	HR Sh Strp, when Cu spec'd	0.02-0.15	0.15 max	0.20 min	0.60 max	0.20 max	0.030 max	0.035 max		Mo <=0.06; Nb <=0.008; Ti <=0.008; V <=0.008; bal Fe		205-345	>=25	
ASTM A569/A569M(98)	C	HR Sh Strp, when Cu not spec'd	0.08 max	0.15 max	0.20 max	0.60 max	0.20 max	0.030 max	0.035 max		Mo <=0.06; Nb <=0.008; Ti <=0.008; V <=0.008; Cu+Ni+Cr+Mo<=0.5; bal Fe		205-345	>=25	
ASTM A569/A569M(98)	C(Cu)	HR Sh Strp, when Cu spec'd	0.08 max	0.15 max	0.20 min	0.60 max	0.20 max	0.030 max	0.035 max		Mo <=0.06; Nb <=0.008; Ti <=0.008; V <=0.008; bal Fe		205-345	>=25	
ASTM A695(95)	A35	HW special qual for fluid power apps						0.040 max	0.050 max	0.15-0.35	bal Fe	415	240	21	
ASTM A695(95)	D35	HW special qual for fluid power apps						0.040 max	0.050 max	0.15-0.35	Pb 0.15-0.35; bal Fe	415	240	21	
ASTM A822(95)	A822	Smls CD tub; hydraulic systems	0.18 max			0.27-0.63		0.048 max	0.058 max		bal Fe	310	170	35	
ASTM A865(97)	A865	Weld or smls threaded couplings						0.14 max	0.35 max		bal Fe				
ASTM A905(93)		Wir, Press ves, 1.02mm thk, Class 1	0.80-0.95			0.30-0.60		0.025 max	0.020 max	0.10-0.30	bal Fe	1965	1725	4.0	
DoD-F-24669/8(91)	Grade A	Frg for Steam Turbines; Supersedes MIL-S-860B(66)	0.45 max			0.90 max		0.025 max	0.025 max	0.15-0.35	V 0.03-0.12; V optional; As 0-0.020; Sb 0-0.010; Sn 0-0.020; bal Fe	517	276	22	
MIL-S-12505B(70)	- - -*	Obs; Shape Plt Bar for weld rivet bolted structures									bal Fe				
SAE J356(96)		Weld flash controlled Tub, no hard req'd if t<1.65mm	0.18 max			0.30-0.60		0.04 max	0.05 max		bal Fe	310	170	35	65 HRB
SAE J524(96)		Smls Ann Tub, no hard req'd if t<1.65mm	0.18 max			0.30-0.60		0.040 max	0.050 max		bal Fe	310	170	35	65 HRB
SAE J525(96)		Weld CD Tub, no hard req'd if t<1.65mm	0.18 max			0.30-0.60		0.040 max	0.050 max		bal Fe	310	170	35	65 HRB

UNS numbers and US grades are provided as a means of cross referencing chemically similar alloys. Exchangability is only possible after independent examination of specifications. Tensile properties are minimum or typical as specified. UTS and YS as MPa. El as %. See Appendix for list of abbreviations used in Notes. * indicates obsolete material.

Specification	Designation	Notes	C	Cr	Cu	Mn	Ni	P	S	Si	Other	UTS	YS	El	Hard

Carbon Steel, Nonresulfurized, No equivalents identified

Yugoslavia

| | C.1402A | Tub w/special req | 0.36 max | | | 0.4 max | | 0.05 max | 0.05 max | 0.1-0.55 | Mo <=0.15; bal Fe | | | | |

Carbon Steel, Nonresulfurized, U

USA

| | UNS K07200 | | 0.65-0.80 | | | 0.60-0.85 | | 0.05 max | 0.05 max | 0.15 min | bal Fe | | | | |

Specification	Designation	Notes	C	Cr	Mn	Ni	P	Pb	S	Si	Other	UTS	YS	El	Hard
Carbon Steel, Resulfurized, 1108															
Bulgaria															
BDS 6886	A10	Free-cutting	0.07-0.13	0.3 max	0.5-0.9		0.06 max		0.15-0.25	0.15-0.4	Cu <=0.30; Mo <=0.15; bal Fe				
France															
AFNOR NFA35551(75)	CC10Pb		0.05-0.15	0.3 max	0.3-0.5	0.4 max	0.04 max	0.15-0.25	0.04 max	0.3 max	Cu <=0.3; bal Fe				
AFNOR NFA35562(81)	10F1	Free-cutting	0.04-0.13	0.3 max	0.6-0.9	0.4 max	0.04 max		0.09-0.13	0.1-0.4	Cu <=0.3; bal Fe				
AFNOR NFA35562(86)	13MF4	Size t<=100mm	0.1-0.16	0.3 max	0.8-1.1	0.4 max	0.04 max		0.1-0.15	0.1-0.4	Cu <=0.3; bal Fe	370-470	250	27	
Hungary															
MSZ 4339(75)	ABPb	Free-cutting; CH	0.07-0.14		0.6-1.1		0.06 max	0.15-0.3	0.15-0.25	0.15-0.4	bal Fe				
MSZ 4339(86)	ABS1	Free-cutting; CH; 0-16mm; HF; bright turned (ground)	0.07-0.13		0.7-1.1		0.06 max		0.15-0.25	0.15-0.35	bal Fe	360-530			159 HB max
MSZ 4339(86)	ABS1	Free-cutting; CH; 40.1-63mm; CD	0.07-0.13		0.7-1.1		0.06 max		0.15-0.25	0.15-0.35	bal Fe	390-640		10L	
International															
ISO 683-9(88)	10S20	Free-cutting, CH, 16mm test	0.07-0.13		0.70-1.10		0.06 max		0.15-0.25	0.15-0.40	bal Fe	450-800	270	12	
ISO 683-9(88)	10SPb20	Free-cutting, CH, 16mm test	0.07-0.13		0.70-1.10		0.06 max	0.15-0.35	0.15-0.25	0.15-0.40	bal Fe	400-700	250	12	
Italy															
UNI 4838(62)	10S22	Free-cutting	0.06-0.12		0.5-0.9		0.07 max		0.18-0.26	0.1-0.4	bal Fe				
UNI 4838(80)	CF10S20	Free-cutting	0.07-0.13		0.6-0.9		0.06 max	0.15-0.3	0.18-0.25	0.1-0.4	bal Fe				
UNI 4838(80)	CF10SPb20	Free-cutting	0.07-0.13		0.6-0.9		0.06 max	0.15-0.3	0.18-0.25	0.1-0.4	bal Fe				
Mexico															
NMX-B-301(86)	1108	Bar	0.08-0.13		0.60-0.80		0.040 max		0.08-0.13		bal Fe				
Poland															
PNH84026	A11	Free-cutting	0.07-0.13	0.3 max	0.5-0.9	0.4 max	0.06 max		0.15-0.25	0.15-0.4	bal Fe				
Spain															
UNE 36021(61)	F.212	Free-cutting	0.07-0.15		2 max		0.11 max	0.15-0.35	0.2-0.35	0.6 max	Mo <=0.15; bal Fe				
UNE 36021(80)	10S20	Free-cutting	0.07-0.13		0.5-0.9		0.06 max		0.15-0.25	0.15-0.4	Mo <=0.15; bal Fe				
UNE 36021(80)	10SPb20	Free-cutting	0.07-0.13		0.5-0.9		0.06 max	0.15-0.35	0.15-0.25	0.15-0.4	Mo <=0.15; bal Fe				
UNE 36021(80)	F.2121*	see 10S20	0.07-0.13		0.5-0.9		0.06 max		0.15-0.25	0.15-0.4	Mo <=0.15; bal Fe				
UNE 36021(80)	F.2122*	see 10SPb20	0.07-0.13		0.5-0.9		0.06 max	0.15-0.35	0.15-0.25	0.15-0.4	Mo <=0.15; bal Fe				
UK															
BS 970/1(83)	210M15	Blm Bil Slab Bar Rod Frg CH	0.12-0.18		0.90-1.30				0.10-0.18		bal Fe	430		18	
BS 970/1(83)	210M15	Blm Bil Slab Bar Rod Frg CH	0.12-0.18		0.90-1.30				0.10-0.18		bal Fe	490		16	
USA															
	AISI 1108	Wir rod	0.08-0.13		0.50-0.80		0.040 max		0.080-0.13		bal Fe				
	UNS G11080		0.08-0.13		0.50-0.80		0.040 max		0.08-0.13		bal Fe				
ASTM A29/A29M(93)	1108	Bar	0.08-0.13		0.60-0.80		0.040 max		0.08-0.13		bal Fe				
ASTM A510(96)	1108	Wir rod	0.08-0.13		0.50-0.80		0.040 max		0.08-0.13		bal Fe				
SAE J1249(95)	1108	Former SAE std valid for Wir rod	0.08-0.13		0.50-0.80		0.040 max		0.08-0.13		bal Fe				
SAE J1397(92)	1108	Bar CD, est mech prop									bal Fe	390	320	20	121 HB
SAE J1397(92)	1108	Bar HR, est mech prop									bal Fe	340	190	30	101 HB

UNS numbers and US grades are provided as a means of cross referencing chemically similar alloys. Exchangability is only possible after independent examination of specifications. Tensile properties are minimum or typical as specified. UTS and YS as MPa. El as %. See Appendix for list of abbreviations used in Notes. * indicates obsolete material.

Specification	Designation	Notes	C	Cr	Mn	Ni	P	Pb	S	Si	Other	UTS	YS	El	Hard

Carbon Steel, Resulfurized, 1108 (Continued from previous page)

Yugoslavia

| | C.0247 | Unalloyed; rivets | 0.1 max | | 0.4-0.7 | | 0.08 max | | 0.12 max | 0.60 max | bal Fe | | | | |
| | C.1190 | Free-cutting | 0.07-0.12 | | 0.6-0.9 | | 0.07 max | | 0.18-0.25 | 0.25 max | Mo <=0.15; bal Fe | | | | |

Carbon Steel, Resulfurized, 1109

Germany

DIN E EN 10087(95)	10S20	Free-cutting HR Bar Rod CD 64-100mm	0.07-0.13		0.70-1.10		0.060 max		0.15-0.25	0.40 max	bal Fe	360-610	235	11	
DIN E EN 10087(95)	10S20	Free-cutting HR Bar Rod CD 11-16mm	0.07-0.13		0.70-1.10		0.060 max		0.15-0.25	0.40 max	bal Fe	490-740	390	8	
DIN E EN 10087(95)	WNr 1.0721	HR Bar Rod free-cutting CD 11-16mm	0.07-0.13		0.70-1.10		0.060 max		0.15-0.25	0.40 max	bal Fe	490-740	390	8	
DIN E EN 10087(95)	WNr 1.0721	HR Bar Rod free-cutting CD 64-100mm	0.07-0.13		0.70-1.10		0.060 max		0.15-0.25	0.40 max	bal Fe	390-610	235	11	

Mexico

| NMX-B-301(86) | 1109 | Bar | 0.08-0.13 | | 0.60-0.90 | | 0.040 max | | 0.08-0.13 | | bal Fe | | | | |

USA

	UNS G11090		0.08-0.13		0.60-0.90		0.040 max		0.08-0.13		bal Fe				
ASTM A29/A29M(93)	1109	Bar	0.08-0.13		0.60-0.80		0.040 max		0.08-0.13		bal Fe				
ASTM A510(96)	1109	Wir rod	0.08-0.13		0.60-0.90		0.040 max		0.08-0.13		bal Fe				
ASTM A576(95)	1109	Special qual HW bar	0.08-0.13		0.60-0.90		0.040 max		0.08-0.13		Si Pb Bi Ca Se Te if spec'd; bal Fe				

Carbon Steel, Resulfurized, 1110

Australia

| AS 1442 | XR1110* | Obs; Bar Bil | 0.06-0.11 | | 0.3-0.6 | | 0.05 | | 0.05-0.1 | | bal Fe | | | | |

Germany

DIN 17111(80)	C10RG1	Bolts Rivets	0.15 max		0.30-0.60		0.050 max		0.08-0.12		trace Si; bal Fe				
DIN 17111(80)	C10RG2	Bolts Nuts Rivets	0.15 max		0.30-0.80		0.050 max		0.08-0.12	0.40 max	bal Fe				
DIN 17111(80)	R10S10	Bolts Rivets	0.15 max		0.30-0.80		0.050 max		0.08-0.12	0.40 max	bal Fe				
DIN 17111(80)	U10S10	Bolts Nuts Rivets	0.15 max		0.30-0.60		0.050 max		0.08-0.12		trace Si; bal Fe				
DIN 17111(80)	WNr 1.0702	Bolts Nuts Rivets	0.15 max		0.30-0.60		0.050 max		0.08-0.12		trace Si; bal Fe				
DIN 17111(80)	WNr 1.0703	Bolts Nuts Rivets	0.15 max		0.30-0.80		0.050 max		0.08-0.12	0.40 max	bal Fe				

Japan

| JIS G4804(83) | SUM11 | Bar Free cutting | 0.08-0.13 | | 0.30-0.60 | | 0.040 max | | 0.08-0.13 | | bal Fe | | | | |
| JIS G4804(83) | SUM12 | Bar Free cutting | 0.08-0.13 | | 0.60-0.90 | | 0.040 max | | 0.08-0.13 | | bal Fe | | | | |

Mexico

| NMX-B-201(68) | 1110 | CD buttweld mech tub | 0.08-0.15 | | 0.30-0.60 | | 0.04 max | | 0.08-0.13 | | bal Fe | | | | |
| NMX-B-301(86) | 1110 | Bar | 0.08-0.13 | | 0.30-0.60 | | 0.040 max | | 0.08-0.13 | | bal Fe | | | | |

Pan America

| COPANT 330 | 1110 | Bar | 0.08-0.13 | | 0.3-0.6 | | 0.04 | | 0.08-0.13 | | bal Fe | | | | |

USA

	AISI 1110	Wir rod	0.08-0.13		0.30-0.60		0.040 max		0.080-0.13		bal Fe				
	AISI 1110	Bar	0.08-0.13		0.30-0.60		0.040 max		0.08-0.12		bal Fe				
	UNS G11100		0.08-0.13		0.30-0.60		0.040 max		0.08-0.13		bal Fe				

UNS numbers and US grades are provided as a means of cross referencing chemically similar alloys. Exchangability is only possible after independent examination of specifications. Tensile properties are minimum or typical as specified. UTS and YS as MPa. El as %. See Appendix for list of abbreviations used in Notes. * indicates obsolete material.

Specification	Designation	Notes	C	Cr	Mn	Ni	P	Pb	S	Si	Other	UTS	YS	El	Hard

Carbon Steel, Resulfurized, 1110 (Continued from previous page)

USA

Specification	Designation	Notes	C	Cr	Mn	Ni	P	Pb	S	Si	Other	UTS	YS	El	Hard
ASTM A29/A29M(93)	1110	Bar	0.08-0.13		0.30-0.60		0.040 max		0.08-0.12		bal Fe				
ASTM A510(96)	1110	Wir rod	0.08-0.13		0.30-0.60		0.040 max		0.08-0.13		bal Fe				
ASTM A512(96)	1110	Buttweld mech tub, CD	0.08-0.15		0.30-0.60		0.040 max		0.08-0.130		bal Fe				
ASTM A576(95)	1110	Special qual HW bar	0.08-0.13		0.30-0.60		0.040 max		0.08-0.13		Si Pb Bi Ca Se Te if spec'd; bal Fe				
FED QQ-S-637A(70)	C1110*	Obs; see ASTM A108; CF Bar std qual, free mach									bal Fe				
SAE J403(95)	1110	Bil Bar Rod Tub	0.08-0.13		0.3-0.6		0.04		0.08-0.13		bal Fe				

Yugoslavia

Specification	Designation	Notes	C	Cr	Mn	Ni	P	Pb	S	Si	Other	UTS	YS	El	Hard
	C.0267	Unalloyed; rivets	0.15 max		0.4-0.7		0.05 max		0.12 max	0.60 max	bal Fe				
	C.3190	Free-cutting	0.14-0.2		1.1-1.4		0.07 max		0.2-0.25	0.1-0.4	Mo <=0.15; bal Fe				

Carbon Steel, Resulfurized, 1116

Hungary

Specification	Designation	Notes	C	Cr	Mn	Ni	P	Pb	S	Si	Other	UTS	YS	El	Hard
MSZ 4339(86)	ABS2	Free-cutting; CH; 40.1-63mm; CD	0.14-0.2		0.7-1.1		0.06 max		0.15-0.25	0.15-0.35	bal Fe	440-690		9L	
MSZ 4339(86)	ABS2	Free-cutting; CH; 0-16mm; HF; bright turned(ground)	0.14-0.2		0.7-1.1		0.06 max		0.15-0.25	0.15-0.35	bal Fe	400-590			174 HB max

Japan

Specification	Designation	Notes	C	Cr	Mn	Ni	P	Pb	S	Si	Other	UTS	YS	El	Hard
JIS G4804(83)	SUM32	Bar Free cutting	0.12-0.20		0.60-1.10		0.040 max		0.10-0.20		bal Fe				

Mexico

Specification	Designation	Notes	C	Cr	Mn	Ni	P	Pb	S	Si	Other	UTS	YS	El	Hard
NMX-B-301(86)	1116	Bar	0.14-0.20		1.10-1.40		0.040 max		0.16-0.23		bal Fe				

USA

Specification	Designation	Notes	C	Cr	Mn	Ni	P	Pb	S	Si	Other	UTS	YS	El	Hard
	UNS G11160		0.14-0.20		1.10-1.40		0.040 max		0.16-0.23		bal Fe				
ASTM A29/A29M(93)	1116	Bar	0.14-0.20		1.10-1.40		0.040 max		0.16-0.23		bal Fe				
ASTM A510(96)	1116	Wir rod	0.14-0.20		1.10-1.40		0.040 max		0.16-0.23		bal Fe				
ASTM A576(95)	1116	Special qual HW bar	0.14-0.20		1.10-1.40		0.040 max		0.16-0.23		Si Pb Bi Ca Se Te if spec'd; bal Fe				

Carbon Steel, Resulfurized, 1117

Australia

Specification	Designation	Notes	C	Cr	Mn	Ni	P	Pb	S	Si	Other	UTS	YS	El	Hard
AS 1442	K1117*	Obs; Bar Bil	0.14-0.2		1-1.3		0.05		0.08-0.13	0.1-0.35	bal Fe				
AS 1443	K1117*	Obs; Bar	0.14-0.2		1-1.3		0.05		0.08-0.13	0.1-0.35	bal Fe				

Bulgaria

Specification	Designation	Notes	C	Cr	Mn	Ni	P	Pb	S	Si	Other	UTS	YS	El	Hard
BDS 6886	A20		0.15-0.25		0.6-0.9		0.06		0.08-0.15	0.15-0.35	bal Fe				

Canada

Specification	Designation	Notes	C	Cr	Mn	Ni	P	Pb	S	Si	Other	UTS	YS	El	Hard
CSA 94-1/1	C1117		0.14-0.2		1-1.3				0.08-0.13		bal Fe				
CSA STAN95-1-1	CL.1.C1117	Bar	0.14-0.2		1-1.3				0.08-0.13		bal Fe				

China

Specification	Designation	Notes	C	Cr	Mn	Ni	P	Pb	S	Si	Other	UTS	YS	El	Hard
GB 8731(88)	Y20	Bar HR/CD	0.17-0.25		0.70-1.00		0.06 max		0.08-0.15	0.15-0.35	bal Fe	450-600		20	

Japan

Specification	Designation	Notes	C	Cr	Mn	Ni	P	Pb	S	Si	Other	UTS	YS	El	Hard
JIS G4804(83)	SUM31	Bar Free cutting	0.14-0.20		1.00-1.30		0.040 max		0.08-0.13		bal Fe				
JIS G4804(83)	SUM31L	Bar Free cutting Leaded	0.14-0.20		1.00-1.30		0.040 max	0.10-0.35	0.08-0.13		bal Fe				

Specification	Designation	Notes	C	Cr	Mn	Ni	P	Pb	S	Si	Other	UTS	YS	El	Hard
Carbon Steel, Resulfurized, 1117 (Continued from previous page)															
Mexico															
NMX-B-301(86)	1117	Bar	0.14-0.20		1.00-1.30		0.040 max		0.08-0.13		bal Fe				
Pan America															
COPANT 330	1117	Bar	0.14-0.2		1-1.3		0.04		0.08-0.13		bal Fe				
COPANT 331	1117	Bar	0.14-0.2		1-1.3		0.04		0.08-0.13		bal Fe				
COPANT 333	1117	Wir rod	0.14-0.2		1-1.3		0.04		0.05		bal Fe				
Romania															
STAS 1350(89)	AUT20	Free-cutting	0.15-0.25	0.3 max	0.6-0.9	0.4 max	0.06 max	0.15 max	0.08-0.15	0.15-0.35	bal Fe				
Russia															
GOST	A15G		0.1-0.2		1-1.4		0.06		0.08-0.15	0.15-0.35	bal Fe				
GOST 1414(75)	A20	Free-cutting	0.17-0.25		0.7-1	0.4 max	0.06 max		0.08-0.15	0.15-0.35	Cu <=0.25; bal Fe				
Spain															
UNE	F.210.F		0.14-0.2		1-1.4		0.045		0.07-0.13	0.13-0.38	bal Fe				
UK															
BS 970/1(83)	210A15	Blm Bil Slab Bar Rod Frg CH	0.13-0.18		0.90-1.20				0.10-0.18		bal Fe				
BS 970/1(83)	210M17	Blm Bil Slab Bar Rod Frg	0.12-0.18		0.90-1.30		0.050 max		0.10-0.18	0.1-0.4	bal Fe				
BS 970/1(96)	210A15	Blm Bil Slab Bar Rod Frg	0.13-0.18		0.90-1.20		0.050 max		0.10-0.18	0.10-0.40	bal Fe				
BS 970/1(96)	210M15	Blm Bil Slab Bar Rod Frg	0.12-0.18		0.90-1.30		0.050 max		0.10-0.18	0.10-0.40	bal Fe				
BS 970/1(96)	214A15	Blm Bil Slab Bar Rod Frg	0.13-0.18		1.20-1.50		0.050 max		0.10-0.18	0.10-0.40	bal Fe				
BS 970/1(96)	214M15	Blm Bil Slab Bar Rod Frg	0.12-0.18		1.20-1.60		0.050 max		0.10-0.18	0.10-0.40	bal Fe	740		12	
BS 970/1(96)	214M15	Wrought CH	0.12-0.18		1.20-1.60		0.050 max		0.10-0.18	0.10-0.40	bal Fe	590		13	
USA															
	AISI 1117	Wir rod	0.14-0.20		1.00-1.30		0.040 max		0.080-0.13		bal Fe				
	AISI 1117	Bar	0.14-0.20		1.00-1.30		0.040 max		0.08-0.13		bal Fe				
	UNS G11170		0.14-0.20		1.00-1.30		0.040 max		0.08-0.13		bal Fe				
AMS 5022L(90)		Mech tub, OD>63.5 and Frg	0.14-0.20		1.00-1.30		0.040 max		0.08-0.13		bal Fe				179 HB max
AMS 5022L(90)		Bar, HF t>63.5mm	0.14-0.20		1.00-1.30		0.040 max		0.08-0.13		bal Fe				179 HB max
ASTM A108(95)	1117	Bar, CF	0.14-0.20		1.00-1.30		0.040 max		0.08-0.13		bal Fe				
ASTM A29/A29M(93)	1117	Bar	0.14-0.20		1.00-1.30		0.040 max		0.08-0.13		bal Fe				
ASTM A311/A311M(95)	1117 Class A	Bar CD SR ann, t<=20mm	0.14-0.20		1.00-1.30		0.040 max		0.08-0.13		bal Fe	520	450	15,R A=40	
ASTM A311/A311M(95)	1117 Class A	Bar CD SR ann, 20<t<=30mm	0.14-0.20		1.00-1.30		0.040 max		0.08-0.13		bal Fe	485	415	15,R A=40	
ASTM A311/A311M(95)	1117 Class A	Bar CD SR ann, 50<t<=75mm	0.14-0.20		1.00-1.30		0.040 max		0.08-0.13		bal Fe	415	345	12,R A=30	
ASTM A311/A311M(95)	1117 Class A	Bar CD SR ann, 30<t<=50mm	0.14-0.20		1.00-1.30		0.040 max		0.08-0.13		bal Fe	450	380	13,R A=35	
ASTM A510(96)	1117	Wir rod	0.14-0.20		1.00-1.30		0.040 max		0.08-0.13		bal Fe				
ASTM A512(96)	1117	Buttweld mech tub, CD	0.14-0.20		1.00-1.30		0.040 max		0.08-0.130		bal Fe				
ASTM A576(95)	1117	Special qual HW bar	0.14-0.20		1.00-1.30		0.040 max		0.08-0.13		Si Pb Bi Ca Se Te if spec'd; bal Fe				
FED QQ-S-637A(70)	C1117*	Obs; see ASTM A108; CF Bar std qual, free mach									bal Fe				

UNS numbers and US grades are provided as a means of cross referencing chemically similar alloys. Exchangability is only possible after independent examination of specifications. Tensile properties are minimum or typical as specified. UTS and YS as MPa. El as %. See Appendix for list of abbreviations used in Notes. * indicates obsolete material.

Specification	Designation	Notes	C	Cr	Mn	Ni	P	Pb	S	Si	Other	UTS	YS	El	Hard

Carbon Steel, Resulfurized, 1117 (Continued from previous page)

USA

Specification	Designation	Notes	C	Cr	Mn	Ni	P	Pb	S	Si	Other	UTS	YS	El	Hard
MIL-S-18411	1117*	Obs; Bar rod wire tube	0.14-0.2		1-1.3		0.04 max		0.08-0.13		bal Fe				
SAE J1397(92)	1117	Bar HR, est mech prop	0.14-0.2		1-1.3		0.04 max		0.08-0.13		bal Fe	430	230	23	121 HB
SAE J1397(92)	1117	Bar CD, est mech prop	0.14-0.2		1-1.3		0.04 max		0.08-0.13		bal Fe	480	400	15	137 HB
SAE J403(95)	1117	Bar Wir rod Smls Tub HR CF	0.14-0.20		1.00-1.30		0.030 max		0.08-0.13		bal Fe				

Carbon Steel, Resulfurized, 1118

Argentina

Specification	Designation	Notes	C	Cr	Mn	Ni	P	Pb	S	Si	Other	UTS	YS	El	Hard
IAS	IRAM 1118		0.14-0.20		1.30-1.60		0.040 max		0.08-0.13		bal Fe	450-540	280-360	23-32	130-161 HB

Canada

Specification	Designation	Notes	C	Cr	Mn	Ni	P	Pb	S	Si	Other	UTS	YS	El	Hard
CSA 95-1/1	C1118		0.14-0.2		1.3-1.6				0.08-0.13		bal Fe				

Mexico

Specification	Designation	Notes	C	Cr	Mn	Ni	P	Pb	S	Si	Other	UTS	YS	El	Hard
NMX-B-301(86)	1118	Bar	0.14-0.20		1.30-1.60		0.040 max		0.08-0.13		bal Fe				

Pan America

Specification	Designation	Notes	C	Cr	Mn	Ni	P	Pb	S	Si	Other	UTS	YS	El	Hard
COPANT 330	1118	Bar	0.14-0.2		1.3-1.6		0.04		0.08-0.13		bal Fe				
COPANT 331	1118	Bar	0.14-0.2		1.3-1.6		0.04		0.08-0.13		bal Fe				
COPANT 514	1118	Tube, Ann	0.14-0.2		1.3-1.6		0.04		0.08-0.13		bal Fe				
COPANT 514	1118	Tube, CF	0.14-0.2		1.3-1.6		0.04		0.08-0.13		bal Fe				
COPANT 514	1118	Tube, Norm	0.14-0.2		1.3-1.6		0.04		0.08-0.13		bal Fe				
COPANT 514	1118	Tube, HF, 324mm diam	0.14-0.2		1.3-1.6		0.04		0.08-0.13		bal Fe				

USA

Specification	Designation	Notes	C	Cr	Mn	Ni	P	Pb	S	Si	Other	UTS	YS	El	Hard
	AISI 1118	Wir rod	0.14-0.20		1.30-1.60		0.040 max		0.080-0.13		bal Fe				
	AISI 1118	Bar	0.14-0.20		1.30-1.60		0.040 max		0.08-0.13		bal Fe				
	UNS G11180		0.14-0.20		1.30-1.60		0.040 max		0.08-0.13		bal Fe				
ASTM A108(95)	1118	Bar, CF	0.14-0.20		1.30-1.60		0.040 max		0.08-0.13		bal Fe				
ASTM A29/A29M(93)	1118	Bar	0.14-0.20		1.30-1.60		0.040 max		0.08-0.13		bal Fe				
ASTM A510(96)	1118	Wir rod	0.14-0.20		1.30-1.60		0.040 max		0.08-0.13		bal Fe				
ASTM A519(96)	1118	Smls mech tub, HR	0.14-0.20		1.30-1.60		0.040 max		0.08-0.130		bal Fe	345	241	25	55 HRB
ASTM A519(96)	1118	Smls mech tub, SR	0.14-0.20		1.30-1.60		0.040 max		0.08-0.130		bal Fe	483	379	8	75 HRB
ASTM A519(96)	1118	Smls mech tub, CW	0.14-0.20		1.30-1.60		0.040 max		0.08-0.130		bal Fe	517	414	5	80 HRB
ASTM A519(96)	1118	Smls mech tub, Norm	0.14-0.20		1.30-1.60		0.040 max		0.08-0.130		bal Fe	379	241	20	60 HRB
ASTM A519(96)	1118	Smls mech tub, Ann	0.14-0.20		1.30-1.60		0.040 max		0.08-0.130		bal Fe	345	207	25	55 HRB
ASTM A519(96)	11L18	Smls mech tub, HR	0.14-0.20		1.30-1.60		0.040 max		0.08-0.130		bal Fe				
ASTM A576(95)	1118	Special qual HW bar	0.14-0.20		1.30-1.60		0.040 max		0.08-0.13		Si Pb Bi Ca Se Te if spec'd; bal Fe				
FED QQ-S-00643(61)	1118*	Obs; Tub smls weld	0.14-0.2		1.3-1.6		0.04 max		0.13 max		bal Fe				
FED QQ-S-633A(63)	1118*	Obs; CF HR Bar	0.14-0.2		1.3-1.6		0.04 max		0.13 max		bal Fe				
FED QQ-S-637A(70)	C1118*	Obs; see ASTM A108; CF Bar std qual, free mach	0.14-0.2		1.3-1.6		0.04 max		0.13 max		bal Fe				

UNS numbers and US grades are provided as a means of cross referencing chemically similar alloys. Exchangability is only possible after independent examination of specifications. Tensile properties are minimum or typical as specified. UTS and YS as MPa. El as %. See Appendix for list of abbreviations used in Notes. * indicates obsolete material.

Specification	Designation	Notes	C	Cr	Mn	Ni	P	Pb	S	Si	Other	UTS	YS	El	Hard
Carbon Steel, Resulfurized, 1118 (Continued from previous page)															
USA															
MIL-S-16124	- - -*	Obs	0.14-0.2		1.3-1.6		0.04 max		0.13 max		bal Fe				
MIL-S-16611	- - -*	Obs	0.14-0.2		1.3-1.6		0.04 max		0.13 max		bal Fe				
SAE J403(95)	1118	Bil Bar Rod Tub	0.14-0.2		1.3-1.6		0.04 max		0.13 max		bal Fe				
SAE J403(95)	1118	Bar Wir rod Smls Tub HR CF	0.14-0.20		1.30-1.60		0.030 max		0.08-0.13		bal Fe				
Carbon Steel, Resulfurized, 1119															
Australia															
AS 1442(92)	X1112	Free-cutting, HR bar, Semifinished	0.08-0.15		1.10-1.40		0.040 max		0.20-0.30	0.10 max	bal Fe				
AS 1443(94)	D11	CR or CR bar, 16-38mm	0.08-0.15		1.10-1.40		0.040 max		0.20-0.30	0.10 max	Pb 0.15-0.35 for lead-bearing steels; bal Fe	430	330	12	
AS 1443(94)	T11	CF bar, not CD or CR;<=260mm	0.08-0.15		1.10-1.40		0.040 max		0.20-0.30	0.10 max	Pb 0.15-0.35 for lead-bearing steels; bal Fe	370	230	17	
AS 1443(94)	X1112	CF bar, Free-cutting	0.08-0.15		1.10-1.40		0.040 max		0.20-0.30	0.10 max	bal Fe				
Mexico															
NMX-B-301(86)	1119	Bar	0.14-0.20		1.00-1.30		0.040 max		0.24-0.33		bal Fe				
USA															
	UNS G11190		0.14-0.20		1.00-1.30		0.040 max		0.24-0.33		bal Fe				
ASTM A183(98)	Grade 1 Bolt	Rail Nuts, load test reqd	0.15 min				0.05 max		0.33 max		bal Fe				
ASTM A183(98)	Grade 1 Bolt	Rail Nuts, load test reqd	0.15 min				0.05 max		0.33 max		bal Fe				
ASTM A183(98)	Grade 1 Nut	Rail Nuts, load test reqd	0.15 min				0.12 max		0.06 max		bal Fe				
ASTM A183(98)	Grade 1 Nut	Rail Nuts, load test reqd	0.15 min				0.12 max		0.06 max		bal Fe				
ASTM A29/A29M(93)	1119	Bar	0.14-0.20		1.00-1.30		0.040 max		0.24-0.33		bal Fe				
ASTM A510(96)	1119	Wir rod	0.14-0.20		1.00-1.30		0.040 max		0.24-0.33		bal Fe				
ASTM A576(95)	1119	Special qual HW bar	0.14-0.20		1.00-1.30		0.040 max		0.24-0.33		Si Pb Bi Ca Se Te if spec'd; bal Fe				
Carbon Steel, Resulfurized, 1123															
USA															
	AISI 1123	Bar	0.20-0.27		1.20-1.50		0.040 max		0.06-0.09		bal Fe				
	UNS G11230		0.20-0.27		1.20-1.50		0.040 max		0.06-0.09		bal Fe				
Carbon Steel, Resulfurized, 1132															
China															
GB 8731(88)	Y30	Bar HR/CD	0.27-0.35		0.70-1.00		0.06 max		0.08-0.15	0.15-0.35	bal Fe	510-655		15	
Mexico															
NMX-B-301(86)	1132	Bar	0.27-0.34		1.35-1.65		0.040 max		0.08-0.13		bal Fe				
Russia															
GOST 1414(75)	A30	Free-cutting	0.26-0.35		0.7-1	0.4 max	0.06 max		0.08-0.15	0.15-0.35	Cu <=0.25; bal Fe				
USA															
	UNS G11320		0.27-0.34		1.35-1.65		0.040 max		0.09-0.13		bal Fe				
ASTM A29/A29M(93)	1132	Bar	0.27-0.34		1.35-1.65		0.040 max		0.08-0.13		bal Fe				
ASTM A510(96)	1132	Wir rod	0.27-0.34		1.35-1.65		0.040 max		0.08-0.13		bal Fe				
ASTM A519(96)	1132	Smls mech tub	0.27-0.34		1.35-1.65		0.040 max		0.08-0.13		bal Fe				

UNS numbers and US grades are provided as a means of cross referencing chemically similar alloys. Exchangability is only possible after independent examination of specifications. Tensile properties are minimum or typical as specified. UTS and YS as MPa. El as %. See Appendix for list of abbreviations used in Notes. * indicates obsolete material.

Specification	Designation	Notes	C	Cr	Mn	Ni	P	Pb	S	Si	Other	UTS	YS	El	Hard
Carbon Steel, Resulfurized, 1132 (Continued from previous page)															
USA															
ASTM A576(95)	1132	Special qual HW bar	0.27-0.34		1.35-1.65		0.040 max		0.08-0.13		Si Pb Bi Ca Se Te if spec'd; bal Fe				
SAE J1397(92)	1132	Bar CD, est mech prop									bal Fe	630	530	12	183 HB
SAE J1397(92)	1132	Bar HR, est mech prop									bal Fe	570	310	16	167 HB
Carbon Steel, Resulfurized, 1137															
Australia															
AS 1442	K1137*	Obs; Bar Bil	0.32-0.39		1.35-1.65		0.05		0.08-0.13		bal Fe				
AS 1442	K1138*	Obs	0.34-0.4		0.7-1		0.05		0.08-0.13	0.1-0.35	bal Fe				
AS 1442(92)	1137	Free-cutting, HR bar, Semifinished	0.32-0.39		1.35-1.65		0.040 max		0.08-0.13	0.10-0.35	bal Fe				
AS 1443	K1137*	Obs; Bar	0.32-0.39		1.35-1.65		0.05		0.08-0.13	0.1-0.35	bal Fe				
AS 1443	K1138*	Obs	0.34-0.4		0.7-1		0.05		0.08-0.13	0.1-0.35	bal Fe				
AS 1443(94)	1137	CF bar, Free-cutting	0.32-0.39		1.35-1.65		0.040 max		0.08-0.13	0.10-0.35	bal Fe				
AS 1443(94)	D14	CR or CR bar, 16-38mm	0.32-0.39		0.35-1.65		0.040 max		0.08-0.13	0.10-0.35	Pb 0.15-0.35 for lead-bearing steels; bal Fe	640	480	7	
AS 1443(94)	T14	CF bar, not CD or CR;<=260mm	0.32-0.39		0.35-1.65		0.040 max		0.08-0.13	0.10-0.34	Pb 0.15-0.35 for lead-bearing steels; bal Fe	600	300	14	
China															
GB/T 15712(95)	YF35MnV	Bar Controlled Rolling/Forging	0.32-0.39		1.00-1.50		0.035 max		0.035-0.075	0.30-0.60	V 0.06-0.13; bal Fe	735	460	17	
France															
AFNOR	35MF4		0.32-0.38		1-1.3		0.06		0.12-0.24	0.1-0.4	bal Fe				
AFNOR NFA35562	35MF6		0.33-0.39		1.3-1.7		0.04 max		0.09-0.13	0.1-0.4	bal Fe				
India															
IS 1570	40Mn2S12		0.35-0.45		1.3-1.7		0.06		0.08-0.15	0.25	bal Fe				
IS 5517	40Mn2S12	Bar, Norm, Ann, Hard Tmp, 60mm	0.35-0.45		1.3-1.7		0.06		0.08-0.15	0.25	bal Fe				
Japan															
JIS G4804(83)	SUM41	Bar Free cutting	0.32-0.39		1.35-1.65		0.040 max		0.08-0.13		bal Fe				
Mexico															
DGN B-296	1137*	Obs	0.32-0.39		1.35-1.65		0.04		0.08-0.13	0.1-0.2	bal Fe				
NMX-B-301(86)	1137	Bar	0.32-0.39		1.35-1.65		0.040 max		0.08-0.13		bal Fe				
Pan America															
COPANT 330	1137	Bar	0.32-0.39		1.35-1.65		0.04		0.08-0.13		bal Fe				
COPANT 331	1137	Bar	0.32-0.39		1.35-1.65		0.04		0.08-0.13		bal Fe				
COPANT 332	1137	Bar, CF Ann, 15mm diam	0.32-0.39		1.35-1.65		0.04		0.08-0.13		bal Fe				
COPANT 514	1137	Tube, Norm	0.32-0.39		1.35-1.65		0.04		0.08-0.13		bal Fe				
COPANT 514	1137	Tube, HF, 324mm diam	0.32-0.39		1.35-1.65		0.04		0.08-0.13		bal Fe				
COPANT 514	1137	Tube, CF	0.32-0.39		1.35-1.65		0.04		0.08-0.13		bal Fe				
COPANT 514	1137	Tube, Ann	0.32-0.39		1.35-1.65		0.04		0.08-0.13		bal Fe				
Russia															
GOST 1414	A35E		0.32-0.4	0.25	0.5-0.8	0.25	0.04		0.06-0.12	0.17-0.37	Cu 0.25; Se 0.04-0.1; bal Fe				
GOST 1414(75)	A35	Free-cutting	0.32-0.40		0.7-1	0.4 max	0.06 max		0.08-0.15	0.15-0.35	Cu <=0.25; bal Fe				

UNS numbers and US grades are provided as a means of cross referencing chemically similar alloys. Exchangability is only possible after independent examination of specifications. Tensile properties are minimum or typical as specified. UTS and YS as MPa. El as %. See Appendix for list of abbreviations used in Notes. * indicates obsolete material.

Specification	Designation	Notes	C	Cr	Mn	Ni	P	Pb	S	Si	Other	UTS	YS	El	Hard

Carbon Steel, Resulfurized, 1137 (Continued from previous page)

Spain

Specification	Designation	Notes	C	Cr	Mn	Ni	P	Pb	S	Si	Other	UTS	YS	El	Hard
UNE	F.210.G		0.32-0.39		1.35-1.65		0.04		0.08-0.13	0.35	bal Fe				
UNE 36021	35MnS6	Free-cutting	0.33-0.39		1.3-1.7		0.04		0.09-0.13	0.1-0.4	bal Fe				
UNE 36021	F.2131*	see 35MnS6	0.33-0.39		1.3-1.7		0.04		0.09-0.13	0.1-0.4	bal Fe				
UNE 36021(80)	35MnSPb6	Free-cutting	0.33-0.39		1.3-1.7		0.04	0.15-0.35	0.09-0.13	0.1-0.4	bal Fe				
UNE 36021(80)	F.2132*	see 35MnSPb6	0.33-0.39		1.3-1.7		0.04	0.15-0.35	0.09-0.13	0.1-0.4	bal Fe				

UK

Specification	Designation	Notes	C	Cr	Mn	Ni	P	Pb	S	Si	Other	UTS	YS	El	Hard
BS 970/1(83)	212M36	Blm Bil Slab Bar Rod Frg	0.32-0.40		1.00-1.40		0.060 max		0.12-0.20	0.25	bal Fe				
BS 970/1(83)	225M36	Blm Bil Slab Bar Rod Frg	0.32-0.40		1.30-1.70		0.060 max		0.12-0.20	0.2-0.25	bal Fe				

USA

Specification	Designation	Notes	C	Cr	Mn	Ni	P	Pb	S	Si	Other	UTS	YS	El	Hard
	AISI 1137	Wir rod	0.32-0.39		1.35-1.65		0.040 max		0.080-0.13		bal Fe				
	AISI 1137	Bar	0.32-0.39		1.35-1.65		0.040 max		0.08-0.13		bal Fe				
	UNS G11370		0.32-0.39		1.35-1.65		0.040 max		0.08-0.13		bal Fe				
AMS 5024F	1137	Bar Rod, Mech tub t<=15.75	0.32-0.39		1.35-1.65		0.040 max		0.08-0.13		bal Fe				207-255 HB
AMS 5024F	1137	Frg	0.32-0.39		1.35-1.65		0.040 max		0.08-0.13		bal Fe				163-229 HB
ASTM A108(95)	1137	Bar, CF	0.32-0.39		1.35-1.65		0.040 max		0.08-0.13		bal Fe				
ASTM A29/A29M(93)	1137	Bar	0.32-0.39		1.35-1.65		0.040 max		0.08-0.13		bal Fe				
ASTM A311/A311M(95)	1137 Class A	Bar CD SR ann, 30<t<=50mm	0.32-0.39		1.35-1.65		0.040 max		0.08-0.13		bal Fe	585	550	10,RA=30	
ASTM A311/A311M(95)	1137 Class A	Bar CD SR ann, 50<t<=75mm	0.32-0.39		1.35-1.65		0.040 max		0.08-0.13		bal Fe	550	520	10,RA=30	
ASTM A311/A311M(95)	1137 Class A	Bar CD SR ann, t<=20mm	0.32-0.39		1.35-1.65		0.040 max		0.08-0.13		bal Fe	655	620	11,RA=35	
ASTM A311/A311M(95)	1137 Class A	Bar CD SR ann, 20<t<=30mm	0.32-0.39		1.35-1.65		0.040 max		0.08-0.13		bal Fe	620	585	11,RA=30	
ASTM A510(96)	1137	Wir rod	0.32-0.39		1.35-1.65		0.040 max		0.08-0.13		bal Fe				
ASTM A519(96)	1137	Smls mech tub, HR	0.32-0.39		1.35-1.65		0.040 max		0.08-0.130		bal Fe	483	276	20	75 HRB
ASTM A519(96)	1137	Smls mech tub, Ann	0.32-0.39		1.35-1.65		0.040 max		0.08-0.130		bal Fe	448	241	22	72 HRB
ASTM A519(96)	1137	Smls mech tub, SR	0.32-0.39		1.35-1.65		0.040 max		0.08-0.130		bal Fe	517	414	8	80 HRB
ASTM A519(96)	1137	Smls mech tub, CW	0.32-0.39		1.35-1.65		0.040 max		0.08-0.130		bal Fe	552	448	5	85 HRB
ASTM A519(96)	1137	Smls mech tub, Norm	0.32-0.39		1.35-1.65		0.040 max		0.08-0.130		bal Fe	483	296	15	75 HRB
ASTM A576(95)	1137	Special qual HW bar	0.32-0.39		1.35-1.65		0.040 max		0.08-0.13		Si Pb Bi Ca Se Te if spec'd; bal Fe				
FED QQ-S-633A(63)	1137*	Obs; CF HR Bar	0.32-0.39		1.35-1.65		0.04 max		0.08-0.13		bal Fe				
FED QQ-S-637A(70)	C1137*	Obs; see ASTM A108; CF Bar std qual, free mach									bal Fe				
MIL-S-16124	- - -*	Obs	0.32-0.39		1.35-1.65		0.04 max		0.08-0.13		bal Fe				
SAE J1397(92)	1137	Bar CD, est mech prop	0.32-0.39		1.35-1.65		0.04 max		0.08-0.13		bal Fe	680	570	10	197 HB
SAE J1397(92)	1137	Bar HR, est mech prop	0.32-0.39		1.35-1.65		0.04 max		0.08-0.13		bal Fe	610	330	15	179 HB
SAE J403(95)	1137	Bar Wir rod Smls Tub HR CF	0.32-0.39		1.35-1.65		0.030 max		0.08-0.13		bal Fe				

Specification	Designation	Notes	C	Cr	Mn	Ni	P	Pb	S	Si	Other	UTS	YS	El	Hard

Carbon Steel, Resulfurized, 1139

Bulgaria

Specification	Designation	Notes	C	Cr	Mn	Ni	P	Pb	S	Si	Other	UTS	YS	El	Hard
BDS 6886	A35		0.32-0.39		0.5-0.9		0.06		0.15-0.25	0.15-0.4	bal Fe				

France

Specification	Designation	Notes	C	Cr	Mn	Ni	P	Pb	S	Si	Other	UTS	YS	El	Hard
AFNOR	35MF4		0.32-0.38		1-1.3		0.06		0.12-0.24	0.1-0.4	bal Fe				
AFNOR NFA35562	35MF6		0.33-0.39		1.11-1.3	0.04 max	0.04-0.09		0.13-0.24	0.1-0.4	bal Fe				

India

Specification	Designation	Notes	C	Cr	Mn	Ni	P	Pb	S	Si	Other	UTS	YS	El	Hard
IS 1570	40S18		0.35-0.45		0.8-1.2	0.04 max	0.06		0.14-0.22	0.25	bal Fe				
IS 5517	40S18		0.35-0.45		0.8-1.2	0.04 max	0.06		0.14-0.22	0.25	bal Fe				

Russia

Specification	Designation	Notes	C	Cr	Mn	Ni	P	Pb	S	Si	Other	UTS	YS	El	Hard
GOST 1414(75)	A40G	Free-cutting	0.35-0.45		1.2-1.55		0.05		0.18-0.3	0.15-0.35	Cu <=0.25; bal Fe				

UK

Specification	Designation	Notes	C	Cr	Mn	Ni	P	Pb	S	Si	Other	UTS	YS	El	Hard
BS 970/1(83)	212A37	Blm Bil Slab Bar Rod Frg	0.35-0.40		1.00-1.30		0.060 max		0.12-0.20	0.25	bal Fe				
BS 970/1(83)	212M36	Blm Bil Slab Bar Rod Frg Hard Tmp; 4 in.	0.32-0.40		1.00-1.40		0.060 max		0.12-0.20	0.25	bal Fe				
BS 970/1(83)	216M36	Blm Bil Slab Bar Rod Frg	0.32-0.40		1.30-1.70		0.060 max		0.12-0.20	0.25	bal Fe				

Carbon Steel, Resulfurized, 1140

Argentina

Specification	Designation	Notes	C	Cr	Mn	Ni	P	Pb	S	Si	Other	UTS	YS	El	Hard
IAS	IRAM 1140		0.37-0.44		0.70-1.00		0.040 max		0.08-0.13		bal Fe	620-720	370-470	10-30	188-215 HB

China

Specification	Designation	Notes	C	Cr	Mn	Ni	P	Pb	S	Si	Other	UTS	YS	El	Hard
GB 8731(88)	Y35	Bar HR/CD	0.32-0.40		0.70-1.00		0.06 max		0.08-0.15	0.15-0.35	bal Fe	510-655		14	

Czech Republic

Specification	Designation	Notes	C	Cr	Mn	Ni	P	Pb	S	Si	Other	UTS	YS	El	Hard
CSN 411140	11140	Free-cutting	0.35-0.45	0.3 max	0.5-1.0		0.1 max		0.11-0.21	0.4 max	bal Fe				

France

Specification	Designation	Notes	C	Cr	Mn	Ni	P	Pb	S	Si	Other	UTS	YS	El	Hard
AFNOR	35MF4		0.32-0.38		1-1.3		0.06		0.12-0.24	0.1-0.4	bal Fe				

Germany

Specification	Designation	Notes	C	Cr	Mn	Ni	P	Pb	S	Si	Other	UTS	YS	El	Hard
DIN E EN 10087(95)	35S20	Free-cutting HR Bar Rod CD 64-100mm	0.32-0.39		0.70-1.10		0.060 max		0.15-0.25	0.40 max	bal Fe	480-680	255	10	
DIN E EN 10087(95)	35S20	Free-cutting HR Bar Rod CD 11-16mm	0.32-0.39		0.70-1.10		0.060 max		0.15-0.25	0.40 max	bal Fe	590-830	400	7	
DIN E EN 10087(95)	WNr 1.0726	HR Bar Rod free-cutting CD Q/T 17-40mm	0.32-0.39		0.70-1.10		0.060 max		0.15-0.25	0.40 max	bal Fe	580-730	365	16	
DIN E EN 10087(95)	WNr 1.0726	HR Bar Rod free-cutting CD 64-100mm	0.32-0.39		0.70-1.10		0.060 max		0.15-0.25	0.40 max	bal Fe	480-680	255	10	
DIN E EN 10087(95)	WNr 1.0726	HR Bar Rod free-cutting CD 11-16mm	0.32-0.39		0.70-1.10		0.060 max		0.15-0.25	0.40 max	bal Fe	590-830	400	7	

Mexico

Specification	Designation	Notes	C	Cr	Mn	Ni	P	Pb	S	Si	Other	UTS	YS	El	Hard
NMX-B-301(86)	1140	Bar	0.37-0.44		0.70-1.00		0.040 max		0.08-0.13		bal Fe				

Romania

Specification	Designation	Notes	C	Cr	Mn	Ni	P	Pb	S	Si	Other	UTS	YS	El	Hard
STAS 1350(89)	AUT30	Free-cutting	0.25-0.35	0.3 max	0.7-1	0.4 max	0.06 max	0.15 max	0.08-0.15	0.15-0.35	bal Fe				

Spain

Specification	Designation	Notes	C	Cr	Mn	Ni	P	Pb	S	Si	Other	UTS	YS	El	Hard
UNE 36021(80)	35MnS6	Free-cutting	0.33-0.39		1.3-1.7		0.04 max		0.09-0.13	0.1-0.4	Mo <=0.15; bal Fe				
UNE 36021(80)	F.2131*	see 35MnS6	0.33-0.39		1.3-1.7		0.04 max		0.09-0.13	0.1-0.4	Mo <=0.15; bal Fe				

Sweden

Specification	Designation	Notes	C	Cr	Mn	Ni	P	Pb	S	Si	Other	UTS	YS	El	Hard
SIS 141957	1957-00	Bar, as roll 63mm diam	0.32-0.39		0.8-1.2		0.06		0.15-0.25	0.1-0.4	bal Fe	540			
SIS 141957	1957-03	Bar, Q/A, Tmp, 16mm diam	0.32-0.39		0.8-1.2		0.06		0.15-0.25	0.1-0.4	bal Fe	650	450	14	
SIS 141957	1957-04	Bar, CW, 16mm diam	0.32-0.39		0.8-1.2		0.06		0.15-0.25	0.1-0.4	bal Fe	640	490	7	

UNS numbers and US grades are provided as a means of cross referencing chemically similar alloys. Exchangability is only possible after independent examination of specifications. Tensile properties are minimum or typical as specified. UTS and YS as MPa. El as %. See Appendix for list of abbreviations used in Notes. * indicates obsolete material.

Specification	Designation	Notes	C	Cr	Mn	Ni	P	Pb	S	Si	Other	UTS	YS	El	Hard

Carbon Steel, Resulfurized, 1140 (Continued from previous page)

Sweden

| SS 141957 | 1957 | | 0.32-0.39 | | 0.8-1.2 | | 0.06 | | 0.15-0.25 | 0.1-0.4 | bal Fe | | | | |

USA

	AISI 1140	Wir rod	0.37-0.44		0.70-1.10		0.040 max		0.080-0.13		bal Fe				
	AISI 1140	Bar	0.37-0.44		0.70-1.00		0.040 max		0.08-0.13		bal Fe				
	UNS G11400		0.37-0.44		0.70-1.00		0.040 max		0.08-0.13		bal Fe				
ASTM A29/A29M(93)	1140	Bar	0.37-0.44		0.70-1.00		0.040 max		0.08-0.13		bal Fe				
ASTM A510(96)	1140	Wir rod	0.37-0.44		0.70-1.10		0.040 max		0.08-0.13		bal Fe				
ASTM A576(95)	1140	Special qual HW bar	0.37-0.44		0.70-1.00		0.040 max		0.08-0.13		Si Pb Bi Ca Se Te if spec'd; bal Fe				
FED QQ-S-637A(70)	C1140*	Obs; see ASTM A108; CF Bar std qual, free mach	0.37-0.44		0.7-1		0.04 max		0.08-0.13		bal Fe				
SAE J1397(92)	1140	Bar CD, est mech prop	0.37-0.44		0.7-1		0.04 max		0.08-0.13		bal Fe	610	510	12	170 HB
SAE J1397(92)	1140	Bar HR, est mech prop	0.37-0.44		0.7-1		0.04 max		0.08-0.13		bal Fe	540	300	16	156 HB
SAE J403(95)	1140	Bar Wir rod Smls Tub HR CF	0.37-0.44		0.70-1.00		0.030 max		0.08-0.13		bal Fe				

Yugoslavia

| | C.1490 | Free-cutting | 0.32-0.39 | | 0.6-0.9 | | 0.07 max | | 0.15-0.25 | 0.1-0.4 | Mo <=0.15; bal Fe | | | | |
| | C.1490 | Free-cutting | 0.32-0.39 | | 0.6-0.9 | | 0.07 max | | 0.15-0.25 | 0.1-0.4 | Mo <=0.15; bal Fe | | | | |

Carbon Steel, Resulfurized, 1141

Argentina

| IAS | IRAM 1141 | | 0.37-0.45 | | 1.35-1.65 | | 0.040 max | | 0.08-0.13 | | bal Fe | 680-780 | 450-530 | 8-17 | 206-233 HB |

China

| GB/T 15712(95) | YF40MnV | Bar Controlled Rolling/Forging | 0.37-0.44 | | 1.00-1.50 | | 0.035 max | | 0.035-0.075 | 0.30-0.60 | V 0.06-0.13; bal Fe | 785 | 490 | 15 | |

France

| AFNOR NFA35562 | 45MF4 | | 0.42-0.49 | | 0.8-1.1 | | 0.04 | | 0.09-0.13 | 0.1-0.4 | bal Fe | | | | |

India

| IS 1570 | 40Mn2S12 | | 0.35-0.45 | | 1.3-1.7 | | 0.06 | | 0.08-0.15 | 0.25 | bal Fe | | | | |
| IS 5517 | 40Mn2S12 | | 0.35-0.45 | | 1.3-1.7 | | 0.06 | | 0.08-0.15 | 0.25 | bal Fe | | | | |

Japan

| JIS G4804(83) | SUM42 | Bar Free cutting | 0.37-0.45 | | 1.35-1.65 | | 0.040 max | | 0.08-0.13 | | bal Fe | | | | |

Mexico

| DGN B-296 | 1141* | Obs | 0.37-0.45 | | 1.35-1.65 | | 0.04 | | 0.08-0.13 | 0.1-0.2 | bal Fe | | | | |
| NMX-B-301(86) | 1141 | Bar | 0.37-0.45 | | 1.35-1.65 | | 0.040 max | | 0.08-0.13 | | bal Fe | | | | |

Pan America

COPANT 330	1141	Bar	0.37-0.45		1.35-1.65		0.04		0.08-0.13		bal Fe				
COPANT 331	1141	Bar	0.37-0.45		1.35-1.65		0.04		0.08-0.13		bal Fe				
COPANT 332	1141	Bar, CF Ann, 15mm diam	0.37-0.45		1.35-1.65		0.04		0.08-0.13		bal Fe				
COPANT 514	1141	Tube	0.37-0.45		1.35-1.65		0.04		0.08-0.13		bal Fe				

UK

| BS 970/1(83) | 216A42 | Blm Bil Slab Bar Rod Frg | 0.40-0.45 | | 1.20-1.50 | | 0.060 max | | 0.12-0.20 | 0.25 | bal Fe | | | | |

UNS numbers and US grades are provided as a means of cross referencing chemically similar alloys. Exchangability is only possible after independent examination of specifications. Tensile properties are minimum or typical as specified. UTS and YS as MPa. El as %. See Appendix for list of abbreviations used in Notes. * indicates obsolete material.

Specification	Designation	Notes	C	Cr	Mn	Ni	P	Pb	S	Si	Other	UTS	YS	El	Hard

Carbon Steel, Resulfurized, 1141 (Continued from previous page)

USA

Specification	Designation	Notes	C	Cr	Mn	Ni	P	Pb	S	Si	Other	UTS	YS	El	Hard
	1141	CD	0.37-0.45		1.35-1.65		0.04 max		0.08-0.13		bal Fe	825	690	10.0	241-321 HB
	AISI 1141	Bar	0.37-0.45		1.35-1.65		0.040 max		0.08-0.13		bal Fe				
	AISI 1141	Wir rod	0.37-0.45		1.35-1.65		0.040 max		0.080-0.13		bal Fe				
	UNS G11410		0.37-0.45		1.35-1.65		0.040 max		0.08-0.13		bal Fe				
ASTM A108(95)	1141	Bar, CF	0.37-0.45		1.35-1.65		0.040 max		0.08-0.13		bal Fe				
ASTM A29/A29M(93)	1141	Bar	0.37-0.45		1.35-1.65		0.040 max		0.08-0.13		bal Fe				
ASTM A311/A311M(95)	1141 Class A	Bar CD SR ann, 50<t<=75mm	0.37-0.45		1.35-1.65		0.040 max		0.08-0.13		bal Fe	550	520	10,RA=30	
ASTM A311/A311M(95)	1141 Class A	Bar CD SR ann, t<=20mm	0.37-0.45		1.35-1.65		0.040 max		0.08-0.13		bal Fe	655	620	11,RA=35	
ASTM A311/A311M(95)	1141 Class A	Bar CD SR ann, 20<t<=30mm	0.37-0.45		1.35-1.65		0.040 max		0.08-0.13		bal Fe	620	585	11,RA=30	
ASTM A311/A311M(95)	1141 Class A	Bar CD SR ann, 30<t<=50mm	0.37-0.45		1.35-1.65		0.040 max		0.08-0.13		bal Fe	585	550	10,RA=30	
ASTM A311/A311M(95)	1141 Class B	Bar CD SR ann, 30<t<=50mm	0.37-0.45		1.35-1.65		0.040 max		0.08-0.13		bal Fe	795	690	9,RA=20	
ASTM A311/A311M(95)	1141 Class B	Bar CD SR ann, t<=50mm	0.37-0.45		1.35-1.65		0.040 max		0.08-0.13		bal Fe	795	690	10,RA=25	
ASTM A311/A311M(95)	1141 Class B	Bar CD SR ann, 75<t<=115mm	0.37-0.45		1.35-1.65		0.040 max		0.08-0.13		bal Fe	725	620	7,RA=20	
ASTM A510(96)	1141	Wir rod	0.37-0.45		1.35-1.65		0.040 max		0.08-0.13		bal Fe				
ASTM A519(96)	1141	Smls mech tub	0.37-0.45		1.35-1.65		0.040 max		0.08-0.130		bal Fe				
ASTM A576(95)	1141	Special qual HW bar	0.37-0.45		1.35-1.65		0.040 max		0.08-0.13		Si Pb Bi Ca Se Te if spec'd; bal Fe				
FED QQ-S-633A(63)	1141*	Obs; CF HR Bar	0.37-0.45		1.35-1.65		0.04 max		0.08-0.13		bal Fe				
FED QQ-S-637A(70)	1141*	Obs; see ASTM A108; CF Bar std qual, free mach	0.37-0.45		1.35-1.65		0.04 max		0.08-0.13		bal Fe				
SAE J1397(92)	1141	Bar HR, est mech prop	0.37-0.45		1.35-1.65		0.04 max		0.08-0.13		bal Fe	650	360	15	187 HB
SAE J1397(92)	1141	Bar CD, est mech prop	0.37-0.45		1.35-1.65		0.04 max		0.08-0.13		bal Fe	720	610	10	212 HB
SAE J403(95)	1141	Bar Wir rod Smls Tub HR CF	0.37-0.45		1.35-1.65		0.030 max		0.08-0.13		bal Fe				

Carbon Steel, Resulfurized, 1144

Australia

Specification	Designation	Notes	C	Cr	Mn	Ni	P	Pb	S	Si	Other	UTS	YS	El	Hard
AS 1442	XK1145*	Obs; Bar Bil	0.42-0.49		1.35-1.65		0.05		0.12-0.2	0.1-0.35	bal Fe				
AS 1442	XK1152*	Obs; Bar Bil	0.48-0.56		1.35-1.65		0.05		0.24-0.33	0.1-0.35	bal Fe				
AS 1442(92)	1144	Free-cutting, HR bar, Semifinished	0.40-0.48		1.35-1.65		0.040 max		0.24-0.33	0.35 max	bal Fe				
AS 1443	XK1152*	Obs; Bar	0.48-0.56		1.35-1.65		0.05		0.24-0.33	0.1-0.35	bal Fe				
AS 1443(94)	1144	CF bar, Free-cutting	0.40-0.48		1.35-1.65		0.040 max		0.24-0.33	0.35 max	bal Fe				

Bulgaria

Specification	Designation	Notes	C	Cr	Mn	Ni	P	Pb	S	Si	Other	UTS	YS	El	Hard
BDS 6886	A45		0.42-0.5		0.5-0.9		0.06		0.15-0.25	0.15-0.4	bal Fe				

China

Specification	Designation	Notes	C	Cr	Mn	Ni	P	Pb	S	Si	Other	UTS	YS	El	Hard
GB 8731(88)	Y40Mn	Bar CD 8-20mm diam	0.37-0.45		1.20-1.55		0.05 max		0.20-0.30	0.15-0.35	bal Fe	590-785		17	

France

Specification	Designation	Notes	C	Cr	Mn	Ni	P	Pb	S	Si	Other	UTS	YS	El	Hard
AFNOR NFA35562	45MF6		0.41-0.48		1.3-1.7		0.04 max		0.24-0.32	0.1-0.4	bal Fe				

UNS numbers and US grades are provided as a means of cross referencing chemically similar alloys. Exchangability is only possible after independent examination of specifications. Tensile properties are minimum or typical as specified. UTS and YS as MPa. El as %. See Appendix for list of abbreviations used in Notes. * indicates obsolete material.

Specification	Designation	Notes	C	Cr	Mn	Ni	P	Pb	S	Si	Other	UTS	YS	El	Hard
Carbon Steel, Resulfurized, 1144 (Continued from previous page)															
Japan															
JIS G4804(83)	SUM43	Bar Free cutting	0.40-0.48		1.35-1.65		0.040 max		0.24-0.33		bal Fe				
Mexico															
NMX-B-301(86)	1144	Bar	0.40-0.48		1.35-1.65		0.040 max		0.24-0.33		bal Fe				
Pan America															
COPANT 330	1444	Bar	0.4-0.48		1.35-1.65		0.04		0.24-0.33		bal Fe				
COPANT 331	1444	Bar	0.4-0.48		1.35-1.65		0.04		0.24-0.33		bal Fe				
COPANT 332	1444	Bar, CF Ann, 15mm diam	0.4-0.48		1.35-1.65		0.04		0.24-0.33		bal Fe				
COPANT 332	1444	Tube	0.4-0.48		1.35-1.65		0.04		0.24-0.33		bal Fe				
Poland															
PNH84026	A45		0.42-0.5		0.5-0.9		0.06		0.15-0.25	0.15-0.4	bal Fe				
Romania															
STAS 1350(89)	AUT40Mn	Free-cutting	0.35-0.45	0.3 max	1.2-1.6	0.4 max	0.06 max	0.15 max	0.18-0.3	0.15-0.35	bal Fe				
Spain															
UNE 36021(61)	45MnS6		0.41-0.48		1.3-1.7		0.04		0.24-0.32	0.1-0.4	bal Fe				
UNE 36021(61)	F.2133*	see 45MnS6	0.41-0.48		1.3-1.7		0.04		0.24-0.32	0.1-0.4	bal Fe				
UK															
BS 970/1(83)	212A42	Blm Bil Slab Bar Rod Frg	0.40-0.45		1.00-1.30		0.060 max		0.12-0.20	0.25	bal Fe				
BS 970/1(83)	216M44	Blm Bil Slab Bar Rod Frg	0.40-0.48		1.20-1.50		0.060 max		0.12-0.20	0.25	bal Fe				
BS 970/3(91)	226M44	Bright bar; free-cutting; 6<t<=13mm; hard CD	0.40-0.48		1.30-1.70		0.060 max		0.22-0.30	0.25 max	Mo <=0.15; bal Fe	850-1000	630	9	248-302 HB
BS 970/3(91)	226M44	Bright bar; free-cutting; 6<t<=100mm; hard	0.40-0.48		1.30-1.70		0.060 max		0.22-0.30	0.25 max	Mo <=0.15; bal Fe	700-850	450	16	201-255 HB
USA															
	1144	CD	0.4-0.48		1.35-1.65		0.04 max		0.24-0.33		bal Fe	965	860	5.0	280 HB
	1144	CD	0.4-0.48		1.35-1.65		0.04 max		0.24-0.33		bal Fe	825	690	10.0	248-321 HB
	AISI 1144	Wir rod	0.40-0.48		1.35-1.65		0.040 max		0.24-0.33		bal Fe				
	AISI 1144	Bar	0.40-0.48		1.35-1.65		0.040 max		0.24-0.33		bal Fe				
	UNS G11440		0.40-0.48		1.35-1.65		0.040 max		0.24-0.33		bal Fe				
ASTM A108(95)	1144	Bar, CF	0.40-0.48		1.35-1.65		0.040 max		0.24-0.33		bal Fe				
ASTM A29/A29M(93)	1144	Bar	0.40-0.48		1.35-1.65		0.040 max		0.24-0.33		bal Fe				
ASTM A311/A311M(95)	1144 Class A	Bar CD SR ann, 50<t<=75mm	0.40-0.48		1.35-1.65		0.040 max		0.24-0.33		bal Fe	620	550	10,RA=20	
ASTM A311/A311M(95)	1144 Class A	Bar CD SR ann, 30<t<=50mm	0.40-0.48		1.35-1.65		0.040 max		0.24-0.33		bal Fe	655	585	10,RA=25	
ASTM A311/A311M(95)	1144 Class A	Bar CD SR ann, 20<t<=30mm	0.40-0.48		1.35-1.65		0.040 max		0.24-0.33		bal Fe	690	620	10,RA=30	
ASTM A311/A311M(95)	1144 Class A	Bar CD SR ann, t<=20mm	0.40-0.48		1.35-1.65		0.040 max		0.24-0.33		bal Fe	725	655	10,RA=30	
ASTM A311/A311M(95)	1144 Class A	Bar CD SR ann, 75<t<=115mm	0.40-0.48		1.35-1.65		0.040 max		0.24-0.33		bal Fe	585	520	10,RA=20	
ASTM A311/A311M(95)	1144 Class B	Bar CD SR ann, t<=50mm	0.40-0.48		1.35-1.65		0.040 max		0.24-0.33		bal Fe	795	690	10,RA=25	
ASTM A311/A311M(95)	1144 Class B	Bar CD SR ann, 75<t<=115mm	0.40-0.48		1.35-1.65		0.040 max		0.24-0.33		bal Fe	725	620	7,RA=20	
ASTM A311/A311M(95)	1144 Class B	Bar CD SR ann, 30<t<=50mm	0.40-0.48		1.35-1.65		0.040 max		0.24-0.33		bal Fe	795	690	9,RA=20	

UNS numbers and US grades are provided as a means of cross referencing chemically similar alloys. Exchangability is only possible after independent examination of specifications. Tensile properties are minimum or typical as specified. UTS and YS as MPa. El as %. See Appendix for list of abbreviations used in Notes. * indicates obsolete material.

Specification	Designation	Notes	C	Cr	Mn	Ni	P	Pb	S	Si	Other	UTS	YS	El	Hard

Carbon Steel, Resulfurized, 1144 (Continued from previous page)

USA

Specification	Designation	Notes	C	Cr	Mn	Ni	P	Pb	S	Si	Other	UTS	YS	El	Hard
ASTM A510(96)	1144	Wir rod	0.40-0.48		1.35-1.65		0.040 max		0.24-0.33		bal Fe				
ASTM A519(96)	1144	Smls mech tub	0.40-0.48		1.35-1.65		0.040 max		0.24-0.330		bal Fe				
ASTM A576(95)	1144	Special qual HW bar	0.40-0.48		1.35-1.65		0.040 max		0.24-0.33		Si Pb Bi Ca Se Te if spec'd; bal Fe				
FED QQ-S-637A(70)	C1144*	Obs; see ASTM A108; CF Bar std qual, free mach	0.4-0.48		1.35-1.65		0.04 max		0.24-0.33		bal Fe				
SAE J1397(92)	1144	Bar CD, est mech prop	0.4-0.48		1.35-1.65		0.04 max		0.24-0.33		bal Fe	740	620	10	217 HB
SAE J1397(92)	1144	Bar HR, est mech prop	0.4-0.48		1.35-1.65		0.04 max		0.24-0.33		bal Fe	670	370	15	197 HB
SAE J403(95)	1144	Bar Wir rod Smls Tub HR CF	0.40-0.48		1.35-1.65		0.030 max		0.24-0.33		bal Fe				

Yugoslavia

Specification	Designation	Notes	C	Cr	Mn	Ni	P	Pb	S	Si	Other	UTS	YS	El	Hard
	C.1590	Free-cutting	0.42-0.51		0.6-0.9		0.07		0.25	0.1-0.4	bal Fe				

Carbon Steel, Resulfurized, 1145

Mexico

Specification	Designation	Notes	C	Cr	Mn	Ni	P	Pb	S	Si	Other	UTS	YS	El	Hard
NMX-B-301(86)	1145	Bar	0.42-0.49		0.70-1.00		0.040 max		0.04-0.07		bal Fe				

USA

Specification	Designation	Notes	C	Cr	Mn	Ni	P	Pb	S	Si	Other	UTS	YS	El	Hard
	UNS G11450		0.41-0.49		0.70-1.00		0.040 max		0.08-0.13		bal Fe				
ASTM A29/A29M(93)	1145	Bar	0.42-0.49		0.70-1.00		0.040 max		0.04-0.07		bal Fe				
ASTM A510(96)	1145	Wir rod	0.42-0.49		0.70-1.00		0.040 max		0.04-0.07		bal Fe				
ASTM A576(95)	1145	Special qual HW bar	0.42-0.49		0.70-1.00		0.040 max		0.04-0.07		Si Pb Bi Ca Se Te if spec'd; bal Fe				

Carbon Steel, Resulfurized, 1146

Australia

Specification	Designation	Notes	C	Cr	Mn	Ni	P	Pb	S	Si	Other	UTS	YS	El	Hard
AS 1442	K1146*	Obs; Bar Bil	0.42-0.49		0.7-1		0.05		0.08-0.13	0.1-0.35	bal Fe				
AS 1442(92)	1146	Free-cutting, HR bar, Semifinished	0.42-0.49		0.70-1.00		0.040 max		0.08-0.13	0.10-0.35	bal Fe				
AS 1443	K1146*	Obs; Bar	0.42-0.49		0.7-1		0.05		0.08-0.13	0.1-0.35	bal Fe				
AS 1443(94)	1146	CF bar, Free-cutting	0.42-0.49		0.70-1.00		0.040 max		0.08-0.13	0.35	bal Fe				

China

Specification	Designation	Notes	C	Cr	Mn	Ni	P	Pb	S	Si	Other	UTS	YS	El	Hard
GB/T 15712(95)	YF40MnV	Bar Controlled Rolling/Forging	0.42-0.49		1.00-1.50		0.035 max		0.035-0.075	0.30-0.60	V 0.06-0.13; bal Fe	835	510	13	

France

Specification	Designation	Notes	C	Cr	Mn	Ni	P	Pb	S	Si	Other	UTS	YS	El	Hard
AFNOR NFA35562	45MF4		0.42-0.49		0.8-1.1		0.04 max		0.09-0.13	0.1-0.4	bal Fe				

Germany

Specification	Designation	Notes	C	Cr	Mn	Ni	P	Pb	S	Si	Other	UTS	YS	El	Hard
DIN E EN 10087(95)	WNr 1.0727	HR Bar Rod free-cutting CD Q/T 17-40mm	0.42-0.50		0.70-1.10		0.060 max		0.15-0.25	0.40 max	bal Fe	660-800	410	13	
DIN E EN 10087(95)	WNr 1.0727	HR Bar Rod free-cutting CD 11-16mm	0.42-0.50		0.70-1.10		0.060 max		0.15-0.25	0.40 max	bal Fe	690-930	470	6	
DIN E EN 10087(95)	WNr 1.0727	HR Bar Rod free-cutting CD 64-100mm	0.42-0.50		0.70-1.10		0.060 max		0.15-0.25	0.40 max	bal Fe	580-770	305	9	
DIN EN 10087(95)	45S20	Free-cutting HR rod bar CD 11-16mm	0.42-0.50		0.70-1.10		0.060 max		0.15-0.25	0.40 max	bal Fe	690-930	470	6	
DIN EN 10087(95)	45S20	Free-cutting HR rod bar CD 64-100mm	0.42-0.50		0.70-1.10		0.060 max		0.15-0.25	0.40 max	bal Fe	580-770	305	9	

Hungary

Specification	Designation	Notes	C	Cr	Mn	Ni	P	Pb	S	Si	Other	UTS	YS	El	Hard
MSZ 4339(86)	ANS2	Free-cutting; Q/T; 40.1-63mm	0.42-0.5		0.7-1.1		0.06 max		0.15-0.25	0.15-0.35	bal Fe	620-770	375	14L	
MSZ 4339(86)	ANS2	Free-cutting; Q/T; 0-10mm	0.42-0.5		0.7-1.1		0.06 max		0.15-0.25	0.15-0.35	bal Fe	700-850	480	10L	

UNS numbers and US grades are provided as a means of cross referencing chemically similar alloys. Exchangability is only possible after independent examination of specifications. Tensile properties are minimum or typical as specified. UTS and YS as MPa. El as %. See Appendix for list of abbreviations used in Notes. * indicates obsolete material.

Specification	Designation	Notes	C	Cr	Mn	Ni	P	Pb	S	Si	Other	UTS	YS	El	Hard

Carbon Steel, Resulfurized, 1146 (Continued from previous page)

International

Specification	Designation	Notes	C	Cr	Mn	Ni	P	Pb	S	Si	Other	UTS	YS	El	Hard
ISO 683-9(88)	46S20	Free-cutting, hard, Q/T, <16mm	0.42-0.50		0.70-1.10		0.06 max		0.15-0.25	0.15-0.40	bal Fe	650-850	450	11	

Italy

UNI 4838(62)	40SMnPb10	Free-cutting	0.37-0.45		1.35-1.65		0.04 max	0.15-0.40	0.08-0.13	0.3 max	bal Fe				
UNI 4838(80)	CF44SMn28	Free-cutting	0.4-0.48		1.35-1.65		0.04 max		0.24-0.32	0.3 max	bal Fe				

Mexico

NMX-B-301(86)	1146	Bar	0.42-0.49		0.70-1.00		0.040 max		0.08-0.13		bal Fe				

Pan America

COPANT 331	1146	Bar	0.42-0.49		7 max		0.04		0.08-0.13		bal Fe				

Poland

PNH84026	A45	Free-cutting	0.42-0.5	0.3 max	0.5-0.9	0.4 max	0.06 max		0.15-0.25	0.15-0.4	bal Fe				

Russia

GOST 1414	A45E		0.42-0.5	0.25	0.5-0.8	0.25	0.04		0.06-0.12	0.17-0.37	Cu 0.25; Se 0.04-0.1; bal Fe				

Spain

UNE 36021(80)	45MnS6	Free-cutting	0.41-0.48		1.3-1.7		0.04 max		0.24-0.32	0.1-0.4	Mo <=0.15; bal Fe				
UNE 36021(80)	F.2133*	see 45MnS6	0.41-0.48		1.3-1.7		0.04 max		0.24-0.32	0.1-0.4	Mo <=0.15; bal Fe				

Sweden

SIS 141973	1973-00	Bar, as rolled 63mm diam	0.46-0.54		0.8 1.2		0.06		0.15-0.25	0.1-0.4	bal Fe	640			
SIS 141973	1973-03	Bar, Q/A, Tmp, 16mm diam	0.46-0.54		0.8-1.2		0.06		0.15-0.25	0.1-0.4	bal Fe	750	520	11	
SIS 141973	1973-04	Bar, CW, 16mm diam	0.46-0.54		0.8-1.2		0.06		0.15-0.25	0.1-0.4	bal Fe	780	590	5	
SS 141973	1973	Free-cutting	0.46-0.54		0.8-1.2		0.06 max		0.15-0.25	0.1-0.4	bal Fe				

UK

BS 970/1(83)	212M44	Blm Bil Slab Bar Rod Frg	0.40-0.48		1.00-1.40		0.060 max		0.12-0.20	0.25	bal Fe				

USA

	AISI 1146	Wir rod	0.42-0.49		0.70-1.00		0.040 max		0.080-0.13		bal Fe				
	AISI 1146	Bar	0.42-0.49		0.70-1.00		0.040 max		0.08-0.13		bal Fe				
	UNS G11460		0.42-0.49		0.70-1.00		0.040 max		0.08-0.13		bal Fe				
ASTM A29/A29M(93)	1146	Bar	0.42-0.49		0.70-1.00		0.040 max		0.08-0.13		bal Fe				
ASTM A510(96)	1146	Wir rod	0.42-0.49		0.70-1.00		0.040 max		0.08-0.13		bal Fe				
ASTM A576(95)	1146	Special qual HW bar	0.42-0.49		0.70-1.00		0.040 max		0.08-0.13		Si Pb Bi Ca Se Te if spec'd; bal Fe				
FED QQ-S-637A(70)	C1146*	Obs; see ASTM A108; CF Bar std qual, free mach	0.42-0.49		0.7-1		0.04 max		0.08-0.13		bal Fe				
SAE J1397(92)	1146	Bar CD, est mech prop	0.42-0.49		0.7-1		0.04 max		0.08-0.13		bal Fe	650	550	12	187 HB
SAE J1397(92)	1146	Bar HR, est mech prop	0.42-0.49		0.7-1		0.04 max		0.08-0.13		bal Fe	590	320	15	170 HB
SAE J403(95)	1146	Bar Wir rod Smls Tub HR CF	0.42-0.49		0.70-1.00		0.030 max		0.08-0.13		bal Fe				

Yugoslavia

	C.1590	Free-cutting	0.42-0.51		0.6-0.9		0.07 max		0.25 max	0.1-0.4	Mo <=0.15; bal Fe				

Carbon Steel, Resulfurized, 1151

Germany

DIN	70S20		0.60-0.72		0.50-0.70		0.070 max		0.15-0.25	0.15-0.25	bal Fe				

UNS numbers and US grades are provided as a means of cross referencing chemically similar alloys. Exchangability is only possible after independent examination of specifications. Tensile properties are minimum or typical as specified. UTS and YS as MPa. El as %. See Appendix for list of abbreviations used in Notes. * indicates obsolete material.

Specification	Designation	Notes	C	Cr	Mn	Ni	P	Pb	S	Si	Other	UTS	YS	El	Hard

Carbon Steel, Resulfurized, 1151 (Continued from previous page)

Germany

Specification	Designation	Notes	C	Cr	Mn	Ni	P	Pb	S	Si	Other	UTS	YS	El	Hard
DIN	WNr 1.0729		0.60-0.72		0.50-0.70		0.070 max		0.15-0.25	0.15-0.25	bal Fe				
DIN E EN 10087(95)	60S20	Free-cutting CD 11-16mm	0.57-0.65		0.70-1.10		0.060 max		0.18-0.25	0.10-0.30	bal Fe	780-1030	540	6	
DIN E EN 10087(95)	60S20	Free-cutting CD 64-100mm	0.57-0.65		0.70-1.10		0.060 max		0.18-0.25	0.10-0.30	bal Fe	640-880	335	9	
DIN E EN 10087(95)	WNr 1.0728	Free-cutting CD 11-16mm	0.57-0.65		0.70-1.10		0.060 max		0.18-0.25	0.10-0.30	bal Fe	780-1030	540	6	
DIN E EN 10087(95)	WNr 1.0728	Free-cutting CD 64-100mm	0.57-0.65		0.70-1.10		0.060 max		0.18-0.25	0.10-0.30	bal Fe	640-880	335	9	

Hungary

Specification	Designation	Notes	C	Cr	Mn	Ni	P	Pb	S	Si	Other	UTS	YS	El	Hard
MSZ 4339(86)	ANS3	Free-cutting; Q/T; 40.1-100mm	0.57-0.65		0.7-1.1		0.06 max		0.15-0.25	0.15-0.35	bal Fe	740-880	450	11L	
MSZ 4339(86)	ANS3	Free-cutting; Q/T; 0-10mm	0.57-0.65		0.7-1.1		0.06 max		0.15-0.25	0.15-0.35	bal Fe	830-980	570	7L	

Mexico

Specification	Designation	Notes	C	Cr	Mn	Ni	P	Pb	S	Si	Other	UTS	YS	El	Hard
DGN B-296	1151*	Bar, CF, 14mm diam; Obs	0.48-0.55		0.7-1		0.04			0.08-0.13	bal Fe				
NMX-B-301(86)	1151	Bar	0.48-0.55		0.70-1.00		0.040 max			0.08-0.13	bal Fe				

Pan America

Specification	Designation	Notes	C	Cr	Mn	Ni	P	Pb	S	Si	Other	UTS	YS	El	Hard
COPANT 330	1151	Bar	0.48-0.55		0.7-1		0.04			0.08-0.13	bal Fe				
COPANT 331	1151	Bar	0.48-0.55		0.7-1		0.04			0.08-0.13	bal Fe				
COPANT 332	1151	Bar, CF Ann, 15mm diam	0.48-0.55		0.7-1		0.04			0.08-0.13	bal Fe				

USA

Specification	Designation	Notes	C	Cr	Mn	Ni	P	Pb	S	Si	Other	UTS	YS	El	Hard
	1151	CD	0.48-0.55		0.7-1		0.04 max			0.08-0.13	bal Fe	825	690	10.0	248-321 HB
	AISI 1151	Wir rod	0.48-0.55		0.70-1.00		0.040 max			0.080-0.13	bal Fe				
	UNS G11510		0.48-0.55		0.70-1.00		0.040 max			0.08-0.13	bal Fe				
ASTM A108(95)	1151	Bar, CF	0.48-0.55		0.70-1.00		0.040 max			0.08-0.13	bal Fe				
ASTM A29/A29M(93)	1151	Bar	0.48-0.55		0.70-1.00		0.040 max			0.08-0.13	bal Fe				
ASTM A510(96)	1151	Wir rod	0.48-0.55		0.70-1.00		0.040 max			0.08-0.13	bal Fe				
ASTM A576(95)	1151	Special qual HW bar	0.48-0.55		0.70-1.00		0.040 max			0.08-0.13	Si Pb Bi Ca Se Te if spec'd; bal Fe				
FED QQ-S-637A(70)	C1151*	Obs; see ASTM A108; CF Bar std qual, free mach	0.48-0.55		0.7-1		0.04 max			0.08-0.13	bal Fe				
MIL-S-20137A	- - -*	Obs; Bar rod bil	0.48-0.55		0.7-1		0.04 max			0.08-0.13	bal Fe				
SAE J1249(95)	1151	Former SAE std valid for Wir rod	0.48-0.55		0.70-1.00		0.040 max			0.08-0.13	bal Fe				
SAE J1397(92)	1151	Bar HR, est mech prop	0.48-0.55		0.7-1		0.04 max			0.08-0.13	bal Fe	630	340	15	187 HB
SAE J1397(92)	1151	Bar CD, est mech prop	0.48-0.55		0.7-1		0.04 max			0.08-0.13	bal Fe	700	590	10	207 HB

Carbon Steel, Resulfurized, 5020

Germany

Specification	Designation	Notes	C	Cr	Mn	Ni	P	Pb	S	Si	Other	UTS	YS	El	Hard
DIN EN 10087(95)	60SPb20	Semi-finished HR Bar Rod CD Q/T 17-40mm	0.57-0.65		0.70-1.10		0.060 max	0.15-0.35	0.18-0.25	0.10-0.30	bal Fe	780-930	490	10	
DIN EN 10087(95)	WNr 1.0758	Semi-finished HR Bar Rod CD Q/T 17-40mm	0.57-0.65		0.70-1.10		0.060 max	0.15-0.35	0.18-0.25	0.10-0.30	bal Fe	780-930	490	10	

Italy

Specification	Designation	Notes	C	Cr	Mn	Ni	P	Pb	S	Si	Other	UTS	YS	El	Hard
UNI 4838(80)	CF35SMnPb10	Free-cutting	0.32-0.39		1.35-1.65		0.04 max	0.15-0.3	0.08-0.13	0.3 max	bal Fe				

USA

Specification	Designation	Notes	C	Cr	Mn	Ni	P	Pb	S	Si	Other	UTS	YS	El	Hard
	UNS G11374	Free-Machining	0.32-0.39		1.35-1.65		0.040 max	0.15-0.35	0.08-0.15		bal Fe				

Specification	Designation	Notes	C	Cr	Mn	Ni	P	Pb	S	Si	Other	UTS	YS	El	Hard
Carbon Steel, Resulfurized, A106(A)															
International															
ISO 2604-2(75)	TS5	Smls Tub	0.17 max	0.3 max	0.4-0.8	0.4 max	0.045 max	0.15 max	0.045 max	0.35 max	Al <=0.1; Co <=0.10; Cu <=0.3; Mo <=0.15; Ti <=0.05; V <=0.1; W <=0.1; bal Fe				
Italy															
UNI 5462(64)	C14	Smls tube	0.17 max		0.4 max		0.035 max		0.035 max	0.1-0.35	bal Fe				
Poland															
PNH84024	K10	High-temp const	0.17 max	0.2 max	0.4-1.60	0.35 max	0.045 max		0.045 max	0.35 max	Cu <=0.25; bal Fe				
USA															
	UNS K02501		0.25 max		0.27-0.93		0.048 max		0.058 max	0.10 min	bal Fe				
ASTM A106(97)	A	Smls Pipe	0.25 max	0.40 max	0.27-0.93	0.40 max	0.035 max		0.035 max	0.10 min	Cu <=0.40; Mo <=0.15; V <=0.08; Cr+Cu+Mo+Ni+V<=1.00; bal Fe	330	205	35L	
Yugoslavia															
	C.1212	Tub w/special req	0.17 max		0.4-1.60		0.05 max		0.05 max	0.35 max	Mo <=0.15; bal Fe				
	C.1214	High-temp const	0.17 max		0.4-1.60		0.05 max		0.05 max	0.1-0.35	Mo <=0.15; bal Fe				
Carbon Steel, Resulfurized, A178(A)															
Mexico															
NMX-B-212(00)		CD Heat exchanger condenser tub	0.06-0.18		0.27-0.63		0.048 max		0.058 max		bal Fe	323	179	35	
USA															
	UNS K01200		0.06-0.18		0.27-0.63		0.050 max		0.060 max		bal Fe				
ASTM A178/A178M	A	ERW Pipe	0.06-0.18		0.27-0.63		0.035 max		0.035 max		bal Fe	345	180	35	
ASTM A179/A179M-90a		Smls CD Low C	0.06-0.18		0.27-0.63		0.035 max		0.035 max		bal Fe	325	180	35	72 HRB max
Carbon Steel, Resulfurized, A179															
Czech Republic															
CSN 412021	12021	High-temp const; Boiler tube	0.07-0.15	0.25 max	0.35-0.6	0.25 max	0.04 max		0.04 max	0.17-0.35	Cu <=0.25; bal Fe				
USA															
	UNS K01201		0.06-0.18		0.27-0.63		0.048 max		0.058 max	0.25 max	bal Fe				
Carbon Steel, Resulfurized, A183															
USA															
	UNS K03015		0.30 max				0.04 max		0.06 max		bal Fe				
Carbon Steel, Resulfurized, A2															
USA															
	UNS K01807		0.18 max		0.27-0.63		0.048 max		0.058 max		bal Fe				
ASTM A214/A214M(96)		ERW heat exchanger/condenser tub	0.18 max		0.27-0.63		0.035 max		0.035 max		bal Fe				
ASTM A556/A556M(96)	A2	Smls CD Feedwater Heater Tub	0.18 max		0.27-0.63		0.035 max		0.035 max		bal Fe	320	180	35	72 HRB max
Carbon Steel, Resulfurized, A325(1)															
USA															
	UNS K02706		0.27 min		0.47 min		0.048 max		0.058 max		bal Fe				

UNS numbers and US grades are provided as a means of cross referencing chemically similar alloys. Exchangability is only possible after independent examination of specifications. Tensile properties are minimum or typical as specified. UTS and YS as MPa. El as %. See Appendix for list of abbreviations used in Notes. * indicates obsolete material.

Specification	Designation	Notes	C	Cr	Mn	Ni	P	Pb	S	Si	Other	UTS	YS	El	Hard
Carbon Steel, Resulfurized, A413															
USA															
	UNS K03700		0.37 max				0.048 max		0.058 max		bal Fe				
Carbon Steel, Resulfurized, A500(A)															
Japan															
JIS G3444(94)	STK400	Tub Gen struct	0.25 max				0.040 max		0.040 max		bal Fe	400	235	23	
USA															
	UNS K03000		0.30 max				0.05 max		0.063 max		Cu 0.18 min (when spec'd); bal Fe				
ASTM A500		CF	0.26 max				0.035 max				Cu >=0.20; bal Fe	310	228	25	
ASTM A500		CF	0.26 max				0.035 max				Cu >=0.20; bal Fe	400	250	23	
ASTM A500		CF	0.26 max				0.035 max				Cu >=0.20; bal Fe	400	290	23	
ASTM A501	A501	HF struct tub	0.26 max				0.035 max		0.035 max		Cu >=0.20; bal Fe	400	250	23	
Carbon Steel, Resulfurized, A502(1)															
USA															
	UNS K01900		0.11-0.27		0.27-0.93		0.048 max		0.058 max		Cu 0.18 min (when spec'd); bal Fe				
Carbon Steel, Resulfurized, A539															
USA															
	UNS K01506		0.15 max		0.63 max		0.050 max		0.060 max		bal Fe				
ASTM A539	A539	ERW Coiled Tub	0.15 max		0.63 max		0.035 max		0.035 max		bal Fe	310	241	21	
Carbon Steel, Resulfurized, A53E-A															
Mexico															
NMX-B-205(86)	Grade A Type E	Pipe for high-press cable circuits, Electric res weld	0.21 max		0.90 max		0.40 max		0.050 max		bal Fe	330	205	35	
NMX-B-205(86)	Grade A Type S	Pipe for high-press cable circuits, Smls	0.22 max		0.90 max		0.040 max		0.050 max		bal Fe	330	205	35	
USA															
	UNS K02504		0.25 max		0.95 max		0.05 max		0.06 max		bal Fe				
ASTM A523(96)	A-ERW	ERW pipe	0.21 max		0.90 max		0.035 max		0.050 max		bal Fe	330	205	35	
ASTM A523(96)	A-S	Smls Weld Pipe	0.22 max		0.90 max		0.035 max		0.050 max		bal Fe	330	205	35	
Carbon Steel, Resulfurized, A556(B2)															
USA															
	UNS K02707		0.27 max		0.93 max		0.048 max		0.058 max	0.10 min	bal Fe				
ASTM A210/A210M(96)	A-1	Tub	0.27 max		0.93 max		0.035 max		0.035 max	0.10 min	bal Fe	485	275	30	89 HRB max
ASTM A556/A556M(96)	B2	Smls CD Feedwater Heater Tub	0.27 max		0.29-0.93		0.035 max		0.035 max	0.10 min	bal Fe	410	260	30	79 HRB max
Carbon Steel, Resulfurized, A557(B2)															
USA															
	UNS K03007		0.30 max		0.27-0.93		0.050 max		0.060 max		bal Fe				
Carbon Steel, Resulfurized, A595(A)															
USA															
	UNS K02004		0.12-0.29		0.26-0.94		0.05 max		0.06 max		bal Fe				
ASTM A595(97)	595A	Tub, tapered; struct use	0.15-0.25		0.30-0.90		0.035 max		0.035 max	0.04 max	bal Fe	380	450	23.0	

UNS numbers and US grades are provided as a means of cross referencing chemically similar alloys. Exchangability is only possible after independent examination of specifications. Tensile properties are minimum or typical as specified. UTS and YS as MPa. El as %. See Appendix for list of abbreviations used in Notes. * indicates obsolete material.

Specification	Designation	Notes	C	Cr	Mn	Ni	P	Pb	S	Si	Other	UTS	YS	El	Hard
Carbon Steel, Resulfurized, A766															
Italy															
UNI 4838(80)	CF17SMnPb10	Free-cutting	0.14-0.2		1-1.3		0.04 max	0.15-0.3	0.08-0.13	0.05 max	bal Fe				
USA															
	UNS K11711	Leaded-Resulfurized	0.14-0.20		1.00-1.30		0.04 max	0.15-0.35	0.08-0.13	0.35 max	bal Fe				
Carbon Steel, Resulfurized, LCST															
USA															
	UNS K01504		0.10-0.20		0.30-0.80		0.048 max		0.058 max	0.25 max	bal Fe				
Carbon Steel, Resulfurized, No equivalents identified															
Hungary															
MSZ 4339(86)	ANSBi2	Free-cutting; Q/T; 0-10mm	0.42-0.5		0.7-1.1		0.06 max		0.15-0.25	0.15-0.35	Bi=0.06-0.15; bal Fe	700-850	480	10L	
MSZ 4339(86)	ANSBi2	Free-cutting; Q/T; 40.1-63mm	0.42-0.5		0.7-1.1		0.06 max		0.15-0.25	0.15-0.35	Bi=0.06-0.15; bal Fe	620-770	375	14L	
Italy															
UNI 4838(80)	CF44SMnPb28	Free-cutting	0.4-0.48		1.35-1.65		0.04 max	0.15-0.3	0.24-0.32	0.3 max	bal Fe				
Mexico															
NMX-B-201(68)	1115	CD buttweld mech tub	0.13-0.20		0.60-0.90		0.04 max		0.08-0.13		bal Fe				
USA															
	AISI 1152	Bar	0.48-0.55		0.70-1.00		0.040 max		0.06-0.09		bal Fe				
ASTM A512(96)	1115	Buttweld mech tub, CD	0.13-0.20		0.60-0.90		0.040 max		0.08-0.130		bal Fe				
ASTM A695(95)	C35	HW special qual for fluid power apps					0.040 max		0.08-0.13	0.35 max	bal Fe	415	240	21	

Specification	Designation	Notes	C	Cr	Mn	Ni	P	Pb	S	Si	Other	UTS	YS	El	Hard
Carbon Steel, Rephosphorized and Resulfurized, 1211															
Bulgaria															
BDS 6886	A12		0.08-0.16		0.6-0.9		0.08-0.15		0.08-0.2	0.15-0.35	bal Fe				
Canada															
CSA STAN95-1-1	CL2.1211	Bar	0.13 max		0.6-0.9		0.07-0.12		0.08-0.15		bal Fe				
Mexico															
NMX-B-301(86)	1211	Bar	0.13 max		0.60-0.90		0.07-0.12		0.10-0.15		bal Fe				
Pan America															
COPANT 330	1211	Bar	0.13 max		0.6-0.9		0.07-0.12		0.1-0.15		bal Fe				
COPANT 330	B1111	Bar	0.13 max		0.6-0.9		0.07-0.12		0.1-0.15		bal Fe				
COPANT 331	B1111	Bar	0.13 max		0.6-0.9		0.07-0.12		0.1-0.15		bal Fe				
COPANT 331	B1211	Bar	0.13 max		0.6-0.9		0.07-0.12		0.1-0.15		bal Fe				
Russia															
GOST 1414(75)	A12	Free-cutting	0.08-0.16		0.6-0.9		0.08-0.15		0.08-0.2	0.15-0.35	Cu <=0.25; bal Fe				
Spain															
UNE	F.210.H		0.17		0.5-0.9		0.07-0.11		0.15-0.25	0.02	bal Fe				
USA															
	AISI 1211	Wir rod	0.13 max		0.60-0.90		0.070-0.12		0.10-0.15		bal Fe				
	UNS G12110		0.13 max		0.60-0.90		0.07-0.12		0.1-0.15		bal Fe				
ASTM A108(95)	1211	Bar, CF	0.13 max		0.60-0.90		0.07-0.12		0.10-0.15		bal Fe				
ASTM A29/A29M(93)	1211	Bar	0.13 max		0.60-0.90		0.07-0.12		0.10-0.15		bal Fe				
ASTM A510(96)	1211	Wir rod	0.13 max		0.60-0.90		0.07-0.12		0.1-0.15		bal Fe				
ASTM A576(95)	1211	Special qual HW bar	0.13 max		0.60-0.90		0.07-0.12		0.10-0.15		Pb Bi Ca Se Te if spec'd; bal Fe				
FED QQ-S-637A(70)	C1211*	Obs; see ASTM A108; CF Bar std qual, free mach									bal Fe				
SAE J1249(95)	1211	Former SAE std valid for Wir rod	0.13 max		0.60-0.90		0.07-0.12		0.10-0.15		bal Fe				
SAE J1397(92)	1211	Bar CD, est mech prop	0.13 max		0.6-0.9		0.07-0.12		0.1-0.15		bal Fe	520	400	10	163 HB
SAE J1397(92)	1211	Bar HR, est mech prop	0.13 max		0.6-0.9		0.07-0.12		0.1-0.15		bal Fe	380	230	25	121 HB
Carbon Steel, Rephosphorized and Resulfurized, 1212															
Argentina															
IAS	IRAM 1212		0.13 max		0.70-1.35		0.04-0.12		0.16-0.40		bal Fe	400-500	240-370	30	116-147 HB
Bulgaria															
BDS 6886	1		0.12		0.5-0.7		0.11		0.15-0.25		bal Fe				
BDS 6886	3		0.07-0.16		0.6-1.2		0.1		0.15-0.25	0.15-0.4	bal Fe				
BDS 6886	A9	Free-cutting	0.13 max	0.3 max	0.6-0.9		0.11 max		0.18-0.25	0.05 max	Cu <=0.30; Mo <=0.15; bal Fe				
Canada															
CSA STAN95-1-1	CL.2.B1111	Bar	0.13 max		0.6-0.9		0.07-0.12		0.16-0.23		bal Fe				
CSA STAN95-1-1	CL.2.B1112	Bar	0.13 max		0.7-1		0.07-0.12		0.16-0.23		bal Fe				
CSA STAN95-1-1	CL2.C1212	Bar	0.13		0.7-1		0.07-0.12		0.16-0.23		bal Fe				

UNS numbers and US grades are provided as a means of cross referencing chemically similar alloys. Exchangability is only possible after independent examination of specifications. Tensile properties are minimum or typical as specified. UTS and YS as MPa. El as %. See Appendix for list of abbreviations used in Notes. * indicates obsolete material.

Carbon Steel, Rephosphorized and Resulfurized, 1212 (Continued from previous page)

Specification	Designation	Notes	C	Cr	Mn	Ni	P	Pb	S	Si	Other	UTS	YS	El	Hard
China															
GB 8731(88)	Y12	Bar HR/CD	0.08-0.16		0.70-1.00		0.08-0.15		0.10-0.20	0.15-0.35	bal Fe	390-540		22	
Czech Republic															
CSN 411107(61)	11107	Free-cutting	0.12 max	0.3 max	0.4-0.9		0.11 max		0.16-0.26	0.4 max	bal Fe				
CSN 411110	11110	Free-cutting	0.07-0.16	0.3 max	0.6-1.1		0.1 max		0.15-0.25	0.4 max	bal Fe				
France															
AFNOR	10F2		0.08-0.14		0.5-0.75		0.06		0.12-0.24	0.1-0.4	bal Fe				
AFNOR	12MF4		0.09-0.15		0.9-1.2		0.06		0.17-0.24	0.1-0.4	bal Fe				
AFNOR NFA35561(92)	S200		0.13		0.7-1.2		0.11		0.2-0.27	0.08	bal Fe				
Germany															
DIN 17405(79)	RFe160	Soft magnetic for DC relays	0.10 max		0.50-0.90		0.080 max		0.18-0.27	0.10 max	Al 0.04-0.10; bal Fe				
DIN 17405(79)	WNr 1.1011	Soft magnetic for DC relays	0.10 max		0.50-0.90		0.080 max		0.18-0.27	0.10 max	Al 0.04-0.10; bal Fe				
DIN E EN 10087(95)	WNr 1.0711	Free-cutting	0.13 max		0.60-1.20		0.100 max		0.18-0.25	0.05 max	bal Fe				
DIN EN 10087(95)	9S20	Free-cutting	0.13 max		0.60-1.20		0.100 max		0.18-0.25	0.05 max	bal Fe				
Hungary															
MSZ 4339(86)	AS1	Free-cutting; gen struct; 0-16mm; HF; bright turned (ground)	0.13 max		0.6-1.2		0.1 max		0.15-0.25	0.05 max	bal Fe	360-550			159 HB max
MSZ 4339(86)	AS1	Free-cutting; gen struct; 63.1-100mm; HF; bright turned (ground)	0.13 max		0.6-1.2		0.1 max		0.15-0.25	0.05 max	bal Fe	350-530			146 HB max
International															
ISO 683-9(88)	9S20	Free-cutting, non HT, <16mm	0.13 max		0.60-1.20		0.11		0.15-0.25	0.05 max	bal Fe	490-790	390	8	163 HB
ISO R683-9(70)	1*	Obs; see 9S20	0.13		0.6-1.2		0.11		0.18-0.25	0.05	bal Fe				
ISO R683-9(70)	1*	Former designation	0.13		0.6-1.2		0.11		0.18-0.25	0.05	bal Fe				
ISO R683-9(70)	3*	Former designation	0.07-0.13		0.5-0.9		0.06		0.15-0.25	0.15-0.4	bal Fe				
ISO R683-9(70)	5*	Former designation	0.09-0.15		0.9-1.2		0.06		0.15-0.25	0.15-0.4	bal Fe				
Italy															
UNI 4838(62)	10S20	Free-cutting	0.06-0.13		0.6-1		0.08 max		0.15-0.25	0.05 max	bal Fe				
UNI 4838(80)	CF9S22	Free-cutting	0.06-0.13		0.7-1.2		0.04-0.1		0.18-0.25	0.05 max	bal Fe				
Japan															
JIS G4804(83)	SUM21	Bar Free cutting	0.13 max		0.70-1.00		0.07-0.12		0.16-0.23		bal Fe				
Mexico															
NMX-B-301(86)	1212	Bar	0.13 max		0.70-1.00		0.07-0.12		0.16-0.23		bal Fe				
Pan America															
COPANT 330	1212	Bar	0.13		0.7-1		0.07-0.12		0.16-0.23		bal Fe				
COPANT 330	B1112	Bar	0.13 max		0.7-1		0.07-0.12		0.16-0.23		bal Fe				
COPANT 331	B1112	Bar	0.13 max		0.7-1		0.07-0.12		0.16-0.23		bal Fe				
COPANT 331	B1212	Bar	0.13 max		0.7-1		0.07-0.12		0.16-0.23		bal Fe				
COPANT 333	1212	Wir rod	0.13 max		0.7-1		0.07-0.12		0.16-0.23		bal Fe				

Specification	Designation	Notes	C	Cr	Mn	Ni	P	Pb	S	Si	Other	UTS	YS	El	Hard

Carbon Steel, Rephosphorized and Resulfurized, 1212 (Continued from previous page)

Poland

Specification	Designation	Notes	C	Cr	Mn	Ni	P	Pb	S	Si	Other	UTS	YS	El	Hard
PNH84026	A11	Free-cutting	0.07-0.13		0.5-0.9		0.06		0.15-0.25	0.15-0.4	bal Fe				
PNH84026	A11X	Free-cutting	0.13 max	0.3 max	0.6-1.2	0.4 max	0.08 max		0.18-0.25	0.05 max	bal Fe				

Romania

Specification	Designation	Notes	C	Cr	Mn	Ni	P	Pb	S	Si	Other	UTS	YS	El	Hard
STAS 1350	AUT12M		0.07-0.16		0.6-1.1		0.1		0.15-0.25	0.15-0.4	bal Fe				
STAS 1350(89)	AUT12	Free-cutting	0.08-0.16	0.3 max	0.6-0.9	0.4 max	0.08-0.15	0.15 max	0.08-0.2	0.15-0.35	bal Fe				
STAS 1350(89)	AUT9	Free-cutting	0.12 max	0.3 max	0.5-0.9	0.4 max	0.035-0.11	0.15 max	0.15-0.25	0.35 max	bal Fe				

Russia

Specification	Designation	Notes	C	Cr	Mn	Ni	P	Pb	S	Si	Other	UTS	YS	El	Hard
GOST 1414(75)	A11	Free-cutting	0.07-0.15	0.25 max	0.8-1.2	0.25 max	0.06-0.12		0.15-0.25	0.1 max	Cu <=0.25; bal Fe				

Spain

Specification	Designation	Notes	C	Cr	Mn	Ni	P	Pb	S	Si	Other	UTS	YS	El	Hard
UNE 36021(61)	10S20	Free-cutting	0.07-0.13		0.5-0.9		0.06		0.15-0.25	0.15-0.4	bal Fe				
UNE 36021(61)	F.210.H		0.17		0.5-0.9		0.07-0.11		0.15-0.25	0.02	bal Fe				
UNE 36021(61)	F.211	Free-cutting	0.2 max		0.6-1.2		0.1 max		0.2-0.3	0.1 max	Mo <=0.15; bal Fe				
UNE 36021(61)	F.2121*	see 10S20	0.07-0.13		0.5-0.9		0.06		0.15-0.25	0.15-0.4	bal Fe				

Sweden

Specification	Designation	Notes	C	Cr	Mn	Ni	P	Pb	S	Si	Other	UTS	YS	El	Hard
SS 141922	1922	Free-cutting	0.12-0.18		0.8-1.2		0.06 max		0.15-0.25	0.1-0.4	bal Fe				

USA

Specification	Designation	Notes	C	Cr	Mn	Ni	P	Pb	S	Si	Other	UTS	YS	El	Hard
	AISI 1212	Bar	0.13 max		0.70-1.00		0.07-0.12		0.16-0.23		bal Fe				
	AISI 1212	Wir rod	0.13 max		0.70-1.00		0.070-0.12		0.16-0.23		bal Fe				
	UNS G12120		0.13 max		0.70-1.00		0.07-0.12		0.16-0.23		bal Fe				
AMS 5010H(88)		Bar, CD, Free Machining, t>=50.8mm	0.13 max		0.70-1.00		0.07-0.12		0.16-0.23		bal Fe				110-201 HB
ASTM A108(95)	1212	Bar, CF	0.13 max		0.70-1.00		0.07-0.12		0.16-0.23		bal Fe				
ASTM A29/A29M(93)	1212	Bar	0.13 max		0.70-1.00		0.07-0.12		0.16-0.23		bal Fe				
ASTM A510(96)	1212	Wir rod	0.13 max		0.70-1.00		0.07-0.12		0.16-0.23		bal Fe				
ASTM A576(95)	1212	Special qual HW bar	0.13 max		0.7-1.00		0.07-0.12		0.16-0.23		Pb Bi Ca Se Te if spec'd; bal Fe				
FED QQ-S-637A(70)	C1212*	Obs; see ASTM A108; CF Bar std qual, free mach	0.13 max		0.7-1		0.07-0.12		0.16-0.23		bal Fe				
SAE J1397(92)	1212	Bar HR, est mech prop	0.13 max		0.7-1		0.07-0.12		0.16-0.23		bal Fe	390	230	25	121 HB
SAE J1397(92)	1212	Bar CD, est mech prop	0.13 max		0.7-1		0.07-0.12		0.16-0.23		bal Fe	540	410	10	167 HB
SAE J403(95)	1212	Bar Wir rod Smls Tub HR CF	0.13 max		0.70-1.00		0.07-0.12		0.16-0.23		bal Fe				
SAE J403(95)	1212	Bil Bar Rod Tub	0.13 max		0.7-1		0.07-0.12		0.16-0.23		bal Fe				

Carbon Steel, Rephosphorized and Resulfurized, 1213

Australia

Specification	Designation	Notes	C	Cr	Mn	Ni	P	Pb	S	Si	Other	UTS	YS	El	Hard
AS 1442(92)	1214	Free-cutting, HR bar, Semifinished	0.15 max		0.80-1.20		0.04-0.09		0.25-0.35	0.10 max	bal Fe				
AS 1443(94)	1214	CF bar, Free-cutting	0.15 max		0.80-1.20		0.04-0.09		0.25-0.35	0.10 max	bal Fe				
AS 1443(94)	D12	CR or CR bar, 16-38mm	0.15 max		0.80-1.20		0.04-0.09		0.25-0.35	0.10 max	Pb 0.15-0.35 for lead-bearing steels; bal Fe	430	330	8	
AS 1443(94)	T12	CF bar, not CD or CR;<=260mm	0.15 max		0.80-1.20		0.04-0.09		0.25-0.35	0.10 max	Pb 0.15-0.35 for lead-bearing steels; bal Fe	370	230	17	

Canada

Specification	Designation	Notes	C	Cr	Mn	Ni	P	Pb	S	Si	Other	UTS	YS	El	Hard
CSA STAN95-1-1	CL.2.B1113	Bar	0.13		0.7-1		0.07-0.12		0.24-0.33		bal Fe				

UNS numbers and US grades are provided as a means of cross referencing chemically similar alloys. Exchangability is only possible after independent examination of specifications. Tensile properties are minimum or typical as specified. UTS and YS as MPa. El as %. See Appendix for list of abbreviations used in Notes. * indicates obsolete material.

Specification	Designation	Notes	C	Cr	Mn	Ni	P	Pb	S	Si	Other	UTS	YS	El	Hard

Carbon Steel, Rephosphorized and Resulfurized, 1213 (Continued from previous page)

Canada

Specification	Designation	Notes	C	Cr	Mn	Ni	P	Pb	S	Si	Other	UTS	YS	El	Hard
CSA STAN95-1-1	CL2.C1213	Bar	0.13		0.7-1		0.07-0.12		0.24-0.33		bal Fe				
China															
GB 8731(88)	Y15	Bar HR/CD	0.10-0.18		0.80-1.20		0.05-0.10		0.23-0.33	0.15 max	bal Fe	390-540		22	
Czech Republic															
CSN 411109	11109	Free-cutting	0.13 max	0.3 max	0.9-1.5		0.1 max		0.21-0.32	0.4 max	bal Fe				
France															
AFNOR NFA35561(92)	S250		0.14		0.9-1.5		0.11		0.25-0.32	0.08	bal Fe				
Germany															
DIN	WNr 1.0740*	Obs	0.14		1-1.5		0.07		0.35-0.45	0.1-0.4	bal Fe				
DIN E EN 10087(95)	WNr 1.0715	HR Bar Rod free-cutting CD 64-100mm	0.14 max		0.90-1.30		0.11 max		0.27-0.33	0.05 max	bal Fe	380-630	245	10	
DIN E EN 10087(95)	WNr 1.0715	HR Bar Rod free-cutting CD 11-16mm	0.14 max		0.90-1.30		0.11 max		0.27-0.33	0.05 max	bal Fe	510-760	410	7	
DIN EN 10087(95)	9SMn28	HR Bar Rod free-cutting CD 11-16mm	0.14 max		0.90-1.30		0.11 max		0.27-0.33	0.05 max	bal Fe	510-760	410	7	
DIN EN 10087(95)	9SMn28	HR Bar Rod free-cutting CD 64-100mm	0.14 max		0.90-1.30		0.11 max		0.27-0.33	0.05 max	bal Fe	380-630	245	10	
Hungary															
MSZ 4339(86)	AS4	Free-cutting; gen struct; 63.1-100mm; HF; bright turned (ground)	0.14 max		0.9-1.3		0.1 max		0.24-0.34	0.05 max	bal Fe	360-520			156 HB max
MSZ 4339(86)	AS4	Free-cutting; gen struct; 0-16mm; HF; bright turned (ground)	0.14 max		0.9-1.3		0.1 max		0.24-0.34	0.05 max	bal Fe	380-570			170 HB max
International															
ISO 683-9(88)	11SMn28	Free-cutting, non HT, <16mm	0.14 max		0.90-1.30		0.11 max		0.24-0.33	0.05 max	bal Fe	510-810	410	7	170 HB
ISO R683-9(70)	2*	Obs; see 11SMn28	0.14		0.9-1.3		0.11		0.24-0.34	0.05	bal Fe				
ISO R683-9(70)	2*	Former designation	0.14		0.9-1.3		0.11		0.24-0.32	0.05	bal Fe				
ISO R683-9(70)	3*	Obs; see 12SMn35	0.15		1-1.5		0.11		0.3-0.4	0.05	bal Fe				
Italy															
UNI 4838(62)	9SMn23	Free-cutting	0.13 max		0.9-1.3		0.1 max		0.2-0.27	0.05 max	bal Fe				
UNI 4838(80)	CF9SMn28	Free-cutting	0.06-0.13		0.9-1.3		0.04-0.1		0.24-0.32	0.05 max	bal Fe				
UNI 4838(80)	CF9SMn32	Free-cutting	0.06-0.13		0.9-1.3		0.04-0.1		0.28-0.35	0.05 max	bal Fe				
Japan															
JIS G4804(83)	SUM22	Bar Free cutting	0.13 max		0.70-1.00		0.07-0.12		0.24-0.33		bal Fe				
Mexico															
NMX-B-301(86)	1213	Bar	0.13 max		0.70-1.00		0.07-0.12		0.24-0.33		bal Fe				
Pan America															
COPANT 330	1213	Bar	0.13		0.7-1		0.07-0.12		0.24-0.33		bal Fe				
COPANT 331	B1213	Bar	0.13		0.7-1		0.07-0.12		0.24-0.33		bal Fe				
COPANT 333	1213	Wir rod	0.13		0.7-1		0.07-0.12		0.24-0.33		bal Fe				
COPANT 514	1213	Tube	0.13		0.7-1		0.072-0.12		0.24-0.33		bal Fe				
Poland															
PNH84026	A10X	Free-cutting	0.12 max	0.3 max	0.9-1.3	0.4 max	0.08 max		0.24-0.34	0.05 max	bal Fe				
PNH84026	A10XN	Free-cutting	0.1		0.9-1.3		0.08		0.24-0.34	0.05	N 0.01-0.016; bal Fe				

UNS numbers and US grades are provided as a means of cross referencing chemically similar alloys. Exchangability is only possible after independent examination of specifications. Tensile properties are minimum or typical as specified. UTS and YS as MPa. El as %. See Appendix for list of abbreviations used in Notes. * indicates obsolete material.

Specification	Designation	Notes	C	Cr	Mn	Ni	P	Pb	S	Si	Other	UTS	YS	El	Hard

Carbon Steel, Rephosphorized and Resulfurized, 1213 (Continued from previous page)

Spain

Specification	Designation	Notes	C	Cr	Mn	Ni	P	Pb	S	Si	Other	UTS	YS	El	Hard
UNE	F.210.A		0.07-0.13		0.9-1.2		0.05-0.09		0.25-0.35	0.1	bal Fe				
UNE 36021(61)	12SMn35	Free-cutting	0.15		1-1.5		0.11		0.3-0.4	0.05	bal Fe				
UNE 36021(61)	F.2113*	see 12SMn35	0.15		1-1.5		0.11		0.3-0.4	0.05	bal Fe				
UNE 36021(80)	11SMn28	Free-cutting	0.14 max		0.9-1.3		0.11 max		0.24-0.32	0.05 max	Mo <=0.15; bal Fe				
UNE 36021(80)	F.2111*	see 11SMn28	0.14 max		0.9-1.3		0.11 max		0.24-0.32	0.05 max	Mo <=0.15; bal Fe				

Sweden

Specification	Designation	Notes	C	Cr	Mn	Ni	P	Pb	S	Si	Other	UTS	YS	El	Hard
SS 141912	1912	Free-cutting	0.14		0.9-1.3		0.11		0.24-0.35	0.05	bal Fe				

Switzerland

Specification	Designation	Notes	C	Cr	Mn	Ni	P	Pb	S	Si	Other	UTS	YS	El	Hard
VSM FH	9Smn36		0.1		0.9-1.3		0.1		0.32-0.4		bal Fe				

UK

Specification	Designation	Notes	C	Cr	Mn	Ni	P	Pb	S	Si	Other	UTS	YS	El	Hard
BS 970/1(83)	220M07	Blm Bil Slab Bar Rod Frg HR, diam<100mm	0.15		0.90-1.30		0.070 max		0.20-0.30		bal Fe				
BS 970/1(83)	230M07	Blm Bil Slab Bar Rod Frg HR, diam<100mm	0.15		0.90-1.30		0.070 max		0.25-0.35	0.05	bal Fe				
BS 970/1(83)	240M07	Blm Bil Slab Bar Rod Frg CD; 0.5/0.625	0.15		1.10-1.50		0.070 max		0.30-0.40		bal Fe				
BS 970/3(91)	230M07	Bright bar; free-cutting; 6<t<=100mm; HR	0.15 max		0.90-1.30		0.090 max		0.25-0.35	0.05 max	Mo <=0.15; bal Fe	360	215	22	
BS 970/3(91)	230M07	Bright bar; free-cutting; 63<t<=76mm; HR CD	0.15 max		0.90-1.30		0.090 max		0.25-0.35	0.05 max	Mo <=0.15; bal Fe	370	240	10	

USA

Specification	Designation	Notes	C	Cr	Mn	Ni	P	Pb	S	Si	Other	UTS	YS	El	Hard
	AISI 1213	Wir rod	0.13 max		0.70-1.00		0.070-0.12		0.24-0.33		bal Fe				
	AISI 1213	Bar	0.13 max		0.70-1.00		0.07-0.12		0.24-0.33		bal Fe				
	UNS G12130		0.13 max		0.70-1.00		0.07-0.12		0.24-0.33		bal Fe				
ASTM A108(95)	1213	Bar, CF	0.13 max		0.70-1.00		0.07-0.12		0.24-0.33		bal Fe				
ASTM A29/A29M(93)	1213	Bar	0.13 max		0.70-1.00		0.07-0.12		0.24-0.33		bal Fe				
ASTM A510(96)	1213	Wir rod	0.13 max		0.70-1.00		0.07-0.12		0.24-0.33		bal Fe				
ASTM A519(96)	1213	Smls mech tub	0.13 max		0.70-1.10		0.07-0.12		0.24-0.330		bal Fe				
ASTM A576(95)	1213	Special qual HW bar	0.13 max		0.70-1.00		0.07-0.12		0.24-0.33		Pb Bi Ca Se Te if spec'd; bal Fe				
FED QQ-S-637A(70)	C1213*	Obs; see ASTM A108; CF Bar std qual, free mach	0.13 max		0.7-1		0.07-0.12		0.24-0.33		bal Fe				
SAE J1397(92)	1213	Bar HR, est mech prop	0.13 max		0.7-1		0.07-0.12		0.24-0.33		bal Fe	390	230	25	121 HB
SAE J1397(92)	1213	Bar CD, est mech prop	0.13 max		0.7-1		0.07-0.12		0.24-0.33		bal Fe	540	410	10	167 HB
SAE J403(95)	1213	Bar Wir rod Smls Tub HR CF	0.13 max		0.70-1.00		0.07-0.12		0.24-0.33		bal Fe				
SAE J403(95)	1213	Bil Bar Rod Tub	0.13 max		0.7-1		0.07-0.12		0.24-0.33		bal Fe				

Yugoslavia

Specification	Designation	Notes	C	Cr	Mn	Ni	P	Pb	S	Si	Other	UTS	YS	El	Hard
	C.1290	Free-cutting	0.12-0.18		0.6-0.9		0.07 max		0.18-0.26	0.1-0.4	Mo <=0.15; bal Fe				
	C.3990	Free-cutting	0.08-0.14		0.9-1.3		0.11 max		0.24-0.32	0.05 max	Mo <=0.15; bal Fe				

Carbon Steel, Rephosphorized and Resulfurized, 1215

France

Specification	Designation	Notes	C	Cr	Mn	Ni	P	Pb	S	Si	Other	UTS	YS	El	Hard
AFNOR NFA35561(92)	S300		0.15		1-1.6		0.11		0.3-0.4	0.09	bal Fe				

Specification	Designation	Notes	C	Cr	Mn	Ni	P	Pb	S	Si	Other	UTS	YS	El	Hard

Carbon Steel, Rephosphorized and Resulfurized, 1215 (Continued from previous page)

Germany

Specification	Designation	Notes	C	Cr	Mn	Ni	P	Pb	S	Si	Other	UTS	YS	El	Hard
DIN E EN 10087(95)	WNr 1.0736	HR Bar Rod free-cutting CD 64-100mm	0.15 max		1.00-1.50		0.11 max		0.34-0.40	0.05 max	bal Fe	390-640	265	10	
DIN E EN 10087(95)	WNr 1.0736	HR Bar Rod free-cutting CD 11-16mm	0.15 max		1.00-1.50		0.11 max		0.34-0.40	0.05 max	bal Fe	540-780	430	7	
DIN EN 10087(95)	9SMn36	HR Bar Rod free-cutting CD 64-100mm	0.15 max		1.00-1.50		0.11 max		0.34-0.40	0.05 max	bal Fe	390-640	265	10	
DIN EN 10087(95)	9SMn36	HR Bar Rod free-cutting CD 11-16mm	0.15 max		1.00-1.50		0.11 max		0.34-0.40	0.05 max	bal Fe	540-780	430	7	

Hungary

Specification	Designation	Notes	C	Cr	Mn	Ni	P	Pb	S	Si	Other	UTS	YS	El	Hard
MSZ 4339(86)	AS5	Free-cutting; gen struct; 63.1-100mm; HF; bright turned (ground)	0.15 max		1-1.5		0.1 max		0.3-0.4	0.05 max	bal Fe	360-520			156 HB max
MSZ 4339(86)	AS5	Free-cutting; gen struct; 0-10mm; HF; bright turned (ground)	0.15 max		1-1.5		0.1 max		0.3-0.4	0.05 max	bal Fe	390-590			174 HB max

International

Specification	Designation	Notes	C	Cr	Mn	Ni	P	Pb	S	Si	Other	UTS	YS	El	Hard
ISO 683-9(88)	12SMn35	Free-cutting, non HT, <16mm	0.15 max		1.00-1.50		0.11 max		0.30-0.40	0.05 max	bal Fe	540-840	430	7	174 HB

Italy

Specification	Designation	Notes	C	Cr	Mn	Ni	P	Pb	S	Si	Other	UTS	YS	El	Hard
UNI 4838(80)	CF9SMn36	Free-cutting	0.06-0.13		1-1.5		0.04-0.1		0.32-0.4	0.05 max	bal Fe				

Japan

Specification	Designation	Notes	C	Cr	Mn	Ni	P	Pb	S	Si	Other	UTS	YS	El	Hard
JIS G4804(83)	SUM23	Bar Free cutting	0.09 max		0.75-1.05		0.04-0.09		0.26-0.35		bal Fe				
JIS G4804(83)	SUM23L	Bar Free cutting Leaded	0.09 max		0.75-1.05		0.04-0.09	0.10-0.35	0.26-0.35		bal Fe				
JIS G4804(83)	SUM25	Bar Free cutting	0.15 max		0.90-1.40		0.07-0.12		0.30-0.40		bal Fe				

Mexico

Specification	Designation	Notes	C	Cr	Mn	Ni	P	Pb	S	Si	Other	UTS	YS	El	Hard
NMX-B-301(86)	1215	Bar	0.09 max		0.75-1.05		0.04-0.09		0.26-0.35		bal Fe				
NMX-B-301(86)	12L15	Bar	0.09 max		0.75-1.05		0.04-0.09	0.15-0.35	0.26-0.35		bal Fe				

Pan America

Specification	Designation	Notes	C	Cr	Mn	Ni	P	Pb	S	Si	Other	UTS	YS	El	Hard
COPANT 330	1215	Bar	0.09		0.75-1.05		0.04-0.09		0.26-0.35		bal Fe				
COPANT 331	1215	Bar	0.09		0.75-1.05		0.04-0.09		0.26-0.35		bal Fe				
COPANT 514	1215	Tube	0.09		0.75-1.05		0.04-0.09		0.26-0.35		bal Fe				

Spain

Specification	Designation	Notes	C	Cr	Mn	Ni	P	Pb	S	Si	Other	UTS	YS	El	Hard
UNE 36021(80)	12SMn35	Free-cutting	0.15 max		1-1.5		0.11 max		0.3-0.4	0.05 max	Mo <=0.15; bal Fe				
UNE 36021(80)	F.2113*	see 12SMn35	0.15 max		1-1.5		0.11 max		0.3-0.4	0.05 max	Mo <=0.15; bal Fe				

USA

Specification	Designation	Notes	C	Cr	Mn	Ni	P	Pb	S	Si	Other	UTS	YS	El	Hard
	AISI 1215	Wir rod	0.09 max		0.75-1.05		0.040-0.090		0.26-0.35		bal Fe				
	AISI 1215	Bar	0.09 max		0.75-1.05		0.04-0.09		0.26-0.35		bal Fe				
	UNS G12150		0.09 max		0.75-1.05		0.04-0.09		0.26-0.35		bal Fe				
AMS 5010H(88)		Bar, CD, Free Machining, t>=50.8mm	0.09 max		0.75-1.05		0.040-0.09		0.26-0.35		bal Fe				110-201 HB
ASTM A108(95)	1215	Bar, CF	0.09 max		0.75-1.05		0.04-0.09		0.26-0.35		bal Fe				
ASTM A29/A29M(93)	1215	Bar	0.09 max		0.75-1.05		0.04-0.09		0.26-0.35		bal Fe				
ASTM A29/A29M(93)	12L15	Bar	0.09 max		0.75-1.05		0.04-0.09	0.15-0.35	0.26-0.35		bal Fe				
ASTM A510(96)	1215	Wir rod	0.09 max		0.75-1.05		0.04-0.09		0.26-0.35		bal Fe				
ASTM A510(96)	12L15	Wir rod	0.09 max		0.75-1.05		0.04-0.09		0.26-0.35		bal Fe				
ASTM A519(96)	1215	Smls mech tub	0.09 max		0.75-1.05		0.04-0.09		0.26-0.350		bal Fe				

UNS numbers and US grades are provided as a means of cross referencing chemically similar alloys. Exchangability is only possible after independent examination of specifications. Tensile properties are minimum or typical as specified. UTS and YS as MPa. El as %. See Appendix for list of abbreviations used in Notes. * indicates obsolete material.

Specification	Designation	Notes	C	Cr	Mn	Ni	P	Pb	S	Si	Other	UTS	YS	El	Hard

Carbon Steel, Rephosphorized and Resulfurized, 1215 (Continued from previous page)

USA

Specification	Designation	Notes	C	Cr	Mn	Ni	P	Pb	S	Si	Other	UTS	YS	El	Hard
ASTM A576(95)	1215	Special qual HW bar	0.09 max		0.75-1.05		0.04-0.09		0.26-0.35		Pb Bi Ca Se Te if spec'd; bal Fe				
FED QQ-S-637A(70)	1215*	Obs; see ASTM A108; CF Bar std qual, free mach	0.09 max		0.75-1.05		0.04-0.09		0.26-0.35		bal Fe				
SAE J403(95)	1215	Bar Wir rod Smls Tub HR CF	0.09 max		0.75-1.05		0.04-0.09		0.26-0.35		bal Fe				
SAE J403(95)	1215	Bil Bar Rod Tub	0.09 max		0.75-1.05		0.04-0.09		0.26-0.35		bal Fe				

Yugoslavia

Specification	Designation	Notes	C	Cr	Mn	Ni	P	Pb	S	Si	Other	UTS	YS	El	Hard
	C.3991	Free-cutting	0.1-0.15		1.0-1.5		0.11 max		0.3-0.4	0.05 max	Mo <=0.15; bal Fe				

Carbon Steel, Rephosphorized and Resulfurized, 12L13

Bulgaria

Specification	Designation	Notes	C	Cr	Mn	Ni	P	Pb	S	Si	Other	UTS	YS	El	Hard
BDS 6886	A9G-Rv	Free-cutting	0.14 max	0.3 max	0.9-1.3		0.11 max	0.15-0.35	0.24-0.32	0.05 max	Cu <=0.30; Mo <=0.15; bal Fe				

Germany

Specification	Designation	Notes	C	Cr	Mn	Ni	P	Pb	S	Si	Other	UTS	YS	El	Hard
DIN E EN 10087(95)	WNr 1.0718	HR Bar Rod free-cutting CD 11-16mm	0.14 max		0.90-1.30		0.11 max	0.15-0.35	0.27-0.33	0.05 max	bal Fe	510-760	410	7	
DIN E EN 10087(95)	WNr 1.0718	HR Bar Rod free-cutting CD 64-100mm	0.14 max		0.90-1.30		0.11 max	0.15-0.35	0.27-0.33	0.05 max	bal Fe	380-630	245	10	

Hungary

Specification	Designation	Notes	C	Cr	Mn	Ni	P	Pb	S	Si	Other	UTS	YS	El	Hard
MSZ 4339(75)	APb1	Free-cutting; gen struct	0.14 max		0.9-1.4		0.1 max	0.15-0.3	0.24-0.34	0.07 max	bal Fe				

International

Specification	Designation	Notes	C	Cr	Mn	Ni	P	Pb	S	Si	Other	UTS	YS	El	Hard
ISO 683-9(88)	11SMnPb28	Free-cutting, non HT, <16mm	0.14 max		0.90-1.30		0.11 max	0.15-0.35	0.24-0.33	0.05 max	bal Fe	510-810	410	7	170 HB

Italy

Specification	Designation	Notes	C	Cr	Mn	Ni	P	Pb	S	Si	Other	UTS	YS	El	Hard
UNI 4838(62)	9SMnPb23	Free-cutting	0.13 max		0.9-1.3		0.1 max	0.15-0.4	0.2-0.27	0.05 max	bal Fe				
UNI 4838(80)	CF9SMnPb28	Free-cutting	0.06-0.13		0.9-1.3		0.04-0.1	0.15-0.3	0.24-0.32	0.05 max	bal Fe				

Mexico

Specification	Designation	Notes	C	Cr	Mn	Ni	P	Pb	S	Si	Other	UTS	YS	El	Hard
NMX-B-301(86)	12L13	Bar	0.13 max		0.70-1.00		0.07-0.12	0.15-0.35	0.24-0.33		bal Fe				

Spain

Specification	Designation	Notes	C	Cr	Mn	Ni	P	Pb	S	Si	Other	UTS	YS	El	Hard
UNE 36021(80)	11SMnPb28	Free-cutting	0.14 max		0.9-1.3		0.11 max	0.15-0.35	0.24-0.32	0.05 max	Mo <=0.15; bal Fe				
UNE 36021(80)	F.2112*	see 11SMnPb28	0.14 max		0.9-1.3		0.11 max	0.15-0.35	0.24-0.32	0.05 max	Mo <=0.15; bal Fe				

Sweden

Specification	Designation	Notes	C	Cr	Mn	Ni	P	Pb	S	Si	Other	UTS	YS	El	Hard
SS 141914	1914	Free-cutting	0.14 max		0.9-1.3		0.11 max	0.15-0.35	0.24-0.35	0.05 max	bal Fe				

USA

Specification	Designation	Notes	C	Cr	Mn	Ni	P	Pb	S	Si	Other	UTS	YS	El	Hard
	UNS G12134	Leaded	0.13 max		0.70-1.00		0.07-0.12	0.15-0.35	0.24-0.33		bal Fe				
ASTM A29/A29M(93)	12L13	Bar	0.13 max		0.70-1.00		0.04-0.09	0.15-0.35	0.26-0.35		bal Fe				
ASTM A510(96)	12L13	Wir rod	0.13 max		0.70-1.00		0.07-0.12		0.24-0.33		bal Fe				

Yugoslavia

Specification	Designation	Notes	C	Cr	Mn	Ni	P	Pb	S	Si	Other	UTS	YS	El	Hard
	C.3993	Free-cutting	0.08-0.14		0.9-1.2		0.11 max	0.15-0.35	0.24-0.32	0.05 max	Mo <=0.15; bal Fe				

Carbon Steel, Rephosphorized and Resulfurized, 12L14

Argentina

Specification	Designation	Notes	C	Cr	Mn	Ni	P	Pb	S	Si	Other	UTS	YS	El	Hard
IAS	IRAM 12L14		0.15 max		0.85-1.35		0.04-0.12	0.15-0.35	0.22-0.40		bal Fe	400-500	240-370	7-16	116-147 HB

Australia

Specification	Designation	Notes	C	Cr	Mn	Ni	P	Pb	S	Si	Other	UTS	YS	El	Hard
AS 1442	S1214*	Obs; Bar Bil	0.15		0.8-1.2		0.04-0.09		0.26-0.35	0.1	bal Fe				
AS 1442(92)	12L14	Free-cutting, HR bar, Semifinished	0.15 max		0.80-1.20		0.04-0.09		0.25-0.35	0.10 max	Pb 0.15-0.35 for lead-bearing steels; bal Fe				
AS 1443(94)	12L14	CF bar, Free-cutting	0.15 max		0.80-1.20		0.04-0.09		0.25-0.35	0.10 max	Pb 0.15-0.35 for lead-bearing steels; bal Fe				

UNS numbers and US grades are provided as a means of cross referencing chemically similar alloys. Exchangability is only possible after independent examination of specifications. Tensile properties are minimum or typical as specified. UTS and YS as MPa. El as %. See Appendix for list of abbreviations used in Notes. * indicates obsolete material.

Carbon Steel, Rephosphorized and Resulfurized, 12L14 (Continued from previous page)

Specification	Designation	Notes	C	Cr	Mn	Ni	P	Pb	S	Si	Other	UTS	YS	El	Hard
Australia															
AS 1443(94)	D13	CR or CR bar, 16-38mm	0.15 max		0.80-1.20		0.04-0.09		0.25-0.35	0.10 max	Pb 0.15-0.35 for lead-bearing steels; bal Fe	430	330	8	
AS 1443(94)	T13	CF bar, not CD or CR;<=260mm	0.15 max		0.80-1.20		0.04-0.09		0.25-0.35	0.10 max	Pb 0.15-0.35 for lead-bearing steels; bal Fe	370	230	17	
France															
AFNOR NFA35561(92)	AD37Pb	t<1000mm; HR	0.08-0.15		0.3-0.6		0.04 max	0.2-0.35	0.04 max	0.1-0.4	bal Fe	340-490	215	28	
AFNOR NFA35561(92)	AD37Pb	4<t<1000mm; CD CW	0.08-0.15		0.3-0.6		0.04 max	0.2-0.35	0.04 max	0.1-0.4	bal Fe	380-580	245	12	
AFNOR NFA35561(92)	AD40Pb	4<t<1000mm; CD CW	0.11-0.18		0.3-0.6		0.04 max	0.2-0.35	0.04 max	0.1-0.4	bal Fe	410-650	255	11	
AFNOR NFA35561(92)	AD40Pb	t<1000mm; HR	0.11-0.18		0.3-0.6		0.04 max	0.2-0.35	0.04 max	0.1-0.4	bal Fe	360-520	225	26	
AFNOR NFA35561(92)	AD42Pb	t<1000mm; HR	0.14-0.21		0.5-0.8		0.04 max	0.2-0.35	0.04 max	0.1-0.4	bal Fe	390-540	235	24	
AFNOR NFA35561(92)	AD42Pb	4<t<1000mm; CD CW	0.14-0.21		0.5-0.8		0.04 max	0.2-0.35	0.04 max	0.1-0.4	bal Fe	430-670	265	10	
AFNOR NFA35561(92)	AD55Pb	t<1000mm; HR	0.31-0.39		0.5-0.8		0.04 max	0.2-0.35	0.04 max	0.1-0.4	bal Fe	510-700	300	18	
AFNOR NFA35561(92)	AD60Pb	t<=30mm; HR	0.37-0.45		0.5-0.8		0.04 max	0.2-0.35	0.04 max	0.1-0.4	bal Fe	590-730	335	16	
AFNOR NFA35561(92)	AD60Pb	t>30mm; HR	0.37-0.45		0.5-0.8		0.04 max	0.2-0.35	0.04 max	0.1-0.4	bal Fe	590-730	315	16	
AFNOR NFA35561(92)	S200Pb		0.14		0.7-1.2		0.11	0.2-0.3	0.25-0.32	0.08	bal Fe				
AFNOR NFA35561(92)	S300Pb		0.15		1-1.6		0.11	0.2-0.3	0.3-0.4	0.08	bal Fe				
Germany															
DIN E EN 10087(95)	WNr 1.0737	HR Bar Rod free-cutting CD 11-16mm	0.15 max		1.00-1.50		0.11 max	0.15-0.35	0.34-0.40	0.05 max	bal Fe	540-780	430	7	
DIN E EN 10087(95)	WNr 1.0737	HR Bar Rod free-cutting CD 64-100mm	0.15 max		1.00-1.50		0.11 max	0.15-0.35	0.34-0.40	0.05 max	bal Fe	390-640	265	10	
DIN EN 10087(95)	9SMnPb28	HR Bar Rod free-cutting CD 64-100mm	0.14 max		0.90-1.30		0.11 max	0.15-0.35	0.27-0.33	0.05 max	bal Fe	380-630	245	10	
DIN EN 10087(95)	9SMnPb28	HR Bar Rod free-cutting CD 11-16mm	0.14 max		0.90-1.30		0.11 max	0.15-0.35	0.27-0.33	0.05 max	bal Fe	510-760	410	7	
DIN EN 10087(95)	9SMnPb36	HR Bar Rod free-cutting CD 64-100mm	0.15 max		1.00-1.50		0.11 max	0.15-0.35	0.34-0.40	0.05 max	bal Fe	390-640	265	10	
DIN EN 10087(95)	9SMnPb36	HR Bar Rod free-cutting CD 11-16mm	0.15 max		1.00-1.50		0.11 max	0.15-0.35	0.34-0.40	0.05 max	bal Fe	540-780	430	7	
Hungary															
MSZ 4339(75)	APb2	Free-cutting; gen struct	0.15 max		1-1.6		0.1 max	0.15-0.3	0.3-0.4	0.07 max	bal Fe				
International															
ISO 683-9(88)	12SMnPb35	Free-cutting, non HT, <16mm	0.15 max		1.00-1.50		0.11 max	0.15-0.35	0.30-0.40	0.05 max	bal Fe	540-840	430	7	174 HB
ISO R683-9(70)	3Pb*	Obs; see 12SMnPb35	0.15		1-1.5		0.11	0.15-0.35	0.3-0.4	0.05	bal Fe				
Italy															
UNI 4838(80)	CF9SMnPb32	Free-cutting	0.06-0.13		0.9-1.3		0.04-0.1	0.15-0.3	0.28-0.35	0.05 max	bal Fe				
UNI 4838(80)	CF9SMnPb36	Free-cutting	0.06-0.13		1-1.5		0.04-0.1	0.15-0.3	0.32-0.4	0.05 max	bal Fe				
Japan															
JIS G4804(83)	SUM22L	Bar Free cutting	0.13 max		0.70-1.00		0.07-0.12	0.10-0.35	0.24-0.33		bal Fe				
JIS G4804(83)	SUM24L	Bar Free cutting	0.15 max		0.85-1.15		0.04-0.09	0.10-0.35	0.26-0.35		bal Fe				
Mexico															
NMX-B-301(86)	12L14	Bar	0.15 max		0.85-1.15		0.04-0.09	0.15-0.35	0.26-0.35		bal Fe				
Pan America															
COPANT 330	12L14		0.15		0.85-1.15		0.04-0.09	0.15-0.35	0.26-0.35		bal Fe				
COPANT 331	12L14		0.15		0.85-1.15		0.04-0.09	0.15-0.35	0.26-0.35		bal Fe				

UNS numbers and US grades are provided as a means of cross referencing chemically similar alloys. Exchangability is only possible after independent examination of specifications. Tensile properties are minimum or typical as specified. UTS and YS as MPa. El as %. See Appendix for list of abbreviations used in Notes. * indicates obsolete material.

Specification	Designation	Notes	C	Cr	Mn	Ni	P	Pb	S	Si	Other	UTS	YS	El	Hard
Carbon Steel, Rephosphorized and Resulfurized, 12L14 (Continued from previous page)															
Pan America															
COPANT 333	12L14		0.15		0.85-1.15		0.04-0.09	0.15-0.35	0.26-0.35		bal Fe				
COPANT 514	12L14	Tube	0.15		0.85-1.15		0.04-0.09	0.15-0.35	0.26-0.35		bal Fe				
Spain															
UNE	F.210.C		0.07-0.13		0.9-1.2		0.05-0.09	0.15-0.35	0.25-0.35	0.1	bal Fe				
UNE	F.210.D		0.07-0.13		0.9-1.2		0.05-0.09	0.15-0.35	0.25-0.35	0.1	Te 0.035; bal Fe				
UNE	F.210.E		0.07-0.13		0.9-1.2		0.05-0.09	0.15-0.35	0.25-0.35	0.1	Se 0.03-0.05; bal Fe				
UNE	F.210.J		0.15		1-1.5		0.11	0.15-0.35	0.3-0.4	0.05	Te 0.035; bal Fe				
UNE	F.210.K		0.15		1-1.5		0.11	0.15-0.35	0.3-0.4	0.05	Se 0.03-0.05; bal Fe				
UNE	F.210.M		0.15		1-1.5		0.11	0.15-0.35	0.3-0.4	0.05	bal Fe				
UNE 36021(80)	12SMnPb35	Free-cutting	0.15 max		1-1.5		0.11 max	0.15-0.35	0.3-0.4	0.05 max	Mo <=0.15; bal Fe				
UNE 36021(80)	F.2114*	see 12SMnPb35	0.15 max		1-1.5		0.11 max	0.15-0.35	0.3-0.4	0.05 max	Mo <=0.15; bal Fe				
Sweden															
SIS 141926	1926-00	Bar, as roll 63mm diam	0.12-0.18		0.8-1.2		0.06	0.15-0.35	0.15-0.25	0.1-0.4	bal Fe	440			
SIS 141926	1926-03	Bar, CH 16mm diam	0.12-0.18		0.8-1.12		0.06	0.15-0.35	0.15-0.25	0.1-0.4	bal Fe	650	400	8	
SIS 141926	1926-04	Bar, CW, 16mm diam	0.12-0.18		0.8-1.2		0.06	0.15-0.35	0.15-0.25	0.1-0.4	bal Fe	550	420	7	
SS 141926	1926	Free-cutting	0.12-0.18		0.8-1.2		0.06 max	0.15-0.35	0.15-0.25	0.1-0.4	bal Fe				
Switzerland															
VSM	PX		0.15		1.3		0.1	0.25	0.38		bal Fe				
USA															
	AISI 12L14	Wir rod	0.15 max		0.85-1.15		0.040-0.090	0.15-0.35	0.26-0.35		bal Fe				
	AISI 12L14	Bar	0.15 max		0.85-1.15		0.04-0.09	0.15-0.35	0.26-0.35		bal Fe				
	UNS G12144	Leaded	0.15 max		0.85-1.15		0.04-0.09	0.15-0.35	0.26-0.35		bal Fe				
ASTM A108(95)	12L14	Bar, CF	0.15 max		0.85-1.15		0.04-0.09	0.15-0.35	0.26-0.35		bal Fe				
ASTM A29/A29M(93)	12L14	Bar	0.15 max		0.85-1.15		0.04-0.09	0.15-0.35	0.26-0.35		bal Fe				
ASTM A510(96)	12L14	Wir rod	0.15 max		0.85-1.15		0.04-0.09	0.15-0.35	0.26-0.35		bal Fe				
ASTM A519(96)	12L14	Smls mech tub	0.15 max		0.85-1.15		0.04-0.09	0.15-0.35	0.26-0.350		bal Fe				
ASTM A576(95)	12L14	Special qual HW bar	0.15 max		0.85-1.15		0.04-0.09	0.15-0.35	0.26-0.35		Pb Bi Ca Se Te if spec'd; bal Fe				
SAE J1397(92)	12L14	Bar CD, est mech prop	0.15 max		0.85-1.15		0.04-0.09	0.15-0.35	0.26-0.35		bal Fe	540	410	10	163 HB
SAE J1397(92)	12L14	Bar HR, est mech prop	0.15 max		0.85-1.15		0.04-0.09	0.15-0.35	0.26-0.35		bal Fe	390	230	22	121 HB
SAE J403(95)	12L14	Bar Wir rod Smls Tub HR CF	0.15 max		0.85-1.15		0.04-0.09	0.15-0.35	0.26-0.35		bal Fe				
SAE J403(95)	12L14	Bil Bar Rod Tub	0.15 max		0.85-1.15		0.04-0.09	0.15-0.35	0.26-0.35		bal Fe				

Carbon Steel, Rephosphorized and Resulfurized, A106(C)

Specification	Designation	Notes	C	Cr	Mn	Ni	P	Pb	S	Si	Other	UTS	YS	El	Hard
Czech Republic															
CSN 410000	10000	Gen struct	0.30 max	0.3 max	1.6 max		0.070 max		0.060 max	0.60 max	bal Fe				
CSN 410004	10004	Gen struct	0.30 max	0.3 max	1.60 max		0.06 max		0.06 max	0.60 max	bal Fe				
CSN 410216	10216	Reinforcing	0.3 max	0.3 max	2 max		0.07 max		0.06 max	0.6 max	bal Fe				

UNS numbers and US grades are provided as a means of cross referencing chemically similar alloys. Exchangability is only possible after independent examination of specifications. Tensile properties are minimum or typical as specified. UTS and YS as MPa. El as %. See Appendix for list of abbreviations used in Notes. * indicates obsolete material.

Specification	Designation	Notes	C	Cr	Mn	Ni	P	Pb	S	Si	Other	UTS	YS	El	Hard
Carbon Steel, Rephosphorized and Resulfurized, A106(C) (Continued from previous page)															
Czech Republic															
CSN 410335	10335	Reinforcing	0.25 max	0.3 max	2 max		0.05 max		0.05 max	0.6 max	bal Fe				
CSN 410338	10338	Reinforcing	0.30 max	0.3 max	1.60 max		0.070 max		0.060 max	0.60 max	bal Fe				
CSN 410425	10425	Ribbed Reinf	0.28 max	0.3 max	2 max		0.05 max		0.05 max	0.6 max	bal Fe				
CSN 410607	10607	Reinforcing	0.3 max	0.3 max	2 max		0.05 max		0.05 max	0.6 max	bal Fe				
CSN 411120	11120	Free-cutting	0.15-0.25	0.3 max	0.6-1.1		0.1 max		0.14-0.24	0.4 max	bal Fe				
Hungary															
MSZ 500(89)	Fe310-0	Gen struct; Base; BS; 16.1-25mm; HR/Frg	0.3 max		2 max		0.07 max		0.06 max	0.6 max	bal Fe	290-510	175	18L; 16T	
MSZ 500(89)	Fe310-0	Gen struct; Base; BS; 0-1mm; HR/Frg	0.3 max		2 max		0.07 max		0.06 max	0.6 max	bal Fe	310-540	185	10L; 8T	
International															
ISO 630(80)	Fe310-0	Struct	0.3 max		2 max		0.07 max		0.06 max	0.6 max	bal Fe				
USA															
	UNS K03501		0.35 max		0.29-1.06		0.048 max		0.058 max	0.10 min	bal Fe				
ASTM A106	C	Smls Pipe	0.35 max	0.40 max	0.29-1.06	0.40 max	0.035 max		0.035 max	0.10 min	Cu <=0.40; Mo <=0.15; V <=0.08; Cr+Cu+Mo+Ni+V<=1.00; bal Fe	485	275	30L	
ASTM A234/A234M(97)	WPC	Frg	0.35 max	0.15 max	0.29-1.06	0.40 max	0.050 max		0.058 max	0.10 min	Cu <=0.40; Mo <=0.15; Nb <=0.02; V <=0.08; Cu+Ni+Cr+Mo<=1.00 Cr+Mo<=0.32; bal Fe	485-655	275	22L	
Yugoslavia															
	C.3105	High-temp const	0.14-0.2	0.3 max	0.9-1.2		0.05 max		0.05 max	0.2-0.4	Mo <=0.15; bal Fe				
Carbon Steel, Rephosphorized and Resulfurized, A178(C)															
USA															
	UNS K03503		0.35 max		0.80 max		0.050 max		0.060 max		bal Fe				
ASTM A178/A178M	C	ERW Pipe	0.35 max		0.80 max		0.035 max		0.035 max		bal Fe	415	255	30	
Carbon Steel, Rephosphorized and Resulfurized, A31(B)															
USA															
	UNS K03100		0.31 max		0.27-0.83		0.048 max		0.058 max		bal Fe				
Carbon Steel, Rephosphorized and Resulfurized, A449															
USA															
	UNS K04200		0.25-0.58		0.57 min		0.048 max		0.058 max		bal Fe				
Carbon Steel, Rephosphorized and Resulfurized, A524															
USA															
	UNS K02104		0.21 max		0.90-1.35		0.048 max		0.058 max	0.10-0.40	bal Fe				
ASTM A524(96)	I	Smls pipe, wall t<=9.52mm	0.21 max		0.90-1.35		0.035 max		0.035 max	0.10-0.40	bal Fe	414-586	240	30L 16.5T	
ASTM A524(96)	II	Smls pipe, wall t>9.52mm	0.21 max		0.90-1.35		0.035 max		0.035 max	0.10-0.40	bal Fe	380-550	205	35L 25T	
Carbon Steel, Rephosphorized and Resulfurized, A563(D)															
USA															
	UNS K05801		0.58 max		0.27 min		0.048 max		0.058 max		bal Fe				

UNS numbers and US grades are provided as a means of cross referencing chemically similar alloys. Exchangability is only possible after independent examination of specifications. Tensile properties are minimum or typical as specified. UTS and YS as MPa. El as %. See Appendix for list of abbreviations used in Notes. * indicates obsolete material.

Specification	Designation	Notes	C	Cr	Mn	Ni	P	Pb	S	Si	Other	UTS	YS	El	Hard

Carbon Steel, Rephosphorized and Resulfurized, A563(DH)

USA

Specification	Designation	Notes	C	Cr	Mn	Ni	P	Pb	S	Si	Other	UTS	YS	El	Hard
	UNS K03800		0.18-0.58		0.57 max		0.048 max		0.058 max		bal Fe				

Carbon Steel, Rephosphorized and Resulfurized, No equivalents identified

Hungary

Specification	Designation	Notes	C	Cr	Mn	Ni	P	Pb	S	Si	Other	UTS	YS	El	Hard
MSZ 4339(86)	ASBi5	Free-cutting; gen struct; 0-10mm; HF; bright turned (ground)	0.15 max		1-1.5		0.1 max		0.3-0.4	0.05 max	Bi=0.06-0.15; bal Fe	390-590			174 HB max
MSZ 4339(86)	ASBi5	Free-cutting; gen struct; 63.1-100mm; HF; bright turned (ground)	0.15 max		1-1.5		0.1 max		0.3-0.4	0.05 max	Bi=0.06-0.15; bal Fe	360-520			156 HB max

Poland

Specification	Designation	Notes	C	Cr	Mn	Ni	P	Pb	S	Si	Other	UTS	YS	El	Hard
PNH84023/05	20P		0.18-0.23	0.3 max	0.4-0.8	0.3 max	0.04-0.07		0.04 max	0.05 max	bal Fe				
PNH84026	A10/XN	Free-cutting	0.1 max	0.3 max	0.9-1.3	0.4 max	0.08 max		0.24-0.34	0.05 max	N 0.01-0.016; bal Fe				

Specification	Designation	Notes	C	Cr	Cu	Mn	Ni	P	S	Si	Other	UTS	YS	El	Hard
Carbon Steel, High-Manganese, 1139															
Hungary															
MSZ 4339(86)	ANS1	Free-cutting; Q/T; 0-10mm	0.32-0.4			0.7-1.1		0.06 max	0.15-0.25	0.15-0.35	bal Fe	620-770	420	13L	
MSZ 4339(86)	ANS1	Free-cutting; Q/T; 40.1-63mm	0.32-0.4			0.7-1.1		0.06 max	0.15-0.25	0.15-0.35	bal Fe	540-690	325	17L	
International															
ISO 683-9(88)	35S20	Free-cutting, hard, Q/T, <16mm	0.32-0.39			0.70-1.10		0.06 max	0.15-0.25	0.15-0.40	bal Fe	570-770	390	14	
ISO R683-9(70)	Type 7*	Obs; see 35S20	0.32-0.39			0.5-0.9		0.06	0.15-0.25	0.15-0.4	bal Fe				
ISO R683-9(70)	Type 8*	Obs; see 35SMn20	0.32-0.39			0.9-1.2		0.06	0.15-0.25	0.15-0.4	bal Fe				
Italy															
UNI 4838(62)	35SMn10	Free-cutting	0.32-0.39			1.35-1.65		0.04 max	0.08-0.13	0.35 max	bal Fe				
UNI 4838(80)	CF35SMn10	Free-cutting	0.32-0.39			1.35-1.65		0.04 max	0.08-0.13	0.3 max	bal Fe				
Japan															
JIS G3507(91)	SWRCH41K	Wir rod FF	0.36-0.44			1.35-1.65		0.030 max	0.035 max	0.10-0.35	bal Fe				
Mexico															
NMX-B-301(86)	1139	Bar	0.35-0.43			1.35-1.65		0.040 max	0.13-0.20		bal Fe				
Poland															
PNH84026	A35	Free-cutting	0.32-0.39	0.3 max		0.5-0.9	0.4 max	0.06 max	0.15-0.25	0.15-0.4	bal Fe				
PNH84026	A35G2	Free-cutting	0.32-0.4	0.25 max		1.4-1.8	0.3 max	0.035 max	0.08-0.2	0.17-0.37	bal Fe				
Spain															
UNE 36021(80)	35MnS6	Free-cutting	0.33-0.39			1.3-1.7		0.04 max	0.09-0.13	0.1-0.4	Mo <=0.15; bal Fe				
Switzerland															
VSM DST-4	35S20		0.32-0.39			0.5-0.9		0.06	0.15-0.25	0.1-0.2	bal Fe				
USA															
	AISI 1139	Wir rod	0.35-0.43			1.35-1.65		0.040 max	0.13-0.20		bal Fe				
	UNS G11390		0.35-0.43			1.35-1.65		0.040 max	0.13-0.20		bal Fe				
ASTM A29/A29M(93)	1139	Bar	0.35-0.43			1.35-1.65		0.040 max	0.13-0.20		bal Fe				
ASTM A510(96)	1139	Wir rod	0.35-0.43			1.35-1.65		0.040 max	0.13-0.20		bal Fe				
ASTM A576(95)	1139	Special qual HW bar	0.35-0.43			1.35-1.65		0.040 max	0.13-0.20		Si Pb Bi Ca Se Te if spec'd; bal Fe				
FED QQ-S-637A(70)	C1139*	Obs; see ASTM A108; CF Bar std qual, free mach	0.35-0.43			1.35-1.65		0.04 max	0.13-0.2		bal Fe				
SAE J1249(95)	1139	Former SAE std valid for Wir rod	0.35-0.43			1.35-1.65		0.040 max	0.13-0.20		bal Fe				
Yugoslavia															
	C.1490	Free-cutting	0.32-0.4			0.6-0.9		0.07	0.25	0.1-0.4	bal Fe				
Carbon Steel, High-Manganese, 1513															
Australia															
AS 1442	K8*	Obs	0.1-0.18			1.3-1.7		0.05	0.05	0.1-0.35	bal Fe				
AS 1442	XK1315*	Obs	0.12-0.18			1.4-1.7		0.05	0.05	0.1-0.35	bal Fe				
AS 1442(92)	8	Bar<=215mm, Bloom/bil/slab<=250mm, HR as rolled or norm	0.10-0.18			1.30-1.70		0.040 max	0.040 max	0.10-0.40	bal Fe	480	270	22	
AS 1443	K8*	Obs	0.1-0.18			1.3-1.7		0.05	0.05	0.1-0.35	bal Fe				

Specification	Designation	Notes	C	Cr	Cu	Mn	Ni	P	S	Si	Other	UTS	YS	El	Hard

Carbon Steel, High-Manganese, 1513 (Continued from previous page)

Australia

Specification	Designation	Notes	C	Cr	Cu	Mn	Ni	P	S	Si	Other	UTS	YS	El	Hard
AS 1443	XK1315*	Obs	0.12-0.18			1.4-1.7		0.05	0.05	0.1-0.35	bal Fe				
Belgium															
NBN 630	E37		0.17			0.4-1.2		0.05	0.05	0.35	bal Fe				
NBN A21-101	AE235		0.2			1.5		0.05	0.05	0.55	bal Fe				
Bulgaria															
BDS 4880(89)	09G2	Struct	0.15 max	0.35 max	0.30 max	1.30-1.90	0.30 max	0.040 max	0.045 max	0.12-0.42	bal Fe				
BDS 4880(89)	09G2B	Struct	0.15 max	0.35 max	0.35 max	1.80 max	0.35 max	0.040 max	0.035 max	0.15-0.45	Nb 0.005-0.050; bal Fe				
BDS 4880(89)	09G2S	Struct	0.15 max	0.35 max	0.35 max	1.20-1.80	0.35 max	0.040 max	0.045 max	0.45-0.85	bal Fe				
BDS 4880(89)	10G2SB	Struct	0.05-0.18	0.35 max	0.35 max	1.10-1.70	0.35 max	0.040 max	0.040 max	0.35-0.70	Nb 0.005-0.060; bal Fe				
BDS 4880(89)	10G2SBF	Struct	0.04-0.16	0.35 max	0.35 max	1.30-1.80	0.35 max	0.040 max	0.040 max	0.40-0.80	Nb 0.005-0.060; V 0.005-0.070; bal Fe				
BDS 4880(89)	10G2SFT	Struct	0.15 max	0.35 max	0.35 max	1.10-1.70	0.35 max	0.040 max	0.040 max	0.35-0.65	Ti 0.005-0.09; V 0.025-0.100; bal Fe				
BDS 4880(89)	11G2SB	Struct	0.08-0.14	0.35 max	0.35 max	1.40-1.75	0.35 max	0.035 max	0.040 max	0.45-0.75	V 0.02-0.060; bal Fe				
BDS 4880(89)	12G2SBF	Struct	0.05-0.18	0.35 max	0.35 max	1.20-1.80	0.35 max	0.040 max	0.040 max	0.55-0.90	Nb 0.005-0.060; V 0.025-0.100; bal Fe				
BDS 5930	09G2	Struct	0.12			1.3-1.7		0.04	0.04	0.5	bal Fe				
BDS 5930	10G2S1	Heat res	0.12 max	0.3 max	0.30 max	1.3-1.65	0.30 max	0.035 max	0.04 max	0.9-1.2	Mo <=0.15; bal Fe				
BDS 6354	5ChGM	Struct	0.12-0.19		0.3	0.8-1.1	0.3	0.035	0.035	0.17-0.37	bal Fe				
BDS 9801	09G2	Struct, weld ships	0.12 max	0.3 max	0.30 max	1.4-1.8	0.3 max	0.035 max	0.035 max	0.17-0.37	Mo <=0.15; bal Fe				
BDS 9801	10G2S1D	Struct weld ships	0.12 max	0.3 max	0.15-0.30	1.3-1.65	0.3 max	0.035 max	0.035 max	0.8-1.1	Mo <=0.15; bal Fe				
BDS 9801	14G2B	Struct, weld ships	0.1-0.16	0.2 max	0.35 max	1.3-1.6	0.4 max	0.040 max	0.04 max	0.2-0.5	Al 0.015-0.06; Mo <=0.15; bal Fe				
BDS 9801	14G2F	Struct, weld ships	0.1-0.16	0.2 max	0.35 max	1.3-1.6	0.4 max	0.040 max	0.04 max	0.2-0.5	Al <=0.1; Mo <=0.15; bal Fe				
China															
GB 1591(88)	12Mn	Plt HR 16-25mm Thk	0.09-0.16			1.10-1.50		0.045 max	0.045 max	0.20-0.55	bal Fe	430-580	275	21	
GB/T 8164(93)	12Mn	Strp HR 4-8mm Thk	0.09-0.16			1.10-1.50		0.045 max	0.045 max	0.20-0.55	bal Fe	430-580	275	21	
YB/T 5132(93)	12Mn2A	Sh HR/CR Ann	0.08-0.17		0.25 max	1.20-1.60		0.030 max	0.030 max	0.20-0.55	bal Fe	395-570		22	
Europe															
EN 10025(90)A1(93) 1.0036		Struct HR QS, t<=1mm	0.17 max			1.40 max		0.045 max	0.045 max		N max na if Al>=0.020; bal Fe	360-510	235	17L 15T	
EN 10025(90)A1(93) 1.0036		Struct HR BS, 16<t<=25mm	0.17 max			1.40 max		0.045 max	0.045 max		N <=0.007; N max na if Al>=0.020; bal Fe	340-470	235	26L 24T	
EN 10025(90)A1(93) 1.0036		Struct HR BS, t<=16mm	0.17 max			1.40 max		0.045 max	0.045 max		N <=0.007; N max na if Al>=0.020; bal Fe	360-510	235	17L 15T	
EN 10025(90)A1(93) 1.0036		Struct HR QS, 3<t<=16mm	0.17 max			1.40 max		0.045 max	0.045 max		N max na if Al>=0.020; bal Fe	340-470	235	26L 24T	
EN 10025(90)A1(93) 1.0037		Struct HR QS, t<=16mm	0.17 max			1.40 max		0.045 max	0.045 max		N <=0.009; N max na if Al>=0.020; bal Fe	360-510	235	17L 15T	
EN 10025(90)A1(93) 1.0037		Struct HR BS, t<=16mm	0.17 max			1.40 max		0.045 max	0.045 max		N <=0.009; N max na if Al>=0.020; bal Fe	360-510	235	17L 15T	
EN 10025(90)A1(93) 1.0038		Struct HR BS, t<=1mm	0.17 max			1.40 max		0.045 max	0.045 max		N <=0.009; N max na if Al>=0.020; bal Fe	360-510	235	17L 15T	
EN 10025(90)A1(93) 1.0038		Struct HR QS, 3<t<=16mm	0.17 max			1.40 max		0.045 max	0.045 max		N max na if Al>=0.020; bal Fe	340-470	235	26L 24T	
EN 10025(90)A1(93) 1.0038		Struct HR QS, t<=1mm	0.17 max			1.40 max		0.045 max	0.045 max		N max na if Al>=0.020; bal Fe	360-510	235	17L 15T	
EN 10025(90)A1(93) 1.0038		Struct HR BS, 3<t<=16mm	0.17 max			1.40 max		0.045 max	0.045 max		N <=0.009; N max na if Al>=0.020; bal Fe	340-470	235	26L 24T	
EN 10025(90)A1(93) 1.0114		Struct HR QS, 3<t<=16mm	0.17 max			1.40 max		0.040 max	0.040 max		N <=0.009; N max na if Al>=0.020; bal Fe	340-470	235	26L 24T	
EN 10025(90)A1(93) 1.0114		Struct HR QS, t<=1mm	0.17 max			1.40 max		0.040 max	0.040 max	·	N <=0.009; N max na if Al>=0.020; bal Fe	360-510	235	17L 15T	

Specification Designation	Notes	C	Cr	Cu	Mn	Ni	P	S	Si	Other	UTS	YS	El	Hard

Carbon Steel, High-Manganese, 1513 (Continued from previous page)

Europe

Specification Designation	Notes	C	Cr	Cu	Mn	Ni	P	S	Si	Other	UTS	YS	El	Hard
EN 10025(90)A1(93) 1.0114	Struct HR QS, 63<t<=80mm	0.17 max			1.40 max		0.040 max	0.040 max		N <=0.009; N max na if Al>=0.020; bal Fe	340-470	215	24L 22T	
EN 10025(90)A1(93) 1.0114	Struct HR QS, 200<t<=250mm	0.17 max			1.40 max		0.040 max	0.040 max		N <=0.009; N max na if Al>=0.020; bal Fe	320-470	175	21L 21T	
EN 10025(90)A1(93) 1.0114	Struct HR QS, t<=1mm	0.17 max			1.40 max		0.040 max	0.040 max		N max na if Al>=0.020; bal Fe	360-510	235	17L 15T	
EN 10025(90)A1(93) 1.0114	Struct HR QS, 3<t<=16mm	0.17 max			1.40 max		0.040 max	0.040 max		N max na if Al>=0.020; bal Fe	340-470	235	26L 24T	
EN 10025(90)A1(93) 1.0116	Struct HR QS, 63<t<=80mm	0.17 max			1.40 max		0.035 max	0.035 max		N max na if Al>=0.020; bal Fe	340-470	215	24L 22T	
EN 10025(90)A1(93) 1.0116	Struct HR QS, 3<t<=16mm	0.17 max			1.40 max		0.035 max	0.035 max		N max na if Al>=0.020; bal Fe	340-470	235	26L 24T	
EN 10025(90)A1(93) 1.0116	Struct HR QS, t<=1mm	0.17 max			1.40 max		0.035 max	0.035 max		N max na if Al>=0.020; bal Fe	360-510	235	17L 15T	
EN 10025(90)A1(93) 1.0116	Struct HR QS, 3<t<=16mm	0.17 max			1.40 max		0.035 max	0.035 max		N max na if Al>=0.020; bal Fe	340-470	235	26L 24T	
EN 10025(90)A1(93) 1.0116	Struct HR QS, t<=1mm	0.17 max			1.40 max		0.035 max	0.035 max		N max na if Al>=0.020; bal Fe	360-510	235	17L 15T	
EN 10025(90)A1(93) 1.0116	Struct HR QS, 200<t<=250mm	0.17 max			1.40 max		0.035 max	0.035 max		N max na if Al>=0.020; bal Fe	320-470	175	21L 21T	
EN 10025(90)A1(93) 1.0117	Struct HR QS, 63<t<=80mm	0.17 max			1.40 max		0.035 max	0.035 max		N max na if Al>=0.020; bal Fe	340-470	215	24L 22T	
EN 10025(90)A1(93) 1.0117	Struct HR QS, 3<t<=16mm	0.17 max			1.40 max		0.035 max	0.035 max		N max na if Al>=0.020; bal Fe	340-470	235	26L 24T	
EN 10025(90)A1(93) 1.0117	Struct HR QS, 200<t<=250mm	0.17 max			1.40 max		0.035 max	0.035 max		N max na if Al>=0.020; bal Fe	320-470	175	21L 21T	
EN 10025(90)A1(93) 1.0117	Struct HR QS, t<=1mm	0.17 max			1.40 max		0.035 max	0.035 max		N max na if Al>=0.020; bal Fe	360-510	235	17L 15T	
EN 10025(90)A1(93) 1.0117	Struct HR QS, 3<t<=16mm	0.17 max			1.40 max		0.035 max	0.035 max		N max na if Al>=0.020; bal Fe	340-470	235	26L 24T	
EN 10025(90)A1(93) 1.0117	Struct HR QS, t<=1mm	0.17 max			1.40 max		0.035 max	0.035 max		N max na if Al>=0.020; bal Fe	360-510	235	17L 15T	
EN 10025(90)A1(93) 1.0143	Struct HR QS, 3<t<=16mm	0.18 max			1.50 max		0.040 max	0.040 max		N max na if Al>=0.020; bal Fe	410-560	275	22L 20T	
EN 10025(90)A1(93) 1.0143	Struct HR QS, 63<t<=80mm	0.18 max			1.50 max		0.040 max	0.040 max		N <=0.009; N max na if Al>=0.020; bal Fe	410-560	245	20L 18T	
EN 10025(90)A1(93) 1.0143	Struct HR QS, t<=1mm	0.18 max			1.50 max		0.040 max	0.040 max		N max na if Al>=0.020; bal Fe	430-580	275	14L 12T	
EN 10025(90)A1(93) 1.0144	Struct HR QS, 3<t<=16mm	0.18 max			1.50 max		0.035 max	0.035 max		N max na if Al>=0.020; bal Fe	410-560	275	22L 20T	
EN 10025(90)A1(93) 1.0144	Struct HR QS, t<=1mm	0.18 max			1.50 max		0.035 max	0.035 max		N max na if Al>=0.020; bal Fe	430-580	275	14L 12T	
EN 10025(90)A1(93) 1.0144	Struct HR QS, 3<t<=16mm	0.18 max			1.50 max		0.035 max	0.035 max		N max na if Al>=0.020; bal Fe	410-560	275	22L 20T	
EN 10025(90)A1(93) 1.0144	Struct HR QS, 63<t<=80mm	0.18 max			1.50 max		0.035 max	0.035 max		N max na if Al>=0.020; bal Fe	410-560	245	20L 18T	
EN 10025(90)A1(93) 1.0145	Struct HR QS, 3<t<=16mm	0.18 max			1.50 max		0.035 max	0.035 max		N max na if Al>=0.020; bal Fe	410-560	275	22L 20T	
EN 10025(90)A1(93) 1.0145	Struct HR QS, 63<t<=80mm	0.18 max			1.50 max		0.035 max	0.035 max		N max na if Al>=0.020; bal Fe	410-560	245	20L 18T	
EN 10025(90)A1(93) 1.0145	Struct HR QS, t<=1mm	0.18 max			1.50 max		0.035 max	0.035 max		N max na if Al>=0.020; bal Fe	430-580	275	14L 12T	
EN 10025(90)A1(93) S235J0	Struct HR QS, 3<t<=16mm	0.17 max			1.40 max		0.040 max	0.040 max		N <=0.009; N max na if Al>=0.020; bal Fe	340-470	235	26L 24T	
EN 10025(90)A1(93) S235J0	Struct HR QS, t<=1mm	0.17 max			1.40 max		0.040 max	0.040 max		N <=0.009; N max na if Al>=0.020; bal Fe	360-510	235	17L 15T	
EN 10025(90)A1(93) S235J0	Struct HR QS, 63<t<=80mm	0.17 max			1.40 max		0.040 max	0.040 max		N <=0.009; N max na if Al>=0.020; bal Fe	340-470	215	24L 22T	
EN 10025(90)A1(93) S235J0	Struct HR QS, 200<t<=250mm	0.17 max			1.40 max		0.040 max	0.040 max		N <=0.009; N max na if Al>=0.020; bal Fe	320-470	175	21L 21T	
EN 10025(90)A1(93) S235J0C	Struct HR QS, t<=1mm	0.17 max			1.40 max		0.040 max	0.040 max		N max na if Al>=0.020; bal Fe	360-510	235	17L 15T	
EN 10025(90)A1(93) S235J0C	Struct HR QS, 3<t<=16mm	0.17 max			1.40 max		0.040 max	0.040 max		N max na if Al>=0.020; bal Fe	340-470	235	26L 24T	
EN 10025(90)A1(93) S235J2G3	Struct HR QS, 3<t<=16mm	0.17 max			1.40 max		0.035 max	0.035 max		N max na if Al>=0.020; bal Fe	340-470	235	26L 24T	
EN 10025(90)A1(93) S235J2G3	Struct HR QS, t<=1mm	0.17 max			1.40 max		0.035 max	0.035 max		N max na if Al>=0.020; bal Fe	360-510	235	17L 15T	

Specification Designation	Notes	C	Cr	Cu	Mn	Ni	P	S	Si	Other	UTS	YS	El	Hard

Carbon Steel, High-Manganese, 1513 (Continued from previous page)

Europe

Specification Designation	Notes	C	Cr	Cu	Mn	Ni	P	S	Si	Other	UTS	YS	El	Hard
EN 10025(90)A1(93) S235J2G3	Struct HR QS, 63<t<=80mm	0.17 max			1.40 max		0.035 max	0.035 max		N max na if Al>=0.020; bal Fe	340-470	215	24L 22T	
EN 10025(90)A1(93) S235J2G3	Struct HR QS, 200<t<=250mm	0.17 max			1.40 max		0.035 max	0.035 max		N max na if Al>=0.020; bal Fe	320-470	175	21L 21T	
EN 10025(90)A1(93) S235J2G3C	Struct HR QS, t<=1mm	0.17 max			1.40 max		0.035 max	0.035 max		N max na if Al>=0.020; bal Fe	360-510	235	17L 15T	
EN 10025(90)A1(93) S235J2G3C	Struct HR QS, 3<t<=16mm	0.17 max			1.40 max		0.035 max	0.035 max		N max na if Al>=0.020; bal Fe	340-470	235	26L 24T	
EN 10025(90)A1(93) S235J2G4	Struct HR QS, 63<t<=80mm	0.17 max			1.40 max		0.035 max	0.035 max		N max na if Al>=0.020; bal Fe	340-470	215	24L 22T	
EN 10025(90)A1(93) S235J2G4	Struct HR QS, t<=1mm	0.17 max			1.40 max		0.035 max	0.035 max		N max na if Al>=0.020; bal Fe	360-510	235	17L 15T	
EN 10025(90)A1(93) S235J2G4	Struct HR QS, 200<t<=250mm	0.17 max			1.40 max		0.035 max	0.035 max		N max na if Al>=0.020; bal Fe	320-470	175	21L 21T	
EN 10025(90)A1(93) S235J2G4	Struct HR QS, 3<t<=16mm	0.17 max			1.40 max		0.035 max	0.035 max		N max na if Al>=0.020; bal Fe	340-470	235	26L 24T	
EN 10025(90)A1(93) S235J2G4C	Struct HR QS, 3<t<=16mm	0.17 max			1.40 max		0.035 max	0.035 max		N max na if Al>=0.020; bal Fe	340-470	235	26L 24T	
EN 10025(90)A1(93) S235J2G4C	Struct HR QS, t<=1mm	0.17 max			1.40 max		0.035 max	0.035 max		N max na if Al>=0.020; bal Fe	360-510	235	17L 15T	
EN 10025(90)A1(93) S235JR	Struct HR BS, t<=16mm	0.17 max			1.40 max		0.045 max	0.045 max		N <=0.009; N max na if Al>=0.020; bal Fe	360-510	235	17L 15T	
EN 10025(90)A1(93) S235JRC	Struct HR QS, t<=16mm	0.17 max			1.40 max		0.045 max	0.045 max		N <=0.009; N max na if Al>=0.020; bal Fe	360-510	235	17L 15T	
EN 10025(90)A1(93) S235JRG1	Struct HR BS, 16<t<=25mm	0.17 max			1.40 max		0.045 max	0.045 max		N <=0.007; N max na if Al>=0.020; bal Fe	340-470	235	26L 24T	
EN 10025(90)A1(93) S235JRG1	Struct HR BS, t<=16mm	0.17 max			1.40 max		0.045 max	0.045 max		N <=0.007; N max na if Al>=0.020; bal Fe	360-510	235	17L 15T	
EN 10025(90)A1(93) S235JRG1C	Struct HR QS, t<=1mm	0.17 max			1.40 max		0.045 max	0.045 max		N max na if Al>=0.020; bal Fe	360-510	235	17L 15T	
EN 10025(90)A1(93) S235JRG1C	Struct HR QS, 3<t<=16mm	0.17 max			1.40 max		0.045 max	0.045 max		N max na if Al>=0.020; bal Fe	340-470	235	26L 24T	
EN 10025(90)A1(93) S235JRG2	Struct HR BS, 3<t<=16mm	0.17 max			1.40 max		0.045 max	0.045 max		N <=0.009; N max na if Al>=0.020; bal Fe	340-470	235	26L 24T	
EN 10025(90)A1(93) S235JRG2	Struct HR BS, t<=1mm	0.17 max			1.40 max		0.045 max	0.045 max		N <=0.009; N max na if Al>=0.020; bal Fe	360-510	235	17L 15T	
EN 10025(90)A1(93) S235JRG2C	Struct HR QS, 3<t<=16mm	0.17 max			1.40 max		0.045 max	0.045 max		N max na if Al>=0.020; bal Fe	340-470	235	26L 24T	
EN 10025(90)A1(93) S235JRG2C	Struct HR QS, t<=1mm	0.17 max			1.40 max		0.045 max	0.045 max		N max na if Al>=0.020; bal Fe	360-510	235	17L 15T	
EN 10025(90)A1(93) S275J0	Struct HR QS, 63<t<=80mm	0.18 max			1.50 max		0.040 max	0.040 max		N <=0.009; N max na if Al>=0.020; bal Fe	410-560	245	20L 18T	
EN 10025(90)A1(93) S275J0	Struct HR QS, t<=1mm	0.18 max			1.50 max		0.040 max	0.040 max		N <=0.009; N max na if Al>=0.020; bal Fe	430-580	275	14L 12T	
EN 10025(90)A1(93) S275J0	Struct HR QS, 3<t<=16mm	0.18 max			1.50 max		0.040 max	0.040 max		N <=0.009; N max na if Al>=0.020; bal Fe	410-560	275	22L 20T	
EN 10025(90)A1(93) S275J0C	Struct HR QS, 3<t<=16mm	0.18 max			1.50 max		0.040 max	0.040 max		N max na if Al>=0.020; bal Fe	410-560	275	22L 20T	
EN 10025(90)A1(93) S275J0C	Struct HR QS, t<=1mm	0.18 max			1.50 max		0.040 max	0.040 max		N max na if Al>=0.020; bal Fe	430-580	275	14L 12T	
EN 10025(90)A1(93) S275J2G3	Struct HR QS, 3<t<=16mm	0.18 max			1.50 max		0.035 max	0.035 max		N max na if Al>=0.020; bal Fe	410-560	275	22L 20T	
EN 10025(90)A1(93) S275J2G3	Struct HR QS, t<=1mm	0.18 max			1.50 max		0.035 max	0.035 max		N max na if Al>=0.020; bal Fe	430-580	275	14L 12T	
EN 10025(90)A1(93) S275J2G3	Struct HR QS, 63<t<=80mm	0.18 max			1.50 max		0.035 max	0.035 max		N max na if Al>=0.020; bal Fe	410-560	245	20L 18T	
EN 10025(90)A1(93) S275J2G3C	Struct HR QS, t<=1mm	0.18 max			1.50 max		0.035 max	0.035 max		N max na if Al>=0.020; bal Fe	430-580	275	14L 12T	
EN 10025(90)A1(93) S275J2G3C	Struct HR QS, 3<t<=16mm	0.18 max			1.50 max		0.035 max	0.035 max		N max na if Al>=0.020; bal Fe	410-560	275	22L 20T	
EN 10025(90)A1(93) S275J2G4	Struct HR QS, t<=1mm	0.18 max			1.50 max		0.035 max	0.035 max		N max na if Al>=0.020; bal Fe	430-580	275	14L 12T	
EN 10025(90)A1(93) S275J2G4	Struct HR QS, 63<t<=80mm	0.18 max			1.50 max		0.035 max	0.035 max		N max na if Al>=0.020; bal Fe	410-560	245	20L 18T	
EN 10025(90)A1(93) S275J2G4	Struct HR QS, 3<t<=16mm	0.18 max			1.50 max		0.035 max	0.035 max		N max na if Al>=0.020; bal Fe	410-560	275	22L 20T	
EN 10025(90)A1(93) S275J2G4C	Struct HR QS, t<=1mm	0.18 max			1.50 max		0.035 max	0.035 max		N max na if Al>=0.020; bal Fe	430-580	275	14L 12T	

Specification	Designation	Notes	C	Cr	Cu	Mn	Ni	P	S	Si	Other	UTS	YS	El	Hard

Carbon Steel, High-Manganese, 1513 (Continued from previous page)

Europe

Specification	Designation	Notes	C	Cr	Cu	Mn	Ni	P	S	Si	Other	UTS	YS	El	Hard
EN 10025(90)A1(93)	S275J2G4C	Struct HR QS, 3<t<=16mm	0.18 max			1.50 max		0.035 max	0.035 max		N max na if Al>=0.020; bal Fe	410-560	275	22L 20T	
EN 10207(91)	1.0112	Press ves 3<t<=16mm	0.16 max			0.40-1.20		0.035 max	0.030 max	0.35 max	Al >=0.020; bal Fe	360-480	235	26	
EN 10207(91)	1.0112	Press ves 40<t<=60mm	0.16 max			0.40-1.20		0.035 max	0.030 max	0.35 max	Al >=0.020; bal Fe	360-480	215	25	
EN 10207(91)	1.0112	Press ves 16<t<=40mm	0.16 max			0.40-1.20		0.035 max	0.030 max	0.35 max	Al >=0.020; bal Fe	360-480	225	26	
EN 10207(91)	1.1100	Press ves 3<t<=16mm	0.16 max			0.50-1.50		0.030 max	0.025 max	0.40 max	Al >=0.020; bal Fe	390-510	275	24	
EN 10207(91)	1.1100	Press ves 16<t<=40mm	0.16 max			0.50-1.50		0.030 max	0.025 max	0.40 max	Al >=0.020; bal Fe	390-510	265	24	
EN 10207(91)	1.1100	Press ves 40<t<=60mm	0.16 max			0.50-1.50		0.030 max	0.025 max	0.40 max	Al >=0.020; bal Fe	390-510	255	24	
EN 10207(91)	SPH235	Press ves 3<t<=16mm	0.16 max			0.40-1.20		0.035 max	0.030 max	0.35 max	Al >=0.020; bal Fe	360-480	235	26	
EN 10207(91)	SPH235	Press ves 16<t<=40mm	0.16 max			0.40-1.20		0.035 max	0.030 max	0.35 max	Al >=0.020; bal Fe	360-480	225	26	
EN 10207(91)	SPH235	Press ves 40<t<=60mm	0.16 max			0.40-1.20		0.035 max	0.030 max	0.35 max	Al >=0.020; bal Fe	360-480	215	25	
EN 10207(91)	SPHL275	Press ves 16<t<=40mm	0.16 max			0.50-1.50		0.030 max	0.025 max	0.40 max	Al >=0.020; bal Fe	390-510	265	24	
EN 10207(91)	SPHL275	Press ves 40<t<=60mm	0.16 max			0.50-1.50		0.030 max	0.025 max	0.40 max	Al >=0.020; bal Fe	390-510	255	24	
EN 10207(91)	SPHL275	Press ves 3<t<=16mm	0.16 max			0.50-1.50		0.030 max	0.025 max	0.40 max	Al >=0.020; bal Fe	390-510	275	24	

Finland

Specification	Designation	Notes	C	Cr	Cu	Mn	Ni	P	S	Si	Other	UTS	YS	El	Hard
SFS 679(86)	CR360	Gen struct	0.12 max			1.2 max		0.03 max	0.03 max	0.5 max	bal Fe				
SFS 679(86)	CR400	Gen struct	0.12 max			1.5 max		0.03 max	0.03 max	0.5 max	bal Fe				

France

Specification	Designation	Notes	C	Cr	Cu	Mn	Ni	P	S	Si	Other	UTS	YS	El	Hard
AFNOR	12M5		0.1-0.15			0.9-1.4		0.04	0.035	0.4	bal Fe				
AFNOR NFA33101(82)	AF50S	t<=450mm; Norm	0.2 max	0.3 max	0.3 max	1.5 max	0.4 max	0.04 max	0.04 max	0.55 max	bal Fe	490-590	355	25	
AFNOR NFA35501	E35-4		0.2			1.6		0.035	0.035	0.55	bal Fe				
AFNOR NFA35501(83)	E36-2*		0.24 max	0.3 max	0.3 max	1.6 max	0.4 max	0.045 max	0.045 max	0.55 max	bal Fe				
AFNOR NFA35501(83)	E36-3*	10t<=350mm; roll	0.22 max	0.3 max	0.3 max	1.6 max	0.4 max	0.04 max	0.04 max	0.55 max	bal Fe	470	315	20	
AFNOR NFA35501(83)	E36-3*	30<t<=100mm; roll	0.22 max	0.3 max	0.3 max	1.6 max	0.4 max	0.04 max	0.04 max	0.55 max	bal Fe	490-630	335	21	
AFNOR NFA35501(83)	E36-3*	t<=30mm; roll	0.22 max	0.3 max	0.3 max	1.6 max	0.4 max	0.04 max	0.04 max	0.55 max	bal Fe	490-630	355	22	
AFNOR NFA36205(82)	A37FP*	Press ves; 80<t<=110mm; Norm	0.16 max	0.25 max	0.30 max	0.40-1.60	0.4 max	0.03 max	0.02 max	0.3 max	Mo <=0.07; Nb <=0.015; V <=0.02; bal Fe	360-430	185	26	
AFNOR NFA36205(82)	A37FP*	Press ves; t<=30mm; Norm	0.16 max	0.25 max	0.30 max	0.40-1.60	0.4 max	0.03 max	0.02 max	0.3 max	Mo <=0.07; Nb <=0.015; V <=0.02; bal Fe	360-430	225	28	
AFNOR NFA36601(80)	A37AP	50<t<=80mm; Norm	0.16 max	0.3 max	0.18 max	0.4-1.60	0.5 max	0.035 max	0.03 max	0.3 max	Cu+6Sn<=0.33; Mn>=0.4; bal Fe	360-430	195	28	
AFNOR NFA36601(80)	A37CP	80<t<=300mm; Q/T	0.16 max	0.15 max	0.1 max	0.4-1.60	0.1 max	0.035 max	0.03 max	0.3 max	Al <=0.18; Mo <=0.5; W <=0.005; Cu+6Sn<=0.33; Mn>=0.4; bal Fe	360-430	185	28L; 26T	
AFNOR NFA36601(80)	A37FP	80<t<=300mm; Q/T	0.16 max	0.3 max	0.18 max	0.40-1.60	0.5 max	0.03 max	0.02 max	0.3 max	Cu+6Sn<=0.33; bal Fe	360-430	185	28L; 26T	
AFNOR NFA36601(80)	A37FP	t<=30mm; Norm	0.16 max	0.3 max	0.18 max	0.40-1.60	0.5 max	0.03 max	0.02 max	0.3 max	Cu+6Sn<=0.33; bal Fe	360-430	225	30L; 28T	

Germany

Specification	Designation	Notes	C	Cr	Cu	Mn	Ni	P	S	Si	Other	UTS	YS	El	Hard
DIN	CS DS*	Obs	0.16 max			1-1.35		0.04 max	0.04 max	0.1-0.35	Al 0.015-0.06; bal Fe				

Specification	Designation	Notes	C	Cr	Cu	Mn	Ni	P	S	Si	Other	UTS	YS	El	Hard

Carbon Steel, High-Manganese, 1513 (Continued from previous page)

Germany

Specification	Designation	Notes	C	Cr	Cu	Mn	Ni	P	S	Si	Other	UTS	YS	El	Hard
DIN	GL-A36		0.18 max	0.20 max	0.35 max	0.90-1.60	0.40 max	0.040 max	0.040 max	0.50 max	Al >=0.020; Mo <=0.08; Nb 0.02-0.05; V 0.05-0.10; V+Nb+Ti<=0.12; bal Fe				
DIN	GL-D36		0.18 max	0.20 max	0.35 max	0.90-1.60	0.40 max	0.040 max	0.040 max	0.50 max	Al >=0.020; Mo <=0.08; Nb 0.02-0.05; V 0.05-0.10; V+Nb+Ti<=0.12; bal Fe				
DIN	GL-E40		0.18 max	0.20 max	0.030 max	0.90-1.60	0.60 max	0.040 max	0.040 max	0.50 max	Al >=0.020; Mo <=0.08; N <=0.007; Nb 0.02-0.05; Ti <=0.02; V <=0.16; Nb+Ti+V<=0.18; bal Fe				
DIN	QStE260N	see S260NC	0.16 max			1.2 max		0.03 max	0.03 max	0.5 max	bal Fe				
DIN	QStE340N	Transverse spec	0.16 max			1.50 max		0.030 max	0.030 max	0.50 max	Al >=0.015; Nb <=0.09; Ti <=0.22; bal Fe	460-580	340	27	
DIN	QStE380N	Transverse spec	0.18 max			1.60 max		0.030 max	0.030 max	0.50 max	Al >=0.015; Nb <=0.09; Ti <=0.22; bal Fe	500-640	380	25	
DIN	WNr 1.0560		0.18 max	0.20 max	0.30 max	0.90-1.60	0.60 max	0.040 max	0.040 max	0.50 max	Al >=0.020; Mo <=0.08; N <=0.007; Nb 0.02-0.05; Ti <=0.02; V <=0.16; Nb+Ti+V<=0.18; bal Fe				
DIN	WNr 1.0583		0.18 max	0.20 max	0.35 max	0.90-1.60	0.40 max	0.040 max	0.040 max	0.50 max	Al >=0.020; Mo <=0.08; Nb 0.02-0.05; V 0.05-0.10; V+Nb+Ti<=0.12; bal Fe				
DIN	WNr 1.0584		0.18 max	0.20 max	0.35 max	0.90-1.60	0.40 max	0.040 max	0.040 max	0.50 max	Al >=0.020; Mo <=0.08; Nb 0.02-0.05; V 0.05-0.10; V+Nb+Ti<=0.12; bal Fe				
DIN	WNr 1.0599		0.15 max			1.70 max		0.030 max	0.030 max	0.50 max	Al Nb Ti or V; bal Fe				
DIN	WNr 1.0975	Transverse spec	0.16 max			1.50 max		0.030 max	0.030 max	0.50 max	Al >=0.015; Nb <=0.09; Ti <=0.22; bal Fe	460-580	340	27	
DIN	WNr 1.0979	Transverse spec	0.18 max			1.60 max		0.030 max	0.030 max	0.50 max	Al >=0.015; Nb <=0.09; Ti <=0.22; bal Fe	500-640	380	25	
DIN	WNr 1.8950*	Obs	0.12 max	0.30-1.25	0.25-0.55	1.00 max		0.03 max	0.03 max	0.5 max	bal Fe				
DIN 17145(80)	12Mn6	Wir rod for weld filler	0.08-0.14	0.12 max	0.17 max	1.35-1.65	0.12 max	0.025 max	0.025 max	0.08-0.22	Al <=0.030; bal Fe				
DIN 17145(80)	13Mn6	Wir rod for weld filler	0.08-0.14	0.12 max	0.17 max	1.35-1.65	0.12 max	0.025 max	0.025 max	0.30-0.45	bal Fe				
DIN 17145(80)	WNr 1.0479	Wir rod for weld filler	0.08-0.14	0.12 max	0.17 max	1.35-1.65	0.12 max	0.025 max	0.025 max	0.30-0.45	bal Fe				
DIN 17145(80)	WNr 1.0496	Wir rod for weld filler	0.08-0.14	0.12 max	0.17 max	1.35-1.65	0.12 max	0.025 max	0.025 max	0.08-0.22	Al <=0.030; bal Fe				
DIN EN 10149(95)	S260NC	Transverse spec	0.16 max			1.20 max		0.025 max	0.020 max	0.50 max	Al >=0.015; Nb <=0.09; Ti <=0.15; V <=0.10; Nb+V+Ti<=0.22; bal Fe	370-490	260	30	
DIN EN 10149(95)	S550MC	HR flat product	0.12 max			1.80 max		0.025 max	0.015 max	0.50 max	Al >=0.015; Nb <=0.09; Ti <=0.15; V <=0.20; Nb+V+Ti<=0.22; bal Fe	600-760	550	15	

Specification	Designation	Notes	C	Cr	Cu	Mn	Ni	P	S	Si	Other	UTS	YS	El	Hard

Carbon Steel, High-Manganese, 1513 (Continued from previous page)

Germany

Specification	Designation	Notes	C	Cr	Cu	Mn	Ni	P	S	Si	Other	UTS	YS	El	Hard
DIN EN 10149(95)	WNr 1.0971	Transverse spec	0.16 max			1.20 max		0.025 max	0.020 max	0.50 max	Al >=0.015; Nb <=0.09; Ti <=0.15; V <=0.10; Nb+Ti+V<=0.22, bal Fe	370-490	260	30	
DIN EN 10149(95)	WNr 1.0986	HR flat product	0.12 max			1.80 max		0.025 max	0.015 max	0.50 max	Al >=0.015; Nb <=0.09; Ti <=0.15; V <=0.20; Nb+Ti+V<=0.22; bal Fe	600-760	550	15	
DIN SEW 028(93)	P265	Press ves, flat products	0.14 max			0.50-1.20		0.030 max	0.025 max	0.30 max	Al 0.015-0.06; bal Fe				
DIN SEW 028(93)	WNr 1.0424	Press ves, flat products	0.14 max			0.50-1.20		0.030 max	0.025 max	0.30 max	Al 0.015-0.06; bal Fe				
DIN SEW 085(88)	FStE 355 OS1	Sect, bar	0.16 max	0.20 max	0.20 max	1.60 max	0.20 max	0.025 max	0.025 max	0.50 max	Al >=0.020; Mo <=0.08; N >=0.015; Nb <=0.040; Ti <=0.05; V <=0.60; bal Fe				

Hungary

Specification	Designation	Notes	C	Cr	Cu	Mn	Ni	P	S	Si	Other	UTS	YS	El	Hard
MSZ 3741(85)	DX42	Spiral weld tube w/special req; 3.5-12mm; HR	0.18 max	0.25 max		1.2 max		0.03 max	0.035 max	0.15-0.5	CE<=0.35D<324mm; CE<=0.38D>=324mm; CCE=C+Mn/6+(Cr+V)/5+Cu/15; bal Fe	410-540	290	24L	
MSZ 3741(85)	DX52	Spiral weld tube w/special req; 3.5-12mm; HR	0.18 max	0.25 max		1.5 max		0.03 max	0.035 max	0.15-0.5	V 0.02-0.06; CE<=0.4D<324mm; CE<=0.42D>=324mm; CE=C+Mn/6+(Cr+V)/5+Cu/15; bal Fe	500-640	360	22L	
MSZ 3770(85)	DX42	Spiral weld tube w/special req; 3.5-12mm; HR	0.18 max	0.25 max		1.2 max		0.03 max	0.035 max	0.15-0.5	CE<=0.35D<324mm; CE<=0.38D>=324mm; CCE=C+Mn/6+(Cr+V)/5+Cu/15; bal Fe	410-540	290	24L	
MSZ 3770(85)	DX52	Spiral weld tube w/special req; 3.5-12mm; HR	0.18 max	0.25 max		1.5 max		0.03 max	0.035 max	0.15-0.5	V 0.02-0.06; CE<=0.4D<324mm; CE<=0.42D>=324mm; CE=C+Mn/6+(Cr+V)/5+Cu/15; bal Fe	500-640	360	22L	
MSZ 3770(85)	DX60	Spiral weld tube w/special req; 5-12mm; HR	0.15 max	0.25 max		1.5 max		0.03 max	0.035 max	0.15-0.5	Nb 0.01-0.08; V 0.02-0.08; V+Nb<=0.12; CE<=0.4D>=324mm; CE=C+Mn/6+(Cr+V)/+5Cu/15; bal Fe	520-690	420	18L	
MSZ 3770(85)	DX65	Spiral weld tube w/special req; 5-12mm; HR	0.15 max	0.25 max		1.6 max		0.03 max	0.03 max	0.15-0.5	Nb 0.01-0.1; V 0.02-0.08; V+Nb<=0.15; CE<=0.4D>=324mm; CE=C+Mn/6+(Cr+V)/5+Cu/15; bal Fe	550-720	450	18L	

India

Specification	Designation	Notes	C	Cr	Cu	Mn	Ni	P	S	Si	Other	UTS	YS	El	Hard
IS 3945	A-N	Plt, Mill 25mm diam	0.17			0.8-1.5		0.05	0.05		bal Fe				
IS 3945	B-N	Plt, Mill 25mm diam	0.19			1.2		0.05	0.05		bal Fe				

Mexico

Specification	Designation	Notes	C	Cr	Cu	Mn	Ni	P	S	Si	Other	UTS	YS	El	Hard
NMX-B-301(86)	1513	Bar	0.10-0.16			1.10-1.40		0.040 max	0.050 max		bal Fe				

Poland

Specification	Designation	Notes	C	Cr	Cu	Mn	Ni	P	S	Si	Other	UTS	YS	El	Hard
PNH83160(88)	L15G		0.12-0.2	0.3	0.3	1.2-1.5	0.3	0.04	0.04	0.2-0.4	bal Fe				
PNH84018(86)	09G2Cu	HS weld const	0.15 max	0.3 max	0.25-0.5	1.2-1.8	0.3 max	0.04 max	0.04 max	0.15-0.4	Al 0.05-0.120; As<=0.08; CEV<=0.44; bal Fe				

UNS numbers and US grades are provided as a means of cross referencing chemically similar alloys. Exchangability is only possible after independent examination of specifications. Tensile properties are minimum or typical as specified. UTS and YS as MPa. El as %. See Appendix for list of abbreviations used in Notes. * indicates obsolete material.

Specification	Designation	Notes	C	Cr	Cu	Mn	Ni	P	S	Si	Other	UTS	YS	El	Hard

Carbon Steel, High-Manganese, 1513 (Continued from previous page)

Romania

Specification	Designation	Notes	C	Cr	Cu	Mn	Ni	P	S	Si	Other	UTS	YS	El	Hard
STAS 11502	O9Mn16		0.12			1.4-1.8	0.3	0.035	0.04	0.17-0.37	Al 0.02; bal Fe				
STAS 11502(89)	9SiMn16	Press ves	0.12 max	0.3 max		1.3-1.7		0.035 max	0.04 max	0.5-0.8	Al 0.02-0.120; Pb <=0.15; bal Fe				
STAS 500/2(88)	OL52.4kf	Gen struct; Specially killed	0.18 max	0.3 max		1.6 max		0.04 max	0.04 max	0.5 max	Al 0.02-0.120; Pb <=0.15; As<=0.08; bal Fe				

Russia

Specification	Designation	Notes	C	Cr	Cu	Mn	Ni	P	S	Si	Other	UTS	YS	El	Hard
GOST 19281	09G2	Weld const	0.12 max	0.3 max	0.3 max	1.4-1.8	0.3 max	0.035 max	0.04 max	0.17-0.37	Al <=0.1; N <=0.012; As<=0.08; bal Fe				
GOST 19281	09G2D	Weld const	0.12 max	0.3 max	0.15-0.3	1.4-1.8	0.3 max	0.035 max	0.04 max	0.17-0.37	Al <=0.1; N <=0.012; As<=0.08; bal Fe				
GOST 19281	09G2S	Weld const	0.12 max	0.3 max	0.3 max	1.3-1.7	0.3 max	0.035 max	0.04 max	0.5-0.8	Al <=0.03; N <=0.012; As<=0.08; bal Fe				
GOST 19281	09G2SD	Weld const	0.12 max	0.3 max	0.15-0.3	1.3-1.7	0.3 max	0.035 max	0.04 max	0.5-0.8	Al <=0.1; N <=0.012; As<=0.08; bal Fe				
GOST 19281	10G2B	Weld const	0.12 max	0.3 max	0.3 max	1.2-1.6	0.3 max	0.035 max	0.04 max	0.17-0.37	Al <=0.1; N <=0.012; Nb 0.02-0.05; As<=0.08; bal Fe				
GOST 19281	10G2BD	Weld const	0.12 max	0.3 max	0.15-0.3	1.2-1.6	0.3 max	0.035 max	0.04 max	0.17-0.37	Al <=0.1; N <=0.012; Nb 0.02-0.05; As<=0.08; bal Fe				
GOST 19281	10G2S1D	Weld const	0.12 max	0.3 max	0.15-0.3	1.3-1.65	0.3 max	0.035 max	0.04 max	0.8-1.1	Al <=0.1; N <=0.012; As<=0.08; bal Fe				
GOST 19282	09G2		0.12	0.3	0.15-0.3	1.4-1.8	0.3	0.035	0.04	0.17-0.37	bal Fe				
GOST 19282	09G2D		0.12	0.3	0.15-0.3	1.4-1.8	0.3	0.035	0.04	0.17-0.37	bal Fe				
GOST 19282	09G2S		0.12	0.3	0.15-0.3	1.3-1.7	0.3	0.035	0.04	0.5-0.8	bal Fe				
GOST 19282	09G2SD		0.12	0.3	0.15-0.3	1.3-1.7	0.3	0.035	0.04	0.5-0.8	bal Fe				
GOST 19282	10G2B		0.12	0.3	0.15-0.3	1.2-1.6	0.3	0.035	0.04	0.17-0.37	Nb 0.02-0.05; bal Fe				
GOST 19282	10G2BD		0.12	0.3	0.15-0.3	1.2-1.6	0.3	0.035	0.04	0.17-0.37	Nb 0.02-0.05; bal Fe				
GOST 19282	10G2S1		0.12	0.3	0.15-0.3	1.3-1.65	0.3	0.035	0.04	0.8-1.1	bal Fe				
GOST 19282	10G2S1D		0.12	0.3	0.15-0.3	1.3-1.65	0.3	0.035	0.04	0.8-1.1	bal Fe				
GOST 19282	14G2		0.12-0.18	0.3	0.3	1.2-1.6	0.3	0.035	0.04	0.17-0.37	bal Fe				
GOST 4543(71)	10G2	CH	0.07-0.15	0.3 max	0.3 max	1.2-1.6	0.3 max	0.035 max	0.035 max	0.17-0.37	Al <=0.1; bal Fe				
GOST 5058	09G2	Weld const; struct	0.12 max	0.3 max	0.3 max	1.4-1.8	0.3 max	0.035 max	0.04 max	0.17-0.37	Al <=0.1; Ti 0.01-0.03; As<=0.08; bal Fe				
GOST 5058	09G2D	Weld const; struct	0.12 max	0.3 max	0.15-0.3	1.4-1.8	0.3 max	0.035 max	0.04 max	0.17-0.37	Al <=0.1; Ti 0.01-0.03; As<=0.08; bal Fe				
GOST 5058(65)	09G2S	Weld const; struct	0.12 max	0.3 max	0.3 max	1.3-1.7	0.3 max	0.035 max	0.04 max	0.5-0.8	Al <=0.1; Ti 0.01-0.03; As<=0.08; bal Fe				
GOST 5058(65)	09G2SD	Weld const; struct	0.12 max	0.3 max	0.15-0.3	1.3-1.7	0.3 max	0.035 max	0.04 max	0.5-0.8	Al <=0.1; Ti 0.01-0.03; As<=0.08; bal Fe				
GOST 5058(65)	10G2S1D	Weld const; struct	0.12 max	0.3 max	0.15-0.3	1.3-1.65	0.3 max	0.035 max	0.04 max	0.8-1.1	Al <=0.1; Ti 0.01-0.03; As<=0.08; bal Fe				
GOST 5521(93)	A36	Struct, weld ships	0.18 max	0.20 max	0.35 max	0.9-1.6	0.40 max	0.035 max	0.035 max	0.15-0.5	Al 0.015-0.06; Mo <=0.08; Nb 0.02-0.05; V 0.05-0.1; bal Fe				
GOST 5521(93)	D36	Struct, weld ships	0.18 max	0.20 max	0.35 max	0.9-1.6	0.40 max	0.035 max	0.035 max	0.15-0.5	Al 0.015-0.06; Mo <=0.08; Nb 0.02-0.05; V 0.05-0.1; bal Fe				

Specification	Designation	Notes	C	Cr	Cu	Mn	Ni	P	S	Si	Other	UTS	YS	El	Hard

Carbon Steel, High-Manganese, 1513 (Continued from previous page)

Russia

Specification	Designation	Notes	C	Cr	Cu	Mn	Ni	P	S	Si	Other	UTS	YS	El	Hard
GOST 5521(93)	E40	Struct, weld ships	0.18 max	0.60-0.90	0.35 max	0.5-0.8	0.40 max	0.035 max	0.035 max	0.8-1.1	Al 0.015-0.06; bal Fe				
GOST 5521(93)	E40S	Struct, weld ships	0.18 max	0.60-0.90	0.40-0.60	0.5-0.8	0.50-0.80	0.035 max	0.035 max	0.8 1.1	Al 0.015-0.06; bal Fe				
GOST 5781(82)	10GT	HR conc-reinf	0.13 max		0.3 max	0.3 max	1-1.4	0.4 max	0.03 max	0.04 max	0.45-0.65	Al 0.02-0.05; N <=0.008; Ti 0.01-0.05; bal Fe				

Spain

Specification	Designation	Notes	C	Cr	Cu	Mn	Ni	P	S	Si	Other	UTS	YS	El	Hard
UNE 36080(80)	A360B	Gen struct	0.2 max			1.60 max		0.05 max	0.05 max	0.60 max	Mo <=0.15; N <=0.009; bal Fe				
UNE 36080(85)	AE235-B	Gen struct	0.2 max			1.60 max		0.045 max	0.045 max	0.60 max	Mo <=0.15; N <=0.009; bal Fe				
UNE 36081(76)	AE315KT	Fine grain Struct	0.17 max	0.25 max		0.7-1.5	0.3 max	0.03 max	0.03 max	0.4 max	Mo <=0.1; +Al+V+Nb; bal Fe				
UNE 36081(76)	F.6409*	see AE315KT	0.17 max	0.25 max		0.7-1.5	0.3 max	0.03 max	0.03 max	0.4 max	Mo <=0.1; +Al+V+Nb; bal Fe				
UNE 36121(85)	AE220-1B	CR; Strp	0.15 max			1.60 max		0.04 max	0.04 max	0.60 max	Mo <=0.15; N <=0.009; bal Fe				
UNE 36121(85)	AE220-1D	CR; Strp	0.15 max			1.60 max		0.035 max	0.035 max	0.60 max	Mo <=0.15; N <=0.015; bal Fe				

UK

Specification	Designation	Notes	C	Cr	Cu	Mn	Ni	P	S	Si	Other	UTS	YS	El	Hard
BS 1449/1(91)	34/20CR	Plt Sh Strp; CR wide based on min strength; 0-16mm	0.15 max			1.20 max		0.050 max	0.050 max		bal Fe	340	200	21-29	
BS 1449/1(91)	34/20CS	Plt Sh Strp; Plt Sh Strp CR t<=16mm	0.15 max			1.20 max		0.050 max	0.050 max		bal Fe	340	200	21-29	
BS 1449/1(91)	34/20HR	Plt Sh Strp HR wide, t<=16mm	0.15 max			1.20 max		0.050 max	0.050 max		bal Fe	340	200	21-29	
BS 1449/1(91)	34/20HS	Plt Sh Strp HR t<=8mm	0.15 max			1.20 max		0.050 max	0.050 max		bal Fe	340	200	21-27	
BS 1449/1(91)	40/30CS	Plt Sh Strp; CR t<=16mm	0.15 max			1.20 max		0.040 max	0.040 max		bal Fe	400	300	18-26	
BS 1449/1(91)	40/30HR	Plt Sh Strp HR wide; t<=16mm	0.15 max			1.20 max		0.040 max	0.040 max		bal Fe	400	300	18-26	
BS 1449/1(91)	40/30HS	Plt Sh Strp HR t<=8mm	0.15 max			1.20 max		0.040 max	0.040 max	0.6 max	bal Fe	400	300	18-26	
BS 1449/1(91)	40F30CS	Plt Sh Strp; CR t<=16mm	0.12 max			1.20 max		0.030 max	0.030 max		bal Fe	400	300	20-28	
BS 1449/1(91)	40F30HR	Plt Sh Strp HR wide; t<=16mm	0.12 max			1.20 max		0.030 max	0.030 max		bal Fe	400	300	20-28	
BS 1449/1(91)	40F30HS	Plt Sh Strp HR t<=8mm	0.12 max			1.20 max		0.030 max	0.030 max		bal Fe	400	300	20-28	
BS 1449/1(91)	43/35CS	Plt Sh Strp; CR t<=16mm	0.15 max			1.20 max		0.040 max	0.040 max		bal Fe	430	350	16-23	
BS 1449/1(91)	43/35HR	Plt Sh Strp HR wide; t<=16mm	0.15 max			1.20 max		0.040 max	0.040 max		bal Fe	430	250	16-23	
BS 1449/1(91)	43/35HS	Plt Sh Strp HR t<=8mm	0.15 max			1.20 max		0.040 max	0.040 max		bal Fe	430	350	16-23	
BS 1449/1(91)	43F35CS	Plt Sh Strp; CR t<=16mm	0.12 max			1.20 max		0.030 max	0.030 max		bal Fe	430	350	18-25	
BS 1449/1(91)	43F35HR	Plt Sh Strp HR wide; t<=16mm	0.12 max			1.20 max		0.030 max	0.030 max		bal Fe	430	350	18-25	
BS 1449/1(91)	43F35HS	Plt Sh Strp HR t<=8mm	0.12 max			1.20 max		0.030 max	0.030 max		bal Fe	430	350	18-25	
BS 1449/1(91)	46/40CS	Plt Sh Strp; CR t<=16mm	0.15 max			1.20 max		0.040 max	0.040 max		bal Fe	460	400	12-20	
BS 1449/1(91)	46/40HR	Plt Sh Strp HR wide; t<=16mm	0.15 max			1.20 max		0.040 max	0.040 max		bal Fe	460	400	12-20	
BS 1449/1(91)	46/40HS	Plt Sh Strp HR t<=8mm	0.15 max			1.20 max		0.040 max	0.040 max		bal Fe	460	400	12-20	
BS 1449/1(91)	46F40CS	Plt Sh Strp; CR t<=16mm	0.12 max			1.20 max		0.030 max	0.030 max		bal Fe	460	400	14-22	
BS 1449/1(91)	46F40HR	Plt Sh Strp HR wide; t<=16mm	0.12 max			1.20 max		0.030 max	0.030 max		bal Fe	460	400	14-22	
BS 1449/1(91)	46F40HS	Plt Sh Strp HR t<=8mm	0.12 max			1.20 max		0.030 max	0.030 max		bal Fe	460	400	14-22	
BS 1449/1(91)	50F45CS	Plt Sh Strp; CR t<=16mm	0.12 max			1.20 max		0.030 max	0.030 max		bal Fe	500	450	14-22	

Specification	Designation	Notes	C	Cr	Cu	Mn	Ni	P	S	Si	Other	UTS	YS	El	Hard

Carbon Steel, High-Manganese, 1513 (Continued from previous page)

UK

Specification	Designation	Notes	C	Cr	Cu	Mn	Ni	P	S	Si	Other	UTS	YS	El	Hard
BS 1449/1(91)	50F45HR	Plt Sh Strp HR wide; t<=16mm	0.12 max			1.20 max		0.030 max	0.030 max		bal Fe	500	450	14-22	
BS 1449/1(91)	50F45HS	Plt sht strp HR t<8mm	0.12 max			1.20 max		0.030 max	0.030 max		bal Fe	500	450	14-22	
BS 1449/1(91)	75F70HS	Plt Sh Strp HR t<=8mm	0.12 max			1.50 max		0.030 max	0.030 max		bal Fe	750	700	8-15	
BS 1453	A2		0.1-0.2			1-1.6		0.04	0.04	0.1-0.35	bal Fe				
BS 1501(86)	223-460	Plate, fully killed	0.20 max	0.25 max	0.30 max	0.80-1.60	0.75 max	0.030 max	0.030 max	0.10-0.40	Mo <=0.10; Nb 0.010-0.080; bal Fe	460-580	340	22	
BS 1501(86)	225-460	Plate, fully killed, Al treated 3<t<=16	0.20 max	0.25 max	0.30 max	0.90-1.60	0.75 max	0.030 max	0.030 max	0.10-0.40	Al >=0.015; Mo <=0.10; Nb 0.010-0.060; bal Fe	490-610	355	21	
BS 1502	28*	Bar Shp Press ves	0.17			0.90-1.50		0.05	0.05	0.1-0.55	bal Fe				
BS 2772	150M12		0.1-0.15			1.3-1.7		0.05	0.05	0.1-0.35	bal Fe				
BS 970/1(83)	130M15	Blm Bil Slab Bar Rod Frg CH	0.12-0.18			1.10-1.50					bal Fe	740		13	
BS 970/1(96)	125A15	Blm Bil Slab Bar Rod Frg	0.13-0.18			1.10-1.40		0.050 max	0.050 max	0.10-0.40	bal Fe				
BS 970/1(96)	130M15	Blm Bil Slab Bar Rod Frg	0.12-0.18			1.10-1.50		0.050 max	0.050 max	0.10-0.40	bal Fe				

USA

Specification	Designation	Notes	C	Cr	Cu	Mn	Ni	P	S	Si	Other	UTS	YS	El	Hard
	AISI 1513	Wir rod	0.10-0.16			1.10-1.40		0.040 max	0.050 max		bal Fe				
	AISI 1513	Bar	0.10-0.16			1.10-1.40		0.040 max	0.050 max		bal Fe				
	UNS G15130		0.10-0.16			1.10-1.40		0.040 max	0.050 max		bal Fe				
ASTM A29/A29M(93)	1513	Bar	0.10-0.16			1.10-1.40		0.040 max	0.050 max		bal Fe				
ASTM A510(96)	1513	Wir rod	0.10-0.16			1.10-1.40		0.040 max	0.050 max		bal Fe				
ASTM A576(95)	1513	Special qual HW bar	0.10-0.16			1.10-1.40		0.040 max	0.050 max		Si Cu Pb B Bi Ca Se Te if spec'd; bal Fe				
SAE J403	1513*	Bil Bar Rod Tub	0.1-0.16			1.1-1.4		0.04 max	0.05 max		bal Fe				

Yugoslavia

Specification	Designation	Notes	C	Cr	Cu	Mn	Ni	P	S	Si	Other	UTS	YS	El	Hard
	Z3		0.08-0.15	0.2	0.22	1.3-1.7	0.15	0.03	0.03	0.05-0.25	bal Fe				

Carbon Steel, High-Manganese, 1518

Czech Republic

Specification	Designation	Notes	C	Cr	Cu	Mn	Ni	P	S	Si	Other	UTS	YS	El	Hard
CSN 413127	13127	High-temp const	0.14-0.2	0.3 max		1.0-1.4		0.04 max	0.04 max	0.2-0.5	Al 0.015-0.10; bal Fe				

Europe

Specification	Designation	Notes	C	Cr	Cu	Mn	Ni	P	S	Si	Other	UTS	YS	El	Hard
EN 10025(90)A1(93) 1.0037		Struct HR BS, 16<t<=25mm	0.20 max			1.40 max		0.045 max	0.045 max		N <=0.009; N max na if Al>=0.020; bal Fe	340-470	225	26L 24T	
EN 10025(90)A1(93) 1.0037		Struct HR QS, 16<t<=25mm	0.20 max			1.40 max		0.045 max	0.045 max		N <=0.009; N max na if Al>=0.020; bal Fe	340-470	235	26L 24T	
EN 10025(90)A1(93) 1.0038		Struct HR BS, 63<t<=80mm	0.20 max			1.40 max		0.045 max	0.045 max		N <=0.009; N max na if Al>=0.020; bal Fe	340-470	215	24L 22T	
EN 10025(90)A1(93) 1.0038		Struct HR BS, 200<t<=250mm	0.20 max			1.40 max		0.045 max	0.045 max		N <=0.009; N max na if Al>=0.020; bal Fe	320-470	175	21L 21T	
EN 10025(90)A1(93) 1.0044		Struct HR BS, 200<t<=250mm	0.22 max			1.50 max		0.045 max	0.045 max		N <=0.009; N max na if Al>=0.020; bal Fe	380-540	205	17L 17T	
EN 10025(90)A1(93) 1.0044		Struct HR QS, 3<t<=16mm	0.21 max			1.50 max		0.045 max	0.045 max		N max na if Al>=0.020; bal Fe	410-560	275	22L 20T	
EN 10025(90)A1(93) 1.0044		Struct HR BS, 63<t<=80mm	0.22 max			1.50 max		0.045 max	0.045 max		N <=0.009; N max na if Al>=0.020; bal Fe	410-560	245	20L 18T	
EN 10025(90)A1(93) 1.0044		Struct HR QS, t<=1mm	0.21 max			1.50 max		0.045 max	0.045 max		N max na if Al>=0.020; bal Fe	430-580	275	14L 12T	
EN 10025(90)A1(93) 1.0044		Struct HR BS, t<=1mm	0.21 max			1.50 max		0.045 max	0.045 max		N <=0.009; N max na if Al>=0.020; bal Fe	430-580	275	14L 12T	
EN 10025(90)A1(93) 1.0044		Struct HR BS, 3<t<=16mm	0.21 max			1.50 max		0.045 max	0.045 max		N <=0.009; N max na if Al>=0.020; bal Fe	410-560	275	22L 20T	

UNS numbers and US grades are provided as a means of cross referencing chemically similar alloys. Exchangability is only possible after independent examination of specifications. Tensile properties are minimum or typical as specified. UTS and YS as MPa. El as %. See Appendix for list of abbreviations used in Notes. * indicates obsolete material.

Specification	Designation	Notes	C	Cr	Cu	Mn	Ni	P	S	Si	Other	UTS	YS	El	Hard

Carbon Steel, High-Manganese, 1518 (Continued from previous page)

Europe

Specification	Designation	Notes	C	Cr	Cu	Mn	Ni	P	S	Si	Other	UTS	YS	El	Hard
EN 10025(90)A1(93)	1.0143	Struct HR QS, 200<t<=250mm	0.20 max			1.50 max		0.040 max	0.040 max		N <=0.009; N max na if Al>=0.020; bal Fe	380-540	205	17L 17T	
FN 10025(90)A1(93)	1.0144	Struct HR QS, 200<t<=250mm	0.20 max			1.50 max		0.035 max	0.035 max		N max na if Al>=0.020; bal Fe	380-540	205	17L 17T	
EN 10025(90)A1(93)	1.0145	Struct HR QS, 200<t<=250mm	0.20 max			1.50 max		0.035 max	0.035 max		N max na if Al>=0.020; bal Fe	380-540	205	17L 17T	
EN 10025(90)A1(93)	S235JR	Struct HR BS, 16<t<=25mm	0.20 max			1.40 max		0.045 max	0.045 max		N <=0.009; N max na if Al>=0.020; bal Fe	340-470	225	26L 24T	
EN 10025(90)A1(93)	S235JRC	Struct HR QS, 16<t<=25mm	0.20 max			1.40 max		0.045 max	0.045 max		N <=0.009; N max na if Al>=0.020; bal Fe	340-470	235	26L 24T	
EN 10025(90)A1(93)	S235JRG2	Struct HR BS, 200<t<=250mm	0.20 max			1.40 max		0.045 max	0.045 max		N <=0.009; N max na if Al>=0.020; bal Fe	320-470	175	21L 21T	
EN 10025(90)A1(93)	S235JRG2	Struct HR BS, 63<t<=80mm	0.20 max			1.40 max		0.045 max	0.045 max		N <=0.009; N max na if Al>=0.020; bal Fe	340-470	215	24L 22T	
EN 10025(90)A1(93)	S275J0	Struct HR QS, 200<t<=250mm	0.20 max			1.50 max		0.040 max	0.040 max		N <=0.009; N max na if Al>=0.020; bal Fe	380-540	205	17L 17T	
EN 10025(90)A1(93)	S275J2G3	Struct HR QS, 200<t<=250mm	0.20 max			1.50 max		0.035 max	0.035 max		N max na if Al>=0.020; bal Fe	380-540	205	17L 17T	
EN 10025(90)A1(93)	S275J2G4	Struct HR QS, 200<t<=250mm	0.20 max			1.50 max		0.035 max	0.035 max		N max na if Al>=0.020; bal Fe	380-540	205	17L 17T	
EN 10025(90)A1(93)	S275JR	Struct HR BS, t<=1mm	0.21 max			1.50 max		0.045 max	0.045 max		N <=0.009; N max na if Al>=0.020; bal Fe	430-580	275	14L 12T	
EN 10025(90)A1(93)	S275JR	Struct HR BS, 63<t<=80mm	0.22 max			1.50 max		0.045 max	0.045 max		N <=0.009; N max na if Al>=0.020; bal Fe	410-560	245	20L 18T	
EN 10025(90)A1(93)	S275JR	Struct HR BS, 200<t<=250mm	0.22 max			1.50 max		0.045 max	0.045 max		N <=0.009; N max na if Al>=0.020; bal Fe	380-540	205	17L 17T	
EN 10025(90)A1(93)	S275JR	Struct HR BS, 3<t<=16mm	0.21 max			1.50 max		0.045 max	0.045 max		N <=0.009; N max na if Al>=0.020; bal Fe	410-560	275	22L 20T	
EN 10025(90)A1(93)	S275JRC	Struct HR QS, t<=1mm	0.21 max			1.50 max		0.045 max	0.045 max		N max na if Al>=0.020; bal Fe	430-580	275	14L 12T	
EN 10025(90)A1(93)	S275JRC	Struct HR QS, 3<t<=16mm	0.21 max			1.50 max		0.045 max	0.045 max		N max na if Al>=0.020; bal Fe	410-560	275	22L 20T	

Mexico

Specification	Designation	Notes	C	Cr	Cu	Mn	Ni	P	S	Si	Other	UTS	YS	El	Hard
NMX-B-301(86)	1518	Bar	0.15-0.21			1.10-1.40		0.040 max	0.050 max		bal Fe				

Spain

Specification	Designation	Notes	C	Cr	Cu	Mn	Ni	P	S	Si	Other	UTS	YS	El	Hard
UNE 36013(76)	20Mn6	CH	0.18-0.25			1.3-1.6		0.035 max	0.035 max	0.15-0.4	Mo <=0.15; bal Fe				
UNE 36087(74)	A37RAII	Sh Plt	0.2 max			0.3-1.3		0.045 max	0.04 max	0.4 max	Al 0.015-0.120; Mo <=0.15; 0-40mm Mn<=1.3; 40mm Mn<=1.1; bal Fe				
UNE 36087(74)	A37RCI	Sh Plt	0.2 max			0.3-1.3		0.055 max	0.05 max	0.4 max	Mo <=0.15; N <=0.009; 0-40mm Mn<=1.3; 40mm Mn<=1.1; bal Fe				
UNE 36087(74)	A37RCII	Sh Plt	0.2 max			0.3-1.3		0.045 max	0.04 max	0.4 max	Al 0.015-0.10; Mo <=0.15; N <=0.009; 0-40mm Mn<=1.3; 40mm Mn<=1.1; bal Fe				

Sweden

Specification	Designation	Notes	C	Cr	Cu	Mn	Ni	P	S	Si	Other	UTS	YS	El	Hard
SS 142132	2132	Fine grain Struct	0.2 max			1.6 max		0.035 max	0.035 max	0.5 max	N <=0.02; Nb <=0.05; V <=0.15; C+Mn/6+(Ni+Cu)/15+(Cr+Mo+V)/5<=0.41; bal Fe				
SS 142172	2172	Gen struct; FF; Non-rimming	0.2 max		0.4 max	1-1.6		0.05 max	0.05 max	0.5 max	N<=0.009/0-0.012; bal Fe				

UK

Specification	Designation	Notes	C	Cr	Cu	Mn	Ni	P	S	Si	Other	UTS	YS	El	Hard
BS 1449/1(91)	50/35HR	Plt Sh Strp HR wide; t<=16mm	0.20 max			1.50 max		0.050 max	0.050 max		bal Fe	500	350	12-20	
BS 1449/1(91)	50/35HS	Plt Sh Strp HR t<=8mm	0.20 max			1.50 max		0.050 max	0.050 max		bal Fe	500	350	12-20	
BS 1449/1(91)	50/45CS	Plt Sh Strp; CR t<=16mm	0.20 max			1.50 max		0.040 max	0.040 max		bal Fe	500	450	12-20	
BS 1449/1(91)	50/45HR	Plt Sh Strp HR wide; t<=16mm	0.20 max			1.50 max		0.040 max	0.040 max		bal Fe	500	450	12-20	

UNS numbers and US grades are provided as a means of cross referencing chemically similar alloys. Exchangability is only possible after independent examination of specifications. Tensile properties are minimum or typical as specified. UTS and YS as MPa. El as %. See Appendix for list of abbreviations used in Notes. * indicates obsolete material.

Specification	Designation	Notes	C	Cr	Cu	Mn	Ni	P	S	Si	Other	UTS	YS	El	Hard

Carbon Steel, High-Manganese, 1518 (Continued from previous page)

UK

Specification	Designation	Notes	C	Cr	Cu	Mn	Ni	P	S	Si	Other	UTS	YS	El	Hard
BS 1449/1(91)	50/45HS	Plt Sh Strp HR t<=8mm	0.20 max			1.50 max		0.040 max	0.040 max		bal Fe	500	450	12-20	
BS 1449/1(91)	60F55CS	Plt Sh Strp; CR t<=16mm	0.12 max			1.20 max		0.030 max	0.030 max		bal Fe	600	550	11-19	
BS 1449/1(91)	60F55HS	Plt Sh Strp HR t<=8mm	0.12 max			1.20 max		0.030 max	0.030 max		bal Fe	600	550	11-19	
BS 1449/1(91)	68F62HS	Plt Sh Strp HR t<=8mm	0.12 max			1.50 max		0.030 max	0.030 max		bal Fe	680	620	10-18	
BS 970/3(91)	150M19	Bright bar; 13<t<=150mm; hard	0.15-0.23			1.30-1.70		0.050 max	0.050 max		Mo <=0.15; bal Fe	550-700	340	18	152-207 HB
BS 970/3(91)	150M19	Bright bar; 150<t<=250mm; norm	0.15-0.23			1.30-1.70		0.050 max	0.050 max		Mo <=0.15; bal Fe	510	295	17	146-197 HB

USA

Specification	Designation	Notes	C	Cr	Cu	Mn	Ni	P	S	Si	Other	UTS	YS	El	Hard
	UNS G15180		0.15-0.21			1.10-1.40		0.040 max	0.050 max		bal Fe				
ASTM A29/A29M(93)	1518	Bar	0.15-0.21			1.10-1.40		0.040 max	0.050 max		bal Fe				
ASTM A510(96)	1518	Wir rod	0.15-0.21			1.10-1.40		0.040 max	0.050 max		bal Fe				
ASTM A519(96)	1518	Smls mech tub	0.15-0.21			1.10-1.40		0.040 max	0.050 max		bal Fe				
ASTM A576(95)	1518	Special qual HW bar	0.15-0.21			1.10-1.40		0.040 max	0.050 max		Si Cu Pb B Bi Ca Se Te if spec'd; bal Fe				

Carbon Steel, High-Manganese, 1522/1522H

Bulgaria

Specification	Designation	Notes	C	Cr	Cu	Mn	Ni	P	S	Si	Other	UTS	YS	El	Hard
BDS 4880(89)	18G2	Struct	0.17-0.22	0.35 max	0.35 max	1.10-1.60	0.35 max	0.035 max	0.045 max	0.17-0.37	bal Fe				

China

Specification	Designation	Notes	C	Cr	Cu	Mn	Ni	P	S	Si	Other	UTS	YS	El	Hard
GB 713(97)	19Mng	Plt HR Norm 16-40mm Thk	0.15-0.22	0.30 max	0.30 max	1.00-1.60	0.30 max	0.030 max	0.025 max	0.30-0.60	bal Fe	510-650	345	20	

Czech Republic

Specification	Designation	Notes	C	Cr	Cu	Mn	Ni	P	S	Si	Other	UTS	YS	El	Hard
CSN 413030	13030	High-temp const	0.14-0.2	0.3 max		1.0-1.4	0.3 max	0.035 max	0.03 max	0.15-0.4	Mo <=0.1; Cr+Mo+Ni+Cu<=0.7; bal Fe				
CSN 413123	13123	Mn-V; Frg; Heavy Plate	0.17-0.23	0.3 max		1.0-1.4	0.3 max	0.04 max	0.04 max	0.15-0.4	V 0.1-0.25; bal Fe				
CSN 413124	13124	Mn-V; Frg	0.16-0.23	0.3 max		1.3-1.6	0.3 max	0.04 max	0.04 max	0.2-0.4	Al 0.02-0.120; V 0.08-0.210; bal Fe				
CSN 413126	13126		0.15-0.22	0.3 max		0.9-1.4	0.3 max	0.04 max	0.045 max	0.15-0.35	Cr+Ni+Cu<=0.75; bal Fe				

Finland

Specification	Designation	Notes	C	Cr	Cu	Mn	Ni	P	S	Si	Other	UTS	YS	El	Hard
SFS 670(86)	Z28	Unalloyed	0.15 max			1.2 max		0.05 max	0.05 max	0.4 max	bal Fe				
SFS 670(86)	Z32	Unalloyed	0.2 max			1.5 max		0.05 max	0.05 max	0.4 max	bal Fe				
SFS 670(86)	Z36	Unalloyed	0.2 max			1.5 max		0.05 max	0.05 max	0.4 max	bal Fe				
SFS 670(86)	Z40	Unalloyed	0.12 max			1.5 max		0.03 max	0.03 max	0.4 max	bal Fe				

France

Specification	Designation	Notes	C	Cr	Cu	Mn	Ni	P	S	Si	Other	UTS	YS	El	Hard
AFNOR NFA35551	20MB5	CH; Bar Rod Wir	0.16-0.22			1.1-1.4		0.035	0.035	0.1-0.4	B 0.0008-0.005; bal Fe				
AFNOR NFA35552	20MB5	Bar Rod Wir	0.16-0.22			1.1-1.4		0.03	0.035	0.1-0.4	B 0.0008-0.005; bal Fe				
AFNOR NFA35552(84)	20M5		0.16-0.22			1.1-1.4		0.035	0.035	0.15-0.35	bal Fe				
AFNOR NFA35553	20MB5	Bar Rod Wir	0.16-0.22			1.1-1.4		0.03	0.035	0.1-0.4	B 0.0008-0.005; bal Fe				
AFNOR NFA35556(83)	20M5		0.16-0.22			1.1-1.4		0.035	0.035	0.15-0.35	bal Fe				
AFNOR NFA35556(84)	20MB5	CH	0.16-0.22			1.1-1.4		0.03 max	0.035 max	0.1-0.4	B 0.0008-0.005; bal Fe				
AFNOR NFA35557	20MB5	Q/T	0.16-0.22			1.1-1.4		0.03	0.035	0.1-0.4	B 0.0008-0.005; bal Fe				
AFNOR NFA35566	20MB5	Q/T; Chains	0.16-0.22			1.1-1.4		0.03	0.035	0.1-0.4	B 0.0008-0.005; bal Fe				

UNS numbers and US grades are provided as a means of cross referencing chemically similar alloys. Exchangability is only possible after independent examination of specifications. Tensile properties are minimum or typical as specified. UTS and YS as MPa. El as %. See Appendix for list of abbreviations used in Notes. * indicates obsolete material.

Carbon Steel, High-Manganese, 1522/1522H (Continued from previous page)

Specification	Designation	Notes	C	Cr	Cu	Mn	Ni	P	S	Si	Other	UTS	YS	El	Hard
Germany															
DIN	20Mn6	Hard Quen	0.17-0.23			1.30-1.60		0.035 max	0.035 max	0.30-0.60	bal Fe	540-690	345	20	
DIN	L690M		0.26 max			1.40 max		0.040 max	0.050 max	0.35 max	bal Fe				
DIN	WNr 1.1169	Hard Quen	0.17-0.23			1.30-1.60		0.035 max	0.035 max	0.30-0.60	bal Fe	540-690	345	20	
DIN	WNr 1.8979		0.26 max			1.40 max		0.040 max	0.020 max	0.35 max	bal Fe				
DIN 10208(96)	L385N		0.23 max			1.00-1.50		0.040 max	0.035 max	0.55 max	Al >=0.020; V <=0.12; bal Fe				
DIN 10208(96)	L415NB	Pipe for combustibles	0.21 max	0.30 max	0.25 max	1.60 max	0.30 max	0.025 max	0.020 max	0.45 max	Al 0.015-0.060; Mo <=0.10; N <=0.012; Nb <=0.05; Ti <=0.04; V <=0.15; bal Fe				
DIN 17115(87)	21MnSi5	Weld round link chain	0.18-0.24			1.10-1.60		0.040 max	0.040 max	0.30-0.35	Al 0.020-0.040; bal Fe				
DIN 17115(87)	WNr 1.0471	Weld round link chains	0.18-0.24			1.10-1.60		0.040 max	0.040 max	0.30-0.35	Al 0.020-0.040; bal Fe				
DIN 17243(90)	20Mn5	HR bar Q/T <=16mm	0.17-0.23	0.30 max		1.00-1.50		0.035 max	0.030 max	0.60 max	Al <=0.015; bal Fe	540-690	390	22	
DIN EN 10147(95)	WNr 1.0529	Hot dip galvanized Sh Strp									bal Fe	420	350	16	
DIN EN 10208(96)	WNr 1.8970	Long distance pipe for combustibles	0.23 max			1.00-1.50		0.040 max	0.035 max	0.55 max	Al >=0.020; V <=0.12; bal Fe				
India															
IS 1570	20Mn2		0.16-0.24			1.3-1.7				0.1-0.35	bal Fe				
Italy															
UNI 6930(71)	20Mn6	Aircraft material	0.17-0.25	0.25 max		1.3-1.7		0.03 max	0.025 max	0.1-0.35	Mo <=0.1; bal Fe				
Japan															
JIS G3445(88)	STKM19A	Tube	0.25 max			1.50 max		0.040 max	0.040 max	0.55 max	bal Fe	490	314	23L	
JIS G3445(88)	STKM19C	Tube	0.25 max			1.50 max		0.040 max	0.040 max	0.55 max	bal Fe	549	412	15L	
JIS G4106(79)	SMn21*	Obs; see SMn420	0.17-0.23		0.3	1.2-1.5	0.25	0.03	0.03	0.15-0.35	bal Fe				
JIS G4106(79)	SMnC21*	Obs; see SMnC420	0.17-0.23	0.35-0.7	0.3	1.2-1.5	0.25	0.03	0.03	0.15-0.35	bal Fe				
JIS G4106(79)	SMnC420	HR Frg	0.17-0.23	0.35-0.70	0.30 max	1.20-1.50	0.25 max	0.030 max	0.030 max	0.15-0.35	bal Fe				
Mexico															
NMX-B-301(86)	1522	Bar	0.18-0.24			1.10-1.40		0.040 max	0.050 max		bal Fe				
NMX-B-301(86)	1522H	Bar	0.17-0.25			1.00-1.50		0.040 max	0.050 max	0.15-0.30	bal Fe				
Romania															
STAS 9382/4	20MnB5q		0.17-0.23	0.3	0.3	1.1-1.4	0.3	0.035	0.035	0.1-0.4	B 0.003-0.005; bal Fe				
Spain															
UNE	F.220.A		0.17-0.23			1-1.4		0.04	0.04	0.13-0.38	bal Fe				
UNE 36087(74)	A42RAI	Sh Plt	0.23 max			0.4-1.4		0.055 max	0.05 max	0.4 max	Mo <=0.15; N <=0.009; bal Fe				
UNE 36087(74)	A42RAII	Sh Plt	0.23 max			0.4-1.4		0.045 max	0.04 max	0.4 max	Al 0.015-0.10; Mo <=0.15; bal Fe				
UNE 36087(74)	A42RBII	Sh Plt	0.23 max			0.4-1.4		0.045 max	0.04 max	0.4 max	Al 0.015-0.10; Mo <=0.15; bal Fe				
UNE 36087(74)	A47RBII	Sh Plt	0.23 max			0.5-1.6		0.045 max	0.04 max	0.4 max	Al 0.015-0.10; Mo <=0.15; bal Fe				
Sweden															
SS 142165	2165		0.24 max			1.6 max		0.06 max	0.05 max	0.6 max	C+Mn/6+(Ni+Cu)/15+(Cr+Mo+V)/5<=0.5; bal Fe				
SS 142168	2168		0.28 max			1.6 max		0.06 max	0.05 max	0.6 max	C<=0.25; Mn<=1.8				

UNS numbers and US grades are provided as a means of cross referencing chemically similar alloys. Exchangability is only possible after independent examination of specifications. Tensile properties are minimum or typical as specified. UTS and YS as MPa. El as %. See Appendix for list of abbreviations used in Notes. * indicates obsolete material.

Specification	Designation	Notes	C	Cr	Cu	Mn	Ni	P	S	Si	Other	UTS	YS	El	Hard

Carbon Steel, High-Manganese, 1522/1522H (Continued from previous page)

UK

Specification	Designation	Notes	C	Cr	Cu	Mn	Ni	P	S	Si	Other	UTS	YS	El	Hard
BS 1503(89)	221-460	Frg press ves; 100<t<=999mm; norm Q/T	0.23 max	0.25 max	0.30 max	0.80-1.40	0.40 max	0.030 max	0.025 max	0.10-0.40	Al <=0.010; Mo <=0.10; W <=0.1; Cr+Mo+Ni+Cu<=0.8; bal Fe	460-580	235	18	
BS 1503(89)	221-460	Frg press ves; t<=100mm; norm Q/T	0.23 max	0.25 max	0.30 max	0.80-1.40	0.40 max	0.030 max	0.025 max	0.10-0.40	Al <=0.010; Mo <=0.10; W <=0.1; Cr+Mo+Ni+Cu<=0.8; bal Fe	460-580	245	18	
BS 1503(89)	223-409*	Frg press ves	0.25			0.90-1.70		0.04	0.04	0.1-0.4	Nb 0.01-0.06; bal Fe				
BS 1503(89)	224-490	Frg press ves; t<=100mm/ norm Q/T	0.25 max	0.25 max	0.30 max	0.90-1.70	0.40 max	0.030 max	0.025 max	0.10-0.40	Al >=0.018; Mo <=0.10; W <=0.1; Cr+Mo+Ni+Cu<=0.8; bal Fe	490-610	305	16	
BS 3146/1(74)	CLA2	Investment Casting	0.18-0.25	0.30 max	0.30 max	1.20-1.70	0.40 max	0.035 max	0.035 max	0.20-0.50	bal Fe				
BS 980	CFS7		0.2-0.3			1.2-1.5		0.05	0.05	0.35	bal Fe				

USA

Specification	Designation	Notes	C	Cr	Cu	Mn	Ni	P	S	Si	Other	UTS	YS	El	Hard
	AISI 1522	Bar	0.18-0.24			1.10-1.40		0.040 max	0.050 max		bal Fe				
	AISI 1522	Wir rod	0.18-0.24			1.10-1.40		0.040 max	0.050 max		bal Fe				
	AISI 1522H	Wir rod, Hard	0.17-0.25			1.00-1.50		0.040 max	0.050 max	0.15-0.35	bal Fe				
	UNS G15220		0.18-0.24			1.10-1.40		0.040 max	0.050 max		bal Fe				
	UNS H15220		0.17-0.25			1.00-1.50		0.040 max	0.050 max	0.15-0.35	bal Fe				
ASTM A29/A29M(93)	1522	Bar	0.18-0.24			1.10-1.40		0.040 max	0.050 max		bal Fe				
ASTM A304(96)	1522H	Bar, hard bands spec	0.17-0.25			1.00-1.50		0.040 max	0.050 max	0.15-0.30	Cu Ni Cr Mo trace allowed; bal Fe				
ASTM A510(96)	1522	Wir rod	0.18-0.24			1.10-1.40		0.040 max	0.050 max		bal Fe				
ASTM A576(95)	1522	Special qual HW bar	0.18-0.24			1.10-1.40		0.040 max	0.050 max		Si Cu Pb B Bi Ca Se Te if spec'd; bal Fe				
SAE J1268(95)	1522H	Multiple forms; See std	0.17-0.25			1.00-1.50		0.030 max	0.050 max	0.15-0.35	bal Fe				
SAE J403(95)	1522	Bar Smls Tub HR CF	0.18-0.24			1.10-1.40		0.030 max	0.050 max		bal Fe				

Yugoslavia

Specification	Designation	Notes	C	Cr	Cu	Mn	Ni	P	S	Si	Other	UTS	YS	El	Hard
	C.3112	Chain	0.16-0.23			1.1-1.5		0.04 max	0.04 max	0.3-0.55	N <=0.007; bal Fe				

Carbon Steel, High-Manganese, 1524/1524H

Argentina

Specification	Designation	Notes	C	Cr	Cu	Mn	Ni	P	S	Si	Other	UTS	YS	El	Hard
IAS	IRAM 1524		0.19-0.25			1.35-1.65		0.040 max	0.050 max		bal Fe	620	400	20	188 HB

Australia

Specification	Designation	Notes	C	Cr	Cu	Mn	Ni	P	S	Si	Other	UTS	YS	El	Hard
AS 1442	KX1320*	Obs	0.18-0.23			1.4-1.7		0.05	0.05	0.1-0.35	bal Fe				
AS 1442(92)	9	Bar<=215mm, Bloom/bil/slab<=250mm, HR as rolled or norm	0.15-0.25			1.30-1.70		0.040 max	0.040 max	0.10-0.40	bal Fe	540	300	18	
AS 1443	KX1320*	Obs	0.18-0.23			1.4-1.7		0.05	0.05	0.1-0.35	bal Fe				
AS 1443(94)	X1320	CF bar (may treat w/microalloying elements)	0.18-0.23			1.40-1.70		0.040 max	0.040 max	0.10-0.35	Si<=0.10 for Al-killed; bal Fe				
AS 1444(96)	X1320H	H series, Hard/Tmp	0.17-0.24			1.30-1.80		0.040 max	0.040 max	0.10-0.35	bal Fe				

Bulgaria

Specification	Designation	Notes	C	Cr	Cu	Mn	Ni	P	S	Si	Other	UTS	YS	El	Hard
BDS 4880(89)	25G2S	Struct	0.20-0.29	0.35 max	0.3	1.20-1.60	0.3	0.045 max	0.045 max	0.60-0.90	bal Fe				

China

Specification	Designation	Notes	C	Cr	Cu	Mn	Ni	P	S	Si	Other	UTS	YS	El	Hard
GB 3077(88)	20Mn2	Bar HR Q/T 15mm diam	0.17-0.24	0.30 max	0.30 max	1.40-1.80	0.30 max	0.035 max	0.035 max	0.17-0.37	bal Fe	785	590	10	

UNS numbers and US grades are provided as a means of cross referencing chemically similar alloys. Exchangability is only possible after independent examination of specifications. Tensile properties are minimum or typical as specified. UTS and YS as MPa. El as %. See Appendix for list of abbreviations used in Notes. * indicates obsolete material.

Specification	Designation	Notes	C	Cr	Cu	Mn	Ni	P	S	Si	Other	UTS	YS	El	Hard
Carbon Steel, High-Manganese, 1524/1524H (Continued from previous page)															
China															
GB/T 3078(94)	20Mn2	Bar CD 15mm diam	0.17-0.24	0.30 max	0.30 max	1.40-1.80	0.30 max	0.035 max	0.035 max	0.17-0.37	bal Fe	785	590	10	
Germany															
DIN	22Mn6		0.18-0.25			1.30-1.70		0.035 max	0.035 max	0.15-0.30	bal Fe				
DIN	WNr 1.1160		0.18-0.25			1.30-1.70		0.035 max	0.035 max	0.15-0.30	bal Fe				
India															
IS 4367	20Mn2	Frg, Hard, Tmp, 63mm diam	0.16-0.24			1.3-1.7		0.05	0.05	0.1-0.35	bal Fe				
IS 4922	St55	Tube, SD/Tmp, or Hard/Tmp	0.16-0.24			1.3-1.7	0.3	0.05	0.05	0.1-0.35	bal Fe				
IS 5517	20Mn2	Bar, Norm, Ann, Hard Tmp, 15mm	0.16-0.24			1.3-1.7				0.1-0.35	bal Fe				
Japan															
JIS G3444(94)	STK490	Tub Gen struct	0.18 max			1.50 max		0.040 max	0.040 max	0.55 max	bal Fe	490	315	23	
JIS G3507(91)	SWRCH24K	Wir rod FF	0.19-0.25			1.35-1.65		0.030 max	0.035 max	0.10-0.35	bal Fe				
Mexico															
NMX-B-301(86)	1524	Bar	0.19-0.25			1.35-1.65		0.040 max	0.050 max		bal Fe				
NMX-B-301(86)	1524H	Bar	0.18-0.26			1.25-1.75		0.040 max	0.050 max	0.15-0.30	bal Fe				
Pan America															
COPANT 333	1024		0.19-0.25			1.35-1.65		0.04	0.05		bal Fe				
Poland															
PNH84023/05	22G2A	Const	0.19-0.25	0.3 max	0.2 max	1.3-1.6	0.25 max	0.035 max	0.035 max	0.17-0.37	bal Fe				
PNH84023/07	20G2	Const	0.17-0.24	0.25 max		1.2-1.6	0.25 max	0.04 max	0.04 max	0.17-0.37	bal Fe				
Russia															
GOST 10702(78)	20G2	for CHd	0.18-0.26	0.25	0.2	1.3-1.6	0.25	0.035	0.035	0.17-0.37	bal Fe				
GOST 19281	17G1S	Weld const	0.15-0.2	0.3 max	0.3 max	1.15-1.6	0.3 max	0.035 max	0.04 max	0.4-0.6	Al <=0.1; N <=0.012; As<=0.08; bal Fe				
GOST 5781(82)	18G2S	Reinforcing	0.14-0.23	0.3 max	0.3 max	1.2-1.6	0.3 max	0.04 max	0.045 max	0.6-0.9	Al <=0.1; Ti 0.01-0.03; bal Fe				
Spain															
UNE	F.120.L		0.19-0.25			1.35-1.75		0.035	0.035	0.13-0.38	bal Fe				
UNE 36013(60)	20Mn6		0.18-0.25			1.3-1.6		0.035	0.035	0.15-0.4	bal Fe				
UNE 36013(60)	F.1515*	see 20Mn6	0.18-0.25			1.3-1.6		0.035	0.035	0.15-0.4	bal Fe				
UNE 36013(76)	20Mn6	CH	0.18-0.25			1.3-1.6		0.035 max	0.035 max	0.15-0.4	Mo <=0.15; bal Fe				
UNE 36013(76)	20Mn6	CH	0.18-0.25			1.3-1.6		0.035 max	0.035 max	0.15-0.4	Mo <=0.15; bal Fe				
UNE 36013(76)	20Mn6-1	CH	0.18-0.25			1.3-1.6		0.035 max	0.02-0.035	0.15-0.4	Mo <=0.15; bal Fe				
UNE 36013(76)	F.1518*	see 20Mn6-1	0.18-0.25			1.3-1.6		0.035 max	0.02-0.035	0.15-0.4	Mo <=0.15; bal Fe				
UNE 36255(79)	AM22Mn5*	Q/T	0.18-0.25			1.2-1.6		0.040 max	0.040 max	0.3-0.5	Mo <=0.15; bal Fe				
UNE 36255(79)	F.8310*	see AM22Mn5*	0.18-0.25			1.2-1.6		0.040 max	0.040 max	0.3-0.5	Mo <=0.15; bal Fe				
UK															
BS 1456	A		0.18-0.25			1.2-1.6		0.05	0.05	0.5	bal Fe				
BS 1503(89)	221-490	Frg press ves; t<=100mm; norm Q/T	0.25 max	0.25 max	0.30 max	0.90-1.70	0.40 max	0.030 max	0.025 max	0.10-0.40	Al <=0.010; Mo <=0.10; W <=0.1; Cr+Mo+Ni+Cu<=0.8; bal Fe	490-610	265	16	

Specification	Designation	Notes	C	Cr	Cu	Mn	Ni	P	S	Si	Other	UTS	YS	El	Hard

Carbon Steel, High-Manganese, 1524/1524H (Continued from previous page)

UK

Specification	Designation	Notes	C	Cr	Cu	Mn	Ni	P	S	Si	Other	UTS	YS	El	Hard
BS 1503(89)	221-490	Frg press ves; 100<t<=999mm; norm Q/T	0.25 max	0.25 max	0.30 max	0.90-1.70	0.40 max	0.030 max	0.025 max	0.10-0.40	Al <=0.010; Mo <=0.10; W <=0.1; Cr+Mo+Ni+Cu<=0.8; bal Fe	490-610	255	16	
BS 1503(89)	221-510	Frg press ves; t<=100mm; Q/T	0.25 max	0.25 max	0.30 max	0.90-1.70	0.40 max	0.030 max	0.025 max	0.10-0.40	Al <=0.010; Mo <=0.10; W <=0.1; Cr+Mo+Ni+Cu<=0.8; bal Fe	510-630	285	16	
BS 1503(89)	223-490	Frg press ves; t<=100mm; norm Q/T	0.25 max	0.25 max	0.30 max	0.90-1.70	0.40 max	0.030 max	0.025 max	0.10-0.40	Mo <=0.10; Nb 0.01-0.06; W <=0.1; Cr+Mo+Ni+Cu<=0.8; bal Fe	490-610	320	16	
BS 1503(89)	223-490	Frg press ves; 100<t<=999mm; norm Q/T	0.25 max	0.25 max	0.30 max	0.90-1.70	0.40 max	0.030 max	0.025 max	0.10-0.40	Mo <=0.10; Nb 0.01-0.06; W <=0.1; Cr+Mo+Ni+Cu<=0.8; bal Fe	490-610	295	16	
BS 970/1(83)	150M19	Blm Bil Slab Bar Rod Frg Hard Tmp; 6 in.	0.15-0.23			1.30-1.70		0.050 max	0.050 max		bal Fe				
BS 980	CDS10	Tube, Drawn or Drawn Tmp, 6 in.									bal Fe				
BS 980	CDS9	Tube, Ann	0.26			1.2-1.7		0.05	0.05	0.35	bal Fe				

USA

Specification	Designation	Notes	C	Cr	Cu	Mn	Ni	P	S	Si	Other	UTS	YS	El	Hard
	AISI 1524	Sh Strp Plt	0.18-0.25			1.30-1.65		0.030 max	0.035 max		bal Fe				
	AISI 1524	Wir rod	0.19-0.25			1.35-1.65		0.040 max	0.050 max		bal Fe				
	AISI 1524	Bar, was 1024	0.19-0.25			1.35-1.65		0.040 max	0.050 max		bal Fe				
	AISI 1524	Struct shp	0.19-0.25			1.30-1.65		0.030 max	0.035 max		bal Fe				
	AISI 1524H	Wir rod, Hard	0.18-0.26			1.25-1.75		0.040 max	0.050 max	0.15-0.35	bal Fe				
	UNS G15240		0.19-0.25			1.35-1.65		0.040 max	0.050 max		Sheets and Plates, C 0.18-0.25; ERW Tubing, C 0.18-0.25 Mn 1.30-1.65; bal Fe				
	UNS H15240		0.18-0.26			1.25-1.75		0.040 max	0.050 max	0.15-0.35	bal Fe				
ASTM A29/A29M(93)	1524	Bar, was 1024	0.19-0.25			1.35-1.65		0.040 max	0.050 max		bal Fe				
ASTM A304(96)	1524H	Bar, hard bands spec	0.18-0.26			1.25-1.75		0.040 max	0.050 max	0.15-0.30	Cu Ni Cr Mo trace allowed; bal Fe				
ASTM A510(96)	1524	Wir rod	0.19-0.25			1.35-1.65		0.040 max	0.050 max		bal Fe				
ASTM A513(97a)	1024	ERW Mech tub	0.18-0.25			1.30-1.65		0.035 max	0.035 max		bal Fe				
ASTM A513(97a)	1524	ERW Mech tub	0.18-0.25			1.35-1.65		0.040 max	0.050 max		bal Fe				
ASTM A519(96)	1524	Smls mech tub	0.19-0.25			1.35-1.65		0.040 max	0.050 max		bal Fe				
ASTM A576(95)	1524	Special qual HW bar	0.19-0.25			1.35-1.65		0.040 max	0.050 max		Si Cu Pb B Bi Ca Se Te if spec'd; bal Fe				
ASTM A635/A635M(98)	1524	Sh Strp, Coil, HR	0.18-0.25		0.20 min	1.30-1.65		0.030 max	0.035 max		bal Fe				
ASTM A830/A830M(98)	1524	Plt	0.18-0.25			1.30-1.65		0.035 max	0.04 max		bal Fe				
FED QQ-S-00643(61)	1524*	Obs; Tub smls weld	0.19-0.25			1.35-1.65		0.04 max	0.05 max		bal Fe				
FED QQ-S-633A(63)	1524*	Obs; CF HR Bar	0.19-0.25			1.35-1.65		0.04 max	0.05 max		bal Fe				
MIL-S-20145	6*	Obs; Bar rod tube bil frg shape	0.19-0.25			1.35-1.65		0.04 max	0.05 max		bal Fe				
SAE J1268(95)	1524H	Bar Rod Wir; See std	0.18-0.26			1.25-1.75		0.030 max	0.050 max	0.15-0.35	bal Fe				
SAE J1397(92)	1524	Bar HR, est mech prop	0.19-0.25			1.35-1.65		0.04 max	0.05 max		bal Fe	510	280	20	149 HB
SAE J1397(92)	1524	Bar CD, est mech prop	0.19-0.25			1.35-1.65		0.04 max	0.05 max		bal Fe	570	480	12	163 HB

UNS numbers and US grades are provided as a means of cross referencing chemically similar alloys. Exchangability is only possible after independent examination of specifications. Tensile properties are minimum or typical as specified. UTS and YS as MPa. El as %. See Appendix for list of abbreviations used in Notes. * indicates obsolete material.

Specification	Designation	Notes	C	Cr	Cu	Mn	Ni	P	S	Si	Other	UTS	YS	El	Hard
Carbon Steel, High-Manganese, 1524/1524H (Continued from previous page)															
USA															
SAE J1397(92)	1524	Bar Ann CD, est mech prop	0.19-0.25			1.35-1.65		0.04 max	0.05 max		bal Fe	650	550	10	184 HB
SAE J403(95)	1524	Struct Shps Plt Strp Sh Weld Tub; formerly 1024	0.18-0.25			1.30-1.65		0.030 max	0.035 max		bal Fe				
SAE J403(95)	1524	Bar Smls Tub HR CF	0.19-0.25			1.35-1.65		0.030 max	0.050 max		bal Fe				
Carbon Steel, High-Manganese, 1524H															
UK															
BS 1449/1(91)	20CS	Plt Sh Strp; CR Ann; t<=16mm	0.15-0.25			1.30-1.70		0.045 max	0.045 max	0.05-0.35	bal Fe	420	230	18-20	
BS 1449/1(91)	20HS	Plt Sh Strp HR; t<=16mm	0.15-0.25			1.30-1.70		0.045 max	0.045 max	0.6 max	bal Fe	540	350	16-18	
Carbon Steel, High-Manganese, 1525															
Mexico															
NMX-B-301(86)	1525	Bar	0.23-0.29			0.80-1.10		0.040 max	0.050 max		bal Fe				
USA															
	UNS G15250		0.23-0.29			0.80-1.10		0.040 max	0.050 max		bal Fe				
ASTM A29/A29M(93)	1525	Bar	0.23-0.29			0.80-1.10		0.040 max	0.050 max		bal Fe				
ASTM A510(96)	1525	Wir rod	0.23-0.29			0.80-1.10		0.040 max	0.050 max		bal Fe				
ASTM A576(95)	1525	Special qual HW bar	0.23-0.29			0.80-1.10		0.040 max	0.050 max		Si Cu Pb B Bi Ca Se Te if spec'd; bal Fe				
Carbon Steel, High-Manganese, 1526/1526H															
Canada															
CSA B193	B	Pipe, Smls, Mill 6.35mm diam	0.27			1.15		0.04	0.05		bal Fe				
CSA B193	G	Pipe, Elect Weld, Mill 6.35mm	0.28			1.25		0.04	0.05		bal Fe				
CSA B193	G	Pipe, Elect Weld, Mill 6.35mm	0.28			1.25		0.04	0.05		bal Fe				
CSA B193	G	Pipe, Elect Weld, Mill 6.35mm	0.28			1.25		0.04	0.05		bal Fe				
CSA B193	G	Pipe, Elect Weld, Mill 6.35mm	0.26			1.15		0.04	0.05		bal Fe				
CSA B193	X42	Pipe, Smls, Mill 6.35mm diam	0.29			1.25		0.04	0.05		bal Fe				
CSA B193	X46	Pipe, Smls, Mill 6.35mm diam	0.29			1.25		0.04	0.05		bal Fe				
CSA B193	X52	Pipe, Smls, Mill 6.35mm diam	0.29			1.25		0.04	0.05		bal Fe				
China															
GB 713(97)	22Mng	Plt HR 25mm Thk	0.30 max	0.30 max	0.30 max	0.90-1.50	0.30 max	0.025 max	0.025 max	0.15-0.40	bal Fe	515-655	275	19	
France															
AFNOR NFA35566	25MS5		0.24-0.3			1.1-1.6		0.035	0.035	0.3-0.55	bal Fe				
Japan															
JIS G3507(91)	SWRCH27K	Wir rod FF	0.22-0.29			1.20-1.50		0.030 max	0.035 max	0.10-0.35	bal Fe				
Mexico															
NMX-B-301(86)	1526	Bar	0.22-0.29			1.10-1.40		0.040 max	0.050 max		bal Fe				
NMX-B-301(86)	1526H	Bar	0.21-0.30			1.00-1.50		0.040 max	0.050 max	0.15-0.30	bal Fe				
Russia															
GOST 977(88)	30GSL		0.25-0.35	0.3		1.1-1.4	0.3	0.04	0.04	0.6-0.8	bal Fe				

Specification	Designation	Notes	C	Cr	Cu	Mn	Ni	P	S	Si	Other	UTS	YS	El	Hard

Carbon Steel, High-Manganese, 1526/1526H (Continued from previous page)

Spain

Specification	Designation	Notes	C	Cr	Cu	Mn	Ni	P	S	Si	Other	UTS	YS	El	Hard
UNE	F.130.D		0.21-0.27			1.6-1.9		0.035	0.035	0.13-0.38	bal Fe				

Sweden

Specification	Designation	Notes	C	Cr	Cu	Mn	Ni	P	S	Si	Other	UTS	YS	El	Hard
SS 142130	2130		0.25-0.3	0.1-0.3		1.1-1.3		0.035	0.035	0.15-0.4	B 0.002; bal Fe				
SS 142131	2131		0.25-0.3	0.3-0.6		1.1-1.4		0.035 max	0.035 max	0.15-0.4	B <=0.002; bal Fe				

UK

Specification	Designation	Notes	C	Cr	Cu	Mn	Ni	P	S	Si	Other	UTS	YS	El	Hard
BS 970/1(83)	120M28	Blm Bil Slab Bar Rod Frg	0.24-0.32			1.00-1.40		0.050 max	0.050 max		bal Fe				

USA

Specification	Designation	Notes	C	Cr	Cu	Mn	Ni	P	S	Si	Other	UTS	YS	El	Hard
	AISI 1526	Wir rod	0.22-0.29			1.10-1.40		0.040 max	0.050 max		bal Fe				
	AISI 1526	Bar	0.22-0.29			1.10-1.40		0.040 max	0.050 max		bal Fe				
	AISI 1526H	Wir rod, Hard	0.21-0.30			1.00-1.50		0.040 max	0.050 max	0.15-0.35	bal Fe				
	UNS G15260		0.22-0.29			1.10-1.40		0.040 max	0.050 max		bal Fe				
	UNS H15260		0.21-0.30			1.00-1.50		0.040 max	0.050 max	0.15-0.35	bal Fe				
ASTM A29/A29M(93)	1526	Bar	0.22-0.29			1.10-1.40		0.040 max	0.050 max		bal Fe				
ASTM A304(96)	1526H	Bar, hard bands spec	0.21-0.30			1.00-1.50		0.040 max	0.040 max	0.15-0.30	Cu Ni Cr Mo trace allowed; bal Fe				
ASTM A510(96)	1526	Wir rod	0.22-0.29			1.10-1.40		0.040 max	0.050 max		bal Fe				
ASTM A576(95)	1526	Special qual HW bar	0.22-0.29			1.10-1.40		0.040 max	0.050 max		Si Cu Pb B Bi Ca Se Te if spec'd; bal Fe				
SAE J1268(95)	1526H	Bar Rod Wir Bil Frg; See std	0.21-0.30			1.00-1.50		0.030 max	0.050 max	0.15-0.35	bal Fe				
SAE J403(95)	1526	Bar Smls Tub HR CF	0.22-0.29			1.10-1.40		0.030 max	0.050 max		bal Fe				

Carbon Steel, High-Manganese, 1526H

India

Specification	Designation	Notes	C	Cr	Cu	Mn	Ni	P	S	Si	Other	UTS	YS	El	Hard
IS 1570/1(78)	Fe290	Sh Plt Bar Frg Tub Pipe	0.3 max	0.3 max	0.3 max	2 max	0.4 max	0.055 max	0.055 max	0.6 max	Co <=0.1; Mo <=0.15; Pb <=0.15; W <=0.1; P:S Varies; bal Fe	290	170	27	
IS 1570/1(78)	Fe310	Sh Plt Bar Frg Tub Pipe	0.3 max	0.3 max	0.3 max	2 max	0.4 max	0.055 max	0.055 max	0.6 max	Co <=0.1; Mo <=0.15; Pb <=0.15; W <=0.1; P:S Varies; bal Fe	310	180	26	
IS 1570/1(78)	Fe330	Sh Plt Bar Frg Tub Pipe	0.3 max	0.3 max	0.3 max	2 max	0.4 max	0.055 max	0.055 max	0.6 max	Co <=0.1; Mo <=0.15; Pb <=0.15; W <=0.1; P:S Varies; bal Fe	330	200	26	
IS 1570/1(78)	Fe360	Sh Plt Bar Frg Tub Pipe	0.3 max	0.3 max	0.3 max	2 max	0.4 max	0.055 max	0.055 max	0.6 max	Co <=0.1; Mo <=0.15; Pb <=0.15; W <=0.1; P:S Varies; bal Fe	360	220	25	
IS 1570/1(78)	Fe410	Sh Plt Bar Frg Tub Pipe	0.3 max	0.3 max	0.3 max	2 max	0.4 max	0.055 max	0.055 max	0.6 max	Co <=0.1; Mo <=0.15; Pb <=0.15; W <=0.1; P:S Varies; bal Fe	410	250	23	
IS 1570/1(78)	Fe490	Sh Plt Bar Frg Tub Pipe	0.3 max	0.3 max	0.3 max	2 max	0.4 max	0.055 max	0.055 max	0.6 max	Co <=0.1; Mo <=0.15; Pb <=0.15; W <=0.1; P:S Varies; bal Fe	490	290	21	
IS 1570/1(78)	Fe540	Sh Plt Bar Frg Tub Pipe	0.3 max	0.3 max	0.3 max	2 max	0.4 max	0.055 max	0.055 max	0.6 max	Co <=0.1; Mo <=0.15; Pb <=0.15; W <=0.1; P:S Varies; bal Fe	540	320	20	
IS 1570/1(78)	Fe620	Sh Plt Bar Frg Tub Pipe	0.3 max	0.3 max	0.3 max	2 max	0.4 max	0.055 max	0.055 max	0.6 max	Co <=0.1; Mo <=0.15; Pb <=0.15; W <=0.1; P:S Varies; bal Fe	620	380	15	

Specification	Designation	Notes	C	Cr	Cu	Mn	Ni	P	S	Si	Other	UTS	YS	El	Hard

Carbon Steel, High-Manganese, 1526H (Continued from previous page)

India

Specification	Designation	Notes	C	Cr	Cu	Mn	Ni	P	S	Si	Other	UTS	YS	El	Hard
IS 1570/1(78)	Fe690	Sh Plt Bar Frg Tub Pipe	0.3 max	0.3 max	0.3 max	2 max	0.4 max	0.055 max	0.055 max	0.6 max	Co <=0.1; Mo <=0.15; Pb <=0.15; W <=0.1; P:S Varies; bal Fe	690	410	12	
IS 1570/1(78)	Fe770	Sh Plt Bar Frg Tub Pipe	0.3 max	0.3 max	0.3 max	2 max	0.4 max	0.055 max	0.055 max	0.6 max	Co <=0.1; Mo <=0.15; Pb <=0.15; W <=0.1; P:S Varies; bal Fe	770	460	10	
IS 1570/1(78)	Fe870	Sh Plt Bar Frg Tub Pipe	0.3 max	0.3 max	0.3 max	2 max	0.4 max	0.055 max	0.055 max	0.6 max	Co <=0.1; Mo <=0.15; Pb <=0.15; W <=0.1; P:S Varies; bal Fe	870	520	8	
IS 1570/1(78)	FeE220	Sh Plt Bar Frg Tub Pipe	0.3 max	0.3 max	0.3 max	2 max	0.4 max	0.055 max	0.055 max	0.6 max	Co <=0.1; Mo <=0.15; Pb <=0.15; W <=0.1; P:S Varies; bal Fe	290	220	27	
IS 1570/1(78)	FeE230	Sh Plt Bar Frg Tub Pipe	0.3 max	0.3 max	0.3 max	2 max	0.4 max	0.055 max	0.055 max	0.6 max	Co <=0.1; Mo <=0.15; Pb <=0.15; W <=0.1; P:S Varies; bal Fe	310	230	26	
IS 1570/1(78)	FeE250	Sh Plt Bar Frg Tub Pipe	0.3 max	0.3 max	0.3 max	2 max	0.4 max	0.055 max	0.055 max	0.6 max	Co <=0.1; Mo <=0.15; Pb <=0.15; W <=0.1; P:S Varies; bal Fe	330	250	26	
IS 1570/1(78)	FeE270	Sh Plt Bar Frg Tub Pipe	0.3 max	0.3 max	0.3 max	2 max	0.4 max	0.055 max	0.055 max	0.6 max	Co <=0.1; Mo <=0.15; Pb <=0.15; W <=0.1; P:S Varies; bal Fe	360	270	25	
IS 1570/1(78)	FeE310	Sh Plt Bar Frg Tub Pipe	0.3 max	0.3 max	0.3 max	2 max	0.4 max	0.055 max	0.055 max	0.6 max	Co <=0.1; Mo <=0.15; Pb <=0.15; W <=0.1; P:S Varies; bal Fe	410	310	23	
IS 1570/1(78)	FeE370	Sh Plt Bar Frg Tub Pipe	0.3 max	0.3 max	0.3 max	2 max	0.4 max	0.055 max	0.055 max	0.6 max	Co <=0.1; Mo <=0.15; Pb <=0.15; W <=0.1; bal Fe	490	370	21	
IS 1570/1(78)	FeE400	Sh Plt Bar Frg Tub Pipe	0.3 max	0.3 max	0.3 max	2 max	0.4 max	0.055 max	0.055 max	0.6 max	Co <=0.1; Mo <=0.15; Pb <=0.15; W <=0.1; bal Fe	540	400	20	
IS 1570/1(78)	FeE460	Sh Plt Bar Frg Tub Pipe	0.3 max	0.3 max	0.3 max	2 max	0.4 max	0.055 max	0.055 max	0.6 max	Co <=0.1; Mo <=0.15; Pb <=0.15; W <=0.1; P:S Varies; bal Fe	620	460	15	
IS 1570/1(78)	FeE520	Sh Plt Bar Frg Tub Pipe	0.3 max	0.3 max	0.3 max	2 max	0.4 max	0.055 max	0.055 max	0.6 max	Co <=0.1; Mo <=0.15; Pb <=0.15; W <=0.1; P:S Varies; bal Fe	690	520	12	
IS 1570/1(78)	FeE580	Sh Plt Bar Frg Tub Pipe	0.3 max	0.3 max	0.3 max	2 max	0.4 max	0.055 max	0.055 max	0.6 max	Co <=0.1; Mo <=0.15; Pb <=0.15; W <=0.1; P:S Varies; bal Fe	770	580	10	
IS 1570/1(78)	FeE650	Sh Plt Bar Frg Tub Pipe	0.3 max	0.3 max	0.3 max	2 max	0.4 max	0.055 max	0.055 max	0.6 max	Co <=0.1; Mo <=0.15; Pb <=0.15; W <=0.1; P:S Varies; bal Fe	870	650	8	
IS 1570/1(78)	St30	Sh Plt Bar Frg Tub Pipe	0.3 max	0.3 max	0.3 max	2 max	0.4 max	0.055 max	0.055 max	0.6 max	Co <=0.1; Mo <=0.15; Pb <=0.15; W <=0.1; P:S Varies; bal Fe	290	170	27	
IS 1570/1(78)	St32	Sh Plt Bar Frg Tub Pipe	0.3 max	0.3 max	0.3 max	2 max	0.4 max	0.055 max	0.055 max	0.6 max	Co <=0.1; Mo <=0.15; Pb <=0.15; W <=0.1; P:S Varies; bal Fe	310	180	26	
IS 1570/1(78)	St34	Sh Plt Bar Frg Tub Pipe	0.3 max	0.3 max	0.3 max	2 max	0.4 max	0.055 max	0.055 max	0.6 max	Co <=0.1; Mo <=0.15; Pb <=0.15; W <=0.1; P:S Varies; bal Fe	330	200	26	
IS 1570/1(78)	St37	Sh Plt Bar Frg Tub Pipe	0.3 max	0.3 max	0.3 max	2 max	0.4 max	0.055 max	0.055 max	0.6 max	Co <=0.1; Mo <=0.15; Pb <=0.15; W <=0.1; P:S Varies; bal Fe	360	220	25	

Specification	Designation	Notes	C	Cr	Cu	Mn	Ni	P	S	Si	Other	UTS	YS	El	Hard

Carbon Steel, High-Manganese, 1526H (Continued from previous page)

India

Specification	Designation	Notes	C	Cr	Cu	Mn	Ni	P	S	Si	Other	UTS	YS	El	Hard
IS 1570/1(78)	St42	Sh Plt Bar Frg Tub Pipe	0.3 max	0.3 max	0.3 max	2 max	0.4 max	0.055 max	0.055 max	0.6 max	Co <=0.1; Mo <=0.15; Pb <=0.15; W <=0.1; P:S Varies; bal Fe	410	250	23	
IS 1570/1(78)	St50	Sh Plt Bar Frg Tub Pipe	0.3 max	0.3 max	0.3 max	2 max	0.4 max	0.055 max	0.055 max	0.6 max	Co <=0.1; Mo <=0.15; Pb <=0.15; W <=0.1; P:S Varies; bal Fe	490	290	21	
IS 1570/1(78)	St55	Sh Plt Bar Frg Tub Pipe	0.3 max	0.3 max	0.3 max	2 max	0.4 max	0.055 max	0.055 max	0.6 max	Co <=0.1; Mo <=0.15; Pb <=0.15; W <=0.1; P:S Varies; bal Fe	540	320	20	
IS 1570/1(78)	St63	Sh Plt Bar Frg Tub Pipe	0.3 max	0.3 max	0.3 max	2 max	0.4 max	0.055 max	0.055 max	0.6 max	Co <=0.1; Mo <=0.15; Pb <=0.15; W <=0.1; P:S Varies; bal Fe	620	380	15	
IS 1570/1(78)	St70	Sh Plt Bar Frg Tub Pipe	0.3 max	0.3 max	0.3 max	2 max	0.4 max	0.055 max	0.055 max	0.6 max	Co <=0.1; Mo <=0.15; Pb <=0.15; W <=0.1; P:S Varies; bal Fe	690	410	12	
IS 1570/1(78)	St78	Sh Plt Bar Frg Tub Pipe	0.3 max	0.3 max	0.3 max	2 max	0.4 max	0.055 max	0.055 max	0.6 max	Co <=0.1; Mo <=0.15; Pb <=0.15; W <=0.1; P:S Varies; bal Fe	770	460	10	
IS 1570/1(78)	St88	Sh Plt Bar Frg Tub Pipe	0.3 max	0.3 max	0.3 max	2 max	0.4 max	0.055 max	0.055 max	0.6 max	Co <=0.1; Mo <=0.15; Pb <=0.15; W <=0.1; P:S Varies; bal Fe	870	520	8	

Carbon Steel, High-Manganese, 1527

Australia

Specification	Designation	Notes	C	Cr	Cu	Mn	Ni	P	S	Si	Other	UTS	YS	El	Hard
AS 1442	K10*	Obs; Bar As rolled or Norm	0.2-0.3			1.3-1.7		0.05	0.05	0.1-0.35	bal Fe				
AS 1442	SK1325*	Obs	0.23-0.28			1.4-1.7		0.05	0.05	0.1-0.35	bal Fe				
AS 1442(92)	10	Bar<=215mm, Bloom/bil/slab<=250mm, HR as rolled or norm	0.20-0.30			1.30-1.70		0.040 max	0.040 max	0.10-0.40	bal Fe	580	330	16	
AS 1443	K10*	Obs; Bar As rolled or Norm	0.2-0.3			1.3-1.7		0.05	0.05	0.1-0.35	bal Fe				
AS 1443	SK1325*	Obs	0.23-0.28			1.4-1.7		0.05	0.05	0.1-0.35	bal Fe				

Belgium

Specification	Designation	Notes	C	Cr	Cu	Mn	Ni	P	S	Si	Other	UTS	YS	El	Hard
NBN 253-02	28Mn6		0.25-0.32			1.3-1.65		0.035	0.035	0.15-0.4	bal Fe				

Bulgaria

Specification	Designation	Notes	C	Cr	Cu	Mn	Ni	P	S	Si	Other	UTS	YS	El	Hard
BDS 4880(89)	25G2S	Struct	0.22-0.29	0.35 max	0.35 max	1.20-1.60	0.35 max	0.045 max	0.045 max	0.60-0.90	bal Fe				
BDS 6354	30G2		0.27-0.35		0.3	1.3-1.7		0.035	0.035	0.17-0.37	bal Fe				

Czech Republic

Specification	Designation	Notes	C	Cr	Cu	Mn	Ni	P	S	Si	Other	UTS	YS	El	Hard
CSN 413140	13140		0.27-0.35			1.2-1.6		0.035 max	0.035 max	0.17-0.37	bal Fe				

Germany

Specification	Designation	Notes	C	Cr	Cu	Mn	Ni	P	S	Si	Other	UTS	YS	El	Hard
DIN	26Mn5		0.22-0.29			1.20-1.50		0.035 max	0.035 max	0.15-0.30	bal Fe				
DIN	WNr 1.1161		0.22-0.29			1.20-1.50		0.035 max	0.035 max	0.15-0.30	bal Fe				
DIN 17115(87)	27MnSi5	Weld round link chain	0.24-0.30		0.25 max	1.10-1.60		0.035 max	0.035 max	0.25-0.45	Al 0.020-0.050; N <=0.012; bal Fe				
DIN 17115(87)	WNr 1.0412	Weld round link chains	0.24-0.30		0.25 max	1.10-1.60		0.035 max	0.035 max	0.25-0.45	Al 0.020-0.050; N <=0.012; bal Fe				
DIN EN 10083(91)	28Mn6	Q/T 17-40mm	0.25-0.32	0.40 max		0.30-1.65	0.40 max	0.035 max	0.035 max	0.40 max	Mo <=0.10; Cr+Mo+Ni<=0.63; bal Fe	700-850	490	15	
DIN EN 10083(91)	WNr 1.1170	Q/T 17-40mm	0.25-0.32	0.40 max		1.30-1.65	0.40 max	0.035 max	0.035 max	0.40 max	Mo <=0.10; Cr+Mo+Ni<=0.63; bal Fe	700-850	490	15	

UNS numbers and US grades are provided as a means of cross referencing chemically similar alloys. Exchangability is only possible after independent examination of specifications. Tensile properties are minimum or typical as specified. UTS and YS as MPa. El as %. See Appendix for list of abbreviations used in Notes. * indicates obsolete material.

Carbon Steel, High-Manganese, 1527 (Continued from previous page)

Specification	Designation	Notes	C	Cr	Cu	Mn	Ni	P	S	Si	Other	UTS	YS	El	Hard
Hungary															
MSZ 61(85)	Mn1E	Q/T; t<=16mm	0.25-0.32			1.3-1.65		0.035 max	0.02-0.035	0.4 max	bal Fe	800-950	600	13L	
MSZ 61(85)	Mn1E	Q/T; 40<t<=100mm	0.25-0.32			1.3-1.65		0.035 max	0.02-0.035	0.4 max	bal Fe	650-800	440	16L	
India															
IS 1570	27Mn2		0.22-0.32			1.3-1.7				0.1-0.35	bal Fe				
IS 5517	27Mn2	Bar, Norm, Ann, Hard Tmp, 30mm	0.22-0.32			1.3-1.7				0.1-0.35	bal Fe				
International															
ISO 683-5	1*		0.25-0.32			1.3-1.65		0.035	0.035	0.15-0.4	bal Fe				
ISO 683-5	1A*		0.25-0.32			1.3-1.65		0.035	0.02-0.035	0.15-0.4	bal Fe				
ISO 683-5	1b*		0.25-0.32			1.3-1.65		0.035	0.03-0.05	0.15-0.4	bal Fe				
Mexico															
NMX-B-301(86)	1527	Bar	0.22-0.29			1.20-1.50		0.040 max	0.050 max		bal Fe				
Pan America															
COPANT 333	1027		0.22-0.29			1.2-1.5		0.04	0.05		bal Fe				
Poland															
PNH84023	26G2S		0.22-0.28	0.3	0.3	1.25-1.5	0.3	0.05	0.05	0.4-0.6	Mo 0.1; bal Fe				
PNH84030	30G2		0.27-0.35	0.25	0.3	1.4-1.8	0.3	0.035	0.035	0.17-0.37	bal Fe				
Russia															
GOST 4543	30G2		0.26-0.35	0.3	0.3	1.4-1.8	0.3	0.035	0.035	0.17-0.37	bal Fe				
GOST 5781(82)	25G2S		0.2-0.29	0.3	0.3	1.2-1.6	0.3	0.04	0.045	0.6-0.9	bal Fe				
GOST 7832	27GL		0.22-0.32	0.3	0.3	1.1-1.5	0.3	0.05	0.05	0.2-0.42	bal Fe				
Spain															
UNE	F.120.D		0.25-0.31			1.3-1.7		0.035	0.035	0.13-0.38	bal Fe				
UK															
BS 1453	A3		0.25-0.3	0.25		1.3-1.6	0.25	0.05	0.05	0.3-0.5	bal Fe				
BS 1456	B1		0.25-0.33			1.2-1.6		0.05	0.05	0.5	bal Fe				
BS 1456	B2		0.25-0.33			1.2-1.6		0.05	0.05	0.5	bal Fe				
BS 1717(83)	CFSC6	Tub	0.29 max			1.50 max		0.050 max	0.050 max	0.35 max	Mo 0.15-0.25; W <=0.1; bal Fe				
BS 970/1(83)	150M28	Blm Bil Slab Bar Rod Frg Norm, diam<150mm	0.25-0.32			1.30-1.70		0.050 max	0.050 max		bal Fe				
USA															
	AISI 1527	Bar, was 1027	0.22-0.29			1.20-1.50		0.040 max	0.050 max		bal Fe				
	AISI 1527	Sh Strp Plt	0.22-0.29			1.20-1.55		0.030 max	0.035 max		bal Fe				
	AISI 1527	Wir rod	0.22-0.29			1.20-1.50		0.040 max	0.050 max		bal Fe				
	AISI 1527	Struct shp	0.22-0.29			1.20-1.55		0.030 max	0.035 max		bal Fe				
	UNS G15270		0.22-0.29			1.20-1.50		0.040 max	0.050 max		Sheets and Plates, Mn 1.20-1.55; ERW Tubing, C 0.21-0.29 Mn 1.20-1.55; bal Fe				
ASTM A29/A29M(93)	1527	Bar, was 1027	0.22-0.29			1.20-1.50		0.040 max	0.050 max		bal Fe				
ASTM A510(96)	1527	Wir rod	0.22-0.29			1.20-1.50		0.040 max	0.050 max		bal Fe				

UNS numbers and US grades are provided as a means of cross referencing chemically similar alloys. Exchangability is only possible after independent examination of specifications. Tensile properties are minimum or typical as specified. UTS and YS as MPa. El as %. See Appendix for list of abbreviations used in Notes. * indicates obsolete material.

Specification	Designation	Notes	C	Cr	Cu	Mn	Ni	P	S	Si	Other	UTS	YS	El	Hard

Carbon Steel, High-Manganese, 1527 (Continued from previous page)

USA

Specification	Designation	Notes	C	Cr	Cu	Mn	Ni	P	S	Si	Other	UTS	YS	El	Hard
ASTM A576(95)	1527	Special qual HW bar	0.22-0.29			1.20-1.50		0.040 max	0.050 max		Si Cu Pb B Bi Ca Se Te if spec'd; bal Fe				
ASTM A830/A830M(98)	1527	Plt	0.22-0.29			1.20-1.55		0.035 max	0.04 max		bal Fe				
SAE J1397(92)	1527	Bar CD, est mech prop									bal Fe	570	480	12	163 HB
SAE J1397(92)	1527	Bar HR, est mech prop									bal Fe	520	280	18	149 HB
SAE J403(95)	1527	Bar Smls Tub HR CF	0.22-0.29			1.20-1.50		0.030 max	0.050 max		bal Fe				
SAE J403(95)	1527	Struct Shps Plt Strp Sh Weld Tub; formerly 1027	0.22-0.29			1.20-1.55		0.030 max	0.035 max		bal Fe				

Yugoslavia

Specification	Designation	Notes	C	Cr	Cu	Mn	Ni	P	S	Si	Other	UTS	YS	El	Hard
	C.3135	Q/T	0.25-0.32			1.3-1.65		0.035 max	0.03 max	0.4 max	Mo <=0.15; bal Fe				

Carbon Steel, High-Manganese, 1533

Czech Republic

Specification	Designation	Notes	C	Cr	Cu	Mn	Ni	P	S	Si	Other	UTS	YS	El	Hard
CSN 413141	13141	Q/T	0.27-0.35	0.3 max		1.2-1.6		0.035 max	0.035 max	0.17-0.37	bal Fe				

USA

Specification	Designation	Notes	C	Cr	Cu	Mn	Ni	P	S	Si	Other	UTS	YS	El	Hard
	AISI 1533	Bar Blm Bil Slab	0.30-0.37			1.10-1.40		0.040 max	0.050 max		bal Fe				
	UNS G15330		0.30-0.37			1.20-1.50		0.040 max	0.050 max		bal Fe				

Carbon Steel, High-Manganese, 1534

USA

Specification	Designation	Notes	C	Cr	Cu	Mn	Ni	P	S	Si	Other	UTS	YS	El	Hard
	AISI 1534	Bar Blm Bil Slab	0.30-0.37			1.20-1.50		0.040 max	0.050 max		bal Fe				
	UNS G15340		0.30-0.37			1.20-1.50		0.040 max	0.050 max		bal Fe				

Carbon Steel, High-Manganese, 1536

Argentina

Specification	Designation	Notes	C	Cr	Cu	Mn	Ni	P	S	Si	Other	UTS	YS	El	Hard
IAS	IRAM 1536		0.30-0.37			1.20-1.50		0.040 max	0.050 max		bal Fe	690	480	16	209 HB

Japan

Specification	Designation	Notes	C	Cr	Cu	Mn	Ni	P	S	Si	Other	UTS	YS	El	Hard
JIS G4052(79)	SMn420H	Struct Hard	0.16-0.23	0.35 max	0.30 max	1.15-1.55	0.25 max	0.030 max	0.030 max	0.15-0.35	bal Fe				
JIS G4106(79)	SMn 1*	Obs; see SMn433	0.30-0.36	0.35 max	0.30 max	1.20-1.50	0.25 max	0.030 max	0.030 max	0.15-0.35	bal Fe				
JIS G4106(79)	SMn433	HR Frg	0.30-0.36	0.35 max	0.30 max	1.20-1.50	0.25 max	0.030 max	0.030 max	0.15-0.35	bal Fe				

Mexico

Specification	Designation	Notes	C	Cr	Cu	Mn	Ni	P	S	Si	Other	UTS	YS	El	Hard
NMX-B-301(86)	1536	Bar	0.30-0.37			1.20-1.50		0.040 max	0.050 max		bal Fe				

USA

Specification	Designation	Notes	C	Cr	Cu	Mn	Ni	P	S	Si	Other	UTS	YS	El	Hard
	AISI 1536	Sh Strp Plt	0.30-0.38			1.20-1.55		0.030 max	0.035 max		bal Fe				
	AISI 1536	Struct shp	0.30-0.37			1.20-1.55		0.030 max	0.035 max		bal Fe				
	UNS G15360		0.30-0.37			1.20-1.50		0.040 max	0.050 max		Sheets and Plates, C 0.30-0.38 Mn 1.20-1.55; bal Fe				
ASTM A29/A29M(93)	1536	Bar, was 1036	0.30-0.37			1.20-1.50		0.040 max	0.050 max		bal Fe				
ASTM A510(96)	1536	Wir rod	0.30-0.37			1.20-1.50		0.040 max	0.050 max		bal Fe				
ASTM A576(95)	1536	Special qual HW bar	0.30-0.37			1.20-1.50		0.040 max	0.050 max		Si Cu Pb B Bi Ca Se Te if spec'd; bal Fe				
ASTM A830/A830M(98)	1536	Plt	0.30-0.38			1.20-1.55		0.035 max	0.04 max		bal Fe				
SAE J1397(92)	1536	Bar CD, est mech prop									bal Fe	630	530	12	187 HB
SAE J1397(92)	1536	Bar HR, est mech prop									bal Fe	570	310	16	163 HB

UNS numbers and US grades are provided as a means of cross referencing chemically similar alloys. Exchangability is only possible after independent examination of specifications. Tensile properties are minimum or typical as specified. UTS and YS as MPa. El as %. See Appendix for list of abbreviations used in Notes. * indicates obsolete material.

Specification	Designation	Notes	C	Cr	Cu	Mn	Ni	P	S	Si	Other	UTS	YS	El	Hard

Carbon Steel, High-Manganese, 1536 (Continued from previous page)

USA

Specification	Designation	Notes	C	Cr	Cu	Mn	Ni	P	S	Si	Other	UTS	YS	El	Hard
SAE J403(95)	1536	Struct Shps Plt Strp Sh Weld Tub; formerly 1036	0.30-0.38			1.20-1.55		0.030 max	0.035 max		bal Fe				

Carbon Steel, High-Manganese, 1541/1541H

Argentina

Specification	Designation	Notes	C	Cr	Cu	Mn	Ni	P	S	Si	Other	UTS	YS	El	Hard
IAS	IRAM 1541		0.36-0.44			1.35-1.65		0.040 max	0.050 max		bal Fe	720	480	12	217 HB

China

Specification	Designation	Notes	C	Cr	Cu	Mn	Ni	P	S	Si	Other	UTS	YS	El	Hard
GB 3077(88)	40Mn2	Bar HR Q/T	0.37-0.44	0.30 max	0.30 max	1.40-1.80	0.30 max	0.035 max	0.035 max	0.17-0.37	bal Fe	885	735	12	
GB 8162(87)	40Mn2	Smls Tub HR/CD Q/T	0.37-0.44	0.30 max		1.40-1.80	0.30 max	0.035 max	0.035 max	0.17-0.37	bal Fe	885	735	12	
GB/T 3078(94)	40Mn2	Bar CD Q/T 25mm diam	0.37-0.44	0.30 max	0.30 max	1.40-1.80	0.30 max	0.035 max	0.035 max	0.17-0.37	bal Fe	885	735	12	
YB/T 5052(93)	40Mn2	Tub HR Norm	0.37-0.44			1.40-1.80		0.040 max	0.040 max	0.20-0.40	bal Fe	685	490	12	

France

Specification	Designation	Notes	C	Cr	Cu	Mn	Ni	P	S	Si	Other	UTS	YS	El	Hard
AFNOR	40M5		0.36-0.44			1-1.35		0.04	0.035	0.1-0.4	bal Fe				
AFNOR	45M5		0.39-0.48			1.2-1.5		0.04	0.035	0.1-0.4	bal Fe				
AFNOR NFA35552	40M6		0.37-0.43			1.3-1.7		0.035	0.035	0.15-0.35	bal Fe				

Germany

Specification	Designation	Notes	C	Cr	Cu	Mn	Ni	P	S	Si	Other	UTS	YS	El	Hard
DIN	36Mn6		0.34-0.42			1.40-1.65		0.035 max	0.035 max	0.15-0.35	bal Fe				
DIN	E75		0.45			1.20		0.040 max	0.060 max	0.10-0.30	N <=0.007; bal Fe				
DIN	GS-40Mn5		0.36-0.44			1.20-1.50		0.035 max	0.035 max	0.30-0.50	bal Fe				
DIN	N-80		0.45			1.20		0.040 max	0.060 max	0.10-0.30	N <=0.007; bal Fe				
DIN	WNr 1.0563		0.45			1.20		0.040 max	0.060 max	0.10-0.30	N <=0.007; bal Fe				
DIN	WNr 1.0564		0.45			1.20		0.040 max	0.060 max	0.10-0.30	N <=0.007; bal Fe				
DIN	WNr 1.1127		0.34-0.42			1.40-1.65		0.035 max	0.035 max	0.15-0.35	bal Fe				
DIN	WNr 1.1168		0.36-0.44			1.20-1.50		0.035 max	0.035 max	0.30-0.50	bal Fe				
DIN 17204(90)	36Mn5	Smls tube Q/T 17-40mm	0.32-0.40			1.20-1.50		0.035 max	0.035 max	0.40 max	bal Fe	830-980	590	10	
DIN 17204(90)	36Mn5	Q/T 17-40mm	0.32-0.40			1.20-1.50		0.035 max	0.035 max	0.40 max	bal Fe	830-980	590	10	
DIN 17204(90)	WNr 1.1167	Tub Q/T 101-160mm	0.32-0.40			1.20-1.50		0.035 max	0.035 max	0.40 max	bal Fe	640-780	440	15	
DIN 17204(90)	WNr 1.1167	Tub, Smls tube Q/T 17-40mm	0.32-0.40			1.20-1.50		0.035 max	0.035 max	0.40 max	bal Fe	830-980	590	10	
DIN SEW 835(95)	36Mn5	For hardening	0.32-0.40			1.20-1.50		0.035 max	0.035 max	0.40 max	bal Fe				
DIN SEW 835(95)	WNr 1.1167	hardening	0.32-0.40			1.20-1.50		0.035 max	0.035 max	0.40 max	bal Fe				

India

Specification	Designation	Notes	C	Cr	Cu	Mn	Ni	P	S	Si	Other	UTS	YS	El	Hard
IS 1570	37Mn2		0.32-0.42			1.3-1.7				0.1-0.35	bal Fe				

Japan

Specification	Designation	Notes	C	Cr	Cu	Mn	Ni	P	S	Si	Other	UTS	YS	El	Hard
JIS G4052(79)	SMn2H*	Obs; see SMn 443H	0.35-0.41			1.35-1.65		0.03	0.03	0.15-0.35	bal Fe				
JIS G4052(79)	SMn3H*	Obs; see SMn 438H	0.4-0.46			1.35-1.65		0.03	0.03	0.15-0.35	bal Fe				
JIS G4052(79)	SMn438H	Struct Hard	0.34-0.41	0.35 max	0.30 max	1.30-1.70	0.25 max	0.030 max	0.030 max	0.15-0.35	bal Fe				
JIS G4052(79)	SMn443H	Struct Hard	0.39-0.46	0.35 max	0.30 max	1.30-1.70	0.25 max	0.030 max	0.030 max	0.15-0.35	bal Fe				
JIS G4106(79)	SMn 2*	Obs; see SMn438	0.35-0.41			1.35-1.65		0.03	0.03	0.15-0.35	bal Fe				

UNS numbers and US grades are provided as a means of cross referencing chemically similar alloys. Exchangability is only possible after independent examination of specifications. Tensile properties are minimum or typical as specified. UTS and YS as MPa. El as %. See Appendix for list of abbreviations used in Notes. * indicates obsolete material.

Specification	Designation	Notes	C	Cr	Cu	Mn	Ni	P	S	Si	Other	UTS	YS	El	Hard

Carbon Steel, High-Manganese, 1541/1541H (Continued from previous page)

Specification	Designation	Notes	C	Cr	Cu	Mn	Ni	P	S	Si	Other	UTS	YS	El	Hard
Japan															
JIS G4106(79)	SMn 3*	Obs; see SMn443	0.4-0.46			1.35-1.65		0.03	0.03	0.15-0.35	bal Fe				
JIS G4106(79)	SMn438	HR Frg	0.35-0.41	0.35 max	0.30 max	1.35-1.65	0.25 max	0.030 max	0.030 max	0.15-0.35	bal Fe				
JIS G4106(79)	SMn443	HR Frg	0.40-0.46	0.35 max	0.30 max	1.35-1.65	0.25 max	0.030 max	0.030 max	0.15-0.35	bal Fe				
Mexico															
NMX-B-301(86)	1541	Bar	0.36-0.44			1.35-1.65		0.040 max	0.050 max	0.15-0.35	bal Fe				
NMX-B-301(86)	1541H	Bar	0.35-0.45			1.25-1.75		0.040 max	0.050 max		Pb 0.15-0.30; bal Fe				
Pan America															
COPANT 333	1041	Wir rod	0.36-0.44			1.35-1.65		0.04	0.05		bal Fe				
Poland															
PNH83156(87)	L45G2		0.4-0.5	0.3	0.3	1.4-1.8	0.3	0.04	0.04	0.2-0.4	bal Fe				
PNH83160(88)	L45G	Wear res	0.4-0.5	0.3 max		1.4-1.8	0.3 max	0.04 max	0.04 max	0.2-0.4	bal Fe				
PNH84030	5G2		0.41-0.49	0.25	0.3	1.4-1.8	0.3	0.035	0.035	0.17-0.37	bal Fe				
Russia															
GOST 4543	35G2		0.31-0.39	0.3 max	0.3 max	1.4-1.8	0.3 max	0.04 max	0.04 max	0.18-0.37	bal Fe				
Sweden															
SS 142128	2128		0.43 max			1.2-1.8		0.04 max	0.04 max	0.1-0.4	P+S<=0.07				
UK															
BS 970/1(83)	135M44	Blm Bil Slab Bar Rod Frg	0.40-0.48			1.20-1.50		0.050 max	0.050 max		bal Fe				
BS 970/1(83)	150M40	Blm Bil Slab Bar Rod Frg	0.36-0.44			1.30-1.70		0.050 max	0.050 max		bal Fe				
USA															
	1541	CD	0.36-0.44			1.35-1.65		0.04 max	0.05 max		bal Fe	825	690	10.0	241-321 HB
	AISI 1541	Bar, was 1041	0.36-0.44			1.35-1.65		0.040 max	0.050 max		bal Fe				
	AISI 1541	Sh Strp Plt	0.36-0.45			1.30-1.65		0.030 max	0.035 max		bal Fe				
	AISI 1541	Struct shp	0.36-0.44			1.30-1.65		0.030 max	0.035 max		bal Fe				
	AISI 1541	Wir rod	0.36-0.44			1.35-1.65		0.040 max	0.050 max		bal Fe				
	AISI 1541H	Wir rod, Hard	0.35-0.45			1.25-1.75		0.040 max	0.050 max	0.15-0.35	bal Fe				
	UNS G15410		0.36-0.44			1.35-1.65		0.040 max	0.050 max		Sheets and Plates, C 0.36-0.45 Mn 1.30-1.65; bal Fe				
	UNS H15410		0.35-0.45			1.25-1.75		0.040 max	0.050 max	0.15-0.35	bal Fe				
ASTM A29/A29M(93)	1541	Bar, was 1041	0.36-0.44			1.35-1.65		0.040 max	0.050 max		bal Fe				
ASTM A304(96)	1541H	Bar, hard bands spec	0.35-0.45			1.25-1.75		0.040 max	0.050 max	0.15-0.35	Cu Ni Cr Mo trace allowed; bal Fe				
ASTM A311/A311M(95)	1541 Class A	Bar CD SR ann, 20<t<=30mm	0.36-0.44			1.35-1.65		0.040 max	0.050 max		bal Fe	655	585	11,RA=30	
ASTM A311/A311M(95)	1541 Class A	Bar CD SR ann, 30<t<=50mm	0.36-0.44			1.35-1.65		0.040 max	0.050 max		bal Fe	620	550	10,RA=30	
ASTM A311/A311M(95)	1541 Class A	Bar CD SR ann, 50<t<=75mm	0.36-0.44			1.35-1.65		0.040 max	0.050 max		bal Fe	585	520	10,RA=30	
ASTM A311/A311M(95)	1541 Class A	Bar CD SR ann, t<=20mm	0.36-0.44			1.35-1.65		0.040 max	0.050 max		bal Fe	690	620	11,RA=35	
ASTM A311/A311M(95)	1541 Class B	Bar CD SR ann, t<=50mm	0.36-0.44			1.35-1.65		0.040 max	0.050 max		bal Fe	795	690	10,RA=25	
ASTM A311/A311M(95)	1541 Class B	Bar CD SR ann, 30<t<=50mm	0.36-0.44			1.35-1.65		0.040 max	0.050 max		bal Fe	795	690	9,RA=20	
ASTM A311/A311M(95)	1541 Class B	Bar CD SR ann, 75<t<=115mm	0.36-0.44			1.35-1.65		0.040 max	0.050 max		bal Fe	725	620	7,RA=20	

UNS numbers and US grades are provided as a means of cross referencing chemically similar alloys. Exchangability is only possible after independent examination of specifications. Tensile properties are minimum or typical as specified. UTS and YS as MPa. El as %. See Appendix for list of abbreviations used in Notes. * indicates obsolete material.

Specification	Designation	Notes	C	Cr	Cu	Mn	Ni	P	S	Si	Other	UTS	YS	El	Hard

Carbon Steel, High-Manganese, 1541/1541H (Continued from previous page)

USA

Specification	Designation	Notes	C	Cr	Cu	Mn	Ni	P	S	Si	Other	UTS	YS	El	Hard
ASTM A510(96)	1541	Wir rod	0.36-0.44			1.35-1.65		0.040 max	0.050 max		bal Fe				
ASTM A519(96)	1541	Smls mech tub	0.36-0.44			1.35-1.65		0.040 max	0.050 max		bal Fe				
ASTM A576(95)	1541	Special qual HW bar	0.36-0.44			1.35-1.65		0.040 max	0.050 max		Si Cu Pb B Bi Ca Se Te if spec'd; bal Fe				
ASTM A830/A830M(98)	1541	Plt	0.36-0.45			1.30-1.65		0.035 max	0.04 max	0.15-0.40	bal Fe				
ASTM A866(94)	1541	Anti-friction bearings	0.36-0.44			1.35-1.65		0.025 max	0.025 max	0.15-0.35	bal Fe				
SAE J1268(95)	1541H	See std	0.35-0.45			1.25-1.75		0.030 max	0.050 max	0.15-0.35	bal Fe				
SAE J1397(92)	1541	Bar CD, est mech prop	0.36-0.44			1.35-1.65		0.04 max	0.05 max		bal Fe	710	600	10	207 HB
SAE J1397(92)	1541	Bar HR, est mech prop	0.36-0.44			1.35-1.65		0.04 max	0.05 max		bal Fe	630	350	15	187 HB
SAE J403(95)	1541	Bar Smls Tub HR CF	0.36-0.44			1.35-1.65		0.030 max	0.050 max		bal Fe				
SAE J403(95)	1541	Struct Shps Plt Strp Sh Weld Tub; formerly 1041	0.36-0.45			1.30-1.65		0.030 max	0.035 max		bal Fe				
SAE J775(93)	1541H	Engine poppet valve, Mart	0.35-0.45			1.25-1.75		0.040 max	0.050 max	0.15-0.35	bal Fe	970	910		
SAE J775(93)	NV1*	Engine poppet valve, Mart; former SAE grade	0.35-0.45			1.25-1.75		0.040 max	0.050 max	0.15-0.35	bal Fe	970	910		

Carbon Steel, High-Manganese, 1547

Mexico

Specification	Designation	Notes	C	Cr	Cu	Mn	Ni	P	S	Si	Other	UTS	YS	El	Hard
NMX-B-301(86)	1547	Bar	0.43-0.51			1.35-1.65		0.040 max	0.050 max		bal Fe				

Russia

Specification	Designation	Notes	C	Cr	Cu	Mn	Ni	P	S	Si	Other	UTS	YS	El	Hard
GOST 1414(75)	AS45G2	Free-cutting	0.40-0.48	0.3 max	0.25 max	1.35-1.65	0.4 max	0.04 max	0.24-0.35	0.15 max	Pb 0.15-0.3; bal Fe				

USA

Specification	Designation	Notes	C	Cr	Cu	Mn	Ni	P	S	Si	Other	UTS	YS	El	Hard
	UNS G15470		0.43-0.51			1.35-1.65		0.040 max	0.050 max		bal Fe				
ASTM A29/A29M(93)	1547	Bar	0.43-0.51			1.35-1.65		0.040 max	0.050 max		bal Fe				
ASTM A510(96)	1547	Wir rod	0.43-0.51			1.35-1.65		0.040 max	0.050 max		bal Fe				
ASTM A576(95)	1547	Special qual HW bar	0.43-0.51			1.35-1.65		0.040 max	0.050 max		Si Cu Pb B Bi Ca Se Te if spec'd; bal Fe				
SAE J775(93)	1547	Engine poppet valve, Mart	0.43-0.51			1.35-1.65		0.040 max	0.050 max		bal Fe	1240	1140		
SAE J775(93)	JIS G4106 SMnC 443	Engine poppet valve	0.40-0.46	0.35-0.70		1.35-1.65		0.030 max	0.030 max	0.15-0.35	bal Fe				
SAE J775(93)	NV2*	Engine poppet valve, Mart; former SAE grade	0.43-0.51			1.35-1.65		0.040 max	0.050 max		bal Fe	1240	1140		

Carbon Steel, High-Manganese, 1548

Argentina

Specification	Designation	Notes	C	Cr	Cu	Mn	Ni	P	S	Si	Other	UTS	YS	El	Hard
IAS	IRAM 1548		0.44-0.52			1.10-1.40		0.040 max	0.050 max		bal Fe	880	580	13	262 HB

China

Specification	Designation	Notes	C	Cr	Cu	Mn	Ni	P	S	Si	Other	UTS	YS	El	Hard
GB 11251(89)	45Mn2	Plt HR Ann	0.42-0.49			1.40-1.80		0.035 max	0.035 max	0.17-0.37	bal Fe	600-850		13	
GB 3077(88)	45Mn2	Bar HR Q/T 25mm diam	0.42-0.49	0.30 max	0.30 max	1.40-1.80	0.30 max	0.035 max	0.035 max	0.17-0.37	bal Fe	885	735	10	
GB 8162(87)	45Mn2	Smls Tub HR/CD Q/T	0.42-0.49	0.30 max		1.40-1.80	0.30 max	0.035 max	0.035 max	0.17-0.37	bal Fe	885	735	10	
GB/T 3078(94)	45Mn2	Bar CD Q/T 25mm diam	0.42-0.49	0.30 max	0.30 max	1.40-1.80	0.30 max	0.035 max	0.035 max	0.17-0.37	bal Fe	885	735	10	
YB/T 5132(93)	45Mn2A	Sh HR/CR Ann	0.42-0.49			1.40-1.80		0.025 max	0.025 max	0.17-0.37	bal Fe	590-835		12	

Specification	Designation	Notes	C	Cr	Cu	Mn	Ni	P	S	Si	Other	UTS	YS	El	Hard

Carbon Steel, High-Manganese, 1548 (Continued from previous page)

Germany

Specification	Designation	Notes	C	Cr	Cu	Mn	Ni	P	S	Si	Other	UTS	YS	El	Hard
DIN	46Mn5		0.42-0.48			1.15-1.35		0.035 max	0.035 max	0.25-0.45	bal Fe				
DIN	WNr 1.1128		0.42-0.48			1.15-1.35		0.035 max	0.035 max	0.25-0.45	bal Fe				

India

Specification	Designation	Notes	C	Cr	Cu	Mn	Ni	P	S	Si	Other	UTS	YS	El	Hard
IS 1570	47Mn2		0.42-0.52			1.3-1.7				0.1-0.35	bal Fe				

Mexico

Specification	Designation	Notes	C	Cr	Cu	Mn	Ni	P	S	Si	Other	UTS	YS	El	Hard
NMX-B-301(86)	1548	Bar	0.44-0.52			1.10-1.40		0.040 max	0.050 max		bal Fe				

USA

Specification	Designation	Notes	C	Cr	Cu	Mn	Ni	P	S	Si	Other	UTS	YS	El	Hard
	AISI 1548	Sh Strp Plt	0.43-0.52			1.05-1.40		0.030 max	0.035 max		bal Fe				
	AISI 1548	Bar, was 1048	0.44-0.52			1.10-1.40		0.040 max	0.050 max		bal Fe				
	AISI 1548	Wir rod	0.44-0.52			1.10-1.40		0.040 max	0.050 max		bal Fe				
	AISI 1548	Struct shp	0.44-0.52			1.05-1.40		0.030 max	0.035 max		bal Fe				
	UNS G15480		0.44-0.52			1.10-1.40		0.040 max	0.050 max		Sheets and Plates, C 0.43-0.52 Mn 1.05-1.40; bal Fe				
ASTM A29/A29M(93)	1548	Bar, was 1048	0.44-0.52			1.10-1.40		0.040 max	0.050 max		bal Fe				
ASTM A510(96)	1548	Wir rod	0.44-0.52			1.10-1.40		0.040 max	0.050 max		bal Fe				
ASTM A576(95)	1548	Special qual HW bar	0.44-0.52			1.10-1.40		0.040 max	0.050 max		Si Cu Pb B Bi Ca Se Te if spec'd; bal Fe				
ASTM A830/A830M(98)	1548	Plt	0.43-0.50			1.05-1.40		0.035 max	0.04 max	0.15-0.40	bal Fe				
SAE J1397(92)	1548	Bar Ann CD, est mech prop	0.44-0.52			1.1-1.4		0.04 max	0.05 max		bal Fe	640	540	10	193 HB
SAE J1397(92)	1548	Bar HR, est mech prop	0.44-0.52			1.1-1.4		0.04 max	0.05 max		bal Fe	660	370	14	197 HB
SAE J1397(92)	1548	Bar CD, est mech prop	0.44-0.52			1.1-1.4		0.04 max	0.05 max		bal Fe	730	620	10	217 HB
SAE J403(95)	1548	Struct Shps Plt Strp Sh Weld Tub; formerly 1048	0.43-0.52			1.05-1.40		0.030 max	0.035 max		bal Fe				
SAE J403(95)	1548	Bar Smls Tub HR CF	0.44-0.52			1.10-1.40		0.030 max	0.050 max		bal Fe				

Carbon Steel, High-Manganese, 1551

China

Specification	Designation	Notes	C	Cr	Cu	Mn	Ni	P	S	Si	Other	UTS	YS	El	Hard
GB 699(88)	50Mn	Bar HR Norm Q/T 25mm diam	0.48-0.56	0.25 max	0.25 max	0.70-1.00	0.25 max	0.035 max	0.035 max	0.17-0.37	bal Fe	645	390	13	
GB/T 3078(94)	50Mn	Bar CD Ann 25mm diam	0.48-0.56	0.25 max	0.25 max	0.70-1.00	0.25 max	0.035 max	0.035 max	0.17-0.37	bal Fe	590		10	
YB/T 5052(93)	50Mn	Tub HR Norm	0.48-0.56			0.70-1.00		0.040 max	0.040 max	0.17-0.37	bal Fe	640	395	14	

France

Specification	Designation	Notes	C	Cr	Cu	Mn	Ni	P	S	Si	Other	UTS	YS	El	Hard
AFNOR	24M4TS		0.49-0.55			0.8-1.1		0.025	0.035	0.1-0.4	bal Fe				

Germany

Specification	Designation	Notes	C	Cr	Cu	Mn	Ni	P	S	Si	Other	UTS	YS	El	Hard
DIN	R0800		0.45-0.65			0.80-1.30		0.040 max	0.040 max	0.10-0.50	bal Fe				
DIN	StSch80		0.45-0.65			0.80-1.30		0.040 max	0.040 max	0.10-0.50	bal Fe				
DIN	WNr 1.0524		0.45-0.65			0.80-1.30		0.040 max	0.040 max	0.10-0.50	bal Fe				

Hungary

Specification	Designation	Notes	C	Cr	Cu	Mn	Ni	P	S	Si	Other	UTS	YS	El	Hard
MSZ 2570(88)	MA1	Non-rimming; FF; 0-75.0mm	0.45-0.6			0.75-1.2		0.04 max	0.04 max	0.15-0.35	bal Fe	750		12L	

Mexico

Specification	Designation	Notes	C	Cr	Cu	Mn	Ni	P	S	Si	Other	UTS	YS	El	Hard
NMX-B-301(86)	1551	Bar	0.45-0.56			0.85-1.15		0.040 max	0.050 max		bal Fe				

UNS numbers and US grades are provided as a means of cross referencing chemically similar alloys. Exchangability is only possible after independent examination of specifications. Tensile properties are minimum or typical as specified. UTS and YS as MPa. El as %. See Appendix for list of abbreviations used in Notes. * indicates obsolete material.

Specification	Designation	Notes	C	Cr	Cu	Mn	Ni	P	S	Si	Other	UTS	YS	El	Hard

Carbon Steel, High-Manganese, 1551 (Continued from previous page)

Pan America
Specification	Designation	Notes	C	Cr	Cu	Mn	Ni	P	S	Si	Other	UTS	YS	El	Hard
COPANT 333	1051	Wir rod	0.45-0.56			0.85-1.15		0.04	0.05		bal Fe				

USA
Specification	Designation	Notes	C	Cr	Cu	Mn	Ni	P	S	Si	Other	UTS	YS	El	Hard
	AISI 1551	Wir rod	0.45-0.56			0.85-1.15		0.040 max	0.050 max		bal Fe				
	UNS G15510		0.45-0.56			0.85-1.15		0.040 max	0.050 max		bal Fe				
ASTM A29/A29M(93)	1551	Bar, was 1051	0.45-0.56			0.85-1.15		0.040 max	0.050 max		bal Fe				
ASTM A510(96)	1551	Wir rod	0.45-0.56			0.85-1.15		0.040 max	0.050 max		bal Fe				
ASTM A576(95)	1551	Special qual HW bar	0.45-0.56			0.85-1.15		0.040 max	0.050 max		Si Cu Pb B Bi Ca Se Te if spec'd; bal Fe				

Carbon Steel, High-Manganese, 1552

Bulgaria
Specification	Designation	Notes	C	Cr	Cu	Mn	Ni	P	S	Si	Other	UTS	YS	El	Hard
BDS 10786	MV2		0.5-0.65			1.3-1.7		0.04	0.045	0.5-0.65	bal Fe				

China
Specification	Designation	Notes	C	Cr	Cu	Mn	Ni	P	S	Si	Other	UTS	YS	El	Hard
GB 3077(88)	50Mn2	Bar HR Q/T 25mm diam	0.47-0.55	0.30	0.30 max	1.40-1.80	0.30	0.035 max	0.035 max	0.17-0.37	bal Fe	930	785	9	
GB/T 3078(94)	50Mn2	Bar CD Ann 25mm diam	0.47-0.55	0.30	0.30 max	1.40-1.80	0.30	0.035 max	0.035 max	0.17-0.37	bal Fe	635		9	

France
Specification	Designation	Notes	C	Cr	Cu	Mn	Ni	P	S	Si	Other	UTS	YS	El	Hard
AFNOR	55M5		0.5-0.6			1.2-1.5		0.04	0.035	0.1-0.4	bal Fe				

Germany
Specification	Designation	Notes	C	Cr	Cu	Mn	Ni	P	S	Si	Other	UTS	YS	El	Hard
DIN	R0900Mn		0.55-0.75			1.30-1.70		0.040 max	0.040 max	0.10-0.50	bal Fe				
DIN	StSch900B		0.55-0.5			1.30-1.70		0.040 max	0.040 max	0.10-0.50	bal Fe				
DIN	WNr 1.0624		0.55-0.75			1.30-1.70		0.040 max	0.040 max	0.10-0.50	bal Fe				

India
Specification	Designation	Notes	C	Cr	Cu	Mn	Ni	P	S	Si	Other	UTS	YS	El	Hard
IS 1570	C50Mn1		0.45-0.55			1.1-1.4					bal Fe				
IS 1570/2(79)	50C12	Sh Plt Strp Tub Pip	0.45-0.55	0.3 max	0.3 max	1.1-1.4	0.4 max	0.055 max	0.055 max	0.6 max	Co <=0.1; Mo <=0.15; Pb <=0.15; W <=0.1; P:S Varies; bal Fe	720	396	11	
IS 1570/2(79)	C50Mn1	Sh Plt Strp Tub Pip	0.45-0.55	0.3 max	0.3 max	1.1-1.4	0.4 max	0.055 max	0.055 max	0.6 max	Co <=0.1; Mo <=0.15; Pb <=0.15; W <=0.1; P:S Varies; bal Fe	720	396	11	

Mexico
Specification	Designation	Notes	C	Cr	Cu	Mn	Ni	P	S	Si	Other	UTS	YS	El	Hard
NMX-B-301(86)	1552	Bar	0.47-0.55			1.20-1.50		0.040 max	0.050 max		bal Fe				

Poland
Specification	Designation	Notes	C	Cr	Cu	Mn	Ni	P	S	Si	Other	UTS	YS	El	Hard
PNH93421	St90PB		0.5-0.65			1.3-1.7		0.05	0.05	0.5	bal Fe				

Russia
Specification	Designation	Notes	C	Cr	Cu	Mn	Ni	P	S	Si	Other	UTS	YS	El	Hard
GOST 4543	50G2		0.46-0.55	0.3	0.3	1.4-1.8	0.3	0.035 max	0.035 max	0.17-0.37	bal Fe				
GOST 4543(71)	50G2	Q/T	0.46-0.55	0.3 max	0.3 max	1.4-1.8	0.3 max	0.035 max	0.035 max	0.17-0.37	Al <=0.1; bal Fe				

Spain
Specification	Designation	Notes	C	Cr	Cu	Mn	Ni	P	S	Si	Other	UTS	YS	El	Hard
UNE	F.120.O		0.47-0.53			1.2-1.6		0.035	0.035	0.13-0.38	bal Fe				

USA
Specification	Designation	Notes	C	Cr	Cu	Mn	Ni	P	S	Si	Other	UTS	YS	El	Hard
	AISI 1552	Bar, was 1052	0.47-0.55			1.20-1.50		0.040 max	0.050 max		bal Fe				
	AISI 1552	Wir rod	0.47-0.55			1.20-1.50		0.040 max	0.050 max		bal Fe				
	AISI 1552	Sh Strp Plt	0.46-0.55			1.20-1.55		0.030 max	0.035 max		bal Fe				
	AISI 1552	Struct shp	0.47-0.55			1.20-1.55		0.030 max	0.035 max		bal Fe				

UNS numbers and US grades are provided as a means of cross referencing chemically similar alloys. Exchangability is only possible after independent examination of specifications. Tensile properties are minimum or typical as specified. UTS and YS as MPa. El as %. See Appendix for list of abbreviations used in Notes. * indicates obsolete material.

Specification	Designation	Notes	C	Cr	Cu	Mn	Ni	P	S	Si	Other	UTS	YS	El	Hard

Carbon Steel, High-Manganese, 1552 (Continued from previous page)

USA

Specification	Designation	Notes	C	Cr	Cu	Mn	Ni	P	S	Si	Other	UTS	YS	El	Hard
	UNS G15520		0.47-0.55			1.20-1.50		0.040 max	0.050 max		Sheets and Plates, C 0.46-0.55 Mn 1.20-1.55; bal Fe				
ASTM A29/A29M(93)	1552	Bar, was 1052	0.47-0.55			1.20-1.50		0.040 max	0.050 max		bal Fe				
ASTM A510(96)	1552	Wir rod	0.47-0.55			1.20-1.50		0.040 max	0.050 max		bal Fe				
ASTM A576(95)	1552	Special qual HW bar	0.47-0.55			1.20-1.50		0.040 max	0.050 max		Si Cu Pb B Bi Ca Se Te if spec'd; bal Fe				
ASTM A830/A830M(98)	1552	Plt	0.46-0.55			1.20-1.55		0.035 max	0.04 max	0.15-0.40	bal Fe				
ASTM A866(94)	1552	Anti-friction bearings	0.47-0.55			1.20-1.50		0.025 max	0.025 max	0.15-0.35	bal Fe				
FED QQ-S-633A(63)	1552*	Obs; CF HR Bar	0.47-0.55			1.2-1.5		0.04 max	0.05 max		bal Fe				
SAE J1397(92)	1552	Bar Ann CD, est mech prop	0.47-0.55			1.2-1.5		0.04 max	0.05 max		bal Fe	680	570	10	193 HB
SAE J1397(92)	1552	Bar HR, est mech prop	0.47-0.55			1.2-1.5		0.04 max	0.05 max		bal Fe	740	410	12	217 HB
SAE J403(95)	1552	Bar Smls Tub HR CF	0.47-0.55			1.20-1.50		0.030 max	0.050 max		bal Fe				
SAE J403(95)	1552	Struct Shps Plt Strp Sh Weld Tub; formerly 1052	0.46-0.55			1.20-1.55		0.030 max	0.035 max		bal Fe				

Carbon Steel, High-Manganese, 1553

USA

Specification	Designation	Notes	C	Cr	Cu	Mn	Ni	P	S	Si	Other	UTS	YS	El	Hard
	AISI 1553	Bar Blm Bil Slab	0.48-0.55			0.80-1.10		0.040 max	0.050 max		bal Fe				
	UNS G15530		0.48-0.56			0.80-1.10		0.040 max	0.050 max		bal Fe				

Carbon Steel, High-Manganese, 1561

China

Specification	Designation	Notes	C	Cr	Cu	Mn	Ni	P	S	Si	Other	UTS	YS	El	Hard
GB 699(88)	60Mn	Bar HR Norm 25mm diam	0.57-0.65	0.25 max	0.25 max	0.70-1.00	0.25 max	0.035 max	0.035 max	0.17-0.37	bal Fe	695	410	11	
GB/T 3078(94)	60Mn	Bar CR Norm 25mm diam	0.57-0.65	0.25 max	0.25 max	0.70-1.00	0.25 max	0.035 max	0.035 max	0.17-0.37	bal Fe	695	410	11	
YB/T 5101(93)	60Mn	Wir CD 0.10-2.00mm diam	0.58-0.64	0.10 max	0.20 max	0.70-1.00	0.15 max	0.025 max	0.020 max	0.17-0.37	bal Fe	1815-3090			
YB/T 5103(93)	60Mn	Wir CD Q/T 2.0-4.0mm diam	0.57-0.65	0.25 max	0.25 max	0.70-1.00	0.25 max	0.035 max	0.035 max	0.17-0.37	bal Fe	1425-1765			

Germany

Specification	Designation	Notes	C	Cr	Cu	Mn	Ni	P	S	Si	Other	UTS	YS	El	Hard
DIN	WNr 1.0908*	Obs	0.55-0.65			0.9-1.1		0.05 max	0.05 max	1-1.3	N 0.007; bal Fe				

Mexico

Specification	Designation	Notes	C	Cr	Cu	Mn	Ni	P	S	Si	Other	UTS	YS	El	Hard
NMX-B-301(86)	1561	Bar	0.55-0.65			0.75-1.05		0.040 max	0.050 max		bal Fe				

Pan America

Specification	Designation	Notes	C	Cr	Cu	Mn	Ni	P	S	Si	Other	UTS	YS	El	Hard
COPANT 333	1061	Wir rod	0.54-0.65			0.75-1.05		0.04	0.05		bal Fe				
COPANT 333	1062	Wir rod	0.54-0.65			0.85-1.15		0.04	0.05		bal Fe				

Russia

Specification	Designation	Notes	C	Cr	Cu	Mn	Ni	P	S	Si	Other	UTS	YS	El	Hard
GOST 1050	60G		0.57-0.65	0.25	0.25	0.7-1	0.25	0.035	0.04	0.17-0.37	bal Fe				
GOST 14959	60G		0.57-0.65	0.25	0.25	0.7-1	0.25	0.035	0.04	0.17-0.37	bal Fe				
GOST 4121	K62		0.5-0.73			0.6-1		0.055	0.05	0.15-0.3	bal Fe				

USA

Specification	Designation	Notes	C	Cr	Cu	Mn	Ni	P	S	Si	Other	UTS	YS	El	Hard
	AISI 1561	Wir rod	0.55-0.65			0.75-1.05		0.040 max	0.050 max		bal Fe				
	UNS G15610		0.55-0.65			0.75-1.05		0.040 max	0.050 max		bal Fe				
ASTM A29/A29M(93)	1561	Bar, was 1061	0.55-0.65			0.75-1.05		0.040 max	0.050 max		bal Fe				

Specification	Designation	Notes	C	Cr	Cu	Mn	Ni	P	S	Si	Other	UTS	YS	El	Hard

Carbon Steel, High-Manganese, 1561 (Continued from previous page)

USA

Specification	Designation	Notes	C	Cr	Cu	Mn	Ni	P	S	Si	Other	UTS	YS	El	Hard
ASTM A510(96)	1561	Wir rod	0.55-0.65			0.75-1.05		0.040 max	0.050 max		bal Fe				
ASTM A576(95)	1561	Special qual HW bar	0.55-0.65			0.75-1.05		0.040 max	0.050 max		Si Cu Pb B Bi Ca Se Te if spec'd; bal Fe				
ASTM A713(93)	1561	Wir, High-C spring, HT	0.55-0.65			0.75-1.05		0.040 max	0.050 max		bal Fe				

Carbon Steel, High-Manganese, 1566

Bulgaria

Specification	Designation	Notes	C	Cr	Cu	Mn	Ni	P	S	Si	Other	UTS	YS	El	Hard
BDS 10786	MV1		0.6-0.75			0.8-1.3		0.04	0.045	0.15-0.5	bal Fe				
BDS 6742	65G	Spring	0.62-0.70	0.25 max	0.30 max	0.90-1.20	0.30 max	0.035 max	0.035 max	0.17-0.37	bal Fe				

China

Specification	Designation	Notes	C	Cr	Cu	Mn	Ni	P	S	Si	Other	UTS	YS	El	Hard
GB 1222(84)	65Mn	Bar Flat HR	0.62-0.70	0.25 max	0.25 max	0.90-1.20	0.25 max	0.035 max	0.035 max	0.17-0.37	bal Fe	980	785	8	
GB 3279(89)	65Mn	Sh HR Ann	0.62-0.70	0.25 max	0.25 max	0.90-1.20	0.25 max	0.035 max	0.035 max	0.17-0.37	bal Fe	850		12	
GB 699(88)	65Mn	Bar HR Norm 25mm diam	0.62-0.70	0.25 max	0.25 max	0.90-1.20	0.25 max	0.035 max	0.035 max	0.17-0.37	bal Fe	735	430	9	
GB/T 3078(94)	65Mn	Bar CD Norm 25mm diam	0.62-0.70	0.25 max	0.25 max	0.90-1.20	0.25 max	0.035 max	0.035 max	0.17-0.37	bal Fe	735	430	9	
GB/T 4358(95)	65Mn	Wir CD 0.10-2.00mm diam	0.62-0.70	0.10 max	0.20 max	0.70-1.00	0.15 max	0.025 max	0.020 max	0.17-0.37	bal Fe	1760-3040			
YB/T 5063(93)	65Mn	Strp HR HT	0.62-0.70	0.25 max	0.25 max	0.90-1.20	0.25 max	0.035 max	0.035 max	0.17-0.37	bal Fe	1275-1570			
YB/T 5101(93)	65Mn	Wir CD 0.10-2.00mm diam	0.63-0.69	0.10 max	0.20 max	0.90-1.20	0.15 max	0.025 max	0.020 max	0.17-0.37	bal Fe	1815-3090			
YB/T 5103(93)	65Mn	Wir CD Q/T 2.0-4.0mm diam	0.62-0.70	0.25 max	0.25 max	0.90-1.20	0.25 max	0.035 max	0.035 max	0.17-0.37	bal Fe	1425-1765			

Germany

Specification	Designation	Notes	C	Cr	Cu	Mn	Ni	P	S	Si	Other	UTS	YS	El	Hard
DIN	65Mn4		0.60-0.70			0.90-1.20		0.035 max	0.035 max	0.25-0.50	bal Fe				
DIN	66Mn4		0.60-0.71			0.85-1.15		0.035 max	0.035 max	0.15-0.30	bal Fe				
DIN	WNr 1.1233		0.60-0.70			0.70-1.20		0.035 max	0.035 max	0.20-0.65	bal Fe				
DIN	WNr 1.1240		0.60-0.70			0.90-1.20		0.035 max	0.035 max	0.25-0.50	bal Fe				
DIN	WNr 1.1260		0.60-0.71			0.85-1.15		0.035 max	0.035 max	0.15-0.30	bal Fe				

Mexico

Specification	Designation	Notes	C	Cr	Cu	Mn	Ni	P	S	Si	Other	UTS	YS	El	Hard
NMX-B-301(86)	1566	Bar	0.60-0.71			0.85-1.15		0.040 max	0.050 max		bal Fe				

Pan America

Specification	Designation	Notes	C	Cr	Cu	Mn	Ni	P	S	Si	Other	UTS	YS	El	Hard
COPANT 333	1066	Wir rod	0.6-0.71			0.85-1.15		0.04	0.05		bal Fe				

Poland

Specification	Designation	Notes	C	Cr	Cu	Mn	Ni	P	S	Si	Other	UTS	YS	El	Hard
PNH84032	65G	Spring	0.6-0.7	0.3 max	0.25 max	0.9-1.2	0.3 max	0.04 max	0.04 max	0.15-0.4	bal Fe				
PNH93421	St90PA		0.6-0.75			0.8-1.3		0.05	0.05	0.5	bal Fe				

Romania

Specification	Designation	Notes	C	Cr	Cu	Mn	Ni	P	S	Si	Other	UTS	YS	El	Hard
STAS 791	65Mn10		0.6-0.7	0.25	0.3	0.9-1.2	0.3	0.035	0.035	0.17-0.37	bal Fe				
STAS 791	65Mn10S		0.6-0.7	0.25	0.3	0.9-1.2	0.3	0.035	0.035	0.17-0.37	bal Fe				
STAS 791	65Mn10X		0.6-0.7	0.25	0.3	0.9-1.2	0.3	0.035	0.035	0.17-0.37	bal Fe				
STAS 791	65Mn10XS		0.6-0.7	0.25	0.3	0.9-1.2	0.3	0.035	0.035	0.17-0.37	bal Fe				

Russia

Specification	Designation	Notes	C	Cr	Cu	Mn	Ni	P	S	Si	Other	UTS	YS	El	Hard
GOST 14959	65G		0.62-0.7	0.25	0.25	0.9-1.2	0.25	0.035	0.035	0.17-0.37	bal Fe				
GOST 18267	M75		0.67-0.78			0.75-1.05		0.035	0.045	0.18-0.33	bal Fe				

UNS numbers and US grades are provided as a means of cross referencing chemically similar alloys. Exchangability is only possible after independent examination of specifications. Tensile properties are minimum or typical as specified. UTS and YS as MPa. El as %. See Appendix for list of abbreviations used in Notes. * indicates obsolete material.

Specification	Designation	Notes	C	Cr	Cu	Mn	Ni	P	S	Si	Other	UTS	YS	El	Hard

Carbon Steel, High-Manganese, 1566 (Continued from previous page)

Russia

Specification	Designation	Notes	C	Cr	Cu	Mn	Ni	P	S	Si	Other	UTS	YS	El	Hard
GOST 9660	M72		0.66-0.77			0.7-1.1		0.035	0.04	0.13-0.28	bal Fe				

USA

Specification	Designation	Notes	C	Cr	Cu	Mn	Ni	P	S	Si	Other	UTS	YS	El	Hard
	AISI 1566	Wir rod	0.60-0.71			0.85-1.15		0.040 max	0.050 max		bal Fe				
	AISI 1566	Bar, was 1066	0.6-0.71			0.85-1.15		0.040 max	0.050 max		bal Fe				
	UNS G15660		0.60-0.71			0.85-1.15		0.040 max	0.050 max		bal Fe				
ASTM A29/A29M(93)	1566	Bar, was 1066	0.6-0.71			0.85-1.15		0.040 max	0.050 max		bal Fe				
ASTM A510(96)	1566	Wir rod	0.60-0.71			0.85-1.15		0.040 max	0.050 max		bal Fe				
ASTM A576(95)	1566	Special qual HW bar	0.60-0.71			0.85-1.15		0.040 max	0.050 max		Si Cu Pb B Bi Ca Se Te if spec'd; bal Fe				
ASTM A713(93)	1566	Wir, High-C spring, HT	0.60-0.71			0.85-1.15		0.040 max	0.050 max		bal Fe				
FED QQ-W-428B(91)	Type I Class 1	Spring wir Q/T 0.207 in.	0.55-0.85			0.80-1.20		0.040 max	0.050 max	0.10-0.35	bal Fe	1310			
FED QQ-W-428B(91)	Type I Class 2	Spring wir Q/T 0.207 in.	0.55-0.85			0.80-1.20		0.040 max	0.050 max	0.10-0.35	bal Fe	1490			
FED QQ-W-428B(91)	Type II Class 1	Spring wir CD 0.207 in.	0.55-0.85			0.80-1.20		0.040 max	0.050 max	0.10-0.35	bal Fe	1310			
FED QQ-W-428B(91)	Type II Class 1	Spring wir CD 0.207 in.	0.45-0.85			0.60-1.30		0.040 max	0.050 max	0.10-0.30	bal Fe	1310			
FED QQ-W-428B(91)	Type II Class 2	Spring wir CD 0.207 in.	0.55-0.85			0.80-1.20		0.040 max	0.050 max	0.10-0.35	bal Fe	1510			
FED QQ-W-428B(91)	Type II Class 2	Spring wir CD 0.207 in.	0.45-0.85			0.60-1.30		0.040 max	0.050 max	0.10-0.30	bal Fe	1510			
FED QQ-W-428B(91)	Type III	Spring wir CF hard	0.55-0.85			0.60-1.20		0.040 max	0.050 max	0.10-0.35	bal Fe				
SAE J403(95)	1566	Bar Smls Tub HR CF	0.60-0.71			0.85-1.15		0.030 max	0.050 max		bal Fe				

Carbon Steel, High-Manganese, 1570

China

Specification	Designation	Notes	C	Cr	Cu	Mn	Ni	P	S	Si	Other	UTS	YS	El	Hard
GB 1222(84)	65Mn	Bar Flat HR Q/T	0.62-0.70			0.90-1.20		0.035 max	0.035 max	0.17-0.37	bal Fe	980	785	8	
GB 3279(89)	65Mn	Sh HR Ann	0.62-0.70			0.90-1.20		0.035 max	0.035 max	0.17-0.37	bal Fe	850		12	
GB 699(88)	65Mn	Bar HR Norm 25mm diam	0.62-0.70			0.90-1.20		0.035 max	0.035 max	0.17-0.37	bal Fe	735	430	9	
GB/T 3078(94)	65Mn	Bar CD Norm 25mm diam	0.62-0.70			0.90-1.20		0.035 max	0.035 max	0.17-0.37	bal Fe	735	430	9	
GB/T 4358(95)	65Mn	Wir CD 0.10-2.00mm diam	0.62-0.70		0.20 max	0.70-1.00		0.025 max	0.020 max	0.17-0.37	bal Fe	1760-3040			
YB/T 5063(93)	65Mn	Strp HR HT	0.62-0.70			0.90-1.20		0.035 max	0.035 max	0.17-0.37	bal Fe	1275-1570			
YB/T 5101(93)	65Mn	Wir CD 0.10-2.00mm diam	0.63-0.69		0.20 max	0.90-1.20		0.025 max	0.020 max	0.17-0.37	bal Fe	1815-3090			
YB/T 5103(93)	65Mn	Wir CD Q/T 2.0-4.0mm diam	0.62-0.70			0.90-1.20		0.035 max	0.035 max	0.17-0.37	bal Fe	1425-1765			

USA

Specification	Designation	Notes	C	Cr	Cu	Mn	Ni	P	S	Si	Other	UTS	YS	El	Hard
	AISI 1570	Bar Blm Bil Slab	0.65-0.75			0.80-1.10		0.040 max	0.050 max		bal Fe				
	UNS G15700		0.65-0.75			0.80-1.10		0.040 max	0.050 max		bal Fe				
SAE J403	1570*		0.65-0.75			0.8-1.1		0.04 max	0.05 max		bal Fe				

Carbon Steel, High-Manganese, 1572

China

Specification	Designation	Notes	C	Cr	Cu	Mn	Ni	P	S	Si	Other	UTS	YS	El	Hard
GB 699(88)	70Mn	Bar HR Norm 25mm diam	0.67-0.75	0.25 max	0.25 max	0.90-1.20	0.25 max	0.035 max	0.035 max	0.17-0.37	bal Fe	785	450	8	
YB/T 5101(93)	70Mn	Wir CD 0.10-2.00mm diam	0.68-0.74	0.10 max	0.20 max	0.90-1.20	0.15 max	0.025 max	0.020 max	0.17-0.37	bal Fe	1815-3090			

Specification	Designation	Notes	C	Cr	Cu	Mn	Ni	P	S	Si	Other	UTS	YS	El	Hard

Carbon Steel, High-Manganese, 1572 (Continued from previous page)

China

Specification	Designation	Notes	C	Cr	Cu	Mn	Ni	P	S	Si	Other	UTS	YS	El	Hard
YB/T 5103(93)	70Mn	Wir CD Q/T 2.0-4.0mm diam	0.67-0.75	0.25 max	0.25 max	0.90-1.20	0.25 max	0.035 max	0.035 max	0.17-0.37	bal Fe	1425-1765			

Mexico

NMX-B-301(86)	1572	Bar	0.65-0.76			1.00-1.30		0.040 max	0.050 max		bal Fe				

USA

	UNS G15720		0.65-0.76			1.00-1.30		0.040 max	0.050 max		bal Fe				
ASTM A29/A29M(93)	1572	Bar, was 1072	0.65-0.76			1.00-1.30		0.040 max	0.050 max		bal Fe				
ASTM A510(96)	1572	Wir rod	0.65-0.76			1.00-1.30		0.040 max	0.050 max		bal Fe				
ASTM A576(95)	1572	Special qual HW bar	0.65-0.76			1.00-1.30		0.040 max	0.050 max		Si Cu Pb B Bi Ca Se Te if spec'd; bal Fe				
ASTM A713(93)	1572	Wir, High-C spring, HT	0.65-0.76			1.00-1.30		0.040 max	0.050 max		bal Fe				

Carbon Steel, High-Manganese, 1580

China

GB 1299(85)	8MnSi	Bar HR Norm	0.75-0.84			0.80-1.10		0.035 max	0.035 max	0.30-0.60	bal Fe	735	430	9	

USA

	AISI 1580	Bar Blm Bil Slab	0.75-0.88			0.80-1.10		0.040 max	0.050 max		bal Fe				
	UNS G15800		0.75-0.85			0.80-1.10		0.040 max	0.050 max		bal Fe				
SAE J403	1580*		0.75-0.85			0.8-1.1		0.04 max	0.05 max		bal Fe				

Carbon Steel, High-Manganese, 1590

USA

	AISI 1590	Bar Blm Bil Slab, UNS no. obs	0.85-0.98			0.80-1.10		0.040 max	0.050 max		bal Fe				
	UNS G15900		0.85-0.98			0.80-1.10		0.040 max	0.050 max		bal Fe				
SAE J403	1590*		0.85-0.98			0.8-1.1		0.04 max	0.05 max		bal Fe				

Carbon Steel, High-Manganese, 15B21H

China

GB 3077(88)	20Mn2B	Bar HR 15mm diam	0.17-0.24	0.30 max	0.30 max	1.50-1.80	0.30 max	0.035 max	0.035 max	0.17-0.37	B 0.0005-0.0035; bal Fe	980	785	10	
GB 5216(85)	20MnVBH	Bar Rod HR/Frg Ann	0.17-0.23	0.35 max		1.05-1.45	0.30 max	0.035 max	0.035 max	0.17-0.37	B 0.0005-0.0035; V 0.07-0.12; bal Fe				207 max HB
GB 8162(87)	20Mn2B	Smls Tub HR/CD Q/T	0.17-0.24	0.30 max		1.50-1.80	0.30 max	0.035 max	0.035 max	0.17-0.37	B 0.0005-0.0035; bal Fe	980	785	10	

Europe

EN 10083/3(95)	1.5530	Boron, Q/T	0.17-0.23	0.30 max		1.10-1.40		0.035 max	0.040 max	0.40 max	B 0.0008-0.0050; bal Fe				
EN 10083/3(95)	20MnB5	Boron, Q/T	0.17-0.23	0.30 max		1.10-1.40		0.035 max	0.040 max	0.40 max	B 0.0008-0.0050; bal Fe				

France

AFNOR NFA35552	21B3	Bar Wir rod	0.18-0.24			0.6-0.9		0.04 max	0.04 max	0.1-0.4	B 0.001-0.005; bal Fe				

Germany

DIN	WNr 1.5523	Cold ext for Q/T	0.17-0.24			0.80-1.15		0.035 max	0.035 max	0.40 max	B 0.0008-0.005; bal Fe				
DIN 1654(89)	19MnB4	Cold ext for Q/T	0.17-0.24			0.80-1.15		0.035 max	0.035 max	0.40 max	B 0.0008-0.0050; bal Fe				

Spain

UNE 36034(85)	20MnB4DF	CHd; Cold ext	0.17-0.23			0.8-1.1		0.035 max	0.035 max	0.15-0.35	B 0.0008-0.005; Mo <=0.15; Al 0.02; bal Fe				
UNE 36034(85)	20MnB5DF	CHd; Cold ext	0.17-0.23			1.1-1.4		0.035 max	0.035 max	0.15-0.35	B 0.0008-0.005; bal Fe				

UNS numbers and US grades are provided as a means of cross referencing chemically similar alloys. Exchangability is only possible after independent examination of specifications. Tensile properties are minimum or typical as specified. UTS and YS as MPa. El as %. See Appendix for list of abbreviations used in Notes. * indicates obsolete material.

Specification	Designation	Notes	C	Cr	Cu	Mn	Ni	P	S	Si	Other	UTS	YS	El	Hard

Carbon Steel, High-Manganese, 15B21H (Continued from previous page)

Spain

Specification	Designation	Notes	C	Cr	Cu	Mn	Ni	P	S	Si	Other	UTS	YS	El	Hard
UNE 36034(85)	F.1292*	see 20MnB4DF	0.17-0.23			0.8-1.1		0.035 max	0.035 max	0.15-0.35	B 0.0008-0.005; Mo <=0.15; Al 0.02; bal Fe				
UNE 36034(85)	F.1293*	see 20MnB5DF	0.17-0.23			1.1-1.4		0.035 max	0.035 max	0.15-0.35	B 0.0008-0.005; bal Fe				

UK

Specification	Designation	Notes	C	Cr	Cu	Mn	Ni	P	S	Si	Other	UTS	YS	El	Hard
BS 970/1(83)	170H15	Blm Bil Slab Bar Rod Frg CH	0.12-0.18			0.80-1.10		0.060 max	0.03-0.06		B 0.0005-0.005; bal Fe				
BS 970/1(83)	170H20	Blm Bil Slab Bar Rod Frg	0.17-0.23			0.80-1.10		0.050 max	0.050 max		B 0.0005-0.005; Mo <=0.15; bal Fe				

USA

Specification	Designation	Notes	C	Cr	Cu	Mn	Ni	P	S	Si	Other	UTS	YS	El	Hard
	AISI 15B21H	Wir rod, Hard	0.17-0.24			0.70-1.20		0.040 max	0.050 max	0.15-0.35	B 0.0005-0.003; bal Fe				
	UNS H15211		0.17-0.24			0.70-1.20		0.040 max	0.050 max	0.15-0.35	B 0.0005-0.003; bal Fe				
ASTM A304(96)	15B21H	Bar, hard bands spec	0.17-0.24			0.70-1.20		0.040 max	0.050 max	0.15-0.30	B >=0.0005; Cu Ni Cr Mo trace allowed; bal Fe				
ASTM A914/914M(92)	15B21RH	Bar restricted end Q hard	0.17-0.22	0.20 max	0.35 max	0.80-1.10	0.25 max	0.035 max	0.040 max	0.15-0.35	B 0.0005-0.003; Mo <=0.06; Electric furnace P, S<=0.025; bal Fe				
SAE J1268(95)	15B21H	Bar Rod Wir; See std	0.17-0.24			0.70-1.20		0.030 max	0.050 max	0.15-0.35	B 0.0005-0.003; bal Fe				
SAE J1868(93)	15B21RH	Restrict hard range	0.17-0.22	0.20 max	0.035 max	0.80-1.10	0.25 max	0.025 max	0.040 max	0.15-0.35	B 0.0005-0.003; Mo <=0.06; Cu, Ni, Cr, Mo not spec'd but acpt; bal Fe				5 HRC max
SAE J776	15B21H*	Bar Rod Wir; obs see J1268	0.17-0.24			0.7-1.2		0.04 max	0.05 max	0.15-0.35	B 0.0005-0.003; bal Fe				

Carbon Steel, High-Manganese, 15B35H

Europe

Specification	Designation	Notes	C	Cr	Cu	Mn	Ni	P	S	Si	Other	UTS	YS	El	Hard
EN 10083/3(95)	1.5531	Boron, Q/T	0.27-0.33	0.30 max		1.15-1.45		0.035 max	0.040 max	0.40 max	B 0.0008-0.0050; bal Fe				
EN 10083/3(95)	30MnB5	Boron, Q/T	0.27-0.33	0.30 max		1.15-1.45		0.035 max	0.040 max	0.40 max	B 0.0008-0.0050; bal Fe				

Italy

Specification	Designation	Notes	C	Cr	Cu	Mn	Ni	P	S	Si	Other	UTS	YS	El	Hard
UNI 7356(74)	C35BKB	Q/T	0.34-0.39			0.5-0.8		0.035 max	0.035 max	0.15-0.4	B 0.001-0.005; bal Fe				

Mexico

Specification	Designation	Notes	C	Cr	Cu	Mn	Ni	P	S	Si	Other	UTS	YS	El	Hard
NMX-B-301(86)	15B35H	Bar	0.31-0.39			0.70-1.20		0.040 max	0.050 max	0.15-0.30	B >=0.0005; bal Fe				

UK

Specification	Designation	Notes	C	Cr	Cu	Mn	Ni	P	S	Si	Other	UTS	YS	El	Hard
BS 970/1(83)	170H36	Blm Bil Slab Bar Rod Frg	0.32-0.39			0.80-1.10		0.050 max	0.050 max		B 0.0005-0.005; Mo <=0.15; P<=0.05 S=0.025-0.05/P=0.025 S=0.015-0.04; bal Fe				

USA

Specification	Designation	Notes	C	Cr	Cu	Mn	Ni	P	S	Si	Other	UTS	YS	El	Hard
	AISI 15B35H	Wir rod, Hard	0.31-0.39			0.70-1.20		0.040 max	0.050 max	0.15-0.35	B 0.0005-0.003; bal Fe				
	UNS H15351		0.31-0.39			0.70-1.20		0.040 max	0.050 max	0.15-0.35	B 0.0005-0.003; bal Fe				
ASTM A304(96)	15B35H	Bar, hard bands spec	0.31-0.39			0.70-1.20		0.040 max	0.050 max	0.15-0.30	B >=0.0005; Cu Ni Cr Mo trace allowed; bal Fe				
ASTM A914/914M(92)	15B35RH	Bar restricted end Q hard	0.33-0.38	0.20 max	0.35 max	0.80-1.10	0.25 max	0.035 max	0.040 max	0.15-0.35	B 0.0005-0.003; Mo <=0.06; Electric furnace P, S<=0.025; bal Fe				
SAE J1268(95)	15B35H	Bar Rod Wir; See std	0.31-0.39			0.70-1.20		0.030 max	0.050 max	0.15-0.35	B 0.0005-0.003; bal Fe				
SAE J1868(93)	15B35RH	Restrict hard range	0.33-0.38	0.20 max	0.035 max	0.80-1.10	0.25 max	0.025 max	0.040 max	0.15-0.35	B 0.0005-0.003; Mo <=0.06; Cu, Ni, Cr, Mo not spec'd but acpt; bal Fe				5 HRC max

Specification	Designation	Notes	C	Cr	Cu	Mn	Ni	P	S	Si	Other	UTS	YS	El	Hard
Carbon Steel, High-Manganese, 15B35H (Continued from previous page)															
USA															
SAE J776	15B35H*	Bar Rod Wir; obs see J1268	0.31-0.39			0.7-1.2		0.04 max	0.05 max	0.15-0.35	B 0.0005-0.003; bal Fe				
Carbon Steel, High-Manganese, 15B37H															
France															
AFNOR NFA35552	38MB5	Bar Rod Wir	0.34-0.4			1.1-1.4		0.04 max	0.04 max	0.1-0.4	B 0.001-0.005; bal Fe				
AFNOR NFA35563(83)	38MB5TS	HT	0.34-0.4			1.1-1.4		0.025 max	0.035 max	0.1-0.4	B 0.0008-0.005; bal Fe				55 HRC
AFNOR NFA35564(83)	38MB5FF	Soft ann	0.34-0.4			1.1-1.4		0.03 max	0.03 max	0.35 max	Al >=0.02; B 0.0008-0.005; bal Fe	0-560			212 HB max
AFNOR NFA35571(84)	38NCD16	t<80mm; soft ann	0.35-0.42	1.6-2	0.3 max	0.15-0.55	3.7-4.2	0.0025 max	0.02 max	0.1-0.4	Mo 0.3-0.55; bal Fe				285 HB max
Mexico															
NMX-B-301(86)	15B37H	Bar	0.30-0.39			1.00-1.50		0.040 max	0.050 max	0.15-0.30	B >=0.0005; bal Fe				
USA															
	AISI 15B37H	Wir rod, Hard	0.30-0.39			1.00-1.50		0.040 max	0.050 max	0.15-0.35	B 0.0005-0.003; bal Fe				
	UNS H15371		0.30-0.39			1.00-1.50		0.040 max	0.050 max	0.15-0.35	B 0.0005-0.003; bal Fe				
ASTM A304(96)	15B37H	Bar, hard bands spec	0.30-0.39			1.00-1.50		0.040 max	0.050 max	0.15-0.30	B >=0.0005; Cu Ni Cr Mo trace allowed; bal Fe				
SAE J1268(95)	15B37H	Bar Rod Wir; See std	0.30-0.39			1.00-1.50		0.030 max	0.050 max	0.15-0.35	B 0.0005-0.003; bal Fe				
Carbon Steel, High-Manganese, 15B41H															
China															
GB 3077(88)	40MnB	Bar HR Q/T 25mm diam	0.37-0.44	0.30 max	0.30 max	1.10-1.40	0.30 max	0.035 max	0.035 max	0.17-0.37	B 0.0005-0.0035; bal Fe	980	785	10	
GB 5216(85)	40MnBH	Bar Rod HR/Frg Ann	0.37-0.44	0.35 max		1.10-1.40	0.30 max	0.035 max	0.035 max	0.17-0.37	B 0.0005-0.0035; bal Fe				207 max HB
GB/T 13795(92)	40MnB	Strp CR Ann	0.37-0.44	0.35 max	0.30 max	1.10-1.40	0.30 max	0.035 max	0.035 max	0.17-0.37	B 0.0005-0.0035; bal Fe	400-700		15	
GB/T 3078(94)	40MnB	Bar CD Q/T 25mm diam	0.37-0.44	0.30 max	0.30 max	1.10-1.40	0.30 max	0.035 max	0.035 max	0.17-0.37	B 0.0005-0.0035; bal Fe	980	785	10	207 max HB
YB/T 5052(93)	40MnVB	Tub HR Norm/Tmp	0.37-0.44			1.10-1.40		0.040 max	0.040 max	0.20-0.40	B 0.001-0.004; V 0.05-0.10; bal Fe	735	539	12	
Germany															
DIN	40MnB4		0.37-0.44			0.80-1.10		0.035 max	0.035 max	0.40 max	B 0.0008-0.005; bal Fe				
DIN	WNr 1.5527		0.37-0.44			0.80-1.10		0.035 max	0.035 max	0.40 max	B 0.0008-0.005; bal Fe				
UK															
BS 970/1(83)	170H41	Blm Bil Slab Bar Rod Frg	0.37-0.44			0.80-1.10		0.050 max	0.050 max		B 0.0005-0.005; bal Fe				
BS 970/1(83)	173H15	Blm Bil Slab Bar Rod Frg CH	0.12-0.18			0.80-1.10		0.060 max	0.03-0.06		B 0.0005-0.005; bal Fe				
BS 970/1(83)	173H16	Blm Bil Slab Bar Rod Frg CH	0.13-0.19			1.10-1.40		0.060 max	0.03-0.06		B 0.0005-0.005; bal Fe				
BS 970/1(83)	174H20	Blm Bil Slab Bar Rod Frg CH	0.17-0.23			1.20-1.50		0.060 max	0.03-0.06		B 0.0005-0.005; bal Fe				
BS 970/1(83)	175H23	Blm Bil Slab Bar Rod Frg CH	0.20-0.25			1.30-1.60		0.060 max	0.03-0.06		B 0.0005-0.005; bal Fe				
BS 970/1(83)	185H40	Blm Bil Slab Bar Rod Frg	0.36-0.43	0.15-0.35		1.25-1.75		0.050 max	0.030-0.060		B 0.0005-0.005; Mo 0.08-0.18; P<=0.05 S=0.025-0.05/P<=0.025 S=0.015-0.04; bal Fe				
USA															
	AISI 15B41H	Wir rod, Hard	0.35-0.45			1.25-1.75		0.040 max	0.050 max	0.15-0.35	B 0.0005-0.003; bal Fe				
	UNS H15411		0.35-0.45			1.25-1.75		0.040 max	0.050 max	0.15-0.35	B 0.0005-0.003; bal Fe				

UNS numbers and US grades are provided as a means of cross referencing chemically similar alloys. Exchangability is only possible after independent examination of specifications. Tensile properties are minimum or typical as specified. UTS and YS as MPa. El as %. See Appendix for list of abbreviations used in Notes. * indicates obsolete material.

Specification	Designation	Notes	C	Cr	Cu	Mn	Ni	P	S	Si	Other	UTS	YS	El	Hard
Carbon Steel, High-Manganese, 15B41H (Continued from previous page)															
USA															
ASTM A304(96)	15B41H	Bar, hard bands spec	0.35-0.45			1.25-1.75		0.040 max	0.050 max	0.15-0.30	B >=0.0005; Cu Ni Cr Mo trace allowed; bal Fe				
SAE J1268(95)	15B41H	Bar Rod Wir; See std	0.35-0.45			1.25-1.75		0.030 max	0.050 max	0.15-0.35	B 0.0005-0.003; bal Fe				
Carbon Steel, High-Manganese, 15B48H															
China															
GB 3077(88)	45MnB	Bar HR Q/T 25mm diam	0.42-0.49	0.30 max	0.30 max	1.10-1.40	0.30 max	0.035 max	0.035 max	0.17-0.37	B 0.0005-0.0035; bal Fe	1030	835	9	
GB 5216(85)	45MnBH	Bar Rod HR/Frg Ann	0.42-0.49	0.35 max		1.10-1.40	0.30 max	0.035 max	0.035 max	0.17-0.37	B 0.0005-0.0035; bal Fe				217 max HB
GB 8162(87)	45MnB	Smls Tub HR/CR Q/T	0.42-0.49	0.30 max		1.10-1.40	0.30 max	0.035 max	0.035 max	0.17-0.37	B 0.0005-0.0035; bal Fe	1030	835	9	
GB/T 3078(94)	45MnB	Bar CD Q/T 25mm diam	0.42-0.49	0.30 max	0.30 max	1.10-1.40	0.30 max	0.035 max	0.035 max	0.17-0.37	B 0.0005-0.0035; bal Fe	1030	835	9	
YB/T 5052(93)	45MnB	Tub HR Norm	0.42-0.49			1.10-1.40		0.040 max	0.040 max	0.20-0.40	B 0.001-0.0035; bal Fe	640	395	14	
Europe															
EN 10083/3(95)	1.5532	Boron, Q/T	0.35-0.42	0.30 max		1.15-1.45		0.035 max	0.040 max	0.40 max	B 0.0008-0.0050; bal Fe				
EN 10083/3(95)	38MnB5	Boron, Q/T	0.35-0.42	0.30 max		1.15-1.45		0.035 max	0.040 max	0.40 max	B 0.0008-0.0050; bal Fe				
USA															
	AISI 15B48H	Wir rod, Hard	0.43-0.53			1.00-1.50		0.040 max	0.050 max	0.15-0.35	B 0.0005-0.003; bal Fe				
	UNS H15481		0.43-0.55			1.00-1.50		0.040 max	0.050 max	0.15-0.35	B 0.0005-0.003; bal Fe				
ASTM A304(96)	15B48H	Bar, hard bands spec	0.43-0.53			1.00-1.50		0.040 max	0.050 max	0.15-0.30	B >=0.0005; Cu Ni Cr Mo trace allowed; bal Fe				
SAE J1268(95)	15B48H	Bar Rod Wir; See std	0.43-0.53			1.00-1.50		0.030 max	0.050 max	0.15-0.35	B 0.0005-0.003; bal Fe				
Carbon Steel, High-Manganese, 15B62H															
USA															
	AISI 15B62H	Wir rod, Hard	0.54-0.67			1.00-1.50		0.040 max	0.050 max	0.40-0.60	B 0.0005-0.003; bal Fe				
	UNS H15621		0.54-0.67			1.00-1.50		0.040 max	0.050 max	0.40-0.60	B 0.0005-0.003; bal Fe				
ASTM A304(96)	15B62H	Bar, hard bands spec	0.54-0.67			1.00-1.50		0.040 max	0.050 max	0.40-0.60	B >=0.0005; Cu Ni Cr Mo trace allowed; bal Fe				
SAE J1268(95)	15B62H	Bar Rod Wir; See std	0.54-0.67			1.00-1.50		0.030 max	0.050 max	0.40-0.60	B 0.0005-0.003; bal Fe				
Carbon Steel, High-Manganese, A105															
Germany															
DIN 2528(91)	C21	Flanges	0.18-0.23			0.60-1.05		0.040 max	0.050 max	0.15-0.35	bal Fe				
DIN 2528(91)	WNr 1.0432	Flanges	0.18-0.23			0.60-1.05		0.040 max	0.050 max	0.15-0.35	bal Fe				
USA															
	UNS K03504		0.35 max			0.60-1.05		0.040 max	0.050 max	0.35 max	bal Fe				
ASTM A105/A105M(98)		Frg, EL is for Test Strp t>7.9	0.35 max	0.30 max	0.40 max	0.60-1.05	0.40 max	0.035 max	0.040 max	0.10-0.35	Mo <=0.12; Nb <=0.02; V <=0.05; Cu+Ni+Cr+Mo<=1.00, Cr+Mo<=0.32; bal Fe	485	250	30	187 HB
ASTM A695(95)	B35	HW special qual for fluid power apps	0.35 max			1.10 max		0.040 max	0.050 max	0.15-0.35	bal Fe	415	240	21	
Carbon Steel, High-Manganese, A106(B)															
Bulgaria															
BDS 4880(89)	08GB	Struct	0.12 max	0.35 max	0.35 max	0.80-1.45	0.35 max	0.040 max	0.040 max	0.10-0.40	Nb 0.015-0.060; bal Fe				

UNS numbers and US grades are provided as a means of cross referencing chemically similar alloys. Exchangability is only possible after independent examination of specifications. Tensile properties are minimum or typical as specified. UTS and YS as MPa. El as %. See Appendix for list of abbreviations used in Notes. * indicates obsolete material.

Carbon Steel, High-Manganese, A106(B) (Continued from previous page)

Specification	Designation	Notes	C	Cr	Cu	Mn	Ni	P	S	Si	Other	UTS	YS	El	Hard
Czech Republic															
CSN 411453	11453	High-temp const; Tube, Pipe	0.24 max	0.3 max		1.60 max		0.05 max	0.05 max	0.60 max	bal Fe				
Hungary															
MSZ 2898/2(80)	A35	Smls CF tube; Smls tubes w/special delivery conds; 0.5-10mm; bright-drawn, hard	0.17 max			2 max		0.05 max	0.05 max	0.6 max	bal Fe	440	350	6L	
MSZ 2898/2(80)	A35	Smls CF tube; Smls tubes w/special delivery conds; 0.5-10mm; bright-drawn, soft	0.17 max			2 max		0.05 max	0.05 max	0.6 max	bal Fe	370	260	10L	
MSZ 2898/2(80)	A45K	Smls CF tube; Smls tubes w/special req; 0.5-10mm; bright ann	0.22 max			0.4-1.60		0.05 max	0.05 max	0.1-0.35	bal Fe	390	195	24L	
MSZ 2898/2(80)	A45K	Smls CF tube; Smls tubes w/special req; 0.5-10mm; bright drawn, soft	0.22 max			0.4-1.60		0.05 max	0.05 max	0.1-0.35	bal Fe	470	330	8L	
MSZ 29/2(64)	A35		0.17 max			2 max		0.05 max	0.05 max	0.6 max	bal Fe				
MSZ 4747(85)	A45.47	High-temp, Tub; Non-rimming; FF; 40.1-60mm; norm	0.21 max			0.4-1.2		0.04 max	0.04 max	0.35 max	bal Fe	410-530	235	22L	
MSZ 4747(85)	A45.47	High-temp, Tub; Non-rimming; FF; 0-16mm; norm	0.21 max			0.4-1.2		0.04 max	0.04 max	0.35 max	bal Fe	410-530	255	22L	
International															
ISO 2604-2(75)	TS9H	Smls Tub	0.21 max	0.3 max	0.3 max	0.4-1.2	0.4 max	0.045 max	0.045 max	0.35 max	Al <=0.1; Co <=0.1; Mo <=0.15; Pb <=0.15; Ti <=0.05; V <=0.1; W <=0.1; bal Fe				
Italy															
UNI 5462(64)	C18	Smls tube	0.21 max			0.5 max		0.035 max	0.035 max	0.1-0.35	bal Fe				
Sweden															
SS 141410	1410		0.15-0.3			0.3-1.1		0.05 max	0.05 max	0.02-0.25	bal Fe				
UK															
BS 3059/2(90)	440	Boiler, Superheater; High-temp; Smls Tub; Tub, Weld; 2-12.5mm	0.12-0.18			0.90-1.20		0.035 max	0.035 max	0.10-0.35	bal Fe	440-580	245	21	
USA															
	UNS K03006		0.30 max			0.29-1.06		0.048 max	0.058 max	0.10 min	bal Fe				
ASTM A106(97)	B	Smls Pipe	0.30 max	0.40 max	0.40 max	0.29-1.06	0.40 max	0.035 max	0.035 max	0.10 min	Mo <=0.15; V <=0.08; Cr+Cu+Mo+Ni+V<=1.00; bal Fe	415	240	30L	
ASTM A234/A234M(97)	WPB	Frg	0.30 max	0.15 max	0.40 max	0.29-1.06	0.40 max	0.050 max	0.058 max	0.10 min	Mo <=0.15; Nb <=0.02; V <=0.08; Cu+Ni+Cr+Mo<=1.00, Cr+Mo<=0.32; bal Fe	415-485	240	22L	
ASTM A556/A556M(96)	C2	Smls CD Feedwater Heater Tub	0.30 max			0.29-1.06		0.035 max	0.035 max	0.10 min	bal Fe	480	280	30	89 HRB max
Yugoslavia															
	C.0000	Struct	0.30 max			1.60 max		0.070 max	0.060 max	0.60 max	bal Fe				
	C.1207	Unalloyed	0.22 max			0.45 max		0.045 max	0.045 max	0.35 max	Mo <=0.15; +Al; bal Fe				
	C.1215	High-temp const	0.22 max			0.45-1.60		0.05 max	0.05 max	0.1-0.35	Mo <=0.15; bal Fe				

Specification	Designation	Notes	C	Cr	Cu	Mn	Ni	P	S	Si	Other	UTS	YS	El	Hard

Carbon Steel, High-Manganese, A131(D)

UK

Specification	Designation	Notes	C	Cr	Cu	Mn	Ni	P	S	Si	Other	UTS	YS	El	Hard
BS 1503(89)	223-410	Frg press ves; 100<t<=999mm; norm Q/T	0.20 max	0.25 max	0.30 max	0.80-1.20	0.40 max	0.030 max	0.025 max	0.10-0.40	Mo <=0.10; Nb 0.01-0.06; W <=0.1; Cr+Mo+Ni+Cu<=0.8; bal Fe	410-530	230	20	
BS 1503(89)	223-410	Frg press ves; t<=100mm; norm Q/T	0.20 max	0.25 max	0.30 max	0.80-1.20	0.40 max	0.030 max	0.025 max	0.10-0.40	Mo <=0.10; Nb 0.01-0.06; W <=0.1; Cr+Mo+Ni+Cu<=0.8; bal Fe	410-530	245	20	

USA

Specification	Designation	Notes	C	Cr	Cu	Mn	Ni	P	S	Si	Other	UTS	YS	El	Hard
	UNS K02101		0.21 max			0.70-1.35		0.04 max	0.04 max	0.10-0.35	bal Fe				
ASTM A131/A131M(94)	D	Ship rivet/cold flanging stl	0.21 max			0.70-1.35		0.035 max	0.04 max	0.10-0.35	bal Fe	380-450	205	26	
ASTM A131/A131M(94)	D	Ship struct	0.21 max			0.70-1.35		0.035 max	0.04 max	0.10-0.35	bal Fe	400-490	235	24	

Carbon Steel, High-Manganese, A139(C)

USA

Specification	Designation	Notes	C	Cr	Cu	Mn	Ni	P	S	Si	Other	UTS	YS	El	Hard
	UNS K03004		0.30 max			1.20 max		0.040 max	0.050 max		bal Fe				
ASTM A139	C	Weld Pipe	0.30 max			1.20 max		0.035 max			bal Fe	415	290	25L	

Carbon Steel, High-Manganese, A139(D)

USA

Specification	Designation	Notes	C	Cr	Cu	Mn	Ni	P	S	Si	Other	UTS	YS	El	Hard
	UNS K03010		0.30 max			1.30 max		0.040 max	0.050 max		bal Fe				
ASTM A139	D	Weld Pipe	0.30 max			1.30 max		0.035 max	0.035 max		bal Fe	415	315	23L	

Carbon Steel, High-Manganese, A139(E)

USA

Specification	Designation	Notes	C	Cr	Cu	Mn	Ni	P	S	Si	Other	UTS	YS	El	Hard
	UNS K03012		0.30 max			1.40 max		0.040 max	0.050 max		bal Fe				
ASTM A139	E	Weld Pipe	0.30 max			1.40 max		0.035 max	0.035 max		bal Fe	455	360	22L	

Carbon Steel, High-Manganese, A181

USA

Specification	Designation	Notes	C	Cr	Cu	Mn	Ni	P	S	Si	Other	UTS	YS	El	Hard
	UNS K03502		0.35 max			1.10 max		0.05 max	0.05 max		bal Fe				
ASTM A181/A181M	Class 60	Frg	0.35 max			1.10 max		0.05 max	0.05 max	0.10-0.35	bal Fe	415	205	22	
ASTM A181/A181M	Class 70	Frg	0.35 max			1.10 max		0.05 max	0.05 max	0.10-0.35	bal Fe	485	250	18	

Carbon Steel, High-Manganese, A227

Norway

Specification	Designation	Notes	C	Cr	Cu	Mn	Ni	P	S	Si	Other	UTS	YS	El	Hard
NS 13601	13611	Spring wir			0.25 max			0.040 max	0.040 max		C varies; bal Fe				
NS 13601	13612	Spring wir			0.25 max			0.040 max	0.040 max		C varies; bal Fe				

USA

Specification	Designation	Notes	C	Cr	Cu	Mn	Ni	P	S	Si	Other	UTS	YS	El	Hard
	UNS K06501		0.45-0.85			0.30-1.30		0.040 max	0.050 max	0.10-0.35	bal Fe				
SAE J441(93)	CW-12	Cut Wir shot; 0.30mm diam	0.45-0.85			0.30-1.30		0.040 max	0.050 max	0.15-0.35	bal Fe	2030-2300			42 HRC
SAE J441(93)	CW-14	Cut Wir shot; 0.35mm diam	0.45-0.85			0.30-1.30		0.040 max	0.050 max	0.15-0.35	bal Fe	2010-2280			42 HRC
SAE J441(93)	CW-17	Cut Wir shot; 0.45mm diam	0.45-0.85			0.30-1.30		0.040 max	0.050 max	0.15-0.35	bal Fe	1980-2250			42 HRC
SAE J441(93)	CW-20	Cut Wir shot; 0.5mm diam	0.45-0.85			0.30-1.30		0.040 max	0.050 max	0.15-0.35	bal Fe	1950-2230			42 HRC
SAE J441(93)	CW-23	Cut Wir shot; 0.6mm diam	0.45-0.85			0.30-1.30		0.040 max	0.050 max	0.15-0.35	bal Fe	1920-2200			42 HRC
SAE J441(93)	CW-28	Cut Wir shot; 0.7mm diam	0.45-0.85			0.30-1.30		0.040 max	0.050 max	0.15-0.35	bal Fe	1870-2140			42 HRC

UNS numbers and US grades are provided as a means of cross referencing chemically similar alloys. Exchangability is only possible after independent examination of specifications. Tensile properties are minimum or typical as specified. UTS and YS as MPa. El as %. See Appendix for list of abbreviations used in Notes. * indicates obsolete material.

Specification	Designation	Notes	C	Cr	Cu	Mn	Ni	P	S	Si	Other	UTS	YS	El	Hard

Carbon Steel, High-Manganese, A227 (Continued from previous page)

USA

Specification	Designation	Notes	C	Cr	Cu	Mn	Ni	P	S	Si	Other	UTS	YS	El	Hard
SAE J441(93)	CW-32	Cut Wir shot; 0.8mm diam	0.45-0.85			0.30-1.30		0.040 max	0.050 max	0.15-0.35	bal Fe	1830-2110			42 HRC
SAE J441(93)	CW-35	Cut Wir shot; 0.9mm diam	0.45-0.85			0.30-1.30		0.040 max	0.050 max	0.15-0.35	bal Fe	1800-2080			42 HRC
SAE J441(93)	CW-41	Cut Wir shot; 1.0mm diam	0.45-0.85			0.30-1.30		0.040 max	0.050 max	0.15-0.35	bal Fe	1760-2020			42 HRC
SAE J441(93)	CW-47	Cut Wir shot; 1.2mm diam	0.45-0.85			0.30-1.30		0.040 max	0.050 max	0.15-0.35	bal Fe	1710-1970			42 HRC
SAE J441(93)	CW-54	Cut Wir shot; 1.4mm diam	0.45-0.85			0.30-1.30		0.040 max	0.050 max	0.15-0.35	bal Fe	1680-1920			42 HRC
SAE J441(93)	CW-62	Cut Wir shot; 1.6mm diam	0.45-0.85			0.30-1.30		0.040 max	0.050 max	0.15-0.35	bal Fe	1630-1880			42 HRC

Carbon Steel, High-Manganese, A229

USA

Specification	Designation	Notes	C	Cr	Cu	Mn	Ni	P	S	Si	Other	UTS	YS	El	Hard
	UNS K07001		0.55-0.85			0.60-1.20		0.040 max	0.050 max	0.10-0.35	bal Fe				
AMS 2431/8(96)		Cond Cut Wir Shot	0.45-0.85			0.60-1.20		0.045 max	0.05 max	0.10-0.30	bal Fe				52-62 HRC

Carbon Steel, High-Manganese, A266

UK

Specification	Designation	Notes	C	Cr	Cu	Mn	Ni	P	S	Si	Other	UTS	YS	El	Hard
BS 1503(89)	221-530	Frg press ves; t<=100mm; Q/T	0.30 max	0.25 max	0.30 max	0.80-1.40	0.40 max	0.030 max	0.025 max	0.10-0.40	Al <=0.010; Mo <=-0.10; W <=0.1; Cr+Mo+Ni+Cu<=0.8; bal Fe	530-650	295	15	

Carbon Steel, High-Manganese, A266(1)

USA

Specification	Designation	Notes	C	Cr	Cu	Mn	Ni	P	S	Si	Other	UTS	YS	El	Hard
	UNS K03506		0.35 max			0.40-1.05		0.040 max	0.040 max	0.15-0.35	bal Fe				
ASTM A266/A266M(96)	1	Frg	0.35 max			0.40-1.05		0.025 max	0.025 max	0.15-0.35	bal Fe	60-85	30	23, RA= 38	121-170 HB
ASTM A266/A266M(96)	2	Frg	0.35 max			0.40-1.05		0.025 max	0.025 max	0.15-0.35	bal Fe	70-95	36	20, RA= 33	137-197 HB
ASTM A541/541M(95)	Grade 1	Q/T frg for press ves	0.35 max	0.25 max		0.40-0.90	0.40 max	0.025 max	0.025 max	0.15-0.35	Mo <=0.10; V <=0.05; Si req vary; bal Fe	480-660	250	20	
ASTM A541/541M(95)	Grade 1A	Q/T frg for press ves	0.30 max	0.25 max		0.70-1.35	0.40 max	0.025 max	0.025 max	0.15-0.40	Mo <=0.10; V <=0.05; Si req vary; bal Fe	480-660	250	20	

Carbon Steel, High-Manganese, A266(4)

USA

Specification	Designation	Notes	C	Cr	Cu	Mn	Ni	P	S	Si	Other	UTS	YS	El	Hard
	UNS K03017		0.30 max			0.80-1.35		0.035 max	0.040 max	0.15-0.35	bal Fe				
ASTM A266/A266M(96)	4	Frg	0.30 max			0.80-1.35		0.025 max	0.025 max	0.15-0.35	bal Fe	70-95	36	20, RA= 33	137-197 HB

Carbon Steel, High-Manganese, A283(A)

Europe

Specification	Designation	Notes	C	Cr	Cu	Mn	Ni	P	S	Si	Other	UTS	YS	El	Hard
EN 10025(90)A1(93)	1.0035	Struct HR BS, t<=1mm									bal Fe	310-540	185	10L 8T	
EN 10025(90)A1(93)	1.0035	Struct HR BS, 3<t<=16mm									bal Fe	290-510	185	18L 16T	
EN 10025(90)A1(93)	S185	Struct HR BS, 3<t<=16mm									bal Fe	290-510	185	18L 16T	
EN 10025(90)A1(93)	S185	Struct HR BS, t<=1mm									bal Fe	310-540	185	10L 8T	

Carbon Steel, High-Manganese, A299

USA

Specification	Designation	Notes	C	Cr	Cu	Mn	Ni	P	S	Si	Other	UTS	YS	El	Hard
	UNS K02803		0.28 max			0.84-1.52		0.035 max	0.040 max	0.13-0.45	bal Fe				

UNS numbers and US grades are provided as a means of cross referencing chemically similar alloys. Exchangability is only possible after independent examination of specifications. Tensile properties are minimum or typical as specified. UTS and YS as MPa. El as %. See Appendix for list of abbreviations used in Notes. * indicates obsolete material.

Specification	Designation	Notes	C	Cr	Cu	Mn	Ni	P	S	Si	Other	UTS	YS	El	Hard
Carbon Steel, High-Manganese, A299 (Continued from previous page)															
USA															
ASTM A299/A299M(97)		Press ves plt, t>25mm	0.30 max			0.90-1.50		0.035 max	0.035 max	0.15-0.40	bal Fe	515-655	275	19	
ASTM A299/A299M(97)		Press ves plt, t<=25mm	0.28 max			0.90-1.40		0.035 max	0.035 max	0.15-0.40	bal Fe	515-655	290	19	
Yugoslavia															
	C.3133	High-temp const	0.17-0.23			1.0-1.3		0.05 max	0.05 max	0.4-0.6	Mo <=0.15; bal Fe				
Carbon Steel, High-Manganese, A333(1)															
USA															
	UNS K03008		0.30 max			0.40-1.06		0.05 max	0.06 max		bal Fe				
Carbon Steel, High-Manganese, A350(LF1)															
USA															
	UNS K03009		0.30 max			1.35 max		0.035 max	0.040 max	0.15-0.30	bal Fe				
Carbon Steel, High-Manganese, A350(LF2)															
USA															
	UNS K03011		0.30 max			1.35 max		0.035 max	0.040 max	0.15-0.30	bal Fe				
ASTM A706(96)	A706	Deformed reinf bar; CEV Req'd	0.30 max			1.50 max		0.035 max	0.045 max	0.50 max	bal Fe	550	420	10-14	
Carbon Steel, High-Manganese, A36															
Italy															
UNI 6363(84)	Fe410	Weld/Smls tub; water main	0.21 max			1.2 max		0.04 max	0.04 max	0.35 max	bal Fe				
UNI 6403(86)	Fe410	Q/T	0.21 max			0.4-1.2		0.04 max	0.04 max	0.1-0.35	bal Fe				
UNI 8913(87)	Fe410	Q/T	0.21 max			0.4-1.2		0.04 max	0.04 max	0.35 max	bal Fe				
USA															
	UNS K02595	Plate t>4 in.	0.29 max			0.85-1.20		0.04 max	0.05 max		bal Fe				
	UNS K02596	Plate 2.5 in.<t<=4 in.	0.27 max			0.85-1.20		0.04 max	0.05 max		bal Fe				
	UNS K02597	Plate 1.5 in.<t<=2.5 in.	0.26 max			0.80-1.20		0.04 max	0.05 max		bal Fe				
	UNS K02598	Plate 0.75 in.<t<=1.5 in.	0.25 max			0.80-1.20		0.04 max	0.05 max		bal Fe				
ASTM A36/A36M(97)	A36	Struct plt, 20<t<=40mm	0.25 max		0.20 min	0.80-1.20		0.04 max	0.05 max	0.40 max	bal Fe	400-550	250	21	
ASTM A36/A36M(97)	A36	Struct plt, 40<t<=65mm	0.26 max		0.20 min	0.80-1.20		0.04 max	0.05 max	0.15-0.40	bal Fe	400-550	250	21	
ASTM A36/A36M(97)	A36	Struct plt, 65<t<=100mm	0.27 max		0.20 min	0.80-1.20		0.04 max	0.05 max	0.15-0.40	bal Fe	400-550	250	21	
ASTM A36/A36M(97)	A36	Struct plt, t>100mm	0.29 max		0.20 min	0.80-1.20		0.04 max	0.05 max	0.15-0.40	bal Fe	400-550	250	21	
Carbon Steel, High-Manganese, A372(II)															
UK															
BS 1503(89)	221-550	Frg press ves; t<=100mm; Q/T	0.35 max	0.25 max	0.30 max	0.80-1.40	0.40 max	0.030 max	0.025 max	0.10-0.40	Al <=0.010; Mo <=0.10; W <=0.1; Cr+Mo+Ni+Cu<=0.8; bal Fe	550-670	305	15	
USA															
	UNS K04001		0.35 max			1.35 max		0.04 max	0.05 max	0.15-0.35	bal Fe				
ASTM A372/372M(95)	Grade B	Thin wall frg press ves	0.35 max			1.35 max		0.025 max	0.025 max	0.15-0.35	bal Fe	515-690	310	18	156 HB
ASTM A372/372M(95)	Type II*	Thin wall frg press ves	0.35 max			1.35 max		0.025 max	0.025 max	0.15-0.35	bal Fe	515-690	310	18	156 HB

UNS numbers and US grades are provided as a means of cross referencing chemically similar alloys. Exchangability is only possible after independent examination of specifications. Tensile properties are minimum or typical as specified. UTS and YS as MPa. El as %. See Appendix for list of abbreviations used in Notes. * indicates obsolete material.

Specification	Designation	Notes	C	Cr	Cu	Mn	Ni	P	S	Si	Other	UTS	YS	El	Hard

Carbon Steel, High-Manganese, A372(III)

USA

Specification	Designation	Notes	C	Cr	Cu	Mn	Ni	P	S	Si	Other	UTS	YS	El	Hard
	UNS K04801		0.48 max			1.65 max		0.04 max	0.05 max	0.15-0.35	bal Fe				
ASTM A372/372M(95)	Grade C	Thin wall frg press ves	0.48 max			1.65 max		0.025 max	0.025 max	0.15-0.35	bal Fe	620-795	380	15	187 HB
ASTM A372/372M(95)	Type III*	Thin wall frg press ves	0.48 max			1.65 max		0.025 max	0.025 max	0.15-0.35	bal Fe	620-795	380	15	187 HB

Carbon Steel, High-Manganese, A381

Japan

Specification	Designation	Notes	C	Cr	Cu	Mn	Ni	P	S	Si	Other	UTS	YS	El	Hard
JIS G3101(95)	SS540	HR Sh Plt Bar	0.30 max			1.60 max		0.040 max	0.040 max		bal Fe	540	390-400	13-17	

USA

Specification	Designation	Notes	C	Cr	Cu	Mn	Ni	P	S	Si	Other	UTS	YS	El	Hard
	UNS K03013		0.30 max			1.50 max		0.050 max	0.060 max		bal Fe				

Carbon Steel, High-Manganese, A414(D)

USA

Specification	Designation	Notes	C	Cr	Cu	Mn	Ni	P	S	Si	Other	UTS	YS	El	Hard
	UNS K02505		0.25 max			1.20 max		0.035 max	0.040 max		Cu 0.20 min (when spec'd); bal Fe				

Carbon Steel, High-Manganese, A414(E)

USA

Specification	Designation	Notes	C	Cr	Cu	Mn	Ni	P	S	Si	Other	UTS	YS	El	Hard
	UNS K02704		0.27 max			1.20 max		0.035 max	0.040 max		Cu 0.20 min (when spec'd); bal Fe				

Carbon Steel, High-Manganese, A414(F)

Italy

Specification	Designation	Notes	C	Cr	Cu	Mn	Ni	P	S	Si	Other	UTS	YS	El	Hard
UNI 5869(75)	Fe510-1KG	Press ves	0.22 max			1.6 max		0.035 max	0.04 max	0.4 max	0-50mm C<=0.2; bal Fe				
UNI 5869(75)	Fe510-1KW	Press ves	0.22 max			1.6 max		0.04 max	0.04 max	0.4 max	0-50mm C<=0.2; bal Fe				
UNI 5869(75)	Fe510-2KG	Press ves	0.22 max			1.6 max		0.03 max	0.035 max	0.4 max	0-50mm C<=0.2; bal Fe				
UNI 5869(75)	Fe510-2KW	Press ves	0.22 max			1.6 max		0.035 max	0.035 max	0.4 max	0-50mm C<=0.2; bal Fe				

Romania

Specification	Designation	Notes	C	Cr	Cu	Mn	Ni	P	S	Si	Other	UTS	YS	El	Hard
STAS 2883/3(88)	K460	Heat res	0.12-0.2	0.3 max		0.9-1.2		0.035 max	0.03 max	0.4 max	Al 0.02-0.035; N <=0.009; Pb <=0.15; V <=0.03; Cr+Ni+Cu<=0.7; As<=0.08; bal Fe				

Spain

Specification	Designation	Notes	C	Cr	Cu	Mn	Ni	P	S	Si	Other	UTS	YS	El	Hard
UNE 36087(74)	A47RAII	Sh Plt	0.23 max			0.5-1.5		0.045 max	0.04 max	0.4 max	Al 0.015-0.120; Mo <=0.15; bal Fe				
UNE 36087(74)	A47RCI	Sh Plt	0.23 max			0.7-1.5		0.055 max	0.05 max	0.4 max	Mo <=0.15; N <=0.009; bal Fe				
UNE 36087(74)	A47RCII	Sh Plt	0.23 max			0.5-1.1		0.045 max	0.04 max	0.4 max	Al 0.015-0.10; Mo <=0.15; bal Fe				

USA

Specification	Designation	Notes	C	Cr	Cu	Mn	Ni	P	S	Si	Other	UTS	YS	El	Hard
	UNS K03102		0.31 max		0.20 min	1.20 max		0.035 max	0.040 max		Cu 0.20 min (when spec'd); bal Fe				

Carbon Steel, High-Manganese, A414(G)

USA

Specification	Designation	Notes	C	Cr	Cu	Mn	Ni	P	S	Si	Other	UTS	YS	El	Hard
	UNS K03103		0.31 max		0.20 min	1.35 max		0.035 max	0.040 max		Cu 0.20 min (when spec'd); bal Fe				

Carbon Steel, High-Manganese, A417

USA

Specification	Designation	Notes	C	Cr	Cu	Mn	Ni	P	S	Si	Other	UTS	YS	El	Hard
	UNS K06201		0.50-0.75			0.60-1.20		0.040 max	0.050 max		bal Fe				

Carbon Steel, High-Manganese, A442(55)

Bulgaria

Specification	Designation	Notes	C	Cr	Cu	Mn	Ni	P	S	Si	Other	UTS	YS	El	Hard
BDS 5930	12K	Heat res	0.08-0.16	0.30 max	0.30 max	0.4-1.60	0.30 max	0.045 max	0.045 max	0.15-0.35	Al <=0.1; Mo 0.4-0.6; bal Fe				

UNS numbers and US grades are provided as a means of cross referencing chemically similar alloys. Exchangability is only possible after independent examination of specifications. Tensile properties are minimum or typical as specified. UTS and YS as MPa. El as %. See Appendix for list of abbreviations used in Notes. * indicates obsolete material.

Specification	Designation	Notes	C	Cr	Cu	Mn	Ni	P	S	Si	Other	UTS	YS	El	Hard

Carbon Steel, High-Manganese, A442(55) (Continued from previous page)

Italy

Specification	Designation	Notes	C	Cr	Cu	Mn	Ni	P	S	Si	Other	UTS	YS	El	Hard
UNI 5869(75)	Fe360-1KG	Press ves	0.17 max			0.4-1.60		0.035 max	0.04 max	0.35 max	bal Fe				
UNI 5869(75)	Fe360-1KW	Press ves	0.17 max			0.4-1.60		0.04 max	0.04 max	0.35 max	bal Fe				
UNI 5869(75)	Fe360-2KG	Press ves	0.17 max			0.4-1.60		0.03 max	0.035 max	0.35 max	bal Fe				
UNI 5869(75)	Fe360-2KW	Press ves	0.17 max			0.4-1.60		0.035 max	0.035 max	0.35 max	bal Fe				

Spain

Specification	Designation	Notes	C	Cr	Cu	Mn	Ni	P	S	Si	Other	UTS	YS	El	Hard
UNE 36087(74)	A37RAII	Sh Plt	0.2 max			0.3-1.3		0.045 max	0.04 max	0.4 max	Al 0.015-0.120; Mo <=0.15; 0-40mm Mn<=1.3; 40mm Mn<=1.1; bal Fe				
UNE 36087(74)	A37RCI	Sh Plt	0.2 max			0.3-1.3		0.055 max	0.05 max	0.4 max	Mo <=0.15; N <=0.009; 0-40mm Mn<=1.3; 40mm Mn<=1.1; bal Fe				

USA

Specification	Designation	Notes	C	Cr	Cu	Mn	Ni	P	S	Si	Other	UTS	YS	El	Hard
	UNS K02202		0.22 max			0.80-1.10		0.035 max	0.04 max	0.13-0.40	bal Fe				

Yugoslavia

Specification	Designation	Notes	C	Cr	Cu	Mn	Ni	P	S	Si	Other	UTS	YS	El	Hard
	C.1202	High-temp const	0.16 max			0.4 max		0.05 max	0.05 max	0.35 max	Mo <=0.15; bal Fe				

Carbon Steel, High-Manganese, A442(60)

Germany

Specification	Designation	Notes	C	Cr	Cu	Mn	Ni	P	S	Si	Other	UTS	YS	El	Hard
DIN 17102(89)	WStE285	see P275NH	0.18 max	0.3 max	0.2 max	0.6-1.4	0.3 max	0.035 max	0.03 max	0.4 max	Mo <=0.08; N <=0.02; Nb <=0.03; bal Fe				

USA

Specification	Designation	Notes	C	Cr	Cu	Mn	Ni	P	S	Si	Other	UTS	YS	El	Hard
	UNS K02402		0.24 max			0.80-1.10		0.35 max	0.04 max	0.15-0.40	bal Fe				

Yugoslavia

Specification	Designation	Notes	C	Cr	Cu	Mn	Ni	P	S	Si	Other	UTS	YS	El	Hard
	C.1204	High-temp const	0.2 max			0.5 max		0.05 max	0.05 max	0.35 max	Mo <=0.15; bal Fe				
	C.1205	Res to ageing	0.2 max			0.45 max		0.045 max	0.045 max	0.35 max	Mo <=0.15; +Al; bal Fe				

Carbon Steel, High-Manganese, A455(I)

USA

Specification	Designation	Notes	C	Cr	Cu	Mn	Ni	P	S	Si	Other	UTS	YS	El	Hard
	UNS K03300		0.33 max			0.81-1.25		0.040 max	0.050 max	0.13	bal Fe				
ASTM A455(96)		Press ves plt, t<=9.5mm	0.33 max			0.85-1.20		0.035 max	0.035 max	0.10 max	bal Fe	515-655	260	22	
ASTM A455(96)		Press ves plt, 9.5<t<=15mm	0.33 max			0.85-1.20		0.035 max	0.035 max	0.10 max	bal Fe	505-640	255	22	
ASTM A455(96)		Press ves plt, 15<t<=20mm	0.33 max			0.85-1.20		0.035 max	0.035 max	0.10 max	bal Fe	485-620	240	22	

Carbon Steel, High-Manganese, A500(C)

USA

Specification	Designation	Notes	C	Cr	Cu	Mn	Ni	P	S	Si	Other	UTS	YS	El	Hard
	UNS K02705		0.27 max			1.40 max		0.05 max	0.063 max		Cu 0.18 min (when spec'd); bal Fe				
ASTM A500		CF	0.23 max		0.20 min	1.35 max		0.035 max			bal Fe	427	317	21	

Carbon Steel, High-Manganese, A502(2)

USA

Specification	Designation	Notes	C	Cr	Cu	Mn	Ni	P	S	Si	Other	UTS	YS	El	Hard
	UNS K02405		0.16-0.33			1.14-1.71		0.048 max	0.058 max	0.08-0.32	Cu 0.18 min (when spec'd); bal Fe				

Carbon Steel, High-Manganese, A516(60)

Bulgaria

Specification	Designation	Notes	C	Cr	Cu	Mn	Ni	P	S	Si	Other	UTS	YS	El	Hard
BDS 4880(89)	08GB	Struct	0.12 max	0.35 max	0.35 max	0.80-1.45	0.35 max	0.040 max	0.040 max	0.10-0.40	Nb 0.015-0.060; bal Fe				

UNS numbers and US grades are provided as a means of cross referencing chemically similar alloys. Exchangability is only possible after independent examination of specifications. Tensile properties are minimum or typical as specified. UTS and YS as MPa. El as %. See Appendix for list of abbreviations used in Notes. * indicates obsolete material.

Specification	Designation	Notes	C	Cr	Cu	Mn	Ni	P	S	Si	Other	UTS	YS	El	Hard
Carbon Steel, High-Manganese, A516(65)															
Bulgaria															
BDS 5930	12K	Heat res	0.08-0.16	0.30 max	0.30 max	0.4-1.60	0.30 max	0.045 max	0.045 max	0.15-0.35	Al <=0.1; Mo 0.4-0.6; bal Fe				
Czech Republic															
CSN 411366	11366	High-temp const	0.15 max	0.3 max		0.65 max	0.3 max	0.045 max	0.04 max	0.35 max	bal Fe				
Italy															
UNI 5869(75)	Fe360-1KG	Press ves	0.17 max			0.4-1.60		0.035 max	0.04 max	0.35 max	bal Fe				
UNI 5869(75)	Fe360-1KW	Press ves	0.17 max			0.4-1.60		0.04 max	0.04 max	0.35 max	bal Fe				
UNI 5869(75)	Fe360-2KG	Press ves	0.17 max			0.4-1.60		0.03 max	0.035 max	0.35 max	bal Fe				
UNI 5869(75)	Fe360-2KW	Press ves	0.17 max			0.4-1.60		0.035 max	0.035 max	0.35 max	bal Fe				
Poland															
PNH84024	St36K	High-temp const	0.08-0.16	0.3 max		0.4-1.6	0.3 max	0.045 max	0.045 max	0.15-0.35	bal Fe				
Spain															
UNE 36080(73)	A37-a	Gen struct	0.2 max			1.60 max		0.06 max	0.05 max	0.60 max	Mo <=0.15; bal Fe				
UNE 36087(74)	A37RAI	Sh Plt	0.2 max			0.3-1.3		0.055 max	0.05 max	0.4 max	Mo <=0.15; N <=0.009; 0-40mm Mn<=1.3; 40mm Mn<=1.1; bal Fe				
UNE 36087(74)	A37RAII	Sh Plt	0.2 max			0.3-1.3		0.045 max	0.04 max	0.4 max	Al 0.015-0.120; Mo <=0.15; 0-40mm Mn<=1.3; 40mm Mn<=1.1; bal Fe				
UNE 36087(74)	A37RCI	Sh Plt	0.2 max			0.3-1.3		0.055 max	0.05 max	0.4 max	Mo <=0.15; N <=0.009; 0-40mm Mn<=1.3; 40mm Mn<=1.1; bal Fe				
Sweden															
SS 141330	1330	High-temp const	0.17 max	0.25 max		0.4-1.2	0.2 max	0.035 max	0.03 max	0.4 max	Mo <=0.1; N <=0.012; Nb <=0.01; Ti <=0.02; V <=0.03; bal Fe				
SS 141331	1331	High-temp const	0.17 max	0.25 max		0.4-1.2	0.2 max	0.035 max	0.03 max	0.4 max	Mo <=0.1; N <=0.012; Nb <=0.01; Ti <=0.02; V <=0.03; bal Fe				
USA															
	UNS K02403		0.24 max			0.79-1.30		0.035 max	0.04 max	0.13-0.45	bal Fe				
ASTM A516(90)	Grade 65	Press ves plt, mod/low-temp, t>50-100mm	0.28 max			0.85-1.20		0.035 max	0.035 max	0.15-0.40	bal Fe	450-585	240	23	
ASTM A516(90)	Grade 65	Press ves plt, mod/low-temp, t<=12.5mm	0.24 max			0.85-1.20		0.035 max	0.035 max	0.15-0.40	bal Fe	450-585	240	23	
ASTM A516(90)	Grade 65	Press ves plt, mod/low-temp, t>100mm	0.29 max			0.85-1.20		0.035 max	0.035 max	0.15-0.40	bal Fe	450-585	240	23	
ASTM A516(90)	Grade 65	Press ves plt, mod/low-temp, t>12.5-50mm	0.26 max			0.85-1.20		0.035 max	0.035 max	0.15-0.40	bal Fe	450-585	240	23	
Carbon Steel, High-Manganese, A516(70)															
Czech Republic															
CSN 411503	11503		0.18 max	0.3 max		1.4 max	0.3 max	0.035 max	0.035 max	0.4 max	Al 0.01-0.110; Nb 0.015-0.08; Ti <=0.2; bal Fe				
Europe															
EN 10028/2(92)	1.0481	High-temp, Press; 60<=t<=100mm	0.08-0.20	0.30 max	0.30 max	0.90-1.50	0.30 max	0.030 max	0.025 max	0.40 max	Al >=0.020; Mo <=0.08; Nb <=0.010; Ti <=0.03; V <=0.02; Cr+Cu+Mo+Ni<=0.70; bal Fe	460-580	260	21T	

Specification	Designation	Notes	C	Cr	Cu	Mn	Ni	P	S	Si	Other	UTS	YS	El	Hard

Carbon Steel, High-Manganese, A516(70) (Continued from previous page)

Europe

Specification	Designation	Notes	C	Cr	Cu	Mn	Ni	P	S	Si	Other	UTS	YS	El	Hard
EN 10028/2(92)	1.0481	High-temp, Press; 16<=t<=40mm	0.08-0.20	0.30 max	0.30 max	0.90-1.50	0.30 max	0.030 max	0.025 max	0.40 max	Al >=0.020; Mo <=0.08; Nb <=0.010; Ti <=0.03; V <=0.02; Cr+Cu+Mo+Ni<=0.70; bal Fe	460-580	290	22T	
EN 10028/2(92)	1.0481	High-temp, Press; 40<=t<=60mm	0.08-0.20	0.30 max	0.30 max	0.90-1.50	0.30 max	0.030 max	0.025 max	0.40 max	Al >=0.020; Mo <=0.08; Nb <=0.010; Ti <=0.03; V <=0.02; Cr+Cu+Mo+Ni<=0.70; bal Fe	460-580	285	22T	
EN 10028/2(92)	1.0481	High-temp, Press; 3<=t<=16mm	0.08-0.20	0.30 max	0.30 max	0.90-1.50	0.30 max	0.030 max	0.025 max	0.40 max	Al >=0.020; Mo <=0.08; Nb <=0.010; Ti <=0.03; V <=0.02; Cr+Cu+Mo+Ni<=0.70; bal Fe	460-580	295	22T	
EN 10028/2(92)	1.0481	High-temp, Press; 100<=t<=150mm	0.08-0.20	0.30 max	0.30 max	0.90-1.50	0.30 max	0.030 max	0.025 max	0.40 max	Al >=0.020; Mo <=0.08; Nb <=0.010; Ti <=0.03; V <=0.02; Cr+Cu+Mo+Ni<=0.70; bal Fe	440-570	235	21T	
EN 10028/2(92)	P295GH	High-temp, Press; 60<=t<=100mm	0.08-0.20	0.30 max	0.30 max	0.90-1.50	0.30 max	0.030 max	0.025 max	0.40 max	Al >=0.020; Mo <=0.08; Nb <=0.010; Ti <=0.03; V <=0.02; Cr+Cu+Mo+Ni<=0.70; bal Fe	460-580	260	21T	
EN 10028/2(92)	P295GH	High-temp, Press; 40<=t<=60mm	0.08-0.20	0.30 max	0.30 max	0.90-1.50	0.30 max	0.030 max	0.025 max	0.40 max	Al >=0.020; Mo <=0.08; Nb <=0.010; Ti <=0.03; V <=0.02; Cr+Cu+Mo+Ni<=0.70; bal Fe	460-580	285	22T	
EN 10028/2(92)	P295GH	High-temp, Press; 16<=t<=40mm	0.08-0.20	0.30 max	0.30 max	0.90-1.50	0.30 max	0.030 max	0.025 max	0.40 max	Al >=0.020; Mo <=0.08; Nb <=0.010; Ti <=0.03; V <=0.02; Cr+Cu+Mo+Ni<=0.70; bal Fe	460-580	290	22T	
EN 10028/2(92)	P295GH	High-temp, Press; 100<=t<=150mm	0.08-0.20	0.30 max	0.30 max	0.90-1.50	0.30 max	0.030 max	0.025 max	0.40 max	Al >=0.020; Mo <=0.08; Nb <=0.010; Ti <=0.03; V <=0.02; Cr+Cu+Mo+Ni<=0.70; bal Fe	440-570	235	21T	
EN 10028/2(92)	P295GH	High-temp, Press; 3<=t<=16mm	0.08-0.20	0.30 max	0.30 max	0.90-1.50	0.30 max	0.030 max	0.025 max	0.40 max	Al >=0.020; Mo <=0.08; Nb <=0.010; Ti <=0.03; V <=0.02; Cr+Cu+Mo+Ni<=0.70; bal Fe	460-580	295	22T	
EN 10113/2(93)	1.0546	HR weld t<=16mm	0.18 max	0.30 max	0.35 max	0.90-1.65	0.50 max	0.030 max	0.025 max	0.50 max	Al >=0.02; Mo <=0.10; N <=0.015; Nb <=0.05; Ti <=0.03; V <=0.12; CEV; bal Fe	470-630	355	22	
EN 10113/2(93)	S355NL	HR weld t<=16mm	0.18 max	0.30 max	0.35 max	0.90-1.65	0.50 max	0.030 max	0.025 max	0.50 max	Al >=0.02; Mo <=0.10; N <=0.015; Nb <=0.05; Ti <=0.03; V <=0.12; CEV; bal Fe	470-630	355	22	

France

Specification	Designation	Notes	C	Cr	Cu	Mn	Ni	P	S	Si	Other	UTS	YS	El	Hard
AFNOR NFA36207(82)	A50Pb	5-25mm; Norm	0.18 max	0.2 max	0.3 max	1.6 max	0.2 max	0.035 max	0.035 max	0.5 max	Mo <=0.1; bal Fe	510-610	335	22	
AFNOR NFA36207(82)	A50Pb	80<t<=100mm	0.2 max	0.2 max	0.3 max	1.6 max	0.2 max	0.035 max	0.035 max	0.5 max	Mo <=0.1; bal Fe	490-590	305		
AFNOR NFA36207(82)	A510FP	80<t<=100mm; Norm	02 max	0.2 max	0.3 max	1.6 max	0.2 max	0.03 max		0.2 max	0.5 max	Mo <=0.1; Nb 0.01-0.06; bal Fe	490-590	305	21	
AFNOR NFA36207(82)	A510FP	5-25mm; Norm	0.18 max	0.2 max	0.3 max	1.6 max	0.2 max	0.03 max		0.2 max	0.5 max	Mo <=0.1; Nb 0.01-0.06; bal Fe	510-610	335	22	
AFNOR NFA36207(82)	A530AP-I*	50<t<=80mm; Norm; EN10028/3	0.22 max	0.2 max	0.3 max	1.6 max	0.2 max	0.035 max	0.035 max	0.5 max	Mo <=0.1; Nb 0.01-0.06; bal Fe	500-600	325	21	
AFNOR NFA36207(82)	A530AP-I*	5-16mm; Norm; EN10028/3	0.2 max	0.2 max	0.3 max	1.6 max	0.2 max	0.035 max	0.035 max	0.5 max	Mo <=0.1; Nb 0.01-0.06; bal Fe	530-630	355	21	
AFNOR NFA36207(82)	A530AP-II*	50<t<=80mm; Norm; EN10028/3	0.2 max	0.2 max	0.3 max	1.6 max	0.2 max	0.035 max	0.035 max	0.5 max	Mo <=0.1; bal Fe	500-600	325	21	

UNS numbers and US grades are provided as a means of cross referencing chemically similar alloys. Exchangability is only possible after independent examination of specifications. Tensile properties are minimum or typical as specified. UTS and YS as MPa. El as %. See Appendix for list of abbreviations used in Notes. * indicates obsolete material.

Specification	Designation	Notes	C	Cr	Cu	Mn	Ni	P	S	Si	Other	UTS	YS	El	Hard

Carbon Steel, High-Manganese, A516(70) (Continued from previous page)

France

Specification	Designation	Notes	C	Cr	Cu	Mn	Ni	P	S	Si	Other	UTS	YS	El	Hard
AFNOR NFA36207(82)	A530AP-II*	5-16mm; Norm; EN10028/3	0.18 max	0.2 max	0.3 max	1.6 max	0.2 max	0.035 max	0.035 max	0.5 max	Mo <=0.1; bal Fe	530-630	355	21	
AFNOR NFA36207(82)	A530FP-I*	50<t<=80mm; Norm; EN10028/3	0.22 max	0.2 max	0.3 max	1.6 max	0.2 max	0.03 max	0.02 max	0.5 max	Mo <=0.1; Nb 0.01-0.06; bal Fe	500-600	325	21	
AFNOR NFA36207(82)	A530FP-I*	5-16mm; Norm; EN10028/3	0.2 max	0.2 max	0.3 max	1.6 max	0.2 max	0.03 max	0.02 max	0.5 max	Mo <=0.1; Nb 0.01-0.06; bal Fe	530-630	355	21	
AFNOR NFA36207(82)	A530FP-II*	5-16mm; Norm; EN10028/3	0.18 max	0.2 max	0.3 max	1.6 max	0.2 max	0.03 max	0.02 max	0.5 max	Mo <=0.1; bal Fe	530-630	355	21	
AFNOR NFA36207(82)	A530FP-II*	50<t<=80mm; Norm; EN10028/3	0.2 max	0.2 max	0.3 max	1.6 max	0.2 max	0.03 max	0.02 max	0.5 max	Mo <=0.1; bal Fe	500-600	325	21	

Hungary

Specification	Designation	Notes	C	Cr	Cu	Mn	Ni	P	S	Si	Other	UTS	YS	El	Hard
MSZ 6280(82)	52E	Weld const; 50<t<=60mm; rolled/norm	0.16 max	0.25 max		1.6 max	1.5 max	0.03 max	0.03 max	0.15-0.5	Al 0.015-0.120; Mo <=0.1; Nb 0.01-0.06; Ti 0.02-0.06; V 0.02-0.1; Zr=0.01-0.06; Al+Nb+V+Zr<=0.15; N<=Al/2+V/4+Nb/7+Ti/3.5<=0.015; bal Fe	490-610	345	23L	
MSZ 6280(82)	52E	Weld const; 3-35mm; rolled/norm	0.16 max	0.25 max		1.6 max	1.5 max	0.03 max	0.03 max	0.15-0.5	Al 0.015-0.120; Mo <=0.1; Nb 0.01-0.06; Ti 0.02-0.06; V 0.02-0.1; Zr=0.01-0.06; Al+Nb+V+Zr<=0.15; N<=Al/2+V/4+Nb/7+Ti/3.5<=0.015; bal Fe	490-610	355	23L	

Italy

Specification	Designation	Notes	C	Cr	Cu	Mn	Ni	P	S	Si	Other	UTS	YS	El	Hard
UNI 5869(75)	Fe510-1KG	Press ves	0.22 max			1.6 max		0.035 max	0.04	0.4 max	0 50mm C<=0.2; bal Fe				
UNI 5869(75)	Fe510-1KW	Press ves	0.22 max			1.6 max		0.04 max	0.04 max	0.4 max	0-50mm C<=0.2; bal Fe				
UNI 5869(75)	Fe510-2KG	Press ves	0.22 max			1.6 max		0.03 max	0.035 max	0.4 max	0-50mm C<=0.2; bal Fe				
UNI 5869(75)	Fe510-2KW	Press ves	0.22 max			1.6 max		0.035 max	0.035 max	0.4 max	0-50mm C<=0.2; bal Fe				

Romania

Specification	Designation	Notes	C	Cr	Cu	Mn	Ni	P	S	Si	Other	UTS	YS	El	Hard
STAS 9021/1(89)	OCS355.5a	Weld const; Fine	0.18 max	0.3 max		1.1-1.6	0.7 max	0.03 max	0.03 max	0.5 max	Al 0.015-0.120; N <=0.02; Nb >=0.02; Pb <=0.15; V 0.03-0.130; bal Fe				
STAS 9021/1(89)	OCS355.5b	Weld const; Fine	0.18 max	0.3 max		1.1-1.6	0.7 max	0.03 max	0.03 max	0.5 max	Al 0.015-0.120; N <=0.02; Nb >=0.02; Pb <=0.15; V 0.03-0.130; bal Fe				
STAS 9021/1(89)	OCS355.6a	Weld const; Fine	0.18 max	0.3 max		1.1-1.6	0.7 max	0.03 max	0.03 max	0.5 max	Al 0.015-0.120; N <=0.02; Nb >=0.02; Pb <=0.15; V 0.03-0.130; bal Fe				
STAS 9021/1(89)	OCS355.6b	Weld const; Fine	0.18 max	0.3 max		1.1-1.6	0.7 max	0.03 max	0.03 max	0.5 max	Al 0.015-0.120; N <=0.02; Nb >=0.02; Pb <=0.15; V 0.03-0.130; bal Fe				
STAS 9021/1(89)	OCS355.7a	Weld const; Fine	0.18 max	0.3 max		1.1-1.6	0.7 max	0.03 max	0.03 max	0.5 max	Al 0.015-0.120; N <=0.02; Nb >=0.02; Pb <=0.15; V 0.03-0.130; bal Fe				
STAS 9021/1(89)	OCS355.7b	Weld const; Fine	0.18 max	0.3 max		1.1-1.6	0.7 max	0.03 max	0.03 max	0.5 max	Al 0.015-0.120; N <=0.02; Nb >=0.02; Pb <=0.15; V 0.03-0.130; bal Fe				

Spain

Specification	Designation	Notes	C	Cr	Cu	Mn	Ni	P	S	Si	Other	UTS	YS	El	Hard
UNE 36081(76)	AE355KT	Fine grain Struct	0.18 max	0.25 max		0.9-1.6	0.3 max	0.03 max	0.03 max	0.5 max	Mo <=0.1; Al+V+Nb; bal Fe				
UNE 36081(76)	AE355KW	Fine grain Struct	0.20 max	0.25 max	0.35 max	0.9-1.6	0.3 max	0.035 max	0.035 max	0.5 max	Al <=0.015; Mo <=0.1; V 0.02-0.2; 0-35mm C<=0.2; 35mm C<=0.18; Nb=0.015-0.06; bal Fe				

Specification	Designation	Notes	C	Cr	Cu	Mn	Ni	P	S	Si	Other	UTS	YS	El	Hard

Carbon Steel, High-Manganese, A516(70) (Continued from previous page)

Spain

Specification	Designation	Notes	C	Cr	Cu	Mn	Ni	P	S	Si	Other	UTS	YS	El	Hard
UNE 36081(76)	F.6411*	see AE355KW	0.20 max	0.25 max	0.35 max	0.9-1.6	0.3 max	0.035 max	0.035 max	0.5 max	Al <=0.015; Mo <=0.1; V 0.02-0.2; 0-35mm C<=0.2; 35mm C<=0.18; Nb=0.015-0.06; bal Fe				
UNE 36081(76)	F.6412*	see AE355KT	0.18 max	0.25 max		0.9-1.6	0.3 max	0.03 max	0.03 max	0.5 max	Mo <=0.1; Al+V+Nb; bal Fe				
UNE 36087(74)	A47RAI	Sh Plt	0.23 max			0.5-1.5		0.055 max	0.05 max	0.4 max	Mo <=0.15; N <=0.009; bal Fe				
UNE 36087(74)	A47RAII	Sh Plt	0.23 max			0.5-1.5		0.045 max	0.04 max	0.4 max	Al 0.015-0.120; Mo <=0.15; bal Fe				
UNE 36087(74)	A47RCI	Sh Plt	0.23 max			0.7-1.5		0.055 max	0.05 max	0.4 max	Mo <=0.15; N <=0.009; bal Fe				
UNE 36121(85)	AE320-1B	CR; Strp	0.20 max			1.5 max		0.04 max	0.04 max	0.60 max	Mo <=0.15; N <=0.009; bal Fe				
UNE 36121(85)	AE320-1D	CR; Strp	0.20 max			1.5 max		0.03 max	0.03 max	0.60 max	Mo <=0.15; N <=0.015; bal Fe				
UNE 36123(90)	AE340HC	HR; Strp	0.10 max			1.0 max		0.03 max	0.5 max	0.01-0.12	Al 0.02-0.08; Mo <=0.15; Nb 0.01-0.06; Pb <=0.03; Ti 0.01-0.08; W <=0.15; bal Fe				

Sweden

Specification	Designation	Notes	C	Cr	Cu	Mn	Ni	P	S	Si	Other	UTS	YS	El	Hard
SS 142135	2135	Fine grain Struct	0.2 max			1.6 max		0.035 max	0.035 max	0.5 max	N <=0.02; Nb <=0.05; V <=0.15; C+Mn/6+(Ni+Cu)/15+(Cr+Mo+V)/5<=0.41; bal Fe				

USA

Specification	Designation	Notes	C	Cr	Cu	Mn	Ni	P	S	Si	Other	UTS	YS	El	Hard
	UNS K02700		0.27 max			0.79-1.30		0.035 max	0.04 max	0.13-0.45	bal Fe				
ASTM A516(90)	Grade 70	Press ves plt, mod/low-temp, t>100mm	0.31 max			0.85-1.20		0.035 max	0.035 max	0.15-0.40	bal Fe	485-620	260	21	
ASTM A516(90)	Grade 70	Press ves plt, mod/low-temp, t<=12.5mm	0.27 max			0.85-1.20		0.035 max	0.035 max	0.15-0.40	bal Fe	485-620	260	21	
ASTM A516(90)	Grade 70	Press ves plt, mod/low-temp, t>50-100mm	0.30 max			0.85-1.20		0.035 max	0.035 max	0.15-0.40	bal Fe	485-620	260	21	
ASTM A516(90)	Grade 70	Press ves plt, mod/low-temp, t>12.5-50mm	0.28 max			0.85-1.20		0.035 max	0.035 max	0.15-0.40	bal Fe	485-620	260	21	

Carbon Steel, High-Manganese, A529

China

Specification	Designation	Notes	C	Cr	Cu	Mn	Ni	P	S	Si	Other	UTS	YS	El	Hard
GB 713(97)	22Mng	Plt HR Norm 25mm Thk	0.30 max		0.30 max	0.90-1.50		0.025 max	0.025 max	0.15-0.40	bal Fe	515-655	275	19	

USA

Specification	Designation	Notes	C	Cr	Cu	Mn	Ni	P	S	Si	Other	UTS	YS	El	Hard
	UNS K02703		0.27 max			1.20 max		0.04 max	0.05 max		Cu 0.20 min (when spec'd); bal Fe				
ASTM A529(96)	Grade 50	Struct qual HS	0.27 max		0.20 max	1.35 max		0.04 max	0.05 max	0.40 max	bal Fe	485-690	345	21	
ASTM A529(96)	Grade 55	Struct qual HS	0.27 max		0.20 max	1.35 max		0.04 max	0.05 max	0.40 max	bal Fe	485-690	380	20	

Carbon Steel, High-Manganese, A53(E-B)

Mexico

Specification	Designation	Notes	C	Cr	Cu	Mn	Ni	P	S	Si	Other	UTS	YS	El	Hard
NMX-B-205(86)	Grade B Type E	Pipe for high-press cable circuits, Electric res weld	0.26 max			1.15 max		0.040 max	0.050 max		bal Fe	415	240	30	
NMX-B-205(86)	Grade B Type S	Pipe for high-press cable circuits, Smls	0.27 max			1.15 max		0.040 max	0.050 max		bal Fe	415	240	30	

USA

Specification	Designation	Notes	C	Cr	Cu	Mn	Ni	P	S	Si	Other	UTS	YS	El	Hard
	UNS K03005		0.30 max			1.20 max		0.05 max	0.06 max		bal Fe				
ASTM A523(96)	B-ERW	ERW pipe	0.26 max			1.15 max		0.035 max	0.050 max		bal Fe	415	240	30	
ASTM A523(96)	B-S	Smls Weld Pipe	0.27 max			1.15 max		0.035 max	0.050 max		bal Fe	415	240	30	

UNS numbers and US grades are provided as a means of cross referencing chemically similar alloys. Exchangability is only possible after independent examination of specifications. Tensile properties are minimum or typical as specified. UTS and YS as MPa. El as %. See Appendix for list of abbreviations used in Notes. * indicates obsolete material.

Specification	Designation	Notes	C	Cr	Cu	Mn	Ni	P	S	Si	Other	UTS	YS	El	Hard

Carbon Steel, High-Manganese, A557(C2)

USA

Specification	Designation	Notes	C	Cr	Cu	Mn	Ni	P	S	Si	Other	UTS	YS	El	Hard
	UNS K03505		0.35 max			0.27-1.06		0.050 max	0.060 max		bal Fe				

Carbon Steel, High-Manganese, A570(45)

UK

Specification	Designation	Notes	C	Cr	Cu	Mn	Ni	P	S	Si	Other	UTS	YS	El	Hard
BS 1502(82)	151	Bar Shp Press ves	0.25 max			0.60-1.40		0.040 max	0.040 max	0.35 max	W <=0.1; bal Fe				
BS 1502(82)	161	Bar Shp Press ves	0.25 max			0.60-1.40		0.040 max	0.040 max	0.1-0.35	W <=0.1; bal Fe				

USA

Specification	Designation	Notes	C	Cr	Cu	Mn	Ni	P	S	Si	Other	UTS	YS	El	Hard
	UNS K02507		0.25 max			1.35 max		0.040 max	0.05 max		Cu 0.20 min (when spec'd); bal Fe				
ASTM A570/A570M(98)	45	HR Sh Strp	0.25 max			1.35 max		0.035 max	0.04 max		If Cu bearing reqd, Cu >=0.20; bal Fe	380	310	13.0-19.0	
ASTM A857/A857M(97)	Grade 30	Sht piling, CF, light gage	0.25 max		0.20 min	1.50 max		0.35 max	0.04 max		bal Fe	340	205	23	

Carbon Steel, High-Manganese, A573(65)

Romania

Specification	Designation	Notes	C	Cr	Cu	Mn	Ni	P	S	Si	Other	UTS	YS	El	Hard
STAS 500/2(88)	OL44.2k	Gen struct; FF; Non-rimming	0.2 max	0.3 max		1.1 max	0.4 max	0.05 max	0.05 max	0.5 max	Pb <=0.15; bal Fe				
STAS 500/2(88)	OL44.3k	Gen struct; FF; Non-rimming	0.2 max	0.3 max		1.1 max		0.045 max	0.045 max	0.5 max	Al 0.025-0.120; Pb <=0.15; As<=0.08; bal Fe				
STAS 500/2(88)	OL44.3kf	Gen struct; Specially killed	0.2 max	0.3 max		1.1 max		0.045 max	0.045 max	0.5 max	Al 0.025-0.120; Pb <=0.15; As<=0.08; bal Fe				
STAS 500/2(88)	OL44.4kf	Gen struct; Specially killed	0.2 max	0.3 max		1.1 max		0.04 max	0.04 max	0.5 max	Al 0.02-0.120; Pb <=0.15; As<=0.08; bal Fe				

USA

Specification	Designation	Notes	C	Cr	Cu	Mn	Ni	P	S	Si	Other	UTS	YS	El	Hard
	UNS K02404		0.24 max			0.85-1.20		0.04 max	0.05 max	0.15-0.40	bal Fe				
ASTM A573/A753M(93)	65	Struct plt, t>=13-40mm	0.26 max			0.85-1.20		0.035 max	0.04 max	0.15-0.40	bal Fe	450-30	240	23	
ASTM A573/A753M(93)	65	Struct plt, t<=13mm	0.24 max			0.85-1.20		0.035 max	0.04 max	0.15-0.40	bal Fe	450-530	240	23	

Carbon Steel, High-Manganese, A573(70)

France

Specification	Designation	Notes	C	Cr	Cu	Mn	Ni	P	S	Si	Other	UTS	YS	El	Hard
AFNOR NFA35501(83)	E28-3*	30<t<=100mm; roll	0.18 max	0.3 max	0.3 max	1.3 max	0.4 max	0.04 max	0.04 max	0.4 max	bal Fe	400-540	255	23	
AFNOR NFA35501(83)	E28-3*	t<=30mm; roll	0.18 max	0.3 max	0.3 max	1.3 max	0.4 max	0.04 max	0.04 max	0.4 max	bal Fe	400-540	275	24	
AFNOR NFA35501(83)	E28-4*	t<3mm; roll	0.18 max	0.3 max		1.3 max	0.4 max	0.035 max	0.035 max	0.4 max	Al >=0.02; bal Fe	420-560	255	21	
AFNOR NFA35501(83)	E28-4*	110<t<=150mm; roll	0.18 max	0.3 max		1.3 max	0.4 max	0.035 max	0.035 max	0.4 max	Al >=0.02; bal Fe	400-540	225	19	
AFNOR NFA35501(83)	E28-4*	t<=30mm; roll	0.18 max	0.3 max		1.3 max	0.4 max	0.035 max	0.035 max	0.4 max	Al >=0.02; bal Fe	400-540	275	24	
AFNOR NFA35501(83)	E28-4*	30<t<=100mm; roll	0.18 max	0.3 max		1.3 max	0.4 max	0.035 max	0.035 max	0.4 max	Al >=0.02; bal Fe	400-540	255	25	
AFNOR NFA35501(83)	E28-4*	10t<=350mm; roll	0.18 max	0.3 max		1.3 max	0.4 max	0.035 max	0.035 max	0.4 max	Al >=0.02; bal Fe	380	235	24	

Hungary

Specification	Designation	Notes	C	Cr	Cu	Mn	Ni	P	S	Si	Other	UTS	YS	El	Hard
MSZ 500(89)	Fe275C	Gen struct; Non-rimming; FF; QS; 0-1mm; HR/Frg	0.2 max			2 max		0.04 max	0.04 max	0.6 max	Al >=0.02; N <=0.009; bal Fe	430-580	275	14L; 12T	
MSZ 500(89)	Fe275C	Gen struct; Non-rimming; FF; QS; 200.1-250mm; HR/Frg	0.2 max			2 max		0.04 max	0.04 max	0.6 max	Al >=0.02; N <=0.009; bal Fe	380-540	205	17L; 17T	
MSZ 500(89)	Fe275D	Gen struct; specially killed; QS; 200.1-250mm; HR/Frg	0.2 max			2 max		0.035 max	0.035 max	0.6 max	bal Fe	380-540	205	17L; 17T	
MSZ 500(89)	Fe275D	Gen struct; specially killed; QS; 0-1mm; HR/Frg	0.2 max			2 max		0.035 max	0.035 max	0.6 max	bal Fe	430-580	275	14L; 12T	

UNS numbers and US grades are provided as a means of cross referencing chemically similar alloys. Exchangability is only possible after independent examination of specifications. Tensile properties are minimum or typical as specified. UTS and YS as MPa. El as %. See Appendix for list of abbreviations used in Notes. * indicates obsolete material.

Specification	Designation	Notes	C	Cr	Cu	Mn	Ni	P	S	Si	Other	UTS	YS	El	Hard

Carbon Steel, High-Manganese, A573(70) (Continued from previous page)

Hungary

Specification	Designation	Notes	C	Cr	Cu	Mn	Ni	P	S	Si	Other	UTS	YS	El	Hard
MSZ 6280(82)	45C	Weld const; 40<t<=60mm; rolled/norm	0.18 max	0.25 max		1.3 max	0.3 max	0.04 max	0.04 max	0.15-0.5	Al 0.015-0.120; Mo <=0.1; Nb 0.01-0.06; Ti 0.02-0.06; V 0.02-0.1; Zr=0.01-0.06; Al+Nb+V+Zr<=0.15; N<=Al/2+V/4+Nb/7+Ti/3.5<=0.015; bal Fe	440-550	255	24L	
MSZ 6280(82)	45C	Weld const; 3-16mm; rolled/norm	0.18 max	0.25 max		1.3 max	0.3 max	0.04 max	0.04 max	0.15-0.5	Al 0.015-0.120; Mo <=0.1; Nb 0.01-0.06; Ti 0.02-0.06; V 0.02-0.1; Zr=0.01-0.06; Al+Nb+V+Zr<=0.15; N<=Al/2+V/4+Nb/7+Ti/3.5<=0.015; bal Fe	440-550	295	24L	
MSZ 6280(82)	45D	Weld const; 40<t<=60mm; rolled/norm	0.18 max	0.25 max		1.3 max	0.3 max	0.035 max	0.035 max	0.15-0.5	Al 0.015-0.120; Mo <=0.1; Nb 0.01-0.06; Ti 0.02-0.06; V 0.02-0.1; Zr=0.01-0.06; Al+Nb+V+Zr<=0.15; N<=Al/2+V/4+Nb/7+Ti/3.5<=0.015; bal Fe	440-550	255	24L	
MSZ 6280(82)	45D	Weld const; 3-16mm; rolled/norm	0.18 max	0.25 max		1.3 max	0.3 max	0.035 max	0.035 max	0.15-0.5	Al 0.015-0.120; Mo <=0.1; Nb 0.01-0.06; Ti 0.02-0.06; V 0.02-0.1; Zr=0.01-0.06; Al+Nb+V+Zr<=0.15; N<=Al/2+V/4+Nb/7+Ti/3.5<=0.015; bal Fe	440-550	295	24L	

International

Specification	Designation	Notes	C	Cr	Cu	Mn	Ni	P	S	Si	Other	UTS	YS	El	Hard
ISO 630(80)	Fe430B	Struct; FF, FN; 0-40mm	0.21 max	0.3 max	0.3 max	2 max	0.4 max	0.05 max	0.05 max	0.6 max	Al <=0.1; Co <=0.1; Mo <=0.15; Pb <=0.15; Ti <=0.05; V <=0.1; W <=0.1; bal Fe				
ISO 630(80)	Fe430B	Struct; FF, FN; Thk>40mm	0.22 max	0.3 max	0.3 max	2 max	0.4 max	0.05 max	0.05 max	0.6 max	Mo <=0.15; N <=0.009; Pb <=0.15; Ti <=0.05; V <=0.1;				
ISO 630(80)	Fe430C	Struct; FN; FF	0.27 max	0.3 max	0.3 max	2 max	0.4 max	0.045 max	0.045 max	0.6 max	Al <=0.1; Co <=0.1; Mo <=0.15; N <=0.009; Pb <=0.15; Ti <=0.05; V <=0.1; W <=0.1; bal Fe				
ISO 630(80)	Fe430C	Struct; FN; FF	0.27 max	0.3 max	0.3 max	2 max	0.4 max	0.045 max	0.045 max	0.6 max	Mo <=0.15; N <=0.009; Pb <=0.15; Ti <=0.05; V <=0.1;				
ISO 630(80)	Fe430C	Struct; FN; FF	0.27 max	0.3 max	0.3 max	2 max	0.4 max	0.045 max	0.045 max	0.6 max	Mo <=0.15; N <=0.009; Pb <=0.15; Ti <=0.05; V <=0.1;				
ISO 630(80)	Fe430D	Struct; FF	0.27 max	0.3 max	0.3 max	2 max	0.4 max	0.04 max	0.04 max	0.6 max	Mo <=0.15; Pb <=0.15; Ti <=0.05; V <=0.1;				
ISO 630(80)	Fe430D	Struct; FF	0.27 max	0.3 max	0.3 max	2 max	0.4 max	0.04 max	0.04 max	0.6 max	Al <=0.1; Co <=0.1; Mo <=0.15; Pb <=0.15; Ti <=0.05; V <=0.1; W <=0.1; bal Fe				
ISO 630(80)	Fe430D	Struct; FF	0.27 max	0.3 max	0.3 max	2 max	0.4 max	0.04 max	0.04 max	0.6 max	Mo <=0.15; Pb <=0.15; Ti <=0.05; V <=0.1;				

Italy

Specification	Designation	Notes	C	Cr	Cu	Mn	Ni	P	S	Si	Other	UTS	YS	El	Hard
UNI 7070(82)	Fe430CFN	Gen struct	0.19 max			2 max		0.04 max	0.045 max	0.6 max	bal Fe				
UNI 7070(82)	Fe430DFF	Gen struct	0.19 max			2 max		0.04 max	0.04 max	0.6 max	Al <=0.015; bal Fe				

Spain

Specification	Designation	Notes	C	Cr	Cu	Mn	Ni	P	S	Si	Other	UTS	YS	El	Hard
UNE 36080(80)	A430C	Gen struct	0.2 max			1.60 max		0.045 max	0.045 max	0.60 max	Mo <=0.15; N <=0.009; bal Fe				
UNE 36080(80)	A430D	Gen struct	0.2 max			1.60 max		0.04 max	0.04 max	0.60 max	Mo <=0.15; +N; bal Fe				
UNE 36080(85)	AE275-C	Gen struct	0.2 max			1.60 max		0.04 max	0.04 max	0.60 max	Mo <=0.15; N <=0.009; bal Fe				

Specification	Designation	Notes	C	Cr	Cu	Mn	Ni	P	S	Si	Other	UTS	YS	El	Hard

Carbon Steel, High-Manganese, A573(70) (Continued from previous page)

Spain

Specification	Designation	Notes	C	Cr	Cu	Mn	Ni	P	S	Si	Other	UTS	YS	El	Hard
UNE 36080(85)	AE275-D	Gen struct	0.2 max			1.60 max		0.035 max	0.035 max	0.60 max	Mo <=0.15; bal Fe				
UNE 36080(85)	F 6211*	see AE275-C	0.2 max			1.60 max		0.04 max	0.04 max	0.60 max	Mo <=0.15; N <=0.009; bal Fe				
UNE 36080(85)	F.6212*	see AE275-D	0.2 max			1.60 max		0.035 max	0.035 max	0.60 max	Mo <=0.15; bal Fe				

Sweden

Specification	Designation	Notes	C	Cr	Cu	Mn	Ni	P	S	Si	Other	UTS	YS	El	Hard
SS 141411	1411	Gen struct; FF; Non-rimming	0.3 max			2 max		0.08 max	0.06 max	0.6 max	C:0.15; Si:0.25; Mn:0.4-1				
SS 141412	1412	Gen struct; FF; Non-rimming	0.2 max		0.4 max	2 max		0.05 max	0.05 max	0.5 max	N<=0.009/N<=0.012; Mn 0.4-1				
SS 141413	1413	Gen struct; FF; Non-rimming	0.18 max		0.4 max	2 max		0.05 max	0.05 max	0.5 max	N<=0.009/N<=0.012; Mn 0.8-1.4				
SS 141414	1414	Gen struct; FF; Non-rimming	0.18 max		0.4 max	1.60 max		0.04 max	0.04 max	0.5 max	N<=0.009/N<=0.012; Mn 0.8-1.4, bal Fe				

USA

Specification	Designation	Notes	C	Cr	Cu	Mn	Ni	P	S	Si	Other	UTS	YS	El	Hard
	UNS K02701		0.27 max			0.85-1.20		0.04 max	0.05 max	0.15-0.40	bal Fe				
ASTM A573/A753M(93)	70	Struct plt, t<=13mm	0.27 max			0.85-1.20		0.035 max	0.04 max	0.15-0.40	bal Fe	485-620	290	21	
ASTM A573/A753M(93)	70	Struct plt, t>=13-40mm	0.28 max			0.85-1.20		0.035 max	0.04 max	0.15-0.40	bal Fe	485-620	290	21	

Carbon Steel, High-Manganese, A595(B)

USA

Specification	Designation	Notes	C	Cr	Cu	Mn	Ni	P	S	Si	Other	UTS	YS	El	Hard
	UNS K02005		0.12-0.29			0.35-1.40		0.05 max	0.06 max		bal Fe				
ASTM A595(97)	595B	Tub, tapered; struct use	0.15-0.25			0.40-1.35		0.035 max	0.035 max	0.04 max	bal Fe	410	480	21.0	

Carbon Steel, High-Manganese, A612

Hungary

Specification	Designation	Notes	C	Cr	Cu	Mn	Ni	P	S	Si	Other	UTS	YS	El	Hard
MSZ 1741(89)	KL7	Boiler, Press ves; 0-20mm; HT	0.14-0.2		0.35 max	0.5-1.5	1.2 max	0.035 max	0.03 max	0.2-0.6	Al 0.02-0.06; Nb 0.01-0.06; Ti 0.02-0.06; V 0.02-0.1; Zr=0.015-0.06; Al+Nb+V+Ti+Zr <=0.15; N<=Al/2+V/4 +Nb/7+Ti/3.5<=0.015; bal Fe	510-650	355	19L	
MSZ 1741(89)	KL7	Boiler, Press ves; 40.1-60mm; HT	0.14-0.2		0.35 max	0.5-1.5	1.2 max	0.035 max	0.03 max	0.2-0.6	Al 0.02-0.06; Nb 0.01-0.06; Ti 0.02-0.06; V 0.02-0.1; Zr=0.015-0.06; Al+Nb+V+Ti+Zr <=0.15; N<=Al/2+V/4 +Nb/7+Ti/3.5<=0.015; bal Fe	510-650	330	19L	
MSZ 1741(89)	KL7C	Boiler, Press ves; 0-20mm; HT	0.14-0.2		0.35 max	1.5 max	1.2 max	0.035 max	0.03 max	0.2-0.6	Al 0.02-0.06; Nb 0.01-0.06; Ti 0.02-0.06; V 0.02-0.1; Zr=0.015-0.06; Al+Nb+V+Ti+Zr <=0.15; N<=Al/2+V/4 +Nb/7+Ti/3.5<=0.015; bal Fe	510-650	355	19L	
MSZ 1741(89)	KL7C	Boiler, Press ves; 40.1-60mm; HT	0.14-0.2		0.35 max	1.5 max	1.2 max	0.035 max	0.03 max	0.2-0.6	Al 0.02-0.06; Nb 0.01-0.06; Ti 0.02-0.06; V 0.02-0.1; Zr=0.015-0.06; Al+Nb+V+Ti+Zr <=0.15; N<=Al/2+V/4 +Nb/7+Ti/3.5<=0.015; bal Fe	510-650	330	19L	
MSZ 1741(89)	KL7D	Boiler, Press ves; 0-20mm; HT	0.14-0.2		0.35 max	1.5 max	1.2 max	0.035 max	0.03 max	0.2-0.6	Al 0.02-0.06; Nb 0.01-0.06; Ti 0.02-0.06; V 0.02-0.1; Zr=0.015-0.06; Al+Nb+V+Ti+Zr <=0.15; N<=Al/2+V/4 +Nb/7+Ti/3.5<=0.015; bal Fe	510-650	355	19L	

Specification	Designation	Notes	C	Cr	Cu	Mn	Ni	P	S	Si	Other	UTS	YS	El	Hard

Carbon Steel, High-Manganese, A612 (Continued from previous page)

Hungary

Specification	Designation	Notes	C	Cr	Cu	Mn	Ni	P	S	Si	Other	UTS	YS	El	Hard
MSZ 1741(89)	KL7D	Boiler, Press ves; 40.1-60mm; HT	0.14-0.2		0.35 max	1.5 max	1.2 max	0.035 max	0.03 max	0.2-0.6	Al 0.02-0.06; Nb 0.01-0.06; Ti 0.02-0.06; V 0.02-0.1; Zr=0.015-0.06; Al+Nb+V+Ti+Zr <=0.15; N<=Al/2+V/4 +Nb/7+Ti3.5<=0.015; bal Fe	510-650	330	19L	
MSZ 1741(89)	KL7E	Boiler, Press ves; 0-20mm; HT	0.14-0.2		0.35 max	1.5 max	1.2 max	0.035 max	0.03 max	0.2-0.6	Al 0.02-0.06; Nb 0.01-0.06; Ti 0.02-0.06; V 0.02-0.1; Zr=0.015-0.06; Al+Nb+V+Ti+Zr <=0.15; N<=Al/2+V/4 +Nb/7+Ti/3.5<=0.015; bal Fe	510-650	355	19L	
MSZ 1741(89)	KL7E	Boiler, Press ves; 40.1-60mm; HT	0.14-0.2		0.35 max	1.5 max	1.2 max	0.035 max	0.03 max	0.2-0.6	Al 0.02-0.06; Nb 0.01-0.06; Ti 0.02-0.06; V 0.02-0.1; Zr=0.015-0.06; Al+Nb+V+Ti+Zr <=0.15; N<=Al/2+V/4 +Nb/7+Ti/3.5<=0.015; bal Fe	510-650	330	19L	

Poland

Specification	Designation	Notes	C	Cr	Cu	Mn	Ni	P	S	Si	Other	UTS	YS	El	Hard
PNH84024	19G2	HS structural	0.16-0.22	0.3 max	0.25 max	1-1.4	0.25 max	0.045 max	0.045 max	0.4-0.6	bal Fe				

Romania

Specification	Designation	Notes	C	Cr	Cu	Mn	Ni	P	S	Si	Other	UTS	YS	El	Hard
STAS 2883/2(80)	R52	Heat res; Press ves	0.18 max	0.3 max		1.05-1.65	0.4 max	0.035 max	0.035 max	0.17-0.45	Al 0.025-0.120; N <=0.009; Pb <=0.15; V <=0.15; As<=0.06; bal Fe				

Spain

Specification	Designation	Notes	C	Cr	Cu	Mn	Ni	P	S	Si	Other	UTS	YS	El	Hard
UNE 36081(76)	AE460KW	Fine grain Struct	0.20 max	0.50 max		1.0-1.7	0.8 max	0.035 max	0.035 max	0.5 max	Mo <=0.4; Ti <=0.2; V 0.02-0.2; bal Fe				
UNE 36081(76)	F.6420*	see AE460KW	0.20 max	0.50 max		1.0-1.7	0.8 max	0.035 max	0.035 max	0.5 max	Mo <=0.4; Ti <=0.2; V 0.02-0.2; bal Fe				
UNE 36087(74)	A52RAII	Sh Plt	0.25 max			0.8-1.7		0.045 max	0.04 max	0.55 max	Al 0.015-0.10; Mo <=0.15; bal Fe				
UNE 36087(74)	A52RCI	Sh Plt	0.25 max			0.8-1.7		0.055 max	0.05 max	0.55 max	Al <=0.01; Mo <=0.15; N <=0.009; bal Fe				
UNE 36087(74)	A52RCII	Sh Plt	0.25 max			0.80-1.70		0.045 max	0.040 max	0.55 max	Al 0.015-0.10; N <=0.009; bal Fe				

Sweden

Specification	Designation	Notes	C	Cr	Cu	Mn	Ni	P	S	Si	Other	UTS	YS	El	Hard
SS 142101	2101	Struct, Const	0.2 max	0.25 max		0.8-1.6	0.2 max	0.035 max	0.03 max	0.25-0.5	Mo <=0.1; N <=0.012; Nb <=0.01; Ti <=0.03; V <=0.03; bal Fe				
SS 142102	2102	Struct, Const	0.2 max	0.25 max		0.8-1.6	0.2 max	0.035 max	0.03 max	0.25-0.5	Mo <=0.1; N <=0.012; Nb <=0.01; Ti <=0.03; V <=0.03; bal Fe				

USA

Specification	Designation	Notes	C	Cr	Cu	Mn	Ni	P	S	Si	Other	UTS	YS	El	Hard
	UNS K02900		0.29 max			0.92-1.46		0.035 max	0.040 max	0.13-0.45	bal Fe				
ASTM A612/A612M(96)		Press ves plt, for mod/low-temp, t>=20-25mm	0.25 max	0.25 max	0.35 max	1.00-1.35	0.25 max	0.035 max	0.035 max	0.15-0.50	Mo <=0.08; V <=0.08; bal Fe	560-695	345	22	
ASTM A612/A612M(96)		Press ves plt, for mod/low-temp, t<=20mm	0.25 max	0.25 max	0.35 max	1.00-1.35	0.25 max	0.035 max	0.035 max	0.15-0.40	Mo <=0.08; V <=0.08; bal Fe	560-725	345	22	

Carbon Steel, High-Manganese, A618(lb)

USA

Specification	Designation	Notes	C	Cr	Cu	Mn	Ni	P	S	Si	Other	UTS	YS	El	Hard
	UNS K02601		0.26 max			1.30 max			0.063 max		bal Fe				
ASTM A618(97)	la	HF Weld and Smls HSLA Struct Tub, Walls<=19.05mm	0.15 max		0.20 min	1.00 max		0.15 max	0.025 max		bal Fe	485	345	22	
ASTM A618(97)	la	HF Weld and Smls HSLA Struct Tub, Walls 19.05-38.1mm	0.15 max		0.20 min	1.00 max		0.15 max	0.025 max		bal Fe	460	315	22	

UNS numbers and US grades are provided as a means of cross referencing chemically similar alloys. Exchangability is only possible after independent examination of specifications. Tensile properties are minimum or typical as specified. UTS and YS as MPa. El as %. See Appendix for list of abbreviations used in Notes. * indicates obsolete material.

Carbon Steel, High-Manganese, A618(Ib) (Continued from previous page)

Specification	Designation	Notes	C	Cr	Cu	Mn	Ni	P	S	Si	Other	UTS	YS	El	Hard
USA															
ASTM A618(97)	Ib	HF Weld and Smls HSLA Struct Tub, Walls 19.05-38.1mm	0.20 max		0.20 min	1.35 max		0.025 max	0.025 max		bal Fe	460	315	22	
ASTM A618(97)	Ib	HF Weld and Smls HSLA Struct Tub, Walls<=19.05mm	0.20 max		0.20 min	1.35 max		0.025 max	0.025 max		bal Fe	485	345	22	

Carbon Steel, High-Manganese, A633

Specification	Designation	Notes	C	Cr	Cu	Mn	Ni	P	S	Si	Other	UTS	YS	El	Hard
Czech Republic															
CSN 413220	13220	Weld Const	0.15-0.2	0.3 max		1.3-1.8		0.04 max	0.04 max	0.25-0.5	Al 0.01-0.110; V 0.1-0.2; bal Fe				
Germany															
DIN 17102(89)	TStE285	see P275NL1	0.16 max	0.3 max	0.2 max	0.6-1.4	0.3 max	0.03 max	0.025 max	0.4 max	N <=0.02; Nb <=0.03; bal Fe				
Hungary															
MSZ 500(89)	Fe355D	Gen struct; specially killed; QS; 0-1mm; HR/Frg	0.22 max			1.6 max		0.035 max	0.035 max	0.55 max	0-30mm C<=0.2; 30-100mm C<=0.22; bal Fe	510-680	355	14L; 12T	
MSZ 500(89)	Fe355D	Gen struct; specially killed; QS; 200.1-250mm; HR/Frg	0.22 max			1.6 max		0.035 max	0.035 max	0.55 max	0-30mm C<=0.2; 30-100mm C<=0.22; bal Fe	450-630	275	17L; 17T	
MSZ 6280(74)	58C	weld const	0.2 max		0.35 max	1.6 max		0.04 max	0.04 max	0.2-0.6	Al <=0.120; Ti <=0.06; +micro; bal Fe				
MSZ 6280(82)	E420C	Weld const; 50<t<=60mm; rolled/norm	0.2 max	0.25 max		1.6 max	0.7 max	0.04 max	0.04 max	0.15-0.5	Al 0.015-0.120; Mo <=0.1; Nb 0.01-0.06; Ti 0.02-0.06; V 0.02-0.2; Zr=0.015-0.06; Al+Nb+V+Ti+Zr <=0.15; N<=Al/2+V/4 +Nb/7+Ti/3.5<=0.015; bal Fe	520-680	380	19L; 17T	
MSZ 6280(82)	E420C	Weld const; 3-16mm; rolled/norm	0.2 max	0.25 max		1.6 max	0.7 max	0.04 max	0.04 max	0.15-0.5	Al 0.015-0.120; Mo <=0.1; Nb 0.01-0.06; Ti 0.02-0.06; V 0.02-0.2; Zr=0.015-0.06; Al+Nb+V+Ti+Zr <=0.15; N<=Al/2+V/4 +Nb/7+Ti/3.5<=0.015; bal Fe	520-680	420	19L; 17T	
MSZ 6280(82)	E420D	Weld const; 50<t<=60mm; rolled/norm	0.18 max	0.25 max		1.7 max	0.7 max	0.035 max	0.035 max	0.15-0.5	Al 0.015-0.120; Mo <=0.1; Nb 0.01-0.06; Ti 0.02-0.06; V 0.02-0.2; Zr=0.015-0.06; Al+Nb+V+Ti+Zr <=0.15; N<=Al/2+V/4 +Nb/7+Ti/3.5<=0.015; bal Fe	520-680	380	19L; 17T	
MSZ 6280(82)	E420D	Weld const; 3-16mm; rolled/norm	0.18 max	0.25 max		1.7 max	0.7 max	0.035 max	0.035 max	0.15-0.5	Al 0.015-0.120; Mo <=0.1; Nb 0.01-0.06; Ti 0.02-0.06; V 0.02-0.2; Zr=0.015-0.06; Al+Nb+V+Ti+Zr <=0.15; N<=Al/2+V/4 +Nb/7+Ti/3.5<=0.015; bal Fe	520-680	420	19L; 17T	
MSZ 6280(82)	E420E	Weld const; 3-16mm; rolled/norm	0.18 max	0.25 max		1.7 max	1.5 max	0.03 max	0.03 max	0.15-0.5	Al 0.015-0.120; Mo <=0.1; Nb 0.01-0.06; Ti 0.02-0.06; V 0.02-0.2; Zr=0.015-0.06; Al+Nb+V+Ti+Zr <=0.15; N<=Al/2+V/4 +Nb/7+Ti/3.5<=0.015; bal Fe	520-680	420	19L; 17T	
MSZ 6280(82)	E420E	Weld const; 50<t<=60mm; rolled/norm	0.18 max	0.25 max		1.7 max	1.5 max	0.03 max	0.03 max	0.15-0.5	Al 0.015-0.120; Mo <=0.1; Nb 0.01-0.06; Ti 0.02-0.06; V 0.02-0.2; Zr=0.015-0.06; Al+Nb+V+Ti+Zr <=0.15; N<=Al/2+V/4 +Nb/7+Ti/3.5<=0.015; bal Fe	520-680	380	19L; 17T	

UNS numbers and US grades are provided as a means of cross referencing chemically similar alloys. Exchangability is only possible after independent examination of specifications. Tensile properties are minimum or typical as specified. UTS and YS as MPa. El as %. See Appendix for list of abbreviations used in Notes. * indicates obsolete material.

Specification	Designation	Notes	C	Cr	Cu	Mn	Ni	P	S	Si	Other	UTS	YS	El	Hard

Carbon Steel, High-Manganese, A633 (Continued from previous page)

Hungary

Specification	Designation	Notes	C	Cr	Cu	Mn	Ni	P	S	Si	Other	UTS	YS	El	Hard
MSZ 6280(82)	E460C	Weld const; 3-16mm; rolled/norm	0.2 max	0.25 max		1.7 max	1 max	0.04 max	0.04 max	0.15-0.5	Al 0.015-0.120; Mo <=0.4; Nb 0.01-0.06; Ti 0.02-0.06; V 0.02-0.2; Zr=0.015-0.06; Al+Nb+V+Ti+Zr <=0.15; N<=Al/2+V/4 +Nb/7+Ti/3.5<=0.015; bal Fe	550-720	460	17L; 15T	
MSZ 6280(82)	E460C	Weld const; 50<t<=60mm; rolled/norm	0.2 max	0.25 max		1.7 max	1 max	0.04 max	0.04 max	0.15-0.5	Al 0.015-0.120; Mo <=0.4; Nb 0.01-0.06; Ti 0.02-0.06; V 0.02-0.2; Zr=0.015-0.06; Al+Nb+V+Ti+Zr <=0.15; N<=Al/2+V/4 +Nb/7+Ti/3.5<=0.015; bal Fe	550-720	420	17L; 15T	
MSZ 6280(82)	E460D	Weld const; 50<t<=60mm; rolled/norm	0.18 max	0.25 max		1.7 max	1 max	0.035 max	0.035 max	0.15-0.5	Al 0.015-0.120; Mo <=0.4; Nb 0.01-0.06; Ti 0.02-0.06; V 0.02-0.2; Zr=0.015-0.06; Al+Nb+V+Ti+Zr <=0.15; N<=Al/2+V/4 +Nb/7+Ti/3.5<=0.015; bal Fe	550-720	420	17L; 15T	
MSZ 6280(82)	E460D	Weld const; 3-16mm; rolled/norm	0.18 max	0.25 max		1.7 max	1 max	0.035 max	0.035 max	0.15-0.5	Al 0.015-0.120; Mo <=0.4; Nb 0.01-0.06; Ti 0.02-0.06; V 0.02-0.2; Zr=0.015-0.06; Al+Nb+V+Ti+Zr <=0.15; N<=Al/2+V/4 +Nb/7+Ti/3.5<=0.015; bal Fe	550-720	460	17L; 15T	
MSZ 6280(82)	E460E	Weld const; 50<t<=60mm; rolled/norm	0.18 max	0.25 max		1.7 max	1.5 max	0.03 max	0.03 max	0.15-0.5	Al 0.015-0.120; Mo <=0.4; Nb 0.01-0.06; Ti 0.02-0.06; V 0.02-0.2; Zr=0.015-0.06; Al+Nb+V+Ti+Zr <=0.15; N<=Al/2+V/4 +Nb/7+Ti/3.5<=0.015; bal Fe	550-720	420	17L; 15T	
MSZ 6280(82)	E460E	Weld const; 3-16mm; rolled/norm	0.18 max	0.25 max		1.7 max	1.5 max	0.03 max	0.03 max	0.15-0.5	Al 0.015-0.120; Mo <=0.4; Nb 0.01-0.06; Ti 0.02-0.06; V 0.02-0.2; Zr=0.015-0.06; Al+Nb+V+Ti+Zr <=0.15; N<=Al/2+V/4 +Nb/7+Ti/3.5<=0.015; bal Fe	550-720	460	17L; 15T	

Poland

Specification	Designation	Notes	C	Cr	Cu	Mn	Ni	P	S	Si	Other	UTS	YS	El	Hard
PNH84018(86)	18G2ANb	HS weld const	0.2 max	0.3 max		1-1.6	0.3 max	0.04 max	0.04 max	0.2-0.55	Al 0.02-0.120; Nb 0.02-0.05; Pb <=0.1; CEV<=0.045; bal Fe				
PNH84018(86)	18G2AV	HS weld const	0.2 max	0.3 max		1.2-1.65	0.3 max	0.04 max	0.04 max	0.2-0.6	Al 0.01-0.120; N 0.01-0.025; V 0.05-0.15; CEV<=0.045; bal Fe				
PNH84018(86)	18G2AVCu	HS weld const	0.2 max	0.3 max	0.25-0.5	1.2-1.65	0.3 max	0.04 max	0.04 max	0.2-0.6	Al 0.01-0.120; N 0.01-0.025; V 0.05-0.2; CEV<=0.045; bal Fe				

Romania

Specification	Designation	Notes	C	Cr	Cu	Mn	Ni	P	S	Si	Other	UTS	YS	El	Hard
STAS 11502(80)	R55	Press ves	0.2 max	0.3 max	0.2 max	1.1-1.65	0.5 max	0.015 max	0.02 max	0.17-0.45	Al 0.02-0.120; Pb <=0.15; V <=0.12; As<=0.06; bal Fe				
STAS 11502(80)	R58	Press ves	0.2 max	0.3 max	0.2 max	1.1-1.65	1 max	0.015 max	0.02 max	0.17-0.45	Al 0.02-0.120; Pb <=0.15; V <=0.16; As<=0.06; bal Fe				
STAS 9021/1(89)	OCS55.5a	Weld const; Fine	0.2 max	0.3 max		1.1-1.65	0.7 max	0.025 max	0.035 max	0.4 max	Al 0.02-0.120; Nb >=0.02; Pb <=0.15; Ti 0.005-0.10; V 0.05-0.110; bal Fe				

UNS numbers and US grades are provided as a means of cross referencing chemically similar alloys. Exchangability is only possible after independent examination of specifications. Tensile properties are minimum or typical as specified. UTS and YS as MPa. El as %. See Appendix for list of abbreviations used in Notes. * indicates obsolete material.

Specification	Designation	Notes	C	Cr	Cu	Mn	Ni	P	S	Si	Other	UTS	YS	El	Hard

Carbon Steel, High-Manganese, A633 (Continued from previous page)

Sweden

Specification	Designation	Notes	C	Cr	Cu	Mn	Ni	P	S	Si	Other	UTS	YS	El	Hard
SS 142142	2142	Fine grain Struct	0.2 max			1.6 max		0.035 max	0.035 max	0.5 max	N <=0.02; Nb <=0.05; V <=0.15; C+Mn/6+(Ni+Cu)/15+ (Cr+Mo+V)/5<=0.41; bal Fe				
SS 142143	2143	Fine grain Struct	0.2 max		0.4 max	1.8 max		0.035 max	0.035 max	0.5 max	N <=0.02; C+Mn/6+(Ni+Cu)/15+ (Cr+Mo+V)/5<=0.45; bal Fe				
SS 142144	2144	Fine grain Struct	0.2 max			1.6 max		0.035 max	0.035 max	0.5 max	N <=0.02; Nb <=0.05; V <=0.15; C+Mn/6+(Ni+Cu)/15+ (Cr+Mo+V)/5<=0.45; bal Fe				
SS 142145	2145	Fine grain Struct	0.2 max			1.6 max		0.035 max	0.035 max	0.5 max	N <=0.02; Nb <=0.05; V <=0.15; C+Mn/6+(Ni+Cu)/15+ (Cr+Mo+V)/5<=0.45; bal Fe				

USA

Specification	Designation	Notes	C	Cr	Cu	Mn	Ni	P	S	Si	Other	UTS	YS	El	Hard
	UNS K01803		0.18 max			1.00-1.35		0.04 max	0.05 max	0.15-0.50	V <=0.10; bal Fe				

Carbon Steel, High-Manganese, A633(A)

China

Specification	Designation	Notes	C	Cr	Cu	Mn	Ni	P	S	Si	Other	UTS	YS	El	Hard
GB 1591(88)	14MnNb	Plt HR 16-25mm Thk	0.12-0.18			0.80-1.20		0.045 max	0.045 max	0.20-0.55	Nb 0.015-0.050; bal Fe	470-620	335	20	
GB/T 1591(94)	Q345	Plt HR 16-35mm Thk	0.20 max			1.00-1.60		0.045 max	0.045 max	0.55 max	Nb 0.015-0.060; Ti 0.02-0.2; V 0.02-0.15; bal Fe	470-630	325	21	
GB/T 8164(93)	14MnNb	Strp HR 4-8mm Thk	0.12-0.18			0.80-1.20		0.045 max	0.045 max	0.20-0.55	Nb 0.015-0.050; bal Fe	470	325	21	

Europe

Specification	Designation	Notes	C	Cr	Cu	Mn	Ni	P	S	Si	Other	UTS	YS	El	Hard
EN 10207(91)	1.0130	Press ves 40<t<=60mm	0.20 max			0.50-1.50		0.035 max	0.030 max	0.40 max	Al >=0.020; bal Fe	410-530	245	22	
EN 10207(91)	1.0130	Press ves 16<t<=40mm	0.20 max			0.50-1.50		0.035 max	0.030 max	0.40 max	Al >=0.020; bal Fe	410-530	255	22	
EN 10207(91)	1.0130	Press ves 3<t<=16mm	0.20 max			0.50-1.50		0.035 max	0.030 max	0.40 max	Al >=0.020; bal Fe	410-530	265	22	
EN 10207(91)	SPH265	Press ves 3<t<=16mm	0.20 max			0.50-1.50		0.035 max	0.030 max	0.40 max	Al >=0.020; bal Fe	410-530	265	22	
EN 10207(91)	SPH265	Press ves 40<t<=60mm	0.20 max			0.50-1.50		0.035 max	0.030 max	0.40 max	Al >=0.020; bal Fe	410-530	245	22	
EN 10207(91)	SPH265	Press ves 16<t<=40mm	0.20 max			0.50-1.50		0.035 max	0.030 max	0.40 max	Al >=0.020; bal Fe	410-530	255	22	

France

Specification	Designation	Notes	C	Cr	Cu	Mn	Ni	P	S	Si	Other	UTS	YS	El	Hard
AFNOR NFA36205(82)	A42AP*	Press ves; t<=30mm; Norm	0.18 max	0.25 max	0.30 max	0.60-2.00	0.3 max	0.035 max	0.03 max	0.3 max	Mo <=0.07; Nb <=0.015; V <=0.02; bal Fe	410-490	245	27	
AFNOR NFA36205(82)	A42AP*	Press ves; 80<t<=110mm; Norm	0.18 max	0.25 max	0.30 max	0.60-2.00	0.3 max	0.035 max	0.03 max	0.3 max	Mo <=0.07; Nb <=0.015; V <=0.02; bal Fe	410-490	215	24	
AFNOR NFA36205(82)	A42CP*	Press ves; t<=30mm; Norm	0.18 max	0.25 max	0.30 max	0.60-2.00	0.3 max	0.035 max	0.03 max	0.3 max	Mo <=0.07; Nb <=0.015; V <=0.02; bal Fe	410-490	245	27	
AFNOR NFA36205(82)	A42CP*	Press ves; 80<t<=110mm; Norm	0.18 max	0.25 max	0.30 max	0.60-2.00	0.3 max	0.035 max	0.03 max	0.3 max	Mo <=0.07; Nb <=0.015; V <=0.02; bal Fe	410-490	215	24	
AFNOR NFA36205(82)	A42FP*	Press ves; t<=30mm; Norm	0.18 max	0.25 max	0.30 max	0.60-1.60	0.4 max	0.03 max	0.02 max	0.3 max	Mo <=0.07; Nb <=0.015; V <=0.02; bal Fe	410-490	245	27	
AFNOR NFA36205(82)	A42FP*	Press ves; 80<t<=110mm; Norm	0.18 max	0.25 max	0.30 max	0.60-1.60	0.4 max	0.03 max	0.02 max	0.3 max	Mo <=0.07; Nb <=0.015; V <=0.02; bal Fe	410-490	215	24	
AFNOR NFA36601(80)	A42AP	80<t<=300mm; Q/T	0.18 max	0.3 max	0.18 max	0.6-1.60	0.5 max	0.035 max	0.03 max	0.3 max	Cu+6Sn<=0.33; bal Fe	410-490	215	26L; 24T	
AFNOR NFA36601(80)	A42AP	t<=30mm; Norm	0.18 max	0.3 max	0.18 max	0.6-1.60	0.5 max	0.035 max	0.03 max	0.3 max	Cu+6Sn<=0.33; bal Fe	410-490	245	29L; 27T	

UNS numbers and US grades are provided as a means of cross referencing chemically similar alloys. Exchangability is only possible after independent examination of specifications. Tensile properties are minimum or typical as specified. UTS and YS as MPa. El as %. See Appendix for list of abbreviations used in Notes. * indicates obsolete material.

Specification	Designation	Notes	C	Cr	Cu	Mn	Ni	P	S	Si	Other	UTS	YS	El	Hard

Carbon Steel, High-Manganese, A633(A) (Continued from previous page)

France

Specification	Designation	Notes	C	Cr	Cu	Mn	Ni	P	S	Si	Other	UTS	YS	El	Hard
AFNOR NFA36601(80)	A42CP	t<=30mm; Norm	0.18 max	0.3 max	0.18 max	0.6-1.60	0.5 max	0.035 max	0.03 max	0.3 max	Cu+6Sn<=0.33; bal Fe	410-490	245	29L; 27T	
AFNOR NFA36601(80)	A42CP	80<t<=300mm; Q/T	0.18 max	0.3 max	0.18 max	0.6-1.60	0.5 max	0.035 max	0.03 max	0.3 max	Cu+6Sn<=0.33; bal Fe	410-490	215	26L; 24T	
AFNOR NFA36601(80)	A42FP	t<=30mm; Norm	0.18 max	0.3 max	0.18 max	0.6-1.60	0.5 max	0.03 max	0.02 max	0.3 max	Cu+6Sn<=0.33; bal Fe	410-490	245	29L; 27T	
AFNOR NFA36601(80)	A42FP	80<t<=300mm; Q/T	0.18 max	0.3 max	0.18 max	0.6-1.60	0.5 max	0.03 max	0.02 max	0.3 max	Cu+6Sn<=0.33; bal Fe	410-490	215	26L; 24T	

Germany

Specification	Designation	Notes	C	Cr	Cu	Mn	Ni	P	S	Si	Other	UTS	YS	El	Hard
DIN 17102(83)	P315N	Norm 125-150mm	0.18 max	0.30 max	0.20 max	0.70-1.50	0.30 max	0.035 max	0.030 max	0.45 max	Al >=0.020; Mo <=0.08; N <=0.020; Nb <=0.03; Nb+Ti+V<=0.05; Cr+Cu+Mo<=0.45; bal Fe	400-520	255	23	
DIN 17102(83)	P315N	Norm 70-85mm	0.18 max	0.30 max	0.20 max	0.70-1.50	0.30 max	0.035 max	0.030 max	0.45 max	Al >=0.020; Mo <=0.08; N <=0.020; Nb <=0.03; Nb+Ti+V<=0.05; Cr+Cu+Mo<=0.45; bal Fe	430-550	285	23	
DIN 17102(83)	StE315	Norm 125-150mm	0.18 max	0.30 max	0.20 max	0.70-1.50	0.30 max	0.035 max	0.030 max	0.45 max	Al >=0.020; Mo <=0.08; N <=0.020; Nb <=0.03; Nb+Ti+V<=0.05; Cr+Cu+Mo<=0.45; bal Fe	400-520	255	23	
DIN 17102(83)	StE315	Norm 70-85mm	0.18 max	0.30 max	0.20 max	0.70-1.50	0.30 max	0.035 max	0.030 max	0.45 max	Al >=0.020; Mo <=0.08; N <=0.020; Nb <=0.03; Nb+Ti+V<=0.05; Cr+Cu+Mo<=0.45; bal Fe	430-550	285	23	
DIN 17102(83)	WNr 1.0505	Norm 70-85mm	0.18 max	0.30 max	0.20 max	0.70-1.50	0.30 max	0.035 max	0.030 max	0.45 max	Al >=0.020; Mo <=0.08; N <=0.020; Nb <=0.03; Nb+Ti+V<=0.05; Cr+Cu+Mo<=0.45; bal Fe	430-550	285	23	
DIN 17102(83)	WNr 1.0505	Norm 125-150mm	0.18 max	0.30 max	0.20 max	0.70-1.50	0.30 max	0.035 max	0.030 max	0.45 max	Al >=0.020; Mo <=0.08; N <=0.020; Nb <=0.03; Nb+Ti+V<=0.05; Cr+Cu+Mo<=0.45; bal Fe	400-520	255	23	
DIN 17102(83)	WNr 1.0506	Norm 70-85mm	0.18 max	0.30 max	0.20 max	0.70-1.50	0.30 max	0.035 max	0.030 max	0.45 max	Al >=0.020; Mo <=0.08; N <=0.020; Nb <=0.03; Nb+Ti+V<=0.05; Cr+Cu+Mo<=0.45; bal Fe	430-550	285	23	
DIN 17102(83)	WStE315	Norm 70-85mm	0.18 max	0.30 max	0.20 max	0.70-1.50	0.30 max	0.035 max	0.030 max	0.45 max	Al >=0.020; Mo <=0.08; N <=0.020; Nb <=0.03; Nb+Ti+V<=0.05; Cr+Cu+Mo<=0.45; bal Fe	430-550	285	23	
DIN 17102(89)	StE285	see P275N	0.18 max	0.30 max	0.30 max	0.50-1.40	0.50 max	0.030 max	0.025 max	0.40 max	Al >=0.020; Mo <=0.08; N <=0.020; Nb <=0.05; Ti <=0.03; V <=0.05; Nb+Ti+V<=0.05; Cr+Cu+Mo<=0.45; bal Fe				

Spain

Specification	Designation	Notes	C	Cr	Cu	Mn	Ni	P	S	Si	Other	UTS	YS	El	Hard
UNE 36081(76)	AE315KG	Fine grain Struct	0.18 max	0.25 max		0.7-1.5	0.3 max	0.035 max	0.035 max	0.4 max	Mo <=0.1; +Al+V+Nb; bal Fe				
UNE 36081(76)	AE315KW	Fine grain Struct	0.18 max	0.25 max		0.7-1.5	0.3 max	0.035 max	0.035 max	0.4 max	Mo <=0.1; +Al+V+Nb; bal Fe				
UNE 36081(76)	F.6407*	see AE315KG	0.18 max	0.25 max		0.7-1.5	0.3 max	0.035 max	0.035 max	0.4 max	Mo <=0.1; +Al+V+Nb; bal Fe				

UNS numbers and US grades are provided as a means of cross referencing chemically similar alloys. Exchangability is only possible after independent examination of specifications. Tensile properties are minimum or typical as specified. UTS and YS as MPa. El as %. See Appendix for list of abbreviations used in Notes. * indicates obsolete material.

Specification	Designation	Notes	C	Cr	Cu	Mn	Ni	P	S	Si	Other	UTS	YS	El	Hard

Carbon Steel, High-Manganese, A633(A) (Continued from previous page)

Spain

Specification	Designation	Notes	C	Cr	Cu	Mn	Ni	P	S	Si	Other	UTS	YS	El	Hard
UNE 36081(76)	F.6408*	see AE315KW	0.18 max	0.25 max		0.7-1.5	0.3 max	0.035 max	0.035 max	0.4 max	Mo <=0.1; +Al+V+Nb; bal Fe				

UK

Specification	Designation	Notes	C	Cr	Cu	Mn	Ni	P	S	Si	Other	UTS	YS	El	Hard
BS 1501(86)	223-490	Plate, fully killed	0.20 max	0.25 max	0.30 max	0.90-1.60	0.75 max	0.030 max	0.030 max	0.10-0.40	Mo <=0.10; Nb 0.010-0.060; bal Fe	490-610	355	21	
BS 1501(86)	224-430	Plate, fully killed, Al treated 3<t<=16	0.20 max	0.25 max	0.30 max	0.9-1.50	0.75 max	0.030 max	0.030 max	0.10-0.40	Al >=0.015; Mo <=0.10; others<=0.50; bal Fe	430-550	305	22	

USA

Specification	Designation	Notes	C	Cr	Cu	Mn	Ni	P	S	Si	Other	UTS	YS	El	Hard
	UNS K01802		0.18 max			1.00-1.35		0.04 max	0.05 max	0.15-0.50	Cb 0.05 max; bal Fe				
ASTM A633/A633M(95)	Class A	Norm struct Plt	0.18 max			1.00-1.35		0.035 max	0.04 max	0.15-0.50	Nb <=0.05; bal Fe	430-570	290	23	

Carbon Steel, High-Manganese, A648(II)

USA

Specification	Designation	Notes	C	Cr	Cu	Mn	Ni	P	S	Si	Other	UTS	YS	El	Hard
	UNS K06700		0.50-0.85			0.60-1.10		0.040 max	0.050 max	0.10-0.35	bal Fe				
ASTM A648/A648(95)	Class II	HD wire for conc pipe, 7.92mm diam	0.50-0.85			0.50-1.10		0.030 max	0.035 max	0.10-0.35	bal Fe	1390			
ASTM A648/A648(95)	Class II	HD wire for conc pipe, 4.88mm diam	0.50-0.85			0.50-1.10		0.030 max	0.035 max	0.10-0.35	bal Fe	1530			
ASTM A648/A648(95)	Class II	HD wire for conc pipe, 6.35mm diam	0.50-0.85			0.50-1.10		0.030 max	0.035 max	0.10-0.35	bal Fe	1450			
ASTM A821(93)	Type A		0.50-0.85			0.60-1.10		0.040 max	0.050 max	0.10-0.35	bal Fe	1500-1780			
ASTM A821(93)	Type B	HD wire-prestress con tank, t=4.11mm	0.50-0.85			0.60-1.10		0.040 max	0.050 max	0.10-0.35	bal Fe	1590-1810			

Carbon Steel, High-Manganese, A648(III)

USA

Specification	Designation	Notes	C	Cr	Cu	Mn	Ni	P	S	Si	Other	UTS	YS	El	Hard
	UNS K07100		0.55-0.85			0.60-1.10		0.040 max	0.050 max	0.10-0.35	bal Fe				
ASTM A648/A648(95)	Class III	HD wire for conc pipe, 7.92mm diam	0.50-0.85			0.50-1.10		0.030 max	0.035 max	0.10-0.35	bal Fe	1520			
ASTM A648/A648(95)	Class III	HD wire for conc pipe, 4.88mm diam	0.50-0.85			0.50-1.10		0.030 max	0.035 max	0.10-0.35	bal Fe	1740			
ASTM A648/A648(95)	Class III	HD wire for conc pipe, 6.35mm diam	0.50-0.85			0.50-1.10		0.030 max	0.035 max	0.10-0.35	bal Fe	1650			

Carbon Steel, High-Manganese, A662(A)

Hungary

Specification	Designation	Notes	C	Cr	Cu	Mn	Ni	P	S	Si	Other	UTS	YS	El	Hard
MSZ 2898/2(80)	A35K	Smls CF tube; Smls tubes w/special req; 0.5-10mm; bright-drawn, soft	0.17 max			0.4-1.60		0.05 max	0.05 max	0.1-0.35	bal Fe	370	260	10L	
MSZ 2898/2(80)	A35K	Smls CF tube; Smls tubes w/special req; 0.5-10mm; bright ann	0.17 max			0.4-1.60		0.05 max	0.05 max	0.1-0.35	bal Fe	310	155	28L	

Spain

Specification	Designation	Notes	C	Cr	Cu	Mn	Ni	P	S	Si	Other	UTS	YS	El	Hard
UNE 36081(76)	AE225KG	Fine grain Struct	0.17 max	0.25 max		0.5-1.4	0.3 max	0.035 max	0.035 max	0.35 max	Mo <=0.1; +Al+V+Nb; bal Fe				
UNE 36081(76)	AE225KT	Fine grain Struct	0.17 max	0.25 max		0.5-1.4	0.3 max	0.035 max	0.035 max	0.35 max	Mo <=0.1; +Al+V+Nb; bal Fe				
UNE 36081(76)	AE225KW	Fine grain Struct	0.17 max	0.25 max		0.5-1.4	0.3 max	0.035 max	0.035 max	0.35 max	Mo <=0.1; +Al+V+Nb; bal Fe				
UNE 36081(76)	F.6401*	see AE225KG	0.17 max	0.25 max		0.5-1.4	0.3 max	0.035 max	0.035 max	0.35 max	Mo <=0.1; +Al+V+Nb; bal Fe				
UNE 36081(76)	F.6402*	see AE225KW	0.17 max	0.25 max		0.5-1.4	0.3 max	0.035 max	0.035 max	0.35 max	Mo <=0.1; +Al+V+Nb; bal Fe				
UNE 36081(76)	F.6403*	see AE225KT	0.17 max	0.25 max		0.5-1.4	0.3 max	0.035 max	0.035 max	0.35 max	Mo <=0.1; +Al+V+Nb; bal Fe				

Sweden

Specification	Designation	Notes	C	Cr	Cu	Mn	Ni	P	S	Si	Other	UTS	YS	El	Hard
SS 141233	1233		0.17 max	0.2 max		2 max		0.05 max	0.05 max	0.1-0.4	N <=0.009; bal Fe				
SS 141234	1234		0.17 max	0.2 max		2 max		0.05 max	0.05 max	0.1-0.4	bal Fe				

UNS numbers and US grades are provided as a means of cross referencing chemically similar alloys. Exchangability is only possible after independent examination of specifications. Tensile properties are minimum or typical as specified. UTS and YS as MPa. El as %. See Appendix for list of abbreviations used in Notes. * indicates obsolete material.

Specification	Designation	Notes	C	Cr	Cu	Mn	Ni	P	S	Si	Other	UTS	YS	El	Hard

Carbon Steel, High-Manganese, A662(A) (Continued from previous page)

USA

Specification	Designation	Notes	C	Cr	Cu	Mn	Ni	P	S	Si	Other	UTS	YS	El	Hard
	UNS K01701		0.17 max			0.84-1.46		0.035 max	0.040 max	0.13-0.45	bal Fe				
ASTM A662/A662M(92)	Grade A	Press ves plt, for mod/low-temp, t<=50mm	0.14 max			0.90-1.35		0.035 max	0.035 max	0.15-0.40	bal Fe	400-540	275	23	

Carbon Steel, High-Manganese, A662(B)

Europe

Specification	Designation	Notes	C	Cr	Cu	Mn	Ni	P	S	Si	Other	UTS	YS	El	Hard
EN 10028/3(92)	1.1104	High-temp, Press; 35<t<=50mm	0.16 max	0.30 max	0.30 max	0.50-1.50	0.50 max	0.025 max	0.015 max	0.40 max	Al >=0.020; Mo <=0.08; N <=0.020; Nb <=0.05; Ti <=0.03; V <=0.05; Nb+Ti+V<=0.05, Cr+Cu+Mo<=0.45; bal Fe	390-510	265	24	
EN 10028/3(92)	1.1104	High-temp, Press; 70<t<=100mm	0.16 max	0.30 max	0.30 max	0.50-1.50	0.50 max	0.025 max	0.015 max	0.40 max	Al >=0.020; Mo <=0.08; N <=0.020; Nb <=0.05; Ti <=0.03; V <=0.05; Nb+Ti+V<=0.05, Cr+Cu+Mo<=0.45; bal Fe	390-510	235	23	
EN 10028/3(92)	1.1104	High-temp, Press; 100<t<=150mm	0.16 max	0.30 max	0.30 max	0.50-1.50	0.50 max	0.025 max	0.015 max	0.40 max	Al >=0.020; Mo <=0.08; N <=0.020; Nb <=0.05; Ti <=0.03; V <=0.00; Nb+Ti+V<=0.05, Cr+Cu+Mo<=0.45; bal Fe	390-510	225	23	
EN 10028/3(92)	1.1104	High-temp, Press; 0<=t<=35mm	0.16 max	0.30 max	0.30 max	0.50-1.50	0.50 max	0.025 max	0.015 max	0.40 max	Al >=0.020; Mo <=0.08; N <=0.020; Nb <=0.05; Ti <=0.03; V <=0.05; Nb+Ti+V<=0.05, Cr+Cu+Mo<=0.45; bal Fe	390-510	275	24	
EN 10028/3(92)	1.1104	High-temp, Press; 50<t<=70mm	0.16 max	0.30 max	0.30 max	0.50-1.50	0.50 max	0.025 max	0.015 max	0.40 max	Al >=0.020; Mo <=0.08; N <=0.020; Nb <=0.05; Ti <=0.03; V <=0.05; Nb+Ti+V<=0.05, Cr+Cu+Mo<=0.45; bal Fe	390-510	255	24	
EN 10028/3(92)	P275NL2	High-temp, Press; 0<=t<=35mm	0.16 max	0.30 max	0.30 max	0.50-1.50	0.50 max	0.025 max	0.015 max	0.40 max	Al >=0.020; Mo <=0.08; N <=0.020; Nb <=0.05; Ti <=0.03; V <=0.05; Nb+Ti+V<=0.05, Cr+Cu+Mo<=0.45; bal Fe	390-510	275	24	
EN 10028/3(92)	P275NL2	High-temp, Press; 35<t<=50mm	0.16 max	0.30 max	0.30 max	0.50-1.50	0.50 max	0.025 max	0.015 max	0.40 max	Al >=0.020; Mo <=0.08; N <=0.020; Nb <=0.05; Ti <=0.03; V <=0.05; Nb+Ti+V<=0.05, Cr+Cu+Mo<=0.45; bal Fe	390-510	265	24	
EN 10028/3(92)	P275NL2	High-temp, Press; 50<t<=70mm	0.16 max	0.30 max	0.30 max	0.50-1.50	0.50 max	0.025 max	0.015 max	0.40 max	Al >=0.020; Mo <=0.08; N <=0.020; Nb <=0.05; Ti <=0.03; V <=0.05; Nb+Ti+V<=0.05, Cr+Cu+Mo<=0.45; bal Fe	390-510	255	24	
EN 10028/3(92)	P275NL2	High-temp, Press; 100<t<=150mm	0.16 max	0.30 max	0.30 max	0.50-1.50	0.50 max	0.025 max	0.015 max	0.40 max	Al >=0.020; Mo <=0.08; N <=0.020; Nb <=0.05; Ti <=0.03; V <=0.00; Nb+Ti+V<=0.05, Cr+Cu+Mo<=0.45; bal Fe	390-510	225	23	

4-280/Carbon Steel

Specification	Designation	Notes	C	Cr	Cu	Mn	Ni	P	S	Si	Other	UTS	YS	El	Hard
Carbon Steel, High-Manganese, A662(B) (Continued from previous page)															
Europe															
EN 10028/3(92)	P275NL2	High-temp, Press; 70<t<=100mm	0.16 max	0.30 max	0.30 max	0.50-1.50	0.50 max	0.025 max	0.015 max	0.40 max	Al >=0.020; Mo <=0.08; N <=0.020; Nb <=0.05; Ti <=0.03; V <=0.05; Nb+Ti+V<=0.05, Cr+Cu+Mo<=0.45; bal Fe	390-510	235	23	
Italy															
UNI 5869(75)	Fe410-1KG	Press ves	0.2 max			0.5-1.60		0.035 max	0.04 max	0.35 max	bal Fe				
UNI 5869(75)	Fe410-1KW	Press ves	0.2 max			0.5-1.60		0.04 max	0.04 max	0.35 max	bal Fe				
UNI 5869(75)	Fe410-2KG	Press ves	0.2 max			0.5-1.60		0.03 max	0.035 max	0.35 max	bal Fe				
UNI 5869(75)	Fe410-2KW	Press ves	0.2 max			0.5-1.60		0.035 max	0.035 max	0.35 max	bal Fe				
UK															
BS 1502(82)	224-490	Bar Shp Press ves	0.22 max			0.90-1.50		0.040 max	0.040 max	0.1-0.40	Al >=0.015; W <=0.1; bal Fe				
BS 1503(89)	224-460	Frg press ves; t<=100mm; norm Q/T	0.23 max	0.25 max	0.30 max	0.80-1.40	0.40 max	0.030 max	0.025 max	0.10-0.40	Al >=0.018; Mo <=0.10; W <=0.1; Cr+Mo+Ni+Cu<=0.8; bal Fe	460-580	275	18	
BS 1503(89)	224-460	Frg press ves; 100<t<=999mm; norm Q/T	0.23 max	0.25 max	0.30 max	0.80-1.40	0.40 max	0.030 max	0.025 max	0.10-0.40	Al >=0.018; Mo <=0.10; W <=0.1; Cr+Mo+Ni+Cu<=0.8; bal Fe	460-580	255	18	
USA															
	UNS K02203		0.22 max			0.80-1.55		0.035 max	0.040 max	0.13-0.33	bal Fe				
ASTM A662/A662M(92)	Grade B	Press ves plt, for mod/low-temp, t<=50mm	0.19 max			0.85-1.50		0.035 max	0.035 max	0.15-0.40	bal Fe	450-585	275	23	
Carbon Steel, High-Manganese, A662(C)															
Spain															
UNE 36121(85)	AE250-1B	CR; Strp	0.20 max			1.60 max		0.04 max	0.04 max	0.60 max	Mo <=0.15; N <=0.009; bal Fe				
UNE 36121(85)	AE250-1D	CR; Strp	0.20 max			1.60 max		0.035 max	0.035 max	0.60 max	Mo <=0.15; N <=0.015; bal Fe				
UNE 36121(85)	AE550-1	CR; Strp	0.20 max			1.50 max		0.040 max	0.040 max	0.60 max	bal Fe				
UK															
BS 1503(89)	224-410	Frg press ves; 100<t<=999mm; norm Q/T	0.20 max	0.25 max	0.30 max	0.80-1.20	0.40 max	0.030 max	0.025 max	0.10-0.40	Al >=0.018; Mo <=0.10; W <=0.1; Cr+Mo+Ni+Cu<=0.8; bal Fe	410-530	220	20	
BS 1503(89)	224-410	Frg press ves; t<=100mm; norm Q/T	0.20 max	0.25 max	0.30 max	0.80-1.20	0.40 max	0.030 max	0.025 max	0.10-0.40	Al >=0.018; Mo <=0.10; W <=0.1; Cr+Mo+Ni+Cu<=0.8; bal Fe	410-530	235	20	
USA															
	UNS K02007		0.20 max			1.00-1.60		0.035 max	0.040 max	0.15-0.50	bal Fe				
ASTM A662/A662M(92)	Grade C	Press ves plt, for mod/low-temp, t<=50mm	0.20 max			1.00-1.60		0.035 max	0.035 max	0.15-0.50	bal Fe	485-620	295	22	
Carbon Steel, High-Manganese, A678(A)															
UK															
BS 1502(82)	224-430	Bar Shp Press ves	0.17 max			0.90-1.50		0.040 max	0.040 max	0.1-0.40	Al >=0.015; W <=0.1; bal Fe				
USA															
	UNS K01600		0.16 max			0.90-1.50		0.04 max	0.05 max	0.15-0.50	Cu 0.20 min (when spec'd); bal Fe				
ASTM A678/A678M(94)	Grade A	Q/T, struct Plt, t<=40mm	0.16 max		0.20 min	0.90-1.50		0.035 max	0.04 max	0.15-0.50	bal Fe	485-620	345	22	

UNS numbers and US grades are provided as a means of cross referencing chemically similar alloys. Exchangability is only possible after independent examination of specifications. Tensile properties are minimum or typical as specified. UTS and YS as MPa. El as %. See Appendix for list of abbreviations used in Notes. * indicates obsolete material.

Specification Designation	Notes	C	Cr	Cu	Mn	Ni	P	S	Si	Other	UTS	YS	El	Hard

Carbon Steel, High-Manganese, A678(C)

Europe

Specification Designation	Notes	C	Cr	Cu	Mn	Ni	P	S	Si	Other	UTS	YS	El	Hard
EN 10025(90)A1(93) 1.0045	Struct HR QS, t<=1mm	0.24 max			1.60 max		0.045 max	0.045 max	0.55 max	N max na if Al>=0.020; bal Fe	510-680	355	14L 12T	
EN 10025(90)A1(93) 1.0045	Struct HR BS, 200<t<=250mm	0.24 max			1.60 max		0.045 max	0.045 max	0.55 max	N <=0.009; N max na if Al>=0.020; bal Fe	450-630	275	17L 17T	
EN 10025(90)A1(93) 1.0045	Struct HR BS, 63<t<=80mm	0.24 max			1.60 max		0.045 max	0.045 max	0.55 max	N <=0.009; N max na if Al>=0.020; bal Fe	490-630	325	20L 18T	
EN 10025(90)A1(93) 1.0045	Struct HR BS, 3<t<=16mm	0.24 max			1.60 max		0.045 max	0.045 max	0.55 max	N <=0.009; N max na if Al>=0.020; bal Fe	490-630	355	22L 20T	
EN 10025(90)A1(93) 1.0045	Struct HR BS, t<=1mm	0.24 max			1.60 max		0.045 max	0.045 max	0.55 max	N <=0.009; N max na if Al>=0.020; bal Fe	510-680	355	14L 12T	
EN 10025(90)A1(93) 1.0533	Struct HR QS, 200<t<=250mm	0.22 max			1.60 max		0.040 max	0.040 max	0.55 max	N <=0.009; N max na if Al>=0.020; bal Fe	450-630	285	17L 17T	
EN 10025(90)A1(93) 1.0533	Struct HR QS, t<=1mm	0.20 max			1.60 max		0.040 max	0.040 max	0.55 max	N max na if Al>=0.020; bal Fe	510-680	355	14L 12T	
EN 10025(90)A1(93) 1.0533	Struct HR QS, 3<t<=16mm	0.20 max			1.60 max		0.040 max	0.040 max	0.55 max	N max na if Al>=0.020; bal Fe	490-630	355	22L 20T	
EN 10025(90)A1(93) 1.0533	Struct HR QS, 3<t<=16mm	0.20 max			1.60 max		0.040 max	0.040 max	0.55 max	N <=0.009; N max na if Al>=0.020; bal Fe	490-630	355	22L 20T	
EN 10025(90)A1(93) 1.0533	Struct HR QS, 63<t<=80mm	0.22 max			1.60 max		0.040 max	0.040 max	0.55 max	N <=0.009; N max na if Al>=0.020; bal Fe	490-630	335	20L 18T	
EN 10025(90)A1(93) 1.0570	Struct HR QS, 3<t<=16mm	0.20 max			1.60 max		0.035 max	0.035 max	0.55 max	CEV may be req'd; bal Fe	490-630	355	22L 20T	
EN 10025(90)A1(93) 1.0570	Struct HR QS, 63<t<=80mm	0.22 max			1.60 max		0.035 max	0.035 max	0.55 max	CEV may be req'd; bal Fe	490-630	325	20L 18T	
EN 10025(90)A1(93) 1.0570	Struct HR QS, 3<t<=16mm	0.20 max			1.60 max		0.035 max	0.035 max	0.55 max	CEV may be req'd; bal Fe	490-630	355	22L 20T	
EN 10025(90)A1(93) 1.0570	Struct HR QS, t<=1mm	0.20 max			1.60 max		0.035 max	0.035 max	0.55 max	CEV may be req'd; bal Fe	510-680	355	14L 12T	
EN 10025(90)A1(93) 1.0570	Struct HR QS, 200<t<=250mm	0.22 max			1.60 max		0.035 max	0.035 max	0.55 max	CEV may be req'd; bal Fe	450-630	275	17L 17T	
EN 10025(90)A1(93) 1.0577	Struct HR QS, t<=1mm	0.20 max			1.60 max		0.035 max	0.035 max	0.55 max	CEV may be req'd; bal Fe	510-680	355	14L 12T	
EN 10025(90)A1(93) 1.0577	Struct HR QS, 3<t<=16mm	0.20 max			1.60 max		0.035 max	0.035 max	0.55 max	CEV may be req'd; bal Fe	490-630	355	22L 20T	
EN 10025(90)A1(93) 1.0595	Struct HR QS, 3<t<=16mm	0.20 max			1.60 max		0.035 max	0.035 max	0.55 max	CEV may be req'd; bal Fe	490-630	355	22L 20T	
EN 10025(90)A1(93) 1.0595	Struct HR QS, t<=1mm	0.20 max			1.60 max		0.035 max	0.035 max	0.55 max	CEV may be req'd; bal Fe	510-680	355	14L 12T	
EN 10025(90)A1(93) 1.0595	Struct HR QS, 63<t<=80mm	0.22 max			1.60 max		0.035 max	0.035 max	0.55 max	CEV may be req'd; bal Fe	490-630	325	20L 18T	
EN 10025(90)A1(93) 1.0595	Struct HR QS, t<=1mm	0.20 max			1.60 max		0.035 max	0.035 max	0.55 max	CEV may be req'd; bal Fe	510-680	355	14L 12T	
EN 10025(90)A1(93) 1.0595	Struct HR QS, 200<t<=250mm	0.22 max			1.60 max		0.035 max	0.035 max	0.55 max	CEV may be req'd; bal Fe	450-630	275	17L 17T	
EN 10025(90)A1(93) 1.0596	Struct HR QS, t<=1mm	0.20 max			1.60 max		0.035 max	0.035 max	0.55 max	CEV may be req'd; bal Fe	510-680	355	14L 12T	
EN 10025(90)A1(93) 1.0596	Struct HR QS, 3<t<=16mm	0.20 max			1.60 max		0.035 max	0.035 max	0.55 max	CEV may be req'd; bal Fe	490-630	355	22L 20T	
EN 10025(90)A1(93) S355J0	Struct HR QS, 200<t<=250mm	0.22 max			1.60 max		0.040 max	0.040 max	0.55 max	N <=0.009; N max na if Al>=0.020; CEV; bal Fe	450-630	285	17L 17T	
EN 10025(90)A1(93) S355J0	Struct HR QS, 3<t<=16mm	0.20 max			1.60 max		0.040 max	0.040 max	0.55 max	N <=0.009; N max na if Al>=0.020; CEV; bal Fe	490-630	355	22L 20T	
EN 10025(90)A1(93) S355J0	Struct HR QS, t<=1mm	0.20 max			1.60 max		0.040 max	0.040 *max	0.55 max	N <=0.009; N max na if Al>=0.020; CEV; bal Fe	510-680	355	14L 12T	
EN 10025(90)A1(93) S355J0	Struct HR QS, 63<t<=80mm	0.22 max			1.60 max		0.040 max	0.040 max	0.55 max	N <=0.009; N max na if Al>=0.020; CEV; bal Fe	490-630	335	20L 18T	
EN 10025(90)A1(93) S355J0C	Struct HR QS, t<=1mm	0.20 max			1.60 max		0.040 max	0.040 max	0.55 max	N max na if Al>=0.020; CEV; bal Fe	510-680	355	14L 12T	
EN 10025(90)A1(93) S355J0C	Struct HR QS, 3<t<=16mm	0.20 max			1.60 max		0.040 max	0.040 max	0.55 max	N max na if Al>=0.020; CEV; bal Fe	490-630	355	22L 20T	
EN 10025(90)A1(93) S355J2G3	Struct HR QS, t<=1mm	0.20 max			1.60 max		0.035 max	0.035 max	0.55 max	N max na if Al>=0.020; CEV; bal Fe	510-680	355	14L 12T	

Specification Designation	Notes	C	Cr	Cu	Mn	Ni	P	S	Si	Other	UTS	YS	El	Hard

Carbon Steel, High-Manganese, A678(C) (Continued from previous page)

Europe

Specification Designation	Notes	C	Cr	Cu	Mn	Ni	P	S	Si	Other	UTS	YS	El	Hard
EN 10025(90)A1(93) S355J2G3	Struct HR QS, 3<t<=16mm	0.20 max			1.60 max		0.035 max	0.035 max	0.55 max	N max na if Al>=0.020; CEV; bal Fe	490-630	355	22L 20T	
EN 10025(90)A1(93) S355J2G3	Struct HR QS, 63<t<=80mm	0.22 max			1.60 max		0.035 max	0.035 max	0.55 max	N max na if Al>=0.020; CEV; bal Fe	490-630	325	20L 18T	
EN 10025(90)A1(93) S355J2G3	Struct HR QS, 200<t<=250mm	0.22 max			1.60 max		0.035 max	0.035 max	0.55 max	N max na if Al>=0.020; CEV; bal Fe	450-630	275	17L 17T	
EN 10025(90)A1(93) S355J2G3C	Struct HR QS, t<=1mm	0.20 max			1.60 max		0.035 max	0.035 max	0.55 max	N max na if Al>=0.020; CEV; bal Fe	510-680	355	14L 12T	
EN 10025(90)A1(93) S355J2G3C	Struct HR QS, 3<t<=16mm	0.20 max			1.60 max		0.035 max	0.035 max	0.55 max	N max na if Al>=0.020; CEV; bal Fe	490-630	355	22L 20T	
EN 10025(90)A1(93) S355J2G4	Struct HR QS, 3<t<=16mm	0.20 max			1.60 max		0.035 max	0.035 max	0.55 max	N max na if Al>=0.020; CEV; bal Fe	490-630	355	22L 20T	
EN 10025(90)A1(93) S355J2G4	Struct HR QS, 200<t<=250mm	0.22 max			1.60 max		0.035 max	0.035 max	0.55 max	N max na if Al>=0.020; CEV; bal Fe	450-630	275	17L 17T	
EN 10025(90)A1(93) S355J2G4	Struct HR QS, t<=1mm	0.20 max			1.60 max		0.035 max	0.035 max	0.55 max	N max na if Al>=0.020; CEV; bal Fe	510-680	355	14L 12T	
EN 10025(90)A1(93) S355J2G4	Struct HR QS, 63<t<=80mm	0.22 max			1.60 max		0.035 max	0.035 max	0.55 max	N max na if Al>=0.020; CEV; bal Fe	490-630	325	20L 18T	
EN 10025(90)A1(93) S355J2G4C	Struct HR QS, 3<t<=16mm	0.20 max			1.60 max		0.035 max	0.035 max	0.55 max	N max na if Al>=0.020; CEV; bal Fe	490-630	355	22L 20T	
EN 10025(90)A1(93) S355J2G4C	Struct HR QS, t<=1mm	0.20 max			1.60 max		0.035 max	0.035 max	0.55 max	N max na if Al>=0.020; CEV; bal Fe	510-680	355	14L 12T	
EN 10025(90)A1(93) S355JR	Struct HR BS, 200<t<=250mm	0.24 max			1.60 max		0.045 max	0.045 max	0.55 max	N <=0.009; N max na if Al>=0.020; bal Fe	450-630	275	17L 17T	
EN 10025(90)A1(93) S355JR	Struct HR BS, 63<t<=80mm	0.24 max			1.60 max		0.045 max	0.045 max	0.55 max	N <=0.009; N max na if Al>=0.020; bal Fe	490-630	325	20L 18T	
EN 10025(90)A1(93) S355JR	Struct HR BS, 3<t<=16mm	0.24 max			1.60 max		0.045 max	0.045 max	0.55 max	N <=0.009; N max na if Al>=0.020; bal Fe	490-630	355	22L 20T	
EN 10025(90)A1(93) S355JR	Struct HR BS, t<=1mm	0.24 max			1.60 max		0.045 max	0.045 max	0.55 max	N <=0.009; N max na if Al>=0.020; bal Fe	510-680	355	14L 12T	
EN 10025(90)A1(93) S355JRC	Struct HR QS, t<=1mm	0.24 max			1.60 max		0.045 max	0.045 max	0.55 max	N max na if Al>=0.020; bal Fe	510-680	355	14L 12T	
EN 10025(90)A1(93) S355K2G3	Struct HR QS, 200<t<=250mm	0.22 max			1.60 max		0.035 max	0.035 max	0.55 max	N max na if Al>=0.020; CEV; bal Fe	450-630	275	17L 17T	
EN 10025(90)A1(93) S355K2G3	Struct HR QS, t<=1mm	0.20 max			1.60 max		0.035 max	0.035 max	0.55 max	N max na if Al>=0.020; CEV; bal Fe	510-680	355	14L 12T	
EN 10025(90)A1(93) S355K2G3	Struct HR QS, 63<t<=80mm	0.22 max			1.60 max		0.035 max	0.035 max	0.55 max	N max na if Al>=0.020; CEV; bal Fe	490-630	325	20L 18T	
EN 10025(90)A1(93) S355K2G3	Struct HR QS, 3<t<=16mm	0.20 max			1.60 max		0.035 max	0.035 max	0.55 max	N max na if Al>=0.020; CEV; bal Fe	490-630	355	22L 20T	
EN 10025(90)A1(93) S355K2G3C	Struct HR QS, t<=1mm	0.20 max			1.60 max		0.035 max	0.035 max	0.55 max	N max na if Al>=0.020; CEV; bal Fe	510-680	355	14L 12T	
EN 10025(90)A1(93) S355K2G3C	Struct HR QS, 3<t<=16mm	0.20 max			1.60 max		0.035 max	0.035 max	0.55 max	N max na if Al>=0.020; CEV; bal Fe	490-630	355	22L 20T	
EN 10025(90)A1(93) S355K2G4	Struct HR QS, 3<t<=16mm	0.20 max			1.60 max		0.035 max	0.035 max	0.55 max	N max na if Al>=0.020; CEV; bal Fe	490-630	355	22L 20T	
EN 10025(90)A1(93) S355K2G4	Struct HR QS, t<=1mm	0.20 max			1.60 max		0.035 max	0.035 max	0.55 max	N max na if Al>=0.020; CEV; bal Fe	510-680	355	14L 12T	
EN 10025(90)A1(93) S355K2G4	Struct HR QS, 63<t<=80mm	0.22 max			1.60 max		0.035 max	0.035 max	0.55 max	N max na if Al>=0.020; CEV; bal Fe	490-630	325	20L 18T	

Specification	Designation	Notes	C	Cr	Cu	Mn	Ni	P	S	Si	Other	UTS	YS	El	Hard

Carbon Steel, High-Manganese, A678(C) (Continued from previous page)

Europe

Specification	Designation	Notes	C	Cr	Cu	Mn	Ni	P	S	Si	Other	UTS	YS	El	Hard
EN 10025(90)A1(93)	S355K2G4	Struct HR QS, 200<t<=250mm	0.22 max			1.60 max		0.035 max	0.035 max	0.55 max	N max na if Al>=0.020; CEV; bal Fe	450-630	275	17L 17T	
EN 10025(90)A1(93)	S355K2G4C	Struct HR QS, t<=1mm	0.20 max			1.60 max		0.035 max	0.035 max	0.55 max	N max na if Al>=0.020; CEV; bal Fe	510-680	355	14L 12T	
EN 10025(90)A1(93)	S355K2G4C	Struct HR QS, 3<t<=16mm	0.20 max			1.60 max		0.035 max	0.035 max	0.55 max	N max na if Al>=0.020; CEV; bal Fe	490-630	355	22L 20T	

Mexico

Specification	Designation	Notes	C	Cr	Cu	Mn	Ni	P	S	Si	Other	UTS	YS	El	Hard
NMX-B-199(86)	Grade C	CF weld Smls struct tub	0.23 max			1.35 max		0.04 max	0.05 max		Cu if spec'd>=0.20; bal Fe	427	317	21	

Romania

Specification	Designation	Notes	C	Cr	Cu	Mn	Ni	P	S	Si	Other	UTS	YS	El	Hard
STAS 2883/3(88)	K510	Heat res	0.15-0.22	0.3 max		1-1.6		0.035 max	0.03 max	0.3-0.6	Al 0.02-0.035; N <=0.009; Pb <=0.15; V <=0.03; Cr+Ni+Cu<=0.7; As<=0.08; bal Fe				
STAS 438/1(89)	PC52	Reinf; HR	0.16-0.22	0.3 max		1.2-1.6		0.045 max	0.045 max	0.55 max	Pb <=0.15; bal Fe				

Spain

Specification	Designation	Notes	C	Cr	Cu	Mn	Ni	P	S	Si	Other	UTS	YS	El	Hard
UNE 36082(84)	AE355W-1A*	Struct; Corr res	0.12 max	0.3-1.25	0.25-0.55	1.0 max	0.65 max	0.06-0.15	0.05 max	0.75 max	Mo <=0.15; bal Fe				
UNE 36082(84)	AE355W-1C*	Struct; Corr res	0.12 max	0.30-1.25	0.25-0.55	1.0 max	0.65 max	0.06-0.15	0.04 max	0.75 max	Mo <=0.15; N <=0.009; bal Fe				
UNE 36082(84)	AE355W-1D*	Struct; Corr res	0.12 max	0.3-1.25	0.25-0.55	1.0 max	0.65 max	0.06-0.15	0.04 max	0.75 max	Mo <=0.15; bal Fe				
UNE 36082(84)	AE355W-2B*	Struct; Corr res	0.19 max	0.4-0.8		0.5-1.5	0.65 max	0.05 max	0.05 max	0.55 max	Mo <=0.3; bal Fe				
UNE 36082(84)	AE355W-2C*	Struct; Corr res	0.19 max	0.4-0.8	0.25-0.55	0.5-1.5	0.65 max	0.04 max	0.04 max	0.55 max	Mo <=0.3; N <=0.009; Zr<=0.15; bal Fe				
UNE 36082(84)	AE355W-2D*	Struct; Corr res	0.19 max	0.4-0.8	0.25-0.55	0.5-1.5	0.65 max	0.04 max	0.04 max	0.55 max	Mo <=0.3; Zr<=0.15; bal Fe				
UNE 36082(84)	F.6434*	see AE355W-1A*	0.12 max	0.3-1.25	0.25-0.55	1.0 max	0.65 max	0.06-0.15	0.05 max	0.75 max	Mo <=0.15; bal Fe				
UNE 36082(84)	F.6435*	see AE355W-1C*	0.12 max	0.30-1.25	0.25-0.55	1.0 max	0.65 max	0.06-0.15	0.04 max	0.75 max	Mo <=0.15; N <=0.009; bal Fe				
UNE 36082(84)	F.6436*	see AE355W-1D*	0.12 max	0.3-1.25	0.25-0.55	1.0 max	0.65 max	0.06-0.15	0.04 max	0.75 max	Mo <=0.15; bal Fe				
UNE 36082(84)	F.6437*	see AE355W-2B*	0.19 max	0.4-0.8		0.5-1.5	0.65 max	0.05 max	0.05 max	0.55 max	Mo <=0.3; bal Fe				
UNE 36082(84)	F.6438*	see AE355W-2C*	0.19 max	0.4-0.8	0.25-0.55	0.5-1.5	0.65 max	0.04 max	0.04 max	0.55 max	Mo <=0.3; N <=0.009; Zr<=0.15; bal Fe				
UNE 36082(84)	F.6439*	see AE355W-2D*	0.19 max	0.4-0.8	0.25-0.55	0.5-1.5	0.65 max	0.04 max	0.04 max	0.55 max	Mo <=0.3; Zr<=0.15; bal Fe				

USA

Specification	Designation	Notes	C	Cr	Cu	Mn	Ni	P	S	Si	Other	UTS	YS	El	Hard
	UNS K02204		0.22 max			1.00-1.60		0.04 max	0.05 max	0.20-0.50	Cu 0.20 min (when spec'd); bal Fe				
ASTM A678/A678M(94)	Grade C	Q/T, struct Plt, t<=20mm	0.22 max		0.20 min	1.00-1.60		0.035 max	0.04 max	0.20-0.50	bal Fe	655-790	515	19	
ASTM A678/A678M(94)	Grade C	Q/T, struct Plt, 20<=t<=40mm	0.22 max		0.20 min	1.00-1.60		0.035 max	0.04 max	0.20-0.50	bal Fe	620-760	485	19	
ASTM A678/A678M(94)	Grade C	Q/T, struct Plt, 40<=t<=50mm	0.22 max		0.20 min	1.00-1.60		0.035 max	0.04 max	0.20-0.50	bal Fe	585-720	450	19	

Carbon Steel, High-Manganese, A679

Norway

Specification	Designation	Notes	C	Cr	Cu	Mn	Ni	P	S	Si	Other	UTS	YS	El	Hard
NS 13601	13613	Spring wir	0.12 max					0.030 max	0.030 max		C varies; bal Fe				

USA

Specification	Designation	Notes	C	Cr	Cu	Mn	Ni	P	S	Si	Other	UTS	YS	El	Hard
	UNS K08200		0.65-1.00			0.20-1.30		0.040 max	0.050 max	0.10-0.40	bal Fe				
ASTM A679(93)	A679	Wir, High-strg 1mm	0.65-1.00			0.20-1.30		0.040 max	0.050 max	0.10-0.40	bal Fe	2150-2400			

Specification	Designation	Notes	C	Cr	Cu	Mn	Ni	P	S	Si	Other	UTS	YS	El	Hard
Carbon Steel, High-Manganese, A687(B)															
Romania															
STAS 2883/3(88)	K410	Heat res	0.2 max	0.3 max		0.5-1.3		0.035 max	0.03 max	0.35 max	Al 0.02-0.035; N <=0.009; Pb <=0.15; V <=0.03; Cr+Ni+Cu<=0.7; As<=0.08; bal Fe				
UK															
BS 1503(89)	221-430	Frg press ves; 500<t<=999mm; Ann norm Q/T	0.20 max	0.25 max	0.30 max	0.80-1.40	0.40 max	0.030 max	0.025 max	0.10-0.40	Al <=0.010; Mo <=0.10; W <=0.1; Cr+Mo+Ni+Cu<=0.8; bal Fe	430-550	215	19	
BS 1503(89)	221-430	Frg press ves; t<=100mm; Ann norm Q/T	0.20 max	0.25 max	0.30 max	0.80-1.40	0.40 max	0.030 max	0.025 max	0.10-0.40	Al <=0.010; Mo <=0.10; W <=0.1; Cr+Mo+Ni+Cu<=0.8; bal Fe	430-550	225	19	
USA															
	UNS K02002		0.20 max			0.70-1.35		0.04 max	0.05 max	0.15-0.50	Cu 0.20 min (when spec'd); bal Fe				
ASTM A678/A678M(94)	Grade B	Q/T, struct Plt, t<=40mm	0.20 max		0.20 min	0.70-1.35		0.035 max	0.04 max	0.15-0.50	bal Fe	550-690	415	22	
Carbon Steel, High-Manganese, A694															
USA															
	UNS K03014		0.30 max			1.50 max		0.050 max	0.060 max	0.13-0.37	bal Fe				
ASTM A694/A694M(98)	A694 F42	Frg or rolled	0.26 max			1.40 max		0.025 max	0.025 max	0.15-0.35	bal Fe	415	290	20	
ASTM A694/A694M(98)	A694 F46	Frg or rolled	0.26 max			1.40 max		0.025 max	0.025 max	0.15-0.35	bal Fe	415	315	20	
ASTM A694/A694M(98)	A694 F48	Frg or rolled	0.26 max			1.40 max		0.025 max	0.025 max	0.15-0.35	bal Fe	425	330	20	
ASTM A694/A694M(98)	A694 F50	Frg or rolled	0.26 max			1.40 max		0.025 max	0.025 max	0.15-0.35	bal Fe	440	345	20	
ASTM A694/A694M(98)	A694 F52	Frg or rolled	0.26 max			1.40 max		0.025 max	0.025 max	0.15-0.35	bal Fe	455	360	20	
ASTM A694/A694M(98)	A694 F56	Frg or rolled	0.26 max			1.40 max		0.025 max	0.025 max	0.15-0.35	bal Fe	470	385	20	
ASTM A694/A694M(98)	A694 F60	Frg or rolled	0.26 max			1.40 max		0.025 max	0.025 max	0.15-0.35	bal Fe	515	415	20	
ASTM A694/A694M(98)	A694 F65	Frg or rolled	0.26 max			1.40 max		0.025 max	0.025 max	0.15-0.35	bal Fe	530	450	20	
ASTM A694/A694M(98)	A694 F70	Frg or rolled	0.26 max			1.40 max		0.025 max	0.025 max	0.15-0.35	bal Fe	565	485	18	
Carbon Steel, High-Manganese, A696(B,C)															
USA															
	UNS K03200		0.32 max			1.04 max		0.035 max	0.045 max	0.15-0.30	bal Fe				
ASTM A696(95)	Grade B	HW CF special qual for press piping	0.32 max			1.04 max		0.035 max	0.045 max	0.15-0.35	Mn depends on C; bal Fe	415	240	20	
ASTM A696(95)	Grade C	HW CF special qual for press piping	0.32 max			1.04 max		0.035 max	0.045 max	0.15-0.35	Mn depends on C; bal Fe	485	275	18	
Carbon Steel, High-Manganese, A707(L1)															
Italy															
UNI 7070(82)	Fe410BFN	Gen struct	0.2 max			2 max		0.045 max	0.045 max	0.6 max	bal Fe				
UNI 7070(82)	Fe410CFN	Gen struct	0.18 max			2 max		0.04 max	0.045 max	0.6 max	bal Fe				
UNI 7070(82)	Fe410DFF	Gen struct	0.18 max			2 max		0.04 max	0.04 max	0.6 max	Al <=0.015; bal Fe				
Japan															
JIS G4052(79)	SMn433H	Struct Hard	0.29-0.36	0.35 max	0.30 max	1.15-1.55	0.25 max	0.030 max	0.030 max	0.15-0.35	bal Fe				
Romania															
STAS 500/2(88)	OL52.2	Gen struct	0.2 max	0.3 max		1.6 max	0.4 max	0.05 max	0.05 max	0.5 max	Pb <=0.15; bal Fe				

Specification	Designation	Notes	C	Cr	Cu	Mn	Ni	P	S	Si	Other	UTS	YS	El	Hard

Carbon Steel, High-Manganese, A707(L1) (Continued from previous page)

Romania

Specification	Designation	Notes	C	Cr	Cu	Mn	Ni	P	S	Si	Other	UTS	YS	El	Hard
STAS 500/2(88)	OL52.3k	Gen struct; FF; Non-rimming	0.2 max	0.3 max		1.6 max		0.045 max	0.045 max	0.5 max	Al 0.025-0.120; Pb <=0.15; As<=0.08; bal Fe				
STAS 500/2(88)	OL52.3kf	Gen struct; Specially killed	0.2 max	0.3 max		1.6 max		0.045 max	0.045 max	0.5 max	Al 0.025-0.120; Pb <=0.15; As<=0.08; bal Fe				

UK

Specification	Designation	Notes	C	Cr	Cu	Mn	Ni	P	S	Si	Other	UTS	YS	El	Hard
BS 1503(89)	223-460	Frg press ves; t<=100mm; norm Q/T	0.23 max	0.25 max	0.30 max	0.80-1.40	0.40 max	0.030 max	0.025 max	0.10-0.40	Mo <=0.10; Nb 0.01-0.06; W <=0.1; Cr+Mo+Ni+Cu<=0.8; bal Fe	460-580	290	18	
BS 1503(89)	223-460	Frg press ves; 100<t<=999mm; norm Q/T	0.23 max	0.25 max	0.30 max	0.80-1.40	0.40 max	0.030 max	0.025 max	0.10-0.40	Mo <=0.10; Nb 0.01-0.06; W <=0.1; Cr+Mo+Ni+Cu<=0.8; bal Fe	460-580	270	18	

USA

Specification	Designation	Notes	C	Cr	Cu	Mn	Ni	P	S	Si	Other	UTS	YS	El	Hard
	UNS K02302		0.23 max			1.60 max		0.035 max	0.040 max	0.37 max	bal Fe				
ASTM A707/A707M(98)	A 707 (L1)	Frg flanges; low-temp; class values given	0.20 max	0.30 max	0.40 max	0.60-1.50	0.40 max	0.030 max	0.030 max	0.35 max	Mo <=0.12; V <=0.05; Cb 0.02 max, Cu+Ni+Cr+Mo<=1.00; bal Fe	415-455	790-360	22	149-217 HB

Carbon Steel, High-Manganese, A707(L2)

USA

Specification	Designation	Notes	C	Cr	Cu	Mn	Ni	P	S	Si	Other	UTS	YS	El	Hard
	UNS K03301		0.33 max			1.45 max		0.035 max	0.040 max	0.37 max	bal Fe				
ASTM A707/A707M(98)	A 707 (L2)	Frg flanges; low-temp; class values given	0.30 max	0.30 max	0.40 max	0.60-1.35	0.40 max	0.030 max	0.030 max	0.35 max	Mo <=0.12; V <=0.05; Cb 0.02 max, Cu+Ni+Cr+Mo<=1.00; bal Fe	415-515	290-415	20-22	149-235 HB

Carbon Steel, High-Manganese, A727

UK

Specification	Designation	Notes	C	Cr	Cu	Mn	Ni	P	S	Si	Other	UTS	YS	El	Hard
BS 1503(89)	164-490	Frg press ves; 100<t<=150mm; norm Q/T	0.25 max	0.25 max	0.30 max	0.80-1.35	0.40 max	0.030 max	0.025 max	0.10-0.40	Al >=0.018; Mo <=0.10; W <=0.1; Cr+Mo+Ni+Cu<=0.8; bal Fe	490-610	280	16	
BS 1503(89)	164-490	Frg press ves; t<=100mm; norm Q/T	0.25 max	0.25 max	0.30 max	0.80-1.35	0.40 max	0.030 max	0.025 max	0.10-0.40	Al >=0.018; Mo <=0.10; W <=0.1; Cr+Mo+Ni+Cu<=0.8; bal Fe	490-610	305	16	

USA

Specification	Designation	Notes	C	Cr	Cu	Mn	Ni	P	S	Si	Other	UTS	YS	El	Hard
	UNS K02506		0.25 max			0.90-1.35		0.035 max	0.025 max	0.15-0.30	bal Fe				
ASTM A727/A727M(97)	A727	Frg; Cu+Ni+Cr+Mo<=1.00 % in heat analysis	0.25 max	0.30 max	0.40 max	0.90-1.35	0.40 max	0.035 max	0.025 max	0.15-0.30	Mo <=0.12; Nb <=0.02; V <=0.05; Cu+Ni+Cr+Mo<=1.00; Cr+Mo<=0.32; bal Fe	415-585	250	22	

Carbon Steel, High-Manganese, A729

Japan

Specification	Designation	Notes	C	Cr	Cu	Mn	Ni	P	S	Si	Other	UTS	YS	El	Hard
JIS G3201(88)	SF340A	Frg Ann Tmp	0.60 max			0.30-1.20		0.030 max	0.035 max	0.15-0.50	bal Fe	340-440	175	27	90 HB
JIS G3201(88)	SF390A	Frg Ann Tmp	0.60 max			0.30-1.20		0.030 max	0.035 max	0.15-0.50	bal Fe	390-490	195	25	105 HB
JIS G3201(88)	SF440A	Frg Ann Tmp	0.60 max			0.30-1.20		0.030 max	0.035 max	0.15-0.50	bal Fe	440-540	225	24	121 HB
JIS G3201(88)	SF490A	Frg Ann Tmp	0.60 max			0.30-1.20		0.030 max	0.035 max	0.15-0.50	bal Fe	490-590	245	22	134 HB
JIS G3201(88)	SF540A	Frg Ann Tmp	0.60 max			0.30-1.20		0.030 max	0.035 max	0.15-0.50	bal Fe	540-640	275	20	152 HB
JIS G3201(88)	SF540B	Q/T Frg	0.60 max			0.30-1.20		0.030 max	0.035 max	0.15-0.50	bal Fe	540-690	295-335	20-21	152 HB
JIS G3201(88)	SF590A	Frg Ann Tmp	0.60 max			0.30-1.20		0.030 max	0.035 max	0.15-0.50	bal Fe	590-690	295	18	167 HB

UNS numbers and US grades are provided as a means of cross referencing chemically similar alloys. Exchangability is only possible after independent examination of specifications. Tensile properties are minimum or typical as specified. UTS and YS as MPa. El as %. See Appendix for list of abbreviations used in Notes. * indicates obsolete material.

Specification	Designation	Notes	C	Cr	Cu	Mn	Ni	P	S	Si	Other	UTS	YS	El	Hard
Carbon Steel, High-Manganese, A729 (Continued from previous page)															
Japan															
JIS G3201(88)	SF590B	Q/T Frg	0.60 max			0.30-1.20		0.030 max	0.035 max	0.15-0.50	bal Fe	590-740	325-360	18-19	167 HB
JIS G3201(88)	SF640B	Q/T Frg	0.60 max			0.30-1.20		0.030 max	0.035 max	0.15-0.50	bal Fe	640-780	345-390	15-16	183 HB
USA															
	UNS K06001		0.60 max			1.30-1.70		0.045 max	0.050 max	0.15 min	bal Fe				
ASTM A729(93)	A729	HT, rail axles	0.60 max			1.30-1.70		0.045 max	0.050 max	0.15 max	bal Fe	690	450	20	
Carbon Steel, High-Manganese, A758															
USA															
	UNS K02741		0.27 max	0.25 max	0.35 max	0.85-1.20	0.25 max	0.035 max	0.035 max	0.15-0.30	Mo <=0.08; Pb <=0.05; V <=0.05; bal Fe				
ASTM A758/A758M(98)	60	Butt-Welding pipe fittings	0.27 max	0.25 max	0.35 max	0.85-1.20		0.035 max	0.035 max	0.15-0.30	Mo <=0.08; N <=0.25; Pb <=0.05; V <=0.05; bal Fe	415-585	240	30L 22T	
ASTM A758/A758M(98)	70	Butt-Welding pipe fittings	0.27 max	0.25 max	0.35 max	0.85-1.20		0.035 max	0.035 max	0.15-0.30	Mo <=0.08; N <=0.25; Pb <=0.05; V <=0.05; bal Fe	485-635	260	27L 20T	
Carbon Steel, High-Manganese, A765(I)															
USA															
	UNS K03046		0.30 max	0.40 max	0.35 max	0.60-1.05	0.50 max	0.020 max	0.020 max	0.15-0.35	Al <=0.05; Mo <=0.25, V <=0.05; bal Fe				
ASTM A765/765M(98)	Grade I	Press ves frg	0.30 max	0.40 max	0.35 max	0.60-1.05	0.50 max	0.020 max	0.020 max	0.15-0.35	Al <=0.05; Mo <=0.25, V <=0.05; bal Fe	415-585	205	25	
Carbon Steel, High-Manganese, A765(II)															
USA															
	UNS K03047		0.30 max	0.40 max	0.35 max	0.60-1.35	0.50 max	0.020 max	0.020 max	0.15-0.35	Al <=0.05; Mo <=0.25, V <=0.05; bal Fe				
ASTM A765/765M(98)	Grade II	Press ves frg	0.30 max	0.40 max	0.35 max	0.60-1.35	0.50 max	0.020 max	0.020 max	0.15-0.35	Al <=0.05; Mo <=0.25, V <=0.05; bal Fe	485-655	250	22	
Carbon Steel, High-Manganese, B															
Sweden															
SS 141306	1306	Gen struct	0.18 max			1.1 max		0.04 max	0.04 max	0.6 max	Mo <=0.15; bal Fe				
USA															
	UNS K02102		0.21 max			0.80-1.10		0.04 max	0.04 max	0.35 max	bal Fe				
ASTM A131/A131M(94)	B	Ship struct	0.21 max			0.80-1.10		0.035 max	0.04 max	0.35 max	bal Fe	400-490	235	24	
ASTM A131/A131M(94)	B	Ship rivet/cold flanging stl	0.21 max			0.80-1.10		0.035 max	0.04 max	0.35 max	bal Fe	380-450	205	26	
Carbon Steel, High-Manganese, DS															
Japan															
JIS G3445(88)	STKM18A	Tube	0.18 max			1.50 max		0.040 max	0.040 max	0.55 max	bal Fe	441	275	25L	
JIS G3445(88)	STKM18B	Tube	0.18 max			1.50 max		0.040 max	0.040 max	0.55 max	bal Fe	490	314	23L	
JIS G3445(88)	STKM18C	Tube	0.18 max			1.50 max		0.040 max	0.040 max	0.55 max	bal Fe	510	382	15L	
UK															
BS 1717(83)	ERWC5	Tub	0.15 max			1.20 max		0.040 max	0.040 max	0.35 max	W <=0.1; bal Fe				
USA															
	UNS K01601		0.16 max			1.00-1.35		0.04 max	0.04 max	0.10-0.35	bal Fe				

UNS numbers and US grades are provided as a means of cross referencing chemically similar alloys. Exchangability is only possible after independent examination of specifications. Tensile properties are minimum or typical as specified. UTS and YS as MPa. El as %. See Appendix for list of abbreviations used in Notes. * indicates obsolete material.

Specification	Designation	Notes	C	Cr	Cu	Mn	Ni	P	S	Si	Other	UTS	YS	El	Hard

Carbon Steel, High-Manganese, DS (Continued from previous page)

USA

Specification	Designation	Notes	C	Cr	Cu	Mn	Ni	P	S	Si	Other	UTS	YS	El	Hard
ASTM A131/A131M(94)	CS	Ship rivet/cold flanging stl	0.16 max			1.00-1.35		0.035 max	0.04 max	0.10-0.35	bal Fe	380-450	205	26	
ASTM A131/A131M(94)	CS	Ship struct	0.16 max			1.00-1.35		0.035 max	0.04 max	0.10-0.35	bal Fe	400-490	235	24	
ASTM A131/A131M(94)	DS	Ship struct	0.16 max			1.00-1.35		0.035 max	0.04 max	0.10-0.35	bal Fe	400-490	235	24	
ASTM A131/A131M(94)	DS	Ship rivet/cold flanging stl	0.16 max			1.00-1.35		0.035 max	0.04 max	0.10-0.50	bal Fe	380-450	205	26	

Carbon Steel, High-Manganese, E

UK

Specification	Designation	Notes	C	Cr	Cu	Mn	Ni	P	S	Si	Other	UTS	YS	El	Hard
BS 1503(89)	221-410	Frg press ves; t<=100mm; Ann norm Q/T	0.20 max	0.25 max	0.30 max	0.80-1.20	0.40 max	0.030 max	0.025 max	0.10-0.40	Al <=0.010; Mo <=0.10; W <=0.1; Cr+Mo+Ni+Cu<=0.8; bal Fe	410-530	215	20	
BS 1503(89)	221-410	Frg press ves; 500<t<=999mm; Ann norm Q/T	0.20 max	0.25 max	0.30 max	0.80-1.20	0.40 max	0.030 max	0.025 max	0.10-0.40	Al <=0.010; Mo <=0.10; W <=0.1; Cr+Mo+Ni+Cu<=0.8; bal Fe	410-530	205	20	

USA

Specification	Designation	Notes	C	Cr	Cu	Mn	Ni	P	S	Si	Other	UTS	YS	El	Hard
	UNS K01801		0.18 max			0.70-1.35		0.04 max	0.04 max	0.10-0.35	bal Fe				
ASTM A131/A131M(94)	E	Ship struct	0.18 max			0.70-1.35		0.035 max	0.04 max	0.10-0.35	bal Fe	400-490	235	24	
ASTM A131/A131M(94)	E	Ship rivet/cold flanging stl	0.18 max			0.70-1.35		0.035 max	0.04 max	0.10-0.35	bal Fe	380-450	205	26	

Carbon Steel, High-Manganese, K02006

UK

Specification	Designation	Notes	C	Cr	Cu	Mn	Ni	P	S	Si	Other	UTS	YS	El	Hard
BS 1449/1(91)	60/55CS	Plt Sh Strp; CR t<=16mm	0.20 max			1.50 max		0.040 max	0.040 max		bal Fe	600	550	10-17	
BS 1449/1(91)	60/55HS	Plt Sh Strp HR t<=8mm	0.20 max			1.50 max		0.040 max	0.040 max		bal Fe				
BS 1503(89)	223-430	Frg press ves; t<=100mm; norm Q/T	0.20 max	0.25 max	0.30 max	0.80-1.40	0.40 max	0.030 max	0.025 max	0.10-0.40	Mo <=0.10; Nb 0.01-0.06; W <=0.1; Cr+Mo+Ni+Cu<=0.8; bal Fe	430-550	260	19	
BS 1503(89)	223-430	Frg press ves; 100<t<=999mm; norm Q/T	0.20 max	0.25 max	0.30 max	0.80-1.40	0.40 max	0.030 max	0.025 max	0.10-0.40	Mo <=0.10; Nb 0.01-0.06; W <=0.1; Cr+Mo+Ni+Cu<=0.8; bal Fe	430-550	245	19	
BS 1503(89)	224-430	Frg press ves; t<=100mm; norm Q/T	0.20 max	0.25 max	0.30 max	0.80-1.40	0.40 max	0.030 max	0.025 max	0.10-0.40	Al >=0.018; Mo <=0.10; W <=0.1; Cr+Mo+Ni+Cu<=0.8; bal Fe	430-550	250	19	
BS 1503(89)	224-430	Frg press ves; 100<t<=999mm; norm Q/T	0.20 max	0.25 max	0.30 max	0.80-1.40	0.40 max	0.030 max	0.025 max	0.10-0.40	Al >=0.018; Mo <=0.10; W <=0.1; Cr+Mo+Ni+Cu<=0.8; bal Fe	430-550	235	19	

USA

Specification	Designation	Notes	C	Cr	Cu	Mn	Ni	P	S	Si	Other	UTS	YS	El	Hard
	UNS K02006		0.20 max			0.90-1.45		0.035 max	0.040 max	0.15-0.35	bal Fe				

Carbon Steel, High-Manganese, K02400

Japan

Specification	Designation	Notes	C	Cr	Cu	Mn	Ni	P	S	Si	Other	UTS	YS	El	Hard
JIS G3445(88)	STKM20A	Tube	0.25 max			1.60 max		0.040 max	0.040 max	0.55 max	Nb <=0.15; Nb or V 0.15 max; bal Fe	539	392	23L	
JIS G4106(79)	SMn420	HR Frg FF	0.17-0.23	0.35 max	0.30 max	1.20-1.50	0.25 max	0.030 max	0.030 max	0.15-0.35	bal Fe				

USA

Specification	Designation	Notes	C	Cr	Cu	Mn	Ni	P	S	Si	Other	UTS	YS	El	Hard
	UNS K02400		0.24 max	0.25 max	0.35 max	0.65-1.40	0.25 max	0.035 max	0.040 max	0.13-0.55	Mo <=0.08; bal Fe				

Carbon Steel, High-Manganese, K02802

USA

Specification	Designation	Notes	C	Cr	Cu	Mn	Ni	P	S	Si	Other	UTS	YS	El	Hard
	UNS K02802		0.28 max			0.81-1.25		0.040 max	0.050 max	0.13-0.33	bal Fe				

UNS numbers and US grades are provided as a means of cross referencing chemically similar alloys. Exchangability is only possible after independent examination of specifications. Tensile properties are minimum or typical as specified. UTS and YS as MPa. El as %. See Appendix for list of abbreviations used in Notes. * indicates obsolete material.

Specification	Designation	Notes	C	Cr	Cu	Mn	Ni	P	S	Si	Other	UTS	YS	El	Hard

Carbon Steel, High-Manganese, K03201

USA

| | UNS K03201 | | 0.25-0.40 | | | 1.15-1.55 | | 0.040 max | 0.08-0.13 | 0.10-0.20 | bal Fe | | | | |

Carbon Steel, High-Manganese, MIL-W-21425(I,III)

UK

| BS 1429(80) | 090A65 | Spring | 0.55-0.75 | | | 0.60-1.20 | | | | 0.30 max | bal Fe | | | | |

USA

| | UNS K06500 | | 0.55-0.75 | | | 0.60-1.20 | | 0.04 max | 0.05 max | 0.10-0.30 | bal Fe | | | | |

Carbon Steel, High-Manganese, MIL-W-21425(II)

USA

| | UNS K06000 | | 0.45-0.75 | | | 0.60-1.10 | | 0.040 max | 0.050 max | 0.10-0.35 | bal Fe | | | | |

Carbon Steel, High-Manganese, No equivalents identified

Canada

Specification	Designation	Notes	C	Cr	Cu	Mn	Ni	P	S	Si	Other	UTS	YS	El	Hard
CSA G40.21(98)	260W	t<65mm; Angles Bars, Rolled	0.20 max			0.50-1.50		0.04 max	0.05 max	0.40 max	Al+Nb+V<=0.10; bal Fe	410-590	260	20L 18T	
CSA G40.21(98)	260WT	t<65mm, Plt, Sht, Bar	0.20 max			0.80-1.50		0.03 max	0.04 max	0.15-0.40	Al+Nb+V<=0.10; bal Fe	410-590	260	20L 18T	
CSA G40.21(98)	300W	t<65mm; Angles Bars, Rolled	0.22 max			0.50-1.50		0.04 max	0.05 max	0.40 max	Al+Nb+V<=0.10; bal Fe	450-620	300	20L 18T	
CSA G40.21(98)	300WT	t<65mm, Plt, Sht, Bar	0.22 max			0.80-1.50		0.03 max	0.04 max	0.15-0.40	Al+Nb+V<=0.10; bal Fe	450-620	300	20L 18T	
CSA G40.21(98)	350A	t<=100mm, Plt Sht Bar	0.20 max	0.70 max	0.20-0.60	0.75-1.35	0.90 max	0.03 max	0.04 max	0.15-0.50	Al+Nb+V<=0.10; bal Fe	480-650	350	19L 17T	
CSA G40.21(98)	350AT	t<65mm, Plt, Sht, Bar	0.20 max	0.70 max	0.20-0.60	0.75-1.35	0.90 max	0.03 max	0.04 max	0.15-0.50	Al+Nb+V<=0.10; bal Fe	480-650	350	19L 17T	
CSA G40.21(98)	350W	t<65mm; Angles Bars, Rolled	0.23 max			0.50-1.50		0.04 max	0.05 max	0.40 max	Al+Nb+V<=0.10; bal Fe	450-650	350	19L 17T	
CSA G40.21(98)	350WT	t<65mm, Plt, Sht, Bar	0.22 max			0.80-1.50		0.03 max	0.04 max	0.15-0.40	Al+Nb+V<=0.10; bal Fe	480-650	350	19L 17T	
CSA G40.21(98)	380W	t<65mm; Angles Bars	0.23 max			0.50-1.50		0.40 max	0.05 max	0.40 max	Al+Nb+V<=0.10; bal Fe	480-650	380	18L	
CSA G40.21(98)	400A	t<65mm, Plt, Sht, Bar	0.20 max	0.70 max	0.20-0.60	0.75-1.35	0.90 max	0.03 max	0.04 max	0.15-0.50	Al+Nb+V<=0.10; bal Fe	520-690	400	18L 15T	
CSA G40.21(98)	400AT	t<65mm, Plt, Sht, Bar	0.20 max	0.70 max	0.20-0.60	0.75-1.35	0.90 max	0.03 max	0.04 max	0.15-0.50	Al+Nb+V<=0.10; bal Fe	520-690	400	18L 15T	
CSA G40.21(98)	400W	t<65mm; Angles Bars	0.23			0.50-1.50		0.40 max	0.05 max	0.40 max	Al+Nb+V<=0.10; bal Fe	520-690	400	16L 13T	
CSA G40.21(98)	400WT	t<65mm, Plt, Sht, Bar	0.22 max			0.80-1.50		0.03 max	0.04 max	0.15-0.40	Al+Nb+V<=0.10; bal Fe	520-690	400	18L 15T	
CSA G40.21(98)	480AT	t<65mm, Plt, Sht, Bar	0.20 max	0.70 max	0.20-0.60	1.00-1.60	0.25-0.50	0.025 max	0.035 max	0.15-0.50	Al+Nb+V<=0.12; bal Fe	590-790	480	15L 12T	
CSA G40.21(98)	480W	t<65mm; Angles Bars	0.26 max			0.50-1.50		0.40 max	0.05 max	0.40 max	Al+Nb+V<=0.10; bal Fe	590-790	480	15L 12T	
CSA G40.21(98)	480WT	t<65mm, Rolled Shapes	0.26 max			0.80-1.50		0.03 max	0.04 max	0.15-0.40	Al+Nb+V<=0.10; bal Fe	590-790	480	15	
CSA G40.21(98)	550AT	t<65mm, Plt, Sht, Bar	0.15 max	0.70 max		1.75 max		0.025 max	0.035 max	0.15-0.40	bal Fe	620-860	550	13L 10T	
CSA G40.21(98)	550W	t<65mm; Angles, Bars	0.15			1.75 max		0.04 max	0.05 max	0.40 max	Al+Nb+V<=0.15; bal Fe	620-860	550	13L 10T	
CSA G40.21(98)	550WT	Hollow Sections	0.15			1.85 max		0.03 max	0.03 max	0.15-0.40	Al+Nb+V=0.15; bal Fe	620-860	550	15	
CSA G40.21(98)	700Q	t<=100mm, Plt Sht Bar	0.20 max			1.50 max		0.03 max	0.04 max	0.15-0.40	B 0.0005-0.0050; bal Fe	800-950	700	18L 16T	
CSA G40.21(98)	700QT	t<=100mm, Plt Sht Bar	0.20 max			1.50 max		0.03 max	0.04 max	0.15-0.40	bal Fe	800-950	700	18L 16T	

Czech Republic

| CSN 411300 | 11300 | Gen struct | 0.09 max | 0.3 max | | 1.60 max | | 0.04 max | 0.04 max | 0.60 max | bal Fe | | | | |

Europe

| EN 10025(90) | Fe510DD1 | Obs EU desig; struct HR FF; QS 16<t<=40mm | 0.20 max | | | 1.6 max | | 0.035 max | 0.035 max | 0.55 max | CEV; bal Fe | 490-630 | 345 | 22 | |

UNS numbers and US grades are provided as a means of cross referencing chemically similar alloys. Exchangability is only possible after independent examination of specifications. Tensile properties are minimum or typical as specified. UTS and YS as MPa. El as %. See Appendix for list of abbreviations used in Notes. * indicates obsolete material.

Specification	Designation	Notes	C	Cr	Cu	Mn	Ni	P	S	Si	Other	UTS	YS	El	Hard

Carbon Steel, High-Manganese, No equivalents identified

Europe

Specification	Designation	Notes	C	Cr	Cu	Mn	Ni	P	S	Si	Other	UTS	YS	El	Hard
EN 10025(90)	Fe510DD1	Obs EU desig; struct HR FF; QS 40<t<=100mm	0.22 max			1.6 max		0.035 max	0.035 max	0.55 max	CEV; bal Fe	490-630	325-335	20-21	
EN 10025(90)	Fe510DD1	Obs EU desig; struct HR FF; QS t<=16mm	0.20 max			1.6 max		0.035 max	0.035 max	0.55 max	CEV; bal Fe	410-580	355	14-22	
EN 10025(90)	Fe510DD1	Obs EU desig; struct HR FF; QS 150<t<=250mm	0.22 max			1.6 max		0.035 max	0.035 max	0.55 max	CEV; bal Fe	440-630	275-285	17	
EN 10025(90)	Fe510DD2	Obs EU desig; struct HR FF; QS 150<t<=250mm	0.22 max			1.6 max		0.035 max	0.035 max	0.55 max	CEV; bal Fe	440-630	275-285	17	
EN 10025(90)	Fe510DD2	Obs EU desig; struct HR FF; QS 40<t<=100mm	0.22 max			1.6 max		0.035 max	0.035 max	0.55 max	CEV; bal Fe	490-630	325-335	20-21	
EN 10025(90)	Fe510DD2	Obs EU desig; struct HR FF; QS 16<t<=40mm	0.20 max			1.6 max		0.035 max	0.035 max	0.55 max	CEV; bal Fe	490-630	345	22	
EN 10025(90)	Fe510DD2	Obs EU desig; struct HR FF; QS t<=16mm	0.20 max			1.6 max		0.035 max	0.035 max	0.55 max	CEV; bal Fe	410-580	355	14-22	
EN 10025(90)A1(93)	1.0577	Struct HR QS, t<=1mm	0.20 max			1.60 max		0.035 max	0.035 max	0.55 max	CEV may be req'd; bal Fe	510-680	355	14L 12T	
EN 10025(90)A1(93)	1.0577	Struct HR QS, 63<t<=80mm	0.22 max			1.60 max		0.035 max	0.035 max	0.55 max	CEV may be req'd; bal Fe	490-630	325	20L 18T	
EN 10025(90)A1(93)	1.0577	Struct HR QS, 200<t<=250mm	0.22 max			1.60 max		0.035 max	0.035 max	0.55 max	CEV may be req'd; bal Fe	450-630	275	17L 17T	

France

Specification	Designation	Notes	C	Cr	Cu	Mn	Ni	P	S	Si	Other	UTS	YS	El	Hard
AFNOR NFA35053(84)	FR22	Wir rod	0.2-0.25	0.15 max	0.2 max	0.5-0.8	0.25 max	0.035 max	0.035 max	0.35 max	Al >=0.02; Cr+Ni+Cu<=0.45; bal Fe				
AFNOR NFA35053(84)	FR38	Wir rod	0.35-0.4	0.15 max	0.2 max	0.5-0.8	0.25 max	0.035 max	0.035 max	0.35 max	Al >=0.02; Cr+Ni+Cu<=0.45; bal Fe				
AFNOR NFA36231(92)	E315D	t<=16mm; HR	0.1 max		0.3 max	1.3 max		0.025 max	0.015 max	0.3 max	Al 0.01-0.08; Nb 0.01-0.06; Ti 0.01-0.10; V 0.01-0.08; bal Fe	400-490	315	28	
AFNOR NFA36231(92)	E315D	1.5-2.99mm; HR	0.1 max		0.3 max	1.3 max		0.025 max	0.015 max	0.3 max	Al 0.01-0.08; Nb 0.01-0.06; Ti 0.01-0.10; V 0.01-0.08; bal Fe	400-490	315	23	
AFNOR NFA36231(92)	E355D	t<=16mm; HR	0.1 max		0.3 max	1.5 max		0.025 max	0.015 max	0.4 max	Al 0.01-0.08; Nb 0.01-0.06; Ti 0.01-0.10; V 0.01-0.08; bal Fe	430-530	355	25	
AFNOR NFA36231(92)	E420D	1.5-2.99mm; HR	0.1 max		0.3 max	1.6 max		0.025 max	0.015 max	0.4 max	Al 0.01-0.08; Nb 0.01-0.06; Ti 0.01-0.10; V 0.01-0.08; bal Fe	490-590	420	17	
AFNOR NFA36231(92)	E420D	t<=16mm; HR	0.1 max		0.3 max	1.6 max		0.025 max	0.015 max	0.4 max	Al 0.01-0.08; Nb 0.01-0.06; Ti 0.01-0.10; V 0.01-0.08; bal Fe	490-590	420	22	
AFNOR NFA36231(92)	E490D	t<=16mm; HR	0.1 max		0.3 max	1.7 max		0.025 max	0.015 max	0.4 max	Al 0.01-0.08; Nb 0.01-0.07; Ti 0.01-0.10; V 0.01-0.10; bal Fe	560-670	490	19	
AFNOR NFA36231(92)	E560D	t<=16mm; HR	0.1 max		0.3 max	1.7 max		0.025 max	0.015 max	0.4 max	Al 0.01-0.08; Nb 0.01-0.08; Ti 0.01-0.10; V 0.01-0.12; bal Fe	630-740	560	16	
AFNOR NFA36231(92)	E560D	1.5-2.9mm; HR	0.1 max		0.3 max	1.7 max		0.025 max	0.015 max	0.4 max	Al 0.01-0.08; Nb 0.01-0.08; Ti 0.01-0.10; V 0.01-0.12; bal Fe	630-740	560	12	
AFNOR NFA36231(92)	E620D	t<=16mm; HR	0.1 max		0.3 max	2.0 max		0.025 max	0.015 max	0.4 max	Al 0.01-0.08; B <=0.005; Mo <=0.50; Nb 0.01-0.08; bal Fe	690-820	620	14	
AFNOR NFA36231(92)	E690D	1.5-2.99mm; HR	0.1 max		0.3 max	2.0 max		0.025 max	0.015 max	0.4 max	Al 0.01-0.08; Mo <=0.50; Ti 0.01-0.10; V 0.01-0.10; bal Fe	750-900	690	10	
AFNOR NFA36231(92)	E690D	t<=16mm; HR	0.1 max		0.3 max	2.0 max		0.025 max	0.015 max	0.4 max	Al 0.01-0.08; Mo <=0.50; Ti 0.01-0.10; V 0.01-0.10; bal Fe	750-900	690	12	
AFNOR NFA36231(92)	S315MC*	t<=16mm; HR; see E315D	0.1 max		0.3 max	1.3 max		0.025 max	0.015 max	0.3 max	Al 0.01-0.08; Nb 0.01-0.06; Ti 0.01-0.10; V 0.01-0.08; bal Fe	400-490	315	28	

Specification	Designation	Notes	C	Cr	Cu	Mn	Ni	P	S	Si	Other	UTS	YS	El	Hard

Carbon Steel, High-Manganese, No equivalents identified

France

Specification	Designation	Notes	C	Cr	Cu	Mn	Ni	P	S	Si	Other	UTS	YS	El	Hard
AFNOR NFA36231(92)	S355MC*	t<=16mm; HR; see E355D	0.1 max		0.3 max	1.5 max		0.025 max	0.015 max	0.4 max	Al 0.01-0.08; Nb 0.01-0.06; Ti 0.01-0.10; V 0.01-0.08; bal Fe	430-530	355	25	
AFNOR NFA36231(92)	S420MC*	t<=16mm; HR; see E420D	0.1 max		0.3 max	1.6 max		0.025 max	0.015 max	0.4 max	Al 0.01-0.08; Nb 0.01-0.06; Ti 0.01-0.10; V 0.01-0.08; bal Fe	490-590	420	22	
AFNOR NFA36231(92)	S490MC*	t<=16mm; HR; see E490D	0.1 max		0.3 max	1.7 max		0.025 max	0.015 max	0.4 max	Al 0.01-0.08; Nb 0.01-0.07; Ti 0.01-0.10; V 0.01-0.10; bal Fe	560-670	490	19	
AFNOR NFA36231(92)	S560MC*	t<=16mm; HR; see E560D	0.1 max		0.3 max	1.7 max		0.025 max	0.015 max	0.4 max	Al 0.01-0.08; Nb 0.01-0.08; Ti 0.01-0.10; V 0.01-0.12; bal Fe	630-740	560	16	
AFNOR NFA36231(92)	S620MC*	t<=16mm; HR; see E620D	0.1 max		0.3 max	2.0 max		0.025 max	0.015 max	0.4 max	Al 0.01-0.08; B <=0.005; Mo <=0.50; Nb 0.01-0.08; bal Fe	690-820	620	14	
AFNOR NFA36231(92)	S690MC*	1.5-2.99mm; HR; see E690D	0.1 max		0.3 max	2.0 max		0.025 max	0.015 max	0.4 max	Al 0.01-0.08; Mo <=0.50; Ti 0.01-0.10; V 0.01-0.10; bal Fe	750-900	690	10	

Germany

Specification	Designation	Notes	C	Cr	Cu	Mn	Ni	P	S	Si	Other	UTS	YS	El	Hard
DIN E EN 10208(93)	L360GA	Pipe for combustible fluids	0.22 max			1.60 max		0.030 max	0.030 max	0.55 max	Al 0.015-0.060; V+Nb+Ti<=0.15; bal Fe				
DIN E EN 10208(93)	WNr 1.0499	Pipe for combustible fluids	0.22 max			1.60 max		0.030 max	0.030 max	0.55 max	Al 0.015-0.060; V+Nb+Ti<=0.15; bal Fe				

Hungary

Specification	Designation	Notes	C	Cr	Cu	Mn	Ni	P	S	Si	Other	UTS	YS	El	Hard
MSZ 23(83)	A341H	Sh Strp; const; 0.5-3mm; CR	0.15 max			2 max	0.3 max	0.045 max	0.05 max	0.6 max	bal Fe	330-460	220	22L	
MSZ 23(83)	A342H	Sh Strp; const; 0.5-3mm; CR	0.15 max			2 max	0.3 max	0.045 max	0.05 max	0.6 max	bal Fe	330-460	220	22L	
MSZ 23(83)	A342P	Sh Strp; const; 0.5-3mm; HR	0.15 max			2 max	0.3 max	0.045 max	0.05 max	0.6 max	bal Fe	330-460	220	20L	
MSZ 23(83)	A343F	Sh Strp; const; 0.5-3mm; HR	0.15 max			2 max	0.3 max	0.045 max	0.05 max	0.6 max	bal Fe	330-460	220	20L	
MSZ 23(83)	A343H	Sh Strp; const; 0.5-3mm; CR	0.15 max			2 max	0.3 max	0.045 max	0.05 max	0.6 max	bal Fe	330-460	220	22L	
MSZ 23(83)	A343P	Sh Strp; const; 0.5-3mm; HR	0.15 max			2 max	0.3 max	0.045 max	0.05 max	0.6 max	bal Fe	330-460	220	20L	
MSZ 23(83)	A381H	Sh Strp; const; 0.5-3mm; CR	0.22 max			2 max	0.3 max	0.045 max	0.05 max	0.6 max	MSZ500; bal Fe	370-510	240	18L	
MSZ 23(83)	A382H	Sh Strp; const; 0.5-3mm; CR	0.22 max			2 max	0.3 max	0.045 max	0.05 max	0.6 max	MSZ500; bal Fe	370-510	240	18L	
MSZ 23(83)	A382P	Sh Strp; const; gen struct; 0.5-3mm; HR	0.22 max			2 max	0.3 max	0.045 max	0.05 max	0.6 max	MSZ500; bal Fe	370-510	240	16L	
MSZ 23(83)	A383F	Sh Strp; const; gen struct; 0.5-3mm; HR	0.22 max			2 max	0.3 max	0.045 max	0.05 max	0.6 max	MSZ500; bal Fe	370-510	240	16L	
MSZ 23(83)	A383H	Sh Strp; const; 0.5-3mm; CR	0.22 max			2 max	0.3 max	0.045 max	0.05 max	0.6 max	MSZ500; bal Fe	370-510	240	18L	
MSZ 23(83)	A383P	Sh Strp; const; gen struct; 0.5-3mm; HR	0.22 max			2 max	0.3 max	0.045 max	0.05 max	0.6 max	MSZ500; bal Fe	370-510	240	16L	
MSZ 23(83)	A442H	Sh Strp; const; gen struct; 0.5-3mm; CR	0.25 max			2 max	0.3 max	0.045 max	0.05 max	0.12 min	MSZ500; bal Fe	430-590	280	16L	
MSZ 23(83)	A442P	Sh Strp; const; gen struct; 0.5-3mm; HR	0.25 max			2 max	0.3 max	0.045 max	0.05 max	0.12 min	MSZ500; bal Fe	430-590	280	14L	
MSZ 23(83)	A443F	Sh Strp; const; gen struct; 0.5-3mm; HR	0.25 max			2 max	0.3 max	0.045 max	0.05 max	0.12 min	MSZ500; bal Fe	430-590	280	14L	
MSZ 23(83)	A443H	Sh Strp; const; gen struct; 0.5-3mm; CR	0.25 max			2 max	0.3 max	0.045 max	0.05 max	0.12 min	MSZ500; bal Fe	430-590	280	16L	
MSZ 23(83)	A443P	Sh Strp; const; gen struct; 0.5-3mm; HR	0.25 max			2 max	0.3 max	0.045 max	0.05 max	0.12 min	MSZ500; bal Fe	430-590	280	14L	
MSZ 23(83)	A503F	Sh Strp; const; 0.5-3mm; HR	0.3 max			2 max		0.045 max	0.05 max	0.12 min	MSZ500; bal Fe	490-640	300	11L	
MSZ 23(83)	A503P	Sh Strp; const; gen struct; 0.5-3mm; HR	0.3 max			2 max		0.045 max	0.05 max	0.12 min	MSZ500; bal Fe	490-640	300	11L	
MSZ 23(83)	A603F	Sh Strp; const; gen struct; 0.5-3mm; HR	0.3 max			2 max		0.045 max	0.05 max	0.12 min	MSZ500; bal Fe	590-740	340	8L	
MSZ 23(83)	A603P	Sh Strp; const; gen struct; 0.5-3mm; HR	0.3 max			2 max		0.045 max	0.05 max	0.12 min	MSZ500; bal Fe	590-740	340	8L	

UNS numbers and US grades are provided as a means of cross referencing chemically similar alloys. Exchangability is only possible after independent examination of specifications. Tensile properties are minimum or typical as specified. UTS and YS as MPa. El as %. See Appendix for list of abbreviations used in Notes. * indicates obsolete material.

Specification	Designation	Notes	C	Cr	Cu	Mn	Ni	P	S	Si	Other	UTS	YS	El	Hard

Carbon Steel, High-Manganese, No equivalents identified

Hungary

Specification	Designation	Notes	C	Cr	Cu	Mn	Ni	P	S	Si	Other	UTS	YS	El	Hard
MSZ 2524(79)	A50	Counterrail; 0-50.0mm; HR	0.3 max			2 max		0.045 max	0.05 max	0.12 min	MSZ500; bal Fe	490-640	285	21L	
MSZ 2524(92)	Fe490-2	HR guiding rail; railway siding	0.3 max			2 max		0.045 max	0.045 max	0.6 max	Al >=0.02; N <=0.009; bal Fe				
MSZ 2570(88)	MA2	Non-rimming; FF; 0-75.0mm	0.6-0.75			0.8-1.3		0.04 max	0.04 max	0.15-0.35	bal Fe	900		9L	
MSZ 339(87)	B38.24	HR; weld conc-reinf; 6-40mm; ageing treated	0.2 max			2 max		0.05 max	0.05 max	0.6 max	bal Fe	370	235	25L	
MSZ 339(87)	B38.24B	HR; weld conc-reinf; 6-40mm; ageing treated	0.2 max			2 max		0.05 max	0.05 max	0.12-0.60	Si>=0.12; bal Fe	370	235	25L	
MSZ 339(87)	B50.36	HR; weld conc-reinf; 8-40mm; ageing treated	0.22 max			1.6 max		0.04 max	0.04 max	0.55 max	Al <=0.15; Nb 0.015-0.1; V 0.02-0.1; 32-40mm C<=0.23; 20-44mm Al>=0.015; Zr=0.015-0.06; Al+V+Nb+Zr<=0.15; N<=Al/2+V/4+Nb/7<=0.015; bal Fe	490	350	23L	
MSZ 339(87)	B75.50	HR Conc-reinf; 8-16mm; ageing treated	0.3 max			2 max		0.06 max	0.06 max	0.6 max	bal Fe	740	490	10L	

International

Specification	Designation	Notes	C	Cr	Cu	Mn	Ni	P	S	Si	Other	UTS	YS	El	Hard
ISO 2604-1(75)	F12	Frg	0.23 max			0.60-1.40		0.040 max	0.040 max	0.10-0.40	bal Fe	460-580			
ISO 2604-1(75)	F13	Frg	0.23 max			0.60-1.40		0.040 max	0.040 max	0.10-0.40	Al <=0.015; bal Fe	460-580			
ISO 2604-1(75)	F17	Frg	0.25 max			0.90-1.70		0.040 max	0.040 max	0.10-0.40	bal Fe	490-610			
ISO 2604-1(75)	F17Q	Frg	0.25 max			0.90-1.70		0.040 max	0.040 max	0.10-0.40	bal Fe	510-630			
ISO 2604-1(75)	F18	Frg	0.25 max			0.90-1.70		0.040 max	0.040 max	0.10-0.40	Al <=0.015; bal Fe	490-610			
ISO 2604-1(75)	F18Q	Frg	0.25 max			0.90-1.70		0.040 max	0.040 max	0.10-0.40	Al <=0.015; bal Fe	510-630			
ISO 2604-1(75)	F22	Frg	0.30 max			0.60-1.40		0.040 max	0.040 max	0.10-0.40	bal Fe	460-580			
ISO 2604-1(75)	F22Q	Frg	0.30 max			0.60-1.40		0.040 max	0.040 max	0.10-0.40	bal Fe	530-650			
ISO 2604-1(75)	F23	Frg	0.35 max			0.60-1.40		0.040 max	0.040 max	0.15-0.40	bal Fe	460-580			
ISO 2604-1(75)	F23Q	Frg	0.35 max			0.60-1.40		0.040 max	0.040 max	0.15-0.40	bal Fe	550-670			
ISO 2604-1(75)	F8	Frg	0.20 max			0.50-1.20		0.040 max	0.040 max	0.10-0.40	bal Fe	410-530			
ISO 2604-1(75)	F9	Frg	0.20 max			0.50-1.20		0.040 max	0.040 max	0.10-0.40	Al <=0.015; bal Fe	410-530			
ISO 2604-6(78)	TSAW18	Tub, weld	0.2 max	0.3 max	0.3 max	0.9-1.6	0.4 max	0.040 max	0.04 max	0.1-0.5	Al 0.015-0.120; Co <=0.1; Mo <=0.15; Pb <=0.15; Ti <=0.05; V <=0.1; W <=0.1; bal Fe				
ISO 2604-6(78)	TSAW9	Tub, weld	0.2 max	0.3 max	0.3 max	0.5-1.3	0.4 max	0.04 max	0.04 max	0.35 max	Al 0.015-0.120; Co <=0.1; Mo <=0.15; Pb <=0.15; Ti <=0.05; V <=0.1; W <=0.1; bal Fe				
ISO 4950-3(95)	E460	HS, flat, Q/T, t<=50mm	0.20 max			0.7-1.7		0.035 max	0.035 max	0.55 max	bal Fe	570-720	460	17	
ISO 4950-3(95)	E550	HS, flat, Q/T, t<=50mm	0.20 max			1.7 max		0.035 max	0.035 max	0.10-0.80	bal Fe	650-830	550	16	
ISO 4950-3(95)	E690	HS, flat, Q/T, t<=50mm	0.20 max			1.7 max		0.035 max	0.035 max	0.10-0.80	bal Fe	770-940	690	14	
ISO 4954(93)	42Mn6E	Wir rod Bar, CH ext	0.39-0.46			1.30-1.65		0.035 max	0.035 max	0.40 max	bal Fe	600			
ISO 683-9(88)	17SMn20	Free-cutting, CH, 16mm test	0.14-0.20			1.20-1.60		0.06 max	0.15-0.25	0.15-0.40	bal Fe	750-1100	500	9	
ISO 683-9(88)	35SMn20	Free-cutting, hard, Q/T, <16mm	0.32-0.39			0.90-1.40		0.06 max	0.15-0.25	0.15-0.40	bal Fe	620-820	420	14	
ISO 683-9(88)	44SMn28	Free-cutting, hard, Q/T, <16mm	0.40-0.48			1.30-1.70		0.06 max	0.24-0.33	0.15-0.40	bal Fe	750-950	530	10	

UNS numbers and US grades are provided as a means of cross referencing chemically similar alloys. Exchangability is only possible after independent examination of specifications. Tensile properties are minimum or typical as specified. UTS and YS as MPa. El as %. See Appendix for list of abbreviations used in Notes. * indicates obsolete material.

Carbon Steel, High-Manganese, No equivalents identified

Specification	Designation	Notes	C	Cr	Cu	Mn	Ni	P	S	Si	Other	UTS	YS	El	Hard
Japan															
JIS G3444(94)	STK540	Tub Gen struct	0.23 max			1.50 max		0.040 max	0.040 max	0.55 max	bal Fe	540	390	20	
JIS G3444(94)	STK540	Tub Gen struct	0.23 max			1.50 max		0.040 max	0.040 max	0.55 max	bal Fe	540	390	20	
Mexico															
NMX-B-206(86)	Grade 1	Smls pipe for atmospheric and low temp	0.21 max			0.90-1.35		0.048 max	0.058 max	0.10-0.40	bal Fe	414-586	240	30	
NMX-B-206(86)	Grade 2	Smls pipe for atmospheric and low temp	0.21 max			0.90-1.35		0.048 max	0.058 max	0.10-0.40	bal Fe	380-550	205	35	
Poland															
PNH74244(79)	G295	Struct; Weld tube	0.18 max	0.3 max		0.7-1.3	0.4 max	0.04 max	0.04 max	0.3-0.55	bal Fe				
PNH74244(79)	G355	Struct; Weld tube	0.22 max	0.3 max		1-1.5	0.4 max	0.04 max	0.04 max	0.2-0.55	CEV<=0.48; bal Fe				
PNH84018(86)	18G2ACu	HS weld const	0.2 max	0.3 max	0.25-0.5	1-1.5	0.3 max	0.04 max	0.04 max	0.2-0.55	Al 0.02-0.120; N <=0.009; CEV<=0.045; bal Fe				
PNH84023/05	22G2SA	Const	0.19-0.25	0.3 max	0.2 max	1.2-1.7	0.3 max	0.035 max	0.04 max	0.4-0.6	bal Fe				
PNH84023/05	25G2NbY	Const	0.23-0.29	0.25 max	0.25 max	1.2-1.6	0.25 max	0.045 max	0.045 max	0.08 max	Nb 0.02-0.04; bal Fe				
PNH84023/05	25G2Y	Const	0.23-0.29	0.3 max		1.2-1.65	0.3 max	0.05 max	0.04 max	0.08 max	Mo <=0.1; bal Fe				
PNH84023/06	18G2-b	Reinforcing	0.15-0.22	0.3 max	0.35 max	1-1.5	0.3 max	0.05 max	0.05 max	0.2-0.55	Mo <=0.1; CEV<=0.045; bal Fe				
PNH84023/06	20G2Y-b	Reinforcing	0.17-0.23	0.3 max		1.1-1.6	0.3 max	0.05 max	0.05 max	0.1 max	Mo <=0.1; CEV<=0.045; bal Fe				
PNH84023/06	25G2S	Reinforcing	0.2-0.29	0.3 max		1.2-1.6	0.3 max	0.04 max	0.045 max	0.6-0.9	Mo <=0.1; CEV<=0.58; bal Fe				
PNH84023/06	35G2	Const	0.29-0.37	0.3 max	0.35 max	1.3-1.7	0.3 max	0.05 max	0.05 max	0.15-0.35	bal Fe				
PNH84023/06	35G2Y	Reinforcing	0.3-0.35	0.3 max	0.35 max	1.2-1.6	0.3 max	0.05 max	0.05 max	0.08 max	Mo <=0.1; CEV<=0.59; bal Fe				
PNH84023/07	16G2Nb	Const	0.15-0.19	0.3 max	0.25 max	1.2-1.45	0.3 max	0.04 max	0.04 max	0.3-0.5	Al 0.02-0.120; N <=0.012; Nb 0.015-0.035; C+Mn/6<=0.43; bal Fe				
PNH84023/07	19G2FA	High-temp tube, pipe	0.15-0.22	0.3 max		1.3-1.7	0.3 max	0.35 max	0.035 max	0.25-0.5	Al 0.01-0.120; N 0.01-0.02; V 0.1-0.17; bal Fe				
PNH84023/08	18G2AA	Chain	0.13-0.2	0.2 max	0.2 max	1.2-1.5	0.2 max	0.04 max	0.04 max	0.3-0.55	Al 0.02-0.120; bal Fe				
PNH84023/09	30G2F	Const	0.27-0.36	0.3 max		1.1-1.4	0.4 max	0.04 max	0.04 max	0.17-0.37	Mo <=0.1; V 0.1-0.2; bal Fe				
PNH84024	St41K	High-temp const	0.12-0.2	0.3 max		0.45-1.60	0.3 max	0.45 max	0.045 max	0.15-0.35	bal Fe				
PNH84024	St44K	High-temp const	0.14-0.22	0.3 max		0.55-1.60	0.3 max	0.045 max	0.045 max	0.15-0.35	bal Fe				
PNH84030/04	35SG	Q/T	0.31-0.39	0.25 max		1.1-1.4	0.3 max	0.035 max	0.035 max	1.1-1.4	bal Fe				
PNH84030/04	45G2	Q/T	0.41-0.49	0.25 max		1.4-1.8	0.3 max	0.035 max	0.035 max	0.17-0.37	bal Fe				
PNH92147	AH32	Struct/welded ships; Sh, Plt	0.18 max	0.2 max	0.25 max	0.9-1.6	0.4 max	0.04 max	0.04 max	0.1-0.5	Al <=0.15; Mo <=0.08; CEV<=0.45; bal Fe				
PNH92147	AH36	Struct/welded ships; Sh, Plt	0.18 max	0.2 max	0.35 max	0.9-1.6	0.4 max	0.04 max	0.04 max	0.1-0.5	Al <=0.15; Mo <=0.08; V 0.05-0.1; CEV<=0.45; bal Fe				
PNH92147	AH40	Struct/welded ships; Sh, Plt	0.18 max	0.2 max	0.35 max	0.9-1.6	0.4 max	0.04 max	0.04 max	0.1-0.5	Al <=0.15; Mo <=0.08; V 0.05-0.1; CEV<=0.45; bal Fe				
PNH92147	D	Struct/welded ships; Sh, Plt	0.21 max	0.3 max		0.6-1.4	0.3 max	0.04 max	0.04 max	0.1-0.35	CEV<=0.4; bal Fe				
PNH92147	DH32	Struct/welded ships; Sh, Plt	0.18 max	0.2 max	0.25 max	0.9-1.6	0.4 max	0.04 max	0.04 max	0.1-0.5	Al <=0.15; Mo <=0.08; CEV<=0.45; bal Fe				

Specification	Designation	Notes	C	Cr	Cu	Mn	Ni	P	S	Si	Other	UTS	YS	El	Hard

Carbon Steel, High-Manganese, No equivalents identified

Poland

Specification	Designation	Notes	C	Cr	Cu	Mn	Ni	P	S	Si	Other	UTS	YS	El	Hard
PNH92147	E	Struct/welded ships; Sh, Plt	0.18 max	0.3 max		0.7-1.5	0.3 max	0.04 max	0.04 max	0.1-0.35	CEV<=0.4; bal Fe				
PNH92147	EH32	Struct/welded ships; Sh, Plt	0.18 max	0.2 max	0.25 max	0.9-1.6	0.4 max	0.04 max	0.04 max	0.1-0.5	Al <=0.15; Mo <=0.08; CEV<=0.45; bal Fe				
PNH92147	EH36	Struct/welded ships; Sh, Plt	0.18 max	0.2 max	0.25 max	0.9-1.6	0.4 max	0.04 max	0.04 max	0.1-0.5	Al <=0.15; Mo <=0.08; V 0.05-0.1; CEV<=0.45; bal Fe				
PNH93215(82)	18G2 A-II	HR reinf; Ribbed	0.15-0.22	0.3 max	0.35 max	1-1.5	0.3 max	0.05 max	0.05 max	0.2-0.55	Mo <=0.1; bal Fe				
PNH93215(82)	20G2VY	HR reinf	0.23 max	0.3 max	0.35 max	1.6 max	0.3 max	0.05 max	0.05 max	0.12 max	Mo <=0.1; bal Fe				
PNH93215(82)	20G2Y; A-II	HR reinf; Ribbed	0.22 max	0.3 max	0.35 max	1.5 max	0.3 max	0.05 max	0.05 max	0.08 max	Mo <=0.1; bal Fe				

Romania

Specification	Designation	Notes	C	Cr	Cu	Mn	Ni	P	S	Si	Other	UTS	YS	El	Hard
STAS 438/1(80)	PC60	Reinf; HR	0.3 max	0.3 max		2 max		0.07 max	0.06 max	0.6 max	Pb <=0.15; No prescription; CE<=0.5; bal Fe				
STAS 500/2(88)	OL30	Gen struct	0.3 max	0.3 max		2 max	0.4 max	0.08 max	0.08 max	0.6 max	Pb <=0.15; bal Fe				

Russia

Specification	Designation	Notes	C	Cr	Cu	Mn	Ni	P	S	Si	Other	UTS	YS	El	Hard
GOST 1050(74)	65G	High-grade struct; Norm; 0-100mm	0.62-0.7	0.25 max	0.25 max	0.9-1.2	0.25 max	0.035 max	0.04 max	0.17-0.37	As<=0.08; bal Fe	740	430	9L	
GOST 1050(74)	65G	High-grade struct; Untreated; 0-250mm	0.62-0.7	0.25 max	0.25 max	0.9-1.2	0.25 max	0.035 max	0.04 max	0.17-0.37	As<=0.08; bal Fe				0-285 HB
GOST 14959(79)	65G	Spring	0.62-0.7	0.25 max	0.2 max	0.9-1.2	0.25 max	0.035 max	0.035 max	0.17-0.37	Al <=0.1; bal Fe				
GOST 4543(71)	30G	Q/T	0.27-0.35	0.3 max	0.3 max	0.7-1.0	0.3 max	0.035 max	0.035 max	0.17-0.37	Al <=0.1; bal Fe				
GOST 5058(65)	17GS	Weld const; struct	0.14-0.2	0.3 max	0.3 max	1-1.4	0.3 max	0.035 max	0.04 max	0.4-0.6	Al <=0.1; Ti 0.01-0.03; As<=0.08; bal Fe				

Spain

Specification	Designation	Notes	C	Cr	Cu	Mn	Ni	P	S	Si	Other	UTS	YS	El	Hard
UNE 36123(90)	AE550HC	HR; Strp; no US equiv	0.12 max			1.7 max		0.03 max	0.03 max	0.5 max	Al 0.02-0.08; Mo <=0.15; Nb 0.01-0.06; Ti 0.01-0.12; V 0.01-0.08; bal Fe				

Sweden

Specification	Designation	Notes	C	Cr	Cu	Mn	Ni	P	S	Si	Other	UTS	YS	El	Hard
SS 142133	2133	Fine grain Struct	0.2 max		0.4 max	1.6 max		0.035 max	0.035 max	0.5 max	N <=0.02; C+Mn/6+(Ni+Cu)/15+(Cr+Mo+V)/5<=0.41; bal Fe				

UK

Specification	Designation	Notes	C	Cr	Cu	Mn	Ni	P	S	Si	Other	UTS	YS	El	Hard
BS 1449/1(91)	43/25HR	Plt Sh Strp HR wide; t<=16mm	0.25 max			1.20 max		0.050 max	0.050 max		bal Fe	430	250	16-25	
BS 1449/1(91)	43/25HS	Plt Sh Strp HR t<=8mm	0.25 max			1.20 max		0.050 max	0.050 max		bal Fe	430	250	16-25	
BS 1502(82)	211	Bar Shp Press ves	0.19 max			0.90-1.50		0.040 max	0.040 max	0.35 max	W <=0.1; bal Fe				
BS 1502(82)	221	Bar Shp Press ves	0.19 max			0.90-1.50		0.040 max	0.040 max	0.1-0.35	W <=0.1; bal Fe				
BS 3059/1(87)	440	Tub Norm	0.12-0.18			0.90-1.20		0.040 max	0.035 max	0.35 max	bal Fe				
BS 3601(87)	ERW430	Pip Tub, press, t<=16mm	0.21 max			0.40-1.20		0.040 max	0.040 max	0.35 max	bal Fe	430-570	275	22	
BS 3602/1(87)	HFS430; CFS430	ERW IW smls Tub; press High-temp; t<=16mm	0.21 max			0.40-1.20		0.035 max	0.035 max	0.35 max	Al <=0.06; bal Fe	430-570	275	22	
BS 3602/1(87)	HFS500Nb; CFS500Nb	ERW IW smls Tub; press High-temp; t<=16mm	0.22 max			1.00-1.50		0.035 max	0.030 max	0.15-0.35	Al <=0.06; Nb 0.015-0.10; bal Fe	500-650	355	21	
BS 3602/2(91)	430	Pip weld Tub High-temp/press; t<=16mm	0.25 max	0.25 max	0.30 max	0.60-1.40	0.30 max	0.030 max	0.030 max	0.10-0.35	Mo <=0.10; Cr+Mo+Ni+Cu<=0.7; CEV req'd; bal Fe	430-550	250	23	
BS 3602/2(91)	490	Pip weld Tub High-temp/press; t<=16mm	0.22 max	0.25 max	0.30 max	0.90-1.60	0.75 max	0.030 max	0.030 max	0.10-0.40	Al >=0.015; Mo <=0.10; Cr+Mo+Ni+Cu<=0.5; CEV req'd; bal Fe	490-610	325	21	

UNS numbers and US grades are provided as a means of cross referencing chemically similar alloys. Exchangability is only possible after independent examination of specifications. Tensile properties are minimum or typical as specified. UTS and YS as MPa. El as %. See Appendix for list of abbreviations used in Notes. * indicates obsolete material.

Carbon Steel, High-Manganese, No equivalents identified

Specification	Designation	Notes	C	Cr	Cu	Mn	Ni	P	S	Si	Other	UTS	YS	El	Hard
USA															
	1052	CD	0.47-0.55			1.20-1.50		0.04 max	0.05 max		bal Fe	825	690	10.0	241-321 HB
	AISI 1544	Bar Blm Bil Slab, UNS no. obs	0.40-0.47			0.80-1.10		0.040 max	0.050 max		bal Fe				
	AISI 1545	Bar Blm Bil Slab, UNS no. obs	0.43-0.50			0.80-1.10		0.040 max	0.050 max		bal Fe				
	AISI 1546	Bar Blm Bil Slab, UNS no. obs	0.44-0.52			1.00		0.040 max	0.050 max		bal Fe				
AMS 2431/3B(96)		Cond Cut Wir Shot	0.45-0.85			0.60-1.20		0.045 max	0.05 max	0.10-0.30	bal Fe				45-52 HRC
ASTM A135	B	Weld Pipe	0.30 max			1.20 max		0.035 max	0.035 max		bal Fe	414	241	30L	
ASTM A178/A178M	D	ERW Pipe	0.27 max			1.00-1.50		0.030 max	0.015 max	0.10 min	bal Fe	485	275	30	
ASTM A304(96)	15L21H	Bar, hard bands spec	0.17-0.24			0.70-1.20		0.040 max	0.050 max	0.15-0.30	Pb 0.15-0.35; Cu Ni Cr Mo trace allowed; bal Fe				
ASTM A513(97a)	1027	ERW Mech tub	0.22-0.29			1.20-1.55		0.035 max	0.035 max		bal Fe				
ASTM A570/A570M(98)	36 Type 2	HR Sh Strp	0.25 max			1.35 max		0.035 max	0.04 max	0.40 max	If Cu bearing reqd, Cu >=0.20; bal Fe	400-550	250	16.0-21.0	
ASTM A570/A570M(98)	50	HR Sh Strp	0.25 max			1.35 max		0.035 max	0.04 max		If Cu bearing reqd, Cu >=0.20; bal Fe	415	345	11.0-17.0	
ASTM A570/A570M(98)	55	HR Sh Strp	0.25 max			1.35 max		0.035 max	0.04 max		If Cu bearing reqd, Cu >=0.20; bal Fe	480	380	9.0-15.0	
ASTM A764(95)		Coated spring Wir, 1.04mm diam class 1	0.45-0.85			0.30-1.30		0.040 max	0.050 max	0.10-0.35	bal Fe	1670-2020			
ASTM A765/765M(98)	Grade IV	Press ves frg	0.20 max	0.40 max	0.35 max	1.00-1.60	0.50 max	0.020 max	0.020 max	0.15-0.50	Al <=0.05; Mo <=0.25; V <=0.06; bal Fe	550-725	345	22	
ASTM A769(94)	Grade 36	ERW struct shps	0.20 max			0.30-1.50		0.035 max	0.04 max		bal Fe	365	250	22	
ASTM A858/A858M(96)	A858M	HT Fittings; low-temp and Corr res	0.20 max	0.30 max	0.35 max	0.90-1.35	0.50 max	0.03 max	0.010 max	0.15-0.40	Mo <=0.20; Ni+Cr+Mo+Cu<=10; bal Fe	485-655	250	22	235 HB
ASTM A907/A907M(96)	Grade 30	Sh Strp, thick, HR	0.25 max			1.50 max		0.035 max	0.04 max		Cu 0.20 if spec; bal Fe	340	205	22.0	
ASTM A907/A907M(96)	Grade 33	Sh Strp, thick, HR	0.25 max			1.50 max		0.035 max	0.04 max		Cu 0.20 if spec; bal Fe	360	230	22.0	
ASTM A907/A907M(96)	Grade 36	Sh Strp, thick, HR	0.25 max			1.50 max		0.035 max	0.04 max		Cu 0.20 if spec; bal Fe	365	250	21.0	
ASTM A907/A907M(96)	Grade 40	Sh Strp, thick, HR	0.25 max			1.50 max		0.035 max	0.04 max		Cu 0.20 if spec; bal Fe	380	275	19.0	
SAE J1268(95)	15B28H	Hard	0.25-0.34			1.00-1.50		0.030 max	0.050 max	0.15-0.35	B 0.0005-0.003; bal Fe				
SAE J1268(95)	15B30H	Hard	0.27-0.35			0.70-1.20		0.030 max	0.050 max	0.15-0.35	B 0.0005-0.003; bal Fe				
SAE J1268(95)	E15B28H	Hard	0.25-0.34			1.00-1.50		0.025 max	0.025 max	0.15-0.35	B 0.0005-0.003; bal Fe				
SAE J1268(95)	E15B30H	Hard	0.27-0.35			0.70-1.20		0.025 max	0.025 max	0.15-0.35	B 0.0005-0.003; bal Fe				
Yugoslavia															
	C.0002	Reinforcing	0.30 max			1.60 max		0.070 max	0.060 max	0.6 max	bal Fe				
	C.0300	Reinforcing	0.3 max			2 max		0.07 max	0.06 max	0.6 max	bal Fe				
	C.0500	Reinforcing	0.6 max			1.60 max		0.05 max	0.05 max	0.6 max	bal Fe				
	C.0551	Reinforcing	0.28 max			1.60 max		0.05 max	0.05 max	0.6 max	bal Fe				
	C.1217	Unalloyed	0.2 max			1.5 max		0.045 max	0.045 max	0.55 max	Mo <=0.15; +Al; bal Fe				
	C.1402	Tub w/special req	0.36 max			0.4-1.60		0.05 max	0.05 max	0.1-0.35	Mo <=0.15; bal Fe				
	CL.0500	Reinforcing	0.30 max			1.60 max		0.070 max	0.060 max	0.60 max	Mo <=0.15; bal Fe				

UNS numbers and US grades are provided as a means of cross referencing chemically similar alloys. Exchangability is only possible after independent examination of specifications. Tensile properties are minimum or typical as specified. UTS and YS as MPa. El as %. See Appendix for list of abbreviations used in Notes. * indicates obsolete material.

Specification	Designation	Notes	C	Cr	Cu	Mn	Ni	P	S	Si	Other	UTS	YS	El	Hard
Carbon Steel, Unclassified, 6321															
USA															
	UNS K03810		0.38-0.43	0.30-0.55	0.35 max	0.75-1.00	0.20-0.40	0.025 max	0.025 max	0.15-0.35	B 0.0005-0.005; Mo 0.08-0.15; bal Fe				
Carbon Steel, Unclassified, 7706															
USA															
	UNS K00095	Commercially Pure Iron			0.15 max			0.010 max	0.030 max		C + Mn + Si + P + S 0.10 max; bal Fe				
AMS 7707B(97)		Commercial pure iron; Bar, Sh, Strp, Plt, HR, unAnn			0.15 max			0.010 max	0.030 max		C+Mn+Si+P+S<=0.10; bal Fe	450 max	380 max	25	75-80 HRB
Carbon Steel, Unclassified, A21(F)															
USA															
	UNS K05200		0.49-0.59			0.60-0.90		0.045 max	0.050 max	0.15 min	bal Fe				
ASTM A21(94)	F	Rail axles, drop test req'd, t 203-305mm	0.45-0.59			0.60-0.90		0.045 max	0.050 max	0.15 min	bal Fe	590	330	21	
ASTM A21(94)	F	Rail axles, drop test req'd, t 305-356mm	0.45-0.59			0.60-0.90		0.045 max	0.050 max	0.15 min	bal Fe	580	320	20	
ASTM A21(94)	F	Rail axles, drop test req'd, t<203mm	0.45-0.59			0.60-0.90		0.045 max	0.050 max	0.15 min	bal Fe	605	345	22	
ASTM A21(94)	G	Rail axles, drop test req'd, t 178-254mm				0.60-0.90		0.045 max	0.050 max	0.15 min	bal Fe	585	345	19	
ASTM A21(94)	G	Rail axles, drop test req'd, t>254mm				0.60-0.90		0.045 max	0.050 max	0.15 min	bal Fe	565	330	19	
ASTM A21(94)	G	Rail axles, drop test req'd, t 102-178mm				0.60-0.90		0.045 max	0.050 max	0.15 min	bal Fe	585	345	20	
ASTM A21(94)	G	Rail axles, drop test req'd, t<=102mm				0.60-0.90		0.045 max	0.050 max	0.15 min	bal Fe	620	427	20	
ASTM A21(94)	H	Rail axles, drop test req'd, t<178mm				0.60-0.90					bal Fe	790	520	16	
ASTM A21(94)	H	Rail axles, drop test req'd, t 178-254mm				0.60-0.90					bal Fe	725	450	18	
ASTM A21(94)	H	Rail axles, drop test req'd, t>254mm				0.60-0.90					bal Fe	690	415	18	
ASTM A617/A617M(96)	Grade 40	axles, bar for conc reinf	0.40-0.90			0.60-0.90		0.045 max	0.050 max	0.15 min	bal Fe	500	300	7-12	
ASTM A617/A617M(96)	Grade 60	axles, bar for conc reinf	0.40-0.59			0.60-0.90		0.045 max	0.050 max	0.15 min	bal Fe	620	420	7-8	
ASTM A730(93)	Grade F	Railway frg; double N/T; t>305-356mm	0.45-0.59			0.60-0.90		0.045 max	0.050 max	0.15 min	bal Fe	580	315	20	
ASTM A730(93)	Grade F	Railway frg; double N/T; t<=203mm	0.45-0.59			0.60-0.90		0.045 max	0.050 max	0.15 min	bal Fe	605	345	22	
ASTM A730(93)	Grade F	Railway frg; double N/T; t>203-305mm	0.45-0.59			0.60-0.90		0.045 max	0.050 max	0.15 min	bal Fe	595	330	21	
Carbon Steel, Unclassified, A21(U)															
USA															
	UNS K04700		0.40-0.55			0.60-0.90		0.045 max	0.050 max	0.15 max	bal Fe				
ASTM A21(94)	U	Rail axles, drop test req'd,	0.40-0.55			0.60-0.90		0.045 max	0.050 max	0.15 min	bal Fe	585	275	14	
ASTM A730(93)	Grade C	Railway frg, ann, norm, N/T; t<=203mm	0.40-0.55			0.60-0.90		0.045 max	0.050 max	0.15 min	bal Fe	515	260	20	
ASTM A730(93)	Grade C	Railway frg, ann, norm, N/T; t>203-356mm	0.40-0.55			0.60-0.90		0.045 max	0.050 max	0.15 min	bal Fe	515	260	19	
ASTM A730(93)	Grade D	Railway frg, ann, norm, N/T; t>203-356mm	0.40-0.55			0.60-0.90		0.045 max	0.050 max	0.15 min	bal Fe	550	275	21	
ASTM A730(93)	Grade D	Railway frg, ann, norm, N/T; t<=203mm	0.40-0.55			0.60-0.90		0.045 max	0.050 max	0.15 min	bal Fe	550	275	22	
ASTM A730(93)	Grade E	Railway frg, N/T; t>203-356mm	0.40-0.55			0.60-0.90		0.045 max	0.050 max	0.15 min	bal Fe	570	295	23	
ASTM A730(93)	Grade E	Railway frg, N/T; t<=203mm	0.40-0.55			0.60-0.90		0.045 max	0.050 max	0.15 min	bal Fe	585	305	25	

UNS numbers and US grades are provided as a means of cross referencing chemically similar alloys. Exchangability is only possible after independent examination of specifications. Tensile properties are minimum or typical as specified. UTS and YS as MPa. El as %. See Appendix for list of abbreviations used in Notes. * indicates obsolete material.

Specification	Designation	Notes	C	Cr	Cu	Mn	Ni	P	S	Si	Other	UTS	YS	El	Hard

Carbon Steel, Unclassified, A283(A)

Europe

Specification	Designation	Notes	C	Cr	Cu	Mn	Ni	P	S	Si	Other	UTS	YS	El	Hard
EN 10025(90)	Fe310-0	Obs EU desig; struct HR, BS, t<=25mm									bal Fe	290-540	175	10-18	

Carbon Steel, Unclassified, A290(B)

USA

Specification	Designation	Notes	C	Cr	Cu	Mn	Ni	P	S	Si	Other	UTS	YS	El	Hard
	UNS K04000		0.35-0.45	0.25 max	0.35	0.60-0.90	0.30 max	0.040 max	0.040 max	0.15-0.30	Mo <=0.10; V <=0.06; bal Fe				
ASTM A290(95)	A	Gear frg	0.35-0.50	0.25 max	0.35	0.60-0.90	0.30	0.040 max	0.040 max	0.35	Mo <=0.10; V <=0.06; bal Fe	550	310	22	163-202 HB
ASTM A290(95)	B	Gear frg	0.35-0.50	0.25 max	0.35	0.60-0.90	0.30	0.040 max	0.040 max	0.35	Mo <=0.10; V <=0.06; bal Fe				163-202 HB

Carbon Steel, Unclassified, A290(C)

USA

Specification	Designation	Notes	C	Cr	Cu	Mn	Ni	P	S	Si	Other	UTS	YS	El	Hard
	UNS K04500		0.40-0.50	0.25 max	0.35	0.60-0.90	0.30 max	0.040 max	0.040 max	0.15-0.35	Mo <=0.10; V <=0.06; bal Fe				
ASTM A290(95)	C	Gear frg	0.40-0.50	0.25 max	0.35	0.60-0.90	0.30	0.040 max	0.040 max	0.35	Mo <=0.10; V <=0.06; bal Fe	655	450	20	197-241 HB
ASTM A290(95)	D	Gear frg	0.40-0.50	0.25 max	0.35	0.60-0.90	0.30	0.040 max	0.040 max	0.35	Mo <=0.10; V <=0.06; bal Fe				197-241 HB

Carbon Steel, Unclassified, A490

USA

Specification	Designation	Notes	C	Cr	Cu	Mn	Ni	P	S	Si	Other	UTS	YS	El	Hard
	UNS K03900		0.28-0.50					0.045 max	0.045 max		bal Fe				

Carbon Steel, Unclassified, A574

USA

Specification	Designation	Notes	C	Cr	Cu	Mn	Ni	P	S	Si	Other	UTS	YS	El	Hard
	UNS K03104		0.31 min					0.045 max	0.040 max		bal Fe				

Carbon Steel, Unclassified, A620

Europe

Specification	Designation	Notes	C	Cr	Cu	Mn	Ni	P	S	Si	Other	UTS	YS	El	Hard
EN 10130(91)	1.0338	Sh Strp, CR t<=3mm	0.08 max			0.40 max		0.030 max	0.030 max		bal Fe	270-350	140	38	
EN 10130(91)	FeP04	Sh Strp, CR t<=3mm	0.08 max			0.40 max		0.030 max	0.030 max		bal Fe	270-350	140	38	

Japan

Specification	Designation	Notes	C	Cr	Cu	Mn	Ni	P	S	Si	Other	UTS	YS	El	Hard
JIS G3141(96)	CR3	CR Sh Coil DDS	0.10 max			0.45 max		0.03 max	0.03 max		bal Fe	350 max		35	53 HRB max
JIS G3141(96)	CR4	CR Sh Coil DDS	0.08 max			0.45 max		0.03 max	0.03 max		bal Fe	340 max		37	50 HRB max
JIS G3141(96)	SPCE	CR Sh Coil DDS	0.08 max			0.40 max		0.030 max	0.030 max		bal Fe	270		36-43	95-170

Russia

Specification	Designation	Notes	C	Cr	Cu	Mn	Ni	P	S	Si	Other	UTS	YS	El	Hard
GOST 4041(71)	08JuA	High-grade Plts for cold pressing	0.10 max	0.1 max	0.2 max	0.20-0.40	0.15 max	0.020 max	0.025 max	0.03 max	Al 0.02-0.08; As<=0.08; bal Fe				

USA

Specification	Designation	Notes	C	Cr	Cu	Mn	Ni	P	S	Si	Other	UTS	YS	El	Hard
	UNS K00040	Special Killed, DDS	0.10 max			0.50 max		0.025 max	0.035 max		bal Fe				
ASTM A620/A620M(97)	A620(A)	Sh, CR, FF, DS	0.08 max	0.15 max	0.20 max	0.50 max	0.20 max	0.020 max	0.030 max		Al >=0.01; Mo <=0.06; Nb <=0.008; Ti <=0.008; V <=0.008; bal Fe		150-240	36 min	
ASTM A620/A620M(97)	A620(B)	Sh, CR, FF, DS	0.02-0.08	0.15 max	0.20 max	0.50 max	0.20 max	0.020 max	0.030 max		Al >=0.01; Mo <=0.06; Nb <=0.008; Ti <=0.008; V <=0.008; bal Fe		150-240	36 min	

Carbon Steel, Unclassified, K03500

USA

Specification	Designation	Notes	C	Cr	Cu	Mn	Ni	P	S	Si	Other	UTS	YS	El	Hard
	UNS K03500		0.35 max					0.040 max	0.045 max		bal Fe				

UNS numbers and US grades are provided as a means of cross referencing chemically similar alloys. Exchangability is only possible after independent examination of specifications. Tensile properties are minimum or typical as specified. UTS and YS as MPa. El as %. See Appendix for list of abbreviations used in Notes. * indicates obsolete material.

Specification	Designation	Notes	C	Cr	Cu	Mn	Ni	P	S	Si	Other	UTS	YS	El	Hard

Carbon Steel, Unclassified, K05003

USA

Specification	Designation	Notes	C	Cr	Cu	Mn	Ni	P	S	Si	Other	UTS	YS	El	Hard
	UNS K05003		0.50 max					0.05 max	0.05 max		bal Fe				

Carbon Steel, Unclassified, K05800

USA

Specification	Designation	Notes	C	Cr	Cu	Mn	Ni	P	S	Si	Other	UTS	YS	El	Hard
	UNS K05800		0.35-0.82					0.050 max			Cu 0.20 min (when spec'd); bal Fe				

Carbon Steel, Unclassified, No equivalents identified

Czech Republic

Specification	Designation	Notes	C	Cr	Cu	Mn	Ni	P	S	Si	Other	UTS	YS	El	Hard
CSN 411305	11305	Unalloyed mild	0.07 max	0.3 max	0.03 max	0.4 max		0.025 max	0.025 max	0.03 max	Al 0.025-0.10; bal Fe				
CSN 411379	11379		0.2 max	0.3 max	0.25-0.5	1.60 max		0.045 max	0.045 max	0.60 max	bal Fe				
CSN 411402	11402	Micro-alloyed Sh, Plt	0.1 max	0.3 max		0.55 max		0.035 max	0.025 max	0.15 max	Ti 0.06-0.2; bal Fe				
CSN 411419	11419		0.2 max	0.3 max		0.8 max	0.3 max	0.04 max	0.04 max	0.35 max	Al 0.02-0.120; Cr+Ni+Cu<=0.7; bal Fe				
CSN 411431	11431	Unalloyed	0.02 max	0.3 max		0.8 max		0.035 max	0.03 max	0.35 max	Al 0.02-0.10; bal Fe				
CSN 411449	11449		0.15 max	0.3 max		1.5 max	0.2 max	0.035 max	0.035 max	0.4 max	Al 0.02-0.120; Nb 0.01-0.05; bal Fe				
CSN 411474	11474	High-temp const	0.22 max	0.3 max		1.1 max	0.3 max	0.045 max	0.04 max	0.35 max	bal Fe				
CSN 411478	11478	High-temp const	0.22 max	0.3 max		1.1 max	0.3 max	0.04 max	0.04 max	0.35 max	bal Fe				
CSN 411481	11481	Unalloyed	0.2 max	0.3 max		1.3 max		0.035 max	0.03 max	0.45	Al 0.02-0.10; bal Fe				
CSN 411483	11483	Weld const; High-Temp	0.2 max	0.3 max		1.4 max	0.3 max	0.045 max	0.045 max	0.55 max	Ti <=0.2; Cr+Ni+Cu<=0.7; bal Fe				
CSN 411484	11484		0.2 max	0.3 max		1.4 max	0.3 max	0.03 max	0.03 max	0.55 max	Ti <=0.2; Cr+Ni+Cu<=0.7; bal Fe				
CSN 411523	11523	Weld Const	0.2 max	0.3 max		1.6 max		0.05 max	0.045 max	0.55 max	Al <=0.10; bal Fe				
CSN 411529	11529		0.2 max	0.3 max	0.25-0.5	1.5 max		0.045 max	0.045 max	0.55 max	bal Fe				
CSN 412011	12011	Smls Tube	0.09 max	0.15 max	0.15 max	0.2-0.45	0.15 max	0.035 max	0.04 max	0.15 max	Al 0.02-0.120; bal Fe				
CSN 412014	12014	Soft Magnetic	0.06 max	0.3 max	0.15 max	0.4 max		0.02 max	0.02 max	0.15 max	Al 0.02-0.120; bal Fe				
CSN 412015	12015	Unalloyed	0.05-0.12	0.15 max		0.2-0.45	0.15 max	0.025 max	0.025 max	0.15 max	Al 0.02-0.120; bal Fe				
CSN 413180	13180	Spring	0.7-0.8	0.3 max		0.9-1.2		0.035 max	0.04 max	0.15-0.35	bal Fe				
CSN 419015	19015	Unalloyed	0.07-0.14	0.2 max		0.35-0.65	0.25 max	0.035 max	0.035 max	0.17-0.37	bal Fe				
CSN 419065	19065	Unalloyed	0.3-0.4	0.2 max		0.3-0.6	0.25 max	0.035 max	0.035 max	0.3 max	bal Fe				
CSN 419083	19083		0.4-0.5	0.3 max		0.55-0.85		0.04 max	0.04 max	0.4 max	bal Fe				
CSN 419096	19096		0.47-0.57	0.2 max		0.4-0.7	0.25 max	0.035 max	0.035 max	0.15-0.3	bal Fe				

France

Specification	Designation	Notes	C	Cr	Cu	Mn	Ni	P	S	Si	Other	UTS	YS	El	Hard
AFNOR NFA35053(84)	FB10	Wir rod	0.08-0.13	0.3 max	0.3 max	0.3-0.6	0.4 max	0.035 max	0.035 max	0.6 max	bal Fe				
AFNOR NFA35053(84)	FB15	Wir rod	0.12-0.17	0.3 max	0.3 max	0.4-0.7	0.4 max	0.035 max	0.035 max	0.6 max	bal Fe				
AFNOR NFA35053(84)	FB18	Wir rod	0.15-0.2	0.3 max	0.3 max	0.4-0.7	0.4 max	0.035 max	0.035 max	0.3 max	Al <=0.08; Al=0.03-0.08 Si<=0.05; Si=0.15-0.3 Al<=0.02				
AFNOR NFA35053(84)	FB5	Wir rod, rolled	0.06 max	0.3 max	0.3 max	0.2-0.45	0.4 max	0.035 max	0.035 max	0.3 max	Al <=0.08; Al=0.03-0.08 Si<=0.05; Si=0.15-0.3 Al<=0.02	0-420			
AFNOR NFA35053(84)	FB8	Wir rod, rolled	0.06-0.1	0.3 max	0.3 max	0.25-0.6	0.4 max	0.035 max	0.035 max	0.6 max	bal Fe	0-460			

UNS numbers and US grades are provided as a means of cross referencing chemically similar alloys. Exchangability is only possible after independent examination of specifications. Tensile properties are minimum or typical as specified. UTS and YS as MPa. El as %. See Appendix for list of abbreviations used in Notes. * indicates obsolete material.

Specification	Designation	Notes	C	Cr	Cu	Mn	Ni	P	S	Si	Other	UTS	YS	El	Hard

Carbon Steel, Unclassified, No equivalents identified

France

Specification	Designation	Notes	C	Cr	Cu	Mn	Ni	P	S	Si	Other	UTS	YS	El	Hard
AFNOR NFA35053(84)	FR10	Wir rod, rolled	0.08-0.13	0.15 max	0.2 max	0.3-0.6	0.25 max	0.035 max	0.035 max	0.6 max	Cr+Ni+Cu<=0.45; bal Fe	0-480			
AFNOR NFA35053(84)	FR12	Wir rod	0.1-0.15	0.15 max	0.2 max	0.3-0.6	0.25 max	0.035 max	0.035 max	0.6 max	Cr+Ni+Cu<=0.45; bal Fe				
AFNOR NFA35053(84)	FR15	Wir rod, rolled	0.12-0.17	0.15 max	0.2 max	0.4-0.7	0.25 max	0.035 max	0.035 max	0.6 max	Cr+Ni+Cu<=0.45; bal Fe	0-530			
AFNOR NFA35053(84)	FR18	Wir rod, rolled	0.15-0.2	0.15 max	0.2 max	0.4-0.7	0.25 max	0.035 max	0.035 max	0.3 max	Al <=0.08; Al=0.03-0.08 Si<=0.05; Si=0.015-0.3 Al<=0.02; Cr+Ni+Cu<=0.45; bal Fe	0-570			
AFNOR NFA35053(84)	FR20	Wir rod	0.18-0.23	0.15 max	0.2 max	0.4-0.7	0.25 max	0.035 max	0.035 max	0.6 max	Cr+Ni+Cu<=0.45; bal Fe				
AFNOR NFA35053(84)	FR28	Wir rod	0.25-0.3	0.15 max	0.2 max	0.5-0.8	0.25 max	0.035 max	0.035 max	0.6 max	Cr+Ni+Cu<=0.45; bal Fe				
AFNOR NFA35053(84)	FR32	Wir rod	0.3-0.35	0.15 max	0.2 max	0.5-0.8	0.25 max	0.035 max	0.035 max	0.6 max	Cr+Ni+Cu<=0.45; bal Fe				
AFNOR NFA35053(84)	FR36	Wir rod	0.33-0.38	0.15 max	0.2 max	0.5-0.8	0.25 max	0.035 max	0.035 max	0.6 max	Cr+Ni+Cu<=0.45; bal Fe				
AFNOR NFA35053(84)	FR5	Wir rod, rolled	0.06 max	0.15 max	0.2 max	0.2-0.45	0.25 max	0.035 max	0.035 max	0.3 max	Al <=0.08; Al 0.03-0.08 Si 0-0.05; Si 0.015-0.3 Al 0-0.02; Cr+Ni+Cu<=0.45; bal Fe	0-400			
AFNOR NFA35053(84)	FR8	Wir rod, rolled	0.06-0.1	0.15 max	0.2 max	0.25-0.6	0.25 max	0.035 max	0.035 max	0.6 max	Cr+Ni+Cu<=0.45; bal Fe	0-430			
AFNOR NFA35054(78)	FMP62-1	Wir rod; Reinf	0.6-0.65	0.15 max	0.2 max	0.6-0.9	0.25 max	0.035 max	0.035 max	0.1-0.35	Mo <=0.1; Cr+Ni+Mo<0.4; bal Fe				
AFNOR NFA35054(78)	FMP62-2	Wir rod; Reinf	0.6-0.65	0.15 max	0.15 max	0.6-0.9	0.25 max	0.035 max	0.03 max	0.1-0.35	Mo <=0.1; Cr+Ni+Mo<0.4; bal Fe				
AFNOR NFA35054(78)	FMP62-3	Wir rod; Reinf	0.6-0.65	0.15 max	0.15 max	0.6-0.9	0.25 max	0.03 max	0.03 max	0.1-0.35	Mo <=0.1; Cr+Ni+Mo<0.4; bal Fe				
AFNOR NFA35054(78)	FMP66-1	Wir rod; Reinf	0.63-0.68	0.15 max	0.2 max	0.6-0.9	0.25 max	0.035 max	0.035 max	0.1-0.35	Mo <=0.1; Cr+Ni+Mo<0.4; bal Fe				
AFNOR NFA35054(78)	FMP66-2	Wir rod; Reinf	0.63-0.68	0.15 max	0.15 max	0.6-0.9	0.25 max	0.035 max	0.03 max	0.1-0.35	Mo <=0.1; Cr+Ni+Mo<0.4; bal Fe				
AFNOR NFA35054(78)	FMP66-3	Wir rod; Reinf	0.63-0.68	0.15 max	0.15 max	0.6-0.9	0.25 max	0.03 max	0.03 max	0.1-0.35	Mo <=0.1; Cr+Ni+Mo<0.4; bal Fe				
AFNOR NFA35054(78)	FMP68-1	Wir rod; Reinf	0.65-0.7	0.15 max	0.2 max	0.6-0.9	0.25 max	0.035 max	0.035 max	0.1-0.35	Mo <=0.1; Cr+Ni+Mo<0.4; bal Fe				
AFNOR NFA35054(78)	FMP68-2	Wir rod; Reinf	0.65-0.7	0.15 max	0.15 max	0.6-0.9	0.25 max	0.035 max	0.03 max	0.1-0.35	Mo <=0.1; Cr+Ni+Mo<0.4; bal Fe				
AFNOR NFA35054(78)	FMP70-1	Wir rod; Reinf	0.68-0.73	0.15 max	0.2 max	0.6-0.9	0.25 max	0.035 max	0.035 max	0.1-0.35	Mo <=0.1; Cr+Ni+Mo<0.4; bal Fe				
AFNOR NFA35054(78)	FMP70-2	Wir rod; Reinf	0.68-0.73	0.15 max	0.15 max	0.6-0.9	0.25 max	0.035 max	0.03 max	0.1-0.35	Mo <=0.1; Cr+Ni+Mo<0.4; bal Fe				
AFNOR NFA35054(78)	FMP70-3	Wir rod; Reinf	0.68-0.73	0.15 max	0.15 max	0.6-0.9	0.25 max	0.03 max	0.03 max	0.1-0.35	Mo <=0.1; Cr+Ni+Mo<0.4; bal Fe				
AFNOR NFA35054(78)	FMP72-1	Wir rod; Reinf	0.7-0.75	0.15 max	0.2 max	0.6-0.9	0.25 max	0.035 max	0.035 max	0.1-0.35	Mo <=0.1; Cr+Ni+Mo<0.4; bal Fe				
AFNOR NFA35054(78)	FMP72-2	Wir rod; Reinf	0.7-0.75	0.15 max	0.15 max	0.6-0.9	0.25 max	0.035 max	0.03 max	0.1-0.35	Mo <=0.1; Cr+Ni+Mo<0.4; bal Fe				
AFNOR NFA35054(78)	FMP72-3	Wir rod; Reinf	0.7-0.75	0.15 max	0.15 max	0.6-0.9	0.25 max	0.03 max	0.03 max	0.1-0.35	Mo <=0.1; Cr+Ni+Mo<0.4; bal Fe				

Specification	Designation	Notes	C	Cr	Cu	Mn	Ni	P	S	Si	Other	UTS	YS	El	Hard

Carbon Steel, Unclassified, No equivalents identified

France

Specification	Designation	Notes	C	Cr	Cu	Mn	Ni	P	S	Si	Other
AFNOR NFA35054(78)	FMP76-1	Wir rod; Reinf	0.73-0.78	0.15 max		0.2 max	0.6-0.9	0.25 max	0.035 max	0.035 max	0.1-0.35	Mo <=0.1; Cr+Ni+Mo<0.4; bal Fe
AFNOR NFA35054(78)	FMP76-2	Wir rod; Reinf	0.73-0.78	0.15 max	0.15 max	0.6-0.9	0.25 max	0.035 max	0.03 max	0.1-0.35	Mo <=0.1; Cr+Ni+Mo<0.4; bal Fe
AFNOR NFA35054(78)	FMP76-3	Wir rod; Reinf	0.73-0.78	0.15 max	0.15 max	0.6-0.9	0.25 max	0.03 max	0.03 max	0.1-0.35	Mo <=0.1; Cr+Ni+Mo<0.4; bal Fe
AFNOR NFA35054(78)	FMP78-1	Wir rod; Reinf	0.75-0.8	0.15 max		0.2 max	0.6-0.9	0.25 max	0.035 max	0.035 max	0.1-0.35	Mo <=0.1; Cr+Ni+Mo<0.4; bal Fe
AFNOR NFA35054(78)	FMP78-2	Wir rod; Reinf	0.75-0.8	0.15 max	0.15 max	0.6-0.9	0.25 max	0.035 max	0.03 max	0.1-0.35	Mo <=0.1; Cr+Ni+Mo<0.4; bal Fe
AFNOR NFA35054(78)	FMP78-3	Wir rod; Reinf	0.75-0.8	0.15 max	0.15 max	0.6-0.9	0.25 max	0.03 max	0.03 max	0.1-0.35	Mo <=0.1; Cr+Ni+Mo<0.4; bal Fe
AFNOR NFA35054(78)	FMP80-1	Wir rod; Reinf	0.78-0.83	0.15 max		0.2 max	0.6-0.9	0.25 max	0.035 max	0.035 max	0.1-0.35	Mo <=0.1; Cr+Ni+Mo<0.4; bal Fe
AFNOR NFA35054(78)	FMP80-2	Wir rod; Reinf	0.78-0.83	0.15 max	0.15 max	0.6-0.9	0.25 max	0.035 max	0.03 max	0.1-0.35	Mo <=0.1; Cr+Ni+Mo<0.4; bal Fe
AFNOR NFA35054(78)	FMP80-3	Wir rod; Reinf	0.78-0.83	0.15 max	0.15 max	0.6-0.9	0.25 max	0.03 max	0.03 max	0.1-0.35	Mo <=0.1; Cr+Ni+Mo<0.4; bal Fe
AFNOR NFA35054(78)	FMP82-1	Wir rod; Reinf	0.8-0.85	0.15 max		0.2 max	0.6-0.9	0.25 max	0.035 max	0.035 max	0.1-0.35	Mo <=0.1; Cr+Ni+Mo<0.4; bal Fe
AFNOR NFA35054(78)	FMP82-2	Wir rod; Reinf	0.8-0.85	0.15 max	0.15 max	0.6-0.9	0.25 max	0.035 max	0.03 max	0.1-0.35	Mo <=0.1; Cr+Ni+Mo<0.4; bal Fe
AFNOR NFA35054(78)	FMP82-3	Wir rod; Reinf	0.8-0.85	0.15 max	0.15 max	0.6-0.9	0.25 max	0.03 max	0.03 max	0.1-0.35	Mo <=0.1; Cr+Ni+Mo<0.4; bal Fe
AFNOR NFA35054(79)	FMR66	Wir rod; Spring, CF	0.63-0.68	0.1 max	0.12 max	0.6-0.9	0.1 max	0.03 max	0.03 max	0.1-0.35	Mo <=0.05; Cr+Ni+Mo<=0.2; Mn<=1; bal Fe
AFNOR NFA35055(84)	FME8-5	Wir rod	0.05-0.1	0.12 max	0.12 max	0.4-0.6	0.12 max	0.025 max	0.025 max	0.6 max	Cr+Ni+Cu<=0.3; Pb+Sb+Sn+As+P+S<=0.15; bal Fe
AFNOR NFA35055(84)	FME8-7	Wir rod	0.05-0.1	0.05 max	0.08 max	0.4-0.6	0.1 max	0.012 max	0.012 max	0.6 max	Co <=0.02; V <=0.005; Cr+Ni+Cu<=0.2; As<=0.06; Sn<=0.01; Pb+Sb+Sn+As+P+S<=0.1; bal Fe
AFNOR NFA35056(84)	FG10M5ATi5	Wir rod, weld	0.04-0.12			0.2 max	1.1-1.4		0.03 max	0.03 max	0.5-0.8	Al 0.05-0.2; Ti 0.05-0.2; bal Fe
AFNOR NFA35056(84)	FG10MS5	Wir rod, weld	0.04-0.12			0.2 max	1-1.3		0.03 max	0.03 max	0.5-0.8	Al <=0.02; bal Fe
AFNOR NFA35056(84)	FG12MS6	Wir rod, weld	0.04-0.12			0.2 max	1.3-1.6		0.03 max	0.03 max	0.7-1	Al <=0.02; bal Fe
AFNOR NFA35056(84)	FG12MS6.4	Wir rod, weld	0.04-0.12			0.2 max	1.3-1.6		0.03 max	0.03 max	0.9-1.2	Al <=0.02; bal Fe
AFNOR NFA35056(84)	FG12MS6.4	Wir rod, weld	0.04-0.12			0.2 max	1.3-1.6		0.03 max	0.03 max	0.9-1.2	Al <=0.02; bal Fe
AFNOR NFA35056(84)	FG12MS7.4	Wir rod, weld	0.04-0.14			0.2 max	1.5-1.8		0.03 max	0.03 max	0.9-1.2	Al <=0.02; bal Fe
AFNOR NFA35056(84)	FS10	Wir rod, weld	0.05-0.12		0.12 max	0.4-0.6		0.025 max	0.025 max	0.6 max	Al <=0.04; Ti <=0.06; Al+Ti<=0.08; bal Fe
AFNOR NFA35056(84)	FS12	Wir rod, weld	0.06-0.13		0.12 max	0.4-0.6		0.025 max	0.025 max	0.1 max	Al <=0.04; Ti <=0.06; Al+Ti<=0.08; bal Fe
AFNOR NFA35056(84)	FS12M4	Wir rod, weld	0.07-0.14		0.12 max	0.9-1.2		0.025 max	0.025 max	0.1 max	Al <=0.04; Ti <=0.06; Al+Ti<=0.08; bal Fe
AFNOR NFA35056(84)	FS12M6	Wir rod, weld	0.07-0.14		0.12 max	1.4-1.7		0.025 max	0.025 max	0.1 max	Al <=0.04; Ti <=0.06; Al+Ti<=0.08; bal Fe
AFNOR NFA35056(84)	FS12M8	Wir rod, weld	0.09-0.16		0.12 max	1.8-2.2		0.025 max	0.025 max	0.1 max	Al <=0.04; Ti <=0.06; Al+Ti<=0.08; bal Fe

UNS numbers and US grades are provided as a means of cross referencing chemically similar alloys. Exchangability is only possible after independent examination of specifications. Tensile properties are minimum or typical as specified. UTS and YS as MPa. El as %. See Appendix for list of abbreviations used in Notes. * indicates obsolete material.

Carbon Steel, Unclassified, No equivalents identified

Specification	Designation	Notes	C	Cr	Cu	Mn	Ni	P	S	Si	Other	UTS	YS	El	Hard
France															
AFNOR NFA35056(84)	FS12MS4	Wir rod, weld	0.07-0.15		0.12 max	0.9-1.2		0.025 max	0.025 max	0.15-0.35	Al <=0.04; Ti <=0.06; Al+Ti<=0.08; bal Fe				
AFNOR NFA36612(82)	F37	t<=300mm; HT	0.16 max	0.3 max	0.3 max	0.3-1.6	0.4 max	0.04 max	0.04 max	0.1-0.35	bal Fe	360-450	195	28	
AFNOR NFA36612(82)	F42	t<=300mm; HT	0.2 max	0.3 max	0.3 max	0.4-1.6	0.4 max	0.04 max	0.04 max	0.1-0.35	bal Fe	410-490	225	26	
AFNOR NFA36612(82)	F48	t<=300mm; HT	0.21 max	0.3 max	0.3 max	0.8-1.5	0.4 max	0.04 max	0.04 max	0.1-0.4	bal Fe	470-570	265	23	
AFNOR NFA36612(82)	F52	t<=300mm; HT	0.22 max	0.3 max	0.3 max	1-1.6	0.4 max	0.04 max	0.04 max	0.1-0.55	bal Fe	510-610	320	22	
Germany															
DIN 1623(86)	E295	Flat product CR Sht Strp 100-150mm thk						0.045 max	0.045 max		N <=0.009; N depends on P; bal Fe	450-610	245	16	
DIN 1623(86)	E335	Flat product CR Sht Strp 100-150mm thk						0.045 max	0.045 max		N <=0.009; N depends on P; bal Fe	550-710	275	12	
DIN 1623(86)	E360	Flat product CR Sht Strp 100-150mm thk						0.045 max	0.045 max		N <=0.009; N depends on P; bal Fe	650-830	305	8	
DIN 1623(86)	WNr 1.0050	Flat product CR Sh Strp 100-150mm thk						0.045 max	0.045 max		N <=0.009; N depends on P; bal Fe	450-610	245	16	
DIN 1623(86)	WNr 1.0060	Flat product CR Sh Strp 100-150mm thk						0.045 max	0.045 max		N <=0.009; N depends on P; bal Fe	550-710	275	12	
DIN 1623(86)	WNr 1.0070	Flat product CR Sh Strp 100-150mm thk						0.045 max	0.045 max		N <=0.009; N depends on P; bal Fe	650-830	305	8	
DIN 1652(90)	E295	100-150mm thk						0.045 max	0.045 max		N <=0.009; N depends on P; bal Fe	450-610	245	16	
DIN 1652(90)	E335	100-150mm thk						0.045 max	0.045 max		N <=0.009; N depends on P; bal Fe	550-710	275	12	
DIN 1652(90)	E360	100-150mm thk						0.045 max	0.045 max		N <=0.009; N depends on P; bal Fe	650-830	305	8	
DIN 1652(90)	WNr 1.0050	100-150mm thk						0.045 max	0.045 max		N <=0.009; N depends on P; bal Fe	450-610	245	16	
DIN 1652(90)	WNr 1.0060	100-150mm thk						0.045 max	0.045 max		N <=0.009; N depends on P; bal Fe	550-710	275	12	
DIN 1652(90)	WNr 1.0070	100-150mm thk						0.045 max	0.045 max		N <=0.009; N depends on P; bal Fe	650-830	305	8	
DIN 17115(87)	C10G1	Weld round link chain	0.06-0.14		0.25	0.40-0.60		0.035 max	0.035 max		N <=0.012; trace Si; bal Fe				
DIN 17115(87)	WNr 1.0207	Weld round link chain	0.06-0.14		0.25	0.40-0.60		0.035 max	0.035 max		N <=0.012; trace Si; bal Fe				
DIN 17405(79)	RFe20	Soft magnetic for DC relays	0.03 max			0.20 max		0.025 max	0.015 max	0.05 max	Al 0.04-0.10; bal Fe				
DIN 17405(79)	WNr 1.1017	Soft magnetic for DC relays	0.03 max			0.20 max		0.025 max	0.015 max	0.05 max	Al 0.04-0.10; bal Fe				
DIN 5512(97)	E295	For rail vehicles, 100-150mm thk						0.045 max	0.045 max		N <=0.009; N depends on P; bal Fe	450-610	245	16	
DIN 5512(97)	E335	For rail vehicles, 100-150mm thk						0.045 max	0.045 max		N <=0.009; N depends on P; bal Fe	550-710	275	12	
DIN 5512(97)	S235J2G3CuC	Rail vehicles, structures	0.17 max		0.25-0.40			0.035 max	0.035 max	1.40 max	bal Fe				
DIN 5512(97)	S235JRG1CuC	Rail vehicles, structures	0.17 max		0.25-0.35	0.20-0.50		0.050 max	0.050 max		N <=0.007; N depends on P;C 0.20 max if >=16 mm thk; trace Si; bal Fe				
DIN 5512(97)	WNr 1.0050	Rail vehicles, 100-150mm thk						0.045 max	0.045 max		N <=0.009; N depends on P; bal Fe	450-610	245	16	
DIN 5512(97)	WNr 1.0060	Rail vehicles, 100-150mm thk						0.045 max	0.045 max		N <=0.009; N depends on P; bal Fe	550-710	275	12	
DIN 5512(97)	WNr 1.0164	Rail vehicles, structures	0.17 max		0.25-0.35	0.20-0.50		0.050 max	0.050 max		N <=0.007; N depends on P; C<=0.20 if >=16mm thk; trace Si; bal Fe				
DIN 5512(97)	WNr 1.0171	Rail vehicles, structures	0.17 max		0.25-0.40			0.035 max	0.035 max	1.40 max	bal Fe				
DIN EN 10025(94)	E295	HR 100-150mm thk						0.045 max	0.045 max		N <=0.009; N depends on P; bal Fe	450-610	245	16	
DIN EN 10025(94)	E335	HR 100-150mm thk						0.045 max	0.045 max		N <=0.009; N depends on P; bal Fe	550-710	275	12	

UNS numbers and US grades are provided as a means of cross referencing chemically similar alloys. Exchangability is only possible after independent examination of specifications. Tensile properties are minimum or typical as specified. UTS and YS as MPa. El as %. See Appendix for list of abbreviations used in Notes. * indicates obsolete material.

Specification	Designation	Notes	C	Cr	Cu	Mn	Ni	P	S	Si	Other	UTS	YS	El	Hard

Carbon Steel, Unclassified, No equivalents identified

Germany

Specification	Designation	Notes	C	Cr	Cu	Mn	Ni	P	S	Si	Other	UTS	YS	El	Hard
DIN EN 10025(94)	E360	HR 100-150mm thk						0.045 max	0.045 max		N <=0.009; N depends on P; bal Fe	650-830	305	8	
DIN EN 10025(94)	WNr 1.0050	HR 100-150mm thk						0.045 max	0.045 max		N <=0.009; N depends on P; bal Fe	450-610	245	16	
DIN EN 10025(94)	WNr 1.0060	HR 100-150mm thk						0.045 max	0.045 max		N <=0.009; N depends on P; bal Fe	550-710	275	12	
DIN EN 10025(94)	WNr 1.0070	HR 100-150mm thk						0.045 max	0.045 max		N <=0.009; N depends on P; bal Fe	650-830	305	8	
DIN EN 10120(97)	P245NB	Sht Strp, welded gas cylinders	0.16 max			0.30 min		0.025 max	0.015 max	0.25 max	N <=0.012; Nb <=0.050; Ti <=0.03; bal Fe				
DIN EN 10120(97)	WNr 1.0111	Sh Strp, welded gas cylinders	0.16 max			0.30 min		0.025 max	0.015 max	0.25 max	N <=0.012; Nb <=0.050; Ti <=0.03; bal Fe				
DIN EN 10202(90)	T57	Cold reduced electrolyt Cr coated									bal Fe				
DIN EN 10202(90)	T61	Cold reduced electrolyt Cr coated									bal Fe				
DIN EN 10202(90)	T65	Cold reduced electrolyt Cr coated									bal Fe				
DIN EN 10202(90)	WNr 1.0375	Cold reduced Cr coated									bal Fe				
DIN EN 10202(90)	WNr 1.0377	Cold reduced Cr coated									bal Fe				
DIN EN 10202(90)	WNr 1.0378	Cold reduced Cr coated									bal Fe				
DIN SEW 094(87)	H300P	CR Sht Strp, P alloyed, WH	0.10 max			0.70 max		0.12 max	0.030 max	0.50 max	Al >=0.020; C+P<=0.16; bal Fe				
DIN SEW 094(87)	WNr 1.0448	CR Sh Strp, P alloyed, WH	0.10 max			0.70 max		0.12 max	0.030 max	0.50 max	Al >=0.020; C+P<=0.16; bal Fe				

Hungary

Specification	Designation	Notes	C	Cr	Cu	Mn	Ni	P	S	Si	Other	UTS	YS	El	Hard
MSZ 17718(77)	MSZ17718	Rolled bell-profile; timbering & walling; HR	0.3 max			2 max		0.07 max	0.06 max	0.6 max	bal Fe	490-690		12L	
MSZ 17718(77)	MSZ17718	Rolled bell-profile; timbering & walling; HR	0.3 max			2 max		0.07 max	0.06 max	0.6 max	bal Fe	490-650	350	20L	
MSZ 2898/2(80)	A52K	Smls CF tube; Smls tubes w/special req; 0.5-10mm; norm free from scale	0.2 max			1.5 max		0.05 max	0.05 max	0.1-0.55	bal Fe	490-630	350	22L	
MSZ 2898/2(80)	A52K	Smls CF tube; Smls tubes w/special req; 0.5-10mm; bright ann	0.2 max			1.5 max		0.05 max	0.05 max	0.1-0.55	bal Fe	490	245	24L	
MSZ 2898/2(80)	A55K	Smls CF tube; Smls tubes w/special req; 0.5-10mm; bright ann	0.36 max			0.4-1.60		0.05 max	0.05 max	0.1-0.35	bal Fe	490	245	18L	
MSZ 2898/2(80)	A55K	Smls CF tube; Smls tubes w/special req; 0.5-10mm; norm free from scale	0.36 max			0.4-1.60		0.05 max	0.05 max	0.1-0.35	bal Fe	540-640	290	17L	
MSZ 2978/2(82)	DKT1	Longitudinal weld & CF tubes; 0.5-3mm; bright-drawn, soft	0.15-0.2			0.35-0.65		0.025 max	0.025 max	0.05-0.15	Al 0.02-0.120; bal Fe	440	310	15L	
MSZ 2978/2(82)	DKT1	Longitudinal weld & CF tubes; 0.5-3mm; bright-drawn, hard	0.15-0.2			0.35-0.65		0.025 max	0.025 max	0.05-0.15	Al 0.02-0.120; bal Fe	490	390	8L	
MSZ 500(89)	Fe355B	Gen struct; Non-rimming; FF; Base; BS; 200.1-250mm; HR/Frg	0.24 max			1.6 max		0.045 max	0.045 max	0.55 max	N<=0.009/Al>0.02; bal Fe	450-630	275	17L; 17T	
MSZ 500(89)	Fe355B	Gen struct; Non-rimming; FF; Base; BS; 0-1mm; HR/Frg	0.24 max			1.6 max		0.045 max	0.045 max	0.55 max	N<=0.009/Al>0.02; bal Fe	510-680	355	14L; 12T	
MSZ 500(89)	Fe355C	Gen struct; Non-rimming; FF; QS; 0-1mm; HR/Frg	0.22 max			1.6 max		0.04 max	0.04 max	0.55 max	0-30mm C<=0.2; 30-100mm C<=0.22; N<=0.009/Al>=0.02; bal Fe	510-680	355	14L; 12T	

Specification	Designation	Notes	C	Cr	Cu	Mn	Ni	P	S	Si	Other	UTS	YS	El	Hard

Carbon Steel, Unclassified, No equivalents identified

Hungary

Specification	Designation	Notes	C	Cr	Cu	Mn	Ni	P	S	Si	Other	UTS	YS	El	Hard
MSZ 500(89)	Fe355C	Gen struct; Non-rimming; FF; QS; 200.1-250mm; HR/Frg	0.22 max			1.6 max		0.04 max	0.04 max	0.55 max	0-30mm C<=0.2; 30-100mm C<=0.22; N<=0.009/Al>=0.02; bal Fe	450-630	275	17L; 17T	
MSZ 500(89)	Fe490-2	Gen struct; Non-rimming; FF; Base; BS; 200.1-250mm; HR/Frg	0.3 max			2 max		0.045 max	0.045 max	0.6 max	Al >=0.02; N <=0.009; bal Fe	450-610	225	15L; 14T	
MSZ 500(89)	Fe490-2	Gen struct; Non-rimming; FF; Base; BS; 0-1mm; HR/Frg	0.3 max			2 max		0.045 max	0.045 max	0.6 max	Al >=0.02; N <=0.009; bal Fe	490-660	295	12L; 10T	
MSZ 500(89)	Fe590-2	Gen struct; Non-rimming; FF; Base BS; 200.1-250mm; HR/Frg	0.3 max			2 max		0.045 max	0.045 max	0.6 max	Al >=0.02; N <=0.009; bal Fe	540-710	255	11L; 10T	
MSZ 500(89)	Fe590-2	Gen struct; Non-rimming; FF; Base BS; 0-1mm; HR/Frg	0.3 max			2 max		0.045 max	0.045 max	0.6 max	Al >=0.02; N <=0.009; bal Fe	590-770	335	8L; 6T	
MSZ 500(89)	Fe690-2	Gen struct; Non-rimming; FF; Base BS; 200.1-250mm; HR/Frg	0.3 max			2 max		0.045 max	0.045 max	0.6 max	Al >=0.02; N <=0.009; bal Fe	640-830	285	7L; 6T	
MSZ 500(89)	Fe690-2	Gen struct; Non-rimming; FF; Base BS; 0-1mm; HR/Frg	0.3 max			2 max		0.045 max	0.045 max	0.6 max	Al >=0.02; N <=0.009; bal Fe	690-900	360	4L; 3T	
MSZ 7262(81)	HA310	Cold bent sect; sect w/special req; 3-4mm; HR; pickled	0.19 max			2 max		0.05 max	0.05 max	0.6 max	bal Fe	310-490	185	18L	
MSZ 7262(81)	HA310	Cold bent sect; sect w/special req; 2-2.99mm; HR	0.19 max			2 max		0.05 max	0.05 max	0.6 max	bal Fe	310-490	185	14L	
MSZ 7262(81)	HA360	Cold bent sect; sect w/special req; 3-4mm; HR; pickled	0.19 max			2 max		0.05 max	0.05 max	0.6 max	bal Fe	360-510	235	23L	
MSZ 7262(81)	HA360	Cold bent sect; sect w/special req; 2-2.99mm; HR	0.19 max			2 max		0.05 max	0.05 max	0.6 max	bal Fe	360-510	235	21L	
MSZ 7262(81)	HA440	Cold bent sect; sect w/special req; 2-2.99mm; HR	0.24 max			2 max		0.06 max	0.06 max	0.6 max	bal Fe	440-580	275	17L	
MSZ 7262(81)	HA440	Cold bent sect; sect w/special req; 3-4mm; HR; pickled	0.24 max			2 max		0.06 max	0.06 max	0.6 max	bal Fe	440-580	275	19L	
MSZ 7262(81)	HL	Cold bent sect; sect w/special req; 2-2.99mm; HR	0.15 max			2 max		0.05 max	0.05 max	0.6 max	bal Fe	270-440		25L	
MSZ 7262(81)	HL	Cold bent sect; sect w/special req; 3-4mm; HR; pickled	0.15 max			2 max		0.05 max	0.05 max	0.6 max	bal Fe	270-440		25L	

India

Specification	Designation	Notes	C	Cr	Cu	Mn	Ni	P	S	Si	Other	UTS	YS	El	Hard
IS 1570/2(79)	113C6	Sh Plt Sect Shp Bar Bil Frg	1.05-1.2	0.3 max	0.3 max	0.5-0.8	0.4 max	0.055 max	0.055 max	0.6 max	Co <=0.1; Mo <=0.15; Pb <=0.15; W <=0.1; P:S Varies; bal Fe				
IS 1570/2(79)	25C4	Sh Plt Sect Shp Bar Bil Frg Tub Pip	0.2-0.3	0.3 max	0.3 max	0.3-0.6	0.4 max	0.055 max	0.055 max	0.6 max	Co <=0.1; Mo <=0.15; Pb <=0.15; W <=0.1; bal Fe	440-540	242	23	
IS 1570/2(79)	C113	Sh Plt Sect Shp Bar Bil Frg	1.05-1.2	0.3 max	0.3 max	0.5-0.8	0.4 max	0.055 max	0.055 max	0.6 max	Co <=0.1; Mo <=0.15; Pb <=0.15; W <=0.1; P:S Varies; bal Fe				
IS 1570/2(79)	C25	Sh Plt Sect Shp Bar Bil Frg Tub Pip	0.2-0.3	0.3 max	0.3 max	0.3-0.6	0.4 max	0.055 max	0.055 max	0.6 max	Co <=0.1; Mo <=0.15; Pb <=0.15; W <=0.1; bal Fe	440-540	242	23	
IS 1570/7(92)	12C7H	smls tube, weld	0.17 max	0.25 max	0.3 max	0.4-1	0.35 max	0.045 max	0.045 max	0.1-0.35	Co <=0.1; Mo <=0.15; Pb <=0.15; W <=0.1; bal Fe	360-480	215	24	
IS 1570/7(92)	3	smls tube, weld	0.17 max	0.25 max	0.3 max	0.4-1	0.35 max	0.045 max	0.045 max	0.1-0.35	Co <=0.1; Mo <=0.15; Pb <=0.15; W <=0.1; bal Fe	360-480	215	24	

UNS numbers and US grades are provided as a means of cross referencing chemically similar alloys. Exchangability is only possible after independent examination of specifications. Tensile properties are minimum or typical as specified. UTS and YS as MPa. El as %. See Appendix for list of abbreviations used in Notes. * indicates obsolete material.

Specification	Designation	Notes	C	Cr	Cu	Mn	Ni	P	S	Si	Other	UTS	YS	El	Hard

Carbon Steel, Unclassified, No equivalents identified

International

Specification	Designation	Notes	C	Cr	Cu	Mn	Ni	P	S	Si	Other	UTS	YS	El	Hard
ISO 2604-4(75)	P15*	Sh Plt, boiler, press ves	0.22 max	0.3 max	0.3 max	0.6-1.5	0.4 max	0.04 max	0.04 max	0.4 max	Al <=0.06; Co <=0.1; Mo <=0.15; Pb <=0.15; Ti <=0.05; V <=0.1; W <=0.1; bal Fe				

Italy

Specification	Designation	Notes	C	Cr	Cu	Mn	Ni	P	S	Si	Other	UTS	YS	El	Hard
UNI 5869(75)	Fe460-1KG	Press ves	0.22 max			1.4 max		0.035 max	0.04 max	0.35 max	0-50mm C<=0.2; bal Fe				
UNI 5869(75)	Fe460-1KW	Press ves	0.22 max			1.4 max		0.04 max	0.04 max	0.35 max	0-50mm C<=0.2; bal Fe				
UNI 5869(75)	Fe460-2KG	Press ves	0.22 max			1.4 max		0.03 max	0.035 max	0.35 max	0-50mm C<=0.2; bal Fe				
UNI 5869(75)	Fe460-2KW	Press ves	0.22 max			1.4 max		0.035 max	0.035 max	0.35 max	0-50mm C<=0.2; bal Fe				
UNI 6102(90)	Fe740	Railway tire	0.58 max	0.25 max		1.2 max	0.5 max	0.03 max	0.03 max	0.5 max	Mo <=0.1; bal Fe				
UNI 6328(92)	Fe680	Railway rail; Railroad rail	0.4-0.6			0.8-1.25		0.05 max	0.05 max	0.05-0.35	bal Fe				
UNI 6328(92)	Fe880-1	Railway rail; Railroad rail	0.6-0.8			0.8-1.3		0.04 max	0.04 max	0.1-0.5	+V,+Mo,+Ti; bal Fe				
UNI 6328(92)	Fe880-2	Railway rail; Railroad rail	0.55-0.75			1.3-1.7		0.04 max	0.04 max	0.1-0.5	+V,+Mo,+Ti; bal Fe				
UNI 6358(90)	Fe690	Railway tire	0.55 max	0.25 max		1 max	0.5 max	0.03 max	0.03 max	0.4 max	Mo <=0.1; P+S<=0.055; bal Fe				
UNI 6363(84)	Fe360	Weld/Smls tub; water main	0.17 max			1.2 max		0.04 max	0.04 max	0.35 max	bal Fe				
UNI 6403(86)	Fe360	Q/T	0.17 max			0.4-0.8		0.04 max	0.04 max	0.1-0.35	bal Fe				
UNI 7070(72)	Fe42CFN	Gen struct	0.2 max			2 max		0.04 max	0.045 max	0.6 max	N <=0.009; bal Fe				
UNI 7070(72)	Fe42DFF	Gen struct	0.2 max			2 max		0.04 max	0.04 max	0.6 max	bal Fe				
UNI 7070(72)	Fe60	Gen struct	0.3 max			2 max		0.045 max	0.05 max	0.6 max	bal Fe				
UNI 7070(72)	Fe70	Gen struct	0.3 max			2 max		0.045 max	0.05 max	0.6 max	bal Fe				
UNI 7070(82)	Fe490	Press ves	0.3 max			2 max		0.05 max	0.05 max	0.6 max	bal Fe				
UNI 7070(82)	Fe590	Gen struct	0.3 max			2 max		0.05 max	0.05 max	0.6 max	bal Fe				
UNI 7070(82)	Fe690	Gen struct	0.3 max			2 max		0.045 max	0.05 max	0.6 max	bal Fe				
UNI 7356(74)	C21BKB	Q/T	0.19-0.25			0.5-0.8		0.035 max	0.035 max	0.4 max	B 0.001-0.005; bal Fe				
UNI 8913(87)	Fe360	Q/T	0.17 max			0.2-0.8		0.04 max	0.04 max	0.35 max	bal Fe				

Japan

Specification	Designation	Notes	C	Cr	Cu	Mn	Ni	P	S	Si	Other	UTS	YS	El	Hard
JIS G3444(94)	STK290	Tub Gen struct						0.050 max	0.050 max		bal Fe	290		30	
JIS G3507(91)	SWRCH19A	Wir rod Al killed	0.15-0.20			0.70-1.00		0.030 max	0.035 max	0.10 max	Al >=0.02; bal Fe				
JIS G3507(91)	SWRCH22A	Wir rod Al killed	0.18-0.23			0.30-0.60		0.030 max	0.035 max	0.10 max	Al >=0.02; bal Fe				
JIS G3507(91)	SWRCH6A	Wir rod Al killed	0.08 max			0.60 max		0.030 max	0.035 max	0.10 max	Al >=0.02; bal Fe				
JIS G3507(91)	SWRCH8A	Wir rod Al killed	0.10 max			0.60 max		0.030 max	0.035 max	0.10 max	Al >=0.02; bal Fe				

Mexico

Specification	Designation	Notes	C	Cr	Cu	Mn	Ni	P	S	Si	Other	UTS	YS	El	Hard
NMX-B-366(90)		CD wire for coiled type spring	0.45-0.70			0.60-1.20					bal Fe				

Poland

Specification	Designation	Notes	C	Cr	Cu	Mn	Ni	P	S	Si	Other	UTS	YS	El	Hard
PNH84019	14P		0.14-0.18	0.3 max		0.35-0.70	0.3 max	0.04-0.07	0.04 max	0.05 max	bal Fe				
PNH84023/02	08XP		0.08 max	0.1 max	0.1 max	0.28-0.5	0.1 max	0.07-0.13	0.025 max	0.04 max	bal Fe				
PNH84023/05	09P		0.08-0.13	0.3 max		0.35-0.7	0.3 max	0.04-0.07	0.04 max	0.05 max	bal Fe				

UNS numbers and US grades are provided as a means of cross referencing chemically similar alloys. Exchangability is only possible after independent examination of specifications. Tensile properties are minimum or typical as specified. UTS and YS as MPa. El as %. See Appendix for list of abbreviations used in Notes. * indicates obsolete material.

Carbon Steel, Unclassified, No equivalents identified

Specification	Designation	Notes	C	Cr	Cu	Mn	Ni	P	S	Si	Other	UTS	YS	El	Hard
Poland															
PNH84032	50S2	Spring	0.47-0.55	0.3 max	0.25 max	0.6-0.9	0.4 max	0.04 max	0.04 max	1.5-1.8	bal Fe				
PNH84032	55S2	Spring	0.52-0.6	0.3 max	0.025 max	0.6-0.9	0.4 max	0.04 max	0.04 max	1.5-1.8	bal Fe				
PNH93027	10E	Struct; Wir Bar	0.07-0.14	0.2 max		0.35-0.55	0.3 max	0.04 max	0.04 max	0.17-0.35	Al 0.02-0.12; bal Fe				
Romania															
STAS 500/3(88)	RCA37	Struct; weather res	0.1 max	0.5-0.8	0.3-0.6	0.4 max	0.4 max	0.06-0.160	0.035 max	0.15-0.3	Al 0.025-0.120; N <=0.015; Pb <=0.15; bal Fe				
STAS 500/3(88)	RCB52	Struct; weather res	0.18 max	0.5-0.8	0.3-0.6	1.15 max	0.4 max	0.04 max	0.04 max	0.6 max	Al 0.025-0.120; N <=0.015; Pb <=0.15; V <=0.12; bal Fe				
Russia															
GOST 14959(79)	65	Spring	0.62-0.7	0.25 max	0.2 max	0.5-0.8	0.25 max	0.035 max	0.035 max	0.17-0.37	Al <=0.1; bal Fe				
GOST 380(71)	St2kp	Gen struct; Semi-Finished; Unkilled; Rimming; R; FU; 0-20mm	0.3 max	0.3 max	0.3 max	2 max	0.4 max	0.07 max	0.06 max	0.6 max	Al <=0.1; bal Fe	320-410		33L	
GOST 380(71)	St2kp	Gen struct; Semi-Finished; Unkilled; Rimming; R; FU; >40mm	0.3 max	0.3 max	0.3 max	2 max	0.4 max	0.07 max	0.06 max	0.6 max	Al <=0.1; bal Fe	320-410		30L	
GOST 380(71)	St2kp2	Gen struct; Unkilled; Rimming; R; FU; 0-20mm	0.3 max	0.3 max	0.3 max	2 max	0.4 max	0.07 max	0.06 max	0.6 max	Al <=0.1; bal Fe	320-410		33L	
GOST 380(71)	St2kp2	Gen struct; Unkilled; Rimming; R; FU; >40mm	0.3 max	0.3 max	0.3 max	2 max	0.4 max	0.07 max	0.06 max	0.6 max	Al <=0.1; bal Fe	320-410		30L	
GOST 380(71)	St2kp3	Gen struct; Unkilled; Rimming; R; FU; 0-20mm	0.3 max	0.3 max	0.3 max	2 max	0.4 max	0.07 max	0.06 max	0.6 max	Al <=0.1; bal Fe	320-410	215	33L	
GOST 380(71)	St2kp3	Gen struct; Unkilled; Rimming; R; FU; >100mm	0.3 max	0.3 max	0.3 max	2 max	0.4 max	0.07 max	0.06 max	0.6 max	Al <=0.1; bal Fe	320-410	185	30L	
GOST 380(71)	St2ps	Gen struct; Semi-Finished; Semi-killed; >40mm	0.3 max	0.3 max	0.3 max	2 max	0.4 max	0.07 max	0.06 max	0.6 max	Al <=0.1; bal Fe	330-430		29L	
GOST 380(71)	St2ps	Gen struct; Semi-Finished; Semi-killed; 0-20mm	0.3 max	0.3 max	0.3 max	2 max	0.4 max	0.07 max	0.06 max	0.6 max	Al <=0.1; bal Fe	330-430		32L	
GOST 380(71)	St2ps2	Gen struct; Semi-killed; 0-20mm	0.3 max	0.3 max	0.3 max	2 max	0.4 max	0.07 max	0.06 max	0.6 max	Al <=0.1; bal Fe	330-430		32L	
GOST 380(71)	St2ps2	Gen struct; Semi-killed; >40mm	0.3 max	0.3 max	0.3 max	2 max	0.4 max	0.07 max	0.06 max	0.6 max	Al <=0.1; bal Fe	330-430		29L	
GOST 380(71)	St2ps3	Gen struct; Semi-killed; >100mm	0.3 max	0.3 max	0.3 max	2 max	0.4 max	0.07 max	0.06 max	0.6 max	Al <=0.1; bal Fe	330-430	195	29L	
GOST 380(71)	St2ps3	Gen struct; Semi-killed; 0-20mm	0.3 max	0.3 max	0.3 max	2 max	0.4 max	0.07 max	0.06 max	0.6 max	Al <=0.1; bal Fe	330-430		32L	
GOST 380(71)	St2sp	Gen struct; Semi-Finished; Non-rimming; FF; >40mm	0.3 max			2 max	0.4 max	0.07 max	0.06 max	0.6 max	bal Fe	330-430		29L	
GOST 380(71)	St2sp	Gen struct; Semi-Finished; Non-rimming; FF; 0-20mm	0.3 max			2 max	0.4 max	0.07 max	0.06 max	0.6 max	bal Fe	330-430		32L	
GOST 380(71)	St2sp2	Gen struct; Non-rimming; FF; 0-20mm	0.3 max	0.3 max	0.3 max	2 max	0.4 max	0.07 max	0.06 max	0.6 max	Al <=0.1; bal Fe	330-430		32L	
GOST 380(71)	St2sp2	Gen struct; Non-rimming; FF; >40mm	0.3 max	0.3 max	0.3 max	2 max	0.4 max	0.07 max	0.06 max	0.6 max	Al <=0.1; bal Fe	330-430		29L	
GOST 380(71)	St2sp3	Gen struct; Non-rimming; FF; 0-20mm	0.3 max	0.3 max	0.3 max	2 max	0.4 max	0.07 max	0.06 max	0.6 max	Al <=0.1; bal Fe	330-430	225	32L	
GOST 380(71)	St2sp3	Gen struct; Non-rimming; FF; >100mm	0.3 max	0.3 max	0.3 max	2 max	0.4 max	0.07 max	0.06 max	0.6 max	Al <=0.1; bal Fe	330-430	195	29L	
GOST 380(71)	St3Gps	Gen struct; Semi-killed; >40mm	0.3 max	0.3 max	0.3 max	2 max	0.4 max	0.07 max	0.06 max	0.6 max	Al <=0.1; bal Fe	370-490		23L	
GOST 380(71)	St3Gps	Gen struct; Semi-killed; 0-20mm	0.3 max	0.3 max	0.3 max	2 max	0.4 max	0.07 max	0.06 max	0.6 max	Al <=0.1; bal Fe	370-490		26L	

Specification	Designation	Notes	C	Cr	Cu	Mn	Ni	P	S	Si	Other	UTS	YS	El	Hard

Carbon Steel, Unclassified, No equivalents identified

Russia

Specification	Designation	Notes	C	Cr	Cu	Mn	Ni	P	S	Si	Other	UTS	YS	El	Hard
GOST 380(71)	St3Gps2	Gen struct; Semi-killed; >40mm	0.3 max	0.3 max	0.3 max	2 max	0.4 max	0.07 max	0.06 max	0.6 max	Al <=0.1; bal Fe	370-490		23L	
GOST 380(71)	St3Gps2	Gen struct; Semi-killed; 0-20mm	0.3 max	0.3 max	0.3 max	2 max	0.4 max	0.07 max	0.06 max	0.6 max	Al <=0.1; bal Fe	370-490		26L	
GOST 380(71)	St3Gps3	Gen struct; Semi-killed; >100mm	0.3 max	0.3 max	0.3 max	2 max	0.4 max	0.07 max	0.06 max	0.6 max	Al <=0.1; bal Fe	370-490	205	23L	
GOST 380(71)	St3Gps3	Gen struct; Semi-killed; 0-20mm	0.3 max	0.3 max	0.3 max	2 max	0.4 max	0.07 max	0.06 max	0.6 max	Al <=0.1; bal Fe	370-490	245	26L	
GOST 380(71)	St3kp2	Gen struct; Unkilled; Rimming; R; FU; 0-20mm	0.3 max	0.3 max	0.3 max	2 max	0.4 max	0.07 max	0.06 max	0.6 max	Al <=0.1; bal Fe	360-460		27L	
GOST 380(71)	St3kp2	Gen struct; Unkilled; Rimming; R; FU; >40mm	0.3 max	0.3 max	0.3 max	2 max	0.4 max	0.07 max	0.06 max	0.6 max	Al <=0.1; bal Fe	360-460		24L	
GOST 380(71)	St3kp3	Gen struct; Unkilled; Rimming; R; FU; >100mm	0.3 max	0.3 max	0.3 max	2 max	0.4 max	0.07 max	0.06 max	0.6 max	Al <=0.1; bal Fe	360-460	195	24L	
GOST 380(71)	St3kp3	Gen struct; Unkilled; Rimming; R; FU; 0-20mm	0.3 max	0.3 max	0.3 max	2 max	0.4 max	0.07 max	0.06 max	0.6 max	Al <=0.1; bal Fe	360-460		27L	
GOST 380(71)	St3ps2	Gen struct; Semi-killed; >40mm	0.3 max	0.3 max	0.3 max	2 max	0.4 max	0.07 max	0.06 max	0.6 max	Al <=0.1; bal Fe	370-480		23L	
GOST 380(71)	St3ps2	Gen struct; Semi-killed; 0-20mm	0.3 max	0.3 max	0.3 max	2 max	0.4 max	0.07 max	0.06 max	0.6 max	Al <=0.1; bal Fe	370-480		26L	
GOST 380(71)	St3ps3	Gen struct; Semi-killed; >100mm	0.3 max	0.3 max	0.3 max	2 max	0.4 max	0.07 max	0.06 max	0.6 max	Al <=0.1; bal Fe	370-480	205	23L	
GOST 380(71)	St3ps3	Gen struct; Semi-killed; 0-20mm	0.3 max	0.3 max	0.3 max	2 max	0.4 max	0.07 max	0.06 max	0.6 max	Al <=0.1; bal Fe	370-480	245	26L	
GOST 380(71)	St3sp	Gen struct; Semi-Finished; Non-rimming; FF; >40mm	0.3 max	0.3 max	0.3 max	2 max	0.4 max	0.07 max	0.06 max	0.6 max	Al <=0.1; bal Fe	370-480		23L	
GOST 380(71)	St3sp	Gen struct; Semi-Finished; Non-rimming; FF; 0-20mm	0.3 max	0.3 max	0.3 max	2 max	0.4 max	0.07 max	0.06 max	0.6 max	Al <=0.1; bal Fe	370-480		26L	
GOST 380(71)	St3sp2	Gen struct; Non-rimming; FF; 0-20mm	0.3 max	0.3 max	0.3 max	2 max	0.4 max	0.07 max	0.06 max	0.6 max	Al <=0.1; bal Fe	370-480		26H	
GOST 380(71)	St3sp2	Gen struct; Non-rimming; FF; >40mm	0.3 max	0.3 max	0.3 max	2 max	0.4 max	0.07 max	0.06 max	0.6 max	Al <=0.1; bal Fe	370-480		23L	
GOST 380(71)	St3sp3	Gen struct; Non-rimming; FF; 0-20mm	0.3 max	0.3 max	0.3 max	2 max	0.4 max	0.07 max	0.06 max	0.6 max	Al <=0.1; bal Fe	370-480	245	26L	
GOST 380(71)	St3sp3	Gen struct; Non-rimming; FF; >100mm	0.3 max	0.3 max	0.3 max	2 max	0.4 max	0.07 max	0.06 max	0.6 max	Al <=0.1; bal Fe	370-480	205	23L	
GOST 380(71)	St4kp2	Gen struct; Unkilled; Rimming; R; FU; >40mm	0.3 max	0.3 max	0.3 max	2 max	0.4 max	0.07 max	0.06 max	0.6 max	Al <=0.1; bal Fe	400-510		22L	
GOST 380(71)	St4kp2	Gen struct; Unkilled; Rimming; R; FU; 0-20mm	0.3 max	0.3 max	0.3 max	2 max	0.4 max	0.07 max	0.06 max	0.6 max	Al <=0.1; bal Fe	400-510		25L	
GOST 380(71)	St4kp3	Gen struct; Unkilled; Rimming; R; FU; 0-20mm	0.3 max	0.3 max	0.3 max	2 max	0.4 max	0.07 max	0.06 max	0.6 max	Al <=0.1; bal Fe	400-510	255	25L	
GOST 380(71)	St4kp3	Gen struct; Unkilled; Rimming; R; FU; >100mm	0.3 max	0.3 max	0.3 max	2 max	0.4 max	0.07 max	0.06 max	0.6 max	Al <=0.1; bal Fe	400-510	225	22L	
GOST 380(71)	St4ps2	Gen struct; Semi-killed; >40mm	0.3 max	0.3 max	0.3 max	2 max	0.4 max	0.07 max	0.06 max	0.6 max	Al <=0.1; bal Fe	410-530		21L	
GOST 380(71)	St4ps2	Gen struct; Semi-killed; 0-20mm	0.3 max	0.3 max	0.3 max	2 max	0.4 max	0.07 max	0.06 max	0.6 max	Al <=0.1; bal Fe	410-530		24L	
GOST 380(71)	St4ps3	Gen struct; Semi-killed; 0-20mm	0.3 max	0.3 max	0.3 max	2 max	0.4 max	0.07 max	0.06 max	0.6 max	Al <=0.1; bal Fe	410-530	265	24L	
GOST 380(71)	St4ps3	Gen struct; Semi-killed; >100mm	0.3 max	0.3 max	0.3 max	2 max	0.4 max	0.07 max	0.06 max	0.6 max	Al <=0.1; bal Fe	410-530	235	21L	
GOST 380(71)	St4sp2	Gen struct; Non-rimming; FF; 0-20mm	0.3 max	0.3 max	0.3 max	2 max	0.4 max	0.07 max	0.06 max	0.6 max	Al <=0.1; bal Fe	410-530		24L	
GOST 380(71)	St4sp2	Gen struct; Non-rimming; FF; >40mm	0.3 max	0.3 max	0.3 max	2 max	0.4 max	0.07 max	0.06 max	0.6 max	Al <=0.1; bal Fe	410-530		21L	
GOST 380(71)	St4sp3	Gen struct; Non-rimming; FF; >100mm	0.3 max	0.3 max	0.3 max	2 max	0.4 max	0.07 max	0.06 max	0.6 max	Al <=0.1; bal Fe	410-530	235	21L	

Carbon Steel, Unclassified, No equivalents identified

Russia

Specification	Designation	Notes	C	Cr	Cu	Mn	Ni	P	S	Si	Other	UTS	YS	El	Hard
GOST 380(71)	St4sp3	Gen struct; Non-rimming; FF; 0-20mm	0.3 max	0.3 max	0.3 max	2 max	0.4 max	0.07 max	0.06 max	0.6 max	Al <=0.1; bal Fe	410-530	265	24L	
GOST 380(71)	St5Gps	Gen struct; Semi-killed; 0-20mm	0.3 max	0.3 max	0.3 max	2 max	0.4 max	0.07 max	0.06 max	0.6 max	Al <=0.1; bal Fe	450-590		20L	
GOST 380(71)	St5Gps	Gen struct; Semi-killed; >40mm	0.3 max	0.3 max	0.3 max	2 max	0.4 max	0.07 max	0.06 max	0.6 max	Al <=0.1; bal Fe	450-590		17L	
GOST 380(71)	St5Gps2	Gen struct; Semi-killed; 0-20mm	0.3 max	0.3 max	0.3 max	2 max	0.4 max	0.07 max	0.06 max	0.6 max	Al <=0.1; bal Fe	450-590		20L	
GOST 380(71)	St5Gps2	Gen struct; Semi-killed; >40mm	0.3 max	0.3 max	0.3 max	2 max	0.4 max	0.07 max	0.06 max	0.6 max	Al <=0.1; bal Fe	450-590		17L	
GOST 380(71)	St5Gps3	Gen struct; Semi-killed; 0-20mm	0.3 max	0.3 max	0.3 max	2 max	0.4 max	0.07 max	0.06 max	0.6 max	Al <=0.1; bal Fe	450-590	285	20L	
GOST 380(71)	St5Gps3	Gen struct; Semi-killed; >100mm	0.3 max	0.3 max	0.3 max	2 max	0.4 max	0.07 max	0.06 max	0.6 max	Al <=0.1; bal Fe	450-590	255	17L	
GOST 380(71)	St5ps2	Gen struct; Semi-killed; 0-20mm	0.3 max	0.3 max	0.3 max	2 max	0.4 max	0.07 max	0.06 max	0.6 max	Al <=0.1; bal Fe	490-630		20L	
GOST 380(71)	St5ps2	Gen struct; Semi-killed; >40mm	0.3 max	0.3 max	0.3 max	2 max	0.4 max	0.07 max	0.06 max	0.6 max	Al <=0.1; bal Fe	490-630		17L	
GOST 380(71)	St5ps3	Gen struct; Semi-killed; >100mm	0.3 max	0.3 max	0.3 max	2 max	0.4 max	0.07 max	0.06 max	0.6 max	Al <=0.1; bal Fe	490-630	255	17L	
GOST 380(71)	St5ps3	Gen struct; Semi-killed; 0-20mm	0.3 max	0.3 max	0.3 max	2 max	0.4 max	0.07 max	0.06 max	0.6 max	Al <=0.1; bal Fe	490-630		20L	
GOST 380(71)	St5sp2	Gen struct; Non-rimming; FF; 0-20mm	0.3 max	0.3 max	0.3 max	2 max	0.4 max	0.07 max	0.06 max	0.6 max	Al <=0.1; bal Fe	490-630		20L	
GOST 380(71)	St5sp2	Gen struct; Non-rimming; FF; >40mm	0.3 max	0.3 max	0.3 max	2 max	0.4 max	0.07 max	0.06 max	0.6 max	Al <=0.1; bal Fe	490-630		17L	
GOST 380(71)	St5sp3	Gen struct; Non-rimming; FF; >100mm	0.3 max	0.3 max	0.3 max	2 max	0.4 max	0.07 max	0.06 max	0.6 max	Al <=0.1; bal Fe	490-630	255	17L	
GOST 380(71)	St5sp3	Gen struct; Non-rimming; FF; 0-20mm	0.3 max	0.3 max	0.3 max	2 max	0.4 max	0.07 max	0.06 max	0.6 max	Al <=0.1; bal Fe	490-630	285	20L	
GOST 380(88)	St2kp	Gen struct; Semi-Finished; Unkilled; R; FU	0.09-0.15		0.3 max	0.3 max	0.25-0.5	0.3 max	0.04 max	0.05 max	0.05 max	Al <=0.1; N <=0.008; Ti <=0.1; As<=0.08; bal Fe				
GOST 380(88)	St2ps	Gen struct; Semi-Finished; Semi-killed	0.09-0.15		0.3 max	0.3 max	0.25-0.5	0.3 max	0.04 max	0.05 max	0.05-0.15	Al <=0.1; N <=0.008; Ti <=0.1; As<=0.08; bal Fe				
GOST 380(88)	St2sp	Gen struct; Semi-Finished; Non-rimming; FF	0.09-0.15				0.25-0.5	0.3 max	0.04 max	0.05 max	0.15-0.3	N <=0.008; Ti <=0.1; As<=0.08; bal Fe				
GOST 380(88)	St3Gps	Gen struct; Semi-Finished; Semi-killed	0.14-0.22		0.3 max	0.1 max	0.8-1.1	0.3 max	0.04 max	0.05 max	0.15 max	Al <=0.1; N <=0.008; Ti <=0.1; As<=0.08; bal Fe				
GOST 380(88)	St3Gsp	Gen struct; Semi-Finished; Non-rimming; FF	0.14-0.2		0.3 max	0.3 max	0.8-1.1	0.3 max	0.04 max	0.05 max	0.15-0.3	Al <=0.1; N <=0.008; Ti <=0.1; As<=0.08; bal Fe				
GOST 380(88)	St3sp	Gen struct; Semi-Finished; Non-rimming; FF	0.14-0.22		0.3 max	0.3 max	0.4-0.65	0.3 max	0.04 max	0.05 max	0.15-0.3	Al 0.02-0.120; N <=0.008; Ti <=0.03; As<=0.08; bal Fe				
GOST 5521(93)	A40	Struct, weld ships	0.18 max	0.60-0.90	0.35 max	0.5-0.8	0.40 max	0.035 max	0.035 max	0.8-1.1	Al 0.015-0.06; Mo <=0.08; bal Fe				
GOST 5521(93)	A40S	Struct, weld ships	0.18 max	0.60-0.90	0.40-0.60	0.5-0.8	0.50-0.80	0.035 max	0.035 max	0.8-1.1	Al 0.015-0.06; Mo <=0.08; bal Fe				
GOST 5521(93)	B	Struct, weld ships	0.21 max	0.30 max	0.35 max	0.4-1.1	0.40 max	0.040 max	0.040 max	0.15-0.37	Al 0.015-0.06; bal Fe				
GOST 5521(93)	D	Struct, weld ships	0.21 max	0.30 max	0.35 max	0.6-1.4	0.40 max	0.040 max	0.040 max	0.15-0.37	Al 0.015-0.06; bal Fe				
GOST 5521(93)	D40	Struct, weld ships	0.18 max	0.60-0.90	0.35 max	0.5-0.8	0.40 max	0.035 max	0.035 max	0.8-1.1	Al 0.015-0.06; bal Fe				
GOST 5521(93)	D40S	Struct, weld ships	0.18 max	0.60-0.90	0.40-0.60	0.5-0.8	0.50-0.80	0.035 max	0.035 max	0.8-1.1	Al 0.015-0.06; bal Fe				
GOST 5521(93)	E	Struct, weld ships	0.18 max	0.30 max	0.35 max	0.6-1.4	0.40 max	0.040 max	0.040 max	0.15-0.37	Al 0.015-0.06; bal Fe				
GOST 5781(82)	80S; A-IV	Reinforcing	0.74-0.82	0.3 max	0.3 max	0.5-0.9	0.3 max	0.04 max	0.045 max	0.6-1.1	Al <=0.1; bal Fe				
GOST 5781(82)	St3kp3; A-I	HR conc reinf	0.3 max	0.3 max	0.3 max	2 max	0.4 max	0.07 max	0.06 max	0.6 max	Al <=0.1; bal Fe				
GOST 5781(82)	St3ps3; A-I	HR conc-reinf	0.3 max	0.3 max	0.3 max	2 max	0.4 max	0.07 max	0.06 max	0.6 max	Al <=0.1; bal Fe				

Carbon Steel, Unclassified, No equivalents identified

Specification	Designation	Notes	C	Cr	Cu	Mn	Ni	P	S	Si	Other	UTS	YS	El	Hard
Russia															
GOST 5781(82)	St3sp3; A-I	HR conc-reinf	0.3 max	0.3 max	0.3 max	2 max	0.4 max	0.07 max	0.06 max	0.6 max	Al <=0.1; bal Fe				
USA															
AMS 7713A(95)		Iron, Electrical Bar and Strp, CF, ANN	0.020 max	0.30 max		0.15 max	0.20 max	0.020 max	0.015 max	0.15 max	V 0.04-0.10; bal Fe				55 or 65 HRB max
ASTM A139	A	Weld Pipe				1.00 max		0.035 max	0.035 max		bal Fe	330	205	35L	
ASTM A252(98)	Grade 1	Weld/Smls pipe; piles						0.050 max			bal Fe	345	205	18-30	
ASTM A252(98)	Grade 2	Weld/Smls pipe; piles						0.050 max			bal Fe	415	240	14-25	
ASTM A252(98)	Grade 3	Weld/Smls pipe; piles						0.050 max			bal Fe	455	310	20	
ASTM A521(96)	CC	die frg; ann, norm, N/T; <12 in.									Reference A322 A576 and other specs for chemisty	415	205	25	
ASTM A521(96)	CF	die frg; N/T; <8 in.									Reference A322 A576 and other specs for chemisty	550	275	22	
ASTM A521(96)	CG	die frg; Q/T, norm Q/T; <4 in. solid, <2 in. bored wall									Reference A322 A576 and other specs for chemisty	620	380	20	
ASTM A589(96)	A 589-A	Smls Weld water-well pipe						0.050 max	0.060 max		bal Fe	330	205		
ASTM A589(96)	A 589-B	Smls Weld water-well pipe						0.050 max	0.060 max		bal Fe	415	240		
ASTM A615/A615M(96)	Grade 40	Deformed/plain Bil Bar for conc reinf						0.06 max			bal Fe	500	300	11-12	
ASTM A615/A615M(96)	Grade 60	Deformed/plain Bil Bar for conc reinf						0.06 max			bal Fe	620	420	7-9	
ASTM A615/A615M(96)	Grade 75	Deformed/plain Bil Bar for conc reinf						0.06 max			bal Fe	690	520	6-7	
ASTM A622/A622M(97)	A622(A)	Sh Strp, HR, FF, DS	0.08 max	0.15 max	0.20 max	0.50 max	0.20 max	0.020 max	0.030 max		Al >=0.01; Mo <=0.06; Nb <=0.008; Ti <=0.008; V <=0.008; bal Fe		205-310	28 min	
ASTM A622/A622M(97)	A622(B)	Sh Strp, HR, FF, DS	0.02-0.08	0.15 max	0.20 max	0.50 max	0.20 max	0.020 max	0.030 max		Al >=0.02; Mo <=0.06; Nb <=0.008; Ti <=0.008; V <=0.008; bal Fe		205-310	28 min	
ASTM A649/649M(98)	Class 4	Frg roll for paperboard machinery	0.35 max			0.60-1.05		0.025 max	0.025 max	0.15-0.35	Si<=0.10 with VCD; bal Fe	415	205	22	
ASTM A663/663M(94)	45 [310]	Merchant qual bar						0.04 max	0.05 max		Cu>=0.20 if spec; bal Fe	310-380	175	33	
ASTM A663/663M(94)	50 [345]	Merchant qual bar						0.04 max	0.05 max		Cu>=0.20 if spec; bal Fe	345-415	195	30	
ASTM A663/663M(94)	55 [380]	Merchant qual bar						0.04 max	0.05 max		Cu>=0.20 if spec; bal Fe	380-450	210	26	
ASTM A663/663M(94)	60 [415]	Merchant qual bar						0.04 max	0.05 max		Cu>=0.20 if spec; bal Fe	415-495	230	22	
ASTM A663/663M(94)	65 [450]	Merchant qual bar						0.04 max	0.05 max		Cu>=0.20 if spec; bal Fe	450-530	250	20	
ASTM A663/663M(94)	70 [485]	Merchant qual bar						0.04 max	0.05 max		Cu>=0.20 if spec; bal Fe	485-585	270	18	
ASTM A663/663M(94)	75 [515]	Merchant qual bar						0.04 max	0.05 max		Cu>=0.20 if spec; bal Fe	51-620	285	18	
ASTM A663/663M(94)	80 [550]	Merchant qual bar						0.04 max	0.05 max		Cu>=0.20 if spec; bal Fe	550	305	17	
ASTM A668/668M(96)	Class A	Frg; untreated; <=20 in.				1.35 max		0.050 max	0.050 max		choice of comp; bal Fe	325			183 HB
ASTM A668/668M(96)	Class B	Frg; ann, norm, N/T; <=20 in.				1.35 max		0.050 max	0.050 max		choice of comp; bal Fe	415	205	24	120-174 HB
ASTM A668/668M(96)	Class C	Frg; ann, norm, N/T; <=20 in.				1.35 max		0.050 max	0.050 max		choice of comp; bal Fe	455	230	22	137-183 HB
ASTM A668/668M(96)	Class D	Frg; ann, norm, N/T; <=20 in.				1.35 max		0.050 max	0.050 max		choice of comp; bal Fe	515	260	20	149-207 HB
ASTM A668/668M(96)	Class E	Frg; N/T, double N/T; <=20 in.				1.35 max		0.050 max	0.050 max		choice of comp; bal Fe	570	295	22	174-217 HB

UNS numbers and US grades are provided as a means of cross referencing chemically similar alloys. Exchangability is only possible after independent examination of specifications. Tensile properties are minimum or typical as specified. UTS and YS as MPa. El as %. See Appendix for list of abbreviations used in Notes. * indicates obsolete material.

Specification	Designation	Notes	C	Cr	Cu	Mn	Ni	P	S	Si	Other	UTS	YS	El	Hard

Carbon Steel, Unclassified, No equivalents identified

USA

Specification	Designation	Notes	C	Cr	Cu	Mn	Ni	P	S	Si	Other	UTS	YS	El	Hard
ASTM A668/668M(96)	Class F	Frg; Q/T, norm Q/T; <=20 in.				1.35 max		0.050 max	0.050 max		choice of comp; bal Fe	565	330	19	174-217 HB
ASTM A675/675M(95)	45 [310]	HW bar special qual				0.040 max		0.040 max	0.050 max		Cu>=0.20 if spec; Pb if spec; bal Fe	310-380	155	33	
ASTM A675/675M(95)	50 [345]	HW bar special qual						0.040 max	0.050 max		Cu>=0.20 if spec; Pb if spec; bal Fe	345-415	170	30	
ASTM A675/675M(95)	55 [380]	HW bar special qual						0.040 max	0.050 max		Cu>=0.20 if spec; Pb if spec; bal Fe	380-450	190	26	
ASTM A675/675M(95)	60 [415]	HW bar special qual						0.040 max	0.050 max		Cu>=0.20 if spec; Pb if spec; bal Fe	415-495	205	22	
ASTM A675/675M(95)	65 [450]	HW bar special qual						0.040 max	0.050 max		Cu>=0.20 if spec; Pb if spec; bal Fe	450-530	225	20	
ASTM A675/675M(95)	70 [485]							0.040 max	0.050 max		Cu>=0.20 if spec; Pb if spec; bal Fe	485-585	240	18	
ASTM A675/675M(95)	75 [515]	HW bar special qual						0.040 max	0.050 max		Cu>=0.20 if spec; Pb if spec; bal Fe	515-620	260	18	
ASTM A675/675M(95)	80 [550]	HW bar special qual						0.040 max	0.050 max		Cu>=0.20 if spec; Pb if spec; bal Fe	550	275	17	
ASTM A675/675M(95)	90 [620]	HW bar special qual						0.040 max	0.050 max		Cu>=0.20 if spec; Pb if spec; bal Fe	620	380	14	
ASTM A730(93)	Grade G	Railway frg; Q/T; t>178-254mm				0.60-0.90		0.045 max	0.050 max	0.15 min	bal Fe	585	345	19	
ASTM A730(93)	Grade G	Railway frg; Q/T; t>102-178mm				0.60-0.90		0.045 max	0.050 max	0.15 min	bal Fe	585	345	20	
ASTM A730(93)	Grade G	Railway frg; Q/T; t<=102mm				0.60-0.90		0.045 max	0.050 max	0.15 min	bal Fe	620	380	20	
ASTM A730(93)	Grade H	Railway frg; N/Q/T; t>178-254mm				0.60-0.90		0.045 max	0.050 max	0.15 min	bal Fe	725	450	18	
ASTM A730(93)	Grade H	Railway frg; N/Q/T; t<=178mm				0.60-0.90		0.045 max	0.050 max	0.15 min	bal Fe	795	515	16	
ASTM A836/A836M(95)	A836	Frg; glass-lined pipe and Press ves	0.20 max			0.90 max		0.05 max	0.05 max	0.35 max	Ti>=4xC<=1.00; bal Fe	380	175	22	
ASTM A909(94)	A909 Class 60	MicroAlloy frg									Mo 0.01-0.30; Nb 0.005-0.07; Ti <=0.030; V 0.02-0.20; see A576 for C Mn P S; bal Fe	515	415	16	167 HB
ASTM A963/A963M(97)		Sh, CR, FF, DDS	0.06 max	0.15 max	0.20 max	0.50 max	0.20 max	0.020 max	0.025 max		Al >=0.01; Mo <=0.06; Nb <=0.008; Ti <=0.008; V <=0.008; Cu+Ni+Cr+Mo<=0.50; bal Fe		115-200	38 min	
ASTM A969/A969M(97)		Sh, CR, FF, extra DDS	0.02 max	0.15 max	0.10 max	0.40 max	0.10 max	0.020 max	0.020 max		Al >=0.01; Mo <=0.03; Nb <=0.10; Ti <=0.15; V <=0.008; Cu+Ni+Cr+Mo<=0.25; bal Fe		105-170	40 min	
ASTM A980(97)	Grade 130	Sh, CR	0.10 max	0.15 max	0.20 max	0.60 max	0.20 max	0.030 max	0.035 max		Al >=0.01; Mo <=0.06; Nb <=0.008; V <=0.008; Cu >=0.20 if spec; bal Fe		896		
ASTM A980(97)	Grade 160	Sh, CR	0.15 max	0.15 max	0.20 max	0.60 max	0.20 max	0.030 max	0.035 max		Al >=0.01; Mo <=0.06; Nb <=0.008; V <=0.008; Cu >=0.20 if spec; bal Fe		1103		
ASTM A980(97)	Grade 190	Sh, CR	0.23 max	0.15 max	0.20 max	0.60 max	0.20 max	0.030 max	0.035 max		Al >=0.01; Mo <=0.06; Nb <=0.008; V <=0.008; Cu >=0.20 if spec; bal Fe		1310		
ASTM A980(97)	Grade 220	Sh, CR	0.28 max	0.15 max	0.20 max	0.60 max	0.20 max	0.030 max	0.035 max		Al >=0.01; Mo <=0.06; Nb <=0.008; V <=0.008; Cu >=0.20 if spec; bal Fe		1517		
FED QQ-S-630A(70) - - -*		Obs; see ASTM A575, A663; HR Bar merchant qual									bal Fe				

UNS numbers and US grades are provided as a means of cross referencing chemically similar alloys. Exchangability is only possible after independent examination of specifications. Tensile properties are minimum or typical as specified. UTS and YS as MPa. El as %. See Appendix for list of abbreviations used in Notes. * indicates obsolete material.

Specification Designation		Notes	C	Cr	Cu	Mn	Ni	P	S	Si	Other	UTS	YS	El	Hard

Carbon Steel, Unclassified, No equivalents identified

USA

Specification Designation		Notes	C	Cr	Cu	Mn	Ni	P	S	Si	Other	UTS	YS	El	Hard
FED QQ-S-631A(71) - - -*		Obs; see ASTM A576, A675; HR Bar spec qual									bal Fe				
FED QQ-S-634A(71) - - -*		Obs; see ASTM A108; CF Bar std qual, free mach									bal Fe				
FED QQ-S-692B(60) - - -*		Obs; CR Sh									bal Fe				
FED QQ-S-693B(63) - - -*		Obs; HR Sh									bal Fe				
FED QQ-S-741D(87) - - -*		Obs; see ASTM A36/36M; Bar plate Shp									bal Fe				
FED QQ-T-830A(61) - - -*		Obs; see ASTM A513, A519; Smls welded tube									bal Fe				
FED QQ-W-409A(60) - - -*		Obs; CHd, cold Frg Wir									bal Fe				
SAE J430(98)	Grade 0	Solid rivet, Ann elong in 200mm						0.040 max	0.050 max		bal Fe	275-379	159	27	65 HRB max
SAE J430(98)	Grade 1	Solid rivet, Ann elong in 200mm						0.040 max	0.050 max		bal Fe	359-427	193	24	85 HRB max
SAE J430(98)	Grade 2	Solid rivet, elong in 200mm; for hot driving only	0.28 max			0.30-0.90		0.040 max	0.050 max	0.25 max	bal Fe	379-483	200	22	
SAE J430(98)	Grade 3	Solid rivet, elong in 200mm; for hot driving only	0.30 max			1.65 max		0.040 max	0.050 max		bal Fe	469-565	262	20	

Yugoslavia

Specification Designation		Notes	C	Cr	Cu	Mn	Ni	P	S	Si	Other	UTS	YS	El	Hard
	C.0145	Unalloyed mild	0.15 max			0.6 max		0.05 max	0.05 max	0.60 max	bal Fe				
	C.0545	Struct	0.30 max			1.60 max		0.06 max	0.06 max	0.60 max	P+S<=0.01; bal Fe				
	C.1216		0.16 max			0.35-0.55		0.035 max	0.035 max	0.1 max	Al 0.2-0.10; Mo <=0.15; bal Fe				
	C.3100	Tub w/special req	0.2 max			1.5 max		0.05 max	0.05 max	0.1-0.55	Mo <=0.15; bal Fe				

Specification	Designation	Notes	C	Cr	Cu	Mn	Mo	P	S	Si	Other	UTS	YS	El	Hard

Alloy Steel, Chromium, 5015

Bulgaria

Specification	Designation	Notes	C	Cr	Cu	Mn	Mo	P	S	Si	Other	UTS	YS	El	Hard
BDS 6354	15Ch	Struct	0.12-0.18	0.70-1.00	0.30 max	0.40-0.70	0.15 max	0.035 max	0.035 max	0.17-0.37	Ni <=0.30; Ti <=0.03; V <=0.05; W <=0.2; bal Fe				

International

Specification	Designation	Notes	C	Cr	Cu	Mn	Mo	P	S	Si	Other	UTS	YS	El	Hard
ISO 683-11(87)	C16E4	CH, 16mm test	0.12-0.18		0.3 max	0.60-0.90		0.035 max	0.035 max	0.15-0.40	Al <=0.1; Co <=0.1; bal Fe	550-900	340		
ISO 683-11(87)	C16M2	CH, 16mm test	0.12-0.18		0.3 max	0.60-0.90		0.035 max	0.020-0.040	0.15-0.40	Al <=0.1; Co <=0.1; bal Fe	550-900	340		

Japan

Specification	Designation	Notes	C	Cr	Cu	Mn	Mo	P	S	Si	Other	UTS	YS	El	Hard
JIS G4052(79)	SCr415H	Struct Hard	0.12-0.18	0.85-1.25	0.30 max	0.55-0.90		0.030 max	0.030 max	0.15-0.35	Ni <=0.25; bal Fe				
JIS G4104(79)	SCr21*	Obs; see SCr415	0.13-0.18	0.90-1.20	0.30 max	0.60-0.85		0.030 max	0.030 max	0.15-0.35	Ni <=0.25; bal Fe				
JIS G4104(79)	SCr415	HR Frg Bar Wir Rod	0.13-0.18	0.90-1.20	0.30 max	0.60-0.85		0.030 max	0.030 max	0.15-0.35	Ni <=0.25; bal Fe				

Mexico

Specification	Designation	Notes	C	Cr	Cu	Mn	Mo	P	S	Si	Other	UTS	YS	El	Hard
NMX-B-300(91)	5010	Bar	0.12-0.17	0.30-0.50		0.30-0.50		0.035 max	0.040 max	0.15-0.35	bal Fe				

Poland

Specification	Designation	Notes	C	Cr	Cu	Mn	Mo	P	S	Si	Other	UTS	YS	El	Hard
PNH84030	14H		0.12-0.18	0.85-1.80		0.4-0.7		0.035 max	0.035 max	0.17-0.37	Ni <=0.3; bal Fe				

Russia

Specification	Designation	Notes	C	Cr	Cu	Mn	Mo	P	S	Si	Other	UTS	YS	El	Hard
GOST 1414(75)	AS14	Free-cutting	0.10-0.17	0.25 max	0.25 max	1-1.3		0.1 max	0.15-0.3	0.12 max	Ni <=0.25; Pb 0.15-0.3; bal Fe				
GOST 4543(71)	15Ch	CH	0.12-0.18	0.7-1.0	0.3 max	0.4-0.7	0.15 max	0.035 max	0.035 max	0.17-0.37	Al <=0.1; Ni <=0.3; Ti <=0.05; bal Fe				

UK

Specification	Designation	Notes	C	Cr	Cu	Mn	Mo	P	S	Si	Other	UTS	YS	El	Hard
BS 970/1(83)	523M15	Blm Bil Slab Bar Rod Frg CH	0.12-0.18	0.30-0.60		0.30-0.60	0.15 max	0.035 max	0.04 max	0.10-0.35	Ni <=0.4; bal Fe	620		12	
BS 970/1(96)	523H15	Blm Bil Slab Bar Rod Frg	0.12-0.18	0.30-0.60		0.30-0.60		0.035 max	0.040 max	0.10-0.35	bal Fe				207 HB max
BS 970/1(96)	523M15	Blm Bil Slab Bar Rod Frg	0.12-0.18	0.30-0.60		0.30-0.60		0.035 max	0.040 max	0.10-0.35	bal Fe				

USA

Specification	Designation	Notes	C	Cr	Cu	Mn	Mo	P	S	Si	Other	UTS	YS	El	Hard
	AISI 5015	Smls mech tub	0.12-0.17	0.30-0.50		0.30-0.50		0.040 max	0.040 max	0.15-0.35	bal Fe				
	UNS G50150		0.12-0.17	0.30-0.50		0.30-0.50		0.035 max	0.040 max	0.15-0.35	bal Fe				
ASTM A29/A29M(93)	5015	Bar	0.12-0.17	0.30-0.50		0.30-0.50		0.035 max	0.040 max	0.15-0.35	bal Fe				
ASTM A519(96)	5015	Smls mech tub	0.12-0.17	0.30-0.50		0.30-0.50		0.040 max	0.040 max	0.15-0.35	bal Fe				
ASTM A752(93)	5015	Rod Wir	0.12-0.17	0.30-0.50		0.30-0.50		0.035 max	0.040 max	0.15-0.30	bal Fe	500			

Yugoslavia

Specification	Designation	Notes	C	Cr	Cu	Mn	Mo	P	S	Si	Other	UTS	YS	El	Hard
	C.4120	CH	0.12-0.18	0.4-0.7		0.4-0.6	0.15 max	0.035 max	0.035 max	0.15-0.4	bal Fe				

Alloy Steel, Chromium, 5046/5046H

France

Specification	Designation	Notes	C	Cr	Cu	Mn	Mo	P	S	Si	Other	UTS	YS	El	Hard
AFNOR NFA35565(94)	44C2*	see 44Cr2	0.42-0.48	0.4-0.6	0.3 max	0.4-0.8	0.15 max	0.025 max	0.035 max	0.4 max	Ni <=0.4; Ti <=0.05; bal Fe				
AFNOR NFA35565(94)	44Cr2	Ball & roller bearing	0.42-0.48	0.4-0.6	0.3 max	0.4-0.8	0.15 max	0.025 max	0.035 max	0.4 max	Ni <=0.4; Ti <=0.05; bal Fe				

Germany

Specification	Designation	Notes	C	Cr	Cu	Mn	Mo	P	S	Si	Other	UTS	YS	El	Hard
DIN 1652(90)	46Cr2	Q/T 17-40mm	0.42-0.50	0.40-0.60		0.50-0.80		0.035 max	0.035 max	0.40 max	bal Fe	800-950	550	14	
DIN 1652(90)	WNr 1.7006	Q/T 17-40mm	0.42-0.50	0.40-0.60		0.50-0.80		0.035 max	0.035 max	0.40 max	bal Fe	800-950	550	14	
DIN 17230(80)	44Cr2	Q/T 17-40mm	0.42-0.48	0.40-0.60	0.30 max	0.50-0.80		0.025 max	0.035 max	0.40 max	bal Fe	780-930	540	14	
DIN 17230(80)	WNr 1.3561	Bearing Q/T 17-40mm	0.42-0.48	0.40-0.60	0.30 max	0.50-0.80		0.025 max	0.035 max	0.40 max	bal Fe	780-930	540	14	
DIN EN 10083(91)	46Cr2	Q/T 17-40mm	0.42-0.50	0.40-0.60		0.50-0.80		0.035 max	0.035 max	0.40 max	bal Fe	800-950	550	14	

Specification	Designation	Notes	C	Cr	Cu	Mn	Mo	P	S	Si	Other	UTS	YS	El	Hard

Alloy Steel, Chromium, 5046/5046H (Continued from previous page)

Germany

Specification	Designation	Notes	C	Cr	Cu	Mn	Mo	P	S	Si	Other	UTS	YS	El	Hard
DIN EN 10083(91)	WNr 1.7006	Q/T 17-40mm	0.42-0.50	0.40-0.60		0.50-0.80		0.035 max	0.035 max	0.40 max	bal Fe	800-950	550	14	

Italy

UNI 7847(79)	45Cr2	SH	0.42-0.48	0.4-0.6		0.5-0.8		0.03 max	0.03 max	0.15-0.4	bal Fe				
UNI 8551(84)	45Cr2	SH	0.42-0.48	0.4-0.6		0.5-0.8		0.03 max	0.025 max	0.15-0.4	bal Fe				

Mexico

NMX-B-268(68)	5046H	Hard	0.43-0.50	0.13-0.43		0.65-1.10				0.20-0.35	bal Fe				
NMX-B-300(91)	5046	Bar	0.43-0.48	0.20-0.35		0.75-1.00		0.035 max	0.040 max	0.15-0.35	bal Fe				
NMX-B-300(91)	5046H	Bar	0.43-0.50	0.13-0.43		0.65-1.10				0.15-0.35	bal Fe				

Pan America

COPANT 514	5046	Tube	0.43-0.5	0.2-0.35		0.75-1		0.04	0.05	0.1	bal Fe				

Russia

GOST 4543(71)	45G2	Q/T	0.41-0.49	0.3 max	0.3 max	1.4-1.8	0.15 max	0.035 max	0.035 max	0.17-0.37	Al <=0.1; Ni <=0.3; Ti <=0.05; bal Fe				

UK

BS 970/3(91)	605M36	Bright bar; Q/T; 29<t<=150mm; hard	0.32-0.40			1.30-1.70	0.22-0.32	0.035 max	0.040 max	0.10-0.40	bal Fe	700-850	525	17	201-255 HB
BS 970/3(91)	605M36	Bright bar; Q/T; 150<t<=250mm; hard	0.32-0.40			1.30-1.70	0.22-0.32	0.035 max	0.040 max	0.10-0.40	bal Fe	700-850	495	15	201-255 HB
BS 970/3(91)	606M36	Bright bar; Q/T; 6<=t<=63mm; hard	0.32-0.40			1.30-1.70	0.22-0.32	0.035 max	0.15-0.25	0.10-0.40	bal Fe	775-925	585	13	223-227 HB
BS 970/3(91)	606M36	Bright bar; Q/T; 29<t<=100mm; hard CD	0.32-0.40			1.30-1.70	0.22-0.32	0.035 max	0.15-0.25	0.10-0.40	bal Fe	700-850	540	11	201-255 HB

USA

	AISI 5046	Smls mech tub	0.43-0.50	0.20-0.35		0.75-1.00		0.040 max	0.040 max	0.15-0.35	bal Fe				
	AISI 5046		0.43-0.5	0.2-0.35		0.75-1		0.04 max	0.04 max	0.15-0.3	bal Fe				
	AISI 5046H	Smls mech tub, Hard	0.44-0.50	0.13-0.43		0.65-1.10		0.035 max	0.040 max	0.15-0.30	bal Fe				
	UNS G50460		0.43-0.50	0.20-0.35		0.75-1.00		0.035 max	0.040 max	0.15-0.35	bal Fe				
	UNS H50460		0.43-0.50	0.13-0.43		0.65-1.10				0.15-0.35	bal Fe				
ASTM A29/A29M(93)	5046	Bar	0.43-0.48	0.20-0.35		0.75-1		0.035 max	0.04 max	0.15-0.35	bal Fe				
ASTM A304(96)	5046H	Bar, hard bands spec	0.43-0.50	0.13-0.43	0.35 max	0.65-1.10	0.06 max	0.035 max	0.040 max	0.15-0.35	Ni <=0.25; Cu Ni Cr Mo trace allowed; bal Fe				
ASTM A331(95)	5046H	Bar	0.43-0.50	0.13-0.43	0.35 max	0.65-1.10	0.06 max	0.025 max	0.025 max	0.15-0.35	Ni <=0.25; bal Fe				
ASTM A519(96)	5046	Smls mech tub	0.43-0.50	0.20-0.35		0.75-1.00		0.040 max	0.040 max	0.15-0.35	bal Fe				
SAE 770(84)	5046*	Obs; see J1397(92)	0.43-0.5	0.2-0.35		0.75-1		0.04 max	0.04 max	0.15-0.3	bal Fe				
SAE J1268(95)	5046H	Bar Tub; See std	0.43-0.50	0.13-0.43	0.35 max	0.65-1.10	0.06 max	0.030 max	0.040 max	0.15-0.35	Ni <=0.25; Cu, Mo, Ni not spec'd but acpt; bal Fe				
SAE J407	5046H*	Obs; see J1268									bal Fe				

Yugoslavia

	C.4133	Q/T	0.42-0.5	0.4-0.6		0.5-0.8	0.15 max	0.035 max	0.03 max	0.4 max	bal Fe				

Alloy Steel, Chromium, 5060

Czech Republic

CSN 414160	14160	Spring	0.5-0.6	0.3-0.5		0.7-1.0		0.04 max	0.04 max	0.3-0.5	bal Fe				

Specification	Designation	Notes	C	Cr	Cu	Mn	Mo	P	S	Si	Other	UTS	YS	El	Hard

Alloy Steel, Chromium, 5060 (Continued from previous page)

USA

Specification	Designation	Notes	C	Cr	Cu	Mn	Mo	P	S	Si	Other	UTS	YS	El	Hard
	UNS G50600		0.56-0.64	0.40-0.60		0.75-1.00		0.035 max	0.040 max	0.15-0.35	bal Fe				

Alloy Steel, Chromium, 50B40/50B40H

China

Specification	Designation	Notes	C	Cr	Cu	Mn	Mo	P	S	Si	Other	UTS	YS	El	Hard
GB 3077(88)	40MnB	Bar HR Q/T 25mm diam	0.37-0.44	0.30 max	0.30 max	1.10-1.40		0.035 max	0.035 max	0.17-0.37	B 0.0005-0.0035; bal Fe	980	785	10	
GB 5216(85)	40MnBH	Bar Rod HR/Frg Ann	0.37-0.44	0.30 max	0.30 max	1.10-1.40		0.035 max	0.035 max	0.17-0.37	B 0.0005-0.0035; bal Fe				
GB/T 1 3795(92)	40MnB	Strp CR Ann	0.37-0.44	0.30 max	0.30 max	1.10-1.40		0.035 max	0.035 max	0.17-0.37	B 0.0005-0.0035; bal Fe	400-700		15	
GB/T 3078(94)	40MnB	Bar CD Q/T 25mm diam	0.37-0.44	0.30 max	0.30 max	1.10-1.40		0.035 max	0.035 max	0.17-0.37	B 0.0005-0.0035; bal Fe	980	785	10	

Europe

Specification	Designation	Notes	C	Cr	Cu	Mn	Mo	P	S	Si	Other	UTS	YS	El	Hard
EN 10083/3(95)	1.7189	Boron, Q/T	0.36-0.45	0.30-0.60		1.40-1.70		0.035 max	0.040 max	0.40 max	B 0.0008-0.0050; bal Fe				
EN 10083/3(95)	39MnCrB6-2	Boron, Q/T	0.36-0.45	0.30-0.60		1.40-1.70		0.035 max	0.040 max	0.40 max	B 0.0008-0.0050; bal Fe				

France

Specification	Designation	Notes	C	Cr	Cu	Mn	Mo	P	S	Si	Other	UTS	YS	El	Hard
AFNOR NFA34552	38C2		0.35-0.4	0.3-0.6		0.6-0.9		0.35	0.35	0.1-0.4	bal Fe				
AFNOR NFA34552	38CBi	Bar Rod Wir	0.34-0.4	0.2-0.4		0.6-0.9		0.04 max	0.04 max	0.1-0.4	B 0.001-0.005; bal Fe				
AFNOR NFA35552	42C2		0.4-0.45	0.3-0.6		0.6-0.9		0.035	0.35	0.1-0.4	bal Fe				
AFNOR NFA35552(82)	38B3	t<4.5mm; Q	0.34-0.4	0.3 max	0.3 max	0.6-0.9	0.15 max	0.035 max	0.035 max	0.1-0.4	B 0.0008-0.005; Ni <=0.4; Ti <=0.05; bal Fe				47 HRC
AFNOR NFA35552(84)	382B1	t<16mm; Q/T	0.34-0.4	0.2-0.4	0.3 max	0.6-0.9	0.15 max	0.035 max	0.035 max	0.1-0.4	B 0.0008-0.005; Ni <=0.4; Ti <=0.05; bal Fe	750-890	560	14	
AFNOR NFA35552(84)	382B1	40<t<=100mm; Q/T	0.34-0.4	0.2-0.4	0.3 max	0.6-0.9	0.15 max	0.035 max	0.035 max	0.1-0.4	B 0.0008-0.005; Ni <=0.4; Ti <=0.05; bal Fe	680-860	510	12	
AFNOR NFA35552(84)	38B3	40<t<=100mm; Q/T	0.34-0.4	0.3 max	0.3 max	0.6-0.9	0.15 max	0.035 max	0.035 max	0.1-0.4	B 0.0008-0.005; Ni <=0.4; Ti <=0.05; bal Fe	650-830	480	12	
AFNOR NFA35552(84)	38B3	t<16mm; soft ann	0.34-0.4	0.3 max	0.3 max	0.6-0.9	0.15 max	0.035 max	0.035 max	0.1-0.4	B 0.0008-0.005; Ni <=0.4; Ti <=0.05; bal Fe	740-880	550	14	
AFNOR NFA35556	38C2		0.35-0.4	0.3-0.6		0.6-0.9		0.35	0.35	0.1-0.4	bal Fe				
AFNOR NFA35556	42C2		0.4-0.45	0.3-0.6		0.6-0.9		0.035	0.35	0.1-0.4	bal Fe				
AFNOR NFA35557	38C2		0.35-0.4	0.3-0.6		0.6-0.9		0.35	0.35	0.1-0.4	bal Fe				
AFNOR NFA35557	42C2		0.4-0.45	0.3-0.6		0.6-0.9		0.035	0.35	0.1-0.4	bal Fe				
AFNOR NFA35557(83)	38B3	16<t<=18mm; Q/T	0.34-0.4	0.3 max	0.3 max	0.6-0.9	0.15 max	0.035 max	0.035 max	0.1-0.4	B 0.0008-0.005; Ni <=0.4; Ti <=0.05; bal Fe	850-1050	650	11	
AFNOR NFA35557(83)	38B3	t<16mm; Q/T	0.34-0.4	0.3 max	0.3 max	0.6-0.9	0.15 max	0.035 max	0.035 max	0.1-0.4	B 0.0008-0.005; Ni <=0.4; Ti <=0.05; bal Fe	860-1070	660	11	
AFNOR NFA35563(83)	38B3TS	HT	0.34-0.4	0.3 max	0.3 max	0.6-0.9	0.15 max	0.025 max	0.035 max	0.1-0.4	B 0.0008-0.005; Ni <=0.4; Ti <=0.05; bal Fe				55 HRC
AFNOR NFA35564(83)	38B3FF	Soft ann	0.34-0.4	0.3 max		0.6-0.9	0.15 max	0.03 max	0.03 max	0.35 max	Al >=0.02; B 0.0008-0.005; Ni <=0.4; Ti <=0.05; bal Fe	0-540			156 HB max

Germany

Specification	Designation	Notes	C	Cr	Cu	Mn	Mo	P	S	Si	Other	UTS	YS	El	Hard
DIN	37CrB1		0.35-0.40	0.30-0.40		0.50-0.80		0.035 max	0.035 max	0.40 max	B >=0.0008; bal Fe				
DIN	WNr 1.7007		0.35-0.40	0.30-0.40		0.50-0.80		0.035 max	0.035 max	0.40 max	B >=0.0008; bal Fe				
DIN 17200	28CrS2*	Obs	0.35-0.42	0.4-0.6		0.5-0.8		0.035 max	0.02-0.035	0.4 max	bal Fe				

Specification	Designation	Notes	C	Cr	Cu	Mn	Mo	P	S	Si	Other	UTS	YS	El	Hard

Alloy Steel, Chromium, 50B40/50B40H (Continued from previous page)

Germany

Specification	Designation	Notes	C	Cr	Cu	Mn	Mo	P	S	Si	Other	UTS	YS	El	Hard
DIN EN 10083(91)	38Cr2	Q/T 17-40mm	0.35-0.42	0.40-0.60		0.50-0.80		0.035 max	0.035 max	0.40 max	bal Fe	700-850	450	15	
DIN EN 10083(91)	38CrS2	Q/T	0.35-0.42	0..40-0.60		0.50-0.80		0.035 max	0.020-0.040	0.40 max	bal Fe				
DIN EN 10083(91)	WNr 1.7003	Q/T 17-40mm	0.35-0.42	0.40-0.60		0.50-0.80		0.035 max	0.035 max	0.40 max	bal Fe	700-850	450	15	
DIN EN 10083(91)	WNr 1.7023	Q/T	0.35-0.42	0.40-0.60		0.50-0.80		0.035 max	0.020-0.040	0.40 max	bal Fe				

Italy

Specification	Designation	Notes	C	Cr	Cu	Mn	Mo	P	S	Si	Other	UTS	YS	El	Hard
UNI 7356(74)	38CrB1KB	Q/T	0.34-0.41	0.2-0.4		0.5-0.8		0.035 max	0.035 max	0.15-0.4	B 0.001-0.005; bal Fe				
UNI 7356(74)	41Cr2KB	Q/T	0.38-0.45	0.4-0.6		0.6-0.9		0.035 max	0.035 max	0.15-0.4	bal Fe				

Japan

Specification	Designation	Notes	C	Cr	Cu	Mn	Mo	P	S	Si	Other	UTS	YS	El	Hard
JIS G4106(79)	SMNC3*	Obs; see SMnC443	0.4-0.46	0.35-0.7		1.35-1.65		0.03	0.03	0.15-0.35	bal Fe				

Mexico

Specification	Designation	Notes	C	Cr	Cu	Mn	Mo	P	S	Si	Other	UTS	YS	El	Hard
DGN B-203	50B40*	Obs	0.38-0.42	0.4-0.6		0.75-1		0.04	0.04	0.2-0.35	B >=0.0005; bal Fe				
NMX-B-268(68)	50B40H	Hard	0.37-0.44	0.30-0.70		0.65-1.10				0.20-0.35	B >=0.0005; bal Fe				
NMX-B-300(91)	50B40H	Bar	0.37-0.44	0.30-0.70		0.65-1.10				0.15-0.35	B >=0.0005; bal Fe				

Pan America

Specification	Designation	Notes	C	Cr	Cu	Mn	Mo	P	S	Si	Other	UTS	YS	El	Hard
COPANT 514	50B40		0.38-0.42	0.4-0.6		0.75-1		0.04	0.05	0.1	B >=0.0005; bal Fe				
COPANT 514	50B40	Tube	0.38-0.42	0.4-0.6		0.75-1		0.04	0.05	0.1	B <=0.003; bal Fe				

Spain

Specification	Designation	Notes	C	Cr	Cu	Mn	Mo	P	S	Si	Other	UTS	YS	El	Hard
UNE 36012(75)	38Cr3	Q/T	0.34-0.41	0.5-0.8		0.6-0.9	0.15 max	0.035 max	0.035 max	0.15-0.4	bal Fe				
UNE 36034(85)	38CrB1DF	CHd; Cold ext	0.34-0.41	0.2-0.4		0.5-0.8	0.15 max	0.035 max	0.035 max	0.15-0.35	Al 0.02; B 0.0008-0.005; bal Fe				
UNE 36034(85)	38MnB5DF	CHd; Cold ext	0.34-0.40			1.1-1.4	0.15 max	0.035 max	0.035 max	0.15-0.35	Al 0.02; B 0.0008-0.005; bal Fe				
UNE 36034(85)	F.1296*	see 38MnB5DF	0.34-0.40			1.1-1.4	0.15 max	0.035 max	0.035 max	0.15-0.35	Al 0.02; B 0.0008-0.005; bal Fe				
UNE 36034(85)	F.1297*	see 38CrB1DF	0.34-0.41	0.2-0.4		0.5-0.8	0.15 max	0.035 max	0.035 max	0.15-0.35	Al 0.02; B 0.0008-0.005; bal Fe				

USA

Specification	Designation	Notes	C	Cr	Cu	Mn	Mo	P	S	Si	Other	UTS	YS	El	Hard
	AISI 50B40	Smls mech Tub; Boron	0.38-0.42	0.40-0.60		0.75-1.00		0.040 max	0.040 max	0.15-0.35	B >=0.0005; bal Fe				
	AISI 50B40H	Smls mech tub, Hard; Boron	0.37-0.44	0.30-0.7		0.65-1.00		0.035 max	0.040 max	0.15-0.30	B 0.0005-0.003; bal Fe				
	UNS G50401		0.38-0.42	0.40-0.60		0.75-1.00		0.035 max	0.040 max	0.15-0.35	B 0.0005-0.003; bal Fe				
	UNS H50401		0.37-0.44	0.30-0.70		0.65-1.10				0.15-0.35	B 0.0005-0.003; bal Fe				
ASTM A304(96)	50B40H	Bar, hard bands spec	0.37-0.44	0.30-0.70	0.35 max	0.65-1.10	0.06 max	0.035 max	0.040 max	0.15-0.35	B >=0.0005; Cu Ni Cr Mo trace allowed; bal Fe				
ASTM A331(95)	50B40H	Bar	0.37-0.44	0.30-0.70	0.35 max	0.65-1.10	0.06 max	0.025 max	0.025 max	0.15-0.35	B >=0.0005; bal Fe				
ASTM A519(96)	50B40	Smls mech tub	0.38-0.42	0.40-0.60		0.75-1.00		0.040 max	0.040 max	0.15-0.35	B >=0.0005; bal Fe				
ASTM A914/914M(92)	50B40RH	Bar restricted end Q hard	0.38-0.43	0.40-0.60	0.35 max	0.75-1.00	0.06 max	0.035 max	0.040 max	0.15-0.35	B 0.0005-0.003; Ni <=0.25; Electric furnace P, S<=0.025; bal Fe				
SAE J1268(95)	50B40H	See std	0.37-0.44	0.30-0.70	0.35 max	0.65-1.10	0.06 max	0.030 max	0.040 max	0.15-0.35	B 0.0005-0.003; Ni <=0.25; Cu, Mo, Ni not spec'd but acpt; bal Fe				
SAE J1868(93)	50B40RH	Restrict hard range	0.38-0.43	0.40-0.60	0.035 max	0.75-1.00	0.06 max	0.025 max	0.040 max	0.15-0.35	B 0.0005-0.003; Ni <=0.25; Cu, Mo, Ni not spec'd but acpt; bal Fe				5 HRC max

UNS numbers and US grades are provided as a means of cross referencing chemically similar alloys. Exchangability is only possible after independent examination of specifications. Tensile properties are minimum or typical as specified. UTS and YS as MPa. El as %. See Appendix for list of abbreviations used in Notes. * indicates obsolete material.

Specification	Designation	Notes	C	Cr	Cu	Mn	Mo	P	S	Si	Other	UTS	YS	El	Hard

Alloy Steel, Chromium, 50B40/50B40H (Continued from previous page)

Yugoslavia

Specification	Designation	Notes	C	Cr	Cu	Mn	Mo	P	S	Si	Other	UTS	YS	El	Hard
	C.4132	Q/T	0.35-0.42	0.4-0.6		0.5-0.8	0.15 max	0.035 max	0.03 max	0.4 max	bal Fe				

Alloy Steel, Chromium, 50B44/50B44H

Belgium

Specification	Designation	Notes	C	Cr	Cu	Mn	Mo	P	S	Si	Other	UTS	YS	El	Hard
NBN 253-06	45Cr2		0.42-0.48	0.4-0.6		0.5-0.8		0.035	0.035	0.15-0.4	bal Fe				

China

Specification	Designation	Notes	C	Cr	Cu	Mn	Mo	P	S	Si	Other	UTS	YS	El	Hard
GB 3077(88)	45MnB	Bar HR Q/T 25mm diam	0.42-0.49	0.030 max	0.030 max	1.10-1.40		0.035 max	0.035 max	0.17-0.37	B 0.0005-0.0035; bal Fe	1030	835	9	
GB 5216(85)	45MnBH	Bar Rod HR/Frg Ann	0.42-0.49	0.35 max	0.030 max	1.10-1.40		0.035 max	0.035 max	0.17-0.37	B 0.0005-0.0035; bal Fe				217 max HB
GB 8162(87)	45MnB	Smls tube HR/CD Q/T	0.42-0.49	0.030 max		1.10-1.40		0.035 max	0.035 max	0.17-0.37	B 0.0005-0.0035; bal Fe	1030	835	9	
GB/T 3078(94)	45MnB	Bar CD Q/T 25mm diam	0.42-0.49	0.030 max	0.030 max	1.10-1.40		0.035 max	0.035 max	0.17-0.37	B 0.0005-0.0035; bal Fe	1030	835	9	
YB/T 5052(93)	45MnB	Tube HR Norm	0.42-0.49			1.10-1.40		0.040 max	0.040 max	0.20-0.40	B 0.0001-0.0035; bal Fe	640	395	14	

France

Specification	Designation	Notes	C	Cr	Cu	Mn	Mo	P	S	Si	Other	UTS	YS	El	Hard
AFNOR	45C2		0.4-0.5	0.4-0.6		0.5-0.8		0.04	0.035	0.35	bal Fe				

Japan

Specification	Designation	Notes	C	Cr	Cu	Mn	Mo	P	S	Si	Other	UTS	YS	El	Hard
JIS G4052(79)	SMnC3H*	Obs; see SMnC 443H	0.39-0.46	0.35-0.7		1.3-1.7		0.03	0.03	0.15-0.35	bal Fe				
JIS G4052(79)	SMnC443H	Struct Hard	0.39-0.46	0.35-0.70	0.30 max	1.30-1.70		0.030 max	0.030 max	0.15-0.35	Ni <=0.25; bal Fe				
JIS G4106(79)	SMnC443	HR Frg	0.40-0.46	0.35-0.70	0.30 max	1.35-1.65		0.030 max	0.030 max	0.15-0.35	Ni <=0.25; bal Fe				

Mexico

Specification	Designation	Notes	C	Cr	Cu	Mn	Mo	P	S	Si	Other	UTS	YS	El	Hard
NMX-B-268(68)	50B44H	Hard	0.42-0.49	0.30-0.70		0.65-1.10				0.20-0.35	B >=0.0005; bal Fe				
NMX-B-300(91)	50B44	Bar	0.43-0.48	0.40-0.60		0.75-1.00		0.035 max	0.040 max	0.15-0.35	B >=0.0005; bal Fe				
NMX-B-300(91)	50B44H	Bar	0.42-0.49	0.30-0.70		0.65-1.10				0.15-0.35	B >=0.0005; bal Fe				

Pan America

Specification	Designation	Notes	C	Cr	Cu	Mn	Mo	P	S	Si	Other	UTS	YS	El	Hard
COPANT 514	50B44	Tube	0.43-0.48	0.4-0.6		0.75-1		0.04	0.05	0.1	bal Fe				

USA

Specification	Designation	Notes	C	Cr	Cu	Mn	Mo	P	S	Si	Other	UTS	YS	El	Hard
	AISI 50B44	Smls mech Tub; Boron	0.43-0.48	0.40-0.60		0.75-1.00		0.040 max	0.040 max	0.15-0.35	B >=0.0005; bal Fe				
	AISI 50B44H	Smls mech tub, Hard; Boron	0.42-0.49	0.30-0.70		0.65-1.10		0.035 max	0.040 max	0.15-0.30	B 0.0005-0.005; bal Fe				
	UNS G50441		0.43-0.48	0.40-0.60		0.75-1.00		0.035 max	0.040 max	0.15-0.35	B 0.0005-0.003; bal Fe				
	UNS H50441		0.42-0.49	0.30-0.70		0.65-1.10		0.035 max	0.04 max	0.15-0.35	B 0.0005-0.003; bal Fe				
ASTM A29/A29M(93)	50B44	Bar	0.43-0.48	0.20-0.60		0.75-1		0.035 max	0.04 max	0.15-0.35	B 0.0005-0.003; bal Fe				
ASTM A304(96)	50B44H	Bar, hard bands spec	0.42-0.49	0.30-0.70	0.35 max	0.65-1.10	0.06 max	0.035 max	0.040 max	0.15-0.35	B >=0.0005; Ni <=0..25; Cu Ni Cr Mo trace allowed; bal Fe				
ASTM A322(96)	50B44	Bar	0.43-0.48	0.20-0.60		0.75-1.00		0.035 max	0.040 max	0.15-0.35	B 0.0005-0.005; bal Fe				
ASTM A331(95)	50B44	Bar	0.43-0.48	0.20-0.60		0.75-1.00		0.035 max	0.040 max	0.15-0.35	bal Fe				
ASTM A331(95)	50B44H	Bar	0.42-0.49	0.30-0.70	0.35 max	0.65-1.10	0.06 max	0.025 max	0.025 max	0.15-0.35	B >=0.0005; Ni <=0.25; bal Fe				
ASTM A519(96)	50B44	Smls mech tub	0.43-0.48	0.40-0.60		0.75-1.00		0.040 max	0.040 max	0.15-0.35	B >=0.0005; bal Fe				
ASTM A752(93)	50B44	Rod Wir	0.43-0.48	0.40-0.60		0.75-1.00		0.035 max	0.040 max	0.15-0.30	B >=0.0005; bal Fe	650			
SAE 770(84)	50B44*	Obs; see J1397(92)	0.43-0.48	0.4-0.6		0.75-1		0.04 max	0.04 max	0.15-0.3	B 0.001-0.005; bal Fe				
SAE J1268(95)	50B44H	Bar Tub; See std	0.42-0.49	0.30-0.70	0.35 max	0.65-1.10	0.06 max	0.030 max	0.040 max	0.15-0.35	B 0.0005-0.003; Ni <=0.25; Cu, Mo, Ni not spec'd but acpt; bal Fe				

Specification	Designation	Notes	C	Cr	Cu	Mn	Mo	P	S	Si	Other	UTS	YS	El	Hard

Alloy Steel, Chromium, 50B44/50B44H (Continued from previous page)

USA

Specification	Designation	Notes	C	Cr	Cu	Mn	Mo	P	S	Si	Other	UTS	YS	El	Hard
SAE J407	50B44H*	Obs; see J1268									bal Fe				

Alloy Steel, Chromium, 50B46/50B46H

China

Specification	Designation	Notes	C	Cr	Cu	Mn	Mo	P	S	Si	Other	UTS	YS	El	Hard
GB 3077(88)	45MnB	Bar HR Q/T 25mm diam	0.42-0.49	0.30 max	0.30 max	1.10-1.40		0.035 max	0.035 max	0.17-0.37	B 0.0005-0.0035; bal Fe	1030	835	9	
GB 5216(85)	45MnBH	Bar Rod HR/Frg Ann	0.42-0.49	0.30 max	0.30 max	1.10-1.40		0.035 max	0.035 max	0.17-0.37	B 0.0005-0.0035; bal Fe				217 max HB
GB 8162(87)	45MnB	Smls tube HR/CD Q/T	0.42-0.49	0.30 max		1.10-1.40		0.035 max	0.035 max	0.17-0.37	B 0.0005-0.0035; bal Fe	1030	835	9	
GB/T 3078(94)	45MnB	Bar CD Q/T 25mm diam	0.42-0.49	0.30 max	0.30 max	1.10-1.40		0.035 max	0.035 max	0.17-0.37	B 0.0005-0.0035; bal Fe	1030	835	9	
YB/T 5052(93)	45MnB	Tube HR Norm	0.42-0.49	0.30 max		1.10-1.40		0.040 max	0.040 max	0.20-0.40	B 0.0001-0.0035; bal Fe	640	395	14	

Europe

Specification	Designation	Notes	C	Cr	Cu	Mn	Mo	P	S	Si	Other	UTS	YS	El	Hard
EN 10083/3(95)	1.7182	Boron, Q/T	0.24-0.30	0.30-0.60		1.10-1.40		0.035 max	0.040 max	0.40 max	B 0.0008-0.0050; bal Fe				
EN 10083/3(95)	1.7185	Boron, Q/T	0.30-0.36	0.30-0.60		1.20-1.50		0.035 max	0.040 max	0.40 max	B 0.0008-0.0050; bal Fe				
EN 10083/3(95)	27MnCrB5-2	Boron, Q/T	0.24-0.30	0.30-0.60		1.10-1.40		0.035 max	0.040 max	0.40 max	B 0.0008-0.0050; bal Fe				
EN 10083/3(95)	33MnCrB5-2	Boron, Q/T	0.30-0.36	0.30-0.60		1.20-1.50		0.035 max	0.040 max	0.40 max	B 0.0008-0.0050; bal Fe				

Mexico

Specification	Designation	Notes	C	Cr	Cu	Mn	Mo	P	S	Si	Other	UTS	YS	El	Hard
NMX-B-268(68)	50B46H	Hard	0.43-0.50	0.13-0.43		0.65-1.10				0.20-0.35	B >=0.0005; bal Fe				
NMX-B-300(91)	50B46	Bar	0.44-0.49	0.20-0.35		0.75-1.00		0.035 max	0.040 max	0.15-0.35	B >=0.0005; bal Fe				
NMX-B-300(91)	50B46H	Bar	0.43-0.50	0.13-0.43		0.65-1.10				0.15-0.35	B >=0.0005; bal Fe				

Pan America

Specification	Designation	Notes	C	Cr	Cu	Mn	Mo	P	S	Si	Other	UTS	YS	El	Hard
COPANT 514	50B46	Tube	0.43-0.5	0.2-0.35		0.75-1		0.04	0.05	0.1	B <=0.003; bal Fe				

USA

Specification	Designation	Notes	C	Cr	Cu	Mn	Mo	P	S	Si	Other	UTS	YS	El	Hard
	AISI 50B46	Smls mech Tub; Boron	0.44-0.49	0.20-0.35		0.75-1.00		0.040 max	0.040 max	0.15-0.35	B >=0.0005; bal Fe				
	AISI 50B46	Bar Blm Bil Slab	0.44-0.49	0.20-0.35		0.75-1.00		0.035 max	0.040 max	0.15-0.35	B 0.0005-0.003; bal Fe				
	AISI 50B46H	Smls mech tub, Hard; Boron	0.43-0.50	0.13-0.43		0.65-1.10		0.035 max	0.040 max	0.15-0.30	B 0.0005-0.003; bal Fe				
	AISI 50B46H	Bar Blm Bil Slab	0.43-0.50	0.13-0.43		0.65-1.10		0.035 max	0.040 max	0.15-0.35	B 0.0005-0.003; bal Fe				
	UNS G50461		0.44-0.49	0.20-0.35		0.75-1.00		0.035 max	0.040 max	0.15-0.35	B 0.0005-0.003; bal Fe				
	UNS H50461		0.43-0.50	0.13-0.43		0.65-1.10		0.035 max	0.040 max	0.15-0.35	B 0.0005-0.003; bal Fe				
ASTM A29/A29M(93)	50B46	Bar	0.44-0.49	0.20-0.35		0.75-1		0.035 max	0.04 max	0.15-0.35	bal Fe				
ASTM A304(96)	50B46H	Bar, hard bands spec	0.43-0.50	0.13-0.43	0.35 max	0.65-1.10	0.06 max	0.035 max	0.040 max	0.15-0.35	B >=0.0005; Ni <=0.25; Cu Ni Cr Mo trace allowed; bal Fe				
ASTM A322(96)	50B46	Bar	0.44-0.49	0.20-0.35		0.75-1.00		0.035 max	0.040 max	0.15-0.35	B 0.0005-0.005; bal Fe				
ASTM A331(95)	50B46	Bar	0.44-0.49	0.20-0.35		0.75-1.00		0.035 max	0.040 max	0.15-0.35	bal Fe				
ASTM A331(95)	50B46H	Bar	0.43-0.50	0.13-0.43	0.35 max	0.65-1.10	0.06 max	0.025 max	0.025 max	0.15-0.35	B >=0.0005; Ni <=0.25; bal Fe				
ASTM A519(96)	50B46	Smls mech tub	0.43-0.50	0.20-0.35		0.75-1.00		0.040 max	0.040 max	0.15-0.35	B >=0.0005; bal Fe				
ASTM A752(93)	50B46	Rod Wir	0.44-0.49	0.20-0.35		0.75-1.00		0.035 max	0.040 max	0.15-0.30	B >=0.0005; bal Fe	630			
SAE 770(84)	50B46*	Obs; see J1397(92)	0.44-0.49	0.2-0.35		0.75-1		0.04 max	0.04 max	0.15-0.3	B 0.001-0.005; bal Fe				
SAE J1268(95)	50B46H	Bar Tub; See std	0.43-0.50	0.13-0.43	0.35 max	0.65-1.10	0.06 max	0.030 max	0.040 max	0.15-0.35	B 0.0005-0.003; Ni <=0.25; Cu, Mo, Ni not spec'd but acpt; bal Fe				

Specification	Designation	Notes	C	Cr	Cu	Mn	Mo	P	S	Si	Other	UTS	YS	El	Hard

Alloy Steel, Chromium, 50B46/50B46H (Continued from previous page)

USA

| SAE J404(94) | 50B46 | Bil Blm Slab Bar HR CF | 0.44-0.49 | 0.20-0.35 | | 0.75-1.00 | | 0.030 max | 0.040 max | 0.15-0.35 | B 0.0005-0.003; bal Fe | | | | |

Alloy Steel, Chromium, 50B50/50B50H

France

| AFNOR | 55C2 | | 0.5-0.6 | 0.4-0.6 | | 0.6-0.9 | | 0.04 | 0.035 | 0.35 | bal Fe | | | | |

Germany

| DIN | 52MnCrB3 | | 0.48-0.55 | 0.40-0.60 | | 0.75-1.00 | | 0.035 max | 0.035 max | 0.15-0.35 | B >=0.0008; bal Fe | | | | |
| DIN | WNr 1.7138 | | 0.48-0.55 | 0.40-0.60 | | 0.75-1.00 | | 0.035 max | 0.035 max | 0.15-0.35 | B >=0.0008; bal Fe | | | | |

Mexico

NMX-B-268(68)	50B50H	Hard	0.47-0.54	0.30-0.70		0.65-1.10				0.20-0.35	B >=0.0005; bal Fe				
NMX-B-300(91)	50B50	Bar	0.48-0.53	0.40-0.60		0.75-1.00		0.035 max	0.040 max	0.20-0.35	B >=0.0005; bal Fe				
NMX-B-300(91)	50B50H	Bar	0.47-0.54	0.30-0.70		0.65-1.10				0.15-0.35	B >=0.0005; bal Fe				

Pan America

| COPANT 514 | 50B50 | Tube | 0.48-0.53 | 0.4-0.6 | | 0.74-1 | | 0.04 | 0.05 | 0.1 | B <=0.003; bal Fe | | | | |

USA

	AISI 50B50	Smls mech Tub; Boron	0.48-0.53	0.40-0.60		0.74-1.00		0.040 max	0.040 max	0.15-0.35	B >=0.0005; bal Fe				
	AISI 50B50H	Smls mech tub, Hard; Boron	0.47-0.65	0.35-0.70		0.65-1.10		0.035 max	0.040 max	0.15-0.30	B 0.0005-0.005; bal Fe				
	UNS G50501		0.48-0.53	0.40-0.60		0.75-1.00		0.035 max	0.040 max	0.15-0.35	B 0.0005-0.003; bal Fe				
	UNS H50501		0.47-0.54	0.30-0.70		0.65-1.10		0.035 max	0.040 max	0.15-0.35	B 0.0005-0.003; bal Fe				
ASTM A229/A229M(93)	50B50	Wir, spring	0.48-0.53	0.4-0.6		0.75-1		0.035 max	0.04 max	0.15-0.35	B 0.0005-0.005; bal Fe				
ASTM A29/A29M(93)	50B50	Bar	0.48-0.53	0.40-0.60		0.75-1.00		0.035 max	0.040 max	0.15-0.35	bal Fe				
ASTM A304(96)	50B50H	Bar, hard bands spec	0.47-0.54	0.30-0.70	0.35 max	0.65-1.10		0.035 max	0.040 max	0.15-0.35	B >=0.0005; Cu Ni Cr Mo trace allowed; bal Fe				
ASTM A322(96)	50B50	Bar	0.48-0.53	0.40-0.60		0.75-1.00		0.035 max	0.040 max	0.15-0.35	B 0.0005-0.005; bal Fe				
ASTM A331(95)	50B50	Bar	0.48-0.53	0.40-0.60		0.75-1.00		0.035 max	0.040 max	0.15-0.35	bal Fe				
ASTM A331(95)	50B50H	Bar	0.47-0.54	0.30-0.70	0.35 max	0.65-1.10	0.06 max	0.025 max	0.025 max	0.15-0.35	B >=0.0005; bal Fe				
ASTM A519(96)	50B50	Smls mech tub	0.48-0.53	0.40-0.60		0.74-1.00		0.040 max	0.040 max	0.15-0.35	B >=0.0005; bal Fe				
ASTM A752(93)	50B50	Rod Wir	0.48-0.53	0.40-0.60		0.75-1.00		0.035 max	0.040 max	0.15-0.30	B >=0.0005; bal Fe	670			
SAE 770(84)	50B50*	Obs; see J1397(92)	0.48-0.53	0.4-0.6		0.75-1		0.04 max	0.04 max	0.15-0.3	B 0.001-0.005; bal Fe				
SAE J1268(95)	50B50H	Bar Tub; See std	0.47-0.54	0.30-0.70	0.35 max	0.65-1.10	0.06 max	0.030 max	0.040 max	0.15-0.35	B 0.0005-0.003; Ni <=0.25; Cu, Mo, Ni not spec'd but acpt; bal Fe				

Alloy Steel, Chromium, 50B60/50B60H

Mexico

NMX-B-268(68)	50B60H	Hard	0.55-0.65	0.30-0.70		0.65-1.10				0.20-0.35	B >=0.0005; bal Fe				
NMX-B-268(68)	51B60H	Hard	0.55-0.65	0.60-1.00		0.65-1.10				0.20-0.35	B >=0.0005; bal Fe				
NMX-B-300(91)	50B60	Bar	0.56-0.64	0.40-0.60		0.75-1.00		0.035 max	0.040 max	0.20-0.35	B >=0.0005; bal Fe				
NMX-B-300(91)	50B60H	Bar	0.55-0.65	0.30-0.70		0.65-1.10				0.15-0.35	B >=0.0005; bal Fe				

Pan America

| COPANT 514 | 50B60 | Tube | 0.55-0.65 | 0.4-0.6 | | 0.75-1 | | 0.04 | 0.05 | 0.1 | B <=0.003; bal Fe | | | | |

UNS numbers and US grades are provided as a means of cross referencing chemically similar alloys. Exchangability is only possible after independent examination of specifications. Tensile properties are minimum or typical as specified. UTS and YS as MPa. El as %. See Appendix for list of abbreviations used in Notes. * indicates obsolete material.

Specification	Designation	Notes	C	Cr	Cu	Mn	Mo	P	S	Si	Other	UTS	YS	El	Hard

Alloy Steel, Chromium, 50B60/50B60H (Continued from previous page)

USA

Specification	Designation	Notes	C	Cr	Cu	Mn	Mo	P	S	Si	Other	UTS	YS	El	Hard
	AISI 50B60	Smls mech Tub; Boron	0.56-0.64	0.40-0.60		0.75-1.00		0.040 max	0.040 max	0.15-0.35	B >=0.0005; bal Fe				
	AISI 50B60	Bar Blm Bil Slab	0.55-0.65	0.40-0.60		0.75-1.00		0.035 max	0.040 max	0.15-0.35	B 0.0005-0.003; bal Fe				
	AISI 50B60H	Bar Blm Bil Slab	0.55-0.64	0.35-0.70		0.70-1.10		0.035 max	0.040 max	0.15-0.35	B 0.0005-0.003; bal Fe				
	AISI 50B60H	Smls mech tub, Hard; Boron	0.55-0.65	0.30-0.65		0.70-1.10		0.035 max	0.040 max	0.15-0.30	B 0.0005-0.005; bal Fe				
	UNS G50601		0.56-0.64	0.40-0.60		0.75-1.00		0.035 max	0.040 max	0.15-0.35	B 0.0005-0.003; bal Fe				
	UNS H50601		0.55-0.65	0.30-0.70		0.65-1.10		0.035 max	0.040 max	0.15-0.35	B 0.0005-0.003; bal Fe				
ASTM A29/A29M(93)	50B60	Bar	0.56-0.64	0.40-0.60		0.75-1		0.035 max	0.04 max	0.15-0.35	B 0.0005-0.005; bal Fe				
ASTM A304(96)	50B60H	Bar, hard bands spec	0.55-0.65	0.30-0.70	0.35 max	0.65-1.10	0.06 max	0.035 max	0.040 max	0.15-0.35	B >=0.0005; Cu Ni Cr Mo trace allowed; bal Fe				
ASTM A322(96)	50B60	Bar	0.56-0.64	0.40-0.60		0.75-1.00		0.035 max	0.040 max	0.15-0.35	B 0.0005-0.005; bal Fe				
ASTM A331(95)	50B60	Bar	0.56-0.64	0.40-0.60		0.75-1.00		0.035 max	0.040 max	0.15-0.35	bal Fe				
ASTM A331(95)	50B60H	Bar	0.55-0.65	0.30-0.70	0.35 max	0.65-1.10	0.06 max	0.025 max	0.025 max	0.15-0.35	B >=0.0005; bal Fe				
ASTM A519(96)	50B60	Smls mech tub	0.55-0.65	0.40-0.60		0.75-1.00		0.040 max	0.040 max	0.15-0.35	B >=0.0005; bal Fe				
ASTM A752(93)	50B60	Rod Wir	0.56-0.64	0.40-0.60		0.75-1.00		0.035 max	0.040 max	0.15-0.30	B >=0.0005; bal Fe	710			
SAE 770(84)	50B60*	Obs; see J1397(92)	0.56-0.64	0.4-0.6		0.75-1		0.04 max	0.04 max	0.15-0.3	B 0.001 0.005; bal Fe				
SAE J1268(95)	50B60H	See std	0.55-0.65	0.35-0.70	0.35 max	0.65-1.10	0.06 max	0.030 max	0.040 max	0.15-0.35	B 0.0005-0.003; Ni <=0.25; Cu, Mo, Ni not spec'd but acpt; bal Fe				

Alloy Steel, Chromium, 5115

Argentina

Specification	Designation	Notes	C	Cr	Cu	Mn	Mo	P	S	Si	Other	UTS	YS	El	Hard
IAS	IRAM 5115		0.13-0.18	0.70-0.90		0.70-0.90		0.035 max	0.040 max	0.20-0.35	bal Fe	510	390	32	143 HB

Bulgaria

Specification	Designation	Notes	C	Cr	Cu	Mn	Mo	P	S	Si	Other	UTS	YS	El	Hard
BDS 6354	15ChF	Struct	0.12-0.18	0.80-1.10	0.30 max	0.40-0.70	0.15 max	0.035 max	0.035 max	0.17-0.37	Ni <=0.30; Ti <=0.03; V 0.06-0.12; W <=0.2; bal Fe				
BDS 6354	15G	Struct	0.12-0.19	0.30 max	0.30 max	0.70-1.00	0.15 max	0.035 max	0.035 max	0.17-0.37	Ni <=0.30; Ti <=0.03; V <=0.05; W <=0.2; bal Fe				

China

Specification	Designation	Notes	C	Cr	Cu	Mn	Mo	P	S	Si	Other	UTS	YS	El	Hard
GB 11251(89)	15Cr	Plt HR/Ann	0.12-0.18	0.70-1.00		0.40-0.70		0.035 max	0.035 max	0.17-0.37	bal Fe	400-600		21	
GB 3077(88)	15Cr	Bar HR Q/T 15mm diam	0.12-0.18	0.70-1.00	0.30 max	0.40-0.70		0.035 max	0.035 max	0.17-0.37	bal Fe	735	490	11	
GB 3077(88)	15CrA	Bar HR Q/T 15mm diam	0.12-0.17	0.70-1.00	0.25 max	0.40-0.70		0.025 max	0.025 max	0.17-0.37	bal Fe	685	490	12	
GB 6478(86)	ML15Cr	Bar HR Q/T	0.12-0.18	0.70-1.00	0.20 max	0.40-0.70		0.035 max	0.035 max	0.30 max	bal Fe	685	490	10	
GB/T 3078(94)	15CrA	Bar CD Q/T 15mm diam	0.12-0.17	0.70-1.00	0.25 max	0.40-0.70		0.025 max	0.025 max	0.17-0.37	bal Fe	685	490	12	
GB/T 3079(93)	15CrA	Wir Q/T	0.12-0.17	0.70-1.00	0.25 max	0.30-0.60		0.025 max	0.025 max	0.17-0.37	bal Fe	590	390	15	
YB/T 5132(93)	15Cr	Sh HR/CR Ann	0.12-0.18	0.70-1.00		0.40-0.70		0.035 max	0.035 max	0.17-0.37	bal Fe	395-590		19	
YB/T 5132(93)	15CrA	Sh HR/CR Ann	0.12-0.17	0.70-1.00		0.40-0.70		0.025 max	0.025 max	0.17-0.37	bal Fe	395-590		19	

Czech Republic

Specification	Designation	Notes	C	Cr	Cu	Mn	Mo	P	S	Si	Other	UTS	YS	El	Hard
CSN 414120	14120	CH	0.12-0.18	0.7-1.0		0.4-0.7		0.035 max	0.035 max	0.17-0.37	bal Fe				

Europe

Specification	Designation	Notes	C	Cr	Cu	Mn	Mo	P	S	Si	Other	UTS	YS	El	Hard
EN 10084(98)	1.7131	CH, Ann	0.14-0.19	0.80-1.10		1.00-1.30		0.035 max	0.035 max	0.40 max	bal Fe				207 HB

UNS numbers and US grades are provided as a means of cross referencing chemically similar alloys. Exchangability is only possible after independent examination of specifications. Tensile properties are minimum or typical as specified. UTS and YS as MPa. El as %. See Appendix for list of abbreviations used in Notes. * indicates obsolete material.

Specification	Designation	Notes	C	Cr	Cu	Mn	Mo	P	S	Si	Other	UTS	YS	El	Hard

Alloy Steel, Chromium, 5115 (Continued from previous page)

Europe

Specification	Designation	Notes	C	Cr	Cu	Mn	Mo	P	S	Si	Other	UTS	YS	El	Hard
EN 10084(98)	1.7139	CH, Ann	0.14-0.19	0.80-1.10		1.00-1.30		0.035 max	0.020-0.040	0.40 max	bal Fe				207 HB
EN 10084(98)	16MnCr5	CH, Ann	0.14-0.19	0.80-1.10		1.00-1.30		0.035 max	0.035 max	0.40 max	bal Fe				207 HB
EN 10084(98)	16MnCrS5	CH, Ann	0.14-0.19	0.80-1.10		1.00-1.30		0.035 max	0.020-0.040	0.40 max	bal Fe				207 HB

France

Specification	Designation	Notes	C	Cr	Cu	Mn	Mo	P	S	Si	Other	UTS	YS	El	Hard
AFNOR NFA35551(86)	16MC5	40<t<=100mm	0.14-0.19	0.8-1.1	0.3 max	1-1.3	0.15 max	0.035 max	0.035 max	0.1-0.4	Ni <=0.4; Ti <=0.05; bal Fe	630-980	450	11	
AFNOR NFA35551(86)	16MC5	t<=16mm	0.14-0.19	0.8-1.1	0.3 max	1-1.3	0.15 max	0.035 max	0.035 max	0.1-0.4	Ni <=0.4; Ti <=0.05; bal Fe	980-1330	700	9	

Hungary

Specification	Designation	Notes	C	Cr	Cu	Mn	Mo	P	S	Si	Other	UTS	YS	El	Hard
MSZ 31(85)	BC3	CH; 63 mm; blank carburized	0.14-0.19	0.8-1.1		1-1.3		0.035 max	0.035 max	0.4 max	bal Fe	700-950	490	12L	
MSZ 31(85)	BC3	CH; 0-11mm; blank carburized	0.14-0.19	0.8-1.1		1-1.3		0.035 max	0.035 max	0.4 max	bal Fe	950-1300	740	9L	
MSZ 31(85)	BC3E	CH; 0-11mm; blank carburized	0.14-0.19	0.8-1.1		1-1.3		0.035 max	0.02-0.035	0.4 max	bal Fe	950-1300	740	9L	
MSZ 31(85)	BC3E	CH; 63 mm; blank carburized	0.14-0.19	0.8-1.1		1-1.3		0.035 max	0.02-0.035	0.4 max	bal Fe	700-950	490	12L	
MSZ 6251(87)	BC3Z	CF; CH; 0-36mm; HR; HR/soft ann; drawn/soft ann; drawn/bright ann; soft ann/ground	0.14-0.19	0.8-1.1		1-1.3		0.035 max	0.035 max	0.4 max	bal Fe	0-560			
MSZ 6251(87)	BC3Z	CF; CH; 0-36mm; drawn, half hard	0.14-0.19	0.8-1.1		1-1.3		0.035 max	0.035 max	0.4 max	bal Fe	0-590			

India

Specification	Designation	Notes	C	Cr	Cu	Mn	Mo	P	S	Si	Other	UTS	YS	El	Hard
IS 1570/4(88)	15Cr3		0.12-0.18	0.5-0.8		0.4-0.6	0.15 max	0.07 max	0.06 max	0.15-0.35	Ni <=0.4; bal Fe	590		13	
IS 1570/4(88)	16Mn5Cr4		0.14-0.19	0.8-1.1		1-1.3	0.15 max	0.07 max	0.06 max	0.1-0.35	Ni <=0.4; bal Fe	790		10	

International

Specification	Designation	Notes	C	Cr	Cu	Mn	Mo	P	S	Si	Other	UTS	YS	El	Hard
ISO 683-11(87)	16MnCr5	CH, 16mm test	0.13-0.19	0.80-1.10	0.3 max	1.00-1.30		0.035 max	0.035 max	0.15-0.40	Al <=0.1; Co <=0.1; bal Fe	880-1230	600		
ISO 683-11(87)	16MnCrS5	CH, 16mm test	0.13-0.19	0.80-1.10	0.3 max	1.00-1.30		0.035 max	0.020-0.040	0.15-0.40	Al <=0.1; Co <=0.1; bal Fe	880-1230	600		

Italy

Specification	Designation	Notes	C	Cr	Cu	Mn	Mo	P	S	Si	Other	UTS	YS	El	Hard
UNI 8550(84)	16MnCr5	CH	0.13-0.19	0.8-1.1		1-1.3		0.035 max	0.035 max	0.15-0.4	bal Fe				
UNI 8788(85)	16MnCr5	CH	0.13-0.19	0.8-1.1		1-1.3		0.035 max	0.035 max	0.15-0.4	P+S<=0.06; bal Fe				

Mexico

Specification	Designation	Notes	C	Cr	Cu	Mn	Mo	P	S	Si	Other	UTS	YS	El	Hard
NMX-B-300(91)	5115	Bar	0.13-0.18	0.70-0.90		0.70-0.90		0.035 max	0.040 max	0.15-0.35	bal Fe				

Poland

Specification	Designation	Notes	C	Cr	Cu	Mn	Mo	P	S	Si	Other	UTS	YS	El	Hard
PNH84030/02	16HG	CH	0.14-0.19	0.8-1.1		1.-1.3		0.035 max	0.035 max	0.17-0.37	Ni <=0.3; bal Fe				

Romania

Specification	Designation	Notes	C	Cr	Cu	Mn	Mo	P	S	Si	Other	UTS	YS	El	Hard
STAS 791(88)	18MnCr11	Struct/const; CH	0.15-0.21	0.9-1.2		0.9-1.2		0.035 max	0.035 max	0.17-0.37	Pb <=0.15; Ti <=0.02; bal Fe				
STAS 791(88)	18MnCr11S	Struct/const; CH	0.15-0.21	0.9-1.2		0.9-1.2		0.035 max	0.02-0.04	0.17-0.37	Pb <=0.15; Ti <=0.02; bal Fe				
STAS 791(88)	18MnCr11X	Struct/const; CH	0.15-0.21	0.9-1.2		0.9-1.2		0.025 max	0.025 max	0.17-0.37	Pb <=0.15; Ti <=0.02; bal Fe				
STAS 791(88)	18MnCr11XS	Struct/const; CH	0.15-0.21	0.9-1.2		0.9-1.2		0.025 max	0.02-0.035	0.17-0.37	Pb <=0.15; Ti <=0.02; bal Fe				

Russia

Specification	Designation	Notes	C	Cr	Cu	Mn	Mo	P	S	Si	Other	UTS	YS	El	Hard
GOST 1414(75)	AS12ChN	Free-cutting	0.09-0.15	0.4-0.7	0.30 max	0.3-0.6		0.035 max	0.035 max	0.17-0.37	Ni 0.5-0.8; Pb 0.15-0.3; P+S<=0.06; bal Fe				
GOST 4543(71)	18ChG	CH	0.15-0.21	0.9-1.2	0.3 max	0.9-1.2	0.15 max	0.035 max	0.035 max	0.17-0.37	Al <=0.1; Ni <=0.3; Ti <=0.05; bal Fe				

Spain

Specification	Designation	Notes	C	Cr	Cu	Mn	Mo	P	S	Si	Other	UTS	YS	El	Hard
UNE	F.155	CH	0.12-0.15	1.0-1.3		0.3-0.6	0.15-0.25	0.04 max	0.04 max	0.1-0.35	bal Fe				

UNS numbers and US grades are provided as a means of cross referencing chemically similar alloys. Exchangability is only possible after independent examination of specifications. Tensile properties are minimum or typical as specified. UTS and YS are MPa. El as %. See Appendix for list of abbreviations used in Notes. * indicates obsolete material.

Specification	Designation	Notes	C	Cr	Cu	Mn	Mo	P	S	Si	Other	UTS	YS	El	Hard

Alloy Steel, Chromium, 5115 (Continued from previous page)

Spain

Specification	Designation	Notes	C	Cr	Cu	Mn	Mo	P	S	Si	Other	UTS	YS	El	Hard
UNE 36013(76)	16MnCr5	CH	0.13-0.19	0.8-1.1		1-1.3	0.15 max	0.035 max	0.035 max	0.15-0.4	bal Fe				
UNE 36013(76)	F.1516*	see 16MnCr5	0.13-0.19	0.8-1.1		1-1.3	0.15 max	0.035 max	0.035 max	0.15-0.4	bal Fe				

Sweden

Specification	Designation	Notes	C	Cr	Cu	Mn	Mo	P	S	Si	Other	UTS	YS	El	Hard
SS 142173	2173	Gen Struct; FF; Non-rimming	0.18 max		0.4 max	1.4 max		0.05 max	0.05 max	0.5 max	N=0.-0.009/0.-0.012; C+Mn/6=0.-0.32; bal Fe				

UK

Specification	Designation	Notes	C	Cr	Cu	Mn	Mo	P	S	Si	Other	UTS	YS	El	Hard
BS 1503(89)	620-440	Frg press ves; t<=999mm; N/T Q/T	0.18 max	0.85-1.15	0.30 max	0.40-0.70	0.45-0.65	0.030 max	0.025 max	0.15-0.40	Al <=0.020; Ni <=0.40; W <=0.1; bal Fe	440-590	275	19	
BS 970/1(96)	527H17	Blm Bil Slab Bar Rod Frg CH	0.14-0.20	0.60-0.90		0.70-1.00				0.10-0.35	bal Fe				217 HB max
BS 970/1(96)	527M17	Wrought CH	0.14-0.20	0.60-0.90		0.70-0.90		0.035 max	0.040 max	0.10-0.35	bal Fe				
BS 970/1(96)	590A15	Blm Bil Slab Bar Rod Frg	0.13-0.18	0.90-1.20		0.90-1.20		0.035 max	0.040 max	0.10-0.35	bal Fe				

USA

Specification	Designation	Notes	C	Cr	Cu	Mn	Mo	P	S	Si	Other	UTS	YS	El	Hard
	AISI 5115	Smls mech tub	0.13-0.18	0.70-0.90		0.70-0.90		0.040 max	0.040 max	0.15-0.35	bal Fe				
	UNS G51150		0.13-0.18	0.70-0.90		0.70-0.90		0.035 max	0.040 max	0.15-0.35	bal Fe				
ASTM A29/A29M(93)	5115	Bar	0.13-0.18	0.70-0.90		0.70-0.90		0.035 max	0.040 max	0.15-0.35	bal Fe				
ASTM A519(96)	5115	Smls mech tub	0.13-0.18	0.70-0.90		0.70-0.90		0.040 max	0.040 max	0.15-0.35	bal Fe				

Yugoslavia

Specification	Designation	Notes	C	Cr	Cu	Mn	Mo	P	S	Si	Other	UTS	YS	El	Hard
	C.4320	CH	0.14-0.19	0.8-1.1		1.0-1.3	0.15 max	0.035 max	0.035 max	0.15-0.4	bal Fe				

Alloy Steel, Chromium, 5117

Belgium

Specification	Designation	Notes	C	Cr	Cu	Mn	Mo	P	S	Si	Other	UTS	YS	El	Hard
NBN 253-03	16MnCr5		0.13-0.19	0.8-1.1		1-1.3		0.035	0.035	0.15-0.4	bal Fe				

Bulgaria

Specification	Designation	Notes	C	Cr	Cu	Mn	Mo	P	S	Si	Other	UTS	YS	El	Hard
BDS 5084	Sv-18ChGS		0.15-0.22	0.80-1.10		0.80-1.10		0.03	0.025	0.70-1.20	Ni 0.3; bal Fe				
BDS 5084	Sw-18ChGS		0.15-0.22	0.80-1.10		0.80-1.10		0.03	0.025	0.70-1.20	Ni 0.3; bal Fe				
BDS 6354	16ChG		0.14-0.19	0.80-1.10	0.3	1.00-1.30		0.035	0.035	0.17-0.37	bal Fe				

China

Specification	Designation	Notes	C	Cr	Cu	Mn	Mo	P	S	Si	Other	UTS	YS	El	Hard
GB 11251(89)	15Cr	Plt HR Ann	0.12-0.18	0.70-1.00		0.40-0.70		0.035 max	0.035 max	0.17-0.37	bal Fe	400-600		21	
GB 3077(88)	15Cr	Bar HR Q/T 15mm diam	0.12-0.18	0.70-1.00	0.30 max	0.40-0.70		0.035 max	0.035 max	0.17-0.37	bal Fe	735	490	11	
GB 6478(86)	ML15Cr	Bar HR Q/T	0.12-0.18	0.70-1.00	0.20 max	0.40-0.70		0.035 max	0.035 max	0.30 max	bal Fe	685	490	10	
YB/T 5132(93)	15Cr	SH HR/CR Ann	0.12-0.18	0.70-1.00		0.40-0.70		0.035 max	0.035 max	0.17-0.37	bal Fe	395-590		19	

Czech Republic

Specification	Designation	Notes	C	Cr	Cu	Mn	Mo	P	S	Si	Other	UTS	YS	El	Hard
CSN 414220	14220	CH	0.14-0.19	0.8-1.1		1.1-1.4		0.035 max	0.035 max	0.17-0.37	bal Fe				

Finland

Specification	Designation	Notes	C	Cr	Cu	Mn	Mo	P	S	Si	Other	UTS	YS	El	Hard
SFS 508(72)	SFS508	CH	0.14-0.19	0.8-1.1		1.0-1.3		0.035 max	0.05 max	0.15-0.4	bal Fe				

France

Specification	Designation	Notes	C	Cr	Cu	Mn	Mo	P	S	Si	Other	UTS	YS	El	Hard
AFNOR	18Cr4		0.16-0.21	0.85-1.15		0.6-0.8		0.04	0.035	0.1-0.4	bal Fe				
AFNOR NFA35551	16MC5		0.14-0.19	0.8-1.1		1-1.3		0.035	0.035	0.1-0.4	bal Fe				
AFNOR NFA36102(93)	15C2*	HR; Strp; CR	0.12-0.17	0.4-0.7		0.4-0.6	0.10 max	0.025 max	0.02 max	0.1-0.4	Ni <=0.10; bal Fe				
AFNOR NFA36102(93)	15Cr2RR	HR; Strp; CR	0.12-0.17	0.4-0.7		0.4-0.6	0.10 max	0.025 max	0.02 max	0.1-0.4	Ni <=0.10; bal Fe				

UNS numbers and US grades are provided as a means of cross referencing chemically similar alloys. Exchangability is only possible after independent examination of specifications. Tensile properties are minimum or typical as specified. UTS and YS as MPa. El as %. See Appendix for list of abbreviations used in Notes. * indicates obsolete material.

Specification	Designation	Notes	C	Cr	Cu	Mn	Mo	P	S	Si	Other	UTS	YS	El	Hard

Alloy Steel, Chromium, 5117 (Continued from previous page)

Germany

Specification	Designation	Notes	C	Cr	Cu	Mn	Mo	P	S	Si	Other	UTS	YS	El	Hard
DIN	16MnCrPb5	CH, 30mm	0.14-0.19	0.80-1.10		1.00-1.30		0.035 max	0.035 max	0.35 max	Pb 0.20-0.35; bal Fe	780-1080	590	10	
DIN	18MnCrB5		0.16-0.20	0.90-1.20		1.00-1.30		0.035 max	0.015-0.035	0.15-0.35	B 0.0008-0.0050; bal Fe				
DIN	WNr 1.7142		0.14-0.19	0.80-1.10		1.00-1.30		0.035 max	0.035 max	0.15-0.35	Pb 0.20-0.35; bal Fe				
DIN	WNr 1.7168		0.16-0.20	0.90-1.20		1.00-1.30		0.035 max	0.015-0.035	0.15-0.35	B 0.0008-0.0050; bal Fe				
DIN 1652(90)	16MnCr5	CH, 30mm	0.14-0.19	0.80-1.10		1.00-1.30		0.035 max	0.035 max	0.40 max	bal Fe	780-1080	590	10	
DIN 1652(90)	16MnCrS5		0.14-0.19	0.80-1.10		1.00-1.30		0.035 max	0.020-0.035	0.15-0.40	bal Fe				
DIN 1652(90)	17Cr3	CH, 30mm	0.14-0.20	0.60-0.90		0.40-0.70		0.035 max	0.035 max	0.40 max	bal Fe	700-900	450	11	
DIN 1652(90)	WNr 1.7016	CH, 30mm	0.14-0.20	0.60-0.90		0.40-0.70		0.035 max	0.035 max	0.40 max	bal Fe	700-900	450	11	
DIN 1652(90)	WNr 1.7131	CH, 30mm	0.14-0.19	0.80-1.10		1.00-1.30		0.035 max	0.035 max	0.40 max	bal Fe	780-1080	590	10	
DIN 1652(90)	WNr 1.7139	CH, 30mm	0.14-0.19	0.80-1.10		1.00-1.30		0.035 max	0.020-0.035	0.40 max	bal Fe	780-1080	590	10	
DIN 17230(80)	17MnCr5	Bearing	0.14-0.19	0.80-1.10	0.30 max	1.00-1.30		0.035 max	0.035 max	0.40 max	bal Fe				
DIN 17230(80)	WNr 1.3521	Bearing	0.14-0.19	0.80-1.10	0.30 max	1.00-1.30		0.035 max	0.035 max	0.40 max	bal Fe				

India

Specification	Designation	Notes	C	Cr	Cu	Mn	Mo	P	S	Si	Other	UTS	YS	El	Hard
IS 1570	17MnCr.95		0.14-0.19	0.8-1.1		1-1.3		0.05	0.05	0.1-0.35	bal Fe				
IS 4367	17MnCr.95		0.14-0.19	0.8-1.1		1-1.3		0.05	0.05	0.1-0.35	bal Fe				

Poland

Specification	Designation	Notes	C	Cr	Cu	Mn	Mo	P	S	Si	Other	UTS	YS	El	Hard
PNH84030	16HG	CH	0.14-0.19	0.8-1.1	0.3	1-1.3		0.035	0.035	0.17-0.37	Ni 0.3; bal Fe				

Romania

Specification	Designation	Notes	C	Cr	Cu	Mn	Mo	P	S	Si	Other	UTS	YS	El	Hard
STAS 791	18MC10		0.15-0.22	0.9-1.2	0.3	0.9-1.2		0.035	0.035	0.17-0.37	Ni 0.3; bal Fe				
STAS 791	18MnCr10		0.15-0.22	0.9-1.2	0.3	0.9-1.2		0.035	0.035	0.17-0.37	Ni 0.3; bal Fe				
STAS 791(88)	15Cr09	Struct/const; CH	0.12-0.18	0.7-1		0.4-0.7		0.035 max	0.035 max	0.17-0.37	Pb <=0.15; Ti <=0.02; bal Fe				
STAS 791(88)	15Cr09S	Struct/const; CH	0.12-0.18	0.7-1		0.4-0.7		0.035 max	0.02-0.04	0.17-0.37	Pb <=0.15; Ti <=0.02; bal Fe				
STAS 791(88)	15Cr09X	Struct/const; CH	0.12-0.18	0.7-1		0.4-0.7		0.025 max	0.025 max	0.17-0.37	Pb <=0.15; Ti <=0.02; bal Fe				
STAS 791(88)	15Cr09XS	Struct/const; CH	0.12-0.18	0.7-1		0.4-0.7		0.025 max	0.02-0.035	0.17-0.37	Pb <=0.15; Ti <=0.02; bal Fe				

Russia

Specification	Designation	Notes	C	Cr	Cu	Mn	Mo	P	S	Si	Other	UTS	YS	El	Hard
GOST 4543	18ChG		0.15-0.21	0.9-1.2	0.3	0.9-1.2		0.035	0.035	0.17-0.37	Ni 0.3; bal Fe				

Spain

Specification	Designation	Notes	C	Cr	Cu	Mn	Mo	P	S	Si	Other	UTS	YS	El	Hard
UNE 36013(76)	16MnCr5-1	CH	0.13-0.19	0.8-1.1		1-1.3	0.15 max	0.035 max	0.02-0.035	0.15-0.4	bal Fe				
UNE 36013(76)	F.1519*	see 16MnCr5-1	0.13-0.19	0.8-1.1		1-1.3	0.15 max	0.035 max	0.02-0.035	0.15-0.4	bal Fe				
UNE 36027(80)	16MnCr5	Ball & roller bearing	0.13-0.19	0.8-1.1		1.0-1.3	0.15 max	0.035 max	0.035 max	0.15-0.4	bal Fe				
UNE 36027(80)	F.1517*	see 16MnCr5	0.13-0.19	0.8-1.1		1.0-1.3	0.15 max	0.035 max	0.035 max	0.15-0.4	bal Fe				

UK

Specification	Designation	Notes	C	Cr	Cu	Mn	Mo	P	S	Si	Other	UTS	YS	El	Hard
BS 970/1(83)	590H17	Blm Bil Slab Bar Rod Frg CH	0.14-0.20	0.80-1.10		1.00-1.30					bal Fe	930		10	217 HB max
BS 970/1(83)	590M17	Blm Bil Slab Bar Rod Frg CH	0.14-0.20	0.80-1.10		1.00-1.30	0.15 max	0.035 max	0.04 max	0.10-0.35	Ni <=0.4; bal Fe	930		10	0-217 HB

USA

Specification	Designation	Notes	C	Cr	Cu	Mn	Mo	P	S	Si	Other	UTS	YS	El	Hard
	AISI 5117		0.15-0.2	0.7-0.9		0.7-0.9		0.035 max	0.04 max	0.15-0.35	bal Fe				
	UNS G51170		0.15-0.20	0.70-0.90		0.70-0.90		0.035 max	0.040 max	0.15-0.35	bal Fe				

UNS numbers and US grades are provided as a means of cross referencing chemically similar alloys. Exchangability is only possible after independent examination of specifications. Tensile properties are minimum or typical as specified. UTS and YS as MPa. El as %. See Appendix for list of abbreviations used in Notes. * indicates obsolete material.

Specification	Designation	Notes	C	Cr	Cu	Mn	Mo	P	S	Si	Other	UTS	YS	El	Hard

Alloy Steel, Chromium, 5117 (Continued from previous page)

USA

Specification	Designation	Notes	C	Cr	Cu	Mn	Mo	P	S	Si	Other	UTS	YS	El	Hard
ASTM A322(96)	5117	Bar	0.15-0.20	0.70-0.90		0.70-0.90		0.035 max	0.040 max	0.15-0.35	bal Fe				
ASTM A331(95)	5117	Bar	0.15-0.20	0.70-0.90		0.70-0.90		0.035 max	0.040 max	0.15-0.35	bal Fe				

Yugoslavia

Specification	Designation	Notes	C	Cr	Cu	Mn	Mo	P	S	Si	Other	UTS	YS	El	Hard
	C.4381	CH	0.14-0.19	0.8-1.1		1.0-1.3	0.15 max	0.035 max	0.035 max	0.15-0.4	bal Fe				

Alloy Steel, Chromium, 5120/5120H

Argentina

Specification	Designation	Notes	C	Cr	Cu	Mn	Mo	P	S	Si	Other	UTS	YS	El	Hard
IAS	IRAM 5121		0.17-0.22	1.00-1.30		1.10-1.40		0.035 max	0.035 max	0.15-0.40	bal Fe	680	400	20	207 HB

Australia

Specification	Designation	Notes	C	Cr	Cu	Mn	Mo	P	S	Si	Other	UTS	YS	El	Hard
AS 1444(96)	5120H	H series, Hard/Tmp	0.17-0.23	0.60-1.00		0.60-1.00		0.040 max	0.040 max	0.10-0.35	bal Fe				

Bulgaria

Specification	Designation	Notes	C	Cr	Cu	Mn	Mo	P	S	Si	Other	UTS	YS	El	Hard
BDS 6354	18ChG		0.17-0.22	1.00-1.30	0.3	1.10-1.40		0.035	0.035	0.17-0.37	bal Fe				
BDS 6354	18ChGT		0.17-0.23	1.00-1.30	0.3	0.80-1.10		0.035	0.035	0.17-0.37	Ti 0.05-0.09; bal Fe				
BDS 6354	20Ch	Struct	0.17-0.23	0.70-1.00	0.3	0.60-0.90		0.035	0.035	0.17-0.37	bal Fe				

China

Specification	Designation	Notes	C	Cr	Cu	Mn	Mo	P	S	Si	Other	UTS	YS	El	Hard
GB 11251(89)	20Cr	Plt HR Ann	0.18-0.24	0.70-1.00		0.50-0.80		0.035 max	0.035 max	0.17-0.37	bal Fe	400-650		20	
GB 3077(88)	20Cr	Bar HR Q/T 15mm diam	0.18-0.24	0.70-1.00	0.30 max	0.50-0.80		0.035 max	0.035 max	0.17-0.37	bal Fe	835	540	10	
GB 3077(88)	20CrMn	Bar HR Q/T 25mm diam	0.17-0.23	0.90-1.20	0.30 max	0.90-1.20		0.035 max	0.035 max	0.17-0.37	bal Fe	930	735	10	
GB 5216(85)	20CrH	Bar Rod HR/Frg Ann	0.17-0.23	0.70-1.00	0.30 max	0.50-0.85		0.035 max	0.035 max	0.17-0.37	bal Fe				
GB 6478(86)	ML20Cr	Bar HR Q/T	0.17-0.24	0.70-1.00	0.20 max	0.50-0.80		0.035 max	0.035 max	0.30 max	bal Fe	785	590	10	
GB 8162(87)	20Cr	Smls tube HR/CD Q/T	0.18-0.24	0.70-1.00		0.50-0.80		0.035 max	0.035 max	0.17-0.37	bal Fe	835	540	10	
GB 8162(87)	20CrMn	Smls tube HR/CD Q/T	0.17-0.23	0.90-1.20		0.90-1.20		0.035 max	0.035 max	0.17-0.37	bal Fe	930	735	10	
GB/T 3078(94)	20Cr	Bar CD Q/T 15mm diam	0.18-0.24	0.70-1.00	0.30 max	0.50-0.80		0.035 max	0.035 max	0.17-0.37	bal Fe	835	540	10	
YB/T 5132(93)	20Cr	Sh HR/CR Ann	0.18-0.24	0.70-1.00		0.50-0.80		0.035 max	0.035 max	0.17-0.37	bal Fe	395-590		18	

Czech Republic

Specification	Designation	Notes	C	Cr	Cu	Mn	Mo	P	S	Si	Other	UTS	YS	El	Hard
CSN 414221	14221	CH	0.17-0.22	1.3-1.5		1.0-1.3		0.035 max	0.035 max	0.17-0.37	bal Fe				
CSN 414223	14223	CH	0.17-0.23	1.0-1.3		0.8-1.1		0.035 max	0.035 max	0.17-0.37	Ti 0.04-0.1; bal Fe				

Europe

Specification	Designation	Notes	C	Cr	Cu	Mn	Mo	P	S	Si	Other	UTS	YS	El	Hard
EN 10084(98)	1.7149	CH, Ann	0.17-0.22	1.00-1.30		1.10-1.40		0.035 max	0.020-0.040	0.40 max	bal Fe				217 HB
EN 10084(98)	1.7243	CH, Ann	0.15-0.21	0.90-1.20		0.60-0.90	0.15-0.25	0.035 max	0.035 max	0.40 max	bal Fe				207 HB
EN 10084(98)	1.7244	CH, Ann	0.15-0.21	0.90-1.20		0.60-0.90	0.15-0.25	0.035 max	0.020-0.040	0.40 max	bal Fe				207 HB
EN 10084(98)	18CrMo4	CH, Ann	0.15-0.21	0.90-1.20		0.60-0.90	0.15-0.25	0.035 max	0.035 max	0.40 max	bal Fe				207 HB
EN 10084(98)	18CrMoS4	CH, Ann	0.15-0.21	0.90-1.20		0.60-0.90	0.15-0.25	0.035 max	0.020-0.040	0.40 max	bal Fe				207 HB
EN 10084(98)	20MnCrS5	CH, Ann	0.17-0.22	1.00-1.30		1.10-1.40		0.035 max	0.020-0.040	0.40 max	bal Fe				217 HB

Finland

Specification	Designation	Notes	C	Cr	Cu	Mn	Mo	P	S	Si	Other	UTS	YS	El	Hard
SFS 510(76)	20MnCr5	CH	0.17-0.22	1.0-1.3		1.1-1.4		0.035 max	0.05 max	0.15-0.4	bal Fe				

France

Specification	Designation	Notes	C	Cr	Cu	Mn	Mo	P	S	Si	Other	UTS	YS	El	Hard
AFNOR NFA35551	20MC5	CH	0.17-0.22	1-1.3		1.1-1.4		0.035 max	0.035 max	0.1-0.4	bal Fe				

Specification	Designation	Notes	C	Cr	Cu	Mn	Mo	P	S	Si	Other	UTS	YS	El	Hard

Alloy Steel, Chromium, 5120/5120H (Continued from previous page)

France

| AFNOR NFA35552 | 20MC5 | CH | 0.17-0.22 | 1-1.3 | | 1.1-1.4 | | 0.035 max | 0.035 max | 0.1-0.4 | bal Fe | | | | |

Germany

DIN	20CrMnS3-3		0.17-0.23	0.60-1.00		0.60-1.00		0.040 max	0.020 min	0.20-0.35	bal Fe				
DIN	20MnCrPb5		0.17-0.22	1.00-1.30		1.10-1.40		0.035 max	0.035 max	0.15-0.35	Pb 0.20-0.35; bal Fe				
DIN	WNr 1.7121		0.17-0.23	0.60-1.00		0.60-1.00		0.040 max	0.020 min	0.20-0.35	bal Fe				
DIN	WNr 1.7146		0.17-0.22	1.00-1.30		1.10-1.40		0.035 max	0.035 max	0.15-0.35	Pb 0.20-0.35; bal Fe				
DIN 1652(90)	20Cr4		0.17-0.23	0.90-1.20		0.6-0.90		0.035 max	0.035 max	0.40 max	bal Fe				
DIN 1652(90)	20CrS4	CH	0.17-0.23	0.90-1.20		0.60-0.90		0.035 max	0.020-0.035	0.40 max	bal Fe				
DIN 1652(90)	20MnCr5	CH	0.17-0.22	1.00-1.30		1.10-1.40		0.035 max	0.035 max	0.40 max	bal Fe	980-1270	685	8	
DIN 1652(90)	20MnCrS5		0.17-0.22	1.00-1.30		1.10-1.40		0.035 max	0.020-0.040	0.40 max	bal Fe				
DIN 1652(90)	WNr 1.7027	CH	0.17-0.23	0.90-1.20		0.60-0.90		0.035 max	0.035 max	0.40 max	bal Fe				
DIN 1652(90)	WNr 1.7028	CH	0.17-0.23	0.90-1.20		0.60-0.90		0.035 max	0.020-0.035	0.40 max	bal Fe				
DIN 1652(90)	WNr 1.7147	CH, 30mm	0.17-0.22	1.00-1.30		1.10-1.40		0.035 max	0.035 max	0.40 max	bal Fe	980-1270	685	8	
DIN 1652(90)	WNr 1.7149	CH, 30mm	0.17-0.22	1.00-1.30		1.10-1.40		0.035 max	0.020-0.040	0.40 max	bal Fe	980-1270	685	8	
DIN 17230(80)	19MnCr5		0.17-0.22	1.10-1.30	0.30 max	1.10-1.40		0.035 max	0.035 max	0.040 max	bal Fe				
DIN 17230(80)	WNr 1.3523	Bearing	0.17-0.22	1.10-1.30	0.30 max	1.10-1.40		0.035 max	0.035 max	0.40 max	bal Fe				

Hungary

MSZ 31(85)	BC2	CH; 0-11mm; blank carburized	0.17-0.23	0.9-1.2		0.6-0.9		0.035 max	0.035 max	0.4 max	bal Fe	850-1200	640	10L	
MSZ 31(85)	BC2	CH; 63 mm; blank carburized	0.17-0.23	0.9-1.2		0.6-0.9		0.035 max	0.035 max	0.4 max	bal Fe	500-750	350	16L	
MSZ 31(85)	BC2E	CH; 63 mm; blank carburized	0.17-0.23	0.9-1.2		0.6-0.9		0.035 max	0.02-0.035	0.4 max	bal Fe	500-750	350	16L	
MSZ 31(85)	BC2E	CH; 0-11mm; blank carburized	0.17-0.23	0.9-1.2		0.6-0.9		0.035 max	0.02-0.035	0.4 max	bal Fe	850-1200	640	10L	
MSZ 6251(87)	BC2Z	CF; CH; 0-36mm; drawn, half-hard	0.17-0.23	0.9-1.2		0.6-0.9		0.035 max	0.035 max	0.4 max	bal Fe	0-590			
MSZ 6251(87)	BC2Z	CF; CH; 0-36mm; HR; HR/soft ann; drawn/soft ann; drawn/bright ann; soft ann/ground	0.17-0.23	0.9-1.2		0.6-0.9		0.035 max	0.035 max	0.4 max	bal Fe	0-560			

India

| IS 1570 | 20MnCr1 | | 0.17-0.22 | 1-1.3 | | 1-1.45 | | 0.05 | 0.05 | 0.1-0.35 | bal Fe | | | | |
| IS 4367 | 20MnCr1 | | 0.17-0.22 | 1-1.3 | | 1-1.45 | | 0.05 | 0.05 | 0.1-0.35 | bal Fe | | | | |

International

| ISO 683-11(87) | 18CrMo4 | CH, 16mm test | 0.15-0.21 | 0.90-1.20 | 0.3 max | 0.60-0.90 | 0.15-0.25 | 0.035 max | 0.035 max | 0.15-0.40 | Al <=0.1; Co <=0.1; bal Fe | 920-1270 | 600 | | |
| ISO 683-11(87) | 18CrMoS4 | CH, 16mm test | 0.15-0.21 | 0.90-1.20 | 0.3 max | 0.60-0.90 | 0.15-0.25 | 0.035 max | 0.020-0.035 | 0.15-0.40 | Al <=0.1; Co <=0.1; bal Fe | 920-1270 | 600 | | |

Italy

UNI 5771(66)	20MnCr4	Chain	0.16-0.24	0.4-0.65		0.9-1.2		0.035 max	0.035 max	0.3 max	bal Fe				
UNI 7846(78)	20CrNi4	CH	0.18-0.23	0.9-1.2		0.8-1.1	0.1 max	0.035 max	0.035 max	0.15-0.4	Ni 0.9-1.2; Pb 0.15-0.3; bal Fe				
UNI 7846(78)	20MnCr5	CH	0.17-0.22	1-1.3		1.1-1.4		0.035 max	0.035 max	0.15-0.4	bal Fe				
UNI 8550(84)	20CrNi4	CH	0.18-0.23	0.9-1.2		0.8-1.1		0.035 max	0.035 max	0.15-0.4	Ni 0.9-1.2; bal Fe				

UNS numbers and US grades are provided as a means of cross referencing chemically similar alloys. Exchangability is only possible after independent examination of specifications. Tensile properties are minimum or typical as specified. UTS and YS as MPa. El as %. See Appendix for list of abbreviations used in Notes. * indicates obsolete material.

Alloy Steel, Chromium, 5120/5120H (Continued from previous page)

Specification	Designation	Notes	C	Cr	Cu	Mn	Mo	P	S	Si	Other	UTS	YS	El	Hard
Italy															
UNI 8550(84)	20MnCr5	CH	0.17-0.22	1-1.3		1.1-1.4		0.035 max	0.035 max	0.15-0.4	bal Fe				
UNI 8788(85)	20CrNi4	CH	0.18-0.23	0.9-1.2		0.8-1.1	0.1 max	0.035 max	0.035 max	0.15-0.4	Ni 0.9-1.2; Pb 0.15-0.3; P+S<=0.06; bal Fe				
UNI 8788(85)	20MnCr5	CH	0.17-0.22	1-1.3		1.1-1.4		0.035 max	0.035 max	0.15-0.4	bal Fe				
Japan															
JIS G3311(88)	SCr420M	CR Strp; Chain	0.18-0.23	0.90-1.20	0.30 max	0.60-0.85		0.030 max	0.030 max	0.15-0.35	Ni <=0.25; bal Fe				180-270 HV
JIS G3441(88)	SCr420TK	Tube	0.18-0.23	0.90-1.20	0.30 max	0.60-0.85		0.030 max	0.030 max	0.15-0.35	Ni <=0.25; bal Fe				
JIS G4052(79)	SCr22H*	Obs; see SCr 420H	0.17-0.23	0.125-0.85		0.55-39		0.03	0.03	0.15-0.35	bal Fe				
JIS G4052(79)	SCr420H	Struct Hard	0.17-0.23	0.85-1.25	0.30 max	0.55-0.90		0.030 max	0.030 max	0.15-0.35	Ni <=0.25; bal Fe				
JIS G4052(79)	SMn21H*	Obs; see SMn 420H	0.16-0.23	0.125-0.85		1.15-1.55		0.03	0.03	0.15-0.35	bal Fe				
JIS G4104(79)	SCr22*	Obs; see SCr420	0.18-0.23	0.90-1.20	0.30 max	0.60-0.85		0.03 max	0.030 max	0.15-0.35	bal Fe				
JIS G4104(79)	SCr420	HR Frg Bar Wir Rod	0.18-0.23	0.90-1.20	0.30 max	0.60-0.85		0.030 max	0.030 max	0.15-0.35	Ni <=0.25; bal Fe				
Mexico															
DGN B-203	5120*	Obs	0.17-0.22	0.7-0.9		0.7-0.9		0.04	0.04	0.23-0.35	bal Fe				
DGN B-297	5120*	Obs	0.17-0.22	0.7-0.9		0.7-0.9		0.035	0.04	0.2-0.35	bal Fe				
NMX-B-268(68)	5120H	Hard	0.17-0.23	0.60-1.00		0.60-1.00				0.20-0.35	bal Fe				
NMX-B-300(91)	5120	Bar	0.17-0.22	0.70-0.90		0.70-0.90		0.035 max	0.040 max	0.15-0.35	bal Fe				
NMX-B-300(91)	5120H	Bar	0.17-0.23	0.60-1.00		0.60-1.00				0.15-0.35	bal Fe				
Pan America															
COPANT 334	5120	Bar	0.17-0.22	0.7-0.9		0.7-0.9		0.04	0.04	0.2-0.35	bal Fe				
COPANT 514	5120	Tube	0.17-0.22	0.7-0.9		0.7-0.9		0.04	0.05	0.1	bal Fe				
Poland															
PNH84030/02	18HGT	CH	0.17-0.23	1.0-1.3		0.8-1.1		0.035 max	0.035 max	0.17-0.37	Ni <=0.3; Ti 0.05-0.12; bal Fe				
PNH84030/02	20H	CH	0.17-0.23	0.7-1		0.5-0.8		0.035 max	0.035 max	0.17-0.37	Ni <=0.3; bal Fe				
PNH84030/02	20HG	CH	0.17-0.22	1.-1.3		1.1-1.4		0.035 max	0.035 max	0.17-0.37	Ni <=0.3; bal Fe				
PNH84030/04	20HGS	Q/T	0.17-0.23	0.8-1.1		0.8-1.1		0.035 max	0.035 max	0.9-1.2	Ni <=0.3; bal Fe				
Romania															
STAS 791	21TiMnCr12		0.18-0.24	1-1.3	0.3	0.8-1.1		0.035	0.035	0.17-0.37	Ni 0.3; Ti 0.04-0.1; bal Fe				
Russia															
GOST 4543	18ChGT		0.17-0.23	1-1.3	0.3	0.8-1.1		0.035	0.035	0.17-0.37	Ni 0.3; Ti 0.03-0.09; bal Fe				
GOST 4543	20Ch		0.17-0.23	0.7-1	0.3	0.5-0.8		0.035	0.035	0.17-0.37	Ni 0.3; bal Fe				
GOST 4543	20ChGR	CH	0.18-0.24	0.75-1.05	0.3 max	0.7-1.0	0.15 max	0.035 max	0.035 max	0.17-0.37	Al <=0.1; B >=0.001; Ni <=0.3; Ti <=0.03; W <=0.3; bal Fe				
GOST 4543	20ChGR	CH	0.18-0.24	0.75-1.05	0.3	0.7-1		0.035	0.035	0.17-0.37	B >=0.001; Ni 0.3; bal Fe				
GOST 4543	20ChGSA	CH	0.17-0.23	0.8-1.1	0.3	0.8-1.1		0.025	0.025	0.9-1.2	Ni 0.3; bal Fe				
GOST 4543(71)	18ChGT	CH	0.17-0.23	1.0-1.3	0.3 max	0.8-1.1	0.15 max	0.035 max	0.035 max	0.17-0.37	Al <=0.1; Ni <=0.3; Ti 0.03-0.09; bal Fe				
GOST 4543(71)	20ChGSA	CH	0.17-0.23	0.8-1.1	0.3 max	0.8-1.1	0.15 max	0.025 max	0.025 max	0.9-1.2	Al <=0.1; Ni <=0.3; Ti <=0.05; bal Fe				

Specification	Designation	Notes	C	Cr	Cu	Mn	Mo	P	S	Si	Other	UTS	YS	El	Hard

Alloy Steel, Chromium, 5120/5120H (Continued from previous page)

Spain
UNE	F.150.D		0.17-0.23	0.85-1.15		1-1.4		0.035	0.035	0.13-0.38	bal Fe				
UNE	F.158	CH	0.15-0.2	0.8-1.2		0.8-1.2	0.15-0.25	0.04 max	0.04 max	0.1-0.35	Ni 0.8-1.2; bal Fe				
UNE 36013(76)	20MoCr5	CH	0.18-0.25	0.3-0.5		0.6-0.9	0.4-0.5	0.035 max	0.035 max	0.15-0.4	bal Fe				
UNE 36027(80)	100CrMnMo7	Ball & roller bearing	0.95-1.1	1.65-1.95		0.6-0.9	0.2-0.4	0.03 max	0.025 max	0.2-0.4	bal Fe				
UNE 36027(80)	F.1314*	see 100CrMnMo7	0.95-1.1	1.65-1.95		0.6-0.9	0.2-0.4	0.03 max	0.025 max	0.2-0.4	bal Fe				

Sweden
| SS 142523 | 2523 | CH | 0.17-0.23 | 0.8-1.2 | | 0.7-1.1 | 0.08-0.16 | 0.035 max | 0.03-0.05 | 0.15-0.4 | Ni 1-1.4; bal Fe | | | | |

UK
| BS 970/3(71) | 527M20* | Hard, 19mm diam | 0.17-0.23 | 0.60-0.90 | | 0.60-0.90 | | | | 0.10-0.35 | bal Fe | | | | |

USA
	AISI 5120	Bar Blm Bil Slab	0.17-0.22	0.70-0.90		0.70-0.90		0.035 max	0.040 max	0.15-0.35	bal Fe				
	AISI 5120	Smls mech tub	0.17-0.22	0.70-0.90		0.70-0.90		0.040 max	0.040 max	0.15-0.35	bal Fe				
	AISI 5120H	Bar Blm Bil Slab	0.17-0.22	0.60-1.00		0.60-1.00		0.035 max	0.040 max	0.15-0.35	bal Fe				
	AISI 5120H	Smls mech tub, Hard	0.17-0.22	0.60-1.00		0.60-1.00		0.040 max	0.040 max	0.15-0.30	bal Fe				
	UNS G51200		0.17-0.22	0.70-0.90		0.70-0.90		0.035 max	0.040 max	0.15-0.35	bal Fe				
	UNS H51200		0.17-0.23	0.60-1.00		0.60-1.00		0.035 max	0.040 max	0.15-0.35	bal Fe				
ASTM A29/A29M(93)	5120	Bar	0.17-0.22	0.70-0.90		0.7-0.9		0.035 max	0.04 max	0.15-0.35	bal Fe				
ASTM A304(96)	5120H	Bar, hard bands spec	0.17-0.23	0.60-1.00	0.35 max	0.60-1.00	0.06 max	0.035 max	0.040 max	0.15-0.35	Cu Ni Cr Mo trace allowed; bal Fe				
ASTM A322(96)	5120	Bar	0.17-0.22	0.70-0.90		0.70-0.90		0.035 max	0.040 max	0.15-0.35	bal Fe				
ASTM A331(95)	5120	Bar	0.17-0.22	0.70-0.90		0.70-0.90		0.035 max	0.040 max	0.15-0.35	bal Fe				
ASTM A331(95)	5120H	Bar	0.17-0.23	0.60-1.00	0.35 max	0.60-1.00	0.06 max	0.025 max	0.025 max	0.15-0.35	bal Fe				
ASTM A519(96)	5120	Smls mech tub	0.17-0.22	0.70-0.90		0.70-0.90		0.040 max	0.040 max	0.15-0.35	bal Fe				
ASTM A534(94)	5120H	Carburizing, anti-friction bearings	0.17-0.23	0.60-1.00		0.60-1.00		0.025 max	0.025 max	0.15-0.35	bal Fe				
ASTM A752(93)	5120	Rod Wir	0.17-0.22	0.70-0.90		0.70-0.90		0.035 max	0.040 max	0.15-0.30	bal Fe	570			
SAE 770(84)	5120*	Obs; see J1397(92)	0.17-0.22	0.7-0.9		0.7-0.9		0.04 max	0.04 max	0.15-0.3	bal Fe				
SAE J1268(95)	5120H	Bar Rod Wir Tub; See std	0.17-0.23	0.60-1.00	0.35 max	0.60-1.00	0.06 max	0.030 max	0.040 max	0.15-0.35	Ni <=0.25; Cu, Mo, Ni not spec'd but acpt; bal Fe				
SAE J404(94)	5120	Bil Blm Slab Bar HR CF	0.17-0.22	0.70-0.90		0.70-0.90		0.030 max	0.040 max	0.15-0.35	bal Fe				

Yugoslavia
	C.4321	CH	0.17-0.22	1.0-1.3		1.1-1.4	0.15 max	0.035 max	0.035 max	0.15-0.4	bal Fe				
	C.4321	Ball & roller bearing	0.17-0.22	1.0-1.3		1.1-1.4	0.15 max	0.035 max	0.035 max	0.4 max	bal Fe				
	C.4382	CH	0.17-0.22	1.0-1.3		1.1-1.4	0.15 max	0.035 max	0.035 max	0.15-0.4	bal Fe				

Alloy Steel, Chromium, 5130/5130H

Argentina
| IAS | IRAM 5130 | | 0.28-0.33 | 0.80-1.10 | | 0.70-0.90 | | 0.035 max | 0.040 max | 0.20-0.35 | bal Fe | 660 | 420 | 20 | 197 HB |

Specification	Designation	Notes	C	Cr	Cu	Mn	Mo	P	S	Si	Other	UTS	YS	El	Hard
Alloy Steel, Chromium, 5130/5130H (Continued from previous page)															
Bulgaria															
BDS 6354	30Ch	Struct	0.24-0.32	0.80-1.10	0.3 max	0.50-0.80	0.15 max	0.035 max	0.035 max	0.17-0.37	Ni <=0.30; Ti <=0.03; V <=0.05; W <=0.2; bal Fe				
BDS 6554	30ChGS		0.28-0.35	0.80-1.10	0.3	0.80-1.10		0.035	0.035	0.90-1.20	bal Fe				
China															
GB 11251(89)	30Cr	Plt HR Ann	0.27-0.34	0.80-1.10		0.50-0.80		0.035 max	0.035 max	0.17-0.37	bal Fe	500-700		19	
GB 3077(88)	30Cr	Bar HR Q/T 25mm diam	0.27-0.34	0.80-1.10	0.30 max	0.50-0.80		0.035 max	0.035 max	0.17-0.37	bal Fe	885	685	11	
GB 8162(87)	30Cr	Smls tube HR/CD Q/T	0.27-0.34	0.80-1.10		0.50-0.80		0.035 max	0.035 max	0.17-0.37	bal Fe	885	685	11	
GB/T 3078(94)	30Cr	Bar CD Q/T 25mm diam	0.27-0.34	0.80-1.10	0.30 max	0.50-0.80		0.035 max	0.035 max	0.17-0.37	bal Fe	885	685	11	
YB/T 5132(93)	30Cr	Sh HR/CR Ann	0.27-0.34	0.80-1.10		0.50-0.80		0.035 max	0.035 max	0.17-0.37	bal Fe	490-685		17	
Czech Republic															
CSN 414331	14331	Q/T	0.28-0.35	0.8-1.1		0.8-1.1		0.035 max	0.035 max	0.9-1.2	bal Fe				
CSN 414340	14340	Nitriding	0.28-0.38	1.3-1.9		0.5-0.9		0.035 max	0.035 max	0.17-0.37	Al 0.9-1.3; bal Fe				
France															
AFNOR	28C4		0.25-0.3	0.85-1.15		0.6-0.9		0.04	0.035	0.4	bal Fe				
Germany															
DIN	30MnCrTi4		0.25-0.35	0.80-1.00		0.90-1.20		0.025 max	0.025 max	0.15-0.35	Al <=0.10; Ti 0.15-0.30; bal Fe				
DIN	WNr 1.8401		0.25-0.35	0.80-1.00		0.90-1.20		0.025 max	0.025 max	0.15-0.35	Al <=0.10; Ti 0.15-0.30; bal Fe				
DIN EN 10083(91)	28Cr4	Q/T 17-40mm	0.24-0.31	0.90-1.20		0.60-0.90		0.035 max	0.035 max	0.40 max	bal Fe	750-900	550	14	
DN EN 10083(91)	WNr 1.7030	CH, Q/T 17-40mm	0.24-0.31	0.90-1.20		0.60-0.90		0.035 max	0.035 max	0.40 max	bal Fe	750-900	550	14	
Japan															
JIS G3311(88)	SCM2M*	Obs; see SCM 430M	0.28-0.33	0.9-1.2	0.3	0.6-0.85	0.15-0.3	0.03	0.03	0.15-0.35	Ni 0.25; bal Fe				
JIS G4052(79)	SCr2H*	Obs; see SCr 430H	0.27-0.34	0.85-1.25		0.55-0.9		0.03	0.03	0.15-0.35	bal Fe				
JIS G4052(79)	SCr430H	Struct Hard	0.27-0.34	0.85-1.25	0.30 max	0.55-0.90		0.030 max	0.030 max	0.15-0.35	Ni <=0.25; bal Fe				
JIS G4104(79)	SCr 2*	Obs; see SCr430	0.28-0.33	0.90-1.20	0.30 max	0.60-0.85		0.03 max	0.030 max	0.15-0.35	bal Fe				
JIS G4104(79)	SCr430	HR Frg Bar Wir Rod	0.28-0.33	0.90-1.20	0.30 max	0.60-0.85		0.030 max	0.030 max	0.15-0.35	Ni <=0.25; bal Fe				
Mexico															
DGN B-203	5130*	Obs	0.28-0.33	0.8-1.1		0.7-0.9		0.04	0.04	0.2-0.35	bal Fe				
NMX-B-268(68)	5130H	Hard	0.27-0.33	0.75-1.20		0.60-1.10				0.20-0.35	bal Fe				
NMX-B-300(91)	5130	Bar	0.28-0.33	0.80-1.10		0.70-0.90		0.035 max	0.040 max	0.15-0.35	bal Fe				
NMX-B-300(91)	5130H	Bar	0.27-0.33	0.75-1.20		0.60-1.00				0.15-0.35	bal Fe				
Pan America															
COPANT 334	5130	Bar	0.28-0.33	0.8-1.1		0.7-0.9		0.04	0.04	0.2-0.35	bal Fe				
COPANT 514	5130	Tube	0.28-0.33	0.8-1.1		0.7-0.9		0.04	0.05	0.1	bal Fe				
Poland															
PNH84030	30H	Q/T	0.27-0.35	0.8-1.1	0.3	0.5-0.8		0.035	0.035	0.17-0.37	Ni 0.3; bal Fe				
PNH84030/04	30HGS	Q/T	0.28-0.35	0.8-1.1		0.8-1.1		0.035 max	0.035 max	0.9-1.2	Ni <=0.3; bal Fe				
PNH84030/04	30HGSA	Q/T	0.28-0.34	0.8-1.1		0.8-1.1		0.025 max	0.025 max	0.9-1.2	Ni <=0.3; bal Fe				

Specification	Designation	Notes	C	Cr	Cu	Mn	Mo	P	S	Si	Other	UTS	YS	El	Hard

Alloy Steel, Chromium, 5130/5130H (Continued from previous page)

Romania

Specification	Designation	Notes	C	Cr	Cu	Mn	Mo	P	S	Si	Other
STAS 791(88)	20MnCrSi11	Struct/const	0.17-0.23	0.8-1.1		0.8-1.1		0.035 max	0.035 max	0.9-1.2	Pb <=0.15; Ti <=0.02; bal Fe
STAS 791(88)	20MnCrSi11S	Struct/const	0.17-0.23	0.8-1.1		0.8-1.1		0.035 max	0.02-0.04	0.9-1.2	Pb <=0.15; Ti <=0.02; bal Fe
STAS 791(88)	20MnCrSi11X	Struct/const	0.17-0.23	0.8-1.1		0.8-1.1		0.025 max	0.025 max	0.9-1.2	Pb <=0.15; Ti <=0.02; bal Fe
STAS 791(88)	20MnCrSi11XS	Struct/const	0.17-0.23	0.8-1.1		0.8-1.1		0.025 max	0.02-0.035	0.9-1.2	Pb <=0.15; Ti <=0.02; bal Fe
STAS 791(88)	28TiMnCr12	Struct/const	0.24-0.32	1-1.3		0.8-1.1		0.035 max	0.035 max	0.17-0.37	Pb <=0.15; Ti 0.03-0.09; bal Fe
STAS 791(88)	28TiMnCr12S	Struct/const	0.24-0.32	1-1.3		0.8-1.1		0.035 max	0.02-0.04	0.17-0.37	Pb <=0.15; Ti 0.03-0.09; bal Fe
STAS 791(88)	28TiMnCr12X	Struct/const	0.24-0.32	1-1.3		0.8-1.1		0.025 max	0.025 max	0.17-0.37	Pb <=0.15; Ti 0.03-0.09; bal Fe
STAS 791(88)	28TiMnCr12XS	Struct/const	0.24-0.32	1-1.3		0.8-1.1		0.025 max	0.02-0.035	0.17-0.37	Pb <=0.15; Ti 0.03-0.09; bal Fe

Russia

Specification	Designation	Notes	C	Cr	Cu	Mn	Mo	P	S	Si	Other
GOST 19277	30ChGSA-VD		0.28-0.34	0.8-1.1		0.9-1.1		0.015	0.012	0.9-1.2	Ni 0.25; bal Fe
GOST 19277	30ChGSA-WD		0.28-0.34	0.8-1.1		0.9-1.1		0.015	0.012	0.9-1.2	Ni 0.25; bal Fe
GOST 4543	27ChGR		0.25-0.31	0.7-1	0.3	0.7-1		0.035	0.035	0.17-0.37	Ni 0.3; bal Fe
GOST 4543	27ChGR	Q/T	0.25-0.31	0.7-1.0	0.3 max	0.7-1.0	0.15 max	0.035 max	0.035 max	0.17-0.37	Al <=0.1; B >=0.001; Ni <=0.3; Ti <=0.03; V <=0.05; W <=0.3; bal Fe
GOST 4543	30ChGS	Q/T	0.28-0.35	0.8-1.1	0.3 max	0.8-1.1	0.15 max	0.035 max	0.035 max	0.9-1.2	Al <=0.1; Ni <=0.3; Ti <=0.05; bal Fe
GOST 4543	30ChGS		0.28-0.35	0.8-1.1	0.3	0.8-1.1		0.035	0.035	0.9-1.2	Ni 0.3; bal Fe
GOST 4543	30ChRA		0.27-0.33	1-1.3	0.3	0.5-0.8		0.025	0.025	0.17-0.37	Ni 0.3; bal Fe
GOST 4543(71)	30Ch	Q/T	0.24-0.32	0.8-1.1	0.3 max	0.5-0.8	0.15 max	0.035 max	0.035 max	0.17-0.37	Al <=0.1; Ni <=0.3; Ti <=0.05; bal Fe
GOST 4543(71)	30ChRA	Q/T	0.27-0.33	1.0-1.3	0.3 max	0.5-0.8	0.15 max	0.025 max	0.025 max	0.17-0.37	Al <=0.1; Ni <=0.3; Ti <=0.05; bal Fe

UK

Specification	Designation	Notes	C	Cr	Cu	Mn	Mo	P	S	Si	Other
BS 970/1	530H30*	Obs; Blm Bil Slab Bar Rod Frg	0.27-0.33	0.80-1.25		0.50-0.90		0.035 max	0.040 max		bal Fe
BS 970/1(83)	530A30	Blm Bil Slab Bar Rod Frg nitriding	0.28-0.33	0.90-1.20		0.60-0.80		0.035 max	0.040 max		bal Fe

USA

Specification	Designation	Notes	C	Cr	Cu	Mn	Mo	P	S	Si	Other
	AISI 5130	Bar Blm Bil Slab	0.28-0.33	0.80-1.10		0.70-0.90		0.035 max	0.040 max	0.15-0.35	bal Fe
	AISI 5130	Smls mech tub	0.28-0.33	0.80-1.10		0.70-0.90		0.040 max	0.040 max	0.15-0.35	bal Fe
	AISI 5130H	Smls mech tub, Hard	0.27-0.33	0.75-1.20		0.60-1.00		0.040 max	0.040 max	0.15-0.30	bal Fe
	AISI 5130H	Bar Blm Bil Slab	0.27-0.33	0.75-1.20		0.60-1.00		0.035 max	0.040 max	0.15-0.35	bal Fe
	UNS G51300		0.28-0.33	0.80-1.10		0.70-0.90		0.035 max	0.040 max	0.15-0.35	bal Fe
	UNS H51300		0.27-0.33	0.75-1.20		0.60-1.00		0.035 max	0.040 max	0.15-0.35	bal Fe
ASTM A29/A29M(93)	5130	Bar	0.28-0.33	0.80-1.10		0.7-0.9		0.035 max	0.04 max	0.15-0.35	bal Fe
ASTM A304(96)	5130H	Bar, hard bands spec	0.27-0.33	0.75-1.20	0.35 max	0.60-1.10	0.06 max	0.035 max	0.040 max	0.15-0.35	Cu Ni Cr Mo trace allowed; bal Fe
ASTM A322(96)	5130	Bar	0.28-0.33	0.80-1.10		0.70-0.90		0.035 max	0.040 max	0.15-0.35	bal Fe
ASTM A331(95)	5130	Bar	0.28-0.33	0.80-1.10		0.70-0.90		0.035 max	0.040 max	0.15-0.35	bal Fe
ASTM A331(95)	5130H	Bar	0.27-0.33	0.75-1.20	0.35 max	0.60-1.10	0.06 max	0.025 max	0.025 max	0.15-0.35	bal Fe
ASTM A513(97a)	5130	ERW Mech Tub	0.23-0.33	0.80-1.10		0.70-0.90		0.035 max	0.040 max	0.15-0.35	bal Fe

UNS numbers and US grades are provided as a means of cross referencing chemically similar alloys. Exchangability is only possible after independent examination of specifications. Tensile properties are minimum or typical as specified. UTS and YS as MPa. El as %. See Appendix for list of abbreviations used in Notes. * indicates obsolete material.

Specification	Designation	Notes	C	Cr	Cu	Mn	Mo	P	S	Si	Other	UTS	YS	El	Hard

Alloy Steel, Chromium, 5130/5130H (Continued from previous page)

USA

Specification	Designation	Notes	C	Cr	Cu	Mn	Mo	P	S	Si	Other	UTS	YS	El	Hard
ASTM A519(96)	5130	Smls mech tub	0.28-0.33	0.80-1.10		0.70-0.90		0.040 max	0.040 max	0.15-0.35	bal Fe				
ASTM A914/914M(92)	5130RH	Bar restricted end Q hard	0.28-0.33	0.80-1.10	0.35 max	0.70-0.90	0.06 max	0.035 max	0.040 max	0.15-0.35	Ni <=0.25; Electric furnace P, S<=0.025; bal Fe				
SAE J1268(95)	5130H	Bar Sh Strp Tub; See std	0.27-0.33	0.75-1.20	0.35 max	0.60-1.00	0.06 max	0.030 max	0.040 max	0.15-0.35	Ni <=0.25; Cu, Mo, Ni not spec'd but acpt; bal Fe				
SAE J1868(93)	5130RH	Restrict hard range	0.28-0.33	0.70-0.90	0.035 max	0.70-0.90	0.06 max	0.025 max	0.040 max	0.15-0.35	Ni <=0.25; Cu, Mo, Ni not spec'd but acpt; bal Fe				5 HRC max
SAE J404(94)	5130	Bil Blm Slab Bar HR CF	0.28-0.33	0.80-1.10		0.70-0.90		0.030 max	0.040 max	0.15-0.35	bal Fe				

Alloy Steel, Chromium, 5132/5132H

Argentina

Specification	Designation	Notes	C	Cr	Cu	Mn	Mo	P	S	Si	Other	UTS	YS	El	Hard
IAS	IRAM 5132		0.30-0.35	0.75-1.00		0.60-0.80		0.035 max	0.040 max	0.15-0.30	bal Fe	720	450	26	212 HB

Australia

Specification	Designation	Notes	C	Cr	Cu	Mn	Mo	P	S	Si	Other	UTS	YS	El	Hard
AS 1444(96)	5132	H series, Hard/Tmp	0.30-0.35	0.75-1.00		0.60-0.80		0.040 max	0.040 max	0.10-0.35	bal Fe				
AS 1444(96)	5132H	H series, Hard/Tmp	0.29-0.35	0.65-1.10		0.50-0.90		0.040 max	0.040 max	0.10-0.35	bal Fe				

Belgium

Specification	Designation	Notes	C	Cr	Cu	Mn	Mo	P	S	Si	Other	UTS	YS	El	Hard
NBN 251	34Cr4		0.3-0.37	0.9-1.2		0.6-0.9		0.035	0.035	0.15-0.4	bal Fe				
NBN 251	37Cr4		0.34-0.41	0.9-1.2		0.6-0.9		0.04 max	0.04 max	0.15-0.4	bal Fe				

Bulgaria

Specification	Designation	Notes	C	Cr	Cu	Mn	Mo	P	S	Si	Other	UTS	YS	El	Hard
BDS 6554	30ChGS		0.28-0.35	0.80-1.10	0.3	0.80-1.10		0.035	0.035	0.90-1.20	B 0.001-0.005; bal Fe				

China

Specification	Designation	Notes	C	Cr	Cu	Mn	Mo	P	S	Si	Other	UTS	YS	El	Hard
GB 11251(89)	35Cr	Plt HR Ann	0.32-0.39	0.80-1.10		0.50-0.80		0.035 max	0.035 max	0.17-0.37	bal Fe	550-750		18	
GB 3077(88)	35Cr	Bar HR Q/T 25mm diam	0.32-0.39	0.80-1.10	0.30 max	0.50-0.80		0.035 max	0.035 max	0.17-0.37	bal Fe	930	735	11	
GB 8162(87)	35Cr	Smls tube HR/CD Q/T	0.32-0.39	0.80-1.10		0.50-0.80		0.035 max	0.035 max	0.17-0.37	bal Fe	930	735	11	
GB/T 3078(94)	35Cr	Bar CD Q/T 25mm diam	0.32-0.39	0.80-1.10	0.30 max	0.50-0.80		0.035 max	0.035 max	0.17-0.37	bal Fe	930	735	11	
YB/T 5132(93)	35Cr	Sh HR/CR Ann	0.32-0.39	0.80-1.10		0.50-0.80		0.035 max	0.035 max	0.17-0.37	bal Fe	540-735		16	

Czech Republic

Specification	Designation	Notes	C	Cr	Cu	Mn	Mo	P	S	Si	Other	UTS	YS	El	Hard
CSN 414230	14230	Q/T	0.28-0.35	0.8-1.1		0.9-1.2		0.035 max	0.035 max	0.35 max	B 0.001-0.005; bal Fe				

Europe

Specification	Designation	Notes	C	Cr	Cu	Mn	Mo	P	S	Si	Other	UTS	YS	El	Hard
EN 10083/1(91)A1(96)	1.7033	Q/T t<=16mm	0.30-0.37	0.90-1.20		0.60-0.90	0.15 max	0.035 max	0.035 max	0.40 max	bal Fe	900-1100	700	12	
EN 10083/1(91)A1(96)	34Cr4	Q/T t<=16mm	0.30-0.37	0.90-1.20		0.60-0.90	0.15 max	0.035 max	0.035 max	0.40 max	bal Fe	900-1100	700	12	

France

Specification	Designation	Notes	C	Cr	Cu	Mn	Mo	P	S	Si	Other	UTS	YS	El	Hard
AFNOR NFA35552	32C4		0.3-0.35	0.85-1.2		0.6-0.9		0.035	0.035	0.1-0.4	bal Fe				
AFNOR NFA35552(84)	38C4	Bar Rod	0.35-0.4	0.9-1.2		0.6-0.9		0.035 max	0.035 max	0.1-0.4	bal Fe				
AFNOR NFA35552(84)	38C4u	40<t<=100mm; Q/T	0.35-0.4	0.9-1.2		0.6-0.9	0.15 max	0.035 max	0.02-0.04	0.1-0.4	bal Fe	730-880	540	14	
AFNOR NFA35552(84)	38C4u	t<=16mm; Q/T	0.35-0.4	0.9-1.2		0.6-0.9	0.15 max	0.035 max	0.02-0.04	0.1-0.4	bal Fe	930-1130	700	12	
AFNOR NFA35553	32C4	Strp	0.3-0.35	0.85-1.15		0.6-0.9		0.04	0.04	0.1-0.4	bal Fe				
AFNOR NFA35553	38C4	Strp	0.35-0.4	0.85-1.15		0.6-0.9		0.035 max	0.035 max	0.1-0.4	bal Fe				
AFNOR NFA35556(84)	32C4		0.3-0.35	0.85-1.2		0.6-0.9		0.035 max	0.035 max	0.1-0.4	bal Fe				
AFNOR NFA35557	32C4		0.3-0.35	0.85-1.2		0.6-0.9		0.035 max	0.035 max	0.1-0.4	bal Fe				

UNS numbers and US grades are provided as a means of cross referencing chemically similar alloys. Exchangability is only possible after independent examination of specifications. Tensile properties are minimum or typical as specified. UTS and YS as MPa. El as %. See Appendix for list of abbreviations used in Notes. * indicates obsolete material.

Specification	Designation	Notes	C	Cr	Cu	Mn	Mo	P	S	Si	Other	UTS	YS	El	Hard

Alloy Steel, Chromium, 5132/5132H (Continued from previous page)

Specification	Designation	Notes	C	Cr	Cu	Mn	Mo	P	S	Si	Other	UTS	YS	El	Hard
Germany															
DIN 1652(90)	34Cr4	Q/T 17-40mm	0.30-0.37	0.90-1.20		0.60-0.90		0.035 max	0.035 max	0.40 max	bal Fe	800-950	590	14	
DIN 1652(90)	WNr 1.7033	Q/T 17-40mm	0.30-0.37	0.90-1.20		0.60-0.90		0.035 max	0.035 max	0.40 max	bal Fe	800-950	590	14	
DIN EN 10083(91)	34Cr4	Q/T 17-40mm	0.30-0.37	0.90-1.20		0.60-0.90		0.035 max	0.035 max	0.40 max	bal Fe	800-950	590	14	
DIN EN 10083(91)	34CrS4	Q/T	0.30-0.37	0.90-1.20		0.60-0.90		0.035 max	0.020-0.040	0.40 max	bal Fe				
DIN EN 10083(91)	WNr 1.7033	Q/T 17-40mm	0.30-0.37	0.90-1.20		0.60-0.90		0.035 max	0.035 max	0.40 max	bal Fe	800-950	590	14	
DIN EN 10083(91)	WNr 1.7037	Q/T	0.30-0.37	0.90-1.20		0.60-0.90		0.035 max	0.020-0.040	0.40 max	bal Fe				
Hungary															
MSZ 61(85)	Cr1	Q/T; 40.1-100mm	0.3-0.37	0.9-1.2		0.6-0.9		0.035 max	0.035 max	0.4 max	bal Fe	700-850	460	15L	
MSZ 61(85)	Cr1	Q/T; 0-16mm	0.3-0.37	0.9-1.2		0.6-0.9		0.035 max	0.035 max	0.4 max	bal Fe	900-1100	700	11L	
MSZ 61(85)	Cr1E	Q/T; 40.1-100mm	0.3-0.37	0.9-1.2		0.6-0.9		0.035 max	0.02-0.035	0.4 max	bal Fe	700-850	460		
MSZ 61(85)	Cr1E	Q/T; 0-16mm	0.3-0.37	0.9-1.2		0.6-0.9		0.035 max	0.02-0.035	0.4 max	bal Fe	900-1100	700	11L	
MSZ 6251(87)	Cr1Z	CF; Q/T; 0-36mm; HR; HR/soft ann; drawn/soft ann; drawn/bright ann; soft ann/ground	0.3-0.37	0.9-1.2		0.6-0.9		0.035 max	0.035 max	0.4 max	bal Fe	0-640			
MSZ 6251(87)	Cr1Z	CF; Q/T; 0-36mm; HR; HR/soft ann; drawn/soft ann; drawn/bright ann; soft ann/ground	0.3-0.37	0.9-1.2		0.6-0.9		0.035 max	0.035 max	0.4 max	bal Fe	0-610			
MSZ 8271	Ao.35MnCr		0.3-0.4	0.6-0.9	0.3	1-1.3		0.04	0.04	0.6-0.8	Ni 0.3; bal Fe				
International															
ISO 683-7	1*		0.3-0.37	0.9-1.2		0.6-0.9		0.035	0.035	0.15-0.4	bal Fe				
ISO 683-7	1a*		0.3-0.37	0.9-1.2		0.6-0.9		0.035	0.02-0.035	0.15-0.4	bal Fe				
ISO 683-7	1b*		0.3-0.37	0.9-1.2		0.6-0.9		0.035	0.03-0.05	0.15-0.4	bal Fe				
ISO R683-7(70)	1	Q/T	0.3-0.37	0.9-1.2	0.3 max	0.6-0.9	0.15 max	0.035 max	0.035 max	0.15-0.4	Al <=0.1; Co <=0.1; Ni <=0.4; Pb <=0.15; Ti <=0.05; V <=0.1; W <=0.1; bal Fe				
Italy															
UNI 7282	38Cr4KB		0.34-0.41	0.9-1.2		0.6-0.9		0.035 max	0.035 max	0.15-0.4	bal Fe				
UNI 7356(74)	34Cr4KB	Q/T	0.3-0.37	0.9-1.2		0.6-0.9		0.035 max	0.035 max	0.15-0.4	bal Fe				
UNI7874(79)	34Cr4	Q/T	0.3-0.37	0.9-1.2		0.6-0.9		0.035 max	0.03 max	0.15-0.4	bal Fe				
Japan															
JIS G4104(79)	SCr 3*	Obs; see SCr435	0.33-0.38	0.90-1.20	0.30 max	0.60-0.85		0.03 max	0.030 max	0.15-0.35	bal Fe				
JIS G4104(79)	SCr435	HR Frg Bar Wir Rod	0.33-0.38	0.90-1.20	0.30 max	0.60-0.85		0.030 max	0.030 max	0.15-0.35	Ni <=0.25; bal Fe				
Mexico															
DGN B-203	5132*	Obs	0.3-0.35	0.75-1		0.6-0.9		0.04	0.04	0.2-0.35	bal Fe				
DGN B-297	5132*	Obs	0.3-0.35	0.75-1		0.6-0.9		0.035	0.04	0.2-0.35	bal Fe				
NMX-B-268(68)	5132H	Hard	0.29-0.35	0.65-1.10		0.50-0.90				0.20-0.35	bal Fe				
NMX-B-300(91)	5132	Bar	0.30-0.35	0.75-1.00		0.60-0.80		0.035 max	0.040 max	0.15-0.35	bal Fe				
NMX-B-300(91)	5132H	Bar	0.29-0.35	0.65-1.10		0.50-0.90				0.15-0.35	bal Fe				

UNS numbers and US grades are provided as a means of cross referencing chemically similar alloys. Exchangability is only possible after independent examination of specifications. Tensile properties are minimum or typical as specified. UTS and YS as MPa. El as %. See Appendix for list of abbreviations used in Notes. * indicates obsolete material.

Specification	Designation	Notes	C	Cr	Cu	Mn	Mo	P	S	Si	Other	UTS	YS	El	Hard

Alloy Steel, Chromium, 5132/5132H (Continued from previous page)

Pan America

Specification	Designation	Notes	C	Cr	Cu	Mn	Mo	P	S	Si	Other	UTS	YS	El	Hard
COPANT 514	5132		0.3-0.35	0.75-1		0.6-0.9		0.04	0.04	0.1	bal Fe				
COPANT 514	5132	Tube	0.3-0.35	0.75-1		0.6-0.9		0.04	0.04	0.1	bal Fe				

Poland

Specification	Designation	Notes	C	Cr	Cu	Mn	Mo	P	S	Si	Other	UTS	YS	El	Hard
PNH83156(87)	635HGs		0.3-0.4	0.6-0.9	0.3	1-1.3		0.04	0.04	0.6-0.8	Ni 0.4; bal Fe				
PNH84030	30HGS	Q/T	0.28-0.35	0.8-1.1	0.3	0.8-1.1		0.035	0.035	0.9-1.2	Ni 0.3; bal Fe				
PNH84030/04	30H	Q/T	0.27-0.35	0.8-1.1		0.5-0.8		0.035 max	0.035 max	0.17-0.37	Ni <=0.3; bal Fe				

Romania

Specification	Designation	Notes	C	Cr	Cu	Mn	Mo	P	S	Si	Other	UTS	YS	El	Hard
STAS 11500/2(89)	32Cr10	Q/T	0.29-0.35	0.85-1.15	0.2 max	0.6-0.9		0.03 max	0.02-0.04	0.15-0.4	Pb <=0.15; bal Fe				
STAS 791	35C10		0.31-0.39	0.8-1.1	0.3	0.5-0.8		0.035	0.035	0.17-0.37	Ni 0.3; bal Fe				

Russia

Specification	Designation	Notes	C	Cr	Cu	Mn	Mo	P	S	Si	Other	UTS	YS	El	Hard
GOST 4543	30ChGS		0.28-0.35	0.8-1.1	0.3	0.8-1.1		0.035	0.035	0.9-1.2	Ni 0.3; bal Fe				
GOST 4543	35Ch		0.31-0.39	0.8-1.1	0.3 max	0.5-0.8		0.04 max	0.04 max	0.17-0.37	Ni <=0.3; bal Fe				
GOST 4543(71)	35Ch	Q/T	0.31-0.39	0.8-1.1	0.3 max	0.5-0.8	0.15 max	0.035 max	0.035 max	0.17-0.37	Al <=0.1; Ni <=0.3; Ti <=0.05; bal Fe				

Spain

Specification	Designation	Notes	C	Cr	Cu	Mn	Mo	P	S	Si	Other	UTS	YS	El	Hard
UNE 36254(79)	35Cr4	Q/T	0.3-0.4	0.8-1.2		0.5-0.8	0.15 max	0.04 max	0.04 max	0.3-0.5	bal Fe				
UNE 36355(79)	AM35Cr4		0.3-0.4	0.8-1.2		0.5-0.8		0.040	0.040	0.3-0.5	bal Fe				
UNE 36355(79)	F.8321*	see AM35Cr4	0.3-0.4	0.8-1.2		0.5-0.8		0.040	0.040	0.3-0.5	bal Fe				

UK

Specification	Designation	Notes	C	Cr	Cu	Mn	Mo	P	S	Si	Other	UTS	YS	El	Hard
BS 970/1(83)	530A32	Blm Bil Slab Bar Rod Frg nitriding	0.30-0.35	0.90-1.20		0.60-0.80		0.035 max	0.040 max		bal Fe				
BS 970/1(83)	530H32	Blm Bil Slab Bar Rod Frg	0.29-0.35	0.80-1.25		0.50-0.90		0.035 max	0.040 max		bal Fe				
BS 970/2(70)	905M31*	Obs; Spring	0.27-0.35	1.4-1.8		0.40-0.65	0.15-0.25	0.025 max	0.025 max	0.1-0.45	Al 0.9-1.3; Pb <=0.15; Ti <=0.05; bal Fe				

USA

Specification	Designation	Notes	C	Cr	Cu	Mn	Mo	P	S	Si	Other	UTS	YS	El	Hard
	AISI 5132	Bar Blm Bil Slab	0.30-0.35	0.75-1.00		0.60-0.80		0.035 max	0.040 max	0.15-0.35	bal Fe				
	AISI 5132	Smls mech tub	0.30-0.35	0.75-1.00		0.60-0.80		0.040 max	0.040 max	0.15-0.35	bal Fe				
	AISI 5132H	Bar Blm Bil Slab	0.29-0.35	0.65-1.10		0.50-0.90		0.035 max	0.040 max	0.15-0.35	bal Fe				
	AISI 5132H	Smls mech tub, Hard	0.29-0.35	0.65-1.10		0.50-0.90		0.040 max	0.040 max	0.15-0.30	bal Fe				
	UNS G51320		0.30-0.35	0.75-1.00		0.60-0.80		0.035 max	0.040 max	0.15-0.35	bal Fe				
	UNS H51320		0.29-0.35	0.65-1.10		0.50-0.90		0.035 max	0.040 max	0.15-0.35	bal Fe				
ASTM A29/A29M(93)	5132	Bar	0.3-0.35	0.75-1.00		0.6-0.8		0.035	0.04	0.15-0.35	bal Fe				
ASTM A304(96)	5132H	Bar, hard bands spec	0.29-0.35	0.65-1.10	0.35 max	0.50-0.90	0.06 max	0.035 max	0.040 max	0.15-0.35	Cu Ni Cr Mo trace allowed; bal Fe				
ASTM A322(96)	5132	Bar	0.30-0.35	0.75-1.00		0.60-0.80		0.035 max	0.040 max	0.15-0.35	bal Fe				
ASTM A331(95)	5132	Bar	0.30-0.35	0.75-1.00		0.60-0.80		0.035 max	0.040 max	0.15-0.35	bal Fe				
ASTM A331(95)	5132H	Bar	0.29-0.35	0.65-1.10	0.35 max	0.50-0.90	0.06 max	0.025 max	0.025 max	0.15-0.35	bal Fe				
ASTM A519(96)	5132	Smls mech tub	0.30-0.35	0.75-1.00		0.60-0.80		0.040 max	0.040 max	0.15-0.35	bal Fe				
ASTM A752(93)	5132	Rod Wir	0.30-0.35	0.75-1.00		0.60-0.80		0.035 max	0.040 max	0.15-0.30	bal Fe		580		
SAE 770(84)	5132*	Obs; see J1397(92)	0.3-0.35	0.75-1		0.6-0.8		0.04 max	0.04 max	0.15-0.3	bal Fe				

Specification	Designation	Notes	C	Cr	Cu	Mn	Mo	P	S	Si	Other	UTS	YS	El	Hard

Alloy Steel, Chromium, 5132/5132H (Continued from previous page)

USA

SAE J1268(95)	5132H	Bar Rod Wir Sh Strp Tub; See std	0.29-0.35	0.65-1.10	0.35 max	0.50-0.90	0.06 max	0.030 max	0.040 max	0.15-0.35	Ni <=0.25; Cu, Mo, Ni not spec'd but acpt; bal Fe				
SAE J404(94)	5132	Bil Blm Slab Bar HR CF	0.30-0.35	0.75-1.00		0.60-0.80		0.030 max	0.040 max	0.15-0.35	bal Fe				
SAE J407	5132H*	Obs; see J1268									bal Fe				

Yugoslavia

	C.4130	Q/T	0.3-0.37	0.9-1.2		0.6-0.9	0.15 max	0.035 max	0.03 max	0.4 max	bal Fe				
	C.4180	Q/T	0.3-0.37	0.9-1.2		0.6-0.9	0.15 max	0.035 max	0.02-0.035	0.4 max	bal Fe				

Alloy Steel, Chromium, 5135/5135H

Australia

AS G18	En18C*	Obs; see AS 1444	0.35-0.38	0.85-1.15		0.65-0.8		0.05		0.1-0.35	Ni 0.05; bal Fe				

Belgium

NBN 253-02	37Cr4		0.34-0.41	0.9-1.2		0.6-0.9		0.035		0.15-0.4	Ni 0.035; bal Fe				
NBN 253-02	41Cr4		0.38-0.45	0.9-1.2		0.5-0.8		0.04 max	0.04 max	0.15-0.4	bal Fe				
NBN 253-06	38Cr4		0.34-0.4	0.9-1.2		0.6-0.9		0.035		0.15-0.4	Ni 0.025; bal Fe				

Bulgaria

BDS 6354	35Ch		0.31-0.38	0.80-1.10	0.3	0.50-0.80		0.035		0.17-0.37	Ni 0.035; bal Fe				
BDS 6354	35ChGS		0.32-0.40	1.10-1.40	0.3	0.80-1.10		0.035		1.10-1.40	Ni 0.035; bal Fe				
BDS 6354	38ChA	Struct	0.35-0.42	1.00-1.30	0.30 max	0.50-0.80		0.025 max	0.025 max	0.17-0.37	Ni <=0.30; Ti <=0.03; V <=0.05; W <=0.2; bal Fe				

China

GB 11251(89)	35Cr	Plt HR Ann	0.32-0.39	0.80-1.10		0.50-0.80		0.035 max	0.035 max	0.17-0.37	bal Fe	550-750		18	
GB 3077(88)	35Cr	Bar HR Q/T 25mm diam	0.32-0.39	0.80-1.10	0.30 max	0.50-0.80		0.035 max	0.035 max	0.17-0.37	bal Fe	930	735	11	
GB 8162(87)	35Cr	Smls tube HR/CD Q/T	0.32-0.39	0.80-1.10		0.50-0.80		0.035 max	0.035 max	0.17-0.37	bal Fe	930	735	11	
GB/T 3078(94)	35Cr	Bar CD Q/T 25mm diam	0.32-0.39	0.80-1.10	0.30 max	0.50-0.80		0.035 max	0.035 max	0.17-0.37	bal Fe	930	735	11	
YB/T 5132(93)	35Cr	Sh HR/CR Ann	0.32-0.39	0.80-1.10		0.50-0.80		0.035 max	0.035 max	0.17-0.37	bal Fe	540-735		16	

Europe

EN 10083/1(91)A1(96)	1.6773	Q/T t<=16mm	0.34-0.41	0.90-1.20		0.60-0.90		0.035 max	0.035 max	0.40 max	bal Fe	950-1150	750	11	
EN 10083/1(91)A1(96)	1.7038	Q/T t<=16mm	0.34-0.41	0.90-1.20		0.60-0.90		0.035 max	0.020-0.040	0.40 max	bal Fe	950-1150	750	11	
EN 10083/1(91)A1(96)	36NiCrMo16	Q/T t<=16mm	0.34-0.41	0.90-1.20		0.60-0.90		0.035 max	0.035 max	0.40 max	bal Fe	950-1150	750	11	
EN 10083/1(91)A1(96)	37CrS4	Q/T t<=16mm	0.34-0.41	0.90-1.20		0.60-0.90		0.035 max	0.020-0.040	0.40 max	bal Fe	950-1150	750	11	

France

AFNOR NFA24452(76)	35CD4TS		0.33-0.39	0.85-1.15	0.3 max	0.6-0.9	0.15-0.3	0.025 max	0.035 max	0.1-0.4	Ni <=0.4; bal Fe				
AFNOR NFA35552	38Cr4		0.35-0.4	0.85-1.2		0.6-0.9		0.035	0.035	0.1-0.4	bal Fe				
AFNOR NFA35552	42C4	Bar Rod	0.4-0.45	0.9-1.2		0.6-0.9		0.035 max	0.035 max	0.1-0.4	bal Fe				
AFNOR NFA35553	38Cr4		0.35-0.4	0.85-1.2		0.6-0.9		0.035	0.035	0.1-0.4	bal Fe				
AFNOR NFA35556(84)	38C4*		0.35-0.4	0.85-1.2		0.6-0.9		0.035 max	0.035 max	0.1-0.4	bal Fe				
AFNOR NFA35556(84)	42C4		0.4-0.45	0.9-1.2		0.6-0.9		0.035 max	0.035 max	0.1-0.4	bal Fe				
AFNOR NFA35557	38Cr4		0.35-0.4	0.85-1.2		0.6-0.9		0.035	0.035	0.1-0.4	bal Fe				

UNS numbers and US grades are provided as a means of cross referencing chemically similar alloys. Exchangability is only possible after independent examination of specifications. Tensile properties are minimum or typical as specified. UTS and YS as MPa. El as %. See Appendix for list of abbreviations used in Notes. * indicates obsolete material.

Alloy Steel, Chromium, 5135/5135H (Continued from previous page)

Specification	Designation	Notes	C	Cr	Cu	Mn	Mo	P	S	Si	Other	UTS	YS	El	Hard
France															
AFNOR NFA35557	42C4		0.4-0.45	0.9-1.2		0.6-0.9		0.035 max	0.035 max	0.1-0.4	bal Fe				
AFNOR NFA35563(83)	38C4TS	Soft ann	0.35-0.4	0.9-1.2		0.6-0.9		0.025 max	0.035 max	0.1-0.4	bal Fe				217 HB max
AFNOR NFA35564(83)	38C4FF	Soft ann	0.35-0.4	0.9-1.2		0.6-0.9		0.03 max	0.03 max	0.35 max	Al >=0.02; bal Fe	0-580			217 HB max
Germany															
DIN 1652(90)	37Cr4	Q/T 17-40mm	0.34-0.41	0.90-1.20		0.60-0.90		0.035 max	0.035 max	0.40 max	bal Fe	850-1100	630	13	
DIN 1652(90)	WNr 1.7034	Q/T 17-40mm	0.34-0.41	0.90-1.20		0.60-0.90		0.035 max	0.035 max	0.40 max	bal Fe	850-1000	630	13	
DIN 17212(72)	38Cr4	Hard	0.34-0.40	0.90-1.20		0.60-0.90		0.025 max	0.035 max	0.15-0.40	bal Fe	830-980	630	13	53-58 HRC
DIN 17212(72)	WNr 1.7043	Hard 17-40mm	0.34-0.40	0.90-1.20		0.60-0.90		0.025 max	0.035 max	0.15-0.40	bal Fe	830-980	630	13	53-58 HRC
DIN EN 10083(91)	37Cr4	Q/T 17-40mm	0.34-0.41	0.90-1.20		0.60-0.90		0.035 max	0.035 max	0.40 max	bal Fe	850-1100	630	13	
DIN EN 10083(91)	37CrS4	Q/T	0.34-0.41	0.90-1.20		0.60-0.90		0.035 max	0.020-0.040	0.40 max	bal Fe				
DIN EN 10083(91)	WNr 1.7034	Q/T 17-40mm	0.34-0.41	0.90-1.20		0.60-0.90		0.035 max	0.035 max	0.40 max	bal Fe	850-1000	630	13	
DIN EN 10083(91)	WNr 1.7038	Q/T	0.34-0.41	0.90-1.20		0.60-0.90		0.035 max	0.020-0.040	0.40 max					
Hungary															
MSZ 61(85)	Cr2	Q/T; 0-16mm	0.34-0.41	0.9-1.2		0.6-0.9		0.035 max	0.035 max	0.4 max	bal Fe	950-1150	750	11L	
MSZ 61(85)	Cr2	Q/T; 40.1-100mm	0.34-0.41	0.9-1.2		0.6-0.9		0.035 max	0.035 max	0.4 max	bal Fe	750-900	510	14L	
MSZ 61(85)	Cr2E	Q/T; 0-16mm	0.34-0.41	0.9-1.2		0.6-0.9		0.035 max	0.02-0.035	0.4 max	bal Fe	950-1150	750	11L	
MSZ 61(85)	Cr2E	Q/T; 40.1-100mm	0.34-0.41	0.9-1.2		0.6-0.9		0.035 max	0.02-0.035	0.4 max	bal Fe	750-900	510	14L	
MSZ 6251(87)	Cr2Z	CF; Q/T; 0-36mm; HR; HR/soft ann; drawn/soft ann; drawn/bright ann; soft ann/ground	0.34-0.41	0.9-1.2		0.6-0.9		0.035 max	0.035 max	0.4 max	bal Fe	0-650			
MSZ 6251(87)	Cr2Z	CF; Q/T; 0-36mm; HR; HR/soft ann; drawn/soft ann; drawn/bright ann; soft ann/ground	0.34-0.41	0.9-1.2		0.6-0.9		0.035 max	0.035 max	0.4 max	bal Fe	0-620			
India															
IS 1570/4(88)	38Cr4BT		0.35-0.4	0.95-1.15		0.3-0.5		0.07 max	0.06 max	0.15-0.3	B 0.0005-0.003; Ni <=0.4; bal Fe				
International															
ISO 683-12(72)	7*	F/IH	0.34-0.4	0.9-1.2		0.6-0.9		0.035 max	0.035 max	0.15-0.4	Ni <=0.4; bal Fe				
ISO 683-7	2*		0.34-0.41	0.9-1.2		0.6-0.9		0.035	0.035	0.15-0.4	bal Fe				
ISO 683-7	2a*		0.34-0.41	0.9-1.2		0.6-0.9		0.035	0.02-0.035	0.15-0.4	bal Fe				
ISO 683-7	2b*		0.34-0.41	0.9-1.2		0.6-0.9		0.035	0.035-0.05	0.15-0.4	bal Fe				
ISO R683-7	2a	Bar Bil, Rod	0.34-0.41	0.9-1.2		0.6-0.9		0.04	0.02-0.04	0.15-0.4	bal Fe				
ISO R683-7	2b	Bar Bil, Rod	0.34-0.41	0.9-1.2		0.6-0.9		0.04	0.04-0.05	0.15-0.4	bal Fe				
ISO R683-7(70)	2	Q/T	0.34-0.41	0.9-1.2		0.6-0.9		0.035 max	0.035 max	0.15-0.4	Ni <=0.4; bal Fe				
Italy															
UNI 5332	35CrMn5		0.32-0.39	1-1.3		0.8-1.1		0.035	0.035	0.4	bal Fe				
UNI 5333	36CrMn4		0.33-0.39	0.9-1.2		0.8-1.1		0.035	0.03	0.4	bal Fe				
UNI 6403(86)	35CrMn5		0.32-0.39	1-1.3		0.8-1.1		0.035	0.035	0.4	bal Fe				

UNS numbers and US grades are provided as a means of cross referencing chemically similar alloys. Exchangability is only possible after independent examination of specifications. Tensile properties are minimum or typical as specified. UTS and YS as MPa. El as %. See Appendix for list of abbreviations used in Notes. * indicates obsolete material.

Specification	Designation	Notes	C	Cr	Cu	Mn	Mo	P	S	Si	Other	UTS	YS	El	Hard

Alloy Steel, Chromium, 5135/5135H (Continued from previous page)

Italy

UNI 6403(86)	36CrMn5	Q/T	0.33-0.4	1-1.3		0.8-1.1		0.035 max	0.035 max	0.15-0.35	bal Fe				
UNI 7356(74)	38Cr4KB	Q/T	0.34-0.41	0.9-1.2		0.6-0.9		0.035 max	0.035 max	0.15-0.4	bal Fe				
UNI 7845(78)	36CrMn5	Q/T	0.33-0.4	1-1.3		0.8-1.1		0.035 max	0.035 max	0.15-0.4	Pb <=0.3; bal Fe				
UNI 7847(79)	36CrMn4	For SH	0.33-0.39	0.9-1.2		0.8-1.1		0.035	0.03	0.4	bal Fe				
UNI 7847(79)	38Cr4	For SH	0.34-0.4	0.9-1.2		0.6-0.9		0.03 max	0.03 max	0.15-0.4	bal Fe				
UNI 7874(79)	36CrMn5		0.33-0.4	1-1.3		0.8-1.1		0.035 max	0.035 max	0.15-0.4	Ni <=0.3; Pb 0.15-0.3; bal Fe				
UNI 8551(84)	37CrMn4	For SH	0.33-0.4	0.9-1.2		0.8-1.1		0.03 max	0.025 max	0.15-0.4	Pb 0.15-0.25; bal Fe				
UNI 8551(84)	38Cr4	For SH	0.34-0.41	0.9-1.2		0.6-0.9		0.03 max	0.025 max	0.15-0.4	bal Fe				

Japan

JIS G3311(88)	SCr435M	CR Strp; Chain	0.33-0.38	0.90-1.20	0.30 max	0.60-0.85		0.030 max	0.030 max	0.15-0.35	Ni <=0.25; bal Fe				190-270 HV
JIS G4052(79)	SCr3H*	Obs; see SCr 435H	0.32-0.38	0.85-1.25		0.55-0.9		0.03	0.03	0.15-0.35	bal Fe				
JIS G4052(79)	SCr435H	Struct Hard	0.32-0.39	0.85-1.25	0.30 max	0.55-0.90		0.030 max	0.030 max	0.15-0.35	Ni <=0.25; bal Fe				

Mexico

DGN B-203	5135*	Obs	0.33-0.38	0.8-1.05		0.6-0.8		0.04	0.04	0.2-0.35	bal Fe				
DGN B-297	5135*	Obs	0.33-0.38	0.8-1.05		0.6-0.9		0.035	0.04	0.2-0.35	bal Fe				
NMX-B-268(68)	5135H	Hard	0.32-0.38	0.70-1.15		0.50-0.90				0.20-0.35	bal Fe				
NMX-B-300(91)	5135	Bar	0.33-0.38	0.80-1.05		0.60-0.80		0.035 max	0.040 max	0.15-0.35	bal Fe				
NMX-B-300(91)	5135H	Bar	0.32-0.38	0.70-1.15		0.50-0.90				0.15-0.35	bal Fe				

Pan America

| COPANT 334 | 5135 | Bar | 0.33-0.38 | 0.8-1.05 | | 0.6-0.8 | | 0.035 | 0.04 | 0.2-0.35 | bal Fe | | | | |

Poland

PNH84030/04	35HGS	Q/T	0.32-0.4	1.1-1.4		0.8-1.1		0.03 max	0.025 max	1.1-1.4	Ni <=0.3; bal Fe				
PNH84030/04	35HGSA	Q/T	0.32-0.39	1.1-1.4		0.8-1.1		0.03 max	0.025 max	1.1-1.4	Ni <=0.3; bal Fe				
PNH84030/04	38HA	Q/T	0.35-0.45	0.8-1.1		0.5-0.8		0.025 max	0.025 max	0.17-0.37	Ni <=0.3; bal Fe				

Romania

| STAS 11500/2(89) | 37Cr10 | Q/T | 0.34-0.4 | 0.85-1.15 | | 0.6-0.9 | | 0.03 max | 0.02-0.04 | 0.15-0.4 | bal Fe | | | | |

Russia

GOST 10702	38ChA		0.35-0.41	0.8-1.1	0.2 max	0.5-0.8		0.025 max	0.025 max	0.17-0.37	Ni <=0.03; bal Fe				
GOST 1414(75)	AS35G2	Free-cutting	0.32-0.39	0.25 max	0.25 max	1.35-1.65		0.04 max	0.08-0.13	0.17-0.37	Ni <=0.25; Pb 0.15-0.3; bal Fe				
GOST 4543	30Ch		0.31-0.39	0.8-1.1	0.3	0.5-0.8		0.035	0.035	0.17-0.37	Ni 0.3; bal Fe				
GOST 4543	35ChGSA		0.32-0.39	1.1-1.4	0.3	0.8-1.1		0.025	0.025	1.1-1.4	Ni 0.3; bal Fe				
GOST 4543	38ChA		0.35-0.42	0.8-1.1	0.3	0.5-0.8		0.025 max	0.025 max	0.17-0.37	Ni <=0.3; bal Fe				
GOST 4543	40Ch		0.34-0.44	0.8-1.1	0.3 max	0.5-0.8		0.04 max	0.04 max	0.17-0.37	Ni <=0.3; bal Fe				
GOST 4543(71)	35ChGSA	Q/T	0.32-0.39	1.1-1.4	0.3 max	0.8-1.0		0.025 max	0.025 max	1.1-1.4	Ni <=0.3; bal Fe				
GOST 4543(71)	38ChA	Q/T	0.35-0.42	0.8-1.1	0.3 max	0.5-0.8		0.025 max	0.025 max	0.17-0.37	Ni <=0.3; bal Fe				
GOST 4543(71)	40Ch	Q/T	0.36-0.44	0.8-1.1	0.3 max	0.5-0.8		0.035 max	0.035 max	0.17-0.37	Ni <=0.3; bal Fe				

Specification	Designation	Notes	C	Cr	Cu	Mn	Mo	P	S	Si	Other	UTS	YS	El	Hard

Alloy Steel, Chromium, 5135/5135H (Continued from previous page)

Russia

Specification	Designation	Notes	C	Cr	Cu	Mn	Mo	P	S	Si	Other	UTS	YS	El	Hard
GOST 808	SChCh10	Soft ann Strp, bent bearing rollers	0.32-0.42	0.8-1.2	0.25 max	0.4-0.7		0.03 max	0.03 max	0.17-0.37	Ni <=0.2; bal Fe				

Spain

Specification	Designation	Notes	C	Cr	Cu	Mn	Mo	P	S	Si	Other	UTS	YS	El	Hard
UNE 36012(75)	38Cr3	Q/T	0.34-0.41	0.5-0.8		0.6-0.9		0.035	0.035	0.15-0.4	bal Fe				
UNE 36012(75)	38Cr4	Q/T	0.34-0.41	0.9-1.2		0.6-0.9		0.035 max	0.035 max	0.15-0.4	bal Fe				
UNE 36012(75)	38Cr4-1		0.34-0.41	0.9-1.2		0.6-0.9		0.035	0.02-0.035	0.15-0.4	bal Fe				
UNE 36012(75)	F.1200*	see 38Cr3	0.34-0.41	0.5-0.8		0.6-0.9		0.035	0.035	0.15-0.4	bal Fe				
UNE 36012(75)	F.1201*	see 38Cr4	0.34-0.41	0.9-1.2		0.6-0.9		0.035 max	0.035 max	0.15-0.4	bal Fe				
UNE 36012(75)	F.1206*	see 38Cr4-1	0.34-0.41	0.9-1.2		0.6-0.9		0.035	0.02-0.035	0.15-0.4	bal Fe				
UNE 36034(85)	38Cr4DF	CHd; Cold ext	0.34-0.41	0.9-1.2		0.6-0.9		0.035 ·	0.035	0.15-0.4	bal Fe				
UNE 36034(85)	F.1210*	see 38Cr4DF	0.34-0.41	0.9-1.2		0.6-0.9		0.035	0.035	0.15-0.4	bal Fe				
UNE 36254(79)	F.8221*	see 35Cr4	0.3-0.4	0.8-1.2		0.5-0.8		0.04	0.04	0.3-0.5	bal Fe				
UNE 36255(79)	AM35Cr4*		0.3-0.4	0.8-1.2		0.5-0.8		0.040	0.040	0.3-0.5	bal Fe				
UNE 36255(79)	F.8321*	see AM35Cr4*	0.3-0.4	0.8-1.2		0.5-0.8		0.040	0.040	0.3-0.5	bal Fe				

UK

Specification	Designation	Notes	C	Cr	Cu	Mn	Mo	P	S	Si	Other	UTS	YS	El	Hard
BS 3111/1(87)	3/1	Q/T Wir	0.35-0.45	0.90-1.20		0.70-0.90		0.035 max	0.035 max	0.15-0.40	bal Fe				
BS 970/1(83)	530A36	Blm Bil Slab Bar Rod Frg nitriding	0.34-0.39	0.90-1.20		0.60-0.80		0.035 max	0.040 max		bal Fe				
BS 970/1(83)	530H36	Blm Bil Slab Bar Rod Frg	0.33-0.40	0.80-1.25		0.50-0.90		0.035 max	0.040 max		bal Fe				
BS 970/1(83)	530H40	Blm Bil Slab Bar Rod Frg	0.37-0.44	0.80-1.25		0.50-0.90		0.035 max	0.040 max		bal Fe				
BS 970/1(83)	530M40	Blm Bil Slab Bar Rod Frg; t<=100mm	0.36-0.44	0.90-1.20		0.60-0.90		0.035 max	0.040 max		bal Fe	700-850	525	17	201-255 HB

USA

Specification	Designation	Notes	C	Cr	Cu	Mn	Mo	P	S	Si	Other	UTS	YS	El	Hard
	AISI 5135	Smls mech tub	0.33-0.38	0.80-1.05		0.60-0.80		0.040 max	0.040 max	0.15-0.35	bal Fe				
	AISI 5135H	Smls mech tub, Hard	0.32-0.38	0.70-1.15		0.50-0.60		0.035 max	0.040 max	0.15-0.30	bal Fe				
	UNS G51350		0.33-0.38	0.80-1.05		0.60-0.80		0.035 max	0.040 max	0.15-0.35	bal Fe				
	UNS H51350		0.32-0.38	0.70-1.15		0.50-0.90		0.035 max	0.040 max	0.15-0.35	bal Fe				
ASTM A29/A29M(93)	5135	Bar	0.33-0.38	0.80-1.05		0.6-0.8		0.035 max	0.04 max	0.15-0.35	bal Fe				
ASTM A304(96)	5135H	Bar, hard bands spec	0.32-0.38	0.70-1.15	0.35 max	0.50-0.90	0.06 max	0.035 max	0.040 max	0.15-0.35	Cu Ni Cr Mo trace allowed; bal Fe				
ASTM A322(96)	5135	Bar	0.33-0.38	0.80-1.05		0.60-0.80		0.035 max	0.040 max	0.15-0.35	bal Fe				
ASTM A331(95)	5135	Bar	0.33-0.38	0.80-1.05		0.60-0.80		0.035 max	0.040 max	0.15-0.35	bal Fe				
ASTM A331(95)	5135H	Bar	0.32-0.38	0.70-1.15	0.35 max	0.50-0.90	0.06 max	0.025 max	0.025 max	0.15-0.35	bal Fe				
ASTM A519(96)	5135	Smls mech tub	0.33-0.38	0.80-1.05		0.60-0.80		0.040 max	0.040 max	0.15-0.35	bal Fe				
ASTM A752(93)	5135	Rod Wir	0.33-0.38	0.80-1.05		0.60-0.80		0.035 max	0.040 max	0.15-0.30	bal Fe	590			
SAE 770(84)	5135*	Obs; see J1397(92)									bal Fe				
SAE J1268(95)	5135H	Bar Rod Wir Tub; See std	0.32-0.38	0.70-1.15	0.35 max	0.50-0.90	0.06 max	0.030 max	0.040 max	0.15-0.35	Ni <=0.25; Cu, Mo, Ni not spec'd but acpt; bal Fe				

Yugoslavia

Specification	Designation	Notes	C	Cr	Cu	Mn	Mo	P	S	Si	Other	UTS	YS	El	Hard
	C.4134	Q/T	0.34-0.41	0.9-1.2		0.6-0.9		0.035 max	0.03 max	0.4 max	bal Fe				
	C.4136	F/IH	0.34-0.4	0.9-1.2		0.6-0.9		0.025 max	0.035 max	0.15-0.4	Ni <=0.25; bal Fe				
	C.4181	Q/T	0.38-0.45	0.9-1.2		0.6-0.9		0.035 max	0.02-0.035	0.4 max	bal Fe				

UNS numbers and US grades are provided as a means of cross referencing chemically similar alloys. Exchangability is only possible after independent examination of specifications. Tensile properties are minimum or typical as specified. UTS and YS as MPa. El as %. See Appendix for list of abbreviations used in Notes. * indicates obsolete material.

Specification	Designation	Notes	C	Cr	Cu	Mn	Mo	P	S	Si	Other	UTS	YS	El	Hard

Alloy Steel, Chromium, 5135/5135H (Continued from previous page)

Yugoslavia

| | C.4184 | Q/T | 0.34-0.41 | 0.9-1.2 | | 0.6-0.9 | | 0.035 max | 0.02-0.035 | 0.4 max | bal Fe | | | | |

Alloy Steel, Chromium, 5140/5140H

Argentina

| IAS | IRAM 5140 | | 0.38-0.43 | 0.70-0.90 | | 0.70-0.90 | | 0.035 max | 0.040 max | 0.20-0.35 | bal Fe | 770 | 520 | 20 | 229 HB |

Australia

AS 1440(73)	5140*	Withdrawn	0.38-0.43	0.7-0.9		0.7-0.9		0.04	0.04	0.1-0.35	bal Fe				
AS 1444(96)	5140	H series, Hard/Tmp	0.38-0.43	0.70-0.90		0.70-0.90		0.040 max	0.040 max	0.10-0.35	bal Fe				
AS G18	En18*	Obs; see AS 1444	0.35-0.45	0.85-1.15		0.6-0.95		0.05	0.05	0.1-0.35	bal Fe				
AS G18	En18D*	Obs; see AS 1444	0.38-0.43	0.85-1.15		0.65-0.8		0.05	0.05	0.1-0.35	bal Fe				

Austria

| ONORM M3167 | 41Cr4SP | | 0.38-0.44 | 0.9-1.2 | | 0.5-0.8 | | 0.040 max | 0.040 max | 0.15-0.35 | bal Fe | | | | |
| ONORM M3167 | 41Cr4SP | Bar | 0.38-0.44 | 0.9-1.2 | | 0.5-0.8 | | 0.040 max | 0.040 max | 0.15-0.35 | bal Fe | | | | |

Belgium

| NBN 251 | 41Cr4 | | 0.38-0.45 | 0.9-1.2 | | 0.5-0.8 | | 0.035 | 0.035 | 0.15-0.4 | bal Fe | | | | |
| NBN 251 | 46Cr2 | | 0.42-0.5 | 0.4-0.6 | | 0.5-0.8 | | 0.04 max | 0.04 max | 0.15-0.4 | bal Fe | | | | |

Bulgaria

| BDS 6354 | 40Ch | | 0.36-0.44 | 0.80-1.10 | 0.2 | 0.50-0.80 | | 0.035 | 0.035 | 0.17-0.37 | Ni 0.25; bal Fe | | | | |

China

GB 11251(89)	40Cr	Plt HR Ann	0.37-0.44	0.80-1.10		0.50-0.80		0.035 max	0.035 max	0.17-0.37	bal Fe	550-800		16	
GB 3077(88)	40Cr	Bar HR Q/T 25mm diam	0.37-0.44	0.80-1.10	0.30 max	0.50-0.80		0.035 max	0.035 max	0.17-0.37	bal Fe	980	785	9	
GB 5216(85)	40CrH	Bar Rod HR/Frg Ann	0.37-0.44	0.80-1.10	0.30 max	0.50-0.80		0.035 max	0.035 max	0.17-0.37	bal Fe				207 max HB
GB 6478(86)	ML40Cr	Bar HR Q/T	0.37-0.44	0.80-1.10	0.20 max	0.50-0.80		0.035 max	0.035 max	0.17-0.37	bal Fe	980	785	9	
GB 8162(87)	40Cr	Smls tube HR/CD Q/T	0.37-0.44	0.80-1.10		0.50-0.80		0.035 max	0.035 max	0.17-0.37	bal Fe	980	785	9	
GB/T 3078(94)	40Cr	Bar CD Q/T 25mm diam	0.37-0.44	0.80-1.10	0.30 max	0.50-0.80		0.035 max	0.035 max	0.17-0.37	bal Fe	980	785	9	
GB/T 3079(93)	40CrA	Wir Q/T	0.37-0.44	0.80-1.10	0.25 max	0.50-0.80		0.035 max	0.035 max	0.17-0.37	bal Fe	980	785	9	
YB/T 5132(93)	40Cr	Sh HR/CR Ann	0.37-0.44	0.80-1.10		0.50-0.80		0.035 max	0.035 max	0.17-0.37	bal Fe	540-785		14	

Czech Republic

| CSN 414140 | 14140 | Q/T | 0.35-0.42 | 0.8-1.1 | | 0.5-0.8 | | 0.035 max | 0.035 max | 0.17-0.37 | bal Fe | | | | |
| CSN 414341 | 14341 | Q/T | 0.34-0.42 | 1.3-1.6 | | 0.3-0.6 | | 0.035 max | 0.035 max | 1.0-1.3 | bal Fe | | | | |

Europe

EN 10083/1(91)A1(96)	1.7003	Q/T t<=16mm	0.35-0.42	0.40-0.60		0.50-0.80	0.15 max	0.035 max	0.035 max	0.40 max	bal Fe	800-950	550	14	
EN 10083/1(91)A1(96)	1.7023	Q/T t<=16mm	0.35-0.42	0.40-0.60		0.50-0.80	0.15 max	0.035 max	0.020-0.040	0.40 max	bal Fe	800-950	550	14	
EN 10083/1(91)A1(96)	1.7035	Q/T t<=16mm	0.38-0.45	0.90-1.20		0.60-0.90	0.15 max	0.035 max	0.035 max	0.40 max	bal Fe	1000-1200	800	11	
EN 10083/1(91)A1(96)	1.7039	Q/T t<=16mm	0.38-0.45	0.90-1.20		0.60-0.90	0.15 max	0.035 max	0.020-0.040	0.40 max	bal Fe	1000-1200	800	11	
EN 10083/1(91)A1(96)	38Cr2	Q/T t<=16mm	0.35-0.42	0.40-0.60		0.50-0.80	0.15 max	0.035 max	0.035 max	0.40 max	bal Fe	800-950	550	14	
EN 10083/1(91)A1(96)	38CrS2	Q/T t<=16mm	0.35-0.42	0.40-0.60		0.50-0.80	0.15 max	0.035 max	0.020-0.040	0.40 max	bal Fe	800-950	550	14	
EN 10083/1(91)A1(96)	41Cr4	Q/T t<=16mm	0.38-0.45	0.90-1.20		0.60-0.90	0.15 max	0.035 max	0.035 max	0.40 max	bal Fe	1000-1200	800	11	

Specification	Designation	Notes	C	Cr	Cu	Mn	Mo	P	S	Si	Other	UTS	YS	El	Hard
Alloy Steel, Chromium, 5140/5140H (Continued from previous page)															
Europe															
EN 10083/1(91)A1(96)	41CrS4	Q/T t<=16mm	0.38-0.45	0.90-1.20		0.60-0.90	0.15 max	0.035 max	0.020-0.040	0.40 max	bal Fe	1000-1200	800	11	
France															
AFNOR NFA35552	42C4		0.4-0.45	0.9-1.2		0.6-0.9		0.035	0.035	0.1-0.4	bal Fe				
AFNOR NFA35552	45C2		0.4-0.5	0.4-0.6		0.5-0.8		0.04 max	0.04 max	0.35 max	bal Fe				
AFNOR NFA35553	45C4	Strp	0.41-0.48	0.85-1.15		0.6-0.9		0.04	0.04	0.1-0.4	bal Fe				
AFNOR NFA35553(82)	45C4	Quen; t<=4.5mm	0.41-0.48	0.85-1.15	0.3 max	0.6-0.9	0.15 max	0.035 max	0.035 max	0.1-0.4	Ni <=0.4; Ti <=0.05; bal Fe				54 HRC
AFNOR NFA35556(84)	42C4		0.39-0.45	0.85-1.15		0.6-0.9		0.025 max	0.035 max	0.1-0.4	bal Fe				
AFNOR NFA35557	42C2		0.4-0.46	0.3-0.6		0.6-0.9		0.04 max	0.04 max	0.1-0.4	bal Fe				
AFNOR NFA35557	42C4		0.4-0.45	0.9-1.2		0.6-0.9		0.035	0.035	0.1-0.4	bal Fe				
AFNOR NFA35563(83)	42C4TS	HT	0.4-0.45	0.9-1.2	0.3 max	0.6-0.9	0.15 max	0.025 max	0.035 max	0.1-0.4	Ni <=0.4; Ti <=0.05; bal Fe				56 HRC
AFNOR NFA35571	45C4	Bar Rod Wir, Q/A Tmp, 80mm	0.41-0.48	0.85-1.15		0.6-0.9		0.04	0.04	0.1-0.4	bal Fe				
Germany															
DIN 1652(90)	41Cr4	Q/T 17-40mm	0.38-0.45	0.90-1.20		0.60-0.90		0.035 max	0.035 max	0.40 max	bal Fe	900-1100	660	12	
DIN 1652(90)	WNr 1.7035	Q/T 17-40mm	0.38-0.45	0.90-1.20		0.60-0.90		0.035 max	0.035 max	0.40 max	bal Fe	900-1100	660	12	
DIN 17212(72)	42Cr4	Hard	0.38-0.44	0.00-1.20		0.50-0.80		0.025 max	0.035 max	0.15-0.40	bal Fe	880-1080	665	12	54-60 HRC
DIN 17212(72)	WNr 1.7045	Hard 17-40mm	0.38-0.44	0.90-1.20		0.50-0.80		0.025 max	0.035 max	0.15-0.40	bal Fe	880-1080	665	12	54-60 HRC
DIN EN 10083(91)	41CrS4	Q/T	0.38-0.45	0.90-1.20		0.60-0.90		0.035 max	0.020-0.040	0.40 max	bal Fe				
DIN EN 10083(91)	WNr 1.7035	Q/T 17-40mm	0.38-0.45	0.90-1.20		0.60-0.90		0.035 max	0.035 max	0.40 max	bal Fe	900-1100	660	12	
DIN EN 10083(91)	WNr 1.7039	Q/T	0.38-0.45	0.90-1.20		0.60-0.90		0.035 max	0.020-0.040	0.40 max	bal Fe				
DIN EN 1083(91)	41Cr4	Q/T 17-40mm	0.38-0.45	0.90-1.20		0.60-0.90		0.035 max	0.035 max	0.40 max	bal Fe	900-1100	660	12	
Hungary															
MSZ 61(85)	Cr3	Q/T; 0-16mm	0.38-0.45	0.9-1.2		0.6-0.9		0.035 max	0.035 max	0.4 max	bal Fe	1000-1200	800	11L	
MSZ 61(85)	Cr3	Q/T; 40.1-100mm	0.38-0.45	0.9-1.2		0.6-0.9		0.035 max	0.035 max	0.4 max	bal Fe	800-950	560	14L	
MSZ 61(85)	Cr3E	Q/T; 0-16mm	0.38-0.45	0.9-1.2		0.6-0.9		0.035 max	0.02-0.035	0.4 max	bal Fe	1000-1200	800	11L	
MSZ 61(85)	Cr3E	Q/T; 40.1-100mm	0.38-0.45	0.9-1.2		0.6-0.9		0.035 max	0.02-0.035	0.4 max	bal Fe	800-950	560	14L	
MSZ 6251(87)	Cr3Z	CF; Q/T; 0-36mm; drawn, half-hard	0.38-0.45	0.9-1.2		0.6-0.9		0.035 max	0.035 max	0.4 max	bal Fe	0-660			
MSZ 6251(87)	Cr3Z	CF; Q/T; 0-36mm; HR; HR/soft ann; drawn/soft ann; drawn/bright ann; soft ann/ground	0.38-0.45	0.9-1.2		0.6-0.9		0.035 max	0.035 max	0.4 max	bal Fe	0-630			
India															
IS 1570	40Cr1		0.35-0.45	0.9-1.2		0.6-0.9		0.05	0.05	0.1-0.35	bal Fe				
IS 1570/4(88)	40Cr4	Bar Frg Tube Pipe	0.35-0.45	0.9-1.2		0.6-0.9	0.15 max	0.07 max	0.06 max	0.1-0.35	Ni <=0.4; bal Fe	1090	890		
IS 4367	40Cr1		0.35-0.45	0.9-1.2		0.6-0.9		0.05	0.05	0.1-0.35	bal Fe				
IS 5517	40Cr1		0.35-0.45	0.9-1.2		0.6-0.9				0.1-0.35	bal Fe				
International															
ISO 683-12(72)	8*	F/IH	0.38-0.44	0.9-1.2	0.3 max	0.6-0.9	0.15 max	0.035 max	0.035 max	0.15-0.4	Al <=0.1; Co <=0.1; Ni <=0.4; Pb <=0.15; Ti <=0.05; V <=0.1; W <=0.1; bal Fe				

UNS numbers and US grades are provided as a means of cross referencing chemically similar alloys. Exchangability is only possible after independent examination of specifications. Tensile properties are minimum or typical as specified. UTS and YS as MPa. El as %. See Appendix for list of abbreviations used in Notes. * indicates obsolete material.

Specification	Designation	Notes	C	Cr	Cu	Mn	Mo	P	S	Si	Other	UTS	YS	El	Hard

Alloy Steel, Chromium, 5140/5140H (Continued from previous page)

International

Specification	Designation	Notes	C	Cr	Cu	Mn	Mo	P	S	Si	Other	UTS	YS	El	Hard
ISO 683-7	3*		0.38-0.45	0.9-1.2		0.6-0.9		0.035	0.035	0.15-0.4	bal Fe				
ISO 683-7	3a*		0.38-0.45	0.9-1.2		0.6-0.9		0.035	0.02-0.035	0.15-0.4	bal Fe				
ISO R683-7(70)	3	Q/T	0.38-0.45	0.9-1.2	0.3 max	0.6-0.9	0.15 max	0.035 max	0.035 max	0.15-0.4	Al <=0.1; Co <=0.1; Ni <=0.4; Pb <=0.15; Ti <=0.05; V <=0.1; W <=0.1; bal Fe				

Italy

Specification	Designation	Notes	C	Cr	Cu	Mn	Mo	P	S	Si	Other	UTS	YS	El	Hard
UNI 5332(64)	40Cr4	Q/T	0.37-0.44	0.9-1.2		0.5-0.8		0.035 max	0.035 max	0.4 max	bal Fe				
UNI 6403(86)	41Cr4	Q/T	0.38-0.45	0.9-1.2		0.5-0.8		0.035 max	0.035 max	0.15-0.35	bal Fe				
UNI 7356(74)	41Cr4KB	Q/T	0.38-0.45	0.9-1.2		0.5-0.8		0.035 max	0.035 max	0.15-0.4	bal Fe				
UNI 7845(78)	41Cr4	Q/T	0.38-0.45	0.9-1.2		0.5-0.8		0.035 max	0.035 max	0.15-0.4	bal Fe				
UNI 7874(79)	41Cr4	Q/T	0.38-0.45	0.9-1.2		0.5-0.9		0.035 max	0.035 max	0.15-0.4	bal Fe				

Japan

Specification	Designation	Notes	C	Cr	Cu	Mn	Mo	P	S	Si	Other	UTS	YS	El	Hard
JIS G3311(88)	SCr440M	CR Strp; Chain	0.38-0.43	0.90-1.20	0.30 max	0.60-0.85		0.030 max	0.030 max	0.15-0.35	Ni <=0.25; bal Fe				200-290 HV
JIS G4052(79)	SCr440H	Struct Hard	0.37-0.44	0.85-1.25	0.30 max	0.55-0.90		0.030 max	0.030 max	0.15-0.35	Ni <=0.25; bal Fe				
JIS G4052(79)	SCr4H*	Obs; see SCr440H	0.37-0.44	0.85-1.25		0.55-0.9		0.03	0.03	0.15-0.35	bal Fe				
JIS G4104(79)	SCr 4*	Obs; see SCr440	0.38-0.43	0.90-1.20	0.30 max	0.60-0.85		0.03 max	0.030 max	0.15-0.35	bal Fe				
JIS G4104(79)	SCr440	HR Frg Bar Wir Rod	0.38-0.43	0.90-1.20	0.30 max	0.60-0.85		0.030 max	0.030 max	0.15-0.35	Ni <=0.25; bal Fe				

Mexico

Specification	Designation	Notes	C	Cr	Cu	Mn	Mo	P	S	Si	Other	UTS	YS	El	Hard
DGN B-203	5140*	Obs	0.38-0.43	0.7-0.9		0.7-0.9		0.04	0.04	0.2-0.35	bal Fe				
DGN B-297	5140*	Obs	0.38-0.43	0.7-0.9		0.7-0.9		0.035	0.04	0.2-0.35	bal Fe				
NMX-B-268(68)	5140H	Hard	0.37-0.44	0.60-1.00		0.60-1.00				0.20-0.35	bal Fe				
NMX-B-300(91)	5140	Bar	0.38-0.43	0.70-0.90		0.70-0.90		0.035 max	0.040 max	0.15-0.35	bal Fe				
NMX-B-300(91)	5140H	Bar	0.37-0.44	0.60-1.00		0.60-1.00				0.15-0.35	bal Fe				

Pan America

Specification	Designation	Notes	C	Cr	Cu	Mn	Mo	P	S	Si	Other	UTS	YS	El	Hard
COPANT 334	5140	Bar	0.38-0.43	0.7-0.9		0.7-0.9		0.04	0.04	0.2-0.35	bal Fe				
COPANT 514	5140		0.38-0.43	0.7-0.9		0.7-0.9		0.04	0.05	0.1	bal Fe				
COPANT 514	5140	Tube	0.38-0.43	0.7-0.9		0.7-0.9		0.04	0.05	0.1	bal Fe				

Poland

Specification	Designation	Notes	C	Cr	Cu	Mn	Mo	P	S	Si	Other	UTS	YS	El	Hard
PNH84030	38HA	Q/T	0.35-0.42	0.8-1.1	0.3	0.5-0.8		0.03	0.03	0.17-0.37	Ni 0.3; bal Fe				
PNH84030/04	40H	Q/T	0.36-0.45	0.8-1.2		0.5-0.9		0.035 max	0.035 max	0.17-0.37	Ni <=0.3; bal Fe				
PNH84030/04	45H	Q/T	0.41-0.49	0.8-1.1		0.5-0.8		0.035 max	0.035 max	0.17-0.37	Ni <=0.3; bal Fe				

Romania

Specification	Designation	Notes	C	Cr	Cu	Mn	Mo	P	S	Si	Other	UTS	YS	El	Hard
STAS 791(88)	40BCr10	Struct/const	0.36-0.44	0.8-1.1		0.5-0.8		0.035 max	0.035 max	0.17-0.37	B 0.001-0.003; Pb <=0.15; Ti <=0.02; bal Fe				
STAS 791(88)	40BCr10S	Struct/const	0.36-0.44	0.8-1.1		0.3-0.8		0.035 max	0.02-0.04	0.17-0.37	B 0.001-0.003; Pb <=0.15; Ti <=0.02; bal Fe				
STAS 791(88)	40BCr10X	Struct/const	0.36-0.44	0.8-1.1		0.3-0.8		0.025 max	0.025 max	0.17-0.37	B 0.001-0.003; Pb <=0.15; Ti <=0.02; bal Fe				

Alloy Steel, Chromium, 5140/5140H (Continued from previous page)

Specification	Designation	Notes	C	Cr	Cu	Mn	Mo	P	S	Si	Other	UTS	YS	El	Hard
Romania															
STAS 791(88)	40BCr10XS	Struct/const	0.36-0.44	0.8-1.1		0.3-0.8		0.025 max	0.02-0.035	0.17-0.37	B 0.001-0.003; Pb <=0.15; Ti <=0.02; bal Fe				
STAS 791(88)	40Cr10	Struct/const; Q/T	0.36-0.44	0.8-1.1		0.5-0.8		0.035 max	0.035 max	0.17-0.37	Pb <=0.15; Ti <=0.02; bal Fe				
STAS 791(88)	40Cr10S	Struct/const; Q/T	0.36-0.44	0.8-1.1		0.5-0.8		0.035 max	0.02-0.04	0.17-0.37	Pb <=0.15; Ti <=0.02; bal Fe				
STAS 791(88)	40Cr10X	Struct/const; Q/T	0.36-0.44	0.8-1.1		0.5-0.8		0.025 max	0.025 max	0.17-0.37	Pb <=0.15; Ti <=0.02; bal Fe				
STAS 791(88)	40Cr10XS	Struct/const; Q/T	0.36-0.44	0.8-1.1		0.5-0.8		0.025 max	0.02-0.035	0.17-0.37	Pb <=0.15; Ti <=0.02; bal Fe				
STAS 9382/4	40BCr10q		0.36-0.44	0.8-1.1	0.3	0.5-0.8		0.035	0.035	0.17-0.37	B 0.001-0.003; Ni 0.3; As 0.05; bal Fe				
STAS 9382/4	40Cr10q		0.36-0.44	0.8-1.1	0.3	0.5-0.8		0.035	0.035	0.17-0.37	Ni 0.3; As 0.05; bal Fe				
Russia															
GOST 1414	A40ChE		0.36-0.44	0.8-1.1	0.3	0.5-0.8		0.035	0.06-0.12	0.17-0.37	Ni 0.3; Se 0.04-0.1; bal Fe				
GOST 1414(75)	AS40	Free-cutting	0.37-0.45	0.25 max	0.25 max	0.5-0.8	0.15 max	0.04 max	0.04 max	0.17-0.37	Ni <=0.25; Pb 0.15-0.3; bal Fe				
GOST 4543	38ChA		0.35-0.42	0.8-1.1	0.3	0.5-0.8		0.025	0.025	0.17-0.37	Ni 0.3; bal Fe				
GOST 4543	40Ch		0.34-0.44	0.8-1.1	0.3	0.5-0.8		0.035	0.035	0.17-0.37	Ni 0.3; bal Fe				
GOST 4543	40ChGTR		0.38-0.45	0.8-1.1	0.3	0.8-1		0.035	0.035	0.17-0.37	B >=0.001; Ni 0.3; Ti 0.03-0.09; bal Fe				
GOST 4543(71)	45Ch	Q/T	0.41-0.49	0.8-1.1	0.3 max	0.5-0.8	0.15 max	0.035 max	0.035 max	0.17-0.37	Al <=0.1; Ni <=0.3; Ti <=0.05; bal Fe				
GOST 7832	40ChL		0.35-0.45	0.8-1.1	0.3	0.4-0.9		0.04	0.04	0.2-0.4	Ni 0.3; bal Fe				
Spain															
UNE 36012(75)	40NiCrMo2	Q/T	0.37-0.44	0.4-0.7		0.7-1.0	0.15-0.3	0.035 max	0.035 max	0.15-0.4	Ni 0.4-0.7; bal Fe				
UNE 36012(75)	42Cr4	Q/T	0.38-0.45	0.9-1.2		0.6-0.9	0.15 max	0.035 max	0.035 max	0.15-0.4	bal Fe				
UNE 36012(75)	42Cr4-1	Q/T	0.38-0.45	0.9-1.2		0.6-0.9	0.15 max	0.035 max	0.02-0.035 max	0.15-0.4	bal Fe				
UNE 36012(75)	F.1202*	see 42Cr4	0.38-0.45	0.9-1.2		0.6-0.9	0.15 max	0.035 max	0.035 max	0.15-0.4	bal Fe				
UNE 36012(75)	F.1207*	see 42Cr4-1	0.38-0.45	0.9-1.2		0.6-0.9	0.15 max	0.035 max	0.02-0.035	0.15-0.4	bal Fe				
UNE 36034(85)	38Cr4DF	CHd; Cold ext	0.38-0.41	0.9-1.2		0.6-0.9		0.035	0.035	0.15-0.4	bal Fe				
UNE 36034(85)	41Cr4DF	CHd; Cold ext	0.38-0.45	0.9-1.2		0.6-0.9	0.15 max	0.035 max	0.035 max	0.15-0.4	bal Fe				
UNE 36034(85)	F.1211*	see 41Cr4DF	0.38-0.45	0.9-1.2		0.6-0.9	0.15 max	0.035 max	0.035 max	0.15-0.4	bal Fe				
Sweden															
SS 142245	2245	for surface hardening	0.38-0.45	0.9-1.2		0.6-0.9		0.035 max	0.02-0.04	0.15-0.4	bal Fe				
UK															
BS 3111/1(87)	3/2	Q/T Wir	0.40-0.45	0.90-1.20		0.70-0.90		0.035 max	0.035 max	0.15-0.40	bal Fe				
BS 970/1(83)	50H40*	Blm Bil Slab Bar Rod Frg	0.37-0.44	0.8-1.25		0.50-0.90		0.04	0.04	0.1-0.35	bal Fe				
BS 970/1(83)	530A40	Blm Bil Slab Bar Rod Frg nitriding	0.38-0.43	0.90-1.20		0.60-0.80		0.035 max	0.040 max		bal Fe				
BS 970/3(91)	530M40	Bright bar; Q/T; 63<t<=100mm; hard	0.36-0.44	0.90-1.20		0.60-0.90	0.15 max	0.035 max	0.040 max	0.10-0.40	bal Fe	700-850	525	17	201-255 HB
BS 970/3(91)	530M40	Bright bar; Q/T; 6<=t<=29mm; hard CD	0.36-0.44	0.90-1.20		0.60-0.90	0.15 max	0.035 max	0.040 max	0.10-0.40	bal Fe	850-1000	700	9	248-302 HB
USA															
	AISI 5140	Sh Strp	0.38-0.43	0.70-0.90		0.70-0.90		0.035	0.035	0.15-0.30	bal Fe				
	AISI 5140	Smls mech tub	0.38-0.43	0.70-0.90		0.70-0.90		0.040 max	0.040 max	0.15-0.35	bal Fe				

UNS numbers and US grades are provided as a means of cross referencing chemically similar alloys. Exchangability is only possible after independent examination of specifications. Tensile properties are minimum or typical as specified. UTS and YS as MPa. El as %. See Appendix for list of abbreviations used in Notes. * indicates obsolete material.

Specification	Designation	Notes	C	Cr	Cu	Mn	Mo	P	S	Si	Other	UTS	YS	El	Hard

Alloy Steel, Chromium, 5140/5140H (Continued from previous page)

USA

Specification	Designation	Notes	C	Cr	Cu	Mn	Mo	P	S	Si	Other	UTS	YS	El	Hard
	AISI 5140	Bar Blm Bil Slab	0.38-0.43	0.70-0.90		0.70-0.90		0.035 max	0.040 max	0.15-0.35	bal Fe				
	AISI 5140H	Bar Blm Bil Slab	0.37-0.44	0.60-1.00		0.60-1.00		0.035 max	0.040 max	0.15-0.35	bal Fe				
	AISI 5140H	Smls mech tub, Hard	0.37-0.44	0.60-1.00		0.60-1.00		0.040 max	0.040 max	0.15-0.30	bal Fe				
	UNS G51400		0.38-0.43	0.70-0.90		0.70-0.90		0.035 max	0.040 max	0.15-0.35	bal Fe				
	UNS H51400		0.37-0.44	0.60-1.00		0.60-1.00		0.035 max	0.040 max	0.15-0.35	bal Fe				
ASTM A29/A29M(93)	5140	Bar	0.38-0.43	0.70-0.90		0.7-0.9		0.035 max	0.04 max	0.15-0.35	bal Fe				
ASTM A304(96)	5140H	Bar, hard bands spec	0.37-0.44	0.60-1.00	0.35 max	0.60-1.00	0.06 max	0.035 max	0.040 max	0.15-0.35	Cu Ni Cr Mo trace allowed; bal Fe				
ASTM A322(96)	5140	Bar	0.38-0.43	0.70-0.90		0.70-0.90		0.035 max	0.040 max	0.15-0.35	bal Fe				
ASTM A331(95)	5140	Bar	0.38-0.43	0.70-0.90		0.70-0.90		0.035 max	0.040 max	0.15-0.35	bal Fe				
ASTM A331(95)	5140H	Bar	0.37-0.44	0.60-1.00	0.35 max	0.60-1.00	0.06 max	0.025 max	0.025 max	0.15-0.35	bal Fe				
ASTM A506(93)	5140	Sh Strp, HR, CR	0.38-0.43	0.70-0.90		0.70-0.90		0.035	0.035	0.15-0.30	bal Fe				
ASTM A519(96)	5140	Smls mech tub	0.38-0.43	0.70-0.90		0.70-0.90		0.040 max	0.040 max	0.15-0.35	bal Fe				
ASTM A752(93)	5140	Rod Wir	0.38-0.43	0.70-0.90		0.70-0.90		0.035 max	0.040 max	0.15-0.30	bal Fe	620			
ASTM A866(94)	5140	Anti-friction bearings	0.38-0.43	0.70-0.90		0.70-0.90		0.025 max	0.025 max	0.15-0.35	bal Fe				
ASTM A914/914M(92)	5140RH	Bar restricted end Q hard	0.38-0.43	0.70-0.90	0.35 max	0.70-0.90	0.06 max	0.035 max	0.040 max	0.15-0.35	Ni <=0.25; Electric furnace P, S<=0.025; bal Fe				
SAE 770(84)	5140*	Obs; see J1397(92)									bal Fe				
SAE J1268(95)	5140H	Bar Rod Wir Sh Strp Tub; See std	0.37-0.44	0.60-1.00	0.35 max	0.60-1.00	0.06 max	0.030 max	0.040 max	0.15-0.35	Ni <=0.25; Cu, Mo, Ni not spec'd but acpt; bal Fe				
SAE J1868(93)	5140RH	Restrict hard range	0.38-0.43	0.70-0.90	0.035 max	0.70-0.90	0.06 max	0.025 max	0.040 max	0.15-0.35	Ni <=0.25; Cu, Mo, Ni not spec'd but acpt; bal Fe				5 HRC max
SAE J404(94)	5140	Bil Blm Slab Bar HR CF	0.38-0.43	0.70-0.90		0.70-0.90		0.030 max	0.040 max	0.15-0.35	bal Fe				

Yugoslavia

Specification	Designation	Notes	C	Cr	Cu	Mn	Mo	P	S	Si	Other	UTS	YS	El	Hard
	C.4131	Q/T	0.38-0.45	0.9-1.2		0.6-0.8	0.15 max	0.035 max	0.03 max	0.4 max	bal Fe				
	C.4137	F/IH	0.38-0.44	0.9-1.2		0.5-0.8		0.025 max	0.035 max	0.15-0.4	Ni <=0.25; bal Fe				

Alloy Steel, Chromium, 5145/5145H

Australia

Specification	Designation	Notes	C	Cr	Cu	Mn	Mo	P	S	Si	Other	UTS	YS	El	Hard
AS 1444(96)	5145	Standard series, Hard/Tmp	0.43-0.48	0.70-0.90		0.70-0.90		0.040 max	0.040 max	0.10-0.35	bal Fe				
AS 1444(96)	5145H	H series, Hard/Tmp	0.42-0.49	0.60-1.00		0.60-1.00		0.040 max	0.040 max	0.10-0.35	bal Fe				

China

Specification	Designation	Notes	C	Cr	Cu	Mn	Mo	P	S	Si	Other	UTS	YS	El	Hard
GB 3077(88)	45Cr	Bar HR Q/T 25mm diam	0.42-0.49	0.80-1.10	0.30 max	0.50-0.80		0.035 max	0.035 max	0.17-0.37	bal Fe	1030	835	9	
GB 5216(85)	45CrH	Bar Rod HR/Frg Ann	0.42-0.49	0.80-1.10	0.30 max	0.50-0.80		0.035 max	0.035 max	0.17-0.37	bal Fe				217 max HB
GB 8162(87)	45Cr	Smls tube HR/CD Q/T	0.42-0.49	0.80-1.10		0.50-0.80		0.035 max	0.035 max	0.17-0.37	bal Fe	1030	835	9	
GB/T 3078(94)	45Cr	Bar CD Q/T 25mm diam	0.42-0.49	0.80-1.10	0.30 max	0.50-0.80		0.035 max	0.035 max	0.17-0.37	bal Fe	1030	835	9	

Japan

Specification	Designation	Notes	C	Cr	Cu	Mn	Mo	P	S	Si	Other	UTS	YS	El	Hard
JIS G4104(79)	SCr 5*	Obs; see SCr445	0.43-0.48	0.90-1.20	0.30 max	0.60-0.85		0.030 max	0.030 max	0.15-0.35	Ni <=0.25; bal Fe				
JIS G4104(79)	SCr445	HR Frg Bar Wir Rod	0.43-0.48	0.90-1.20	0.30 max	0.60-0.85		0.030 max	0.030 max	0.15-0.35	Ni <=0.25; bal Fe				

Specification	Designation	Notes	C	Cr	Cu	Mn	Mo	P	S	Si	Other	UTS	YS	El	Hard
Alloy Steel, Chromium, 5145/5145H (Continued from previous page)															
Mexico															
NMX-B-268(68)	5145H	Hard	0.42-0.49	0.60-1.00		0.60-1.00				0.20-0.35	bal Fe				
NMX-B-300(91)	5145	Bar	0.43-0.48	0.70-0.90		0.70-0.90		0.035 max	0.040 max	0.15-0.35	bal Fe				
NMX-B-300(91)	5145H	Bar	0.42-0.49	0.60-1.00		0.60-1.00				0.15-0.35	bal Fe				
USA															
	AISI 5145	Smls mech tub	0.43-0.48	0.70-0.90		0.70-0.90		0.040 max	0.040 max	0.15-0.35	bal Fe				
	UNS G51450		0.43-0.48	0.70-0.90		0.70-0.90		0.035 max	0.040 max	0.15-0.35	bal Fe				
	UNS H51450		0.42-0.49	0.60-1.00		0.60-1.00		0.035 max	0.040 max	0.15-0.35	bal Fe				
ASTM A29/A29M(93)	5145	Bar	0.43-0.48	0.70-0.90		0.70-0.90		0.035 max	0.040 max	0.15-0.35	bal Fe				
ASTM A519(96)	5145	Smls mech tub	0.43-0.48	0.70-0.90		0.70-0.90		0.040 max	0.040 max	0.15-0.35	bal Fe				
ASTM A752(93)	5145	Rod Wir	0.43-0.48	0.70-0.90		0.70-0.90		0.035 max	0.040 max	0.15-0.30	bal Fe	650			
Alloy Steel, Chromium, 5147/5147H															
Mexico															
NMX-B-268(68)	5147H	Hard	0.45-0.52	0.80-1.25		0.60-1.05				0.20-0.35	bal Fe				
NMX-B-300(91)	5147	Bar	0.46-0.51	0.85-1.15		0.70-0.95		0.035 max	0.040 max	0.15-0.35	bal Fe				
NMX-B-300(91)	5147H	Bar	0.45-0.52	0.80-1.25		0.60-1.05				0.15-0.35	bal Fe				
USA															
	AISI 5147	Smls mech tub	0.45-0.52	0.85-1.15		0.70-0.95		0.040 max	0.040 max	0.15-0.35	bal Fe				
	UNS G51470		0.46-0.51	0.85-1.15		0.70-0.95		0.035 max	0.040 max	0.15-0.35	bal Fe				
	UNS H51470		0.45-0.52	0.80-1.25		0.60-1.05		0.035 max	0.040 max	0.15-0.35	bal Fe				
ASTM A29/A29M(93)	5147	Bar	0.46-0.51	0.85-1.15		0.70-0.95		0.035 max	0.040 max	0.15-0.35	bal Fe				
ASTM A519(96)	5147	Smls mech tub	0.45-0.52	0.85-1.15		0.70-0.95		0.040 max	0.040 max	0.15-0.35	bal Fe				
ASTM A752(93)	5147	Rod Wir	0.46-0.51	0.85-1.15		0.70-0.90		0.035 max	0.040 max	0.15-0.30	bal Fe	680			
SAE J1268(95)	5147H	Hard	0.45-0.52	0.80-1.25	0.35 max	0.60-1.05	0.06 max	0.030 max	0.040 max	0.15-0.35	Ni <=0.25; Cu, Mo, Ni not spec'd but acpt; bal Fe				
SAE J1268(95)	E5147H	Hard	0.45-0.52	0.80-1.25	0.35 max	0.60-1.05	0.06 max	0.025 max	0.025 max	0.15-0.35	Ni <=0.25; Cu, Mo, Ni not spec'd but acpt; bal Fe				
Alloy Steel, Chromium, 5150/5150H															
Australia															
AS 1444	5150*	Obs; Bar	0.48-0.55	0.7-0.9		0.7-1		0.05	0.05	0.1-0.35	bal Fe				
Bulgaria															
BDS 6742	50ChGA	Spring	0.46-0.54	0.90-1.20	0.30 max	0.80-1.10		0.025 max	0.025 max	0.17-0.37	Ni <=0.30; bal Fe				
China															
GB 3077(88)	50Cr	Bar HR Q/T 25mm diam	0.47-0.54	0.80-1.10	0.30 max	0.50-0.80		0.035 max	0.035 max	0.17-0.37	bal Fe	1080	930	9	
GB 8162(87)	50Cr	Smls tube HR/CD Q/T	0.47-0.54	0.80-1.10		0.50-0.80		0.035 max	0.035 max	0.17-0.37	bal Fe	1080	930	9	
YB/T 5132(93)	50Cr	Sh HR/CR	0.47-0.54	0.80-1.10		0.50-0.80		0.035 max	0.035 max	0.17-0.37	bal Fe				
France															
AFNOR NFA35552	A5C2		0.4-0.5	0.4-0.6		0.5-0.8		0.04 max	0.04 max	0.35 max	bal Fe				
AFNOR NFA35557	42C2		0.4-0.46	0.3-0.6		0.6-0.9		0.04 max	0.04 max	0.1-0.4	bal Fe				

UNS numbers and US grades are provided as a means of cross referencing chemically similar alloys. Exchangability is only possible after independent examination of specifications. Tensile properties are minimum or typical as specified. UTS and YS as MPa. El as %. See Appendix for list of abbreviations used in Notes. * indicates obsolete material.

Specification	Designation	Notes	C	Cr	Cu	Mn	Mo	P	S	Si	Other	UTS	YS	El	Hard

Alloy Steel, Chromium, 5150/5150H (Continued from previous page)

Germany

Specification	Designation	Notes	C	Cr	Cu	Mn	Mo	P	S	Si	Other	UTS	YS	El	Hard
DIN	60MnCrTi4		0.45-0.60	0.80-1.00		0.90-1.20		0.025 max	0.025 max	0.15-0.35	Al <=0.10; Ti 0.15-0.30; bal Fe				
DIN	WNr 1.7145*	Obs	0.46-0.54	0.8-1.1		0.8-1.2		0.035 max	0.035	0.3-0.5	bal Fe				
DIN	WNr 1.8404		0.45-0.60	0.80-1.00		0.90-1.20		0.025 max	0.025 max	0.15-0.35	Al <=0.10; Ti 0.15-0.30; bal Fe				

Mexico

Specification	Designation	Notes	C	Cr	Cu	Mn	Mo	P	S	Si	Other	UTS	YS	El	Hard
DGN B-203	5150*	Obs	0.48-0.53	0.7-0.9		0.7-0.9		0.04	0.04	0.2-0.35	bal Fe				
DGN B-297	5150*	Obs	0.48-0.53	0.7-0.9		0.7-0.9		0.035	0.04	0.2-0.35	bal Fe				
NMX-B-268(68)	5150H	Hard	0.47-0.54	0.60-1.00		0.60-1.00				0.20-0.35	bal Fe				
NMX-B-300(91)	5150	Bar	0.48-0.53	0.70-0.90		0.70-0.90		0.035 max	0.040 max	0.15-0.35	bal Fe				
NMX-B-300(91)	5150H	Bar	0.47-0.54	0.60-1.00		0.60-1.00				0.15-0.35	bal Fe				

Pan America

Specification	Designation	Notes	C	Cr	Cu	Mn	Mo	P	S	Si	Other	UTS	YS	El	Hard
COPANT 334	5150	Bar	0.48-0.53	0.7-0.9		0.7-0.9		0.035	0.04	0.2-0.35	bal Fe				
COPANT 514	5150	Tube	0.48-0.53	0.7-0.9		0.7-0.9		0.04	0.05	0.1	bal Fe				

Poland

Specification	Designation	Notes	C	Cr	Cu	Mn	Mo	P	S	Si	Other	UTS	YS	El	Hard
PNH84030	50HG	Spring	0.46-0.54	0.9-1.2	0.25	0.8-1.1		0.03	0.03	0.15-0.4	Ni 0.4; bal Fe				
PNH84030/04	50H	Q/T	0.47-0.55	0.8-1.1		0.5-0.8		0.035 max	0.035 max	0.17-0.37	Ni <=0.3; bal Fe				
PNH84032	50HG	Spring	0.46-0.54	0.9-1.2	0.25 max	0.8-1.1		0.03 max	0.03 max	0.15-0.4	Ni <=0.4; bal Fe				

Russia

Specification	Designation	Notes	C	Cr	Cu	Mn	Mo	P	S	Si	Other	UTS	YS	El	Hard
GOST 14959	50ChG		0.46-0.54	0.9-1.2	0.2	0.7-1		0.035	0.035	0.17-0.37	Ni 0.25; bal Fe				
GOST 14959	50ChGA		0.47-0.52	0.95-1.2	0.2	0.8-1		0.025	0.025	0.17-0.37	Ni 0.25; bal Fe				
GOST 14959(79)	50ChG	Spring	0.46-0.54	0.9-1.2	0.2 max	0.7-1.0	0.15 max	0.035 max	0.035 max	0.17-0.37	Al <=0.1; Ni <=0.25; Ti <=0.05; bal Fe				
GOST 4543	50Ch		0.46-0.54	0.8-1.1	0.3	0.5-0.8		0.035	0.035	0.17-0.37	Ni 0.3; bal Fe				

UK

Specification	Designation	Notes	C	Cr	Cu	Mn	Mo	P	S	Si	Other	UTS	YS	El	Hard
BS 3146/1(74)	CLA12(A)	Inv Casting	0.45-0.55	0.80-1.20	0.30 max	0.50-1.00	0.10 max	0.035 max	0.035 max	0.30-0.80	Ni <=0.40; bal Fe				
BS 3146/1(74)	CLA12(B)	Inv Casting	0.45-0.55	0.80-1.20	0.36 max	0.50-1.00	0.10 max	0.035 max	0.035 max	0.30-0.80	Ni <=0.40; bal Fe				

USA

Specification	Designation	Notes	C	Cr	Cu	Mn	Mo	P	S	Si	Other	UTS	YS	El	Hard
	AISI 5150	Bar Blm Bil Slab	0.48-0.53	0.70-0.90		0.70-0.90		0.035 max	0.040 max	0.15-0.35	bal Fe				
	AISI 5150	Sh Strp	0.48-0.53	0.70-0.90		0.70-0.90		0.035	0.035	0.15-0.30	bal Fe				
	AISI 5150	Smls mech tub	0.48-0.53	0.70-0.90		0.70-0.90		0.040 max	0.040 max	0.15-0.35	bal Fe				
	AISI 5150H	Smls mech tub, Hard	0.47-0.54	0.60-1.00		0.60-1.00		0.040 max	0.040 max	0.15-0.30	bal Fe				
	AISI 5150H	Bar Blm Bil Slab	0.47-0.54	0.60-1.00		0.60-1.00		0.035 max	0.040 max	0.15-0.35	bal Fe				
	UNS G51500		0.48-0.53	0.70-0.90		0.70-0.90		0.035 max	0.040 max	0.15-0.35	bal Fe				
	UNS H51500		0.47-0.54	0.60-1.00		0.60-1.00		0.035 max	0.040 max	0.15-0.35	bal Fe				
ASTM A29/A29M(93)	5150	Bar	0.48-0.53	0.70-0.90		0.7-0.9		0.035 max	0.04 max	0.15-0.35	bal Fe				
ASTM A304(96)	5150H	Bar, hard bands spec	0.47-0.54	0.60-1.00	0.35 max	0.60-1.00	0.06 max	0.035 max	0.040 max	0.15-0.35	Cu Ni Cr Mo trace allowed; bal Fe				
ASTM A322(96)	5150	Bar	0.48-0.53	0.70-0.90		0.70-0.90		0.035 max	0.040 max	0.15-0.35	bal Fe				
ASTM A331(95)	5150	Bar	0.48-0.53	0.70-0.90		0.70-0.90		0.035 max	0.040 max	0.15-0.35	bal Fe				

UNS numbers and US grades are provided as a means of cross referencing chemically similar alloys. Exchangability is only possible after independent examination of specifications. Tensile properties are minimum or typical as specified. UTS and YS as MPa. El as %. See Appendix for list of abbreviations used in Notes. * indicates obsolete material.

Specification	Designation	Notes	C	Cr	Cu	Mn	Mo	P	S	Si	Other	UTS	YS	El	Hard

Alloy Steel, Chromium, 5150/5150H (Continued from previous page)

USA

Specification	Designation	Notes	C	Cr	Cu	Mn	Mo	P	S	Si	Other	UTS	YS	El	Hard
ASTM A331(95)	5150H	Bar	0.47-0.54	0.60-1.00	0.35 max	0.60-1.00	0.06 max	0.025 max	0.025 max	0.15-0.35	bal Fe				
ASTM A506(93)	5150	Sh Strp, HR, CR	0.48-0.53	0.70-0.90		0.70-0.90		0.035	0.035	0.15-0.30	bal Fe				
ASTM A519(96)	5150	Smls mech tub	0.48-0.53	0.70-0.90		0.70-0.90		0.040 max	0.040 max	0.15-0.35	bal Fe				
ASTM A752(93)	5150	Rod Wir	0.48-0.53	0.70-0.90		0.70-0.90		0.035 max	0.035 max	0.15-0.30	bal Fe	670			
ASTM A866(94)	5150	Anti-friction bearings	0.48-0.53	0.70-0.90		0.70-0.90		0.025 max	0.025 max	0.15-0.35	bal Fe				
SAE 770(84)	5150*	Obs; see J1397(92)									bal Fe				
SAE J1268(95)	5150H	Bar Rod Wir Sh Strp Tub; See std	0.47-0.54	0.60-1.00	0.35 max	0.60-1.00	0.06 max	0.030 max	0.040 max	0.15-0.35	Ni <=0.25; Cu, Mo, Ni not spec'd but acpt; bal Fe				
SAE J404(94)	5150	Bil Blm Slab Bar HR CF	0.48-0.53	0.70-0.90		0.70-0.90		0.030 max	0.040 max	0.15-0.35	bal Fe				
SAE J775(93)	5150H	Engine poppet valve, Mart	0.47-0.54	0.60-1.00		0.60-1.00		0.035 max	0.040 max	0.15-0.35	bal Fe	990	910		
SAE J775(93)	NV6*	Engine poppet valve, Mart; former SAE grade	0.47-0.54	0.60-1.00		0.60-1.00		0.035 max	0.040 max	0.15-0.35	bal Fe	990	910		

Alloy Steel, Chromium, 5155/5155H

Australia

Specification	Designation	Notes	C	Cr	Cu	Mn	Mo	P	S	Si	Other	UTS	YS	El	Hard
AS 1444(96)	5155	H series, Hard/Tmp	0.50-0.60	0.70-0.90		0.70-1.00		0.050 max	0.050 max	0.10-0.35	bal Fe				
AS G18	En11*	Obs; see AS 1444	0.5-0.6	0.5-0.8		0.5-0.8		0.05	0.05	0.1-0.35	bal Fe				

Belgium

Specification	Designation	Notes	C	Cr	Cu	Mn	Mo	P	S	Si	Other	UTS	YS	El	Hard
NBN 235-05	55Cr3		0.52-0.59	0.6-0.9		0.7-1		0.035	0.035	0.15-0.4	bal Fe				

China

Specification	Designation	Notes	C	Cr	Cu	Mn	Mo	P	S	Si	Other	UTS	YS	El	Hard
GB 1222(84)	55CrMnA	Bar Flat HR Q/T	0.52-0.60	0.65-0.95	0.25 max	0.65-0.95		0.030 max	0.030 max	0.17-0.37	bal Fe	1225	1080	9	

France

Specification	Designation	Notes	C	Cr	Cu	Mn	Mo	P	S	Si	Other	UTS	YS	El	Hard
AFNOR NFA35571	55C3	Bar Wir rod, Q/A Tmp, 80mm	0.52-0.59	0.6-0.9		0.7-1		0.035	0.035	0.1-0.4	bal Fe				
AFNOR NFA35571(84)	55C3	t<=80mm; Q/T	0.52-0.59	0.6-0.9	0.3 max	0.7-1	0.15 max	0.035 max	0.035 max	0.1-0.4	Ni <=0.4; Ti <=0.05; bal Fe	1370-1620	1180	6	
AFNOR NFA35571(96)	55Cr3	t<1000mm; Q/T	0.52-0.59	0.7-1	0.3 max	0.7-1	0.15 max	0.025 max	0.025 max	0.1-0.4	Ni <=0.4; Ti <=0.05; bal Fe	1400-1700	1250	3	

Germany

Specification	Designation	Notes	C	Cr	Cu	Mn	Mo	P	S	Si	Other	UTS	YS	El	Hard
DIN 17221(88)	55Cr3	Spring hard	0.52-0.59	0.70-1.00		0.70-1.00		0.030 max	0.030 max	0.25-0.50	bal Fe	1320-1720	1175	6	
DIN 17221(88)	WNr 1.7176	HR spring hard	0.52-0.59	0.70-1.00		0.70-1.00		0.030 max	0.030 max	0.25-0.50	bal Fe	1320-1720	1175	6	

India

Specification	Designation	Notes	C	Cr	Cu	Mn	Mo	P	S	Si	Other	UTS	YS	El	Hard
IS 1570	55Cr70		0.5-0.6	0.6-0.8		0.6-0.8				0.1-0.35	bal Fe				
IS 3479	T55Cr70	Bar	0.5-0.6	0.6-0.8		0.6-0.8		0.04	0.04	0.1-0.35	bal Fe				
IS 5517	55Cr70	Bar, Norm, Ann, Hard Tmp, 63mm	0.5-0.6	0.6-0.8		0.6-0.8				0.1-0.35	bal Fe				

International

Specification	Designation	Notes	C	Cr	Cu	Mn	Mo	P	S	Si	Other	UTS	YS	El	Hard
ISO 683-14(92)	55Cr3	Bar Rod Wir; Q/T, springs, Ann	0.52-0.59	0.70-1.00		0.70-1.00		0.030 max	0.030 max	0.15-0.40	bal Fe				255 HB

Italy

Specification	Designation	Notes	C	Cr	Cu	Mn	Mo	P	S	Si	Other	UTS	YS	El	Hard
UNI 3545(80)	55Cr3	Spring	0.52-0.59	0.6-0.8		0.7-1		0.035 max	0.035 max	0.15-0.4	bal Fe				

Japan

Specification	Designation	Notes	C	Cr	Cu	Mn	Mo	P	S	Si	Other	UTS	YS	El	Hard
JIS G3311(88)	SUP9M	CR Strp; Spring	0.52-0.60	0.65-0.95	0.30 max	0.65-0.95		0.035 max	0.035 max	0.15-0.35	bal Fe				200-290 HV
JIS G4801(84)	SUP 9	Bar Spring Q/T	0.52-0.60	0.65-0.95	0.30 max	0.65-0.95		0.035 max	0.035 max	0.15-0.35	bal Fe	1226	1079	9	363-429 HB

Specification	Designation	Notes	C	Cr	Cu	Mn	Mo	P	S	Si	Other	UTS	YS	El	Hard

Alloy Steel, Chromium, 5155/5155H (Continued from previous page)

Specification	Designation	Notes	C	Cr	Cu	Mn	Mo	P	S	Si	Other	UTS	YS	El	Hard
Mexico															
NMX-B-268(68)	5155H	Hard	0.50-0.60	0.60-1.00		0.60-1.00				0.20-0.35	bal Fe				
NMX-B-300(91)	5155	Bar	0.51-0.59	0.70-0.90		0.70-0.90		0.035 max	0.040 max	0.15-0.35	bal Fe				
NMX-B-300(91)	5155H	Bar	0.50-0.60	0.60-1.00		0.60-1.00				0.15-0.35	bal Fe				
Pan America															
COPANT 514	5155	Tube	0.5-0.6	0.7-0.9		0.7-0.9		0.04	0.05	0.1	bal Fe				
Russia															
GOST 14959	55ChGR		0.52-0.6	0.9-1.2	0.2	0.9-1.2		0.035	0.035	0.17-0.37	B 0.001-0.003; Ni 0.25; bal Fe				
GOST 14959(79)	50ChGA	Spring	0.47-0.52	0.95-1.2	0.2 max	0.8-1.0	0.15 max	0.025 max	0.025 max	0.17-0.37	Al <=0.1; Ni <=0.25; Ti <=0.05; bal Fe				
GOST 14959(79)	55ChGR	Spring	0.52-0.6	0.9-1.2	0.2 max	0.9-1.2	0.15 max	0.035 max	0.035 max	0.17-0.37	Al <=0.1; B 0.001-0.003; Ni <=0.25; Ti <=0.05; bal Fe				
Spain															
UNE 36015(60)	55Cr3	Spring	0.52-0.59	0.6-0.9		0.7-1		0.035	0.035	0.15-0.4	bal Fe				
UNE 36015(60)	F.1431*	see 55Cr3	0.52-0.59	0.6-0.9		0.7-1		0.035	0.035	0.15-0.4	bal Fe				
UNE 36015(77)	55Cr3	Spring	0.52-0.59	0.6-0.9		0.7-1	0.15 max	0.035 max	0.035 max	0.15-0.4	bal Fe				
UNE 36015(77)	F.1431*	see 55Cr3	0.52-0.59	0.6-0.9		0.7-1	0.15 max	0.035 max	0.035 max	0.15-0.4	bal Fe				
Sweden															
SS 142253	2253	Spring	0.52-0.59	0.6-0.9		0.7-1		0.035 max	0.035 max	0.15-0.4	bal Fe				
Turkey															
TS 2288(97)	55Cr3-1	Ref. ISO 683-14(92), wire rods, flts, HT	0.52-0.59	0.6-0.9		0.7-1		0.035	0.035	0.15-0.4	bal Fe				
TS 7176	55Cr3-1*		0.52-0.59	0.6-0.9		0.7-1		0.035	0.035	0.15-0.4	bal Fe				
UK															
BS 970/2(88)	525A58	Spring	0.55-0.60	0.70-0.85		0.80-0.95	0.10 max	0.035 max	0.035 max	0.20-0.35	bal Fe				
BS 970/2(88)	525A60	Spring	0.57-0.62	0.80-0.95		0.85-1.00	0.06 max	0.035 max	0.035 max	0.20-0.35	bal Fe				
BS 970/2(88)	525A61	Spring	0.57-0.63	0.85-1.00		0.85-1.00	0.08-0.15	0.035 max	0.035 max	0.20-0.35	bal Fe				
BS 970/2(88)	525H60	Spring	0.55-0.64	0.60-1.00		0.65-1.00		0.035 max	0.035 max	0.15-0.40	bal Fe				
BS 970/5(72)	527A60*	Obs; Spring	0.55-0.65	0.6-0.9		0.70-1.00	0.15 max	0.04 max	0.05 max	0.10-0.35	Pb <=0.15; Ti <=0.05; bal Fe				
USA															
	AISI 5155	Smls mech tub	0.50-0.60	0.70-0.90		0.70-0.90		0.040 max	0.040 max	0.15-0.35	bal Fe				
	AISI 5155		0.51-0.59	0.70-0.90		0.70-0.90		0.035 max	0.04 max	0.15-0.3	bal Fe				
	AISI 5155H	Smls mech tub, Hard	0.50-0.60	0.60-1.00		0.60-1.00		0.035 max	0.040 max	0.15-0.30	bal Fe				
	UNS G51550		0.51-0.59	0.70-0.90		0.70-0.90		0.035 max	0.040 max	0.15-0.35	bal Fe				
	UNS H51550		0.50-0.60	0.60-1.00		0.60-1.00		0.035 max	0.040 max	0.15-0.35	bal Fe				
ASTM A29/A29M(93)	5155	Bar	0.51-0.59	0.70-0.90		0.7-0.9		0.035 max	0.04 max	0.15-0.35	bal Fe				
ASTM A304(96)	5155H	Bar, hard bands spec	0.50-0.60	0.60-1.00	0.35 max	0.60-1.00	0.06 max	0.035 max	0.040 max	0.15-0.35	Cu Ni Cr Mo trace allowed; bal Fe				
ASTM A322(96)	5155	Bar	0.51-0.59	0.70-0.90		0.70-0.90		0.035 max	0.040 max	0.15-0.35					
ASTM A331(95)	5155	Bar	0.51-0.59	0.70-0.90		0.70-0.90		0.035 max	0.040 max	0.15-0.35	bal Fe				
ASTM A331(95)	5155H	Bar	0.50-0.60	0.60-1.00	0.35 max	0.60-1.00	0.06 max	0.025 max	0.025 max	0.15-0.35	bal Fe				
ASTM A519(96)	5155	Smls mech tub	0.50-0.60	0.70-0.90		0.70-0.90		0.040 max	0.040 max	0.15-0.35	bal Fe				

UNS numbers and US grades are provided as a means of cross referencing chemically similar alloys. Exchangability is only possible after independent examination of specifications. Tensile properties are minimum or typical as specified. UTS and YS as MPa. El as %. See Appendix for list of abbreviations used in Notes. * indicates obsolete material.

Specification	Designation	Notes	C	Cr	Cu	Mn	Mo	P	S	Si	Other	UTS	YS	El	Hard

Alloy Steel, Chromium, 5155/5155H (Continued from previous page)

USA

Specification	Designation	Notes	C	Cr	Cu	Mn	Mo	P	S	Si	Other	UTS	YS	El	Hard
ASTM A752(93)	5155	Rod Wir	0.51-0.59	0.70-0.90		0.70-0.90		0.035 max	0.040 max	0.15-0.30	bal Fe	710			
SAE 770(84)	5155*	Obs; see J1397(92)									bal Fe				
3AE J1268(95)	5155H	Bar Rod Wir Tub; See std	0.50-0.60	0.60-1.00	0.35 max	0.60-1.00	0.06 max	0.030 max	0.040 max	0.15-0.35	Ni <=0.25; Cu, Mo, Ni not spec'd but acpt; bal Fe				

Yugoslavia

Specification	Designation	Notes	C	Cr	Cu	Mn	Mo	P	S	Si	Other	UTS	YS	El	Hard
	C.4332	Spring	0.52-0.59	0.6-0.9		0.7-1.0	0.15 max	0.035 max	0.035 max	0.15-0.4	bal Fe				

Alloy Steel, Chromium, 5160/5160H

Argentina

Specification	Designation	Notes	C	Cr	Cu	Mn	Mo	P	S	Si	Other	UTS	YS	El	Hard
IAS	IRAM 5160		0.56-0.64	0.70-0.90		0.75-1.00		0.035 max	0.040 max	0.20-0.35	bal Fe	1000	680	15	302 HB

Australia

Specification	Designation	Notes	C	Cr	Cu	Mn	Mo	P	S	Si	Other	UTS	YS	El	Hard
AS 1444(96)	5160	H series, Hard/Tmp	0.55-0.65	0.70-0.90		0.70-1.00		0.050 max	0.050 max	0.10-0.35	bal Fe				

Bulgaria

Specification	Designation	Notes	C	Cr	Cu	Mn	Mo	P	S	Si	Other	UTS	YS	El	Hard
BDS 6742	60S2ChA	Spring	0.56-0.64	0.70-1.00	0.30 max	0.40-0.70		0.025 max	0.025 max	1.40-1.80	Ni <=0.30; bal Fe				

China

Specification	Designation	Notes	C	Cr	Cu	Mn	Mo	P	S	Si	Other	UTS	YS	El	Hard
GB 1222(84)	60CrMnA	Bar Flat HR Q/T	0.56-0.64	0.70-1.00	0.25 max	0.70-1.00		0.030 max	0.030 max	0.17-0.37	bal Fe	1225	1080	9	

Hungary

Specification	Designation	Notes	C	Cr	Cu	Mn	Mo	P	S	Si	Other	UTS	YS	El	Hard
MSZ 2666(88)	60Cr3	HF spring; 0-100.0mm; Q/T	0.56-0.64	0.7-1		0.7-1		0.03 max	0.03 max	0.15-0.4	bal Fe	1370	1180	5L	
MSZ 2666(00)	60Cr3	HF spring; 0-999mm	0.56-0.64	0.7-1		0.7-1		0.03 max	0.03 max	0.15-0.4	bal Fe				340 HB max
MSZ 2666(88)	60CrB3	HF spring; 0-100.0mm; Q/T	0.56-0.64	0.7-1		0.7-1		0.03 max	0.03 max	0.15-0.4	B >=0.0005; bal Fe	1370	1180	6L	
MSZ 2666(88)	60CrB3	HF spring; 0-999mm	0.56-0.64	0.7-1		0.7-1		0.03 max	0.03 max	0.15-0.4	B >=0.0005; bal Fe				340 HB max
MSZ 2666(88)	60CrMo3	HF spring; 0-999mm	0.56-0.64	0.7-1		0.7-1	0.25-0.35	0.03 max	0.03 max	0.15-0.4	bal Fe				340 HB max
MSZ 2666(88)	60CrMo3	HF spring; 0-100.0mm; Q/T	0.56-0.64	0.7-1		0.7-1	0.25-0.35	0.03 max	0.03 max	0.15-0.4	bal Fe	1370	1180	6L	
MSZ 2666(88)	60SiMn5	HF spring; 0-100.0mm; Q/T	0.55-0.65			1-1.4		0.04 max	0.04 max	1-1.5	bal Fe	1370	1180	6L	
MSZ 2666(88)	60SiMn5	HF spring; 0-999mm	0.55-0.65			1-1.4		0.04 max	0.04 max	1-1.5	bal Fe				340 HB max

International

Specification	Designation	Notes	C	Cr	Cu	Mn	Mo	P	S	Si	Other	UTS	YS	El	Hard
ISO 683-14(92)	55Cr3	Bar Rod Wir; Q/T, springs, Ann	0.52-0.59	0.70-1.00		0.70-1.00		0.030 max	0.030 max	0.15-0.40					255 HB

Japan

Specification	Designation	Notes	C	Cr	Cu	Mn	Mo	P	S	Si	Other	UTS	YS	El	Hard
JIS G4801(84)	SUP 9A	Bar Spring Q/T	0.56-0.64	0.70-1.00	0.30 max	0.70-1.00		0.035 max	0.035 max	0.15-0.35	bal Fe	1226	1079	9	363-429 HB
JIS G4801(84)	SUP11A	Bar Spring Q/T	0.56-0.64	0.70-1.00	0.30 max	0.70-1.00		0.035 max	0.035 max	0.15-0.35	B >=0.0005; bal Fe	1226	1079	9	363-429 HB

Mexico

Specification	Designation	Notes	C	Cr	Cu	Mn	Mo	P	S	Si	Other	UTS	YS	El	Hard
DGN B-203	5160*	Obs	0.55-0.65	0.7-0.9		0.75-1		0.04	0.04	0.2-0.35	bal Fe				
DGN B-297	5160*	Obs	0.56-0.64	0.7-0.9		0.75-1		0.035	0.04	0.2-0.35	bal Fe				
NMX-B-268(68)	5160H	Hard	0.55-0.65	0.60-1.00		0.65-1.10				0.20-0.35	bal Fe				
NMX-B-300(91)	5160	Bar	0.56-0.61	0.70-0.90		0.75-1.00		0.035 max	0.040 max	0.15-0.35	bal Fe				
NMX-B-300(91)	5160H	Bar	0.55-0.65	0.60-1.00		0.65-1.10				0.15-0.35	bal Fe				

Pan America

Specification	Designation	Notes	C	Cr	Cu	Mn	Mo	P	S	Si	Other	UTS	YS	El	Hard
COPANT 334	5160	Bar	0.56-0.64	0.7-1		0.75-1		0.04	0.04	0.2-0.35	bal Fe				
COPANT 514	5160	Tube	0.55-0.64	0.7-0.9		0.75-1		0.04	0.05	0.1	bal Fe				
COPANT 514	5160		0.55-0.64	0.7-0.9		0.75-1		0.04	0.05	0.1	bal Fe				

Specification	Designation	Notes	C	Cr	Cu	Mn	Mo	P	S	Si	Other	UTS	YS	El	Hard

Alloy Steel, Chromium, 5160/5160H (Continued from previous page)

Russia

Specification	Designation	Notes	C	Cr	Cu	Mn	Mo	P	S	Si	Other	UTS	YS	El	Hard
GOST 14959	60S2ChA		0.56-0.64	0.7-1	0.2	0.4-0.7		0.025	0.025	1.4-1.8	Ni 0.25; bal Fe				
GOST 14959(79)	60S2ChA	Spring	0.56-0.64	0.7-1.0	0.2 max	0.4-0.7	0.15 max	0.025 max	0.025 max	1.4-1.8	Al <=0.1; Ni <=0.25; Ti <=0.05; bal Fe				
Spain															
UNE 36015(77)	55Cr3	Spring	0.52-0.59	0.6-0.9		0.7-1	0.15 max	0.035 max	0.035 max	0.15-0.4	bal Fe				
Sweden															
SS 142254	2254	Q/T	0.56-0.64	0.6-0.9		0.7-1		0.035 max	0.035 max	0.15-0.4	bal Fe				
UK															
BS 970/1(83)	527A17*	Wrought CH	0.14-0.19	0.70-0.90		0.70-0.90		0.035 max	0.040 max	0.10-0.35	bal Fe	770		12	
BS 970/1(83)	527A60*	Blm Bil Slab Bar Rod Frg	0.55-0.64	0.6-0.9		0.70-1.00		0.04	0.05	0.1-0.35	bal Fe				
BS 970/2(70)	526M60*	Obs; Spring	0.55-0.65	0.5-0.8		0.50-0.80	0.15 max	0.04 max	0.05 max	0.1-0.35	Pb <=0.15; Ti <=0.05; bal Fe				
BS 970/5(72)	527H60*	Blm Bil Slab Bar Rod Frg	0.55-0.64	0.55-0.95		0.65-1.05		0.04	0.05	0.1-0.35	bal Fe				
BS 970/5(72)	527H60*	Obs; Spring	0.55-0.65	0.55-0.95		0.65-1.05	0.15 max	0.04 max	0.05 max	0.1-0.35	Pb <=0.15; Ti <=0.05; bal Fe				
USA															
	AISI 5160	Bar Blm Bil Slab	0.56-0.64	0.70-0.90		0.75-1.00		0.035 max	0.040 max	0.15-0.3	bal Fe				
	AISI 5160	Smls mech tub	0.55-0.65	0.70-0.90		0.75-1.00		0.040 max	0.040 max	0.15-0.35	bal Fe				
	AISI 5160	Plt	0.54-0.65	0.60-0.90		0.70-1.00		0.035 max	0.040 max	0.15-0.40	bal Fe				
	AISI 5160	Sh Strp	0.55-0.65	0.70-0.90		0.75-1.00		0.035	0.035	0.15-0.30	bal Fe				
	AISI 5160H	Bar Blm Bil Slab	0.55-0.65	0.60-1.00		0.65-1.10		0.035 max	0.040 max	0.15-0.35	bal Fe				
	AISI 5160H	Smls mech tub, Hard	0.55-0.65	0.60-1.00		0.65-1.10		0.040 max	0.040 max	0.15-0.30	bal Fe				
	UNS G51600		0.56-0.64	0.70-0.90		0.75-1.00		0.035 max	0.040 max	0.15-0.35	bal Fe				
	UNS H51600		0.55-0.65	0.60-1.00		0.65-1.10		0.035 max	0.040 max	0.15-0.35	bal Fe				
ASTM A29/A29M(93)	5160	Bar	0.56-0.61	0.70-0.90		0.75-1		0.035 max	0.04	0.15-0.35	bal Fe				
ASTM A295(94)	5160	Bearing, coil bar tub, CD ann	0.56-0.64	0.70-0.90	0.35 max	0.75-1.00	0.10 max	0.025 max	0.025 max	0.15-0.35	Ni <=0.25; bal Fe				250 HB max
ASTM A304(96)	5160H	Bar, hard bands spec	0.55-0.65	0.60-1.00	0.35 max	0.65-1.10	0.06 max	0.035 max	0.040 max	0.15-0.35	Cu Ni Cr Mo trace allowed; bal Fe				
ASTM A322(96)	5160	Bar	0.56-0.64	0.70-0.90		0.75-1.00		0.035 max	0.040 max	0.15-0.35	bal Fe				
ASTM A331(95)	5160	Bar	0.56-0.64	0.70-0.90		0.75-1.00		0.035 max	0.040 max	0.15-0.35	bal Fe				
ASTM A331(95)	5160H	Bar	0.55-0.65	0.60-1.00	0.35 max	0.65-1.10	0.06 max	0.025 max	0.025 max	0.15-0.35	bal Fe				
ASTM A506(93)	5160	Sh Strp, HR, CR	0.55-0.65	0.70-0.90		0.75-1.00		0.035	0.035	0.15-0.30	bal Fe				
ASTM A519(96)	5160	Smls mech tub	0.55-0.65	0.70-0.90		0.75-1.00		0.040 max	0.040 max	0.15-0.35	bal Fe				
ASTM A752(93)	5160	Rod Wir	0.56-0.64	0.70-0.90		0.75-1.00		0.035 max	0.040 max	0.15-0.30	bal Fe	720			
ASTM A829/A829M(95)	5160	Plt	0.54-0.65	0.60-0.90		0.70-1.00		0.035 max	0.040 max	0.15-0.40	bal Fe	480-965			
ASTM A914/914M(92)	5160RH	Bar restricted end Q hard	0.56-0.64	0.70-0.90	0.35 max	0.75-1.00	0.06 max	0.035 max	0.040 max	0.15-0.35	Ni <=0.25; Electric furnace P, S<=0.025; bal Fe				
SAE 770(84)	5160*	Obs; see J1397(92)									bal Fe				
SAE J1268(95)	5160H	Bar Rod Wir Sh Strp Tub; See std	0.55-0.65	0.60-1.00	0.35 max	0.65-1.10	0.06 max	0.030 max	0.040 max	0.15-0.35	Ni <=0.25; Cu, Mo, Ni not spec'd but acpt; bal Fe				

UNS numbers and US grades are provided as a means of cross referencing chemically similar alloys. Exchangability is only possible after independent examination of specifications. Tensile properties are minimum or typical as specified. UTS and YS as MPa. El as %. See Appendix for list of abbreviations used in Notes. * indicates obsolete material.

Specification	Designation	Notes	C	Cr	Cu	Mn	Mo	P	S	Si	Other	UTS	YS	El	Hard
Alloy Steel, Chromium, 5160/5160H (Continued from previous page)															
USA															
SAE J1868(93)	5160RH	Restrict hard range	0.56-0.64	0.70-0.90	0.035 max	0.75-1.00	0.06 max	0.025 max	0.040 max	0.15-0.35	Ni <=0.25; Cu, Mo, Ni not spec'd but acpt; bal Fe				5 HRC max
SAE J404(94)	5160	Bil Blm Slab Bar HR CF	0.56-0.64	0.70-0.90		0.75-1.00		0.030 max	0.040 max	0.15-0.35	bal Fe				
SAE J404(94)	5160	Plt	0.54-0.65	0.60-0.90		0.70-1.00		0.035 max	0.040 max	0.15-0.35	bal Fe				
Alloy Steel, Chromium, 51B60/51B60H															
China															
GB 1222(84)	60CrMnBA	Bar Flat HR Q/T	0.56-0.64	0.70-1.00	0.25 max	0.70-1.00		0.030 max	0.030 max	0.17-0.37	B 0.0005-0.0040; bal Fe	1225	1080	9	
International															
ISO 683-14(92)	60CrB3	Bar Rod Wir; Q/T, springs, Ann	0.56-0.64	0.60-0.90		0.70-1.00		0.030 max	0.030 max	0.15-0.40	B >=0.0008; bal Fe				255 HB
Mexico															
DGN B-203	51B60*	Obs	0.55-0.65	0.7-0.9		0.75-1		0.04	0.04	0.2-0.35	B >=0.0005; bal Fe				
DGN B-297	51B60*	Obs	0.56-0.64	0.7-0.9		0.75-1		0.035	0.04	0.2-0.35	B >=0.0005; bal Fe				
NMX-B-300(91)	51B60	Bar	0.56-0.64	0.70-0.90		0.75-1.00		0.035 max	0.040 max	0.20-0.35	B >=0.0005; bal Fe				
NMX-B-300(91)	51B60H	Bar	0.55-0.65	0.60-1.00		0.65-1.10				0.15-0.35	B >=0.0005; bal Fe				
Pan America															
COPANT 514	51B60		0.55-0.65	0.7-0.9		0.75-1		0.04	0.05	0.1	B >=0.0005; bal Fe				
COPANT 514	51B60	Tube	0.55-0.65	0.7-0.9		0.75-1		0.04	0.05	0.1	B <=0.003; bal Fe				
USA															
	AISI 51B60	Smls mech Tub; Boron	0.56-0.64	0.70-0.90		0.75-1.00		0.035 max	0.040 max	0.15-0.35	B 0.0005-0.003; bal Fe				
	AISI 51B60	Bar Blm Bil Slab	0.55-0.65	0.70-0.90		0.75-1.00		0.040 max	0.040 max	0.15-0.35	B >=0.0005; bal Fe				
	AISI 51B60H	Smls mech tub, Hard; Boron	0.55-0.65	0.60-1.00		0.65-1.10		0.035 max	0.040 max	0.15-0.30	B 0.0005-0.005; bal Fe				
	AISI 51B60H	Bar Blm Bil Slab	0.55-0.65	0.60-1.00		0.65-1.10		0.035 max	0.040 max	0.15-0.30	B 0.0005-0.003; bal Fe				
	UNS G51601		0.56-0.64	0.70-0.90		0.75-1.00		0.035 max	0.040 max	0.15-0.35	B >=0.0005; bal Fe				
	UNS H51601		0.55-0.65	0.60-1.00		0.65-1.10				0.15-0.35	B 0.0005-0.003; bal Fe				
ASTM A29/A29M(93)	51B60	Bar	0.56-0.64	0.70-0.90		0.75-1		0.035 max	0.04 max	0.15-0.35	B 0.0005-0.005; bal Fe				
ASTM A304(96)	51B60H	Bar, hard bands spec	0.55-0.65	0.60-1.00	0.35 max	0.65-1.10	0.06 max	0.035 max	0.040 max	0.15-0.35	B >=0.0005; Cu Ni Cr Mo trace allowed; bal Fe				
ASTM A322(96)	51B60	Bar	0.56-0.64	0.70-0.90		0.75-1.00		0.035 max	0.040 max	0.15-0.35	B 0.0005-0.005; bal Fe				
ASTM A331(95)	51B60	Bar	0.56-0.64	0.70-0.90		0.75-1.00		0.035 max	0.040 max	0.15-0.35	bal Fe				
ASTM A331(95)	51B60H	Bar	0.55-0.65	0.60-1.00	0.35 max	0.65-1.10	0.06 max	0.025 max	0.025 max	0.15-0.35	B >=0.0005; bal Fe				
ASTM A519(96)	51B60	Smls mech tub	0.55-0.65	0.70-0.90		0.75-1.00		0.040 max	0.040 max	0.15-0.35	B >=0.0005; bal Fe				
ASTM A752(93)	51B60	Rod Wir	0.56-0.64	0.70-0.90		0.75-1.00		0.035 max	0.040 max	0.15-0.30	B >=0.0005; bal Fe	750			
SAE 770(84)	51B60*	Obs; see J1397(92)	0.56-0.64	0.7-0.9		0.75-1		0.04 max	0.04 max	0.15-0.3	B 0.001-0.005; bal Fe				
SAE J1268(95)	51B60H	Bar Rod Wir Sh Strp Tub; See std	0.55-0.65	0.60-1.00	0.35 max	0.65-1.10	0.06 max	0.030 max	0.040 max	0.15-0.35	B 0.0005-0.003; Ni <=0.25; Cu, Mo, Ni not spec'd but acpt; bal Fe				
SAE J404(94)	51B60	Bil Blm Slab Bar HR CF	0.56-0.64	0.70-0.90		0.75-1.00		0.030 max	0.040 max	0.15-0.35	bal Fe				
SAE J407	51B60H*	Obs; see J1268									bal Fe				

Specification	Designation	Notes	C	Cr	Cu	Mn	Mo	P	S	Si	Other	UTS	YS	El	Hard
Alloy Steel, Chromium, 6445															
USA															
	UNS K22097		0.92-1.02	0.90-1.15	0.35 max	0.95-1.25	0.08 max	0.015 max	0.015 max	0.50-0.70	Ni <=0.25; bal Fe				
Alloy Steel, Chromium, A202(A)															
USA															
	UNS K11742		0.17 max	0.31-0.64		0.97-1.52		0.035 max	0.040 max	0.54-0.96	bal Fe				
ASTM A202/A202M(93)	A202(A)	Press ves plt	0.17 max	0.35-0.60		1.05-1.40		0.035 max	0.035 max	0.060-0.090	bal Fe	75-95 ksi	45 ksi	19	
Alloy Steel, Chromium, A202(B)															
USA															
	UNS K12542		0.25 max	0.31-0.64		0.97-1.52		0.035 max	0.040 max	0.54-0.96	bal Fe				
ASTM A202/A202M(93)	A202(B)	Press ves plt	0.25 max	0.35-0.60		1.05-1.40		0.035 max	0.035 max	0.060-0.090	bal Fe	85-110 ksi	47 ksi	18	
Alloy Steel, Chromium, A288(2)															
USA															
	UNS K14542		0.45 max	0.70-1.25		0.60-1.00	0.15 min	0.025 max	0.020 max	0.15-0.35	bal Fe				
ASTM A288(98)	Class 2	Frg	0.45 max	0.70-1.25		0.80-1.00	0.15 min	0.025 max	0.020 max	0.15-0.35	bal Fe	620	450	20	
ASTM A288(98)	Class 3	Frg	0.45 max	0.70-1.25		0.80-1.00	0.15 min	0.025 max	0.020 max	0.15-0.35	bal Fe	760	550	18	
Alloy Steel, Chromium, A325(A)															
USA															
	UNS K13643		0.31-0.42	0.42-0.68	0.22-0.48	0.86-1.24		0.045 max	0.055 max	0.13-0.32	Ni 0.22-0.48; bal Fe				
Alloy Steel, Chromium, A325(B)															
USA															
	UNS K14358		0.36-0.50	0.47-0.83	0.17-0.43	0.67-0.93	0.07 max	0.06-0.125	0.055 max	0.25-0.55	Ni 0.47-0.83; bal Fe				
Alloy Steel, Chromium, A423(1)															
Germany															
DIN EN 10155(93)	S355J0WP		0.12 max	0.30-1.25	0.25-0.55	1.00 max		0.06-0.15	0.040 max	0.75 max	N <=0.009; Ni <=0.65; bal Fe				
DIN EN 10155(93)	WNr 1.8945		0.12 max	0.30-1.25	0.25-0.55	1.00 max		0.06-0.15	0.040 max	0.75 max	N <=0.009; Ni <=0.65; bal Fe				
USA															
	UNS K11535		0.15 max	0.24-1.31	0.20-0.60	0.55 max		0.06-0.16	0.060 max	0.10 min	Ni 0.20-0.70; bal Fe				
Alloy Steel, Chromium, A485(1)															
USA															
	UNS K19667	Bearing	0.90-1.05	0.90-1.20	0.35 max	0.95-1.25	0.10 max	0.025 max	0.025 max	0.45-0.75	Ni <=0.25; bal Fe				
ASTM A485(94)	Grade 1	Bearing, hard	0.90-1.05	0.90-1.20	0.35 max	0.95-1.25	0.10 max	0.025 max	0.025 max	0.45-0.75	N <=0.25; bal Fe				46 HRC
Alloy Steel, Chromium, A485(2)															
Bulgaria															
BDS 12731(75)	SchCh15SG	Bearing; smls tub	0.95-1.05	1.30-1.65	0.25 max	0.90-1.20		0.027 max	0.020 max	0.40-0.65	Ni <=0.30; bal Fe				
France															
AFNOR NFA35565(94)	100CM6*	Ball & roller bearing	0.9-1.05	1.4-1.65	0.3 max	1-1.2	0.1 max	0.025 max	0.015 max	0.5-0.7	Al <=0.05; Ni <=0.25; bal Fe				223 HB max
AFNOR NFA35565(94)	100CrMn6	Ball & roller bearing	0.9-1.05	1.4-1.65		1-1.2	0.1 max	0.025 max	0.015 max	0.5-0.7	Ni <=0.25; bal Fe				223 HB max
AFNOR NFA35590(78)	100CM6*	CW, soft ann	0.9-1.05	1.35-1.6	0.3 max	0.95-1.25	0.15 max	0.025 max	0.025 max	0.4-0.7	Ni <=0.4; bal Fe				217 HB max

UNS numbers and US grades are provided as a means of cross referencing chemically similar alloys. Exchangability is only possible after independent examination of specifications. Tensile properties are minimum or typical as specified. UTS and YS as MPa. El as %. See Appendix for list of abbreviations used in Notes. * indicates obsolete material.

Specification	Designation	Notes	C	Cr	Cu	Mn	Mo	P	S	Si	Other	UTS	YS	El	Hard
Alloy Steel, Chromium, A485(2) (Continued from previous page)															
France															
AFNOR NFA35590(92)	100CM6*	CW, wear res; soft An	0.9-1.05	1.35-1.6	0.3 max	0.95-1.25	0.15 max	0.025 max	0.025 max	0.4-0.7	Ni <=0.4; bal Fe				223 HB max
AFNOR NFA35590(92)	100CrMn6	CW, wear res; soft Ann	0.9-1.05	1.35-1.6		0.95-1.25	0.15 max	0.025 max	0.025 max	0.4-0.7	Ni <-0.4; bal Fe				223 HB max
Germany															
DIN 17230(80)	WNr 1.3520	Bearing Q/T 200C	0.90-1.05	1.40-1.65	0.30 max	1.00-1.20		0.030 max	0.025 max	0.50-0.70	Ni <=0.30; bal Fe				61 HRC
DIN 17350(80)	100CrMn6	Q/T to 200C	0.90-1.05	1.40-1.65	0.30 max	1.00-1.20		0.030 max	0.025 max	0.50-0.70	Ni <=0.30; bal Fe				61 HRC
International															
ISO 683-17(76)	3	Ball & roller bearing, HT	0.95-1.10	1.40-1.65		0.95-1.25		0.030 max	0.025 max	0.45-0.75	bal Fe				
Italy															
UNI 3097(50)	100CM4	Ball & roller bearing	0.95-1.1	0.9-1.1		0.9-1.1		0.03 max	0.03 max	0.5-0.85	bal Fe				
UNI 3097(75)	100CrMo7	Ball & roller bearing	0.95-1.1	1.65-1.95		0.25-0.45	0.3-0.4	0.025 max	0.025 max	0.2-0.4	bal Fe				
Romania															
STAS 1456/1(89)	RUL2	Ball & roller bearing	0.95-1.1	1.3-1.65	0.25 max	0.9-1.2	0.8 max	0.027 max	0.02 max	0.4-0.65	Pb <=0.15; Cu+Ni=0-0.4; bal Fe				
Russia															
GOST 801(60)	SHKH15SG	Ball & roller bearing	0.95-1.05	1.3-1.65	0.3 max	0.9-1.2	0.15 max	0.027 max	0.02 max	0.4-0.65	Al <=0.1; Ni <=0.3; Ti <=0.05; bal Fe				
GOST 801(78)	SChCh15SG	Ball & roller bearing	0.95-1.05	1.3-1.65	0.25 max	0.9-1.2	0.15 max	0.027 max	0.02 max	0.4-0.65	Al <=0.1; Ni <=0.3; Ti <=0.05; Ni+Cu<=0.5; bal Fe				
Spain															
UNE 36027(80)	100CrMn4	Ball & roller bearing	0.95-1.1	0.9-1.2		0.95-1.25	0.15 max	0.03 max	0.025 max	0.45-0.75	bal Fe				
UNE 36027(80)	100CrMn6	Ball & roller bearing	0.95-1.1	1.4-1.65		0.95-1.25	0.15 max	0.03 max	0.025 max	0.45-0.75	bal Fe				
UNE 36027(80)	F.1311*	see 100CrMn4	0.95-1.1	0.9-1.2		0.95-1.25	0.15 max	0.03 max	0.025 max	0.45-0.75	bal Fe				
UNE 36027(80)	F.1312*	see 100CrMn6	0.95-1.1	1.4-1.65		0.95-1.25	0.15 max	0.03 max	0.025 max	0.45-0.75	bal Fe				
USA															
	UNS K19195	Bearing	0.85-1.00	1.40-1.80	0.35 max	1.40-1.70	0.10 max	0.025 max	0.025 max	0.50-0.80	Ni <=0.25; bal Fe				
ASTM A485(94)	Grade 2	Bearing, hard	0.85-1.00	1.40-1.80	0.35 max	1.40-1.70	0.10 max	0.025 max	0.025 max	0.50-0.80	N <=0.25; bal Fe				32 HRC
Yugoslavia															
	C.4340	Ball & roller bearing	0.9-1.05	1.4-1.65		0.9-1.2	0.15 max	0.03 max	0.025 max	0.45-0.7	bal Fe				
Alloy Steel, Chromium, A502(3B)															
USA															
	UNS K12244		0.21 max	0.37-0.73	0.17-0.43	0.71-1.29		0.045 max	0.055 max	0.13-0.32	Ni 0.22-0.53; V <=0.11; bal Fe				
Alloy Steel, Chromium, A563(DH3)															
USA															
	UNS K13650		0.20-0.53	0.45 min	0.20 min	0.40 min	0.15 min	0.046 max	0.050 max		Ni >=0.20; bal Fe				
Alloy Steel, Chromium, A595(C)															
USA															
	UNS K11526		0.15 max	0.24-1.31	0.22-0.58	0.17-0.53		0.06-0.16	0.06 max	0.19-0.81	Ni <=0.68; bal Fe				
ASTM A595(97)	595C	Tube, tapered; struct use	0.12 max	0.30-1.25	0.25-0.55	0.20-0.50		0.07-0.15	0.025 max	0.25-0.75	Ni <=0.65; bal Fe	410	480	21.0	
Alloy Steel, Chromium, A714(IV)															
Mexico															
NMX-B-069(86)	Grade IV	Weld and Smls pipe	0.10 max	0.50-1.20	0.25-0.45	0.60 max		0.03-0.04	0.05 max		Ni 0.20-0.50; bal Fe	400	250		

UNS numbers and US grades are provided as a means of cross referencing chemically similar alloys. Exchangability is only possible after independent examination of specifications. Tensile properties are minimum or typical as specified. UTS and YS as MPa. El as %. See Appendix for list of abbreviations used in Notes. * indicates obsolete material.

Specification	Designation	Notes	C	Cr	Cu	Mn	Mo	P	S	Si	Other	UTS	YS	El	Hard
Alloy Steel, Chromium, A714(IV) (Continued from previous page)															
USA															
	UNS K11356		0.13 max	0.74-1.26	0.22-0.48	0.65 max			0.06 max		Ni 0.17-0.53; bal Fe				
ASTM A714(96)	IV	Weld, Smls pipe	0.10 max	0.80-1.20	0.25-0.45	0.60 max		0.03-0.08	0.05 max		Ni 0.20-0.50; bal Fe	400	250		
Alloy Steel, Chromium, B(IIB)															
USA															
	UNS K92400	Electrical Heating Element		20-24							Al 4.00-5.25; bal Fe				
Alloy Steel, Chromium, B603(I)															
USA															
	UNS K92500	Electrical Heating Element		20-24							Al 5.00-6.00; bal Fe				
Alloy Steel, Chromium, E50100															
China															
YB/T 1(80)	GCr6	Bar Tub HR/Frg/CD Ann	1.05-1.15	0.40-0.70	0.25 max	0.20-0.40		0.025 max	0.025 max	0.15-0.35	Ni+Cu<=0.50; bal Fe				179-207 HB
Mexico															
NMX-B-300(91)	E50100	Bar	0.98-1.10	0.40-0.60		0.25-0.45		0.025 max	0.025 max	0.15-0.35	bal Fe				
USA															
	AISI E50100	Smls mech Tub; BEF	0.95-1.10	0.40-0.60		0.25-0.45		0.025 max	0.025 max	0.15-0.35	bal Fe				
	UNS G50986		0.95-1.10	0.40-0.60		0.25-0.45		0.025 max	0.025 max	0.15-0.35	bal Fe				
ASTM A29/A29M(93)	E50100	Bar	0.98-1.10	0.40-0.60		0.25-0.45		0.025 max	0.025 max	0.15-0.35	bal Fe				
ASTM A295(94)	50100	Bearing, coil bar tub, CD ann	0.98-1.10	0.40-0.60	0.35 max	0.25-0.45	0.10 max	0.025 max	0.025 max	0.15-0.35	Ni <=0.25; bal Fe				250 HB max
ASTM A519(96)	E50100	Smls mech tub	0.95-1.10	0.40-0.60		0.25-0.45		0.025 max	0.025 max	0.15-0.35	bal Fe				
Alloy Steel, Chromium, E51100															
China															
YB/T 1(80)	GCr9	Bar Tub HR/Frg/CD Ann	1.00-1.10	0.90-1.20	0.25 max	0.25-0.45		0.025 max	0.025 max	0.15-0.35	Ni+Cu<=0.50; bal Fe				179-207 HB
Czech Republic															
CSN 414208	14208	Q/T; Ball & roller bearing	0.9-1.1	0.8-1.2	0.25 max	0.9-1.2		0.027 max	0.03 max	0.4-0.7	Ni <=0.3; bal Fe				
Germany															
DIN	105Cr4	Bearing Q/T to 200C	1.00-1.10	0.90-1.15		0.25-0.40		0.030 max	0.025 max	0.15-0.35	bal Fe				62 HRC
DIN	WNr 1.3503	Q/T 200C	1.00-1.10	0.90-1.15		0.25-0.40		0.030 max	0.025 max	0.15-0.35	bal Fe				62 HRC
India															
IS 1570	103Cr1		0.95-1.1	0.9-1.2		0.25-0.45				0.15-0.3	bal Fe				
Mexico															
NMX-B-300(91)	E51100	Bar	0.98-1.10	0.90-1.15		0.25-0.45		0.025 max	0.025 max	0.15-0.35	bal Fe				
Pan America															
COPANT 514	E51100	Tube	0.95-1.1	0.9-1.15		0.25-0.45		0.04	0.05	0.1	bal Fe				
Spain															
UNE 36254(79)	90Cr4		0.8-1	0.8-1.2		0.5-0.8		0.04	0.04	0.3-0.5	bal Fe				
UNE 36254(79)	F.8222*	see 90Cr4	0.8-1	0.8-1.2		0.5-0.8		0.04	0.04	0.3-0.5	bal Fe				
USA															
	AISI E51100	Smls mech Tub; BEF	0.98-1.10			0.25-0.45		0.025 max	0.025 max	0.15-0.35	bal Fe				
	AISI E51100	Bar Blm Bil Slab	0.98-1.10	0.90-1.15		0.25-0.45		0.025 max	0.025 max	0.15-0.35	bal Fe				
	UNS G51986		0.98-1.10	0.90-1.15		0.25-0.45		0.025 max	0.025 max	0.15-0.35	bal Fe				

UNS numbers and US grades are provided as a means of cross referencing chemically similar alloys. Exchangability is only possible after independent examination of specifications. Tensile properties are minimum or typical as specified. UTS and YS as MPa. El as %. See Appendix for list of abbreviations used in Notes. * indicates obsolete material.

Specification	Designation	Notes	C	Cr	Cu	Mn	Mo	P	S	Si	Other	UTS	YS	El	Hard

Alloy Steel, Chromium, E51100 (Continued from previous page)

USA

Specification	Designation	Notes	C	Cr	Cu	Mn	Mo	P	S	Si	Other	UTS	YS	El	Hard
AMS 6443		Bar Rod Wir; Tub, Bil	0.98-1.1	0.90-1.15		0.25-0.45		0.03 max	0.03 max	0.15-0.30	bal Fe				
AMS 6446		Bar Rod Wir; Tub, Bil	0.98-1.1	0.90-1.15		0.25-0.45		0.03 max	0.03 max	0.15-0.30	bal Fe				
AMS 6449		Bar Rod Wir; Tub, Bil	0.98-1.1	0.90-1.15		0.25-0.45		0.03 max	0.03 max	0.15-0.30	bal Fe				
ASTM A29/A29M(93)	E51100	Bar	0.98-1.10	0.90-1.15		0.25-0.45		0.025	0.025	0.15-0.35	bal Fe				
ASTM A295(94)	51100	Bearing, coil bar tub, CD ann	0.98-1.10	0.9-1.15	0.35 max	0.25-0.45	0.10 max	0.025 max	0.025 max	0.15-0.35	Ni <=0.25; bal Fe				250 HB max
ASTM A322(96)	E51100	Bar	0.98-1.10	0.9-1.15		0.25-0.45		0.025 max	0.025 max	0.15-0.35	bal Fe				
ASTM A331(95)	E51100	Bar	0.98-1.10	0.9-1.15		0.25-0.45		0.025 max	0.025 max	0.15-0.35	bal Fe				
ASTM A519(96)	E51100	Smls mech tub	0.95-1.10	0.90-1.15		0.25-0.45		0.025 max	0.025 max	0.15-0.35	bal Fe				
ASTM A752(93)	E51100	Rod Wir	0.98-1.10	0.90-1.15		0.25-0.45		0.025 max	0.025 max	0.15-0.30	bal Fe				
SAE 770(84)	E51100*	Obs; see J1397(92)	0.98-1.1	0.9-1.15		0.25-0.45		0.03 max	0.03 max	0.15-0.3	bal Fe				

Alloy Steel, Chromium, E52100

Argentina

Specification	Designation	Notes	C	Cr	Cu	Mn	Mo	P	S	Si	Other	UTS	YS	El	Hard
IAS	IRAM 52100		0.98-1.10	1.30-1.60		0.25-0.45		0.025 max	0.025 max	0.20-0.35	bal Fe	1080	780	20	325 HB

Australia

Specification	Designation	Notes	C	Cr	Cu	Mn	Mo	P	S	Si	Other	UTS	YS	El	Hard
AS G18	En31*	Obs; see AS 1444	0.9-1.2	1.2-1.6		0.3-0.75		0.05	0.05	0.1-0.35	bal Fe				

China

Specification	Designation	Notes	C	Cr	Cu	Mn	Mo	P	S	Si	Other	UTS	YS	El	Hard
YB/T 1(80)	GCr15	Bar Tub HR/Frg/CD Ann	0.95-1.05	1.40-1.65	0.25 max	0.25-0.45		0.025 max	0.025 max	0.15-0.35	Ni+Cu<=0.50; bal Fe				179-207 HB

Czech Republic

Specification	Designation	Notes	C	Cr	Cu	Mn	Mo	P	S	Si	Other	UTS	YS	El	Hard
CSN 414100	14100	Q/T	0.95-1.05	1.25-1.5		0.35-0.6		0.035 max	0.035 max	0.17-0.37	bal Fe				
CSN 414109	14109	Ball & roller bearing	0.9-1.1	1.3-1.65	0.25 max	0.3-0.5		0.027 max	0.03 max	0.15-0.35	Ni <=0.3; bal Fe				
CSN 414209	14209	Ball & roller bearing	0.9-1.1	1.3-1.65	0.25 max	0.9-1.2		0.027 max	0.03 max	0.35-0.65	Ni <=0.3; bal Fe				

France

Specification	Designation	Notes	C	Cr	Cu	Mn	Mo	P	S	Si	Other	UTS	YS	El	Hard
AFNOR NFA35552(84)	100C6*	Ball & roller bearing; bar rod	0.95-1.1	1.35-1.6		0.2-0.4	0.15 max	0.03 max	0.025 max	0.15-0.35	bal Fe				
AFNOR NFA35553(82)	100C6*	Ball & roller bearing; bar rod	0.95-1.1	1.3-1.6		0.25-0.4	0.15 max	0.025 max	0.025 max	0.1-0.35	bal Fe				62 HRC
AFNOR NFA35565(94)	100C6*	see 100Cr6	0.95-1.1	1.35-1.6		0.2-0.4	0.1	0.03	0.025	0.15-0.35	bal Fe				
AFNOR NFA36102(93)	100Cr6RR	HR; Strp CR	0.95-1.1	1.35-1.6		0.2-0.4	0.10 max	0.025 max	0.02 max	0.15-0.35	Ni <=0.20; bal Fe				

Germany

Specification	Designation	Notes	C	Cr	Cu	Mn	Mo	P	S	Si	Other	UTS	YS	El	Hard
DIN 17230(80)	100Cr6	Bearing Q/T 200C	0.90-1.05	1.35-1.65	0.30 max	0.25-0.45		0.030 max	0.025 max	0.15-0.35	Ni <=0.30; bal Fe				62 HRC
DIN 17230(80)	WNr 1.3505	Bearing Q/T 200C	0.90-1.05	1.35-1.65	0.30 max	0.25-0.45		0.030 max	0.025 max	0.15-0.35	Ni <=0.30; bal Fe				62 HRC
DIN 17350(80)	100Cr6	Q/T to 200C	0.95-1.10	1.35-1.65		0.25-0.45		0.030 max	0.030 max	0.15-0.35	bal Fe				63 HRC
DIN(Aviation Hdbk)	101Cr6		0.95-1.10			0.25-0.45		0.015 max	0.015 max	0.15-0.35	Ni <=0.40; bal Fe				
DIN(Aviation Hdbk)	WNr 1.3514		0.95-1.10	1.35-1.65		0.25-0.45		0.015 max	0.015 max	0.15-0.35	Ni <=0.40; bal Fe				

Hungary

Specification	Designation	Notes	C	Cr	Cu	Mn	Mo	P	S	Si	Other	UTS	YS	El	Hard
MSZ 17789(83)	GO3	Ball & roller bearing; 0-999mm; soft ann to shearing	0.95-1.05	1.35-1.65	0.25 max	0.25-0.45		0.027 max	0.025 max	0.17-0.37	Ni <=0.3; Ni+Cu<=0.5; bal Fe				277-321 HB
MSZ 17789(83)	GO3	Ball & roller bearing; 0-999mm; soft ann to mach	0.95-1.05	1.35-1.65	0.25 max	0.25-0.45		0.027 max	0.025 max	0.17-0.37	Ni <=0.3; Ni+Cu<=0.5; bal Fe				217 HB max

UNS numbers and US grades are provided as a means of cross referencing chemically similar alloys. Exchangability is only possible after independent examination of specifications. Tensile properties are minimum or typical as specified. UTS and YS as MPa. El as %. See Appendix for list of abbreviations used in Notes. * indicates obsolete material.

Specification	Designation	Notes	C	Cr	Cu	Mn	Mo	P	S	Si	Other	UTS	YS	El	Hard

Alloy Steel, Chromium, E52100 (Continued from previous page)

Specification	Designation	Notes	C	Cr	Cu	Mn	Mo	P	S	Si	Other	UTS	YS	El	Hard
Hungary															
MSZ 17789(83)	GO4	Ball & roller bearing; 0-999mm; soft ann to shearing	0.95-1.05	1.35-1.65	0.25 max	0.9-1.2		0.027 max	0.025 max	0.4-0.65	Ni <=0.3; Ni+Cu<=0.5; bal Fe				217 HB max
MSZ 17789(83)	GO4	Ball & roller bearing; 0-999mm; soft ann to shearing	0.95-1.05	1.35-1.65	0.25 max	0.9-1.2		0.027 max	0.025 max	0.4-0.65	Ni <=0.3; Ni+Cu<=0.5; bal Fe				277-321 HB
India															
IS 1570	103Cr2		0.95-1.1	1.3-1.6		0.25-0.45				0.15-0.3	bal Fe				
IS 1570	105Cr1Mn60		0.9-1.2	1-1.6		0.4-0.8				0.1-0.35	bal Fe				
International															
ISO 683-17(76)	1	Ball & roller bearing, HT	0.95-1.10	1.35-1.65		0.25-0.45		0.030 max	0.025 max	0.15-0.35	bal Fe				
ISO 683-17(76)	2	Ball & roller bearing, HT	0.95-1.10	0.90-1.20		0.95-1.25		0.030 max	0.025 max	0.45-0.75	bal Fe				
Italy															
UNI 3097(50)	100C6	Ball & roller bearing	0.95-1.1	1.4-1.6		0.3-0.5		0.03 max	0.03 max	.035 max	P+S=0.05; bal Fe				
UNI 3097(75)	100Cr6	Ball & roller bearing	0.95-1.1	1.4-1.6		0.25-0.45		0.025 max	0.025 max	0.15-0.35	bal Fe				
Mexico															
DGN B-297	E52100*	Obs	0.98-1.1	1.3-1.6		0.25-0.45		0.025 max	0.025 max	0.2-0.35	bal Fe				
NMX-B-300(91)	E52100	Bar	0.98-1.10	1.30-1.60		0.25-0.45		0.025 max	0.025 max	0.15-0.35	bal Fe				
Pan America															
COPANT 334	52100	Bar	0.98-1.1	1.3-1.6		0.25-0.45		0.04	0.03	0.2-0.35	bal Fe				
COPANT 514	52100		0.98-1.1	1.3-1.6		0.25-0.45		0.04	0.05	0.1	bal Fe				
Poland															
PNH84041(74)	LH15	Ball & roller bearing	0.95-1.1	1.3-1.65	0.25	0.25-0.45		0.027	0.02	0.15-0.35	Ni 0.3; bal Fe				
PNH84041(74)	LH15SG	Ball & roller bearing	0.95-1.1	1.2-1.65	0.25	0.95-1.25		0.027	0.02	0.4-0.65	Ni 0.3; bal Fe				
Romania															
STAS 11250(89)	RUL1v	Ball & roller bearing	0.95-1.1	1.3-1.65	0.25 max	0.2-0.45	0.8 max	0.027 max	0.02 max	0.17-0.37	Pb <=0.15; bal Fe				
STAS 11250(89)	RUL2v	Ball & roller bearing	0.95-1.1	1.3-1.65	0.25 max	0.9-1.2	0.8 max	0.027 max	0.02 max	0.4-0.65	Pb <=0.15; bal Fe				
Russia															
GOST 21022	SChc15-SChD		0.95-1.05	1.3-1.65	0.25 max	0.3-0.5		0.025	0.01	0.2-0.37	Ni 0.3; Ni+Cu=0.50; bal Fe				
GOST 21022	SChCh15-SChD		0.95-1.05	1.3-1.65	0.25 max	0.3-0.5		0.025	0.01	0.2-0.37	Ni 0.3; Ni+Cu=0.50; bal Fe				
GOST 21022(75)	SChCh15-SChD	Ball & roller bearing	0.95-1.05	1.3-1.65	0.25 max	0.3-0.5	0.15 max	0.025 max	0.01 max	0.2-0.37	Al <=0.1; Ni <=0.3; Ti <=0.05; Ni+Cu<=0.5; bal Fe				
GOST 801	SChCh15		0.95-1.05	1.3-1.65	0.25 max	0.2-0.4		0.03 max	0.02 max	0.17-0.37	Ni <=0.3; bal Fe				
GOST 801	SChCh20SG		0.9-1	1.4-1.7	0.25	1.4-1.7		0.027	0.02	0.55-0.85	Ni 0.3; Ni+Cu=0.50; bal Fe				
GOST 801	ShCh15		0.95-1.05	1.3-1.65	0.25 max	0.2-0.4		0.027	0.02	0.17-0.37	Ni <=0.3; bal Fe				
GOST 801	ShCh20SG		0.9-1	1.4-1.7	0.25	1.4-1.7		0.027	0.02	0.55-0.85	Ni 0.3; Ni+Cu=0.50; bal Fe				
GOST 801(78)	SChCh20SG	Ball & roller bearing	0.9-1.0	1.4-1.7	0.25 max	1.4-1.7	0.15 max	0.027 max	0.02 max	0.55-0.85	Al <=0.1; Ni <=0.3; Ti <=0.05; Ni+Cu<=0.5; bal Fe				
Spain															
UNE	F.131		0.95-1.2	1.4-1.8		0.25-0.4		0.04	0.05	0.1-0.35	bal Fe				
Sweden															
SS 142258	2258		0.95-1.1	1.35-1.65		0.25-0.45		0.03	0.025	0.15-0.35	bal Fe				

Specification	Designation	Notes	C	Cr	Cu	Mn	Mo	P	S	Si	Other	UTS	YS	El	Hard

Alloy Steel, Chromium, E52100 (Continued from previous page)

UK

Specification	Designation	Notes	C	Cr	Cu	Mn	Mo	P	S	Si	Other	UTS	YS	El	Hard
BS 970/1(83)	535A99	Blm Bil Slab Bar Rod Frg	0.95-1.10	1.20-1.60		0.40-0.70		0.035 max	0.040 max		bal Fe				229 HB max

USA

Specification	Designation	Notes	C	Cr	Cu	Mn	Mo	P	S	Si	Other	UTS	YS	El	Hard
	AISI E52100	Bar Blm Bil Slab; BEF	0.98-1.10	1.30-1.60		0.25-0.45		0.025 max	0.025 max	0.15-0.35	bal Fe				
	AISI E52100	Smls mech Tub; BEF	0.98-1.10			0.25-0.45		0.025 max	0.025 max	0.15-0.35	bal Fe				
	UNS G52986		0.98-1.10	1.30-1.60		0.25-0.45		0.025 max	0.025 max	0.15-0.35	bal Fe				
AMS 6440		Bar Frg Tub	0.98-1.1	1.3-1.6	0.35	0.25-0.45	0.1	0.025	0.025	0.15-0.35	Ni 0.25; bal Fe				
AMS 6444		Bar Frg Tub	0.98-1.1	1.3-1.6	0.35	0.25-0.45	0.08	0.015	0.015	0.15-0.35	Ni 0.25; bal Fe				
AMS 6447		Bar Frg Tub	0.98-1.1	1.3-1.6	0.35	0.25-0.45	0.08	0.015	0.015	0.15-0.35	Ni 0.25; bal Fe				
ASTM A274	E52100*	Obs, 1975; see A711									bal Fe				
ASTM A29/A29M(93)	E52100	Bar	0.98-1.10	1.30-1.60		0.25-0.45		0.025 max	0.025 max	0.15-0.35	bal Fe				
ASTM A295(94)	52100	Bearing, coil bar tub, CD ann	0.98-1.10	1.3-1.6	0.35 max	0.25-0.45	0.10 max	0.025 max	0.025 max	0.15-0.35	Ni <=0.25; bal Fe				250 HB max
ASTM A322(96)	E52100	Bar	0.98-1.10	1.30-1.60		0.25-0.45		0.025 max	0.025 max	0.15-0.35	bal Fe				
ASTM A331(95)	E52100	Bar	0.98-1.10	1.30-1.60		0.25-0.45		0.025 max	0.025 max	0.15-0.35	bal Fe				
ASTM A519(96)	E52100	Smls mech tub	0.95-1.10	1.30-1.60		0.25-0.45		0.025 max	0.025 max	0.15-0.35	bal Fe				
ASTM A535(92)	E52100	Ball & roller bearing	0.95-1.10	1.30-1.60	0.35 max	0.25-0.45	0.10 max	0.015 max	0.015 max	0.15-0.35	Ni <=0.25; bal Fe				
ASTM A535(92)	E52100 Mod 1	Ball & roller bearing	0.90-1.05	0.90-1.20	0.35 max	0.95-1.25	0.10 max	0.015 max	0.015 max	0.15-0.35	Ni <=0.25; bal Fe				
ASTM A535(92)	E52100 Mod 2	Ball & roller bearing	0.85-1.00	1.40-1.80	0.35 max	1.40-1.70	0.10 max	0.015 max	0.015 max	0.15-0.35	Ni <=0.25; bal Fe				
ASTM A535(92)	E52100 Mod 3	Ball & roller bearing	0.95-1.10	1.10-1.50	0.35 max	0.65-0.90	0.20-0.30	0.015 max	0.015 max	0.15-0.35	Ni <=0.25; bal Fe				
ASTM A535(92)	E52100 Mod 4	Ball & roller bearing	0.95-1.10	1.10-1.50	0.35 max	1.05-1.35	0.45-0.60	0.015 max	0.015 max	0.15-0.35	Ni <=0.25; bal Fe				
ASTM A646(95)	Grade 15	Blm Bil for aircraft aerospace frg, ann	0.98-1.10	1.30-1.60		0.25-0.45		0.025 max	0.010 max	0.20-0.35	bal Fe				302 HB
ASTM A752(93)	E52100	Rod Wir	0.98-1.10	1.30-1.60		0.25-0.45		0.025 max	0.025 max	0.15-0.30	bal Fe				
MIL-B-81793	E52100	Bearing, ball ring	0.98-1.10	1.30-1.60		0.25-0.45		0.025 max	0.025 max	0.15-0.35	bal Fe				
MIL-B-913(93)	E52100	Bearing, ball ring	0.98-1.10	1.30-1.60		0.25-0.45		0.025 max	0.025 max	0.15-0.35	bal Fe				
MIL-S-7420B(58)	E51200*	Obs for new design; Bar, aircraft qual	0.95-1.1	1.3-1.6		0.25-0.45		0.025	0.025	0.2-0.35	bal Fe				
MIL-S-980	51200*	Obs	0.98-1.1	1.3-1.6	0.25	0.25-0.45	0.08	0.025	0.025	0.2-0.35	Ni 0.35; bal Fe				
SAE 770(84)	E52100*	Obs; see J1397(92)									bal Fe				
SAE J404(94)	E52100	Electric Furnace	0.98-1.10	1.30-1.60		0.25-0.45		0.025 max	0.025 max	0.15-0.35	bal Fe				

Yugoslavia

Specification	Designation	Notes	C	Cr	Cu	Mn	Mo	P	S	Si	Other	UTS	YS	El	Hard
	C.4146	Ball & roller bearing	0.9-1.05	1.35-1.65		0.25-0.45	0.15 max	0.03 max	0.025 max	0.15-0.35	bal Fe				

Alloy Steel, Chromium, F256(II)

Bulgaria

Specification	Designation	Notes	C	Cr	Cu	Mn	Mo	P	S	Si	Other	UTS	YS	El	Hard
BDS 9801	09G2S	Struct weld ships	0.12 max	0.20 max	0.30 max	1.30-1.70	0.15 max	0.035 max	0.035 max	0.50-0.80	Ni <=0.30; bal Fe				
BDS 9801	10ChSND	Struct, weld ships	0.12 max	0.60-0.90	0.4-0.5	0.50-0.80	0.15 max	0.035 max	0.035 max	0.80-1.10	Al 0.015-0.06; Ni 0.50-0.80; bal Fe				
BDS 9801	10ChSND	Struct weld ships	0.12 max	0.60-0.90	0.4-0.65	0.50-0.80	0.15 max	0.035 max	0.035 max	0.80-1.10	Al 0.015-0.06; Ni 0.50-0.80; bal Fe				

USA

Specification	Designation	Notes	C	Cr	Cu	Mn	Mo	P	S	Si	Other	UTS	YS	El	Hard
	UNS K92801	Cr Sealing Alloy	0.12 max	28		1.00 max		0.04 max	0.03 max	0.75 max	N <=0.20; Ni <=0.50; bal Fe				

UNS numbers and US grades are provided as a means of cross referencing chemically similar alloys. Exchangability is only possible after independent examination of specifications. Tensile properties are minimum or typical as specified. UTS and YS as MPa. El as %. See Appendix for list of abbreviations used in Notes. * indicates obsolete material.

Specification	Designation	Notes	C	Cr	Cu	Mn	Mo	P	S	Si	Other	UTS	YS	El	Hard

Alloy Steel, Chromium, K12040

USA

| | UNS K12040 | | 0.20 max | 0.50-1.00 | 0.30-0.50 | 1.20 max | 0.10 max | 0.04 max | 0.05 max | 0.25-0.70 | Ni <=0.80; Ti <=0.07; bal Fe | | | | |

Alloy Steel, Chromium, K14245

USA

| | UNS K14245 | | 0.35-0.50 | 0.75-1.50 | | 0.60-1.00 | 0.15 min | 0.025 max | 0.025 max | 0.15-0.35 | bal Fe | | | | |

Alloy Steel, Chromium, K19526

USA

| ASTM A295(94) | K19526 | Bearing, coil bar tub, CD ann | 0.89-1.01 | 0.40-0.60 | 0.35 max | 0.50-0.80 | 0.08-0.15 | 0.025 max | 0.025 max | 0.15-0.35 | Ni <=0.25; bal Fe | | | | 250 HB max |

Alloy Steel, Chromium, No equivalents identified

Australia

| AS 1444(96) | 51B40 | Standard series, Hard/Tmp | 0.38-0.43 | 0.70-0.90 | | 0.70-0.90 | | 0.040 max | 0.040 max | 0.10-0.35 | B >=0.0005; bal Fe | | | | |

Canada

| CSA G40.21(98) | 350R | t<65mm,PLT,SHT,Bar | 0.16 max | 0.30-1.25 | 0.20-0.60 | 0.75 max | | 0.05-0.15 | 0.04 max | 0.75 max | Ni <=0.90; Al+Nb+V<=0.10; bal Fe | 480-650 | 350 | 19L 16T | |

Czech Republic

| CSN 414231 | 14231 | Nitriding | 0.23-0.29 | 1.0-1.3 | | 0.8-1.1 | | 0.035 max | 0.035 max | 0.17-0.37 | Ti 0.04-0.1; bal Fe | | | | |
| CSN 414240 | 14240 | Q/T; Frg | 0.32-0.4 | 0.2-0.4 | | 1.5-1.9 | | 0.035 max | 0.035 max | 0.17-0.37 | bal Fe | | | | |

Europe

EN 10083/1(91)A1(96)	1.7006	Q/T t<=16mm	0.42-0.50	0.90-1.20		0.50-0.80	0.15 max	0.035 max	0.035 max	0.40 max	bal Fe	900-1100	650	12	
EN 10083/1(91)A1(96)	1.7025	Q/T t<=16mm	0.42-0.50	0.90-1.20		0.50-0.80	0.15 max	0.035 max	0.020-0.040	0.40 max	bal Fe	900-1100	650	12	
EN 10083/1(91)A1(96)	46Cr2	Q/T t<=16mm	0.42-0.50	0.90-1.20		0.50-0.80	0.15 max	0.035 max	0.035 max	0.40 max	bal Fe	900-1100	650	12	
EN 10083/1(91)A1(96)	46CrS2	Q/T t<=16mm	0.42-0.50	0.90-1.20		0.50-0.80	0.15 max	0.035 max	0.020-0.040	0.40 max	bal Fe	900-1100	650	12	

France

AFNOR NFA35564(83)	32C4FF	Bar Wir rod CF; Soft ann	0.3-0.35	0.9-1.2		0.6-0.9	0.15 max	0.03 max	0.03 max	0.35 max	Al >=0.02; Ni <=0.4; Ti <=0.05; bal Fe	0-560			207 HB max
AFNOR NFA35565(94)	100Cr6	Ball & roller bearing	0.95-1.1	1.35-1.6	0.35 max	0.25-0.45	0.1 max	0.025 max	0.015 max	0.15-0.35	Al <=0.05; Ni <=0.3; bal Fe				
AFNOR NFA35565(94)	100CrMo7-3	Ball & roller bearing	0.9-1.1	1.65-1.95	0.3 max	0.2-0.4	0.2-0.4	0.025 max	0.015 max	0.2-0.4	Al <=0.05; Ni <=0.3; bal Fe				
AFNOR NFA35571(96)	45Cr4	t<1000mm; Q/T	0.41-0.48	0.85-1.15	0.3 max	0.6-0.9	0.15 max	0.025 max	0.025 max	0.1-0.4	Ni <=0.4; Ti <=0.05; bal Fe	1200-1500	1100	6	

Hungary

MSZ 6259(82)	LK37B	Struct; Weather res; 40.1-100mm; rolled/norm	0.13 max	0.5-0.8	0.2-0.5	0.6 max		0.04 max	0.04 max	0.15-0.5	Ni 0.2-0.4; 0-16mm P<=0.15; bal Fe	360-490	215	26L; 24T	
MSZ 6259(82)	LK37B	Struct; Weather res; 0-16mm; rolled/norm	0.13 max	0.5-0.8	0.2-0.5	0.6 max		0.04 max	0.04 max	0.15-0.5	Ni 0.2-0.4; 0-16mm P<=0.15; bal Fe	360-490	235	26L; 24T	
MSZ 6259(82)	LK37C	Struct; Weather res; 0-16mm; rolled/norm	0.13 max	0.5-0.8	0.2-0.5	0.6 max		0.04 max	0.04 max	0.15-0.5	Al 0.015-0.120; Nb 0.01-0.06; Ni 0.2-0.4; Ti <=0.06; V 0.02-0.1; 0-16mm P<=0.15; Ti=0.02-0.06; Zr=0.01-0.06; Al+V>=0.02; Al+Nb+V+Ti+Zr<=0.15; bal Fe	360-490	235	26L; 24T	
MSZ 6259(82)	LK37C	Struct; Weather res; 40.1-100mm; rolled/norm	0.13 max	0.5-0.8	0.2-0.5	0.6 max		0.04 max	0.04 max	0.15-0.5	Al 0.015-0.120; Nb 0.01-0.06; Ni 0.2-0.4; Ti <=0.06; V 0.02-0.1; 0-16mm P<=0.15; Ti=0.02-0.06; Zr=0.01-0.06; Al+V>=0.02; Al+Nb+V+Ti+Zr<=0.15; bal Fe	360-490	215	26L; 24T	

UNS numbers and US grades are provided as a means of cross referencing chemically similar alloys. Exchangability is only possible after independent examination of specifications. Tensile properties are minimum or typical as specified. UTS and YS as MPa. El as %. See Appendix for list of abbreviations used in Notes. * indicates obsolete material.

Specification	Designation	Notes	C	Cr	Cu	Mn	Mo	P	S	Si	Other	UTS	YS	El	Hard

Alloy Steel, Chromium, No equivalents identified

Hungary

Specification	Designation	Notes	C	Cr	Cu	Mn	Mo	P	S	Si	Other	UTS	YS	El	Hard
MSZ 6259(82)	LK37D	Struct; Weather res; 40.1-100mm; rolled/norm	0.13 max	0.5-0.8	0.2-0.5	0.6 max		0.04 max	0.04 max	0.15-0.5	Al 0.015-0.120; Nb 0.01-0.06; Ni 0.2-0.4; Ti <=0.06; V 0.02-0.1; 0-16mm P<=0.15; Ti=0.02-0.06; Zr=0.01-0.06; Al+V>=0.02; Al+Nb+V+Ti+Zr<=0.15; bal Fe	360-490	215	26L; 24T	
MSZ 6259(82)	LK37D	Struct; Weather res; 0-16mm; rolled/norm	0.13 max	0.5-0.8	0.2-0.5	0.6 max		0.04 max	0.04 max	0.15-0.5	Al 0.015-0.120; Nb 0.01-0.06; Ni 0.2-0.4; Ti <=0.06; V 0.02-0.1; 0-16mm P<=0.15; Ti=0.02-0.06; Zr=0.01-0.06; Al+V>=0.02; Al+Nb+V+Ti+Zr<=0.15; bal Fe	360-490	235	26L; 24T	
MSZ 6259(82)	LK45B	Struct; Weather res; 40.1-100mm; rolled/norm	0.14 max	0.5-1	0.2-0.5	0.8 max		0.04 max	0.04 max	0.15-0.5	Ni 0.3-0.6; 0-16mm P<=0.15; bal Fe	440-590	255	24L; 22T	
MSZ 6259(82)	LK45B	Struct; Weather res; 0-16mm; rolled/norm	0.14 max	0.5-1	0.2-0.5	0.8 max		0.04 max	0.04 max	0.15-0.5	Ni 0.3-0.6; 0-16mm P<=0.15; bal Fe	440-590	295	24L; 22T	
MSZ 6259(82)	LK45C	Struct; Weather res; 0-16mm; rolled/norm	0.14 max	0.5-1	0.2-0.5	0.8 max		0.04 max	0.04 max	0.15-0.5	Al 0.015-0.120; Nb 0.01-0.06; Ni 0.3-0.6; Ti <=0.06; V 0.02-0.1; 0-16mm P<=0.15; Ti=0.02-0.06; Zr=0 01-0.06; Al+Nb+V+Ti+Zr<=0.15; bal Fe	440-590	295	24L; 22T	
MSZ 6259(82)	LK45C	Struct; Weather res; 40.1-100mm; rolled/norm	0.14 max	0.5-1	0.2-0.5	0.8 max		0.04 max	0.04 max	0.15-0.5	Al 0.015-0.120; Nb 0.01-0.06; Ni 0.3-0.6; Ti <=0.06; V 0.02-0.1; 0-16mm P<=0.15; Ti=0.02-0.06; Zr=0.01-0.06; Al+Nb+V+Ti+Zr<=0.15; bal Fe	440-590	255	24L; 22T	
MSZ 6259(82)	LK45D	Struct; Weather res; 40.1-100mm; rolled/norm	0.14 max	0.5-1	0.2-0.5	0.8 max		0.04 max	0.04 max	0.15-0.5	Al 0.015-0.120; Nb 0.01-0.06; Ni 0.3-0.6; Ti <=0.06; V 0.02-0.1; 0-16mm P<=0.15; Ti=0.02-0.06; Zr=0.01-0.06; Al+Nb+V+Ti+Zr<=0.15; bal Fe	440-590	255	24L; 22T	
MSZ 6259(82)	LK45D	Struct; Weather res; 0-16mm; rolled/norm	0.14 max	0.5-1	0.2-0.5	0.8 max		0.04 max	0.04 max	0.15-0.5	Al 0.015-0.120; Nb 0.01-0.06; Ni 0.3-0.6; Ti <=0.06; V 0.02-0.1; 0-16mm P<=0.15; Ti=0.02-0.06; Zr=0.01-0.06; Al+Nb+V+Ti+Zr<=0.15; bal Fe	440-590	295	24L; 22T	

India

Specification	Designation	Notes	C	Cr	Cu	Mn	Mo	P	S	Si	Other	UTS	YS	El	Hard
IS 1570/4(88)	103Cr4	Const, Spring	0.95-1.1	0.9-1.2		0.25-0.45	0.15 max	0.07 max	0.06 max	0.1-0.35	Ni <=0.4; bal Fe				
IS 1570/4(88)	103Cr6	Const, Spring	0.95-1.1	1.4-1.6		0.25-0.45	0.15 max	0.07 max	0.06 max	0.1-0.35	Ni <=0.4; bal Fe				
IS 1570/4(88)	105Cr5	Const, Spring	0.9-1.2	1.-1.6		0.4-0.8	0.15 max	0.07 max	0.06 max	0.1-0.35	Ni <=0.4; bal Fe				
IS 1570/4(88)	25Cr13Mo6	0-63mm; Bar Frg	0.2-0.3	2.9-3.4		0.4-0.7	0.45-0.65 max	0.07 max	0.06 max	0.1-0.35	Ni <=0.3; bal Fe	1540	1240	8	444 HB
IS 1570/4(88)	42Cr6V1	0-15mm; Bar Frg	0.38-0.46	1.4-1.7		0.5-0.8	0.15 max	0.07 max	0.06 max	0.1-0.35	Ni <=0.4; V 0.07-0.12; bal Fe	1080-1280	880	10	320-380 HB
IS 1570/4(88)	42Cr6V1	30.1-100mm; Bar Frg	0.38-0.46	1.4-1.7		0.5-0.8	0.15 max	0.07 max	0.06 max	0.1-0.35	Ni <=0.4; V 0.07-0.12; bal Fe	880-1030	690	12	265-310 HB
IS 1570/4(88)	47C15	Sh Plt Strp	0.42-0.5	0.3 max		1.3-1.7	0.15 max	0.07 max	0.06 max	0.1-0.35	Ni <=0.4; bal Fe				230 HB max
IS 1570/4(88)	50Cr4	Bar Frg Tub Pipe	0.45-0.55	0.9-1.2		0.6-0.9	0.15 max	0.07 max	0.06 max	0.1-0.35	Ni <=0.4; bal Fe	1090	890		

UNS numbers and US grades are provided as a means of cross referencing chemically similar alloys. Exchangability is only possible after independent examination of specifications. Tensile properties are minimum or typical as specified. UTS and YS as MPa. El as %. See Appendix for list of abbreviations used in Notes. * indicates obsolete material.

Specification	Designation	Notes	C	Cr	Cu	Mn	Mo	P	S	Si	Other	UTS	YS	El	Hard

Alloy Steel, Chromium, No equivalents identified

India

Specification	Designation	Notes	C	Cr	Cu	Mn	Mo	P	S	Si	Other	UTS	YS	El	Hard
IS 1570/4(88)	50Cr4V2		0.45-0.55	0.9-1.2		0.5-0.8	0.15 max	0.07 max	0.06 max	0.1-0.35	Ni <=0.4; V 0.15-0.3; bal Fe				
IS 1570/4(88)	55Cr3	0-63mm	0.5-0.6	0.6-0.8		0.6-0.8	0.15 max	0.07 max	0.06 max	0.1-0.35	Ni <=0.4; bal Fe	890-1040	650	12	255-311 HB
IS 1570/4(88)	XT160Cr48	Const, Spring	1.5-1.7	11-13		0.25-0.5	0.8 max	0.07 max	0.06 max	0.1-0.35	Ni <=0.4; V <=0.8; bal Fe				

International

Specification	Designation	Notes	C	Cr	Cu	Mn	Mo	P	S	Si	Other	UTS	YS	El	Hard
ISO 4954(93)	16MnCr5E	Wir Rod Bar, CH ext	0.13-0.19	0.90-1.10		1.00-1.30		0.035 max	0.035 max	0.40 max	bal Fe	550			
ISO 4954(93)	20Cr4E	Wir Rod Bar, CH ext	0.17-0.23	0.90-1.20		0.60-0.90		0.035 max	0.035 max	0.40 max	bal Fe	560			
ISO 4954(93)	34Cr4E	Wir Rod Bar, CH ext	0.30-0.37	0.90-1.20		0.60-0.90		0.035 max	0.035 max	0.40 max	bal Fe	600			
ISO 4954(93)	37Cr2E	Wir Rod Bar, CH ext	0.34-0.41	0.40-0.60		0.50-0.80		0.035 max	0.035 max	0.40 max	bal Fe	600			
ISO 4954(93)	37Cr4E	Wir Rod Bar, CH ext	0.34-0.41	0.90-1.20		0.60-0.90		.0.035 max	0.035 max	0.40 max	bal Fe	610			
ISO 4954(93)	37CrB1E	Wir Rod Bar, CH ext, boron	0.34-0.41	0.20-0.40		0.50-0.80		0.035 max	0.035 max	0.40 max	Al >=0.020; B 0.0008-0.005; bal Fe	600			
ISO 4954(93)	41Cr4E	Wir Rod Bar, CH ext	0.38-0.45	0.90-1.20		0.60-0.90		0.035 max	0.035 max	0.40 max	bal Fe	620			
ISO 4954(93)	46Cr2E	Wir Rod Bar, CH ext	0.42-0.50	0.40-0.60		0.50-0.80		0.035 max	0.035 max	0.40 max	bal Fe	620			
ISO 683-1(87)	34Cr4	Bar Frg Plt Wir Rod Bolt Slab; Q/T, t<16mm	0.30-0.37	0.90-1.20		0.60-0.90		0.035 max	0.035 max	0.10-0.40	bal Fe	900-1100	700		
ISO 683-1(87)	34CrS4	Bar Frg Plt Wir Rod Bolt Slab; Q/T, t<16mm	0.30-0.37	0.90-1.20		0.60-0.90		0.035 max	0.020-0.040	0.10-0.40	bal Fe	900-1100	700		
ISO 683-1(87)	37Cr4	Bar Frg Plt Wir Rod Bolt Slab; Q/T, t<16mm	0.34-0.41	0.90-1.20		0.60-0.90		0.035 max	0.035 max	0.10-0.40	bal Fe	960-1150	750		
ISO 683-1(87)	37CrS4	Bar Frg Plt Wir Rod Bolt Slab; Q/T, t<16mm	0.34-0.41	0.90-1.20		0.60-0.90		0.035 max	0.020-0.040	0.10-0.40	bal Fe	960-1150	750		
ISO 683-1(87)	41Cr4	Bar Frg Plt Wir Rod Bolt Slab; Q/T, t<16mm	0.38-0.45	0.90-1.20		0.60-0.90		0.035 max	0.035 max	0.10-0.40	bal Fe	100-1200	800		
ISO 683-1(87)	41CrS4	Bar Frg Plt Wir Rod Bolt Slab; Q/T, t<16mm	0.38-0.45	0.90-1.20		0.60-0.90		0.035 max	0.020-0.040	0.10-0.40	bal Fe	100-1200	800		
ISO 683-11(87)	20MnCrS5	CH, 16mm test	0.17-0.23	1.00-1.30		1.10-1.40		0.035 max	0.020-0.040	0.15-0.40	bal Fe	100-1350	670		
ISO 683-18(96)	16MnCr5	Flat, CH, t<=16mm	0.13-0.19	0.80-1.10		1.00-1.30		0.035 max	0.035 max	0.15-0.40	bal Fe	820 max			207 HB max
ISO 683-18(96)	16MnCrS5	Flat, CH, t<=16mm	0.13-0.19	0.80-1.10		1.00-1.30		0.035 max	0.020-0.040	0.15-0.40	bal Fe	820 max			207 HB max
ISO 683-18(96)	20Cr4	Flat, CH, t<=16mm	0.17-0.23	0.90-1.20		0.60-0.90		0.035 max	0.035 max	0.15-0.40	bal Fe	820 max			197 HB max
ISO 683-18(96)	20CrS4	Flat, CH, t<=16mm	0.17-0.23	0.90-1.20		0.60-0.90		0.035 max	0.020-0.040	0.15-0.40	bal Fe	820 max			197 HB max
ISO 683-18(96)	20MnCr5	Flat, CH, t<=16mm	0.17-0.23	1.00-1.30		1.10-1.40		0.035 max	0.035 max	0.15-0.40	bal Fe	850 max			217 HB max
ISO 683-18(96)	20MnCrS5	Flat, CH, t<=16mm	0.17-0.23	1.00-1.30		1.10-1.40		0.035 max	0.020-0.040	0.15-0.40	bal Fe	850 max			217 HB max
ISO 683-18(96)	34Cr4	Flat, CD, Ann, t<=16mm	0.30-0.37	0.90-1.20		0.60-0.90		0.035 max	0.035 max	0.10-0.40	bal Fe	940 max			223 HB max
ISO 683-18(96)	34CrS4	Flat, CD, Ann, t<=16mm	0.30-0.37	0.90-1.20		0.60-0.90		0.035 max	0.020-0.040	0.10-0.40	bal Fe	940 max			223 HB max
ISO 683-18(96)	37Cr4	Flat, CD, Ann, t<=16mm	0.34-0.41	0.90-1.20		0.60-0.90		0.035 max	0.035 max	0.10-0.40	bal Fe	960 max			235 HB max
ISO 683-18(96)	37CrS4	Flat, CD, Ann, t<=16mm	0.34-0.41	0.90-1.20		0.60-0.90		0.035 max	0.020-0.040	0.10-0.40	bal Fe	960 max			235 HB max
ISO 683-18(96)	41Cr4	Flat, CD, Ann, t<=16mm	0.38-0.45	0.90-1.20		0.60-0.90		0.035 max	0.035 max	0.10-0.40	bal Fe	980 max			241 HB max
ISO 683-18(96)	41CrS4	Flat, CD, Ann, t<=16mm	0.38-0.45	0.90-1.20		0.60-0.90		0.035 max	0.020-0.040	0.10-0.40	bal Fe	980 max			241 HB max

UNS numbers and US grades are provided as a means of cross referencing chemically similar alloys. Exchangability is only possible after independent examination of specifications. Tensile properties are minimum or typical as specified. UTS and YS as MPa. El as %. See Appendix for list of abbreviations used in Notes. * indicates obsolete material.

Specification	Designation	Notes	C	Cr	Cu	Mn	Mo	P	S	Si	Other	UTS	YS	El	Hard

Alloy Steel, Chromium, No equivalents identified

Italy

Specification	Designation	Notes	C	Cr	Cu	Mn	Mo	P	S	Si	Other	UTS	YS	El	Hard
UNI 7356(74)	38Cr1KB	Q/T	0.34-0.41	0.2-0.4		0.5-0.8		0.035 max	0.035 max	0.15-0.4	bal Fe				
UNI 7874(79)	38Cr2	Q/T	0.34-0.41	0.4-0.6		0.5-0.8		0.035 max	0.035 max	0.15-0.4	bal Fe				

Japan

| JIS G4107(94) | SNB5 | Class 1 diam<=100mm Bar | 0.10 min | 4.00-6.00 | | 1.00 max | 0.40-0.65 | 0.040 max | 0.030 max | 1.00 max | bal Fe | 690 | 550 | 16 | |

Poland

PNH84023/05	16HSN	Const	0.13-0.2	0.8-1.1	0.2 max	0.3-0.6		0.035 max	0.035 max	0.6-0.9	Ni 0.6-0.9; bal Fe				
PNH84023/07	32HA	Tube, Pipe	0.28-0.35	0.85-1.15	0.2 max	0.55-0.85		0.025 max	0.025 max	0.17-0.37	Al 0.02-0.120; Ni <=0.2; bal Fe				
PNH84030/02	15H	CH	0.12-0.18	0.7-1		0.5-0.8		0.035 max	0.035 max	0.17-0.37	Ni <=0.3; bal Fe				
PNH84030/04	37HS	Q/T	0.34-0.42	1.3-1.6		0.3-0.6		0.035 max	0.035 max	1-1.4	Ni <=0.3; bal Fe				
PNH84030/04	40H2MF	Q/T	0.38-0.45	1.6-1.9		0.5-0.8	0.3-0.4	0.035 max	0.035 max	0.17-0.37	Ni <=0.3; V 0.15-0.25; bal Fe				
PNH84030/04	40HA	Q/T	0.36-0.44	0.8-1.1		0.5-0.8		0.025 max	0.025 max	0.17-0.37	Ni <=0.3; bal Fe				
PNH84030/04	45HNMFA	Q/T	0.42-0.5	0.8-1.1		0.5-0.8	0.2-0.3	0.025 max	0.025 max	0.17-0.37	Ni 1.3-1.8; V 0.1-0.2; bal Fe				
PNH84032	50HS	Spring	0.45-0.55	0.9-1.2	0.25 max	0.3-0.6		0.03 max	0.03 max	0.8-1.2	Ni <=0.4; Pb <=.15; bal Fe				
PNH84032	60SGH	Spring	0.55-0.65	0.4-0.6	0.25 max	0.9-1.1		0.035 max	0.035 max	1-1.3	Ni <=0.4; bal Fe				
PNH84035	25HGS	Q/T	0.22-0.28	0.8-1.1	0.2 max	0.8-1.1		0.035 max	0.035 max	0.9-1.2	Ni <=0.3; bal Fe				

Romania

| STAS 11500/2(89) | 37Cr5 | Q/T | 0.34-0.4 | 0.3-0.6 | | 0.6-0.8 | | 0.03 max | 0.02-0.04 | 0.15-0.4 | Ni <=0.4; Pb <=0.15; bal Fe | | | | |
| STAS 1456/1(89) | RUL1 | Ball & roller bearing | 0.95-1.1 | 1.3-1.65 | 0.25 max | 0.2-0.45 | 0.8 max | 0.027 max | 0.02 max | 0.17-0.37 | Pb <=0.15; Cu+Ni=0-0.4; bal Fe | | | | |

Russia

GOST 4543(71)	15ChA	CH	0.12-0.17	0.7-1.0	0.3 max	0.4-0.7	0.15 max	0.025 max	0.025 max	0.17-0.37	Al <=0.1; Ni <=0.3; Ti <=0.05; bal Fe				
GOST 4543(71)	15ChF	CH	0.12-0.18	0.8-1.1	0.3 max	0.4-0.7	0.15 max	0.035 max	0.035 max	0.17-0.37	Al <=0.1; Ni <=0.3; Ti <=0.05; V 0.06-0.12; bal Fe				
GOST 4543(71)	25ChGSA	Q/T	0.22-0.28	0.8-1.1	0.3 max	0.8-1.1	0.15 max	0.025 max	0.025 max	0.9-1.2	Al <=0.1; Ni <=0.3; Ti <=0.05; bal Fe				
GOST 4543(71)	25ChGT	Q/T	0.22-0.29	1.0-1.3	0.3 max	0.8-1.1	0.15 max	0.035 max	0.035 max	0.17-0.37	Al <=0.1; Ni <=0.3; Ti 0.03-0.09; bal Fe				
GOST 4543(71)	30ChGSA	Q/T	0.28-0.34	0.8-1.1	0.3 max	0.8-1.1	0.15 max	0.025 max	0.025 max	0.9-1.2	Al <=0.1; Ni <=0.3; Ti <=0.05; bal Fe				
GOST 4543(71)	30ChGSN2A	Q/T	0.27-0.34	0.9-1.2	0.3 max	1.0-1.3	0.15 max	0.025 max	0.025 max	0.9-1.2	Al <=0.1; Ni 1.4-1.8; Ti <=0.05; bal Fe				
GOST 4543(71)	30ChGT	Q/T	0.24-0.32	1.0-1.3	0.3 max	0.8-1.1	0.15 max	0.035 max	0.035 max	0.17-0.37	Al <=0.1; Ni <=0.3; Ti 0.03-0.09; bal Fe				
GOST 4543(71)	33ChS	Q/T	0.29-0.37	1.3-1.6	0.3 max	0.3-0.6	0.15 max	0.035 max	0.035 max	1.0-1.4	Al <=0.1; Ni <=0.3; Ti <=0.05; bal Fe				
GOST 4543(71)	35ChGF	Q/T	0.31-0.38	1.0-1.3	0.3 max	0.95-1.25	0.15 max	0.035 max	0.035 max	0.17-0.37	Al <=0.1; Ni <=0.3; Ti <=0.05; V 0.06-0.12; bal Fe				
GOST 4543(71)	38ChS	Q/T	0.34-0.42	1.3-1.6	0.3 max	0.3-0.6	0.15 max	0.035 max	0.035 max	1.0-1.4	Al <=0.1; Ni <=0.3; Ti <=0.05; bal Fe				
GOST 4543(71)	38ChW	Q/T	0.35-0.42	0.9-1.3	0.3 max	0.35-0.6	0.15 max	0.035 max	0.035 max	0.17-0.37	Al <=0.1; Ni <=0.3; Ti <=0.05; W 0.5-0.8; bal Fe				

USA

	UNS G15216*	Obs; see G52986									bal Fe				
ASTM A234/A234M(97)	WP11 CL2	Frg	0.05-0.20	1.00-1.50		0.30-0.80	0.44-0.65	0.040 max	0.040 max	0.50-1.00	bal Fe	485-655	275	22L	
ASTM A234/A234M(97)	WP11 CL3	Frg	0.05-0.20	1.00-1.50		0.30-0.80	0.44-0.65	0.040 max	0.040 max	0.50-1.00	bal Fe	520-690	310	22L	
ASTM A295(94)	5195	Bearing, coil bar tub, CD ann	0.90-1.03	0.70-0.90	0.35 max	0.75-1.00	0.10 max	0.025 max	0.025 max	0.15-0.35	Ni <=0.25; bal Fe				250 HB max

UNS numbers and US grades are provided as a means of cross referencing chemically similar alloys. Exchangability is only possible after independent examination of specifications. Tensile properties are minimum or typical as specified. UTS and YS as MPa. El as %. See Appendix for list of abbreviations used in Notes. * indicates obsolete material.

Specification	Designation	Notes	C	Cr	Cu	Mn	Mo	P	S	Si	Other	UTS	YS	El	Hard

Alloy Steel, Chromium, No equivalents identified

USA

Specification	Designation	Notes	C	Cr	Cu	Mn	Mo	P	S	Si	Other	UTS	YS	El	Hard
ASTM A768(95)	Grade 1 Class 1	Turbine rotor shaft frg	0.15 max	11.5-13.0		1.0 max	0.50 max	0.018 max	0.015 max	0.35 max	Ni 0.40-0.75; bal Fe	620-758	483	16	
ASTM A768(95)	Grade 1 Class 2	Turbine rotor shaft frg	0.15 max	11.5-13.0		1.0 max	0.50 max	0.018 max	0.015 max	0.35 max	Ni 0.40-0.75; bal Fe	586	434	18	
ASTM A768(95)	Grade 2	Turbine rotor shaft frg	0.08-0.15	11.0-13.0		0.50-0.90	1.5-2.0	0.02 max	0.015 max	0.30 max	N <=0.06; Ni 2.0-3.0; V 0.25-0.40; bal Fe	1034	758	15	
ASTM A768(95)	Grade 4 Class 1	Turbine rotor shaft frg	0.05-0.07	11.25-12.25		0.70-1.00	0.30-0.50	0.015 max	0.012 max	0.30-0.50	Al <=0.03; Ni 3.5-4.25; V <=0.03; bal Fe	758-965	586	16	
ASTM A768(95)	Grade 4 Class 2	Turbine rotor shaft frg	0.05-0.07	11.25-12.25		0.70-1.00	0.30-0.50	0.015 max	0.012 max	0.30-0.50	Al <=0.03; Ni 3.5-4.25; V <=0.03; bal Fe	827-1034	655	15	

Yugoslavia

Specification	Designation	Notes	C	Cr	Cu	Mn	Mo	P	S	Si	Other	UTS	YS	El	Hard
	C.4135	F/IH	0.42-0.48	0.4-0.6		0.5-0.8		0.025 max	0.035 max	0.15-0.4	Ni <=0.25; bal Fe				

Alloy Steel, Chromium, PS59

Japan

Specification	Designation	Notes	C	Cr	Cu	Mn	Mo	P	S	Si	Other	UTS	YS	El	Hard
JIS G4052(79)	SMnC420H	Struct Hard	0.16-0.23	0.35-0.70	0.30 max	1.15-1.55		0.030 max	0.030 max	0.15-0.35	Ni <=0.25; bal Fe				

USA

Specification	Designation	Notes	C	Cr	Cu	Mn	Mo	P	S	Si	Other	UTS	YS	El	Hard
	UNS G51210		0.18-0.23	0.70-0.90		1.00-1.30		0.035 max	0.040 max	0.15-0.35	bal Fe				

Alloy Steel, Chromium, PS64

Mexico

Specification	Designation	Notes	C	Cr	Cu	Mn	Mo	P	S	Si	Other	UTS	YS	El	Hard
NMX-B-300(91)	PS64	Bar	0.16-0.21	0.70-0.90		1.00-1.30		0.035 max	0.040 max	0.15-0.35	bal Fe				

USA

Specification	Designation	Notes	C	Cr	Cu	Mn	Mo	P	S	Si	Other	UTS	YS	El	Hard
	UNS G51190		0.16-0.21	0.70-0.90		1.00-1.30		0.035 max	0.040 max	0.15 max	bal Fe				

Alloy Steel, Chromium, PS65

Mexico

Specification	Designation	Notes	C	Cr	Cu	Mn	Mo	P	S	Si	Other	UTS	YS	El	Hard
NMX-B-300(91)	PS65	Bar	0.21-0.26	0.70-0.90		1.00-1.30		0.035 max	0.040 max	0.15-0.35	bal Fe				

USA

Specification	Designation	Notes	C	Cr	Cu	Mn	Mo	P	S	Si	Other	UTS	YS	El	Hard
	UNS G51240		0.21-0.26	0.70-0.90		1.00-1.30		0.035 max	0.040 max	0.15-0.35	bal Fe				

Alloy Steel, Chromium, W-FeCrC

USA

Specification	Designation	Notes	C	Cr	Cu	Mn	Mo	P	S	Si	Other	UTS	YS	El	Hard
	UNS T87510	Thermal Spray Wire	1.0 max	1.35-1.65		0.25-0.40		0.020 max	0.02 max	0.50 max	bal Fe				

Specification	Designation	Notes	C	Cr	Mn	Mo	Ni	P	S	Si	Other	UTS	YS	El	Hard

Alloy Steel, Chromium Molybdenum, 4118/4118H

Specification	Designation	Notes	C	Cr	Mn	Mo	Ni	P	S	Si	Other	UTS	YS	El	Hard
Argentina															
IAS	IRAM 4117		0.15-0.22	0.85-1.1	0.60-0.90	0.10-0.40		0.035 max	0.035 max	0.10-0.40	bal Fe	600	380	20	179 HB
Belgium															
NBN 250	18CrMo4		0.15-0.21	0.85-1.15	0.6-0.9	0.15-0.25		0.035	0.035	0.15-0.4	bal Fe				
Bulgaria															
BDS 5084	Sw-18ChMa		0.15-0.22	0.80-1.10	0.40-0.70	0.15-0.30	0.3	0.025	0.025	0.12-0.35	bal Fe				
BDS 6354	18ChGa		0.16-0.23	1.00-1.30	0.90-1.20	0.20-0.30		0.025	0.025	0.17-0.37	bal Fe				
China															
GB 3077(88)	20CrMo	Bar HR Q/T 25mm diam	0.17-0.24	0.80-1.10	0.40-0.70	0.15-0.25		0.035 max	0.035 max	0.17-0.37	Cu <=0.30; bal Fe	885	685	12	
GB 3203(82)	G20CrMo	Bar HR/CD	0.17-0.23	0.35-0.65	0.65-0.95	0.08-0.15		0.030 max	0.030 max	0.20-0.35	Cu <=0.25; bal Fe				
GB 8162(87)	20CrMo	Smls tube HR/CD Q/T	0.17-0.24	0.80-1.10	0.40-0.70	0.15-0.25		0.035 max	0.035 max	0.17-0.37	Cu <=0.30; bal Fe	885	685	11	
GB/T 13796(92)	20CrMo	Bar Wir CD Ann	0.17-0.24	0.80-1.10	0.40-0.70	0.15-0.25		0.035 max	0.035 max	0.17-0.37	Cu <=0.30; bal Fe	490-740			
GB/T 3078(94)	20CrMo	Bar CD Q/T 25mm diam	0.17-0.24	0.80-1.10	0.40-0.70	0.15-0.25		0.035 max	0.035 max	0.17-0.37	Cu <=0.30; bal Fe	885	685	12	
YB/T 5132(93)	20CrMo	Sh HR/CR	0.17-0.24	0.80-1.10	0.40-0.70	0.15-0.25		0.035 max	0.035 max	0.17-0.37	bal Fe				
Czech Republic															
CSN 415124	15124	Q/T	0.17-0.24	0.8-1.1	0.4-0.8	0.15-0.25		0.035 max	0.035 max	0.17-0.37	bal Fe				
Finland															
SFS 507(72)	SFS507	CH	0.17-0.22	0.3-0.5	0.6-0.9	0.4-0.5		0.035 max	0.05 max	0.15-0.4	bal Fe				
France															
AFNOR	15CD3.5		0.14-0.18	0.85-1.15	0.3-0.8	0.15-0.3		0.04	0.035	0.35	bal Fe				
AFNOR	20CD4		0.17-0.23	0.85-1.15	0.6-0.9	0.15-0.3		0.04	0.035	0.1-0.4	bal Fe				
AFNOR NFA35551	18CD4		0.16-0.22	0.9-1.2	0.6-0.9	0.15-0.35		0.035	0.035	0.1-0.4	bal Fe				
AFNOR NFA35553	18CD4		0.15-0.22	0.85-1.15	0.6-0.9	0.15-0.3		0.035	0.035	0.1-0.4	bal Fe				
AFNOR NFA35556	18CD4		0.16-0.22	0.9-1.2	0.6-0.9	0.15-0.35		0.035	0.035	0.1-0.4	bal Fe				
Germany															
DIN	20CrMo5	CH, 30mm	0.18-0.23	1.10-1.40	0.90-1.20	0.20-0.30		0.035 max	0.035 max	0.15-0.35	bal Fe	980-1270	685	8	
DIN	23CrMoB4		0.20-0.25	0.90-1.20	0.50-0.80	0.10-0.20		0.035 max	0.035 max	0.15-0.35	B >=0.0008; bal Fe				
DIN	WNr 1.7211		0.20-0.25	0.90-1.20	0.50-0.80	0.10-0.20		0.035 max	0.035 max	0.15-0.35	B 0.0008-0.0050; bal Fe				
DIN	WNr 1.7264	CH, 30mm	0.18-0.23	1.10-1.40	0.90-1.20	0.20-0.30		0.035 max	0.035 max	0.15-0.35	bal Fe	980-1270	685	8	
DIN 1652(90)	20MoCr4	Bright CH 30mm	0.17-0.23	0.30-0.60	0.70-1.00	0.40-0.50		0.035 max	0.035 max	0.40 max	bal Fe	780-1080	590	10	
DIN 1652(90)	WNr 1.7321	Bright CH 30mm	0.17-0.23	0.30-0.60	0.70-1.00	0.40-0.50		0.035 max	0.035 max	0.40 max	bal Fe	780-1080	590	10	
DIN 1654(89)	20MoCr4	CHd, cold ext, CH 30mm	0.17-0.23	0.30-0.60	0.70-1.00	0.40-0.50		0.035 max	0.035 max	0.40 max	bal Fe	780-1080	590	10	
DIN 1654(89)	WNr 1.7321	CHd, cold ext, CH 30mm	0.17-0.23	0.30-0.60	0.70-1.00	0.40-0.50		0.035 max	0.035 max	0.40 max	bal Fe	780-1080	590	10	
DIN E EN 10084(95)	20MoCr4	CH, 30mm	0.17-0.23	0.30-0.60	0.70-1.00	0.40-0.50		0.035 max	0.035 max	0.40 max	bal Fe	780-1080	590	10	
DIN E EN 10084(95)	WNr 1.7321	CH, 30mm	0.17-0.23	0.30-0.60	0.70-1.00	0.40-0.50		0.035 max	0.035 max	0.40 max	bal Fe	780-1080	590	10	
Hungary															
MSZ 31(85)	BCMo2	CH; 63 mm; blank carburized	0.17-0.23	1.1-1.4	0.9-1.2	0.2-0.3		0.035 max	0.035 max	0.4 max	bal Fe	800-1100	690	10L	
MSZ 31(85)	BCMo2	CH; 0-11mm; blank carburized	0.17-0.23	1.1-1.4	0.9-1.2	0.2-0.3		0.035 max	0.035 max	0.4 max	bal Fe	1100-1400	880	8L	

Specification	Designation	Notes	C	Cr	Mn	Mo	Ni	P	S	Si	Other	UTS	YS	El	Hard

Alloy Steel, Chromium Molybdenum, 4118/4118H (Continued from previous page)

Specification	Designation	Notes	C	Cr	Mn	Mo	Ni	P	S	Si	Other	UTS	YS	El	Hard
Hungary															
MSZ 31(85)	BCMo2E	CH; 0-11mm; blank carburized	0.17-0.23	1.1-1.4	0.9-1.2	0.2-0.3		0.035 max	0.02-0.035	0.4 max	bal Fe	1100-1400	880	8L	
MSZ 31(85)	BCMo2E	CH; 63 mm; blank carburized	0.17-0.23	1.1-1.4	0.9-1.2	0.2-0.3		0.035 max	0.02-0.035	0.4 max	bal Fe	800-1100	690	10L	
Italy															
UNI 6403(86)	18CrMo4	Q/T	0.15-0.21	0.85-1.15	0.6-0.9	0.15-0.25		0.035 max	0.035 max	0.15-0.35	bal Fe				
UNI 7846(78)	18CrMo4	CH	0.15-0.21	0.85-1.15	0.6-0.9	0.15-0.25		0.035 max	0.035 max	0.15-0.4	bal Fe				
UNI 8550(84)	18CrMo4	CH	0.15-0.21	0.85-1.15	0.6-0.9	0.15-0.25		0.035 max	0.035 max	0.15-0.35	bal Fe				
UNI 8788(85)	18CrMo4	CH	0.15-0.21	0.85-1.15	0.6-0.9	0.15-0.25		0.035 max	0.035 max	0.15-0.4	P+S<=0.06; bal Fe				
Japan															
JIS G4052(79)	SCM418H	Struct Hard	0.15-0.21	0.85-1.25	0.55-0.90	0.15-0.35	0.25 max	0.030 max	0.030 max	0.15-0.35	Cu <=0.30; bal Fe				
JIS G4052(79)	SCM420H	Struct Hard	0.17-0.23	0.85-1.25	0.55-0.90	0.15-0.35	0.25 max	0.030 max	0.030 max	0.15-0.35	Cu <=0.30; bal Fe				
Mexico															
DGN B-203	4118*	Obs	0.18-0.23	0.4-0.6	0.7-0.9	0.08-0.15		0.04	0.04	0.2-0.35	bal Fe				
NMX-B-268(68)	4118H	Hard	0.17-0.23	0.30-0.70	0.60-1.00	0.08-0.15				0.20-0.35	bal Fe				
NMX-B-300(91)	4118	Bar	0.18-0.23	0.40-0.60	0.70-0.90	0.08-0.15		0.035 max	0.040 max	0.15-0.35	bal Fe				
NMX-B-300(91)	4118H	Bar	0.17-0.23	0.30-0.70	0.60-1.00	0.08-0.15				0.15-0.35	bal Fe				
Pan America															
COPANT 514	4118	Tube	0.18-0.23	0.4-0.6	0.7-0.9	0.08-0.15		0.04	0.04	0.2-0.35	bal Fe				
Poland															
PNH84030/02	15HGM	CH	0.12-0.19	0.8-1.1	0.8-1.1	0.15-0.25	0.3 max	0.035 max	0.035 max	0.17-0.37	bal Fe				
PNH84030/02	18HGM	CH	0.16-0.23	0.9-1.2	0.9-1.2	0.2-0.3	0.3 max	0.035 max	0.035 max	0.17-0.37	bal Fe				
Romania															
STAS 3127	21MoMnCr12q		0.18-0.24	1-1.4	0.8-1.2	0.2-0.3	0.3	0.035	0.035	0.17-0.37	Cu 0.3; As 0.05; bal Fe				
STAS 791(88)	21MoMnCr12	Struct/const; CH	0.18-0.24	1-1.4	0.8-1.2	0.2-0.3		0.035 max	0.035 max	0.17-0.37	Pb <=0.15; Ti <=0.02; bal Fe				
STAS 791(88)	21MoMnCr12S	Struct/const; CH	0.18-0.24	1-1.4	0.8-1.2	0.2-0.3		0.035 max	0.02-0.04	0.17-0.37	Pb <=0.15; Ti <=0.02; bal Fe				
STAS 791(88)	21MoMnCr12X	Struct/const; CH	0.18-0.24	1-1.4	0.8-1.2	0.2-0.3		0.025 max	0.025 max	0.17-0.37	Pb <=0.15; Ti <=0.02; bal Fe				
STAS 791(88)	21MoMnCr12XS	Struct/const; CH	0.18-0.24	1-1.4	0.8-1.2	0.2-0.3		0.025 max	0.02-0.035	0.17-0.37	Pb <=0.15; Ti <=0.02; bal Fe				
Spain															
UNE 36013(60)	18CrMo4		0.15-0.21	0.85-1.15	0.6-0.9	0.15-0.25		0.035	0.035	0.15-0.4	bal Fe				
UNE 36013(60)	18CrMo4-1		0.15-0.21	0.85-1.15	0.6-0.9	0.15-0.25		0.035	0.035	0.15-0.4	bal Fe				
UNE 36013(60)	F.1550*	see 18CrMo4	0.15-0.21	0.85-1.15	0.6-0.9	0.15-0.25		0.035	0.035	0.15-0.4	bal Fe				
UNE 36013(60)	F.1559*	see 18CrMo4-1	0.15-0.21	0.85-1.15	0.6-0.9	0.15-0.25		0.035	0.035	0.15-0.4	bal Fe				
UNE 36013(76)	18CrMo4	CH	0.15-0.21	0.85-1.15	0.6-0.9	0.15-0.25		0.035 max	0.035 max	0.15-0.4	bal Fe				
UNE 36013(76)	F.1550*	see 18CrMo4	0.15-0.21	0.85-1.15	0.6-0.9	0.15-0.25		0.035 max	0.035 max	0.15-0.4	bal Fe				
UK															
BS 970/1(83)	708H20	Blm Bil Slab Bar Rod Frg CH	0.17-0.23	0.85-1.15	0.60-0.90	0.15-0.25					bal Fe				
BS 970/1(96)	708M20	Blm Bil Slab Bar Rod Frg	0.17-0.23	0.85-1.15	0.60-0.90	0.15-0.25		0.035 max	0.040 max	0.10-0.35	Pb <=0.15; bal Fe				

UNS numbers and US grades are provided as a means of cross referencing chemically similar alloys. Exchangability is only possible after independent examination of specifications. Tensile properties are minimum or typical as specified. UTS and YS as MPa. El as %. See Appendix for list of abbreviations used in Notes. * indicates obsolete material.

Specification	Designation	Notes	C	Cr	Mn	Mo	Ni	P	S	Si	Other	UTS	YS	El	Hard

Alloy Steel, Chromium Molybdenum, 4118/4118H (Continued from previous page)

USA

Specification	Designation	Notes	C	Cr	Mn	Mo	Ni	P	S	Si	Other	UTS	YS	El	Hard
	AISI 4118	Plt	0.17-0.23	0.40-0.60	0.60-0.90	0.08-0.15		0.035 max	0.040 max	0.15-0.40	bal Fe				
	AISI 4118	Sh Strp	0.18-0.23	0.40-0.60	0.70-0.90	0.08-0.15		0.035	0.035	0.15-0.30	bal Fe				
	AISI 4118	Smls mech tub	0.18-0.23	0.40-0.60	0.70-0.90	0.08-0.15		0.040 max	0.040 max	0.15-0.35	bal Fe				
	AISI 4118	Bar Blm Bil Slab	0.18-0.23	0.40-0.60	0.70-0.90	0.08-0.15		0.035 max	0.040 max	0.15-0.35	bal Fe				
	AISI 4118H	Bar Blm Bil Slab	0.17-0.23	0.30-0.70	0.60-1.00	0.08-0.15		0.035 max	0.040 max	0.15-0.35	bal Fe				
	AISI 4118H	Smls mech tub, Hard	0.17-0.23	0.30-0.70	0.60-1.00	0.08-0.15		0.040 max	0.040 max	0.15-0.30	bal Fe				
	UNS G41180		0.18-0.23	0.40-0.60	0.70-0.90	0.08-0.15		0.035 max	0.040 max	0.15-0.35	bal Fe				
	UNS H41180		0.17-0.23	0.30-0.70	0.60-1.00	0.08-0.15		0.035 max	0.040 max	0.15-0.35	bal Fe				
ASTM A29/A29M(93)	4118	Bar	0.18-0.23	0.40-0.60	0.70-0.90	0.08-0.15		0.035 max	0.040 max	0.15-0.35	bal Fe				
ASTM A304(96)	4118H	Bar, hard bands spec	0.17-0.23	0.30-0.70	0.60-1.00	0.08-0.15		0.040 max	0.040 max	0.15-0.35	Cu <=0.35; Cu Ni Cr Mo trace allowed; bal Fe				
ASTM A322(96)	4118	Bar	0.18-0.23	0.40-0.60	0.70-0.90	0.08-0.15		0.035 max	0.040 max	0.15-0.35	bal Fe				
ASTM A331(95)	4118	Bar	0.18-0.23	0.40-0.60	0.70-0.90	0.08-0.15		0.035 max	0.040 max	0.15-0.35	bal Fe				
ASTM A331(95)	4118H	Bar	0.17-0.23	0.30-0.70	0.60-1.00	0.08-0.15		0.025 max	0.025 max	0.15-0.35	Cu <=0.35; bal Fe				
ASTM A506(93)	4118	Sh Strp, HR, CR	0.18-0.23	0.40-0.60	0.70-0.90	0.08-0.15		0.035	0.035	0.15-0.30	bal Fe				
ASTM A507(93)	4118	Sh Strp, HR, CR	0.18-0.23	0.40-0.60	0.70-0.90	0.08-0.15		0.025	0.030	0.15-0.30	bal Fe				
ASTM A513(97a)	4118	ERW Mech Tub	0.18-0.23	0.40-0.60	0.70-0.90	0.08-0.15		0.035 max	0.040 max	0.15-0.35	bal Fe				
ASTM A519(96)	4118	Smls mech tub	0.18-0.23	0.40-0.60	0.70-0.90	0.08-0.15		0.040 max	0.040 max	0.15-0.30	bal Fe				
ASTM A534(94)	4118H	Carburizing, anti-friction bearings	0.17-0.23	0.30-0.70	0.60-1.00	0.08-0.15		0.025 max	0.025 max	0.15-0.35	bal Fe				
ASTM A752(93)	4118	Rod Wir	0.18-0.23	0.40-0.60	0.70-0.90	0.08-0.15		0.035 max	0.040 max	0.15-0.30	bal Fe	570			
ASTM A829/A829M(95)	4118	Plt	0.17-0.23	0.40-0.65	0.60-0.90	0.08-0.15		0.035 max	0.040 max	0.15-0.40	bal Fe	480-965			
ASTM A914/914M(92)	4118RH	Bar restricted end Q hard	0.18-0.23	0.40-0.60	0.70-0.90	0.08-0.15	0.25 max	0.035 max	0.040 max	0.15-0.35	Cu <=0.35; Electric furnace P, S<=0.025; bal Fe				
SAE 770(84)	4118*	Obs; see J1397(92)	0.18-0.23	0.4-0.6	0.7-0.9	0.08-0.15		0.04 max	0.04 max	0.15-0.3	bal Fe				
SAE J1268(95)	4118H	See std	0.17-0.23	0.30-0.70	0.60-1.00	0.08-0.15	0.25 max	0.030 max	0.040 max	0.15-0.35	Cu <=0.35; Cu, Ni not spec'd but acpt; bal Fe				
SAE J1868(93)	4118RH	Restrict hard range	0.18-0.23	0.40-0.60	0.70-0.90	0.08-0.15	0.25 max	0.025 max	0.040 max	0.15-0.35	Cu <=0.035; Cu, Ni not spec'd but acpt; bal Fe				5 HRC max
SAE J404(94)	4118	Plt	0.17-0.23	0.40-0.65	0.60-0.90	0.08-0.15		0.035 max	0.040 max	0.15-0.35	bal Fe				
SAE J404(94)	4118	Bil Blm Slab Bar HR CF	0.18-0.23	0.40-0.60	0.70-0.90	0.08-0.15		0.030 max	0.040 max	0.15-0.35	bal Fe				

Alloy Steel, Chromium Molybdenum, 4120/4120H

France

Specification	Designation	Notes	C	Cr	Mn	Mo	Ni	P	S	Si	Other	UTS	YS	El	Hard
AFNOR NFA36102(93)	25M4*	HR; Strp; CR	0.23-0.28	0.2 max	0.95-1.15	0.10 max	0.20 max	0.025 max	0.02 max	0.15-0.3	bal Fe				
AFNOR NFA36102(93)	25Mn4RR	HR; Strp; CR	0.23-0.28	0.2 max	0.95-1.15	0.10 max	0.20 max	0.025 max	0.02 max	0.15-0.3	bal Fe				

USA

Specification	Designation	Notes	C	Cr	Mn	Mo	Ni	P	S	Si	Other	UTS	YS	El	Hard
	AISI 4120	Bar Blm Bil Slab	0.18-0.23	0.40-0.60	0.90-1.20	0.13-0.20		0.035 max	0.040 max	0.15-0.35	bal Fe				
	UNS G41200		0.18-0.23	0.40-0.60	0.80-1.20	0.15-0.25		0.035 max	0.040 max	0.15-0.35	bal Fe				

UNS numbers and US grades are provided as a means of cross referencing chemically similar alloys. Exchangability is only possible after independent examination of specifications. Tensile properties are minimum or typical as specified. UTS and YS as MPa. El as %. See Appendix for list of abbreviations used in Notes. * indicates obsolete material.

Specification	Designation	Notes	C	Cr	Mn	Mo	Ni	P	S	Si	Other	UTS	YS	El	Hard

Alloy Steel, Chromium Molybdenum, 4120/4120H (Continued from previous page)

USA

Specification	Designation	Notes	C	Cr	Mn	Mo	Ni	P	S	Si	Other	UTS	YS	El	Hard
ASTM A29/A29M(93)	4120	Bar	0.18-0.23	0.40-0.60	0.90-1.20	0.13-0.20		0.035 max	0.040 max	0.15-0.35	bal Fe				
ASTM A322(96)	G41200	Bar	0.18-0.23	0.40-0.60	0.70-0.90	0.13-0.20		0.035 max	0.04 max	0.15-0.35	bal Fe				
ASTM A331(95)	G41200	Bar	0.18-0.23	0.40-0.60	0.70-0.90	0.13-0.20		0.035 max	0.04 max	0.15-0.35	bal Fe				
ASTM A914/914M(92)	4120RH	Bar restricted end Q hard	0.18-0.23	0.40-0.60	0.90-1.20	0.13-0.20	0.25 max	0.035 max	0.040 max	0.15-0.35	Cu <=0.35; Electric furnace P, S<=0.025; bal Fe				
SAE J1268(95)	4120H	Hard	0.18-0.23	0.40-0.60	0.90-1.20	0.13-0.20	0.25 max	0.030 max	0.040 max	0.15-0.35	Cu <=0.35; Cu, Ni not spec'd but acpt; bal Fe				
SAE J1268(95)	E4120H	Hard	0.18-0.23	0.40-0.60	0.90-1.20	0.13-0.20	0.25 max	0.025 max	0.025 max	0.15-0.35	Cu <=0.35; Cu, Ni not spec'd but acpt; bal Fe				
SAE J1868(93)	4120RH	Restrict hard range	0.18-0.23	0.40-0.60	0.90-1.20	0.13-0.20	0.25 max	0.025 max	0.040 max	0.15-0.35	Cu <=0.035; Cu, Ni not spec'd but acpt; bal Fe				5 HRC max
SAE J404(94)	4120	Bil Blm Slab Bar HR CF	0.18-0.23	0.40-0.60	0.90-1.20	0.13-0.20		0.030 max	0.040 max	0.15-0.35	bal Fe				

Alloy Steel, Chromium Molybdenum, 4121

USA

Specification	Designation	Notes	C	Cr	Mn	Mo	Ni	P	S	Si	Other	UTS	YS	El	Hard
	AISI 4121	Bar Blm Bil Slab	0.18-0.23	0.45-0.65	0.75-1.00	0.20-0.30		0.035 max	0.040 max	0.15-0.35	bal Fe				
	UNS G41210		0.18-0.23	0.45-0.65	0.75-1.00	0.15-0.25		0.035 max	0.040 max	0.15-0.35	bal Fe				
ASTM A29/A29M(93)	4121	Bar	0.18-0.23	0.45-0.65	0.75-1.00	0.20-0.30		0.035 max	0.040 max	0.15-0.35	bal Fe				
ASTM A322(96)	G41210	Bar	0.18-0.23	0.45-0.65	0.75-1.00	0.20-0.30		0.035 max	0.04 max	0.15-0.35	bal Fe				
ASTM A331(95)	G41210	Bar	0.18-0.23	0.45-0.65	0.75-1.00	0.20-0.30		0.035 max	0.04 max	0.15-0.35	bal Fe				

Alloy Steel, Chromium Molybdenum, 4130/4130H

Argentina

Specification	Designation	Notes	C	Cr	Mn	Mo	Ni	P	S	Si	Other	UTS	YS	El	Hard
IAS	IRAM 4130		0.28-0.33	0.80-1.10	0.40-0.60	0.15-0.25		0.035 max	0.040 max	0.15-0.35	bal Fe	800	483	22	229 HB

Australia

Specification	Designation	Notes	C	Cr	Mn	Mo	Ni	P	S	Si	Other	UTS	YS	El	Hard
AS 1444(96)	4130	H series, Hard/Tmp	0.28-0.33	0.80-1.10	0.40-0.60	0.15-0.25		0.040 max	0.040 max	0.10-0.35	bal Fe				
AS 1444(96)	4130H	H series, Hard/Tmp	0.27-0.33	0.75-1.20	0.30-0.70	0.15-0.25		0.040 max	0.040 max	0.10-0.35	bal Fe				

Belgium

Specification	Designation	Notes	C	Cr	Mn	Mo	Ni	P	S	Si	Other	UTS	YS	El	Hard
NBN 253-02	25CrMo4		0.22-0.29	0.9-1.2	0.5-0.8	0.15-0.3		0.04 max	0.04 max	0.15-0.4	bal Fe				
NBN 253-02	34CrMo4		0.3-0.37	0.9-1.2	0.5-0.8	0.15-0.3		0.035 max	0.035 max	0.15-0.4	bal Fe				

Bulgaria

Specification	Designation	Notes	C	Cr	Mn	Mo	Ni	P	S	Si	Other	UTS	YS	El	Hard
BDS 6354	30ChMA		0.26-0.34	0.90-1.20	0.50-0.80	0.15-0.25	0.30 max	0.025 max	0.025 max	0.90-1.20	Cu <=0.30; Ti <=0.3; bal Fe				
BDS 6354	35ChM		0.3-0.37	0.90-1.20	0.50-0.80	0.15-0.25	0.30 max	0.035 max	0.035 max	0.17-0.37	Cu <=0.30; Ti <=0.03; V <=0.05; W <=0.2; bal Fe				

China

Specification	Designation	Notes	C	Cr	Mn	Mo	Ni	P	S	Si	Other	UTS	YS	El	Hard
GB 3077(88)	30CrMo	Bar HR Q/T 25mm diam	0.26-0.34	0.80-1.10	0.40-0.70	0.15-0.25		0.035 max	0.035 max	017-0.37	Cu <=0.30; bal Fe	930	785	12	
GB 3077(88)	30CrMoA	Bar HR Q/T 25mm diam	0.26-0.33	0.80-1.10	0.40-0.70	0.15-0.25		0.025 max	0.025 max	017-0.37	Cu <=0.30; bal Fe	930	735	12	
GB 6478(86)	ML30CrMo	Bar HR Q/T	0.26-0.34	0.80-1.10	0.40-0.70	0.15-0.25		0.035 max	0.035 max	0.30 max	bal Fe	930	785	12	
GB/T 3078(94)	30CrMo	Bar CD Q/T 25mm diam	0.26-0.34	0.80-1.10	0.40-0.70	0.15-0.25		0.035 max	0.035 max	017-0.37	Cu <=0.30; bal Fe	930	785	12	
YB/T 5132(93)	30CrMo	Sh HR/CR	0.26-0.34	0.80-1.10	0.40-0.70	0.15-0.25		0.035 max	0.035 max	017-0.37	bal Fe				

UNS numbers and US grades are provided as a means of cross referencing chemically similar alloys. Exchangability is only possible after independent examination of specifications. Tensile properties are minimum or typical as specified. UTS and YS as MPa. El as %. See Appendix for list of abbreviations used in Notes. * indicates obsolete material.

Alloy Steel, Chromium Molybdenum, 4130/4130H (Continued from previous page)

Specification	Designation	Notes	C	Cr	Mn	Mo	Ni	P	S	Si	Other	UTS	YS	El	Hard
Czech Republic															
CSN 415130	15130	Q/T	0.22-0.29	0.9-1.2	0.5-0.8	0.15-0.25		0.035 max	0.035 max	0.17-0.37	bal Fe				
CSN 415131	15131	Q/T	0.26-0.34	0.8-1.1	0.4-0.7	0.15-0.25		0.035 max	0.035 max	0.17-0.37	bal Fe				
Europe															
EN 10083/1(91)A1(96)	1.7213	Q/T t<=16mm	0.22-0.29	0.90-1.20	0.60-0.90	0.15-0.30		0.035 max	0.020-0.040	0.40 max	bal Fe	900-1000	700	12	
EN 10083/1(91)A1(96)	1.7218	Q/T t<=16mm	0.22-0.29	0.90-1.20	0.60-0.90	0.15-0.30		0.035 max	0.035 max	0.40 max	bal Fe	900-1000	700	12	
EN 10083/1(91)A1(96)	25CrMo4	Q/T t<=16mm	0.22-0.29	0.90-1.20	0.60-0.90	0.15-0.30		0.035 max	0.035 max	0.40 max	bal Fe	900-1000	700	12	
EN 10083/1(91)A1(96)	25CrMoS4	Q/T t<=16mm	0.22-0.29	0.90-1.20	0.60-0.90	0.15-0.30		0.035 max	0.020-0.040	0.40 max	bal Fe	900-1000	700	12	
Finland															
SFS 457(73)	SFS457	Q/T	0.38-0.45		1.1-1.4			0.035 max	0.035 max	0.15-0.4	bal Fe				
SFS 459	Co.35Cr1.1Mo0.20		0.3-0.37	0.9-1.2	0.5-0.8	0.15-0.25		0.035 max	0.035 max	0.15-0.4	bal Fe				
France															
AFNOR NFA35552	30CD4		0.27-0.33	0.9-1.2	0.6-0.9	0.15-0.25		0.035 max	0.035 max	0.1-0.4	bal Fe				
AFNOR NFA35553	25CD4	Strp	0.22-0.29	0.85-1.15	0.6-0.9	0.15-0.3		0.035 max	0.035 max	0.1-0.4	bal Fe				
AFNOR NFA35554	25CD4S	Plt HR Norm, 16mm diam	0.22-0.29	0.85-1.15	0.5-0.8	0.15-0.3		0.03	0.03	0.1-0.25	bal Fe				
AFNOR NFA35556(84)	30CD4		0.27-0.33	0.9-1.2	0.6-0.9	0.15-0.25		0.035 max	0.035 max	0.1-0.4	bal Fe				
AFNOR NFA35557	30CD4		0.27-0.33	0.9-1.2	0.6-0.9	0.15-0.25		0.035 max	0.035 max	0.1-0.4	bal Fe				
AFNOR NFA35558(83)	25CD4	t<=160mm	0.23-0.29	0.9-1.2	0.6-0.9	0.15-0.25	0.4 max	0.035 max	0.035 max	0.1-0.4	Cu <=0.3; Ti <=0.05; bal Fe	600-750	440	18	
Germany															
DIN 1652(90)	25CrMo4	Q/T 17-40mm	0.22-0.29	0.90-1.20	0.60-0.90	0.15-0.30		0.035 max	0.035 max	0.40 max	bal Fe	800-950	600	14	
DIN 1652(90)	WNr 1.7218	Q/T 17-40mm	0.22-0.29	0.90-1.20	0.60-0.90	0.15-0.30		0.035 max	0.035 max	0.40 max	bal Fe	800-950	600	14	
Hungary															
MSZ 61(85)	CMo1	Q/T; 100.1-160mm	0.22-0.29	0.9-1.2	0.6-0.9	0.15-0.3		0.035 max	0.035 max	0.4 max	bal Fe	650-800	400	16L	
MSZ 61(85)	CMo1	Q/T; 0-16mm	0.22-0.29	0.9-1.2	0.6-0.9	0.15-0.3		0.035 max	0.035 max	0.4 max	bal Fe	900-1100	700	12L	
MSZ 61(85)	CMo1E	Q/T; 0-16mm	0.22-0.29	0.9-1.2	0.6-0.9	0.15-0.3		0.035 max	0.02-0.035	0.4 max	bal Fe	900-1100	700	12L	
MSZ 61(85)	CMo1E	Q/T; 100.1-160mm	0.22-0.29	0.9-1.2	0.6-0.9	0.15-0.3		0.035 max	0.02-0.035	0.4 max	bal Fe	650-800	400	16L	
MSZ 6251(87)	CMo1Z	CF; Q/T; 0-36mm; HR; HR/soft ann; drawn/soft ann; drawn/bright ann; soft ann/ground	0.22-0.29	0.9-1.2	0.6-0.9	0.15-0.3		0.035 max	0.035 max	0.4 max	bal Fe	0-600			
MSZ 6251(87)	CMo1Z	CF; Q/T; 0-36mm; drawn, half-hard	0.22-0.29	0.9-1.2	0.6-0.9	0.15-0.3		0.035 max	0.035 max	0.4 max	bal Fe	0-630			
International															
ISO 683-2(68)	Type 1*	Q/T	0.22-0.29	0.9-1.2	0.5-0.8	0.15-0.3	0.4 max	0.035 max	0.035 max	0.15-0.4	Al <=0.1; Co <=0.1; Cu <=0.3; Pb <=0.15; Ti <=0.05; V <=0.1; W <=0.1; bal Fe				
ISO 683-4	Type 2a*		0.3-0.37	0.9-1.2	0.5-0.8	0.15-0.3		0.035 max	0.02-0.035	0.15-0.4	bal Fe				
ISO 683-4	Type 2b*		0.3-0.37	0.9-1.2	0.5-0.8	0.15-0.3		0.035 max	0.03-0.035	0.15-0.4	bal Fe				
Italy															
UNI 6403(86)	25CrMo4	Q/T	0.22-0.29	0.9-1.2	0.5-0.8	0.15-0.25		0.035 max	0.035 max	0.15-0.35	bal Fe				
UNI 6403(86)	30CrMo4	Q/T	0.27-0.34	0.8-1.1	0.4-0.7	0.15-0.25		0.035 max	0.035 max	0.15-0.35	bal Fe				

UNS numbers and US grades are provided as a means of cross referencing chemically similar alloys. Exchangability is only possible after independent examination of specifications. Tensile properties are minimum or typical as specified. UTS and YS as MPa. El as %. See Appendix for list of abbreviations used in Notes. * indicates obsolete material.

Specification	Designation	Notes	C	Cr	Mn	Mo	Ni	P	S	Si	Other	UTS	YS	El	Hard

Alloy Steel, Chromium Molybdenum, 4130/4130H (Continued from previous page)

Specification	Designation	Notes	C	Cr	Mn	Mo	Ni	P	S	Si	Other	UTS	YS	El	Hard
Italy															
UNI 6928(71)	25CrMo4F	Aircraft material	0.22-0.29	0.9-1.2	0.5-0.8	0.15-0.25	0.3 max	0.02 max	0.015 max	0.15-0.35	bal Fe				
UNI 7356	35CrMo4KB		0.3-0.37	0.9-1.2	0.5-0.8	0.15-0.3		0.035 max	0.035 max	0.15-0.4	bal Fe				
UNI 7356(74)	25CrMo4KB	Q/T	0.22-0.29	0.9-1.2	0.5-0.8	0.15-0.3		0.035 max	0.035 max	0.15-0.4	bal Fe				
UNI 7845(78)	25CrMo4	Q/T	0.22-0.29	0.9-1.2	0.5-0.8	0.15-0.25		0.035 max	0.035 max	0.15-0.4	bal Fe				
UNI 7845(78)	30CrMo4	Q/T	0.17-0.34	0.8-1.1	0.4-0.7	0.15-0.25		0.035 max	0.035 max	0.15-0.4	bal Fe				
UNI 7874(79)	25CrMo4	Q/T	0.22-0.29	0.9-1.2	0.5-0.9	0.15-0.3		0.035 max	0.035 max	0.15-0.4	bal Fe				
UNI 7874(79)	30CrMo4	Q/T	0.27-0.34	0.9-1.2	0.5-0.9	0.15-0.3		0.035 max	0.035 max	0.15-0.4	bal Fe				
UNI 8913(87)	25CrMo4	Q/T	0.22-0.29	0.8-1.1	0.5-0.8	0.15-0.25		0.035 max	0.035 max	0.35 max	bal Fe				
Japan															
JIS G3311(88)	SCM430M	CR Strp; Chain	0.28-0.33	0.90-1.20	0.60-0.85	0.15-0.30	0.25 max	0.030 max	0.030 max	0.15-0.35	Cu <=0.30; bal Fe				180-250 HV
JIS G3441(88)	SCM430TK	Tube	0.28-0.33	0.90-1.20	0.60-0.85	0.15-0.30	0.25 max	0.030 max	0.030 max	0.25-0.35	Cu <=0.30; bal Fe				
JIS G3441(88)	STKS1*	Obs; see SCM430TK	0.26-0.33	0.8-1.2	0.4-0.85	0.15-0.25		0.03	0.03	0.15-0.35	bal Fe				
JIS G3441(88)	STKS3*	Obs; see SCM435TK	0.26-0.33	0.8-1.2	0.4-0.85	0.15-0.25		0.03	0.03	0.15-0.35	bal Fe				
JIS G4105(79)	SCM 1*	Obs; see SCM432	0.27-0.37	1-1.5	0.3-0.6	0.15-0.3		0.03	0.03	0.15-0.35	bal Fe				
JIS G4105(79)	SCM 2*	Obs; see SCM430	0.28-0.33	0.9-1.2	0.6-0.85	0.15-0.3		0.03 max	0.03 max	0.15-0.35	bal Fe				
JIS G4105(79)	SCM24*	Obs; see SCM822	0.20-0.25	0.90-1.20	0.60-0.85	0.35-0.45	0.25 max	0.030 max	0.030 max	0.15-0.35	Cu <=0.30; bal Fe				
JIS G4105(79)	SCM430	HR Frg Bar Rod	0.28-0.33	0.90-1.20	0.60-0.85	0.15-0.30	0.25 max	0.030 max	0.030 max	0.15-0.35	Cu <=0.30; bal Fe				
JIS G4105(79)	SCM432	HR Frg Bar Rod	0.27-0.37	1.00-1.50	0.30-0.60	0.15-0.30	0.25 max	0.030 max	0.030 max	0.15-0.35	Cu <=0.30; bal Fe				
JIS G4105(79)	SCM822	HR Frg Bar Rod	0.20-0.25	0.90-1.20	0.60-0.85	0.35-0.45	0.25 max	0.030 max	0.030 max	0.15-0.35	Cu <=0.30; bal Fe				
Mexico															
DGN B-203	4130*	Obs	0.28-0.33	0.8-1.1	0.4-0.6	0.15-0.25		0.04	0.04	0.2-0.35	bal Fe				
DGN B-297	4130*	Obs	0.28-0.33	0.8-1.1	0.4-0.6	0.15-0.35		0.035	0.04	0.2-0.35	bal Fe				
NMX-B-268(68)	4130H	Hard	0.27-0.33	0.75-1.20	0.30-0.70	0.15-0.25				0.20-0.35	bal Fe				
NMX-B-300(91)	4130	Bar	0.28-0.33	0.80-1.10	0.40-0.60	0.15-0.25		0.035 max	0.040 max	0.15-0.35	bal Fe				
NMX-B-300(91)	4130H	Bar	0.27-0.33	0.75-1.20	0.30-0.70	0.15-0.25				0.15-0.35	bal Fe				
Pan America															
COPANT 334	4130	Bar	0.28-0.35	0.8-1.1	0.4-0.6	0.15-0.25		0.04	0.04	0.2-0.35	bal Fe				
COPANT 514	4130	Tube, HF	0.28-0.33	0.8-1.1	0.4-0.6	0.15-0.25		0.04	0.04	0.2-0.35	bal Fe				
Poland															
PNH84030/04	25HM	Q/T	0.22-0.29	0.8-1.1	0.4-0.7	0.15-0.25	0.3 max	0.035 max	0.035 max	0.17-0.37	bal Fe				
PNH84030/04	30HM	Q/T	0.26-0.34	0.8-1.1	0.4-0.7	0.15-0.25	0.3 max	0.035 max	0.035 max	0.17-0.37	bal Fe				
Romania															
STAS	25MoCr11		0.22-0.29	0.9-1.3	0.4-0.8	0.15-0.3		0.035 max	0.035 max	0.17-0.37	bal Fe				
STAS 11500/2	30MoCr10		0.27-0.34	0.95-1.2	0.6-0.85	0.2-0.3	0.25 max	0.025 max	0.035 max	0.4	Cu <=0.2; bal Fe				
STAS 11500/2	33MoCr10		0.3-0.36	0.95-1.2	0.6-0.85	0.2-0.3	0.25 max	0.025 max	0.035 max	0.4	Cu <=0.2; bal Fe				
STAS 11500/2(89)	30MoCr11	Q/T	0.27-0.34	0.95-1.2	0.6-0.85	0.2-0.3		0.025 max	0.02-0.035	0.15-0.4	Pb <=0.15; bal Fe				

UNS numbers and US grades are provided as a means of cross referencing chemically similar alloys. Exchangability is only possible after independent examination of specifications. Tensile properties are minimum or typical as specified. UTS and YS as MPa. El as %. See Appendix for list of abbreviations used in Notes. * indicates obsolete material.

Specification	Designation	Notes	C	Cr	Mn	Mo	Ni	P	S	Si	Other	UTS	YS	El	Hard

Alloy Steel, Chromium Molybdenum, 4130/4130H (Continued from previous page)

Romania

Specification	Designation	Notes	C	Cr	Mn	Mo	Ni	P	S	Si	Other	UTS	YS	El	Hard
STAS 791(88)	19MoCr11	Struct/const	0.16-0.22	0.9-1.2	0.6-0.9	0.15-0.25		0.035 max	0.035 max	0.17-0.37	Pb <=0.15; Ti <=0.02; bal Fe				
STAS 791(88)	19MoCr11S	Struct/const	0.16-0.22	0.9-1.2	0.6-0.9	0.15-0.25		0.035 max	0.02-0.04	0.17-0.37	Pb <=0.15; Ti <=0.02; bal Fe				
STAS 791(88)	19MoCr11X	Struct/const	0.16-0.22	0.9-1.2	0.6-0.9	0.15-0.25		0.025 max	0.025 max	0.17-0.37	Pb <=0.15; Ti <=0.02; bal Fe				
STAS 791(88)	19MoCr11XS	Struct/const	0.16-0.22	0.9-1.2	0.6-0.9	0.15-0.25		0.025 max	0.02-0.035	0.17-0.37	Pb <=0.15; Ti <=0.02; bal Fe				
STAS 791(88)	26MoCr11	Struct/const; Q/T	0.22-0.29	0.9-1.2	0.6-0.9	0.15-0.3		0.035 max	0.035 max	0.17-0.37	Pb <=0.15; Ti <=0.02; bal Fe				
STAS 791(88)	26MoCr11S	Struct/const; Q/T	0.22-0.29	0.9-1.2	0.6-0.9	0.15-0.3		0.035 max	0.02-0.04	0.17-0.37	Pb <=0.15; Ti <=0.02; bal Fe				
STAS 791(88)	26MoCr11X	Struct/const; Q/T	0.22-0.29	0.9-1.2	0.6-0.9	0.15-0.3		0.025 max	0.025 max	0.17-0.37	Pb <=0.15; Ti <=0.02; bal Fe				
STAS 791(88)	26MoCr11XS	Struct/const; Q/T	0.22-0.29	0.9-1.2	0.6-0.9	0.15-0.3		0.025 max	0.02-0.035	0.17-0.37	Pb <=0.15; Ti <=0.02; bal Fe				
STAS 9382/4	33MoCr11q		0.3-0.37	0.9-1.3	0.4-0.8	0.15-0.3		0.035 max	0.035 max	0.17-0.37	As<=0.05; bal Fe				

Russia

Specification	Designation	Notes	C	Cr	Mn	Mo	Ni	P	S	Si	Other	UTS	YS	El	Hard
GOST 10702	30ChMA		0.26-0.33	0.8-1.1	0.4-0.7	0.15-0.25	0.25 max	0.03 max	0.03 max	0.2 max	Cu <=0.3; bal Fe				
GOST 4543	30ChM		0.26-0.34	0.8-1.1	0.4-0.7	0.15-0.25	0.3	0.025 max	0.025 max	0.17-0.37	Cu 0.3; bal Fe				
GOST 4543	30ChMA		0.26-0.34	0.8-1.1	0.4-0.7	0.15-0.25	0.3	0.025 max	0.025 max	0.17-0.37	Cu 0.3; bal Fe				
GOST 4543(71)	20ChM	Q/T	0.15-0.25	0.8-1.1	0.4-0.7	0.15-0.25	0.3 max	0.035 max	0.03 max	0.17-0.37	Al <=0.1; Cu <=0.3; Ti <=0.05; bal Fe				
GOST 4543(71)	30ChM	Q/T	0.26-0.34	0.8-1.1	0.4-0.7	0.15-0.25	0.3 max	0.035 max	0.035 max	0.17-0.37	Al <=0.1; Cu <=0.3; Ti <=0.05; bal Fe				
GOST 4543(71)	30ChMA	Q/T	0.26-0.33	0.8-1.1	0.4-0.7	0.15-0.25	0.3 max	0.025 max	0.025 max	0.17-0.37	Al <=0.1; Cu <=0.3; Ti <=0.05; bal Fe				

Spain

Specification	Designation	Notes	C	Cr	Mn	Mo	Ni	P	S	Si	Other	UTS	YS	El	Hard
UNE 36012(75)	30CrMo4	Q/T	0.27-0.33	0.85-1.15	0.6-0.9	0.15-0.25		0.035 max	0.035 max	0.15-0.4	bal Fe				
UNE 36012(75)	30CrMo4-1	Q/T	0.27-0.33	0.85-1.15	0.6-0.9	0.15-0.25		0.035 max	0.02-0.035	0.15-0.4	bal Fe				
UNE 36012(75)	F.1251*	see 30CrMo4	0.27-0.33	0.85-1.15	0.6-0.9	0.15-0.25		0.035 max	0.035 max	0.15-0.4	bal Fe				
UNE 36012(75)	F.1256*	see 30CrMo4-1	0.27-0.33	0.85-1.15	0.6-0.9	0.15-0.25		0.035 max	0.02-0.035	0.15-0.4	bal Fe				
UNE 36034(85)	35CrMo4DF	CHd; Cold ext	0.3-0.37	0.9-1.2	0.5-0.8	0.15-0.3		0.035 max	0.035	0.15-0.4	bal Fe				

Sweden

Specification	Designation	Notes	C	Cr	Mn	Mo	Ni	P	S	Si	Other	UTS	YS	El	Hard
SIS 142225	2225-01	Frg Plt Sh Tub, Norm	0.22-0.29	0.9-1.2	0.5-0.8	0.15-0.25	0.3	0.03	0.02	0.15-0.4	bal Fe	660	490	13	
SIS 142225	2225-03	Bar Frg Q/A, Tmp, 100mm diam	0.22-0.29	0.9-1.2	0.5-0.8	0.15-0.25	0.3	0.03	0.02	0.15-0.4	bal Fe	690	491	17	
SIS 142225	2225-04	Bar Frg Tube, Q/A, Tmp	0.22-0.29	0.9-1.2	0.5-0.8	0.15-0.25	0.3	0.03	0.02	0.15-0.4	bal Fe	780	590	15	
SIS 142225	2225-05	Bar Frg Plt Sh Tube	0.22-0.29	0.9-1.2	0.5-0.8	0.15-0.25	0.3	0.03	0.02	0.15-0.4	bal Fe	880	690	10	
SIS 142225	2225-06	Bar Frg Q/A, Tmp, 100-16mm diam	0.22-0.29	0.9-1.2	0.5-0.8	0.15-0.25	0.3	0.03	0.02	0.15-0.4	bal Fe	640	410	16	
SIS 142225	2225-07	Bar, Tube, Q/A, Tmp, 12mm diam	0.22-0.29	0.9-1.2	0.5-0.8	0.15-0.25	0.3	0.03	0.02	0.15-0.4	bal Fe	1080	880	8	
SS 142225	2225	Q/T	0.22-0.29	0.9-1.2	0.6-0.9	0.15-0.3		0.035 max	0.035 max	0.1-0.4	bal Fe				
SS 142233	2233		0.25-0.35	0.1-0.8	0.4-0.9	0.15-0.3		0.04 max	0.04 max	0.15-0.4	bal Fe				

UK

Specification	Designation	Notes	C	Cr	Mn	Mo	Ni	P	S	Si	Other	UTS	YS	El	Hard
BS 1717(83)	CDS110	Tube	0.26 max	0.80-1.20	0.80 max	0.15-0.30		0.050 max	0.050 max	0.35 max	bal Fe				
BS 970/1(83)	708A30	Blm Bil Slab Bar Rod Frg	0.28-0.33	0.90-1.20	0.40-0.60	0.15-0.25		0.035 max	0.040 max		bal Fe				

USA

Specification	Designation	Notes	C	Cr	Mn	Mo	Ni	P	S	Si	Other	UTS	YS	El	Hard
	AISI 4130	Sh Strp	0.28-0.33	0.80-1.10	0.40-0.60	0.15-0.25		0.035	0.035	0.15-0.30	bal Fe				

UNS numbers and US grades are provided as a means of cross referencing chemically similar alloys. Exchangability is only possible after independent examination of specifications. Tensile properties are minimum or typical as specified. UTS and YS as MPa. El as %. See Appendix for list of abbreviations used in Notes. * indicates obsolete material.

Specification	Designation	Notes	C	Cr	Mn	Mo	Ni	P	S	Si	Other	UTS	YS	El	Hard

Alloy Steel, Chromium Molybdenum, 4130/4130H (Continued from previous page)

USA

Specification	Designation	Notes	C	Cr	Mn	Mo	Ni	P	S	Si	Other	UTS	YS	El	Hard
	AISI 4130	Plt	0.27-0.34	0.80-1.15	0.35-0.60	0.15-0.25		0.035 max	0.040 max	0.15-0.30	bal Fe				
	AISI 4130	Bar Blm Bil Slab	0.28-0.33	0.80-1.10	0.40-0.60	0.15-0.25		0.035 max	0.040 max	0.15-0.35	bal Fe				
	AISI 4130	Smls mech tub	0.28-0.33	0.80-1.10	0.40-0.60	0.15-0.25		0.040 max	0.040 max	0.15-0.35	bal Fe				
	AISI 4130H	Smls mech tub, Hard	0.27-0.33	0.75-1.20	0.30-0.70	0.15-0.25		0.040 max	0.040 max	0.15-0.30	bal Fe				
	AISI 4130H	Bar Blm Bil Slab	0.27-0.33	0.75-1.20	0.30-0.70	0.15-0.25		0.035 max	0.040 max	0.15-0.35	bal Fe				
	UNS G41300		0.28-0.33	0.80-1.10	0.40-0.60	0.15-0.25		0.035 max	0.040 max	0.15-0.35	bal Fe				
	UNS H41300		0.27-0.33	0.75-1.20	0.30-0.70	0.15-0.25		0.035 max	0.040 max	0.15-0.35	bal Fe				
AMS 6348		Bar	0.28-0.33	0.8-1.1	0.4-0.60	0.15-0.25	0.25 min	0.025	0.025	0.15-0.35	Cu 0.35; bal Fe				
AMS 6350		Sh, Strp, Plt	0.28-0.33	0.8-1.1	0.4-0.60	0.15-0.25	0.25 min	0.025	0.025	0.15-0.35	Cu 0.35; bal Fe				
AMS 6351		Sh, Strp, Plt	0.28-0.33	0.8-1.1	0.4-0.60	0.15-0.25	0.25 min	0.025	0.025	0.15-0.35	Cu 0.35; bal Fe				
AMS 6360		Tube	0.28-0.33	0.8-1.1	0.4-0.60	0.15-0.25	0.25 min	0.025	0.025	0.15-0.35	Cu 0.35; bal Fe				
AMS 6361		Tube	0.28-0.33	0.8-1.1	0.4-0.60	0.15-0.25	0.25 min	0.025	0.025	0.15-0.35	Cu 0.35; bal Fe				
AMS 6362		Tube	0.28-0.33	0.8-1.1	0.4-0.60	0.15-0.25	0.25 min	0.025	0.025	0.15-0.35	Cu 0.35; bal Fe				
AMS 6370		Bar Frg	0.28-0.33	0.8-1.1	0.4-0.60	0.15-0.25	0.25 min	0.025	0.025	0.15-0.35	Cu 0.35; bal Fe				
AMS 6371		Tube	0.28-0.33	0.8-1.1	0.4-0.60	0.15-0.25	0.25 min	0.025	0.025	0.15-0.35	Cu 0.35; bal Fe				
AMS 6373		Tube	0.28-0.33	0.8-1.1	0.4-0.60	0.15-0.25	0.25 min	0.025	0.025	0.15-0.35	Cu 0.35; bal Fe				
AMS 6374											bal Fe				
AMS 6528											bal Fe				
AMS 7496											bal Fe				
ASTM A29/A29M(93)	4130	Bar	0.28-0.33	0.80-1.10	0.40-0.60	0.15-0.25		0.035 max	0.040 max	0.15-0.35	bal Fe				
ASTM A304(96)	4130H	Bar, hard bands spec	0.27-0.33	0.75-1.20	0.30-0.70	0.15-0.25		0.035 max	0.040 max	0.15-0.35	Cu <=0.35; Cu Ni Cr Mo trace allowed; bal Fe				
ASTM A322(96)	4130	Bar	0.28-0.33	0.80-1.10	0.40-0.60	0.15-0.25		0.035 max	0.040 max	0.15-0.35	bal Fe				
ASTM A331(95)	4130	Bar	0.28-0.33	0.80-1.10	0.40-0.60	0.15-0.25		0.035 max	0.040 max	0.15-0.35	bal Fe				
ASTM A331(95)	4130H	Bar	0.27-0.33	0.75-1.20	0.30-0.70	0.15-0.25		0.025 max	0.025 max	0.15-0.35	Cu <=0.35; bal Fe				
ASTM A372	Grade 1*	Thin wall frg press ves	0.25-0.35	0.8-1.15	0.4-0.9	0.15-0.25		0.025 max	0.025 max	0.15-0.35	bal Fe				
ASTM A506(93)	4130	Sh Strp, HR, CR	0.28-0.33	0.80-1.10	0.40-0.60	0.15-0.25		0.035	0.035	0.15-0.30	bal Fe				
ASTM A507(93)	4130	Sh Strp, HR, CR	0.28-0.33	0.80-1.10	0.40-0.60	0.15-0.25		0.025	0.030	0.15-0.30	bal Fe				
ASTM A513(97a)	4130	ERW Mech Tub	0.28-0.33	0.80-1.10	0.40-0.60	0.15-0.25		0.035 max	0.040 max	0.15-0.35	bal Fe				
ASTM A519(96)	4130	Smls mech tub, Norm	0.28-0.33	0.80-1.10	0.40-0.60	0.15-0.25		0.040 max	0.040 max	0.15-0.35	bal Fe	621	414	20	89 HRB
ASTM A519(96)	4130	Smls mech tub, Ann	0.28-0.33	0.80-1.10	0.40-0.60	0.15-0.25		0.040 max	0.040 max	0.15-0.35	bal Fe	517	379	30	81 HRB
ASTM A519(96)	4130	Smls mech tub, SR	0.28-0.33	0.80-1.10	0.40-0.60	0.15-0.25		0.040 max	0.040 max	0.15-0.35	bal Fe	724	586	10	95 HRB
ASTM A519(96)	4130	Smls mech tub, HR	0.28-0.33	0.80-1.10	0.40-0.60	0.15-0.25		0.040 max	0.040 max	0.15-0.35	bal Fe	621	483	20	89 HRB
ASTM A646(95)	Grade 11	Blm Bil for aircraft aerospace frg, ann	0.28-0.33	0.80-1.10	0.40-0.60	0.15-0.25		0.025 max	0.025 max	0.20-0.35	bal Fe				229 HB
ASTM A752(93)	4130	Rod Wir	0.28-0.33	0.80-1.10	0.40-0.60	0.15-0.25		0.035 max	0.040 max	0.15-0.30	bal Fe	590			
ASTM A752(93)	5130	Rod Wir	0.28-0.33	0.80-1.10	0.70-0.90			0.035 max	0.040 max	0.15-0.30	bal Fe	580			

UNS numbers and US grades are provided as a means of cross referencing chemically similar alloys. Exchangability is only possible after independent examination of specifications. Tensile properties are minimum or typical as specified. UTS and YS as MPa. El as %. See Appendix for list of abbreviations used in Notes. * indicates obsolete material.

Specification	Designation	Notes	C	Cr	Mn	Mo	Ni	P	S	Si	Other	UTS	YS	El	Hard

Alloy Steel, Chromium Molybdenum, 4130/4130H (Continued from previous page)

USA

Specification	Designation	Notes	C	Cr	Mn	Mo	Ni	P	S	Si	Other	UTS	YS	El	Hard
ASTM A829/A829M(95)	4130	Plt	0.27-0.34	0.80-1.15	0.35-0.60	0.15-0.25		0.035 max	0.040 max	0.15-0.40	bal Fe	480-965			
ASTM A866(94)	4130	Anti-friction bearings	0.28-0.33	0.80-1.10	0.40-0.60	0.15-0.25		0.025 max	0.025 max	0.15-0.35	bal Fe				
ASTM A914/914M(92)	4130RH	Bar restricted end Q hard	0.28-0.33	0.80-1.10	0.40-0.60	0.15-0.25	0.25 max	0.035 max	0.040 max	0.15-0.35	Cu <=0.35; Electric furnace P, S<=0.025; bal Fe				
ASTM A983/983M(98)	Grade 3	Continuous grain flow frg for crankshafts	0.28-0.33	0.80-1.20	0.40-1.00	0.15-0.25		0.025 max	0.025 max	0.15-0.40	V <=0.10; bal Fe				
DoD-F-24669/1(86)	4130	Bar Bil; Supersedes MIL-S-866 & MIL-S-16974	0.28-0.33	0.8-1.1	0.4-0.6	0.15-0.25		0.035	0.04	0.15-0.35	bal Fe				
MIL-S-16974E(86)	4130*	Obs; Bar Bil Blm slab for refrg									bal Fe				
MIL-S-18729C(66)	4130*	Obs; see SAE-AMS 6345, 6350, 6351; Aircraft Plt Sh Strp	0.27-0.33	0.8-1.1	0.4-0.6	0.15-0.35	0.25	0.025	0.025	0.2-0.35	bal Fe				
MIL-S-46059	G41300*	Obs; Sh	0.28-0.33	0.8-1.1	0.4-0.6	0.15-0.25		0.035	0.04	0.15-0.3	bal Fe				
MIL-S-6758	- - -*	Obs; Bar	0.28-0.33	0.8-1.1	0.4-0.6	0.15-0.25	0.25	0.025	0.025	0.2-0.35	bal Fe				
MIL-T-6736B	- - -*	Obs; Tube	0.27-0.33	0.8-1.1	0.4-0.6	0.15-0.25		0.025	0.025	0.2-0.35	bal Fe				
SAE 770(84)	4130*	Obs; see J1397(92)									bal Fe				
SAE J1268(95)	4130H	Bar Tub, See std	0.27-0.33	0.75-1.20	0.30-0.70	0.15-0.25	0.25 max	0.030 max	0.040 max	0.15-0.35	Cu <=0.35; Cu, Ni not spec'd but acpt; bal Fe				
SAE J1868(93)	4130RI I	Restrict hard range	0.28-0.33	0.80-1.10	0.40-0.60	0.15-0.25	0.25 max	0.025 max	0.040 max	0.15-0.35	Cu <=0.035; Cu, Ni not spec'd but acpt; bal Fe				5 HRC max
SAE J404(94)	4130	Bil Blm Slab Bar HR CF	0.28-0.33	0.80-1.10	0.40-0.60	0.15-0.25		0.030 max	0.040 max	0.15-0.35	bal Fe				
SAE J404(94)	4130	Plt	0.27-0.34	0.80-1.15	0.35-0.60	0.15-0.25		0.035 max	0.040 max	0.15-0.35	bal Fe				

Yugoslavia

Specification	Designation	Notes	C	Cr	Mn	Mo	Ni	P	S	Si	Other	UTS	YS	El	Hard
	C.4730	Q/T	0.22-0.29	0.9-1.2	0.6-0.9	0.15-0.3		0.035 max	0.03 max	0.4 max	bal Fe				
	C.4781	Q/T	0.3-0.37	0.9-1.2	0.5-0.8	0.15-0.3		0.035 max	0.035 max	0.15-0.4	bal Fe				

Alloy Steel, Chromium Molybdenum, 4135/4135H

Argentina

Specification	Designation	Notes	C	Cr	Mn	Mo	Ni	P	S	Si	Other	UTS	YS	El	Hard
IAS	IRAM 4133		0.30-0.36	0.80-1.10	0.70-0.90	0.15-0.25		0.035 max	0.040 max	0.20-0.35	bal Fe	850	530	55	270 HB

Belgium

Specification	Designation	Notes	C	Cr	Mn	Mo	Ni	P	S	Si	Other	UTS	YS	El	Hard
NBN 253-02	34CrMo4		0.3-0.37	0.9-1.2	0.5-0.8	0.15-0.3		0.035 max	0.035 max	0.15-0.4	bal Fe				

Bulgaria

Specification	Designation	Notes	C	Cr	Mn	Mo	Ni	P	S	Si	Other	UTS	YS	El	Hard
BDS 6354	35ChM		0.3-0.37	0.90-1.20	0.50-0.80	0.15-0.25		0.035	0.035	0.17-0.37	bal Fe				

China

Specification	Designation	Notes	C	Cr	Mn	Mo	Ni	P	S	Si	Other	UTS	YS	El	Hard
GB 3077(88)	35CrMo	Bar HR Q/T 25mm diam	0.32-0.40	0.80-1.10	0.40-0.70	0.15-0.25		0.035 max	0.035 max	0.17-0.37	Cu <=0.30; bal Fe	980	835	12	
GB 6478(86)	ML35CrMo	Bar HR Q/T	0.32-0.40	0.80-1.10	0.40-0.70	0.15-0.25		0.035 max	0.035 max	0.17-0.37	Cu <=0.30; bal Fe	980	835	12	
GB 8162(87)	35CrMo	Smls tube HR/CD Q/T	0.32-0.40	0.80-1.10	0.40-0.70	0.15-0.25		0.035 max	0.035 max	0.17-0.37	bal Fe	980	835	12	
GB/T 3078(94)	35CrMo	Bar CD Q/T 25mm diam	0.32-0.40	0.80-1.10	0.40-0.70	0.15-0.25		0.035 max	0.035 max	0.17-0.37	Cu <=0.30; bal Fe	980	835	12	
YB/T 5132(93)	35CrMo	Sh HR/CR	0.32-0.40	0.80-1.10	0.40-0.70	0.15-0.25		0.035 max	0.035 max	0.17-0.37	bal Fe				

Finland

Specification	Designation	Notes	C	Cr	Mn	Mo	Ni	P	S	Si	Other	UTS	YS	El	Hard
SFS 458(73)	SFS458	Q/T	0.22-0.29	0.9-1.2	0.5-0.8	0.15-0.25	0.3 max	0.035 max	0.035 max	0.15-0.4	bal Fe				
SFS 459	Co.35Cr1.1Mo0.20	Bar Frg Q/A Tmp, 180mm diam	0.3-0.37	0.9-1.2	0.5-0.8	0.15-0.25		0.035	0.035	0.15-0.4	bal Fe				

Specification	Designation	Notes	C	Cr	Mn	Mo	Ni	P	S	Si	Other	UTS	YS	El	Hard

Alloy Steel, Chromium Molybdenum, 4135/4135H (Continued from previous page)

France

Specification	Designation	Notes	C	Cr	Mn	Mo	Ni	P	S	Si	Other	UTS	YS	El	Hard
AFNOR NFA35552	34CD4	Bar Rod Wir	0.31-0.37	0.9-1.2	0.6-0.9	0.15-0.25		0.035	0.035	0.1-0.4	bal Fe				
AFNOR NFA35552	35CD4		0.3-0.37	0.85-1.2	0.6-0.9	0.15-0.25		0.035	0.035	0.1-0.4	bal Fe				
AFNOR NFA35553	35CD4	Strp	0.3-0.37	0.85-1.15	0.6-0.9	0.15-0.3		0.035	0.035	0.1-0.4	bal Fe				
AFNOR NFA35556(84)	34CD4		0.31-0.37	0.9-1.2	0.6-0.9	0.15-0.25		0.035 max	0.035 max	0.1-0.4	bal Fe				
AFNOR NFA35556(84)	35CD4		0.3-0.37	0.85-1.2	0.6-0.9	0.15-0.25		0.035 max	0.035 max	0.1-0.4	bal Fe				
AFNOR NFA35557(83)	34CD4	t<=16mm	0.31-0.37	0.9-1.2	0.6-0.9	0.15-0.25	0.4 max	0.035 max	0.035 max	0.1-0.4	Cu <=0.3; Ti <=0.05; bal Fe	1080-1280	870	10	
AFNOR NFA36102(93)	30MnB5RR	HR; Strp; CR; t<1000mm	0.27-0.33	0.15-0.25	1.1-1.4	0.10 max	0.20 max	0.025 max	0.02 max	0.1-0.4	B 0.0008-0.005; bal Fe	620	450	14	
AFNOR NFA36102(93)	34CrMo4RR	HR; Strp; CR	0.31-0.38	0.9-1.2	0.6-0.9	0.15-0.30	0.20 max	0.025 max	0.02 max	0.15-0.4	bal Fe				
AFNOR NFA36102(93)	35B3RR	HR; Strp; CR; t<1000mm	0.32-0.39	0.155-0.25	0.6-0.9	0.10 max	0.20 max	0.025 max	0.02 max	0.1-0.4	B 0.0008-0.005; bal Fe	900	700	9	

Germany

Specification	Designation	Notes	C	Cr	Mn	Mo	Ni	P	S	Si	Other	UTS	YS	El	Hard
	33CrMo4		0.30-0.37	0.90-1.20	0.50-0.80	0.15-0.30		0.035 max	0.035 max	0.40 max	bal Fe				
DIN	WNr 1.7231		0.30-0.37	0.90-1.20	0.50-0.80	0.15-0.30		0.035 max	0.035 max	0.40 max	bal Fe				
DIN 1652(90)	34CrMoS4	Q/T	0.30-0.37	0.90-1.20	0.60-0.90	0.15-0.30		0.035 max	0.020-0.040	0.40 max	bal Fe				
DIN 1652(90)	WNr 1.7220	Q/T 17-40mm	0.30-0.37	0.90-1.20	0.60-0.90	0.15-0.30		0.035 max	0.035 max	0.40 max	bal Fe	900-1100	650	12	
DIN 1652(90)	WNr 1.7226	Q/T	0.30-0.37	0.90-1.20	0.60-0.90	0.15-0.30		0.035 max	0.020-0.040	0.40 max	bal Fe				
DIN EN 10083(91)	34CrMo4	Q/T 17-40mm	0.30-0.37	0.90-1.20	0.60-0.90	0.15-0.30		0.035 max	0.035 max	0.40 max	bal Fe	900-1100	650	12	
DIN EN 10083(91)	WNr 1.7220	Q/T 17-40mm	0.30-0.37	0.90-1.20	0.60-0.90	0.15-0.30		0.035 max	0.035 max	0.40 max	bal Fe	900-1100	650	12	

Hungary

Specification	Designation	Notes	C	Cr	Mn	Mo	Ni	P	S	Si	Other	UTS	YS	El	Hard
MSZ 61(85)	CMo3	Q/T; 0-16mm	0.3-0.37	0.9-1.2	0.6-0.9	0.15-0.3		0.035 max	0.035 max	0.4 max	bal Fe	1000-1200	800	11L	
MSZ 61(85)	CMo3	Q/T; 160.1-250mm	0.3-0.37	0.9-1.2	0.6-0.9	0.15-0.3		0.035 max	0.035 max	0.4 max	bal Fe	700-850	460	15L	
MSZ 61(85)	CMo3E	Q/T; 160.1-250mm	0.3-0.37	0.9-1.2	0.6-0.9	0.15-0.3		0.035 max	0.02-0.035	0.4 max	bal Fe	700-850	460	15L	
MSZ 61(85)	CMo3E	Q/T; 0-16mm	0.3-0.37	0.9-1.2	0.6-0.9	0.15-0.3		0.035 max	0.02-0.035	0.4 max	bal Fe	1000-1200	800	11L	
MSZ 6251(87)	CMo3Z	CF; Q/T, 0-36mm; HR; HR/soft ann; drawn/soft ann; drawn/bright ann; soft ann/ground	0.3-0.37	0.9-1.2	0.6-0.9	0.15-0.3		0.035 max	0.035 max	0.4 max	bal Fe	0-620			
MSZ 6251(87)	CMo3Z	CF; Q/T, 0-36mm; drawn, half-hard	0.3-0.37	0.9-1.2	0.6-0.9	0.15-0.3		0.035 max	0.035 max	0.4 max	bal Fe	0-650			

International

Specification	Designation	Notes	C	Cr	Mn	Mo	Ni	P	S	Si	Other	UTS	YS	El	Hard
ISO R683-4	26	Bar Bil, Rod, Q/A, Tmp, 16mm	0.3-0.37	0.9-1.2	0.5-0.8	0.15-0.3		0.04	0.03-0.05	0.15-0.4	bal Fe				

Italy

Specification	Designation	Notes	C	Cr	Mn	Mo	Ni	P	S	Si	Other	UTS	YS	El	Hard
UNI 6403(86)	35CrMo4	Q/T	0.32-0.39	0.9-1.2	0.6-0.9	0.15-0.25		0.035 max	0.035 max	0.15-0.4	bal Fe				
UNI 6929(71)	35CrMo4F		0.3-0.37	0.9-1.2	0.5-0.8	0.15-0.25	0.3	0.02	0.015	0.15-0.35	bal Fe				
UNI 7356(74)	34CrMo4KB	Q/T	0.3-0.37	0.9-1.2	0.5-0.8	0.15-0.3		0.035 max	0.035 max	0.15-0.4	bal Fe				
UNI 7845(78)	35CrMo4	Q/T	0.32-0.39	0.9-1.2	0.6-0.9	0.15-0.25		0.035 max	0.035 max	0.15-0.4	Pb 0.15-0.3; bal Fe				
UNI 7874(79)	35CrMo4	Q/T	0.32-0.39	0.9-1.2	0.5-0.9	0.15-0.3	0.3 max	0.035 max	0.035 max	0.15-0.4	Pb 0.15-0.3; bal Fe				

Japan

Specification	Designation	Notes	C	Cr	Mn	Mo	Ni	P	S	Si	Other	UTS	YS	El	Hard
JIS G3311(88)	SCM3M*	Obs; see SCM 435M	0.33-0.38	0.9-1.2	0.6-0.85	0.15-0.3	0.25	0.03	0.03	0.15-0.35	Cu 0.3; bal Fe				

UNS numbers and US grades are provided as a means of cross referencing chemically similar alloys. Exchangability is only possible after independent examination of specifications. Tensile properties are minimum or typical as specified. UTS and YS as MPa. El as %. See Appendix for list of abbreviations used in Notes. * indicates obsolete material.

Alloy Steel, Chromium Molybdenum, 4135/4135H (Continued from previous page)

Specification	Designation	Notes	C	Cr	Mn	Mo	Ni	P	S	Si	Other	UTS	YS	El	Hard
Japan															
JIS G4052(79)	SCM3H*	Obs; see SCM435H	0.32-0.39	0.85-1.25	0.55-0.9	0.15-0.35		0.03	0.03	0.15-0.35	Cu 0.3; bal Fe				
JIS G4052(79)	SCM435H	Struct Hard	0.32-0.39	0.85-1.25	0.55-0.90	0.15-0.35	0.25 max	0.030 max	0.030 max	0.15-0.35	Cu <=0.30; bal Fe				
JIS G4105(79)	SCM 3*	Obs; see SCM435	0.33-0.38	0.9-1.2	0.6-0.85	0.15-0.3		0.03	0.03	0.15-0.35	bal Fe				
JIS G4105(79)	SCM435	HR Frg Bar Rod	0.33-0.38	0.90-1.20	0.60-0.85	0.15-0.30	0.25 max	0.030 max	0.030 max	0.15-0.35	Cu <=0.30; bal Fe				
Mexico															
DGN B-203	4135*	Obs	0.33-0.38	0.8-1.1	0.7-0.9	0.15-0.25		0.04	0.04	0.2-0.35	bal Fe				
NMX-B-268(68)	4135H	Hard	0.32-0.38	0.75-1.20	0.60-1.00	0.15-0.25				0.20-0.35	bal Fe				
NMX-B-300(91)	4135	Bar	0.33-0.38	0.80-1.10	0.70-0.90	0.15-0.25		0.035 max	0.040 max	0.15-0.35	bal Fe				
NMX-B-300(91)	4135H	Bar	0.32-0.38	0.75-1.20	0.60-1.00	0.15-0.25				0.15-0.35	bal Fe				
Norway															
NS	CrMoII		0.34	1.1	0.75	0.2		0.03	0.03	0.25	bal Fe				
Pan America															
COPANT 334	4135	Bar	0.33-0.38	0.8-1.1	0.7-0.9	0.15-0.25		0.035	0.04	0.2-0.35	bal Fe				
COPANT 514	4135	Tube	0.33-0.38	0.8-1.1	0.7-0.9	0.15-0.25		0.04	0.04	0.2-0.35	bal Fe				
Poland															
PNH84030	35HM	Q/T	0.34-0.4	0.9-1.2	0.4-0.7	0.15-0.25	0.3	0.03	0.03	0.17-0.37	Cu 0.3; bal Fe				
PNH84030	35HMA	Q/T	0.34-0.4	0.9-1.2	0.4-0.7	0.15-0.25	0.3	0.03	0.03	0.17-0.37	Cu 0.3; bal Fe				
Romania															
STAS 791	33MoCr11		0.3-0.37	0.9-1.2	0.4-0.8	0.15-0.3	0.3	0.035	0.035	0.17-0.37	Cu 0.3; bal Fe				
STAS 791	35MoCr11		0.3-0.37	0.9-1.2	0.4-0.8	0.15-0.3	0.3	0.035	0.035	0.17-0.37	Cu 0.3; bal Fe				
STAS 791	36MoCr10		0.32-0.4	0.8-1.1	0.4-0.7	0.15-0.3	0.3	0.035	0.035	0.17-0.37	Cu 0.3; bal Fe				
STAS 791(88)	34MoCr11	Struct/const; Q/T	0.3-0.37	0.9-1.3	0.6-0.9	0.15-0.3		0.035 max	0.035 max	0.17-0.37	Pb <=0.15; Ti <=0.02; bal Fe				
STAS 791(88)	34MoCr11S	Struct/const; Q/T	0.3-0.37	0.9-1.3	0.6-0.9	0.15-0.3		0.035 max	0.02-0.04	0.17-0.37	Pb <=0.15; Ti <=0.02; bal Fe				
STAS 791(88)	34MoCr11X	Struct/const; Q/T	0.3-0.37	0.9-1.3	0.6-0.9	0.15-0.3		0.025 max	0.025 max	0.17-0.37	Pb <=0.15; Ti <=0.02; bal Fe				
STAS 791(88)	34MoCr11XS	Struct/const; Q/T	0.3-0.37	0.9-1.3	0.6-0.9	0.15-0.3		0.025 max	0.02-0.035	0.17-0.37	Pb <=0.15; Ti <=0.02; bal Fe				
Russia															
GOST 4543	35ChM		0.32-0.4	0.8-1.1	0.4-0.7	0.15-0.25	0.3 max	0.035 max	0.035 max	0.17-0.37	Cu <=0.3; bal Fe				
Spain															
UNE	F.125		0.3-0.4	0.9-1.5	0.4-0.7	0.2-0.4		0.04	0.04	0.1-0.35	bal Fe				
UNE	F.125A		0.32-0.38	0.9-1.5	0.6-0.9	0.2-0.4		0.035	0.035	0.15-0.4	bal Fe				
UNE 36012(75)	35CrMo4	Q/T	0.32-0.38	0.85-1.15	0.6-0.9	0.15-0.25		0.035 max	0.035 max	0.15-0.4	bal Fe				
UNE 36012(75)	35CrMo4-1	Q/T	0.32-0.38	0.85-1.15	0.6-0.9	0.15-0.25		0.035 max	0.02-0.035	0.15-0.4	bal Fe				
UNE 36012(75)	38CrMo4		0.34-0.4	0.85-1.15	0.6-0.9	0.15-0.25		0.035	0.035	0.15-0.4	bal Fe				
UNE 36012(75)	F.1250*	see 35CrMo4	0.32-0.38	0.85-1.15	0.6-0.9	0.15-0.25		0.035 max	0.035 max	0.15-0.4	bal Fe				
UNE 36012(75)	F.1255*	see 35CrMo4-1	0.32-0.38	0.85-1.15	0.6-0.9	0.15-0.25		0.035 max	0.02-0.035	0.15-0.4	bal Fe				
UNE 36034(85)	35CrMo4DF	CHd; Cold ext	0.3-0.37	0.9-1.2	0.5-0.8	0.15-0.3		0.035 max	0.035 max	0.15-0.4	bal Fe				
UNE 36034(85)	F.1254*	see 35CrMo4DF	0.3-0.37	0.9-1.2	0.5-0.8	0.15-0.3		0.035 max	0.035 max	0.15-0.4	bal Fe				

UNS numbers and US grades are provided as a means of cross referencing chemically similar alloys. Exchangability is only possible after independent examination of specifications. Tensile properties are minimum or typical as specified. UTS and YS as MPa. El as %. See Appendix for list of abbreviations used in Notes. * indicates obsolete material.

Specification	Designation	Notes	C	Cr	Mn	Mo	Ni	P	S	Si	Other	UTS	YS	El	Hard

Alloy Steel, Chromium Molybdenum, 4135/4135H (Continued from previous page)

Spain

Specification	Designation	Notes	C	Cr	Mn	Mo	Ni	P	S	Si	Other	UTS	YS	El	Hard
UNE 36254(79)	F.8231*	see 34CrMo4	0.3-0.38	0.8-1.2	0.5-0.8	0.15-0.3		0.04 max	0.04 max	0.3-0.5	bal Fe				

Sweden

Specification	Designation	Notes	C	Cr	Mn	Mo	Ni	P	S	Si	Other	UTS	YS	El	Hard
SIS 142234	2234-03	Bar Frg Q/A, Tmp, 250mm diam	0.3-0.37	0.9-1.2	0.5-0.8	0.15-0.3		0.04	0.04	0.15-0.4	bal Fe	690	490	15	
SIS 142234	2234-04	Bar Frg Q/A, Tmp, 100mm diam	0.3-0.37	0.9-1.2	0.5-0.8	0.15-0.3		0.04	0.04	0.15-0.4	bal Fe	780	590	14	
SIS 142234	2234-05	Bar Frg Q/A, Tmp, 40mm diam	0.3-0.37	0.9-1.2	0.5-0.8	0.15-0.3		0.04	0.04	0.15-0.4	bal Fe	880	690	12	
SS 142234	2234	Q/T	0.3-0.37	0.9-1.2	0.6-0.9	0.15-0.3		0.035 max	0.035 max	0.1-0.4	bal Fe				

UK

Specification	Designation	Notes	C	Cr	Mn	Mo	Ni	P	S	Si	Other	UTS	YS	El	Hard
BS 970/1(83)	708A37	Blm Bil Slab Bar Rod Frg	0.35-0.40	0.90-1.20	0.70-0.90	0.15-0.25		0.035 max	0.040 max		bal Fe				
BS 970/1(83)	708H37	Blm Bil Slab Bar Rod Frg	0.34-0.41	0.80-1.25	0.65-1.05	0.15-0.25		0.035 max	0.040 max		bal Fe				

USA

Specification	Designation	Notes	C	Cr	Mn	Mo	Ni	P	S	Si	Other	UTS	YS	El	Hard
	AISI 4135	Smls mech tub	0.33-0.38	0.80-1.10	0.70-0.90	0.15-0.25		0.040 max	0.040 max	0.15-0.35	bal Fe				
	AISI 4135	Plt	0.33-0.38	0.80-1.10	0.70-0.90	0.15-0.25		0.035 max	0.040 max	0.15-0.35	bal Fe				
	AISI 4135H	Bar Blm Bil Slab	0.32-0.38	0.75-1.20	0.60-1.00	0.15-0.25		0.035 max	0.040 max	0.15-0.35	bal Fe				
	AISI 4135H	Smls mech tub, Hard	0.32-0.38	0.75-1.20	0.60-1.00	0.15-0.25		0.040 max	0.040 max	0.15-0.30	bal Fe				
	UNS G41350		0.33-0.38	0.80-1.10	0.70-0.90	0.15-0.25		0.035 max	0.040 max	0.15-0.35	bal Fe				
	UNS H41350		0.32-0.38	0.75-1.20	0.60-1.00	0.15-0.25				0.15-0.35	bal Fe				
AMS 6352	4135	Sh, Strp, Plt	0.33-0.38	0.8-1.1	0.7-0.90	0.15-0.25	0.25	0.025	0.025	0.15-0.35	Cu 0.35; bal Fe				
AMS 6365	4135	Tube	0.33-0.38	0.8-1.1	0.7-0.90	0.15-0.25	0.25	0.025	0.025	0.15-0.35	Cu 0.35; bal Fe				
AMS 6365C	4135	Wire, Frg	0.33-0.38	0.8-1.1	0.7-0.90	0.15-0.25		0.040 max	0.040 max	0.15-0.30	bal Fe				
AMS 6372	4135	Tube	0.33-0.38	0.8-1.1	0.7-0.90	0.15-0.25	0.25	0.025	0.025	0.15-0.35	Cu 0.35; bal Fe				
AMS 6372C	4135	Wire, Frg	0.33-0.38	0.8-1.1	0.7-0.90	0.15-0.25		0.040 max	0.040 max	0.15-0.30	bal Fe				
ASTM A29/A29M(93)	4135	Bar	0.33-0.38	0.80-1.10	0.70-0.90	0.15-0.25		0.035 max	0.040 max	0.15-0.35	bal Fe				
ASTM A304(96)	4135H	Bar, hard bands spec	0.32-0.38	0.75-1.20	0.6-1.00	0.15-0.25		0.035 max	0.040 max	0.15-0.35	Cu <=0.35; Cu Ni Cr Mo trace allowed; bal Fe				
ASTM A331(95)	4135H	Bar	0.32-0.38	0.75-1.20	0.6-1.00	0.15-0.25		0.025 max	0.025 max	0.15-0.35	Cu <=0.35; bal Fe				
ASTM A372/372M(95)	Grade F	Thin wall frg press ves	0.30-0.40	0.80-1.15	0.70-1.00	0.15-0.25		0.025 max	0.025 max	0.15-0.35	bal Fe	545-1000	380-485	15-18	179-248 HB
ASTM A372/372M(95)	Type V Grade 2*	Thin wall frg press ves	0.30-0.40	0.80-1.15	0.70-1.00	0.15-0.25		0.025 max	0.025 max	0.15-0.35	bal Fe				
ASTM A519(96)	4135	Smls mech tub	0.33-0.38	0.80-1.10	0.70-0.90	0.15-0.25		0.040 max	0.040 max	0.15-0.35	bal Fe				
ASTM A829/A829M(95)	4135	Plt	0.32-0.39	0.80-1.15	0.65-0.95	0.15-0.25		0.035 max	0.040 max	0.15-0.40	bal Fe	480-965			
DoD-F-24669/1(86)	4135	Bar Bil; Supersedes MIL-S-866 & MIL-S-16974	0.33-0.38	0.8-1.1	0.7-0.9	0.15-0.25		0.035	0.04	0.15-0.35	bal Fe				
MIL-S-16974E(86)	4135*	Obs; Bar Bil Blm slab for refrg	0.33-0.38	0.8-1.1	0.7-0.9	0.15-0.25		0.04 max	0.04 max	0.15-0.3	bal Fe				
MIL-S-18733A(56)	4135*	Obs; see AMS 6352; Aircraft Plt Sh Strp	0.33-0.38	0.8-1.1	0.7-0.9	0.15-0.25		0.04 max	0.04 max	0.15-0.3	bal Fe				
MIL-T-6735A	---*	Obs; tube	0.32-0.39	0.8-1.1	0.7-0.9	0.15-0.25		0.025	0.025	0.2-0.35	bal Fe				
SAE 770(84)	4135*	Obs; see J1397(92)	0.33-0.38	0.8-1.1	0.7-0.9	0.15-0.25		0.04 max	0.04 max	0.15-0.3	bal Fe				
SAE J1268(95)	4135H	Wir Frg; See std	0.32-0.38	0.75-1.20	0.60-1.00	0.15-0.25	0.25 max	0.030 max	0.040 max	0.15-0.35	Cu <=0.35; Cu, Ni not spec'd but acpt; bal Fe				

UNS numbers and US grades are provided as a means of cross referencing chemically similar alloys. Exchangability is only possible after independent examination of specifications. Tensile properties are minimum or typical as specified. UTS and YS as MPa. El as %. See Appendix for list of abbreviations used in Notes. * indicates obsolete material.

Specification	Designation	Notes	C	Cr	Mn	Mo	Ni	P	S	Si	Other	UTS	YS	El	Hard

Alloy Steel, Chromium Molybdenum, 4135/4135H (Continued from previous page)

USA

Specification	Designation	Notes	C	Cr	Mn	Mo	Ni	P	S	Si	Other	UTS	YS	El	Hard
SAE J404(94)	4135	Plt	0.32-0.39	0.80-1.15	0.65-0.95	0.15-0.25		0.035 max	0.040 max	0.15-0.35	bal Fe				
SAE J407	4135H*	Obs; see J1268									bal Fe				

Yugoslavia

Specification	Designation	Notes	C	Cr	Mn	Mo	Ni	P	S	Si	Other	UTS	YS	El	Hard
	C.4731	Q/T	0.3-0.37	0.9-1.2	0.6-0.9	0.15-0.3		0.035 max	0.03 max	0.4 max	bal Fe				
	C.4783	Q/T	0.3-0.37	0.9-1.2	0.6-0.9	0.15-0.3		0.035 max	0.02-0.035	0.4 max	bal Fe				

Alloy Steel, Chromium Molybdenum, 4137/4137H

Argentina

Specification	Designation	Notes	C	Cr	Mn	Mo	Ni	P	S	Si	Other	UTS	YS	El	Hard
IAS	IRAM 4137		0.35-0.40	0.80-1.10	0.70-0.90	0.15-0.25		0.035 max	0.040 max	0.15-0.30	bal Fe	850	585	18	250 HB

Australia

Specification	Designation	Notes	C	Cr	Mn	Mo	Ni	P	S	Si	Other	UTS	YS	El	Hard
AS G18	En19B*	Obs; see AS 1444	0.35-0.4	0.9-1.2	0.5-0.8	0.2-0.35		0.05	0.05	0.1-0.35	bal Fe				

China

Specification	Designation	Notes	C	Cr	Mn	Mo	Ni	P	S	Si	Other	UTS	YS	El	Hard
GB 3077(88)	35CrMo	Bar HR Q/T 25mm diam	0.32-0.40	0.80-1.10	0.40-0.70	0.15-0.25		0.035 max	0.035 max	0.17-0.37	Cu <=0.30; bal Fe	980	835	12	
GB 6478(86)	ML35CrMo	Smls tube HR/CD Q/T	0.32-0.40	0.80-1.10	0.40-0.70	0.15-0.25		0.035 max	0.035 max	0.17-0.37	Cu <=0.20; bal Fe	980	835	12	
GB 8162(87)	35CrMo	Bar HR Q/T	0.32-0.40	0.80-1.10	0.40-0.70	0.15-0.25		0.035 max	0.035 max	0.17-0.37	bal Fe	980	835	12	
GB/T 3078(94)	35CrMo	Bar CD Q/T 25mm diam	0.32-0.40	0.80-1.10	0.40-0.70	0.15-0.25		0.035 max	0.035 max	0.17-0.37	Cu <=0.30; bal Fe	980	835	12	
YB/T 5132(93)	35CrMo	Sh HR/CR	0.32-0.40	0.80-1.10	0.40-0.70	0.15-0.25		0.035 max	0.035 max	0.17-0.37	bal Fe				

Europe

Specification	Designation	Notes	C	Cr	Mn	Mo	Ni	P	S	Si	Other	UTS	YS	El	Hard
EN 10083/1(91)A1(96)	1.7220	Q/T t<=16mm	0.30-0.37	0.90-1.20	0.60-0.90	0.15-0.30	0.40 max	0.035 max	0.035 max	0.40 max	bal Fe	1100-1200	800	11	
EN 10083/1(91)A1(96)	34CrMo4	Q/T t<=16mm	0.30-0.37	0.90-1.20	0.60-0.90	0.15-0.30	0.40 max	0.035 max	0.035 max	0.40 max	bal Fe	1100-1200	800	11	

France

Specification	Designation	Notes	C	Cr	Mn	Mo	Ni	P	S	Si	Other	UTS	YS	El	Hard
AFNOR NFA35552	38CD4		0.35-0.41	0.9-1.2	0.6-0.9	0.15-0.25		0.035	0.035	0.1-0.4	bal Fe				
AFNOR NFA35557	38CD4		0.35-0.41	0.9-1.2	0.6-0.9	0.15-0.25		0.035	0.035	0.1-0.4	bal Fe				
AFNOR NFA35557(83)	38CD4	t<16mm; Q/T	0.35-0.41	0.9-1.2	0.6-0.9	0.15-0.25	0.4 max	0.035 max	0.035 max	0.1-0.4	Cu <=0.3; Ti <=0.05; bal Fe	1130-1330	910	9	
AFNOR NFA35563(83)	38CD4TS	HT	0.35-0.41	0.9-1.2	0.6-0.9	0.15-0.25	0.4 max	0.025 max	0.035 max	0.1-0.4	Cu <=0.3; Ti <=0.05; bal Fe				55 HRC
AFNOR NFA35564(83)	38CD4FF	Soft ann	0.35-0.41	0.9-1.2	0.6-0.9	0.15-0.25	0.4 max	0.03 max	0.03 max	0.35 max	Al >=0.02; Ti <=0.05; bal Fe	0-600			229 HB max
AFNOR NFA35564(83)	42C2FF	Soft ann	0.4-0.45	0.4-0.6	0.6-0.9	0.15 max	0.4 max	0.03 max	0.03 max	0.35 max	Al >=0.02; Ti <=0.05; bal Fe	0-580			217 HB max
AFNOR NFA35564(83)	42C4FF	Soft ann	0.4-0.45	0.9-1.2	0.6-0.9	0.15 max	0.4 max	0.03 max	0.03 max	0.35 max	Al >=0.02; Ti <=0.05; bal Fe	0-600			223 HB max

International

Specification	Designation	Notes	C	Cr	Mn	Mo	Ni	P	S	Si	Other	UTS	YS	El	Hard
ISO 683-2(68)	Type 2*	Q/T	0.35-0.40	0.8-1.1	0.7-0.9	0.15-0.25	0.4 max	0.035 max	0.035 max	0.15-0.35	Al <=0.1; Co <=0.1; Cu <=0.3; Pb <=0.15; Ti <=0.05; V <=0.1; W <=0.1; bal Fe				

Italy

Specification	Designation	Notes	C	Cr	Mn	Mo	Ni	P	S	Si	Other	UTS	YS	El	Hard
UNI 7356(74)	38CrMo4KB	Q/T	0.34-0.41	0.9-1.2	0.5-0.8	0.15-0.3		0.035 max	0.035 max	0.15-0.4	bal Fe				

Japan

Specification	Designation	Notes	C	Cr	Mn	Mo	Ni	P	S	Si	Other	UTS	YS	El	Hard
JIS G3311(88)	SCM435M	CR Strp; Chain	0.33-0.38	0.90-1.20	0.60-0.85	0.15-0.30	0.25 max	0.030 max	0.030 max	0.15-0.35	Cu <=0.30; bal Fe				190-270 HV
JIS G3441(88)	SCM435TK	Tube	0.33-0.38	0.90-1.20	0.60-0.85	0.15-0.30	0.25 max	0.030 max	0.030 max	0.15-0.35	Cu <=0.30; bal Fe				

Mexico

Specification	Designation	Notes	C	Cr	Mn	Mo	Ni	P	S	Si	Other	UTS	YS	El	Hard
DGN B-203	4137*	Obs	0.35-0.4	0.8-1.1	0.7-0.9	0.15-0.25		0.04	0.04	0.2-0.35	bal Fe				
DGN B-297	4137*	Obs	0.35-0.4	0.8-1.1	0.7-0.9	0.15-0.25		0.035	0.04	0.2-0.35	bal Fe				

Specification	Designation	Notes	C	Cr	Mn	Mo	Ni	P	S	Si	Other	UTS	YS	El	Hard

Alloy Steel, Chromium Molybdenum, 4137/4137H (Continued from previous page)

Mexico

Specification	Designation	Notes	C	Cr	Mn	Mo	Ni	P	S	Si	Other	UTS	YS	El	Hard
NMX-B-300(91)	4137	Bar	0.35-0.40	0.80-1.10	0.70-0.90	0.15-0.25		0.035 max	0.040 max	0.15-0.35	bal Fe				
NMX-B-300(91)	4137H	Bar	0.34-0.41	0.75-1.20	0.60-1.00	0.15-0.25				0.15-0.35	bal Fe				

Pan America

| COPANT 514 | 4137 | Tube | 0.35-0.4 | 0.8-1.1 | 0.7-0.9 | 0.15-0.25 | | 0.04 | 0.04 | 0.2-0.35 | bal Fe | | | | |

Poland

| PNH84030/04 | 35HM | Q/T | 0.34-0.4 | 0.9-1.2 | 0.4-0.7 | 0.15-0.25 | 0.3 max | 0.035 max | 0.035 max | 0.17-0.37 | bal Fe | | | | |
| PNH84030/04 | 35HMA | Q/T | 0.34-0.4 | 0.9-1.2 | 0.4-0.7 | 0.15-0.25 | 0.3 max | 0.03 max | 0.03 max | 0.17-0.37 | bal Fe | | | | |

Russia

GOST 1414(75)	AS38ChGM	Free-cutting	0.34-0.40	0.8-1.1	0.6-0.9	0.15-0.25	0.3 max	0.035 max	0.03 max	0.17-0.37	Cu <=0.30; Pb 0.15-0.3; P+S<=0.06; bal Fe				
GOST 4543	38ChM		0.35-0.42	0.9-1.3	0.35-0.65	0.2-0.3	0.3	0.035	0.035	0.17-0.37	Cu 0.3; bal Fe				
GOST 4543(71)	35ChM	Q/T	0.32-0.4	0.8-1.1	0.4-0.7	0.15-0.25	0.3 max	0.035 max	0.035 max	0.17-0.37	Al <=0.1; Cu <=0.3; Ti <=0.05; bal Fe				

Spain

UNE	F.125B		0.34-0.4	0.8-1.1	0.6-0.9	0.15-0.25		0.035	0.035	0.4	bal Fe				
UNE 36012(75)	35CrMo4	Q/T	0.32-0.38	0.85-1.15	0.6-0.9	0.15-0.25		0.035 max	0.035 max	0.15-0.4	bal Fe				
UNE 36012(75)	38CrMo4		0.34-0.4	0.85-1.15	0.6-0.9	0.15-0.25		0.035	0.035	0.15-0.4	bal Fe				
UNE 36012(75)	F.1253*	see 38CrMo4	0.34-0.4	0.85-1.15	0.6-0.9	0.15-0.25		0.035	0.035	0.15-0.4	bal Fe				

UK

| | Type 5 | | 0.35-0.43 | 0.9-1.1 | 0.7-0.9 | 0.15-0.25 | | 0.04 | 0.04 | 0.15-0.4 | bal Fe | | | | |

USA

	AISI 4137	Bar Blm Bil Slab	0.35-0.40	0.80-1.10	0.70-0.90	0.15-0.25		0.035 max	0.040 max	0.15-0.35	bal Fe				
	AISI 4137	Plt	0.35-0.40	0.80-1.10	0.70-0.90	0.15-0.25		0.035 max	0.040 max	0.15-0.35	bal Fe				
	AISI 4137	Smls mech tub	0.35-0.40	0.80-1.10	0.70-0.90	0.15-0.25		0.040 max	0.040 max	0.15-0.35	bal Fe				
	AISI 4137H	Bar Blm Bil Slab	0.34-0.41	0.75-1.20	0.60-1.00	0.15-0.25		0.035 max	0.040 max	0.15-0.35	bal Fe				
	AISI 4137H	Smls mech tub, Hard	0.34-0.41	0.75-1.20	0.60-1.00	0.15-0.25		0.040 max	0.040 max	0.15-0.30	bal Fe				
	UNS G41370		0.35-0.4	0.80-1.10	0.70-0.90	0.15-0.25		0.035 max	0.040 max	0.15-0.35	bal Fe				
	UNS H41370		0.34-0.41	0.75-1.20	0.60-1.00	0.15-0.25		0.035 max	0.040 max	0.15-0.35	bal Fe				
ASTM A29/A29M(93)	4137	Bar	0.35-0.40	0.80-1.10	0.70-0.90	0.15-0.25		0.035 max	0.040 max	0.15-0.35	bal Fe				
ASTM A304(96)	4137H	Bar, hard bands spec	0.34-0.41	0.75-1.20	0.60-1.00	0.15-0.25		0.035 max	0.040 max	0.15-0.35	Cu <=0.35; Cu Ni Cr Mo trace allowed; bal Fe				
ASTM A322(96)	4137	Bar	0.35-0.40	0.80-1.10	0.70-0.90	0.15-0.25		0.035 max	0.040 max	0.15-0.35	bal Fe				
ASTM A331(95)	4137	Bar	0.35-0.40	0.80-1.10	0.70-0.90	0.15-0.25		0.035 max	0.040 max	0.15-0.35	bal Fe				
ASTM A331(95)	4137H	Bar	0.34-0.41	0.75-1.20	0.60-1.00	0.15-0.25		0.025 max	0.025 max	0.15-0.35	Cu <=0.35; bal Fe				
ASTM A372	Grade 2*	Thin wall frg press ves	0.30-0.40	0.80-1.15	0.70-1.00	0.15-0.25		0.025 max	0.025 max	0.15-0.35	bal Fe				
ASTM A519(96)	4137	Smls mech tub	0.35-0.40	0.80-1.10	0.70-0.90	0.15-0.25		0.040 max	0.040 max	0.15-0.35	bal Fe				
ASTM A752(93)	4137	Rod Wir	0.35-0.40	0.80-1.10	0.70-0.90	0.15-0.25		0.035 max	0.040 max	0.15-0.30	bal Fe	630			
ASTM A829/A829M(95)	4137	Plt	0.33-0.40	0.80-1.15	0.65-0.95	0.15-0.25		0.035 max	0.040 max	0.15-0.40	bal Fe	480-965			

UNS numbers and US grades are provided as a means of cross referencing chemically similar alloys. Exchangability is only possible after independent examination of specifications. Tensile properties are minimum or typical as specified. UTS and YS as MPa. El as %. See Appendix for list of abbreviations used in Notes. * indicates obsolete material.

Specification	Designation	Notes	C	Cr	Mn	Mo	Ni	P	S	Si	Other	UTS	YS	El	Hard

Alloy Steel, Chromium Molybdenum, 4137/4137H (Continued from previous page)

USA

Specification	Designation	Notes	C	Cr	Mn	Mo	Ni	P	S	Si	Other	UTS	YS	El	Hard
SAE 770(84)	4137*	Obs; see J1397(92)	0.35-0.4	0.8-1.1	0.7-0.9	0.15-0.25		0.04 max	0.04 max	0.15-0.3	bal Fe				
SAE J1268(95)	4137H	Bar Wir Sh Strp; See std	0.34-0.41	0.75-1.20	0.60-1.00	0.15-0.25	0.25 max	0.030 max	0.040 max	0.15-0.35	Cu <=0.35; Cu, Ni not spec'd but acpt; bal Fe				
SAE J404(94)	4137	Bil Blm Slab Bar HR CF	0.35-0.40	0.80-1.10	0.70-0.90	0.15-0.25		0.030 max	0.040 max	0.15-0.35	bal Fe				
SAE J404(94)	4137	Plt	0.33-0.40	0.80-1.15	0.65-0.95	0.15-0.25		0.035 max	0.040 max	0.15-0.35	bal Fe				

Alloy Steel, Chromium Molybdenum, 4140/4140H

Argentina

Specification	Designation	Notes	C	Cr	Mn	Mo	Ni	P	S	Si	Other	UTS	YS	El	Hard
IAS	IRAM 4140		0.38-0.43	0.80-1.10	0.75-1.00	0.15-0.25		0.035 max	0.040 max	0.20-0.35	bal Fe	1030	680	15	311 HB

Australia

Specification	Designation	Notes	C	Cr	Mn	Mo	Ni	P	S	Si	Other	UTS	YS	El	Hard
AS 1444(96)	4140	H series, Hard/Tmp	0.38-0.43	0.80-1.10	0.75-1.00	0.15-0.25		0.040 max	0.040 max	0.10-0.35	bal Fe				
AS 1444(96)	4140H	H series, Hard/Tmp	0.37-0.44	0.75-1.20	0.65-1.10	0.15-0.25		0.040 max	0.040 max	0.10-0.35	bal Fe				
AS G18	En19*	Obs; see AS 1444	0.35-0.45	0.9-1.5	0.5-0.8	0.2-0.4		0.05	0.05	0.1-0.35	bal Fe				
AS G18	En19A*	Obs; see AS 1444	0.35-0.45	0.9-1.2	0.5-0.8	0.2-0.35		0.05	0.05	0.1-0.35	bal Fe				

Austria

Specification	Designation	Notes	C	Cr	Mn	Mo	Ni	P	S	Si	Other	UTS	YS	El	Hard
ONORM M3167	42CrMo4SP		0.38-0.45	0.9-1.2	0.5-0.8	0.15-0.25		0.035	0.035	0.15-0.35	bal Fe				
ONORM M3170	StL5		0.38-0.45	0.9-1.2	0.5-0.8	0.15-0.30		0.025	0.025	0.15-0.40	bal Fe				

Belgium

Specification	Designation	Notes	C	Cr	Mn	Mo	Ni	P	S	Si	Other	UTS	YS	El	Hard
NBN 253-02	41CrMo4		0.32-0.44	0.9-1.2	0.5-0.8	0.15-0.3		0.025 max	0.035 max	0.15-0.4	bal Fe				
NBN 253-02	42CrMo4		0.38-0.45	0.9-1.2	0.5-0.8	0.15-0.3		0.035	0.035	0.15-0.4	bal Fe				

Bulgaria

Specification	Designation	Notes	C	Cr	Mn	Mo	Ni	P	S	Si	Other	UTS	YS	El	Hard
BDS 6354	40ChMA		0.38-0.45	0.90-1.20	0.50-0.80	0.15-0.25		0.025	0.025	0.17-0.37	Cu 0.3; bal Fe				

China

Specification	Designation	Notes	C	Cr	Mn	Mo	Ni	P	S	Si	Other	UTS	YS	El	Hard
GB 3077(88)	42CrMnMo	Bar HR Q/T 25mm diam	0.37-0.45	0.90-1.20	0.90-1.20	0.20-0.30		0.035 max	0.035 max	0.17-0.37	Cu <=0.30; bal Fe	980	785	10	
GB 3077(88)	42CrMo	Bar HR Q/T 25mm diam	0.38-0.45	0.90-1.20	0.50-0.80	0.15-0.25		0.035 max	0.035 max	0.17-0.37	Cu <=0.30; bal Fe	1080	930	12	
GB 8162(87)	42CrMo	Smls tube HR/CD Q/T	0.38-0.45	0.90-1.20	0.50-0.80	0.15-0.25		0.035 max	0.035 max	0.17-0.37	bal Fe	1080	930	12	
GB/T 3078(94)	42CrMnMo	Bar CD Q/T	0.37-0.45	0.90-1.20	0.90-1.20	0.20-0.30		0.035 max	0.035 max	0.17-0.37	Cu <=0.30; bal Fe	980	785	10	
GB/T 3078(94)	42CrMo	Bar CD Q/T 25mm diam	0.38-0.45	0.90-1.20	0.50-0.80	0.15-0.25		0.035 max	0.035 max	0.17-0.37	Cu <=0.30; bal Fe	1080	930	12	

Europe

Specification	Designation	Notes	C	Cr	Mn	Mo	Ni	P	S	Si	Other	UTS	YS	El	Hard
EN 10083/1(91)A1(96)	1.7225	Q/T t<=16mm	0.38-0.45	0.90-1.20	0.60-0.90	0.15-0.30		0.035 max	0.035 max	0.40 max	bal Fe	1100-1300	900	10	
EN 10083/1(91)A1(96)	42CrMo4	Q/T t<=16mm	0.38-0.45	0.90-1.20	0.60-0.90	0.15-0.30		0.035 max	0.035 max	0.40 max	bal Fe	1100-1300	900	10	

Finland

Specification	Designation	Notes	C	Cr	Mn	Mo	Ni	P	S	Si	Other	UTS	YS	El	Hard
SFS 459(73)	SFS459	Q/T	0.3-0.37	0.9-1.2	0.5-0.8	0.15-0.25	0.3 max	0.035 max	0.035 max	0.15-0.4	bal Fe				
SFS 460	Co.42Cr1.1Mo0.20	Bar Frg Q/A Tmp, 160mm diam	0.38-0.45	0.9-1.2	0.6-0.9	0.15-0.25		0.035	0.035	0.15-0.4	bal Fe				
SFS 460(73)	SFS460	Q/T	0.38-0.45	0.9-1.2	0.6-0.9	0.15-0.25	0.3 max	0.035 max	0.035 max	0.15-0.4	bal Fe				

France

Specification	Designation	Notes	C	Cr	Mn	Mo	Ni	P	S	Si	Other	UTS	YS	El	Hard
AFNOR NFA35552	42CD4		0.39-0.45	0.85-1.2	0.6-0.9	0.15-0.25		0.035	0.035	0.1-0.4	bal Fe				
AFNOR NFA35552	42CDTS		0.39-0.45	0.85-1.2	0.6-0.9	0.15-0.25		0.035	0.035	0.1-0.4	bal Fe				
AFNOR NFA35553	42CD4	Strp	0.38-0.45	0.85-1.15	0.6-0.9	0.15-0.3		0.035	0.035	0.1-0.4	bal Fe				

UNS numbers and US grades are provided as a means of cross referencing chemically similar alloys. Exchangability is only possible after independent examination of specifications. Tensile properties are minimum or typical as specified. UTS and YS as MPa. El as %. See Appendix for list of abbreviations used in Notes. * indicates obsolete material.

Specification	Designation	Notes	C	Cr	Mn	Mo	Ni	P	S	Si	Other	UTS	YS	El	Hard

Alloy Steel, Chromium Molybdenum, 4140/4140H (Continued from previous page)

Specification	Designation	Notes	C	Cr	Mn	Mo	Ni	P	S	Si	Other	UTS	YS	El	Hard
France															
AFNOR NFA35553	42CDTS		0.38-0.45	0.85-1.2	0.6-0.9	0.15-0.25		0.035	0.035	0.15-0.4	bal Fe				
AFNOR NFA35556	42CDTS		0.39-0.45	0.9-1.2	0.6-0.9	0.15-0.25		0.035	0.035	0.1-0.4	bal Fe				
AFNOR NFA35556(84)	42CD4		0.39-0.45	0.9-1.2	0.6-0.9	0.15-0.25		0.035 max	0.035 max	0.1-0.4	bal Fe				
AFNOR NFA35557	42CD4		0.39-0.45	0.85-1.2	0.6-0.9	0.15-0.25		0.035	0.035	0.1-0.4	bal Fe				
AFNOR NFA35557	42CDTS		0.39-0.45	0.85-1.2	0.6-0.9	0.15-0.25		0.035	0.035	0.1-0.4	bal Fe				
AFNOR NFA35564(83)	42CD4FF	Soft ann	0.39-0.45	0.9-1.2	0.6-0.9	0.15-0.25	0.4 max	0.03 max	0.03 max	0.35 max	Al >=0.02; Ti <=0.05; bal Fe	0-620			229 HB max
Germany															
DIN 1652(90)	42CrMo4	Q/T 17-40mm	0.38-0.45	0.90-1.20	0.60-0.90	0.15-0.30		0.035 max	0.035 max	0.40 max	bal Fe	1000-1200	750	11	
DIN 1652(90)	42CrMoS4	Q/T	0.38-0.45	0.90-1.20	0.60-0.90	0.15-0.30		0.035 max	0.020-0.040	0.40 max	bal Fe				
DIN 1652(90)	WNr 1.7225	Q/T 17-40mm	0.38-0.45	0.90-1.20	0.60-0.90	0.15-0.30		0.035 max	0.035 max	0.40 max	bal Fe	1000-1200	750	11	
DIN 1652(90)	WNr 1.7225	Q/T 17-40mm	0.38-0.45	0.90-1.20	0.60-0.90	0.15-0.30		0.035 max	0.035 max	0.40 max	bal Fe	1000-1200	750	11	
DIN 1652(90)	WNr 1.7227	Q/T	0.38-0.45	0.90-1.20	0.60-0.90	0.15-0.30		0.035 max	0.020-0.040	0.40 max	bal Fe				
DIN EN 10083(91)	42CrMo4	Q/T 17-40mm	0.38-0.45	0.90-1.20	0.60-0.90	0.15-0.30		0.035 max	0.035 max	0.40 max	bal Fe	1000	1200	750	11
Hungary															
MSZ 61(85)	CMo4	Q/T; 0-16mm	0.38-0.45	0.9-1.2	0.6-0.9	0.15-0.3		0.035 max	0.035 max	0.4 max	bal Fe	1100-1300	900	10L	
MSZ 61(85)	CMo4	Q/T; 160.1-250mm	0.38-0.45	0.9-1.2	0.6-0.9	0.15-0.3		0.035 max	0.035 max	0.4 max	bal Fe	750-900	510	14L	
MSZ 61(85)	CMo4E	Q/T; 0-16mm	0.38-0.45	0.9-1.2	0.6-0.9	0.15-0.3		0.035 max	0.02-0.035	0.4 max	bal Fe	1100-1300	900	10L	
MSZ 61(85)	CMo4E	Q/T; 160.1-250mm	0.38-0.45	0.9-1.2	0.6-0.9	0.15-0.3		0.035 max	0.02-0.035	0.4 max	bal Fe	750-900	510	14L	
MSZ 6251(87)	CMo4Z	CF; Q/T; 0-36mm; HR; HR/soft ann; drawn/soft ann; drawn/bright ann; soft ann/ground	0.38-0.45	0.9-1.2	0.6-0.9	0.15-0.3		0.035 max	0.035 max	0.4 max	bal Fe	0-640			
MSZ 6251(87)	CMo4Z	CF, Q/T; 0-36mm; HR/soft ann; drawn/soft ann; drawn/bright ann; soft ann/ground	0.38-0.45	0.9-1.2	0.6-0.9	0.15-0.3		0.035 max	0.035 max	0.4 max	bal Fe	0-670			
India															
IS 1570	40Cr1Mo28		0.35-0.45	0.9-1.2	0.5-0.8	0.2-0.35				0.1-0.35	bal Fe				
IS 4367	40Cr1Mo28	Frg, Hard, Tmp, 100mm diam	0.35-0.45	0.9-1.2	0.5-0.8	0.2-0.35		0.05	0.05	0.1-0.35	bal Fe				
IS 5517	40Cr1Mo28	Bar, Norm, Ann, Tmp, Hard Tmp	0.35-0.45	0.9-1.2	0.5-0.8	0.2-0.35				0.1-0.35	bal Fe				
International															
ISO 683-12-72)	9*	F/IH	0.38-0.44	0.9-1.2	0.5-0.8	0.15-0.3	0.4 max	0.035 max	0.035 max	0.15-0.4	Al <=0.1; Co <=0.1; Cu <=0.3; Pb <=0.15; Ti <=0.05; V <=0.1; W <=0.1; bal Fe				
ISO 683-2(68)	Type 3*	Q/T	0.38-0.43	0.8-1.1	0.75-1	0.15-0.25	0.4 max	0.035 max	0.035 max	0.15-0.35	Al <=0.1; Co <=0.1; Cu <=0.3; Pb <=0.15; Ti <=0.05; V <=0.1; W <=0.1; bal Fe				
ISO 683-4	Type 3a*		0.38-0.45	0.9-1.2	0.5-1	0.15-0.3		0.035	0.02-0.035	0.15-0.4	bal Fe				
ISO 683-4	Type 3b*		0.38-0.45	0.9-1.2	0.5-1	0.15-0.3		0.035	0.03-0.05	0.15-0.4	bal Fe				
ISO R683-4	3A	Bar Bil, Rod, Q/A, Tmp, 16mm	0.38-0.45	0.9-1.2	0.5-1	0.15-0.3		0.04	0.02-0.04	0.15-0.4	bal Fe				
ISO R683-4	3b	Bar Bil, Rod	0.38-0.45	0.9-1.2	0.5-1	0.15-0.3		0.04	0.03-0.05	0.15-0.4	bal Fe				

UNS numbers and US grades are provided as a means of cross referencing chemically similar alloys. Exchangability is only possible after independent examination of specifications. Tensile properties are minimum or typical as specified. UTS and YS as MPa. El as %. See Appendix for list of abbreviations used in Notes. * indicates obsolete material.

Specification	Designation	Notes	C	Cr	Mn	Mo	Ni	P	S	Si	Other	UTS	YS	El	Hard

Alloy Steel, Chromium Molybdenum, 4140/4140H (Continued from previous page)

Italy

Specification	Designation	Notes	C	Cr	Mn	Mo	Ni	P	S	Si	Other	UTS	YS	El	Hard
UNI 5332(64)	40CrMo4	Q/T	0.37-0.44	0.9-1.2	0.7-1	0.15-0.25		0.035 max	0.035 max	0.4 max	bal Fe				
UNI 7847(79)	41CrMo4	SH	0.38-0.44	0.9-1.2	0.5-0.8	0.15-0.25		0.03 max	0.03 max	0.15-0.4	bal Fe				
UNI 8551(84)	41CrMo4	SH	0.38-0.44	0.9-1.2	0.75-0.95	0.15-0.25		0.03 max	0.025 max	0.15-0.4	Pb 0.15-0.25; bal Fe				

Japan

Specification	Designation	Notes	C	Cr	Mn	Mo	Ni	P	S	Si	Other	UTS	YS	El	Hard
JIS G3311(88)	SCM440M	CR Strp; Chain	0.38-0.43	0.90-1.20	0.60-0.85	0.15-0.30	0.25 max	0.030 max	0.030 max	0.15-0.35	Cu <=0.30; bal Fe				200-280 HV
JIS G3311(88)	SCM4M*	Obs; see SCM 440M	0.38-0.43	0.9-1.2	0.6-0.85	0.15-0.3	0.25	0.03	0.03	0.15-0.35	Cu 0.3; bal Fe				
JIS G3441(88)	SCM440TK	Tube	0.38-0.43	0.90-1.20	0.60-0.85	0.15-0.30	0.25 max	0.030 max	0.030 max	0.15-0.35	Cu <=0.30; bal Fe				
JIS G4052(79)	SCM440H	Struct Hard	0.37-0.44	0.85-1.25	0.55-0.9	0.15-0.35	0.25 max	0.030 max	0.030 max	0.15-0.35	Cu <=0.30; bal Fe				
JIS G4052(79)	SCM4H*	Obs; see SCM 440H	0.37-0.44	0.85-1.25	0.55-0.9	0.15-0.35		0.03	0.03	0.15-0.35	bal Fe				
JIS G4105(79)	SCM 4*	Obs; see SCM440	0.38-0.43	0.9-1.2	0.6-0.85	0.15-0.3		0.03	0.03	0.15-0.35	bal Fe				
JIS G4105(79)	SCM440	HR Frg Bar Rod	0.38-0.43	0.90-1.20	0.60-0.85	0.15-0.30	0.25 max	0.030 max	0.030 max	0.15-0.35	Cu <=0.30; bal Fe				
JIS G4107(94)	SNB7	Class 2 diam<=120mm Bar	0.38-0.48	0.80-1.10	0.75-1.00	0.15-0.25		0.040 max	0.040 max	0.20-0.35	bal Fe	690-860	520-725	16-18	

Mexico

Specification	Designation	Notes	C	Cr	Mn	Mo	Ni	P	S	Si	Other	UTS	YS	El	Hard
DGN B-203	4140*	Obs	0.38-0.43	0.8-1.1	0.75-1	0.15-0.25		0.04	0.04	0.2-0.35	bal Fe				
DGN B-297	4140*	Obs	0.38-0.43	0.8-1.1	0.75-1	0.15-0.25		0.035	0.04	0.2-0.35	bal Fe				
NMX-B-268(68)	4140H	Hard	0.37-0.44	0.75-1.20	0.65-1.10	0.15-0.25				0.20-0.35	bal Fe				
NMX-B-300(91)	4140	Bar	0.38-0.43	0.80-1.10	0.75-1.00	0.15-0.25		0.035 max	0.040 max	0.15-0.35	bal Fe				
NMX-B-300(91)	4140H	Bar	0.37-0.44	0.75-1.20	0.65-1.10	0.15-0.25				0.15-0.35	bal Fe				

Norway

Specification	Designation	Notes	C	Cr	Mn	Mo	Ni	P	S	Si	Other	UTS	YS	El	Hard
NS	CrMolV		0.41	1.05	0.75	0.2		0.03	0.03	0.25	bal Fe				

Pan America

Specification	Designation	Notes	C	Cr	Mn	Mo	Ni	P	S	Si	Other	UTS	YS	El	Hard
COPANT 334	4140	Bar	0.38-0.43	0.8-1.1	0.75-1	0.15-0.25		0.04	0.04	0.2-0.35	bal Fe				
COPANT 514	4140	Tube, HF, 324mm diam	0.38-0.43	0.8-1.1	0.75-1	0.15-0.25		0.04	0.04	0.2-0.35	bal Fe				
COPANT 514	4140	Ann	0.38-0.43	0.8-1.1	0.75-1	0.15-0.25		0.04	0.04	0.2-0.35	bal Fe				

Poland

Specification	Designation	Notes	C	Cr	Mn	Mo	Ni	P	S	Si	Other	UTS	YS	El	Hard
PNH84030/04	40HM	Q/T	0.38-0.45	0.9-1.2	0.4-0.7	0.15-0.25	0.3 max	0.035 max	0.035 max	0.17-0.37	bal Fe				

Romania

Specification	Designation	Notes	C	Cr	Mn	Mo	Ni	P	S	Si	Other	UTS	YS	El	Hard
STAS 10677(84)	41MoCr11CS	F/IH	0.37-0.43	0.9-1.2	0.4-0.8	0.15-0.3		0.025 max	0.025 max	0.17-0.37	Pb <=0.15; Ti <=0.02; As<=0.05; bal Fe				
STAS 8185(88)	41MoCr11	Q/T	0.38-0.45	0.9-1.3	0.4-0.8	0.15-0.3		0.035 max	0.035 max	0.17-0.37	Pb <=0.15; bal Fe				
STAS 9382/4	41MoCr11q		0.38-0.45	0.9-1.3	0.4-0.8	0.15-0.3	0.3	0.035	0.035	0.17-0.37	Cu 0.3; As 0.05; bal Fe				

Russia

Specification	Designation	Notes	C	Cr	Mn	Mo	Ni	P	S	Si	Other	UTS	YS	El	Hard
GOST 4543	38ChM		0.35-0.42	0.9-1.3	0.35-0.65	0.2-0.3	0.3	0.035	0.035	0.17-0.37	Cu 0.3; bal Fe				
GOST 4543	40ChMFA		0.37-0.44	0.8-1.1	0.4-0.7	0.2-0.3	0.3 max	0.03 max	0.03 max	0.17-0.37	Cu <=0.3; V 0.1-0.18; bal Fe				
GOST 4543(71)	38ChM	Q/T	0.35-0.42	0.9-1.3	0.35-0.65	0.2-0.3	0.3 max	0.035 max	0.035 max	0.17-0.37	Al <=0.1; Cu <=0.3; Ti <=0.05; bal Fe				

Spain

Specification	Designation	Notes	C	Cr	Mn	Mo	Ni	P	S	Si	Other	UTS	YS	El	Hard
UNE 36012(75)	40CrMo4	Q/T	0.37-0.43	0.85-1.15	0.6-0.9	0.15-0.25		0.035 max	0.035 max	0.15-0.4	bal Fe				
UNE 36012(75)	40CrMo4-1		0.37-0.43	0.85-1.15	0.6-0.9	0.15-0.25		0.035	0.02-0.035	0.4-15	bal Fe				

UNS numbers and US grades are provided as a means of cross referencing chemically similar alloys. Exchangability is only possible after independent examination of specifications. Tensile properties are minimum or typical as specified. UTS and YS as MPa. El as %. See Appendix for list of abbreviations used in Notes. * indicates obsolete material.

Specification	Designation	Notes	C	Cr	Mn	Mo	Ni	P	S	Si	Other	UTS	YS	El	Hard

Alloy Steel, Chromium Molybdenum, 4140/4140H (Continued from previous page)

Spain

Specification	Designation	Notes	C	Cr	Mn	Mo	Ni	P	S	Si	Other	UTS	YS	El	Hard
UNE 36012(75)	F.1252*	see 40CrMo4	0.37-0.43	0.85-1.15	0.6-0.9	0.15-0.25		0.035 max	0.035 max	0.15-0.4	bal Fe				
UNE 36254(79)	42CrMo4	Q/T	0.38-0.45	0.8-1.2	0.5-0.8	0.15-0.3				0.3-0.5	bal Fe				
UNE 36254(79)	F.8232*	see 42CrMo4	0.38-0.45	0.8-1.2	0.5-0.8	0.15-0.3				0.3-0.5	bal Fe				

UK

Specification	Designation	Notes	C	Cr	Mn	Mo	Ni	P	S	Si	Other	UTS	YS	El	Hard
	Type 5		0.35-0.43	0.9-1.1	0.7-0.9	0.15-0.25		0.04	0.04	0.15-0.4	bal Fe				
BS 4670	711M40		0.36-0.44	0.9-1.5	0.6-1.1	0.25-0.4	0.4	0.04	0.04	0.1-0.35	bal Fe				
BS 970/1(83)	708A40	Blm Bil Slab Bar Rod Frg	0.38-0.43	0.90-1.20	0.75-1.00	0.15-0.25		0.035 max	0.040 max		bal Fe				
BS 970/1(83)	709A40	Blm Bil Slab Bar Rod Frg	0.38-0.43	0.90-1.20	0.75-1.00	0.25-0.35		0.035 max	0.040 max		bal Fe				
BS 970/3(91)	708M40	Bright bar; Hard, Tmp; Q/T; 6<=t<=63mm	0.36-0.44	0.90-1.20	0.70-1.00	0.15-0.25		0.035 max	0.040 max	0.10-0.04	bal Fe	850-1000	680	13	248-302 HB
BS 970/3(91)	708M40	Bright bar; Q/T; 150<t<=250mm; hard	0.36-0.44	0.90-1.20	0.70-1.00	0.15-0.25		0.035 max	0.040 max	0.10-0.40	bal Fe	700-850	495	15	201-255 HB
BS 970/3(91)	709M40	Bright bar; Q/T; 6<t<=19mm; hard CD	0.36-0.44	0.90-1.20	0.70-1.00	0.25-0.36		0.035 max	0.040 max	0.10-0.40	bal Fe	1075-1225	955	8	311-375 HB
BS 970/3(91)	709M40	Bright bar; Q/T; 100<t<=250mm; hard	0.36-0.44	0.90-1.20	0.70-1.00	0.25-0.36		0.035 max	0.040 max	0.10-0.40	bal Fe	700-850	495	15	201-255 HB

USA

Specification	Designation	Notes	C	Cr	Mn	Mo	Ni	P	S	Si	Other	UTS	YS	El	Hard
	AISI 4140	Smls mech tub	0.38-0.43	0.80-1.10	0.75-1.00	0.15-0.25		0.040 max	0.040 max	0.15-0.35	bal Fe				
	AISI 4140	Bar Blm Bil Slab	0.38-0.43	0.80-1.10	0.75-1.00	0.15-0.25		0.035 max	0.040 max	0.15-0.35	bal Fe				
	AISI 4140	Sh Strp	0.38-0.43	0.80-1.10	0.75-1.00	0.15-0.25		0.035	0.035	0.15-0.30	bal Fe				
	AISI 4140	Plt	0.36-0.44	0.80-1.15	0.70-1.00	0.15-0.25		0.035 max	0.040 max	0.15-0.40	bal Fe				
	AISI 4140H	Smls mech tub, Hard	0.37-0.44	0.75-1.20	0.65-1.10	0.15-0.25		0.040 max	0.040 max	0.15-0.30	bal Fe				
	AISI 4140H	Bar Blm Bil Slab	0.37-0.44	0.75-1.20	0.65-1.10	0.15-0.25		0.035 max	0.040 max	0.15-0.35	bal Fe				
	UNS G41400		0.38-0.43	0.80-1.10	0.75-1.00	0.15-0.25		0.035 max	0.040 max	0.15-0.35	bal Fe				
	UNS H41400		0.37-0.44	0.75-1.20	0.65-1.10	0.15-0.25		0.035 max	0.040 max	0.15-0.35	bal Fe				
	UNS K11546	4140 Modified	0.40-0.53	0.80-1.10	0.75-1.00	0.15-0.25	0.25 max	0.040 max	0.040 max	0.20-0.35	Cu <=0.35; Te 0.035-0.060; bal Fe				
AMS 6349	4140	Bar	0.38-0.43	0.8-1.1	0.75-1	0.15-0.25	0.25	0.025	0.025	0.15-0.35	Cu <=0.35; bal Fe				
AMS 6381	4140	Tube	0.38-0.43	0.8-1.1	0.75-1	0.15-0.25	0.25	0.025	0.025	0.15-0.35	Cu <=0.35; bal Fe				
AMS 6382	4140	Tube	0.38-0.43	0.8-1.1	0.75-1	0.15-0.25	0.25	0.025	0.025	0.15-0.35	Cu <=0.35; bal Fe				
AMS 6390	4140	Tube	0.38-0.43	0.8-1.1	0.75-1	0.15-0.25	0.25	0.025	0.025	0.15-0.35	Cu <=0.35; bal Fe				
AMS 6395	4140										bal Fe				
AMS 6396	4140	Sh, Strp, Plt	0.38-0.43	0.8-1.1	0.75-1	0.15-0.25	0.25	0.025	0.025	0.15-0.35	Cu <=0.35; bal Fe				
AMS 6529	4140										bal Fe				
ASTM A193/A193M(98)	B7	Bolt, high-temp	0.37-0.49	0.75-1.20	0.65-1.10	0.15-0.25		0.035 max	0.040 max	0.15-0.35	bal Fe	860	720	16	321 HB
ASTM A193/A193M(98)	B7M	Bolt, high-temp	0.37-0.49	0.75-1.20	0.65-1.10	0.15-0.25		0.035 max	0.040 max	0.15-0.35	bal Fe	690	550	18	235 HB
ASTM A194/A194M(98)	7	Nuts, high-temp press	0.37-0.49	0.75-1.20	0.65-1.10	0.15-0.25		0.04 max	0.04 max	0.15-0.35	bal Fe				248-352 HB
ASTM A194/A194M(98)	7M	Nuts, high-temp press	0.37-0.49	0.75-1.20	0.65-1.10	0.15-0.25		0.04 max	0.04 max	0.15-0.35	bal Fe				159-237 HB
ASTM A29/A29M(93)	4140	Bar	0.38-0.43	0.80-1.10	0.75-1.00	0.15-0.25		0.035 max	0.040 max	0.15-0.35	bal Fe				
ASTM A304(96)	4140H	Bar, hard bands spec	0.37-0.44	0.75-1.20	0.65-1.10	0.15-0.25		0.035 max	0.040 max	0.15-0.35	Cu <=0.35; Cu Ni Cr Mo trace allowed; bal Fe				

UNS numbers and US grades are provided as a means of cross referencing chemically similar alloys. Exchangability is only possible after independent examination of specifications. Tensile properties are minimum or typical as specified. UTS and YS as MPa. El as %. See Appendix for list of abbreviations used in Notes. * indicates obsolete material.

Specification	Designation	Notes	C	Cr	Mn	Mo	Ni	P	S	Si	Other	UTS	YS	El	Hard

Alloy Steel, Chromium Molybdenum, 4140/4140H (Continued from previous page)

USA

Specification	Designation	Notes	C	Cr	Mn	Mo	Ni	P	S	Si	Other	UTS	YS	El	Hard
ASTM A322(96)	4140	Bar	0.38-0.43	0.80-1.10	0.75-1.00	0.15-0.25		0.035 max	0.040 max	0.15-0.35	bal Fe				
ASTM A331(95)	4140	Bar	0.38-0.43	0.80-1.10	0.75-1.00	0.15-0.25		0.035 max	0.040 max	0.15-0.35	bal Fe				
ASTM A331(95)	4140H	Bar	0.37-0.44	0.75-1.20	0.65-1.10	0.15-0.25		0.025 max	0.025 max	0.15-0.35	Cu <=0.35; bal Fe				
ASTM A506(93)	4140	Sh Strp, HR, CR	0.38-0.43	0.80-1.10	0.75-1.00	0.15-0.25		0.035	0.035	0.15-0.30	bal Fe				
ASTM A513(97a)	4140	ERW Mech Tub	0.38-0.43	0.80-1.10	0.75-1.00	0.15-0.25		0.035 max	0.040 max	0.15-0.35	bal Fe				
ASTM A519(96)	4140	Smls mech tub, Ann	0.38-0.43	0.80-1.10	0.75-1.00	0.15-0.25		0.040 max	0.040 max	0.15-0.35	bal Fe	552	414	25	85 HRB
ASTM A519(96)	4140	Smls mech tub, Norm	0.38-0.43	0.80-1.10	0.75-1.00	0.15-0.25		0.040 max	0.040 max	0.15-0.35	bal Fe	855	621	20	100 HRB
ASTM A519(96)	4140	Smls mech tub, HR	0.38-0.43	0.80-1.10	0.75-1.00	0.15-0.25		0.040 max	0.040 max	0.15-0.35	bal Fe	855	621	15	100 HRB
ASTM A519(96)	4140	Smls mech tub, SR	0.38-0.43	0.80-1.10	0.75-1.00	0.15-0.25		0.040 max	0.040 max	0.15-0.35	bal Fe	855	689	10	100 HRB
ASTM A646(95)	Grade 12	Blm Bil for aircraft aerospace frg, ann	0.38-0.43	0.80-1.10	0.75-1.00	0.15-0.25		0.025 max	0.025 max	0.20-0.35	bal Fe				235 HB
ASTM A752(93)	4140	Rod Wir	0.38-0.43	0.80-1.10	0.75-1.00	0.15-0.25		0.035 max	0.040 max	0.15-0.30	bal Fe	650			
ASTM A829/A829M(95)	4140	Plt	0.36-0.44	0.80-1.15	0.70-1.00	0.15-0.25		0.035 max	0.040 max	0.15-0.40	bal Fe	480-965			
ASTM A866(94)	4140	Anti-friction bearings	0.38-0.43	0.80-1.10	0.75-1.00	0.15-0.25		0.025 max	0.025 max	0.15-0.35	bal Fe				
ASTM A914/914M(92)	4140RH	Bar restricted end Q hard	0.38-0.43	0.80-1.10	0.75-1.00	0.15-0.25	0.25 max	0.035 max	0.040 max	0.15-0.35	Cu <=0.35; Electric furnace P, S<=0.025; bal Fe				
ASTM A983/983M(98)	Grade 4	Continuous grain flow frg for crankshafts	0.38-0.48	0.80-1.20	0.75-1.10	0.15-0.25		0.025 max	0.025 max	0.15-0.40	V <=0.10; bal Fe				
ASTM A983/983M(98)	Grade 5	Continuous grain flow frg for crankshafts	0.30-0.45	0.80-1.20	0.65-1.00	0.15-0.25		0.025 max	0.025 max	0.15-0.40	V <=0.10; bal Fe				
DoD-F-24669/1(86)	4140	Bar Bil; Supersedes MIL-S-866 & MIL-S-16974	0.38-0.43	0.8-1.1	0.75-1	0.15-0.25		0.035	0.04	0.15-0.35	bal Fe				
FED QQ-S-626C(91)	4140*	Obs; see ASTM A829; Plt	0.38-0.4	0.8-1.1	0.75-1	0.15-0.25		0.035	0.04	0.15-0.35	bal Fe				
MIL-S-19434B(SH)(90)	Class 3	Gear, pinion frg HT; shipboard propulsion and turbine	0.35-0.45	0.75-1.15	0.70-1.05	0.13-0.27		0.040 max	0.040 max	0.15-0.35	V <=0.10; Si 0.10 max if VCD used; bal Fe	758	552		229-277 HB
MIL-S-46059	G41400*	Obs	0.38-0.43	0.8-1.1	0.75-1	0.15-0.25		0.035	0.04	0.15-0.3	bal Fe				
MIL-S-5626C(72)	4140*	Obs; see ASM 6382 6349; Bar rod frg for aircraft apps	0.38-0.43	0.8-1.1	0.75-1	0.15-0.25	0.25	0.025	0.025	0.2-0.35	Cu 0.35; bal Fe				
SAE 770(84)	4140*	Obs; see J1397(92)	0.38-0.43	0.8-1.1	0.75-1	0.15-0.25		0.035 max	0.04 max	0.15-0.35	bal Fe				
SAE J1268(95)	4140H	Bar Sh Strp Tub; See std	0.37-0.44	0.75-1.20	0.65-1.10	0.15-0.25	0.25 max	0.030 max	0.040 max	0.15-0.35	Cu <=0.35; Cu, Ni not spec'd but acpt; bal Fe				
SAE J1868(93)	4140RH	Restrict hard range	0.38-0.43	0.80-1.10	0.75-1.00	0.15-0.25	0.25 max	0.025 max	0.040 max	0.15-0.35	Cu <=0.035; Cu, Ni not spec'd but acpt; bal Fe				5 HRC max
SAE J404(94)	4140	Plt	0.36-0.44	0.80-1.15	0.70-1.00	0.15-0.25		0.035 max	0.040 max	0.15-0.35	bal Fe				
SAE J404(94)	4140	Bil Blm Slab Bar HR CF	0.38-0.43	0.80-1.10	0.75-1.00	0.15-0.25		0.030 max	0.040 max	0.15-0.35	bal Fe				
SAE J775(93)	4140H	Engine poppet valve, Mart	0.37-0.44	0.75-1.20	0.65-1.10	0.15-0.25		0.035 max	0.040 max	0.15-0.35	bal Fe	900	800		
SAE J775(93)	NV7*	Engine poppet valve, Mart; former SAE grade	0.37-0.44	0.75-1.20	0.65-1.10	0.15-0.25		0.035 max	0.040 max	0.15-0.35	bal Fe	900	800		

Yugoslavia

Specification	Designation	Notes	C	Cr	Mn	Mo	Ni	P	S	Si	Other	UTS	YS	El	Hard
	C.4732	Q/T	0.38-0.45	0.9-1.2	0.6-0.9	0.15-0.3		0.035 max	0.03 max	0.4 max	bal Fe				

Alloy Steel, Chromium Molybdenum, 4142/4142H

Specification	Designation	Notes	C	Cr	Mn	Mo	Ni	P	S	Si	Other	UTS	YS	El	Hard
Austria															
ONORM M3167	42CrMo4SP		0.38-0.45	0.9-1.2	0.5-0.8	0.15-0.25		0.035	0.035	0.15-0.35	bal Fe				
ONORM M3170	StL5		0.38-0.45	0.9-1.2	0.5-0.8	0.15-0.30		0.025	0.025	0.15-0.40	bal Fe				
Belgium															
NBN 253-02	42CrMo4		0.38-0.45	0.9-1.2	0.5-0.8	0.15-0.3		0.035	0.035	0.15-0.4	bal Fe				
NBN 253-06	41CrMo4		0.38-0.44	0.9-1.2	0.5-0.8	0.15-0.3		0.025	0.035	0.15-0.4	bal Fe				
Bulgaria															
BDS 6354	40ChMA		0.38-0.45	0.90-1.20	0.50-0.80	0.15-0.25		0.025	0.025	0.17-0.37	Cu 0.3; bal Fe				
China															
GB 3077(88)	40CrMnMo	Bar HR Q/T 25mm diam	0.37-0.45	0.90-1.20	0.90-1.20	0.20-0.30		0.035 max	0.035 max	0.17-0.37	Cu <=0.30; bal Fe	980	785	10	
GB 3077(88)	42CrMo	Smls tube HR/CD	0.38-0.45	0.90-1.20	0.50-0.80	0.15-0.25		0.035 max	0.035 max	0.17-0.37	bal Fe	1080	930	12	
GB 3077(88)	42CrMo	Bar HR Q/T 25mm diam	0.38-0.45	0.90-1.20	0.50-0.80	0.15-0.25		0.035 max	0.035 max	0.17-0.37	Cu <=0.30; bal Fe	1080	930	12	
GB 3077(88)	42CrMo	Bar CD Q/T 25mm diam	0.38-0.45	0.90-1.20	0.50-0.80	0.15-0.25		0.035 max	0.035 max	0.17-0.37	Cu <=0.30; bal Fe	1080	930	12	
GB/T 3078(94)	40CrMnMo	Bar CD Q/T 25mm diam	0.37-0.45	0.90-1.20	0.90-1.20	0.20-0.30		0.035 max	0.035 max	0.17-0.37	Cu <=0.30; bal Fe	980	785	10	
Czech Republic															
CSN 415142	15142	Q/T	0.38-0.45	0.9-1.2	0.5-0.8	0.15-0.3		0.035 max	0.035 max	0.17-0.37	bal Fe				
Finland															
SFS 460	Co.42Cr1.1Mo0.20		0.38-0.45	0.9-1.2	0.6-0.9	0.15-0.25		0.035	0.035	0.15-0.4	bal Fe				
France															
AFNOR	40CD4		0.39-0.46	0.95-1.3	0.5-0.8	0.15-0.3		0.03	0.025	0.2-0.5	bal Fe				
AFNOR NFA35552	42CD4		0.39-0.45	0.85-1.2	0.6-0.9	0.15-0.25		0.035	0.035	0.1-0.4	bal Fe				
AFNOR NFA35552	42CDTS		0.39-0.45	0.85-1.2	0.6-0.9	0.15-0.25		0.035	0.035	0.1-0.4	bal Fe				
AFNOR NFA35553	42CD4		0.38-0.45	0.85-1.2	0.6-0.9	0.15-0.25		0.035	0.035	0.1-0.4	bal Fe				
AFNOR NFA35553	42CDTS		0.38-0.45	0.85-1.2	0.6-0.9	0.15-0.25		0.035	0.035	0.1-0.4	bal Fe				
AFNOR NFA35556	42CDTS		0.39-0.45	0.9-1.2	0.6-0.9	0.15-0.25		0.035	0.035	0.1-0.4	bal Fe				
AFNOR NFA35556(84)	42CD4		0.39-0.45	0.9-1.2	0.6-0.9	0.15-0.25		0.035 max	0.035 max	0.1-0.4	bal Fe				
AFNOR NFA35557	42CD4		0.39-0.45	0.85-1.2	0.6-0.9	0.15-0.25		0.035	0.035	0.1-0.4	bal Fe				
AFNOR NFA35557	42CDTS		0.39-0.45	0.85-1.2	0.6-0.9	0.15-0.25		0.035	0.035	0.1-0.4	bal Fe				
Germany															
DIN 17212(72)	41CrMo4	HT hard 17-40mm	0.38-0.44	0.90-1.20	0.50-0.80	0.15-0.30		0.025 max	0.035 max	0.15-0.40	bal Fe	980-1080	765	11	
DIN 17212(72)	WNr 1.7223	HT hard 17-40mm	0.38-0.44	0.90-1.20	0.50-0.80	0.15-0.30		0.025 max	0.035 max	0.15-0.40	bal Fe	980-1080	765	11	
DIN 17230(80)	43CrMo4	Bearing Q/T 17-40mm	0.40-0.46	0.90-1.20	0.60-0.90	0.15-0.30		0.025 max	0.035 max	0.040 max	Cu <=0.30; bal Fe	980-1180	760	11	
DIN 17230(80)	WNr 1.3563	Bearing Q/T 17-40mm	0.40-0.46	0.90-1.20	0.60-0.90	0.15-0.30		0.025 max	0.035 max	0.40 max	Cu <=0.30; bal Fe	980-1180	760	11	
Italy															
UNI 5333(64)	38CrMo4	F/IH	0.34-0.4	0.8-1.1	0.6-0.9	0.15-0.25		0.035 max	0.03 max	0.4 max	bal Fe				
UNI 6403(86)	42CrMo4	Q/T	0.38-0.45	0.9-1.2	0.6-0.9	0.15-0.3		0.035 max	0.035 max	0.15-0.35	bal Fe				
UNI 7768	38CrMo4		0.38-0.44	0.9-1.2	0.5-0.8	0.15-0.25		0.03 max	0.03 max	0.15-0.4	bal Fe				
UNI 7845(78)	42CrMo4	Q/T	0.38-0.45	0.9-1.2	0.6-0.9	0.15-0.25		0.035 max	0.035 max	0.15-0.4	Pb 0.15-0.3; bal Fe				

UNS numbers and US grades are provided as a means of cross referencing chemically similar alloys. Exchangability is only possible after independent examination of specifications. Tensile properties are minimum or typical as specified. UTS and YS as MPa. El as %. See Appendix for list of abbreviations used in Notes. * indicates obsolete material.

Alloy Steel, Chromium Molybdenum, 4142/4142H (Continued from previous page)

Specification	Designation	Notes	C	Cr	Mn	Mo	Ni	P	S	Si	Other	UTS	YS	El	Hard
Italy															
UNI 7847(79)	41CrMo4	SH	0.38-0.44	0.9-1.2	0.5-0.8	0.15-0.25		0.03 max	0.03 max	0.15-0.4	bal Fe				
UNI 7874(79)	42CrMo4	Q/T	0.38-0.45	0.9-1.2	0.5-0.9	0.15-0.3		0.035 max	0.035 max	0.15-0.4	Pb 0.15-0.3; bal Fe				
Japan															
JIS G4108(94)	SNB22-1	Bar diam<=38mm	0.39-0.46	0.75-1.20	0.65-1.10	0.15-0.25		0.025 max	0.025 max	0.20-0.35	bal Fe	1140	1030	10	321-401 HB
JIS G4108(94)	SNB22-2	Bar diam<=75mm	0.39-0.46	0.75-1.20	0.65-1.10	0.15-0.25		0.025 max	0.025 max	0.20-0.35	bal Fe	1070	960	11	311-401 HB
JIS G4108(94)	SNB22-3	Bar 50<diam<=100mm	0.39-0.46	0.75-1.20	0.65-1.10	0.15-0.25		0.025 max	0.025 max	0.20-0.35	bal Fe	1000	890	12	302-375 HB
JIS G4108(94)	SNB22-4	Bar 25<diam<=100mm	0.39-0.46	0.75-1.20	0.65-1.10	0.15-0.25		0.025 max	0.025 max	0.20-0.35	bal Fe	930	825	13	277-363 HB
JIS G4108(94)	SNB22-5	Bar 50<diam<=100mm	0.39-0.46	0.75-1.20	0.65-1.10	0.15-0.25		0.025 max	0.025 max	0.20-0.35	bal Fe	790	685	15	255-302 HB
Mexico															
DGN B-297	4142*	Obs	0.4-0.45	0.8-1.1	0.75-1	0.15-0.25		0.035	0.04	0.2-0.35	bal Fe				
NMX-B-268(68)	4142H	Hard	0.39-0.46	0.75-1.20	0.65-1.10	0.15-0.25				0.20-0.35	bal Fe				
NMX-B-300(91)	4142	Bar	0.40-0.45	0.80-1.10	0.75-1.00	0.15-0.25		0.035 max	0.040 max	0.15-0.35	bal Fe				
NMX-B-300(91)	4142H	Bar	0.39-0.46	0.75-1.20	0.65-1.10	0.15-0.25				0.15-0.35	bal Fe				
Pan America															
COPANT 514	4142	Tube	0.4-0.45	0.8-1.1	0.75-1	0.15-0.25		0.04	0.04	0.2-0.35	bal Fe				
COPANT 514	4142		0.4-0.45	0.8-1.15	0.75-1	0.15-0.25		0.04	0.04	0.2-0.35	bal Fe				
Poland															
PNH84030	40HM	Q/T	0.38-0.45	0.8-1.1	0.4-0.7	0.15-0.25	0.3	0.035	0.035	0.17-0.37	Cu 0.3; bal Fe				
Romania															
STAS 791	41MoCr11		0.38-0.45	0.9-1.3	0.4-0.8	0.15-0.3	0.3	0.035	0.035	0.17-0.37	Cu 0.3; bal Fe				
STAS 791(88)	42MoCr11	Struct/const; Q/T	0.38-0.45	0.9-1.2	0.6-0.9	0.15-0.3		0.035 max	0.035 max	0.17-0.37	Pb <=0.15; Ti <=0.02; bal Fe				
STAS 791(88)	42MoCr11S	Struct/const; Q/T	0.38-0.45	0.9-1.2	0.6-0.9	0.15-0.3		0.035 max	0.02-0.04	0.17-0.37	Pb <=0.15; Ti <=0.02; bal Fe				
STAS 791(88)	42MoCr11X	Struct/const; Q/T	0.38-0.45	0.9-1.2	0.6-0.9	0.15-0.3		0.025 max	0.025 max	0.17-0.37	Pb <=0.15; Ti <=0.02; bal Fe				
STAS 791(88)	42MoCr11XS	Struct/const; Q/T	0.38-0.45	0.9-1.2	0.6-0.9	0.15-0.3		0.025 max	0.02-0.035	0.17-0.37	Pb <=0.15; Ti <=0.02; bal Fe				
STAS 9382/4(89)	42MoCr11q	Q/T	0.38-0.45	0.9-1.2	0.4-0.8	0.15-0.3		0.035 max	0.035 max	0.17-0.37	Pb <=0.15; As<=0.05; bal Fe				
Russia															
GOST 4543(71)	40ChFA	Surface hardening	0.37-0.44	0.8-1.1	0.5-0.8	0.15 max	0.3 max	0.025 max	0.025 max	0.17-0.37	Al <=0.1; Cu <=0.3; Ti <=0.05; V 0.1-0.18; bal Fe				
Spain															
UNE 36012(75)	40CrMo4	Q/T	0.37-0.43	0.85-1.15	0.6-0.9	0.15-0.25		0.035 max	0.035 max	0.15-0.4	bal Fe				
UNE 36254(79)	42CrMo4	Q/T	0.38-0.45	0.8-1.2	0.5-0.8	0.15-0.3		0.04 max	0.04 max	0.3-0.5	bal Fe				
Sweden															
SIS 142244	2244-04	Tube, Frg As Rolled As Frg 160mm diam	0.38-0.45	0.9-1.2	0.6-0.9	0.15-0.3		0.04	0.04	0.15-0.4	bal Fe	780	590	14	
SIS 142244	2244-05	Bar Frg Q/A, Tmp, 100mm diam	0.38-0.45	0.9-1.2	0.6-0.9	0.15-0.3		0.04	0.04	0.15-0.4	bal Fe	880	690	12	
SIS 142244	2244-06	Bar Frg Tube, Q/A, Tmp	0.38-0.45	0.9-1.2	0.6-0.9	0.15-0.3		0.04	0.04	0.15-0.4	bal Fe	1080	880	10	
SS 142244	2244	Q/T	0.38-0.45	0.9-1.2	0.6-0.9	0.15-0.3		0.035 max	0.035 max	0.1-0.4	bal Fe				
UK															
BS 970/1(83)	708A42	Blm Bil Slab Bar Rod Frg	0.40-0.45	0.90-1.20	0.70-1.00	0.15-0.25		0.035 max	0.040 max		bal Fe				

UNS numbers and US grades are provided as a means of cross referencing chemically similar alloys. Exchangability is only possible after independent examination of specifications. Tensile properties are minimum or typical as specified. UTS and YS as MPa. El as %. See Appendix for list of abbreviations used in Notes. * indicates obsolete material.

Specification	Designation	Notes	C	Cr	Mn	Mo	Ni	P	S	Si	Other	UTS	YS	El	Hard

Alloy Steel, Chromium Molybdenum, 4142/4142H (Continued from previous page)

UK

Specification	Designation	Notes	C	Cr	Mn	Mo	Ni	P	S	Si	Other	UTS	YS	El	Hard
BS 970/1(83)	708H42	Blm Bil Slab Bar Rod Frg	0.39-0.46	0.80-1.25	0.65-1.05	0.15-0.25		0.035 max	0.040 max		bal Fe				
BS 970/1(83)	709A42	Blm Bil Slab Bar Rod Frg	0.40-0.45	0.90-1.20	0.75-1.00	0.25-0.35		0.035 max	0.040 max		bal Fe				

USA

Specification	Designation	Notes	C	Cr	Mn	Mo	Ni	P	S	Si	Other	UTS	YS	El	Hard
	AISI 4142	Smls mech tub	0.40-0.45	0.80-1.10	0.75-1.00	0.15-0.25		0.040 max	0.040 max	0.15-0.35	bal Fe				
	AISI 4142	Plt	0.40-0.45	0.80-1.10	0.75-1.00	0.15-0.25		0.035 max	0.040 max	0.15-0.3	bal Fe				
	AISI 4142	Bar Blm Bil Slab	0.40-0.45	0.80-1.10	0.75-1.00	0.15-0.25		0.035 max	0.040 max	0.15-0.3	bal Fe				
	AISI 4142H	Bar Blm Bil Slab	0.39-0.46	0.75-1.20	0.65-1.10	0.15-0.25		0.035 max	0.040 max	0.15-0.35	bal Fe				
	AISI 4142H	Smls mech tub, Hard	0.39-0.46	0.75-1.20	0.65-1.10	0.15-0.25		0.040 max	0.040 max	0.15-0.30	bal Fe				
	UNS G41420		0.4-0.45	0.80-1.10	0.75-1.00	0.15-0.25		0.035 max	0.040 max	0.15-0.35	bal Fe				
	UNS H41420		0.39-0.46	0.75-1.20	0.65-1.10	0.15-0.25		0.035 max	0.040 max	0.15-0.35	bal Fe				
ASTM A29/A29M(93)	4142	Bar	0.40-0.45	0.80-1.10	0.75-1.00	0.15-0.25		0.035 max	0.040 max	0.15-0.35	bal Fe				
ASTM A304(96)	4142H	Bar, hard bands spec	0.39-0.46	0.75-1.20	0.65-1.10	0.15-0.25		0.035 max	0.040 max	0.15-0.35	Cu <=0.35; Cu Ni Cr Mo trace allowed; bal Fe				
ASTM A322(96)	4142	Bar	0.40-0.45	0.80-1.10	0.75-1.00	0.15-0.25		0.035 max	0.040 max	0.15-0.35	bal Fe				
ASTM A331(95)	4142	Bar	0.40-0.45	0.80-1.10	0.75-1.00	0.15-0.25		0.035 max	0.040 max	0.15-0.35	bal Fe				
ASTM A331(95)	4142H	Bar	0.39-0.46	0.75-1.20	0.65-1.10	0.15-0.25		0.025 max	0.025 max	0.15-0.35	Cu <=0.35; bal Fe				
ASTM A519(96)	4142	Smls mech tub	0.40-0.45	0.80-1.10	0.75-1.00	0.15-0.25		0.040 max	0.040 max	0.15-0.35	bal Fe				
ASTM A540/A540M(98)	A540 (B22)	Bolting matl	0.39-0.46	0.75-1.20	0.65-1.10	0.15-0.25		0.025 max	0.025 max	0.15-0.35	bal Fe	827-1138	724-1034	15-10	248-321/293-401 HB
ASTM A752(93)	4142	Rod Wir	0.40-0.45	0.80-1.10	0.75-1.00	0.15-0.25		0.035 max	0.040 max	0.15-0.30	bal Fe	670			
ASTM A829/A829M(95)	4142	Plt	0.38-0.46	0.80-1.15	0.70-1.00	0.15-0.25		0.035 max	0.040 max	0.15-0.40	bal Fe	480-965			
SAE 770(84)	4142*	Obs; see J1397(92)									bal Fe				
SAE J1268(95)	4142H	See std	0.39-0.46	0.75-1.20	0.65-1.10	0.15-0.25	0.25 max	0.030 max	0.040 max	0.15-0.35	Cu <=0.35; Cu, Ni not spec'd but acpt; bal Fe				
SAE J404(94)	4142	Bil Blm Slab Bar HR CF	0.40-0.45	0.80-1.10	0.75-1.00	0.15-0.25		0.030 max	0.040 max	0.15-0.35	bal Fe				
SAE J404(94)	4142	Plt	0.38-0.46	0.80-1.15	0.70-1.00	0.15-0.25		0.035 max	0.040 max	0.15-0.35	bal Fe				

Yugoslavia

Specification	Designation	Notes	C	Cr	Mn	Mo	Ni	P	S	Si	Other	UTS	YS	El	Hard
	C.4732	Ball & roller bearing	0.38-0.45	0.9-1.2	0.6-0.9	0.15-0.3		0.035 max	0.03 max	0.4 max	bal Fe				
	C.4735	F/IH	0.38-0.44	0.9-1.2	0.5-0.8	0.15-0.3	0.25 max	0.025 max	0.035 max	0.15-0.4	bal Fe				
	C.4782	Q/T	0.38-0.45	0.9-1.2	0.6-0.9	0.15-0.3		0.035 max	0.02-0.035	0.4 max	bal Fe				

Alloy Steel, Chromium Molybdenum, 4145/4145H

France

Specification	Designation	Notes	C	Cr	Mn	Mo	Ni	P	S	Si	Other	UTS	YS	El	Hard
AFNOR NFA35552	45SCD6		0.42-0.5	0.5-0.75	0.5-0.8	0.15-0.3		0.035	0.035	1.3-1.7	bal Fe				
AFNOR NFA35553	45SCD6		0.42-0.5	0.5-0.75	0.5-0.8	0.15-0.3		0.035	0.035	1.3-1.7	bal Fe				

Japan

Specification	Designation	Notes	C	Cr	Mn	Mo	Ni	P	S	Si	Other	UTS	YS	El	Hard
JIS G4052(79)	SCM445H	Struct Hard	0.42-0.49	0.85-1.25	0.55-0.90	0.15-0.35	0.25 max	0.030 max	0.030 max	0.15-0.35	Cu <=0.30; bal Fe				
JIS G4052(79)	SCM5H*	Obs; see SCM 445H	0.42-0.49	0.85-1.25	0.55-0.9	0.15-0.35		0.03	0.03	0.15-0.35	bal Fe				

UNS numbers and US grades are provided as a means of cross referencing chemically similar alloys. Exchangability is only possible after independent examination of specifications. Tensile properties are minimum or typical as specified. UTS and YS as MPa. El as %. See Appendix for list of abbreviations used in Notes. * indicates obsolete material.

Specification	Designation	Notes	C	Cr	Mn	Mo	Ni	P	S	Si	Other	UTS	YS	El	Hard

Alloy Steel, Chromium Molybdenum, 4145/4145H (Continued from previous page)

Japan

Specification	Designation	Notes	C	Cr	Mn	Mo	Ni	P	S	Si	Other	UTS	YS	El	Hard
JIS G4105(79)	SCM 5*	Obs; see SCM445	0.43-0.48	0.9-1.2	0.6-0.85	0.15-0.3		0.03	0.03	0.15-0.35	bal Fe				
JIS G4105(79)	SCM445	HR Frg Bar Rod	0.43-0.48	0.90-1.20	0.60-0.85	0.15-0.30	0.25 max	0.030 max	0.030 max	0.15-0.35	Cu <=0.30; bal Fe				

Mexico

Specification	Designation	Notes	C	Cr	Mn	Mo	Ni	P	S	Si	Other	UTS	YS	El	Hard
DGN B-297	4145*	Obs	0.43-0.48	0.8-1.1	0.75-1	0.15-0.25		0.035	0.04	0.2-0.35	bal Fe				
NMX-B-268(68)	4145H	Hard	0.42-0.49	0.75-1.20	0.65-1.10	0.15-0.25				0.20-0.35	bal Fe				
NMX-B-300(91)	4145	Bar	0.43-0.48	0.80-1.10	0.75-1.00	0.15-0.25		0.035 max	0.040 max	0.15-0.35	bal Fe				
NMX-B-300(91)	4145H	Bar	0.42-0.49	0.75-1.20	0.65-1.10	0.15-0.25				0.15-0.35	bal Fe				

Pan America

Specification	Designation	Notes	C	Cr	Mn	Mo	Ni	P	S	Si	Other	UTS	YS	El	Hard
COPANT 514	4145	Tube	0.43-0.48	0.8-1.1	0.75-1	0.15-0.25		0.04	0.04	0.2-0.35	bal Fe				

Russia

Specification	Designation	Notes	C	Cr	Mn	Mo	Ni	P	S	Si	Other	UTS	YS	El	Hard
GOST 7832	35ChML		0.3-0.4	0.8-1.1	0.4-0.9	0.2-0.3	0.3 max	0.04 max	0.04 max	0.2-0.42	Cu <=0.3; bal Fe				

UK

Specification	Designation	Notes	C	Cr	Mn	Mo	Ni	P	S	Si	Other	UTS	YS	El	Hard
BS 970/1(83)	708H45	Blm Bil Slab Bar Rod Frg	0.42-0.49	0.80-1.25	0.65-1.05	0.15-0.25		0.035 max	0.040 max		bal Fe				

USA

Specification	Designation	Notes	C	Cr	Mn	Mo	Ni	P	S	Si	Other	UTS	YS	El	Hard
	AISI 4145	Plt	0.43-0.48	0.80-1.10	0.75-1.00	0.15-0.25		0.035 max	0.040 max	0.15-0.35	bal Fe				
	AISI 4145	Bar Blm Bil Slab	0.43-0.48	0.80-1.10	0.75-1.00	0.15-0.25		0.035 max	0.040 max	0.15-0.35	bal Fe				
	AISI 4145	Smls mech tub	0.43-0.48	0.80-1.10	0.75-1.00	0.15-0.25		0.040 max	0.040 max	0.15-0.35	bal Fe				
	AISI 4145H	Bar Blm Bil Slab	0.42-0.48	0.75-1.20	0.75-1.20	0.15-0.25		0.035 max	0.040 max	0.15-0.35	bal Fe				
	AISI 4145H	Smls mech tub, Hard	0.42-0.48	0.75-1.20	0.75-1.20	0.15-0.25		0.040 max	0.040 max	0.15-0.30	bal Fe				
	UNS G41450		0.43-0.48	0.80-1.10	0.75-1.00	0.15-0.25		0.035 max	0.040 max	0.15-0.35	bal Fe				
	UNS H41450		0.42-0.49	0.75-1.20	0.65-1.10	0.15-0.25		0.035 max	0.040 max	0.15-0.35	bal Fe				
ASTM A29/A29M(93)	4145	Bar	0.43-0.48	0.80-1.10	0.75-1.00	0.15-0.25		0.035 max	0.040 max	0.15-0.35	bal Fe				
ASTM A304(96)	4145H	Bar, hard bands spec	0.42-0.48	0.75-1.20	0.65-1.10	0.15-0.25		0.035 max	0.040 max	0.15-0.35	Cu <=0.35; Cu Ni Cr Mo trace allowed; bal Fe				
ASTM A322(96)	4145	Bar	0.43-0.48	0.80-1.10	0.75-1.00	0.15-0.25		0.035 max	0.040 max	0.15-0.35	bal Fe				
ASTM A331(95)	4145	Bar	0.43-0.48	0.80-1.10	0.75-1.00	0.15-0.25		0.035 max	0.040 max	0.15-0.35	bal Fe				
ASTM A331(95)	4145H	Bar	0.42-0.48	0.75-1.20	0.65-1.10	0.15-0.25		0.025 max	0.025 max	0.15-0.35	Cu <=0.35; bal Fe				
ASTM A519(96)	4145	Smls mech tub	0.43-0.48	0.80-1.10	0.75-1.00	0.15-0.25		0.040 max	0.040 max	0.15-0.35	bal Fe				
ASTM A752(93)	4145	Rod Wir	0.43-0.48	0.80-1.10	0.75-1.00	0.15-0.25		0.035 max	0.040 max	0.15-0.30	bal Fe	680			
ASTM A829/A829M(95)	4145	Plt	0.41-0.49	0.80-1.15	0.70-1.00	0.15-0.25		0.035 max	0.040 max	0.15-0.40	bal Fe	480-965			
ASTM A914/914M(92)	4145RH	Bar restricted end Q hard	0.43-0.48	0.80-1.10	0.75-1.00	0.15-0.25	0.25 max	0.035 max	0.040 max	0.15-0.35	Cu <=0.35; Electric furnace P, S<=0.025; bal Fe				
MIL-S-16974E(86)	4145*	Obs; Bar Bil Blm slab for refrg	0.43-0.48	0.8-1.1	0.75-1	0.15-0.25		0.04 max	0.04 max	0.15-0.3	bal Fe				
SAE 770(84)	4145*	Obs; see J1397(92)	0.43-0.48	0.8-1.1	0.75-1	0.15-0.25		0.04 max	0.04 max	0.15-0.3	bal Fe				
SAE J1268(95)	4145H	Bar Sh Strp Tub; See std	0.42-0.49	0.75-1.20	0.65-1.10	0.15-0.25	0.25 max	0.030 max	0.040 max	0.15-0.35	Cu <=0.35; Cu, Ni not spec'd but acpt; bal Fe				
SAE J1868(93)	4145RH	Restrict hard range	0.43-0.48	0.80-1.10	0.75-1.00	0.15-0.25	0.25 max	0.025 max	0.040 max	0.15-0.35	Cu <=0.035; Cu, Ni not spec'd but acpt; bal Fe				5 HRC max

Specification	Designation	Notes	C	Cr	Mn	Mo	Ni	P	S	Si	Other	UTS	YS	El	Hard

Alloy Steel, Chromium Molybdenum, 4145/4145H (Continued from previous page)

USA

Specification	Designation	Notes	C	Cr	Mn	Mo	Ni	P	S	Si	Other	UTS	YS	El	Hard
SAE J404(94)	4145	Bil Blm Slab Bar HR CF	0.43-0.48	0.80-1.10	0.75-1.00	0.15-0.25		0.030 max	0.040 max	0.15-0.35	bal Fe				
SAE J404(94)	4145	Plt	0.41-0.49	0.80-1.15	0.70-1.00	0.15-0.25		0.035 max	0.040 max	0.15-0.35	bal Fe				

Alloy Steel, Chromium Molybdenum, 4147/4147H

Australia

Specification	Designation	Notes	C	Cr	Mn	Mo	Ni	P	S	Si	Other	UTS	YS	El	Hard
AS 1444(96)	4150H	H series, Hard/Tmp	0.47-0.54	0.75-1.20	0.65-1.10	0.15-0.25		0.040 max	0.040 max	0.10-0.35	bal Fe				

France

Specification	Designation	Notes	C	Cr	Mn	Mo	Ni	P	S	Si	Other	UTS	YS	El	Hard
AFNOR NFA35552	45SCD6		0.42-0.5	0.5-0.75	0.5-0.8	0.15-0.3		0.035	0.035	1.3-1.7	bal Fe				
AFNOR NFA35553	45SCD6		0.42-0.5	0.5-0.75	0.5-0.8	0.15-0.3		0.035	0.035	1.3-1.7	bal Fe				
AFNOR NFA35571	50SCD6		0.46-0.54	0.8-1.1	0.7-1.1	0.2-0.35		0.025	0.02	1.4-1.8	bal Fe				

Germany

Specification	Designation	Notes	C	Cr	Mn	Mo	Ni	P	S	Si	Other	UTS	YS	El	Hard
DIN	WNr 1.7230*	Obs	0.46-0.54	0.9-1.2	0.5-0.8	0.15-0.3	0.6	0.035 max	0.035 max	0.15-0.4	Pb 0.15-0.3; bal Fe				

Mexico

Specification	Designation	Notes	C	Cr	Mn	Mo	Ni	P	S	Si	Other	UTS	YS	El	Hard
DGN B-203	4147*	Obs *	0.45-0.5	0.8-1.1	0.75-1	0.15-0.25		0.04	0.04	0.2-0.35	bal Fe				
DGN B-297	4147*	Obs	0.45-0.5	0.8-1.1	0.75-1	0.15-0.25		0.035	0.04	0.2-0.35	bal Fe				
NMX-B-268(68)	4147H	Hard	0.44-0.51	0.75-1.20	0.65-1.10	0.15-0.25				0.20-0.35	bal Fe				
NMX-B-300(91)	4147	Bar	0.45-0.50	0.80-1.10	0.75-1.00	0.15-0.25		0.035 max	0.040 max	0.15-0.35	bal Fe				
NMX-B-300(91)	4147H	Bar	0.44-0.51	0.75-1.20	0.65-1.10	0.15-0.25				0.15-0.35	bal Fe				

Pan America

Specification	Designation	Notes	C	Cr	Mn	Mo	Ni	P	S	Si	Other	UTS	YS	El	Hard
COPANT 514	4147	Tube	0.45-0.5	0.8-1.1	0.75-1	0.15-0.25		0.04	0.04	0.2-0.35	bal Fe				

Russia

Specification	Designation	Notes	C	Cr	Mn	Mo	Ni	P	S	Si	Other	UTS	YS	El	Hard
GOST 14959	50ChFA		0.46-0.54	0.8-1.1	0.5-0.8		0.25 max	0.03 max	0.03 max	0.17-0.37	Cu <=0.2; V 0.1-0.2; bal Fe				

UK

Specification	Designation	Notes	C	Cr	Mn	Mo	Ni	P	S	Si	Other	UTS	YS	El	Hard
BS 970/1(83)	708A47	Blm Bil Slab Bar Rod Frg	0.45-0.50	0.90-1.20	0.75-1.00	0.15-0.25		0.035 max	0.040 max		bal Fe				

USA

Specification	Designation	Notes	C	Cr	Mn	Mo	Ni	P	S	Si	Other	UTS	YS	El	Hard
	AISI 4147	Smls mech tub	0.45-0.50	0.80-1.10	0.75-1.00	0.15-0.30		0.040 max	0.040 max	0.15-0.35	bal Fe				
	AISI 4147	Bar Blm Bil Slab	0.45-0.50	0.80-1.10	0.75-1.00	0.15-0.30		0.035 max	0.040 max	0.15-0.35	bal Fe				
	AISI 4147H	Smls mech tub, Hard	0.44-0.51	0.75-1.20	0.65-1.10	0.15-0.30		0.040 max	0.040 max	0.15-0.30	bal Fe				
	AISI 4147H	Bar Blm Bil Slab	0.44-0.51	0.75-1.20	0.65-1.10	0.15-0.30		0.035 max	0.040 max	0.15-0.35	bal Fe				
	UNS G41470		0.45-0.5	0.80-1.10	0.75-1.00	0.15-0.25		0.035 max	0.040 max	0.15-0.35	bal Fe				
	UNS H41470		0.44-0.51	0.75-1.20	0.65-1.10	0.15-0.25		0.035 max	0.040 max	0.15-0.35	bal Fe				
ASTM A29/A29M(93)	4147	Bar	0.45-0.50	0.80-1.10	0.75-1.00	0.15-0.25		0.035 max	0.040 max	0.15-0.35	bal Fe				
ASTM A304(96)	4147H	Bar, hard bands spec	0.44-0.51	0.75-1.20	0.65-1.10	0.15-0.25		0.035 max	0.040 max	0.15-0.35	Cu <=0.35; Cu Ni Cr Mo trace allowed; bal Fe				
ASTM A322(96)	4147	Bar	0.45-0.50	0.80-1.10	0.75-1.00	0.15-0.25		0.035 max	0.040 max	0.15-0.35	bal Fe				
ASTM A331(95)	4147	Bar	0.45-0.50	0.80-1.10	0.75-1.00	0.15-0.25		0.035 max	0.040 max	0.15-0.35	bal Fe				
ASTM A331(95)	4147H	Bar	0.44-0.51	0.75-1.20	0.65-1.10	0.15-0.25		0.025 max	0.025 max	0.15-0.35	Cu <=0.35; bal Fe				
ASTM A519(96)	4147	Smls mech tub	0.45-0.50	0.80-1.10	0.75-1.00	0.15-0.25		0.040 max	0.040 max	0.15-0.35	bal Fe				
ASTM A752(93)	4147	Rod Wir	0.45-0.50	0.80-1.10	0.75-1.00	0.15-0.25		0.035 max	0.040 max	0.15-0.30	bal Fe	690			

UNS numbers and US grades are provided as a means of cross referencing chemically similar alloys. Exchangability is only possible after independent examination of specifications. Tensile properties are minimum or typical as specified. UTS and YS as MPa. El as %. See Appendix for list of abbreviations used in Notes. * indicates obsolete material.

Specification	Designation	Notes	C	Cr	Mn	Mo	Ni	P	S	Si	Other	UTS	YS	El	Hard

Alloy Steel, Chromium Molybdenum, 4147/4147H (Continued from previous page)

USA

Specification	Designation	Notes	C	Cr	Mn	Mo	Ni	P	S	Si	Other	UTS	YS	El	Hard
MIL-S-24093A(SH)(91)	Type II Class B	Frg for shipboard apps; 10 in. max section thick	0.44 max	0.80-1.10	1.10 max	0.15-0.25		0.025 max	0.025 max	0.20-0.35	bal Fe	965	827	14	
MIL-S-24093A(SH)(91)	Type II Class C	Frg for shipboard apps; 10 in. max section thick	0.44 max	0.80-1.10	1.10 max	0.15-0.25		0.025 max	0.025 max	0.20-0.35	bal Fe	827	690	16	
MIL-S-24093A(SH)(91)	Type II Class D	Frg for shipboard apps; 10 in. max section thick	0.44 max	0.80-1.10	1.10 max	0.15-0.25		0.025 max	0.025 max	0.20-0.35	bal Fe	690	552	18	
MIL-S-24093A(SH)(91)	Type II Class E	Frg for shipboard apps; 10 in. max section thick	0.44 max	0.80-1.10	1.10 max	0.15-0.25		0.025 max	0.025 max	0.20-0.35	bal Fe	655	448	20	
MIL-S-24093A(SH)(91)	Type II Class F	Frg for shipboard apps; 10 in. max section thick	0.44 max	0.80-1.10	1.10 max	0.15-0.25		0.025 max	0.025 max	0.20-0.35	bal Fe	621 max	310	22	
SAE 770(84)	4147*	Obs; see J1397(92)	0.45-0.5	0.8-1.1	0.75-1	0.15-0.3		0.04 max	0.04 max	0.15-0.3	bal Fe				
SAE J1268(95)	4147H	See std	0.44-0.51	0.75-1.20	0.65-1.10	0.15-0.35	0.25 max	0.030 max	0.040 max	0.15-0.35	Cu <=0.35; Cu, Ni not spec'd but acpt; bal Fe				

Alloy Steel, Chromium Molybdenum, 4150/4150H

Argentina

Specification	Designation	Notes	C	Cr	Mn	Mo	Ni	P	S	Si	Other	UTS	YS	El	Hard
IAS	IRAM 4150		0.48-0.53	0.80-1.10	0.75-1.00	0.15-0.25		0.035 max	0.040 max	0.15-0.35	bal Fe	1070	725	13	311 HB

Australia

Specification	Designation	Notes	C	Cr	Mn	Mo	Ni	P	S	Si	Other	UTS	YS	El	Hard
AS 1444	X4150*	Obs	0.47-0.55	0.4-0.8	1-1 4	0.1-0.2		0.04	0.04	0.1-0.4	bal Fe				
AS 1444(96)	4150	H series, Hard/Tmp	0.48-0.53	0.80-1.10	0.75-1.00	0.15-0.25		0.040 max	0.040 max	0.10-0.35	bal Fe				

Europe

Specification	Designation	Notes	C	Cr	Mn	Mo	Ni	P	S	Si	Other	UTS	YS	El	Hard
EN 10083/1(91)A1(96)	1.7228	Q/T t<=16mm	0.46-0.54	0.90-1.20	0.50-0.80	0.15-0.30		0.035 max	0.035 max	0.40 max	bal Fe	1100-1300	900	9	
EN 10083/1(91)A1(96)	50CrMo4	Q/T t<=16mm	0.46-0.54	0.90-1.20	0.50-0.80	0.15-0.30		0.035 max	0.035 max	0.40 max	bal Fe	1100-1300	900	9	

France

Specification	Designation	Notes	C	Cr	Mn	Mo	Ni	P	S	Si	Other	UTS	YS	El	Hard
AFNOR NFA35565(94)	48CD4*	see 48CrMo4	0.46-0.52	0.9-1.2	0.5-0.8	0.15-0.3	0.4 max	0.025 max	0.035 max	0.4 max	Cu <=0.3; Ti <=0.05; bal Fe				
AFNOR NFA35565(94)	48CrMo4	Ball & roller bearing	0.46-0.52	0.9-1.2	0.5-0.8	0.15-0.3	0.4 max	0.025 max	0.035 max	0.4 max	Cu <=0.3; Ti <=0.05; bal Fe				
AFNOR NFA35571	50SCD5		0.46-0.54	0.8-1.1	0.7-1.1	0.2-0.35		0.025	0.02	1.4-1.8	bal Fe				

Germany

Specification	Designation	Notes	C	Cr	Mn	Mo	Ni	P	S	Si	Other	UTS	YS	El	Hard
DIN 1652(90)	50CrMo4	Q/T 17-40mm	0.46-0.54	0.90-1.20	0.50-0.80	0.15-0.30		0.035 max	0.035 max	0.40 max	bal Fe	1000-1200	780	10	
DIN 1652(90)	WNr 1.7228	Q/T 17-40mm	0.46-0.54	0.90-1.20	0.50-0.80	0.15-0.30		0.035 max	0.035 max	0.40 max	bal Fe	1000-1200	780	10	
DIN 1652(90)	WNr 1.7228	Q/T 17-40mm	0.46-0.54	0.90-1.20	0.50-0.80	0.15-0.30		0.035 max	0.035 max	0.40 max	bal Fe	1000-1200	780	10	
DIN 17201(89)	WNr 1.7228	Frg Q/T 17-40mm	0.46-0.54	0.90-1.20	0.50-0.80	0.15-0.30		0.035 max	0.035 max	0.40 max	bal Fe	1000-1200	780	10	
DIN 17212(72)	49CrMo4	HT hard 17-40mm	0.46-0.52	0.90-1.20	0.50-0.80	0.15-0.30		0.025 max	0.035 max	0.15-0.40	bal Fe	980-1180	780	10	
DIN 17212(72)	WNr 1.7238	HT hard 17-40mm	0.46-0.52	0.90-1.20	0.50-0.80	0.15-0.30		0.025 max	0.035 max	0.15-0.40	bal Fe	980-1180	780	10	
DIN 17212(72)	WNr 1.7238	HT hard 101-250mm	0.46-0.52	0.90-1.20	0.50-0.80	0.15-0.30		0.025 max	0.035 max	0.15-0.40	bal Fe	780-980	510	13	
DIN 17230(80)	48CrMo4	Bearing Q/T 17-40mm	0.46-0.52	0.90-1.20	0.50-0.80	0.15-0.30		0.025 max	0.035 max	0.40 max	bal Fe	980-1180	780	10	
DIN 17230(80)	WNr 1.3565	Bearing Q/T 17-40mm	0.46-0.52	0.90-1.20	0.50-0.80	0.15-0.30		0.025 max	0.035 max	0.40 max	bal Fe	980-1180	780	10	
DIN EN 10083(91)	50CrMo4	Q/T 17-40mm	0.46-0.54	0.90-1.20	0.50-0.80	0.15-0.30		0.035 max	0.035 max	0.40 max	bal Fe	1000-1200	780	10	

Italy

Specification	Designation	Notes	C	Cr	Mn	Mo	Ni	P	S	Si	Other	UTS	YS	El	Hard
UNI 3545(80)	51CrMoV4	Spring	0.48-0.56	0.9-1.2	0.7-1	0.15-0.25		0.035 max	0.035 max	0.15-0.4	V 0.07-0.12; bal Fe				

Specification	Designation	Notes	C	Cr	Mn	Mo	Ni	P	S	Si	Other	UTS	YS	El	Hard

Alloy Steel, Chromium Molybdenum, 4150/4150H (Continued from previous page)

Mexico

Specification	Designation	Notes	C	Cr	Mn	Mo	Ni	P	S	Si	Other	UTS	YS	El	Hard
DGN B-203	4150*	Obs	0.48-0.53	0.8-1.1	0.75-1	0.15-0.25		0.04	0.04	0.25-0.35	bal Fe				
DGN B-297	4150*	Obs	0.48-0.53	0.8-1.1	0.75-1	0.15-0.25		0.035	0.04	0.2-0.35	bal Fe				
NMX-B-268(68)	4150H	Hard	0.47-0.54	0.75-1.20	0.65-1.10	0.15-0.25				0.20-0.35	bal Fe				
NMX-B-300(91)	4150	Bar	0.48-0.53	0.80-1.10	0.75-1.00	0.15-0.25		0.035 max	0.040 max	0.15-0.35	bal Fe				
NMX-B-300(91)	4150H	Bar	0.47-0.54	0.75-1.20	0.65-1.10	0.15-0.25				0.15-0.35	bal Fe				

Pan America

Specification	Designation	Notes	C	Cr	Mn	Mo	Ni	P	S	Si	Other	UTS	YS	El	Hard
COPANT 334	4150	Bar	0.48-0.53	0.8-1.1	0.75-1	0.15-0.25		0.035	0.04	0.2-0.35	bal Fe				
COPANT 514	4150	Tube	0.48-0.53	0.8-1.1	0.75-1	0.15-0.25		0.04	0.04	0.2-0.35	bal Fe				

Russia

Specification	Designation	Notes	C	Cr	Mn	Mo	Ni	P	S	Si	Other	UTS	YS	El	Hard
GOST 14959	50ChFA	Q/T; Spring	0.46-0.54	0.8-1.1	0.5-0.8	0.15 max	0.25 max	0.025 max	0.025 max	0.17-0.37	Al <=0.1; Cu <=0.2; Ti <=0.05; V 0.1-0.2; bal Fe				

Turkey

Specification	Designation	Notes	C	Cr	Mn	Mo	Ni	P	S	Si	Other	UTS	YS	El	Hard
TS 2288(97)	51CrMoV4-17701	Ref. ISO 683-14(92), wire rods, flts, HT	0.48-0.56	0.9-1.2	0.7-1.1	0.15-0.25		0.04	0.04	0.15-0.4	bal Fe				

USA

Specification	Designation	Notes	C	Cr	Mn	Mo	Ni	P	S	Si	Other	UTS	YS	El	Hard
	AISI 4150	Plt	0.48-0.53	0.80-1.10	0.75-1.00	0.15-0.25		0.035 max	0.040 max	0.15-0.35	bal Fe				
	AISI 4150	Smls mech tub	0.48-0.53	0.80-1.10	0.75-1.00	0.15-0.25		0.040 max	0.040 max	0.15-0.35	bal Fe				
	AISI 4150	Bar Blm Bil Slab	0.48-0.53	0.80-1.10	0.75-1.00	0.15-0.25		0.035 max	0.040 max	0.15-0.35	bal Fe				
	AISI 4150H	Smls mech tub, Hard	0.47-0.54	0.75-1.20	0.65-1.10	0.15-0.25		0.040 max	0.040 max	0.15-0.30	bal Fe				
	AISI 4150H	Bar Blm Bil Slab	0.47-0.54	0.75-1.20	0.65-1.10	0.15-0.25		0.035 max	0.040 max	0.15-0.35	bal Fe				
	UNS G41500		0.48-0.53	0.80-1.10	0.75-1.00	0.15-0.25		0.035 max	0.040 max	0.15-0.35	bal Fe				
	UNS H41500		0.47-0.54	0.75-1.20	0.65-1.10	0.15-0.25		0.035 max	0.040 max	0.15-0.35	bal Fe				
ASTM A29/A29M(93)	4150	Bar	0.48-0.53	0.80-1.10	0.75-1.00	0.15-0.25		0.035 max	0.040 max	0.15-0.35	bal Fe				
ASTM A304(96)	4150H	Bar, hard bands spec	0.47-0.54	0.75-1.20	0.65-1.10	0.15-0.25		0.035 max	0.040 max	0.15-0.35	Cu <=0.35; Cu Ni Cr Mo trace allowed; bal Fe				
ASTM A322(96)	4150	Bar	0.48-0.53	0.80-1.10	0.75-1.00	0.15-0.25		0.035 max	0.040 max	0.15-0.35	bal Fe				
ASTM A331(95)	4150	Bar	0.48-0.53	0.80-1.10	0.75-1.00	0.15-0.25		0.035 max	0.040 max	0.15-0.35	bal Fe				
ASTM A331(95)	4150H	Bar	0.47-0.54	0.75-1.20	0.65-1.10	0.15-0.25		0.025 max	0.025 max	0.15-0.35	Cu <=0.35; bal Fe				
ASTM A519(96)	4150	Smls mech tub	0.48-0.53	0.80-1.10	0.75-1.00	0.15-0.25		0.040 max	0.040 max	0.15-0.35	bal Fe				
ASTM A752(93)	4150	Rod Wir	0.48-0.53	0.80-1.10	0.75-1.00	0.15-0.25		0.035 max	0.040 max	0.15-0.30	bal Fe	690			
ASTM A829/A829M(95)	4150	Plt	0.48-0.53	0.80-1.15	0.70-1.00	0.15-0.25		0.035 max	0.040 max	0.15-0.40	bal Fe	480-965			
ASTM A866(94)	4150	Anti-friction bearings	0.48-0.53	0.80-1.10	0.75-1.00	0.15-0.25		0.025 max	0.025 max	0.15-0.35	bal Fe				
MIL-B-11595E(88)	ORD 4150	Small arms barrels	0.48-0.55	0.80-1.10	0.75-1.00	0.15-0.25		0.040 max	0.040 max	0.20-0.35	Al <=0.040; Cu <=0.35; bal Fe				
MIL-B-11595E(88)	ORD 4150 ReS	Small arms barrels	0.47-0.55	0.80-1.15	0.70-1.00	0.15-0.25		0.040 max	0.05-0.09	0.20-0.35	Al <=0.040; Cu <=0.35; bal Fe				
MIL-S-11595	ORD4150*	Obs; see MIL-B-11595E; Bar bil	0.48-0.55	0.8-1.1	0.75-1	0.15-0.25		0.04	0.04	0.2-0.35	bal Fe				
SAE 770(84)	4150*	Obs; see J1397(92)									bal Fe				
SAE J1268(95)	4150H	Bar; See std	0.47-0.54	0.75-1.20	0.65-1.10	0.15-0.25	0.25 max	0.030 max	0.040 max	0.15-0.35	Cu <=0.35; Cu, Ni not spec'd but acpt; bal Fe				

UNS numbers and US grades are provided as a means of cross referencing chemically similar alloys. Exchangability is only possible after independent examination of specifications. Tensile properties are minimum or typical as specified. UTS and YS as MPa. El as %. See Appendix for list of abbreviations used in Notes. * indicates obsolete material.

Specification	Designation	Notes	C	Cr	Mn	Mo	Ni	P	S	Si	Other	UTS	YS	El	Hard

Alloy Steel, Chromium Molybdenum, 4150/4150H (Continued from previous page)

USA

Specification	Designation	Notes	C	Cr	Mn	Mo	Ni	P	S	Si	Other	UTS	YS	El	Hard
SAE J404(94)	4150	Bil Blm Slab Bar HR CF	0.48-0.53	0.80-1.10	0.75-1.00	0.15-0.25		0.030 max	0.040 max	0.15-0.35	bal Fe				

Yugoslavia

	C.4733	Q/T	0.48-0.54	0.9-1.2	0.5-0.8	0.15-0.3		0.035 max	0.03 max	0.4 max	bal Fe				
	C.4733	Ball & roller bearing	0.46-0.54	0.9-1.2	0.5-0.8	0.15-0.3		0.035 max	0.03 max	0.4 max	bal Fe				
	C.4736	F/IH	0.46-0.52	0.9-1.2	0.5-0.8	0.15-0.3	0.25 max	0.025 max	0.035 max	0.15-0.4	bal Fe				

Alloy Steel, Chromium Molybdenum, 4161/4161H

China

Specification	Designation	Notes	C	Cr	Mn	Mo	Ni	P	S	Si	Other	UTS	YS	El	Hard
GB 1222(84)	60CrMnMoA	BAR Flat HR/CD	0.56-0.64	0.70-0.90	0.70-1.00	0.25-0.35		0.035 max	0.035 max	0.17-0.37	Cu <=0.25; bal Fe				

Germany

DIN	61CrMo4		0.57-0.65	0.90-1.20	0.40-0.60	0.15-0.25		0.035 max	0.035 max	0.15-0.35	bal Fe				
DIN	WNr 1.7229		0.57-0.65	0.90-1.20	0.40-0.60	0.15-0.25		0.035 max	0.035 max	0.15-0.35	bal Fe				
DIN	WNr 1.7266*	Obs	0.54-0.62	0.8-1.2	0.8-1.2	0.2-0.3		0.035	0.035	0.3-0.5	bal Fe				

International

| ISO 683-14(92) | 60CrMo33 | Bar Rod Wir; Q/T, springs, Ann | 0.56-0.64 | 0.70-1.00 | 0.70-1.00 | 0.25-0.35 | | 0.030 max | 0.030 max | 0.15-0.40 | bal Fe | | | | 255 HB |

Japan

| JIS G4801(84) | SUP13 | Bar Spring Q/T | 0.56-0.64 | 0.70-0.90 | 0.70-1.00 | 0.25-0.35 | | 0.035 max | 0.035 max | 0.15-0.35 | Cu <=0.30; bal Fe | 1226 | 1079 | 10 | 363-429 HB |

Mexico

NMX-B-268(68)	4161H	Hard	0.55-0.65	0.65-0.95	0.65-1.10	0.25-0.35				0.20-0.35	bal Fe				
NMX-B-300(91)	4161	Bar	0.56-0.64	0.70-0.90	0.75-1.00	0.25-0.35		0.035 max	0.040 max	0.15-0.35	bal Fe				
NMX-B-300(91)	4161H	Bar	0.55-0.65	0.65-0.95	0.65-1.10	0.25-0.35				0.15-0.35	bal Fe				

UK

BS 3146/1(74)	CLA12(C)	Inv Casting	0.55-0.65	0.80-1.50	0.50-1.00	0.20-0.40	0.40 max	0.035 max	0.035 max	0.30-0.80	Cu <=0.30; bal Fe				
BS 970/2(88)	704A60	Spring	0.57-0.62	0.80-0.95	0.85-1.00	0.15-0.25		0.035 max	0.035 max	0.20-0.35	bal Fe				
BS 970/2(88)	704H60	Spring	0.55-0.64	0.60-1.00	0.65-1.10	0.15-0.25		0.035 max	0.035 max	0.15-0.40	bal Fe				
BS 970/2(88)	705A60	Spring	0.57-0.62	0.85-1.00	0.85-1.00	0.25-0.35		0.035 max	0.035 max	0.20-0.35	bal Fe				
BS 970/2(88)	705H60	Spring	0.55-0.64	0.60-1.00	0.65-1.10	0.25-0.35		0.035 max	0.035 max	0.15-0.40	bal Fe				

USA

	AISI 4161		0.55-0.65	0.7-0.9	0.75-1.1	0.25-0.35		0.04 max	0.04 max	0.25-0.35	bal Fe				
	AISI 4161H	Smls mech tub, Hard	0.55-0.65	0.65-0.95	0.65-1.10	0.25-0.35		0.035 max	0.040 max	0.15-0.30	bal Fe				
	UNS G41610		0.56-0.64	0.70-0.90	0.75-1.00	0.25-0.35		0.035 max	0.040 max	0.15-0.35	bal Fe				
	UNS H41610		0.55-0.65	0.65-0.95	0.65-1.10	0.25-0.35		0.035 max	0.040 max	0.15-0.35	bal Fe				
ASTM A29/A29M(93)	4161	Bar	0.56-0.64	0.70-0.90	0.75-1.00	0.25-0.35		0.035 max	0.040 max	0.15-0.35	bal Fe				
ASTM A304(96)	4161H	Bar, hard bands spec	0.55-0.65	0.65-0.95	0.65-1.10	0.25-0.35		0.035 max	0.040 max	0.15-0.35	Cu <=0.35; Cu Ni Cr Mo trace allowed; bal Fe				
ASTM A322(96)	4161	Bar	0.56-0.64	0.70-0.90	0.75-1.00	0.25-0.35		0.035 max	0.040 max	0.15-0.35	bal Fe				
ASTM A331(95)	4161	Bar	0.56-0.64	0.70-0.90	0.75-1.00	0.25-0.35		0.035 max	0.040 max	0.15-0.35	bal Fe				
ASTM A331(95)	4161H	Bar	0.55-0.65	0.65-0.95	0.65-1.10	0.25-0.35		0.025 max	0.025 max	0.15-0.35	Cu <=0.35; bal Fe				
ASTM A752(93)	4161	Rod Wir	0.56-0.65	0.70-0.90	0.75-1.00	0.25-0.35		0.035 max	0.040 max	0.15-0.30	bal Fe	770			

Specification	Designation	Notes	C	Cr	Mn	Mo	Ni	P	S	Si	Other	UTS	YS	El	Hard

Alloy Steel, Chromium Molybdenum, 4161/4161H (Continued from previous page)

USA

ASTM A914/914M(92)	4161RH	Bar restricted end Q hard	0.56-0.64	0.70-0.90	0.75-1.00	0.25-0.35	0.25 max	0.035 max	0.040 max	0.15-0.35	Cu <=0.35; Electric furnace P, S<=0.025; bal Fe				
SAE 770(84)	4161*	Obs; see J1397(92)	0.55-0.65	0.7-0.9	0.75-1.1	0.25-0.35		0.04 max	0.04 max	0.25-0.35	bal Fe				
SAE J1268(95)	4161H	See std	0.55-0.65	0.65-0.95	0.65-1.10	0.25-0.35	0.25 max	0.030 max	0.040 max	0.15-0.35	Cu <=0.35; Cu, Ni not spec'd but acpt; bal Fe				
SAE J1868(93)	4161RH	Restrict hard range	0.56-0.64	0.70-0.90	0.75-1.00	0.25-0.35	0.25 max	0.025 max	0.040 max	0.15-0.35	Cu <=0.035; Cu, Ni not spec'd but acpt; bal Fe				5 HRC max

Alloy Steel, Chromium Molybdenum, 6356

USA

| | UNS K13247 | | 0.30-0.35 | 0.80-1.10 | 0.40-0.60 | 0.15-0.25 | | 0.025 max | 0.025 max | 0.20-0.50 | Cu <=0.35; N <=0.25; bal Fe | | | | |
| AMS 6356 | | Sh, Strp, Plt | 0.30-0.35 | 0.80-1.10 | 0.40-0.60 | 0.15-0.25 | 0.25 max | 0.025 max | 0.025 max | 0.20-0.50 | Cu <=0.35; bal Fe | | | | |

Alloy Steel, Chromium Molybdenum, 6378

USA

| | UNS K11542 | | 0.38-0.45 | 0.80-1.10 | 0.75-1.00 | 0.15-0.25 | 0.25 max | 0.040 max | 0.040 max | 0.20-0.35 | Cu <=0.35; Te 0.035-0.060; bal Fe | | | | |
| AMS 6378E(91) | | Bar | 0.39-0.48 | 0.75-1.20 | 0.70-1.10 | 0.15-0.25 | 0.25 max | 0.040 max | 0.040 max | 0.15-0.35 | Cu <=0.35; bal Fe | 1034 | 896 | 5 | 302-341 HB |

Alloy Steel, Chromium Molybdenum, 6386(4)

Japan

JIS G3441(88)	SCM420TK	Tube	0.18-0.23	0.90-1.20	0.60-0.85	0.15-0.30	0.25 max	0.030 max	0.030 max	0.15-0.35	Cu <=0.30; bal Fe				
JIS G4052(79)	SCM415H	Struct Hard	0.12-0.18	0.85-1.25	0.55-0.90	0.15-0.35	0.25 max	0.030 max	0.030 max	0.15-0.35	Cu <=0.30; bal Fe				
JIS G4105(79)	SCM418	HR Frg Bar Rod	0.16-0.21	0.90-1.20	0.60-0.85	0.15-0.30	0.25 max	0.030 max	0.030 max	0.15-0.35	Cu <=0.30; bal Fe				

USA

| | UNS K11662 | | 0.13-0.20 | 0.85-1.20 | 0.40-0.70 | 0.15-0.25 | | 0.035 max | 0.04 max | 0.20-0.35 | B 0.0015-0.005; Cu 0.20-0.40; Ti 0.04-0.10; bal Fe | | | | |

Alloy Steel, Chromium Molybdenum, 6459

USA

| | UNS K22720 | | 0.18-0.23 | 0.80-1.20 | 0.40-0.60 | 0.80-1.20 | | 0.015 max | 0.008 max | 0.60-0.90 | Cu <=0.50; N <=0.005; V 0.08-0.15; H 0.0010 max; O 0.0025 max; bal Fe | | | | |

Alloy Steel, Chromium Molybdenum, A182(F2)

USA

| | UNS K12122 | | 0.21 max | 0.50-0.81 | 0.30-0.80 | 0.44-0.65 | | 0.040 max | 0.040 max | 0.10-0.60 | bal Fe | | | | |
| ASTM A182/A182M(98) | F2 | Frg/roll pipe flange valve, former F2 is now F12 | 0.05-0.21 | 0.50-0.81 | 0.30-0.80 | 0.44-0.65 | | 0.040 max | 0.40 max | 0.10-0.60 | bal Fe | 485 | 275 | 20.0 | 143-192 HB |

Alloy Steel, Chromium Molybdenum, A182(F9)

USA

	UNS K90941		0.15 max	8.00-10.00	0.30-0.60	0.90-1.10		0.030 max	0.030 max	0.50-1.00	bal Fe				
ASTM A182/A182M(98)	F9	Frg/roll pipe flange valve	0.15 max	8.0-10.0	0.30-0.60	0.90-1.10		0.030 max	0.030 max	0.50-1.00	bal Fe	585	380	20.0	179-217 HB
ASTM A213/A213M(95)	T9	Smls tube boiler, superheater, heat exchanger	0.15 max	8.00-10.00	0.30-0.60	0.90-1.10		0.025 max	0.025 max	0.25-1.00	bal Fe	415	205	30	
ASTM A234/A234M(97)	WP9	Frg	0.15 max	8.0-10.0	0.30-0.60	0.90-1.10		0.030 max	0.030 max	0.25-1.00	bal Fe	415-585	205	22L	
ASTM A336/A336M(98)	F9	Ferr Frg	0.15 max	8.0-10.0	0.30-0.60	0.90-1.10		0.025 max	0.025 max	0.50-1.00	bal Fe	585-760	380	20	

UNS numbers and US grades are provided as a means of cross referencing chemically similar alloys. Exchangability is only possible after independent examination of specifications. Tensile properties are minimum or typical as specified. UTS and YS as MPa. El as %. See Appendix for list of abbreviations used in Notes. * indicates obsolete material.

Specification	Designation	Notes	C	Cr	Mn	Mo	Ni	P	S	Si	Other	UTS	YS	El	Hard
Alloy Steel, Chromium Molybdenum, A182(F9) (Continued from previous page)															
USA															
ASTM A989(98)	K90941	HIP Flanges, Fittings, Valves/parts; Heat res	0.15 max	8.0-10.0	0.30-0.60	0.90-1.10		0.030 max	0.030 max	0.50-1.00	bal Fe	585	380	20.0	179-217 HB
Alloy Steel, Chromium Molybdenum, A182/A182M															
India															
IS 1570/4(88)	16Ni6Cr7Mo3		0.14-0.19	1.5-1.8	0.4-0.6	0.25-0.35	1.4-1.7	0.07 max	0.06 max	0.15-0.35	bal Fe				
Spain															
UNE 36087(76/78)	13MoCrV6	Sh Plt Strp CR	0.08-0.18	0.30-0.60	0.40-0.80	0.50-0.70	0.40 max	0.035 max	0.035 max	0.15-0.35	Cu <=0.30; V 0.25-0.35; bal Fe				
UNE 36087(76/78)	F.2621*	see 13MoCrV6	0.08-0.18	0.30-0.60	0.40-0.80	0.50-0.70	0.40 max	0.035 max	0.035 max	0.15-0.35	Cu <=0.30; V 0.25-0.35; bal Fe				
UK															
BS 1503(89)	620-540	Frg press ves; t<=200mm; Q/T	0.18 max	1.10-1.40	0.40-0.70	0.45-0.65	0.40 max	0.030 max	0.025 max	0.15-0.40	Al <=0.020; Cu <=0.30; W <=0.1; bal Fe	540-690	375	18	
BS 1503(89)	621-460	Frg press ves; t<=999mm; Q/T	0.18 max	1.10-1.40	0.40-0.70	0.45-0.65	0.40 max	0.030 max	0.025 max	0.15-0.40	Al <=0.020; Cu <=0.30; W <=0.1; bal Fe	460-610	275	18	
USA															
	UNS K31835	2.25 Cr, 1 Mo, 0.25 V	0.09-0.18	2.00-2.50	0.30-0.60	0.90-1.10	0.25 max	0.015 max	0.010 max	0.10 max	B <=0.0020; Cu <=0.20; Nb <=0.07; Ti <=0.030; V 0.25-0.35; Ca 0.015 max; Other REM 0.020 max; bal Fe				
ASTM A182/A182M(98)	F22V	Frg/roll pipe flange valve	0.11-0.15	2.00-2.50	0.30-0.60	0.90-1.10	0.25 max	0.015 max	0.010 max	0.10 max	B <=0.002; Cu <=0.20; Nb <=0.007; Ti <=0.030; V 0.25-0.35; Ca<=0.015; bal Fe	585-780	415	18.0	174-237
ASTM A336/A336M(98)	F22V	Ferr Frg	0.11-0.15	2.00-2.50	0.30-0.60	0.90-1.10	0.25 max	0.015 max	0.010 max	0.10 max	B <=0.0020; Cu <=0.20; Nb <=0.07; Ti <=0.030; V 0.25-0.35; Ca<=0.015; bal Fe	585-760	415	18	
Alloy Steel, Chromium Molybdenum, A199(T4)															
USA															
	UNS K31509		0.15 max	2.15-2.85	0.30-0.60	0.44-0.65		0.030 max	0.030 max	0.50-1.00	bal Fe				
Alloy Steel, Chromium Molybdenum, A213(T5b)															
UK															
BS 3606(92)	625	Heat exch Tub; t<=3.2mm	0.15 max	4.00-6.00	0.30-0.60	0.45-0.65	0.30 max	0.030 max	0.030 max	0.50 max	Al <=0.02; Sn<=0.03; bal Fe	450-600	170	20	
USA															
	UNS K51545		0.15 max	4.00-6.00	0.30-0.60	0.45-0.65		0.03 max	0.03 max	1.00-2.00	bal Fe				
ASTM A213/A213M(95)	T5b	Smls tube boiler, superheater, heat exchanger	0.15 max	4.00-6.00	0.30-0.60	0.45-0.65		0.025 max	0.025 max	1.00-2.00	bal Fe	415	205	30	
Alloy Steel, Chromium Molybdenum, A213(T5c)															
USA															
	UNS K41245		0.12 max	4.00-6.00	0.30-0.60	0.45-0.65		0.03 max	0.03 max	0.50 max	bal Fe				
ASTM A213/A213M(95)	T5c	Smls tube boiler, superheater, heat exchanger	0.12 max	4.00-6.00	0.30-0.60	0.45-0.65		0.025 max	0.025 max	0.50 max	Ti 4xC-0.70; bal Fe	415	205	30	
Alloy Steel, Chromium Molybdenum, A290(E,F)															
USA															
	UNS K14048		0.35-0.45	0.80-1.15	0.70-1.00	0.15-0.25		0.040 max	0.040 max	0.20-0.35	V <=0.06; bal Fe				
ASTM A290(95)	E	Gear frg	0.35-0.45	0.80-1.15	0.70-1.00	0.15-0.25	0.50 max	0.040 max	0.040 max	0.35 max	Cu <=0.35; V <=0.06; bal Fe	725	515	20	223-269 HB

UNS numbers and US grades are provided as a means of cross referencing chemically similar alloys. Exchangability is only possible after independent examination of specifications. Tensile properties are minimum or typical as specified. UTS and YS as MPa. El as %. See Appendix for list of abbreviations used in Notes. * indicates obsolete material.

Specification	Designation	Notes	C	Cr	Mn	Mo	Ni	P	S	Si	Other	UTS	YS	El	Hard

Alloy Steel, Chromium Molybdenum, A290(E,F) (Continued from previous page)

USA

Specification	Designation	Notes	C	Cr	Mn	Mo	Ni	P	S	Si	Other	UTS	YS	El	Hard
ASTM A290(95)	F	Gear frg	0.35-0.45	0.80-1.15	0.70-1.00	0.15-0.25	0.50 max	0.040 max	0.040 max	0.35 max	Cu <=0.35; V <=0.06; bal Fe				223-269 HB

Alloy Steel, Chromium Molybdenum, A369(FP3b)

Italy

Specification	Designation	Notes	C	Cr	Mn	Mo	Ni	P	S	Si	Other	UTS	YS	El	Hard
UNI 5462(64)	12CrMo910	Smls tube	0.15 max	2-2.5	0.4-0.6	0.9-1.1		0.035 max	0.035 max	0.15-0.5	bal Fe				

USA

Specification	Designation	Notes	C	Cr	Mn	Mo	Ni	P	S	Si	Other	UTS	YS	El	Hard
	UNS K21509		0.15 max	1.65-2.35	0.30-0.60	0.44-0.65		0.030 max	0.030 max	0.50 max	bal Fe				

Alloy Steel, Chromium Molybdenum, A372(5E, 8)

USA

Specification	Designation	Notes	C	Cr	Mn	Mo	Ni	P	S	Si	Other	UTS	YS	El	Hard
	UNS K14248		0.35-0.50	0.80-1.15	0.75-1.05	0.15-0.25		0.035 max	0.04 max	0.15-0.35	bal Fe				
ASTM A372/372M(95)	Grade J	Thin wall frg press ves, Class 110	0.35-0.50	0.80-1.15	0.75-1.05	0.15-0.25		0.025 max	0.025 max	0.15-0.35	bal Fe	930-1100	760	15	277 HB
ASTM A372/372M(95)	Type VIII*	Thin wall frg press ves, Class 110	0.35-0.50	0.80-1.15	0.75-1.05	0.15-0.25		0.025 max	0.025 max	0.15-0.35	bal Fe	930-1100	760	15	277 HB

Alloy Steel, Chromium Molybdenum, A372(V-1)

USA

Specification	Designation	Notes	C	Cr	Mn	Mo	Ni	P	S	Si	Other	UTS	YS	El	Hard
	UNS K13047		0.26-0.34	0.80-1.15	0.40-0.70	0.15-0.25		0.035 max	0.04 max	0.15-0.35	bal Fe				
ASTM A372/372M(95)	Grade E	Thin wall frg press ves	0.25-0.35	0.80-1.15	0.40-0.90	0.15-0.25		0.025 max	0.025 max	0.15-0.35	bal Fe	545-1000	380-485	18-20	179-248 HB
ASTM A372/372M(95)	Type V Grade 1*	Thin wall frg press ves	0.25-0.35	0.80-1.15	0.40-0.90	0.15-0.25		0.025 max	0.025 max	0.15-0.35	bal Fe	545-1000	380-485	18-20	179-248 HB
ASTM A649/649M(98)	Class 3	Frg roll for paperboard mach	0.35 max	0.80-1.15	0.40-0.70	0.15-0.25		0.025 max	0.025 max	0.15-0.35	Si<=0.10 with VCD; bal Fe	415	205	22	

Alloy Steel, Chromium Molybdenum, A372(V-4)

USA

Specification	Designation	Notes	C	Cr	Mn	Mo	Ni	P	S	Si	Other	UTS	YS	El	Hard
	UNS K13547		0.31-0.39	0.40-0.65	0.75-1.05	0.15-0.25		0.035 max	0.04 max	0.15-0.35	bal Fe				
ASTM A372/372M(95)	Grade H	Thin wall frg press ves	0.30-0.40	0.40-0.65	0.75-1.05	0.15-0.25		0.025 max	0.025 max	0.15-0.35	bal Fe	545-1000	380-485	18-20	179-248 HB
ASTM A372/372M(95)	Type V Grade 4*	Thin wall frg press ves	0.30-0.40	0.40-0.65	0.75-1.05	0.15-0.25		0.025 max	0.025 max	0.15-0.35	bal Fe	545-1000	380-485	18-20	179-248 HB

Alloy Steel, Chromium Molybdenum, A372(V-5)

USA

Specification	Designation	Notes	C	Cr	Mn	Mo	Ni	P	S	Si	Other	UTS	YS	El	Hard
	UNS K13548		0.31-0.39	0.80-1.15	0.70-1.00	0.15-0.25		0.035 max	0.04 max	0.15-0.35	bal Fe				
ASTM A372/372M(95)	Grade J	Thin wall frg press ves	0.35-0.50	0.80-1.15	0.75-1.05	0.15-0.25		0.025 max	0.025 max	0.15-0.35	bal Fe	545-1000	380-485	18-20	179-248 HB
ASTM A372/372M(95)	Type V Grade 5*	Thin wall frg press ves	0.35-0.50	0.80-1.15	0.75-1.05	0.15-0.25		0.025 max	0.025 max	0.15-0.35	bal Fe	545-1000	380-485	18-20	179-248 HB

Alloy Steel, Chromium Molybdenum, A387(11)

France

Specification	Designation	Notes	C	Cr	Mn	Mo	Ni	P	S	Si	Other	UTS	YS	El	Hard
AFNOR NFA36206(83)	10CD12.10	Boiler, Press ves, Q/T; 60<t<=100mm	0.15 max	2.75-3.25	0.3-0.6	0.9-1.1	0.3 max	0.03 max	0.03 max	0.35 max	Cu <=0.25; Ti <=0.05; V <=0.04; bal Fe	520-670	310	18	
AFNOR NFA36206(83)	10CD12.10	Boiler, Press ves, Q/T; t<30mm	0.15 max	2.75-3.25	0.3-0.6	0.9-1.1	0.3 max	0.03 max	0.03 max	0.35 max	Cu <=0.25; Ti <=0.05; V <=0.04; bal Fe	520-670	310	21	
AFNOR NFA36206(83)	10CD9.10	Boiler, Press ves; t<=30mm	0.15 max	2-2.5	0.3-0.6	0.9-1.1	0.3 max	0.03 max	0.03 max	0.35 max	Cu <=0.25; Ti <=0.05; V <=0.04; bal Fe	520-670	310	21	
AFNOR NFA36602(88)	10CD12.10	Weld Frg, Boiler & Press ves, Q/T; t<30mm	0.15 max	2.75-3.25	0.3-0.6	0.9-1.1	0.3 max	0.02 max	0.015 max	0.35 max	Cu <=0.25; Ti <=0.05; V <=0.04; bal Fe	520-670	310	23	
AFNOR NFA36602(88)	10CD12.10	Weld Frg, Boiler & Press ves, Q/T; 60<t<=300mm	0.15 max	2.75-3.25	0.3-0.6	0.9-1.1	0.3 max	0.02 max	0.015 max	0.35 max	Cu <=0.25; Ti <=0.05; V <=0.04; bal Fe	520-670	310	20	

UNS numbers and US grades are provided as a means of cross referencing chemically similar alloys. Exchangability is only possible after independent examination of specifications. Tensile properties are minimum or typical as specified. UTS and YS as MPa. El as %. See Appendix for list of abbreviations used in Notes. * indicates obsolete material.

Specification	Designation	Notes	C	Cr	Mn	Mo	Ni	P	S	Si	Other	UTS	YS	El	Hard

Alloy Steel, Chromium Molybdenum, A387(11) (Continued from previous page)

France

Specification	Designation	Notes	C	Cr	Mn	Mo	Ni	P	S	Si	Other	UTS	YS	El	Hard
AFNOR NFA36602(88)	10CD9.10	Weld Frg, Boiler & Press ves, Q/T; 60<t<=100mm	0.15 max	2-2.5	0.3-0.6	0.9-1.1	0.3 max	0.02 max	0.015 max	0.35 max	Cu <=0.25; Ti <=0.05; V <=0.04; bal Fe	520-670	310	18	

UK

Specification	Designation	Notes	C	Cr	Mn	Mo	Ni	P	S	Si	Other	UTS	YS	El	Hard
BS 1503(89)	271-560	Frg press ves; t<=999mm; Norm Q/T	0.17 max	0.50-1.00	1.00-1.50	0.2-0.35	0.30-0.70	0.030 max	0.025 max	0.15-0.40	Al <=0.020; Cu <=0.30; W <=0.1; bal Fe	560-710	370	17	
BS 1503(89)	660-460	Frg Press ves; 0-999mm; N/T Q/T	0.1-0.18	0.3-0.6	0.4-0.7	0.5-0.7	0.4 max	0.03 max	0.025 max	0.15-0.4	Al <=0.02; Co <=0.1; Cu <=0.3; Pb <=0.15; Ti <=0.05; V 0.22-0.28; W <=0.1; Sn<=0.025; bal Fe	460-610	300	18L	

USA

Specification	Designation	Notes	C	Cr	Mn	Mo	Ni	P	S	Si	Other	UTS	YS	El	Hard
	UNS K11789		0.17 max	0.94-1.56	0.35-0.73	0.40-0.70		0.035 max	0.040 max	0.44-0.86	bal Fe				
ASTM A387/A387(97)	Grade 11	Press ves plt, Class 2	0.05-0.17	1.00-1.50	0.40-0.65	0.45-0.65		0.035 max	0.035 max	0.50-0.80	bal Fe	515-690	310	22	

Alloy Steel, Chromium Molybdenum, A387(12)

Bulgaria

Specification	Designation	Notes	C	Cr	Mn	Mo	Ni	P	S	Si	Other	UTS	YS	El	Hard
BDS 6354	15ChM	Struct	0.11-0.18	0.80-1.10	0.40-0.70	0.40-0.55	0.30 max	0.035 max	0.035 max	0.17-0.37	Cu <=0.30; Ti <=0.03; V <=0.05; W <=0.2; bal Fe				
BDS 6354	15ChM	Struct	0.11-0.18	0.80-1.10	0.40-0.70	0.40-0.55	0.30 max	0.035 max	0.035 max	0.17-0.37	Cu <=0.30; Ti <=0.03; V <=0.05; W <=0.2; bal Fe				
BDS 6354	15ChM	Struct	0.11-0.18	0.80-1.10	0.40-0.70	0.40-0.55	0.30 max	0.035 max	0.035 max	0.17-0.37	Cu <=0.30; Ti <=0.03; V <=0.05; W <=0.2; bal Fe				
BDS 6609(73)	14ChM	Heat res boiler tube	0.10-0.18	0.70-1.00	0.40-0.70	0.40-0.60	0.30 max	0.040 max	0.040 max	0.15-0.35	Cu <=0.30; bal Fe				
BDS 6609(73)	14ChM	Heat res boiler tube	0.10-0.18	0.70-1.00	0.40-0.70	0.40-0.60	0.30 max	0.040 max	0.040 max	0.15-0.35	Cu <=0.30; bal Fe				
BDS 6609(73)	14ChM	Heat res boiler tube	0.10-0.18	0.70-1.00	0.40-0.70	0.40-0.60	0.30 max	0.040 max	0.040 max	0.15-0.35	Cu <=0.30; bal Fe				

Czech Republic

Specification	Designation	Notes	C	Cr	Mn	Mo	Ni	P	S	Si	Other	UTS	YS	El	Hard
CSN 415121	15121	High-temp const	0.1-0.18	0.7-1.0	0.4-0.7	0.4-0.5		0.04 max	0.04 max	0.15-0.35	bal Fe				

Europe

Specification	Designation	Notes	C	Cr	Mn	Mo	Ni	P	S	Si	Other	UTS	YS	El	Hard
EN 10028/2(92)	1.7335	High-temp, Press; 100<=t<=150mm	0.08-0.18	0.70-1.15	0.40-1.00	0.40-0.60		0.030 max	0.025 max	0.35 max	Cu <=0.30; bal Fe	430-580	255	19T	
EN 10028/2(92)	1.7335	High-temp, Press; 60<=t<=100mm	0.08-0.18	0.70-1.15	0.40-1.00	0.40-0.60		0.030 max	0.025 max	0.35 max	Cu <=0.30; bal Fe	440-590	275	19T	
EN 10028/2(92)	1.7335	High-temp, Press; 16<=t<=60mm	0.08-0.18	0.70-1.15	0.40-1.00	0.40-0.60		0.030 max	0.025 max	0.35 max	Cu <=0.30; bal Fe	450-600	295	20T	
EN 10028/2(92)	1.7335	High-temp, Press; t<16mm	0.08-0.18	0.70-1.15	0.40-1.00	0.40-0.60		0.030 max	0.025 max	0.35 max	Cu <=0.30; bal Fe	450-600	300	20T	
EN 10028/2(92)	13CrMo4-5	High-temp, Press; 100<=t<=150mm	0.08-0.18	0.70-1.15	0.40-1.00	0.40-0.60		0.030 max	0.025 max	0.35 max	Cu <=0.30; bal Fe	430-580	255	19T	
EN 10028/2(92)	13CrMo4-5	High-temp, Press; 16<=t<=60mm	0.08-0.18	0.70-1.15	0.40-1.00	0.40-0.60		0.030 max	0.025 max	0.35 max	Cu <=0.30; bal Fe	450-600	295	20T	
EN 10028/2(92)	13CrMo4-5	High-temp, Press; t<16mm	0.08-0.18	0.70-1.15	0.40-1.00	0.40-0.60		0.030 max	0.025 max	0.35 max	Cu <=0.30; bal Fe	450-600	300	20T	
EN 10028/2(92)	13CrMo4-5	High-temp, Press; 60<=t<=100mm	0.08-0.18	0.70-1.15	0.40-1.00	0.40-0.60		0.030 max	0.025 max	0.35 max	Cu <=0.30; bal Fe	440-590	275	19T	

France

Specification	Designation	Notes	C	Cr	Mn	Mo	Ni	P	S	Si	Other	UTS	YS	El	Hard
AFNOR NFA35558(83)	15CD4.05	t<=160mm	0.18 max	0.8-1.2	0.4-0.8	0.4-0.6	0.3 max	0.035 max	0.035 max	0.15-0.35	Cu <=0.3; Ti <=0.05; bal Fe	490-640	350	20	
AFNOR NFA36205/2(92)	13CrMo4-5	Press ves; Press; High-temp; 60<=t<=100mm	0.08-0.18	0.7-1.15	0.40-1.00	0.4-0.60	0.4 max	0.03 max	0.025 max	0.35 max	Cu <=0.30; Ti <=0.03; Cu, Sn; bal Fe	440-590	275	19T	
AFNOR NFA36206(83)	15CD2.05	t<=30mm	0.18 max	0.4-0.6	0.5-0.9	0.4-0.6	0.3 max	0.03 max	0.03 max	0.15-0.3	Cu <=0.25; Ti <=0.05; V <=0.04; bal Fe	450-570	275	25	
AFNOR NFA36206(83)	15CD2.05	80<t<=100mm	0.18 max	0.4-0.6	0.5-0.9	0.4-0.6	0.3 max	0.03 max	0.03 max	0.15-0.3	Cu <=0.25; Ti <=0.05; V <=0.04; bal Fe	450-570	255	22	

Alloy Steel, Chromium Molybdenum, A387(12) (Continued from previous page)

Specification	Designation	Notes	C	Cr	Mn	Mo	Ni	P	S	Si	Other	UTS	YS	El	Hard
France															
AFNOR NFA36206(83)	15CD4.05	80<t<=100mm	0.18 max	0.8-1.2	0.4-0.8	0.4-0.6	0.3 max	0.03 max	0.03 max	0.15-0.35	Cu <=025; Ti <=0.05; V <=0.04; W <=01; bal Fe	80.1-100	275	21	
AFNOR NFA36206(83)	15CD4.05	t<=30mm	0.18 max	0.8-1.2	0.4-0.8	0.4-0.6	0.3 max	0.03 max	0.03 max	0.15-0.35	Cu <=025; Ti <=0.05; V <=0.04; W <=01; bal Fe	470-610	295	23	
AFNOR NFA36602(88)	15CD4.05	t<=30mm	0.18 max	0.8-1.2	0.4-0.8	0.4-0.6	0.3 max	0.02 max	0.015 max	0.15-0.35	Cu <=0.25; Ti <=0.05; V <=0.04; bal Fe	470-610	295	25L, 23T	
AFNOR NFA36602(88)	15CD4.05	80<t<=300mm	0.18 max	0.8-1.2	0.4-0.8	0.4-0.6	0.3 max	0.02 max	0.015 max	0.15-0.35	Cu <=0.25; Ti <=0.05; V <=0.04; bal Fe	470-610	275	23L, 21T	
Hungary															
MSZ 2295(79)	14HCM	High-press hydrogenation ves; 0-100mm; Q/T	0.1-0.18	0.8-1.1	0.4-0.7	0.4-0.6		0.03 max	0.03 max	0.17-0.37	bal Fe	450-600	300	21L	180 HB max
MSZ 4747(85)	Cr5Mo45.47	High-temp, Tub; Non-rimming; K; FF; 40.1-60mm; Q/T	0.1-0.18	0.7-1.1	0.4-0.7	0.45-0.65		0.035 max	0.035 max	0.35 max	bal Fe	440-590	280	22L; 20T	
MSZ 4747(85)	Cr5Mo45.47	High-temp, Tub; Non-rimming; K; FF; 0-30mm; Q/T	0.1-0.18	0.7-1.1	0.4-0.7	0.45-0.65		0.035 max	0.035 max	0.35 max	bal Fe	440-590	290	22L; 20T	
Italy															
UNI 5462(64)	14CrMo3	Smls tube	0.1-0.18	0.7-1	0.4-0.7	0.45-0.65		0.035 max	0.035 max	0.15-0.35	bal Fe				
UNI 5869(75)	14CrMo45	Press ves	0.1-0.18	0.8-1.15	0.4-0.8	0.4-0.6		0.03 max	0.035 max	0.15-0.35	bal Fe				
Poland															
PNH84024	15HM	High-temp const	0.11-0.18	0.7-1	0.4-0.7	0.4-0.55	0.35 max	0.04 max	0.04 max	0.15-0.35	Al 0.02-0.120; Cu <=0.25; bal Fe				
Romania															
STAS 2883/3(88)	14CrMo4	Heat res	0.1-0.18	0.7-1	0.4-0.7	0.4-0.6		0.035 max	0.03 max	0.15-0.35	Al 0.015-0.035; Pb <=0.15; As<=0.07; bal Fe				
STAS 8184(87)	14CrMo4	High-temp const	0.1-0.18	0.7-1	0.4-0.7	0.4-0.55		0.04 max	0.04 max	0.15-0.35	Pb <=0.15; Ti <=0.02; As<=0.05; bal Fe				
Russia															
GOST 4543(71)	15ChM	High-temp const	0.11-0.18	0.8-1.1	0.4-0.7	0.4-0.55	0.3 max	0.035 max	0.035 max	0.17-0.37	Al <=0.1; Cu <=0.3; Ti <=0.05; bal Fe				
GOST 5520(79)	12ChM	Boiler, Press ves	0.16 max	0.80-1.10	0.40-0.70	0.40-0.55	0.30 max	0.025 max	0.025 max	0.17-0.37	Al <=0.02; Cu <=0.30; bal Fe				
Spain															
UNE 36087(76/78)	14CrMo45		0.1-0.18	0.8-1.15	0.4-0.8	0.4-0.6		0.035 max	0.035 max	0.15-0.35	bal Fe				
UNE 36087(76/78)	F.2631*	see 14CrMo45	0.1-0.18	0.8-1.15	0.4-0.8	0.4-0.6		0.035 max	0.035 max	0.15-0.35	bal Fe				
Sweden															
SS 142216	2216	High-temp const	0.1-0.18	0.7-1.1	0.4-1	0.4-0.6		0.035 max	0.03 max	0.1-0.35	bal Fe				
UK															
BS 1502(82)	620-440	Bar Shp Press ves	0.10-0.18	0.80-1.20	0.40-0.70	0.45-0.65		0.040 max	0.040 max	0.15-0.40	Al <=0.020; Co <=0.1; W <=0.1; bal Fe				
BS 1502(82)	620-470	Bar Shp Press ves	0.10-0.18	0.80-1.20	0.40-0.70	0.45-0.65		0.040 max	0.040 max	0.15-0.40	Al <=0.030; Co <=0.1; W <=0.1; bal Fe				
BS 1502(82)	620-540	Bar Shp Press ves	0.10-0.18	0.80-1.20	0.40-0.70	0.45-0.65		0.040 max	0.040 max	0.15-0.40	Al <=0.020; Co <=0.1; W <=0.1; bal Fe				
USA															
	UNS K11757		0.17 max	0.74-1.21	0.35-0.73	0.40-0.65		0.035 max	0.040 max	0.13-0.32	bal Fe				
ASTM A387/A387(97)	Grade 12	Press ves plt, Class 2	0.05-0.17	0.80-1.15	0.40-0.65	0.45-0.60		0.035 max	0.035 max	0.15-0.40	bal Fe	450-585	275	22	

Specification	Designation	Notes	C	Cr	Mn	Mo	Ni	P	S	Si	Other	UTS	YS	El	Hard
Alloy Steel, Chromium Molybdenum, A387(2)															
USA															
	UNS K12143		0.21 max	0.46-0.85	0.50-0.88	0.40-0.65		0.035 max	0.040 max	0.13-0.45	bal Fe				
ASTM A387/A387(97)	Grade 2	Press ves plt, Class 2	0.05-0.21	0.50-0.80	0.55-0.80	0.45-0.60		0.035 max	0.035 max	0.15-0.40	bal Fe	485-620	310	22	
Alloy Steel, Chromium Molybdenum, A405(P24)															
USA															
	UNS K11591		0.15 max	0.80-1.25	0.30-0.60	0.87-1.13		0.030 max	0.030 max	0.10-0.35	V 0.15-0.25; bal Fe				
Alloy Steel, Chromium Molybdenum, A485(3)															
France															
AFNOR NFA35565(94)	100CD7*	see 100CrMo7-2; Ball & roller bearing	0.9-1.05	1.65-1.95	0.25-0.45	0.15-0.3	0.3 max	0.025 max	0.015 max	0.2-0.4	Cu <=0.3; bal Fe				
International															
ISO 683-17(76)	4	Ball & roller bearing, HT	0.95-1.10	1.65-1.95	0.25-0.45	0.20-0.40		0.030 max	0.025 max	0.20-0.40	bal Fe				
Spain															
UNE 36027(80)	100CrMn7	Ball & roller bearing	0.95-1.1	1.65-1.95	0.25-0.45	0.2-0.4		0.03 max	0.025 max	0.2-0.4	bal Fe				
UNE 36027(80)	F.1313*	see 100CrMn7	0.95-1.1	1.65-1.95	0.25-0.45	0.2-0.4		0.03 max	0.025 max	0.2-0.4	bal Fe				
USA															
	UNS K19965	Bearing	0.95-1.0	1.10-1.50	0.65-0.90	0.20-0.30	0.25 max	0.025 max	0.025 max	0.15-0.35	Cu <=0.35; bal Fe				
ASTM A485(94)	Grade 3	Bearing, hard	0.95-1.10	1.10-1.50	0.65-0.90	0.20-0.30		0.025 max	0.025 max	0.15-0.35	Cu <=0.35; N <=0.25; bal Fe				46 HRC
Alloy Steel, Chromium Molybdenum, A485(4)															
USA															
	UNS K19990	Bearing	0.95-1.10	1.10-1.50	1.05-1.35	0.45-0.60	0.25 max	0.025 max	0.025 max	0.15-0.35	Cu <=0.35; bal Fe				
ASTM A485(94)	Grade 4	Bearing, hard	0.95-1.10	1.10-1.50	1.05-1.35	0.45-0.60		0.025 max	0.025 max	0.15-0.35	Cu <=0.35; N <=0.25; bal Fe				35 HRC
Alloy Steel, Chromium Molybdenum, A514(E)															
USA															
	UNS K21604		0.12-0.20	1.40-2.00	0.40-0.70	0.40-0.60		0.035 max	0.04 max	0.20-0.35	B 0.0015-0.005; Cu 0.20-0.40; Ti 0.04-0.10; bal Fe				
ASTM A514(94)	Grade E	Q/T, Plt, t=150mm, weld	0.12-0.20	1.40-2.00	0.40-0.70	0.40-0.60		0.035 max	0.035 max	0.20-0.40	B 0.001-0.005; Ti 0.01-0.10; bal Fe	690-895	620	16	
ASTM A517(93)	Grade E	Q/T, Press ves plt, t<=65mm	0.12-0.20	1.40-2.00	0.40-0.70	0.40-0.60		0.035 max	0.035 max	0.10-0.40	B 0.001-0.005; Ti 0.01-0.10; bal Fe	795-930	690	16	
ASTM A709(97)	Grade 100, Type E	Q/T, Plt, t<=100mm	0.12-0.20	1.40-2.00	0.40-0.70	0.40-0.60		0.035 max	0.035 max	0.20-0.40	B 0.001-0.005; Ti 0.01-0.10; bal Fe	690-895	620	16	
Alloy Steel, Chromium Molybdenum, A540(B21)															
Japan															
JIS G4108(94)	SNB21-1	Bar diam<=100mm	0.36-0.44	0.80-1.15	0.45-0.70	0.50-0.65		0.025 max	0.025 max	0.20-0.35	V 0.25-0.35; bal Fe	1140	1030	10	321-429 HB
JIS G4108(94)	SNB21-2	Bar 75<diam<=150mm	0.36-0.44	0.80-1.15	0.45-0.70	0.50-0.65		0.025 max	0.025 max	0.20-0.35	V 0.25-0.35; bal Fe	1070	960	11	311-429 HB
JIS G4108(94)	SNB21-3	Bar 75<diam<=150mm	0.36-0.44	0.80-1.15	0.45-0.70	0.50-0.65		0.025 max	0.025 max	0.20-0.35	V 0.25-0.35; bal Fe	1000	890	12	302-375 HB
JIS G4108(94)	SNB21-4	Bar 75<diam<=150mm	0.36-0.44	0.80-1.15	0.45-0.70	0.50-0.65		0.025 max	0.025 max	0.20-0.35	V 0.25-0.35; bal Fe	930	825	13	277-352 HB
JIS G4108(94)	SNB21-5	Bar 150<diam<=200mm	0.36-0.44	0.80-1.15	0.45-0.70	0.50-0.65		0.025 max	0.025 max	0.20-0.35	V 0.25-0.35; bal Fe	790	685	15	255-311 HB
USA															
	UNS K14073		0.36-0.44	0.80-1.15	0.45-0.70	0.50-0.65		0.025 max	0.025 max	0.15-0.35	V 0.25-0.35; bal Fe				
ASTM A540/A540M(98)	A540 (B21) *	Bolting matl	0.36-0.44	0.80-1.15	0.45-0.70	0.50-0.65		0.025 max	0.025 max	0.15-0.35	V 0.25-0.35; bal Fe	827-1138	724-1034	15-10	241-321/285-429 HB

UNS numbers and US grades are provided as a means of cross referencing chemically similar alloys. Exchangability is only possible after independent examination of specifications. Tensile properties are minimum or typical as specified. UTS and YS as MPa. El as %. See Appendix for list of abbreviations used in Notes. * indicates obsolete material.

Specification	Designation	Notes	C	Cr	Mn	Mo	Ni	P	S	Si	Other	UTS	YS	El	Hard

Alloy Steel, Chromium Molybdenum, A541(22B)

Romania

| STAS 2883/3(88) | 12MoCr22 | Heat res | 0.08-0.15 | 2-2.5 | 0.4-0.7 | 0.9-1.1 | | 0.035 max | 0.03 max | 0.15-0.5 | Al 0.015-0.035; N <=0.009; Pb <=0.15; As<=0.05; bal Fe | | | | |

USA

	UNS K21390		0.15 max	2.00-2.50	0.30-0.60	0.90-1.10		0.035 max	0.040 max	0.50 max	bal Fe				
ASTM A541/541M(95)	Grade 22 Class 3	Q/T frg for press ves components	0.11-0.15	2.00-2.50	0.30-0.60	0.90-1.10	0.25 max	0.015 max	0.015 max	0.50 max	V <=0.02; Si req vary; bal Fe	585-760	380	18	
ASTM A541/541M(95) *	Grade 22 Class 4	Q/T frg for press ves components	0.05-0.15	2.00-2.50	0.30-0.60	0.90-1.10	0.50 max	0.025 max	0.025 max	0.50 max	V <=0.05; Si req vary; bal Fe	720-900	590	16	
ASTM A541/541M(95)	Grade 22 Class 5	Q/T frg for press ves components	0.05-0.15	2.00-2.50	0.30-0.60	0.90-1.10	0.50 max	0.025 max	0.025 max	0.50 max	V <=0.05; Si req vary; bal Fe	790-1000	690	15	
ASTM A541/541M(95)	Grade 22V	Q/T frg for press ves components	0.11-0.15	2.00-2.50	0.30-0.60	0.90-1.10	0.25 max	0.015 max	0.010 max	0.10 max	B <=0.0020; Cu <=0.20; Nb <=0.07; Ti <=0.030; V 0.25-0.35; Si req vary;Ca<=0.015 or REM; bal Fe	585-760	415	18	
ASTM A739(95)	B22	Bar HW for elev temp or press containing	0.05-0.15	2.00-2.50	0.30-0.60	0.90-1.10		0.035 max	0.040 max	0.50 max	bal Fe	517-655	310	18	

Alloy Steel, Chromium Molybdenum, A592(E)

Czech Republic

| CSN 415422 | 15422 | Boron | 0.14-0.19 | 1.4-1.7 | 1.4-1.7 | 0.4-0.6 | 0.3 max | 0.035 max | 0.035 max | 0.2-0.4 | B 0.002-0.006; Ti <=0.04; bal Fe | | | | |

Japan

JIS G4105(79)	SCM22*	Obs; see SCM420	0.18-0.23	0.90-1.20	0.60-0.85	0.15-0.30	0.25 max	0.030 max	0.030 max	0.15-0.35	Cu <=0.30; bal Fe				
JIS G4105(79)	SCM23*	Obs; see SCM421	0.17-0.23	0.90-1.20	0.70-1.00	0.15-0.30	0.25 max	0.030 max	0.030 max	0.15-0.35	Cu <=0.30; bal Fe				
JIS G4105(79)	SCM420	HR Frg Bar Rod	0.18-0.23	0.90-1.20	0.60-0.85	0.15-0.30	0.25 max	0.030 max	0.030 max	0.15-0.35	Cu <=0.30; bal Fe				
JIS G4105(79)	SCM421	HR Frg Bar Rod	0.17-0.23	0.90-1.20	0.70-1.00	0.15-0.30	0.25 max	0.030 max	0.030 max	0.15-0.35	Cu <=0.30; bal Fe				

USA

| | UNS K11695 | | 0.10-0.22 | 1.34-2.06 | 0.37-0.74 | 0.36-0.64 | | 0.035 max | 0.040 max | 0.18-0.37 | B 0.0015-0.005; Cu 0.17-0.43; Ti 0.03-0.11; bal Fe | | | | |
| ASTM A592/592M(94) | Grade E | High-strength Q/T frg for press ves, <=2.5 in. | 0.12-0.20 | 1.40-2.00 | 0.40-0.70 | 0.40-0.60 | | 0.025 max | 0.025 max | 0.20-0.35 | B 0.0015-0.005; Cu 0.20-0.40; Ti 0.04-0.10; V may be substituted for Ti; bal Fe | 795-930 | 690 | 18 | |

Alloy Steel, Chromium Molybdenum, A649(1A)

USA

| | UNS K14247 | | 0.35-0.55 | 0.80-1.15 | 0.55-1.05 | 0.15-0.50 | | 0.035 max | 0.040 max | 0.15-0.35 | bal Fe | | | | |
| ASTM A649/649M(98) | Class 1A | Frg roll for paperboard mach | 0.45-0.60 | 0.80-1.15 | 0.55-1.05 | 0.15-0.50 | | 0.025 max | 0.025 max | 0.15-0.35 | Si<=0.10 with VCD; bal Fe | 690-1030 | 450-890 | 12-14 | |

Alloy Steel, Chromium Molybdenum, A687(II)

USA

| | UNS K14044 | | 0.36-0.45 | 0.75-1.15 | 0.71-1.04 | 0.13-0.27 | | 0.040 max | 0.045 max | 0.18-0.37 | bal Fe | | | | |

Alloy Steel, Chromium Molybdenum, A739(B11)

USA

| | UNS K11797 | | 0.20 max | 1.00-1.50 | 0.40-0.65 | 0.45-0.65 | | 0.035 max | 0.040 max | 0.50-0.80 | bal Fe | | | | |
| ASTM A739(95) | B11 | Bar HW for elev temp or press containing | 0.05-0.20 | 1.00-1.50 | 0.40-0.65 | 0.45-0.65 | | 0.035 max | 0.040 max | 0.50-0.80 | bal Fe | 483-655 | 310 | 18 | |

Alloy Steel, Chromium Molybdenum, AMS 6255

USA

| | UNS K21940 | NbS600 | 0.16-0.22 | 1.25-1.65 | 0.45-0.75 | 0.90-1.10 | 0.25 max | 0.010 max | 0.010 max | 0.90-1.25 | Al 0.03-0.12; Cu <=0.35; bal Fe | | | | |

UNS numbers and US grades are provided as a means of cross referencing chemically similar alloys. Exchangability is only possible after independent examination of specifications. Tensile properties are minimum or typical as specified. UTS and YS as MPa. El as %. See Appendix for list of abbreviations used in Notes. * indicates obsolete material.

Specification	Designation	Notes	C	Cr	Mn	Mo	Ni	P	S	Si	Other	UTS	YS	El	Hard

Alloy Steel, Chromium Molybdenum, B16

Japan

Specification	Designation	Notes	C	Cr	Mn	Mo	Ni	P	S	Si	Other	UTS	YS	El	Hard
JIS G4107(94)	SNB16	Class 3 diam<=180mm Bar	0.36-0.44	0.80-1.15	0.45-0.70	0.50-0.65		0.040 max	0.040 max	0.20-0.35	V 0.25-0.35; bal Fe	690-860	590-725	16-18	

USA

Specification	Designation	Notes	C	Cr	Mn	Mo	Ni	P	S	Si	Other	UTS	YS	El	Hard
	UNS K14072		0.36-0.44	0.80-1.15	0.45-0.70	0.50-0.65		0.04 max	0.04 max	0.15-0.35	V 0.25-0.35; bal Fe				
ASTM A193/A193M(98)	B16	Bolt, high-temp	0.36-0.47	0.80-1.15	0.45-0.70	0.50-0.65		0.035 max	0.040 max	0.15-0.35	Al <=0.015; V 0.25-0.35; bal Fe	860	725	18	321 HB
ASTM A194/A194M(98)	16	Nuts, high-temp press	0.36-0.47	0.80-1.15	0.45-0.70	0.50-0.65		0.035 max	0.040 max	0.15-0.35	Al <=0.015; V 0.25-0.35; bal Fe				248-352 HB
ASTM A437/437M(98)	B4D	High-temp bolt stud, t<=65mm	0.36-0.44	0.80-1.15	0.45-0.70	0.50-0.65		0.04 max	0.04 max	0.15-0.35	V 0.25-0.35; bal Fe	860	720	18	302 HB
ASTM A437/437M(98)	B4D	High-temp nut washer mat, t<=65mm	0.36-0.44	0.80-1.15	0.45-0.70	0.50-0.65		0.04 max	0.04 max	0.15-0.35	V 0.25-0.35; bal Fe	860	720	18	263-311 HB
MIL-B-11595E(88)	Cr-Mo-V	Small arms barrels	0.41-0.49	0.80-1.15	0.60-0.90	0.30-0.40		0.040 max	0.040 max	0.20-0.35	Al <=0.040; Cu <=0.35; V 0.20-0.30; bal Fe				
SAE J775(93)	B16	Engine poppet valve, Mart	0.36-0.44	0.80-1.15	0.45-0.70			0.040 max	0.040 max	0.15-0.35	bal Fe				

Alloy Steel, Chromium Molybdenum, F11-1

USA

Specification	Designation	Notes	C	Cr	Mn	Mo	Ni	P	S	Si	Other	UTS	YS	El	Hard
	UNS K11597		0.15 max	1.00-1.50	0.30-0.60	0.44-0.65		0.030 max	0.030 max	0.50-1.00	bal Fe				
ASTM A182/A182M(98)	F11 Class 1	Frg/roll pipe flange valve	0.05-0.15	1.00-1.50	0.30-0.60	0.44-0.65		0.030 max	0.030 max	0.50-1.00	bal Fe	415	205	20	121-174 HB
ASTM A213/A213M(95)	T11	Smls tube boiler, superheater, heat exchanger	0.05-0.15	1.00-1.50	0.30-0.60	0.44-0.65		0.025 max	0.025 max	0.50-1.00	bal Fe	415	205	30	
ASTM A234/A234M(97)	WP11 CL1	Frg	0.05-0.15	1.00-1.50	0.30-0.60	0.44-0.65		0.030 max	0.030 max	0.50-1.00	bal Fe	415-585	205	22L	
ASTM A336/A336M(98)	F11 Class 2	Ferr Frg	0.10-0.20	1.00-1.50	0.30-0.80	0.45-0.65		0.025 max	0.025 max	0.50-1.00	bal Fe	485-660	275	20	

Alloy Steel, Chromium Molybdenum, F11-2

USA

Specification	Designation	Notes	C	Cr	Mn	Mo	Ni	P	S	Si	Other	UTS	YS	El	Hard
	UNS K11572		0.10-0.20	1.00-1.50	0.30-0.80	0.44-0.65		0.040 max	0.040 max	0.50-1.00	bal Fe				
ASTM A182/A182M(98)	F11 Class 2	Frg/roll pipe flange valve	0.01-0.20	1.00-1.50	0.30-0.60	0.44-0.65		0.040 max	0.040 max	0.50-1.00	bal Fe	485	275	20.0	143-207 HB
ASTM A182/A182M(98)	F11 Class 3	Frg/roll pipe flange valve	0.01-0.20	1.00-1.50	0.30-0.60	0.44-0.65		0.040 max	0.040 max	0.50-1.00	bal Fe	515	310	20	156-207 HB
ASTM A336/A336M(98)	F11 Class 1	Ferr Frg	0.05-0.15	1.00-1.50	0.30-0.60	0.45-0.65		0.025 max	0.025 max	0.50-1.00	bal Fe	415-585	205	20	
ASTM A336/A336M(98)	F11 Class 3	Ferr Frg	0.10-0.20	1.00-1.50	0.30-0.80	0.45-0.65		0.025 max	0.025 max	0.50-1.00	bal Fe	515-690	310	18	
ASTM A541/541M(95)	Grade 11 Class 4	Q/T frg for press ves components	0.10-0.20	1.00-1.50	0.30-0.80	0.45-0.65	0.50 max	0.025 max	0.025 max	0.50-1.00	V <=0.05; Si req vary; bal Fe	550-720	340	18	

Alloy Steel, Chromium Molybdenum, F12

Japan

Specification	Designation	Notes	C	Cr	Mn	Mo	Ni	P	S	Si	Other	UTS	YS	El	Hard
JIS G4105(79)	SCM21*	Obs; see SCM415	0.13-0.18	0.90-1.20	0.60-0.85	0.15-0.30	0.25 max	0.030 max	0.030 max	0.15-0.35	Cu <=0.30; bal Fe				
JIS G4105(79)	SCM415	HR Frg Bar Rod	0.13-0.18	0.90-1.20	0.60-0.85	0.15-0.30	0.25 max	0.030 max	0.030 max	0.15-0.35	Cu <=0.30; bal Fe				

USA

Specification	Designation	Notes	C	Cr	Mn	Mo	Ni	P	S	Si	Other	UTS	YS	El	Hard
	UNS K11564		0.10-0.20	0.80-1.10	0.30-0.80	0.45-0.65		0.040 max	0.040 max	0.10-0.60	bal Fe				
ASTM A182/A182M(98)	F12 Class 2	Frg/roll pipe flange valve	0.10-0.20	0.80-1.25	0.30-0.60	0.44-0.65		0.040 max	0.040 max	0.10-0.60	bal Fe	485	275	20.0	143-207 HB
ASTM A336/A336M(98)	F12	Ferr Frg	0.10-0.20	0.80-1.10	0.30-0.80	0.45-0.65		0.025 max	0.025 max	0.10-0.60	bal Fe	485-660	275	20	

Alloy Steel, Chromium Molybdenum, F21

USA

Specification	Designation	Notes	C	Cr	Mn	Mo	Ni	P	S	Si	Other	UTS	YS	El	Hard
	UNS K31545		0.15 max	2.65-3.35	0.30-0.60	0.80-1.06		0.030 max	0.030 max	0.50 max	bal Fe				

UNS numbers and US grades are provided as a means of cross referencing chemically similar alloys. Exchangability is only possible after independent examination of specifications. Tensile properties are minimum or typical as specified. UTS and YS as MPa. El as %. See Appendix for list of abbreviations used in Notes. * indicates obsolete material.

Specification	Designation	Notes	C	Cr	Mn	Mo	Ni	P	S	Si	Other	UTS	YS	El	Hard

Alloy Steel, Chromium Molybdenum, F21 (Continued from previous page)

USA

Specification	Designation	Notes	C	Cr	Mn	Mo	Ni	P	S	Si	Other	UTS	YS	El	Hard
ASTM A182/A182M(98)	F21	Frg/roll pipe flange valve	0.05-0.15	2.7-3.3	0.30-0.60	0.80-1.06		0.040 max	0.040 max	0.50 max	bal Fe	515	310	20.0	156-207 HB
ASTM A213/A213M(95)	T21	Smls tube boiler, superheater, heat exchanger	0.05-0.15	2.65-3.35	0.30-0.60	0.80-1.06		0.025 max	0.025 max	0.50 max	bal Fe	415	205	30	
ASTM A336/A336M(98)	F21 Class 3	Ferr Frg	0.05-0.15	2.7-3.3	0.30-0.60	0.80-1.06		0.025 max	0.025 max	0.50 max	bal Fe	515-690	310	19	
ASTM A336/A336M(98)	F21 Class1	Ferr Frg	0.05-0.15	2.7-3.3	0.30-0.60	0.80-1.06		0.025 max	0.025 max	0.50 max	bal Fe	415-585	205	20	
ASTM A387/A387(97)	Grade 21	Press ves plt, Class 2	0.05-0.15	2.75-3.25	0.30-0.60	0.90-1.10		0.035 max	0.035 max	0.50 max		515-690	310	18	
ASTM A387/A387(97)	Grade 21L	Press ves plt, Class 1	0.10 max	2.75-3.25	0.30-0.60	0.90-1.10		0.035 max	0.035 max	0.50 max	bal Fe	415-485	207	18	
ASTM A989(98)	K31545	HIP Flanges, Fittings, Valves/parts; Heat res	0.05-0.15	2.7-3.3	0.30-0.60	0.80-1.06		0.040 max	0.040 max	0.50 max	bal Fe	515	310	20.0	156-207 HB

Alloy Steel, Chromium Molybdenum, F21b

USA

Specification	Designation	Notes	C	Cr	Mn	Mo	Ni	P	S	Si	Other	UTS	YS	El	Hard
	UNS K31830		0.18 max	2.75-3.25	0.30-0.60	0.90-1.10		0.02 max	0.02 max	0.12 max	B 0.001-0.003; Ti 0.015-0.035; V 0.20-0.30; bal Fe				
ASTM A182/A182M(98)	F3V	Frg/roll pipe flange valve	0.05-0.18	2.8-3.2	0.30-0.60	0.90-1.10		0.020 max	0.020 max	0.10 max	B 0.001-0.003; Ti 0.015-0.035; V 0.20-0.30; bal Fe	585-760	415	18	174-237 HB
ASTM A336/A336M(98)	F3V	Ferr Frg	0.05-0.15	2.7-3.3	0.30-0.60	0.90-1.10		0.020 max	0.020 max	0.10 max	B 0.001-0.003; Ti 0.015-0.035; V 0.20-0.30; bal Fe	585-760	415	18	
ASTM A541/541M(95)	Grade 3V	Q/T frg for press ves components	0.10-0.15	2.8-3.3	0.30-0.60	0.90-1.10		0.020 max	0.020 max	0.10 max	B 0.001-0.003; Ti 0.015-0.035; V 0.20-0.30; Si req vary; bal Fe	585-760	415	18	
ASTM A541/541M(95)	Grade 3VCb	Q/T frg for press ves components	0.10-0.15	2.7-3.3	0.30-0.60	0.90-1.10	0.25 max	0.020 max	0.010 max	0.10 max	Cu <=0.25; Nb 0.015-0.070; Ti <=0.015; V 0.20-0.30; Si req vary;Ca 0.0005-0.0150; bal Fe	585-760	415	18	
ASTM A542(95)	Type C	Q/T, Press ves plt, Class 4, t>=5mm	0.10-0.15	2.75-3.25	0.30-0.60	0.90-1.10	0.25 max	0.025 max	0.025 max	0.13 max	B 0.001-0.003; Cu <=0.25; Ti 0.015-0.035; V 0.20-0.30; bal Fe	585-790	380	20	
ASTM A542(95)	Type C	Q/T, Press ves plt, Class 4a, t>=5mm	0.10-0.15	2.75-3.25	0.30-0.60	0.90-1.10	0.25 max	0.025 max	0.025 max	0.13 max	B 0.001-0.003; Cu <=0.25; Ti 0.015-0.035; V 0.20-0.30; bal Fe	585-760	415	18	
ASTM A542(95)	Type C	Q/T, Press ves plt, Class 2, t>=5mm	0.10-0.15	2.75-3.25	0.30-0.60	0.90-1.10	0.25 max	0.025 max	0.025 max	0.13 max	B 0.001-0.003; Cu <=0.25; Ti 0.015-0.035; V 0.20-0.30; bal Fe	795-930	690	13	
ASTM A542(95)	Type C	Q/T, Press ves plt, Class 1, t>=5mm	0.10-0.15	2.75-3.25	0.30-0.60	0.90-1.10	0.25 max	0.025 max	0.025 max	0.13 max	B 0.001-0.003; Cu <=0.25; Ti 0.015-0.035; V 0.20-0.30; bal Fe	725-860	585	14	
ASTM A542(95)	Type C	Q/T, Press ves plt, Class 3, t>=5mm	0.10-0.15	2.75-3.25	0.30-0.60	0.90-1.10	0.25 max	0.025 max	0.025 max	0.13 max	B 0.001-0.003; Cu <=0.25; Ti 0.015-0.035; V 0.20-0.30; bal Fe	655-795	515	20	
ASTM A832/A832M(97)	Grade 21V	Press ves plt	0.10-0.15	2.75-3.25	0.30-0.60	0.90-1.10		0.025 max	0.025 max	0.10 max	B 0.001-0.003; Ti 0.015-0.035; V 0.20-0.30; bal Fe	585-760	415	18	

Alloy Steel, Chromium Molybdenum, F5

Hungary

Specification	Designation	Notes	C	Cr	Mn	Mo	Ni	P	S	Si	Other	UTS	YS	El	Hard
MSZ 2295(88)	12CrMo205	High-press hydrogenation ves; 0-100mm; air Q/T	0.08-0.15	4.5-5.5	0.3-0.6	0.45-0.55		0.035 max	0.035 max	0.3-0.5	bal Fe	440-590	245	20L	130-175 HB
MSZ 2295(88)	12CrMo205	High-press hydrogenation ves; 0-100mm; soft ann	0.08-0.15	4.5-5.5	0.3-0.6	0.45-0.55		0.035 max	0.035 max	0.3-0.5	bal Fe	410	180	21L	190 HB max

UNS numbers and US grades are provided as a means of cross referencing chemically similar alloys. Exchangability is only possible after independent examination of specifications. Tensile properties are minimum or typical as specified. UTS and YS as MPa. El as %. See Appendix for list of abbreviations used in Notes. * indicates obsolete material.

Specification	Designation	Notes	C	Cr	Mn	Mo	Ni	P	S	Si	Other	UTS	YS	El	Hard

Alloy Steel, Chromium Molybdenum, F5 (Continued from previous page)

Romania

Specification	Designation	Notes	C	Cr	Mn	Mo	Ni	P	S	Si	Other	UTS	YS	El	Hard
STAS 2883/3(88)	12MoCr50	Heat res	0.08-0.15	4-6	0.3-0.6	0.45-0.65		0.035 max	0.03 max	0.15-0.5	Al 0.015-0.035; N <=0.009; Pb <=0.15; V 0.15-0,3; As<=0.05; bal Fe				

Russia

Specification	Designation	Notes	C	Cr	Mn	Mo	Ni	P	S	Si	Other	UTS	YS	El	Hard
GOST 20072(74)	15Ch5M	High-temp const	0.15 max	4.50-6.00	0.50 max	0.45-0.60	0.60 max	0.030 max	0.025 max	0.50 max	Cu <=0.20; Ti <=0.03; V <=0.05; W <=0.30; bal Fe				

USA

Specification	Designation	Notes	C	Cr	Mn	Mo	Ni	P	S	Si	Other	UTS	YS	El	Hard
	UNS K41545		0.15 max	4.00-6.00	0.30-0.60	0.45-0.65		0.030 max	0.030 max	0.50 max	bal Fe				
ASTM A182/A182M(98)	F5	Frg/roll pipe flange valve	0.15 max	4.0-6.0	0.30-0.60	0.44-0.65	0.50 max	0.030 max	0.030 max	0.50 max	bal Fe	485	275	20.0	143-217 HB
ASTM A213/A213M(95)	T5	Smls tube boiler, superheater, heat exchanger	0.15 max	4.00-6.00	0.30-0.60	0.45-0.65		0.025 max	0.025 max	0.50 max	bal Fe	415	205	30	
ASTM A234/A234M(97)	WP5	Frg	0.15 max	4.0-6.0	0.30-0.60	0.44-0.65		0.040 max	0.030 max	0.50 max	bal Fe	415-585	205	22L	
ASTM A336/A336M(98)	F5	Ferr Frg	0.15 max	4.0-5.0	0.30-0.60	0.45-0.65	0.50 max	0.025 max	0.025 max	0.50 max	bal Fe	415-585	250	20	
ASTM A387/A387(97)	Grade 5	Press ves plt, Class 2	0.15 max	4.00-6.00	0.30-0.60	0.45-0.65		0.035 max	0.035 max	0.50 max	bal Fe	515-690	310	18	

Alloy Steel, Chromium Molybdenum, K11598

USA

Specification	Designation	Notes	C	Cr	Mn	Mo	Ni	P	S	Si	Other	UTS	YS	El	Hard
	UNS K11598		0.10-0.20	1.00-1.50	0.30-0.80	0.45-0.65		0.035 max	0.040 max	0.50-1.00	bal Fe				

Alloy Steel, Chromium Molybdenum, K11682

USA

Specification	Designation	Notes	C	Cr	Mn	Mo	Ni	P	S	Si	Other	UTS	YS	El	Hard
	UNS K11682		0.13-0.20	1.15-1.65	0.40-0.70	0.25-0.40		0.035 max	0.04 max	0.20-0.35	B 0.0015-0.005; Cu 0.20-0.40; Ti 0.04-0.10; bal Fe				

Alloy Steel, Chromium Molybdenum, K12125

Italy

Specification	Designation	Notes	C	Cr	Mn	Mo	Ni	P	S	Si	Other	UTS	YS	El	Hard
UNI 6403(86)	25MnCr6	Q/T	0.2-0.29	0.3-0.6	1.2-1.6			0.035 max	0.035 max	0.15-0.35	bal Fe				
UNI 8913(87)	25MnCr6	Q/T	0.2-0.29	0.3-0.7	1.2-1.6			0.035 max	0.035 max	0.35 max	bal Fe				

USA

Specification	Designation	Notes	C	Cr	Mn	Mo	Ni	P	S	Si	Other	UTS	YS	El	Hard
	UNS K12125	Boron treated	0.16-0.26	0.40-0.60	1.10-1.60	0.10-0.25		0.025 max	0.025 max	0.10-0.40	bal Fe				

Alloy Steel, Chromium Molybdenum, K14047

UK

Specification	Designation	Notes	C	Cr	Mn	Mo	Ni	P	S	Si	Other	UTS	YS	El	Hard
BS 3111/1(87)	5/1	Q/T Wir	0.35-0.40	0.90-1.10	0.70-0.90	0.15-0.25		0.035 max	0.035 max	0.15-0.40	bal Fe				
BS 970/1(96)	897M39	Blm Bil Slab Bar Rod Frg CH	0.35-0.43	3.00-3.50	0.45-0.70	0.80-1.10		0.025 max	0.025 max	0.10-0.35	V 0.15-0.25; 4xP+Sn<=0.10; bal Fe				

USA

Specification	Designation	Notes	C	Cr	Mn	Mo	Ni	P	S	Si	Other	UTS	YS	El	Hard
	UNS K14047		0.38-0.43	0.80-1.10	0.75-1.00	0.15-0.25		0.025 max	0.025 max	0.20-0.35	bal Fe				

Alloy Steel, Chromium Molybdenum, K22094

USA

Specification	Designation	Notes	C	Cr	Mn	Mo	Ni	P	S	Si	Other	UTS	YS	El	Hard
	UNS K22094		0.20 max	1.00-1.50	0.30-0.80	0.44-0.65		0.040 max	0.040 max	0.50-1.00	bal Fe				

Alloy Steel, Chromium Molybdenum, MIL-S-46047

USA

Specification	Designation	Notes	C	Cr	Mn	Mo	Ni	P	S	Si	Other	UTS	YS	El	Hard
	UNS K14185		0.38-0.45	0.95-1.35	0.75-1.00	0.55-0.70		0.025 max	0.020 max	0.20-0.35	V 0.20-0.30; bal Fe				

UNS numbers and US grades are provided as a means of cross referencing chemically similar alloys. Exchangability is only possible after independent examination of specifications. Tensile properties are minimum or typical as specified. UTS and YS as MPa. El as %. See Appendix for list of abbreviations used in Notes. * indicates obsolete material.

Specification	Designation	Notes	C	Cr	Mn	Mo	Ni	P	S	Si	Other	UTS	YS	El	Hard

Alloy Steel, Chromium Molybdenum, No equivalents identified

Bulgaria

Specification	Designation	Notes	C	Cr	Mn	Mo	Ni	P	S	Si	Other	UTS	YS	El	Hard
BDS 5930	12Ch2M	Boiler	0.07-0.15	2.00-2.50	0.40-0.70	0.15 max	0.30 max	0.040 max	0.040 max	0.20-0.50	Cu <=0.30; bal Fe				

Czech Republic

Specification	Designation	Notes	C	Cr	Mn	Mo	Ni	P	S	Si	Other	UTS	YS	El	Hard
CSN 415222	15222	Cr-Mo-B	0.14-0.18	0.5-0.8	1.0-1.4	0.5-0.65	0.3 max	0.04 max	0.04 max	0.2-0.4	B 0.002-0.005; Nb 0.02-0.06; bal Fe				
CSN 415313	15313	High-temp const	0.08-0.15	2.0-2.5	0.4-0.8	0.9-1.1		0.035 max	0.04 max	0.15-0.4	bal Fe				
CSN 415342	15342	Cr-Mo Frg; bar	0.32-0.4	2.0-2.5	0.5-0.8	0.25-0.35		0.04 max	0.04 max	0.15-0.4	bal Fe				
CSN 415412	15412	High-press hydro ves	0.12 max	2.5-3.0	0.25-0.5	0.25-0.5		0.035 max	0.04 max	0.15-0.4	bal Fe				
CSN 415423	15423	High-press hydro ves	0.17-0.23	3.0-3.5	0.3-0.5	0.5-0.65		0.035 max	0.03 max	0.2-0.35	V 0.45-0.6; bal Fe				

Europe

Specification	Designation	Notes	C	Cr	Mn	Mo	Ni	P	S	Si	Other	UTS	YS	El	Hard
EN 10028/2(92)	10CrMo9-10	High-temp, Press; 100<=t<=150mm	0.08-0.14	2.00-2.50	0.40-0.80	0.90-1.10		0.030 max	0.025 max	0.50 max	Cu <=0.30; bal Fe	460-610	250	17T	
EN 10028/2(92)	10CrMo9-10	High-temp, Press; 60<=t<=100mm	0.06-0.15	2.00-2.50	0.40-0.80	0.90-1.10		0.030 max	0.025 max	0.50 max	Cu <=0.30; bal Fe	470-620	270	17T	
EN 10028/2(92)	10CrMo9-10	High-temp, Press; 40<=t<=60mm	0.06-0.15	2.00-2.50	0.40-0.80	0.90-1.10		0.030 max	0.025 max	0.50 max	Cu <=0.30; bal Fe	480-630	290	18T	
EN 10028/2(92)	10CrMo9-10	High-temp, Press; 16<= t<=40mm	0.08-0.14	2.00-2.50	0.40-0.80	0.90-1.10		0.030 max	0.025 max	0.50 max	Cu <=0.30; bal Fe	480-630	300	18T	
EN 10028/2(92)	10CrMo9-10	High-temp, Press; t<16mm	0.08-0.14	2.00-2.50	0.40-0.80	0.90-1.10		0.030 max	0.025 max	0.50 max	Cu <=0.30; bal Fe	480-630	310	18T	
EN 10028/2(92)	11CrMo9-10	High-temp, Press; t<60mm	0.08-0.15	2.00-2.50	0.40-0.80	0.90-1.10		0.030 max	0.025 max	0.50 max	Cu <=0.30; bal Fe	520-670	310	18T	
EN 10028/2(92)	11CrMo9-10	High-temp, Press; 60<= t<=100mm	0.08-0.15	2.00-2.50	0.40-0.80	0.90-1.10		0.030 max	0.025 max	0.50 max	Cu <=0.30; bal Fe	520-670	310	17T	

France

Specification	Designation	Notes	C	Cr	Mn	Mo	Ni	P	S	Si	Other	UTS	YS	El	Hard
AFNOR NFA35552(84)	25CD4u	t<=16mm	0.23-0.29	0.9-1.2	0.6-0.9	0.15-0.25	0.4 max	0.035 max	0.02-0.04	0.1-0.4	Cu <=0.3; Ti <=0.05; bal Fe	880-1080	700	12	
AFNOR NFA35552(84)	25CD4u	100<t<=160mm	0.23-0.29	0.9-1.2	0.6-0.9	0.15-0.25	0.4 max	0.035 max	0.02-0.04	0.1-0.4	Cu <=0.3; Ti <=0.05; bal Fe	630-780	490	15	
AFNOR NFA35552(84)	30CD12	t<=16mm	0.28-0.35	2.8-3.3	0.4-0.7	0.3-0.5	0.4 max	0.035 max	0.035 max	0.1-0.4	Cu <=0.3; Ti <=0.05; bal Fe	1080-1280	880	10	
AFNOR NFA35552(84)	30CD12	160<t<=250mm	0.28-0.35	2.8-3.3	0.4-0.7	0.3-0.5	0.4 max	0.035 max	0.035 max	0.1-0.4	Cu <=0.3; Ti <=0.05; bal Fe	880-1180	700	12	
AFNOR NFA35552(84)	30CD4u	t<=16mm	0.27-0.33	0.9-1.2	0.6-0.9	0.15-0.25	0.4 max	0.035 max	0.02-0.04	0.1-0.4	Cu <=0.3; Ti <=0.05; bal Fe	930-1130	730	12	
AFNOR NFA35552(84)	30CD4u	100<t<=160mm	0.27-0.33	0.9-1.2	0.6-0.9	0.15-0.25	0.4 max	0.035 max	0.02-0.04	0.1-0.4	Cu <=0.3; Ti <=0.05; bal Fe	680-830	530	15	
AFNOR NFA35552(84)	34CD4u	160<t<=250mm	0.31-0.37	0.9-1.2	0.6-0.9	0.15-0.25	0.4 max	0.035 max	0.02-0.04	0.1-0.4	Cu <=0.3; Ti <=0.05; bal Fe	680-830	530	15	
AFNOR NFA35552(84)	34CD4u	t<=16mm	0.31-0.37	0.9-1.2	0.6-0.9	0.15-0.25	0.4 max	0.035 max	0.02-0.04	0.1-0.4	Cu <=0.3; Ti <=0.05; bal Fe	980-1180	770	12	
AFNOR NFA35552(84)	42CD4u	t<=16mm; Q/T	0.39-0.45	0.9-1.2	0.6-0.9	0.15-0.25	0.4 max	0.035 max	0.02-0.04	0.1-0.4	Cu <=0.3; Ti <=0.05; bal Fe	1080-1280	850	10	
AFNOR NFA35552(84)	42CD4u	160<t<=250mm; Q/T	0.39-0.45	0.9-1.2	0.6-0.9	0.15-0.25	0.4 max	0.035 max	0.02-0.04	0.1-0.4	Cu <=0.3; Ti <=0.05; bal Fe	780-930	600	13	
AFNOR NFA35558(83)	20CDV5.07	t<=250mm	0.17-0.24	1.1-1.5	0.3-0.6	0.5-0.8	0.5 max	0.015 max	0.015 max	0.5 max	Cu <=0.3; Ti <=0.05; V 0.15-0.35; bal Fe	700-850	600	16	
AFNOR NFA35558(83)	42CDV4	t<=180mm; Q/T	0.36-0.44	0.8-1.15	0.45-0.7	0.5-0.65	0.4 max	0.03 max	0.03 max	0.2-0.35	Cu <=0.3; Ti <=0.05; V 0.25-0.35; bal Fe	700-900	590	16	
AFNOR NFA35564(83)	12CD4FF	Bar Wir rod; CF; t<=16mm	0.1-0.15	0.9-1.2	0.6-0.9	0.15-0.25	0.4 max	0.03 max	0.03 max	0.1-0.25	Cu <=0.3; Ti <=0.05; bal Fe	900-1250	630	9	
AFNOR NFA35564(83)	30CD4FF	Soft ann	0.27-0.33	0.9-1.2	0.6-0.9	0.15-0.25	0.4 max	0.03 max	0.03 max	0.35 max	Al >=0.02; Ti <=0.05; bal Fe	0-570			217 HB max
AFNOR NFA35564(83)	34CD4FF	Soft ann	0.31-0.37	0.9-1.2	0.6-0.9	0.15-0.25	0.4 max	0.03 max	0.03 max	0.35 max	Al 0.02-0.12; Cu <=0.3; Ti <=0.05; bal Fe	0-580			223 HB max
AFNOR NFA35564(83)	34CD4FF	isothermal Ann	0.31-0.37	0.9-1.2	0.6-0.9	0.15-0.25	0.4 max	0.03 max	0.03 max	0.35 max	Al 0.02-0.12; Cu <=0.3; Ti <=0.05; bal Fe				174-229 HB
AFNOR NFA35565(94)	100CD8*	Ball & roller bearing	0.9-1.05	1.7-1.95	0.6-0.8	0.2-0.4	0.3 max	0.03 max	0.015 max	0.2-0.4	Al <=0.05; Cu <=0.3; bal Fe				
AFNOR NFA36206(83)	10CD9.10	Boiler, Press ves, Q/T 30<t<=60mm	0.15 max	2-2.5	0.3-0.6	0.9-1.1	0.3 max	0.03 max	0.03 max	0.35 max	Cu <=0.25; Ti <=0.05; V <=0.04; bal Fe	520-670	310	20	

UNS numbers and US grades are provided as a means of cross referencing chemically similar alloys. Exchangability is only possible after independent examination of specifications. Tensile properties are minimum or typical as specified. UTS and YS as MPa. El as %. See Appendix for list of abbreviations used in Notes. * indicates obsolete material.

Alloy Steel, Chromium Molybdenum, No equivalents identified

Specification	Designation	Notes	C	Cr	Mn	Mo	Ni	P	S	Si	Other	UTS	YS	El	Hard
France															
AFNOR NFA36210(88)	12CD12-10	Sh Plt; Boiler, Press ves, HS, HT, Weld, Q/T, t<=125mm	0.16 max	2.75-3.25	0.3-0.6	0.9-1.1	0.3 max	0.015 max	0.012 max	0.1-0.4	Cu <=0.3; Ti <=0.05; V <=0.03; bal Fe	560-680	400	19	
Hungary															
MSZ 2295(79)	16HCM	High-press hydrogenation ves	0.12-0.2	2-2.5	0.3-0.6	0.3-0.4		0.03 max	0.03 max	0.17-0.37	bal Fe				
MSZ 2295(79)	17HCMV	High-press hydrogenation ves	0.14-0.22	2.5-3.0	0.3-0.6	0.2-0.3		0.03 max	0.03 max	0.17-0.37	V 0.1-0.2; bal Fe				
MSZ 2295(79)	20HCMV	High-press hydrogenation ves	0.16-0.24	3-3.5	0.3-0.6	0.4-0.6		0.03 max	0.03 max	0.17-0.37	V 0.4-0.55; bal Fe				
MSZ 2295(79)	24HCMN	High-press hydrogenation ves	0.2-0.28	2.2-2.7	0.3-0.6	0.4-0.6	0.5-0.8	0.03 max	0.03 max	0.17-0.37	bal Fe				
MSZ 2295(88)	10CrMo910	High-press hydrogenation ves; 0-100mm; soft ann	0.08-0.15	2-2.5	0.4-0.7	0.9-1.1		0.035 max	0.035 max	0.2-0.5	bal Fe				175 HB max
MSZ 2295(88)	10CrMo910	High-press hydrogenation ves; 0-100mm; Q/T	0.08-0.15	2-2.5	0.4-0.7	0.9-1.1		0.035 max	0.035 max	0.2-0.5	bal Fe	440-590	260	20L	130-175 HB
MSZ 2295(88)	16CrMo93	High-press hydrogenation ves; 0-100mm; Q/T	0.12-0.2	2-2.5	0.3-0.5	0.3-0.4		0.035 max	0.035 max	0.15-0.35	bal Fe	540-640	345	20L	160-190 HB
MSZ 2295(88)	16CrMo93	High-press hydrogenation ves; 0-100mm; soft ann	0.12-0.2	2-2.5	0.3-0.5	0.3-0.4		0.035 max	0.035 max	0.15-0.35	bal Fe				190 HB max
MSZ 2295(88)	17CrMoV10	High-press hydrogenation ves; 0-100mm; soft ann	0.15-0.2	2.7-3.0	0.3-0.5	0.2-0.3		0.035 max	0.035 max	0.15-0.35	V 0.1-0.2; bal Fe				230 HB max
MSZ 2295(88)	17CrMoV10	High-press hydrogenation ves; 0-100mm; Q/T	0.15-0.2	2.7-3.0	0.3-0.5	0.2-0.3		0.035 max	0.035 max	0.15-0.35	V 0.1-0.2; bal Fe	640-780	440	17L	190-240 HB
MSZ 2295(88)	20CrMoV135	High-press hydrogenation ves; 0-100mm; soft ann	0.17-0.23	3-3.3	0.3-0.5	0.5-0.6		0.035 max	0.035 max	0.15-0.35	V 0.45-0.55; bal Fe				235 HB max
MSZ 2295(88)	20CrMoV135	High-press hydrogenation ves; 0-100mm; Q/T	0.17-0.23	3-3.3	0.3-0.5	0.5-0.6		0.035 max	0.035 max	0.15-0.35	V 0.45-0.55; bal Fe	780-930	620	14L	240-280 HB
MSZ 2295(88)	21CrVMoW12	High-press hydrogenation ves; 0-100mm; Q/T	0.18-0.25	2.7-3.0	0.3-0.5	0.35-0.45		0.035 max	0.035 max	0.15-0.35	V 0.75-0.85; W 0.3-0.45; bal Fe	690-830	540	16L	205-250 HB
MSZ 2295(88)	21CrVMoW12	High-press hydrogenation ves; 0-100mm; soft ann	0.18-0.25	2.7-3.0	0.3-0.5	0.35-0.45		0.035 max	0.035 max	0.15-0.35	V 0.75-0.85; W 0.3-0.45; bal Fe				250 HB max
MSZ 2295(88)	24CrMo10	High-press hydrogenation ves; 0-100mm; Q/T	0.2-0.28	2.3-2.6	0.5-0.8	0.2-0.3	0.8 max	0.035 max	0.035 max	0.15-0.35	bal Fe	640-780	440	17L	190-235 HB
MSZ 2295(88)	24CrMo10	High-press hydrogenation ves; 0-100mm; soft ann	0.2-0.28	2.3-2.6	0.5-0.8	0.2-0.3	0.8 max	0.035 max	0.035 max	0.15-0.35	bal Fe				235 HB max
MSZ 31(85)	BCMo1	CH; 63 mm; blank carburized	0.12-0.18	1-1.3	0.8-1.1	0.2-0.3		0.035 max	0.035 max	0.4 max	bal Fe	750-1050	590	11L	
MSZ 31(85)	BCMo1	CH; 0-11mm; blank carburized	0.12-0.18	1-1.3	0.8-1.1	0.2-0.3		0.035 max	0.035 max	0.4 max	bal Fe	1000-1300	790	9L	
MSZ 31(85)	BCMo1E	CH; 0-11mm; blank carburized	0.12-0.18	1-1.3	0.8-1.1	0.2-0.3		0.035 max	0.02-0.035	0.4 max	bal Fe	1000-1300	790	9L	
MSZ 31(85)	BCMo1E	CH; 63 mm; blank carburized	0.12-0.18	1-1.3	0.8-1.1	0.2-0.3		0.035 max	0.02-0.035	0.4 max	bal Fe	750-1050	590	11L	
MSZ 4747(63)	2Cr10Mo45.47	High-temp, Tub; Non-rimming; K; FF	0.15 max	2-2.5	0.4-0.6	0.9-1.1		0.04 max	0.04 max	0.15-0.5	P+S<=0.07; bal Fe				
MSZ 4747(85)	2Cr10Mo45.47	High-temp, Tub; Non-rimming; K; FF; 40.1-60mm; Q/T	0.08-0.15	2-2.5	0.4-0.7	0.9-1.2		0.035 max	0.035 max	0.5 max	bal Fe	450-600	270	20L	
MSZ 4747(85)	2Cr10Mo45.47	High-temp, Tub; Non-rimming; K; FF; 0-30mm; Q/T	0.08-0.15	2-2.5	0.4-0.7	0.9-1.2		0.035 max	0.035 max	0.5 max	bal Fe	450-600	280	20L	
MSZ 6251(87)	BCMo1Z	CF; CH; 0-36mm; drawn, half-hard	0.12-0.18	1-1.3	0.8-1.1	0.2-0.3		0.035 max	0.035 max	0.4 max	bal Fe	0-590			

Specification	Designation	Notes	C	Cr	Mn	Mo	Ni	P	S	Si	Other	UTS	YS	El	Hard

Alloy Steel, Chromium Molybdenum, No equivalents identified

Hungary

Specification	Designation	Notes	C	Cr	Mn	Mo	Ni	P	S	Si	Other	UTS	YS	El	Hard
MSZ 6251(87)	BCMo1Z	CF; CH; 0-36mm; HR; HR/soft ann; drawn/soft ann; drawn/bright ann; soft ann/ground	0.12-0.18	1-1.3	0.8-1.1	0.2-0.3		0.035 max	0.035 max	0.4 max	bal Fe	0-560			

India

Specification	Designation	Notes	C	Cr	Mn	Mo	Ni	P	S	Si	Other	UTS	YS	El	Hard
IS 1570/4(88)	07Cr4Mo6	Const, Spring	0.12 max	0.7-1.1	0.4-0.7	0.45-0.65	0.3 max	0.07 max	0.06 max	0.15-0.6	bal Fe				
IS 1570/4(88)	10Cr20Mo6		0.15 max	4-6	0.4-0.7	0.45-0.65	0.3 max	0.07 max	0.06 max	0.5 max	bal Fe				
IS 1570/4(88)	30Cr4Mo2		0.28-0.33	0.8-1.1	0.4-0.8	0.15-0.28	0.4 max	0.07 max	0.06 max	0.1-0.35	bal Fe				
IS 1570/4(88)	35Cr5Mo6V2		0.25-0.45	1.-1.5	0.4-0.7	0.5-0.8	0.3 max	0.07 max	0.06 max	0.1-0.35	V 0.2-0.3; bal Fe				
IS 1570/4(88)	40Cr13Mo10V2	30.1-63mm; Bar Frg	0.35-0.45	3-3.5	0.4-0.7	0.9-1.1	0.3 max	0.07 max	0.06 max	0.1-0.35	V 0.15-0.25; bal Fe	1340	1050	8	363 HB
IS 1570/4(88)	40Cr13Mo10V2	0-30mm; Bar Frg	0.35-0.45	3-3.5	0.4-0.7	0.9-1.1	0.3 max	0.07 max	0.06 max	0.1-0.35	V 0.15-0.25; bal Fe	1540	1240	8	444 HB
IS 1570/4(88)	40Cr5Mo6		0.35-0.45	1-1.5	0.4-0.7	0.5-0.7	0.4 max	0.07 max	0.06 max	0.1-0.35	bal Fe				
IS 1570/4(88)	40Cr7Al10Mo2	0-63mm; Bar Frg	0.35-0.45	1.5-1.8	0.4-0.7	0.1-0.25	0.3 max	0.07 max	0.06 max	0.1-0.45	Al 0.9-1.3; bal Fe	890-1040	650	15	255-311 HB
IS 1570/4(88)	42Cr4Mo2	Bar Frg Tub Pipe	0.38-0.45	0.9-1.2	0.6-0.9	0.15-0.3	0.4 max	0.07 max	0.06 max	0.1-0.35	bal Fe	710	590		
IS 1570/7(92)	10	Smls tube Sh Plt Bar Frg	0.1-0.18	0.5-0.7	0.4-0.7	0.15 max	0.22-0.32	0.04 max	0.3-0.6	0.1-0.35	Al <=0.02; Almet<=0.02; bal Fe	460-610	275	16	
IS 1570/7(92)	10Cr36Mo10H	Soft Ann Smls tube	0.15 max	8-10	0.3-0.6	0.9-1.1	0.4 max	0.04 max	0.03 max	0.25-1	Al <=0.02; bal Fe	410-560	135	20	
IS 1570/7(92)	10Mo6H	Tube Pipe	0.15 max	1-1.5	0.3-0.6	0.45-0.6	0.4 max	0.03 max	0.03 max	0.5-1	bal Fe	415	205	30	
IS 1570/7(92)	11	Smls tube Bar Frg	0.08-0.15	1.9-2.6	0.3-0.7	0.9-1.15	0.4 max	0.04 max	0.04 max	0.1-0.5	Al <=0.02; Almet<=0.02; bal Fe	410-560	135	20	
IS 1570/7(92)	12	Soft Ann Smls tube	0.15 max	8-10	0.3-0.6	0.9-1.1	0.4 max	0.04 max	0.03 max	0.25-1	Al <=0.02; bal Fe	410-560	135	20	
IS 1570/7(92)	12Cr2Mo5V2H	Smls tube Sh Plt Bar Frg	0.1-0.18	0.5-0.7	0.4-0.7	0.15 max	0.22-0.32	0.04 max	0.3-0.6	0.1-0.35	Al <=0.02; Almet<=0.02; bal Fe	460-610	275	16	
IS 1570/7(92)	12Cr4Mo5H	Smls tube Sh Plt Bar Frg	0.1-0.18	0.7-1.1	0.4-0.7	0.45-0.65	0.4 max	0.04 max	0.04 max	0.1-0.35	Al <=0.02; Almet<=0.02; bal Fe	540-690	375	15	
IS 1570/7(92)	12Cr9Mo10H	Smls tube Bar Frg	0.08-0.15	1.9-2.6	0.3-0.7	0.9-1.15	0.4 max	0.04 max	0.04 max	0.1-0.5	Al <=0.02; Almet<=0.02; bal Fe	410-560	135	20	
IS 1570/7(92)	40CrMoH	0-200mm; Bar Frg	0.35-0.45	1-1.5	0.4-0.7	0.5-0.8	0.4 max	0.035 max	0.035 max	0.1-0.35	bal Fe	850-1000	635	14	
IS 1570/7(92)	40CrMoVH	0-100mm; Bar Frg	0.36-0.44	0.9-1.2	0.45-0.85	0.55-0.75	0.4 max	0.03 max	0.03 max	0.1-0.35	V 0.25-0.35; bal Fe	850-1000	700	14	
IS 1570/7(92)	7	Tube Pipe	0.15 max	1-1.5	0.3-0.6	0.45-0.6	0.4 max	0.03 max	0.03 max	0.5-1	bal Fe	415	205	30	
IS 1570/7(92)	9	Smls tube Sh Plt Bar Frg	0.1-0.18	0.7-1.1	0.4-0.7	0.45-0.65	0.4 max	0.04 max	0.04 max	0.1-0.35	Al <=0.02; Almet<=0.02; bal Fe	540-690	375	15	

International

Specification	Designation	Notes	C	Cr	Mn	Mo	Ni	P	S	Si	Other	UTS	YS	El	Hard
ISO 2604-1(75)	F31	Frg	0.20-0.28	0.90-1.20	0.50-0.80	0.20-0.35		0.040 max	0.040 max	0.15-0.40	Al <=0.020; bal Fe	640-780			
ISO 2604-1(75)	F32	Frg	0.20 max	0.85-1.15	0.40-0.70	0.45-0.65		0.040 max	0.040 max	0.15-0.40	Al <=0.020; bal Fe	410-560			
ISO 2604-1(75)	F32Q	Frg	0.20 max	0.85-1.15	0.40-0.70	0.45-0.65		0.040 max	0.040 max	0.15-0.40	Al <=0.020; bal Fe	540-690			
ISO 2604-1(75)	F33	Frg	0.10-0.18	0.30-0.60	0.40-0.70	0.50-0.70		0.040 max	0.040 max	0.15-0.40	Al <=0.020; bal Fe	460-510			
ISO 2604-1(75)	F34	Frg	0.15 max	2.00-2.50	0.40-0.70	0.90-1.20		0.040 max	0.040 max	0.15-0.40	Al <=0.020; bal Fe	490-640			
ISO 2604-1(75)	F34Q	Frg	0.15 max	2.00-2.50	0.40-0.70	0.90-1.20		0.040 max	0.040 max	0.15-0.40	Al <=0.020; bal Fe	540-690			
ISO 2604-1(75)	F35	Frg	0.22 max	2.75-3.50	0.30-0.80	0.45-0.65		0.040 max	0.040 max	0.15-0.40	bal Fe	590-740			
ISO 2604-1(75)	F36	Frg	0.30 max	2.75-3.50	0.30-0.80	0.45-0.65		0.040 max	0.040 max	0.15-0.40	bal Fe	740-880			
ISO 2604-1(75)	F37	Frg	0.18 max	4.00-6.00	0.30-0.80	0.45-0.65		0.040 max	0.040 max	0.15-0.40	Al <=0.020; bal Fe	410-510			

Alloy Steel, Chromium Molybdenum, No equivalents identified

Specification	Designation	Notes	C	Cr	Mn	Mo	Ni	P	S	Si	Other	UTS	YS	El	Hard
International															
ISO 2604-1(75)	F40	Frg	0.23 max	11.00-12.50	0.30-1.00	0.70-1.20	0.30-1.0	0.040 max	0.040 max	0.15-0.40	bal Fe	780-930			
ISO 4954(93)	18CrMo4E	Wir Rod Bar, CH ext	0.15-0.21	0.90-1.20	0.60-0.90	0.15-0.25		0.035 max	0.035 max	0.40 max	bal Fe	560			
ISO 4954(93)	25CrMo4E	Wir Rod Bar, CH ext	0.22-0.29	0.90-1.20	0.60-0.90	0.15-0.30		0.035 max	0.035 max	0.40 max	bal Fe	580			
ISO 4954(93)	34CrMo4E	Wir Rod Bar, CH ext	0.30-0.37	0.90-1.20	0.60-0.90	0.15-0.30		0.035 max	0.035 max	0.40 max	bal Fe	610			
ISO 4954(93)	42CrMo4E	Wir Rod Bar, CH ext	0.38-0.45	0.90-1.20	0.60-0.90	0.15-0.30		0.035 max	0.035 max	0.40 max	bal Fe	630			
ISO 683-1(87)	25CrMo4 O	Bar Frg Plt Wir Rod Bolt Slab; Q/T, t<16mm	0.22-0.29	0.90-1.20	0.60-0.90	0.15-0.30		0.035 max	0.035 max	0.10-0.40	bal Fe	900-1100	700		
ISO 683-1(87)	25CrMoS4 O	Bar Frg Plt Wir Rod Bolt Slab; Q/T, t<16mm	0.22-0.29	0.90-1.20	0.60-0.90	0.15-0.30		0.035 max	0.020-0.040	0.10-0.40	bal Fe	900-1100	700		
ISO 683-1(87)	34CrMo4	Bar Frg Plt Wir Rod Bolt Slab; Q/T, t<16mm	0.30-0.37	0.90-1.20	0.60-0.90	0.15-0.30		0.035 max	0.035 max	0.10-0.40	bal Fe	1000-1200	800		
ISO 683-1(87)	34CrMoS4	Bar Frg Plt Wir Rod Bolt Slab; Q/T, t<16mm	0.30-0.37	0.90-1.20	0.60-0.90	0.15-0.30		0.035 max	0.020-0.040	0.10-0.40	bal Fe	1000-1200	800		
ISO 683-1(87)	42CrMo4	Bar Frg Plt Wir Rod Bolt Slab; Q/T, t<16mm	0.38-0.45	0.90-1.20	0.60-0.90	0.15-0.30		0.035 max	0.035 max	0.10-0.40	bal Fe	1100-1300	900		
ISO 683-1(87)	42CrMoS4	Bar Frg Plt Wir Rod Bolt Slab; Q/T, t<16mm	0.38-0.45	0.90-1.20	0.60-0.90	0.15-0.30		0.035 max	0.020-0.040	0.10-0.40	bal Fe	1100-1300	900		
ISO 683-1(87)	50CrMo4	Bar Frg Plt Wir Rod Bolt Slab; Q/T, t<16mm	0.46-0.54	0.90-1.20	0.50-0.80	0.15-0.30		0.035 max	0.035 max	0.10-0.40	bal Fe	1100-1300	900		
ISO 683-12(72)	6*	F/IH	0.42-0.48	0.4-0.6	0.5-0.8	0.15 max	0.4 max	0.035 max	0.035 max	0.15-0.4	Al <=0.1; Co <=0.1; Cu <=0.3; Pb <=0.15; Ti <=0.05; V <=0.1; W <=0.1; bal Fe				
ISO 683-14(92)	60CrMo31	Bar Rod Wir; Q/T, springs, Ann	0.56-0.64	0.70-1.00	0.70-1.00	0.08-0.15		0.030 max	0.030 max	0.15-0.40	bal Fe				255 HB
ISO 683-17(76)	5	Ball & roller bearing, HT	0.95-1.10	1.65-1.95	0.60-0.90	0.20-0.40		0.030 max	0.025 max	0.20-0.40	bal Fe				
ISO 683-18(96)	25CrMo4	Flat, CD, Ann, t<=16mm	0.22-0.29	0.90-1.20	0.60-0.90	0.15-0.30		0.035 max	0.035 max	0.10-0.40	bal Fe	880 max			212 HB max
ISO 683-18(96)	25CrMoS4	Flat, CD, Ann, t<=16mm	0.22-0.29	0.90-1.20	0.60-0.90	0.15-0.30		0.035 max	0.020-0.040	0.10-0.40	bal Fe	880 max			212 HB max
ISO 683-18(96)	34CrMo4	Flat, CD, Ann, t<=16mm	0.30-0.37	0.90-1.20	0.60-0.90	0.15-0.30		0.035 max	0.035 max	0.10-0.40	bal Fe	940 max			223 HB max
ISO 683-18(96)	34CrMoS4	Flat, CD, Ann, t<=16mm	0.30-0.37	0.90-1.20	0.60-0.90	0.15-0.30		0.035 max	0.020-0.040	0.10-0.40	bal Fe	940 max			223 HB max
ISO 683-18(96)	42CrMo4	Flat, CD, Ann, t<=16mm	0.38-0.45	0.90-1.20	0.60-0.90	0.15-0.30		0.035 max	0.035 max	0.10-0.40	bal Fe	980 max			241 HB max
ISO 683-18(96)	42CrMoS4	Flat, CD, Ann, t<=16mm	0.38-0.45	0.90-1.20	0.60-0.90	0.15-0.30		0.035 max	0.020-0.040	0.10-0.40	bal Fe	980 max			241 HB max
ISO 683-18(96)	50CrMo4	Flat, CD, Ann, t<=16mm	0.46-0.54	0.90-1.20	0.50-0.80	0.15-0.30		0.035 max	0.035 max	0.10-0.40	bal Fe	1050 max			248 HB max
ISO 683-18(96)	51CrV6	Flat, CD, Ann, t<=16mm	0.47-0.55	0.80-1.10	0.60-1.00			0.035 max	0.035 max	0.10-0.40	V 0.10-0.25; bal Fe	1050 max			248 HB max
ISO R683-6(70)	1	Q/T	0.28-0.35	2.8-3.3	0.4-0.7	0.35-0.5	0.3 max	0.035 max	0.035 max	0.15-0.4	Al <=0.1; Co <=0.1; Cu <=0.3; Pb <=0.15; Ti <=0.05; V <=0.1; W <=0.1; bal Fe				
Italy															
UNI 7874(79)	32CrMo12	Q/T	0.28-0.35	2.8-3.3	0.5-0.9	0.3-0.5	0.3 max	0.035 max	0.035 max	0.15-0.4	bal Fe				
UNI 8077(80)	31CrMo12	Nitriding	0.29-0.34	2.8-3.2	0.4-0.7	0.3-0.4		0.035 max	0.035 max	0.15-0.4	bal Fe				
UNI 8077(80)	31CrMoV10	Nitriding	0.29-0.34	2.3-2.8	0.4-0.7	0.3-0.4		0.03 max	0.035 max	0.15-0.4	V 0.1-0.2; bal Fe				
UNI 8552(84)	31CrMo10	Nitriding	0.29-0.34	2.3-2.8	0.4-0.7	0.3-0.4	0.3 max	0.035 max		0.15-0.4	V 0.1-0.2; bal Fe				

UNS numbers and US grades are provided as a means of cross referencing chemically similar alloys. Exchangability is only possible after independent examination of specifications. Tensile properties are minimum or typical as specified. UTS and YS as MPa. El as %. See Appendix for list of abbreviations used in Notes. * indicates obsolete material.

Specification	Designation	Notes	C	Cr	Mn	Mo	Ni	P	S	Si	Other	UTS	YS	El	Hard

Alloy Steel, Chromium Molybdenum, No equivalents identified

Italy

Specification	Designation	Notes	C	Cr	Mn	Mo	Ni	P	S	Si	Other	UTS	YS	El	Hard
UNI 8552(84)	31CrMo12	Nitriding	0.29-0.34	2.8-3.2	0.4-0.7	0.3-0.4		0.03 max	0.035 max	0.15-0.4	bal Fe				

Japan

| JIS G3441(88) | SCM418TK | Tube | 0.16-0.21 | 0.90-1.20 | 0.60-0.85 | 0.15-0.30 | 0.25 max | 0.030 max | 0.030 max | 0.15-0.35 | Cu <=0.30; bal Fe | | | | |

Poland

Specification	Designation	Notes	C	Cr	Mn	Mo	Ni	P	S	Si	Other	UTS	YS	El	Hard
PNH84024	10H2M	High-temp const	0.08-0.15	2-2.5	0.4-0.6	0.9-1.1	0.3 max	0.03 max	0.03 max	0.15-0.5	Al 0.02-0.120; Cu <=0.25; bal Fe				
PNH84024	12HMF	High-temp const	0.08-0.15	0.9-1.2	0.4-0.7	0.25-0.4	0.25 max	0.03 max	0.03 max	0.15-0.4	Al 0.02-0.120; Cu <=0.25; V 0.15-0.35; bal Fe				
PNH84024	13HMF	High-temp const	0.1-0.18	0.3-0.6	0.4-0.7	0.5-0.65	0.3 max	0.04 max	0.04 max	0.15-0.35	Al 0.02-0.120; Cu <=0.25; V 0.22-0.35; bal Fe				
PNH84024	15HM	High-temp const	0.11-0.18	0.7-1	0.4-0.7	0.4-0.55	0.35 max	0.04 max	0.04 max	0.15-0.35	Al 0.02-0.120; Cu <=0.25; bal Fe				
PNH84024	15HM	High-temp const	0.11-0.18	0.7-1	0.4-0.7	0.4-0.55	0.35 max	0.04 max	0.04 max	0.15-0.35	Al 0.02-0.120; Cu <=0.25; bal Fe				
PNH84024	15HMF	High-temp const	0.12-0.19	1.2-1.6	0.4-0.7	0.9-1.1	0.3 max	0.03 max	0.025 max	0.17-0.37	V 0.25-0.35; bal Fe				
PNH84024	20HM	High-temp const	0.19-0.26	0.9-1.2	0.5-0.8	0.4-0.5	0.6 max	0.035 max	0.035 max	0.15-0.35	Al 0.02-0.120; bal Fe				
PNH84024	20HMFTB	High-temp const	0.17-0.24	0.9-1.4	0.5 max	0.8-1.1	0.5 max	0.03 max	0.03 max	0.35 max	B <=0.005; Ti 0.05-0.12; V 0.7-1; bal Fe				
PNH84024	20MF	High-temp const	0.15-0.22	0.3 max	0.4-0.6	0.5-0.7	0.3 max	0.04 max	0.04 max	0.3 max	V 0.25-0.35; bal Fe				
PNH84024	21HMF	High-temp const	0.17-0.25	1.2-1.5	0.6-0.5	1.-1.2	0.6 max	0.035 max	0.035 max	0.3-0.6	V 0.25-0.35; bal Fe				
PNH84024	24H2MF	High-temp const	0.2-0.3	2.1-2.5	0.5-0.8	0.9-1.2	0.4 max	0.03 max	0.03 max	0.17-0.37	V 0.3-0.6; bal Fe				
PNH84024	26H2MF	High-temp const	0.22-0.3	1.5-1.8	0.3-0.6	0.6-0.8	0.3 max	0.035 max	0.035 max	0.3-0.5	Cu <=0.25; V 0.2-0.3; bal Fe				
PNH84024	30H2MF	High-temp const	0.26-0.34	2.3-2.7	0.4-0.7	0.15-0.25	0.4 max	0.035 max	0.035 max	0.15-0.4	V 0.1-0.3; bal Fe				
PNH84030/02	15HGMA	CH	0.12-0.19	0.8-1.1	0.8-1.1	0.15-0.25	0.3 max	0.025 max	0.025 max	0.17-0.37	bal Fe				
PNH84030/03	25H3M	Nitriding	0.2-0.3	2.9-3.5	0.4-0.65	0.4-0.55	0.4 max	0.035 max	0.035 max	0.17-0.37	bal Fe				
PNH84030/03	33H3MF	Nitriding	0.29-0.36	2.4-2.8	0.5-0.8	0.35-0.45	0.3 max	0.035 max	0.035 max	0.17-0.37	V 0.2-0.3; bal Fe				
PNH84030/03	38HMJ	Nitriding	0.35-0.42	1.35-1.65	0.3-0.6	0.15-0.25	0.25 max	0.025 max	0.025 max	0.17-0.37	Al 0.7-1.1; bal Fe				
PNH84030/04	20HGSA	Q/T	0.17-0.23	0.8-1.1	0.8-1.1		0.3 max	0.025 max	0.025 max	0.9-1.2	bal Fe				
PNH84030/04	25HMA	Q/T	0.22-0.29	0.8-1.1	0.4-0.7	0.15-0.25	0.3 max	0.03 max	0.03 max	0.17-0.37	bal Fe				
PNH84030/04	30HMA	Q/T	0.28-0.33	0.8-1.1	0.4-0.7	0.15-0.25	0.3 max	0.025 max	0.025 max	0.17-0.37	bal Fe				
PNH84035	20H3MWF	High-temp const	0.16-0.24	2.4-3.3	0.25-0.6	0.35-0.55	0.5 max	0.035 max	0.03 max	0.4 max	V 0.6-0.85; W 0.3-0.5; bal Fe				
PNH86022(71)	H10S2M	Heat res	0.35-0.45	9-10.5	0.7 max	0.7-0.9	0.5 max	0.035 max		1.9-2.6	bal Fe				

Romania

| STAS 8184(87) | 10CrMo10 | High-temp const | 0.15 max | 2-2.5 | 0.4-0.7 | 0.9-1.1 | | 0.04 max | 0.04 max | 0.15-0.5 | Pb <=0.15; Ti <=0.02; As<=0.05; bal Fe | | | | |

Russia

| GOST 4543(71) | 25ChGM | Q/T | 0.23-0.29 | 0.9-1.2 | 0.9-1.2 | 0.2-0.3 | 0.3 max | 0.035 max | 0.035 max | 0.17-0.37 | Al <=0.1; Cu <=0.3; Ti <=0.05; bal Fe | | | | |
| GOST 550 | 12Ch8 | Smls tube, pipelines | 0.12 max | 7.5-9.0 | 0.3-0.6 | 0.15 max | 0.4 max | 0.025 max | 0.03 max | 0.17-0.37 | Al <=0.1; Cu <=0.25; Ti <=0.05; bal Fe | | | | |

Spain

| UNE 36012(75) | 38CrMo4 | Q/T | 0.34-0.4 | 0.85-1.15 | 0.6-0.9 | 0.15-0.25 | | 0.035 max | 0.035 max | 0.15-0.4 | bal Fe | | | | |
| UNE 36013(76) | 12CrMo4 | CH | 0.1-0.15 | 0.85-1.15 | 0.6-0.9 | 0.15-0.25 | | 0.035 max | 0.035 max | 0.15-0.4 | bal Fe | | | | |

UNS numbers and US grades are provided as a means of cross referencing chemically similar alloys. Exchangability is only possible after independent examination of specifications. Tensile properties are minimum or typical as specified. UTS and YS as MPa. El as %. See Appendix for list of abbreviations used in Notes. * indicates obsolete material.

Alloy Steel, Chromium Molybdenum, No equivalents identified

Specification	Designation	Notes	C	Cr	Mn	Mo	Ni	P	S	Si	Other	UTS	YS	El	Hard
Spain															
UNE 36013(76)	F.1551*	see 12CrMo4	0.1-0.15	0.85-1.15	0.6-0.9	0.15-0.25		0.035 max	0.035 max	0.15-0.4	bal Fe				
UNE 36014(76)	25CrMo12	Nitriding	0.22-0.29	2.75-3.25	0.4-0.7	0.4-0.6		0.035 max	0.035 max	0.15-0.4	bal Fe				
UNE 36014(76)	31CrMo12	Nitriding	0.28-0.35	2.8-3.3	0.4-0.7	0.3-0.5	0.3 max	0.035 max	0.035 max	0.15-0.4	bal Fe				
UNE 36014(76)	31CrMoV10	Nitriding	0.28-0.35	2.3-2.8	0.4-0.7	0.3-0.5		0.035 max	0.035 max	0.15-0.4	V 0.2-0.3; bal Fe				
UNE 36014(76)	F.1711*	see 25CrMo12	0.22-0.29	2.75-3.25	0.4-0.7	0.4-0.6		0.035 max	0.035 max	0.15-0.4	bal Fe				
UNE 36014(76)	F.1712*	see 31CrMo12	0.28-0.35	2.8-3.3	0.4-0.7	0.3-0.5	0.3 max	0.035 max	0.035 max	0.15-0.4	bal Fe				
UNE 36014(76)	F.1721*	see 31CrMoV10	0.28-0.35	2.3-2.8	0.4-0.7	0.3-0.5		0.035 max	0.035 max	0.15-0.4	V 0.2-0.3; bal Fe				
UNE 36087(76/78)	12CrMo910	Sh Plt Strp CR	0.08-0.15	2.0-2.5	0.4-0.8	0.9-1.1		0.035 max	0.035 max	0.15-0.35	bal Fe				
UNE 36087(76/78)	F.2632*	see 12CrMo910	0.08-0.15	2.0-2.5	0.4-0.8	0.9-1.1		0.035 max	0.035 max	0.15-0.35	bal Fe				
UK															
BS 3604/1(90)	620-440	Pip Tub High-press/temp smls weld t<=200mm	0.10-0.15	0.70-1.10	0.40-0.70	0.45-0.65	0.30 max	0.030 max	0.030 max	0.10-0.35	Al <=0.02; Cu <=0.25; Sn<=0.03; bal Fe	440-590	290	22	
BS 3604/1(90)	621	Pip Tub High-press/temp smls weld t<=200mm	0.15 max	1.00-1.50	0.30-0.60	0.45-0.65	0.30 max	0.030 max	0.030 max	0.50-1.00	Al <=0.02; Cu <=0.25; Sn<=0.03; bal Fe	420-570	275	22	
BS 3604/1(90)	622	Pip Tub High-press/temp smls weld t<=200mm	0.08-0.15	2.00-2.50	0.40-0.70	0.90-1.20	0.30 max	0.030 max	0.030 max	0.50 max	Al <=0.02; Cu <=0.25; Sn<=0.03; bal Fe	490-620	275	20	
BS 3604/1(90)	625	Pip Tub High-press/temp smls weld t<=200mm	0.15 max	4.00-6.00	0.30-0.60	0.45-0.65	0.30 max	0.030 max	0.030 max	0.50 max	Al <=0.02; Cu <=0.25; Sn<=0.03; bal Fe	450-600	170	20	
BS 3604/1(90)	629-470	Pip Tub High-press/temp smls weld t<=200mm	0.15 max	8.00-10.0	0.30-0.60	0.90-1.10	0.30 max	0.030 max	0.030 max	0.25-1.00	Al <=0.02; Cu <=0.25; Sn<=0.03; bal Fe	470-620	185	20	
BS 3604/1(90)	629-590	Pip Tub High-press/temp smls weld t<=200mm	0.15 max	8.00-10.0	0.30-0.60	0.90-1.10	0.30 max	0.030 max	0.030 max	0.25-1.00	Al <=0.02; Cu <=0.25; Sn<=0.03; bal Fe	590-740	400	18	
USA															
ASTM A182/A182M(98)	F3VCb	Frg/roll pipe flange valve	0.10-0.15	2.7-3.3	0.30-0.60	0.90-1.10	0.25 max	0.020 max	0.010 max	0.10 max	Cu <=0.25; Nb 0.015-0.070; Ti <=0.015; V 0.20-0.30; Ca 0.0005-0.0150; bal Fe	585-760	415	18	174-237 HB
ASTM A290(95)	M	Gear frg	0.35-0.45	1.40-1.80	0.40-0.70	0.30-0.45	0.30 max	0.040 max	0.040 max	0.40 max	Al 0.85-1.30; Cu <=0.35; V <=0.03; bal Fe	825	585	15	255-302 HB
ASTM A290(95)	P	Gear frg	0.35-0.45	1.40-1.80	0.40-0.70	0.30-0.45	0.30 max	0.040 max	0.040 max	0.40 max	Al 0.85-1.30; Cu <=0.35; V <=0.03; bal Fe				255-302 HB
ASTM A336/A336M(98)	F3VCb	Ferr Frg	0.05-0.15	2.7-3.3	0.30-0.60	0.90-1.10	0.25 max	0.020 max	0.010 max	0.10 max	Cu <=0.25; Nb 0.015-0.070; Ti <=0.015; V 0.20-0.30; Ca=0.0005-0.0150; bal Fe	585-760	415	18	
ASTM A336/A336M(98)	F6	Ferr Frg	0.12 max	11.5-13.5	1.00 max		0.50 max	0.025 max	0.025 max	1.00 max	bal Fe	585-760	380	18	
ASTM A372/372M(95)	Grade G	Thin wall frg press ves	0.25-0.35	0.40-0.65	0.70-1.00	0.15-0.25		0.025 max	0.025 max	0.15-0.35	bal Fe	545-1000	380-485	18-20	179-248 HB
ASTM A387/A387(97)	Grade 9	Press ves plt, Class 2	0.15 max	8.00-10.00	0.30-0.60	0.90-1.10		0.035 max	0.035 max	1.00 max	bal Fe	515-690	310	18	
ASTM A542(95)	Type D	Q/T, Press ves plt, Class 2, t>=5mm	0.11-0.15	2.00-2.50	0.30-0.60	0.90-1.10	0.25 max	0.015 max	0.010 max	0.10 max	B 0.0020; Cu <=0.20; Nb <=0.07; Ti 0.030; V 0.25-0.35; Ca=0-0.015, bal Fe	795-930	690	13	
ASTM A542(95)	Type D	Q/T, Press ves plt, Class 4, t>=5mm	0.11-0.15	2.00-2.50	0.30-0.60	0.90-1.10	0.25 max	0.015 max	0.010 max	0.10 max	B 0.0020; Cu <=0.20; Nb <=0.07; Ti 0.030; V 0.25-0.35; Ca=0-0.015, bal Fe	585-790	380	20	

UNS numbers and US grades are provided as a means of cross referencing chemically similar alloys. Exchangability is only possible after independent examination of specifications. Tensile properties are minimum or typical as specified. UTS and YS as MPa. El as %. See Appendix for list of abbreviations used in Notes. * indicates obsolete material.

Specification	Designation	Notes	C	Cr	Mn	Mo	Ni	P	S	Si	Other	UTS	YS	El	Hard

Alloy Steel, Chromium Molybdenum, No equivalents identified

USA

Specification	Designation	Notes	C	Cr	Mn	Mo	Ni	P	S	Si	Other	UTS	YS	El	Hard
ASTM A542(95)	Type D	Q/T, Press ves plt, Class 1, t>=5mm	0.11-0.15	2.00-2.50	0.30-0.60	0.90-1.10	0.25 max	0.015 max	0.010 max	0.10 max	B 0.0020; Cu <=0.20; Nb <=0.07; Ti 0.030; V 0.25-0.35; Ca=0-0.015, bal Fe	725-860	585	14	
ASTM A542(95)	Type D	Q/T, Press ves plt, Class 3, t>=5mm	0.11-0.15	2.00-2.50	0.30-0.60	0.90-1.10	0.25 max	0.015 max	0.010 max	0.10 max	B 0.0020; Cu <=0.20; Nb <=0.07; Ti 0.030; V 0.25-0.35; Ca=0-0.015, bal Fe	655-795	515	20	
ASTM A542(95)	Type D	Q/T, Press ves plt, Class 4a, t>=5mm	0.11-0.15	2.00-2.50	0.30-0.60	0.90-1.10	0.25 max	0.015 max	0.010 max	0.10 max	B 0.0020; Cu <=0.20; Nb <=0.07; Ti 0.030; V 0.25-0.35; Ca=0-0.015, bal Fe	585-760	415	18	
ASTM A542(95)	Type E	Q/T, Press ves plt, Class 4, t>=5mm	0.10-0.15	2.75-3.25	0.30-0.60	0.90-1.10	0.25 max	0.025 max	0.010 max	0.15 max	Cu <=0.25; Nb 0.015-0.070; V 0.20-0.30; Ca=0.0005-0.0150, bal Fe	585-760	380	20	
ASTM A542(95)	Type E	Q/T, Press ves plt, Class 4a, t>=5mm	0.10-0.15	2.75-3.25	0.30-0.60	0.90-1.10	0.25 max	0.025 max	0.010 max	0.15 max	Cu <=0.25; Nb 0.015-0.070; V 0.20-0.30; Ca=0.0005-0.0150, bal Fe	585-760	415	18	
ASTM A579(96)	41	Superstrength frg	0.38-0.43	4.75-5.25	0.20-0.40	1.20-1.40		0.015 max	0.015 max	0.80-1.00	V 0.40-0.60; bal Fe	1790-1930	1380-1550	8-9	235 HB
ASTM A782/A783M(90)	Class 1	Press ves plt, Q/T	0.20 max	0.50-1.00	0.70-1.20	0.20-0.60		0.035 max	0.035 max	0.40-0.80	Zr 0.04-0.12	670-820	550	18	
ASTM A832/A832M(97)	Grade 23V	Press ves plt	0.10-0.15	2.75-3.25	0.30-0.60	0.90-1.10		0.025 max	0.010 max	0.10 max	Nb 0.015-0.070; V 0.20-0.30; Ca=0.0005-0.0150, bal Fe	585-760	415	18	
ASTM A983/983M(98)	Grade 2	Continuous grain flow frg for crankshafts	0.43-0.52	0.20-0.35	0.75-1.10	0.10 max		0.025 max	0.025 max	0.15-0.40	V <=0.10; bal Fe				
ASTM A983/983M(98)	Grade 6	Continuous grain flow frg for crankshafts	0.35-0.45	0.80-1.20	0.65-1.00	0.30-0.50		0.025 max	0.025 max	0.15-0.40	V <=0.10; bal Fe				
ASTM A983/983M(98)	Grade 7	Continuous grain flow frg for crankshafts	0.28-0.35	2.8-3.3	0.40-1.00	0.30-0.50		0.025 max	0.025 max	0.15-0.40	V <=0.10; bal Fe				
DoD-F-24669/2(86)	F11	HT, (H for Ann); Supersedes MIL-S-18410 & MIL-S-872A	0.15 max	2.00-2.50	0.30-0.60	0.90-1.10		0.040 max	0.040 max	0.50 max	bal Fe	485	275	20	180 HB
DoD-F-24669/2(86)	F22	HT, (H for Ann); Supersedes MIL-S-18410 & MIL-S-872A	0.10-0.20	1.00-1.50	0.30-0.80	0.45-0.65		0.040 max	0.040 max	0.50-1.00	bal Fe	485	275	20	180 HB
DoD-F-24669/8(91)	Grade B	Frg for Steam Turbines; Supersedes MIL-S-860B(66)	0.25 max	0.75 max	0.20-0.60	0.25 min	2.50 min	0.012 max	0.015 max	0.15-0.30	V >=0.03; As 0-0.020; Sb 0-0.010; Sn 0-0.020; bal Fe	551	379	22	
DoD-F-24669/8(91)	Grade C	Frg for Steam Turbines; Supersedes MIL-S-860B(66)	0.28 max	0.75 max	0.20-0.60	0.25 min	2.50 min	0.012 max	0.015 max	0.15-0.30	V >=0.03; As 0-0.020; Sb 0-0.010; Sn 0-0.020; bal Fe	620	482	20	
DoD-F-24669/8(91)	Grade D	Frg for Steam Turbines; Supersedes MIL-S-860B(66)	0.28 max	1.25-2.00	0.20-0.60	0.30-0.60	3.25-4.00	0.012 max	0.015 max	0.15-0.30	V 0.05-0.15; As 0-0.020; Sb 0-0.010; Sn 0-0.020; bal Fe	723-861	586	18	
DoD-F-24669/8(91)	Grade E	Frg for Steam Turbines; Supersedes MIL-S-860B(66)	0.28 max	1.25-2.00	0.20-0.60	0.30-0.60	3.25-4.00	0.012 max	0.015 max	0.15-0.30	V 0.05-0.15; As 0-0.020; Sb 0-0.010; Sn 0-0.020; bal Fe	792-896	655	18	
DoD-F-24669/8(91)	Grade F	Frg for Steam Turbines; Supersedes MIL-S-860B(66)	0.25-0.35	0.90-1.50	1.00 max	1.00-1.50	0.75 max	0.012 max	0.015 max	0.15-0.35	V 0.20-0.30; As 0-0.020; Sb 0-0.010; Sn 0-0.020; bal Fe	723-861	586	17	

Yugoslavia

Specification	Designation	Notes	C	Cr	Mn	Mo	Ni	P	S	Si	Other	UTS	YS	El	Hard
	C.4721	CH	0.18-0.23	1.1-1.4	0.9-1.2	0.2-0.3		0.035 max	0.035 max	0.15-0.4	bal Fe				
	C.4738	Q/T	0.28-0.35	2.8-3.3	0.4-0.7	0.3-0.5	0.3 max	0.035 max	0.035 max	0.15-0.4	bal Fe				
	C.7480	CH	0.17-0.22	0.3-0.5	0.6-0.9	0.4-0.5		0.035 max	0.035 max	0.15-0.4	bal Fe				
	C.7481	CH	0.23-0.29	0.4-0.6	0.6-0.9	0.4-0.5		0.035 max	0.035 max	0.15-0.4	bal Fe				

Alloy Steel, Chromium Molybdenum, T1

USA

Specification	Designation	Notes	C	Cr	Mn	Mo	Ni	P	S	Si	Other	UTS	YS	El	Hard
	UNS K11522		0.10-0.20		0.30-0.80	0.44-0.65		0.045 max	0.045 max	0.10-0.50	bal Fe				

UNS numbers and US grades are provided as a means of cross referencing chemically similar alloys. Exchangability is only possible after independent examination of specifications. Tensile properties are minimum or typical as specified. UTS and YS as MPa. El as %. See Appendix for list of abbreviations used in Notes. * indicates obsolete material.

Specification	Designation	Notes	C	Cr	Mn	Mo	Ni	P	S	Si	Other	UTS	YS	El	Hard

Alloy Steel, Chromium Molybdenum, T1 (Continued from previous page)

USA

| ASTM A161(94) | T1 | C-Mo Still tub | 0.10-0.20 | | 0.30-0.80 | 0.44-0.65 | | 0.025 max | 0.025 max | 0.10-0.50 | bal Fe | 379 | 207 | 30 | 137 or 150 HB max |
| ASTM A209/A209M(95) | T1 | Smls Boiler/Super Heater tubs, H for >5.1mm Wall | 0.10-0.20 | | 0.30-0.80 | 0.44-0.65 | | 0.025 max | 0.025 max | 0.10-0.50 | bal Fe | 380 | 205 | 30 | 146 HB |

Alloy Steel, Chromium Molybdenum, T11

UK

| BS 3604/2(91) | 622 | Pip Tub High-press/temp t<=40mm | 0.09-0.15 | 2.00-2.50 | 0.30-0.60 | 0.90-1.10 | 0.30 max | 0.025 max | 0.015 max | 0.50 max | Al <=0.020; Cu <=0.30; bal Fe | 515-690 | 310 | 16 | |

Alloy Steel, Chromium Molybdenum, T12

International

| ISO 2604-2(75) | TS32 | Smls tube | 0.1-0.18 | 0.7-1.1 | 0.4-0.7 | 0.45-0.65 | 0.4 max | 0.04 max | 0.04 max | 0.1-0.35 | Al <=0.02; Co <=0.1; Cu <=0.3; Pb <=0.15; Ti <=0.05; V <=0.1; W <=0.1; bal Fe | | | | |

UK

BS 3059/2(90)	620-460	Boiler, Superheater; High-temp; Smls Tub; Tub, Weld; 2-12.5mm	0.10-0.15	0.70-1.10	0.40-0.70	0.45-0.65	0.30 max	0.030 max	0.030 max	0.10-0.35	Al <=0.020; Cu <=0.25; Sn<=0.03; bal Fe	460-610	180	22	
BS 3606(92)	620	Heat exch Tub; t<=3.2mm	0.15 max	0.70-1.10	0.40-0.70	0.45-0.65	0.30 max	0.040 max	0.040 max	0.10-0.35	Al <=0.02; Sn<=0.03; bal Fe	460-610	180	22	
BS 3606(92)	621	Heat exch Tub; t<=3.2mm	0.15 max	1.00-1.50	0.30-0.60	0.45-0.65	0.30 max	0.040 max	0.040 max	0.50-1.00	Al <=0.02; Sn<=0.03; bal Fe	420-570	275	22	

USA

	UNS K11562	Ferritic Cr-Mo T12	0.15 max	0.80-1.25	0.30-0.61	0.44-0.65		0.045 max	0.045 max	0.50 max	bal Fe				
ASTM A182/A182M(98)	F12 Class 1	Frg/roll pipe flange valve	0.05-0.15	0.80-1.25	0.30-0.60	0.44-0.65		0.045 max	0.045 max	0.50 max	bal Fe	415	220	20	121-174 HB
ASTM A213/A213M(95)	T12	Smls tube boiler, superheater, heat exchanger	0.05-0.15	0.80-1.25	0.30-0.61	0.44-0.65		0.025 max	0.025 max	0.50 max	bal Fe	415	220	30	

Alloy Steel, Chromium Molybdenum, T2

India

| IS 1570/7(92) | 15Mo6H | High-temp; Tub pipe | 0.1-0.2 | 0.3 max | 0.3-0.8 | 0.44-0.65 | 0.4 max | 0.045 max | 0.045 max | 0.1-0.5 | bal Fe | 380 | 205 | 22 | |
| IS 1570/7(92) | 6 | High-temp; Tub pipe | 0.1-0.2 | 0.3 max | 0.3-0.8 | 0.44-0.65 | 0.4 max | 0.045 max | 0.045 max | 0.1-0.5 | bal Fe | 380 | 205 | 22 | |

UK

| BS 3604/2(91) | 620 | Pip Tub High-press/temp t<=40mm | 0.09-0.18 | 0.80-1.15 | 0.40-0.65 | 0.45-0.60 | 0.30 max | 0.025 max | 0.015 max | 0.15-0.40 | Al <=0.020; Cu <=0.30; bal Fe | 480-600 | 340 | 18 | |
| BS 3604/2(91) | 621 | Pip Tub High-press/temp t<=40mm | 0.09-0.17 | 1.00-1.50 | 0.40-0.65 | 0.45-0.60 | 0.30 max | 0.025 max | 0.015 max | 0.50-0.80 | Al <=0.020; Cu <=0.30; bal Fe | 515-690 | 340 | 18 | |

USA

| | UNS K11547 | | 0.10-0.20 | 0.50-0.81 | 0.30-0.61 | 0.44-0.65 | | 0.045 max | 0.045 max | 0.10-0.30 | bal Fe | | | | |
| ASTM A213/A213M(95) | T2 | Smls tube boiler, superheater, heat exchanger | 0.10-0.20 | 0.50-0.81 | 0.30-0.61 | 0.44-0.65 | | 0.025 max | 0.025 max | 0.10-0.30 | bal Fe | 415 | 205 | 30 | |

Alloy Steel, Chromium Molybdenum, T22

Bulgaria

| BDS 6609(73) | 12Ch2M | Heat res; boiler tube | 0.08-0.15 | 2.00-2.5 | 0.40-0.70 | 0.90-1.10 | 0.30 max | 0.040 max | 0.040 max | 0.15-0.50 | Cu <=0.30; bal Fe | | | | |

Europe

EN 10028/2(92)	1.7380	High-temp, Press; 100<=t<=150mm	0.08-0.14	2.00-2.50	0.40-0.80	0.90-1.10		0.030 max	0.025 max	0.50 max	Cu <=0.30; bal Fe	460-610	250	17T	
EN 10028/2(92)	1.7380	High-temp, Press; 60<=t<=100mm	0.06-0.15	2.00-2.50	0.40-0.80	0.90-1.10		0.030 max	0.025 max	0.50 max	Cu <=0.30; bal Fe	470-620	270	17T	
EN 10028/2(92)	1.7380	High-temp, Press; t<16mm	0.08-0.14	2.00-2.50	0.40-0.80	0.90-1.10		0.030 max	0.025 max	0.50 max	Cu <=0.30; bal Fe	480-630	310	18T	
EN 10028/2(92)	1.7380	High-temp, Press; 16<= t<=40mm	0.08-0.14	2.00-2.50	0.40-0.80	0.90-1.10		0.030 max	0.025 max	0.50 max	Cu <=0.30; bal Fe	480-630	300	18T	

UNS numbers and US grades are provided as a means of cross referencing chemically similar alloys. Exchangability is only possible after independent examination of specifications. Tensile properties are minimum or typical as specified. UTS and YS as MPa. El as %. See Appendix for list of abbreviations used in Notes. * indicates obsolete material.

Alloy Steel, Chromium Molybdenum, T22 (Continued from previous page)

Specification	Designation	Notes	C	Cr	Mn	Mo	Ni	P	S	Si	Other	UTS	YS	El	Hard
Europe															
EN 10028/2(92)	1.7380	High-temp, Press; 40<=t<=60mm	0.06-0.15	2.00-2.50	0.40-0.80	0.90-1.10		0.030 max	0.025 max	0.50 max	Cu <=0.30; bal Fe	480-630	290	18T	
EN 10028/2(92)	1.7383	High-temp, Press; 60<= t<=100mm	0.08-0.15	2.00-2.50	0.40-0.80	0.90-1.10		0.030 max	0.025 max	0.50 max	Cu <=0.30; bal Fe	520-670	310	17T	
EN 10028/2(92)	1.7383	High-temp, Press; t<60mm	0.08-0.15	2.00-2.50	0.40-0.80	0.90-1.10		0.030 max	0.025 max	0.50 max	Cu <=0.30; bal Fe	520-670	310	18T	
France															
AFNOR NFA36205/2(92)	10CrMo9-10	Press ves; High-temp; 150mm	0.08-0.17	2-2.5	0.40-0.80	0.9-1.10	0.4 max	0.03 max	0.025 max	0.5 max	Cu <=0.30; Ti <=0.03; bal Fe	470-620	270	17T	
AFNOR NFA36205/2(92)	11CrMo9-10	Press ves; Sh Plt Strp; Press; High-temp; Size 2-60mm	0.06-0.15	2-2.5	0.40-0.80	0.9-1.10	0.4 max	0.03 max	0.025 max	0.5 max	Cu <=0.30; Ti <=0.03; bal Fe	520-670	310	18T	
AFNOR NFA36210(88)	12CD9-10	Size t<=70mm	0.16 max	2-2.5	0.3-0.6	0.9-1.1	0.3 max	0.015 max	0.012 max	0.1-0.4	Cu <=0.3; Ti <=0.05; bal Fe	590-700	420	17	
AFNOR NFA36210(88)	12CD9-10	Sh Plt; Boiler, Press Ves; 125<t<=200mm; Q/T	0.16 max	2-2.5	0.3-0.6	0.9-1.1	0.3 max	0.015 max	0.012 max	0.1-0.4	Cu <=0.3; Ti <=0.05; bal Fe	540-660	380	17	
International															
ISO 2604-2(75)	TS34	Smls tube	0.08-0.15	2-2.5	0.4-0.7	0.9-1.2	0.4 max	0.04 max	0.04 max	0.5 max	Al <=0.02; Co <=0.1; Cu <=0.3; Pb <=0.15; V 0.05-0.1; W <=0.1; bal Fe				
Italy															
UNI 7660(77)	12CrMo910KG	Press ves	0.15 max	2-2.5	0.4-0.7	0.9-1.2		0.03 max	0.03 max	0.15-0.4	Al <=0.02; bal Fe				
Sweden															
SS 142218	2218	High-temp const	0.06-0.15	2-2.5	0.4-0.7	0.9-1.1	0.3 max	0.035 max	0.03 max	0.15-0.5	bal Fe				
UK															
BS 1502(82)	622	Bar Shp Press ves	0.08-0.15	2.00-2.50	0.40-0.70	0.90-1.20		0.040 max	0.040 max	0.15-0.50	Al <=0.020; Co <=0.1; W <=0.1; bal Fe				
BS 3059/2(90)	622-490	Boiler, Superheater; High-temp; Smls Tub; 2-12.5mm	0.08-0.15	2.00-2.50	0.40-0.70	0.9-1.20	0.30 max	0.030 max	0.030 max	0.50 max	Al <=0.020; Cu <=0.25; Sn<=0.03; bal Fe	490-640	275	20	
BS 3606(92)	622	Heat exch Tub; t<=3.2mm	0.08-0.15	2.00-2.50	0.40-0.70	0.90-1.20	0.30 max	0.040 max	0.040 max	0.50 max	Al <=0.02; Sn<=0.03; bal Fe	490-640	275	16	
USA															
	UNS K21590		0.15 max	2.00-2.50	0.30-0.60	0.90-1.10		0.030 max	0.030 max	0.50 max	bal Fe				
ASTM A182/A182M(98)	F22 Class1	Frg/roll pipe flange valve	0.05-0.15	2.00-2.50	0.30-0.60	0.87-1.13		0.040 max	0.040 max	0.50 max	bal Fe	415	205	20.0	170 HB max
ASTM A182/A182M(98)	F22 Class3	Frg/roll pipe flange valve	0.05-0.15	2.00-2.50	0.30-0.60	0.87-1.13		0.040 max	0.040 max	0.50 max	bal Fe	515	310	20.0	156-207
ASTM A213/A213M(95)	T22	Smls tube boiler, superheater, heat exchanger	0.05-0.15	1.90-2.60	0.30-0.60	0.087-1.13		0.025 max	0.025 max	0.50 max	bal Fe	415	205	30	
ASTM A234/A234M(97)	WP22 CL1	Frg	0.05-0.15	1.90-2.60	0.30-0.60	0.87-1.13		0.040 max	0.040 max	0.50 max	bal Fe	415-585	205	22L	
ASTM A234/A234M(97)	WP22 CL3	Frg	0.05-0.15	1.90-2.60	0.30-0.60	0.87-1.13		0.040 max	0.040 max	0.50 max	bal Fe	520-690	310	22L	
ASTM A336/A336M(98)	F22 Class 1	Ferr Frg	0.05-0.15	2.00-2.50	0.30-0.60	0.90-1.10		0.025 max	0.025 max	0.50 max	bal Fe	415-585	205	20	
ASTM A336/A336M(98)	F22 Class 3	Ferr Frg	0.05-0.15	2.00-2.50	0.30-0.60	0.90-1.10		0.025 max	0.025 max	0.50 max	bal Fe	515-690	310	19	
ASTM A387/A387(97)	Grade 22	Press ves plt, Class 2	0.05-0.15	2.00-2.50	0.30-0.60	0.90-1.10		0.035 max	0.035 max	0.50 max	bal Fe	515-690	310	18	
ASTM A387/A387(97)	Grade 22L	Press ves plt, Class 1	0.10 max	2.00-2.50	0.30-0.60	0.90-1.10		0.035 max	0.035 max	0.50 max	bal Fe	415-485	207	18	
ASTM A508/508M(95)	Grade 22 Class 3	Q/T vacuum-treated frg for press ves	0.11-0.15	2.00-2.50	0.30-0.60	0.90-1.10	0.25 max	0.015 max	0.015 max	0.50 max	bal Fe	585-790	380	18	
ASTM A508/508M(95)	Grade 3V	Q/T vacuum-treated frg for press ves	0.10-0.15	2.8-3.3	0.30-0.60	0.90-1.10		0.020 max	0.020 max	0.10 max	B 0.001-0.003; Ti 0.015-0.035; bal Fe	585-760	415	18	
ASTM A508/508M(95)	Grade 3VCb	Q/T vacuum-treated frg for press ves	0.10-0.15	2.7-3.3	0.30-0.60	0.90-1.10	0.25 max	0.020 max	0.010 max	0.10 max	Cu <=0.25; Nb 0.015-0.070; Ti <=0.015; Ca 0.0005-0.0150; bal Fe	585-760	415	18	

UNS numbers and US grades are provided as a means of cross referencing chemically similar alloys. Exchangability is only possible after independent examination of specifications. Tensile properties are minimum or typical as specified. UTS and YS as MPa. El as %. See Appendix for list of abbreviations used in Notes. * indicates obsolete material.

Specification	Designation	Notes	C	Cr	Mn	Mo	Ni	P	S	Si	Other	UTS	YS	El	Hard

Alloy Steel, Chromium Molybdenum, T22 (Continued from previous page)

USA

Specification	Designation	Notes	C	Cr	Mn	Mo	Ni	P	S	Si	Other	UTS	YS	El	Hard
ASTM A542(95)	Type A	Q/T, Press ves plt, Class 1, t>=5mm	0.15 max	2.00-2.50	0.30-0.60	0.90-1.10	0.40 max	0.025 max	0.025 max	0.50 max	Cu <=0.40; V <=0.03; bal Fe	725-860	585	14	
ASTM A542(95)	Type A	Q/T, Press ves plt, Class 3, t>=5mm	0.15 max	2.00-2.50	0.30-0.60	0.90-1.10	0.40 max	0.025 max	0.025 max	0.50 max	Cu <=0.40; V <=0.03; bal Fe	655-795	515	20	
ASTM A542(95)	Type A	Q/T, Press ves plt, Class 4a, t=5mm	0.15 max	2.00-2.50	0.30-0.60	0.90-1.10	0.40 max	0.025 max	0.025 max	0.50 max	Cu <=0.40; V <=0.03; bal Fe	585-760	415	18	
ASTM A542(95)	Type A	Q/T, Press ves plt, Class 2, t>=5mm	0.15 max	2.00-2.50	0.30-0.60	0.90-1.10	0.40 max	0.025 max	0.025 max	0.50 max	Cu <=0.40; V <=0.03; bal Fe	795-930	690	13	
ASTM A542(95)	Type A	Q/T, Press ves plt, Class 4, t>=5mm	0.15 max	2.00-2.50	0.30-0.60	0.90-1.10	0.40 max	0.025 max	0.025 max	0.50 max	Cu <=0.40; V <=0.03; bal Fe	585-790	380	20	
ASTM A542(95)	Type B	Q/T, Press ves plt, Class 2, t>=5mm	0.11-0.15	2.00-2.50	0.30-0.60	0.90-1.10	0.25 max	0.015 max	0.015 max	0.50 max	Cu <=0.25; V <=0.02; bal Fe	795-930	690	13	
ASTM A542(95)	Type B	Q/T, Press ves plt, Class 1, t>=5mm	0.11-0.15	2.00-2.50	0.30-0.60	0.90-1.10	0.25 max	0.015 max	0.015 max	0.50 max	Cu <=0.25; V <=0.02; bal Fe	725-860	585	14	
ASTM A542(95)	Type B	Q/T, Press ves plt, Class 3, t>=5mm	0.11-0.15	2.00-2.50	0.30-0.60	0.90-1.10	0.25 max	0.015 max	0.015 max	0.50 max	Cu <=0.25; V <=0.02; bal Fe	655-795	515	20	
ASTM A542(95)	Type B	Q/T, Press ves plt, Class 4, t>=5mm	0.11-0.15	2.00-2.50	0.30-0.60	0.90-1.10	0.25 max	0.015 max	0.015 max	0.50 max	Cu <=0.25; V <=0.02; bal Fe	585-790	380	20	
ASTM A542(95)	Type B	Q/T, Press ves plt, Class 4a, t=5mm	0.11-0.15	2.00-2.50	0.30-0.60	0.90-1.10	0.25 max	0.015 max	0.015 max	0.50 max	Cu <=0.25; V <=0.02; bal Fe	585-760	415	18	
ASTM A832/A832M(97)	Grade 22V	Press ves plt	0.11-0.15	2.00-2.50	0.30-0.60	0.90-1.10	0.25 max	0.015 max	0.010 max	0.10 max	B <=0.0020; Cu <=0.20; Nb <=0.07; Ti <=0.030; V 0.25-0.35; Ca=0-0.015, bal Fe	585-760	415	18	
ASTM A989(98)	K21590(1)	HIP Flanges, Fittings, Valves/parts; Heat res	0.05-0.15	2.00-2.50	0.30-0.60	0.87-1.13		0.040 max	0.040 max	0.50 max	bal Fe	415	205	20.0	170 HB max
ASTM A989(98)	K21500(2)	HIP Flanges, Fittings, Valves/parts; Heat res	0.05-0.15	2.00-2.50	0.30-0.60	0.87-1.13		0.040 max	0.040 max	0.50 max	bal Fe	515	310	20.0	156-207 HB

Yugoslavia

Specification	Designation	Notes	C	Cr	Mn	Mo	Ni	P	S	Si	Other	UTS	YS	El	Hard
	C.7401	High-temp const	0.15 max	2.0-2.5	0.4-0.8	0.9-1.1		0.04 max	0.04 max	0.15-0.5	bal Fe				

Alloy Steel, Chromium Molybdenum, W-FeCrMn

USA

Specification	Designation	Notes	C	Cr	Mn	Mo	Ni	P	S	Si	Other	UTS	YS	El	Hard
	UNS T87515	Thermal Spray Wire	1.0 max	1.6-2.0	1.7-2.0	0.15-0.25		0.010 max	0.040 max	0.35 max	Ti 0.11-0.15; bal Fe				

Alloy Steel, Chromium Molybdenum, WP12

Japan

Specification	Designation	Notes	C	Cr	Mn	Mo	Ni	P	S	Si	Other	UTS	YS	El	Hard
JIS G3311(88)	SCM415M	CR Strp; Chain Blades	0.13-0.18	0.90-1.20	0.60-0.85	0.15-0.30	0.25 max	0.030 max	0.030 max	0.15-0.35	Cu <=0.30; bal Fe				170-240 HV
JIS G3441(88)	SCM415TK	Tube	0.13-0.18	0.90-1.20	0.60-0.85	0.15-0.30	0.25 max	0.030 max	0.030 max	0.15-0.35	Cu <=0.30; bal Fe				

USA

Specification	Designation	Notes	C	Cr	Mn	Mo	Ni	P	S	Si	Other	UTS	YS	El	Hard
	UNS K12062		0.20 max	0.80-1.25	0.30-0.80	0.44-0.65		0.045 max	0.045 max	0.60 max	bal Fe				
ASTM A234/A234M(97)	WP12 CL1	Frg	0.05-0.20	0.80-1.25	0.30-0.80	0.44-0.65		0.045 max	0.045 max	0.60 max	bal Fe	415-585	220	22L	
ASTM A234/A234M(97)	WP12 CL2	Frg	0.05-0.20	0.80-1.25	0.30-0.80	0.44-0.65		0.045 max	0.045 max	0.60 max	bal Fe	485-655	275	22L	

Yugoslavia

Specification	Designation	Notes	C	Cr	Mn	Mo	Ni	P	S	Si	Other	UTS	YS	El	Hard
	C.7400	High-temp const	0.1-0.18	0.7-1.0	0.4-0.7	0.4-0.5		0.04 max	0.04 max	0.15-0.35	bal Fe				

Specification	Designation	Notes	C	Cr	Mn	Mo	P	S	Si	V	Other	UTS	YS	El	Hard

Alloy Steel, Chromium Vanadium, 6118/6118H

China

Specification	Designation	Notes	C	Cr	Mn	Mo	P	S	Si	V	Other	UTS	YS	El	Hard
GB 3077(88)	20CrV	Bar HR Q/T 15mm diam	0.17-0.23	0.80-1.10	0.50-0.80		0.035 max	0.035 max	0.17-0.37	0.10-0.20	Cu <=0.30; bal Fe	835	590	12	

Czech Republic

Specification	Designation	Notes	C	Cr	Mn	Mo	P	S	Si	V	Other	UTS	YS	El	Hard
CSN 415110	15110	High-temp const	0.08-0.16	0.5-0.75	0.4-0.7		0.04 max	0.04 max	0.15-0.4	0.2-0.35	bal Fe				
CSN 415111	15111	High-temp const	0.08-0.16	0.5-0.75	0.4-0.7	0.2-0.4	0.04 max	0.04 max	0.15-0.4	0.2-0.35	bal Fe				
CSN 415112	15112	High-temp const	0.08-0.16	0.5-0.75	0.4-0.7	0.08-0.14	0.035 max	0.035 max	0.15-0.4	0.2-0.35	bal Fe				

Germany

Specification	Designation	Notes	C	Cr	Mn	Mo	P	S	Si	V	Other	UTS	YS	El	Hard
DIN	22CrV3		0.20-0.24	0.65-0.80	0.60-0.75		0.035 max	0.035 max	0.35 max	0.10-0.15	bal Fe				
DIN	WNr 1.7511		0.20-0.24	0.65-0.80	0.60-0.75		0.035 max	0.035 max	0.35 max	0.10-0.15	bal Fe				

Italy

Specification	Designation	Notes	C	Cr	Mn	Mo	P	S	Si	V	Other	UTS	YS	El	Hard
UNI 6551(69)	22MnCrV5	axles	0.17-0.27	0.2-0.7	1-1.4		0.025 max	0.03 max	0.4 max	0.05-0.15	Cu <=0.25; Ni <=0.15; bal Fe				

Mexico

Specification	Designation	Notes	C	Cr	Mn	Mo	P	S	Si	V	Other	UTS	YS	El	Hard
DGN B-203	6118*	Obs	0.16-0.21	0.5-0.7	0.5-0.7		0.04	0.04	0.2-0.35	0.1-0.15	bal Fe				
DGN B-297	6118*	Obs	0.16-0.21	0.5-0.7	0.5-0.7		0.035	0.04	0.2-0.35	0.1-0.15	bal Fe				
NMX-B-268(68)	6118H	Hard	0.15-0.21	0.40-0.80	0.40-0.80				0.20-0.35	0.10-0.15	bal Fe				
NMX-B-300(91)	6118	Bar	0.16-0.21	0.50-0.70	0.50-0.70		0.035 max	0.040 max	0.15-0.35	0.10-0.15	bal Fe				
NMX-B-300(91)	6118H	Bar	0.15-0.21	0.40-0.80	0.40-0.80				0.15-0.35	0.10-0.15	bal Fe				

Pan America

Specification	Designation	Notes	C	Cr	Mn	Mo	P	S	Si	V	Other	UTS	YS	El	Hard
COPANT 514	6118	Tube	0.16-0.21	0.5-0.7	0.5-0.7		0.04	0.05	0.1	0.1-0.15	bal Fe				
COPANT 514	6118		0.16-0.21	0.5-0.7	0.5-0.7		0.04	0.05	0.1	0.1-0.15	bal Fe				

Poland

Specification	Designation	Notes	C	Cr	Mn	Mo	P	S	Si	V	Other	UTS	YS	El	Hard
PNH93215(82)	20HG2V		0.18-0.25	0.7-1	1.6-1.9		0.04	0.04	0.4-0.6	0.05-0.1	Cu 0.3; Ni 0.3; bal Fe				

USA

Specification	Designation	Notes	C	Cr	Mn	Mo	P	S	Si	V	Other	UTS	YS	El	Hard
	AISI 6118	Smls mech tub	0.16-0.21	0.50-0.70	0.50-0.70		0.040 max	0.040 max	0.15-0.35	0.10-0.15	bal Fe				
	AISI 6118H	Smls mech tub, Hard	0.15-0.21	0.40-0.80	0.40-0.80		0.035 max	0.040 max	0.15-0.30	0.10-0.15	bal Fe				
	UNS G61180		0.16-0.21	0.50-0.70	0.50-0.70		0.035 max	0.040 max	0.15-0.35	0.10-0.15	bal Fe				
	UNS H61180		0.15-0.21	0.40-0.80	0.40-0.80		0.035 max	0.040 max	0.15-0.35	0.10-0.15	bal Fe				
ASTM A29/A29M(93)	6118	Bar	0.16-0.21	0.50-0.70	0.5-0.7		0.035 max	0.04 max	0.15-0.35	0.1-0.15	bal Fe				
ASTM A304(96)	6118H	Bar, hard bands spec	0.15-0.21	0.40-0.80	0.40-0.80	0.06 max	0.035 max	0.040 max	0.15-0.35	0.1-0.15	Cu <=0.35; Cu Ni Cr Mo trace allowed; bal Fe				
ASTM A322(96)	6118	Bar	0.16-0.21	0.50-0.70	0.50-0.70		0.035 max	0.040 max	0.15-0.35	0.1-0.15	bal Fe				
ASTM A331(95)	6118	Bar	0.16-0.21	0.50-0.70	0.50-0.70		0.035 max	0.040 max	0.15-0.35	0.10-0.15	bal Fe				
ASTM A331(95)	6118H	Bar	0.15-0.21	0.40-0.80	0.40-0.80	0.06 max	0.025 max	0.025 max	0.15-0.35	0.10-0.15	Cu <=0.35; bal Fe				
ASTM A519(96)	6118	Smls mech tub	0.16-0.21	0.50-0.70	0.50-0.70		0.040 max	0.040 max	0.15-0.35	0.10-0.15	bal Fe				
ASTM A752(93)	6118	Rod Wir	0.16-0.21	0.50-0.70	0.50-0.70		0.035 max	0.040 max	0.15-0.30	0.10-0.15	bal Fe	550			
SAE 770(84)	6118*	Obs; see J1397(92)	0.16-0.21	0.5-0.7	0.5-0.7		0.04 max	0.04 max	0.15-0.3	0.1-0.15	bal Fe				
SAE J1268(95)	6118H	See std	0.15-0.21	0.40-0.80	0.40-0.80	0.06 max	0.030 max	0.040 max	0.15-0.35	0.10-0.15	Cu <=0.35; Ni <=0.25; Cu, Mo, Ni not spec'd but acpt; bal Fe				

Specification	Designation	Notes	C	Cr	Mn	Mo	P	S	Si	V	Other	UTS	YS	El	Hard
Alloy Steel, Chromium Vanadium, 6120															
China															
GB 3077(88)	20CrV	Bar HR Q/T 15mm diam	0.17-0.23	0.80-1.10	0.50-0.80		0.035 max	0.035 max	0.17-0.37	0.10-0.20	Cu <=0.30; bal Fe	835	590	12	
USA															
	AISI 6120	Smls mech tub	0.17-0.22	0.70-0.90	0.70-0.90		0.040 max	0.040 max	0.15-0.35	0.10 min	bal Fe				
	UNS G61200		0.17-0.22	0.70-0.90	0.70-0.90		0.035 max	0.040 max	0.15-0.35	0.10 min	bal Fe				
ASTM A519(96)	6120	Smls mech tub	0.17-0.22	0.70-0.90	0.70-0.90		0.040 max	0.040 max	0.15-0.35	0.10 min	bal Fe				
Alloy Steel, Chromium Vanadium, 6150/6150H															
Australia															
AS 1444(96)	6150	H series, Hard/Tmp	0.48-0.53	0.80-1.10	0.70-0.90		0.040 max	0.040 max	0.10-0.35	0.15-0.25	bal Fe				
AS 1444(96)	6150H	H series, Hard/Tmp	0.47-0.54	0.75-1.20	0.60-1.00		0.040 max	0.040 max	0.10-0.35	0.15-0.25	bal Fe				
AS G18	En47*	Obs; see AS 1444	0.45-0.55	0.8-1.2	0.5-0.8		0.05	0.05	0.5	0.15 min	bal Fe				
Belgium															
NBN 253-02	50CrV4		0.47-0.55	0.9-1.2	0.7-1.1		0.035	0.035	0.15-0.4	0.1-0.2	bal Fe				
NBN 253-05	50CrV4	Wire, Ann or Q/A Tmp 40 mm diam	0.48-0.55	0.9-1.2	0.7-1		0.04	0.04	0.15-0.4	0.1-0.2	bal Fe				
Bulgaria															
BDS 6742	50ChFA	Spring	0.46-0.54	0.80-1.10	0.50-0.80		0.025 max	0.025 max	0.17-0.37	0.10-0.20	Cu <=0.30; Ni <=0.30; bal Fe				
China															
GB 1222(84)	50CrVA	Bar Flat Q/T	0.46-0.54	0.80-1.10	0.50-0.80		0.030 max	0.030 max	0.17-0.37	0.10-0.20	Cu <=0.25; bal Fe	1275	1130	10	
GB 3077(88)	50CrVA	Bar HR Q/T 25mm diam	0.47-0.54	0.80-1.10	0.50-0.80		0.025 max	0.025 max	0.17-0.37	0.10-0.20	Cu <=0.30; bal Fe	1275	1130	10	
GB 3279(89)	50CrVA	Sh HR Ann	0.46-0.54	0.80-1.10	0.50-0.80		0.030 max	0.030 max	0.17-0.37	0.10-0.20	Cu <=0.25; bal Fe	950		12	
GB 8162(87)	50CrVA	Smls Pip HR/CD Q/T	0.47-0.54	0.80-1.10	0.50-0.80		0.025 max	0.025 max	0.17-0.37	0.10-0.20	bal Fe	1275	1130	10	
GB/T 3079(93)	50CrVA	Wir Q/T	0.47-0.54	0.80-1.10	0.50-0.80		0.025 max	0.025 max	0.17-0.37	0.10-0.20	Cu <=0.25; bal Fe	1275	1130	10	
YB/T 5136(93)	50CrVA	Wir Q/T	0.46-0.54	0.80-1.10	0.50-0.80		0.025 max	0.025 max	0.17-0.37	0.10-0.20	Cu <=0.25; bal Fe	1470-1765			
Czech Republic															
CSN 415260	15260	Q/T	0.47-0.55	0.9-1.2	0.7-1.0		0.035 max	0.035 max	0.15-0.4	0.1-0.2	Ni <=0.3; bal Fe				
Europe															
EN 10083/1(91)A1(96)	1.8159	Q/T t<=16mm	0.47-0.55	0.90-1.20	0.70-1.10	0.15 max	0.035 max	0.035 max	0.40 max	0.10-0.25	bal Fe	1100-1300	900	9	
EN 10083/1(91)A1(96)	51CrV4	Q/T t<=16mm	0.47-0.55	0.90-1.20	0.70-1.10	0.15 max	0.035 max	0.035 max	0.40 max	0.10-0.25	bal Fe	1100-1300	900	9	
France															
AFNOR NFA35552	50CV4		0.47-0.55	0.85-1.2	0.7-1.1		0.035	0.035	0.1-0.4	0.1-0.2	bal Fe				
AFNOR NFA35553	50CV4	Strp	0.47-0.55	0.85-1.2	0.7-1.1		0.035	0.035	0.1-0.4	0.1-0.2	bal Fe				
AFNOR NFA35553(82)	50CV4	t<=4.5mm; Q	0.47-0.55	0.85-1.15	0.7-1	0.15 max	0.035 max	0.035 max	0.1-0.4	0.1-0.2	Cu <=0.3; Ni <=0.4; Ti <=0.05; bal Fe				54 HRC
AFNOR NFA35571	50CV4	Bar Wir rod, Q/A Tmp, 80mm	0.47-0.55	0.85-1.2	0.7-1.1		0.035	0.035	0.1-0.4	0.1-0.2	bal Fe				
AFNOR NFA35571(96)	51CrV4	t<1000mm; Q/T	0.47-0.55	0.9-1.2	0.7-1.1	0.15 max	0.025 max	0.025 max	0.1-0.4	0.07-0.25	Cu <=0.3; Ni <=0.4; Ti <=0.05; bal Fe	1350-1650	1200	6	
AFNOR NFA36102(93)	50CrV4RR	HR; Strp; CR	0.47-0.55	0.9-1.2	0.7-1.1	0.10 max	0.025 max	0.02	0.15-0.4	0.1-0.25	Ni <=0.20; bal Fe				
AFNOR NFA36102(93)	50CV4*	HR; Strp; CR	0.47-0.55	0.9-1.2	0.7-1.1	0.10 max	0.025 max	0.02 max	0.15-0.4	0.1-0.25	Ni <=0.20; bal Fe				
Germany															
DIN 1652(90)	WNr 1.8159	Q/T 17-40mm	0.47-0.55	0.90-1.20	0.70-1.10		0.035 max	0.035 max	0.40 max	0.10-0.25	bal Fe	1000-1200	800	10	

Alloy Steel, Chromium Vanadium, 6150/6150H (Continued from previous page)

Specification	Designation	Notes	C	Cr	Mn	Mo	P	S	Si	V	Other	UTS	YS	El	Hard
Germany															
DIN 17221(88)	WNr 1.8159	HR QY 17-40mm	0.47-0.55	0.90-1.20	0.70-1.10		0.035 max	0.035 max	0.40 max	0.10-0.25	bal Fe	1000-1200	800	10	
DIN EN 10083(91)	50CrV4	Q/T 17-40mm	0.47-0.55	0.90-1.20	0.70-1.10		0.035 max	0.035 max	0.40 max	0.10-0.25	bal Fe	1000-1200	800	10	
DIN EN 10083(91)	WNr 1.8159	Q/T 17-40mm	0.47-0.55	0.90-1.20	0.70-1.10		0.035 max	0.035 max	0.40 max	0.10-0.25	bal Fe	1000-1200	800	10	
Hungary															
MSZ 2666(76)	50CV2	Spring HF	0.47-0.55	0.9-1.2	0.8-1.1		0.03 max	0.03 max	0.15-0.4	0.1-0.2	Cu <=0.25; Ni <=0.3; bal Fe				
MSZ 2666(88)	51CrV4	HF spring; 0-100.0mm; Q/T	0.47-0.55	0.8-1.1	0.6-1		0.03 max	0.03 max	0.15-0.4	0.1-0.25	bal Fe	1320	1125	7L	
MSZ 2666(88)	51CrV4	HF spring; 0-999mm	0.47-0.55	0.8-1.1	0.6-1		0.03 max	0.03 max	0.15-0.4	0.1-0.25	bal Fe				340 HB max
MSZ 4217(85)	50CV2	CR; Q/T or spring Strp; 0-2mm	0.47-0.55	0.9-1.2	0.8-1.1		0.03 max	0.03 max	0.15-0.4	0.1-0.2	bal Fe	1860-2350			
MSZ 4217(85)	50CV2	CR; Q/T or spring Strp; 0-2mm	0.47-0.55	0.9-1.2	0.8-1.1		0.03 max	0.03 max	0.15-0.4	0.1-0.2	bal Fe	1280-1580		4L	
MSZ 61(74)	CrV3E	Q/T	0.46-0.55	0.8-1.1	0.5-0.8		0.035 max	0.02-0.035	0.17-0.37	0.1-0.2	bal Fe				
MSZ 61(85)	CrV3	Q/T; 160.1-250mm	0.47-0.55	0.8-1.1	0.6-1		0.035 max	0.035 max	0.4 max	0.1-0.2	bal Fe	800-950	550	13L	
MSZ 61(85)	CrV3	Q/T; 0-16mm	0.47-0.55	0.8-1.1	0.6-1		0.035 max	0.035 max	0.4 max	0.1-0.2	bal Fe	1100-1300	900	9L	
MSZ 6251(87)	CrV3Z	CF; Q/T; 0-36mm; drawn, half-hard	0.47-0.55	0.8-1.1	0.6-1		0.035 max	0.035 max	0.4 max	0.1-0.2	bal Fe	0-690			
MSZ 6251(87)	CrV3Z	CF; Q/T; 0-36mm; HR; HR/soft ann; drawn/soft ann; drawn/bright ann; soft ann/ground	0.47-0.55	0.8-1.1	0.6-1		0.035 max	0.035 max	0.4 max	0.1-0.2	bal Fe	0-660			
India															
IS 1570	50Cr1V23		0.45-0.55	0.9-1.2	0.5-0.8				0.1-0.35	0.15-0.3	bal Fe				
IS 4367	50Cr1V23	Frg, Ann, 63mm diam	0.45-0.55	0.9-1.2	0.5-0.8		0.05	0.05	0.1-0.35	0.15-0.3	bal Fe				
International															
ISO 683-14(92)	51CrV4	Bar Rod Wir; Q/T, springs, Ann	0.47-0.55	0.80-1.10	0.60-1.00		0.030 max	0.030 max	0.10-0.40	0.10-0.25	bal Fe				255 HB
Italy															
UNI 3545(80)	50CrV4	Spring	0.47-0.55	0.8-1.2	0.7-1.1		0.035 max	0.035 max	0.15-0.4	0.1-0.2	bal Fe				
UNI 7064(82)	50CrV4	Spring	0.48-0.55	0.9-1.2	0.7-1		0.035 max	0.035 max	0.15-0.4	0.1-0.2	bal Fe				
UNI 7065	50CrV4		0.47-0.55	0.8-1.2	0.7-0.9		0.035	0.035	0.2-0.4	0.1-0.2	bal Fe				
UNI 7845(78)	50CrV4	Q/T	0.47-0.55	0.8-1.2	0.7-1.1		0.035 max	0.035 max	0.15-0.4	0.1-0.2	bal Fe				
UNI 7874(79)	50CrV4	Q/T	0.47-0.55	0.9-1.2	0.7-1.1		0.035 max	0.035 max	0.15-0.4	0.1-0.2	bal Fe				
UNI 8893(86)	50CrV4	Spring	0.48-0.55	0.9-1.2	0.7-1		0.035 max	0.035 max	0.15-0.4	0.1-0.2	bal Fe				
Japan															
JIS G3311(88)	SUP10M	CR Strp; Spring	0.47-0.55	0.80-1.10	0.65-0.95		0.035 max	0.035 max	0.15-0.35	0.15-0.25	Cu <=0.30; bal Fe				200-290 HV
JIS G4801(84)	SUP10	Bar Spring Q/T	0.47-0.55	0.80-1.10	0.65-0.95		0.035 max	0.035 max	0.15-0.35	0.15-0.25	Cu <=0.30; bal Fe	1226	1079	10	363-429 HB
Mexico															
DGN B-203	6150*	Obs	0.48-0.53	0.8-1.1	0.7-0.9		0.04	0.04	0.2-0.35	0.15 min	bal Fe				
DGN B-297	6150*	Obs	0.48-0.53	0.8-1.1	0.7-0.9		0.035	0.04	0.2-0.35	0.15 min	bal Fe				
NMX-B-268(68)	6150H	Hard	0.47-0.54	0.75-1.20	0.60-1.00				0.20-0.35	0.15	bal Fe				
NMX-B-300(91)	6150	Bar	0.48-0.53	0.80-1.10	0.70-0.90		0.035 max	0.040 max	0.15-0.35	0.15 min	bal Fe				
NMX-B-300(91)	6150H	Bar	0.47-0.54	0.75-1.20	0.60-1.00				0.15-0.35	0.15 max	bal Fe				

UNS numbers and US grades are provided as a means of cross referencing chemically similar alloys. Exchangability is only possible after independent examination of specifications. Tensile properties are minimum or typical as specified. UTS and YS as MPa. El as %. See Appendix for list of abbreviations used in Notes. * indicates obsolete material.

Specification	Designation	Notes	C	Cr	Mn	Mo	P	S	Si	V	Other	UTS	YS	El	Hard

Alloy Steel, Chromium Vanadium, 6150/6150H (Continued from previous page)

Specification	Designation	Notes	C	Cr	Mn	Mo	P	S	Si	V	Other	UTS	YS	El	Hard
Pan America															
COPANT 334	6150	Bar	0.48-0.53	0.8-1.1	0.7-0.9		0.04	0.04	0.2-0.35	0.15	bal Fe				
COPANT 514	6150	Tube	0.48-0.53	0.8-1.1	0.7-0.9		0.04	0.05	0.1	0.15	bal Fe				
COPANT 514	6150		0.48-0.53	0.8-1.1	0.7-0.9		0.04	0.05	0.1	0.15 min	bal Fe				
Poland															
PNH84032	50HF	Spring	0.46-0.54	0.8-1.1	0.5-0.8		0.03 max	0.03 max	0.15-0.4	0.1-0.2	Cu <=0.25; Ni <=0.4; bal Fe				
Romania															
STAS 10677(76)	50VCr11CS	F/IH	0.49-0.55	0.9-1.25	0.55-0.9		0.025 max	0.025 max	0.17-0.37	0.1-0.25	Pb <=0.15; Ti <=0.02; As<=0.05; bal Fe				
STAS 791	50Cr11		0.45-0.55	0.9-1.25	0.55-0.9		0.035	0.035	0.17-0.37	0.1-0.25	Cu 0.3; Ni 0.3; bal Fe				
STAS 791(80)	50VCr11	Struct/const; Q/T	0.45-0.55	0.9-1.25	0.55-0.9		0.035 max	0.035 max	0.17-0.37	0.1-0.25	Pb <=0.15; bal Fe				
STAS 791(88)	51VMnCr11	Struct/const	0.47-0.55	0.9-1.2	0.7-1.1		0.035 max	0.035 max	0.17-0.37	0.1-0.2	Pb <=0.15; Ti <=0.02; bal Fe				
STAS 791(88)	51VMnCr11S	Struct/const	0.47-0.55	0.9-1.2	0.7-1.1		0.035 max	0.02-0.04	0.17-0.37	0.1-0.2	Pb <=0.15; Ti <=0.02; bal Fe				
STAS 791(88)	51VMnCr11X	Struct/const	0.47-0.55	0.9-1.2	0.7-1.1		0.025 max	0.025 max	0.17-0.37	0.1-0.2	Pb <=0.15; Ti <=0.02; bal Fe				
STAS 791(88)	51VMnCr11XS	Struct/const	0.47-0.55	0.9-1.2	0.7-1.1		0.025 max	0.02-0.035	0.17-0.37	0.1-0.2	Pb <=0.15; Ti <=0.02; bal Fe				
STAS 795	ARC1		0.46-0.54	0.8-1.1	0.5-0.8		0.035	0.035	0.17-0.37	0.1-0.2	Cu 0.25; Ni 0.3; bal Fe				
STAS 795(92)	51VCr11A	Spring; HF	0.47-0.55	0.9-1.2	0.8-1.1		0.035 max	0.035 max	0.15-0.35	0.07-0.12	Cu <=0.25; Pb <=0.15; bal Fe				
Russia															
GOST 14958	50ChF		0.46-0.54	0.8-1.1	0.5-0.8		0.025	0.025	0.17-0.37	0.1-0.2	Cu 0.2; Ni 0.25; bal Fe				
GOST 14959	50ChGFA	Spring; Q/T	0.48-0.55	0.95-1.2	0.8-1	0.15 max	0.025 max	0.025 max	0.17-0.37	0.15-0.25	Al <=0.1; Cu <=0.2; Ni <=0.25; Ti <=0.05; bal Fe				
GOST 14959	50ChGFA		0.48-0.55	0.95-1.2	0.8-1		0.03 max	0.03 max	0.17-0.37	0.15-0.25	Cu <=0.2; Ni <=0.25; bal Fe				
GOST 14963	50ChFA		0.47-0.55	0.75-1.1	0.3-0.6		0.03	0.03	0.15-0.3	0.15-0.25	bal Fe				
GOST 4543(71)	50Ch	Q/T	0.46-0.54	0.8-1.1	0.5-0.8	0.15 max	0.035 max	0.035 max	0.17-0.37		Al <=0.1; Cu <=0.3; Ni <=0.3; Ti <=0.05; bal Fe				
Spain															
UNE	F.143	Spring; Q/T	0.45-0.55	0.8-1.1	0.5-0.7	0.15 max	0.04 max	0.04 max	0.1-0.35	0.15-0.25	bal Fe				
UNE	F.143	Spring; Q/T	0.45-0.55	0.8-1.1	0.5-0.7	0.15 max	0.04 max	0.04 max	0.1-0.35	0.15-0.25	bal Fe				
UNE	F.143.A		0.47-0.54	0.9-1.2	0.7-1		0.035	0.035	0.15-0.4	0.1-0.2	bal Fe				
UNE 36015(77)	51CrV4	Spring; Q/T	0.48-0.55	0.9-1.2	0.7-1	0.15 max	0.035 max	0.035 max	0.15-0.4	0.1-0.2	bal Fe				
UNE 36015(77)	51CrV4	Spring; Q/T	0.48-0.55	0.9-1.2	0.7-1	0.15 max	0.035 max	0.035 max	0.15-0.4	0.1-0.2	bal Fe				
UNE 36015(77)	52CrMoV4	Spring	0.48-0.56	0.9-1.2	0.7-1	0.15-0.25	0.035 max	0.035 max	0.15-0.4	0.07-0.12	bal Fe				
UNE 36015(77)	F.1430*	see 51CrV4	0.48-0.55	0.9-1.2	0.7-1	0.15 max	0.035 max	0.035 max	0.15-0.4	0.1-0.2	bal Fe				
UNE 36015(77)	F.1460*	see 52CrMoV4	0.48-0.56	0.9-1.2	0.7-1	0.15-0.25	0.035 max	0.035 max	0.15-0.4	0.07-0.12	bal Fe				
Sweden															
SIS 142230	2230-03	Bar, Wire, Frg Q/A, Tmp 30mm	0.48-0.55	0.9-1.2	0.7-1		0.04	0.04	0.15-0.4	0.1-0.2	bal Fe	1300	1150	8	
SIS 142230	2230-04	Bar, Wire, Frg Q/A, Tmp	0.48-0.55	0.9-1.2	0.7-1		0.04	0.04	0.15-0.4	0.1-0.2	bal Fe	1500	1300	6	
SIS 142230	2230-05	Bar, Wire, Q/A, Tmp, 4 5mm diam	0.48-0.55	0.9-1.2	0.7-1		0.04	0.04	0.15-0.4	0.1-0.2	bal Fe	1450	1250		

UNS numbers and US grades are provided as a means of cross referencing chemically similar alloys. Exchangability is only possible after independent examination of specifications. Tensile properties are minimum or typical as specified. UTS and YS as MPa. El as %. See Appendix for list of abbreviations used in Notes. * indicates obsolete material.

Specification	Designation	Notes	C	Cr	Mn	Mo	P	S	Si	V	Other	UTS	YS	El	Hard

Alloy Steel, Chromium Vanadium, 6150/6150H (Continued from previous page)

Sweden

Specification	Designation	Notes	C	Cr	Mn	Mo	P	S	Si	V	Other	UTS	YS	El	Hard
SIS 142230	2230-06	Bar, Wire, Ann, CW	0.48-0.55	0.9-1.2	0.7-1		· 0.04	0.04	0.15-0.4	0.1-0.2	bal Fe	800			
SS 142230	2230	Spring	0.48-0.55	0.9-1.2	0.7-1		0.035 max	0.035 max	0.15-0.4		bal Fe				

Turkey

Specification	Designation	Notes	C	Cr	Mn	Mo	P	S	Si	V	Other	UTS	YS	El	Hard
TS 2288(97)	50CrV4-1,8159	Ref. ISO 683-14(92), wire rods, flts, HT	0.47-0.55	0.9-1.2	0.7-1.1		0.04	0.04	0.15-0.4	0.1-0.2	bal Fe				

UK

Specification	Designation	Notes	C	Cr	Mn	Mo	P	S	Si	V	Other	UTS	YS	El	Hard
BS 1429(80)	735A50	Spring	0.46-0.54	0.80-1.10	0.60-0.90				0.10-0.35	0.15-0.30	bal Fe				
BS 970/1(83)	735A50*	Blm Bil Slab Bar Rod Frg	0.46-0.54	0.8-1.1	0.60-0.90		0.04	0.04	0.1-0.35	0.15 min	bal Fe				
BS 970/2(88)	735A51	Spring	0.48-0.54	0.90-1.20	0.70-1.00	0.15 max	0.035 max	0.035 max	0.20-0.35	0.10-0.20	bal Fe				
BS 970/2(88)	735A54	Spring	0.52-0.57	1.05-1.20	0.90-1.15	0.15 max	0.035 max	0.035 max	0.20-0.35	0.12-0.20	bal Fe				
BS 970/2(88)	735H51	Spring	0.47-0.55	0.90-1.20	0.70-1.10	0.15 max	0.035 max	0.035 max	0.15-0.40	0.10-0.25	bal Fe				
BS 970/5(72)	735A50*	Blm Bil Slab Bar Rod Frg	0.46-0.54	0.8-1.1	0.35-0.60				0.1-0.35	0.15	bal Fe				

USA

Specification	Designation	Notes	C	Cr	Mn	Mo	P	S	Si	V	Other	UTS	YS	El	Hard
	AISI 6150	Plt	0.46-0.54	0.80-1.15	0.60-0.90		0.035 max	0.040 max	0.15-0.40	0.15 min	bal Fe				
	AISI 6150	Bar Blm Bil Slab	0.48-0.53	0.80-1.10	0.70-0.90		0.035 max	0.040 max	0.15-0.35	0.15 min	bal Fe				
	AISI 6150	Smls mech tub	0.48-0.53	0.80-1.10	0.70-0.90		0.040 max	0.040 max	0.15-0.35	0.15 min	bal Fe				
	AISI 6150H	Smls mech tub, Hard	0.47-0.54	0.75-1.20	0.60-1.00		0.040 max	0.040 max	0.15-0.30	0.15 min	bal Fe				
	AISI 6150H	Bar Blm Bil Slab	0.47-0.54	0.75-1.20	0.60-1.00		0.035 max	0.040 max	0.15-0.35	0.15 min	bal Fe				
	UNS G61500		0.48-0.53	0.80-1.10	0.70-0.90		0.035 max	0.040 max	0.15-0.35	0.15 min	bal Fe				
	UNS H61500		0.47-0.54	0.75-1.20	0.60-1.00		0.035 max	0.040 max	0.15-0.35	0.15 min	bal Fe				
	UNS K15047		0.48-0.53	0.80-1.10	0.70-0.90		0.020 max	0.035 max	0.20-0.35	0.15 min	bal Fe				
	UNS K15048		0.48-0.53	0.80-1.10	0.70-0.90		0.040 max	0.040 max	0.20-0.35	0.15 min	bal Fe				
AMS 6448		Bar Frg Tub	0.48-0.53	0.8-1.1	0.7-0.90	0.06	0.025	0.025	0.15-0.35	0.15-0.30	Cu 0.35; Ni 0.25; bal Fe				
AMS 6450		Wire	0.48-0.53	0.8-1.1	0.7-0.90	0.06	0.025	0.025	0.15-0.35	0.15-0.30	Cu 0.35; Ni 0.25; bal Fe				
AMS 6450		Sh, Strp, Plt	0.48-0.53	0.8-1.1	0.7-0.90	0.06	0.025	0.025	0.15-0.35	0.15-0.30	Cu 0.35; Ni 0.25; bal Fe				
AMS 7301											bal Fe				
ASTM A29/A29M(93)	6150	Bar	0.48-0.53	0.80-1.10	0.7-0.9		0.035 max	0.04 max	0.15-0.35	0.15 min	bal Fe				
ASTM A304(96)	6150H	Bar, hard bands spec	0.47-0.54	0.75-1.20	0.60-1.00	0.06 max	0.035 max	0.040 max	0.15-0.35	0.15 min	Cu <=0.35; Cu Ni Cr Mo trace allowed; bal Fe				
ASTM A322(96)	6150	Bar	0.48-0.53	0.80-1.10	0.70-0.90		0.035 max	0.040 max	0.15-0.35	0.15 min	bal Fe				
ASTM A331(95)	6150	Bar	0.48-0.53	0.80-1.10	0.70-0.90		0.035 max	0.040 max	0.15-0.35		bal Fe				
ASTM A331(95)	6150H	Bar	0.47-0.54	0.75-1.20	0.60-1.00	0.06 max	0.025 max	0.025 max	0.15-0.35	0.15 min	Cu <=0.35; bal Fe				
ASTM A519(96)	6150	Smls mech tub	0.48-0.53	0.80-1.10	0.70-0.90		0.040 max	0.040 max	0.15-0.35	0.15 min	bal Fe				
ASTM A646(95)	Grade 14	Blm Bil for aircraft aerospace frg, ann	0.48-0.53	0.80-1.10	0.70-0.90		0.025 max	0.025 max	0.20-0.35	0.15 min	bal Fe				235 HB
ASTM A752(93)	6150	Rod Wir	0.48-0.53	0.80-1.10	0.70-0.90		0.035 max	0.040 max	0.15-0.30	0.15 min	bal Fe	670			
ASTM A829/A829M(95)	6150	Plt	0.46-0.54	0.80-1.15	0.60-0.90		0.035 max	0.040 max	0.15-0.40	0.15 min	bal Fe	480-965			
ASTM A866(94)	6150	Anti-friction bearings	0.48-0.53	0.80-1.10	0.70-0.90		0.025 max	0.025 max	0.15-0.35	0.15 min	bal Fe				

UNS numbers and US grades are provided as a means of cross referencing chemically similar alloys. Exchangability is only possible after independent examination of specifications. Tensile properties are minimum or typical as specified. UTS and YS as MPa. El as %. See Appendix for list of abbreviations used in Notes. * indicates obsolete material.

Alloy Steel, Chromium Vanadium, 6150/6150H (Continued from previous page)

Specification	Designation	Notes	C	Cr	Mn	Mo	P	S	Si	V	Other	UTS	YS	El	Hard
USA															
ASTM A878/A878M(93)		Wir, valve spring qual, 1mm diam	0.60-0.75	0.35-0.60	0.50-0.90		0.025 max	0.025 max	0.15-0.30	0.10-0.25	bal Fe	1930-2100			
MIL-S-8503	6150*	Obs; Bar	0.48-0.53	0.75-1.2	0.7-0.9		0.035	0.035	0.2-0.35	0.15 min	bal Fe				
SAE 770(84)	6150*	Obs; see J1397(92)									bal Fe				
SAE J1268(95)	6150H	See std	0.47-0.54	0.75-1.20	0.60-1.00	0.06 max	0.030 max	0.040 max	0.15-0.35	0.15 min	Cu <=0.35; Ni <=0.25; Cu, Mo, Ni not spec'd but acpt; bal Fe				
SAE J404(94)	6150	Bil Blm Slab Bar HR CF	0.48-0.53	0.80-1.10	0.70-0.90		0.030 max	0.040 max	0.15-0.35	0.15 min	bal Fe				
SAE J404(94)	6150	Plt	0.46-0.54	0.80-1.15	0.60-0.90		0.035 max	0.040 max	0.15-0.35	0.15 min	bal Fe				
Yugoslavia															
	C.4830	Q/T	0.47-0.55	0.9-1.2	0.7-1.1	0.15 max	0.035 max	0.035 max	0.15-0.4	0.07-0.2	bal Fe				

Alloy Steel, Chromium Vanadium, 6481

Specification	Designation	Notes	C	Cr	Mn	Mo	P	S	Si	V	Other	UTS	YS	El	Hard
Italy															
UNI 6120(67)	36CrMoV12	Nitriding	0.33-0.4	2.7-3.3	0.4-0.7	0.7-1.2	0.035 max	0.035 max	0.4 max	0.15-0.25	bal Fe				
USA															
	UNS K24340	Cr-O-Mo-V Martensitic	0.29-0.36	2.80-3.30	0.40-0.70	0.70-1.20	0.015 max	0.005 max	0.10-0.40	0.15-0.35	Cu <=0.10; Ni <=0.30; bal Fe				

Alloy Steel, Chromium Vanadium, A213(T17)

Specification	Designation	Notes	C	Cr	Mn	Mo	P	S	Si	V	Other	UTS	YS	El	Hard
India															
IS 1570/4(88)	21Cr4Mo2	Sh Plt Strp Tub Pipe	0.26 max	0.9-1.2	0.6-0.9	0.15-0.3	0.07 max	0.06 max	0.1-0.35		Ni <=0.4; bal Fe	790-990	610		
IS 1570/7(92)	14	0-250mm; Bar Frg	0.17-0.25	1.2-1.5	0.4-0.8	0.65-0.8	0.03 max	0.03 max	0.4 max	0.25-0.35	Ni <=0.6; bal Fe	700-850	550	16	205-250 HB
IS 1570/7(92)	21CrMoVH	0-250mm; Bar Frg	0.17-0.25	1.2-1.5	0.4-0.8	0.65-0.8	0.03 max	0.03 max	0.4 max	0.25-0.35	Ni <=0.6; bal Fe	700-850	550	16	205-250 HB
USA															
	UNS K12047		0.15-0.25	0.80-1.25	0.30-0.61		0.045 max	0.045 max	0.15-0.35	0.15 min	bal Fe				
ASTM A213/A213M(95)	T17	Smls tube boiler, superheater, heat exchanger	0.15-0.25	0.80-1.25	0.30-0.61		0.025 max	0.025 max	0.15-0.35	0.15 min	bal Fe	415	205	30	

Alloy Steel, Chromium Vanadium, F91

Specification	Designation	Notes	C	Cr	Mn	Mo	P	S	Si	V	Other	UTS	YS	El	Hard
USA															
	UNS K90901	Ferritic Cr-Mo-V (Grade 91, T91)	0.08-0.12	8.00-9.50	0.30-0.60	0.85-1.05	0.020 max	0.010 max	0.20-0.50	0.18-0.25	Al <=0.04; N 0.030-0.070; Nb 0.06-0.10; Ni <=0.40; bal Fe				
	UNS K91560	9Cr-1Mo-V-Nb-N	0.08-0.12	8.0-9.5	0.30-0.60	0.85-1.05	0.020 max	0.010 max	0.20-0.50	0.18-0.25	Al <=0.04; N 0.03-0.07; Nb 0.06-0.10; Ni <=0.40; bal Fe				
ASTM A182/A182M(98)	F91	Frg/roll pipe flange valve	0.08-0.12	8.0-9.5	0.30-0.60	0.85-1.05	0.020 max	0.010 max	0.20-0.50	0.18-0.25	Al <=0.04; N 0.03-0.07; Nb 0.06-0.10; Ni <=0.40; bal Fe	585	415	20.0	248 HB max
ASTM A213/A213M(95)	T91	Smls tube boiler, superheater, heat exchanger	0.08-0.12	8.00-9.50	0.30-0.60	0.85-1.05	0.020 max	0.010 max	0.20-0.50	0.18-0.25	Al <=0.04; N 0.030-0.070; Nb 0.06-0.10; Ni <=0.40; bal Fe	585	415	20	
ASTM A234/A234M(97)	WP91	Frg	0.08-0.12	8.0-9.5	0.30-0.60	0.85-1.05	0.020 max	0.010 max	0.20-0.50	0.18-0.25	Al <=0.04; Nb 0.06-0.10; Ni <=0.40; bal Fe	585-760	415	20L	
ASTM A336/A336M(98)	F91	Ferr Frg	0.08-0.12	8.0-9.5	0.30-0.60	0.85-1.05	0.025 max	0.025 max	0.20-0.50	0.18-0.25	Al <=0.04; N 0.03-0.07; Nb 0.06-0.10; Ni <=0.40; bal Fe	585-760	415	20	
ASTM A387/A387(97)	Grade 91	Press ves plt, Class 2	0.08-0.12	8.00-9.50	0.30-0.60	0.85-1.05	0.020 max	0.010 max	0.20-0.50	0.18-0.25	Al <=0.04; Nb 0.06-0.10; Ni <=0.40; bal Fe	585-760	415	18	
ASTM A989(98)	K91560	HIP Flanges, Fittings, Valves/parts; Heat res; 9Cr-1Mo-V-Cb-N	0.08-0.12	8.0-9.5	0.30-0.60	0.85-1.05	0.020 max	0.010 max	0.20-0.50	0.18-0.25	Al <=0.04; N 0.03-0.07; Nb 0.06-0.10; Ni <=0.40; bal Fe	585	415	20.0	248 HB max

UNS numbers and US grades are provided as a means of cross referencing chemically similar alloys. Exchangability is only possible after independent examination of specifications. Tensile properties are minimum or typical as specified. UTS and YS as MPa. El as %. See Appendix for list of abbreviations used in Notes. * indicates obsolete material.

Specification	Designation	Notes	C	Cr	Mn	Mo	P	S	Si	V	Other	UTS	YS	El	Hard
Alloy Steel, Chromium Vanadium, No equivalents identified															
Czech Republic															
CSN 415221	15221	Mn-Cr-V	0.12-0.20	0.8-1.2	0.8-1.2		0.04 max	0.04 max	0.2-0.4	0.1-0.2	Ni <=0.04; bal Fe				
CSN 415230	15230	Q/T; Nitriding	0.24-0.34	2.2-2.5	0.4-0.8		0.035 max	0.035 max	0.17-0.37	0.1-0.2	bal Fe				
CSN 415231	15231	Q/T	0.24-0.3	0.6-0.9	1.0-1.3		0.035 max	0.035 max	0.17-0.37	0.1-0.2	bal Fe				
CSN 415233	15233	High-temp const	0.18-0.28	1.2-1.5	0.3-0.6		0.035 max	0.035 max	0.3-0.6	0.45-0.65	bal Fe				
CSN 415235	15235	High-temp const	0.25-0.35	1.5-2.0	0.5-0.9		0.035 max	0.035 max	0.15-0.4	0.2-0.35	bal Fe				
CSN 415240	15240	Q/T	0.3-0.4	0.7-1.1	0.7-1.0		0.035 max	0.035 max	0.17-0.37	0.1-0.2	bal Fe				
CSN 415241	15241	Q/T	0.35-0.43	1.7-2.0	0.6-0.8		0.035 max	0.035 max	0.17-0.37	0.1-0.2	Ni <=0.6; bal Fe				
CSN 415261	15261	Q/T	0.55-0.62	0.9-1.2	0.8-1.1		0.035 max	0.035 max	0.17-0.37	0.1-0.2	bal Fe				
CSN 415331	15331	High-temp const	0.17-0.27	1.1-1.5	1 max		0.04 max	0.04 max	0.2-0.4	0.45-0.65	W 0.8-1.3; bal Fe				
CSN 415334	15334	Frg	0.2-0.27	1.4-1.8	1.1-1.55		0.04 max	0.04 max	0.15-0.4	0.1-0.25	bal Fe				
CSN 415341	15341	Frg	0.35-0.43	0.95-1.35	0.95-1.35		0.04 max	0.04 max	0.6-1.0	0.15-0.3	bal Fe				
Hungary															
MSZ 2666(76)	51CMoV	Spring HF	0.48-0.56	0.9-1.2	0.7-1	0.15-0.25	0.035 max	0.035 max	0.15-0.4	0.07-0.12	Cu <=0.25; Ni <=0.3; bal Fe				
MSZ 2666(88)	52CrMoV4	HF spring; 0-100.0mm; Q/T	0.48-0.56	0.9-1.2	0.7-1	0.15-0.25	0.03 max	0.03 max	0.15-0.4	0.07-0.15	bal Fe	1450	1300	8L	
MSZ 2666(88)	52CrMoV4	HF spring; 0-999mm	0.48-0.56	0.9-1.2	0.7-1	0.15-0.25	0.03 max	0.03 max	0.15-0.4	0.07-0.15	bal Fe				310 HB max
MSZ 61(74)	CrV1E	Q/T	0.3-0.37	0.8-1.1	0.5-0.8		0.035 max	0.02-0.035	0.17-0.37	0.1-0.2	bal Fe				
MSZ 61(74)	CrV2E	Q/T	0.37-0.46	0.8-1.1	0.5-0.8		0.035 max	0.02-0.035	0.17-0.37	0.1-0.2	bal Fe				
MSZ 61(85)	CrV1	Q/T; 0-16mm	0.3-0.37	0.8-1.1	0.6-1		0.035 max	0.035 max	0.4 max	0.1-0.2	bal Fe	900-1100	700	12L	
MSZ 61(85)	CrV1	Q/T; 100.1-160mm	0.3-0.37	0.8-1.1	0.6-1		0.035 max	0.035 max	0.4 max	0.1-0.2	bal Fe	650-800	450	15L	
MSZ 61(85)	CrV2	Q/T; 0-16mm	0.38-0.45	0.8-1.1	0.6-1		0.035 max	0.035 max	0.4 max	0.1-0.2	bal Fe	1000-1200	800	11L	
MSZ 61(85)	CrV2	Q/T; 100.1-160mm	0.38-0.45	0.8-1.1	0.6-1		0.035 max	0.035 max	0.4 max	0.1-0.2	bal Fe	750-900	550	14L	
MSZ 6251(87)	CrV2Z	CF; Q/T; 0-36mm; HR; HR/soft ann; drawn/soft ann; drawn/bright ann; soft ann/ground	0.38-0.45	0.8-1.1	0.6-1		0.035 max	0.035 max	0.4 max	0.1-0.2	bal Fe	0-640			
MSZ 6251(87)	CrV2Z	CF; Q/T; 0-36mm; drawn, half-hard	0.38-0.45	0.8-1.1	0.6-1		0.035 max	0.035 max	0.4 max	0.1-0.2	bal Fe	0-650			
India															
IS 1570/4(88)	60Cr4V2		0.55-0.65	0.9-1.2	0.8-1.1	0.15 max	0.07 max	0.06 max	0.1-0.35	0.15-1.150	Ni <=0.4; bal Fe				
International															
ISO 2604-2(75)	TS40	Smls tube	0.17-0.23	10-12.5	1 max	0.8-1.2	0.03 max	0.03 max	0.5 max	0.25-0.35	Al <=0.1; Co <=0.1; Cu <=0.3; Ni 0.3-0.8; Pb <=0.15; Ti <=0.05; W <=0.7; bal Fe		430		
ISO 683-1(87)	51CrV4	Bar Frg Plt Wir Rod Bolt Slab; Q/T, t<16mm	0.47-0.55	0.80-1.10	0.60-1.00		0.035 max	0.035 max	0.10-0.40	0.10-0.25	bal Fe	1100-1300	900		
ISO 683-14(92)	52CrMoV4	Bar Rod Wir; Q/T, springs, Ann	0.48-0.56	0.90-1.20	0.70-1.00	0.15-0.25	0.030 max	0.030 max	0.15-0.40	0.07-0.15	bal Fe				255 HB
Italy															
UNI 6927(71)	15CrMoV6	Aircraft material	0.12-0.18	1.25-1.5	0.8-1.1	0.8-1	0.02 max	0.015 max	0.2 max	0.2-0.3	bal Fe				

Specification	Designation	Notes	C	Cr	Mn	Mo	P	S	Si	V	Other	UTS	YS	El	Hard

Alloy Steel, Chromium Vanadium, No equivalents identified

Romania

Specification	Designation	Notes	C	Cr	Mn	Mo	P	S	Si	V	Other	UTS	YS	El	Hard
STAS 2883/3(88)	12VMoCr10	Heat res	0.08-0.15	0.9-1.2	0.4-0.7	0.25-0.35	0.03 max	0.025 max	0.17-0.37	0.15-0.3	Al 0.015-0.035; N <=0.009; Pb <=0.15; As<=0.05; bal Fe				

Russia

Specification	Designation	Notes	C	Cr	Mn	Mo	P	S	Si	V	Other	UTS	YS	El	Hard
GOST 20072(74)	12Ch1MF	High-temp const	0.08-0.15	0.90-1.20	0.40-0.70	0.25-0.35	0.030 max	0.025 max	0.17-0.37	0.15-0.30	Cu <=0.20; Ni <=0.30; Ti <=0.03; W <=0.20; bal Fe				
GOST 20072(74)	20Ch1M1F1BR	High-temp const	0.18-0.25	1.00-1.50	0.50-0.80	0.80-1.10	0.030 max	0.030 max	0.37 max	0.70-1.00	B <=0.005; Cu <=0.20; Nb 0.05-0.15; Ni <=0.30; Ti <=0.06; W <=0.20; Ce 0.05-0.10; bal Fe				
GOST 20072(74)	20Ch1M1F1TR	High-temp const	0.17-0.24	0.90-1.40	0.50 max	0.80-1.10	0.030 max	0.030 max	0.37 max	0.70-1.00	B <=0.005; Cu <=0.20; Ni <=0.30; Ti 0.05-0.12; W <=0.20; bal Fe				
GOST 20072(74)	20Ch3MWF	High-temp const	0.15-0.23	2.80-3.30	0.25-0.50	0.35-0.55	0.030 max	0.025 max	0.17-0.37	0.60-0.85	Cu <=0.20; Ni <=0.30; Ti <=0.05; W 0.30-0.50; bal Fe				
GOST 20072(74)	25Ch1M1F	High-temp const	0.22-0.29	1.50-1.80	0.40-0.70	0.60-0.80	0.030 max	0.025 max	0.17-0.37	0.15-0.30	Cu <=0.20; Ni <=0.30; Ti <=0.05; W <=0.20; bal Fe				
GOST 20072(74)	25Ch1MF	High-temp const	0.22-0.29	1.50-1.80	0.40-0.70	0.25-0.35	0.030 max	0.025 max	0.17-0.37	0.15-0.30	Cu <=0.20; Ni <=0.30; Ti <=0.05; W <=0.20; bal Fe				
GOST 20072(74)	25Ch2M1F	High-temp const	0.22-0.29	2.10-2.60	0.40-0.70	0.90-1.10	0.030 max	0.025 max	0.17-0.37	0.30-0.50	Cu <=0.20; Ni <=0.30; Ti <=0.05; W <=0.20; bal Fe				

Alloy Steel, Chromium Vanadium, T92

USA

Specification	Designation	Notes	C	Cr	Mn	Mo	P	S	Si	V	Other	UTS	YS	El	Hard
	UNS K92460	Ferritic Cr-Mo-W-V-Nb T92	0.06-0.13	8.00-9.50	0.30-0.60	0.30-0.60	0.020 max	0.010 max	0.50 max	0.15-0.25	Al <=0.04; B <=0.006; N 0.030-0.070; Nb 0.03-0.10; Ni <=0.40; W 1.50-2.20; bal Fe				
ASTM A213/A213M(95)	T92	Smls tube boiler, superheater, heat exchanger	0.07-0.13	8.50-9.50	0.30-0.60	0.30-0.60	0.020 max	0.010 max	0.50 max	0.15-0.25	Al <=0.04; B 0.001-0.006; N 0.03-0.07; Nb 0.04-0.09; Ni <=0.40; W 1.5-2.00; bal Fe	620	440	20	

Alloy Steel, Manganese, 1330/1330H

Specification	Designation	Notes	C	Cr	Cu	Mn	Mo	P	S	Si	Other	UTS	YS	El	Hard
Australia															
AS 1442(92)	X1315	HR bar, Semifinished (may treat w/microalloying elements)	0.12-0.18			1.40-1.70		0.040 max	0.040 max	0.10-0.35	Si<=0.10 for Al-killed steels; bal Fe				
AS 1442(92)	X1320	HR bar, Semifinished (may treat w/microalloying elements)	0.18-0.23			1.40-1.70		0.040 max	0.040 max	0.10-0.35	Si<=0.10 for Al-killed steels; bal Fe				
AS 1442(92)	X1325	HR bar, Semifinished (may treat w/microalloying elements)	0.23-0.28			1.40-1.70		0.040 max	0.040 max	0.10-0.35	Si<=0.10 for Al-killed steels; bal Fe				
China															
GB 3077(88)	30Mn2	Bar HR Q/T 25mm diam	0.27-0.34	0.30 max	0.30 max	1.40-1.80		0.035 max	0.035 max	0.17-0.37	bal Fe	785	635	12	
Europe															
EN 10083/1(91)A1(96)	1.1170	Q/T t<=16mm	0.25-0.32	0.40 max		1.30-1.65	0.10 max	0.035 max	0.035 max	0.40 max	Ni <=0.40; Cr+Mo+Ni<=0.63; bal Fe	800-950	590	13	
EN 10083/1(91)A1(96)	28Mn6	Q/T t<=16mm	0.25-0.32	0.40 max		1.30-1.65	0.10 max	0.035 max	0.035 max	0.40 max	Ni <=0.40; Cr+Mo+Ni<=0.63; bal Fe	800-950	590	13	
Germany															
DIN 17205(92)	30Mn5	Q/T 17-40mm	0.27-0.34	0.30 max		0.20-1.50		0.035 max	0.035 max	0.15-0.40	bal Fe	690-830	440	15	
Hungary															
MSZ 61(85)	Mn1	Q/T; 0-16mm;	0.25-0.32			1.3-1.65		0.035 max	0.035 max	0.4 max	bal Fe	800-950	600	13L	
MSZ 61(85)	Mn1	Q/T; 40.1-100mm	0.25-0.32			1.3-1.65		0.035 max	0.035 max	0.4 max	bal Fe	650-800	440	16L	
International															
ISO R683-5(70)	1	Q/T	0.25-0.32	0.3 max	0.3 max	1.3-1.65	0.15 max	0.035 max	0.035 max	0.15-0.4	Al <=0.1; Co <=0.1; Ni <=0.4; Pb <=0.15; Ti <=0.05; V <=0.1; W <=0.1; bal Fe				
Italy															
UNI 7874(79)	C28Mn	Q/T	0.25-0.32			1.3-1.65		0.035 max	0.035 max	0.15-0.4	bal Fe				
Japan															
JIS G4052(79)	SMn1H*	Obs; see SMn 433H	0.29-0.36			1.15-1.55		0.03 max	0.03 max	0.15-0.35	bal Fe				
Mexico															
DGN B-203	1330*	Obs	0.28-0.33			1.6-1.9		0.04	0.04	0.2-0.35	bal Fe				
DGN B-297	1330*	Obs	0.28-0.33			1.6-1.9		0.035	0.035	0.2-0.35	bal Fe				
NMX-B-268(68)	1330H	Hard	0.27-0.33			1.45-2.05				0.20-0.35	bal Fe				
NMX-B-300(91)	1330	Bar	0.28-0.33			1.60-1.90		0.035 max	0.040 max	0.15-0.35	bal Fe				
NMX-B-300(91)	1330H	Bar	0.27-0.33			1.45-2.05				0.15-0.35	bal Fe				
NOM-061-SCFI(94)	Type II	Plt for gasoline containers	0.28 max			1.40 max		0.035 max	0.04 max	0.4 max	bal Fe	451-520	226	24	
NOM-123-SCFI(96)		Corrugated rod for conc reinf	0.30-0.33			1.50-1.55		0.035-0.043	0.045-0.053	0.50-0.55	bal Fe				
Pan America															
COPANT 334	1330	Bar	0.28-0.33			1.6-1.9		0.04	0.04	0.2-0.35	bal Fe				
COPANT 514	1330	Tube	0.28-0.33			1.6-1.9		0.04 max	0.04	0.2-0.35	bal Fe				
Poland															
PNH84030/04	30G2	Q/T	0.27-0.35	0.25 max		1.4-1.8		0.035 max	0.035 max	0.17-0.37	Ni <=0.3; bal Fe				
Russia															
GOST 4543(71)	30G2	Q/T	0.26-0.35	0.3 max	0.3 max	1.4-1.8	0.15 max	0.035 max	0.035 max	0.17-0.37	Al <=0.1; Ni <=0.3; Ti <=0.05; bal Fe				

UNS numbers and US grades are provided as a means of cross referencing chemically similar alloys. Exchangability is only possible after independent examination of specifications. Tensile properties are minimum or typical as specified. UTS and YS as MPa. El as %. See Appendix for list of abbreviations used in Notes. * indicates obsolete material.

Specification	Designation	Notes	C	Cr	Cu	Mn	Mo	P	S	Si	Other	UTS	YS	El	Hard

Alloy Steel, Manganese, 1330/1330H (Continued from previous page)

Specification	Designation	Notes	C	Cr	Cu	Mn	Mo	P	S	Si	Other	UTS	YS	El	Hard
Russia															
GOST 4543(71)	35G2	Q/T	0.31-0.39	0.3 max	0.3 max	1.4-1.8	0.15 max	0.035 max	0.035 max	0.17-0.37	Al <=0.1; Ni <=0.3; Ti <=0.05; bal Fe				
GOST 7832	30GSL	Bar	0.25-0.35	0.3 max	0.3 max	1.1-1.4		0.04 max	0.04 max	0.6-0.8	Ni <=0.3, bal Fe				
GOST 7832	30GSL		0.25-0.35	0.3 max	0.3 max	1.1-1.4		0.04 max	0.04 max	0.6-0.8	Ni <=0.3; bal Fe				
GOST 924	26G2SA		0.25-0.31	0.4	0.3	1.45-1.85		0.03	0.03	0.5-0.75	Ni 0.4; bal Fe				
Spain															
UNE 36012(75)	36Mn6	Q/T	0.33-0.4			1.3-1.65	0.15 max	0.035 max	0.035 max	0.15-0.4	bal Fe				
UNE 36012(75)	36Mn6-1	Q/T	0.33-0.4			1.3-1.65	0.15 max	0.035 max	0.02-0.035	0.15-0.4	bal Fe				
UNE 36012(75)	F.1208*	see 36Mn6-1	0.33-0.4			1.3-1.65	0.15 max	0.035 max	0.02-0.035	0.15-0.4	bal Fe				
USA															
	AISI 1330	Bar Blm Bil Slab	0.28-0.33			1.60-1.90		0.035 max	0.040 max	0.15-0.35	bal Fe				
	AISI 1330	Smls mech tub	0.28-0.33			1.60-1.90		0.040 max	0.040 max	0.15-0.35	bal Fe				
	AISI 1330	Plt	0.27-0.34			1.50-1.90		0.035 max	0.040 max	0.15-0.40	bal Fe				
	AISI 1330H	Bar Blm Bil Slab	0.27-0.33			1.45-2.05		0.035 max	0.040 max	0.15-0.35	bal Fe				
	AISI 1330H	Smls mech tub, Hard	0.27-0.33			1.45-2.05		0.035 max	0.040 max	0.15-0.30	bal Fe				
	UNS G13300		0.28-0.33			1.60-1.90		0.035 max	0.040 max	0.15-0.35	bal Fe				
	UNS H13300		0.27-0.33			1.45-2.05		0.035 max	0.040 max	0.15-0.35	bal Fe				
ASTM A29/A29M(93)	1330	Bar	0.28-0.33			1.60-1.90		0.040 max	0.040 max	0.15-0.35	bal Fe				
ASTM A304(96)	1330H	Bar, hard bands spec	0.27-0.33	0.20 max	0.35 max	1.45-2.05	0.06 max	0.035 max	0.040 max	0.15-0.35	Cu Ni Cr Mo trace allowed; bal Fe				
ASTM A322(96)	1330	Bar	0.28-0.33			1.60-1.90		0.035 max	0.040 max	0.15-0.35	bal Fe				
ASTM A331(95)	1330	Bar	0.28-0.33			1.60-1.90		0.035 max	0.040 max	0.15-0.35	bal Fe				
ASTM A331(95)	1330H	Bar	0.27-0.33	0.20 max	0.35 max	1.45-2.05	0.06 max	0.025 max	0.025 max	0.15-0.35	bal Fe				
ASTM A519(96)	1330	Smls mech tub	0.28-0.33			1.60-1.90		0.040 max	0.040 max	0.15-0.35	bal Fe				
ASTM A752(93)	1330	Rod Wir	0.28-0.33			1.60-1.90		0.035 max	0.040 max	0.15-0.30	bal Fe	610			
ASTM A829/A829M(95)	1330	Plt	0.27-0.34			1.50-1.90		0.035 max	0.040 max	0.15-0.40	bal Fe	480-965			
MIL-S-16974E(86)	1330*	Obs; Bar Bil Blm slab for refrg	0.28-0.33			1.6-1.9		0.04 max	0.04 max	0.15-0.45	bal Fe				
SAE 770(84)	1330*	Bar	0.28-0.33			1.6-1.9		0.04 max	0.04 max	0.15-0.45	bal Fe				
SAE J1268(95)	1330H	Bar; See std	0.27-0.33	0.20 max	0.35 max	1.45-2.05	0.06 max	0.030 max	0.040 max	0.15-0.35	Ni <=0.25; Cu, Ni, Cr, Mo not spec'd but acpt; bal Fe				
SAE J404(94)	1330	Plt	0.27-0.34			1.50-1.90		0.035 max	0.040 max	0.15-0.35	bal Fe				
Yugoslavia															
	C.3139	Q/T	0.25-0.32	0.3 max		1.3-1.65	0.15 max	0.035 max	0.035 max	0.15-0.4	bal Fe				

Alloy Steel, Manganese, 1335/1335H

Specification	Designation	Notes	C	Cr	Cu	Mn	Mo	P	S	Si	Other	UTS	YS	El	Hard
Bulgaria															
BDS 6354	35G2	Struct	0.32-0.40	0.30 max	0.30 max	1.40-1.80	0.15 max	0.035 max	0.035 max	0.17-0.37	Ni <=0.30; Ti <=0.03; V <=0.05; W <=0.2; bal Fe				
BDS 6550(86)	35GL	Casting	0.30-0.40	0.30 max	0.30 max	1.20-1.60	0.15 max	0.040 max	0.040 max	0.20-0.50	Ni <=0.30; bal Fe				

UNS numbers and US grades are provided as a means of cross referencing chemically similar alloys. Exchangability is only possible after independent examination of specifications. Tensile properties are minimum or typical as specified. UTS and YS as MPa. El as %. See Appendix for list of abbreviations used in Notes. * indicates obsolete material.

Specification	Designation	Notes	C	Cr	Cu	Mn	Mo	P	S	Si	Other	UTS	YS	El	Hard

Alloy Steel, Manganese, 1335/1335H (Continued from previous page)

Specification	Designation	Notes	C	Cr	Cu	Mn	Mo	P	S	Si	Other	UTS	YS	El	Hard
China															
GB 3077(88)	35Mn2	Bar HR Q/T 25mm diam	0.32-0.39	0.30 max	0.30 max	1.40-1.80		0.035 max	0.035 max	0.17-0.37	bal Fe	835	685	12	
GB/T 3078(94)	35Mn2	Bar HR Q/T 25mm diam	0.32-0.39	0.30 max	0.30 max	1.40-1.80		0.035 max	0.035 max	0.17-0.37	bal Fe	835	685	12	
Czech Republic															
CSN 413242	13242	Q/T	0.36-0.46	0.3 max		1.5-2.0		0.035 max	0.035 max	0.15-0.4	V 0.07-0.15; bal Fe				
France															
AFNOR	40M5		0.36-0.44			1-1.35		0.04 max	0.035 max	0.1-0.4	bal Fe				
Germany															
DIN	36Mn7		0.35			1.60-1.90		0.025 max	0.025 max	0.50	bal Fe				
DIN	WNr 1.5069		0.35			1.60		0.025 max	0.025 max	0.50	bal Fe				
Hungary															
MSZ 61(85)	Mn2	Q/T; 100.1-160mm	0.33-0.4			1.3-1.65		0.035 max	0.035 max	0.4 max	bal Fe	650-800	410	16L	
MSZ 61(85)	Mn2	Q/T; 0-16mm	0.33-0.4			1.3-1.65		0.035 max	0.035 max	0.4 max	bal Fe	850-1000	640	12L	
MSZ 61(85)	Mn2E	Q/T; 100.1-160mm	0.33-0.4			1.3-1.65		0.035 max	0.02-0.035	0.4 max	bal Fe	650-800	410	16L	
MSZ 61(85)	Mn2E	Q/T; 0-16mm	0.33-0.4			1.3-1.65		0.035 max	0.02-0.035	0.4 max	bal Fe	850-1000	640	12L	
Japan															
JIS G3311(88)	SMn438M	CR Strp; Chain	0.35-0.41	0.35 max	0.30 max	1.35-1.65		0.030 max	0.030 max	0.15-0.35	Ni <=0.25; bal Fe				200-290 HV
Mexico															
NMX-B-268(68)	1335H	Hard	0.32-0.38			1.45-2.05				0.20-0.35	bal Fe				
NMX-B-300(91)	1335	Bar	0.33-0.38			1.60-1.90		0.035 max	0.040 max	0.15-0.35	bal Fe				
NMX-B-300(91)	1335H	Bar	0.32-0.38			1.45-2.05				0.15-0.35	bal Fe				
Pan America															
COPANT 514	1335	Tube	0.33-0.38			1.6-1.9		0.04	0.04	0.2-0.35	bal Fe				
Russia															
GOST 977(88)	35GL	Q/T	0.3-0.4	0.3 max		1.2-1.6		0.04 max	0.04 max	0.2-0.4	Ni <=0.3; Ti <=0.05; bal Fe				
Spain															
UNE 36012(75)	36Mn6	Q/T	0.33-0.4			1.3-1.65	0.15 max	0.035 max	0.035 max	0.15-0.4	bal Fe				
UNE 36012(75)	F.1203*	see 36Mn6	0.33-0.4			1.3-1.65	0.15 max	0.035 max	0.035 max	0.15-0.4	bal Fe				
Sweden															
SS 142120	2120	Q/T	0.38-0.45			1.1-1.4		0.035 max	0.035 max	0.1-0.4	bal Fe				
UK															
BS 970/3(91)	150M36	Bright bar; 19<t<=150mm; hard	0.32-0.40			1.30-1.70	0.15 max	0.050 max	0.050 max		bal Fe	625-775	400	18	179-229 HB
BS 970/3(91)	150M36	Bright bar; 150<t<=250mm; Norm	0.32-0.40			1.30-1.70	0.15 max	0.050 max	0.050 max		bal Fe	600	355	15	170-223 HB
USA															
	AISI 1335	Smls mech tub	0.33-0.38			1.60-1.90		0.040 max	0.040 max	0.15-0.35	bal Fe				
	AISI 1335	Plt	0.32-0.39			1.50-1.90		0.035 max	0.040 max	0.15-0.40	bal Fe				
	AISI 1335	Bar Blm Bil Slab	0.33-0.38			1.60-1.90		0.035 max	0.040 max	0.15-0.35	bal Fe				
	AISI 1335H	Smls mech tub, Hard	0.32-0.38			1.45-2.05		0.035 max	0.040 max	0.15-0.30	bal Fe				
	AISI 1335H	Bar Blm Bil Slab	0.32-0.38			1.45-2.05		0.035 max	0.040 max	0.15-0.35	bal Fe				
	UNS G13350		0.33-0.38			1.60-1.90		0.035 max	0.040 max	0.15-0.35	bal Fe				

UNS numbers and US grades are provided as a means of cross referencing chemically similar alloys. Exchangability is only possible after independent examination of specifications. Tensile properties are minimum or typical as specified. UTS and YS as MPa. El as %. See Appendix for list of abbreviations used in Notes. * indicates obsolete material.

Specification	Designation	Notes	C	Cr	Cu	Mn	Mo	P	S	Si	Other	UTS	YS	El	Hard

Alloy Steel, Manganese, 1335/1335H (Continued from previous page)

USA

Specification	Designation	Notes	C	Cr	Cu	Mn	Mo	P	S	Si	Other	UTS	YS	El	Hard
	UNS H13350		0.32-0.38			1.45-2.05		0.035 max	0.040 max	0.15-0.35	bal Fe				
ASTM A29/A29M(93)	1335	Bar	0.33-0.38			1.60-1.90		0.035 max	0.040 max	0.15-0.35	bal Fe				
ASTM A304(96)	1335H	Bar, hard bands spec	0.32-0.38	0.35 max		1.45-2.05	0.06 max	0.035 max	0.040 max	0.15-0.35	Cu Ni Cr Mo trace allowed; bal Fe				
ASTM A322(96)	1335	Bar	0.33-0.38			1.60-1.90		0.035 max	0.040 max	0.15-0.35	bal Fe				
ASTM A331(95)	1335	Bar	0.33-0.38			1.60-1.90		0.035 max	0.040 max	0.15-0.35	bal Fe				
ASTM A331(95)	1335H	Bar	0.32-0.38	0.20 max	0.35 max	1.45-2.05	0.06 max	0.025 max	0.025 max	0.15-0.35	bal Fe				
ASTM A519(96)	1335	Smls mech tub	0.33-0.38			1.60-1.90		0.040 max	0.040 max	0.15-0.35	bal Fe				
ASTM A752(93)	1335	Rod Wir	0.33-0.38			1.60-1.90		0.035 max	0.040 max	0.15-0.30	bal Fe	620			
ASTM A829/A829M(95)	1335	Plt	0.32-0.39			1.50-1.90		0.035 max	0.040 max	0.15-0.40	bal Fe	480-965			
DoD-F-24669/1(86)(86)	1335	Bar Bil; Supersedes MIL-S-866 & MIL-S-16974	0.33-0.38			1.6-1.9		0.035	0.04	0.15-0.35	bal Fe				
MIL-S-16974E(86)	1335*	Obs; Bar Bil Blm slab for refrg	0.33-0.38			1.6-1.9		0.04 max	0.04 max	0.15-0.3	bal Fe				
SAE 770(84)	1335*	Obs; see J1397(92)	0.33-0.38			1.6-1.9		0.04 max	0.04 max	0.15-0.3	bal Fe				
SAE J1268(95)	1335H	Bar Wir; See std	0.32-0.38	0.20 max	0.35 max	1.45-2.05	0.06 max	0.030 max	0.040 max	0.15-0.35	Ni <=0.25; Cu, Ni, Cr, Mo not spec'd but acpt; bal Fe				
SAE J404(94)	1335	Plt	0.32-0.39			1.50-1.90		0.035 max	0.040 max	0.15-0.35	bal Fe				
SAE J404(94)	1335	Bil Blm Slab Bar HR CF	0.33-0.38			1.60-1.90		0.030 max	0.040 max	0.15-0.35	bal Fe				

Alloy Steel, Manganese, 1340/1340H

Australia

Specification	Designation	Notes	C	Cr	Cu	Mn	Mo	P	S	Si	Other	UTS	YS	El	Hard
AS 1442	K1340*	Obs; Bar Bil	0.38-0.43			1.6-1.9		0.05	0.05	0.1-0.35	bal Fe				
AS 1442(92)	X1340	HR bar, Semifinished (may treat w/microalloying elements)	0.38-0.43			1.40-1.70		0.040 max	0.040 max	0.10-0.35	Si<=0.10 for Al-killed steels; bal Fe				
AS 1443(94)	X1340	CF bar (may treat w/microalloying elements)	0.38-0.43			1.40-1.70		0.040 max	0.040 max	0.10-0.35	Si<=0.10 for Al-killed steels; bal Fe				

Bulgaria

Specification	Designation	Notes	C	Cr	Cu	Mn	Mo	P	S	Si	Other	UTS	YS	El	Hard
BDS 6354	40G2F		0.38-0.45		0.3	1.60-1.90		0.035	0.035	0.17-0.37	V 0.06-0.1; bal Fe				

China

Specification	Designation	Notes	C	Cr	Cu	Mn	Mo	P	S	Si	Other	UTS	YS	El	Hard
GB 3077(88)	40Mn2	Bar HR Q/T 25mm diam	0.37-0.44	0.30 max	0.30 max	1.40-1.80		0.035 max	0.035 max	0.17-0.37	bal Fe	885	735	12	
GB 8162(87)	40Mn2	Smls tube HR/CD Q/T	0.37-0.44			1.40-1.80		0.035 max	0.035 max	0.17-0.37	bal Fe	885	735	12	
GB/T 3078(94)	40Mn2	Bar CD Q/T 25mm diam	0.37-0.44	0.30 max	0.30 max	1.40-1.80		0.035 max	0.035 max	0.17-0.37	bal Fe	885	735	12	
YB/T 5052(93)	40Mn2	Tube HR Norm	0.37-0.44			1.40-1.80		0.040 max	0.040 max	0.20-0.40	bal Fe	685	490	12	

Germany

Specification	Designation	Notes	C	Cr	Cu	Mn	Mo	P	S	Si	Other	UTS	YS	El	Hard
DIN	42MnV7	Q/T 17-40mm	0.38-0.45			1.60-1.90		0.035 max	0.035 max	0.15-0.35	V 0.07-0.12; bal Fe	980-1180	785	11	
DIN	WNr 1.5223	Q/T 17-40mm	0.38-0.45			1.60-1.90		0.035 max	0.035 max	0.15-0.35	V 0.07-0.12; bal Fe	980-1180	785	11	

Mexico

Specification	Designation	Notes	C	Cr	Cu	Mn	Mo	P	S	Si	Other	UTS	YS	El	Hard
DGN B-203	1340*	Obs	0.38-0.43			1.6-1.9		0.04	0.04	0.2-0.35	bal Fe				
DGN B-297	1340*	Obs	0.38-0.43			1.6-1.9		0.035	0.04	0.2-0.35	bal Fe				
NMX-B-268(68)	1340H	Hard	0.37-0.44			0.45-2.05				0.20-0.35	bal Fe				

UNS numbers and US grades are provided as a means of cross referencing chemically similar alloys. Exchangability is only possible after independent examination of specifications. Tensile properties are minimum or typical as specified. UTS and YS as MPa. El as %. See Appendix for list of abbreviations used in Notes. * indicates obsolete material.

Specification	Designation	Notes	C	Cr	Cu	Mn	Mo	P	S	Si	Other	UTS	YS	El	Hard
Alloy Steel, Manganese, 1340/1340H (Continued from previous page)															
Mexico															
NMX-B-300(91)	1340	Bar	0.38-0.43			1.60-1.90		0.035 max	0.040 max	0.15-0.35	bal Fe				
NMX-B-300(91)	1340H	Bar	0.37-0.44			1.45-2.05				0.15-0.35	bal Fe				
Pan America															
COPANT 334	1340	Bar	0.38-0.43			1.6-1.9		0.04	0.04	0.2-0.35	bal Fe				
COPANT 514	1340	Tube	0.38-0.43			1.6-1.9		0.04	0.04	0.2-0.35	bal Fe				
USA															
	AISI 1340	Bar Blm Bil Slab	0.38-0.43			1.60-1.90		0.035 max	0.040 max	0.15-0.35	bal Fe				
	AISI 1340	Plt	0.36-0.44			1.50-1.90		0.035 max	0.040 max	0.15-0.40	bal Fe				
	AISI 1340	Smls mech tub	0.38-0.43			1.60-1.90		0.040 max	0.040 max	0.15-0.35	bal Fe				
	AISI 1340H	Bar Blm Bil Slab	0.37-0.43			1.45-2.05		0.035 max	0.040 max	0.15-0.35	bal Fe				
	AISI 1340H	Smls mech tub, Hard	0.37-0.43			1.45-2.05		0.040 max	0.040 max	0.15-0.30	bal Fe				
	UNS G13400		0.38-0.43			1.60-1.90		0.035 max	0.040 max	0.15-0.35	bal Fe				
	UNS H13400		0.37-0.44			1.45-2.05		0.035 max	0.040 max	0.15-0.35	bal Fe				
ASTM A29/A29M(93)	1340	Bar	0.38-0.43			1.60-1.90		0.035 max	0.040 max	0.15-0.35	bal Fe				
ASTM A304(96)	1340H	Bar, hard bands spec	0.37-0.43	0.20 max	0.35 max	1.45-2.05	0.06 max	0.04-0.035	0.040 max	0.15-0.35	Cu Ni Cr Mo trace allowed; bal Fe				
ASTM A322(96)	1340	Bar	0.38-0.43			1.60-1.90		0.035 max	0.040 max	0.15-0.35	bal Fe				
ASTM A331(95)	1340	Bar	0.38-0.43			1.60-1.90		0.035 max	0.040 max	0.15-0.35	bal Fe				
ASTM A331(95)	1340H	Bar	0.37-0.43	0.20 max	0.35 max	1.45-2.05	0.06 max	0.025 max	0.025 max	0.15-0.35					
ASTM A513(97a)	1340	ERW Mech Tub	0.38-0.43			1.60-1.90		0.035 max	0.040 max	0.15-0.35	bal Fe				
ASTM A519(96)	1340	Smls mech tub	0.38-0.43			1.60-1.90		0.040 max	0.040 max	0.15-0.35	bal Fe				
ASTM A547	1340	Wir	0.38-0.43			1.6-1.9		0.04 max	0.04 max	0.15-0.35	bal Fe				
ASTM A752(93)	1340	Rod Wir	0.38-0.43			1.60-1.90		0.035 max	0.040 max	0.15-0.30	bal Fe	630			
ASTM A829/A829M(95)	1340	Plt	0.36-0.44			1.50-1.90		0.035 max	0.040 max	0.15-0.40	bal Fe	480-965			
DoD-F-24669/1(86)(86)	1340	Bar Bil; Supersedes MIL-S-866 & MIL-S-16974	0.38-0.43			1.6-1.9		0.035	0.04	0.15-0.3	bal Fe				
MIL-S-16974E(86)	1340*	Obs; Bar Bil Blm slab for refrg	0.38-0.43			1.6-1.9		0.04 max	0.04 max	0.15-0.3	bal Fe				
SAE 770(84)	1340*	Obs; see J1397(92)	0.38-0.43			1.6-1.9		0.04 max	0.04 max	0.15-0.3	bal Fe				
SAE J1268(95)	1340H	Bar Wir; See std	0.37-0.44	0.20 max	0.35 max	1.45-2.05	0.06 max	0.030 max	0.040 max	0.15-0.35	Ni <=0.25; Cu, Ni, Cr, Mo not spec'd but acpt; bal Fe				
SAE J404(94)	1340	Bil Blm Slab Bar HR CF	0.38-0.43			1.60-1.90		0.030 max	0.040 max	0.15-0.35	bal Fe				
SAE J404(94)	1340	Plt	0.36-0.44			1.50-1.90		0.035 max	0.040 max	0.15-0.35	bal Fe				
Alloy Steel, Manganese, 1345/1345H															
Australia															
AS 1442(92)	X1147	Free-cutting, HR bar, Semifinished	0.40-0.47			1.60-1.90		0.040 max	0.07-0.12	0.10-0.35	bal Fe				
AS 1442(92)	X1345	HR bar, Semifinished (may treat w/microalloying elements)	0.43-0.48			1.40-1.70		0.040 max	0.040 max	0.10-0.35	Si<=0.10 for Al-killed steels; bal Fe				

Specification	Designation	Notes	C	Cr	Cu	Mn	Mo	P	S	Si	Other	UTS	YS	El	Hard
Alloy Steel, Manganese, 1345/1345H (Continued from previous page)															
Australia															
AS 1443(94)	X1147	CF bar, Free-cutting	0.40-0.47			1.60-1.90		0.040 max	0.07-0.12	0.10-0.35	bal Fe				
China															
GB 11251(89)	45Mn2	Plt HR Ann	0.42-0.49			1.40-1.80		0.035 max	0.035 max	0.17-0.37	bal Fe	600-850		13	
GB 3077(88)	45Mn2	Bar HR Q/T 25mm diam	0.42-0.49	0.30 max	0.30 max	1.40-1.80		0.035 max	0.035 max	0.17-0.37	bal Fe	885	735	10	
GB 3077(88)	50Mn2	Bar HR Q/T 25mm diam	0.47-0.55	0.30 max	0.30 max	1.40-1.80		0.035 max	0.035 max	0.17-0.37	bal Fe	930	785	9	
GB 8162(87)	45Mn2	Smls tube HR/CD Q/T	0.42-0.49			1.40-1.80		0.035 max	0.035 max	0.17-0.37	bal Fe	885	735	10	
GB/T 3078(94)	45Mn2	Bar CD Q/T 25mm diam	0.42-0.49	0.30 max	0.30 max	1.40-1.80		0.035 max	0.035 max	0.17-0.37	bal Fe	885	735	10	
GB/T 3078(94)	50Mn2	Bar CD Ann	0.47-0.55	0.30 max	0.30 max	1.40-1.80		0.035 max	0.035 max	0.17-0.37	bal Fe	635		9	
YB/T 5132(93)	45Mn2	Sh HR/CR Ann	0.42-0.49			1.40-1.80		0.035 max	0.035 max	0.17-0.37	bal Fe	590-835		12	
Germany															
DIN	46Mn7		0.42-0.50			1.60-1.90		0.050 max	0.050 max	0.15-0.35	N <=0.007; bal Fe				
DIN	50Mn7		0.45-0.55			1.60-2.00		0.040 max	0.040 max	0.40 max	N <=0.007; bal Fe				
DIN	51Mn7		0.55			1.90		0.025 max	0.025 max	0.60	Al >=0.025; bal Fe				
DIN	51MnV7	Spring hard	0.48-0.55			1.60-1.90		0.035 max	0.035 max	0.15-0.35	V 0.07-0.12; bal Fe	1230-1420	1080	8	
DIN	R1100Cr		0.60-0.82	0.80-1.30		0.80-1.30		0.030 max	0.030 max	0.30-0.90	bal Fe				
DIN	WNr 1.0912		0.42-0.50			1.60-1.90		0.050 max	0.050 max	0.15-0.35	N <=0.007; bal Fe				
DIN	WNr 1.0913		0.45-0.55			1.60-2.00		0.040 max	0.040 max	0.40 max	N <=0.007; bal Fe				
DIN	WNr 1.0915		0.60-0.82	0.80-1.30		0.80-1.30		0.030 max	0.030 max	0.30-0.90	bal Fe				
DIN	WNr 1.5085		0.55			1.90		0.025 max	0.025 max	0.60	Al >=0.025; bal Fe				
DIN	WNr 1.5225	Spring hard	0.48-0.55			1.60-1.90		0.035 max	0.035 max	0.15-0.35	V 0.07-0.12; bal Fe	1230-1420	1080	8	
Japan															
JIS G3311(88)	SMn443M	CR Strp; Chain	0.40-0.46	0.35 max	0.30 max	1.35-1.65		0.030 max	0.030 max	0.15-0.35	Ni <=0.25; bal Fe				200-290 HV
Mexico															
DGN B-203	1345*	Obs	0.43-0.48			1.6-1.9		0.04	0.04	0.2-0.35	bal Fe				
DGN B-297	1345*	Obs	0.43-0.48			1.6-1.9		0.035	0.04	0.2-0.35	bal Fe				
NMX-B-268(68)	1345H	Hard	0.42-0.49			1.45-2.05				0.20-0.35	bal Fe				
NMX-B-300(91)	1345	Bar	0.43-0.48			1.60-1.90		0.035 max	0.040 max	0.15-0.35	bal Fe				
NMX-B-300(91)	1345H	Bar	0.42-0.49			1.45-2.05				0.15-0.35	bal Fe				
Pan America															
COPANT 514	1345	Tube	0.43-0.48			1.6-1.9		0.04	0.04	0.2-0.35	bal Fe				
Poland															
PNH93421	St90PC		0.45-0.6			1.7-2.1		0.03	0.03	0.3	bal Fe				
USA															
	AISI 1345	Plt	0.41-0.49			1.50-1.90		0.035 max	0.040 max	0.15-0.40	bal Fe				
	AISI 1345	Smls mech tub	0.43-0.48			1.60-1.90		0.040 max	0.040 max	0.15-0.35	bal Fe				
	AISI 1345	Bar Blm Bil Slab	0.43-0.48			1.60-1.90		0.035 max	0.040 max	0.15-0.35	bal Fe				
	AISI 1345H	Bar Blm Bil Slab	0.42-0.49			1.45-2.05		0.04 max	0.04 max	0.15-0.35	bal Fe				

UNS numbers and US grades are provided as a means of cross referencing chemically similar alloys. Exchangability is only possible after independent examination of specifications. Tensile properties are minimum or typical as specified. UTS and YS as MPa. El as %. See Appendix for list of abbreviations used in Notes. * indicates obsolete material.

Specification	Designation	Notes	C	Cr	Cu	Mn	Mo	P	S	Si	Other	UTS	YS	El	Hard
Alloy Steel, Manganese, 1345/1345H (Continued from previous page)															
USA															
	AISI 1345H	Smls mech tub, Hard	0.42-0.49			1.45-2.05		0.035 max	0.040 max	0.15-0.30	bal Fe				
	UNS G13450		0.43-0.48			1.60-1.90		0.035 max	0.040 max	0.15-0.35	bal Fe				
	UNS H13450		0.42-0.49			1.45-2.05		0.035 max	0.040 max	0.15-0.35	bal Fe				
ASTM A29/A29M(93)	1345	Bar	0.43-0.48			1.60-1.90		0.035 max	0.040 max	0.15-0.35	bal Fe				
ASTM A304(96)	1345H	Bar, hard bands spec	0.42-0.49	0.20 max	0.35 max	1.45-2.05	0.06 max	0.035 max	0.040 max	0.15-0.35	Cu Ni Cr Mo trace allowed; bal Fe				
ASTM A322(96)	1345	Bar	0.43-0.48			1.60-1.90		0.035 max	0.040 max	0.15-0.35	bal Fe				
ASTM A331(95)	1345	Bar	0.43-0.48			1.60-1.90		0.035 max	0.040 max	0.15-0.35	bal Fe				
ASTM A331(95)	1345H	Bar	0.42-0.49	0.20 max	0.35 max	1.45-2.05	0.06 max	0.025 max	0.025 max	0.15-0.35	bal Fe				
ASTM A372	Type IV*	Thin wall frg press ves	0.4-0.5			1.4-1.8		0.025 max	0.025 max	0.15-0.35	bal Fe				
ASTM A519(96)	1345	Smls mech tub	0.43-0.48			1.60-1.90		0.040 max	0.040 max	0.15-0.35	bal Fe				
ASTM A752(93)	1345	Rod Wir	0.43-0.48			1.60-1.90		0.035 max	0.040 max	0.15-0.30	bal Fe	670			
ASTM A829/A829M(95)	1345	Plt	0.41-0.49			1.50-1.90		0.035 max	0.040 max	0.15-0.40	bal Fe	480-965			
FED QQ-S-626C(91)	1345*	Obs; see ASTM A829; Plt	0.43-0.48			1.6-1.9		0.035	0.04	0.15-0.35	bal Fe				
SAE 770(84)	1345*	Obs; see J1397(92)	0.43-0.48			1.6-1.9		0.04 max	0.04 max	0.15-0.3	bal Fe				
SAE J1268(95)	1345H	See std	0.42-0.49	0.20 max	0.35 max	1.45-2.05	0.06 max	0.030 max	0.040 max	0.15-0.35	Ni <=0.25; Cu, Ni, Cr, Mo not spec'd but acpt; bal Fe				
SAE J404(94)	1345	Plt	0.41-0.49			1.50-1.90		0.035 max	0.040 max	0.15-0.35	bal Fe				
Alloy Steel, Manganese, A372(IV)															
USA															
	UNS K14508		0.40-0.50			1.40-1.80	0.17-0.27	0.035 max	0.04 max	0.15-0.35	bal Fe				
ASTM A372/372M(95)	Grade D	Thin wall frg press ves	0.40-0.50			1.40-1.80	0.17-0.27	0.025 max	0.025 max	0.15-0.35	bal Fe	725-895	450	15	217 HB
ASTM A372/372M(95)	Grade D	Thin wall frg press ves	0.48-0.53			1.40-1.80	0.17-0.27	0.025 max	0.025 max	0.15-0.35	bal Fe	725-895	450	15	217 HB
ASTM A372/372M(95)	Type IV*	Thin wall frg press ves	0.48-0.53			1.40-1.80	0.17-0.27	0.025 max	0.025 max	0.15-0.35	bal Fe	725-895	450	15	217 HB
ASTM A372/372M(95)	Type IV*	Thin wall frg press ves	0.40-0.50			1.40-1.80	0.17-0.27	0.025 max	0.025 max	0.15-0.35	bal Fe	725-895	450	15	217 HB
Alloy Steel, Manganese, A562															
USA															
	UNS K11224		0.12 max		0.15 max	1.20 max		0.035 max	0.04 max	0.15-0.50	Ti 4xC min; bal Fe				
ASTM A562/A562M(96)		Press ves plt, for glass or diffused metal coating t<50mm	0.12 max		0.15 max	1.20 max		0.035 max	0.035 max	0.15-0.50	Ti 4xC min; bal Fe	380-515	205	26	
Alloy Steel, Manganese, A707(L6)															
USA															
	UNS K20902		0.09 max			1.75-2.30	0.22-0.38	0.030 max	0.035 max	0.17 max	Nb 0.05-0.11; bal Fe				
ASTM A707/A707M(98)	A 707 (L6)	Frg flanges; low-temp; class values given	0.07 max	0.30 max	0.40 max	1.85-2.20	0.25-0.35	0.025 max	0.025 max	0.15 max	Nb 0.06-0.10; Ni <=0.40; V <=0.05; bal Fe	415-620	290-515	20-22	149-265 HB
Alloy Steel, Manganese, A714(I)															
Mexico															
NMX-B-069(86)	Grade I	Weld and Smls pipe	0.22 max		0.20 min	1.25 max			0.05 max		bal Fe	485	345	22	

UNS numbers and US grades are provided as a means of cross referencing chemically similar alloys. Exchangability is only possible after independent examination of specifications. Tensile properties are minimum or typical as specified. UTS and YS as MPa. El as %. See Appendix for list of abbreviations used in Notes. * indicates obsolete material.

Specification	Designation	Notes	C	Cr	Cu	Mn	Mo	P	S	Si	Other	UTS	YS	El	Hard
Alloy Steel, Manganese, A714(I) (Continued from previous page)															
USA															
	UNS K12608		0.26 max		0.18 min	1.30 max			0.063 max		bal Fe				
ASTM A714(96)	I	Weld, Smls pipe	0.22 max		0.20 min	1.25 max			0.05 max		bal Fe	485	345	22	
Alloy Steel, Manganese, A735															
USA															
	UNS K10623		0.06 max	0.20-0.35		1.20-1.90	0.23-0.47	0.04 max	0.025 max	0.40	Nb 0.03-0.09; bal Fe				
ASTM A735(96)	Class 1	Press ves plt, mod-temp, roll or Q/T; t <=16mm	0.06 max	0.20-0.35		1.20-1.90	0.23-0.47	0.035 max	0.025 max	0.40 max	Nb 0.03-0.09; bal Fe	550-690	450	18	
ASTM A735(96)	Class 1	Press ves plt, mod-temp, roll t>16-25mm; Q/T t>16mm	0.06 max	0.20-0.35		1.50-2.20	0.23-0.47	0.035 max	0.025 max	0.40 max	Nb 0.03-0.09; bal Fe	550-690	450	18	
Alloy Steel, Manganese, C110															
Italy															
UNI 5771(66)	20MnSi5	Chain	0.16-0.24			1.1-1.5		0.035 max	0.035 max	0.3-0.5	bal Fe				
USA															
	UNS K01907		0.16-0.22			1.40-1.80		0.025 max	0.025 max	0.25-0.45	V 0.07-0.18; bal Fe				
Alloy Steel, Manganese, K10614															
USA															
	UNS K10614		0.06 max			1.20-1.90	0.25-0.35	0.04 max	0.025 max	0.40 max	Nb 0.03-0.09; bal Fe				
Alloy Steel, Manganese, K11201															
UK															
BS 3606(92)	400	Heat exch Tub; t<=3.2mm	0.12 max	0.20 max	0.25 max	0.90-1.20		0.020 max	0.020 max	0.10-0.35	Al <=0.04; Ni <=0.30; Sn<=0.025; bal Fe	400-520	230	21	
BS 3606(92)	440	Heat exch Tub; t<=3.2mm	0.12-0.18			0.90-1.20	0.15 max	0.035 max	0.035 max	0.10-0.35	Ni <=0.40; bal Fe	440-560	265	21	
USA															
	UNS K11201		0.12 max			1.30 max		0.035 max	0.040 max	0.15-0.35	V >=0.020; bal Fe				
Alloy Steel, Manganese, K12810															
India															
IS 1570/4(88)	26C10BT		0.23-0.29	0.3 max		0.9-1.2	0.15 max	0.07 max	0.06 max	0.15-0.3	B 0.0005-0.003; Ni <=0.4; bal Fe				
USA															
	UNS K12810		0.28 max		0.20 min	1.10-1.60		0.04 max	0.05 max	0.30 max	bal Fe				
Alloy Steel, Manganese, MIL-A-13259															
USA															
	UNS K91209	Non-magnetic	1.20-1.50	0.60 max		12.00-15.00	0.10 max	0.08 max	0.04 max	0.55 max	Ni <=0.75; bal Fe				
Alloy Steel, Manganese, MIL-S-12504(MnMo)															
USA															
	UNS K17145		0.65-0.77			0.75-1.05	0.90-1.10	0.04 max	0.04 max	0.20-0.35	bal Fe				
MIL-B-12504E(90)	Mn-Mo	CD Bar wire for bullets	0.65-0.77			0.75-1.05	0.90-1.10	0.04 max	0.04 max	0.20-0.35	bal Fe				24-26 HRC
Alloy Steel, Manganese, No equivalents identified															
Europe															
EN 10028/4(94)	1.5637	High-temp, Press; 0<=t<=30mm	0.15 max			0.30-0.80		0.020 max	0.010 max	0.35 max	Al >=0.020; Ni 3.25-3.75; V <=0.05; Cr+Cu+Mo<=0.50; bal Fe	490-640	355	22	

Specification	Designation	Notes	C	Cr	Cu	Mn	Mo	P	S	Si	Other	UTS	YS	El	Hard

Alloy Steel, Manganese, No equivalents identified

Europe

Specification	Designation	Notes	C	Cr	Cu	Mn	Mo	P	S	Si	Other	UTS	YS	El	Hard
EN 10028/4(94)	1.5637	High-temp, Press; 30<=t<=50mm	0.15 max			0.30-0.80		0.020 max	0.010 max	0.35 max	Al >=0.020; Ni 3.25-3.75; V <=0.05; Cr+Cu+Mo<=0.50; bal Fe	490-640	345	22	
EN 10028/4(94)	12Ni14	High-temp, Press; 0<=t<=30mm	0.15 max			0.30-0.80		0.020 max	0.010 max	0.35 max	Al >=0.020; Ni 3.25-3.75; V <=0.05; Cr+Cu+Mo<=0.50; bal Fe	490-640	355	22	
EN 10028/4(94)	12Ni14	High-temp, Press; 30<=t<=50mm	0.15 max			0.30-0.80		0.020 max	0.010 max	0.35 max	Al >=0.020; Ni 3.25-3.75; V <=0.05; Cr+Cu+Mo<=0.50; bal Fe	490-640	345	22	

International

Specification	Designation	Notes	C	Cr	Cu	Mn	Mo	P	S	Si	Other	UTS	YS	El	Hard
ISO 683-1(87)	22Mn6	Bar Frg Plt Wir Rod Bolt Slab; Q/T, t<16mm	0.19-0.26			1.30-1.65		0.035 max	0.035 max	0.10-0.40	bal Fe	700-850	550		
ISO 683-1(87)	28Mn6	Bar Frg Plt Wir Rod Bolt Slab; Q/T, t<16mm	0.26-0.32			1.30-1.65		0.035 max	0.035 max	0.10-0.40	bal Fe	800-950	590		
ISO 683-1(87)	36Mn6	Bar Frg Plt Wir Rod Bolt Slab; Q/T, t<16mm	0.33-0.40			1.30-1.65		0.035 max	0.035 max	0.10-0.40	bal Fe	850-1000	640		
ISO 683-1(87)	42Mn6	Bar Frg Plt Wir Rod Bolt Slab; Q/T, t<16mm	0.39-0.46			1.30-1.65		0.035 max	0.035 max	0.10-0.40	bal Fe	900-1050	690		
ISO 683-18(96)	28Mn6	Flat, CD, Ann, t<=16mm	0.25-0.32			1.30-1.65		0.035 max	0.035 max	0.10-0.40	bal Fe	920 max			223 HB max
ISO 683-18(96)	36Mn6	Flat, CD, Ann, t<=16mm	0.33-0.40			1.30-1.65		0.035 max	0.035 max	0.10-0.40	bal Fe	930 max			229 HB max
ISO 683-18(96)	42Mn6	Flat, CD, Ann, t<=16mm	0.39-0.49			1.30-1.65		0.035 max	0.035 max	0.10-0.40	bal Fe	940 max			229 HB max

Russia

Specification	Designation	Notes	C	Cr	Cu	Mn	Mo	P	S	Si	Other	UTS	YS	El	Hard
GOST 4543	40G2		0.36-0.44	0.3	0.3	1.4-1.8		0.035	0.035	0.17-0.37	Ni 0.3; bal Fe				
GOST 4543(71)	40G2	Q/T	0.36-0.44	0.3 max	0.3 max	1.4-1.8	0.15 max	0.035 max	0.035 max	0.17-0.37	Al <=0.1; Ni <=0.3; Ti <=0.05; bal Fe				

Spain

Specification	Designation	Notes	C	Cr	Cu	Mn	Mo	P	S	Si	Other	UTS	YS	El	Hard
UNE 36034(85)	35Mn5DF	CH	0.32-0.39			1.1-1.4	0.15 max	0.035 max	0.035 max	0.15-0.4	bal Fe				
UNE 36034(85)	F.1209*	see 35Mn5DF	0.32-0.39			1.1-1.4	0.15 max	0.035 max	0.035 max	0.15-0.4	bal Fe				
UNE 36087(76/78)	12Ni14	Sh Plt Strp CR	0.15 max			0.8 max	0.15 max	0.035 max	0.035 max	0.15-0.35	Ni 3.25-3.75; bal Fe				
UNE 36087(76/78)	F.2643*	see 12Ni14	0.15 max			0.8 max	0.15 max	0.035 max	0.035 max	0.15-0.35	Ni 3.25-3.75; bal Fe				

Specification	Designation	Notes	C	Cr	Cu	Mn	Mo	P	S	Si	Other	UTS	YS	El	Hard

Alloy Steel, Molybdenum/Molybdenum Sulfide, 4012

Bulgaria

Specification	Designation	Notes	C	Cr	Cu	Mn	Mo	P	S	Si	Other	UTS	YS	El	Hard
BDS 5930	14ChMF	Heat res	0.1-0.18	0.50-0.75	0.30 max	0.45-0.70	0.15-0.25	0.040 max	0.040 max	0.15-0.40	Ni <=0.40; bal Fe				
BDS 6600(73)	12ChMF	Heat res boiler tube	0.08-0.15	0.90-1.20	0.30 max	0.40-0.70	0.25-0.40	0.030 max	0.03 max	0.15-0.40	Ni <=0.30; V 0.15-0.35; bal Fe				
BDS 6609(73)	14ChMF	Heat res boiler tube	0.10-0.18	0.40-0.70	0.30 max	0.40-0.70	0.40-0.60	0.040 max	0.040 max	0.15-0.40	Ni <=0.30; V 0.22-0.35; bal Fe				

Mexico

Specification	Designation	Notes	C	Cr	Cu	Mn	Mo	P	S	Si	Other	UTS	YS	El	Hard
NMX-B-300(91)	4012	Bar	0.09-0.14			0.75-1.00	0.15-0.25	0.035 max	0.040 max	0.15-0.35	bal Fe				

USA

Specification	Designation	Notes	C	Cr	Cu	Mn	Mo	P	S	Si	Other	UTS	YS	El	Hard
	AISI 4012	Smls mech tub	0.09-0.14			0.75-1.00	0.15-0.25	0.040 max	0.040 max	0.15-0.35	bal Fe				
	UNS G40120		0.09-0.14			0.75-1.00	0.15-0.25	0.035 max	0.040 max	0.15-0.35	bal Fe				
ASTM A29/A29M(93)	4012	Bar	0.09-0.14			0.75-1.00	0.15-0.25	0.035 max	0.040 max	0.15-0.35	bal Fe				
ASTM A519(96)	4012	Smls mech tub	0.09-0.14			0.75-1.00	0.15-0.25	0.040 max	0.040 max	0.15-0.35	bal Fe				
ASTM A752(93)	4012	Rod Wir	0.09-0.14			0.75-1.00	0.15-0.25	0.035 max	0.040 max	0.15-0.30	bal Fe	490			

Alloy Steel, Molybdenum/Molybdenum Sulfide, 4023

Belgium

Specification	Designation	Notes	C	Cr	Cu	Mn	Mo	P	S	Si	Other	UTS	YS	El	Hard
NBN A25-102	15Mo3	Tube, HW CW Norm	0.11-0.21			0.47-0.84	0.22-0.39	0.04	0.04	0.12-0.38	bal Fe				

Germany

Specification	Designation	Notes	C	Cr	Cu	Mn	Mo	P	S	Si	Other	UTS	YS	El	Hard
DIN	WNr 1.5416*	Obs	0.16-0.24			0.5-0.8	0.25-0.35	0.04	0.04	0.15-0.35	bal Fe				

International

Specification	Designation	Notes	C	Cr	Cu	Mn	Mo	P	S	Si	Other	UTS	YS	El	Hard
ISO 2603	F27	Frg Norm, Tmp, Q/A	0.18-0.25			0.5-0.8	0.25-0.35	0.04	0.04	0.15-0.4	Al 0.01; bal Fe				

Mexico

Specification	Designation	Notes	C	Cr	Cu	Mn	Mo	P	S	Si	Other	UTS	YS	El	Hard
DGN B-203	4023*	Obs	0.2-0.25			0.7-0.9	0.2-0.3	0.04	0.04	0.2-0.35	bal Fe				
DGN B-297	4023*	Obs	0.2-0.25			0.7-0.9	0.2-0.3	0.035	0.04	0.2-0.35	bal Fe				
NMX-B-300(91)	4023	Bar	0.20-0.25			0.70-0.90	0.20-0.30	0.035 max	0.040 max	0.15-0.35	bal Fe				

Pan America

Specification	Designation	Notes	C	Cr	Cu	Mn	Mo	P	S	Si	Other	UTS	YS	El	Hard
COPANT 514	4023	Tube	0.2-0.25			0.7-0.9	0.2-0.3	0.04	0.04-0.05	0.2-0.35	bal Fe				

Poland

Specification	Designation	Notes	C	Cr	Cu	Mn	Mo	P	S	Si	Other	UTS	YS	El	Hard
PNH84024	2M		0.17-0.23	0.3	0.3	0.8-1.1	0.25-0.4	0.045	0.045	0.15-0.35	Al 0.012; Ni 0.3; bal Fe				

Russia

Specification	Designation	Notes	C	Cr	Cu	Mn	Mo	P	S	Si	Other	UTS	YS	El	Hard
GOST 21357	25MLS		0.22-0.3	0.3 max	0.3 max	0.35-0.8	0.1-0.2	0.02	0.02	0.2-0.4	Ni 0.3; bal Fe				

USA

Specification	Designation	Notes	C	Cr	Cu	Mn	Mo	P	S	Si	Other	UTS	YS	El	Hard
	AISI 4023	Bar Blm Bil Slab	0.20-0.25			0.70-0.90	0.20-0.30	0.035 max	0.040 max	0.15-0.35	bal Fe				
	AISI 4023	Smls mech tub	0.20-0.25			0.70-0.90	0.20-0.30	0.040 max	0.036-0.040	0.15-0.35	bal Fe				
	UNS G40230		0.20-0.25			0.70-0.90	0.20-0.30	0.035 max	0.040 max	0.15-0.35	bal Fe				
ASTM A29/A29M(93)	4023	Bar	0.20-0.25			0.70-0.90	0.20-0.30	0.035 max	0.040 max	0.15-0.35	bal Fe				
ASTM A322(96)	4023	Bar	0.20-0.25			0.70-0.90	0.20-0.30	0.035 max	0.040 max	0.15-0.35	bal Fe				
ASTM A331(95)	4023	Bar	0.20-0.25			0.70-0.90	0.20-0.30	0.035 max	0.040 max	0.15-0.35	bal Fe				
ASTM A519(96)	4023	Smls mech tub	0.20-0.25			0.70-0.90	0.2-0.3	0.040 max	0.040 max	0.15-0.35	bal Fe				
ASTM A752(93)	4023	Rod Wir	0.20-0.25			0.70-0.90	0.20-0.30	0.035 max	0.040 max	0.15-0.30	bal Fe	500			
SAE 770(84)	4023*	Obs; see J1397(92)	0.2-0.25			0.7-0.9	0.2-0.3	0.04 max	0.04 max	0.15-0.3	bal Fe				

Specification	Designation	Notes	C	Cr	Cu	Mn	Mo	P	S	Si	Other	UTS	YS	El	Hard
Alloy Steel, Molybdenum/Molybdenum Sulfide, 4023 (Continued from previous page)															
USA															
SAE J404(94)	4023	Bil Blm Slab Bar HR CF	0.20-0.25			0.70-0.90	0.20-0.30	0.030 max	0.040 max	0.15-0.35	bal Fe				
Alloy Steel, Molybdenum/Molybdenum Sulfide, 4024															
Belgium															
NBN 629	16Mo3	Strip Tmp 3-40 mm diam	0.2			0.5-0.9	0.25-0.35	0.04	0.04	0.15-0.35	bal Fe				
China															
GB 5310(95)	20Mog	Smls tube Pip HR/CD Norm	0.15-0.25	0.30 max	0.20 max	0.40-0.80	0.44-0.65	0.030 max	0.030 max	0.17-0.37	bal Fe	415	220	22	
Mexico															
DGN B-203	4024*	Obs	0.2-0.25			0.7-0.9	0.2-0.3	0.04	0.05	0.2-0.35	bal Fe				
DGN B-297	4024*	Obs	0.2-0.25			0.7-0.9	0.2-0.3	0.035	0.035-0.05	0.2-0.35	bal Fe				
NMX-B-300(91)	4024	Bar	0.20-0.25			0.70-0.90	0.20-0.30	0.035 max	0.035-0.050	0.15-0.35	bal Fe				
Pan America															
COPANT 514	4024	Tube	0.2-0.25			0.7-0.9	0.2-0.3	0.04	0.05	0.2-0.35	bal Fe				
USA															
	AISI 4024	Smls mech tub	0.20-0.25			0.70-0.90	0.20-0.30	0.040 max	0.050 max	0.15-0.35	bal Fe				
	UNS G40240		0.20-0.25			0.70-0.90	0.20-0.30	0.035 max	0.035-0.050	0.15-0.35	bal Fe				
ASTM A29/A29M(93)	4024	Bar	0.20-0.25			0.70-0.90	0.20-0.30	0.035 max	0.035-0.050	0.15-0.35	bal Fe				
ASTM A322(96)	4024	Bar	0.20-0.25			0.70-0.90	0.20-0.30	0.035 max	0.035-0.050	0.15-0.35	bal Fe				
ASTM A331(95)	4024	Bar	0.20-0.25			0.70-0.90	0.20-0.30	0.035 max	0.035-0.050	0.15-0.35	bal Fe				
ASTM A519(96)	4024	Smls mech tub	0.20-0.25			0.70-0.90	0.20-0.30	0.040 max	0.035-0.050	0.15-0.35	bal Fe				
ASTM A752(93)	4024	Rod Wir	0.20-0.25			0.70-0.90		0.035 max	0.035-0.050	0.15-0.30	bal Fe	500			
SAE 770(84)	4024*	Obs; see J1397(92)	0.2-0.25			0.7-0.9	0.2-0.3	0.04	0.04-0.05	0.15-0.3	bal Fe				
Alloy Steel, Molybdenum/Molybdenum Sulfide, 4027/4027H															
Australia															
AS G18	En16A*	Obs; see AS 1444	0.25-0.3			1.3-1.8	0.2-0.35	0.05	0.05	0.1-0.35	bal Fe				
Mexico															
DGN B-203	4027*	Obs	0.25-0.3			0.7-0.9	0.2-0.3	0.04	0.04	0.2-0.35	bal Fe				
DGN B-297	4027*	Obs	0.25-0.3			0.7-0.9	0.2-0.3	0.035	0.04	0.2-0.35	bal Fe				
NMX-B-268(68)	4027H	Hard	0.24-0.30			0.60-1.00	0.20-0.30			0.20-0.35	bal Fe				
NMX-B-300(91)	4027	Bar	0.25-0.30			0.70-0.90	0.20-0.30	0.035 max	0.040 max	0.15-0.35	bal Fe				
NMX-B-300(91)	4027H	Bar	0.24-0.30			0.60-1.00	0.20-0.30			0.15-0.35	bal Fe				
Pan America															
COPANT 334	4027	Bar	0.25-0.3			0.7-0.9	0.2-0.3	0.04	0.04	0.2-0.35	bal Fe				
COPANT 514	4027	Tube	0.25-0.3			0.7-0.9	0.2-0.3	0.04	0.04	0.2-0.35	bal Fe				
Spain															
UNE	F.120.H		0.24-0.3			0.65-0.95	0.2-0.3	0.035	0.035	0.13-0.38	bal Fe				
USA															
	AISI 4027	Bar Blm Bil Slab	0.25-0.30			0.70-0.90	0.20-0.30	0.035 max	0.040 max	0.15-0.35	bal Fe				
	AISI 4027	Smls mech tub	0.25-0.30			0.70-0.90	0.20-0.30	0.040 max	0.040 max	0.15-0.35	bal Fe				

UNS numbers and US grades are provided as a means of cross referencing chemically similar alloys. Exchangability is only possible after independent examination of specifications. Tensile properties are minimum or typical as specified. UTS and YS as MPa. El as %. See Appendix for list of abbreviations used in Notes. * indicates obsolete material.

Specification	Designation	Notes	C	Cr	Cu	Mn	Mo	P	S	Si	Other	UTS	YS	El	Hard

Alloy Steel, Molybdenum/Molybdenum Sulfide, 4027/4027H (Continued from previous page)

USA

Specification	Designation	Notes	C	Cr	Cu	Mn	Mo	P	S	Si	Other	UTS	YS	El	Hard
	AISI 4027H	Smls mech tub, Hard	0.24-0.30			0.60-1.00	0.20-0.30	0.040 max	0.040 max	0.15-0.35	bal Fe				
	AISI 4027H	Bar Blm Bil Slab	0.24-0.30			0.60-1.00	0.20-0.30	0.035 max	0.040 max	0.15-0.35	bal Fe				
	UNS G40270		0.25-0.30			0.70-0.90	0.20-0.30	0.035 max	0.040 max	0.15-0.35	bal Fe				
	UNS H40270		0.24-0.30			0.60-1.00	0.20-0.30	0.035 max	0.040 max	0.15-0.35	bal Fe				
ASTM A29/A29M(93)	4027	Bar	0.25-0.30			0.70-0.90	0.20-0.30	0.035 max	0.040 max	0.15-0.35	bal Fe				
ASTM A304(96)	4027H	Bar, hard bands spec	0.24-0.3	0.20 max	0.35 max	0.60-1.00	0.20-0.30	0.035 max	0.040 max	0.15-0.35	Cu Ni Cr Mo trace allowed; bal Fe				
ASTM A322(96)	4027	Bar	0.25-0.30			0.70-0.90	0.20-0.30	0.035 max	0.040 max	0.15-0.35	bal Fe				
ASTM A331(95)	4027	Bar	0.25-0.30			0.70-0.90	0.20-0.30	0.035 max	0.040 max	0.15-0.35	bal Fe				
ASTM A331(95)	4027H	Bar	0.24-0.3	0.20 max	0.35 max	0.60-1.00	0.20-0.30	0.025 max	0.025 max	0.15-0.35	bal Fe				
ASTM A519(96)	4027	Smls mech tub	0.25-0.30			0.70-0.90	0.20-0.30	0.040 max	0.040 max	0.15-0.35	bal Fe				
ASTM A752(93)	4027	Rod Wir	0.25-0.30			0.70-0.90	0.20-0.30	0.035 max	0.040 max	0.15-0.30	bal Fe	570			
ASTM A914/914M(92)	4027RH	Bar restricted end Q hard	0.25-0.30	0.20 max	0.35 max	0.70-0.90	0.20-0.30	0.035 max	0.040 max	0.15-0.35	Ni <=0.25; Electric furnace P, S<=0.025; bal Fe				
SAE 770(84)	4027*	Obs; see J1397(92)	0.25-0.3			0.7-0.9	0.2-0.3	0.04 max	0.04	0.15-0.3	bal Fe				
SAE J1268(95)	4027H	Bar; See std	0.24-0.30	0.20 max	0.35 max	0.60-1.00	0.20-0.30	0.030 max	0.040 max	0.15-0.35	Ni <=0.25; Cu, Ni, Cr not spec'd but acpt; bal Fe				
SAE J1868(93)	4027RH	Restrict hard range	0.25-0.30	0.20 max	0.035 max	0.70-0.90	0.20-0.30	0.025 max	0.040 max	0.15-0.35	Ni <=0.25; Cu, Ni, Cr not spec'd but acpt; bal Fe				5 HRC max
SAE J404(94)	4027	Bil Blm Slab Bar HR CF	0.25-0.30			0.70-0.90	0.20-0.30	0.030 max	0.040 max	0.15-0.35	bal Fe				

Alloy Steel, Molybdenum/Molybdenum Sulfide, 4028/4028H

Argentina

Specification	Designation	Notes	C	Cr	Cu	Mn	Mo	P	S	Si	Other	UTS	YS	El	Hard
IAS	IRAM 4028		0.25-0.30			0.70-0.90	0.20-0.30	0.035 max	0.035-0.050	0.20-0.35	bal Fe	660	400	25	183 HB

Mexico

Specification	Designation	Notes	C	Cr	Cu	Mn	Mo	P	S	Si	Other	UTS	YS	El	Hard
DGN B-203	4028*	Obs	0.25-0.3			0.7-0.9	0.2-0.3	0.04	0.035-0.05	0.2-0.35	bal Fe				
DGN B-297	4028*	Obs	0.25-0.3			0.7-0.9		0.035	0.035-0.05	0.2-0.35	bal Fe				
NMX-B-268(68)	4028H	Hard	0.24-0.30			0.60-1.00	0.20-0.30		0.035-0.050	0.20-0.35	bal Fe				
NMX-B-300(91)	4028	Bar	0.25-0.30			0.70-0.90	0.20-0.30	0.035 max	0.035-0.050	0.15-0.35	bal Fe				
NMX-B-300(91)	4028H	Bar	0.24-0.30			0.60-1.00	0.20-0.30			0.15-0.35	bal Fe				

Pan America

Specification	Designation	Notes	C	Cr	Cu	Mn	Mo	P	S	Si	Other	UTS	YS	El	Hard
COPANT 514	4028	Tube	0.25-0.3			0.7-0.9	0.2-0.3	0.04	0.04-0.05	0.2-0.35	bal Fe				

USA

Specification	Designation	Notes	C	Cr	Cu	Mn	Mo	P	S	Si	Other	UTS	YS	El	Hard
	AISI 4028	Smls mech tub	0.25-0.30			0.70-0.90	0.20-0.30	0.040 max	0.035-0.050	0.15-0.35	bal Fe				
	AISI 4028	Bar Blm Bil Slab	0.25-0.30			0.70-0.90	0.20-0.30	0.035 max	0.035-0.050	0.15-0.35	bal Fe				
	AISI 4028H	Bar Blm Bil Slab	0.24-1.00			0.60-1.00	0.20-0.30	0.035 max	0.035-0.050	0.15-0.35	bal Fe				
	AISI 4028H	Smls mech tub, Hard	0.24-0.30			0.60-1.00	0.20-0.30	0.040 max	0.035-0.050	0.15-0.30	bal Fe				
	UNS G40280		0.25-0.30			0.70-0.90	0.20-0.30	0.035 max	0.035-0.050	0.15-0.35	bal Fe				
	UNS H40280		0.24-0.30			0.60-1.00	0.20-0.30	0.035 max	0.035-0.050	0.15-0.35	bal Fe				

Specification	Designation	Notes	C	Cr	Cu	Mn	Mo	P	S	Si	Other	UTS	YS	El	Hard

Alloy Steel, Molybdenum/Molybdenum Sulfide, 4028/4028H (Continued from previous page)

USA

Specification	Designation	Notes	C	Cr	Cu	Mn	Mo	P	S	Si	Other	UTS	YS	El	Hard
ASTM A29/A29M(93)	4028	Bar	0.25-0.30			0.70-0.90	0.20-0.30	0.035 max	0.035-0.050	0.15-0.35	bal Fe				
ASTM A304(96)	4028H	Bar, hard bands spec	0.24-0.30	0.20 max	0.35 max	0.60-1.00	0.20-0.30	0.035 max	0.050 max	0.15-0.35	Cu Ni Cr Mo trace allowed; bal Fe				
ASTM A322(96)	4028	Bar	0.25-0.30			0.70-0.90	0.20-0.30	0.035 max	0.035-0.050	0.15-0.35	bal Fe				
ASTM A331(95)	4028	Bar	0.25-0.30			0.70-0.90	0.20-0.30	0.035 max	0.035-0.050	0.15-0.35	bal Fe				
ASTM A331(95)	4028H	Bar	0.24-0.30	0.20 max	0.35 max	0.60-1.00	0.20-0.30	0.025 max	0.025 max	0.15-0.35	bal Fe				
ASTM A519(96)	4028	Smls mech tub	0.25-0.30			0.70-0.90	0.20-0.30	0.040 max	0.035-0.050	0.15-0.35	bal Fe				
ASTM A752(93)	4028	Rod Wir	0.25-0.30			0.70-0.90	0.20-0.30	0.035 max	0.035-0.050	0.15-0.30	bal Fe	570			
SAE 770(84)	4028*	Obs; see J1397(92)	0.25-0.3			0.7-0.9	0.2-0.3	0.04 max	0.04-0.05	0.15-0.3	bal Fe				
SAE J1268(95)	4028H	Bar; See std	0.24-0.30	0.20 max	0.35 max	0.60-1.00	0.20-0.30	0.030 max	0.035-0.050	0.15-0.35	Ni <=0.25; Cu, Ni, Cr not spec'd but acpt; bal Fe				

Alloy Steel, Molybdenum/Molybdenum Sulfide, 4032/4032H

Germany

Specification	Designation	Notes	C	Cr	Cu	Mn	Mo	P	S	Si	Other	UTS	YS	El	Hard
DIN	WNr 1.5411*	Obs	0.32-0.38			1.1-1.4	0.15-0.25	0.035	0.035	0.3-0.5	Ni 0.3; bal Fe				

India

Specification	Designation	Notes	C	Cr	Cu	Mn	Mo	P	S	Si	Other	UTS	YS	El	Hard
IS 5517	35Mn2Mo28	Bar, Norm, Ann, Hard Tmp, 30mm	0.3-0.4			1.3-1.8	0.2-0.35	0.05	0.05	0.1-0.35	bal Fe				
IS G18	EN16		0.3-0.4			1.3-1.8	0.2-0.35	0.05	0.05	0.1-0.35	bal Fe				
IS G18	EN16D		0.3-0.4			1.3-1.8	0.2-0.35	0.05	0.05	0.1-0.35	bal Fe				

Mexico

Specification	Designation	Notes	C	Cr	Cu	Mn	Mo	P	S	Si	Other	UTS	YS	El	Hard
NMX-B-268(68)	4032H	Hard	0.29-0.35			0.60-1.00	0.20-0.30			0.20-0.35	bal Fe				
NMX-B-300(91)	4032	Bar	0.30-0.35			0.70-0.90	0.20-0.30	0.035 max	0.040 max	0.15-0.35	bal Fe				
NMX-B-300(91)	4032H	Bar	0.29-0.35			0.60-1.00	0.20-0.30			0.15-0.35	bal Fe				

Russia

Specification	Designation	Notes	C	Cr	Cu	Mn	Mo	P	S	Si	Other	UTS	YS	El	Hard
GOST 21357	35GMLS		0.3-0.4	0.3 max	0.3 max	1.2-1.5	0.15-0.25	0.02	0.02	0.2-0.4	Ni 0.3; bal Fe				
GOST 21357	35MLS		0.32-0.4	0.3 max	0.3 max	0.4-0.9	0.1-0.2	0.02	0.02	0.2-0.4	Ni 0.3; bal Fe				

Spain

Specification	Designation	Notes	C	Cr	Cu	Mn	Mo	P	S	Si	Other	UTS	YS	El	Hard
UNE	F.130.B		0.32-0.38			1.3-1.7	0.2-0.4	0.035	0.035	0.15-0.38	bal Fe				

UK

Specification	Designation	Notes	C	Cr	Cu	Mn	Mo	P	S	Si	Other	UTS	YS	El	Hard
BS 970/1(83)	605A32	Blm Bil Slab Bar Rod Frg	0.30-0.35			1.30-1.70	0.22-0.32	0.035 max	0.040 max		bal Fe				
BS 970/1(83)	605H32	Blm Bil Slab Bar Rod Frg	0.29-0.35			1.25-1.75	0.22-0.32	0.035 max	0.040 max		bal Fe				
BS 970/1(83)	605M36	Blm Bil Slab Bar Rod Frg	0.32-0.40			1.30-1.70	0.22-0.32	0.035 max	0.040 max		bal Fe				

USA

Specification	Designation	Notes	C	Cr	Cu	Mn	Mo	P	S	Si	Other	UTS	YS	El	Hard
	AISI 4032		0.3-0.35			0.7-0.9	0.2-0.3	0.035 max	0.04 max	0.2-0.35	bal Fe				
	AISI 4032H	Smls mech tub, Hard	0.29-0.35			0.60-1.00	0.20-0.30	0.035 max	0.040 max	0.15-0.30	bal Fe				
	UNS G40320		0.30-0.35			0.70-0.90	0.20-0.30	0.035 max	0.040 max	0.15-0.35	bal Fe				
	UNS H40320		0.29-0.35			0.60-1.00	0.20-0.30			0.15-0.35	bal Fe				
ASTM A29/A29M(93)	4032	Bar	0.30-0.35			0.70-0.90	0.20-0.30	0.035 max	0.040 max	0.15-0.35	bal Fe				
ASTM A304(96)	4032H	Bar, hard bands spec	0.29-0.35	0.20 max	0.35 max	0.60-1.00	0.20-0.30	0.035 max	0.400 max	0.15-0.35	Cu Ni Cr Mo trace allowed; bal Fe				

Specification	Designation	Notes	C	Cr	Cu	Mn	Mo	P	S	Si	Other	UTS	YS	El	Hard

Alloy Steel, Molybdenum/Molybdenum Sulfide, 4032/4032H (Continued from previous page)

USA

Specification	Designation	Notes	C	Cr	Cu	Mn	Mo	P	S	Si	Other	UTS	YS	El	Hard
ASTM A322(96)	4032*	Bar	0.20-0.25			0.70-0.90	0.20-0.30	0.035 max	0.040 max	0.15-0.35	bal Fe				
ASTM A331(95)	4032	Bar	0.20-0.25			0.70-0.90	0.20-0.30	0.035 max	0.040 max	0.15-0.35	bal Fe				
ASTM A331(95)	4032H	Bar	0.29-0.35	0.20 max	0.35 max	0.60-1.00	0.20-0.30	0.025 max	0.025 max	0.15-0.35	bal Fe				
FED QQ-S-00629(63)	FS4032*	Obs; Tube smls weld									bal Fe				
SAE 770(84)	4032*	Obs; see J1397(92)									bal Fe				
SAE J1268(95)	4032H	See std	0.29-0.35	0.20 max	0.35 max	0.60-1.00	0.20-0.30	0.030 max	0.040 max	0.15-0.35	Ni <=0.25; Cu, Ni, Cr not spec'd but acpt; bal Fe				
SAE J407	4032H*	Obs; see J1268									bal Fe				

Alloy Steel, Molybdenum/Molybdenum Sulfide, 4037/4037H

Australia

Specification	Designation	Notes	C	Cr	Cu	Mn	Mo	P	S	Si	Other	UTS	YS	El	Hard
AS 1444(96)	4037	H series, Hard/Tmp	0.35-0.40			0.70-0.90	0.20-0.30	0.035 max	0.040 max	0.15-0.35	bal Fe				
AS G18	En16C*	Obs; see AS 1444	0.35-0.4			0.7-0.9	0.2-0.3	0.04	0.04	0.1-0.35	bal Fe				

Germany

Specification	Designation	Notes	C	Cr	Cu	Mn	Mo	P	S	Si	Other	UTS	YS	El	Hard
DIN	WNr 1.5412*	Obs									bal Fe				

India

Specification	Designation	Notes	C	Cr	Cu	Mn	Mo	P	S	Si	Other	UTS	YS	El	Hard
IS 5517	35Mn2Mo28		0.3-0.4			1.3-1.8	0.2-0.35	0.05	0.05	0.1-0.35	bal Fe				
IS G18	EN16		0.3-0.4			1.3-1.8	0.2-0.35	0.05	0.05	0.1-0.35	bal Fe				
IS G18	EN16D		0.3-0.4			1.3-1.8	0.2-0.35	0.05	0.05	0.1-0.35	bal Fe				

Mexico

Specification	Designation	Notes	C	Cr	Cu	Mn	Mo	P	S	Si	Other	UTS	YS	El	Hard
DGN B-203	4037*	Obs	0.35-0.4			0.7-0.9	0.2-0.3	0.04	0.04	0.2-0.35	bal Fe				
DGN B-297	4037*	Obs	0.35-0.4			0.7-0.9	0.2-0.3	0.035	0.04	0.2-0.35	bal Fe				
NMX-B-268(68)	4037H	Hard	0.34-0.41			0.60-1.00	0.20-0.30			0.20-0.35	bal Fe				
NMX-B-268(68)	4137H	Hard	0.34-0.41	0.75-1.20		0.60-1.00	0.15-0.25			0.20-0.35	bal Fe				
NMX-B-300(91)	4037	Bar	0.35-0.40			0.70-0.90	0.20-0.30	0.035 max	0.040 max	0.15-0.35	bal Fe				
NMX-B-300(91)	4037H	Bar	0.34-0.41			0.60-1.00	0.20-0.30			0.15-0.35	bal Fe				

Pan America

Specification	Designation	Notes	C	Cr	Cu	Mn	Mo	P	S	Si	Other	UTS	YS	El	Hard
COPANT 334	4037	Bar	0.35-0.4			0.7-0.9	0.2-0.3	0.04	0.04	0.2-0.35	bal Fe				

Russia

Specification	Designation	Notes	C	Cr	Cu	Mn	Mo	P	S	Si	Other	UTS	YS	El	Hard
GOST 21357	35MLS		0.32-0.4	0.3 max	0.3 max	0.4-0.9	0.1-0.2	0.02	0.02	0.2-0.4	Ni 0.3; bal Fe				

Spain

Specification	Designation	Notes	C	Cr	Cu	Mn	Mo	P	S	Si	Other	UTS	YS	El	Hard
UNE	F.120.I		0.34-0.4			0.65-0.9	0.2-0.3	0.035	0.035	0.13-0.38	bal Fe				

UK

Specification	Designation	Notes	C	Cr	Cu	Mn	Mo	P	S	Si	Other	UTS	YS	El	Hard
BS 3111/1(87)	2/1	Q/T Wir	0.35-0.40			0.70-0.90	0.20-0.30	0.035 max	0.035 max	0.15-0.40	bal Fe				
BS 3111/1(87)	2/2	Q/T Wir	0.40-0.45			0.80-1.00	0.25-0.35	0.035 max	0.035 max	0.15-0.40	bal Fe				
BS 970/1(83)	605A37	Blm Bil Slab Bar Rod Frg	0.35-0.40			1.30-1.70	0.22-0.32	0.035 max	0.040 max		bal Fe				
BS 970/1(83)	605H37	Blm Bil Slab Bar Rod Frg	0.34-0.41			1.25-1.75	0.22-0.32	0.035 max	0.040 max		bal Fe				

USA

Specification	Designation	Notes	C	Cr	Cu	Mn	Mo	P	S	Si	Other	UTS	YS	El	Hard
	AISI 4037	Smls mech tub	0.35-0.40			0.70-0.90	0.20-0.30	0.040 max	0.040 max	0.15-0.35	bal Fe				
	AISI 4037	Bar Blm Bil Slab	0.35-0.40			0.70-0.90	0.20-0.30	0.035 max	0.040 max	0.15-0.35	bal Fe				

UNS numbers and US grades are provided as a means of cross referencing chemically similar alloys. Exchangability is only possible after independent examination of specifications. Tensile properties are minimum or typical as specified. UTS and YS as MPa. El as %. See Appendix for list of abbreviations used in Notes. * indicates obsolete material.

Specification	Designation	Notes	C	Cr	Cu	Mn	Mo	P	S	Si	Other	UTS	YS	El	Hard

Alloy Steel, Molybdenum/Molybdenum Sulfide, 4037/4037H (Continued from previous page)

USA

Specification	Designation	Notes	C	Cr	Cu	Mn	Mo	P	S	Si	Other	UTS	YS	El	Hard
	AISI 4037H	Bar Blm Bil Slab	0.34-0.41			0.60-1.00	0.20-0.30	0.035 max	0.040 max	0.15-0.35	bal Fe				
	AISI 4037H	Smls mech tub, Hard	0.34-0.41			0.60-1.00	0.20-0.30	0.040 max	0.040 max	0.15-0.30	bal Fe				
	UNS G40370		0.35-0.40			0.70-0.90	0.20-0.30	0.035 max	0.040 max	0.15-0.35	bal Fe				
	UNS H40370		0.34-0.41			0.60-1.00	0.20-0.30	0.035 max	0.040 max	0.15-0.35	bal Fe				
AMS 6300	4037	Bar Frg	0.35-0.4	0.2	0.35	0.7-0.90	0.2-0.30	0.040	0.040	0.15-0.35	Ni 0.25; bal Fe				
ASTM A29/A29M(93)	4037	Bar	0.35-0.40			0.70-0.90	0.20-0.30	0.035 max	0.040 max	0.15-0.35	bal Fe				
ASTM A304(96)	4037H	Bar, hard bands spec	0.34-0.41	0.20 max	0.35 max	0.60-1.00	0.20-0.30	0.035 max	0.040 max	0.15-0.35	Cu Ni Cr Mo trace allowed; bal Fe				
ASTM A322(96)	4037	Bar	0.35-0.40			0.70-0.90	0.20-0.30	0.035 max	0.040 max	0.15-0.35	bal Fe				
ASTM A331(95)	4037	Bar	0.35-0.40			0.70-0.90	0.20-0.30	0.035 max	0.040 max	0.15-0.35	bal Fe				
ASTM A331(95)	4037H	Bar	0.34-0.41	0.20 max	0.35 max	0.60-1.00	0.20-0.30	0.025 max	0.025 max	0.15-0.35	bal Fe				
ASTM A519(96)	4037	Smls mech tub	0.35-0.40			0.70-0.90	0.20-0.30	0.040 max	0.040 max	0.15-0.35	bal Fe				
ASTM A752(93)	4037	Rod Wir	0.35-0.40			0.70-0.90	0.20-0.30	0.035 max	0.040 max	0.15-0.30	bal Fe	610			
SAE 770(84)	4037*	Obs; see J1397(92)	0.35-0.4			0.7-0.9	0.2-0.3	0.04 max	0.04 max	0.15-0.3	bal Fe				
SAE J1268(95)	4037H	Bar Wir; See std	0.34-0.41	0.20 max	0.35 max	0.60-1.00	0.20-0.30	0.030 max	0.040 max	0.15-0.35	Ni <=0.25; Cu, Ni, Cr not spec'd but acpt; bal Fe				
SAE J404(94)	4037	Bil Blm Slab Bar HR CF	0.35-0.40			0.70-0.90	0.20-0.30	0.030 max	0.040 max	0.15-0.35	bal Fe				

Alloy Steel, Molybdenum/Molybdenum Sulfide, 4042/4042H

Germany

Specification	Designation	Notes	C	Cr	Cu	Mn	Mo	P	S	Si	Other	UTS	YS	El	Hard
DIN	42MnMo7		0.38-0.45			1.55-1.85	0.15-0.25	0.040 max	0.040 max	0.20-0.35	bal Fe				
DIN	WNr 1.5432		0.38-0.45			1.55-1.85	0.15-0.25	0.040 max	0.040 max	0.20-0.35	bal Fe				

Mexico

Specification	Designation	Notes	C	Cr	Cu	Mn	Mo	P	S	Si	Other	UTS	YS	El	Hard
DGN B-203	4042*	Obs	0.4-0.45			0.7-0.9	0.2-0.3	0.04	0.04	0.2-0.35	bal Fe				
NMX-B-268(68)	4042H	Hard	0.39-0.46			0.60-1.00	0.20-0.30			0.20-0.35	bal Fe				
NMX-B-300(91)	4042	Bar	0.40-0.45			0.70-0.90	0.20-0.30	0.035 max	0.040 max	0.15-0.35	bal Fe				
NMX-B-300(91)	4042H	Bar	0.39-0.46			0.60-1.00	0.20-0.30			0.15-0.35	bal Fe				

Pan America

Specification	Designation	Notes	C	Cr	Cu	Mn	Mo	P	S	Si	Other	UTS	YS	El	Hard
COPANT 514	4042	Tube	0.4-0.45			0.7-0.9	0.2-0.3	0.04	0.04	0.2-0.35	bal Fe				

USA

Specification	Designation	Notes	C	Cr	Cu	Mn	Mo	P	S	Si	Other	UTS	YS	El	Hard
	AISI 4042	Smls mech tub	0.40-0.45			0.70-0.90	0.20-0.30	0.040 max	0.040 max	0.15-0.35	bal Fe				
	AISI 4042		0.40-0.45			0.70-0.90	0.20-0.30	0.035 max	0.040 max	0.15-0.35	bal Fe				
	AISI 4042H	Smls mech tub, Hard	0.39-0.46			0.60-1.00	0.20-0.30	0.035 max	0.040 max	0.15-0.30	bal Fe				
	UNS G40420		0.40-0.45			0.70-0.90	0.20-0.30	0.035 max	0.040 max	0.15-0.30	bal Fe				
	UNS H40420		0.39-0.46			0.60-1.00	0.20-0.30			0.15-0.35	bal Fe				
ASTM A29/A29M(93)	4042	Bar	0.40-0.45			0.70-0.90	0.20-0.30	0.035 max	0.040 max	0.15-0.35	bal Fe				
ASTM A304(96)	4042H	Bar, hard bands spec	0.39-0.46	0.20 max	0.35 max	0.60-1.00	0.20-0.30	0.035 max	0.040 max	0.15-0.35	Cu Ni Cr Mo trace allowed; bal Fe				
ASTM A322(96)	4042*	Bar	0.40-0.45			0.70-0.90	0.20-0.30	0.035 max	0.040 max	0.15-0.35	bal Fe				

Specification	Designation	Notes	C	Cr	Cu	Mn	Mo	P	S	Si	Other	UTS	YS	El	Hard

Alloy Steel, Molybdenum/Molybdenum Sulfide, 4042/4042H (Continued from previous page)

USA

Specification	Designation	Notes	C	Cr	Cu	Mn	Mo	P	S	Si	Other	UTS	YS	El	Hard
ASTM A331(95)	4042	Bar	0.40-0.45			0.70-0.90	0.20-0.30	0.035 max	0.040 max	0.15-0.35	bal Fe				
ASTM A331(95)	4042H	Bar	0.39-0.46	0.20 max	0.35 max	0.60-1.00	0.20-0.30	0.025 max	0.025 max	0.15-0.35	bal Fe				
ASTM A519(96)	4042	Smls mech tub	0.40-0.45			0.70-0.90	0.20-0.30	0.040 max	0.040 max	0.15-0.35	bal Fe				
SAE 770(84)	4042*	Obs; see J1397(92)	0.4-0.45			0.7-0.9	0.2-0.3	0.04 max	0.04 max	0.15-0.3	bal Fe				
SAE J1268(95)	4042H	Bar; See std	0.39-0.46	0.20 max	0.35 max	0.60-1.00	0.20-0.30	0.030 max	0.040 max	0.15-0.35	Ni <=0.25; Cu, Ni, Cr not spec'd but acpt; bal Fe				

Alloy Steel, Molybdenum/Molybdenum Sulfide, 4047/4047H

Mexico

Specification	Designation	Notes	C	Cr	Cu	Mn	Mo	P	S	Si	Other	UTS	YS	El	Hard
DGN B-203	4047*	Obs	0.45-0.5			0.7-0.9	0.2-0.3	0.04	0.04	0.2-0.35	bal Fe				
DGN B-297	4047*	Obs	0.45-0.5			0.7-0.9	0.2-0.3	0.035	0.04	0.2-0.35	bal Fe				
NMX-B-268(68)	4047H	Hard	0.44-0.51			0.60-1.00	0.20-0.30			0.20-0.35	bal Fe				
NMX-B-300(91)	4047	Bar	0.45-0.50			0.70-0.90	0.20-0.30	0.035 max	0.040 max	0.15-0.35	bal Fe				
NMX-B-300(91)	4047H	Bar	0.44-0.51			0.60-1.00	0.20-0.30			0.15-0.35	bal Fe				

Pan America

Specification	Designation	Notes	C	Cr	Cu	Mn	Mo	P	S	Si	Other	UTS	YS	El	Hard
COPANT 514	4047	Tube	0.45-0.5			0.7-0.9	0.2-0.3	0.04	0.04	0.2-0.35	bal Fe				

USA

Specification	Designation	Notes	C	Cr	Cu	Mn	Mo	P	S	Si	Other	UTS	YS	El	Hard
	AISI 4047	Bar Blm Bil Slab	0.45-0.50			0.70-0.90	0.20-0.30	0.035 max	0.040 max	0.15-0.35	bal Fe				
	AISI 4047	Smls mech tub	0.45-0.50			0.70-0.90	0.20-0.30	0.040 max	0.040 max	0.15-0.35	bal Fe				
	AISI 4047H	Bar Blm Bil Slab	0.44-0.51			0.60-1.00	0.20-0.30	0.035 max	0.040 max	0.15-0.35	bal Fe				
	AISI 4047H	Smls mech tub, Hard	0.44-0.51			0.60-1.00	0.20-0.30	0.040 max	0.040 max	0.15-0.30	bal Fe				
	UNS G40470		0.45-0.50			0.70-0.90	0.20-0.30	0.035 max	0.040 max	0.15-0.35	bal Fe				
	UNS H40470		0.44-0.51			0.60-1.00	0.20-0.30	0.035 max	0.040 max	0.15-0.35	bal Fe				
ASTM A29/A29M(93)	4047	Bar	0.45-0.50			0.70-0.90	0.20-0.30	0.035 max	0.040 max	0.15-0.35	bal Fe				
ASTM A304(96)	4047H	Bar, hard bands spec	0.44-0.51	0.20 max	0.35 max	0.60-1.00	0.20-0.30	0.035 max	0.040 max	0.15-0.35	Cu Ni Cr Mo trace allowed; bal Fe				
ASTM A322(96)	4047	Bar	0.45-0.50			0.70-0.90	0.20-0.30	0.035 max	0.040 max	0.15-0.35	bal Fe				
ASTM A331(95)	4047	Bar	0.45-0.50			0.70-0.90	0.20-0.30	0.035 max	0.040 max	0.15-0.35	bal Fe				
ASTM A331(95)	4047H	Bar	0.44-0.51	0.20 max	0.35 max	0.60-1.00	0.20-0.30	0.025 max	0.025 max	0.15-0.35	bal Fe				
ASTM A519(96)	4047	Smls mech tub	0.45-0.50			0.70-0.90	0.20-0.30	0.040 max	0.040 max	0.15-0.35	bal Fe				
ASTM A752(93)	4047	Rod Wir	0.45-0.50			0.70-0.90	0.20-0.30	0.035 max	0.040 max	0.15-0.30	bal Fe	670			
SAE 770(84)	4047*	Obs; see J1397(92)	0.45-0.5			0.7-0.9	0.2-0.3	0.04 max	0.04 max	0.15-0.3	bal Fe				
SAE J1268(95)	4047H	Bar; See std	0.44-0.51	0.20 max	0.35 max	0.60-1.00	0.20-0.30	0.030 max	0.040 max	0.15-0.35	Ni <=0.25; Cu, Ni, Cr not spec'd but acpt; bal Fe				
SAE J404(94)	4047	Bil Blm Slab Bar HR CF	0.45-0.50			0.70-0.90	0.20-0.30	0.030 max	0.040 max	0.15-0.35	bal Fe				

Alloy Steel, Molybdenum/Molybdenum Sulfide, 4063

USA

Specification	Designation	Notes	C	Cr	Cu	Mn	Mo	P	S	Si	Other	UTS	YS	El	Hard
	AISI 4063	Smls mech tub	0.60-0.67			0.75-1.00	0.20-0.30	0.040 max	0.040 max	0.15-0.35	bal Fe				

UNS numbers and US grades are provided as a means of cross referencing chemically similar alloys. Exchangability is only possible after independent examination of specifications. Tensile properties are minimum or typical as specified. UTS and YS as MPa. El as %. See Appendix for list of abbreviations used in Notes. * indicates obsolete material.

Specification	Designation	Notes	C	Cr	Cu	Mn	Mo	P	S	Si	Other	UTS	YS	El	Hard
Alloy Steel, Molybdenum/Molybdenum Sulfide, 4063 (Continued from previous page)															
USA															
	UNS G40630		0.60-0.67			0.75-1.00	0.20-0.30	0.035 max	0.040 max	0.15-0.35	bal Fe				
ASTM A519(96)	4063	Smls mech tub	0.60-0.67			0.75-1.00	0.20-0.30	0.040 max	0.040 max	0.15-0.35	bal Fe				
Alloy Steel, Molybdenum/Molybdenum Sulfide, 4419/4419H															
Mexico															
NMX-B-268(68)	4419H	Hard	0.17-0.23			0.35-0.75	0.45-0.60			0.20-0.35	bal Fe				
NMX-B-300(91)	4419	Bar	0.18-0.23			0.45-0.65	0.45-0.60	0.035 max	0.040 max	0.15-0.35	bal Fe				
NMX-B-300(91)	4419H	Bar	0.17-0.23			0.35-0.75	0.45-0.60			0.15-0.35	bal Fe				
Sweden															
SS 142512	2512	CH	0.18-0.23	0.6-1		0.7-1.1	0.1 max	0.035 max	0.03-0.05	0.15-0.4	Ni 0.8-1.2; bal Fe				
UK															
BS 1503(89)	243-430	Frg press ves; t<=200mm; Norm Q/T	0.12-0.20	0.25 max	0.30 max	0.50-0.80	0.25-0.35	0.030 max	0.025 max	0.15-0.40	Al <=0.012; Ni <=0.40; W <=0.1; bal Fe	430-580	250	19	
USA															
	UNS G44190		0.18-0.23			0.45-0.65	0.45-0.60	0.035 max	0.040 max	0.15-0.35	bal Fe				
	UNS H44190		0.17-0.23			0.35-0.75	0.45-0.60	0.035 max	0.040 max	0.15-0.35	bal Fe				
ASTM A29/A29M(93)	4419	Bar	0.18-0.23			0.45-0.65	0.45-0.60	0.035 max	0.040 max	0.15-0.35	bal Fe				
ASTM A752(93)	4419	Rod Wir	0.18-0.23			0.45-0.65	0.45-0.60	0.035 max	0.040 max	0.15-0.30	bal Fe	570			
Alloy Steel, Molybdenum/Molybdenum Sulfide, 4422															
Mexico															
NMX-B-300(91)	4422	Bar	0.20-0.25			0.70-0.90	0.35-0.45	0.035 max	0.040 max	0.15-0.35	bal Fe				
USA															
	AISI 4422	Smls mech tub	0.20-0.25			0.70-0.90	0.35-0.45	0.040 max	0.040 max	0.15-0.35	bal Fe				
	UNS G44220		0.20-0.25			0.70-0.90	0.35-0.45	0.035 max	0.040 max	0.15-0.35	bal Fe				
ASTM A29/A29M(93)	4422	Bar	0.20-0.25			0.70-0.90	0.35-0.45	0.035 max	0.040 max	0.15-0.35	bal Fe				
ASTM A519(96)	4422	Smls mech tub	0.20-0.25			0.70-0.90	0.35-0.45	0.040 max	0.040 max	0.15-0.35	bal Fe				
Alloy Steel, Molybdenum/Molybdenum Sulfide, 4427															
Mexico															
NMX-B-300(91)	4427	Bar	0.24-0.29			0.70-0.90	0.35-0.45	0.035 max	0.040 max	0.15-0.35	bal Fe				
Russia															
GOST 1414(75)	AS30ChM	Free-cutting	0.27-0.33	0.8-1.1	0.30 max	0.4-0.7	0.15-0.25	0.035 max	0.035 max	0.17-0.37	Ni <=0.3; Pb 0.15-0.3; P+S<=0.06; bal Fe				
USA															
	AISI 4427	Smls mech tub	0.24-0.29			0.70-0.90	0.35-0.45	0.040 max	0.040 max	0.15-0.35	bal Fe				
	UNS G44270		0.24-0.29			0.70-0.90	0.35-0.45	0.035 max	0.040 max	0.15-0.35	bal Fe				
ASTM A29/A29M(93)	4427	Bar	0.24-0.29			0.70-0.90	0.35-0.45	0.035 max	0.040 max	0.15-0.35	bal Fe				
ASTM A519(96)	4427	Smls mech tub	0.24-0.29			0.70-0.90	0.35-0.45	0.040 max	0.040 max	0.15-0.35	bal Fe				
Alloy Steel, Molybdenum/Molybdenum Sulfide, 4520															
Italy															
UNI 5462(64)	16Mo5	Smls tube	0.12-0.2			0.5-0.8	0.45-0.65	0.035 max	0.035 max	0.15-0.35	bal Fe				

UNS numbers and US grades are provided as a means of cross referencing chemically similar alloys. Exchangability is only possible after independent examination of specifications. Tensile properties are minimum or typical as specified. UTS and YS as MPa. El as %. See Appendix for list of abbreviations used in Notes. * indicates obsolete material.

Specification	Designation	Notes	C	Cr	Cu	Mn	Mo	P	S	Si	Other	UTS	YS	El	Hard
Alloy Steel, Molybdenum/Molybdenum Sulfide, 4520 (Continued from previous page)															
Romania															
STAS 2883/3(88)	16Mo5	Heat res	0.12-0.2	0.3 max		0.5-0.8	0.45-0.65	0.035 max	0.03 max	0.15-0.35	Al 0.01-0.110; N <=0.009; Pb <=0.15; Cr+Ni+Cu<=0.7; As<=0.07; bal Fe				
Spain															
UNE 36087(76/78)	16Mo5	Sh Plt Strp CR	0.12-0.2			0.5-0.8	0.45-0.65	0.035 max	0.035 max	0.15-0.35	bal Fe				
UNE 36087(76/78)	F.2602*	see 16Mo5	0.12-0.2			0.5-0.8	0.45-0.65	0.035 max	0.035 max	0.15-0.35	bal Fe				
USA															
	AISI 4520	Smls mech tub	0.18-0.23			0.45-0.65	0.45-0.60	0.040 max	0.040 max	0.15-0.35	bal Fe				
	UNS G45200		0.18-0.23			0.45-0.65	0.45-0.60	0.035 max	0.040 max	0.15-0.35	bal Fe				
ASTM A519(96)	4520	Smls mech tub	0.18-0.23			0.45-0.65	0.45-0.60	0.040 max	0.040 max	0.15-0.35	bal Fe				
Alloy Steel, Molybdenum/Molybdenum Sulfide, A182(F1)															
USA															
	UNS K12822		0.28 max			0.60-0.90	0.44-0.65	0.045 max	0.045 max	0.15-0.35	bal Fe				
ASTM A182/A182M(98)	F1	Frg/roll pipe flange valve	0.28 max			0.60-0.90	0.44-0.65	0.045 max	0.045 max	0.15-0.35	bal Fe	485	275	20.0	143-192 HB
Alloy Steel, Molybdenum/Molybdenum Sulfide, A194(4)															
USA															
	UNS K14510		0.40-0.50			0.70-0.90	0.20-0.30	0.035 max	0.040 max	0.15-0.35	bal Fe				
ASTM A194/A194M(98)	4	Nuts, high-temp press	0.40-0.50			0.70-0.90	0.20-0.30	0.035 max	0.040 max	0.15-0.35	bal Fe				248-352 HB
Alloy Steel, Molybdenum/Molybdenum Sulfide, A204(B)															
UK															
BS 3059/2(90)	243	Boiler, Superheater; High-temp; Smls Tub; Tub, Weld; 2-12.5mm	0.12-0.20		0.25 max	0.40-0.80	0.25-0.35	0.035 max	0.035 max	0.10-0.35	Al <=0.012; Ni <=0.30; Sn<=0.03; bal Fe	480-630	275	22	
USA															
	UNS K12020		0.20 max			0.90 max	0.41-0.64	0.035 max	0.040 max	0.13-0.45	bal Fe				
ASTM A204/A204M(95)	A204(B)	Press ves plt, t<=25mm	0.20 max			0.90 max	0.45-0.60	0.035 max	0.035 max	0.15-0.40	bal Fe	70-90	40	21	
ASTM A204/A204M(95)	A204(B)	Press ves plt, t 25-50mm	0.23 max			0.90 max	0.45-0.60	0.035 max	0.035 max	0.15-0.40	bal Fe	70-90	40	21	
ASTM A204/A204M(95)	A204(B)	Press ves plt, t>100mm	0.27 max			0.90 max	0.45-0.60	0.035 max	0.035 max	0.15-0.40	bal Fe	70-90	40	21	
ASTM A204/A204M(95)	A204(B)	Press ves plt, t 50-100mm	0.25 max			0.90 max	0.45-0.60	0.035 max	0.035 max	0.15-0.40	bal Fe	70-90	40	21	
Alloy Steel, Molybdenum/Molybdenum Sulfide, A204(C)															
USA															
	UNS K12320		0.23 max			0.90 max	0.41-0.64	0.035 max	0.040 max	0.13-0.45	bal Fe				
ASTM A204/A204M(95)	A204(C)	Press ves plt, t 25-50mm	0.26 max			0.90 max	0.45-0.60	0.035 max	0.035 max	0.15-0.40	bal Fe	75-95	43	20	
ASTM A204/A204M(95)	A204(C)	Press ves plt, t<25mm	0.23 max			0.90 max	0.45-0.60	0.035 max	0.035 max	0.15-0.40	bal Fe	75-95	43	20	
ASTM A204/A204M(95)	A204(C)	Press ves plt, t>50mm	0.28 max			0.90 max	0.45-0.60	0.035 max	0.035 max	0.15-0.40	bal Fe	75-95	43	20	
ASTM A225A225M(93)	C	Press ves plt	0.25 max			1.60 max		0.035 max	0.035 max	0.15-0.40	Ni 0.40-0.70; V 0.13-0.18; bal Fe	105-135	70	20	
Alloy Steel, Molybdenum/Molybdenum Sulfide, A209(T1a)															
USA															
	UNS K12023		0.15-0.25			0.30-0.80	0.44-0.65	0.045 max	0.045 max	0.10-0.50	bal Fe				

UNS numbers and US grades are provided as a means of cross referencing chemically similar alloys. Exchangability is only possible after independent examination of specifications. Tensile properties are minimum or typical as specified. UTS and YS as MPa. El as %. See Appendix for list of abbreviations used in Notes. * indicates obsolete material.

Specification	Designation	Notes	C	Cr	Cu	Mn	Mo	P	S	Si	Other	UTS	YS	El	Hard

Alloy Steel, Molybdenum/Molybdenum Sulfide, A209(T1a) (Continued from previous page)

USA

Specification	Designation	Notes	C	Cr	Cu	Mn	Mo	P	S	Si	Other	UTS	YS	El	Hard
ASTM A209/A209M(95)	T1a	Smls Boiler/Super Heater tubs, H for >5.1mm Wall	0.15-0.25			0.30-0.80	0.44-0.65	0.025 max	0.025 max	0.10-0.50	bal Fe	365	195	30	153 HB

Alloy Steel, Molybdenum/Molybdenum Sulfide, A209(T1b)

USA

Specification	Designation	Notes	C	Cr	Cu	Mn	Mo	P	S	Si	Other	UTS	YS	El	Hard
	UNS K11422		0.14 max			0.30-0.80	0.44-0.65	0.045 max	0.045 max	0.10-0.50	bal Fe				
ASTM A209/A209M(95)	T1b	Smls Boiler/Super Heater tubs, H for >5.1mm Wall	0.14 max			0.30-0.80	0.44-0.65	0.025 max	0.025 max	0.1-0.50	bal Fe	415	220	30	137 HB

Alloy Steel, Molybdenum/Molybdenum Sulfide, A234(WP1)

USA

Specification	Designation	Notes	C	Cr	Cu	Mn	Mo	P	S	Si	Other	UTS	YS	El	Hard
	UNS K12821		0.28 max			0.30-0.90	0.44-0.65	0.045 max	0.045 max	0.10-0.50	bal Fe				
ASTM A234/A234M(97)	WP1	Frg	0.28 max			0.30-0.90	0.44-0.65	0.045 max	0.045 max	0.10-0.50	bal Fe	380-550	205	22L	

Alloy Steel, Molybdenum/Molybdenum Sulfide, A302(A)

USA

Specification	Designation	Notes	C	Cr	Cu	Mn	Mo	P	S	Si	Other	UTS	YS	El	Hard
	UNS K12021		0.20 max			0.87-1.41	0.41-0.64	0.035 max	0.040 max	0.13-0.45	bal Fe				
ASTM A302/302M(97)	A	Press ves plt, t<=25mm	0.20 max			0.95-1.30	0.45-0.60	0.035 max	0.035 max	0.15-0.40	bal Fe	515-655	310	19	
ASTM A302/302M(97)	A	Press ves plt, 25<t<=50mm	0.23 max			0.95-1.30	0.45-0.60	0.035 max	0.035 max	0.15-0.40	bal Fe	515-655	310	19	
ASTM A302/302M(97)	A	Press ves plt, t>50mm	0.25 max			0.95-1.30	0.45-0.60	0.035 max	0.035 max	0.15-0.40	bal Fe	515-655	310	19	

Alloy Steel, Molybdenum/Molybdenum Sulfide, A302(B)

USA

Specification	Designation	Notes	C	Cr	Cu	Mn	Mo	P	S	Si	Other	UTS	YS	El	Hard
	UNS K12022		0.20 max			1.07-1.62	0.41-0.64	0.035 max	0.040 max	0.13-0.45	bal Fe				
ASTM A302/302M(97)	B	Press ves plt, t>50mm	0.25 max			1.15-1.50	0.45-0.60	0.035 max	0.035 max	0.15-0.40	bal Fe	550-690	345	18	
ASTM A302/302M(97)	B	Press ves plt, 25<t<=50mm	0.23 max			1.15-1.50	0.45-0.60	0.035 max	0.035 max	0.15-0.40	bal Fe	550-690	345	18	
ASTM A302/302M(97)	B	Press ves plt, t<=25mm	0.20 max			1.15-1.50	0.45-0.60	0.035 max	0.035 max	0.15-0.40	bal Fe	550-690	345	18	

Alloy Steel, Molybdenum/Molybdenum Sulfide, A336(F1)

USA

Specification	Designation	Notes	C	Cr	Cu	Mn	Mo	P	S	Si	Other	UTS	YS	El	Hard
	UNS K12520		0.20-0.30			0.60-0.80	0.40-0.60	0.040 max	0.040 max	0.20-0.35	bal Fe				
ASTM A336/A336M(98)	F1	Ferr Frg	0.20-0.30			0.60-0.80	0.40-0.60	0.025 max	0.025 max	0.20-0.35	bal Fe	485-660	275	20	

Alloy Steel, Molybdenum/Molybdenum Sulfide, A533(A)

USA

Specification	Designation	Notes	C	Cr	Cu	Mn	Mo	P	S	Si	Other	UTS	YS	El	Hard
	UNS K12521		0.25 max			1.07-1.62	0.41-0.64	0.035 max	0.040 max	0.13-0.45	bal Fe				
ASTM A533	Type A	Q/T, Press ves plt, Class 3, t=6.5-65mm	0.25 max			1.15-1.50	0.45-0.60	0.035 max	0.035 max	0.15-0.40	bal Fe	690-860	570	16	
ASTM A533	Type A	Q/T, Press ves plt, Class 2, t=6.5-300mm	0.25 max			1.15-1.50	0.45-0.60	0.035 max	0.035 max	0.15-0.40	bal Fe	620-795	485	16	
ASTM A533	Type A	Q/T, Press ves plt, Class 1, t=6.5-300mm	0.25 max			1.15-1.50	0.45-0.60	0.035 max	0.035 max	0.15-0.40	bal Fe	550-690	345	18	

Alloy Steel, Molybdenum/Molybdenum Sulfide, A579(33)

USA

Specification	Designation	Notes	C	Cr	Cu	Mn	Mo	P	S	Si	Other	UTS	YS	El	Hard
	UNS K14394		0.41-0.46			0.75-1.00	0.45-0.60	0.025 max	0.025 max	1.40-1.75	V 0.03-0.08; bal Fe				
ASTM A579(96)	33	Superstrength frg	0.41-0.46	1.90-2.25		0.75-1.00	0.45-0.60	0.025 max	0.025 max	1.40-1.75	V 0.03-0.08; bal Fe	1720	1550	6	302 HB

UNS numbers and US grades are provided as a means of cross referencing chemically similar alloys. Exchangability is only possible after independent examination of specifications. Tensile properties are minimum or typical as specified. UTS and YS as MPa. El as %. See Appendix for list of abbreviations used in Notes. * indicates obsolete material.

Specification	Designation	Notes	C	Cr	Cu	Mn	Mo	P	S	Si	Other	UTS	YS	El	Hard
Alloy Steel, Molybdenum/Molybdenum Sulfide, K12220															
USA															
	UNS K12220		0.22 max	0.25 max	0.25 max	0.80 max	0.40-0.60	0.04 max	0.04 max	0.15-0.40	Ni <=0.25; bal Fe				
Alloy Steel, Molybdenum/Molybdenum Sulfide, K14520															
USA															
	UNS K14520		0.45 max			0.50-0.90	0.30-0.60	0.040 max	0.040 max	0.15-0.45	V 0.10-0.25; bal Fe				
Alloy Steel, Molybdenum/Molybdenum Sulfide, No equivalents identified															
India															
IS 1570/4(88)	10Mo6		0.15 max	0.25 max		0.4-0.7	0.45-0.65	0.07 max	0.06 max	0.15-0.25	Ni <=0.3; bal Fe				
IS 1570/4(88)	33Mo6		0.25-0.4	0.25 max		0.4-0.7	0.45-0.65	0.07 max	0.06 max	0.1-0.35	Ni <=0.3; bal Fe				
International															
ISO 2604-1(75)	F26	Frg	0.12-0.20			0.50-0.80	0.25-0.35	0.040 max	0.040 max	0.15-0.40	Al <=0.012; bal Fe	440-590			
ISO 2604-1(75)	F27	Frg	0.18-0.25			0.50-0.80	0.25-0.35	0.040 max	0.040 max	0.15-0.40	Al <=0.012; bal Fe	440-590			
ISO 2604-1(75)	F28	Frg	0.12-0.20			0.50-0.80	0.45-0.65	0.040 max	0.040 max	0.15-0.40	Al <=0.012; bal Fe	450-600			
ISO 2604-1(75)	F29	Frg	0.16-0.25			0.50-0.80	0.45-0.65	0.040 max	0.040 max	0.15-0.40	Al <=0.012; bal Fe	450-600			
ISO 4954(93)	36Mo3E	Wir Rod Bar, CH ext	0.33-0.40			0.70-1.00	0.20-0.30	0.035 max	0.035 max	0.40 max	bal Fe	620			
Poland															
PNH84023/07	17G2MFA	Const	0.13-0.19	0.3 max	0.25 max	1.4-1.7	0.2-0.5	0.035 max	0.035 max	0.2-0.5	Ni <=0.4; V <=0.15; bal Fe				
PNH84024	20M	High-temp const	0.1 max	0.3 max	0.8-1.1	0.25-0.4	0.3 max	0.045 max	0.045 max	0.15-0.35	Al <=0.012; Co <=0.3; Ni <=0.045; bal Fe				
Alloy Steel, Molybdenum/Molybdenum Sulfide, T1															
Mexico															
NMX-B-142-SCFI(94)	Grade T1	Smls tube for refinery service	0.10-0.20			0.30-0.80	0.44-0.65	0.025 max	0.25 max	0.10-0.50	bal Fe	379	207	30	

Specification	Designation	Notes	C	Cr	Mn	Mo	Ni	P	S	Si	Other	UTS	YS	El	Hard

Alloy Steel, Nickel Chromium Molybdenum, 4320/4320H

Argentina

Specification	Designation	Notes	C	Cr	Mn	Mo	Ni	P	S	Si	Other	UTS	YS	El	Hard
IAS	IRAM 4320		0.17-0.22	0.40-0.60	0.45-0.65	0.20-0.30	1.65-2.00	0.035 max	0.040 max	0.15-0.35	bal Fe	825	535	20	248 HB

Australia

Specification	Designation	Notes	C	Cr	Mn	Mo	Ni	P	S	Si	Other	UTS	YS	El	Hard
AS 1444(96)	X4317	Standard series, Hard/Tmp	0.15-0.20	1.50-1.80	0.40-0.60	0.25-0.35	1.40-1.70	0.040 max	0.040 max	0.10-0.35	bal Fe				
AS 1444(96)	X4317H	H series, Hard/Tmp	0.15-0.20	1.50-1.80	0.40-0.60	0.25-0.35	1.40-1.70	0.040 max	0.040 max	0.10-0.35	bal Fe				
AS 1444(96)	X9931	Standard series, Hard/Tmp	0.27-0.35	0.50-0.80	0.45-0.70	0.45-0.65	2.30-2.80	0.040 max	0.040 max	0.10-0.35	bal Fe				

Bulgaria

Specification	Designation	Notes	C	Cr	Mn	Mo	Ni	P	S	Si	Other	UTS	YS	El	Hard
BDS 6354(74)	15Ch2N2	Struct	0.11-0.17	1.40-1.70	0.40-0.60	0.15 max	1.40-1.70	0.035 max	0.035 max	0.17-0.37	Cu <=0.30; Ti <=0.05; V <=0.1; W <=0.1; bal Fe				

China

Specification	Designation	Notes	C	Cr	Mn	Mo	Ni	P	S	Si	Other	UTS	YS	El	Hard
GB 3203(82)	G20CrNi2Mo	Bar HR/CD Q/T 25mm diam	0.17-0.23	0.35-0.65	0.40-0.70	0.20-0.30	1.60-2.00	0.030 max	0.030 max	0.15-0.40	Cu <=0.25; bal Fe	980		13	

Finland

Specification	Designation	Notes	C	Cr	Mn	Mo	Ni	P	S	Si	Other	UTS	YS	El	Hard
SFS 506	21NiCrMo2(506)	Bar Frg	0.17-0.23	0.35-0.65	0.6-0.95	0.15-0.25	0.35-0.75	0.04	0.05	0.15-0.4	bal Fe				
SFS 509(76)	20NiCrMo5	CH	0.17-0.23	0.8-1.2	0.7-1.1	0.08-0.16	1.0-1.4	0.035 max	0.05 max	0.15-0.4	bal Fe				
SFS 511(76)	17CrNiMo6	CH	0.14-0.19	1.5-1.8	0.4-0.6	0.25-0.35	1.4-1.7	0.035 max	0.05 max	0.15-0.4	bal Fe				

France

Specification	Designation	Notes	C	Cr	Mn	Mo	Ni	P	S	Si	Other	UTS	YS	El	Hard
AFNOR NFA35551(86)	16NC6	t<=16mm	0.12-0.17	0.9-1.2	0.6-0.9	0.15 max	1.2-1.5	0.035 max	0.035 max	0.1-0.4	Cu <=0.3; Ti <=0.05; bal Fe	1100-1400	800	9	
AFNOR NFA35565	20NCD7*	see 20NiCrMo7	0.16-0.22	0.2-0.6	0.45-0.65	0.25-0.3	1.65-2	0.03 max	0.03 max	0.2-0.35	Cu <=0.35; bal Fe				
AFNOR NFA35565(84)	18NCD4		0.16-0.22	0.35-0.55	0.5-0.8	0.15-0.3	0.9-1.2	0.03	0.025	0.2-0.35	Cu 0.35; bal Fe				
AFNOR NFA35565(94)	20NCD2*	see 20NiCrMo2	0.18-0.23	0.4-0.6	0.7-0.9	0.15-0.25	0.4-0.7	0.025 max	0.015 max	0.15-0.35	Al <=0.05; Cu <=0.3; Ti <=0.05; bal Fe				
AFNOR NFA35565(94)	20NCD7*	see 20NiCrMo7	0.17-0.22	0.4-0.6	0.45-0.65	0.2-0.3	1.65-2	0.025 max	0.015 max	0.15-0.35	Al <=0.05; Cu <=0.35; Ti <=0.05; bal Fe				
AFNOR NFA35565(94)	20NiCrMo2	Ball & roller bearing	0.18-0.23	0.4-0.6	0.7-0.9	0.15-0.25	0.4-0.7	0.025 max	0.015 max	0.15-0.35	Al <=0.05; Cu <=0.3; Ti <=0.05; bal Fe				
AFNOR NFA35565(94)	20NiCrMo7	Ball & roller bearing	0.17-0.22	0.4-0.6	0.45-0.65	0.2-0.3	1.65-2	0.025 max	0.015 max	0.15-0.35	Al <=0.05; Cu <=0.35; Ti <=0.05; bal Fe				

Hungary

Specification	Designation	Notes	C	Cr	Mn	Mo	Ni	P	S	Si	Other	UTS	YS	El	Hard
MSZ 31(85)	BNC5	CH; 63 mm; blank carburized	0.14-0.2	1.4-1.7	0.3-0.6		1.4-1.7	0.035 max	0.035 max	0.4 max	bal Fe	800-1100	590	10L	
MSZ 31(85)	BNC5	CH; 0-11mm; blank carburized	0.14-0.2	1.4-1.7	0.3-0.6		1.4-1.7	0.035 max	0.035 max	0.4 max	bal Fe	1000-1300	740	8L	
MSZ 31(85)	BNC5E	CH; 63 mm; blank carburized	0.14-0.2	1.4-1.7	0.3-0.6		1.4-1.7	0.035 max	0.02-0.035	0.4 max	bal Fe	800-1100	590	10L	
MSZ 31(85)	BNC5E	CH; 0-11mm; blank carburized	0.14-0.2	1.4-1.7	0.3-0.6		1.4-1.7	0.035 max	0.02-0.035	0.4 max	bal Fe	1000-1300	740	8L	
MSZ 31(85)	BNCMo1	CH; 0-11mm; blank carburized	0.14-0.2	0.8-1.1	0.6-0.9	0.15-0.25	1.2-1.6	0.035 max	0.035 max	0.4 max	bal Fe	1100-1450	740	8L	
MSZ 31(85)	BNCMo1	CH; 63 mm; blank carburized	0.14-0.2	0.8-1.1	0.6-0.9	0.15-0.25	1.2-1.6	0.035 max	0.035 max	0.4 max	bal Fe	850-1150	540	10L	
MSZ 31(85)	BNCMo1E	CH; 63 mm; blank carburized	0.14-0.2	0.8-1.1	0.6-0.9	0.15-0.25	1.2-1.6	0.035 max	0.02-0.035	0.4 max	bal Fe	850-1150	540	10L	
MSZ 31(85)	BNCMo1E	CH; 0-11mm; blank carburized	0.14-0.2	0.8-1.1	0.6-0.9	0.15-0.25	1.2-1.6	0.035 max	0.02-0.035	0.4 max	bal Fe	1100-1450	740	8L	

International

Specification	Designation	Notes	C	Cr	Mn	Mo	Ni	P	S	Si	Other	UTS	YS	El	Hard
ISO 683-17(76)	13	Ball & roller bearing, HT	0.17-0.23	0.35-0.65	0.40-0.70	0.15-0.25	0.90-1.20	0.035 max	0.035 max	0.15-0.40	bal Fe				
ISO 683-17(76)	14	Ball & roller bearing, HT	0.17-0.23	0.35-0.65	0.40-0.70	0.20-0.30	1.60-2.00	0.035 max	0.035 max	0.15-0.40	bal Fe				

Italy

Specification	Designation	Notes	C	Cr	Mn	Mo	Ni	P	S	Si	Other	UTS	YS	El	Hard
UNI 3097(75)	20NiCrMo7	Ball & roller bearing	0.17-0.22	0.4-0.6	0.45-0.65	0.2-0.3	1.6-2	0.035 max	0.035 max	0.15-0.4	bal Fe				
UNI 5331(64)	16NiCr11	CH	0.12-0.18	0.6-0.9	0.3-0.6	0.1 max	2.5-3	0.035 max	0.035 max	0.35 max	bal Fe				

UNS numbers and US grades are provided as a means of cross referencing chemically similar alloys. Exchangability is only possible after independent examination of specifications. Tensile properties are minimum or typical as specified. UTS and YS are MPa. El as %. See Appendix for list of abbreviations used in Notes. * indicates obsolete material.

Specification	Designation	Notes	C	Cr	Mn	Mo	Ni	P	S	Si	Other	UTS	YS	El	Hard

Alloy Steel, Nickel Chromium Molybdenum, 4320/4320H (Continued from previous page)

Specification	Designation	Notes	C	Cr	Mn	Mo	Ni	P	S	Si	Other	UTS	YS	El	Hard
Italy															
UNI 8550(84)	16NiCr11	CH	0.12-0.18	0.6-0.9	0.3-0.6		2.5-3	0.035 max	0.035 max	0.15-0.4	bal Fe				
UNI 8550(04)	10NiCrMo7	Cl I	0.17-0.22	0.4-0.7	0.45-0.65	0.2-0.3	1.0-2	0.035 max	0.035 max	0.15-0.4	Pb <=0.1; bal Fe				
Japan															
JIS G3311(88)	SNCM21M*	Obs	0.17-0.23	0.4-0.65	0.6-0.9	0.15-0.3	1.6-2	0.03	0.03	0.15-0.35	Cu 0.3; bal Fe				
JIS G3311(88)	SNCM415M	CR Strp; Chain	0.12-0.18	0.40-0.65	0.40-0.70	0.15-0.30	1.60-2.00	0.030 max	0.030 max	0.15-0.35	Cu <=0.30; bal Fe				170-240 HV
JIS G4052(79)	SNCM420H	Struct Hard	0.17-0.23	0.35-0.65	0.40-0.70	0.15-0.30	1.55-2.00	0.030 max	0.030 max	0.15-0.35	Cu <=0.30; bal Fe				
JIS G4103(79)	SNCM23*	Obs; see SNCM420	0.17-0.23	0.4-0.65	0.4-0.6	0.15-0.3	1.6-2	0.03 max	0.03 max	0.15-0.35	bal Fe				
JIS G4103(79)	SNCM415	HR Frg	0.12-0.18	0.40-0.65	0.40-0.70	0.15-0.30	1.60-2.00	0.030 max	0.030 max	0.15-0.35	Cu <=0.30; bal Fe				
JIS G4103(79)	SNCM420	HR Frg Bar Rod	0.17-0.23	0.40-0.65	0.40-0.70	0.15-0.30	1.60-2.00	0.030 max	0.030 max	0.15-0.35	Cu <=0.30; bal Fe				
Mexico															
DGN B-297	K4320*	Obs	0.17-0.22	0.4-0.6	0.45-0.65	0.2-0.3	1.65-2	0.035	0.04	0.2-0.3	bal Fe				
NMX-B-268(68)	4320H	Hard	0.17-0.23	0.35-0.65	0.40-0.70	0.20-0.30	1.55-2.00			0.20-0.35	bal Fe				
NMX-B-300(91)	4320	Bar	0.17-0.22	0.40-0.60	0.45-0.60	0.20-0.30	1.65-2.00	0.035 max	0.040 max	0.15-0.35	bal Fe				
NMX-B-300(91)	4320H	Bar	0.17-0.23	0.35-0.65	0.40-0.70	0.20-0.30	1.55-2.00		0.035	0.15-0.35	bal Fe				
Norway															
NS 13101	13124	Struct const cement	0.12-0.17	1.40-1.70	0.40-0.60		1.20-1.40	0.035 max	0.035 max	0.15-0.40	bal Fe				
Pan America															
COPANT 334	4320	Bar	0.17-0.22	0.4-0.6	0.45-0.65	0.2-0.3	1.65-2	0.035	0.04	0.2-0.35	bal Fe				
Russia															
GOST 4543	20ChN2M		0.15-0.22	0.4-0.6	0.4-0.7	0.2-0.3	1.6-2	0.035 max	0.035 max	0.17-0.37	Cu <=0.3; bal Fe				
GOST 4543(71)	12ChN3A	CH	0.09-0.16	0.6-0.9	0.3-0.6	0.15 max	2.75-3.15	0.025 max	0.025 max	0.17-0.37	Al <=0.1; Cu <=0.3; Ti <=0.03; W <=0.2; bal Fe				
GOST 4543(71)	20ChN2M	CH	0.15-0.22	0.4-0.6	0.4-0.7	0.2-0.3	1.6-2.0	0.035 max	0.035 max	0.17-0.37	Al <=0.1; Cu <=0.3; Ti <=0.05; bal Fe				
Spain															
UNE	F.159	CH	0.12-0.16	0.8-1.2	0.55-0.85	0.15-0.25	0.8-1.2	0.04 max	0.04 max	0.1-0.35	bal Fe				
UNE 36013(76)	16NiCr4	CH	0.13-0.18	0.8-1.2	0.8-1	0.15 max	0.8-1.2	0.035 max	0.035 max	0.15-0.4	bal Fe				
UNE 36013(76)	20NiCrMo6	CH	0.18-0.23	0.4-0.6	0.6-0.8	0.3-0.4	1.4-1.7	0.035 max	0.035 max	0.15-0.4	bal Fe				
UNE 36013(76)	20NiCrMo6-1	CH	0.18-0.23	0.4-0.6	0.6-0.8	0.3-0.4	1.4-1.7	0.035 max	0.02-0.035	0.15-0.4	bal Fe				
UNE 36013(76)	F.1525*	see 20NiCrMo6	0.18-0.23	0.4-0.6	0.6-0.8	0.3-0.4	1.4-1.7	0.035 max	0.035 max	0.15-0.4	bal Fe				
UNE 36013(76)	F.1535*	see 20NiCrMo6-1	0.18-0.23	0.4-0.6	0.6-0.8	0.3-0.4	1.4-1.7	0.035 max	0.02-0.035	0.15-0.4	bal Fe				
UNE 36013(76)	F.1581*	see 16NiCr4	0.13-0.18	0.8-1.2	0.8-1	0.15 max	0.8-1.2	0.035 max	0.035 max	0.15-0.4	bal Fe				
UNE 36027(80)	20NiCrMo4	Ball & roller bearing	0.17-0.23	0.35-0.65	0.4-0.7	0.15-0.25	0.9-1.2	0.035 max	0.035 max	0.15-0.4	bal Fe				
UNE 36027(80)	20NiCrMo7	Ball & roller bearing	0.17-0.23	0.35-0.65	0.4-0.7	0.2-0.3	1.6-2.0	0.035 max	0.035 max	0.15-0.4	bal Fe				
UNE 36027(80)	F.1528*	see 20NiCrMo4	0.17-0.23	0.35-0.65	0.4-0.7	0.15-0.25	0.9-1.2	0.035 max	0.035 max	0.15-0.4	bal Fe				
UNE 36027(80)	F.1529*	see 20NiCrMo7	0.17-0.23	0.35-0.65	0.4-0.7	0.2-0.3	1.6-2.0	0.035 max	0.035 max	0.15-0.4	bal Fe				
UK															
BS 970/1(83)	815H17	Blm Bil Slab Bar Rod Frg CH	0.14-0.20	0.80-1.20	0.60-0.90	0.10-0.20	1.2-1.7	0.035 max	0.04 max	0.10-0.35	bal Fe				241 HB max

UNS numbers and US grades are provided as a means of cross referencing chemically similar alloys. Exchangability is only possible after independent examination of specifications. Tensile properties are minimum or typical as specified. UTS and YS as MPa. El as %. See Appendix for list of abbreviations used in Notes. * indicates obsolete material.

Specification	Designation	Notes	C	Cr	Mn	Mo	Ni	P	S	Si	Other	UTS	YS	El	Hard

Alloy Steel, Nickel Chromium Molybdenum, 4320/4320H (Continued from previous page)

UK

Specification	Designation	Notes	C	Cr	Mn	Mo	Ni	P	S	Si	Other	UTS	YS	El	Hard
BS 970/1(83)	815M17	Blm Bil Slab Bar Rod Frg CH	0.14-0.20	0.80-1.20	0.60-0.90	0.10-0.20	1.2-1.7	0.040 max	0.05 max	0.10-0.40	bal Fe	1080		8	
BS 970/1(83)	820H17	Blm Bil Slab Bar Rod Frg CH	0.14-0.20	0.80-1.20	0.60-0.90	0.10-0.20	1.5-2.0	0.035 max	0.04 max	0.10-0.35	bal Fe				0-248 HB
BS 970/1(83)	820M17	Blm Bil Slab Bar Rod Frg CH	0.14-0.20	0.80-1.20	0.60-0.90	0.10-0.20	1.5-2.0	0.035 max	0.04 max	0.10-0.35	bal Fe	1160		8	
BS 970/1(83)	822H17	Blm Bil Slab Bar Rod Frg CH	0.14-0.20	1.30-1.70	0.40-0.70	0.15-0.25	1.75-2.25	0.035 max	0.04 max	0.10-0.35	bal Fe				0-255 HB
BS 970/1(83)	822M17	Blm Bil Slab Bar Rod Frg CH	0.14-0.20	1.30-1.70	0.40-0.70	0.15-0.25	1.75-2.25	0.040 max	0.05 max	0.10-0.40		1310		8	

USA

Specification	Designation	Notes	C	Cr	Mn	Mo	Ni	P	S	Si	Other	UTS	YS	El	Hard
	AISI 4320	Smls mech tub	0.17-0.22	0.40-0.60	0.45-0.65	0.20-0.30	1.65-2.00	0.040 max	0.040 max	0.15-0.35	bal Fe				
	AISI 4320	Bar Blm Bil Slab	0.17-0.22	0.40-0.60	0.45-0.65	0.20-0.30	1.65-2.00	0.035 max	0.040 max	0.15-0.35	bal Fe				
	AISI 4320H	Bar Blm Bil Slab	0.17-0.23	0.35-0.65	0.40-0.70	0.20-0.30	1.55-2.00	0.035 max	0.040 max	0.15-0.35	bal Fe				
	AISI 4320H	Smls mech tub, Hard	0.17-0.23	0.35-0.65	0.40-0.70	0.20-0.30	1.55-2.00	0.035 max	0.040 max	0.15-0.30	bal Fe				
	UNS G43200		0.17-0.22	0.40-0.60	0.45-0.65	0.20-0.30	1.65-2.00	0.035 max	0.040 max	0.15-0.35	bal Fe				
	UNS H43200		0.17-0.23	0.35-0.65	0.40-0.70	0.20-0.30	1.55-2.00	0.035 max	0.040 max	0.15-0.35	bal Fe				
AMS 6299											bal Fe				
ASTM A29/A29M(93)	4320	Bar	0.17-0.22	0.40-0.60	0.45-0.65	0.20-0.30	1.65-2.00	0.035 max	0.040 max	0.15-0.35	bal Fe				
ASTM A304(96)	4320H	Bar, hard bands spec	0.17-0.23	0.35-0.65	0.40-0.70	0.20-0.30	1.55-2.00	0.035 max	0.040 max	0.15-0.35	Cu <=0.35; Cu Ni Cr Mo trace allowed; bal Fe				
ASTM A322(96)	4320	Bar	0.17-0.22	0.40-0.60	0.45-0.65	0.20-0.30	1.65-2.00	0.035 max	0.040 max	0.15-0.35	bal Fe				
ASTM A331(95)	4320	Bar	0.17-0.22	0.40-0.60	0.45-0.65	0.20-0.30	1.65-2.00	0.035 max	0.040 max	0.15-0.35	bal Fe				
ASTM A331(95)	4320H	Bar	0.17-0.23	0.35-0.65	0.40-0.70	0.20-0.30	1.55-2.00	0.025 max	0.025 max	0.15-0.35	Cu <=0.35; bal Fe				
ASTM A519(96)	4320	Smls mech tub	0.17-0.22	0.40-0.60	0.45-0.65	0.20-0.30	1.65-2.00	0.040 max	0.040 max	0.15-0.35	bal Fe				
ASTM A534(94)	4320H	Carburizing, anti-friction bearings	0.17-0.23	0.35-0.65	0.40-0.70	0.20-0.30	1.55-2.00	0.025 max	0.025 max	0.15-0.35	bal Fe				
ASTM A535(92)	4320	Ball & roller bearing	0.17-0.22	0.40-0.60	0.45-0.65	0.20-0.30	1.65-2.00	0.015 max	0.015 max	0.15-0.35	Cu <=0.35; bal Fe				
ASTM A752(93)	4320	Rod Wir	0.17-0.22	0.40-0.60	0.45-0.65	0.20-0.30	1.65-2.00	0.035 max	0.040 max	0.15-0.30	bal Fe	650			
ASTM A837(96)	4320	Frg for carburizing	0.17-0.22	0.40-0.60	0.45-0.65	0.20-0.30	1.65-2.00	0.035 max	0.040 max	0.15-0.35	Cu <=0.35; Si<=0.10 if VCD used; bal Fe				
ASTM A914/914M(92)	4320RH	Bar restricted end Q hard	0.17-0.22	0.40-0.60	0.45-0.65	0.20-0.30	1.65-2.00	0.035 max	0.040 max	0.15-0.35	Cu <=0.35; Electric furnace P, S<=0.025; bal Fe				
SAE 770(84)	4320*	Obs; see J1397(92)									bal Fe				
SAE J1268(95)	4320H	See std	0.17-0.23	0.35-0.65	0.40-0.70	0.20-0.30	1.55-2.00	0.030 max	0.040 max	0.15-0.35	Cu <=0.35; Cu not spec'd but acpt; bal Fe				
SAE J1868(93)	4320RH	Restrict hard range	0.17-0.22	0.40-0.60	0.45-0.65	0.20-0.30	1.65-2.00	0.025 max	0.040 max	0.15-0.35	Cu <=0.035; Cu not spec'd but acpt; bal Fe				5 HRC max
SAE J404(94)	4320	Bil Blm Slab Bar HR CF	0.17-0.22	0.40-0.60	0.45-0.65	0.20-0.30	1.65-2.00	0.030 max	0.040 max	0.15-0.35	bal Fe				

Yugoslavia

Specification	Designation	Notes	C	Cr	Mn	Mo	Ni	P	S	Si	Other	UTS	YS	El	Hard
	C.4520	CH	0.14-0.19	1.5-1.8	0.4-0.6	0.25-0.35	1.4-1.7	0.035 max	0.035 max	0.15-0.4	bal Fe				
	C.5420	CH	0.12-0.17	1.4-1.7	0.4-0.6	0.15 max	1.4-1.7	0.035 max	0.035 max	0.15-0.4	bal Fe				

Alloy Steel, Nickel Chromium Molybdenum, 4337

Australia

Specification	Designation	Notes	C	Cr	Mn	Mo	Ni	P	S	Si	Other	UTS	YS	El	Hard
AS 1444(96)	X4330	Standard series, Hard/Tmp	0.28-0.33	0.80-1.10	0.85-1.15	0.35-0.55	1.10-1.40	0.040 max	0.040 max	0.10-0.35	bal Fe				

UNS numbers and US grades are provided as a means of cross referencing chemically similar alloys. Exchangability is only possible after independent examination of specifications. Tensile properties are minimum or typical as specified. UTS and YS as MPa. El as %. See Appendix for list of abbreviations used in Notes. * indicates obsolete material.

Alloy Steel, Nickel Chromium Molybdenum, 4337 (Continued from previous page)

Specification	Designation	Notes	C	Cr	Mn	Mo	Ni	P	S	Si	Other	UTS	YS	El	Hard
Czech Republic															
CSN 416343	16343	High-temp const	0.32-0.4	1.3-1.7	0.5-0.8	0.2-0.3	1.3-1.7	0.035 max	0.035 max	0.15-0.4	bal Fe				
France															
AFNOR NFA35552(84)	30CND8	160<t<=250mm	0.26-0.33	1.8-2.2	0.3-0.6	0.3-0.5	1.8-2.2	0.03 max	0.025 max	0.1-0.4	Cu <=0.3; Ti <=0.05; bal Fe	930-1130	750	12	
AFNOR NFA35552(84)	30CND8	t<=40mm	0.26-0.33	1.8-2.2	0.3-0.6	0.3-0.5	1.8-2.2	0.03 max	0.025 max	0.1-0.4	Cu <=0.3; Ti <=0.05; bal Fe	1030-1230	850	12	
AFNOR NFA35557(83)	30CND8	t<=100mm	0.26-0.33	1.8-2.2	0.3-0.6	0.3-0.5	1.8-2.2	0.03 max	0.025 max	0.1-0.4	Cu <=0.3; Ti <=0.05; bal Fe	1130-1330	950	10	
AFNOR NFA35557(83)	30CND8	t<=50mm	0.26-0.33	1.8-2.2	0.3-0.6	0.3-0.5	1.8-2.2	0.03 max	0.025 max	0.1-0.4	Cu <=0.3; Ti <=0.05; bal Fe	1130-1330	950	10	
Germany															
DIN 1652(90)	WNr 1.6580	Bright Q/T 17-40mm	0.26-0.34	1.80-2.20	0.30-0.60	0.30-0.50	1.80-2.20	0.035 max	0.035 max	0.40 max	bal Fe	1250-1450	1050	9	
DIN 1652(990)	30CrNiMo8	Bright Q/T 17-40mm	0.26-0.34	1.80-2.20	0.30-0.60	0.30-0.50	1.80-2.20	0.035 max	0.035 max	0.40 max	bal Fe	1250-1450	1050	9	
DIN 1654(89)	30CrNiMo8	CHd, cold ext, Q/T 17-40mm	0.26-0.34	1.80-2.20	0.30-0.60	0.30-0.50	1.80-2.20	0.035 max	0.035 max	0.40 max	bal Fe	1250-1450	1050	9	
DIN 1654(89)	WNr 1.6580	CHd, cold ext, Q/T 17-40mm	0.26-0.34	1.80-2.20	0.30-0.60	0.30-0.50	1.80-2.20	0.035 max	0.035 max	0.40 max	bal Fe	1250-1450	1050	9	
DIN EN 10083(91)	30CrNiMo8	Q/T 17-40mm	0.26-0.34	1.80-2.20	0.30-0.60	0.30-0.50	1.80-2.20	0.035 max	0.035 max	0.40 max	bal Fe	1250-1450	1050	9	
DIN EN 10083(91)	WNr 1.6580	Q/T 17-40mm	0.26-0.34	1.80-2.20	0.30-0.60	0.30-0.50	1.80-2.20	0.035 max	0.035 max	0.40 max	bal Fe	1250-1450	1050	9	
International															
ISO R683-8(70)	3	Q/T	0.32-0.39	1.3-1.7	0.5-0.8	0.15-0.3	1.3-1.7	0.035 max	0.035 max	0.15-0.4	Cu <=0.3; Pb <=0.15; Tl <=0.05; V <=0.1;				
Italy															
UNI 7356(74)	35NiCrMo6KB	Q/T	0.32-0.39	1.3-1.7	0.5-0.8	0.2-0.3	1.3-1.7	0.035 max	0.035 max	0.15-0.4	bal Fe				
Poland															
PNH84030/04	34HNM	Q/T	0.32-0.4	1.3-1.7	0.4-0.7	0.15-0.25	1.3-1.7	0.035 max	0.035 max	0.17-0.37	bal Fe				
Romania															
STAS 791(88)	34MoCrNi16	Struct/const; Q/T	0.3-0.38	1.4-1.7	0.4-0.7	0.15-0.3	1.4-1.7	0.035 max	0.035 max	0.17-0.37	Pb <=0.15; Ti <=0.02; bal Fe				
STAS 791(88)	34MoCrNi16S	Struct/const; Q/T	0.3-0.38	1.4-1.7	0.4-0.7	0.15-0.3	1.4-1.7	0.035 max	0.02-0.04	0.17-0.37	Pb <=0.15; Ti <=0.02; bal Fe				
STAS 791(88)	34MoCrNi16X	Struct/const; Q/T	0.3-0.38	1.4-1.7	0.4-0.7	0.15-0.3	1.4-1.7	0.025 max	0.025 max	0.17-0.37	Pb <=0.15; Ti <=0.02; bal Fe				
STAS 791(88)	34MoCrNi16XS	Struct/const; Q/T	0.3-0.38	1.4-1.7	0.4-0.7	0.15-0.3	1.4-1.7	0.025 max	0.02-0.035	0.17-0.37	Pb <=0.15; Ti <=0.02; bal Fe				
STAS 8185(88)	34MoCrNi15	Q/T	0.3-0.38	1.4-1.7	0.4-0.7	0.15-0.25	1.4-1.7	0.035 max	0.035 max	0.17-0.37	Pb <=0.15; bal Fe				
STAS 9382/4(89)	34MoCrNi16q	Q/T	0.3-0.38	1.4-1.7	0.4-0.7	0.15-0.3	1.4-1.7	0.035 max	0.035 max	0.17-0.37	Pb <=0.15; As<=0.05; bal Fe				
Spain															
UNE 36012(75)	40NiCrMo7	Q/T	0.37-0.43	0.65-0.95	0.55-0.85	0.15-0.3	1.6-2.0	0.035 max	0.035 max	0.15-0.4	bal Fe				
Sweden															
SS 142541	2541	Q/T	0.32-0.39	1.3-1.7	0.5-0.8	0.15-0.3	1.3-1.7	0.035 max	0.035 max	0.1-0.4	bal Fe				
UK															
BS 970/3(91)	826M40	Bright bar; Q/T; 150<t<=250mm; hard	0.36-0.44	0.50-0.80	0.45-0.70	0.45-0.65	2.30-2.80	0.035 max	0.040 max	0.10-0.40	bal Fe	925-1075	740	12	269-331 HB
BS 970/3(91)	826M40	Bright bar; Q/T; 29<t<=150mm; hard CD	0.36-0.44	0.50-0.80	0.45-0.70	0.45-0.65	2.30-2.80	0.035 max	0.040 max	0.10-0.40	bal Fe	1225-1375	1110	7	363-429 HB
USA															
	AISI 4337	Smls mech tub	0.35-0.40	0.70-0.90	0.60-0.80	0.20-0.30	1.65-2.00	0.040 max	0.040 max	0.15-0.35	bal Fe				
	UNS G43370		0.35-0.40	0.70-0.90	0.60-0.80	0.20-0.30	1.65-2.00	0.035 max	0.040 max	0.15-0.35	bal Fe				
AMS 6413H(90)		Mech tub, CF	0.35-0.40	0.70-0.90	0.65-0.85	0.20-0.30	1.65-2.00	0.025 max	0.025 max	0.15-0.35	Cu <=0.35; bal Fe				25 HRC max

UNS numbers and US grades are provided as a means of cross referencing chemically similar alloys. Exchangability is only possible after independent examination of specifications. Tensile properties are minimum or typical as specified. UTS and YS as MPa. El as %. See Appendix for list of abbreviations used in Notes. * indicates obsolete material.

Specification	Designation	Notes	C	Cr	Mn	Mo	Ni	P	S	Si	Other	UTS	YS	El	Hard
Alloy Steel, Nickel Chromium Molybdenum, 4337 (Continued from previous page)															
USA															
AMS 6413H(90)		Mech tub, HF and Ann	0.35-0.40	0.70-0.90	0.65-0.85	0.20-0.30	1.65-2.00	0.025 max	0.025 max	0.15-0.35	Cu <=0.35; bal Fe				99 HRB max
ASTM A519(96)	4337	Smls mech tub	0.35-0.40	0.70-0.90	0.60-0.80	0.20-0.30	1.65-2.00	0.040 max	0.040 max	0.15-0.35	bal Fe				
Yugoslavia															
	C.5432	Q/T	0.26-0.33	1.8-2.2	0.3-0.6	0.3-0.5	1.8-2.2	0.035 max	0.03 max	0.4 max	bal Fe				
Alloy Steel, Nickel Chromium Molybdenum, 4340/4340H															
Argentina															
IAS	IRAM 4340		0.38-0.43	0.70-0.90	0.60-0.80	0.20-0.30	1.65-2.00	0.035 max	0.040 max	0.20-0.35	bal Fe	1300	860	10	388 HB
Australia															
AS 1444(96)	4340	H series, Hard/Tmp	0.38-0.43	0.70-0.90	0.60-0.80	0.20-0.30	1.65-2.00	0.040 max	0.040 max	0.10-0.35	bal Fe				
AS 1444(96)	4340H	H series, Hard/Tmp	0.37-0.44	0.65-0.95	0.55-0.90	0.20-0.30	1.55-2.00	0.040 max	0.040 max	0.10-0.35	bal Fe				
Bulgaria															
BDS 6354	40ChN2M		0.37-0.44	0.60-0.90	0.50-0.80	0.15-0.25	1.25-1.65	0.035	0.035	0.17-0.37	Cu 0.3; bal Fe				
BDS 6354(74)	35Ch2N2M	Struct	0.32-0.39	1.30-1.70	0.50-0.80	0.15-0.30	1.30-1.70	0.035 max	0.035 max	0.17-0.37	Cu <=0.30; Ti <=0.05; V <=0.1; W <=0.1; bal Fe				
BDS 6354(74)	35Ch2N2M	Struct	0.32-0.39	1.30-1.70	0.50-0.80	0.15-0.30	1.30-1.70	0.035 max	0.035 max	0.17-0.37	Cu <=0.30; Ti <=0.05; V <=0.1; W <=0.1; bal Fe				
China															
GB 3077(88)	40CrNiMoA	Bar HR Q/T 25mm diam	0.37-0.44	0.60-0.90	0.50-0.80	0.15-0.25	1.25-1.65	0.025 max	0.025 max	0.17-0.37	Cu <=0.25; bal Fe	980	835	12	
GB 8162(87)	40CrNiMoA	Smls tube HR/CD Q/T	0.37-0.44	0.60-0.90	0.50-0.80	0.15-0.25	1.25-1.65	0.025 max	0.025 max	0.17-0.37	bal Fe	980	835	12	
GB/T 3078(94)	40CrNiMoA	Bar CD Q/T 25mm diam	0.37-0.44	0.60-0.90	0.50-0.80	0.15-0.25	1.25-1.65	0.025 max	0.025 max	0.17-0.37	Cu <=0.25; bal Fe	980	835	12	
GB/T 3079(93)	40CrNiMoA	Wir Q/T	0.37-0.44	0.60-0.90	0.50-0.80	0.15-0.25	1.25-1.65	0.025 max	0.025 max	0.17-0.37	Cu <=0.25; bal Fe	980	835	12	
Czech Republic															
CSN 416341	16341	Q/T	0.35-0.43	0.8-1.2	0.6-0.8	0.1-0.2	1.6-2.1	0.035 max	0.035 max	0.17-0.37	bal Fe				
Europe															
EN 10083/1(91)A1(96)	1.6582	Q/T t<=16mm	0.30-0.38	1.30-1.70	0.50-0.80	0.15-0.30	1.30-1.70	0.035 max	0.035 max	0.40 max	bal Fe	1200-1400	1000	9	
EN 10083/1(91)A1(96)	1.7037	Q/T t<=16mm	0.30-0.37	0.90-1.20	0.60-0.90	0.15 max		0.035 max	0.020-0.040	0.40 max	bal Fe	900-1100	700	12	
EN 10083/1(91)A1(96)	34CrNiMo6	Q/T t<=16mm	0.30-0.38	1.30-1.70	0.50-0.80	0.15-0.30	1.30-1.70	0.035 max	0.035 max	0.40 max	bal Fe	1200-1400	1000	9	
EN 10083/1(91)A1(96)	34CrS4	Q/T t<=16mm	0.30-0.37	0.90-1.20	0.60-0.90	0.15 max		0.035 max	0.020-0.040	0.40 max	bal Fe	900-1100	700	12	
France															
AFNOR NFA35551(75)	35NCD6		0.37 max	0.85-1.15	0.6-0.9	0.15-0.3	1.2-1.6	0.035 max	0.035 max	0.1-0.4	Cu <=0.3; Ti <=0.05; bal Fe				
Germany															
DIN	40NiCrMo6		0.35-0.45	0.90-1.40	0.5-0.7	0.20-0.30	1.40-1.70	0.035 max	0.035 max	0.15-0.35	bal Fe				
DIN	40NiCrMo8-4		0.37-0.44	0.70-0.95	0.70-0.90	0.30-0.40	1.65-2.00	0.020 max	0.015 max	0.20-0.35	Al 0.005-0.050; bal Fe				
DIN	WNr 1.6562		0.37-0.44	0.70-0.95	0.70-0.90	0.30-0.40	1.65-2.00	0.020 max	0.015 max	0.20-0.35	Al 0.005-0.050; bal Fe				
DIN	WNr 1.6565		0.35-0.45	0.90-1.40	0.50-0.70	0.20-0.30	1.40-1.70	0.035 max	0.035 max	0.15-0.35	bal Fe				
Hungary															
MSZ 61(85)	NCMo5	Q/T; 160.1-250mm	0.32-0.39	1.3-1.7	0.5-0.8	0.15-0.3	1.3-1.7	0.035 max	0.035 max	0.4 max	bal Fe	800-950	600	13L	
MSZ 61(85)	NCMo5	Q/T; 0-16mm	0.32-0.39	1.3-1.7	0.5-0.8	0.15-0.3	1.3-1.7	0.035 max	0.035 max	0.4 max	bal Fe	1200-1400	1000	9L	

Specification	Designation	Notes	C	Cr	Mn	Mo	Ni	P	S	Si	Other	UTS	YS	El	Hard

Alloy Steel, Nickel Chromium Molybdenum, 4340/4340H (Continued from previous page)

India

Specification	Designation	Notes	C	Cr	Mn	Mo	Ni	P	S	Si	Other	UTS	YS	El	Hard
IS 1570	40Ni2Cr1Mo28		0.35-0.45	0.9-1.3	0.4-0.7	0.2-0.35	1.25-1.75			0.1-0.35	bal Fe				
IS 1570	40NiCr1Mo15		0.35-0.45	0.9-1.3	0.4-0.7	0.1-0.2	1.2-1.6			0.1-0.35	bal Fe				

International

Specification	Designation	Notes	C	Cr	Mn	Mo	Ni	P	S	Si	Other	UTS	YS	El	Hard
ISO 683-8	4*		0.37-0.44	0.65-0.95	0.55-0.85	0.15-0.3	1.6-2	0.035	0.035	0.15-0.4	bal Fe				
ISO 683-8	4A*		0.37-0.44	0.65-0.95	0.55-0.85	0.15-0.3	1.6-2	0.035	0.02-0.035	0.15-0.4	bal Fe				
ISO 683-8	4b*		0.37-0.44	0.65-0.95	0.55-0.85	0.15-0.3	1.6-2	0.035	0.03-0.05	0.15-0.4	bal Fe				
ISO R683-8	4A	Bar Bil, Rod, Q/A, Tmp, 16mm diam	0.37-0.44	0.65-0.95	0.55-0.85	0.15-0.3	1.6-2	0.04	0.02-0.04	0.15-0.4	bal Fe				
ISO R683-8	4b	Bar Bil, Rod, Q/A, Tmp, 16mm diam	0.37-0.44	0.65-0.95	0.55-0.85	0.15-0.3	1.6-2	0.04	0.03-0.05	0.15-0.4	bal Fe				
ISO R683-8(70)	3	Q/T	0.32-0.39	1.3-1.7	0.5-0.8	0.15-0.3	1.3-1.7	0.035 max	0.035 max	0.15-0.4	Al <=0.1; Co <=0.1; Cu <=0.3; Pb <=0.15; Ti <=0.05; V <=0.1; W <=0.1; bal Fe				

Italy

Specification	Designation	Notes	C	Cr	Mn	Mo	Ni	P	S	Si	Other	UTS	YS	El	Hard
UNI 5332(64)	40NiCrMo7	Q/T	0.37-0.43	0.6-0.9	0.5-0.8	0.2-0.3	1.6-1.9	0.035 max	0.035 max	0.4 max	bal Fe				
UNI 6926(71)	40NiCrMo7		0.37-0.43	0.6-0.9	0.5-0.8	0.2-0.3	1.6-1.9	0.035	0.035	0.4	bal Fe				
UNI 7356(74)	40NiCrMo7KB	Q/T	0.37-0.44	0.7-1	0.5-0.8	0.2-0.3	1.6-2	0.035 max	0.035 max	0.15-0.4	bal Fe				
UNI 7845(78)	40NiCrMo7	Q/T	0.37-0.44	0.6-0.9	0.5-0.8	0.2-0.3	1.6-1.9	0.035 max	0.035 max	0.15-0.4	bal Fe				
UNI 7874(79)	40NiCrMo7	Q/T	0.37-0.44	0.6-0.9	0.5-0.8	0.2-0.35	1.6-1.9	0.035 max	0.035 max	0.15-0.4	bal Fe				

Japan

Specification	Designation	Notes	C	Cr	Mn	Mo	Ni	P	S	Si	Other	UTS	YS	El	Hard
JIS G4103(79)	SNCM439	HR Frg Bar Rod	0.36-0.43	0.60-1.00	0.60-0.90	0.15-0.30	1.60-2.00	0.030 max	0.030 max	0.15-0.35	Cu <=0.30; bal Fe				
JIS G4103(79)	SNCM8*	Obs; see SNCM439	0.36-0.43	0.6-1	0.6-0.9	0.15-0.3	1.6-2	0.03 max	0.03 max	0.15-0.35	bal Fe				
JIS G4108(94)	SNB24-1	Bar 150<diam<=200mm	0.37-0.44	0.70-0.95	0.70-0.90	0.30-0.40	1.65-2.00	0.025 max	0.025 max	0.20-0.35	bal Fe	1140	1030	10	331-429 HB
JIS G4108(94)	SNB24-2	Bar 175<diam<=240mm	0.37-0.44	0.70-0.95	0.70-0.90	0.30-0.40	1.65-2.00	0.025 max	0.025 max	0.20-0.35	bal Fe	1070	960	11	321-415 HB
JIS G4108(94)	SNB24-3	Bar 200<diam<=240mm	0.37-0.44	0.70-0.95	0.70-0.90	0.30-0.40	1.65-2.00	0.025 max	0.025 max	0.20-0.35	bal Fe	1000	890	12	311-388 HB
JIS G4108(94)	SNB24-4	Bar 200<diam<=240mm	0.37-0.44	0.70-0.95	0.70-0.90	0.30-0.40	1.65-2.00	0.025 max	0.025 max	0.20-0.35	bal Fe	930	825	13	293-363 HB
JIS G4108(94)	SNB24-5	Bar 200<diam<=240mm	0.37-0.44	0.70-0.95	0.70-0.90	0.30-0.40	1.65-2.00	0.025 max	0.025 max	0.20-0.35	bal Fe	790	685	15	262-321 HB

Mexico

Specification	Designation	Notes	C	Cr	Mn	Mo	Ni	P	S	Si	Other	UTS	YS	El	Hard
DGN B-203	4340*	Obs	0.38-0.43	0.7-0.9	0.6-0.8	0.2-0.3	1.65-2	0.04	0.04	0.2-0.35	bal Fe				
DGN B-297	4340*	Obs	0.38-0.43	0.7-0.9	0.6-0.8	0.2-0.3	1.65-2	0.035	0.04	0.2-0.35	bal Fe				
NMX-B-268(68)	4340H	Hard	0.37-0.44	0.65-0.95	0.55-0.90	0.20-0.30	1.55-2.00			0.20-0.35	bal Fe				
NMX-B-300(91)	4340	Bar	0.38-0.43	0.70-0.90	0.60-0.80	0.20-0.30	1.65-2.00	0.035 max	0.040 max	0.15-0.35	bal Fe				
NMX-B-300(91)	4340H	Bar	0.37-0.44	0.65-0.95	0.55-0.90	0.20-0.30	1.55-2.00			0.15-0.35	bal Fe				

Pan America

Specification	Designation	Notes	C	Cr	Mn	Mo	Ni	P	S	Si	Other	UTS	YS	El	Hard
COPANT 334	4340	Bar	0.38-0.43	0.7-0.9	0.6-0.8	0.2-0.3	1.65-2	0.04	0.04	0.02-0.35	bal Fe				
COPANT 514	4340	Tube	0.38-0.43	0.7-0.9	0.6-0.8	0.2-0.3	1.65-2	0.04	0.04	0.04-0.35	bal Fe				

Poland

Specification	Designation	Notes	C	Cr	Mn	Mo	Ni	P	S	Si	Other	UTS	YS	El	Hard
PNH84030/04	36HNM	Q/T	0.32-0.4	0.9-1.2	0.5-0.8	0.15-0.25	0.9-1.2	0.035 max	0.035 max	0.17-0.37	bal Fe				
PNH84030/04	40HNMA	Q/T	0.37-0.44	0.6-0.9	0.5-0.8	0.15-0.25	1.25-1.65	0.03 max	0.025 max	0.17-0.37	bal Fe				

UNS numbers and US grades are provided as a means of cross referencing chemically similar alloys. Exchangability is only possible after independent examination of specifications. Tensile properties are minimum or typical as specified. UTS and YS as MPa. El as %. See Appendix for list of abbreviations used in Notes. * indicates obsolete material.

Alloy Steel, Nickel Chromium Molybdenum, 4340/4340H (Continued from previous page)

Specification	Designation	Notes	C	Cr	Mn	Mo	Ni	P	S	Si	Other	UTS	YS	El	Hard
Russia															
GOST 4543	40Ch2N2MA		0.35-0.42	1.25-1.65	0.3-0.6	0.2-0.3	1.35-1.75	0.025	0.025	0.17-0.37	Cu 0.3; bal Fe				
GOST 4543(61)	38ChNWA	Q/T	0.34-0.42	1.3-1.7	0.3-0.6	0.15 max	1.25-1.65	0.025 max	0.025 max	0.17-0.37	Al <=0.1; Cu <=0.3; Ti <=0.05; W 0.5-0.8; bal Fe				
GOST 4543(71)	36Ch2N2MFA	Q/T	0.33-0.4	1.3-1.7	0.25-0.5	0.3-0.4	1.3-1.7	0.025 max	0.025 max	0.17-0.37	Al <=0.1; Cu <=0.3; Ti <=0.05; V 0.1-0.18; bal Fe				
GOST 4543(71)	38Ch2N2MA	Q/T	0.33-0.4	1.3-1.7	0.25-0.5	0.2-0.3	1.3-1.7	0.025 max	0.025 max	0.17-0.37	Al <=0.1; Cu <=0.3; Ti <=0.05; bal Fe				
GOST 4543(71)	40Ch2N2MA	Q/T	0.35-0.42	1.25-1.65	0.3-0.6	0.2-0.3	1.35-1.75	0.025 max	0.025 max	0.17-0.37	Al <=0.1; Cu <=0.3; Ti <=0.05; bal Fe				
Spain															
UNE 36012(75)	35NiCrMo4	Q/T	0.32-0.38	0.6-0.9	0.5-0.8	0.15-0.3	0.7-1	0.035 max	0.035 max	0.15-0.4	bal Fe				
UNE 36012(75)	40NiCrMo4	Q/T	0.37-0.42	0.6-0.9	0.5-0.8	0.15-0.3	0.7-1.0	0.035 max	0.035 max	0.15-0.4	bal Fe				
UNE 36012(75)	40NiCrMo7	Q/T	0.37-0.43	0.65-0.95	0.55-0.85	0.15-0.3	1.6-2	0.035	0.035	0.15-0.4	bal Fe				
UNE 36012(75)	F.1272*	see 40NiCrMo7	0.37-0.43	0.65-0.95	0.55-0.85	0.15-0.3	1.6-2	0.035	0.035	0.15-0.4	bal Fe				
UK															
BS 4670	818M40		0.36-0.44	1-1.5	0.45-0.85	0.2-0.4	1.3-1.8	0.04	0.04	0.1-0.35	bal Fe				
BS 970/3(91)	817M40	Bright bar; Q/T; 6<=t<=29mm; hard	0.36-0.44	1.00-1.40	0.45-0.70	0.2-0.35	1.30-1.70	0.035 max	0.040 max	0.10-0.40	bal Fe	1075-1225	940	11	311-375 HB
BS 970/3(91)	817M40	Bright bar; Q/T; 150<t<=250mm; hard	0.36-0.44	1.00-1.40	0.45-0.70	0.2-0.35	1.30-1.70	0.035 max	0.040 max	0.10-0.40	bal Fe	850-1000	650	13	248-302 HB
USA															
	AISI 4340	Smls mech tub	0.38-0.43	0.70-0.90	0.60-0.80	0.20-0.30	1.65-2.00	0.040 max	0.040 max	0.15-0.35	bal Fe				
	AISI 4340	Bar Blm Bil Slab	0.38-0.43	0.70-0.90	0.60-0.80	0.20-0.30	1.65-2.00	0.035 max	0.040 max	0.15-0.35	bal Fe				
	AISI 4340	Plt	0.38-0.43	0.70-0.90	0.60-0.80	0.20-0.30	1.65-2.00	0.035 max	0.040 max	0.15-0.35	bal Fe				
	AISI 4340	Sh Strp	0.38-0.43	0.70-0.90	0.60-0.80	0.20-0.30	1.65-2.00	0.035	0.035	0.15-0.30	bal Fe				
	AISI 4340H	Bar Blm Bil Slab	0.37-0.44	0.65-0.95	0.65-0.85	0.20-0.30	1.65-2.00	0.025 max	0.040 max	0.15-0.35	bal Fe				
	AISI 4340H	Smls mech tub, Hard	0.37-0.44	0.65-0.95	0.65-0.85	0.20-0.30	1.65-2.00	0.025 max	0.040 max	0.15-0.30	bal Fe				
	UNS G43400		0.38-0.43	0.70-0.90	0.60-0.80	0.20-0.30	1.65-2.00	0.035 max	0.040 max	0.15-0.30	bal Fe				
	UNS H43400		0.37-0.44	0.65-0.95	0.55-0.90	0.20-0.30	1.55-2.00	0.035 max	0.040 max	0.15-0.35	bal Fe				
AMS 6359		Sh, Strp, Plt	0.38-0.43	0.7-0.90	0.60-0.8	0.2-0.30	1.65-2	0.025	0.025	0.15-0.35	Cu 0.35; bal Fe				
AMS 6409											bal Fe				
AMS 6414			0.38-0.43	0.7-0.90	0.60-0.90	0.2-0.30	1.65-2	0.015	0.015	0.15-0.35	Cu 0.35; bal Fe				
AMS 6415		Bar Frg Tub	0.38-0.43	0.7-0.90	0.60-0.8	0.2-0.30	1.65-2	0.025	0.025	0.15-0.35	Cu 0.35; bal Fe				
AMS 6454			0.38-0.43	0.7-0.90	0.60-0.90	0.2-0.30	1.65-2	0.015	0.015	0.15-0.35	Cu 0.35; bal Fe				
ASTM A29/A29M(93)	4340	Bar	0.38-0.43	0.70-0.90	0.60-0.80	0.20-0.30	1.65-2.00	0.035 max	0.040 max	0.15-0.35	bal Fe				
ASTM A304(96)	4340H	Bar, hard bands spec	0.37-0.44	0.65-0.95	0.55-0.90	0.20-0.30	1.55-2.00	0.035 max	0.040 max	0.15-0.35	Cu <=0.35; Cu Ni Cr Mo trace allowed; bal Fe				
ASTM A322(96)	4340	Bar	0.38-0.43	0.70-0.90	0.60-0.80	0.20-0.30	1.65-2.00	0.035 max	0.040 max	0.15-0.35	bal Fe				
ASTM A331(95)	4340	Bar	0.38-0.43	0.70-0.90	0.60-0.80	0.20-0.30	1.65-2.00	0.035 max	0.040 max	0.15-0.35	bal Fe				
ASTM A372	Type VII*	Thin wall frg press ves	0.38-0.43	0.7-0.9	0.6-0.8	0.2-0.3	1.65-2	0.035	0.04	0.15-0.35	bal Fe				
ASTM A506(93)	4340	Sh Strp, HR, CR	0.38-0.43	0.70-0.90	0.60-0.80	0.20-0.30	1.65-2.00	0.035	0.035	0.15-0.30	bal Fe				

UNS numbers and US grades are provided as a means of cross referencing chemically similar alloys. Exchangability is only possible after independent examination of specifications. Tensile properties are minimum or typical as specified. UTS and YS as MPa. El as %. See Appendix for list of abbreviations used in Notes. * indicates obsolete material.

Alloy Steel, Nickel Chromium Molybdenum, 4340/4340H (Continued from previous page)

Specification	Designation	Notes	C	Cr	Mn	Mo	Ni	P	S	Si	Other	UTS	YS	El	Hard
USA															
ASTM A519(96)	4340	Smls mech tub	0.38-0.43	0.70-0.90	0.60-0.80	0.20-0.30	1.65-2.00	0.040 max	0.040 max	0.15-0.35	bal Fe				
ASTM A540/A540M(98)	A540 (B23)	Bolting matl	0.37-0.44	0.65-0.95	0.60-0.95	0.20-0.30	1.55-2.00	0.025 max	0.025 max	0.15-0.35	bal Fe	827-1138	724-1034	15-10	248-341/311-444 HB
ASTM A646(95)	Grade 7	Blm Bil for aircraft aerospace frg, ann	0.38-0.43	0.70-0.90	0.65-0.85	0.20-0.30	1.65-2.00	0.025 max	0.025 max	0.20-0.35	bal Fe				285 HB
ASTM A752(93)	4340	Rod Wir	0.38-0.43	0.70-0.90	0.60-0.80	0.20-0.30	1.65-2.00	0.035 max	0.040 max	0.15-0.30	bal Fe	720			
ASTM A829/A829M(95)	4340	Plt	0.36-0.44	0.60-0.90	0.55-0.80	0.20-0.30	1.65-2.00	0.035 max	0.040 max	0.15-0.40	bal Fe	480-965			
DoD-F-24669/1(86)	4340	Bar Bil; Supersedes MIL-S-866 & MIL-S-16974	0.38-0.43	0.7-0.9	0.6-0.8	0.2-0.3	1.65-2	0.035	0.04	0.15-0.3	bal Fe				
FED QQ-S-626C(91)	4340*	Obs; see ASTM A829; Plt	0.38-0.43	0.7-0.95	0.6-0.8	0.2-0.3	1.65-2	0.035	0.04	0.15-0.3	bal Fe				
MIL-S-16974E(86)	4340*	Obs; Bar Bil Blm slab for refrg									bal Fe				
MIL-S-19434B(SH)(90)	Class 2	Gear, pinion frg HT; shipboard propulsion and turbine	0.45 max	1.25 max	0.55-0.90	0.50 max	2.25 max	0.040 max	0.040 max	0.15-0.35	V <=0.10; Si 0.10 max if VCD used; bal Fe	655	483		201-241 HB
MIL-S-24093A(SH)(91)	Type I Class A	Frg for shipboard apps; 10 in. max section thick	0.44 max	0.70-0.95	0.90 max	0.20-0.30	1.65-2.00	0.025 max	0.025 max	0.20-0.35	bal Fe	1138	965	10	
MIL-S-24093A(SH)(91)	Type I Class B	Frg for shipboard apps; 10 in. max section thick	0.44 max	0.70-0.95	0.90 max	0.20-0.30	1.65-2.00	0.025 max	0.025 max	0.20-0.35	bal Fe	965	827	14	
MIL-S-24093A(SH)(91)	Type I Class C	Frg for shipboard apps; 10 in. max section thick	0.44 max	0.70-0.95	0.90 max	0.20-0.30	1.65-2.00	0.025 max	0.025 max	0.20-0.35	bal Fe	827	690	16	
MIL-S-46059	G43400*	Obs; Shape	0.38-0.45	0.7-0.9	0.6-0.8	0.2-0.3	1.65-2	0.035	0.04	0.15-0.3	bal Fe				
MIL-S-5000E(82)	4340*	Obs; see AMS 6415, 6484; Bar frg	0.38-0.43	0.7-0.9	0.65-0.85	0.2-0.3	1.65-2	0.025	0.025	0.15-0.35	Cu 0.35; bal Fe				
MIL-S-83135USAF3(95)	4340M*	Obs for new design; see AMS spec; Bar refrg stock tube	0.38-0.43	0.7-0.95	0.65-0.9	0.35-0.45	1.65-2	0.012	0.012	1.45-1.8	V >=0.05; bal Fe				
MIL-S-8844D(90)	4340*	Obs; see AMS specs; Bar refrg stock tube, premium qual									bal Fe				
SAE 770(84)	4340*	Obs; see J1397(92)									bal Fe				
SAE J1268(95)	4340H	Rod; See std	0.37-0.44	0.65-0.95	0.55-0.90	0.20-0.30	1.55-2.00	0.030 max	0.040 max	0.15-0.35	Cu <=0.35; Cu not spec'd but acpt; bal Fe				
SAE J404(94)	4340	Plt	0.36-0.44	0.60-0.90	0.55-0.80	0.20-0.30	1.65-2.00	0.035 max	0.040 max	0.15-0.35	bal Fe				
SAE J404(94)	4340	Bil Blm Slab Bar HR CF	0.38-0.43	0.70-0.90	0.60-0.80	0.20-0.30	1.65-2.00	0.030 max	0.040 max	0.15-0.35	bal Fe				
SAE J407	4330H*	Obs; see J1268									bal Fe				
Yugoslavia															
	C.5431	Q/T	0.3-0.38	1.4-1.7	0.4-0.7	0.15-0.3	1.4-1.7	0.035 max	0.03 max	0.4 max	bal Fe				

Alloy Steel, Nickel Chromium Molybdenum, 4715

Specification	Designation	Notes	C	Cr	Mn	Mo	Ni	P	S	Si	Other	UTS	YS	El	Hard
Bulgaria															
BDS 6354	15ChGN2TA	Struct	0.13-0.18	0.70-1.00	0.70-1.00	0.15 max	1.40-1.80	0.025 max	0.025 max	0.17-0.37	Cu <=0.30; Ti 0.05-0.09; V <=0.05; W <=0.2; bal Fe				
Italy															
UNI 3097(75)	16CrNi4	Ball & roller bearing	0.13-0.18	0.8-1.1	0.7-1		0.8-1.1	0.035 max	0.035 max	0.15-0.4	bal Fe				
UNI 7846(78)	16CrNi4	CH	0.13-0.18	0.8-1.1	0.7-1	0.1 max	0.8-1.1	0.035 max	0.035 max	0.15-0.4	Pb <=0.30; bal Fe				
UNI 8550(84)	16CrNi4	CH	0.13-0.18	0.8-1.1	0.7-1		0.8-1.1	0.035 max	0.035 max	0.15-0.4	bal Fe				
UNI 8788(85)	16CrNi4	CH	0.13-0.18	0.8-1.1	0.7-1	0.1 max	0.8-1.1	0.035 max	0.035 max	0.15-0.4	Pb 0.15-0.3; P+S<=0.06; bal Fe				

UNS numbers and US grades are provided as a means of cross referencing chemically similar alloys. Exchangability is only possible after independent examination of specifications. Tensile properties are minimum or typical as specified. UTS and YS as MPa. El as %. See Appendix for list of abbreviations used in Notes. * indicates obsolete material.

Specification	Designation	Notes	C	Cr	Mn	Mo	Ni	P	S	Si	Other	UTS	YS	El	Hard

Alloy Steel, Nickel Chromium Molybdenum, 4715 (Continued from previous page)

UK

Specification	Designation	Notes	C	Cr	Mn	Mo	Ni	P	S	Si	Other	UTS	YS	El	Hard
BS 970/1(83)	635H15	Blm Bil Slab Bar Rod Frg CH	0.12-0.18	0.40-0.80	0.60-0.90	0.15 max	0.7-1.1	0.035 max	0.04 max	0.10-0.35	bal Fe				0-207 HB
BS 970/1(83)	635H17	Blm Bil Slab Bar Rod Frg CH	0.14-0.20	0.60-1.00	0.60-0.90	0.15 max	0.85-1.25	0.035 max	0.04 max	0.10-0.35	bal Fe				0-217 HB
BS 970/1(83)	635M15	Blm Bil Slab Bar Rod Frg CH	0.12-0.18	0.40-0.80	0.60-0.90	0.15 max	0.7-1.1	0.040 max	0.05 max	0.10-0.40	bal Fe	770		12	
BS 970/1(83)	637M17	Blm Bil Slab Bar Rod Frg CH	0.14-0.20	0.60-1.00	0.60-0.90	0.15 max	0.85-1.25	0.040 max	0.05 max	0.10-0.40	bal Fe	930		10	
BS 970/1(96)	637A16	Blm Bil Slab Bar Rod Frg	0.14-0.19	0.70-1.00	0.70-0.90	0.10 max	0.90-1.20	0.035 max	0.040 max	0.10-0.35	bal Fe				

USA

Specification	Designation	Notes	C	Cr	Mn	Mo	Ni	P	S	Si	Other	UTS	YS	El	Hard
	AISI 4715	Bar Blm Bil Slab	0.13-0.18	0.45-0.65	0.70-0.90	0.45-0.65	0.70-1.00	0.035 max	0.040 max	0.15-0.35	bal Fe				
	UNS G47150		0.13-0.18	0.45-0.65	0.70-0.90	0.45-0.65	0.70-1.00	0.035 max	0.040 max	0.15-0.35	bal Fe				
ASTM A29/A29M(93)	4715	Bar	0.13-0.18	0.45-0.65	0.70-0.90	0.45-0.60	0.70-1.00	0.035 max	0.040 max	0.15-0.35	bal Fe				
ASTM A322(96)	G47150	Bar	0.03-0.18	0.45-0.65	0.70-0.90	0.45-0.60	0.70-1.00	0.035 max	0.04 max	0.15-0.35	bal Fe				
ASTM A331(95)	G47150	Bar	0.03-0.18	0.45-0.65	0.70-0.90	0.45-0.60	0.70-1.00	0.035 max	0.04 max	0.15-0.35	bal Fe				

Alloy Steel, Nickel Chromium Molybdenum, 4718/4718H

Europe

Specification	Designation	Notes	C	Cr	Mn	Mo	Ni	P	S	Si	Other	UTS	YS	El	Hard
EN 10084(98)	1.6566	CH, Ann	0.14-0.20	0.80-1.10	0.60-0.90	0.15-0.25	1.20-1.50	0.035 max	0.035 max	0.40 max	bal Fe				229 HB
EN 10084(98)	1.6569	CH, Ann	0.14-0.20	0.80-1.10	0.60-0.90	0.15-0.25	1.20-1.50	0.035 max	0.020-0.040	0.40 max	bal Fe				229 HB
EN 10084(98)	17NiCrMo6-4	CH, Ann	0.14-0.20	0.80-1.10	0.60-0.90	0.15-0.25	1.20-1.50	0.035 max	0.035 max	0.40 max	bal Fe				229 HB
EN 10084(98)	17NiCrMoS6-4	CH, Ann	0.14-0.20	0.80-1.10	0.60-0.90	0.15-0.25	1.20-1.50	0.035 max	0.020-0.040	0.40 max	bal Fe				229 HB

Mexico

Specification	Designation	Notes	C	Cr	Mn	Mo	Ni	P	S	Si	Other	UTS	YS	El	Hard
NMX-B-268(68)	4718H	Hard	0.13-0.21	0.30-0.60	0.60-0.95	0.30-0.40	0.85-1.25			0.20-0.35	bal Fe				
NMX-B-300(91)	4718	Bar	0.16-0.21	0.35-0.55	0.70-0.90	0.30-0.40	0.90-1.20	0.035 max	0.040 max	0.15-0.35	bal Fe				
NMX-B-300(91)	4718H	Bar	0.15-0.21	0.30-0.60	0.60-0.95	0.30-0.40	0.85-1.25			0.15-0.35	bal Fe				

USA

Specification	Designation	Notes	C	Cr	Mn	Mo	Ni	P	S	Si	Other	UTS	YS	El	Hard
	AISI 4718	Smls mech tub	0.16-0.21	0.35-0.55	0.70-0.90	0.30-0.40	0.90-1.20	0.040 max	0.040 max	0.15-0.35	bal Fe				
	AISI 4718H	Ni-Cr-Mo H-Alloy	0.15-0.21	0.30-0.60	0.60-0.95	0.30-0.40	0.85-1.25	0.035 max	0.040 max	0.15-0.35	bal Fe				
	UNS G47180		0.16-0.21	0.35-0.55	0.70-0.90	0.30-0.40	0.90-1.20	0.035 max	0.040 max	0.15-0.35	bal Fe				
	UNS H47180		0.15-0.21	0.30-0.60	0.60-0.95	0.30-0.40	0.85-1.25	0.035 max	0.040 max	0.15-0.35	bal Fe				
ASTM A29/A29M(93)	4718	Bar	0.16-0.21	0.35-0.55	0.70-0.90	0.30-0.40	0.90-1.20	0.035 max	0.040 max	0.15-0.35	bal Fe				
ASTM A519(96)	4718	Smls mech tub	0.16-0.21	0.35-0.55	0.70-0.90	0.30-0.40	0.90-1.20	0.040 max	0.040 max	0.15-0.35	bal Fe				
ASTM A752(93)	4718	Rod Wir	0.16-0.21		0.70-0.90		0.90-1.20	0.035 max	0.040 max	0.15-0.30	bal Fe	580			
SAE J1268(95)	4718H	Hard	0.15-0.21	0.30-0.60	0.60-0.95	0.30-0.40	0.85-1.25	0.030 max	0.040 max	0.15-0.35	Cu <=0.35; Cu not spec'd, but acpt; bal Fe				
SAE J1268(95)	E4718H	Hard	0.15-0.21	0.30-0.60	0.60-0.95	0.30-0.40	0.85-1.25	0.025 max	0.025 max	0.15-0.35	Cu <=0.35; Cu not spec'd but acpt; bal Fe				

Alloy Steel, Nickel Chromium Molybdenum, 4720/4720H

France

Specification	Designation	Notes	C	Cr	Mn	Mo	Ni	P	S	Si	Other	UTS	YS	El	Hard
AFNOR NFA35565(84)	18NCD4		0.16-0.22	0.35-0.55	0.5-0.8	0.15-0.3	0.9-1.2	0.03	0.025	0.2-0.35	Cu 0.35; bal Fe				

UNS numbers and US grades are provided as a means of cross referencing chemically similar alloys. Exchangability is only possible after independent examination of specifications. Tensile properties are minimum or typical as specified. UTS and YS as MPa. El as %. See Appendix for list of abbreviations used in Notes. * indicates obsolete material.

Specification	Designation	Notes	C	Cr	Mn	Mo	Ni	P	S	Si	Other	UTS	YS	El	Hard

Alloy Steel, Nickel Chromium Molybdenum, 4720/4720H (Continued from previous page)

Mexico

Specification	Designation	Notes	C	Cr	Mn	Mo	Ni	P	S	Si	Other
DGN B-203	4720*	Obs	0.17-0.22	0.35-0.55	0.5-0.7	0.15-0.25	0.9-1.2	0.04	0.04	0.2-0.35	bal Fe
DGN B-297	4720*	Obs	0.17-0.22	0.35-0.55	0.5-0.7	0.15-0.25	0.9-1.2	0.035	0.04	0.2-0.35	bal Fe
NMX-B-268(68)	4720H	Hard	0.17-0.23	0.30-0.60	0.45-0.75	0.15-0.25	0.85-1.25			0.20-0.35	bal Fe
NMX-B-300(91)	4720	Bar	0.17-0.22	0.35-0.55	0.50-0.70	0.15-0.25	0.90-1.20	0.035 max	0.040 max	0.15-0.35	bal Fe
NMX-B-300(91)	4720H	Bar	0.17-0.23	0.30-0.60	0.45-0.75	0.15-0.25	0.85-1.25			0.15-0.35	bal Fe

Pan America

Specification	Designation	Notes	C	Cr	Mn	Mo	Ni	P	S	Si	Other
COPANT 514	4720	Tube	0.17-0.22	0.35-0.55	0.5-0.7	0.15-0.25	0.9-1.2	0.04	0.04	0.2-0.35	bal Fe

USA

Specification	Designation	Notes	C	Cr	Mn	Mo	Ni	P	S	Si	Other	UTS
	AISI 4720	Smls mech tub	0.17-0.22	0.35-0.55	0.50-0.70	0.15-0.25	0.90-1.20	0.040 max	0.040 max	0.15-0.35	bal Fe	
	AISI 4720	Bar Blm Bil Slab	0.17-0.22	0.35-0.55	0.50-0.70	0.15-0.25	0.90-1.20	0.035 max	0.040 max	0.15-0.35	bal Fe	
	AISI 4720H	Smls mech tub, Hard	0.17-0.23	0.30-0.60	0.45-0.75	0.15-0.25	0.85-1.25	0.040 max	0.040 max	0.15-0.30	bal Fe	
	AISI 4720H	Bar Blm Bil Slab	0.17-0.23	0.30-0.60	0.45-0.75	0.15-0.25	0.85-1.25	0.035 max	0.040 max	0.15-0.35	bal Fe	
	UNS G47200		0.17-0.22	0.35-0.55	0.50-0.70	0.15-0.25	0.90-1.20	0.035 max	0.040 max	0.15-0.35	bal Fe	
	UNS H47200		0.17-0.23	0.30-0.60	0.45-0.75	0.15-0.25	0.85-1.25	0.035 max	0.040 max	0.15-0.35	bal Fe	
ASTM A29/A29M(93)	4720	Bar	0.17-0.22	0.35-0.55	0.50-0.70	0.15-0.25	0.90-1.20	0.035 max	0.040 max	0.15-0.35	bal Fe	
ASTM A304(96)	4720H	Bar, hard bands spec	0.17-0.23	0.30-0.60	0.45-0.75	0.15-0.25	0.85-1.25	0.035 max	0.040 max	0.15-0.35	Cu <=0.35; Cu Ni Cr Mo trace allowed; bal Fe	
ASTM A322(96)	4720	Bar	0.17-0.22	0.35-0.55	0.50-0.70	0.15-0.25	0.90-1.20	0.035 max	0.040 max	0.15-0.35	bal Fe	
ASTM A331(95)	4720	Bar	0.17-0.22	0.35-0.55	0.50-0.70	0.15-0.25	0.90-1.20	0.035 max	0.040 max	0.15-0.35	bal Fe	
ASTM A331(95)	4720H	Bar	0.17-0.23	0.30-0.60	0.45-0.75	0.15-0.25	0.85-1.25	0.025 max	0.025 max	0.15-0.35	Cu <=0.35; bal Fe	
ASTM A519(96)	4720	Smls mech tub	0.17-0.22	0.35-0.55	0.50-0.70	0.15-0.25	0.90-1.20	0.040 max	0.040 max	0.15-0.35	bal Fe	
ASTM A534(94)	4720H	Carburizing, anti-friction bearings	0.17-0.23	0.30-0.60	0.45-0.75	0.15-0.25	0.85-1.25	0.025 max	0.025 max	0.15-0.35	bal Fe	
ASTM A535(92)	4720	Ball & roller bearing	0.17-0.22	0.35-0.55	0.50-0.70	0.15-0.25	0.90-1.20	0.015 max	0.015 max	0.15-0.35	Cu <=0.35; bal Fe	
ASTM A752(93)	4720	Rod Wir	0.17-0.22	0.35-0.55	0.50-0.70	0.15-0.25	0.90-1.20	0.035 max	0.040 max	0.15-0.30	bal Fe	570
SAE 770(84)	4720*	Obs; see J1397(92)	0.17-0.22	0.35-0.55	0.5-0.7	0.15-0.25	0.9-1.2	0.04 max	0.04 max	0.15-0.3	bal Fe	
SAE J1268(95)	4720H	See std	0.17-0.23	0.30-0.60	0.45-0.75	0.15-0.25	0.85-1.25	0.030 max	0.040 max	0.15-0.35	Cu <=0.35; Cu not spec'd but acpt; bal Fe	

Alloy Steel, Nickel Chromium Molybdenum, 6250

Italy

Specification	Designation	Notes	C	Cr	Mn	Mo	Ni	P	S	Si	Other
UNI 3097(75)	10NiCr	Ball & roller bearing	0.08-0.13	1.4-1.75	0.4-0.6		3.25-3.75	0.035 max	0.035 max	0.15-0.4	bal Fe

USA

Specification	Designation	Notes	C	Cr	Mn	Mo	Ni	P	S	Si	Other
	UNS K44910		0.07-0.13	1.25-1.75	0.40-0.70	0.06 max	3.25-3.75	0.025 max	0.025 max	0.20-0.35	Cu <=0.35; bal Fe
MIL-S-83030A2(95)	3310*	Obs for new design; see AMS specs; Bar Bil premium qual									bal Fe

Alloy Steel, Nickel Chromium Molybdenum, 6256

Bulgaria

Specification	Designation	Notes	C	Cr	Mn	Mo	Ni	P	S	Si	Other
BDS 6354	12Ch2N4A	Struct	0.09-0.16	1.25-1.65	0.30-0.60	0.15 max	3.25-3.65	0.025 max	0.025 max	0.17-0.37	Cu <=0.30; Ti <=0.03; V <=0.05; W <=0.2; bal Fe

UNS numbers and US grades are provided as a means of cross referencing chemically similar alloys. Exchangability is only possible after independent examination of specifications. Tensile properties are minimum or typical as specified. UTS and YS as MPa. El as %. See Appendix for list of abbreviations used in Notes. * indicates obsolete material.

Specification	Designation	Notes	C	Cr	Mn	Mo	Ni	P	S	Si	Other	UTS	YS	El	Hard

Alloy Steel, Nickel Chromium Molybdenum, 6256 (Continued from previous page)

Bulgaria

Specification	Designation	Notes	C	Cr	Mn	Mo	Ni	P	S	Si	Other	UTS	YS	El	Hard
BDS 6354	12ChN2	Struct	0.09-0.16	0.60-0.90	0.30-0.60	0.15 max	1.50-1.90	0.035 max	0.035 max	0.17-0.37	Cu <=0.30; Ti <=0.03; V <=0.05; W <=0.2; bal Fe				

UK

Specification	Designation	Notes	C	Cr	Mn	Mo	Ni	P	S	Si	Other	UTS	YS	El	Hard
BS 970/1(96)	832H13	Blm Bil Slab Bar Rod Frg; sub-critically Ann	0.10-0.16	0.70-1.00	0.35-0.60	0.10-0.25	3.00-3.75	0.035 max	0.040 max	0.10-0.35	bal Fe				255 HB max

USA

Specification	Designation	Notes	C	Cr	Mn	Mo	Ni	P	S	Si	Other	UTS	YS	El	Hard
	UNS K71350		0.10-0.16	0.90-1.20	0.40-0.70	4.00-5.00	2.75-3.25	0.010 max	0.010 max	0.40-0.60	Al 0.03-0.12; Cu <=0.35; V 0.25-0.50; bal Fe				

Alloy Steel, Nickel Chromium Molybdenum, 6263

Bulgaria

Specification	Designation	Notes	C	Cr	Mn	Mo	Ni	P	S	Si	Other	UTS	YS	El	Hard
BDS 6354	14Ch2N3MA	Struct	0.12-0.17	1.50-1.75	0.30-0.60	0.20-0.30	2.75-3.15	0.025 max	0.025 max	0.17-0.37	Cu <=0.03; Ti <=0.03; V <=0.05; W <=0.2; bal Fe				

Italy

Specification	Designation	Notes	C	Cr	Mn	Mo	Ni	P	S	Si	Other	UTS	YS	El	Hard
UNI 6935(71)	10NiCrMo13	Aircraft material	0.08-0.13	1-1.4	0.4-0.7	0.08	3-3.5	0.03 max	0.025 max	0.2-0.35	Cu <=0.35; bal Fe				

USA

Specification	Designation	Notes	C	Cr	Mn	Mo	Ni	P	S	Si	Other	UTS	YS	El	Hard
	UNS K44414		0.11-0.17	1.00-1.40	0.40-0.70	0.08-0.15	3.00-3.50	0.025 max	0.025 max	0.15-0.35	Cu <=0.35; bal Fe				

Alloy Steel, Nickel Chromium Molybdenum, 6266

Bulgaria

Specification	Designation	Notes	C	Cr	Mn	Mo	Ni	P	S	Si	Other	UTS	YS	El	Hard
BDS 4758	10GT	HR, con reinf	0.13 max	0.30 max	1.00-1.40		0.30 max	0.030 max	0.040 max	0.45-0.65	Al 0.02-0.05; Ti 0.015-0.05; As<=0.08				

USA

Specification	Designation	Notes	C	Cr	Mn	Mo	Ni	P	S	Si	Other	UTS	YS	El	Hard
	UNS K21028		0.08-0.13	0.40-0.60	0.75-1.00	0.20-0.30	1.65-2.00	0.025 max	0.025 max	0.20-0.40	B <=0.007; Cu <=0.35; V 0.03-0.08; bal Fe				

Alloy Steel, Nickel Chromium Molybdenum, 6302

USA

Specification	Designation	Notes	C	Cr	Mn	Mo	Ni	P	S	Si	Other	UTS	YS	El	Hard
	UNS K23015		0.27-0.33	1.00-1.50	0.45-0.65	0.40-0.60	0.25 max	0.025 max	0.025 max	0.55-0.75	Cu <=0.35; V 0.20-0.30; bal Fe				

Alloy Steel, Nickel Chromium Molybdenum, 6303

UK

Specification	Designation	Notes	C	Cr	Mn	Mo	Ni	P	S	Si	Other	UTS	YS	El	Hard
BS 970/3(91)	722M24	Bright bar; Q/T; 6<t<=150mm; hard	0.20-0.28	3.00-3.50	0.45-0.70	0.45-0.65		0.035 max	0.040 max	0.10-0.40	bal Fe	850-1000	680	13	248-302 HB
BS 970/3(91)	722M24	Bright bar; Q/T; 150<t<=250mm; hard CD	0.20-0.28	3.00-3.50	0.45-0.70	0.45-0.65		0.035 max	0.040 max	0.10-0.40	bal Fe	850-1000	650	13	248-302 HB

USA

Specification	Designation	Notes	C	Cr	Mn	Mo	Ni	P	S	Si	Other	UTS	YS	El	Hard
	UNS K22770		0.25-0.30	1.00-1.50	0.60-0.90	0.40-0.60	0.50 max	0.025 max	0.025 max	0.55-0.75	Cu <=0.50; V 0.75-0.95; bal Fe				

Alloy Steel, Nickel Chromium Molybdenum, 6304

USA

Specification	Designation	Notes	C	Cr	Mn	Mo	Ni	P	S	Si	Other	UTS	YS	El	Hard
	UNS K14675	17-22A	0.42-0.50	0.80-1.10	0.40-0.70	0.45-0.65	0.25 max	0.025 max	0.025 max	0.20-0.35	Cu <=0.35; V 0.25-0.35; bal Fe				
SAE J467(68)	17-22-A		0.45	1.25	0.55	0.55				0.65	V 0.30; bal Fe	1165	1110	13	341 HB
SAE J467(68)	17-22-A S		0.30	1.25	0.55	0.50				0.65	V 0.25; bal Fe	1055	924	18	320 HB
SAE J467(68)	17-22-A V		0.28	1.25	0.75	0.50				0.65	V 0.85; bal Fe	1103	1000	17	341 HB

Alloy Steel, Nickel Chromium Molybdenum, 6308

USA

Specification	Designation	Notes	C	Cr	Mn	Mo	Ni	P	S	Si	Other	UTS	YS	El	Hard
	UNS K71040	Pyrowear Alloy 53	0.07-0.13	0.75-1.25	0.25-0.50	3.00-3.50	1.60-2.40	0.015 max	0.010 max	0.60-1.20	Cu 1.80-2.30; V 0.05-0.15; bal Fe				

Specification	Designation	Notes	C	Cr	Mn	Mo	Ni	P	S	Si	Other	UTS	YS	El	Hard
Alloy Steel, Nickel Chromium Molybdenum, 6324															
France															
AFNOR NFA36205/4(94)	11MnNi5-3	Press ves; Sh Plt Strp; Press, low-temp, Norm t<=30mm	0.14 max	0.3 max	0.70-1.50	0.15 max	0.15-0.8	0.025 max	0.015 max	0.5 max *	Al >=0.02; Cu <=0.30; Nb <=0.05; Ti <=0.05; V <=0.05; Cr+Cu+Mo<=0.5; bal Fe	420-530	285	24	
AFNOR NFA36205/4(94)	13MnNi6-3	Press ves; t<=30mm	0.16 max	0.3 max	0.85-1.70	0.15 max	0.15-0.85	0.025 max	0.015 max	0.5 max	Al >=0.02; Cu <=0.30; Nb <=0.05; Ti <=0.05; V <=0.05; Cr+Cu+Mo<=0.5; bal Fe	490-610	355	22	
USA															
	UNS K11640		0.38-0.43	0.55-0.75	0.75-1.00	0.20-0.30	0.55-0.85	0.025 max	0.025 max	0.20-0.35	Cu <=0.35; bal Fe				
Alloy Steel, Nickel Chromium Molybdenum, 6328															
USA															
	UNS K13550		0.48-0.53	0.40-0.60	0.75-1.00	0.20-0.30	0.40-1.00	0.025 max	0.025 max	0.15-0.35	Cu <=0.35; bal Fe				
Alloy Steel, Nickel Chromium Molybdenum, 6396															
USA															
	UNS K22950		0.49-0.55	0.70-0.90	0.65-0.85	0.20-0.30	1.65-2.00	0.025 max	0.025 max	0.15-0.35	Cu <=0.35; bal Fe				
Alloy Steel, Nickel Chromium Molybdenum, 6406															
USA															
	UNS K34378		0.41-0.46	1.90-2.25	0.75-1.00	0.45-0.60	0.25 max	0.015 max	0.015 max	1.40-1.75	Cu <=0.35; V 0.03-0.08; bal Fe				
Alloy Steel, Nickel Chromium Molybdenum, 6407															
Japan															
JIS G4103(79)	SNCM625	HR Frg Bar Rod	0.20-0.30	1.00-1.50	0.35-0.60	0.15-0.30	3.00-3.50	0.030 max	0.030 max	0.15-0.35	Cu <=0.30; bal Fe				
UK															
BS 970/2(70)	823M30	Q/T	0.26-0.34	1.8-2.2	0.35-0.60	0.3-0.5	1.8-2.2	0.025 max	0.025 max	0.1-0.35	Pb <=0.15; Ti <=0.05; bal Fe				
USA															
	UNS K33020		0.27-0.33	1.00-1.35	0.60-0.80	0.35-0.55	1.85-2.25	0.025 max	0.025 max	0.40-0.70	Cu <=0.35; bal Fe				
Alloy Steel, Nickel Chromium Molybdenum, 6411															
Japan															
JIS G4103(79)	SNCM431	HR Frg	0.27-0.35	0.60-1.00	0.60-0.90	0.15-0.30	1.60-2.00	0.030 max	0.030 max	0.15-0.35	Cu <=0.30; bal Fe				
USA															
	UNS K23080		0.28-0.33	0.75-1.00	0.75-1.00	0.35-0.50	1.65-2.00	0.015 max	0.015 max	0.20-0.35	Cu <=0.35; V 0.05-0.10; bal Fe				
Alloy Steel, Nickel Chromium Molybdenum, 6422															
USA															
	UNS K11940		0.38-0.43	0.70-0.90	0.65-0.85	0.15-0.25	0.70-1.00	0.025 max	0.025 max	0.15-0.35	B 0.0005-0.005; Cu <=0.35; V 0.01-0.06; bal Fe				
Alloy Steel, Nickel Chromium Molybdenum, 6426															
USA															
	UNS K18597		0.80-0.90	0.85-1.15	0.20-0.50	0.50-0.65	0.15 max	0.015 max	0.015 max	0.60-0.90	Cu <=0.15; bal Fe				
Alloy Steel, Nickel Chromium Molybdenum, 6475															
Japan															
JIS G4052(79)	SCM822H	Struct Hard	0.19-0.25	0.85-1.25	0.55-0.90	0.35-0.45	0.25 max	0.030 max	0.030 max	0.15-0.35	Cu <=0.30; bal Fe				
USA															
	UNS K52355		0.21-0.26	1.00-1.25	0.50-0.70	0.20-0.30	3.25-3.75	0.025 max	0.025 max	0.20-0.40	Al 1.10-1.40; Cu <=0.35; bal Fe				

UNS numbers and US grades are provided as a means of cross referencing chemically similar alloys. Exchangability is only possible after independent examination of specifications. Tensile properties are minimum or typical as specified. UTS and YS as MPa. El as %. See Appendix for list of abbreviations used in Notes. * indicates obsolete material.

Specification	Designation	Notes	C	Cr	Mn	Mo	Ni	P	S	Si	Other	UTS	YS	El	Hard

Alloy Steel, Nickel Chromium Molybdenum, 6550

USA

| | UNS K13048 | | 0.28-0.33 | 0.40-0.60 | 0.70-0.90 | 0.15-0.25 | 0.40-0.70 | 0.025 max | 0.025 max | 0.20-0.35 | Cu <=0.35; bal Fe | | | | |

Alloy Steel, Nickel Chromium Molybdenum, 8115/8115H

Australia

| AS 1444(96) | 8115 | Standard series, Hard/Tmp | 0.13-0.18 | 0.30-0.50 | 0.70-0.90 | 0.08-0.15 | 0.20-0.40 | 0.040 max | 0.040 max | 0.10-0.35 | bal Fe | | | | |
| AS 1444(96) | 8115H | H series, Hard/Tmp | 0.12-0.18 | 0.30-0.55 | 0.60-0.95 | 0.08-0.15 | 0.20-0.40 | 0.040 max | 0.040 max | 0.10-0.35 | bal Fe | | | | |

Mexico

| NMX-B-300(91) | 8115 | Bar | 0.13-0.18 | 0.30-0.50 | 0.70-0.90 | 0.08-0.15 | 0.20-0.40 | 0.035 max | 0.040 max | 0.15-0.35 | bal Fe | | | | |

USA

	AISI 8115	Smls mech tub	0.13-0.18	0.30-0.50	0.70-0.90	0.08-0.15	0.20-0.40	0.040 max	0.040 max	0.15-0.35	bal Fe				
	UNS G81150		0.13-0.18	0.30-0.50	0.70-0.90	0.08-0.15	0.20-0.40	0.035 max	0.040 max	0.15-0.35	bal Fe				
ASTM A29/A29M(93)	8115	Bar	0.13-0.18	0.30-0.50	0.70-0.90	0.08-0.15	0.20-0.40	0.035 max	0.040 max	0.15-0.35	bal Fe				
ASTM A519(96)	8115	Smls mech tub	0.13-0.18	0.30-0.50	0.70-0.90	0.08-0.15	0.20-0.40	0.040 max	0.040 max	0.15-0.35	bal Fe				

Alloy Steel, Nickel Chromium Molybdenum, 81B45/81B45H

Mexico

NMX-B-268(68)	81B45H	Hard	0.42-0.49	0.30-0.60	0.70-1.05	0.08-0.15	0.15-0.45			0.20-0.35	B >=0.0005; bal Fe				
NMX-B-300(91)	81B45	Bar	0.43-0.48	0.35-0.55	0.70-0.90	0.08-0.15	0.20-0.40	0.035 max	0.040 max	0.20-0.35	B >=0.0005; bal Fe				
NMX-B-300(91)	81B45H	Bar	0.42-0.49	0.30-0.60	0.70-1.05	0.08-0.15	0.15-0.45			0.15-0.35	B >=0.0005; bal Fe				

Pan America

| COPANT 514 | 81B45 | Tube | 0.43-0.48 | 0.35-0.55 | 0.75-1 | 0.08-0.15 | 0.2-0.4 | 0.04 | 0.05 | 0.1 | B <=0.003; bal Fe | | | | |

USA

	AISI 81B45	Smls mech Tub; Boron	0.43-0.48	0.35-0.55	0.75-1.00	0.08-0.15	0.20-0.40	0.040 max	0.040 max	0.15-0.35	B >=0.0005; bal Fe				
	AISI 81B45H	Smls mech tub, Hard; Boron	0.42-0.49	0.30-0.60	0.70-1.05	0.08-0.15	0.15-0.45	0.035 max	0.040 max	0.15-0.30	B 0.0005-0.005; bal Fe				
	UNS G81451		0.43-0.48	0.35-0.55	0.75-1.00	0.08-0.15	0.20-0.40	0.035 max	0.040 max	0.15-0.35	B >=0.0005; bal Fe				
	UNS H81451		0.42-0.49	0.30-0.60	0.70-1.05	0.08-0.15	0.15-0.45	0.035 max	0.040 max	0.15-0.35	B 0.0005-0.003; bal Fe				
ASTM A29/A29M(93)	81B45	Bar	0.43-0.48	0.35-0.55	0.75-1	0.08-0.15	0.2-0.4	0.035 max	0.04 max	0.15-0.35	bal Fe				
ASTM A304(96)	81B45H	Bar, hard bands spec	0.42-0.49	0.30-0.60	0.70-1.05	0.08-0.15	0.15-0.45	0.035 max	0.040 max	0.15-0.35	B >=0.0005; Cu <=0.35; Cu Ni Cr Mo trace allowed; bal Fe				
ASTM A322(96)	81B45	Bar	0.43-0.48	0.35-0.55	0.75-1.00	0.08-0.15	0.20-0.40	0.035 max	0.040 max	0.15-0.35	B 0.0005-0.005; bal Fe				
ASTM A331(95)	81B45	Bar	0.43-0.48	0.35-0.55	0.75-1.00	0.08-0.15	0.20-0.40	0.035 max	0.040 max	0.15-0.35	bal Fe				
ASTM A331(95)	81B45H	Bar	0.42-0.49	0.30-0.60	0.70-1.05	0.08-0.15	0.15-0.45	0.025 max	0.025 max	0.15-0.35	B >=0.0005; Cu <=0.35; bal Fe				
ASTM A519(96)	81B45	Smls mech tub	0.43-0.48	0.35-0.55	0.75-1.00	0.08-0.15	0.20-0.40	0.040 max	0.040 max	0.15-0.35	B >=0.0005; bal Fe				
ASTM A752(93)	81B45	Rod Wir, Boron	0.43-0.48	0.35-0.55	0.75-1.00	0.08-0.15	0.20-0.40	0.035 max	0.040 max	0.15-0.30	B >=0.0005; bal Fe	630			
SAE 770(84)	81B45*	Obs; see J1397(92)	0.43-0.48	0.35-0.55	0.75-1	0.08-0.15	0.2-0.4	0.04 max	0.04 max	0.15-0.3	B 0.001-0.005; bal Fe				
SAE J1268(95)	81B45H	Bar; See std	0.42-0.49	0.30-0.60	0.70-1.05	0.08-0.15	0.15-0.45	0.030 max	0.040 max	0.15-0.35	B 0.0005-0.003; Cu <=0.35; Cu not spec'd but acpt; bal Fe				

Alloy Steel, Nickel Chromium Molybdenum, 8615

Specification	Designation	Notes	C	Cr	Mn	Mo	Ni	P	S	Si	Other	UTS	YS	El	Hard
Argentina															
IAS	IRAM 8615		0.13-0.18	0.40-0.60	0.70-0.90	0.15-0.25	0.40-0.70	0.035 max	0.040 max	0.20-0.35	bal Fe	600	320	25	170 HB
Australia															
AS G18	EN361*	Obs; see AS 1444	0.13-0.17	0.55-0.8	0.7-1	0.08-0.15	0.4-0.7	0.05	0.05	0.35	bal Fe				
China															
GB 3077(88)	20CrNiMo	Bar HR Q/T 15mm diam	0.17-0.23	0.40-0.70	0.60-0.95	0.20-0.30	0.35-0.75	0.035 max	0.035 max	0.17-0.37	Cu <=0.30; bal Fe	980	785	9	
GB 3203(82)	G20CrNiMo	Bar HR/CD Q/T 15mm diam	0.17-0.23	0.35-0.65	0.60-0.90	0.15-0.30	0.40-0.70	0.030 max	0.030 max	0.15-0.40	Cu <=0.25; bal Fe	1175		9	
France															
AFNOR	15NCD2		0.13-0.18	0.4-0.6	0.7-0.9	0.15-0.25	0.4-0.7	0.04	0.035	0.1-0.4	bal Fe				
AFNOR	15NCD4		0.12-0.19	0.4-0.7	0.5-0.9	0.1-0.2	1-1.3	0.04	0.035	0.1-0.4	bal Fe				
Italy															
UNI 3097(75)	12NiCr3	Ball & roller bearing	0.09-0.15	0.4-0.7	0.3-0.6		0.5-0.8	0.035 max	0.035 max	0.15-0.4	bal Fe				
UNI 6403(86)	12NiCr3	Q/T; Smls Tub	0.09-0.15	0.4-0.7	0.3-0.6	0.1 max	0.5-0.8	0.035 max	0.035 max	0.15-0.35	bal Fe				
UNI 7846(78)	12NiCr3	CH	0.09-0.15	0.4-0.7	0.3-0.6	0.1 max	0.5-0.8	0.035 max	0.035 max	0.15-0.4	Pb 0.15-0.3; bal Fe				
UNI 8550(84)	12NiCr3	CH	0.09-0.15	0.4-0.7	0.3-0.6		0.5-0.8	0.035 max	0.035 max	0.15-0.4	bal Fe				
UNI 8788(85)	12NiCr3	CH	0.09-0.15	0.4-0.7	0.3-0.6	0.1 max	0.5-0.8	0.035 max	0.035 max	0.15-0.4	Pb 0.15-0.3; P+S<=0.06; bal Fe				
Mexico															
DGN B-203	8615*	Obs	0.13-0.18	0.4-0.6	0.7-0.9	0.15-0.25	0.4-0.7	0.04	0.04	0.2-0.35	bal Fe				
DGN B-297	8615*	Obs	0.13-0.18	0.4-0.6	0.7-0.9	0.15-0.25	0.4-0.7	0.035	0.04	0.2-0.35	bal Fe				
NMX-B-300(91)	8615	Bar	0.13-0.18	0.40-0.60	0.70-0.90	0.15-0.25	0.40-0.70	0.035 max	0.040 max	0.15-0.35	bal Fe				
Pan America															
COPANT 334	8615	Bar	0.13-0.18	0.4-0.6	0.7-0.9	0.15-0.25	0.4-0.7	0.04	0.04	0.2-0.35	bal Fe				
COPANT 514	8615		0.13-0.18	0.4-0.6	0.7-0.9	0.15-0.25	0.4-0.7	0.04	0.05	0.1	bal Fe				
COPANT 514	8615	Tube	0.13-0.18	0.4-0.6	0.7-0.9	0.15-0.25	0.4-0.7	0.04	0.05	0.1	bal Fe				
Russia															
GOST 1414(75)	AS14ChGN	Free-cutting	0.13-0.18	0.8-1.1	0.7-1	0.10 max	0.8-1.1	0.035 max	0.035 max	0.17-0.37	Cu <=0.30; Pb 0.15-0.3; P+S<=0.06; bal Fe				
GOST 1414(75)	AS19ChGN	Free-cutting	0.16-0.21	0.8-1.1	0.7-1.1	0.10 max	0.8-1.1	0.035 max	0.035 max	0.17-0.37	Cu <=0.30; Pb 0.15-0.3; P+S<=0.06; bal Fe				
UK															
BS 970/1(83)	805A15*	Blm Bil Slab Bar Rod Frg	0.13-0.18	0.4-0.6	0.70-0.90	0.15-0.25	0.4-0.7	0.04	0.05	0.1-0.35	bal Fe				
BS 970/1(83)	808H17	Blm Bil Slab Bar Rod Frg CH	0.14-0.20	0.35-0.65	0.70-1.05	0.30-0.40	0.35-0.75	0.035 max	0.04 max	0.10-0.35	bal Fe	930		10	
BS 970/1(83)	808M17	Blm Bil Slab Bar Rod Frg CH	0.14-0.20	0.35-0.65	0.70-1.05	0.30-0.40	0.35-0.75	0.035 max	0.04 max	0.10-0.35	bal Fe	930		10	
BS 970/1(96)	635A14	Blm Bil Slab Bar Rod Frg	0.12-0.17	0.50-0.75	0.70-0.90	0.10 max	0.70-1.00	0.035 max	0.040 max	0.10-0.35	bal Fe				
USA															
	AISI 8615	Smls mech tub	0.13-0.18	0.40-0.60	0.70-0.90	0.15-0.25	0.40-0.70	0.040 max	0.040 max	0.15-0.35	bal Fe				
	AISI 8615	Sh Strp	0.13-0.18	0.04-0.60	0.70-0.90	0.15-0.25	0.40-0.70	0.035	0.035	0.15-0.30	bal Fe				
	AISI 8615	Plt	0.12-0.18	0.35-0.60	0.60-0.90	0.15-0.25	0.40-0.70	0.035 max	0.040 max	0.15-0.40	bal Fe				
	AISI 8615	Bar Blm Bil Slab	0.13-0.18	0.40-0.60	0.70-0.90	0.15-0.25	0.40-0.70	0.035 max	0.040 max	0.15-0.35	bal Fe				

Specification	Designation	Notes	C	Cr	Mn	Mo	Ni	P	S	Si	Other	UTS	YS	El	Hard

Alloy Steel, Nickel Chromium Molybdenum, 8615 (Continued from previous page)

USA

Specification	Designation	Notes	C	Cr	Mn	Mo	Ni	P	S	Si	Other	UTS	YS	El	Hard
	UNS G86150		0.13-0.18	0.40-0.60	0.70-0.90	0.15-0.25	0.40-0.70	0.035 max	0.040 max	0.15-0.35	bal Fe				
AMS 6270		Bar Frg Tub	0.11-0.17	0.4-0.60	0.7-1	0.15-0.25	0.4-0.7	0.025	0.025	0.15-0.35	Cu 0.35; bal Fe				
ASTM A29/A29M(93)	8615	Bar	0.13-0.18	0.40-0.60	0.7-0.9	0.15-0.25	0.4-0.7	0.035 max	0.04 max	0.15-0.35	bal Fe				
ASTM A322(96)	8615	Bar	0.13-0.18	0.40-0.60	0.70-0.90	0.15-0.25	0.40-0.70	0.035 max	0.040 max	0.15-0.35	bal Fe				
ASTM A331(95)	8615	Bar	0.13-0.18	0.40-0.60	0.70-0.90	0.15-0.25	0.40-0.70	0.035 max	0.040 max	0.15-0.35	bal Fe				
ASTM A506(93)	8615	Sh Strp, HR, CR	0.13-0.18	0.040-0.60	0.70-0.90	0.15-0.25	0.40-0.70	0.035	0.035	0.15-0.30	bal Fe				
ASTM A507(93)	8615	Sh Strp, HR, CR	0.13-0.18	0.40-0.60	0.70-0.90	0.15-0.25	0.40-0.70	0.025	0.030	0.15-0.30	bal Fe				
ASTM A519(96)	8615	Smls mech tub	0.13-0.18	0.40-0.60	0.70-0.90	0.15-0.25	0.40-0.70	0.040 max	0.040 max	0.15-0.35	bal Fe				
ASTM A752(93)	8615	Rod Wir	0.13-0.18	0.40-0.60	0.70-0.90	0.15-0.25	0.40-0.70	0.035 max	0.040 max	0.15-0.30	bal Fe	550			
ASTM A829/A829M(95)	8615	Plt	0.12-0.18	0.35-0.60	0.60-0.90	0.15-0.25	0.40-0.70	0.035 max	0.040 max	0.15-0.40	bal Fe	480-965			
DoD-F-24669/1(86)	8615	Bar Bil; Supersedes MIL-S-866 & MIL-S-16974	0.13-0.18	0.4-0.6	0.7-0.9	0.15-0.25	0.4-0.7	0.035	0.04	0.15-0.3	bal Fe				
MIL-S-866C	- - -*	Obs									bal Fe				
SAE 770(84)	8615*	Obs; see J1397(92)	0.13-0.18	0.4-0.6	0.7-0.9	0.15-0.25	0.4-0.7	0.04 max	0.04 max	0.15-0.3	bal Fe				
SAE J404(94)	8615	Bil Blm Slab Bar HR CF	0.13-0.18	0.40-0.60	0.70-0.90	0.15-0.25	0.40-0.70	0.030 max	0.040 max	0.15-0.35	bal Fe				
SAE J404(94)	8615	Plt	0.12-0.18	0.35-0.60	0.60-0.90	0.15-0.25	0.40-0.70	0.035 max	0.040 max	0.15-0.35	bal Fe				

Alloy Steel, Nickel Chromium Molybdenum, 8617/8617H

Australia

Specification	Designation	Notes	C	Cr	Mn	Mo	Ni	P	S	Si	Other	UTS	YS	El	Hard
AS 1444(96)	8617	Standard series, Hard/Tmp	0.15-0.20	0.40-0.60	0.70-0.90	0.15-0.25	0.40-0.70	0.040 max	0.040 max	0.10-0.35	bal Fe				
AS 1444(96)	8617H	H series, Hard/Tmp	0.14-0.20	0.35-0.65	0.60-0.95	0.15-0.25	0.35-0.75	0.040 max	0.040 max	0.10-0.35	bal Fe				*

France

Specification	Designation	Notes	C	Cr	Mn	Mo	Ni	P	S	Si	Other	UTS	YS	El	Hard
AFNOR	15NCD2		0.13-0.18	0.4-0.6	0.7-0.9	0.15-0.25	0.4-0.7	0.04 max	0.04 max	0.1-0.4	bal Fe				
AFNOR NFA35551(86)	18NCD6		0.14-0.2	0.85-1.15	0.6-0.9	0.15-0.3	1.2-1.6	0.035	0.035	0.1-0.4	bal Fe				
AFNOR NFA35553	20NCD2	Strp	0.18-0.23	0.4-0.6	0.7-0.9	0.15-0.3	0.4-0.7	0.04	0.04	0.1-0.4	bal Fe				
AFNOR NFA35565(84)	18NCD4		0.16-0.22	0.35-0.55	0.5-0.8	0.15-0.3	0.9-1.2	0.03	0.025	0.2-0.35	Cu 0.35; bal Fe				
AFNOR NFA35566	22NCD2		0.2-0.25	0.4-0.65	0.65-0.95	0.15-0.25	0.4-0.7	0.03 max	0.025 max	0.1-0.35	Al >=0.02; bal Fe				

India

Specification	Designation	Notes	C	Cr	Mn	Mo	Ni	P	S	Si	Other	UTS	YS	El	Hard
IS 1570/4(88)	16Ni3Cr2		0.12-0.2	0.4-0.8	0.6-1	0.15 max	0.6-1	0.07 max	0.06 max	0.15-0.35	bal Fe	690		15	
IS 1570/4(88)	16Ni4Cr3		0.12-0.2	0.6-1	0.6-1	0.15 max	0.8-1.2	0.07 max	0.06 max	0.15-0.35	bal Fe	840		12	

Japan

Specification	Designation	Notes	C	Cr	Mn	Mo	Ni	P	S	Si	Other	UTS	YS	El	Hard
JIS G4052(79)	SNCM21H*	Obs; see SNCM 220H	0.17-0.23	0.35-0.65	0.6-0.95	0.15-0.3	0.35-0.75	0.03	0.03	0.15-0.35	Cu 0.3; bal Fe				

Mexico

Specification	Designation	Notes	C	Cr	Mn	Mo	Ni	P	S	Si	Other	UTS	YS	El	Hard
NMX-B-268(68)	8617H	Hard	0.14-0.20	0.35-0.65	0.60-0.95	0.15-0.25	0.35-0.75			0.20-0.35	bal Fe				
NMX-B-300(91)	8617	Bar	0.15-0.20	0.40-0.60	0.70-0.90	0.15-0.25	0.40-0.70	0.035 max	0.040 max	0.15-0.35	bal Fe				
NMX-B-300(91)	8617H	Bar	0.14-0.20	0.35-0.65	0.60-0.95	0.15-0.25	0.35-0.75			0.15-0.35	bal Fe				

Pan America

Specification	Designation	Notes	C	Cr	Mn	Mo	Ni	P	S	Si	Other	UTS	YS	El	Hard
COPANT 334	8617	Bar	0.15-0.2	0.4-0.6	0.7-0.9	0.15-0.25	0.4-0.7	0.04	0.04	0.2-0.35	bal Fe				

UNS numbers and US grades are provided as a means of cross referencing chemically similar alloys. Exchangability is only possible after independent examination of specifications. Tensile properties are minimum or typical as specified. UTS and YS as MPa. El as %. See Appendix for list of abbreviations used in Notes. * indicates obsolete material.

Alloy Steel, Nickel Chromium Molybdenum, 8617/8617H (Continued from previous page)

Specification	Designation	Notes	C	Cr	Mn	Mo	Ni	P	S	Si	Other	UTS	YS	El	Hard
Pan America															
COPANT 514	8617	Tube	0.15-0.2	0.4-0.6	0.7-0.9	0.15-0.25	0.4-0.7	0.04	0.05	0.1	bal Fe				
Russia															
GOST 10702(78)	15ChGNM	for CHd	0.13-0.18	0.4-0.7	0.7-1.1	0.15-0.25	0.4-0.7	0.035 max	0.04 max	0.17-0.37	Al <=0.1; Cu <=0.3; Ti <=0.05; bal Fe				
UK															
BS 4670:1971	785M19	Frg Hard Tmp, 500mm diam	0.15-0.23	0.4	1.4-1.8	0.15-0.35	0.4-0.7	0.04	0.04	0.1-0.35	bal Fe				
BS 970/1(83)	805H17	Blm Bil Slab Bar Rod Frg CH	0.14-0.20	0.35-0.65	0.60-0.95	0.15-0.25	0.35-0.75				bal Fe				207 HB max
BS 970/1(83)	805M17	Blm Bil Slab Bar Rod Frg CH	0.14-0.20	0.35-0.65	0.60-0.95	0.15-0.25	0.35-0.75	0.040 max	0.05 max	0.10-0.40	bal Fe	770		12	
BS 970/1(96)	805A17	Blm Bil Slab Bar Rod Frg	0.15-0.20	0.40-0.60	0.70-0.90	0.15-0.25	0.40-0.70	0.035 max	0.040 max	0.10-0.35	bal Fe				
BS 970/1(96)	805M17	Blm Bil Slab Bar Rod Frg	0.14-0.20	0.35-0.65	0.60-0.95	0.15-0.25	0.35-0.75	0.035	0.035 max	0.10-0.35	bal Fe				
BS 970/3(91)	805A17	Bright bar	0.15-0.20	0.40-0.60	0.70-0.90	0.15-0.25	0.40-0.70				bal Fe				
USA															
	AISI 8617	Plt	0.15-0.21	0.40-0.70	0.60-0.90	0.15-0.25	0.40-0.70	0.035 max	0.040 max	0.15-0.40	bal Fe				
	AISI 8617	Smls mech tub	0.15-0.20	0.40-0.60	0.70-0.90	0.15-0.25	0.40-0.70	0.040 max	0.040 max	0.15-0.35	bal Fe				
	AISI 8617	Bar Blm Bil Slab	0.15-0.20	0.40-0.60	0.70-0.90	0.15-0.25	0.40-0.70	0.035 max	0.040 max	0.15-0.35	bal Fe				
	AISI 8617H	Smls mech tub, Hard	0.14-0.2	0.35-0.65	0.65-0.95	0.15-0.25	0.35-0.75	0.040 max	0.040 max	0.15-0.30	bal Fe				
	AISI 8617H	Bar Blm Bil Slab	0.14-0.2	0.35-0.65	0.65-0.95	0.15-0.25	0.35-0.75	0.035 max	0.040 max	0.15-0.35	bal Fe				
	UNS G86170		0.15-0.20	0.40-0.60	0.70-0.90	0.15-0.25	0.40-0.70	0.035 max	0.040 max	0.15-0.35	bal Fe				
	UNS H86170		0.14-0.20	0.35-0.65	0.60-0.95	0.15-0.25	0.35-0.75	0.035 max	0.040 max	0.15-0.3	bal Fe				
AMS 6272		Bar Frg Tub	0.15-0.2	0.4-0.60	0.7-1	0.15-0.25	0.4-0.7	0.025	0.025	0.15-0.35	Cu 0.35; bal Fe				
AMS 6272		Bar Rod Wir	0.15-0.2	0.4-0.60	0.7-0.90	0.15-0.25	0.4-0.7	0.040 max	0.040 max	0.15-0.30	bal Fe				
ASTM A29/A29M(93)	8617	Bar	0.15-0.20	0.40-0.60	0.7-0.9	0.15-0.25	0.4-0.7	0.035 max	0.04	0.15-0.35	bal Fe				
ASTM A304(96)	8617H	Bar, hard bands spec	0.14-0.20	0.35-0.65	0.60-0.95	0.15-0.25	0.35-0.75	0.035 max	0.040 max	0.15-0.35	Cu <=0.35; Cu Ni Cr Mo trace allowed; bal Fe				
ASTM A322(96)	8617	Bar	0.15-0.20	0.40-0.60	0.70-0.90	0.15-0.25	0.40-0.70	0.035 max	0.040 max	0.15-0.35	bal Fe				
ASTM A331(95)	8617	Bar	0.15-0.20	0.40-0.60	0.70-0.90	0.15-0.25	0.40-0.70	0.035 max	0.040 max	0.15-0.35	bal Fe				
ASTM A331(95)	8617H	Bar	0.14-0.20	0.35-0.65	0.60-0.95	0.15-0.25	0.35-0.75	0.025 max	0.025 max	0.15-0.35	Cu <=0.35; bal Fe				
ASTM A519(96)	8617	Smls mech tub	0.15-0.20	0.40-0.60	0.70-0.90	0.15-0.25	0.40-0.70	0.040 max	0.040 max	0.15-0.35	bal Fe				
ASTM A534(94)	8617H	Carburizing, anti-friction bearings	0.14-0.20	0.35-0.65	0.60-0.95	0.15-0.25	0.35-0.75	0.025 max	0.025 max	0.15-0.35	bal Fe				
ASTM A752(93)	8617	Rod Wir	0.15-0.20	0.40-0.60	0.70-0.90	0.15-0.25	0.40-0.70	0.035 max	0.040 max	0.15-0.30	bal Fe	550			
ASTM A829/A829M(95)	8617	Plt	0.15-0.21	0.35-0.60	0.60-0.90	0.15-0.25	0.40-0.70	0.035 max	0.040 max	0.15-0.40	bal Fe	480-965			
SAE 770(84)	8617*	Obs; see J1397(92)	0.15-0.2	0.4-0.6	0.7-0.9	0.15-0.25	0.4-0.7	0.04 max	0.04 max	0.15-0.3	bal Fe				
SAE J1268(95)	8617H	Bar Rod Wir; See std	0.14-0.20	0.35-0.65	0.60-0.95	0.15-0.25	0.35-0.75	0.030 max	0.040 max	0.15-0.35	Cu <=0.35; Cu not spec'd but acpt; bal Fe				
SAE J404(94)	8617	Plt	0.15-0.21	0.35-0.60	0.60-0.90	0.15-0.25	0.40-0.70	0.035 max	0.040 max	0.15-0.35	bal Fe				
SAE J404(94)	8617	Bil Blm Slab Bar HR CF	0.15-0.20	0.40-0.60	0.70-0.90	0.15-0.25	0.40-0.70	0.030 max	0.040 max	0.15-0.35	bal Fe				

UNS numbers and US grades are provided as a means of cross referencing chemically similar alloys. Exchangability is only possible after independent examination of specifications. Tensile properties are minimum or typical as specified. UTS and YS as MPa. El as %. See Appendix for list of abbreviations used in Notes. * indicates obsolete material.

Alloy Steel, Nickel Chromium Molybdenum, 8620/8620H

Specification	Designation	Notes	C	Cr	Mn	Mo	Ni	P	S	Si	Other	UTS	YS	El	Hard
Argentina															
IAS	IRAM 8620		0.18-0.23	0.40-0.60	0.70-0.90	0.15-0.25	0.40-0.70	0.035 max	0.040 max	0.20-0.35	bal Fe	650	380	20	192 HB
Australia															
AS 1444(96)	8620	H series, Hard/Tmp	0.18-0.23	0.40-0.60	0.70-0.90	0.15-0.25	0.40-0.70	0.040 max	0.040 max	0.10-0.35	bal Fe				
AS 1444(96)	8620H	H series, Hard/Tmp	0.17-0.23	0.35-0.65	0.60-0.95	0.15-0.25	0.35-0.75	0.040 max	0.040 max	0.10-0.35	bal Fe				
Belgium															
NBN 253-03	20NiCrMo2		0.17-0.23	0.35-0.65	0.6-0.9	0.15-0.25	0.4-0.7	0.035	0.035	0.15-0.4	bal Fe				
Bulgaria															
BDS 6354(74)	20ChM	Struct	0.17-0.22	0.30-0.50	0.50-0.80	0.45-0.55	0.40 max	0.035 max	0.035 max	0.17-0.37	Cu <=0.30; Ti <=0.05; V <=0.1; W <=0.1; bal Fe				
China															
GB 3077(88)	20CrNiMo	Bar HR Q/T 15mm diam	0.17-0.23	0.40-0.70	0.60-0.95	0.20-0.30	0.35-0.75	0.035 max	0.035 max	0.17-0.37	Cu <=0.30; bal Fe	980	785	9	
GB 3203(82)	G20CrNiMo	Bar HR/CD Q/T 15mm diam	0.17-0.23	0.35-0.65	0.60-0.90	0.15-0.30	0.40-0.70	0.030 max	0.030 max	0.15-0.40	Cu <=0.25; bal Fe	1175		9	
GB 5216(85)	20CrNiMoH	Bar Rod HR/Frg Ann	0.17-0.23	0.35-0.65	0.60-0.95	0.15-0.25	0.35-0.75	0.035 max	0.035 max	0.17-0.37	bal Fe				
Europe															
EN 10084(98)	1.6523	CH, Ann	0.17-0.23	0.35-0.70	0.65-0.95	0.15-0.25	0.40-0.70	0.035 max	0.035 max	0.40 max	bal Fe				212 HB
EN 10084(98)	1.6526	CH, Ann	0.17-0.23	0.35-0.70	0.65-0.95	0.15-0.25	0.40-0.70	0.035 max	0.020-0.040	0.40 max	bal Fe				212 HB
EN 10084(98)	1.7147	CH, Ann	0.17-0.22	1.00-1.30	1.10-1.40			0.035 max	0.035 max	0.40 max	bal Fe				217 HB
EN 10084(98)	20MnCr5	CH, Ann	0.17-0.22	1.00-1.30	1.10-1.40			0.035 max	0.035 max	0.40 max	bal Fe				217 HB
EN 10084(98)	20NiCrMo2-2	CH, Ann	0.17-0.23	0.35-0.70	0.65-0.95	0.15-0.25	0.40-0.70	0.035 max	0.035 max	0.40 max	bal Fe				212 HB
EN 10084(98)	20NiCrMoS2-2	CH, Ann	0.17-0.23	0.35-0.70	0.65-0.95	0.15-0.25	0.40-0.70	0.035 max	0.020-0.040	0.40 max	bal Fe				212 HB
Finland															
SFS 506(76)	21NiCrMo2	CH	0.17-0.23	0.35-0.65	0.6-0.95	0.15-0.25	0.35-0.75	0.035 max	0.05 max	0.15-0.4	bal Fe				
France															
AFNOR	20NCD2		0.18-0.23	0.4-0.6	0.7-0.9	0.15-0.3	0.4-0.7	0.03	0.025	0.1-0.4	Cu 0.35; bal Fe				
AFNOR NFA35551	20NCD2	CH	0.17-0.25	0.4-0.65	0.65-0.95	0.15-0.25	0.4-1	0.035	0.035	0.1-0.4	bal Fe				
AFNOR NFA35551(86)	19NCDB2		0.17-0.23	0.4-0.65	0.65-0.95	0.15-0.25	0.4-0.7	0.035	0.035	0.1-0.4	B 0.0008-0.005; bal Fe				
AFNOR NFA35552(84)	19NCDB2	Bar Rod Wir; CH	0.17-0.23	0.4-0.65	0.65-0.95	0.15-0.25	0.4-0.7	0.035	0.035	0.1-0.4	B 0.0008-0.005; bal Fe				
AFNOR NFA35553	20NCD2	Strp	0.18-0.23	0.4-0.6	0.7-0.9	0.15-0.3	0.4-0.7	0.04	0.04	0.1-0.4	bal Fe				
AFNOR NFA35565	20NCD2*	see 20NiCrMo2	0.17-0.25	0.4-0.65	0.65-0.95	0.15-0.25	0.4-1	0.035	0.035	0.1-0.2	bal Fe				
AFNOR NFA35565(84)	18NCD4		0.16-0.22	0.35-0.55	0.5-0.8	0.15-0.3	0.9-1.2	0.03	0.025	0.2-0.35	Cu 0.35; bal Fe				
AFNOR NFA35566	20NCD2		0.17-0.25	0.4-0.65	0.65-0.95	0.15-0.25	0.4-1	0.035	0.035	0.1-0.2	bal Fe				
Germany															
DIN 1652(90)	20NiCrMo2-2	CH, 30mm	0.17-0.23	0.40-0.70	0.65-0.95	0.15-0.25	0.40-0.70	0.035 max	0.035 max	0.40 max	bal Fe	780-1080	590	10	
DIN 1652(90)	21NiCrMoS2	CH	0.17-0.23	0.40-0.70	0.65-0.95	0.15-0.25	0.40-0.70	0.035 max	0.020-0.035	0.40 max	bal Fe				
DIN 1652(90)	WNr 1.6523	CH, 30mm	0.17-0.23	0.40-0.70	0.65-0.95	0.15-0.25	0.40-0.70	0.035 max	0.035 max	0.40 max	bal Fe	780-1080	590	10	
DIN 1652(90)	WNr 1.6526	CH	0.17-0.23	0.40-0.70	0.65-0.85	0.15-0.25	0.40-0.70	0.035 max	0.020-0.035	0.40 max	bal Fe				
DIN 1654(89)	WNr 1.6523	CH, 30mm	0.17-0.23	0.40-0.70	0.65-0.95	0.15-0.25	0.40-0.70	0.035 max	0.035 max	0.40 max	bal Fe	780-1080	590	10	

UNS numbers and US grades are provided as a means of cross referencing chemically similar alloys. Exchangability is only possible after independent examination of specifications. Tensile properties are minimum or typical as specified. UTS and YS as MPa. El as %. See Appendix for list of abbreviations used in Notes. * indicates obsolete material.

Specification	Designation	Notes	C	Cr	Mn	Mo	Ni	P	S	Si	Other	UTS	YS	El	Hard

Alloy Steel, Nickel Chromium Molybdenum, 8620/8620H (Continued from previous page)

Germany

| DIN 17115(87) | 20NiCrMo2 | Welded round link chain | 0.17-0.23 | 0.35-0.65 | 0.60-0.90 | 0.15-0.25 | 0.40-0.70 | 0.020 max | 0.020 max | 0.25 max | Al 0.02-0.05; Cu <=0.25; P+S 0.035 max; bal Fe | | | | |
| DIN 17115(87) | WNr 1.6522 | Welded round link chain | 0.17-0.23 | 0.35-0.65 | 0.60-0.90 | 0.15-0.25 | 0.40-0.70 | 0.020 max | 0.020 max | 0.25 max | Al 0.02-0.05; Cu <=0.25; N <=0.012; P+S 0.035 max; bal Fe | | | | |

India

| IS 1570 | 20N55Cr50Mo20 | | 0.18-0.23 | 0.4-0.6 | 0.7-0.9 | 0.15-0.25 | 0.4-0.7 | | | 0.2-0.35 | bal Fe | | | | |

International

ISO 683-11(87)	20Cr4	CH, 16mm test	0.17-0.23	0.90-1.20	0.60-0.90			0.035 max	0.035 max	0.15-0.40	Al <=0.1; Co <=0.1; Cu <=0.3; bal Fe	820-1170	550		
ISO 683-11(87)	20CrS4	CH, 16mm test	0.17-0.23	0.90-1.20	0.60-0.90			0.035 max	0.020-0.040	0.15-0.40	Al <=0.1; Co <=0.1; Cu <=0.3; bal Fe	820-1170	550		
ISO 683-11(87)	20MnCr5	CH, 16mm test	0.17-0.23	1.00-1.30	1.10-1.40			0.035 max	0.035 max	0.15-0.40	Al <=0.1; Co <=0.1; Cu <=0.3; bal Fe	100-1350	670		
ISO 683-11(87)	20NiCrMo2	CH, 16mm test	0.17-0.23	0.30-0.65	0.65-0.95	0.15-0.25	0.40-0.70	0.035 max	0.035 max	0.15-0.40	Al <=0.1; Co <=0.1; Cu <=0.3; bal Fe	810-1160	560		
ISO 683-11(87)	20NiCrMoS2	CH, 16mm test	0.17-0.23	0.30-0.65	0.65-0.95	0.15-0.25	0.40-0.70	0.035 max	0.020-0.040	0.15-0.40	Al <=0.1; Co <=0.1; Cu <=0.3; bal Fe	810-1160	560		
ISO 683-17(76)	12	Ball & roller bearing, HT	0.17-0.23	0.35-0.65	0.60-0.90	0.15-0.25	0.40-0.70	0.035 max	0.035 max	0.15-0.40	bal Fe				
ISO R683-11(70)	12*	Bar Carburized Hardness	0.17-0.23	0.35-0.65	0.6-0.9	0.15-0.24	0.4-0.7	0.04	0.04	0.15-0.4	bal Fe				
ISO R683-11(70)	12a*	Bar Carburized Hardness	0.17-0.23	0.35-0.65	0.6-0.9	0.15-0.24	0.4-0.7	0.04	0.02-0.04	0.15-0.4	bal Fe				

Italy

UNI 3097(75)	16NiCrMo2	Ball & roller bearing	0.13-0.18	0.4-0.6	0.6-0.9	0.15-0.25	0.35-0.65	0.035 max	0.035 max	0.15-0.4	bal Fe				
UNI 6403(86)	16NiCrMo2	Q/T	0.13-0.18	0.4-0.6	0.6-0.9	0.15-0.25	0.4-0.7	0.035 max	0.035 max	0.15-0.35	bal Fe				
UNI 7846(78)	16NiCrMo2	CH	0.13-0.18	0.35-0.65	0.6-0.9	0.15-0.25	0.4-0.7	0.035 max	0.035 max	0.15-0.4	Pb 0.15-0.3; bal Fe				
UNI 7846(78)	20NiCrMo2	CH	0.17-0.23	0.35-0.65	0.6-0.9	0.15-0.25	0.4-0.7	0.035 max	0.035 max	0.15-0.4	Pb 0.15-0.3; bal Fe				
UNI 8550(84)	16NiCrMo2	CH	0.13-0.18	0.35-0.65	0.6-0.9	0.15-0.25	0.4-0.7	0.035 max	0.035 max	0.15-0.4	bal Fe				
UNI 8550(84)	20NiCrMo2	CH	0.17-0.23	0.35-0.65	0.6-0.9	0.15-0.25	0.4-0.7	0.035 max	0.035 max	0.15-0.4	bal Fe				
UNI 8788(85)	16NiCrMo2	CH	0.13-0.18	0.35-0.69	0.6-0.9	0.15-0.25	0.4-0.7	0.035 max	0.035 max	0.15-0.4	Pb <=0.30; P+S<=0.06; bal Fe				
UNI 8788(85)	20NiCrMo2	CH	0.17-0.23	0.35-0.65	0.6-0.9	0.15-0.25	0.4-0.7	0.03 max	0.035 max	0.15-0.4	Pb 0.15-0.3; P+S<=0.06; bal Fe				

Japan

JIS G3311(88)	SNCM220M	CR Strp; Chain	0.17-0.23	0.40-0.65	0.60-0.90		0.15-0.30	0.40-0.70	0.030 max	0.15-0.35	Cu <=0.30; bal Fe				180-240 HV
JIS G4052(79)	SNCM220H	Ni-Cr Struct Hard	0.17-0.23	0.35-0.65	0.60-0.95	0.15-0.30	0.35-0.75	0.030 max	0.030 max	0.15-0.35	Cu <=0.30; bal Fe				
JIS G4103(79)	SNCM21*	Obs; see SNCM220	0.17-0.23	0.4-0.65	0.6-0.9	0.15-0.3	0.4-0.7	0.03	0.03	0.15-0.35	bal Fe				
JIS G4103(79)	SNCM220	HR Frg Bar Rod	0.17-0.23	0.40-0.65	0.60-0.90	0.15-0.30	0.40-0.70	0.030 max	0.030 max	0.15-0.35	Cu <=0.30; bal Fe				

Mexico

DGN B-203	8620*	Obs	0.18-0.23	0.4-0.6	0.7-0.9	0.15-0.25	0.4-0.7	0.04	0.04	0.2-0.35	bal Fe				
NMX-B-268(68)	8620H	Hard	0.17-0.23	0.35-0.65	0.60-0.95	0.15-0.25	0.35-0.75			0.20-0.35	bal Fe				
NMX-B-300(91)	8620	Bar	0.18-0.23	0.40-0.60	0.70-0.90	0.15-0.25	0.40-0.70	0.035 max	0.040 max	0.15-0.35	bal Fe				
NMX-B-300(91)	8620H	Bar	0.17-0.23	0.35-0.65	0.60-0.95	0.15-0.25	0.35-0.75			0.15-0.35	bal Fe				

Pan America

| COPANT 514 | 8620 | Tube | 0.18-0.23 | 0.4-0.6 | 0.7-0.9 | 0.15-0.25 | 0.4-0.7 | 0.04 | 0.05 | 0.1 | bal Fe | | | | |
| COPANT 514 | 8620 | | 0.18-0.23 | 0.4-0.6 | 0.7-0.9 | 0.15-0.25 | 0.4-0.7 | 0.04 | 0.05 | 0.1 | bal Fe | | | | |

Specification	Designation	Notes	C	Cr	Mn	Mo	Ni	P	S	Si	Other	UTS	YS	El	Hard
Alloy Steel, Nickel Chromium Molybdenum, 8620/8620H (Continued from previous page)																
Poland																
PNH84023/08	20HNMA	Chain	0.17-0.23	0.35-0.65	0.6-0.9	0.15-0.25	0.4-0.7	0.03 max	0.03 max	0.1-0.25	Al 0.02-0.05; Cu <=0.2; bal Fe			
PNH84030	19HM	Q/T; CH	0.17-0.22	0.3-0.5	0.6-0.9	0.4-0.5	0.3 max	0.035 max	0.035 max	0.15-0.4	bal Fe			
Russia																
GOST 1414(75)	AS20ChGNM	Free-cutting	0.18-0.23	0.4-0.7	0.7-1.1	0.15-0.35	0.4-0.7	0.035	0.035	0.17-0.37	Cu <=0.30; Pb 0.15-0.3; bal Fe			
GOST 4543(71)	20Ch	CH	0.17-0.23	0.7-1.0	0.5-0.8	0.15 max	0.3 max	0.035 max	0.035 max	0.17-0.37	Al <=0.1; Cu <=0.3; Ti <=0.05; bal Fe			
Spain																
UNE 36013(60)	20NiCrMo2	Ball & roller bearing	0.17-0.22	0.35-0.65	0.6-0.9	0.15-0.25	0.4-0.7	0.035	0.035	0.15-0.4	bal Fe			
UNE 36013(60)	20NiCrMo3		0.18-0.23	0.4-0.6	0.6-0.8	0.3-0.4	0.7-0.9	0.035	0.035	0.15-0.4	bal Fe			
UNE 36013(60)	20NiCrMo3-1		0.18-0.23	0.4-0.6	0.6-0.8	0.3-0.4	0.7-0.9	0.035	0.02-0.035	0.15-0.4	bal Fe			
UNE 36013(60)	F.1522*	see 20NiCrMo2	0.17-0.22	0.35-0.65	0.6-0.9	0.15-0.25	0.4-0.7	0.035	0.035	0.15-0.4	bal Fe			
UNE 36013(60)	F.1524*	see 20NiCrMo3	0.18-0.23	0.4-0.6	0.6-0.8	0.3-0.4	0.7-0.9	0.035	0.035	0.15-0.4	bal Fe			
UNE 36013(60)	F.1534*	see 20NiCrMo3-1	0.18-0.23	0.4-0.6	0.6-0.8	0.3-0.4	0.7-0.9	0.035	0.02-0.035	0.15-0.4	bal Fe			
UNE 36013(76)	20MoCr5	CH	0.18-0.25	0.3-0.5	0.6-0.9	0.4-0.5		0.035 max	0.035 max	0.15-0.4	bal Fe			
UNE 36013(76)	20MoCr5-1	CH	0.18-0.23	0.3-0.5	0.6-0.9	0.4-0.5		0.035 max	0.02-0.035	0.15-0.4	bal Fe			
UNE 36013(76)	20NiCrMo2-1	CH	0.17-0.23	0.35-0.65	0.6-0.9	0.15-0.25	0.4-0.7	0.035 max	0.02-0.035	0.15-0.4	bal Fe			
UNE 36013(76)	20NiCrMo3-1	CH	0.18-0.23	0.4-0.6	0.6-0.8	0.3-0.4	0.7-0.9	0.035 max	0.02-0.035	0.15-0.4	bal Fe			
UNE 36013(76)	F.1523*	see 20MoCr5	0.18-0.25	0.3-0.5	0.6-0.9	0.4-0.5		0.035 max	0.035 max	0.15-0.4	bal Fe			
UNE 36013(76)	F.1532*	see 20NiCrMo2-1	0.17-0.23	0.35-0.65	0.6-0.9	0.15-0.25	0.4-0.7	0.035 max	0.02-0.035	0.15-0.4	bal Fe			
UNE 36013(76)	F.1533*	see 20MoCr5-1	0.18-0.23	0.3-0.5	0.6-0.9	0.4-0.5		0.035 max	0.02-0.035	0.15-0.4	bal Fe			
UNE 36013(76)	F.1534*	see 20NiCrMo3-1	0.18-0.23	0.4-0.6	0.6-0.8	0.3-0.4	0.7-0.9	0.035 max	0.02-0.035	0.15-0.4	bal Fe			
Sweden																
SIS 142506	2506-03	Bar Frg, CH Q/A, Ann 11mm	0.17-0.23	0.35-0.65	0.6-0.95	0.15-0.25	0.35-0.75	0.04	0.03-0.05	0.15-0.4	bal Fe	980	640	8	
SS 142506	2506	CH	0.17-0.23	0.35-0.65	0.6-0.95	0.15-0.25	0.35-0.75	0.035 max	0.03-0.05	0.15-0.4	bal Fe			
UK																
BS 2772	806M20		0.17-0.23	0.35-0.65	0.6-0.95	0.15-0.25	0.35-0.75	0.04	0.05	0.1-0.35	bal Fe			
BS 970/1(83)	805H20	Blm Bil Slab Bar Rod Frg CH	0.17-0.23	0.35-0.65	0.60-0.95	0.15-0.25	0.35-0.75				bal Fe				207 HB max
BS 970/1(96)	805A20	Blm Bil Slab Bar Rod Frg	0.18-0.23	0.40-0.60	0.70-0.90	0.15-0.25	0.40-0.70	0.035 max	0.040 max	0.10-0.35	bal Fe			
BS 970/1(96)	805M20	Blm Bil Slab Bar Rod Frg	0.17-0.23	0.35-0.65	0.60-0.95	0.15-0.25	0.35-0.75	0.035 max	0.040 max	0.10-0.35	bal Fe			
BS 970/3(91)	805A20	Bright bar	0.18-0.23	0.40-0.60	0.70-0.90	0.15-0.25	0.40-0.70				bal Fe			
BS 970/3(91)	805H20	Bright bar	0.17-0.23	0.35-0.65	0.60-0.95	0.15-0.25	0.35-0.75			0.10-0.35	bal Fe			
USA																
	AISI 8620	Plt	0.17-0.23	0.35-0.60	0.60-0.90	0.15-0.25	0.40-0.70	0.035 max	0.040 max	0.15-0.40	bal Fe			
	AISI 8620	Smls mech tub	0.18-0.23	0.40-0.60	0.70-0.90	0.15-0.25	0.40-0.70	0.040 max	0.040 max	0.15-0.35	bal Fe			
	AISI 8620	Sh Strp	0.18-0.23	0.04-0.60	0.70-0.90	0.15-0.25	0.40-0.70	0.035	0.035	0.15-0.30	bal Fe			
	AISI 8620	Bar Blm Bil Slab	0.18-0.23	0.40-0.60	0.70-0.90	0.15-0.25	0.40-0.70	0.035 max	0.040 max	0.15-0.35	bal Fe			
	AISI 8620H	Smls mech tub, Hard	0.19-0.25	0.35-0.65	0.60-0.95	0.15-0.25	0.35-0.75	0.040 max	0.040 max	0.15-0.30	bal Fe			

UNS numbers and US grades are provided as a means of cross referencing chemically similar alloys. Exchangability is only possible after independent examination of specifications. Tensile properties are minimum or typical as specified. UTS and YS as MPa. El as %. See Appendix for list of abbreviations used in Notes. * indicates obsolete material.

Specification	Designation	Notes	C	Cr	Mn	Mo	Ni	P	S	Si	Other	UTS	YS	El	Hard

Alloy Steel, Nickel Chromium Molybdenum, 8620/8620H (Continued from previous page)

USA

Specification	Designation	Notes	C	Cr	Mn	Mo	Ni	P	S	Si	Other	UTS	YS	El	Hard
	AISI 8620H	Bar Blm Bil Slab	0.19-0.25	0.35-0.65	0.60-0.95	0.15-0.25	0.35-0.75	0.035 max	0.040 max	0.15-0.35	bal Fe				
	UNS G86200		0.18-0.23	0.40-0.60	0.70-0.90	0.15-0.25	0.40-0.70	0.035 max	0.040 max	0.15-0.35	bal Fe				
	UNS H86200		0.17-0.23	0.35-0.65	0.60-0.95	0.15-0.25	0.35-0.75	0.035 max	0.040 max	0.15-0.35	bal Fe				
AMS 6274		Bar Rod Wir	0.18-0.23	0.4-0.60	0.7-0.90	0.15-0.25	0.4-0.7	0.4 max	0.040 max	0.15-0.30	bal Fe				
AMS 6276		Bar Rod Wir	0.18-0.23	0.4-0.60	0.7-0.90	0.15-0.25	0.4-0.7	0.4 max	0.040 max	0.15-0.30	bal Fe				
AMS 6277		Bar Rod Wir	0.18-0.23	0.4-0.60	0.7-0.90	0.15-0.25	0.4-0.7	0.4 max	0.040 max	0.15-0.30	bal Fe				
ASTM A29/A29M(93)	8620	Bar	0.18-0.23	0.40-0.60	0.7-0.9	0.15-0.25	0.4-0.7	0.035 max	0.04 max	0.15-0.35	bal Fe				
ASTM A304(96)	8620H	Bar, hard bands spec	0.17-0.23	0.35-0.65	0.60-0.95	0.15-0.25	0.35-0.75	0.035 max	0.040 max	0.15-0.35	Cu <=0.35; Cu Ni Cr Mo trace allowed; bal Fe				
ASTM A322(96)	8620	Bar	0.18-0.23	0.40-0.60	0.70-0.90	0.15-0.25	0.40-0.70	0.035 max	0.040 max	0.15-0.35	bal Fe				
ASTM A331(95)	8620	Bar	0.18-0.23	0.40-0.60	0.70-0.90	0.15-0.25	0.40-0.70	0.035 max	0.040 max	0.15-0.35	bal Fe				
ASTM A331(95)	8620H	Bar	0.17-0.23	0.35-0.65	0.60-0.95	0.15-0.25	0.35-0.75	0.025 max	0.025 max	0.15-0.35	Cu <=0.35; bal Fe				
ASTM A506(93)	8620	Sh Strp, HR, CR	0.18-0.23	0.040-0.60	0.70-0.90	0.15-0.25	0.40-0.70	0.035	0.035	0.15-0.30	bal Fe				
ASTM A507(93)	8620	Sh Strp, HR, CR	0.18-0.23	0.40-0.60	0.70-0.90	0.15-0.25	0.40-0.70	0.025	0.030	0.15-0.30	bal Fe				
ASTM A513(97a)	8620	ERW Mech Tub	0.18-0.23	0.40-0.60	0.70-0.90	0.15-0.25	0.40-0.70	0.035 max	0.040 max	0.15-0.35	bal Fe				
ASTM A519(96)	8620	Smls mech tub	0.18-0.23	0.40-0.60	0.70-0.90	0.15-0.25	0.40-0.70	0.040 max	0.040 max	0.15-0.35	bal Fe				
ASTM A534(94)	8620H	Carburizing, anti-friction bearings	0.17-0.23	0.35-0.65	0.60-0.95	0.15-0.25	0.35-0.75	0.025 max	0.025 max	0.15-0.35	bal Fe				
ASTM A535(92)	8620	Ball & roller bearing	0.18-0.23	0.40-0.60	0.70-0.90	0.15-0.25	0.40-0.70	0.015 max	0.015 max	0.15-0.35	Cu <=0.35; bal Fe				
ASTM A646(95)	Grade 4	Blm Bil for aircraft aerospace frg, ann	0.18-0.23	0.40-0.60	0.70-0.90	0.15-0.25	0.40-0.70	0.025 max	0.025 max	0.20-0.35	bal Fe				229 HB
ASTM A752(93)	8620	Rod Wir	0.18-0.23	0.40-0.60	0.70-0.90	0.15-0.25	0.40-0.70	0.035 max	0.040 max	0.15-0.30	bal Fe	570			
ASTM A829/A829M(95)	8620	Plt	0.17-0.23	0.35-0.60	0.60-0.90	0.15-0.25	0.40-0.70	0.035 max	0.040 max	0.15-0.40	bal Fe	480-965			
ASTM A837(96)	8620	Frg for carburizing	0.18-0.23	0.40-0.60	0.70-0.90	0.15-0.25	0.40-0.70	0.035 max	0.040 max	0.15-0.35	Cu <=0.35; Si<=0.10 if VCD used; bal Fe				
ASTM A914/914M(92)	8620RH	Bar restricted end Q hard	0.18-0.23	0.40-0.60	0.70-0.90	0.15-0.25	0.40-0.70	0.035 max	0.040 max	0.15-0.35	Cu <=0.35; Electric furnace P, S<=0.025; bal Fe				
DoD-F-24669/1(86)	8620	Bar Bil; Supersedes MIL-S-866 & MIL-S-16974	0.18-0.23	0.4-0.6	0.7-0.9	0.15-0.25	0.4-0.7	0.035	0.04	0.15-0.3	bal Fe				
FED QQ-S-626C(91)	8620*	Obs; see ASTM A829; Plt	0.18-0.23	0.4-0.6	0.7-0.9	0.15-0.25	0.4-0.7	0.035	0.04	0.15-0.35	bal Fe				
MIL-S-16974E(86)	8620*	Obs; Bar Bil Blm slab for refrg	0.18-0.23	0.4-0.6	0.7-0.9	0.15-0.25	0.4-0.7	0.04 max	0.04 max	0.15-0.3	bal Fe				
MIL-S-8690B(78)	8620*	Obs; see AMS spec; Bar aircraft qual	0.18-0.23	0.4-0.7	0.7-1	0.12-0.25	0.4-0.7	0.025	0.025	0.2-0.35	Cu 0.35; bal Fe				
SAE 770(84)	8620*	Obs; see J1397(92)	0.18-0.23	0.4-0.6	0.7-0.9	0.15-0.25	0.4-0.7	0.04 max	0.04 max	0.15-0.3	bal Fe				
SAE J1268(95)	8620H	Bar Rod Wir; See std	0.17-0.23	0.35-0.65	0.60-0.95	0.15-0.25	0.35-0.75	0.030 max	0.040 max	0.15-0.35	Cu <=0.35; Cu not spec'd but acpt; bal Fe				
SAE J1868(93)	8620RH	Restrict hard range	0.18-0.23	0.40-0.60	0.70-0.90	0.15-0.25	0.40-0.70	0.025 max	0.040 max	0.15-0.35	Cu <=0.035; Cu not spec'd but acpt; bal Fe				5 HRC max
SAE J404(94)	8620	Bil Blm Slab Bar HR CF	0.18-0.23	0.40-0.60	0.70-0.90	0.15-0.25	0.40-0.70	0.030 max	0.040 max	0.15-0.35	bal Fe				
SAE J404(94)	8620	Plt	0.17-0.23	0.35-0.60	0.60-0.90	0.15-0.25	0.40-0.70	0.035 max	0.040 max	0.15-0.35	bal Fe				

UNS numbers and US grades are provided as a means of cross referencing chemically similar alloys. Exchangability is only possible after independent examination of specifications. Tensile properties are minimum or typical as specified. UTS and YS as MPa. El as %. See Appendix for list of abbreviations used in Notes. * indicates obsolete material.

Specification	Designation	Notes	C	Cr	Mn	Mo	Ni	P	S	Si	Other	UTS	YS	El	Hard

Alloy Steel, Nickel Chromium Molybdenum, 8620/8620H (Continued from previous page)

Yugoslavia

Specification	Designation	Notes	C	Cr	Mn	Mo	Ni	P	S	Si	Other	UTS	YS	El	Hard
	C.7420	CH	0.17-0.22	0.3-0.5	0.6-0.9	0.4-0.5		0.035 max	0.035 max	0.15-0.4	bal Fe				

Alloy Steel, Nickel Chromium Molybdenum, 8622/8622H

China

Specification	Designation	Notes	C	Cr	Mn	Mo	Ni	P	S	Si	Other	UTS	YS	El	Hard
GB 3077(88)	20CrNiMo	Bar HR Q/T 15mm diam	0.17-0.23	0.40-0.70	0.60-0.95	0.20-0.30	0.35-0.75	0.035 max	0.035 max	0.17-0.37	Cu <=0.30; bal Fe	980	785	9	
GB 3203(82)	G20CrNiMo	Bar HR/CD Q/T 15mm diam	0.17-0.23	0.35-0.65	0.60-0.90	0.15-0.30	0.40-0.70	0.030 max	0.030 max	0.15-0.40	Cu <=0.25; bal Fe	1175		9	
GB 5216(85)	20CrNiMoH	Bar Rod HR/Frg Ann	0.17-0.23	0.35-0.65	0.60-0.95	0.15-0.25	0.35-0.75	0.035 max	0.035 max	0.17-0.37	bal Fe				179 max HB

France

Specification	Designation	Notes	C	Cr	Mn	Mo	Ni	P	S	Si	Other	UTS	YS	El	Hard
AFNOR	23NCDB4		0.2-0.25	0.4-0.65	0.65-0.95	0.15-0.25	0.4-0.7	0.03	0.025	0.1-0.25	B >=0.0008; bal Fe				
AFNOR NFA35556	22NCD2		0.2-0.25	0.4-0.65	0.65-0.95	0.15-0.25	0.4-0.7	0.03	0.025	0.1-0.35	bal Fe				
AFNOR NFA35556	23MNCD5		0.2-0.26	0.4-0.6	1.1-1.4	0.2-0.3	0.4-0.7	0.03	0.025	0.1-0.35	Al 0.02; bal Fe				
AFNOR NFA35556	23NCDB2		0.2-0.25	0.4-0.65	0.65-0.95	0.15-0.25	0.4-0.7	0.03	0.025	0.1-0.35	Al 0.02; B >=0.0008; bal Fe				

Germany

Specification	Designation	Notes	C	Cr	Mn	Mo	Ni	P	S	Si	Other	UTS	YS	El	Hard
DIN	21NiCrMo22		0.18-0.23	0.40-0.60	0.7-0.90	0.20-0.30	0.40-0.70	0.035 max	0.035 max	0.20-0.35	bal Fe				
DIN	WNr 1.6543		0.18-0.23	0.40-0.60	0.70-0.90	0.20-0.30	0.40-0.70	0.035 max	0.035 max	0.20-0.35	bal Fe				
DIN 17115(87)	23MnNiCrMo5-2	Welded round link chain	0.20-0.26	0.40-0.60	1.10-1.40	0.20-0.30	0.40-0.70	0.020 max	0.020 max	0.25 max	Al 0.02-0.05; Cu <=0.25; N <=0.012; P+S 0.035 max; bal Fe				
DIN 17115(87)	WNr 1.6541	Welded round link chain	0.20-0.26	0.40-0.60	1.10-1.40	0.20-0.30	0.40-0.70	0.020 max	0.020 max	0.25 max	Al 0.02-0.05; Cu <=0.25; N <=0.012; P+S 0.035 max; bal Fe				

Mexico

Specification	Designation	Notes	C	Cr	Mn	Mo	Ni	P	S	Si	Other	UTS	YS	El	Hard
DGN B-203	8622*	Obs	0.2-0.25	0.4-0.6	0.7-0.9	0.15-0.25	0.4-0.7	0.04	0.04	0.2-0.35	bal Fe				
DGN B-297	8622*	Obs	0.2-0.25	0.4-0.6	0.7-0.9	0.15-0.25	0.4-0.7	0.035	0.04	0.2-0.35	bal Fe				
NMX-B-268(68)	8622H	Hard	0.19-0.25	0.35-0.65	0.60-0.95	0.15-0.25	0.35-0.75			0.20-0.35	bal Fe				
NMX-B-300(91)	8622	Bar	0.20-0.25	0.40-0.60	0.70-0.90	0.15-0.25	0.40-0.70	0.035 max	0.040 max	0.15-0.35	bal Fe				
NMX-B-300(91)	8622H	Bar	0.19-0.25	0.35-0.65	0.60-0.95	0.15-0.25	0.35-0.75			0.15-0.35	bal Fe				

Pan America

Specification	Designation	Notes	C	Cr	Mn	Mo	Ni	P	S	Si	Other	UTS	YS	El	Hard
COPANT 514	8622		0.2-0.25	0.4-0.6	0.7-0.9	0.15-0.25	0.4-0.6	0.04	0.05	0.1	bal Fe				

Poland

Specification	Designation	Notes	C	Cr	Mn	Mo	Ni	P	S	Si	Other	UTS	YS	El	Hard
PNH84023	23GHNMA		0.2-0.26	0.4-0.6	1-1.25	0.2-0.3	0.4-0.7	0.025	0.025	0.15-0.35	Al 0.02-0.05; Cu 0.02; P+ 0.04; bal Fe				

UK

Specification	Designation	Notes	C	Cr	Mn	Mo	Ni	P	S	Si	Other	UTS	YS	El	Hard
BS 2772	806M22		0.19-0.25	0.35-0.65	0.6-0.95	0.15-0.25	0.35-0.75	0.04	0.05	0.1-0.35	bal Fe				
BS 970/1(83)	805M22	Blm Bil Slab Bar Rod Frg CH	0.19-0.25	0.35-0.65	0.60-0.95	0.15-0.25	0.35-0.75	0.035 max	0.04 max	0.10-0.35	bal Fe	930		10	
BS 970/1(96)	805M22	Blm Bil Slab Bar Rod Frg	0.19-0.25	0.35-0.65	0.60-0.95	0.15-0.25	0.35-0.75	0.035 max	0.040 max	0.10-0.35	bal Fe				
BS 970/3(91)	805A22	Bright bar	0.20-0.25	0.40-0.60	0.70-0.90	0.15-0.25	0.40-0.70				bal Fe				
BS 970/3(91)	805H22	Blm Bil Slab Bar Rod Frg CH	0.19-0.25	0.35-0.65	0.60-0.95	0.15-0.25	0.35-0.75			0.10-0.35	bal Fe				217 HB max

USA

Specification	Designation	Notes	C	Cr	Mn	Mo	Ni	P	S	Si	Other	UTS	YS	El	Hard
	AISI 8622	Plt	0.19-0.25	0.35-0.60	0.60-0.90	0.15-0.25	0.40-0.70	0.035 max	0.040 max	0.15-0.35	bal Fe				
	AISI 8622	Smls mech tub	0.20-0.25	0.40-0.60	0.70-0.90	0.15-0.25	0.40-0.70	0.040 max	0.040 max	0.15-0.35	bal Fe				

UNS numbers and US grades are provided as a means of cross referencing chemically similar alloys. Exchangability is only possible after independent examination of specifications. Tensile properties are minimum or typical as specified. UTS and YS as MPa. El as %. See Appendix for list of abbreviations used in Notes. * indicates obsolete material.

Specification	Designation	Notes	C	Cr	Mn	Mo	Ni	P	S	Si	Other	UTS	YS	El	Hard

Alloy Steel, Nickel Chromium Molybdenum, 8622/8622H (Continued from previous page)

USA

Specification	Designation	Notes	C	Cr	Mn	Mo	Ni	P	S	Si	Other	UTS	YS	El	Hard
	AISI 8622	Bar Blm Bil Slab	0.20-0.25	0.40-0.60	0.70-0.90	0.15-0.25	0.40-0.70	0.035 max	0.040 max	0.15-0.35	bal Fe				
	AISI 8622H	Bar Blm Bil Slab	0.19-0.25	0.35-0.65	0.60-0.95	0.15-0.25	0.35-0.75	0.035 max	0.040 max	0.15-0.35	bal Fe				
	AISI 8622H	Smls mech tub, Hard	0.19-0.25	0.35-0.65	0.60-0.95	0.15-0.25	0.35-0.75	0.040 max	0.040 max	0.15-0.30	bal Fe				
	UNS G86220		0.20-0.25	0.40-0.60	0.70-0.90	0.15-0.25	0.40-0.70	0.035 max	0.040 max	0.15-0.35	bal Fe				
	UNS H86220		0.19-0.25	0.35-0.65	0.60-0.95	0.15-0.25	0.35-0.75	0.035 max	0.040 max	0.15-0.35	bal Fe				
ASTM A29/A29M(93)	8622	Bar	0.2-0.25	0.40-0.60	0.7-0.9	0.15-0.25	0.4-0.7	0.035 max	0.04 max	0.15-0.35	bal Fe				
ASTM A304(96)	8622H	Bar, hard bands spec	0.19-0.25	0.35-0.65	0.60-0.95	0.15-0.25	0.35-0.75	0.035 max	0.040 max	0.15-0.35	Cu <=0.35; Cu Ni Cr Mo trace allowed; bal Fe				
ASTM A322(96)	8622	Bar	0.20-0.25	0.40-0.60	0.70-0.90	0.15-0.25	0.40-0.70	0.035 max	0.040 max	0.15-0.35	bal Fe				
ASTM A331(95)	8622	Bar	0.20-0.25	0.40-0.60	0.70-0.90	0.15-0.25	0.40-0.70	0.035 max	0.040 max	0.15-0.35	bal Fe				
ASTM A331(95)	8622H	Bar	0.19-0.25	0.35-0.65	0.60-0.95	0.15-0.25	0.35-0.75	0.025 max	0.025 max	0.15-0.35	Cu <=0.35; bal Fe				
ASTM A519(96)	8622	Smls mech tub	0.20-0.25	0.40-0.60	0.70-0.90	0.15-0.25	0.40-0.70	0.040 max	0.040 max	0.15-0.35	bal Fe				
ASTM A752(93)	8622	Rod Wir	0.20-0.25	0.40-0.62	0.70-0.90	0.15-0.25	0.40-0.70	0.035 max	0.040 max	0.15-0.30	bal Fe	580			
ASTM A829/A829M(95)	8622	Plt	0.19-0.25	0.35-0.60	0.60-0.90	0.15-0.25	0.40-0.70	0.035 max	0.040 max	0.15-0.40	bal Fe	480-965			
ASTM A914/914M(92)	8622RH	Bar restricted end Q hard	0.20-0.25	0.40-0.60	0.70-0.90	0.15-0.25	0.40-0.70	0.035 max	0.040 max	0.15-0.35	Cu <=0.35; Electric furnace P, S<=0.025; bal Fe				
DoD-F-24669/1(86)	8622	Bar Bil; Supersedes MIL-S-866 & MIL-S-16974	0.2-0.25	0.4-0.6	0.7-0.9	0.15-0.25	0.4-0.7	0.035 max	0.04 max	0.15-0.35	bal Fe				
SAE 770(84)	8622*	Obs; see J1397(92)	0.2-0.25	0.4-0.6	0.7-0.9	0.15-0.25	0.4-0.7	0.04 max	0.04 max	0.15-0.3	bal Fe				
SAE J1268(95)	8622H	Bar Rod Wir; See std	0.19-0.25	0.35-0.65	0.60-0.95	0.15-0.25	0.35-0.75	0.030 max	0.040 max	0.15-0.35	Cu <=0.35; Cu not spec'd but acpt; bal Fe				
SAE J1868(93)	8622RH	Restrict hard range	0.20-0.25	0.40-0.60	0.70-0.90	0.15-0.25	0.40-0.70	0.025 max	0.040 max	0.15-0.35	Cu <=0.035; Cu not spec'd but acpt; bal Fe				5 HRC max
SAE J404(94)	8622	Plt	0.19-0.25	0.35-0.60	0.60-0.90	0.15-0.25	0.40-0.70	0.035 max	0.040 max	0.15-0.35	bal Fe				*
SAE J404(94)	8622	Bil Blm Slab Bar HR CF	0.20-0.25	0.40-0.60	0.70-0.90	0.15-0.25	0.40-0.70	0.030 max	0.040 max	0.15-0.35	bal Fe				

Alloy Steel, Nickel Chromium Molybdenum, 8625/8625H

France

Specification	Designation	Notes	C	Cr	Mn	Mo	Ni	P	S	Si	Other	UTS	YS	El	Hard
AFNOR	25NCD4		0.22-0.28	0.4-0.7	0.5-0.9	0.1-0.2	1-1.3	0.04	0.035	0.1-0.4	bal Fe				
AFNOR NFA35556	25MNCD6		0.23-0.28	0.4-0.7	1.4-1.7	0.2-0.3	0.4-0.7	0.02	0.02	0.1-0.35	Al 0.02; bal Fe				
AFNOR NFA35556	25MNDC6		0.23-0.26	0.4-0.6	1.4-1.7	0.2-0.3	0.4-0.7	0.02	0.02	0.1-0.35	Al 0.02; bal Fe				

Mexico

Specification	Designation	Notes	C	Cr	Mn	Mo	Ni	P	S	Si	Other	UTS	YS	El	Hard
DGN B-203	8625*	Obs	0.23-0.28	0.4-0.6	0.7-0.9	0.15-0.25	0.4-0.7	0.04	0.04	0.2-0.35	bal Fe				
NMX-B-268(68)	8625H	Hard	0.22-0.28	0.35-0.65	0.60-0.95	0.15-0.25	0.35-0.75			0.20-0.35	bal Fe				
NMX-B-300(91)	8625	Bar	0.23-0.28	0.40-0.60	0.70-0.90	0.15-0.25	0.40-0.70	0.035 max	0.040 max	0.15-0.35	bal Fe				
NMX-B-300(91)	8625H	Bar	0.22-0.28	0.35-0.65	0.60-0.95	0.15-0.25	0.35-0.75			0.15-0.35	bal Fe				

Pan America

Specification	Designation	Notes	C	Cr	Mn	Mo	Ni	P	S	Si	Other	UTS	YS	El	Hard
COPANT 514	8625	Tube	0.23-0.28	0.4-0.6	0.7-0.9	0.15-0.25	0.4-0.7	0.04	0.05	0.1	bal Fe				

Specification	Designation	Notes	C	Cr	Mn	Mo	Ni	P	S	Si	Other	UTS	YS	El	Hard

Alloy Steel, Nickel Chromium Molybdenum, 8625/8625H (Continued from previous page)

Russia

Specification	Designation	Notes	C	Cr	Mn	Mo	Ni	P	S	Si	Other	UTS	YS	El	Hard
GOST 924	25ChGMNTBA		0.23-0.29	0.5-0.8	0.8-1.4	0.15-0.25	0.5-0.8	0.025	0.025	0.2-0.4	Cu 0.3; Nb 0.01-0.05; Ti 0.01-0.04; bal Fe				

UK

Specification	Designation	Notes	C	Cr	Mn	Mo	Ni	P	S	Si	Other	UTS	YS	El	Hard
BS 970/1(83)	805H25*	Blm Bil Slab Bar Rod Frg	0.22-0.28	0.35-0.65	0.60-0.95	0.15-0.25	0.35-0.75	0.04	0.05	0.1-0.35	bal Fe				
BS 970/1(83)	805M25*	Blm Bil Slab Bar Rod Frg	0.22-0.28	0.35-0.65	0.60-0.95	0.15-0.25	0.35-0.75	0.04	0.05	0.1-0.35	bal Fe				
BS 970/2(88)	805H60	Spring	0.55-0.64	0.35-0.65	0.65-1.05	0.15-0.25	0.35-0.75	0.035 max	0.035 max	0.15-0.40	bal Fe				
BS 970/3(71)	805M25*	Hard, 19mm diam	0.22-0.28	0.35-0.65	0.60-0.95	0.15-0.25	0.35-0.75			0.10-0.35	bal Fe				
BS 970/3(91)	805A24	Bright bar	0.22-0.27	0.40-0.60	0.70-0.90	0.15-0.25	0.40-0.70				bal Fe				

USA

Specification	Designation	Notes	C	Cr	Mn	Mo	Ni	P	S	Si	Other	UTS	YS	El	Hard
	AISI 8625	Plt	0.22-0.29	0.35-0.60	0.60-0.90	0.15-0.25	0.40-0.70	0.035 max	0.040 max	0.15-0.40	bal Fe				
	AISI 8625	Smls mech tub	0.23-0.28	0.40-0.60	0.70-0.90	0.15-0.25	0.40-0.70	0.040 max	0.040 max	0.15-0.35	bal Fe				
	AISI 8625H	Smls mech tub, Hard	0.22-0.28	0.35-0.65	0.70-0.90	0.15-0.25	0.35-0.75	0.035 max	0.040 max	0.15-0.30	bal Fe				
	UNS G86250		0.23-0.28	0.40-0.60	0.70-0.90	0.15-0.25	0.40-0.70	0.035 max	0.040 max	0.15-0.35	bal Fe				
	UNS H86250		0.22-0.28	0.35-0.65	0.60-0.95	0.15-0.25	0.35-0.75	0.035 max	0.040 max	0.15-0.35	bal Fe				
ASTM A29/A29M(93)	8625	Bar	0.23-0.28	0.40-0.60	0.7-0.9	0.15-0.25	0.4-0.7	0.035 max	0.04 max	0.15-0.35	bal Fe				
ASTM A304(96)	8625H	Bar, hard bands spec	0.22-0.28	0.35-0.65	0.60-0.95	0.15-0.25	0.35-0.75	0.035 max	0.040 max	0.15-0.35	Cu <=0.35; Cu Ni Cr Mo trace allowed; bal Fe				
ASTM A322(96)	8625	Bar	0.23-0.28	0.40-0.60	0.70-0.90	0.15-0.25	0.40-0.70	0.035 max	0.040 max	0.15-0.35					
ASTM A331(95)	8625	Bar	0.23-0.28	0.40-0.60	0.70-0.90	0.15-0.25	0.40-0.70	0.035 max	0.040 max	0.15-0.35	bal Fe				
ASTM A331(95)	8625H	Bar	0.22-0.28	0.35-0.65	0.60-0.95	0.15-0.25	0.35-0.75	0.025 max	0.025 max	0.15-0.35	Cu <=0.35; bal Fe				
ASTM A519(96)	8625	Smls mech tub	0.23-0.28	0.40-0.60	0.70-0.90	0.15-0.25	0.40-0.70	0.040 max	0.040 max	0.15-0.35	bal Fe				
ASTM A752(93)	8625	Rod Wir	0.23-0.28	0.40-0.60	0.70-0.90	0.15-0.25	0.40-0.70	0.035 max	0.040 max	0.15-0.30	bal Fe	580			
ASTM A829/A829M(95)	8625	Plt	0.22-0.29	0.35-0.60	0.60-0.90	0.15-0.25	0.40-0.70	0.035 max	0.040 max	0.15-0.40	bal Fe	480-965			
DoD-F-24669/1(86)	8625	Bar Bil; Supersedes MIL-S-866 & MIL-S-16974	0.23-0.28	0.4-0.6	0.7-0.9	0.15-0.25	0.4-0.7	0.035	0.04	0.15-0.35	bal Fe				
MIL-S-16974E(86)	8625*	Obs; Bar Bil Blm slab for refrg	0.23-0.28	0.4-0.6	0.7-0.9	0.15-0.25	0.4-0.7	0.04 max	0.04 max	0.15-0.3	bal Fe				
SAE 770(84)	8625*	Obs; see J1397(92)	0.23-0.28	0.4-0.6	0.7-0.9	0.15-0.25	0.4-0.7	0.04 max	0.04 max	0.15-0.3	bal Fe				
SAE J1268(95)	8625H	Bar Rod Wir; See std	0.22-0.28	0.35-0.65	0.60-0.95	0.15-0.25	0.35-0.75	0.030 max	0.040 max	0.15-0.35	Cu <=0.35; Cu not spec'd but acpt; bal Fe				
SAE J404(94)	8625	Plt	0.22-0.29	0.35-0.60	0.60-0.90	0.15-0.25	0.40-0.70	0.035 max	0.040 max	0.15-0.35	bal Fe				

Yugoslavia

Specification	Designation	Notes	C	Cr	Mn	Mo	Ni	P	S	Si	Other	UTS	YS	El	Hard
	C.7421	CH	0.23-0.29	0.4-0.6	0.6-0.9	0.4-0.5		0.035 max	0.035 max	0.15-0.4	bal Fe				

Alloy Steel, Nickel Chromium Molybdenum, 8627/8627H

Mexico

Specification	Designation	Notes	C	Cr	Mn	Mo	Ni	P	S	Si	Other	UTS	YS	El	Hard
DGN B-203	8627*	Obs	0.25-0.3	0.4-0.6	0.7-0.9	0.15-0.25	0.4-0.7	0.04	0.04	0.2-0.35	bal Fe				
DGN B-297	8627*	Obs	0.25-0.3	0.4-0.6	0.7-0.9	0.15-0.25	0.4-0.7	0.035	0.04	0.2-0.35	bal Fe				
NMX-B-268(68)	8627H	Hard	0.24-0.30	0.35-0.65	0.60-0.95	0.15-0.25	0.35-0.75			0.20-0.35	bal Fe				
NMX-B-300(91)	8627	Bar	0.25-0.30	0.40-0.60	0.70-0.90	0.15-0.25	0.40-0.70	0.035 max	0.040 max	0.15-0.35	bal Fe				

UNS numbers and US grades are provided as a means of cross referencing chemically similar alloys. Exchangability is only possible after independent examination of specifications. Tensile properties are minimum or typical as specified. UTS and YS as MPa. El as %. See Appendix for list of abbreviations used in Notes. * indicates obsolete material.

Alloy Steel, Nickel Chromium Molybdenum, 8627/8627H (Continued from previous page)

Specification	Designation	Notes	C	Cr	Mn	Mo	Ni	P	S	Si	Other	UTS	YS	El	Hard
Mexico															
NMX-B-300(91)	8627H	Bar	0.24-0.30	0.35-0.65	0.60-0.95	0.15-0.25	0.35-0.75			0.15-0.35	bal Fe				
Pan America															
COPANT 514	8627	Tube	0.25-0.3	0.4-0.6	0.7-0.9	0.15-0.25	0.4-0.7	0.04	0.05	0.1	bal Fe				
COPANT 514	8627		0.25-0.3	0.4-0.6	0.7-0.9	0.15-0.25	0.4-0.7	0.04	0.05	0.1	bal Fe				
Russia															
GOST 924	25ChGMNTBA		0.23-0.29	0.5-0.8	0.8-1.4	0.15-0.25	0.5-0.8	0.025	0.025	0.2-0.4	Cu 0.3; Nb 0.01-0.05; Ti 0.01-0.04; bal Fe				
USA															
	AISI 8627	Plt	0.24-0.31	0.35-0.60	0.60-0.90	0.15-0.25	0.40-0.70	0.035 max	0.040 max	0.15-0.40	bal Fe				
	AISI 8627	Smls mech tub	0.25-0.30	0.40-0.60	0.70-0.90	0.15-0.25	0.40-0.70	0.040 max	0.040 max	0.15-0.35	bal Fe				
	AISI 8627H	Smls mech tub, Hard	0.24-0.30	0.35-0.65	0.60-0.95	0.15-0.25	0.35-0.75	0.035 max	0.040 max	0.15-0.30	bal Fe				
	UNS G86270		0.25-0.30	0.40-0.60	0.70-0.90	0.15-0.25	0.40-0.70	0.035 max	0.040 max	0.15-0.35	bal Fe				
	UNS H86270		0.24-0.30	0.35-0.65	0.60-0.95	0.15-0.25	0.35-0.75	0.035 max	0.040 max	0.15-0.35	bal Fe				
ASTM A29/A29M(93)	8627	Bar	0.25-0.30	0.40-0.60	0.7-0.9	0.15-0.25	0.4-0.7	0.035 max	0.04 max	0.15-0.35	bal Fe				
ASTM A304(96)	8627H	Bar, hard bands spec	0.24-0.30	0.35-0.65	0.60-0.95	0.15-0.25	0.35-0.75	0.035 max	0.040 max	0.15-0.35	Cu <=0.35; Cu Ni Cr Mo trace allowed; bal Fe				
ASTM A322(96)	8627	Bar	0.25-0.30	0.40-0.60	0.70-0.90	0.15-0.25	0.40-0.70	0.035 max	0.040 max	0.15-0.35	bal Fe				
ASTM A331(95)	8627	Bar	0.25-0.30	0.40-0.60	0.70-0.90	0.15-0.25	0.40-0.70	0.035 max	0.040 max	0.15-0.35	bal Fe				
ASTM A331(95)	8627H	Bar	0.24-0.30	0.35-0.65	0.60-0.95	0.15-0.25	0.35-0.75	0.025 max	0.025 max	0.15-0.35	Cu <=0.35; bal Fe				
ASTM A519(96)	8627	Smls mech tub	0.25-0.30	0.40-0.60	0.70-0.90	0.15-0.25	0.40-0.70	0.040 max	0.040 max	0.15-0.35	bal Fe				
ASTM A752(93)	8627	Rod Wir	0.25-0.30	0.40-0.60	0.70-0.90	0.15-0.25	0.40-0.70	0.035 max	0.040 max	0.15-0.30	bal Fe		580		
ASTM A829/A829M(95)	8627	Plt	0.24-0.31	0.35-0.60	0.60-0.90	0.15-0.25	0.40-0.70	0.035 max	0.040 max	0.15-0.40	bal Fe		480-965		
DoD-F-24669/1(86)	8627	Bar Bil; Supersedes MIL-S-866 & MIL-S-16974	0.25-0.3	0.4-0.6	0.7-0.9	0.15-0.25	0.4-0.7	0.035	0.04	0.15-0.35	bal Fe				
SAE 770(84)	8627*	Obs; see J1397(92)	0.25-0.3	0.4-0.6	0.7-0.9	0.15-0.25	0.4-0.7	0.04 max	0.05 max	0.15-0.3	bal Fe				
SAE J1268(95)	8627H	Bar Rod Wir; See std	0.24-0.30	0.35-0.65	0.60-0.95	0.15-0.25	0.35-0.75	0.030 max	0.040 max	0.15-0.35	Cu <=0.35; Cu not spec'd but acpt; bal Fe				
SAE J404(94)	8627	Plt	0.24-0.31	0.35-0.60	0.60-0.90	0.15-0.25	0.40-0.70	0.035 max	0.040 max	0.15-0.35	bal Fe				

Alloy Steel, Nickel Chromium Molybdenum, 8630/8630H

Specification	Designation	Notes	C	Cr	Mn	Mo	Ni	P	S	Si	Other	UTS	YS	El	Hard
France															
AFNOR	30NCD2		0.3-0.35	0.4-0.6	0.7-0.9	0.15-0.3	0.5-0.8	0.04	0.04	0.1-0.4	bal Fe				
Germany															
DIN	30NiCrMo2-2		0.27-0.34	0.40-0.60	0.70-1.00	0.15-0.30	0.40-0.70	0.035 max	0.035 max	0.15-0.40	bal Fe				
DIN	WNr 1.6545		0.27-0.34	0.40-0.60	0.70-1.00	0.15-0.30	0.40-0.70	0.035 max	0.035 max	0.15-0.40	bal Fe				
Italy															
UNI 7356(74)	30NiCrMo2KB	Q/T	0.27-0.34	0.4-0.6	0.7-1	0.15-0.3	0.4-0.7	0.035 max	0.035 max	0.15-0.4	bal Fe				
Mexico															
DGN B-203	8630*	Obs	0.28-0.33	0.4-0.6	0.7-0.9	0.15-0.25	0.4-0.7	0.04	0.04	0.2-0.35	bal Fe				
DGN B-297	8630*	Obs	0.28-0.33	0.4-0.6	0.7-0.9	0.15-0.25	0.4-0.7	0.035	0.04	0.2-0.35	bal Fe				

UNS numbers and US grades are provided as a means of cross referencing chemically similar alloys. Exchangability is only possible after independent examination of specifications. Tensile properties are minimum or typical as specified. UTS and YS as MPa. El as %. See Appendix for list of abbreviations used in Notes. * indicates obsolete material.

Specification	Designation	Notes	C	Cr	Mn	Mo	Ni	P	S	Si	Other	UTS	YS	El	Hard

Alloy Steel, Nickel Chromium Molybdenum, 8630/8630H (Continued from previous page)

Mexico

Specification	Designation	Notes	C	Cr	Mn	Mo	Ni	P	S	Si	Other	UTS	YS	El	Hard
NMX-B-268(68)	8630H	Hard	0.27-0.33	0.35-0.65	0.60-0.95	0.15-0.25	0.35-0.75			0.20-0.35	bal Fe				
NMX-B-300(91)	8630	Bar	0.28-0.33	0.40-0.60	0.70-0.90	0.15-0.25	0.40-0.70	0.035 max	0.040 max	0.15-0.35	bal Fe				
NMX-B-300(91)	8630H	Bar	0.27-0.33	0.35-0.65	0.60-0.95	0.15-0.25	0.35-0.75			0.15-0.35	bal Fe				

Pan America

Specification	Designation	Notes	C	Cr	Mn	Mo	Ni	P	S	Si	Other	UTS	YS	El	Hard
COPANT 334	8630	Bar	0.28-0.33	0.4-0.6	0.7-0.9	0.15-0.25	0.4-0.7	0.04	0.04	0.2-0.35	bal Fe				
COPANT 514	8630	Tube	0.28-0.33	0.4-0.6	0.7-0.9	0.15-0.25	0.4-0.7	0.04	0.05	0.1	bal Fe				
COPANT 514	8630		0.28-0.33	0.4-0.6	0.7-0.9	0.15-0.25	0.4-0.7	0.04	0.05	0.1	bal Fe				

Spain

Specification	Designation	Notes	C	Cr	Mn	Mo	Ni	P	S	Si	Other	UTS	YS	El	Hard
UNE	F.129		0.25-0.35	0.55-0.85	0.55-0.85	0.1-0.3	0.55-0.85	0.04	0.04	0.1-0.35	bal Fe				

USA

Specification	Designation	Notes	C	Cr	Mn	Mo	Ni	P	S	Si	Other	UTS	YS	El	Hard
	AISI 8630	Bar Blm Bil Slab	0.28-0.33	0.40-0.60	0.70-0.90	0.15-0.25	0.40-0.70	0.035 max	0.040 max	0.15-0.35	bal Fe				
	AISI 8630	Plt	0.27-0.34	0.35-0.60	0.60-0.90	0.15-0.25	0.40-0.70	0.035 max	0.040 max	0.15-0.40	bal Fe				
	AISI 8630	Smls mech tub	0.28-0.33	0.40-0.60	0.70-0.90	0.15-0.25	0.40-0.70	0.040 max	0.040 max	0.15-0.35	bal Fe				
	AISI 8630H	Bar Blm Bil Slab	0.27-0.33	0.35-0.65	0.65-0.95	0.15-0.25	0.35-0.75	0.035 max	0.040 max	0.15-0.35	bal Fe				
	AISI 8630H	Smls mech tub, Hard	0.27-0.33	0.35-0.65	0.65-0.95	0.15-0.25	0.35-0.75	0.040 max	0.040 max	0.15-0.30	bal Fe				
	UNS G86300		0.28-0.33	0.40-0.60	0.70-0.90	0.15-0.25	0.40-0.70	0.035 max	0.040 max	0.15-0.35	bal Fe				
	UNS H86300		0.27-0.33	0.35-0.65	0.60-0.95	0.15-0.25	0.35-0.75	0.035 max	0.040 max	0.15-0.35	bal Fe				
AMS 6280		Bar Frg	0.28-0.33	0.4-0.60	0.7-0.90	0.15-0.25	0.4-0.7	0.025	0.025	0.15-0.35	Cu 0.35; bal Fe				
AMS 6281		Tube	0.28-0.33	0.4-0.60	0.7-0.90	0.15-0.25	0.4-0.7	0.025	0.025	0.15-0.35	Cu 0.35; bal Fe				
AMS 6530		Tube	0.28-0.33	0.4-0.60	0.7-0.90	0.15-0.25	0.40-0.70	0.025	0.025	0.15-0.35	Cu 0.35; bal Fe				
AMS 6535		Sh, Strp, Plt	0.28-0.33	0.40-0.60	0.70-0.90	0.15-0.25	0.40-0.70	0.025	0.040	0.15-0.35	bal Fe				
AMS 6550		Tube	0.28-0.33	0.4-0.60	0.7-0.90	0.15-0.25	0.4-0.7	0.025	0.025	0.15-0.35	Cu 0.35; bal Fe				
AMS 7496											bal Fe				
ASTM A29/A29M(93)	8630	Bar	0.28-0.33	0.40-0.60	0.7-0.9	0.15-0.25	0.4-0.7	0.035 max	0.04 max	0.15-0.35	bal Fe				
ASTM A304(96)	8630H	Bar, hard bands spec	0.27-0.33	0.35-0.65	0.60-0.95	0.15-0.25	0.35-0.75	0.035 max	0.040 max	0.15-0.35	Cu <=0.35; Cu Ni Cr Mo trace allowed; bal Fe				
ASTM A322(96)	8630	Bar	0.28-0.33	0.40-0.60	0.70-0.90	0.15-0.25	0.40-0.70	0.035 max	0.040 max	0.15-0.35	bal Fe				
ASTM A331(95)	8630	Bar	0.28-0.33	0.40-0.60	0.70-0.90	0.15-0.25	0.40-0.70	0.035 max	0.040 max	0.15-0.35	bal Fe				
ASTM A331(95)	8630H	Bar	0.27-0.33	0.35-0.65	0.60-0.95	0.15-0.25	0.35-0.75	0.025 max	0.025 max	0.15-0.35	Cu <=0.35; bal Fe				
ASTM A513(97a)	8630	ERW Mech Tub	0.28-0.33	0.40-0.60	0.70-0.90	0.15-0.25	0.40-0.70	0.035 max	0.040 max	0.15-0.35	bal Fe				
ASTM A519(96)	8630	Smls mech tub	0.28-0.33	0.40-0.60	0.70-0.90	0.15-0.25	0.40-0.70	0.040 max	0.040 max	0.15-0.35	bal Fe				
ASTM A752(93)	8630	Rod Wir	0.28-0.33	0.40-0.60	0.70-0.90	0.15-0.25	0.40-0.70	0.035 max	0.040 max	0.15-0.30	bal Fe	610			
ASTM A829/A829M(95)	8630	Plt	0.27-0.34	0.35-0.60	0.60-0.90	0.15-0.25	0.40-0.70	0.035 max	0.040 max	0.15-0.40	bal Fe	480-965			
DoD-F-24669/1(86)	8630	Bar Bil; Supersedes MIL-S-866 & MIL-S-16974									bal Fe				
MIL-S-16974E(86)	8630*	Obs; Bar Bil Blm slab for refrg									bal Fe				

UNS numbers and US grades are provided as a means of cross referencing chemically similar alloys. Exchangability is only possible after independent examination of specifications. Tensile properties are minimum or typical as specified. UTS and YS as MPa. El as %. See Appendix for list of abbreviations used in Notes. * indicates obsolete material.

Specification	Designation	Notes	C	Cr	Mn	Mo	Ni	P	S	Si	Other	UTS	YS	El	Hard

Alloy Steel, Nickel Chromium Molybdenum, 8630/8630H (Continued from previous page)

USA

Specification	Designation	Notes	C	Cr	Mn	Mo	Ni	P	S	Si	Other	UTS	YS	El	Hard
MIL-S-18728D(83)	8630*	Obs; see SAE-AMS 6345, 6350, 6351; Aircraft Plt Sh Strp	0.27-0.33	0.4-0.6	0.7-0.9	0.15-0.25	0.4-0.7	0.04	0.04	0.2-0.35	bal Fe				
MIL-S-46059	G86300*	Obs; Shape	0.28-0.33	0.4-0.6	0.7-0.9	0.15-0.25	0.4-0.7	0.035	0.04	0.15-0.3	bal Fe				
MIL-S-6050A(94)	8630*	Obs for new design; Bar refrg stock, aircraft qual	0.28-0.33	0.4-0.6	0.7-0.9	0.15-0.25	0.4-0.7	0.025	0.025	.	Cu 0.35; bal Fe				
SAE 770(84)	8630*	Obs; see J1397(92)									bal Fe				
SAE J1268(95)	8630H	Bar Rod Wir Tub; See std	0.27-0.33	0.35-0.65	0.60-0.95	0.15-0.25	0.35-0.75	0.030 max	0.040 max	0.15-0.35	Cu <=0.35; Cu not spec'd but acpt; bal Fe				
SAE J404(94)	8630	Plt	0.27-0.34	0.35-0.60	0.60-0.90	0.15-0.25	0.40-0.70	0.035 max	0.040 max	0.15-0.35	bal Fe				
SAE J404(94)	8630	Bil Blm Slab Bar HR CF	0.28-0.33	0.40-0.60	0.70-0.90	0.15-0.25	0.40-0.70	0.030 max	0.040 max	0.15-0.35	bal Fe				

Alloy Steel, Nickel Chromium Molybdenum, 8637/8637H

Belgium

Specification	Designation	Notes	C	Cr	Mn	Mo	Ni	P	S	Si	Other	UTS	YS	El	Hard
NBN 253-02	39NiCrMo3		0.36-0.43	0.6-0.9	0.5-0.8	0.15-0.3	0.7-1	0.035	0.035	0.15-0.4	bal Fe				

France

Specification	Designation	Notes	C	Cr	Mn	Mo	Ni	P	S	Si	Other	UTS	YS	El	Hard
AFNOR	40NCD3		0.36-0.43	0.6-0.9	0.5-0.8	0.15-0.3	0.7-1	0.035	0.035	0.1-0.4	bal Fe				

Hungary

Specification	Designation	Notes	C	Cr	Mn	Mo	Ni	P	S	Si	Other	UTS	YS	El	Hard
MSZ	61NCMo4		0.36-0.43	0.6-0.9	0.5-0.8	0.15-0.3	0.7-1	0.035	0.035	0.17-0.37	bal Fe				
MSZ	NCMo4E		0.36-0.43	0.6-0.9	0.5-0.8	0.15-0.3	0.7-1	0.035	0.035	0.17-0.37	bal Fe				
MSZ 61(85)	NCMo4	Q/T; 160.1-250mm	0.36-0.43	0.6-0.9	0.5-0.8	0.15-0.3	0.7-1	0.035 max	0.035 max	0.4 max	bal Fe	750-900	540	13L	
MSZ 61(85)	NCMo4	Q/T; 0-16mm	0.36-0.43	0.6-0.9	0.5-0.8	0.15-0.3	0.7-1	0.035 max	0.035 max	0.4 max	bal Fe	1000-1200	840	10L	

International

Specification	Designation	Notes	C	Cr	Mn	Mo	Ni	P	S	Si	Other	UTS	YS	El	Hard
ISO 683-8	2*		0.36-0.43	0.6-0.9	0.5-0.8	0.15-0.3	0.7-1	0.035	0.035	0.15-0.4	bal Fe				
ISO 683-8	2a*		0.36-0.43	0.6-0.9	0.5-0.8	0.15-0.3	0.7-1	0.035	0.02-0.035	0.15-0.4	bal Fe				
ISO 683-8	2b*		0.36-0.43	0.6-0.9	0.5-0.8	0.15-0.3	0.7-1	0.035	0.03-0.05	0.15-0.4	bal Fe				

Italy

Specification	Designation	Notes	C	Cr	Mn	Mo	Ni	P	S	Si	Other	UTS	YS	El	Hard
UNI 5332	38NiCrMo4		0.34-0.42	0.7-1	0.5-0.8	0.15-0.25	0.7-1	0.035	0.035	0.4	bal Fe				
UNI 6403(86)	39NiCrMo3	Q/T	0.35-0.43	0.6-1	0.5-0.8	0.15-0.25	0.7-1	0.035 max	0.035 max	0.15-0.35	bal Fe				
UNI 7845(78)	39NiCrMo3	Q/T	0.35-0.43	0.6-1	0.5-0.8	0.15-0.25	0.7-1	0.035 max	0.035 max	0.15-0.4	Pb 0.15-0.3; bal Fe				
UNI 7874(79)	39NiCrMo3	Q/T	0.36-0.43	0.6-0.9	0.5-0.8	0.15-0.3	0.7-1	0.035 max	0.035 max	0.15-0.4	Pb 0.15-0.3; bal Fe				

Mexico

Specification	Designation	Notes	C	Cr	Mn	Mo	Ni	P	S	Si	Other	UTS	YS	El	Hard
DGN B-203	8637*	Obs	0.35-0.4	0.4-0.6	0.75-1	0.15-0.25	0.4-0.7	0.04	0.04	0.2-0.35	bal Fe				
DGN B-297	8637*	Obs	0.35-0.4	0.4-0.6	0.75-1	0.15-0.25	0.4-0.7	0.035	0.04	0.2-0.35	bal Fe				
NMX-B-268(68)	8637H	Hard	0.34-0.41	0.35-0.65	0.70-1.05	0.15-0.25	0.35-0.75		.	0.20-0.35	bal Fe				
NMX-B-300(91)	8637	Bar	0.35-0.40	0.40-0.60	0.75-1.00	0.15-0.25	0.40-0.70	0.035 max	0.040 max	0.15-0.35	bal Fe				
NMX-B-300(91)	8637H	Bar	0.34-0.41	0.35-0.65	0.70-1.05	0.15-0.25	0.35-0.75			0.15-0.35	bal Fe				

Pan America

Specification	Designation	Notes	C	Cr	Mn	Mo	Ni	P	S	Si	Other	UTS	YS	El	Hard
COPANT 334	8637	Bar	0.35-0.4	0.4-0.6	0.75-1	0.15-0.25	0.4-0.7	0.04	0.04	0.2-0.35	bal Fe				
COPANT 514	8637	Tube	0.35-0.4	0.4-0.6	0.75-1	0.15-0.25	0.4-0.7	0.04	0.05	0.1	bal Fe				
COPANT 514	8637		0.35-0.4	0.4-0.6	0.75-1	0.15-0.25	0.4-0.7	0.04	0.05	0.1	bal Fe				

UNS numbers and US grades are provided as a means of cross referencing chemically similar alloys. Exchangability is only possible after independent examination of specifications. Tensile properties are minimum or typical as specified. UTS and YS as MPa. El as %. See Appendix for list of abbreviations used in Notes. * indicates obsolete material.

Specification	Designation	Notes	C	Cr	Mn	Mo	Ni	P	S	Si	Other	UTS	YS	El	Hard
Alloy Steel, Nickel Chromium Molybdenum, 8637/8637H (Continued from previous page)															
Poland															
PNH84030/04	37HGNM	Q/T	0.35-0.43	0.4-0.7	0.8-1.1	0.15-0.25	0.4-0.7	0.035 max	0.035 max	0.17-0.37	bal Fe				
UK															
BS 970/1(83)	945M38*	Blm Bil Slab Bar Rod Frg	0.34-0.42	0.4-0.6	1.20-1.60	0.15-0.25	0.6-0.9	0.04	0.05	0.1-0.35	bal Fe				
BS 970/3(91)	945M38	Bright bar; Q/T; 150<t<=250mm; hard	0.34-0.42	0.40-0.60	1.20-1.60	0.15-0.25	0.60-0.90	0.035 max	0.040 max	0.10-0.40	bal Fe	700-850	495	15	201-255 HB
BS 970/3(91)	945M38	Bright bar; Q/T; 6<=t<=29mm; hard CD	0.34-0.42	0.40-0.60	1.20-1.60	0.15-0.25	0.60-0.90	0.035 max	0.040 max	0.10-0.40	bal Fe	1000-1150	865	9	293-352 HB
USA															
	AISI 8637	Smls mech tub	0.35-0.40	0.40-0.60	0.75-1.00	0.15-0.25	0.40-0.70	0.040 max	0.040 max	0.15-0.35	bal Fe				
	AISI 8637	Bar Blm Bil Slab	0.35-0.40	0.40-0.60	0.75-1.00	0.15-0.25	0.40-0.70	0.035 max	0.040 max	0.15-0.35	bal Fe				
	AISI 8637	Plt	0.33-0.40	0.35-0.60	0.70-1.00	0.15-0.25	0.40-0.70	0.035 max	0.040 max	0.15-0.40	bal Fe				
	AISI 8637H	Bar Blm Bil Slab	0.34-0.41	0.35-0.65	0.70-1.05	0.15-0.25	0.35-0.75	0.035 max	0.040 max	0.15-0.35	bal Fe				
	AISI 8637H	Smls mech tub, Hard	0.34-0.41	0.35-0.65	0.70-1.05	0.15-0.25	0.35-0.75	0.040 max	0.040 max	0.15-0.30	bal Fe				
	UNS G86370		0.35-0.40	0.40-0.60	0.75-1.00	0.15-0.25	0.40-0.70	0.035 max	0.040 max	0.15-0.35	bal Fe				
	UNS H86370		0.34-0.41	0.35-0.65	0.70-1.05	0.15-0.25	0.35-0.75	0.035 max	0.040 max	0.15-0.35	bal Fe				
ASTM A29/A29M(93)	8637	Bar	0.35-0.40	0.40-0.60	0.75-1	0.15-0.25	0.4-0.7	0.035 max	0.04 max	0.15-0.35	bal Fe				
ASTM A304(96)	8637H	Bar, hard bands spec	0.34-0.41	0.35-0.65	0.70-1.05	0.15-0.25	0.35-0.75	0.035 max	0.040 max	0.15-0.35	Cu <=0.35; Cu Ni Cr Mo trace allowed; bal Fe				
ASTM A322(96)	8637	Bar	0.35-0.40	0.40-0.60	0.75-1.00	0.15-0.25	0.40-0.70	0.035 max	0.040 max	0.15-0.35	bal Fe				
ASTM A331(95)	8637	Bar	0.35-0.40	0.40-0.60	0.75-1.00	0.15-0.25	0.40-0.70	0.035 max	0.040 max	0.15-0.35	bal Fe				
ASTM A331(95)	8637H	Bar	0.34-0.41	0.35-0.65	0.70-1.05	0.15-0.25	0.35-0.75	0.025 max	0.025 max	0.15-0.35	Cu <=0.35; bal Fe				
ASTM A519(96)	8637	Smls mech tub	0.35-0.40	0.40-0.60	0.75-1.00	0.15-0.25	0.40-0.70	0.040 max	0.040 max	0.15-0.35	bal Fe				
ASTM A752(93)	8637	Rod Wir	0.35-0.40	0.40-0.60	0.75-1.00	0.15-0.25	0.40-0.70	0.035 max	0.040 max	0.15-0.30	bal Fe	630			
ASTM A829/A829M(95)	8637	Plt	0.33-0.40	0.35-0.60	0.70-1.00	0.15-0.25	0.40-0.70	0.035 max	0.040 max	0.15-0.40	bal Fe	480-965			
SAE 770(84)	8637*	Obs; see J1397(92)	0.35-0.4	0.4-0.6	0.75-1	0.15-0.25	0.4-0.7	0.04 max	0.04 max	0.15-0.3	bal Fe				
SAE J1268(95)	8637H	Bar Rod Wir Tub; See std	0.34-0.41	0.35-0.65	0.70-1.05	0.15-0.25	0.35-0.75	0.030 max	0.040 max	0.15-0.35	Cu <=0.35; Cu not spec'd but acpt; bal Fe				
SAE J404(94)	8637	Plt	0.33-0.40	0.35-0.60	0.70-1.00	0.15-0.25	0.40-0.70	0.035 max	0.040 max	0.15-0.35	bal Fe				
Alloy Steel, Nickel Chromium Molybdenum, 8640/8640H															
Argentina															
IAS	IRAM 8640		0.38-0.43	0.40-0.60	0.75-1.00	0.15-0.25	0.40-0.70	0.035 max	0.040 max	0.20-0.35	bal Fe	850	600	16	225 HB
Australia															
AS 1444	8640*	Obs; Bar	0.38-0.43	0.4-0.6	0.75-1	0.15-0.25	0.4-0.7	0.04	0.04	0.1-0.35	bal Fe				
Belgium															
NBN 253-02	40NiCrMo2		0.37-0.44	0.4-0.6	0.7-1	0.15-0.3	0.4-0.7	0.04 max	0.04 max	0.15-0.4	bal Fe				
NBN 253-06	40NiCrMo3		0.37-0.43	0.6-0.9	0.5-0.8	0.15-0.3	0.75-1	0.025	0.035	0.15-0.4	bal Fe				
France															
AFNOR	40NCD2		0.37-0.44	0.4-0.6	0.6-0.9	0.15-0.3	0.4-0.7	0.04	0.035	0.1-0.4	bal Fe				
AFNOR	40NCD2TS		0.38-0.44	0.4-0.6	0.7-1	0.15-0.3	0.4-0.7	0.025	0.03	0.1-0.4	bal Fe				

UNS numbers and US grades are provided as a means of cross referencing chemically similar alloys. Exchangability is only possible after independent examination of specifications. Tensile properties are minimum or typical as specified. UTS and YS as MPa. El as %. See Appendix for list of abbreviations used in Notes. * indicates obsolete material.

Alloy Steel, Nickel Chromium Molybdenum, 8640/8640H (Continued from previous page)

Specification	Designation	Notes	C	Cr	Mn	Mo	Ni	P	S	Si	Other	UTS	YS	El	Hard
France															
AFNOR	40NCD3		0.36-0.43	0.6-0.9	0.5-0.8	0.15-0.3	0.7-1	0.035	0.035	0.1-0.4	bal Fe				
AFNOR	40NCD3TS		0.38-0.44	0.4-0.6	0.7-1	0.15-0.3	0.4-0.7	0.025	0.03	0.1-0.4	bal Fe				
Germany															
DIN	40NiCrMo2-2		0.37-0.44	0.40-0.60	0.70-1.00	0.15-0.3	0.40-0.70	0.035 max	0.035 max	0.15-0.40	bal Fe				
DIN	WNr 1.6546		0.37-0.44	0.40-0.60	0.70-1.00	0.15-0.30	0.40-0.70	0.035 max	0.035 max	0.15-0.40	bal Fe				
Hungary															
MSZ 61	NCMo3E		0.37-0.44	0.4-0.6	0.7-1	0.15-0.3	0.4-0.7	0.035	0.035	0.17-0.37	bal Fe				
MSZ 61(85)	NCMo3	Q/T; 100.1-160mm	0.37-0.44	0.4-0.6	0.7-1	0.15-0.3	0.4-0.7	0.035 max	0.035 max	0.4 max	bal Fe	750-900	540	13L	
MSZ 61(85)	NCMo3	Q/T; 0-16mm	0.37-0.44	0.4-0.6	0.7-1	0.15-0.3	0.4-0.7	0.035 max	0.035 max	0.4 max	bal Fe	1000-1200	840	10L	
MSZ 6251(87)	NCMo3Z	CF; Q/T; 0-36mm; HR; HR/ann; drawn/soft ann; drawn/bright ann; soft ann/ground	0.37-0.44	0.4-0.6	0.7-1	0.15-0.3	0.4-0.7	0.035 max	0.035 max	0.4 max	bal Fe	0-650			
MSZ 6251(87)	NCMo3Z	CF; Q/T; 0-36mm; drawn, half-hard	0.37-0.44	0.4-0.6	0.7-1	0.15-0.3	0.4-0.7	0.035 max	0.035 max	0.4 max	bal Fe	0-680			
International															
ISO 683-12(72)	10*	Bar Plt, Q/A, Tmp, 16mm	0.38-0.44	0.4-0.6	0.7-1	0.15-0.3	0.4-0.7	0.04	0.04	0.15-0.4	bal Fe				
ISO 683-8	1*		0.37-0.44	0.4-0.6	0.7-1	0.15-0.3	0.4-0.7	0.035	0.035	0.15-0.4	bal Fe				
ISO 683-8	1A*		0.37-0.44	0.4-0.6	0.7-1	0.15-0.3	0.4-0.7	0.035	0.02-0.035	0.15-0.4	bal Fe				
ISO 683-8	1b*		0.37-0.44	0.4-0.6	0.7-1	0.15-0.3	0.4-0.7	0.035	0.03-0.05	0.15-0.4	bal Fe				
ISO R683-8	1A	Bar Bil Rod, Q/A, Tmp, 16mm	0.37-0.44	0.4-0.6	0.7-1	0.15-0.3	0.4-0.7	0.04	0.02-0.04	0.15-0.4	bal Fe				
ISO R683-8	1b	Bar Bil Rod, Q/A, Tmp, 16mm	0.37-0.44	0.4-0.6	0.7-1	0.15-0.3	0.4-0.7	0.04	0.03-0.05	0.15-0.4	bal Fe				
Italy															
UNI 5333	40NiCrMo4		0.37-0.43	0.6-0.9	0.5-0.8	0.15-0.25	0.7-1.1	0.035	0.03	0.4	bal Fe				
UNI 7356(74)	40NiCrMo2KB	Q/T	0.37-0.44	0.4-0.6	0.7-1	0.15-0.3	0.4-0.7	0.035 max	0.035 max	0.15-0.4	bal Fe				
UNI 7845(78)	40NiCrMo2	Q/T	0.37-0.44	0.4-0.6	0.7-1	0.15-0.25	0.4-0.7	0.035 max	0.035 max	0.15-0.4	bal Fe				
UNI 7847(79)	40NiCrMo3	SH	0.37-0.43	0.6-1	0.5-0.8	0.15-0.25	0.7-1	0.03 max	0.03 max	0.15-0.4	bal Fe				
UNI 7874(79)	40NiCrMo2	Q/T	0.37-0.44	0.4-0.6	0.7-1	0.15-0.3	0.4-0.7	0.035 max	0.035 max	0.15-0.4	bal Fe				
UNI 8551(84)	40NiCrMo3	SH	0.37-0.43	0.7-1	0.5-0.8	0.15-0.25	0.7-1	0.03 max	0.025 max	0.15-0.4	Pb 0.15-0.25; bal Fe				
Japan															
JIS G4103(79)	SNCM6*	Obs; see SNCM240	0.38-0.43	0.4-0.65	0.7-1	0.15-0.3	0.4-0.7	0.035	0.03	0.15-0.35	bal Fe				
Mexico															
DGN B-203	8640*	Obs	0.38-0.43	0.4-0.6	0.75-1	0.15-0.25	0.4-0.7	0.04	0.04	0.2-0.35	bal Fe				
DGN B-297	8640*	Obs	0.38-0.43	0.4-0.6	0.75-1	0.15-0.25	0.4-0.7	0.035	0.04	0.2-0.35	bal Fe				
NMX-B-268(68)	8640H	Hard	0.37-0.44	0.35-0.65	0.70-1.05	0.15-0.25	0.35-0.75			0.20-0.35	bal Fe				
NMX-B-300(91)	8640	Bar	0.38-0.43	0.40-0.60	0.75-1.00	0.15-0.25	0.40-0.70	0.035 max	0.040 max	0.15-0.35	bal Fe				
NMX-B-300(91)	8640H	Bar	0.37-0.44	0.35-0.65	0.70-1.05	0.15-0.25	0.35-0.75			0.15-0.35	bal Fe				
Pan America															
COPANT 334	8640	Bar	0.38-0.43	0.4-0.6	0.7-1	0.15-0.25	0.4-0.7	0.04	0.04	0.2-0.35	bal Fe				
COPANT 514	8640	Tube	0.38-0.43	0.4-0.6	0.7-1	0.15-0.25	0.4-0.7	0.04	0.05	0.1	bal Fe				

UNS numbers and US grades are provided as a means of cross referencing chemically similar alloys. Exchangability is only possible after independent examination of specifications. Tensile properties are minimum or typical as specified. UTS and YS as MPa. El as %. See Appendix for list of abbreviations used in Notes. * indicates obsolete material.

Specification	Designation	Notes	C	Cr	Mn	Mo	Ni	P	S	Si	Other	UTS	YS	El	Hard

Alloy Steel, Nickel Chromium Molybdenum, 8640/8640H (Continued from previous page)

Pan America

Specification	Designation	Notes	C	Cr	Mn	Mo	Ni	P	S	Si	Other	UTS	YS	El	Hard
COPANT 514	8640		0.38-0.43	0.4-0.6	0.75-1	0.15-0.25	0.4-0.7	0.04	0.05	0.1	bal Fe				

Poland

PNH84030	37HGNM	Q/T	0.35-0.43	0.4-0.7	0.8-1	0.15-0.25	0.4-0.7	0.035	0.035	0.17-0.37	bal Fe				

Russia

GOST 10702	38ChGNM		0.37-0.43	0.4-0.6	0.7-1	0.15-0.25	0.4-0.7	0.035	0.035	0.17-0.37	bal Fe				
GOST 10702	38ChGNM	Cold upsetting; CHd	0.37-0.43	0.4-0.6	0.5-0.8	0.15-0.25	0.4-0.7	0.035 max	0.035 max	0.17-0.37	Al <=0.1; Cu <=0.3; Ti <=0.05; bal Fe				
GOST 1414(75)	AS40Ch	Free-cutting	0.36-0.44	0.8-0.1	0.6-0.9	0.15 max	0.3 max	0.035 max	0.035 max	0.17-0.37	Cu <=0.30; Pb <=0.3; bal Fe				
GOST 1414(75)	AS40ChGNM	Free-cutting	0.37-0.43	0.6-0.9	0.5-0.8	0.15-0.25	0.7-1.1	0.035 max	0.03 max	0.17-0.37	Cu <=0.30; Pb 0.15-0.3; P+S<=0.06; bal Fe				

Spain

UNE 36012(75)	40NiCrMo2	Q/T	0.37-0.44	0.4-0.7	0.7-1.0	0.15-0.3	0.4-0.7	0.035 max	0.035 max	0.15-0.4	bal Fe				
UNE 36012(75)	F.1204*	see 40NiCrMo2	0.37-0.44	0.4-0.7	0.7-1.0	0.15-0.3	0.4-0.7	0.035 max	0.035 max	0.15-0.4	bal Fe				
UNE 36034(85)	40NiCrMo2DF	CHd; Cold ext	0.37-0.44	0.4-0.6	0.7-1	0.15-0.3	0.4-0.7	0.035 max	0.035 max	0.15-0.4	bal Fe				
UNE 36034(85)	F.1205*	see 40NiCrMo2DF	0.37-0.44	0.4-0.6	0.7-1	0.15-0.3	0.4-0.7	0.035 max	0.035 max	0.15-0.4	bal Fe				

UK

BS 3111/1(87)	5/2	Q/T Wir	0.38-0.43	0.90-1.10	0.70-0.90	0.15-0.25		0.035 max	0.035 max	0.15-0.40	bal Fe				
BS 3111/1(87)	7	Q/T Wir	0.38-0.43	0.40-0.60	0.75-1.00	0.20-0.30	0.40-0.70	0.035 max	0.035 max	0.15-0.40	bal Fe				
BS 970/1(83)	945A40*	Blm Bil Slab Bar Rod Frg	0.38-0.43	0.4-0.6	1.20-1.60	0.15-0.25	0.6-0.9	0.04	0.05	0.1-0.35	bal Fe				

USA

	AISI 8640	Bar Blm Bil Slab	0.38-0.43	0.40-0.60	0.75-1.00	0.15-0.25	0.40-0.70	0.035 max	0.040 max	0.15-0.35	bal Fe				
	AISI 8640	Plt	0.36-0.44	0.35-0.60	0.70-1.00	0.15-0.25	0.40-0.70	0.035 max	0.040 max	0.15-0.40	bal Fe				
	AISI 8640	Smls mech tub	0.38-0.43	0.40-0.60	0.75-1.00	0.15-0.25	0.40-0.70	0.040 max	0.040 max	0.15-0.35	bal Fe				
	AISI 8640H	Bar Blm Bil Slab	0.37-0.44	0.35-0.65	0.70-1.05	0.15-0.25	0.35-0.75	0.035 max	0.040 max	0.15-0.35	bal Fe				
	AISI 8640H	Smls mech tub, Hard	0.37-0.44	0.35-0.65	0.70-1.05	0.15-0.25	0.35-0.75	0.040 max	0.040 max	0.15-0.30	bal Fe				
	UNS G86400		0.38-0.43	0.40-0.60	0.75-1.00	0.15-0.25	0.40-0.70	0.035 max	0.040 max	0.15-0.35	bal Fe				
	UNS H86400		0.37-0.44	0.35-0.65	0.70-1.05	0.15-0.25	0.35-0.75	0.035 max	0.040 max	0.15-0.35	bal Fe				
ASTM A29/A29M(93)	8640	Bar	0.38-0.43	0.40-0.60	0.75-1	0.15-0.25	0.4-0.7	0.035 max	0.04 max	0.15-0.35	bal Fe				
ASTM A304(96)	8640H	Bar, hard bands spec	0.37-0.44	0.35-0.65	0.70-1.05	0.15-0.25	0.35-0.75	0.035 max	0.040 max	0.15-0.35	Cu <=0.35; Cu Ni Cr Mo trace allowed; bal Fe				
ASTM A322(96)	8640	Bar	0.38-0.43	0.40-0.60	0.75-1.00	0.15-0.25	0.40-0.70	0.035 max	0.040 max	0.15-0.35	bal Fe				
ASTM A331(95)	8640	Bar	0.38-0.43	0.40-0.60	0.75-1.00	0.15-0.25	0.40-0.70	0.035 max	0.040 max	0.15-0.35	bal Fe				
ASTM A331(95)	8640H	Bar	0.37-0.44	0.35-0.65	0.70-1.05	0.15-0.25	0.35-0.75	0.025 max	0.025 max	0.15-0.35	Cu <=0.35; bal Fe				
ASTM A519(96)	8640	Smls mech tub	0.38-0.43	0.40-0.60	0.75-1.00	0.15-0.25	0.40-0.70	0.040 max	0.040 max	0.15-0.35	bal Fe				
ASTM A752(93)	8640	Rod Wir	0.38-0.43	0.40-0.60	0.75-1.00	0.15-0.25	0.40-0.70	0.035 max	0.040 max	0.15-0.30	bal Fe	650			
ASTM A829/A829M(95)	8640	Plt	0.36-0.44	0.35-0.60	0.70-1.00	0.15-0.25	0.40-0.70	0.035 max	0.040 max	0.15-0.40	bal Fe	480-965			
MIL-S-16974E(86)	8640*	Obs; Bar Bil Blm slab for refrg													
SAE 770(84)	8640*	Obs; see J1397(92)													

UNS numbers and US grades are provided as a means of cross referencing chemically similar alloys. Exchangability is only possible after independent examination of specifications. Tensile properties are minimum or typical as specified. UTS and YS as MPa. El as %. See Appendix for list of abbreviations used in Notes. * indicates obsolete material.

Specification	Designation	Notes	C	Cr	Mn	Mo	Ni	P	S	Si	Other	UTS	YS	El	Hard

Alloy Steel, Nickel Chromium Molybdenum, 8640/8640H (Continued from previous page)

USA

Specification	Designation	Notes	C	Cr	Mn	Mo	Ni	P	S	Si	Other	UTS	YS	El	Hard
SAE J1268(95)	8640H	Bar Rod Wir Sh Strp Tub; See std	0.37-0.44	0.35-0.65	0.70-1.05	0.15-0.25	0.35-0.75	0.030 max	0.040 max	0.15-0.35	Cu <=0.35; Cu not spec'd but acpt; bal Fe				
SAE J404(94)	8640	Bil Blm Slab Bar HR CF	0.38-0.43	0.40-0.60	0.75-1.00	0.15-0.25	0.40-0.70	0.030 max	0.040 max	0.15-0.35	bal Fe				
SAE J404(94)	8640	Plt	0.36-0.44	0.35-0.60	0.70-1.00	0.15-0.25	0.40-0.70	0.035 max	0.040 max	0.15-0.35	bal Fe				

Alloy Steel, Nickel Chromium Molybdenum, 8642/8642H

Mexico

Specification	Designation	Notes	C	Cr	Mn	Mo	Ni	P	S	Si	Other	UTS	YS	El	Hard
DGN B-297	8642*	Obs	0.4-0.45	0.4-0.6	0.75-1	0.15-0.25	0.4-0.7	0.035	0.04	0.2-0.35	bal Fe				
NMX-B-268(68)	8642H	Hard	0.39-0.46	0.35-0.65	0.70-1.05	0.15-0.25	0.35-0.75			0.20-0.35	bal Fe				
NMX-B-300(91)	8642	Bar	0.40-0.45	0.40-0.60	0.75-1.00	0.15-0.25	0.40-0.70	0.035 max	0.040 max	0.15-0.35	bal Fe				
NMX-B-300(91)	8642H	Bar	0.39-0.46	0.35-0.65	0.70-1.05	0.15-0.25	0.35-0.75			0.15-0.35	bal Fe				

Pan America

Specification	Designation	Notes	C	Cr	Mn	Mo	Ni	P	S	Si	Other	UTS	YS	El	Hard
COPANT 514	8642		0.4-0.45	0.4-0.6	0.75-1	0.15-0.25	0.4-0.7	0.04	0.05	0.1	bal Fe				
COPANT 514	8642	Tube	0.4-0.45	0.4-0.6	0.75-1	0.15-0.25	0.4-0.7	0.04	0.05	0.1	bal Fe				

USA

Specification	Designation	Notes	C	Cr	Mn	Mo	Ni	P	S	Si	Other	UTS	YS	El	Hard
	AISI 8642	Smls mech tub	0.40-0.45	0.40-0.60	0.75-1.00	0.15-0.25	0.40-0.70	0.040 max	0.040 max	0.15-0.35	bal Fe				
	AISI 8642H	Smls mech tub, Hard	0.39-0.46	0.35-0.65	0.70-1.05	0.15-0.25	0.35-0.75	0.035 max	0.040 max	0.15-0.30	bal Fe				
	UNS G86420		0.40-0.45	0.40-0.60	0.75-1.00	0.15-0.25	0.40-0.70	0.035 max	0.040 max	0.15-0.35	bal Fe				
	UNS H86420		0.39-0.46	0.35-0.65	0.70-1.05	0.15-0.25	0.35-0.75	0.035 max	0.040 max	0.15-0.35	bal Fe				
ASTM A29/A29M(93)	8642	Bar	0.40-0.45	0.40-0.60	0.75-1	0.15-0.25	0.4-0.7	0.035 max	0.04 max	0.15-0.35	bal Fe				
ASTM A304(96)	8642H	Bar, hard bands spec	0.39-0.46	0.35-0.65	0.70-1.05	0.15-0.25	0.35-0.75	0.035 max	0.040 max	0.15-0.35	Cu <=0.35; Cu Ni Cr Mo trace allowed; bal Fe				
ASTM A322(96)	8642	Bar	0.40-0.45	0.40-0.60	0.75-1.00	0.15-0.25	0.40-0.70	0.035 max	0.040 max	0.15-0.35	bal Fe				
ASTM A331(95)	8642	Bar	0.40-0.45	0.40-0.60	0.75-1.00	0.15-0.25	0.40-0.70	0.035 max	0.040 max	0.15-0.35	bal Fe				
ASTM A331(95)	8642H	Bar	0.39-0.46	0.35-0.65	0.70-1.05	0.15-0.25	0.35-0.75	0.025 max	0.025 max	0.15-0.35	Cu <=0.35; bal Fe				
ASTM A519(96)	8642	Smls mech tub	0.40-0.45	0.40-0.60	0.75-1.00	0.15-0.25	0.40-0.70	0.040 max	0.040 max	0.15-0.35	bal Fe				
ASTM A752(93)	8642	Rod Wir	0.40-0.45	0.40-0.60	0.70-1.00	0.15-0.25	0.40-0.70	0.035 max	0.040 max	0.15-0.30	bal Fe	670			
SAE J1268(95)	8642H	Bar Rod Wir Sh Strp Tub; See std	0.39-0.46	0.35-0.65	0.70-1.05	0.15-0.25	0.35-0.75	0.030 max	0.040 max	0.15-0.35	Cu <=0.35; Cu not spec'd but acpt; bal Fe				

Alloy Steel, Nickel Chromium Molybdenum, 8645/8645H

Australia

Specification	Designation	Notes	C	Cr	Mn	Mo	Ni	P	S	Si	Other	UTS	YS	El	Hard
AS 1444	8645*	Obs; Bar	0.43-0.48	0.4-0.6	0.75-1	0.15-0.25	0.4-0.7	0.04	0.04	0.1-0.35	bal Fe				
AS 1444	8645H*	Obs; Bar	0.42-0.49	0.35-0.65	0.7-1.05	0.15-0.25	0.35-0.75	0.04	0.04	0.1-0.35	bal Fe				

Mexico

Specification	Designation	Notes	C	Cr	Mn	Mo	Ni	P	S	Si	Other	UTS	YS	El	Hard
DGN B-203	8645*	Obs	0.43-0.48	0.4-0.6	0.75-1	0.15-0.25	0.4-0.7	0.04	0.04	0.2-0.35	bal Fe				
DGN B-297	8645*	Obs	0.43-0.48	0.4-0.6	0.75-1	0.15-0.25	0.4-0.7	0.035	0.04	0.2-0.35	bal Fe				
NMX-B-268(68)	8645H	Hard	0.42-0.49	0.35-0.65	0.70-1.05	0.15-0.25	0.35-0.75			0.20-0.35	bal Fe				
NMX-B-300(91)	8645	Bar	0.43-0.48	0.40-0.60	0.75-1.00	0.15-0.25	0.40-0.70	0.035 max	0.040 max	0.15-0.35	bal Fe				

Specification	Designation	Notes	C	Cr	Mn	Mo	Ni	P	S	Si	Other	UTS	YS	El	Hard

Alloy Steel, Nickel Chromium Molybdenum, 8645/8645H (Continued from previous page)

Mexico

Specification	Designation	Notes	C	Cr	Mn	Mo	Ni	P	S	Si	Other	UTS	YS	El	Hard
NMX-B-300(91)	8645H	Bar	0.42-0.49	0.35-0.65	0.70-1.05	0.15-0.25	0.35-0.75			0.15-0.35	bal Fe				

Pan America

Specification	Designation	Notes	C	Cr	Mn	Mo	Ni	P	S	Si	Other	UTS	YS	El	Hard
COPANT 514	8645		0.43-0.48	0.4-0.6	0.75-1	0.15-0.25	0.4-0.7	0.04	0.05	0.1	bal Fe				

USA

Specification	Designation	Notes	C	Cr	Mn	Mo	Ni	P	S	Si	Other	UTS	YS	El	Hard
	AISI 8645	Bar Blm Bil Slab	0.43-0.48	0.40-0.60	0.75-1.00	0.15-0.25	0.40-0.70	0.035 max	0.040 max	0.15-0.35	bal Fe				
	AISI 8645	Smls mech tub	0.43-0.48	0.40-0.60	0.75-1.00	0.15-0.25	0.40-0.70	0.040 max	0.040 max	0.15-0.35	bal Fe				
	AISI 8645H	Smls mech tub, Hard	0.42-0.49	0.35-0.65	0.70-1.05	0.15-0.25	0.35-0.75	0.040 max	0.040 max	0.15-0.30	bal Fe				
	AISI 8645H	Bar Blm Bil Slab	0.42-0.49	0.35-0.65	0.70-1.05	0.15-0.25	0.35-0.75	0.035 max	0.040 max	0.15-0.35	bal Fe				
	UNS G86450		0.43-0.48	0.40-0.60	0.75-1.00	0.15-0.25	0.40-0.70	0.035 max	0.040 max	0.15-0.35	bal Fe				
	UNS H86450		0.42-0.49	0.35-0.65	0.70-1.05	0.15-0.25	0.35-0.75	0.035 max	0.040 max	0.15-0.35	bal Fe				
ASTM A29/A29M(93)	8645	Bar	0.43-0.48	0.40-0.60	0.75-1	0.15-0.25	0.4-0.7	0.035 max	0.04 max	0.15-0.35	bal Fe				
ASTM A304(96)	8645H	Bar, hard bands spec	0.42-0.49	0.35-0.65	0.70-1.05	0.15-0.25	0.35-0.75	0.035 max	0.040 max	0.15-0.35	Cu <=0.35; Cu Ni Cr Mo trace allowed; bal Fe				
ASTM A322(96)	8645	Bar	0.43-0.48	0.40-0.60	0.75-1.00	0.15-0.25	0.40-0.70	0.035 max	0.040 max	0.15-0.35	bal Fe				
ASTM A331(95)	8645	Bar	0.43-0.48	0.40-0.60	0.75-1.00	0.15-0.25	0.40-0.70	0.035 max	0.040 max	0.15-0.35	bal Fe				
ASTM A331(95)	8645H	Bar	0.42-0.49	0.35-0.65	0.70-1.05	0.15-0.25	0.35-0.75	0.025 max	0.025 max	0.15-0.35	Cu <=0.35; bal Fe				
ASTM A519(96)	8645	Smls mech tub	0.43-0.48	0.40-0.60	0.75-1.00	0.15-0.25	0.40-0.70	0.040 max	0.040 max	0.15-0.35	bal Fe				
ASTM A752(93)	8645	Rod Wir	0.43-0.48	0.40-0.60	0.75-1.00	0.15-0.25	0.40-0.70	0.035 max	0.040 max	0.15-0.30	bal Fe	680			
DoD-F-24669/1(86)	8645	Bar Bil; Supersedes MIL-S-866 & MIL-S-16974	0.43-0.48	0.4-0.6	0.75-1	0.15-0.25	0.4-0.7	0.035	0.04	0.15-0.35	bal Fe				
MIL-S-16974E(86)	8645*	Obs; Bar Bil Blm slab for refrg	0.43-0.48	0.4-0.6	0.75-1	0.15-0.25	0.4-0.7	0.04 max	0.04 max	0.15-0.3	bal Fe				
SAE 770(84)	8645*	Obs; see J1397(92)	0.43-0.48	0.4-0.6	0.75-1	0.15-0.25	0.4-0.7	0.04 max	0.04 max	0.15-0.3	bal Fe				
SAE J1268(95)	8645H	Bar Rod Wir Sh Strp Tub; See std	0.42-0.49	0.35-0.65	0.70-1.05	0.15-0.25	0.35-0.75	0.030 max	0.040 max	0.15-0.35	Cu <=0.35; Cu not spec'd but acpt; bal Fe				
SAE J404(94)	8645	Bil Blm Slab Bar HR CF	0.43-0.48	0.40-0.60	0.75-1.00	0.15-0.25	0.40-0.70	0.030 max	0.040 max	0.15-0.35	bal Fe				
SAE J775(93)	8645	Engine poppet valve, Mart	0.43-0.48	0.40-0.60	0.75-1.00	0.15-0.25	0.40-0.70	0.035 max	0.040 max	0.15-0.35	bal Fe	1240	1140		
SAE J775(93)	JIS G4103 SNCM 240	Engine poppet valve	0.38-0.43	0.40-0.65	0.70-1.00	0.15-0.30	0.40-0.70	0.030 max	0.030 max	0.15-0.35	bal Fe				
SAE J775(93)	NV5*	Engine poppet valve, Mart; former SAE grade	0.43-0.48	0.40-0.60	0.75-1.00	0.15-0.25	0.40-0.70	0.035 max	0.040 max	0.15-0.35	bal Fe	1240	1140		

Alloy Steel, Nickel Chromium Molybdenum, 8650/8650H

Mexico

Specification	Designation	Notes	C	Cr	Mn	Mo	Ni	P	S	Si	Other	UTS	YS	El	Hard
DGN B-203	8650*	Obs	0.48-0.53	0.4-0.6	0.75-1	0.15-0.25	0.4-0.7	0.04	0.04	0.2-0.35	bal Fe				
NMX-B-268(68)	8650H	Hard	0.47-0.54	0.35-0.65	0.70-1.05	0.15-0.25	0.35-0.75			0.20-0.35	bal Fe				
NMX-B-300(91)	8650	Bar	0.48-0.53	0.40-0.60	0.75-1.00	0.15-0.25	0.40-0.70	0.035 max	0.040 max	0.15-0.35	bal Fe				
NMX-B-300(91)	8650H	Bar	0.47-0.54	0.35-0.65	0.70-1.05	0.15-0.25	0.35-0.75			0.15-0.35	bal Fe				

Pan America

Specification	Designation	Notes	C	Cr	Mn	Mo	Ni	P	S	Si	Other	UTS	YS	El	Hard
COPANT 334	8650	Bar	0.48-0.53	0.4-0.6	0.75-1	0.15-0.25	0.4-0.7	0.04	0.04	0.2-0.35	bal Fe				
COPANT 514	8650	Tube	0.48-0.53	0.4-0.6	0.75-1	0.15-0.25	0.4-0.7	0.04	0.05	0.1	bal Fe				

UNS numbers and US grades are provided as a means of cross referencing chemically similar alloys. Exchangability is only possible after independent examination of specifications. Tensile properties are minimum or typical as specified. UTS and YS as MPa. El as %. See Appendix for list of abbreviations used in Notes. * indicates obsolete material.

Specification	Designation	Notes	C	Cr	Mn	Mo	Ni	P	S	Si	Other	UTS	YS	El	Hard

Alloy Steel, Nickel Chromium Molybdenum, 8650/8650H (Continued from previous page)

Pan America

| COPANT 514 | 8650 | | 0.48-0.53 | 0.4-0.6 | 0.75-1 | 0.15-0.25 | 0.4-0.7 | 0.04 | 0.05 | 0.1 | bal Fe | | | | |

USA

	AISI 8650	Smls mech tub	0.48-0.53	0.40-0.60	0.75-1.00	0.15-0.25	0.40-0.70	0.040 max	0.040 max	0.15-0.35	bal Fe				
	AISI 8650H	Smls mech tub, Hard	0.47-0.54	0.35-0.65	0.70-1.05	0.15-0.25	0.35-0.75	0.035 max	0.040 max	0.15-0.30	bal Fe				
	UNS G86500		0.48-0.53	0.40-0.60	0.75-1.00	0.15-0.25	0.40-0.70	0.035 max	0.040 max	0.15-0.35	bal Fe				
	UNS H86500		0.47-0.54	0.35-0.65	0.70-1.05	0.15-0.25	0.35-0.75			0.15-0.35	bal Fe				
ASTM A29/A29M(93)	8650	Bar	0.48-0.53	0.40-0.60	0.75-1	0.15-0.25	0.4-0.7	0.035 max	0.04 max	0.15-0.35	bal Fe				
ASTM A304(96)	8650H	Bar, hard bands spec	0.47-0.54	0.35-0.65	0.70-1.05	0.15-0.25	0.35-0.75	0.035 max	0.040 max	0.15-0.35	Cu <=0.35; Cu Ni Cr Mo trace allowed; bal Fe				
ASTM A322(96)	8650*	Bar	0.48-0.53	0.40-0.60	0.75-1.00	0.15-0.25	0.40-0.70	0.035 max	0.040 max	0.15-0.35	bal Fe				
ASTM A331(95)	8650	Bar	0.48-0.53	0.40-0.60	0.75-1.00	0.15-0.25	0.40-0.70	0.035 max	0.040 max	0.15-0.35	bal Fe				
ASTM A331(95)	8650H	Bar	0.47-0.54	0.35-0.65	0.70-1.05	0.15-0.25	0.35-0.75	0.025 max	0.025 max	0.15-0.35	Cu <=0.35; bal Fe				
ASTM A519(96)	8650	Smls mech tub	0.48-0.53	0.40-0.60	0.75-1.00	0.15-0.25	0.40-0.70	0.040 max	0.040 max	0.15-0.35	bal Fe				
SAE 770(84)	8650*	Obs; see J1397(92)	0.48-0.53	0.4-0.6	0.75-1	0.15-0.25	0.4-0.7	0.04 max	0.05 max	0.15-0.3	bal Fe				
SAE J1268(95)	8650H	See std	0.47-0.54	0.35-0.65	0.70-1.05	0.15-0.25	0.35-0.70	0.030 max	0.040 max	0.15-0.35	Cu <=0.35; Cu not spec'd but acpt; bal Fe				

Alloy Steel, Nickel Chromium Molybdenum, 8655/8655H

Mexico

DGN B-203	8655*	Obs	0.5-0.6	0.4-0.6	0.75-1	0.15-0.25	0.4-0.7	0.04	0.04	0.2-0.35	bal Fe				
NMX-B-268(68)	8655H	Hard	0.50-0.60	0.35-0.65	0.70-1.05	0.15-0.25	0.35-0.75			0.20-0.35	bal Fe				
NMX-B-300(91)	8655	Bar	0.51-0.59	0.40-0.60	0.75-1.00	0.15-0.25	0.40-0.70	0.035 max	0.040 max	0.15-0.35	bal Fe				
NMX-B-300(91)	8655H	Bar	0.50-0.60	0.35-0.65	0.70-1.05	0.15-0.25	0.35-0.75			0.15-0.35	bal Fe				

Pan America

| COPANT 514 | 8655 | | 0.5-0.6 | 0.4-0.6 | 0.75-1 | 0.15-0.25 | 0.4-0.7 | 0.04 | 0.05 | 0.1 | bal Fe | | | | |

USA

	AISI 8655	Smls mech tub	0.51-0.59	0.40-0.60	0.75-1.00	0.15-0.25	0.40-0.70	0.040 max	0.040 max	0.15-0.35	bal Fe				
	AISI 8655	Plt	0.49-0.60	0.35-0.60	0.70-1.00	0.15-0.25	0.40-0.70	0.035 max	0.040 max	0.15-0.40	bal Fe				
	AISI 8655H	Smls mech tub, Hard	0.50-0.60	0.35-0.65	0.70-1.05	0.15-0.25	0.35-0.75	0.035 max	0.040 max	0.15-0.30	bal Fe				
	UNS G86550		0.51-0.59	0.40-0.60	0.75-1.00	0.15-0.25	0.40-0.70	0.035 max	0.040 max	0.15-0.35	bal Fe				
	UNS H86550		0.50-0.60	0.35-0.65	0.70-1.05	0.15-0.25	0.35-0.75	0.035 max	0.040 max	0.15-0.35	bal Fe				
ASTM A29/A29M(93)	8655	Bar	0.51-0.59	0.40-0.60	0.75-1	0.15-0.25	0.4-0.7	0.035 max	0.04 max	0.15-0.35	bal Fe				
ASTM A304(96)	8655H	Bar, hard bands spec	0.50-0.60	0.35-0.65	0.70-1.05	0.15-0.25	0.35-0.75	0.035 max	0.040 max	0.15-0.35	Cu <=0.35; Cu Ni Cr Mo trace allowed; bal Fe				
ASTM A322(96)	8655	Bar	0.51-0.59	0.40-0.60	0.75-1.00	0.15-0.25	0.40-0.70	0.035 max	0.040 max	0.15-0.35	bal Fe				
ASTM A331(95)	8655	Bar	0.51-0.59	0.40-0.60	0.75-1.00	0.15-0.25	0.40-0.70	0.035 max	0.040 max	0.15-0.35	bal Fe				
ASTM A331(95)	8655H	Bar	0.50-0.60	0.35-0.65	0.70-1.05	0.15-0.25	0.35-0.75	0.025 max	0.025 max	0.15-0.35	Cu <=0.35; bal Fe				
ASTM A519(96)	8655	Smls mech tub	0.50-0.60	0.40-0.60	0.75-1.00	0.15-0.25	0.40-0.70	0.040 max	0.040 max	0.15-0.35	bal Fe				

UNS numbers and US grades are provided as a means of cross referencing chemically similar alloys. Exchangability is only possible after independent examination of specifications. Tensile properties are minimum or typical as specified. UTS and YS as MPa. El as %. See Appendix for list of abbreviations used in Notes. * indicates obsolete material.

Specification	Designation	Notes	C	Cr	Mn	Mo	Ni	P	S	Si	Other	UTS	YS	El	Hard

Alloy Steel, Nickel Chromium Molybdenum, 8655/8655H (Continued from previous page)

USA

Specification	Designation	Notes	C	Cr	Mn	Mo	Ni	P	S	Si	Other	UTS	YS	El	Hard
ASTM A752(93)	8655	Rod Wir	0.51-0.59	0.40-0.60	0.75-1.00	0.15-0.25	0.40-0.70	0.035 max	0.040 max	0.15-0.30	bal Fe	720			
ASTM A829/A829M(95)	8655	Plt	0.49-0.60	0.35-0.60	0.70-1.00	0.15-0.25	0.40-0.70	0.035 max	0.040 max	0.15-0.40	bal Fe	480-965			
SAE 770(84)	8655*	Obs; see J1397(92)	0.51-0.59	0.4-0.6	0.75-1	0.15-0.25	0.4-0.7	0.04 max	0.05 max	0.15-0.3	bal Fe				
SAE J1268(95)	8655H	Bar Sh Strp Tub; See std	0.50-0.60	0.35-0.65	0.70-1.05	0.15-0.25	0.35-0.75	0.030 max	0.040 max	0.15-0.35	Cu <=0.35; Cu not spec'd but acpt; bal Fe				
SAE J404(94)	8655	Plt	0.49-0.60	0.35-0.60	0.70-1.00	0.15-0.25	0.40-0.70	0.035 max	0.040 max	0.15-0.35	bal Fe				

Alloy Steel, Nickel Chromium Molybdenum, 8660/8660H

Australia

Specification	Designation	Notes	C	Cr	Mn	Mo	Ni	P	S	Si	Other	UTS	YS	El	Hard
AS 1444(96)	8660	H series, Hard/Tmp	0.55-0.65	0.40-0.60	0.75-1.00	0.15-0.25	0.40-0.70	0.040 max	0.040 max	0.10-0.35	bal Fe				
AS 1444(96)	8660H	H series, Hard/Tmp	0.55-0.65	0.35-0.65	0.70-1.05	0.15-0.25	0.35-0.75	0.040 max	0.040 max	0.10-0.35	bal Fe				

Mexico

Specification	Designation	Notes	C	Cr	Mn	Mo	Ni	P	S	Si	Other	UTS	YS	El	Hard
DGN B-203	8660*	Obs	0.55-0.65	0.4-0.6	0.75-1	0.15-0.25	0.4-0.7	0.04	0.04	0.2-0.35	bal Fe				
NMX-B-268(68)	8660H	Hard	0.55-0.65	0.35-0.65	0.70-1.05	0.15-0.25	0.35-0.75			0.20-0.35	bal Fe				
NMX-B-300(91)	8660	Bar	0.56-0.64	0.40-0.60	0.75-1.00	0.15-0.25	0.40-0.70	0.035 max	0.040 max	0.15-0.35	bal Fe				
NMX-B-300(91)	8660H	Bar	0.55-0.65	0.35-0.65	0.70-1.05	0.15-0.25	0.35-0.75			0.15-0.35	bal Fe				

Pan America

Specification	Designation	Notes	C	Cr	Mn	Mo	Ni	P	S	Si	Other	UTS	YS	El	Hard
COPANT 334	8660	Bar	0.56-0.64	0.4-0.6	0.75-1	0.15-0.25	0.4-0.7	0.04	0.04	0.2-0.35	bal Fe				
COPANT 514	8660	Tube	0.55-0.65	0.4-0.6	0.75-1	0.15-0.25	0.4-0.7	0.04	0.05	0.1	bal Fe				
COPANT 514	8660		0.55-0.65	0.4-0.6	0.75-1	0.15-0.25	0.4-0.7	0.04	0.05	0.1	bal Fe				

Spain

Specification	Designation	Notes	C	Cr	Mn	Mo	Ni	P	S	Si	Other	UTS	YS	El	Hard
UNE	F.140.E		0.55-0.65	0.4-0.7	0.7-1	0.15-0.25	0.4-0.7	0.035	0.035	0.13-0.38	bal Fe				

UK

Specification	Designation	Notes	C	Cr	Mn	Mo	Ni	P	S	Si	Other	UTS	YS	El	Hard
BS 970/1(83)	805A60*	Blm Bil Slab Bar Rod Frg	0.55-0.65	0.4-0.6	0.70-1.00	0.15-0.25	0.4-0.7	0.04	0.05	0.1-0.35	bal Fe				
BS 970/1(83)	805H60*	Blm Bil Slab Bar Rod Frg	0.55-0.65	0.35-0.65	0.65-1.05	0.15-0.25	0.35-0.75	0.04	0.05	0.1-0.35	bal Fe				

USA

Specification	Designation	Notes	C	Cr	Mn	Mo	Ni	P	S	Si	Other	UTS	YS	El	Hard
	AISI 8660	Smls mech tub	0.55-0.65	0.40-0.60	0.75-1.00	0.15-0.25	0.40-0.70	0.040 max	0.040 max	0.15-0.35	bal Fe				
	AISI 8660H	Smls mech tub, Hard	0.55-0.65	0.35-0.65	0.70-1.05	0.15-0.25	0.35-0.75	0.035 max	0.040 max	0.15-0.30	bal Fe				
	UNS G86600		0.55-0.65	0.40-0.60	0.75-1.00	0.15-0.25	0.40-0.70	0.035 max	0.040 max	0.15-0.35	bal Fe				
	UNS H86600		0.55-0.65	0.35-0.65	0.70-1.05	0.15-0.25	0.35-0.75			0.15-0.35	bal Fe				
ASTM A29/A29M(93)	8660	Bar	0.55-0.65	0.40-0.60	0.1-0.75	0.15-0.25	0.4-0.7	0.035 max	0.04 max	0.15-0.35	bal Fe				
ASTM A304(96)	8660H	Bar, hard bands spec	0.55-0.65	0.35-0.65	0.70-1.05	0.15-0.25	0.35-0.75	0.035 max	0.040 max	0.15-0.35	Cu <=0.35; Cu Ni Cr Mo trace allowed; bal Fe				
ASTM A331(95)	8660H	Bar	0.55-0.65	0.35-0.65	0.70-1.05	0.15-0.25	0.35-0.75	0.025 max	0.025 max	0.15-0.35	Cu <=0.35; bal Fe				
ASTM A332	8660	Bar Rod Wir Sh Strp Tub	0.55-0.65	0.4-0.6	0.1-0.75	0.15-0.25	0.4-0.7	0.04 max	0.04 max	0.15-0.3	bal Fe				
ASTM A519(96)	8660	Smls mech tub	0.55-0.65	0.40-0.60	0.75-1.00	0.15-0.25	0.40-0.70	0.040 max	0.040 max	0.15-0.35	bal Fe				
SAE 770(84)	8660*	Obs; see J1397(92)	0.55-0.65	0.4-0.6	0.75-1	0.15-0.25	0.4-0.7	0.04 max	0.04 max	0.15-0.3	bal Fe				
SAE J1268(95)	8660H	Bar Rod Wir Sh Strp Tub; See std	0.55-0.65	0.35-0.65	0.70-1.05	0.15-0.25	0.35-0.75	0.030 max	0.040 max	0.15-0.35	Cu <=0.35; Cu not spec'd but acpt; bal Fe				

UNS numbers and US grades are provided as a means of cross referencing chemically similar alloys. Exchangability is only possible after independent examination of specifications. Tensile properties are minimum or typical as specified. UTS and YS as MPa. El as %. See Appendix for list of abbreviations used in Notes. * indicates obsolete material.

Specification	Designation	Notes	C	Cr	Mn	Mo	Ni	P	S	Si	Other	UTS	YS	El	Hard
Alloy Steel, Nickel Chromium Molybdenum, 8660/8660H (Continued from previous page)															
USA															
SAE J407	8660H*	Obs; see J1268									bal Fe				
Alloy Steel, Nickel Chromium Molybdenum, 86B30/86B30H															
Australia															
AS 1444(96)	86B30H	H series, Hard/Tmp	0.27-0.33	0.35-0.65	0.60-0.95	0.15-0.25	0.35-0.75	0.040 max	0.040 max	0.10-0.35	B >=0.0005; bal Fe				
Mexico															
NMX-B-300(91)	86B30H	Bar	0.27-0.33	0.35-0.65	0.60-0.95	0.15-0.25	0.35-0.75			0.15-0.35	B >=0.0005; bal Fe				
USA															
	AISI 86B30H	Smls mech tub, Hard; Boron	0.27-0.33	0.35-0.65	0.65-0.95	0.15-0.25	0.35-0.75	0.035 max	0.040 max	0.15-0.30	B 0.0005-0.005; bal Fe				
	UNS H86301		0.27-0.33	0.35-0.65	0.60-0.95	0.15-0.25	0.35-0.75			0.15-0.35	B 0.0005-0.003; bal Fe				
ASTM A304(96)	86B30H	Bar, hard bands spec	0.27-0.33	0.35-0.65	0.60-0.95	0.15-0.25	0.35-0.75	0.035 max	0.040 max	0.15-0.35	Cu <=0.35; Cu Ni Cr Mo trace allowed; bal Fe				
ASTM A331(95)	86B30H	Bar	0.27-0.33	0.35-0.65	0.60-0.95	0.15-0.25	0.35-0.75	0.025 max	0.025 max	0.15-0.35	Cu <=0.35; bal Fe				
SAE J1268(95)	86B30H	Bar Rod Wir Tub; See std	0.27-0.33	0.35-0.65	0.60-0.95	0.15-0.25	0.35-0.75	0.030 max	0.040 max	0.15-0.35	B 0.0005-0.003; Cu <=0.35; Cu not spec'd but acpt; bal Fe				
Alloy Steel, Nickel Chromium Molybdenum, 86B45/86B45H															
Mexico															
NMX-B-268(68)	86B45H	Hard	0.42-0.49	0.35-0.65	0.70-1.05	0.15-0.25	0.35-0.75			0.20-0.35	B >=0.0005; bal Fe				
NMX-B-300(91)	86B45H	Bar	0.42-0.49	0.35-0.65	0.70-1.05	0.15-0.25	0.35-0.75			0.15-0.35	B >=0.0005; bal Fe				
Pan America															
COPANT 514	86B45	Tube	0.43-0.48	0.4-0.6	0.75-1	0.15-0.25	0.4-0.7	0.04	0.05	0.1	bal Fe				
USA															
	AISI 86B45	Smls mech Tub; Boron	0.43-0.48	0.40-0.60	0.75-1.00	0.15-0.25	0.40-0.70	0.040 max	0.040 max	0.15-0.35	B >=0.0005; bal Fe				
	AISI 86B45H	Smls mech tub, Hard; Boron	0.42-0.49	0.35-0.65	0.70-1.05	0.15-0.25	0.35-0.75	0.035 max	0.040 max	0.15-0.30	B 0.0005-0.005; bal Fe				
	UNS G86451		0.43-0.48	0.40-0.60	0.75-1.00	0.15-0.25	0.40-0.70	0.035 max	0.040 max	0.15-0.35	B 0.0005-0.003; bal Fe				
	UNS H86451		0.42-0.49	0.35-0.65	0.70-1.05	0.15-0.25	0.35-0.75			0.15-0.35	B 0.0005-0.003; bal Fe				
ASTM A304(96)	86B45H	Bar, hard bands spec	0.42-0.49	0.35-0.65	0.70-1.05	0.15-0.25	0.35-0.75	0.035 max	0.040 max	0.15-0.35	Cu <=0.35; Cu Ni Cr Mo trace allowed; bal Fe				
ASTM A331(95)	86B45H	Bar	0.42-0.49	0.35-0.65	0.70-1.05	0.15-0.25	0.35-0.75	0.025 max	0.025 max	0.15-0.35	Cu <=0.35; bal Fe				
ASTM A519(96)	86B45	Smls mech tub	0.43-0.48	0.40-0.60	0.75-1.00	0.15-0.25	0.40-0.70	0.040 max	0.040 max	0.15-0.35	B >=0.0005; bal Fe				
SAE 770(84)	86B45*	Obs; see J1397(92)									bal Fe				
SAE J1268(95)	86B45H	Bar Rod Wir Sh Strp Tub; See std	0.42-0.49	0.35-0.65	0.70-1.05	0.15-0.25	0.35-0.75	0.030 max	0.040 max	0.15-0.35	B 0.0005-0.003; Cu <=0.35; Cu not spec'd but acpt; bal Fe				
Alloy Steel, Nickel Chromium Molybdenum, 8720/8720H															
Mexico															
NMX-B-268(68)	8720H	Hard	0.17-0.23	0.35-0.65	0.60-0.95	0.20-0.30	0.35-0.75			0.20-0.35	bal Fe				
NMX-B-300(91)	8720	Bar	0.18-0.23	0.40-0.60	0.70-0.90	0.20-0.30	0.40-0.70	0.035 max	0.040 max	0.15-0.35	bal Fe				
NMX-B-300(91)	8720H	Bar	0.17-0.23	0.35-0.65	0.60-0.95	0.20-0.30	0.35-0.75			0.15-0.35	bal Fe				
Pan America															
COPANT 334	8720	Bar	0.18-0.23	0.4-0.6	0.7-0.9	0.2-0.3	0.4-0.7	0.04	0.04	0.2-0.35	bal Fe				

UNS numbers and US grades are provided as a means of cross referencing chemically similar alloys. Exchangability is only possible after independent examination of specifications. Tensile properties are minimum or typical as specified. UTS and YS as MPa. El as %. See Appendix for list of abbreviations used in Notes. * indicates obsolete material.

Specification	Designation	Notes	C	Cr	Mn	Mo	Ni	P	S	Si	Other	UTS	YS	El	Hard

Alloy Steel, Nickel Chromium Molybdenum, 8720/8720H (Continued from previous page)

Pan America

| COPANT 514 | 8720 | Tube | 0.18-0.23 | 0.4-0.6 | 0.7-0.9 | 0.2-0.3 | 0.4-0.7 | 0.04 | 0.05 | 0.1 | bal Fe | | | | |

UK

| BS 970/3(91) | 805A20 | Bright bar | 0.18-0.23 | 0.40-0.60 | 0.70-0.90 | 0.15-0.25 | 0.40-0.70 | | | | bal Fe | | | | |

USA

	AISI 8720	Bar Blm Bil Slab	0.18-0.23	0.40-0.60	0.70-0.90	0.20-0.30	0.40-0.70	0.035 max	0.040 max	0.15-0.35	bal Fe				
	AISI 8720	Smls mech tub	0.18-0.23	0.40-0.60	0.70-0.90	0.20-0.30	0.40-0.70	0.040 max	0.040 max	0.15-0.35	bal Fe				
	AISI 8720H	Smls mech tub, Hard	0.17-0.23	0.35-0.65	0.60-0.95	0.20-0.30	0.35-0.75	0.040 max	0.040 max	0.15-0.30	bal Fe				
	AISI 8720H	Bar Blm Bil Slab	0.17-0.23	0.35-0.65	0.60-0.95	0.20-0.30	0.35-0.75	0.035 max	0.040 max	0.15-0.35	bal Fe				
	UNS G87200		0.18-0.23	0.40-0.60	0.70-0.90	0.20-0.30	0.40-0.70	0.035 max	0.040 max	0.15-0.35	bal Fe				
	UNS H87200		0.17-0.23	0.35-0.65	0.60-0.95	0.20-0.30	0.35-0.75	0.035 max	0.040 max	0.15-0.35	bal Fe				
ASTM A29/A29M(93)	8720	Bar	0.18-0.23	0.40-0.60	0.7-0.9	0.2-0.3	0.4-0.7	0.035 max	0.04 max	0.15-0.35	bal Fe				
ASTM A304(96)	8720H	Bar, hard bands spec	0.17-0.23	0.35-0.65	0.60-0.95	0.20-0.30	0.35-0.75	0.035 max	0.040 max	0.15-0.35	Cu <=0.35; Cu Ni Cr Mo trace allowed; bal Fe				
ASTM A322(96)	8720	Bar	0.18-0.23	0.40-0.60	0.70-0.90	0.2-0.30	0.40-0.70	0.035 max	0.040 max	0.15-0.35	bal Fe				
ASTM A331(95)	8720	Bar	0.18-0.23	0.40-0.60	0.70-0.90	0.2-0.30	0.40-0.70	0.035 max	0.040 max	0.15-0.35	bal Fe				
ASTM A331(95)	8720H	Bar	0.17-0.23	0.35-0.65	0.60-0.95	0.20-0.30	0.35-0.75	0.025 max	0.025 max	0.15-0.35	Cu <=0.35; bal Fe				
ASTM A519(96)	8720	Smls mech tub	0.18-0.23	0.40-0.60	0.70-0.90	0.20-0.30	0.40-0.70	0.040 max	0.040 max	0.15-0.35	bal Fe				
ASTM A752(93)	8720	Rod Wir	0.18-0.23	0.40-0.60	0.70-0.90	0.20-0.30	0.40-0.70	0.035 max	0.040 max	0.15-0.30	bal Fe	570			
ASTM A914/914M(92)	8720RH	Bar restricted end Q hard	0.18-0.23	0.40-0.60	0.70-0.90	0.20-0.30	0.40-0.70	0.035 max	0.040 max	0.15-0.35	Cu <=0.35; Electric furnace P, S<=0.025; bal Fe				
SAE 770(84)	8720*	Obs; see J1397(92)	0.18-0.23	0.4-0.6	0.7-0.9	0.2-0.3	0.4-0.7	0.04 max	0.04 max	0.15-0.3	bal Fe				
SAE J1268(95)	8720H	Bar Rod Wir Sh Strp Tub; See std	0.17-0.23	0.35-0.65	0.60-0.95	0.20-0.30	0.35-0.75	0.030 max	0.040 max	0.15-0.35	Cu <=0.35; Cu not spec'd but acpt; bal Fe				
SAE J1868(93)	8720RH	Restrict hard range	0.18-0.23	0.40-0.60	0.70-0.90	0.20-0.30	0.40-0.70	0.025 max	0.040 max	0.15-0.35	Cu <=0.035; Cu not spec'd but acpt; bal Fe				5 HRC max
SAE J404(94)	8720	Bil Blm Slab Bar HR CF	0.18-0.23	0.40-0.60	0.70-0.90	0.20-0.30	0.40-0.70	0.030 max	0.040 max	0.15-0.35	bal Fe				

Alloy Steel, Nickel Chromium Molybdenum, 8735

Russia

| GOST 977(88) | 35ChGSL | | 0.3-0.4 | 0.6-0.9 | 1-1.3 | | 0.3 | 0.04 | 0.04 | 0.6-0.8 | bal Fe | | | | |

USA

	AISI 8735	Smls mech tub	0.33-0.38	0.40-0.60	0.75-1.00	0.20-0.30	0.40-0.70	0.040 max	0.040 max	0.15-0.35	bal Fe				
	UNS G87350		0.33-0.38	0.40-0.60	0.75-1.00	0.20-0.30	0.40-0.70	0.035 max	0.040 max	0.15-0.35	bal Fe				
ASTM A519(96)	8735	Smls mech tub	0.33-0.38	0.40-0.60	0.75-1.00	0.20-0.30	0.40-0.70	0.040 max	0.040 max	0.15-0.35	bal Fe				

Alloy Steel, Nickel Chromium Molybdenum, 8740/8740H

Australia

| AS 1444(96) | 8740 | H series, Hard/Tmp | 0.38-0.43 | 0.40-0.60 | 0.75-1.00 | 0.20-0.30 | 0.40-0.70 | 0.040 max | 0.040 max | 0.10-0.35 | bal Fe | | | | |
| AS 1444(96) | 8740H | H series, Hard/Tmp | 0.37-0.44 | 0.35-0.65 | 0.70-1.05 | 0.20-0.30 | 0.35-0.75 | 0.040 max | 0.040 max | 0.10-0.35 | bal Fe | | | | |

Belgium

| NBN 253-02 | 40NiCrMo2 | | 0.37-0.44 | 0.4-0.6 | 0.7-1 | 0.15-0.3 | 0.4-0.7 | 0.035 | 0.035 | 0.15-0.4 | bal Fe | | | | |

UNS numbers and US grades are provided as a means of cross referencing chemically similar alloys. Exchangability is only possible after independent examination of specifications. Tensile properties are minimum or typical as specified. UTS and YS as MPa. El as %. See Appendix for list of abbreviations used in Notes. * indicates obsolete material.

Specification	Designation	Notes	C	Cr	Mn	Mo	Ni	P	S	Si	Other	UTS	YS	El	Hard

Alloy Steel, Nickel Chromium Molybdenum, 8740/8740H (Continued from previous page)

Specification	Designation	Notes	C	Cr	Mn	Mo	Ni	P	S	Si	Other	UTS	YS	El	Hard
Belgium															
NBN 253-06	40NiCrMo3		0.37-0.43	0.6-0.9	0.5-0.8	0.15-0.3	0.7-1	0.025	0.035	0.15-0.4	bal Fe				
France															
AFNOR	40NCD2		0.37-0.44	0.4-0.6	0.6-0.9	0.15-0.3	0.4-0.7	0.04	0.035	0.1-0.4	bal Fe				
AFNOR	40NCD2TS		0.38-0.44	0.4-0.6	0.7-1	0.15-0.3	0.4-0.7	0.025	0.03	0.1-0.4	bal Fe				
AFNOR	40NCD3TS		0.38-0.44	0.4-0.6	0.7-1	0.15-0.3	0.4-0.7	0.025	0.03	0.1-0.4	bal Fe				
International															
ISO 683-8	1*		0.37-0.44	0.4-0.6	0.7-1	0.15-0.3	0.4-0.7	0.035	0.035	0.15-0.4	bal Fe				
ISO 683-8	1A*		0.37-0.44	0.4-0.6	0.7-1	0.15-0.3	0.4-0.7	0.035	0.02-0.035	0.15-0.4	bal Fe				
ISO 683-8	1b*		0.37-0.44	0.4-0.6	0.7-1	0.15-0.3	0.4-0.7	0.035	0.03-0.05	0.15-0.4	bal Fe				
Italy															
UNI 7845(78)	40NiCrMo2	Q/T	0.37-0.44	0.4-0.6	0.7-1	0.15-0.25	0.4-0.7	0.035 max	0.035 max	0.15-0.4	bal Fe				
Japan															
JIS G4103(79)	SNCM240	HR Frg Bar Rod	0.38-0.43	0.40-0.65	0.70-1.00	0.15-0.30	0.40-0.70	0.030 max	0.030 max	0.15-0.35	Cu <=0.30; bal Fe				
Mexico															
NMX-B-268(68)	8740H	Hard	0.37-0.44	0.35-0.65	0.70-1.05	0.20-0.30	0.35-0.75			0.20-0.35	bal Fe				
NMX-B-300(91)	8740	Bar	0.38-0.43	0.40-0.60	0.75-1.00	0.20-0.30	0.40-0.70	0.035 max	0.040 max	0.15-0.35	bal Fe				
NMX-B-300(91)	8740H	Bar	0.37-0.44	0.35-0.65	0.70-1.05	0.20-0.30	0.35-0.75			0.15-0.35	bal Fe				
Pan America															
COPANT 334	8740	Bar	0.38-0.43	0.4-0.6	0.75-1	0.2-0.3	0.4-0.7	0.04	0.04	0.2-0.35	bal Fe				
COPANT 514	8740	Tube	0.38-0.43	0.4-0.6	0.75-1	0.2-0.3	0.4-0.7	0.04	0.05	0.1	bal Fe				
Spain															
UNE 36012(75)	40NiCrMo2	Q/T	0.37-0.44	0.4-0.7	0.7-1.0	0.15-0.3	0.4-0.7	0.035 max	0.035 max	0.15-0.4	bal Fe				
UNE 36034(85)	40NiCrMo2DF	CHd; Cold ext	0.37-0.44	0.4-0.6	0.7-1	0.15-0.3	0.4-0.7	0.035 max	0.035 max	0.15-0.4	bal Fe				
USA															
	AISI 8740	Smls mech tub	0.38-0.43	0.40-0.60	0.75-1.00	0.20-0.30	0.40-0.70	0.040 max	0.040 max	0.15-0.35	bal Fe				
	AISI 8740H	Smls mech tub, Hard	0.37-0.44	0.35-0.65	0.70-1.05	0.20-0.30	0.35-0.75	0.035 max	0.040 max	0.15-0.30	bal Fe				
	UNS G87400		0.38-0.43	0.40-0.60	0.75-1.00	0.20-0.30	0.40-0.70	0.035 max	0.040 max	0.15-0.35	bal Fe				
	UNS H87400		0.37-0.44	0.35-0.65	0.70-1.00	0.20-0.30	0.35-0.75	0.035 max	0.040 max	0.15-0.35	bal Fe				
AMS 6322		Bar Frg	0.38-0.43	0.4-0.60	0.75-1	0.2-0.30	0.4-0.7	0.025	0.025	0.15-0.35	Cu 0.35; bal Fe				
AMS 6323		Tube	0.38-0.43	0.4-0.60	0.75-1	0.2-0.30	0.4-0.7	0.025	0.025	0.15-0.35	Cu 0.35; bal Fe				
AMS 6325		Bar Frg	0.38-0.43	0.4-0.60	0.75-1	0.2-0.30	0.4-0.7	0.025	0.025	0.15-0.35	Cu 0.35; bal Fe				
AMS 6327		Bar Frg	0.38-0.43	0.4-0.60	0.75-1	0.2-0.30	0.4-0.7	0.025	0.025	0.15-0.35	Cu 0.35; bal Fe				
AMS 6358		Sh, Strp, Plt	0.38-0.43	0.4-0.60	0.75-1	0.2-0.30	0.4-0.7	0.025	0.025	0.15-0.35	Cu 0.35; bal Fe				
AMS 7496											bal Fe				
ASTM A29/A29M(93)	8740	Bar	0.38-0.43	0.40-0.60	0.75-1	0.2-0.3	0.4-0.7	0.035 max	0.04 max	0.15-0.35	bal Fe				
ASTM A304(96)	8740H	Bar, hard bands spec	0.37-0.44	0.35-0.65	0.70-1.05	0.20-0.30	0.35-0.75	0.035 max	0.040 max	0.15-0.35	Cu <=0.35; Cu Ni Cr Mo trace allowed; bal Fe				
ASTM A322(96)	8740	Bar	0.38-0.43	0.40-0.60	0.75-1.00	0.20-0.30	0.40-0.70	0.035 max	0.040 max	0.15-0.35	bal Fe				

UNS numbers and US grades are provided as a means of cross referencing chemically similar alloys. Exchangability is only possible after independent examination of specifications. Tensile properties are minimum or typical as specified. UTS and YS as MPa. El as %. See Appendix for list of abbreviations used in Notes. * indicates obsolete material.

Specification	Designation	Notes	C	Cr	Mn	Mo	Ni	P	S	Si	Other	UTS	YS	El	Hard

Alloy Steel, Nickel Chromium Molybdenum, 8740/8740H (Continued from previous page)

USA

Specification	Designation	Notes	C	Cr	Mn	Mo	Ni	P	S	Si	Other	UTS	YS	El	Hard
ASTM A331(95)	8740	Bar	0.38-0.43	0.40-0.60	0.75-1.00	0.20-0.30	0.40-0.70	0.035 max	0.040 max	0.15-0.35	bal Fe				
ASTM A331(95)	8740H	Bar	0.37-0.44	0.35-0.65	0.70-1.05	0.20-0.30	0.35-0.75	0.025 max	0.025 max	0.15-0.35	Cu <=0.35; bal Fe				
ASTM A519(96)	8740	Smls mech tub	0.38-0.43	0.40-0.60	0.75-1.00	0.20-0.30	0.40-0.70	0.040 max	0.040 max	0.15-0.35	bal Fe				
ASTM A752(93)	8740	Rod Wir	0.38-0.43	0.40-0.60	0.75-1.00	0.20-0.30	0.40-0.70	0.035 max	0.040 max	0.15-0.30	bal Fe	670			
MIL-S-6049A(67)	8740*	Obs for new design; see AMS specs; Bar refrg stock, aircraft qual	0.38-0.43	0.4-0.6	0.75-1	0.2-0.3	0.4-0.7	0.025	0.025		Cu 0.35; bal Fe				
SAE 770(84)	8740*	Obs; see J1397(92)									bal Fe				
SAE J1268(95)	8740H	Bar Rod Wir Sh Strp Tub; See std	0.37-0.44	0.35-0.65	0.70-1.05	0.20-0.30	0.35-0.75	0.030 max	0.040 max	0.15-0.35	Cu <=0.35; Cu not spec'd but acpt; bal Fe				

Alloy Steel, Nickel Chromium Molybdenum, 8742

USA

Specification	Designation	Notes	C	Cr	Mn	Mo	Ni	P	S	Si	Other	UTS	YS	El	Hard
	AISI 8742	Smls mech tub	0.40-0.45	0.40-0.60	0.75-1.00	0.20-0.30	0.40-0.70	0.040 max	0.040 max	0.15-0.35	bal Fe				
	AISI 8742	Plt	0.38-0.46	0.35-0.60	0.70-1.00	0.20-0.30	0.40-0.70	0.035 max	0.040 max	0.15-0.40	bal Fe				
	UNS G87420		0.40-0.45	0.40-0.60	0.75-1.00	0.20-0.30	0.40-0.70	0.035 max	0.040 max	0.15-0.35	bal Fe				
ASTM A519(96)	8742	Smls mech tub	0.40-0.45	0.40-0.60	0.75-1.00	0.20-0.30	0.40-0.70	0.040 max	0.040 max	0.15-0.35	bal Fe				
ASTM A829/A829M(95)	8742	Plt	0.38-0.46	0.35-0.60	0.70-1.00	0.20-0.30	0.40-0.70	0.035 max	0.040 max	0.15-0.40	bal Fe	480-965			
SAE J404(94)	8742	Plt	0.38-0.46	0.35-0.60	0.70-1.00	0.20-0.30	0.40-0.70	0.035 max	0.040 max	0.15-0.35	bal Fe				

Alloy Steel, Nickel Chromium Molybdenum, 8822/8822H

Mexico

Specification	Designation	Notes	C	Cr	Mn	Mo	Ni	P	S	Si	Other	UTS	YS	El	Hard
DGN B-203	8822*	Obs	0.2-0.25	0.4-0.6	0.75-1	0.3-0.4	0.4-0.7	0.04	0.04	0.2-0.35	bal Fe				
DGN B-297	8822*	Obs	0.2-0.25	0.4-0.6	0.75-1	0.3-0.4	0.4-0.7	0.035	0.04	0.2-0.35	bal Fe				
NMX-B-268(68)	8822H	Hard	0.19-0.25	0.35-0.65	0.70-1.05	0.30-0.40	0.35-0.75			0.20-0.35	bal Fe				
NMX-B-300(91)	8822	Bar	0.20-0.25	0.40-0.60	0.75-1.00	0.30-0.40	0.40-0.70	0.035 max	0.040 max	0.15-0.35	bal Fe				
NMX-B-300(91)	8822H	Bar	0.19-0.25	0.35-0.65	0.70-1.05	0.30-0.40	0.35-0.75			0.15-0.35	bal Fe				

Pan America

Specification	Designation	Notes	C	Cr	Mn	Mo	Ni	P	S	Si	Other	UTS	YS	El	Hard
COPANT 334	8822	Bar	0.2-0.25	0.4-0.6	0.75-1	0.3-0.4	0.4-0.7	0.04	0.04	0.2-0.35	bal Fe				
COPANT 514	8822	Tube	0.2-0.25	0.4-0.6	0.75-1	0.3-0.4	0.4-0.7	0.04	0.05	0.1	bal Fe				
COPANT 514	8822		0.2-0.25	0.4-0.6	0.75-1	0.3-0.4	0.4-0.7	0.04	0.05	0.1	bal Fe				

Romania

Specification	Designation	Notes	C	Cr	Mn	Mo	Ni	P	S	Si	Other	UTS	YS	El	Hard
STAS 11504	22MoCrNi05		0.19-0.25	0.35-0.65	0.7-1.05	0.3-0.4	0.35-0.75	0.035	0.035	0.2-0.35	bal Fe				

USA

Specification	Designation	Notes	C	Cr	Mn	Mo	Ni	P	S	Si	Other	UTS	YS	El	Hard
	AISI 8822	Smls mech tub	0.20-0.25	0.40-0.60	0.75-1.00	0.30-0.40	0.40-0.70	0.040 max	0.040 max	0.15-0.35	bal Fe				
	AISI 8822	Bar Blm Bil Slab	0.20-0.25	0.40-0.60	0.75-1.00	0.30-0.40	0.40-0.70	0.035 max	0.040 max	0.15-0.35	bal Fe				
	AISI 8822H	Smls mech tub, Hard	0.19-0.25	0.35-0.65	0.70-1.05	0.30-0.40	0.35-0.75	0.040 max	0.040 max	0.15-0.30	bal Fe				
	AISI 8822H	Bar Blm Bil Slab	0.19-0.25	0.35-0.65	0.70-1.05	0.30-0.40	0.35-0.75	0.035 max	0.040 max	0.15-0.35	bal Fe				
	UNS G88220		0.20-0.25	0.40-0.60	0.75-1.00	0.30-0.40	0.40-0.70	0.035 max	0.040 max	0.15-0.35	bal Fe				
	UNS H88220		0.19-0.25	0.35-0.65	0.70-1.05	0.30-0.40	0.35-0.75			0.15-0.35	bal Fe				

Specification	Designation	Notes	C	Cr	Mn	Mo	Ni	P	S	Si	Other	UTS	YS	El	Hard

Alloy Steel, Nickel Chromium Molybdenum, 8822/8822H (Continued from previous page)

USA

Specification	Designation	Notes	C	Cr	Mn	Mo	Ni	P	S	Si	Other	UTS	YS	El	Hard
ASTM A29/A29M(93)	8822	Bar	0.20-0.25	0.40-0.60	0.75-1	0.3-0.4	0.4-0.7	0.035 max	0.04 max	0.15-0.35	bal Fe				
ASTM A304(96)	8822H	Bar, hard bands spec	0.19-0.25	0.35-0.65	0.70-1.05	0.30-0.40	0.35-0.75	0.035 max	0.040 max	0.15-0.35	Cu <=0.35; Cu Ni Cr Mo trace allowed; bal Fe				
ASTM A322(96)	8822	Bar	0.20-0.25	0.40-0.60	0.75-1.00	0.30-0.40	0.40-0.70	0.035 max	0.040 max	0.15-0.35	bal Fe				
ASTM A331(95)	8822	Bar	0.20-0.25	0.40-0.60	0.75-1.00	0.30-0.40	0.40-0.70	0.035 max	0.040 max	0.15-0.35	bal Fe				
ASTM A331(95)	8822H	Bar	0.19-0.25	0.35-0.65	0.70-1.05	0.30-0.40	0.35-0.75	0.025 max	0.025 max	0.15-0.35	Cu <=0.35; bal Fe				
ASTM A519(96)	8822	Smls mech tub	0.20-0.25	0.40-0.60	0.75-1.00	0.30-0.40	0.40-0.70	0.040 max	0.040 max	0.15-0.35	bal Fe				
ASTM A752(93)	8822	Rod Wir	0.2-0.25	0.40-0.60	0.75-1.00	0.30-0.40	0.40-0.70	0.035 max	0.040 max	0.15-0.30	bal Fe	610			
ASTM A914/914M(92)	8822RH	Bar restricted end Q hard	0.20-0.25	0.40-0.60	0.75-1.00	0.30-0.40	0.40-0.70	0.035 max	0.040 max	0.15-0.35	Cu <=0.35; Electric furnace P, S<=0.025; bal Fe				
SAE 770(84)	8822*	Obs; see J1397(92)	0.2-0.25	0.4-0.6	0.75-1	0.3-0.4	0.4-0.7	0.04 max	0.04 max	0.15-0.3	bal Fe				
SAE J1268(95)	8822H	Bar Rod Wir Tub; See std	0.19-0.25	0.35-0.65	0.70-1.05	0.30-0.40	0.35-0.75	0.030 max	0.040 max	0.15-0.35	Cu <=0.35; Cu not spec'd but acpt; bal Fe				
SAE J1868(93)	8822RH	Restrict hard range	0.20-0.25	0.40-0.60	0.75-1.00	0.30-0.40	0.40-0.70	0.025 max	0.040 max	0.15-0.35	Cu <=0.035; Cu not spec'd but acpt; bal Fe				5 HRC max
SAE J404(94)	8822	Bil Blm Slab Bar HR CF	0.20-0.25	0.40-0.60	0.75-1.00	0.30-0.40	0.40-0.70	0.030 max	0.040 max	0.15-0.35	bal Fe				

Alloy Steel, Nickel Chromium Molybdenum, 9310H

China

Specification	Designation	Notes	C	Cr	Mn	Mo	Ni	P	S	Si	Other	UTS	YS	El	Hard
GB 3203(82)	G10CrNi3Mo	Bar HR/CD Q/T 15mm diam	0.08-0.13	1.00-1.40	0.40-0.70	0.08-0.15	3.00-3.50	0.030 max	0.030 max	0.15-0.40	Cu <=0.25; bal Fe	1080		9	

Mexico

Specification	Designation	Notes	C	Cr	Mn	Mo	Ni	P	S	Si	Other	UTS	YS	El	Hard
NMX-B-268(68)	9310H	Hard	0.07-0.13	1.00-1.45	0.40-0.70	0.08-0.15	2.95-3.55			0.20-0.35	bal Fe				
NMX-B-300(91)	9310H	Bar	0.07-0.13	1.00-1.45	0.40-0.70	0.08-0.15	2.95-3.55			0.15-0.35	bal Fe				

USA

Specification	Designation	Notes	C	Cr	Mn	Mo	Ni	P	S	Si	Other	UTS	YS	El	Hard
	AISI 9310H	Smls mech tub, Hard	0.07-0.13	1.00-1.45	0.40-0.70	0.08-0.15	2.95-3.55	0.035 max	0.040 max	0.15-0.30	bal Fe				
	UNS H93100		0.07-0.13	1.00-1.45	0.40-0.70	0.08-0.15	2.95-3.55			0.15-0.35	bal Fe				
ASTM A304(96)	9310H	Bar, hard bands spec	0.07-0.13	1.00-1.45	0.40-0.70	0.08-0.15	2.95-3.55	0.035 max	0.040 max	0.15-0.35	Cu <=0.35; Cu Ni Cr Mo trace allowed; bal Fe				
ASTM A331(95)	9310H	Bar	0.07-0.13	1.00-1.45	0.40-0.70	0.08-0.15	2.95-3.55	0.025 max	0.025 max	0.15-0.35	Cu <=0.35; bal Fe				
ASTM A534(94)	9310H	Carburizing, anti-friction bearings	0.07-0.13	1.00-1.45	0.40-0.70	0.08-0.15	2.95-3.55	0.025 max	0.025 max	0.15-0.35	bal Fe				
ASTM A837(96)	9310	Frg for carburizing	0.07-0.13	1.00-1.45	0.40-0.70	0.08-0.15	2.95-3.55	0.035 max	0.040 max	0.15-0.35	Cu <=0.35; Si<=0.10 if VCD used; bal Fe				
ASTM A914/914M(92)	9310RH	Bar restricted end Q hard	0.08-0.13	1.00-1.40	0.45-0.65	0.08-0.15	3.00-3.50	0.035 max	0.040 max	0.15-0.35	Cu <=0.35; Electric furnace P, S<=0.025; bal Fe				
SAE J1268(95)	E9310H	Bar Rod Wir Tub; See std	0.07-0.13	1.00-1.45	0.40-0.70	0.08-0.15	2.95-3.55	0.025 max	0.025 max	0.15-0.35	Cu <=0.35; Cu not spec'd but acpt; bal Fe				
SAE J1868(93)	9310RH	Restrict hard range	0.08-0.13	1.00-1.40	0.45-0.65	0.08-0.15	3.00-3.50	0.025 max	0.040 max	0.15-0.35	Cu <=0.035; Cu not spec'd but acpt; bal Fe				5 HRC max

Alloy Steel, Nickel Chromium Molybdenum, 94B15/94B15H

Mexico

Specification	Designation	Notes	C	Cr	Mn	Mo	Ni	P	S	Si	Other	UTS	YS	El	Hard
DGN B-203	94B15*	Obs	0.13-0.18		0.75-1	0.08-0.15		0.3-0.6	0.04	0.04	0.2-0.35	B >=0.0005; bal Fe				
NMX-B-268(68)	94B15H	Hard	0.12-0.18	0.25-0.55	0.70-1.05	0.08-0.15	0.25-0.65			0.20-0.35	bal Fe				

UNS numbers and US grades are provided as a means of cross referencing chemically similar alloys. Exchangability is only possible after independent examination of specifications. Tensile properties are minimum or typical as specified. UTS and YS as MPa. El as %. See Appendix for list of abbreviations used in Notes. * indicates obsolete material.

Specification	Designation	Notes	C	Cr	Mn	Mo	Ni	P	S	Si	Other	UTS	YS	El	Hard

Alloy Steel, Nickel Chromium Molybdenum, 94B15/94B15H (Continued from previous page)

Mexico

Specification	Designation	Notes	C	Cr	Mn	Mo	Ni	P	S	Si	Other	UTS	YS	El	Hard
NMX-B-300(91)	94B15H	Bar	0.12-0.18	0.25-0.55	0.70-1.05	0.08-0.15	0.25-0.65			0.15-0.35	B >=0.0005; bal Fe				

Pan America

Specification	Designation	Notes	C	Cr	Mn	Mo	Ni	P	S	Si	Other	UTS	YS	El	Hard
COPANT 514	94B15		0.13-0.18	0.3-0.5	0.75-1	0.08-0.15	0.3-0.6	0.04	0.05	0.1	B >=0.0005; bal Fe				
COPANT 514	94B15	Tube	0.13-0.18	0.3-0.5	0.75-1	0.08-0.15	0.3-0.6	0.04	0.05	0.1	B <=0.003; bal Fe				

USA

Specification	Designation	Notes	C	Cr	Mn	Mo	Ni	P	S	Si	Other	UTS	YS	El	Hard
	AISI 94B15	Smls mech Tub; Boron	0.13-0.18	0.30-0.50	0.75-1.00	0.09-0.15	0.30-0.60	0.040 max	0.040 max	0.15-0.35	B >=0.0005; bal Fe				
	AISI 94B15H	Smls mech tub, Hard; Boron	0.12-0.18	0.25-0.55	0.70-1.05	0.08-0.15	0.25-0.65	0.035 max	0.040 max	0.15-0.30	B 0.0005-0.005; bal Fe				
	UNS G94151		0.13-0.18	0.30-0.50	0.75-1.00	0.08-0.15	0.30-0.60	0.035 max	0.040 max	0.15-0.35	B >=0.0005; bal Fe				
	UNS H94151		0.12-0.18	0.25-0.55	0.70-1.05	0.08-0.15	0.25-0.65			0.15-0.35	B 0.0005-0.003; bal Fe				
AMS 6275		Bar, Tub	0.13-0.18	0.30-0.50	0.75-1	0.08-0.15	0.30-0.60	0.040 max	0.040 max	0.15-0.30	B 0.001-0.005; bal Fe				
ASTM A304(96)	94B15H	Bar, hard bands spec	0.12-0.18	0.25-0.55	0.70-1.05	0.08-0.15	0.25-0.65	0.035 max	0.040 max	0.15-0.35	B >=0.0005; Cu <=0.35; Cu Ni Cr Mo trace allowed; bal Fe				
ASTM A331(95)	94B15H	Bar	0.12-0.18	0.25-0.55	0.70-1.05	0.08-0.15	0.25-0.65	0.025 max	0.025 max	0.15-0.35	B >=0.0005; Cu <=0.35; bal Fe				
ASTM A519(96)	94B15	Smls mech tub	0.13-0.18	0.30-0.50	0.75-1.00	0.08-0.15	0.30-0.60	0.040 max	0.040 max	0.15-0.35	B >=0.0005; bal Fe				
SAE 770(84)	94B15*	Obs; see J1397(92)	0.13-0.18	0.3-0.5	0.75-1	0.08-0.15	0.3-0.6	0.04 max	0.04 max	0.15-0.3	B 0.001-0.005; bal Fe				
SAE J1268(95)	94B15H	Bar Tub; See std	0.12-0.18	0.25-0.55	0.70-1.05	0.08-0.15	0.25-0.65	0.030 max	0.040 max	0.15-0.35	B 0.0005-0.003; Cu <=0.35; Cu not spec'd but acpt; bal Fe				
SAE J407	94B15H*	Obs; see J1268									bal Fe				

Alloy Steel, Nickel Chromium Molybdenum, 94B17/94B17H

China

Specification	Designation	Notes	C	Cr	Mn	Mo	Ni	P	S	Si	Other	UTS	YS	El	Hard
GB 3077(88)	20MnMoB	Bar HR Q/T 15mm diam	0.16-0.22	0.30 max	0.90-1.20	0.20-0.30	0.30 max	0.035 max	0.035 max	0.17-0.37	B 0.0005-0.0035; Cu <=0.30; bal Fe	1080	885	10	

France

Specification	Designation	Notes	C	Cr	Mn	Mo	Ni	P	S	Si	Other	UTS	YS	El	Hard
AFNOR NFA35551(86)	19NCDB2		0.17-0.23	0.4-0.65	0.65-0.95	0.15-0.25	0.4-0.7	0.04 max	0.04 max	0.1-0.4	B 0.001-0.005; bal Fe				
AFNOR NFA35552(84)	19NCDB2	Bar Rod Wir	0.17-0.23	0.4-0.65	0.65-0.95	0.15-0.25	0.4-0.7	0.04 max	0.04 max	0.1-0.4	B 0.001-0.005; bal Fe				
AFNOR NFA36102(93)	20MB5*	HR; Strp; CR	0.16-0.22	0.15-0.25	1.1-1.4	0.10 max	0.20 max	0.025 max		0.2 max	0.1-0.4 B 0.0008-0.005; bal Fe				
AFNOR NFA36102(93)	20MnB5RR	HR; Strp; CR	0.16-0.22	0.15-0.25	1.1-1.4	0.10 max	0.20 max	0.025 max		0.2 max	0.1-0.4 B 0.0008-0.005; bal Fe				

India

Specification	Designation	Notes	C	Cr	Mn	Mo	Ni	P	S	Si	Other	UTS	YS	El	Hard
IS 1570/4(88)	18C10BT	Boron	0.15-0.2	0.1-0.3	0.8-1.1	0.15 max	0.4 max	0.07 max	0.06 max	0.15-0.3	B 0.0005-0.003; bal Fe				
IS 1570/7(92)	18C10BT	Boron	0.15-0.2	0.1-0.3	0.8-1.1	0.15 max	0.4 max	0.07 max	0.06 max	0.15-0.3	B 0.0005-0.003; bal Fe				

Mexico

Specification	Designation	Notes	C	Cr	Mn	Mo	Ni	P	S	Si	Other	UTS	YS	El	Hard
DGN B-203	94B17*	Obs	0.15-0.2	0.3-0.5	0.75-1	0.08-0.15	0.3-0.6	0.04	0.04	0.2-0.35	B >=0.0005; bal Fe				
DGN B-297	94B17*	Obs	0.15-0.2	0.3-0.5	0.75-1	0.08-0.15	0.3-0.6	0.035	0.04	0.2-0.35	B >=0.0005; bal Fe				
NMX-B-268(68)	94B17H	Hard	0.12-0.20	0.25-0.55	0.70-1.05	0.08-0.15	0.25-0.65			0.20-0.35	bal Fe				
NMX-B-300(91)	94B17	Bar	0.15-0.20	0.30-0.50	0.75-1.00	0.08-0.15	0.30-0.60	0.035 max	0.040 max	0.20-0.35	B >=0.0005; bal Fe				
NMX-B-300(91)	94B17H	Bar	0.14-0.20	0.25-0.55	0.70-1.05	0.08-0.15	0.25-0.65			0.15-0.35	B >=0.0005; bal Fe				

Pan America

Specification	Designation	Notes	C	Cr	Mn	Mo	Ni	P	S	Si	Other	UTS	YS	El	Hard
COPANT 514	94B17	Tube	0.15-0.2	0.3-0.5	0.75-1	0.08-0.15	0.3-0.6	0.04	0.05	0.1	B <=0.003; bal Fe				

Specification	Designation	Notes	C	Cr	Mn	Mo	Ni	P	S	Si	Other	UTS	YS	El	Hard

Alloy Steel, Nickel Chromium Molybdenum, 94B17/94B17H (Continued from previous page)

USA

Specification	Designation	Notes	C	Cr	Mn	Mo	Ni	P	S	Si	Other	UTS	YS	El	Hard
	AISI 94B17	Smls mech Tub; Boron	0.15-0.20	0.30-0.50	0.75-1.00	0.09-0.15	0.30-0.60	0.040 max	0.040 max	0.15-0.35	B >=0.0005; bal Fe				
	AISI 94B17H	Smls mech tub, Hard; Boron	0.14-0.20	0.25-0.55	0.70-1.05	0.08-0.15	0.25-0.65	0.035 max	0.040 max	0.15-0.30	B 0.0005-0.005; bal Fe				
	UNS G94171		0.15-0.20	0.30-0.50	0.75-1.00	0.08-0.15	0.30-0.60	0.035 max	0.040 max	0.15-0.35	B 0.0005-0.003; bal Fe				
	UNS H94171		0.14-0.20	0.25-0.55	0.70-1.05	0.08-0.15	0.25-0.65	0.035 max	0.040 max	0.15-0.35	B 0.0005-0.003; bal Fe				
AMS 6275											bal Fe				
ASTM A29/A29M(93)	94B17	Bar	0.15-0.20	0.30-0.50	0.75-1	0.08-0.15	0.3-0.6	0.035 max	0.04 max	0.15-0.35	bal Fe				
ASTM A304(96)	94B17H	Bar, hard bands spec	0.14-0.20	0.25-0.55	0.70-1.05	0.08-0.15	0.25-0.65	0.035 max	0.040 max	0.15-0.35	B >=0.0005; Cu <=0.35; Cu Ni Cr Mo trace allowed; bal Fe				
ASTM A322(96)	94B17	Bar	0.15-0.20	0.30-0.50	0.75-1.00	0.08-0.15	0.30-0.60	0.035 max	0.040 max	0.15-0.35	B 0.0005-0.005; bal Fe				
ASTM A331(95)	94B17	Bar	0.15-0.20	0.30-0.50	0.75-1.00	0.08-0.15	0.30-0.60	0.035 max	0.040 max	0.15-0.35	bal Fe				
ASTM A331(95)	94B17H	Bar	0.14-0.20	0.25-0.55	0.70-1.05	0.08-0.15	0.25-0.65	0.025 max	0.025 max	0.15-0.35	B >=0.0005; Cu <=0.35; bal Fe				
ASTM A519(96)	94B17	Smls mech tub	0.15-0.20	0.30-0.50	0.75-1.00	0.08-0.15	0.30-0.60	0.040 max	0.040 max	0.15-0.35	B >=0.0005; bal Fe				
ASTM A752(93)	94B17	Rod Wir, Boron	0.15-0.20	0.30-0.50	0.75-1.00	0.08-0.15	0.20-0.40	0.035 max	0.040 max	0.15-0.30	B >=0.0005; bal Fe	500			
SAE 770(84)	94B17*	Obs; see J1397(92)	0.15-0.2	0.3-0.5	0.75-1	0.08-0.15	0.3-0.6	0.04 max	0.04 max	0.15-0.3	B 0.001-0.005; bal Fe				
SAE J1268(95)	94B17H	See std	0.14-0.20	0.25-0.55	0.70-1.05	0.08-0.15	0.25-0.65	0.030 max	0.040 max	0.15-0.35	B 0.0005-0.003; Cu <=0.35; Cu not spec'd but acpt; bal Fe				

Alloy Steel, Nickel Chromium Molybdenum, 94B30/94B30H

Mexico

Specification	Designation	Notes	C	Cr	Mn	Mo	Ni	P	S	Si	Other	UTS	YS	El	Hard
DGN B-203	94B30*	Obs	0.28-0.33	0.3-0.5	0.75-1	0.08-0.15	0.3-0.6	0.04	0.04	0.2-0.35	B >=0.0005; bal Fe				
DGN B-297	94B30*	Obs	0.28-0.33	0.3-0.5	0.75-1	0.08-0.15	0.3-0.6	0.035	0.04	0.2-0.35	B >=0.0005; bal Fe				
NMX-B-268(68)	94B30H	Hard	0.27-0.33	0.25-0.55	0.70-1.05	0.08-0.15	0.25-0.65			0.20-0.35	bal Fe				
NMX-B-300(91)	94B30	Bar	0.28-0.33	0.30-0.50	0.75-1.00	0.08-0.15	0.30-0.60	0.035 max	0.040 max	0.20-0.35	B >=0.0005; bal Fe				
NMX-B-300(91)	94B30H	Bar	0.27-0.33	0.25-0.55	0.70-1.05	0.08-0.15	0.25-0.65			0.15-0.35	B >=0.0005; bal Fe				

Pan America

Specification	Designation	Notes	C	Cr	Mn	Mo	Ni	P	S	Si	Other	UTS	YS	El	Hard
COPANT 514	94B30		0.28-0.33	0.3-0.5	0.75-1	0.08-0.15	0.3-0.6	0.04	0.04	0.1	B >=0.0005; bal Fe				

USA

Specification	Designation	Notes	C	Cr	Mn	Mo	Ni	P	S	Si	Other	UTS	YS	El	Hard
	AISI 94B30	Smls mech Tub; Boron	0.28-0.33	0.30-0.50	0.75-1.00	0.09-0.15	0.30-0.60	0.040 max	0.040 max	0.15-0.35	B >=0.0005; bal Fe				
	AISI 94B30H	Smls mech tub, Hard; Boron	0.27-0.33	0.25-0.55	0.70-1.05	0.08-0.15	0.25-0.65	0.035 max	0.040 max	0.15-0.30	B 0.0005-0.005; bal Fe				
	UNS G94301		0.28-0.33	0.30-0.50	0.75-1.00	0.08-0.15	0.30-0.60	0.035 max	0.040 max	0.15-0.35	B 0.0005-0.003; bal Fe				
	UNS H94301		0.27-0.33	0.25-0.55	0.70-1.05	0.08-0.15	0.25-0.65	0.035 max	0.040 max	0.15-0.35	B 0.0005-0.003; bal Fe				
ASTM A29/A29M(93)	94B30	Bar	0.28-0.33	0.30-0.50	0.75-1	0.08-0.15	0.3-0.6	0.035 max	0.04 max	0.15-0.35	bal Fe				
ASTM A304(96)	94B30H	Bar, hard bands spec	0.27-0.33	0.25-0.55	0.70-1.05	0.08-0.15	0.25-0.65	0.035 max	0.040 max	0.15-0.35	B >=0.0005; Cu <=0.35; Cu Ni Cr Mo trace allowed; bal Fe				
ASTM A322(96)	94B30	Bar	0.28-0.33	0.30-0.50	0.75-1.00	0.08-0.15	0.30-0.60	0.035 max	0.040 max	0.15-0.35	B 0.0005-0.005; bal Fe				
ASTM A331(95)	94B30	Bar	0.28-0.33	0.30-0.50	0.75-1.00	0.08-0.15	0.30-0.60	0.035 max	0.040 max	0.15-0.35	bal Fe				
ASTM A331(95)	94B30H	Bar	0.27-0.33	0.25-0.55	0.70-1.05	0.08-0.15	0.25-0.65	0.025 max	0.025 max	0.15-0.35	B >=0.0005; Cu <=0.35; bal Fe				
ASTM A519(96)	94B30	Smls mech tub	0.28-0.33	0.30-0.50	0.75-1.00	0.08-0.15	0.30-0.60	0.040 max	0.040 max	0.15-0.35	B >=0.0005; bal Fe				

UNS numbers and US grades are provided as a means of cross referencing chemically similar alloys. Exchangability is only possible after independent examination of specifications. Tensile properties are minimum or typical as specified. UTS and YS as MPa. El as %. See Appendix for list of abbreviations used in Notes. * indicates obsolete material.

Specification	Designation	Notes	C	Cr	Mn	Mo	Ni	P	S	Si	Other	UTS	YS	El	Hard

Alloy Steel, Nickel Chromium Molybdenum, 94B30/94B30H (Continued from previous page)

USA

Specification	Designation	Notes	C	Cr	Mn	Mo	Ni	P	S	Si	Other	UTS	YS	El	Hard
ASTM A752(93)	94B30	Rod Wir, Boron	0.28-0.33	0.30-0.50	0.75-1.00	0.08-0.15	0.20-0.40	0.035 max	0.040 max	0.15-0.30	B >=0.0005; bal Fe	580			
SAE 770(84)	94B30*	Obs; see J1397(92)	0.28-0.33	0.3-0.5	0.75-1	0.08-0.15	0.3-0.6	0.04 max	0.04 max	0.15-0.3	B 0.001-0.005; bal Fe				
SAE J1268(95)	94B30H	Bar Rod Wir Tub; See std	0.27-0.33	0.25-0.55	0.70-1.05	0.08-0.15	0.25-0.65	0.030 max	0.040 max	0.15-0.35	B 0.0005-0.003; Cu <=0.35; Cu not spec'd but acpt; bal Fe				

Alloy Steel, Nickel Chromium Molybdenum, 94B40

USA

Specification	Designation	Notes	C	Cr	Mn	Mo	Ni	P	S	Si	Other	UTS	YS	El	Hard
	AISI 94B40	Smls mech Tub; Boron	0.38-0.43	0.30-0.50	0.75-1.00	0.09-0.15	0.30-0.60	0.040 max	0.040 max	0.15-0.35	B >=0.0005; bal Fe				
	UNS G94401		0.38-0.43	0.30-0.50	0.75-1.00	0.08-0.15	0.30-0.60	0.035 max	0.040 max	0.15-0.35	B 0.0005-0.003; bal Fe				
ASTM A519(96)	94B40	Smls mech tub	0.38-0.43	0.30-0.50	0.75-1.00	0.08-0.15	0.30-0.60	0.040 max	0.040 max	0.15-0.35	B >=0.0005; bal Fe				

Alloy Steel, Nickel Chromium Molybdenum, 9840

Argentina

Specification	Designation	Notes	C	Cr	Mn	Mo	Ni	P	S	Si	Other	UTS	YS	El	Hard
IAS	IRAM 9840		0.38-0.43	0.70-0.90	0.70-0.90	0.20-0.30	0.85-1.15	0.040 max	0.040 max	0.20-0.35	bal Fe	920	650	14	302 HB

Bulgaria

Specification	Designation	Notes	C	Cr	Mn	Mo	Ni	P	S	Si	Other	UTS	YS	El	Hard
BDS 6354(74)	36ChNM	Struct	0.32-0.40	0.90-1.20	0.50-0.80	0.15-0.25	0.80-1.20	0.035 max	0.035 max	0.17-0.37	Cu <=0.30; Ti <=0.05; V <=0.1; W <=0.1; bal Fe				
BDS 6354(74)	36ChNM	Struct	0.32-0.40	0.90-1.20	0.50-0.80	0.15-0.25	0.80-1.20	0.035 max	0.035 max	0.17-0.37	Cu <=0.30; Ti <=0.05; V <=0.1; W <=0.1; bal Fe				

Europe

Specification	Designation	Notes	C	Cr	Mn	Mo	Ni	P	S	Si	Other	UTS	YS	El	Hard
EN 10083/1(91)A1(96)	1.6511	Q/T t<=16mm	0.32-0.40	0.90-1.20	0.50-0.80	0.15-0.30	0.90-1.20	0.035 max	0.035 max	0.40 max	bal Fe	1100-1300	900	10	
EN 10083/1(91)A1(96)	36CrNiMo4	Q/T t<=16mm	0.32-0.40	0.90-1.20	0.50-0.80	0.15-0.30	0.90-1.20	0.035 max	0.035 max	0.40 max	bal Fe	1100-1300	900	10	

Finland

Specification	Designation	Notes	C	Cr	Mn	Mo	Ni	P	S	Si	Other	UTS	YS	El	Hard
SFS 461(73)	SFS461	Q/T	0.32-0.39	1.2-1.6	0.5-0.8	0.15-0.25	1.2-1.6	0.035 max	0.035 max	0.15-0.4	bal Fe				

Germany

Specification	Designation	Notes	C	Cr	Mn	Mo	Ni	P	S	Si	Other	UTS	YS	El	Hard
DIN EN 10083(91)	36CrNiMo4	Q/T 41-100mm	0.32-0.4	0.90-1.20	0.50-0.80	0.15-0.30	0.90-1.20	0.035 max	0.035 max	0.40 max	bal Fe	900-1100	700	12	
DIN EN 10083(91)	WNr 1.6511	Q/T 17-40mm	0.32-0.40	0.90-1.20	0.50-0.80	0.15-0.30	0.90-1.20	0.035 max	0.035 max	0.40 max	bal Fe	1100-1200	800	11	
DIN EN 10083(91)	WNr 1.6511	Q/T 101-160mm	0.32-0.40	0.90-1.20	0.50-0.80	0.15-0.30	0.90-1.20	0.035 max	0.035 max	0.40 max	bal Fe	800-950	600	13	

International

Specification	Designation	Notes	C	Cr	Mn	Mo	Ni	P	S	Si	Other	UTS	YS	El	Hard
ISO R683-8(70)	2	Q/T	0.36-0.43	0.6-0.9	0.5-0.8	0.15-0.3	0.7-1	0.035 max	0.035 max	0.15-0.4	Cu <=0.3; Pb <=0.15; Ti <=0.05; V <=0.1;				
ISO R683-8(70)	2	Q/T	0.36-0.43	0.6-0.9	0.5-0.8	0.15-0.3	0.7-1	0.035 max	0.035 max	0.15-0.4	Al <=0.1; Co <=0.1; Cu <=0.3; Pb <=0.15; Ti <=0.05; V <=0.1; W <=0.1; bal Fe				

Italy

Specification	Designation	Notes	C	Cr	Mn	Mo	Ni	P	S	Si	Other	UTS	YS	El	Hard
UNI 7356(74)	38NiCrMo4KB	Q/T	0.34-0.41	0.7-1	0.5-0.8	0.15-0.3	0.7-1	0.035 max	0.035 max	0.15-0.4	bal Fe				

Mexico

Specification	Designation	Notes	C	Cr	Mn	Mo	Ni	P	S	Si	Other	UTS	YS	El	Hard
NMX-B-300(91)	9840	Bar	0.38-0.43	0.70-0.90	0.45-0.90	0.20-0.30	0.85-1.15	0.035 max	0.040 max	0.15-0.35	bal Fe				

Pan America

Specification	Designation	Notes	C	Cr	Mn	Mo	Ni	P	S	Si	Other	UTS	YS	El	Hard
COPANT 334	9840	Bar	0.38-0.43	0.7-0.9	0.7-0.9	0.2-0.3	0.85-1.15	0.035	0.04	0.2-0.35	bal Fe				
COPANT 514	9840		0.38-0.43	0.7-0.9	0.7-0.9	0.2-0.3	0.85-1.15	0.04	0.04	0.1	bal Fe				

Poland

Specification	Designation	Notes	C	Cr	Mn	Mo	Ni	P	S	Si	Other	UTS	YS	El	Hard
PNH84030	36HNM	Q/T	0.32-0.4	0.9-1.2	0.5-0.8	0.15-0.25	0.9-1.2	0.035	0.035	0.17-0.37	Cu 0.3; bal Fe				

UNS numbers and US grades are provided as a means of cross referencing chemically similar alloys. Exchangability is only possible after independent examination of specifications. Tensile properties are minimum or typical as specified. UTS and YS as MPa. El as %. See Appendix for list of abbreviations used in Notes. * indicates obsolete material.

Specification	Designation	Notes	C	Cr	Mn	Mo	Ni	P	S	Si	Other	UTS	YS	El	Hard

Alloy Steel, Nickel Chromium Molybdenum, 9840 (Continued from previous page)

Romania

Specification	Designation	Notes	C	Cr	Mn	Mo	Ni	P	S	Si	Other	UTS	YS	El	Hard
STAS 791	36MCN10		0.32-0.4	0.9-1.2	0.5-0.8	0.15-0.3	0.9-1.2	0.035	0.035	0.17-0.37	Cu 0.3; bal Fe				

Russia

Specification	Designation	Notes	C	Cr	Mn	Mo	Ni	P	S	Si	Other	UTS	YS	El	Hard
GOST 4543(71)	40ChN2MA	Q/T	0.37-0.44	0.6-0.9	0.5-0.8	0.15-0.25	1.25-1.65	0.025 max	0.025 max	0.17-0.37	Al <=0.1; Cu <=0.3; Ti <=0.05; bal Fe				

Spain

Specification	Designation	Notes	C	Cr	Mn	Mo	Ni	P	S	Si	Other	UTS	YS	El	Hard
UNE 36012(75)	35NiCrMo4	Q/T	0.32-0.38	0.6-0.9	0.5-0.8	0.15-0.3	0.7-1	0.035 max	0.035 max	0.15-0.4	bal Fe				
UNE 36012(75)	40NiCrMo4	Q/T	0.37-0.42	0.6-0.9	0.5-0.8	0.15-0.3	0.7-1.0	0.035 max	0.035 max	0.15-0.4	bal Fe				
UNE 36012(75)	F.1280*	see 35NiCrMo4	0.32-0.38	0.6-0.9	0.5-0.8	0.15-0.3	0.7-1	0.035 max	0.035 max	0.15-0.4	bal Fe				
UNE 36012(75)	F.1282*	see 40NiCrMo4	0.37-0.42	0.6-0.9	0.5-0.8	0.15-0.3	0.7-1.0	0.035 max	0.035 max	0.15-0.4	bal Fe				

UK

Specification	Designation	Notes	C	Cr	Mn	Mo	Ni	P	S	Si	Other	UTS	YS	El	Hard
BS 970/1(83)	817A37	Blm Bil Slab Bar Rod Frg	0.35-0.40	1.00-1.40	0.45-0.70	0.20-0.35	1.30-1.70	0.035 max	0.040 max		bal Fe				
BS 970/1(83)	817A42	Blm Bil Slab Bar Rod Frg	0.40-0.45	1.00-1.40	0.45-0.70	0.20-0.35	1.30-1.70	0.035 max	0.040 max		bal Fe				
BS 970/2(70)	816M40*	Q/T	0.36-0.44	1.0-1.4	0.45-0.70	0.1-0.2	1.3-1.7	0.040 max	0.050 max	0.1-0.35	Pb <=0.15; bal Fe				

USA

Specification	Designation	Notes	C	Cr	Mn	Mo	Ni	P	S	Si	Other	UTS	YS	El	Hard
	AISI 9840	Smls mech tub	0.38-0.43	0.70-0.90	0.70-0.90	0.20-0.30	0.85-1.15	0.040 max	0.040 max	0.15-0.35	bal Fe				
	UNS G98400		0.38-0.43	0.70-0.90	0.70-0.90	0.20-0.30	0.85-1.15	0.035 max	0.040 max	0.15-0.35	bal Fe				
ASTM A519(96)	9840	Smls mech tub	0.38-0.43	0.70-0.90	0.70-0.90	0.20-0.30	0.85-1.15	0.040 max	0.040 max	0.15-0.35	bal Fe				
SAE J778	9840*	Obs; see J1249									bal Fe				

Yugoslavia

Specification	Designation	Notes	C	Cr	Mn	Mo	Ni	P	S	Si	Other	UTS	YS	El	Hard
	C.5430	Q/T	0.32-0.4	0.9-1.2	0.5-0.8	0.15-0.3	0.9-1.2	0.035 max	0.03 max	0.4 max	bal Fe				

Alloy Steel, Nickel Chromium Molybdenum, 9850

Germany

Specification	Designation	Notes	C	Cr	Mn	Mo	Ni	P	S	Si	Other	UTS	YS	El	Hard
DIN 17350(80)	WNr 1.2767	Tmp to 200C	0.40-0.50	1.20-1.50	0.15-0.45	0.15-0.35	3.80-4.30	0.030 max	0.030 max	0.10-0.40	bal Fe				54 HRC
DIN 17350(80)	X45NiCrMo4	Tmp to 200C	0.40-0.50	1.20-1.50	0.15-0.45	0.15-0.35	3.80-4.30	0.030 max	0.030 max	0.10-0.40	bal Fe				54 HRC

Mexico

Specification	Designation	Notes	C	Cr	Mn	Mo	Ni	P	S	Si	Other	UTS	YS	El	Hard
DGN B-203	9850*	Obs	0.48-0.53	0.7-0.9	0.7-0.9	0.2-0.35	0.85-1.15	0.04	0.04	0.2-0.35	bal Fe				

Pan America

Specification	Designation	Notes	C	Cr	Mn	Mo	Ni	P	S	Si	Other	UTS	YS	El	Hard
COPANT 514	9850	Tube	0.48-0.53	0.7-0.9	0.7-0.9	0.2-0.3	0.85-1.15	0.04	0.05	0.1	bal Fe				

Russia

Specification	Designation	Notes	C	Cr	Mn	Mo	Ni	P	S	Si	Other	UTS	YS	El	Hard
GOST 4543	40Ch2MA		0.35-0.42	1.25-1.65	0.3-0.6	0.2-0.3	1.35-1.75	0.03 max	0.03 max	0.17-0.37	bal Fe				

USA

Specification	Designation	Notes	C	Cr	Mn	Mo	Ni	P	S	Si	Other	UTS	YS	El	Hard
	AISI 9850	Smls mech tub	0.48-0.53	0.70-0.90	0.70-0.90	0.20-0.30	0.85-1.15	0.040 max	0.040 max	0.15-0.35	bal Fe				
	UNS G98500		0.48-0.53	0.70-0.90	0.70-0.90	0.20-0.30	0.85-1.15	0.035 max	0.040 max	0.15-0.35	bal Fe				
ASTM A519(96)	9850	Smls mech tub	0.48-0.53	0.70-0.90	0.70-0.90	0.20-0.30	0.85-1.15	0.040 max	0.040 max	0.15-0.35	bal Fe				
MIL-S-19434B(SH)(90)	Class 4	Gear, pinion frg HT; shipboard propulsion and turbine	0.35-0.45	0.50 min	0.60-0.90	0.20 min	1.65 max	0.015 max	0.015 max	0.15-0.35	V <=0.10; Si 0.10 max if VCD used; bal Fe	862	690		248-293 HB
MIL-S-19434B(SH)(90)	Class 5	Gear, pinion frg HT; shipboard propulsion and turbine	0.50 max	0.50 min	0.60-0.90	0.20 min	1.65 max	0.015 max	0.015 max	0.15-0.35	V <=0.10; Si 0.10 max if VCD used; bal Fe	1000	827		302-352 HB
SAE J778	9850*	Obs; see J1249									bal Fe				

Specification	Designation	Notes	C	Cr	Mn	Mo	Ni	P	S	Si	Other	UTS	YS	El	Hard

Alloy Steel, Nickel Chromium Molybdenum, A182(F5a)

USA

	UNS K42544		0.25 max	4.0-6.0	0.60 max	0.44-0.65	0.50 max	0.040 max	0.030 max	0.50 max	bal Fe				
ASTM A182/A182M(98)	F5a	Frg/roll pipe flange valve	0.25 max	4.0-6.0	0.60 max	0.44-0.65	0.50 max	0.040 max	0.030 max	0.50 max	bal Fe	620	450	22.0	187-248 HB
ASTM A336/A336M(98)	F5A	Ferr Frg, called F5 before 1955	0.25 max	4.0-5.0	0.60 max	0.45-0.65	0.50 max	0.025 max	0.025 max	0.50 max	bal Fe	550-725	345	19	

Alloy Steel, Nickel Chromium Molybdenum, A288(4)

Italy

| UNI 6786(71) | 30NiCrMoV12 | axles | 0.26-0.32 | 0.6-1 | 0.4-0.7 | 0.4-0.6 | 2.7-3.3 | 0.02 max | 0.02 max | 0.4 max | Cu <=0.2; V 0.08-0.13; bal Fe | | | | |

USA

	UNS K24562		0.45 max	0.70-1.25	1.00 max	0.20 min	1.65-3.50	0.025 max	0.020 max	0.15-0.35	V 0.07-0.12; bal Fe				
ASTM A288(98)	Class 4	Frg	0.45 max	0.70-1.25	1.00 max	0.20 min	1.65-3.50	0.025 max	0.020 max	0.15-0.30	V 0.07-0.12; bal Fe	825	655	18	
ASTM A288(98)	Class 5	Frg	0.45 max	0.70-1.25	1.00 max	0.20 min	1.65-3.50	0.025 max	0.020 max	0.15-0.30	V 0.07-0.12; bal Fe	895	760	16	
ASTM A288(98)	Class 6	Frg	0.45 max	0.70-1.25	1.00 max	0.20 min	1.65-3.50	0.025 max	0.020 max	0.15-0.30	V 0.07-0.12; bal Fe	965	860	14	
ASTM A288(98)	Class 7	Frg	0.45 max	0.70-1.25	1.00 max	0.20 min	1.65-3.50	0.025 max	0.020 max	0.15-0.30	V 0.07-0.12; bal Fe	1035	930	13	
ASTM A288(98)	Class 8	Frg	0.45 max	0.70-1.25	1.00 max	0.20 min	1.65-3.50	0.025 max	0.020 max	0.15-0.30	V 0.07-0.12; bal Fe	1140	1035	12	

Alloy Steel, Nickel Chromium Molybdenum, A290(G)

USA

	UNS K24045		0.35-0.45	0.60-0.90	0.60-0.90	0.20-0.50	1.65-2.00	0.040 max	0.040 max	0.20-0.35	V <=0.10; bal Fe				
ASTM A290(95)	G	Gear frg	0.35-0.45	0.60-0.90	0.60-0.90	0.20-0.50	1.65-2.00	0.040 max	0.040 max	0.35 max	Cu <=0.35; V <=0.10; bal Fe	860	690	15	262-311 HB
ASTM A290(95)	H	Gear frg	0.35-0.45	0.60-0.90	0.60-0.90	0.20-0.50	1.65-2.00	0.040 max	0.040 max	0.35 max	Cu <=0.35; V <=0.10; bal Fe				262-311 HB
ASTM A290(95)	I	Gear frg	0.35-0.45	0.60-0.90	0.60-0.90	0.20-0.50	1.65-2.00	0.040 max	0.040 max	0.35 max	Cu <=0.35; V <=0.10; bal Fe	1000	825	14	302-352 HB
ASTM A290(95)	J	Gear frg	0.35-0.45	0.60-0.90	0.60-0.90	0.20-0.50	1.65-2.00	0.040 max	0.040 max	0.35 max	Cu <=0.35; V <=0.10; bal Fe				302-352 HB
ASTM A290(95)	K	Gear frg	0.35-0.45	0.60-0.90	0.60-0.90	0.20-0.50	1.65-2.00	0.040 max	0.040 max	0.35 max	Cu <=0.35; V <=0.10; bal Fe	1175	1000	10	341-401 HB
ASTM A290(95)	L	Gear frg	0.35-0.45	0.60-0.90	0.60-0.90	0.20-0.50	1.65-2.00	0.040 max	0.040 max	0.35 max	Cu <=0.35; V <=0.10; bal Fe				341-401 HB

Alloy Steel, Nickel Chromium Molybdenum, A291(3A)

USA

	UNS K14557		0.45 max	1.50 max	0.40-0.90	0.15 min	1.00-3.00	0.040 max	0.040 max	0.15-0.30	Cu 0.35; V <=0.10; bal Fe				
ASTM A291(95)	3A	Gear pinion frg, 250<t<=510mm	0.45 max	1.50 max	0.40-0.90	0.15 min	1.00-3.00	0.040 max	0.040 max	0.35 max	Cu <=0.35; V <=0.10; bal Fe	725	550	19 L 16T	223-262 HB
ASTM A291(95)	3A	Gear pinion frg, t>510mm	0.45 max	1.50 max	0.40-0.90	0.15 min	1.00-3.00	0.040 max	0.040 max	0.35 max	Cu <=0.35; V <=0.10; bal Fe	725	550	19 L 14T	223-262 HB
ASTM A291(95)	3A	Gear pinion frg, t<=250mm	0.45 max	1.50 max	0.40-0.90	0.15 min	1.00-3.00	0.040 max	0.040 max	0.35 max	Cu <=0.35; V <=0.10; bal Fe	725	550	19 L	223-262 HB

Alloy Steel, Nickel Chromium Molybdenum, A291(4 to 7)

Japan

| JIS G4103(79) | SNCM447 | HR Frg Bar Rod | 0.44-0.50 | 0.60-1.00 | 0.60-0.90 | 0.15-0.30 | 1.60-2.00 | 0.030 max | 0.030 max | 0.15-0.35 | Cu <=0.30; bal Fe | | | | |

USA

| | UNS K24245 | | 0.35-0.50 | 0.60 min | 0.40-0.90 | 0.20-0.60 | 1.65 min | 0.040 max | 0.040 max | 0.15-0.35 | Cu 0.35; V <=0.10; bal Fe | | | | |
| ASTM A291(95) | 4-7 | Gear pinion frg | | | | | | | | | bal Fe | | | | |

Alloy Steel, Nickel Chromium Molybdenum, A325(D)

USA

| | UNS K12059 | | 0.14-0.26 | 0.45-1.05 | 0.36-1.24 | 0.11 max | 0.47-0.83 | 0.045 max | 0.055 max | 0.20-0.55 | Cu 0.27-0.53; Ti <=0.05; bal Fe | | | | |

UNS numbers and US grades are provided as a means of cross referencing chemically similar alloys. Exchangability is only possible after independent examination of specifications. Tensile properties are minimum or typical as specified. UTS and YS as MPa. El as %. See Appendix for list of abbreviations used in Notes. * indicates obsolete material.

Specification	Designation	Notes	C	Cr	Mn	Mo	Ni	P	S	Si	Other	UTS	YS	El	Hard

Alloy Steel, Nickel Chromium Molybdenum, A372(VI)

India

Specification	Designation	Notes	C	Cr	Mn	Mo	Ni	P	S	Si	Other	UTS	YS	El	Hard
IS 1570/4(88)	15Cr6Ni6		0.12-0.18	1.4-1.7	0.4-0.6	0.15 max	1.4-1.7	0.07 max	0.06 max	0.15-0.35	bal Fe				

USA

Specification	Designation	Notes	C	Cr	Mn	Mo	Ni	P	S	Si	Other	UTS	YS	El	Hard
	UNS K31820	HY80	0.18 max	1.00-1.80	0.10-0.40	0.20-0.60	2.00-3.25	0.025 max	0.025 max	0.15-0.35	bal Fe				
ASTM A372/372M(95)	Grade K	Thin wall frg press ves, HY80	0.18 max	1.00-1.80	0.10-0.40	0.20-0.60	2.00-3.30	0.025 max	0.025 max	0.15-0.35	bal Fe	690-860	550	20	207 HB
ASTM A372/372M(95)	Type IV*	Thin wall frg press ves, HY80	0.18 max	1.00-1.80	0.10-0.40	0.20-0.60	2.00-3.30	0.025 max	0.025 max	0.15-0.35	bal Fe	690-860	550	20	207 HB
MIL-S-16216K(SH)(87)	HY-80	Plt 1.25-3 in.	0.13-0.18	1.40-1.80	0.10-0.40	0.35-0.60	2.50-3.50	0.015 max	0.008 max	0.15-0.38	Cu <=0.25; Ti <=0.02; V <=0.03; Sb 0.025 max; As 0.025 max; Sn 0.030 max; bal Fe	552	552-686	20	
MIL-S-16216K(SH)(87)	HY-80	Plt >3 in.	0.13-0.18	1.50-1.90	0.10-0.40	0.50-0.65	3.00-3.50	0.015 max	0.008 max	0.15-0.38	Cu <=0.25; Ti <=0.02; V <=0.03; Sb 0.025 max; As 0.025 max; Sn 0.030 max; bal Fe	552	552-686	20	
MIL-S-16216K(SH)(87)	HY-80	Plt <=1.25 in.	0.12-0.18	1.00-1.80	0.10-0.40	0.20-0.60	2.00-3.25	0.015 max	0.008 max	0.15-0.38	Cu <=0.25; Ti <=0.02; V <=0.03; Sb 0.025 max; As 0.025 max; Sn 0.030 max; bal Fe	552			
MIL-S-23009C(SH)(87)	HY-80	Frg struct apps	0.12-0.18	1.35-1.80	0.10-0.40	0.30-0.60	2.50-3.25	0.015 max	0.008 max	0.15-0.35	Cu <=0.25; Ti <=0.02; V <=0.03; Sb 0.025 max; As 0.025 max; Sn 0.030 max; bal Fe	552	552-686	20	

Alloy Steel, Nickel Chromium Molybdenum, A372(VII)

USA

Specification	Designation	Notes	C	Cr	Mn	Mo	Ni	P	S	Si	Other	UTS	YS	El	Hard
	UNS K24055		0.38-0.43	0.70-0.90	0.60-0.80	0.20-0.30	1.65-2.00	0.035 max	0.04 max	0.15-0.35	bal Fe				
ASTM A372/372M(95)	Grade L	Thin wall frg press ves	0.38-0.43	0.70-0.90	0.60-0.80	0.20-0.30	1.65-2.00	0.025 max	0.04 max	0.15-0.35	bal Fe	1070-1240	930	12	311 HB
ASTM A372/372M(95)	Type VII*	Thin wall frg press ves	0.38-0.43	0.70-0.90	0.60-0.80	0.20-0.30	1.65-2.00	0.025 max	0.025 max	0.15-0.35	bal Fe	1070-1240	930	12	311 HB

Alloy Steel, Nickel Chromium Molybdenum, A469(2)

USA

Specification	Designation	Notes	C	Cr	Mn	Mo	Ni	P	S	Si	Other	UTS	YS	El	Hard
	UNS K22573		0.25 max	0.50 max	0.60 max	0.20-0.50	2.50 min	0.015 max	0.018 max	0.15-0.30	V >=0.03; bal Fe				

Alloy Steel, Nickel Chromium Molybdenum, A469(3)

USA

Specification	Designation	Notes	C	Cr	Mn	Mo	Ni	P	S	Si	Other	UTS	YS	El	Hard
	UNS K22773		0.27 max	0.50 max	0.60 max	0.20-0.50	2.50 min	0.015 max	0.018 max	0.15-0.30	N >=2.50; V >=0.03; bal Fe				

Alloy Steel, Nickel Chromium Molybdenum, A469(4)

USA

Specification	Designation	Notes	C	Cr	Mn	Mo	Ni	P	S	Si	Other	UTS	YS	El	Hard
	UNS K32723		0.27 max	0.50 max	0.70 max	0.20-0.60	3.00 min	0.015 max	0.018 max	0.15-0.30	V >=0.03; bal Fe				

Alloy Steel, Nickel Chromium Molybdenum, A469(5)

USA

Specification	Designation	Notes	C	Cr	Mn	Mo	Ni	P	S	Si	Other	UTS	YS	El	Hard
	UNS K33125		0.31 max	0.50 max	0.70 max	0.20-0.70	3.00 min	0.015 max	0.018 max	0.15-0.30	V 0.05-0.15; bal Fe				

Alloy Steel, Nickel Chromium Molybdenum, A469(6)

Italy

Specification	Designation	Notes	C	Cr	Mn	Mo	Ni	P	S	Si	Other	UTS	YS	El	Hard
UNI 7845(78)	34NiCrMo16	Q/T	0.31-0.38	1.6-2	0.3-0.6	0.25-0.45	3.7-4.2	0.035 max	0.035 max	0.15-0.4	bal Fe				
UNI 7874(79)	34NiCrMo16	Q/T	0.31-0.38	1.6-2	0.3-0.6	0.3-0.6	3.7-4.2	0.035 max	0.035 max	0.15-0.4	V <=0.15; bal Fe				

UNS numbers and US grades are provided as a means of cross referencing chemically similar alloys. Exchangability is only possible after independent examination of specifications. Tensile properties are minimum or typical as specified. UTS and YS as MPa. El as %. See Appendix for list of abbreviations used in Notes. * indicates obsolete material.

Specification	Designation	Notes	C	Cr	Mn	Mo	Ni	P	S	Si	Other	UTS	YS	El	Hard
Alloy Steel, Nickel Chromium Molybdenum, A469(6) (Continued from previous page)															
USA															
	UNS K42885		0.28 max	1.25-2.00	0.60 max	0.30-0.60	3.25-4.00	0.015 max	0.018 max	0.15-0.30	V 0.05-0.15; bal Fe				
Alloy Steel, Nickel Chromium Molybdenum, A470(2)															
USA															
	UNS K22578		0.25 max	0.75 max	0.020-0.60	0.25 min	2.50 min	0.015 max	0.018 max	0.15-0.30	V >=0.03; bal Fe				
Alloy Steel, Nickel Chromium Molybdenum, A470(3,4)															
USA															
	UNS K22878		0.28 max	0.75 max	0.20-0.60	0.25 min	2.50 min	0.015 max	0.018 max	0.15-0.30	V >=0.03; bal Fe				
Alloy Steel, Nickel Chromium Molybdenum, A470(8)															
USA															
	UNS K23010		0.25-0.35	0.90-1.50	1.00 max	1.00-1.50	0.75 max	0.015 max	0.018 max	0.15-0.35	V 0.20-0.30; bal Fe				
Alloy Steel, Nickel Chromium Molybdenum, A471(1-6)															
USA															
	UNS K32800		0.28 max	0.75-2.00	0.70 max	0.20-0.70	2.00-4.00	0.015 max	0.015 max	0.15-0.35	V >=0.05; bal Fe				
Alloy Steel, Nickel Chromium Molybdenum, A471(10)															
USA															
	UNS K23205		0.27-0.37	0.85-1.25	0.70-1.00	1.00-1.50	0.50 max	0.015 max	0.015 max	0.20 min	V 0.20-0.30; bal Fe				
Alloy Steel, Nickel Chromium Molybdenum, A508(2,2a)															
USA															
	UNS K12766		0.27 max	0.25-0.45	0.50-1.00	0.55-0.70	0.50-1.00	0.025 max	0.025 max	0.15-0.40	V <=0.05; bal Fe				
ASTM A508/508M(95)	Grade 2 Class 1	Q/T vacuum-treated frg for press ves	0.27 max	0.25-0.45	0.50-1.00	0.55-0.70	0.50-1.00	0.025 max	0.025 max	0.15-0.40	bal Fe	550-725	345	18	
ASTM A508/508M(95)	Grade 2 Class 2	Q/T vacuum-treated frg for press ves	0.27 max	0.25-0.45	0.50-1.00	0.55-0.70	0.50-1.00	0.025 max	0.025 max	0.15-0.40	bal Fe	620-795	450	16	
Alloy Steel, Nickel Chromium Molybdenum, A508(4)															
USA															
	UNS K22375		0.23 max	1.50-2.00	0.20-0.40	0.40-0.60	2.75-3.90	0.020 max	0.020 max	0.15-0.40	V <=0.03; bal Fe				
ASTM A508/508M(95)	Grade 4N Class 1	Q/T vacuum-treated frg for press ves	0.23 max	1.50-2.00	0.20-0.40	0.40-0.60	2.8-3.9	0.020 max	0.020 max	0.15-0.40	bal Fe	725-895	585	18	
ASTM A508/508M(95)	Grade 4N Class 2	Q/T vacuum-treated frg for press ves	0.23 max	1.50-2.00	0.20-0.40	0.40-0.60	2.8-3.9	0.020 max	0.020 max	0.15-0.40	bal Fe	795-965	690	16	
ASTM A508/508M(95)	Grade 4N Class 3	Q/T vacuum-treated frg for press ves	0.23 max	1.50-2.00	0.20-0.40	0.40-0.60	2.8-3.9	0.020 max	0.020 max	0.15-0.40	bal Fe	620-795	485	20	
Alloy Steel, Nickel Chromium Molybdenum, A508(5)															
USA															
	UNS K42365		0.23 max	1.50-2.00	0.20-0.40	0.40-0.60	2.75-3.90	0.020 max	0.020 max	0.30 max	V <=0.08; bal Fe				
ASTM A508/508M(95)	Grade 5 Class 1	Q/T vacuum-treated frg for press ves	0.23 max	1.50-2.00	0.20-0.40	0.40-0.60	2.8-3.9	0.020 max	0.020 max	0.30 max	bal Fe	725-895	585	18	
ASTM A508/508M(95)	Grade 5 Class 2	Q/T vacuum-treated frg for press ves	0.23 max	1.50-2.00	0.20-0.40	0.40-0.60	2.8-3.9	0.020 max	0.020 max	0.30 max	bal Fe	795-965	690	16	
Alloy Steel, Nickel Chromium Molybdenum, A514(H)															
USA															
	UNS K11646		0.12-0.21	0.40-0.65	0.95-1.30	0.20-0.30	0.30-0.70	0.035 max	0.04	0.20-0.35	B 0.0005-0.005; V 0.03-0.08; bal Fe				
ASTM A514(94)	Grade H	Q/T, Plt, t=50mm, weld	0.12-0.21	0.40-0.65	0.95-1.30	0.20-0.30	0.30-0.70	0.035 max	0.035 max	0.20-0.35	B 0.0005-0.005; V 0.03-0.08; bal Fe	760-895	690	18	
ASTM A517(93)	Grade H	Q/T, Press ves plt, t<=50mm	0.12-0.21	0.40-0.65	0.95-1.30	0.20-0.30	0.30-0.70	0.035 max	0.035 max	0.15-0.35	B >=0.0005; V 0.03-0.08; bal Fe	795-930	690	16	

UNS numbers and US grades are provided as a means of cross referencing chemically similar alloys. Exchangability is only possible after independent examination of specifications. Tensile properties are minimum or typical as specified. UTS and YS as MPa. El as %. See Appendix for list of abbreviations used in Notes. * indicates obsolete material.

Specification	Designation	Notes	C	Cr	Mn	Mo	Ni	P	S	Si	Other	UTS	YS	El	Hard
Alloy Steel, Nickel Chromium Molybdenum, A514(H) (Continued from previous page)															
USA															
ASTM A709(97)	Grade 100, Type H	Q/T, Plt, t<=50mm	0.12-0.21	0.40-0.65	0.95-1.30	0.20-0.30	0.30-0.70	0.035 max	0.035 max	0.20-0.35	B 0.0005-0.005; V 0.03-0.08; bal Fe	760-895	690	18	
Alloy Steel, Nickel Chromium Molybdenum, A514(P)															
Europe															
EN 10084(98)	1.6587	CH, Ann	0.15-0.21	1.50-1.80	0.50-0.90	0.25-0.35	1.40-1.70	0.035 max	0.035 max	0.40 max	bal Fe				229 HB
EN 10084(98)	18CrNiMo7-6	CH, Ann	0.15-0.21	1.50-1.80	0.50-0.90	0.25-0.35	1.40-1.70	0.035 max	0.035 max	0.40 max	bal Fe				229 HB
Italy															
UNI 3097(75)	17NiCrMo5	Ball & roller bearing	0.14-0.2	0.8-1.1	0.6-0.9	0.15-0.25	1.2-1.5	0.035 max	0.035 max	0.15-0.4	bal Fe				
UNI 7846(78)	18NiCrMo5	CH	0.15-0.21	0.7-1	0.6-0.9	0.15-0.25	1.2-1.5	0.035 max	0.035 max	0.15-0.4	Pb 0.15-0.3; bal Fe				
UNI 7846(78)	18NiCrMo7	CH	0.17-0.22	0.4-0.7	0.45-0.65	0.2-0.3	1.6-2	0.035 max	0.035 max	0.15-0.4	Pb 0.15-0.3; bal Fe				
UNI 8550(84)	18NiCrMo5	CH	0.15-0.21	0.7-1	0.6-0.9	0.15-0.25	1.2-1.5	0.035 max	0.035 max	0.15-0.4	bal Fe				
UNI 8788(85)	18NiCrMo5	CH	0.15-0.21	0.7-1	0.6-0.9	0.15-0.25	1.2-1.5	0.035 max	0.035 max	0.15-0.4	Pb 0.15-0.3; P+S<=0.06; bal Fe				
UNI 8788(85)	18NiCrMo7	CH	0.17-0.22	0.8-1.1	0.45-0.65	0.3-0.4	1.6-2	0.035 max	0.035 max	0.15-0.4	Pb 0.15-0.3; P+S<=0.06; bal Fe				
USA															
	UNS K21650		0.12-0.21	0.85-1.20	0.45-0.70	0.45-0.60	1.20-1.50	0.035 max	0.04 max	0.20-0.35	B 0.001-0.005; bal Fe				
ASTM A514(94)	Grade P	Q/T, Plt, t=150mm, weld	0.12-0.21	0.85-1.20	0.45-0.70	0.45-0.60	1.20-1.50	0.035 max	0.035 max	0.20-0.35	B 0.001-0.005; bal Fe	690-895	620	16	
ASTM A517(93)	Grade P	Q/T, Press ves plt, t<=65mm	0.12-0.21	0.85-1.20	0.45-0.70	0.45-0.60	1.20-1.50	0.035 max	0.035 max	0.20-0.35	B 0.001-0.005; bal Fe	795-930	690	16	
ASTM A517(93)	Grade P	Q/T, Press ves plt, t>=65-100mm	0.12-0.21	0.85-1.20	0.45-0.70	0.45-0.60	1.20-1.50	0.035 max	0.035 max	0.20-0.35	B 0.001-0.005; bal Fe	725-930	620	14	
ASTM A709(97)	Grade 100, Type P	Q/T, Plt, t<=100mm	0.12-0.21	0.85-1.20	0.45-0.70	0.45-0.60	1.20-1.50	0.035 max	0.035 max	0.20-0.35	B 0.001-0.005; bal Fe	690-895	620	16	
Alloy Steel, Nickel Chromium Molybdenum, A540(B24)															
USA															
	UNS K24064		0.37-0.44	0.70-0.95	0.70-0.90	0.30-0.40	1.65-2.00	0.025 max	0.025 max	0.20-0.35	bal Fe				
ASTM A540/A540M(98)	A540 (B24)	Bolting matl	0.37-0.44	0.70-0.95	0.70-0.90	0.30-0.40	1.65-2.00	0.025 max	0.025 max	0.15-0.35	bal Fe	827-1138	724-1034	15-10	248-331/311-429 HB
Alloy Steel, Nickel Chromium Molybdenum, A540(B24V)															
USA															
	UNS K24070		0.38-0.43	0.70-0.90	0.60-0.90	0.30-0.60	1.65-2.00	0.025 max	0.025 max	0.20-0.35	V 0.05-0.10; bal Fe				
ASTM A540/A540M(98)	A540 (B24V)	Bolting matl	0.37-0.44	0.60-0.95	0.60-0.95	0.40-0.60	1.55-2.00	0.025 max	0.025 max	0.15-0.35	V 0.04-0.10; bal Fe	1000	896	12	293-311/363-388 HB
ASTM A579(96)	22	Superstrength frg	0.38-0.43	0.70-0.90	0.60-0.90	0.30-0.60	1.65-2.00	0.025 max	0.025 max	0.20-0.35	V 0.05-0.10; bal Fe	1035-1720	965-1550	6-13	302 HB
Alloy Steel, Nickel Chromium Molybdenum, A541(2,2A)															
USA															
	UNS K12765		0.27 max	0.25-0.45	0.50-0.90	0.55-0.70	0.50-1.00	0.035 max	0.040 max	0.15-0.35	V <=0.05; bal Fe				
ASTM A541/541M(95)	Grade 2 Class 1	Q/T frg for press ves components	0.27 max	0.25-0.45	0.50-0.90	0.55-0.70	0.50-1.00	0.025 max	0.025 max	0.15-0.35	V <=0.05; Si req vary; bal Fe	550-720	340	18	
ASTM A541/541M(95)	Grade 2 Class 2	Q/T frg for press ves components	0.27 max	0.25-0.45	0.50-0.90	0.55-0.70	0.50-1.00	0.025 max	0.025 max	0.15-0.35	V <=0.05; Si req vary; bal Fe	620-790	450	16	
Alloy Steel, Nickel Chromium Molybdenum, A541(7)															
USA															
	UNS K42343		0.23 max	1.25-2.00	0.20-0.40	0.40-0.60	2.75-3.90	0.035 max	0.040 max	0.30 max	V <=0.03; bal Fe				
ASTM A541/541M(95)	Grade 4N Class 1	Q/T frg for press ves components	0.23 max	1.25-2.00	0.20-0.40	0.40-0.60	2.8-3.9	0.025 max	0.025 max	0.30 max	V <=0.03; Si req vary; bal Fe	720-900	590	18	

UNS numbers and US grades are provided as a means of cross referencing chemically similar alloys. Exchangability is only possible after independent examination of specifications. Tensile properties are minimum or typical as specified. UTS and YS as MPa. El as %. See Appendix for list of abbreviations used in Notes. * indicates obsolete material.

Specification	Designation	Notes	C	Cr	Mn	Mo	Ni	P	S	Si	Other	UTS	YS	El	Hard

Alloy Steel, Nickel Chromium Molybdenum, A541(7) (Continued from previous page)

USA

Specification	Designation	Notes	C	Cr	Mn	Mo	Ni	P	S	Si	Other	UTS	YS	El	Hard
ASTM A541/541M(95)	Grade 4N Class 2	Q/T frg for press ves components	0.23 max	1.25-2.00	0.20-0.40	0.40-0.60	2.8-3.9	0.025 max	0.025 max	0.30 max	V <=0.03; Si req vary; bal Fe	790-1000	690	16	
ASTM A541/541M(95)	Grade 4N Class 3	Q/T frg for press ves components	0.23 max	1.25-2.00	0.20-0.40	0.40-0.60	2.8-3.9	0.025 max	0.025 max	0.30 max	V <=0.03; Si req vary; bal Fe	620-790	480	20	

Alloy Steel, Nickel Chromium Molybdenum, A541(8)

USA

Specification	Designation	Notes	C	Cr	Mn	Mo	Ni	P	S	Si	Other	UTS	YS	El	Hard
	UNS K42348		0.23 max	1.25-2.00	0.20-0.40	0.40-0.60	2.75-3.90	0.035 max	0.040 max	0.30 max	V <=0.08; bal Fe				
ASTM A541/541M(95)	Grade 5 Class 1	Q/T frg for press ves components	0.23 max	1.25-2.00	0.20-0.40	0.40-0.60	2.8-3.9	0.025 max	0.025 max	0.30 max	V <=0.08; Si req vary; bal Fe	720-900	590	18	
ASTM A541/541M(95)	Grade 5 Class 2	Q/T frg for press ves components	0.23 max	1.25-2.00	0.20-0.40	0.40-0.60	2.8-3.9	0.025 max	0.025 max	0.30 max	V <=0.08; Si req vary; bal Fe	790-1000	690	16	

Alloy Steel, Nickel Chromium Molybdenum, A543/A543M(B)

USA

Specification	Designation	Notes	C	Cr	Mn	Mo	Ni	P	S	Si	Other	UTS	YS	El	Hard
	UNS K42339		0.23 max	1.44-2.06	0.40 max	0.41-0.64	2.53-3.32	0.020 max	0.020 max		V <=0.03; bal Fe				
ASTM A543(93)	Type B	Q/T, Press ves plt, t>=5mm, Class 3	0.20 max	1.00-1.90	0.40 max	0.20-0.65	2.25-4.00	0.020 max	0.020 max	0.15-0.40	V <=0.03; bal Fe	620-795	485	16(in 50mm)	
ASTM A543(93)	Type B	Q/T, Press ves plt, t>=5mm, Class 2	0.20 max	1.00-1.90	0.40 max	0.20-0.65	2.25-4.00	0.020 max	0.020 max	0.15-0.40	V <=0.03; bal Fe	795-930	690	14(in 50mm)	
ASTM A543(93)	Type B	Q/T, Press ves plt, t>=5mm, Class 1	0.20 max	1.00-1.90	0.40 max	0.20-0.65	2.25-4.00	0.020 max	0.020 max	0.15-0.40	V <=0.03; bal Fe	725-860	585	14	

Alloy Steel, Nickel Chromium Molybdenum, A579(11)

USA

Specification	Designation	Notes	C	Cr	Mn	Mo	Ni	P	S	Si	Other	UTS	YS	El	Hard
	UNS K42598		0.23-0.28	1.40-1.65	0.20 max	0.8-1.0	2.75-3.25	0.01 max	0.01 max	0.10 max	Nb 0.03-0.07; bal Fe				
ASTM A579(96)	11	Superstrength frg	0.23-0.28	1.40-1.65	0.20 max	0.8-1.0	2.75-3.25	0.01 max	0.01 max	0.10 max	Nb 0.03-0.07; bal Fe				321 HB

Alloy Steel, Nickel Chromium Molybdenum, A579(12)

USA

Specification	Designation	Notes	C	Cr	Mn	Mo	Ni	P	S	Si	Other	UTS	YS	El	Hard
	UNS K51255	HY30	0.12 max	0.40-0.70	0.60-0.90	0.30-0.65	4.75-5.25	0.010 max	0.010 max	0.20-0.35	V 0.05-0.10; bal Fe				
ASTM A579(96)	12	Superstrength frg	0.12 max	0.40-0.70	0.60-0.90	0.30-0.65	4.75-5.25	0.010 max	0.010 max	0.20-0.35	V 0.05-0.10; bal Fe				
ASTM A579(96)	12a	Superstrength frg	0.20 max	0.40-0.70	0.60-0.90	0.30-0.65	4.75-5.25	0.015 max	0.015 max	0.20-0.35	V 0.05-0.10; bal Fe				

Alloy Steel, Nickel Chromium Molybdenum, A579(21)

USA

Specification	Designation	Notes	C	Cr	Mn	Mo	Ni	P	S	Si	Other	UTS	YS	El	Hard
	UNS K23477		0.31-0.38	0.65-0.90	0.60-0.90	0.30-0.60	1.65-2.00	0.025 max	0.025 max	0.20-0.35	V 0.17-0.23; bal Fe				
ASTM A579(96)	21	Superstrength frg	0.31-0.38	0.65-0.90	0.60-0.90	0.30-0.60	1.65-2.00	0.025 max	0.025 max	0.20-0.35	V 0.17-0.23; bal Fe	1035-1450	965-1380	9-13	285 HB
ASTM A646(95)	Grade 6	Blm Bil for aircraft aerospace frg, ann	0.33-0.38	0.65-0.90	0.60-0.90	0.30-0.40	1.65-2.00	0.025 max	0.025 max	0.40-0.60	V 0.17-0.23; bal Fe				285 HB

Alloy Steel, Nickel Chromium Molybdenum, A579(31)

USA

Specification	Designation	Notes	C	Cr	Mn	Mo	Ni	P	S	Si	Other	UTS	YS	El	Hard
	UNS K32550		0.23-0.28	0.20-0.40	1.20-1.50	0.35-0.45	1.65-2.00	0.025 max	0.025 max	1.30-1.70	bal Fe				
ASTM A579(96)	31	Superstrength frg	0.23-0.28	0.20-0.40	1.20-1.50	0.35-0.45	1.65-2.00	0.025 max	0.025 max	1.30-1.70	bal Fe				262 HB

Alloy Steel, Nickel Chromium Molybdenum, A649(1B)

USA

Specification	Designation	Notes	C	Cr	Mn	Mo	Ni	P	S	Si	Other	UTS	YS	El	Hard
	UNS K24040		0.37-0.44	0.65-0.95	0.60-0.95	0.20-0.30	1.55-2.00	0.025 max	0.025 max	0.15-0.35	bal Fe				
ASTM A649/649M(98)	Class 1B	Frg roll for paperboard mach	0.40-0.60	0.65-0.93	0.60-0.95	0.20-0.30		0.025 max	0.025 max	0.15-0.35	Si<=0.10 with VCD; bal Fe	690-1030	450-890	12-14	

UNS numbers and US grades are provided as a means of cross referencing chemically similar alloys. Exchangability is only possible after independent examination of specifications. Tensile properties are minimum or typical as specified. UTS and YS as MPa. El as %. See Appendix for list of abbreviations used in Notes. * indicates obsolete material.

Specification	Designation	Notes	C	Cr	Mn	Mo	Ni	P	S	Si	Other	UTS	YS	El	Hard
Alloy Steel, Nickel Chromium Molybdenum, A687(I)															
USA															
	UNS K13521		0.31-0.40	0.22-0.43	0.91-1.29	0.04-0.11	0.17-0.43	0.040 max	0.045 max	0.18-0.37	bal Fe				
Alloy Steel, Nickel Chromium Molybdenum, A707(L5)															
USA															
	UNS K20934		0.09 max	0.56-0.94	0.35-0.75	0.14-0.28	0.67-1.03	0.030 max	0.035 max	0.37 max	Cu 0.95-1.35; Nb >=0.02; bal Fe				
ASTM A707/A707M(98)	A 707 (L5)	Frg flanges; low-temp; class values given	0.07 max	0.60-0.90	0.40-0.70	0.15-0.25	0.70-1.00	0.025 max	0.025 max	0.35 max	Cu 1.00-1.30; Nb >=0.03; V <=0.05; bal Fe	415-620	290-515	20-22	149-265 HB
Alloy Steel, Nickel Chromium Molybdenum, A707(L8)															
USA															
	UNS K42247		0.22 max	1.44-2.06	0.15-0.45	0.35-0.65	2.68-3.97	0.025 max	0.025 max	0.37 max	V <=0.05; bal Fe				
ASTM A707/A707M(98)	A 707 (L8)	Frg flanges; low-temp; class values given	0.20 max	1.5-2.00	0.20-0.40	0.40-0.60	2.8-3.9	0.020 max	0.020 max	0.35 max	Cu <=0.40; Nb <=0.02; V <=0.05; bal Fe	415-620	290-515	20-22	149-265 HB
Alloy Steel, Nickel Chromium Molybdenum, A723(1)															
USA															
	UNS K23550		0.35 max	0.80-2.00	0.90 max	0.20-0.40	1.5-2.25	0.015 max	0.015 max	0.35 max	V <=0.20; bal Fe				
ASTM A723/723M(94)	Grade 1 Class 1	Frg for high-strength press components	0.35 max	0.80-2.00	0.90 max	0.20-0.40	1.5-2.25	0.015 max	0.015 max	0.35 max	V <=0.20; bal Fe	795	690	16	
Alloy Steel, Nickel Chromium Molybdenum, A723(2)															
Italy															
UNI 3097(75)	16NiCrMo12	Ball & roller bearing	0.13-0.19	0.8-1.1	0.4-0.7	0.3-0.4	2.7-3.2	0.035 max	0.035 max	0.15-0.4	bal Fe				
UNI 7846(78)	16NiCrMo12	CH	0.13-0.19	0.8-1.1	0.4-0.7	0.3-0.4	2.7-3.2	0.035 max	0.035 max	0.15-0.4	bal Fe				
UNI 8550(84)	16NiCrMo12	CH	0.13-0.19	0.8-1.1	0.4-0.7	0.3-0.4	2.7-3.2	0.035 max	0.035 max	0.15-0.4	bal Fe				
UNI 8788(85)	16NiCrMo12	CH	0.13-0.19	0.8-1.1	0.4-0.7	0.3-0.4	2.7-3.2	0.035 max	0.035 max	0.15-0.4	P+S<=0.06; bal Fe				
USA															
	UNS K34035		0.40 max	0.80-2.00	0.90 max	0.30-0.50	2.25-3.25	0.015 max	0.015 max	0.35 max	V <=0.20; bal Fe				
ASTM A723/723M(94)	Grade 2 Class 1	Frg for high-strength press components	0.40 max	0.80-2.00	0.90 max	0.30-0.50	2.3-3.3	0.015 max	0.015 max	0.35 max	V <=0.20; bal Fe	795	690	16	
Alloy Steel, Nickel Chromium Molybdenum, A723(3)															
Italy															
UNI 6925(71)	35NiCrMo15	Aircraft material	0.3-0.38	1.6-1.9	0.3-0.6	0.25-0.45	3.5-4	0.025 max	0.02 max	0.15-0.35	bal Fe				
USA															
	UNS K44045		0.40 max	0.80-2.00	0.90 max	0.40-0.80	3.25-4.50	0.015 max	0.015 max	0.35 max	V <=0.20; bal Fe				
ASTM A723/723M(94)	Grade 3 Class 1	Frg for high-strength press components	0.40 max	0.80-2.00	0.90 max	0.40-0.80	3.3-4.5	0.015 max	0.015 max	0.35 max	V <=0.20; bal Fe	795	690	16	
Alloy Steel, Nickel Chromium Molybdenum, A734(A)															
USA															
	UNS K21205		0.12 max	0.90-1.20	0.45-0.75	0.25-0.40	0.90-1.20	0.035 max	0.015 max	0.40 max	Al <=0.06; bal Fe				
ASTM A734(97)	Type A	Press ves plt, Q/T	0.12 max	0.90-1.20	0.45-0.75	0.25-0.40	0.90-1.20	0.035 max	0.015 max	0.40 max	Al <=0.06; bal Fe	530-670	450	20	
Alloy Steel, Nickel Chromium Molybdenum, A736															
USA															
	UNS K20747		0.07 max	0.60-0.90	0.40-0.70	0.15-0.25	0.70-1.00	0.025 max	0.025 max	0.40 max	Cu 1.00-1.30; Nb >=0.02; bal Fe				
	UNS K20747		0.07 max	0.60-0.90	0.40-0.70	0.15-0.25	0.70-1.00	0.025 max	0.025 max	0.40 max	Cu 1.00-1.30; Nb >=0.02; bal Fe				

UNS numbers and US grades are provided as a means of cross referencing chemically similar alloys. Exchangability is only possible after independent examination of specifications. Tensile properties are minimum or typical as specified. UTS and YS as MPa. El as %. See Appendix for list of abbreviations used in Notes. * indicates obsolete material.

Specification	Designation	Notes	C	Cr	Mn	Mo	Ni	P	S	Si	Other	UTS	YS	El	Hard

Alloy Steel, Nickel Chromium Molybdenum, A736 (Continued from previous page)

USA

Specification	Designation	Notes	C	Cr	Mn	Mo	Ni	P	S	Si	Other	UTS	YS	El	Hard
ASTM A710(95)	Grade A, Class 1	AH struct Plt	0.07 max	0.60-0.90	0.40-0.70	0.15-0.25	0.70-1.00	0.025 max	0.025 max	0.40 max	Cu 1.00-1.30; Nb >=0.02; bal Fe	620	585	20	
ASTM A710(95)	Grade A, Class 1	AH struct Plt	0.07 max	0.60-0.90	0.40-0.70	0.15-0.25	0.70-1.00	0.025 max	0.025 max	0.40 max	Cu 1.00-1.30; Nb >=0.02; bal Fe	620	585	20	
ASTM A710(95)	Grade A, Class 2	AH struct Plt	0.07 max	0.60-0.90	0.40-0.70	0.15-0.25	0.70-1.00	0.025 max	0.025 max	0.40 max	Cu 1.00-1.30; Nb >=0.02; bal Fe	495	450	20	
ASTM A710(95)	Grade A, Class 2	AH struct Plt	0.07 max	0.60-0.90	0.40-0.70	0.15-0.25	0.70-1.00	0.025 max	0.025 max	0.40 max	Cu 1.00-1.30; Nb >=0.02; bal Fe	495	450	20	
ASTM A710(95)	Grade A, Class 3	AH struct Plt	0.07 max	0.60-0.90	0.40-0.70	0.15-0.25	0.70-1.00	0.025 max	0.025 max	0.40 max	Cu 1.00-1.30; Nb >=0.02; bal Fe	585	515	20	
ASTM A710(95)	Grade A, Class 3	AH struct Plt	0.07 max	0.60-0.90	0.40-0.70	0.15-0.25	0.70-1.00	0.025 max	0.025 max	0.40 max	Cu 1.00-1.30; Nb >=0.02; bal Fe	585	515	20	
ASTM A736(94)	Grade A	Press ves plt, rolled or PHT	0.07 max	0.60-0.90	0.40-0.70	0.15-0.25	0.70-1.00	0.025 max	0.025 max	0.40 max	Cu 1.00-1.30; Nb >=0.02; bal Fe	620-760	550	20	
ASTM A736(94)	Grade C	Press ves plt, rolled or PHT	0.07 max		1.30-1.65	0.15-0.25	0.70-1.00	0.025 max	0.025 max	0.40 max	Cu 1.00-1.30; Nb >=0.02; bal Fe	690-825	620	20	

Alloy Steel, Nickel Chromium Molybdenum, AMS 6423

USA

Specification	Designation	Notes	C	Cr	Mn	Mo	Ni	P	S	Si	Other	UTS	YS	El	Hard
	UNS K24336		0.40-0.46	0.80-1.05	0.75-1.00	0.45-0.60	0.60-0.90	0.025 max	0.025 max	0.50-0.80	B 0.0005-0.005; Cu <=0.35; V 0.01-0.06; bal Fe				

Alloy Steel, Nickel Chromium Molybdenum, E4337

USA

Specification	Designation	Notes	C	Cr	Mn	Mo	Ni	P	S	Si	Other	UTS	YS	El	Hard
	AISI E4337	Smls mech Tub; BEF	0.35-0.40	0.70-0.90	0.65-0.85	0.20-0.30	1.65-2.00	0.025 max	0.025 max	0.15-0.35	bal Fe				
	UNS G43376		0.35-0.40	0.70-0.90	0.65-0.85	0.20-0.30	1.65-2.00	0.025 max	0.025 max	0.15-0.35	bal Fe				
ASTM A519(96)	E3310	Smls mech tub	0.08-0.13	1.40-1.75	0.45-0.60		3.25-3.75	0.025 max	0.025 max	0.15-0.35	bal Fe				
ASTM A519(96)	E4337	Smls mech tub	0.35-0.40	0.70-0.90	0.65-0.85	0.20-0.30	1.65-2.00	0.025 max	0.025 max	0.15-0.35	bal Fe				

Alloy Steel, Nickel Chromium Molybdenum, E4340/E4340H

China

Specification	Designation	Notes	C	Cr	Mn	Mo	Ni	P	S	Si	Other	UTS	YS	El	Hard
GB 3077(88)	40CrNiMoA	Bar HR Q/T 15mm diam	0.37-0.44	0.60-0.90	0.50-0.80	0.15-0.25	1.25-1.65	0.025 max	0.025 max	0.17-0.37	Cu <=0.25; bal Fe	980	835	12	
GB 8162(87)	40CrNiMoA	Smls tube HR/CD Q/T	0.37-0.44	0.60-0.90	0.50-0.80	0.15-0.25	1.25-1.65	0.025 max	0.025 max	0.17-0.37	bal Fe	980	835	12	
GB/T 3078(94)	40CrNiMoA	Bar CD Q/T 15mm diam	0.37-0.44	0.60-0.90	0.50-0.80	0.15-0.25	1.25-1.65	0.025 max	0.025 max	0.17-0.37	Cu <=0.25; bal Fe	980	835	12	
GB/T 3079(93)	40CrNiMoA	Wir Q/T	0.37-0.44	0.60-0.90	0.50-0.80	0.15-0.25	1.25-1.65	0.025 max	0.025 max	0.17-0.37	Cu <=0.25; bal Fe	980	835	12	

Germany

Specification	Designation	Notes	C	Cr	Mn	Mo	Ni	P	S	Si	Other	UTS	YS	El	Hard
DIN	40NiCrMo73		0.37-0.44	0.7-0.95	0.7-0.9	0.3-0.4	1.65-2	0.02	0.015	0.4 max	bal Fe				

Japan

Specification	Designation	Notes	C	Cr	Mn	Mo	Ni	P	S	Si	Other	UTS	YS	El	Hard
JIS G4108(94)	SNB23-1	Bar 150<diam<=200mm	0.37-0.44	0.65-0.95	0.60-0.95	0.20-0.30	1.55-2.00	0.025 max	0.025 max	0.20-0.35	bal Fe	1140	1030	10	341-444 HB
JIS G4108(94)	SNB23-2	Bar 150<diam<=240mm	0.37-0.44	0.65-0.95	0.60-0.95	0.20-0.30	1.55-2.00	0.025 max	0.025 max	0.20-0.35	bal Fe	1070	960	11	321-415 HB
JIS G4108(94)	SNB23-3	Bar 150<diam<=240mm	0.37-0.44	0.65-0.95	0.60-0.95	0.20-0.30	1.55-2.00	0.025 max	0.025 max	0.20-0.35	bal Fe	1000	890	12	311-388 HB
JIS G4108(94)	SNB23-4	Bar 200<diam<=240mm	0.37-0.44	0.65-0.95	0.60-0.95	0.20-0.30	1.55-2.00	0.025 max	0.025 max	0.20-0.35	bal Fe	930	825	13	285-363 HB
JIS G4108(94)	SNB23-5	Bar 150<diam<=240mm	0.37-0.44	0.65-0.95	0.60-0.95	0.20-0.30	1.55-2.00	0.025 max	0.025 max	0.20-0.35	bal Fe	790	685	15	262-321 HB

Mexico

Specification	Designation	Notes	C	Cr	Mn	Mo	Ni	P	S	Si	Other	UTS	YS	El	Hard
DGN B-203	E4340*	Obs	0.38-0.43	0.7-0.9	0.65-0.85	0.2-0.3	1.65-2	0.025	0.025	0.2-0.35	bal Fe				
DGN B-297	E4340*	Obs	0.38-0.43	0.7-0.9	0.65-0.85	0.2-0.3	1.65-2	0.025	0.025	0.2-0.35	bal Fe				
NMX-B-268(68)	E4340H	Hard	0.37-0.44	0.65-0.95	0.60-0.95	0.20-0.30	1.55-2.00			0.20-0.35	bal Fe				
NMX-B-300(91)	E4340	Bar	0.38-0.43	0.70-0.90	0.65-0.85	0.20-0.30	1.65-2.00	0.035 max	0.040 max	0.15-0.35	bal Fe				

UNS numbers and US grades are provided as a means of cross referencing chemically similar alloys. Exchangability is only possible after independent examination of specifications. Tensile properties are minimum or typical as specified. UTS and YS as MPa. El as %. See Appendix for list of abbreviations used in Notes. * indicates obsolete material.

Alloy Steel, Nickel Chromium Molybdenum, E4340/E4340H (Continued from previous page)

Specification	Designation	Notes	C	Cr	Mn	Mo	Ni	P	S	Si	Other	UTS	YS	El	Hard
Mexico															
NMX-B-300(91)	E4340H	Bar	0.37-0.44	0.65-0.95	0.60-0.95	0.20-0.30	1.55-2.00			0.15-0.35	bal Fe				
Pan America															
COPANT 514	E4340	Tube	0.38-0.43	0.7-0.9	0.65-0.85	0.2-0.3	1.65-2	0.03	0.03	0.2-0.35	bal Fe				
USA															
	AISI E4340	Smls mech Tub; BEF	0.38-0.43	0.7-0.9	0.65-0.85	0.20-0.30	1.65-2.00	0.025 max	0.025 max	0.15-0.35	bal Fe				
	AISI E4340H	Smls mech tub, Hard; BEF	0.37-0.44	0.65-0.95	0.60-0.95	0.20-0.30	1.55-2.00	0.025 max	0.025 max	0.15-0.30	bal Fe				
	UNS G43406		0.38-0.43	0.70-0.90	0.65-0.85	0.20-0.30	1.65-2.00	0.025 max	0.025 max	0.15-0.35	bal Fe				
	UNS H43406		0.37-0.44	0.65-0.95	0.60-0.95	0.20-0.30	1.55-2.00	0.025 max	0.025 max	0.15-0.35	bal Fe				
AMS 6415											bal Fe				
ASTM A29/A29M(93)	E4340	Bar	0.38-0.43	0.70-0.90	0.65-0.85	0.2-0.3	1.65-2	0.025 max	0.025 max	0.15-0.35	bal Fe				
ASTM A304(96)	E4340H	Bar, hard bands spec	0.37-0.44	0.65-0.95	0.60-0.95	0.20-0.30	1.55-2.00	0.035 max	0.040 max	0.15-0.35	Cu <=0.35; Cu Ni Cr Mo trace allowed; bal Fe				
ASTM A322(96)	E4340	Bar	0.38-0.43	0.70-0.90	0.65-0.85	0.20-0.30	1.65-2.00	0.025 max	0.025 max	0.15-0.35	bal Fe				
ASTM A331(95)	4340H	Bar	0.37-0.44	0.65-0.95	0.55-0.90	0.20-0.30	1.55-2.00	0.025 max	0.025 max	0.15-0.35	Cu <=0.35; bal Fe				
ASTM A331(95)	E4340	Bar	0.38-0.43	0.70-0.90	0.65-0.85	0.20-0.30	1.55-2.00	0.025 max	0.025 max	0.15-0.35	bal Fe				
ASTM A331(95)	E4340H	Bar	0.37-0.44	0.65-0.95	0.60-0.95	0.20-0.30	1.55-2.00	0.025 max	0.025 max	0.15-0.35	Cu <=0.35; bal Fe				
ASTM A519(96)	E4340	Smls mech tub	0.38-0.43	0.70-0.90	0.65-0.85	0.20-0.30	1.65-2.00 *	0.025 max	0.025 max	0.15-0.35	bal Fe				
ASTM A752(93)	E4340	Rod Wir	0.38-0.43	0.70-0.90	0.65-0.85	0.20-0.30	1.65-2.00	0.025 max	0.025 max	0.15-0.30	bal Fe				
ASTM A829/A829M(95)	E4340	Plt	0.37-0.44	0.65-0.90	0.60-0.85	0.20-0.30	1.65-2.00	0.025 max	0.025 max	0.15-0.40	bal Fe	480-965			
SAE 770(84)	E4340*	Obs; see J1397(92)	0.38-0.43	0.7-0.9	0.65-0.85	0.2-0.3	1.65-2	0.03 max	0.02 max	0.15-0.3	bal Fe				
SAE J1268(95)	E4340H	Rod; See std	0.37-0.44	0.65-0.95	0.60-0.95	0.20-0.30	1.55-2.00	0.030 max	0.040 max	0.15-0.35	Cu <=0.35; Cu, Ni not spec'd but acpt; bal Fe				
SAE J404(94)	E4340	Bil Blm Slab Bar HR CF	0.37-0.44	0.65-0.90	0.60-0.85	0.20-0.30	1.65-2.00	0.025 max	0.025 max	0.15-0.35	bal Fe				
SAE J404(94)	E4340	Electric Furnace	0.38-0.43	0.70-0.90	0.65-0.85	0.20-0.30	1.65-2.00	0.025 max	0.025 max	0.15-0.35	bal Fe				
SAE J407	E4340H*	Obs; see J1268									bal Fe				

Alloy Steel, Nickel Chromium Molybdenum, E9310

Specification	Designation	Notes	C	Cr	Mn	Mo	Ni	P	S	Si	Other	UTS	YS	El	Hard
Bulgaria															
BDS 6354	15ChA	Struct	0.12-0.18	0.70-1.00	0.40-0.70	0.20-0.30	3.00-3.50	0.025 max	0.025 max	0.17-0.37	Cu <=0.30; Ti <=0.05; V <=0.1; W <=0.1; bal Fe				
BDS 6354(74)	15ChN3M	Struct	0.12-0.17	0.80-1.10	0.30-0.60	0.20-0.30	3.00-3.50	0.035 max	0.035 max	0.17-0.37	Cu <=0.30; Ti <=0.05; V <=0.1; W <=0.1; bal Fe				
Japan															
JIS G4103(79)	SNCM815	HR Frg Bar Rod	0.12-0.18	0.70-1.00	0.30-0.60	0.15-0.30	4.00-4.50	0.030 max	0.030 max	0.15-0.35	Cu <=0.30; bal Fe				
Mexico															
NMX-B-300(91)	E9310	Bar	0.08-0.13	1.00-1.40	0.45-0.65	0.08-0.15	3.0-3.5	0.035 max	0.040 max	0.15-0.35	bal Fe				
USA															
	AISI E9310	Smls mech Tub; BEF	0.08-0.13	1.00-1.40	0.45-0.65	0.08-0.15	3.00-3.50	0.025 max	0.025 max	0.15-0.35	bal Fe				
	UNS G93106		0.08-0.13	1.00-1.40	0.45-0.65	0.08-0.15	3.00-3.50	0.025 max	0.025 max	0.15-0.35	bal Fe				
ASTM A29/A29M(93)	E9310	Bar	0.08-0.13	1.00-1.40	0.45-0.65	0.08-0.15	3.00-3.50	0.025 max	0.025 max	0.15-0.30	bal Fe				

UNS numbers and US grades are provided as a means of cross referencing chemically similar alloys. Exchangability is only possible after independent examination of specifications. Tensile properties are minimum or typical as specified. UTS and YS as MPa. El as %. See Appendix for list of abbreviations used in Notes. * indicates obsolete material.

Specification	Designation	Notes	C	Cr	Mn	Mo	Ni	P	S	Si	Other	UTS	YS	El	Hard
Alloy Steel, Nickel Chromium Molybdenum, E9310 (Continued from previous page)															
USA															
ASTM·A519(96)	E9310	Smls mech tub	0.08-0.13	1.00-1.40	0.45-0.65	0.08-0.15	3.00-3.50	0.025 max	0.025 max	0.15-0.35	bal Fe				
ASTM A535(92)	9310	Ball & roller bearing	0.08-0.13	1.0-1.40	0.45-0.65	0.08-0.15	3.00-3.50	0.015 max	0.015 max	0.15-0.35	Cu <=0.35; bal Fe				
ASTM A646(95)	Grade 2	Blm Bil for aircraft aerospace frg, ann	0.08-0.13	1.00-1.40	0.45-0.65	0.08-0.15	3.00-3.50	0.025 max	0.025 max	0.20-0.35	bal Fe				262 HB
MIL-S-7393C(94)	9310	Carb Bar bil, aircraft qual									bal Fe				
MIL-S-83030A2(95)	3316*	Obs for new design; see AMS specs; Bar Bil premium qual									bal Fe				
MIL-S-83030A2(95)	9310*	Obs for new design; see AMS specs; Bar Bil premium qual									bal Fe				
Alloy Steel, Nickel Chromium Molybdenum, HY100															
Japan															
JIS G4103(79)	SNCM616	HR Frg Bar Rod	0.13-0.20	1.40-1.80	0.80-1.20	0.40-0.60	2.80-3.20	0.030 max	0.030 max	0.15-0.35	Cu <=0.30; bal Fe				
UK															
BS 970/1(96)	822M17	Blm Bil Slab Bar Rod Frg	0.14-0.20	1.30-1.70	0.40-0.70	0.15-0.25	1.75-2.25	0.035 max	0.040 max	0.10-0.35	Pb <=0.15; bal Fe	1310		8	
USA															
	UNS K32045	HY 100	0.20 max	1.00-1.80	0.10-0.40	0.20-0.60	2.25-3.50	0.25 max	0.25 max	0.15-0.35	bal Fe				
MIL-S-16216K(SH)(87)	HY-100	Plt >3 in.	0.14-0.20	1.50-1.90	0.10-0.40	0.50-0.65	3.00-3.50	0.015 max	0.008 max	0.15-0.38	Cu <=0.25; Ti <=0.02; V <=0.03; Sb 0.025 max; As 0.025 max; Sn 0.030 max; bal Fe	690	690-827	18	
MIL-S-16216K(SH)(87)	HY-100	Plt 1.25-3 in.	0.14-0.20	1.40-1.80	0.10-0.40	0.35-0.60	2.75-3.50	0.015 max	0.008 max	0.15-0.38	Cu <=0.25; Ti <=0.02; V <=0.03; Sb 0.025 max; As 0.025 max; Sn 0.030 max; bal Fe	690	690-827	18	
MIL-S-16216K(SH)(87)	HY-100	Plt <=1.25 in.	0.12-0.18	1.00-1.80	0.10-0.40	0.20-0.60	2.25-3.50	0.015 max	0.008 max	0.15-0.38	Cu <=0.25; Ti <=0.02; V <=0.03; Sb 0.025 max; As 0.025 max; Sn 0.030 max; bal Fe	690			
MIL-S-23009C(SH)(87)	HY-100	Frg struct apps	0.12-0.20	1.35-1.80	0.10-0.40	0.30-0.60	2.75-3.50	0.015 max	0.008 max	0.15-0.35	Cu <=0.25; Ti <=0.02; V <=0.03; Sb 0.025 max; As 0.025 max; Sn 0.030 max; bal Fe	690	690-793	18	
Alloy Steel, Nickel Chromium Molybdenum, K11541															
India															
IS 1570/4(88)	16Ni8Cr6Mo2		0.12-0.2	1.4-1.7	0.4-0.7	0.15-0.25	1.8-2.2	0.07 max	0.06 max	0.15-0.35	bal Fe	1340		9	
IS 1570/7(92)	5	Smls tube Bar Frg Sh Plt; high-temp	0.12-0.2	0.3 max	0.4-0.8	0.25-0.35	0.35 max	0.04 max	0.04 max	0.1-0.35	Al <=0.02; Almet<=0.02; bal Fe	440-590	250	17	
USA															
	UNS K11541		0.10-0.20	0.30 max	0.50-1.00	0.10-0.20	0.40-1.10	0.04 max	0.05 max	0.30 max	Cu 0.30-1.00; V 0.01-0.10; bal Fe				
Alloy Steel, Nickel Chromium Molybdenum, K12220															
Europe															
EN 10084(98)	1.7321	CH, Ann	0.17-0.23	0.30-0.60	0.70-1.00			0.035 max	0.035 max	0.40 max	bal Fe				207 HB
EN 10084(98)	1.7323	CH, Ann	0.17-0.23	0.30-0.60	0.70-1.00			0.035 max	0.020-0.040	0.40 max	bal Fe				207 HB
EN 10084(98)	20MoCr4	CH, Ann	0.17-0.23	0.30-0.60	0.70-1.00			0.035 max	0.035 max	0.40 max	bal Fe				207 HB
EN 10084(98)	20MoCrS4	CH, Ann	0.17-0.23	0.30-0.60	0.70-1.00			0.035 max	0.020-0.040	0.40 max	bal Fe				207 HB

UNS numbers and US grades are provided as a means of cross referencing chemically similar alloys. Exchangability is only possible after independent examination of specifications. Tensile properties are minimum or typical as specified. UTS and YS as MPa. El as %. See Appendix for list of abbreviations used in Notes. * indicates obsolete material.

Specification	Designation	Notes	C	Cr	Mn	Mo	Ni	P	S	Si	Other	UTS	YS	El	Hard

Alloy Steel, Nickel Chromium Molybdenum, K13020

USA

| | UNS K13020 | | 0.30 max | 0.25 max | 0.80 max | 0.40-0.60 | 0.25 max | 0.04 max | 0.04 max | 0.15-0.40 | Cu <=0.25; bal Fe | | | | |

Alloy Steel, Nickel Chromium Molybdenum, K13049

USA

| | UNS K13049 | | 0.26-0.34 | 0.40-0.65 | 0.70-1.00 | 0.15-0.25 | 0.40-0.70 | 0.035 max | 0.04 max | 0.15-0.35 | bal Fe | | | | |

Alloy Steel, Nickel Chromium Molybdenum, K13262

USA

| | UNS K13262 | | 0.32 max | 0.40 min | 1.00 max | 0.15 min | 0.70 min | 0.025 max | 0.025 max | 0.75 max | bal Fe | | | | |

Alloy Steel, Nickel Chromium Molybdenum, K13586

USA

| | UNS K13586 | | 0.35 max | 0.70 max | 0.60-0.90 | 0.20 min | 1.50-3.50 | 0.025 max | 0.025 max | 0.15-0.35 | V 0.03-0.12; bal Fe | | | | |

Alloy Steel, Nickel Chromium Molybdenum, K23028

India

| IS 1570/7(92) | 17 | 0-900mm; Frg Turbine Shaft | 0.2-0.3 | 1.5-1.8 | 0.3-0.6 | 0.9-1.2 | 0.4 max | 0.03 max | 0.025 max | 0.2-0.5 | V 0.2-0.3; bal Fe | 630 | 420 | 16 | |
| IS 1570/7(92) | 25Cr2MoVH | 0-900mm; Frg Turbine Shaft | 0.2-0.3 | 1.5-1.8 | 0.3-0.6 | 0.9-1.2 | 0.4 max | 0.03 max | 0.025 max | 0.2-0.5 | V 0.2-0.3; bal Fe | 630 | 420 | 16 | |

USA

| | UNS K23028 | | 0.30 max | 0.75 max | 0.70 max | 0.25 min | 2.00 min | 0.035 max | 0.035 max | 0.15-0.35 | V 0.03-0.12; bal Fe | | | | |

Alloy Steel, Nickel Chromium Molybdenum, K23545

USA

| | UNS K23545 | | 0.35 max | | 0.50-0.90 | 0.20-0.50 | 2.25-3.00 | 0.040 max | 0.040 max | 0.10-0.40 | V <=0.15; bal Fe | | | | |

Alloy Steel, Nickel Chromium Molybdenum, K23578

USA

| | UNS K23578 | | 0.35 max | 0.75 max | 0.70 max | 0.25 min | 2.50 min | 0.035 max | 0.035 max | 0.15-0.35 | V 0.03-0.12; bal Fe | | | | |

Alloy Steel, Nickel Chromium Molybdenum, K23579

USA

| | UNS K23579 | | 0.35 max | 1.25 max | 0.70 max | 0.25 min | 2.50 min | 0.035 max | 0.035 max | 0.15-0.35 | V >=0.03; bal Fe | | | | |

Alloy Steel, Nickel Chromium Molybdenum, K23705

Italy

UNI 7874(79)	21CrMoV511	High-temp const	0.18-0.25	0.9-1.35	0.4-0.8	0.9-1.2	0.7 max	0.035 max	0.035 max	0.15-0.4	V 0.28-0.35; bal Fe				
UNI 7874(79)	24CrMoV55	High-temp const	0.2-0.28	0.9-1.35	0.4-0.8	0.4-0.7	0.6 max	0.035 max	0.035 max	0.15-0.4	V 0.28-0.35; bal Fe				
UNI 7874(79)	30CrMoV511	High-temp const	0.27-0.33	0.9-1.35	0.4-0.8	0.9-1.2	0.7 max	0.035 max	0.035 max	0.15-0.4	V 0.2-0.35; bal Fe				

USA

| | UNS K23705 | | 0.37 max | 0.85-1.25 | 1.00 max | 1.00-1.50 | 0.50 max | 0.035 max | 0.035 max | 0.15-0.35 | V 0.20-0.30; bal Fe | | | | |

Alloy Steel, Nickel Chromium Molybdenum, K24535

USA

| | UNS K24535 | | 0.45 max | 0.50-1.25 | 0.60-1.00 | 0.20 min | 1.65-3.50 | 0.025 max | 0.025 max | 0.15-0.35 | bal Fe | | | | |

Alloy Steel, Nickel Chromium Molybdenum, K33585

Japan

| JIS G4103(79) | SNCM630 | HR Frg Bar Rod | 0.25-0.35 | 2.50-3.50 | 0.35-0.60 | 0.50-0.70 | 2.50-3.50 | 0.030 max | 0.030 max | 0.15-0.35 | Cu <=0.30; bal Fe | | | | |

UNS numbers and US grades are provided as a means of cross referencing chemically similar alloys. Exchangability is only possible after independent examination of specifications. Tensile properties are minimum or typical as specified. UTS and YS as MPa. El as %. See Appendix for list of abbreviations used in Notes. * indicates obsolete material.

Specification	Designation	Notes	C	Cr	Mn	Mo	Ni	P	S	Si	Other	UTS	YS	El	Hard

Alloy Steel, Nickel Chromium Molybdenum, K33585 (Continued from previous page)

Sweden

Specification	Designation	Notes	C	Cr	Mn	Mo	Ni	P	S	Si	Other	UTS	YS	El	Hard
SS 142240	2240	Q/T	0.28-0.35	2.8-3.3	0.4-0.7	0.4-0.6	0.3 max	0.035 max	0.035 max	0.15-0.4	bal Fe				

USA

Specification	Designation	Notes	C	Cr	Mn	Mo	Ni	P	S	Si	Other	UTS	YS	El	Hard
	UNS K33585		0.35 max	3.00-3.60	0.50-0.90	0.30-0.50	0.50-1.00	0.040 max	0.040 max	0.15-0.45	V 0.05-0.15; bal Fe				

Alloy Steel, Nickel Chromium Molybdenum, K42338

USA

Specification	Designation	Notes	C	Cr	Mn	Mo	Ni	P	S	Si	Other	UTS	YS	El	Hard
	UNS K42338		0.23 max	1.44-2.06	0.40 max	0.41-0.64	2.53-3.32	0.035 max	0.040 max	0.18-0.37	V <=0.03; bal Fe				

Alloy Steel, Nickel Chromium Molybdenum, K42570

USA

Specification	Designation	Notes	C	Cr	Mn	Mo	Ni	P	S	Si	Other	UTS	YS	El	Hard
	UNS K42570		0.23-0.27	1.25-1.75	0.70-1.00	0.20-0.30	3.25-3.75	0.020 max	0.020 max	0.20-0.35	bal Fe				

Alloy Steel, Nickel Chromium Molybdenum, K91401

USA

Specification	Designation	Notes	C	Cr	Mn	Mo	Ni	P	S	Si	Other	UTS	YS	El	Hard
	UNS K91401		0.16-0.23	0.61-0.89	0.20-0.40	0.86-1.14	8.40-9.60	0.010 max	0.010 max	0.12 max	Co 4.15-5.10; V 0.04-0.14; bal Fe				

Alloy Steel, Nickel Chromium Molybdenum, K91890

USA

Specification	Designation	Notes	C	Cr	Mn	Mo	Ni	P	S	Si	Other	UTS	YS	El	Hard
	UNS K91890		0.03 max	4.50-5.50	0.10 max	2.75-3.25	11.50-12.50	0.010 max	0.010 max	0.10 max	Al <=0.40; Ti 0.20-0.35; bal Fe				

Alloy Steel, Nickel Chromium Molybdenum, MIL-A-46173

Italy

Specification	Designation	Notes	C	Cr	Mn	Mo	Ni	P	S	Si	Other	UTS	YS	El	Hard
UNI 7845(78)	30NiCrMo12	Q/T	0.28-0.35	0.6-1	0.5-0.8	0.3-0.5	2.6-3.2	0.035 max	0.035 max	0.15-0.4	bal Fe				
UNI 7874(79)	30NiCrMo12	Q/T	0.27-0.34	0.6-1	0.5-0.8	0.3-0.6	2.6-3.2	0.035 max	0.035 max	0.15-0.4	V <=0.15; bal Fe				

UK

Specification	Designation	Notes	C	Cr	Mn	Mo	Ni	P	S	Si	Other	UTS	YS	El	Hard
BS 970/3(91)	826M31	Bright bar; Q/T; 100<t<=150mm; hard	0.27-0.35	0.50-0.80	0.45-0.70	0.45-0.65	2.30-2.80	0.035 max	0.040 max	0.10-0.40	bal Fe	850-1000	680	13	248-302 HB
BS 970/3(91)	826M31	Bright bar; Q/T; 6<=t<=63mm; hard CD	0.27-0.35	0.50-0.80	0.45-0.70	0.45-0.65	2.30-2.80	0.035 max	0.040 max	0.10-0.40	bal Fe	1150-1300	1035	7	341-401 HB

USA

Specification	Designation	Notes	C	Cr	Mn	Mo	Ni	P	S	Si	Other	UTS	YS	El	Hard
	UNS K33370		0.31-0.35	0.65-0.90	0.60-0.90	0.30-0.50	2.75 min	0.005 max	0.005 max	0.20-0.30	O 25ppm max; bal Fe				

Alloy Steel, Nickel Chromium Molybdenum, MIL-S-12504(MnCrMo)

USA

Specification	Designation	Notes	C	Cr	Mn	Mo	Ni	P	S	Si	Other	UTS	YS	El	Hard
	UNS K15747		0.52-0.62	0.80-1.15	0.75-1.05	0.15-0.25		0.04 max	0.04 max	0.20-0.35	bal Fe				
MIL-B-12504E(90)	Mn-Cr-Mo	CD Bar wire for bullets	0.52-0.62	0.80-1.15	0.75-1.05	0.15-0.25		0.04 max	0.04 max	0.20-0.35	bal Fe				24-26 HRC

Alloy Steel, Nickel Chromium Molybdenum, MIL-S-980

USA

Specification	Designation	Notes	C	Cr	Mn	Mo	Ni	P	S	Si	Other	UTS	YS	El	Hard
	UNS K19964	Bearing	0.95-1.10	1.30-1.60	0.25-0.45	0.08 max	0.35 max	0.025 max	0.025 max	0.20-0.35	Cu <=0.25; bal Fe				

Alloy Steel, Nickel Chromium Molybdenum, No equivalents identified

Czech Republic

Specification	Designation	Notes	C	Cr	Mn	Mo	Ni	P	S	Si	Other	UTS	YS	El	Hard
CSN 415421	15421	High-temp const; Frg	0.18-0.28	2.7-3.5	0.5-0.9	0.25-0.6	0.8 max	0.04 max	0.04 max	0.15-0.4	bal Fe				
CSN 416224	16224		0.12-0.18	0.7-1.1	0.7-1.0	0.3-0.5	0.7-1.0	0.03 max	0.03 max	0.12-0.35	B 0.002-0.005; Ti <=0.04; V 0.03-0.08; bal Fe				
CSN 416310	16310		0.18-0.25	1.4-2.0	0.4-0.85		0.6-0.8	0.8-1.2	0.03 max	0.03 max	0.2-0.4	V 0.03-0.08; bal Fe				
CSN 416431	16431		0.25-0.32	1.2-1.6	0.4-0.6	0.35-0.45	1.8-2.3	0.035 max	0.035 max	0.37 max	Cu <=0.2; V 0.05; bal Fe				
CSN 416444	16444	High-temp const	0.32-0.4	1.7-2.0	0.4-0.6	0.15-0.25	1.5-1.8	0.035 max	0.035 max	0.15-0.4	V 0.1-0.2; bal Fe				

UNS numbers and US grades are provided as a means of cross referencing chemically similar alloys. Exchangability is only possible after independent examination of specifications. Tensile properties are minimum or typical as specified. UTS and YS as MPa. El as %. See Appendix for list of abbreviations used in Notes. * indicates obsolete material.

Specification	Designation	Notes	C	Cr	Mn	Mo	Ni	P	S	Si	Other	UTS	YS	El	Hard

Alloy Steel, Nickel Chromium Molybdenum, No equivalents identified

Czech Republic

Specification	Designation	Notes	C	Cr	Mn	Mo	Ni	P	S	Si	Other	UTS	YS	El	Hard
CSN 416540	16540	High-temp const	0.3-0.4	0.7-1.1	0.5-0.8	0.25-0.4	2.75-3.75	0.035 max	0.035 max	0.15-0.4	bal Fe				

France

Specification	Designation	Notes	C	Cr	Mn	Mo	Ni	P	S	Si	Other	UTS	YS	El	Hard
AFNOR NFA35551(86)	10NC6	CH, 16t<=40; Bar Wir rod	0.07-0.12	0.9-1.2	0.6-0.9	0.15 max	1.2-1.5	0.035 max	0.035 max	0.1-0.4	Cu <=0.3; Ti <=0.05; bal Fe	700-1000	500	11	
AFNOR NFA35551(86)	14NC11	Bar Rod; CH; t<=16mm	0.11-0.17	0.6-0.9	0.25-0.6	0.15 max	2.5-3	0.035 max	0.035 max	0.1-0.4	Cu <=0.3; Ti <=0.05; bal Fe	1100-1400	800	9	
AFNOR NFA35551(86)	14NC11	Bar Rod; CH; 40<t<=100mm	0.11-0.17	0.6-0.9	0.25-0.6	0.15 max	2.5-3	0.035 max	0.035 max	0.1-0.4	Cu <=0.3; Ti <=0.05; bal Fe	650-950	470	11	
AFNOR NFA35551(86)	20NC6	CH; 16<t<=40mm	0.16-0.21	0.9-1.2	0.6-0.9	0.15 max	1.2-1.5	0.035 max	0.035 max	0.1-0.4	Cu <=0.3; Ti <=0.05; bal Fe	950-1300	700	9	
AFNOR NFA35552(84)	35NCD16	t<=100mm	0.32-0.39	1.6-2	0.3-0.6	0.25-0.45	3.6-4.1	0.03 max	0.025 max	0.1-0.4	Cu <=0.3; Ti <=0.05; bal Fe	1080-1280	880	10	
AFNOR NFA35553(82)	10NC6	CH; Strp, CR	0.07-0.12	0.8-1.1	0.6-0.9	0.15 max	1.2-1.6	0.035 max	0.035 max	0.1-0.4	Cu <=0.3; Ti <=0.05; bal Fe				
AFNOR NFA35553(82)	14NC11		0.11-0.17	0.6-0.9	0.35-0.6	0.15 max	2.5-3	0.035 max	0.035 max	0.1-0.4	Cu <=0.3; Ti <=0.05; bal Fe				
AFNOR NFA35556(84)	30NC11	t<=16mm	0.27-0.34	0.6-0.9	0.35-0.6		2.5-3	0.035 max	0.035 max	0.1-0.4	Cu <=0.3; Ti <=0.05; bal Fe	1030-1230	850	11	
AFNOR NFA35556(84)	35NCD16	t<=100mm	0.32-0.39	1.6-2	0.3-0.6	0.25-0.45	3.6-4.2	0.03 max	0.025 max	0.1-0.4	Cu <=0.3; Ti <=0.05; bal Fe	1180-1380	980	8	
AFNOR NFA35564(83)	10NC6FF	Bar, Wir rod, CF, soft ann	0.07-0.12	0.9-1.2	0.6-0.9	0.15 max	1.2-1.5	0.03 max	0.03 max	0.35 max	Al >=0.02; Ti <=0.05; bal Fe	0-500			192 HB max
AFNOR NFA35564(83)	16NC6FF		0.12-0.17	0.9-1.2	0.6-0.9	0.15 max	1.2-1.5	0.03 max	0.03 max	0.35 max	Al >=0.02; Cu <=0.03; Ti <=0.05; bal Fe				156-207 HB
AFNOR NFA35564(83)	19NCDB2FF		0.17-0.23	0.4-0.65	0.65-0.95	0.15-0.25	0.4-0.7	0.03 max	0.03 max	0.35 max	Al >=0.02; B 0.0008-0.005; Ti <=0.05; bal Fe	0-530			201 HB max
AFNOR NFA35564(83)	20MB5FF	Bar Wir rod CF	0.16-0.22	0.3 max	1.1-1.4	0.15 max	0.4 max	0.03 max	0.03 max	0.35 max	Al >=0.02; B 0.0008-0.005; Ti <=0.05; bal Fe	0-510			174 HB max
AFNOR NFA35564(83)	20MC5FF	Bar Wir rod CF	0.17-0.22	1-1.3	1.1-1.4	0.15 max	0.4 max	0.03 max	0.03 max	0.35 max	Al >=0.02; Ti <=0.05; bal Fe	0-550			207 HB max
AFNOR NFA35564(83)	20NC6FF		0.16-0.21	0.9-1.2	0.6-0.9	0.15 max	1.2-1.5	0.03 max	0.03 max	0.35 max	Al >=0.02; Ti <=0.05; bal Fe	0-550			207 HB max
AFNOR NFA35564(83)	20NCD2FF		0.17-0.23	0.4-0.65	0.65-0.95	0.15-0.25	0.4-0.7	0.03 max	0.03 max	0.35 max	Al >=0.02; Ti <=0.05; bal Fe	0-530			201 HB max
AFNOR NFA35564(83)	21B3FF	Soft ann	0.18-0.24	0.3 max	0.6-0.9	0.15 max	0.4 max	0.03 max	0.03 max	0.35 max	Al >=0.02; B 0.0008-0.005; Ti <=0.05; bal Fe	0-470			149 HB max
AFNOR NFA35565(94)	20MnCrNi4	Ball & roller bearing; CH	0.17-0.23	0.45-0.75	1.05-1.4	0.2 max	0.2-0.45	0.025 max	0.015 max	0.15-0.35	Al <=0.05; Cu <=0.3; Ti <=0.05; bal Fe				
AFNOR NFA36208(82)	5Ni390	t<=30mm; Norm	0.12 max	0.25 max	0.3-0.8	0.1 max	4.75-5.25	0.025 max	0.02 max	0.35 max	Al >=0.015; Ti <=0.05; V <=0.4; Cr+Cu+Mo<=0.5; bal Fe	590-720	390	23	

Germany

Specification	Designation	Notes	C	Cr	Mn	Mo	Ni	P	S	Si	Other	UTS	YS	El	Hard
DIN E 17201(89)	33NiCrMo14-5	Frg Q/T 41-100mm	0.28-0.36	1.00-1.70	0.15-0.40	0.30-0.60	3.20-4.00	0.035 max	0.035 max	0.40 max	V <=0.15; bal Fe	1270-1470	1030	7	
DIN E 17201(89)	WNr 1.6956	Frg Q/T 41-100mm	0.28-0.36	1.00-1.70	0.15-0.40	0.30-0.60	3.20-4.00	0.035 max	0.035 max	0.40 max	V <=0.15; bal Fe	1270-1470	1030	7	

Hungary

Specification	Designation	Notes	C	Cr	Mn	Mo	Ni	P	S	Si	Other	UTS	YS	El	Hard
MSZ 1741(89)	KL10	Broiler, Press ves; 0-20mm; HT	0.09-0.17	0.5-0.75	0.45-0.75	0.4-0.6	0.35 max	0.035 max	0.03 max	0.15-0.4	Al 0.02-0.06; Cu <=0.35; Nb 0.01-0.06; Ti 0.02-0.06; V 0.02-0.2; W 0.35-0.1; Zr=0.015-0.06; Al+Nb+V+Ti+Zr<=0.15; N<=Al/2+V/4+Nb/7+Ti/3.5<=0.015; bal Fe	470-660	300	16L	

Specification	Designation	Notes	C	Cr	Mn	Mo	Ni	P	S	Si	Other	UTS	YS	El	Hard

Alloy Steel, Nickel Chromium Molybdenum, No equivalents identified

Hungary

Specification	Designation	Notes	C	Cr	Mn	Mo	Ni	P	S	Si	Other	UTS	YS	El	Hard
MSZ 1741(89)	˙KL10	Broiler, Press ves; 40.1-60mm; HT	0.09-0.17	0.5-0.75	0.45-0.75	0.4-0.6	0.35 max	0.035 max	0.03 max	0.15-0.4	Al 0.02-0.06; Cu <=0.35; Nb 0.01-0.06; Ti 0.02-0.06; V 0.02-0.2; W 0.35-0.1; Zr=0.015-0.06; Al+Nb+V+Ti+Zr<=0.15; N<=Al/2+V/4+Nb/7+Ti/3.5<=0.015; bal Fe	470-660	280	16L	
MSZ 1741(89)	KL9	Boiler, Press ves; 0-20mm; HT	0.1-0.18	0.7-1	0.4-0.9	0.45-0.55	0.35 max	0.035 max	0.03 max	0.15-0.4	Al 0.02-0.06; Cu <=0.35; Nb 0.01-0.06; Ti 0.02-0.06; V 0.02-0.1; Zr=0.015-0.06; Al+Nb+V+Ti+Zr<=0.15; N<=Al/2+V/4+Nb/7+Ti/3.5<=0.015; bal Fe	440-580	290	20L	
MSZ 1741(89)	KL9	Boiler, Press ves; 40.1-60mm; HT	0.1-0.18	0.7-1	0.4-0.9	0.45-0.55	0.35 max	0.035 max	0.03 max	0.15-0.4	Al 0.02-0.06; Cu <=0.35; Nb 0.01-0.06; Ti 0.02-0.06; V 0.02-0.1; Zr=0.015-0.06; Al+Nb+V+Ti+Zr<=0.15; N<=Al/2+V/4+Nb/7+Ti/3.5<=0.015; bal Fe	440-580	270	20L	
MSZ 1745(79)	MCrMo	High-temp const; 0-80mm; Q/T	0.2-0.28	0.9-1.2	0.5-0.8	0.2-0.3	0.3-0.6	0.035 max	0.035 max	0.17-0.37	bal Fe	590-740	440	16L	
MSZ 1745(79)	MCrMo	High-temp const; 80.1-160mm; Q/T	0.2-0.28	0.9-1.2	0.5-0.8	0.2-0.3	0.3-0.6	0.035 max	0.035 max	0.17-0.37	bal Fe	590-740	440	16L	
MSZ 1745(79)	MCrMoNiV	High-temp const; 0-80mm; Q/T	0.17-0.25	11.0-13.0	0.5-0.8	0.8-1	0.5-0.8	0.035 max	0.035 max	0.17-0.37	V 0.3-0.45; bal Fe	800-950	590	12L	
MSZ 1745(79)	MCrMoNiV	High-temp const; 80.1-160mm; Q/T	0.17-0.25	11.0-13.0	0.5-0.8	0.8-1	0.5-0.8	0.035 max	0.035 max	0.17-0.37	V 0.3-0.45; bal Fe	800-950	590	12L	
MSZ 1745(79)	MCrMoV	High-temp const; 80.1-160mm; Q/T	0.17-0.25	1.2-1.5	0.5-0.8	0.8-1	0.3-0.6	0.035 max	0.035 max	0.17-0.37	V 0.3-0.45; bal Fe	690-830	540	15L	
MSZ 1745(79)	MCrMoV	High-temp const; 0-80mm; Q/T	0.17-0.25	1.2-1.5	0.5-0.8	0.8-1	0.3-0.6	0.035 max	0.035 max	0.17-0.37	V 0.3-0.45; bal Fe	690-830	540	15L	
MSZ 31(85)	BNCMo2	CH; 0-11mm; blank carburized	0.12-0.17	0.8-1.1	0.3-0.6	0.2-0.3	3-3.5	0.035 max	0.035 max	0.4 max	bal Fe	1150-1500	790	8L	
MSZ 31(85)	BNCMo2	CH; 63 mm; blank carburized	0.12-0.17	0.8-1.1	0.3-0.6	0.2-0.3	3-3.5	0.035 max	0.035 max	0.4 max	bal Fe	950-1250	640	9L	
MSZ 31(85)	BNCMo2E	CH; 63 mm; blank carburized	0.12-0.17	0.8-1.1	0.3-0.6	0.2-0.3	3-3.5	0.035 max	0.02-0.035	0.4 max	bal Fe	950-1250	640	9L	
MSZ 31(85)	BNCMo2E	CH; 0-11mm; blank carburized	0.12-0.17	0.8-1.1	0.3-0.6	0.2-0.3	3-3.5	0.035 max	0.02-0.035	0.4 max	bal Fe	1150-1500	790	8L	
MSZ 4747(85)	12Cr10MoVNi70.47	High-temp, Tub; Non-rimming; K; FF; 0-10mm; Q/T	0.17-0.23	10-12.5	1 max	0.8-1.2	0.3-0.8	0.03 max	0.03 max	0.5 max	V 0.25-0.35; bal Fe	690-840	490	17L; 14T	
MSZ 4747(85)	12Cr10MoVNi70.47	High-temp, Tub; Non-rimming; K; FF; 10.1-60mm; Q/T	0.17-0.23	10-12.5	1 max	0.8-1.2	0.3-0.8	0.03 max	0.03 max	0.5 max	V 0.25-0.35; bal Fe	690-840	490	17L; 14T	
MSZ 61(85)	NCMo6	Q/T; 160.1-250mm	0.26-0.34	1.8-2.2	0.3-0.6	0.3-0.5	1.8-2.2	0.035 max	0.035 max	0.4 max	bal Fe	900-1100	700	12L	
MSZ 61(85)	NCMo6	Q/T; 0-16mm	0.26-0.34	1.8-2.2	0.3-0.6	0.3-0.5	1.8-2.2	0.035 max	0.035 max	0.4 max	bal Fe	1250-1450	1050	9L	
MSZ 61(85)	NCMoV	Q/T; 0-16mm	0.33-0.4	1.2-1.5	0.25-0.5	0.3-0.5	3-3.5	0.035 max	0.035 max	0.4 max	V 0.1-0.2; bal Fe	1250-1450	1050	9L	
MSZ 61(85)	NCMoV	Q/T; 160.1-250mm	0.33-0.4	1.2-1.5	0.25-0.5	0.3-0.5	3-3.5	0.035 max	0.035 max	0.4 max	V 0.1-0.2; bal Fe	1000-1150	800	11L	
MSZ 6251(87)	NCMo6Z	CF; Q/T; 0-36mm; HR; HR/soft ann; drawn/soft ann; drawn/bright ann; soft ann/ground	0.26-0.34	1.8-2.2	0.3-0.6	0.3-0.5	1.8-2.2	0.035 max	0.035 max	0.4 max	bal Fe	0-700			
MSZ 6251(87)	NCMo6Z	CF; Q/T; 0-36mm; drawn, half-hard	0.26-0.34	1.8-2.2	0.3-0.6	0.3-0.5	1.8-2.2	0.035 max	0.035 max	0.4 max	bal Fe	0-730			
MSZ 6259(82)	LK52B	Struct; Weather res; 40.1-100mm; rolled/norm	0.15 max	0.5-1	1.3 max	0.3 max	0.3-0.6	0.04 max	0.04 max	0.15-0.5	Cu 0.2-0.5; 0-16mm P<=0.15; bal Fe	490-630	325	23L; 21T	

Specification	Designation	Notes	C	Cr	Mn	Mo	Ni	P	S	Si	Other	UTS	YS	El	Hard

Alloy Steel, Nickel Chromium Molybdenum, No equivalents identified

Hungary

Specification	Designation	Notes	C	Cr	Mn	Mo	Ni	P	S	Si	Other	UTS	YS	El	Hard
MSZ 6259(82)	LK52B	Struct; Weather res; 0-16mm; rolled/norm	0.15 max	0.5-1	1.3 max	0.3 max	0.3-0.6	0.04 max	0.04 max	0.15-0.5	Cu 0.2-0.5; 0-16mm P<=0.15; bal Fe	490-630	355	23L; 21T	
MSZ 6259(82)	LK52C	Struct; Weather res; 50.1-100mm; rolled/norm	0.15 max	0.5-1	1.3 max	0.3 max	0.3-0.6	0.04 max	0.04 max	0.15-0.5	Al 0.015-0.120; Cu 0.2-0.5; Nb 0.01-0.06; Ti <=0.06; V 0.02-0.1; 0-16mm P<=0.15; Ti=0.02-0.06; Zr=0.01-0.06; Al+Nb+V+Ti+Zr<=0.15; bal Fe	490-630	325	23L; 21T	
MSZ 6259(82)	LK52C	Struct; Weather res; 0-16mm; rolled/norm	0.15 max	0.5-1	1.3 max	0.3 max	0.3-0.6	0.04 max	0.04 max	0.15-0.5	Al 0.015-0.120; Cu 0.2-0.5; Nb 0.01-0.06; Ti <=0.06; V 0.02-0.1; 0-16mm P<=0.15; Ti=0.02-0.06; Zr=0.01-0.06; Al+Nb+V+Ti+Zr<=0.15; bal Fe	490-630	355	23L; 21T	
MSZ 6259(82)	LK52D	Struct; Weather res; 50.1-100mm; rolled/norm	0.15 max	0.5-1	1.3 max	0.3 max	0.3-0.6	0.04 max	0.04 max	0.15-0.5	Al 0.015-0.120; Cu 0.2-0.5; Nb 0.01-0.06; Ti <=0.06; V 0.02-0.1; 0-16mm P<=0.15; Ti=0.02-0.06; Zr=0.01-0.06; Al+Nb+V+Ti+Zr<=0.15; bal Fe	490-630	325	23L; 21T	
MSZ 6259(82)	LK52D	Struct; Weather res; 0-16mm; rolled/norm	0.15 max	0.5-1	1.3 max	0.3 max	0.3-0.6	0.04 max	0.04 max	0.15-0.5	Al 0.015-0.120; Cu 0.2-0.5; Nb 0.01-0.06; Ti <=0.06; V 0.02-0.1; 0-16mm P<=0.15; Ti=0.02-0.06; Zr=0.01-0.06; Al+Nb+V+Ti+Zr<=0.15; bal Fe	490-630	355	23L; 21T	

India

Specification	Designation	Notes	C	Cr	Mn	Mo	Ni	P	S	Si	Other	UTS	YS	El	Hard
IS 1570/4(88)	13Ni13Cr3		0.1-0.15	3-3.5	0.4-0.7	0.15 max	0.6-1	0.07 max	0.06 max	0.15-0.35	bal Fe	840		12	
IS 1570/4(88)	31Ni10Cr3Mo6	Bar Frg Tub Pipe	0.27-0.35	0.5-0.8	0.4-0.7	0.4-0.7	2.25-2.75	0.07 max	0.06 max	0.1-0.35	bal Fe	1190	990		
IS 1570/4(88)	40Ni10Cr3Mo6	0-100mm; Bar Frg	0.36-0.44	0.5-0.8	0.4-0.7	0.4-0.7	2.25-2.75	0.07 max	0.06 max	0.1-0.35	bal Fe	1540	1240	8	444 HB
IS 1570/4(88)	40Ni6Cr4Mo2	0-30mm; Bar Frg	0.35-0.45	0.9-1.3	0.4-0.7	0.1-0.2	1.2-1.6	0.07 max	0.06 max	0.1-0.35	bal Fe	1090-1240	830	11	311-363 HB
IS 1570/4(88)	40Ni6Cr4Mo2	100.1-150mm; Bar Frg	0.35-0.45	0.9-1.3	0.4-0.7	0.1-0.2	1.2-1.6	0.07 max	0.06 max	0.1-0.35	bal Fe	790-940	550	16	229-277 HB
IS 1570/4(88)	40Ni6Cr4Mo3	0-30mm; Bar Frg	0.35-0.45	0.9-1.3	0.4-0.7	0.2-0.35	1.25-1.75	0.07 max	0.06 max	0.1-0.35	bal Fe	1540	1240	6	444 HB
IS 1570/4(88)	40Ni6Cr4Mo3	0-100mm; Bar Frg	0.35-0.45	0.9-1.3	0.4-0.7	0.2-0.35	1.25-1.75	0.07 max	0.06 max	0.1-0.35	bal Fe	990-1140	750	13	285-341 HB

International

Specification	Designation	Notes	C	Cr	Mn	Mo	Ni	P	S	Si	Other	UTS	YS	El	Hard
ISO 4954(93)	20NiCrMo2E	Wir Rod Bar, CH ext	0.17-0.23	0.30-0.65	0.65-0.95	0.15-0.25	0.40-0.70	0.035 max	0.035 max	0.40 max	bal Fe	590			
ISO 4954(93)	31NiCrMo8E	Wir Rod Bar, CH ext	0.27-0.34	1.80-2.20	0.30-0.60	0.30-0.50	1.80-2.20	0.035 max	0.035 max	0.40 max	bal Fe	700			
ISO 4954(93)	41CrNiMo2E	Wir Rod Bar, CH ext	0.37-0.44	0.40-0.60	0.70-1.00	0.15-0.30	0.40-0.70	0.035 max	0.035 max	0.40 max	bal Fe	650			
ISO 4954(93)	41NiCrMo7E	Wir Rod Bar, CH ext	0.37-0.44	0.65-0.95	0.55-0.85	0.15-0.30	1.60-2.00	0.035 max	0.035 max	0.40 max	bal Fe	680			
ISO 683-1(87)	31CrNiMo8	Bar Frg Plt Wir Rod Bolt Slab; Q/T, t<16mm	0.27-0.34	1.80-2.20	0.30-0.60	0.30-0.50	1.80-2.20	0.035 max	0.035 max	0.10-0.40	bal Fe	1030-1230	850		
ISO 683-1(87)	36CrNiMo4	Bar Frg Plt Wir Rod Bolt Slab; Q/T, t<16mm	0.32-0.40	0.90-1.20	0.50-0.80	0.15-0.30	0.90-1.20	0.035 max	0.035 max	0.10-0.40	bal Fe	1100-1300	900		
ISO 683-1(87)	36CrNiMo6	Bar Frg Plt Wir Rod Bolt Slab; Q/T, t<16mm	0.32-0.39	1.30-1.70	0.50-0.80	0.15-0.30	1.30-1.70	0.035 max	0.035 max	0.10-0.40	bal Fe	1200-1400	1000		
ISO 683-1(87)	41CrNiMo4	Bar Frg Plt Wir Rod Bolt Slab; Q/T, t<16mm	0.37-0.41	0.40-0.60	0.70-1.00	0.15-0.30	0.40-0.70	0.035 max	0.035 max	0.10-0.40	bal Fe	1000-1200	840		

Specification	Designation	Notes	C	Cr	Mn	Mo	Ni	P	S	Si	Other	UTS	YS	El	Hard

Alloy Steel, Nickel Chromium Molybdenum, No equivalents identified

International

Specification	Designation	Notes	C	Cr	Mn	Mo	Ni	P	S	Si	Other	UTS	YS	El	Hard
ISO 683-1(87)	41CrNiMoS4	Bar Frg Plt Wir Rod Bolt Slab; Q/T, t<16mm	0.37-0.41	0.40-0.60	0.70-1.00	0.15-0.30	0.40-0.70	0.035 max	0.020-0.040	0.10-0.40	bal Fe	1000-1200	840		
ISO 683-11(87)	17NiCrMo6	CH, 16mm test	0.14-0.20	0.80-1.10	0.60-0.90	0.15-0.25	1.20-1.60	0.035 max	0.035 max	0.15-0.40	bal Fe	1030-1380	700		
ISO 683-11(87)	18CrNiMo7	CH, 16mm test	0.15-0.21	1.50-1.80	0.35-0.65	0.25-0.35	1.40-1.70	0.035 max	0.035 max	0.15-0.40	bal Fe	1130-1480	820		
ISO 683-18(96)	18CrNi Mo7	Flat, CH, t<=16mm	0.15-0.21	1.50-1.80	0.35-0.65	0.25-0.35	1.40-1.70	0.035 max	0.035 max	0.15-0.40	bal Fe	900 max			229 HB max
ISO 683-18(96)	20NiCrMo2	Flat, CH, t<=16mm	0.17-0.23	0.30-0.65	0.65-0.95	0.15-0.25	0.40-0.70	0.035 max	0.035 max	0.15-0.40	bal Fe	820 max			197 HB max
ISO 683-18(96)	20NiCrMoS2	Flat, CH, t<=16mm	0.17-0.23	0.30-0.65	0.65-0.95	0.15-0.25	0.40-0.70	0.035 max	0.020-0.040	0.15-0.40	bal Fe	820 max			197 HB max
ISO 683-18(96)	31CrNiMo8	Flat, CD, Ann, t<=16mm	0.27-0.34	1.80-2.20	0.30-0.90	0.30-0.50	1.80-2.20	0.035 max	0.035 max	0.10-0.40	bal Fe	1000 max			248 HB max
ISO 683-18(96)	36CrNiMo4	Flat, CD, Ann, t<=16mm	0.32-0.40	0.90-1.20	0.50-0.80	0.15-0.30	0.90-1.20	0.035 max	0.035 max	0.10-0.40	bal Fe	1000 max			248 HB max
ISO 683-18(96)	36CrNiMo6	Flat, CD, Ann, t<=16mm	0.32-0.39	1.30-1.70	0.50-0.80	0.15-0.30	1.30-1.70	0.035 max	0.035 max	0.10-0.40	bal Fe	1000 max			248 HB max
ISO R683-8(70)	5	Q/T	0.26-0.33	1.8-2.2	0.3-0.6	0.3-0.5	1.8-2.2	0.035 max	0.035 max	0.15-0.4	Al <=0.1; Co <=0.1; Cu <=0.3; Pb <=0.15; Ti <=0.05; V <=0.1; W <=0.1; bal Fe				

Italy

Specification	Designation	Notes	C	Cr	Mn	Mo	Ni	P	S	Si	Other	UTS	YS	El	Hard
UNI 6786(71)	31NiCrMo12	axles	0.27-0.35	0.6-1	0.4-0.7	0.3-0.6	2.6-3.2	0.025 max	0.025 max	0.4 max	Cu <=0.2; bal Fe				
UNI 6932(71)	15NiCrMo13	Aircraft material	0.12-0.17	0.8-1.1	0.3-0.6	0.2-0.3	3-3.5	0.03 max	0.025 max	0.15-0.4	bal Fe				
UNI 7874(79)	30NiCrMo8	Q/T	0.26-0.33	1.8-2.2	0.3-0.6	0.3-0.5	1.8-2.2	0.035 max	0.035 max	0.15-0.4	V <=0.15; bal Fe				

Norway

Specification	Designation	Notes	C	Cr	Mn	Mo	Ni	P	S	Si	Other	UTS	YS	El	Hard
NS 13101	13136	Struct const cement	0.14-0.19	1.50-1.80	0.40-0.60	0.25-0.35	1.40-1.70	0.035 max	0.035 max	0.15-0.40	bal Fe				

Poland

Specification	Designation	Notes	C	Cr	Mn	Mo	Ni	P	S	Si	Other	UTS	YS	El	Hard
PNH84023/08	22GHNMA	Chain	0.2-0.26	0.4-0.6	1-1.25	0.2-0.3	0.4-0.7	0.025 max	0.025 max	0.15-0.35	Al 0.02-0.07; Cu <=0.2; P+S<=0.04; bal Fe				
PNH84023/08	25HGNMA	Chain	0.22-0.28	0.45-0.65	1-1.3	0.2-0.3	0.45-0.65	0.025 max	0.025 max	0.17-0.37	Al 0.02-0.12; Cu <=0.2; P+S<=0.4; bal Fe				
PNH84024	22H2NM	High-temp const	0.18-0.25	1.2-2	0.25-0.8	0.5-0.8	0.9-1.1	0.035 max	0.035 max	0.1-0.4	V <=0.05; bal Fe				
PNH84024	23H12MNF	High-temp const	0.2-0.26	11-12.5	0.3-0.8	0.8-1.2	0.3-0.8	0.035 max	0.035 max	0.1-0.5	Nb <=0.05; V 0.25-0.35; W <=0.6; bal Fe				
PNH84024	23H2MF	High-temp const	0.21-0.29	1.5-1.8	0.3-0.6	0.9-1.1	0.4 max	0.025 max	0.025 max	0.25-0.5	Cu <=0.2; V 0.22-0.32; bal Fe				
PNH84024	32H2NMJ	High-temp const; Nitriding	0.3-0.37	1.5-1.8	0.4-0.6	0.15-0.25	0.9-1.1	0.025 max	0.025 max	0.15-0.35	Al 0.8-1.1; bal Fe				
PNH84024	32HN3M	High-temp const	0.28-0.35	0.6-0.9	0.3-0.6	0.3-0.4	2.75-3.25	0.04 max	0.03 max	0.17-0.37	bal Fe				
PNH84024	34HN3M	High-temp const	0.3-0.4	0.7-1.1	0.5-0.8	0.25-0.4	2.75-3.25	0.03 max	0.03 max	0.17-0.37	Cu <=0.15; bal Fe				
PNH84030/02	17HNM	CH	0.14-0.19	1.5-1.8	0.4-0.7	0.25-0.35	1.4-1.7	0.035 max	0.035 max	0.17-0.37	bal Fe				
PNH84030/02	20HNM	CH	0.17-0.23	0.35-0.65	0.6-0.9	0.15-0.25	0.35-0.75	0.035 max	0.035 max	0.17-0.37	bal Fe				
PNH84030/02	22HNM	CH	0.19-0.25	0.35-0.65	0.6-0.9	0.15-0.25	0.35-0.75	0.035 max	0.035 max	0.17-0.37	bal Fe				
PNH84030/04	34HNMA	Q/T	0.32-0.4	1.3-1.7	0.4-0.7	0.15-0.25	1.3-1.7	0.025 max	0.025 max	0.17-0.37	bal Fe				
PNH84030/04	38HNM	Q/T	0.34-0.43	0.6-0.9	0.5-0.8	0.15-0.25	0.7-1	0.035 max	0.035 max	0.17-0.37	bal Fe				
PNH84035	30H2N2M	Q/T	0.26-0.34	1.8-2.1	0.3-0.6	0.25-0.35	1.8-2.1	0.035 max	0.035 max	0.17-0.37	Cu <=0.2; bal Fe				

Romania

Specification	Designation	Notes	C	Cr	Mn	Mo	Ni	P	S	Si	Other	UTS	YS	El	Hard
STAS 11512(80)	17CrNiMo6	CH	0.15-0.21	1.44-1.86	0.36-0.64	0.21-0.39	1.35-1.75	0.035 max	0.035 max	0.12-0.43	Al 0.02-0.120; Pb <=0.15; bal Fe				

UNS numbers and US grades are provided as a means of cross referencing chemically similar alloys. Exchangability is only possible after independent examination of specifications. Tensile properties are minimum or typical as specified. UTS and YS as MPa. El as %. See Appendix for list of abbreviations used in Notes. * indicates obsolete material.

Specification	Designation	Notes	C	Cr	Mn	Mo	Ni	P	S	Si	Other	UTS	YS	El	Hard

Alloy Steel, Nickel Chromium Molybdenum, No equivalents identified

Romania

Specification	Designation	Notes	C	Cr	Mn	Mo	Ni	P	S	Si	Other	UTS	YS	El	Hard
STAS 11522(80)	24VMoCr12	Heat res	0.2-0.28	1.2-1.5	0.3-0.6	0.5-0.6	0.6 max	0.035 max	0.035 max	0.15-0.35	Pb <=0.15; V 0.15-0.25; bal Fe				
STAS 791(80)	18MoCrNi13	Struct/const; CH	0.15-0.21	0.8-1.1	0.5-0.8	0.15-0.3	1.2-1.5	0.035 max	0.035 max	0.17-0.37	Pb <=0.15; bal Fe				
STAS 791(88)	17MoCrNi14	Struct/const; CH	0.14-0.2	0.85-*1.15	0.6-0.9	0.15-0.25	1.2-1.6	0.035 max	0.035 max	0.17-0.37	Pb <=0.15; Ti <=0.02; bal Fe				
STAS 791(88)	17MoCrNi14S	Struct/const; CH	0.14-0.2	0.85-1.15	0.6-0.9	0.15-0.25	1.2-1.6	0.035 max	0.02-0.04	0.17-0.37	Pb <=0.15; Ti <=0.02; bal Fe				
STAS 791(88)	17MoCrNi14X	Struct/const; CH	0.14-0.2	0.85-1.15	0.6-0.9	0.15-0.25	1.2-1.6	0.025 max	0.025 max	0.17-0.37	Pb <=0.15; Ti <=0.02; bal Fe				
STAS 791(88)	17MoCrNi14XS	Struct/const; CH	0.14-0.2	0.85-1.15	0.6-0.9	0.15-0.25	1.2-1.6	0.025 max	0.02-0.035	0.17-0.37	Pb <=0.15; Ti <=0.02; bal Fe				
STAS 791(88)	30MoCrNi20	Struct/const; Q/T	0.26-0.34	1.8-2.1	0.3-0.6	0.3-0.5	1.8-2.2	0.035 max	0.035 max	0.17-0.37	Pb <=0.15; Ti <=0.02; bal Fe				
STAS 791(88)	30MoCrNi20S	Struct/const; Q/T	0.26-0.34	1.8-2.1	0.3-0.6	0.3-0.5	1.8-2.2	0.035 max	0.02-0.04	0.17-0.37	Pb <=0.15; Ti <=0.02; bal Fe				
STAS 791(88)	30MoCrNi20X	Struct/const; Q/T	0.26-0.34	1.8-2.1	0.3-0.6	0.3-0.5	1.8-2.2	0.025 max	0.025 max	0.17-0.37	Pb <=0.15; Ti <=0.02; bal Fe				
STAS 791(88)	30MoCrNi20XS	Struct/const; Q/T	0.26-0.34	1.8-2.1	0.3-0.6	0.3-0.5	1.8-2.2	0.025 max	0.02-0.035	0.17-0.37	Pb <=0.15; Ti <=0.02; bal Fe				

Russia

Specification	Designation	Notes	C	Cr	Mn	Mo	Ni	P	S	Si	Other	UTS	YS	El	Hard
GOST 4543(71)	12Ch2N4A	CH	0.09-0.15	1.25-1.65	0.3-0.6	0.15 max	3.25-3.65	0.025 max	0.025 max	0.17-0.37	Al <=0.1; Cu <=0.3; Ti <=0.03; W <=0.2; bal Fe				
GOST 4543(71)	12ChN2	CH	0.09-0.16	0.6-0.9	0.3-0.6	0.15 max	1.5-1.9	0.035 max	0.035 max	0.17-0.37	Al <=0.1; Cu <=0.3; Ti <=0.05; bal Fe				
GOST 4543(71)	14Ch2N3MA	CH	0.12-0.17	1.5-1.75	0.3-0.6	0.2-0.3	2.75-3.25	0.025 max	0.025 max	0.17-0.37	Al <=0.1; Cu <=0.3; Ti <=0.05; bal Fe				
GOST 4543(71)	15ChGN2TA	CH	0.13-0.18	0.7-1.0	0.7-1.0	0.15 max	1.4-1.8	0.025 max	0.025 max	0.17-0.37	Al <=0.1; Cu <=0.3; Ti 0.03-0.09; bal Fe				
GOST 4543(71)	18Ch2N4MA	CH	0.14-0.2	1.35-1.65	0.25-0.55	0.3-0.4	4.0-4.4	0.025 max	0.025 max	0.17-0.37	Al <=0.1; Cu <=0.3; Ti <=0.03; V <=0.05; W <=0.2; bal Fe				
GOST 4543(71)	18Ch2N4WA	CH	0.14-0.2	1.35-1.65	0.25-0.55	0.15 max	4.0-4.4	0.025 max	0.025 max	0.17-0.37	Al <=0.1; Cu <=0.3; Ti <=0.03; W 0.8-1.2; bal Fe				
GOST 4543(71)	20Ch2N4A	CH	0.16-0.22	1.25-1.65	0.3-0.6	0.15 max	3.25-3.65	0.025 max	0.025 max	0.17-0.37	Al 0.01-0.1; Cu <=0.3; Ti <=0.05; bal Fe				
GOST 4543(71)	20ChGNTR	CH	0.18-0.24	0.4-0.7	0.8-1.1	0.15 max	0.4-0.7	0.035 max	0.035 max	0.17-0.37	Al <=0.1; B >=0.001; Cu <=0.3; Ti 0.03-0.09; W <=0.3; bal Fe				
GOST 4543(71)	20ChN	CH	0.17-0.23	0.45-0.75	0.4-0.7	0.15 max	1.0-1.4	0.035 max	0.035 max	0.17-0.37	Al <=0.1; Cu <=0.3; Ti <=0.05; bal Fe				
GOST 4543(71)	20ChN4FA	CH	0.17-0.24	0.7-1.1	0.25-0.55	0.15 max	3.75-4.15	0.025 max	0.025 max	0.17-0.37	Al <=0.1; Cu <=0.3; Ti <=0.03; V 0.1-0.18; W <=0.2; bal Fe				
GOST 4543(71)	20ChNR	CH	0.16-0.23	0.7-1.1	0.6-0.9	0.15 max	0.8-1.1	0.035 max	0.035 max	0.17-0.37	Al <=0.1; B >=0.001; Cu <=0.3; Ti <=0.03; V <=0.5; W <=0.2; bal Fe				
GOST 4543(71)	25Ch2N4MA	Q/T	0.21-0.28	1.35-1.65	0.25-0.55	0.3-0.4	4.0-4.4	0.025 max	0.025 max	0.17-0.37	Al <=0.1; Cu <=0.3; Ti <=0.05; bal Fe				
GOST 4543(71)	25Ch2N4WA	Q/T	0.21-0.28	1.35-1.65	0.25-0.55	0.15 max	4.0-4.4	0.025 max	0.025 max	0.17-0.37	Al <=0.1; Cu <=0.3; Ti <=0.05; W 0.8-1.2; bal Fe				
GOST 4543(71)	30ChN2MA	Q/T	0.27-0.34	0.6-0.9	0.3-0.6	0.2-0.3	1.25-1.65	0.025 max	0.025 max	0.17-0.37	Al <=0.1; Cu <=0.3; Ti <=0.05; bal Fe				
GOST 4543(71)	30ChN2MFA	Q/T	0.27-0.34	0.6-0.9	0.3-0.6	0.2-0.3	2.0-2.4	0.025 max	0.025 max	0.17-0.37	Al <=0.1; Cu <=0.3; Ti <=0.05; V 0.1-0.18; bal Fe				
GOST 4543(71)	30ChN2WA	Q/T	0.27-0.34	0.6-0.9	0.3-0.6	0.15 max	1.25-1.65	0.025 max	0.025 max	0.17-0.37	Al <=0.1; Cu <=0.3; Ti <=0.05; W 0.5-0.8; bal Fe				
GOST 4543(71)	30ChN2WFA	Q/T	0.27-0.34	0.6-0.9	0.3-0.6	0.15 max	2.0-2.4	0.025 max	0.025 max	0.17-0.37	Al <=0.1; Cu <=0.3; Ti <=0.05; V 0.1-0.18; W 0.5-0.8; bal Fe				

Specification	Designation	Notes	C	Cr	Mn	Mo	Ni	P	S	Si	Other	UTS	YS	El	Hard

Alloy Steel, Nickel Chromium Molybdenum, No equivalents identified

Russia

Specification	Designation	Notes	C	Cr	Mn	Mo	Ni	P	S	Si	Other	UTS	YS	El	Hard
GOST 4543(71)	30ChN3A	Q/T	0.27-0.33	0.6-0.9	0.3-0.6	0.15 max	2.75-3.15	0.025 max	0.025 max	0.17-0.37	Al <=0.1; Cu <=0.3; Ti <=0.03; W <=0.2; bal Fe				
GOST 4543(71)	38Ch2N2WA	Q/T	0.33-0.4	1.3-1.7	0.25-0.5	0.15 max	1.3-1.7	0.025 max	0.025 max	0.17-0.37	Al <=0.1; Cu <=0.3; Ti <=0.05; W 0.5-0.8; bal Fe				
GOST 4543(71)	38ChGN	Q/T	0.35-0.43	0.5-0.8	0.8-1.1	0.15 max	0.7-1.0	0.035 max	0.035 max	0.17-0.37	Al <=0.1; Cu <=0.3; Ti <=0.05; bal Fe				
GOST 4543(71)	38ChN3MA	Q/T	0.33-0.4	0.8-1.2	0.25-0.5	0.2-0.3	2.75-3.25	0.025 max	0.025 max	0.17-0.37	Al <=0.1; Cu <=0.3; Ti <=0.05; bal Fe				
GOST 4543(71)	38ChN3MFA	Q/T	0.33-0.4	1.2-1.5	0.25-0.5	0.35-0.45	3.0-3.5	0.025 max	0.025 max	0.17-0.37	Al <=0.1; Cu <=0.3; Ti <=0.05; V 0.1-0.18; bal Fe				
GOST 4543(71)	38ChN3WA	Q/T	0.33-0.4	0.8-1.2	0.25-0.5	0.15 max	2.75-3.25	0.025 max	0.025 max	0.17-0.37	Al <=0.1; Cu <=0.3; Ti <=0.05; W 0.5-0.8; bal Fe				
GOST 4543(71)	40Ch2N2WA	Q/T	0.34-0.42	1.25-1.65	0.3-0.6	0.15 max	1.35-1.75	0.025 max	0.025 max	0.17-0.37	Al <=0.1; Cu <=0.3; Ti <=0.05; W 0.6-0.9; bal Fe				
GOST 4543(71)	45ChN	Q/T	0.41-0.49	0.45-0.75	0.5-0.8	0.15 max	1.0-1.4	0.035 max	0.035 max	0.17-0.37	Al <=0.1; Cu <=0.3; Ti <=0.05; bal Fe				
GOST 4543(71)	45ChN2MFA	Q/T	0.42-0.5	0.8-1.1	0.5-0.8	0.2-0.3	1.3-1.8	0.025 max	0.025 max	0.17-0.37	Al <=0.1; Cu <=0.3; Ti <=0.05; V 0.1-0.18; bal Fe				
GOST 4543(71)	50ChN	Q/T	0.46-0.54	0.45-0.75	0.5-0.8	0.15 max	1.0-1.4	0.035 max	0.035 max	0.17-0.37	Al <=0.1; Cu <=0.3; Ti <=0.05; bal Fe				

Spain

Specification	Designation	Notes	C	Cr	Mn	Mo	Ni	P	S	Si	Other	UTS	YS	El	Hard
UNE 36012(75)	32NiCrMo12	Q/T	0.3-0.36	0.7-0.9	0.6-0.8	0.3-0.4	2.75-3.25	0.035 max	0.035 max	0.15-0.4	bal Fe				
UNE 36012(75)	32NiCrMo16	Q/T	0.3-0.37	1.1-1.4	0.3-0.6	0.25-0.4	3.7-4.2	0.035 max	0.035 max	0.15-0.4	bal Fe				
UNE 36012(75)	35NiCrMo7	Q/T	0.32-0.38	0.65-0.95	0.55-0.85	0.15-0.3	1.6-2.0	0.035 max	0.035 max	0.15-0.4	bal Fe				
UNE 36012(75)	F.1260*	see 32NiCrMo16	0.3-0.37	1.1-1.4	0.3-0.6	0.25-0.4	3.7-4.2	0.035 max	0.035 max	0.15-0.4	bal Fe				
UNE 36012(75)	F.1262*	see 32NiCrMo12	0.3-0.36	0.7-0.9	0.6-0.8	0.3-0.4	2.75-3.25	0.035 max	0.035 max	0.15-0.4	bal Fe				
UNE 36012(75)	F.1270*	see 35NiCrMo7	0.32-0.38	0.65-0.95	0.55-0.85	0.15-0.3	1.6-2.0	0.035 max	0.035 max	0.15-0.4	bal Fe				
UNE 36013(76)	14NiCrMo13	CH	0.11-0.17	0.8-1.1	0.3-0.6	0.2-0.3	3-3.5	0.035 max	0.035 max	0.15-0.4	bal Fe				
UNE 36013(76)	14NiCrMo13-1	CH	0.11-0.17	0.8-1.1	0.3-0.6	0.2-0.3	3-3.5	0.035 max	0.02-0.035	0.15-0.4	bal Fe				
UNE 36013(76)	15NiCr11	CH	0.12-0.17	0.6-0.9	0.3-0.6	0.15 max	2.5-3	0.035 max	0.035 max	0.15-0.4	bal Fe				
UNE 36013(76)	20NiCr4	CH	0.17-0.22	0.8-1.2	0.8-1	0.15 max	0.8-1.2	0.035 max	0.035 max	0.15-0.4	bal Fe				
UNE 36013(76)	20NiCr4-1	CH	0.17-0.22	0.8-1.2	0.8-1	0.15 max	0.8-1.2	0.035 max	0.02-0.035	0.15-0.4	bal Fe				
UNE 36013(76)	20NiCrMo3	CH	0.18-0.23	0.4-0.6	0.6-0.8	0.3-0.4	0.7-0.9	0.035 max	0.035 max	0.15-0.4	bal Fe				
UNE 36013(76)	F.1524*	see 20NiCrMo3	0.18-0.23	0.4-0.6	0.6-0.8	0.3-0.4	0.7-0.9	0.035 max	0.035 max	0.15-0.4	bal Fe				
UNE 36013(76)	F.1540*	see 15NiCr11	0.12-0.17	0.6-0.9	0.3-0.6	0.15 max	2.5-3	0.035 max	0.035 max	0.15-0.4	bal Fe				
UNE 36013(76)	F.1560*	see 14NiCrMo13	0.11-0.17	0.8-1.1	0.3-0.6	0.2-0.3	3-3.5	0.035 max	0.035 max	0.15-0.4	bal Fe				
UNE 36013(76)	F.1569*	see 14NiCrMo13-1	0.11-0.17	0.8-1.1	0.3-0.6	0.2-0.3	3-3.5	0.035 max	0.02-0.035	0.15-0.4	bal Fe				
UNE 36013(76)	F.1580*	see 20NiCr4	0.17-0.22	0.8-1.2	0.8-1	0.15 max	0.8-1.2	0.035 max	0.035 max	0.15-0.4	bal Fe				
UNE 36013(76)	F.1589*	see 20NiCr4-1	0.17-0.22	0.8-1.2	0.8-1	0.15 max	0.8-1.2	0.035 max	0.02-0.035	0.15-0.4	bal Fe				
UNE 36027(80)	18NiCrMo6	Ball & roller bearing	0.14-0.2	0.8-1.1	0.6-0.9	0.15-0.25	1.2-1.6	0.035 max	0.035 max	0.15-0.4	bal Fe				
UNE 36027(80)	F.1555*	see 18NiCrMo6	0.14-0.2	0.8-1.1	0.6-0.9	0.15-0.25	1.2-1.6	0.035 max	0.035 max	0.15-0.4	bal Fe				

UNS numbers and US grades are provided as a means of cross referencing chemically similar alloys. Exchangability is only possible after independent examination of specifications. Tensile properties are minimum or typical as specified. UTS and YS as MPa. El as %. See Appendix for list of abbreviations used in Notes. * indicates obsolete material.

Specification	Designation	Notes	C	Cr	Mn	Mo	Ni	P	S	Si	Other	UTS	YS	El	Hard

Alloy Steel, Nickel Chromium Molybdenum, No equivalents identified

UK

Specification	Designation	Notes	C	Cr	Mn	Mo	Ni	P	S	Si	Other	UTS	YS	El	Hard
BS 3604/1(90)	591	Pip Tub High-press/temp smls weld t<=200mm	0.10-0.17	0.30 max	0.80-1.20	0.25-0.50	1.00-1.30	0.030 max	0.030 max	0.25-0.50	Al <=0.045; Cu 0.50-0.80; N <=0.20; Nb 0.015-0.045; Sn<=0.03; bal Fe	610-760	440	20	
BS 3604/1(90)	762	Pip Tub High-press/temp smls weld t<=200mm	0.17-0.23	10.0-12.5	1.00 max	0.80-1.20	0.30-0.80	0.030 max	0.030 max	0.50 max	Al <=0.02; Cu <=0.25; V 0.25-0.35; Sn<=0.03; bal Fe	690-840	490	15	
BS 970/1(83)	832H13	Blm Bil Slab Bar Rod Frg CH	0.10-0.16	0.70-1.00	0.35-0.60	0.10-0.25	3.0-3.75	0.035 max	0.04 max	0.10-0.35	bal Fe				248 HB max
BS 970/1(83)	832M13	Blm Bil Slab Bar Rod Frg CH	0.10-0.16	0.70-1.00	0.35-0.60	0.10-0.25	3.0-3.75	0.035 max	0.04 max	0.10-0.35	bal Fe	1080		8	
BS 970/1(83)	835H15	Blm Bil Slab Bar Rod Frg CH	0.12-0.18	1.00-1.40	0.25-0.50	0.15-0.30	3.9-4.3	0.035 max	0.04 max	0.10-0.35	bal Fe				269 HB max
BS 970/1(83)	835M15	Blm Bil Slab Bar Rod Frg CH	0.12-0.18	1.00-1.40	0.25-0.50	0.15-0.30	3.9-4.3	0.040 max	0.05 max	0.10-0.40	bal Fe	1310		8	
BS 970/1(83)	945M38 (En100)	Blm Bil Slab Bar Rod Frg CH	0.14-0.20	0.80-1.10	1.00-1.30						bal Fe	930		10	217 HB max
BS 970/1(96)	835H15	Blm Bil Slab Bar Rod Frg CH	0.12-0.18	1.00-1.40	0.25-0.50	0.15-0.30	3.90-4.30	0.035 max	0.040 max	0.10-0.35	bal Fe				277 HB max
BS 970/1(96)	835M15	Blm Bil Slab Bar Rod Frg CH	0.12-0.18	1.00-1.40	0.25-0.50	0.15-0.30	3.90-4.30	0.035 max	0.040 max	0.10-0.35	bal Fe	1310		8	

USA

Specification	Designation	Notes	C	Cr	Mn	Mo	Ni	P	S	Si	Other	UTS	YS	El	Hard
ASTM A290(95)	T	Gear frg	0.25-0.39	1.25-1.75	0.20-0.60	0.30-0.70	3.25-4.00	0.015 max	0.015 max	0.35 max	Al 0.85-1.30; Cu <=0.35; V 0.05-0.15; bal Fe	1175	965	10	352-401 HB
ASTM A322(96)	9259	Bar	0.56-0.64	0.45-0.65	0.75-1.00			0.035 max	0.04 max	0.70-1.10	bal Fe				
ASTM A372/372M(95)	Grade M	Thin wall frg press ves	0.23 max	1.50-2.00	0.20-0.40	0.40-0.60	2.8-3.9	0.020 max	0.020 max	0.30 max	V <=0.08; bal Fe	725-1000	585-690	16-18	217-248 HB
ASTM A372/372M(95)	Type IX*	Thin wall frg press ves	0.23 max	1.50-2.00	0.20-0.40	0.40-0.60	2.8-3.9	0.020 max	0.020 max	0.30 max	V <=0.08; bal Fe	725-1000	585-690	16-18	217-248 HB
ASTM A543(93)	Type C	Q/T, Press ves plt, t>=5mm, Class 1	0.18 max	1.00-1.90	0.40 max	0.20-0.65	2.00-3.50	0.020 max	0.020 max	0.15-0.40	V <=0.03; bal Fe	725-860	585	14	
ASTM A543(93)	Type C	Q/T, Press ves plt, t>=5mm, Class 2	0.18 max	1.00-1.90	0.40 max	0.20-0.65	2.00-3.50	0.020 max	0.020 max	0.15-0.40	V <=0.03; bal Fe	795-930	690	14(in 50mm)	
ASTM A543(93)	Type C	Q/T, Press ves plt, t>=5mm, Class 3	0.18 max	1.00-1.90	0.40 max	0.20-0.65	2.00-3.50	0.020 max	0.020 max	0.15-0.40	V <=0.03; bal Fe	620-795	485	16(in 50mm)	
ASTM A646(95)	Grade 5	Blm Bil for aircraft aerospace frg, ann	0.28-0.33	0.70-0.95	0.75-1.00	0.35-0.50	1.65-2.00	0.025 max	0.025 max	0.20-0.35	V 0.05-0.10; bal Fe				285 HB
ASTM A768(95)	Grade 3 Class 1	Turbine rotor shaft frg	0.10-0.16	11.0-13.0	0.25-1.00	0.20 max	0.75 max	0.015 max	0.012 max	0.15-0.45	Nb <=0.15; bal Fe	620-758	483	15	
ASTM A768(95)	Grade 3 Class 2	Turbine rotor shaft frg	0.10-0.16	11.0-13.0	0.25-1.00	0.20 max	0.75 max	0.015 max	0.012 max	0.15-0.45	Nb <=0.15; bal Fe	724-862	552	16	
ASTM A940(96)	A940 Grade 1	Frg differentially vacuum HT for turbine rotors	0.23-0.31	0.90-1.50	0.50-1.00	1.10-1.50	0.80-1.10	0.012 max	0.015 max	0.10 max	Al <=0.010; Nb 0.01-0.05; V 0.20-0.30; bal Fe	725-860	585	17	
ASTM A983/983M(98)	Grade 8	Continuous grain flow frg for crankshafts	0.30-0.35	1.00-1.40	0.40-0.80	0.20-0.35	1.30-1.70	0.025 max	0.025 max	0.15-0.40	V <=0.15; bal Fe				

Yugoslavia

Specification	Designation	Notes	C	Cr	Mn	Mo	Ni	P	S	Si	Other	UTS	YS	El	Hard
	C.4520	Ball & roller bearing	0.15-0.2	1.5-1.8	0.4-0.6	0.25-0.35	1.4-1.7	0.035 max	0.035 max	0.4 max	bal Fe				
	C.5420	Ball & roller bearing	0.14-0.19	1.4-1.7	0.4-0.6	0.15 max	1.4-1.7	0.035 max	0.035 max	0.4 max	bal Fe				
	C.5421	CH	0.15-0.2	1.8-2.1	0.4-0.6	0.15 max	1.8-2.1	0.035 max	0.035 max	0.15-0.4	bal Fe				

Specification	Designation	Notes	C	Cr	Mn	Mo	Ni	P	S	Si	Other	UTS	YS	El	Hard

Alloy Steel, Nickel Molybdenum, 4615

Australia

Specification	Designation	Notes	C	Cr	Mn	Mo	Ni	P	S	Si	Other	UTS	YS	El	Hard
AS 1444	4615*	Obs	0.13-0.18		0.45-0.65	0.2-0.3	1.65-2.00	0.040 max	0.040 max	0.10-0.35	bal Fe				

France

Specification	Designation	Notes	C	Cr	Mn	Mo	Ni	P	S	Si	Other	UTS	YS	El	Hard
AFNOR	15ND8		0.13-0.18		0.2-0.5	0.15-0.3	1.8-2.3	0.04	0.035	0.1-0.35	bal Fe				

India

Specification	Designation	Notes	C	Cr	Mn	Mo	Ni	P	S	Si	Other	UTS	YS	El	Hard
IS 1570/4(88)	15Ni13Cr3Mo2		0.12-0.18	0.6-1.1	0.3-0.6	0.1-0.25	3-3.75	0.07 max	0.06 max	0.15-0.35	bal Fe	1080		8	

Mexico

Specification	Designation	Notes	C	Cr	Mn	Mo	Ni	P	S	Si	Other	UTS	YS	El	Hard
DGN B-203	4615*	Obs	0.13-0.18		0.45-0.65	0.2-0.3	1.65-2	0.04	0.04 •	0.2-0.35	bal Fe				
DGN B-297	4615*	Obs	0.13-0.18		0.45-0.65	0.2-0.3	1.65-2	0.035	0.04	0.2-0.35	bal Fe				
NMX-B-300(91)	4615	Bar	0.13-0.18		0.45-0.65	0.20-0.30	1.65-2.00	0.035 max	0.040 max	0.15-0.35	bal Fe				

Pan America

Specification	Designation	Notes	C	Cr	Mn	Mo	Ni	P	S	Si	Other	UTS	YS	El	Hard
COPANT 514	4615	Tube	0.13-0.18		0.45-0.65	0.2-0.3	1.65-2	0.04	0.04	0.2-0.35	bal Fe				

Russia

Specification	Designation	Notes	C	Cr	Mn	Mo	Ni	P	S	Si	Other	UTS	YS	El	Hard
GOST 4543	15N2M		0.1-0.18	0.3 max	0.4-0.7	0.2-0.3	1.5-1.9	0.035	0.035	0.17-0.37	Cu <=0.3; bal Fe				
GOST 4543(71)	15N2M	CH	0.1-0.18	0.3 max	0.4-0.7	0.2-0.3	1.5-1.9	0.035 max	0.035 max	0.17-0.37	Al <=0.1; Cu <=0.3; Ti <=0.05; bal Fe				

UK

Specification	Designation	Notes	C	Cr	Mn	Mo	Ni	P	S	Si	Other	UTS	YS	El	Hard
BS 970/1(83)	665H17	Blm Bil Slab Bar Rod Frg CH	0.14-0.20		0.35-0.75	0.20-0.30	1.5-2.0	0.035 max	0.04 max	0.10-0.35	bal Fe				0-207 HB
BS 970/1(83)	665M17	Blm Bil Slab Bar Rod Frg CH	0.14-0.20		0.35-0.75	0.20-0.30	1.5-2.0	0.040 max	0.05 max	0.10-0.40	bal Fe	770		12	

USA

Specification	Designation	Notes	C	Cr	Mn	Mo	Ni	P	S	Si	Other	UTS	YS	El	Hard
	AISI 4615	Plt	0.13-0.18		0.45-0.65	0.20-0.30	1.65-2.00	0.040 max	0.040 max	0.15-0.35	bal Fe				
	AISI 4615	Smls mech tub	0.13-0.18		0.45-0.65	0.20-0.30	1.65-2.00	0.040 max	0.040 max	0.15-0.35	bal Fe				
	UNS G46150		0.13-0.18		0.45-0.65	0.20-0.30	1.65-2.00	0.035 max	0.040 max	0.15-0.35	bal Fe				
AMS 6290		Bar Frg	0.11-0.17	0.2	0.45-0.65	0.2-0.30	1.65-2	0.025	0.025	0.15-0.35	Cu 0.35; bal Fe				
ASTM A29/A29M(93)	4615	Bar	0.13-0.18		0.45-0.65	0.20-0.30	1.65-2.00	0.035 max	0.040 max	0.15-0.35	bal Fe				
ASTM A322(96)	4615	Bar	0.13-0.18		0.45-0.65	0.20-0.30	1.65-2.00	0.035 max	0.040 max	0.15-0.35	bal Fe				
ASTM A331(95)	4615	Bar	0.13-0.18		0.45-0.65	0.20-0.30	1.65-2.00	0.035 max	0.040 max	0.15-0.35	bal Fe				
ASTM A519(96)	4615	Smls mech tub	0.13-0.18		0.45-0.65	0.20-0.30	1.65-2.00	0.040 max	0.040 max	0.15-0.35	bal Fe				
ASTM A752(93)	4615	Rod Wir	0.13-0.18		0.45-0.65	0.20-0.30	1.65-2.00	0.035 max	0.040 max	0.15-0.30	bal Fe	550			
ASTM A829/A829M(95)	4615	Plt	0.12-0.18		0.40-0.65	0.20-0.30	1.65-2.00	0.035 max	0.040 max	0.15-0.40	bal Fe	480-965			
MIL-S-7493B	- - -*	Obs									bal Fe				
SAE 770(84)	4615*	Obs; see J1397(92)									bal Fe				
SAE J404(94)	4615	Bil Blm Slab Bar HR CF	0.12-0.18		0.40-0.65	0.20-0.30	1.65-2.00	0.030 max	0.040 max	0.15-0.35	bal Fe				

Alloy Steel, Nickel Molybdenum, 4617

USA

Specification	Designation	Notes	C	Cr	Mn	Mo	Ni	P	S	Si	Other	UTS	YS	El	Hard
	AISI 4617	Plt	0.15-0.20		0.45-0.65	0.20-0.30	1.65-2.00	0.040 max	0.040 max	0.15-0.35	bal Fe				
	AISI 4617	Smls mech tub	0.15-0.20		0.45-0.65	0.20-0.30	1.65-2.00	0.040 max	0.040 max	0.15-0.35	bal Fe				
	UNS G46170		0.15-0.20		0.45-0.65	0.20-0.30	1.65-2.00	0.035 max	0.040 max	0.15-0.35	bal Fe				
ASTM A519(96)	4617	Smls mech tub	0.15-0.20		0.45-0.65	0.20-0.30	1.65-2.00	0.040 max	0.040 max	0.15-0.35	bal Fe	*			
ASTM A829/A829M(95)	4617	Plt	0.15-0.21		0.40-0.65	0.20-0.30	1.65-2.00	0.035 max	0.040 max	0.15-0.40	bal Fe	480-965			

UNS numbers and US grades are provided as a means of cross referencing chemically similar alloys. Exchangability is only possible after independent examination of specifications. Tensile properties are minimum or typical as specified. UTS and YS as MPa. El as %. See Appendix for list of abbreviations used in Notes. * indicates obsolete material.

Specification	Designation	Notes	C	Cr	Mn	Mo	Ni	P	S	Si	Other	UTS	YS	El	Hard

Alloy Steel, Nickel Molybdenum, 4620/4620H

Australia

AS 1444(96)	4620	H series, Hard/Tmp	0.17-0.22		0.45-0.65	0.20-0.30	1.65-2.00	0.040 max	0.040 max	0.10-0.35	bal Fe				
AS 1444(96)	4620H	H series, Hard/Tmp	0.17-0.23		0.35-0.75	0.20-0.30	1.55-2.00	0.040 max	0.040 max	0.10-0.35	bal Fe				

France

AFNOR	20ND8		0.16-0.23		0.2-0.5	0.15-0.3	1.8-2.3	0.04	0.025	0.1-0.35	bal Fe				

India

IS 1570	20Ni2Mo2		0.17-0.22		0.45-0.65	0.2-0.3	1.65-2			0.1-0.35	bal Fe				
IS 1570/4(88)	20Ni7Cr2Mo2		0.17-0.22	0.4-0.6	0.45-0.65	0.2-0.3	1.65-2	0.07 max	0.06 max	0.15-0.35	bal Fe				
IS 1570/4(88)	20Ni7Mo2		0.17-0.22	0.3 max	0.45-0.65	0.2-0.3	1.65-2	0.07 max	0.06 max	0.15-0.35	bal Fe				

International

ISO 683-17(76)	10	Ball & roller bearing, HT	0.13-0.19	0.80-1.10	1.00-1.30			0.035 max	0.035 max	0.15-0.40	bal Fe				
ISO 683-17(76)	11	Ball & roller bearing, HT	0.17-0.23		0.40-0.70	0.20-0.30	1.60-2.00	0.035 max	0.035 max	0.15-0.40	bal Fe				
ISO R683-9(70)	6A*	Former designation	0.17-0.23		0.4-0.7	0.2-0.3	1.6-2	0.04	0.02-0.04	0.15-0.4	bal Fe				

Mexico

DGN B-203	4620*	Obs	0.17-0.22		0.45-0.65	0.2-0.3	1.65-2	0.04	0.04	0.2-0.35	bal Fe				
DGN B-297	4620*	Obs	0.17-0.22		0.45-0.65	0.2-0.3	1.65-2	0.035	0.04	0.2-0.35	bal Fe				
NMX-B-268(68)	4620H	Hard	0.17-0.23		0.35-0.75	0.20-0.30	1.55-2.00			0.20-0.35	bal Fe				
NMX-B-300(91)	4620	Bar	0.17-0.22		0.45-0.60	0.20-0.30	1.65-2.00	0.035 max	0.040 max	0.15-0.35	bal Fe				
NMX-B-300(91)	4620H	Bar	0.17-0.23		0.35-0.75	0.20-0.30	1.55-2.00			0.15-0.35	bal Fe				

Pan America

COPANT 334	4620	Bar	0.17-0.22		0.45-0.65	0.2-0.3	1.65-2	0.04	0.04	0.2-0.35	bal Fe				
COPANT 514	4620	Tube	0.17-0.22		0.45-0.65	0.2-0.3	1.65-2	0.04	0.04	0.2-0.35	bal Fe				

Russia

GOST 4543	20N2M		0.17-0.25	0.3 max	0.4-0.7	0.2-0.3	1.5-1.9	0.035	0.035	0.17-0.37	Cu <=0.3; bal Fe				
GOST 4543(71)	20N2M	Q/T	0.17-0.25	0.3 max	0.4-0.7	0.2-0.3	1.5-1.9	0.035 max	0.035 max	0.17-0.37	Al <=0.1; Cu <=0.3; Ti <=0.05; bal Fe				

Spain

UNE	F.310.F		0.19-0.25		0.3-0.6	0.2-0.3	1.55-1.85	0.035	0.035	0.13-0.38	bal Fe				

UK

BS 970/1(83)	665A19*	Blm Bil Slab Bar Rod Frg	0.17-0.22	0.25	0.45-0.65	0.20-0.30	1.60-2.0	0.035 max	0.040 max		bal Fe				
BS 970/1(83)	665H20	Blm Bil Slab Bar Rod Frg CH	0.17-0.23		0.35-0.75	0.20-0.30	1.50-2.0				bal Fe				0-207 HB
BS 970/1(83)	665M20	Blm Bil Slab Bar Rod Frg CH	0.17-0.23		0.35-0.75	0.20-0.30	1.50-2.0	0.035 max	0.04	0.10-0.35	bal Fe	850		11	

USA

	AISI 4620	Smls mech tub	0.17-0.22		0.45-0.65	0.20-0.30	1.65-2.00	0.040 max	0.040 max	0.15-0.35	bal Fe				
	AISI 4620	Bar Blm Bil Slab	0.17-0.22		0.45-0.65	0.20-0.30	1.65-2.00	0.035 max	0.040 max	0.15-0.35	bal Fe				
	AISI 4620	Plt	0.17-0.22		0.45-0.65	0.20-0.30	1.65-2.00	0.035 max	0.040 max	0.15-0.35	bal Fe				
	AISI 4620H	Bar Blm Bil Slab	0.17-0.23		0.35-0.75	0.20-0.30	1.55-2.00	0.035 max	0.040 max	0.15-0.35	bal Fe				
	AISI 4620H	Smls mech tub, Hard	0.17-0.23		0.35-0.75	0.20-0.30	1.55-2.00	0.040 max	0.040 max	0.15-0.30	bal Fe				
	UNS G46200		0.17-0.22		0.45-0.65	0.20-0.30	1.65-2.00	0.035 max	0.040 max	0.15-0.35	bal Fe				

UNS numbers and US grades are provided as a means of cross referencing chemically similar alloys. Exchangability is only possible after independent examination of specifications. Tensile properties are minimum or typical as specified. UTS and YS as MPa. El as %. See Appendix for list of abbreviations used in Notes. * indicates obsolete material.

Specification	Designation	Notes	C	Cr	Mn	Mo	Ni	P	S	Si	Other	UTS	YS	El	Hard

Alloy Steel, Nickel Molybdenum, 4620/4620H (Continued from previous page)

USA

Specification	Designation	Notes	C	Cr	Mn	Mo	Ni	P	S	Si	Other	UTS	YS	El	Hard
	UNS H46200		0.17-0.23		0.35-0.75	0.20-0.30	1.55-2.00	0.035 max	0.040 max	0.15-0.35	bal Fe				
AMS 6294			0.17-0.22	0.2	0.45-0.65	0.2-0.30	1.65-2	0.025	0.025	0.15-0.35	Cu 0.35; bal Fe				
ASTM A29/A29M(93)	4620	Bar	0.17-0.22		0.45-0.65	0.20-0.30	1.65-2.00	0.035 max	0.040 max	0.15-0.35	bal Fe				
ASTM A304(96)	4620H	Bar, hard bands spec	0.17-0.23	0.20 max	0.35-0.75	0.20-0.30	1.55-2.00	0.035 max	0.040 max	0.15-0.35	Cu <=0.35; Cu Ni Cr Mo trace allowed; bal Fe				
ASTM A322(96)	4620	Bar	0.17-0.22		0.45-0.65	0.20-0.30	1.65-2.00	0.035 max	0.040 max	0.15-0.35	bal Fe				
ASTM A331(95)	4620	Bar	0.17-0.22		0.45-0.65	0.20-0.30	1.65-2.00	0.035 max	0.040 max	0.15-0.35	bal Fe				
ASTM A331(95)	4620H	Bar	0.17-0.23	0.20 max	0.35-0.75	0.20-0.30	1.55-2.00	0.025 max	0.025 max	0.15-0.35	Cu <=0.35; bal Fe				
ASTM A519(96)	4620	Smls mech tub	0.17-0.22		0.45-0.65	0.20-0.30	1.65-2.00	0.040 max	0.040 max	0.15-0.35	bal Fe				
ASTM A534(94)	4620H	Carburizing, anti-friction bearings	0.17-0.23		0.35-0.75	0.20-0.30	1.55-2.00	0.025 max	0.025 max	0.15-0.35	bal Fe				
ASTM A535(92)	4620	Ball & roller bearing	0.17-0.22	0.20 max	0.45-0.65	0.20-0.30	1.65-2.00	0.015 max	0.015 max	0.15-0.35	Cu <=0.35; bal Fe				
ASTM A646(95)	Grade 3	Blm Bil for aircraft aerospace frg, ann	0.17-0.22		0.45-0.65	0.20-0.30	1.65-2.00	0.025 max	0.025 max	0.20-0.35	bal Fe				229 HB
ASTM A752(93)	4620	Rod Wir	0.17-0.22		0.45-0.65	0.20-0.30	1.65-2.00	0.035 max	0.040 max	0.15-0.30	bal Fe	570			
ASTM A829/A829M(95)	4620	Plt	0.16-0.22		0.40-0.65	0.20-0.30	1.65-2.00	0.035 max	0.040 max	0.15-0.40	bal Fe	480-965			
ASTM A837(96)	4620	Frg for carburizing	0.17-0.22	0.25 max	0.45-0.65	0.20-0.30	1.65-2.00	0.035 max	0.040 max	0.15-0.35	Cu <=0.35; Si<=0.10 if VCD used; bal Fe				
ASTM A914/914M(92)	4620RH	Bar restricted end Q hard	0.17-0.22	0.20 max	0.45-0.65	0.20-0.30	1.65-2.00	0.035 max	0.040 max	0.15-0.35	Cu <=0.35; Electric furnace P, S<=0.025; bal Fe				
FED QQ-S-626C(91)	4620*	Obs; see ASTM A829; Plt	0.17-0.22		0.45-0.65	0.2-0.3	1.65-2	0.035	0.04	0.15-0.35	bal Fe				
MIL-S-7493B	- - -*	Obs									bal Fe				
SAE 770(84)	4620*	Obs; see J1397(92)									bal Fe				
SAE J1268(95)	4620H	Bar Rod Wir Sh Strp; See std	0.17-0.23	0.20 max	0.35-0.75	0.20-0.30	1.55-2.00	0.030 max	0.040 max	0.15-0.35	Cu <=0.35; Cu, Cr not spec'd but acpt; bal Fe				
SAE J1868(93)	4620RH	Restrict hard range	0.17-0.22	0.20 max	0.45-0.65	0.20-0.30	1.65-2.00	0.025 max	0.040 max	0.15-0.35	Cu <=0.035; Cu, Cr not spec'd but acpt; bal Fe				5 HRC max
SAE J404(94)	4620	Bil Blm Slab Bar HR CF	0.16-0.22		0.40-0.65	0.20-0.30	1.65-2.00	0.030 max	0.040 max	0.15-0.35	bal Fe				
SAE J407	4620H*	Obs; see J1268									bal Fe				

Alloy Steel, Nickel Molybdenum, 4621/4621H

Mexico

Specification	Designation	Notes	C	Cr	Mn	Mo	Ni	P	S	Si	Other	UTS	YS	El	Hard
NMX-B-268(68)	4621H	Hard	0.17-0.23		0.60-1.00	0.20-0.30	1.55-2.00			0.20-0.35	bal Fe				
NMX-B-300(91)	4621H	Bar	0.17-0.23		0.60-1.00	0.20-0.30	1.55-2.00			0.15-0.35	bal Fe				

UK

Specification	Designation	Notes	C	Cr	Mn	Mo	Ni	P	S	Si	Other	UTS	YS	El	Hard
BS 970/1(83)	665H23	Blm Bil Slab Bar Rod Frg CH	0.20-0.26		0.35-0.75	0.20-0.30	1.5-2.0	0.035 max	0.04 max	0.10-0.35	bal Fe				0-229 HB
BS 970/1(83)	665M23	Blm Bil Slab Bar Rod Frg CH	0.20-0.26		0.35-0.75	0.20-0.30	1.5-2.0	0.035 max	0.04 max	0.10-0.35	bal Fe		930	10	

USA

Specification	Designation	Notes	C	Cr	Mn	Mo	Ni	P	S	Si	Other	UTS	YS	El	Hard
	AISI 4621	Smls mech tub	0.18-0.23		0.70-0.90	0.20-0.30	1.65-2.00	0.040 max	0.040 max	0.15-0.35	bal Fe				
	UNS G46210		0.18-0.23		0.70-0.90	0.20-0.30	1.65-2.00	0.035 max	0.040 max	0.15-0.35	bal Fe				
	UNS H46210		0.17-0.23		0.60-1.00	0.20-0.30	1.55-2.00	0.035 max	0.040 max	0.15-0.35	bal Fe				
ASTM A29/A29M(93)	4621	Bar	0.18-0.23		0.70-0.90	0.20-0.30	1.65-2.00	0.035 max	0.040 max	0.15-0.35	bal Fe				

Specification	Designation	Notes	C	Cr	Mn	Mo	Ni	P	S	Si	Other	UTS	YS	El	Hard

Alloy Steel, Nickel Molybdenum, 4621/4621H (Continued from previous page)

USA

Specification	Designation	Notes	C	Cr	Mn	Mo	Ni	P	S	Si	Other	UTS	YS	El	Hard
ASTM A322(96)	4621	Bar	0.18-0.23		0.90-0.20	0.20-0.30	1.65-2.00	0.035 max	0.04 max	0.15-0.35	bal Fe				
ASTM A331(95)	4621	Bar	0.18-0.23		0.90-0.20	0.20-0.30	1.65-2.00	0.035 max	0.04 max	0.15-0.35	bal Fe				
ASTM A519(96)	4621	Smls mech tub	0.18-0.23		0.70-0.90	0.20-0.30	1.65-2.00	0.040 max	0.040 max	0.15-0.35	bal Fe				
ASTM A752(93)	4621	Rod Wir	0.18-0.23		0.70-0.90	0.20-0.30	1.65-2.00	0.035 max	0.040 max	0.15-0.30	bal Fe	570			

Alloy Steel, Nickel Molybdenum, 4626/4626H

Australia

Specification	Designation	Notes	C	Cr	Mn	Mo	Ni	P	S	Si	Other	UTS	YS	El	Hard
AS G18	En35B*	Obs; see AS 1444	0.23-0.28		0.3-0.6	0.2-0.3	1.5-2	0.05	0.05	0.1-0.35	bal Fe				

Mexico

Specification	Designation	Notes	C	Cr	Mn	Mo	Ni	P	S	Si	Other	UTS	YS	El	Hard
NMX-B-300(91)	4626	Bar	0.24-0.29		0.45-0.65	0.15-0.25	0.70-1.00	0.035 max	0.040 max	0.15-0.35	bal Fe				
NMX-B-300(91)	4626H	Bar	0.23-0.29		0.40-0.70	0.15-0.25	0.65-1.05			0.15-0.35	bal Fe				

UK

Specification	Designation	Notes	C	Cr	Mn	Mo	Ni	P	S	Si	Other	UTS	YS	El	Hard
BS 970/1(83)	665A24*	Blm Bil Slab Bar Rod Frg	0.22-0.27	0.25	0.45-0.65	0.20-0.30	1.60-2.0	0.035 max	0.040 max		bal Fe				

USA

Specification	Designation	Notes	C	Cr	Mn	Mo	Ni	P	S	Si	Other	UTS	YS	El	Hard
	AISI 4626		0.24-0.29		0.45-0.65	0.15-0.25	0.7-1	0.04 max	0.04 max	0.15-0.3	bal Fe				
	AISI 4626H	Smls mech tub, Hard	0.23-0.29		0.40-0.70	0.15-0.25	0.65-1.05	0.035 max	0.040 max	0.15-0.30	bal Fe				
	UNS G46260		0.24-0.29		0.45-0.65	0.15-0.25	0.70-1.00	0.035 max	0.040 max	0.15-0.35	bal Fe				
	UNS H46260		0.23-0.29		0.40-0.70	0.15-0.25	0.65-1.05	0.035 max	0.040 max	0.15-0.35	bal Fe				
ASTM A29/A29M(93)	4626	Bar	0.24-0.29		0.45-0.65	0.15-0.25	0.70-1.00	0.035 max	0.040 max	0.15-0.35	bal Fe				
ASTM A304(96)	4626H	Bar, hard bands spec	0.23-0.29	0.20 max	0.40-0.70	0.15-0.25	0.65-1.05	0.035 max	0.040 max	0.15-0.35	Cu <=0.35; Cu Ni Cr Mo trace allowed; bal Fe				
ASTM A322(96)	4626	Bar	0.24-0.29		0.45-0.65	0.15-0.25	0.70-1.00	0.035 max	0.040 max	0.15-0.35	bal Fe				
ASTM A331(95)	4626	Bar	0.24-0.29		0.45-0.65	0.15-0.25	0.70-1.00	0.035 max	0.040 max	0.15-0.35	bal Fe				
ASTM A331(95)	4626H	Bar	0.23-0.29	0.20 max	0.40-0.70	0.15-0.25	0.65-1.05	0.025 max	0.025 max	0.15-0.35	Cu <=0.35; bal Fe				
ASTM A752(93)	4626	Rod Wir	0.24-0.29		0.45-0.65	0.15-0.25	0.70-1.00	0.035 max	0.040 max	0.15-0.30	bal Fe	580			
SAE 770(84)	4626*	Obs; see J1397(92)	0.24-0.29		0.45-0.65	0.15-0.25	0.7-1	0.04 max	0.04 max	0.15-0.3	bal Fe				

Alloy Steel, Nickel Molybdenum, 4815/4815H

Argentina

Specification	Designation	Notes	C	Cr	Mn	Mo	Ni	P	S	Si	Other	UTS	YS	El	Hard
IAS	IRAM 4815		0.13-0.18		0.40-0.60	0.20-0.30	3.25-3.75	0.035 max	0.040 max	0.20-0.35	bal Fe	650	440	20	210 HB

France

Specification	Designation	Notes	C	Cr	Mn	Mo	Ni	P	S	Si	Other	UTS	YS	El	Hard
AIR 9160C83	9160C301	Bar Frg Q/A Tmp, 16mm diam	0.12-0.17	0.8-1.1	0.3-0.6	0.2-0.3	3-3.5	0.03	0.02	0.15-0.4	bal Fe				

Mexico

Specification	Designation	Notes	C	Cr	Mn	Mo	Ni	P	S	Si	Other	UTS	YS	El	Hard
DGN B-203	4815*	Obs	0.13-0.18		0.4-0.6	0.2-0.3	3.25-3.75	0.04	0.04	0.2-0.3	bal Fe				
DGN B-297	4815*	Obs	0.13-0.18		0.4-0.6	0.2-0.3	3.25-3.75	0.035	0.04	0.2-0.35	bal Fe				
NMX-B-268(68)	4815H	Hard	0.12-0.18		0.30-0.70	0.20-0.30	3.20-3.80			0.20-0.35	bal Fe				
NMX-B-300(91)	4815	Bar	0.13-0.18		0.40-0.60	0.20-0.30	3.25-3.75	0.035 max	0.040 max	0.15-0.35	bal Fe				
NMX-B-300(91)	4815H	Bar	0.12-0.18		0.30-0.70	0.20-0.30	3.20-3.80			0.15-0.35	bal Fe				

Alloy Steel, Nickel Molybdenum, 4815/4815H (Continued from previous page)

Specification	Designation	Notes	C	Cr	Mn	Mo	Ni	P	S	Si	Other	UTS	YS	El	Hard
Pan America															
COPANT 514	4815	Tube	0.13-0.18		0.4-0.6	0.2-0.3	3.25-3.75	0.04	0.04	0.1	bal Fe				
Spain															
UNE 36087(76/78)	15Ni14	Sh Plt Strp CR	0.18 max		0.8 max	0.15 max	3.25-3.75	0.035 max	0.035 max	0.15-0.35	bal Fe				
UNE 36087(76/78)	F.2644*	see 15Ni14	0.18 max		0.8 max	0.15 max	3.25-3.75	0.035 max	0.035 max	0.15-0.35	bal Fe				
USA															
	AISI 4815	Smls mech tub	0.13-0.18		0.40-0.60	0.20-0.30	3.25-3.75	0.040 max	0.040 max	0.15-0.35	bal Fe				
	AISI 4815	Bar Blm Bil Slab	0.13-0.18		0.40-0.60	0.20-0.30	3.25-3.75	0.035 max	0.040 max	0.15-0.35	bal Fe				
	AISI 4815H	Smls mech tub, Hard	0.12-0.18		0.30-0.70	0.20-0.30	3.20-3.80	0.040 max	0.040 max	0.15-0.30					
	AISI 4815H	Bar Blm Bil Slab	0.12-0.18		0.30-0.70	0.20-0.30	3.20-3.80	0.035 max	0.040 max	0.15-0.35	bal Fe				
	UNS G48150		0.13-0.18		0.40-0.60	0.20-0.30	3.25-3.75	0.035 max	0.040 max	0.15-0.35	bal Fe				
	UNS H48150		0.12-0.18		0.30-0.70	0.20-0.30	3.20-3.80	0.035 max	0.04 max	0.15-0.35	bal Fe				
ASTM A29/A29M(93)	4815	Bar	0.13-0.18		0.40-0.60	0.20-0.30	3.25-3.75	0.035 max	0.040 max	0.15-0.35	bal Fe				
ASTM A304(96)	4815H	Bar, hard bands spec	0.12-0.18	0.20 max	0.30-0.70	0.20-0.30	3.20-3.80	0.035 max	0.040 max	0.15-0.35	Cu <=0.35; Cu Ni Cr Mo trace allowed; bal Fe				
ASTM A322(96)	4815	Bar	0.13-0.18		0.40-0.60	0.20-0.30	3.25-3.75	0.035 max	0.040 max	0.15-0.35	bal Fe				
ASTM A331(95)	4815	Bar	0.13-0.18		0.40-0.60	0.20-0.30	3.25-3.75	0.035 max	0.040 max	0.15-0.35	bal Fe				
ASTM A331(95)	4815H	Bar	0.12-0.18	0.20 max	0.30-0.70	0.20-0.30	3.20-3.80	0.025 max	0.025 max	0.15-0.35	Cu <=0.35; bal Fe				
ASTM A519(96)	4815	Smls mech tub	0.13-0.18		0.40-0.60	0.20-0.30	3.25-3.75	0.040 max	0.040 max	0.15-0.35	bal Fe				
ASTM A752(93)	4815	Rod Wir	0.13-0.18		0.40-0.60	0.20-0.30	3.25-3.75	0.035 max	0.040 max	0.15-0.30	bal Fe	630			
ASTM A837(96)	4815	Frg for carburizing	0.13-0.18	0.25 max	0.40-0.60	0.20-0.30	3.25-3.75	0.035 max	0.040 max	0.15-0.35	Cu <=0.35; Si<=0.10 if VCD used; bal Fe				
SAE 770(84)	4815*	Obs; see J1397(92)	0.13-0.18		0.4-0.6	0.2-0.3	3.25-3.75	0.04 max	0.04 max	0.15-0.3	bal Fe				
SAE J1268(95)	4815H	Bar Rod Wir Sh Strp; See std	0.12-0.18	0.20 max	0.30-0.70	0.20-0.30	3.20-3.80	0.030 max	0.040 max	0.15-0.35	Cu <=0.35; Cu, Cr not spec'd but acpt; bal Fe				

Alloy Steel, Nickel Molybdenum, 4817/4817H

Specification	Designation	Notes	C	Cr	Mn	Mo	Ni	P	S	Si	Other	UTS	YS	El	Hard
Mexico															
DGN B-203	4817*	Obs	0.15-0.2		0.4-0.6	0.2-0.3	3.25-3.75	0.04	0.04	0.2-0.35	bal Fe				
DGN B-297	4817*	Obs	0.15-0.2		0.4-0.6	0.2-0.3	3.25-3.75	0.035	0.04	0.2-0.35	bal Fe				
NMX-B-268(68)	4817H	Hard	0.14-0.20		0.30-0.70	0.20-0.30	3.20-3.80			0.20-0.35	bal Fe				
NMX-B-300(91)	4817	Bar	0.15-0.20		0.40-0.60	0.20-0.30	3.25-3.75	0.035 max	0.040 max	0.15-0.35	bal Fe				
NMX-B-300(91)	4817H	Bar	0.14-0.20		0.30-0.70	0.20-0.30	3.20-3.80			0.15-0.35	bal Fe				
Pan America															
COPANT 334	4817	Bar	0.15-0.2		0.4-0.6	0.2-0.3	3.25-3.75	0.04	0.04	0.2-0.35	bal Fe				
COPANT 514	4817	Tube	0.15-0.2		0.4-0.6	0.2-0.3	3.25-3.75	0.04	0.05	0.1	bal Fe				
UK															
BS 3606(92)	243	Heat exch Tub; t<=3.2mm	0.12-0.20		0.40-0.80	0.25-0.35		0.040 max	0.040 max	0.10-0.35	Al <=0.012; Sn<=0.03; bal Fe	480-630	275	22	
USA															
	AISI 4817	Smls mech tub	0.15-0.20		0.40-0.60	0.20-0.30	3.25-3.75	0.040 max	0.040 max	0.15-0.35	bal Fe				

UNS numbers and US grades are provided as a means of cross referencing chemically similar alloys. Exchangability is only possible after independent examination of specifications. Tensile properties are minimum or typical as specified. UTS and YS as MPa. El as %. See Appendix for list of abbreviations used in Notes. * indicates obsolete material.

Specification	Designation	Notes	C	Cr	Mn	Mo	Ni	P	S	Si	Other	UTS	YS	El	Hard

Alloy Steel, Nickel Molybdenum, 4817/4817H (Continued from previous page)

USA

Specification	Designation	Notes	C	Cr	Mn	Mo	Ni	P	S	Si	Other	UTS	YS	El	Hard
	AISI 4817H	Smls mech tub, Hard	0.14-0.20		0.30-0.70	0.20-0.30	3.20-3.80	0.035 max	0.040 max	0.15-0.30	bal Fe				
	UNS G48170		0.15-0.20		0.40-0.60	0.20-0.30	3.25-3.75	0.035 max	0.040 max	0.15-0.35	bal Fe				
	UNS H48170		0.14-0.20		0.30-0.70	0.20-0.30	3.20-3.80	0.035 max	0.04 max	0.15-0.35	bal Fe				
ASTM A29/A29M(93)	4817	Bar	0.15-0.20		0.40-0.60	0.20-0.30	3.25-3.75	0.035 max	0.040 max	0.15-0.35	bal Fe				
ASTM A304(96)	4817H	Bar, hard bands spec	0.14-0.20	0.20 max	0.30-0.70	0.20-0.30	3.20-3.80	0.035 max	0.040 max	0.15-0.35	Cu <=0.35; Cu Ni Cr Mo trace allowed; bal Fe				
ASTM A322(96)	4817	Bar	0.15-0.20		0.40-0.60	0.20-0.30	3.25-3.75	0.035 max	0.040 max	0.15-0.35	bal Fe				
ASTM A331(95)	4817	Bar	0.15-0.20		0.40-0.60	0.20-0.30	3.25-3.75	0.035 max	0.040 max	0.15-0.35	bal Fe				
ASTM A331(95)	4817H	Bar	0.14-0.20	0.20 max	0.30-0.70	0.20-0.30	3.20-3.80	0.025 max	0.025 max	0.15-0.35	Cu <=0.35; bal Fe				
ASTM A519(96)	4817	Smls mech tub	0.15-0.20		0.40-0.60	0.20-0.30	3.25-3.75	0.040 max	0.040 max	0.15-0.35	bal Fe				
ASTM A534(94)	4817H	Carburizing, anti-friction bearings	0.14-0.20		0.30-0.70	0.20-0.30	3.20-3.80	0.025 max	0.025 max	0.15-0.35	bal Fe				
ASTM A752(93)	4817	Rod Wir	0.15-0.20		0.40-0.60	0.20-0.30	3.25-3.75	0.035 max	0.040 max	0.15-0.30	bal Fe	650			
SAE 770(84)	4817*	Obs; see J1397(92)	0.15-0.2		0.4-0.6	0.2-0.3	3.25-3.75	0.04 max	0.04 max	0.15-0.3	bal Fe				
SAE J1268(95)	4817H	Bar Rod Wir Pipe; See std	0.14-0.20	0.20 max	0.30-0.70	0.20-0.30	3.20-3.80	0.030 max	0.040 max	0.15-0.35	Cu <=0.35; Cu, Cr not spec'd but acpt; bal Fe				

Alloy Steel, Nickel Molybdenum, 4820/4820H

Mexico

Specification	Designation	Notes	C	Cr	Mn	Mo	Ni	P	S	Si	Other	UTS	YS	El	Hard
DGN B-203	4820*	Obs	0.18-0.23		0.5-0.7	0.2-0.3	3.25-3.75	0.035	0.04	0.2-0.35	bal Fe				
NMX-B-268(68)	4820H	Hard	0.17-0.23		0.40-0.80	0.20-0.30	3.20-3.80			0.20-0.35	bal Fe				
NMX-B-300(91)	4820	Bar	0.18-0.23		0.50-0.70	0.20-0.30	3.25-3.75	0.035 max	0.040 max	0.15-0.35	bal Fe				
NMX-B-300(91)	4820H	Bar	0.17-0.23		0.40-0.80	0.20-0.30	3.20-3.80			0.15-0.35	bal Fe				

Pan America

Specification	Designation	Notes	C	Cr	Mn	Mo	Ni	P	S	Si	Other	UTS	YS	El	Hard
COPANT 514	4820	Tube	0.18-0.23		0.5-0.7	0.2-0.3	3.25-3.75	0.04	0.05	0.1	bal Fe				

Romania

Specification	Designation	Notes	C	Cr	Mn	Mo	Ni	P	S	Si	Other	UTS	YS	El	Hard
STAS 791(88)	20MoNi35	Struct/const	0.17-0.23	0.3 max	0.4-0.8	0.2-0.3	3.2-3.8	0.035 max	0.035 max	0.17-0.37	Pb <=0.15; Ti <=0.02; bal Fe				
STAS 791(88)	20MoNi35S	Struct/const	0.17-0.23	0.3 max	0.4-0.8	0.2-0.3	3.2-3.8	0.035 max	0.02-0.04	0.17-0.37	Pb <=0.15; Ti <=0.02; bal Fe				
STAS 791(88)	20MoNi35X	Struct/const	0.17-0.23	0.3 max	0.4-0.8	0.2-0.3	3.2-3.8	0.025 max	0.025 max	0.17-0.37	Pb <=0.15; Ti <=0.02; bal Fe				
STAS 791(88)	20MoNi35XS	Struct/const	0.17-0.23	0.3 max	0.4-0.8	0.2-0.3	3.2-3.8	0.025 max	0.02-0.035	0.17-0.37	Pb <=0.15; Ti <=0.02; bal Fe				

USA

Specification	Designation	Notes	C	Cr	Mn	Mo	Ni	P	S	Si	Other	UTS	YS	El	Hard
	AISI 4820	Bar Blm Bil Slab	0.18-0.23		0.50-0.70	0.20-0.30	3.25-3.75	0.035 max	0.040 max	0.15-0.35	bal Fe				
	AISI 4820	Smls mech tub	0.18-0.23		0.50-0.70	0.20-0.30	3.25-3.75	0.035 max	0.040 max	0.15-0.35	bal Fe				
	AISI 4820H	Smls mech tub, Hard	0.17-0.23		0.40-0.80	0.20-0.30	3.20-3.80	0.040 max	0.040 max	0.15-0.30	bal Fe				
	AISI 4820H	Bar Blm Bil Slab	0.17-0.23		0.40-0.80	0.20-0.30	3.20-3.80	0.035 max	0.040 max	0.15-0.35	bal Fe				
	UNS G48200		0.18-0.23		0.50-0.70	0.20-0.30	3.25-3.75	0.035 max	0.040 max	0.15-0.35	bal Fe				
	UNS H48200		0.17-0.23		0.40-0.80	0.20-0.30	3.20-3.80	0.035 max	0.04 max	0.15-0.35	bal Fe				
ASTM A29/A29M(93)	4820	Bar	0.18-0.23		0.50-0.70	0.20-0.30	3.25-3.75	0.035 max	0.040 max	0.15-0.35	bal Fe				

Specification	Designation	Notes	C	Cr	Mn	Mo	Ni	P	S	Si	Other	UTS	YS	El	Hard

Alloy Steel, Nickel Molybdenum, 4820/4820H (Continued from previous page)

USA

Specification	Designation	Notes	C	Cr	Mn	Mo	Ni	P	S	Si	Other	UTS	YS	El	Hard
ASTM A304(96)	4820H	Bar, hard bands spec	0.17-0.23	0.20 max	0.40-0.80	0.20-0.30	3.20-3.80	0.035 max	0.040 max	0.15-0.35	Cu <=0.35; Cu Ni Cr Mo trace allowed; bal Fe				
ASTM A322(96)	4820	Bar	0.18-0.23		0.50-0.70	0.20-0.30	3.25-3.75	0.035 max	0.040 max	0.15-0.35	bal Fe				
ASTM A331(95)	4820	Bar	0.18-0.23		0.50-0.70	0.20-0.30	3.25-3.75	0.035 max	0.040 max	0.15-0.35	bal Fe				
ASTM A331(95)	4820H	Bar	0.17-0.23	0.20 max	0.40-0.80	0.20-0.30	3.20-3.80	0.025 max	0.025 max	0.15-0.35	Cu <=0.35; bal Fe				
ASTM A519(96)	4820	Smls mech tub	0.18-0.23		0.50-0.70	0.20-0.30	3.25-3.75	0.040 max	0.040 max	0.15-0.35	bal Fe				
ASTM A534(94)	4820H	Carburizing, anti-friction bearings	0.17-0.23		0.40-0.80	0.20-0.30	3.20-3.80	0.025 max	0.025 max	0.15-0.35	bal Fe				
ASTM A535(92)	4820	Ball & roller bearing	0.18-0.23	0.20 max	0.50-0.70	0.20-0.30	3.25-3.75	0.015 max	0.015 max	0.15-0.35	Cu <=0.35; bal Fe				
ASTM A752(93)	4820	Rod Wir	0.18-0.23		0.50-0.70	0.20-0.30	3.25-3.75	0.035 max	0.040 max	0.15-0.30	bal Fe	650			
ASTM A914/914M(92)	4820RH	Bar restricted end Q hard	0.18-0.23	0.20 max	0.50-0.70	0.20-0.30	3.25-3.75	0.035 max	0.040 max	0.15-0.35	Cu <=0.35; Electric furnace P, S<=0.025; bal Fe				
SAE 770(84)	4820*	Obs; see J1397(92)									bal Fe				
SAE J1268(95)	4820H	See std	0.17-0.23	0.20 max	0.40-0.80	0.20-0.30	3.20-3.80	0.030 max	0.040 max	0.15-0.35	Cu <=0.35; Cu not spec'd but acpt; bal Fe				
SAE J1868(93)	4820RH	Restrict hard range	0.18-0.23	0.20 max	0.50-0.70	0.20-0.30	3.25-3.75	0.025 max	0.040 max	0.15-0.35	Cu <=0.035; Cu, Cr not spec'd but acpt; bal Fe				5 HRC max
SAE J404(94)	4820	Bil Blm Slab Bar HR CF	0.18-0.23		0.50-0.70	0.20-0.30	3.25-3.75	0.030 max	0.040 max	0.15-0.35	bal Fe				

Alloy Steel, Nickel Molybdenum, 6312

USA

Specification	Designation	Notes	C	Cr	Mn	Mo	Ni	P	S	Si	Other	UTS	YS	El	Hard
	UNS K22440		0.38-0.43	0.20 max	0.60-0.80	0.20-0.30	1.65-2.00	0.025 max	0.025 max	0.15-0.35	Cu <=0.35; bal Fe				

Alloy Steel, Nickel Molybdenum, A302(D)

USA

Specification	Designation	Notes	C	Cr	Mn	Mo	Ni	P	S	Si	Other	UTS	YS	El	Hard
	UNS K12054		0.20 max		1.07-1.62	0.41-0.64	0.67-1.03	0.035 max	0.040 max	0.13-0.45	bal Fe				
ASTM A302/302M(97)	D	Press ves plt, 25<t<=50mm	0.23 max		1.15-1.50	0.45-0.60	0.70-1.00	0.035 max	0.035 max	0.15-0.40	bal Fe	550-690	345	20	
ASTM A302/302M(97)	D	Press ves plt, t>50mm	0.25 max		1.15-1.50	0.45-0.60	0.70-1.00	0.035 max	0.035 max	0.15-0.40	bal Fe	550-690	345	20	
ASTM A302/302M(97)	D	Press ves plt, t<=25mm	0.20 max		1.15-1.50	0.45-0.60	0.70-1.00	0.035 max	0.035 max	0.15-0.40	bal Fe	550-690	345	20	

Alloy Steel, Nickel Molybdenum, A514(M)

USA

Specification	Designation	Notes	C	Cr	Mn	Mo	Ni	P	S	Si	Other	UTS	YS	El	Hard
	UNS K11683		0.12-0.21		0.45-0.70	0.45-0.60	1.20-1.50	0.035 max	0.04 max	0.20-0.35	B 0.001-0.005; bal Fe				
ASTM A514(94)	Grade M	Q/T, Plt, t=50mm, weld	0.12-0.21		0.45-0.70	0.45-0.60	1.20-1.50	0.035 max	0.035 max	0.20-0.35	B 0.001-0.005; bal Fe	760-895	690	18	
ASTM A517(93)	Grade M	Q/T, Press ves plt, t<=50mm	0.12-0.21		0.45-0.70	0.45-0.60	1.20-1.50	0.035 max	0.035 max	0.20-0.35	B 0.001-0.005; bal Fe	795-930	690	16	
ASTM A709(97)	Grade 100, Type M	Q/T, Plt, t<=50mm	0.12-0.21		0.45-0.70	0.45-0.60	1.20-1.50	0.035 max	0.035 max	0.20-0.35	B 0.001-0.005; bal Fe	760-895	690	18	

Alloy Steel, Nickel Molybdenum, A533(B)

USA

Specification	Designation	Notes	C	Cr	Mn	Mo	Ni	P	S	Si	Other	UTS	YS	El	Hard
	UNS K12539		0.25 max		1.07-1.62	0.41-0.64	0.37-0.73	0.035 max	0.040 max	0.13-0.45	bal Fe				
ASTM A533(93)	Type B	Q/T, Press ves plt, Class 2, t=6.5-300mm	0.25 max		1.15-1.50	0.45-0.60	0.40-0.70	0.035 max	0.035 max	0.15-0.40	bal Fe	620-795	485	16	
ASTM A533(93)	Type B	Q/T, Press ves plt, Class 1, t=6.5-300mm	0.25 max		1.15-1.50	0.45-0.60	0.40-0.70	0.035 max	0.035 max	0.15-0.40	bal Fe	550-690	345	18	
ASTM A533(93)	Type B	Q/T, Press ves plt, Class 3, t=6.5-65mm	0.25 max		1.15-1.50	0.45-0.60	0.40-0.70	0.035 max	0.035 max	0.15-0.40	bal Fe	690-860	570	16	

UNS numbers and US grades are provided as a means of cross referencing chemically similar alloys. Exchangability is only possible after independent examination of specifications. Tensile properties are minimum or typical as specified. UTS and YS as MPa. El as %. See Appendix for list of abbreviations used in Notes. * indicates obsolete material.

Specification	Designation	Notes	C	Cr	Mn	Mo	Ni	P	S	Si	Other	UTS	YS	El	Hard
Alloy Steel, Nickel Molybdenum, A533(C)															
USA															
	UNS K12554		0.25 max		1.07-1.62	0.41-0.64	0.67-1.03	0.035 max	0.040 max	0.13-0.45	bal Fe				
ASTM A533(93)	Type C	Q/T, Press ves plt, Class 2, t=6.5-300mm	0.25 max		1.15-1.50	0.45-0.60	0.70-1.00	0.035 max	0.035 max	0.15-0.40	bal Fe	620-795	485	16	
ASTM A533(93)	Type C	Q/T, Press ves plt, Class 1, t=6.5-300mm	0.25 max		1.15-1.50	0.45-0.60	0.70-1.00	0.035 max	0.035 max	0.15-0.40	bal Fe	550-690	345	18	
ASTM A533(93)	Type C	Q/T, Press ves plt, Class 3, t=6.5-65mm	0.25 max		1.15-1.50	0.45-0.60	0.70-1.00	0.035 max	0.035 max	0.15-0.40	bal Fe	690-860	570	16	
Alloy Steel, Nickel Molybdenum, A533(D)															
UK															
BS 1503(89)	224-490	Frg Press ves; 100.1-999mm	0.25 max	0.25 max	0.9-1.7	0.1 max	0.4 max	0.03 max	0.025 max	0.1-0.4	Al 0.018-0.10; Co <=0.1; Cu <=0.3; Pb <=0.15; Ti <=0.05; V <=0.1; W <=0.1; Cr+Mo+Ni+Cu<=0.8; bal Fe	490-610	280	16	
USA															
	UNS K12529		0.25 max		1.07-1.62	0.41-0.64	0.17-0.43	0.035 max	0.040 max	0.13-0.45	bal Fe				
ASTM A533(93)	Type D	Q/T, Press ves plt, Class 3, t=6.5-65mm	0.25 max		1.15-1.50	0.45-0.60	0.20-0.40	0.035 max	0.035 max	0.15-0.40	bal Fe	690-860	570	16	
ASTM A533(93)	Type D	Q/T, Press ves plt, Class 1, t=6.5-300mm	0.25 max		1.15-1.50	0.45-0.60	0.20-0.40	0.035 max	0.035 max	0.15-0.40	bal Fe	550-690	345	18	
ASTM A533(93)	Type D	Q/T, Press ves plt, Class 2, t=6.5-300mm	0.25 max		1.15-1.50	0.45-0.60	0.20-0.40	0.035 max	0.035 max	0.15-0.40	bal Fe	620-795	485	16	
Alloy Steel, Nickel Molybdenum, A541(3,3a)															
USA															
	UNS K12045		0.25 max		1.20-1.50	0.45-0.60	0.40-1.00	0.035 max	0.040 max	0.15-0.35	V <=0.05; bal Fe				
ASTM A541/541M(95)	Grade 3 Class 1	Q/T frg for press ves components	0.25 max	0.25 max	1.20-1.50	0.45-0.60	0.40-1.00	0.025 max	0.025 max	0.15-0.35	V <=0.05; Si req vary; bal Fe	550-720	340	18	
ASTM A541/541M(95)	Grade 3 Class 2	Q/T frg for press ves components	0.25 max	0.25 max	1.20-1.50	0.45-0.60	0.40-1.00	0.025 max	0.025 max	0.15-0.35	V <=0.05; Si req vary; bal Fe	620-790	450	16	
Alloy Steel, Nickel Molybdenum, A645															
France															
AFNOR NFA36205/4(94)	12Ni19	Press ves; Sh Plt Strp; Press; Low-temp; t<=30mm; N/T; Q/T	0.15 max	0.3 max	0.30-0.80	0.15 max	4.75-5.25	0.02 max	0.01 max	0.35 max	Cu <=0.30; Ti <=0.05; V <=0.05; Cr+Cu+Mo<=0.5; bal Fe	530-710	390	20	
UK															
BS 1503(89)	503-490	Frg press ves; t<=100mm; Q/T	0.15 max	0.25 max	0.80 max	0.10 max	3.25-3.75	0.025 max	0.020 max	0.15-0.40	Al >=0.018; Cu <=0.30; W <=0.1; bal Fe	490-640	300	17	
BS 1503(89)	509-690	Frg press ves; t<=150mm; single or double refined	0.10 max	0.25 max	0.80 max	0.10 max	8.50-10.0	0.025 max	0.020 max	0.15-0.40	Al >=0.018; Cu <=0.30; W <=0.1; bal Fe	690-840	490	16	
USA															
	UNS K41583		0.15 max		0.25-0.66	0.17-0.38	4.65-5.35	0.035 max	0.035 max	0.18-0.37	Al 0.01-0.16; N <=0.025; bal Fe				
ASTM A645/A645M(97)		HT, Press ves plt	0.13 max		0.30-0.60	0.20-0.35	4.75-5.25	0.025 max	0.025 max	0.20-0.40	Al 0.02-0.12; N <=0.020; bal Fe	655-795	450	20.0	
Alloy Steel, Nickel Molybdenum, A707(L4)															
USA															
	UNS K12089		0.20 max		0.40-0.70	0.19-0.33	1.60-2.05	0.030 max	0.035 max	0.37 max	bal Fe				
ASTM A707/A707M(98)	A 707 (L4)	Frg flanges; low-temp; class values given	0.18 max	0.30 max	0.45-0.65	0.20-0.30	1.65-2.00	0.025 max	0.025 max	0.35 max	Cu <=0.40; Nb <=0.02; V <=0.05; bal Fe	415-515	290-415	20-22	149-235 HB
Alloy Steel, Nickel Molybdenum, A710(B)															
USA															
	UNS K20622		0.06 max		0.40-0.65		1.20-1.50	0.025 max	0.025 max	0.15-0.40	Cu 1.00-1.30; Nb >=0.02; bal Fe				

UNS numbers and US grades are provided as a means of cross referencing chemically similar alloys. Exchangability is only possible after independent examination of specifications. Tensile properties are minimum or typical as specified. UTS and YS as MPa. El as %. See Appendix for list of abbreviations used in Notes. * indicates obsolete material.

Specification	Designation	Notes	C	Cr	Mn	Mo	Ni	P	S	Si	Other	UTS	YS	El	Hard
Alloy Steel, Nickel Molybdenum, A714(VI)															
Mexico															
NMX-B-069(86)	Grade VI	Weld and Smls pipe	0.15 max	0.30 max	0.50-1.00	0.10-0.20	0.40-1.10	0.035 max	0.045 max		Cu 0.30-1.00; bal Fe	450	315		
USA															
	UNS K11835		0.18 max	0.33 max	0.45-1.05	0.09-0.21	0.35-1.15	0.045 max	0.055 max		Cu 0.27-1.03; bal Fe				
ASTM A714(96)	VI	Weld, Smls pipe, Grade VI Type E , Type S	0.15 max	0.30 max	0.50-1.00	0.10-0.20	0.40-1.10	0.035 max	0.045 max		Cu 0.30-1.00; bal Fe	450	315		
Alloy Steel, Nickel Molybdenum, K11268															
USA															
	UNS K11268	High Strength, Low Alloy, Precipitation Hardening	0.12 max		0.50-1.00	0.25 max	0.50-1.00	0.05 max	0.05 max	0.15 max	Al 0.02-0.23; Cu 0.85-1.30; bal Fe				
Alloy Steel, Nickel Molybdenum, K11567															
USA															
	UNS K11567		0.15 max		1.20 max	0.08-0.25	0.75-1.25	0.04 max	0.05 max	0.30 max	Cu 0.50-0.80; V <=0.05; bal Fe				
Alloy Steel, Nickel Molybdenum, K43170															
USA															
	UNS K43170		0.29-0.33		0.70-1.00	0.20-0.30	3.25-3.75	0.020 max	0.020 max	0.20-0.35	bal Fe				
Alloy Steel, Nickel Molybdenum, No equivalents identified															
France															
AFNOR NFA36102(93)	75N8*	HR; Strp; CR	0.72-0.78	0.15 max	0.3-0.5	0.10 max	1.90-2.10	0.025 max	0.02 max	0.15-0.3	B 0.0008-0.005; bal Fe				
AFNOR NFA36102(93)	75Ni8RR	HR; Strp; CR	0.72-0.78	0.15 max	0.3-0.5	0.10 max	1.90-2.10	0.025 max	0.02 max	0.15-0.3	B 0.0008-0.005; bal Fe				
India															
IS 1570/4(88)	35Ni5Cr2	0-63mm; Bar Frg	0.3-0.4	0.45-0.75	0.6-0.9	0.15 max	1-1.5	0.07 max	0.06 max	0.1-0.35	bal Fe	890-1040	650	10	255-311 HB
IS 1570/4(88)	40Ni14	63.1-100mm; Bar Frg	0.35-0.45	0.3 max	0.5-0.8	0.15 max	3.2-3.6	0.07 max	0.06 max	0.1-0.35	bal Fe	790-940	550	16	229-277 HB
IS 1570/4(88)	40Ni14	0-63mm; Bar Frg	0.35-0.45	0.3 max	0.5-0.8	0.15 max	3.2-3.6	0.07 max	0.06 max	0.1-0.35	bal Fe	890-1040	650	15	255-311 HB
Poland															
PNH84024	15NCuMNb	High-temp const	0.17 max	0.3 max	0.8-1.2	0.25-0.4	1-1.3	0.035 max	0.035 max	0.25-0.5	Al 0.015-0.120; Cu 0.5-0.8; Nb 0.02-0.06; bal Fe				
Spain															
UNE 36027(80)	20NiMo5	Ball & roller bearing	0.17-0.23		0.4-0.7	0.2-0.3	1.6-2.0	0.035 max	0.035 max	0.15-0.4	bal Fe				
UNE 36027(80)	F.1526*	see 20NiMo5	0.17-0.23		0.4-0.7	0.2-0.3	1.6-2.0	0.035 max	0.035 max	0.15-0.4	bal Fe				

UNS numbers and US grades are provided as a means of cross referencing chemically similar alloys. Exchangability is only possible after independent examination of specifications. Tensile properties are minimum or typical as specified. UTS and YS as MPa. El as %. See Appendix for list of abbreviations used in Notes. * indicates obsolete material.

Specification	Designation	Notes	C	Cr	Cu	Mn	Mo	P	S	Si	Other	UTS	YS	El	Hard
Alloy Steel, Silicon, A335(P15)															
USA															
	UNS K11578		0.15 max			0.30-0.60	0.44-0.65	0.030 max	0.030 max	1.15-1.65	bal Fe				
Alloy Steel, Silicon, A469(1)															
USA															
	UNS K14501		0.45 max			0.90 max		0.035 max	0.035 max	0.15-0.35	V 0.03-0.12; bal Fe				
Alloy Steel, Silicon, No equivalents identified															
Czech Republic															
CSN 413251	13251	Spring	0.42-0.52	0.3 max		0.5-0.8		0.035 max	0.035 max	1.5-1.9	bal Fe				
Hungary															
MSZ 2666(88)	38Si7	HF spring; 0-999mm; soft ann	0.34-0.42			0.5-0.8		0.04 max	0.04 max	1.4-1.7	bal Fe				250 HB max
MSZ 2666(88)	38Si7	HF spring; 0-100.0mm; Q/T	0.34-0.42			0.5-0.8		0.04 max	0.04 max	1.4-1.7	bal Fe	1250	1030	6L	
MSZ 2666(88)	61Si7	HF spring; 0-999mm	0.57-0.65			0.6-0.9		0.04 max	0.04 max	1.5-2	bal Fe				300 HB max
MSZ 2666(88)	61Si7	HF spring; 0-100.0mm; Q/T	0.57-0.65			0.6-0.9		0.04 max	0.04 max	1.5-2	bal Fe	1370	1180	6L	
India															
IS 1570/4(88)	36Si7		0.33-0.4	0.3 max		0.8-1	0.15 max	0.07 max	0.06 max	1.5-2	Ni <=0.4; bal Fe				
IS 1570/4(88)	55Si7		0.5-0.6	0.3 max		0.8-1	0.15 max	0.07 max	0.06 max	1.5-2	Ni <=0.4; bal Fe				
IS 1570/4(88)	60Si7		0.55-0.65	0.3 max		0.8-1	0.15 max	0.07 max	0.06 max	1.5-2	Ni <=0.4; bal Fe				
IS 1570/4(88)	65Si7	Smls tube Bar Frg Sh Plt	0.6-0.7	0.3 max		0.8-1	0.15 max	0.07 max	0.06 max	1.5-2.0	Ni <=0.4; bal Fe				
Romania															
STAS 795(92)	40Si17A	Spring; HF	0.35-0.45	0.3 max	0.25 max	0.55-0.95		0.04 max	0.04 max	1.5-1.9	Pb <=0.15; bal Fe				
USA															
SAE J1868(93)	3310RH	Restrict hard range	0.08-0.13	1.40-1.75	0.035 max	0.40-0.60	0.06 max	0.025 max	0.040 max	0.15-0.35	Ni 3.25-3.75; Cu, Mo not spec'd but acpt; bal Fe				5 HRC max

Specification	Designation	Notes	C	Cr	Mn	Mo	Ni	P	S	Si	Other	UTS	YS	El	Hard

Alloy Steel, Silicon Chromium, No equivalents identified

International

Specification	Designation	Notes	C	Cr	Mn	Mo	Ni	P	S	Si	Other	UTS	YS	El	Hard
ISO 683-14(92)	55SiCr63	Bar Rod Wir; Q/T, springs, Ann	0.51-0.59	0.55-0.85	0.50-0.80			0.030 max	0.030 max	1.20-1.60	bal Fe				255 HB
ISO 683-14(92)	56SiCr7	Bar Rod Wir; Q/T, springs, Ann	0.52-0.59	0.20-0.40	0.70-1.00			0.030 max	0.030 max	1.60-2.00	bal Fe				255 HB

Japan

Specification	Designation	Notes	C	Cr	Mn	Mo	Ni	P	S	Si	Other	UTS	YS	El	Hard
JIS G3125(87)	SPA-C	CR Plt Sh Strp 0.6-2.3mm thk	0.12 max	0.30-1.25	0.20-0.50		0.65 max	0.070-0.150	0.040 max	0.25-0.75	Cu 0.25-0.60; bal Fe			26	
JIS G3125(87)	SPA-H	HR Plt >6.0mm	0.12 max	0.30-1.25	0.20-0.60		0.65 max	0.070-0.150	0.040 max	0.25-0.75	Cu 0.25-0.60; bal Fe			15	
JIS G3125(87)	SPA-H	HR Plt <=6.0mm	0.12 max	0.30-1.25	0.20-0.50		0.65 max	0.070-0.150	0.040 max	0.25-0.75	Cu 0.25-0.60; bal Fe			22	

Romania

Specification	Designation	Notes	C	Cr	Mn	Mo	Ni	P	S	Si	Other	UTS	YS	El	Hard
STAS 791(88)	25MnCrSi11	Struct/const	0.22-0.28	0.8-1.1	0.8-1.1			0.035 max	0.035 max	0.9-1.2	Pb <=0.15; Ti <=0.02; bal Fe				
STAS 791(88)	25MnCrSi11S	Struct/const	0.22-0.28	0.8-1.1	0.8-1.1			0.035 max	0.02-0.04	0.9-1.2	Pb <=0.15; Ti <=0.02; bal Fe				
STAS 791(88)	25MnCrSi11X	Struct/const	0.22-0.28	0.8-1.1	0.8-1.1			0.025 max	0.025 max	0.9-1.2	Pb <=0.15; Ti <=0.02; bal Fe				
STAS 791(88)	25MnCrSi11XS	Struct/const	0.22-0.28	0.8-1.1	0.8-1.1			0.025 max	0.02-0.035	0.9-1.2	Pb <=0.15; Ti <=0.02; bal Fe				
STAS 795(92)	60CrMnSi12A	Spring; HF	0.55-0.65	0.4-0.7	0.9-1.1			0.035 max	0.035 max	1-1.3	Cu <=0.25; Pb <=0.15; bal Fe				

Alloy Steel, Silicon Manganese, 9254

Specification	Designation	Notes	C	Cr	Cu	Mn	Mo	P	S	Si	Other	UTS	YS	El	Hard
Australia															
AS 1444	9254*	Obs; Bar	0.51-0.59	0.6-0.8		0.6-0.8		0.05	0.05	1.2-1.6	bal Fe				
Belgium															
NBN 253-05	55Cr3	Wire, Ann or Q/A Tmp 32 mm	0.52-0.59	0.6-0.9		0.7-1		0.04	0.04	0.15-0.4	bal Fe				
Czech Republic															
CSN 414260	14260	Spring	0.5-0.6	0.5-0.7		0.5-0.8		0.035 max	0.035 max	1.3-1.6	Ni <=0.5; bal Fe				
France															
AFNOR NFA35571(84)	51S7	t<=80mm; Q/T	0.48-0.54	0.3 max	0.3 max	0.5-0.8	0.15 max	0.035 max	0.035 max	1.6-2	Ni <=0.4; Ti <=0.05; bal Fe	1500-1750	1350	6	
AFNOR NFA35571(96)	54SiCr6	t<1000mm; Q/T	0.5-0.58	0.5-0.8	0.3 max	0.5-0.8	0.15 max	0.025 max	0.025 max	1.2-1.6	Ni <=0.4; Ti <=0.05; bal Fe	1450-1750	1300	6	
Germany															
DIN	55SiCr7		0.52-0.60	0.20-0.40		0.70-1.00		0.050 max	0.050 max	1.50-1.80	N <=0.007; bal Fe				
DIN	WNr 1.0958*	Obs	0.52-0.6	0.2-0.4		0.7-1		0.05	0.05 max	1.5-1.8	N 0.007; bal Fe				
DIN	WNr 1.7106		0.52-0.60	0.20-0.40		0.70-1.00		0.050 max	0.050 max	1.50-1.80	N <=0.007; bal Fe				
Italy															
UNI 3545(80)	45SiCrMo6	Spring	0.42-0.5	0.5-0.75		0.5-0.8	0.15-0.25	0.035 max	0.035 max	1.3-1.7	bal Fe				
UNI 3545(80)	52SiCrNi5	Spring	0.49-0.56	0.7-1		0.7-0.9		0.035 max	0.035 max	1.2-1.5	Ni 0.5 0.7; bal Fe				
UNI 8893(86)	52SiCrNi5	Spring	0.49-0.56	0.7-1		0.7-0.9		0.035 max	0.035 max	1.2-1.5	Ni 0.5-0.7; bal Fe				
Japan															
JIS G4801(84)	SUP12	Spring	0.51-0.59	0.60-0.90	0.30 max	0.60-0.90		0.035 max	0.035 max	1.20-1.60	bal Fe	1226	1079	9	363-429 HB
Mexico															
NMX-B-300(91)	9254	Bar	0.51-0.59	0.60-0.80		0.60-0.80		0.035 max	0.040 max	1.20-1.60	bal Fe				
Spain															
UNE 36015(77)	46Si7	Spring	0.43-0.5			0.5-0.8	0.15 max	0.035 max	0.035 max	1.5-2	bal Fe				
UNE 36015(77)	50Si7	Spring	0.47-0.53			0.5-0.8	0.15 max	0.035 max	0.035 max	1.5-2	bal Fe				
UNE 36015(77)	F.1451*	see 46Si7	0.43-0.5			0.5-0.8	0.15 max	0.035 max	0.035 max	1.5-2	bal Fe				
UK															
BS 1429(80)	685A55	Spring	0.50-0.60	0.50-0.80		0.50-0.80				1.20-1.60	bal Fe				
BS 970/2(88)	685A57	Spring	0.55-0.60	0.60-0.85		0.70-0.90		0.035 max	0.035 max	1.20-1.60	bal Fe				
BS 970/2(88)	685H57	Spring	0.54-0.62	0.50-0.80		0.50-0.80		0.035 max	0.035 max	1.20-1.60	bal Fe				
USA															
	AISI 9254		0.51-0.59	0.6-0.8		0.6-0.8		0.035 max	0.04 max	1.2-1.6	bal Fe				
	UNS G92540		0.51-0.59	0.60-0.80		0.60-0.80		0.035 max	0.040 max	1.20-1.60	bal Fe				
AMS 6451	9254														
ASTM A401/A401M(96)	9254	Wir	0.51-0.59	0.6-0.8		0.6-0.8		0.035 max	0.04 max	1.2-1.6	bal Fe				
ASTM A752(93)	9254	Rod Wir	0.51-0.59	0.60-0.80		0.60-0.80		0.035 max	0.040 max	1.20-1.60	bal Fe	770			
ASTM A877/A877M(93)		Wir, valve spring qual, 1mm diam	0.51-0.59	0.60-0.80		0.50-0.80		0.025 max	0.025 max	1.20-1.60	bal Fe	2070-2240			
SAE 770(84)	9254*	Obs; see J1397(92)													
SAE J1249(95)	9254	Former SAE std valid for Wir Rod	0.51-0.59	0.60-0.80		0.60-0.80		0.035 max	0.040 max	1.20-1.60	bal Fe				

UNS numbers and US grades are provided as a means of cross referencing chemically similar alloys. Exchangability is only possible after independent examination of specifications. Tensile properties are minimum or typical as specified. UTS and YS as MPa. El as %. See Appendix for list of abbreviations used in Notes. * indicates obsolete material.

Specification	Designation	Notes	C	Cr	Cu	Mn	Mo	P	S	Si	Other	UTS	YS	El	Hard

Alloy Steel, Silicon Manganese, 9255

Australia

Specification	Designation	Notes	C	Cr	Cu	Mn	Mo	P	S	Si	Other	UTS	YS	El	Hard
AS 1444(96)	9255	H series, Hard/Tmp	0.5-0.6			0.70-1.05		0.050 max	0.050 max	1.60-2.20	bal Fe				
Belgium															
NBN 253-05	55Si7	Wire, Ann or Q/A Tmp 12 mm	0.52-0.6			0.6-0.9		0.04	0.04	1.5-2	bal Fe				
Bulgaria															
BDS 6742	55S2	Spring	0.52-0.60	0.30 max	0.30 max	0.60-0.90		0.035 max	0.035 max	1.50-2.00	Ni <=0.30; bal Fe				
China															
GB 1222(84)	55Si2Mn	Bar Flat Q/T	0.52-0.60	0.35 max	0.25 max	0.60-0.90		0.035 max	0.035 max	1.50-2.00	bal Fe				
GB 1222(84)	55Si2Mn	Bar Flat Q/T	0.52-0.60	0.35 max	0.25 max	0.60-0.90		0.035 max	0.035 max	1.50-2.00	Ni <=0.35; bal Fe	1275	1175	6	
Czech Republic															
CSN 413270	13270	Spring	0.58-0.68	0.3 max		0.65-0.9		0.04 max	0.04 max	1.5-1.9	bal Fe				
France															
AFNOR	55S7		0.53-0.59	0.45		0.6-0.9		0.035	0.035	1.6-2	bal Fe				
AFNOR NFA35553(82)	55S7	t<=4.5mm; Q	0.51-0.6	0.45 max	0.3 max	0.6-1	0.15 max	0.04 max	004 max	1.6-2	Ni <=0.4; Ti <=0.05; bal Fe				54 HRC
AFNOR NFA35571(84)	56SC7	t<=80mm; Q/T	0.53-0.59	0.2-0.45	0.3 max	0.6-0.9	0.15 max	0.035 max	0.035 max	1.6-2	Ni <=0.4; Ti <=0.05; bal Fe	1520-1770	1370	6	
AFNOR NFA35571(96)	51Si7	t<1000mm; Q/T	0.48-0.54	0.3 max	0.3 max	0.5-0.8	0.15 max	0.025 max	0.025 max	1.5-2	Ni <=0.4; Ti <=0.05; bal Fe	1450-1750	1300	6	
AFNOR NFA35571(96)	56Si7	t<1000mm; Q/T	0.52-0.6	0.3 max	0.3 max	0.6-0.9	0.15 max	0.025 max	0.025 max	1.5-2	Ni <=0.4; Ti <=0.05; bal Fe	1450-1750	1300	6	
AFNOR NFA35590(92)	51Si7	Soft ann	0.48-0.54	0.3 max	0.3 max	0.5-0.8	0.15 max	0.025 max	0.025 max	1.6-2	Ni <=0.4; Ti <=0.05; bal Fe				248 HB max
AFNOR NFA35590(92)	Y51S7	Soft ann	0.48-0.54	0.3 max	0.3 max	0.5-0.8	0.15 max	0.025 max	0.025 max	1.6-2	Ni <=0.4; Ti <=0.05; bal Fe				248 HB max
AFNOR NFA36102(93)	55S7*	HR; Strp; CR	0.51-0.6	0.45 max		0.6-0.9	0.10 max	0.025 max	0.02 max	1.6-2	Ni <=0.20; bal Fe				
AFNOR NFA36102(93)	55Si7RR	HR; Strp; CR	0.51-0.6	0.45 max		0.6-0.9	0.10 max	0.025 max	0.02 max	1.6-2	Ni <=0.20; bal Fe				
Germany															
DIN	WNr 1.0904*	Obs	0.52-0.6			0.7-1		0.05	0.05	1.5-1.8	bal Fe				
DIN 17222(79)	55Si7	CR spring hard	0.52-0.60			0.70-1.0		0.045 max	0.045 max	1.50-1.80	bal Fe	1470-1670	1080	6	
DIN 17222(79)	WNr 1.5026	CR spring hard	0.52-0.60			0.70-1.00		0.045 max	0.045 max	1.50-1.80	bal Fe	1470-1670	1080	6	
Hungary															
MSZ 2666(76)	55S	Spring HF	0.52-0.6		0.25 max	0.6-0.9		0.04 max	0.04 max	1.5-2	Ni <=0.3; bal Fe				
MSZ 2666(88)	55Si7	HF spring; 0-999mm	0.52-0.6			0.6-0.9		0.04 max	0.04 max	1.5-2	bal Fe				285 HB max
MSZ 2666(88)	55Si7	HF spring; 0-100.0mm; Q/T	0.52-0.6			0.6-0.9		0.04 max	0.04 max	1.5-2	bal Fe	1320	1130	6L	
India															
IS 1570	55Si2Mo90		0.5-0.6			0.8-1		0.05	0.05	1.5-2	bal Fe				
IS 4367	55Si2Mo90	Frg, Hard, Tmp	0.5-0.6			0.8-1		0.05	0.05	1.5-2	bal Fe				
Italy															
UNI 3545(68)	50Si7	Spring	0.42-0.55			0.4-0.8		0.035 max	0.035 max	1.5-2	bal Fe				
UNI 3545(80)	48Si7	Spring	0.42-0.52			0.5-0.8		0.035 max	0.035 max	1.5-2	bal Fe				
UNI 3545(80)	55Si7	Spring	0.52-0.6	0.15-0.45		0.6-0.9		0.035 max	0.035 max	1.5-2	bal Fe				
UNI 7064	55Si8		0.5-0.6	0.15-0.45		0.7-1		0.035	0.035	1.8-2.2	bal Fe				
UNI 7064(82)	48Si7	Spring	0.42-0.52			0.6-0.9		0.035 max	0.035 max	1.5-2	bal Fe				
UNI 8893(86)	48Si7	Spring	0.42-0.52			0.6-0.9		0.035 max	0.035 max	1.5-2	bal Fe				

UNS numbers and US grades are provided as a means of cross referencing chemically similar alloys. Exchangability is only possible after independent examination of specifications. Tensile properties are minimum or typical as specified. UTS and YS as MPa. El as %. See Appendix for list of abbreviations used in Notes. * indicates obsolete material.

Alloy Steel, Silicon Manganese, 9255 (Continued from previous page)

Specification	Designation	Notes	C	Cr	Cu	Mn	Mo	P	S	Si	Other	UTS	YS	El	Hard
Italy															
UNI 8893(86)	55Si7	Spring	0.52-0.6	0.15-0.45		0.6-0.9		0.035 max	0.035 max	1.5-2	bal Fe				
Mexico															
DGN B-203	9255*	Obs	0.5-0.6			0.7-0.95		0.04	0.04	1.8-2.2	bal Fe				
NMX-B-300(91)	9255	Bar	0.51-0.59			0.70-0.95		0.035 max	0.040 max	1.80-2.20	bal Fe				
Pan America															
COPANT 334	9255	Bar	0.51-0.59			0.7-0.95		0.035	0.04	1.8-2.2	bal Fe				
COPANT 514	9255	Tube	0.5-0.6			0.7-0.95		0.04	0.05	1.8-2.2	bal Fe				
Romania															
STAS 795	ARC4		0.52-0.6	0.3	0.25	0.6-0.9		0.04	0.04	1.5-2	Ni 0.3; bal Fe				
STAS 795(92)	51Si17A	Spring; HF	0.47-0.55	0.3 max	0.25 max	0.6-0.9		0.04 max	0.04 max	1.5-2	Pb <=0.15; bal Fe				
STAS 795(92)	56Si17A	Spring; HF	0.52-0.6	0.3 max	0.25 max	0.6-1		0.04 max	0.04 max	1.4-2	Pb <=0.15; bal Fe				
Russia															
GOST 14959	55S2		0.52-0.6	0.3	0.2	0.6-0.9		0.035	0.035	1.5-2	Ni 0.25; bal Fe				
GOST 14959	55S2A		0.53-0.58	0.3	0.2	0.6-0.9		0.025	0.025	1.5-2	Ni 0.25; bal Fe				
GOST 14959	55SGF		0.52-0.6	0.3	0.2	0.95-1.25		0.035	0.035	1.5-2	Ni 0.25; V 0.1-0.15; bal Fe				
GOST 14959(79)	55S2	Spring	0.52-0.6	0.3 max	0.2 max	0.6-0.9	0.15 max	0.035 max	0.035 max	1.5-2.0	Al <=0.1; Ni <=0.25; Ti <=0.05; bal Fe				
GOST 14959(79)	55S2A	Spring	0.53-0.58	0.3 max	0.2 max	0.6-0.9	0.15 max	0.025 max	0.025 max	1.5-2.0	Al <=0.1; Ni <=0.25; Ti <=0.05; bal Fe				
Spain															
UNE	F.144		0.5-0.6			0.6-1		0.04	0.04	1.5-2	bal Fe				
UNE	F.145	Spring	0.45-0.55			0.6-0.9	0.15 max	0.04 max	0.04 max	1.5-2	bal Fe				
UNE 36015(77)	50Si7	Spring	0.47-0.53			0.5-0.8	0.15 max	0.035 max	0.035 max	1.5-2	bal Fe				
UNE 36015(77)	56Si7	Spring	0.52-0.6			0.6-0.9	0.15 max	0.035 max	0.035 max	1.5-2	bal Fe				
UNE 36015(77)	F.1440*	see 56Si7	0.52-0.6			0.6-0.9	0.15 max	0.035 max	0.035 max	1.5-2	bal Fe				
UNE 36015(77)	F.1450*	see 50Si7	0.47-0.53			0.5-0.8	0.15 max	0.035 max	0.035 max	1.5-2	bal Fe				
Sweden															
SS 142085	2085	Spring	0.5-0.6	0.3		0.6-0.9		0.05	0.05	1.5-2	bal Fe				
SS 142090	2090	Spring	0.52-0.6			0.6-0.9		0.035 max	0.035 max	1.5-2	bal Fe				
UK															
BS 970/1(83)	250A53*	Blm Bil Slab Bar Rod Frg	0.50-0.57			0.70-1.00		0.05	0.05	1.7-2.1	bal Fe				
USA															
	AISI 9255	Smls mech tub	0.50-0.60			0.70-0.95		0.040 max	0.040 max	1.80-2.20	bal Fe				
	AISI 9255		0.51-0.59			0.7-0.95		0.035 max	0.04 max	1.8-2.2	bal Fe				
	UNS G92550		0.51-0.59			0.70-0.95		0.035 max	0.040 max	1.80-2.20	bal Fe				
ASTM A29/A29M(93)	9255	Bar	0.51-0.59			0.7-0.95		0.035 max	0.04 max	1.8-2.2	bal Fe				
ASTM A519(96)	9255	Smls mech tub	0.50-0.60			0.70-0.95		0.040 max	0.040 max	1.80-2.20	bal Fe				
ASTM A752(93)	9255	Rod Wir	0.51-0.59			0.70-0.95		0.035 max	0.040 max	1.80-2.20	bal Fe	770			
SAE 770(84)	9255*	Obs; see J1397(92)									bal Fe				
Yugoslavia															
	C.2132	Spring	0.5-0.55	0.4 max		0.5-0.8	0.15 max	0.05 max	0.05 max	1.5-1.8	bal Fe				

UNS numbers and US grades are provided as a means of cross referencing chemically similar alloys. Exchangability is only possible after independent examination of specifications. Tensile properties are minimum or typical as specified. UTS and YS as MPa. El as %. See Appendix for list of abbreviations used in Notes. * indicates obsolete material.

Specification	Designation	Notes	C	Cr	Cu	Mn	Mo	P	S	Si	Other	UTS	YS	El	Hard

Alloy Steel, Silicon Manganese, 9255 (Continued from previous page)

Yugoslavia

Specification	Designation	Notes	C	Cr	Cu	Mn	Mo	P	S	Si	Other	UTS	YS	El	Hard
	C.2133	Spring	0.52-0.6			0.7-1.0	0.15 max	0.04 max	0.04 max	1.5-1.8	bal Fe				
	C.2430	Spring	0.55-0.56	0.2-0.4		0.7-1.0	0.15 max	0.04 max	0.04 max	1.5-1.8	bal Fe				

Alloy Steel, Silicon Manganese, 9259

USA

Specification	Designation	Notes	C	Cr	Cu	Mn	Mo	P	S	Si	Other	UTS	YS	El	Hard
	AISI 9259	Bar Blm Bil Slab	0.56-0.64	0.45-0.65		0.75-1.00		0.035 max	0.040 max	0.70-1.10	bal Fe				
ASTM A331(95)	9259	Bar	0.56-0.64	0.45-0.65		0.75-1.00		0.035 max	0.04 max	0.70-1.10	bal Fe				
SAE J404(94)	9259	Bil Blm Slab Bar HR CF	0.56-0.64	0.45-0.65		0.75-1.00		0.030 max	0.040 max	0.70-1.10	bal Fe				

Alloy Steel, Silicon Manganese, 9260/9260H

Argentina

Specification	Designation	Notes	C	Cr	Cu	Mn	Mo	P	S	Si	Other	UTS	YS	El	Hard
IAS	IRAM 9260		0.56-0.64			0.75-1.00		0.035 max	0.040 max	1.80-2.20	bal Fe	1000	630	16	293 HB

Australia

Specification	Designation	Notes	C	Cr	Cu	Mn	Mo	P	S	Si	Other	UTS	YS	El	Hard
AS 1444(96)	9260	H series, Hard/Tmp	0.55-0.65			0.70-1.00		0.050 max	0.050 max	1.80-2.20	bal Fe				
AS 1444(96)	9260H	H series, Hard/Tmp	0.55-0.65			0.65-1.10		0.050 max	0.050 max	1.70-2.20	bal Fe				

Belgium

Specification	Designation	Notes	C	Cr	Cu	Mn	Mo	P	S	Si	Other	UTS	YS	El	Hard
NBN 253-05	60Si7	Wire, Ann or Q/A Tmp 20 mm	0.57-0.64			0.6-0.9		0.04	0.04	1.5-2	bal Fe				

Bulgaria

Specification	Designation	Notes	C	Cr	Cu	Mn	Mo	P	S	Si	Other	UTS	YS	El	Hard
BDS 6742	60S2	Spring	0.57-0.62	0.30 max	0.30 max	0.60-0.90		0.035 max	0.035 max	1.50-2.00	Ni <=0.30; bal Fe				
BDS 6742	60SGA	Spring	0.56-0.64	0.30 max	0.30 max	0.80-1.10		0.025 max	0.025 max	1.30-1.80	Ni <=0.30; bal Fe				

France

Specification	Designation	Notes	C	Cr	Cu	Mn	Mo	P	S	Si	Other	UTS	YS	El	Hard
AFNOR	60S7		0.55-0.65			0.7-1		0.05	0.05	1.5-2	bal Fe				
AFNOR	61S7		0.57-0.64	0.45 max		0.6-0.9		0.04 max	0.04 max	1.6-2	bal Fe				
AFNOR	760S7		0.55-0.65			0.7-1		0.05 max	0.05 max	1.5-2	bal Fe				
AFNOR NFA35571	61SC7*		0.57-0.64	0.2-0.45		0.6-0.9		0.035	0.035	1.6-2	bal Fe				
AFNOR NFA35571(96)	60Si7	t<1000mm; Q/T	0.56-0.64	0.3 max	0.3 max	0.7-1.1	0.15 max	0.025 max	0.025 max	1.5-2	Ni <=0.4; Ti <=0.05; bal Fe	1500-1800	1350	6	

Germany

Specification	Designation	Notes	C	Cr	Cu	Mn	Mo	P	S	Si	Other	UTS	YS	El	Hard
DIN	WNr 1.0909*	Obs	0.56-0.64			0.7-1		0.045 max	0.045 max	1.5-1.8	bal Fe				

Hungary

Specification	Designation	Notes	C	Cr	Cu	Mn	Mo	P	S	Si	Other	UTS	YS	El	Hard
MSZ 2666(76)	60S	Spring HF	0.57-0.65		0.25 max	0.6-0.9		0.04 max	0.04 max	1.5-2	Ni <=0.3; bal Fe				
MSZ 2666(88)	61SiCr7	HF spring; 0-100.0mm; Q/T	0.57-0.65	0.2-0.4		0.7-1		0.03 max	0.03 max	1.5-1.8	bal Fe	1370	1180	6L	
MSZ 2666(88)	61SiCr7	HF spring; 0-999mm	0.57-0.65	0.2-0.4		0.7-1		0.03 max	0.03 max	1.5-1.8	bal Fe				300 HB max
MSZ 4217(85)	60S	CR, Q/T or spring Strp; 0-2mm; roll, hard	0.57-0.65			0.6-0.9		0.04 max	0.04 max	1.5-2	bal Fe	780-1180			
MSZ 4217(85)	60S	CR, Q/T or spring Strp; 0-2mm	0.57-0.65			0.6-0.9		0.04 max	0.04 max	1.5-2	bal Fe	1280-1580		4L	
MSZ 4217(85)	60SM1	CR, Q/T or spring Strp; 0-2mm; roll, hard	0.55-0.65			1-1.4		0.04 max	0.04 max	1-1.5	bal Fe	780-1180			
MSZ 4217(85)	60SM1	CR, Q/T or spring Strp; 0-2mm	0.55-0.65			1-1.4		0.04 max	0.04 max	1-1.5	bal Fe	1280-1580		4L	

International

Specification	Designation	Notes	C	Cr	Cu	Mn	Mo	P	S	Si	Other	UTS	YS	El	Hard
ISO 683-14(92)	59Si7	Bar Rod Wir; Q/T, springs, Ann	0.55-0.63			0.60-1.00		0.030 max	0.030 max	1.60-2.00	bal Fe				255 HB
ISO 683-14(92)	61SiCr7	Bar Rod Wir; Q/T, springs, Ann	0.57-0.65	0.20-0.40		0.70-1.00		0.030 max	0.030 max	1.60-2.00					255 HB

UNS numbers and US grades are provided as a means of cross referencing chemically similar alloys. Exchangability is only possible after independent examination of specifications. Tensile properties are minimum or typical as specified. UTS and YS as MPa. El as %. See Appendix for list of abbreviations used in Notes. * indicates obsolete material.

Alloy Steel, Silicon Manganese, 9260/9260H (Continued from previous page)

Specification	Designation	Notes	C	Cr	Cu	Mn	Mo	P	S	Si	Other	UTS	YS	El	Hard
Italy															
UNI 3545(80)	60Si7	Spring	0.57-0.64			0.6-0.9		0.035 max	0.035 max	1.5-2	bal Fe				
UNI 3545(80)	60SiCr8	Spring	0.57-0.64	0.25-0.4		0.7-1		0.035 max	0.035 max	1.7-2.2	bal Fe				
Japan															
JIS G3311(88)	SUP6M	CR Strp; Spring	0.56-0.64	0.30 max		0.70-1.00		0.035 max	0.035 max	1.50-1.80	bal Fe				210-310 HV
JIS G4801(84)	SUP 6	Spring Q/T	0.56-0.64	0.30 max		0.70-1.00		0.035 max	0.035 max	1.50-1.80	bal Fe	1226	1079	9	363-429 HB
JIS G4801(84)	SUP 7	Bar Spring Q/T	0.56-0.64	0.30 max		0.70-1.00		0.035 max	0.035 max	1.80-2.20	bal Fe	1226	1079	9	363-429 HB
Mexico															
DGN B-203	9260*	Obs	0.55-0.65			0.75-1		0.04	0.04	1.8-2.2	bal Fe				
DGN B-297	9260*	Obs	0.56-0.64			0.75-1		0.035	0.04	1.8-2.2	bal Fe				
NMX-B-268(68)	9260H	Hard	0.55-0.65			0.65-1.10				1.70-2.20	bal Fe				
NMX-B-300(91)	9260	Bar	0.56-0.64			0.75-1.00		0.035 max	0.040 max	1.80-2.20	bal Fe				
NMX-B-300(91)	9260H	Bar	0.55-0.65			0.65-1.10				1.70-2.20	bal Fe				
Pan America															
COPANT 334	9260	Bar	0.56-0.64			0.75-1		0.04	0.04	1.8-2.2	bal Fe				
Poland															
PNH84032	60GS	Spring	0.56-0.64	0.3 max	0.25 max	0.8-1.1		0.03 max	0.03 max	1.3-1.8	Ni <=0.4; bal Fe				
PNH84032	60S2A	Spring	0.57-0.63	0.3 max	0.25 max	0.6-0.9		0.03 max	0.03 max	1.6-2	Ni <=0.4; bal Fe				
Romania															
STAS 795(92)	60Si15A	Spring; HF	0.55-0.65	0.3 max	0.25 max	0.8-1.1		0.04 max	0.04 max	1.3-1.8	Pb <=0.15; bal Fe				
Russia															
GOST 14959	6052A		0.58-0.63	0.3	0.2	0.6-0.9		0.025	0.025	1.6-2	Ni 0.25; bal Fe				
GOST 14959	60S2		0.57-0.65	0.3	0.2	0.6-0.9		0.035	0.035	1.5-2	Ni 0.25; bal Fe				
GOST 14959	60SGA		0.56-0.64	0.3 max	0.2 max	0.8-1		0.03 max	0.03 max	1.3-1.8	Ni <=0.25; bal Fe				
GOST 14959(79)	60S2G	Spring	0.55-0.65	0.3 max	0.2 max	0.7-1.0	0.15 max	0.035 max	0.035 max	1.8-2.2	Al <=0.1; Ni <=0.25; Ti <=0.05; S+P<=0.06; bal Fe				
Spain															
UNE	F.144	Spring	0.5-0.6			0.7-1	0.15 max	0.04 max	0.04 max	1.5-2	bal Fe				
UNE	F.144.B		0.57-0.63			0.6-0.9		0.035	0.035	1.6-2	bal Fe				
UNE 36015(60)	60Si7	Spring	0.57-0.64			0.6-0.9		0.035	0.035	1.5-2	bal Fe				
UNE 36015(60)	F.1441*	see 60Si7	0.57-0.64			0.6-0.9		0.035	0.035	1.5-2	bal Fe				
UNE 36015(77)	50Si7	Spring	0.47-0.53			0.5-0.8	0.15 max	0.035 max	0.035 max	1.5-2	bal Fe				
UNE 36015(77)	60Si7	Spring	0.57-0.64			0.6-0.9	0.15 max	0.035 max	0.035 max	1.5-2	bal Fe				
UNE 36015(77)	F.1441*	see 60Si7	0.57-0.64			0.6-0.9	0.15 max	0.035 max	0.035 max	1.5-2	bal Fe				
UK															
BS 970/1(83)	250A58*	Blm Bil Slab Bar Rod Frg	0.55-0.62			0.70-1.00		0.05	0.05	1.7-2.1	bal Fe				
BS 970/1(83)	250A61*	Blm Bil Slab Bar Rod Frg	0.58-0.65			0.70-1.00		0.05	0.05	1.7-2.1	bal Fe				
BS 970/2(88)	251A58	Spring	0.55-0.60	0.15-0.30		0.80-1.00	0.10 max	0.035 max	0.035 max	1.80-2.10	bal Fe				
BS 970/2(88)	251A60	Spring	0.57-0.62	0.25-0.40		0.80-1.00	0.12 max	0.035 max	0.035 max	1.80-2.10	bal Fe				

UNS numbers and US grades are provided as a means of cross referencing chemically similar alloys. Exchangability is only possible after independent examination of specifications. Tensile properties are minimum or typical as specified. UTS and YS as MPa. El as %. See Appendix for list of abbreviations used in Notes. * indicates obsolete material.

Specification	Designation	Notes	C	Cr	Cu	Mn	Mo	P	S	Si	Other	UTS	YS	El	Hard

Alloy Steel, Silicon Manganese, 9260/9260H (Continued from previous page)

UK

Specification	Designation	Notes	C	Cr	Cu	Mn	Mo	P	S	Si	Other	UTS	YS	El	Hard
BS 970/2(88)	251H60	Spring	0.57-0.62	0.25-0.40		0.80-1.00	0.12 max	0.035 max	0.035 max	1.80-2.10	bal Fe				
BS 970/5(72)	250A58*	Blm Bil Slab Bar Rod Frg	0.55-0.62			0.70-1.00				1.7-2.1	bal Fe				

USA

Specification	Designation	Notes	C	Cr	Cu	Mn	Mo	P	S	Si	Other	UTS	YS	El	Hard
	AISI 9260	Smls mech tub	0.55-0.65			0.70-1.00		0.040 max	0.040 max	1.80-2.20	bal Fe				
	AISI 9260	Bar Blm Bil Slab	0.56-0.64			0.75-1.00		0.035 max	0.040 max	1.80-2.20	bal Fe				
	AISI 9260H	Bar Blm Bil Slab	0.55-0.65			0.65-1.10		0.035 max	0.040 max	1.70-2.20	bal Fe				
	AISI 9260H	Smls mech tub, Hard	0.55-0.65			0.65-1.10		0.040 max	0.040 max	1.7-2.20	bal Fe				
	UNS G92600		0.56-0.64			0.75-1.00		0.035 max	0.040 max	1.80-2.20	bal Fe				
	UNS H92600		0.55-0.65			0.65-1.10				1.70-2.2	bal Fe				
ASTM A29/A29M(93)	9260	Bar	0.56-0.64			0.75-1		0.035 max	0.04 max	1.8-2.2	bal Fe				
ASTM A304(96)	9260H	Bar, hard bands spec	0.55-0.65	0.20 max	0.35 max	0.65-1.10	0.06 max	0.035 max	0.040 max	1.70-2.20	Ni <=0.25; Cu Ni Cr Mo trace allowed; bal Fe				
ASTM A322(96)	9260	Bar	0.56-0.64			0.75-1.00		0.035 max	0.040 max	1.80-2.20	bal Fe				
ASTM A331(95)	9260	Bar	0.56-0.64			0.75-1.00		0.035 max	0.040 max	1.80-2.20	bal Fe				
ASTM A331(95)	9260H	Bar	0.55-0.65	0.20 max	0.35 max	0.65-1.10	0.06 max	0.025 max	0.025 max	1.70-2.20	Ni <=0.25; bal Fe				
ASTM A519(96)	9260	Smls mech tub	0.55-0.65			0.70-1.00		0.040 max	0.040 max	1.80-2.20	bal Fe				
ASTM A752(93)	9260	Rod Wir	0.56-0.64			0.75-1.00		0.035 max	0.040 max	1.80-2.20	bal Fe	770			
SAE 770(84)	9260*	Obs; see J1397(92)									bal Fe				
SAE J1268(95)	9260H	Bar Rod Wir Sh Strp Tub; See std	0.55-0.65	0.20 max	0.35 max	0.65-1.10	0.06 max	0.030 max	0.040 max	1.70-2.20	Ni <=0.25; Cr, Ni, Cu, Mo not spec'd but acpt; bal Fe				
SAE J404(94)	9260	Bil Blm Slab Bar HR CF	0.56-0.64			0.75-1.00		0.030 max	0.040 max	1.80-2.20	bal Fe				
SAE J778	9260*	Obs; see J1249									bal Fe				

Yugoslavia

Specification	Designation	Notes	C	Cr	Cu	Mn	Mo	P	S	Si	Other	UTS	YS	El	Hard
	C.2331	Spring	0.55-0.65	0.4 max		0.7-1.0	0.15 max	0.05 max	0.05 max	1.5-1.8	bal Fe				

Alloy Steel, Silicon Manganese, 9262

Australia

Specification	Designation	Notes	C	Cr	Cu	Mn	Mo	P	S	Si	Other	UTS	YS	El	Hard
AS 1444(96)	9261	Standard series, Hard/Tmp	0.55-0.65	0.10-0.25		0.70-1.00		0.050 max	0.050 max	1.80-2.20	bal Fe				
AS 1444(96)	9261H	H series, Hard/Tmp	0.55-0.65	0.05-0.35		0.65-1.10		0.050 max	0.050 max	1.70-2.20	bal Fe				

Belgium

Specification	Designation	Notes	C	Cr	Cu	Mn	Mo	P	S	Si	Other	UTS	YS	El	Hard
NBN 253-05	60SiCr8		0.57-0.64	0.25-0.4		0.7-1		0.035	0.035	1.7-2.2	bal Fe				

France

Specification	Designation	Notes	C	Cr	Cu	Mn	Mo	P	S	Si	Other	UTS	YS	El	Hard
AFNOR NFA35552(84)	60SC7	t<=16mm; Q/T	0.55-0.65	0.45-0.7	0.3 max	0.6-0.9	0.15 max	0.035 max	0.035 max	1.3-1.8	Ni <=0.4; Ti <=0.05; bal Fe	1150-1350	950	8	
AFNOR NFA35552(84)	60SC7	160<t<=250mm; Q/T	0.55-0.65	0.45-0.7	0.3 max	0.6-0.9	0.15 max	0.035 max	0.035 max	1.3-1.8	Ni <=0.4; Ti <=0.05; bal Fe	750-980	600	12	
AFNOR NFA35571(96)	56SiCr7	t<1000mm; Q/T	0.53-0.59	0.2-0.45	0.3 max	0.6-0.9	0.15 max	0.025 max	0.025 max	1.6-2	Ni <=0.4; Ti <=0.05; bal Fe	1500-1800	1350	6	
AFNOR NFA35571(96)	61SiCr7	t<1000mm; Q/T	0.57-0.64	0.2-0.45	0.3 max	0.7-1	0.15 max	0.025 max	0.025 max	1.6-2	Ni <=0.4; Ti <=005; bal Fe	1550-1850	1400	5	

Germany

Specification	Designation	Notes	C	Cr	Cu	Mn	Mo	P	S	Si	Other	UTS	YS	El	Hard
DIN	60SiCr7		0.55-0.66	0.20-0.40		0.70-1.00		0.035 max	0.035 max	1.50-1.80	bal Fe				
DIN	WNr 1.5092		0.55-0.66	0.20-0.40		0.70-1.00		0.035 max	0.035 max	1.50-1.80	bal Fe				

Specification	Designation	Notes	C	Cr	Cu	Mn	Mo	P	S	Si	Other	UTS	YS	El	Hard

Alloy Steel, Silicon Manganese, 9262 (Continued from previous page)

Germany

Specification	Designation	Notes	C	Cr	Cu	Mn	Mo	P	S	Si	Other	UTS	YS	El	Hard
DIN 17221	WNr 1.0961*	Obs	0.55-0.65	0.2-0.4		0.7-1		0.045	0.045	1.5-1.8	bal Fe				

International

Specification	Designation	Notes	C	Cr	Cu	Mn	Mo	P	S	Si	Other	UTS	YS	El	Hard
ISO 683-14(92)	61SiCr7	Bar Rod Wir; Q/T, springs, Ann	0.57-0.65	0.20-0.40		0.70-1.00		0.030 max	0.030 max	1.60-2.00	bal Fe				255 HB

Italy

Specification	Designation	Notes	C	Cr	Cu	Mn	Mo	P	S	Si	Other	UTS	YS	El	Hard
UNI 7064(82)	67SiCr5	Spring	0.62-0.72	0.2-0.4		0.4-0.6		0.035 max	0.035 max	1.2-1.4	bal Fe				
UNI 8893(86)	67SiCr5	Spring	0.62-0.72	0.2-0.4		0.4-0.6		0.035 max	0.035 max	1.2-1.4	bal Fe				

Mexico

Specification	Designation	Notes	C	Cr	Cu	Mn	Mo	P	S	Si	Other	UTS	YS	El	Hard
DGN B-203	9262*	Obs	0.55-0.65	0.25-0.4		0.75-1		0.04	0.04	1.8-2.2	bal Fe				

Pan America

Specification	Designation	Notes	C	Cr	Cu	Mn	Mo	P	S	Si	Other	UTS	YS	El	Hard
COPANT 514	9262	Tube	0.55-0.65	0.25-0.4		0.75-1		0.04	0.05	1.8-2.2	bal Fe				

Romania

Specification	Designation	Notes	C	Cr	Cu	Mn	Mo	P	S	Si	Other	UTS	YS	El	Hard
STAS 11500/2(89)	X	Q/T	0.55-0.65	0.25-0.45		0.6-1		0.04 max	0.035 max	1.5-1.9	Ni <=0.4; Pb <=0.15; bal Fe				

Russia

Specification	Designation	Notes	C	Cr	Cu	Mn	Mo	P	S	Si	Other	UTS	YS	El	Hard
GOST 14959(79)	60S2	Spring	0.57-0.65	0.3 max	0.2 max	0.6-0.9	0.15 max	0.035 max	0.035 max	1.5-2.0	Al <=0.1; Ni <=0.25; Ti <=0.05; bal Fe				

Spain

Specification	Designation	Notes	C	Cr	Cu	Mn	Mo	P	S	Si	Other	UTS	YS	El	Hard
UNE	F.144	Spring	0.5-0.6			0.7-1	0.15 max	0.04 max	0.04 max	1.5-2	bal Fe				
UNE 36015(77)	60SiCr8	Spring	0.57-0.64	0.25-0.4		0.7-1	0.15 max	0.035 max	0.035 max	1.7-2.2	bal Fe				
UNE 36015(77)	F.1442*	see 60SiCr8	0.57-0.64	0.25-0.4		0.7-1	0.15 max	0.035 max	0.035 max	1.7-2.2	bal Fe				

UK

Specification	Designation	Notes	C	Cr	Cu	Mn	Mo	P	S	Si	Other	UTS	YS	El	Hard
BS 970/2(88)	925A60	Spring	0.55-0.65	0.20-0.40		0.70-1.00	0.20-0.30	0.035 max	0.035 max	1.70-2.10	bal Fe				

USA

Specification	Designation	Notes	C	Cr	Cu	Mn	Mo	P	S	Si	Other	UTS	YS	El	Hard
	AISI 9262	Smls mech tub	0.55-0.65	0.25-0.40		0.75-1.00		0.040 max	0.040 max	1.80-2.20	bal Fe				
	UNS G92620		0.55-0.65	0.25-0.40		0.75-1.00		0.035 max	0.040 max	1.80-2.20	bal Fe				
ASTM A519(96)	9262	Smls mech tub	0.55-0.65	0.25-0.40		0.75-1.00		0.040 max	0.040 max	1.80-2.20	bal Fe				
SAE J778	9262*	Obs; see J1249									bal Fe				

Alloy Steel, Silicon Manganese, A204(A)

Czech Republic

Specification	Designation	Notes	C	Cr	Cu	Mn	Mo	P	S	Si	Other	UTS	YS	El	Hard
CSN 415020	15020	High-temp const	0.12-0.2	0.3 max		0.5-0.8	0.25-0.35	0.04 max	0.04 max	0.15-0.35	bal Fe				

Europe

Specification	Designation	Notes	C	Cr	Cu	Mn	Mo	P	S	Si	Other	UTS	YS	El	Hard
EN 10028/2(92)	1.5415	High-temp, Press; 16<=t<=40mm	0.12-0.20	0.30 max	0.30 max	0.40-0.90	0.25-0.35	0.030 max	0.025 max	0.35 max	Ni <=0.30; bal Fe	440-590	270	24T	
EN 10028/2(92)	1.5415	High-temp, Press; 100<=t<=150mm	0.12-0.20	0.30 max	0.30 max	0.40-0.90	0.25-0.35	0.030 max	0.025 max	0.35 max	Ni <=0.30; bal Fe	420-570	220	19T	
EN 10028/2(92)	1.5415	High-temp, Press; 60<=t<=100mm	0.12-0.20	0.30 max	0.30 max	0.40-0.90	0.25-0.35	0.030 max	0.025 max	0.35 max	Ni <=0.30; bal Fe	430-560	240	22T	
EN 10028/2(92)	1.5415	High-temp, Press; 40<=t<=60mm	0.12-0.20	0.30 max	0.30 max	0.40-0.90	0.25-0.35	0.030 max	0.025 max	0.35 max	Ni <=0.30; bal Fe	440-590	260	23T	
EN 10028/2(92)	1.5415	High-temp, Press; 3<=t<=16mm	0.12-0.20	0.30 max	0.30 max	0.40-0.90	0.25-0.35	0.030 max	0.025 max	0.35 max	Ni <=0.30; bal Fe	440-590	275	24T	
EN 10028/2(92)	16Mo3	High-temp, Press; 60<=t<=100mm	0.12-0.20	0.30 max	0.30 max	0.40-0.90	0.25-0.35	0.030 max	0.025 max	0.35 max	Ni <=0.30; bal Fe	430-560	240	22T	
EN 10028/2(92)	16Mo3	High-temp, Press; 16<=t<=40mm	0.12-0.20	0.30 max	0.30 max	0.40-0.90	0.25-0.35	0.030 max	0.025 max	0.35 max	Ni <=0.30; bal Fe	440-590	270	24T	
EN 10028/2(92)	16Mo3	High-temp, Press; 40<=t<=60mm	0.12-0.20	0.30 max	0.30 max	0.40-0.90	0.25-0.35	0.030 max	0.025 max	0.35 max	Ni <=0.30; bal Fe	440-590	260	23T	
EN 10028/2(92)	16Mo3	High-temp, Press; 100<=t<=150mm	0.12-0.20	0.30 max	0.30 max	0.40-0.90	0.25-0.35	0.030 max	0.025 max	0.35 max	Ni <=0.30; bal Fe	420-570	220	19T	
EN 10028/2(92)	16Mo3	High-temp, Press; 3<=t<=16mm	0.12-0.20	0.30 max	0.30 max	0.40-0.90	0.25-0.35	0.030 max	0.025 max	0.35 max	Ni <=0.30; bal Fe	440-590	275	24T	

UNS numbers and US grades are provided as a means of cross referencing chemically similar alloys. Exchangability is only possible after independent examination of specifications. Tensile properties are minimum or typical as specified. UTS and YS as MPa. El as %. See Appendix for list of abbreviations used in Notes. * indicates obsolete material.

Specification	Designation	Notes	C	Cr	Cu	Mn	Mo	P	S	Si	Other	UTS	YS	El	Hard
Alloy Steel, Silicon Manganese, A204(A) (Continued from previous page)															
France															
AFNOR NFA36206(83)	15D3	t<=30mm	0.18 max	0.3 max	0.25 max	0.5-0.8	0.25-0.35	0.035 max	0.03 max	0.15-0.3	Ni <=0.3; Ti <=0.05; V <=0.04; bal Fe	430-550	265	25	
AFNOR NFA36206(83)	15D3	80<t<=120mm	0.18 max	0.3 max	0.25 max	0.5-0.8	0.25-0.35	0.035 max	0.03 max	0.15-0.3	Ni <=0.3; Ti <=0.05; V <=0.04; bal Fe	430-550	245	22	
AFNOR NFA36602(88)	15D3	80<t<=300mm	0.18 max	0.3 max	0.25 max	0.5-0.8	0.25-0.35	0.02 max	0.015 max	0.15-0.3	Ni <=0.3; Ti <=0.05; V <=0.04; bal Fe	430-550	245	24L, 22T	
AFNOR NFA36602(88)	15D3	t<=30mm	0.18 max	0.3 max	0.25 max	0.5-0.8	0.25-0.35	0.02 max	0.015 max	0.15-0.3	Ni <=0.3; Ti <=0.05; V <=0.04; bal Fe	430-550	265	27L, 25T	
Hungary															
MSZ 1741(89)	KL8	Boiler, Press ves; 40.1-60mm; HT	0.12-0.2		0.35 max	0.5-0.8	0.25-0.35	0.035 max	0.03 max	0.15-0.4	Al 0.02-0.06; Nb 0.01-0.06; Ni <=0.35; Ti 0.02-0.06; V 0.02-0.1; Zr=0.015-0.06; Al+Nb+V+Ti+Zr<=0.15; N<=Al/2+V/4+Nb/7+Ti/3.5<=0.015; bal Fe	440-560	250	20L	
MSZ 1741(89)	KL8	Boiler, Press ves; 0-20mm; HT	0.12-0.2		0.35 max	0.5-0.8	0.25-0.35	0.035 max	0.03 max	0.15-0.4	Al 0.02-0.06; Nb 0.01-0.06; Ni <=0.35; Ti 0.02-0.06; V 0.02-0.1; Zr=0.015-0.06; Al+Nb+V+Ti+Zr<=0.15; N<=Al/2+V/4+Nb/7+Ti/3.5<=0.015; bal Fe	440-560	270	20L	
India															
IS 1570/4(88)	15Ni5Cr4Mo1		0.12-0.18	0.75-1.25		0.6-1	0.08-0.15	0.07 max	0.06 max	0.15-0.35	Ni 1-1.5; bal Fe	990		9	
IS 1570/4(88)	15Ni7Cr4Mo2		0.12-0.18	0.75-1.25		0.6-1	0.1-0.2	0.07 max	0.06 max	0.15-0.35	Ni 1.5-2; bal Fe	1080		9	
Italy															
UNI 5869(75)	16Mo3	Press ves	0.12-0.2			0.5-0.8	0.25-0.35	0.03 max	0.03 max	0.15-0.35	bal Fe				
UNI 7660(77)	16Mo3KG	Press ves	0.12-0.2			0.5-0.8	0.25-0.35	0.03 max	0.03 max	0.15-0.4	Al <=0.012; bal Fe				
Poland															
PNH84024	16M	High-temp const	0.12-0.2	0.3 max	0.25 max	1.4-1.7	0.2-0.5	0.035 max	0.035 max	0.2-0.5	Al <=0.012; Ni <=0.4; V <=0.15; bal Fe				
Romania															
STAS 2883/3(88)	16Mo3	Heat res	0.12-0.2	0.3 max		0.5-0.8	0.25-0.4	0.035 max	0.03 max	0.15-0.35	Al 0.01-0.03; Pb <=0.15; Cr+Ni+Cu<=0.7; As<=0.07; bal Fe				
STAS 8184(87)	16Mo3	High-temp const	0.12-0.2	0.3 max		0.5-0.8	0.25-0.4	0.04 max	0.04 max	0.15-0.35	Pb <=0.15; Ti <=0.02; As<=0.05; bal Fe				
Spain															
UNE 36087(76/78)	16Mo3	Sh Plt Strp CR	0.12-0.2			0.5-0.8	0.25-0.35	0.035 max	0.035 max	0.15-0.35	bal Fe				
UNE 36087(76/78)	F.2601*	see 16Mo3	0.12-0.2			0.5-0.8	0.25-0.35	0.035 max	0.035 max	0.15-0.35	bal Fe				
Sweden															
SS 142912	2912	High-temp const	0.12-0.2	0.25 max		0.4-0.9	0.25-0.35	0.035 max	0.03 max	0.1-0.35	Al <=0.012; Ni <=0.3; bal Fe				
USA															
	UNS K11820		0.18 max			0.90 max	0.41-0.64	0.035 max	0.040 max	0.13-0.32	bal Fe				
ASTM A204/A204M(95)	A	Press ves plt, t>100mm	0.25 max			0.90 max	0.45-0.60	0.035 max	0.035 max	0.15-0.40	bal Fe	65-85	37	23	
ASTM A204/A204M(95)	A	Press ves plt, t 50-100mm	0.23 max			0.90 max	0.45-0.60	0.035 max	0.035 max	0.15-0.40	bal Fe	65-85	37	23	
ASTM A204/A204M(95)	A	Press ves plt, t 25-50mm	0.21 max			0.90 max	0.45-0.60	0.035 max	0.035 max	0.15-0.40	bal Fe	65-85	37	23	
ASTM A204/A204M(95)	A	Press ves plt, t<=25mm	0.18 max			0.90 max	0.45-0.60	0.035 max	0.035 max	0.15-0.40	bal Fe	65-85	37	23	
Yugoslavia															
	C.7100	High-temp const	0.12-0.2			0.5-0.8	0.25-0.35	0.04 max	0.04 max	0.15-0.35	bal Fe				

UNS numbers and US grades are provided as a means of cross referencing chemically similar alloys. Exchangability is only possible after independent examination of specifications. Tensile properties are minimum or typical as specified. UTS and YS as MPa. El as %. See Appendix for list of abbreviations used in Notes. * indicates obsolete material.

Specification	Designation	Notes	C	Cr	Cu	Mn	Mo	P	S	Si	Other	UTS	YS	El	Hard

Alloy Steel, Silicon Manganese, A204(A) (Continued from previous page)

Yugoslavia

| | C.7100 | High-temp const | 0.12-0.2 | | | 0.5-0.7 | 0.25-0.35 | 0.04 max | 0.04 max | 0.15-0.35 | bal Fe | | | | |

Alloy Steel, Silicon Manganese, A656(1)

USA

| | UNS K11804 | | 0.18 max | | | 1.60 max | | 0.025 max | 0.035 max | 0.60 max | N 0.005-0.030; V 0.030-0.20; bal Fe | | | | |

Alloy Steel, Silicon Manganese, A738(A)

Czech Republic

| CSN 411531 | 11531 | | 0.2 max | 0.3 max | | 1.5 max | | 0.035 max | 0.03 max | 0.55 max | Al 0.02-0.120; bal Fe | | | | |

Europe

EN 10025(90)	Fe510D2	Obs EU desig; struct HR FF; QS 150<t<=250mm	0.22 max			1.6 max		0.035 max	0.035 max	0.55 max	CEV; bal Fe	440-630	275-285	17	
EN 10025(90)	Fe510D2	Obs EU desig; struct HR FF; QS 40<t<=100mm	0.22 max			1.6 max		0.035 max	0.035 max	0.55 max	CEV; bal Fe	490-630	325-335	20-21	
EN 10025(90)	Fe510D2	Obs EU desig; struct HR FF; QS 16<t<=40mm	0.20 max			1.6 max		0.035 max	0.035 max	0.55 max	CEV; bal Fe	490-630	345	22	
EN 10025(90)	Fe510D2	Obs EU desig; struct HR FF; QS t<=16mm	0.20 max			1.6 max		0.035 max	0.035 max	0.55 max	CEV; bal Fe	410-580	355	14-22	

France

AFNOR NFA36205(82)	A52FP*	Press ves; t<=25mm; Norm	0.20 max	0.25 max	0.30 max	1.00-1.60	0.10 max	0.03 max	0.02 max	0.5 max	Nb <=0.010; Ni <=0.4; Ti <=0.05; V <=0.05; bal Fe	510-620	335	22	
AFNOR NFA36205(82)	A52FP*	Press ves; 80<t<=110mm; Norm	0.20 max	0.25 max	0.30 max	1.00-1.60	0.10 max	0.03 max	0.02 max	0.5 max	Nb <=0.010; Ni <=0.4; Ti <=0.05; V <=0.05; bal Fe	510-620	305	20	
AFNOR NFA36205(82)	A52FPR*	Press ves; 80<t<=110mm; Norm	0.20 max	0.25 max	0.30 max	1.00-1.60	0.15 max	0.03 max	0.02 max	0.5 max	Nb <=0.010; Ni <=0.4; Ti <=0.05; V <=0.05; bal Fe	510-620	305	20	
AFNOR NFA36205(82)	A52FPR*	Press ves; t<=25mm; Norm	0.20 max	0.25 max	0.30 max	1.00-1.60	0.15 max	0.03 max	0.02 max	0.5 max	Nb <=0.010; Ni <=0.4; Ti <=0.05; V <=0.05; bal Fe	510-620	335	22	
AFNOR NFA36601(80)	A52FP	800<t<=300mm; Q/T	0.2 max	0.25 max	0.18 max	1-1.6	0.1 max	0.03 max	0.02 max	0.5 max	Nb <=0.04; Ni <=0.5; Ti <=0.05; V <=0.05; Cu+6Sn<=0.33; bal Fe	510-620	305	22L; 20T	
AFNOR NFA36601(80)	A52FP	t<=30mm; Norm	0.2 max	0.25 max	0.18 max	1-1.6	0.1 max	0.03 max	0.02 max	0.5 max	Nb <=0.04; Ni <=0.5; Ti <=0.05; V <=0.05; Cu+6Sn<=0.33; bal Fe	510-620	325	24L; 22T	

Spain

UNE 36080(85)	AE355-D	Gen Struct	0.22 max			1.6 max	0.15 max	0.035 max	0.035 max	0.55 max	N <=0.009; bal Fe				
UNE 36080(85)	F.6215*	see AE355-D	0.22 max			1.6 max	0.15 max	0.035 max	0.035 max	0.55 max	N <=0.009; bal Fe				
UNE 36087(74)	A52RBII	Sh Plt	0.25 max			0.8-1.7	0.15 max	0.045 max	0.04 max	0.55 max	Al 0.015-0.10; bal Fe				

Sweden

| SS 142174 | 2174 | Gen Struct; FF; Non-rimming | 0.18 max | 0.2 max | | 1.4 max | | 0.04 max | 0.04 max | 0.5 max | N=0.-0.009/0.-0.012; C+Mn/6=0.-0.32; bal Fe | | | | |

USA

	UNS K12447		0.24 max	0.25 max	0.35 max	1.50 max	0.08 max	0.035 max	0.040 max	0.15-0.50	Ni <=0.50; bal Fe				
ASTM A738(96)	Grade A	HT, Press ves plt, mod/low-temp; t>65mm	0.24 max	0.25 max	0.35 max	1.60 max	0.08 max	0.035 max	0.035 max	0.15-0.50	Nb <=0.04; Ni <=0.50; V <=0.07; Nb+V<=0.08	515-655	310	20	
ASTM A738(96)	Grade A	HT, Press ves plt, mod/low-temp; t<=65mm	0.24 max	0.25 max	0.35 max	1.50 max	0.08 max	0.035 max	0.035 max	0.15-0.50	Nb <=0.04; Ni <=0.50; V <=0.07; Nb+V<=0.08	515-655	310	20	

UNS numbers and US grades are provided as a means of cross referencing chemically similar alloys. Exchangability is only possible after independent examination of specifications. Tensile properties are minimum or typical as specified. UTS and YS as MPa. El as %. See Appendix for list of abbreviations used in Notes. * indicates obsolete material.

Specification	Designation	Notes	C	Cr	Cu	Mn	Mo	P	S	Si	Other	UTS	YS	El	Hard
Alloy Steel, Silicon Manganese, K11802															
USA															
	UNS K11802		0.18 max	0.15 max	0.35 max	1.30 max	0.06 max	0.035 max	0.040 max	0.15-0.35	Ti <=0.005; V 0.02-0.13; bal Fe				
Alloy Steel, Silicon Manganese, K11803															
Germany															
DIN 17102(89)	WStE355	see P355NH	0.2 max	0.3 max	0.2 min	0.9-1.65	0.08 max	0.035 max	0.03 max	0.1-0.5	N <=0.02; Nb <=0.05; Ni <=0.3; V <=0.1; bal Fe				
Spain															
UNE 36087(76/78)	14MnMo55	Sh Plt Strp CR	0.1-0.18			0.9-1.4	0.4-0.6	0.035 max	0.035 max	0.15-0.35	V 0.04-0.08; bal Fe				
UNE 36087(76/78)	F.2611*	see 14MnMo55	0.1-0.18			0.9-1.4	0.4-0.6	0.035 max	0.035 max	0.15-0.35	V 0.04-0.08; bal Fe				
USA															
	UNS K11803		0.18 max			1.45 max		0.035 max	0.040 max	0.13-0.32	V 0.07-0.16; bal Fe				
Alloy Steel, Silicon Manganese, K11805															
USA															
	UNS K11805		0.18 max			1.30 max		0.035 max	0.040 max	0.15-0.35	V >=0.02; bal Fe				
Alloy Steel, Silicon Manganese, K91955															
Czech Republic															
CSN 417436	17436	Valve	0.46-0.56	2.9-3.6		17.0-19.0		0.1 max	0.035 max	0.3-0.8	Ni <=1.3; bal Fe				
USA															
	UNS K91955	Non-magnetic	0.40-0.60	3.50-6.00		16.00-20.00		0.08 max	0.025 max	0.20-0.65	Ni <=2.00; bal Fe				
Alloy Steel, Silicon Manganese, No equivalents identified															
Czech Republic															
CSN 413240	13240	Q/T	0.33-0.41	0.3 max		1.1-1.4		0.035 max	0.035 max	1.1-1.4	bal Fe				
CSN 413320	13320	Wear res; Mn	0.12-0.2	0.3 max		2.0-2.4		0.035 max	0.035 max	0.3-0.6	bal Fe				
Europe															
EN 10028/4(94)	1.6217	High-temp, Press; 30<=t<=50mm	0.16 max			0.85-1.70		0.025 max	0.015 max	0.50 max	Al >=0.020; Nb <=0.05; Ni 0.30-0.85; V <=0.05; Cr+Cu+Mo<=0.50; bal Fe	490-610	345	22	
EN 10028/4(94)	1.6217	High-temp, Press; 0<=t<=30mm	0.16 max			0.85-1.70		0.025 max	0.015 max	0.50 max	Al >=0.020; Nb <=0.05; Ni 0.15-0.85; V <=0.05; Cr+Cu+Mo<=0.50; bal Fe	490-610	355	22	
EN 10028/4(94)	13MnNi6-3	High-temp, Press; 30<=t<=50mm	0.16 max			0.85-1.70		0.025 max	0.015 max	0.50 max	Al >=0.020; Nb <=0.05; Ni 0.30-0.85; V <=0.05; Cr+Cu+Mo<=0.50; bal Fe	490-610	345	22	
EN 10028/4(94)	13MnNi6-3	High-temp, Press; 0<=t<=30mm	0.16 max			0.85-1.70		0.025 max	0.015 max	0.50 max	Al >=0.020; Nb <=0.05; Ni 0.15-0.85; V <=0.05; Cr+Cu+Mo<=0.50; bal Fe	490-610	355	22	
France															
AFNOR NFA35552(84)	45S7	40<t<=100mm; Q/T	0.42-0.5	0.3 max		0.5-0.8	0.15 max	0.035 max	0.035 max	1.6-2	Ni <=0.4; Ti <=0.05; bal Fe	640-780	510	15	
AFNOR NFA35552(84)	45S7	t<=16mm; Q/T	0.42-0.5	0.3 max		0.5-0.8	0.15 max	0.035 max	0.035 max	1.6-2	Ni <=0.4; Ti <=0.05; bal Fe	980-1180	780	11	
AFNOR NFA35553(82)	45S7	t<=4.5mm; Q	0.42-0.5	0.3 max		0.5-0.8	0.15 max	0.04 max	0.04 max	1.6-2	Ni <=0.4; Ti <=0.05; bal Fe				52 HRC
AFNOR NFA35571(84)	46S7	t<=80mm; Q/T	0.43-0.49	0.3 max		0.5-0.8	0.15 max	0.035 max	0.035 max	1.6-2	Ni <=0.4; Ti <=0.05; bal Fe	1450-1700	1300	7	

UNS numbers and US grades are provided as a means of cross referencing chemically similar alloys. Exchangability is only possible after independent examination of specifications. Tensile properties are minimum or typical as specified. UTS and YS as MPa. El as %. See Appendix for list of abbreviations used in Notes. * indicates obsolete material.

Specification	Designation	Notes	C	Cr	Cu	Mn	Mo	P	S	Si	Other	UTS	YS	El	Hard

Alloy Steel, Silicon Manganese, No equivalents identified

France

Specification	Designation	Notes	C	Cr	Cu	Mn	Mo	P	S	Si	Other	UTS	YS	El	Hard
AFNOR NFA35571(96)	46Si7	t<1000mm; Q/T	0.43-0.49	0.3 max		0.5-0.8	0.15 max	0.025 max	0.025 max	1.6-2	Ni <=0.4; Ti <=0.05; bal Fe	1400-1700	1250	7	
AFNOR NFA35590(92)	46Si7	Soft ann	0.43-0.49	0.3 max		0.5-0.8	0.15 max	0.025 max	0.025 max	1.6-2	Ni <=0.4; Ti <=0.05; bal Fe				241 HB max

Hungary

Specification	Designation	Notes	C	Cr	Cu	Mn	Mo	P	S	Si	Other	UTS	YS	El	Hard
MSZ 2666(76)	60SM1	Spring HF	0.55-0.65		0.25 max	1-1.4		0.04 max	0.04 max	1-1.5	Ni <=0.3; bal Fe				

Romania

Specification	Designation	Notes	C	Cr	Cu	Mn	Mo	P	S	Si	Other	UTS	YS	El	Hard
STAS 791(88)	35MnSi13	Struct/const; Q/T	0.31-0.39	0.25 max		1.1-1.4		0.035 max	0.035 max	1.1-1.4	Pb <=0.15; Ti <=0.02; bal Fe				
STAS 791(88)	35MnSi13S	Struct/const; Q/T	0.31-0.39	0.25 max		1.1-1.4		0.035 max	0.02-0.04	1.1-1.4	Pb <=0.15; Ti <=0.02; bal Fe				
STAS 791(88)	35MnSi13X	Struct/const; Q/T	0.31-0.39	0.25 max		1.1-1.4		0.025 max	0.025 max	1.1-1.4	Pb <=0.15; Ti <=0.02; bal Fe				
STAS 791(88)	35MnSi13XS	Struct/const; Q/T	0.31-0.39	0.25 max		1.1-1.4		0.025 max	0.02-0.035	1.1-1.4	Pb <=0.15; Ti <=0.02; bal Fe				

USA

Specification	Designation	Notes	C	Cr	Cu	Mn	Mo	P	S	Si	Other	UTS	YS	El	Hard
	UNS K15590*	Obs; see G92540									bal Fe				
ASTM A29/A29M(93)	9254	Bar	0.51-0.59	0.60-0.80		0.6-0.8		0.035 max	0.04 max	1.2-1.6	bal Fe				
ASTM A738(96)	Grade C	HT, Press ves plt, mod/low-temp; t>100-150mm	0.20 max	0.25 max	0.35 max	1.60 max	0.08 max	0.025 max	0.025 max	0.15-0.50	Ni <=0.50; V <=0.05; bal Fe	480-620	315	20	
ASTM A738(96)	Grade C	HT, Press ves plt, mod/low-temp; t>65-100mm	0.20 max	0.25 max	0.35 max	1.60 max	0.08 max	0.025 max	0.025 max	0.15-0.50	Ni <=0.50; V <=0.05; bal Fe	515-655	380	22	
ASTM A738(96)	Grade C	HT, Press ves plt, mod/low-temp; t<=65mm	0.20 max	0.25 max	0.35 max	1.50 max	0.08 max	0.025 max	0.025 max	0.15-0.50	Ni <=0.50; V <=0.05; bal Fe	550-690	415	22	
SAE J1268(95)	9259H	Hard	0.56-0.64	0.45-0.65	0.35 max	0.65-1.10	0.06 max	0.030 max	0.040 max	0.70-1.20	Ni <=0.25; Cu, Mo, Ni not spec'd, but acpt, bal Fe				
SAE J1268(95)	E9259H	Hard	0.56-0.64	0.45-0.65	0.35 max	0.65-1.10	0.06 max	0.025 max	0.025 max	0.70-1.20	Ni <=0.25; Cu, Mo, Ni not spec'd but acpt; bal Fe				

Yugoslavia

Specification	Designation	Notes	C	Cr	Cu	Mn	Mo	P	S	Si	Other	UTS	YS	El	Hard
	C.2330	Spring	0.55-0.65			0.7-1.0	0.15 max	0.04 max	0.04 max	1.3-1.5	bal Fe				
	C.2332	Spring	0.6-0.68			0.7-1.0	0.15 max	0.04 max	0.04 max	1.5-1.8	bal Fe				

Specification	Designation	Notes	C	Cr	Mn	Mo	Ni	P	S	Si	Other	UTS	YS	El	Hard
Alloy Steel, Unclassified, 3140															
Bulgaria															
BDS 6354	40ChN	Struct	0.36-0.44	0.45-0.75	0.50-0.80	0.15 max	1.00-1.40	0.035 max	0.035 max	0.17-0.37	Cu <=0.30; Ti <=0.03; V <=0.05; W <=0.2; bal Fe				
Japan															
JIS G4052(79)	SNC631H	Ni-Cr Struct Hard	0.26-0.35	0.55-1.05	0.30-0.70		2.45-3.00	0.030 max	0.030 max	0.15-0.35	Cu <=0.30; bal Fe				
Romania															
STAS 791(80)	41CrNi12	Struct/const; Q/T	0.37-0.45	0.45-0.75	0.4-0.8		1-1.4	0.035 max	0.035 max	0.17-0.37	Pb <=0.15; bal Fe				
Russia															
GOST 4543(71)	40ChN	Q/T	0.36-0.44	0.45-0.75	0.5-0.8	0.15 max	1.0-1.4	0.035 max	0.035 max	0.17-0.37	Al <=0.1; Cu <=0.3; Ti <=0.05; bal Fe				
USA															
	AISI 3140	Smls mech tub	0.38-0.43	0.55-0.75	0.70-0.90		1.10-1.40	0.040 max	0.040 max	0.15-0.35	bal Fe				
	UNS G31400	Ni-Cr	0.38-0.43	0.55-0.75	0.70-0.90		1.10-1.40	0.040 max	0.040 max	0.15-0.35	bal Fe				
ASTM A519(96)	3140	Smls mech tub, Ni-Cr	0.38-0.43	0.55-0.75	0.70-0.90		1.10-1.40	0.040 max	0.040 max	0.15-0.35	bal Fe				
SAE J775(93)	3140	Engine poppet valve, Mart	0.38-0.43	0.55-0.75	0.70-0.90		1.10-1.40	0.040 max	0.040 max	0.15-0.35	bal Fe	790	520		
SAE J775(93)	NV4*	Engine poppet valve, Mart; former SAE grade	0.38-0.43	0.55-0.75	0.70-0.90		1.10-1.40	0.040 max	0.040 max	0.15-0.35	bal Fe	790	520		
Alloy Steel, Unclassified, 5623															
USA															
	UNS K91456	High Expansion	0.55-0.65	5.00-6.00			8.50-10.50	0.040 max	0.030 max	1.00 max	bal Fe				
AMS 5625C(90)		Bar, High exp, cw, t<=25.4mm	0.55-0.65	5.00-6.00			8.50-10.50	0.040 max	0.030 max	1.00 max	bal Fe	862	689	16	255-331 HB
AMS 5625C(90)		Bar, High exp, cw, 25.4<t<=27mm	0.55-0.65	5.00-6.00			8.50-10.50	0.040 max	0.030 max	1.00 max	bal Fe	827	621	16	248-331 HB
Alloy Steel, Unclassified, 5624															
USA															
	UNS K91505	High Expansion	0.50-0.60	3.00-5.00	3.50-5.50	0.50 max	11.0-14.0	0.040 max	0.030 max	0.50 max	Cu <=0.50; bal Fe				
AMS 5624D(89)		Bar, High exp, cw	0.50-0.60	3.00-5.00	3.50-5.50	0.50 max	11.00-14.00	0.040 max	0.030 max	0.50 max	Cu <=0.50; bal Fe				269-352 HB
Alloy Steel, Unclassified, 6354															
USA															
	UNS K11914	NAX 9115-AC	0.10-0.17	0.50-0.75	0.50-0.80	0.15-0.25	0.25 max	0.025 max	0.035 max	0.60-0.90	Cu <=0.35; Zr 0.05-0.10; bal Fe				
Alloy Steel, Unclassified, 6523															
USA															
	UNS K91472	HP 9-4-20	0.17-0.23	0.65-0.85	0.20-0.40	0.90-1.10	8.50-9.50	0.010 max	0.010 max	0.20 max	Co 4.25-4.75; Cu <=0.35; V 0.06-0.12; bal Fe				
Alloy Steel, Unclassified, A203(A)															
India															
IS 1570/4(88)	11C15	0-100mm	0.16 max	0.3 max	1.3-1.7	0.15 max	0.4 max	0.07 max	0.06 max	0.1-0.35	bal Fe	460-560	270	26	
Japan															
JIS G4052(79)	SNC415H	Ni-Cr Struct Hard	0.11-0.18	0.20-0.55	0.30-0.70		1.95-2.50	0.030 max	0.030 max	0.15-0.35	Cu <=0.30; bal Fe				
USA															
	UNS K21703		0.17 max		0.70 max		2.03-2.57	0.035 max	0.040 max	0.13-0.32	bal Fe				
ASTM A203/A203M(97)	A	Press ves plt, t<=50mm	0.17 max		0.70 max		2.10-2.50	0.035 max	0.035 max	0.15-0.40	bal Fe	65-85 ksi	37 ksi	23	
ASTM A203/A203M(97)	A	Press ves plt, t=50-100mm	0.20 max		0.80 max		2.10-2.50	0.035 max	0.035 max	0.15-0.40	bal Fe	65-85 ksi	37 ksi	23	

UNS numbers and US grades are provided as a means of cross referencing chemically similar alloys. Exchangability is only possible after independent examination of specifications. Tensile properties are minimum or typical as specified. UTS and YS as MPa. El as %. See Appendix for list of abbreviations used in Notes. * indicates obsolete material.

Specification	Designation	Notes	C	Cr	Mn	Mo	Ni	P	S	Si	Other	UTS	YS	El	Hard
Alloy Steel, Unclassified, A203(A) (Continued from previous page)															
USA															
ASTM A203/A203M(97)	A	Press ves plt, t>100mm	0.23 max		0.80 max		2.10-2.50	0.035 max	0.035 max	0.15-0.40	bal Fe	65-85 ksi	37 ksi	23	
Alloy Steel, Unclassified, A203(B)															
USA															
	UNS K22103		0.21 max		0.70 max		2.03-2.57	0.035 max	0.040 max	0.13-0.45	bal Fe				
Alloy Steel, Unclassified, A203(D)															
Italy															
UNI 5949(67)	18Ni14	Smls tube w/low-temp impact test	0.18 max		0.65 max		3.2-3.8	0.035 max	0.035 max	0.2-0.35	bal Fe				
USA															
	UNS K31718		0.17 max		0.70 max		3.18-3.82	0.035 max	0.040 max	0.13-0.32	bal Fe				
ASTM A203/A203M(97)	D	Press ves plt, t=50-100mm	0.20 max		0.80 max		3.25-3.75	0.035 max	0.035 max	0.15-0.40	bal Fe	65-85 ksi	37 ksi	23	
ASTM A203/A203M(97)	D	Press ves plt, t<=50mm	0.17 max		0.70 max		3.25-3.75	0.035 max	0.035 max	0.15-0.40	bal Fe	65-85 ksi	37 ksi	23	
Alloy Steel, Unclassified, A203(E)															
India															
IS 1570/7(92)	1	3-16mm; Sh Plt	0.2 max	0.25 max	0.4-1.2	0.1 max	0.3 max	0.03 max	0.03 max	0.1-0.35	bal Fe	360-480	205	26	
IS 1570/7(92)	1	100.1-150mm; Sh Plt	0.2 max	0.25 max	0.4-1.2	0.1 max	0.3 max	0.03 max	0.03 max	0.1-0.35	bal Fe	360-480	170	24	
IS 1570/7(92)	Fe360H	100.1-150mm; Sh Plt	0.2 max	0.25 max	0.4-1.2	0.1 max	0.3 max	0.03 max	0.03 max	0.1-0.35	bal Fe	360-480	170	24	
IS 1570/7(92)	Fe360H	3-16mm; Sh Plt	0.2 max	0.25 max	0.4-1.2	0.1 max	0.3 max	0.03 max	0.03 max	0.1-0.35	bal Fe	360-480	205	26	
USA															
	UNS K32018		0.20 max		0.70 max		3.18-3.82	0.035 max	0.040 max	0.13-0.32	bal Fe				
ASTM A203/A203M(97)	B	Press ves plt, t<=50mm	0.21 max		0.70 max		2.10-2.50	0.035 max	0.035 max	0.15-0.40	bal Fe	70-90 ksi	40 ksi	21	
ASTM A203/A203M(97)	B	Press ves plt, t>100mm	0.25 max		0.80 max		2.10-2.50	0.035 max	0.035 max	0.15-0.40	bal Fe	70-90 ksi	40 ksi	21	
ASTM A203/A203M(97)	B	Press ves plt, t=50-100mm	0.24 max		0.80 max		2.10-2.50	0.035 max	0.035 max	0.15-0.40	bal Fe	70-90 ksi	40 ksi	21	
ASTM A203/A203M(97)	E	Press ves plt, t<=50mm	0.20 max		0.70 max		3.25-3.75	0.035 max	0.035 max	0.15-0.40	bal Fe	70-90 ksi	40 ksi	21	
ASTM A203/A203M(97)	E	Press ves plt, t=50-100mm	0.23 max		0.80 max		3.25-3.75	0.035 max	0.035 max	0.15-0.40	bal Fe	70-90 ksi	40 ksi	21	
ASTM A203/A203M(97)	F	Press ves plt, t<=50mm	0.20 max		0.70 max		3.25-3.75	0.035 max	0.035 max	0.15-0.40	bal Fe	80-100 ksi	55 ksi	20	
ASTM A203/A203M(97)	F	Press ves plt, t=50-100mm	0.23 max		0.80 max		3.25-3.75	0.035 max	0.035 max	0.15-0.40	bal Fe	75-95 ksi	55 ksi	20	
Alloy Steel, Unclassified, A225(C)															
USA															
	UNS K12524		0.25 max		1.60 max		0.37-0.73	0.035 max	0.040 max	0.13-0.45	V 0.11-0.20; bal Fe				
ASTM A225A225M(93)	D	Press ves plt, t<=75mm	0.20 max		1.70 max		0.40-0.70	0.035	0.035	0.15-0.40	V 0.10-0.18; bal Fe	80-105	60	19	
ASTM A225A225M(93)	D	Press ves plt, t>75mm	0.20 max		1.70 max		0.40-0.70	0.035	0.035	0.15-0.40	V 0.10-0.18; bal Fe	75-100	55	19	
Alloy Steel, Unclassified, A289(A)															
Czech Republic															
CSN 417455	17455	Corr res	0.63-0.73	2.7-3.7	8.0-10.0		7.0-9.0	0.06 max	0.04 max	0.8 max	bal Fe				
USA															
	UNS K91555	Non-magnetic	0.40-0.75	3.50-6.00	6.00-10.00		6.00-10.00	0.05 max	0.045 max	0.20-0.65	bal Fe				

UNS numbers and US grades are provided as a means of cross referencing chemically similar alloys. Exchangability is only possible after independent examination of specifications. Tensile properties are minimum or typical as specified. UTS and YS as MPa. El as %. See Appendix for list of abbreviations used in Notes. * indicates obsolete material.

Specification	Designation	Notes	C	Cr	Mn	Mo	Ni	P	S	Si	Other	UTS	YS	El	Hard
Alloy Steel, Unclassified, A291(3)															
USA															
	UNS K14507		0.45 max	1.25 max	0.40-0.90	0.15 min	0.50	0.040 max	0.040 max	0.15-0.30	Cu <=0.35; V <=0.10; bal Fe				
ASTM A291(95)	3	Gear pinion frg, t>510mm	0.45 max	1.25 max	0.40-0.90	0.15 min	0.50	0.040 max	0.040 max	0.35 max	Cu 0.35; V <=0.50; bal Fe	725	550	18 L 14T	223-262 HB
ASTM A291(95)	3	Gear pinion frg, 250<t<=510mm	0.45 max	1.25 max	0.40-0.90	0.15 min	0.50	0.040 max	0.040 max	0.35 max	Cu 0.35; V <=0.50; bal Fe	725	550	19 L 16T	223-262 HB
ASTM A291(95)	3	Gear pinion frg, t<=250mm	0.45 max	1.25 max	0.40-0.90	0.15 min	0.50	0.040 max	0.040 max	0.35 max	Cu 0.35; V <=0.50; bal Fe	725	550	19 L	223-262 HB
Alloy Steel, Unclassified, A302(C)															
USA															
	UNS K12039		0.20 max		1.07-1.62	0.41-0.64	0.37-0.73	0.035 max	0.040 max	0.13-0.45	bal Fe				
ASTM A302/302M(97)	C	Press ves plt, t>50mm	0.25 max		1.15-1.50	0.45-0.60	0.40-0.70	0.035 max	0.035 max	0.15-0.40	bal Fe	550-690	345	20	
ASTM A302/302M(97)	C	Press ves plt, 25<t<=50mm	0.23 max		1.15-1.50	0.45-0.60	0.40-0.70	0.035 max	0.035 max	0.15-0.40	bal Fe	550-690	345	20	
ASTM A302/302M(97)	C	Press ves plt, t<=25mm	0.20 max		1.15-1.50	0.45-0.60	0.40-0.70	0.035 max	0.035 max	0.15-0.40	bal Fe	550-690	345	20	
Alloy Steel, Unclassified, A325(2)															
USA															
	UNS K11900		0.13-0.25		0.67 min			0.048 max	0.058 max		B >=0.0005; bal Fe				
Alloy Steel, Unclassified, A325(C)															
USA															
	UNS K12033		0.14-0.26	0.27-0.53	0.76-1.39		0.22-0.53	0.040 max	0.045 max	0.13-0.32	Cu 0.17-0.53; V >=0.010; bal Fe				
Alloy Steel, Unclassified, A325(E)															
USA															
	UNS K12254		0.18-0.27	0.55-0.95	0.56-1.04		0.27-0.63	0.045 max	0.045 max	0.13-0.32	Cu 0.27-0.63; bal Fe				
Alloy Steel, Unclassified, A333(3)															
Mexico															
NMX-B-197(85)	Grade 3	Smls weld tub for low temp service	0.19 max		0.31-0.64		3.18-3.82	0.05 max	0.05 max	0.18-0.37	bal Fe	448	241	30	
USA															
	UNS K31918		0.19 max		0.31-0.64		3.18-3.82	0.05 max	0.05 max	0.18-0.37	bal Fe				
Alloy Steel, Unclassified, A333(4)															
Argentina															
IAS	IRAM 3112		0.09-0.15	0.40-0.70	0.30-0.60		0.50-0.80	0.035 max	0.035 max	0.35 max	bal Fe	460	340	25	137 HB
USA															
	UNS K11267	High Strength, Low Alloy, Precipitation Hardening	0.12 max	0.44-1.01	0.50-1.05		0.47-0.98	0.04 max	0.04 max	0.08-0.37	Al 0.04-0.30; Cu 0.40-0.75; bal Fe				
Alloy Steel, Unclassified, A333(7)															
Italy															
UNI 5949(67)	18Ni9	Smls tube w/low-temp impact test	0.18 max		0.9 max		2.1-2.6	0.035 max	0.035 max	0.15-0.3	bal Fe				
Mexico															
NMX-B-197(85)	Grade 7	Smls weld tub for low temp service	0.19 max		0.90 max		2.03-2.57	0.04 max	0.05 max	0.13-0.32	bal Fe	448	241	30	
USA															
	UNS K21903		0.19 max		0.90 max		2.03-2.57	0.04 max	0.05 max	0.13-0.32	bal Fe				

Specification	Designation	Notes	C	Cr	Mn	Mo	Ni	P	S	Si	Other	UTS	YS	El	Hard
Alloy Steel, Unclassified, A333(8)															
Europe															
EN 10028/4(94)	1.5663	High-temp, Press; 30<=t<=50mm Q/T	0.10 max		0.30-0.80	0.10 max	8.50-10.0	0.015 max	0.005 max	0.35 max	Al >=0.020; V <=0.01; Cr+Cu+Mo<=0.50; bal Fe	680-820	575	18	
EN 10028/4(94)	1.5663	High-temp, Press; 0<=t<=14.99mm NNT	0.10 max		0.30-0.80	0.10 max	8.50-10.0	0.015 max	0.005 max	0.35 max	Al >=0.020; V <=0.01; Cr+Cu+Mo<=0.50; bal Fe	680-820	585	18	
EN 10028/4(94)	1.5663	High-temp, Press; 15<=t<=30mm Q/T	0.10 max		0.30-0.80	0.10 max	8.50-10.0	0.015 max	0.005 max	0.35 max	Al >=0.020; V <=0.01; Cr+Cu+Mo<=0.50; bal Fe	680-820	585	18	
EN 10028/4(94)	X7Ni9	High-temp, Press; 15<=t<=30mm Q/T	0.10 max		0.30-0.80	0.10 max	8.50-10.0	0.015 max	0.005 max	0.35 max	Al >=0.020; V <=0.01; Cr+Cu+Mo<=0.50; bal Fe	680-820	585	18	
EN 10028/4(94)	X7Ni9	High-temp, Press; 0<=t<=14.99mm NNT	0.10 max		0.30-0.80	0.10 max	8.50-10.0	0.015 max	0.005 max	0.35 max	Al >=0.020; V <=0.01; Cr+Cu+Mo<=0.50; bal Fe	680-820	585	18	
EN 10028/4(94)	X7Ni9	High-temp, Press; 30<=t<=50mm Q/T	0.10 max		0.30-0.80	0.10 max	8.50-10.0	0.015 max	0.005 max	0.35 max	Al >=0.020; V <=0.01; Cr+Cu+Mo<=0.50; bal Fe	680-820	575	18	
Italy															
UNI 5949(67)	X12Ni09	Smls tube w/low-temp impact test	0.13 max		0.9 max		8.4-9.6	0.035 max	0.035 max	0.15-0.3	bal Fe				
Mexico															
NMX-B-197(85)	Grade 8	Smls weld tub for low temp service	0.13 max		0.90 max		8.40-9.60	0.045 max	0.045 max	0.13-0.32	bal Fe	689	517	22	
UK															
BS 1506(90)	509-650	Bolt matl Pres/Corr res; t<=75mm; Q/T	0.10 max	0.25 max	0.30-0.80		8.50-10.0	0.025 max	0.020 max	0.15-0.35	Al >=0.015; bal Fe	650	480	18	
BS 1506(90)	509-690	Bolt matl Pres/Corr res; t<=100mm; Q/T	0.10 max	0.25 max	0.30-0.80		8.50-10.0	0.025 max	0.020 max	0.15-0.35	Al >=0.015; bal Fe	690-850	580	18	
USA															
	UNS K81340		0.13 max		0.90 max		8.40-9.60	0.045 max	0.045 max	0.13-0.32	bal Fe				
ASTM A353/A353M(93)	A353	Press ves plt	0.13 max		0.90 max		8.50-9.50	0.035 max	0.035 max	0.15-0.40	bal Fe	690-825	515	20	
ASTM A522/A522M(95b)	A522(1)		0.13 max		0.90 max		8.5-9.5	0.025 max	0.025 max	0.15-0.30	bal Fe	690	515	22	
ASTM A553/A553M(95)	Type I	Q/T, Press ves plt	0.13 max		0.90 max		8.50-9.50	0.035 max	0.035 max	0.15-0.40	bal Fe	690-825	585	20.0	
ASTM A844/A844M(93)		Press ves plt, Quen	0.13 max		0.90 max		8.50-9.50	0.020 max	0.020 max	0.15-0.40	bal Fe	690-825	585	20.0	
Alloy Steel, Unclassified, A350(LF3)															
France															
AFNOR NFA36205/4(94)	12Ni14	Press ves; Sh Plt Strp; Ni-Alloy; low-temp; t<=30mm; N/T; Q/T	0.15 max	0.3 max	0.30-0.80	0.15 max	3.25-3.75	0.02 max	0.01 max	0.35 max	Cu <=0.30; V <=0.05; Cr+Cu+Mo<=0.5; bal Fe	490-640	355	22	
Hungary															
MSZ 4400(78)	AH80	Tough at subzero; 0-100mm; Norm	0.14 max		0.5-0.8	0.15-0.25	3-4	0.035 max	0.03 max	0.15-0.4	V 0.05-0.15; 100mm V=0.05-0.25; bal Fe	590	440	20L	
USA															
	UNS K32025		0.20 max		0.90 max		3.25-3.75	0.035 max	0.040 max	0.20-0.35	bal Fe				
Alloy Steel, Unclassified, A350(LF5)															
Hungary															
MSZ 4400(78)	AH60	Tough at subzero; 0-100mm; Norm	0.18 max		0.5-0.8		1.5-2.2	0.035 max	0.03 max	0.15-0.4	V 0.05-0.15; 100mm V=0.05-0.25; bal Fe	510	345	20L	
Italy															
UNI 7660(77)	14Ni6KG	Press ves	0.18 max	0.25 max	0.8-1.5	0.1 max	1.3-1.7	0.03 max	0.03 max	0.35 max	Al 0.015-0.120; V <=0.08; bal Fe				
UNI 7660(77)	14Ni6KT	Press ves	0.18 max	0.25 max	0.8-1.5	0.1 max	1.3-1.7	0.03 max	0.03 max	0.35 max	Al 0.015-0.120; V <=0.05; bal Fe				

UNS numbers and US grades are provided as a means of cross referencing chemically similar alloys. Exchangability is only possible after independent examination of specifications. Tensile properties are minimum or typical as specified. UTS and YS as MPa. El as %. See Appendix for list of abbreviations used in Notes. * indicates obsolete material.

Specification	Designation	Notes	C	Cr	Mn	Mo	Ni	P	S	Si	Other	UTS	YS	El	Hard
Alloy Steel, Unclassified, A350(LF5) (Continued from previous page)															
USA															
	UNS K13050		0.30 max		1.35 max		1.0-2.0	0.035 max	0.040 max	0.20-0.35	bal Fe				
Alloy Steel, Unclassified, A350(LF9)															
USA															
	UNS K22036		0.20 max		0.40-1.06		1.60-2.24	0.035 max	0.040 max		Cu 0.75-1.25; bal Fe				
Alloy Steel, Unclassified, A355(D)															
France															
AFNOR NFA35551(75)	30CAD612		0.28-0.35	1.5-1.8	0.5-0.8	0.25-0.4	0.4 max	0.035 max	0.035 max	0.2-0.4	Al 1-1.3; Cu <=0.3; Ti <=0.05; bal Fe				
Hungary															
MSZ 7779(79)	34CAMo	Nitriding; 0-100mm; Q/T	0.28-0.35	2.8-3.3	0.4-0.7	0.3-0.5	0.3 max	0.03 max	0.035 max	0.2-0.5	bal Fe	780-930	590	14L	
MSZ 7779(79)	34CAMo	Nitriding; 0-999mm; soft ann	0.28-0.35	2.8-3.3	0.4-0.7	0.3-0.5	0.3 max	0.03 max	0.035 max	0.2-0.5	bal Fe				248 HB max
MSZ 7779(88)	34CrA1Mo54	Nitriding; 0-70mm; Q/T/drawn and SR	0.3-0.37	1-1.3	0.5-0.8	0.15-0.25		0.03 max	0.03 max	0.5 max	Al 0.8-1.2; bal Fe	800-1000	600	14L	
MSZ 7779(88)	34CrA1Mo54	Nitriding; 0-250mm; soft ann	0.3-0.37	1-1.3	0.5-0.8	0.15-0.25		0.03 max	0.03 max	0.5 max	Al 0.8-1.2; bal Fe				248 HB max
International															
ISO 683-10(87)	33CrAlMo54	Nitriding, <70mm	0.30-0.37	1.00-1.30	0.50-0.80	0.15-0.25		0.030 max	0.035 max	0.50 max	Al 0.80-1.20; bal Fe	800-1000	600		950 HV
ISO R683-10(75)	X/3*	Nitriding	0.3-0.37	1-1.3	0.5-0.8	0.15-0.25	0.4 max	0.03 max	0.035 max	0.2-0.5	Al 0.8-1.2; Co <=0.1; Cu <=0.3; Pb <=0.15; Ti <=0.05; V <=0.1; W <=0.1; bal Fe				
Italy															
UNI 7356(74)	34CrAlMo7	Nitriding	0.31-0.36	1.5-1.8	0.5-0.8	0.25-0.4		0.03 max	0.035 max	0.2-0.5	Al 0.3-0.5; V 0.1-0.2; bal Fe				
UNI 8077(80)	34CrAlMo7	Nitriding	0.31-0.36	1.5-1.8	0.5-0.8	0.25-0.4		0.03 max	0.035 max	0.2-0.5	Al 0.8-1.2; V 0.1-0.2; bal Fe				
UNI 8552(84)	34CrAlMo7	Nitriding	0.31-0.36	1.5-1.8	0.5-0.8	0.25-0.4		0.03 max	0.035 max	0.2-0.5	Al 0.8-1.2; V 0.1-0.2; bal Fe				
Spain															
UNE 36014(76)	34CrAlMo5	Nitriding	0.3-0.37	1-1.3	0.5-0.8	0.15-0.3		0.035 max	0.035 max	0.2-0.5	Al 0.8-1.2; bal Fe				
UNE 36014(76)	F.1741*	see 34CrAlMo5	0.3-0.37	1-1.3	0.5-0.8	0.15-0.3		0.035 max	0.035 max	0.2-0.5	Al 0.8-1.2; bal Fe				
USA															
	UNS K23510	Nitriding	0.33-0.38	1.00-1.30	0.50-0.70	0.15-0.25		0.035 max	0.040 max	0.15-0.35	Al 0.95-1.30; bal Fe				
Alloy Steel, Unclassified, A423(2)															
USA															
	UNS K11540		0.15 max		0.50-1.00	0.10 min	0.40-1.10	0.04 max	0.05 max		Cu 0.30-1.00; bal Fe				
Alloy Steel, Unclassified, A441															
Germany															
DIN 17102(89)	StE355	see P355N	0.2 max	0.3 max	0.9-1.65	0.08 max	0.3 max	0.035 max	0.03 max	0.1-0.5	Cu <=0.2; N <=0.02; Nb <=0.05; V <=0.1; bal Fe				
India															
IS 1570/4(88)	20Ni2Cr2Mo2	Const, Spring	0.18-0.23	0.4-0.6	0.7-0.9	0.15-0.25	0.4-0.7	0.07 max	0.06 max	0.15-0.35	bal Fe				
IS 1570/4(88)	21C10BT		0.18-0.23	0.3 max	0.8-1.1	0.15 max	0.4 max	0.07 max	0.06 max	0.15-0.3	B 0.0005-0.003; bal Fe				
USA															
	UNS K12211		0.22 max		0.85-1.25			0.04 max	0.05 max	0.40 max	Cu >=0.20; V >=0.02; bal Fe				

Specification	Designation	Notes	C	Cr	Mn	Mo	Ni	P	S	Si	Other	UTS	YS	El	Hard
Alloy Steel, Unclassified, A508(1,1A)															
USA															
	UNS K13502		0.35 max	0.25 max	0.40-1.05	0.10 max	0.40 max	0.025 max	0.025 max	0.15-0.40	V <=0.05; bal Fe				
ASTM A508/508M(95)	Grade 1	Q/I vacuum-treated frg for press ves	0.35 max	0.25 max	0.40-1.05	0.10 max	0.40 max	0.025 max	0.025 max	0.15-0.40	bal Fe	485-655	250	20	
ASTM A508/508M(95)	Grade 1A	Q/T vacuum-treated frg for press ves	0.30 max	0.25 max	0.70-1.35	0.10 max	0.40 max	0.025 max	0.025 max	0.15-0.40	bal Fe	485-655	250	20	
Alloy Steel, Unclassified, A508(3)															
USA															
	UNS K12042		0.25 max	0.25 max	1.20-1.50	0.45-0.60	0.40-1.00	0.025 max	0.025 max	0.15-0.40	V <=0.05; bal Fe				
ASTM A508/508M(95)	Grade 3 Class 1	Q/T vacuum-treated frg for press ves	0.25 max	0.25 max	1.20-1.50	0.45-0.60	0.40-1.00	0.025 max	0.025 max	0.15-0.40	bal Fe	550-725	345	18	
ASTM A508/508M(95)	Grade 3 Class 2	Q/T vacuum-treated frg for press ves	0.25 max	0.25 max	1.20-1.50	0.45-0.60	0.40-1.00	0.025 max	0.025 max	0.15-0.40	bal Fe	620-795	450	16	
Alloy Steel, Unclassified, A522(II)															
USA															
	UNS K71340		0.13 max		0.90 max		7.40-8.60	0.035 max	0.040 max	0.13-0.32	bal Fe				
ASTM A522/A522M(95b)	2		0.13 max		0.90 max		7.5-8.5	0.025 max	0.025 max	0.15-0.30	bal Fe	690	515	22	
ASTM A553/A553M(95)	Type II	Q/T, Press ves plt	0.13 max		0.90 max		7.50-8.50	0.035 max	0.035 max	0.15-0.40	bal Fe	690-825	585	20.0	
Alloy Steel, Unclassified, A541(4)															
USA															
	UNS K11800		0.18 max	0.15 max	1.30 max	0.05 max	0.25 max	0.035 max	0.040 max	0.15-0.35	V 0.02-0.12; bal Fe				
ASTM A541/541M(95)	Grade 1C	Q/T frg for press ves components	0.18 max	0.15 max	1.30 max	0.05 max	0.25 max	0.025 max	0.025 max	0.15-0.35	V 0.02-0.12; Si req vary; bal Fe	550-720	340	18	
Alloy Steel, Unclassified, A579(72)															
USA															
	UNS K92940	Superstrength	0.03 max		0.10 max	4.6-5.2	17.0-19.0	0.01 max	0.01 max	0.10 max	Al 0.05-0.15; Co 7.5-8.5; Ti 0.30-0.50; bal Fe				
ASTM A579(96)	72	Superstrength frg	0.03 max		0.10 max	4.6-5.2	17.0-19.0	0.01 max	0.01 max	0.10 max	Al 0.05-0.15; B 0.003; Co 7.0-8.5; Ti 0.30-0.50; Ca 0.06; Zr 0.02; bal Fe	1760	1720	10	321 HB
Alloy Steel, Unclassified, A65															
USA															
	UNS K11210		0.12 min								Cu >=0.20; bal Fe				
Alloy Steel, Unclassified, A692															
USA															
	UNS K12121		0.17-0.26		0.46-0.94	0.42-0.68		0.045 max	0.045 max	0.18-0.37	bal Fe				
Alloy Steel, Unclassified, A707(L3)															
USA															
	UNS K12510		0.25 max		1.05-1.60			0.030 max	0.035 max	0.32 max	Cu >=0.18; N 0.005-0.035; V 0.03-0.13; bal Fe				
ASTM A707/A707M(98)	A 707 (L3)	Frg flanges; low-temp; class values given	0.22 max	0.30 max	1.15-1.50	0.12 max	0.40 max	0.025 max	0.025 max	0.30 max	Cu >=0.20; N 0.010-0.030; Nb <=0.02; V 0.04-0.11; bal Fe	415-515	290-415	20-22	149-235 HB
Alloy Steel, Unclassified, A707(L7)															
USA															
	UNS K32218		0.22 max		1.00 max		3.18-3.82	0.030 max	0.035 max	0.37 max	bal Fe				

UNS numbers and US grades are provided as a means of cross referencing chemically similar alloys. Exchangability is only possible after independent examination of specifications. Tensile properties are minimum or typical as specified. UTS and YS as MPa. El as %. See Appendix for list of abbreviations used in Notes. * indicates obsolete material.

Specification	Designation	Notes	C	Cr	Mn	Mo	Ni	P	S	Si	Other	UTS	YS	El	Hard

Alloy Steel, Unclassified, A707(L7) (Continued from previous page)

USA

| ASTM A707/A707M(98) | A 707 (L7) | Frg flanges; low-temp; class values given | 0.20 max | 0.30 max | 0.90 max | 0.12 max | 3.2-3.7 | 0.025 max | 0.025 max | 0.35 max | Cu <=0.40; Nb <=0.02; V <=0.05; Cr+Mo+V<=0.32; bal Fe | 415-455 | 290-360 | 22 | 149-217 HB |

Alloy Steel, Unclassified, A753(Alloy 1)

USA

| | UNS K94490 | Fe-Ni Ferromagnetic | 0.05 max | 0.30 max | 0.80 max | 0.30 max | 43.5-46.5 | 0.03 max | 0.01 max | 0.50 max | Co <=0.50; Cu <=0.30; bal Fe | | | | |

Alloy Steel, Unclassified, A753(Alloy 2)

USA

| | UNS K94840 | Fe-Ni Ferromagnetic | 0.05 max | 0.30 max | 0.80 max | 0.30 max | 47.0-49.0 | 0.03 max | 0.01 max | 0.50 max | Co <=0.50; Cu <=0.30; bal Fe | | | | |

Alloy Steel, Unclassified, A765(III)

Romania

| STAS 10382(88) | 10Ni35 | Tough at subzero | 0.12 max | 0.3 max | 0.4 max | 0.05 max | 3.2-3.8 | 0.035 max | 0.035 max | 0.15-0.5 | Al 0.02-0.06; Cu <=0.2; Pb <=0.15; Ti <=0.02; As<=0.05; bal Fe | | | | |

USA

| | UNS K32026 | | 0.20 max | 0.20 max | 0.90 max | 0.06 max | 3.25-3.75 | 0.020 max | 0.020 max | 0.15-0.35 | Al <=0.05; Cu <=0.35; V <=0.05; bal Fe | | | | |
| ASTM A765/765M(98) | Grade III | Press ves frg | 0.20 max | 0.20 max | 0.90 max | 0.06 max | 3.3-3.8 | 0.020 max | 0.020 max | 0.15-0.35 | Al <=0.05; Cu <=0.35; V <=0.05; bal Fe | 485-655 | 260 | 22 | |

Alloy Steel, Unclassified, A841(A)

USA

ASTM A841/A841M(95)	Grade A, Class 1	TMCP Press ves plt; t<=40mm	0.20 max	0.25 max	0.70-1.35	0.08 max	0.25 max	0.030 max	0.030 max	0.15-0.50	Al >=0.020; Cu <=0.35; Nb <=0.03; V <=0.06; bal Fe	485-620	345	22	
ASTM A841/A841M(95)	Grade A, Class 1	TMCP Press ves plt; t=40-65mm	0.20 max	0.25 max	1.00-1.60	0.08 max	0.25 max	0.030 max	0.030 max	0.15-0.50	Al >=0.020; Cu <=0.35; Nb <=0.03; V <=0.06; bal Fe	485-620	345	22	
ASTM A841/A841M(95)	Grade A, Class 1	TMCP Press ves plt; t 65-100mm	0.20 max	0.25 max	1.00-1.60	0.08 max	0.25 max	0.030 max	0.030 max	0.15-0.50	Al >=0.020; Cu <=0.35; Nb <=0.03; V <=0.06; bal Fe	450-585	310	22	
ASTM A841/A841M(95)	Grade A, Class 2	TMCP Press ves plt; t=65-100mm	0.20 max	0.25 max	1.00-1.60	0.08 max	0.25 max	0.030 max	0.030 max	0.15-0.50	Al >=0.020; Cu <=0.35; Nb <=0.03; V <=0.06; bal Fe	515-655	380	22	
ASTM A841/A841M(95)	Grade A, Class 2	TMCP Press ves plt; t <=40mm	0.20 max	0.25 max	0.70-1.35	0.08 max	0.25 max	0.030 max	0.030 max	0.15-0.50	Al >=0.020; Cu <=0.35; Nb <=0.03; V <=0.06; bal Fe	550-690	415	22	
ASTM A841/A841M(95)	Grade A, Class 2	TMCP Press ves plt; t 40-65mm	0.20 max	0.25 max	1.00-1.60	0.08 max	0.25 max	0.030 max	0.030 max	0.15-0.50	Al >=0.020; Cu <=0.35; Nb <=0.03; V <=0.06; bal Fe	550-690	415	22	

Alloy Steel, Unclassified, A841(B)

Europe

| EN 10028/5(96) | 1.8821 | High-temp, Press; t<=40mm | 0.14 max | | 1.60 max | | 0.50 max | 0.025 max | 0.020 max | 0.50 max | Al >=0.020; N <=0.020; Nb <=0.05; Ti <=0.05; V <=0.10; Cr+Cu+Mo<=0.60, V+Nb+Ti<=0.15; bal Fe | 450-610 | 355 | 22 | |
| EN 10028/5(96) | 1.8821 | High-temp, Press; 40<t<=63mm | 0.14 max | | 1.60 max | | 0.50 max | 0.025 max | 0.020 max | 0.50 max | Al >=0.020; N <=0.020; Nb <=0.05; Ti <=0.05; V <=0.10; Cr+Cu+Mo<=0.60, V+Nb+Ti<=0.15; bal Fe | 450-610 | 345 | 22 | |

UNS numbers and US grades are provided as a means of cross referencing chemically similar alloys. Exchangability is only possible after independent examination of specifications. Tensile properties are minimum or typical as specified. UTS and YS as MPa. El as %. See Appendix for list of abbreviations used in Notes. * indicates obsolete material.

Specification	Designation	Notes	C	Cr	Mn	Mo	Ni	P	S	Si	Other	UTS	YS	El	Hard

Alloy Steel, Unclassified, A841(B) (Continued from previous page)

Europe

Specification	Designation	Notes	C	Cr	Mn	Mo	Ni	P	S	Si	Other	UTS	YS	El	Hard
EN 10028/5(96)	1.8824	High-temp, Press; 40<t<=63mm	0.16 max		1.70 max	0.20 max	0.50 max	0.025 max	0.020 max	0.50 max	Al >=0.020; N <=0.015; Nb <=0.05; Ti <=0.05; V <=0.10; Cr+Cu+Mo<=0.60, V+Nb+Ti<=0.15; bal Fe	500-660	390	19	
EN 10028/5(96)	1.8824	High-temp, Press; t<=16mm	0.16 max		1.70 max	0.20 max	0.50 max	0.025 max	0.020 max	0.50 max	Al >=0.020; N <=0.015; Nb <=0.05; Ti <=0.05; V <=0.10; Cr+Cu+Mo<=0.60, V+Nb+Ti<=0.15; bal Fe	500-660	420	19	
EN 10028/5(96)	1.8824	High-temp, Press; 16<=t<=40mm	0.16 max		1.70 max	0.20 max	0.50 max	0.025 max	0.020 max	0.50 max	Al >=0.020; N <=0.015; Nb <=0.05; Ti <=0.05; V <=0.10; Cr+Cu+Mo<=0.60, V+Nb+Ti<=0.15; bal Fe	500-660	400	19	
EN 10028/5(96)	1.8826	High-temp, Press; 40<t<=63mm	0.16 max		1.70 max	0.20 max	0.50 max	0.025 max	0.020 max	0.60 max	Al >=0.020; N <=0.015; Nb <=0.05; Ti <=0.05; V <=0.10; Cr+Cu+Mo<=0.60, V+Nb+Ti<=0.15; bal Fe	530-720	430	17	
EN 10028/5(96)	1.8826	High-temp, Press; t<=16mm	0.16 max		1.70 max	0.20 max	0.50 max	0.025 max	0.020 max	0.60 max	Al >=0.020; N <=0.015; Nb <-0.05; Ti <=0.05; V <=0.10; Cr+Cu+Mo<=0.60, V+Nb+Ti<=0.15; bal Fe	530-720	460	17	
EN 10028/5(96)	1.8826	High-temp, Press; 16<=t<=40mm	0.16 max		1.70 max	0.20 max	0.50 max	0.025 max	0.020 max	0.60 max	Al >=0.020; N <=0.015; Nb <=0.05; Ti <=0.05; V <=0.10; Cr+Cu+Mo<=0.60, V+Nb+Ti<=0.15; bal Fe	530-720	440	17	
EN 10028/5(96)	1.8828	High-temp, Press; t<=16mm	0.16 max		1.70 max	0.20 max	0.50 max	0.020 max	0.015 max	0.50 max	Al >=0.020; N <=0.015; Nb <=0.05; Ti <=0.05; V <=0.10; Cr+Cu+Mo<=0.60, V+Nb+Ti<=0.15; bal Fe	500-660	420	19	
EN 10028/5(96)	1.8828	High-temp, Press; 40<t<=63mm	0.16 max		1.70 max	0.20 max	0.50 max	0.020 max	0.015 max	0.50 max	Al >=0.020; N <=0.015; Nb <=0.05; Ti <=0.05; V <=0.10; Cr+Cu+Mo<=0.60, V+Nb+Ti<=0.15; bal Fe	500-660	390	19	
EN 10028/5(96)	1.8828	High-temp, Press; 16<=t<=40mm	0.16 max		1.70 max	0.20 max	0.50 max	0.020 max	0.015 max	0.50 max	Al >=0.020; N <=0.015; Nb <=0.05; Ti <=0.05; V <=0.10; Cr+Cu+Mo<=0.60, V+Nb+Ti<=0.15; bal Fe	500-660	400	19	
EN 10028/5(96)	1.8831	High-temp, Press; 40<t<=63mm	0.16 max		1.70 max	0.20 max	0.50 max	0.020 max	0.015 max	0.60 max	Al >=0.020; N <=0.015; Nb <=0.05; Ti <=0.05; V <=0.10; Cr+Cu+Mo<=0.60, V+Nb+Ti<=0.15; bal Fe	530-720	430	17	
EN 10028/5(96)	1.8831	High-temp, Press; 16<=t<=40mm	0.16 max		1.70 max	0.20 max	0.50 max	0.020 max	0.015 max	0.60 max	Al >=0.020; N <=0.015; Nb <=0.05; Ti <=0.05; V <=0.10; Cr+Cu+Mo<=0.60, V+Nb+Ti<=0.15; bal Fe	530-720	440	17	
EN 10028/5(96)	1.8831	High-temp, Press; t<=16mm	0.16 max		1.70 max	0.20 max	0.50 max	0.020 max	0.015 max	0.60 max	Al >=0.020; N <=0.015; Nb <=0.05; Ti <=0.05; V <=0.10; Cr+Cu+Mo<=0.60, V+Nb+Ti<=0.15; bal Fe	530-720	460	17	

Specification	Designation	Notes	C	Cr	Mn	Mo	Ni	P	S	Si	Other	UTS	YS	El	Hard

Alloy Steel, Unclassified, A841(B) (Continued from previous page)

Europe

Specification	Designation	Notes	C	Cr	Mn	Mo	Ni	P	S	Si	Other	UTS	YS	El	Hard
EN 10028/5(96)	1.8832	High-temp, Press; 40<t<=63mm	0.14 max		1.60 max		0.50 max	0.020 max	0.015 max	0.50 max	Al >=0.020; N <=0.020; Nb <=0.05; Ti <=0.05; V <=0.10; Cr+Cu+Mo<=0.60, V+Nb+Ti<=0.15; bal Fe	450-610	345	22	
EN 10028/5(96)	1.8832	High-temp, Press; t<=40mm	0.14 max		1.60 max		0.50 max	0.020 max	0.015 max	0.50 max	Al >=0.020; N <=0.020; Nb <=0.05; Ti <=0.05; V <=0.10; Cr+Cu+Mo<=0.60, V+Nb+Ti<=0.15; bal Fe	450-610	355	22	
EN 10028/5(96)	1.8833	High-temp, Press; t<=40mm	0.14 max		1.60 max		0.50 max	0.020 max	0.015 max	0.50 max	Al >=0.020; N <=0.020; Nb <=0.05; Ti <=0.05; V <=0.10; Cr+Cu+Mo<=0.60, V+Nb+Ti<=0.15; bal Fe	450-610	355	22	
EN 10028/5(96)	1.8833	High-temp, Press; 40<t<=63mm	0.14 max		1.60 max		0.50 max	0.020 max	0.015 max	0.50 max	Al >=0.020; N <=0.020; Nb <=0.05; Ti <=0.05; V <=0.10; Cr+Cu+Mo<=0.60, V+Nb+Ti<=0.15; bal Fe	450-610	345	22	
EN 10028/5(96)	1.8835	High-temp, Press; t<=16mm	0.16 max		1.70 max	0.20 max	0.50 max	0.020 max	0.015 max	0.50 max	Al >=0.020; N <=0.015; Nb <=0.05; Ti <=0.05; V <=0.10; Cr+Cu+Mo<=0.60, V+Nb+Ti<=0.15; bal Fe	500-660	420	19	
EN 10028/5(96)	1.8835	High-temp, Press; 16<=t<=40mm	0.16 max		1.70 max	0.20 max	0.50 max	0.020 max	0.015 max	0.50 max	Al >=0.020; N <=0.015; Nb <=0.05; Ti <=0.05; V <=0.10; Cr+Cu+Mo<=0.60, V+Nb+Ti<=0.15; bal Fe	500-660	400	19	
EN 10028/5(96)	1.8835	High-temp, Press; 40<t<=63mm	0.16 max		1.70 max	0.20 max	0.50 max	0.020 max	0.015 max	0.50 max	Al >=0.020; N <=0.015; Nb <=0.05; Ti <=0.05; V <=0.10; Cr+Cu+Mo<=0.60, V+Nb+Ti<=0.15; bal Fe	500-660	390	19	
EN 10028/5(96)	1.8837	High-temp, Press; t<=16mm	0.16 max		1.70 max	0.20 max	0.50 max	0.020 max	0.015 max	0.60 max	Al >=0.020; N <=0.015; Nb <=0.05; Ti <=0.05; V <=0.10; Cr+Cu+Mo<=0.60, V+Nb+Ti<=0.15; bal Fe	530-720	460	17	
EN 10028/5(96)	1.8837	High-temp, Press; 16<=t<=40mm	0.16 max		1.70 max	0.20 max	0.50 max	0.020 max	0.015 max	0.60 max	Al >=0.020; N <=0.015; Nb <=0.05; Ti <=0.05; V <=0.10; Cr+Cu+Mo<=0.60, V+Nb+Ti<=0.15; bal Fe	530-720	440	17	
EN 10028/5(96)	1.8837	High-temp, Press; 40<t<=63mm	0.16 max		1.70 max	0.20 max	0.50 max	0.020 max	0.015 max	0.60 max	Al >=0.020; N <=0.015; Nb <=0.05; Ti <=0.05; V <=0.10; Cr+Cu+Mo<=0.60, V+Nb+Ti<=0.15; bal Fe	530-720	430	17	
EN 10028/5(96)	P355M	High-temp, Press; t<=40mm	0.14 max		1.60 max		0.50 max	0.025 max	0.020 max	0.50 max	Al >=0.020; N <=0.020; Nb <=0.05; Ti <=0.05; V <=0.10; Cr+Cu+Mo<=0.60, V+Nb+Ti<=0.15; bal Fe	450-610	355	22	
EN 10028/5(96)	P355M	High-temp, Press; 40<t<=63mm	0.14 max		1.60 max		0.50 max	0.025 max	0.020 max	0.50 max	Al >=0.020; N <=0.020; Nb <=0.05; Ti <=0.05; V <=0.10; Cr+Cu+Mo<=0.60, V+Nb+Ti<=0.15; bal Fe	450-610	345	22	

UNS numbers and US grades are provided as a means of cross referencing chemically similar alloys. Exchangability is only possible after independent examination of specifications. Tensile properties are minimum or typical as specified. UTS and YS as MPa. El as %. See Appendix for list of abbreviations used in Notes. * indicates obsolete material.

Alloy Steel, Unclassified, A841(B) (Continued from previous page)

Europe

Specification	Designation	Notes	C	Cr	Mn	Mo	Ni	P	S	Si	Other	UTS	YS	El	Hard
EN 10028/5(96)	P355ML1	High-temp, Press; t<=40mm	0.14 max		1.60 max		0.50 max	0.020 max	0.015 max	0.50 max	Al >=0.020; N <=0.020; Nb <=0.05; Ti <=0.05; V <=0.10; Cr+Cu+Mo<=0.60, V+Nb+Ti<=0.15; bal Fe	450-610	355	22	
EN 10028/5(96)	P355ML1	High-temp, Press; 40<t<=63mm	0.14 max		1.60 max		0.50 max	0.020 max	0.015 max	0.50 max	Al >=0.020; N <=0.020; Nb <=0.05; Ti <=0.05; V <=0.10; Cr+Cu+Mo<=0.60, V+Nb+Ti<=0.15; bal Fe	450-610	345	22	
EN 10028/5(96)	P355ML2	High-temp, Press; 40<t<=63mm	0.14 max		1.60 max		0.50 max	0.020 max	0.015 max	0.50 max	Al >=0.020; N <=0.020; Nb <=0.05; Ti <=0.05; V <=0.10; Cr+Cu+Mo<=0.60, V+Nb+Ti<=0.15; bal Fe	450-610	345	22	
EN 10028/5(96)	P355ML2	High-temp, Press; t<=40mm	0.14 max		1.60 max		0.50 max	0.020 max	0.015 max	0.50 max	Al >=0.020; N <=0.020; Nb <=0.05; Ti <=0.05; V <=0.10; Cr+Cu+Mo<=0.60, V+Nb+Ti<=0.15; bal Fe	450-610	355	22	
EN 10028/5(96)	P420M	High-temp, Press; 16<=t<=40mm	0.16 max		1.70 max	0.20 max	0.50 max	0.025 max	0.020 max	0.50 max	Al >=0.020; N <=0.015; Nb <=0.05; Ti <=0.05; V <=0.10; Cr+Cu+Mo<=0.60, V+Nb+Ti<=0.15; bal Fe	500-660	400	19	
EN 10028/5(96)	P420M	High-temp, Press; 40<t<=63mm	0.16 max		1.70 max	0.20 max	0.50 max	0.025 max	0.020 max	0.50 max	Al >=0.020; N <=0.015; Nb <=0.05; Ti <=0.05; V <=0.10; Cr+Cu+Mo<=0.60, V+Nb+Ti<=0.15; bal Fe	500-660	390	19	
EN 10028/5(96)	P420M	High-temp, Press; t<=16mm	0.16 max		1.70 max	0.20 max	0.50 max	0.025 max	0.020 max	0.50 max	Al >=0.020; N <=0.015; Nb <=0.05; Ti <=0.05; V <=0.10; Cr+Cu+Mo<=0.60, V+Nb+Ti<=0.15; bal Fe	500-660	420	19	
EN 10028/5(96)	P420ML1	High-temp, Press; 40<t<=63mm	0.16 max		1.70 max	0.20 max	0.50 max	0.020 max	0.015 max	0.50 max	Al >=0.020; N <=0.015; Nb <=0.05; Ti <=0.05; V <=0.10; Cr+Cu+Mo<=0.60, V+Nb+Ti<=0.15; bal Fe	500-660	390	19	
EN 10028/5(96)	P420ML1	High-temp, Press; 16<=t<=40mm	0.16 max		1.70 max	0.20 max	0.50 max	0.020 max	0.015 max	0.50 max	Al >=0.020; N <=0.015; Nb <=0.05; Ti <=0.05; V <=0.10; Cr+Cu+Mo<=0.60, V+Nb+Ti<=0.15; bal Fe	500-660	400	19	
EN 10028/5(96)	P420ML1	High-temp, Press; t<=16mm	0.16 max		1.70 max	0.20 max	0.50 max	0.020 max	0.015 max	0.50 max	Al >=0.020; N <=0.015; Nb <=0.05; Ti <=0.05; V <=0.10; Cr+Cu+Mo<=0.60, V+Nb+Ti<=0.15; bal Fe	500-660	420	19	
EN 10028/5(96)	P420ML2	High-temp, Press; 40<t<=63mm	0.16 max		1.70 max	0.20 max	0.50 max	0.020 max	0.015 max	0.50 max	Al >=0.020; N <=0.015; Nb <=0.05; Ti <=0.05; V <=0.10; Cr+Cu+Mo<=0.60, V+Nb+Ti<=0.15; bal Fe	500-660	390	19	
EN 10028/5(96)	P420ML2	High-temp, Press; 16<=t<=40mm	0.16 max		1.70 max	0.20 max	0.50 max	0.020 max	0.015 max	0.50 max	Al >=0.020; N <=0.015; Nb <=0.05; Ti <=0.05; V <=0.10; Cr+Cu+Mo<=0.60, V+Nb+Ti<=0.15; bal Fe	500-660	400	19	

Specification	Designation	Notes	C	Cr	Mn	Mo	Ni	P	S	Si	Other	UTS	YS	El	Hard

Alloy Steel, Unclassified, A841(B) (Continued from previous page)

Europe

Specification	Designation	Notes	C	Cr	Mn	Mo	Ni	P	S	Si	Other	UTS	YS	El	Hard
EN 10028/5(96)	P420ML2	High-temp, Press; t<=16mm	0.16 max		1.70 max	0.20 max	0.50 max	0.020 max	0.015 max	0.50 max	Al >=0.020; N <=0.015; Nb <=0.05; Ti <=0.05; V <=0.10; Cr+Cu+Mo<=0.60, V+Nb+Ti<=0.15; bal Fe	500-660	420	19	
EN 10028/5(96)	P460M	High-temp, Press; 40<t<=63mm	0.16 max		1.70 max	0.20 max	0.50 max	0.025 max	0.020 max	0.60 max	Al >=0.020; N <=0.015; Nb <=0.05; Ti <=0.05; V <=0.10; Cr+Cu+Mo<=0.60, V+Nb+Ti<=0.15; bal Fe	530-720	430	17	
EN 10028/5(96)	P460M	High-temp, Press; 16<=t<=40mm	0.16 max		1.70 max	0.20 max	0.50 max	0.025 max	0.020 max	0.60 max	Al >=0.020; N <=0.015; Nb <=0.05; Ti <=0.05; V <=0.10; Cr+Cu+Mo<=0.60, V+Nb+Ti<=0.15; bal Fe	530-720	440	17	
EN 10028/5(96)	P460M	High-temp, Press; t<=16mm	0.16 max		1.70 max	0.20 max	0.50 max	0.025 max	0.020 max	0.60 max	Al >=0.020; N <=0.015; Nb <=0.05; Ti <=0.05; V <=0.10; Cr+Cu+Mo<=0.60, V+Nb+Ti<=0.15; bal Fe	530-720	460	17	
EN 10028/5(96)	P460ML1	High-temp, Press; 16<=t<=40mm	0.16 max		1.70 max	0.20 max	0.50 max	0.020 max	0.015 max	0.60 max	Al >=0.020; N <=0.015; Nb <=0.05; Ti <=0.05; V <=0.10; Cr+Cu+Mo<=0.60, V+Nb+Ti<=0.15; bal Fe	530-720	440	17	
EN 10028/5(96)	P460ML1	High-temp, Press; t<=16mm	0.16 max		1.70 max	0.20 max	0.50 max	0.020 max	0.015 max	0.60 max	Al >=0.020; N <=0.015; Nb <=0.05; Ti <=0.05; V <=0.10; Cr+Cu+Mo<=0.60, V+Nb+Ti<=0.15; bal Fe	530-720	460	17	
EN 10028/5(96)	P460ML1	High-temp, Press; 40<t<=63mm	0.16 max		1.70 max	0.20 max	0.50 max	0.020 max	0.015 max	0.60 max	Al >=0.020; N <=0.015; Nb <=0.05; Ti <=0.05; V <=0.10; Cr+Cu+Mo<=0.60, V+Nb+Ti<=0.15; bal Fe	530-720	430	17	
EN 10028/5(96)	P460ML2	High-temp, Press; 40<t<=63mm	0.16 max		1.70 max	0.20 max	0.50 max	0.020 max	0.015 max	0.60 max	Al >=0.020; N <=0.015; Nb <=0.05; Ti <=0.05; V <=0.10; Cr+Cu+Mo<=0.60, V+Nb+Ti<=0.15; bal Fe	530-720	430	17	
EN 10028/5(96)	P460ML2	High-temp, Press; 16<=t<=40mm	0.16 max		1.70 max	0.20 max	0.50 max	0.020 max	0.015 max	0.60 max	Al >=0.020; N <=0.015; Nb <=0.05; Ti <=0.05; V <=0.10; Cr+Cu+Mo<=0.60, V+Nb+Ti<=0.15; bal Fe	530-720	440	17	
EN 10028/5(96)	P460ML2	High-temp, Press; t<=16mm	0.16 max		1.70 max	0.20 max	0.50 max	0.020 max	0.015 max	0.60 max	Al >=0.020; N <=0.015; Nb <=0.05; Ti <=0.05; V <=0.10; Cr+Cu+Mo<=0.60, V+Nb+Ti<=0.15; bal Fe	530-720	460	17	

USA

Specification	Designation	Notes	C	Cr	Mn	Mo	Ni	P	S	Si	Other	UTS	YS	El	Hard
ASTM A841/A841M(95)	Grade B, Class 1	TMCP Press ves plt; t<=40mm	0.15 max	0.25 max	0.70-1.35	0.30 max	0.60 max	0.030 max	0.025 max	0.15-0.50	Cu <=0.35; Nb <=0.03; V <=0.06; bal Fe	485-620	345	22	
ASTM A841/A841M(95)	Grade B, Class 1	TMCP Press ves plt; t=65-100mm	0.15 max	0.25 max	1.00-1.60	0.30 max	0.60 max	0.030 max	0.025 max	0.15-0.50	Cu <=0.35; Nb <=0.03; V <=0.06; bal Fe	450-585	310	22	
ASTM A841/A841M(95)	Grade B, Class 1	TMCP Press ves plt; t=40-65mm	0.15 max	0.25 max	1.00-1.60	0.30 max	0.60 max	0.030 max	0.025 max	0.15-0.50	Cu <=0.35; Nb <=0.03; V <=0.06; bal Fe	485-620	345	22	

UNS numbers and US grades are provided as a means of cross referencing chemically similar alloys. Exchangability is only possible after independent examination of specifications. Tensile properties are minimum or typical as specified. UTS and YS as MPa. El as %. See Appendix for list of abbreviations used in Notes. * indicates obsolete material.

Specification	Designation	Notes	C	Cr	Mn	Mo	Ni	P	S	Si	Other	UTS	YS	El	Hard

Alloy Steel, Unclassified, A841(B) (Continued from previous page)

USA

ASTM A841/A841M(95)	Grade B, Class 2	TMCP Press ves plt; t<=40mm	0.15 max	0.25 max	0.70-1.35	0.30 max	0.60 max	0.030 max	0.025 max	0.15-0.50	Cu <=0.35; Nb <=0.03; V <=0.06; bal Fe	550-690	415	22	
ASTM A841/A841M(95)	Grade B, Class 2	TMCP Press ves plt; t=40-65mm	0.15 max	0.25 max	1.00-1.60	0.30 max	0.60 max	0.030 max	0.025 max	0.15-0.50	Cu <=0.35; Nb <=0.03; V <=0.06; bal Fe	550-690	415	22	
ASTM A841/A841M(95)	Grade B, Class 2	TMCP Press ves plt; t=65-100mm	0.15 max	0.25 max	1.00-1.60	0.30 max	0.60 max	0.030 max	0.025 max	0.15-0.50	Cu <=0.35; Nb <=0.03; V <=0.06; bal Fe	515-655	380	22	

Alloy Steel, Unclassified, AMS 6330

USA

| | UNS K22033 | | 0.33-0.38 | 0.55-0.75 | 0.60-0.80 | 0.06 max | 1.10-1.40 | 0.025 max | 0.025 max | 0.20-0.35 | Cu <=0.35; bal Fe | | | | |

Alloy Steel, Unclassified, B388

USA

	UNS K92100	Thermostat Alloy		2			19				Fe 79; bal Fe				
	UNS K92350	Thermostat Alloy		8.5			25				Fe 66.5; bal Fe				
	UNS K92510	Thermostat		3			22				Fe 75; bal Fe				
	UNS K92850	Thermostat									Al 5 Fe 71.5 Mn 9.5 Ni 14				
	UNS K93600	Thermostat (Invar)									Fe 64 Ni 36				
	UNS K93800	Thermostat					31-1				Co 7; Fe 62; bal Fe				
	UNS K94000	Thermostat									Fe 60 Ni 40				
	UNS K94200	Thermostat					42				Fe 58				
	UNS K94500	Thermostat									Fe 55				
	UNS K95000	Ni Thermostat									Fe 50 Ni 50				

Alloy Steel, Unclassified, B603(IIA)

USA

| | UNS K91870 | Electrical Heating Element | | 14-17 | | | | | | | Al 4.75-5.75; bal Fe | | | | |

Alloy Steel, Unclassified, B603(III)

USA

| | UNS K91670* | Obs; Electrical Heating Element | | 13-16 | | | | | | | Al 3.75-4.75; bal Fe | | | | |

Alloy Steel, Unclassified, B603(IV)

USA

| | UNS K91470 | Electrical Heating Element | | 12-15 | | | | | | | Al 2.75-3.75; bal Fe | | | | |

Alloy Steel, Unclassified, C

India

| IS 1570/7(92) | 16Mo3H | Smls tube Bar Frg Sh Plt; high-temp | 0.12-0.2 | 0.3 max | 0.4-0.8 | 0.25-0.35 | 0.35 max | 0.04 max | 0.04 max | 0.1-0.35 | Al <=0.02; Almet<=0.02; bal Fe | 440-590 | 250 | 17 | |

USA

	UNS K11511		0.10-0.20		1.10-1.50	0.20-0.30		0.035 max	0.04 max	0.15-0.30	B 0.001-0.005; bal Fe				
ASTM A514(94)	Grade C	Q/T, Plt, t=32mm, weld	0.10-0.20		1.10-1.50	0.15-0.30		0.035 max	0.035 max	0.15-0.30	B 0.001-0.005; bal Fe	760-895	690	18	
ASTM A517(93)	Grade C	Q/T, Press ves plt, t<=32mm	0.10-0.20		1.10-1.50	0.20-0.30		0.035 max	0.035 max	0.15-0.30	B 0.001-0.005; bal Fe	795-930	690	16	
ASTM A709(97)	Grade 100, Type C	Q/T, Plt, t<=32mm	0.10-0.20		1.10-1.50	0.15-0.30		0.035 max	0.035 max	0.15-0.30	B 0.001-0.005; bal Fe	760-895	690	18	

Alloy Steel, Unclassified, E3310

Bulgaria

| BDS 6354 | 12ChN3A | Struct | 0.09-0.16 | 0.60-0.90 | 0.30-0.60 | 0.15 max | 2.75-3.15 | 0.025 max | 0.025 max | 0.17-0.37 | Cu <=0.30; Ti <=0.03; V <=0.05; W <=0.2; bal Fe | | | | |

UNS numbers and US grades are provided as a means of cross referencing chemically similar alloys. Exchangability is only possible after independent examination of specifications. Tensile properties are minimum or typical as specified. UTS and YS as MPa. El as %. See Appendix for list of abbreviations used in Notes. * indicates obsolete material.

Specification	Designation	Notes	C	Cr	Mn	Mo	Ni	P	S	Si	Other	UTS	YS	El	Hard
Alloy Steel, Unclassified, E3310 (Continued from previous page)															
France															
AFNOR NFA35565(84)	16NCD13	100<t<=160mm	0.12-0.17	0.9-1.2	0.3-0.6	0.15-0.3	3-3.5	0.035 max	0.025 max	0.15-0.4	Cu <=0.35; Ti <=0.05; bal Fe	900-1200	700	12	
AFNOR NFA35567(84)	16NCD13	40<t<=100mm	0.12-0.17	0.9-1.2	0.3-0.6	0.15-0.3	3-3.5	0.035 max	0.035 max	0.15-0.4	Cu <=0.35; Ti <=0.05; bal Fe	950-1250	750	11	
Hungary															
MSZ 31(85)	BNC2	CH; 0-11mm; blank carburized	0.14-0.2	0.6-0.9	0.3-0.6		3-3.5	0.035 max	0.035 max	0.4 max	bal Fe	950-1300	690	9L	
MSZ 31(85)	BNC2	CH; 63 mm; blank carburized	0.14-0.2	0.6-0.9	0.3-0.6		3-3.5	0.035 max	0.035 max	0.4 max	bal Fe	800-1100	490	12L	
MSZ 31(85)	BNC2E	CH; 0-11mm; blank carburized	0.14-0.2	0.6-0.9	0.3-0.6		3-3.5	0.035 max	0.02-0.035	0.4 max	bal Fe	950-1300	690	9L	
MSZ 31(85)	BNC2E	CH; 63 mm; blank carburized	0.14-0.2	0.6-0.9	0.3-0.6		3-3.5	0.035 max	0.02-0.035	0.4 max	bal Fe	800-1100	490	12L	
International															
ISO 683-17(76)	15	Ball & roller bearing, HT	0.14-0.20	0.80-1.10	0.60-0.90	0.15-0.25	1.20-1.60	0.035 max	0.035 max	0.15-0.40	bal Fe				
ISO 683-17(76)	16	Ball & roller bearing, HT	0.14-0.20	1.30-1.60	0.40-0.70	0.15-0.25	3.25-3.75	0.035 max	0.035 max	0.15-0.40	bal Fe				
Japan															
JIS G4052(79)	SNC815H	Ni-Cr Struct Hard	0.11-0.18	0.65-1.05	0.30-0.70		2.95-3.50	0.030 max	0.030 max	0.15-0.35	Cu <=0.30; bal Fe				
Poland															
PNH84035	12HN3A	CH	0.09-0.16	0.6-0.9	0.3-0.6		2.75-3.15	0.03 max	0.025 max	0.17-0.37	Cu <=0.2; bal Fe				
Romania															
STAS 791(80)	13CrNi30	Struct/const; CH	0.09-0.16	0.6-0.9	0.3-0.6		2.75-3.15	0.035 max	0.035 max	0.17-0.37	Pb <=0.15; bal Fe				
Russia															
GOST 4543(71)	20ChN3A	CH	0.17-0.24	0.6-0.9	0.3-0.6	0.15 max	2.75-3.15	0.025 max	0.025 max	0.17-0.37	Al <=0.1; Cu <=0.3; Ti <=0.03; W <=0.2; bal Fe				
Spain															
UNE 36027(80)	18NiCrMo14	Ball & roller bearing	0.14-0.2	1.3-1.6	0.4-0.7	0.15-0.25	3.25-3.75	0.035 max	0.035 max	0.15-0.4	bal Fe				
UNE 36027(80)	F.1556*	see 18NiCrMo14	0.14-0.2	1.3-1.6	0.4-0.7	0.15-0.25	3.25-3.75	0.035 max	0.035 max	0.15-0.4	bal Fe				
UK															
BS 970/1(83)	655H13	Blm Bil Slab Bar Rod Frg CH	0.10-0.16	0.70-1.00	0.35-0.60	0.15 max	3.0-3.75	0.035 max	0.04 max	0.10-0.35	bal Fe				0-223 HB
BS 970/3(91)	655M13	Blm Bil Slab Bar Rod Frg CH	0.10-0.16	0.70-1.00	0.35-0.60	0.15 max	3.0-3.75	0.040 max	0.05 max	0.10-0.40	bal Fe	1000		9	
USA															
	AISI E3310	Smls mech Tub; BEF	0.08-0.13	1.40-1.75	0.45-0.60		3.25-3.75	0.025 max	0.025 max	0.15-0.35	bal Fe				
	UNS G33106	Ni-Cr	0.08-0.13	1.40-1.75	0.45-0.60		3.25-3.75	0.025 max	0.025 max	0.15-0.30	bal Fe				
ASTM A535(92)	3310	Ball & roller bearing	0.08-0.13	1.40-1.75	0.45-0.60	0.10 max	3.25-3.75	0.015 max	0.015 max	0.15-0.35	Cu <=0.35; bal Fe				
ASTM A646(95)	Grade 1	Blm Bil for aircraft aerospace frg, ann	0.08-0.13	1.40-1.75	0.45-0.60		3.25-3.75	0.025 max	0.025 max	0.20-0.35	bal Fe				262 HB
ASTM A837(96)	E3310	Frg for carburizing	0.08-0.13	1.40-1.75	0.45-0.60	0.10 max	3.25-3.75	0.025 max	0.025 max	0.15-0.35	Cu <=0.35; Si<=0.10 if VCD used; bal Fe				
ASTM A914/914M(92)	3310RH	Bar restricted end Q hard	0.08-0.13	1.40-1.75	0.40-0.60	0.06 max	3.25-3.75	0.035 max	0.040 max	0.15-0.35	Cu <=0.35; Electric furnace P, S<=0.025; bal Fe				
MIL-S-7393C(94)	3310	Carb Bar bil, aircraft qual									bal Fe				
MIL-S-7393C(94)	3316	Carb Bar bil, aircraft qual									bal Fe				
Alloy Steel, Unclassified, F															
USA															
	UNS K12238		0.19-0.26	0.42-0.68	0.86-1.24		0.17-0.43	0.045 max	0.045 max	0.13-0.32	Cu 0.17-0.43; bal Fe				

Specification	Designation	Notes	C	Cr	Mn	Mo	Ni	P	S	Si	Other	UTS	YS	El	Hard

Alloy Steel, Unclassified, F15

Germany

Specification	Designation	Notes	C	Cr	Mn	Mo	Ni	P	S	Si	Other	UTS	YS	El	Hard
DIN SEW 385(91)	NiCo29 18	Thermal expansion	0.05 max		0.50 max		28.0-30.0	0.030 max	0.030 max	0.30 max	Co 17.0-18.0; bal Fe				
DIN SEW 385(91)	WNr 1.3981	Thermal expansion	0.05 max		0.50 max		28.0-30.0	0.030 max	0.030 max	0.30 max	Co 17.0-18.0; bal Fe				

USA

Specification	Designation	Notes	C	Cr	Mn	Mo	Ni	P	S	Si	Other	UTS	YS	El	Hard
	UNS K94610	Ni-Co Sealing KOVAR	0.04 max	0.20 max	0.50 max	0.20 max	29			0.20 max	Al <=0.10; Co 17; Cu <=0.20; Mg <=0.10; Ti <=0.10; Fe 53 nom; Zr 0.10 max; Al+Mg+Zr+Ti 0.20 max				

Alloy Steel, Unclassified, F1684

USA

Specification	Designation	Notes	C	Cr	Mn	Mo	Ni	P	S	Si	Other	UTS	YS	El	Hard
	UNS K93050	Fe-Ni Low Expansion, Free Machining (Invar FC)	0.15 max	0.25 max	1.00 max		34.0-38.0	0.020 max	0.020 max	0.35 max	Co <=0.50; Se 0.15-0.30; bal Fe				
	UNS K93500	Fe-Ni-Co Low Expansion Alloy (Super Invar)	0.05 max	0.25 max	0.60 max		30.5-33.5	0.015 max	0.015 max	0.25 max	Al <=0.10; Co 4.0-6.5; Mg <=0.10; Ti <=0.10; Zr 0.10 max; bal Fe				
	UNS K93603	Fe-Ni Low Expansion (Invar)	0.05 max	0.25 max	0.60 max		34.0-38.0	0.015 max	0.015 max	0.40 max	Al <=0.10; Co <=0.50; Mg <=0.10; Ti <=0.10; Zr 0.10 max; bal Fe				

Alloy Steel, Unclassified, F256(I)

USA

Specification	Designation	Notes	C	Cr	Mn	Mo	Ni	P	S	Si	Other	UTS	YS	El	Hard
	UNS K91800	Cr Sealing Alloy	0.08 max	18	1.00 max		0.50 max	0.04 max	0.03 max	0.75 max	Ti 5xC-0.60; bal Fe				

Alloy Steel, Unclassified, F30(42)

USA

Specification	Designation	Notes	C	Cr	Mn	Mo	Ni	P	S	Si	Other	UTS	YS	El	Hard
	UNS K94100	Ni Sealing	0.05 max	0.25 max	0.80 max			0.025 max	0.025 max	0.30 max	Al <=0.10; Ni 41 nom; bal Fe				

Alloy Steel, Unclassified, F30(46)

USA

Specification	Designation	Notes	C	Cr	Mn	Mo	Ni	P	S	Si	Other	UTS	YS	El	Hard
	UNS K94600	Ni Sealing	0.05 max	0.25 max	0.80 max			0.025 max	0.025 max	0.30 max	Al <=0.10; Ni 46 nom; bal Fe				

Alloy Steel, Unclassified, F30(48)

USA

Specification	Designation	Notes	C	Cr	Mn	Mo	Ni	P	S	Si	Other	UTS	YS	El	Hard
	UNS K94800	Ni Sealing	0.05 max	0.25 max	0.80 max			0.025 max	0.025 max	0.30 max	Al <=0.10; Ni 48 nom; bal Fe				

Alloy Steel, Unclassified, F31

USA

Specification	Designation	Notes	C	Cr	Mn	Mo	Ni	P	S	Si	Other	UTS	YS	El	Hard
	UNS K94760	Ni-Cr Sealing	0.07 max	5.6	0.25 max		42	0.025 max	0.025 max	0.30 max	Al <=0.20; bal Fe				

Alloy Steel, Unclassified, FR

India

Specification	Designation	Notes	C	Cr	Mn	Mo	Ni	P	S	Si	Other	UTS	YS	El	Hard
IS 1570/7(92)	2	3-16mm; Sh Plt Smls pipe	0.2 max	0.25 max	0.5-1.3	0.1 max	0.3 max	0.03 max	0.03 max	0.1-0.35	bal Fe	410-520	225	24	
IS 1570/7(92)	2	100.1-150mm; Sh Plt Smls pipe	0.2 max	0.25 max	0.5-1.3	0.1 max	0.3 max	0.03 max	0.03 max	0.1-0.35	bal Fe	410-520	195	22	
IS 1570/7(92)	Fe410H	3-16mm; Sh Plt Smls pipe	0.2 max	0.25 max	0.5-1.3	0.1 max	0.3 max	0.03 max	0.03 max	0.1-0.35	bal Fe	410-520	225	24	
IS 1570/7(92)	Fe410H	100.1-150mm; Sh Plt Smls pipe	0.2 max	0.25 max	0.5-1.3	0.1 max	0.3 max	0.03 max	0.03 max	0.1-0.35	bal Fe	410-520	195	22	

USA

Specification	Designation	Notes	C	Cr	Mn	Mo	Ni	P	S	Si	Other	UTS	YS	El	Hard
	UNS K22035		0.20 max		0.40-1.06		1.60-2.24	0.045 max	0.050 max		Cu 0.75-1.25; bal Fe				
ASTM A182/A182M(98)	FR	Frg/roll pipe flange valve	0.20 max		0.40-1.06		1.60-2.24	0.045 max	0.050 max		Cu 0.75-1.25; bal Fe	435	315	25.0	197 HB max

UNS numbers and US grades are provided as a means of cross referencing chemically similar alloys. Exchangability is only possible after independent examination of specifications. Tensile properties are minimum or typical as specified. UTS and YS as MPa. El as %. See Appendix for list of abbreviations used in Notes. * indicates obsolete material.

Specification	Designation	Notes	C	Cr	Mn	Mo	Ni	P	S	Si	Other	UTS	YS	El	Hard

Alloy Steel, Unclassified, FR (Continued from previous page)

USA

Specification	Designation	Notes	C	Cr	Mn	Mo	Ni	P	S	Si	Other	UTS	YS	El	Hard
ASTM A234/A234M(97)	WPR	Frg	0.20 max		0.40-1.06		1.60-2.24	0.045 max	0.050 max		Cu 0.75-1.25; bal Fe	435-605	315	20L	
ASTM A714(96)	V	Weld, Smls pipe, Grade V Type F	0.16 max		0.40-1.01		1.65 min	0.035 max	0.040 max		Cu <=0.80; bal Fe	380	275		
ASTM A714(96)	V	Weld, Smls pipe, Grade V Type E , Type S	0.16 max		0.40-1.01		1.65 min	0.035 max	0.040 max		Cu <=0.80; bal Fe	450	315	21	

Alloy Steel, Unclassified, HY100

Europe

Specification	Designation	Notes	C	Cr	Mn	Mo	Ni	P	S	Si	Other	UTS	YS	El	Hard
EN 10084(98)	1.5752	CH, Ann	0.14-0.20	0.60-0.90	0.35-0.65		3.00-3.50	0.035 max	0.035 max	0.40 max	bal Fe				229 HB
EN 10084(98)	15NiCr13	CH, Ann	0.14-0.20	0.60-0.90	0.35-0.65		3.00-3.50	0.035 max	0.035 max	0.40 max	bal Fe				229 HB

Alloy Steel, Unclassified, K11523

USA

Specification	Designation	Notes	C	Cr	Mn	Mo	Ni	P	S	Si	Other	UTS	YS	El	Hard
	UNS K11523		0.10-0.20		1.10-1.50	0.45-0.55		0.035 max	0.04 max	0.15-0.30	B 0.001-0.005; bal Fe				

Alloy Steel, Unclassified, K11801

USA

Specification	Designation	Notes	C	Cr	Mn	Mo	Ni	P	S	Si	Other	UTS	YS	El	Hard
	UNS K11801		0.18 max	0.15 max	1.30 max	0.06 max	0.25 max	0.040 max	0.050 max	0.15-0.35	Cu <=0.35; Ti >=0.005; V >=0.02; bal Fe				

Alloy Steel, Unclassified, K11847

USA

Specification	Designation	Notes	C	Cr	Mn	Mo	Ni	P	S	Si	Other	UTS	YS	El	Hard
	UNS K11847		0.15-0.21	0.50-0.80	0.80-1.10	0.25 max		0.035 max	0.04 max	0.40-0.90	B 0.0005-0.0025; Zr 0.05-0.15; bal Fe				

Alloy Steel, Unclassified, K12032

Germany

Specification	Designation	Notes	C	Cr	Mn	Mo	Ni	P	S	Si	Other	UTS	YS	El	Hard
DIN	9CrNiCuP324	Long sample 40-60mm	0.12 max	0.50-1.25	0.20-0.50		0.65 max	0.07-0.15	0.035 max	0.25-0.75	Cu 0.25-0.55; bal Fe	510-610	335	22	
DIN	WNr 1.8962	16-40mm, long sample	0.12 max	0.50-1.25	0.20-0.50		0.65 max	0.07-0.15	0.035 max	0.25-0.75	Cu 0.25-0.55; bal Fe	510-610	335	22	
DIN EN 10155(93)	S355J2G1W	Long sample 40-60mm	0.16 max	0.40-0.80	0.50-1.50		0.65 max	0.035 max	0.035 max	0.50 max	Cu 0.25-0.55; V 0.02-0.12; bal Fe	510-610	335	22	
DIN EN 10155(93)	WNr 1.8963	16-40mm, long sample	0.15 max	0.40-0.80	0.50-1.50		0.65 max	0.035 max	0.035 max	0.50 max	Cu 0.25-0.55; V 0.02-0.12; Zr<=0.15; bal Fe	510-610	345	22	
DIN EN 10155(93)	WTSt52-3	Long sample 40-60mm	0.16 max	0.40-0.80	0.50-1.50		0.65 max	0.035 max	0.035 max	0.50 max	Cu 0.25-0.55; V 0.02-0.12; Zr<=0.15; bal Fe	510-610	335	22	

USA

Specification	Designation	Notes	C	Cr	Mn	Mo	Ni	P	S	Si	Other	UTS	YS	El	Hard
	UNS K12032		0.20 max	0.10-0.25	1.25 max	0.15 max	0.30-0.60	0.035 max	0.040 max	0.25-0.75	Cu 0.20-0.35; Ti 0.005-0.030; V 0.02-0.10; bal Fe				

Alloy Steel, Unclassified, K12044

USA

Specification	Designation	Notes	C	Cr	Mn	Mo	Ni	P	S	Si	Other	UTS	YS	El	Hard
	UNS K12044		0.20 max		0.60-1.00		0.50-0.70	0.04 max	0.05 max	0.30-0.50	Cu >=0.30; Ti 0.05; bal Fe				

Alloy Steel, Unclassified, K12103

USA

Specification	Designation	Notes	C	Cr	Mn	Mo	Ni	P	S	Si	Other	UTS	YS	El	Hard
	UNS K12103		0.10-0.32	0.30-0.40	0.30-0.40	0.07-0.15	0.50 max	0.04 max	0.04 max	0.20-0.40	V <=0.10; bal Fe				

Alloy Steel, Unclassified, K41650

USA

Specification	Designation	Notes	C	Cr	Mn	Mo	Ni	P	S	Si	Other	UTS	YS	El	Hard
	UNS K41650	HCM2S	0.04-0.10	1.90-2.60	0.10-0.60	0.05-0.30		0.030 max	0.010 max	0.50 max	Al <=0.030; B <=0.006; N <=0.030; Nb 0.02-0.08; V 0.20-0.30; W 1.45-1.75; bal Fe				

UNS numbers and US grades are provided as a means of cross referencing chemically similar alloys. Exchangability is only possible after independent examination of specifications. Tensile properties are minimum or typical as specified. UTS and YS as MPa. El as %. See Appendix for list of abbreviations used in Notes. * indicates obsolete material.

Specification	Designation	Notes	C	Cr	Mn	Mo	Ni	P	S	Si	Other	UTS	YS	El	Hard
Alloy Steel, Unclassified, K51210															
Italy															
UNI 6933(71)	18NiCr16	Aircraft material	0.15-0.2	0.8-1.1	0.3-0.6		3.75-4.25	0.03 max	0.025 max	0.35 max	bal Fe				
USA															
	UNS K51210	Krupp	0.10-0.15	1.35-1.75	0.45-0.65		3.75-4.25			0.15-0.35	bal Fe				
Alloy Steel, Unclassified, K93120															
USA															
	UNS K93160*	Obs; see K93120									bal Fe				
Alloy Steel, Unclassified, K93601															
USA															
	UNS K93601	Ni Steel, 36% Ni	0.10 max	0.50 max	1.00 max	0.50 max	35.0-37.0	0.025 max	0.025 max	0.45 max	Co <=0.50; bal Fe				
Alloy Steel, Unclassified, K94620															
USA															
	UNS K94620	Fe-Ni-Co Metal-to-Ceramic Sealing Alloy	0.02 max	0.03 max	0.35 max	0.06 max	27	0.006 max	0.006 max	0.15 max	Al <=0.01; Co 25; Mg <=0.01; Ti <=0.01; Zr 0.01 max; bal Fe				
Alloy Steel, Unclassified, K94630															
USA															
	UNS K94630	Fe-Ni-Co Metal-to-Ceramic Sealing Alloy	0.02 max	0.03 max	0.35 max	0.06 max	29	0.006 max	0.006 max	0.15 max	Al <=0.01; Co 17; Cu <=0.20; Mg <=0.01; Ti <=0.01; Zr 0.01 max; bal Fe				
Alloy Steel, Unclassified, MIL-A-12560															
USA															
	UNS K11918*	Obs	0.10-0.28	0.30-0.40	0.30-0.40	0.07-0.15	0.50 max			0.20-0.40	V <=0.10; bal Fe				
Alloy Steel, Unclassified, MIL-S-16598															
USA															
	UNS K93602	Ni Steel, Free Cutting, 36% Ni	0.15 max		1.00 max		35.0-36.5			0.35 max	Se 0.15-0.25; bal Fe				
Alloy Steel, Unclassified, MIL-S-23093A(Type III)															
USA															
	UNS K34025		0.40 max		0.85 max		3.25-3.75	0.025 max	0.025 max	0.20-0.35	bal Fe				
MIL-S-24093A(SH)(91)	Type III Class D	Frg for shipboard apps; 10 in. max section thick	0.40 max		0.85 max		3.25-3.75	0.025 max	0.025 max	0.20-0.35	bal Fe	690	552	18	
MIL-S-24093A(SH)(91)	Type III Class E	Frg for shipboard apps; 10 in. max section thick	0.40 max		0.85 max		3.25-3.75	0.025 max	0.025 max	0.20-0.35	bal Fe	655	448	20	
MIL-S-24093A(SH)(91)	Type III Class F	Frg for shipboard apps	0.40 max		0.85 max		3.25-3.75	0.025 max	0.025 max	0.20-0.35	bal Fe	621 max	310	22	
Alloy Steel, Unclassified, No equivalents identified															
Argentina															
IAS	IRAM 3115		0.13-0.18	0.55-0.75	0.40-0.60		1.10-1.40	0.040 max	0.040 max	0.20-0.35	bal Fe	550	380	25	167 HB
Australia															
AS 1444(96)	X1320	Standard series, Hard/Tmp	0.18-0.23		1.40-1.70			0.040 max	0.040 max	0.10-0.35	bal Fe				
AS 1444(96)	X3312	Standard series, Hard/Tmp	0.10-0.16	0.70-1.00	0.35-0.60		3.00-3.75	0.040 max	0.040 max	0.10-0.35	bal Fe				
AS 1444(96)	X3312H	H series, Hard/Tmp	0.10-0.16	0.70-1.00	0.35-0.60		3.00-3.75	0.040 max	0.040 max	0.10-0.35	bal Fe				
AS 1444(96)	X9315	Standard series, Hard/Tmp	0.12-0.18	1.00-1.40	0.25-0.50	0.15-0.30	3.90-4.30	0.040 max	0.040 max	0.10-0.35	bal Fe				
AS 1444(96)	X9315H	H series, Hard/Tmp	0.12-0.18	1.00-1.40	0.25-0.50	0.15-0.30	3.90-4.30	0.040 max	0.040 max	0.10-0.35	bal Fe				

UNS numbers and US grades are provided as a means of cross referencing chemically similar alloys. Exchangability is only possible after independent examination of specifications. Tensile properties are minimum or typical as specified. UTS and YS as MPa. El as %. See Appendix for list of abbreviations used in Notes. * indicates obsolete material.

Specification	Designation	Notes	C	Cr	Mn	Mo	Ni	P	S	Si	Other	UTS	YS	El	Hard

Alloy Steel, Unclassified, No equivalents identified

Australia

Specification	Designation	Notes	C	Cr	Mn	Mo	Ni	P	S	Si	Other	UTS	YS	El	Hard
AS 1444(96)	X9940	Standard series, Hard/Tmp	0.36-0.44	0.50-0.80	0.45-0.70	0.45-0.65	2.30-2.80	0.040 max	0.040 max	0.10-0.35	bal Fe				

Bulgaria

Specification	Designation	Notes	C	Cr	Mn	Mo	Ni	P	S	Si	Other	UTS	YS	El	Hard
BDS 6354	12ChN3A	Struct	0.09-0.16	0.60-0.90	0.30-0.60	0.15 max	2.75-3.15	0.025 max	0.025 max	0.17-0.37	Cu <=0.30; Ti <=0.03; V <=0.05; W <=0.2; bal Fe				
BDS 6354	12ChN3A	Struct	0.09-0.16	0.60-0.90	0.30-0.60	0.15 max	2.75-3.15	0.025 max	0.025 max	0.17-0.37	Cu <=0.30; Ti <=0.03; V <=0.05; W <=0.2; bal Fe				

Canada

Specification	Designation	Notes	C	Cr	Mn	Mo	Ni	P	S	Si	Other	UTS	YS	El	Hard
CSA G40.21(98)	480A	t<65mm,PLT,SHT,Bar	0.20 max	0.70 max	1.00-1.60		0.25-0.50	0.025 max	0.035 max	0.15-0.50	Cu 0.20-0.60; Al+Nb+V<=0.12; bal Fe	590-790	480	15L 12T	
CSA G40.21(98)	550A	t<65mm,PLT,SHT,Bar	0.15 max	0.70 max	1.75 max		0.25-0.50	0.025 max	0.035 max	0.15-0.50	Cu 0.20-0.60; Al+Nb+V<=0.15; bal Fe	620-860	550	13L 10T	

Czech Republic

Specification	Designation	Notes	C	Cr	Mn	Mo	Ni	P	S	Si	Other	UTS	YS	El	Hard
CSN 413221	13221	Mn-Ni-V-N	0.15-0.2	0.3 max	1.4-1.7		0.4-0.6	0.03 max	0.03 max	0.25-0.5	Al 0.015-0.07; N 0.01-0.02; V 0.1-0.2; bal Fe				
CSN 415127	15127	Weather res; Struct	0.1-0.17	0.4-0.8	0.9-1.2		0.3-0.6	0.04 max	0.04 max	0.2-0.45	Al 0.015-0.120; Cu 0.3-0.55; Nb 0.02-0.06; V 0.02-0.06; bal Fe				
CSN 415128	15128	Q/T	0.1-0.18	0.5-0.75	0.45-0.7	0.4-0.6		0.04 max	0.04 max	0.15-0.4	V 0.22-0.35; bal Fe				
CSN 415217	15217	Weather res; Struct	0.12 max	0.5-1.2	0.3-0.8		0.3-0.6	0.07-0.15	0.04 max	0.25-0.7	Al 0.01-0.10; Cu <=0.55; bal Fe				
CSN 415223	15223	High-temp Mn-Mo	0.17-0.23	0.2 max	1.2-1.6	0.3-0.5	0.25 max	0.04 max	0.04 max	0.15-0.4	bal Fe				
CSN 415236	15236	High-temp const	0.17-0.27	1.2-1.5	0.3-0.6	0.25-0.5		0.035 max	0.035 max	0.15-0.4	V 0.45-0.65; bal Fe				
CSN 415320	15320	High-temp const	0.2-0.28	1.1-1.4	0.5-0.8	0.55-0.75		0.035 max	0.035 max	0.17-0.37	V 0.15-0.3; bal Fe				
CSN 415323	15323	High-press hydro ves	0.15-0.2	2.7-3.2	0.3-0.5	0.2-0.3		0.035 max	0.03 max	0.2-0.35	V 0.1-0.2; bal Fe				
CSN 415330	15330	Q/T; Nitriding	0.24-0.34	2.3-2.7	0.4-0.8	0.2-0.3		0.035 max	0.035 max	0.17-0.37	V 0.15-0.3; bal Fe				
CSN 415335	15335	High-temp const	0.2-0.27	1.0-1.5	0.25-0.5	0.45-0.65	0.3 max	0.035 max	0.035 max	0.25-0.5	V 0.65-0.85; W 0.4-0.7; bal Fe				
CSN 416121	16121	CH	0.14-0.19	0.8-1.1	0.7-1.0		0.4-0.7	0.035 max	0.035 max	0.15-0.4	bal Fe				
CSN 416220	16220	CH	0.14-0.19	0.8-1.1	0.7-1.0		1.3-1.6	0.035 max	0.035 max	0.17-0.37	bal Fe				
CSN 416221	16221	High-temp const	0.18-0.24	0.2 max	0.7-1.0		1.0-1.5	0.035 max	0.035 max	0.15-0.4	V 0.1-0.25; bal Fe				
CSN 416222	16222	Tough at subzero	0.18 max	0.3 max	1.0-1.5		0.7-1.1	0.04 max	0.04 max	0.35 max	Al 0.01-0.110; Nb 0.04-0.08; bal Fe				
CSN 416231	16231	CH	0.19-0.24	0.8-1.1	0.7-1.0		1.3-1.6	0.035 max	0.035 max	0.17-0.37	bal Fe				
CSN 416240	16240	Q/T	0.32-0.4	0.5-0.9	0.35-0.7		1.2-1.7	0.035 max	0.035 max	0.17-0.37	bal Fe				
CSN 416250	16250	Q/T Ni-Cr; Frg	0.4-0.5	0.5-0.9	0.35-0.7		1.2-1.7	0.035 max	0.035 max	0.17-0.37	bal Fe				
CSN 416320	16320	Q/T	0.1-0.17	0.3 max	0.3-0.6		2.8-3.3	0.04 max	0.04 max	0.15-0.35	bal Fe				
CSN 416322	16322	High-temp const	0.18-0.24	0.2 max	0.8-1.1		1.2-1.8	0.035 max	0.035 max	0.15-0.4	V 0.1-0.25; W 0.25-0.5; bal Fe				
CSN 416342	16342	High-temp const	0.32-0.42	1.6-2.0	0.4-0.7		1.0-1.3	0.04 max	0.04 max	0.15-0.4	V 0.1-0.25; bal Fe				
CSN 416420	16420	CH	0.1-0.17	0.6-0.9	0.3-0.6		3.2-3.7	0.035 max	0.035 max	0.17-0.37	bal Fe				
CSN 416440	16440	Q/T	0.3-0.4	0.6-1.0	0.4-0.8		3.2-3.7	0.035 max	0.035 max	0.17-0.37	bal Fe				
CSN 416520	16520	CH	0.1-0.17	0.9-1.3	0.3-0.6		3.9-4.7	0.04 max	0.04 max	0.15-0.35	bal Fe				
CSN 416523	16523		0.12-0.19	1.2-1.75	0.3-0.6		3.25-4.0	0.035 max	0.035 max	0.17-0.37	bal Fe				

UNS numbers and US grades are provided as a means of cross referencing chemically similar alloys. Exchangability is only possible after independent examination of specifications. Tensile properties are minimum or typical as specified. UTS and YS as MPa. El as %. See Appendix for list of abbreviations used in Notes. * indicates obsolete material.

Specification	Designation	Notes	C	Cr	Mn	Mo	Ni	P	S	Si	Other	UTS	YS	El	Hard

Alloy Steel, Unclassified, No equivalents identified

Czech Republic

Specification	Designation	Notes	C	Cr	Mn	Mo	Ni	P	S	Si	Other	UTS	YS	El	Hard
CSN 416532	16532	Q/T	0.27-0.34	0.9-1.2	1.0-1.3		1.4-1.8	0.035 max	0.035 max	0.9-1.2	bal Fe				
CSN 416640	16640	Q/T	0.3-0.38	0.8-1.2	0.35 0.6		4.7-5.2	0.035 max	0.035 max	0.17-0.37	bal Fe				
CSN 416720	16720	Q/T	0.14-0.21	1.35-1.65	0.25-0.55		4.0-4.5	0.035 max	0.035 max	0.17-0.37	W 0.8-1.2; bal Fe				

Europe

Specification	Designation	Notes	C	Cr	Mn	Mo	Ni	P	S	Si	Other	UTS	YS	El	Hard
EN 10028/4(94)	1.5662	High-temp, Press; 30<=t<=50mm N/T or Q/T	0.10 max		0.30-0.80	0.10 max	8.50-10.0	0.020 max	0.010 max	0.35 max	Al >=0.020; V <=0.05; Cr+Cu+Mo<=0.50; bal Fe	640-840	480	18	
EN 10028/4(94)	1.5662	High-temp, Press; t<=30mm N/T or Q/T	0.10 max		0.30-0.80	0.10 max	8.50-10.0	0.020 max	0.010 max	0.35 max	Al >=0.020; V <=0.05; Cr+Cu+Mo<=0.50; bal Fe'	640-840	490	18	
EN 10028/4(94)	1.5662	High-temp, Press; 30<=t<=50mm Q/T	0.10 max		0.30-0.80	0.10 max	8.50-10.0	0.020 max	0.010 max	0.35 max	Al >=0.020; V <=0.05; Cr+Cu+Mo<=0.50; bal Fe	680-820	575	18	
EN 10028/4(94)	1.5662	High-temp, Press; 15<=t<=30mm Q/T	0.10 max		0.30-0.80	0.10 max	8.50-10.0	0.020 max	0.010 max	0.35 max	Al >=0.020; V <=0.05; Cr+Cu+Mo<=0.50; bal Fe	680-820	585	18	
EN 10028/4(94)	1.5662	High-temp, Press; 0<=t<=14.99mm NNT or Q/T	0.10 max		0.30-0.80	0.10 max	8.50-10.0	0.020 max	0.010 max	0.35 max	Al >=0.020; V <=0.05; Cr+Cu+Mo<=0.50; bal Fe	680-820	585	18	
EN 10028/4(94)	X8Ni9	High-temp, Press; 30<=t<=50mm N/T or Q/T	0.10 max		0.30-0.80	0.10 max	8.50-10.0	0.020 max	0.010 max	0.35 max	Al >=0.020; V <=0.05; Cr+Cu+Mo<=0.50; bal Fe	640-840	480	18	
EN 10028/4(94)	X8Ni9	High-temp, Press; 30<=t<=50mm Q/T	0.10 max		0.30-0.80	0.10 max	8.50-10.0	0.020 max	0.010 max	0.35 max	Al >=0.020; V <=0.05; Cr+Cu+Mo<=0.50; bal Fe	680-820	575	18	
EN 10028/4(94)	X8Ni9	High-temp, Press; 0<=t<=14.99mm NNT or Q/T	0.10 max		0.30-0.80	0.10 max	8.50-10.0	0.020 max	0.010 max	0.35 max	Al >=0.020; V <=0.05; Cr+Cu+Mo<=0.50; bal Fe	680-820	585	18	
EN 10028/4(94)	X8Ni9	High-temp, Press; 15<=t<=30mm Q/T	0.10 max		0.30-0.80	0.10 max	8.50-10.0	0.020 max	0.010 max	0.35 max	Al >=0.020; V <=0.05; Cr+Cu+Mo<=0.50; bal Fe	680-820	585	18	
EN 10028/4(94)	X8Ni9	High-temp, Press; t<=30mm N/T or Q/T	0.10 max		0.30-0.80	0.10 max	8.50-10.0	0.020 max	0.010 max	0.35 max	Al >=0.020; V <=0.05; Cr+Cu+Mo<=0.50; bal Fe	640-840	490	18	

France

Specification	Designation	Notes	C	Cr	Mn	Mo	Ni	P	S	Si	Other	UTS	YS	El	Hard
AFNOR NFA35056(84)	FG12MSD6	Wir rod, weld	0.04-0.14		1.7-2.1	0.4-0.6		0.03 max	0.03 max	0.5-0.8	Al <=0.02; Cu <=0.2; Ti <=0.05; bal Fe				
AFNOR NFA35056(84)	FS12MD4	Wir rod, weld	0.07-0.14		0.9-1.2	0.45-0.6		0.025 max	0.025 max	0.1 max	Al <=0.04; Cu <=0.12; Ti <=0.06; Al+Ti<=0.08; bal Fe				
AFNOR NFA35056(84)	FS12MD8	Wir rod, weld	0.09-0.16		1.8-2.2	0.45-0.6		0.025 max	0.025 max	0.1 max	Al <=0.04; Cu <=0.12; Ti <=0.06; Al+Ti<=0.08; bal Fe				
AFNOR NFA35565(94)	15NCD16*	see 15NiCrMo16-5	0.14-0.18	1-1.4	0.25-0.55	0.2-0.3	3.8-4.3	0.015 max	0.008 max	0.15-0.35	Al <=0.05; Cu <=0.35; Ti <=0.05; bal Fe				
AFNOR NFA35565(94)	15NiCrMo16-5	Ball & roller bearing; CH	0.14-0.18	1-1.4	0.25-0.55	0.2-0.3	3.8-4.3	0.025 max	0.015 max	0.15-0.35	Al <=0.05; Cu <=0.35; Ti <=0.05; bal Fe				
AFNOR NFA35565(94)	15NiCrMo16-5	Ball & roller bearing; CH	0.14-0.18	1-1.4	0.25-0.55	0.2-0.3	3.8-4.3	0.015 max	0.008 max	0.15-0.35	Al <=0.05; Cu <=0.35; Ti <=0.05; bal Fe				
AFNOR NFA36208(82)	9Ni490	t<=30mm; Norm	0.10 max	0.25 max	0.3-0.8	0.1 max	8.5-10	0.025 max	0.02 max	0.35 max	Al >=0.015; Ti <=0.05; V <=0.04; Cr+Cu+Mo<=0.5; bal Fe	670-800	490	18	

Specification	Designation	Notes	C	Cr	Mn	Mo	Ni	P	S	Si	Other	UTS	YS	El	Hard
Alloy Steel, Unclassified, No equivalents identified																
France																
AFNOR NFA36208(82)	9Ni490	30<t<=50mm; Norm	0.10 max	0.25 max	0.3-0.8	0.1 max	8.5-10	0.025 max	0.02 max	0.35 max	Al >=0.015; Ti <=0.05; V <=0.04; Cr+Cu+Mo<=0.5; bal Fe	670-800	480	18	
AFNOR NFA36208(82)	9Ni585	t<=30mm; Q/T	0.10 max	0.25 max	0.3-0.8	0.1 max	8.5-10	0.025 max	0.02 max	0.35 max	Al >=0.015; Ti <=0.05; V <=0.04; Cr+Cu+Mo<=0.5; bal Fe	690-820	585	18	
AFNOR NFA36208(82)	9Ni585	30<t<=50mm; Q/T	0.10 max	0.25 max	0.3-0.8	0.1 max	8.5-10	0.025 max	0.02 max	0.35 max	Al >=0.015; Ti <=0.05; V <=0.04; Cr+Cu+Mo<=0.5; bal Fe	690-820	575	18	
Hungary																
MSZ 1741(89)	KL2	Broiler, Press ves; 0-20mm; HT	0.2 max		0.5-1			0.35 max	0.035 max	0.03 max	0.15-0.4	Al 0.02-0.06; Cu <=0.35; Nb 0.01-0.06; Ti 0.02-0.06; V 0.02-0.1; Zr=0.015-0.06; Al+Nb+V+Ti+Zr<=0.15; N<=Al/2+V/4+Nb/7+Ti/3.5<=0.015; bal Fe	410-530	260	22L	
MSZ 1741(89)	KL2	Broiler, Press ves; 40.1-60mm; HT	0.2 max		0.5-1			0.35 max	0.035 max	0.03 max	0.15-0.4	Al 0.02-0.06; Cu <=0.35; Nb 0.01-0.06; Ti 0.02-0.06; V 0.02-0.1; Zr=0.015-0.06; Al+Nb+V+Ti+Zr<=0.15; N<=Al/2+V/4+Nb/7+Ti/3.5<=0.015; bal Fe	410-530	240	22L	
MSZ 1741(89)	KL2C	Broiler, Press ves; 40.1-60mm; HT	0.2 max		0.5-1			0.35 max	0.035 max	0.03 max	0.15-0.4	Al 0.02-0.06; Cu <=0.35; Nb 0.01-0.06; Ti 0.02-0.06; V 0.02-0.1; Zr=0.015-0.06; Al+Nb+V+Ti+Zr<=0.15; N<=Al/2+V/4+Nb/7+Ti/3.5<=0.015; bal Fe	410-530	240	22L	
MSZ 1741(89)	KL2C	Broiler, Press ves; 0-20mm; HT	0.2 max		0.5-1			0.35 max	0.035 max	0.03 max	0.15-0.4	Al 0.02-0.06; Cu <=0.35; Nb 0.01-0.06; Ti 0.02-0.06; V 0.02-0.1; Zr=0.015-0.06; Al+Nb+V+Ti+Zr<=0.15; N<=Al/2+V/4+Nb/7+Ti/3.5<=0.015; bal Fe	410-530	260	22L	
MSZ 1741(89)	KL2D	Broiler, Press ves; 40.1-60mm; HT	0.2 max		0.5-1			0.35 max	0.035 max	0.03 max	0.15-0.4	Al 0.02-0.06; Cu <=0.35; Nb 0.01-0.06; Ti 0.02-0.06; V 0.02-0.1; Zr=0.015-0.06; Al+Nb+V+Ti+Zr<=0.15; N<=Al/2+V/4+Nb/7+Ti/3.5<=0.015; bal Fe	410-530	240	22L	
MSZ 1741(89)	KL2D	Broiler, Press ves; 0-20mm; HT	0.2 max		0.5-1			0.35 max	0.035 max	0.03 max	0.15-0.4	Al 0.02-0.06; Cu <=0.35; Nb 0.01-0.06; Ti 0.02-0.06; V 0.02-0.1; Zr=0.015-0.06; Al+Nb+V+Ti+Zr<=0.15; N<=Al/2+V/4+Nb/7+Ti/3.5<=0.015; bal Fe	410-530	260	22L	
MSZ 2666(76)	38S	Spring HF	0.34-0.42		0.5-0.8		0.3 max	0.04 max	0.04 max	1.4-1.7	Cu <=0.25; bal Fe				
MSZ 2666(76)	75	Spring HF; 0-999mm	0.72-0.8		0.5-0.8		0.3 max	0.04 max	0.04 max	0.17-0.37	Cu <=0.25; bal Fe				285 HB max
MSZ 2666(76)	75	Spring HF; 0-999mm; Q/T	0.72-0.8		0.5-0.8		0.3 max	0.04 max	0.04 max	0.17-0.37	Cu <=0.25; bal Fe	1080	880	9L	

UNS numbers and US grades are provided as a means of cross referencing chemically similar alloys. Exchangability is only possible after independent examination of specifications. Tensile properties are minimum or typical as specified. UTS and YS as MPa. El as %. See Appendix for list of abbreviations used in Notes. * indicates obsolete material.

Specification	Designation	Notes	C	Cr	Mn	Mo	Ni	P	S	Si	Other	UTS	YS	El	Hard
Alloy Steel, Unclassified, No equivalents identified															
Hungary															
MSZ 31(74)	BNC7	CH; 0-11mm; blank carburized	0.12-0.18	0.8-1.1	0.6-0.9		1.3-1.6	0.035 max	0.035 max	0.17-0.37	bal Fe	883-1226	637	8L	
MSZ 31(74)	BNC7	CH; 63 mm; blank carburized	0.12-0.18	0.8-1.1	0.6-0.9		1.3-1.6	0.035 max	0.035 max	0.17-0.37	bal Fe	686-981	490	11L	
MSZ 31(74)	BNCMo3	CH; 63 mm; blank carburized	0.12-0.18	1.1-1.4	0.25-0.55	0.2-0.3	3.8-4.3	0.035 max	0.035 max	0.17-0.37	bal Fe	1177-1471	785	8L	
MSZ 31(74)	BNCMo3	CH; 0-11mm; blank carburized	0.12-0.18	1.1-1.4	0.25-0.55	0.2-0.3	3.8-4.3	0.035 max	0.035 max	0.17-0.37	bal Fe	1275-1618	883	7L	
MSZ 31(74)	BNCMo3E	CH; 63 mm; blank carburized	0.12-0.18	1.1-1.4	0.25-0.55	0.2-0.3	3.8-4.3	0.035 max	0.02-0.035	0.17-0.37	bal Fe	1177-1471	785	8L	
MSZ 31(74)	BNCMo3E	CH; 0-11mm; blank carburized	0.12-0.18	1.1-1.4	0.25-0.55	0.2-0.3	3.8-4.3	0.035 max	0.02-0.035	0.17-0.37	bal Fe	1275-1618	883	7L	
MSZ 4400(78)	AH120	Tough at subzero; 0-100mm; Norm	0.12 max		0.6-1	0.15-0.25	4-5	0.035 max	0.03 max	0.15-0.4	V 0.05-0.15; 100mm V=0.05-0.25; bal Fe	590	440	20L	
MSZ 4400(78)	AH195	Tough at subzero; 0-100mm; Norm	0.08 max		1-1.4	0.15-0.25	8-10	0.035 max	0.03 max	0.15-0.4	V 0.05-0.15; 100mm V=0.05-0.25; bal Fe	640	490	18L	
MSZ 4400(78)	AHC195	Tough at subzero; 0-100mm; SA	0.1 max	17-19	2 max		8-10	0.035 max	0.03 max	1 max	bal Fe	490	195	40L	
MSZ 4400(78)	AHCN195	Tough at subzero; 0-100mm; SA	0.1 max	17-19	2 max		8-10	0.035 max	0.03 max	1 max	Nb=8C-1.1; bal Fe	490	195	40L	
MSZ 4400(78)	AHCT195	Tough at subzero; 0-100mm; SA	0.1 max	17-19	2 max		8-10	0.035 max	0.03 max	1 max	Ti <=0.8; Ti+5C-0.8; bal Fe	490	195	40L	
MSZ 4747(85)	Mo45.47	High-temp, Tub; Non-rimming; K; FF; 0-20mm; Norm	0.12-0.2		0.4-0.8	0.25-0.35		0.035 max	0.035 max	0.35 max	bal Fe	450-600	270	22L	
MSZ 4747(85)	Mo45.47	High-temp, Tub; Non-rimming; K; FF; 40.1-60mm; Norm	0.12-0.2		0.4-0.8	0.25-0.35		0.035 max	0.035 max	0.35 max	bal Fe	450-600	260	22L	
MSZ 7779(79)	31CMo	Nitriding; 160.1-250mm; Q/T	0.28-0.35	2.8-3.3	0.4-0.7	0.3-0.5		0.03 max	0.035 max	0.15-0.4	bal Fe	880-1080	690	12L	
MSZ 7779(79)	31CMo	Nitriding; 0-16mm; Q/T	0.28-0.35	2.8-3.3	0.4-0.7	0.3-0.5		0.03 max	0.035 max	0.15-0.4	bal Fe	1080-1280	880	10L	
MSZ 7779(79)	39CMoV	Nitriding; 0-100mm; Q/T	0.35-0.42	3-3.5	0.4-0.7	0.8-1.1		0.03 max	0.035 max	0.15-0.4	V 0.15-0.25; bal Fe	1270-1470	1080	8L	
MSZ 7779(79)	39CMoV	Nitriding; 0-999mm; soft ann	0.35-0.42	3-3.5	0.4-0.7	0.8-1.1		0.03 max	0.035 max	0.15-0.4	V 0.15-0.25; bal Fe				262 HB max
MSZ 7779(88)	31CrMo12	Nitriding; Q/T; 100.1-250mm	0.28-0.35	2.8-3.3	0.4-0.7	0.3-0.5	0.3 max	0.03 max	0.03 max	0.4 max	bal Fe	900-1000	700	12L	
MSZ 7779(88)	31CrMo12	Nitriding; Q/T; 0-100mm	0.28-0.35	2.8-3.3	0.4-0.7	0.3-0.5	0.3 max	0.03 max	0.03 max	0.4 max	bal Fe	1000-1200	800	11L	
International															
ISO 2604-1(75)	F44	Frg	0.20 max		0.80 max		3.25-3.75	0.040 max	0.040 max	0.15-0.40	Al <=0.015; bal Fe	490-540			
ISO 2604-1(75)	F45	Frg	0.13 max		0.80 max		8.50-10.00	0.040 max	0.040 max	0.15-0.40	Al <=0.015; bal Fe	690-830			
ISO 4952(81)	Fe235W	Corr res, Struct, Bars, Shp, t<=16mm	0.13 max	0.40-0.80	0.20-0.60		0.65 max	0.040 max	0.035 max	0.10-0.40	Cu 0.20-0.50; bal Fe	360	235	25	
ISO 4952(81)	Fe355W	Corr res, Struct, Bars, Shp, t<=16mm	0.12 max	0.30-1.25	1.00 max		0.65 max	0.06-0.15	0.050 max	0.20-0.75	Cu 0.25-0.55; bal Fe	480	355	20	
ISO 683-10	X/1*	Nitriding	0.28-0.35	2.8-3.3	0.4-0.7	0.3-0.5	0.3 max	0.03 max	0.035 max	0.15-0.4	Al <=0.1; Co <=0.1; Cu <=0.3; Pb <=0.15; Ti <=0.05; V <=0.1; W <=0.1; bal Fe				
ISO 683-10	X/2*	Nitriding	0.35-0.42	3-3.5	0.4-0.7	0.8-1.1	0.4 max	0.03 max	0.035 max	0.15-0.4	Al <=0.1; Co <=0.1; Cu <=0.3; Pb <=0.15; Ti <=0.05; V 0.15-0.25; W <=0.1; bal Fe				
ISO 683-10(87)	31CrMo12	Nitriding, <100mm	0.28-0.35	2.80-3.30	0.40-0.70	0.30-0.5	0.30 max	0.030 max	0.035 max	0.40 max	bal Fe	1000-1200	800		800 HV
ISO 683-11(87)	15NiCr13	CH, 16mm test	0.12-0.18	0.60-0.90	0.35-0.65		3.00-3.50	0.035 max	0.035 max	0.15-0.40	Al <=0.1; Co <=0.1; Cu <=0.3; bal Fe	1010-1360	650		
Italy															
UNI 3097(75)	X80MoCrV44	Ball & roller bearing	0.77-0.85	3.75-4.25	0.35 max	4-4.5		0.03 max	0.03 max	0.25 max	V 0.9-1.1; bal Fe				
UNI 6120(67)	30CrMo12	Nitriding	0.27-0.34	2.7-3.3	0.4-0.7	0.3-0.4		0.035 max	0.035 max	0.4 max	bal Fe				

UNS numbers and US grades are provided as a means of cross referencing chemically similar alloys. Exchangability is only possible after independent examination of specifications. Tensile properties are minimum or typical as specified. UTS and YS as MPa. El as %. See Appendix for list of abbreviations used in Notes. * indicates obsolete material.

Specification	Designation	Notes	C	Cr	Mn	Mo	Ni	P	S	Si	Other	UTS	YS	El	Hard

Alloy Steel, Unclassified, No equivalents identified

Japan

Specification	Designation	Notes	C	Cr	Mn	Mo	Ni	P	S	Si	Other	UTS	YS	El	Hard
JIS G3311(88)	SNC415M	CR Strp; Chain	0.12-0.18	0.20-0.50	0.35-0.65		2.00-2.50	0.030 max	0.030 max	0.15-0.35	Cu <=0.30; bal Fe				170-240 HV
JIS G3311(88)	SNC631M	CR Strp; Mach parts	0.27-0.35	0.60-1.00	0.35-0.65		2.50-3.00	0.030 max	0.030 max	0.15-0.35	Cu <=0.30; bal Fe				180-240 hv
JIS G3311(88)	SNC836M	CR Strp; Mach parts	0.32-0.40	0.60-1.00	0.35-0.65		3.00-3.50	0.030 max	0.030 max	0.15-0.35	Cu <=0.30; bal Fe				190-250 HV

Mexico

Specification	Designation	Notes	C	Cr	Mn	Mo	Ni	P	S	Si	Other	UTS	YS	El	Hard
NMX-B-197(85)	Grade 9	Smls weld tub for low temp service	0.20 max		0.40-1.06		1.60-2.24	0.045 max	0.050 max		Cu 0.75-1.25; bal Fe	434	317	28	

Poland

Specification	Designation	Notes	C	Cr	Mn	Mo	Ni	P	S	Si	Other	UTS	YS	El	Hard
PNH84018(86)	15G2ANb	HS weld const	0.18 max	0.3 max	1.1-1.6		0.3 max	0.04 max	0.04 max	0.2-0.55	Al 0.02-0.120; Nb 0.02-0.05; CEV<=0.045; bal Fe				
PNH84018(86)	15G2ANNb	HS weld const	0.17 max	0.2 max	1.2-1.5		0.5-0.7	0.04 max	0.04 max	0.3-0.5	Al 0.02-0.120; CEV<=0.045; bal Fe				
PNH84030/02	15HGN	CH	0.13-0.2	0.8-1.1	0.7-1		1.3-1.6	0.035 max	0.035 max	0.17-0.37	bal Fe				
PNH84030/02	15HN	CH	0.12-0.18	1.4-1.7	0.4-0.7		1.4-1.7	0.035 max	0.035 max	0.17-0.37	bal Fe				
PNH84030/02	15HNA	CH	0.12-0.18	1.4-1.7	0.15 max	0.03 max	0.17-0.37	0.03 max	0.03 max	0.17-0.37	Cu <=0.03; bal Fe				
PNH84030/02	17HGN	CH	0.15-0.21	0.8-1.1	1-1.3		0.6-0.9	0.035 max	0.035 max	0.17-0.37	bal Fe				
PNH84030/02	18H2N2	CH	0.15-0.22	1.8-2.1	0.4-0.7		1.8-2.1	0.035 max	0.035 max	0.17-0.37	bal Fe				
PNH84035	12H2N4A	CH	0.09-0.16	1.25-1.65	0.3-0.6		3.25-3.65	0.03 max	0.025 max	0.17-0.37	Cu <=0.2; bal Fe				
PNH84035	12HN3A	CH	0.09-0.16	0.6-0.9	0.3-0.6		2.75-3.15	0.03 max	0.025 max	0.17-0.37	Cu <=0.2; bal Fe				
PNH84035	12HN3A	CH	0.09-0.16	0.6-0.9	0.3-0.6		2.75-3.15	0.03 max	0.025 max	0.17-0.37	Cu <=0.2; bal Fe				
PNH84035	18H2N4WA	CH	0.14-0.2	1.35-1.65	0.25-0.55		4-4.4	0.03 max	0.025 max	0.17-0.37	Cu <=0.2; W 0.8-1.2; bal Fe				
PNH84035	20H2N4A	CH	0.16-0.22	1.25-1.65	0.3-0.6		3.25-3.65	0.03 max	0.025 max	0.17-0.37	Cu <=0.2; bal Fe				
PNH84035	20HN3A	CH	0.18-0.24	0.6-0.9	0.3-0.6		2.8-3.2	0.03 max	0.03 max	0.17-0.37	Cu <=0.2; bal Fe				
PNH84035	25H2N4WA	Q/T	0.21-0.28	1.35-1.65	0.25-0.55		4-4.4	0.03 max	0.25 max	0.17-0.37	Cu <=0.2; W 0.8-1.2; bal Fe				
PNH84035	30HGSNA	Q/T	0.27-0.34	0.9-1.2	1-1.3		1.4-1.8	0.03 max	0.025 max	0.9-1.2	Cu <=0.2; bal Fe				
PNH84035	30HN3A	Q/T	0.27-0.34	0.6-0.9	0.3-0.6		2.8-3.2	0.03 max	0.03 max	0.17-0.37	Cu <=0.2; V <=0.15; bal Fe				
PNH84035	37HN3A	Q/T	0.33-0.41	1.2-1.6	0.25-0.55		3-3.5	0.3 max	0.3 max	0.17-0.37	Cu <=0.2; bal Fe				

Romania

Specification	Designation	Notes	C	Cr	Mn	Mo	Ni	P	S	Si	Other	UTS	YS	El	Hard
STAS 791(66)	20C08	Struct/const; CH	0.17-0.23	0.7-1	0.5-0.8			0.035 max	0.035 max	0.17-0.37	Pb <=0.15; bal Fe				
STAS 791(88)	17CrNi16	Struct/const; CH	0.14-0.19	1.4-1.7	0.4-0.6		1.4-1.7	0.035 max	0.035 max	0.17-0.37	Pb <=0.15; Ti <=0.02; bal Fe				
STAS 791(88)	17CrNi16S	Struct/const; CH	0.14-0.19	1.4-1.7	0.4-0.6			0.035 max	0.02-0.04	0.17-0.37	Pb <=0.15; Ti <=0.02; bal Fe				
STAS 791(88)	17CrNi16X	Struct/const; CH	0.14-0.19	1.4-1.7	0.4-0.6			0.025 max	0.025 max	0.17-0.37	Pb <=0.15; Ti <=0.02; bal Fe				
STAS 791(88)	17CrNi16XS	Struct/const; CH	0.14-0.19	1.4-1.7	0.4-0.6			0.025 max	0.02-0.035	0.17-0.37	Pb <=0.15; Ti <=0.02; bal Fe				
STAS 791(88)	17MnCr10	Struct/const; CH	0.14-0.19	0.8-1.1	1-1.3			0.035 max	0.035 max	0.17-0.37	Pb <=0.15; Ti <=0.02; bal Fe				
STAS 791(88)	17MnCr10S	Struct/const; CH	0.14-0.19	0.8-1.1	1-1.3			0.035 max	0.02-0.04	0.17-0.37	Pb <=0.15; Ti <=0.02; bal Fe				
STAS 791(88)	17MnCr10X	Struct/const; CH	0.14-0.19	0.8-1.1	1-1.3			0.025 max	0.025 max	0.17-0.37	Pb <=0.15; Ti <=0.02; bal Fe				
STAS 791(88)	17MnCr10XS	Struct/const; CH	0.14-0.19	0.8-1.1	1-1.3			0.025 max	0.02-0.035	0.17-0.37	Pb <=0.15; Ti <=0.02; bal Fe				
STAS 791(88)	18CrNi20	Struct/const; CH	0.15-0.2	1.8-2.1	0.4-0.6		1.8-2.1	0.035 max	0.035 max	0.17-0.37	Pb <=0.15; Ti <=0.02; bal Fe				
STAS 791(88)	18CrNi20S	Struct/const; CH	0.15-0.2	1.8-2.1	0.4-0.6		1.8-2.1	0.035 max	0.02-0.04	0.17-0.37	Pb <=0.15; Ti <=0.02; bal Fe				

Specification	Designation	Notes	C	Cr	Mn	Mo	Ni	P	S	Si	Other	UTS	YS	El	Hard

Alloy Steel, Unclassified, No equivalents identified

Romania

Specification	Designation	Notes	C	Cr	Mn	Mo	Ni	P	S	Si	Other	UTS	YS	El	Hard
STAS 791(88)	18CrNi20X	Struct/const; CH	0.15-0.2	1.8-2.1	0.4-0.6		1.8-2.1	0.025 max	0.025 max	0.17-0.37	Pb <=0.15; Ti <=0.02; bal Fe				
STAS 791(88)	18CrNi20XS	Struct/const; CH	0.15-0.2	1.8-2.1	0.4-0.6		1.8-2.1	0.025 max	0.02-0.035	0.17-0.37	Pb <=0.15; Ti <=0.02; bal Fe				
STAS 791(88)	20MnCr12	Struct/const; CH	0.17-0.22	1-1.3	1.1-1.4			0.035 max	0.035 max	0.17-0.37	Pb <=0.15; Ti <=0.02; bal Fe				
STAS 791(88)	20MnCr12S	Struct/const; CH	0.17-0.22	1-1.3	1.1-1.4			0.035 max	0.02-0.04	0.17-0.37	Pb <=0.15; Ti <=0.02; bal Fe				
STAS 791(88)	20MnCr12X	Struct/const; CH	0.17-0.22	1-1.3	1.1-1.4			0.025 max	0.025 max	0.17-0.37	Pb <=0.15; Ti <=0.02; bal Fe				
STAS 791(88)	20MnCr12XS	Struct/const; CH	0.17-0.22	1-1.3	1.1-1.4			0.025 max	0.02-0.035	0.17-0.37	Pb <=0.15; Ti <=0.02; bal Fe				
STAS 791(88)	20TiMnCr12	Struct/const	0.17-0.23	1-1.3	0.8-1.1			0.035 max	0.035 max	0.17-0.37	Pb <=0.15; Ti 0.03-0.09; bal Fe				
STAS 791(88)	20TiMnCr12S	Struct/const	0.17-0.23	1-1.3	0.8-1.1			0.035 max	0.02-0.04	0.17-0.37	Pb <=0.15; Ti 0.03-0.09; bal Fe				
STAS 791(88)	20TiMnCr12X	Struct/const	0.17-0.23	1-1.3	0.8-1.1			0.025 max	0.025 max	0.17-0.37	Pb <=0.15; Ti 0.03-0.09; bal Fe				
STAS 791(88)	20TiMnCr12XS	Struct/const	0.17-0.23	1-1.3	0.8-1.1			0.025 max	0.02-0.035	0.17-0.37	Pb <=0.15; Ti 0.03-0.09; bal Fe				
STAS 791(88)	31MnCrSi11	Struct/const	0.28-0.34	0.8-1.1	0.8-1.1			0.035 max	0.035 max	0.9-1.2	Pb <=0.15; Ti <=0.02; bal Fe				
STAS 791(88)	31MnCrSi11S	Struct/const	0.28-0.34	0.8-1.1	0.8-1.1			0.035 max	0.02-0.04	0.9-1.2	Pb <=0.15; Ti <=0.02; bal Fe				
STAS 791(88)	31MnCrSi11X	Struct/const	0.28-0.34	0.8-1.1	0.8-1.1			0.025 max	0.025 max	0.9-1.2	Pb <=0.15; Ti <=0.02; bal Fe				
STAS 791(88)	31MnCrSi11XS	Struct/const	0.28-0.34	0.8-1.1	0.8-1.1			0.025 max	0.02-0.035	0.9-1.2	Pb <=0.15; Ti <=0.02; bal Fe				
STAS 791(88)	36MnCrSi13	Struct/const	0.32-0.39	1.1-1.4	0.8-1.1			0.035 max	0.035 max	1.1-1.4	Pb <=0.15; Ti <=0.02; bal Fe				
STAS 791(88)	36MnCrSi13S	Struct/const	0.32-0.39	1.1-1.4	0.8-1.1			0.035 max	0.02-0.04	1.1-1.4	Pb <=0.15; Ti <=0.02; bal Fe				
STAS 791(88)	36MnCrSi13X	Struct/const	0.32-0.39	1.1-1.4	0.8-1.1			0.025 max	0.025 max	1.1-1.4	Pb <=0.15; Ti <=0.02; bal Fe				
STAS 791(88)	36MnCrSi13XS	Struct/const	0.32-0.39	1.1-1.4	0.8-1.1			0.025 max	0.02-0.035	1.1-1.4	Pb <=0.15; Ti <=0.02; bal Fe				
STAS 791(88)	39MoAlCr15	Struct/const	0.35-0.42	1.35-1.65	0.3-0.6	0.15-0.25		0.035 max	0.035 max	0.17-0.37	Al 0.7-1.1; Pb <=0.15; Ti <=0.02; bal Fe				
STAS 791(88)	39MoAlCr15S	Struct/const	0.35-0.42	1.35-1.65	0.3-0.6	0.15-0.25		0.035 max	0.02-0.035	0.2-0.45	Al 0.7-1.1; Pb <=0.15; Ti <=0.02; bal Fe				
STAS 791(88)	39MoAlCr15X	Struct/const	0.35-0.42	1.35-1.65	0.3-0.6	0.15-0.25		0.025 max	0.025 max	0.2-0.45	Al 0.7-1.1; Pb <=0.15; Ti <=0.02; bal Fe				
STAS 791(88)	39MoAlCr15XS	Struct/const	0.35-0.42	1.35-1.65	0.3-0.6	0.15-0.25		0.025 max	0.02-0.035	00.2-0.45	Al 0.7-1.1; Pb <=0.15; Ti <=0.02; bal Fe				
STAS 791(88)	40CrNi12	Struct/const; Q/T	0.36-0.44	0.45-0.75	0.5-0.8		1-1.4	0.035 max	0.035 max	0.17-0.37	Pb <=0.15; Ti <=0.02; bal Fe				
STAS 791(88)	40CrNi12S	Struct/const; Q/T	0.36-0.44	0.45-0.75	0.5-0.8		1-1.4	0.035 max	0.02-0.04	0.17-0.37	Pb <=0.15; Ti <=0.02; bal Fe				
STAS 791(88)	40CrNi12X	Struct/const; Q/T	0.36-0.44	0.45-0.75	0.5-0.8		1-1.4	0.025 max	0.025 max	0.17-0.37	Pb <=0.15; Ti <=0.02; bal Fe				
STAS 791(88)	40CrNi12XS	Struct/const; Q/T	0.36-0.44	0.45-0.75	0.5-0.8		1-1.4	0.025 max	0.02-0.035	0.17-0.37	Pb <=0.15; Ti <=0.02; bal Fe				
STAS 795(92)	65WSi18A	Spring; HF	0.61-0.69	0.3 max	0.7-1			0.025 max	0.025 max	1.5-2	Cu <=0.25; Pb <=0.15; W 0.8-1.2; bal Fe				
STAS 9382/4(89)	30MoCrNi20q	Q/T	0.26-0.34	1.8-2.1	0.3-0.6	0.3-0.5	1.8-2.1	0.035 max	0.035 max	0.17-0.37	Pb <=0.15; As<=0.05; bal Fe				

Russia

Specification	Designation	Notes	C	Cr	Mn	Mo	Ni	P	S	Si	Other	UTS	YS	El	Hard
GOST 977(88)	40ChL		0.35-0.45	0.8-1.1	0.4-0.9		0.3	0.04	0.04	0.2-0.4	bal Fe				

Spain

Specification	Designation	Notes	C	Cr	Mn	Mo	Ni	P	S	Si	Other	UTS	YS	El	Hard
UNE 36087(76/78)	F.2645*	see X8Ni09	0.1 max		0.8 max	0.15 max	8.5-10.0	0.035 max	0.035 max	0.15-0.35	bal Fe				

UNS numbers and US grades are provided as a means of cross referencing chemically similar alloys. Exchangability is only possible after independent examination of specifications. Tensile properties are minimum or typical as specified. UTS and YS as MPa. El as %. See Appendix for list of abbreviations used in Notes. * indicates obsolete material.

Specification	Designation	Notes	C	Cr	Mn	Mo	Ni	P	S	Si	Other	UTS	YS	El	Hard

Alloy Steel, Unclassified, No equivalents identified

Spain

Specification	Designation	Notes	C	Cr	Mn	Mo	Ni	P	S	Si	Other	UTS	YS	El	Hard
UNE 36087(76/78)	X8Ni09		0.1 max		0.8 max	0.15 max	8.5-10.0	0.035 max	0.035 max	0.15-0.35	bal Fe				
UNE 36123(90)	AE390HC	HR; Strp	0.10 max		1.2 max	0.15 max		0.03 max	0.5 max	0.01-0.12	Al 0.02-0.08; Nb 0.01-0.06; Pb <=0.03; Ti 0.01-0.08; W <=0.15; bal Fe				
UNE 36123(90)	AE440HC	HR; Strp	0.12 max		1.3 max	0.15 max		0.03 max	0.03 max	0.5 max	Al 0.02-0.08; Nb 0.01-0.06; Ti 0.01-0.12; V 0.01-0.08; bal Fe				
UNE 36123(90)	AE490HC	HR; Strp	0.12 max		1.5 max	0.15 max		0.03 max	0.03 max	0.5 max	Al 0.02-0.08; Nb 0.01-0.06; Ti 0.01-0.12; V 0.01-0.08; bal Fe				

Sweden

Specification	Designation	Notes	C	Cr	Mn	Mo	Ni	P	S	Si	Other	UTS	YS	El	Hard
SS 142127	2127		0.13-0.19	0.8-1.1	1-1.3			0.035 max	0.03-0.05	0.15-0.4	bal Fe				
SS 142511	2511	CH	0.13-0.18	0.6-1	0.7-1.1	0.1 max	0.8-1.2	0.035 max	0.03-0.05	0.15-0.4	bal Fe				
SS 142534	2534	Q/T	0.28-0.35	0.9-1.2	0.4-0.7	0.2-0.3	3-3.5	0.035 max	0.035 max	0.15-0.4	bal Fe				

UK

Specification	Designation	Notes	C	Cr	Mn	Mo	Ni	P	S	Si	Other	UTS	YS	El	Hard
BS 1502(82)	509-690	Bar Shp Press ves	0.10 max	0.25 max	0.30-0.80	0.10 max	8.50-10.0	0.025 max	0.020 max	0.15-0.35	Al >=0.015; Co <=0.1; W <=0.1; bal Fe				
BS 1503(89)	225-490	Frg press ves; t<=100mm; Norm Q/T	0.20 max	0.25 max	0.90-1.70	0.10 max	0.40 max	0.030 max	0.025 max	0.10-0.40	Al >=0.018; Cu <=0.30; Nb 0.01-0.04; W <=0.1; Cr+Mo+Ni+Cu<=0.8; bal Fe	490-610	340	16	
BS 1503(89)	225-490	Frg press ves; 100<t<=999mm; Norm Q/T	0.20 max	0.25 max	0.90-1.70	0.10 max	0.40 max	0.030 max	0.025 max	0.10-0.40	Al >=0.018; Cu <=0.30; Nb 0.01-0.04; W <=0.1; Cr+Mo+Ni+Cu<=0.8; bal Fe	490-610	300	16	
BS 3603(91)	HFS430LT; CFS430LT; ERW430LT; CEW430LT	Pip Tub press low temp t<=200mm	0.20 max		0.60-1.20			0.035 max	0.035 max	0.35 max	Al >=0.020; bal Fe	430-570	275	22	
BS 3603(91)	HFS503LT; CFS503LT	Pip Tub press low temp t<=200mm	0.15 max		0.30-0.80		3.25-3.75	0.025 max	0.020 max	0.15-0.35	Al >=0.020; bal Fe	440-590	245	16	

USA

Specification	Designation	Notes	C	Cr	Mn	Mo	Ni	P	S	Si	Other	UTS	YS	El	Hard
	AISI E7140	Smls mech Tub; BEF, see K24065	0.38-0.43	1.40-1.80	0.50-0.70	0.30-0.40		0.025 max	0.025 max	0.15-0.35 *	Al 0.95-1.30; bal Fe				
	UNS G15116*	Obs; see G51986									bal Fe				
	UNS G71406*	Obs; see K24065; W-Cr									bal Fe				
	UNS K44210*	Obs; see K44220									bal Fe				
	UNS K44315*	Obs; see K44220; 300M									bal Fe				
	UNS K61595*	Obs; see S50300									bal Fe				
	UNS K63005*	Obs; see S63005; Valve Steel (CNS)									bal Fe				
	UNS K63007*	Obs; see S63007; Valve Steel (21-55 N)									bal Fe				
	UNS K63008*	Obs; see S63008; Valve Steel (21-4 N)									bal Fe				
	UNS K63011*	Obs; see S63011; Valve Steel (746)									bal Fe				
	UNS K63012*	Obs; see S63012; Valve Steel (21-2 N)									bal Fe				
	UNS K63013*	Obs; see S63013; Valve Steel (Gaman H)									bal Fe				
	UNS K63014*	Obs; see S63014; Valve Steel (10)									bal Fe				
	UNS K63015*	Obs; see S63015; Valve Steel (10 N)									bal Fe				
	UNS K63016*	Obs; see S63016; Valve Steel (21-12)									bal Fe				

Specification	Designation	Notes	C	Cr	Mn	Mo	Ni	P	S	Si	Other	UTS	YS	El	Hard

Alloy Steel, Unclassified, No equivalents identified

USA

Specification	Designation	Notes	C	Cr	Mn	Mo	Ni	P	S	Si	Other	UTS	YS	El	Hard
	UNS K63017*	Obs; see S63017; Valve Steel (21-12 N)									bal Fe				
	UNS K63198*	Obs; see S63198; Fe Base Superalloy (19-9-DL)									bal Fe				
	UNS K63199*	Obs; see S63199; Fe Base Superalloy (19-9-DX or 19-0-W-Mo)									bal Fe				
	UNS K64005*	Obs; see S64005; Valve Steel (2)									bal Fe				
	UNS K64006*	Obs; see S64006; Valve Steel (F)									bal Fe				
	UNS K64152*	Obs; see S64152; High-Strength (M152)									bal Fe				
	UNS K64299*	Obs; see S64299; Fe Base Superalloy (29-9)									bal Fe				
	UNS K65006*	Obs; see S65006; Valve Steel (XB)									bal Fe				
	UNS K65007*	Obs; see S65007; Valve Steel (1)									bal Fe				
	UNS K65150*	Obs; see S65150; Fe Base Superalloy (Pyromet X-15)									bal Fe				
	UNS K65770*	Obs; see S65770; Fe Base Superalloy (AFC77)									bal Fe				
	UNS K66009*	Obs; see S66009; Valve Steel (TPA)									bal Fe				
	UNS K66220*	Obs; see S66220; Fe Base Superalloy (Discaloy)									bal Fe				
	UNS K66286*	Obs; see S66286; Fe Base Superalloy (A286)									bal Fe				
	UNS K66545*	Obs; see S66545; Fe Base Superalloy (W545)									bal Fe				
	UNS K66979*	Obs; see N09979; Fe-Ni Base Superalloy (D-979)									bal Fe				
	UNS K70640*	Obs; see N09902; Cr-Ni-Ti Alloy									bal Fe				
	UNS K74015*	Obs; see T20811									bal Fe				
	UNS K81590*	Obs; see S50400									bal Fe				
	UNS K88165*	Obs; see T11350; Bearing Steel									bal Fe				
	UNS K91151*	Obs; see S41000									bal Fe				
	UNS K91161*	Obs; see S41001; Superstrength									bal Fe				
	UNS K91313*	Obs; Superstrength	0.29-0.34	0.90-1.10	0.10-0.35	0.90-1.10	7.00-8.00	0.010 max	0.010 max	0.10 max	Co 4.25-4.75; V 0.06-0.12; bal Fe				
	UNS K91342*	Obs; see S42201; Superstrength									bal Fe				
	UNS K91970*	Obs; see K92571									bal Fe				
	UNS K95050*	Obs; see N14052; Ni Sealing Alloy									bal Fe				
	UNS K95100*	Obs; see R30005; Fe-Co-V Ferromagnetic Alloy									bal Fe				
ASTM A521(96)	AA	die frg; Ann, norm, N/T; <12 in.									Reference A322 A576 and other specs for chemisty	550	345	24	
ASTM A521(96)	AH	die frg; Norm Q/T; <4 in. solid, <2 in. bored wall									Reference A322 A576 and other specs for chemisty	1175	965	13	
ASTM A646(95)	Grade 10	Blm Bil for aircraft aerospace frg, ann	0.38-0.43	4.75-5.25	0.20-0.40	1.20-1.40		0.015 max	0.015 max	0.80-1.00	V 0.40-0.60; bal Fe				235 HB

UNS numbers and US grades are provided as a means of cross referencing chemically similar alloys. Exchangability is only possible after independent examination of specifications. Tensile properties are minimum or typical as specified. UTS and YS as MPa. El as %. See Appendix for list of abbreviations used in Notes. * indicates obsolete material.

Specification	Designation	Notes	C	Cr	Mn	Mo	Ni	P	S	Si	Other	UTS	YS	El	Hard

Alloy Steel, Unclassified, No equivalents identified

USA

Specification	Designation	Notes	C	Cr	Mn	Mo	Ni	P	S	Si	Other	UTS	YS	El	Hard
ASTM A646(95)	Grade 13	Blm Bil for aircraft aerospace frg, ann	0.40-0.46	0.80-1.05	0.75-1.00	0.45-0.60	0.60-0.90	0.025 max	0.025 max	0.50-0.80	B >=0.0005; V 0.01-0.06; bal Fe				285 HB
ASTM A646(95)	Grade 16	Blm Bil for aircraft aerospace frg, ann	0.17-0.23	0.65-0.85	0.20-0.40	0.90-1.10	8.5-9.5	0.010 max	0.010 max	0.10 max	Co 4.25-4.75; V 0.06-0.12; bal Fe				341 HB
ASTM A646(95)	Grade 17	Blm Bil for aircraft aerospace frg, ann	0.29-0.34	0.90-1.10	0.10-0.35	0.90-1.10	7.0-8.0	0.010 max	0.010 max	0.10 max	Co 4.25-4.75; V 0.06-0.12; bal Fe				341 HB
ASTM A646(95)	Grade 18	Blm Bil for aircraft aerospace frg, ann	0.03 max		0.10 max	3.0-3.50	17.0-19.0	0.010 max	0.010 max	0.10 max	Al 0.05-0.105; Co 8.0-9.0; Ti 0.10-0.25; B Zr Ca added; bal Fe				321 HB
ASTM A646(95)	Grade 19	Blm Bil for aircraft aerospace frg, ann	0.03 max		0.10 max	4.6-5.2	17.0-19.0	0.010 max	0.010 max	0.10 max	Al 0.05-0.15; Co 7.0-8.5; Ti 0.30-0.50; B Zr Ca added; bal Fe				321 HB
ASTM A646(95)	Grade 20	Blm Bil for aircraft aerospace frg, ann	0.03 max		0.10 max	4.7-5.2	18.0-19.0	0.010 max	0.010 max	0.10 max	Al 0.05-0.15; Co 8.5-9.5; Ti 0.50-0.80; B Zr Ca added; bal Fe				321 HB
ASTM A646(95)	Grade 21	Blm Bil for aircraft aerospace frg, ann	0.38-0.43	1.40-1.80	0.50-0.70	0.30-0.40		0.025 max	0.025 max	0.20-0.40	Al 0.95-1.30; bal Fe				285 HB
ASTM A646(95)	Grade 9	Blm Bil for aircraft aerospace frg, ann	0.45-0.50	0.90-1.20	0.60-0.90	0.90-1.10	0.40-0.70	0.010 max	0.010 max	0.15-0.30	V 0.08-0.15; bal Fe				285 HB
ASTM A649/649M(98)	Class 5	Frg roll for paperboard mach	0.50-0.60	0.30 max	0.90-1.50	0.15 max	0.60	0.025 max	0.025 max	0.15-0.35	Si<=0.10 with VCD; bal Fe	690-1030	450-890	12-14	
ASTM A668/668M(96)	Class G	Frg; Ann, norm, N/T; <=20 in.						0.040 max	0.040 max		choice of comp; bal Fe	550	345	22	163-207 HB
ASTM A668/668M(96)	Class H	Frg; N/T; <=20 in.						0.040 max	0.040 max		choice of comp; bal Fe	620	400	18	187-235 HB
ASTM A668/668M(96)	Class J	Frg; N/T, norm Q/T; <=20 in.						0.040 max	0.040 max		choice of comp; bal Fe	620	450	18	207-255 HB
ASTM A668/668M(96)	Class K	Frg; Norm Q/T; <=20 in.						0.040 max	0.040 max		choice of comp; bal Fe	690	515	18	207-269 HB
ASTM A668/668M(96)	Class L	Frg; Norm Q/T; <=20 in.						0.040 max	0.040 max		choice of comp; bal Fe	760	585	14	223-293 HB
ASTM A668/668M(96)	Class M	Frg; Norm Q/T; <=20 in.						0.040 max	0.040 max		choice of comp; bal Fe	930	758	12	269-341 HB
ASTM A668/668M(96)	Class N	Frg; Norm Q/T; <=20 in.						0.040 max	0.040 max		choice of comp; bal Fe	1105	905	11	321-402 HB
ASTM A730(93)	Grade I	Railway frg; N/T; t<=203-508mm						0.045 max	0.050 max		bal Fe	550	380	28	
ASTM A730(93)	Grade J	Railway frg; N/T; t>330-508mm						0.045 max	0.050 max		bal Fe	605	385	20	
ASTM A730(93)	Grade J	Railway frg; N/T; t<=127mm						0.045 max	0.050 max		bal Fe	620	415	24	
ASTM A730(93)	Grade J	Railway frg; N/T; t>127-229mm						0.045 max	0.050 max		bal Fe	620	415	22	
ASTM A730(93)	Grade J	Railway frg; N/T; t>229-330mm						0.045 max	0.050 max		bal Fe	620	400	21	
ASTM A730(93)	Grade K	Railway frg; N/T; t<=127mm						0.045 max	0.050 max		bal Fe	655	495	23	
ASTM A730(93)	Grade K	Railway frg; N/T; t>229-330mm						0.045 max	0.050 max		bal Fe	640	460	22	
ASTM A730(93)	Grade K	Railway frg; N/T; t>330-508mm						0.045 max	0.050 max		bal Fe	625	450	21	
ASTM A730(93)	Grade K	Railway frg; N/T; t>127-229mm						0.045 max	0.050 max		bal Fe	655	485	22	
ASTM A730(93)	Grade L	Railway frg; N/Q/T; t>178-254mm						0.045 max	0.050 max		bal Fe	620	450	20	
ASTM A730(93)	Grade L	Railway frg; N/Q/T; t<=178mm						0.045 max	0.050 max		bal Fe	655	485	23	
ASTM A730(93)	Grade M	Railway frg; N/Q/T; t>178-254mm						0.045 max	0.050 max		bal Fe	690	515	19	
ASTM A730(93)	Grade M	Railway frg; N/Q/T; t<=178mm						0.045 max	0.050 max		bal Fe	725	550	20	
ASTM A730(93)	Grade N	Railway frg; N/Q/T; t>178-254mm						0.045 max	0.050 max		bal Fe	760	585	16	
ASTM A730(93)	Grade N	Railway frg; N/Q/T; t>102-178mm						0.045 max	0.050 max		bal Fe	795	655	16	
ASTM A730(93)	Grade N	Railway frg; N/Q/T; t<=102mm						0.045 max	0.050 max		bal Fe	860	725	16	

Alloy Steel, Unclassified, No equivalents identified

USA

Specification	Designation	Notes	C	Cr	Mn	Mo	Ni	P	S	Si	Other	UTS	YS	El	Hard
ASTM A859/859M(95)	A859 Class 1	Age-hard frg for press ves	0.07 max	0.60-0.90	0.40-0.70	0.15-0.25	0.70-1.00	0.025 max	0.025 max	0.40 max	Cu 1.00-1.30; Nb >=0.02; bal Fe	450-585	380	20	
ASTM A859/859M(95)	A859 Class 2	Age-hard frg for press ves	0.07 max	0.60-0.90	0.40-0.70	0.15-0.25	0.70-1.00	0.025 max	0.025 max	0.40 max	Cu 1.00-1.30; Nb >=0.02; bal Fe	515-655	450	20	
ASTM A860/A860M	WPHY 42	Fittings	0.20 max	0.30 max	1.00-1.45	0.25 max	0.50 max	0.030 max	0.010 max	0.15-0.40	Al <=0.06; Cu <=0.35; Nb <=0.04; Ti <=0.05; V <=0.10; V+Nb<=0.12; Ni+Cr+Mo+Cu<=1.0; bal Fe	415-585	290	25	235 HB max
ASTM A860/A860M	WPHY 46	Fittings	0.20 max	0.30 max	1.00-1.45	0.25 max	0.50 max	0.030 max	0.010 max	0.15-0.40	Al <=0.06; Cu <=0.35; Nb <=0.04; Ti <=0.05; V <=0.10; V+Nb<=0.12; Ni+Cr+Mo+Cu<=1.0; bal Fe	435-605	315	25	235 HB max
ASTM A860/A860M	WPHY 52	Fittings	0.20 max	0.30 max	1.00-1.45	0.25 max	0.50 max	0.030 max	0.010 max	0.15-0.40	Al <=0.06; Cu <=0.35; Nb <=0.04; Ti <=0.05; V <=0.10; V+Nb<=0.12; Ni+Cr+Mo+Cu<=1.0; bal Fe	455-625	360	25	235 HB max
ASTM A860/A860M	WPHY 60	Fittings	0.20 max	0.30 max	1.00-1.45	0.25 max	0.50 max	0.030 max	0.010 max	0.15-0.40	Al <=0.06; Cu <=0.35; Nb <=0.04; Ti <=0.05; V <=0.10; V+Nb<=0.12; Ni+Cr+Mo+Cu<=1.0; bal Fe	515-690	415	20	235 HB max
ASTM A860/A860M	WPHY 65	Fittings	0.20 max	0.30 max	1.00-1.45	0.25 max	0.50 max	0.030 max	0.010 max	0.15-0.40	Al <=0.06; Cu <=0.35; Nb <=0.04; Ti <=0.05; V <=0.10; V+Nb<=0.12; Ni+Cr+Mo+Cu<=1.0; bal Fe	530-705	450	20	235 HB max
ASTM A860/A860M	WPHY 70	Fittings	0.20 max	0.30 max	1.00-1.45	0.25 max	0.50 max	0.030 max	0.010 max	0.15-0.40	Al <=0.06; Cu <=0.35; Nb <=0.04; Ti <=0.05; V <=0.10; V+Nb<=0.12; Ni+Cr+Mo+Cu<=1.0; bal Fe	550-725	485	20	235 HB max
ASTM A871/A871M(97)	Type I, Grade 60	Corr res, struct plt	0.19 max	0.40-0.70	0.80-1.35		0.40 max	0.04 max	0.05 max	0.30-0.65	Cu 0.25-0.40; V 0.02-0.10; bal Fe	520	415	18	
ASTM A871/A871M(97)	Type I, Grade 65	Corr res, struct plt	0.19 max	0.40-0.70	0.80-1.35		0.40 max	0.04 max	0.05 max	0.30-0.65	Cu 0.25-0.40; V 0.02-0.10; bal Fe	550	450	17	
ASTM A871/A871M(97)	Type II, Grade 60	Corr res, struct plt	0.20 max	0.40-0.70	0.75-1.35		0.50 max	0.04 max	0.05 max	0.15-0.50	Cu 0.20-0.40; V 0.01-0.10; bal Fe	520	415	18	
ASTM A871/A871M(97)	Type II, Grade 65	Corr res, struct plt	0.20 max	0.40-0.70	0.75-1.35		0.50 max	0.04 max	0.05 max	0.15-0.50	Cu 0.20-0.40; V 0.01-0.10; bal Fe	550	450	17	
ASTM A871/A871M(97)	Type III, Grade 60	Corr res, struct plt	0.15 max	0.30-0.50	0.80-1.35		0.25-0.50	0.04 max	0.05 max	0.15-0.40	Cu 0.20-0.50; V 0.01-0.10; bal Fe	520	415	18	
ASTM A871/A871M(97)	Type III, Grade 65	Corr res, struct plt	0.15 max	0.30-0.50	0.80-1.35		0.25-0.50	0.04 max	0.05 max	0.15-0.40	Cu 0.20-0.50; V 0.01-0.10; bal Fe	550	450	17	
ASTM A871/A871M(97)	Type IV, Grade 60	Corr res, struct plt	0.17 max	0.40-0.70	0.50-1.20	0.10 max	0.40 max	0.04 max	0.05 max	0.25-0.50	Cu 0.30-0.50; Nb 0.005-0.05; bal Fe	520	415	18	
ASTM A871/A871M(97)	Type IV, Grade 65	Corr res, struct plt	0.17 max	0.40-0.70	0.50-1.20	0.10 max	0.40 max	0.04 max	0.05 max	0.25-0.50	Cu 0.30-0.50; Nb 0.005-0.05; bal Fe	550	450	17	
ASTM A913/A913M(97)	Grade 50	Quen self tmp struct shp	0.12 max	0.25 max	1.60 max	0.07 max	0.25 max	0.040 max	0.030 max	0.40 max	Cu <=0.45; Nb <=0.05; V <=0.06; bal Fe	450	345	21	
ASTM A913/A913M(97)	Grade 60	Quen self tmp struct shp	0.14 max	0.25 max	1.60 max	0.07 max	0.25 max	0.030 max	0.030 max	0.40 max	Cu <=0.35; Nb <=0.04; V <=0.06; bal Fe	520	415	18	
ASTM A913/A913M(97)	Grade 65	Quen self tmp struct shp	0.16 max	0.25 max	1.60 max	0.07 max	0.25 max	0.030 max	0.030 max	0.40 max	Cu <=0.35; Nb <=0.05; V <=0.06; bal Fe	550	450	17	
ASTM A913/A913M(97)	Grade 70	Quen self tmp struct shp	0.16 max	0.25 max	1.60 max	0.07 max	0.25 max	0.040 max	0.030 max	0.40 max	Cu <=0.45; Nb <=0.05; V <=0.09; bal Fe	620	485	16	
ASTM A945(95)	Grade 50	Struct plt; t<=50mm	0.10 max	0.20 max	1.10-1.65	0.08 max	0.40 max	0.025 max	0.010 max	0.10-0.50	Cu <=0.35; Nb <=0.05; V <=0.10; bal Fe	485-620	345	24	

UNS numbers and US grades are provided as a means of cross referencing chemically similar alloys. Exchangability is only possible after independent examination of specifications. Tensile properties are minimum or typical as specified. UTS and YS as MPa. El as %. See Appendix for list of abbreviations used in Notes. * indicates obsolete material.

Specification	Designation	Notes	C	Cr	Mn	Mo	Ni	P	S	Si	Other	UTS	YS	El	Hard

Alloy Steel, Unclassified, No equivalents identified

USA

Specification	Designation	Notes	C	Cr	Mn	Mo	Ni	P	S	Si	Other	UTS	YS	El	Hard
ASTM A945(95)	Grade 65	Struct plt; t<=32mm	0.10 max	0.20 max	1.10-1.65	0.08 max	0.40 max	0.025 max	0.010 max	0.10-0.50	Cu <=0.35; Nb <=0.05; V <=0.10; bal Fe	540-690	450	22	
ASTM A992(98)		Struct Shp for bldg	0.23 max	0.35 max	0.50-1.50	0.15 max	0.45 max	0.035 max	0.045 max	0.40 max	Cu <=0.60; Nb <=0.05; V <=0.11; Nb+V<=0.15; Bal Fe	450	345-450	21	
FED QQ-S-624C(71)	- - -*	Obs; see ASTM A304, A322; HR CF Bar									bal Fe				
FED QQ-S-627B(61)	- - -*	Obs; see ASTM A505, A506, A507; HR Sh Strp									bal Fe				
FED QQ-W-412(71)	- - -*	Obs; Wir for mech spring									bal Fe				
MIL-C-24527B(92)		Bar Rod Frg; Corr res, PH									NiCr alloys, AISI 31xx, 33xx; bal Fe				
MIL-F-24669/4(86)	- - -*	Obs; Bar Bil mod low magnetic perm									bal Fe				
MIL-F-24669/8(91)	- - -*	Obs for new design; Frg for steam turbine rotor									bal Fe				
MIL-S-19434B(SH)(90)	Class 6	Gear, pinion frg HT; shipboard propulsion and turbine; <- 10 in.	0.55 max	0.50 min	0.60-0.90	0.13-0.50	1.65 max	0.015 max	0.015 max	0.15-0.35	V <=0.10; Si 0.10 max if VCD used; bal Fe	1138	931	12	341-388 HB

Yugoslavia

Specification	Designation	Notes	C	Cr	Mn	Mo	Ni	P	S	Si	Other	UTS	YS	El	Hard
	C.3811	Chain	0.2-0.25	0.2-0.3	1.3-1.6		0.2-0.3	0.04 max	0.04 max	0.2-0.3	N <=0.007; V <=0.07; bal Fe				
	C.4140	Ball & roller bearing	0.95-1.1	0.4-0.6	0.25-0.45	0.15 max		0.03 max	0.025 max	0.15-0.35	bal Fe				
	C.4230	Spring	0.62-0.72	0.4-0.6	0.4-0.6	0.15 max		0.035 max	0.035 max	1.2-1.4	bal Fe				
	C.4734	Q/T	0.26-0.34	2.3-2.7	0.4-0.7	0.15-0.25		0.035 max	0.03 max	0.4 max	V 0.1-0.2; bal Fe				
	C.4811	Chain	0.18-0.24	0.9-1.2	0.8-1.0	0.3 max		0.035 max	0.035 max	0.1-0.2	V 0.07-0.12; bal Fe				
	CL.0300	Reinforcing	0.30 max		1.60 max	0.15 max		0.070 max	0.060 max	0.60 max	bal Fe				

Specification	Designation	Notes	C	Cr	Cu	Mn	P	S	Si	V	Other	UTS	YS	El	Hard

High Strength Steel, Low-Alloy (HSLA), A242(1)

China

Specification	Designation	Notes	C	Cr	Cu	Mn	P	S	Si	V	Other	UTS	YS	El	Hard
GB 4171(84)	09CuP	Sh HR/Norm 6mm Thk	0.12 max		0.25-0.45	0.20-0.50	0.07-0.12	0.040 max	0.20-0.40		bal Fe	415	295	24	

USA

Specification	Designation	Notes	C	Cr	Cu	Mn	P	S	Si	V	Other	UTS	YS	El	Hard
	UNS K11510		0.15 max		0.20 min	1 max	0.15 max	0.05 max			bal Fe				
ASTM A242/A242M(97)	Bar/Plate<=1-1/2>3/4	Struct bar/plt, t 20-40mm	0.15 max		0.20 min	1.00 max	0.15 max	0.05 max			Cr,Ni,Si,V,Ti,Zr; bal Fe	460	315	21	
ASTM A242/A242M(97)	Bar/Plate<=3/4	Struct bar/plt, t <=20mm	0.15 max		0.20 min	1.00 max	0.15 max	0.05 max			Cr,Ni,Si,V,Ti,Zr; bal Fe	480	345	21	
ASTM A242/A242M(97)	Bar/Plate<=4>1-1/2	Struct bar/plt, t 40-100mm	0.15 max		0.20 min	1.00 max	0.15 max	0.05 max			Cr,Ni,Si,V,Ti,Zr; bal Fe	435	290	21	
ASTM A242/A242M(97)	Shapes Group 1&2	Struct Shp, Group 1&2	0.15 max		0.20 min	1.00 max	0.15 max	0.05 max			Cr,Ni,Si,V,Ti,Zr; bal Fe	485	345	21	
ASTM A242/A242M(97)	Shapes Group 3	Struct Shp, Group 3	0.15 max		0.20 min	1.00 max	0.15 max	0.05 max			Cr,Ni,Si,V,Ti,Zr; bal Fe	460	315	21	
ASTM A242/A242M(97)	Shapes Group 4&5	Struct Shp, Group 4&5	0.15 max		0.20 min	1.00 max	0.15 max	0.05 max			Cr,Ni,Si,V,Ti,Zr; bal Fe	435	290	21	
SAE J410c	*	Obs; see J1392	0.15 max		0.2 min	1 max	0.15 max	0.05 max			bal Fe				

High Strength Steel, Low-Alloy (HSLA), A514(A)

USA

Specification	Designation	Notes	C	Cr	Cu	Mn	P	S	Si	V	Other	UTS	YS	El	Hard
	UNS K11856		0.15-0.21	0.50-0.80		0.80-1.10	0.035 max	0.04 max	0.40-0.80		B <=0.0025; Mo 0.18-0.28; Zr 0.05-0.15; bal Fe				
ASTM A514(94)	Class A	Q/T, Plt, t=32mm, weld	0.15-0.21	0.50-0.80		0.80-1.10	0.035 max	0.035 max	0.40-0.80		B <=0.0025; Mo 0.18-0.28; Zr 0.05-0.15; bal Fe	780-895	690	18	
ASTM A517(93)	Grade A	Q/T, Press ves plt, t<=32mm	0.15-0.21	0.50-0.80		0.80-1.10	0.035 max	0.035 max	0.40-0.80		B <=0.0025; Mo 0.18-0.28; Zr 0.05-0.15; bal Fe	795-930	690	16	
ASTM A592/592M(94)	Grade A	High-strength QT frg for press ves, <=2.5 in.	0.15-0.21	0.50-0.80		0.80-1.10	0.025 max	0.025 max	0.40-0.80		B <=0.0025; Mo 0.18-0.28; Zr 0.05-0.15; bal Fe	795-930	690	18	
ASTM A709(97)	Grade 100, Type A	Q/T, Plt, t<=32mm	0.15-0.21	0.50-0.80		0.80-1.10	0.035 max	0.035 max	0.40-0.80		B <=0.0025; Mo 0.18-0.28; Zr 0.05-0.15; bal Fe	760-895	690	18	

High Strength Steel, Low-Alloy (HSLA), A514(F)

Europe

Specification	Designation	Notes	C	Cr	Cu	Mn	P	S	Si	V	Other	UTS	YS	El	Hard
EN 10028/3(92)	1.8918	High-temp, Press; 50<t<=70mm	0.20 max	0.30 max	0.70 max	1.00-1.70	0.025 max	0.015 max	0.60 max	0.20 max	Al >=0.020; Mo <=0.10; N <=0.025; Ni <=0.80; Nb <=0.05; Ti <=0.03; Nb+Ti+ V<=0.12; bal Fe	390-510	420	17	
EN 10028/3(92)	1.8918	High-temp, Press; 70<t<=100mm	0.20 max	0.30 max	0.70 max	1.00-1.70	0.025 max	0.015 max	0.60 max	0.20 max	Al >=0.020; Mo <=0.10; N <=0.025; Ni <=0.80; Nb <=0.05; Ti <=0.03; Nb+Ti+ V<=0.12; bal Fe	540-710	400	16	
EN 10028/3(92)	1.8918	High-temp, Press; 0<=t<=16mm	0.20 max	0.30 max	0.70 max	1.00-1.70	0.025 max	0.015 max	0.60 max	0.20 max	Al >=0.020; Mo <=0.10; N <=0.025; Ni <=0.80; Nb <=0.05; Ti <=0.03; Nb+Ti+ V<=0.12; bal Fe	570-730	460	17	
EN 10028/3(92)	1.8918	High-temp, Press; 35<t<=50mm	0.20 max	0.30 max	0.70 max	1.00-1.70	0.025 max	0.015 max	0.60 max	0.20 max	Al >=0.020; Mo <=0.10; N <=0.025; Ni <=0.80; Nb <=0.05; Ti <=0.03; Nb+Ti+ V<=0.12; bal Fe	390-510	440	17	
EN 10028/3(92)	1.8918	High-temp, Press; 16<=t<=35mm	0.20 max	0.30 max	0.70 max	1.00-1.70	0.025 max	0.015 max	0.60 max	0.20 max	Al >=0.020; Mo <=0.10; N <=0.025; Ni <=0.80; Nb <=0.05; Ti <=0.03; Nb+Ti+ V<=0.12; bal Fe	570-720	450	17	

UNS numbers and US grades are provided as a means of cross referencing chemically similar alloys. Exchangability is only possible after independent examination of specifications. Tensile properties are minimum or typical as specified. UTS and YS as MPa. El as %. See Appendix for list of abbreviations used in Notes. * indicates obsolete material.

6-2/High Strength Steel

High Strength Steel, Low-Alloy (HSLA), A514(F) (Continued from previous page)

Europe

Specification	Designation	Notes	C	Cr	Cu	Mn	P	S	Si	V	Other	UTS	YS	El	Hard
EN 10028/3(92)	1.8918	High-temp, Press; 100<t<=150mm	0.20 max	0.30 max	0.70 max	1.00-1.70	0.025 max	0.015 max	0.60 max	0.20 max	Al >=0.020; Mo <=0.10; N <=0.025; Ni <=0.80; Nb <=0.05; Ti <=0.03; Nb+Ti+ V<=0.12; bal Fe	520-690	380	16	
EN 10028/3(92)	P460NL2	High-temp, Press; 0<=t<=16mm	0.20 max	0.30 max	0.70 max	1.00-1.70	0.025 max	0.015 max	0.60 max	0.20 max	Al >=0.020; Mo <=0.10; N <=0.025; Ni <=0.80; Nb <=0.05; Ti <=0.03; Nb+Ti+ V<=0.12; bal Fe	570-730	460	17	
EN 10028/3(92)	P460NL2	High-temp, Press; 16<=t<=35mm	0.20 max	0.30 max	0.70 max	1.00-1.70	0.025 max	0.015 max	0.60 max	0.20 max	Al >=0.020; Mo <=0.10; N <=0.025; Ni <=0.80; Nb <=0.05; Ti <=0.03; Nb+Ti+ V<=0.12; bal Fe	570-720	450	17	
EN 10028/3(92)	P460NL2	High-temp, Press; 35<t<=50mm	0.20 max	0.30 max	0.70 max	1.00-1.70	0.025 max	0.015 max	0.60 max	0.20 max	Al >=0.020; Mo <=0.10; N <=0.025; Ni <=0.80; Nb <=0.05; Ti <=0.03; Nb+Ti+ V<=0.12; bal Fe	390-510	440	17	
EN 10028/3(92)	P460NL2	High-temp, Press; 50<t<=70mm	0.20 max	0.30 max	0.70 max	1.00-1.70	0.025 max	0.015 max	0.60 max	0.20 max	Al >=0.020; Mo <=0.10; N <=0.025; Ni <=0.80; Nb <=0.05; Ti <=0.03; Nb+Ti+ V<=0.12; bal Fe	390-510	420	17	
EN 10028/3(92)	P460NL2	High-temp, Press; 100<t<=150mm	0.20 max	0.30 max	0.70 max	1.00-1.70	0.025 max	0.015 max	0.60 max	0.20 max	Al >=0.020; Mo <=0.10; N <=0.025; Ni <=0.80; Nb <=0.05; Ti <=0.03; Nb+Ti+ V<=0.12; bal Fe	520-690	380	16	
EN 10028/3(92)	P460NL2	High-temp, Press; 70<t<=100mm	0.20 max	0.30 max	0.70 max	1.00-1.70	0.025 max	0.015 max	0.60 max	0.20 max	Al >=0.020; Mo <=0.10; N <=0.025; Ni <=0.80; Nb <=0.05; Ti <=0.03; Nb+Ti+ V<=0.12; bal Fe	540-710	400	16	
EN 10028/6(96)	1.8864 *	100<t<=150mm	0.18 max	1.50 max	0.50 max	1.70 max	0.020 max	0.020 max	0.80 max	0.12 max	B <=0.0050; Mo <=0.70; N <=0.015; Ni <=2.00; Nb <=0.06; Ti <=0.05; Zr<=0.15; Nb+Ti+V+Zr>=0.015; bal Fe	500-670	400	19	
EN 10028/6(96)	1.8864	High-temp, Press; 50<t<=100mm	0.18 max	1.50 max	0.50 max	1.70 max	0.020 max	0.020 max	0.80 max	0.12 max	B <=0.0050; Mo <=0.70; N <=0.015; Ni <=2.00; Nb <=0.06; Ti <=0.05; Zr<=0.15; Nb+Ti+V+Zr>=0.015; bal Fe	550-720	440	19	
EN 10028/6(96)	1.8864	High-temp, Press; t<=50mm	0.18 max	1.50 max	0.50 max	1.70 max	0.020 max	0.020 max	0.80 max	0.12 max	B <=0.0050; Mo <=0.70; N <=0.015; Ni <=2.00; Nb <=0.06; Ti <=0.05; Zr<=0.15; Nb+Ti+V+Zr>=0.015; bal Fe	550-720	460	19	
EN 10028/6(96)	1.8865	100<t<=150mm	0.18 max	1.50 max	0.50 max	1.70 max	0.020 max	0.020 max	0.80 max	0.12 max	B <=0.0050; Mo <=0.70; N <=0.015; Ni <=2.00; Nb <=0.06; Ti <=0.05; Zr<=0.15; Nb+Ti+V+Zr>=0.015; bal Fe	540-720	440	17	

UNS numbers and US grades are provided as a means of cross referencing chemically similar alloys. Exchangability is only possible after independent examination of specifications. Tensile properties are minimum or typical as specified. UTS and YS as MPa. El as %. See Appendix for list of abbreviations used in Notes. * indicates obsolete material.

High Strength Steel, Low-Alloy (HSLA), A514(F) (Continued from previous page)

Europe

Specification	Designation	Notes	C	Cr	Cu	Mn	P	S	Si	V	Other	UTS	YS	El	Hard
EN 10028/6(96)	1.8865	High-temp, Press; t<=50mm	0.18 max	1.50 max	0.50 max	1.70 max	0.020 max	0.020 max	0.80 max	0.12 max	B <=0.0050; Mo <=0.70; N <=0.015; Ni <=2.00; Nb <=0.06; Ti <=0.05; Zr<=0.15; Nb+Ti+V+Zr>=0.015; bal Fe	590-770	500	17	
EN 10028/6(96)	1.8865	High-temp, Press; 50<t<=100mm	0.18 max	1.50 max	0.50 max	1.70 max	0.020 max	0.020 max	0.80 max	0.12 max	B <=0.0050; Mo <=0.70; N <=0.015; Ni <=2.00; Nb <=0.06; Ti <=0.05; Zr<=0.15; Nb+Ti+V+Zr>=0.015; bal Fe	590-770	480	17	
EN 10028/6(96)	1.8873	High-temp, Press; 50<t<=100mm	0.18 max	1.50 max	0.50 max	1.70 max	0.025 max	0.015 max	0.80 max	0.12 max	B <=0.0050; Mo <=0.70; N <=0.015; Ni <=2.00; Nb <=0.06; Ti <=0.05; Zr<=0.15; Nb+Ti+V+Zr>=0.015; bal Fe	590-770	480	17	
EN 10028/6(96)	1.8873	100<t<=150mm	0.18 max	1.50 max	0.50 max	1.70 max	0.025 max	0.015 max	0.80 max	0.12 max	B <=0.0050; Mo <=0.70; N <=0.015; Ni <=2.00; Nb <=0.06; Ti <=0.05; Zr<=0.15; Nb+Ti+V+Zr>=0.015; bal Fe	540-720	440	17	
EN 10028/6(96)	1.8873	High-temp, Press; t<=50mm	0.18 max	1.50 max	0.50 max	1.70 max	0.025 max	0.015 max	0.80 max	0.12 max	B <=0.0050; Mo <=0.70; N <=0.015; Ni <=2.00; Nb <=0.06; Ti <=0.05; Zr<=0.15; Nb+Ti+V+Zr>=0.015; bal Fe	590-770	500	17	
EN 10028/6(96)	1.8874	100<t<=150mm	0.18 max	1.50 max	0.50 max	1.70 max	0.025 max	0.015 max	0.80 max	0.12 max	B <=0.0050; Mo <=0.70; N <=0.015; Ni <=2.00; Nb <=0.06; Ti <=0.05; Zr<=0.15; Nb+Ti+V+Zr>=0.015; bal Fe	540-720	440	17	
EN 10028/6(96)	1.8874	High-temp, Press; 50<t<=100mm	0.18 max	1.50 max	0.50 max	1.70 max	0.025 max	0.015 max	0.80 max	0.12 max	B <=0.0050; Mo <=0.70; N <=0.015; Ni <=2.00; Nb <=0.06; Ti <=0.05; Zr<=0.15; Nb+Ti+V+Zr>=0.015; bal Fe	590-770	480	17	
EN 10028/6(96)	1.8874	High-temp, Press; t<=50mm	0.18 max	1.50 max	0.50 max	1.70 max	0.025 max	0.015 max	0.80 max	0.12 max	B <=0.0050; Mo <=0.70; N <=0.015; Ni <=2.00; Nb <=0.06; Ti <=0.05; Zr<=0.15; Nb+Ti+V+Zr>=0.015; bal Fe	590-770	500	17	
EN 10028/6(96)	1.8875	100<t<=150mm	0.18 max	1.50 max	0.50 max	1.70 max	0.020 max	0.020 max	0.80 max	0.12 max	B <=0.0050; Mo <=0.70; N <=0.015; Ni <=2.00; Nb <=0.06; Ti <=0.05; Zr<=0.15; Nb+Ti+V+Zr>=0.015; bal Fe	540-720	440	17	
EN 10028/6(96)	1.8875	High-temp, Press; 50<t<=100mm	0.18 max	1.50 max	0.50 max	1.70 max	0.020 max	0.020 max	0.80 max	0.12 max	B <=0.0050; Mo <=0.70; N <=0.015; Ni <=2.00; Nb <=0.06; Ti <=0.05; Zr<=0.15; Nb+Ti+V+Zr>=0.015; bal Fe	590-770	480	17	

Specification	Designation	Notes	C	Cr	Cu	Mn	P	S	Si	V	Other	UTS	YS	El	Hard

High Strength Steel, Low-Alloy (HSLA), A514(F) (Continued from previous page)

Europe

Specification	Designation	Notes	C	Cr	Cu	Mn	P	S	Si	V	Other	UTS	YS	El	Hard
EN 10028/6(96)	1.8875	High-temp, Press; t<=50mm	0.18 max	1.50 max	0.50 max	1.70 max	0.020 max	0.020 max	0.80 max	0.12 max	B <=0.0050; Mo <=0.70; N <=0.015; Ni <=2.00; Nb <=0.06; Ti <=0.05; Zr<=0.15; Nb+Ti+V+Zr>=0.015; bal Fe	590-770	500	17	
EN 10028/6(96)	1.8879	High-temp, Press; t<=50mm	0.20 max	1.50 max	0.50 max	1.70 max	0.025 max	0.015 max	0.80 max	0.12 max	B <=0.0050; Mo <=0.70; N <=0.015; Ni <=2.00; Nb <=0.06; Ti <=0.05; Zr<=0.15; Nb+Ti+V+Zr>=0.015; bal Fe	770-940	690	14	
EN 10028/6(96)	1.8879	High-temp, Press; 50<t<=100mm	0.20 max	1.50 max	0.50 max	1.70 max	0.025 max	0.015 max	0.80 max	0.12 max	B <=0.0050; Mo <=0.70; N <=0.015; Ni <=2.00; Nb <=0.06; Ti <=0.05; Zr<=0.15; Nb+Ti+V+Zr>=0.015; bal Fe	770-940	670	14	
EN 10028/6(96)	1.8879	100<t<=150mm	0.20 max	1.50 max	0.50 max	1.70 max	0.025 max	0.015 max	0.80 max	0.12 max	B <=0.0050; Mo <=0.70; N <=0.015; Ni <=2.00; Nb <=0.06; Ti <=0.05; Zr<=0.15; Nb+Ti+V+Zr>=0.015; bal Fe	720-900	630	14	
EN 10028/6(96)	1.8880	High-temp, Press; 50<t<=100mm	0.20 max	1.50 max	0.50 max	1.70 max	0.025 max	0.015 max	0.80 max	0.12 max	B <=0.0050; Mo <=0.70; N <=0.015; Ni <=2.00; Nb <=0.06; Ti <=0.05; Zr<=0.15; Nb+Ti+V+Zr>=0.015; bal Fe	770-940	670	14	
EN 10028/6(96)	1.8880	100<t<=150mm	0.20 max	1.50 max	0.50 max	1.70 max	0.025 max	0.015 max	0.80 max	0.12 max	B <=0.0050; Mo <=0.70; N <=0.015; Ni <=2.00; Nb <=0.06; Ti <=0.05; Zr<=0.15; Nb+Ti+V+Zr>=0.015; bal Fe	720-900	630	14	
EN 10028/6(96)	1.8880	High-temp, Press; t<=50mm	0.20 max	1.50 max	0.50 max	1.70 max	0.025 max	0.015 max	0.80 max	0.12 max	B <=0.0050; Mo <=0.70; N <=0.015; Ni <=2.00; Nb <=0.06; Ti <=0.05; Zr<=0.15; Nb+Ti+V+Zr>=0.015; bal Fe	770-940	690	14	
EN 10028/6(96)	1.8881	High-temp, Press; t<=50mm	0.20 max	1.50 max	0.50 max	1.70 max	0.020 max	0.020 max	0.80 max	0.12 max	B <=0.0050; Mo <=0.70; N <=0.015; Ni <=2.00; Nb <=0.06; Ti <=0.05; Zr<=0.15; Nb+Ti+V+Zr>=0.015; bal Fe	770-940	690	14	
EN 10028/6(96)	1.8881	High-temp, Press; 50<t<=100mm	0.20 max	1.50 max	0.50 max	1.70 max	0.020 max	0.020 max	0.80 max	0.12 max	B <=0.0050; Mo <=0.70; N <=0.015; Ni <=2.00; Nb <=0.06; Ti <=0.05; Zr<=0.15; Nb+Ti+V+Zr>=0.015; bal Fe	770-940	670	14	
EN 10028/6(96)	1.8881	100<t<=150mm	0.20 max	1.50 max	0.50 max	1.70 max	0.020 max	0.020 max	0.80 max	0.12 max	B <=0.0050; Mo <=0.70; N <=0.015; Ni <=2.00; Nb <=0.06; Ti <=0.05; Zr<=0.15; Nb+Ti+V+Zr>=0.015; bal Fe	720-900	630	14	

High Strength Steel, Low-Alloy (HSLA), A514(F) (Continued from previous page)

Specification	Designation	Notes	C	Cr	Cu	Mn	P	S	Si	V	Other	UTS	YS	El	Hard
Europe															
EN 10028/6(96)	1.8888	100<t<=150mm	0.20 max	1.50 max	0.50 max	1.70 max	0.020 max	0.020 max	0.80 max	0.12 max	B <=0.0050; Mo <=0.70; N <=0.015; Ni <=2.00; Nb <=0.06; Ti <=0.05; Zr<=0.15; Nb+Ti+V+Zr>=0.015; bal Fe	720-900	630	14	
EN 10028/6(96)	1.8888	High-temp, Press; t<=50mm	0.20 max	1.50 max	0.50 max	1.70 max	0.020 max	0.020 max	0.80 max	0.12 max	B <=0.0050; Mo <=0.70; N <=0.015; Ni <=2.00; Nb <=0.06; Ti <=0.05; Zr<=0.15; Nb+Ti+V+Zr>=0.015; bal Fe	770-940	690	14	
EN 10028/6(96)	1.8888	High-temp, Press; 50<t<=100mm	0.20 max	1.50 max	0.50 max	1.70 max	0.020 max	0.020 max	0.80 max	0.12 max	B <=0.0050; Mo <=0.70; N <=0.015; Ni <=2.00; Nb <=0.06; Ti <=0.05; Zr<=0.15; Nb+Ti+V+Zr>=0.015; bal Fe	770-940	670	14	
EN 10028/6(96)	P460QL2	High-temp, Press; t<=50mm	0.18 max	1.50 max	0.50 max	1.70 max	0.020 max	0.020 max	0.80 max	0.12 max	B <=0.0050; Mo <=0.70; N <=0.015; Ni <=2.00; Nb <=0.06; Ti <=0.05; Zr<=0.15; Nb+Ti+V+Zr>=0.015; bal Fe	550-720	460	19	
EN 10028/6(96)	P460QL2	High-temp, Press; 50<t<=100mm	0.18 max	1.50 max	0.50 max	1.70 max	0.020 max	0.020 max	0.80 max	0.12 max	B <=0.0050; Mo <=0.70; N <=0.015; Ni <=2.00; Nb <=0.06; Ti <=0.05; Zr<=0.15; Nb+Ti+V+Zr>=0.015; bal Fe	550-720	440	19	
EN 10028/6(96)	P460QL2	100<t<=150mm	0.18 max	1.50 max	0.50 max	1.70 max	0.020 max	0.020 max	0.80 max	0.12 max	B <=0.0050; Mo <=0.70; N <=0.015; Ni <=2.00; Nb <=0.06; Ti <=0.05; Zr<=0.15; Nb+Ti+V+Zr>=0.015; bal Fe	500-670	400	19	
EN 10028/6(96)	P500Q	100<t<=150mm	0.18 max	1.50 max	0.50 max	1.70 max	0.025 max	0.015 max	0.80 max	0.12 max	B <=0.0050; Mo <=0.70; N <=0.015; Ni <=2.00; Nb <=0.06; Ti <=0.05; Zr<=0.15; Nb+Ti+V+Zr>=0.015; bal Fe	540-720	440	17	
EN 10028/6(96)	P500Q	High-temp, Press; 50<t<=100mm	0.18 max	1.50 max	0.50 max	1.70 max	0.025 max	0.015 max	0.80 max	0.12 max	B <=0.0050; Mo <=0.70; N <=0.015; Ni <=2.00; Nb <=0.06; Ti <=0.05; Zr<=0.15; Nb+Ti+V+Zr>=0.015; bal Fe	590-770	480	17	
EN 10028/6(96)	P500Q	High-temp, Press; t<=50mm	0.18 max	1.50 max	0.50 max	1.70 max	0.025 max	0.015 max	0.80 max	0.12 max	B <=0.0050; Mo <=0.70; N <=0.015; Ni <=2.00; Nb <=0.06; Ti <=0.05; Zr<=0.15; Nb+Ti+V+Zr>=0.015; bal Fe	590-770	500	17	
EN 10028/6(96)	P500QH	High-temp, Press; 50<t<=100mm	0.18 max	1.50 max	0.50 max	1.70 max	0.025 max	0.015 max	0.80 max	0.12 max	B <=0.0050; Mo <=0.70; N <=0.015; Ni <=2.00; Nb <=0.06; Ti <=0.05; Zr<=0.15; Nb+Ti+V+Zr>=0.015; bal Fe	590-770	480	17	

High Strength Steel, Low-Alloy (HSLA), A514(F) (Continued from previous page)

Europe

Specification	Designation	Notes	C	Cr	Cu	Mn	P	S	Si	V	Other	UTS	YS	El	Hard
EN 10028/6(96)	P500QH	High-temp, Press; t<=50mm	0.18 max	1.50 max	0.50 max	1.70 max	0.025 max	0.015 max	0.80 max	0.12 max	B <=0.0050; Mo <=0.70; N <=0.015; Ni <=2.00; Nb <=0.06; Ti <=0.05; Zr<=0.15; Nb+Ti+V+Zr>=0.015; bal Fe	590-770	500	17	
EN 10028/6(96)	P500QH	100<t<=150mm	0.18 max	1.50 max	0.50 max	1.70 max	0.025 max	0.015 max	0.80 max	0.12 max	B <=0.0050; Mo <=0.70; N <=0.015; Ni <=2.00; Nb <=0.06; Ti <=0.05; Zr<=0.15; Nb+Ti+V+Zr>=0.015; bal Fe	540-720	440	17	
EN 10028/6(96)	P500QL1	High-temp, Press; 50<t<=100mm	0.18 max	1.50 max	0.50 max	1.70 max	0.020 max	0.020 max	0.80 max	0.12 max	B <=0.0050; Mo <=0.70; N <=0.015; Ni <=2.00; Nb <=0.06; Ti <=0.05; Zr<=0.15; Nb+Ti+V+Zr>=0.015; bal Fe	590-770	480	17	
EN 10028/6(96)	P500QL1	100<t<=150mm	0.18 max	1.50 max	0.50 max	1.70 max	0.020 max	0.020 max	0.80 max	0.12 max	B <=0.0050; Mo <=0.70; N <=0.015; Ni <=2.00; Nb <=0.06; Ti <=0.05; Zr<=0.15; Nb+Ti+V+Zr>=0.015; bal Fe	540-720	440	17	
EN 10028/6(96)	P500QL1	High-temp, Press; t<=50mm	0.18 max	1.50 max	0.50 max	1.70 max	0.020 max	0.020 max	0.80 max	0.12 max	B <=0.0050; Mo <=0.70; N <=0.015; Ni <=2.00; Nb <=0.06; Ti <=0.05; Zr<=0.15; Nb+Ti+V+Zr>=0.015; bal Fe	590-770	500	17	
EN 10028/6(96)	P500QL2	High-temp, Press; t<=50mm	0.18 max	1.50 max	0.50 max	1.70 max	0.020 max	0.020 max	0.80 max	0.12 max	B <=0.0050; Mo <=0.70; N <=0.015; Ni <=2.00; Nb <=0.06; Ti <=0.05; Zr<=0.15; Nb+Ti+V+Zr>=0.015; bal Fe	590-770	500	17	
EN 10028/6(96)	P500QL2	High-temp, Press; 50<t<=100mm	0.18 max	1.50 max	0.50 max	1.70 max	0.020 max	0.020 max	0.80 max	0.12 max	B <=0.0050; Mo <=0.70; N <=0.015; Ni <=2.00; Nb <=0.06; Ti <=0.05; Zr<=0.15; Nb+Ti+V+Zr>=0.015; bal Fe	590-770	480	17	
EN 10028/6(96)	P500QL2	100<t<=150mm	0.18 max	1.50 max	0.50 max	1.70 max	0.020 max	0.020 max	0.80 max	0.12 max	B <=0.0050; Mo <=0.70; N <=0.015; Ni <=2.00; Nb <=0.06; Ti <=0.05; Zr<=0.15; Nb+Ti+V+Zr>=0.015; bal Fe	540-720	440	17	
EN 10028/6(96)	P690Q	100<t<=150mm	0.20 max	1.50 max	0.50 max	1.70 max	0.025 max	0.015 max	0.80 max	0.12 max	B <=0.0050; Mo <=0.70; N <=0.015; Ni <=2.00; Nb <=0.06; Ti <=0.05; Zr<=0.15; Nb+Ti+V+Zr>=0.015; bal Fe	720-900	630	14	
EN 10028/6(96)	P690Q	High-temp, Press; 50<t<=100mm	0.20 max	1.50 max	0.50 max	1.70 max	0.025 max	0.015 max	0.80 max	0.12 max	B <=0.0050; Mo <=0.70; N <=0.015; Ni <=2.00; Nb <=0.06; Ti <=0.05; Zr<=0.15; Nb+Ti+V+Zr>=0.015; bal Fe	770-940	670	14	

Specification	Designation	Notes	C	Cr	Cu	Mn	P	S	Si	V	Other	UTS	YS	El	Hard
High Strength Steel, Low-Alloy (HSLA), A514(F) (Continued from previous page)															
Europe															
EN 10028/6(96)	P690Q	High-temp, Press; t<=50mm	0.20 max	1.50 max	0.50 max	1.70 max	0.025 max	0.015 max	0.80 max	0.12 max	B <=0.0050; Mo <=0.70; N <=0.015; Ni <=2.00; Nb <=0.06; Ti <=0.05; Zr<=0.15; Nb+Ti+V+Zr>=0.015; bal Fe	770-940	690	14	
EN 10028/6(96)	P690QH	100<t<=150mm	0.20 max	1.50 max	0.50 max	1.70 max	0.025 max	0.015 max	0.80 max	0.12 max	B <=0.0050; Mo <=0.70; N <=0.015; Ni <=2.00; Nb <=0.06; Ti <=0.05; Zr<=0.15; Nb+Ti+V+Zr>=0.015; bal Fe	720-900	630	14	
EN 10028/6(96)	P690QH	High-temp, Press; 50<t<=100mm	0.20 max	1.50 max	0.50 max	1.70 max	0.025 max	0.015 max	0.80 max	0.12 max	B <=0.0050; Mo <=0.70; N <=0.015; Ni <=2.00; Nb <=0.06; Ti <=0.05; Zr<=0.15; Nb+Ti+V+Zr>=0.015; bal Fe	770-940	670	14	
EN 10028/6(96)	P690QH	High-temp, Press; t<=50mm	0.20 max	1.50 max	0.50 max	1.70 max	0.025 max	0.015 max	0.80 max	0.12 max	B <=0.0050; Mo <=0.70; N <=0.015; Ni <=2.00; Nb <=0.06; Ti <=0.05; Zr<=0.15; Nb+Ti+V+Zr>=0.015; bal Fe	770-940	690	14	
EN 10028/6(96)	P690QL1	100<t<=150mm	0.20 max	1.50 max	0.50 max	1.70 max	0.020 max	0.020 max	0.80 max	0.12 max	B <=0.0050; Mo <=0.70; N <=0.015; Ni <=2.00; Nb <=0.06; Ti <=0.05; Zr<=0.15; Nb+Ti+V+Zr>=0.015; bal Fe	720-900	630	14	
EN 10028/6(96)	P690QL1	High-temp, Press; 50<t<=100mm	0.20 max	1.50 max	0.50 max	1.70 max	0.020 max	0.020 max	0.80 max	0.12 max	B <=0.0050; Mo <=0.70; N <=0.015; Ni <=2.00; Nb <=0.06; Ti <=0.05; Zr<=0.15; Nb+Ti+V+Zr>=0.015; bal Fe	770-940	670	14	
EN 10028/6(96)	P690QL1	High-temp, Press; t<=50mm	0.20 max	1.50 max	0.50 max	1.70 max	0.020 max	0.020 max	0.80 max	0.12 max	B <=0.0050; Mo <=0.70; N <=0.015; Ni <=2.00; Nb <=0.06; Ti <=0.05; Zr<=0.15; Nb+Ti+V+Zr>=0.015; bal Fe	770-940	690	14	
EN 10028/6(96)	P690QL2	High-temp, Press; t<=50mm	0.20 max	1.50 max	0.50 max	1.70 max	0.020 max	0.020 max	0.80 max	0.12 max	B <=0.0050; Mo <=0.70; N <=0.015; Ni <=2.00; Nb <=0.06; Ti <=0.05; Zr<=0.15; Nb+Ti+V+Zr>=0.015; bal Fe	770-940	690	14	
EN 10028/6(96)	P690QL2	100<t<=150mm	0.20 max	1.50 max	0.50 max	1.70 max	0.020 max	0.020 max	0.80 max	0.12 max	B <=0.0050; Mo <=0.70; N <=0.015; Ni <=2.00; Nb <=0.06; Ti <=0.05; Zr<=0.15; Nb+Ti+V+Zr>=0.015; bal Fe	720-900	630	14	
EN 10028/6(96)	P690QL2	High-temp, Press; 50<t<=100mm	0.20 max	1.50 max	0.50 max	1.70 max	0.020 max	0.020 max	0.80 max	0.12 max	B <=0.0050; Mo <=0.70; N <=0.015; Ni <=2.00; Nb <=0.06; Ti <=0.05; Zr<=0.15; Nb+Ti+V+Zr>=0.015; bal Fe	770-940	670	14	

Specification	Designation	Notes	C	Cr	Cu	Mn	P	S	Si	V	Other	UTS	YS	El	Hard

High Strength Steel, Low-Alloy (HSLA), A514(F) (Continued from previous page)

France

Specification	Designation	Notes	C	Cr	Cu	Mn	P	S	Si	V	Other	UTS	YS	El	Hard
AFNOR NFA36204(83)	E460TFP-I	5<=t<=50mm; N/T	0.08 max	2 max	1.5 max	0.3-1.3	0.025 max	0.025 max	0.1-0.5	0.2 max	B <=0.05; Mo <=1; Ni <=2; Nb <=0.06; Ti <=0.2; Zr<=0.12; Cu+6Sn<=0.33; bal Fe	570-720	460	17	
AFNOR NFA36204(83)	E460TFP-I	50<t<=70mm; N/T	0.08 max	2 max	1.5 max	0.3-1.3	0.025 max	0.025 max	0.1-0.5	0.2 max	B <=0.05; Mo <=1; N <=0.015; Ni <=2; Nb <=0.06; Ti <=0.2; Zr<=0.12; Cu+6Sn<=0.33; bal Fe	570-720	440	17	
AFNOR NFA36204(92)	E420T-II-K2	50<t<=70mm; Q/T	0.18 max	2 max	1.5 max	1.5 max	0.025 max	0.015 max	0.5 max	0.2 max	B <=0.005; Mo <=1; N <=0.015; Ni <=2; Nb <=0.06; Ti <=0.2; CEV; bal Fe	510-680	400	19T	
AFNOR NFA36204(92)	E420T-II-K2	t<50mm; Q/T	0.18 max	2 max	1.5 max	1.5 max	0.025 max	0.015 max	0.5 max	0.2 max	B <=0.005; Mo <=1; N <=0.015; Ni <=2; Nb <=0.06; Ti <=0.2; CEV; bal Fe	510-680	420	19T	
AFNOR NFA36204(92)	E420T-II-K4	t<50mm; Q/T	0.18 max	2 max	1.5 max	1.5 max	0.02 max	0.01 max	0.5 max	0.2 max	B <=0.005; Mo <=1; N <=0.015; Ni <=2; Nb <=0.06; Ti <=0.2; CEV; bal Fe	510-680	420	19T	
AFNOR NFA36204(92)	E420T-II-K4	50<t<=70mm; Q/T	0.18 max	2 max	1.5 max	1.5 max	0.02 max	0.01 max	0.5 max	0.2 max	B <=0.005; Mo <=1; N <=0.015; Ni <=2; Nb <=0.06; Ti <=0.2; CEV; bal Fe	510-680	400	19T	
AFNOR NFA36204(92)	E460T-II-K2	t<50mm; Q/T	0.18 max	2 max	1.5 max	1.6 max	0.025 max	0.015 max	0.5 max	0.2 max	B <=0.005; Mo <=1; N <=0.015; Ni <=2; Nb <=0.06; Ti <=0.2; CEV; bal Fe	550-720	460	17T	
AFNOR NFA36204(92)	E460T-II-K2	50<t<=70mm; Q/T	0.18 max	2 max	1.5 max	1.6 max	0.025 max	0.015 max	0.5 max	0.2 max	B <=0.005; Mo <=1; N <=0.015; Ni <=2; Nb <=0.06; Ti <=0.2; CEV; bal Fe	550-720	440	17T	
AFNOR NFA36204(92)	E460T-II-K4	t<50mm; Q/T	0.18 max	2 max	1.5 max	1.6 max	0.02 max	0.01 max	0.5 max	0.2 max	B <=0.005; Mo <=1; N <=0.015; Ni <=2; Nb <=0.06; Ti <=0.2; CEV; bal Fe	550-720	460	17T	
AFNOR NFA36204(92)	E460T-II-K4	50<t<=70mm; Q/T	0.18 max	2 max	1.5 max	1.6 max	0.02 max	0.01 max	0.5 max	0.2 max	B <=0.005; Mo <=1; N <=0.015; Ni <=2; Nb <=0.06; Ti <=0.2; CEV; bal Fe	550-720	440	17T	
AFNOR NFA36204(92)	E500T-II-K2	50<t<=70mm; Q/T	0.18 max	2 max	1.5 max	1.6 max	0.025 max	0.015 max	0.5 max	0.2 max	B <=0.005; Mo <=1; N <=0.015; Ni <=2; Nb <=0.06; Ti <=0.2; CEV; bal Fe	600-770	480	17T	
AFNOR NFA36204(92)	E500T-II-K2	t<50mm; Q/T	0.18 max	2 max	1.5 max	1.6 max	0.025 max	0.015 max	0.5 max	0.2 max	B <=0.005; Mo <=1; N <=0.015; Ni <=2; Nb <=0.06; Ti <=0.2; CEV; bal Fe	600-770	500	17T	
AFNOR NFA36204(92)	E500T-II-K4	t<50mm; Q/T	0.18 max	2 max	1.5 max	1.6 max	0.02 max	0.01 max	0.5 max	0.2 max	B <=0.005; Mo <=1; N <=0.015; Ni <=2; Nb <=0.06; Ti <=0.2; CEV; bal Fe	600-770	500	17T	
AFNOR NFA36204(92)	E500T-II-K4	50<t<=70mm; Q/T	0.18 max	2 max	1.5 max	1.6 max	0.02 max	0.01 max	0.5 max	0.2 max	B <=0.005; Mo <=1; N <=0.015; Ni <=2; Nb <=0.06; Ti <=0.2; CEV; bal Fe	600-770	480	17T	
AFNOR NFA36204(92)	E550T-II-K2	t<50mm; Q/T	0.18 max	2 max	1.5 max	1.6 max	0.025 max	0.15 max	0.5 max	0.2 max	B <=0.005; Mo <=1; N <=0.015; Ni <=2; Nb <=0.06; Ti <=0.2; CEV; bal Fe	650-820	550	16T	
AFNOR NFA36204(92)	E550T-II-K2	50<t<=70mm; Q/T	0.18 max	2 max	1.5 max	1.6 max	0.025 max	0.15 max	0.5 max	0.2 max	B <=0.005; Mo <=1; N <=0.015; Ni <=2; Nb <=0.06; Ti <=0.2; CEV; bal Fe	650-820	530	16T	
AFNOR NFA36204(92)	E550T-II-K4	t<50mm; Q/T	0.18 max	2 max	1.5 max	1.6 max	0.02 max	0.01 max	0.5 max	0.2 max	B <=0.005; Mo <=1; N <=0.015; Ni <=2; Nb <=0.06; Ti <=0.2; CEV; bal Fe	650-820	550	16T	

High Strength Steel, Low-Alloy (HSLA), A514(F) (Continued from previous page)

France

Specification	Designation	Notes	C	Cr	Cu	Mn	P	S	Si	V	Other	UTS	YS	El	Hard
AFNOR NFA36204(92)	E550T-II-K4	50<t<=70mm; Q/T	0.18 max	2 max	1.5 max	1.6 max	0.02 max	0.01 max	0.5 max	0.2 max	B <=0.005; Mo <=1; N <=0.015; Ni <=2; Nb <=0.06; Ti <=0.2; CEV; bal Fe	650-820	530	16T	
AFNOR NFA36204(92)	E620T-II-K2	t<50mm; Q/T	0.18 max	2 max	1.5 max	1.6 max	0.025 max	0.015 max	0.5 max	0.2 max	B <=0.005; Mo <=1; N <=0.015; Ni <=2; Nb <=0.06; Ti <=0.2; CEV; bal Fe	720-890	620	15T	
AFNOR NFA36204(92)	E620T-II-K2	50<t<=70mm; Q/T	0.18 max	2 max	1.5 max	1.6 max	0.025 max	0.015 max	0.5 max	0.2 max	B <=0.005; Mo <=1; N <=0.015; Ni <=2; Nb <=0.06; Ti <=0.2; CEV; bal Fe	720-890	600	15T	
AFNOR NFA36204(92)	E620T-II-K4	50<t<=70mm; Q/T	0.18 max	2 max	1.5 max	1.6 max	0.02 max	0.01 max	0.5 max	0.2 max	B <=0.005; Mo <=1; N <=0.015; Ni <=2; Nb <=0.06; Ti <=0.2; CEV; bal Fe	720-890	600	15T	
AFNOR NFA36204(92)	E620T-II-K4	t<50mm; Q/T	0.18 max	2 max	1.5 max	1.6 max	0.02 max	0.01 max	0.5 max	0.2 max	B <=0.005; Mo <=1; N <=0.015; Ni <=2; Nb <=0.06; Ti <=0.2; CEV; bal Fe	720-890	620	15T	
AFNOR NFA36204(92)	E690T-II-K2	50<t<=70mm; Q/T	0.18 max	2 max	1.5 max	1.6 max	0.025 max	0.015 max	0.5 max	0.2 max	B <=0.005; Mo <=1; N <=0.015; Ni <=2; Nb <=0.06; Ti <=0.2; CEV; bal Fe	770-940	670	14T	
AFNOR NFA36204(92)	E690T-II-K2	t<50mm; Q/T	0.18 max	2 max	1.5 max	1.6 max	0.025 max	0.015 max	0.5 max	0.2 max	B <=0.005; Mo <=1; N <=0.015; Ni <=2; Nb <=0.06; Ti <=0.2; CEV; bal Fe	770-940	690	14T	
AFNOR NFA36204(92)	E690T-II-K4	50<t<=70mm; Q/T	0.18 max	2 max	1.5 max	1.6 max	0.02 max	0.01 max	0.5 max	0.2 max	B <=0.005; Mo <=1; N <=0.015; Ni <=2; Nb <=0.06; Ti <=0.2; CEV; bal Fe	770-940	670	14T	
AFNOR NFA36204(92)	E690T-II-K4	t<50mm; Q/T	0.18 max	2 max	1.5 max	1.6 max	0.02 max	0.01 max	0.5 max	0.2 max	B <=0.005; Mo <=1; N <=0.015; Ni <=2; Nb <=0.06; Ti <=0.2; CEV; bal Fe	770-940	690	14T	
AFNOR NFA36204(92)	S420Q-II-K2	t<50mm; Q/T	0.18 max	2 max	1.5 max	1.5 max	0.025 max	0.015 max	0.5 max	0.2 max	B <=0.005; Mo <=1; N <=0.015; Ni <=2; Nb <=0.06; Ti <=0.2; CEV; bal Fe	510-680	420	19T	
AFNOR NFA36204(92)	S420Q-II-K4	t<50mm; Q/T	0.18 max	2 max	1.5 max	1.5 max	0.02 max	0.01 max	0.5 max	0.2 max	B <=0.005; Mo <=1; N <=0.015; Ni <=2; Nb <=0.06; Ti <=0.2; CEV; bal Fe	510-680	420	19T	
AFNOR NFA36204(92)	S420Q-II-K4	50<t<=70mm; Q/T	0.18 max	2 max	1.5 max	1.5 max	0.02 max	0.01 max	0.5 max	0.2 max	B <=0.005; Mo <=1; N <=0.015; Ni <=2; Nb <=0.06; Ti <=0.2; CEV; bal Fe	510-680	400	19T	
AFNOR NFA36204(92)	S460Q-II-K2	t<50mm; Q/T	0.18 max	2 max	1.5 max	1.6 max	0.025 max	0.015 max	0.5 max	0.2 max	B <=0.005; Mo <=1; N <=0.015; Ni <=2; Nb <=0.06; Ti <=0.2; CEV; bal Fe	550-720	460	17T	
AFNOR NFA36204(92)	S460Q-II-K2	50<t<=70mm; Q/T	0.18 max	2 max	1.5 max	1.6 max	0.025 max	0.015 max	0.5 max	0.2 max	B <=0.005; Mo <=1; N <=0.015; Ni <=2; Nb <=0.06; Ti <=0.2; CEV; bal Fe	550-720	440	17T	
AFNOR NFA36204(92)	S460Q-II-K4	t<50mm; Q/T	0.18 max	2 max	1.5 max	1.6 max	0.02 max	0.01 max	0.5 max	0.2 max	B <=0.005; Mo <=1; N <=0.015; Ni <=2; Nb <=0.06; Ti <=0.2; CEV; bal Fe	550-720	460	17T	
AFNOR NFA36204(92)	S460Q-II-K4	50<t<=70mm; Q/T	0.18 max	2 max	1.5 max	1.6 max	0.02 max	0.01 max	0.5 max	0.2 max	B <=0.005; Mo <=1; N <=0.015; Ni <=2; Nb <=0.06; Ti <=0.2; CEV; bal Fe	550-720	440	17T	
AFNOR NFA36204(92)	S500-Q-II-K2	50<t<=70mm; Q/T	0.18 max	2 max	1.5 max	1.6 max	0.025 max	0.015 max	0.5 max	0.2 max	B <=0.005; Mo <=1; N <=0.015; Ni <=2; Nb <=0.06; Ti <=0.2; CEV; bal Fe	600-770	480	17T	
AFNOR NFA36204(92)	S500-Q-II-K2	t<50mm; Q/T	0.18 max	2 max	1.5 max	1.6 max	0.025 max	0.015 max	0.5 max	0.2 max	B <=0.005; Mo <=1; N <=0.015; Ni <=2; Nb <=0.06; Ti <=0.2; CEV; bal Fe	600-770	500	17T	

UNS numbers and US grades are provided as a means of cross referencing chemically similar alloys. Exchangability is only possible after independent examination of specifications. Tensile properties are minimum or typical as specified. UTS and YS as MPa. El as %. See Appendix for list of abbreviations used in Notes. * indicates obsolete material.

High Strength Steel, Low-Alloy (HSLA), A514(F) (Continued from previous page)

Specification	Designation	Notes	C	Cr	Cu	Mn	P	S	Si	V	Other	UTS	YS	El	Hard
France															
AFNOR NFA36204(92)	S500-Q-II-K4	t<50mm; Q/T	0.18 max	2 max	1.5 max	1.6 max	0.02 max	0.01 max	0.5 max	0.2 max	B <=0.005; Mo <=1; N <=0.015; Ni <=2; Nb <=0.06; Ti <=0.2; CEV; bal Fe	600-770	500	17T	
AFNOR NFA36204(92)	S500-Q-II-K4	50<t<=70mm; Q/T	0.18 max	2 max	1.5 max	1.6 max	0.02 max	0.01 max	0.5 max	0.2 max	B <=0.005; Mo <=1; N <=0.015; Ni <=2; Nb <=0.06; Ti <=0.2; CEV; bal Fe	600-770	480	17T	
AFNOR NFA36204(92)	S550-Q-II-K2	t<50mm; Q/T	0.18 max	2 max	1.5 max	1.6 max	0.025 max	0.15 max	0.5 max	0.2 max	B <=0.005; Mo <=1; N <=0.015; Ni <=2; Nb <=0.06; Ti <=0.2; CEV; bal Fe	650-820	550	16T	
AFNOR NFA36204(92)	S550-Q-II-K2	50<t<=70mm; Q/T	0.18 max	2 max	1.5 max	1.6 max	0.025 max	0.15 max	0.5 max	0.2 max	B <=0.005; Mo <=1; N <=0.015; Ni <=2; Nb <=0.06; Ti <=0.2; CEV; bal Fe	650-820	530	16T	
AFNOR NFA36204(92)	S550-Q-II-K4	50<t<=70mm; Q/T	0.18 max	2 max	1.5 max	1.6 max	0.02 max	0.01 max	0.5 max	0.2 max	B <=0.005; Mo <=1; N <=0.015; Ni <=2; Nb <=0.06; Ti <=0.2; CEV; bal Fe	650-820	530	16T	
AFNOR NFA36204(92)	S550-Q-II-K4	t<50mm; Q/T	0.18 max	2 max	1.5 max	1.6 max	0.02 max	0.01 max	0.5 max	0.2 max	B <=0.005; Mo <=1; N <=0.015; Ni <=2; Nb <=0.06; Ti <=0.2; CEV; bal Fe	650-820	550	16T	
AFNOR NFA36204(92)	S620-Q-II-K2	50<t<=70mm; Q/T	0.18 max	2 max	1.5 max	1.6 max	0.025 max	0.015 max	0.5 max	0.2 max	B <=0.005; Mo <=1; N <=0.015; Ni <=2; Nb <=0.06; Ti <=0.2; CEV; bal Fe	720-890	600	15T	
AFNOR NFA36204(92)	S620-Q-II-K2	t<50mm; Q/T	0.18 max	2 max	1.5 max	1.6 max	0.025 max	0.015 max	0.5 max	0.2 max	B <=0.005; Mo <=1; N <=0.015; Ni <=2; Nb <=0.06; Ti <=0.2; CEV; bal Fe	720-890	620	15T	
AFNOR NFA36204(92)	S620-Q-II-K4	50<t<=70mm; Q/T	0.18 max	2 max	1.5 max	1.6 max	0.02 max	0.01 max	0.5 max	0.2 max	B <=0.005; Mo <=1; N <=0.015; Ni <=2; Nb <=0.06; Ti <=0.2; CEV; bal Fe	720-890	600	15T	
AFNOR NFA36204(92)	S620-Q-II-K4	t<50mm; Q/T	0.18 max	2 max	1.5 max	1.6 max	0.02 max	0.01 max	0.5 max	0.2 max	B <=0.005; Mo <=1; N <=0.015; Ni <=2; Nb <=0.06; Ti <=0.2; CEV; bal Fe	720-890	620	15T	
AFNOR NFA36204(92)	S690Q-11-K4	50<t<=70mm; Q/T	0.18 max	2 max	1.5 max	1.6 max	0.02 max	0.01 max	0.5 max	0.2 max	B <=0.005; Mo <=1; N <=0.015; Ni <=2; Nb <=0.06; Ti <=0.2; CEV; bal Fe	770-940	670	14T	
AFNOR NFA36204(92)	S690Q-11-K4	t<50mm; Q/T	0.18 max	2 max	1.5 max	1.6 max	0.02 max	0.01 max	0.5 max	0.2 max	B <=0.005; Mo <=1; N <=0.015; Ni <=2; Nb <=0.06; Ti <=0.2; CEV; bal Fe	770-940	690	14T	
AFNOR NFA36204(92)	S690Q-II-K2	50<t<=70mm; Q/T	0.18 max	2 max	1.5 max	1.6 max	0.025 max	0.015 max	0.5 max	0.2 max	B <=0.005; Mo <=1; N <=0.015; Ni <=2; Nb <=0.06; Ti <=0.2; CEV; bal Fe	770-940	670	14T	
AFNOR NFA36204(92)	S690Q-II-K2	t<50mm; Q/T	0.18 max	2 max	1.5 max	1.6 max	0.025 max	0.015 max	0.5 max	0.2 max	B <=0.005; Mo <=1; N <=0.015; Ni <=2; Nb <=0.06; Ti <=0.2; CEV; bal Fe	770-940	690	14T	
Germany															
DIN EN 10137(95)	S960Q	Plt flat Q/T or PH	0.20 max	1.50 max	0.50 max	1.70 max	0.025 max	0.015 max	0.80 max	0.12 max	Al >=0.018; B <=0.0050; Mo <=0.70; N <=0.015; Ni <=2.00; Nb <=0.06; Ti <=0.05; Zr<=0.15; bal Fe				
DIN EN 10137(95)	WNr 1.8941	Plt flat product Q/T or PH	0.20 max	1.50 max	0.50 max	1.70 max	0.025 max	0.015 max	0.80 max	0.12 max	Al >=0.018; B <=0.0050; Mo <=0.70; N <=0.015; Ni <=2.00; Nb <=0.06; Ti <=0.05; Zr<=0.15; bal Fe				

UNS numbers and US grades are provided as a means of cross referencing chemically similar alloys. Exchangability is only possible after independent examination of specifications. Tensile properties are minimum or typical as specified. UTS and YS as MPa. El as %. See Appendix for list of abbreviations used in Notes. * indicates obsolete material.

Specification	Designation	Notes	C	Cr	Cu	Mn	P	S	Si	V	Other	UTS	YS	El	Hard

High Strength Steel, Low-Alloy (HSLA), A514(F) (Continued from previous page)

USA

Specification	Designation	Notes	C	Cr	Cu	Mn	P	S	Si	V	Other	UTS	YS	El	Hard
	UNS K11576		0.10-0.20	0.40-0.65	0.15-0.50	0.60-1.00	0.035 max	0.040 max	0.15-0.35	0.03-0.08	B 0.0005-0.006; Mo 0.40-0.60; Ni 0.70-1.00; bal Fe				
ASTM A514(94)	Grade F	Q/T, Plt, t=65mm, weld	0.10-0.20	0.40-0.65	0.15-0.50	0.60-1.00	0.035 max	0.035 max	0.15-0.35	0.03-0.08	B 0.0005-0.006; Mo 0.40-0.60; Ni 0.70-1.00; bal Fe	760-895	690	18	
ASTM A517(93)	Grade F	Q/T, Press ves plt, t<=65mm	0.10-0.20	0.40-0.65	0.15-0.50	0.60-1.00	0.035 max	0.035 max	0.15-0.35	0.03-0.08	B 0.0005-0.006; Mo 0.40-0.60; Ni 0.70-1.00; bal Fe	795-930	690	16	
ASTM A592/592M(94)	Grade F	High-strength QT frg for press ves, <=2.5 in.	0.10-0.22	0.40-0.65	0.15-0.50	0.60-1.00	0.025 max	0.025 max	0.15-0.35	0.03-0.08	B 0.002-0.006; Mo 0.40-0.60; Ni 0.70-1.00; bal Fe	795-930	690	18	
ASTM A709(97)	Grade 100, Type F	Q/T, Plt, t<=65mm	0.10-0.20	0.40-0.65	0.15-0.50	0.60-1.00	0.035 max	0.035 max	0.15-0.35	0.03-0.08	B 0.0005-0.006; Mo 0.40-0.60; Ni 0.70-1.00; bal Fe	760-895	690	18	

High Strength Steel, Low-Alloy (HSLA), A537

Bulgaria

Specification	Designation	Notes	C	Cr	Cu	Mn	P	S	Si	V	Other	UTS	YS	El	Hard
BDS 4880	09G2BF	Struct	0.15 max	0.35 max	0.35 max	1.75 max	0.030 max	0.020 max	0.65 max	0.025-0.100	Ni <=0.35; Nb 0.005-0.060; bal Fe				

Europe

Specification	Designation	Notes	C	Cr	Cu	Mn	P	S	Si	V	Other	UTS	YS	El	Hard
EN 10028/2(92)	1.0473	High-temp, Press; 3<=t<=16mm	0.10-0.22	0.30 max	0.30 max	1.00-1.70	0.030 max	0.025 max	0.60 max	0.02 max	Al >=0.020; Mo <=0.08; Ni <=0.30; Nb <=0.010; Ti <=0.03; Cr+Cu+Mo+Ni<=0.70; bal Fe	510-650	355	21T	
EN 10028/2(92)	1.0473	High-temp, Press; 40<=t<=60mm	0.10-0.22	0.30 max	0.30 max	1.00-1.70	0.030 max	0.025 max	0.60 max	0.02 max	Al >=0.020; Mo <=0.08; Ni <=0.30; Nb <=0.010; Ti <=0.03; Cr+Cu+Mo+Ni<=0.70; bal Fe	510-650	335	21T	
EN 10028/2(92)	1.0473	High-temp, Press; 60<=t<=100mm	0.10-0.22	0.30 max	0.30 max	1.00-1.70	0.030 max	0.025 max	0.60 max	0.02 max	Al >=0.020; Mo <=0.08; Ni <=0.30; Nb <=0.010; Ti <=0.03; Cr+Cu+Mo+Ni<=0.70; bal Fe	490-630	315	20T	
EN 10028/2(92)	1.0473	High-temp, Press; 100<=t<=150mm	0.10-0.22	0.30 max	0.30 max	1.00-1.70	0.030 max	0.025 max	0.60 max	0.02 max	Al >=0.020; Mo <=0.08; Ni <=0.30; Nb <=0.010; Ti <=0.03; Cr+Cu+Mo+Ni<=0.70; bal Fe	480-630	295	20T	
EN 10028/2(92)	1.0473	High-temp, Press; 16<=t<=40mm	0.10-0.22	0.30 max	0.30 max	1.00-1.70	0.030 max	0.025 max	0.60 max	0.02 max	Al >=0.020; Mo <=0.08; Ni <=0.30; Nb <=0.010; Ti <=0.03; Cr+Cu+Mo+Ni<=0.70; bal Fe	510-650	345	21T	
EN 10028/2(92)	P355GH	High-temp, Press; 16<=t<=40mm	0.10-0.22	0.30 max	0.30 max	1.00-1.70	0.030 max	0.025 max	0.60 max	0.02 max	Al >=0.020; Mo <=0.08; Ni <=0.30; Nb <=0.010; Ti <=0.03; Cr+Cu+Mo+Ni<=0.70; bal Fe	510-650	345	21T	
EN 10028/2(92)	P355GH	High-temp, Press; 40<=t<=60mm	0.10-0.22	0.30 max	0.30 max	1.00-1.70	0.030 max	0.025 max	0.60 max	0.02 max	Al >=0.020; Mo <=0.08; Ni <=0.30; Nb <=0.010; Ti <=0.03; Cr+Cu+Mo+Ni<=0.70; bal Fe	510-650	335	21T	
EN 10028/2(92)	P355GH	High-temp, Press; 60<=t<=100mm	0.10-0.22	0.30 max	0.30 max	1.00-1.70	0.030 max	0.025 max	0.60 max	0.02 max	Al >=0.020; Mo <=0.08; Ni <=0.30; Nb <=0.010; Ti <=0.03; Cr+Cu+Mo+Ni<=0.70; bal Fe	490-630	315	20T	

UNS numbers and US grades are provided as a means of cross referencing chemically similar alloys. Exchangability is only possible after independent examination of specifications. Tensile properties are minimum or typical as specified. UTS and YS as MPa. El as %. See Appendix for list of abbreviations used in Notes. * indicates obsolete material.

High Strength Steel, Low-Alloy (HSLA), A537 (Continued from previous page)

Specification	Designation	Notes	C	Cr	Cu	Mn	P	S	Si	V	Other	UTS	YS	El	Hard
Europe															
EN 10028/2(92)	P355GH	High-temp, Press; 100<=t<=150mm	0.10-0.22	0.30 max	0.30 max	1.00-1.70	0.030 max	0.025 max	0.60 max	0.02 max	Al >=0.020; Mo <=0.08; Ni <=0.30; Nb <=0.010; Ti <=0.03; Cr+Cu+Mo+Ni<=0.70; bal Fe	480-630	295	20T	
EN 10028/2(92)	P355GH	High-temp, Press; 3<=t<=16mm	0.10-0.22	0.30 max	0.30 max	1.00-1.70	0.030 max	0.025 max	0.60 max	0.02 max	Al >=0.020; Mo <=0.08; Ni <=0.30; Nb <=0.010; Ti <=0.03; Cr+Cu+Mo+Ni<=0.70; bal Fe	510-650	355	21T	
EN 10028/6(96)	1.8866	100<t<=150mm	0.16 max	0.3 max	0.3 max	1.5 max	0.025 max	0.015 max	0.4 max	0.06 max	Al >=0.020; B <=0.0050; Mo <=0.25; N <=0.015; Nb <=0.05; Ti <=0.03; Zr<=0.05; Nb+Ti+V+Zr>=0.015; bal Fe	450-590	315	22	
EN 10028/6(96)	1.8866	High-temp, Press; t<=50mm	0.16 max	0.3 max	0.3 max	1.5 max	0.025 max	0.015 max	0.4 max	0.06 max	Al >=0.020; B <=0.0050; Mo <=0.25; N <=0.015; Nb <=0.05; Ti <=0.03; Zr<=0.05; Nb+Ti+V+Zr>=0.015; bal Fe	490-630	355	22	
EN 10028/6(96)	1.8866	High-temp, Press; 50<t<=100mm	0.16 max	0.3 max	0.3 max	1.5 max	0.025 max	0.015 max	0.4 max	0.06 max	Al >=0.020; B <=0.0050; Mo <=0.25; N <=0.015; Nb <=0.05; Ti <=0.03; Zr<=0.05; Nb+Ti+V+Zr>=0.015; bal Fe	490-630	335	22	
EN 10028/6(96)	1.8867	High-temp, Press; 50<t<=100mm	0.16 max	0.3 max	0.3 max	1.5 max	0.025 max	0.015 max	0.4 max	0.06 max	Al >=0.020; B <=0.0050; Mo <=0.25; N <=0.015; Nb <=0.05; Ti <=0.03; Zr<=0.05; Nb+Ti+V+Zr>=0.015; bal Fe	490-630	335	22	
EN 10028/6(96)	1.8867	100<t<=150mm	0.16 max	0.3 max	0.3 max	1.5 max	0.025 max	0.015 max	0.4 max	0.06 max	Al >=0.020; B <=0.0050; Mo <=0.25; N <=0.015; Nb <=0.05; Ti <=0.03; Zr<=0.05; Nb+Ti+V+Zr>=0.015; bal Fe	450-590	315	22	
EN 10028/6(96)	1.8867	High-temp, Press; t<=50mm	0.16 max	0.3 max	0.3 max	1.5 max	0.025 max	0.015 max	0.4 max	0.06 max	Al >=0.020; B <=0.0050; Mo <=0.25; N <=0.015; Nb <=0.05; Ti <=0.03; Zr<=0.05; Nb+Ti+V+Zr>=0.015; bal Fe	490-630	355	22	
EN 10028/6(96)	1.8868	High-temp, Press; t<=50mm	0.16 max	0.3 max	0.3 max	1.5 max	0.02 max	0.01 max	0.4 max	0.06 max	Al >=0.020; B <=0.0050; Mo <=0.25; N <=0.015; Nb <=0.05; Ti <=0.03; Zr<=0.05; Nb+Ti+V+Zr>=0.015; bal Fe	490-630	355	22	
EN 10028/6(96)	1.8868	High-temp, Press; 50<t<=100mm	0.16 max	0.3 max	0.3 max	1.5 max	0.02 max	0.01 max	0.4 max	0.06 max	Al >=0.020; B <=0.0050; Mo <=0.25; N <=0.015; Nb <=0.05; Ti <=0.03; Zr<=0.05; Nb+Ti+V+Zr>=0.015; bal Fe	490-630	335	22	

Specification	Designation	Notes	C	Cr	Cu	Mn	P	S	Si	V	Other	UTS	YS	El	Hard

High Strength Steel, Low-Alloy (HSLA), A537 (Continued from previous page)

Europe

Specification	Designation	Notes	C	Cr	Cu	Mn	P	S	Si	V	Other	UTS	YS	El	Hard
EN 10028/6(96)	1.8868	100<t<=150mm	0.16 max	0.3 max	0.3 max	1.5 max	0.02 max	0.01 max	0.4 max	0.06 max	Al >=0.020; B <=0.0050; Mo <=0.25; N <=0.015; Nb <=0.05; Ti <=0.03; Zr<=0.05; Nb+Ti+V+Zr>=0.015; bal Fe	450-590	315	22	
EN 10028/6(96)	1.8869	High-temp, Press; 50<t<=100mm	0.16 max	0.3 max	0.3 max	1.5 max	0.02 max	0.01 max	0.4 max	0.06 max	Al >=0.020; B <=0.0050; Mo <=0.25; N <=0.015; Nb <=0.05; Ti <=0.03; Zr<=0.05; Nb+Ti+V+Zr>=0.015; bal Fe	490-630	335	22	
EN 10028/6(96)	1.8869	High-temp, Press; t<=50mm	0.16 max	0.3 max	0.3 max	1.5 max	0.02 max	0.01 max	0.4 max	0.06 max	Al >=0.020; B <=0.0050; Mo <=0.25; N <=0.015; Nb <=0.05; Ti <=0.03; Zr<=0.05; Nb+Ti+V+Zr>=0.015; bal Fe	490-630	355	22	
EN 10028/6(96)	1.8869	100<t<=150mm	0.16 max	0.3 max	0.3 max	1.5 max	0.02 max	0.01 max	0.4 max	0.06 max	Al >=0.020; B <=0.0050; Mo <=0.25; N <=0.015; Nb <=0.05; Ti <=0.03; Zr<=0.05; Nb+Ti+V+Zr>=0.015; bal Fe	450-590	315	22	
EN 10028/6(96)	1.8870	100<t<=150mm	0.18 max	1.50 max	0.50 max	1.70 max	0.025 max	0.015 max	0.80 max	0.12 max	Al >=0.020; B <=0.0050; Mo <=0.70; N <=0.015; Ni <=2.00; Nb <=0.05; Ti <=0.05; Zr<=0.15; Nb+Ti+V+Zr>=0.015; bal Fe	500-670	400	19	
EN 10028/6(96)	1.8870	High-temp, Press; 50<t<=100mm	0.18 max	1.50 max	0.50 max	1.70 max	0.025 max	0.015 max	0.80 max	0.12 max	Al >=0.020; B <=0.0050; Mo <=0.70; N <=0.015; Ni <=2.00; Nb <=0.05; Ti <=0.05; Zr<=0.15; Nb+Ti+V+Zr>=0.015; bal Fe	550-720	440	19	
EN 10028/6(96)	1.8870	High-temp, Press; t<=50mm	0.18 max	1.50 max	0.50 max	1.70 max	0.025 max	0.015 max	0.80 max	0.12 max	Al >=0.020; B <=0.0050; Mo <=0.70; N <=0.015; Ni <=2.00; Nb <=0.05; Ti <=0.05; Zr<=0.15; Nb+Ti+V+Zr>=0.015; bal Fe	550-720	460	19	
EN 10028/6(96)	1.8871	High-temp, Press; 50<t<=100mm	0.18 max	1.50 max	0.50 max	1.70 max	0.025 max	0.015 max	0.80 max	0.12 max	B <=0.0050; Mo <=0.70; N <=0.015; Ni <=2.00; Nb <=0.06; Ti <=0.05; Zr<=0.15; Nb+Ti+V+Zr>=0.015; bal Fe	550-720	440	19	
EN 10028/6(96)	1.8871	100<t<=150mm	0.18 max	1.50 max	0.50 max	1.70 max	0.025 max	0.015 max	0.80 max	0.12 max	B <=0.0050; Mo <=0.70; N <=0.015; Ni <=2.00; Nb <=0.06; Ti <=0.05; Zr<=0.15; Nb+Ti+V+Zr>=0.015; bal Fe	500-670	400	19	
EN 10028/6(96)	1.8871	High-temp, Press; t<=50mm	0.18 max	1.50 max	0.50 max	1.70 max	0.025 max	0.015 max	0.80 max	0.12 max	Al >=0.020; B <=0.0050; Mo <=0.70; N <=0.015; Ni <=2.00; Nb <=0.06; Ti <=0.05; Zr<=0.15; Nb+Ti+V+Zr>=0.015; bal Fe	550-720	460	19	

UNS numbers and US grades are provided as a means of cross referencing chemically similar alloys. Exchangability is only possible after independent examination of specifications. Tensile properties are minimum or typical as specified. UTS and YS as MPa. El as %. See Appendix for list of abbreviations used in Notes. * indicates obsolete material.

High Strength Steel, Low-Alloy (HSLA), A537 (Continued from previous page)

Europe

Specification	Designation	Notes	C	Cr	Cu	Mn	P	S	Si	V	Other	UTS	YS	El	Hard
EN 10028/6(96)	1.8872	High-temp, Press; t<=50mm	0.18 max	1.50 max	0.50 max	1.70 max	0.020 max	0.020 max	0.80 max	0.12 max	B <=0.0050; Mo <=0.70; N <=0.015; Ni <=2.00; Nb <=0.06; Ti <=0.05; Zr<=0.15; Nb+Ti+V+Zr>=0.015; bal Fe	550-720	460	19	
EN 10028/6(96)	1.8872	High-temp, Press; 50<t<=100mm	0.18 max	1.50 max	0.50 max	1.70 max	0.020 max	0.020 max	0.80 max	0.12 max	B <=0.0050; Mo <=0.70; N <=0.015; Ni <=2.00; Nb <=0.06; Ti <=0.05; Zr<=0.15; Nb+Ti+V+Zr>=0.015; bal Fe	550-720	440	19	
EN 10028/6(96)	1.8872	100<t<=150mm	0.18 max	1.50 max	0.50 max	1.70 max	0.020 max	0.020 max	0.80 max	0.12 max	B <=0.0050; Mo <=0.70; N <=0.015; Ni <=2.00; Nb <=0.06; Ti <=0.05; Zr<=0.15; Nb+Ti+V+Zr>=0.015; bal Fe	500-670	400	19	
EN 10028/6(96)	P355Q	High-temp, Press; 50<t<=100mm	0.16 max	0.3 max	0.3 max	1.5 max	0.025 max	0.015 max	0.4 max	0.06 max	Al >=0.020; B <=0.0050; Mo <=0.25; N <=0.015; Nb <=0.05; Ti <=0.03; Zr<=0.05; Nb+Ti+V+Zr>=0.015; bal Fe	490-630	335	22	
EN 10028/6(96)	P355Q	100<t<=150mm	0.16 max	0.3 max	0.3 max	1.5 max	0.025 max	0.015 max	0.4 max	0.06 max	Al >=0.020; B <=0.0050; Mo <=0.25; N <=0.015; Nb <=0.05; Ti <=0.03; Zr<=0.05; Nb+Ti+V+Zr>=0.015; bal Fe	450-590	315	22	
EN 10028/6(96)	P355Q	High-temp, Press; t<=50mm	0.16 max	0.3 max	0.3 max	1.5 max	0.025 max	0.015 max	0.4 max	0.06 max	Al >=0.020; B <=0.0050; Mo <=0.25; N <=0.015; Nb <=0.05; Ti <=0.03; Zr<=0.05; Nb+Ti+V+Zr>=0.015; bal Fe	490-630	355	22	
EN 10028/6(96)	P355QH	High-temp, Press; t<=50mm	0.16 max	0.3 max	0.3 max	1.5 max	0.025 max	0.015 max	0.4 max	0.06 max	Al >=0.020; B <=0.0050; Mo <=0.25; N <=0.015; Nb <=0.05; Ti <=0.03; Zr<=0.05; Nb+Ti+V+Zr>=0.015; bal Fe	490-630	355	22	
EN 10028/6(96)	P355QH	High-temp, Press; 50<t<=100mm	0.16 max	0.3 max	0.3 max	1.5 max	0.025 max	0.015 max	0.4 max	0.06 max	Al >=0.020; B <=0.0050; Mo <=0.25; N <=0.015; Nb <=0.05; Ti <=0.03; Zr<=0.05; Nb+Ti+V+Zr>=0.015; bal Fe	490-630	335	22	
EN 10028/6(96)	P355QH	100<t<=150mm	0.16 max	0.3 max	0.3 max	1.5 max	0.025 max	0.015 max	0.4 max	0.06 max	Al >=0.020; B <=0.0050; Mo <=0.25; N <=0.015; Nb <=0.05; Ti <=0.03; Zr<=0.05; Nb+Ti+V+Zr>=0.015; bal Fe	450-590	315	22	
EN 10028/6(96)	P355QL1	100<t<=150mm	0.16 max	0.3 max	0.3 max	1.5 max	0.02 max	0.01 max	0.4 max	0.06 max	Al >=0.020; B <=0.0050; Mo <=0.25; N <=0.015; Nb <=0.05; Ti <=0.03; Zr<=0.05; Nb+Ti+V+Zr>=0.015; bal Fe	450-590	315	22	

Specification	Designation	Notes	C	Cr	Cu	Mn	P	S	Si	V	Other	UTS	YS	El	Hard

High Strength Steel, Low-Alloy (HSLA), A537 (Continued from previous page)

Europe

Specification	Designation	Notes	C	Cr	Cu	Mn	P	S	Si	V	Other	UTS	YS	El	Hard
EN 10028/6(96)	P355QL1	High-temp, Press; 50<t<=100mm	0.16 max	0.3 max	0.3 max	1.5 max	0.02 max	0.01 max	0.4 max	0.06 max	Al >=0.020; B <=0.0050; Mo <=0.25; N <=0.015; Nb <=0.05; Ti <=0.03; Zr<=0.05; Nb+Ti+V+Zr>=0.015; bal Fe	490-630	335	22	
EN 10028/6(96)	P355QL1	High-temp, Press; t<=50mm	0.16 max	0.3 max	0.3 max	1.5 max	0.02 max	0.01 max	0.4 max	0.06 max	Al >=0.020; B <=0.0050; Mo <=0.25; N <=0.015; Nb <=0.05; Ti <=0.03; Zr<=0.05; Nb+Ti+V+Zr>=0.015; bal Fe	490-630	355	22	
EN 10028/6(96)	P355QL2	High-temp, Press; 50<t<=100mm	0.16 max	0.3 max	0.3 max	1.5 max	0.02 max	0.01 max	0.4 max	0.06 max	Al >=0.020; B <=0.0050; Mo <=0.25; N <=0.015; Nb <=0.05; Ti <=0.03; Zr<=0.05; Nb+Ti+V+Zr>=0.015; bal Fe	490-630	335	22	
EN 10028/6(96)	P355QL2	100<t<=150mm	0.16 max	0.3 max	0.3 max	1.5 max	0.02 max	0.01 max	0.4 max	0.06 max	Al >=0.020; B <=0.0050; Mo <=0.25; N <=0.015; Nb <=0.05; Ti <=0.03; Zr<=0.05; Nb+Ti+V+Zr>=0.015; bal Fe	450-590	315	22	
EN 10028/6(96)	P355QL2	High-temp, Press; t<=50mm	0.16 max	0.3 max	0.3 max	1.5 max	0.02 max	0.01 max	0.4 max	0.06 max	Al >=0.020; B <=0.0050; Mo <=0.25; N <=0.015; Nb <=0.05; Ti <=0.03; Zr<=0.05; Nb+Ti+V+Zr>=0.015; bal Fe	490-630	355	22	
EN 10028/6(96)	P460Q	High-temp, Press; 50<t<=100mm	0.18 max	1.50 max	0.50 max	1.70 max	0.025 max	0.015 max	0.80 max	0.12 max	Al >=0.020; B <=0.0050; Mo <=0.70; N <=0.015; Ni <=2.00; Nb <=0.05; Ti <=0.05; Zr<=0.15; Nb+Ti+V+Zr>=0.015; bal Fe	550-720	440	19	
EN 10028/6(96)	P460Q	High-temp, Press; t<=50mm	0.18 max	1.50 max	0.50 max	1.70 max	0.025 max	0.015 max	0.80 max	0.12 max	Al >=0.020; B <=0.0050; Mo <=0.70; N <=0.015; Ni <=2.00; Nb <=0.05; Ti <=0.05; Zr<=0.15; Nb+Ti+V+Zr>=0.015; bal Fe	550-720	460	19	
EN 10028/6(96)	P460Q	100<t<=150mm	0.18 max	1.50 max	0.50 max	1.70 max	0.025 max	0.015 max	0.80 max	0.12 max	Al >=0.020; B <=0.0050; Mo <=0.70; N <=0.015; Ni <=2.00; Nb <=0.05; Ti <=0.05; Zr<=0.15; Nb+Ti+V+Zr>=0.015; bal Fe	500-670	400	19	
EN 10028/6(96)	P460QH	High-temp, Press; 50<t<=100mm	0.18 max	1.50 max	0.50 max	1.70 max	0.025 max	0.015 max	0.80 max	0.12 max	B <=0.0050; Mo <=0.70; N <=0.015; Ni <=2.00; Nb <=0.06; Ti <=0.05; Zr<=0.15; Nb+Ti+V+Zr>=0.015; bal Fe	550-720	440	19	
EN 10028/6(96)	P460QH	High-temp, Press; t<=50mm	0.18 max	1.50 max	0.50 max	1.70 max	0.025 max	0.015 max	0.80 max	0.12 max	Al >=0.020; B <=0.0050; Mo <=0.70; N <=0.015; Ni <=2.00; Nb <=0.06; Ti <=0.05; Zr<=0.15; Nb+Ti+V+Zr>=0.015; bal Fe	550-720	460	19	

UNS numbers and US grades are provided as a means of cross referencing chemically similar alloys. Exchangability is only possible after independent examination of specifications. Tensile properties are minimum or typical as specified. UTS and YS as MPa. El as %. See Appendix for list of abbreviations used in Notes. * indicates obsolete material.

High Strength Steel, Low-Alloy (HSLA), A537 (Continued from previous page)

Specification	Designation	Notes	C	Cr	Cu	Mn	P	S	Si	V	Other	UTS	YS	El	Hard
Europe															
EN 10028/6(96)	P460QH	100<t<=150mm	0.18 max	1.50 max	0.50 max	1.70 max	0.025 max	0.015 max	0.80 max	0.12 max	B <=0.0050; Mo <=0.70; N <=0.015; Ni <=2.00; Nb <=0.06; Ti <=0.05; Zr<=0.15; Nb+Ti+V+Zr>=0.015; bal Fe	500-670	400	19	
EN 10028/6(96)	P460QL1	100<t<=150mm	0.18 max	1.50 max	0.50 max	1.70 max	0.020 max	0.020 max	0.80 max	0.12 max	B <=0.0050; Mo <=0.70; N <=0.015; Ni <=2.00; Nb <=0.06; Ti <=0.05; Zr<=0.15; Nb+Ti+V+Zr>=0.015; bal Fe	500-670	400	19	
EN 10028/6(96)	P460QL1	High-temp, Press; t<=50mm	0.18 max	1.50 max	0.50 max	1.70 max	0.020 max	0.020 max	0.80 max	0.12 max	B <=0.0050; Mo <=0.70; N <=0.015; Ni <=2.00; Nb <=0.06; Ti <=0.05; Zr<=0.15; Nb+Ti+V+Zr>=0.015; bal Fe	550-720	460	19	
EN 10028/6(96)	P460QL1	High-temp, Press; 50<t<=100mm	0.18 max	1.50 max	0.50 max	1.70 max	0.020 max	0.020 max	0.80 max	0.12 max	B <=0.0050; Mo <=0.70; N <=0.015; Ni <=2.00; Nb <=0.06; Ti <=0.05; Zr<=0.15; Nb+Ti+V+Zr>=0.015; bal Fe	550-720	440	19	
EN 10113/2(93)	1.8912	HR weld t<=16mm	0.20 max	0.30 max	0.35-0.70	1.00-1.70	0.030 max	0.025 max	0.60 max	0.20 max	Al >=0.02; Mo <=0.10; N <=0.025; Ni <=0.80; Nb <=0.05; Ti <=0.03; CEV; bal Fe	520-680	420	19	
EN 10113/2(93)	S420NL	HR weld t<=16mm	0.20 max	0.30 max	0.35-0.70	1.00-1.70	0.030 max	0.025 max	0.60 max	0.20 max	Al >=0.02; Mo <=0.10; N <=0.025; Ni <=0.80; Nb <=0.05; Ti <=0.03; CEV; bal Fe	520-680	420	19	
France															
AFNOR NFA36205(82)	A52AP*	Press ves; t<=25mm; Norm	0.20 max	0.25 max	0.30 max	1.00-1.60	0.035 max	0.03 max	0.5 max	0.05 max	Mo <=0.10; Ni <=0.3; Nb <=0.010; Ti <=0.05; bal Fe	510-620	335	22	
AFNOR NFA36205(82)	A52AP*	Press ves; 80<t<=100mm; Norm	0.20 max	0.25 max	0.30 max	1.00-1.60	0.035 max	0.03 max	0.5 max	0.05 max	Mo <=0.10; Ni <=0.3; Nb <=0.010; Ti <=0.05; bal Fe	510-620	305	20	
AFNOR NFA36205(82)	A52APR*	Press ves; t<=25mm; Norm	0.20 max	0.25 max	0.30 max	1.00-1.60	0.035 max	0.03 max	0.5 max	0.05 max	Mo <=0.15; Ni <=0.3; Nb <=0.010; Ti <=0.05; bal Fe	510-620	335	22	
AFNOR NFA36205(82)	A52APR*	Press ves; 80<t<=100mm; Norm	0.20 max	0.25 max	0.30 max	1.00-1.60	0.035 max	0.03 max	0.5 max	0.05 max	Mo <=0.15; Ni <=0.3; Nb <=0.010; Ti <=0.05; bal Fe	510-620	305	20	
AFNOR NFA36205(82)	A52CP*	Press ves; t<=25mm; Norm	0.20 max	0.25 max	0.30 max	1.00-1.60	0.035 max	0.03 max	0.5 max	0.05 max	Mo <=0.10; Ni <=0.3; Nb <=0.010; Ti <=0.05; bal Fe	510-620	335	22	
AFNOR NFA36205(82)	A52CP*	Press ves; 80<t<=100mm; Norm	0.20 max	0.25 max	0.30 max	1.00-1.60	0.035 max	0.03 max	0.5 max	0.05 max	Mo <=0.10; Ni <=0.3; Nb <=0.010; Ti <=0.05; bal Fe	510-620	305	20	
AFNOR NFA36205(82)	A52CPR*	Press ves; t<=25mm; Norm	0.20 max	0.25 max	0.30 max	1.00-1.60	0.035 max	0.03 max	0.5 max	0.05 max	Mo <=0.15; Ni <=0.3; Nb <=0.010; Ti <=0.05; bal Fe	510-620	335	22	
AFNOR NFA36205(82)	A52CPR*	Press ves; 80<t<=110mm; Norm	0.20 max	0.25 max	0.30 max	1.00-1.60	0.035 max	0.03 max	0.5 max	0.05 max	Mo <=0.15; Ni <=0.3; Nb <=0.010; Ti <=0.05; bal Fe	510-620	305	20	
AFNOR NFA36601(80)	A52AP	80<t<=300mm; Q/T	0.2 max	0.25 max	0.18 max	1-1.6	0.035 max	0.03 max	0.5 max	0.05 max	Mo <=0.1; Ni <=0.5; Nb <=0.04; Ti <=0.05; Cu+(6xSn) <=0.33; bal Fe	510-620	305	22L; 20T	
AFNOR NFA36601(80)	A52AP	t<=30mm; Norm	0.2 max	0.25 max	0.18 max	1-1.6	0.035 max	0.03 max	0.5 max	0.05 max	Mo <=0.1; Ni <=0.5; Nb <=0.04; Ti <=0.05; Cu+(6xSn) <=0.33; bal Fe	510-620	325	24L; 22T	

Specification	Designation	Notes	C	Cr	Cu	Mn	P	S	Si	V	Other	UTS	YS	El	Hard

High Strength Steel, Low-Alloy (HSLA), A537 (Continued from previous page)

France

Specification	Designation	Notes	C	Cr	Cu	Mn	P	S	Si	V	Other	UTS	YS	El	Hard
AFNOR NFA36601(80)	A52CP	t<=30mm; Norm	0.2 max	0.25 max	0.18 max	1-1.6	0.035 max	0.03 max	0.5 max	0.05 max	Mo <=0.1; Ni <=0.5; Nb <=0.04; Ti <=0.05; Cu+(6xSn) <=0.33; bal Fe	510-620	325	24L; 22T	
AFNOR NFA36601(80)	A52CP	80<t<=300mm; Norm	0.2 max	0.25 max	0.18 max	1-1.6	0.035 max	0.03 max	0.5 max	0.05 max	Mo <=0.1; Ni <=0.5; Nb <=0.04; Ti <=0.05; Cu+(6xSn) <=0.33; bal Fe	510-620	305	22L; 20T	

Germany

Specification	Designation	Notes	C	Cr	Cu	Mn	P	S	Si	V	Other	UTS	YS	El	Hard
DIN 17103(89)	P420NH	85-100mm	0.2 max	0.30 max	0.70 max	1.00-1.70	0.035 max	0.030 max	0.10-0.60	0.20 max	Al >=0.020; Mo <=0.10; N <=0.020; Ni <=1.00; Nb <=0.05; Ti <=0.20; bal Fe	510-660	365	19	
DIN 17103(89)	WNr 1.8932	85-100mm	0.20 max	0.30 max	0.70 max	1.00-1.70	0.035 max	0.030 max	0.10-0.60	0.20 max	Al >=0.020; Mo <=0.10; N <=0.020; Ni <=1.00; Nb <=0.05; Ti <=0.20; bal Fe	510-660	365	19	
DIN EN 10113(93)	S420N	HR 85-100mm	0.20 max	0.30 max	0.70 max	1.00-1.70	0.035 max	0.030 max	0.60 max	0.20 max	Al >=0.020; Mo <=0.10; Ni <=0.80; Nb <=0.05; bal Fe	510-660	365	19	
DIN EN 10113(93)	WNr 1.8902	HR 85-100mm	0.20 max	0.30 max	0.70 max	1.00-1.70	0.035 max	0.030 max	0.60 max	0.20 max	Al >=0.020; Mo <=0.10; N <=0.025; Ni <=0.80; Nb <=0.05; Ti <=0.03; Ni+Ti+V<=0.22; N depends on Cu; bal Fe	510-660	365	19	

Italy

Specification	Designation	Notes	C	Cr	Cu	Mn	P	S	Si	V	Other	UTS	YS	El	Hard
UNI 7382(75)	FeE420KG	weld const	0.2 max	0.4 max	0.6 max	1-1.6	0.035 max	0.035 max	0.5 max	0.02-0.2	Al <=0.015; Mo <=0.4; Ni <=0.7; Nb 0.015-0.06; Ti <=0.2; bal Fe				

Japan

Specification	Designation	Notes	C	Cr	Cu	Mn	P	S	Si	V	Other	UTS	YS	El	Hard
JIS G3444(94)	STK540	Tube; Gen Struct	0.23 max			1.50 max	0.040 max	0.040 max	0.55 max		bal Fe	540	390	20	

Poland

Specification	Designation	Notes	C	Cr	Cu	Mn	P	S	Si	V	Other	UTS	YS	El	Hard
PNH84024	19G2	HS structural	0.16-0.22	0.3 max	0.25 max	1-1.4	0.045 max	0.045 max	0.4-0.6		Ni <=0.25; bal Fe				

Spain

Specification	Designation	Notes	C	Cr	Cu	Mn	P	S	Si	V	Other	UTS	YS	El	Hard
UNE 36081(76)	AE420KT	Fine grain Struct	0.20 max	0.40 max		1.0-1.6	0.03 max	0.03 max	0.5 max		Mo <=0.4; Ni <=0.7; +Al+V+Nb; bal Fe				
UNE 36081(76)	AE420KW	Fine grain Struct	0.20 max	0.40 max		1.0-1.6	0.035 max	0.035 max	0.5 max	0.02-0.2	Al <=0.015; Mo <=0.4; Ni <=0.7; Nb 0.015-0.06; Ti <=0.2; bal Fe				
UNE 36081(76)	F.6417*	see AE420KW	0.20 max	0.40 max	0.6 max	1.0-1.6	0.035 max	0.035 max	0.5 max	0.02-0.2	Al <=0.015; Mo <=0.4; Ni <=0.7; Nb 0.015-0.06; Ti <=0.2; bal Fe				
UNE 36081(76)	F.6418*	see AE420KT	0.20 max	0.40 max		1.0-1.6	0.03 max	0.03 max	0.5 max		Mo <=0.4; Ni <=0.7; +Al+V+Nb; bal Fe				
UNE 36087(74)	A52RAI	Sh Plt	0.25 max			0.80-1.70	0.055 max	0.050 max	0.55 max		N <=0.009; bal Fe				
UNE 36087(74)	A52RAII	Sh Plt	0.25 max			0.8-1.7	0.045 max	0.04 max	0.55 max		Al 0.015-0.10; Mo <=0.15; bal Fe				
UNE 36087(74)	A52RCI	Sh Plt	0.25 max			0.8-1.7	0.055 max	0.05 max	0.55 max		Al <=0.01; Mo <=0.15; N <=0.009; bal Fe				

USA

Specification	Designation	Notes	C	Cr	Cu	Mn	P	S	Si	V	Other	UTS	YS	El	Hard
	UNS K12437		0.24 max	0.25 max	0.35 max	0.70-1.35	0.035 max	0.040 max	0.13-0.55		Mo <=0.08; Ni <=0.25; bal Fe				
ASTM A537(95)	Class 1	HT, Press ves plt, t>=65-100mm	0.24 max	0.25 max	0.35 max	1.00-1.60	0.035 max	0.035 max	0.15-0.50		Mo <=0.08; Ni <=0.25; bal Fe	450-585	310	22	
ASTM A537(95)	Class 1	HT, Press ves plt, t>=40-65mm	0.24 max	0.25 max	0.35 max	1.00-1.60	0.035 max	0.035 max	0.15-0.50		Mo <=0.08; Ni <=0.25; bal Fe	485-620	345	22	
ASTM A537(95)	Class 1	HT, Press ves plt, t<=40mm	0.24 max	0.25 max	0.35 max	0.70-1.35	0.035 max	0.035 max	0.15-0.50		Mo <=0.08; Ni <=0.25; bal Fe	485-620	345	22	

UNS numbers and US grades are provided as a means of cross referencing chemically similar alloys. Exchangability is only possible after independent examination of specifications. Tensile properties are minimum or typical as specified. UTS and YS as MPa. El as %. See Appendix for list of abbreviations used in Notes. * indicates obsolete material.

Specification	Designation	Notes	C	Cr	Cu	Mn	P	S	Si	V	Other	UTS	YS	El	Hard
High Strength Steel, Low-Alloy (HSLA), A58(D)															
USA															
	UNS K11552		0.10-0.20	0.50-0.90	0.30 max	0.75-1.25	0.04 max	0.05 max	0.50-0.90		Nb <=0.04; Zr 0.05-0.15; bal Fe				
High Strength Steel, Low-Alloy (HSLA), A588(A)															
China															
GB 4172(84)	15MnCuCr	Plt HR/Norm 16-40mm Thk	0.10-0.19	0.30-0.65	0.20-0.40	0.90-1.30	0.040 max	0.040 max	0.15-0.35		bal Fe	490	335	22	
GB 4172(84)	15MnCuCrQT	Plt HR Q/T 16-40mm Thk	0.10-0.19	0.30-0.65	0.20-0.40	0.90-1.30	0.040 max	0.040 max	0.15-0.35		bal Fe	550-695	430	22	
USA															
	UNS K11430		0.09-0.20	0.37-0.68	0.22-0.43	0.86-1.24	0.045 max	0.055 max	0.13-0.32	0.01-0.11	bal Fe				
ASTM A502	3A														
ASTM A588/A588M(97)	Grade A	Struct Shp, t<=100mm for plt/bar	0.19 max	0.40-0.65	0.25-0.40	0.80-1.25	0.04 max	0.05 max	0.30-0.65	0.02-0.10	Ni <=0.40; bal Fe	485	345	21	
ASTM A588/A588M(97)	Grade A	t>=100-125mm for plt/bar	0.19 max	0.40-0.65	0.25-0.40	0.80-1.25	0.04 max	0.05 max	0.30-0.65	0.02-0.10	Ni <=0.40; bal Fe	460	315	21	
ASTM A588/A588M(97)	Grade A	t>=125-200mm for plt/bar	0.19 max	0.40-0.65	0.25-0.40	0.80-1.25	0.04 max	0.05 max	0.30-0.65	0.02-0.10	Ni <=0.40; bal Fe	435	290	21	
ASTM A709(97)	Grade 50W, Type A	Bar, Plt, Shp, struct for bridges, t<=100mm	0.19 max	0.40-0.65	0.25-0.40	0.90-1.25	0.04 max	0.04 max	0.15-0.65	0.02-0.10	Ni <=0.40; bal Fe	485	345	21	
High Strength Steel, Low-Alloy (HSLA), A588(B)															
China															
GB 4172(84)	15MnCuCr	Plt HR/Norm 16-40mm Thk	0.10-0.19	0.30-0.65	0.20-0.40	0.90-1.30	0.040 max	0.040 max	0.15-0.35		Ni <=0.65; bal Fe	490	335	22	
GB 4172(84)	15MnCuCrQT	Plt HR Q/T 16-40mm Thk	0.10-0.19	0.30-0.65	0.20-0.40	0.90-1.30	0.040 max	0.040 max	0.15-0.35		Ni <=0.65; bal Fe	550-695	430	22	
USA															
	UNS K12043		0.20 max	0.40-0.70	0.20-0.40	0.75-1.25	0.04 max	0.05 max	0.15-0.30	0.01-0.10	Ni <=0.50; bal Fe				
ASTM A588/A588M(97)	Grade B	t>=125-200mm for plt/bar	0.20 max	0.40-0.70	0.20-0.40	0.75-1.35	0.04 max	0.05 max	0.15-0.50	0.01-0.10	Ni <=0.50; bal Fe	435	290	21	
ASTM A588/A588M(97)	Grade B	t>=100-125mm for plt/bar	0.20 max	0.40-0.70	0.20-0.40	0.75-1.35	0.04 max	0.05 max	0.15-0.50	0.01-0.10	Ni <=0.50; bal Fe	460	315	21	
ASTM A588/A588M(97)	Grade B	Struct Shp, t<=100mm for plt/bar	0.20 max	0.40-0.70	0.20-0.40	0.75-1.35	0.04 max	0.05 max	0.15-0.50	0.01-0.10	Ni <=0.50; bal Fe	485	345	21	
ASTM A709(97)	Grade 50W, Type B	Bar, Plt, Shp, struct for bridges, t<=100mm	0.20 max	0.40-0.70	0.20-0.40	0.75-1.25	0.04 max	0.05 max	0.15-0.50	0.01-0.10	Ni <=0.50; bal Fe	485	345	21	
ASTM A852/A852M(97)		Q/T, struct plt; t<=100mm	0.19 max	0.40-0.70	0.20-0.40	0.80-1.35	0.035 max	0.04 max	0.20-0.65	0.02-0.10	Ni <=0.50; bal Fe	620-760	485	19	
High Strength Steel, Low-Alloy (HSLA), A588(C)															
China															
GB 4172(84)	12MnCuCr	Plt HR/Norm 16-40mm Thk	0.08-0.15	0.30-0.65	0.20-0.40	0.60-1.00	0.040 max	0.040 max	0.15-0.35		Ni <=0.65; bal Fe	420	285	24	
USA															
	UNS K11538		0.15 max	0.30-0.50	0.20-0.50	0.80-1.35	0.04 max	0.05 max	0.15-0.3	0.01-0.10	Ni 0.25-0.50; bal Fe				
ASTM A588/A588M(97)	Grade C	t>=125-200mm for plt/bar	0.15 max	0.30-0.50	0.20-0.50	0.80-1.35	0.04 max	0.05 max	0.15-0.40	0.01-0.10	Ni 0.25-0.50; bal Fe	435	290	21	
ASTM A588/A588M(97)	Grade C	Struct Shp, t<=100mm for plt/bar	0.15 max	0.30-0.50	0.20-0.50	0.80-1.35	0.04 max	0.05 max	0.15-0.40	0.01-0.10	Ni 0.25-0.50; bal Fe	485	345	21	
ASTM A588/A588M(97)	Grade C	t>=100-125mm for plt/bar	0.15 max	0.30-0.50	0.20-0.50	0.80-1.35	0.04 max	0.05 max	0.15-0.40	0.01-0.10	Ni 0.25-0.50; bal Fe	460	315	21	
SAE J410c	*	Obs; see J1392	0.15 max	0.3-0.5	0.2-0.5	0.8-1.35	0.04 max	0.05 max	0.15-0.3	0.01-0.1	Ni 0.25-0.5; bal Fe				
High Strength Steel, Low-Alloy (HSLA), A618(II)															
Germany															
DIN 17102(89)	TStE355	see P355NL1	0.18 max	0.3 max	0.2 min	0.9-1.65	0.03 max	0.025 max	0.1-0.5	0.1 max	Mo <=0.08; N <=0.02; Ni <=0.3; Nb <=0.05; bal Fe				
USA															
	UNS K12609		0.26 max		0.18 min	1.30 max	0.05 max	0.063 max	0.33 max	0.01 min	bal Fe				

UNS numbers and US grades are provided as a means of cross referencing chemically similar alloys. Exchangability is only possible after independent examination of specifications. Tensile properties are minimum or typical as specified. UTS and YS as MPa. El as %. See Appendix for list of abbreviations used in Notes. * indicates obsolete material.

Specification	Designation	Notes	C	Cr	Cu	Mn	P	S	Si	V	Other	UTS	YS	El	Hard
High Strength Steel, Low-Alloy (HSLA), A618(II) (Continued from previous page)															
USA															
ASTM A618(97)	II	HF Weld and Smls HSLA Struct Tub, Walls<=19.05 mm	0.22 max		0.20 min	0.85-1.25	0.025 max	0.025 max	0.30 max	0.02 min	bal Fe	485	345	22	
ASTM A618(97)	II	HF Weld and Smls HSLA Struct Tub, Walls 19.05-38.1 mm	0.22 max		0.20 min	0.85-1.25	0.025 max	0.025 max	0.30 max	0.02 min	bal Fe	460	315	22	
ASTM A714(96)	II	Weld, smls pipe	0.22 max		0.20 min	0.85-1.25	0.04 max	0.05 max	0.30 max	0.02 min	bal Fe	485	345	22	
High Strength Steel, Low-Alloy (HSLA), A618(III)															
China															
GB 3077(88)	20MnV	Bar HR Q/T 15mm diam	0.17-0.24		0.30 max	1.30-1.60	0.035 max	0.035 max	0.17-0.37	0.07-0.12	bal Fe	785	590	10	
GB/T 3078(94)	20MnV	Bar CD Q/T 15mm diam	0.17-0.24		0.30 max	1.30-1.60	0.035 max	0.035 max	0.17-0.37	0.07-0.12	bal Fe	785	590	10	
Germany															
DIN EN 10208(96)	L555MB		0.16 max	0.30 max	0.25 max	1.80 max	0.025 max	0.020 max	0.45 max	0.10 max	Al 0.015-0.060; Mo <=0.10; N <=0.012; Ni <=0.30; Nb <=0.06; Ti <=0.06; V+Nb+Ti<=0.15; bal Fe				
DIN EN 10208(96)	WNr 1.8972	Pipe for combustibles	0.21 max	0.30 max	0.25 max	1.60 max	0.25 max	0.020 max	0.45 max	0.15 max	Al 0.015-0.060; Mo <=0.10; N <=0.012; Ni <=0.30; Nb <=0.05; Ti <=0.04; bal Fe				
DIN EN 10208(96)	WNr 1.8978	Pipe for combustibles	0.16 max	0.30 max	0.25 max	1.80 max	0.25 max	0.020 max	0.45 max	0.10 max	Al 0.015-0.060; Mo <=0.10; N <=0.012; Ni <=0.30; Nb <=0.06; Ti <=0.06; bal Fe				
USA															
	UNS K12700		0.27 max			1.40 max	0.05 max	0.06 max	0.35 max	0.01 min	bal Fe				
ASTM A618(97)	III	HF Weld and Smls HSLA Struct Tub	0.23 max			1.35 max	0.025 max	0.025 max	0.30 max	0.02 min	Nb may be used >=0.005; bal Fe	450	345	20	
SAE J410c	*	Obs; see J1392	0.23 max			1.35 max	0.04 max	0.05 max	0.3 max	0.02 min	Nb 0.01-0.05; bal Fe				
High Strength Steel, Low-Alloy (HSLA), A633(C)															
Bulgaria															
BDS 4880(79)	15GF*	Struct	0.12-0.18	0.30 max	0.30 max	0.90-1.20	0.035 max	0.040 max	0.17-0.37	0.05-0.100	Al 0.02-0.120; Ni <=0.30; bal Fe				
China															
GB 1591(88)	16MnNb	Plt HR 16-25mm Thk	0.12-0.20			1.00-1.40	0.045 max	0.045 max	0.20-0.55		Nb 0.015-0.050; bal Fe	510-660	375	19	
GB/T 1591(94)	Q390	Plt HR 16-35mm Thk	0.20 max			1.00-1.60	0.045 max	0.045 max	0.55 max	0.02-0.15	Nb 0.015-0.060; Ti 0.02-0.2; bal Fe	490-650	370	19	
GB/T 8164(93)	16MnNb	Strp HR 4-8mm Thk	0.12-0.20			1.00-1.40	0.045 max	0.045 max	0.20-0.55		Nb 0.015-0.050; bal Fe	510	375	19	
Europe															
EN 10028/3(92)	1.0562	High-temp, Press; 100<t<=150mm	0.2 max	0.30 max	0.30 max	0.90-1.70	0.030 max	0.025 max	0.50 max	0.10 max	Al >=0.020; Mo <=0.08; N <=0.020; Ni <=0.50; Nb <=0.05; Ti <=0.03; Nb+Ti+V<=0.12; Cr+Cu+Mo<=0.45; bal Fe	450-590	295	21	
EN 10028/3(92)	1.0562	High-temp, Press; 0<=t<=35mm	0.2 max	0.30 max	0.30 max	0.90-1.70	0.030 max	0.025 max	0.50 max	0.10 max	Al >=0.020; Mo <=0.08; N <=0.020; Ni <=0.50; Nb <=0.05; Ti <=0.03; Nb+Ti+V<=0.12; Cr+Cu+Mo<=0.45; bal Fe	490-630	355	22	

UNS numbers and US grades are provided as a means of cross referencing chemically similar alloys. Exchangability is only possible after independent examination of specifications. Tensile properties are minimum or typical as specified. UTS and YS as MPa. El as %. See Appendix for list of abbreviations used in Notes. * indicates obsolete material.

High Strength Steel, Low-Alloy (HSLA), A633(C) (Continued from previous page)

Europe

Specification	Designation	Notes	C	Cr	Cu	Mn	P	S	Si	V	Other	UTS	YS	El	Hard
EN 10028/3(92)	1.0562	High-temp, Press; 35<t<=50mm	0.2 max	0.30 max	0.30 max	0.90-1.70	0.030 max	0.025 max	0.50 max	0.10 max	Al >=0.020; Mo <=0.08; N <=0.020; Ni <=0.50; Nb <=0.05; Ti <=0.03; Nb+Ti+V<=0.12; Cr+Cu+Mo<=0.45; bal Fe	490-630	345	22	
EN 10028/3(92)	1.0562	High-temp, Press; 50<t<=70mm	0.2 max	0.30 max	0.30 max	0.90-1.70	0.030 max	0.025 max	0.50 max	0.10 max	Al >=0.020; Mo <=0.08; N <=0.020; Ni <=0.50; Nb <=0.05; Ti <=0.03; Nb+Ti+V<=0.12; Cr+Cu+Mo<=0.45; bal Fe	490-630	325	22	
EN 10028/3(92)	1.0562	High-temp, Press; 70<t<=100mm	0.2 max	0.30 max	0.30 max	0.90-1.70	0.030 max	0.025 max	0.50 max	0.10 max	Al >=0.020; Mo <=0.08; N <=0.020; Ni <=0.50; Nb <=0.05; Ti <=0.03; Nb+Ti+V<=0.12; Cr+Cu+Mo<=0.45; bal Fe	470-610	315	21	
EN 10028/3(92)	1.0565	High-temp, Press; 100<t<=150mm	0.2 max	0.30 max	0.30 max	0.90-1.70	0.030 max	0.025 max	0.50 max	0.10 max	Al >=0.020; Mo <=0.08; N <=0.020; Ni <=0.50; Nb <=0.05; Ti <=0.03; Nb+Ti+V<=0.12; Cr+Cu+Mo<=0.45; bal Fe	450-590	295	21	
EN 10028/3(92)	1.0565	High-temp, Press; 35<t<=50mm	0.2 max	0.30 max	0.30 max	0.90-1.70	0.030 max	0.025 max	0.50 max	0.10 max	Al >=0.020; Mo <=0.08; N <=0.020; Ni <=0.50; Nb <=0.05; Ti <=0.03; Nb+Ti+V<=0.12; Cr+Cu+Mo<=0.45; bal Fe	490-630	345	22	
EN 10028/3(92)	1.0565	High-temp, Press; 50<t<=70mm	0.2 max	0.30 max	0.30 max	0.90-1.70	0.030 max	0.025 max	0.50 max	0.10 max	Al >=0.020; Mo <=0.08; N <=0.020; Ni <=0.50; Nb <=0.05; Ti <=0.03; Nb+Ti+V<=0.12; Cr+Cu+Mo<=0.45; bal Fe	490-630	325	22	
EN 10028/3(92)	1.0565	High-temp, Press; 0<=t<=35mm	0.2 max	0.30 max	0.30 max	0.90-1.70	0.030 max	0.025 max	0.50 max	0.10 max	Al >=0.020; Mo <=0.08; N <=0.020; Ni <=0.50; Nb <=0.05; Ti <=0.03; Nb+Ti+V<=0.12; Cr+Cu+Mo<=0.45; bal Fe	490-630	355	22	
EN 10028/3(92)	1.0565	High-temp, Press; 70<t<=100mm	0.2 max	0.30 max	0.30 max	0.90-1.70	0.030 max	0.025 max	0.50 max	0.10 max	Al >=0.020; Mo <=0.08; N <=0.020; Ni <=0.50; Nb <=0.05; Ti <=0.03; Nb+Ti+V<=0.12; Cr+Cu+Mo<=0.45; bal Fe	470-610	315	21	
EN 10028/3(92)	P355N	High-temp, Press; 0<=t<=35mm	0.2 max	0.30 max	0.30 max	0.90-1.70	0.030 max	0.025 max	0.50 max	0.10 max	Al >=0.020; Mo <=0.08; N <=0.020; Ni <=0.50; Nb <=0.05; Ti <=0.03; Nb+Ti+V<=0.12; Cr+Cu+Mo<=0.45; bal Fe	490-630	355	22	
EN 10028/3(92)	P355N	High-temp, Press; 35<t<=50mm	0.2 max	0.30 max	0.30 max	0.90-1.70	0.030 max	0.025 max	0.50 max	0.10 max	Al >=0.020; Mo <=0.08; N <=0.020; Ni <=0.50; Nb <=0.05; Ti <=0.03; Nb+Ti+V<=0.12; Cr+Cu+Mo<=0.45; bal Fe	490-630	345	22	

High Strength Steel, Low-Alloy (HSLA), A633(C) (Continued from previous page)

Specification	Designation	Notes	C	Cr	Cu	Mn	P	S	Si	V	Other	UTS	YS	El	Hard
Europe															
EN 10028/3(92)	P355N	High-temp, Press; 50<t<=70mm	0.2 max	0.30 max	0.30 max	0.90-1.70	0.030 max	0.025 max	0.50 max	0.10 max	Al >=0.020; Mo <=0.08; N <=0.020; Ni <=0.50; Nb <=0.05; Ti <=0.03; Nb+Ti+V<=0.12; Cr+Cu+Mo<=0.45; bal Fe	490-630	325	22	
EN 10028/3(92)	P355N	High-temp, Press; 100<t<=150mm	0.2 max	0.30 max	0.30 max	0.90-1.70	0.030 max	0.025 max	0.50 max	0.10 max	Al >=0.020; Mo <=0.08; N <=0.020; Ni <=0.50; Nb <=0.05; Ti <=0.03; Nb+Ti+V<=0.12; Cr+Cu+Mo<=0.45; bal Fe	450-590	295	21	
EN 10028/3(92)	P355N	High-temp, Press; 70<t<=100mm	0.2 max	0.30 max	0.30 max	0.90-1.70	0.030 max	0.025 max	0.50 max	0.10 max	Al >=0.020; Mo <=0.08; N <=0.020; Ni <=0.50; Nb <=0.05; Ti <=0.03; Nb+Ti+V<=0.12; Cr+Cu+Mo<=0.45; bal Fe	470-610	315	21	
EN 10028/3(92)	P355NH	High-temp, Press; 35<t<=50mm	0.2 max	0.30 max	0.30 max	0.90-1.70	0.030 max	0.025 max	0.50 max	0.10 max	Al >=0.020; Mo <=0.08; N <=0.020; Ni <=0.50; Nb <=0.05; Ti <=0.03; Nb+Ti+V<=0.12; Cr+Cu+Mo<=0.45; bal Fe	490-630	345	22	
EN 10028/3(92)	P355NH	High-temp, Press; 100<t<=150mm	0.2 max	0.30 max	0.30 max	0.90-1.70	0.030 max	0.025 max	0.50 max	0.10 max	Al >=0.020; Mo <=0.08; N <=0.020; Ni <=0.50; Nb <=0.05; Ti <=0.03; Nb+Ti+V<=0.12; Cr+Cu+Mo<=0.45; bal Fe	450-590	295	21	
EN 10028/3(92)	P355NH	High-temp, Press; 0<=t<=35mm	0.2 max	0.30 max	0.30 max	0.90-1.70	0.030 max	0.025 max	0.50 max	0.10 max	Al >=0.020; Mo <=0.08; N <=0.020; Ni <=0.50; Nb <=0.05; Ti <=0.03; Nb+Ti+V<=0.12; Cr+Cu+Mo<=0.45; bal Fe	490-630	355	22	
EN 10028/3(92)	P355NH	High-temp, Press; 70<t<=100mm	0.2 max	0.30 max	0.30 max	0.90-1.70	0.030 max	0.025 max	0.50 max	0.10 max	Al >=0.020; Mo <=0.08; N <=0.020; Ni <=0.50; Nb <=0.05; Ti <=0.03; Nb+Ti+V<=0.12; Cr+Cu+Mo<=0.45; bal Fe	470-610	315	21	
EN 10028/3(92)	P355NH	High-temp, Press; 50<t<=70mm *	0.2 max	0.30 max	0.30 max	0.90-1.70	0.030 max	0.025 max	0.50 max	0.10 max	Al >=0.020; Mo <=0.08; N <=0.020; Ni <=0.50; Nb <=0.05; Ti <=0.03; Nb+Ti+V<=0.12; Cr+Cu+Mo<=0.45; bal Fe	490-630	325	22	
EN 10113/2(93)	1.0545	HR weld t<=16mm	0.20 max	0.30 max	0.35 max	0.90-1.65	0.035 max	0.030 max	0.50 max	0.12 max	Al >=0.02; Mo <=0.10; N <=0.015; Ni <=0.50; Nb <=0.05; Ti <=0.03; CEV; bal Fe	470-630	355	22	
EN 10113/2(93)	S355N	HR weld t<=16mm	0.20 max	0.30 max	0.35 max	0.90-1.65	0.035 max	0.030 max	0.50 max	0.12 max	Al >=0.02; Mo <=0.10; N <=0.015; Ni <=0.50; Nb <=0.05; Ti <=0.03; CEV; bal Fe	470-630	355	22	
Finland															
SFS 1150	Fe355P	Fine grain Struct	0.18 max			0.9-1.6	0.03 max	0.03 max	0.15-0.5	0.03-0.12	Al 0.015-0.08; N <=0.012; Nb 0.015-0.06; Ti 0.02-0.2; bal Fe				

High Strength Steel, Low-Alloy (HSLA), A633(C) (Continued from previous page)

Specification	Designation	Notes	C	Cr	Cu	Mn	P	S	Si	V	Other	UTS	YS	El	Hard
Finland															
SFS 1150	Fe390P	Fine grain Struct; boiler, press vessel	0.2 max			1.0-1.8	0.03 max	0.03 max	0.15-0.5	0.03-0.15	Al 0.015-0.08; N <=0.015; Nb 0.015-0.06; Ti 0.02-0.2; bal Fe				
SFS 255(77)	Fe355	Fine grain Struct	0.18 max			0.9-1.6	0.035 max	0.035 max	0.55 max	0.02-0.15	Al 0.015-0.08; N <=0.015; Nb 0.015-0.06; Ti 0.02-0.20; bal Fe				
SFS 256(77)	Fe390	Fine grain Struct	0.2 max			1.0-1.8	0.035 max	0.035 max	0.55 max	0.02-0.2	Al 0.015-0.08; N <=0.02; Nb 0.015-0.06; Ti 0.02-0.2; bal Fe				
France															
AFNOR NFA36201(84)	E355R	>35mm	0.2 max	0.2 max	0.3 max	1.6 max	0.035 max	0.035 max	0.5 max	0.1 max	Mo <=0.1; Ni <=0.2; Nb 0.01-0.06; Ti <=0.05; 0-35mm C<=0.18; Cu+6Sn<=0.33; bal Fe				
Germany															
DIN EN 10025(94)	WNr 1.0579	HR 150-250mm	0.20 max			1.60 max	0.035 max	0.035 max	0.55 max		Cu<=0.22 if thk>30mm; bal Fe	450-630	285	17	
DIN EN 10025(94)	WNr 1.0579	HR 100-150mm	0.20 max			1.60 max	0.035 max	0.035 max	0.55 max		Cu<=0.22 if thk>30mm; bal Fe	470-630	295	18	
Hungary															
MSZ 6280(82)	52D	weld const; 50.1-60mm; rolled; norm	0.18 max	0.25 max		1.5 max	0.035 max	0.035 max	0.15-0.5	0.02-0.1	Al 0.015-0.120; Mo <=0.1; Ni <=0.3; Nb 0.01-0.06; Ti 0.02-0.06; Zr=0.015-0.06; Al+Nb+V+Zr<=0.15; N<=Al/2+V/4+Nb/7+Ti/3.5<=0.015; bal Fe	490-610	335	23L	
MSZ 6280(82)	52D	weld const; 3-35mm; rolled/norm	0.18 max	0.25 max		1.5 max	0.035 max	0.035 max	0.15-0.5	0.02-0.1	Al 0.015-0.120; Mo <=0.1; Ni <=0.3; Nb 0.01-0.06; Ti 0.02-0.06; Zr=0.015-0.06; Al+Nb+V+Zr<=0.15; N<=Al/2+V/4+Nb/7+Ti/3.5<=0.015; bal Fe	490-610	355	23L	
Italy															
UNI 7382(75)	FeE355KG	Fine grain Struct	0.2 max	0.35 max	0.35 max	0.7-1.6	0.035 max	0.035 max	0.5 max	0.02-0.2	Al <=0.015; Mo <=0.1; Ni <=0.3; Nb 0.015-0.06; bal Fe				
Poland															
PNH84018(86)	18G2A	HS weld const	0.2 max	0.3 max		1-1.5	0.04 max	0.04 max	0.2-0.55		Al 0.02-0.120; N <=0.009; Ni <=0.3; CEV>=0.045; bal Fe				
Russia															
GOST 19281	15GF	Weld constr	0.12-0.18		0.3 max	0.3 max	0.9-1.2	0.035 max	0.04 max	0.17-0.37	0.05-0.12	Al <=0.1; Mo <=0.15; N <=0.012; Ni <=0.3; Ti <=0.05; As<=0.08; bal Fe				
GOST 5058(65)	15GF	struct, constr; Weld	0.12-0.18		0.3 max	0.3 max	0.9-1.2	0.035 max	0.04 max	0.17-0.37	0.05-0.1	Al <=0.1; Mo <=0.15; Ni <=0.3; Ti 0.01-0.03; As<=0.08; bal Fe				
Spain															
UNE 36080(85)	AE355-DD	Gen Struct	0.22 max			1.6 max	0.035 max	0.035 max	0.55 max		Mo <=0.15; bal Fe				
UNE 36080(85)	F.6220*	see AE355-DD	0.22 max			1.6 max	0.035 max	0.035 max	0.55 max		Mo <=0.15; bal Fe				
UNE 36081(76)	AE355KG	Fine grain Struct; 0-35mm	0.20 max	0.25 max		0.9-1.6	0.035 max	0.035 max	0.5 max		Mo <=0.1; Ni <=0.3; >35mm C<=0.18; +Al+V+Nb; bal Fe				
UNE 36081(76)	F.6410*	see AE355KG; 0-35mm	0.20 max	0.25 max		0.9-1.6	0.035 max	0.035 max	0.5 max		Mo <=0.1; Ni <=0.3; >35mm C<=0.18; +Al+V+Nb; bal Fe				

Specification	Designation	Notes	C	Cr	Cu	Mn	P	S	Si	V	Other	UTS	YS	El	Hard

High Strength Steel, Low-Alloy (HSLA), A633(C) (Continued from previous page)

Sweden

Specification	Designation	Notes	C	Cr	Cu	Mn	P	S	Si	V	Other	UTS	YS	El	Hard
SS 141430	1430	High-temp Const	0.18 max	0.25 max		0.6-1.4	0.035 max	0.03 max	0.25-0.4	0.03 max	N <=0.012; Nb <=0.01; Ti <=0.03; bal Fe				
SS 141431	1431	High-temp Const	0.18 max	0.25 max		0.6-1.4	0.035 max	0.03 max	0.25-0.4	0.03 max	Mo <=0.1; N <=0.012; Ni <=0.2; Nb <=0.01; Ti <=0.03; bal Fe				
SS 141432	1432	High-temp Const	0.16 max	0.25 max		0.6-1.4	0.03 max	0.03 max	0.25-0.4	0.03 max	Mo <=0.1; N <=0.012; Ni <=0.2; Nb <=0.01; Ti <=0.03; bal Fe				
SS 141434	1434	High-temp Const	0.22 max	0.25 max		0.6-1	0.035 max	0.03 max	0.25-0.4	0.03 max	N <=0.012; Nb <=0.01; Ti <=0.03; bal Fe				
SS 141435	1435	High-temp Const	0.22 max	0.25 max		0.6-1	0.035 max	0.03 max	0.25-0.4	0.03 max	N <=0.012; Nb <=0.01; Ti <=0.03; bal Fe				
SS 142103	2103		0.16 max	0.25 max		0.9-1.6	0.035 max	0.03 max	0.25-0.5	0.03 max	Mo <=0.1; N <=0.012; Ni <=0.2; Nb <=0.01; Ti <=0.03; C>=0.15 Mn=0.9-1.7; C>=0.14 Mn=0.9-1.8; bal Fe				
SS 142106	2106	Fine grain Struct	0.2 max			1.6 max	0.035 max	0.035 max	0.5 max		Al 0.015-0.06; N <=0.015; Nb <=0.05; Ti <=0.2; Nb+V<=0.1; Almet=0.015-0.06				
SS 142107	2107		0.2 max			1.6 max	0.035 max	0.035 max	0.5 max		Al 0.015-0.06; N <=0.015; Nb <=0.05; Ti <=0.2; Nb+V<=0.1; Almet=0.015-0.06				
SS 142108	2108		0.2 max	0.1 max	0.2 max	0.9-1.6	0.04 max	0.04 max	0.5 max		N<=0.009/N<=0.012				
SS 142116	2116	Fine grain Struct	0.2 max			1.8 max	0.035 max	0.035 max	0.5 max		Al 0.015-0.06; N <=0.015; Nb <=0.05; Ti <=0.2; Nb+V<=0.1; Almet=0.015-0.06; C+Mn/6+(Ni+Cu)/15+ (Cr+Mo+V)/5<=0.45				
SS 142117	2117	Fine grain Struct	0.2 max			1.8 max	0.035 max	0.035 max	0.5 max		Al 0.015-0.06; N <=0.015; Nb <=0.05; Ti <=0.2; Nb+V<=0.1; Almet=0.015-0.06; C+Mn/6+(Ni+Cu)/15+ (Cr+Mo+V)/5<=0.45				
SS 142134	2134	Fine grain Struct	0.2 max			1.6 max	0.035 max	0.035 max	0.5 max	0.15 max	N <=0.02; Nb <=0.05; C+Mn/6+(Ni+Cu)/15+ (Cr+Mo+V)/5>=0.41; bal Fe				

USA

Specification	Designation	Notes	C	Cr	Cu	Mn	P	S	Si	V	Other	UTS	YS	El	Hard
	UNS K12000		0.20 max			1.15-1.50	0.04 max	0.05 max	0.15-0.50		Nb 0.01-0.05; bal Fe				
ASTM A633/A633M(95)	Class C	Norm struct Plt, t<=65mm	0.20 max			1.15-1.50	0.035 max	0.04 max	0.15-0.50		Nb 0.01-0.05; bal Fe	485-620	345	23	
ASTM A633/A633M(95)	Class C	Norm struct Plt, t>=65-100mm	0.20 max			1.15-1.50	0.035 max	0.04 max	0.15-0.50		Nb 0.01-0.05; bal Fe	450-590	315	23	

High Strength Steel, Low-Alloy (HSLA), A633(D)

USA

Specification	Designation	Notes	C	Cr	Cu	Mn	P	S	Si	V	Other	UTS	YS	El	Hard
	UNS K02003*	Obs; see K12037													
	UNS K12037		0.20 max	0.25 max	0.35 max	0.70-1.35	0.04 max	0.05 max	0.15-0.50		Mo <=0.08; Ni <=0.25; bal Fe				
ASTM A633/A633M(95)	Class D	Norm struct Plt, t<=40mm	0.20 max	0.25 max	0.35 max	0.70-1.35	0.035 max	0.04 max	0.15-0.50		Mo <=0.08; Ni <=0.25; bal Fe	485-620	345	23	

UNS numbers and US grades are provided as a means of cross referencing chemically similar alloys. Exchangability is only possible after independent examination of specifications. Tensile properties are minimum or typical as specified. UTS and YS as MPa. El as %. See Appendix for list of abbreviations used in Notes. * indicates obsolete material.

Specification	Designation	Notes	C	Cr	Cu	Mn	P	S	Si	V	Other	UTS	YS	El	Hard

High Strength Steel, Low-Alloy (HSLA), A633(D) (Continued from previous page)

USA

Specification	Designation	Notes	C	Cr	Cu	Mn	P	S	Si	V	Other	UTS	YS	El	Hard
ASTM A633/A633M(95)	Class D	Norm struct Plt, t>=65-100mm	0.20 max	0.25 max		1.00-1.60	0.035 max	0.04 max			Mo <=0.08; Ni <=0.25; bal Fe	450-590	315	23	
ASTM A633/A633M(95)	Class D	Norm struct Plt, t>=40-65mm	0.20 max	0.25 max	0.35 max	1.00-1.60	0.035 max	0.04 max	0.15-0.50		Mo <=0.08; Ni <=0.25; bal Fe	485-620	345	23	
ASTM A724(96)	Grade C	Q/T, Press ves plt, weld	0.22 max	0.25 max	0.35 max	1.10-1.60	0.035 max	0.035 max	0.20-0.60	0.08 max	B <=0.005; Mo <=0.08; Ni <=0.25; bal Fe	620-760	485	19	

High Strength Steel, Low-Alloy (HSLA), A656(2)

China

Specification	Designation	Notes	C	Cr	Cu	Mn	P	S	Si	V	Other	UTS	YS	El	Hard
GB 1591(88)	15MnTi	Plt HR 25-40mm Thk	0.12-0.18			1.20-1.60	0.045 max	0.045 max	0.20-0.55		Ti 0.12-0.20; bal Fe	510-660	375	20	
GB/T 1591(94)	Q390	Plt HR 16-35mm Thk	0.20 max			1.00-1.60	0.045 max	0.045 max	0.55 max	0.02-0.15	Nb 0.015-0.060; Ti 0.02-0.2; bal Fe	490-650	370	19	
GB/T 8164(93)	15MnTi	Strp HR 4-8mm Thk	0.12-0.18			1.20-1.60	0.045 max	0.045 max	0.20-0.55		Ti 0.12-0.20; bal Fe	490	370	19	

USA

Specification	Designation	Notes	C	Cr	Cu	Mn	P	S	Si	V	Other	UTS	YS	El	Hard
	UNS K11509		0.18 max			1.65 max	0.025 max	0.035 max	0.30 max		Ti 0.05-0.40; bal Fe				
ASTM A715	70 3*	Obs; see A715(98); Sh Strp, HR, CR	0.15			1.65	0.03	0.04	0.6	0.08	N 0.02; Nb 0.005; bal Fe				

High Strength Steel, Low-Alloy (HSLA), A690

USA

Specification	Designation	Notes	C	Cr	Cu	Mn	P	S	Si	V	Other	UTS	YS	El	Hard
	UNS K12249		0.22 max		0.50 min	0.60-0.90	0.08-0.15	0.05 max	0.10 max		Ni 0.40-0.75; bal Fe				
ASTM A690/A690M(94)		H piles/sheet piling for marine environments	0.22 max		0.50 min	0.60-0.90	0.08-0.15	0.04 max	0.40 max		Ni 0.40-0.75; bal Fe	485	345	21	

High Strength Steel, Low-Alloy (HSLA), A709(C)

USA

Specification	Designation	Notes	C	Cr	Cu	Mn	P	S	Si	V	Other	UTS	YS	El	Hard
ASTM A709(97)	Grade 50W, Type C	Bar, Plt, Shp, struct for bridges, t<=100mm	0.15 max	0.30-0.50	0.20-0.50	0.80-1.35	0.04 max	0.05 max	0.15-0.40	0.01-0.10	Ni 0.25-0.50; bal Fe	485	345	21	

High Strength Steel, Low-Alloy (HSLA), A714(III)

Mexico

Specification	Designation	Notes	C	Cr	Cu	Mn	P	S	Si	V	Other	UTS	YS	El	Hard
NMX-B-069(86)	Grade III	Pipe, weld and Smls	0.23 max		0.20 min	1.35 max	0.04 max	0.05 max	0.30 max	0.02 min	bal Fe	450	345	20	

USA

Specification	Designation	Notes	C	Cr	Cu	Mn	P	S	Si	V	Other	UTS	YS	El	Hard
	UNS K12709		0.27 max		0.18 min	1.40 max	0.05 max	0.06 max	0.35 max	0.01 min	bal Fe				
ASTM A714(96)	III	Weld, smls pipe	0.23 max		0.20 min	1.35 max	0.04 max	0.05 max	0.30 max	0.02 min	bal Fe	450	345	20	

High Strength Steel, Low-Alloy (HSLA), A715(1)

China

Specification	Designation	Notes	C	Cr	Cu	Mn	P	S	Si	V	Other	UTS	YS	El	Hard
GB 1591(88)	15MnTi	Plt HR 25-40mm Thk	0.12-0.18			1.20-1.60	0.045 max	0.045 max	0.20-0.55		Ti 0.12-0.20; bal Fe	510-660	375	20	
GB/T 1591(94)	Q390	Plt HR 16-35mm Thk	0.20 max			1.00-1.60	0.045 max	0.045 max	0.55 max	0.02-0.15	Nb 0.015-0.060; Ti 0.02-0.2; bal Fe	490-650	370	19	
GB/T 8164(93)	15MnTi	Strp HR 4-8mm Thk	0.12-0.18			1.20-1.60	0.045 max	0.045 max	0.20-0.55		Ti 0.12-0.20; bal Fe	490	370	19	

Europe

Specification	Designation	Notes	C	Cr	Cu	Mn	P	S	Si	V	Other	UTS	YS	El	Hard
EN 10149/3(95)	1.0971	HR 1.5<=t<3mm	0.16 max			1.20 max	0.025 max	0.020 max	0.50 max	0.10 max	Al >=0.015; Nb <=0.09; Ti <=0.15; Nb+Ti+V<=0.22; bal Fe	370-490	260	24	
EN 10149/3(95)	1.0971	HR 3<=t<20mm	0.16 max			1.20 max	0.025 max	0.020 max	0.50 max	0.10 max	Al >=0.015; Nb <=0.09; Ti <=0.15; Nb+Ti+V<=0.22; bal Fe	370-490	260	30	
EN 10149/3(95)	1.0973	HR 3<=t<20mm	0.16 max			1.40 max	0.025 max	0.020 max	0.50 max	0.10 max	Al >=0.015; Nb <=0.09; Ti <=0.15; Nb+Ti+V<=0.22; bal Fe	430-550	315	27	

Specification	Designation	Notes	C	Cr	Cu	Mn	P	S	Si	V	Other	UTS	YS	El	Hard

High Strength Steel, Low-Alloy (HSLA), A715(1) (Continued from previous page)

Europe

Specification	Designation	Notes	C	Cr	Cu	Mn	P	S	Si	V	Other	UTS	YS	El	Hard
EN 10149/3(95)	1.0973	HR 1.5<=t<3mm	0.16 max			1.40 max	0.025 max	0.020 max	0.50 max	0.10 max	Al >=0.015; Nb <=0.09; Ti <=0.15; Nb+Ti+V<=0.22; bal Fe	430-550	315	22	
EN 10149/3(95)	1.0977	HR 3<=t<20mm	0.18 max			1.60 max	0.025 max	0.015 max	0.50 max	0.10 max	Al >=0.015; Nb <=0.09; Ti <=0.15; Nb+Ti+V<=0.22; bal Fe	470-610	355	25	
EN 10149/3(95)	1.0977	HR 1.5<=t<3mm	0.18 max			1.60 max	0.025 max	0.015 max	0.50 max	0.10 max	Al >=0.015; Nb <=0.09; Ti <=0.15; Nb+Ti+V<=0.22; bal Fe	470-610	355	20	
EN 10149/3(95)	1.0981	HR 3<=t<20mm	0.2 max			1.60 max	0.025 max	0.015 max	0.50 max	0.10 max	Al >=0.015; Nb <=0.09; Ti <=0.15; Nb+Ti+V<=0.22; bal Fe	530-670	420	23	
EN 10149/3(95)	1.0981	HR 1.5<=t<3mm	0.2 max			1.60 max	0.025 max	0.015 max	0.50 max	0.10 max	Al >=0.015; Nb <=0.09; Ti <=0.15; Nb+Ti+V<=0.22; bal Fe	530-670	420	18	
EN 10149/3(95)	S260NC	HR 1.5<=t<3mm	0.16 max			1.20 max	0.025 max	0.020 max	0.50 max	0.10 max	Al >=0.015; Nb <=0.09; Ti <=0.15; Nb+Ti+V<=0.22; bal Fe	370-490	260	24	
EN 10149/3(95)	S260NC	HR 3<=t<20mm	0.16 max			1.20 max	0.025 max	0.020 max	0.50 max	0.10 max	Al >=0.015; Nb <=0.09; Ti <=0.15; Nb+Ti+V<=0.22; bal Fe	370-490	260	30	
EN 10149/3(95)	S315NC	HR 3<=t<20mm	0.16 max			1.40 max	0.025 max	0.020 max	0.50 max	0.10 max	Al >=0.015; Nb <=0.09; Ti <=0.15; Nb+Ti+V<=0.22; bal Fe	430-550	315	27	
EN 10149/3(95)	S315NC	HR 1.5<=t<3mm	0.16 max			1.40 max	0.025 max	0.020 max	0.50 max	0.10 max	Al >=0.015; Nb <=0.09; Ti <=0.15; Nb+Ti+V<=0.22; bal Fe	430-550	315	22	
EN 10149/3(95)	S355NC	HR 1.5<=t<3mm	0.18 max			1.60 max	0.025 max	0.015 max	0.50 max	0.10 max	Al >=0.015; Nb <=0.09; Ti <=0.15; Nb+Ti+V<=0.22; bal Fe	470-610	355	20	
EN 10149/3(95)	S355NC	HR 3<=t<20mm	0.18 max			1.60 max	0.025 max	0.015 max	0.50 max	0.10 max	Al >=0.015; Nb <=0.09; Ti <=0.15; Nb+Ti+V<=0.22; bal Fe	470-610	355	25	
EN 10149/3(95)	S420NC	HR 3<=t<20mm	0.2 max			1.60 max	0.025 max	0.015 max	0.50 max	0.10 max	Al >=0.015; Nb <=0.09; Ti <=0.15; Nb+Ti+V<=0.22; bal Fe	530-670	420	23	
EN 10149/3(95)	S420NC	HR 1.5<=t<3mm	0.2 max			1.60 max	0.025 max	0.015 max	0.50 max	0.10 max	Al >=0.015; Nb <=0.09; Ti <=0.15; Nb+Ti+V<=0.22; bal Fe	530-670	420	18	

USA

Specification	Designation	Notes	C	Cr	Cu	Mn	P	S	Si	V	Other	UTS	YS	El	Hard
	UNS K11501		0.15 max			1.65 max	0.025 max	0.035 max	0.10 max		Ti >=0.05; bal Fe				
ASTM A715	1*	Obs; see A715(98); Sh Strp, HR, CR													
ASTM A715	70 1*	Obs; see A715(98); Sh Strp, HR, CR	0.15			1.65	0.03	0.04	0.1		Ti 0.05; bal Fe				
ASTM A715	80 1*	Obs; see A715(98); Sh Strp, HR, CR	0.15			1.65	0.03	0.04	0.1		Ti 0.05; bal Fe				
ASTM A715(98)	Grade 50	Sh Strp, HR, CR	0.15 max			1.65 max	0.020 max	0.025 max			bal Fe	415	345	22-24	
ASTM A715(98)	Grade 60	Sh Strp, HR, CR	0.15 max			1.65 max	0.020 max	0.025 max			bal Fe	485	415	18-22	
ASTM A715(98)	Grade 70	Sh Strp, HR, CR	0.15 max			1.65 max	0.020 max	0.025 max			bal Fe	550	485	16-20	
ASTM A715(98)	Grade 80	Sh Strp, HR, CR	0.15 max			1.65 max	0.020 max	0.025 max			bal Fe	620	550	14-18	

UNS numbers and US grades are provided as a means of cross referencing chemically similar alloys. Exchangability is only possible after independent examination of specifications. Tensile properties are minimum or typical as specified. UTS and YS as MPa. El as %. See Appendix for list of abbreviations used in Notes. * indicates obsolete material.

High Strength Steel, Low-Alloy (HSLA), A715(2)

Specification	Designation	Notes	C	Cr	Cu	Mn	P	S	Si	V	Other	UTS	YS	El	Hard
China															
GB 1591(88)	15MnVN	Plt HR 25-38mm Thk	0.12-0.20			1.30-1.70	0.045 max	0.045 max	0.20-0.55	0.10-0.20	N 0.010-0.020; bal Fe	550-700	410	18	
GB 6654(96)	15MnVNR	Plt HR Norm 16-36mm Thk	0.20 max			1.30-1.70	0.035 max	0.030 max	0.20-0.55	0.10-0.20	N 0.010-0.020; bal Fe	550-690	420	18	
GB/T 8164(93)	15MnVN	Strp HR 4-8mm Thk	0.12-0.20			1.30-1.70	0.045 max	0.045 max	0.20-0.55	0.10-0.20	N 0.010-0.020; bal Fe	520	420	18	
USA															
	UNS K11502		0.15 max			1.65 max	0.025 max	0.035 max	0.60 max	0.02 min	N >=0.005; bal Fe				
ASTM A715	70 2*	Obs; see A715(98); Sh Strp, HR, CR	0.15			1.65	0.03	0.04	0.6	0.02	N 0.005; bal Fe				
ASTM A715	80 2*	Obs; see A715(98); Sh Strp, HR, CR	0.15			1.65	0.03	0.04	0.6	0.02	N 0.005; bal Fe				

High Strength Steel, Low-Alloy (HSLA), A715(4)

Specification	Designation	Notes	C	Cr	Cu	Mn	P	S	Si	V	Other	UTS	YS	El	Hard
France															
AFNOR NFA36204(83)	E420TFP-II	50<t<=70mm; N/T	0.18 max	2 max	1.5 max	1.5 max	0.025 max	0.025 max	0.1-0.5	0.2 max	B <=0.05; Mo <=1; N <=0.015; Ni <=2; Nb <=0.06; Ti <=0.2; Zr<=0.12; Cu+6xSn<=0.33; bal Fe	530-680	400	19	
AFNOR NFA36204(83)	E420TFP-II	5<=t<=50mm; N/T	0.18 max	2 max	1.5 max	1.5 max	0.025 max	0.025 max	0.1-0.5	0.2 max	B <=0.05; Mo <=1; N <=0.015; Ni <=2; Nb <=0.06; Ti <=0.2; Zr<=0.12; Cu+6xSn<=0.33; bal Fe	530-680	420	19	
AFNOR NFA36204(83)	E420TR-II	5<=t<=50mm; N/T	0.18 max	2 max	1.5 max	1.5 max	0.03 max	0.03 max	0.1-0.5	0.2 max	B <=0.05; Mo <=1; N <=0.015; Ni <=2; Nb <=0.06; Ti <=0.2; Zr<=0.12; Cu+6xSn<=0.33; bal Fe	530-680	420	19	
AFNOR NFA36204(83)	E420TR-II	50<t<=70mm; N/T	0.18 max	2 max	1.5 max	1.5 max	0.03 max	0.03 max	0.1-0.5	0.2 max	B <=0.05; Mo <=1; N <=0.015; Ni <=2; Nb <=0.06; Ti <=0.2; Zr<=0.12; Cu+6xSn<=0.33; bal Fe	530-680	400	19	
AFNOR NFA36204(83)	E460TFP-II	5<=t<=50mm; N/T	0.18 max	2 max	1.5 max	1.5 max	0.025 max	0.025 max	0.1-0.5	0.2 max	B <=0.05; Mo <=1; N <=0.015; Ni <=2; Nb <=0.06; Ti <=0.2; Zr<=0.12; Cu+6xSn<=0.33; bal Fe	570-720	460	17	
AFNOR NFA36204(83)	E460TFP-II	50<t<=70mm; N/T	0.18 max	2 max	1.5 max	1.5 max	0.025 max	0.025 max	0.1-0.5	0.2 max	B <=0.05; Mo <=1; N <=0.015; Ni <=2; Nb <=0.06; Ti <=0.2; Zr<=0.12; Cu+6xSn<=0.33; bal Fe	570-720	440	17	
AFNOR NFA36204(83)	E460TR-II	5<=t<=50mm; N/T	0.18 max	2 max	1.5 max	1.5 max	0.03 max	0.03 max	0.1-0.5	0.2 max	B <=0.05; Mo <=1; N <=0.015; Ni <=2; Nb <=0.06; Ti <=0.2; Zr<=0.12; Cu+6xSn<=0.33; bal Fe	570-720	460	17	
AFNOR NFA36204(83)	E460TR-II	50<t<=70mm; N/T	0.18 max	2 max	1.5 max	1.5 max	0.03 max	0.03 max	0.1-0.5	0.2 max	B <=0.05; Mo <=1; N <=0.015; Ni <=2; Nb <=0.06; Ti <=0.2; Zr<=0.12; Cu+6xSn<=0.33; bal Fe	570-720	440	17	
AFNOR NFA36204(83)	E500TR-II	t<=50mm; N/T	0.18 max	2 max	1.5 max	1.5 max	0.03 max	0.03 max	0.1-0.5	0.2 max	B <=0.05; Mo <=1; N <=0.015; Ni <=2; Nb <=0.06; Ti <=0.2; Zr<=0.12; Cu+6xSn<=0.33; bal Fe	620-770	500	17	

High Strength Steel, Low-Alloy (HSLA), A715(4) (Continued from previous page)

Specification	Designation	Notes	C	Cr	Cu	Mn	P	S	Si	V	Other	UTS	YS	El	Hard
France															
AFNOR NFA36204(83)	E500TR-II	50<t<=70mm; N/T	0.18 max	2 max	1.5 max	1.5 max	0.03 max	0.03 max	0.1-0.5	0.2 max	B <=0.05; Mo <=1; N <=0.015; Ni <=2; Nb <=0.06; Ti <=0.2; Zr<=0.12; Cu+6xSn <=0.33; bal Fe	620-770	480	17	
AFNOR NFA36204(83)	E550TFP-II	5<=t<=50mm; N/T	0.18 max	2 max	1.5 max	1.5 max	0.025 max	0.025 max	0.1-0.5	0.2 max *	B <=0.05; Mo <=1; N <=0.015; Ni <=2; Nb <=0.06; Ti <=0.2; Zr<=0.12; Cu+6xSn<=0.33; bal Fe	670-820	550	16T	
AFNOR NFA36204(83)	E550TFP-II	50<t<=70mm; N/T	0.18 max	2 max	1.5 max	1.5 max	0.025 max	0.025 max	0.1-0.5	0.2 max	B <=0.05; Mo <=1; N <=0.015; Ni <=2; Nb <=0.06; Ti <=0.2; Zr<=0.12; Cu+6xSn<=0.33; bal Fe	670-820	530	16T	
AFNOR NFA36204(83)	E550TR-II	50<t<=70mm; N/T	0.18 max	2 max	1.5 max	1.5 max	0.03 max	0.03 max	0.1-0.5	0.2 max	B <=0.05; Mo <=1; N <=0.015; Ni <=2; Nb <=0.06; Ti <=0.2; Zr<=0.12; Cu+6xSn<=0.33; bal Fe	670-820	530	16T	
AFNOR NFA36204(83)	E550TR-II	5<=t<=50mm; N/T	0.18 max	2 max	1.5 max	1.5 max	0.03 max	0.03 max	0.1-0.5	0.2 max	B <=0.05; Mo <=1; N <=0.015; Ni <=2; Nb <=0.06; Ti <=0.2; Zr<=0.12; Cu+6xSn<=0.33; bal Fe	670-820	550	16T	
AFNOR NFA36204(83)	E620TFP-II	50<t<=70mm; Norm	0.18 max	2 max	1.5 max	1.6 max	0.025 max	0.025 max	0.1-0.5	0.2 max	B <=0.05; Mo <=1; N <=0.015; Ni <=2; Nb <=0.06; Ti <=0.2; Zr<=0.12; Cu+6xSn<=0.33; bal Fe	740-890	600	15	
AFNOR NFA36204(83)	E620TFP-II	5<=t<=50mm; Norm	0.18 max	2 max	1.5 max	1.6 max	0.025 max	0.025 max	0.1-0.5	0.2 max	B <=0.05; Mo <=1; N <=0.015; Ni <=2; Nb <=0.06; Ti <=0.2; Zr<=0.12; Cu+6xSn<=0.33; bal Fe	740-890	620	15	
AFNOR NFA36204(83)	E620TR-II	50<t<=70mm; Norm	0.18 max	2 max	1.5 max	1.6 max	0.03 max	0.03 max	0.1-0.5	0.2 max	B <=0.05; Mo <=1; N <=0.015; Ni <=2; Nb <=0.06; Ti <=0.2; Zr<=0.12; Cu+6xSn<=0.33; bal Fe	740-890	600	15	
AFNOR NFA36204(83)	E620TR-II	5<=t<=50mm; Norm	0.18 max	2 max	1.5 max	1.6 max	0.03 max	0.03 max	0.1-0.5	0.2 max	B <=0.05; Mo <=1; N <=0.015; Ni <=2; Nb <=0.06; Ti <=0.2; Zr<=0.12; Cu+6xSn<=0.33; bal Fe	740-890	620	15	
AFNOR NFA36204(83)	E690TFP	50<t<=70mm; Norm	0.18 max	2 max	1.5 max	1.6 max	0.025 max	0.025 max	0.1-0.5	0.2 max	B <=0.05; Mo <=1; N <=0.015; Ni <=2; Nb <=0.06; Ti <=0.2; Zr<=0.12; Cu+6xSn<=0.33; bal Fe	790-940	670	14	
AFNOR NFA36204(83)	E690TFP	5<=t<=50mm; Norm	0.18 max	2 max	1.5 max	1.6 max	0.025 max	0.025 max	0.1-0.5	0.2 max	B <=0.05; Mo <=1; N <=0.015; Ni <=2; Nb <=0.06; Ti <=0.2; Zr<=0.12; Cu+6xSn<=0.33; bal Fe	790-940	690	14	
AFNOR NFA36204(83)	E690TR	5<=t<=50mm; Norm	0.18 max	2 max	1.5 max	1.6 max	0.03 max	0.03 max	0.1-0.5	0.2 max	B <=0.05; Mo <=1; N <=0.015; Ni <=2; Nb <=0.06; Ti <=0.2; Zr<=0.12; Cu+6xSn<=0.33; bal Fe	790-940	690	14	

UNS numbers and US grades are provided as a means of cross referencing chemically similar alloys. Exchangability is only possible after independent examination of specifications. Tensile properties are minimum or typical as specified. UTS and YS as MPa. El as %. See Appendix for list of abbreviations used in Notes. * indicates obsolete material.

Specification	Designation	Notes	C	Cr	Cu	Mn	P	S	Si	V	Other	UTS	YS	El	Hard

High Strength Steel, Low-Alloy (HSLA), A715(4) (Continued from previous page)

France

Specification	Designation	Notes	C	Cr	Cu	Mn	P	S	Si	V	Other	UTS	YS	El	Hard
AFNOR NFA36204(83)	E690TR	50<t<=70mm; Norm	0.18 max	2 max	1.5 max	1.6 max	0.03 max	0.03 max	0.1-0.5	0.2 max	B <=0.05; Mo <=1; N <=0.015; Ni <=2; Nb <=0.06; Ti <=0.2; Zr<=0.12; Cu+6xSn<=0.33; bal Fe	790-940	670	14	

USA

Specification	Designation	Notes	C	Cr	Cu	Mn	P	S	Si	V	Other	UTS	YS	El	Hard
	UNS K11504		0.15 max	0.80 max		1.65 max	0.025 max	0.035 max	0.90 max		B <=0.0025; Nb 0.005-0.06; Ti <=0.1; Zr 0.05 min; bal Fe				
ASTM A715	4*	Obs; see A715(98); Sh Strp, HR, CR													
ASTM A715	70 4*	Obs; see A715(98); Sh Strp, HR, CR	0.15	0.8		1.65	0.03	0.04	0.9		B 0.003; Nb 0.005-0.06; Ti 0.1; Zn 0.05; bal Fe				
ASTM A715	80 4*	Obs; see A715(98); Sh Strp, HR, CR	0.15	0.8		1.65	0.03	0.04	0.9		B 0.005; Nb 0.005-0.06; Ti 0.1; Zn 0.05; bal Fe				

High Strength Steel, Low-Alloy (HSLA), A715(5)

USA

Specification	Designation	Notes	C	Cr	Cu	Mn	P	S	Si	V	Other	UTS	YS	El	Hard
	UNS K11505		0.15 max			1.65 max	0.025 max	0.035 max	0.30 max		Mo >=0.20; Nb <=0.03; bal Fe				
ASTM A715	5*	Obs; see A715(98); Sh Strp, HR, CR													
ASTM A715	80 5*	Obs; see A715(98); Sh Strp, HR, CR	0.15			1.65	0.03	0.04	0.3		Mo 0.2; Nb 0.03; bal Fe				

High Strength Steel, Low-Alloy (HSLA), A715(6)

China

Specification	Designation	Notes	C	Cr	Cu	Mn	P	S	Si	V	Other	UTS	YS	El	Hard
GB 1591(88)	14MnNb	Plt HR 16-25mm Thk	0.12-0.18			0.80-1.20	0.045 max	0.045 max	0.20-0.55		Nb 0.015-0.05; bal Fe	470-620	335	20	
GB/T 1591(94)	Q345	Plt HR 16-35mm Thk	0.20 max			1.00-1.60	0.045 max	0.045 max	0.55 max	0.02-0.15	Nb 0.015-0.060; Ti 0.02-0.2; bal Fe	470-630	325	21	
GB/T 8164(93)	14MnNb	Strp HR 4-8mm Thk	0.12-0.18			0.80-1.20	0.045 max	0.045 max	0.20-0.55		Nb 0.015-0.050; bal Fe	470	325	21	

USA

Specification	Designation	Notes	C	Cr	Cu	Mn	P	S	Si	V	Other	UTS	YS	El	Hard
	UNS K11506		0.15 max			1.65 max	0.025 max	0.035 max	0.90 max		Nb 0.005-0.10; bal Fe				
ASTM A715	50 6*	Obs; see A715(98); Sh Strp, HR, CR	0.15			1.65	0.03	0.04	0.9		Nb 0.005-0.1; bal Fe				
ASTM A715	6*	Obs; see A715(98); Sh Strp, HR, CR													
ASTM A715	60 6*	Obs; see A715(98); Sh Strp, HR, CR	0.15			1.65	0.03	0.04	0.9		Nb 0.005-0.1; bal Fe				
ASTM A715	70 6*	Obs; see A715(98); Sh Strp, HR, CR	0.15			1.65	0.03	0.04	0.9		Nb 0.005-0.1; bal Fe				
ASTM A715	80 6*	Obs; see A715(98); Sh Strp, HR, CR	0.15			1.65	0.03	0.04	0.9		Nb 0.005-0.1; bal Fe				

High Strength Steel, Low-Alloy (HSLA), A715(7)

Bulgaria

Specification	Designation	Notes	C	Cr	Cu	Mn	P	S	Si	V	Other	UTS	YS	El	Hard
BDS 6354	10G2	Struct	0.07-0.15	0.30 max	0.30 max	1.20-1.60	0.035 max	0.035 max	0.17-0.37	0.05 max	Mo <=0.15; Ni <=0.30; Ti <=0.03; W <=0.20; bal Fe				

Europe

Specification	Designation	Notes	C	Cr	Cu	Mn	P	S	Si	V	Other	UTS	YS	El	Hard
EN 10113/3(93)	1.8818	HR weld t<=16mm	0.13 max			1.50 max	0.035 max	0.030 max	0.50 max	0.08 max	Al >=0.02; Mo <=0.20; N <=0.050; Ni <=0.30; Nb <=0.05; Ti <=0.05; Cr+Cu+Mo<=0.6; CEV; bal Fe	360-510	275	24	
EN 10113/3(93)	1.8819	HR weld t<=16mm	0.13 max			1.50 max	0.030 max	0.025 max	0.50 max	0.08 max	Al >=0.02; Mo <=0.20; N <=0.015; Ni <=0.30; Nb <=0.05; Ti <=0.05; Cr+Cu+Mo<=0.6; CEV; bal Fe	360-510	275	24	

Specification	Designation	Notes	C	Cr	Cu	Mn	P	S	Si	V	Other	UTS	YS	El	Hard

High Strength Steel, Low-Alloy (HSLA), A715(7) (Continued from previous page)

Europe

Specification	Designation	Notes	C	Cr	Cu	Mn	P	S	Si	V	Other	UTS	YS	El	Hard
EN 10113/3(93)	1.8823	HR weld t<=16mm	0.14 max			1.60 max	0.035 max	0.030 max	0.50 max	0.10 max	Al >=0.02; Mo <=0.20; N <=0.015; Ni <=0.30; Nb <=0.05; Ti <=0.05; Cr+Cu+Mo<=0.6; CEV; bal Fe	450-610	355	22	
EN 10113/3(93)	1.8834	HR weld t<=16mm	0.14 max			1.60 max	0.030 max	0.025 max	0.50 max	0.10 max	Al >=0.02; Mo <=0.20; N <=0.015; Ni <=0.30; Nb <=0.05; Ti <=0.05; Cr+Cu+Mo<=0.6; CEV; bal Fe	450-610	355	22	
EN 10113/3(93)	S275M	HR weld t<=16mm	0.13 max			1.50 max	0.035 max	0.030 max	0.50 max	0.08 max	Al >=0.02; Mo <=0.20; N <=0.050; Ni <=0.30; Nb <=0.05; Ti <=0.05; Cr+Cu+Mo<=0.6; CEV; bal Fe	360-510	275	24	
EN 10113/3(93)	S275ML	HR weld t<=16mm	0.13 max			1.50 max	0.030 max	0.025 max	0.50 max	0.08 max	Al >=0.02; Mo <=0.20; N <=0.015; Ni <=0.30; Nb <=0.05; Ti <=0.05; Cr+Cu+Mo<=0.6; CEV; bal Fe	360-510	275	24	
EN 10113/3(93)	S355M	HR weld t<=16mm	0.14 max			1.60 max	0.035 max	0.030 max	0.50 max	0.10 max	Al >=0.02; Mo <=0.20; N <=0.015; Ni <=0.30; Nb <=0.05; Ti <=0.05; Cr+Cu+Mo<=0.6; CEV; bal Fe	450-610	355	22	
EN 10113/3(93)	S355ML	HR weld t<=16mm	0.14 max			1.60 max	0.030 max	0.025 max	0.50 max	0.10 max	Al >=0.02; Mo <=0.20; N <=0.015; Ni <=0.30; Nb <=0.05; Ti <=0.05; Cr+Cu+Mo<=0.6; CEV; bal Fe	450-610	355	22	

USA

Specification	Designation	Notes	C	Cr	Cu	Mn	P	S	Si	V	Other	UTS	YS	El	Hard
	UNS K11507		0.15 max			1.65 max	0.025 max	0.035 max	0.60 max		N <=0.020; Nb+V>=0.005; bal Fe				
ASTM A715	50 7*	Obs; see A715(98); Sh Strp, HR, CR	0.15			1.65	0.03	0.04	0.6	0.005	N 0.02; Nb 0.005; bal Fe				
ASTM A715	60 7*	Obs; see A715(98); Sh Strp, HR, CR	0.15			1.65	0.03	0.04	0.6	0.005	N 0.02; Nb 0.005; bal Fe				
ASTM A715	7*	Obs; see A715(98); Sh Strp, HR, CR													
ASTM A715	70 7*	Obs; see A715(98); Sh Strp, HR, CR	0.15			1.65	0.03	0.04	0.6	0.005	N 0.02; Nb 0.005; bal Fe				
ASTM A715	80 7*	Obs; see A715(98); Sh Strp, HR, CR	0.15			1.65	0.03	0.04	0.6	0.005	N 0.02; Nb 0.005; bal Fe				

High Strength Steel, Low-Alloy (HSLA), A715(8)

USA

Specification	Designation	Notes	C	Cr	Cu	Mn	P	S	Si	V	Other	UTS	YS	El	Hard
	UNS K11508		0.15 max			1.65 max	0.025 max	0.035 max			Nb 0.005-0.15; bal Fe				
ASTM A715	8*	Obs; see A715(98); Sh Strp, HR, CR													

High Strength Steel, Low-Alloy (HSLA), A724(A)

USA

Specification	Designation	Notes	C	Cr	Cu	Mn	P	S	Si	V	Other	UTS	YS	El	Hard
	UNS K11831		0.18 max	0.25 max	0.35 max	1.00-1.60	0.035 max	0.040 max	0.55 max	0.08 max	Mo <=0.08; Ni <=0.25; bal Fe				
ASTM A724(96)	Grade A	Q/T, Press ves plt, weld	0.18 max	0.25 max	0.35 max	1.00-1.60	0.035 max	0.035 max	0.55 max	0.08 max	Mo <=0.08; Ni <=0.25; bal Fe	620-760	485	19	

High Strength Steel, Low-Alloy (HSLA), A724(B)

USA

Specification	Designation	Notes	C	Cr	Cu	Mn	P	S	Si	V	Other	UTS	YS	El	Hard
	UNS K12031		0.20 max	0.25 max	0.35 max	1.00-1.60	0.035 max		0.50 max	0.08 max	Mo <=0.08; Ni <=0.25; bal Fe				
ASTM A724(96)	Grade B	Q/T, Press ves plt, weld	0.20 max	0.25 max	0.35 max	1.00-1.60	0.035 max	0.035 max	0.50 max	0.08 max	Mo <=0.08; Ni <=0.25; bal Fe	655-795	515	17	

UNS numbers and US grades are provided as a means of cross referencing chemically similar alloys. Exchangability is only possible after independent examination of specifications. Tensile properties are minimum or typical as specified. UTS and YS as MPa. El as %. See Appendix for list of abbreviations used in Notes. * indicates obsolete material.

Specification	Designation	Notes	C	Cr	Cu	Mn	P	S	Si	V	Other	UTS	YS	El	Hard

High Strength Steel, Low-Alloy (HSLA), A734(B)

Europe

Specification	Designation	Notes	C	Cr	Cu	Mn	P	S	Si	V	Other	UTS	YS	El	Hard
EN 10113/3(93)	1.8825	HR weld t<=16mm	0.16 max			1.70 max	0.035 max	0.030 max	0.50 max	0.12 max	Al >=0.02; Mo <=0.20; N <=0.020; Ni <=0.30; Nb <=0.05; Ti <=0.05; Cr+Cu+Mo<=0.6; CEV; bal Fe	500-660	420	19	
EN 10113/3(93)	1.8827	HR weld t<=16mm	0.16 max			1.70 max	0.035 max	0.030 max	0.50 max	0.12 max	Al >=0.02; Mo <=0.20; N <=0.025; Ni <=0.45; Nb <=0.05; Ti <=0.05; Cr+Cu+Mo<=0.6; CEV; bal Fe	530-720	460	17	
EN 10113/3(93)	1.8836	HR weld t<=16mm	0.16 max			1.70 max	0.030 max	0.025 max	0.50 max	0.12 max	Al >=0.02; Mo <=0.20; N <=0.020; Ni <=0.30; Nb <=0.05; Ti <=0.05; Cr+Cu+Mo<=0.6; CEV; bal Fe	500-660	420	19	
EN 10113/3(93)	1.8838	HR weld t<=16mm	0.16 max			1.70 max	0.030 max	0.025 max	0.50 max	0.12 max	Al >=0.02; Mo <=0.20; N <=0.025; Ni <=0.45; Nb <=0.05; Ti <=0.05; Cr+Cu+Mo<=0.6; CEV; bal Fe	530-720	460	17	
EN 10113/3(93)	S420M	HR weld t<=16mm	0.16 max			1.70 max	0.035 max	0.030 max	0.50 max	0.12 max	Al >=0.02; Mo <=0.20; N <=0.020; Ni <=0.30; Nb <=0.05; Ti <=0.05; Cr+Cu+Mo<=0.6; CEV; bal Fe	500-660	420	19	
EN 10113/3(93)	S420ML	HR weld t<=16mm	0.16 max			1.70 max	0.030 max	0.025 max	0.50 max	0.12 max	Al >=0.02; Mo <=0.20; N <=0.020; Ni <=0.30; Nb <=0.05; Ti <=0.05; Cr+Cu+Mo<=0.6; CEV; bal Fe	500-660	420	19	
EN 10113/3(93)	S460M	HR weld t<=16mm	0.16 max			1.70 max	0.035 max	0.030 max	0.50 max	0.12 max	Al >=0.02; Mo <=0.20; N <=0.025; Ni <=0.45; Nb <=0.05; Ti <=0.05; Cr+Cu+Mo<=0.6; CEV; bal Fe	530-720	460	17	
EN 10113/3(93)	S460ML	HR weld t<=16mm	0.16 max			1.70 max	0.030 max	0.025 max	0.50 max	0.12 max	Al >=0.02; Mo <=0.20; N <=0.025; Ni <=0.45; Nb <=0.05; Ti <=0.05; Cr+Cu+Mo<=0.6; CEV; bal Fe	530-720	460	17	

USA

Specification	Designation	Notes	C	Cr	Cu	Mn	P	S	Si	V	Other	UTS	YS	El	Hard
	UNS K11720		0.17 max		0.35 max	1.60 max	0.035 max	0.015 max	0.40 max	0.11 max	Al <=0.06; Nb <=0.050; bal Fe				
ASTM A734(97)	Type B	Press ves plt, Q/T	0.17 max	0.25 max	0.35 max	1.60 max	0.035 max	0.015 max	0.40 max	0.11 max	Al <=0.06; N <=0.030; bal Fe	530-670*	450	20	

High Strength Steel, Low-Alloy (HSLA), AH32

Bulgaria

Specification	Designation	Notes	C	Cr	Cu	Mn	P	S	Si	V	Other	UTS	YS	El	Hard
BDS 9801	09G2B-K	Struct, weld ships	0.12 max	0.20 max	0.35 max	1.30-1.60	0.040 max	0.040 max	0.30-0.50		Al 0.015-0.06; Ni <=0.40; bal Fe				
BDS 9801	10G2-K	Struct, weld ships	0.14 max	0.20 max	0.35 max	1.20-1.60	0.040 max	0.040 max	0.17-0.37		Al 0.015-0.06; Ni <=0.40; bal Fe				

Germany

Specification	Designation	Notes	C	Cr	Cu	Mn	P	S	Si	V	Other	UTS	YS	El	Hard
DIN	GL-A32		0.18 max	0.20 max	0.35 max	0.90-1.60	0.040 max	0.040 max	0.50 max	0.05-0.10	Al >=0.015; Mo <=0.08; Ni <=0.40; Nb 0.02-0.05; Ti <=0.02; Nb V subs for Al; bal Fe				
DIN	GL-D32		0.18 max	0.20 max	0.35 max	0.90-1.60	0.040 max	0.040 max	0.50 max	0.05-0.10	Al >=0.015; Mo <=0.08; Ni <=0.40; Nb 0.02-0.05; Nb V subs for Al; bal Fe				

UNS numbers and US grades are provided as a means of cross referencing chemically similar alloys. Exchangability is only possible after independent examination of specifications. Tensile properties are minimum or typical as specified. UTS and YS as MPa. El as %. See Appendix for list of abbreviations used in Notes. * indicates obsolete material.

Specification	Designation	Notes	C	Cr	Cu	Mn	P	S	Si	V	Other	UTS	YS	El	Hard

High Strength Steel, Low-Alloy (HSLA), AH32 (Continued from previous page)

Germany

Specification	Designation	Notes	C	Cr	Cu	Mn	P	S	Si	V	Other	UTS	YS	El	Hard
DIN	GL-E32		0.18 max	0.20 max	0.35 max	0.90-1.60	0.040 max	0.040 max	0.50 max	0.05-0.10	Al >=0.015; Mo <=0.08; Ni <=0.40; Nb 0.02-0.05; Nb V subs for Al; bal Fe				
DIN	WNr 1.0513		0.18 max	0.20 max	0.35 max	0.90-1.60	0.040 max	0.040 max	0.50 max	0.05-0.10	Al >=0.015; Mo <=0.08; Ni <=0.40; Nb 0.02-0.05; Ti <=0.02; May subst Nb or V for Al; bal Fe				
DIN	WNr 1.0514		0.18 max	0.20 max	0.35 max	0.90-1.60	0.040 max	0.040 max	0.50 max	0.05-0.10	Al >=0.015; Mo <=0.08; Ni <=0.40; Nb 0.02-0.05; May subst Nb or V for Al; bal Fe				
DIN	WNr 1.0515		0.18 max	0.20 max	0.35 max	0.90-1.60	0.040 max	0.040 max	0.50 max	0.05-0.10	Al >=0.015; Mo <=0.08; Ni <=0.40; Nb 0.02-0.05; May subst Nb or V for Al; bal Fe				

Russia

Specification	Designation	Notes	C	Cr	Cu	Mn	P	S	Si	V	Other	UTS	YS	El	Hard
GOST 5521(93)	A32	struct, weld, ships	0.18 max	0.20 max	0.35 max	0.9-1.6	0.035 max	0.035 max	0.15-0.5	0.1 max	Al 0.015-0.06; Mo <=0.08; Ni <=0.40; Ti <=0.05; bal Fe				
GOST 5521(93)	D32	struct, weld, ships	0.18 max	0.20 max	0.35 max	0.9-1.6	0.035 max	0.035 max	0.15-0.5	0.1 max	Al 0.015-0.06; Mo <=0.08; Ni <=0.40; Ti <=0.05; bal Fe				
GOST 5521(93)	E32	struct, weld, ships	0.18 max	0.20 max	0.35 max	0.9-1.6	0.035 max	0.035 max	0.15-0.5	0.1 max	Al 0.015-0.06; Mo <=0.08; Ni <=0.40; Ti <=0.05; bal Fe				

USA

Specification	Designation	Notes	C	Cr	Cu	Mn	P	S	Si	V	Other	UTS	YS	El	Hard
	UNS K11846		0.18 max	0.25 max	0.35 max	0.90-1.60	0.04 max	0.04 max	0.10-0.50		Al <=0.065; Mo <=0.08; Ni <=0.40; bal Fe				
ASTM A131/A131M(94)	AH32	Ship struct	0.18 max	0.25 max	0.35 max	0.90-1.60	0.035 max	0.04 max	0.10-0.50	0.10 max	Mo <=0.08; Ni <=0.40; Nb <=0.05; bal Fe	470-585	315	22	
ASTM A131/A131M(94)	AH36	Ship struct	0.18 max	0.25 max	0.35 max	0.90-1.60	0.035 max	0.04 max	0.10-0.50	0.10 max	Mo <=0.08; Ni <=0.40; Nb <=0.05; bal Fe	490-620	360	22	
ASTM A131/A131M(94)	AH40	Ship struct	0.18 max	0.25 max	0.35 max	0.90-1.60	0.035 max	0.04 max	0.10-0.50	0.10 max	Mo <=0.08; Ni <=0.40; Nb <=0.05; bal Fe	510-650	390	22	
ASTM A131/A131M(94)	DH32	Ship struct	0.18 max	0.25 max	0.35 max	0.90-1.60	0.035 max	0.04 max	0.10-0.50	0.10 max	Mo <=0.08; Ni <=0.40; Nb <=0.05; bal Fe	470-585	315	22	
ASTM A131/A131M(94)	DH36	Ship struct	0.18 max	0.25 max	0.35 max	0.90-1.60	0.035 max	0.04 max	0.10-0.50	0.10 max	Mo <=0.08; Ni <=0.40; Nb <=0.05; bal Fe	490-620	360	22	
ASTM A131/A131M(94)	DH40	Ship struct	0.18 max	0.25 max	0.35 max	0.90-1.60	0.035 max	0.04 max	0.10-0.50	0.10 max	Mo <=0.08; Ni <=0.40; Nb <=0.05; bal Fe	510-650	390	22	
ASTM A131/A131M(94)	EH32	Ship struct	0.18 max	0.25 max	0.35 max	0.90-1.60	0.035 max	0.04 max	0.10-0.50	0.10 max	Mo <=0.08; Ni <=0.40; Nb <=0.05; bal Fe	470-585	315	22	
ASTM A131/A131M(94)	EH36	Ship struct	0.18 max	0.25 max	0.35 max	0.90-1.60	0.035 max	0.04 max	0.10-0.50	0.10 max	Mo <=0.08; Ni <=0.40; Nb <=0.05; bal Fe	490-620	360	22	
ASTM A131/A131M(94)	EH40	Ship struct	0.18 max	0.25 max	0.35 max	0.90-1.60	0.035 max	0.04 max	0.10-0.50	0.10 max	Mo <=0.08; Ni <=0.40; Nb <=0.05; bal Fe	510-650	390	22	

High Strength Steel, Low-Alloy (HSLA), AH36

Germany

Specification	Designation	Notes	C	Cr	Cu	Mn	P	S	Si	V	Other	UTS	YS	El	Hard
DIN	GL-E36		0.18 max	0.20 max	0.35 max	0.90-1.60	0.040 max	0.040 max	0.50 max	0.05-0.10	Al >=0.020; Mo <=0.08; Ni <=0.40; Nb 0.02-0.05; V+Nb+Ti<=0.12; bal Fe				

Specification	Designation	Notes	C	Cr	Cu	Mn	P	S	Si	V	Other	UTS	YS	El	Hard
High Strength Steel, Low-Alloy (HSLA), AH36 (Continued from previous page)															
Germany															
DIN	WNr 1.0589		0.18 max	0.20 max	0.35 max	0.90-1.60	0.040 max	0.040 max	0.50 max	0.05-0.10	Al >=0.020; Mo <=0.08; Ni <=0.40; Nb 0.02-0.05; V+Nb+Ti<=0.12; bal Fe				
Russia															
GOST 5521(93)	E36	struct, weld, ships	0.18 max	0.20 max	0.35 max	0.9-1.6	0.035 max	0.035 max	0.15-0.5	0.05-0.1	Al 0.015-0.06; Mo <=0.08; Ni <=0.40; Nb 0.02-0.05; Ti <=0.05; bal Fe				
USA															
	UNS K11852		0.18 max	0.25 max	0.35 max	0.90-1.60	0.04 max	0.04 max	0.10-0.50	0.10 max	Mo <=0.08; Ni <=0.40; Nb <=0.05; bal Fe				
ASTM A808/A808M(98)		Struct qual, t<=40mm	0.12 max			1.65 max	0.035 max	0.04 max	0.15-0.50	0.10 max	Nb 0.02-0.10; Nb+V<=0.15; bal Fe	450	345	22	
High Strength Steel, Low-Alloy (HSLA), B															
Europe															
EN 10028/3(92)	1.1106	High-temp, Press; 70<t<=100mm	0.18 max	0.30 max	0.30 max	0.90-1.70	0.025 max	0.015 max	0.50 max	0.10 max	Al >=0.020; Mo <=0.08; N <=0.020; Ni <=0.50; Nb <=0.05; Ti <=0.03; Nb+Ti+V<=0.12; Cr+Cu+Mo<=0.45; bal Fe	470-610	315	21	
EN 10028/3(92)	1.1106	High-temp, Press; 100<t<=150mm	0.18 max	0.30 max	0.30 max	0.90-1.70	0.025 max	0.015 max	0.50 max	0.10 max	Al >=0.020; Mo <=0.08; N <=0.020; Ni <=0.50; Nb <=0.05; Ti <=0.03; Nb+Ti+V<=0.12; Cr+Cu+Mo<=0.45; bal Fe	450-590	295	21	
EN 10028/3(92)	1.1106	High-temp, Press; 35<t<=50mm	0.18 max	0.30 max	0.30 max	0.90-1.70	0.025 max	0.015 max	0.50 max	0.10 max	Al >=0.020; Mo <=0.08; N <=0.020; Ni <=0.50; Nb <=0.05; Ti <=0.03; Nb+Ti+V<=0.12; Cr+Cu+Mo<=0.45; bal Fe	490-630	345	22	
EN 10028/3(92)	1.1106	High-temp, Press; 0<=t<=35mm	0.18 max	0.30 max	0.30 max	0.90-1.70	0.025 max	0.015 max	0.50 max	0.10 max	Al >=0.020; Mo <=0.08; N <=0.020; Ni <=0.50; Nb <=0.05; Ti <=0.03; Nb+Ti+V<=0.12; Cr+Cu+Mo<=0.45; bal Fe	490-630	355	22	
EN 10028/3(92)	1.1106	High-temp, Press; 50<t<=70mm	0.18 max	0.30 max	0.30 max	0.90-1.70	0.025 max	0.015 max	0.50 max	0.10 max	Al >=0.020; Mo <=0.08; N <=0.020; Ni <=0.50; Nb <=0.05; Ti <=0.03; Nb+Ti+V<=0.12; Cr+Cu+Mo<=0.45; bal Fe	490-630	325	22	
EN 10028/3(92)	P355NL2	High-temp, Press; 0<=t<=35mm	0.18 max	0.30 max	0.30 max	0.90-1.70	0.025 max	0.015 max	0.50 max	0.10 max	Al >=0.020; Mo <=0.08; N <=0.020; Ni <=0.50; Nb <=0.05; Ti <=0.03; Nb+Ti+V<=0.12; Cr+Cu+Mo<=0.45; bal Fe	490-630	355	22	
EN 10028/3(92)	P355NL2	High-temp, Press; 35<t<=50mm	0.18 max	0.30 max	0.30 max	0.90-1.70	0.025 max	0.015 max	0.50 max	0.10 max	Al >=0.020; Mo <=0.08; N <=0.020; Ni <=0.50; Nb <=0.05; Ti <=0.03; Nb+Ti+V<=0.12; Cr+Cu+Mo<=0.45; bal Fe	490-630	345	22	

High Strength Steel, Low-Alloy (HSLA), B (Continued from previous page)

Specification	Designation	Notes	C	Cr	Cu	Mn	P	S	Si	V	Other	UTS	YS	El	Hard
Europe															
EN 10028/3(92)	P355NL2	High-temp, Press; 50<t<=70mm	0.18 max	0.30 max	0.30 max	0.90-1.70	0.025 max	0.015 max	0.50 max	0.10 max	Al >=0.020; Mo <=0.08; N <=0.020; Ni <=0.50; Nb <=0.05; Ti <=0.03; Nb+Ti+V<=0.12; Cr+Cu+Mo<=0.45; bal Fe	490-630	325	22	
EN 10028/3(92)	P355NL2	High-temp, Press; 100<t<=150mm	0.18 max	0.30 max	0.30 max	0.90-1.70	0.025 max	0.015 max	0.50 max	0.10 max	Al >=0.020; Mo <=0.08; N <=0.020; Ni <=0.50; Nb <=0.05; Ti <=0.03; Nb+Ti+V<=0.12; Cr+Cu+Mo<=0.45; bal Fe	450-590	295	21	
EN 10028/3(92)	P355NL2	High-temp, Press; 70<t<=100mm	0.18 max	0.30 max	0.30 max	0.90-1.70	0.025 max	0.015 max	0.50 max	0.10 max	Al >=0.020; Mo <=0.08; N <=0.020; Ni <=0.50; Nb <=0.05; Ti <=0.03; Nb+Ti+V<=0.12; Cr+Cu+Mo<=0.45; bal Fe	470-610	315	21	
EN 10028/4(94)	1.6212	High-temp, Press; 30<=t<=50mm	0.14 max			0.70-1.50	0.025 max	0.015 max	0.50 max	0.05 max	Al >=0.020; Ni 0.30-0.80; Nb <=0.05; Cr+Cu+Mo<=0.50; bal Fe	420-530	275	24	
EN 10028/4(94)	1.6212	High-temp, Press; 0<=t<=30mm	0.14 max			0.70-1.50	0.025 max	0.015 max	0.50 max	0.05 max	Al >=0.020; Ni 0.30-0.80; Nb <=0.05; Cr+Cu+Mo<=0.50; bal Fe	420-530	285	24	
EN 10028/4(94)	1.6228	High-temp, Press; 0<=t<=30mm	0.18 max			0.80-1.50	0.025 max	0.015 max	0.35 max	0.05 max	Al >=0.020; Ni 1.30-1.70; Cr+Cu+Mo<=0.50; bal Fe	490-640	355	22	
EN 10028/4(94)	1.6228	High-temp, Press; 30<=t<=50mm	0.18 max			0.80-1.50	0.025 max	0.015 max	0.35 max	0.05 max	Al >=0.020; Ni 1.30-1.70; Cr+Cu+Mo<=0.50; bal Fe	490-640	345	22	
EN 10028/4(94)	11MnNi5-3	High-temp, Press; 0<=t<=30mm	0.14 max			0.70-1.50	0.025 max	0.015 max	0.50 max	0.05 max	Al >=0.020; Ni 0.30-0.80; Nb <=0.05; Cr+Cu+Mo<=0.50; bal Fe	420-530	285	24	
EN 10028/4(94)	11MnNi5-3	High-temp, Press; 30<=t<=50mm	0.14 max			0.70-1.50	0.025 max	0.015 max	0.50 max	0.05 max	Al >=0.020; Ni 0.30-0.80; Nb <=0.05; Cr+Cu+Mo<=0.50; bal Fe	420-530	275	24	
EN 10028/4(94)	15NiMn6	High-temp, Press; 30<=t<=50mm	0.18 max			0.80-1.50	0.025 max	0.015 max	0.35 max	0.05 max	Al >=0.020; Ni 1.30-1.70; Cr+Cu+Mo<=0.50; bal Fe	490-640	345	22	
EN 10028/4(94)	15NiMn6	High-temp, Press; 0<=t<=30mm	0.18 max			0.80-1.50	0.025 max	0.015 max	0.35 max	0.05 max	Al >=0.020; Ni 1.30-1.70; Cr+Cu+Mo<=0.50; bal Fe	490-640	355	22	
Spain															
UNE 36087(76/78)	15NiMn6	Sh Plt Strp CR	0.18 max			1.5 max	0.035 max	0.035 max	0.15-0.35		Mo <=0.15; Ni 1.3-1.7; bal Fe				
UNE 36087(76/78)	F.2642*	see 15NiMn6	0.18 max			1.5 max	0.035 max	0.035 max	0.15-0.35		Mo <=0.15; Ni 1.3-1.7; bal Fe				
UK															
BS 1506(90)	681-820	Bolt matl Pres/Corr res Q/T t<=63mm	0.17-0.23	0.90-1.20	0.20 max	0.35-0.75	0.020 max	0.020 max	0.1-0.35	0.60-0.80	Al <=0.08; B 0.001-0.010; Mo 0.90-1.10; Ni <=0.20; Ti 0.07-0.15; As<=0.020; Sn<=0.020; bal Fe	820-1000	660	15	241-302 HB
USA															
	UNS K11630		0.12-0.21	0.40-0.65		0.70-1.00	0.035 max	0.04 max	0.20-0.35	0.03-0.08	B 0.0005-0.005; Mo 0.15-0.25; Ti 0.01-0.03; bal Fe				
	UNS K12001		0.20 max			1.15-1.50	0.035 max	0.030 max	0.15-0.50		Nb <=0.05; bal Fe				

UNS numbers and US grades are provided as a means of cross referencing chemically similar alloys. Exchangability is only possible after independent examination of specifications. Tensile properties are minimum or typical as specified. UTS and YS as MPa. El as %. See Appendix for list of abbreviations used in Notes. * indicates obsolete material.

Specification	Designation	Notes	C	Cr	Cu	Mn	P	S	Si	V	Other	UTS	YS	El	Hard

High Strength Steel, Low-Alloy (HSLA), B (Continued from previous page)

USA

Specification	Designation	Notes	C	Cr	Cu	Mn	P	S	Si	V	Other	UTS	YS	El	Hard
ASTM A514(94)	Grade B	Q/T, Plt, t=32mm, weld	0.12-0.21	0.40-0.65		0.70-1.00	0.035 max	0.035 max	0.35 max	0.03-0.08	B 0.0005-0.005; Mo 0.15-0.25; Ti 0.01-0.03; bal Fe	760-895	690	18	
ASTM A517(93)	Grade B	Q/T, Press ves plt, t<=32mm	0.15-0.21	0.40-0.65		0.70-1.00	0.035 max	0.035 max	0.15-0.35	0.03-0.08	B <=0.0005; Mo 0.15-0.25; Ti 0.01-0.03; bal Fe	795-930	690	16	
ASTM A709(97)	Grade 100, Type B	Q/T, Plt, t<=32mm	0.12-0.21	0.40-0.65		0.70-1.00	0.035 max	0.035 max	0.35 max	0.03-0.08	B 0.0005-0.005; Mo 0.15-0.25; Ti 0.01-0.03; bal Fe	760-895	690	18	
ASTM A737(97)	Grade B	Press ves plt	0.20 max			1.15-1.50	0.035 max	0.030 max	0.15-0.50		Nb <=0.05; bal Fe	485-620	345	23	
ASTM A738(96)	Grade B	HT, Press ves plt, mod/low-temp; t<=40mm	0.20 max	0.30 max	0.35 max	0.90-1.50	0.030 max	0.030 max	0.15-0.55	0.07 max	Mo <=0.20; Ni <=0.60; Nb <=0.04; Nb+V<=0.08	585-705	415	20	
ASTM A738(96)	Grade B	HT, Press ves plt, mod/low-temp; t>40-65mm	0.20 max	0.30 max	0.35 max	0.90-1.50	0.030 max	0.030 max	0.15-0.55	0.07 max	Mo <=0.30; Ni <=0.60; Nb <=0.04; Nb+V<=0.08	585-705	415	20	

High Strength Steel, Low-Alloy (HSLA), J

USA

Specification	Designation	Notes	C	Cr	Cu	Mn	P	S	Si	V	Other	UTS	YS	El	Hard
	UNS K11625		0.12-0.21			0.45-0.70	0.035 max	0.04 max	0.20-0.35		B 0.001-0.005; Mo 0.50-0.65; bal Fe				
ASTM A514(94)	Grade J	Q/T, Plt, t=32mm, weld	0.12-0.21			0.45-0.70	0.035 max	0.035 max	0.20-0.35		B 0.001-0.005; Mo 0.50-0.65; bal Fe	760-895	690	18	
ASTM A517(93)	Grade J	Q/T, Press ves plt, t<=32mm	0.12-0.21			0.45-0.70	0.035 max	0.035 max	0.20-0.35		B 0.001-0.005; Mo 0.50-0.65; bal Fe	795-930	690	16	
ASTM A709(97)	Grade 100, Type J	Q/T, Plt, t<=32mm	0.12-0.21			0.45-0.70	0.035 max	0.035 max	0.20-0.35		B 0.001-0.005; Mo 0.50-0.65; bal Fe	760-895	690	18	

High Strength Steel, Low-Alloy (HSLA), K11803

Bulgaria

Specification	Designation	Notes	C	Cr	Cu	Mn	P	S	Si	V	Other	UTS	YS	El	Hard
BDS 4880	10G2SAF	Struct	0.15 max	0.35 max	0.35 max	1.10-1.70	0.040 max	0.040 max	0.35-0.65	0.025-0.120	Ni <=0.35; Nb 0.005-0.030; bal Fe				

High Strength Steel, Low-Alloy (HSLA), K11872

USA

Specification	Designation	Notes	C	Cr	Cu	Mn	P	S	Si	V	Other	UTS	YS	El	Hard
	UNS K11872		0.15-0.21	0.50-0.90		0.80-1.10	0.035 max	0.04 max	0.50-0.90		B <=0.0025; Mo 0.40-0.60; Zr 0.05-0.15; bal Fe				

High Strength Steel, Low-Alloy (HSLA), K12003

China

Specification	Designation	Notes	C	Cr	Cu	Mn	P	S	Si	V	Other	UTS	YS	El	Hard
GB 1591(88)	15MnV	Plt HR 25-36mm Thk	0.12-0.18			1.20-1.60	0.045 max	0.045 max	0.20-0.55	0.04-0.12	bal Fe	490-640	355	18	
GB 6479(86)	15MnV	Smls Tub HR/CD Norm	0.12-0.18			1.20-1.60	0.040 max	0.040 max	0.20-0.60	0.04-0.12	bal Fe	510-690	350	19	
GB 6654(96)	15MnVR	Plt HR Norm 16-36mm Thk	0.18 max			1.20-1.60	0.035 max	0.030 max	0.20-0.55	0.04-0.12	bal Fe	510-645	370	19	
GB/T 8164(93)	15MnV	Strp HR 4-8mm Thk	0.12-0.18			1.20-1.60	0.045 max	0.045 max	0.20-0.55	0.04-0.12	bal Fe	490	370	19	

Germany

Specification	Designation	Notes	C	Cr	Cu	Mn	P	S	Si	V	Other	UTS	YS	El	Hard
DIN SEW 028(93)	18MnMoV5-2	Press ves	0.20 max			1.00-1.50	0.030 max	0.025 max	0.20-0.50	0.05-0.10	Mo 0.10-0.30; N <=0.020; Al; bal Fe				
DIN SEW 028(93)	WNr 1.8812	Press ves	0.20 max			1.00-1.50	0.030 max	0.025 max	0.20-0.50	0.05-0.10	Mo 0.10-0.30; N <=0.020; Al; bal Fe				

USA

Specification	Designation	Notes	C	Cr	Cu	Mn	P	S	Si	V	Other	UTS	YS	El	Hard
	UNS K12003		0.20 max			1.45 max	0.035 max	0.040 max	0.13-0.32	0.07-0.16	bal Fe				

High Strength Steel, Low-Alloy (HSLA), K12010

USA

Specification	Designation	Notes	C	Cr	Cu	Mn	P	S	Si	V	Other	UTS	YS	El	Hard
	UNS K12010		0.20 max		0.20 min	1.35 max	0.04 max	0.05 max			bal Fe				

High Strength Steel, Low-Alloy (HSLA), LF6

Specification	Designation	Notes	C	Cr	Cu	Mn	P	S	Și	V	Other	UTS	YS	El	Hard
China															
GB 1591(88)	15MnVN	Plt HR 25-38mm Thk	0.12-0.20			1.30-1.70	0.045 max	0.045 max	0.20-0.55	0.10-0.20	N 0.010-0.020; bal Fe	550-700	410	18	
GB 6654(96)	15MnVNR	Plt HR Norm 16-36mm Thk	0.20 max			1.30-1.70	0.035 max	0.030 max	0.20-0.55	0.10-0.20	N 0.010-0.020; bal Fe	550-690	420	18	
GB/T 8164(93)	15MnVN	Strp HR 4-8mm Thk	0.12-0.20			1.30-1.70	0.045 max	0.045 max	0.20-0.55	0.10-0.20	N 0.010-0.020; bal Fe	520	420	18	
Europe															
EN 10028/3(92)	1.8905	High-temp, Press; 100<t<=150mm	0.20 max	0.30 max	0.70 max	1.00-1.70	0.030 max	0.025 max	0.60 max	0.20 max	Al >=0.020; Mo <=0.10; N <=0.025; Ni <=0.80; Nb <=0.05; Ti <=0.03; Nb+Ti+ V<=0.12; bal Fe	520-690	380	16	
EN 10028/3(92)	1.8905	High-temp, Press; 0<=t<=16mm	0.20 max	0.30 max	0.70 max	1.00-1.70	0.030 max	0.025 max	0.60 max	0.20 max	Al >=0.020; Mo <=0.10; N <=0.025; Ni <=0.80; Nb <=0.05; Ti <=0.03; Nb+Ti+ V<=0.12; bal Fe	570-730	460	17	
EN 10028/3(92)	1.8905	High-temp, Press; 16<=t<=35mm	0.20 max	0.30 max	0.70 max	1.00-1.70	0.030 max	0.025 max	0.60 max	0.20 max	Al >=0.020; Mo <=0.10; N <=0.025; Ni <=0.80; Nb <=0.05; Ti <=0.03; Nb+Ti+ V<=0.12; bal Fe	570-720	450	17	
EN 10028/3(92)	1.8905	High-temp, Press; 35<t<=50mm	0.20 max	0.30 max	0.70 max	1.00-1.70	0.030 max	0.025 max	0.60 max	0.20 max	Al >=0.020; Mo <=0.10; N <=0.025; Ni <=0.80; Nb <=0.05; Ti <=0.03; Nb+Ti+ V<=0.12; bal Fe	390-510	440	17	
EN 10028/3(92)	1.8905	High-temp, Press; 50<t<=70mm	0.20 max	0.30 max	0.70 max	1.00-1.70	0.030 max	0.025 max	0.60 max	0.20 max	Al >=0.020; Mo <=0.10; N <=0.025; Ni <=0.80; Nb <=0.05; Ti <=0.03; Nb+Ti+ V<=0.12; bal Fe	390-510	420	17	
EN 10028/3(92)	1.8905	High-temp, Press; 70<t<=100mm	0.20 max	0.30 max	0.70 max	1.00-1.70	0.030 max	0.025 max	0.60 max	0.20 max	Al >=0.020; Mo <=0.10; N <=0.025; Ni <=0.80; Nb <=0.05; Ti <=0.03; Nb+Ti+ V<=0.12; bal Fe	540-710	400	16	
EN 10028/3(92)	1.8915	High-temp, Press; 16<=t<=35mm	0.20 max	0.30 max	0.70 max	1.00-1.70	0.030 max	0.020 max	0.60 max	0.20 max	Al >=0.020; Mo <=0.10; N <=0.025; Ni <=0.80; Nb <=0.05; Ti <=0.03; Nb+Ti+ V<=0.12; bal Fe	570-720	450	17	
EN 10028/3(92)	1.8915	High-temp, Press; 50<t<=70mm	0.20 max	0.30 max	0.70 max	1.00-1.70	0.030 max	0.020 max	0.60 max	0.20 max	Al >=0.020; Mo <=0.10; N <=0.025; Ni <=0.80; Nb <=0.05; Ti <=0.03; Nb+Ti+ V<=0.12; bal Fe	390-510	420	17	
EN 10028/3(92)	1.8915	High-temp, Press; 70<t<=100mm	0.20 max	0.30 max	0.70 max	1.00-1.70	0.030 max	0.020 max	0.60 max	0.20 max	Al >=0.020; Mo <=0.10; N <=0.025; Ni <=0.80; Nb <=0.05; Ti <=0.03; Nb+Ti+ V<=0.12; bal Fe	540-710	400	16	
EN 10028/3(92)	1.8915	High-temp, Press; 35<t<=50mm	0.20 max	0.30 max	0.70 max	1.00-1.70	0.030 max	0.020 max	0.60 max	0.20 max	Al >=0.020; Mo <=0.10; N <=0.025; Ni <=0.80; Nb <=0.05; Ti <=0.03; Nb+Ti+ V<=0.12; bal Fe	390-510	440	17	

Specification	Designation	Notes	C	Cr	Cu	Mn	P	S	Si	V	Other	UTS	YS	El	Hard

High Strength Steel, Low-Alloy (HSLA), LF6 (Continued from previous page)

Europe

Specification	Designation	Notes	C	Cr	Cu	Mn	P	S	Si	V	Other	UTS	YS	El	Hard
EN 10028/3(92)	1.8915	High-temp, Press; 0<=t<=16mm	0.20 max	0.30 max	0.70 max	1.00-1.70	0.030 max	0.020 max	0.60 max	0.20 max	Al >=0.020; Mo <=0.10; N <=0.025; Ni <=0.80; Nb <=0.05; Ti <=0.03; Nb+Ti+ V<=0.12; bal Fe	570-730	460	17	
EN 10028/3(92)	1.8915	High-temp, Press; 100<t<=150mm	0.20 max	0.30 max	0.70 max	1.00-1.70	0.030 max	0.020 max	0.60 max	0.20 max	Al >=0.020; Mo <=0.10; N <=0.025; Ni <=0.80; Nb <=0.05; Ti <=0.03; Nb+Ti+ V<=0.12; bal Fe	520-690	380	16	
EN 10028/3(92)	1.8935	High-temp, Press; 16<=t<=35mm	0.20 max	0.30 max	0.30 max	1.00-1.70	0.030 max	0.020 max	0.60 max	0.20 max	Al >=0.020; Mo <=0.10; N <=0.025; Ni <=0.80; Nb <=0.05; Ti <=0.03; Nb+Ti+ V<=0.12; bal Fe	570-720	450	17	
EN 10028/3(92)	1.8935	High-temp, Press; 35<t<=50mm	0.20 max	0.30 max	0.30 max	1.00-1.70	0.030 max	0.020 max	0.60 max	0.20 max	Al >=0.020; Mo <=0.10; N <=0.025; Ni <=0.80; Nb <=0.05; Ti <=0.03; Nb+Ti+ V<=0.12; bal Fe	390-510	440	17	
EN 10028/3(92)	1.8935	High-temp, Press; 70<t<=100mm	0.20 max	0.30 max	0.30 max	1.00-1.70	0.030 max	0.020 max	0.60 max	0.20 max	Al >=0.020; Mo <=0.10; N <=0.025; Ni <=0.80; Nb <=0.05; Ti <=0.03; Nb+Ti+ V<=0.12; bal Fe	540-710	400	16	
EN 10028/3(92)	1.8935	High-temp, Press; 50<t<=70mm	0.20 max	0.30 max	0.30 max	1.00-1.70	0.030 max	0.020 max	0.60 max	0.20 max	Al >=0.020; Mo <=0.10; N <=0.025; Ni <=0.80; Nb <=0.05; Ti <=0.03; Nb+Ti+ V<=0.12; bal Fe	390-510	420	17	
EN 10028/3(92)	1.8935	High-temp, Press; 0<=t<=16mm	0.20 max	0.30 max	0.30 max	1.00-1.70	0.030 max	0.020 max	0.60 max	0.20 max	Al >=0.020; Mo <=0.10; N <=0.025; Ni <=0.80; Nb <=0.05; Ti <=0.03; Nb+Ti+ V<=0.12; bal Fe	570-730	460	17	
EN 10028/3(92)	1.8935	High-temp, Press; 100<t<=150mm	0.20 max	0.30 max	0.30 max	1.00-1.70	0.030 max	0.020 max	0.60 max	0.20 max	Al >=0.020; Mo <=0.10; N <=0.025; Ni <=0.80; Nb <=0.05; Ti <=0.03; Nb+Ti+ V<=0.12; bal Fe	520-690	380	16	
EN 10028/3(92)	P460N	High-temp, Press; 16<=t<=35mm	0.20 max	0.30 max	0.70 max	1.00-1.70	0.030 max	0.025 max	0.60 max	0.20 max	Al >=0.020; Mo <=0.10; N <=0.025; Ni <=0.80; Nb <=0.05; Ti <=0.03; Nb+Ti+ V<=0.12; bal Fe	570-720	450	17	
EN 10028/3(92)	P460N	High-temp, Press; 0<=t<=16mm	0.20 max	0.30 max	0.70 max	1.00-1.70	0.030 max	0.025 max	0.60 max	0.20 max	Al >=0.020; Mo <=0.10; N <=0.025; Ni <=0.80; Nb <=0.05; Ti <=0.03; Nb+Ti+ V<=0.12; bal Fe	570-730	460	17	
EN 10028/3(92)	P460N	High-temp, Press; 35<t<=50mm	0.20 max	0.30 max	0.70 max	1.00-1.70	0.030 max	0.025 max	0.60 max	0.20 max	Al >=0.020; Mo <=0.10; N <=0.025; Ni <=0.80; Nb <=0.05; Ti <=0.03; Nb+Ti+ V<=0.12; bal Fe	390-510	440	17	
EN 10028/3(92)	P460N	High-temp, Press; 100<t<=150mm	0.20 max	0.30 max	0.70 max	1.00-1.70	0.030 max	0.025 max	0.60 max	0.20 max	Al >=0.020; Mo <=0.10; N <=0.025; Ni <=0.80; Nb <=0.05; Ti <=0.03; Nb+Ti+ V<=0.12; bal Fe	520-690	380	16	

Specification	Designation	Notes	C	Cr	Cu	Mn	P	S	Si	V	Other	UTS	YS	El	Hard

High Strength Steel, Low-Alloy (HSLA), LF6 (Continued from previous page)

Europe

Specification	Designation	Notes	C	Cr	Cu	Mn	P	S	Si	V	Other	UTS	YS	El	Hard
EN 10028/3(92)	P460N	High-temp, Press; 70<t<=100mm	0.20 max	0.30 max	0.70 max	1.00-1.70	0.030 max	0.025 max	0.60 max	0.20 max	Al >=0.020; Mo <=0.10; N <=0.025; Ni <=0.80; Nb <=0.05; Ti <=0.03; Nb+Ti+ V<=0.12; bal Fe	540-710	400	16	
EN 10028/3(92)	P460N	High-temp, Press; 50<t<=70mm	0.20 max	0.30 max	0.70 max	1.00-1.70	0.030 max	0.025 max	0.60 max	0.20 max	Al >=0.020; Mo <=0.10; N <=0.025; Ni <=0.80; Nb <=0.05; Ti <=0.03; Nb+Ti V<=0.12; bal Fe	390-510	420	17	
EN 10028/3(92)	P460NH	High-temp, Press; 35<t<=50mm	0.20 max	0.30 max	0.30 max	1.00-1.70	0.030 max	0.020 max	0.60 max	0.20 max	Al >=0.020; Mo <=0.10; N <=0.025; Ni <=0.80; Nb <=0.05; Ti <=0.03; Nb+Ti V<=0.12; bal Fe	390-510	440	17	
EN 10028/3(92)	P460NH	High-temp, Press; 50<t<=70mm	0.20 max	0.30 max	0.30 max	1.00-1.70	0.030 max	0.020 max	0.60 max	0.20 max	Al >=0.020; Mo <=0.10; N <=0.025; Ni <=0.80; Nb <=0.05; Ti <=0.03; Nb+Ti V<=0.12; bal Fe	390-510	420	17	
EN 10028/3(92)	P460NH	High-temp, Press; 70<t<=100mm	0.20 max	0.30 max	0.30 max	1.00-1.70	0.030 max	0.020 max	0.60 max	0.20 max	Al >=0.020; Mo <=0.10; N <=0.025; Ni <=0.80; Nb <=0.05; Ti <=0.03; Nb+Ti V<=0.12; bal Fe	540-710	400	16	
EN 10028/3(92)	P460NH	High-temp, Press; 0<=t<=16mm	0.20 max	0.30 max	0.30 max	1.00-1.70	0.030 max	0.020 max	0.60 max	0.20 max	Al >=0.020; Mo <=0.10; N <=0.025; Ni <=0.80; Nb <=0.05; Ti <=0.03; Nb+Ti V<=0.12; bal Fe	570-730	460	17	
EN 10028/3(92)	P460NH	High-temp, Press; 16<=t<=35mm	0.20 max	0.30 max	0.30 max	1.00-1.70	0.030 max	0.020 max	0.60 max	0.20 max	Al >=0.020; Mo <=0.10; N <=0.025; Ni <=0.80; Nb <=0.05; Ti <=0.03; Nb+Ti V<=0.12; bal Fe	570-720	450	17	
EN 10028/3(92)	P460NH	High-temp, Press; 100<t<=150mm	0.20 max	0.30 max	0.30 max	1.00-1.70	0.030 max	0.020 max	0.60 max	0.20 max	Al >=0.020; Mo <=0.10; N <=0.025; Ni <=0.80; Nb <=0.05; Ti <=0.03; Nb+Ti V<=0.12; bal Fe	520-690	380	16	
EN 10028/3(92)	P460NL1	High-temp, Press; 0<=t<=16mm	0.20 max	0.30 max	0.70 max	1.00-1.70	0.030 max	0.020 max	0.60 max	0.20 max	Al >=0.020; Mo <=0.10; N <=0.025; Ni <=0.80; Nb <=0.05; Ti <=0.03; Nb+Ti V<=0.12; bal Fe	570-730	460	17	
EN 10028/3(92)	P460NL1	High-temp, Press; 50<t<=70mm	0.20 max	0.30 max	0.70 max	1.00-1.70	0.030 max	0.020 max	0.60 max	0.20 max	Al >=0.020; Mo <=0.10; N <=0.025; Ni <=0.80; Nb <=0.05; Ti <=0.03; Nb+Ti V<=0.12; bal Fe	390-510	420	17	
EN 10028/3(92)	P460NL1	High-temp, Press; 100<t<=150mm	0.20 max	0.30 max	0.70 max	1.00-1.70	0.030 max	0.020 max	0.60 max	0.20 max	Al >=0.020; Mo <=0.10; N <=0.025; Ni <=0.80; Nb <=0.05; Ti <=0.03; Nb+Ti V<=0.12; bal Fe	520-690	380	16	
EN 10028/3(92)	P460NL1	High-temp, Press; 70<t<=100mm	0.20 max	0.30 max	0.70 max	1.00-1.70	0.030 max	0.020 max	0.60 max	0.20 max	Al >=0.020; Mo <=0.10; N <=0.025; Ni <=0.80; Nb <=0.05; Ti <=0.03; Nb+Ti V<=0.12; bal Fe	540-710	400	16	

High Strength Steel, Low-Alloy (HSLA), LF6 (Continued from previous page)

Specification	Designation	Notes	C	Cr	Cu	Mn	P	S	Si	V	Other	UTS	YS	El	Hard
Europe															
EN 10028/3(92)	P460NL1	High-temp, Press; 35<t<=50mm	0.20 max	0.30 max	0.70 max	1.00-1.70	0.030 max	0.020 max	0.60 max	0.20 max	Al >=0.020; Mo <=0.10; N <=0.025; Ni <=0.80; Nb <=0.05; Ti <=0.03; Nb+Ti+ V <=0.12; bal Fe	390-510	440	17	
EN 10028/3(92)	P460NL1	High-temp, Press; 16<=t<=35mm	0.20 max	0.30 max	0.70 max	1.00-1.70	0.030 max	0.020 max	0.60 max	0.20 max	Al >=0.020; Mo <=0.10; N <=0.025; Ni <=0.80; Nb <=0.05; Ti <=0.03; Nb+Ti+ V <=0.12; bal Fe	570-720	450	17	
EN 10113/2(93)	1.8901	HR weld t<=16mm	0.20 max	0.30 max	0.35-0.70	1.00-1.70	0.035 max	0.030 max	0.60 max	0.20 max	Al >=0.02; Mo <=0.10; N <=0.025; Ni <=0.80; Nb <=0.05; Ti <=0.03; Nb+Ti+V<=0.22; Cr+Mo<=0.3; CEV; bal Fe	550-720	460	17	
EN 10113/2(93)	1.8902	HR weld t<=16mm	0.20 max	0.30 max	0.35-0.70	1.00-1.70	0.035 max	0.030 max	0.60 max	0.20 max	Al >=0.02; Mo <=0.10; N <=0.025; Ni <=0.80; Nb <=0.05; Ti <=0.03; CEV; bal Fe	520-680	420	19	
EN 10113/2(93)	1.8903	HR weld t<=16mm	0.20 max	0.30 max	0.35-0.70	1.00-1.70	0.030 max	0.025 max	0.60 max	0.20 max	Al >=0.02; Mo <=0.10; N <=0.025; Ni <=0.80; Nb <=0.05; Ti <=0.03; Nb+Ti+V<=0.22; Cr+Mo<=0.3; CEV; bal Fe	550-720	460	17	
EN 10113/2(93)	S420N	HR weld t<=16mm	0.20 max	0.30 max	0.35-0.70	1.00-1.70	0.035 max	0.030 max	0.60 max	0.20 max	Al >=0.02; Mo <=0.10; N <=0.025; Ni <=0.80; Nb <=0.05; Ti <=0.03; CEV; bal Fe	520-680	420	19	
EN 10113/2(93)	S460N	HR weld t<=16mm	0.20 max	0.30 max	0.35-0.70	1.00-1.70	0.035 max	0.030 max	0.60 max	0.20 max	Al >=0.02; Mo <=0.10; N <=0.025; Ni <=0.80; Nb <=0.05; Ti <=0.03; Nb+Ti+V<=0.22; Cr+Mo<=0.3; CEV; bal Fe	550-720	460	17	
EN 10113/2(93)	S460NL	HR weld t<=16mm	0.20 max	0.30 max	0.35-0.70	1.00-1.70	0.030 max	0.025 max	0.60 max	0.20 max	Al >=0.02; Mo <=0.10; N <=0.025; Ni <=0.80; Nb <=0.05; Ti <=0.03; Nb+Ti+V<=0.22; Cr+Mo<=0.3; CEV; bal Fe	550-720	460	17	
France															
AFNOR NFA36201(84)	E420R-I	>35mm	0.2 max	0.2 max	0.3 max	1.6 max	0.035 max	0.035 max	0.5 max	0.02-0.15	Mo <=0.1; Ni <=0.2; Nb 0.01-0.06; Ti <=0.05; 0-35mm V=0.02-0.12; bal Fe				
AFNOR NFA36207(82)	A550AP-I*	5-16mm; Norm; EN10028/3	0.2 max	0.2 max	0.3 max	1.6 max	0.035 max	0.035 max	0.5 max	0.02-0.12	Mo <=0.1; Ni <=0.2; Nb 0.01-0.06; Ti <=0.05; bal Fe	550-670	400	20	
AFNOR NFA36207(82)	A550AP-I*	50<t<=80mm; Norm; EN10028/3	0.2 max	0.2 max	0.3 max	1.6 max	0.035 max	0.035 max	0.5 max	0.02-0.15	Mo <=0.1; Ni <=0.2; Nb 0.01-0.06; Ti <=0.05; bal Fe	540-660	360	20	
AFNOR NFA36207(82)	A550AP-II*	5-16mm; Norm; EN10028/3	0.22 max	0.25 max	0.35 max	1.6 max	0.035 max	0.035 max	0.55 max	0.1 max	Mo <=0.15; Ni 0.2-0.7; Ti <=0.05; Cr+Ni+Mo+Cu<=0.8; bal Fe	550-670	400	20	
AFNOR NFA36207(82)	A550AP-II*	50<t<=80mm; Norm; EN10028/3	0.22 max	0.25 max	0.35 max	1.6 max	0.035 max	0.035 max	0.55 max	0.1 max	Mo <=0.15; Ni 0.2-0.7; Ti <=0.05; Cr+Ni+Mo+Cu<=0.8; bal Fe	540-660	360	20	
AFNOR NFA36207(82)	A550FP-I*	5-16mm; Norm; EN10028/3	0.2 max	0.2 max	0.3 max	1.6 max	0.03 max	0.02 max	0.5 max	0.02-0.12	Mo <=0.1; Ni <=0.2; Nb 0.01-0.06; Ti <=0.05; bal Fe	550-670	400	20	

UNS numbers and US grades are provided as a means of cross referencing chemically similar alloys. Exchangability is only possible after independent examination of specifications. Tensile properties are minimum or typical as specified. UTS and YS as MPa. El as %. See Appendix for list of abbreviations used in Notes. * indicates obsolete material.

Specification	Designation	Notes	C	Cr	Cu	Mn	P	S	Si	V	Other	UTS	YS	El	Hard

High Strength Steel, Low-Alloy (HSLA), LF6 (Continued from previous page)

France

Specification	Designation	Notes	C	Cr	Cu	Mn	P	S	Si	V	Other	UTS	YS	El	Hard
AFNOR NFA36207(82)	A550FP-I*	50<t<=80mm; Norm; EN10028/3	0.2 max	0.2 max	0.3 max	1.6 max	0.03 max	0.02 max	0.5 max	0.02-0.15	Mo <=0.1; Ni <=0.2; Nb 0.01-0.06; Ti <=0.05; bal Fe	540-660	360	20	
AFNOR NFA36207(82)	A550FP-II*	50<t<=80mm; Norm; EN10028/3	0.22 max	0.25 max	0.35 max	1.6 max	0.03 max	0.02 max	0.55 max	0.1 max	Mo <=0.15; Ni 0.2-0.7; Ti <=0.05; Cr+Ni+Mo+Cu<=0.8; bal Fe	540-660	360	20	
AFNOR NFA36207(82)	A550FP-II*	5-16mm; Norm; EN10028/3	0.22 max	0.25 max	0.35 max	1.6 max	0.03 max	0.02 max	0.55 max	0.1 max	Mo <=0.15; Ni 0.2-0.7; Ti <=0.05; Cr+Ni+Mo+Cu<=0.8; bal Fe	550-670	400	20	
AFNOR NFA36207(82)	A590AP-I*	16<t<=35mm; Norm; EN10028/3	0.2 max	0.2 max	0.3 max	1.7 max	0.035 max	0.035 max	0.5 max	0.02-0.15	Mo <=0.1; Ni <=0.2; Nb 0.01-0.06; Ti <=0.05; bal Fe	590-700	430	17	
AFNOR NFA36207(82)	A590AP-I*	5-16mm; Norm; EN10028/3	0.2 max	0.2 max	0.3 max	1.7 max	0.035 max	0.035 max	0.5 max	0.02-0.15	Mo <=0.1; Ni <=0.2; Nb 0.01-0.06; Ti <=0.05; bal Fe	590-700	440	17	
AFNOR NFA36207(82)	A590AP-II*	5-16mm; Norm; EN10028/3	0.18 max	0.4 max	0.6 max	1.7 max	0.035 max	0.035 max	0.4 max	0.02-0.15	Mo <=0.15; Ni 0.2-0.7; Ti <=0.05; Cr+Mo+Cu<=0.7; bal Fe	590-700	440	17	
AFNOR NFA36207(82)	A590AP-II*	50<t<=80mm; Norm; EN10028/3	0.18 max	0.4 max	0.6 max	1.7 max	0.035 max	0.035 max	0.4 max	0.02-0.18	Mo <=0.15; Ni 0.2-0.7; Ti <=0.05; Cr+Mo+Cu<=0.7; bal Fe	570-700	400	17	
AFNOR NFA36207(82)	A590FP-I*	16<t<=35mm; Norm; EN10028/3	0.2 max	0.2 max	0.3 max	1.7 max	0.03 max	0.02 max	0.5 max	0.02-0.15	Mo <=0.1; Ni <=0.2; Nb 0.01-0.06; Ti <=0.05; bal Fe	590-700	430	17	
AFNOR NFA36207(82)	A590FP-I*	5-16mm; Norm; EN10028/3	0.2 max	0.2 max	0.3 max	1.7 max	0.03 max	0.02 max	0.5 max	0.02-0.15	Mo <=0.1; Ni <=0.2; Nb 0.01-0.06; Ti <=0.05; bal Fe	590-700	440	17	
AFNOR NFA36207(82)	A590FP-II*	50<t<=80mm; Norm; EN10028/3	0.18 max	0.4 max	0.6 max	1.7 max	0.03 max	0.02 max	0.4 max	0.02-0.18	Mo <=0.15; Ni 0.2-0.7; Ti <=0.05; Cr+Mo+Cu<=0.7; bal Fe	570-700	400	17	
AFNOR NFA36207(82)	A590FP-II*	5-16mm; Norm; EN10028/3	0.18 max	0.4 max	0.6 max	1.7 max	0.03 max	0.02 max	0.4 max	0.02-0.15	Mo <=0.15; Ni 0.2-0.7; Ti <=0.05; Cr+Mo+Cu<=0.7; bal Fe	590-700	440	17	

Germany

Specification	Designation	Notes	C	Cr	Cu	Mn	P	S	Si	V	Other	UTS	YS	El	Hard
DIN	EStE500	Weld 85-100mm	0.20 max	0.30 max	0.70 max	1.00-1.70	0.025 max	0.015 max	0.10-0.60	0.22 max	Al >=0.020; Mo <=0.10; N <=0.020; Ni <=1.00; Nb <=0.05; Ti <=0.20; Nb+Ti+V<=0.22; bal Fe	590-760	430	16	
DIN	WNr 1.8919	85-100mm	0.20 max	0.30 max	0.70 max	1.00-1.70	0.025 max	0.015 max	0.10-0.60	0.22 max	Al >=0.020; Mo <=0.10; N <=0.020; Ni <=1.00; Nb <=0.05; Nb+Ti+V<=0.22; bal Fe	590-760	430	16	
DIN 17102	StE460*	Obs; see P460N	0.2 max	0.3 max	0.2 max	1-1.7	0.035 max	0.03 max	0.1-0.6	0.2 max	Mo <=0.1; N <=0.02; Ni <=1; Nb <=0.05; bal Fe				
DIN 17102	TStE460*	Obs; see P460NL1	0.2 max	0.3 max	0.2 max	1-1.7	0.03 max	0.025 max	0.1-0.6	0.2 max	Mo <=0.1; N <=0.02; Ni <=1; Nb <=0.05; bal Fe				
DIN 17102	WStE460	see P460NH	0.2 max	0.3 max	0.2 max	1-1.7	0.035 max	0.03 max	0.1-0.6	0.2 max	Mo <=0.1; N <=0.02; Ni <=1; Nb <=0.05; bal Fe				
DIN 17103(89)	WNr 1.8912	Frg	0.20 max	0.30 max	0.70 max	1.00-1.70	0.030 max	0.025 max	0.60 max	0.20 max	Al >=0.020; Mo <=0.10; N <=0.025; Ni <=0.80; Nb <=0.05; Ti <=0.03; bal Fe				
DIN 17119(84)	WNr 1.8912	Tube	0.20 max	0.30 max	0.70 max	1.00-1.70	0.030 max	0.025 max	0.60 max	0.20 max	Al >=0.020; Mo <=0.10; N <=0.025; Ni <=0.80; Nb <=0.05; Ti <=0.03; bal Fe				

Specification	Designation	Notes	C	Cr	Cu	Mn	P	S	Si	V	Other	UTS	YS	El	Hard

High Strength Steel, Low-Alloy (HSLA), LF6 (Continued from previous page)

Germany

Specification	Designation	Notes	C	Cr	Cu	Mn	P	S	Si	V	Other	UTS	YS	El	Hard
DIN EN 10113(93)	S420NL	HR 100-125mm	0.20 max	0.30 max	0.70 max	1.00-1.70	0.030 max	0.025 max	0.60 max	0.20 max	Al >=0.020; Mo <=0.10; N <=0.025; Ni <=0.80; Nb <=0.05; bal Fe	500-650	355	19	
DIN EN 10113(93)	WNr 1.8912	HR 100-125mm	0.20 max	0.30 max	0.70 max	1.00-1.70	0.030 max	0.025 max	0.60 max	0.20 max	Al >=0.020; Mo <=0.10; N <=0.025; Ni <=0.80; Nb <=0.05; Ti <=0.03; bal Fe	500-650	355	19	

Russia

Specification	Designation	Notes	C	Cr	Cu	Mn	P	S	Si	V	Other	UTS	YS	El	Hard
GOST 19281	16G2AF	Weld constr	0.14-0.2	0.4 max	0.3 max	1.3-1.7	0.035 max	0.04 max	0.3-0.6	0.08-0.14	Al <=0.1; Mo <=0.15; N 0.015-0.025; Ni <=0.3; Ti <=0.05; As<=0.08; bal Fe				
GOST 19281	18G2AF	Weld constr	0.14-0.22	0.3 max	0.3 max	1.3-1.7	0.035 max	0.04 max	0.17 max	0.08-0.15	Al <=0.1; Mo <=0.15; N 0.015-0.03; Ni <=0.3; Ti <=0.05; As<=0.08; bal Fe				

Spain

Specification	Designation	Notes	C	Cr	Cu	Mn	P	S	Si	V	Other	UTS	YS	El	Hard
UNE 36081(76)	AE390KG	Fine grain Struct	0.20 max			1.0-1.6	0.035 max	0.035 max	0.5 max		Mo <=0.3; Ni <=0.7; +Al+V+Nb; bal Fe				
UNE 36081(76)	AE390KT	Fine grain Struct	0.20 max			1.0-1.6	0.03 max	0.03 max	0.5 max		Mo <=0.3; Ni <=0.7; +Al+V+Nb; bal Fe				
UNE 36081(76)	AE390KW	Fine grain Struct	0.20 max			1.0-1.6	0.035 max	0.035 max	0.5 max	0.02-0.2	Al <=0.015; Mo <=0.3; Ni <=0.7; Nb 0.015-0.06; Ti <=0.2; bal Fe				
UNE 36081(76)	AE420KG	Fine grain Struct	0.20 max	0.40 max		1.0-1.6	0.035 max	0.035 max	0.5 max		Mo <=0.4; Ni <=0.7; +Al+V+Nb; bal Fe				
UNE 36081(76)	AE420KT	Fine grain Struct	0.20 max	0.40 max		1.0-1.6	0.03 max	0.03 max	0.5 max		Mo <=0.4; Ni <=0.7; +Al+V+Nb; bal Fe				
UNE 36081(76)	AE460KG	Fine grain Struct	0.20 max	0.50 max		1.0-1.7	0.035 max	0.035 max	0.5 max		Mo <=0.4; Ni <=0.8; Nb 0.015-0.06; +Al+V; bal Fe				
UNE 36081(76)	AE460KT	Fine grain Struct	0.20 max	0.50 max		1.0-1.7	0.03 max	0.03 max	0.5 max		Al <=0.015; Mo <=0.4; Ni <=0.8; +Al+V+Nb; bal Fe				
UNE 36081(76)	AE460KW	Fine grain Struct	0.20 max	0.50 max		1.0-1.7	0.035 max	0.035 max	0.5 max	0.02-0.2	Mo <=0.4; Ni <=0.8; Ti <=0.2; bal Fe				
UNE 36081(76)	F.6413*	see AE390KG	0.20 max			1.0-1.6	0.035 max	0.035 max	0.5 max		Mo <=0.3; Ni <=0.7; +Al+V+Nb; bal Fe				
UNE 36081(76)	F.6414*	see AE390KW	0.20 max			1.0-1.6	0.035 max	0.035 max	0.5 max	0.02-0.2	Al <=0.015; Mo <=0.3; Ni <=0.7; Nb 0.015-0.06; Ti <=0.2; bal Fe				
UNE 36081(76)	F.6415*	see AE390KT	0.20 max			1.0-1.6	0.03 max	0.03 max	0.5 max		Mo <=0.3; Ni <=0.7; +Al+V+Nb; bal Fe				
UNE 36081(76)	F.6416*	see AE420KG	0.20 max	0.40 max		1.0-1.6	0.035 max	0.035 max	0.5 max		Mo <=0.4; Ni <=0.7; +Al+V+Nb; bal Fe				
UNE 36081(76)	F.6419*	see AE460KG	0.20 max	0.50 max		1.0-1.7	0.035 max	0.035 max	0.5 max		Mo <=0.4; Ni <=0.8; Nb 0.015-0.06; +Al=V; bal Fe				
UNE 36081(76)	F.6420*	see AE460KW	0.20 max	0.50 max		1.0-1.7	0.035 max	0.035 max	0.5 max	0.02-0.2	Mo <=0.4; Ni <=0.8; Ti <=0.2; bal Fe				
UNE 36081(76)	F.6421*	see AE460KT	0.20 max	0.50 max		1.0-1.7	0.03 max	0.03 max	0.5 max		Al <=0.015; Mo <=0.4; Ni <=0.8; +Al+V+Nb; bal Fe				

USA

Specification	Designation	Notes	C	Cr	Cu	Mn	P	S	Si	V	Other	UTS	YS	El	Hard
	UNS K12202		0.22 max			1.15-1.50	0.04 max	0.05 max	0.15-0.50	0.04-0.11	N 0.01-0.03; bal Fe				
ASTM A633/A633M(95)	Class E	Norm struct Plt, t>=100-150mm	0.22 max			1.15-1.50	0.035 max	0.04 max	0.15-0.50	0.04-0.11	N 0.01-0.03; bal Fe	515-655	380	23	
ASTM A633/A633M(95)	Class E	Norm struct Plt, t<=100mm	0.22 max			1.15-1.50	0.035 max	0.04 max	0.15-0.50	0.04-0.11	N 0.01-0.03; bal Fe	550-690	415	23	
ASTM A678/A678M(94)	Grade D	Q/T, struct Plt, t<=75mm	0.22 max		0.20 min	1.15-1.50	0.035 max	0.04 max	0.15-0.50		N 0.01-0.03; bal Fe	620-760	515	18	
ASTM A737(97)	Grade C	Press ves plt; Nb<=0.05 may be present	0.22 max			1.15-1.50	0.035 max	0.030 max	0.15-0.50	0.04-0.11	N <=0.03; bal Fe	550-690	415	23	

Specification	Designation	Notes	C	Cr	Cu	Mn	P	S	Si	V	Other	UTS	YS	El	Hard

High Strength Steel, Low-Alloy (HSLA), No equivalents identified

Europe

Specification	Designation	Notes	C	Cr	Cu	Mn	P	S	Si	V	Other	UTS	YS	El	Hard
EN 10149/2(95)	1.0972	HR 1.5<=t<3mm	0.12 max			1.30 max	0.025 max	0.020 max	0.5 max	0.20 max	Al >=0.015; Nb <=0.09; Ti <=0.15; Nb+Ti+V<=0.22; bal Fe	390-510	315	20	
EN 10149/2(95)	1.0972	HR 3<=t<20mm	0.12 max			1.30 max	0.025 max	0.020 max	0.5 max	0.20 max	Al >=0.015; Nb <=0.09; Ti <=0.15; Nb+Ti+V<=0.22; bal Fe	390-510	315	24	
EN 10149/2(95)	1.0976	HR 3<=t<20mm	0.12 max			1.50 max	0.025 max	0.020 max	0.5 max	0.20 max	Al >=0.015; Nb <=0.09; Ti <=0.15; Nb+Ti+V<=0.22; bal Fe	430-550	355	23	
EN 10149/2(95)	1.0976	HR 1.5<=t<3mm	0.12 max			1.50 max	0.025 max	0.020 max	0.5 max	0.20 max	Al >=0.015; Nb <=0.09; Ti <=0.15; Nb+Ti+V<=0.22; bal Fe	430-550	355	19	
EN 10149/2(95)	1.0980	HR 1.5<=t<3mm	0.12 max			1.60 max	0.025 max	0.015 max	0.5 max	0.20 max	Al >=0.015; Nb <=0.09; Ti <=0.15; Nb+Ti+V<=0.22; bal Fe	480-620	420	16	
EN 10149/2(95)	1.0980	HR 3<=t<20mm	0.12 max			1.60 max	0.025 max	0.015 max	0.50 max	0.20 max	Al >=0.015; Nb <=0.09; Ti <=0.15; Nb+Ti+V<=0.22; bal Fe	480-620	420	19	
EN 10149/2(95)	1.0982	HR 1.5<=t<3mm	0.12 max			1.60 max	0.025 max	0.015 max	0.50 max	0.20 max	Al >=0.015; Nb <=0.09; Ti <=0.15; Nb+Ti+V<=0.22; bal Fe	520-670	460	14	
EN 10149/2(95)	1.0982	HR 3<=t<20mm	0.12 max			1.60 max	0.025 max	0.015 max	0.50 max	0.20 max	Al >=0.015; Nb <=0.09; Ti <=0.15; Nb+Ti+V<=0.22; bal Fe	520-670	460	17	
EN 10149/2(95)	1.0984	HR 1.5<=t<3mm	0.12 max			1.70 max	0.025 max	0.015 max	0.50 max	0.20 max	Al >=0.015; Nb <=0.09; Ti <=0.15; Nb+Ti+V<=0.22; bal Fe	550-700	500	12	
EN 10149/2(95)	1.0984	HR 3<t<=16mm	0.12 max			1.70 max	0.025 max	0.015 max	0.50 max	0.20 max	Al >=0.015; Nb <=0.09; Ti <=0.15; Nb+Ti+V<=0.22; bal Fe	550-700	500	14	
EN 10149/2(95)	1.0986	HR 1.5<=t<3mm	0.12 max			1.80 max	0.025 max	0.015 max	0.50 max	0.20 max	Al >=0.015; Nb <=0.09; Ti <=0.15; Nb+Ti+V<=0.22; bal Fe	600-760	550	12	
EN 10149/2(95)	1.0986	HR 3<t<=16mm	0.12 max			1.80 max	0.025 max	0.015 max	0.50 max	0.20 max	Al >=0.015; Nb <=0.09; Ti <=0.15; Nb+Ti+V<=0.22; bal Fe	600-760	550	14	
EN 10149/2(95)	1.8969	HR 3<t<=16mm	0.12 max			1.90 max	0.025 max	0.015 max	0.50 max	0.20 max	Al >=0.015; B <=0.005; Mo <=0.50; Nb <=0.09; Ti <=0.22; Nb+Ti+V<=0.22; bal Fe	650-820	600	13	
EN 10149/2(95)	1.8969	HR 1.5<=t<3mm	0.12 max			1.90 max	0.025 max	0.015 max	0.50 max	0.20 max	Al >=0.015; B <=0.005; Mo <=0.50; Nb <=0.09; Ti <=0.22; Nb+Ti+V<=0.22; bal Fe	650-820	600	11	
EN 10149/2(95)	1.8974	HR 3<=t<8mm	0.12 max			2.10 max	0.025 max	0.015 max	0.60 max	0.20 max	Al >=0.015; B <=0.005; Mo <=0.50; Nb <=0.09; Ti <=0.22; Nb+Ti+V<=0.22; bal Fe	750-950	700	12	
EN 10149/2(95)	1.8974	HR 8<t<=16mm	0.12 max			2.10 max	0.025 max	0.015 max	0.60 max	0.20 max	Al >=0.015; B <=0.005; Mo <=0.50; Nb <=0.09; Ti <=0.22; Nb+Ti+V<=0.22; bal Fe	750-950	680	12	

UNS numbers and US grades are provided as a means of cross referencing chemically similar alloys. Exchangability is only possible after independent examination of specifications. Tensile properties are minimum or typical as specified. UTS and YS as MPa. El as %. See Appendix for list of abbreviations used in Notes. * indicates obsolete material.

Specification	Designation	Notes	C	Cr	Cu	Mn	P	S	Si	V	Other	UTS	YS	El	Hard

High Strength Steel, Low-Alloy (HSLA), No equivalents identified

Europe

Specification	Designation	Notes	C	Cr	Cu	Mn	P	S	Si	V	Other	UTS	YS	El	Hard
EN 10149/2(95)	1.8974	HR 1.5<=t<3mm	0.12 max			2.10 max	0.025 max	0.015 max	0.60 max	0.20 max	Al >=0.015; B <=0.005; Mo <=0.50; Nb <=0.09; Ti <=0.22; Nb+Ti+V<=0.22; bal Fe	750-950	700	10	
EN 10149/2(95)	1.8976	HR 8<t<=16mm	0.12 max			2.00 max	0.025 max	0.015 max	0.60 max	0.20 max	Al >=0.015; B <=0.005; Mo <=0.50; Nb <=0.09; Ti <=0.22; Nb+Ti+V<=0.22; bal Fe	700-880	630	12	
EN 10149/2(95)	1.8976	HR 1.5<=t<3mm	0.12 max			2.00 max	0.025 max	0.015 max	0.60 max	0.20 max	Al >=0.015; B <=0.005; Mo <=0.50; Nb <=0.09; Ti <=0.22; Nb+Ti+V<=0.22; bal Fe	700-880	650	10	
EN 10149/2(95)	1.8976	HR 3-8mm	0.12 max			2.00 max	0.025 max	0.015 max	0.60 max	0.20 max	Al >=0.015; B <=0.005; Mo <=0.50; Nb <=0.09; Ti <=0.22; Nb+Ti+V<=0.22; bal Fe	700-880	650	12	
EN 10149/2(95)	S315MC	HR 1.5<=t<3mm	0.12 max			1.30 max	0.025 max	0.020 max	0.5 max	0.20 max	Al >=0.015; Nb <=0.09; Ti <=0.15; Nb+Ti+V<=0.22; bal Fe	390-510	315	20	
EN 10149/2(95)	S315MC	HR 3<=t<20mm	0.12 max			1.30 max	0.025 max	0.020 max	0.5 max	0.20 max	Al >=0.015; Nb <=0.09; Ti <=0.15; Nb+Ti+V<=0.22; bal Fe	390-510	315	24	
EN 10149/2(95)	S355MC	HR 3<=t<20mm	0.12 max			1.50 max	0.025 max	0.020 max	0.5 max	0.20 max	Al >=0.015; Nb <=0.09; Ti <=0.15; Nb+Ti+V<=0.22; bal Fe	430-550	355	23	
EN 10149/2(95)	S355MC	HR 1.5<=t<3mm	0.12 max			1.50 max	0.025 max	0.020 max	0.5 max	0.20 max	Al >=0.015; Nb <=0.09; Ti <=0.15; Nb+Ti+V<=0.22; bal Fe	430-550	355	19	
EN 10149/2(95)	S420MC	HR 1.5<=t<3mm	0.12 max			1.60 max	0.025 max	0.015 max	0.5 max	0.20 max	Al >=0.015; Nb <=0.09; Ti <=0.15; Nb+Ti+V<=0.22; bal Fe	480-620	420	16	
EN 10149/2(95)	S420MC	HR 3<=t<20mm	0.12 max			1.60 max	0.025 max	0.015 max	0.50 max	0.20 max	Al >=0.015; Nb <=0.09; Ti <=0.15; Nb+Ti+V<=0.22; bal Fe	480-620	420	19	
EN 10149/2(95)	S460MC	HR 1.5<=t<3mm	0.12 max			1.60 max	0.025 max	0.015 max	0.50 max	0.20 max	Al >=0.015; Nb <=0.09; Ti <=0.15; Nb+Ti+V<=0.22; bal Fe	520-670	460	14	
EN 10149/2(95)	S460MC	HR 3<=t<20mm	0.12 max			1.60 max	0.025 max	0.015 max	0.50 max	0.20 max	Al >=0.015; Nb <=0.09; Ti <=0.15; Nb+Ti+V<=0.22; bal Fe	520-670	460	17	
EN 10149/2(95)	S500MC	HR 1.5<=t<3mm	0.12 max			1.70 max	0.025 max	0.015 max	0.50 max	0.20 max	Al >=0.015; Nb <=0.09; Ti <=0.15; Nb+Ti+V<=0.22; bal Fe	550-700	500	12	
EN 10149/2(95)	S500MC	HR 3<t<=16mm	0.12 max			1.70 max	0.025 max	0.015 max	0.50 max	0.20 max	Al >=0.015; Nb <=0.09; Ti <=0.15; Nb+Ti+V<=0.22; bal Fe	550-700	500	14	
EN 10149/2(95)	S550MC	HR 3<t<=16mm	0.12 max			1.80 max	0.025 max	0.015 max	0.50 max	0.20 max	Al >=0.015; Nb <=0.09; Ti <=0.15; Nb+Ti+V<=0.22; bal Fe	600-760	550	14	
EN 10149/2(95)	S550MC	HR 1.5<=t<3mm	0.12 max			1.80 max	0.025 max	0.015 max	0.50 max	0.20 max	Al >=0.015; Nb <=0.09; Ti <=0.15; Nb+Ti+V<=0.22; bal Fe	600-760	550	12	

UNS numbers and US grades are provided as a means of cross referencing chemically similar alloys. Exchangability is only possible after independent examination of specifications. Tensile properties are minimum or typical as specified. UTS and YS as MPa. El as %. See Appendix for list of abbreviations used in Notes. * indicates obsolete material.

High Strength Steel, Low-Alloy (HSLA), No equivalents identified

Specification	Designation	Notes	C	Cr	Cu	Mn	P	S	Si	V	Other	UTS	YS	El	Hard
Europe															
EN 10149/2(95)	S600MC	HR 1.5<=t<3mm	0.12 max			1.90 max	0.025 max	0.015 max	0.50 max	0.20 max	Al >=0.015; B <=0.005; Mo <=0.50; Nb <=0.09; Ti <=0.22; Nb+Ti+V<=0.22; bal Fe	650-820	600	11	
EN 10149/2(95)	S600MC	HR 3<t<=16mm	0.12 max			1.90 max	0.025 max	0.015 max	0.50 max	0.20 max	Al >=0.015; B <=0.005; Mo <=0.50; Nb <=0.09; Ti <=0.22; Nb+Ti+V<=0.22; bal Fe	650-820	600	13	
EN 10149/2(95)	S650MC	HR 1.5<=t<3mm	0.12 max			2.00 max	0.025 max	0.015 max	0.60 max	0.20 max	Al >=0.015; B <=0.005; Mo <=0.50; Nb <=0.09; Ti <=0.22; Nb+Ti+V<=0.22; bal Fe	700-880	650	10	
EN 10149/2(95)	S650MC	HR 3<t<=8mm	0.12 max			2.00 max	0.025 max	0.015 max	0.60 max	0.20 max	Al >=0.015; B <=0.005; Mo <=0.50; Nb <=0.09; Ti <=0.22; Nb+Ti+V<=0.22; bal Fe	700-880	650	12	
EN 10149/2(95)	S650MC	HR 8<t<=16mm	0.12 max			2.00 max	0.025 max	0.015 max	0.60 max	0.20 max	Al >=0.015; B <=0.005; Mo <=0.50; Nb <=0.09; Ti <=0.22; Nb+Ti+V<=0.22; bal Fe	700-880	630	12	
EN 10149/2(95)	S700MC	HR 8<t<=16mm	0.12 max			2.10 max	0.025 max	0.015 max	0.60 max	0.20 max	Al >=0.015; B <=0.005; Mo <=0.50; Nb <=0.09; Ti <=0.22; Nb+Ti+V<=0.22; bal Fe	750-950	680	12	
EN 10149/2(95)	S700MC	HR 1.5<=t<3mm	0.12 max			2.10 max	0.025 max	0.015 max	0.60 max	0.20 max	Al >=0.015; B <=0.005; Mo <=0.50; Nb <=0.09; Ti <=0.22; Nb+Ti+V<=0.22; bal Fe	750-950	700	10	
EN 10149/2(95)	S700MC	HR 3<=t<8mm	0.12 max			2.10 max	0.025 max	0.015 max	0.60 max	0.20 max	Al >=0.015; B <=0.005; Mo <=0.50; Nb <=0.09; Ti <=0.22; Nb+Ti+V<=0.22; bal Fe	750-950	700	12	
France															
AFNOR NFA35502(84)	E36WA3*	t<=3mm; roll	0.12 max	0.30-1.25	0.25-0.55	1.00 max	0.06-0.15	0.040 max	0.75 max		Mo <=0.15; Ni <=0.65; bal Fe	480-580	315	20	
AFNOR NFA35502(84)	E36WA3*	t<=12mm; roll	0.12 max	0.30-1.25	0.25-0.55	1.00 max	0.06-0.15	0.040 max	0.75 max		Mo <=0.15; Ni <=0.65; bal Fe	480-580	355	22	
AFNOR NFA35502(84)	E36WA3*	3<t<=12mm; roll	0.12 max	0.30-1.25	0.25-0.55	1.00 max	0.06-0.15	0.040 max	0.75 max		Mo <=0.15; Ni <=0.65; bal Fe	480-580	355	20	
AFNOR NFA35502(84)	E36WA4*	t<=12mm; roll	0.12 max	0.30-1.25	0.25-0.55	1 max	0.06-0.15	0.04 max	0.2-0.75	0.1 max	Mo <=0.15; Ni <=0.65; Ti <=0.05; bal Fe	480-580	355	22	
AFNOR NFA35502(84)	E36WA4*	t<=3mm; roll	0.12 max	0.30-1.25	0.25-0.55	1.00 max	0.06-0.15	0.040 max	0.75 max		Al >=0.015; Mo <=0.15; Ni <=0.65; Nb 0.015-0.06; bal Fe	480-580	315	20	
AFNOR NFA35502(84)	E36WA4*	3<t<=12mm; roll	0.12 max	0.30-1.25	0.25-0.55	1.00 max	0.06-0.15	0.040 max	0.75 max		Al >=0.015; Mo <=0.15; Ni <=0.65; Nb 0.015-0.06; bal Fe	480-580	355	21	
AFNOR NFA36204(83)	E420TFP-I	5<=t<=50mm; N/T	0.08 max	2 max	1.5 max	0.3-1.3	0.025 max	0.025 max	0.1-0.5	0.2 max	B <=0.05; Mo <=1; N <=0.015; Ni <=2; Nb <=0.06; Ti <=0.2; Zr<=0.12; Cu+6Sn<=0.33; bal Fe	530-680	420	19	

Specification	Designation	Notes	C	Cr	Cu	Mn	P	S	Si	V	Other	UTS	YS	El	Hard

High Strength Steel, Low-Alloy (HSLA), No equivalents identified

France

Specification	Designation	Notes	C	Cr	Cu	Mn	P	S	Si	V	Other	UTS	YS	El	Hard
AFNOR NFA36204(83)	E420TFP-I	50<t<=70mm; N/T	0.08 max		2 max	1.5 max	0.3-1.3	0.025 max	0.025 max	0.1-0.5	0.2 max	B <=0.05; Mo <=1; N <=0.015; Ni <=2; Nb <=0.06; Ti <=0.2; Zr<=0.12; Cu+6Sn<=0.33; bal Fe	530-680	400	19	
AFNOR NFA36204(83)	E420TR-I	50<t<=70mm; N/T	0.08 max		2 max	1.5 max	0.3-1.3	0.03 max	0.03 max	0.1-0.5	0.2 max	B <=0.05; Mo <=1; N <=0.015; Ni <=2; Nb <=0.06; Ti <=0.2; Zr<=0.12; Cu+6Sn<=0.33; bal Fe	530-680	400	19	
AFNOR NFA36204(83)	E420TR-I	5<=t<=50mm; N/T	0.08 max		2 max	1.5 max	0.3-1.3	0.03 max	0.03 max	0.1-0.5	0.2 max	B <=0.05; Mo <=1; N <=0.015; Ni <=2; Nb <=0.06; Ti <=0.2; Zr<=0.12; Cu+6Sn<=0.33; bal Fe	530-680	420	19	
AFNOR NFA36204(83)	E460TR-I	5<=t<=50mm; N/T	0.08 max		2 max	1.5 max	0.3-1.3	0.03 max	0.03 max	0.1-0.5	0.2 max	B <=0.05; Mo <=1; N <=0.015; Ni <=2; Nb <=0.06; Ti <=0.2; Zr<=0.12; Cu+6Sn<=0.33; bal Fe	570-720	460	17	
AFNOR NFA36204(83)	E460TR-I	50<t<=70mm; N/T	0.08 max		2 max	1.5 max	0.3-1.3	0.03 max	0.03 max	0.1-0.5	0.2 max	B <=0.05; Mo <=1; N <=0.015; Ni <=2; Nb <=0.06; Ti <=0.2; Zr<=0.12; Cu+6Sn<=0.33; bal Fe	570-720	440	17	
AFNOR NFA36204(83)	E500FTP-I	50<t<=70mm; N/T	0.08 max		2 max	1.5 max	0.3-1.3	0.025 max	0.025 max	0.1-0.5	0.2 max	B <=0.05; Mo <=1; N <=0.015; Ni <=2; Nb <=0.06; Ti <=0.2; Zr<=0.12; Cu+6Sn<=0.33; bal Fe	620-770	480	17	
AFNOR NFA36204(83)	E500FTP-I	t<=50mm; N/T	0.08 max		2 max	1.5 max	0.3-1.3	0.025 max	0.025 max	0.1-0.5	0.2 max	B <=0.05; Mo <=1; N <=0.015; Ni <=2; Nb <=0.06; Ti <=0.2; Zr<=0.12; Cu+6Sn<=0.33; bal Fe	620-770	500	17	
AFNOR NFA36204(83)	E500TR-I	50<t<=70mm; N/T	0.08 max		2 max	1.5 max	0.3-1.3	0.03 max	0.03 max	0.1-0.5	0.2 max	B <=0.05; Mo <=1; N <=0.015; Ni <=2; Nb <=0.06; Ti <=0.2; Zr<=0.12; Cu+6Sn<=0.33; bal Fe	620-770	480	17	
AFNOR NFA36204(83)	E500TR-I	t<=50mm; N/T	0.08 max		2 max	1.5 max	0.3-1.3	0.03 max	0.03 max	0.1-0.5	0.2 max	B <=0.05; Mo <=1; N <=0.015; Ni <=2; Nb <=0.06; Ti <=0.2; Zr<=0.12; Cu+6Sn<=0.33; bal Fe	620-770	500	17	
AFNOR NFA36204(83)	E550TFP-I	50<t<=70mm; N/T	0.08 max		2 max	1.5 max	0.3-1.3	0.025 max	0.025 max	0.1-0.5	0.2 max	B <=0.05; Mo <=1; N <=0.015; Ni <=2; Nb <=0.06; Ti <=0.2; Zr<=0.12; Cu+6Sn<=0.33; bal Fe	670-820	530	16T	
AFNOR NFA36204(83)	E550TFP-I	5<=t<=50mm; N/T	0.08 max		2 max	1.5 max	0.3-1.3	0.025 max	0.025 max	0.1-0.5	0.2 max	B <=0.05; Mo <=1; N <=0.015; Ni <=2; Nb <=0.06; Ti <=0.2; Zr<=0.12; Cu+6Sn<=0.33; bal Fe	670-820	550	16T	
AFNOR NFA36204(83)	E550TR-I	5<=t<=50mm; N/T	0.08 max		2 max	1.5 max	0.3-1.3	0.03 max	0.03 max	0.1-0.5	0.2 max	B <=0.05; Mo <=1; N <=0.015; Ni <=2; Nb <=0.06; Ti <=0.2; Zr<=0.12; Cu+6Sn<=0.33; bal Fe	670-820	550	16T	

Specification	Designation	Notes	C	Cr	Cu	Mn	P	S	Si	V	Other	UTS	YS	El	Hard

High Strength Steel, Low-Alloy (HSLA), No equivalents identified

France

Specification	Designation	Notes	C	Cr	Cu	Mn	P	S	Si	V	Other	UTS	YS	El	Hard
AFNOR NFA36204(83)	E550TR-I	50<t<=70mm; N/T	0.08 max	2 max		1.5 max	0.3-1.3	0.03 max	0.03 max	0.1-0.5	0.2 max	B <=0.05; Mo <=1; N <=0.015; Ni <=2; Nb <=0.06; Ti <=0.2; Zr<=0.12; Cu+6Sn<=0.33; bal Fe	670-820	530	16	
AFNOR NFA36204(83)	E620TFP-I	50<t<=70mm; Norm	0.08 max	2 max		1.5 max	0.3-1.3	0.025 max	0.025 max	0.1-0.5	0.2 max	B <=0.05; Mo <=1; N <=0.015; Ni <=2; Nb <=0.06; Ti <=0.2; Zr<=0.12; Cu+6Sn<=0.33; bal Fe	740-890	600	15	
AFNOR NFA36204(83)	E620TFP-I	5<=t<=50mm; Norm	0.08 max	2 max		1.5 max	0.3-1.3	0.025 max	0.025 max	0.1-0.5	0.2 max	B <=0.05; Mo <=1; N <=0.015; Ni <=2; Nb <=0.06; Ti <=0.2; Zr<=0.12; Cu+6Sn<=0.33; bal Fe	740-890	620	15	
AFNOR NFA36204(83)	E620TR-I	5<=t<=50mm; Norm	0.08 max	2 max		1.5 max	0.3-1.3	0.03 max	0.03 max	0.1-0.5	0.2 max	B <=0.05; Mo <=1; N <=0.015; Ni <=2; Nb <=0.06; Ti <=0.2; Zr<=0.12; Cu+6Sn<=0.33; bal Fe	740-890	620	15	
AFNOR NFA36204(83)	E620TR-I	50<t<=70mm; Norm	0.08 max	2 max		1.5 max	0.3-1.3	0.03 max	0.03 max	0.1-0.5	0.2 max	B <=0.05; Mo <=1; N <=0.015; Ni <=2; Nb <=0.06; Ti <=0.2; Zr<=0.12; Cu+6Sn<=0.33; bal Fe	740-890	600	15	
AFNOR NFA36204(92)	E420T-I-K2	50<t<=70mm; Q/T	0.08 max	2 max		1.5 max	0.3-1.3	0.025 max	0.015 max	0.5 max	0.2 max	B <=0.005; Mo <=1; Ni <=2; Nb <=0.06; Ti <=0.2; N2<=0.015; CEV maybe used; bal Fe	510-680	400	19	
AFNOR NFA36204(92)	E420T-I-K2	t<50mm; Q/T	0.08 max	2 max		1.5 max	0.3-1.3	0.025 max	0.015 max	0.5 max	0.2 max	B <=0.005; Mo <=1; Ni <=2; Nb <=0.06; Ti <=0.2; N2<=0.015; CEV maybe used; bal Fe	510-680	420	19	
AFNOR NFA36204(92)	E420T-I-K4	50<t<=70mm; Q/T	0.08 max	2 max		1.5 max	0.3-1.3	0.02 max	0.01 max	0.5 max	0.2 max	B <=0.005; Mo <=1; Ni <=2; Nb <=0.06; Ti <=0.2; N2<=0.015; CEV maybe used; bal Fe	510-680	400	19T	
AFNOR NFA36204(92)	E420T-I-K4	t<50mm; Q/T	0.08 max	2 max		1.5 max	0.3-1.3	0.02 max	0.01 max	0.5 max	0.2 max	B <=0.005; Mo <=1; Ni <=2; Nb <=0.06; Ti <=0.2; N2<=0.015; CEV maybe used; bal Fe	510-680	420	19T	
AFNOR NFA36204(92)	E460T-I-K2	50<t<=70mm; Q/T	0.08 max	2 max		1.5 max	0.3-1.3	0.025 max	0.015 max	0.5 max	0.2 max	B <=0.005; Mo <=1; Ni <=2; Nb <=0.06; Ti <=0.2; N2<=0.015; CEV maybe used; bal Fe	550-720	440	17T	
AFNOR NFA36204(92)	E460T-I-K2	t<50mm; Q/T	0.08 max	2 max		1.5 max	0.3-1.3	0.025 max	0.015 max	0.5 max	0.2 max	B <=0.005; Mo <=1; Ni <=2; Nb <=0.06; Ti <=0.2; N2<=0.015; CEV maybe used; bal Fe	550-720	460	17T	
AFNOR NFA36204(92)	E460T-I-K4	50<t<=70mm; Q/T	0.08 max	2 max		1.5 max	0.3-1.3	0.02 max	0.01 max	0.5 max	0.2 max	B <=0.005; Mo <=1; Ni <=2; Nb <=0.06; Ti <=0.2; N2<=0.015; CEV maybe used; bal Fe	550-720	440	17T	
AFNOR NFA36204(92)	E460T-I-K4	t<50mm; Q/T	0.08 max	2 max		1.5 max	0.3-1.3	0.02 max	0.01 max	0.5 max	0.2 max	B <=0.005; Mo <=1; Ni <=2; Nb <=0.06; Ti <=0.2; N2<=0.015; CEV maybe used; bal Fe	550-720	460	17T	

High Strength Steel, Low-Alloy (HSLA), No equivalents identified

France

Specification	Designation	Notes	C	Cr	Cu	Mn	P	S	Si	V	Other	UTS	YS	El	Hard
AFNOR NFA36204(92)	E500T-I-K2	t<50mm; Q/T	0.08 max	2 max	1.5 max	0.3-1.3	0.025 max	0.015 max	0.5 max	0.2 max	B <=0.005; Mo <=1; Ni <=2; Nb <=0.06; Ti <=0.2; N2<=0.015; CEV maybe used; bal Fe	600-770	500	17T	
AFNOR NFA36204(92)	E500T-I-K2	50<t<=70mm; Q/T	0.08 max	2 max	1.5 max	0.3-1.3	0.025 max	0.015 max	0.5 max	0.2 max	B <=0.005; Mo <=1; Ni <=2; Nb <=0.06; Ti <=0.2; N2<=0.015; CEV maybe used; bal Fe	600-770	480	17T	
AFNOR NFA36204(92)	E500T-I-K4	t<50mm; Q/T	0.08 max	2 max	1.5 max	0.3-1.3	0.02 max	0.01 max	0.5 max	0.2 max	B <=0.005; Mo <=1; Ni <=2; Nb <=0.06; Ti <=0.2; N2<=0.015; CEV maybe used; bal Fe	600-770	500	17T	
AFNOR NFA36204(92)	E500T-I-K4	50<t<=70mm; Q/T	0.08 max	2 max	1.5 max	0.3-1.3	0.02 max	0.01 max	0.5 max	0.2 max	B <=0.005; Mo <=1; Ni <=2; Nb <=0.06; Ti <=0.2; N2<=0.015; CEV maybe used; bal Fe	600-770	480	17T	
AFNOR NFA36204(92)	E550T-I-K2	t<50mm; Q/T	0.08 max	2 max	1.5 max	0.3-1.3	0.025 max	0.015 max	0.5 max	0.2 max	B <=0.005; Mo <=1; Ni <=2; Nb <=0.06; Ti <=0.2; N2<=0.015; CEV maybe used; bal Fe	650-820	550	16T	
AFNOR NFA36204(92)	E550T-I-K2	50<t<=70mm; Q/T	0.08 max	2 max	1.5 max	0.3-1.3	0.025 max	0.015 max	0.5 max	0.2 max	B <=0.005; Mo <=1; Ni <=2; Nb <=0.06; Ti <=0.2; N2<=0.015; CEV maybe used; bal Fe	650-820	530	16T	
AFNOR NFA36204(92)	E550T-I-K4	t<50mm; Q/T	0.08 max	2 max	1.5 max	0.3-1.3	0.02 max	0.01 max	0.5 max	0.2 max	B <=0.005; Mo <=1; Ni <=2; Nb <=0.06; Ti <=0.2; N2<=0.015; CEV maybe used; bal Fe	650-820	550	16T	
AFNOR NFA36204(92)	E550T-I-K4	50<t<=70mm; Q/T	0.08 max	2 max	1.5 max	0.3-1.3	0.02 max	0.01 max	0.5 max	0.2 max	B <=0.005; Mo <=1; Ni <=2; Nb <=0.06; Ti <=0.2; N2<=0.015; CEV maybe used; bal Fe	650-820	530	16T	
AFNOR NFA36204(92)	E620T-I-K2	50<t<=70mm; Q/T	0.08 max	2 max	1.5 max	0.3-1.3	0.025 max	0.015 max	0.5 max	0.2 max	B <=0.005; Mo <=1; Ni <=2; Nb <=0.06; Ti <=0.2; N2<=0.015; CEV maybe used; bal Fe	720-890	600	15T	
AFNOR NFA36204(92)	E620T-I-K2	t<50mm; Q/T	0.08 max	2 max	1.5 max	0.3-1.3	0.025 max	0.015 max	0.5 max	0.2 max	B <=0.005; Mo <=1; Ni <=2; Nb <=0.06; Ti <=0.2; N2<=0.015; CEV maybe used; bal Fe	720-890	620	15T	
AFNOR NFA36204(92)	E620T-I-K4	t<50mm; Q/T	0.08 max	2 max	1.5 max	0.3-1.3	0.02 max	0.1 max	0.5 max	0.2 max	B <=0.005; Mo <=1; Ni <=2; Nb <=0.06; Ti <=0.2; N2<=0.015; CEV maybe used; bal Fe	720-890	620	15T	
AFNOR NFA36204(92)	E620T-I-K4	50<t<=70mm; Q/T	0.08 max	2 max	1.5 max	0.3-1.3	0.02 max	0.1 max	0.5 max	0.2 max	B <=0.005; Mo <=1; Ni <=2; Nb <=0.06; Ti <=0.2; N2<=0.015; CEV maybe used; bal Fe	720-890	600	15T	
AFNOR NFA36204(92)	E960T-II-K2	t<50mm; Q/T	0.2 max	2 max	1.5 max	1.7 max	0.025 max	0.015 max	0.5 max	0.2 max	B <=0.005; Mo <=1; Ni <=2; Nb <=0.06; Ti <=0.2; N2<=0.015; CEV maybe used; bal Fe	1000-1200	960	10T	
AFNOR NFA36204(92)	E960T-II-K2	50<t<=70mm; Q/T	0.2 max	2 max	1.5 max	1.7 max	0.025 max	0.015 max	0.5 max	0.2 max	B <=0.005; Mo <=1; Ni <=2; Nb <=0.06; Ti <=0.2; N2<=0.015; CEV maybe used; bal Fe	1000-1200	940	10T	

Specification	Designation	Notes	C	Cr	Cu	Mn	P	S	Si	V	Other	UTS	YS	El	Hard

High Strength Steel, Low-Alloy (HSLA), No equivalents identified

France

Specification	Designation	Notes	C	Cr	Cu	Mn	P	S	Si	V	Other	UTS	YS	El	Hard
AFNOR NFA36204(92)	E960T-II-K4	50<t<=70mm; Q/T	0.2 max	2 max	1.5 max	1.7 max	0.02 max	0.01 max	0.5 max	0.2 max	B <=0.005; Mo <=1; Ni <=2; Nb <=0.06; Ti <=0.2; N2<=0.015; CEV maybe used; bal Fe	1000-1200	940	10T	
AFNOR NFA36204(92)	S420-Q-I-K4	t<50mm; Q/T	0.08 max	2 max	1.5 max	0.3-1.3	0.02 max	0.01 max	0.5 max	0.2 max	B <=0.005; Mo <=1; Ni <=2; Nb <=0.06; Ti <=0.2; N2<=0.015; CEV maybe used; bal Fe	510-680	420	19T	
AFNOR NFA36204(92)	S420-Q-I-K4	50<t<=70mm; Q/T	0.08 max	2 max	1.5 max	0.3-1.3	0.02 max	0.01 max	0.5 max	0.2 max	B <=0.005; Mo <=1; Ni <=2; Nb <=0.06; Ti <=0.2; N2<=0.015; CEV maybe used; bal Fe	510-680	400	19T	
AFNOR NFA36204(92)	S420Q-1-K2	t<50mm; Q/T	0.08 max	2 max	1.5 max	0.3-1.3	0.025 max	0.015 max	0.5 max	0.2 max	B <=0.005; Mo <=1; Ni <=2; Nb <=0.06; Ti <=0.2; N2<=0.015; CEV maybe used; bal Fe	510-680	420	19	
AFNOR NFA36204(92)	S420Q-1-K2	50<t<=70mm; Q/T	0.08 max	2 max	1.5 max	0.3-1.3	0.025 max	0.015 max	0.5 max	0.2 max	B <=0.005; Mo <=1; Ni <=2; Nb <=0.06; Ti <=0.2; N2<=0.015; CEV maybe used; bal Fe	510-680	400	19	
AFNOR NFA36204(92)	S460Q-I-K2	t<50mm; Q/T	0.08 max	2 max	1.5 max	0.3-1.3	0.025 max	0.015 max	0.5 max	0.2 max	B <=0.005; Mo <=1; Ni <=2; Nb <=0.06; Ti <=0.2; N2<=0.015; CEV maybe used; bal Fe	550-720	460	17T	
AFNOR NFA36204(92)	S460Q-I-K2	50<t<=70mm; Q/T	0.08 max	2 max	1.5 max	0.3-1.3	0.025 max	0.015 max	0.5 max	0.2 max	B <=0.005; Mo <=1; Ni <=2; Nb <=0.06; Ti <=0.2; N2<=0.015; CEV maybe used; bal Fe	550-720	440	17T	
AFNOR NFA36204(92)	S460Q-I-K4	t<50mm; Q/T	0.08 max	2 max	1.5 max	0.3-1.3	0.02 max	0.01 max	0.5 max	0.2 max	B <=0.005; Mo <=1; Ni <=2; Nb <=0.06; Ti <=0.2; N2<=0.015; CEV maybe used; bal Fe	550-720	460	17T	
AFNOR NFA36204(92)	S460Q-I-K4	50<t<=70mm; Q/T	0.08 max	2 max	1.5 max	0.3-1.3	0.02 max	0.01 max	0.5 max	0.2 max	B <=0.005; Mo <=1; Ni <=2; Nb <=0.06; Ti <=0.2; N2<=0.015; CEV maybe used; bal Fe	550-720	440	17T	
AFNOR NFA36204(92)	S500-Q-I-K2	50<t<=70mm; Q/T	0.08 max	2 max	1.5 max	0.3-1.3	0.025 max	0.015 max	0.5 max	0.2 max	B <=0.005; Mo <=1; Ni <=2; Nb <=0.06; Ti <=0.2; N2<=0.015; CEV maybe used; bal Fe	600-770	480	17T	
AFNOR NFA36204(92)	S500-Q-I-K4	50<t<=70mm; Q/T	0.08 max	2 max	1.5 max	0.3-1.3	0.02 max	0.01 max	0.5 max	0.2 max	B <=0.005; Mo <=1; Ni <=2; Nb <=0.06; Ti <=0.2; N2<=0.015; CEV maybe used; bal Fe	600-770	480	17T	
AFNOR NFA36204(92)	S500-Q-I-K4	t<50mm; Q/T	0.08 max	2 max	1.5 max	0.3-1.3	0.02 max	0.01 max	0.5 max	0.2 max	B <=0.005; Mo <=1; Ni <=2; Nb <=0.06; Ti <=0.2; N2<=0.015; CEV maybe used; bal Fe	600-770	500	17T	
AFNOR NFA36204(92)	S550-Q-I-K2	50<t<=70mm; Q/T	0.08 max	2 max	1.5 max	0.3-1.3	0.025 max	0.015 max	0.5 max	0.2 max	B <=0.005; Mo <=1; Ni <=2; Nb <=0.06; Ti <=0.2; N2<=0.015; CEV maybe used; bal Fe	650-820	530	16T	
AFNOR NFA36204(92)	S550-Q-I-K2	t<50mm; Q/T	0.08 max	2 max	1.5 max	0.3-1.3	0.025 max	0.015 max	0.5 max	0.2 max	B <=0.005; Mo <=1; Ni <=2; Nb <=0.06; Ti <=0.2; N2<=0.015; CEV maybe used; bal Fe	650-820	550	16T	

Specification	Designation	Notes	C	Cr	Cu	Mn	P	S	Si	V	Other	UTS	YS	El	Hard

High Strength Steel, Low-Alloy (HSLA), No equivalents identified

France

Specification	Designation	Notes	C	Cr	Cu	Mn	P	S	Si	V	Other	UTS	YS	El	Hard
AFNOR NFA36204(92)	S550-Q-I-K4	t<50mm; Q/T	0.08 max	2 max		1.5 max	0.3-1.3	0.02 max	0.01 max	0.5 max	0.2 max	B <=0.005; Mo <=1; Ni <=2; Nb <=0.06; Ti <=0.2; N2<=0.015; CEV maybe used; bal Fe	650-820	550	16T	
AFNOR NFA36204(92)	S550-Q-I-K4	50<t<=70mm; Q/T	0.08 max	2 max		1.5 max	0.3-1.3	0.02 max	0.01 max	0.5 max	0.2 max	B <=0.005; Mo <=1; Ni <=2; Nb <=0.06; Ti <=0.2; N2<=0.015; CEV maybe used; bal Fe	650-820	530	16T	
AFNOR NFA36204(92)	S620-Q-I-K2	50<t<=70mm; Q/T	0.08 max	2 max		1.5 max	0.3-1.3	0.025 max	0.015 max	0.5 max	0.2 max	B <=0.005; Mo <=1; Ni <=2; Nb <=0.06; Ti <=0.2; N2<=0.015; CEV maybe used; bal Fe	720-890	600	15T	
AFNOR NFA36204(92)	S620-Q-I-K2	t<50mm; Q/T	0.08 max	2 max		1.5 max	0.3-1.3	0.025 max	0.015 max	0.5 max	0.2 max	B <=0.005; Mo <=1; Ni <=2; Nb <=0.06; Ti <=0.2; N2<=0.015; CEV maybe used; bal Fe	720-890	620	15T	
AFNOR NFA36204(92)	S620-Q-I-K4	t<50mm; Q/T	0.08 max	2 max		1.5 max	0.3-1.3	0.02 max	0.1 max	0.5 max	0.2 max	B <=0.005; Mo <=1; Ni <=2; Nb <=0.06; Ti <=0.2; N2<=0.015; CEV maybe used; bal Fe	720-890	620	15T	
AFNOR NFA36204(92)	S620-Q-I-K4	50<t<=70mm; Q/T	0.08 max	2 max		1.5 max	0.3-1.3	0.02 max	0.1 max	0.5 max	0.2 max	B <=0.005; Mo <=1; Ni <=2; Nb <=0.06; Ti <=0.2; N2<=0.015; CEV maybe used; bal Fe	720-890	600	15T	

International

Specification	Designation	Notes	C	Cr	Cu	Mn	P	S	Si	V	Other	UTS	YS	El	Hard
ISO 1052(82)	Fe590	Struct, t<=16mm					0.050 max	0.050 max			bal Fe	590-740	335		
ISO 1052(82)	Fe690	Struct, t<=16mm					0.050 max	0.050 max			bal Fe	690-840	365		
ISO 4950-2(95)	E355	Flat, norm, t<=16mm	0.18 max	0.25 max	0.35 max	0.9-1.6	0.030 max	0.030 max	0.50 max	0.02-0.10	Al >=0.020; Mo <=0.10; Ni <=0.30; Nb 0.015-0.060; Ti 0.02-0.20; bal Fe	470-630	355	22	
ISO 4950-2(95)	E460	Flat, norm, t<=16mm	0.20 max	0.70 max	0.70 max	1.0-1.7	0.040 max	0.040 max	0.50 max	0.02-0.20	Al >=0.020; Mo <=0.40; Ni <=1.0; Nb 0.015-0.060; Ti 0.02-0.20; bal Fe	550-720	460	17	

Mexico

Specification	Designation	Notes	C	Cr	Cu	Mn	P	S	Si	V	Other	UTS	YS	El	Hard
NMX-B-069(86)	Grade II	Pipe, weld and Smls	0.22 max		0.20 min	0.85-1.25	0.04 max	0.05 max	0.30 max	0.02 min	bal Fe	485	345	11	
NMX-B-069(86)	Grade V (Typ E, S)	Pipe, weld and Smls	0.16 max		0.80 min	0.40-1.01	0.035 max	0.040 max			Ni >=1.65; bal Fe	450	315		
NMX-B-069(86)	Grade VII	Pipe, weld and Smls	0.12 max	0.30-1.25	0.25-0.55	0.20-0.50	0.07-0.15	0.05 max	0.25-0.75		Ni >=0.65; bal Fe	450	310	22	
NMX-B-069(86)	Grade VIII	Pipe, weld and Smls	0.19 max	0.40-0.65	0.25-0.40	0.80-1.25	0.04 max	0.05 max	0.30-0.65	0.02-0.10	Ni <=0.40; bal Fe	485	345	21	

Spain

Specification	Designation	Notes	C	Cr	Cu	Mn	P	S	Si	V	Other	UTS	YS	El	Hard
UNE 36087(76/78)	15Ni6	Sh Plt Strp CR	0.18 max			0.8 max	0.035 max	0.035 max	0.15-0.35		Mo <=0.15; Ni 1.3-1.7; bal Fe				
UNE 36087(76/78)	F.2641*	see 15Ni6	0.18 max			0.8 max	0.035 max	0.035 max	0.15-0.35		Mo <=0.15; Ni 1.3-1.7; bal Fe				

UK

Specification	Designation	Notes	C	Cr	Cu	Mn	P	S	Si	V	Other	UTS	YS	El	Hard
BS 1503(89)	223-510	Frg press ves; t<=100mm; Q/T	0.25 max	0.25 max	0.30 max	0.90-1.70	0.030 max	0.025 max	0.10-0.40		Mo <=0.10; Ni <=0.40; Nb 0.01-0.06; W <=0.1; Cr+Mo+Ni+Cu<=0.8; bal Fe	510-630	340	16	

USA

Specification	Designation	Notes	C	Cr	Cu	Mn	P	S	Si	V	Other	UTS	YS	El	Hard
ASTM A514(94)	Grade K	Q/T, Plt, t=50mm, weld	0.10-0.20			1.10-1.50	0.035 max	0.035 max	0.15-0.30		B 0.001-0.005; Mo 0.45-0.55; bal Fe	760-895	690	18	
ASTM A514(94)	Grade Q	Q/T, Plt, t=150mm, weld	0.14-0.21	1.00-1.50		0.95-1.30	0.035 max	0.035 max	0.15-0.35	0.03-0.08	Mo 0.40-0.60; Ni 1.20-1.50; bal Fe	690-895	620	16	
ASTM A514(94)	Grade R	Q/T, Plt, t=65mm, weld	0.15-0.20	0.35-0.65		0.85-1.15	0.035 max	0.035 max	0.20-0.35	0.03-0.08	Mo 0.15-0.25; Ni 0.90-1.10; bal Fe	760-895	690	18	

UNS numbers and US grades are provided as a means of cross referencing chemically similar alloys. Exchangability is only possible after independent examination of specifications. Tensile properties are minimum or typical as specified. UTS and YS as MPa. El as %. See Appendix for list of abbreviations used in Notes. * indicates obsolete material.

Specification	Designation	Notes	C	Cr	Cu	Mn	P	S	Si	V	Other	UTS	YS	El	Hard

High Strength Steel, Low-Alloy (HSLA), No equivalents identified

USA

Specification	Designation	Notes	C	Cr	Cu	Mn	P	S	Si	V	Other	UTS	YS	El	Hard
ASTM A514(94)	Grade S	Q/T, Plt, t=65mm, weld	0.11-0.21			0.85-1.15	0.035 max	0.020 max	0.15-0.45	0.06 max	B 0.001-0.005; Mo 0.10-0.60; Nb <=0.06; Ti <=0.06; Ti may be present to protect Boron; bal Fe	760-895	690	18	
ASTM A514(94)	Grade T	Q/T, Plt, t=50mm, weld	0.08-0.14			1.20-1.50	0.035 max	0.010 max	0.40-0.60	0.03-0.08	B 0.001-0.005; Mo 0.45-0.60; bal Fe	760-895	690	18	
ASTM A517(93)	Grade K	Q/T, Press ves plt, t<=50mm	0.10-0.20			0.10-1.50	0.035 max	0.035 max	0.15-0.30		B 0.001-0.005; Mo 0.45-0.55; bal Fe	795-930	690	16	
ASTM A517(93)	Grade Q	Q/T, Press ves plt, t<=65mm	0.14-0.21	1.00-1.50		0.95-1.30	0.035 max	0.035 max	0.15-0.35	0.03-0.08	Mo 0.40-0.60; Ni 1.20-1.50; bal Fe	795-930	690	16	
ASTM A517(93)	Grade S	Q/T, Press ves plt, t<=50mm	0.10-0.20			1.10-1.50	0.035 max	0.035 max	0.15-0.40		Mo 0.10-0.35; Nb <=0.06; Ti <=0.06; bal Fe	795-930	690	16	
ASTM A517(93)	Grade T	Q/T, Press ves plt; t<=50mm	0.08-0.14			1.20-1.50	0.035 max	0.010 max	0.40-0.60	0.03-0.08	B 0.001-0.005; Mo 0.45-0.60; bal Fe	795-930	690	16	
ASTM A588/A588M(97)	Grade K	t>=100-125mm for plt/bar	0.17 max	0.40-0.70	0.30-0.50	0.50-1.20	0.04 max	0.05 max	0.25-0.50		Mo <=0.10; Ni <=0.40; Nb 0.005-0.05; bal Fe	460	315	21	
ASTM A588/A588M(97)	Grade K	Struct Shp, t<=100mm for plt/bar	0.17 max	0.40-0.70	0.30-0.50	0.50-1.20	0.04 max	0.05 max	0.25-0.50		Mo <=0.10; Ni <=0.40; Nb 0.005-0.05; bal Fe	485	345	21	
ASTM A588/A588M(97)	Grade K	t>=125-200mm for plt/bar	0.17 max	0.40-0.70	0.30-0.50	0.50-1.20	0.04 max	0.05 max	0.25-0.50		Mo <=0.10; Ni <=0.40; Nb 0.005-0.05; bal Fe	435	290	21	
ASTM A606(98)	A606	HR, Sh Strp	0.22 max			1.25 max		0.04 max			bal Fe	480	340	22	
ASTM A606(98)	A606	CR, Sh Strp	0.22 max			1.25 max		0.04 max			bal Fe	450	310	22	
ASTM A607(98)	45	HR, CR, Sh Strp	0.22 max			1.35 max	0.04 max	0.04 max		0.01 min	Nb >=0.005; Nb or V min; bal Fe	410	310	22-25	
ASTM A607(98)	50	HR, CR, Sh Strp	0.23 max			1.35 max	0.04 max	0.04 max		0.01 min	Nb >=0.005; Nb or V min; bal Fe	450	340	20-22	
ASTM A607(98)	55	HR, CR, Sh Strp	0.25 max			1.35 max	0.04 max	0.04 max		0.01 min	Nb >=0.005; Nb or V min; bal Fe	480	380	18-20	
ASTM A607(98)	60	HR, CR, Sh Strp	0.26 max			1.50 max	0.04 max	0.04 max		0.01 min	Nb >=0.005; Nb or V min; bal Fe	520	410	16-18	
ASTM A607(98)	65	HR, CR, Sh Strp	0.26 max			1.50 max	0.04 max	0.04 max		0.01 min	N <=0.012; Nb >=0.005; Nb or V min; bal Fe	550	450	14-16	
ASTM A607(98)	70	HR, CR, Sh Strp	0.26 max			1.65 max	0.04 max	0.04 max		0.01 min	N <=0.012; Nb >=0.005; Nb or V min; bal Fe	590	480	12-14	
ASTM A656/A656M(97)	Type 3	HR struct, plt	0.18 max			1.65 max	0.025 max	0.035 max	0.60 max	0.08 max	N <=0.020; Nb 0.008-0.15; bal Fe	415	345	23	
ASTM A656/A656M(97)	Type 7	HR struct, plt	0.18 max			1.65 max	0.025 max	0.035 max	0.60 max	0.15 max	N <=0.020; Nb <=0.10; bal Fe	415	345	23	
ASTM A709(97)	Grade 100, Type Q	Q/T, Plt, t<=100mm	0.14-0.21	1.00-1.50		0.95-1.30	0.035 max	0.035 max	0.15-0.35	0.03-0.08	Mo 0.40-0.60; Ni 1.20-1.50; bal Fe	690-895	620	16	
ASTM A709(97)	Grade 70W	HT, t<=100mm	0.19 max	0.40-0.70	0.20-0.40	0.80-1.35	0.035 max	0.04 max	0.20-0.65	0.02-0.10	Ni <=0.50; bal Fe	620-760	485	19	
ASTM A714(96)	VII	Weld, smls pipe, Grade VII Type E, Type S	0.12 max	0.30-1.25	0.25-0.55	0.20-0.50	0.07-0.15	0.05	0.25-0.75		Ni <=0.65; bal Fe	450	310		
ASTM A714(96)	VIII	Weld, smls pipe, Grade VIII Type E, Type S	0.19 max	0.40-0.65	0.25-0.40	0.80-1.25	0.04 max	0.05 max	0.30-0.65	0.02-0.10	Ni <=0.40; bal Fe	485	345	22	
ASTM A847(96)	A847	CF, weld/smls struct tub; Corr res	0.20 max		0.20 min	1.35 max	0.15 max	0.05 max			bal Fe	483	345	19	
ASTM A935/A935M(97)	Grade 45	Sh Strp, thick, HR, Class 1	0.22 max			1.50 max	0.04 max	0.04 max		0.01 min	Nb >=0.005; Nb or V min; Cu 0.20 if spec; bal Fe*	410	310	22.0	
ASTM A935/A935M(97)	Grade 50	Sh Strp, thick, HR, Class 1	0.23 max			1.50 max	0.04 max	0.04 max		0.01 min	Nb >=0.005; Nb or V min; Cu 0.20 if spec; bal Fe	450	340	20.0	
ASTM A935/A935M(97)	Grade 55	Sh Strp, thick, HR, Class 1	0.25 max			1.50 max	0.04 max	0.04 max		0.01 min	Nb >=0.005; Nb or V min; Cu 0.20 if spec; bal Fe	480	380	18.0	

UNS numbers and US grades are provided as a means of cross referencing chemically similar alloys. Exchangability is only possible after independent examination of specifications. Tensile properties are minimum or typical as specified. UTS and YS as MPa. El as %. See Appendix for list of abbreviations used in Notes. * indicates obsolete material.

High Strength Steel, Low-Alloy (HSLA), No equivalents identified

USA

Specification	Designation	Notes	C	Cr	Cu	Mn	P	S	Si	V	Other	UTS	YS	El	Hard
ASTM A935/A935M(97)	Grade 60	Sh Strp, thick, HR, Class 1	0.26 max			1.50 max	0.04 max	0.04 max		0.01 min	Nb >=0.005; Nb or V min; Cu 0.20 if spec; bal Fe	520	410	16.0	
ASTM A935/A935M(97)	Grade 65	Sh Strp, thick, HR, Class 1	0.26 max			1.50 max	0.04 max	0.04 max		0.01 min	N <=0.012; Nb >=0.005; Nb or V min; Cu 0.20 if spec; bal Fe	550	450	14.0	
ASTM A935/A935M(97)	Grade 70	Sh Strp, thick, HR, Class 1	0.26 max			1.65 max	0.04 max	0.04 max		0.01 min	N <=0.012; Nb >=0.005; Nb or V min; Cu 0.20 if spec; bal Fe	590	480	12.0	
ASTM A936/A936M(97)	Grade 50	Sh Strp, thick, HR	0.15 max	0.15 max	0.20 max	1.65 max	0.025 max	0.035 max			Mo <=0.06; Ni <=0.20; Nb,V or Ti req'd; Cu+Ni+Cr+Mo<=0.50; bal Fe	415	345	22.0	
ASTM A936/A936M(97)	Grade 60	Sh Strp, thick, HR	0.15 max	0.15 max	0.20 max	1.65 max	0.025 max	0.035 max			Mo <=0.06; Ni <=0.20; Nb,V or Ti req'd; Cu+Ni+Cr+Mo<=0.50; bal Fe	485	415	16.0	
ASTM A936/A936M(97)	Grade 70	Sh Strp, thick, HR	0.15 max	0.15 max	0.20 max	1.65 max	0.025 max	0.035 max			Mo <=0.06; Ni <=0.20; Nb,V or Ti req'd; Cu+Ni+Cr+Mo<=0.50; bal Fe	550	485	12.0	
ASTM A936/A936M(97)	Grade 80	Sh Strp, thick, HR	0.15 max	0.15 max	0.20 max	1.65 max	0.025 max	0.035 max			Mo <=0.06; Ni <=0.20; Nb,V or Ti req'd; Cu+Ni+Cr+Mo<=0.50; bal Fe	620	550	12.0	
MIL-S-12505B(70)	- - -*	Obs; Shape Plt Bar for weld rivet bolted structures													
SAE J1392(84)	035A	High-C, Sh Strp, CR coated	0.20 max			0.60 max					bal Fe		240	22	
SAE J1392(84)	035A	High-C, Sh Strp, HR	0.25 max			0.60 max					bal Fe		240	21	
SAE J1392(84)	035B	High-C, Sh Strp, HR	0.25 max			0.60 max					N added; bal Fe		240	21	
SAE J1392(84)	035B	High-C, Sh Strp, CR coated	0.20 max			0.60 max					N added; bal Fe		240	22	
SAE J1392(84)	035C	High-C, Sh Strp, CR coated	0.20 max			0.60 max					P added; bal Fe		240	22	
SAE J1392(84)	035C	High-C, Sh Strp, HR	0.25 max			0.60 max					P added; bal Fe		240	21	
SAE J1392(84)	035S	High-C, Sh Strp, CR coated	0.20 max			0.60 max					Opt add; bal Fe		240	22	
SAE J1392(84)	035S	High-C, Sh Strp, HR	0.25 max			0.60 max					Opt add; bal Fe		240	21	
SAE J1392(84)	035X,Y,Z	Low-C, Sh Strp, HR	0.13 max			0.60 max					Microalloy; bal Fe		240	28	
SAE J1392(84)	035X,Y,Z	High-C, Sh Strp, CR coated	0.18 max			0.60 max					Microalloy; bal Fe		240	27	
SAE J1392(84)	040A	High-C, Sh Strp, CR coated	0.24 max			0.90 max					bal Fe		280	20	
SAE J1392(84)	040A	High-C, Sh Strp, HR	0.25 max			0.90 max					bal Fe		280	20	
SAE J1392(84)	040B	High-C, Sh Strp, CR coated	0.24 max			0.90 max					N added; bal Fe		280	20	
SAE J1392(84)	040B	High-C, Sh Strp, HR	0.25 max			0.90 max					N added; bal Fe		280	20	
SAE J1392(84)	040C	High-C, Sh Strp, CR coated	0.24 max			0.90 max					P added; bal Fe		280	20	
SAE J1392(84)	040C	High-C, Sh Strp, HR	0.25 max			0.90 max					P; bal Fe		280	20	
SAE J1392(84)	040S	High-C, Sh Strp, CR coated	0.24 max			0.90 max					Opt add; bal Fe		280	20	
SAE J1392(84)	040S	High-C, Sh Strp, HR	0.25 max			0.90 max					Opt add; bal Fe		280	20	

High Strength Steel, Low-Alloy (HSLA), No equivalents identified

USA

Specification	Designation	Notes	C	Cr	Cu	Mn	P	S	Si	V	Other	UTS	YS	El	Hard
SAE J1392(84)	040X,Y,Z	High-C, Sh Strp, CR coated	0.20 max			0.90 max					Microalloy; bal Fe		280	25	
SAE J1392(84)	040X,Y,Z	Low-C, Sh Strp, HR	0.13 max			0.60 max					Microalloy; bal Fe		280	27	
SAE J1392(84)	045A	High-C, Sh Strp, CR coated	0.25 max			1.20 max					bal Fe		310	18	
SAE J1392(84)	045A	High-C, Sh Strp, HR	0.25 max			0.90 max					bal Fe		310	18	
SAE J1392(84)	045B	High-C, Sh Strp, CR coated	0.25 max			1.20 max					N added; bal Fe		310	18	
SAE J1392(84)	045B	High-C, Sh Strp, HR	0.25 max			0.90 max					N added; bal Fe		310	18	
SAE J1392(84)	045C	High-C, Sh Strp, CR coated	0.25 max			1.20 max					P added; bal Fe		310	18	
SAE J1392(84)	045C	High-C, Sh Strp, HR	0.25 max			0.90 max					P; bal Fe		310	18	
SAE J1392(84)	045S	High-C, Sh Strp, CR coated	0.25 max			1.20 max					Opt add; bal Fe		310	18	
SAE J1392(84)	045S	High-C, Sh Strp, HR	0.25 max			0.90 max					Opt add; bal Fe		310	18	
SAE J1392(84)	045W	High-C, Sh Strp, HR	0.22 max			1.25 max					Si P Cu Ni Cr added; bal Fe	450	310	25	
SAE J1392(84)	045W	High-C, Sh Strp, CR coated	0.22 max			1.35 max					Si P Cu Ni Cr added; bal Fe	450	310	22	
SAE J1392(84)	045X	High-C, Sh Strp, CR coated	0.22 max			1.20 max					Microalloy; bal Fe	380	310	22	
SAE J1392(84)	045X	Low-C, Sh Strp, HR	0.13 max			1.35 max					Microalloy; bal Fe	380	310	25	
SAE J1392(84)	045Y	High-C, Sh Strp, CR coated	0.22 max			1.20 max					Microalloy; bal Fe	410	310	22	
SAE J1392(84)	045Y	High-C, Sh Strp, HR	0.22 max			1.35 max					Microalloy; bal Fe	410	310	25	
SAE J1392(84)	045Z	High-C, Sh Strp, CR coated	0.22 max			1.20 max					Microalloy; bal Fe	450	310	22	
SAE J1392(84)	045Z	High-C, Sh Strp, HR	0.22 max			1.35 max					Microalloy; bal Fe	450	310	25	
SAE J1392(84)	050A	High-C, Sh Strp, CR coated	0.25 max			1.35 max					bal Fe		340	16	
SAE J1392(84)	050A	High-C, Sh Strp, HR	0.25 max			1.35 max					bal Fe		340	16	
SAE J1392(84)	050B	High-C, Sh Strp, HR	0.25 max			1.35 max					N added; bal Fe		340	16	
SAE J1392(84)	050B	High-C, Sh Strp, CR coated	0.25 max			1.35 max					N added; bal Fe		340	16	
SAE J1392(84)	050C	High-C, Sh Strp, CR coated	0.25 max			1.35 max					P added; bal Fe		340	16	
SAE J1392(84)	050C	High-C, Sh Strp, HR	0.25 max			1.35 max					P added; bal Fe		340	16	
SAE J1392(84)	050S	High-C, Sh Strp, CR coated	0.25 max			1.35 max					Opt add; bal Fe		340	16	
SAE J1392(84)	050S	High-C, Sh Strp, HR	0.25 max			1.35 max					Opt add; bal Fe		340	16	
SAE J1392(84)	050W	High-C, Sh Strp, HR	0.22 max			1.25 max					Si P Cu Ni Cr added; bal Fe	480	340	22	
SAE J1392(84)	050X	High-C, Sh Strp, CR coated	0.23 max			1.35 max					Microalloy; bal Fe	410	340	20	
SAE J1392(84)	050X	Low-C, Sh Strp, HR	0.13 max			0.90 max					Microalloy; bal Fe	410	340	22	
SAE J1392(84)	050Y	High-C, Sh Strp, HR	0.23 max			1.35 max					Microalloy; bal Fe	450	340	22	
SAE J1392(84)	050Y	High-C, Sh Strp, CR coated	0.23 max			1.35 max					Microalloy; bal Fe	450	340	20	
SAE J1392(84)	050Z	High-C, Sh Strp, HR	0.23 max			1.35 max					Microalloy; bal Fe	480	340	22	
SAE J1392(84)	050Z	High-C, Sh Strp, CR coated	0.23 max			1.35 max					Microalloy; bal Fe	480	340	20	
SAE J1392(84)	060X	Low-C, Sh Strp, HR	0.13 max			0.90 max					Microalloy; bal Fe	480	410	20	

High Strength Steel, Low-Alloy (HSLA), No equivalents identified

USA

Specification	Designation	Notes	C	Cr	Cu	Mn	P	S	Si	V	Other	UTS	YS	El	Hard
SAE J1392(84)	060Y	High-C, Sh Strp, HR	0.26 max			1.50 max					Microalloy; bal Fe	520	410	20	
SAE J1392(84)	060Z	High-C, Sh Strp, HR	0.26 max			1.50 max					Microalloy; bal Fe	550	410		
SAE J1392(84)	070X	Low-C, Sh Strp, HR	0.13 max			1.65 max					Microalloy; bal Fe	550	480	17	
SAE J1392(84)	070Y	High-C, Sh Strp, HR	0.26 max			1.65 max					Microalloy; bal Fe	590	480	17	
SAE J1392(84)	070Z	High-C, Sh Strp, HR	0.26 max			1.65 max					Microalloy; bal Fe				
SAE J1392(84)	080X	Low-C, Sh Strp, HR	0.13 max			1.65 max					Microalloy; bal Fe	620	550	14	
SAE J1392(84)	080Y	High-C, Sh Strp, HR	0.18 max			1.65 max					Microalloy; bal Fe	650	550	14	
SAE J1442(93)	290A	HR Plt Bar Shps, t<=150mm	0.21 max			1.50 max	0.040 max	0.050 max	0.40 max		Cr,Cu,Mo,Ni,Nb,V,Ti, Zr can be added; bal Fe	415-620	290	24	
SAE J1442(93)	345A	HR Plt Bar Shps, t<=100mm	0.23 max			1.50 max	0.040 max	0.050 max	0.40 max		Cr,Cu,Mo,Ni,Nb,V,Ti, Zr can be added; bal Fe	450-655	345	21	
SAE J1442(93)	345F	HR Plt, t<=50mm	0.18 max			1.65 max	0.025 max	0.035 max	0.60 max		bal Fe	415-620	345	23	
SAE J1442(93)	345W	HR Plt Bar Shps, t<=100mm	0.20 max			1.35 max	0.040 max	0.050 max	0.90 max		bal Fe	485-655	345	21	
SAE J1442(93)	415A	HR Plt Bar Shps, t<=32mm	0.26 max			1.50 max	0.040 max	0.050 max	0.40 max		Cr,Cu,Mo,Ni,Nb,V,Ti, Zr can be added; bal Fe	515-690	415	18	
SAE J1442(93)	415F	HR Plt, t<=40mm	0.18 max			1.65 max	0.025 max	0.035 max	0.60 max		bal Fe	485-655	415	20	
SAE J1442(93)	450A	HR Plt Bar Shps, t<=32mm	0.26 max			1.65 max	0.040 max	0.050 max	0.40 max		bal Fe	550-725	450	17	
SAE J1442(93)	485F	HR Plt, t<=25mm	0.18 max			1.65 max	0.025 max	0.035 max	0.60 max		bal Fe	550-725	485	17	
SAE J1442(93)	550F	HR Plt, t<=20mm	0.18 max			1.65 max	0.025 max	0.035 max	0.60 max		bal Fe	620-795	550	15	
SAE J467(68)	Chromoloy		0.20	1.00		0.50			0.75	0.10	Mo 1.00; bal Fe	952	807	7	30 HRC
SAE J467(68)	UCX2		0.39	1.10		0.70			1.00	0.15	Co 1.00; Mo 0.25; bal Fe	1875	1620	6	51 HRC

Specification	Designation	Notes	C	Co	Mn	Mo	Ni	P	S	Si	Other	UTS	YS	El	Hard

High Strength Steel, Maraging, A538(A)

China

Specification	Designation	Notes	C	Co	Mn	Mo	Ni	P	S	Si	Other	UTS	YS	El	Hard
	18Ni(200)	Plt Frg SHT Aged	0.03 max	8.0-9.0	0.20 max	3.0-3.5	17.5-18.5	0.01 max	0.01 max	0.10 max	Al 0.05-0.15; Ti 0.15-0.25; bal Fe	1480	1430	9	

USA

Specification	Designation	Notes	C	Co	Mn	Mo	Ni	P	S	Si	Other	UTS	YS	El	Hard
	UNS K92810		0.03 max	7.0-8.5	0.10 max	4.0-4.5	17.0-19.0	0.010 max	0.010 max	0.10 max	Al 0.05-0.15; Ti 0.10-0.25; bal Fe				
ASTM A538/A538M	A														
MIL-S-46850D(91)	- - -*	Obs for new design; Bar Plt Sh Strp frg ext													

High Strength Steel, Maraging, A538(B)

China

Specification	Designation	Notes	C	Co	Mn	Mo	Ni	P	S	Si	Other	UTS	YS	El	Hard
	18Ni(250)	Plt Frg SHT Aged	0.03 max	8.5-9.5	0.20 max	4.6-5.2	17.5-18.5	0.01 max	0.01 max	0.10 max	Al 0.05-0.15; Ti 0.30-0.50; bal Fe	1785	1725	12	

Germany

Specification	Designation	Notes	C	Co	Mn	Mo	Ni	P	S	Si	Other	UTS	YS	El	Hard
DIN(Aviation Hdbk)	WNr 1.6359		0.03 max	7.00-8.50	0.10 max	4.60-5.20	17.0-19.0	0.010 max	0.010 max	0.10 max	Al 0.05-0.15; B <=0.003; Ti 0.30-0.60; Zn <=0.02; Ca<=0.05; bal Fe				
DIN(Aviation Hdbk)	X2NiCoMo18-8-5		0.03 max	7.00-8.50	0.10 max	4.60-5.20	17.0-19.0	0.010 max	0.010 max	0.10 max	Al 0.05-0.15; B <=0.003; Ti 0.30-0.60; Zn <=0.02; Ca<=0.05; bal Fe				

USA

Specification	Designation	Notes	C	Co	Mn	Mo	Ni	P	S	Si	Other	UTS	YS	El	Hard
	UNS K92890		0.03 max	7.0-8.5	0.10 max	4.6-5.1	17.0-19.0	0.010 max	0.010 max	0.1 max	Al 0.05-0.15; Ti 0.30-0.50; bal Fe				
AMS 6501															
AMS 6512															
AMS 6520															
ASTM A538/A538M	B														
MIL-S-46850D(91)	- - -*	Obs for new design; Bar Plt Sh Strp frg ext													

High Strength Steel, Maraging, A538(C)

China

Specification	Designation	Notes	C	Co	Mn	Mo	Ni	P	S	Si	Other	UTS	YS	El	Hard
	18Ni(300)	Plt Frg SHT Aged	0.03 max	8.0-9.0	0.20 max	4.6-5.2	17.5-18.5	0.01 max	0.01 max	0.10 max	Al 0.05-0.15; Ti 0.55-0.80; bal Fe	2050	1970	12	

Germany

Specification	Designation	Notes	C	Co	Mn	Mo	Ni	P	S	Si	Other	UTS	YS	El	Hard
DIN	WNr 1.6358		0.03 max	8.00-10.0	0.10 max	4.50-5.50	17.0-19.0	0.010 max	0.010 max	0.10 max	Al 0.05-0.15; Ti 0.50-0.80; B, Zr; bal Fe				

USA

Specification	Designation	Notes	C	Co	Mn	Mo	Ni	P	S	Si	Other	UTS	YS	El	Hard
	UNS K93120		0.03 max	8.0-9.5	0.10 max	4.6-5.2	18.0-19.0	0.010 max	0.010 max	0.10 max	Al 0.05-0.15; Ti 0.55-0.80; bal Fe				
AMS 6514															
AMS 6521															
ASTM A538/A538M	C														
ASTM A579(96)	73	Superstrength frg	0.03 max	8.5-9.5	0.10 max	4.6-5.2	18.0-19.0	0.01 max	0.01 max	0.10 max	Al 0.05-0.15; B 0.003; Ti 0.50-0.80; Ca 0.06; Zr 0.02; bal Fe	1930	1900	9	321 HB
MIL-S-46850D(91)	- - -*	Obs for new design; Bar Plt Sh Strp frg ext													

Specification	Designation	Notes	C	Cu	Mn	N	P	S	Si	V	Other	UTS	YS	El	Hard
High Strength Steel, Structural, 50-1															
Bulgaria															
BDS 2592(71)	ASt5	Struct													
BDS 2592(71)	ASt6	Struct													
BDS 2592(71)	WSt5ps	Struct	0.37 max	0.30 max	1.10 max		0.045 max	0.055 max	0.05-0.17		Cr <=0.30; Ni <=0.30; bal Fe				
BDS 2592(71)	WSt5sp	Struct	0.37 max	0.30 max	1.10 max		0.045 max	0.055 max	0.15-0.35		Cr <=0.30; Ni <=0.30; bal Fe				
BDS 2592(71)	WSt6sp	Struct	0.49 max	0.30 max	1.10 max		0.045 max	0.055 max	0.15-0.35		Cr <=0.30; Ni <=0.30; bal Fe				
China															
GB 1591(88)	16MnNb	Plt HR 16-25mm Thk	0.12-0.20		1.00-1.40		0.045 max	0.045 max	0.20-0.55		Nb 0.015-0.050; bal Fe	510-660	375	19	
GB/T 1591(94)	Q390	Plt HR 16-35mm Thk	0.20 max		1.00-1.60		0.045 max	0.045 max	0.55 max	0.02-0.15	Nb 0.015-0.060; Ti 0.02-0.2; bal Fe	490-650	370	19	
GB/T 8164(93)	16MnNb	Strp HR 4-8mm Thk	0.12-0.20		1.00-1.40		0.045 max	0.045 max	0.20-0.55		Nb 0.015-0.050; bal Fe	510	375	19	
Europe															
EN 10025(90)	Fe490-2	Obs EU desig; struct HR FN; BS 16<t<=40mm			1.6 max	0.009 max	0.045 max	0.045 max			N max na if Al>=0.020; CEV; bal Fe	470-610	285	20	
EN 10025(90)	Fe490-2	Obs EU desig; struct HR FN; BS 40<t<=100mm			1.6 max	0.009 max	0.045 max	0.045 max			N max na if Al>=0.020; CEV; bal Fe	470-610	255-275	18-19	
EN 10025(90)	Fe490-2	Obs EU desig; struct HR FN; BS 150<t<=250mm			1.6 max	0.009 max	0.045 max	0.045 max			N max na if Al>=0.020; CEV; bal Fe	540-710	225-235	15	
EN 10025(90)	Fe490-2	Obs EU desig; struct HR FN; BS t<=16mm			1.6 max	0.009 max	0.045 max	0.045 max			N max na if Al>=0.020; CEV; bal Fe	470-660	295	12-20	
EN 10025(90)	Fe510B	Obs EU desig; struct HR FN; BS t<=16mm	0.24 max		1.6 max	0.009 max	0.045 max	0.045 max	0.55 max		N max na if Al>=0.020; bal Fe	410-580	355	14-22	
EN 10025(90)	Fe510B	Obs EU desig; struct HR FN; BS 16<t<=40mm	0.24 max		1.6 max	0.009 max	0.045 max	0.045 max	0.55 max		N max na if Al>=0.020; bal Fe	490-630	345	22	
EN 10025(90)	Fe510B	Obs EU desig; struct HR FN; BS 40<t<=100mm	0.24 max		1.6 max	0.009 max	0.045 max	0.045 max	0.55 max		N max na if Al>=0.020; bal Fe	490-630	325-335	20-21	
EN 10025(90)	Fe510B	Obs EU desig; struct HR FN; BS 150<t<=250mm	0.24 max		1.6 max	0.009 max	0.045 max	0.045 max	0.55 max		N max na if Al>=0.020; bal Fe	440-630	275-285	17	
EN 10025(90)	Fe510C	Obs EU desig; struct HR FN; QS t<=16mm	0.20 max		1.6 max	0.009 max	0.040 max	0.040 max	0.55 max		N max na if Al>=0.020; CEV; bal Fe	410-580	355	14-22	
EN 10025(90)	Fe510C	Obs EU desig; struct HR FN; QS 16<t<=40mm	0.20 max		1.6 max	0.009 max	0.040 max	0.040 max	0.55 max		N max na if Al>=0.020; CEV; bal Fe	490-630	345	22	
EN 10025(90)	Fe510C	Obs EU desig; struct HR FN; QS 40<t<=100mm	0.22 max		1.6 max	0.009 max	0.040 max	0.040 max	0.55 max		N max na if Al>=0.020; CEV; bal Fe	490-630	325-335	20-21	
EN 10025(90)	Fe510C	Obs EU desig; struct HR FN; QS 150<t<=250mm	0.22 max		1.6 max	0.009 max	0.040 max	0.040 max	0.55 max		N max na if Al>=0.020; CEV; bal Fe	440-630	275-285	17	
EN 10025(90)	Fe510D1	Obs EU desig; struct HR FF; QS t<=16mm	0.20 max		1.6 max		0.035 max	0.035 max	0.55 max		N max na if Al>=0.020; CEV; bal Fe	410-580	355	14-22	
EN 10025(90)	Fe510D1	Obs EU desig; struct HR FF; QS 16<t<=40mm	0.20 max		1.6 max		0.035 max	0.035 max	0.55 max		N max na if Al>=0.020; CEV; bal Fe	490-630	345	22	
EN 10025(90)	Fe510D1	Obs EU desig; struct HR FF; QS 40<t<=100mm	0.22 max		1.6 max		0.035 max	0.035 max	0.55 max		N max na if Al>=0.020; CEV; bal Fe	490-630	325-335	20-21	
EN 10025(90)	Fe510D1	Obs EU desig; struct HR FF; QS 150<t<=250mm	0.22 max		1.6 max		0.035 max	0.035 max	0.55 max		N max na if Al>=0.020; CEV; bal Fe	440-630	275-285	17	
EN 10025(90)A1(93)	1.0050	Struct HR BS, t<=16mm					0.045 max	0.045 max			N max na if Al>=0.020; bal Fe	470-610	295	20L 18T	
EN 10025(90)A1(93)	1.0050	Struct HR QS, t<=1mm					0.045 max	0.045 max			N max na if Al>=0.020; bal Fe	490-660	295	12L 10T	
EN 10025(90)A1(93)	1.0050	Struct HR BS, 200<t<=250mm					0.045 max	0.045 max			N max na if Al>=0.020; bal Fe	440-610	225	15L 14T	

UNS numbers and US grades are provided as a means of cross referencing chemically similar alloys. Exchangability is only possible after independent examination of specifications. Tensile properties are minimum or typical as specified. UTS and YS as MPa. El as %. See Appendix for list of abbreviations used in Notes. * indicates obsolete material.

Specification	Designation	Notes	C	Cu	Mn	N	P	S	Si	V	Other	UTS	YS	El	Hard

High Strength Steel, Structural, 50-1 (Continued from previous page)

Europe

Specification	Designation	Notes	C	Cu	Mn	N	P	S	Si	V	Other	UTS	YS	El	Hard
EN 10025(90)A1(93)	1.0050	Struct HR, t<=1mm					0.045 max	0.045 max			N max na if Al>=0.020; bal Fe	490-660	295	12L 10T	
EN 10025(90)A1(93)	1.0050	Struct HR BS, 63<t<=80mm					0.045 max	0.045 max			N max na if Al>=0.020; bal Fe	470-610	265	18L 16T	
EN 10025(90)A1(93)	1.0060	Struct HR BS, 63<t<=80mm					0.045 max	0.045 max			N max na if Al>=0.020; bal Fe	570-710	305	14L 12T	
EN 10025(90)A1(93)	1.0060	Struct HR BS, t<=1mm					0.045 max	0.045 max			N max na if Al>=0.020; bal Fe	590-770	335	8L 6T	
EN 10025(90)A1(93)	1.0060	Struct HR BS, 200<t<=250mm					0.045 max	0.045 max			N max na if Al>=0.020; bal Fe	540-710	255	11L 10T	
EN 10025(90)A1(93)	1.0060	Struct HR BS, 3<t<=16mm					0.045 max	0.045 max			N max na if Al>=0.020; bal Fe	570-710	335	16L 14T	
EN 10025(90)A1(93)	1.0060	Struct HR QS, t<=1mm					0.045 max	0.045 max			N max na if Al>=0.020; bal Fe	590-770	335	8L 6T	
EN 10025(90)A1(93)	1.0070	Struct HR BS, 200<t<=250mm					0.045 max	0.045 max			N max na if Al>=0.020; bal Fe	640-830	285	7L 6T	
EN 10025(90)A1(93)	1.0070	Struct HR BS, t<=1mm					0.045 max	0.045 max			N max na if Al>=0.020; bal Fe	690-900	360	4L 3T	
EN 10025(90)A1(93)	1.0070	Struct HR QS, t<=1mm					0.045 max	0.045 max			N max na if Al>=0.020; bal Fe	690-900	360	4L 3T	
EN 10025(90)A1(93)	1.0070	Struct HR BS, 63<t<=80mm					0.045 max	0.045 max			N max na if Al>=0.020; bal Fe	670-830	335	9L 8T	
EN 10025(90)A1(93)	1.0070	Struct HR BS, 3<t<=16mm					0.045 max	0.045 max			N max na if Al>=0.020; bal Fe	670-830	360	11L 10T	
EN 10025(90)A1(93)	E295	Struct HR BS, 200<t<=250mm					0.045 max	0.045 max			N max na if Al>=0.020; bal Fe	440-610	225	15L 14T	
EN 10025(90)A1(93)	E295	Struct HR BS, 63<t<=80mm					0.045 max	0.045 max			N max na if Al>=0.020; bal Fe	470-610	265	18L 16T	
EN 10025(90)A1(93)	E295	Struct HR, t<=1mm					0.045 max	0.045 max			N max na if Al>=0.020; bal Fe	490-660	295	12L 10T	
EN 10025(90)A1(93)	E295	Struct HR BS, t<=16mm					0.045 max	0.045 max			N max na if Al>=0.020; bal Fe	470-610	295	20L 18T	
EN 10025(90)A1(93)	E295GC	Struct HR QS, t<=1mm					0.045 max	0.045 max			N max na if Al>=0.020; bal Fe	490-660	295	12L 10T	
EN 10025(90)A1(93)	E335	Struct HR BS, t<=1mm					0.045 max	0.045 max			N max na if Al>=0.020; bal Fe	590-770	335	8L 6T	
EN 10025(90)A1(93)	E335	Struct HR BS, 3<t<=16mm					0.045 max	0.045 max			N max na if Al>=0.020; bal Fe	570-710	335	16L 14T	
EN 10025(90)A1(93)	E335	Struct HR BS, 63<t<=80mm					0.045 max	0.045 max			N max na if Al>=0.020; bal Fe	570-710	305	14L 12T	
EN 10025(90)A1(93)	E335	Struct HR BS, 200<t<=250mm					0.045 max	0.045 max			N max na if Al>=0.020; bal Fe	540-710	255	11L 10T	
EN 10025(90)A1(93)	E335GC	Struct HR QS, t<=1mm					0.045 max	0.045 max			N max na if Al>=0.020; bal Fe	590-770	335	8L 6T	
EN 10025(90)A1(93)	E360	Struct HR BS, 3<t<=16mm					0.045 max	0.045 max			N max na if Al>=0.020; bal Fe	670-830	360	11L 10T	
EN 10025(90)A1(93)	E360	Struct HR BS, 200<t<=250mm					0.045 max	0.045 max			N max na if Al>=0.020; bal Fe	640-830	285	7L 6T	
EN 10025(90)A1(93)	E360	Struct HR BS, t<=1mm					0.045 max	0.045 max			N max na if Al>=0.020; bal Fe	690-900	360	4L 3T	
EN 10025(90)A1(93)	E360	Struct HR BS, 63<t<=80mm					0.045 max	0.045 max			N max na if Al>=0.020; bal Fe	670-830	335	9L 8T	
EN 10025(90)A1(93)	E360GC	Struct HR QS, t<=1mm					0.045 max	0.045 max			N max na if Al>=0.020; bal Fe	690-900	360	4L 3T	

France

Specification	Designation	Notes	C	Cu	Mn	N	P	S	Si	V	Other	UTS	YS	El	Hard
AFNOR NFA35501(83)	A50-2*	30<t<=100mm; roll	0.3 max	0.3 max	2 max		0.045 max	0.045 max	0.6 max	0.1 max	Cr <=0.3; Mo <=0.15; Ni <=0.4; Ti <=0.05; bal Fe	490-610	275	20	
AFNOR NFA35501(83)	A50-2*	t<=30mm; roll	0.3 max	0.3 max	2 max		0.045 max	0.045 max	0.6 max	0.1 max	Cr <=0.3; Mo <=0.15; Ni <=0.4; Ti <=0.05; bal Fe	490-610	295	21	
AFNOR NFA35501(83)	A50-2*	110<t<=150mm; roll	0.3 max	0.3 max	2 max		0.045 max	0.045 max	0.6 max	0.1 max	Cr <=0.3; Mo <=0.15; Ni <=0.4; Ti <=0.05; bal Fe	470-610	255	13	
AFNOR NFA35501(83)	A50-2*	t<=2.99mm; roll	0.3 max	0.3 max	2 max		0.045 max	0.045 max	0.6 max	0.1 max	Cr <=0.3; Mo <=0.15; Ni <=0.4; Ti <=0.05; bal Fe	490-630	275	17	

UNS numbers and US grades are provided as a means of cross referencing chemically similar alloys. Exchangability is only possible after independent examination of specifications. Tensile properties are minimum or typical as specified. UTS and YS as MPa. El as %. See Appendix for list of abbreviations used in Notes. * indicates obsolete material.

Specification	Designation	Notes	C	Cu	Mn	N	P	S	Si	V	Other	UTS	YS	El	Hard

High Strength Steel, Structural, 50-1 (Continued from previous page)

France

Specification	Designation	Notes	C	Cu	Mn	N	P	S	Si	V	Other	UTS	YS	El	Hard
AFNOR NFA35501(83)	A50-2*	10t<=350mm; roll	0.3 max	0.3 max	2 max		0.045 max	0.045 max	0.6 max	0.1 max	Cr <=0.3; Mo <=0.15; Ni <=0.4; Ti <=0.05; bal Fe	470	255	19	
AFNOR NFA35501(83)	A60-2*	t<=30mm; roll	0.3 max	0.3 max	2 max		0.045 max	0.045 max	0.6 max		Cr <=0.3; Ni <=0.4; bal Fe	590-710	335	16	
AFNOR NFA35501(83)	A60-2*	30<t<=100mm; roll	0.3 max	0.3 max	2 max		0.045 max	0.045 max	0.6 max		Cr <=0.3; Ni <=0.4; bal Fe	590-710	315	15	
AFNOR NFA35501(83)	A60-2*	10t<=350mm; roll	0.3 max	0.3 max	2 max		0.045 max	0.045 max	0.6 max		Cr <=0.3; Ni <=0.4; bal Fe	570	295	14	
AFNOR NFA35501(83)	A70-2*	30<t<=100mm; roll	0.3 max	0.3 max	2 max		0.045 max	0.045 max	0.6 max		Cr <=0.3; Ni <=0.4; bal Fe	690-830	345	10	
AFNOR NFA35501(83)	A70-2*	10t<=350mm; roll	0.3 max	0.3 max	2 max		0.045 max	0.045 max	0.6 max		Cr <=0.3; Ni <=0.4; bal Fe	670	325	9	
AFNOR NFA35501(83)	A70-2*	t<=30mm; roll	0.3 max	0.3 max	2 max		0.045 max	0.045 max	0.6 max		Cr <=0.3; Ni <=0.4; bal Fe	690-830	365	11	
AFNOR NFA35501(83)	E28-3*	10t<=350mm; roll	0.18 max		0.3 max	1.3 max	0.04 max	0.04 max	0.4 max	0.1 max	Cr <=0.3; Mo <=0.15; Ni <=0.4; Ti <=0.05; bal Fe	380	235	22	
AFNOR NFA35501(83)	E28-3*	11t<=150mm; roll	0.18 max		0.3 max	1.3 max	0.04 max	0.04 max	0.4 max	0.1 max	Cr <=0.3; Mo <=0.15; Ni <=0.4; Ti <=0.05; bal Fe	400-540	225	18	
AFNOR NFA35501(83)	E28-3*	t<=30mm; roll	0.18 max		0.3 max	1.3 max	0.04 max	0.04 max	0.4 max	0.1 max	Cr <=0.3; Mo <=0.15; Ni <=0.4; Ti <=0.05; bal Fe	400-540	275	23	
AFNOR NFA35501(83)	E36-3*	110<t<=150mm; roll	0.22 max		0.3 max	1.6 max	0.04 max	0.04 max	0.55 max	0.1 max	Cr <=0.3; Mo <=0.15; Ni <=0.4; Ti <=0.05; bal Fe	490-630	305	15	
AFNOR NFA35501(83)	E36-3*	t<=2.99mm; roll	0.22 max		0.3 max	1.6 max	0.04 max	0.04 max	0.55 max	0.1 max	Cr <=0.3; Mo <=0.15; Ni <=0.4; Ti <=0.05; bal Fe	510-650	325	17	
AFNOR NFA35501(83)	E36-4*	30<t<=100mm; roll	0.22 max		1.6 max		0.035 max	0.035 max	0.55 max	0.1 max	Al >=0.02; Cr <=0.3; Mo <=0.15; Ni <=0.4; Ti <=0.05; bal Fe	490-630	335	22	
AFNOR NFA35501(83)	E36-4*	3<t<=30mm; roll	0.22 max		1.6 max		0.035 max	0.035 max	0.55 max	0.1 max	Al >=0.02; Cr <=0.3; Mo <=0.15; Ni <=0.4; Ti <=0.05; bal Fe	490-630	355	23	
AFNOR NFA35501(83)	E36-4*	t<3mm; roll	0.22 max		1.6 max		0.035 max	0.035 max	0.55 max	0.1 max	Al >=0.02; Cr <=0.3; Mo <=0.15; Ni <=0.4; Ti <=0.05; bal Fe	510-650	325	19	
AFNOR NFA35561(78)	A60Pb		0.3 max	0.3 max	2 max		0.045 max	0.045 max	0.6 max		Cr <=0.3; Ni <=0.4; Pb 0.2-0.3; bal Fe				
AFNOR NFA35561(78)	A70Pb	t<=30mm	0.08-0.15	0.3 max	0.3-0.6		0.04 max	0.04 max	0.3 max		Cr <=0.3; Ni <=0.4; Pb 0.2-0.3; bal Fe	360-480	235		
AFNOR NFA35561(78)	A70Pb	30<t<=100mm	0.08-0.15	0.3 max	0.3-0.6		0.04 max	0.04 max	0.3 max		Cr <=0.3; Ni <=0.4; Pb 0.2-0.3; bal Fe	360-480	215		

Hungary

Specification	Designation	Notes	C	Cu	Mn	N	P	S	Si	V	Other	UTS	YS	El	Hard
MSZ 6280(82)	52C	weld const; 3-35mm; rolled/norm	0.2 max		1.5 max		0.04 max	0.04 max	0.15-0.5	0.02-0.1	Al 0.015-0.120; Cr <=0.25; Mo <=0.1; Ni <=0.3; Nb 0.01-0.06; Ti 0.02-0.06; Zr=0.01-0.06; Al+Nb+V+Zr<=0.15; N<=Al/2+V/4+Nb/7+Ti/3.5<=0.015; bal Fe	490-610	355	23L	
MSZ 6280(82)	52C	weld const; 50.1-60mm; rolled/norm	0.2 max		1.5 max		0.04 max	0.04 max	0.15-0.5	0.02-0.1	Al 0.015-0.120; Cr <=0.25; Mo <=0.1; Ni <=0.3; Nb 0.01-0.06; Ti 0.02-0.06; Zr=0.01-0.06; Al+Nb+V+Zr<=0.15; N<=Al/2+V/4+Nb/7+Ti/3.5<=0.015; bal Fe	490-610	335	23L	

International

Specification	Designation	Notes	C	Cu	Mn	N	P	S	Si	V	Other	UTS	YS	El	Hard
ISO 1052(82)	Fe490	Struct, t<=16mm					0.050 max	0.050 max			bal Fe	490-640	295		
ISO 630(80)	Fe510D	Struct; Specially killed	0.23 max	0.3 max	2 max		0.04 max	0.04 max	0.6 max	0.1 max	Al <=0.1; Co <=0.1; Cr <=0.3; Mo <=0.15; Ni <=0.4; Pb <=0.15; Ti <=0.05; W <=0.1; bal Fe				

UNS numbers and US grades are provided as a means of cross referencing chemically similar alloys. Exchangability is only possible after independent examination of specifications. Tensile properties are minimum or typical as specified. UTS and YS as MPa. El as %. See Appendix for list of abbreviations used in Notes. * indicates obsolete material.

Specification	Designation	Notes	C	Cu	Mn	N	P	S	Si	V	Other	UTS	YS	El	Hard

High Strength Steel, Structural, 50-1 (Continued from previous page)

International

Specification	Designation	Notes	C	Cu	Mn	N	P	S	Si	V	Other	UTS	YS	El	Hard
ISO R630(67)	Fe42C	Struct	0.2 max	0.3 max	1.60 max		0.05 max	0.05 max	0.60 max	0.1 max	Al <=0.1; Co <=0.1; Cr <=0.3; Mo <=0.15; Ni <=0.4; Pb <=0.15; Ti <=0.05; W <=0.1; bal Fe				
ISO R630(67)	Fe42D	Struct	0.2 max	0.3 max	1.60 max		0.045 max	0.045 max	0.60 max	0.1 max	Al <=0.1; Co <=0.1; Cr <=0.3; Mo <=0.15; Ni <=0.4; Pb <=0.15; Ti <=0.05; W <=0.1; bal Fe				
ISO R630(67)	Fe52B	Struct	0.22 max	0.3 max	1.5 max		0.06 max	0.05 max	0.55 max	0.1 max	Al <=0.1; Co <=0.1; Cr <=0.3; Mo <=0.15; Ni <=0.4; Pb <=0.15; Ti <=0.05; W <=0.1; bal Fe				

Italy

Specification	Designation	Notes	C	Cu	Mn	N	P	S	Si	V	Other	UTS	YS	El	Hard
UNI 7070(82)	Fe510BFN	Gen struct	0.22 max		1.6 max		0.045 max	0.045 max	0.55 max		bal Fe				
UNI 7070(82)	Fe510CFN	Gen struct	0.22 max		1.6 max		0.04 max	0.045 max	0.55 max		0-16mm C<=0.2; bal Fe				
UNI 7070(82)	Fe510DDFF	Gen struct	0.22 max		1.6 max		0.04 max	0.04 max	0.55 max		Al <=0.015; 0-30mm C<=0.2; bal Fe				
UNI 7070(82)	Fe510DFF	Gen struct	0.22 max		1.6 max		0.04 max	0.04 max	0.55 max		Al <=0.015; 0-30mm C<=0.2; bal Fe				

Russia

Specification	Designation	Notes	C	Cu	Mn	N	P	S	Si	V	Other	UTS	YS	El	Hard
GOST 19282(73)	14G2	Weld constr	0.12-0.18	0.3 max	1.2-1.6		0.035 max	0.04 max	0.17-0.37	0.1 max	Al <=0.1; Cr <=0.3; Mo <=0.15; Ni <=0.3; Ti <=0.05; bal Fe				
GOST 380(71)	St6ps	Gen struct; Semi-Finished; Semi-killed; 0-20 mm	0.3 max	0.3 max	2 max		0.07 max	0.06 max	0.6 max	0.1 max	Al <=0.1; Cr <=0.3; Mo <=0.15; Ni <=0.4; Ti <=0.05; bal Fe	590		15L	
GOST 380(71)	St6ps	Gen struct; Semi-Finished; Semi-killed; 40.1-999 mm	0.3 max	0.3 max	2 max		0.07 max	0.06 max	0.6 max	0.1 max	Al <=0.1; Cr <=0.3; Mo <=0.15; Ni <=0.4; Ti <=0.05; bal Fe	590		12L	
GOST 380(71)	St6sp	Gen struct; Non-rimming; K; FF; 0-20 mm	0.3 max	0.3 max	2 max		0.07 max	0.06 max	0.6 max	0.1 max	Al <=0.1; Cr <=0.3; Mo <=0.15; Ni <=0.4; Ti <=0.05; bal Fe	590		15L	
GOST 380(71)	St6sp	Gen struct; Non-rimming; K; FF; 40.1-999 mm	0.3 max	0.3 max	2 max		0.07 max	0.06 max	0.6 max	0.1 max	Al <=0.1; Cr <=0.3; Mo <=0.15; Ni <=0.4; Ti <=0.05; bal Fe	590		12L	
GOST 380(88)	St6ps	Gen struct; Semi-Finished; Semi-killed	0.38-0.49	0.3 max	0.5-0.8	0.008 max	0.04 max	0.05 max	0.05-0.15	0.1 max	Al <=0.1; Cr <=0.3; Mo <=0.15; Ni <=0.3; Ti <=0.05; As<=0.08; bal Fe				
GOST 380(88)	St6sp	Gen struct; Semi-Finished; Non-rimming; K; FF	0.38-0.49	0.3 max	0.5-0.8	0.008 max	0.04 max	0.05 max	0.15-0.3	0.1 max	Al <=0.1; Cr <=0.3; Mo <=0.15; Ni <=0.3; Ti <=0.05; As<=0.08; bal Fe				

Spain

Specification	Designation	Notes	C	Cu	Mn	N	P	S	Si	V	Other	UTS	YS	El	Hard
UNE 36080(73)	A50-2	Gen Struct	0.30 max		1.60 max		0.05 max	0.05 max	0.60 max		Mo <=0.15; bal Fe				
UNE 36080(73)	A60-2	Gen Struct	0.30 max		1.60 max		0.05 max	0.05 max	0.60 max		Mo <=0.15; bal Fe				
UNE 36080(73)	A70-2	Gen Struct	0.30 max		1.60 max	0.009 max	0.05 max	0.05 max	0.60 max		Mo <=0.15; bal Fe				
UNE 36080(80)	A410B	Gen Struct	0.24 max		1.60 max	0.009 max	0.05 max	0.05 max	0.60 max		Mo <=0.15; bal Fe				
UNE 36080(80)	A410C	Gen Struct	0.22 max		1.60 max	0.009 max	0.045 max	0.045 max	0.60 max		Mo <=0.15; bal Fe				
UNE 36080(80)	A410D	Gen Struct	0.22 max		1.60 max		0.04 max	0.04 max	0.60 max		Mo <=0.15; +N; bal Fe				
UNE 36080(80)	A430B	Gen Struct	0.22 max		1.60 max	0.009 max	0.05 max	0.05 max	0.60 max		Mo <=0.15; bal Fe				
UNE 36080(80)	A430C	Gen Struct	0.2 max		1.60 max	0.009 max	0.045 max	0.045 max	0.60 max		Mo <=0.15; bal Fe				
UNE 36080(80)	A430D	Gen Struct	0.2 max		1.60 max		0.04 max	0.04 max	0.60 max		Mo <=0.15; +N; bal Fe				
UNE 36080(80)	A510C	Gen Struct	0.22 max		1.6 max	0.009 max	0.045 max	0.045 max	0.55 max		Mo <=0.15; bal Fe				

UNS numbers and US grades are provided as a means of cross referencing chemically similar alloys. Exchangability is only possible after independent examination of specifications. Tensile properties are minimum or typical as specified. UTS and YS as MPa. El as %. See Appendix for list of abbreviations used in Notes. * indicates obsolete material.

Specification	Designation	Notes	C	Cu	Mn	N	P	S	Si	V	Other	UTS	YS	El	Hard

High Strength Steel, Structural, 50-1 (Continued from previous page)

Spain

Specification	Designation	Notes	C	Cu	Mn	N	P	S	Si	V	Other	UTS	YS	El	Hard
UNE 36080(80)	A510D	Gen Struct	0.22 max		1.6 max		0.04 max	0.04 max	0.55 max		Mo <=0.15; +N; bal Fe				
UNE 36080(85)	A490-2	Gen Struct	0.30 max		1.60 max	0.009 max	0.045 max	0.045 max	0.60 max		Mo <=0.15; bal Fe				
UNE 36080(85)	A590-2	Gen Struct	0.30 max		1.60 max	0.009 max	0.045 max	0.045 max	0.60 max		Mo <=0.15; bal Fe				
UNE 36080(85)	A690-2	Gen Struct	0.30 max		1.60 max	0.009 max	0.045 max	0.045 max	0.60 max		Mo <=0.15; bal Fe				
UNE 36080(85)	AE275-B	Gen Struct	0.22 max		1.60 max	0.009 max	0.045 max	0.045 max	0.60 max		Mo <=0.15; bal Fe				
UNE 36080(85)	AE275-C	Gen Struct	0.2 max		1.60 max	0.009 max	0.04 max	0.04 max	0.60 max		Mo <=0.15; bal Fe				
UNE 36080(85)	AE275-D	Gen Struct	0.2 max		1.60 max		0.035 max	0.035 max	0.60 max		Mo <=0.15; bal Fe				
UNE 36080(85)	AE355-C	Gen Struct	0.22 max		1.6 max	0.009 max	0.04 max	0.04 max	0.55 max		Mo <=0.15; bal Fe				
UNE 36080(85)	AE355-D	Gen Struct	0.22 max		1.6 max	0.009 max	0.035 max	0.035 max	0.55 max		Mo <=0.15; bal Fe				
UNE 36080(85)	F.6210*	see AE275-B	0.22 max		1.60 max	0.009 max	0.045 max	0.045 max	0.60 max		Mo <=0.15; bal Fe				
UNE 36080(85)	F.6211*	see AE275-C	0.2 max		1.60 max	0.009 max	0.04 max	0.04 max	0.60 max		Mo <=0.15; bal Fe				
UNE 36080(85)	F.6212*	see AE275-D	0.2 max		1.60 max		0.035 max	0.035 max	0.60 max		Mo <=0.15; bal Fe				
UNE 36080(85)	F.6214*	see AE355-C	0.22 max		1.6 max	0.009 max	0.04 max	0.04 max	0.55 max		Mo <=0.15; bal Fe				
UNE 36080(85)	F.6215*	see AE355-D	0.22 max		1.6 max	0.009 max	0.035 max	0.035 max	0.55 max		Mo <=0.15; bal Fe				
UNE 36080(85)	F.6216*	see A490-2	0.30 max		1.60 max	0.009 max	0.045 max	0.045 max	0.60 max		Mo <=0.15; bal Fe				
UNE 36080(85)	F.6217*	see A590-2	0.30 max		1.60 max	0.009 max	0.045 max	0.045 max	0.60 max		Mo <=0.15; bal Fe				
UNE 36080(85)	F.6218*	see A690-2	0.30 max		1.60 max	0.009 max	0.045 max	0.045 max	0.60 max		Mo <=0.15; bal Fe				

USA

Specification	Designation	Notes	C	Cu	Mn	N	P	S	Si	V	Other	UTS	YS	El	Hard
	UNS K02303		0.23 max	0.20 min	1.35 max		0.04 max	0.05 max	0.40 max		Nb 0.005-0.05; bal Fe				
ASTM A572/A572M(97)	50-1	t<=100mm	0.23 max		1.35 max		0.04 max	0.05 max	0.40 max		Si min depends on t; bal Fe	450	345	21	
ASTM A709(97)	Grade 50, Type 1	t<=100mm	0.23 max		1.35 max		0.04 max	0.05 max	0.15-0.40		Nb 0.005-0.05; bal Fe	450	345	21	

High Strength Steel, Structural, 50-2

China

Specification	Designation	Notes	C	Cu	Mn	N	P	S	Si	V	Other	UTS	YS	El	Hard
GB 3077(88)	20MnV	Bar HR Q/T 15mm diam	0.17-0.24	0.30 max	1.30-1.60		0.035 max	0.035 max	0.17-0.37	0.07-0.12	bal Fe	785	590	10	
GB/T 3078(94)	20MnV	Bar CD Q/T 15mm diam	0.17-0.24	0.30 max	1.30-1.60		0.035 max	0.035 max	0.17-0.37	0.07-0.12	bal Fe	785	590	10	

International

Specification	Designation	Notes	C	Cu	Mn	N	P	S	Si	V	Other	UTS	YS	El	Hard
ISO 630(80)	Fe510C	Struct; FF; FN; 0-16mm	0.23 max	0.3 max	1.6 max		0.045 max	0.045 max	0.55 max	0.1 max	Al <=0.1; Co <=0.1; Cr <=0.3; Mo <=0.15; Ni <=0.4; Pb <=0.15; Ti <=0.05; W <=0.1; bal Fe				

USA

Specification	Designation	Notes	C	Cu	Mn	N	P	S	Si	V	Other	UTS	YS	El	Hard
	UNS K02304		0.23 max	0.20 min	1.35 max		0.04 max	0.05 max	0.04 max	0.01-0.15	bal Fe				
ASTM A572/A572M(97)	50-2	t<=100mm	0.23 max		1.35 max		0.04 max	0.05 max	0.40 max	0.01-0.15	Si min depends on t; bal Fe	450	345	21	
ASTM A709(97)	Grade 50, Type 2	t<=100mm	0.23 max		1.35 max		0.04 max	0.05 max	0.15-0.40	0.01-0.15	bal Fe	450	345	21	

High Strength Steel, Structural, 50-3

China

Specification	Designation	Notes	C	Cu	Mn	N	P	S	Si	V	Other	UTS	YS	El	Hard
GB 1499(91)	20MnNb-b	Bar HR 840mm diam	0.17-0.25	0.30 max	1.00-1.50		0.045 max	0.045 max	0.17 max		Nb <=0.05; bal Fe	490-510	335	16	

UNS numbers and US grades are provided as a means of cross referencing chemically similar alloys. Exchangability is only possible after independent examination of specifications. Tensile properties are minimum or typical as specified. UTS and YS as MPa. El as %. See Appendix for list of abbreviations used in Notes. * indicates obsolete material.

Specification	Designation	Notes	C	Cu	Mn	N	P	S	Si	V	Other	UTS	YS	El	Hard

High Strength Steel, Structural, 50-3 (Continued from previous page)

International

Specification	Designation	Notes	C	Cu	Mn	N	P	S	Si	V	Other	UTS	YS	El	Hard
ISO 630(80)	Fe510C	Struct; FF; FN; Thk>16mm	0.23 max	0.3 max	1.6 max		0.045 max	0.045 max	0.55 max	0.1 max	Al <=0.1; Co <=0.1; Cr <=0.3; Mo <=0.15; Ni <=0.4; Pb <=0.15; Ti <=0.05; W <=0.1; bal Fe				

USA

Specification	Designation	Notes	C	Cu	Mn	N	P	S	Si	V	Other	UTS	YS	El	Hard
	UNS K02305		0.23 max	0.20 min	1.35 max		0.04 max	0.05 max	0.40 max		Nb <=0.05; Nb+V 0.02-0.15; bal Fe				
ASTM A572/A572M(97)	50-3	t<=100mm	0.23 max		1.35 max		0.04 max	0.05 max	0.40 max	0.01-0.15	Nb 0.005-0.05; Si min depends on t; bal Fe	450	345	21	
ASTM A709(97)	Grade 50, Type 3	t<=100mm	0.23 max		1.35 max		0.04 max	0.05 max	0.15-0.40	0.01-0.15	Nb 0.005-0.05; Nb+V=0.02-0.15; bal Fe	450	345	21	

High Strength Steel, Structural, 50-4

Europe

Specification	Designation	Notes	C	Cu	Mn	N	P	S	Si	V	Other	UTS	YS	El	Hard
EN 10025(90)	Fe590-2	Obs EU desig; struct HR FN; BS 40<t<=100mm			1.6 max	0.009 max	0.045 max	0.045 max			N max na if Al>=0.020; bal Fe	570-710	295-315	14-15	
EN 10025(90)	Fe590-2	Obs EU desig; struct HR FN; BS 150<t<=250mm			1.6 max	0.009 max	0.045 max	0.045 max			N max na if Al>=0.020; bal Fe	540-710	255-265	11	
EN 10025(90)	Fe590-2	Obs EU desig; struct HR FN; BS 16<t<=40mm			1.6 max	0.009 max	0.045 max	0.045 max			N max na if Al>=0.020; bal Fe	570-710	325	16	
EN 10025(90)	Fe590-2	Obs EU desig; struct HR FN; BS t<=16mm			1.6 max	0.009 max	0.045 max	0.045 max			N max na if Al>=0.020; bal Fe	570-770	335	8-16	
EN 10025(90)	Fe690-2	Obs EU desig; struct HR FN; BS t<=16mm			1.6 max	0.009 max	0.045 max	0.045 max			N max na if Al>=0.020; bal Fe	670-900	360	4-11	
EN 10025(90)	Fe690-2	Obs EU desig; struct HR FN; BS 16<t<=40mm			1.6 max	0.009 max	0.045 max	0.045 max			N max na if Al>=0.020; bal Fe	670-830	355	11	
EN 10025(90)	Fe690-2	Obs EU desig; struct HR FN; BS 40<t<=100mm			1.6 max	0.009 max	0.045 max	0.045 max			N max na if Al>=0.020; bal Fe	670-830	325-345	9-10	
EN 10025(90)	Fe690-2	Obs EU desig; struct HR FN; BS 150<t<=250mm			1.6 max	0.009 max	0.045 max	0.045 max			N max na if Al>=0.020; bal Fe	640-830	285-295	7	

USA

Specification	Designation	Notes	C	Cu	Mn	N	P	S	Si	V	Other	UTS	YS	El	Hard
	UNS K02306		0.23 max	0.20 min	1.35 max	0.015 max	0.04 max	0.05 max	0.40 max		V 4xN min; bal Fe				
ASTM A572/A572M(97)	50-4	t<=100mm	0.23 max		1.35 max		0.04 max	0.05 max	0.40 max	0.01-0.15	Si min depends on t; bal Fe	450	345	21	
ASTM A709(97)	Grade 50, Type 4	t<=100mm	0.23 max		1.35 max	0.015 max	0.04 max	0.05 max	0.15-0.40	0.01-0.15	V/N>=4; bal Fe	450	345	21	

High Strength Steel, Structural, No equivalents identified

France

Specification	Designation	Notes	C	Cu	Mn	N	P	S	Si	V	Other	UTS	YS	El	Hard
AFNOR NFA35502(84)	E24W2*	t<=16mm; roll	0.13 max	0.25-0.55	0.20-0.60		0.040 max	0.035 max	0.40 max		Cr 0.40-0.80; Mo <=0.15; Ni <=0.65; bal Fe	360-460	235	26	
AFNOR NFA35502(84)	E24W2*	50<t<=80mm; roll	0.13 max	0.25-0.55	0.20-0.60		0.040 max	0.035 max	0.40 max		Cr 0.40-0.80; Mo <=0.15; Ni <=0.65; bal Fe	360	215	25	
AFNOR NFA35502(84)	E24W2*	80<t<=110mm; roll	0.13 max	0.25-0.55	0.20-0.60		0.040 max	0.035 max	0.40 max		Cr 0.40-0.80; Mo <=0.15; Ni <=0.65; bal Fe	360	195	22	
AFNOR NFA35502(84)	E24W2*	t<=3mm; roll	0.13 max	0.25-0.55	0.20-0.60		0.040 max	0.035 max	0.40 max		Cr 0.40-0.80; Mo <=0.15; Ni <=0.65; bal Fe	360-460	215	21	
AFNOR NFA35502(84)	E24W3*	t<=30mm; roll	0.13 max	0.25-0.55	0.20-0.60		0.040 max	0.035 max	0.40 max		Cr 0.40-0.80; Mo <=0.15; Ni <=0.65; bal Fe	360-460	215	21	
AFNOR NFA35502(84)	E24W3*	80<t<=110mm; roll	0.13 max	0.25-0.55	0.20-0.60		0.040 max	0.035 max	0.40 max		Cr 0.40-0.80; Mo <=0.15; Ni <=0.65; bal Fe	360-460	195	22	

High Strength Steel, Structural, No equivalents identified

Specification	Designation	Notes	C	Cu	Mn	N	P	S	Si	V	Other	UTS	YS	El	Hard
France															
AFNOR NFA35502(84)	E24W3*	t<=30mm; roll	0.13 max	0.25-0.55	0.20-0.60		0.040 max	0.035 max	0.40 max		Cr 0.40-0.80; Mo <=0.15; Ni <=0.65; bal Fe	360-460	235	26	
AFNOR NFA35502(84)	E24W3*	50<t<=80mm; roll	0.13 max	0.25-0.55	0.20-0.60		0.040 max	0.035 max	0.40 max		Cr 0.40-0.80; Mo <=0.15; Ni <=0.65; bal Fe	360	215	25	
AFNOR NFA35502(84)	E24W4*	t<=3mm; roll	0.13 max	0.25-0.55	0.20-0.60		0.040 max	0.035 max	0.40 max		Al >=0.015; Cr 0.40-0.80; Mo <=0.15; Ni <=0.65; Nb 0.015-0.06; bal Fe	360-460	215	21	
AFNOR NFA35502(84)	E24W4*	80<t<=110mm; roll	0.13 max	0.25-0.55	0.20-0.60		0.040 max	0.035 max	0.40 max		Al >=0.015; Cr 0.40-0.80; Mo <=0.15; Ni <=0.65; Nb 0.015-0.06; bal Fe	360	195	22	
AFNOR NFA35502(84)	E36WA2*	t<=3mm; roll	0.12 max	0.25-0.55	1.00 max		0.06-0.15	0.040 max	0.75 max		Cr 0.30-1.25; Mo <=0.15; Ni <=0.65; bal Fe	480-580	315	20	
AFNOR NFA35502(84)	E36WA2*	3<t<=12mm; roll	0.12 max	0.25-0.55	1.00 max		0.06-0.15	0.040 max	0.75 max		Cr 0.30-1.25; Mo <=00; Ni <=0.65; bal Fe	480-580	355	20	
AFNOR NFA35502(84)	E36WA2*	t<=12mm; roll	0.12 max	0.25-0.55	1 max		0.06-0.15	0.04 max	0.2-0.75	0.1 max	Cr 0.3-1.25; Mo <=0.15; Ni <=0.65; Ti <=0.05; bal Fe	480-580m m	355	22	
AFNOR NFA35502(84)	E36WB3*	3<t<=25mm; roll	0.19 max	0.25-0.55	1.50 max		0.040 max	0.040 max	0.50 max		Al >=0.015; Cr 0.40-0.80; Mo <=0.3; Ni <=0.65; Nb 0.015-0.06; Zr<=0.15; bal Fe	480-580	355	20	
AFNOR NFA35502(84)	E36WB3*	80<t<=110mm; roll	0.19 max	0.25-0.55	1.50 max		0.040 max	0.040 max	0.50 max		Al >=0.015; Cr 0.40-0.80; Mo <=0.3; Ni <=0.65; Nb 0.015-0.06; Zr<=0.15; bal Fe	480	315	17	
AFNOR NFA35502(84)	E36WB3*	t<=30mm; roll	0.19 max	0.25-0.55	1.50 max		0.040 max	0.040 max	0.50 max		Al >=0.015; Cr 0.40-0.80; Mo <=0.3; Ni <=0.65; Nb 0.015-0.06; Zr<=0.15; bal Fe	480-580	355	22	
AFNOR NFA35502(84)	E36WB3*	50<t<=80mm; roll	0.19 max	0.25-0.55	1.50 max		0.040 max	0.040 max	0.50 max		Al >=0.015; Cr 0.40-0.80; Mo <=0.3; Ni <=0.65; Nb 0.015-0.06; Zr<=0.15; bal Fe	480	335	21	
AFNOR NFA35502(84)	E36WB4*	3<t<=25mm; roll	0.19 max	0.25-0.55	1.50 max		0.040 max	0.040 max	0.50 max		Al >=0.015; Cr 0.40-0.80; Mo <=0.3; Ni <=0.65; Nb 0.015-0.06; Zr<=0.15; bal Fe	480-580	355	22	
AFNOR NFA35502(84)	E36WB4*	80<t<=110mm; roll	0.19 max	0.25-0.55	1.50 max		0.040 max	0.040 max	0.50 max		Al >=0.015; Cr 0.40-0.80; Mo <=0.3; Ni <=0.65; Nb 0.015-0.06; Zr<=0.15; bal Fe	480	315	19	
Germany															
DIN EN 10137(95)	S690QL1	Plt flat product Q/T	0.20 max	0.50 max	1.70 max	0.015 max	0.020 max	0.010 max	0.80 max	0.12 max	Al >=0.018; B <=0.005; Cr <=1.50; Mo <=0.70; Ni <=2.00; Nb <=0.06; Ti <=0.05; grain refining elements>=0.015; Zr 0.015 max; bal Fe				
DIN EN 10137(95)	WNr 1.8988	Plt flat product Q/T	0.20 max	0.50 max	1.70 max	0.015 max	0.020 max	0.010 max	0.80 max	0.12 max	Al >=0.018; B <=0.005; Cr <=1.50; Mo <=0.70; Ni <=2.00; Nb <=0.06; Ti <=0.05; grain refining elements>=0.015; Zr 0.015 max; bal Fe				

UNS numbers and US grades are provided as a means of cross referencing chemically similar alloys. Exchangability is only possible after independent examination of specifications. Tensile properties are minimum or typical as specified. UTS and YS as MPa. El as %. See Appendix for list of abbreviations used in Notes. * indicates obsolete material.

High Strength Steel, Structural, No equivalents identified

Specification	Designation	Notes	C	Cu	Mn	N	P	S	Si	V	Other	UTS	YS	El	Hard
Germany															
DIN SEW 085(88)	FStE 420 OS2	Weld section, bar	0.14 max	0.40 max	1.60 max	0.015 max	0.025 max	0.010 max	0.50 max	0.10 max	Al >=0.020; Cr <=0.20; Mo <=0.08; Ni <=0.60; Nb <=0.04; Ti <=0.05; bal Fe				
DIN SEW 085(88)	WNr 1.8855	Weld section, bar	0.14 max	0.40 max	1.60 max	0.015 max	0.025 max	0.010 max	0.50 max	0.10 max	Al >=0.020; Cr <=0.20; Mo <=0.08; Ni <=0.60; Nb <=0.04; Ti <=0.05; bal Fe				
DIN SEW 090(93)	S690QL1	Tube; Q/T	0.20 max	0.50 max	1.70 max	0.015 max	0.020 max	0.010 max	0.80 max	0.12 max	Al >=0.018; B <=0.005; Cr <=1.50; Mo <=0.70; Ni <=2.00; Nb <=0.06; Ti <=0.05; grain refining elements>=0.015; Zr 0.015 max; bal Fe				
DIN SEW 090(93)	WNr 1.8988	Tube; Q/T	0.20 max	0.50 max	1.70 max	0.015 max	0.020 max	0.010 max	0.80 max	0.12 max	Al >=0.018; B <=0.005; Cr <=1.50; Mo <=0.70; Ni <=2.00; Nb <=0.06; Ti <=0.05; grain refining elements>=0.015; Zr 0.015 max; bal Fe				
Hungary															
MSZ 17718(77)	52C	Rolled bell-profile; timbering & walling; 0-999mm; HR	0.18 max	0.35 max	1.5 max		0.035 max	0.035 max	0.2-0.5	0.02-0.1	Al 0.015-0.150; Ni <=0.3; Nb 0.015-0.1; Ti 0.02-0.06; Zr=0.015-0.06; Al+V+Nb+Ti+Zr<=0.15; N<=Al/2+V/4+Nb/7+Ti/3.5<=0.015; bal Fe	640-830		10L	
Italy															
UNI 7070(72)	Fe34BFN	Gen struct	0.17 max		2 max	0.09 max	0.045 max	0.045 max	0.6 max		bal Fe				
UNI 7070(72)	Fe34BFU	Gen struct	0.17 max		2 max	0.07 max	0.045 max	0.045 max	0.6 max		bal Fe				
UNI 7070(82)	Fe330BFN	Gen struct	0.17 max		2 max		0.045 max	0.045 max	0.6 max		bal Fe				
UNI 7070(82)	Fe330BFU	Gen struct	0.17 max		2 max	0.07 max	0.045 max	0.045 max	0.6 max		bal Fe				
UNI 7070(82)	Fe330CFN	Gen struct	0.15 max		2 max		0.04 max	0.045 max	0.6 max		bal Fe				
UNI 7070(82)	Fe330DFF	Gen struct	0.15 max		2 max		0.04 max	0.04 max	0.6 max		Al <=0.015; bal Fe				
Russia															
GOST 380(71)	St4kp	Gen struct; Semi-Finished; Unkilled; Rimming; R; FU; 0-20 mm	0.3 max	0.3 max	2 max		0.07 max	0.06 max	0.6 max	0.1 max	Al <=0.1; Cr <=0.3; Mo <=0.15; Ni <=0.4; Ti <=0.05; bal Fe	400-510		25L	
GOST 380(71)	St4kp	Gen struct; Semi-Finished; Unkilled; Rimming; R; FU; 40.1-999 mm	0.3 max	0.3 max	2 max		0.07 max	0.06 max	0.6 max	0.1 max	Al <=0.1; Cr <=0.3; Mo <=0.15; Ni <=0.4; Ti <=0.05; bal Fe	400-510		22L	
GOST 380(71)	St4ps	Gen struct; Semi-Finished; Semi-killed; 40.1-999 mm	0.3 max	0.3 max	2 max		0.07 max	0.06 max	0.6 max	0.1 max	Al <=0.1; Cr <=0.3; Mo <=0.15; Ni <=0.4; Ti <=0.05; bal Fe	410-530		21L	
GOST 380(71)	St4ps	Gen struct; Semi-Finished; Semi-killed; 0-20 mm	0.3 max	0.3 max	2 max		0.07 max	0.06 max	0.6 max	0.1 max	Al <=0.1; Cr <=0.3; Mo <=0.15; Ni <=0.4; Ti <=0.05; bal Fe	410-530		24L	
GOST 380(71)	St4sp	Gen struct; Semi-Finished; Non-rimming; K; FF; 0-20 mm	0.3 max	0.3 max	2 max		0.07 max	0.06 max	0.6 max	0.1 max	Al <=0.1; Cr <=0.3; Mo <=0.15; Ni <=0.4; Ti <=0.05; bal Fe	410-530		24L	
GOST 380(71)	St4sp	Gen struct; Semi-Finished; Non-rimming; K; FF; 40.1-999 mm	0.3 max	0.3 max	2 max		0.07 max	0.06 max	0.6 max	0.1 max	Al <=0.1; Cr <=0.3; Mo <=0.15; Ni <=0.4; Ti <=0.05; bal Fe	410-530		21L	

UNS numbers and US grades are provided as a means of cross referencing chemically similar alloys. Exchangability is only possible after independent examination of specifications. Tensile properties are minimum or typical as specified. UTS and YS as MPa. El as %. See Appendix for list of abbreviations used in Notes. * indicates obsolete material.

Specification	Designation	Notes	C	Cu	Mn	N	P	S	Si	V	Other	UTS	YS	El	Hard

High Strength Steel, Structural, No equivalents identified

Russia

Specification	Designation	Notes	C	Cu	Mn	N	P	S	Si	V	Other
GOST 380(88)	St4kp	Gen struct; Semi-Finished; Unkilled; R; FU	0.18-0.27	0.3 max	0.4-0.7	0.008 max	0.04 max	0.05 max	0.05 max	0.1 max	Al <=0.1; Cr <=0.3; Mo <=0.15; Ni <=0.3; Ti <=0.05; As<=0.08; bal Fe
GOST 380(88)	St4ps	Gen struct; Semi-Finished; Semi-killed	0.18-0.27	0.3 max	0.4-0.7	0.008 max	0.04 max	0.05 max	0.05-0.15	0.1 max	Al <=0.1; Cr <=0.3; Mo <=0.15; Ni <=0.3; Ti <=0.05; As<=0.08; bal Fe
GOST 380(88)	St4sp	Gen struct; Semi-Finished; Non-rimmimg; K; FF	0.18-0.27	0.3 max	0.4-0.7	0.008 max	0.04 max	0.05 max	0.15-0.3	0.1 max	Al <=0.1; Cr <=0.3; Mo <=0.15; Ni <=0.3; Ti <=0.05; As<=0.15; bal Fe
GOST 5058	14G2	Weld constr; struct, constr	0.12-0.18	0.3 max	1.2-1.6		0.035 max	0.04 max	0.17-0.37	0.1 max	Al <=0.1; Cr <=0.3; Mo <=0.15; Ni <=0.3; Ti 0.01-0.03; As<=0.08; bal Fe

Sweden

Specification	Designation	Notes	C	Cu	Mn	N	P	S	Si	V	Other
SS 142614	2614	Fine grain Struct	0.2 max		1.7 max	0.2 max	0.03 max	0.03 max	0.55 max		B <=0.005; C+Mn/6<=0.43-0.49; bal Fe
SS 142615	2615	Fine grain Struct	0.2 max		1.7 max	0.2 max	0.03 max	0.03 max	0.55 max		B <=0.005; C+Mn/6<=0.43-0.49; bal Fe
SS 142624	2624	Fine grain Struct	0.2 max		1.7 max	0.02 max	0.03 max	0.03 max	0.1-0.8		B <=0.005; C+Mn/6<=0.43-0.6; bal Fe
SS 142625	2625	Fine grain Struct	0.2 max		1.7 max	0.02 max	0.03 max	0.03 max	0.1-0.8		B <=0.005; C+Mn/6<=0.43-0.6; bal Fe
SS 142632	2632	High-temp Const	0.18 max		1.3 max	0.02 max	0.03 max	0.03 max	0.5 max		bal Fe
SS 142642	2642	Fine grain Struct	0.18 max	1.65 max	0.015 max	0.03 max	0.03 max	0.5 max		bal Fe
SS 142652	2652	Fine grain Struct	0.18 max	1.65 max	0.02 max	0.03 max	0.03 max	0.5 max		bal Fe
SS 142662	2662	Fine grain Struct	0.18 max		1.8 max	0.015 max	0.03 max	0.03 max	0.5 max		bal Fe

Specification	Designation	Notes	C	Cr	Mn	Mo	Ni	P	S	Si	Other	UTS	YS	El	Hard

High Strength Steel, Ultrahigh-Strength, 300M

China

Specification	Designation	Notes	C	Cr	Mn	Mo	Ni	P	S	Si	Other	UTS	YS	El	Hard
	40CrNi2MoVA	Frg Q/T	0.38-0.43	0.70-0.95	0.60-0.90	0.30-0.50	1.65-2.00	0.010 max	0.010 max	1.45-1.80	V 0.05-0.10; bal Fe	1925	1630	12.5	

UK

Specification	Designation	Notes	C	Cr	Mn	Mo	Ni	P	S	Si	Other	UTS	YS	El	Hard
BS S.155	S-155A	Bil, Bar, Norm, Ann, Austenitic, Q/A, Tmp	0.39-0.44	0.7-0.95	0.6-0.9	0.3-0.45	1.65-2	0.01 max	0.01 max	1.5-1.8	V 0.05-0.1; P+S = 0.025; bal Fe				
BS S.155	S-155B	Bar, Norm, Ann, Austenitic, Q/A, Tmp	0.39-0.44	0.7-0.95	0.6-0.9	0.3-0.45	1.65-2	0.01 max	0.01 max	1.5-1.8	V 0.05-0.1; P+S = 0.025; bal Fe				
BS S.155	S-155C	Frg Norm, Ann, Austenitic, Q/A, Tmp	0.39-0.44	0.7-0.95	0.6-0.9	0.3-0.45	1.65-2	0.01 max	0.01 max	1.5-1.8	V 0.05-0.1; P+S = 0.025; bal Fe				

USA

Specification	Designation	Notes	C	Cr	Mn	Mo	Ni	P	S	Si	Other	UTS	YS	El	Hard
	UNS K44220	300M	0.38-0.46	0.70-0.95	0.60-0.90	0.30-0.65	1.65-2.00	0.010 max	0.010 max	1.45-1.80	V >=0.05; bal Fe				
AMS 6417(96)		Bar	0.38-0.43	0.70-0.95	0.60-0.90	0.30-0.50	1.65-2.00	0.010 max	0.010 max	1.45-1.80	Cu <=0.35; V 0.05-0.10; bal Fe	1862	1517	8	
AMS 6419		Bar, Frg, Tub	0.40-0.45	0.70-0.95	0.60-0.90	0.30-0.50	1.65-2.00	0.010 max	0.010 max	1.45-1.80	Cu <=0.35; V 0.05-0.10; bal Fe	1931	1586	7	
ASTM A579(96)	32	Superstrength frg	0.40-0.45	0.65-0.90	0.65-0.90	0.30-0.45	1.65-2.00	0.025 max	0.025 max	1.45-1.80	bal Fe	1720	1550	6	302 HB
ASTM A646(95)	Grade 8	Bloom bil for aircraft aerospace frg, ann	0.38-0.43	0.70-0.95	0.65-0.90	0.35-0.45	1.65-2.00	0.012 max	0.012 max	1.45-1.80	V 0.05-0.10; bal Fe				285 HB
MIL-S-8844D(90)	300M*	Obs; see AMS specs; Bar refrg stock tube, premium qual													
SAE J467(68)	300M		0.40	0.85	0.75	0.40	1.85			1.60	V 0.08; bal Fe	1993	1689	9	525 HB

High Strength Steel, Ultrahigh-Strength, 6434

USA

Specification	Designation	Notes	C	Cr	Mn	Mo	Ni	P	S	Si	Other	UTS	YS	El	Hard
	UNS K33517		0.32-0.38	0.65-0.90	0.60-0.90	0.30-0.40	1.65-2.00	0.040 max	0.040 max	0.20-0.60	V 0.17-0.23; bal Fe				
AMS 6429												1655	1448	10 L, 7T	
AMS 6430															
AMS 6433															
AMS 6434															
AMS 6435			0.33-0.38	0.65-0.90	0.60-0.90	0.30-0.40	1.65-2.00	0.010 max	0.010 max	0.40-0.60	Cu <=0.35; V 0.17-0.23; bal Fe				

High Strength Steel, Ultrahigh-Strength, 6522

Spain

Specification	Designation	Notes	C	Cr	Mn	Mo	Ni	P	S	Si	Other	UTS	YS	El	Hard
UNE 36017(85)	F.3151*	see X10CrAl7	0.12 max	6-7	1 max	0.15 max		0.04 max	0.03 max	0.5-1	Al 0.5-1; bal Fe				
UNE 36017(85)	X10CrAl7	Heat res	0.12 max	6-7	1 max	0.15 max		0.04 max	0.03 max	0.5-1	Al 0.5-1; bal Fe				

USA

Specification	Designation	Notes	C	Cr	Mn	Mo	Ni	P	S	Si	Other	UTS	YS	El	Hard
	UNS K92571	AF 1410	0.13-0.17	1.80-2.20	0.10 max	0.90-1.10	9.50-10.50	0.008 max	0.005 max	0.10 max	Al <=0.015; Co 13.5-14.5; N <=0.0015; Ti <=0.015; O 0.0015 max; bal Fe				
AMS 6522A(89)		Plt, Vacuum melted, Norm, Ann	0.13-0.17	1.80-2.20	0.10 max	0.90-1.10	9.50-10.50	0.008 max	0.005 max	0.10 max	Al <=0.015; Co 13.50-14.50; N <=0.0015; Ti <=0.015; P+S<=0.010; O<=0.0020; bal Fe	1620	1482	12	
AMS 6544B(91)		Plt, Maraging, Dbl vacuum melted, SHT, HR, or Frg	0.10-0.14	1.80-2.20	0.05-0.25	0.90-1.10	9.50-10.50	0.010 max	0.006 max	0.10 max	Al <=0.025; Co 7.50-8.50; N <=0.0075; Ti <=0.015; O<=0.0025; bal Fe	1310	1172-1241	14-15	42 HRC min

High Strength Steel, Ultrahigh-Strength, A579(13)

France

Specification	Designation	Notes	C	Cr	Mn	Mo	Ni	P	S	Si	Other	UTS	YS	El	Hard
AFNOR NFA35565(84)	16NCD13		0.12-0.17	0.9-1.15	0.3-0.6	0.15-0.3	3-3.5	0.025 max	0.015 max	0.15-0.35	Al <=0.05; Cu <=0.35; Ti <=0.05; V <=0.1; bal Fe				
AFNOR NFA35565(94)	16NiCrMo13	ball & roller bearing; CH	0.12-0.17	0.9-1.15	0.3-0.6	0.15-0.3	3-3.5	0.015 max	0.008 max	0.15-0.35	Al <=0.05; Cu <=0.35; Ti <=0.05; V <=0.1; bal Fe				

UNS numbers and US grades are provided as a means of cross referencing chemically similar alloys. Exchangability is only possible after independent examination of specifications. Tensile properties are minimum or typical as specified. UTS and YS as MPa. El as %. See Appendix for list of abbreviations used in Notes. * indicates obsolete material.

Specification	Designation	Notes	C	Cr	Mn	Mo	Ni	P	S	Si	Other	UTS	YS	El	Hard
High Strength Steel, Ultrahigh-Strength, A579(13) (Continued from previous page)															
UK															
BS 1506(90)	670-860	Bolt matl Pres/Corr res Q/T t<=63mm	0.36-0.44	0.80-1.15	0.45-0.70	0.50-0.65		0.035 max	0.040 max	0.15-0.35	Ti <=0.05; V 0.25-0.35; bal Fe	860-1000	725	16	248-302 HB
BS 1506(90)	671-850;En661	Bolt matl Pres/Corr res Q/T t<=63mm	0.30-0.45	1.00-1.50	0.40-0.70	0.50-0.70		0.040 max	0.040 max	0.1-0.35	Ti <=0.05; V 0.20-0.30; bal Fe	850-1000	635	14	248-302 HB
BS 1506(90)	671-850;En661	Bolt matl Pres/Corr res Q/T t<=63mm	0.30-0.45	1.00-1.50	0.40-0.70	0.50-0.70		0.040 max	0.040 max	0.1-0.35	Ti <=0.05; V 0.20-0.30; bal Fe	850-1000	665	14	248-302 HB
USA															
	UNS K13051		0.27-0.33	0.80-1.10	0.40-0.60	0.15-0.25		0.025 max	0.025 max	0.20-0.35	V 0.05-0.10; bal Fe				
ASTM A579(96)	13	Superstrength frg	0.27-0.33	0.80-1.10	0.40-0.60	0.15-0.25		0.025 max	0.025 max	0.20-0.35	V 0.05-0.10; bal Fe	1035-1450	965-1380	9-13	229 HB
High Strength Steel, Ultrahigh-Strength, A579(81)															
USA															
	UNS K91122	Superstrength (HP9-4-25)	0.24-0.30	0.35-0.60	0.10-0.35	0.35-0.60	7.00-9.00	0.010 max	0.010 max	0.10 max	Co 3.50-4.50; V 0.06-0.12; bal Fe				
AMS 6546															
ASTM A579(96)	81	Superstrength frg	0.30 max	0.35-0.60	0.10-0.35	0.35-0.60	7.0-9.0	0.01 max	0.01 max	0.10 max	B 0.003; Co 3.5-4.5; V 0.06-0.12; Ca 0.06; Zr 0.02; bal Fe	1310	1240	13	341 HB
High Strength Steel, Ultrahigh-Strength, A579(82)															
USA															
	UNS K91283	Superstrength (HP9-4-30)	0.28-0.34		0.10-0.35	0.90-1.10	7.0-8.5	0.01 max	0.01 max	0.10 max	Co 4.0-5.0; Cu 0.90-1.10; V 0.06-0.12; bal Fe				
AMS 6524															
AMS 6526															
ASTM A579(96)	82	Superstrength frg	0.28-0.34	0.90-1.10	0.10-0.35	0.90-1.10	7.0-8.5	0.01 max	0.01 max	0.10 max	B 0.003; Co 4.0-5.0; V 0.06-0.12; Ca 0.06; Zr 0.02; bal Fe	1450	1380	10	341 HB
High Strength Steel, Ultrahigh-Strength, A579(83)															
USA															
	UNS K91094	Superstrength	0.42-0.47	0.20-0.35	0.10-0.35	0.20-0.35	7.0-8.5	0.01 max	0.01 max	0.10 max	Co 3.5-4.5; V 0.06-0.12; bal Fe				
ASTM A579(96)	83	Superstrength frg	0.42-0.47	0.20-0.35	0.10-0.35	0.20-0.35	7.0-8.5	0.01 max	0.01 max	0.10 max	B 0.003; Co 3.5-4.5; V 0.06-0.12; Ca 0.06; Zr 0.02; bal Fe	1790-1930	1550-1720	4-7	341 HB
High Strength Steel, Ultrahigh-Strength, K92580															
USA															
	UNS K92580	Martensitic Age-Hardenable Alloy	0.21-0.25	2.85-3.35	0.10 max		10.5-12.5	0.008 max	0.005 max	0.10 max	Al <=0.015; Co 12.5-14.5; N <=0.0015; Ti <=0.015; O 0.0015 max; bal Fe				
High Strength Steel, Ultrahigh-Strength, No equivalents identified															
Germany															
DIN	41CrAlMo7	Q/T <=40mm	0.38-0.45	1.50-1.80	0.50-0.80	0.25-0.40	0.30 max		0.035 max	0.40 max	Al 0.80-1.20; bal Fe	980	735	12	
DIN	WNr 1.8509	Q/T <=40mm	0.38-0.45	1.50-1.80	0.50-0.80	0.25-0.40		0.030 max	0.035 max	0.40 max	Al 0.80-1.20; bal Fe	980	735	12	
Russia															
GOST 4543(71)	38Ch2MJuA	Nitriding	0.35-0.42	1.35-1.65	0.3-0.6	0.15-0.25	0.3 max	0.025 max	0.025 max	0.2-0.45	Al 0.7-1.1; Cu <=0.3; Ti <=0.05; V <=0.1; bal Fe				

UNS numbers and US grades are provided as a means of cross referencing chemically similar alloys. Exchangability is only possible after independent examination of specifications. Tensile properties are minimum or typical as specified. UTS and YS as MPa. El as %. See Appendix for list of abbreviations used in Notes. * indicates obsolete material.

Specification	Designation	Notes	C	Cr	Mn	Mo	Ni	P	S	Si	Other	UTS	YS	El	Hard
High Strength Steel, Unclassified, A355(A)															
Bulgaria															
BDS 6354	38Ch2MJuA	Struct	0.35-0.42	1.35-1.65	0.30-0.60	0.15-0.25	0.30 max	0.025 max	0.025 max	0.20-0.45	Al 0.70-1.10; Cu <=0.30; Ti <=0.03; V <=0.05; W <=0.20; bal Fe				
Czech Republic															
CSN 415340	15340	Nitriding	0.35-0.42	1.35-1.65	0.3-0.6	0.15-0.25		0.035 max	0.035 max	0.17-0.37	Al 0.7-1.1; bal Fe				
France															
AFNOR NFA35552(84)	40CAD6.12	t<=16mm; Q/T	0.36-0.43	1.5-1.8	0.5-0.8	0.2-0.4	0.4 max	0.035 max	0.035 max	0.1-0.4	Al 0.8-1.3; Cu <=0.3; Ti <=0.05; V <=0.1; bal Fe	1000-1200	800	11	
AFNOR NFA35552(84)	40CAD6.12	100<t<=160mm; Q/T	0.36-0.43	1.5-1.8	0.5-0.8	0.2-0.4	0.4 max	0.035 max	0.035 max	0.1-0.4	Al 0.8-1.3; Cu <=0.3; Ti <=0.05; V <=0.1; bal Fe	850-1000	670	14	
Hungary															
MSZ 17779(88)	41CrAlMo74	Nitriding; 100.1-160mm; Q/T; drawn	0.38-0.45	1.5-1.8	0.5-0.8	0.25-0.4		0.03 max	0.03 max	0.5 max	Al 0.8-1.2; bal Fe	800-1000	600	14L	
MSZ 17779(88)	41CrAlMo74	Nitriding; 0-100mm; Q/T; drawn	0.38-0.45	1.5-1.8	0.5-0.8	0.25-0.4		0.03 max	0.03 max	0.5 max	Al 0.8-1.2; bal Fe	900-1000	700	12L	
MSZ 7779(79)	41CAMo	Nitriding; 0-100mm; Q/T	0.38-0.45	1.5-1.8	0.5-0.8	0.25-0.4	0.04 max	0.03 max	0.035 max	0.2-0.5	Al 0.8-1.2; bal Fe	930-1130	740	12L	
MSZ 7779(79)	41CAMo	Nitriding; 100.1-160mm; Q/T	0.38-0.45	1.5-1.8	0.5-0.8	0.25-0.4	0.04 max	0.03 max	0.035 max	0.2-0.5	Al 0.8-1.2; bal Fe	830-980	640	14L	
Italy															
UNI 6120(67)	38CrAlMo7	Nitriding	0.35-0.42	1.5-1.8	0.5-0.7	0.25-0.4		0.035 max	0.035 max	0.4 max	Al 0.8-1.3; bal Fe				
UNI 8077(80)	41CrAlMo7	Nitriding	0.38-0.43	1.5-1.8	0.5-0.8	0.25-0.4		0.03 max	0.035 max	0.2-0.5	Al 0.8-1.2; V 0.1-0.2; bal Fe				
UNI 8552(84)	41CrAlMo7	Nitriding	0.38-0.43	1.5-1.8	0.5-0.8	0.25-0.4		0.03 max	0.035 max	0.2-0.5	Al 0.3-0.5; V 0.1-0.2; bal Fe				
Romania															
STAS 791(80)	38MoCrAl09	Struct/const	0.35-0.42	1.35-1.65	0.3-0.6	0.15-0.3		0.035 max	0.035 max	0.17-0.37	Al 0.7-1.1; Pb <=0.15; bal Fe				
Spain															
UNE 36014(76)	41CrAlMo7	Nitriding	0.38-0.45	1.5-1.8	0.5-0.8	0.25-0.4		0.035 max	0.035 max	0.2-0.5	Al 0.8-1.2; bal Fe				
UNE 36014(76)	F.1740*	see 41CrAlMo7	0.38-0.45	1.5-1.8	0.5-0.8	0.25-0.4		0.035 max	0.035 max	0.2-0.5	Al 0.8-1.2; bal Fe				
Sweden															
SS 142940	2940	Nitriding	0.38-0.45	1.5-1.8	0.5-0.8	0.25-0.35		0.03 max	0.035 max	0.2-0.5	Al 0.9-1.2; bal Fe				
UK															
BS 1506(90)	631-850;En621B	Bolt matl Pres/Corr res Q/T t<=63mm	0.35-0.45	1.00-1.50	0.40-0.70	0.50-0.70	0.40 max	0.035 max	0.040 max	0.15-0.35	bal Fe	850-1000	695	14	248-302 HB
BS 1506(90)	631-850;En621B	Bolt matl Pres/Corr res Q/T t<=63mm	0.35-0.45	1.00-1.50	0.40-0.70	0.50-0.70	0.40 max	0.035 max	0.040 max	0.15-0.35	bal Fe	850-1000	635	14	248-302 HB
USA															
	UNS K24065	Nitriding Steel (135)	0.38-0.43	1.40-1.80	0.50-0.80	0.30-0.40	0.25 max	0.025 max	0.025 max	0.20-0.40	Al 0.95-1.30; Cu <=0.35; bal Fe				
ASTM A355(94)	Class A Range 1	Bar, nitriding	0.38-0.43	1.40-1.80	0.50-0.70	0.30-0.40		0.035 max	0.040 max	0.15-0.35	Al 0.95-1.30; bal Fe				223-269 HB
ASTM A355(94)	Class A Range 2	Bar, nitriding	0.38-0.43	1.40-1.80	0.50-0.70	0.30-0.40		0.035 max	0.040 max	0.15-0.35	Al 0.95-1.30; bal Fe				248-302 HB
ASTM A519(96)	E7140	Smls mech tub	0.38-0.43	1.40-1.80	0.50-0.70	0.30-0.40		0.025 max	0.025 max	0.15-0.40	Al 0.95-1.30; bal Fe				
DoD-F-24669/3(86)	Class A	Frg, Bar, Bil for Nitriding; Supersedes MIL-S-869C(58)	0.38-0.45	1.35-1.85	0.40-0.70	0.30-0.45		0.040 max	0.040 max	0.20-0.40	Al 0.85-1.20; bal Fe	930	690	16	277-331 HB
DoD-F-24669/3(86)	Class B	Frg, Bar, Bil for Nitriding; Supersedes MIL-S-869C(58)	0.30-0.40	1.00-1.50	0.50-1.10	0.15-0.25		0.040 max	0.060 max	0.20-0.40	Al 0.85-1.20; Se 0.15-0.25; bal Fe	730	525	20	217-262 HB
DoD-F-24669/3(86)	Class C	Frg, Bar, Bil for Nitriding; Supersedes MIL-S-869C(58)	0.20-0.27	0.95-1.35	0.40-0.70	0.20-0.30	3.25-3.75	0.040 max	0.040 max	0.20-0.40	Al 0.85-1.20; bal Fe	860	725	19	255-302 HB
DoD-F-24669/3(86)	Class D	Frg, Bar, Bil for Nitriding; Supersedes MIL-S-869C(58)	0.30-0.40	0.90-1.40	0.40-0.70	0.15-0.25		0.040 max	0.040 max	0.20-0.40	Al 0.85-1.20; bal Fe	860	690	18	255-302 HB

UNS numbers and US grades are provided as a means of cross referencing chemically similar alloys. Exchangability is only possible after independent examination of specifications. Tensile properties are minimum or typical as specified. UTS and YS as MPa. El as %. See Appendix for list of abbreviations used in Notes. * indicates obsolete material.

Specification	Designation	Notes	C	Cr	Mn	Mo	Ni	P	S	Si	Other	UTS	YS	El	Hard

High Strength Steel, Unclassified, A355(B)

UK

Specification	Designation	Notes	C	Cr	Mn	Mo	Ni	P	S	Si	Other	UTS	YS	El	Hard
BS 1506(90)	630-690;En621A	Bolt matl Pres/Corr res Q/T t<=63mm	0.37-0.49	0.75-1.20	0.65-1.10	0.15-0.25	0.40 max	0.035 max	0.040 max	0.15-0.35	Al >=0.015; bal Fe	690-800	550	16	201-235 HB
BS 1506(90)	630-790;En621A	Bolt matl Pres/Corr res Q/T t<=63mm	0.37-0.49	0.75-1.20	0.65-1.10	0.15-0.25	0.40 max	0.035 max	0.040 max	0.15-0.35	Al >=0.015; bal Fe	790-1000	655	14	223-302 HB
BS 1506(90)	630-860;En621A	Bolt matl Pres/Corr res Q/T t<=63mm	0.37-0.49	0.75-1.20	0.65-1.10	0.15-0.25	0.40 max	0.035 max	0.040 max	0.15-0.35	Al >=0.015; bal Fe	860-1070	725	14	248-331 HB

USA

Specification	Designation	Notes	C	Cr	Mn	Mo	Ni	P	S	Si	Other	UTS	YS	El	Hard
	UNS K23745	Nitriding Steel	0.35-0.40	1.20-1.50	0.70-0.95	0.15-0.25		0.035 max	0.060 max	0.15-0.35	Al 0.95-1.30; Se 0.15-0.25; bal Fe				
ASTM A355(94)	Class B Range 1	Bar, nitriding	0.35-0.40	1.20-1.50	0.70-0.95	0.15-0.25		0.035 max	0.060 max	0.15-0.35	Al 0.95-1.30; Se 0.15-0.25; bal Fe				223-269 HB
ASTM A355(94)	Class B Range 2	Bar, nitriding	0.35-0.40	1.20-1.50	0.70-0.95	0.15-0.25		0.035 max	0.060 max	0.15-0.35	Al 0.95-1.30; Se 0.15-0.25; bal Fe				248-302 HB

High Strength Steel, Unclassified, A355(C)

USA

Specification	Designation	Notes	C	Cr	Mn	Mo	Ni	P	S	Si	Other	UTS	YS	El	Hard
	UNS K52440	Nitriding Steel	0.22-0.27	1.00-1.35	0.50-0.70	0.20-0.30	3.25-3.75	0.035 max	0.040 max	0.15-0.35	Al 0.95-1.30; bal Fe				
ASTM A355(94)	Class C Range 1	Bar, nitriding	0.22-0.27	1.00-1.35	0.50-0.70	0.20-0.30	3.25-3.75	0.035 max	0.040 max	0.15-0.35	Al 0.95-1.30; bal Fe				223-269 HB
ASTM A355(94)	Class C Range 2	Bar, nitriding	0.22-0.27	1.00-1.35	0.50-0.70	0.20-0.30	3.25-3.75	0.035 max	0.040 max	0.15-0.35	Al 0.95-1.30; bal Fe				248-302 HB

High Strength Steel, Unclassified, A579(71)

USA

Specification	Designation	Notes	C	Cr	Mn	Mo	Ni	P	S	Si	Other	UTS	YS	El	Hard
	UNS K92820	Superstrength	0.03 max		0.10 max	3.0-3.5	17.0-19.0	0.01 max	0.01 max	0.10 max	Al 0.05-0.15; Co 8.0-9.0; Ti 0.15-0.25; bal Fe				
ASTM A579(96)	71	Superstrength frg	0.03 max		0.10 max	3.0-3.5	17.0-19.0	0.01 max	0.01 max	0.10 max	Al 0.05-0.15; B 0.003; Co 8.0-9.0; Ti 0.15-0.25; Ca 0.06; Zr 0.02; bal Fe	1450	1380	12	321 HB
MIL-S-46850D(91)	---*	Obs for new design; Bar Plt Sh Strp frg ext													

High Strength Steel, Unclassified, A579(74)

USA

Specification	Designation	Notes	C	Cr	Mn	Mo	Ni	P	S	Si	Other	UTS	YS	El	Hard
	UNS K91930	Superstrength	0.03 max	4.75-5.25	0.10 max	2.75-3.25	11.5-12.5	0.01 max	0.01 max	0.12 max	Al 0.25-0.40; Ti 0.05-0.15; bal Fe				
ASTM A579(96)	74	Superstrength frg	0.03 max	4.75-5.25	0.10 max	2.75-3.25	11.5-12.5	0.01 max	0.01 max	0.12 max	Al 0.25-0.40; B 0.003; Ti 0.05-0.15; Ca 0.06; Zr 0.02; bal Fe	1170	1100	15	321 HB

High Strength Steel, Unclassified, A579(75)

USA

Specification	Designation	Notes	C	Cr	Mn	Mo	Ni	P	S	Si	Other	UTS	YS	El	Hard
	UNS K91940	Superstrength	0.03 max	4.75-5.25	0.10 max	2.75-3.25	11.5-12.5	0.01 max	0.01 max	0.12 max	Al 0.35-0.50; Ti 0.10-0.25; bal Fe				
ASTM A579(96)	75	Superstrength frg	0.03 max	4.75-5.25	0.10 max	2.75-3.25	11.5-12.5	0.01 max	0.01 max	0.12 max	Al 0.35-0.50; B 0.003; Ti 0.10-0.25; Ca 0.06; Zr 0.02; bal Fe	1310	1240	14	321 HB

High Strength Steel, Unclassified, A801(Alloy 2)

USA

Specification	Designation	Notes	C	Cr	Mn	Mo	Ni	P	S	Si	Other	UTS	YS	El	Hard
	UNS K92650	Fe-Co Soft Magnetic Alloy	0.025 max	0.75 max	0.35 max		0.75 max	0.015 max	0.015 max	0.35 max	Co 26.50-28.50; V <=0.35; bal Fe				

High Strength Steel, Unclassified, B4B

USA

Specification	Designation	Notes	C	Cr	Mn	Mo	Ni	P	S	Si	Other	UTS	YS	El	Hard
	UNS K91352	Superstrength	0.20-0.25	11.00-12.50	0.50-1.00	0.90-1.25	0.50-1.00	0.025 max	0.025 max	0.20-0.50	Al <=0.05; Ti <=0.05; V 0.20-0.30; W 0.90-1.25; Sn 0.04 max; bal Fe				

UNS numbers and US grades are provided as a means of cross referencing chemically similar alloys. Exchangability is only possible after independent examination of specifications. Tensile properties are minimum or typical as specified. UTS and YS as MPa. El as %. See Appendix for list of abbreviations used in Notes. * indicates obsolete material.

Specification	Designation	Notes	C	Cr	Mn	Mo	Ni	P	S	Si	Other	UTS	YS	El	Hard

High Strength Steel, Unclassified, B4B (Continued from previous page)

USA

Specification	Designation	Notes	C	Cr	Mn	Mo	Ni	P	S	Si	Other	UTS	YS	El	Hard
ASTM A437/437M(98)	B4B	high-temp nut washer	0.20-0.25	11.0-12.5	0.50-1.00	0.90-1.25		0.025 max	0.025 max	0.20-0.50	Al <=0.05; Ti <=0.05; V 0.20-0.30; W 0.90-1.25; Sn<=0.04; bal Fe	1000	720	13	293-341 HB
ASTM A437/437M(98)	B4B	high-temp bolt stud	0.20-0.25	11.0-12.5	0.50-1.00	0.90-1.25		0.025 max	0.025 max	0.20-0.50	Al <=0.05; Ti <=0.05; V 0.20-0.30; W 0.90-1.25; Sn<=0.04; bal Fe	1000	720	13	331 HB
ASTM A437/437M(98)	B4C	high-temp nut washer	0.20-0.25	11.0-12.5	0.50-1.00	0.90-1.25		0.025 max	0.025 max	0.20-0.50	Ti <=0.05; V 0.20-0.30; W 0.90-1.25; Sn<=0.04; bal Fe	790	585	18	229-277 HB
ASTM A437/437M(98)	B4C	high-temp bolt stud	0.20-0.25	11.0-12.5	0.50-1.00	0.90-1.25		0.025 max	0.025 max	0.20-0.50	Ti <=0.05; V 0.20-0.30; W 0.90-1.25; Sn<=0.04; bal Fe	790	585	18	277 HB

High Strength Steel, Unclassified, D6

China

Specification	Designation	Notes	C	Cr	Mn	Mo	Ni	P	S	Si	Other	UTS	YS	El	Hard
	45CrNiMo1VA	Frg Q/T	0.42-0.48	0.90-1.20	0.60-0.90	0.90-1.10	0.40-0.70	0.020 max	0.020 max	0.15-0.35	V 0.05-0.15; bal Fe	1595	1470	12.6	

USA

Specification	Designation	Notes	C	Cr	Mn	Mo	Ni	P	S	Si	Other	UTS	YS	El	Hard
	UNS K24728		0.45-0.50	0.90-1.20	0.60-0.90	0.90-1.10	0.40-0.70	0.010 max	0.010 max	0.15-0.30	V 0.08-0.15; bal Fe				
AMS 6431															
AMS 6432															
AMS 6439															
ASTM A579(96)	23	Superstrength frg	0.45-0.50	0.90-1.20	0.60-0.90	0.90-1.10	0.40-0.70	0.015 max	0.015 max	0.15-0.30	V 0.08-0.15; bal Fe	1035-1720	965-1550	6-13	302 HB
MIL-S-8949(65)	D6*	Obs; Bar Plt Sh bil refrg stock													
SAE J467(68)	D6A		0.47	1.05	0.75	1.00	0.55			0.22	V 0.10; bal Fe	1841	1703	10	53 HRC

High Strength Steel, Unclassified, No equivalents identified

France

Specification	Designation	Notes	C	Cr	Mn	Mo	Ni	P	S	Si	Other	UTS	YS	El	Hard
AFNOR NFA35566(83)	23D5	Q/T t<=16mm	0.2-0.26	0.3 max	0.5-0.8	0.45-0.6	0.4 max	0.03 max	0.025 max	0.1-0.35	Al >=0.02; Ti <=0.05; V <=0.1; bal Fe	1320	1125	9	
AFNOR NFA35566(83)	25MNCDV5	t<=40mm	0.23-0.28	0.4-0.6	1.1-1.4	0.2-0.3	0.4-0.7	0.02 max	0.02 max	0.1-0.35	Al >=0.02; Ti <=0.05; V 0.15-0.25; bal Fe	1520	1325	7	
AFNOR NFA35566(83)	25MNCDV6	t<=40mm	0.23-0.28	0.4-0.6	1.4-1.7	0.2-0.3	0.4-0.7	0.02 max	0.02 max	0.1-0.35	Al >=0.02; Ti <=0.05; V 0.15-0.25; bal Fe	1520	1325	7	
AFNOR NFA36210(88)	14MNDV5	Sh Plt; Boiler, Press ves; t<=70mm; Q/T	0.16 max	0.3 max	1.15-1.6	0.4-0.65	0.4-0.9	0.015 max	0.012 max	0.1-0.4	Cu <=0.3; Ti <=0.05; V 0.04-0.08; bal Fe	610-730	460	18	
AFNOR NFA36210(88)	14MNDV5	Sh Plt; Boiler, Press ves; 125<t<=200mm; Q/T	0.16 max	0.3 max	1.15-1.6	0.4-0.65	0.4-0.9	0.015 max	0.012 max	0.1-0.4	Cu <=0.3; Ti <=0.05; V 0.04-0.08; bal Fe	590-710	440	17	

Italy

Specification	Designation	Notes	C	Cr	Mn	Mo	Ni	P	S	Si	Other	UTS	YS	El	Hard
UNI 7382(75)	FeE390KG	Fine grain Struct	0.2 max	0.4 max	1-1.6	0.3 max	0.7 max	0.035 max	0.035 max	0.5 max	Al <=0.015; Cu <=0.5; Nb 0.015-0.06; Ti <=0.2; V 0.02-0.2; bal Fe				
UNI 7382(75)	FeE390KT	Fine grain Struct	0.2 max	0.4 max	1-1.6	0.3 max	0.7 max	0.03 max	0.03 max	0.5 max	Al <=0.015; Cu <=0.5; Nb 0.015-0.06; Ti <=0.2; V 0.02-0.2; bal Fe				

Russia

Specification	Designation	Notes	C	Cr	Mn	Mo	Ni	P	S	Si	Other	UTS	YS	El	Hard
GOST 380(71)	St6ps3	Gen struct; Semi-killed; 0-20 mm	0.3 max	0.3 max	2 max	0.15 max	0.4 max	0.07 max	0.06 max	0.6 max	Al <=0.1; Cu <=0.3; Ti <=0.05; V <=0.1; bal Fe	590		15L	
GOST 380(71)	St6ps3	Gen struct; Semi-killed; 40.1-999 mm	0.3 max	0.3 max	2 max	0.15 max	0.4 max	0.07 max	0.06 max	0.6 max	Al <=0.1; Cu <=0.3; Ti <=0.05; V <=0.1; bal Fe	590	295	12L	
GOST 380(71)	St6sp3	Gen struct; Non-rimming; K; FF; 40.1-999 mm	0.3 max	0.3 max	2 max	0.15 max	0.4 max	0.07 max	0.06 max	0.6 max	Al <=0.1; Cu <=0.3; Ti <=0.05; V <=0.1; bal Fe	590	295	12L	
GOST 380(71)	St6sp3	Gen struct; Non-rimming; K; FF; 0-20 mm	0.3 max	0.3 max	2 max	0.15 max	0.4 max	0.07 max	0.06 max	0.6 max	Al <=0.1; Cu <=0.3; Ti <=0.05; V <=0.1; bal Fe	590	315	15L	

Specification	Designation	Notes	C	Cr	Mn	Mo	Ni	P	S	Si	Other	UTS	YS	El	Hard

High Strength Steel, Unclassified, No equivalents identified

UK

Specification	Designation	Notes	C	Cr	Mn	Mo	Ni	P	S	Si	Other	UTS	YS	El	Hard
BS 1506(90)	162	Bolt matl Pres/Corr res Q/T t<=63mm	0.4-0.6	0.3 max	0.6-1.0	0.15 max	0.4 max	0.04 max	0.05 max	0.15-0.35	bal Fe				248-352 HB
BS 1506(90)	253	Bolt matl Pres/Corr res; t<=22mm; Q/T	0.4-0.5	0.3 max	0.4-0.8	0.2-0.3	0.4 max	0.035 max	0.04 max	0.15-0.35	Al 0.015-0.10; bal Fe				248-352 HB

USA

Specification	Designation	Notes	C	Cr	Mn	Mo	Ni	P	S	Si	Other	UTS	YS	El	Hard
ASTM A355(94)	Class D Range 1	Bar, nitriding	0.33-0.38	1.00-1.30	0.50-0.70	0.15-0.25		0.035 max	0.040 max	0.15-0.35	Al 0.95-1.30; bal Fe				223-269 HB
ASTM A355(94)	Class D Range 2	Bar, nitriding	0.33-0.38	1.00-1.30	0.50-0.70	0.15-0.25		0.035 max	0.040 max	0.15-0.35	Al 0.95-1.30; bal Fe				248-302 HB
ASTM A722(98)	A722	Uncoated bar for prestressing conc						0.040 max	0.050 max		bal Fe	1035			
ASTM A891(95)	A891 Tyoe 1	PH frg for turbine rotor disks wheels	0.05 max	13.50-16.00	0.50 max	1.00-1.50	24.00-27.00	0.025 max	0.015 max	0.50 max	Al <=0.35; B 0.003-0.010; Ti 1.90-2.35; V 0.10-0.50; bal Fe	965	655	12	277-363 HB
ASTM A891(95)	A891 Type 2	PH frg for turbine rotor disks wheels	0.05 max	13.50-16.00	0.50 max	1.00-1.50	24.00-27.00	0.025 max	0.015 max	0.50 max	Al <=0.35; B 0.003-0.010; Ti 1.90-2.35; V 0.10-0.50; bal Fe	895	585	15	248-341 HB

Specification	Designation	Notes	C	Cr	Mn	Mo	Ni	P	S	Si	Other	UTS	YS	El	Hard

Stainless Steel, Austenitic, 18-18 PLUS

USA

Specification	Designation	Notes	C	Cr	Mn	Mo	Ni	P	S	Si	Other	UTS	YS	El	Hard
	UNS S28200	18-18-Plus	0.15 max	17.00-19.00	17.00-19.00	0.50-1.50		0.045 max	0.030 max	1.00 max	Cu 0.50-1.50; N 0.40-0.60; bal Fe				
ASTM A580/A580M(98)	S28200	Wir, Ann	0.15 max	17.0-19.0	17.0-19.0	0.75-1.25		0.045 max	0.030 max	1.00 max	Cu 0.75-1.25; N 0.40-0.60; bal Fe	760	415	35	

Stainless Steel, Austenitic, 19-9 DX

USA

Specification	Designation	Notes	C	Cr	Mn	Mo	Ni	P	S	Si	Other	UTS	YS	El	Hard
	UNS S63199	Fe Base Superalloy 19-9-DX or 19-9-W-Mo	0.28-0.35	18.0-21.0	0.75-1.50	1.25-2.00	8.00-11.00	0.040 max	0.030 max	0.30-0.80	Cu <=0.50; Ti 0.40-0.75; W 1.00-1.75; bal Fe				
SAE J467(68)	19-9DX		0.32	18.5	1.15	1.60	9.00			0.55	Ti 0.55; W 1.35; bal Fe	814	476	55	216 HB

Stainless Steel, Austenitic, 201

Australia

Specification	Designation	Notes	C	Cr	Mn	Mo	Ni	P	S	Si	Other	UTS	YS	El	Hard
AS 1449(94)	201-2	Plt Sh Strp	0.15 max	16.00-18.00	5.50-7.50		3.50-5.50	0.060 max	0.030 max	1.00 max	N <=0.25; bal Fe				

China

Specification	Designation	Notes	C	Cr	Mn	Mo	Ni	P	S	Si	Other	UTS	YS	El	Hard
GB 1220(92)	1Cr17Mn6Ni5N	Bar SA	0.15 max	16.00-18.00	5.50-7.50		3.50-5.50	0.060 max	0.030 max	1.00 max	N <=0.25; bal Fe	520	275	40	
GB 3280(92)	1Cr17Mn6Ni5N	Sh Plt, CR SA	0.15 max	16.00-18.00	5.50-7.50		3.50-5.50	0.060 max	0.030 max	1.00 max	N <=0.25; bal Fe	635	245	40	
GB 4237(92)	1Cr17Mn6Ni5N	Sh Plt, HR SA	0.15 max	16.00-18.00	5.50-7.50		3.50-5.50	0.060 max	0.030 max	1.00 max	N <=0.25; bal Fe	635	245	40	
GB 4239(91)	1Cr17Mn6Ni5N	Sh Strp, CR SA	0.15 max	16.00-18.00	5.50-7.50		3.50-5.50	0.060 max	0.030 max	1.00 max	N <=0.25; bal Fe	635	245	40	

Japan

Specification	Designation	Notes	C	Cr	Mn	Mo	Ni	P	S	Si	Other	UTS	YS	El	Hard
JIS G4303(91)	SUS201	Bar; SA; <=180mm diam	0.15 max	16.00-18.00	5.50-7.50		3.50-5.50	0.060 max	0.030 max	1.00 max	N <=0.25; bal Fe	520	275	40	241 HB
JIS G4308	SUS201	Wir rod	0.15 max	16.00-18.00	5.50-7.50		3.50-5.50	0.060 max	0.030 max	1.00 max	N <=0.25; bal Fe				
JIS G4309	SUS201	Wir	0.15 max	16.00-18.00	5.50-7.50		3.50-5.50	0.060 max	0.030 max	1.00 max	N <=0.25; bal Fe				

USA

Specification	Designation	Notes	C	Cr	Mn	Mo	Ni	P	S	Si	Other	UTS	YS	El	Hard
	AISI 201	Tube	0.15 max	16.00-18.00	5.50-7.50		3.50-5.50	0.060 max	0.030 max	1.00 max	N <=0.25; bal Fe				
	UNS S20100	Cr-Mn-Ni	0.15 max	16.00-18.00	5.50-7.50		3.50-5.50	0.060 max	0.030 max	1.00 max	N <=0.25; bal Fe				
ASTM A213/A213M(95)	TP201	Smls tube boiler, superheater, heat exchanger	0.15 max	16.0-18.0	5.50-7.50		3.50-5.50	0.060 max	0.030 max	1.00 max	N <=0.25; bal Fe	655	260	35	
ASTM A240/A240M(98)	S20100	Cr-Mn-Ni; Rich type 201-1	0.15 max	16.0-18.0	5.5-7.5		3.5-5.5	0.060 max	0.030 max	1.00 max	N <=0.25; bal Fe	515	260	40.0	
ASTM A249/249M(96)	TP201	Weld tube; boiler, superheater, heat exch	0.15 max	16.0-18.0	5.50-7.50		3.50-5.50	0.060 max	0.030 max	1.00 max	N <=0.25; bal Fe	655	260	35	95 HRB
ASTM A666(96)	201	Sh Strp Plt Bar, Ann	0.15 max	16.00-18.00	5.50-7.50		3.50-5.50	0.060 max	0.030 max	0.75 max	N <=0.25; bal Fe	655	230-310	40	217-241 HB
SAE J405(98)	S20100	Cr-Mn-Ni	0.15 max	16.00-18.00	5.50-7.50		3.50-5.50	0.060 max	0.030 max	1.00 max	N <=0.25; bal Fe				

Stainless Steel, Austenitic, 201L

USA

Specification	Designation	Notes	C	Cr	Mn	Mo	Ni	P	S	Si	Other	UTS	YS	El	Hard
	UNS S20103	Cr-Mn-Ni Low-C 201L	0.03 max	16.0-18.0	5.5-7.5		3.5-5.5	0.045 max	0.030 max	0.75 max	N <=0.25; bal Fe				
ASTM A240/A240M(98)	S20103	Cr-Mn-Ni Low-C 201L	0.03 max	16.0-18.0	5.5-7.5		3.5-5.5	0.045 max	0.030 max	0.75 max	N <=0.25; bal Fe	655	260	40.0	217 HB
SAE J405(98)	S20103	Cr-Mn-Ni Low-C 201L	0.03 max	16.00-18.00	5.50-7.50		3.50-5.50	0.045 max	0.030 max	0.75 max	N <=0.25; bal Fe				

Stainless Steel, Austenitic, 201LN

USA

Specification	Designation	Notes	C	Cr	Mn	Mo	Ni	P	S	Si	Other	UTS	YS	El	Hard
	UNS S20153	Cr-Mn-Ni Low-C, High Nitrogen T201LN	0.03 max	16.0-17.5	6.4-7.5		4.0-5.0	0.045 max	0.015 max	0.75 max	N 0.10-0.25; bal Fe				
ASTM A240/A240M(98)	S20153	Cr-Mn-Ni Low-C, High-N T201LN	0.03 max	16.0-17.5	6.4-7.5		4.0-5.0	0.045 max	0.015 max	0.75 max	Cu <=1.00; N 0.10-0.25; bal Fe	655	310	45.0	241 HB
SAE J405(98)	S20153	Cr-Mn-Ni Low-C, High Nitrogen T201LN	0.03 max	16.00-17.50	*6.40-7.50		4.00-5.00	0.045 max	0.015 max	0.75 max	N 0.10-0.25; bal Fe				

UNS numbers and US grades are provided as a means of cross referencing chemically similar alloys. Exchangability is only possible after independent examination of specifications. Tensile properties are minimum or typical as specified. UTS and YS as MPa. El as %. See Appendix for list of abbreviations used in Notes. * indicates obsolete material.

Stainless Steel, Austenitic, 202

Specification	Designation	Notes	C	Cr	Mn	Mo	Ni	P	S	Si	Other	UTS	YS	El	Hard
Bulgaria															
BDS 6738(72)	2Ch13N4G9	Corr res; bar	0.15-0.30	12.0-14.0	8.80-10.0	0.30 max	3.70-4.70	0.050 max	0.025 max	0.80 max	bal Fe				
BDS 6738(72)	Ch17G9AN4	Sh Strp Plt Bar	0.12 max	16.0-18.0	8.00-10.5	0.30 max	3.50-4.50	0.050 max	0.025 max	0.80 max	N 0.15-0.25; bal Fe				
Canada															
CSA G110.3	202	Sh Strp Plt Bar	0.15 max	17-19	7.5-10		4-6	0.06 max	0.03 max	1 max	N <=0.25; bal Fe				
CSA G110.6	202	Sh Strp Plt Bar, HR, Ann, HR, Q/A, Tmp	0.15 max	17-19	7.5-10		4-6	0.06 max	0.03 max	1 max	N <=0.25; bal Fe				
China															
GB 1220(92)	1Cr18Mn8Ni5N	Bar SA	0.15 max	17.00-19.00	7.50-10.00		4.00-6.00	0.060 max	0.030 max	1.00 max	N <=0.25; bal Fe	520	275	40	
GB 3280(92)	1Cr18Mn8Ni5N	Sh Plt, HR SA	0.15 max	17.00-19.00	7.50-10.00		4.00-6.00	0.060 max	0.030 max	1.00 max	N <=0.25; bal Fe	590	245	40	
GB 4237(92)	1Cr18Mn8Ni5N	Sh Plt, HR SA	0.15 max	17.00-19.00	7.50-10.00		4.00-6.00	0.060 max	0.030 max	1.00 max	N <=0.25; bal Fe	590	245	40	
GB 4239(91)	1Cr18Mn8Ni5N	Sh Strp, CR SA	0.15 max	17.00-19.00	7.50-10.00		4.00-6.00	0.060 max	0.030 max	1.00 max	N <=0.25; bal Fe	590	245	40	
Czech Republic															
CSN 417460	17460	Corr res	0.12 max	17.0-20.0	7.0-10.0		4.0-6.0	0.06 max	0.035 max	0.9 max	N 0.1-0.25; bal Fe				
France															
AFNOR NFA35583	Z8CNM19.8	Sh Strp Plt Bar	0.1 max	17-20	6.5-8	0.5 max	7.5-9.5	0.03 max	0.02 max	0.3-0.65	Cu <=0.5; bal Fe				
Germany															
DIN 17480(92)	WNr 1.4871	Valve, HT age	0.48-0.58	20.0-22.0	8.00-10.0		3.25-4.50	0.050 max	0.030 max	0.25 max	N 0.35-0.50; bal Fe	950-1200	580	8	
DIN 17480(92)	X53CrMnNiN21-9	Valve, HT age	0.48-0.58	20.0-22.0	8.00-10.0		3.25-4.50	0.050 max	0.030 max	0.25 max	N 0.35-0.50; bal Fe	950-1200	580	8	
TGL 39672	X12CrNiMn19.9*	Obs	0.15 max	18.5-20	6.5-7.5		8.5-9.5	0.03 max	0.02 max	0.3-0.8	bal Fe				
Hungary															
MSZ 4360(80)	KO23	Corr res	0.15-0.3	17-19	8-10.5	0.5 max	4-5.5	0.06 max	0.03 max	1 max	N 0.15-0.25; bal Fe				
MSZ 4360(80)	KO31	Corr res	0.12 max	16-18	8-10.5	0.5 max	3.5-4.5	0.06 max	0.03 max	1 max	N 0.15-0.25; bal Fe				
International															
COMECON PC4-70	13	Sh Strp Plt Bar	0.12 max	16-18	8-10		3.5-4.5	0.05 max	0.025 max	0.8 max	N 0.15-0.25; bal Fe				
ISO 683-13(74)	A-3*	Sh Strp Bar, ST	0.15 max	17-19	7.5-10.5		4-6	0.06 max	0.03 max	1 max	N 0.05-0.25; bal Fe				
Japan															
JIS G4303(91)	SUS202	Bar; SA; <=180mm diam	0.15 max	17.00-19.00	7.50-10.00		4.00-6.00	0.060 max	0.030 max	1.00 max	N <=0.25; bal Fe	520	275	40	207 HB
JIS G4304(91)	SUS202	Plt Sh, HR SA	0.15 max	17.00-19.00	7.50-10.00		4.0-6.00	0.060 max	0.030 max	1.00 max	N <=0.25; bal Fe	590	245	40	95 HRB max
JIS G4305(91)	SUS202	Plt Sh, CR SA	0.15 max	17.00-19.00	7.50-10.00		4.00-6.00	0.060 max	0.030 max	1.00 max	N <=0.25; bal Fe	590	245	40	207 HB max
JIS G4306	SUS202	Strp HR SA	0.15 max	17-19	7.5-10		4-6	0.06 max	0.03 max	1 max	N <=0.25; bal Fe				
JIS G4307	SUS202	Strp CR SA	0.15 max	17-19	7.5-10		4-6	0.06 max	0.03 max	1 max	N <=0.25; bal Fe				
Poland															
PNH86020	0H17N4G8	Corr res	0.07 max	16-18	7-9		4-6	0.05 max	0.03 max	0.8 max	N 0.12-0.25; bal Fe				
PNH86020	1H17N4G9	Corr res; Sh Strp Plt Bar	0.12 max	16-18	8-10.5		3.5-4.5	0.05 max	0.03 max	0.8 max	N 0.15-0.25; bal Fe				
PNH86020	H13N4G9	Corr res	0.15-0.3	12-14	8-10		3.7-4.7	0.05 max	0.03 max	0.8 max	bal Fe				
Romania															
STAS 3583(64)	10AzMNC170	Sh Strp Plt Bar; Corr/Heat res	0.12 max	16-18	8-10.5		3.5-4.5	0.035 max	0.03 max	0.8 max	N <=0.25; Pb <=0.15; bal Fe				
STAS 3583(64)	22MNC130	Corr res; Heat res	0.15-0.3	12-14	8-10		3.7-4.7	0.035 max	0.03 max	0.8 max	Pb <=0.15; bal Fe				
STAS 3583(87)	12NNiMnCr180	Corr res; Heat res	0.15 max	17-19	1 max	0.2 max	4-6	0.045 max	0.03 max	1 max	N <=0.25; Pb <=0.15; bal Fe				

UNS numbers and US grades are provided as a means of cross referencing chemically similar alloys. Exchangability is only possible after independent examination of specifications. Tensile properties are minimum or typical as specified. UTS and YS as MPa. El as %. See Appendix for list of abbreviations used in Notes. * indicates obsolete material.

Specification	Designation	Notes	C	Cr	Mn	Mo	Ni	P	S	Si	Other	UTS	YS	El	Hard

Stainless Steel, Austenitic, 202 (Continued from previous page)

Russia

Specification	Designation	Notes	C	Cr	Mn	Mo	Ni	P	S	Si	Other	UTS	YS	El	Hard
GOST 5632	12Ch17G9AN4	Sh Strp Plt Bar	0.12 max	16-18	8-10.5	0.3 max	3.5-4.5	0.035 max	0.02 max	0.8 max	N 0.15-0.25; bal Fe				
GOST 5632	20Ch13N4G9	Corr res	0.15-0.3	12.0-14.0	8.0-10.0	0.3 max	3.7-4.7	0.05 max	0.025 max	0.8 max	Cu <=0.3; Ti <=0.2; W <=0.2; bal Fe				
GOST 5632(72)	12Ch17G9AN4	Corr res	0.12 max	16.0-18.0	8.0-10.5	0.3 max	3.5-4.5	0.035 max	0.02 max	0.8 max	Cu <=0.3; N 0.15-0.25; Ti <=0.2; W <=0.2; bal Fe				

UK

Specification	Designation	Notes	C	Cr	Mn	Mo	Ni	P	S	Si	Other	UTS	YS	El	Hard
BS 1449/2(83)	284S16	Sh Strp Plt	0.07 max	16.5-18.5	7.00-10.0		4-6.5	0.06 max	0.03 max	1 max	N 0.15-0.25; bal Fe	630	300	40	0-220 HB
BS 1554(90)	202S16	Wir Corr/Heat res; diam<0.49mm; Ann	0.15 max	17.0-19.0	7.50-10.0		4.00-6.00	0.060 max	0.030 max	1.00 max	N <=0.25; W <=0.1; bal Fe	0-900			
BS 1554(90)	202S16	Wir Corr/Heat res; 6<diam<=13mm; Ann	0.15 max	17.0-19.0	7.50-10.0		4.00-6.00	0.060 max	0.030 max	1.00 max	N <=0.25; W <=0.1; bal Fe	0-750			

USA

Specification	Designation	Notes	C	Cr	Mn	Mo	Ni	P	S	Si	Other	UTS	YS	El	Hard
	AISI 202	Tube	0.15 max	17.00-19.00	7.50-10.00		4.00-6.00	0.060 max	0.030 max	1.00 max	N <=0.25; bal Fe				
	UNS S20200	Cr-Mn-Ni	0.15 max	17.00-19.00	7.50-10.00		4.00-6.00	0.060 max	0.030 max	1.00 max	N <=0.25; bal Fe				
ASTM A213/A213M(95)	TP202	Smls tube boiler, superheater, heat exchanger	0.15 max	17.0-19.0	7.50-10.0		4.00-6.00	0.060 max	0.030 max	1.00 max	N <=0.25; bal Fe	620	310	35	
ASTM A240/A240M(98)	S20200	Cr-Mn-Ni	0.15 max	17.0-19.0	7.5-10.0		4.0-6.0	0.060 max	0.030 max	1.00 max	N <=0.25; bal Fe	620	260	40.0	241 HB
ASTM A249/249M(96)	TP202	Weld tube; boiler, superheater, heat exch	0.15 max	17.0-19.0	7.50-10.0		4.00-6.00	0.060 max	0.030 max	1.00 max	N <=0.25; bal Fe	620	260	35	95 HRB
ASTM A314	202	Bil	0.15 max	17-19	7.5-10		4-6	0.06 max	0.03 max	1 max	N <=0.25; bal Fe				
ASTM A412	202	Sh Strp Plt	0.15 max	17-19	7.5-10		4-6	0.06 max	0.03 max	1 max	N <=0.25; bal Fe				
ASTM A473	202	Frg, Heat res SHT	0.15 max	17.00-19.00	7.50-10.00		4.00-6.00	0.060 max	0.030 max	1.00 max	N <=0.25; bal Fe	620	310	40	
ASTM A666(96)	202	Sh Strp Plt Bar, Ann	0.15 max	17.00-19.00	7.50-10.00		4.00-6.00	0.060 max	0.030 max	0.75 max	N <=0.25; bal Fe	620	260	40	241 HB
FED QQ-S-763F(96)	202*	Obs; Bar Wir Shp Frg; Corr res	0.15 max	17-19	7.5-10		4-6	0.06 max	0.03 max	1 max	N <=0.25; bal W				
FED QQ-S-766D(93)	202*	Obs; see ASTM A240, A666, A693; Sh Strp Plt	0.15 max	17-19	7.5-10		4-6	0.06 max	0.03 max	1 max	N <=0.25; bal Fe				
SAE J405(98)	S20200	Cr-Mn-Ni	0.15 max	17.00-19.00	7.50-10.00		4.00-6.00	0.060 max	0.030 max	1.00 max	N <=0.25; bal Fe				

Stainless Steel, Austenitic, 205

USA

Specification	Designation	Notes	C	Cr	Mn	Mo	Ni	P	S	Si	Other	UTS	YS	El	Hard
	AISI 205	Tube	0.12-0.25	16.50-18.00	14.00-15.50		1.00-1.75	0.060 max	0.030 max	1.00 max	N 0.32-0.40; bal Fe				
	UNS S20500	Cr-Mn-Ni	0.12-0.25	16.00-18.00	14.00-15.50		1.00-1.75	0.060 max	0.030 max	1.00 max	N 0.32-0.40; bal Fe				
ASTM A666(96)	205	Sh Strp Plt Bar, Ann	0.12-0.25	16.50-18.00	14.00-15.00		1.00-1.75	0.060 max	0.030 max	0.75 max	N 0.32-0.40; bal Fe	790	450	40	255 HB

Stainless Steel, Austenitic, 301

Australia

Specification	Designation	Notes	C	Cr	Mn	Mo	Ni	P	S	Si	Other	UTS	YS	El	Hard
AS 1449(94)	301	Sh Strp Plt	0.15 max	16.00-18.00	2.00 max		6.00-8.00	0.045 max	0.030 max	0.75 max	bal Fe				

Canada

Specification	Designation	Notes	C	Cr	Mn	Mo	Ni	P	S	Si	Other	UTS	YS	El	Hard
CSA G110.6	301	Sh Strp Plt, HR, Ann, or HR, Q/A, Tmp	0.15 max	16-18	2 max		6-8	0.05 max	0.03 max	1 max	bal Fe				

China

Specification	Designation	Notes	C	Cr	Mn	Mo	Ni	P	S	Si	Other	UTS	YS	El	Hard
GB 1220(92)	1Cr17Ni7	Bar SA	0.15 max	16.00-18.00	2.00 max		6.00-8.00	0.035 max	0.030 max	1.00 max	bal Fe	520	205	40	
GB 3280(92)	1Cr17Ni7	Sh Plt, HR SA	0.15-0.25	16.00-18.00	2.00 max		6.00-8.00	0.035 max	0.030 max	1.00 max	bal Fe	520	205	40	
GB 4231(93)	1Cr17Ni7	Strp CW	0.15 max	16.00-18.00	2.00 max		6.00-8.00	0.035 max	0.030 max	1.00 max	bal Fe	930	510	10	
GB 4239(91)	1Cr17Ni7	Sh Strp, SA	0.15 max	16.00-18.00	2.00 max		6.00-8.00	0.035 max	0.030 max	1.00 max	bal Fe	520	205	40	

UNS numbers and US grades are provided as a means of cross referencing chemically similar alloys. Exchangability is only possible after independent examination of specifications. Tensile properties are minimum or typical as specified. UTS and YS as MPa. El as %. See Appendix for list of abbreviations used in Notes. * indicates obsolete material.

Stainless Steel, Austenitic, 301 (Continued from previous page)

Specification	Designation	Notes	C	Cr	Mn	Mo	Ni	P	S	Si	Other	UTS	YS	El	Hard
Europe															
EN 10088/2(95)	1.4310	Strp, CR Corr res; t<=6mm, Ann	0.05-0.15	16.00-19.00	2.00 max	0.80 max	6.00-9.50	0.045 max	0.015 max	2.00 max	N <=0.11; bal Fe	600-950	250	40	
EN 10088/2(95)	X10CrNi18-8	Bar Rod Sect, Corr res; t<=6mm	0.05-0.15	16.00-19.00	2.00 max	0.80 max	6.00-9.50	0.045 max	0.015 max	2.00 max	N <=0.11; bal Fe	600-950	250	40	
EN 10088/2(95)	X10CrNi18-8	Strp, CR Corr res; t<=6mm, Ann	0.05-0.15	16.00-19.00	2.00 max	0.80 max	6.00-9.50	0.045 max	0.015 max	2.00 max	N <=0.11; bal Fe	600-950	250	40	
EN 10088/3(95)	1.4310	Bar Rod Sect, t<=40mm	0.05-0.15	16.00-19.00	2.00 max	0.80 max	6.00-9.50	0.045 max	0.015 max	2.00 max	N <=0.11; bal Fe	500-750	195	40	230 HB
EN 10088/3(95)	X10CrNi18-8	Bar Rod Sect, t<=40mm	0.05-0.15	16.00-19.00	2.00 max	0.80 max	6.00-9.50	0.045 max	0.015 max	2.00 max	N <=0.11; bal Fe	500-750	195	40	230 HB
France															
AFNOR NFA35572	Z12CN17.07	Bar, 25mm diam	0.08-0.15	16-18	2 max		6-8	0.04 max	0.03 max	1 max	bal Fe				
AFNOR NFA35573	Z12CN17.07	Sh Strp Plt, CR, 5mm diam	0.08-0.15	16-18	2 max		6-8	0.04 max	0.03 max	1 max	Cu <=0.05; bal Fe				
AFNOR NFA35575	Z12CN17.07	Wir rod	0.08-0.15	16-18	2 max		6-8	0.04 max	0.03 max	1 max	bal Fe				
AFNOR NFA36209	Z12CN17.07	Plt	0.08-0.15	16-18	2 max		6.5-8.5	0.04 max	0.03 max	1 max	bal Fe				
Germany															
DIN EN 10088(95)	WNr 1.4310	Sh Plt Strp Bar Rod	0.05-0.15	16.0-19.0	2.00 max	0.80 max	6.00-9.50	0.045 max	0.015 max	2.00 max	N <=0.11; bal Fe	500-750	195	40	
Hungary															
MSZ 4360(87)	KO32	Corr res; 0-160mm; SA	0.12 max	17-19	2 max	0.5 max	8-10	0.045 max	0.03 max	1 max	bal Fe	500-700	200	40L; 35T	190 HB max
MSZ 4360(87)	X12CrNi189	Corr res; 0-160mm; SA	0.12 max	17-19	2 max	0.5 max	8-10	0.045 max	0.03 max	1 max	bal Fe	500-700	200	40L; 35T	190 HB max
India															
IS 6527	10Cr17Ni7	Wir rod	0.15 max	16-18	2 max		6-8	0.05 max	0.03 max	1 max	bal Fe				
IS 6528	10Cr17Ni7	Wir	0.15 max	16-18	2 max		6-8	0.05 max	0.03 max	1 max	bal Fe				
IS 6529	10Cr17Ni7	Bil	0.15 max	16-18	2 max		6-8	0.05 max	0.03 max	1 max	bal Fe				
IS 6603	10Cr17Ni7	Bar, CD, 45mm diam	0.15 max	16-18	2 max		6-8	0.05 max	0.03 max	1 max	bal Fe				
IS 6911	10Cr17Ni7	Sh Strp Plt Wir	0.15 max	16-18	2 max		6-8	0.05 max	0.03 max	1 max	bal Fe				
International															
ISO 683-13(74)	14*	Sh Strp Plt Bar Frg ST	0.15 max	16-18	2 max		6-8	0.045 max	0.03 max	1 max	bal Fe				
Italy															
UNI 6901(71)	X12CrNi1707	Corr res	0.15 max	16-18	2 max		6.-8	0.045 max	0.03 max	1 max	bal Fe				
UNI 8317(81)	X12CrNi1707	Sh Strp Plt	0.15 max	16-18	2 max		6-8	0.045 max	0.03 max	1 max	bal Fe				
Japan															
JIS G4303(91)	SUS301	Bar; SA; <=180mm diam	0.15 max	16.00-18.00	2.00 max		6.00-8.00	0.045 max	0.045 max	1.00 max	bal Fe	520	205	40	207 HB
JIS G4304(91)	SUS301	Sh Plt, HR	0.15 max	16.00-18.00	2.00 max		6.0-8.0	0.045 max	0.030 max	1.00 max	bal Fe	520	205	40	95 HRB max
JIS G4305(91)	SUS301	Sh Plt, CR SA	0.15 max	16.00-18.00	2.00 max		6.00-8.00	0.045 max	0.03 max	1.00 max	bal Fe	520	205	40	207 HB
JIS G4305(91)	SUS301	Sh Plt 1/2 hard 0.4mm	0.15 max	16.00-18.00	2.00 max		6.00-8.00	0.045 max	0.030 max	1.00 max	bal Fe	1030	755	10m m	207 HB max
JIS G4306	SUS301	Strp HR SA	0.15 max	16.00-18.00	2.00 max		6.00-8.00	0.045 max	0.030 max	1.00 max	bal Fe				
JIS G4307	SUS301	Strp CR SA	0.15 max	16-18	2 max		6-8	0.04 max	0.03 max	1 max	bal Fe				
JIS G4313(96)	SUS301-CSP	CR Strp for springs Htemp	0.15 max	16.00-18.00	2.00 max		6.00-8.00	0.045 max	0.030 max	1.00 max	bal Fe	1320	1030		430 HV
Poland															
PNH86020	1H18N9	Corr res	0.12 max	17-19	2 max		8-10	0.045 max	0.03 max	0.8 max	bal Fe				

Specification	Designation	Notes	C	Cr	Mn	Mo	Ni	P	S	Si	Other	UTS	YS	El	Hard
Stainless Steel, Austenitic, 301 (Continued from previous page)															
Romania															
STAS 10322(80)	12NiCr180	Corr res; Heat res	0.12 max	17-19	2 max		8-10	0.04 max	0.03 max	1 max	Cu <=0.5; Pb <=0.15; bal Fe				
Spain															
UNE 36016(75)	F.3517*	see X12CrNi17-07	0.15 max	16.0-18.0	2.0 max	0.15 max	6.0-8.0	0.045 max	0.03 max	1.0 max	bal Fe				
UNE 36016(75)	X12CrNi17-07	Sh Strp Plt Bar Wir; Corr res	0.15 max	16.0-18.0	2.0 max	0.15 max	6.0-8.0	0.045 max	0.03 max	1.0 max	bal Fe				
UNE 36016/1(89)	E-301	Corr res	0.15 max	16.0-18.0	2.0 max	0.15 max	6.0-8.0	0.045 max	0.03 max	1.0 max	N <=0.1; bal Fe				
UNE 36016/1(89)	F.3517*	Obs; E-301; Corr res	0.15 max	16.0-18.0	2.0 max	0.15 max	6.0-8.0	0.045 max	0.03 max	1.0 max	N <=0.1; bal Fe				
Sweden															
SIS 142331	2331-02	Sh Strp Bar Wir Frg, SA	0.12 max	17-19	2 max		7-9.5	0.05 max	0.03 max	1 max	bal Fe	490	210	45	
SIS 142331	2331-11	Strp, CW, 2.5mm diam	0.12 max	17-19	2 max		7-9.5	0.05 max	0.03 max	1 max	bal Fe	620	260		
SIS 142331	2331-12	Strp, CW, 2mm diam	0.12 max	17-19	2 max		7-9.5	0.05 max	0.03 max	1 max	bal Fe	690	290		
SIS 142331	2331-14	Strp, CW, 2mm diam	0.12 max	17-19	2 max		7-9.5	0.05 max	0.03 max	1 max	bal Fe	890	590		
SIS 142331	2331-17	Strp, CW, 2mm diam	0.12 max	17-19	2 max		7-9.5	0.05 max	0.03 max	1 max	bal Fe	230	130		
SIS 142331	2331-18	Strp, CW, 1.5mm	0.12 max	17-19	2 max		7-9.5	0.05 max	0.03 max	1 max	bal Fe	370	230		
SIS 142331	2331-19	Strp, CW, 1.4mm diam	0.12 max	17-19	2 max		7-9.5	0.05 max	0.03 max	1 max	bal Fe	570	320		
SS 14231	2331	Corr res	0.12 max	16-19	2 max	0.8 max	6.5-9.5	0.045 max	0.03 max	1.5 max	bal Fe				
UK															
BS 1449/2(83)	301S21	Sh Strp, 1/2 Hard	0.15 max	16-18	0.50-2.00		6-8	0.05 max	0.03 max	0.2-1	bal Fe	540	215	40	0-220 HB
BS 1554(90)	301S22	Wir Corr/Heat res; 0.5-1.49mm; Ann	0.15 max	16.0-18.0	2.00 max	0.80 max	6.50-9.00	0.045 max	0.030 max	1.00 max	W <=0.1; bal Fe	0-900			
BS 1554(90)	301S22	Wir Corr/Heat res; 6<diam<=13mm; Ann	0.15 max	16.0-18.0	2.00 max	0.80 max	6.50-9.00	0.045 max	0.030 max	1.00 max	W <=0.1; bal Fe	0-750			
BS 2056(83)	301S81	Corr res; spring	0.09 max	16.0-18.0	1.00 max		6.50-7.75	0.045 max	0.030 max	1.00 max	Al 0.75-1.50; W <=0.1; bal Fe				
USA															
	AISI 301	Tube	0.15 max	16.00-18.00	2.00 max		6.00-8.00	0.045 max	0.030 max	1.00 max	bal Fe				
	UNS S30100		0.15 max	16.00-18.00	2.00 max		6.00-8.00	0.045 max	0.030 max	1.00 max	bal Fe				
AMS 5517		Sh Strp Plt Bar Wir	0.15 max	16-18	2 max		6-8	0.045 max	0.03 max	1 max	bal Fe				
AMS 5518		Sh Strp Plt Bar Wir	0.15 max	16-18	2 max		6-8	0.045 max	0.03 max	1 max	bal Fe				
AMS 5519		Sh Strp Plt Bar Wir	0.15 max	16-18	2 max		6-8	0.045 max	0.03 max	1 max	bal Fe				
ASTM A167(96)	301*	Sh Strp Plt	0.15 max	16-18	2 max		6-8	0.045 max	0.03 max	1 max	bal Fe				
ASTM A177	301*	Obs; see A666; Sht strp	0.15 max	16-18	2 max		6-8	0.045 max	0.03 max	1 max	bal Fe				
ASTM A240/A240M(98)	S30100		0.15 max	16.0-18.0	2.00 max		6.0-8.0	0.045 max	0.030 max	1.00 max	N <=0.10; bal Fe	515	205	40.0	217 HB
ASTM A554(94)	MT301	Weld mech tub, Rnd Ann	0.15 max	16.0-18.0	2.00 max		6.0-8.0	0.040 max	0.030 max	1.00 max	bal Fe	517	207	35	
ASTM A666(96)	301	Sh Strp Plt Bar, Ann	0.15 max	16.00-18.00	2.00 max		6.00-8.00	0.045 max	0.030 max	1.00 max	N <=0.10; bal Fe	515	205	40	217 HB
FED QQ-S-766D(93)	301*	Obs; see ASTM A240, A666, A693; Sh Strp Plt	0.15 max	16-18	2 max		6-8	0.045 max	0.03 max	1 max	bal Fe				
MIL-S-5059D(90)	301*	Obs see AMS specs; Plt Sh Strp; Corr res	0.15 max	16-18	2 max		6-8	0.045 max	0.03 max	1 max	bal Fe				
SAE J405(98)	S30100		0.15 max	16.00-18.00	2.00 max		6.00-8.00	0.045 max	0.030 max	1.00 max	N <=0.10; bal Fe				

UNS numbers and US grades are provided as a means of cross referencing chemically similar alloys. Exchangability is only possible after independent examination of specifications. Tensile properties are minimum or typical as specified. UTS and YS as MPa. El as %. See Appendix for list of abbreviations used in Notes. * indicates obsolete material.

Specification	Designation	Notes	C	Cr	Mn	Mo	Ni	P	S	Si	Other	UTS	YS	El	Hard

Stainless Steel, Austenitic, 301L

USA

Specification	Designation	Notes	C	Cr	Mn	Mo	Ni	P	S	Si	Other	UTS	YS	El	Hard
	UNS S30103		0.030 max	16.0-18.0	2.00 max		6.0-8.0	0.045 max	0.030 max	1.0 max	N <=0.20; bal Fe				

Stainless Steel, Austenitic, 301LN

USA

Specification	Designation	Notes	C	Cr	Mn	Mo	Ni	P	S	Si	Other	UTS	YS	El	Hard
	UNS S30153		0.030 max	16.0-18.0	2.00 max		6.0-8.0	0.045 max	0.030 max	1.0 max	N 0.07-0.20; bal Fe				

Stainless Steel, Austenitic, 302

Australia

Specification	Designation	Notes	C	Cr	Mn	Mo	Ni	P	S	Si	Other	UTS	YS	El	Hard
AS 1449(94)	302	Sh Strp Plt	0.15 max	17.00-19.00	2.00 max		8.00-10.00	0.045 max	0.030 max	0.75 max	bal Fe				
AS 2837(86)	302	Bar, Semi-finished product	0.12 max	17.0-19.0	2.00 max		8.0-10.0	0.045 max	0.030 max	1.00 max	bal Fe				
AS 2837(86)	302HQ	Bar, Semi-finished product	0.08 max	17.0-19.0	2.00 max		9.0-11.0	0.045 max	0.030 max	1.00 max	Cu 3.0-4.0; bal Fe				

Bulgaria

Specification	Designation	Notes	C	Cr	Mn	Mo	Ni	P	S	Si	Other	UTS	YS	El	Hard
BDS 6738(72)	Ch18N9	Sh Strp Plt Bar Wir	0.12 max	17.0-19.0	2.00 max	0.30 max	8.0-10.0	0.035 max	0.025 max	0.80 max	bal Fe				

Canada

Specification	Designation	Notes	C	Cr	Mn	Mo	Ni	P	S	Si	Other	UTS	YS	El	Hard
CSA G110.3	302	Bar Bil	0.15 max	17-19	2 max		8-10	0.05 max	0.03 max	1 max	bal Fe				
CSA G110.6	302	Sh Strp Plt, HR, Ann, or HR, Q/A, Tmp	0.15 max	17-19	2 max		8-10	0.05 max	0.03 max	1 max	bal Fe				
CSA G110.9	302	Sh Strp Plt, HR, CR, Ann, Q/A, Tmp	0.15 max	17-19	2 max		8-10	0.05 max	0.03 max	1 max	bal Fe				

China

Specification	Designation	Notes	C	Cr	Mn	Mo	Ni	P	S	Si	Other	UTS	YS	El	Hard
GB 1220(92)	1Cr18Ni9	Bar SA	0.15 max	17.00-19.00	2.00 max		8.00-10.00	0.035 max	0.030 max	1.00 max	bal Fe	520	205	40	
GB 12770(91)	1Cr18Ni9	Weld tube; SA	0.15 max	17.00-19.00	2.00 max		8.00-10.00	0.035 max	0.030 max	1.00 max	bal Fe	520	210	35	
GB 12771(91)	1Cr18Ni9	Weld Pipe SA	0.15 max	17.00-19.00	2.00 max		8.00-10.00	0.035 max	0.030 max	1.00 max	bal Fe	520	210	35	
GB 13296(91)	1Cr19Ni9	Smls tube SA	0.04-0.10	18.00-20.00	2.00 max		8.00-11.00	0.035 max	0.030 max	1.00 max	bal Fe	520	205	35	
GB 3280(92)	1Cr18Ni9	Sh Plt, CR SA	0.15 max	17.00-19.00	2.00 max		8.00-10.00	0.035 max	0.030 max	1.00 max	bal Fe	520	205	40	
GB 4237(92)	1Cr18Ni9	Sh Plt, HR SA	0.15 max	17.00-19.00	2.00 max		8.00-10.00	0.035 max	0.030 max	1.00 max	bal Fe	520	205	40	
GB 4239(91)	1Cr18Ni9	Sh Strp, CR SA	0.15 max	17.00-19.00	2.00 max		8.00-10.00	0.035 max	0.030 max	1.00 max	bal Fe	520	205	40	
GB 4240(93)	1Cr18Ni9(-L,-Q,-R)	Wir Ann 0.6-1mm diam, L-CD; Q-TLC	0.15 max	17.00-19.00	2.00 max		8.00-10.00	0.035 max	0.030 max	1.00 max	bal Fe	540-880		25	
GB 5310(95)	1Cr18Ni9	Smls tube Pip SA	0.15 max	17.00-19.00	2.00 max		8.00-10.00	0.035 max	0.030 max	1.00 max	Cu <=0.20; bal Fe	520	205	35	

Czech Republic

Specification	Designation	Notes	C	Cr	Mn	Mo	Ni	P	S	Si	Other	UTS	YS	El	Hard
CSN 417241	17241	Corr res	0.12 max	17.0-20.0	2.0 max		8.0-11.0	0.045 max	0.03 max	1.0 max	bal Fe				

France

Specification	Designation	Notes	C	Cr	Mn	Mo	Ni	P	S	Si	Other	UTS	YS	El	Hard
AFNOR NFA35209	Z10CN18.09	Plt	0.12 max	17-19	2 max		8-10	0.04 max	0.03 max	1 max	bal Fe				
AFNOR NFA35572	Z10CN18.09	Bar	0.12 max	17-19	2 max		7.5-9.5	0.04 max	0.03 max	1 max	bal Fe				
AFNOR NFA35573	Z10CN18.09	Sh Strp Plt, CR, 5mm diam	0.12 max	17-19	2 max		7.5-9.5	0.04 max	0.03 max	1 max	bal Fe				
AFNOR NFA35575	Z10CN18.09	Wir rod, 60% Hard	0.12 max	17-19	2 max		7.5-9.5	0.04 max	0.03 max	1 max	bal Fe				

Germany

Specification	Designation	Notes	C	Cr	Mn	Mo	Ni	P	S	Si	Other	UTS	YS	El	Hard
DIN	WNr 1.4300		0.12 max	17.0-19.0	2.00 max		8.00-10.0	0.045 max	0.030 max	1.00 max	bal Fe				
DIN	X10CrNi18 8		0.12 max	17.0-19.0	2.00 max		8.00-10.0	0.045 max	0.030 max	1.00 max	bal Fe				
DIN	X12CrNi188		0.12 max	17.0-19.0	2.00 max		8.00-10.0	0.045 max	0.030 max	1.00 max	bal Fe				

Specification	Designation	Notes	C	Cr	Mn	Mo	Ni	P	S	Si	Other	UTS	YS	El	Hard

Stainless Steel, Austenitic, 302 (Continued from previous page)

International

Specification	Designation	Notes	C	Cr	Mn	Mo	Ni	P	S	Si	Other	UTS	YS	El	Hard
COMECON PC4-70	14	Sh Strp Plt Bar Tub Frg	0.12 max	17-19	2 max		8-10	0.035 max	0.025 max	1 max	bal Fe				
ISO 4954(93)	X10CrNi189E	Wir rod Bar, CH ext	0.12 max	17.0-19.0	2.00 max		8.0-10.0	0.045 max	0.030 max	1.00 max	bal Fe	650			
ISO 683-13(74)	12*	ST	0.12 max	17-19	2 max		8-10	0.045 max	0.03 max	1 max	bal Fe				
ISO 6931-2(89)	X9CrNi188	Spring	0.12 max	16-19	2 max		6.5-9.5			1.5 max	bal Fe				

Italy

Specification	Designation	Notes	C	Cr	Mn	Mo	Ni	P	S	Si	Other	UTS	YS	El	Hard
UNI 6901(71)	X10CrNi1809	Corr res	0.12 max	17-19	2 max		8-10	0.045 max	0.03 max	1 max	bal Fe				
UNI 8317(81)	X10CrNi1809	Sh Strp Plt	0.12 max	17-19	2 max		8-10	0.045 max	0.03 max	1 max	bal Fe				

Japan

Specification	Designation	Notes	C	Cr	Mn	Mo	Ni	P	S	Si	Other	UTS	YS	El	Hard
JIS G4303(91)	SUS302	Bar; SA; <=180mm diam	0.15 max	17.00-19.00	2.00 max		8.00-10.00	0.045 max	0.030 max	1.00 max	bal Fe	520	205	40	187 HB
JIS G4304(91)	SUS302	Sh Plt, HR SA	0.15 max	17.00-19.00	2.00 max		8.0-10.00	0.045 max	0.030 max	1.00 max	bal Fe	520	205	40	90 HRB max
JIS G4305(91)	SUS302	Sh Plt, CR SA	0.15 max	17.00-19.00	2.00 max		8.00-10.00	0.045 max	0.030 max	1.00 max	bal Fe	520	205	40	187 HB max
JIS G4306	SUS302	Strp HR SA	0.15 max	17-19	2 max		8-10	0.04 max	0.03 max	1 max	bal Fe				
JIS G4307	SUS302	Strp CR SA	0.15 max	17-19	2 max		8-10	0.04 max	0.03 max	1 max	bal Fe				
JIS G4308(98)	SUS302	Wir rod	0.15 max	17.00-19.00	2.00 max		8.00-10.00	0.045 max	0.030 max	1.00 max	bal Fe				

Mexico

Specification	Designation	Notes	C	Cr	Mn	Mo	Ni	P	S	Si	Other	UTS	YS	El	Hard
DGN B-83	302*	Obs; Bar, HR, CR	0.15 max	17-19	2 max		8-10	0.045 max	0.03 max	1 max	bal Fe				
NMX-B-171(91)	MT302	mech tub	0.08-0.20	17.0-19.0	2.00 max		8.0-10.0	0.040 max	0.030 max	1.00 max	bal Fe				

Poland

Specification	Designation	Notes	C	Cr	Mn	Mo	Ni	P	S	Si	Other	UTS	YS	El	Hard
PNH86020	1H18N9		0.12 max	17-19	2 max		8-10	0.045 max	0.03 max	0.8 max	bal Fe				

Romania

Specification	Designation	Notes	C	Cr	Mn	Mo	Ni	P	S	Si	Other	UTS	YS	El	Hard
STAS 3583	13NC180	Sh Strp Plt Bar Wir Tub Frg	0.12 max	17-19	2 max	0.2 max	8-10	0.035 max	0.03 max	0.8 max	bal Fe				
STAS 3583(64)	10TNC180	Corr res; Heat res	0.12 max	17-19	2 max	0.2 max	8-9.5	0.035 max	0.03 max	0.8 max	Pb <=0.15; Ti <=0.7; 5C<=Ti<=0.7; bal Fe				

Russia

Specification	Designation	Notes	C	Cr	Mn	Mo	Ni	P	S	Si	Other	UTS	YS	El	Hard
GOST 5632	12Ch18N9	Sh Strp Plt Bar Tub Frg	0.12 max	17-19	2 max	0.3 max	8-10	0.035 max	0.02 max	0.8 max	Cu <=0.3; Ti <=0.5; W <=0.2; bal Fe				
GOST 5632(72)	12Ch18N9	Corr res; Heat res	0.12 max	17.0-19.0	2.0 max	0.3 max	8.0-10.0	0.035 max	0.02 max	0.8 max	Cu <=0.3; Ti <=0.5; W <=0.2; bal Fe				

Spain

Specification	Designation	Notes	C	Cr	Mn	Mo	Ni	P	S	Si	Other	UTS	YS	El	Hard
UNE 36016/1(89)	E-302	Corr res	0.12 max	17.0-19.0	2.0 max	0.15 max	8.0-10.0	0.045 max	0.03 max	1.0 max	N <=0.1; bal Fe				
UNE 36016/1(89)	F.3507*	Obs; E-302; Corr res	0.12 max	17.0-19.0	2.0 max	0.15 max	8.0-10.0	0.045 max	0.03 max	1.0 max	N <=0.1; bal Fe				
UNE 36017(85)	F.3311*	see X9CrNi18-09	0.12 max	17-19	2 max	0.15 max	8-10	0.045 max	0.03 max	1 max	bal Fe				
UNE 36017(85)	X9CrNi18-09	Heat res	0.12 max	17-19	2 max	0.15 max	8-10	0.045 max	0.03 max	1 max	bal Fe				

UK

Specification	Designation	Notes	C	Cr	Mn	Mo	Ni	P	S	Si	Other	UTS	YS	El	Hard
BS 1449/2(83)	302S25	Sh Strp, Ann	0.12 max	17-19	0.50-2.00		8-11	0.05 max	0.03 max	0.2-1	bal Fe				
BS 1554(90)	302S31	Wir Corr/Heat res; 6<diam<=13mm; Ann	0.12 max	17.0-19.0	2.00 max		8.00-10.0	0.045 max	0.030 max	1.00 max	W <=0.1; bal Fe	0-750			
BS 1554(90)	302S31	Wir Corr/Heat res; diam<0.49mm; Ann	0.12 max	17.0-19.0	2.00 max		8.00-10.0	0.045 max	0.030 max	1.00 max	W <=0.1; bal Fe	0-900			
BS 2056(83)	302S26	Corr res; spring	0.12 max	17.0-19.0	2.00 max		7.50-10.0	0.045 max	0.030 max	1.00 max	W <=0.1; bal Fe				
BS 970/1(96)	302S31	Wrought Corr/Heat res; t<=160mm	0.12 max	17.0-19.0	2.00 max		8.00-10.0	0.045 max	0.030 max	1.00 max	bal Fe	510	190	40	183 HB max

Stainless Steel, Austenitic, 302 (Continued from previous page)

USA

Specification	Designation	Notes	C	Cr	Mn	Mo	Ni	P	S	Si	Other	UTS	YS	El	Hard
	AISI 302	Tube	0.15 max	17.00-19.00	2.00 max		8.00-10.00	0.045 max	0.030 max	1.00 max	bal Fe				
	UNS S30200		0.15 max	17.00-19.00	2.00 max		8.00-10.00	0.045 max	0.030 max	1.00 max	bal Fe				
AMS 2431/4A(96)	302	Cond Cut Wire Shot	0.15 max	17.00-20.00	2.00 max		8.00-12.00	0.045 max	0.030 max	1.00 max	bal Fe				45 HRC min
AMS 5515		Sh Strp Plt Bar Wir	0.15	17-19	2 max		8-10	0.045 max	0.03 max	1 max	bal Fe				
AMS 5516		Sh Strp Plt Bar Wir	0.15 max	17-19	2 max		8-10	0.045 max	0.03 max	1 max	bal Fe				
AMS 5600		Sh Strp Plt Bar Wir	0.15 max	17-19	2 max		8-10	0.045 max	0.03 max	1 max	bal Fe				
AMS 5636		Sh Strp Plt Bar Wir	0.15 max	17-19	2 max		8-10	0.045 max	0.03 max	1 max	bal Fe				
AMS 5637		Sh Strp Plt Bar Wir	0.15 max	17-19	2 max		8-10	0.045 max	0.03 max	1 max	bal Fe				
AMS 5688		Sh Strp Plt Bar Wir	0.15 max	17-19	2 max		8-10	0.045 max	0.03 max	1 max	bal Fe				
AMS 5693		Sh Strp Plt Bar Wir	0.15 max	17-19	2 max		8-10	0.045 max	0.03 max	1 max	bal Fe				
AMS 5866B(95)		Flat Wire, Spring tmp, 0.20<t<=0.56	0.15 max	17.00-19.00	2.00 max	0.75 max	8.00-10.00	0.040 max	0.030 max	1.00 max	Cu <=0.75; bal Fe	1379-1655		3	72.0-74.9 HRA
AMS 5866B(95)		Flat Wire, Spring tmp, 2.26<t<=2.41	0.15 max	17.00-19.00	2.00 max	0.75 max	8.00-10.00	0.040 max	0.030 max	1.00 max	Cu <=0.75; bal Fe	1069-1413		3	34.5-44.0 HRC
AMS 7210		Sh Strp Plt Bar Wir	0.15 max	17-19	2 max		8-10	0.045 max	0.03 max	1 max	bal Fe				
AMS 7241		Sh Strp Plt Bar Wir	0.15 max	17-19	2 max		8-10	0.045 max	0.03 max	1 max	bal Fe				
ASME SA240	302	Refer to ASTM A240(95)													
ASME SA479	302	Refer to ASTM A479/A479M(95)													
ASTM A167(96)	302*	Sh Strp Plt	0.15 max	17-19	2 max		8-10	0.045 max	0.03 max	1 max	bal Fe				
ASTM A240/A240M(98)	S30200		0.15 max	17.0-19.0	2.00 max		8.0-10.0	0.045 max	0.030 max	0.75 max	N <=0.10; bal Fe	515	205	40.0	201 HB
ASTM A276(98)	302	Bar Shp	0.15 max	17-19	2 max		8-10	0.045 max	0.03 max	1 max	bal Fe				
ASTM A313/A313M(95)	302 Class 1	Spring wire, t<=0.23mm	0.15 max	17.00-19.00	2.00 max		8.00-10.00	0.045 max	0.030 max	1.00 max	bal Fe	2240-2450			
ASTM A313/A313M(95)	302 Class 2	Spring wire, CD, 1.27<t<=4.06mm	0.15 max	17.00-19.00	2.00 max		8.00-10.00	0.045 max	0.030 max	1.00 max	bal Fe	1998			
ASTM A314	302	Bil Bar, for Frg	0.15 max	17-19	2 max		8-10	0.045 max	0.03 max	1 max	bal Fe				
ASTM A368(95)	302	Wir, Heat res	0.15 max	17.00-19.00	2.00 max		8.00-10.00	0.045 max	0.030 max	1.00 max	bal Fe				
ASTM A473	302	Frg, Heat res SHT	0.15 max	17.00-19.00	2.00 max		8.00-10.00	0.045 max	0.030 max	1.00 max	bal Fe	515	205	40	
ASTM A478	302	Wir CD 0.76<t<=3.18mm	0.15 max	17.00-19.00	2.00 max		8.00-10.00	0.045 max	0.030 max	1.00 max	bal Fe	830-1030		15	
ASTM A479	302	Wir Bar Shp	0.15 max	17-19	2 max		8-10	0.045 max	0.03 max	1 max	bal Fe				
ASTM A492	302	Wir, 2.29<t<=2.54	0.15 max	17.00-19.00	2.00 max		8.00-10.00	0.045 max	0.030 max	1.00 max	bal Fe	1620-1830			
ASTM A493	302	Wir rod, CHd, cold frg	0.15 max	17.0-19.0	2.00 max		8.0-10.0	0.045 max	0.030 max	1.00 max	bal Fe	655			
ASTM A511(96)	MT302	Smls mech tub, Ann	0.08-0.20	17.0-19.0	2.00 max		8.0-10.0	0.040 max	0.030 max	1.00 max	bal Fe	517	207	35	192 HB
ASTM A554(94)	MT302	Weld mech tub, Rnd Ann	0.15 max	17.0-19.0	2.00 max		8.0-10.0	0.040 max	0.030 max	1.00 max	bal Fe	517	207	35	
ASTM A580/A580M(98)	302	Wir, Ann	0.15 max	17.0-19.0	2.00 max		8.0-10.0	0.045 max	0.030 max	1.00 max	N <=0.10; bal Fe	520	210	35	
ASTM A666(96)	302	Sh Strp Plt Bar, Ann	0.15 max	17.00-19.00	2.00 max		8.00-10.00	0.045 max	0.030 max	0.75 max	bal Fe	515	205	40	201 HB
FED QQ-S-763F(96)	302*	Obs; Bar Wir Shp Frg; Corr res	0.15 max	17-19	2 max		8-10	0.045 max	0.03 max	1 max	bal Fe				

UNS numbers and US grades are provided as a means of cross referencing chemically similar alloys. Exchangability is only possible after independent examination of specifications. Tensile properties are minimum or typical as specified. UTS and YS as MPa. El as %. See Appendix for list of abbreviations used in Notes. * indicates obsolete material.

Specification	Designation	Notes	C	Cr	Mn	Mo	Ni	P	S	Si	Other	UTS	YS	El	Hard

Stainless Steel, Austenitic, 302 (Continued from previous page)

USA

Specification	Designation	Notes	C	Cr	Mn	Mo	Ni	P	S	Si	Other	UTS	YS	El	Hard
FED QQ-S-766D(93)	302*	Obs; see ASTM A240, A666, A693; Sh Strp Plt	0.15 max	17-19	2 max		8-10	0.045 max	0.03 max	1 max	bal Fe				
FED QQ-W-423B(85)	302*	Obs; see ASTM A313, A580; Wir Corr res	0.15 max	17-19	2 max		8-10	0.045 max	0.03 max	1 max	bal Fe				
MIL-S-5059D(90)	302*	Obs see AMS specs; Plt Sh Strp; Corr res	0.15 max	17-19	2 max		8-10	0.045 max	0.03 max	1 max	bal Fe				
MIL-S-7720A	302*	Obs	0.15 max	17-19	2 max		8-10	0.045 max	0.03 max	1 max	bal Fe				
MIL-S-862B	302*	Obs	0.15 max	17-19	2 max		8-10	0.045 max	0.03 max	1 max	bal Fe				
SAE J405(98)	S30200		0.15 max	17.00-19.00	2.00 max		8.00-10.00	0.045 max	0.030 max	0.75 max	N <=0.10; bal Fe				
SAE J467(68)	302		0.08	18.0	1.00		9.0			0.50	bal Fe	634	262	68	85 HRB

Stainless Steel, Austenitic, 302B

Bulgaria

Specification	Designation	Notes	C	Cr	Mn	Mo	Ni	P	S	Si	Other	UTS	YS	El	Hard
BDS 6738(72)	1CH18N95	Sh Strp Plt Bar Wir	0.10-0.20	17.0-20.0	2.00 max	0.30 max	8.0-11.0	0.035 max	0.025 max	0.80-2.00	bal Fe				

Canada

Specification	Designation	Notes	C	Cr	Mn	Mo	Ni	P	S	Si	Other	UTS	YS	El	Hard
CSA G110.3	302B	Bar Bil	0.15 max	17-19	2 max		8-10	0.05 max	0.03 max	2-3	bal Fe				

Japan

Specification	Designation	Notes	C	Cr	Mn	Mo	Ni	P	S	Si	Other	UTS	YS	El	Hard
JIS G4304(91)	SUS302B	Plt Sh, HR	0.15 max	17.00-19.00	2.00 max		8.0-10.00	0.045 max	0.030 max	2.00-3.00	bal Fe	520	205	40	95 HRB max
JIS G4305(91)	SUS302B	Plt Sh, CR	0.15 max	17.00-19.00	2.00 max		8.00-10.00	0.045 max	0.030 max	2-3.00	bal Fe	520	205	40	207 HB max
JIS G4306	SUS302B	Strp HR	0.15 max	17-19	2 max		8-10	0.045 max	0.03 max	2-3	bal Fe				
JIS G4307	SUS302B	Strp CR	0.15 max	17-19	2 max		8-10	0.045 max	0.03 max	2-3	bal Fe				

Mexico

Specification	Designation	Notes	C	Cr	Mn	Mo	Ni	P	S	Si	Other	UTS	YS	El	Hard
DGN B-83	302B*	Obs; Bar, HR, CR	0.15 max	17-19	2 max		8-10	0.05 max	0.03 max	2-3	bal Fe				

USA

Specification	Designation	Notes	C	Cr	Mn	Mo	Ni	P	S	Si	Other	UTS	YS	El	Hard
	AISI 302B	Tube	0.15 max	17.00-19.00	2.00 max		8.00-10.00	0.045 max	0.030 max	2.00-3.00	bal Fe				
	UNS S30215		0.15 max	17.00-19.00	2.00 max		8.00-10.00	0.045 max	0.030 max	2.00-3.00	bal Fe				
ASTM A167(96)	302B	Sh Strp Plt	0.15 max	17.00-19.00	2.00 max		8.00-10.00	0.045 max	0.030 max	2.00-3.00	Al <=0.025; N <=0.10; bal Fe	515	205	40.0	217 HB
ASTM A276(98)	302B	Bar Shp	0.15 max	17-19	2 max		8-10	0.045 max	0.03 max	2-3	bal Fe				
ASTM A314	302B	Bil Bar, for Frg	0.15 max	17-19	2 max		8-10	0.045 max	0.03 max	2-3	bal Fe				
ASTM A473	302B	Frg, Heat res SHT	0.15 max	17.00-19.00	2.00 max		8.00-10.00	0.045 max	0.030 max	2.00-3.00	bal Fe	515	205	40	
ASTM A580/A580M(98)	302B	Wir, Ann	0.15 max	17.0-19.0	2.00 max		8.0-10.0	0.045 max	0.030 max	2.00-3.00	bal Fe	520	210	35	

Stainless Steel, Austenitic, 303

Australia

Specification	Designation	Notes	C	Cr	Mn	Mo	Ni	P	S	Si	Other	UTS	YS	El	Hard
AS 2837(86)	303	Bar, Semi-finished product	0.12 max	17.0-19.0	2.00 max		8.0-10.0	0.060 max	0.15-0.35	1.00 max	bal Fe				

Canada

Specification	Designation	Notes	C	Cr	Mn	Mo	Ni	P	S	Si	Other	UTS	YS	El	Hard
CSA G110.3	303	Bar Bil	0.15 max	17-19	2 max	0.6 max	8-10	0.2 max	0.15 max	1 max	bal Fe				

China

Specification	Designation	Notes	C	Cr	Mn	Mo	Ni	P	S	Si	Other	UTS	YS	El	Hard
GB 1220(92)	Y1Cr18Ni9	Bar SA	0.15 max	17.00-19.00	2.00 max	0.60 max	8.00-10.00	0.20 max	0.15 min	1.00 max	bal Fe	520	205	40	

Europe

Specification	Designation	Notes	C	Cr	Mn	Mo	Ni	P	S	Si	Other	UTS	YS	El	Hard
EN 10088/2(95)	1.4305	Plt, HR Corr res; t<=75mm, Ann	0.10 max	17.00-19.00	2.00 max		8.00-10.00	0.045 max	0.15-0.35	1.00 max	Cu <=1.00; N <=0.11; bal Fe	500-700	190	35	
EN 10088/2(95)	X8CrNiS18-9	Plt, HR Corr res; t<=75mm, Ann	0.10 max	17.00-19.00	2.00 max		8.00-10.00	0.045 max	0.15-0.35	1.00 max	Cu <=1.00; N <=0.11; bal Fe	500-700	190	35	

Specification	Designation	Notes	C	Cr	Mn	Mo	Ni	P	S	Si	Other	UTS	YS	El	Hard

Stainless Steel, Austenitic, 303 (Continued from previous page)

Specification	Designation	Notes	C	Cr	Mn	Mo	Ni	P	S	Si	Other	UTS	YS	El	Hard
Europe															
EN 10088/3(95)	1.4305	Bar Rod Sect, Corr res; t<=160mm, SA	0.10 max	17.00-19.00	2.00 max		8.00-10.00	0.045 max	0.15-0.35	1.00 max	N <=0.11; bal Fe	500-750	190	35	230 HB
EN 10088/3(95)	X8CrNiS18-9	Bar Rod Sect, Corr res; t<=160mm, SA	0.10 max	17.00-19.00	2.00 max		8.00-10.00	0.045 max	0.15-0.35	1.00 max	N <=0.11; bal Fe	500-750	190	35	230 HB
Germany															
DIN EN 10088(95)	WNr 1.4305	Sh Plt Strp Bar Rod, HT	0.10 max	17.0-19.0	2.00 max		8.00-10.0	0.045 max	0.15-0.35	1.00 max	bal Fe	500-700	190	35	
DIN EN 10088(95)	X8CrNiS18-9	SHT >=160mm	0.10 max	17.0-19.0	2.00 max		8.00-10.0	0.045 max	0.35 max	1.00 max	Cu <=1.00; N <=0.11; bal Fe	500-750	190	35	160 HB
Hungary															
MSZ 4360(87)	KO36S	Corr res; 0-160mm; SA	0.12 max	17-19	2 max	0.5 max	8-11	0.045 max	0.15-0.35	1 max	bal Fe	500-700	195	35L max	190 HB max
MSZ 4360(87)	X12CrNiS189	Corr res; 0-160mm; SA	0.12 max	17-19	2 max	0.5 max	8-11	0.045 max	0.15-0.35	1 max	bal Fe	500-700	195	35L max	190 HB max
India															
IS 1570/5(85)	X07Cr18Ni9	Sh Plt Strp Bar Flat Band	0.15 max	17-19	2 max	0.15 max	8-10	0.045 max	0.03 max	1 max	W <=0.1; bal Fe	515	205	40	183 HB max
International															
ISO 683-13(74)	17*	ST		17-19	2 max		8-10	0.06 max	0.15-0.35	1 max	bal Fe				
Italy															
UNI 6901(71)	X10CrNiS1809	Corr res	0.12 max	17-19	2 max	0.6 max	8-11	0.2 max	0.15-0.35	1 max	bal Fe				
Japan															
JIS G4303(91)	SUS303	Bar; SA; <=180mm diam	0.15 max	17.00-19.00	2.00 max	0.60 max	8.00-10.00	0.20 max	0.15 min	1.00 max	Mo optional; bal Fe	520	205	40	187 HB
JIS G4308	SUS303Cu	Wir rod	0.15 max	17.00-19.00	3.00 max		8.00-10.00	0.20 max	0.15 max	1.00 max	bal Fe				
JIS G4308(98)	SUS303	Wir rod	0.15 max	17.00-19.00	2.00 max	0.60 max	8.00-10.00	0.20 max	0.15 max	1.00 max	bal Fe				
JIS G4309	SUS303	Wir	0.15 max	17.00-19.00	2.00 max	0.60 max	8.00-10.00	0.20 max	0.15 min	1.00 max	bal Fe				
Mexico															
DGN B-83	303*	Obs; Bar, HR, CR	0.15 max	17-19	2 max	0.6 max	8-10	0.2 max	0.15 max	1 max	bal Fe				
Romania															
STAS 3583(87)	10NC180	Corr res; Heat res	0.12 max	17-19	2 max	0.2 max	8-10	0.035 max	0.03 max	0.8 max	Pb <=0.15; bal Fe				
Russia															
GOST 5632(61)	0KH18N10E	Corr res; Heat res; High-temp constr	0.12 max	17.0-19.0	1.0-2.0	0.15 max	9.0-11.0	0.035 max	0.02 max	0.8 max	Cu <=0.3; Ti <=0.05; bal Fe				
GOST 5632(72)	12Ch18N10E	Corr res	0.12 max	17.0-19.0	2.0 max	0.3 max	9.0-11.0	0.035 max	0.02 max	0.8 max	Ti <=0.05; W <=0.2; Se 0.18-0.35; bal Fe				
Spain															
UNE 36016(75)	F.3508*	see X10CrNiS18-09	0.12 max	17.0-19.0	2.0 max	0.6 max	8.0-10.0	0.2 max	0.15-0.35	1.0 max	bal Fe				
UNE 36016(75)	X10CrNiS18-09	Bar Rod; Corr res	0.12 max	17.0-19.0	2.0 max	0.6 max	8.0-10.0	0.2 max	0.15-0.35	1.0 max	bal Fe				
UNE 36016/1(89)	E-303	Corr res	0.12 max	17.0-19.0	2.0 max	0.7 max	8.0-10.0	0.06 max	0.15-0.35	1.0 max	bal Fe				
UNE 36016/1(89)	F.3508*	Obs; E-303; Corr res	0.12 max	17.0-19.0	2.0 max	0.7 max	8.0-10.0	0.06 max	0.15-0.35	1.0 max	bal Fe				
Sweden															
SS 142346	2346	Corr res	0.12 max	17-19	2 max	0.6 max	8-10	0.06 max	0.15-0.35	1 max	bal Fe				
UK															
BS 1506(90)	303S22	Bolt matl Pres/Corr res; t<=160mm; Hard; was En80(AM)	0.12 max	17.0-19.0	2.00 max	0.70 max	8.00-10.0	0.060 max	0.15-0.35	1.00 max	bal Fe	510	190	40	
BS 1506(90)	303S22	Bolt matl Pres/Corr res; Corr res; 38<t<=44mm; SHT CD was En80(AM)	0.12 max	17.0-19.0	2.00 max	0.70 max	8.00-10.0	0.060 max	0.15-0.35	1.00 max	bal Fe	650	310	28	0-320 HB
BS 1554(90)	303S31	Wir Corr/Heat res; 6<diam<=13mm; Ann	0.12 max	17.0-19.0	2.00 max	1.00 max	8.00-10.0	0.060 max	0.15-0.35	1.00 max	W <=0.1; bal Fe	0-700			

UNS numbers and US grades are provided as a means of cross referencing chemically similar alloys. Exchangability is only possible after independent examination of specifications. Tensile properties are minimum or typical as specified. UTS and YS as MPa. El as %. See Appendix for list of abbreviations used in Notes. * indicates obsolete material.

Specification	Designation	Notes	C	Cr	Mn	Mo	Ni	P	S	Si	Other	UTS	YS	El	Hard

Stainless Steel, Austenitic, 303 (Continued from previous page)

UK

| BS 1554(90) | 303S31 | Wir Corr/Heat res; diam<0.49mm; Ann | 0.12 max | 17.0-19.0 | 2.00 max | 1.00 max | 8.00-10.0 | 0.060 max | 0.15-0.35 | 1.00 max | W <=0.1; bal Fe | 0-900 | | | |
| BS 970/1(96) | 303S31 | Wrought Corr/Heat res; t<=160mm | 0.12 max | 17.0-19.0 | 2.00 max | 1.00 max | 8.00-10.0 | 0.060 max | 0.15-0.35 | 1.00 max | bal Fe | 510 | 190 | 40 | 183 HB max |

USA

	AISI 303	Tube	0.15 max	17.00-19.00	2.00 max	0.60 max	8.00-10.00	0.20 max	0.15 min	1.00 max	bal Fe				
	UNS S30300	Free Machining	0.15 max	17.00-19.00	2.00 max	0.60 max	8.00-10.00	0.20 max	0.15 min	1.00 max	bal Fe				
AMS 5640	Type 1	Bar Wir Bil	0.15 max	17-19	2 max	0.60 max	8-10	0.2 max	0.15 min	1 max	bal Fe				
ASTM A194/A194M(98)	303*	Nuts	0.15 max	17-19	2 max	0.6 max	8-10	0.2 max	0.15-99.9	1 max	bal Fe				
ASTM A194/A194M(98)	8F	Nuts, high-temp press	0.15 max	17.0-19.0	2.00 max		8.0-10.0	0.20 max	0.15 min	1.00 max	bal Fe				126-300 HB
ASTM A194/A194M(98)	8FA	Nuts, high-temp press	0.15 max	17.0-19.0	2.00 max		8.0-10.0	0.20 max	0.15 min	1.00 max	bal Fe				126-192 HB
ASTM A314	303	Bil	0.15 max	17-19	2 max	0.6 max	8-10	0.2 max	0.15-99.9	1 max	bal Fe				
ASTM A320	303	Bolt	0.15 max	17-19	2 max	0.6 max	8-10	0.2 max	0.15-99.9	1 max	bal Fe				
ASTM A473	303	Frg, Heat res SHT	0.15 max	17.00-19.00	2.00 max	0.60 max	8.00-10.00	0.20 max	0.15 min	1.00 max	bal Fe	515	205	40	
ASTM A581/A581M(95)	303	Wir rod, Ann	0.15 max	17.0-19.0	2.00 max		8.0-10.0	0.20 max	0.15 min	1.00 max	bal Fe	585-860			
ASTM A582/A582M(95)	303	Bar, HF, CF, Ann	0.15 max	17.00-19.00	2.00 max		8.00-10.00	0.20 max	0.15 min	1.00 max	bal Fe				262 HB max
ASTM A895(94)	303	Sh Strp Plt, free mach, Ann	0.15 max	17.0-19.0	2.00 max		8.0-10.0	0.20 max	0.15 min	1.00 max	bal Fe				202 HB max
MIL-S-862B	303*	Obs	0.15 max	17-19	2 max	0.6 max	8-10	0.2 max	0.15-99.9	1 max	Zn <=0.6; bal Fe				

Stainless Steel, Austenitic, 303Cu

USA

| | UNS S30330 | Free Machining Cu-bearing | 0.15 max | 17.00-19.00 | 2.00 max | | 6.00-10.00 | 0.15 max | 0.10 max | 1.00 max | Cu 2.50-4.00; Se<=0.10; bal Fe | | | | |

Stainless Steel, Austenitic, 303Se

Canada

| CSA G110.3 | 303Se | Bar Bil | 0.15 max | 17-19 | 2 max | | 8-10 | 0.2 max | 0.06 max | 1 max | Se>=0.15; bal Fe | | | | |

China

| GB 1220(92) | Y1Cr18Ni9Se | Bar SA | 0.15 max | 17.00-19.00 | 2.00 max | | 8.00-10.00 | 0.20 max | 0.060 max | 1.00 max | Se>=0.15; bal Fe | 520 | 205 | 40 | |

International

| ISO 683-13(74) | 17a* | ST | 0.12 max | 17-19 | 2 max | | 8-10 | 0.2 max | 0.06 max | 1 max | Se>=0.15; bal Fe | | | | |

Japan

JIS G4303(91)	SUS303Se	Bar; SA; <=180mm diam	0.15 max	17.00-19.00	2.00 max		8.00-10.00	0.20 max	0.060 max	1.00 max	Se>=0.15; bal Fe	520	205	40	187 HB
JIS G4308(98)	SUS303Se	Wir rod	0.15 max	17.00-19.00	2.00 max		8.00-10.00	0.20 max	0.060 max	1.00 max	Se>=0.15; bal Fe				
JIS G4309	SUS303Se	Wir	0.15 max	17.00-19.00	2.00 max		8.00-10.00	0.20 max	0.060 max	1.00 max	Se>=0.15; bal Fe				

Mexico

| DGN B-83 | 303Se* | Obs; Bar, HR, CR | 0.15 max | 18-20 | 2 max | | 8-12 | 0.05 max | 0.03 max | 1 max | Se>=0.15; bal Fe | | | | |
| NMX-B-171(91) | MT303SE | mech tub | 0.15 max | 17.0-19.0 | 2.00 max | | 8.0-11.0 | 0.040 max | 0.040 max | 1.00 max | Se 0.12-0.20; bal Fe | | | | |

Russia

| GOST 5632 | 12Ch18N10E | Bar Wire | 0.12 max | 17-19 | 2 max | 0.3 max | 9-11 | 0.035 max | 0.02 max | 0.8 max | Cu <=0.3; W <=0.2; Se 0.18-0.35; bal Fe | | | | |

UK

| BS 1554(90) | 303S42 | Wir Corr/Heat res; diam<0.49mm; Ann | 0.12 max | 17.0-19.0 | 2.00 max | 1.00 max | 8.00-10.0 | 0.060 max | 0.060 max | 1.00 max | W <=0.1; 0.15<=Se<=0.35; bal Fe | 0-900 | | | |

UNS numbers and US grades are provided as a means of cross referencing chemically similar alloys. Exchangability is only possible after independent examination of specifications. Tensile properties are minimum or typical as specified. UTS and YS as MPa. El as %. See Appendix for list of abbreviations used in Notes. * indicates obsolete material.

Specification	Designation	Notes	C	Cr	Mn	Mo	Ni	P	S	Si	Other	UTS	YS	El	Hard

Stainless Steel, Austenitic, 303Se (Continued from previous page)

UK

Specification	Designation	Notes	C	Cr	Mn	Mo	Ni	P	S	Si	Other	UTS	YS	El	Hard
BS 1554(90)	303S42	Wir Corr/Heat res; 6<diam<=13mm; Ann	0.12 max	17.0-19.0	2.00 max	1.00 max	8.00-10.0	0.060 max	0.060 max	1.00 max	W <=0.1; 0.15<=Se<=0.35; bal Fe	0-700			
BS 1554(90)	326S42	Wir Corr/Heat res; diam<0.49mm; Ann	0.12 max	16.5-18.5	2.00 max	2.25-3.00	10.0-13.0	0.060 max	0.060 max	1.00 max	W <=0.1; 0.15<=Se<=0.35; bal Fe	0-900			
BS 1554(90)	326S42	Wir Corr/Heat res; 6<diam<=13mm; Ann	0.12 max	16.5-18.5	2.00 max	2.25-3.00	10.0-13.0	0.060 max	0.060 max	1.00 max	W <=0.1; 0.15<=Se<=0.35; bal Fe	0-750			
BS 970/1(96)	303S42	Wrought Corr/Heat res; t<=160mm	0.12 max	17.0-19.0	2.00 max	1.00 max	8.00-10.0	0.060 max	0.060 max	1.00 max	0.15<=Se<=0.35; bal Fe	510	190	40	183 HB max

USA

Specification	Designation	Notes	C	Cr	Mn	Mo	Ni	P	S	Si	Other	UTS	YS	El	Hard
	AISI 303Se	Tube	0.15 max	17.00-19.00	2.00 max		8.00-10.00	0.20 max	0.060 max	1.00 max	Se>=0.15; bal Fe				
	UNS S30323	Free Machining Se-bearing	0.15 max	17.00-19.00	2.00 max		8.00-10.00	0.20 max	0.060 max	1.00 max	Se>=0.15; bal Fe				
AMS 5640	Type 2	Bar Wir Bil	0.15 max	17-19	2 max		8-10	0.2 max	0.06 max	1 max	Se>=0.15; bal Fe				
AMS 5641		Bar Wir Bil	0.15 max	17-19	2 max		8-10	0.2 max	0.06 max	1 max	Se>=0.15; bal Fe				
AMS 5733		Bar Wir Bil	0.15 max	17-19	2 max		8-10	0.2 max	0.06 max	1 max	Al <=0.35; B 0.0010-0.010; Se>=0.15; bal Fe				
AMS 5738B(89)		Bar Wir; Corr res; 38.10<t<=44.45	0.12 max	17.00-19.00	0.20-2.00	0.75 max	8.00-10.00	0.17 max	0.10 max	1.00 max	Cu <=0.75; Se 0.15-0.35; bal Fe	655	310	28	
AMS 5738B(89)		Bar Wir; Corr res; t<=19.05mm	0.12 max	17.00-19.00	0.20-2.00	0.75 max	8.00-10.00	0.17 max	0.10 max	1.00 max	Cu <=0.75; Se 0.15-0.35; bal Fe	862	689	12	
ASTM A194	303Se*	Nuts	0.15 max	17-19	2 max		8-10	0.2 max	0.06 max	1 max	Se>=0.15; bal Fe				
ASTM A194/A194M(98)	8F	Nuts, high-temp press	0.15 max	17.0-19.0	2.00 max		8.0-10.0	0.20 max	0.06 max	1.00 max	Se>=0.15; bal Fe				126-300 HB
ASTM A194/A194M(98)	8FA	Nuts, high-temp press	0.15 max	17.0-19.0	2.00 max		8.0-10.0	0.20 max	0.06 max	1.00 max	Se>=0.15; bal Fe				126-192 HB
ASTM A314	303Se	Bil Bar, for Frg	0.15 max	17-19	2 max		8-10	0.2 max	0.06 max	1 max	Se>=0.15; bal Fe				
ASTM A320	303Se	Bolt	0.15 max	17-19	2 max		8-10	0.2 max	0.06 max	1 max	Se>=0.15; bal Fe				
ASTM A473	303Se	Frg, Heat res SHT	0.15 max	17.00-19.00	2.00 max		8.00-10.00	0.20 max	0.06 max	1.00 max	Se>=0.15; bal Fe	515	205	40	
ASTM A581/A581M(95)	303Se	Wir rod, Ann	0.15 max	17.0-19.0	2.00 max		8.0-10.0	0.20 max	0.06 max	1.00 max	Se>=0.15; bal Fe	585-860			
ASTM A582/A582M(95)	303Se	Bar, HF, CF, Ann	0.15 max	17.00-19.00	2.00 max		8.00-10.00	0.20 max	0.06 max	1.00 max	Se>=0.15; bal Fe				262 HB max
ASTM A895(94)	303Se	Sh Strp Plt, free mach, Ann	0.15 max	17.0-19.0	2.00 max		8.0-10.0	0.20 max	0.06 max	1.00 max	Se>=0.15; bal Fe				202 HB max
MIL-S-862B	303Se*	Obs	0.15 max	17-19	2 max		8-10	0.2 max	0.06 max	1 max	Se>=0.15; bal Fe				

Stainless Steel, Austenitic, 304

Australia

Specification	Designation	Notes	C	Cr	Mn	Mo	Ni	P	S	Si	Other	UTS	YS	El	Hard
AS 1449(94)	304	Sh Strp Plt	0.08 max	18.00-20.00	2.00 max		8.00-10.50	0.045 max	0.030 max	0.75 max	bal Fe				
AS 2837(86)	304	Bar, Semi-finished product	0.08 max	18.0-20.0	2.00 max		8.0-10.5	0.045 max	0.030 max	1.00 max	bal Fe				

Austria

Specification	Designation	Notes	C	Cr	Mn	Mo	Ni	P	S	Si	Other	UTS	YS	El	Hard
ONORM M3120	X5CrNi18105	Sh Strp Plt Bar Tub Frg	0.07 max	17-19	2 max		8.5-11.0	0.045 max	0.030 max	1 max	bal Fe				

Bulgaria

Specification	Designation	Notes	C	Cr	Mn	Mo	Ni	P	S	Si	Other	UTS	YS	El	Hard
BDS 6738(72)	0Ch18N10	Corr res	0.08 max	17.0-19.0	2.00 max	0.30 max	9.0-11.0	0.035 max	0.025 max	0.80 max	bal Fe				

Canada

Specification	Designation	Notes	C	Cr	Mn	Mo	Ni	P	S	Si	Other	UTS	YS	El	Hard
CSA G110.3	304	Bar Bil	0.08 max	18-20	2 max		8-12	0.05 max	0.03 max	1 max	bal Fe				
CSA G110.6	304	Sh Strp Plt	0.08 max	18-20	2 max		8-12	0.05 max	0.03 max	1 max	bal Fe				
CSA G110.9	304	Plt, HR, Ann, Q/A, Tmp	0.08 max	18-20	2 max		8-12	0.05 max	0.03 max	1 max	bal Fe				

UNS numbers and US grades are provided as a means of cross referencing chemically similar alloys. Exchangability is only possible after independent examination of specifications. Tensile properties are minimum or typical as specified. UTS and YS as MPa. El as %. See Appendix for list of abbreviations used in Notes. * indicates obsolete material.

Stainless Steel, Austenitic, 304 (Continued from previous page)

Specification	Designation	Notes	C	Cr	Mn	Mo	Ni	P	S	Si	Other	UTS	YS	El	Hard
China															
GB 1220(92)	0Cr18Ni9	Bar SA	0.07 max	17.00-19.00	2.00 max		8.00-11.00	0.035 max	0.030 max	1.00 max	bal Fe	520	205	40	
GB 1221(92)	0Cr18Ni9	Bar SA	0.07 max	17.00-19.00	2.00 max		8.00-11.00	0.035 max	0.030 max	1.00 max	bal Fe	520	205	40	
GB 13296(91)	0Cr18Ni9	Smls tube SA	0.07 max	17.00-19.00	2.00 max		8.00-11.00	0.035 max	0.030 max	1.00 max	bal Fe	520	205	35	
GB 4232(93)	ML0Cr18Ni9	Wir HT 1-3mm diam	0.07 max	17.00-19.00	2.00 max		8.00-11.00	0.035 max	0.030 max	1.00 max	bal Fe	590-740		30	
GB 4237(92)	0Cr18Ni9	Sh Plt, HR SA	0.07 max	17.00-19.00	2.00 max		8.00-11.00	0.035 max	0.030 max	1.00 max	bal Fe	520	205	40	
GB 4238(92)	0Cr18Ni9	Plt, HR SA	0.07 max	17.00-19.00	2.00 max		8.00-11.00	0.035 max	0.030 max	1.00 max	bal Fe	520	205	40	
GB 4239(91)	0Cr18Ni9	Sh Strp, CR SA	0.07 max	17.00-19.00	2.00 max		8.00-11.00	0.035 max	0.030 max	1.00 max	bal Fe	520	205	40	
GB 4240(93)	0Cr18Ni9(-L,-Q,-R)	Wir Ann 0.6-1mm diam, L-CD; Q-TLC	0.07 max	17.00-19.00	2.00 max		8.00-11.00	0.035 max	0.030 max	1.00 max	bal Fe	540-880		25	
Czech Republic															
CSN 417240	17240	Corr res	0.07 max	17.0-20.0	2.0 max		9.0-11.5	0.045 max	0.03 max	1.0 max	bal Fe				
Europe															
EN 10088/3(95)	1.4301	Bar Rod Sect, t<=160mm, Ann	0.07 max	17.00-19.50	2.00 max		8.00-10.50	0.045 max	0.030 max	1.00 max	N <=0.11; bal Fe	500-700	190	45	215 HB
EN 10088/3(95)	1.4301	Bar Rod Sect; Corr res; 160<t<=250mm, Ann	0.07 max	17.00-19.50	2.00 max		8.00-10.50	0.045 max	0.030 max	1.00 max	N <=0.11; bal Fe	500-700	190	35	215 HB
EN 10088/3(95)	X5CrNi18-10	Bar Rod Sect; Corr res; 160<t<=250mm, Ann	0.07 max	17.00-19.50	2.00 max		8.00-10.50	0.045 max	0.030 max	1.00 max	N <=0.11; bal Fe	500-700	190	35	215 HB
EN 10088/3(95)	X5CrNi18-10	Bar Rod Sect, t<=160mm, Ann	0.07 max	17.00-19.50	2.00 max		8.00-10.50	0.045 max	0.030 max	1.00 max	N <=0.11; bal Fe	500-700	190	45	215 HB
Finland															
SFS 700	X4CrNi189	Sh Strp Plt Bar Wir Tub Frg	0.05 max	17-19	2 max		8-11	0.045 max	0.03 max	1 max	bal Fe				
SFS 725(86)	X4CrNi189	Corr res	0.05 max	17.0-19.0	2.0 max		8.0-11.0	0.045 max	0.03 max	1.0 max	bal Fe				
France															
AFNOR NFA35573	Z6CN18.09	Sh Strp Plt	0.07 max	17-19	2 max		8-10	0.04 max	0.03 max	1 max	bal Fe				
AFNOR NFA35574	Z6CN18.09	Bar	0.07 max	17-19	2 max		8-10	0.04 max	0.03 max	1 max	bal Fe				
AFNOR NFA35577	Z6CN18.09	Bar Wir rod	0.07 max	17-19	2 max		8-10	0.04 max	0.03 max	1 max	bal Fe				
AFNOR NFA36209	Z5CN18.09	Sh Strp Plt	0.06 max	17-20	2 max		8-11	0.04 max	0.03 max	1 max	bal Fe				
AFNOR NFA36607	Z5CN18.09	Frg	0.06 max	17-20	2 max		8-11	0.04 max	0.03 max	1 max	bal Fe				
Germany															
DIN 17440(96)	WNr 1.4301	Plt Sh, HR Strp Bar Wir Frg, HT	0.07 max	17.0-19.5	2.00 max		8.00-10.5	0.045 max	0.030 max	1.00 max	N <=0.11; bal Fe	500-700	190	35-45	
DIN 17441(97)	WNr 1.4301	CR Strp Plt Sh	0.07 max	17.0-19.5	2.00 max		8.00-10.5	0.045 max	0.030 max	1.00 max	N <=0.11; bal Fe				
DIN EN 10088(95)	WNr 1.4301	Sh Plt Strp Bar Rod, HT	0.07 max	17.0-19.5	2.00 max		8.00-10.5	0.045 max	0.030 max	1.00 max	N <=0.11; bal Fe	500-700	190	35-45	
DIN EN 10088(95)	X5CrNi18-10	Sh Plt Strp Bar Rod, HT	0.07 max	17.0-19.5	2.00 max		8.00-10.5	0.045 max	0.030 max	1.00 max	N <=0.11; bal Fe	500-700	190	35-45	
Hungary															
MSZ 4360(87)	KO33	Corr res; 0-160mm; SA	0.08	17-19	2 max	0.5 max	9-11.5	0.045 max	0.03 max	1 max	bal Fe	500-700	195	40L; 35T	190 HB
MSZ 4360(87)	X8CrNi1810	Corr res; 0-160mm; SA	0.08 max	17-19	2 max	0.5 max	9-11.5	0.045 max	0.03 max	1 max	bal Fe	500-700	195	40L; 35T	190 HB max
MSZ 4398(86)	KO33	Heat res; Corr res; longitudinal weld tube; 0.8-3mm; SA	0.07 max	17-19	2 max	0.5 max	8.5-10.5	0.045 max	0.03 max	1 max	bal Fe	500-720	195	40L	
India															
IS 1570/5(85)	X04Cr19Ni9	Sh Plt Strp Bar Flat Band	0.08 max	17.5-20	2 max	0.15 max	8-10.5	0.045 max	0.03 max	1 max	Co <=0.1; Cu <=0.3; Pb <=0.15; W <=0.1; bal Fe	515	205	40	183 HB max

UNS numbers and US grades are provided as a means of cross referencing chemically similar alloys. Exchangability is only possible after independent examination of specifications. Tensile properties are minimum or typical as specified. UTS and YS as MPa. El as %. See Appendix for list of abbreviations used in Notes. * indicates obsolete material.

Stainless Steel, Austenitic, 304 (Continued from previous page)

Specification	Designation	Notes	C	Cr	Mn	Mo	Ni	P	S	Si	Other	UTS	YS	El	Hard
India															
IS 6527	04Cr18Ni10	Wir rod	0.08 max	17-20	2 max		8-12	0.05 max	0.03 max	1 max	bal Fe				
IS 6528	04Cr18Ni10	Wir, Ann, SA	0.08 max	17-20	2 max		8-12	0.05 max	0.03 max	1 max	bal Fe				
IS 6529	04Cr18Ni10	Bil	0.08 max	17-20	2 max		8-12	0.05 max	0.03 max	1 max	bal Fe				
IS 6603	04Cr18Ni10	Bar, CD, 45mm diam	0.08 max	17-20	2 max		8-12	0.05 max	0.03 max	1 max	bal Fe				
IS 6911	04Cr18Ni10	Sh Strp Plt, CR, 0.5-0.8mm	0.08 max	17-20	2 max		8-12	0.05 max	0.03 max	1 max	bal Fe				
International															
COMECON PC4-70 15		Sh Strp Plt Bar Wir Tub Frg	0.08 max	17-19	2 max		9-11	0.035 max	0.025 max	0.8 max	bal Fe				
ISO 4954(93)	X5CrNi189E	Wir rod Bar, CH ext	0.07 max	17.0-19.0	2.00 max		8.0-11.0	0.045 max	0.030 max	1.00 max	bal Fe	650			
ISO 683-13(74)	11*	Sh Plt Bar, ST	0.07 max	17-19	2 max		8-11	0.045 max	0.03 max	1 max	bal Fe				
Italy															
UNI 6901(71)	X5CrNi1810	Corr res	0.06 max	17-19	2 max		8-11	0.045 max	0.03 max	1 max	bal Fe				
UNI 6904(71)	X5CrNi1810	Heat res	0.06 max	18-20	2 max		8-12	0.04 max	0.03 max	0.75 max	bal Fe				
UNI 7500(75)	X5CrNi1810	Press ves	0.06 max	17-19	2 max		8-11	0.045 max	0.03 max	1 max	bal Fe				
Japan															
JIS G3214(91)	SUSF304	Frg press ves 130-200mm diam	0.08 max	18.00-20.00	2.00 max		8.00-11.00	0.040 max	0.030 max	1.00 max	bal Fe	480	205	29	187 HB max
JIS G4303(91)	SUS304	Bar; SA; <=180mm diam	0.08 max	18.00-20.00	2.00 max		8.00-10.50	0.045 max	0.030 max	1.00 max	bal Fe	520	205	40	187 HB
JIS G4303(91)	SUS304J3	Bar; SA; <=180mm diam	0.08 max	17.00-19.00	2.00 max		8.00-10.50	0.045 max	0.030 max	1.00 max	Cu 1.00-3.00; bal Fe	480	175	40	187 HB
JIS G4304(91)	SUS304	Plt Sh, HR SA	0.08 max	18.00-20.00	2.00 max		8.0-10.50	0.045 max	0.030 max	1.00 max	bal Fe	520	205	40	90 HRB max
JIS G4305(91)	SUS304	Plt Sh, CR SA	0.08 max	18.00-20.00	2.00 max		8.00-10.50	0.045 max	0.030 max	1.00 max	bal Fe	520	205	40	187 HB max
JIS G4305(91)	SUS304J1		0.08 max	15.00-18.00	3.00 max		6.00-9.00	0.045 max	0.030 max	1.70 max	Cu 1.00-3.00; bal Fe	450	155	40	
JIS G4305(91)	SUS304J2		0.08 max	15.00-18.00	3.00-5.00		6.00-9.00	0.045 max	0.030 max	1.70 max	Cu 1.00-3.00; bal Fe	450	155	40	
JIS G4306	SUS304	Strp HR SA	0.08 max	18-20	2 max		8-10.5	0.04 max	0.03 max	1 max	bal Fe				
JIS G4307	SUS304	Strp CR SA	0.08 max	18-20	2 max		8-10.5	0.04 max	0.03 max	1 max	bal Fe				
JIS G4308	SUS304J3	Wir rod	0.08 max	17.00-19.00	2.00 max		8.00-10.50	0.045 max	0.030 max	1.00 max	Cu 1.00-3.00; bal Fe				
JIS G4308(98)	SUS304	Wir rod	0.080 max	18.00-20.00	2.00 max		8.00-10.50	0.045 max	0.030 max	1.00 max	bal Fe				
JIS G4309	SUS304	Wir	0.08 max	18.00-20.00	2.00 max		8.00-10.50	0.045 max	0.030 max	1.00 max	bal Fe				
JIS G4309	SUS304J3	Wir	0.08 max	17.00-19.00	2.00 max		8.00-10.50	0.045 max	0.030 max	1.00 max	Cu 1.00-3.00; bal Fe				
JIS G4313(96)	SUS304-CSP	CR Strp for springs Htemp	0.08 max	18.00-20.00	2.00 max		8.00-10.50	0.045 max	0.030 max	1.00 max	bal Fe	1130	880		370 HV
JIS G4315	SUS304	Wir rod	0.08 max	18.00-20.00	2.00 max		8.00-10.50	0.045 max	0.030 max	1.00 max	bal Fe				
JIS G4315	SUS304J3	Wir rod	0.08 max	17.00-19.00	2.00 max		8.00-10.50	0.045 max	0.030 max	1.00 max	Cu 1.00-3.00; bal Fe				
Mexico															
DGN B-218	TP304*	Obs; Tube, Pipe	0.08 max	18-20	2 max		8-11	0.03 max	0.03 max	0.75 max	bal Fe				
DGN B-224	TP304*	Obs; Tube, Pipe	0.08 max	18-20	2 max		8-11	0.03 max	0.03 max	0.75 max	bal Fe				
DGN B-83	304*	Obs; Bar, HR, CR	0.08 max	18-20	2 max		8-12	0.05 max	0.03 max	1 max	bal Fe				
NMX-B-171(91)	MT304	CF HF Smls tub	0.08 max	18.0-20.0	2.00 max		8.0-11.0	0.040 max	0.030 max	1.00 max	bal Fe				

UNS numbers and US grades are provided as a means of cross referencing chemically similar alloys. Exchangability is only possible after independent examination of specifications. Tensile properties are minimum or typical as specified. UTS and YS as MPa. El as %. See Appendix for list of abbreviations used in Notes. * indicates obsolete material.

Specification	Designation	Notes	C	Cr	Mn	Mo	Ni	P	S	Si	Other	UTS	YS	El	Hard

Stainless Steel, Austenitic, 304 (Continued from previous page)

Mexico

Specification	Designation	Notes	C	Cr	Mn	Mo	Ni	P	S	Si	Other	UTS	YS	El	Hard
NMX-B-176(91)	TP304	Smls, weld sanitary tub	0.08 max	18.00-20.00	2.00 max		8.00-11.00	0.040 max	0.030 max	0.75 max	bal Fe				
NMX-B-186-SCFI(94)	TP304	Smls pipe for high-temp central station service	0.08 max	18.0-20.0	2.00 max		8.00-11.0	0.040 max	0.030 max	0.75 max	bal Fe	517	207		
NMX-B-196(68)	TP304	Smls tube for refinery	0.08 max	18.0-20.0	2.00 max		8.0-11.0	0.040 max	0.030 max	0.75 max	bal Fe				

Norway

Specification	Designation	Notes	C	Cr	Mn	Mo	Ni	P	S	Si	Other	UTS	YS	El	Hard
NS 14350	14350	Bar	0.05 max	17-19	2 max		8-11	0.045 max	0.03 max	1 max	bal Fe				

Pan America

Specification	Designation	Notes	C	Cr	Mn	Mo	Ni	P	S	Si	Other	UTS	YS	El	Hard
COPANT 513	TP304	Tube, HF, CF, 8mm diam	0.08 max	18-20	2 max		8-11	0.04 max	0.03 max	0.75 max	bal Fe				
COPANT R195	TP 304	HF, CF, 5.6mm diam	0.08 max	18-20	2 max		8-11	0.04 max	0.03 max	0.75 max	bal Fe				

Poland

Specification	Designation	Notes	C	Cr	Mn	Mo	Ni	P	S	Si	Other	UTS	YS	El	Hard
PNH86020	0H18N9	Corr res	0.07 max	17-19	2 max		9-11	0.045 max	0.03 max	0.8 max	bal Fe				

Romania

Specification	Designation	Notes	C	Cr	Mn	Mo	Ni	P	S	Si	Other	UTS	YS	El	Hard
STAS 3583(87)	5NiCr180	Corr res; Heat res	0.07 max	17-19	2 max	0.2 max	8.5-10.5	0.045 max	0.03 max	1 max	Pb <=0.15; bal Fe				

Russia

Specification	Designation	Notes	C	Cr	Mn	Mo	Ni	P	S	Si	Other	UTS	YS	El	Hard
GOST	O8Ch18N10	Sh Strp Plt Bar Wir Frg	0.08 max	17-19	2 max		9-11	0.035 max	0.02 max	0.8 max	bal Fe				
GOST 5632(61)	0KH18N10	Corr res; Heat res; High-temp constr	0.08 max	17.0-19.0	1.0-2.0	0.15 max	9.0-11.0	0.035 max	0.02 max	0.8 max	Cu <=0.3; Ti <=0.05; bal Fe				
GOST 5632(72)	08Ch18N10	Corr res; Heat res	0.08 max	17.0-19.0	2.0 max	0.3 max	9.0-11.0	0.035 max	0.02 max	0.8 max	Cu <=0.3; Ti <=0.5; W <=0.2; bal Fe				

Spain

Specification	Designation	Notes	C	Cr	Mn	Mo	Ni	P	S	Si	Other	UTS	YS	El	Hard
UNE 36016(75)	F.3504*	see X6CrNi19-10	0.08 max	18.0-20.0	2.0 max	0.15 max	8.0-10.5	0.045 max	0.03 max	1.0 max	bal Fe				
UNE 36016(75)	X6CrNi19-10	Sh Strp Plt Bar Rod; Corr res	0.08 max	18.0-20.0	2.0 max	0.15 max	8.0-10.5	0.045 max	0.03 max	1.0 max	bal Fe				
UNE 36016/1(89)	E-304	Corr res	0.07 max	17.0-19.0	2.0 max	0.15 max	8.0-11.0	0.045 max	0.03 max	1.0 max	N <=0.1; bal Fe				
UNE 36016/1(89)	F.3504*	Obs; E-304; Corr res	0.07 max	17.0-19.0	2.0 max	0.15 max	8.0-11.0	0.045 max	0.03 max	1.0 max	N <=0.1; bal Fe				
UNE 36087(78)	F.3541*	see X5CrNi18-10	0.03-0.07	17.0-19.0	2.0 max	0.15 max	8.0-11.0	0.045 max	0.03 max	1.0 max	bal Fe				
UNE 36087(78)	F.3551*	see X5CrNi18-11	0.07 max	17.0-19.0	2.0 max	0.5 max	9.0-11.5	0.045 max	0.03 max	1.0 max	bal Fe				
UNE 36087(78)	X5CrNi18-10	Corr res	0.03-0.07	17.0-19.0	2.0 max	0.15 max	8.0-11.0	0.045 max	0.03 max	1.0 max	bal Fe				
UNE 36087(78)	X5CrNi18-11	Corr res	0.07 max	17.0-19.0	2.0 max	0.5 max	9.0-11.5	0.045 max	0.03 max	1.0 max	bal Fe				

Sweden

Specification	Designation	Notes	C	Cr	Mn	Mo	Ni	P	S	Si	Other	UTS	YS	El	Hard
SS 142332	2332	Corr res	0.07 max	17-19	2 max		8-11	0.045 max	0.03 max	1 max	bal Fe				
SS 142333	2333	Corr res	0.05 max	17-19	2 max		8-11	0.045 max	0.03 max	1 max	bal Fe				

UK

Specification	Designation	Notes	C	Cr	Mn	Mo	Ni	P	S	Si	Other	UTS	YS	El	Hard
BS 1449/2(83)	304S15	Corr/Heat res; t<=100mm	0.06 max	17.5-19.0	2.00 max		8.00-11.00	0.045 max	0.030 max	1.00 max	bal Fe	500	195	40	0-190 HB
BS 1449/2(83)	304S16	Corr/Heat res; t<=100mm	0.06 max	17.5-19.0	2.00 max		9.00-11.00	0.045 max	0.030 max	1.00 max	bal Fe	500	195	40	0-190 HB
BS 1449/2(83)	304S31	Sh Strp Plt Bar Wir Frg, 0.05-100mm	0.07 max	17-19	0.50-2.00		8-11	0.045 max	0.03 max	0.2-1	bal Fe	500	195	40	0-190 HB
BS 1501/3(73)	304S15	Press ves; Corr res; was En801	0.06 max	17.5-19.0	0.5-2.0	0.15 max	8.0-11.0	0.045 max	0.03 max	0.2-1.0	Cu <=0.3; Ti <=0.05; W <=0.1; bal Fe				
BS 1501/3(73)	304S29	Press ves; Corr res	0.04-0.09	17.5-19.0	0.5-2.0	0.15 max	8.0-11.0	0.04 max	0.03 max	0.2-1.0	Cu <=0.3; Ti <=0.05; W <=0.1; Nb<=0.05; bal Fe				
BS 1501/3(90)	304S31	Press ves; Corr/Heat res; 0.05-100mm	0.07 max	17.0-19.0	2.0 max	0.15 max	8.0-11.0	0.045 max	0.025 max	1.0 max	Cu <=0.3; Ti <=0.05; W <=0.1; bal Fe	500-700	195	40	
BS 1501/3(90)	304S51	Press ves; Corr/Heat res; 0.05-100mm	0.04-0.1	17.0-19.0	2.0 max	0.15 max	8.0-11.0	0.045 max	0.025 max	1.0 max	Cu <=0.3; Ti <=0.05; W <=0.1; bal Fe	490-690	195	40	

UNS numbers and US grades are provided as a means of cross referencing chemically similar alloys. Exchangability is only possible after independent examination of specifications. Tensile properties are minimum or typical as specified. UTS and YS as MPa. El as %. See Appendix for list of abbreviations used in Notes. * indicates obsolete material.

Specification	Designation	Notes	C	Cr	Mn	Mo	Ni	P	S	Si	Other	UTS	YS	El	Hard
Stainless Steel, Austenitic, 304 (Continued from previous page)															
UK															
BS 1501/3(90)	304S61	Press ves; Corr/Heat res; 0.05-100mm	0.03 max	17.0-19.0	2.0 max	0.15 max	8.5-11.5	0.045 max	0.025 max	1.0 max	Cu <=0.3; N 0.12-0.22; Ti <=0.05; W <=0.1; bal Fe	550-750	270	35	
BS 1502	304S31	Bar Shp; Press ves	0.07 max	17.0-19.0	2.00 max		8.00-11.0	0.045 max	0.030 max	0.2-1.00	bal Fe				
BS 1503(89)	304S31	Frg press ves; t<=999mm; HT	0.07 max	17.0-19.0	2.00 max	0.70 max	8.00-11.0	0.040 max	0.025 max	1.00 max	B <=0.005; Cu <=0.50; Nb <=0.20; Ti <=0.10; W <=0.1; bal Fe	490-690	230	30	
BS 1506(90)	304S31	Bolt matl Pres/Corr res; t<=160mm; Hard	0.07 max	17.0-19.0	2.00 max		8.00-11.0	0.045 max	0.030 max	1.00 max	bal Fe	520	205	40	
BS 1506(90)	304S31	Bolt matl Pres/Corr res; Corr res; 38<t<=44mm; SHT CD	0.07 max	17.0-19.0	2.00 max		8.00-11.0	0.045 max	0.030 max	1.00 max	bal Fe	650	310	28	0-320 HB
BS 1554(90)	304S15	Wir Corr/Heat res; diam<0.49mm; Ann	0.06 max	17.5-19.0	2.00 max		8.00-11.0	0.045 max	0.030 max	1.00 max	W <=0.1; bal Fe	0-900			
BS 1554(90)	304S15	Wir Corr/Heat res; 6<diam<=13mm; Ann	0.06 max	17.5-19.0	2.00 max		8.00-11.0	0.045 max	0.030 max	1.00 max	W <=0.1; bal Fe	0-700			
BS 1554(90)	304S31	Wir Corr/Heat res; 6<diam<=13mm; Ann	0.07 max	17.0-19.0	2.00 max		8.00-11.0	0.045 max	0.030 max	1.00 max	W <=0.1; bal Fe	0-700			
BS 1554(90)	304S31	Wir Corr/Heat res; diam<0.49mm; Ann	0.07 max	17.0-19.0	2.00 max		8.00-11.0	0.045 max	0.030 max	1.00 max	W <=0.1; bal Fe	0-900			
BS 3059/2(90) *	304S51	Boiler, Superheater; High-temp; Smls tub; 2-11mm	0.04-0.10	17.0-19.0	2.00 max	0.15 max	8.0-11.0	0.040 max	0.030 max	1.00 max	bal Fe	490-690	230	35	
BS 3605	304S18	Tube, Pipe	0.060 max	17.5-19.0	0.5-2.0		9.0-12.0	0.040 max	0.030 max	0.20-1.00	bal Fe				
BS 3605	304S25	Tube, Pipe	0.060 max	17.5-19.0	0.5-2.0		8.0-11.0	0.040 max	0.030 max	0.20-1.00	bal Fe				
BS 3605/1(91)	304S31	Pip Tube press smls; t<=200mm	0.070 max	17.0-19.0	2.00 max		8.0-11.0	0.040 max	0.030 max	1.00 max	bal Fe	490-690		35	
BS 3605/1(91)	304S51	Pip Tube press smls; t<=200mm	0.04-0.10	17.0-19.0	2.0 max		8.0-11.0	0.040 max	0.030 max	1.00 max	bal Fe	490-690		35	
BS 3606(78)	304S22	Corr res; Heat exch Tub	0.030 max	17.0-19.0	0.5-2.0		9.0-12.0	0.045 max	0.030 max	1.00 max	bal Fe				
BS 3606(78)	304S25	Corr res; Heat exch Tub	0.060 max	17.0-19.0	0.5-2.0		8.0-11.0	0.045 max	0.030 max	1.00 max	bal Fe				
BS 3606(92)	304S31	Heat exch Tub; Smls tub, Tub, Weld; 1.2-3.2mm	0.070 max	17.0-19.0	2.00 max		8.0-11.0	0.040 max	0.030 max	1.00 max	bal Fe	490-690	235	30	
BS 970/1(96)	304S15	Wrought Corr/Heat res; t<=160mm	0.06 max	17.5-19.0	2.00 max		8.00-11.0	0.045 max	0.030 max	1.00 max	bal Fe	480	195	40	183 HB max
BS 970/1(96)	304S31	Wrought Corr/Heat res; t<=160mm	0.07 max	17.0-19.0	2.00 max		8.00-11.0	0.045 max	0.030 max	1.00 max	bal Fe	490	195	40	183 HB max
USA															
	AISI 304	Tube	0.08 max	18.00-20.00	2.00 max		8.00-10.50	0.045 max	0.030 max	1.00 max	bal Fe				
	UNS S30400		0.08 max	18.00-20.00	2.00 max		8.00-10.50	0.045 max	0.030 max	1.00 max	bal Fe				
AMS 5501		Sh Strp Plt Bar Wir	0.08 max	18-20	2 max		8-10.5	0.045 max	0.03 max	1 max	bal Fe				
AMS 5513		Sh Strp Plt Bar Wir	0.08 max	18-20	2 max		8-10.5	0.045 max	0.03 max	1 max	bal Fe				
AMS 5560H(92)		Smls tube OD<=4.78	0.08 max	18.00-20.00	2.00 max	0.75 max	8.00-12.00	0.040 max	0.030 max	1.00 max	Cu <=0.75; bal Fe	793		35	
AMS 5560H(92)		Smls tube OD<=12.70	0.08 max	18.00-20.00	2.00 max	0.75 max	8.00-12.00	0.040 max	0.030 max	1.00 max	Cu <=0.75; bal Fe	689		30	
AMS 5563		Sh Strp Plt Bar Wir	0.08 max	18-20	2 max		8-10.5	0.045 max	0.03 max	1 max	bal Fe				
AMS 5564		Sh Strp Plt Bar Wir	0.08 max	18-20	2 max		8-10.5	0.045 max	0.03 max	1 max	bal Fe				
AMS 5565		Sh Strp Plt Bar Wir	0.08 max	18-20	2 max		8-10.5	0.045 max	0.03 max	1 max	bal Fe				
AMS 5566		Sh Strp Plt Bar Wir	0.08 max	18-20	2 max		8-10.5	0.045 max	0.03 max	1 max	bal Fe				
AMS 5567		Sh Strp Plt Bar Wir	0.08 max	18-20	2 max		8-10.5	0.045 max	0.03 max	1 max	bal Fe				

UNS numbers and US grades are provided as a means of cross referencing chemically similar alloys. Exchangability is only possible after independent examination of specifications. Tensile properties are minimum or typical as specified. UTS and YS as MPa. El as %. See Appendix for list of abbreviations used in Notes. * indicates obsolete material.

Specification	Designation	Notes	C	Cr	Mn	Mo	Ni	P	S	Si	Other	UTS	YS	El	Hard

Stainless Steel, Austenitic, 304 (Continued from previous page)

USA

Specification	Designation	Notes	C	Cr	Mn	Mo	Ni	P	S	Si	Other	UTS	YS	El	Hard
AMS 5639		Sh Strp Plt Bar Wir	0.08 max	18-20	2 max		8-10.5	0.045 max	0.03 max	1 max	bal Fe				
AMS 5697		Sh Strp Plt Bar Wir	0.08 max	18-20	2 max		8-10.5	0.045 max	0.03 max	1 max	bal Fe				
AMS 5857(90)		Bar Wir; Corr res; t<=19.05mm	0.08 max	18.00-20.00	2.00 max	0.75 max	8.00-12.00	0.040 max	0.030 max	1.00 max	Cu <=0.75; bal Fe	862	689	12	
AMS 5857(90)		Bar Wir; Corr res; 38.10<t<=44.45mm	0.08 max	18.00-20.00	2.00 max	0.75 max	8.00-12.00	0.040 max	0.030 max	1.00 max	Cu <=0.75; bal Fe	655	310	30	
AMS 5868(93)		Corr res; Smls/weld aircraft tub, CD, Half Hard	0.08 max	18.00-20.00	2.00 max	0.75 max	8.00-11.00	0.045 max	0.030 max	1.00 max	Cu <=0.75; bal Fe	1034	758	7	
AMS 7228		Sh Strp Plt Bar Wir	0.08 max	18-20	2 max		8-10.5	0.045 max	0.03 max	1 max	bal Fe				
AMS 7245		Sh Strp Plt Bar Wir	0.08 max	18-20	2 max		8-10.5	0.045 max	0.03 max	1 max	bal Fe				
ASME SA182	304	Refer to ASTM A182/A182M													
ASME SA213	304	Refer to ASTM A213/A213M(95)													
ASME SA240	304	Refer to ASTM A240(95)													
ASME SA249	304	Refer to ASTM A249/A249M(96)													
ASME SA312	304	Refer to ASTM A312/A312M(95)													
ASME SA358	304	Refer to ASTM A358/A358M(95)													
ASME SA376	304	Refer to ASTM A376/A376M(93)													
ASME SA403	304	Refer to ASTM A403/A403M(95)													
ASME SA409	304	Refer to ASTM A409/A409M(95)													
ASME SA430	304	Refer to ASTM A430/A430M(91)													
ASME SA479	304	Refer to ASTM A479/A479M(95)													
ASME SA688	304	Refer to ASTM A688/A688M(96)													
ASTM A167(96)	304*	Sh Strp Plt	0.08 max	18-20	2 max		8-10.5	0.045 max	0.03 max	1 max	bal Fe				
ASTM A182	304*	Bar Frg	0.08 max	18-20	2 max		8-10.5	0.045 max	0.03 max	1 max	bal Fe				
ASTM A182/A182M(98)	F304	Frg/roll pipe flange valve	0.08 max	18.0-20.0	2.00 max		8.00-11.00	0.045 max	0.030 max	1.00 max	N <=0.10; bal Fe	515	205	30	
ASTM A193/A193M(98)	304*	Bolt, high-temp	0.08 max	18.0-20.0	2.00 max		8.0-10.5	0.045 max	0.030 max	1 max	bal Fe				
ASTM A193/A193M(98)	B8	Bolt, high-temp, HT	0.08 max	18.0-20.0	2.00 max		8.0-11.0	0.045 max	0.030 max	1.00 max	bal Fe	515	205	30	223 HB
ASTM A193/A193M(98)	B8A	Bolt, high-temp, HT	0.08 max	18.0-20.0	2.00 max		8.0-11.0	0.045 max	0.030 max	1.00 max	bal Fe	515	205	30	192 HB
ASTM A194	304*	Nuts	0.08 max	18-20	2 max		8-10.5	0.045 max	0.03 max	1 max	bal Fe				
ASTM A194/A194M(98)	8	Nuts, high-temp press	0.08 max	18.0-20.0	2.00 max		8.0-11.0	0.045 max	0.030 max	1.00 max	bal Fe				126-300 HB
ASTM A194/A194M(98)	8A	Nuts, high-temp press	0.08 max	18.0-20.0	2.00 max		8.0-11.0	0.045 max	0.030 max	1.00 max	bal Fe				126-192 HB
ASTM A213	304*	Tube	0.08 max	18-20	2 max		8-10.5	0.045 max	0.03 max	1 max	bal Fe				
ASTM A213/A213M(95)	TP304	Smls tube boiler, superheater, heat exchanger	0.08 max	18.0-20.0	2.00 max		8.00-11.00	0.040 max	0.030 max	0.75 max	bal Fe	515	205	35	
ASTM A240/A240M(98)	S30400		0.08 max	18.0-20.0	2.00 max		8.0-10.5	0.045 max	0.030 max	0.75 max	N <=0.10; bal Fe	515	205	40.0	201 HB
ASTM A249/249M(96)	TP304	Weld tube; boiler, superheater, heat exch	0.08 max	18.0-20.0	2.00 max		8.00-11.00	0.040 max	0.030 max	0.75 max	bal Fe	515	205	35	90 HRB
ASTM A269	304	Smls Weld, Tube	0.08 max	18-20	2 max		8-10.5	0.045 max	0.03 max	1 max	bal Fe				

UNS numbers and US grades are provided as a means of cross referencing chemically similar alloys. Exchangability is only possible after independent examination of specifications. Tensile properties are minimum or typical as specified. UTS and YS as MPa. El as %. See Appendix for list of abbreviations used in Notes. * indicates obsolete material.

Specification	Designation	Notes	C	Cr	Mn	Mo	Ni	P	S	Si	Other	UTS	YS	El	Hard

Stainless Steel, Austenitic, 304 (Continued from previous page)

USA

Specification	Designation	Notes	C	Cr	Mn	Mo	Ni	P	S	Si	Other	UTS	YS	El	Hard
ASTM A270(95)	304	Smls Weld, Tube	0.08 max	18-20	2 max		8-10.5	0.045 max	0.03 max	1 max	bal Fe				
ASTM A271(96)	304	Smls tube	0.08 max	18-20	2 max		8-10.5	0.045 max	0.03 max	1 max	bal Fe				
ASTM A276(98)	304	Bar Shp	0.08 max	18-20	2 max		8-10.5	0.045 max	0.03 max	1 max	bal Fe				
ASTM A312/A312M(95)	304	Smls Weld, Pipe	0.08 max	18-20	2 max		8-10.5	0.045 max	0.03 max	1 max	bal Fe				
ASTM A313/A313M(95)	304	Spring wire, t<=0.23mm	0.08 max	18.00-20.00	2.00 max		8.00-10.50	0.045 max	0.030 max	1.00 max	bal Fe	2240-2450			
ASTM A314	304	Bil Bar, for Frg	0.08 max	18-20	2 max		8-10.5	0.045 max	0.03 max	1 max	bal Fe				
ASTM A320	304	Bolt	0.08 max	18-20	2 max		8-10.5	0.045 max	0.03 max	1 max	bal Fe				
ASTM A336/A336M(98)	F304	Frg, press/high-temp	0.08 max	18.0-20.0	2.00 max		8.00-11.00	0.040 max	0.030 max	1.00 max	bal Fe	485	205	30	90 HRB
ASTM A358/A358M(95)	304	Elect fusion weld pipe, high-temp	0.08 max	18-20	2 max		8-10.5	0.045 max	0.03 max	1 max	bal Fe				
ASTM A368(95)	304	Wir, Heat res	0.08 max	18.00-20.00	2.00 max		8.00-10.50	0.045 max	0.030 max	1.00 max	bal Fe				
ASTM A376	304	Smls pipe	0.08 max	18-20	2 max		8-10.5	0.045 max	0.03 max	1 max	bal Fe				
ASTM A409	304	Weld, Pipe	0.08 max	18-20	2 max		8-10.5	0.045 max	0.03 max	1 max	bal Fe				
ASTM A430	304*	Obs 1995; see A312; Pipe	0.08 max	18-20	2 max		8-10.5	0.045 max	0.03 max	1 max	bal Fe				
ASTM A473	304	Frg, Heat res SHT, t<127mm	0.08 max	18.00-20.00	2.00 max		8.00-10.50	0.045 max	0.030 max	1.00 max	bal Fe	515	205	40	
ASTM A479	304	Bar	0.08 max	18-20	2 max		8-10.5	0.045 max	0.03 max	1 max	bal Fe				
ASTM A492	304	Wir, 2.29<t<=2.54	0.08 max	18.00-20.00	2.00 max		8.00-10.50	0.045 max	0.030 max	1.00 max	bal Fe	1620-1830			
ASTM A493	304	Wir rod, CHd, cold frg	0.08 max	18.0-20.0	2.00 max		8.0-10.5	0.045 max	0.030 max	1.00 max	bal Fe	620			
ASTM A511(96)	MT304	Smls mech tub, Ann	0.08 max	18.0-20.0	2.00 max		8.0-11.0	0.040 max	0.030 max	1.00 max	bal Fe	517	207	35	192 HB
ASTM A554(94)	MT304	Weld mech tub, Rnd Ann	0.08 max	18.0-20.0	2.00 max		8.0-11.0	0.040 max	0.030 max	1.00 max	bal Fe	517	207	35	
ASTM A580/A580M(98)	304	Wir, Ann	0.08 max	18.0-20.0	2.00 max		8.0-10.5	0.045 max	0.030 max	1.00 max	N <=0.10; bal Fe	520	210	35	
ASTM A632(90)	TP304	Smls Weld tub	0.08 max	18.0-20.0	2.00 max		8.0-11.0	0.040 max	0.030 max	0.75 max	bal Fe	515	205	35	
ASTM A666(96)	304	Sh Strp Plt Bar, Ann	0.08 max	18.00-20.00	2.00 max		8.00-10.50	0.045 max	0.030 max	0.75 max	N <=0.10; bal Fe	515	205	40	201 HB
ASTM A688/A688M(96)	TP304	Weld Feedwater Heater Tub	0.08 max	18.00-20.00	2.00 max		8.00-11.00	0.040 max	0.03 max	0.75 max	bal Fe	515	205	35	
ASTM A793(96)	304	Rolled floor plt	0.08 max	18.00-20.00	2.00 max		8.00-10.50	0.045 max	0.030 max	0.75 max	N <=0.10; bal Fe				
ASTM A813/A813M(95)	TP304	Weld Pipe	0.08 max	18.0-20.0	2.00 max		8.00-11.00	0.045 max	0.030 max	0.75 max	B <=1; bal Fe	515	205		
ASTM A814/A814M(96)	TP304	CW Weld Pipe	0.08 max	18.0-20.0	2.00 max		8.0-11.0	0.045 max	0.030 max	0.75 max	bal Fe	515	205		
ASTM A851(96)	TP304	HFIW UnAnn Condenser Tub	0.08 max	18.0-20.0	2.00 max		8.0-11.00	0.040 max	0.030 max	0.75 max	bal Fe	515	205	35	
ASTM A908(95)	304	Needle tub	0.08 max	18.0-20.0	2.00 max		8.0-11.0	0.040 max	0.030 max	0.75 max	bal Fe	1030-1370			
ASTM A943/A943M(95)	TP304	Spray formed smls pipe	0.08 max	18.0-20.0	2.00 max		8.00-11.00	0.040 max	0.030 max	0.75 max	bal Fe	515	205		
ASTM A965/965M(97)	F304	Frg for press high-temp parts	0.08 max	18.0-20.0	2.00 max		8.0-11.0	0.040 max	0.030 max	1.00 max	bal Fe	485	205	30	
ASTM A988(98)	S30400	HIP Flanges, Fittings, Valves/parts; Heat res	0.08 max	18.0-20.0	2.00 max		8.0-11.0	0.045 max	0.030 max	1.00 max	N <=0.10; bal Fe	515	205	30	
MIL-S-23195(A)(65)	304	Corr res; Bar frg; Rev E controlled distribution	0.08 max	18.00-20.00	2.00 max		8.00-12.00	0.035 max	0.030 max	1.00 max	Co <=0.10; Nb+Ta 1.00 max; bal Fe	517	207-379	40	
MIL-S-23196	304	Controlled distribution													
MIL-S-27419(USAF)(68)	304*	Obs; Corr res ann bil	0.08 max	18-20	2 max		8-10.5	0.045 max	0.03 max	1 max	bal Fe				

UNS numbers and US grades are provided as a means of cross referencing chemically similar alloys. Exchangability is only possible after independent examination of specifications. Tensile properties are minimum or typical as specified. UTS and YS as MPa. El as %. See Appendix for list of abbreviations used in Notes. * indicates obsolete material.

Specification	Designation	Notes	C	Cr	Mn	Mo	Ni	P	S	Si	Other	UTS	YS	El	Hard
Stainless Steel, Austenitic, 304 (Continued from previous page)															
USA															
MIL-S-5059D(90)	304*	Obs see AMS specs; Plt Sh Strp; Corr res													
MIL-T-8504B(98)	304*	Obs for new design; Tube; aerospace; ann smls weld	0.08 max	18-20	2 max		8-10.5	0.045 max	0.03 max	1 max	bal Fe				
MIL-T-8506A	304*	Obs	0.08 max	18-20	2 max		8-10.5	0.045 max	0.03 max	1 max	bal Fe				
SAE J405(98)	S30400		0.08 max	18.00-20.00	2.00 max		8.00-10.50	0.045 max	0.030 max	0.75 max	N <=0.10; bal Fe				
SAE J467(68)	304		0.04	19.0	1.00		10.0			0.50	bal Fe	586	207	60	80 HRB
Stainless Steel, Austenitic, 304B															
USA															
	UNS S30460		0.08 max	18.0-20.0	2.00 max		12.0-15.0	0.045 max	0.030 max	0.75 max	B 0.20-0.29; N <=0.10; bal Fe				
ASTM A887(94)	S30460	Sh Strp Plt, Boron, for nuclear, Grade A	0.08 max	18.0-20.0	2.00 max		12.0-15.0	0.045 max	0.030 max	0.75 max	B 0.20-0.29; N <=0.10; bal Fe	515	205	40.0	201 HB
Stainless Steel, Austenitic, 304B1															
USA															
	UNS S30461		0.08 max	18.0-20.0	2.00 max		12.0-15.0	0.045 max	0.030 max	0.75 max	B 0.30-0.49; N <=0.10; bal Fe				
ASTM A887(94)	S30461	Sh Strp Plt, Boron, for nuclear, Grade A	0.08 max	18.0-20.0	2.00 max		12.0-15.0	0.045 max	0.030 max	0.75 max	B 0.30-0.49; N <=0.10; bal Fe	515	205	40.0	201 HB
Stainless Steel, Austenitic, 304B2															
USA															
	UNS S30462		0.08 max	18.0-20.0	2.00 max		12.0-15.0	0.045 max	0.030 max	0.75 max	B 0.50-0.74; N <=0.10; bal Fe				
ASTM A887(94)	S30462	Sh Strp Plt, Boron, for nuclear, Grade A	0.08 max	18.0-20.0	2.00 max		12.0-15.0	0.045 max	0.030 max	0.75 max	B 0.50-0.74; N <=0.10; bal Fe	515	205	35.0	201 HB
Stainless Steel, Austenitic, 304B3															
USA															
	UNS S30463		0.08 max	18.0-20.0	2.00 max		12.0-15.0	0.045 max	0.030 max	0.75 max	B 0.75-0.99; N <=0.10; bal Fe				
ASTM A887(94)	S30463	Sh Strp Plt, Boron, for nuclear, Grade A	0.08 max	18.0-20.0	2.00 max		12.0-15.0	0.045 max	0.030 max	0.75 max	B 0.75-0.99; N <=0.10; bal Fe	515	205	31.0	201 HB
Stainless Steel, Austenitic, 304B4															
USA															
	UNS S30464		0.08 max	18.0-20.0	2.00 max		12.0-15.0	0.045 max	0.030 max	0.75 max	B 1.00-1.24; N <=0.10; bal Fe				
ASTM A887(94)	S30464	Sh Strp Plt, Boron, for nuclear, Grade A	0.08 max	18.0-20.0	2.00 max		12.0-15.0	0.045 max	0.030 max	0.75 max	B 1.00-1.24; N <=0.10; bal Fe	515	205	27.0	217 HB
Stainless Steel, Austenitic, 304B5															
USA															
	UNS S30465		0.08 max	18.0-20.0	2.00 max		12.0-15.0	0.045 max	0.030 max	0.75 max	B 1.25-1.49; N <=0.10; bal Fe				
ASTM A887(94)	S30465	Sh Strp Plt, Boron, for nuclear, Grade A	0.08 max	18.0-20.0	2.00 max		12.0-15.0	0.045 max	0.030 max	0.75 max	B 1.25-1.49; N <=0.10; bal Fe	515	205	24.0	217 HB
Stainless Steel, Austenitic, 304B6															
USA															
	UNS S30466		0.08 max	18.0-20.0	2.00 max		12.0-15.0	0.045 max	0.030 max	0.75 max	B 1.50-1.74; N <=0.10; bal Fe				
ASTM A887(94)	S30466	Sh Strp Plt, Boron, for nuclear, Grade A	0.08 max	18.0-20.0	2.00 max		12.0-15.0	0.045 max	0.030 max	0.75 max	B 1.50-1.74; N <=0.10; bal Fe	515	205	20.0	241 HB
Stainless Steel, Austenitic, 304B7															
USA															
	UNS S30467		0.08 max	18.0-20.0	2.00 max		12.0-15.0	0.045 max	0.030 max	0.75 max	B 1.75-2.25; N <=0.10; bal Fe				
ASTM A887(94)	S30467	Sh Strp Plt, Boron, for nuclear, Grade A	0.08 max	18.0-20.0	2.00 max		12.0-15.0	0.045 max	0.030 max	0.75 max	B 1.75-2.25; N <=0.10; bal Fe	515	205	17.0	241 HB

UNS numbers and US grades are provided as a means of cross referencing chemically similar alloys. Exchangability is only possible after independent examination of specifications. Tensile properties are minimum or typical as specified. UTS and YS as MPa. El as %. See Appendix for list of abbreviations used in Notes. * indicates obsolete material.

Specification	Designation	Notes	C	Cr	Mn	Mo	Ni	P	S	Si	Other	UTS	YS	El	Hard
Stainless Steel, Austenitic, 304H															
Bulgaria															
BDS 6738(72)	0Ch18N10	Corr res	0.08 max	17.0-19.0	2.00 max	0.30 max	9.0-11.0	0.035 max	0.025 max	0.80 max	bal Fe				
China															
GB 12770(91)	0Cr19Ni9	Weld tube; SA	0.08 max	18.00-20.00	2.00 max		8.00-10.50	0.035 max	0.030 max	1.00 max	bal Fe	520	210	35	
GB 12771(91)	0Cr19Ni9	Weld Pipe SA	0.08 max	18.00-20.00	2.00 max		8.00-10.50	0.035 max	0.030 max	1.00 max	bal Fe	520	210	35	
GB 4231(93)	0Cr19Ni9	Strp CW	0.08 max	18.00-20.00	2.50 max		7.00-10.50	0.035 max	0.030 max	1.00 max	bal Fe	780	470	6	
Europe															
EN 10088/2(95)	1.4301	Strp, CR Corr res; t<=6mm, Ann	0.07 max	17.00-19.50	2.00 max		8.00-10.50	0.045 max	0.030 max	1.00 max	N <=0.11; bal Fe	540-750	230	45	
EN 10088/2(95)	X5CrNi18-10	Strp, CR Corr res; t<=6mm, Ann	0.07 max	17.00-19.50	2.00 max		8.00-10.50	0.045 max	0.030 max	1.00 max	N <=0.11; bal Fe	540-750	230	45	
EN 10088/3(95)	1.4301	Bar Rod Sect, t<=25mm, strain Hard	0.07 max	17.00-19.50	2.00 max		8.00-10.50	0.045 max	0.030 max	1.00 max	N <=0.11; bal Fe	800-1000	500	12	
EN 10088/3(95)	1.4301	Bar Rod Sect, t<=35mm, strain Hard	0.07 max	17.00-19.50	2.00 max		8.00-10.50	0.045 max	0.030 max	1.00 max	N <=0.11; bal Fe	700-850	350	20	
EN 10088/3(95)	X5CrNi18-10	Bar Rod Sect, t<=35mm, strain Hard	0.07 max	17.00-19.50	2.00 max		8.00-10.50	0.045 max	0.030 max	1.00 max	N <=0.11; bal Fe	700-850	350	20	
EN 10088/3(95)	X5CrNi18-10	Bar Rod Sect, t<=25mm, strain Hard	0.07 max	17.00-19.50	2.00 max		8.00-10.50	0.045 max	0.030 max	1.00 max	N <=0.11; bal Fe	800-1000	500	12	
International															
ISO 2604-1(75)	F48	Frg	0.04-0.09	17.00-19.00	2.00 max		8.00-12.00	0.045 max	0.030 max	1.00 max	bal Fe	490-690			
Italy															
UNI 6901(71)	X5CrNi1810	Corr res	0.06 max	17-19	2 max		8-11	0.045 max	0.03 max	1 max	bal Fe				
UNI 6904(71)	X8CrNi1910	Heat res	0.04-0.1	18-20	2 max		8-12	0.04 max	0.03 max	0.75 max	bal Fe				
UNI 8317(81)	X8CrNi1910	Sh Strp Plt	0.04-0.1	18-20	2 max		8-12	0.04 max	0.03 max	0.75 max	bal Fe				
Japan															
JIS G3214(91)	SUSF304H	Frg press ves 130-200mm diam	0.04-0.10	18.00-20.00	2.00 max		8.00-12.00	0.040 max	0.030 max	1.00 max	bal Fe	480	205	29	187 HB max
JIS G3459(94)	SUS304HTP	Pipe SA 8mm diam	0.04-0.1	18-20	2 max		8-11	0.04 max	0.03 max	0.75 max	bal Fe	520	205	35	
JIS G3463(94)	SUS304HTB	Tube, SA 8mm diam	0.04-0.10	18.00-20.00	2.00 max		8.00-11.00	0.040 max	0.030 max	0.75 max	bal Fe	520	205		90 HRB max
Mexico															
NMX-B-186-SCFI(94)	TP304H	Smls pipe for high-temp central station service	0.04-1.10	18.0-20.0	2.00 max		8.00-11.0	0.040 max	0.030 max	0.75 max	bal Fe	517	207		
NMX-B-196(68)	TP304H	Smls tube for refinery	0.04-0.10	18.0-20.0	2.00 max		8.0-11.0	0.040 max	0.030 max	0.75 max	bal Fe	520	205	35	
Pan America															
COPANT 195	TP304H	Tube, HF, CF, 5.65mm diam	0.04-0.1	18-20	2 max		8-11	0.04 max	0.03 max	0.75 max	bal Fe				
COPANT 513	TP304H	Tube, SA, Q/A, 8mm diam	0.04-0.1	18-20	2 max		8-11	0.04 max	0.03 max	0.75 max	bal Fe				
Spain															
UNE 36016(75)	X6CrNi19-10	Sh Strp Plt Bar Rod; Corr res	0.08 max	18.0-20.0	2.0 max	0.15 max	8.0-10.5	0.045 max	0.03 max	1.0 max	bal Fe				
UNE 36016/1(89)	E-304	Corr res	0.07 max	17.0-19.0	2.0 max	0.15 max	8.0-11.0	0.045 max	0.03 max	1.0 max	N <=0.1; bal Fe				
UNE 36016/1(89)	F.3504*	Obs; EN10088-3(96); X5CrNi1810	0.07 max	17.0-19.0	2.0 max	0.15 max	8.0-11.0	0.045 max	0.03 max	1.0 max	N <=0.1; bal Fe				
UNE 36087(78)	X5CrNi18-10	Corr res	0.03-0.07	17.0-19.0	2.0 max	0.15 max	8.0-11.0	0.045 max	0.03 max	1.0 max	bal Fe				
UNE 36087(78)	X5CrNi18-11	Corr res	0.07 max	17.0-19.0	2.0 max	0.5 max	9.0-11.5	0.045 max	0.03 max	1.0 max	bal Fe				
UK															
BS 1501	304S49*		0.04-0.09	17.5-19	0.5-2		8-11	0.04 max	0.03 max	0.2-1	Nb <=0.05; bal Fe				
BS 1502(82)	304S51	Bar Shp; Press ves	0.04-0.10	17.0-19.0	2.00 max		8.00-11.00	0.045 max	0.030 max	1.00 max	W <=0.1; bal Fe				

UNS numbers and US grades are provided as a means of cross referencing chemically similar alloys. Exchangability is only possible after independent examination of specifications. Tensile properties are minimum or typical as specified. UTS and YS as MPa. El as %. See Appendix for list of abbreviations used in Notes. * indicates obsolete material.

Stainless Steel, Austenitic, 304H (Continued from previous page)

Specification	Designation	Notes	C	Cr	Mn	Mo	Ni	P	S	Si	Other	UTS	YS	El	Hard
UK															
BS 1503(89)	304S50		0.04-0.1	17-19	2 max		8-11	0.045 max	0.03 max	1 max	bal Fe				
BS 1503(89)	304S51	Frg press ves; t<=999mm; HT	0.04-0.10	17.0-19.0	2.00 max	0.70 max	8.00-11.0	0.040 max	0.025 max	1.00 max	B <=0.005; Cu <=0.50; Nb <=0.20; Ti <=0.10; W <=0.1; bal Fe	490-690	230	30	
BS 1506(90)	304S51	Bolt matl Pres/Corr res; t<=160mm; Hard; was En80(A)	0.04-0.10	17.0-19.0	2.00 max		8.00-11.0	0.045 max	0.030 max	1.00 max	bal Fe	520	205	40	
BS 3059/2(78)	304S59	Boiler, Superheater; Tub; Pip	0.04-0.09	17.0-19.0	0.50-2.00	0.15 max	9.0-12.0	0.040 max	0.030 max	0.20-1.00	bal Fe				
BS 3605(73)	304S59	Pip Tube press smls weld; t<=200mm	0.04-0.09	17.0-19.0	0.5-2.0		9.0-12.0	0.040 max	0.030 max	0.2-1.0	bal Fe				
BS 5059	304S59	Tube, Pipe	0.04-0.09	17-19	0.5-2		9-12	0.04 max	0.03 max	0.2-1	bal Fe				
USA															
	AISI 304H*	Obs; Tube	0.04-0.1	18.00-20.00	2.00 max		8.00-11.00	0.040 max	0.030 max	1.00 max	bal Fe				
	UNS S30409	High-C	0.04-0.10	18.00-20.00	2.00 max		8.00-11.00	0.040 max	0.030 max	1.00 max	bal Fe				
ASME SA182	304H	Refer to ASTM A182/A182M													
ASME SA213	304H	Refer to ASTM A213/A213M(95)													
ASME SA240	304H	Refer to ASTM A240(95)													
ASME SA249	304H	Refer to ASTM A249/A249M(96)													
ASME SA312	304H	Refer to ASTM A312/A312M(95)													
ASME SA376	304H	Refer to ASTM A376/A376M(93)													
ASME SA403	304H	Refer to ASTM A403/A403M(95)													
ASME SA430	304H	Refer to ASTM A430/A430M(91)													
ASME SA479	304H	Refer to ASTM A479/A479M(95)													
ASTM A182	304H*	Bar Frg	0.04-0.1	18-20	2 max		8-11	0.04 max	0.03 max	1 max	bal Fe				
ASTM A182/A182M(98)	F304H	Frg/roll pipe flange valve	0.040-0.10	18.0-20.0	2.00 max		8.00-11.00	0.045 max	0.030 max	1.00 max	bal Fe	515	205	30	
ASTM A213	304H*	Tube	0.04-0.1	18-20	2 max		8-11	0.04 max	0.03 max	1 max	bal Fe				
ASTM A213/A213M(95)	TP304H	Smls tube boiler, superheater, heat exchanger	0.04-0.10	18.0-20.0	2.00 max		8.00-11.00	0.040 max	0.030 max	0.75 max	bal Fe	515	205	35	
ASTM A240/A240M(98)	S30409	High-C	0.04-0.10	18.0-20.0	2.00 max		8.0-11.0	0.040 max	0.030 max	0.75 max	bal Fe	515	205	40.0	201 HB
ASTM A249/249M(96)	TP304H	Weld tube; boiler, superheater, heat exch	0.04-0.10	18.0-20.0	2.00 max		8.00-11.00	0.040 max	0.030 max	0.75 max	bal Fe	515	205	35	90 HRB
ASTM A271(96)	304H	Smls tube	0.04-0.1	18-20	2 max		8-11	0.04 max	0.03 max	1 max	bal Fe				
ASTM A312/A312M(95)	304H	Smls Weld, Pipe	0.04-0.1	18-20	2 max		8-11	0.04 max	0.03 max	1 max	bal Fe				
ASTM A336/A336M(98)	F304H	Frg, press/high-temp	0.04-0.10	19.0-20.0	2.00 max		8.00-12.00	0.045 max	0.030 max	1.00 max	bal Fe	485	205	30	90 HRB
ASTM A358/A358M(95)	304H	Elect fusion weld pipe, high-temp	0.04-0.1	18-20	2 max		8-11	0.04 max	0.03 max	1 max	bal Fe				
ASTM A376	304H	Smls pipe	0.04-0.1	18-20	2 max		8-11	0.04 max	0.03 max	1 max	bal Fe				
ASTM A403	304H	Pipe	0.04-0.1	18-20	2 max		8-11	0.04 max	0.03 max	1 max	bal Fe				
ASTM A430	304H*	Obs 1995; see A312; Pipe	0.04-0.1	18-20	2 max		8-11	0.04 max	0.03 max	1 max	bal Fe				
ASTM A479	304H	Wir Bar Shp	0.04-0.1	18-20	2 max		8-11	0.04 max	0.03 max	1 max	bal Fe				

UNS numbers and US grades are provided as a means of cross referencing chemically similar alloys. Exchangability is only possible after independent examination of specifications. Tensile properties are minimum or typical as specified. UTS and YS as MPa. El as %. See Appendix for list of abbreviations used in Notes. * indicates obsolete material.

Specification	Designation	Notes	C	Cr	Mn	Mo	Ni	P	S	Si	Other	UTS	YS	El	Hard

Stainless Steel, Austenitic, 304H (Continued from previous page)

USA

Specification	Designation	Notes	C	Cr	Mn	Mo	Ni	P	S	Si	Other	UTS	YS	El	Hard
ASTM A813/A813M(95)	TP304H	Weld Pipe, High-C	0.04-0.10	18.0-20.0	2.00 max		8.00-11.00	0.045 max	0.030 max	0.75 max	B <=2; bal Fe	515	205		
ASTM A814/A814M(96)	TP304H	CW Weld Pipe, High-C	0.04-0.10	18.0-20.0	2.00 max		8.00-11.0	0.045 max	0.030 max	0.75 max	bal Fe	515	205		
ASTM A943/A943M(95)	TP304H	Spray formed smls pipe	0.40-0.10	18.0-20.0	2.00 max		8.00-11.00	0.040 max	0.030 max	0.75 max	bal Fe	515	205		
ASTM A965/965M(97)	F304H	Frg for press high-temp parts	0.04-0.10	19.0-20.0	2.00 max		8.0-12.0	0.045 max	0.030 max	1.00 max	bal Fe	485	205	30	
SAE J405(98)	S30409	High-C	0.04-0.10	18.00-20.00	2.00 max		8.00-10.50	0.045 max	0.030 max	0.75 max	bal Fe				

Stainless Steel, Austenitic, 304L

Australia

Specification	Designation	Notes	C	Cr	Mn	Mo	Ni	P	S	Si	Other	UTS	YS	El	Hard
AS 1449(94)	304L	Sh Strp Plt	0.030 max	18.00-20.00	2.00 max		8.00-12.00	0.045 max	0.030 max	0.75 max	bal Fe				
AS 2837(86)	304L	Bar, Semi-finished product	0.030 max	18.0-20.0	2.00 max		9.0-12.0	0.045 max	0.030 max	1.00 max	bal Fe				

Austria

Specification	Designation	Notes	C	Cr	Mn	Mo	Ni	P	S	Si	Other	UTS	YS	El	Hard
ONORM M3121	X2CrNi1911KKW	Sh Strp Plt Bar Wir Tub	0.03 max	18-20	2 max		10.0-12.5	0.045 max	0.030 max	1 max	bal Fe				

Canada

Specification	Designation	Notes	C	Cr	Mn	Mo	Ni	P	S	Si	Other	UTS	YS	El	Hard
CSA G110.3	304L	Bar Bil	0.03 max	18-20	2 max		8-12	0.05 max	0.03 max	1 max	bal Fe				
CSA G110.6	304L	Sh Strp Plt, HR, Ann or HR, Q/A, Tmp	0.03 max	18-20	2 max		8-12	0.05 max	0.03 max	1 max	bal Fe				
CSA G110.9	304L	Sh Strp Plt, HR, CR, Ann, Q/A, Tmp	0.03 max	18-20	2 max		8-12	0.05 max	0.03 max	1 max	bal Fe				

China

Specification	Designation	Notes	C	Cr	Mn	Mo	Ni	P	S	Si	Other	UTS	YS	El	Hard
GB 1220(92)	00Cr19Ni10	Bar SA	0.030 max	18.00-20.00	2.00 max		8.00-12.00	0.035 max	0.030 max	1.00 max	bal Fe	480	177	40	
GB 12770(91)	00Cr19Ni11	Weld tube; SA	0.030 max	18.00-20.00	2.00 max		9.00-13.00	0.035 max	0.030 max	1.00 max	bal Fe	480	180	35	
GB 12771(91)	00Cr19Ni11	Weld Pipe SA	0.030 max	18.00-20.00	2.00 max		9.00-13.00	0.035 max	0.030 max	1.00 max	bal Fe	480	180	35	
GB 13296(91)	00Cr19Ni10	Smls tube SA	0.030 max	18.00-20.00	2.00 max		8.00-12.00	0.035 max	0.030 max	1.00 max	bal Fe	480	175	35	
GB 3280(92)	00Cr19Ni10	Sh Plt, CR SA	0.030 max	18.00-20.00	2.00 max		8.00-12.00	0.035 max	0.030 max	1.00 max	bal Fe	480	177	40	
GB 4237(92)	00Cr19Ni10	Sh Plt, HR SA	0.030 max	18.00-20.00	2.00 max		8.00-12.00	0.035 max	0.030 max	1.00 max	bal Fe	480	177	40	
GB 4239(91)	00Cr19Ni10	Sh Strp, CR SA	0.030 max	18.00-20.00	2.00 max		8.00-12.00	0.035 max	0.030 max	1.00 max	bal Fe	480	175	40	
GB 4240(93)	00Cr19Ni11(-R)	Wir Ann 1-3mm diam, R-As Ann	0.030 max	18.00-20.00	2.00 max		9.00-13.00	0.035 max	0.030 max	1.00 max	bal Fe	490-830		25	
GB/T 14975(94)	00Cr19Ni10	Smls tube SA	0.030 max	18.00-20.00	2.00 max		8.00-12.00	0.035 max	0.030 max	1.00 max	bal Fe	480	175	35	
GB/T 14976(94)	00Cr19Ni10	Smls pipe SA	0.030 max	18.00-20.00	2.00 max		8.00-12.00	0.035 max	0.030 max	1.00 max	bal Fe	480	175	35	

Czech Republic

Specification	Designation	Notes	C	Cr	Mn	Mo	Ni	P	S	Si	Other	UTS	YS	El	Hard
CSN 417249	17249	Corr res	0.03 max	17.0-20.0	2.0 max		10.0-12.5	0.045 max	0.03 max	1.0 max	bal Fe				

Europe

Specification	Designation	Notes	C	Cr	Mn	Mo	Ni	P	S	Si	Other	UTS	YS	El	Hard
EN 10088/2(95)	1.4306	Strp, CR Corr res; t<=6mm, Ann	0.030 max	18.00-20.00	2.00 max		10.00-12.00	0.045 max	0.015 max	1.00 max	N <=0.11; bal Fe	520-670	220	45	
EN 10088/2(95)	X2CrNi19-11	Strp, CR Corr res; t<=6mm, Ann	0.030 max	18.00-20.00	2.00 max		10.00-12.00	0.045 max	0.015 max	1.00 max	N <=0.11; bal Fe	520-670	220	45	
EN 10088/3(95)	1.4306	Bar Rod Sect; Corr res; 160<t<=250mm, Ann	0.030 max	18.00-20.00	2.00 max		10.00-12.00	0.045 max	0.030 max	1.00 max	N <=0.11; bal Fe	460-680	180	35	215 HB
EN 10088/3(95)	1.4306	Bar Rod Sect, t<=25mm, strain Hard	0.030 max	18.00-20.00	2.00 max		10.00-12.00	0.045 max	0.030 max	1.00 max	N <=0.11; bal Fe	800-1000	500	12	
EN 10088/3(95)	1.4306	Bar Rod Sect, t<=160mm, Ann	0.030 max	18.00-20.00	2.00 max		10.00-12.00	0.045 max	0.030 max	1.00 max	N <=0.11; bal Fe	460-680	180	45	215 HB
EN 10088/3(95)	1.4306	Bar Rod Sect, t<=35mm, strain Hard	0.030 max	18.00-20.00	2.00 max		10.00-12.00	0.045 max	0.030 max	1.00 max	N <=0.11; bal Fe	700-850	350	20	
EN 10088/3(95)	X2CrNi19-11	Bar Rod Sect, t<=25mm, strain Hard	0.030 max	18.00-20.00	2.00 max		10.00-12.00	0.045 max	0.030 max	1.00 max	N <=0.11; bal Fe	800-1000	500	12	

UNS numbers and US grades are provided as a means of cross referencing chemically similar alloys. Exchangability is only possible after independent examination of specifications. Tensile properties are minimum or typical as specified. UTS and YS as MPa. El as %. See Appendix for list of abbreviations used in Notes. * indicates obsolete material.

Specification	Designation	Notes	C	Cr	Mn	Mo	Ni	P	S	Si	Other	UTS	YS	El	Hard

Stainless Steel, Austenitic, 304L (Continued from previous page)

Europe

Specification	Designation	Notes	C	Cr	Mn	Mo	Ni	P	S	Si	Other	UTS	YS	El	Hard
EN 10088/3(95)	X2CrNi19-11	Bar Rod Sect; Corr res; 160<t<=250mm, Ann	0.030 max	18.00-20.00	2.00 max		10.00-12.00	0.045 max	0.030 max	1.00 max	N <=0.11; bal Fe	460-680	180	35	215 HB
EN 10088/3(95)	X2CrNi19-11	Bar Rod Sect, t<=160mm, Ann	0.030 max	18.00-20.00	2.00 max		10.00-12.00	0.045 max	0.030 max	1.00 max	N <=0.11; bal Fe	460-680	180	45	215 HB
EN 10088/3(95)	X2CrNi19-11	Bar Rod Sect, t<=35mm, strain Hard	0.030 max	18.00-20.00	2.00 max		10.00-12.00	0.045 max	0.030 max	1.00 max	N <=0.11; bal Fe	700-850	350	20	

Finland

Specification	Designation	Notes	C	Cr	Mn	Mo	Ni	P	S	Si	Other	UTS	YS	El	Hard
SFS 720(86)	X2CrNi1810	Sh Strp Plt Bar Wir	0.03 max	17.0-19.0	2.0 max		9.0-12.0	0.045 max	0.03 max	1.0 max	bal Fe				

France

Specification	Designation	Notes	C	Cr	Mn	Mo	Ni	P	S	Si	Other	UTS	YS	El	Hard
AFNOR NFA35572	Z2CN18.10	Bar, 25mm diam	0.03 max	17-19	2 max		9-11	0.04 max	0.03 max	1 max	bal Fe				
AFNOR NFA35573	Z2CN18.10	Sh Strp Plt, CR, 5mm diam	0.03 max	17-19	2 max		9-11	0.04 max	0.03 max	1 max	bal Fe				
AFNOR NFA35575	Z2CN18.10	Wir rod, 60% Hard	0.03 max	17-19	2 max		9-11	0.04 max	0.03 max	1 max	bal Fe				

Germany

Specification	Designation	Notes	C	Cr	Mn	Mo	Ni	P	S	Si	Other	UTS	YS	El	Hard
DIN 17440(96)	WNr 1.4306	Plt Sh, HR Strp Bar Wir Frg, HT	0.030 max	18.0-20.0	2.00 max		10.0-12.5	0.045 max	0.030 max	1.00 max	N <=0.11; bal Fe	460-680	180	35-45	
DIN 17441(97)	WNr 1.4306	CR Strp Plt Sh	0.030 max	18.0-20.0	2.00 max		10.0-12.5	0.045 max	0.030 max	1.00 max	N <=0.11; bal Fe				
DIN EN 10088(95)	GX2CrNiN18-9	Sht Plt Strp Bar Rod	0.030 max	18.0-20.0	2.00 max		10.0-12.5	0.045 max	0.030 max	1.00 max	N <=0.11; bal Fe	460-680	180	35-45	
DIN EN 10088(95)	WNr 1.4306	Sh Plt Strp Bar Rod, HT	0.030 max	18.0-20.0	2.00 max		10.0-12.5	0.045 max	0.030 max	1.00 max	N <=0.11; bal Fe	460-680	180	35-45	

Hungary

Specification	Designation	Notes	C	Cr	Mn	Mo	Ni	P	S	Si	Other	UTS	YS	El	Hard
MSZ 4360	KO41LC	Sh Strp Plt Bar Wir Tub	0.03 max	18-20	2 max		10-12.5	0.045 max	0.03 max	1 max	bal Fe				
MSZ 4360(80)	KO41	Corr res	0.03 max	17-19	2 max	0.5 max	10.5-13.0	0.04 max	0.03 max	1 max	bal Fe				
MSZ 4360(87)	KO41LC	Corr res; 0-160mm; SA	0.03 max	18-20	2 max	0.5 max	10.0-12.5	0.045 max	0.03 max	1 max	bal Fe	480-680	180	40L; 35T	190 HB max
MSZ 4360(87)	X3CrNi1911	Corr res; 0-160mm; SA	0.03 max	18-20	2 max	0.5 max	10.0-12.5	0.045 max	0.03 max	1 max	bal Fe	480-680	180	40L; 35T	190 HB max

India

Specification	Designation	Notes	C	Cr	Mn	Mo	Ni	P	S	Si	Other	UTS	YS	El	Hard
IS 6527	02Cr18Ni11	Wir rod	0.03 max	17-20	2 max		9-13	0.05 max	0.03 max	1 max	bal Fe				
IS 6528	02Cr18Ni11	Wir, Ann, SA	0.03 max	17-20	2 max		9-13	0.05 max	0.03 max	1 max	bal Fe				
IS 6529	02Cr18Ni11	Bil	0.03 max	17-20	2 max		9-13	0.05 max	0.03 max	1 max	bal Fe				
IS 6603	02Cr18Ni11	Bar, Soft	0.03 max	17-20	2 max		9-13	0.05 max	0.03 max	1 max	bal Fe				
IS 6911	02Cr18Ni11	Sh Strp Plt, Soft	0.03 max	17-20	2 max		9-13	0.05 max	0.03 max	1 max	bal Fe				

International

Specification	Designation	Notes	C	Cr	Mn	Mo	Ni	P	S	Si	Other	UTS	YS	El	Hard
COMECON PC4-70	16	Sh Strp Plt Bar Wir Bil	0.03 max	17-19	2 max		10-12.5	0.035 max	0.025 max	0.8 max	bal Fe				
ISO 4954(93)	X2CrNi1810E	Wir rod Bar, CH ext	0.030 max	17.0-19.0	2.00 max		9.0-12.0	0.045 max	0.030 max	1.00 max	bal Fe	630			
ISO 683-13(74)	10*	Sh Plt Bar, ST	0.03 max	17-19	2 max		9-12	0.045 max	0.03 max	1 max	bal Fe				

Italy

Specification	Designation	Notes	C	Cr	Mn	Mo	Ni	P	S	Si	Other	UTS	YS	El	Hard
UNI 6901(71)	X2CrNi1811	Corr res	0.03 max	17-19	2 max		9-12	0.045 max	0.03 max	1 max	bal Fe				
UNI 6904(71)	X2CrNi1811	Heat res	0.03 max	18-20	2 max		8-13	0.04 max	0.03 max	0.75 max	bal Fe				
UNI 7500(75)	X2CrNi1811	Press ves	0.03 max	17-19	2 max		9-12	0.045 max	0.03 max	1 max	bal Fe				
UNI 8317(81)	X2CrNi1811	Sh Strp Plt	0.03 max	17-19	2 max		9-12	0.035 max	0.03 max	1 max	bal Fe				

Japan

Specification	Designation	Notes	C	Cr	Mn	Mo	Ni	P	S	Si	Other	UTS	YS	El	Hard
JIS G3214(91)	SUSF304L	Frg press ves 130-200mm diam	0.030 max	18.00-20.00	2.00 max		9.00-13.00	0.040 max	0.030 max	1.00 max	bal Fe	450	175	29	187 HB max
JIS G3447(94)	SUS304LTBS	Tube	0.030 max	18.00-20.00	2.00 max		9.00-13.00	0.040 max	0.030 max	1.00 max	bal Fe	480		35	

UNS numbers and US grades are provided as a means of cross referencing chemically similar alloys. Exchangability is only possible after independent examination of specifications. Tensile properties are minimum or typical as specified. UTS and YS as MPa. El as %. See Appendix for list of abbreviations used in Notes. * indicates obsolete material.

Specification	Designation	Notes	C	Cr	Mn	Mo	Ni	P	S	Si	Other	UTS	YS	El	Hard

Stainless Steel, Austenitic, 304L (Continued from previous page)

Japan

Specification	Designation	Notes	C	Cr	Mn	Mo	Ni	P	S	Si	Other	UTS	YS	El	Hard
JIS G3459(94)	SUS304L	Pipe	0.03 max	18-20	2 max		9-13	0.04 max	0.03 max	1 max	bal Fe				
JIS G3463(94)	SUS304LTB	Tube, SA 8mm diam	0.030 max	18.00-20.00	2.00 max		9.00-13.00	0.040 max	0.030 max	1.00 max	bal Fe	520	205	35	90 HRB max
JIS G4303(91)	SUS304L	Bar; SA; <=180mm diam	0.030 max	18.00-20.00	2.00 max		9.00-13.00	0.045 max	0.030 max	1.00 max	bal Fe	480	175	40	187 HB
JIS G4304(91)	SUS304L	Sh Plt, HR SA	0.030 max	18.00-20.00	2.00 max		9.0-13.00	0.045 max	0.030 max	1.00 max	bal Fe	480	175	40	90 HRB max
JIS G4305(91)	SUS304L	Sh Plt, CR SA	0.030 max	18.00-20.00	2.00 max		9.00-13.00	0.045 max	0.030 max	1.00 max	bal Fe	480	175	40	187 HB max
JIS G4306	SUS304L	Strp HR SA	0.03 max	18-20	2 max		9-13	0.04 max	0.03 max	1 max	bal Fe				
JIS G4307	SUS304L	Strp CR SA	0.03 max	18-20	2 max		9-13	0.04 max	0.03 max	1 max	bal Fe				
JIS G4308(98)	SUS304L	Wir rod	0.030 max	18.00-20.00	2.00 max		9.00-13.00	0.045 max	0.030 max	1.00 max	bal Fe				
JIS G4309	SUS304L	Wir	0.030 max	18.00-20.00	2.00 max		9.00-13.00	0.045 max	0.030 max	1.00 max	bal Fe				
JIS G4315	SUS304L	Wir rod	0.030 max	18.00-20.00	2.00 max		9.00-13.00	0.045 max	0.030 max	1.00 max	bal Fe				

Mexico

Specification	Designation	Notes	C	Cr	Mn	Mo	Ni	P	S	Si	Other	UTS	YS	El	Hard
DGN B-229	TP304L*	Obs; Tube	0.04 max	18-20	2 max		8-13	0.04 max	0.03 max	0.75 max	bal Fe				
NMX-B-171(91)	MT304L	mech tub	0.035 max	18.0-20.0	2.00 max		8.0-13.0	0.040 max	0.030 max	1.00 max	bal Fe				
NMX-B-176(91)	TP304L	Smls, weld sanitary tub	0.08 max	18.00-20.00	2.00 max		8.00-13.00	0.040 max	0.030 max	0.75 max	bal Fe				

Norway

Specification	Designation	Notes	C	Cr	Mn	Mo	Ni	P	S	Si	Other	UTS	YS	El	Hard
NS 14360	14360	Sh Strp Plt Bar	0.03 max	17-19	2 max		9-12	0.045 max	0.03 max	1 max	bal Fe				

Pan America

Specification	Designation	Notes	C	Cr	Mn	Mo	Ni	P	S	Si	Other	UTS	YS	El	Hard
COPANT 513	TP304L	Tube, HF, CF	0.04 max	18-20	2 max		8-11	0.04 max	0.03 max	1 max	bal Fe				

Poland

Specification	Designation	Notes	C	Cr	Mn	Mo	Ni	P	S	Si	Other	UTS	YS	El	Hard
PNH86020	00H18N10	Corr res	0.3 max	17-19	2 max		10-12.5	0.045 max	0.03 max	0.8 max	bal Fe				
PNH86020	00M18N10		0.03 max	17-19	2 max		10-12.5	0.045 max	0.03 max	0.8 max	bal Fe				

Romania

Specification	Designation	Notes	C	Cr	Mn	Mo	Ni	P	S	Si	Other	UTS	YS	El	Hard
STAS 3583(87)	2NiCr185	Corr res; Heat res	0.03 max	18-20	2 max	0.2 max	10-12	0.045 max	0.03 max	1 max	Pb <=0.15; bal Fe				

Russia

Specification	Designation	Notes	C	Cr	Mn	Mo	Ni	P	S	Si	Other	UTS	YS	El	Hard
GOST	O3Ch18N11	Sh Strp Plt Bar Wir Tub	0.03 max	17-19	2 max	0.1 max	10.5-12.5	0.035 max	0.02 max	0.8 max	Cu <=0.3; Ti <=0.2; W <=0.2; bal Fe				
GOST 5632(61)	00KH18N10	Corr res; Heat res; High-temp constr	0.04 max	17.0-19.0	1.0-2.0	0.15 max	9.0-11.0	0.035 max	0.02 max	0.8 max	Cu <=0.3; Ti <=0.05; bal Fe				
GOST 5632(72)	03Ch18N11	Corr res	0.03 max	17.0-19.0	0.7-2.0	0.1 max	10.5-12.5	0.035 max	0.02 max	0.8 max	Cu <=0.3; Ti <=0.2; W <=0.2; bal Fe				
GOST 5632(72)	03Ch18N12	Corr res	0.03 max	17.0-19.0	0.4 max	0.3 max	11.5-13.0	0.03 max	0.02 max	0.4 max	Cu <=0.3; Ti <=0.005; W <=0.2; bal Fe				
GOST 5632(72)	04Ch18N10	Corr res	0.04 max	17.0-19.0	2.0 max	0.3 max	9.0-11.0	0.035 max	0.02 max	0.8 max	Cu <=0.3; Ti <=0.2; W <=0.2; bal Fe				

Spain

Specification	Designation	Notes	C	Cr	Mn	Mo	Ni	P	S	Si	Other	UTS	YS	El	Hard
UNE 36016(75)	F.3503*	see X2CrNi19-10	0.03 max	18.0-20.0	2.0 max	0.15 max	8.0-12.0	0.045 max	0.03 max	1.0 max	bal Fe				
UNE 36016(75)	X2CrNi19-10	Sh Strp Bar; Corr res	0.03 max	18.0-20.0	2.0 max	0.15 max	8.0-12.0	0.045 max	0.03 max	1.0 max	bal Fe				
UNE 36016/1(89)	E-304L	Corr res	0.03 max	17.0-19.0	2.0 max	0.15 max	9.0-12.0	0.045 max	0.03 max	1.0 max	N <=0.1; bal Fe				
UNE 36016/1(89)	F.3503*	Obs; E-304 L; Corr res	0.03 max	17.0-19.0	2.0 max	0.15 max	9.0-12.0	0.045 max	0.03 max	1.0 max	N <=0.1; bal Fe				
UNE 36087(78)	F.3503*	see X2CrNi19-10	0.03 max	18-20	2 max		8-12	0.045 max	0.03 max	1 max	bal Fe				
UNE 36087(78)	X2CrNi19-10	Sh Strp Bar	0.03 max	18-20	2 max		8-12	0.045 max	0.03 max	1 max	bal Fe				

UNS numbers and US grades are provided as a means of cross referencing chemically similar alloys. Exchangability is only possible after independent examination of specifications. Tensile properties are minimum or typical as specified. UTS and YS as MPa. El as %. See Appendix for list of abbreviations used in Notes. * indicates obsolete material.

Stainless Steel, Austenitic, 304L (Continued from previous page)

Specification	Designation	Notes	C	Cr	Mn	Mo	Ni	P	S	Si	Other	UTS	YS	El	Hard
Spain															
UNE 36087/4(89)	F.3503*	see X2CrNi18 10	0.03 max	17-19	2 max	0.15 max	9-12	0.045 max	0.03 max	1 max	N <=0.1; bal Fe				
UNE 36087/4(89)	X2CrNi18 10	Corr res; Sh Strp	0.03 max	17-19	2 max	0.15 max	9-12	0.045 max	0.03 max	1 max	N <=0.1; bal Fe				
Sweden															
SS 142352	2352	Corr res	0.03 max	17-19	2 max		9-125	0.045 max	0.03 max	1 max	bal Fe				
UK															
BS 1449/2(83)	304S11	Sh Strp Plt Bar Rod, 0.05-100mm	0.03 max	17-19	2.00 max		9-12	0.045 max	0.03 max	1 max	bal Fe	480	180	40	0-185 HB
BS 1501/3(73)	304S12	Bar Rod Press ves; Corr res	0.03 max	17.5-19.0	0.5-2.0	0.15 max	9.0-12.0	0.045 max	0.03 max	0.2-1.0	Cu <=0.3; Ti <=0.05; W <=0.1; bal Fe				
BS 1501/3(90)	304S11	Press ves; Corr/Heat res; 0.05-100mm	0.03 max	17.0-19.0	2.0 max		9.0-12.0	0.045 max	0.025 max	1.0 max	Cu <=0.3; Ti <=0.05; W <=0.1; bal Fe	480-680	180	40	
BS 1503(89)	304S11	Frg press ves; t<=999mm; HT	0.030 max	17.0-19.0	2.00 max	0.70 max	9.00-12.0	0.040 max	0.025 max	1.00 max	B <=0.005; Cu <=0.50; Nb <=0.20; Ti <=0.10; W <=0.1; bal Fe	480-680	215	30	
BS 1506(90)	304S11	Bolt matl Pres/Corr res; Corr res; 38<t<=44mm; SHT CD	0.03 max	17.0-19.0	2.00 max		9.00-12.0	0.045 max	0.030 max	1.00 max	bal Fe	650	310	28	0-320 HB
BS 1506(90)	304S11	Bolt matl Pres/Corr res; t<=160mm; Hard	0.03 max	17.0-19.0	2.00 max		9.00-12.0	0.045 max	0.030 max	1.00 max	bal Fe	480-680	180	40	
BS 1554(90)	304S11	Wir Corr/Heat res; diam<0.49mm; Ann	0.03 max	17.0-19.0	2.00 max		9.00-12.0	0.045 max	0.030 max	1.00 max	W <=0.1; bal Fe	0-850			
BS 1554(90)	304S11	Wir Corr/Heat res; 6<diam<=13mm; Ann	0.03 max	17.0-19.0	2.00 max		9.00-12.0	0.045 max	0.030 max	1.00 max	W <=0.1; bal Fe	0-650			
BS 1554(90)	305S11	Wir Corr/Heat res; 6<diam<=13mm; Ann	0.030 max	17.0-19.0	2.00 max		11.0-13.0	0.045 max	0.030 max	1.00 max	W <=0.1; bal Fe	0-650			
BS 1554(90)	305S11	Wir Corr/Heat res; diam<0.49mm; Ann	0.030 max	17.0-19.0	2.00 max		11.0-13.0	0.045 max	0.030 max	1.00 max	W <=0.1; bal Fe	0-850			
BS 3605	304S22	Tube, Pipe	0.030 max	17.0-19.0	0.5-2.0		9.0-12.0	0.040 max	0.030 max	0.20-1.00	bal Fe				
BS 3605/1(91)	304S11	Pip Tube press smls; t<=200mm	0.030 max	17.0-19.0	2.00 max		9.0-12.0	0.040 max	0.030 max	1.00 max	bal Fe	480-680		35	
BS 3606(92)	304S11	Heat exch Tub; t= 1.2-3.2mm	0.030 max	17.0-19.0	2.00 max		9.0-12.0	0.040 max	0.030 max	1.00 max	bal Fe	490-690	205	30	
BS 970/1(96)	304S11	Wrought Corr/Heat res; t<=160mm	0.030 max	17.0-19.0	2.00 max		9.00-12.0	0.045 max	0.030 max	1.00 max	bal Fe	480	180	40	183 HB max
BS 970/4(70)	304S12*	Obs; Corr res	0.03 max	17.5-19.0	0.50-2.00	0.7 max	9.0-12.0	0.045 max	0.03 max	0.2-1.0	Cu <=0.5; Pb <=0.15; Nb<=0.2; bal Fe				
USA															
	AISI 304L	Tube	0.030 max	18.00-20.00	2.00 max		8.00-12.00	0.045 max	0.030 max	1.00 max	bal Fe				
	UNS S30403	Low-C	0.03 max	18.00-20.00	2.00 max		8.00-12.00	0.045 max	0.030 max	1.00 max	bal Fe				
AMS 5511		Sh Strp Plt Bar Wir	0.03 max	18-20	2 max		8-12	0.045 max	0.03 max	1 max	bal Fe				
AMS 5569(93)		Corr/Heat res; Smls/weld hydraulic tub	0.03 max	18.0-20.0	2.00 max	0.50 max	8.0-11.0	0.04 max	0.03 max	1.00 max	Cu <=0.75; bal Fe	724-965	517-689	20	
AMS 5584(93)		Corr/Heat res; Smls/Weld tub, CD, 1/8 hard temp	0.03 max	16.0-18.0	2.00 max	2.0-3.0	10.0-14.0	0.04 max	0.03 max	1.00 max	Cu <=0.75; bal Fe	724-965	517-689	20	
AMS 5647		Sh Strp Plt Bar Wir	0.03 max	18-20	2 max		8-12	0.045 max	0.03 max	1 max	bal Fe				
ASME SA182	304L	Refer to ASTM A182/A182M													
ASME SA213	304L	Refer to ASTM A213/A213M(95)													
ASME SA240	304L	Refer to ASTM A240(95)													
ASME SA249	304L	Refer to ASTM A249/A249M(96)													
ASME SA312	304L	Refer to ASTM A312/A312M(95)													
ASME SA403	304L	Refer to ASTM A403/A403M(95)													

UNS numbers and US grades are provided as a means of cross referencing chemically similar alloys. Exchangability is only possible after independent examination of specifications. Tensile properties are minimum or typical as specified. UTS and YS as MPa. El as %. See Appendix for list of abbreviations used in Notes. * indicates obsolete material.

Stainless Steel, Austenitic, 304L (Continued from previous page)

USA

Specification	Designation	Notes	C	Cr	Mn	Mo	Ni	P	S	Si	Other	UTS	YS	El	Hard
ASME SA479	304L	Refer to ASTM A479/A479M(95)													
ASME SA688	304L	Refer to ASTM A688/A688M(96)													
ASTM A167(96)	304L*	Sh Strp Plt	0.03 max	18-20	2 max		8-12	0.045 max	0.03 max	1 max	bal Fe				
ASTM A182	304L*	Bar Frg	0.03 max	18-20	2 max		8-12	0.045 max	0.03 max	1 max	bal Fe				
ASTM A182/A182M(98)	F304L	Frg/roll pipe flange valve	0.035 max	18.0-20.0	2.00 max		8.00-13.00	0.045 max	0.030 max	1.00 max	N <=0.10; bal Fe	485	170	30	
ASTM A213	304L*	Tube	0.03 max	18-20	2 max		8-12	0.045 max	0.03 max	1 max	bal Fe				
ASTM A213/A213M(95)	TP304L	Smls tube boiler, superheater, heat exchanger	0.035 max	18.0-20.0	2.00 max		8.00-13.0	0.040 max	0.030 max	0.75 max	bal Fe	485	170	35	
ASTM A240/A240M(98)	S30403	Low-C	0.030 max	18.0-20.0	2.00 max		8.0-12.0	0.045 max	0.030 max	0.75 max	N <=0.10; bal Fe	485	170	40.0	201 HB
ASTM A249/249M(96)	TP304L	Weld tube; boiler, superheater, heat exch	0.035 max	18.0-20.0	2.00 max		8.00-13.0	0.040 max	0.030 max	0.75 max	bal Fe	485	170	35	90 HRB
ASTM A269	304L	Smls Weld, Tube	0.03 max	18-20	2 max		8-12	0.045 max	0.03 max	1 max	bal Fe				
ASTM A276(98)	304L	Bar	0.03 max	18-20	2 max		8-12	0.045 max	0.03 max	1 max	bal Fe				
ASTM A312/A312M(95)	304L	Smls Weld, Pipe	0.03 max	18-20	2 max		8-12	0.045 max	0.03 max	1 max	bal Fe				
ASTM A314	304L	Bil	0.03 max	18-20	2 max		8-12	0.045 max	0.03 max	1 max	bal Fe				
ASTM A336/A336M(98)	F304L	Frg, press/high-temp	0.035 max	18.0-20.0	2.00 max		8.00-13.00	0.040 max	0.030 max	1.00 max	bal Fe	450	170	30	90 HRB
ASTM A403	304L	Pipe	0.03 max	18-20	2 max		8-12	0.045 max	0.03 max	1 max	bal Fe				
ASTM A473	304L	Frg, Heat res SHT	0.030 max	18.00-20.00	2.00 max		8.00-12.00	0.045 max	0.030 max	1.00 max	bal Fe	515	205	40	
ASTM A478	304L	Wir CD 0.76<t<=3.18mm	0.03 max	18.00-20.00	2.00 max		8.00-12.00	0.045 max	0.030 max	1.00 max	bal Fe	830-1030		15	
ASTM A479	304L	Bar	0.03 max	18-20	2 max		8-12	0.045 max	0.03 max	1 max	bal Fe				
ASTM A511(96)	MT304L	Smls mech tub, Ann	0.035 max	18.0-20.0	2.00 max		8.0-11.0	0.040 max	0.030 max	1.00 max	bal Fe	517	207	35	192 HB
ASTM A554(94)	MT304L	Weld mech tub, Rnd Ann, Low-C	0.035 max	18.0-20.0	2.00 max		8.0-13.0	0.040 max	0.030 max	1.00 max	bal Fe	483	172	35	
ASTM A580/A580M(98)	304L	Wir, Ann	0.030 max	18.0-20.0	2.00 max		8.0-12.0	0.045 max	0.030 max	1.00 max	N <=0.10; bal Fe	485	170	35	
ASTM A632(90)	TP304L	Smls Weld tub, Low-C	0.04 max	18.0-20.0	2.00 max		8.0-13.0	0.040 max	0.030 max	0.75 max	bal Fe	485	170	35	
ASTM A666(96)	304L	Sh Strp Plt Bar, Ann	0.030 max	18.00-20.00	2.00 max		8.00-12.00	0.045 max	0.030 max	0.75 max	N <=0.10; bal Fe	485	170	40	201 HB
ASTM A688/A688M(96)	TP304L	Weld Feedwater Heater Tub, Low-C	0.035 max	18.00-20.00	2.00 max		8.00-13.00	0.040 max	0.03 max	0.75 max	bal Fe	485	175	35	
ASTM A774/A774M(98)	TP304L	Fittings; Corr service at low/mod-temp, Low-C	0.030 max	18.0-20.0	2.00 max		8.0-12.0	0.045 max	0.030 max	1.00 max	N <=0.10; bal Fe	485-655	170	40.0	183 HB
ASTM A778(90)	TP304L	Low-C	0.030 max	18.00-20.00	2.00 max		8.00-13.00	0.045 max	0.030 max	1.00 max	N <=0.10; bal Fe	485	170	40	
ASTM A793(96)	304L	Rolled floor plt	0.030 max	18.00-20.00	2.00 max		8.00-12.00	0.045 max	0.030 max	0.75 max	N <=0.10; bal Fe				
ASTM A813/A813M(95)	TP304L	Weld Pipe, Low-C	0.035 max	18.0-20.0	2.00 max		8.00-13.00	0.045 max	0.030 max	0.75 max	B <=3; bal Fe	485	170		
ASTM A814/A814M(96)	TP304L	CW Weld Pipe, Low-C	0.035 max	18.0-20.0	2.00 max		8.00-13.0	0.045 max	0.030 max	0.75 max	bal Fe	485	170		
ASTM A851(96)	TP304L	HFIW UnAnn Condenser Tub; Low-C	0.035 max	18.0-20.0	2.00 max		8.00-13.00	0.040 max	0.030 max	0.75 max	bal Fe	485	170	35	
ASTM A943/A943M(95)	TP304L	Spray formed smls pipe	0.030 max	18.0-20.0	2.00 max		8.00-13.00	0.040 max	0.030 max	0.75 max	bal Fe	485	170		
ASTM A965/965M(97)	F304L	Frg for press high-temp parts	0.035 max	18.0-20.0	2.00 max		8.0-13.0	0.040 max	0.030 max	1.00 max	bal Fe	450	170	30	

UNS numbers and US grades are provided as a means of cross referencing chemically similar alloys. Exchangability is only possible after independent examination of specifications. Tensile properties are minimum or typical as specified. UTS and YS as MPa. El as %. See Appendix for list of abbreviations used in Notes. * indicates obsolete material.

Specification	Designation	Notes	C	Cr	Mn	Mo	Ni	P	S	Si	Other	UTS	YS	El	Hard

Stainless Steel, Austenitic, 304L (Continued from previous page)

USA

Specification	Designation	Notes	C	Cr	Mn	Mo	Ni	P	S	Si	Other	UTS	YS	El	Hard
ASTM A988(98)	S30403	HIP Flanges, Fittings, Valves/parts; Heat res; Low-C	0.035 max	18.0-20.0	2.00 max		8.0-13.0	0.045 max	0.030 max	1.00 max	N <=0.10; bal Fe	485	170	30	
FED QQ-S-763F(96)	304L*	Obs; Bar Wir Shp Frg; Corr res	0.03 max	18-20	2 max		8-12	0.045 max	0.03 max	1 max	bal Fe				
FED QQ-S-766D(93)	304L*	Obs; see ASTM A240, A666, A693; Sh Strp Plt	0.03 max	18-20	2 max		8-12	0.045 max	0.03 max	1 max	bal Fe				
MIL-P-1144D	304L*	Obs; Sh Strp Plt Bar Bil Tub; SA	0.03 max	18-20	2 max		8-12	0.045 max	0.03 max	1 max	bal Fe				
MIL-S-23195(A)(65)	304L	Corr res; Bar frg; Rev E controlled distribution	0.03 max	18.00-20.00	2.00 max		8.00-12.00	0.035 max	0.030 max	1.00 max	Co <=0.10; Nb+Ta 1.00 max; bal Fe	483	172-345	40	
MIL-S-23196	304L	Controlled distribution													
MIL-S-27419(USAF)(68)	304L*	Obs; Corr res ann bil	0.03 max	18-20	2 max		8-12	0.045 max	0.03 max	1 max	bal Fe				
MIL-S-4043B(87)	304L*	Obs for new design; Plt Sh Strp; Corr res	0.03 max	18-20	2 max		8-12	0.045 max	0.03 max	1 max	bal Fe				
MIL-S-862B	304L*	Obs	0.03 max	18-20	2 max		8-12	0.045 max	0.03 max	1 max	bal Fe				
MIL-T-8606C(79)	304L*	Obs for new design; Tube; Corr res	0.03 max	18-20	2 max		8-12	0.045 max	0.03 max	1 max	bal Fe				
MIL-T-8973(69)	304L*	Obs for new design; Tube; Corr/Heat res; aerospace	0.03 max	18.0-20.0	2.00 max	0.50 max	8.0-11.0	0.040 max	0.030 max	1.00 max	Cu <=0.50; bal Fe				
SAE J405(98)	S30403	Low-C	0.030 max	18.00-20.00	2.00 max		8.00-12.00	0.045 max	0.030 max	0.75 max	N <=0.10; bal Fe				
SAE J467(68)	304L		0.02	19.0	1.00		10.0			0.40	bal Fe	552	207	60	80 HRB

Stainless Steel, Austenitic, 304LH

India

Specification	Designation	Notes	C	Cr	Mn	Mo	Ni	P	S	Si	Other	UTS	YS	El	Hard
IS 1570/5(85)	X02Cr19Ni10	Sh Plt Strp Bar Flat Band	0.03 max	17.5-20	2 max	0.15 max	8-12	0.045 max	0.03 max	1 max	Co <=0.1; Cu <=0.3; Pb <=0.15; W <=0.1; bal Fe	485	170	40	183 HB max

USA

Specification	Designation	Notes	C	Cr	Mn	Mo	Ni	P	S	Si	Other	UTS	YS	El	Hard
	UNS S30454	N-bearing	0.03 max	18.00-20.00	2.00 max		8.00-12.00	0.045 max	0.030 max	1.00 max	N 0.16-0.30; bal Fe				

Stainless Steel, Austenitic, 304LN

Austria

Specification	Designation	Notes	C	Cr	Mn	Mo	Ni	P	S	Si	Other	UTS	YS	El	Hard
ONORM M3121	X2CrNiN1810KKW	Bar Bil	0.03 max	17-19	2 max		8.5-11.5	0.045 max	0.030 max	1 max	N 0.12-0.22; bal Fe				

China

Specification	Designation	Notes	C	Cr	Mn	Mo	Ni	P	S	Si	Other	UTS	YS	El	Hard
GB 1220(92)	00Cr18Ni10N	Bar SA	0.030 max	17.00-19.00	2.00 max		8.50-11.50	0.035 max	0.030 max	1.00 max	N 0.12-0.22; bal Fe	550	245	40	
GB 3280(92)	00Cr18Ni10N	Sh Plt, CR SA	0.030 max	17.00-19.00	2.00 max		8.50-11.50	0.035 max	0.030 max	1.00 max	N 0.12-0.22; bal Fe	550	245	40	
GB 4237(92)	00Cr18Ni10N	Sh Plt, HR SA	0.030 max	17.00-19.00	2.00 max		8.50-11.50	0.035 max	0.030 max	1.00 max	N 0.12-0.22; bal Fe	550	245	40	
GB 4239(91)	00Cr18Ni10N	Sh Strp, CR SA	0.030 max	17.00-19.00	2.00 max		8.50-11.50	0.035 max	0.030 max	1.00 max	N 0.12-0.22; bal Fe	550	245	40	

Europe

Specification	Designation	Notes	C	Cr	Mn	Mo	Ni	P	S	Si	Other	UTS	YS	El	Hard
EN 10088/2(95)	1.4311	Strp, CR Corr res; t<=6mm, Ann	0.030 max	17.00-19.50	2.00 max		8.50-11.50	0.045 max	0.015 max	1.00 max	N 0.12-0.22; bal Fe	550-750	290	40	
EN 10088/2(95)	X2CrNiN18-10	Strp, CR Corr res; t<=6mm, Ann	0.030 max	17.00-19.50	2.00 max		8.50-11.50	0.045 max	0.015 max	1.00 max	N 0.12-0.22; bal Fe	550-750	290	40	
EN 10088/3(95)	1.4311	Bar Rod Sect, t<=160mm, Ann	0.030 max	17.00-19.50	2.00 max		8.50-11.50	0.045 max	0.030 max	1.00 max	N 0.12-0.22; bal Fe	550-760	270	40	230 HB
EN 10088/3(95)	1.4311	Bar Rod Sect, t<=160mm, Ann	0.030 max	17.00-19.50	2.00 max		8.50-11.50	0.045 max	0.030 max	1.00 max	N 0.12-0.22; bal Fe	550-760	270	30	230 HB
EN 10088/3(95)	1.4560	Bar Rod Sect, Corr res; t<=160mm, Ann	0.035 max	18.00-19.00	1.5-2.00		8.00-9.00	0.045 max	0.015 max	1.00 max	Co <=0.035; Cu 1.50-2.00; N <=0.11; bal Fe	450-650	175	45	215 HB
EN 10088/3(95)	X2CrNiN18-10	Bar Rod Sect, t<=160mm, Ann	0.030 max	17.00-19.50	2.00 max		8.50-11.50	0.045 max	0.030 max	1.00 max	N 0.12-0.22; bal Fe	550-760	270	30	230 HB
EN 10088/3(95)	X2CrNiN18-10	Bar Rod Sect, t<=160mm, Ann	0.030 max	17.00-19.50	2.00 max		8.50-11.50	0.045 max	0.030 max	1.00 max	N 0.12-0.22; bal Fe	550-760	270	40	230 HB

Specification	Designation	Notes	C	Cr	Mn	Mo	Ni	P	S	Si	Other	UTS	YS	El	Hard

Stainless Steel, Austenitic, 304LN (Continued from previous page)

Europe

Specification	Designation	Notes	C	Cr	Mn	Mo	Ni	P	S	Si	Other	UTS	YS	El	Hard
EN 10088/3(95)	X3CrNiCu19-9-2	Bar Rod Sect, Corr res; t<=160mm, Ann	0.035 max	18.00-19.00	1.5-2.00		8.00-9.00	0.045 max	0.015 max	1.00 max	Co <=0.035; Cu 1.50-2.00; N <=0.11; bal Fe	450-650	175	45	215 HB

Finland

Specification	Designation	Notes	C	Cr	Mn	Mo	Ni	P	S	Si	Other	UTS	YS	El	Hard
SFS 721(86)	X2CrNiN1810	Corr res	0.03 max	17.0-19.0	2.0 max		8.0-11.0	0.045 max	0.03 max	1.0 max	N 0.12-0.22; bal Fe				

France

Specification	Designation	Notes	C	Cr	Mn	Mo	Ni	P	S	Si	Other	UTS	YS	El	Hard
AFNOR NFA35582	Z2CN18.10AZ	Plt Bar Bil	0.03 max	17-19	2 max		9-11	0.04 max	0.03 max	1 max	N 0.1-0.2; bal Fe				
AFNOR NFA36209	Z2CN18.10AZ		0.03 max	17-20	2 max		9-12	0.04 max	0.03 max	1 max	N 0.12-0.2; bal Fe				

Germany

Specification	Designation	Notes	C	Cr	Mn	Mo	Ni	P	S	Si	Other	UTS	YS	El	Hard
DIN EN 10088(95)	WNr 1.4311	Sh Plt Strp Bar Rod	0.030 max	17.0-19.5	2.00 max		8.50-11.5	0.045 max	0.030 max	1.00 max	N 0.12-0.22; bal Fe	540-740	255	40	
DIN EN 10088(95)	X2CrNiN1810	Sh Plt Strp Bar Rod, HT	0.030 max	17.0-19.5	2.00 max		8.50-11.5	0.045 max	0.030 max	1.00 max	N 0.12-0.22; bal Fe	540-740	255	40	

International

Specification	Designation	Notes	C	Cr	Mn	Mo	Ni	P	S	Si	Other	UTS	YS	El	Hard
ISO 683-13(74)	10N*	ST	0.03 max	17-19	2 max		8.5-11.5	0.045 max	0.03 max	1 max	N 0.12-0.22; bal Fe				

Italy

Specification	Designation	Notes	C	Cr	Mn	Mo	Ni	P	S	Si	Other	UTS	YS	El	Hard
UNI 7500(75)	X2CrNiN1811	Press ves	0.03 max	17-19	2 max		9-12	0.045 max	0.03 max	1 max	N 0.12-0.25; bal Fe				

Japan

Specification	Designation	Notes	C	Cr	Mn	Mo	Ni	P	S	Si	Other	UTS	YS	El	Hard
JIS G3214(91)	SUSF304LN	Frg press ves 130-200mm diam	0.030 max	18.00-20.00	2.00 max		8.00-11.00	0.040 max	0.030 max	1.00 max	N 0.10-0.16; bal Fe	480	205	29	187 HB max
JIS G4303(91)	SUS304LN	Bar; SA; <=180mm diam	0.030 max	17.00-19.00	2.00 max		8.50-11.50	0.045 max	0.030 max	1.00 max	N 0.12-0.22; bal Fe	550	245	40	217 HB
JIS G4304(91)	SUS304LN	Plt Sh, HR	0.030 max	17.00-19.00	2.00 max		8.50-11.50	0.045 max	0.030 max	1.00 max	N 0.12-0.22; bal Fe	550	245	40	95 HRB max
JIS G4305(91)	SUS304LN	Plt Sh, CR	0.030 max	17.00-19.00	2.00 max		8.50-11.50	0.045 max	0.030 max	1.00 max	N 0.12-0.22; bal Fe	550	245	40	217 HB max
JIS G4306	SUS304LN	Strp HR	0.03 max	17-19	2 max		8.5-11.5	0.045 max	0.03 max	1 max	N 0.12-0.22; bal Fe				
JIS G4307	SUS304LN	Strp CR	0.03 max	17-19	2 max		8.5-11.5	0.045 max	0.03 max	1 max	N 0.12-0.22; bal Fe				

Mexico

Specification	Designation	Notes	C	Cr	Mn	Mo	Ni	P	S	Si	Other	UTS	YS	El	Hard
NMX-B-186-SCFI(94)	TP304LN	Pipe; High-temp	0.035 max	18.0-20.0	2.00 max		8.00-11.0	0.040 max	0.030 max	0.75 max	N 0.10-0.16; bal Fe	517	207		

Spain

Specification	Designation	Notes	C	Cr	Mn	Mo	Ni	P	S	Si	Other	UTS	YS	El	Hard
UNE 36016/1(89)	F.3541*	Obs; EN10088-3(96); X2CrNiN18-10	0.03 max	17.0-19.0	2.0 max	0.15 max	8.5-11.5	0.045 max	0.03 max	1.0 max	N 0.12-0.22; bal Fe				
UNE 36087/4(89)	F.3541*	see X2CrNiN18 10	0.03 max	17-19	2 max	0.15 max	8.5-11.5	0.045 max	0.03 max	1 max	N 0.12-0.22; bal Fe				
UNE 36087/4(89)	X2CrNiN18 10	Corr res; Sh Strp	0.03 max	17-19	2 max	0.15 max	8.5-11.5	0.045 max	0.03 max	1 max	N 0.12-0.22; bal Fe				

Sweden

Specification	Designation	Notes	C	Cr	Mn	Mo	Ni	P	S	Si	Other	UTS	YS	El	Hard
SS 142371	2371	Corr res	0.03 max	17-19	2 max		8-11	0.045 max	0.03 max	1 max	N 0.12-0.22; bal Fe				

UK

Specification	Designation	Notes	C	Cr	Mn	Mo	Ni	P	S	Si	Other	UTS	YS	El	Hard
BS 1501/3(73)	304S62	Press ves; Corr res	0.03 max	17.5-19.0	0.5-2.0	0.15 max	9.0-12.0	0.045 max	0.03 max	0.2-1.0	Cu <=0.3; N 0.15-0.25; Ti <=0.05; W <=0.1; bal Fe				
BS 1502(82)	304S61	Bar Shp; Press ves	0.030 max	17.0-19.0	2.00 max		8.50-11.5	0.045 max	0.030 max	1.00 max	N 0.12-0.22; W <=0.1; bal Fe				
BS 1503(89)	304S61	Frg press ves; t<=999mm; HT	0.03 max	17.0-19.0	2.00 max	0.70 max	8.50-11.5	0.045 max	0.025 max	1.00 max	B <=0.005; Cu <=0.50; N 0.12-0.22; Nb <=0.20; Ti <=0.10; W <=0.1; bal Fe	550-750	305	35	
BS 1506(90)	304S61	Bolt matl Pres/Corr res; t<=160mm; Hard	0.03 max	17.0-19.0	2.00 max		8.50-11.5	0.045 max	0.030 max	1.00 max	N 0.12-0.22; bal Fe	550	270	35	

USA

Specification	Designation	Notes	C	Cr	Mn	Mo	Ni	P	S	Si	Other	UTS	YS	El	Hard
	UNS S30453	N-bearing	0.030 max	18.00-20.00	2.00 max		8.00-12.00	0.045 max	0.030 max	1.00 max	N 0.10-0.16; bal Fe				
ASTM A182/A182M(98)	F304NL	Frg/roll pipe flange valve	0.030 max	18.0-20.0	2.00 max		8.00-10.5	0.045 max	0.030 max	0.75 max	N 0.10-0.16; bal Fe	515	205	30	

UNS numbers and US grades are provided as a means of cross referencing chemically similar alloys. Exchangability is only possible after independent examination of specifications. Tensile properties are minimum or typical as specified. UTS and YS as MPa. El as %. See Appendix for list of abbreviations used in Notes. * indicates obsolete material.

Specification	Designation	Notes	C	Cr	Mn	Mo	Ni	P	S	Si	Other	UTS	YS	El	Hard

Stainless Steel, Austenitic, 304LN (Continued from previous page)

USA

Specification	Designation	Notes	C	Cr	Mn	Mo	Ni	P	S	Si	Other	UTS	YS	El	Hard
ASTM A193/A193M(98)	B8LN	Bolt, high-temp, HT	0.030 max	18.0-20.0	2.00 max		8.0-11.00	0.045 max	0.030 max	1.00 max	N 0.10-0.16; bal Fe	515	205	30	223 HB
ASTM A193/A193M(98)	B8LNA	Bolt, high-temp, HT	0.030 max	18.0-20.0	2.00 max		8.0-11.00	0.045 max	0.030 max	1.00 max	N 0.10-0.16; bal Fe	515	205	30	192 HB
ASTM A194/A194M(98)	8LN	Nuts, high-temp press	0.030 max	18.0-20.0	2.00 max		8.0-11.0	0.045 max	0.030 max	1.00 max	N 0.10-0.16; bal Fe				126-300 HB
ASTM A194/A194M(98)	8LNA	Nuts, high-temp press	0.030 max	18.0-20.0	2.00 max		8.0-11.0	0.045 max	0.030 max	1.00 max	N 0.10-0.16; bal Fe				126-192 HB
ASTM A213/A213M(95)	TP304LN	Smls tube boiler, superheater, heat exchanger	0.035 max	18.0-20.0	2.00 max		8.00-11.0	0.040 max	0.030 max	0.75 max	N 0.10-0.16; bal Fe	515	205	35	
ASTM A240/A240M(98)	S30453	N-bearing	0.030 max	18.0-20.0	2.00 max		8.0-12.0	0.045 max	0.030 max	0.75 max	N 0.10-0.16; bal Fe	515	205	40.0	201 HB
ASTM A249/249M(96)	TP304LN	Weld tube; boiler, superheater, heat exch	0.035 max	18.0-20.0	2.00 max		8.00-13.0	0.040 max	0.030 max	0.75 max	N 0.10-0.16; bal Fe	515	205	35	90 HRB
ASTM A276(98)	304LN	Bar Shp	0.03 max	18-20	2 max		8-12	0.045 max	0.03 max	1 max	N 0.1-0.16; bal Fe				
ASTM A336/A336M(98)	F304LN	Frg, press/high-temp	0.030 max	18.0-20.0	2.00 max		8.00-11.0	0.040 max	0.030 max	1.00 max	N 0.10-0.16; bal Fe	585	205	30	90 HRB
ASTM A666(96)	304LN	Sh Strp Plt Bar, Ann	0.030 max	18.00-20.00	2.00 max		8.00-10.50	0.045 max	0.030 max	0.75 max	N 0.10-0.16; bal Fe	515	205	40	217 HB
ASTM A688/A688M(96)	TP304LN	Weld Feedwater Heater Tub, N-bearing	0.035 max	18.00-20.00	2.00 max		8.00-13.00	0.040 max	0.03 max	0.75 max	N 0.10-0.16; bal Fe	515	205	35	
ASTM A813/A813M(95)	TP304LN	Weld Pipe, N-bearing	0.035 max	18.0-20.0	2.00 max		8.00-11.00	0.045 max	0.030 max	0.75 max	N 0.10-0.16; bal Fe	515	205		
ASTM A814/A814M(96)	TP304LN	CW Weld Pipe, N-bearing	0.035 max	18.0-20.0	2.00 max		8.00-11.0	0.045 max	0.030 max	0.75 max	N 0.10-0.16; bal Fe	515	205		
ASTM A943/A943M(95)	TP304LN	Spray formed smls pipe	0.030 max	18.0-20.0	2.00 max		8.00-11.00	0.040 max	0.030 max	0.75 max	N 0.10-0.16; bal Fe	515	205		
ASTM A965/965M(97)	F304LN	Frg for press high-temp parts	0.030 max	18.0-20.0	2.00 max		8.0-11.00	0.040 max	0.030 max	1.00 max	N 0.10-0.16; bal Fe	485	205	30	
ASTM A988(98)	S30453	HIP Flanges, Fittings, Valves/parts; Heat res; N-bearing	0.030 max	18.0-20.0	2.00 max		8.0-11.0	0.045 max	0.030 max	1.00 max	N 0.10-0.16; bal Fe	515	205	30	
SAE J405(98)	S30453	N-bearing	0.030 max	18.00-20.00	2.00 max		8.00-12.00	0.045 max	0.030 max	0.75 max	N 0.10-0.16; bal Fe				

Stainless Steel, Austenitic, 304N

China

Specification	Designation	Notes	C	Cr	Mn	Mo	Ni	P	S	Si	Other	UTS	YS	El	Hard
GB 1220(92)	0Cr19Ni9N	Bar SA	0.08 max	18.00-20.00	2.00 max		7.00-10.50	0.035 max	0.030 max	1.00 max	N 0.10-0.25; bal Fe	550	275	35	
GB 12770(91)	0Cr19Ni9N	Weld tube; SA	0.08 max	18.00-20.00	2.00 max		7.00-10.50	0.035 max	0.030 max	1.00 max	N 0.10-0.25; bal Fe	520	210	35	
GB 3280(92)	0Cr19Ni9N	Sh Plt, CR SA	0.08 max	18.00-20.00	2.00 max		7.00-10.50	0.035 max	0.030 max	1.00 max	N 0.10-0.25; bal Fe	550	275	35	
GB 4237(92)	0Cr19Ni9N	Sh Plt, HR SA	0.08 max	18.00-20.00	2.00 max		7.00-10.50	0.035 max	0.030 max	1.00 max	N 0.10-0.25; bal Fe	550	275	35	
GB 4239(91)	0Cr19Ni9N	Sh Strp, CR SA	0.08 max	18.00-20.00	2.00 max		7.00-10.50	0.035 max	0.030 max	1.00 max	N 0.10-0.25; bal Fe	550	275	35	
GB 4240(93)	0Cr19Ni9N	Wir Ann 0.6-1mm diam	0.08 max	18.00-20.00	2.00 max		7.00-10.50	0.035 max	0.030 max	1.00 max	N 0.10-0.25; bal Fe	540-880		25	

France

Specification	Designation	Notes	C	Cr	Mn	Mo	Ni	P	S	Si	Other	UTS	YS	El	Hard
AFNOR NFA36209	Z5CN18.09Az		0.06 max	17-20	2 max		8-11	0.04 max	0.03 max	1 max	N 0.12-0.2; bal Fe				

Germany

Specification	Designation	Notes	C	Cr	Mn	Mo	Ni	P	S	Si	Other	UTS	YS	El	Hard
TGL 7143	X5CrNiN19.7*	Obs	0.06 max	18-20	2.5 max		6-8	0.045 max	0.03 max	0.8 max	N 0.15-0.25; bal Fe				

Italy

Specification	Designation	Notes	C	Cr	Mn	Mo	Ni	P	S	Si	Other	UTS	YS	El	Hard
UNI 7500(75)	X5CrNiN1810	Press ves	0.06 max	17-19	2 max		8-11	0.045 max	0.03 max	1 max	N 0.12-0.25; bal Fe				

Japan

Specification	Designation	Notes	C	Cr	Mn	Mo	Ni	P	S	Si	Other	UTS	YS	El	Hard
JIS G3214(91)	SUSF304N	Frg press ves 130-200mm diam	0.08 max	18.00-20.00	2.00 max		8.00-11.00	0.040 max	0.030 max	0.75 max	N 0.10-0.16; bal Fe	550	240	24	217 HB max
JIS G4303(91)	SUS304N1	Bar; SA; <=180mm diam	0.08 max	18.00-20.00	2.50 max		7.00-10.50	0.045 max	0.030 max	1.00 max	N 0.10-0.25; bal Fe	550	275	35	217 HB
JIS G4303(91)	SUS304N2	Bar; SA; <=180mm diam	0.08 max	18.00-20.00	2.50 max		7.00-10.50	0.045 max	0.030 max	1.00 max	N 0.15-0.30; Nb <=0.15; bal Fe	690	345	35	250 HB

UNS numbers and US grades are provided as a means of cross referencing chemically similar alloys. Exchangability is only possible after independent examination of specifications. Tensile properties are minimum or typical as specified. UTS and YS as MPa. El as %. See Appendix for list of abbreviations used in Notes. * indicates obsolete material.

Specification	Designation	Notes	C	Cr	Mn	Mo	Ni	P	S	Si	Other	UTS	YS	El	Hard
Stainless Steel, Austenitic, 304N (Continued from previous page)															
Japan															
JIS G4304(91)	SUS304N1	Plt Sh, HR	0.08 max	18.00-20.00	2.50 max		7.0-10.50	0.045 max	0.030 max	1.00 max	N 0.10-0.25; bal Fe	550	275	35	95 HRB max
JIS G4305(91)	SUS304N1	Plt Sh, CR	0.08 max	18.00-20.00	2.50 max		7.00-10.50	0.045 max	0.030 max	1.00 max	N 0.10-0.25; bal Fe	550	275	35	217 HB max
JIS G4305(91)	SUS304N2		0.08 max	18.00-20.00	2.50 max		7.50-10.50	0.045 max	0.030 max	1.00 max	N 0.15-0.30; Nb <=0.15; bal Fe	690	345	35	
JIS G4306	SUS304N1	Strp HR	0.08 max	18-20	2.5 max		7-10.5	0.045 max	0.03 max	1 max	N 0.1-0.25; bal Fe				
JIS G4307	SUS304N1	Strp CR	0.08 max	18-20	2.5 max		7-10.5	0.045 max	0.03 max	1 max	N 0.1-0.25; bal Fe				
JIS G4308(98)	SUS304N1	Wir rod	0.08 max	18.00-20.00	2.50 max		7.00-10.50	0.045 max	0.030 max	1.00 max	N 0.10-0.25; bal Fe				
JIS G4309	SUS304NI	Wir	0.08 max	18.00-20.00	2.50 max		7.00-10.50	0.045 max	0.030 max	1.00 max	N 0.10-0.25; bal Fe				
Mexico															
NMX-B-186-SCFI(94)	TP304N	Pipe; High-temp	0.08 max	18.0-20.0	2.00 max		8.00-11.0	0.040 max	0.030 max	0.75 max	N 0.10-0.16; bal Fe				
UK															
BS 1501/3(73)	304S65	Press ves; Corr res	0.06 max	17.5-19.0	0.5-2.0	0.15 max	8.0-11.0	0.045 max	0.03	0.2-1.0	Cu <=0.3; N 0.15-0.25; Ti <=0.05; W <=0.1; bal Fe				
BS 1502(82)	304S71	Bar Shp; Press ves	0.07 max	17.0-19.0	2.00 max		8.00-11.0	0.045 max	0.030 max	1.00 max	N 0.12-0.22; W <=0.1; bal Fe				
BS 1506(90)	304S71	Bolt matl Pres/Corr res; t<=160mm; Hard	0.07 max	17.0-19.0	2.00 max		8.00-11.0	0.045 max	0.030 max	1.00 max	N 0.12-0.22; bal Fe	550	270	35	
USA															
	AISI 304N	Tube	0.08 max	18.00-20.00	2.00 max		8.00-10.50	0.045 max	0.030 max	1.00 max	N 0.10-0.16; bal Fe				
	UNS S30451	N-bearing	0.08 max	18.00-20.00	2.00 max		8.00-10.50	0.045 max	0.030 max	1.00 max	N 0.10-0.16; bal Fe				
AMS 5511		Sh Strp Plt Bar	0.08 max	18-20	2 max		8-10.5	0.045 max	0.03 max	1 max	N 0.1-0.16; bal Fe				
AMS 5647		Sh Strp Plt Bar	0.08 max	18-20	2 max		8-10.5	0.045 max	0.03 max	1 max	N 0.1-0.16; bal Fe				
ASME SA182	304N	Refer to ASTM A182/A182M													
ASME SA213	304N	Refer to ASTM A213/A213M(95)													
ASME SA240	304N	Refer to ASTM A240(95)													
ASME SA249	304N	Refer to ASTM A249/A249M(96)													
ASME SA312	304N	Refer to ASTM A312/A312M(95)													
ASME SA358	304N	Refer to ASTM A358/A358M(95)													
ASME SA376	304N	Refer to ASTM A376/A376M(93)													
ASME SA430	304N	Refer to ASTM A430/A430M(91)													
ASME SA479	304N	Refer to ASTM A479/A479M(95)													
ASTM A182	304N*	Bar Frg	0.08 max	18-20	2 max		8-10.5	0.045 max	0.03 max	1 max	N 0.1-0.16; bal Fe				
ASTM A182/A182M(98)	F304N	Frg/roll pipe flange valve	0.08 max	18.0-20.0	2.00 max		8.00-10.5	0.045 max	0.030 max	0.75 max	N 0.10-0.16; bal Fe	550	240	30	
ASTM A193/A193M(98)	B8N	Bolt, high-temp, HT	0.08 max	18.0-20.0	2.00 max		8.0-11.0	0.045 max	0.030 max	1.00 max	N 0.10-0.16; bal Fe	550	240	30	223 HB
ASTM A193/A193M(98)	B8NA	Bolt, high-temp, HT	0.08 max	18.0-20.0	2.00 max		8.0-11.0	0.045 max	0.030 max	1.00 max	N 0.10-0.16; bal Fe	515	205	30	192 HB
ASTM A194/A194M(98)	8N	Nuts, high-temp press	0.08 max	18.0-20.0	2.00 max		8.0-11.0	0.045 max	0.030 max	1.00 max	N 0.10-0.16; bal Fe				126-300 HB
ASTM A194/A194M(98)	8NA	Nuts, high-temp press	0.08 max	18.0-20.0	2.00 max		8.0-11.0	0.045 max	0.030 max	1.00 max	N 0.10-0.16; bal Fe				126-192 HB
ASTM A213	304N*	Tube	0.08 max	18-20	2 max		8-10.5	0.045 max	0.03 max	1 max	N 0.1-0.16; bal Fe				

UNS numbers and US grades are provided as a means of cross referencing chemically similar alloys. Exchangability is only possible after independent examination of specifications. Tensile properties are minimum or typical as specified. UTS and YS as MPa. El as %. See Appendix for list of abbreviations used in Notes. * indicates obsolete material.

Specification	Designation	Notes	C	Cr	Mn	Mo	Ni	P	S	Si	Other	UTS	YS	El	Hard

Stainless Steel, Austenitic, 304N (Continued from previous page)

USA

Specification	Designation	Notes	C	Cr	Mn	Mo	Ni	P	S	Si	Other	UTS	YS	El	Hard
ASTM A213/A213M(95)	TP304N	Smls tube boiler, superheater, heat exchanger	0.08 max	18.0-20.0	2.00 max		8.00-11.0	0.040 max	0.030 max	0.75 max	N 0.10-0.16; bal Fe	550	240	35	
ASTM A240/A240M(98)	S30451	N-bearing	0.08 max	18.0-20.0	2.00 max		8.0-10.5	0.045 max	0.030 max	0.75 max	N 0.10-0.16; bal Fe	550	240	30.0	201 HB
ASTM A249/249M(96)	TP304N	Weld tube; boiler, superheater, heat exch	0.08 max	18.0-20.0	2.00 max		8.00-11.0	0.040 max	0.030 max	0.75 max	N 0.10-0.16; bal Fe	550	240	35	90 HRB
ASTM A312/A312M(95)	304N	Smls Weld, Pipe	0.08 max	18-20	2 max		8-10.5	0.045 max	0.03 max	1 max	N 0.1-0.16; bal Fe				
ASTM A336/A336M(98)	F304N	Frg, press/high-temp	0.08 max	18.0-20.0	2.00 max		8.00-11.0	0.030 max	0.030 max	0.75 max	N 0.10-0.16; bal Fe	550	240	25	90 HRB
ASTM A358/A358M(95)	304N	Elect fusion weld pipe, high-temp	0.08 max	18-20	2 max		8-10.5	0.045 max	0.03 max	1 max	N 0.1-0.16; bal Fe				
ASTM A376	304N	Smls pipe	0.08 max	18-20	2 max		8-10.5	0.045 max	0.03 max	1 max	N 0.1-0.16; bal Fe				
ASTM A403	304N	Pipe	0.08 max	18-20	2 max		8-10.5	0.045 max	0.03 max	1 max	N 0.1-0.16; bal Fe				
ASTM A430	304N*	Obs 1995; see A312; Pipe	0.08 max	18-20	2 max		8-10.5	0.045 max	0.03 max	1 max	N 0.1-0.16; bal Fe				
ASTM A479	304N	Wir Bar Shp	0.08 max	18-20	2 max		8-10.5	0.045 max	0.03 max	1 max	N 0.1-0.16; bal Fe				
ASTM A666(96)	304N	Sh Strp Plt Bar, Ann	0.08 max	18.00-20.00	2.00 max		8.00-12.00	0.045 max	0.030 max	0.75 max	N 0.10-0.16; bal Fe	550	240	30	217 HB
ASTM A688/A688M(96)	TP304N	Weld Feedwater Heater Tub, N-bearing	0.08 max	18.00-20.00	2.00 max		8.00-11.0	0.040 max	0.03 max	0.75 max	N 0.10-0.16; bal Fe	550	240	35	
ASTM A813/A813M(95)	TP304N	Weld Pipe, N-bearing	0.08 max	18.0-20.0	2.00 max		8.00-11.00	0.045 max	0.030 max	0.75 max	B <=4; N 0.10-0.16; bal Fe	550	240		
ASTM A814/A814M(96)	TP304N	CW Weld Pipe, N-bearing	0.08 max	18.0-20.0	2.00 max		8.00-11.0	0.045 max	0.030 max	0.75 max	N 0.10-0.16; bal Fe	550	240		
ASTM A943/A943M(95)	TP304N	Spray formed smls pipe	0.08 max	18.0-20.0	2.00 max		8.00-11.00	0.040 max	0.030 max	0.75 max	N 0.10-0.16; bal Fe	550	240		
ASTM A965/965M(97)	F304N	Frg for press high-temp parts	0.08 max	18.0-20.0	2.00 max		8.0-11.0	0.030 max	0.030 max	0.75 max	N 0.10-0.16; bal Fe	550	240	25	
ASTM A988(98)	S30451	HIP Flanges, Fittings, Valves/parts; Heat res; N-bearing	0.08 max	18.0-20.0	2.00 max		8.0-11.0	0.045 max	0.030 max	1.00 max	N 0.10-0.16; bal Fe	550	240	30	
SAE J405(98)	S30451	N-bearing	0.08 max	18.00-20.00	2.00 max		8.00-10.50	0.045 max	0.030 max	0.75 max	N 0.10-0.16; bal Fe				

Stainless Steel, Austenitic, 305

Canada

Specification	Designation	Notes	C	Cr	Mn	Mo	Ni	P	S	Si	Other	UTS	YS	El	Hard
CSA G110.3	305	Bar Bil	0.12 max	17-19	2 max		10-13	0.05 max	0.03 max	1 max	bal Fe				
CSA G110.6	305	Sh Strp Plt, HR, Ann, or HR, Q/A, Tmp	0.12 max	17-19	2 max		10-13	0.05 max	0.03 max	1 max	bal Fe				
CSA G110.9	305	Sh Strp Plt, HR, CR, Ann, Q/A, Tmp	0.12 max	17-19	2 max		10-13	0.05 max	0.03 max	1 max	bal Fe				

China

Specification	Designation	Notes	C	Cr	Mn	Mo	Ni	P	S	Si	Other	UTS	YS	El	Hard
GB 1220(92)	1Cr18Ni12	Bar SA	0.12 max	17.00-19.00	2.00 max		10.50-13.00	0.035 max	0.030 max	1.00 max	bal Fe	480	177	40	
GB 3280(92)	1Cr18Ni12	Sh Plt, CR SA	0.12 max	17.00-19.00	2.00 max		10.50-13.00	0.035 max	0.030 max	1.00 max	bal Fe	480	177	40	
GB 4232(93)	ML1Cr18Ni12	Wir Ann 1-3mm diam	0.12 max	17.00-19.00	2.00 max		10.50-13.00	0.035 max	0.030 max	1.00 max	bal Fe	490-640		30	
GB 4237(92)	1Cr18Ni12	Sh Plt, HR SA	0.12 max	17.00-19.00	2.00 max		10.50-13.00	0.035 max	0.030 max	1.00 max	bal Fe	480	177	40	
GB 4239(91)	1Cr18Ni12	Sh Strp, CR SA	0.12 max	17.00-19.00	2.00 max		10.50-13.00	0.035 max	0.030 max	1.00 max	bal Fe	480	175	40	
GB 4240(93)	1Cr18Ni12(-Q,-R)	Wir Ann 1-3mm diam, Q-TLC; R-As Ann	0.12 max	17.00-19.00	2.00 max		10.50-13.00	0.035 max	0.030 max	1.00 max	bal Fe	490-830		25	

Europe

Specification	Designation	Notes	C	Cr	Mn	Mo	Ni	P	S	Si	Other	UTS	YS	El	Hard
EN 10088/2(95)	1.4303	Strp, CR Corr res; t<=6mm, Ann	0.06 max	17.00-19.00	2.00 max		11.00-13.00	0.045 max	0.030 max	1.00 max	N <=0.11; bal Fe	500-650	220	45	
EN 10088/2(95)	X4CrNi18-12	Strp, CR Corr res; t<=6mm, Ann	0.06 max	17.00-19.00	2.00 max		11.00-13.00	0.045 max	0.030 max	1.00 max	N <=0.11; bal Fe	500-650	220	45	
EN 10088/3(95)	1.4303	Bar Rod Sect; Corr res; 160<t<=250mm, Ann	0.6 max	17.00-19.00	2.00 max		11.00-13.00	0.045 max	0.030 max	1.00 max	N <=0.11; bal Fe	500-700	190	35	215 HB

UNS numbers and US grades are provided as a means of cross referencing chemically similar alloys. Exchangability is only possible after independent examination of specifications. Tensile properties are minimum or typical as specified. UTS and YS as MPa. El as %. See Appendix for list of abbreviations used in Notes. * indicates obsolete material.

Stainless Steel, Austenitic, 305 (Continued from previous page)

Specification	Designation	Notes	C	Cr	Mn	Mo	Ni	P	S	Si	Other	UTS	YS	El	Hard
Europe															
EN 10088/3(95)	1.4303	Bar Rod Sect, t<=160mm, Ann	0.6 max	17.00-19.00	2.00 max		11.00-13.00	0.045 max	0.030 max	1.00 max	N <=0.11; bal Fe	500-700	190	45	215 HB
EN 10088/3(95)	X4CrNi18-12	Bar Rod Sect; Corr res; 160<t<=250mm, Ann	0.6 max	17.00-19.00	2.00 max		11.00-13.00	0.045 max	0.030 max	1.00 max	N <=0.11; bal Fe	500-700	190	35	215 HB
EN 10088/3(95)	X4CrNi18-12	Bar Rod Sect, t<=160mm, Ann	0.6 max	17.00-19.00	2.00 max		11.00-13.00	0.045 max	0.030 max	1.00 max	N <=0.11; bal Fe	500-700	190	45	215 HB
France															
AFNOR NFA35575	Z8CN18.12	Wir rod, 60% Hard	0.1 max	17-19	2 max		11-13	0.04 max	0.03 max	1 max	bal Fe				
AFNOR NFA35577	Z8CN18.12	Bar Wir rod	0.1 max	17-19	2 max		11-13	0.04 max	0.03 max	1 max	bal Fe				
AFNOR NFA36209	Z8CN18.12	Plt	0.1 max	17-19	2 max		11-13	0.04 max	0.03 max	1 max	bal Fe				
International															
ISO 4954(93)	X5CrNi1812E	Wir rod Bar, CH ext	0.07 max	17.0-19.0	2.00 max		11.0-13.0	0.045 max	0.030 max	1.00 max	bal Fe	650			
ISO 683-13(74)	13*	ST	0.1 max	17-19	2 max		11-13	0.045 max	0.03 max	1 max	bal Fe				
Italy															
UNI 6901(71)	X8CrNi1910	Corr res	0.04-0.1	18-20	2 max		8-12	0.04 max	0.03 max	0.75 max	bal Fe				
UNI 8317(81)	X8CrNi1910	Sh Strp Plt	0.04-0.1	18-20	2 max		8-12	0.04 max	0.03 max	0.75 max	bal Fe				
Japan															
JIS G4303(91)	SUS305	Bar; SA; <=180mm diam	0.12 max	17.00-19.00	2.00 max		10.50-13.00	0.045 max	0.030 max	1.00 max	bal Fe	480	175	40	187 HB
JIS G4304(91)	SUS305	Sh Plt, HR SA	0.12 max	17.00-19.00	2.00 max		10.50-13.00	0.045 max	0.030 max	1.00 max	bal Fe	480	175	40	90 HRB max
JIS G4305(91)	SUS305	Sh Plt, CR SA	0.12 max	17.00-19.00	2.00 max		10.50-13.00	0.045 max	0.030 max	1.00 max	bal Fe	480	175	40	187 HB max
JIS G4306	SUS305	Strp HR SA	0.12 max	17-19	2 max		10.5-13	0.04 max	0.03 max	1 max	bal Fe				
JIS G4307	SUS305	Strp CR SA	0.12 max	17-19	2 max		10.5-13	0.04 max	0.03 max	1 max	bal Fe				
JIS G4308	SUS305J1	Wir rod	0.08 max	16.50-19.00	2.00 max		11.00-13.50	0.045 max	0.030 max	1.00 max	bal Fe				
JIS G4308(98)	SUS305	Wir rod	0.12 max	17.00-19.00	2.00 max		10.50-13.00	0.045 max	0.030 max	1.00 max	bal Fe				
JIS G4309	SUS305	Wir	0.12 max	17.00-19.00	2.00 max		10.50-13.00	0.045 max	0.030 max	1.00 max	bal Fe				
JIS G4309	SUS305J1	Wir	0.08 max	16.50-19.00	2.00 max		11.00-13.50	0.045 max	0.030 max	1.00 max	bal Fe				
JIS G4315	SUS305	Wir rod	0.12 max	17.00-19.00	2.00 max		10.50-13.00	0.045 max	0.030 max	1.00 max	bal Fe				
JIS G4315	SUS305J1	Wir rod	0.08 max	16.50-19.00	2.00 max		11.00-13.50	0.045 max	0.030 max	1.00 max	bal Fe				
Mexico															
DGN B-83	SUS305*	Obs; Bar, HR, CR	0.12 max	17-19	2 max		10.5-13	0.05 max	0.03 max	1 max	bal Fe				
NMX-B-171(91)	MT305	mech tub	0.12	17.0-19.0	2.00 max		10.0-13.0	0.040 max	0.030 max	1.00 max	bal Fe				
Russia															
GOST	O6Ch18N11	Sh Strp Plt Bar Wir Bil	0.06 max	17-19	2 max	0.3 max	10-12	0.035 max	0.02 max	0.8 max	Cu <=0.3; Ti <=0.2; W <=0.2; bal Fe				
GOST 5632(61)	0KH18N11	Corr res; Heat res; High-temp constr	0.06 max	17.0-19.0	1.0-2.0	0.15 max	10.0-12.0	0.025 max	0.02 max	0.8 max	Cu <=0.3; Ti <=0.05; bal Fe				
GOST 5632(72)	06Ch18N11	Corr res	0.06 max	17.0-19.0	2.0 max	0.3 max	10.0-12.0	0.035 max	0.02 max	0.8 max	Cu <=0.3; Ti <=0.2; W <=0.2; bal Fe				
Spain															
UNE 36016(75)	X8CrNi18-12	Sh Strp Plt; Corr res	0.1 max	17.0-19.0	2.0 max	0.15 max	11.0-13.0	0.045 max	0.03 max	1.0 max	bal Fe				
UNE 36016/1(89)	E-305	Corr res	0.1 max	17.0-19.0	2.0 max	0.15 max	11.0-13.0	0.045 max	0.03 max	1.0 max	N <=0.1; bal Fe				
UNE 36016/1(89)	F.3513*	Obs; E-305; Corr res	0.1 max	17.0-19.0	2.0 max	0.15 max	11.0-13.0	0.045 max	0.03 max	1.0 max	N <=0.1; bal Fe				

UNS numbers and US grades are provided as a means of cross referencing chemically similar alloys. Exchangability is only possible after independent examination of specifications. Tensile properties are minimum or typical as specified. UTS and YS as MPa. El as %. See Appendix for list of abbreviations used in Notes. * indicates obsolete material.

Specification	Designation	Notes	C	Cr	Mn	Mo	Ni	P	S	Si	Other	UTS	YS	El	Hard

Stainless Steel, Austenitic, 305 (Continued from previous page)

Spain

Specification	Designation	Notes	C	Cr	Mn	Mo	Ni	P	S	Si	Other	UTS	YS	El	Hard
UNE 36087/4(89)	F.3513*	see X8CrNi18 12	0.1 max	17-19	2 max	0.15 max	11-13	0.045 max	0.03 max	1 max	N <=0.1; bal Fe				
UNE 36087/4(89)	X8CrNi18 12	Corr res; Sh Strp	0.1 max	17-19	2 max	0.15 max	11-13	0.045 max	0.03 max	1 max	N <=0.1; bal Fe				

UK

Specification	Designation	Notes	C	Cr	Mn	Mo	Ni	P	S	Si	Other	UTS	YS	El	Hard
BS 1449/2(83)	305S19	Sh Strp, SA, 3mm diam	0.1 max	17-19	2.00 max		11-13	0.045 max	0.03 max	1 max	bal Fe				
BS 1554(90)	325S31	Wir Corr/Heat res; diam<0.49mm; Ann	0.12 max	17.0-19.0	2.00 max	0.70 max	8.00-11.0	0.060 max	0.15-0.35	1.00 max	Ti <=0.09; W <=0.1; 5xC<=Ti<=0.90; bal Fe	0-900			
BS 1554(90)	325S31	Wir Corr/Heat res; 6<diam<=13mm; Ann	0.12 max	17.0-19.0	2.00 max	0.70 max	8.00-11.0	0.060 max	0.15-0.35	1.00 max	Ti <=0.09; W <=0.1; 5xC<=Ti<=0.90; bal Fe	0-750			
BS 970/1(96)	325S31	Wrought Corr/Heat res; t<=160mm	0.12 max	17.0-19.0	2.00 max		8.00-11.0	0.045 max	0.15-0.35	1.00 max	5xC<=Ti<=0.90; bal Fe	510	200	35	183 HB max

USA

Specification	Designation	Notes	C	Cr	Mn	Mo	Ni	P	S	Si	Other	UTS	YS	El	Hard
	AISI 305	Tube	0.12 max	17.00-19.00	2.00 max		10.50-13.00	0.045 max	0.030 max	1.00 max	bal Fe				
	UNS S30500	Bare Filler Metal, Low-C, WH	0.12 max	17.00-19.00	2.00 max		10.00-13.00	0.045 max	0.030 max	1.00 max	bal Fe				
AMS 5514		Sh Strp Plt Bar Wir	0.12 max	17-19	2 max		10.5-13	0.045 max	0.03 max	1 max	bal Fe				
AMS 5685		Sh Strp Plt Bar Wir	0.12 max	17-19	2 max		10.5-13	0.045 max	0.03 max	1 max	bal Fe				
AMS 5686		Sh Strp Plt Bar Wir	0.12 max	17-19	2 max		10.5-13	0.045 max	0.03 max	1 max	bal Fe				
ASME SA193	305	Refer to ASTM A193/A193M(95)													
ASME SA240	305	Refer to ASTM A240(95)													
ASTM A167(96)	305*	Sh Strp Plt	0.12 max	17-19	2 max		10.5-13	0.045 max	0.03 max	1 max	bal Fe				
ASTM A193/A193M(98)	B8P	Bolt, high-temp, HT	0.12 max	17.0-19.0	2.00 max		11.0-13.0	0.045 max	0.030 max	1.00 max	bal Fe	515	205	30	223 HB
ASTM A193/A193M(98)	B8PA	Bolt, high-temp, HT	0.12 max	17.0-19.0	2.00 max		11.0-13.0	0.045 max	0.030 max	1.00 max	bal Fe	515	205	30	192 HB
ASTM A194/A194M(98)	8P	Nuts, high-temp press	0.08 max	17.0-19.0	2.00 max		11.0-13.0	0.045 max	0.030 max	1.00 max	bal Fe				126-300 HB
ASTM A194/A194M(98)	8PA	Nuts, high-temp press	0.08 max	17.0-19.0	2.00 max		11.0-13.0	0.045 max	0.030 max	1.00 max	bal Fe				126-192 HB
ASTM A240/A240M(98)	S30500	Bare Filler Metal, Low-C, WH	0.12 max	17.0-19.0	2.00 max		10.0-13.0	0.045 max	0.030 max	0.75 max	bal Fe	485	170	40.0	183 HB
ASTM A249/249M(96)	TP305	Weld tube; boiler, superheater, heat exch	0.12 max	17.0-19.0	2.00 max		10.0-13.0	0.045 max	0.030 max	1.00 max	bal Fe	515	205	35	90 HRB
ASTM A276(98)	305	Bar	0.12 max	17-19	2 max		10.5-13	0.045 max	0.03 max	1 max	bal Fe				
ASTM A313/A313M(95)	305	Spring wire, t<=0.25mm	0.12 max	17.00-19.00	2.00 max		10.5-13.00	0.045 max	0.030 max	1.00 max	bal Fe	1690-1895			
ASTM A314	305	Bi, Bar, for Frg	0.12 max	17-19	2 max		10.5-13	0.045 max	0.03 max	1 max	bal Fe				
ASTM A368(95)	305	Wir, Heat res	0.12 max	17.00-19.00	2.00 max		10.50-13.00	0.045 max	0.030 max	1.00 max	bal Fe				
ASTM A473	305	Frg, Heat res SHT	0.12 max	17.00-19.00	2.00 max		10.50-13.00	0.045 max	0.030 max	1.00 max	bal Fe	515	205	40	
ASTM A478	305	Wir CD 0.76<t<=3.18mm	0.12 max	17.00-19.00	2.00 max		10.5.-13.00	0.045 max	0.030 max	1.00 max	bal Fe	830-1030		15	
ASTM A492	305	Wir, 2.29<t<=2.54	0.12 max	17.00-19.00	2.00 max		10.50-13.00	0.045 max	0.030 max	1.00 max	bal Fe	1410			
ASTM A493	305	Wir rod, CHd, cold frg	0.12 max	17.0-19.0	2.00 max		10.5-13.0	0.045 max	0.030 max	1.00 max	bal Fe	585			
ASTM A511(96)	MT305	Smls mech tub, Ann	0.12 max	17.0-19.0	2.00 max		10.0-13.0	0.040 max	0.030 max	1.00 max	bal Fe	517	207	35	192 HB
ASTM A554(94)	MT305	Weld mech tub, Rnd Ann	0.12 max	17.0-19.0	2.00 max		10.0-13.0	0.040 max	0.030 max	1.00 max	bal Fe	517	207	35	
ASTM A580/A580M(98)	305	Wir, Ann	0.12 max	17.0-19.0	2.00 max		10.5-13.0	0.045 max	0.030 max	1.00 max	bal Fe	520	210	35	
FED QQ-S-763F(96)	305*	Obs; Bar Wir Shp Frg; Corr res	0.12 max	17-19	2 max		10.5-13	0.045 max	0.03 max	1 max	bal Fe				

UNS numbers and US grades are provided as a means of cross referencing chemically similar alloys. Exchangability is only possible after independent examination of specifications. Tensile properties are minimum or typical as specified. UTS and YS as MPa. El as %. See Appendix for list of abbreviations used in Notes. * indicates obsolete material.

Specification	Designation	Notes	C	Cr	Mn	Mo	Ni	P	S	Si	Other	UTS	YS	El	Hard

Stainless Steel, Austenitic, 305 (Continued from previous page)

USA

Specification	Designation	Notes	C	Cr	Mn	Mo	Ni	P	S	Si	Other	UTS	YS	El	Hard
FED QQ-S-766D(93)	305*	Obs; see ASTM A240, A666, A693; Sh Strp Plt	0.12 max	17-19	2 max		10.5-13	0.045 max	0.03 max	1 max	bal Fe				
SAE J405(98)	S30500	Bare Filler Metal, Low-C, WH	0.12 max	17.00-19.00	2.00 max		10.50-13.00	0.045 max	0.030 max	0.75 max	bal Fe				

Stainless Steel, Austenitic, 308

Canada

Specification	Designation	Notes	C	Cr	Mn	Mo	Ni	P	S	Si	Other	UTS	YS	El	Hard
CSA G110.3	308	Bar Bil	0.08 max	19-21	2 max		10-12	0.045 max	0.03 max	1 max	bal Fe				
CSA G110.6	308	Sh Strp Plt, HR/Ann or HR/Q/A/Tmp	0.08 max	19-21	2 max		10-12	0.045 max	0.03 max	1 max	bal Fe				

Germany

Specification	Designation	Notes	C	Cr	Mn	Mo	Ni	P	S	Si	Other	UTS	YS	El	Hard
DIN EN 10088(95)	WNr 1.4303	Sh Plt Strp Bar Rod, HT	0.06 max	17.0-19.0	2.00 max		11.0-13.0	0.045 max	0.030 max	1.00 max	bal Fe	500-700	190	35-45	
DIN EN 10088(95)	X4CrNi1812	SHT	0.06 max	17.0-19.0	2.00 max		11.0-13.0	0.045 max	0.030 max	1.00 max	N <=0.11; bal Fe	500-700	190	35-45	215 HB

Mexico

Specification	Designation	Notes	C	Cr	Mn	Mo	Ni	P	S	Si	Other	UTS	YS	El	Hard
DGN B-83	308*	Obs; Bar, HR, CR	0.08 max	19-21	2 max		10-12	0.05 max	0.03 max	1 max	bal Fe				

Spain

Specification	Designation	Notes	C	Cr	Mn	Mo	Ni	P	S	Si	Other	UTS	YS	El	Hard
UNE 36016(75)	F.3513*	see X8CrNi18-12	0.1 max	17.0-19.0	2.0 max	0.15 max	11.0-13.0	0.045 max	0.03 max	1.0 max	bal Fe				
UNE 36016(75)	X8CrNi18-12	Sh Strp Plt; Corr res	0.1 max	17.0-19.0	2.0 max	0.15 max	11.0-13.0	0.045 max	0.03 max	1.0 max	bal Fe				

USA

Specification	Designation	Notes	C	Cr	Mn	Mo	Ni	P	S	Si	Other	UTS	YS	El	Hard
	AISI 308	Tube	0.08 max	19.00-21.00	2.00 max		10.00-12.00	0.045 max	0.030 max	1.00 max	bal Fe				
	UNS S30800	Heat res	0.08 max	19.00-21.00	2.00 max		10.00-12.00	0.045 max	0.030 max	1.00 max	bal Fe				
ASTM A167(96)	308	Sh Strp Plt	0.08 max	19.00-21.00	2.00 max		10.00-12.00	0.045 max	0.030 max	0.75 max	bal Fe	515	205	40.0	183 HB
ASTM A276(98)	308	Bar	0.08 max	19-21	2 max		10-12	0.04 max	0.03 max	1 max	bal Fe				
ASTM A314	308	Bil	0.08 max	19-21	2 max		10-12	0.04 max	0.03 max	1 max	bal Fe				
ASTM A473	308	Frg, Heat res SHT	0.08 max	19.00-21.00	2.00 max		10.00-12.00	0.045 max	0.030 max	1.00 max	bal Fe	515	205	40	
ASTM A580/A580M(98)	308	Wir, Ann	0.08 max	19.0-21.0	2.00 max		10.0-12.0	0.040 max	0.030 max	1.00 max	bal Fe	520	210	35	

Stainless Steel, Austenitic, 309

Bulgaria

Specification	Designation	Notes	C	Cr	Mn	Mo	Ni	P	S	Si	Other	UTS	YS	El	Hard
BDS 6738(72)	Ch20N14S2	Corr res; bar	0.20 max	19.0-22.0	1.50 max	0.30 max	12.0-15.0	0.035 max	0.025 max	2.00-3.00	bal Fe				

Canada

Specification	Designation	Notes	C	Cr	Mn	Mo	Ni	P	S	Si	Other	UTS	YS	El	Hard
CSA G110.3	309	Bar Bil	0.2 max	22-24	2 max		12-15	0.05 max	0.03 max	1 max	bal Fe				
CSA G110.6	309	Sh Strp Plt, HR/Ann or HR/Q/A/Tmp	0.2 max	22-24	2 max		12-15	0.05 max	0.03 max	1 max	bal Fe				

China

Specification	Designation	Notes	C	Cr	Mn	Mo	Ni	P	S	Si	Other	UTS	YS	El	Hard
GB 1221(92)	2Cr23NI13	Bar SA	0.20 max	22.00-24.00	2.00 max		12.00-15.00	0.035 max	0.030 max	1.00 max	bal Fe	560	205	45	
GB 13296(91)	2Cr23NI13	Smls tube SA	0.20 max	22.00-24.00	2.00 max		12.00-15.00	0.035 max	0.030 max	1.00 max	bal Fe	520	205	35	
GB 4238(92)	2Cr23NI13	Plt, HR SA	0.20 max	22.00-24.00	2.00 max		12.00-15.00	0.035 max	0.030 max	1.00 max	bal Fe	560	205	40	

Czech Republic

Specification	Designation	Notes	C	Cr	Mn	Mo	Ni	P	S	Si	Other	UTS	YS	El	Hard
CSN 417251	17251	Heat res	0.2 max	18.0-21.0	1.5 max		8.0-11.0	0.045 max	0.03 max	2.0 max	bal Fe				

France

Specification	Designation	Notes	C	Cr	Mn	Mo	Ni	P	S	Si	Other	UTS	YS	El	Hard
AFNOR NFA35586	Z15CN24.13	Sh Strp Plt Bar Bil Tub	0.2 max	22-25	2 max		11-14	0.04 max	0.03 max	1 max	bal Fe				

Germany

Specification	Designation	Notes	C	Cr	Mn	Mo	Ni	P	S	Si	Other	UTS	YS	El	Hard
DIN EN 10095(95)	WNr 1.4828	Roll frg	0.20 max	19.0-21.0	2.00 max		11.0-13.0	0.045 max	0.030 max	1.50-2.50	N <=0.11; bal Fe	500-750	230	30	

UNS numbers and US grades are provided as a means of cross referencing chemically similar alloys. Exchangability is only possible after independent examination of specifications. Tensile properties are minimum or typical as specified. UTS and YS as MPa. El as %. See Appendix for list of abbreviations used in Notes. * indicates obsolete material.

Specification	Designation	Notes	C	Cr	Mn	Mo	Ni	P	S	Si	Other	UTS	YS	El	Hard

Stainless Steel, Austenitic, 309 (Continued from previous page)

Germany

Specification	Designation	Notes	C	Cr	Mn	Mo	Ni	P	S	Si	Other	UTS	YS	El	Hard
DIN SEW 470(76)	X15CrNiSi20-12	Heat res roll frg	0.20 max	19.0-21.0	2.00 max		11.0-13.0	0.045 max	0.030 max	1.50-2.50	bal Fe	500-750	230	30	

Hungary

Specification	Designation	Notes	C	Cr	Mn	Mo	Ni	P	S	Si	Other	UTS	YS	El	Hard
MSZ 4359(82)	H8	Heat res; 0-60mm; SA	0.2 max	17-20	2 max		8-11	0.04 max	0.03 max	0.8-2	bal Fe	550	230	30L	223 HB max

India

Specification	Designation	Notes	C	Cr	Mn	Mo	Ni	P	S	Si	Other	UTS	YS	El	Hard
IS 1570/5(85)	X15Cr24Ni13	Sh Plt Strp Bar Flat Band	0.2 max	22-25	2 max	0.15 max	11-15	0.045 max	0.03 max	1.5 max	Co <=0.1; Cu <=0.3; Pb <=0.15; W <=0.1; bal Fe	490	210	40	223 HB max

International

Specification	Designation	Notes	C	Cr	Mn	Mo	Ni	P	S	Si	Other	UTS	YS	El	Hard
ISO 4955(94)	H13	Heat res	0.2 max	19.0-21.0	2.0 max	0.15 max	11.0-13.0	0.045 max	0.03 max	1.5-2.5	Al <=0.1; Co <=0.1; Cu <=0.3; Pb <=0.15; Ti <=0.05; V <=0.1; W <=0.1; bal Fe				

Italy

Specification	Designation	Notes	C	Cr	Mn	Mo	Ni	P	S	Si	Other	UTS	YS	El	Hard
UNI 6901(71)	X16CrNi2314	Corr res	0.2 max	22-24	2 max		12-15	0.045 max	0.03 max	1 max	bal Fe				
UNI 6904(71)	X16CrNi2314	Heat res	0.2 max	22-24	2 max		12-15	0.04 max	0.03 max	0.75 max	bal Fe				
UNI 8317(81)	X16CrNi2314	Sh Strp Plt	0.2 max	22-24	2 max		12-15	0.045 max	0.03 max	1 max	bal Fe				

Japan

Specification	Designation	Notes	C	Cr	Mn	Mo	Ni	P	S	Si	Other	UTS	YS	El	Hard
JIS G4311	SUH309	Heat res Bar <=180mm	0.20 max	22.00-24.00	2.00 max		12.00-15.00	0.040 max	0.030 max	1.00 max	bal Fe	560	205	45	201 HB
JIS G4312	SUH309	Heat res Plt Sh	0.20 max	22.00-24.00	2.00 max		12.00-15.00	0.040 max	0.030 max	1.00 max	bal Fe	560	205	40	201 HB max

Mexico

Specification	Designation	Notes	C	Cr	Mn	Mo	Ni	P	S	Si	Other	UTS	YS	El	Hard
DGN B-83	309*	Obs; Bar, HR, CR	0.2 max	22-24	2 max		12-15	0.05 max	0.03 max	1 max	bal Fe				

Poland

Specification	Designation	Notes	C	Cr	Mn	Mo	Ni	P	S	Si	Other	UTS	YS	El	Hard
PNH86022(71)	H18N9S	Heat res	0.1-0.2	17-20	2 max		8-11	1-0.045	0.03 max	0.8-2	bal Fe				
PNH86022(71)	H20N12S2	Heat res	0.2 max	19-22	1.5 max		11-13	0.045 max	0.03 max	1.8-2.5	bal Fe				

Romania

Specification	Designation	Notes	C	Cr	Mn	Mo	Ni	P	S	Si	Other	UTS	YS	El	Hard
STAS 3583(64)	15SNC200	Corr res; Heat res	0.2 max	19-22	1.5 max		12-15	0.035 max	0.03 max	2-3	Pb <=0.15; bal Fe				

Russia

Specification	Designation	Notes	C	Cr	Mn	Mo	Ni	P	S	Si	Other	UTS	YS	El	Hard
GOST	20Ch23N13	Sh Strp Plt Bar Wir	0.2 max	22-25	2 max		12-15	0.035 max	0.025 max	1 max	bal Fe				
GOST 5632(72)	20Ch20N14S2	Heat res	0.2 max	19.0-22.0	1.5 max	0.3 max	12.0-15.0	0.035 max	0.025 max	2.0-3.0	Cu <=0.3; Ti <=0.2; W <=0.2; bal Fe				

Spain

Specification	Designation	Notes	C	Cr	Mn	Mo	Ni	P	S	Si	Other	UTS	YS	El	Hard
UNE 36017(85)	F.3312*	see X15CrNiSi20-12	0.2 max	19-21	2 max	0.15 max	11-13	0.045 max	0.03 max	1.5-2.5	bal Fe				
UNE 36017(85)	X15CrNiSi20-12	Heat res	0.2 max	19-21	2 max	0.15 max	11-13	0.045 max	0.03 max	1.5-2.5	bal Fe				

UK

Specification	Designation	Notes	C	Cr	Mn	Mo	Ni	P	S	Si	Other	UTS	YS	El	Hard
BS 1449/2(83)	309S24	Sh Strp Plt, SA, 3mm diam	0.15 max	22-25	2.00 max		13-16	0.045 max	0.03 max	1 max	bal Fe				
BS 1501/3(90)	309S16	Press ves; Corr/Heat res; 0.05-100mm	0.08 max	22.0-25.0	2.0 max	0.15 max	13.0-16.0	0.045 max	0.025 max	1.0 max	Cu <=0.3; Ti <=0.05; W <=0.1; bal Fe	510-710	205	40	
BS 1554(90)	309S20	Wir Corr/Heat res; diam<0.49mm; Ann	0.12 max	22.0-25.0	2.00 max		12.0-15.0	0.045 max	0.030 max	1.00 max	W <=0.1; bal Fe	0-900			
BS 1554(90)	309S20	Wir Corr/Heat res; 6<diam<=13mm; Ann	0.12 max	22.0-25.0	2.00 max		12.0-15.0	0.045 max	0.030 max	1.00 max	W <=0.1; bal Fe	0-700			
BS 2901	309S24	Wire, Rod	0.12 max	23-25	1-2.5		12-14	0.05 max	0.03 max	0.25-0.65	bal Fe				
BS 970/4(70)	302S25*	Obs; Corr res	0.12 max	17.0-19.0	0.50-2.00	0.7 max	8.0-11.0	0.045 max	0.03 max	0.2-1.0	Cu <=0.5; Pb <=0.15; Nb<=0.2; bal Fe				

USA

Specification	Designation	Notes	C	Cr	Mn	Mo	Ni	P	S	Si	Other	UTS	YS	El	Hard
	AISI 309	Tube	0.20 max	22.00-24.00	2.00 max		12.00-15.00	0.045 max	0.030 max	1.00 max	bal Fe				
	UNS S30900	Heat res	0.20 max	22.00-24.00	2.00 max		12.00-15.00	0.045 max	0.030 max	1.00 max	bal Fe				

UNS numbers and US grades are provided as a means of cross referencing chemically similar alloys. Exchangability is only possible after independent examination of specifications. Tensile properties are minimum or typical as specified. UTS and YS as MPa. El as %. See Appendix for list of abbreviations used in Notes. * indicates obsolete material.

Specification	Designation	Notes	C	Cr	Mn	Mo	Ni	P	S	Si	Other	UTS	YS	El	Hard

Stainless Steel, Austenitic, 309 (Continued from previous page)

USA

Specification	Designation	Notes	C	Cr	Mn	Mo	Ni	P	S	Si	Other	UTS	YS	El	Hard
ASME SA249	309	Refer to ASTM A249/A249M(96)									*				
ASME SA312	309	Refer to ASTM A312/A312M(95)													
ASME SA358	309	Refer to ASTM A358/A358M(95)													
ASME SA403	309	Refer to ASTM A403/A403M(95)													
ASME SA409	309	Refer to ASTM A409/A409M(95)													
ASTM A167(96)	309	Sh Strp Plt	0.20 max	22.00-24.00	2.00 max		12.00-15.00	0.045 max	0.030 max	0.75 max	bal Fe	515	205	40.0	217 HB
ASTM A249	309*	Weld tube	0.2 max	22-24	2 max		12-15	0.045 max	0.03 max	1 max	bal Fe				
ASTM A312/A312M(95)	309	Smls Weld, Pipe	0.2 max	22-24	2 max		12-15	0.045 max	0.03 max	1 max	bal Fe				
ASTM A314	309	Bil	0.2 max	22-24	2 max		12-15	0.045 max	0.03 max	1 max	bal Fe				
ASTM A358/A358M(95)	309	Elect fusion weld pipe, high-temp	0.2 max	22-24	2 max		12-15	0.045 max	0.03 max	1 max	bal Fe				
ASTM A403	309	Pipe	0.2 max	22-24	2 max		12-15	0.045 max	0.03 max	1 max	bal Fe				
ASTM A409	309	Weld, Pipe	0.2 max	22-24	2 max		12-15	0.045 max	0.03 max	1 max	bal Fe				
ASTM A473	309	Frg, Heat res SHT	0.20 max	22.00-24.00	2.00 max		12.00-15.00	0.045 max	0.030 max	1.00 max	bal Fe	515	205	40	
ASTM A554(94)	MT309S	Weld mech tub, Rnd Ann	0.08 max	22.0-24.0	2.00 max		12.0-15.0	0.040 max	0.030 max	1.00 max	bal Fe	517	207	35	
ASTM A580/A580M(98)	309	Wir, Ann	0.2 max	22.0-24.0	2.00 max		12.0-15.0	0.045 max	0.030 max	1.00 max	bal Fe	520	210	35	
FED QQ-S-763F(96)	309*	Obs; Bar Wir Shp Frg; Corr res	0.2 max	22-24	2 max		12-15	0.045 max	0.03 max	1 max	bal Fe				
FED QQ-S-766D(93)	309*	Obs; see ASTM A240, A666, A693; Sh Strp Plt	0.2 max	22-24	2 max		12-15	0.045 max	0.03 max	1 max	bal Fe				
MIL-S-862B	309*	Obs	0.2 max	22-24	2 max		12-15	0.045 max	0.03 max	1 max	bal Fe				

Stainless Steel, Austenitic, 309Cb

USA

Specification	Designation	Notes	C	Cr	Mn	Mo	Ni	P	S	Si	Other	UTS	YS	El	Hard
	UNS S30940	Heat res	0.08 max	22.00-24.00	2.00 max		12.00-15.00	0.040 max	0.030 max	1.00 max	Nb 10xC-1.00; bal Fe				
ASTM A213/A213M(95)	TP309Cb	Smls tube boiler, superheater, heat exchanger	0.08 max	22.00-24.00	2.00 max	0.75 max	12.00-16.00	0.045 max	0.030 max	0.75 max	10xC<=Nb+Ta<=1.10; bal Fe	515	205	35	
ASTM A240/A240M(98)	S30940	Heat res	0.08 max	22.0-24.0	2.00 max		12.0-15.0	0.040 max	0.030 max	0.75 max	10xC<=Nb<=1.10; bal Fe	515	205	40.0	217 HB
ASTM A249/249M(96)	TP309Cb	Weld tube; boiler, superheater, heat exch	0.08 max	22.00-24.00	2.00 max	0.75 max	12.00-16.00	0.045 max	0.030 max	0.75 max	10xC<=Nb+Ta<=1.10; bal Fe	515	205	35	90 HRB
ASTM A580/A580M(98)	309Cb	Wir, Ann	0.08 max	22.00-24.00	2.00 max		12.00-16.00	0.045 max	0.030 max	1.00 max	N <=0.10; 10xC<=Ta+Nb<=1.10; bal Fe	520	210	35	90 HBR
ASTM A813/A813M(95)	TP309Cb	Weld Pipe, Heat res	0.08 max	22.00-24.00	2.00 max	0.75 max	12.00-16.00	0.045 max	0.030 max	0.75 max	Cu <=0.75; Nb+Ta>=10xC<=1.10, bal Fe	515	205		
ASTM A814/A814M(96)	TP309Cb	CW Weld Pipe, Heat res	0.08 max	22.00-24.00	2.00 max	0.75 max	12.00-16.00	0.045 max	0.030 max	0.75 max	Cu <=0.75; Nb+Ta>=10xC<=1.10, bal Fe	515	205		
ASTM A943/A943M(95)	TP309Cb	Spray formed smls pipe	0.08 max	22.0-24.0	2.00 max	0.75 max	12.0-16.0	0.045 max	0.030 max	0.75 max	Nb+Ta>=10xC<=1.10; bal Fe	515	205		
SAE J405(98)	S30940	Heat res	0.08 max	22.00-24.00	2.00 max		12.00-15.00	0.045 max	0.030 max	0.75 max	10xC<=Nb<=1.10; bal Fe				

Stainless Steel, Austenitic, 309H

USA

Specification	Designation	Notes	C	Cr	Mn	Mo	Ni	P	S	Si	Other	UTS	YS	El	Hard
	UNS S30909	Heat res	0.04-0.10	22.0-24.0	2.00 max		12.0-16.0	0.040 max	0.030 max	0.75 max	bal Fe				

UNS numbers and US grades are provided as a means of cross referencing chemically similar alloys. Exchangability is only possible after independent examination of specifications. Tensile properties are minimum or typical as specified. UTS and YS as MPa. El as %. See Appendix for list of abbreviations used in Notes. * indicates obsolete material.

Specification	Designation	Notes	C	Cr	Mn	Mo	Ni	P	S	Si	Other	UTS	YS	El	Hard

Stainless Steel, Austenitic, 309H (Continued from previous page)

USA

Specification	Designation	Notes	C	Cr	Mn	Mo	Ni	P	S	Si	Other	UTS	YS	El	Hard
ASTM A182/A182M(98)	F309H	Frg/roll pipe flange valve	0.04-0.10	22.0-24.0	2.00 max		12.0-15.0	0.045 max	0.030 max	1.00 max	N 0.10-0.16; bal Fe	515	205	30	
ASTM A213/A213M(95)	TP309H	Smls tube boiler, superheater, heat exchanger	0.04-0.10	22.00-24.00	2.00 max	0.75 max	12.00-15.00	0.045 max	0.030 max	0.75 max	bal Fe	515	205	35	
ASTM A240/A240M(98)	S30909	Heat res	0.04-0.10	22.0-24.0	2.00 max		12.0-16.0	0.040 max	0.030 max	0.75 max	bal Fe	515	205	40.0	217 HB
ASTM A249/249M(96)	TP309H	Weld tube; boiler, superheater, heat exch	0.04-0.10	22.00-24.00	2.00 max	0.75 max	12.00-15.00	0.040 max	0.030 max	0.75 max	bal Fe	515	205	35	90 HRB
ASTM A336/A336M(98)	F309H	Frg, press/high-temp	0.04-0.10	22.00-24.00	2.00		12.00-15.00	0.040 max	0.030 max	1.00 max	bal Fe	485	205	30	90 HRB
ASTM A943/A943M(95)	TP309H	Spray formed smls pipe	0.04-0.10	22.0-24.0	2.00 max		12.0-15.0	0.040 max	0.030 max	0.75 max	bal Fe	515	205		
ASTM A965/965M(97) *	F309H	Frg for press high-temp parts	0.04-0.10	22.0-24.0	2.00		12.0-15.0	0.040 max	0.030 max	1.00 max	bal Fe	485	205	30	
SAE J405(98)	S30909	Heat res	0.04-0.10	22.00-24.00	2.00 max		12.00-16.00	0.045 max	0.030 max	0.75 max	bal Fe	515	205		

Stainless Steel, Austenitic, 309HCb

USA

Specification	Designation	Notes	C	Cr	Mn	Mo	Ni	P	S	Si	Other	UTS	YS	El	Hard
	UNS S30941	Heat res	0.04-0.10	22.0-24.0	2.00 max	0.75 max	12.0-16.0	0.040 max	0.030 max	0.75 max	Cu <=0.75; Nb+ a 10xC min-1.10 max; bal Fe				
ASTM A213/A213M(95)	TP309HCb	Smls tube boiler, superheater, heat exchanger	0.04-0.10	22.00-24.00	2.00 max	0.75 max	12.00-16.00	0.045 max	0.030 max	0.75 max	Nb+Ta 10xC-1.10; bal Fe	515	205	35	
ASTM A240/A240M(98)	S30941	Heat res	0.04-0.10	22.0-24.0	2.00 max		12.0-16.0	0.040 max	0.030 max	0.75 max	Cu <=0.7; 10xC<=Nb<=1.10; bal Fe	515	205	40.0	217 HB
ASTM A249/249M(96)	TP309HCb	Weld tube; boiler, superheater, heat exch	0.04-0.10	22.00-24.00	2.00 max	0.75 max	12.00-16.00	0.045 max	0.030 max	0.75 max	10xC<=Nb+Ta<=1.10; bal Fe	515	205	35	90 HRB
ASTM A943/A943M(95)	TP309HCb	Spray formed smls pipe	0.04-0.10	22.0-24.0	2.00 max	0.75 max	12.0-16.0	0.045 max	0.030 max	0.75 max	Nb+Ta>=10xC<=1.10; bal Fe	515	205		
SAE J405(98)	S30941	Heat res	0.04-0.10	22.00-24.00	2.00 max		12.00-16.00	0.045 max	0.030 max	0.75 max	Cu <=0.75; 10xC<=Nb<=1.10; bal Fe				

Stainless Steel, Austenitic, 309S

Canada

Specification	Designation	Notes	C	Cr	Mn	Mo	Ni	P	S	Si	Other	UTS	YS	El	Hard
CSA G110.3	309S	Bar Bil	0.08 max	22-24	2 max		12-15	0.05 max	0.03 max	1 max	bal Fe				
CSA G110.6	309S	Sh Strp Plt, HR, Ann or HR/Ann or HR/Q/A/Tmp	0.08 max	22-24	2 max		12-15	0.05 max	0.03 max	1 max	bal Fe				
CSA G110.9	309S	Sh Strp Plt, HR, Ann Q/A, Tmp, CR	0.08 max	22-24	2 max		12-15	0.05 max	0.03 max	1 max	bal Fe				

China

Specification	Designation	Notes	C	Cr	Mn	Mo	Ni	P	S	Si	Other	UTS	YS	El	Hard
GB 1220(92)	0Cr23Ni13	Bar SA	0.08 max	22.00-24.00	2.00 max		12.00-15.00	0.035 max	0.030 max	1.00 max	bal Fe	520	205	40	
GB 1221(92)	0Cr23Ni13	Bar SA	0.08 max	22.00-24.00	2.00 max		12.00-15.00	0.035 max	0.030 max	1.00 max	bal Fe	520	205	40	
GB 13296(91)	0Cr23Ni13	Smls tube SA	0.08 max	22.00-24.00	2.00 max		12.00-15.00	0.035 max	0.030 max	1.00 max	bal Fe	520	205	35	
GB 3280(92)	0Cr23Ni13	Sh Plt, CR SA	0.08 max	22.00-24.00	2.00 max		12.00-15.00	0.035 max	0.030 max	1.00 max	bal Fe	520	205	40	
GB 4237(92)	0Cr23Ni13	Sh Plt, HR SA	0.08 max	22.00-24.00	2.00 max		12.00-15.00	0.035 max	0.030 max	1.00 max	bal Fe	520	205	40	
GB 4238(92)	0Cr23Ni13	Plt, HR SA	0.08 max	22.00-24.00	2.00 max		12.00-15.00	0.035 max	0.030 max	1.00 max	bal Fe	520	205	40	
GB 4239(91)	0Cr23Ni13	Sh Strp, CR SA	0.08 max	22.00-24.00	2.00 max		12.00-15.00	0.035 max	0.030 max	1.00 max	bal Fe	520	205	40	
GB 4240(93)	0Cr23Ni13(-R)	Wir Ann 0.6-1mm diam, R-As Ann	0.08 max	22.00-24.00	2.00 max		12.00-15.00	0.035 max	0.030 max	1.00 max	bal Fe	540-880		25	
GB/T 14976(94)	0Cr23Ni13	Smls pipe SA	0.08 max	22.00-24.00	2.00 max		12.00-15.00	0.035 max	0.030 max	1.00 max	bal Fe	520	205	35	

France

Specification	Designation	Notes	C	Cr	Mn	Mo	Ni	P	S	Si	Other	UTS	YS	El	Hard
AFNOR NFA35583	Z10CN24.13	Wir rod	0.12 max	23-25	1-2.5	0.5 max	12-14	0.03 max	0.02 max	0.3-0.65	Cu <=0.5; bal Fe				

UNS numbers and US grades are provided as a means of cross referencing chemically similar alloys. Exchangability is only possible after independent examination of specifications. Tensile properties are minimum or typical as specified. UTS and YS as MPa. El as %. See Appendix for list of abbreviations used in Notes. * indicates obsolete material.

Specification	Designation	Notes	C	Cr	Mn	Mo	Ni	P	S	Si	Other	UTS	YS	El	Hard

Stainless Steel, Austenitic, 309S (Continued from previous page)

Germany

Specification	Designation	Notes	C	Cr	Mn	Mo	Ni	P	S	Si	Other	UTS	YS	El	Hard
DIN EN 10095(95)	WNr 1.4833		0.15 max	22.0-24.0	2.00 max		12.0-14.0	0.035 max		0.75 max	N <=0.11; bal Fe	500-750	210	26	
DIN EN 10095(95)	X7CrNi2314	Heat res quench	0.15 max	22.0-24.0	2.00 max		12.0-14.0	0.035 max		0.75 max	N <=0.11; bal Fe	560-750	210	26	
DIN SEW 470(76)	WNr 1.4833	Roll frg	0.15 max	22.0-24.0	2.00 max		12.0-14.0	0.035 max		0.75 max	N <=0.11; bal Fe	500-750	210	26	
DIN SEW 470(76)	X7CrNi2314	Roll frg Q	0.15 max	22.0-24.0	2.00 max		12.0-14.0	0.035 max		0.75 max	N <=0.11; bal Fe	500-750	210	26	

International

Specification	Designation	Notes	C	Cr	Mn	Mo	Ni	P	S	Si	Other	UTS	YS	El	Hard
ISO 4955(94)	H14	Heat res	0.08 max	22.0-24.0	2.0 max	0.15 max	12.0-15.0	0.045 max	0.03 max	1.0 max	Al <=0.1; Co <=0.1; Cu <=0.3; Pb <=0.15; Ti <=0.05; V <=0.1; W <=0.1; bal Fe				

Italy

Specification	Designation	Notes	C	Cr	Mn	Mo	Ni	P	S	Si	Other	UTS	YS	El	Hard
UNI 6901(71)	X6CrNi2314	Heat res	0.08 max	22-24	2 max		12-15	0.045 max	0.03 max	1 max	bal Fe				
UNI 6904(71)	X6CrNi2314	Heat res	0.08 max	22-24	2 max		12-15	0.04 max	0.03 max	0.75 max	bal Fe				
UNI 7500(75)	X6CrNi2314	Press ves	0.08 max	22-24	2 max		12-15	0.045 max	0.03 max	1 max	bal Fe				
UNI 8317(81)	X6CrNi2314	Sh Strp Plt	0.08 max	22-24	2 max		12-15	0.045 max	0.03 max	1 max	bal Fe				

Japan

Specification	Designation	Notes	C	Cr	Mn	Mo	Ni	P	S	Si	Other	UTS	YS	El	Hard
JIS G4303(91)	SUS309S	Bar; SA; <=180mm diam	0.08 max	22.00-24.00	2.00 max		12.00-15.00	0.045 max	0.030 max	1.00 max	bal Fe	520	205	40	187 HB
JIS G4304(91)	SUS309S	Sh Plt, HR SA	0.08 max	22.00-24.00	2.00 max		12.0-15.00	0.045 max	0.030 max	1.00 max	bal Fe	520	205	40	90 HRB max
JIS G4305(91)	SUS309S	Sh Plt, CR SA	0.08 max	22.00-24.00	2.00 max		12.0-15.00	0.045 max	0.030 max	1.00 max	bal Fe	520	205	40	187 HB max
JIS G4306	SUS309S	Strp HR SA	0.08 max	22-24	2 max		12-15	0.04 max	0.03 max	1 max	bal Fe				
JIS G4307	SUS309S	Strp CR SA	0.08 max	22-24	2 max		12-15	0.04 max	0.03 max	1 max	bal Fe				
JIS G4308(98)	SUS309S	Wir rod	0.08 max	22.00-24.00	2.00 max		12.00-15.00	0.045 max	0.030 max	1.00 max	bal Fe				
JIS G4309	SUS309S	Wir	0.08 max	22.00-24.00	2.00 max		12.00-15.00	0.045 max	0.030 max	1.00 max	bal Fe				

Mexico

Specification	Designation	Notes	C	Cr	Mn	Mo	Ni	P	S	Si	Other	UTS	YS	El	Hard
DGN B-83	309S*	Obs; Bar, HR, CR	0.08 max	22-24	2 max		12-15	0.05 max	0.03 max	1 max	bal Fe				
NMX-B-171(91)	MT309S	mech tub	0.08 max	22.0-24.0	2.00 max		12.0-15.0	0.040 max	0.030 max	1.00 max	bal Fe				

USA

Specification	Designation	Notes	C	Cr	Mn	Mo	Ni	P	S	Si	Other	UTS	YS	El	Hard
	AISI 309S	Tube	0.08 max	22.00-24.00	2.00 max		12.00-15.00	0.045 max	0.030 max	1.00 max	bal Fe				
	UNS S30908	Heat res	0.08 max	22.00-24.00	2.00 max		12.00-15.00	0.045 max	0.030 max	1.00 max	bal Fe				
AMS 5523		Sh Strp Plt Bar Wir	0.08 max	22-24	2 max		12-15	0.045 max	0.03 max	1 max	bal Fe				
AMS 5574		Sh Strp Plt Bar Wir	0.08 max	22-24	2 max		12-15	0.045 max	0.03 max	1 max	bal Fe				
AMS 5650		Sh Strp Plt Bar Wir	0.08 max	22-24	2 max		12-15	0.045 max	0.03 max	1 max	bal Fe				
AMS 7490		Sh Strp Plt Bar Wir	0.08 max	22-24	2 max		12-15	0.045 max	0.03 max	1 max	bal Fe				
ASME SA240	309S	Refer to ASTM A240(95)													
ASTM A167(96)	309S*	Sh Strp Plt	0.08 max	22-24	2 max		12-15	0.045 max	0.03 max	1 max	bal Fe				
ASTM A213/A213M(95)	TP309S	Smls tube boiler, superheater, heat exchanger	0.08 max	22.00-24.00	2.00 max	0.75 max	12.00-15.00	0.045 max	0.030 max	0.75 max	bal Fe	515	205	35	
ASTM A240/A240M(98)	S30908	Heat res	0.08 max	22.0-24.0	2.00 max		12.0-15.0	0.045 max	0.030 max	0.75 max	bal Fe	515	205	40.0	217 HB
ASTM A249/249M(96)	TP309S	Weld tube; boiler, superheater, heat exch	0.08 max	22.00-24.00	2.00 max	0.75 max	12.00-15.00	0.045 max	0.030 max	0.75 max	bal Fe	515	205	35	90 HRB

UNS numbers and US grades are provided as a means of cross referencing chemically similar alloys. Exchangability is only possible after independent examination of specifications. Tensile properties are minimum or typical as specified. UTS and YS as MPa. El as %. See Appendix for list of abbreviations used in Notes. * indicates obsolete material.

Specification	Designation	Notes	C	Cr	Mn	Mo	Ni	P	S	Si	Other	UTS	YS	El	Hard

Stainless Steel, Austenitic, 309S (Continued from previous page)

USA

Specification	Designation	Notes	C	Cr	Mn	Mo	Ni	P	S	Si	Other	UTS	YS	El	Hard
ASTM A276(98)	309S	Bar	0.08 max	22-24	2 max		12-15	0.045 max	0.03 max	1 max	bal Fe				
ASTM A314	309S	Bil	0.08 max	22-24	2 max		12-15	0.045 max	0.03 max	1 max	bal Fe				
ASTM A473	309S	Frg, Heat res SHT	0.08 max	22.00-24.00	2.00 max		12.00-15.00	0.045 max	0.030 max	1.00 max	bal Fe	515	205	40	
ASTM A511(96)	MT309S	Smls mech tub, Ann	0.08 max	22.0-24.0	2.00 max		12.0-15.0	0.040 max	0.030 max	1.00 max	bal Fe	517	207	35	192 HB
ASTM A554(94)	MT309S-Cb	Weld mech tub, Rnd Ann	0.08 max	22.0-24.0	2.00 max		12.0-15.0	0.040 max	0.030 max	1.00 max	Nb+Ta>=10xC<=1.00; bal Fe	517	207	35	
ASTM A580/A580M(98)	309S	Wir, Ann	0.08 max	22.0-24.0	2.00 max		12.0-15.0	0.045 max	0.030 max	1.00 max	bal Fe	520	210	35	
ASTM A813/A813M(95)	TP309S	Weld Pipe, Heat res	0.08 max	22.00-24.00	2.00 max	0.75 max	12.00-15.00	0.045 max	0.030 max	0.75 max	Cu <=0.75; bal Fe	515	205		
ASTM A814/A814M(96)	TP309S	CW Weld Pipe, Heat res	0.08 max	22.00-24.00	2.00 max	0.75 max	12.00-15.00	0.045 max	0.030 max	0.75 max	Cu <=0.75; bal Fe	515	205		
ASTM A943/A943M(95)	TP309S	Spray formed smls pipe	0.08 max	22.0-24.0	2.00 max	0.75 max	12.0-15.0	0.045 max	0.030 max	0.75 max	bal Fe	515	205		
SAE J405(98)	S30908	Heat res	0.08 max	22.00-24.00	2.00 max		12.00-15.00	0.045 max	0.030 max	1.00 max	bal Fe				
SAE J467(68)	309S		0.04	23.0	1.00	0.25	13.5			0.50	Cu 0.25; bal Fe	620	276	50	83 HRB

Stainless Steel, Austenitic, 310

Australia

Specification	Designation	Notes	C	Cr	Mn	Mo	Ni	P	S	Si	Other	UTS	YS	El	Hard
AS 1449(94)	310	Sh Strp Plt	0.25 max	24.00-26.00	2.00 max		19.00-22.00	0.045 max	0.030 max	1.50 max	bal Fe				
AS 2837(86)	310	Bar, Semi-finished product	0.08 max	24.0-26.0	2.00 max		19.0-22.0	0.045 max	0.030 max	1.50 max	bal Fe				

Canada

Specification	Designation	Notes	C	Cr	Mn	Mo	Ni	P	S	Si	Other	UTS	YS	El	Hard
CSA G110.3	310	Bar Bil	0.25 max	24-26	2 max		19-22	0.05 max	0.03 max	1.5 max	bal Fe				
CSA G110.6	310	Sh Strp Plt, HR/Ann or HR/Q/A/Tmp	0.25 max	24-26	2 max		19-22	0.05 max	0.03 max	1.5 max	bal Fe				

China

Specification	Designation	Notes	C	Cr	Mn	Mo	Ni	P	S	Si	Other	UTS	YS	El	Hard
GB 1221(92)	2Cr25Ni20	Bar SA	0.25 max	24.00-26.00	2.00 max		19.00-22.00	0.035 max	0.030 max	1.50 max	bal Fe	590	205	40	
GB 13296(91)	2Cr25Ni20	Smls tube SA	0.25 max	24.00-26.00	2.00 max		19.00-22.00	0.035 max	0.030 max	1.50 max	bal Fe	520	205	35	
GB 4238(92)	2Cr25Ni20	Plt, HR SA	0.25 max	24.00-26.00	2.00 max		19.00-22.00	0.035 max	0.030 max	1.50 max	bal Fe	520	205	40	

Czech Republic

Specification	Designation	Notes	C	Cr	Mn	Mo	Ni	P	S	Si	Other	UTS	YS	El	Hard
CSN 417255	17255	Heat res	0.25 max	23.0-27.0	1.5 max		18.0-22.0	0.045 max	0.03 max	2.0 max	bal Fe				

France

Specification	Designation	Notes	C	Cr	Mn	Mo	Ni	P	S	Si	Other	UTS	YS	El	Hard
AFNOR NFA35578	Z12CN25.20	Sh Strp Plt Bar	0.15 max	23-26	2 max		18-21	0.04 max	0.03 max	1 max	bal Fe				
AFNOR NFA35583	Z12CN25.20	Wir rod	0.08-0.15	25-27	1-2.5	0.5 max	20-22	0.02 max	0.02 max	0.5 max	Cu <=0.5; bal Fe				

Germany

Specification	Designation	Notes	C	Cr	Mn	Mo	Ni	P	S	Si	Other	UTS	YS	El	Hard
DIN SEW 470(76)	WNr 1.4845	Roll frg Q	0.15 max	24.0-26.0	2.00 max		20.0-23.0	0.045 max	0.015 max	0.75 max	bal Fe	500-750	210	35	
DIN SEW 470(76)	X8CrNi25-21	Heat res roll frg	0.15 max	24.0-26.0	2.00 max		20.0-23.0	0.045 max	0.015 max	0.75 max	bal Fe	500-750	210	35	

Hungary

Specification	Designation	Notes	C	Cr	Mn	Mo	Ni	P	S	Si	Other	UTS	YS	El	Hard
MSZ 4359(82)	H10	Sh Plt Bar Bil; Heat res; 0-60mm; SA	0.2 max	24-27	1.5 max		18-21	0.04 max	0.03 max	2-3.0	bal Fe	550	230	30L	223 HB max
MSZ 4359(82)	H9	Heat res; 0-60mm; SA	0.2 max	22-25	1.5 max		17-20	0.04 max	0.03 max	1 max	bal Fe	500	210	35L	192 HB max
MSZ 4398(86)	H10	Corr res; Heat res; longitudinal weld tube; 0.8-3mm; SA	0.2 max	24-26	2 max	0.5 max	19-22	0.045 max	0.03 max	1.5-2.5	bal Fe	550-800	230	30L	
MSZ 4398(86)	H9	Corr res; Heat res; longitudinal weld tube; 0.8-3mm; SA	0.15 max	24-26	2 max	0.5 max	19-22	0.045 max	0.03 max	0.75 max	bal Fe	500-750	210	35L	

UNS numbers and US grades are provided as a means of cross referencing chemically similar alloys. Exchangability is only possible after independent examination of specifications. Tensile properties are minimum or typical as specified. UTS and YS as MPa. El as %. See Appendix for list of abbreviations used in Notes. * indicates obsolete material.

Stainless Steel, Austenitic, 310 (Continued from previous page)

Specification	Designation	Notes	C	Cr	Mn	Mo	Ni	P	S	Si	Other	UTS	YS	El	Hard
India															
IS 1570/5(85)	X20Cr25Ni20	Sh Plt Strp Bar Flat Band	0.25 max	24-26	2 max	0.15 max	18-21	0.045 max	0.03 max	2.5 max	Co <=0.1; Cu <=0.3; Pb <=0.15; W <=0.1; bal Fe	515	210	40	223 HB max
International															
ISO 2604-1(75)	F68	Frg	0.15 max	24.00-26.00	2.00 max		19.00-23.00	0.045 max	0.030 max	1.50 max	bal Fe	490-690			
ISO 4955(94)	H16	Heat res	0.25 max	24.0-26.0	2.0 max	0.15 max	19.0-22.0	0.045 max	0.03 max	1.5-2.5	Cu <=0.3; Pb <=0.15; Ti <=0.05; V <=0.1; bal Fe				
Italy															
UNI 6901(71)	X22CrNi2520	Corr res	0.25 max	24-26	2 max		19-22	0.045 max	0.03 max	1.5 max	bal Fe				
UNI 6904(71)	X22CrNi2520	Heat res	0.25 max	24-26	2 max		19-22	0.04 max	0.03 max	0.75 max	bal Fe				
Japan															
JIS G3214(91)	SUSF310	Frg press ves 130-200mm diam	0.15 max	24.00-26.00	2.00 max		19.00-22.00	0.040 max	0.030 max	1.00 max	bal Fe	480	205	29	187 HB max
JIS G4311(91)	SUH310	Bar; SA; 180mm diam	0.25 max	24.00-26.00	2.00 max		19.00-22.00	0.040 max	0.030 max	1.50 max	bal Fe	590	205	40	201 HB
JIS G4312(91)	SUH310	Sh Plt, SA Heat res	0.25 max	24.00-26.00	2.00 max		19.00-22.00	0.040 max	0.030 max	1.50 max	bal Fe	590	205	35	201 HB
JIS G4316	SUH310	Wir	0.25 max	24-26	2 max		19-22	0.04 max	0.03 max	1.5 max	bal Fe				
Mexico															
DGN B-218	TP310*	Obs; Tube, CF, Q/A or Tmp 8mm diam	0.15 max	24-26	2 max		19-22	0.04 max	0.03 max	0.75 max	bal Fe				
DGN B-83	310*	Obs; Bar, HR, CR	0.25 max	24-26	2 max		19-22	0.05 max	0.03 max	1.5 max	bal Fe				
Pan America															
COPANT 513	TP310	Tube, HF, CF, 8mm diam	0.15 max	24-26	2 max		19-22	0.04 max	0.03 max	0.75 max	bal Fe				
Poland															
PNH86022(71)	H25N20S2	Heat res	0.2 max	24-27	1.5 max		18-21	0.045 max	0.03 max	2-3	bal Fe				
Romania															
STAS 11510(80)	Cr25Ni20	Sh Strp Plt Bar Bil; Heat res	0.14 max	23-27	2 max		18-25	0.04 max	0.03 max	2 max	Pb <=0.15; bal Fe				
STAS 11523(87)	15SiNiCr200	Corr res; Heat res	0.2 max	19-21	2 max	0.2 max	11-13	0.045 max	0.03 max	1.5-2.5	Pb <=0.15; bal Fe				
STAS 11523(87)	40SiNiCr250	Corr res; Heat res	0.3-0.5	24-26	0.5-1.5	0.2 max	19-21	0.045 max	0.03 max	1-2.5	Pb <=0.15; bal Fe				
STAS 3583(64)	15NC230	Corr res; Heat res	0.2 max	22-25	2 max		17-20	0.035 max	0.03 max	1 max	Pb <=0.15; bal Fe				
Russia															
GOST 10994	Ch25N20	Sh Strp Plt Bar Tub Frg	0.15 max	24-27	2 max		17-20	0.035 max	0.03 max	1 max	Ti <=0.3; Zn <=0.5; bal Fe				
GOST 10994(74)	Ch25N20	Precision alloy	0.15 max	24.0-27.0	2.0 max	0.15 max	17.0-20.0	0.03 max	0.02 max	1.0 max	Cu <=0.3; Ti <=0.05; Zr<=0.5; bal Fe				
Spain															
UNE 36017(85)	X15CrNiSi25-20	Sh Plt Bar Rod; Heat res	0.2 max	24-26	2 max	0.15 max	19-21	0.045 max	0.03 max	1.5-2.5	bal Fe				
UK															
BS 1449/2(83)	310S24	Sh Strp Plt, SA, t= 0.05-100mm	0.15 max	24-26	2.00 max		19-22	0.045 max	0.03 max	1 max	bal Fe	510	205	40	0-205 HB
BS 970/1(96)	310S31	Wrought Corr/Heat res; t<=160mm	0.15 max	24.0-26.0	2.00 max		19.0-22.0	0.045 max	0.030 max	1.00 max	bal Fe	510	205	40	207 HB max
USA															
	AISI 310	Tube	0.25 max	24.00-26.00	2.00 max		19.00-22.00	0.045 max	0.030 max	1.50 max	bal Fe				
	UNS S31000	Heat res	0.25 max	24.00-26.00	2.00 max		19.00-22.00	0.045 max	0.030 max	1.50 max	bal Fe				
	UNS S31002	Low-C, 2RE10	0.015 max	24.0-26.0	2.00 max	0.10 max	19.0-22.0	0.020 max	0.015 max	0.15 max	N <=0.10; bal Fe				
ASME SA182	310	Refer to ASTM A182/A182M													

Stainless Steel, Austenitic, 310 (Continued from previous page)

USA

Specification	Designation	Notes	C	Cr	Mn	Mo	Ni	P	S	Si	Other	UTS	YS	El	Hard
ASME SA213	310	Refer to ASTM A213/A213M(95)													
ASME SA249	310	Refer to ASTM A249/A249M(96)													
ASME SA312	310	Refer to ASTM A312/A312M(95)													
ASME SA358	310	Refer to ASTM A358/A358M(95)													
ASME SA403	310	Refer to ASTM A403/A403M(95)													
ASME SA409	310	Refer to ASTM A409/A409M(95)													
ASTM A167(96)	310	Sh Strp Plt	0.25 max	24.00-26.00	2.00 max		19.00-22.00	0.045 max	0.030 max	1.50 max	bal Fe	515	205	40.0	217 HB
ASTM A182	310*	Bar Frg	0.25 max	24-26	2 max		19-22	0.045 max	0.03 max	1.5 max	bal Fe				
ASTM A182/A182M(98)	F310	Frg/roll pipe flange valve	0.25 max	24.0-26.0	2.00 max		19.0-22.0	0.045 max	0.030 max	1.00 max	bal Fe	515	205	30	
ASTM A213	310*	Tube	0.25 max	24-26	2 max		19-22	0.045 max	0.03 max	1.5 max	bal Fe				
ASTM A249	310*	Weld tube	0.25 max	24-26	2 max		19-22	0.045 max	0.03 max	1.5 max	bal Fe				
ASTM A276(98)	310	Bar	0.25 max	24-26	2 max		19-22	0.045 max	0.03 max	1.5 max	bal Fe				
ASTM A312/A312M(95)	310	Smls Weld, Pipe	0.25 max	24-26	2 max		19-22	0.045 max	0.03 max	1.5 max	bal Fe				
ASTM A314	310	Bil	0.25 max	24-26	2 max		19-22	0.045 max	0.03 max	1.5 max	bal Fe				
ASTM A336/A336M(98)	F310	Frg, press/high-temp	0.15 max	24.00-26.00	2.00 max		19.0-22.0	0.040 max	0.030 max	1.00 max	bal Fe	515	205	30	90 HRB
ASTM A358	310*	Elect fusion weld pipe, high-temp	0.25 max	24-26	2 max		19-22	0.045 max	0.03 max	1.5 max	bal Fe				
ASTM A403	310	Pipe	0.25 max	24-26	2 max		19-22	0.045 max	0.03 max	1.5 max	bal Fe				
ASTM A409	310	Weld, Pipe	0.25 max	24-26	2 max		19-22	0.045 max	0.03 max	1.5 max	bal Fe				
ASTM A473	310	Frg, Heat res SHT	0.25 max	24.00-26.00	2.00 max		19.00-22.00	0.045 max	0.030 max	1.50 max	bal Fe	515	205	40	
ASTM A580/A580M(98)	310	Wir, Ann	0.25 max	24.00-26.00	2.00 max		19.0-22.0	0.045 max	0.030 max	1.50 max	bal Fe	520	210	35	
ASTM A632(90)	TP310	Smls Weld tub, Heat res	0.15 max	24.0-26.0	2.00 max		19.0-22.0	0.040 max	0.030 max	0.75 max	bal Fe	515	205	35	
ASTM A965/965M(97)	F310	Frg for press high-temp parts	0.15 max	24.0-26.0	2.00 max		19.0-22.0	0.040 max	0.030 max	1.00 max	bal Fe	515	205	30	
FED QQ-S-763F(96)	310*	Obs; Bar Wir Shp Frg; Corr res	0.25 max	24-26	2 max		19-22	0.045 max	0.03 max	1.5 max	bal Fe				
FED QQ-S-766D(93)	310*	Obs; see ASTM A240, A666, A693; Sh Strp Plt	0.25 max	24-26	2 max		19-22	0.045 max	0.03 max	1.5 max	bal Fe				
MIL-S-862B	310*	Obs	0.25 max	24-26	2 max		19-22	0.045 max	0.03 max	1.5 max	bal Fe				
SAE J467(68)	310		0.12	25.0	1.00		20.5			0.40	bal Fe	634	276	47	89 HRB

Stainless Steel, Austenitic, 310Cb

USA

Specification	Designation	Notes	C	Cr	Mn	Mo	Ni	P	S	Si	Other	UTS	YS	El	Hard
	UNS S31040	Heat res	0.08 max	24.0-26.00	2.00 max		19.00-22.00	0.045 max	0.030 max	1.50 max	N <=0.10; Nb+Ta 10xC min-1.10 max; bal Fe				
ASTM A213/A213M(95)	TP310Cb	Smls tube boiler, superheater, heat exchanger	0.08 max	24.00-26.00	2.00 max	0.75 max	19.00-22.00	0.045 max	0.030 max	0.75 max	Nb+Ta 10xC-1.10; bal Fe	515	205	35	
ASTM A240/A240M(98)	S31040	Heat res	0.08 max	24.0-26.0	2.00 max		19.0-22.0	0.045 max	0.030 max	1.50 max	10xC<=Nb<=1.10; bal Fe	515	205	40.0	217 HB
ASTM A249/249M(96)	TP310Cb	Weld tube; boiler, superheater, heat exch	0.08 max	24.00-26.00	2.00 max	0.75 max	19.00-22.00	0.045 max	0.030 max	0.75 max	10xC<=Nb+Ta<=1.10; bal Fe	515	205	35	90 HRB
ASTM A336/A336M(98)	F310H	Frg, press/high-temp	0.04-0.10	24.00-26.00	2.00 max		19.0-22.0	0.040 max	0.030 max	1.00 max	bal Fe	485	205	30	90 HRB

UNS numbers and US grades are provided as a means of cross referencing chemically similar alloys. Exchangability is only possible after independent examination of specifications. Tensile properties are minimum or typical as specified. UTS and YS as MPa. El as %. See Appendix for list of abbreviations used in Notes. * indicates obsolete material.

Specification	Designation	Notes	C	Cr	Mn	Mo	Ni	P	S	Si	Other	UTS	YS	El	Hard

Stainless Steel, Austenitic, 310Cb (Continued from previous page)

USA

Specification	Designation	Notes	C	Cr	Mn	Mo	Ni	P	S	Si	Other	UTS	YS	El	Hard
ASTM A813/A813M(95)	TP310Cb	Weld Pipe, Heat res	0.08 max	24.00-26.00	2.00 max	0.75 max	19.00-22.00	0.045 max	0.030 max	0.75 max	Cu <=0.75; Nb+Ta>=10xC<=1.10, bal Fe	515	205		
ASTM A814/A814M(96)	TP310Cb	CW Weld Pipe, Heat res	0.08 max	24.00-26.00	2.00 max	0.75 max	19.00-22.00	0.045 max	0.030 max	0.75 max	Cu <=0.75; Nb+Ta>=10xC<=1.10, bal Fe	515	205		
ASTM A943/A943M(95)	TP310Cb	Spray formed smls pipe	0.08 max	24.0-26.0	2.00 max	0.75 max	19.0-22.0	0.045 max	0.030 max	0.75 max	Nb+Ta>=10xC<=1.10; bal Fe	515	205		
SAE J405(98)	S31040	Heat res	0.08 max	24.00-26.00	2.00 max		19.00-22.00	0.045 max	0.030 max	1.50 max	N <=0.10; 10xC<=Nb<=1.10; bal Fe				

Stainless Steel, Austenitic, 310H

USA

Specification	Designation	Notes	C	Cr	Mn	Mo	Ni	P	S	Si	Other	UTS	YS	El	Hard
	UNS S31009	Heat res	0.04-0.10	24.0-26.0	2.00 max		19.0-22.0	0.040 max	0.030 max	0.75 max	bal Fe				
ASTM A182/A182M(98)	F310H	Frg/roll pipe flange valve	0.04-0.10	24.0-26.0	2.00 max		19.0-22.0	0.045 max	0.030 max	1.00 max	bal Fe	515	205	30	
ASTM A213/A213M(95)	TP310H	Smls tube boiler, superheater, heat exchanger	0.04-0.10	24.00-26.00	2.00 max		19.00-22.00	0.040 max	0.030 max	0.75 max	bal Fe	515	205	35	
ASTM A240/A240M(98)	S31009	Heat res	0.04-0.10	24.0-26.0	2.00 max		19.0-22.0	0.040 max	0.030 max	0.75 max	bal Fe	515	205	40.0	217 HB
ASTM A249/249M(96)	TP310H	Weld tube; boiler, superheater, heat exch	0.04-0.10	24.00-26.00	2.00 max		19.00-22.00	0.040 max	0.030 max	0.75 max	bal Fe	515	205	35	90 HRB
ASTM A943/A943M(95)	TP310H	Spray formed smls pipe	0.04-0.10	24.0-26.0	2.00 max		19.0-22.0	0.040 max	0.030 max	0.75 max	bal Fe	515	205		
ASTM A965/965M(97)	F310H	Frg for press high-temp parts	0.04-0.10	24.0-26.0	2.00		19.00-22.0	0.040 max	0.30	1.50 max	bal Fe	485	205	30	
SAE J405(98)	S31009	Heat res	0.04-0.10	24.00-26.00	2.00 max		19.00-22.00	0.040 max	0.030 max	0.75 max	bal Fe				

Stainless Steel, Austenitic, 310S

Australia

Specification	Designation	Notes	C	Cr	Mn	Mo	Ni	P	S	Si	Other	UTS	YS	El	Hard
AS 1449(94)	310S	Plt Sh Strp	0.08 max	24.00-26.00	2.00 max		19.00-22.00	0.045 max	0.030 max	1.50 max	bal Fe				

Bulgaria

Specification	Designation	Notes	C	Cr	Mn	Mo	Ni	P	S	Si	Other	UTS	YS	El	Hard
BDS 6738(72)	Ch23N18	Corr res; bar	0.20 max	22.0-25.0	1.50 max	0.30 max	17.0-20.0	0.035 max	0.025 max	1.00 max	bal Fe				

Canada

Specification	Designation	Notes	C	Cr	Mn	Mo	Ni	P	S	Si	Other	UTS	YS	El	Hard
CSA G110.3		Bar Bil	0.08 max	24-26	2 max		19-22	0.045 max	0.03 max	1.5 max	bal Fe				
CSA G110.6		Sh Strp Plt, HR/Ann or HR/Q/A/Tmp	0.08 max	24-26	2 max		19-22	0.045 max	0.03 max	1.5 max	bal Fe				
CSA G110.9		Sh Strp Plt, HR, Ann, Q/A, Tmp	0.08 max	24-26	2 max		19-22	0.045 max	0.03 max	1.5 max	bal Fe				

China

Specification	Designation	Notes	C	Cr	Mn	Mo	Ni	P	S	Si	Other	UTS	YS	El	Hard
GB 1220(92)	0Cr25Ni20	Bar SA	0.08 max	24.00-26.00	2.00 max		19.00-22.00	0.035 max	0.030 max	1.00 max	bal Fe	520	205	40	
GB 1221(92)	0Cr25Ni20	Bar SA	0.08 max	24.00-26.00	2.00 max		19.00-22.00	0.035 max	0.030 max	1.50 max	bal Fe	520	205	40	
GB 12770(91)	0Cr25Ni20	Weld tube; SA	0.08 max	24.00-26.00	2.00 max		19.00-22.00	0.035 max	0.030 max	1.50 max	bal Fe	520	210	35	
GB 12771(91)	0Cr25Ni20	Weld Pipe SA	0.08 max	24.00-26.00	2.00 max		19.00-22.00	0.035 max	0.030 max	1.50 max	bal Fe	520	210	35	
GB 13296(91)	0Cr25Ni20	Smls tube SA	0.08 max	24.00-26.00	2.00 max		19.00-22.00	0.035 max	0.030 max	1.00 max	bal Fe	520	205	35	
GB 3280(92)	0Cr25Ni20	Sh Plt, CR SA	0.08 max	24.00-26.00	2.00 max		19.00-22.00	0.035 max	0.030 max	1.50 max	bal Fe	520	205	40	
GB 4237(92)	0Cr25Ni20	Sh Plt, HR SA	0.08 max	24.00-26.00	2.00 max		19.00-22.00	0.035 max	0.030 max	1.00 max	bal Fe	520	205	40	
GB 4238(92)	0Cr25Ni20	Plt, HR SA	0.08 max	24.00-26.00	2.00 max		19.00-22.00	0.035 max	0.030 max	1.50 max	bal Fe	520	205	40	
GB 4239(91)	0Cr25Ni20	Sh Strp, CR SA	0.08 max	24.00-26.00	2.00 max		19.00-22.00	0.035 max	0.030 max	1.50 max	bal Fe	520	205	40	
GB 4240(93)	0Cr25Ni20(-R)	Wir Ann 0.6-1mm diam, R-As Ann	0.08 max	24.00-26.00	2.00 max		19.00-22.00	0.035 max	0.030 max	1.00 max	bal Fe	540-880			

UNS numbers and US grades are provided as a means of cross referencing chemically similar alloys. Exchangability is only possible after independent examination of specifications. Tensile properties are minimum or typical as specified. UTS and YS as MPa. El as %. See Appendix for list of abbreviations used in Notes. * indicates obsolete material.

Specification	Designation	Notes	C	Cr	Mn	Mo	Ni	P	S	Si	Other	UTS	YS	El	Hard

Stainless Steel, Austenitic, 310S (Continued from previous page)

China

| GB/T 14976(94) | 0Cr25Ni20 | Smls pipe SA | 0.08 max | 24.00-26.00 | 2.00 max | | 19.00-22.00 | 0.035 max | 0.030 max | 1.00 max | bal Fe | 520 | 205 | 35 | |

India

| IS 1570/5(85) | X04Cr25Ni20 | Sh Plt Strp Bar Flat Band | 0.08 max | 24-26 | 2 max | 0.15 max | 19-22 | 0.045 max | .03 max | 1.5 max | Co <=0.1; Cu <=0.3; Pb <=0.15; W <=0.1; bal Fe | 515 | 205 | 40 | 217 HB max |

International

| ISO 4955(94) | H15 | Heat res | 0.15 max | 24.0-26.0 | 2.0 max | 0.15 max | 19.0-22.0 | 0.045 max | 0.03 max | 1.5 max | Al <=0.1; Co <=0.1; Cu <=0.3; Pb <=0.15; Ti <=0.05; V <=0.1; W <=0.1; bal Fe | | | | |

Italy

UNI 6901(71)	X6CrNi2520	Heat res	0.08 max	24-26	2 max		19-22	0.045 max	0.03 max	1.5 max	bal Fe				
UNI 6904(71)	X6CrNi2520	Heat res	0.08 max	24-26	2 max		19-22	0.04 max	0.03 max	0.75 max	bal Fe				
UNI 7500(75)	X6CrNi2520	Press ves	0.08 max	24-26	2 max		19-22	0.045 max	0.03 max	1.5 max	bal Fe				
UNI 8317(81)	X6CrNi2520	Sh Strp Plt	0.08 max	24-26	2 max		19-22	0.045 max	0.03 max	1.5 max	bal Fe				

Japan

JIS G4303(91)	SUS310S	Bar; SA; <=180mm diam	0.08 max	24.00-26.00	2.00 max		19.00-22.00	0.045 max	0.030 max	1.50 max	bal Fe	520	205	40	187 HB
JIS G4304(91)	SUS310S	Sh Plt, HR SA	0.08 max	24.00-26.00	2.00 max		19.0-22.00	0.045 max	0.030 max	1.50 max	bal Fe	520	205	40	90 HRB max
JIS G4305(91)	SUS310S	Sh Plt, CR SA	0.08 max	24.00-26.00	2.00 max		19.00-22.00	0.045 max	0.030 max	1.50 max	bal Fe	520	205	40	187 HB max
JIS G4306	SUS310S	Strp HR SA	0.08 max	24-26	2 max		19-22	0.04 max	0.03 max	1.5 max	bal Fe				
JIS G4307	SUS310S	Strp CR SA	0.08 max	24-26	2 max		19-22	0.04 max	0.03 max	1.5 max	bal Fe				
JIS G4308(98)	SUS310S	Wir rod	0.08 max	24.00-26.00	2.00 max		19.00-22.00	0.045 max	0.030 max	1.50 max	bal Fe				
JIS G4309	SUS310S	Wir	0.08 max	24.00-26.00	2.00 max		19.00-22.00	0.045 max	0.030 max	1.50 max	bal Fe				

Mexico

| DGN B-83 | 310S* | Obs; Bar, HR, CR | 0.08 max | 24-26 | 2 max | | 19-22 | 0.05 max | 0.03 max | 1.5 max | bal Fe | | | | |
| NMX-B-171(91) | MT310S | CF HF Smls tub | 0.08 max | 24.0-26.0 | 2.00 max | | 19.0-22.0 | 0.040 max | 0.030 max | 1.00 max | bal Fe | | | | |

Norway

| NS 14480 | 14480 | Sh Strp Plt Bar Bil | 0.08 max | 24-26 | 2 max | | 19-22 | 0.045 max | 0.03 max | 1 max | bal Fe | | | | |

Poland

| PNH86022(71) | H23N18 | Heat res | 0.2 max | 22-25 | 1.5 max | | 17-20 | 0.015 max | 0.03 max | 1 max | bal Fe | | | | |

Romania

| STAS 11523(87) | 12NiCr250 | Sh Strp Plt Bar; Corr/Heat res | 0.15 max | 24-26 | 2 max | 0.2 max | 19-22 | 0.045 max | 0.03 max | 0.75 max | Pb <=0.15; bal Fe | | | | |
| STAS 3583(87) | 2NbNiCr250 | Corr res; Heat res | 0.03 max | 23-26 | 1 max | 0.2 max | 19-22 | 0.035 max | 0.025 max | 0.4 max | Nb 0.2-0.3; Pb <=0.15; bal Fe | | | | |

Russia

| GOST 5632(72) | 20Ch23N18 | Heat res; High-temp constr | 0.2 max | 22.0-25.0 | 2.0 max | 0.3 max | 17.0-20.0 | 0.035 max | 0.02 max | 1.0 max | Cu <=0.3; Ti <=0.2; W <=0.2; bal Fe | | | | |

Sweden

| SS 142361 | 2361 | Heat res | 0.08 max | 24-26 | 2 max | | 19-22 | 0.045 max | 0.03 max | 1.5 max | bal Fe | | | | |

UK

BS 1501/3(73)	310S24	Press ves; Corr res	0.15 max	23.0-26.0	0.5-2.0	0.15 max	19.0-22.0	0.045 max	0.03 max	0.2-1.0	Cu <=0.3; Ti <=0.05; W <=0.1; bal Fe				
BS 1501/3(90)	310S16	Press ves; Corr/Heat res; 0.05-100mm	0.08 max	23.0-26.0	2.0 max	0.15 max	19.0-22.0	0.045 max	0.025 max	1.0 max	Cu <=0.3; Ti <=0.05; W <=0.1; bal Fe	510-710	205	40	
BS 1503(89)	310S31	Frg press ves; t<=999mm; HT	0.15 max	24.0-26.0	2.00 max	0.70 max	19.0-22.0	0.040 max	0.025 max	1.50 max	B <=0.005; Cu <=0.50; Nb <=0.20; Ti <=0.10; W <=0.1; bal Fe	510-710	240	30	

UNS numbers and US grades are provided as a means of cross referencing chemically similar alloys. Exchangability is only possible after independent examination of specifications. Tensile properties are minimum or typical as specified. UTS and YS as MPa. El as %. See Appendix for list of abbreviations used in Notes. * indicates obsolete material.

Stainless Steel, Austenitic, 310S (Continued from previous page)

Specification	Designation	Notes	C	Cr	Mn	Mo	Ni	P	S	Si	Other	UTS	YS	El	Hard
UK															
BS 1554(90)	310S17	Wir Corr/Heat res; 6<diam<=13mm; Ann	0.08 max	24.0-26.0	2.00 max		19.0-22.0	0.045 max	0.030 max	1.50 max	W <=0.1; bal Fe	0-750			
BS 1554(90)	310S17	Wir Corr/Heat res; diam<0.49mm; Ann	0.08 max	24.0-26.0	2.00 max		19.0-22.0	0.045 max	0.030 max	1.50 max	W <=0.1; bal Fe	0-900			
BS 1554(90)	310S25	Wir Corr/Heat res; diam<0.49mm; Ann	0.15 max	25.0-28.0	2.50 max		20.0-22.5	0.045 max	0.030 max	1.50 max	W <=0.1; bal Fe	0-900			
BS 1554(90)	310S25	Wir Corr/Heat res; 6<diam<=13mm; Ann	0.15 max	25.0-28.0	2.50 max		20.0-22.5	0.045 max	0.030 max	1.50 max	W <=0.1; bal Fe	0-750			
USA															
	AISI 310S	Tube	0.08 max	24.00-26.00	2.00 max		19.00-22.00	0.045 max	0.030 max	1.50 max	bal Fe				
	UNS S31008	Heat res	0.08 max	24.00-26.00	2.00 max		19.00-22.00	0.045 max	0.030 max	1.50 max	bal Fe				
AMS 5521	310S	Sh Strp Plt Bar Wir	0.08 max	24-26	2 max		19-22	0.045 max	0.03 max	1.5 max	bal Fe				
AMS 5572	310S	Sh Strp Plt Bar Wir	0.08 max	24-26	2 max		19-22	0.045 max	0.03 max	1.5 max	bal Fe				
AMS 5577G(95)	310S	Corr/Heat res; Weld tub; OD<7.92	0.08 max	24.00-26.00	2.00 max	0.75 max	19.00-22.00	0.040 max	0.030 max	0.75 max	Cu <=0.75; bal Fe	724		40	
AMS 5577G(95)	310S	Corr/Heat res; Weld tub; OD>=7.92	0.08 max	24.00-26.00	2.00 max	0.75 max	19.00-22.00	0.040 max	0.030 max	0.75 max	Cu <=0.75; bal Fe	689		40	
AMS 5651	310S	Sh Strp Plt Bar Wir	0.08 max	24-26	2 max		19-22	0.045 max	0.03 max	1.5 max	bal Fe				
AMS 7490	310S	Sh Strp Plt Bar Wir	0.08 max	24-26	2 max		19-22	0.045 max	0.03 max	1.5 max	bal Fe				
ASME SA240	310S	Refer to ASTM A240(95)													
ASME SA479	310S	Refer to ASTM A479/A479M(95)													
ASTM A167(96)	310S*	Sh Strp Plt	0.08 max	24-26	2 max		19-22	0.045 max	0.03 max	1.5 max	bal Fe				
ASTM A213/A213M(95)	TP310S	Smls tube boiler, superheater, heat exchanger	0.08 max	24.00-26.00	2.00 max	0.75 max	19.00-22.00	0.045 max	0.030 max	0.75 max	bal Fe	515	205	35	
ASTM A240/A240M(98)	S31008	Heat res	0.08 max	24.0-26.0	2.00 max		19.0-22.0	0.045 max	0.030 max	1.50 max	bal Fe	515	205	40.0	217 HB
ASTM A249/249M(96)	TP310S	Weld tube; boiler, superheater, heat exch	0.08 max	24.00-26.00	2.00 max	0.75 max	19.00-22.00	0.045 max	0.030 max	0.75 max	bal Fe	515	205	35	90 HRB
ASTM A276(98)	310S	Bar	0.08 max	24-26	2 max		19-22	0.045 max	0.03 max	1.5 max	bal Fe				
ASTM A314	310S	Bil	0.08 max	24-26	2 max		19-22	0.045 max	0.03 max	1.5 max	bal Fe				
ASTM A473	310S	Frg, Heat res SHT	0.08 max	24.00-26.00	2.00 max		19.00-22.00	0.045 max	0.030 max	1.50 max	bal Fe	515	205	40	
ASTM A479	310S	Bar	0.08 max	24-26	2 max		19-22	0.045 max	0.03 max	1.5 max	bal Fe				
ASTM A511(96)	MT310S	Smls mech tub, Ann	0.08 max	24.0-26.0	2.00 max		19.0-22.0	0.040 max	0.030 max	1.5 max	bal Fe	517	207	35	192 HB
ASTM A554(94)	MT310S	Weld mech tub, Rnd Ann	0.08 max	24.0-26.0	2.00 max		19.0-22.0	0.040 max	0.030 max	1.00 max	bal Fe	517	207	35	
ASTM A580/A580M(98)	310S	Wir, Ann	0.08 max	24.0-26.0	2.00 max		19.0-22.0	0.045 max	0.030 max	1.50 max	bal Fe	520	210	35	
ASTM A813/A813M(95)	TP310S	Weld Pipe, Heat res	0.08 max	24.00-26.00	2.00 max	0.75 max	19.00-22.00	0.045 max	0.030 max	0.75 max	Cu <=0.75; bal Fe	515	205		
ASTM A814/A814M(96)	TP310S	CW Weld Pipe, Heat res	0.08 max	24.00-26.00	2.00 max	0.75 max	19.00-22.00	0.045 max	0.030 max	0.75 max	Cu <=0.75; bal Fe	515	205		
ASTM A943/A943M(95)	TP310S	Spray formed smls pipe	0.08 max	24.0-26.0	2.00 max	0.75 max	19.0-22.0	0.045 max	0.030 max	0.75 max	bal Fe	515	205		
FED QQ-S-763F(96)	310S*	Obs; Bar Wir Shp Frg; Corr res	0.08 max	24-26	2 max		19-22	0.045 max	0.03 max	1.5 max	bal Fe				
SAE J405(98)	S31008	Heat res	0.08 max	24.00-26.00	2.00 max		19.00-22.00	0.045 max	0.030 max	1.50 max	bal Fe				

Stainless Steel, Austenitic, 314

Specification	Designation	Notes	C	Cr	Mn	Mo	Ni	P	S	Si	Other	UTS	YS	El	Hard
Bulgaria															
BDS 6738(72)	Ch25N20S2	Sh Plt Bar Bil	0.20 max	24.0-27.0	1.50 max	0.30 max	18.0-21.0	0.035 max	0.025 max	2.00-3.00	bal Fe				

Specification	Designation	Notes	C	Cr	Mn	Mo	Ni	P	S	Si	Other	UTS	YS	El	Hard

Stainless Steel, Austenitic, 314 (Continued from previous page)

Canada

Specification	Designation	Notes	C	Cr	Mn	Mo	Ni	P	S	Si	Other	UTS	YS	El	Hard
CSA G110.3	314	Bar Bil	0.25 max	23-26	2 max		19-22	0.045 max	0.03 max	1.5-3	bal Fe				

China

GB 1221(92)	1Cr25Ni20Si2	Bar SA	0.20 max	24.00-27.00	1.50 max		18.00-21.00	0.035 max	0.030 max	1.50-2.50	bal Fe	590	295	35	
GB 4238(92)	1Cr25Ni20Si2	Plt, HR SA	0.20 max	24.00-27.00	1.50 max		18.00-21.00	0.035 max	0.030 max	1.50-2.50	bal Fe	540		35	

France

AFNOR NFA35578	Z12CNS2520	Plt Bar Bil	0.15 max	23-26	2 max		18-21	0.04 max	0.035 max	1.5-2.5	bal Fe				

Germany

DIN EN 10095(95)	WNr 1.4841		0.20 max	24.0-26.0	2.00 max		19.0-22.0	0.045 max	0.030 max	1.50-2.50	N <=0.11; bal Fe				
DIN EN 10095(95)	X15CrNiSi25-20	Heat res	0.20 max	24.0-26.0	2.00 max		19.0-22.0	0.045 max	0.030 max	1.50-2.50	bal Fe				

International

COMECON PC4-70	34	Sh Plt Bar Wir Bil	0.2 max	24-27	1.5 max		18-21	0.035 max	0.025 max	2-3	bal Fe				
ISO 4955(94)	H16	Heat res	0.25 max	24.0-26.0	2.0 max	0.15 max	19.0-22.0	0.045 max	0.03 max	1.5-2.5	Al <=0.1; Co <=0.1; Cu <=0.3; Pb <=0.15; Ti <=0.05; V <=0.1; W <=0.1; bal Fe				

Italy

UNI 6901(71)	X16CrNiSi2520	Heat res	0.2 max	24-26	2 max		19-22	0.045 max	0.03 max	1.5-2.5	bal Fe				

Mexico

DGN B-83	314*	Obs; Bar, HR, CR	0.25 max	23-26	2 max		19-22	0.05 max	0.03 max	1.5-3	bal Fe				

Poland

PNH86022(71)	H25N20S2	Heat res	0.2 max	24-27	1.5 max		18-21	0.045 max	0.03 max	2-3	bal Fe				

Romania

STAS 11523(87)	15SiNiCr250	Sh Plt Bar Bil; Corr res; Heat res	0.2 max	24-26	2 max	0.2 max	19-21	0.045 max	0.03 max	1.5-2.5	Pb <=0.15; bal Fe				
STAS 3583(87)	15SNC250	Corr res; Heat res	0.2 max	24-27	1.5 max		18-21	0.035 max	0.03 max	2-3	Pb <=0.15; bal Fe				

Russia

GOST 5632	20Ch25N20S2	Sh Plt Bar Bil	0.2 max	24-27	1.5 max	0.3 max	18-21	0.035 max	0.03 max	2-3	Cu <=0.3; W <=0.2; bal Fe				
GOST 5632(72)	20Ch25N20S2	Heat res	0.2 max	24.0-27.0	1.5 max	0.3 max	18.0-21.0	0.035 max	0.02 max	2.0-3.0	Cu <=0.3; Ti <=0.2; W <=0.2; bal Fe				
GOST 977	20Ch25N19S2L	Sh Plt Bar Bil	0.2 max	23-27	0.5-1.5		18-20	0.035 max	0.03 max	2-3	Cu <=0.3; bal Fe				

Spain

UNE 36017(85)	F.3310*	see X15CrNiSi25-20	0.2 max	24-26	2 max	0.15 max	19-21	0.045 max	0.03 max	1.5-2.5	bal Fe				
UNE 36017(85)	X15CrNiSi25-20	Sh Plt Bar Rod; Heat res	0.2 max	24-26	2 max	0.15 max	19-21	0.045 max	0.03 max	1.5-2.5	bal Fe				

UK

BS 1554(90)	314S25	Wir Corr/Heat res; 6<diam<=13mm; Ann	0.25 max	23.0-26.0	2.50 max		19.0-22.0	0.045 max	0.030 max	1.5-3.00	W <=0.1; bal Fe	0-750			
BS 1554(90)	314S25	Wir Corr/Heat res; diam<0.49mm; Ann	0.25 max	23.0-26.0	2.50 max		19.0-22.0	0.045 max	0.030 max	1.5-3.00	W <=0.1; bal Fe	0-900			

USA

	AISI 314	Tube	0.25 max	23.00-26.00	2.00 max		19.00-22.00	0.045 max	0.030 max	1.50-3.00	bal Fe				
	UNS S31400	Heat res	0.25 max	23.00-26.00	2.00 max		19.00-22.00	0.045 max	0.030 max	1.50-3.00	bal Fe				
AMS 5522		Sh Plt Bar Wir	0.25 max	23-26	2 max		19-22	0.045 max	0.03 max	1.5-3	bal Fe				
AMS 5652		Sh Plt Bar Wir	0.25 max	23-26	2 max		19-22	0.045 max	0.03 max	1.5-3	bal Fe				
AMS 7490		Sh Plt Bar Wir	0.25 max	23-26	2 max		19-22	0.045 max	0.03 max	1.5-3	bal Fe				
ASTM A276(98)	314	Bar Shp	0.25 max	23-26	2 max		19-22	0.045 max	0.03 max	1.5-3	bal Fe				

UNS numbers and US grades are provided as a means of cross referencing chemically similar alloys. Exchangability is only possible after independent examination of specifications. Tensile properties are minimum or typical as specified. UTS and YS as MPa. El as %. See Appendix for list of abbreviations used in Notes. * indicates obsolete material.

Specification	Designation	Notes	C	Cr	Mn	Mo	Ni	P	S	Si	Other	UTS	YS	El	Hard

Stainless Steel, Austenitic, 314 (Continued from previous page)

USA

Specification	Designation	Notes	C	Cr	Mn	Mo	Ni	P	S	Si	Other	UTS	YS	El	Hard
ASTM A314	314	Bil	0.25 max	23-26	2 max		19-22	0.045 max	0.03 max	1.5-3	bal Fe				
ASTM A473	314	Frg, Heat res SHI	0.25 max	23.00-26.00	2.00 max		19.00-22.00	0.045 max	0.030 max	1.50-3.00	bal Fe	515	205	40	
ASTM A580/A580M(98)	314	Wir, Ann	0.25 max	23.0-26.0	2.00 max		19.0-22.0	0.045 max	0.030 max	1.50-3.00	bal Fe	520	210	35	
SAE J467(68)	314		0.12	24.5	1.00		20.5			2.25	bal Fe	690	345	45	89 HRB

Stainless Steel, Austenitic, 316

Australia

Specification	Designation	Notes	C	Cr	Mn	Mo	Ni	P	S	Si	Other	UTS	YS	El	Hard
AS 1449(94)	316	Sh Strp Plt	0.08 max	16.00-18.00	2.00 max	2.00-3.00	10.00-14.00	0.045 max	0.030 max	0.75 max	bal Fe				
AS 2837(86)	316	Bar, Semi-finished product	0.08 max	16.0-18.0	2.00 max	2.0-3.0	10.0-14.0	0.045 max	0.030 max	1.00 max	bal Fe				

Austria

Specification	Designation	Notes	C	Cr	Mn	Mo	Ni	P	S	Si	Other	UTS	YS	El	Hard
ONORM M3120	X5CrNiMo17122S	Sh Strp Plt Bar Wir	0.07 max	16.5-18.5	2 max	2.0-2.5	10.5-13.5	0.045 max	0.030 max	1 max	bal Fe				

Bulgaria

Specification	Designation	Notes	C	Cr	Mn	Mo	Ni	P	S	Si	Other	UTS	YS	El	Hard
BDS 9631	0Ch18N10M2SL	Corr res	0.07 max	17-19	2 max	2-2.5	9-12	0.04 max	0.04 max	2 max	bal Fe				

Canada

Specification	Designation	Notes	C	Cr	Mn	Mo	Ni	P	S	Si	Other	UTS	YS	El	Hard
CSA G110.3	316	Bar Bil	0.08 max	16-18	2 max	2-3	10-14	0.05 max	0.03 max	1 max	bal Fe				
CSA G110.6	316	Sh Strp Plt, HR/Ann or HR/Q/A/Tmp	0.08 max	16-18	2 max	2-3	10-14	0.05 max	0.03 max	1 max	bal Fe				
CSA G110.9	316	Sh Strp Plt, HR, CR, Ann	0.08 max	16-18	2 max	2-3	10-14	0.05 max	0.03 max	1 max	bal Fe				

China

Specification	Designation	Notes	C	Cr	Mn	Mo	Ni	P	S	Si	Other	UTS	YS	El	Hard
GB 1220(92)	0Cr17Ni12Mo2	Bar SA	0.08 max	16.00-18.00	2.00 max	2.00-3.00	10.00-14.00	0.035 max	0.030 max	1.00 max	bal Fe	520	205	40	
GB 1221(92)	0Cr17Ni12Mo2	Bar SA	0.08 max	16.00-18.00	2.00 max	2.00-3.00	10.00-14.00	0.035 max	0.030 max	1.00 max	bal Fe	520	205	40	
GB 12770(91)	0Cr17Ni12Mo2	Weld tube; SA	0.080 max	16.00-18.00	2.00 max	2.00-3.00	10.00-14.00	0.035 max	0.030 max	1.00 max	bal Fe	520	210	35	
GB 12771(91)	0Cr17Ni12Mo2	Weld Pipe SA	0.080 max	16.00-18.00	2.00 max	2.00-3.00	10.00-14.00	0.035 max	0.030 max	1.00 max	bal Fe	520	210	35	
GB 13296(91)	0Cr17Ni12Mo2	Smls tube SA	0.080 max	16.00-18.00	2.00 max	2.00-3.00	10.00-14.00	0.035 max	0.030 max	1.00 max	bal Fe	520	205	35	
GB 3280(92)	0Cr17Ni12Mo2	Sh Plt, CR SA	0.08 max	16.00-18.00	2.00 max	2.00-3.00	10.00-14.00	0.035 max	0.030 max	1.00 max	bal Fe	520	205	40	
GB 4237(92)	0Cr17Ni12Mo2	Sh Plt, HR SA	0.08 max	16.00-18.00	2.00 max	2.00-3.00	10.00-14.00	0.035 max	0.030 max	1.00 max	bal Fe	520	205	40	
GB 4238(92)	0Cr17Ni12Mo2	Plt, HR SA	0.080 max	16.00-18.00	2.00 max	2.00-3.00	10.00-14.00	0.035 max	0.030 max	1.00 max	bal Fe	520	205	40	
GB 4239(91)	0Cr17Ni12Mo2	Sh Strp, CR SA	0.080 max	16.00-18.00	2.00 max	2.00-3.00	10.00-14.00	0.035 max	0.030 max	1.00 max	bal Fe	520	205	40	
GB 4240(93)	0Cr17Ni12Mo2(-L,-Q,-R)	Wir Ann 0.6-1mm diam, L-CD; Q-TLC	0.080 max	16.00-18.00	2.00 max	2.00-3.00	10.00-14.00	0.035 max	0.030 max	1.00 max	bal Fe	540-880			
GB/T 14975(94)	0Cr17Ni12Mo2	Smls tube SA	0.080 max	16.00-18.00	2.00 max	2.00-3.00	10.00-14.00	0.035 max	0.030 max	1.00 max	bal Fe	520	205	35	
GB/T 14976(94)	0Cr17Ni12Mo2	Smls pipe SA	0.080 max	16.00-18.00	2.00 max	2.00-3.00	10.00-14.00	0.035 max	0.030 max	1.00 max	bal Fe	520	205	35	

Czech Republic

Specification	Designation	Notes	C	Cr	Mn	Mo	Ni	P	S	Si	Other	UTS	YS	El	Hard
CSN 417346	17346	Corr res	0.07 max	16.5-18.5	2.0 max	2.0-2.5	10.5-13.5	0.045 max	0.03 max	1.0 max	bal Fe				
CSN 417352	17352	Corr res	0.07 max	16.5-18.5	2.0 max	2.5-3.0	11.0-14.0	0.045 max	0.03 max	1.0 max	bal Fe	*			

Europe

Specification	Designation	Notes	C	Cr	Mn	Mo	Ni	P	S	Si	Other	UTS	YS	El	Hard
EN 10088/2(95)	1.4401	Strp, CR Corr res; t<=6mm, Ann	0.07 max	16.50-18.50	2.00 max	2.00-2.50	10.00-13.00	0.045 max	0.015 max	1.00 max	N <=0.11; bal Fe	530-680	240	40	
EN 10088/2(95)	1.4436	Strp, CR Corr res; t<=6mm, Ann	0.05 max	16.50-18.50	2.00 max	2.50-3.00	10.50-13.00	0.045 max	0.015 max	1.00 max	N <=0.11; bal Fe	550-700	240	40	
EN 10088/2(95)	X3CrNiMo17-13-3	Strp, CR Corr res; t<=6mm, Ann	0.05 max	16.50-18.50	2.00 max	2.50-3.00	10.50-13.00	0.045 max	0.015 max	1.00 max	N <=0.11; bal Fe	550-700	240	40	
EN 10088/2(95)	X5CrNiMo17-12-2	Strp, CR Corr res; t<=6mm, Ann	0.07 max	16.50-18.50	2.00 max	2.00-2.50	10.00-13.00	0.045 max	0.015 max	1.00 max	N <=0.11; bal Fe	530-680	240	40	

Specification	Designation	Notes	C	Cr	Mn	Mo	Ni	P	S	Si	Other	UTS	YS	El	Hard

Stainless Steel, Austenitic, 316 (Continued from previous page)

Europe

Specification	Designation	Notes	C	Cr	Mn	Mo	Ni	P	S	Si	Other	UTS	YS	El	Hard
EN 10088/3(95)	1.4401	Bar Rod Sect, t<=25mm, strain Hard	0.07 max	16.50-18.50	2.00 max	2.00-2.50	10.00-13.00	0.045 max	0.030 max	1.00 max	N <=0.11; bal Fe	800-1000	500	12	
EN 10088/3(95)	1.4401	Bar Rod Sect, t<=35mm, strain Hard	0.07 max	16.50-18.50	2.00 max	2.00-2.50	10.00-13.00	0.045 max	0.030 max	1.00 max	N <=0.11; bal Fe	700-850	350	20	
EN 10088/3(95)	1.4401	Bar Rod Sect, t<=160mm, Ann	0.07 max	16.50-18.50	2.00 max	2.00-2.50	10.00-13.00	0.045 max	0.030 max	1.00 max	N <=0.11; bal Fe	500-700	200	40	215 HB
EN 10088/3(95)	1.4401	Bar Rod Sect; Corr res; 160<t<=250mm, Ann	0.07 max	16.50-18.50	2.00 max	2.00-2.50	10.00-13.00	0.045 max	0.030 max	1.00 max	N <=0.11; bal Fe	500-700	200	30	215 HB
EN 10088/3(95)	1.4436	Bar Rod Sect; Corr res; 0-160mm, Ann	0.05 max	16.50-18.50	2.00 max	2.50-3.00	10.50-13.00	0.045 max	0.030 max	1.00 max	N <=0.11; bal Fe	500-700	200	40	215 HB
EN 10088/3(95)	1.4436	Bar Rod Sect; Corr res; 160<t<=250mm, Ann	0.05 max	16.50-18.50	2.00 max	2.50-3.00	10.50-13.00	0.045 max	0.030 max	1.00 max	N <=0.11; bal Fe	500-700	200	30	215 HB
EN 10088/3(95)	X3CrNiMo17-13-3	Bar Rod Sect; Corr res; 0-160mm, Ann	0.05 max	16.50-18.50	2.00 max	2.50-3.00	10.50-13.00	0.045 max	0.030 max	1.00 max	N <=0.11; bal Fe	500-700	200	40	215 HB
EN 10088/3(95)	X3CrNiMo17-13-3	Bar Rod Sect; Corr res; 160<t<=250mm, Ann	0.05 max	16.50-18.50	2.00 max	2.50-3.00	10.50-13.00	0.045 max	0.030 max	1.00 max	N <=0.11; bal Fe	500-700	200	30T	215 HB
EN 10088/3(95)	X5CrNiMo17-12-2	Bar Rod Sect, t<=160mm, Ann	0.07 max	16.50-18.50	2.00 max	2.00-2.50	10.00-13.00	0.045 max	0.030 max	1.00 max	N <=0.11; bal Fe	500-700	200	40	215 HB
EN 10088/3(95)	X5CrNiMo17-12-2	Bar Rod Sect; Corr res; 160<t<=250mm, Ann	0.07 max	16.50-18.50	2.00 max	2.00-2.50	10.00-13.00	0.045 max	0.030 max	1.00 max	N <=0.11; bal Fe	500-700	200	30	215 HB
EN 10088/3(95)	X5CrNiMo17-12-2	Bar Rod Sect, t<=35mm, strain Hard	0.07 max	16.50-18.50	2.00 max	2.00-2.50	10.00-13.00	0.045 max	0.030 max	1.00 max	N <=0.11; bal Fe	700-850	350	20	
EN 10088/3(95)	X5CrNiMo17-12-2	Bar Rod Sect, t<=25mm, strain Hard	0.07 max	16.50-18.50	2.00 max	2.00-2.50	10.00-13.00	0.045 max	0.030 max	1.00 max	N <=0.11; bal Fe	800-1000	500	12	

France

Specification	Designation	Notes	C	Cr	Mn	Mo	Ni	P	S	Si	Other	UTS	YS	El	Hard
AFNOR NFA35572	Z6CND17.11	Bar, 25mm diam	0.07 max	16-18	2 max	2-2.5	10-12.5	0.04 max	0.03 max	1 max	bal Fe				
AFNOR NFA35573	Z6CND17.11	Sh Strp Plt, CR, 5mm diam	0.07 max	16-18	2 max	2-2.5	10-12.5	0.04 max	0.03 max	1 max	bal Fe ·				
AFNOR NFA35574	Z6CND17.11	Bar Rod	0.07 max	16-18	2 max	2-2.5	10-12.5	0.04 max	0.03 max	1 max	bal Fe				
AFNOR NFA35575	Z6CND17.11	Wir rod, 60% Hard	0.07 max	16-18	2 max	2-2.5	10-12.5	0.04 max	0.03 max	1 max	bal Fe				
AFNOR NFA35577	Z6CND17.11	Bar Wir	0.07 max	16-18	2 max	2-2.5	10-12	0.04 max	0.03 max	1 max	bal Fe				
AFNOR NFA36209	Z6CND17.11	Plt	0.07 max	16-18	2 max	2-2.4	10-12.5	0.04 max	0.03 max	1 max	bal Fe				
AFNOR NFA36607	Z6CND17.11	Frg	0.07 max	16-18	2 max	2-2.4	10-12.5	0.04 max	0.03 max	1 max	Cu <=1; bal Fe				

Germany

Specification	Designation	Notes	C	Cr	Mn	Mo	Ni	P	S	Si	Other	UTS	YS	El	Hard
DIN 17224(82)	WNr 1.4401	Spring wire, Strp	0.07 max	16.5-18.5	2.00 max	2.00-2.50	10.5-13.5	0.045 max	0.030 max	1.00 max	N <=0.11; bal Fe				
DIN 17440(96)	WNr 1.4401	Plt Sh, HR Strp Bar Wir Frg	0.07 max	16.5-18.5	2.00 max	2.00-2.50	10.5-13.5	0.045 max	0.030 max	1.00 max	N <=0.11; bal Fe				
DIN 17441(97)	WNr 1.4401	CR Strp Plt Sh	0.07 max	16.5-18.5	2.00 max	2.00-2.50	10.5-13.5	0.045 max	0.030 max	1.00 max	N <=0.11; bal Fe				
DIN 17455(85)	WNr 1.4401	Weld tube	0.07 max	16.5-18.5	2.00 max	2.00-2.50	10.5-13.5	0.045 max	0.030 max	1.00 max	N <=0.11; bal Fe				
DIN 17456(85)	WNr 1.4401	Smls tube	0.07 max	16.5-18.5	2.00 max	2.00-2.50	10.5-13.5	0.045 max	0.030 max	1.00 max	N <=0.11; bal Fe				
DIN EN 10028(96)	WNr 1.4401	Flat product for Press, high-temp	0.07 max	16.5-18.5	2.00 max	2.00-2.50	10.5-13.5	0.045 max	0.030 max	1.00 max	N <=0.11; bal Fe				
DIN EN 10088(95)	WNr 1.4401	Sh Plt Strp Bar Rod, long SHT	0.07 max	16.5-18.5	2.00 max	2.00-2.50	10.5-13.5	0.045 max	0.030 max	1.00 max	N <=0.11; bal Fe	500-700	235	30-40	
DIN EN 10088(95)	X5CrNiMo17-12-2	Sh Plt Strp Bar Rod, long SHT	0.07 max	16.5-18.5	2.00 max	2.00-2.50	10.5-13.5	0.045 max	0.030 max	1.00 max	N <=0.11; bal Fe	500-700	215	30-40	
DIN EN 10222(94)	WNr 1.4401	Frg	0.07 max	16.5-18.5	2.00 max	2.00-2.50	10.5-13.5	0.045 max	0.030 max	1.00 max	N <=0.11; bal Fe				
TGL 7143	X5CrNiMo1811*	Obs	0.06 max	16.5-18.5	2 max	2-2.5	10.5-13.5	0.045 max	0.03 max	0.8 max	bal Fe				

India

Specification	Designation	Notes	C	Cr	Mn	Mo	Ni	P	S	Si	Other	UTS	YS	El	Hard
IS 6527	04Cr17Ni12Mo2	Wir rod	0.08 max	16-18.5	2 max	2-3	10-14	0.04 max	0.03 max	1 max	bal Fe				

UNS numbers and US grades are provided as a means of cross referencing chemically similar alloys. Exchangability is only possible after independent examination of specifications. Tensile properties are minimum or typical as specified. UTS and YS as MPa. El as %. See Appendix for list of abbreviations used in Notes. * indicates obsolete material.

Stainless Steel, Austenitic, 316 (Continued from previous page)

Specification	Designation	Notes	C	Cr	Mn	Mo	Ni	P	S	Si	Other	UTS	YS	El	Hard
India															
IS 6528	04Cr17Ni12Mo2	Wir, Ann, SA	0.08 max	16-18.5	2 max	2-3	10-14	0.04 max	0.03 max	1 max	bal Fe				
IS 6529	04Cr17Ni12Mo2	Bil	0.08 max	16-18.5	2 max	2-3	10-14	0.04 max	0.03 max	1 max	bal Fe				
IS 6603	04Cr17Ni12Mo2	Bar	0.08 max	16-18.5	2 max	2-3	10-14	0.04 max	0.03 max	1 max	bal Fe				
IS 6911	04Cr17Ni12Mo2	Sh Strp Plt	0.08 max	16-18.5	2 max	2-3	10-14	0.04 max	0.03 max	1 max	bal Fe				
International															
ISO 2604-1(75)	F62	Frg	0.07 max	16.00-18.00	2.00 max	2.00-3.00	10.00-14.00	0.045 max	0.030 max	1.00 max	bal Fe	490-690			
ISO 2604-2(75)	TS60	Smls tube	0.07 max	16-18.5	2 max	2-2.5	11-14	0.05 max	0.03 max	1 max	bal Fe				
ISO 2604-2(75)	TS61	Smls tube	0.07 max	16-18.5	2 max	2.5-3	11-14.5	0.05 max	0.03 max	1 max	bal Fe				
ISO 2604-4	P60*	Plt	0.07 max	16-18.5	2 max	2-2.5	10.5-14	0.05 max	0.03 max	1 max	bal Fe				
ISO 2604-4	P61*	Plt	0.07 max	16-18.5	2 max	2.5-3	11-14.5	0.05 max	0.03 max	1 max	bal Fe				
ISO 4954(93)	X5CrNiMo17122E	Wir rod Bar, CH ext	0.07 max	16.5-18.5	2.00 max	2.0-2.5	10.5-13.5	0.045 max	0.030 max	1.00 max	bal Fe	660			
ISO 683-13(74)	20*	Sh Plt Bar, ST	0.07 max	16.5-18.5	2 max	2-2.5	10.5-13.5	0.045 max	0.03 max	1 max	bal Fe				
ISO 683-13(74)	20a*	Sh Plt Bar, ST	0.07 max	16.5-18.5	2 max	2.5-3	11-14	0.045 max	0.03 max	1 max	bal Fe				
ISO 6931-2(89)	X5CrNiMo17122	Spring	0.07 max	16.5-18.5	2 max	2-2.5	10.5-13.5			1 max	bal Fe				
Italy															
UNI 6901(71)	X5CrNiMo1712	Bar Rod; Corr res	0.06 max	16-18.5	2 max	2-2.5	10.5-13.5	0.045 max	0.03 max	1 max	bal Fe				
UNI 6901(71)	X5CrNiMo1713	Corr res	0.06 max	16-18.5	2 max	2.5-3	11-14	0.045 max	0.03 max	1 max	bal Fe				
UNI 6901(71)	X8CrNiMo1713	Corr res	0.04-0.1	16-18	2 max	2.5-3	11-14	0.03 max	0.03 max	0.75 max	bal Fe				
UNI 6904(71)	X5CrNiMo1712	Tube; Heat res	0.08 max	16-18	2 max	2-2.5	11-13.5	0.03 max	0.03 max	0.75 max	V <=0.1; bal Fe				
UNI 6904(71)	X5CrNiMo1713	Heat res	0.08 max	16-18	2 max	2.5-3	11-14	0.03 max	0.03 max	0.75 max	bal Fe				
UNI 6904(71)	X8CrNiMo1713	Heat res	0.04-0.1	16-18	2 max	2.5-3	11-14	0.03 max	0.03 max	0.75 max	bal Fe				
UNI 7500(75)	X5CrNiMo1712	Press ves	0.06 max	16-18.5	2 max	2-2.5	10.5-13.5	0.045 max	0.03 max	1 max	bal Fe				
UNI 7500(75)	X5CrNiMo1713	Press ves	0.06 max	16-18.5	2 max	2.5-3	11-14	0.045 max	0.03 max	1 max	bal Fe				
UNI 8317(81)	X5CrNiMo1712	Sh Strp Plt	0.06 max	16-18.5	2 max	2-2.5	10.5-13.5	0.045 max	0.03 max	1 max	bal Fe				
Japan															
JIS G3214(91)	SUSF316	Frg press ves 130-200mm diam	0.08 max	16.00-18.00	2.00 max	2.00-3.00	10.00-14.00	0.040 max	0.030 max	1.00 max	bal Fe	480	205	29	187 HB max
JIS G3459(94)	SUS316TP	Pipe	0.08 max	16-18	2 max	2-3	10-14	0.04 max	0.03 max	1 max	bal Fe	520	205	35	
JIS G4303(91)	SUS316	Bar; SA; <=180mm diam	0.08 max	16.00-18.00	2.00 max	2.00-3.00	10.00-14.00	0.045 max	0.030 max	1.00 max	bal Fe	520	205	40	187 HB
JIS G4303(91)	SUS316J1	Bar; SA; <=180mm diam	0.08 max	17.00-19.00	2.00 max	1.00-2.50	10.00-14.00	0.045 max	0.030 max	1.00 max	Cu 1.00-2.50; bal Fe	520	205	40	187 HB
JIS G4303(91)	SUS316J1L	Bar; SA; <=180mm diam	0.030 max	17.00-19.00	2.00 max	1.20-2.75	12.00-16.00	0.045 max	0.030 max	1.00 max	Cu 1.00-2.50; bal Fe	480	175	40	187 HB
JIS G4303(91)	SUS316Ti	Bar; SA; <=180mm diam	0.08 max	16.00-18.00	2.00 max	2.00-3.00	10.00-14.00	0.045 max	0.030 max	1.00 max	Ti>=5xC; bal Fe	520	205	40	187 HB
JIS G4304(91)	SUS316	Sh Plt, HR SA	0.08 max	16.00-18.00	2.00 max	2.00-3.0	10.0-14.00	0.045 max	0.030 max	1.00 max	bal Fe	520	205	40	90 HRB max
JIS G4305(91)	SUS316	Sh Plt, CR SA	0.08 max	16.00-18.00	2.00 max	2.00-3.00	10.00-14.00	0.045 max	0.030 max	1.00 max	bal Fe	520	205	40	187 HB max
JIS G4305(91)	SUS316J1		0.08 max	17.00-19.00	2.00 max	1.20-2.75	10.00-14.00	0.045 max	0.030 max	1.00 max	Cu 1.00-2.50; bal Fe	520	205	40	
JIS G4305(91)	SUS316J1L		0.030 max	17.00-19.00	2.00 max	1.20-2.75	12.00-16.00	0.045 max	0.030 max	1.00 max	Cu 1.00-2.50; bal Fe	480	175	40	

UNS numbers and US grades are provided as a means of cross referencing chemically similar alloys. Exchangability is only possible after independent examination of specifications. Tensile properties are minimum or typical as specified. UTS and YS as MPa. El as %. See Appendix for list of abbreviations used in Notes. * indicates obsolete material.

Specification	Designation	Notes	C	Cr	Mn	Mo	Ni	P	S	Si	Other	UTS	YS	El	Hard

Stainless Steel, Austenitic, 316 (Continued from previous page)

Specification	Designation	Notes	C	Cr	Mn	Mo	Ni	P	S	Si	Other	UTS	YS	El	Hard
Japan															
JIS G4306	SUS316	Strp HR SA	0.08 max	16-18	2 max	2-3	10-14	0.04 max	0.03 max	1 max	bal Fe				
JIS G4307	SUS316	Strp CR SA	0.08 max	16-18	2 max	2-3	10-14	0.04 max	0.03 max	1 max	bal Fe				
JIS G4308	SUS316F	Wir rod	0.08 max	16.00-18.00	2.00 max	2.00-3.00	10.00-14.00	0.045 max	0.10 max	1.00	bal Fe				
JIS G4308(98)	SUS316	Wir rod	0.08 max	16.00-18.00	2.00 max	2.00-3.00	10.00-14.00	0.045 max	0.030 max	1.00 max	bal Fe				
JIS G4309	SUS316	Wir	0.08 max	16.00-18.00	2.00 max		10.00-14.00	0.045 max	0.030 max	1.00 max	bal Fe				
JIS G4315	SUS316	Wir rod	0.08 max	16.00-18.00	2.00 max	2.00-3.00	10.00-14.00	0.045 max	0.030 max	1.00 max	bal Fe				
Mexico															
DGN B-150	AE-316*	Obs; Wire	0.08 max	18-20	1-2.5	2-3	11-14	0.03 max	0.03 max	0.25-0.6	Cu <=0.25; N <=0.7; bal Fe				
DGN B-83	316*	Obs; Bar, HR, CR	0.08 max	16-18	2 max	2-3	10-14	0.05 max	0.03 max	1 max	bal Fe				
NMX-B-171(91)	MT316	CF HF Smls tub	0.08 max	16.0-18.0	2.00 max	2.0-3.0	11.0-14.0	0.040 max	0.030 max	1.00 max	bal Fe				
NMX-B-176(91)	TP316	Smls, weld sanitary tub	0.08 max	16.00-18.00	2.00 max	2.00-3.00	10.00-14.00	0.040 max	0.030 max	0.75 max	bal Fe				
NMX-B-186-SCFI(94)	TP316	Pipe; High-temp	0.08 max	16.0-18.0	2.00 max	2.00-3.00	11.0-14.0	0.040 max	0.030 max	0.75 max	bal Fe	517	207		
Pan America															
COPANT 513	TP316	Tube, HF, CF	0.08 max	16-18	2 max	2-3	11-14	0.03 max	0.03 max	0.75 max	bal Fe				
Spain															
UNE 36016(75)	F.3534*	see X6CrNiMo17-12-03	0.08 max	16.0-18.0	2.0 max	2.0-3.0	10.0-14.0	0.045 max	0.03 max	1.0 max	bal Fe				
UNE 36016(75)	X6CrNiMo17-12-03	Corr res	0.08 max	16.0-18.0	2.0 max	2.0-3.0	10.0-14.0	0.045 max	0.03 max	1.0 max	bal Fe				
UNE 36016(75)	XCrNiMo17-12-03	Sh Strp Plt Bar Wir Bil	0.08 max	16-18	2 max	2-3	10-14	0.045 max	0.03 max	1 max	bal Fe				
UNE 36016/1(89)	E-316	Corr res	0.07 max	16.5-18.5	2.0 max	2.0-2.5	10.5-13.5	0.045 max	0.03 max	1.0 max	N <=0.1; bal Fe				
UNE 36016/1(89)	F.3534*	Obs; E-316; Corr res	0.07 max	16.5-18.5	2.0 max	2.0-2.5	10.5-13.5	0.045 max	0.03 max	1.0 max	N <=0.1; bal Fe				
UNE 36016/1(89)	F.3538*	Obs; EN10088-3(96); X5CrNiMo17-13-3	0.07 max	16.5-18.5	2.0 max	2.5-3.0	11.0-14.0	0.045 max	0.03 max	1.0 max	N <=0.1; bal Fe				
UNE 36087(78)	F.3543*	see X5CrNiMo17-12	0.03-0.07	16.0-18.5	2.0 max	2.0-2.5	10.5-14.0	0.045 max	0.03 max	1.0 max	bal Fe				
UNE 36087(78)	X5CrNiMo17-12	Corr res	0.03-0.07	16.0-18.5	2.0 max	2.0-2.5	10.5-14.0	0.045 max	0.03 max	1.0 max	bal Fe				
UNE 36087(78)	XCrNiMo17-12-03	Sh Strip	0.07 max	16-18	2 max	2-3	10-14	0.045 max	0.03 max	1 max	bal Fe				
UNE 36087/4(89)	F.3534*	see X5CrNiMo17 12 2	0.07 max	16.5-18.5	2 max	2-2.5	10.5-13.5	0.045 max	0.03 max	1 max	N <=0.1; bal Fe				
UNE 36087/4(89)	F.3538*	see X5CrNiMo17 13 3	0.07 max	16.5-18.5	2 max	2.5-3	11-14	0.045 max	0.03 max	1 max	N <=0.1; bal Fe				
UNE 36087/4(89)	X5CrNiMo17 12 2	Corr res; Sh Strp	0.07 max	16.5-18.5	2 max	2-2.5	10.5-13.5	0.045 max	0.03 max	1 max	N <=0.1; bal Fe				
UNE 36087/4(89)	X5CrNiMo17 13 3	Corr res; Sh Strp	0.07 max	16.5-18.5	2 max	2.5-3	11-14	0.045 max	0.03 max	1 max	N <=0.1; bal Fe				
Sweden															
SS 142343	2343	Corr res	0.05 max	16-18.5	2 max	2.5-3	10.5-14	0.045 max	0.03 max	1 max	bal Fe				
SS 142347	2347	Corr res	0.05 max	16.5-18.5	2 max	2-2.5	10.5-14	0.045 max	0.03 max	1 max	bal Fe				
UK															
BS 1449/2(83)	316S31	Corr/Heat res; t<=100mm	0.07 max	16.5-18.5	2.00 max	2.00-2.50	10.5-13.5	0.045 max	0.030 max	1.00 max	bal Fe	510	205	40	0-205 HB
BS 1449/2(83)	316S33	Corr/Heat res; t<=100mm	0.07 max	16.5-18.5	2.00 max	2.50-3.00	11.0-14.0	0.045 max	0.030 max	1.00 max	bal Fe	510	205	40	0-205 HB
BS 1501/3(73)	316S16	Plate	0.07 max	16.5-18.5	0.5-2.0	2.25-3.0	10.0-13.0	0.045 max	0.03 max	0.2-1.0	bal Fe				
BS 1501/3(90)	316S31	Press ves; Corr/Heat res; 0.05-100mm	0.07 max	16.5-18.5	2.0 max	2.0-2.5	10.5-13.5	0.045 max	0.025 max	1.0 max	Cu <=0.3; Ti <=0.05; W <=0.1; bal Fe	510-710	205	40	

UNS numbers and US grades are provided as a means of cross referencing chemically similar alloys. Exchangability is only possible after independent examination of specifications. Tensile properties are minimum or typical as specified. UTS and YS as MPa. El as %. See Appendix for list of abbreviations used in Notes. * indicates obsolete material.

Specification	Designation	Notes	C	Cr	Mn	Mo	Ni	P	S	Si	Other	UTS	YS	El	Hard

Stainless Steel, Austenitic, 316 (Continued from previous page)

UK

Specification	Designation	Notes	C	Cr	Mn	Mo	Ni	P	S	Si	Other	UTS	YS	El	Hard
BS 1501/3(90)	316S33	Press ves; Corr/Heat res	0.07 max	16.5-18.5	2.0 max	2.5-3.0	11.0-14.0	0.045 max	0.025 max	1.0 max	Cu <=0.3; Ti <=0.05; W <=0.1; bal Fe	510-710	205	40	
BS 1501/3(90)	316S51	Press ves; Corr/Heat res	0.04-0.1	16.5-18.5	2.0 max	2.0-2.5	10.0-13.0	0.045 max	0.025 max	1.0 max	Cu <=0.3; Ti <=0.05; W <=0.1; bal Fe	510-710	205	40	
BS 1501/3(90)	316S61	Press ves; Corr/Heat res	0.03 max	16.5-18.5	2.0 max	2.0-2.5	10.5-13.5	0.045 max	0.025 max	1.0 max	Cu <=0.3; N 0.12-0.22; Ti <=0.05; W <=0.1; bal Fe	580-780	280	35	
BS 1502	316S31	Bar Shp; Press ves	0.07 max	16.5-18.5	2.00 max	2.00-2.50	10.5-13.5	0.045 max	0.030 max	1.00 max	bal Fe				
BS 1502	316S32	Bar Shp; Press ves	0.07 max	16.5-18.5	2.00 max	2.50-3.00	11.0-14.0	0.045 max	0.030 max	1.00 max	bal Fe				
BS 1502(82)	316S51	Bar Shp; Press ves	0.04-0.10	16.5-18.5	2.00 max	2.00-2.50	10.5-13.5	0.045 max	0.030 max	1.00 max	W <=0.1; bal Fe				
BS 1503(89)	316S31	Frg press ves; t<=999mm; HT	0.07 max	16.5-18.5	2.00 max	2.00-2.50	10.5-13.5	0.040 max	0.025 max	1.00 max	B <=0.005; Cu <=0.70; Nb <=0.20; Ti <=0.10; W <=0.1; bal Fe	510-710	240	30	
BS 1503(89)	316S33	Frg press ves; t<=999mm; HT	0.07 max	16.5-18.5	2.00 max	2.50-3.00	11.0-14.0	0.040 max	0.025 max	1.00 max	B <=0.005; Cu <=0.70; Nb <=0.20; Ti <=0.10; W <=0.1; bal Fe	510-710	240	30	
BS 1506(90)	316S31	Bolt matl Pres/Corr res; t<=160mm; Hard	0.07 max	16.5-18.5	2.00 max	2.00-2.50	10.5-13.5	0.045 max	0.030 max	1.00 max	bal Fe	520	205	40	
BS 1506(90)	316S31	Bolt matl Pres/Corr res; Corr res; 38<t<=44mm; SHT CD	0.07 max	16.5-18.5	2.00 max	2.00-2.50	10.5-13.5	0.045 max	0.030 max	1.00 max	bal Fe	650	310	28	0-320 HB
BS 1506(90)	316S33	Bolt matl Pres/Corr res; Corr res; 38<t<=44mm; SHT CD; was En 845	0.07 max	16.5-18.5	2.00 max	2.50-3.00	11.0-14.0	0.045 max	0.030 max	1.00 max	bal Fe	650	310	28	0-320 HB
BS 1506(90)	316S33	Bolt matl Pres/Corr res; t<=160mm; Hard; was En845	0.07 max	16.5-18.5	2.00 max	2.50-3.00	11.0-14.0	0.045 max	0.030 max	1.00 max	bal Fe	520	205	40	
BS 1554(90)	316S19	Wir Corr/Heat res; diam<0.49mm; Ann	0.07 max	16.0-18.5	2.00 max	2.00-3.00	10.0-14.0	0.045 max	0.030 max	1.00 max	W <=0.1; bal Fe	0-900			
BS 1554(90)	316S19	Wir Corr/Heat res; 6<diam<=13mm; Ann	0.07 max	16.0-18.5	2.00 max	2.00-3.00	10.0-14.0	0.045 max	0.030 max	1.00 max	W <=0.1; bal Fe	0-700			
BS 1554(90)	316S33	Wir Corr/Heat res; diam<0.49mm; Ann	0.07 max	16.0-18.5	2.00 max	2.00-3.00	10.0-14.0	0.045 max	0.030 max	1.00 max	W <=0.1; bal Fe	0-900			
BS 1554(90)	316S33	Wir Corr/Heat res; 6<diam<=13mm; Ann	0.07 max	16.0-18.5	2.00 max	2.00-3.00	10.0-14.0	0.045 max	0.030 max	1.00 max	W <=0.1; bal Fe	0-700			
BS 3059/2(90)	316S52	Boiler, Superheater; High-temp; Smls tub; 2-11mm	0.04-0.10	16.5-18.5	2.00 max	2.00-2.50	10.5-13.5	0.040 max	0.030 max	1.00 max	B 0.0015-0.006; bal Fe	510-710	240	35	
BS 3111/2(87)	316S17	Wir	0.07 max	16.5-18.5	2.00 max	2.00-3.00	12.0-14.0	0.045 max	0.030 max	1.00 max	bal Fe				
BS 3605(73)	316S18	Pip Tube press smls weld	0.07 max	16.0-18.5	0.5-2.0	2.0-3.0	11.0-14.0	0.040 max	0.030 max	0.2-1.0	bal Fe				
BS 3605(73)	316S26	Corr res; Tub, Weld; Smls tub	0.07 max	16.0-18.5	0.5-2.0	2.0-3.0	10.0-13.0	0.040 max	0.030 max	0.2-1.0	bal Fe				
BS 3605/1(91)	316S13	Pip Tube press smls; t<=200mm	0.030 max	16.5-18.5	2.00 max	2.50-3.00	11.5-14.5	0.040 max	0.030 max	1.00 max	bal Fe	490-690		35	
BS 3605/1(91)	316S31	Pip Tube press smls; t<=200mm	0.070 max	16.5-18.5	2.00 max	2.00-2.50	10.5-13.5	0.040 max	0.030 max	1.00 max	bal Fe	510-710		35	
BS 3605/1(91)	316S33	Pip Tube press smls; t<=200mm	0.070 max	16.5-18.5	2.00 max	2.50-3.00	11.0-14.0	0.040 max	0.030 max	1.00 max	bal Fe	510-710		35	
BS 3605/1(91)	316S51	Pip Tube press smls; t<=200mm	0.04-0.10	16.5-18.5	2.00 max	2.00-2.5	11.0-14.0	0.040 max	0.030 max	1.00 max	bal Fe	510-710		35	
BS 3605/1(91)	316S52	Pip Tube press smls; t<=200mm	0.04-0.10	16.5-18.5	2.00 max	2.00-2.5	10.5-13.5	0.040 max	0.030 max	1.00 max	B 0.0015-0.006; bal Fe	510-710		35	
BS 3606(78)	316S24	Corr res; Heat exch Tub	0.030 max	16.5-18.5	0.5-2.0	2.00-2.50	11.0-14.0	0.045 max	0.030 max	1.00 max	bal Fe				
BS 3606(78)	316S25	Corr res; Heat exch Tub	0.070 max	16.5-18.5	0.5-2.0	2.00-2.50	10.5-13.5	0.045 max	0.030 max	1.00 max	bal Fe				
BS 3606(78)	316S30	Corr res; Heat exch Tub	0.070 max	16.5-18.5	0.5-2.0	2.50-3.00	11.0-14.0	0.045 max	0.030 max	1.00 max	bal Fe				
BS 3606(92)	316S13	Heat exch Tub; Smls tub, Tub, Weld; 1.2-3.2mm	0.030 max	16.5-18.5	2.00 max	2.50-3.00	11.5-14.5	0.040 max	0.030 max	1.00 max	bal Fe	490-690	215	30	

UNS numbers and US grades are provided as a means of cross referencing chemically similar alloys. Exchangability is only possible after independent examination of specifications. Tensile properties are minimum or typical as specified. UTS and YS as MPa. El as %. See Appendix for list of abbreviations used in Notes. * indicates obsolete material.

Specification	Designation	Notes	C	Cr	Mn	Mo	Ni	P	S	Si	Other	UTS	YS	El	Hard

Stainless Steel, Austenitic, 316 (Continued from previous page)

UK

Specification	Designation	Notes	C	Cr	Mn	Mo	Ni	P	S	Si	Other	UTS	YS	El	Hard
BS 3606(92)	316S31	Heat exch Tub; Smls tub, Tub, Weld; 1.2-3.2mm	0.070 max	16.5-18.5	2.00 max	2.00-2.50	10.5-13.5	0.040 max	0.030 max	1.00 max	bal Fe	510-710	245	30	
BS 3606(92)	316S33	Heat exch Tub; Smls tub, Tub, Weld; 1.2-3.2mm	0.070 max	16.5-18.5	2.00 max	2.50-3.00	11.0-14.0	0.040 max	0.030 max	1.00 max	bal Fe	510-710	245	30	
BS 970/1(96)	316S31	Wrought Corr/Heat res; t<=160mm	0.07 max	16.5-18.5	2.00 max	2.00-2.50	10.5-13.5	0.045 max	0.030 max	1.00 max	bal Fe	510	205	40	183 HB max
BS 970/1(96)	316S33	Wrought Corr/Heat res; t<=160mm	0.07 max	16.5-18.5	2.00 max	2.50-3.00	11.0-14.0	0.045 max	0.030 max	1.00 max	bal Fe	510	205	40	183 HB max
BS 970/4(70)	316S16*	Obs; Corr res	0.07 max	16.5-18.5	0.50-2.00	2.25-3.0	10.0-13.0	0.045 max	0.03 max	0.2-1.0	Cu <=0.7; Pb <=0.15; Nb<=0.2; bal Fe				

USA

Specification	Designation	Notes	C	Cr	Mn	Mo	Ni	P	S	Si	Other	UTS	YS	El	Hard
	AISI 316	Tube	0.08 max	16.00-18.00	2.00 max	2.00-3.00	10.00-14.00	0.045 max	0.030 max	1.00 max	bal Fe				
	UNS S31600		0.08 max	16.00-18.00	2.00 max	2.00-3.00	10.00-14.00	0.045 max	0.030 max	1.00 max	bal Fe				
	UNS S31670	Surgical Implant 316	0.08 max	17.00-19.00	2.00 max	2.00-3.00	13.00-15.50	0.025 max	0.010 max	0.75 max	Cu <=0.50; N <=0.10; bal Fe				
AMS 5524		Sh Strp Plt Bar Wir	0.08 max	16-18	2 max	2-3	10-14	0.045 max	0.03 max	1 max	bal Fe				
AMS 5573		Sh Strp Plt Bar Wir	0.08 max	16-18	2 max	2-3	10-14	0.045 max	0.03 max	1 max	bal Fe				
AMS 5648		Sh Strp Plt Bar Wir	0.08 max	16-18	2 max	2-3	10-14	0.045 max	0.03 max	1 max	bal Fe				
AMS 5690		Sh Strp Plt Bar Wir	0.08 max	16-18	2 max	2-3	10-14	0.045 max	0.03 max	1 max	bal Fe				
AMS 5696		Sh Strp Plt Bar Wir	0.08 max	16-18	2 max	2-3	10-14	0.045 max	0.03 max	1 max	bal Fe				
AMS 7490		Sh Strp Plt Bar Wir	0.08 max	16-18	2 max	2-3	10-14	0.045 max	0.03 max	1 max	bal Fe				
ASME SA182	316	Refer to ASTM A182/A182M													
ASME SA213	316	Refer to ASTM A213/A213M(95)													
ASME SA240	316	Refer to ASTM A240(95)													
ASME SA249	316	Refer to ASTM A249/A249M(96)													
ASME SA312	316	Refer to ASTM A312/A312M(95)													
ASME SA358	316	Refer to ASTM A358/A358M(95)													
ASME SA376	316	Refer to ASTM A376/A376M(93)													
ASME SA403	316	Refer to ASTM A403/A403M(95)													
ASME SA409	316	Refer to ASTM A409/A409M(95)													
ASME SA430	316	Refer to ASTM A430/A430M(91)													
ASME SA479	316	Refer to ASTM A479/A479M(95)													
ASME SA688	316	Refer to ASTM A688/A688M(96)													
ASTM A167(96)	316*	Sh Strp Plt	0.08 max	16-18	2 max	2-3	10-14	0.045 max	0.03 max	1 max	bal Fe				
ASTM A182	316*	Bar Frg	0.08 max	16-18	2 max	2-3	10-14	0.045 max	0.03 max	1 max	bal Fe				
ASTM A182/A182M(98)	F316	Frg/roll pipe flange valve	0.08 max	16.0-18.0	2.00 max	2.00-3.00	10.0-14.0	0.045 max	0.030 max	1.00 max	N <=0.10; bal Fe	515	205	30	
ASTM A193/A193M(98)	316*	Bolt, high-temp	0.08 max	16-18	2 max	2-3	10-14	0.045 max	0.03 max	1 max	bal Fe				
ASTM A193/A193M(98)	B8M	Bolt, high-temp, HT	0.08 max	16.0-18.0	2.00 max	2.00-3.00	10.0-14.0	0.045 max	0.030 max	1.00 max	bal Fe	515	205	30	223 HB
ASTM A193/A193M(98)	B8M2	Bolt, high-temp, HT	0.08 max	16.0-18.0	2.00 max	2.00-3.00	10.0-14.0	0.045 max	0.030 max	1.00 max	bal Fe	655	515	25	321 HB

UNS numbers and US grades are provided as a means of cross referencing chemically similar alloys. Exchangability is only possible after independent examination of specifications. Tensile properties are minimum or typical as specified. UTS and YS as MPa. El as %. See Appendix for list of abbreviations used in Notes. * indicates obsolete material.

Stainless Steel, Austenitic, 316 (Continued from previous page)

USA

Specification	Designation	Notes	C	Cr	Mn	Mo	Ni	P	S	Si	Other	UTS	YS	El	Hard
ASTM A193/A193M(98)	B8M3	Bolt, high-temp, HT	0.08 max	16.0-18.0	2.00 max	2.00-3.00	10.0-14.0	0.045 max	0.030 max	1.00 max	bal Fe	585	450	30	321 HB
ASTM A193/A193M(98)	B8MA	Bolt, high-temp, HT	0.08 max	16.0-18.0	2.00 max	2.00-3.00	10.0-14.0	0.045 max	0.030 max	1.00 max	bal Fe	515	205	30	192 HB
ASTM A194	316*	Nuts	0.08 max	16-18	2 max	2-3	10-14	0.045 max	0.03 max	1 max	bal Fe				
ASTM A194/A194M(98)	8M	Nuts, high-temp press	0.08 max	16.0-18.0	2.00 max	2.00-3.00	10.0-14.0	0.045 max	0.030 max	1.00 max	bal Fe				126-300 HB
ASTM A194/A194M(98)	8MA	Nuts, high-temp press	0.08 max	16.0-18.0	2.00 max	2.00-3.00	10.0-14.0	0.045 max	0.030 max	1.00 max	bal Fe				126-192 HB
ASTM A213	316*	Tube	0.08 max	16-18	2 max	2-3	10-14	0.045 max	0.03 max	1 max	bal Fe				
ASTM A213/A213M(95)	TP316	Smls tube boiler, superheater, heat exchanger	0.08 max	16.0-18.0	2.00 max	2.00-3.00	11.0-14.0	0.040 max	0.030 max	0.75 max	bal Fe	515	205	35	
ASTM A240/A240M(98)	S31600		0.08 max	16.0-18.0	2.00 max	2.00-3.00	10.0-14.0	0.045 max	0.030 max	1.00 max	N <=0.10; bal Fe	515	205	40.0	217 HB
ASTM A249/249M(96)	TP316	Weld tube; boiler, superheater, heat exch	0.08 max	16.0-18.0	2.00 max	2.00-3.00	10.0-14.0	0.040 max	0.030 max	0.75 max	bal Fe	515	205	35	90 HRB
ASTM A269	316	Smls Weld, Tube	0.08 max	16-18	2 max	2-3	10-14	0.045 max	0.03 max	1 max	bal Fe				
ASTM A276(98)	316	Bar	0.08 max	16-18	2 max	2-3	10-14	0.045 max	0.03 max	1 max	bal Fe				
ASTM A312/A312M(95)	316	Smls Weld, Pipe	0.08 max	16-18	2 max	2-3	10-14	0.045 max	0.03 max	1 max	bal Fe				
ASTM A313/A313M(95)	316	Spring wire, t<=0.25mm	0.08 max	16.00-18.00	2.00 max	2.00-3.00	10.00-14.00	0.045 max	0.030 max	1.00 max	bal Fe	1690-1895			
ASTM A313/A313M(95)	XM-16	Spring wire, CD, t<=0.23mm	0.50 max	11.00-12.50	0.50 max	0.50 max	7.50-9.50	0.040 max	0.030 max	0.50 max	Cu 1.50-2.50; Ti 0.80-1.40; 0.10<=Nb+Ta<=0.50; bal Fe	1690			
ASTM A314	316	Bil	0.08 max	16-18	2 max	2-3	10-14	0.045 max	0.03 max	1 max	bal Fe				
ASTM A320	316	Bolt	0.08 max	16-18	2 max	2-3	10-14	0.045 max	0.03 max	1 max	bal Fe				
ASTM A336/A336M(98)	F316	Frg, press/high-temp	0.08 max	16.0-18.0	2.00 max	2.00-3.00	10.0-14.0	0.040 max	0.030 max	1.00 max	bal Fe	485	205	30	90 HRB
ASTM A358/A358M(95)	316	Elect fusion weld pipe, high-temp	0.08 max	16-18	2 max	2-3	10-14	0.045 max	0.03 max	1 max	bal Fe				
ASTM A368(95)	316	Wir, Heat res	0.08 max	16.00-18.00	2.00 max	2.00-3.00	10.00-14.00	0.045 max	0.030 max	1.00 max	bal Fe				
ASTM A376	316	Smls pipe	0.08 max	16-18	2 max	2-3	10-14	0.045 max	0.03 max	1 max	bal Fe				
ASTM A403	316	Pipe	0.08 max	16-18	2 max	2-3	10-14	0.045 max	0.03 max	1 max	bal Fe				
ASTM A409	316	Weld, Pipe	0.08 max	16-18	2 max	2-3	10-14	0.045 max	0.03 max	1 max	bal Fe				
ASTM A430	316*	Obs,1995; see A312; Pipe	0.08 max	16-18	2 max	2-3	10-14	0.045 max	0.03 max	1 max	bal Fe				
ASTM A473	316	Frg, Heat res SHT, t<127mm	0.08 max	16.00-18.00	2.00 max	2.00-3.00	10.00-14.00	0.045 max	0.030 max	1.00 max	bal Fe	515	205	40	
ASTM A478	316	Wir CD 0.76<t<=3.18mm	0.08 max	16.00-18.00	2.00 max	2.00-3.00	10.00-14.00	0.045 max	0.030 max	1.00 max	bal Fe	830-1030		15	
ASTM A479	316	Bar	0.08 max	16-18	2 max	2-3	10-14	0.045 max	0.03 max	1 max	bal Fe				
ASTM A511(96)	MT316	Smls mech tub, Ann	0.08 max	16.0-18.0	2.00 max	2.0-3.0	11.0-14.0	0.040 max	0.030 max	1.00 max	bal Fe	517	207	35	192 HB
ASTM A554(94)	MT316	Weld mech tub, Rnd Ann	0.08 max	16.0-18.0	2.00 max	2.0-3.0	10.0-14.0	0.040 max	0.030 max	1.00 max	bal Fe	517	207	35	
ASTM A580/A580M(98)	316	Wir, Ann	0.08 max	16.0-18.0	2.00 max	2.00-3.00	10.0-14.0	0.045 max	0.030 max	1.00 max	N <=0.10; bal Fe	520	210	35	
ASTM A632(90)	TP316	Smls Weld tub	0.08 max	16.0-18.0	2.00 max	2.00-3.00	11.0-14.0	0.040 max	0.030 max	0.75 max	bal Fe	515	205	35	
ASTM A666(96)	316	Sh Strp Plt Bar, Ann	0.08 max	16.00-18.00	2.00 max	2.00-3.00	10.00-14.00	0.045 max	0.030 max	0.75 max	bal Fe	515	205	40	217 HB
ASTM A688/A688M(96)	TP316	Weld Feedwater Heater Tub	0.08 max	16.00-18.00	2.00 max	2.00-3.00	10.00-14.00	0.040 max	0.03 max	0.75 max	bal Fe	515	205	35	

UNS numbers and US grades are provided as a means of cross referencing chemically similar alloys. Exchangability is only possible after independent examination of specifications. Tensile properties are minimum or typical as specified. UTS and YS as MPa. El as %. See Appendix for list of abbreviations used in Notes. * indicates obsolete material.

Specification	Designation	Notes	C	Cr	Mn	Mo	Ni	P	S	Si	Other	UTS	YS	El	Hard

Stainless Steel, Austenitic, 316 (Continued from previous page)

USA

Specification	Designation	Notes	C	Cr	Mn	Mo	Ni	P	S	Si	Other	UTS	YS	El	Hard
ASTM A771/A771M(95)	316	Smls tube; Reactor Core, CW	0.040-0.060	17.0-18.0	1.00-2.00	2.00-3.00	13.0-14.0	0.040 max	0.010 max	0.50-0.75	Al <=0.050; B <=0.0020; Co <=0.050; Cu <=0.04; N <=0.010; Nb <=0.050; V <=0.05; Ta:0.020; As<=0.030; bal Fe	758-862	552-758	15	
ASTM A771/A771M(95)	316	Smls tube; Reactor Core	0.040-0.060	17.0-18.0	1.00-2.00	2.00-3.00	13.0-14.0	0.040 max	0.010 max	0.50-0.75	Al <=0.050; B <=0.0020; Co <=0.050; Cu <=0.04; N <=0.010; Nb <=0.050; V <=0.05; Ta<=0.020; As<=0.030; bal Fe	517-689	207-345	40	
ASTM A793(96)	316	Rolled floor plt	0.08 max	16.00-18.00	2.00 max	2.00-3.00	10.00-14.00	0.045 max	0.030 max	0.75 max	N <=0.10; bal Fe				
ASTM A813/A813M(95)	TP316	Weld Pipe	0.08 max	16.0-18.0	2.00 max	2.00-3.00	10.0-14.0	0.045 max	0.030 max	0.75 max	bal Fe	515	205		
ASTM A814/A814M(96)	TP316	CW Weld Pipe	0.08 max	16.0-18.0	2.00 max	2.00-3.00	10.0-14.0	0.045 max	0.030 max	0.75 max	bal Fe	515	205		
ASTM A826/A826M(95)	TP 316	Smls duct tub; Reactor grade	0.040-0.060	17.0-18.0	1.00-2.00	2.00-3.00	13.0-14.0	0.040 max	0.010 max	0.50-0.75	Al <=0.050; B <=0.0020; Co <=0.050; Cu <=0.04; N <=0.010; Nb <=0.050; V <=0.05; Ta<=0.020; As<=0.030; bal Fe	689	586	15	
ASTM A831/A831M(95)	316	Bar Bil Frg; reactor core, Ann	0.040-0.060	17.0-18.0	1.00-2.00	2.00-3.00	13.0-14.0	0.040 max	0.010 max	0.50-0.75	Al <=0.050; B <=0.0020; Co <=0.050; Cu <=0.04; N <=0.010; Nb <=0.050; V <=0.05; Ta 0.020, As<=0.030; bal Fe	515	205	30	
ASTM A943/A943M(95)	TP316	Spray formed smls pipe	0.08 max	16.0-18.0	2.00 max	2.00-3.00	11.0-14.0	0.040 max	0.030 max	0.75 max	bal Fe	515	205		
ASTM A965/965M(97)	F316	Frg for press high-temp parts	0.08 max	16.0-18.0	2.00 max	2.00-3.00	10.0-14.0	0.040 max	0.030 max	1.00 max	bal Fe	485	205	30	
ASTM A988(98)	S31600	HIP Flanges, Fittings, Valves/parts; Heat res	0.08 max	16.0-18.0	2.00 max	2.00-3.00	10.0-14.0	0.045 max	0.030 max	1.00 max	N <=0.10; bal Fe	515	205	30	
FED QQ-S-763F(96)	316*	Obs; Bar Wir Shp Frg; Corr res	0.08 max	16-18	2 max	2-3	10-14	0.045 max	0.03 max	1 max	bal Fe				
FED QQ-S-766D(93)	316*	Obs; see ASTM A240, A666, A693; Sh Strp Plt	0.08 max	16-18	2 max	2-3	10-14	0.045 max	0.03 max	1 max	bal Fe				
MIL-P-1144D	316*	Obs; Sh Strp Plt Bar Wir Pipe	0.08 max	16-18	2 max	2-3	10-14	0.045 max	0.03 max	1 max	bal Fe				
MIL-S-5059D(90)	316*	Obs see AMS specs; Plt Sh Strp; Corr res	0.08 max	16-18	2 max	2-3	10-14	0.045 max	0.03 max	1 max	bal Fe				
MIL-S-7720A	316*	Obs	0.08 max	16-18	2 max	2-3	10-14	0.045 max	0.03 max	1 max	bal Fe				
MIL-S-862B	316*	Obs	0.08 max	16-18	2 max	2-3	10-14	0.045 max	0.03 max	1 max	bal Fe				
SAE J405(98)	S31600		0.08 max	16.00-18.00	2.00 max	2.00-3.00	10.00-14.00	0.045 max	0.030 max	0.75 max	N <=0.10; bal Fe				
SAE J467(68)	316		0.05	17.0	1.00	2.50	12.5			0.40	bal Fe	586	262	60	80 HRB

Stainless Steel, Austenitic, 316Cb

Austria

Specification	Designation	Notes	C	Cr	Mn	Mo	Ni	P	S	Si	Other	UTS	YS	El	Hard
ONORM M3121	X6CrNiMoNb17122K KW	Sh Strp Plt Bar Bil	0.08 max	16.5-18.5	2 max	2.0-2.5	10.5-13.5	0.045 max	0.030 max	1 max	Nb/Ta>=10xC<=1.00; bal Fe				

Europe

Specification	Designation	Notes	C	Cr	Mn	Mo	Ni	P	S	Si	Other	UTS	YS	El	Hard
EN 10088/2(95)	1.4580	Plt, HR Corr res; t<=75mm, Ann	0.08 max	16.50-18.50	2.00 max	2.00-2.50	10.50-13.50	0.045 max	0.015 max	1.00 max	bal Fe	520-720	220	40	
EN 10088/2(95)	X6CrNiMoNb17-12-2	Plt, HR Corr res; t<=75mm, Ann	0.08 max	16.50-18.50	2.00 max	2.00-2.50	10.50-13.50	0.045 max	0.015 max	1.00 max	bal Fe	520-720	220	40	
EN 10088/3(95)	1.4580	Bar Rod Sect, t<=160mm, Ann	0.08 max	16.50-18.50	2.00 max	2.00-2.50	10.50-13.50	0.045 max	0.015 max	1.00 max	bal Fe	510-740	215	35	230 HB

UNS numbers and US grades are provided as a means of cross referencing chemically similar alloys. Exchangability is only possible after independent examination of specifications. Tensile properties are minimum or typical as specified. UTS and YS as MPa. El as %. See Appendix for list of abbreviations used in Notes. * indicates obsolete material.

Specification	Designation	Notes	C	Cr	Mn	Mo	Ni	P	S	Si	Other	UTS	YS	El	Hard

Stainless Steel, Austenitic, 316Cb (Continued from previous page)

Europe

Specification	Designation	Notes	C	Cr	Mn	Mo	Ni	P	S	Si	Other	UTS	YS	El	Hard
EN 10088/3(95)	1.4580	Bar Rod Sect; Corr res; 160<t<=250mm, Ann	0.08 max	16.50-18.50	2.00 max	2.00-2.50	10.50-13.50	0.045 max	0.015 max	1.00 max	bal Fe	510-740	215	30	230 HB
EN 10088/3(95)	X6CrNiMoNb17-12-2	Bar Rod Sect; Corr res; 160<t<=250mm, Ann	0.08 max	16.50-18.50	2.00 max	2.00-2.50	10.50-13.50	0.045 max	0.015 max	1.00 max	bal Fe	510-740	215	30	230 HB
EN 10088/3(95)	X6CrNiMoNb17-12-2	Bar Rod Sect, t<=160mm, Ann	0.08 max	16.50-18.50	2.00 max	2.00-2.50	10.50-13.50	0.045 max	0.015 max	1.00 max	bal Fe	510-740	215	35	230 HB

France

Specification	Designation	Notes	C	Cr	Mn	Mo	Ni	P	S	Si	Other	UTS	YS	El	Hard
AFNOR NFA35573	Z6CNDNb17.12	Sh Strp Plt, CR, 5mm diam	0.08 max	16-18	2 max	2-2.5	10.5-13	0.04 max	0.03 max	1 max	Nb+Ta>=10xC<=1.00; bal Fe				
AFNOR NFA35574	Z6CNDNb17.12	Bar Bil Rod	0.08 max	16-18	2 max	2-2.5	10.5-13	0.04 max	0.03 max	1 max	Nb+Ta>=10xC<=1.00; bal Fe				
AFNOR NFA35583	Z6CNDNb17.12	Wir rod	0.08 max	18-20	1-2.5	2-3	11-14	0.03 max	0.02 max	0.3-0.65	Cu <=0.5; Nb>=10xC<=1.00; bal Fe				

Germany

Specification	Designation	Notes	C	Cr	Mn	Mo	Ni	P	S	Si	Other	UTS	YS	El	Hard
DIN 17440(96)	WNr 1.4580	Plt Sh, HR Strp Bar, SHT	0.08 max	16.5-18.5	2.00 max	2.00-2.50	10.5-13.5	0.045 max	0.030 max	1.00 max	Nb 10xC>=1.00; bal Fe	510-740	215	30-35	
DIN 17458(85)	WNr 1.4580	Smls tube SHT	0.08 max	16.5-18.5	2.00 max	2.00-2.50	10.5-13.5	0.045 max	0.030 max	1.00 max	Nb 10xC>=1.00; bal Fe	510-740	215	30-35	
DIN EN 10088(95)	G-X10CrNiMoNb1810	Sht Plt Strp Bar Rod SHT	0.08 max	16.5-18.5	2.00 max	2.00-2.50	10.5-13.5	0.045 max	0.030 max	1.00 max	Nb 10xC>=1.00; bal Fe	510-740	215	30-35	
DIN EN 10088(95)	X6CrNiMoNb17-12-2	Sh Plt Strp Bar Sect, SHT	0.08 max	16.5-18.5	2.00 max	2.00-2.50	10.5-13.5	0.045 max	0.030 max	1.00 max	Nb 10xC<=1.00; bal Fe	510-740	215	30-35	
TGL 39672	X5CrNiMoNb19.11*	Obs	0.07 max	18.5-19.5	1-1.5	2.2-2.7	10.5-11.5	0.025 max	0.015 max	0.7 max	Nb 12xC<=1.00; bal Fe				

International

Specification	Designation	Notes	C	Cr	Mn	Mo	Ni	P	S	Si	Other	UTS	YS	El	Hard
ISO 683-13(74)	23*	Sh Strp Bar, ST	0.08 max	16.5-18.5	2 max	2-2.5	11-14	0.045 max	0.03 max	1 max	Nb>= 10xC <= 1.00; bal Fe				

Italy

Specification	Designation	Notes	C	Cr	Mn	Mo	Ni	P	S	Si	Other	UTS	YS	El	Hard
UNI 6901(71)	X6CrNiMoNb1712	Corr res	0.08 max	16-18.5	2 max	2-2.5	10.5-13.5	0.045 max	0.03 max	1 max	Nb+Ta=10C-1; bal Fe				
UNI 6901(71)	X6CrNiMoNb1713	Corr res	0.08 max	16-18.5	2 max	2.5-3	11.5-14.5	0.045 max	0.03 max	1 max	Nb+Ta=10C-1; bal Fe				
UNI 6904(71)	X6CrNiMoNb1712	Heat res	0.08 max	16-18.5	2 max	2-2.5	10.5-13.5	0.04 max	0.03 max	0.75 max	Nb+Ta=10C-1; bal Fe				
UNI 6904(71)	X6CrNiMoNb1713	Heat res	0.08 max	16-18.5	2 max	2.5-3	11.5-14.5	0.04 max	0.03 max	0.75 max	Nb+Ta=10C-1; bal Fe				
UNI 7500(75)	X6CrNiMoNb1712	Press ves	0.08 max	16-18.5	2 max	2-2.5	10.5-13.5	0.045 max	0.03 max	1 max	Nb+Ta=10C-1; bal Fe				
UNI 8317(81)	X6CrNiMoNb1712	Sh Strp Plt	0.08 max	16-18.5	2 max	2-2.5	10.5-13.5	0.045 max	0.03 max	1 max	Nb+Ta>=10xC<=1.00; bal Fe				
UNI 8317(81)	X6CrNiMoNb1713	Sh Strp Plt	0.08 max	16-18.5	2 max	2.5-3	11.5-14.5	0.045 max	0.03 max	1 max	Nb+Ta>=10xC<=1.00; bal Fe				

Russia

Specification	Designation	Notes	C	Cr	Mn	Mo	Ni	P	S	Si	Other	UTS	YS	El	Hard
GOST	O8Ch16N13M2B	Sh Strp Plt Bar Wir Frg Tub	0.08 max	15-17	1 max	2-2.5	12.5-14.5	0.035 max	0.02 max	0.8 max	Cu <=0.3; Nb 0.90-1.30; Ti <=0.2; W <=0.2; bal Fe				
GOST 5632(61)	1KH16N13M2B	Heat res; Corr res; High-temp constr	0.06-0.12	15.0-17.0	1.0 max	2.0-2.5	12.5-14.5	0.035 max	0.02 max	0.8 max	Cu <=0.3; Nb 0.9-1.3; Ti <=0.05; bal Fe				
GOST 5632(72)	08Ch16N13M2B	High-temp const	0.06-0.12	15.0-17.0	1.0 max	2.0-2.5	12.5-14.5	0.035 max	0.02 max	0.8 max	Cu <=0.3; Nb 0.9-1.3; Ti <=0.2; W <=0.2; bal Fe				

Spain

Specification	Designation	Notes	C	Cr	Mn	Mo	Ni	P	S	Si	Other	UTS	YS	El	Hard
UNE 36016/1(89)	F.3536*	Obs; EN10088-3(96); X6CrNiMoNb17-12-2	0.08 max	16.5-18.5	2.0 max	2.0-2.5	10.5-13.5	0.045 max	0.03 max	1.0 max	N <=0.1; 10xC<=Nb<=1.00; bal Fe				
UNE 36016/2(89)	F.3123*	Obs; En10088-2(96); X2CrMoTiNb18-2	0.025 max	17.5-19.5	1.0 max	1.75-2.5	1.0 max	0.04 max	0.03 max	1.0 max	N <=0.035; Ti <=0.80; Ti+Nb=0.2+4(C+N)-0.8; bal Fe				

UK

Specification	Designation	Notes	C	Cr	Mn	Mo	Ni	P	S	Si	Other	UTS	YS	El	Hard
BS 1554(90)	318S17	Wir Corr/Heat res; 6<diam<=13mm; Ann	0.08 max	18.0-20.0	2.00 max	2.00-3.00	11.0-14.0	0.045 max	0.030 max	1.00 max	W <=0.1; 10xC<=Nb<=1.00; bal Fe	0-700			

UNS numbers and US grades are provided as a means of cross referencing chemically similar alloys. Exchangability is only possible after independent examination of specifications. Tensile properties are minimum or typical as specified. UTS and YS as MPa. El as %. See Appendix for list of abbreviations used in Notes. * indicates obsolete material.

Stainless Steel, Austenitic, 316Cb (Continued from previous page)

Specification	Designation	Notes	C	Cr	Mn	Mo	Ni	P	S	Si	Other	UTS	YS	El	Hard
UK															
BS 1554(90)	318S17	Wir Corr/Heat res; diam<0.49mm; Ann	0.08 max	18.0-20.0	2.00 max	2.00-3.00	11.0-14.0	0.045 max	0.030 max	1.00 max	W <=0.1; 10xC<=Nb<=1.00; bal Fe	0-900			
BS 2901	318S96	Wire, Rod	0.08 max	18-20	1-2.5	2-3	11-14	0.03 max	0.03 max	0.25-0.65	Cu <=0.5; Nb>=10xC<=1.00; bal Fe				
USA															
	UNS S31640	Nb stabilized	0.08 max	16.00-18.00	2.00 max	2.00-3.00	10.00-14.00	0.045 max	0.030 max	1.50 max	Nb+Ta 10xC min 1.10 max; bal Fe				
ASTM A240/A240M(98)	S31640	Nb-stabilized	0.08 max	16.0-18.0	2.00 max	2.00-3.00	10.0-14.0	0.045 max	0.030 max	1.50 max	N <=0.10; 10xC<=Nb<=1.10; bal Fe	515	205	30.0	217 HB
ASTM A313	316Cb*	Wir	0.08 max	16.00-18.00	2.00 max	2.00-3.00	10.00-14.00	0.045 max	0.030 max	1.00 max	10xC<=Nb+Ta<=1.10; bal Fe				
ASTM A368(95)	316Cb	Wir, Heat res	0.08 max	16.00-18.00	2.00 max	2.00-3.00	10.00-14.00	0.045 max	0.030 max	1.00 max	10xC<=(Nb+Ta)<=1.10; bal Fe				
SAE J405(98)	S31640	Nb stabilized	0.08 max	16.00-18.00	2.00 max	2.0-3.0	10.00-14.00	0.045 max	0.030 max	0.75 max	10xC<=Nb+Ta<=1.10; bal Fe				

Stainless Steel, Austenitic, 316F

Specification	Designation	Notes	C	Cr	Mn	Mo	Ni	P	S	Si	Other	UTS	YS	El	Hard
USA															
	AISI 316F	Tube	0.08 max	16.00-18.00	2.00 max	1.75-2.50	10.00-14.00	0.20 max	0.10 min	1.00 max	bal Fe				
	UNS S31620	Free Machining	0.08 max	17.00-19.00	2.00 max	1.75-2.50	12.00-14.00	0.20 max	0.10 max	1.00 max	bal Fe				

Stainless Steel, Austenitic, 316H

Specification	Designation	Notes	C	Cr	Mn	Mo	Ni	P	S	Si	Other	UTS	YS	El	Hard
Italy															
UNI 6901(71)	X8CrNiMo1712	Corr res	0.04-0.1	16-18	2 max	2-2.5	11-13.5	0.03 max	0.03 max	0.75 max	bal Fe				
UNI 6904(71)	X8CrNiMo1712	Heat res	0.04-0.1	16-18	2 max	2-2.5	11-13.5	0.03 max	0.03 max	0.75 max	bal Fe				
Japan															
JIS G3214(91)	SUSF316H	Frg press ves 130-200mm diam	0.04-0.10	16.00-18.00	2.00 max	2.00-3.00	10.00-14.00	0.040 max	0.030 max	1.00 max	bal Fe	480	205	29	187 HB max
Mexico															
NMX-B-186-SCFI(94)	TP316H	Pipe; High-temp	0.04-0.10	16.0-18.0	2.00 max	2.00-3.00	11.0-14.0	0.040 max	0.030 max	0.75 max	bal Fe	517	207		
Romania															
STAS 3583(64)	8TMoNC170	Corr res; Heat res	0.1 max	16-18	1-2	1.8-2.5	12-14	0.035 max	0.03 max	0.8 max	Pb <=0.15; Ti 0.3-0.6; bal Fe				
UK															
BS 1502(82)	316S53	Bar Shp; Press ves	0.04-0.10	16.5-18.5	2.00 max	2.50-3.00	11.0-14.0	0.045 max	0.030 max	1.00 max	W <=0.1; bal Fe				
BS 1503(89)	316S51	Frg press ves; t<=999mm; HT	0.04-0.10	16.5-18.5	2.00 max	2.00-2.50	10.5-13.5	0.040 max	0.025 max	1.00 max	B <=0.005; Cu <=0.70; Nb <=0.20; Ti <=0.10; W <=0.1; bal Fe	510-710	240	30	
BS 1506(90)	316S51	Bolt matl Pres/Corr res; t<=160mm; Hard	0.04-0.10	16.5-18.5	2.00 max	2.00-2.50	10.5-13.5	0.045 max	0.030 max	1.00 max	bal Fe	520	205	40	
BS 1506(90)	316S53	Bolt matl Pres/Corr res; t<=160mm; Hard	0.04-0.10	16.5-18.5	2.00 max	2.50-3.00	11.0-14.0	0.045 max	0.030 max	1.00 max	bal Fe	520	205	40	
BS 3059/2(78)	316S59	Boiler, Superheater; Tub; Pip	0.04-0.09	16.0-18.0	0.50-2.00	2.00-2.75	11.0-14.0	0.040 max	0.030 max	0.20-1.00	bal Fe				
BS 3605(73)	316S59	Corr res; Tub, Weld; Smls tub	0.04-0.09	16.0-18.0	0.5-2.0	2.0-2.75	12.0-14.0	0.040 max	0.030 max	0.2-1.0	B 0.0015-0.006; bal Fe				
USA															
	UNS S31609	High-C	0.04-0.10	16.00-18.00	2.00 max	2.00-3.00	10.00-14.00	0.040 max	0.030 max	1.00 max	bal Fe				
ASTM A182/A182M(98)	F316H	Frg/roll pipe flange valve	0.04-0.10	16.0-18.0	2.00 max	2.00-3.00	10.0-14.0	0.045 max	0.030 max	1.00 max	bal Fe	515	205	30	
ASTM A213/A213M(95)	TP316H	Smls tube boiler, superheater, heat exchanger	0.04-0.10	16.0-18.0	2.00 max	2.00-3.00	11.0-14.0	0.040 max	0.030 max	0.75 max	bal Fe	515	205	35	
ASTM A240/A240M(98)	S31609	High-C	0.04-0.10	16.0-18.0	2.00 max	2.00-3.00	10.0-14.0	0.040 max	0.030 max	1.00 max	bal Fe	515	205	40.0	217 HB

UNS numbers and US grades are provided as a means of cross referencing chemically similar alloys. Exchangability is only possible after independent examination of specifications. Tensile properties are minimum or typical as specified. UTS and YS as MPa. El as %. See Appendix for list of abbreviations used in Notes. * indicates obsolete material.

Specification	Designation	Notes	C	Cr	Mn	Mo	Ni	P	S	Si	Other	UTS	YS	El	Hard

Stainless Steel, Austenitic, 316H (Continued from previous page)

USA

Specification	Designation	Notes	C	Cr	Mn	Mo	Ni	P	S	Si	Other	UTS	YS	El	Hard
ASTM A249/249M(96)	TP316H	Weld tube; boiler, superheater, heat exch	0.04-0.10	16.0-18.0	2.00 max	2.00-3.00	10.0-14.0	0.040 max	0.030 max	0.75 max	bal Fe	515	205	35	90 HRB
ASTM A336/A336M(98)	F316H	Frg, press/high-temp	0.04-0.10	16.0-18.0	2.00 max	2.00-3.00	10.0-14.0	0.045 max	0.030 max	1.00 max	bal Fe	485	205	30	90 HRB
ASTM A813/A813M(95)	TP316H	Weld Pipe, High-C	0.04-0.10	16.0-18.0	2.00 max	2.00-3.00	10.0-14.0	0.045 max	0.030 max	0.75 max	bal Fe	515	205		
ASTM A814/A814M(96)	TP316H	CW Weld Pipe, High-C	0.04-0.10	16.0-18.0	2.00 max	2.00-3.00	10.0-14.0	0.045 max	0.030 max	0.75 max	bal Fe	515	205		
ASTM A943/A943M(95)	TP316H	Spray formed smls pipe	0.04-0.10	16.0-18.0	2.00 max	2.00-3.00	11.0-14.0	0.040 max	0.030 max	0.75 max	bal Fe	515	205		
ASTM A965/965M(97)	F316H	Frg for press high-temp parts	0.04-0.10	16.0-18.0	2.00 max	2.00-3.00	10.0-14.0	0.045 max	0.030 max	1.00 max	bal Fe	485	205	30	
SAE J405(98)	S31609	High-C	0.04-0.10	16.00-18.00	2.00 max	2.00-3.00	10.00-14.00	0.040 max	0.030 max	0.75 max	N <=0.10; bal Fe				

Stainless Steel, Austenitic, 316L

Australia

Specification	Designation	Notes	C	Cr	Mn	Mo	Ni	P	S	Si	Other	UTS	YS	El	Hard
AS 1449(94)	316L	Sh Strp Plt	0.030 max	16.00-18.00	2.00 max	2.00-3.00	10.00-14.00	0.045 max	0.030 max	0.75 max	bal Fe				
AS 2837(86)	316L	Bar, Semi-finished product	0.030 max	16.0-18.0	2.00 max	2.0-3.0	10.0-14.0	0.045 max	0.030 max	1.00 max	bal Fe				

Austria

Specification	Designation	Notes	C	Cr	Mn	Mo	Ni	P	S	Si	Other	UTS	YS	El	Hard
ONORM M3121	X2CrNiMo17132KKW	Sh Strp Plt Bar Wir	0.03 max	16.5-18.5	2 max	2.0-2.5	11.0-14.0	0.045 max	0.030 max	1 max	bal Fe				

Canada

Specification	Designation	Notes	C	Cr	Mn	Mo	Ni	P	S	Si	Other	UTS	YS	El	Hard
CSA G110.3	316L	Bar Bil	0.03 max	16-18	2 max	2-3	10-14	0.045 max	0.03 max	1 max	bal Fe				
CSA G110.6	316L	Sh Strp Plt, HR/Ann or HR/Q/A/Tmp	0.03 max	16-18	2 max	2-3	10-14	0.045 max	0.03 max	1 max	bal Fe				
CSA G110.9	316L	Sh Strp Plt, HR, CR, Ann	0.03 max	16-18	2 max	2-3	10-14	0.045 max	0.03 max	1 max	bal Fe				

China

Specification	Designation	Notes	C	Cr	Mn	Mo	Ni	P	S	Si	Other	UTS	YS	El	Hard
GB 1220(92)	00Cr17Ni14Mo2	Bar SA	0.030 max	16.00-18.00	2.00 max	2.00-3.00	12.00-15.00	0.035 max	0.030 max	1.00 max	bal Fe	480	177	40	
GB 12770(91)	00Cr17Ni14Mo2	Weld tube; SA	0.030 max	16.00-18.00	2.00 max	2.00-3.00	12.00-15.00	0.035 max	0.030 max	1.00 max	bal Fe	480	180	35	
GB 12771(91)	00Cr17Ni14Mo2	Weld Pipe SA	0.030 max	16.00-18.00	2.00 max	2.00-3.00	12.00-15.00	0.035 max	0.030 max	1.00 max	bal Fe	480	180	35	
GB 13296(91)	00Cr17Ni14Mo2	Smls tube SA	0.030 max	16.00-18.00	2.00 max	2.00-3.00	12.00-15.00	0.035 max	0.030 max	1.00 max	bal Fe	480	175	35	
GB 3280(92)	00Cr17Ni14Mo2	Sh Plt, CR SA	0.030 max	16.00-18.00	2.00 max	2.00-3.00	12.00-15.00	0.035 max	0.030 max	1.00 max	bal Fe	480	177	40	
GB 4237(92)	00Cr17Ni14Mo2	Sh Plt, HR SA	0.030 max	16.00-18.00	2.00 max	2.00-3.00	12.00-15.00	0.035 max	0.030 max	1.00 max	bal Fe	480	177	40	
GB 4239(91)	00Cr17Ni14Mo2	Sh Strp, CR SA	0.030 max	16.00-18.00	2.00 max	2.00-3.00	12.00-15.00	0.035 max	0.030 max	1.00 max	bal Fe	480	175	40	
GB 4240(93)	00Cr17Ni14Mo2	Wir Ann 1-3mm diam	0.030 max	16.00-18.00	2.00 max	2.00-3.00	12.00-15.00	0.035 max	0.030 max	1.00 max	bal Fe	490-830		25	
GB/T 14975(94)	00Cr17Ni14Mo2	Smls tube SA	0.030 max	16.00-18.00	2.00 max	2.00-3.00	12.00-15.00	0.035 max	0.030 max	1.00 max	bal Fe	480	175	35	
GB/T 14976(94)	00Cr17Ni14Mo2	Smls pipe SA	0.030 max	16.00-18.00	2.00 max	2.00-3.00	12.00-15.00	0.035 max	0.030 max	1.00 max	bal Fe	480	175	35	

Czech Republic

Specification	Designation	Notes	C	Cr	Mn	Mo	Ni	P	S	Si	Other	UTS	YS	El	Hard
CSN 417349	17349	Corr res	0.03 max	16.5-18.5	2.0 max	2.0-2.5	11.0-14.0	0.045 max	0.03 max	1.0 max	bal Fe				
CSN 417350	17350	Corr res	0.03 max	16.5-18.5	2.0 max	2.5-3.0	12.0-15.0	0.045 max	0.03 max	1.0 max	bal Fe				

Europe

Specification	Designation	Notes	C	Cr	Mn	Mo	Ni	P	S	Si	Other	UTS	YS	El	Hard
EN 10088/2(95)	1.4307	Strp, CR Corr res; t<=6mm, Ann	0.030 max	17.50-19.50	2.00 max		8.00-10.00	0.045 max	0.015 max	1.00 max	N <=0.11; bal Fe	520-670	220	45	
EN 10088/2(95)	1.4404	Strp, CR Corr res; t<=6mm, Ann	0.030 max	16.50-18.50	2.00 max	2.00-2.50	10.00-13.00	0.045 max	0.015 max	1.00 max	N <=0.11; bal Fe	530-680	240	40	
EN 10088/2(95)	1.4432	Strp, CR Corr res; t<=6mm, Ann	0.030 max	16.50-18.50	2.00 max	2.50-3.00	10.50-13.00	0.045 max	0.015 max	1.00 max	N <=0.11; bal Fe	550-700	240	40	
EN 10088/2(95)	1.4435	Strp, CR Corr res; t<=6mm, Ann	0.030 max	17.00-19.00	2.00 max	2.50-3.00	12.50-15.00	0.045 max	0.015 max	1.00 max	N <=0.11; bal Fe	550-700	240	40	

UNS numbers and US grades are provided as a means of cross referencing chemically similar alloys. Exchangability is only possible after independent examination of specifications. Tensile properties are minimum or typical as specified. UTS and YS as MPa. El as %. See Appendix for list of abbreviations used in Notes. * indicates obsolete material.

Stainless Steel, Austenitic, 316L (Continued from previous page)

Specification	Designation	Notes	C	Cr	Mn	Mo	Ni	P	S	Si	Other	UTS	YS	El	Hard
Europe															
EN 10088/2(95)	1.4439	Strp, CR Corr res; t<=6mm, Ann	0.030 max	16.50-18.50	2.00 max	4.00-5.00	12.50-14.50	0.045 max	0.015 max	1.00 max	N 0.12-0.22; bal Fe	580-780	290	35	
EN 10088/2(95)	X2CrNi18-9	Strp, CR Corr res; t<=6mm, Ann	0.030 max	17.50-19.50	2.00 max		8.00-10.00	0.045 max	0.015 max	1.00 max	N <=0.11; bal Fe	520-670	220	45	
EN 10088/2(95)	X2CrNiMo17-12-2	Strp, CR Corr res; t<=6mm, Ann	0.030 max	16.50-18.50	2.00 max	2.00-2.50	10.00-13.00	0.045 max	0.015 max	1.00 max	N <=0.11; bal Fe	530-680	240	40	
EN 10088/2(95)	X2CrNiMo17-12-3	Strp, CR Corr res; t<=6mm, Ann	0.030 max	16.50-18.50	2.00 max	2.50-3.00	10.50-13.00	0.045 max	0.015 max	1.00 max	N <=0.11; bal Fe	550-700	240	40	
EN 10088/2(95)	X2CrNiMo18-14-3	Strp, CR Corr res; t<=6mm, Ann	0.030 max	17.00-19.00	2.00 max	2.50-3.00	12.50-15.00	0.045 max	0.015 max	1.00 max	N <=0.11; bal Fe	550-700	240	40	
EN 10088/2(95)	X2CrNiMoN17-13-5	Strp, CR Corr res; t<=6mm, Ann	0.030 max	16.50-18.50	2.00 max	4.00-5.00	12.50-14.50	0.045 max	0.015 max	1.00 max	N 0.12-0.22; bal Fe	580-780	290	35	
EN 10088/3(95)	1.4307	Bar Rod Sect; Corr res; 160<t<=250mm Ann	0.030 max	17.50-19.50	2.00 max		8.00-10.00	0.045 max	0.030 max	1.00 max	N <=0.11; bal Fe	450-680	175	35	215 HB
EN 10088/3(95)	1.4307	Bar Rod Sect, t<=25mm, strain Hard	0.030 max	17.50-19.50	2.00 max		8.00-10.00	0.045 max	0.030 max	1.00 max	N <=0.11; bal Fe	800-1000	500	12	
EN 10088/3(95)	1.4307	Bar Rod Sect, t<=35mm, strain Hard	0.030 max	17.50-19.50	2.00 max		8.00-10.00	0.045 max	0.030 max	1.00 max	N <=0.11; bal Fe	700-850	350	20	
EN 10088/3(95)	1.4307	Bar Rod Sect, t<=160mm Ann	0.030 max	17.50-19.50	2.00 max		8.00-10.00	0.045 max	0.030 max	1.00 max	N <=0.11; bal Fe	450-680	175	45	215 HB
EN 10088/3(95)	1.4404	Bar Rod Sect, t<=35mm, strain Hard	0.030 max	16.50-18.50	2.00 max	2.00-2.50	10.00-13.00	0.045 max	0.030 max	1.00 max	N <=0.11; bal Fe	700-850	350	20	
EN 10088/3(95)	1.4404	Bar Rod Sect, t<=160mm, Ann	0.030 max	16.50-18.50	2.00 max	2.00-2.50	10.00-13.00	0.045 max	0.030 max	1.00 max	N <=0.11; bal Fe	500-700	200	40	215 HB
EN 10088/3(95)	1.4404	Bar Rod Sect; Corr res; 160<t<=250mm, Ann	0.030 max	16.50-18.50	2.00 max	2.00-2.50	10.00-13.00	0.045 max	0.030 max	1.00 max	N <=0.11; bal Fe	500-700	200	30	215 HB
EN 10088/3(95)	1.4404	Bar Rod Sect, t<=25mm, strain Hard	0.030 max	16.50-18.50	2.00 max	2.00-2.50	10.00-13.00	0.045 max	0.030 max	1.00 max	N <=0.11; bal Fe	800-1000	500	12	
EN 10088/3(95)	1.4432	Bar Rod Sect; Corr res; 160<t<=250mm, Ann	0.030 max	16.50-18.50	2.00 max	2.50-3.00	10.50-13.00	0.045 max	0.030 max	1.00 max	N <=0.11; bal Fe	500-700	200	30	215 HB
EN 10088/3(95)	1.4432	Bar Rod Sect, t<=160mm, Ann	0.030 max	16.50-18.50	2.00 max	2.50-3.00	10.50-13.00	0.045 max	0.030 max	1.00 max	N <=0.11; bal Fe	500-700	200	40	215 HB
EN 10088/3(95)	1.4435	Bar Rod Sect; Corr res; 160<t<=250mm, Ann	0.030 max	17.00-19.00	2.00 max	2.50-3.00	12.50-15.00	0.045 max	0.030 max	1.00 max	N <=0.11; bal Fe	500-700	200	30	215 HB
EN 10088/3(95)	1.4435	Bar Rod Sect, t<=160mm, Ann	0.030 max	17.00-19.00	2.00 max	2.50-3.00	12.50-15.00	0.045 max	0.030 max	1.00 max	N <=0.11; bal Fe	500-700	200	40	215 HB
EN 10088/3(95)	1.4439	Bar Rod Sect; Corr res; 160<t<=250mm, Ann	0.030 max	16.50-18.50	2.00 max	4.00-5.00	12.50-14.50	0.045 max	0.015 max	1.00 max	N 0.12-0.22; bal Fe	580-800	280	30	250 HB
EN 10088/3(95)	1.4439	Bar Rod Sect, t<=160mm, Ann	0.030 max	16.50-18.50	2.00 max	4.00-5.00	12.50-14.50	0.045 max	0.015 max	1.00 max	N 0.12-0.22; bal Fe	580-800	280	35	250 HB
EN 10088/3(95)	X2CrNi18-9	Bar Rod Sect, t<=25mm, strain Hard	0.030 max	17.50-19.50	2.00 max		8.00-10.00	0.045 max	0.030 max	1.00 max	N <=0.11; bal Fe	800-1000	500	12	
EN 10088/3(95)	X2CrNi18-9	Bar Rod Sect; Corr res; 160<t<=250mm Ann	0.030 max	17.50-19.50	2.00 max		8.00-10.00	0.045 max	0.030 max	1.00 max	N <=0.11; bal Fe	450-680	175	35	215 HB
EN 10088/3(95)	X2CrNi18-9	Bar Rod Sect, t<=160mm Ann	0.030 max	17.50-19.50	2.00 max		8.00-10.00	0.045 max	0.030 max	1.00 max	N <=0.11; bal Fe	450-680	175	45	215 HB
EN 10088/3(95)	X2CrNi18-9	Bar Rod Sect, t<=35mm, strain Hard	0.030 max	17.50-19.50	2.00 max		8.00-10.00	0.045 max	0.030 max	1.00 max	N <=0.11; bal Fe	700-850	350	20	
EN 10088/3(95)	X2CrNiMo17-12-2	Bar Rod Sect, t<=35mm, strain Hard	0.030 max	16.50-18.50	2.00 max	2.00-2.50	10.00-13.00	0.045 max	0.030 max	1.00 max	N <=0.11; bal Fe	700-850	350	20	
EN 10088/3(95)	X2CrNiMo17-12-2	Bar Rod Sect, t<=160mm, Ann	0.030 max	16.50-18.50	2.00 max	2.00-2.50	10.00-13.00	0.045 max	0.030 max	1.00 max	N <=0.11; bal Fe	500-700	200	40	215 HB
EN 10088/3(95)	X2CrNiMo17-12-2	Bar Rod Sect; Corr res; 160<t<=250mm, Ann	0.030 max	16.50-18.50	2.00 max	2.00-2.50	10.00-13.00	0.045 max	0.030 max	1.00 max	N <=0.11; bal Fe	500-700	200	30	215 HB
EN 10088/3(95)	X2CrNiMo17-12-2	Bar Rod Sect, t<=25mm, strain Hard	0.030 max	16.50-18.50	2.00 max	2.00-2.50	10.00-13.00	0.045 max	0.030 max	1.00 max	N <=0.11; bal Fe	800-1000	500	12	
EN 10088/3(95)	X2CrNiMo17-12-3	Bar Rod Sect, t<=160mm, Ann	0.030 max	16.50-18.50	2.00 max	2.50-3.00	10.50-13.00	0.045 max	0.030 max	1.00 max	N <=0.11; bal Fe	500-700	200	40	215 HB
EN 10088/3(95)	X2CrNiMo17-12-3	Bar Rod Sect; Corr res; 160<t<=250mm, Ann	0.030 max	16.50-18.50	2.00 max	2.50-3.00	10.50-13.00	0.045 max	0.030 max	1.00 max	N <=0.11; bal Fe	500-700	200	30	215 HB

UNS numbers and US grades are provided as a means of cross referencing chemically similar alloys. Exchangability is only possible after independent examination of specifications. Tensile properties are minimum or typical as specified. UTS and YS as MPa. El as %. See Appendix for list of abbreviations used in Notes. * indicates obsolete material.

Specification	Designation	Notes	C	Cr	Mn	Mo	Ni	P	S	Si	Other	UTS	YS	El	Hard

Stainless Steel, Austenitic, 316L (Continued from previous page)

Europe

Specification	Designation	Notes	C	Cr	Mn	Mo	Ni	P	S	Si	Other	UTS	YS	El	Hard
EN 10088/3(95)	X2CrNiMo18-14-3	Bar Rod Sect, t<=160mm, Ann	0.030 max	17.00-19.00	2.00 max	2.50-3.00	12.50-15.00	0.045 max	0.030 max	1.00 max	N <=0.11; bal Fe	500-700	200	40	215 HB
EN 10088/3(95)	X2CrNiMo18-14-3	Bar Rod Sect; Corr res; 160<t<=250mm, Ann	0.030 max	17.00-19.00	2.00 max	2.50-3.00	12.50-15.00	0.045 max	0.030 max	1.00 max	N <=0.11; bal Fe	500-700	200	30	215 HB
EN 10088/3(95)	X2CrNiMoN17-13-5	Bar Rod Sect; Corr res; 160<t<=250mm, Ann	0.030 max	16.50-18.50	2.00 max	4.00-5.00	12.50-14.50	0.045 max	0.015 max	1.00 max	N 0.12-0.22; bal Fe	580-800	280	30	250 HB
EN 10088/3(95)	X2CrNiMoN17-13-5	Bar Rod Sect, t<=160mm, Ann	0.030 max	16.50-18.50	2.00 max	4.00-5.00	12.50-14.50	0.045 max	0.015 max	1.00 max	N 0.12-0.22; bal Fe	580-800	280	35	250 HB

Finland

Specification	Designation	Notes	C	Cr	Mn	Mo	Ni	P	S	Si	Other	UTS	YS	El	Hard
SFS 750(86)	X2CrNiMo17122	Corr res	0.03 max	16.0-18.5	2.0 max	2.0-2.5	11.0-14.0	0.045 max	0.03 max	1.0 max	bal Fe				
SFS 752	XCrNiMo1812	Sh Strp Plt Bar Wir Tub Frg	0.03 max	16-18.5	2 max	2.5-3	11.5-14.5	0.045 max	0.03 max	1 max	bal Fe				
SFS 752(86)	X2CrNiMo17133	Corr res	0.03 max	16.0-18.5	2.0 max	2.5-3.0	11.5-14.5	0.045 max	0.03 max	1.0 max	bal Fe				

France

Specification	Designation	Notes	C	Cr	Mn	Mo	Ni	P	S	Si	Other	UTS	YS	El	Hard
AFNOR NFA35573	Z2CND17.12	Sh Strp Plt, CR, 5mm diam	0.03 max	16-18	2 max	2-2.5	10.5-13	0.04 max	0.03 max	1 max	bal Fe				
AFNOR NFA35574	Z2CND17.12	Bar Rod	0.03 max	16-18	2 max	2-2.5	10.5-13	0.04 max	0.03 max	1 max	bal Fe				
AFNOR NFA35575	Z2CND17.12	Wir rod, 60% Hard	0.03 max	16-18	2 max	2-2.5	10.5-13	0.04 max	0.03 max	1 max	bal Fe				
AFNOR NFA35577	Z2CND17.12	Bar Wir rod	0.03 max	16-18	2 max	2-2.5	10.5-13	0.04 max	0.03 max	1 max	bal Fe				
AFNOR NFA36209	Z2CND17.12	Plt	0.03 max	16-18	2 max	2-2.4	10.5-13	0.04 max	0.03 max	1 max	bal Fe				
AFNOR NFA36607	Z2CND17.12	Frg	0.03 max	16-18	2 max	2-2.5	10.5-13	0.04 max	0.03 max	1 max	Cu <=1; bal Fe				

Germany

Specification	Designation	Notes	C	Cr	Mn	Mo	Ni	P	S	Si	Other	UTS	YS	El	Hard
DIN 17440(96)	GX2CrNiMoN18-10	HR Strp Bar Wir Frg Ann	0.030 max	16.5-18.5	2.00 max	2.00-2.50	10.0-13.0	0.045 max	0.030 max	1.00 max	bal Fe	440-640	185	25	
DIN 17440(96)	WNr 1.4404	HR Strp Bar Wir Frg Ann	0.030 max	16.5-18.5	2.00 max	2.00-2.50	10.0-13.0	0.045 max	0.030 max	1.00 max	N <=0.11; bal Fe	440-640	185	25	
DIN 17440(96)	WNr 1.4435	Plt Sh, HR Strp Bar Wir Frg, SHT	0.030 max	17.0-19.0	2.00 max	2.50-3.00	12.5-15.0	0.045 max	0.030 max	1.00 max	N <=0.11; bal Fe	500-700	200	30-40	
DIN 17440(96)	X2CrNiMo18-14-3	Plt Sh, HR Strp Bar Wir Frg, SHT	0.030 max	17.0-19.0	2.00 max	2.50-3.00	12.5-15.0	0.045 max	0.030 max	1.00 max	bal Fe	500-700	200	30-40	
DIN 17441(97)	GX2CrNiMoN18-10	CR Strp Plt Sh Ann	0.030 max	16.5-18.5	2.00 max	2.00-2.50	10.0-13.0	0.045 max	0.030 max	1.00 max	bal Fe	440-640	185	25	
DIN 17441(97)	WNr 1.4404	CR Strp Plt Sh Ann	0.030 max	16.5-18.5	2.00 max	2.00-2.50	10.0-13.0	0.045 max	0.030 max	1.00 max	N <=0.11; bal Fe	440-640	185	25	
DIN 17441(97)	WNr 1.4435	CR Strp Plt Sh SHT	0.030 max	17.0-19.0	2.00 max	2.50-3.00	12.5-15.0	0.045 max	0.030 max	1.00 max	N <=0.11; bal Fe	500-700	200	30-40	
DIN 17441(97)	X2CrNiMo18-14-3	CR Strp Plt Sh SHT	0.030 max	17.0-19.0	2.00 max	2.50-3.00	12.5-15.0	0.045 max	0.030 max	1.00 max	bal Fe	500-700	200	30-40	

Hungary

Specification	Designation	Notes	C	Cr	Mn	Mo	Ni	P	S	Si	Other	UTS	YS	El	Hard
MSZ 4360(80)	KO38	Corr res	0.03 max	16-18	2 max	1-2.5	12.5-15.0	0.04 max	0.03 max	1 max	bal Fe				
MSZ 4360(87)	KO38LC	Corr res; 0-160mm; SA	0.03 max	16.5-18.5	2 max	2.5-3	12.5-15.0	0.045 max	0.025 max	1 max	bal Fe	490-690	190	40L; 30T	190 HB max
MSZ 4360(87)	X3CrNiMo17143	Corr res; 0-160mm; SA	0.03 max	16.5-18.5	2 max	2.5-3	12.5-15.0	0.045 max	0.025 max	1 max	bal Fe	490-690	190	40L; 30T	190 HB max

India

Specification	Designation	Notes	C	Cr	Mn	Mo	Ni	P	S	Si	Other	UTS	YS	El	Hard
IS 1570/5(85)	X02Cr17Ni12Mo2	Sh Plt Strp Bar Flat Band	0.03 max	16-18	2 max	2-3	10-14	0.045 max	0.03 max	1 max	Co <=0.1; Cu <=0.3; Pb <=0.15; W <=0.1; bal Fe	485	170	40	217 HB max
IS 6527	02Cr17Ni12Mo2	Wir rod	0.03 max	16-18.5	2 max	2-3	10.5-14	0.05 max	0.03 max	1 max	bal Fe				
IS 6528	02Cr17Ni12Mo2	Wir, Ann, SA	0.03 max	16-18.5	2 max	2-3	10.5-14	0.05 max	0.03 max	1 max	bal Fe				
IS 6529	02Cr17Ni12Mo2	Bil	0.03 max	16-18.5	2 max	2-3	10.5-14	0.05 max	0.03 max	1 max	bal Fe				
IS 6603	02Cr17Ni12Mo2	Bar, Soft 5-100mm diam	0.03 max	16-18.5	2 max	2-3	10.5-14	0.05 max	0.03 max	1 max	bal Fe				

UNS numbers and US grades are provided as a means of cross referencing chemically similar alloys. Exchangability is only possible after independent examination of specifications. Tensile properties are minimum or typical as specified. UTS and YS as MPa. El as %. See Appendix for list of abbreviations used in Notes. * indicates obsolete material.

Specification	Designation	Notes	C	Cr	Mn	Mo	Ni	P	S	Si	Other	UTS	YS	El	Hard

Stainless Steel, Austenitic, 316L (Continued from previous page)

India

| IS 6911 | 02Cr17Ni12Mo2 | Sh Strp Plt | 0.03 max | 16-18.5 | 2 max | 2-3 | 10.5-14 | 0.05 max | 0.03 max | 1 max | bal Fe | | | | |

International

COMECON PC4-70	22	Sh Strp Plt Bar Wir Tub	0.03 max	16-18	2 max	2-2.5	12-15	0.035 max	0.02 max	0.8 max	bal Fe				
ISO 2604-1(75)	F59	Frg	0.03 max	16.00-18.00	2.00 max	2.00-3.00	11.00-15.00	0.045 max	0.030 max	1.00 max	bal Fe	440-640			
ISO 2604-4	P57*	Plt	0.03 max	16-18.5	2 max	2-2.5	11-15	0.05 max	0.03 max	1 max	bal Fe				
ISO 2604-4	P58*	Plt	0.03 max	16-18.5	2 max	2.5-3	11.5-14.5	0.05 max	0.03 max	1 max	bal Fe				
ISO 4954(93)	X2CrNiMo17133E	Wir rod Bar, CH ext	0.030 max	16.5-18.5	2.00 max	2.5-3.0	11.5-14.5	0.045 max	0.030 max	1.00 max	bal Fe	680			
ISO 683-13(74)	19*	ST	0.03 max	16.5-18.5	2 max	2-2.5	11-14	0.045 max	0.03 max	1 max	bal Fe				
ISO 683-13(74)	19a*	ST	0.03 max	16.5-18.5	2 max	2.5-3	11.5-14.5	0.04 max	0.03 max	1 max	bal Fe				

Italy

UNI 6901(71)	X2CrNiMo1712	Corr res	0.03 max	16-18.5	2 max	2-2.5	11-14	0.045 max	0.03 max	1 max	bal Fe				
UNI 6901(71)	X2CrNiMo1713	Corr res	0.03 max	16-18.5	2 max	2.5-3	11.5-14.5	0.045 max	0.03 max	1 max	bal Fe				
UNI 6904(71)	X2CrNiMo1712	Heat res	0.03 max	16-18	2 max	2-2.5	11-14	0.03 max	0.03 max	0.75 max	bal Fe				
UNI 6904(71)	X2CrNiMo1713	Heat res	0.03 max	16-18	2 max	2.5-3	11.5-14.5	0.03 max	0.03 max	0.75 max	bal Fe				
UNI 7500(75)	X2CrNiMo1712	Press ves	0.03 max	16-18.5	2 max	2-2.5	11-14	0.045 max	0.03 max	1 max	bal Fe				
UNI 7500(75)	X2CrNiMo1713	Press ves	0.03 max	16-18.5	2 max	2.5-3	11.5-14.5	0.045 max	0.03 max	1 max	bal Fe				
UNI 8317(81)	X2CrNiMo1712	Sh Strp Plt	0.03 max	16-18.5	2 max	2-2.5	11-14	0.045 max	0.03 max	1 max	bal Fe				

Japan

JIS G3214(91)	SUSF316L	Frg press ves 130-200mm diam	0.030 max	16.00-18.00	2.00 max	2.00-3.00	12.00-15.00	0.040 max	0.030 max	1.00 max	bal Fe	450	175	29	187 HB max
JIS G3447(94)	SUS316LTBS	Tube	0.030 max	16.00-18.00	2.00 max	2.00-3.00	12.00-16.00	0.040 max	0.030 max	1.00 max	bal Fe	480		35	
JIS G3459(94)	SUS316LLTP	Pipe	0.03 max	16-18	2 max	2-3	12-16	0.04 max	0.03 max	1 max	bal Fe				
JIS G4303(91)	SUS316L	Bar; SA; <=180mm diam	0.030 max	16.00-18.00	2.00 max	2.00-3.00	12.00-15.00	0.045 max	0.030 max	1.00 max	bal Fe	480	175	40	187 HB
JIS G4304(91)	SUS316L	Sh Plt, HR SA	0.030 max	16.00-18.00	2.00 max	2.00-3.0	12.0-15.00	0.045 max	0.030 max	1.00 max	bal Fe	480	175	40	90 HRB max
JIS G4305(91)	SUS316L	Sh Plt, CR SA	0.030 max	16.00-18.00	2.00 max	2.00-3.00	12.00-15.00	0.045 max	0.030 max	1.00 max	bal Fe	480	175	40	187 HB max
JIS G4306	SUS316L	Strp HR SA	0.03 max	16-18	2 max	2-3	12-15	0.045 max	0.03 max	1 max	bal Fe				
JIS G4307	SUS316L	Strp CR SA	0.03 max	16-18	2 max	2-3	12-15	0.045 max	0.03 max	1 max	bal Fe				
JIS G4308(98)	SUS316L	Wir rod	0.030 max	16.00-18.00	2.00 max	2.00-3.00	12.00-15.00	0.045 max	0.030 max	1.00 max	bal Fe				
JIS G4309	SUS316L	Wir	0.030 max	16.00-18.00	2.00 max	2.00-3.00	12.00-15.00	0.045 max	0.030 max	1.00 max	bal Fe				
JIS G4315	SUS316L	Wir rod	0.030 max	16.00-18.00	2.00 max	2.00-3.00	12.00-15.00	0.045 max	0.030 max	1.00 max	bal Fe				

Mexico

DGN B-229	TP316L*	Obs; Tube	0.04 max	16-18	2 max	2-3	10-15	0.04 max	0.03 max	0.75 max	bal Fe				
NMX-B-171(91)	MT316L	mech tub	0.035 max	16.0-18.0	2.00 max	2.0-3.0	10.0-15.0	0.040 max	0.030 max	1.00 max	bal Fe				
NMX-B-176(91)	TP316L	Smls, weld sanitary tub	0.08 max	16.00-18.00	2.00 max	2.00-3.00	10.00-15.00	0.040 max	0.030 max	0.75 max	bal Fe				

Norway

| NS | 14455 | Sh Strp Plt Bar Wir | 0.03 max | 16-18.5 | 2 max | 2-2.5 | 11-14 | 0.045 max | 0.03 max | 1 max | bal Fe | | | | |

UNS numbers and US grades are provided as a means of cross referencing chemically similar alloys. Exchangability is only possible after independent examination of specifications. Tensile properties are minimum or typical as specified. UTS and YS as MPa. El as %. See Appendix for list of abbreviations used in Notes. * indicates obsolete material.

Stainless Steel, Austenitic, 316L (Continued from previous page)

Specification	Designation	Notes	C	Cr	Mn	Mo	Ni	P	S	Si	Other	UTS	YS	El	Hard
Pan America															
COPANT 513	TP316L	Tube, HF, CF, 8mm diam	0.04 max	16-18	2 max	2-3	10-15	0.04 max	0.03 max	0.75 max	bal Fe				
Poland															
PNH86020	00H17N14M2	Corr res	0.03 max	16-18	2 max	2-2.5	12-15	0.045 max	0.03 max	0.8 max	bal Fe				
Romania															
STAS 3583(87)	2MoNiCr175	Corr res; Heat res	0.03 max	17-18.5	2 max	2.5-3	12.5-15	0.045 max	0.025 max	1 max	Pb <=0.15; bal Fe				
Russia															
GOST 5632(72)	03Ch17N14M2	Corr res	0.03 max	16.0-18.0	1.0-2.0	2.0-2.8	13.0-15.0	0.035 max	0.02 max	0.8 max	Cu <=0.3; Ti <=0.2; W <=0.2; bal Fe				
Spain															
UNE 36016(75)	F.3533*	see X2CrNiMo17-12-03	0.03 max	16-18	2 max	2-3	10-14	0.045 max	0.03 max	1 max	bal Fe				
UNE 36016(75)	X2CrNiMo17-12-03	Corr res	0.03 max	16-18	2 max	2-3	10-14	0.045 max	0.03 max	1 max	bal Fe				
UNE 36016/1(89)	E-316L	Corr res	0.03 max	16.5-18.5	2.0 max	2.0-2.5	11.0-14.0	0.045 max	0.03 max	1.0 max	N <=0.1; bal Fe				
UNE 36016/1(89)	F.3533*	Obs; E-316 L; Corr res	0.03 max	16.5-18.5	2.0 max	2.0-2.5	11.0-14.0	0.045 max	0.03 max	1.0 max	N <=0.1; bal Fe				
UNE 36016/1(89)	F.3537*	Obs; EN10088-3(96); X2CrNiMo17-13-3	0.03 max	16.5-18.5	2.0 max	2.5-3.0	11.5-14.5	0.045 max	0.03 max	1.0 max	N <=0.1; bal Fe				
UNE 36016/1(89)	F.3539*	Obs; EN10088-3(96); X2CrNiMo18-16-4	0.03 max	17.5-19.5	2.0 max	3.0-4.0	14.0-17.0	0.045 max	0.03 max	1.0 max	N <=0.1; bal Fe				
UNE 36016/1(89)	F.3544*	Obs; EN10088-3(96); X2CrNiMoN17-13-5	0.03 max	16.5-18.5	2.0 max	4.0-5.0	12.5-14.5	0.045 max	0.03 max	1.0 max	N 0.12-0.22; bal Fe				
UNE 36087/4(89)	F.3533*	see X2CrNiMo17 13 2	0.03 max	16.5-18.5	2 max	2-2.5	11-14	0.045 max	0.03 max	1 max	N <=0.1; bal Fe				
UNE 36087/4(89)	F.3537*	see X2CrNiMo17 13 3	0.03 max	16.5-18.5	2 max	2.5-3	11.5-14.5	0.045 max	0.03 max	1 max	N <=0.1; bal Fe				
UNE 36087/4(89)	F.3544*	see X2CrNiMoN17 13 5	0.03 max	16.5-18.5	2 max	4-5	12.5-14.5	0.045 max	0.03 max	1 max	N 00.12-0.22; bal Fe				
UNE 36087/4(89)	X2CrNiMo17 13 2	Corr res; Sh Strp	0.03 max	16.5-18.5	2 max	2-2.5	11-14	0.045 max	0.03 max	1 max	N <=0.1; bal Fe				
UNE 36087/4(89)	X2CrNiMo17 13 3	Corr res; Sh Strp	0.03 max	16.5-18.5	2 max	2.5-3	11.5-14.5	0.045 max	0.03 max	1 max	N <=0.1; bal Fe				
UNE 36087/4(89)	X2CrNiMoN17 13 5	Corr res; Sh Strp	0.03 max	16.5-18.5	2 max	4-5	12.5-14.5	0.045 max	0.03 max	1 max	N 00.12-0.22; bal Fe				
Sweden															
SS 142348	2348	Corr res	0.03 max	16.5-18.5	2 max	2-2.5	11-14	0.045 max	0.3 max	1 max	bal Fe				
SS 142353	2353	Corr res	0.03 max	16.5-18.5	2 max	2.5-3	11.5-14.5	0.045 max	0.03 max	1 max	bal Fe				
UK															
BS 1449/2(83)	316S11	Corr/Heat res; t<=100mm	0.03 max	16.5-18.5	2.00 max	2.00-2.50	11.0-14.0	0.045 max	0.030 max	1.00 max	bal Fe	490	190	40	0-195 HB
BS 1449/2(83)	316S13	Corr/Heat res; t<=100mm	0.03 max	16.5-18.5	2.00 max	2.50-3.00	11.5-14.5	0.045 max	0.030 max	1.00 max	bal Fe	490	190	40	0-195 HB
BS 1501	316S37	Plate	0.03 max	16.5-18	0.5-2	2.25-3	13-15	0.04 max	0.03 max	0.2-1	bal Fe				
BS 1501/3(71)	316S82	Plate; Press ves; Corr res	0.03 max	16.5-18.0	0.5-2.0	2.25-3.0	12.0-14.0	0.045 max	0.03 max	0.2-1.0	Cu <=0.3; Ti <=0.05; W <=0.1; bal Fe				
BS 1501/3(73)	316S12	Plate; Press ves; Corr res	0.03 max	16.5-18.5	0.5-2.0	2.25-3.0	11.0-14.0	0.045 max	0.03 max	0.2-1.0	Cu <=0.3; Ti <=0.05; W <=0.1; bal Fe				
BS 1501/3(73)	316S49	Press ves; Corr res	0.04-0.09	16.0-18.0	0.5-2.0	2.0-2.75	10.0-13.0	0.04 max	0.03 max	0.2-1.0	B 0.001-0.006; Cu <=0.3; Ti <=0.05; W <=0.1; Nb<=0.05; bal Fe				
BS 1501/3(90)	316S11	Press ves; Corr/Heat res; 0.05-100mm	0.03 max	16.5-18.5	2.0 max	2.0-2.5	11.0-14.0	0.045 max	0.025 max	1.0 max	Cu <=0.3; Ti <=0.05; W <=0.1; bal Fe	490-690	190	40	
BS 1501/3(90)	316S13	Press ves; Corr/Heat res; 0.05-100mm; wasEn845B	0.03 max	16.5-18.5	2.0 max	2.5-3.0	11.5-14.5	0.045 max	0.025 max	1.0 max	Cu <=0.3; Ti <=0.05; W <=0.1; bal Fe	490-690	190	40	
BS 1503(89)	316S11	Frg press ves; t<=999mm; HT	0.03 max	16.5-18.5	2.00 max	2.00-2.50	11.0-14.0	0.040 max	0.025 max	1.00 max	B <=0.005; Cu <=0.70; Nb <=0.20; Ti <=0.10; W <=0.1; bal Fe	490-690	225	30	

UNS numbers and US grades are provided as a means of cross referencing chemically similar alloys. Exchangability is only possible after independent examination of specifications. Tensile properties are minimum or typical as specified. UTS and YS as MPa. El as %. See Appendix for list of abbreviations used in Notes. * indicates obsolete material.

Stainless Steel, Austenitic, 316L (Continued from previous page)

Specification	Designation	Notes	C	Cr	Mn	Mo	Ni	P	S	Si	Other	UTS	YS	El	Hard
UK															
BS 1503(89)	316S13	Frg press ves; t<=999mm; HT	0.03 max	16.5-18.5	2.00 max	2.50-3.00	11.5-14.5	0.040 max	0.025 max	1.00 max	B <=0.005; Cu <=0.70; Nb <=0.20; Ti <=0.10; W <=0.1; bal Fe	490-690	225	30	
BS 1503(89)	316S63	Frg press ves; t<=999mm; HT	0.030 max	16.5-18.5	2.00 max	2.50-3.00	11.5-14.5	0.045 max	0.025 max	1.00 max	B <=0.005; Cu <=0.70; N 0.12-0.22; Nb <=0.20; Ti <=0.10; W <=0.1; bal Fe	490-690	315	30	
BS 1506(90)	316S11	Bolt matl Pres/Corr res; Corr res; 38<t<=44mm; SHT CD	0.03 max	16.5-18.5	2.00 max	2.00-2.50	11.0-14.0	0.045 max	0.030 max	1.00 max	bal Fe	650	310	28	0-320 HB
BS 1506(90)	316S11	Bolt matl Pres/Corr res; t<=160mm; Hard	0.03 max	16.5-18.5	2.00 max	2.00-2.50	11.0-14.0	0.045 max	0.030 max	1.00 max	bal Fe	490-690	190	40	
BS 1506(90)	316S13	Bolt matl Pres/Corr res; Corr res; 38<t<=44mm; SHT CD	0.03 max	16.5-18.5	2.00 max	2.50-3.00	11.5-14.5	0.045 max	0.030 max	1.00 max	bal Fe	650	310	28	0-320 HB
BS 1506(90)	316S13	Bolt matl Pres/Corr res; t<=160mm; Hard	0.03 max	16.5-18.5	2.00 max	2.50-3.00	11.5-14.5	0.045 max	0.030 max	1.00 max	bal Fe	490-690	190	40	
BS 1506(90)	316S63	Bolt matl Pres/Corr res; t<=160mm; Hard	0.07 max	16.5-18.5	2.00 max	2.00-2.50	10.0-13.0	0.045 max	0.030 max	1.00 max	N 0.12-0.22; bal Fe	580	280	35	
BS 1554(90)	316S14	Wir Corr/Heat res; diam<0.49mm; Ann	0.03 max	16.0-18.5	2.00 max	2.00-3.00	10.0-14.0	0.045 max	0.030 max	1.5-1.00	W <=0.1; bal Fe	0-850			
BS 1554(90)	316S14	Wir Corr/Heat res; 6<diam<=13mm; Ann	0.03 max	16.0-18.5	2.00 max	2.00-3.00	10.0-14.0	0.045 max	0.030 max	1.5-1.00	W <=0.1; bal Fe	0-650			
BS 2056(83)	316S42	Corr res; spring	0.07 max	16.0-18.5	2.00 max	2.00-2.50	9.50-13.5	0.045 max	0.030 max	1.00 max	W <=0.1; bal Fe				
BS 3605(73)	316S14	Pip Tube press smls weld	0.03 max	16.0-18.5	0.5-2.0	2.0-3.0	12.0-15.0	0.040 max	0.030 max	0.2-1.0	bal Fe				
BS 3605(73)	316S22	Corr res; Tub, Weld; Smls tub	0.03 max	16.0-18.5	0.5-2.0	2.0-3.0	11.0-14.0	0.040 max	0.030 max	0.2-1.0	bal Fe				
BS 3605/1(91)	316S11	Pip Tube press smls; t<=200mm	0.030 max	16.5-18.5	2.00 max	2.00-2.50	11.0-14.0	0.040 max	0.030 max	1.00 max	bal Fe	490-690		35	
BS 3606(78)	316S29	Tube	0.030 max	16.5-18.5	0.5-2.0	2.50-3.00	11.5-14.5	0.045 max	0.030 max	1.00 max	bal Fe				
BS 3606(92)	316S11	Heat exch Tub; Smls tub, Tub, Weld; 1.2-3.2mm	0.030 max	16.5-18.5	2.00 max	2.00-2.50	11.0-14.0	0.040 max	0.030 max	1.00 max	bal Fe	490-690	215	30	
BS 970/1(96)	316S11	Wrought Corr/Heat res; t<=160mm	0.030 max	16.5-18.5	2.00 max	2.00-2.50	11.0-14.0	0.045 max	0.030 max	1.00 max	bal Fe	490	190	40	183 HB max
BS 970/1(96)	316S13	Wrought Corr/Heat res; t<=160mm	0.030 max	16.5-18.5	2.00 max	2.50-3.00	11.5-14.5	0.045 max	0.030 max	1.00 max	bal Fe	490	190	40	183 HB max
BS 970/4(70)	316A12*	Obs; Corr res	0.030 max	16.5-18.5	0.50-2.00	2.25-3.0	11.0-14.0	0.045 max	0.03 max	0.2-1.0	Cu <=0.7; Pb <=0.15; Nb<=0.2; bal Fe				
USA															
	AISI 316L	Tube	0.030 max	16.00-18.00	2.00 max	2.00-3.00	10.00-14.00	0.045 max	0.030 max	1.00 max	bal Fe				
	UNS S31603	Low-C	0.030 max	16.00-18.00	2.00 max	2.00-3.00	10.00-14.00	0.045 max	0.030 max	1.00 max	bal Fe				
	UNS S31673	Surgical Implant 316L	0.030 max	17.00-19.00	2.00 max	2.00-3.00	13.00-15.50	0.025 max	0.010 max	0.75 max	Cu <=0.50; N <=0.10; bal Fe				
AMS 5507		Sh Strp Plt Bar Wir	0.03 max	16-18	2 max	2-3	10-14	0.045 max	0.03 max	1 max	bal Fe				
AMS 5653		Sh Strp Plt Bar Wir	0.03 max	16-18	2 max	2-3	10-14	0.045 max	0.03 max	1 max	bal Fe				
ASME SA182	316L	Refer to ASTM A182/A182M													
ASME SA213	316L	Refer to ASTM A213/A213M(95)													
ASME SA240	316L	Refer to ASTM A240(95)													
ASME SA249	316L	Refer to ASTM A249/A249M(96)													
ASME SA312	316L	Refer to ASTM A312/A312M(95)													
ASME SA403	316L	Refer to ASTM A403/A403M(95)													
ASME SA479	316L	Refer to ASTM A479/A479M(95)													

UNS numbers and US grades are provided as a means of cross referencing chemically similar alloys. Exchangability is only possible after independent examination of specifications. Tensile properties are minimum or typical as specified. UTS and YS as MPa. El as %. See Appendix for list of abbreviations used in Notes. * indicates obsolete material.

Stainless Steel, Austenitic, 316L (Continued from previous page)

USA

Specification	Designation	Notes	C	Cr	Mn	Mo	Ni	P	S	Si	Other	UTS	YS	El	Hard
ASME SA688	316L	Refer to ASTM A688/A688M(96)													
ASTM A167(96)	316L*	Sh Strp Plt	0.03 max	16-18	2 max	2-3	10-14	0.045 max	0.03 max	1 max	bal Fe				
ASTM A182	316L*	Bar Frg	0.03 max	16-18	2 max	2-3	10-14	0.045 max	0.03 max	1 max	bal Fe				
ASTM A182/A182M(98)	F316L	Frg/roll pipe flange valve	0.035 max	16.0-18.0	2.00 max	2.00-3.00	10.0-15.0	0.045 max	0.030 max	1.00 max	N <=0.10; bal Fe	485	170	30	
ASTM A213	316L*	Tube	0.03 max	16-18	2 max	2-3	10-14	0.045 max	0.03 max	1 max	bal Fe				
ASTM A213/A213M(95)	TP316L	Smls tube boiler, superheater, heat exchanger	0.035 max	16.0-18.0	2.00 max	2.00-3.00	10.0-15.0	0.040 max	0.030 max	0.75 max	bal Fe	485	170	35	
ASTM A240/A240M(98)	S31603	Low-C	0.030 max	16.0-18.0	2.00 max	2.00-3.00	10.0-14.0	0.045 max	0.030 max	1.00 max	N <=0.10; bal Fe	485	170	40.0	217 HB
ASTM A249/249M(96)	TP316L	Weld tube; boiler, superheater, heat exch	0.035 max	16.0-18.0	2.00 max	2.00-3.00	10.0-15.0	0.040 max	0.030 max	0.75 max	bal Fe	485	170	35	90 HRB
ASTM A269	316L	Smls Weld, Pipe	0.03 max	16-18	2 max	2-3	10-14	0.045 max	0.03 max	1 max	bal Fe				
ASTM A276(98)	316L	Bar	0.03 max	16-18	2 max	2-3	10-14	0.045 max	0.03 max	1 max	bal Fe				
ASTM A312/A312M(95)	316L	Smls Weld, Pipe	0.03 max	16-18	2 max	2-3	10-14	0.045 max	0.03 max	1 max	bal Fe				
ASTM A314	316L	Bil	0.03 max	16-18	2 max	2-3	10-14	0.045 max	0.03 max	1 max	bal Fe				
ASTM A336/A336M(98)	F316L	Frg, press/high-temp	0.035 max	16.0-18.0	2.00 max	2.00-3.00	10.0-15.0	0.040 max	0.030 max	1.00 max	bal Fe	450	170	30	90 HRB
ASTM A403	316L	Pipe	0.03 max	16-18	2 max	2-3	10-14	0.045 max	0.03 max	1 max	bal Fe				
ASTM A473	316L	Frg, Heat res SHT	0.030 max	16.00-18.00	2.00 max	2.00-3.00	10.00-14.00	0.045 max	0.030 max	1.00 max	bal Fe	515	205	40	
ASTM A478	316L	Wir CD 0.76<t<=3.18mm	0.03 max	16.00-18.00	2.00 max	2.00-3.00	10.00-14.00	0.045 max	0.030 max	1.00 max	bal Fe	830-1030		15	
ASTM A479	316L	Wir Bar Shp	0.03 max	16-18	2 max	2-3	10-14	0.045 max	0.03 max	1 max	bal Fe				
ASTM A511(96)	MT316L	Smls mech tub, Ann	0.035 max	16.0-18.0	2.00 max	2.0-3.0	10.0-15.0	0.040 max	0.030 max	1.00 max	bal Fe	517	207	35	192 HB
ASTM A554(94)	MT316L	Weld mech tub, Rnd Ann, Low-C	0.035 max	16.0-18.0	2.00 max	2.0-3.0	10.0-15.0	0.040 max	0.030 max	1.00 max	bal Fe	483	172	35	
ASTM A580/A580M(98)	316L	Wir, Ann	0.030 max	16.0-18.0	2.00 max	2.00-3.00	10.0-14.0	0.045 max	0.030 max	1.00 max	N <=0.10; bal Fe	485	170	35	
ASTM A632(90)	TP316L	Smls Weld tub, Low-C	0.040 max	16.0-18.0	2.00 max	2.00-3.00	10.0-15.0	0.040 max	0.030 max	0.75	bal Fe	485	170	35	
ASTM A666(96)	316L	Sh Strp Plt Bar, Ann	0.030 max	16.00-18.00	2.00 max	2.00-3.00	10.00-14.00	0.045 max	0.030 max	0.75 max	bal Fe	485	170	40	217 HB
ASTM A688/A688M(96)	TP316L	Weld Feedwater Heater Tub, Low-C	0.035 max	16.00-18.00	2.00 max	2.00-3.00	10.00-15.00	0.040 max	0.03 max	0.75 max	bal Fe	485	175	35	
ASTM A774/A774M(98)	TP316L	Fittings; Corr service at low/mod-temp, Low-C	0.030 max	16.0-18.0	2.00 max	2.00-3.00	10.0-14.0	0.045 max	0.030 max	1.00 max	N <=0.10; bal Fe	485-655	170	40.0	217 HB
ASTM A778(90)	TP316L	Low-C	0.030 max	16.00-18.00	2.00 max	2.00-3.00	10.00-15.00	0.045 max	0.030 max	1.00 max	N <=0.10; bal Fe	485	170	40	
ASTM A793(96)	316L	Rolled floor plt	0.030 max	16.00-18.00	2.00 max	2.00-3.00	10.00-14.00	0.045 max	0.030 max	0.75 max	N <=0.10; bal Fe				
ASTM A813/A813M(95)	TP316L	Weld Pipe, Low-C	0.035 max	16.0-18.0	2.00 max	2.00-3.00	10.0-15.0	0.045 max	0.030 max	0.75 max	bal Fe	485	170		
ASTM A814/A814M(96)	TP316L	CW Weld Pipe, Low-C	0.035 max	16.0-18.0	2.00 max	2.00-3.00	10.0-15.0	0.045 max	0.030 max	0.75 max	bal Fe	485	170		
ASTM A943/A943M(95)	TP316L	Spray formed smls pipe	0.030 max	16.0-18.0	2.00 max	2.00-3.00	10.0-15.0	0.040 max	0.030 max	0.75 max	bal Fe	485	170		
ASTM A965/965M(97)	F316L	Frg for press high-temp parts	0.035 max	16.0-18.0	2.00 max	2.00-3.00	10.0-15.0	0.040 max	0.030 max	1.00 max	bal Fe	450	170	30	
ASTM A988(98)	S31603	HIP Flanges, Fittings, Valves/parts; Heat res; Low-C	0.030 max	16.0-18.0	2.00 max	2.00-3.00	10.0-14.0	0.045 max	0.030 max	1.00 max	N <=0.10; bal Fe	485	170	30	
FED QQ-S-763F(96)	316L*	Obs; Bar Wir Shp Frg; Corr res	0.03 max	16-18	2 max	2-3	10-14	0.045 max	0.03 max	1 max	bal Fe				

Specification	Designation	Notes	C	Cr	Mn	Mo	Ni	P	S	Si	Other	UTS	YS	El	Hard

Stainless Steel, Austenitic, 316L (Continued from previous page)

USA

Specification	Designation	Notes	C	Cr	Mn	Mo	Ni	P	S	Si	Other	UTS	YS	El	Hard
FED QQ-S-766D(93)	316L*	Obs; see ASTM A240, A666, A693; Sh Strp Plt	0.03 max *	16-18	2 max	2-3	10-14	0.045 max	0.03 max	1 max	bal Fe				
MIL-T-8973(69)	316L*	Obs for new design; Tube; Corr/Heat res; aerospace	0.03 max	16.0-18.0	2.00 max	2.00-3.00	10.0-14.0	0.040 max	0.030 max	1.00 max	Cu <=0.50; bal Fe				
SAE J405(98)	S31603	Low-C	0.030 max	16.00-18.00	2.00 max	2.00-3.00	10.00-14.00	0.045 max	0.030 max	0.75 max	N <=0.10; bal Fe				

Stainless Steel, Austenitic, 316L(Hi)N

Japan

Specification	Designation	Notes	C	Cr	Mn	Mo	Ni	P	S	Si	Other	UTS	YS	El	Hard
JIS G4305(91)	SUS316Ti		0.08 max	16.00-18.00	2.00 max	2.00-3.00	10.00-14.00	0.045 max	0.030 max	1.00 max	Ti>=5xC; bal Fe	520	205	40	

USA

Specification	Designation	Notes	C	Cr	Mn	Mo	Ni	P	S	Si	Other	UTS	YS	El	Hard
	UNS S31654	N-bearing	0.03 max	16.00-18.00	2.00 max	2.00-3.00	10.00-14.00	0.045 max	0.030 max	1.00 max	N 0.16-0.30; bal Fe				

Stainless Steel, Austenitic, 316LN

Austria

Specification	Designation	Notes	C	Cr	Mn	Mo	Ni	P	S	Si	Other	UTS	YS	El	Hard
ONORM M3121	X2CrNiMoN17122KK W	Sh Strp Plt Bar	0.03 max	16.5-18.5	2 max	2.0-2.5	10.5-13.5	0.045 max	0.030 max	1 max	N 0.12-0.22; bal Fe				
ONORM M3121	X2CrNiMoN17133KK W	Sh Strp Plt Bar	0.03 max	16.5-18.5	2 max	2.5-3.0	11.5-14.5	0.045 max	0.030 max	1 max	N 0.14-0.22; bal Fe				

China

Specification	Designation	Notes	C	Cr	Mn	Mo	Ni	P	S	Si	Other	UTS	YS	El	Hard
GB 1220(92)	00C17Ni13Mo2N	Bar SA	0.030 max	16.00-18.00	2.00 max	2.00-3.00	10.50-14.50	0.035 max	0.030 max	1.00 max	N 0.12-0.22; bal Fe	550	245	40	
GB 3280(92)	00C17Ni13Mo2N	Sh Plt, CR SA	0.030 max	16.00-18.00	2.00 max	2.00-3.00	10.50-14.50	0.035 max	0.030 max	1.00 max	N 0.12-0.22; bal Fe	550	245	40	
GB 4237(92)	00C17Ni13Mo2N	Sh Plt, HR SA	0.030 max	16.00-18.00	2.00 max	2.00-3.00	10.50-14.50	0.035 max	0.030 max	1.00 max	N 0.12-0.22; bal Fe	550	245	40	
GB 4239(91)	00C17Ni13Mo2N	Sh Strp, CR SA	0.030 max	16.00-18.00	2.00 max	2.00-3.00	10.50-14.50	0.035 max	0.030 max	1.00 max	N 0.12-0.22; bal Fe	550	245	40	

Europe

Specification	Designation	Notes	C	Cr	Mn	Mo	Ni	P	S	Si	Other	UTS	YS	El	Hard
EN 10088/2(95)	1.4406	Strp, CR Corr res; t<=6mm, Ann	0.030 max	16.50-18.50	2.00 max	2.00-2.50	10.00-12.00	0.045 max	0.015 max	1.00 max	N 0.12-0.22; bal Fe	580-780	300	40	
EN 10088/2(95)	1.4429	Strp, CR Corr res; t<=6mm, Ann	0.030 max	16.50-18.50	2.00 max	2.50-3.00	11.00-14.00	0.045 max	0.015 max	1.00 max	N 0.12-0.22; bal Fe	580-780	300	35	
EN 10088/2(95)	X2CrNiMoN17-11-2	Strp, CR Corr res; t<=6mm, Ann	0.030 max	16.50-18.50	2.00 max	2.00-2.50	10.00-12.00	0.045 max	0.015 max	1.00 max	N 0.12-0.22; bal Fe	580-780	300	40	
EN 10088/2(95)	X2CrNiMoN17-13-3	Strp, CR Corr res; t<=6mm, Ann	0.030 max	16.50-18.50	2.00 max	2.50-3.00	11.00-14.00	0.045 max	0.015 max	1.00 max	N 0.12-0.22; bal Fe	580-780	300	35	
EN 10088/3(95)	1.4406	Bar Rod Sect; Corr res; 160<t<=250mm, Ann	0.030 max	16.50-18.50	2.00 max	2.00-2.50	10.00-12.00	0.045 max	0.030 max	1.00 max	N 0.12-0.22; bal Fe	580-800	280	30	250 HB
EN 10088/3(95)	1.4406	Bar Rod Sect, t<=160mm, Ann	0.030 max	16.50-18.50	2.00 max	2.00-2.50	10.00-12.00	0.045 max	0.030 max	1.00 max	N 0.12-0.22; bal Fe	580-800	280	40	250 HB
EN 10088/3(95)	1.4429	Bar Rod Sect, t<=160mm, Ann	0.030 max	16.50-18.50	2.00 max	2.50-3.00	11.00-14.00	0.045 max	0.015 max	1.00 max	N 0.12-0.22; bal Fe	580-800	280	40	250 HB
EN 10088/3(95)	1.4429	Bar Rod Sect; Corr res; 160<t<=250mm, Ann	0.030 max	16.50-18.50	2.00 max	2.50-3.00	11.00-14.00	0.045 max	0.015 max	1.00 max	N 0.12-0.22; bal Fe	580-800	280	30	250 HB
EN 10088/3(95)	X2CrNiMoN17-11-2	Bar Rod Sect; Corr res; 160<t<=250mm, Ann	0.030 max	16.50-18.50	2.00 max	2.00-2.50	10.00-12.00	0.045 max	0.030 max	1.00 max	N 0.12-0.22; bal Fe	580-800	280	30	250 HB
EN 10088/3(95)	X2CrNiMoN17-11-2	Bar Rod Sect, t<=160mm, Ann	0.030 max	16.50-18.50	2.00 max	2.00-2.50	10.00-12.00	0.045 max	0.030 max	1.00 max	N 0.12-0.22; bal Fe	580-800	280	40	250 HB
EN 10088/3(95)	X2CrNiMoN17-13-3	Bar Rod Sect; Corr res; 160<t<=250mm, Ann	0.030 max	16.50-18.50	2.00 max	2.50-3.00	11.00-14.00	0.045 max	0.015 max	1.00 max	N 0.12-0.22; bal Fe	580-800	280	30	250 HB
EN 10088/3(95)	X2CrNiMoN17-13-3	Bar Rod Sect, t<=160mm, Ann	0.030 max	16.50-18.50	2.00 max	2.50-3.00	11.00-14.00	0.045 max	0.015 max	1.00 max	N 0.12-0.22; bal Fe	580-800	280	40	250 HB

Finland

Specification	Designation	Notes	C	Cr	Mn	Mo	Ni	P	S	Si	Other	UTS	YS	El	Hard
SFS 753(86)	X2CrNiMoN17113	Corr res	0.03 max	16.0-18.5	2.0 max	2.5-3.0	9.5-13.0	0.045 max	0.03 max	1.0 max	N 0.12-0.22; bal Fe				
SFS 757(86)	X4CrNiMo17123	Corr res	0.05 max	16.0-18.5	2.0 max	2.5-3.0	10.5-14.0	0.045 max	0.03 max	1.0 max	bal Fe				

UNS numbers and US grades are provided as a means of cross referencing chemically similar alloys. Exchangability is only possible after independent examination of specifications. Tensile properties are minimum or typical as specified. UTS and YS as MPa. El as %. See Appendix for list of abbreviations used in Notes. * indicates obsolete material.

Specification	Designation	Notes	C	Cr	Mn	Mo	Ni	P	S	Si	Other	UTS	YS	El	Hard

Stainless Steel, Austenitic, 316LN (Continued from previous page)

France

Specification	Designation	Notes	C	Cr	Mn	Mo	Ni	P	S	Si	Other	UTS	YS	El	Hard
AFNOR NFA35582	Z2CND17.12Az	Sh Strp Plt Bar	0.03 max	16-18	2 max	2-2.5	11-13	0.045 max	0.03 max	1 max	N 0.1-0.2; bal Fe				
AFNOR NFA35582	Z2CND17.13Az	Sh Strp Plt Bar	0.03 max	16-18	2 max	2.5-3	11.5-13.5	0.045 max	0.03 max	1 max	N 0.1-0.2; bal Fe				

Germany

Specification	Designation	Notes	C	Cr	Mn	Mo	Ni	P	S	Si	Other	UTS	YS	El	Hard
DIN 17440(96)	GX2CrNiMoN17-12-2	Plt Sht HR Strp Bar Wir Frg Ann	0.030 max	16.5-18.5	2.00 max	2.00-2.50	10.0-13.0	0.045 max	0.030 max	1.00 max	N <=0.11; bal Fe	440-640	185	25	
DIN 17440(96)	WNr 1.4406	HR Strp Bar Wir Frg Hard Q	0.030 max	16.5-18.5	2.00 max	2.00-2.50	10.0-12.0	0.045 max	0.030 max	1.00 max	N 0.12-0.22; bal Fe	590-780	295	40	
DIN 17440(96)	WNr 1.4429	HR Strp Bar Wir Frg HT	0.030 max	16.5-18.5	2.00 max	2.50-3.00	11.0-14.0	0.045 max	0.015 max	1.00 max	N 0.12-0.22; bal Fe	580-800	280	30-40	
DIN 17441(97)	WNr 1.4406	CR Strp Plt Sh, hard Q	0.030 max	16.5-18.5	2.00 max	2.00-2.50	10.0-12.0	0.045 max	0.030 max	1.00 max	N 0.12-0.22; bal Fe	590-780	295	40	
DIN 17441(97)	WNr 1.4429	CR Strp Plt Sh HT	0.030 max	16.5-18.5	2.00 max	2.50-3.00	11.0-14.0	0.045 max	0.015 max	1.00 max	N 0.12-0.22; bal Fe	580-800	280	30-40	
DIN 17459(92)	WNr 1.4910	Smls tube SHT	0.04 max	16.0-18.0	2.00 max	2.00-2.80	12.0-14.0	0.035 max	0.015 max	0.75	B 0.0015-0.0050; N 0.10-0.18; bal Fe	550-750	260	35	
DIN 17459(92)	X3CrNiMoN17-13	Smls tube SHT	0.04 max	16.0-18.0	2.00 max	2.00-2.80	12.0-14.0	0.035 max	0.015 max	0.75	B 0.0015-0.0050; N 0.10-0.18; bal Fe	550-750	260	35	
DIN EN 10028(96)	WNr 1.4910	Flat product for Press purposes SHT	0.04 max	16.0-18.0	2.00 max	2.00-2.80	12.0-14.0	0.035 max	0.015 max	0.75	B 0.0015-0.0050; N 0.10-0.18; bal Fe	550-750	260	35	
DIN EN 10028(96)	X3CrNiMoN17-13	Flat product for Press purposes SHT	0.04 max	16.0-18.0	2.00 max	2.00-2.80	12.0-14.0	0.035 max	0.015 max	0.75	B 0.0015-0.0050; N 0.10-0.18; bal Fe	550-750	260	35	
DIN EN 10088(95)	X2CrNiMoN17-13-3	Sh Plt Strp Bar Rod Sect, HT	0.030 max	16.5-18.5	2.00 max	2.50-3.00	11.0-14.0	0.045 max	0.015 max	1.00 max	N 0.12-0.22; bal Fe	580-800	280	30-40	
TGL 7143	X2CrNiMoN18.12*	Obs	0.03 max	16.5-18.5	2 max	2-3	11-14	0.045 max	0.03 max	1 max	N 0.12-0.22; bal Fe				

International

Specification	Designation	Notes	C	Cr	Mn	Mo	Ni	P	S	Si	Other	UTS	YS	El	Hard
ISO 4954(93)	X2CrNiMoN17133E	Wir rod Bar, CH ext	0.030 max	16.5-18.5	2.00 max	2.5-3.0	11.5-14.5	0.045 max	0.030 max	1.00 max	N 0.12-0.22; bal Fe	780			
ISO 683-13(74)	19aN*	ST	0.03 max	16.5-18.5	2 max	2.5-3	11.5-14.5	0.045 max	0.03 max	1 max	N 0.12-0.22; bal Fe				
ISO 683-13(74)	19N*	ST	0.03 max	16.5-18.5	2 max	2-2.5	10.5-13.5	0.045 max	0.03 max	1 max	N 0.12-0.22; bal Fe				

Italy

Specification	Designation	Notes	C	Cr	Mn	Mo	Ni	P	S	Si	Other	UTS	YS	El	Hard
UNI 7500(75)	X2CrNiMoN1712	Press ves	0.03 max	16-18.5	2 max	2-2.5	11-14	0.045 max	0.03 max	1 max	N 0.12-0.25; bal Fe				
UNI 7500(75)	X2CrNiMoN1713	Press ves	0.03 max	16-18.5	2 max	2.5-3	11.5-14	0.045 max	0.03 max	1 max	N 0.12-0.25; bal Fe				
UNI 8317(81)	X2CrNiMoN1712	Sh Strp Plt	0.03 max	16-18.5	2 max	2-2.5	11-14	0.045 max	0.03 max	1 max	N 0.1-0.2; bal Fe				
UNI 8317(81)	X2CrNiMoN1713	Sh Strp Plt	0.03 max	16-18.5	2 max	2.5-3	11.5-14.5	0.045 max	0.03 max	1 max	N 0.1-0.2; bal Fe				

Japan

Specification	Designation	Notes	C	Cr	Mn	Mo	Ni	P	S	Si	Other	UTS	YS	El	Hard
JIS G3214(91)	SUSF316LN	Frg press ves 130-200mm diam	0.030 max	16.00-18.00	2.00 max	2.00-3.00	10.00-14.00	0.040 max	0.030 max	1.00 max	bal Fe	480	205	29	187 HB max
JIS G4303(91)	SUS316LN	Bar; SA; <=180mm diam	0.030 max	16.50-18.50	2.00 max	2.00-3.00	10.50-14.50	0.045 max	0.030 max	1.00 max	N 0.12-0.22; bal Fe	550	245	40	217 HB
JIS G4304(91)	SUS316LN	Sh Plt, HR	0.030 max	16.50-18.50	2.00 max	2.00-3.0	10.50-14.50	0.045 max	0.030 max	1.00 max	N 0.12-0.22; bal Fe	550	245	40	95 HRB max
JIS G4305(91)	SUS316LN	Sh Plt, CR	0.030 max	16.50-18.50	2.00 max	2.00-3.00	10.50-14.50	0.045 max	0.030 max	1.00 max	N 0.12-0.22; bal Fe	550	245	40	217 HB max
JIS G4306	SUS316LN	Strp HR	0.03 max	16.5-18.5	2 max	2-3	10.5-14.5	0.045 max	0.03 max	1 max	N 0.12-0.22; bal Fe				
JIS G4307	SUS316LN	Strp CR	0.03 max	16.5-18.5	2 max	2-3	10.5-14.5	0.045 max	0.03 max	1 max	N 0.12-0.22; bal Fe				

Mexico

Specification	Designation	Notes	C	Cr	Mn	Mo	Ni	P	S	Si	Other	UTS	YS	El	Hard
NMX-B-186-SCFI(94)	TP316LN	Pipe; High-temp	0.035 max	16.0-18.0	2.00 max	2.00-3.00	11.0-14.0	0.040 max	0.030 max	0.75 max	N 0.10-0.16; bal Fe	517	207		

Spain

Specification	Designation	Notes	C	Cr	Mn	Mo	Ni	P	S	Si	Other	UTS	YS	El	Hard
UNE 36016/1(89)	F.3542*	Obs; EN10088-3(96); X2CrNiMoN17-12-2	0.03 max	16.5-18.5	2.0 max	2.0-2.5	10.5-13.5	0.045 max	0.03 max	1.0 max	N 0.12-0.22; bal Fe				
UNE 36016/1(89)	F.3543*	Obs; EN10088-3(96); X2CrNiMoN17-13-3	0.03 max	16.5-18.5	2.0 max	2.5-3.0	11.5-14.5	0.045 max	0.03 max	1.0 max	N 0.12-0.22; bal Fe				
UNE 36087/4(89)	F.3542*	see X2CrNiMoN17 12 2	0.03 max	16.5-18.5	2 max	2-2.5	10.5-13.5	0.045 max	0.03 max	1 max	N 0.12-0.22; bal Fe				

UNS numbers and US grades are provided as a means of cross referencing chemically similar alloys. Exchangability is only possible after independent examination of specifications. Tensile properties are minimum or typical as specified. UTS and YS as MPa. El as %. See Appendix for list of abbreviations used in Notes. * indicates obsolete material.

Specification	Designation	Notes	C	Cr	Mn	Mo	Ni	P	S	Si	Other	UTS	YS	El	Hard
Stainless Steel, Austenitic, 316LN (Continued from previous page)															
Spain															
UNE 36087/4(89)	F.3543*	see X2CrNiMoN17 13 3	0.03 max	16.5-18.5	2 max	2.5-3	11.5-14.5	0.045 max	0.03 max	1 max	N 0.12-0.22; bal Fe				
UNE 36087/4(89)	X2CrNiMoN17 12 2	Corr res; Sh Strp	0.03 max	16.5-18.5	2 max	2-2.5	10.5-13.5	0.045 max	0.03 max	1 max	N 0.12-0.22; bal Fe				
UNE 36087/4(89)	X2CrNiMoN17 13 3	Corr res; Sh Strp	0.03 max	16.5-18.5	2 max	2.5-3	11.5-14.5	0.045 max	0.03 max	1 max	N 0.12-0.22; bal Fe				
Sweden															
SS 142375	2375	Corr res	0.03 max	16.5-18.5	2 max	2.5-3	9.5-13	0.045 max	0.03 max	1 max	N 0.12-0.22; bal Fe				
UK															
BS 1501/3(73)	316S62	Plate	0.03 max	16.5-18.5	0.5-2.0	2.25-3.0	11.0-14.0	0.045 max	0.03 max	0.2-1.0	N 0.15-0.25; bal Fe				
BS 1501/3(90)	316S63	Press ves; Corr/Heat res	0.03 max	16.5-18.5	2.0 max	2.5-3.0	11.5-14.5	0.045 max	0.025 max	1.0 max	Cu <=0.3; N 0.12-0.22; Ti <=0.05; W <=0.1; bal Fe	580-780	280	35	
BS 1502(82)	316S61	Bar Shp; Press ves	0.030 max	16.5-18.5	2.00 max	2.00-2.50	10.5-13.5	0.045 max	0.030 max	1.00 max	N 0.12-0.22; W <=0.1; bal Fe				
BS 1502(82)	316S63	Bar Shp; Press ves	0.030 max	16.5-18.5	2.00 max	2.50-3.00	11.5-14.5	0.045 max	0.030 max	1.00 max	N 0.12-0.22; W <=0.1; bal Fe				
BS 1503(89)	316S61	Frg press ves; t<=999mm; HT	0.030 max	16.5-18.5	2.00 max	2.00-2.50	10.5-13.5	0.045 max	0.025 max	1.00 max	B <=0.005; Cu <=0.70; N 0.12-0.22; Nb <=0.20; Ti <=0.10; W <=0.1; bal Fe	580-780	315	35	
BS 1506(90)	316S61	Bolt matl Pres/Corr res; t<=160mm; Hard	0.030 max	16.5-18.5	2.00 max	2.00-2.50	10.5-13.5	0.045 max	0.030 max	1.00 max	N 0.12-0.22; bal Fe	580	280	35	
USA															
	UNS S31653	N-bearing	0.03 max	16.00-18.00	2.00 max	2.00-3.00	10.00-14.00	0.045 max	0.030 max	1.00 max	N 0.10-0.16; bal Fe				
ASTM A182/A182M(98)	F316LN	Frg/roll pipe flange valve	0.030 max	16.0-18.0	2.00 max	2.00-3.00	11.0-14.0	0.045 max	0.030 max	0.75 max	N 0.10-0.16; bal Fe	515	205	30	
ASTM A193/A193M(98)	B8MLN	Bolt, high-temp, HT	0.030 max	16.0-18.0	2.00 max	2.00-3.00	10.0-13.0	0.045 max	0.030 max	1.00 max	N 0.10-0.16; bal Fe	760	665	15	321 HB
ASTM A193/A193M(98)	B8MLNA	Bolt, high-temp, HT	0.030 max	16.0-18.0	2.00 max	2.00-3.00	10.0-13.0	0.045 max	0.030 max	1.00 max	N 0.10-0.16; bal Fe	515	205	30	192 HB
ASTM A194/A194M(98)	8MLN	Nuts, high-temp press	0.030 max	16.0-18.0	2.00 max	2.00-3.00	10.0-13.0	0.045 max	0.030 max	1.00 max	N 0.10-0.16; bal Fe				126-300 HB
ASTM A194/A194M(98)	8MLNA	Nuts, high-temp press	0.030 max	16.0-18.0	2.00 max	2.00-3.00	10.0-13.0	0.045 max	0.030 max	1.00 max	N 0.10-0.16; bal Fe				126-192 HB
ASTM A213/A213M(95)	TP316LN	Smls tube boiler, superheater, heat exchanger	0.035 max	16.0-18.0	2.00 max	2.00-3.00	11.0-14.0	0.040 max	0.030 max	0.75 max	N 0.10-0.16; bal Fe	515	205	35	
ASTM A240/A240M(98)	S31653	N-bearing	0.030 max	16.0-18.0	2.00 max	2.00-3.00	10.0-14.0	0.045 max	0.030 max	1.00 max	N 0.10-0.16; bal Fe	515	205	40.0	217 HB
ASTM A249/249M(96)	TP316LN	Weld tube; boiler, superheater, heat exch	0.035 max	16.0-18.0	2.00 max	2.00-3.00	10.0-15.0	0.040 max	0.030 max	0.75 max	N 0.10-0.16; bal Fe	515	205	35	90 HRB
ASTM A276(98)	316LN	Bar Shp	0.03 max	16-18	2 max	2-3	10-14	0.045 max	0.03 max	1 max	N 0.1-0.16; bal Fe				
ASTM A336/A336M(98)	F316LN	Frg, press/high-temp	0.030 max	16.0-18.0	2.00 max	2.00-3.00	10.0-14.0	0.040 max	0.030 max	1.00 max	N 0.10-0.16; bal Fe	485	205	30	90 HRB
ASTM A688/A688M(96)	TP316LN	Weld Feedwater Heater Tub, N-bearing	0.035 max	10.00-18.00	2.00 max	2.00-3.00	10.00-15.00	0.040 max	0.03 max	0.75 max	N 0.10-0.16; bal Fe	515	205	35	
ASTM A813/A813M(95)	TP316LN	Weld Pipe, N-bearing	0.035 max	16.0-18.0	2.00 max	2.00-3.00	10.0-14.0	0.045 max	0.030 max	0.75 max	N 0.10-0.16; bal Fe	515	205		
ASTM A814/A814M(96)	TP316LN	CW Weld Pipe, N-bearing	0.035 max	16.0-18.0	2.00 max	2.00-3.00	10.0-14.0	0.045 max	0.030 max	0.75 max	N 0.10-0.16; bal Fe	515	205		
ASTM A943/A943M(95)	TP316LN	Spray formed smls pipe	0.030 max	16.0-18.0	2.00 max	2.00-3.00	11.0-14.0	0.040 max	0.030 max	0.75 max	N 0.10-0.16; bal Fe	515	205		
ASTM A965/965M(97)	F316LN	Frg for press high-temp parts	0.030 max	16.0-18.0	2.00 max	2.00-3.00	10.0-14.0	0.040 max	0.030 max	1.00 max	N 0.10-0.16; bal Fe	485	205	30	
ASTM A988(98)	S31653	HIP Flanges, Fittings, Valves/parts; Heat res; N-bearing	0.030 max	16.0-18.0	2.00 max	2.00-3.00	10.0-13.0	0.045 max	0.030 max	1.00 max	N 0.10-0.16; bal Fe	515	205	30	
SAE J405(98)	S31653	N-bearing	0.030 max	16.00-18.00	2.00 max	2.00-3.00	10.00-14.00	0.045 max	0.030 max	0.75 max	N 0.10-0.16; bal Fe				

UNS numbers and US grades are provided as a means of cross referencing chemically similar alloys. Exchangability is only possible after independent examination of specifications. Tensile properties are minimum or typical as specified. UTS and YS as MPa. El as %. See Appendix for list of abbreviations used in Notes. * indicates obsolete material.

Specification	Designation	Notes	C	Cr	Mn	Mo	Ni	P	S	Si	Other	UTS	YS	El	Hard

Stainless Steel, Austenitic, 316N

China

Specification	Designation	Notes	C	Cr	Mn	Mo	Ni	P	S	Si	Other	UTS	YS	El	Hard
GB 1220(92)	0Cr17Ni12Mo2N	Bar SA	0.08 max	16.00-18.00	2.00 max	2.00-3.00	10.00-14.00	0.035 max	0.030 max	1.00 max	N 0.10-0.22; bal Fe	550	275	35	
GB 3280(92)	0Cr17Ni12Mo2N	Sh Plt, CR SA	0.08 max	16.00-18.00	2.00 max	2.00-3.00	10.00-14.00	0.035 max	0.030 max	1.00 max	N 0.10-0.22; bal Fe	550	275	35	
GB 4237(92)	0Cr17Ni12Mo2N	Sh Plt, HR SA	0.08 max	16.00-18.00	2.00 max	2.00-3.00	10.00-14.00	0.035 max	0.030 max	1.00 max	N 0.10-0.22; bal Fe	550	275	35	
GB 4239(91)	0Cr17Ni12Mo2N	Sh Strp, CR SA	0.08 max	16.00-18.00	2.00 max	2.00-3.00	10.00-14.00	0.035 max	0.030 max	1.00 max	N 0.10-0.22; bal Fe	550	275	35	

Japan

Specification	Designation	Notes	C	Cr	Mn	Mo	Ni	P	S	Si	Other	UTS	YS	El	Hard
JIS G3214(91)	SUSF316N	Frg press ves 130-200mm diam	0.08 max	16.00-18.00	2.00 max	2.00-3.00	11.00-14.00	0.040 max	0.030 max	0.75 max	N 0.10-0.16; bal Fe	550	240	24	217 HB max
JIS G4303(91)	SUS316N	Bar; SA; <=180mm diam	0.08 max	16.00-18.00	2.00 max	2.00-3.00	10.00-14.00	0.045 max	0.030 max	1.00 max	N 0.10-0.22; bal Fe	550	275	35	217 HB
JIS G4304(91)	SUS316N	Sh Plt, HR	0.08 max	16.00-18.00	2.00 max	2.00-3.0	10.0-14.00	0.045 max	0.030 max	1.00 max	N 0.12-0.22; bal Fe	550	275	35	95 HRB max
JIS G4305(91)	SUS316N	Sh Plt, CR	0.08 max	16.00-18.0	2.00 max	2.00-3.00	10.00-14.00	0.045 max	0.030 max	1.00 max	N 0.10-0.22; bal Fe	550	275	35	217 HB max
JIS G4306	SUS316N	Strp HR	0.08 max	16-18	2 max	2-3	10-14	0.045 max	0.03 max	1 max	N 0.1-0.22; bal Fe				
JIS G4307	SUS316N	Strp CR	0.08 max	16-18	2 max	2-3	10-14	0.045 max	0.03 max	1 max	N 0.1-0.22; bal Fe				

Mexico

Specification	Designation	Notes	C	Cr	Mn	Mo	Ni	P	S	Si	Other	UTS	YS	El	Hard
NMX-B-186-SCFI(94)	TP316N	Pipe; High-temp	0.08 max	16.0-18.0	2.00 max	2.00-3.00	11.0-14.0	0.040 max	0.030 max	0.75 max	N 0.10-0.16; bal Fe	552	241		

UK

Specification	Designation	Notes	C	Cr	Mn	Mo	Ni	P	S	Si	Other	UTS	YS	El	Hard
BS 1501/3(73)	316S66	Press ves; Corr res	0.07 max	16.5-18.5	0.5-2.0	2.25-3.0	10.0-13.0	0.045 max	0.03 max	0.2-1.0	Cu <=0.3; N 0.15-0.25; Ti <=0.05; W <=0.1; bal Fe				
BS 1502(82)	316S65	Bar Shp; Press ves	0.07 max	16.5-18.5	2.00 max	2.00-2.50	10.0-13.0	0.045 max	0.030 max	1.00 max	N 0.12-0.22; W <=0.1; bal Fe				
BS 1502(82)	316S67	Bar Shp; Press ves	0.07 max	16.5-18.5	2.00 max	2.50-3.00	10.5-13.5	0.045 max	0.030 max	1.00 max	N 0.12-0.22; W <=0.1; bal Fe				
BS 1506(90)	316S67	Bolt matl Pres/Corr res; t<=160mm; Hard	0.07 max	16.5-18.5	2.00 max	2.50-3.00	10.5-13.5	0.045 max	0.030 max	1.00 max	N 0.12-0.22; bal Fe	580	280	35	

USA

Specification	Designation	Notes	C	Cr	Mn	Mo	Ni	P	S	Si	Other	UTS	YS	El	Hard
	AISI 316N	Tube	0.08 max	16.00-18.00	2.00 max	2.00-3.00	10.00-14.00	0.045 max	0.030 max	1.00 max	N 0.10-0.16; bal Fe				
	UNS S31651	N-bearing	0.08 max	16.00-18.00	2.00 max	2.00-3.00	10.00-14.00	0.045 max	0.030 max	1.00 max	N 0.10-0.16; bal Fe				
ASME SA182	316N	Refer to ASTM A182/A182M													
ASME SA213	316N	Refer to ASTM A213/A213M(95)													
ASME SA240	316N	Refer to ASTM A240(95)													
ASME SA249	316N	Refer to ASTM A249/A249M(96)													
ASME SA312	316N	Refer to ASTM A312/A312M(95)													
ASME SA358	316N	Refer to ASTM A358/A358M(95)													
ASME SA376	316N	Refer to ASTM A376/A376M(93)													
ASME SA403	316N	Refer to ASTM A403/A403M(95)													
ASME SA430	316N	Refer to ASTM A430/A430M(91)													
ASME SA479	316N	Refer to ASTM A479/A479M(95)													
ASTM A182	316N*	Bar Frg	0.08 max	16-18	2 max	2-3	10-14	0.045 max	0.03 max	1 max	N 0.1-0.16; bal Fe				
ASTM A182/A182M(98)	F316N	Frg/roll pipe flange valve	0.08 max	16.0-18.0	2.00 max	2.00-3.00	11.0-14.0	0.045 max	0.030 max	0.75 max	N 0.10-0.16; bal Fe	550	240	30	
ASTM A193/A193M(98)	B8MN	Bolt, high-temp, HT	0.08 max	16.0-18.0	2.00 max	2.00-3.00	10.0-13.0	0.045 max	0.030 max	1.00 max	N 0.10-0.16; bal Fe	550	240	30	223 HB
ASTM A193/A193M(98)	B8MNA	Bolt, high-temp, HT	0.08 max	16.0-18.0	2.00 max	2.00-3.00	10.0-13.0	0.045 max	0.030 max	1.00 max	N 0.10-0.16; bal Fe	515	205	30	192 HB

UNS numbers and US grades are provided as a means of cross referencing chemically similar alloys. Exchangability is only possible after independent examination of specifications. Tensile properties are minimum or typical as specified. UTS and YS as MPa. El as %. See Appendix for list of abbreviations used in Notes. * indicates obsolete material.

Specification	Designation	Notes	C	Cr	Mn	Mo	Ni	P	S	Si	Other	UTS	YS	El	Hard

Stainless Steel, Austenitic, 316N (Continued from previous page)

USA

Specification	Designation	Notes	C	Cr	Mn	Mo	Ni	P	S	Si	Other	UTS	YS	El	Hard
ASTM A194/A194M(98)	8MN	Nuts, high-temp press	0.08 max	16.0-18.0	2.00 max	2.00-3.00	10.0-13.0	0.045 max	0.030 max	1.00 max	N 0.10-0.16; bal Fe				126-300 HB
ASTM A194/A194M(98)	8MNA	Nuts, high-temp press	0.08 max	16.0-18.0	2.00 max	2.00-3.00	10.0-13.0	0.045 max	0.030 max	1.00 max	N 0.10-0.16; bal Fe				126-192 HB
ASTM A213	316N*	Tube	0.08 max	16-18	2 max	2-3	10-14	0.045 max	0.03	1 max	N 0.1-0.16; bal Fe				
ASTM A213/A213M(95)	TP316N	Smls tube boiler, superheater, heat exchanger	0.08 max	16.0-18.0	2.00 max	2.00-3.00	11.0-14.0	0.040 max	0.030 max	0.75 max	N 0.10-0.16; bal Fe	550	240	35	
ASTM A240/A240M(98)	S31651	N-bearing	0.08 max	16.0-18.0	2.00 max	2.00-3.00	10.0-14.0	0.045 max	0.030 max	1.00 max	N 0.10-0.16; bal Fe	550	240	35.0	217 HB
ASTM A249/249M(96)	TP316N	Weld tube; boiler, superheater, heat exch	0.08 max	16.0-18.0	2.00 max	2.00-3.00	10.0-14.0	0.040 max	0.030 max	0.75 max	N 0.10-0.16; bal Fe	550	240	35	90 HRB
ASTM A276(98)	316N	Bar	0.08 max	16-18	2 max	2-3	10-14	0.045 max	0.03 max	1 max	N 0.1-0.16; bal Fe				
ASTM A312/A312M(95)	316N	Smls Weld, Pipe	0.08 max	16-18	2 max	2-3	10-14	0.045 max	0.03 max	1 max	N 0.1-0.16; bal Fe				
ASTM A336/A336M(98)	F316N	Frg, press/high-temp	0.08 max	16.0-18.0	2.00 max	2.00-3.00	11.0-14.0	0.030 max	0.030 max	0.75 max	N 0.10-0.16; bal Fe	550	240	25	90 HRB
ASTM A358/A358M(95)	316N	Elect fusion weld pipe, high-temp	0.08 max	16-18	2 max	2-3	10-14	0.045 max	0.03 max	1 max	N 0.1-0.16; bal Fe				
ASTM A376	316N	Smls pipe	0.08 max	16-18	2 max	2-3	10-14	0.045 max	0.03 max	1 max	N 0.1-0.16; bal Fe				
ASTM A403	316N	Pipe	0.08 max	16-18	2 max	2-3	10-14	0.045 max	0.03 max	1 max	N 0.1-0.16; bal Fe				
ASTM A430	316N*	Obs 1995; see A312; Pipe	0.08 max	16-18	2 max	2-3	10-14	0.045 max	0.03 max	1 max	N 0.1-0.16; bal Fe				
ASTM A479	316N	Bar	0.08 max	16-18	2 max	2-3	10-14	0.045 max	0.03 max	1 max	N 0.1-0.16; bal Fe				
ASTM A666(96)	316N	Sh Strp Plt Bar, Ann	0.08 max	16.00-18.00	2.00 max	2.00-3.00	10.00-14.00	0.045 max	0.030 max	0.75 max	N 0.10-0.16; bal Fe	550	240	35	217 HB
ASTM A688/A688M(96)	TP316N	Weld Feedwater Heater Tub, N-bearing	0.08 max	16.00-18.00	2.00 max	2.00-3.00	10.00-14.00	0.040 max	0.03 max	0.75 max	N 0.10-0.16; bal Fe	550	240	35	
ASTM A813/A813M(95)	TP316N	Weld Pipe, N-bearing	0.08 max	16.0-18.0	2.00 max	2.00-3.00	10.0-14.0	0.045 max	0.030 max	0.75 max	N 0.10-0.16; bal Fe	550	240		
ASTM A814/A814M(96)	TP316N	CW Weld Pipe, N-bearing	0.08 max	16.0-18.0	2.00 max	2.00-3.00	10.0-14.0	0.045 max	0.030 max	0.75 max	N 0.10-0.16; bal Fe	550	240		
ASTM A943/A943M(95)	TP316N	Spray formed smls pipe	0.08 max	16.0-18.0	2.00 max	2.00-3.00	11.0-14.0	0.040 max	0.030 max	0.75 max	N 0.10-0.16; bal Fe	550	240		
ASTM A965/965M(97)	F316N	Frg for press high-temp parts	0.08 max	16.0-18.0	2.00 max	2.00-3.00	11.0-14.0	0.030 max	0.030 max	0.75 max	N 0.10-0.16; bal Fe	550	240	25	
ASTM A988(98)	S31651	HIP Flanges, Fittings, Valves/parts; Heat res; N-bearing	0.08 max	16.0-18.0	2.00 max	2.00-3.00	10.0-13.0	0.045 max	0.030 max	1.00 max	N 0.10-0.16; bal Fe	550	240	30	
SAE J405(98)	S31651	N-bearing	0.08 max	16.00-18.00	2.00 max	2.00-3.00	10.00-14.00	0.045 max	0.030 max	0.75 max	N 0.10-0.16; bal Fe				

Stainless Steel, Austenitic, 316Ti

Australia

Specification	Designation	Notes	C	Cr	Mn	Mo	Ni	P	S	Si	Other	UTS	YS	El	Hard
AS 1449(94)	316Ti	Sh Strp Plt	0.08 max	16.00-18.00	2.00 max	2.00-3.00	10.00-14.00	0.045 max	0.030 max	0.75 max	Ti>=4xC; bal Fe				

Austria

Specification	Designation	Notes	C	Cr	Mn	Mo	Ni	P	S	Si	Other	UTS	YS	El	Hard
ONORM M3121	X6CrNiMoTi17122KKW	Sh Strp Plt Bar Wir	0.08 max	16.5-18.5	2 max	2.0-2.5	10.5-13.5	0.045 max	0.030 max	1 max	Ti>=5xC<=0.80; bal Fe				

Bulgaria

Specification	Designation	Notes	C	Cr	Mn	Mo	Ni	P	S	Si	Other	UTS	YS	El	Hard
BDS 6738(72)	0Ch17N13M2T	Corr res	0.08 max	16.0-18.0	2.00 max	2.00-2.50	11.0-14.0	0.035 max	0.025 max	0.80 max	5xC<=Ti<=0.70; bal Fe				
BDS 6738(72)	0Ch17N15M3T	Corr res	0.08 max	16.0-18.0	2.00 max	3.00-4.00	14.0-16.0	0.035 max	0.025 max	0.80 max	Ti 0.30-0.60; bal Fe				

China

Specification	Designation	Notes	C	Cr	Mn	Mo	Ni	P	S	Si	Other	UTS	YS	El	Hard
GB 1220(92)	0Cr18Ni12Mo2Ti	Bar SA	0.08 max	16.00-19.00	2.00 max	1.80-2.50	11.00-14.00	0.035 max	0.030 max	1.00 max	Ti>=5xC<=0.70; bal Fe	530	205	40	
GB 13296(91)	0Cr18Ni12Mo2Ti	Smls tube SA	0.08 max	16.00-19.00	2.00 max	1.80-2.50	11.00-14.00	0.035 max	0.030 max	1.00 max	Ti>=5xC<=0.70; bal Fe	530	205	35	
GB 3280(92)	0Cr18Ni12Mo2Ti	Sh Plt, CR SA	0.08 max	16.00-19.00	2.00 max	1.80-2.50	11.00-14.00	0.035 max	0.030 max	1.00 max	Ti>=5xC<=0.70; bal Fe	530	205	35	

Specification	Designation	Notes	C	Cr	Mn	Mo	Ni	P	S	Si	Other	UTS	YS	El	Hard

Stainless Steel, Austenitic, 316Ti (Continued from previous page)

China

Specification	Designation	Notes	C	Cr	Mn	Mo	Ni	P	S	Si	Other	UTS	YS	El	Hard
GB 4237(92)	0Cr18Ni12Mo2Ti	Sh Plt, HR SA	0.08 max	16.00-19.00	2.00 max	1.80-2.50	11.00-14.00	0.035 max	0.030 max	1.00 max	Ti>=5xC<=0.70; bal Fe	530	205	37	
GB/T 14975(94)	0Cr18Ni12Mo2Ti	Weld tube; SA	0.08 max	16.00-19.00	2.00 max	1.80-2.50	11.00-14.00	0.035 max	0.030 max	1.00 max	Ti>=5xC<=0.70; bal Fe	530	205	35	
GB/T 14976(94)	0Cr18Ni12Mo2Ti	Smls pipe SA	0.08 max	16.00-19.00	2.00 max	1.80-2.50	11.00-14.00	0.035 max	0.030 max	1.00 max	Ti>=5xC<=0.70; bal Fe	530	205	35	

Czech Republic

Specification	Designation	Notes	C	Cr	Mn	Mo	Ni	P	S	Si	Other	UTS	YS	El	Hard
CSN 417347	17347	Corr res	0.12 max	16.0-19.0	2.0 max	1.5-2.5	9.0-12.0	0.045 max	0.03 max	1.5 max	Ti <=1.50; Ti>=5xC; bal Fe				
CSN 417348	17348	Corr res	0.1 max	16.5-18.5	2.0 max	2.0-2.5	11.0-14.0	0.045 max	0.03 max	1.0 max	Ti>=5xC; bal Fe				

Europe

Specification	Designation	Notes	C	Cr	Mn	Mo	Ni	P	S	Si	Other	UTS	YS	El	Hard
EN 10088/2(95)	1.4571	Strp, CR Corr res; t<=6mm, Ann	0.08 max	16.50-18.50	2.00 max	2.00-2.50	10.50-13.50	0.045 max	0.015 max	1.00 max	5xC<=Ti<=0.70; bal Fe	540-690	240	40	
EN 10088/2(95)	X6CrNiMoTi17-12-2	Strp, CR Corr res; t<=6mm, Ann	0.08 max	16.50-18.50	2.00 max	2.00-2.50	10.50-13.50	0.045 max	0.015 max	1.00 max	5xC<=Ti<=0.70; bal Fe	540-690	240	40	
EN 10088/3(95)	1.4571	Bar Rod Sect, t<=160mm, Ann	0.08 max	16.50-18.50	2.00 max	2.00-2.50	10.50-13.50	0.045 max	0.030 max	1.00 max	Ti <=0.7; 5xC<=Ti<=0.70; bal Fe	500-700	200	40	215 HB
EN 10088/3(95)	1.4571	Bar Rod Sect; Corr res; 160<t<=250mm, Ann	0.08 max	16.50-18.50	2.00 max	2.00-2.50	10.50-13.50	0.045 max	0.030 max	1.00 max	Ti <=0.7; 5xC<=Ti<=0.70; bal Fe	500-700	200	30	215 HB
EN 10088/3(95)	1.4571	Bar Rod Sect, t<=25mm, strain Hard	0.08 max	16.50-18.50	2.00 max	2.00-2.50	10.50-13.50	0.045 max	0.030 max	1.00 max	Ti <=0.7; 5xC<=Ti<=0.70; bal Fe	800-1000	500	12	
EN 10088/3(95)	1.4571	Bar Rod Sect, t<=35mm, strain Hard	0.08 max	16.50-18.50	2.00 max	2.00-2.50	10.50-13.50	0.045 max	0.030 max	1.00 max	Ti <=0.7; 5xC<=Ti<=0.70; bal Fe	700-850	350	20	
EN 10088/3(95)	X5CrNiMoTi17-12-2	Bar Rod Sect, t<=160mm, Ann	0.08 max	16.50-18.50	2.00 max	2.00-2.50	10.50-13.50	0.045 max	0.030 max	1.00 max	Ti <=0.7; 5xC<=Ti<=0.70; bal Fe	500-700	200	40	215 HB
EN 10088/3(95)	X5CrNiMoTi17-12-2	Bar Rod Sect, t<=35mm, strain Hard	0.08 max	16.50-18.50	2.00 max	2.00-2.50	10.50-13.50	0.045 max	0.030 max	1.00 max	Ti <=0.7; 5xC<=Ti<=0.70; bal Fe	700-850	350	20	
EN 10088/3(95)	X5CrNiMoTi17-12-2	Bar Rod Sect, t<=25mm, strain Hard	0.08 max	16.50-18.50	2.00 max	2.00-2.50	10.50-13.50	0.045 max	0.030 max	1.00 max	Ti <=0.7; 5xC<=Ti<=0.70; bal Fe	800-1000	500	12	
EN 10088/3(95)	X5CrNiMoTi17-12-2	Bar Rod Sect; Corr res; 160<t<=250mm, Ann	0.08 max	16.50-18.50	2.00 max	2.00-2.50	10.50-13.50	0.045 max	0.030 max	1.00 max	Ti <=0.7; 5xC<=Ti<=0.70; bal Fe	500-700	200	30	215 HB

France

Specification	Designation	Notes	C	Cr	Mn	Mo	Ni	P	S	Si	Other	UTS	YS	El	Hard
AFNOR NFA35573	Z6CNDT1712	Sh Strp Plt	0.08 max	16-18	2 max	2-2.5	10.5-13	0.04 max	0.03 max	1 max	Ti>=5xC<=0.60; bal Fe				
AFNOR NFA35574	Z6CNDT1712	Bar Bil	0.08 max	16-18	2 max	2-2.5	10.5-13	0.04 max	0.03 max	1 max	Ti>=5xC<=0.60; bal Fe				
AFNOR NFA35575	Z6CNDT1712	Wir rod	0.08 max	16-18	2 max	2-2.5	10.5-13	0.04 max	0.03 max	1 max	Ti>=5xC<=0.60; bal Fe				
AFNOR NFA36209	Z6CNDT1712	Plt	0.08 max	16-18	2 max	2-2.4	10.5-13	0.04 max	0.03 max	1 max	Ti>=5xC<=0.60; bal Fe				
AFNOR NFA36607	Z6CNDT1712	Frg	0.08 max	16-18	2 max	2-2.5	10.5-13	0.04 max	0.03 max	1 max	Cu <=1; Ti>=5xC<=0.60; bal Fe				

Germany

Specification	Designation	Notes	C	Cr	Mn	Mo	Ni	P	S	Si	Other	UTS	YS	El	Hard
DIN 17440(96)	WNr 1.4571	Plt Sh, HR Strp Bar, SHT	0.08 max	16.5-18.5	2.00 max	2.00-2.50	10.5-13.5	0.045 max	0.030 max	1.00 max	Ti 5xC>=0.70; bal Fe	500-700	200	30-40	
DIN 17441(97)	WNr 1.4571	CR Strp Plt Sh SHT	0.08 max	16.5-18.5	2.00 max	2.00-2.50	10.5-13.5	0.045 max	0.030 max	1.00 max	Ti 5xC>=0.70; bal Fe	500-700	200	30-40	
DIN EN 10088(95)	X6CrNiMoTi17-12-2	Sh Plt Strp Bar Rod Sect, SHT	0.08 max	16.5-18.5	2.00 max	2.00-2.50	10.5-13.5	0.045 max	0.030 max	1.00 max	Ti 5xC<=0.80; bal Fe	500-700	200	30-40	
TGL 7143	X8CrNiMoTi1811*	Obs	0.1 max	16.5-18.5	2 max	2-2.5	10.5-14	0.045 max	0.03 max	0.8 max	Ti>=5xC; bal Fe				

Hungary

Specification	Designation	Notes	C	Cr	Mn	Mo	Ni	P	S	Si	Other	UTS	YS	El	Hard
MSZ 4360(80)	KO35	Corr res	0.08 max	17-19	2 max	2-2.5	11-14	0.04 max	0.03 max	1 max	Ti <=0.7; Ti+5C-0.7; bal Fe				
MSZ 4360(87)	KO35Ti	Corr res; 0-160mm; SA	0.08 max	16.5-18.5	2 max	2-2.5	11-13.5	0.045 max	0.03 max	1 max	Ti <=0.8; Ti=5C-0.8; bal Fe	500-730	210	35L	190 HB max
MSZ 4360(87)	X8CrNiMoTi17122	Corr res; 0-160mm; SA	0.08 max	16.5-18.5	2 max	2-2.5	11-13.5	0.045 max	0.03 max	1 max	Ti <=0.8; Ti=5C-0.8; bal Fe	500-730	210	35L	190 HB max

UNS numbers and US grades are provided as a means of cross referencing chemically similar alloys. Exchangability is only possible after independent examination of specifications. Tensile properties are minimum or typical as specified. UTS and YS as MPa. El as %. See Appendix for list of abbreviations used in Notes. * indicates obsolete material.

Specification	Designation	Notes	C	Cr	Mn	Mo	Ni	P	S	Si	Other	UTS	YS	El	Hard

Stainless Steel, Austenitic, 316Ti (Continued from previous page)

Hungary

MSZ 4398(86)	KO35	Corr res; Heat res; longitudinal weld tube; 0.8-3mm; SA	0.08 max	16.5-18.5	2 max	2-2.5	10.5-13.5	0.045 max	0.03 max	1 max	Ti <=0.8; Ti=5C-0.8; bal Fe	500-730	210	35L	

India

IS 1570/5(85)	X04Cr17Ni12Mo2	Sh Plt Strp Bar Flat Band	0.08 max	16-18	2 max	2-3	10-14	0.045 max	0.03 max	1 max	Co <=0.1; Cu <=0.3; Pb <=0.15; W <=0.1; bal Fe	515	205	40	217 HB max
IS 1570/5(85)	X04Cr17Ni12Mo2Ti	Sh Plt Strp Bar Flat Band	0.08 max	16-18	2 max	2-3	10-14	0.045 max	0.03 max	1 max	Co <=0.1; Cu <=0.3; Pb <=0.15; Ti <=0.8; W <=0.1; 5C<=Ti<=0.8; bal Fe	515	205	40	217 HB max
IS 6529	04Cr17Ni12Mo2Ti20	Bil	0.08 max	16-18.5	2 max	2-3	10.5-14	0.045 max	0.03 max	1 max	Ti>=5xC<=0.80; bal Fe				
IS 6603	04Cr17Ni12Mo2Ti20	Bar, Soft 5 100mm diam	0.08 max	16-18.5	2 max	2-3	10.5-14	0.045 max	0.03 max	1 max	Ti>=5xC<=0.80; bal Fe				
IS 6911	04Cr17Ni12Mo2Ti20	Sh Strp Plt, Soft 0.5-3mm diam	0.08 max	16-18.5	2 max	2-3	10.5-14	0.045 max	0.03 max	1 max	Ti>=5xC<=0.80; bal Fe				

International

COMECON PC4-70	21	Sh Strp Plt Bar Wir	0.08 max	16-18	2 max	2-2.5	10-14	0.035 max	0.025 max	0.8 max	Ti>=5xC<=0.70; bal Fe				
ISO 4954(93)	X6CrNiMoTi17122E	Wir rod Bar, CH ext	0.08 max	16.5-18.5	2.00 max	2.0-2.5	11.0-14.0	0.045 max	0.030 max	1.00 max	Ti=5xC<= 0.80; bal Fe	680			
ISO 683-13(74)	21*	Sh Plt Bar, ST	0.08 max	16.5-18.5	2 max	2-2.5	11-14	0.045 max	0.03 max	1 max	Ti>=5xC<= 0.80; bal Fe				

Italy

UNI 6901(71)	X6CrNiMoTi1712	Corr res	0.08 max	16-18.5	2 max	2-2.5	10.5-13.5	0.045 max	0.03 max	1 max	Ti <=0.8; Ti=5C-0.8; bal Fe				
UNI 6901(71)	X6CrNiMoTi1713	Corr res	0.08 max	16-18.5	2 max	2.5-3	11.5-14.5	0.045 max	0.03 max	1 max	Ti <=0.8; Ti=5C-0.8; bal Fe				
UNI 6904(71)	X6CrNiMoTi1712	Heat res	0.08 max	16-18.5	2 max	2-2.5	10.5-13.5	0.04 max	0.03 max	0.75 max	Ti <=0.6; Ti=5C-0.6; bal Fe				
UNI 6904(71)	X6CrNiMoTi1713	Heat res	0.08 max	16-18.5	2 max	2.5-3	11.5-14.5	0.04 max	0.03 max	0.75 max	Ti <=0.6; Ti=5C-0.6; bal Fe				
UNI 7500(75)	X6CrNiMoTi1712	Press ves	0.08 max	16-18.5	2 max	2-2.5	10.5-13.5	0.045 max	0.03 max	1 max	Ti <=0.8; Ti=5C-0.8; bal Fe				
UNI 8317(81)	X6CrNiMoTi1712	Sh Strp Plt	0.08 max	16-18.5	2 max	2-2.5	10.5-13.5	0.045 max	0.03 max	1 max	Ti>=5xC<=0.80; bal Fe				

Poland

PNH86020	H17N13M2T	Corr res	0.08 max	16-18	2 max	2-2.5	11-14	0.045 max	0.03 max	0.8 max	Ti <=0.7; Ti=5C-0.7; bal Fe				
PNH86020	H18N10MT		0.1 max	17-20	2 max	1.5-2.2	9-11	0.045 max	0.03 max	0.8 max	Ti>=5xC<=0.80; bal Fe				

Romania

STAS 10322	8TiMoNiCr175	Sh Strp Plt Bar Wir Tub	0.12 max	16.5-18.5	2 max	2-2.5	10.5-13.5	0.04 max	0.03 max	0.8 max	Ti>=5xC; bal Fe				
STAS 3583(87)	10TiMoNiCr175	Corr res; Heat res	0.08 max	16.5-18.5	2 max	2-2.5	10.5-13.5	0.045 max	0.03 max	1 max	Pb <=0.15; Ti <=0.8; 5C<=Ti<=0.8; bal Fe				

Russia

GOST	O8Ch17N13M2T	Sh Strp Plt Bar Wir Tub Frg	0.08 max	16-18	2 max	2-3	12-14	0.035 max	0.02 max	0.8 max	Cu <=0.3; W <=0.2; Ti>=5xC<=0.70; bal Fe				
GOST 5632(61)	KH17N13M3T	Corr res; Heat res; High-temp constr	0.1 max	16.0-18.0	1.0-2.0	3.0-4.0	12.0-14.0	0.035 max	0.02 max	0.8 max	Cu <=0.3; Ti 0.3-0.6; bal Fe				
GOST 5632(72)	08Ch17N13M2T	Corr res	0.08 max	16.0-18.0	2.0 max	2.0-3.0	12.0-14.0	0.035 max	0.02 max	0.8 max	Cu <=0.3; Ti <=0.7; W <=0.2; Ti=5C-0.7; bal Fe				
GOST 5632(72)	10Ch17N13M2T	Corr res	0.1 max	16.0-18.0	2.0 max	2.0-3.0	12.0-14.0	0.035 max	0.02 max	0.8 max	Cu <=0.3; Ti <=0.7; W <=0.2; Ti=5C-0.7; bal Fe				
GOST 5632(72)	10Ch17N13M3T	Corr res	0.1 max	16.0-18.0	2.0 max	3.0-4.0	12.0-14.0	0.035 max	0.02 max	0.8 max	Cu <=0.3; Ti <=0.7; W <=0.2; Ti=5C-0.7; bal Fe				

Spain

UNE 36016(75)	F.3535*	see X6CrNiMoTi17-12-03	0.08 max	16.0-18.0	2.0 max	2.0-3.0	10.0-14.0	0.045 max	0.03 max	1.0 max	Ti <=0.8; 5xC<=Ti<=0.8; bal Fe				
UNE 36016(75)	X6CrNiMoTi17-12-03	Sh Strp Plt Bar Wir Bil; Corr res	0.08 max	16.0-18.0	2.0 max	2.0-3.0	10.0-14.0	0.045 max	0.03 max	1.0 max	Ti <=0.8; 5xC<=Ti<=0.8; bal Fe				

UNS numbers and US grades are provided as a means of cross referencing chemically similar alloys. Exchangability is only possible after independent examination of specifications. Tensile properties are minimum or typical as specified. UTS and YS as MPa. El as %. See Appendix for list of abbreviations used in Notes. * indicates obsolete material.

Specification	Designation	Notes	C	Cr	Mn	Mo	Ni	P	S	Si	Other	UTS	YS	El	Hard

Stainless Steel, Austenitic, 316Ti (Continued from previous page)

Spain

Specification	Designation	Notes	C	Cr	Mn	Mo	Ni	P	S	Si	Other	UTS	YS	El	Hard
UNE 36016/1(89)	E-316Ti	Corr res	0.08 max	16.5-18.5	2.0 max	2.0-2.5	10.5-13.5	0.045 max	0.03 max	1.0 max	N <=0.1; Ti <=0.8; Ti=5C-0.8; bal Fe				
UNE 36016/1(89)	F.3535*	Obs; E-316Ti; Corr res	0.08 max	16.5-18.5	2.0 max	2.0-2.5	10.5-13.5	0.045 max	0.03 max	1.0 max	N <=0.1; Ti <=0.8; 5xC<=Ti<=0.8; bal Fe				
UNE 36087/4(89)	F.3535*	see X6CrNiMoTi17 12 2	0.08 max	16.5-18.5	2 max	2-2.5	10.5-13.5	0.045 max	0.03 max	1 max	N <=0.1; Ti <=0.8; 5xC<=Ti<=0.8; bal Fe				
UNE 36087/4(89)	X6CrNiMoTi17 12 2	Corr res; Sh Strp	0.08 max	16.5-18.5	2 max	2-2.5	10.5-13.5	0.045 max	0.03 max	1 max	N <=0.1; Ti <=0.8; 5xC<=Ti<=0.8; bal Fe				

Sweden

Specification	Designation	Notes	C	Cr	Mn	Mo	Ni	P	S	Si	Other	UTS	YS	El	Hard
SS 142350	2350	Corr res	0.08 max	12.5-18.5	2 max	2.-2.5	10.5-14	0.045 max		0.3 max	1 max	Ti <=0.8; Ti=5(C+N)-0.8				

UK

Specification	Designation	Notes	C	Cr	Mn	Mo	Ni	P	S	Si	Other	UTS	YS	El	Hard
BS 1449/2(83)	320S31	Corr/Heat res; t<=100mm	0.08 max	16.5-18.5	2.00 max	2.00-2.50	11.0-14.0	0.045 max	0.030 max	1.00 max	5xC<=Ti<=0.80; bal Fe	510	210	40	0-205 HB
BS 1449/2(83)	320S33	Corr/Heat res; t<=100mm	0.08 max	16.5-18.5	2.00 max	2.50-3.00	11.5-14.5	0.045 max	0.030 max	1.00 max	5xC<=Ti<=0.80; bal Fe	510	210	40	0-205 HB
BS 1501/3(73)	320S17	Plate; Press ves; Corr res	0.08 max	16.5-18.5	0.5-2.0	2.25-3.0	11.0-14.0	0.045 max	0.03 max	0.2-1.0	Cu <=0.3; Ti <=0.6; W <=0.1; Ti=4C-0.6; bal Fe				
BS 1501/3(90)	320S31	Press ves; Corr/Heat res; 0.05-100mm	0.08 max	16.5-18.5	2.0 max	2.0-2.5	11.0-14.0	0.045 max	0.025 max	1.0 max	Cu <=0.3; Ti <=0.8; W <=0.1; Ti=5C-0.8; bal Fe	510-710	210	35	
BS 1503(89)	320S33	Frg press ves; t<=999mm; HT	0.08 max	16.5-18.5	2.00 max	2.50-3.00	11.5-14.5	0.040 max	0.025 max	1.00 max	B <=0.005; Cu <=0.70; W <=0.1; 5xC<=Ti<=0.80; bal Fe	510-710	245	30	
BS 1554(90)	320S18	Wir Corr/Heat res; 6<diam<=13mm; Ann	0.08 max	16.5-18.5	2.00 max	2.00-2.50	10.5-13.5	0.045 max	0.030 max	1.00 max	Ti <=0.08; W <=0.1; 5xC<=Ti<=0.80; bal Fe	0-700			
BS 1554(90)	320S18	Wir Corr/Heat res; diam<0.49mm; Ann	0.08 max	16.5-18.5	2.00 max	2.00-2.50	10.5-13.5	0.045 max	0.030 max	1.00 max	Ti <=0.08; W <=0.1; 5xC<=Ti<=0.80; bal Fe	0-900			
BS 970/1(96)	320S31	Wrought Corr/Heat res; t<=160mm	0.08 max	16.5-18.5	2.00 max	2.00-2.50	11.0-14.0	0.045 max	0.030 max	1.00 max	5xC<=Ti<=0.80; bal Fe	510	210	35	183 HB max

USA

Specification	Designation	Notes	C	Cr	Mn	Mo	Ni	P	S	Si	Other	UTS	YS	El	Hard
	UNS S31635	Ti-stabilized	0.08 max	16.00-18.00	2.00 max	2.00-3.00	10.00-14.00	0.045 max	0.030 max	1.00 max	N <=0.10; Ti 5x(C+N) min, 0.70 max; bal Fe				
ASTM A240/A240M(98)	S31635	Ti-stabilized	0.08 max	16.0-18.0	2.00 max	2.00-3.00	10.0-14.0	0.045 max	0.030 max	1.00 max	N <=0.10; 5x(C+N)<=Ti<=0.70; bal Fe	515	205	40.0	217 HB
ASTM A313	316Ti*	Wir	0.08 max	16.00-18.00	2.00 max	2.00-3.00	10.00-14.00	0.045 max	0.030 max	1.00 max	N <=0.1; Ti>=5(C+N)<=0.70; bal Fe				
ASTM A368(95)	316Ti	Wir, Heat res	0.08 max	16.00-18.00	2.00 max	2.00-3.00	10.00-14.00	0.045 max	0.030 max	1.00 max	5x(C+N)<=Ti<=0.70; bal Fe				
SAE J405(98)	S31635	Ti-stabilized	0.08 max	16.00-18.00	2.00 max	2.0-3.0	10.00-14.00	0.045 max	0.030 max	0.75 max	N <=0.10; 5x(C+N)<=Ti<=0.70; bal Fe				

Stainless Steel, Austenitic, 317

Australia

Specification	Designation	Notes	C	Cr	Mn	Mo	Ni	P	S	Si	Other	UTS	YS	El	Hard
AS 1444	317*	Obs; Bar	0.08 max	18-20	2 max	3-4	11-15	0.045 max	0.04 max	1 max	bal Fe				
AS 1449(94)	317	Sh Strp Plt	0.08 max	18.00-20.00	2.00 max	3.00-4.00	11.00-15.00	0.045 max	0.030 max	0.75 max	bal Fe				

Canada

Specification	Designation	Notes	C	Cr	Mn	Mo	Ni	P	S	Si	Other	UTS	YS	El	Hard
CSA G110.3	317	Bar Bil	0.08 max	18-20	2 max	3-4	11-15	0.045 max	0.03 max	1 max	bal Fe				
CSA G110.6	317	Sh Strp Plt, HR/Ann or HR/Q/A/Tmp	0.08 max	18-20	2 max	3-4	11-15	0.045 max	0.03 max	1 max	bal Fe				
CSA G110.9	317	Sh Strp Plt, HR, CR, Q/A, Tmp, Ann	0.08 max	18-20	2 max	3-4	11-15	0.045 max	0.03 max	1 max	bal Fe				

China

Specification	Designation	Notes	C	Cr	Mn	Mo	Ni	P	S	Si	Other	UTS	YS	El	Hard
GB 1220(92)	0Cr19Ni13Mo3	Bar SA	0.08 max	18.00-20.00	2.00 max	3.00-4.00	11.00-15.00	0.035 max	0.030 max	1.00 max	bal Fe	520	205	40	
GB 1221(92)	0Cr19Ni13Mo3	Bar SA	0.08 max	18.00-20.00	2.00 max	3.00-4.00	11.00-15.00	0.035 max	0.030 max	1.00 max	bal Fe	520	205	40	

UNS numbers and US grades are provided as a means of cross referencing chemically similar alloys. Exchangability is only possible after independent examination of specifications. Tensile properties are minimum or typical as specified. UTS and YS as MPa. El as %. See Appendix for list of abbreviations used in Notes. * indicates obsolete material.

Specification	Designation	Notes	C	Cr	Mn	Mo	Ni	P	S	Si	Other	UTS	YS	El	Hard

Stainless Steel, Austenitic, 317 (Continued from previous page)

China

Specification	Designation	Notes	C	Cr	Mn	Mo	Ni	P	S	Si	Other	UTS	YS	El	Hard
GB 13296(91)	0Cr19Ni13Mo3	Smls tube SA	0.08 max	18.00-20.00	2.00 max	3.00-4.00	11.00-15.00	0.035 max	0.030 max	1.00 max	bal Fe	520	205	35	
GB 3280(92)	0Cr19Ni13Mo3	Sh Plt, CR SA	0.08 max	18.00-20.00	2.00 max	3.00-4.00	11.00-15.00	0.035 max	0.030 max	1.00 max	bal Fe	520	205	40	
GB 4237(92)	0Cr19Ni13Mo3	Sh Plt, HR SA	0.08 max	18.00-20.00	2.00 max	3.00-4.00	11.00-15.00	0.035 max	0.030 max	1.00 max	bal Fe	520	205	35	
GB 4238(92)	0Cr19Ni13Mo3	Plt, HR SA	0.08 max	18.00-20.00	2.00 max	3.00-4.00	11.00-15.00	0.035 max	0.030 max	1.00 max	bal Fe	520	205	40	
GB 4239(91)	0Cr19Ni13Mo3	Sh Strp, SA	0.08 max	18.00-20.00	2.00 max	3.00-4.00	11.00-15.00	0.035 max	0.030 max	1.00 max	bal Fe	520	205	40	
GB/T 14975(94)	0Cr19Ni13Mo3	Smls tube SA	0.08 max	18.00-20.00	2.00 max	3.00-4.00	11.00-15.00	0.035 max	0.030 max	1.00 max	bal Fe	520	205	35	
GB/T 14976(94)	0Cr19Ni13Mo3	Smls pipe SA	0.08 max	18.00-20.00	2.00 max	3.00-4.00	11.00-15.00	0.035 max	0.030 max	1.00 max	bal Fe	520	205	35	

Finland

Specification	Designation	Notes	C	Cr	Mn	Mo	Ni	P	S	Si	Other	UTS	YS	El	Hard
SFS 773(86)	X2CrNiMo17145	Corr res	0.03 max	16.0-18.0	2.0 max	4.0-5.0	12.5-16.0	0.045 max	0.03 max	1.0 max	bal Fe				

Germany

Specification	Designation	Notes	C	Cr	Mn	Mo	Ni	P	S	Si	Other	UTS	YS	El	Hard
DIN	WNr 1.4449		0.07 max	16.0-18.0	2.00 max	4.00-5.00	12.5-14.5	0.045 max	0.030 max	1.00 max	bal Fe				
DIN	X5CrMiMo1713		0.07 max	16.0-18.0	2.00 max	4.00-5.00	12.5-14.5	0.045 max	0.030 max	1.00 max	bal Fe				

Hungary

Specification	Designation	Notes	C	Cr	Mn	Mo	Ni	P	S	Si	Other	UTS	YS	El	Hard
MSZ 4360(87)	X3CrNiMo18164	Corr res; 0-160mm; SA	0.03 max	17.5-19.5	2 max	3-4	14-17	0.045 max	0.025 max	1 max	bal Fe	490-690	195	35L	190 HB max

India

Specification	Designation	Notes	C	Cr	Mn	Mo	Ni	P	S	Si	Other	UTS	YS	El	Hard
IS 1570/5(85)	X04Cr19Ni13Mo3	Sh Plt Strp Bar Flat Band	0.08 max	18-20	2 max	3-4	11-15	0.045 max	0.03 max	1 max	Co <=0.1; Cu <=0.3; Pb <=0.15; W <=0.1; bal Fe	515	205	35	217 HB max

Italy

Specification	Designation	Notes	C	Cr	Mn	Mo	Ni	P	S	Si	Other	UTS	YS	El	Hard
UNI 7500(75)	X5CrMiMo1815	Sh Plt Bar Bil	0.06 max	17.5-19.5	2 max	3-4	13-16	0.045 max	0.03 max	1 max	bal Fe				

Japan

Specification	Designation	Notes	C	Cr	Mn	Mo	Ni	P	S	Si	Other	UTS	YS	El	Hard
JIS G3214(91)	SUSF317	Frg press ves 130-200mm diam	0.08 max	18.00-20.00	2.00 max	3.00-4.00	11.00-15.00	0.040 max	0.030 max	1.00 max	bal Fe	480	205	29	187 HB max
JIS G4303(91)	SUS317	Bar; SA; <=180mm diam	0.08 max	18.00-20.00	2.00 max	3.00-4.00	11.00-15.00	0.045 max	0.030 max	1.00 max	bal Fe	520	205	40	187 HB
JIS G4304(91)	SUS317	Sh Plt, HR SA	0.08 max	18.00-20.00	2.00 max	3.00-4.00	11.0-15.00	0.045 max	0.030 max	1.00 max	bal Fe	520	205	40	90 HRB max
JIS G4305(91)	SUS317	Sh Plt, CR SA	0.08 max	18.00-20.00	2.00 max	3.00-4.00	11.00-15.00	0.045 max	0.030 max	1.00 max	bal Fe	520	205	40	187 HB max
JIS G4305(91)	SUS317J1		0.040 max	16.00-19.00	2.50 max	4.00-6.00	15.00-17.00	0.045 max	0.030 max	1.00 max	bal Fe	480	175	40	
JIS G4305(91)	SUS317J2		0.06 max	23.00-26.00	2.00 max	0.50-1.20	12.00-16.00	0.045 max	0.030 max	1.50 max	N 0.25-0.40; bal Fe	690	345	40	
JIS G4308	SUS317	Wir rod	0.08 max	18.00-20.00	2.00 max	3.00-4.00	11.00-15.00	0.045 max	0.030 max	1.00 max	bal Fe				
JIS G4309	SUS317	Wir	0.08 max	18.00-20.00	2.00 max	3.00-4.00	11.00-15.00	0.045 max	0.030 max	1.00 max	bal Fe				
JIS G4359	SUS317	Pipe	0.08 max	18-20	2 max	3-4	11-15	0.04 max	0.03 max	1 max	bal Fe				

Mexico

Specification	Designation	Notes	C	Cr	Mn	Mo	Ni	P	S	Si	Other	UTS	YS	El	Hard
DGN B-83	317*	Obs; Bar, HR, CR	0.08 max	18-20	2 max	3-4	11-15	0.045 max	0.03 max	1 max	bal Fe				
NMX-B-171(91)	MT317	mech tub	0.08 max	18.0-20.0	2.00 max	3.0-4.0	11.0-14.0	0.040 max	0.030 max	1.00 max	bal Fe				

Spain

Specification	Designation	Notes	C	Cr	Mn	Mo	Ni	P	S	Si	Other	UTS	YS	El	Hard
UNE 36016/1(89)	F.3539*	Obs; EN10088-3(96); X2CrNiMo18-16-4	0.03 max	17.5-19.5	2.0 max	3.0-4.0	14.0-17.0	0.045 max	0.03 max	1.0 max	N <=0.1; bal Fe				

Sweden

Specification	Designation	Notes	C	Cr	Mn	Mo	Ni	P	S	Si	Other	UTS	YS	El	Hard
SS 142366	2366	Corr res	0.07 max	17-20	2 max	3-4	13-16	0.045 max	0.03 max	1.5 max	bal Fe				
SS 142367	2367	Corr res	0.03 max	17.5-19.5	2 max	3-4	13-17	0.045 max	0.03 max	1 max	bal Fe				

UNS numbers and US grades are provided as a means of cross referencing chemically similar alloys. Exchangability is only possible after independent examination of specifications. Tensile properties are minimum or typical as specified. UTS and YS as MPa. El as %. See Appendix for list of abbreviations used in Notes. * indicates obsolete material.

Stainless Steel, Austenitic, 317 (Continued from previous page)

Specification	Designation	Notes	C	Cr	Mn	Mo	Ni	P	S	Si	Other	UTS	YS	El	Hard
UK															
BS 1449/2(83)	317S16	Corr/Heat res; t<=100mm	0.06 max	17.5-19.5	2.00 max	3.00-4.00	12.0-15.0	0.045 max	0.030 max	1.00 max	bal Fe	510	205	40	0-205 HB
BS 1554(90)	317S17	Wir Corr/Heat res; 6<diam<=13mm; Ann	0.08 max	18.0-20.0	2.00 max	3.00-4.00	12.0-15.0	0.045 max	0.030 max	1.00 max	W <=0.1; bal Fe	0-700			
BS 1554(90)	317S17	Wir Corr/Heat res; diam<0.49mm; Ann	0.08 max	18.0-20.0	2.00 max	3.00-4.00	12.0-15.0	0.045 max	0.030 max	1.00 max	W <=0.1; bal Fe	0-900			
BS 2901	317S96	Wire, Rod	0.08 max	18.5-20.5	1-2.5	3-4	13-15		0.03 max	0.25-0.65	bal Fe				
USA															
	AISI 317	Tube	0.08 max	18.00-20.00	2.00 max	3.00-4.00	11.00-15.00	0.045 max	0.030 max	1.00 max	bal Fe				
	UNS S31700		0.08 max	18.00-20.00	2.00 max	3.00-4.00	11.00-15.00	0.045 max	0.030 max	1.00 max	bal Fe				
ASME SA240	317	Refer to ASTM A240(95)													
ASME SA249	317	Refer to ASTM A249/A249M(96)													
ASME SA312	317	Refer to ASTM A312/A312M(95)													
ASME SA403	317	Refer to ASTM A403/A403M(95)													
ASME SA409	317	Refer to ASTM A409/A409M(95)													
ASTM A167(96)	317*	Sh Strp Plt	0.08 max	18-20	2 max	3-4	11-15	0.045 max	0.03 max	1 max	bal Fe				
ASTM A182/A182M(98)	F317	Frg/roll pipe flange valve	0.08 max	18.0-20.0	2.00 max	3.0-4.0	11.0-15.0	0.045 max	0.030 max	1.00 max	bal Fe	515	205	30	
ASTM A213/A213M(95)	TP317	Smls tube boiler, superheater, heat exchanger	0.08 max	18.0-20.0	2.00 max	3.00-4.00	11.0-14.0	0.040 max	0.030 max	0.75 max	bal Fe	515	205	35	
ASTM A240/A240M(98)	S31700		0.08 max	18.0-20.0	2.00 max	3.0-4.0	11.0-15.0	0.045 max	0.030 max	1.00 max	N <=0.10; bal Fe	515	205	35.0	217 HB
ASTM A249/249M(96)	TP317	Weld tube; boiler, superheater, heat exch	0.08 max	18.0-20.0	2.00 max	3.00-4.00	11.0-14.0	0.04 max	0.03 max	0.75 max	bal Fe	515	205	35	90 HRB
ASTM A269	317	Smls Weld, Tube	0.08 max	18-20	2 max	3-4	11-15	0.045 max	0.03 max	1 max	bal Fe				
ASTM A276(98)	317	Bar	0.08 max	18-20	2 max	3-4	11-15	0.045 max	0.03 max	1 max	bal Fe				
ASTM A312/A312M(95)	317	Smls Weld, Pipe	0.08 max	18-20	2 max	3-4	11-15	0.045 max	0.03 max	1 max	bal Fe				
ASTM A314	317	Bil	0.08 max	18-20	2 max	3-4	11-15	0.045 max	0.03 max	1 max	bal Fe				
ASTM A403	317	Pipe	0.08 max	18-20	2 max	3-4	11-15	0.045 max	0.03 max	1 max	bal Fe				
ASTM A409	317	Weld, Pipe	0.08 max	18-20	2 max	3-4	11-15	0.045 max	0.03 max	1 max	bal Fe				
ASTM A473	317	Frg, Heat res SHT	0.08 max	18.00-20.00	2.00 max	3.00-4.00	11.00-15.00	0.045 max	0.030 max	1.00 max	bal Fe	515	205	40	
ASTM A478	317	Wir CD 0.76<t<=3.18mm	0.08 max	18.00-20.00	2.00 max	3.00-4.00	11.00-15.00	0.045 max	0.030 max	1.00 max	bal Fe	830-1030		15	
ASTM A511(96)	MT317	Smls mech tub, Ann	0.08 max	18.0-20.0	2.00 max	3.0-4.0	11.0-14.0	0.040 max	0.030 max	1.00 max	bal Fe	517	207	35	192 HB
ASTM A554(94)	MT317	Weld mech tub, Rnd Ann	0.08 max	18.0-20.0	2.00 max	3.0-4.0	11.0-14.0	0.040 max	0.030 max	1.00 max	bal Fe	517	207	35	
ASTM A580/A580M(98)	317	Wir, Ann	0.08 max	18.0-20.0	2.00 max	3.0-4.0	11.0-15.0	0.045 max	0.030 max	1.00 max	N <=0.10; bal Fe	520	210	35	
ASTM A632(90)	TP317	Smls Weld tub	0.08 max	18.0-20.0	2.00 max	3.00-4.00	11.0-14.0	0.040 max	0.030 max	0.75 max	bal Fe	515	205	35	
ASTM A813/A813M(95)	TP317	Weld Pipe	0.08 max	18.0-20.0	2.00 max	3.00-4.00	11.0-14.0	0.045 max	0.030 max	0.75 max	bal Fe	515	205		
ASTM A814/A814M(96)	TP317	CW Weld Pipe	0.08 max	18.0-20.0	2.00 max	3.00-4.00	11.0-14.0	0.045 max	0.030 max	0.75 max	bal Fe	515	205		
ASTM A943/A943M(95)	TP317	Spray formed smls pipe	0.08 max	18.0-20.0	2.00 max	3.00-4.00	11.0-14.0	0.040 max	0.030 max	0.75 max	bal Fe	515	205		
ASTM A988(98)	S31700	HIP Flanges, Fittings, Valves/parts; Heat res	0.08 max	18.0-20.0	2.00 max	3.00-4.00	11.0-15.0	0.045 max	0.030 max	1.00 max	bal Fe	515	205	30	
FED QQ-S-763F(96)	317*	Obs; Bar Wir Shp Frg; Corr res	0.08 max	18-20	2 max	3-4	11-15	0.045 max	0.03 max	1 max	bal Fe				

UNS numbers and US grades are provided as a means of cross referencing chemically similar alloys. Exchangability is only possible after independent examination of specifications. Tensile properties are minimum or typical as specified. UTS and YS as MPa. El as %. See Appendix for list of abbreviations used in Notes. * indicates obsolete material.

Specification	Designation	Notes	C	Cr	Mn	Mo	Ni	P	S	Si	Other	UTS	YS	El	Hard

Stainless Steel, Austenitic, 317 (Continued from previous page)

USA

Specification	Designation	Notes	C	Cr	Mn	Mo	Ni	P	S	Si	Other	UTS	YS	El	Hard
SAE J405(98)	S31700		0.08 max	18.00-20.00	2.00 max	3.00-4.00	11.00-15.00	0.045 max	0.030 max	0.75 max	N <=0.10; bal Fe				

Stainless Steel, Austenitic, 317L

Australia

Specification	Designation	Notes	C	Cr	Mn	Mo	Ni	P	S	Si	Other	UTS	YS	El	Hard
AS 1444	317L*	Obs; Bar	0.03 max	18-20	2 max	3-4	11-15	0.045 max	0.03 max	1 max	bal Fe				

Canada

Specification	Designation	Notes	C	Cr	Mn	Mo	Ni	P	S	Si	Other	UTS	YS	El	Hard
CSA G110.6	317L	Sh Strp Plt, HR, Ann or HR/Q/A/Tmp	0.03 max	18-20	2 max	3-4	11-15	0.045 max	0.03 max	1 max	bal Fe				
CSA G110.9	317L	Plt, HR, CR, Ann, Q/A, Tmp	0.03 max	18-20	2 max	3-4	11-15	0.045 max	0.03 max	1 max	bal Fe				

China

Specification	Designation	Notes	C	Cr	Mn	Mo	Ni	P	S	Si	Other	UTS	YS	El	Hard
GB 1220(92)	00Cr19Ni13Mo3	Bar SA	0.030 max	18.00-20.00	2.00 max	3.00-4.00	11.00-15.00	0.035 max	0.030 max	1.00 max	bal Fe	480	177	40	
GB 13296(91)	00Cr19Ni13Mo3	Smls tube SA	0.030 max	18.00-20.00	2.00 max	3.00-4.00	11.00-15.00	0.035 max	0.030 max	1.00 max	bal Fe	480	175	40	
GB 3280(92)	00Cr19Ni13Mo3	Sh Plt, CR SA	0.030 max	18.00-20.00	2.00 max	3.00-4.00	11.00-15.00	0.035 max	0.030 max	1.00 max	bal Fe	480	177	40	
GB 4237(92)	00Cr19Ni13Mo3	Sh Plt, HR SA	0.030 max	18.00-20.00	2.00 max	3.00-4.00	11.00-15.00	0.035 max	0.030 max	1.00 max	bal Fe	480	175	35	
GB 4239(91)	00Cr19Ni13Mo3	Sh Strp, CR SA	0.030 max	18.00-20.00	2.00 max	3.00-4.00	11.00-15.00	0.035 max	0.030 max	1.00 max	bal Fe	480	175	40	
GB/T 14976(94)	00Cr19Ni13Mo3	Smls pipe SA	0.030 max	18.00-20.00	2.00 max	3.00-4.00	11.00-15.00	0.035 max	0.030 max	1.00 max	bal Fe	480	175	35	

Europe

Specification	Designation	Notes	C	Cr	Mn	Mo	Ni	P	S	Si	Other	UTS	YS	El	Hard
EN 10088/2(95)	1.4438	Strp, CR Corr res; t<=6mm, Ann	0.030 max	17.50-19.50	2.00 max	3.00-4.00	13.00-16.00	0.045 max	0.015 max	1.00 max	N <=0.11; bal Fe	550-700	240	35	
EN 10088/2(95)	X2CrNiMo18-15-4	Strp, CR Corr res; t<=6mm, Ann	0.030 max	17.50-19.50	2.00 max	3.00-4.00	13.00-16.00	0.045 max	0.015 max	1.00 max	N <=0.11; bal Fe	550-700	240	35	
EN 10088/3(95)	1.4438	Bar Rod Sect; Corr res; 160<t<=250mm, Ann	0.030 max	17.50-19.50	2.00 max	3.00-4.00	13.00-16.00	0.045 max	0.030 max	1.00 max	N <=0.11; bal Fe	500-700	200	30	215 HB
EN 10088/3(95)	1.4438	Bar Rod Sect, t<=160mm, Ann	0.030 max	17.50-19.50	2.00 max	3.00-4.00	13.00-16.00	0.045 max	0.030 max	1.00 max	N <=0.11; bal Fe	500-700	200	40	215 HB
EN 10088/3(95)	X2CrNiMo18-15-4	Bar Rod Sect; Corr res; 160<t<=250mm, Ann	0.030 max	17.50-19.50	2.00 max	3.00-4.00	13.00-16.00	0.045 max	0.030 max	1.00 max	N <=0.11; bal Fe	500-700	200	30	215 HB
EN 10088/3(95)	X2CrNiMo18-15-4	Bar Rod Sect, t<=160mm, Ann	0.030 max	17.50-19.50	2.00 max	3.00-4.00	13.00-16.00	0.045 max	0.030 max	1.00 max	N <=0.11; bal Fe	500-700	200	40	215 HB

Finland

Specification	Designation	Notes	C	Cr	Mn	Mo	Ni	P	S	Si	Other	UTS	YS	El	Hard
SFS 770	X2CrNiMo19124	Sh Strp Plt Bar Wir Tub	0.03 max	18-20	2 max	3-4	11-15	0.045 max	0.03 max	1 max	bal Fe				
SFS 770(86)	X2CrNiMo19134	Corr res	0.03 max	18.0-20.0	2.0 max	3.0-4.0	11.0-15.0	0.045 max	0.03 max	1.0 max	bal Fe				
SFS 772(86)	X2CrNiMoN18145	Corr res	0.03 max	16.5-18.5	2.0 max	4.0-5.0	12.5-14.5	0.045 max	0.03 max	1.0 max	N 0.12-0.22; bal Fe				

France

Specification	Designation	Notes	C	Cr	Mn	Mo	Ni	P	S	Si	Other	UTS	YS	El	Hard
AFNOR NFA35573	Z2CND19.15	Sh Strp Plt, CR, 5mm diam	0.03 max	17.5-19.5	2 max	3-4	14-16	0.04 max	0.03 max	1 max	bal Fe				
AFNOR NFA35574	Z2CND19.15	Bar Bil	0.03 max	17.5-19.5	2 max	3-4	14-16	0.04 max	0.03 max	1 max	bal Fe				
AFNOR NFA35583	Z2CND19.14	Wir rod	0.025 max	18.5-20.5	1-2.5	3-4	13-15	0.03 max	0.02 max	0.3-0.65	Cu <=0.5; bal Fe				
AFNOR NFA36209	Z2CND19.15	Plt	0.03 max	17.5-19.5	2 max	3-4	14-16	0.04 max	0.03 max	1 max	bal Fe				

Germany

Specification	Designation	Notes	C	Cr	Mn	Mo	Ni	P	S	Si	Other	UTS	YS	El	Hard
DIN 17440(96)	WNr 1.4438	Plt Sh, HR Strp Bar Wir Frg, SHT	0.030 max	17.5-19.5	2.00 max	3.00-4.00	13.0-16.0	0.045 max	0.030 max	1.00 max	N <=0.11; bal Fe	500-700	200	30-40	
DIN 17440(96)	X2CrNiMo18-15-4	SHT	0.030 max	17.5-19.5	2.00 max	3.00-4.00	13.0-16.0	0.045 max	0.030 max	1.00 max	N <=0.11; bal Fe	500-700	200	30-40	215 HB
DIN 17441(97)	WNr 1.4438	CR Strp Plt Sh SHT	0.030 max	17.5-19.5	2.00 max	3.00-4.00	13.0-16.0	0.045 max	0.030 max	1.00 max	N <=0.11; bal Fe	500-700	200	30-40	

Hungary

Specification	Designation	Notes	C	Cr	Mn	Mo	Ni	P	S	Si	Other	UTS	YS	El	Hard
MSZ 4360(87)	KO42LC	Corr res; 0-160mm; SA	0.03 max	17.5-19.5	2 max	3-4	14-17	0.045 max	0.025 max	1 max	bal Fe	490-690	195	35L	190 HB max

UNS numbers and US grades are provided as a means of cross referencing chemically similar alloys. Exchangability is only possible after independent examination of specifications. Tensile properties are minimum or typical as specified. UTS and YS as MPa. El as %. See Appendix for list of abbreviations used in Notes. * indicates obsolete material.

Specification	Designation	Notes	C	Cr	Mn	Mo	Ni	P	S	Si	Other	UTS	YS	El	Hard

Stainless Steel, Austenitic, 317L (Continued from previous page)

International

Specification	Designation	Notes	C	Cr	Mn	Mo	Ni	P	S	Si	Other	UTS	YS	El	Hard
ISO 683-13(74)	24*	Sh Plt Bar, ST	0.03 max	17.5-19.5	2 max	3-4	14-17	0.045 max	0.03 max	1 max	bal Fe				

Italy

UNI 6901(71)	X2CrNiMo1816	Corr res	0.03 max	17.5-19.5	2 max	3-4	14-17	0.045 max	0.03 max	1 max	bal Fe				
UNI 7500(75)	X2CrNiMo1815	Press ves	0.03 max	17.5-19.5	2 max	3-4	13-16	0.045 max	0.03 max	1 max	bal Fe				

Japan

JIS G3214(91)	SUSF317L	Frg press ves 130-200mm diam	0.030 max	18.00-20.00	2.00 max	3.00-4.00	11.00-15.00	0.040 max	0.030 max	1.00 max	bal Fe	450	175	29	187 HB max
JIS G3459(94)	SUS317LTP	Pipe	0.03 max	18-20	2 max	3-4	11-15	0.04 max	0.03 max	1 max	bal Fe	480	175	35	
JIS G4303(91)	SUS317L	Bar; SA; <=180mm diam	0.030 max	18.00-20.00	2.00 max	3.00-4.00	11.00-15.00	0.045 max	0.030 max	1.00 max	bal Fe	480	175	40	187 HB
JIS G4304(91)	SUS317L	Plt Sh, HR SA	0.030 max	18.00-20.00	2.00 max	3.00-4.0	11.0-15.00	0.045 max	0.030 max	1.00 max	bal Fe	480	175	40	90 HRB max
JIS G4305(91)	SUS317J3L		0.030 max	20.50-22.50	2.00 max	2.00-3.00	11.00-13.00	0.045 max	0.030 max	1.00 max	N 0.18-0.30; bal Fe	640	275	40	
JIS G4305(91)	SUS317J4L		0.030 max	19.00-24.00	2.00 max	5.00-7.00	24.00-26.00	0.045 max	0.030 max	1.00 max	N <=0.25; bal Fe	520	205	35	
JIS G4305(91)	SUS317J5L		0.020 max	19.00-23.00	2.00 max	4.00-5.00	23.00-28.00	0.045 max	0.030 max	1.00 max	Cu 1.00-2.00; bal Fe	490	215	35	
JIS G4305(91)	SUS317L	Plt Sh, CR SA	0.030 max	18.00-20.00	2.00 max	3.00-4.00	11.00-15.00	0.045 max	0.030 max	1.00 max	bal Fe	480	175	40	187 HB max
JIS G4305(91)	SUS317LN		0.030 max	18.00-20.00	2.00 max	3.00-4.00	11.00-15.00	0.045 max	0.030 max	1.00 max	N 0.10-0.22; bal Fe	550	245	40	
JIS G4308	SUS317L	Wir rod	0.030 max	18.00-20.00	2.00 max	3.00-4.00	11.00-15.00	0.045 max	0.030 max	1.00 max	bal Fe				
JIS G4309	SUS317L	Wir	0.08 max	18.00-20.00	2.00 max	3.00-4.00	11.00-15.00	0.045 max	0.030 max	1.00 max	bal Fe				

Spain

UNE 36087/4(89)	F.3539*	see X2CrNiMo18 16 4	0.03 max	17.5-19.5	2 max	3-4	14-17	0.045 max	0.03 max	1 max	N <=0.1; bal Fe				
UNE 36087/4(89)	X2CrNiMo18 16 4	Corr res; Sh Strp	0.03 max	17.5-19.5	2 max	3-4	14-17	0.045 max	0.03 max	1 max	N <=0.1; bal Fe				
UNE 6901(71)	X2CrNiMo18-16	Bar Rod	0.03 max	17.5-19.5	2 max	3-4	14-17	0.045 max	0.03 max	1 max	bal Fe				
UNE 7500(75)	X2CrNiMo18-15	Bar Frg	0.03 max	17.5-19.5	2 max	3-4	13-16	0.045 max	0.03 max	1 max	bal Fe				
UNE 8317(81)	X2CrNiMo18-16	Sh Strp Plt	0.03 max	17.5-19.5	2 max	3-4	14-17	0.045 max	0.03 max	1 max	bal Fe				

UK

BS 1449/2(83)	317S12	Corr/Heat res; t<=100mm	0.03 max	17.5-19.5	2.00 max	3.00-4.00	14.0-17.0	0.045 max	0.030 max	1.00 max	bal Fe	490	195	40	0-195 HB
BS 1554(90)	317S11	Wir Corr/Heat res; diam<0.49mm; Ann	0.03 max	18.0-20.0	2.00 max	3.00-4.00	12.0-15.0	0.045 max	0.030 max	1.00 max	W <=0.1; bal Fe	0-850			
BS 1554(90)	317S11	Wir Corr/Heat res; 6<diam<=13mm; Ann	0.03 max	18.0-20.0	2.00 max	3.00-4.00	12.0-15.0	0.045 max	0.030 max	1.00 max	W <=0.1; bal Fe	0-650			

USA

	AISI 317L	Tube	0.030 max	18.00-20.00	2.00 max	3.00-4.00	11.00-15.00	0.045 max	0.030 max	1.00 max	bal Fe				
	UNS S31703	Low-C	0.03 max	18.00-20.00	2.00 max	3.00-4.00	11.00-15.00	0.045 max	0.030 max	1.00 max	bal Fe				
ASME SA240	317L	Refer to ASTM A240(95)													
ASTM A167(96)	317L*	Sh Strp Plt	0.03 max	18-20	2 max	3-4	11-15	0.045 max	0.03 max	1 max	bal Fe				
ASTM A182/A182M(98)	F317L	Frg/roll pipe flange valve	0.03 max	18.0-20.0	2.00 max	3.0-4.0	11.0-15.0	0.045 max	0.030 max	1.00 max	bal Fe	485	170	30	
ASTM A213/A213M(95)	TP317L	Smls tube boiler, superheater, heat exchanger	0.035 max	18.0-20.0	2.00 max	3.00-4.00	11.0-15.00	0.040 max	0.030 max	0.75 max	bal Fe	515	205	35	
ASTM A240/A240M(98)	S31703	Low-C	0.030 max	18.0-20.0	2.00 max	3.0-4.0	11.0-15.0	0.045 max	0.030 max	1.00 max	N <=0.10; bal Fe	515	205	40.0	217 HB
ASTM A249/249M(96)	TP317L	Weld tube; boiler, superheater, heat exch	0.035 max	18.0-20.0	2.00 max	3.00-4.00	11.0-15.00	0.04 max	0.03 max	0.75 max	bal Fe	515	205	35	90 HRB

UNS numbers and US grades are provided as a means of cross referencing chemically similar alloys. Exchangability is only possible after independent examination of specifications. Tensile properties are minimum or typical as specified. UTS and YS as MPa. El as %. See Appendix for list of abbreviations used in Notes. * indicates obsolete material.

Stainless Steel, Austenitic, 317L (Continued from previous page)

Specification	Designation	Notes	C	Cr	Mn	Mo	Ni	P	S	Si	Other	UTS	YS	El	Hard
USA															
ASTM A774/A774M(98)	TP317L	Fittings; Corr service at low/mod-temp, Low-C	0.030 max	18.0-20.0	2.00 max	3.0-4.0	11.0-15.0	0.045 max	0.030 max	1.00 max	N <=0.10; bal Fe	515-690	205	35.0	217 HB
ASTM A778(90)	TP317L	Low-C	0.030 max	18.00-20.00	2.00 max	3.00-4.00	11.00-14.00	0.045 max	0.030 max	1.00 max	N <=0.10; bal Fe	515	205	35	
ASTM A813/A813M(95)	TP317L	Weld Pipe, Low-C	0.035 max	18.0-20.0	2.00 max	3.00-4.00	11.0-15.0	0.045 max	0.030 max	0.75 max	bal Fe	515	205		
ASTM A814/A814M(96)	TP317L	CW Weld Pipe, Low-C	0.035 max	18.0-20.0	2.00 max	3.00-4.00	11.0-15.0	0.045 max	0.030 max	0.75 max	bal Fe	515	205		
ASTM A943/A943M(95)	TP317L	Spray formed smls pipe	0.030 max	18.0-20.0	2.00 max	3.00-4.00	11.0-15.0	0.040 max	0.030 max	0.75 max	bal Fe	515	205		
ASTM A988(98)	S31703	HIP Flanges, Fittings, Valves/parts; Heat res; Low-C	0.030 max	18.0-20.0	2.00 max	3.00-4.00	11.0-15.0	0.045 max	0.030 max	1.00 max	bal Fe	485	170	30	
SAE J405(98)	S31703	Low-C	0.030 max	18.00-20.00	2.00 max	3.00-4.00	11.00-15.00	0.045 max	0.030 max	0.75 max	N <=0.10; bal Fe				

Stainless Steel, Austenitic, 317LM

Specification	Designation	Notes	C	Cr	Mn	Mo	Ni	P	S	Si	Other	UTS	YS	El	Hard
Mexico															
NMX-B-186-SCFI(94)	S31725	Pipe; High-temp	0.03 max	18.0-20.0	2.00 max	4.0-5.0	13.5-17.5	0.040 max	0.030 max	0.75 max	Cu <=0.75; N <=0.10; bal Fe	517	207		
USA															
	UNS S31725		0.03 max	18.0-20.0	2.00 max	4.00-5.00	13.0-17.0	0.045 max	0.030 max	0.75 max	Cu <=0.75; N <=0.10; bal Fe				
ASTM A182/A182M(98)	F47	Frg/roll pipe flange valve	0.030 max	18.0-20.0	2.00 max	4.0-5.0	13.0-17.5	0.045 max	0.030 max	0.75 max	N <=0.10; bal Fe	525	205	40.0	
ASTM A213/A213M(95)	S31725	Smls tube boiler, superheater, heat exchanger	0.03 max	18.0-20.0	2.00 max	4.0-5.00	13.5-17.5	0.040 max	0.030 max	0.75 max	Cu <=0.75; N <=0.10; bal Fe	515	205	35	
ASTM A240/A240M(98)	S31725		0.030 max	18.0-20.0	2.00 max	4.0-5.0	13.0-17.0	0.045 max	0.030 max	0.75 max	Cu <=0.75; N <=0.20; bal Fe	515	205	40.0	217 HB
ASTM A249/249M(96)	S31725	Weld tube; boiler, superheater, heat exch	0.03 max	18.0-20.0	2.00 max	4.0-5.00	13.5-17.5	0.045 max	0.03 max	0.75 max	Cu <=0.75; N <=0.10; bal Fe	515	205	35	90 HRB
ASTM A943/A943M(95)		Spray formed smls pipe	0.03 max	18.0-20.0	2.0 max	4.0-5.0	13.5-17.5	0.040 max	0.030 max	0.75 max	Cu <=0.75; N <=0.10; bal Fe	515	205		
ASTM A988(98)	S31725	HIP Flanges, Fittings, Valves/parts; Heat res	0.030 max	18.0-20.0	2.00 max	4.0-5.0	13.5-17.5	0.045 max	0.030 max	1.00 max	N <=0.20; bal Fe	525	205	40.0	
SAE J405(98)	S31725		0.030 max	18.00-20.00	2.00 max	4.0-5.0	13.50-17.50	0.045 max	0.030 max	0.75 max	Cu <=0.75; N <=0.10; bal Fe				

Stainless Steel, Austenitic, 317LMN

Specification	Designation	Notes	C	Cr	Mn	Mo	Ni	P	S	Si	Other	UTS	YS	El	Hard
Mexico															
NMX-B-186-SCFI(94)	S31726	Pipe; High-temp	0.03 max	17.0-20.0	2.00 max	4.0-5.0	13.5-17.5	0.040 max	0.030 max	0.75 max	Cu <=0.75; N 0.10-0.20; bal Fe	552	241		
USA															
	UNS S31726		0.03 max	17.0-20.0	2.00 max	4.00-5.00	13.5-17.5	0.045 max	0.030 max	0.75 max	Cu <=0.75; N 0.10-0.20; bal Fe				
ASTM A182/A182M(98)	F48	Frg/roll pipe flange valve	0.030 max	17.0-20.0	2.00 max	4.0-5.0	13.5-17.5	0.045 max	0.030 max	0.75 max	N 0.10-0.20; bal Fe	550	240	40.0	
ASTM A213/A213M(95)	S31726	Smls tube boiler, superheater, heat exchanger	0.03 max	17.0-20.0	2.00 max	4.0-5.00	13.5-17.5	0.040 max	0.030 max	0.75 max	Cu <=0.75; N 0.10-0.20; bal Fe	550	240	35	
ASTM A240/A240M(98)	S31726		0.030 max	17.0-20.0	2.00 max	4.0-5.0	13.5-17.5	0.045 max	0.030 max	0.75 max	Cu <=0.75; N 0.10-0.20; bal Fe	550	240	40.0	223 HB
ASTM A249/249M(96)	S31726	Weld tube; boiler, superheater, heat exch	0.03 max	17.0-20.0	2.00 max	4.0-5.00	13.5-17.5	0.045 max	0.03 max	0.75 max	Cu <=0.75; N 0.10-0.20; bal Fe	550	240	35	90 HRB
ASTM A943/A943M(95)		Spray formed smls pipe	0.03 max	17.0-20.0	2.0 max	4.0-5.0	13.5-17.5	0.040 max	0.030 max	0.75 max	Cu <=0.75; N 0.10-0.20; bal Fe	550	240		
ASTM A988(98)	S31726	HIP Flanges, Fittings, Valves/parts; Heat res	0.030 max	17.0-20.0	2.00 max	4.0-5.0	14.5-17.5	0.045 max	0.030 max	1.00 max	N 0.10-0.20; bal Fe	550	240	40.0	
SAE J405(98)	S31726		0.030 max	17.00-20.00	2.00 max	4.0-5.0	13.50-17.50	0.045 max	0.030 max	0.75 max	Cu <=0.75; N 0.10-0.20; bal Fe				

Stainless Steel, Austenitic, 317LN

Specification	Designation	Notes	C	Cr	Mn	Mo	Ni	P	S	Si	Other	UTS	YS	El	Hard
Japan															
JIS G4303(91)	SUS317J1	Bar; SA; <=180mm diam	0.040 max	16.00-19.00	2.50 max	4.00-6.00	15.00-17.00	0.045 max	0.030 max	1.00 max	bal Fe	480	175	40	187 HB

UNS numbers and US grades are provided as a means of cross referencing chemically similar alloys. Exchangability is only possible after independent examination of specifications. Tensile properties are minimum or typical as specified. UTS and YS as MPa. El as %. See Appendix for list of abbreviations used in Notes. * indicates obsolete material.

Specification	Designation	Notes	C	Cr	Mn	Mo	Ni	P	S	Si	Other	UTS	YS	El	Hard

Stainless Steel, Austenitic, 317LN (Continued from previous page)

Japan

Specification	Designation	Notes	C	Cr	Mn	Mo	Ni	P	S	Si	Other	UTS	YS	El	Hard
JIS G4303(91)	SUS317J4L	Bar; SA; <=180mm diam	0.030 max	19.00-24.00	2.00 max	5.00-7.00	24.00-26.00	0.045 max	0.030 max	1.00 max	bal Fe	520	205	35	217 HB
JIS G4303(91)	SUS317J5L	Bar; SA; <=180mm diam	0.020 max	19.00-23.00	2.00 max	4.00-5.00	23.00-28.00	0.045 max	0.030 max	1.00 max	Cu 1.00-2.00; bal Fe	490	215	35	187 HB
JIS G4303(91)	SUS317LN	Bar; SA; <=180mm diam	0.030 max	18.00-20.00	2.00 max	3.00-4.00	11.00-15.00	0.045 max	0.030 max	1.00 max	N 0.10-0.22; bal Fe	550	245	40	217 HB

USA

Specification	Designation	Notes	C	Cr	Mn	Mo	Ni	P	S	Si	Other	UTS	YS	El	Hard
	UNS S31753	N-bearing	0.030 max	18.0-20.0	2.00 max	3.00-4.00	11.0-15.0	0.045 max	0.30 max	1.00 max	N 0.10-0.22; bal Fe				
ASTM A240/A240M(98)	S31753	N-bearing	0.030 max	18.0-20.0	2.00 max	3.00-4.00	11.0-15.0	0.045 max	0.30 max	1.00 max	N 0.10-0.22; bal Fe	550	240	40.0	217 HB
SAE J405(98)	S31753	N-bearing	0.030 max	18.00-20.00	2.00 max	3.00-4.00	11.00-15.00	0.045 max	0.30 max	0.75 max	N 0.10-0.22; bal Fe				

Stainless Steel, Austenitic, 321

Australia

Specification	Designation	Notes	C	Cr	Mn	Mo	Ni	P	S	Si	Other	UTS	YS	El	Hard
AS 1449(94)	321	Plt Sh Strp	0.08 max	17.00-19.00	2.00 max		9.00-12.00	0.045 max	0.030 max	0.75 max	N <=0.10; Ti 5x(C+N) min, 0.70 max; bal Fe				
AS 2837(86)	321	Bar, Semi-finished product	0.08 max	17.0-19.0	2.00 max		9.0-12.0	0.045 max	0.030 max	1.00 max	Ti 5xC min, 0.80 max; bal Fe				

Bulgaria

Specification	Designation	Notes	C	Cr	Mn	Mo	Ni	P	S	Si	Other	UTS	YS	El	Hard
BDS 6738(72)	0Ch18N10T	Corr res	0.08 max	17.0-19.0	2.00 max	0.30 max	9.0-11.0	0.035 max	0.025 max	0.80 max	5xC<=Ti<=0.70; bal Fe				
BDS 6738(72)	Ch18N9T	Corr res	0.12 max	17.0-19.0	0.20 max	0.30 max	8.0-10.0	0.035 max	0.025 max	0.80 max	Ti <=0.8; 5xC<=Ti<=0.80; bal Fe				

China

Specification	Designation	Notes	C	Cr	Mn	Mo	Ni	P	S	Si	Other	UTS	YS	El	Hard
GB 1220(92)	0Cr18Ni10Ti	Bar SA	0.08 max	17.00-19.00	2.00 max		9.00-12.00	0.035 max	0.03 max	1.00 max	Ti>=5xC; bal Fe	520	205	40	
GB 1220(92)	1Cr18Ni9Ti	Bar SA	0.12 max	17.00-19.00	2.00 max		8.00-11.00	0.035 max	0.03 max	1.00 max	Ti>=5(C-0.02)<=0.80; bal Fe	520	205	40	
GB 1221(92)	0Cr18Ni10Ti	Bar SA	0.08 max	17.00-19.00	2.00 max		9.00-12.00	0.035 max	0.03 max	1.00 max	Ti>=5xC; bal Fe	520	205	40	
GB 1221(92)	1Cr18Ni9Ti	Bar SA	0.12 max	17.00-19.00	2.00 max		8.00-11.00	0.035 max	0.03 max	1.00 max	Ti>=5(C-0.02)<=0.80; bal Fe	520	205	40	
GB 12770(91)	1Cr18Ni9Ti	Weld tube; SA	0.12 max	17.00-19.00	2.00 max		8.00-11.00	0.035 max	0.03 max	1.00 max	Ti>=5(C-0.02)<=0.80; bal Fe	520	210	35	
GB 12771(91)	0Cr18Ni11Ti	Weld Pipe SA	0.08 max	17.00-19.00	2.00 max		9.00-13.00	0.035 max	0.03 max	1.00 max	Ti>=5xC; bal Fe	520	210	35	
GB 12771(91)	1Cr18Ni9Ti	Weld Pipe SA	0.12 max	17.00-19.00	2.00 max		8.00-11.00	0.035 max	0.03 max	1.00 max	Ti>=5(C-0.02)<=0.80; bal Fe	520	210	35	
GB 13296(91)	0Cr18Ni10Ti	Smls tube SA	0.08 max	17.00-19.00	2.00 max		9.00-12.00	0.035 max	0.03 max	1.00 max	Ti>=5xC; bal Fe	520	205	35	
GB 13296(91)	1Cr18Ni9Ti	Smls tube SA	0.12 max	17.00-19.00	2.00 max		8.00-11.00	0.035 max	0.03 max	1.00 max	Ti>=5(C-0.02)<=0.80; bal Fe	520	205	40	
GB 3280(92)	0Cr18Ni10Ti	Sh Plt, CR SA	0.08 max	17.00-19.00	2.00 max		9.00-12.00	0.035 max	0.03 max	1.00 max	Ti>=5xC; bal Fe	520	205	40	
GB 3280(92)	1Cr18Ni9Ti	Sh Plt, CR SA	0.12 max	17.00-19.00	2.00 max		8.00-11.00	0.035 max	0.03 max	1.00 max	Ti>=5(C-0.02)<=0.80; bal Fe	520	205	40	
GB 4237(92)	0Cr18Ni10Ti	Sh Plt, HR SA	0.08 max	17.00-19.00	2.00 max		9.00-12.00	0.035 max	0.03 max	1.00 max	Ti>=5xC; bal Fe	520	205	40	
GB 4237(92)	1Cr18Ni9Ti	Sh Plt, HR SA	0.12 max	17.00-19.00	2.00 max		8.00-11.00	0.035 max	0.03 max	1.00 max	Ti>=5(C-0.02)<=0.80; bal Fe	520	205	40	
GB 4238(92)	0Cr18Ni10Ti	Plt, HR SA	0.08 max	17.00-19.00	2.00 max		9.00-12.00	0.035 max	0.03 max	1.00 max	Ti>=5xC; bal Fe	520	205	40	
GB 4238(92)	1Cr18Ni9Ti	Plt, HR SA	0.12 max	17.00-19.00	2.00 max		8.00-11.00	0.035 max	0.03 max	1.00 max	Ti>=5(C-0.02)<=0.80; bal Fe	520	205	40	
GB 4239(91)	0Cr18Ni10Ti	Sh Strp, CR SA	0.08 max	17.00-19.00	2.00 max		9.00-12.00	0.035 max	0.03 max	1.00 max	Ti>=5xC; bal Fe	520	205	40	
GB 4239(91)	1Cr18Ni9Ti	Sh Strp, CR SA	0.12 max	17.00-19.00	2.00 max		8.00-11.00	0.035 max	0.03 max	1.00 max	Ti>=5(C-0.02)<=0.80; bal Fe	520	205	40	
GB 4240(93)	0Cr18Ni11Ti(-Q, -R)	Wir Ann 0.6-1mm diam, Q-TLC; R-As Ann	0.08 max	17.00-19.00	2.00 max		9.00-13.00	0.035 max	0.03 max	1.00 max	Ti>=5xC; bal Fe	540-880			
GB 4240(93)	1Cr18Ni9Ti(-L, -Q, -R)	Wir Ann 0.6-1mm diam, L-CD; Q-TLC	0.12 max	17.00-19.00	2.00 max		8.00-11.00	0.035 max	0.03 max	1.00 max	Ti>=5(C-0.02)<=0.80; bal Fe	540-880			

Stainless Steel, Austenitic, 321 (Continued from previous page)

Specification	Designation	Notes	C	Cr	Mn	Mo	Ni	P	S	Si	Other	UTS	YS	El	Hard
China															
GB/T 14975(94)	0Cr18Ni10Ti	Smls tube SA	0.08 max	17.00-19.00	2.00 max		9.00-12.00	0.035 max	0.03 max	1.00 max	Ti>=5xC; bal Fe	520	205	35	
GB/T 14975(94)	1Cr18Ni9Ti	Smls tube SA	0.12 max	17.00-19.00	2.00 max		8.00-11.00	0.035 max	0.03 max	1.00 max	Ti>=5(C-0.02)<=0.80; bal Fe	520	205	35	
GB/T 14976(94)	0Cr18Ni10Ti	Smls pipe SA	0.08 max	17.00-19.00	2.00 max		9.00-12.00	0.035 max	0.03 max	1.00 max	Ti>=5xC; bal Fe	520	205	35	
GB/T 14976(94)	1Cr18Ni9Ti	Smls pipe SA	0.12 max	17.00-19.00	2.00 max		8.00-11.00	0.035 max	0.03 max	1.00 max	Ti>=5(C-0.02)<=0.80; bal Fe	520	205	35	
Czech Republic															
CSN 417246	17246	Corr res; tough at subzero	0.12 max	17.0-20.0	2.0 max		8.0-11.0	0.045 max	0.03 max	1.0 max	Ti <=0.90; Ti>=5xC; bal Fe				
CSN 417247	17247	Corr res	0.08 max	17.0-19.0	2.0 max		9.5-12.0	0.045 max	0.03 max	1.0 max	Ti <=0.90; Ti>=5xC; bal Fe				
CSN 417248	17248	Corr res	0.1 max	17.0-19.0	2.0 max		9.5-12.0	0.045 max	0.03 max	1.0 max	Ti <=0.90; Ti>=5xC; bal Fe				
Europe															
EN 10088/2(95)	1.4541	Strp, CR Corr res; t<=6mm, Ann	0.08 max	17.00-19.00	2.00 max		9.00-12.00	0.045 max	0.015 max	1.00 max	5xC<=Ti<=0.70; bal Fe	520-720	220	40	
EN 10088/2(95)	X6CrNiTi18-10	Strp, CR Corr res; t<=6mm, Ann	0.08 max	17.00-19.00	2.00 max		9.00-12.00	0.045 max	0.015 max	1.00 max	5xC<=Ti<=0.70; bal Fe	520-720	220	40	
EN 10088/3(95)	1.4541	Bar Rod Sect, t<=25mm, strain Hard	0.08 max	17.00-19.00	2.00 max		9.00-12.00	0.045 max	0.030 max	1.00 max	Ti <=0.7; 5xC<=Ti<=0.70; bal Fe	800-1000	500	12	
EN 10088/3(95)	1.4541	Bar Rod Sect, t<=35mm, strain Hard	0.08 max	17.00-19.00	2.00 max		9.00-12.00	0.045 max	0.030 max	1.00 max	Ti <=0.7; 5xC<=Ti<=0.70; bal Fe	700-850	350	20	
EN 10088/3(95)	1.4541	Bar Rod Sect, t<=160mm, Ann	0.08 max	17.00-19.00	2.00 max		9.00-12.00	0.045 max	0.030 max	1.00 max	Ti <=0.7; 5xC<=Ti<=0.70; bal Fe	500-700	190	40	215 HB
EN 10088/3(95)	1.4541	Bar Rod Sect; Corr res; 160<t<=250mm, Ann	0.08 max	17.00-19.00	2.00 max		9.00-12.00	0.045 max	0.030 max	1.00 max	Ti <=0.7; 5xC<=Ti<=0.70; bal Fe	500-700	190	30	215 HB
EN 10088/3(95)	X6CrNiTi18-10	Bar Rod Sect, t<=25mm, strain Hard	0.08 max	17.00-19.00	2.00 max		9.00-12.00	0.045 max	0.030 max	1.00 max	Ti <=0.7; 5xC<=Ti<=0.70; bal Fe	800-1000	500	12	
EN 10088/3(95)	X6CrNiTi18-10	Bar Rod Sect; Corr res; 160<t<=250mm, Ann	0.08 max	17.00-19.00	2.00 max		9.00-12.00	0.045 max	0.030 max	1.00 max	Ti <=0.7; 5xC<=Ti<=0.70; bal Fe	500-700	190	30	215 HB
EN 10088/3(95)	X6CrNiTi18-10	Bar Rod Sect, t<=160mm, Ann	0.08 max	17.00-19.00	2.00 max		9.00-12.00	0.045 max	0.030 max	1.00 max	Ti <=0.7; 5xC<=Ti<=0.70; bal Fe	500-700	190	40	215 HB
EN 10088/3(95)	X6CrNiTi18-10	Bar Rod Sect, t<=35mm, strain Hard	0.08 max	17.00-19.00	2.00 max		9.00-12.00	0.045 max	0.030 max	1.00 max	Ti <=0.7; 5xC<=Ti<=0.70; bal Fe	700-850	350	20	
Germany															
DIN 1654(89)	WNr 1.4541	CHd, cold ext	0.08 max	17.0-19.0	2.00 max		9.00-12.0	0.045 max	0.030 max	1.00 max	Ti 5xC<=0.70; bal Fe				
DIN 1654(89)	X6CrNiTi18-10	CHd, cold ext	0.08 max	17.0-19.0	2.00 max		9.00-12.0	0.045 max	0.030 max	1.00 max	Ti 5xC<=0.70; bal Fe				
DIN 17440(96)	WNr 1.4541	CR Strp Plt Sh	0.08 max	17.0-19.0	2.00 max		9.00-12.0	0.045 max	0.030 max	1.00 max	Ti 5xC<=0.70; bal Fe				
DIN 17440(96)	X6CrNiTi18-10	CR Strp Plt Sh	0.08 max	17.0-19.0	2.00 max		9.00-12.0	0.045 max	0.030 max	1.00 max	Ti 5xC<=0.70; bal Fe				
DIN 17441(97)	WNr 1.4541	Frg for Press purposes	0.08 max	17.0-19.0	2.00 max		9.00-12.0	0.045 max	0.030 max	1.00 max	Ti 5xC<=0.70; bal Fe				
DIN 17441(97)	X6CrNiTi18-10	Frg for Press purposes	0.08 max	17.0-19.0	2.00 max		9.00-12.0	0.045 max	0.030 max	1.00 max	Ti 5xC<=0.70; bal Fe				
DIN EN 10028(96)	WNr 1.4541	Flat product for Press purposes	0.08 max	17.0-19.0	2.00 max		9.00-12.0	0.045 max	0.030 max	1.00 max	Ti 5xC<=0.70; bal Fe				
DIN EN 10028(96)	X6CrNiTi18-10	Flat product for Press purposes	0.08 max	17.0-19.0	2.00 max		9.00-12.0	0.045 max	0.030 max	1.00 max	Ti 5xC<=0.70; bal Fe				
DIN EN 10088(95)	WNr 1.4541	Sh Plt Strp Bar Rod Sect, SHT	0.08 max	17.0-19.0	2.00 max		9.00-12.0	0.045 max	0.030 max	1.00 max	Ti 5xC<=0.70; bal Fe	500-700	190	30-40	215 HB
DIN EN 10088(95)	X6CrNiTi18-10	Sh Plt Strp Bar Rod Sect, SHT	0.08 max	17.0-19.0	2.00 max		9.00-12.0	0.045 max	0.030 max	1.00 max	Ti 5xC<=0.70; bal Fe	500-700	190	30-40	215 HB
DIN EN 10222(94)	WNr 1.4541	Plt Sh, HR Strp Bar Wir Frg	0.08 max	17.0-19.0	2.00 max		9.00-12.0	0.045 max	0.030 max	1.00 max	Ti 5xC<=0.70; bal Fe				

Specification	Designation	Notes	C	Cr	Mn	Mo	Ni	P	S	Si	Other	UTS	YS	El	Hard

Stainless Steel, Austenitic, 321 (Continued from previous page)

Germany

Specification	Designation	Notes	C	Cr	Mn	Mo	Ni	P	S	Si	Other	UTS	YS	El	Hard
DIN EN 10222(94)	X6CrNiTi18-10	Plt Sh, HR Strp Bar Wir Frg	0.08 max	17.0-19.0	2.00 max		9.00-12.0	0.045 max	0.030 max	1.00 max	Ti 5xC<=0.70; bal Fe				

Hungary

Specification	Designation	Notes	C	Cr	Mn	Mo	Ni	P	S	Si	Other	UTS	YS	El	Hard
MSZ 4360(72)	KO36	Corr res; Heat res; 60.1-100mm; Quen; Q	0.12 max	17-19	2 max	0.5 max	8-11	0.04 max	0.03 max	1 max	Ti <=0.8; V <=0.2; W <=0.3; Ti=5C-0.8; bal Fe	490	206	39L	
MSZ 4360(72)	KO36	Corr res; 0-60mm; Quen; Q	0.12 max	17-19	2 max	0.5 max	8-11	0.04 max	0.03 max	1 max	Ti <=0.8; V <=0.2; W <=0.3; Ti=5C-0.8; bal Fe	490	206	40L	
MSZ 4360(80)	KO37	Corr res	0.08 max	17-19	2 max	0.5 max	9-11.5	0.04 max	0.03 max	1 max	Ti <=0.7; Ti=5C-0.7; bal Fe				
MSZ 4360(87)	KO36Ti	Corr res; 0-160mm; SA	0.12 max	17-19	2 max	0.5 max	8-11	0.045 max	0.03 max	1 max	Ti <=0.8; Ti=5C-0.8; bal Fe	490-720	200	40L; 30T	190 HB max
MSZ 4360(87)	KO37Nb	Corr res; 0-160mm; SA	0.08 max	17-19	2 max	0.5 max	9-12	0.045 max	0.03 max	1 max	Nb+Ta=10C-1.1; bal Fe	500-740	205	40L; 30T	190 HB max
MSZ 4360(87)	KO37Ti	Corr res; 0-160mm; SA	0.08 max	17-19	2 max	0.5 max	9-12	0.045 max	0.03 max	1 max	Ti <=0.8; Ti=5C-0.8; bal Fe	500-730	200	40L; 30T	190 HB max
MSZ 4360(87)	X12CrNiTi189	Corr res; 0-160mm; SA	0.12 max	17-19	2 max	0.5 max	8-11	0.045 max	0.03 max	1 max	Ti <=0.8; Ti=5C-0.8; bal Fe	490-720	200	40L; 30T	190 HB max
MSZ 4360(87)	X8CrNiTi1810	Corr res; 0-160mm; SA	0.08 max	17-19	2 max	0.5 max	9-12	0.045 max	0.03 max	1 max	Ti <=0.8; Ti=5C-0.8; bal Fe	500-730	200	40L; 30T	190 HB max
MSZ 4398(86)	KO36Ti	Corr res; Heat res; longitudinal weld tube; 0.8-3mm; SA	0.08 max	17-19	2 max	0.5 max	9-12	0.045 max	0.03 max	1 max	Ti <=0.8; Ti=5C-0.8; bal Fe	500-730	200	35L	

International

Specification	Designation	Notes	C	Cr	Mn	Mo	Ni	P	S	Si	Other	UTS	YS	El	Hard
ISO 4955(94)	H1	Heat res	0.08 max	17.0-19.0	1.0 max	0.15 max	0.4 max	0.04 max	0.03 max	1.0 max	Al <=0.1; Co <=0.1; Cu <=0.3; Pb <=0.15; V <=0.1; W <=0.1; 6C<=Ti<=1.0				
ISO 4955(94)	H11	Heat res	0.-0.12	17.0-19.0	2.0 max	0.15 max	9.0-12.0	0.045 max	0.03 max	1.0 max	Al <=0.1; Co <=0.1; Cu <=0.3; Pb <=0.15; V <=0.1; W <=0.1; 5c<=Ti<=0.8, bal Fe				

Italy

Specification	Designation	Notes	C	Cr	Mn	Mo	Ni	P	S	Si	Other	UTS	YS	El	Hard
UNI 6901(71)	X6CrNiTi1811	Corr res	0.08 max	17-19	2 max		9-12	0.045 max	0.03 max	1 max	Ti <=0.8; Ti=5C-0.8; bal Fe				
UNI 6904(71)	X6CrNiTi1811	Heat res	0.08 max	17-19	2 max		9-13	0.03 max	0.03 max	0.75 max	Ti <=0.6; Ti=5C-0.6; bal Fe				
UNI 7500(75)	X6CrNiTi1811	Press ves	0.08 max	17-19	2 max		9-12	0.045 max	0.03 max	1 max	Ti <=0.8; Ti=5C-0.8; bal Fe				

Japan

Specification	Designation	Notes	C	Cr	Mn	Mo	Ni	P	S	Si	Other	UTS	YS	El	Hard
JIS G3214(91)	SUSF321	Frg press ves 130-200mm diam	0.08 max	17.00 min	2.00 max		9.00-12.00	0.040 max	0.030 max	1.00 max	Ti 5xC to 0.60; bal Fe	480	205	29	187 HB max
JIS G4303(91)	SUS321	Bar; SA; <=180mm diam	0.08 max	17.00-19.00	2.00 max		9.00-13.00	0.045 max	0.030 max	1.00 max	Ti>=5xC; bal Fe	520	205	40	187 HB
JIS G4305(91)	SUS321		0.08 max	17.00-19.00	2.00 max		9.00-13.00	0.045 max	0.030 max	1.00 max	Ti>=5xC; bal Fe	520	205	40	
JIS G4308	SUS321	Wir rod	0.08 max	17.00-19.00	2.00 max		9.00-13.00	0.045 max	0.030 max	1.00 max	Ti>=5xC; bal Fe				
JIS G4309	SUS321	Wir	0.08 max	17.00-19.00	2.00 max		9.00-13.00	0.045 max	0.030 max	1.00 max	Ti>=5xC; bal Fe				

Mexico

Specification	Designation	Notes	C	Cr	Mn	Mo	Ni	P	S	Si	Other	UTS	YS	El	Hard
NMX-B-171(91)	MT321	mech tub	0.08 max	17.0-20.0	2.00 max		9.0-13.0	0.040 max	0.030 max	1.00 max	Ti=(5xC) min, 0.60 max; bal Fe				
NMX-B-186-SCFI(94)	TP321	Pipe; High-temp; <=9.5mm	0.08 max	17.0-20.0	2.00 max		9.00-13.0	0.040 max	0.030 max	0.75 max	Ti=5xC min, 0.60 max; bal Fe	517	207		
NMX-B-196(68)	TP321	Smls tube for refinery	0.08 max	17.0-20.0	2.00 max		9.0-13.0	0.040 max	0.030 max	0.75 max	Ti=5xC min, 0.60 max; bal Fe				

Poland

Specification	Designation	Notes	C	Cr	Mn	Mo	Ni	P	S	Si	Other	UTS	YS	El	Hard
PNH74245/09	1H18N9Tselekt	Corr res; hot-formed Smls tube	0.1 max	17-18.5	2 max		10-11	0.035 max	0.03 max	0.8 max	Ti <=0.8; Ti=5C-0.8; bal Fe				
PNH86020	0H18N10T	Corr res	0.08 max	17-19	2 max		9-11	0.045 max	0.03 max	0.8 max	Ti <=0.7; Ti=5C-0.7; bal Fe				
PNH86020	1H18N9T	Corr res	0.1 max	17-19	2 max		8-10	0.045 max	0.03 max	0.8 max	Ti <=0.8; Ti=5C-0.8; bal Fe				

Specification	Designation	Notes	C	Cr	Mn	Mo	Ni	P	S	Si	Other	UTS	YS	El	Hard

Stainless Steel, Austenitic, 321 (Continued from previous page)

Romania

Specification	Designation	Notes	C	Cr	Mn	Mo	Ni	P	S	Si	Other	UTS	YS	El	Hard
STAS 10322(80)	10TiNiCr180	Corr res; Heat res	0.12 max	17-19	2 max		8-9.5	0.04 max	0.03 max	0.8 max	Cu <=0.5; Pb <=0.15; Ti <=0.90; Ti>=5xC; bal Fe				
STAS 3583(87)	10TiNiCr180	Corr res; Heat res	0.08 max	17-19	2 max	0.2 max	9-12	0.045 max	0.03 max	1 max	Pb <=0.15; Ti <=0.8; 5C<=Ti<=0.8; bal Fe				

Russia

Specification	Designation	Notes	C	Cr	Mn	Mo	Ni	P	S	Si	Other	UTS	YS	El	Hard
GOST 10498(63)	06Ch18N10T	Corr res	0.06 max	17.0-19.0	1.0-2.0		9.0-11.0	0.035 max	0.02 max	0.8 max	Ti <=0.6; Ti=5xC; bal Fe				
GOST 10498(63)	09Ch18N10T	Corr res	0.07-0.10	17.0-19.0	1.0-2.0		9.0-11.0	0.035 max	0.02 max	0.8 max	Ti <=0.7; Ti=5xC; bal Fe				
GOST 10500(63)	KH18N10T	Heat res	0.12 max	17.0-19.0	1.0-2.0		9.0-11.0	0.035 max	0.02 max	0.8 max	Ti <=0.7; Ti=5(C-0.02)-0.7; bal Fe				
GOST 19277	Ch18N10T-WD	Corr res	0.08-0.12	17.0-19.0	1.0 max	0.15 max	9.0-11.0	0.015 max	0.015 max	0.8 max	Cu <=0.25; Ti <=0.7; Ti=5(C-0.02)-0.7; bal Fe				
GOST 5632(61)	0KH18N10T	Corr res; Heat res; High-temp constr	0.08 max	17.0-19.0	1.0-2.0	0.15 max	9.0-11.0	0.035 max	0.02 max	0.8 max	Cu <=0.3; Ti <=0.6; Ti=5C-0.6; bal Fe				
GOST 5632(72)	08Ch18N10T	Corr res; Heat res	0.08 max	17.0-19.0	2.0 max	0.3 max	9.0-11.0	0.035 max	0.02 max	0.8 max	Cu <=0.3; Ti <=0.7; W <=0.2; Ti=5C-0.7; bal Fe				
GOST 5632(72)	12Ch18N10T	Corr res; Heat res; High-temp constr	0.12 max	17.0-19.0	2.0 max	0.3 max	9.0-11.0	0.035 max	0.02 max	0.8 max	Cu <=0.3; Ti <=0.8; W <=0.2; Ti=5C-0.8; bal Fe				

Spain

Specification	Designation	Notes	C	Cr	Mn	Mo	Ni	P	S	Si	Other	UTS	YS	El	Hard
UNE 36016(75)	F.3523*	see X6CrNiTi18-11	0.08 max	17.0-19.0	2.0 max	0.15 max	9.0-12.0	0.045 max	0.03 max	1.0 max	Ti <=0.8; 5xC<=Ti<=0.8; bal Fe				
UNE 36016(75)	X6CrNiTi18-11	Corr res	0.08 max	17.0-19.0	2.0 max	0.15 max	9.0-12.0	0.045 max	0.03 max	1.0 max	Ti <=0.8; 5xC<=Ti<=0.8; bal Fe				
UNE 36016/1(89)	E-321	Corr res	0.08 max	17.0-19.0	2.0 max	0.15 max	9.0-12.0	0.045 max	0.03 max	1.0 max	N <=0.1; Ti <=0.8; Ti=5C-0.8; bal Fe				
UNE 36016/1(89)	F.3523*	Obs; E-321; Corr res	0.08 max	17.0-19.0	2.0 max	0.15 max	9.0-12.0	0.045 max	0.03 max	1.0 max	N <=0.1; Ti <=0.8; 5xC<=Ti<=0.8; bal Fe				
UNE 36087(78)	F.3553*	see X7CrNiTi18-11	0.1 max	17.0-19.0	2.0 max	0.5 max	10.0-12.0	0.045 max	0.03 max	1.0 max	Ti <=0.8; 5xC<=Ti<=0.8; bal Fe				
UNE 36087(78)	X7CrNiTi18-11	Corr res	0.1 max	17.0-19.0	2.0 max	0.5 max	10.0-12.0	0.045 max	0.03 max	1.0 max	Ti <=0.8; 5xC<=Ti<=0.8; bal Fe				
UNE 36087/4(89)	F.3523*	see X6CrNiTi18 10	0.08 max	17-19	2 max	0.15 max	9-12	0.045 max	0.03 max	1 max	N <=0.1; Ti <=0.8; 5xC<=Ti<=0.8; bal Fe				
UNE 36087/4(89)	X6CrNiTi18 10	Corr res; Sh Strp	0.08 max	17-19	2 max	0.15 max	9-12	0.045 max	0.03 max	1 max	N <=0.1; Ti <=0.8; 5xC<=Ti<=0.8; bal Fe				

Sweden

Specification	Designation	Notes	C	Cr	Mn	Mo	Ni	P	S	Si	Other	UTS	YS	El	Hard
SS 142337	2337	Corr res	0.08 max	17-19	2 max		9-12	0.045 max	0.03 max	1 max	Ti <=0.8; Ti=5C-0.8; bal Fe				

UK

Specification	Designation	Notes	C	Cr	Mn	Mo	Ni	P	S	Si	Other	UTS	YS	El	Hard
BS 1449/2(83)	321S31	Corr/Heat res; t<=100mm	0.08 max	17.0-19.0	2.00 max		9.00-12.0	0.045 max	0.030 max	1.00 max	5xC<=Ti<=0.80; bal Fe	500	200	35	0-200 HB
BS 1501/3(73)	321S12	Press ves; Corr res	0.08 max	17.0-19.0	0.5-2.0	0.15 max	9.0-12.0	0.045 max	0.03 max	0.2-1.0	Cu <=0.3; Ti <=0.7; W <=0.1; Ti=5C-0.7; bal Fe				
BS 1501/3(73)	321S87	Press ves; Corr res	0.08 max	17.0-19.0	0.5-2.0	0.15 max	9.0-12.0	0.045 max	0.03 max	0.2-1.0	Cu <=0.3; Ti <=0.05; W <=0.1; Ti=5C-0.7; bal Fe				
BS 1501/3(90)	321S31	Press ves; Corr/Heat res; 0.05-100mm	0.08 max	17.0-19.0	2.0 max	0.15 max	9.0-12.0	0.045 max	0.025 max	1.0 max	Cu <=0.3; Ti <=0.8; W <=0.1; Ti=5C-0.8; bal Fe	510-710	200	35	
BS 1502(82)	321S51-510	Bar Shp; Press ves	0.04-0.10	17.0-19.0	2.00 max		9.00-12.0	0.045 max	0.030 max	1.00 max	W <=0.1; 5xC<=Ti<=0.8; bal Fe				
BS 1503(89)	321S51-490	Frg press ves; t<=350mm; Quen	0.04-0.10	17.0-19.0	2.00 max	0.70 max	9.00-12.0	0.040 max	0.025 max	1.00 max	B <=0.005; Cu <=0.50; W <=0.1; 5xC<=Ti<=0.80; bal Fe	490-690	190	30	
BS 1503(89)	321S51-510	Frg press ves; t<=350mm; Quen	0.04-0.10	17.0-19.0	2.00 max	0.70 max	9.00-12.0	0.040 max	0.025 max	1.00 max	B <=0.005; Cu <=0.50; W <=0.1; 5xC<=Ti<=0.80; bal Fe	510-710	235	30	
BS 1506(90)	321S51-490	Bolt matl Pres/Corr res; t<=160mm; Hard	0.04-0.10	17.0-19.0	2.00 max		9.00-12.0	0.045 max	0.030 max	1.00 max	Ti <=0.8; 5xC<=Ti<=0.80; bal Fe	490	155	35	

UNS numbers and US grades are provided as a means of cross referencing chemically similar alloys. Exchangability is only possible after independent examination of specifications. Tensile properties are minimum or typical as specified. UTS and YS as MPa. El as %. See Appendix for list of abbreviations used in Notes. * indicates obsolete material.

Stainless Steel, Austenitic, 321 (Continued from previous page)

Specification	Designation	Notes	C	Cr	Mn	Mo	Ni	P	S	Si	Other	UTS	YS	El	Hard
UK															
BS 1506(90)	321S51-520	Bolt matl Pres/Corr res; t<=160mm; Hard	0.04-0.10	17.0-19.0	2.00 max		9.00-12.0	0.045 max	0.030 max	1.00 max	Ti <=0.8; 5xC<=Ti<=0.80; bal Fe	520	205	35	
BS 3059/2(90)	316S51	Boiler, Superheater; High-temp; Smls tub; 2-11mm	0.04-0.10	16.5-18.5	2.00 max	2.00-2.50	10.5-13.5	0.040 max	0.030 max	1.00 max	bal Fe	510-710	240	35	
BS 3059/2(90)	321S51(1010)	Boiler, Superheater; High-temp; Smls tub; 2-11mm	0.04-0.10	17.0-19.0	2.00 max	0.15 max	9.0-12.0	0.040 max	0.030 max	1.00 max	5xC<=Ti<=0.80; bal Fe	510-710	235	35	
BS 3059/2(90)	321S51(1105)	Boiler, Superheater; High-temp; Smls tub; 2-11mm	0.04-0.10	17.0-19.0	2.00 max	0.15 max	9.0-12.0	0.040 max	0.030 max	1.00 max	5xC<=Ti<=0.80; bal Fe	490-690	190	35	
BS 3605/1(91)	321S51(1105)	Pip Tube press smls; t<=200mm	0.04-0.10	17.0-19.0	2.00 max		9.00-12.0	0.040 max	0.030 max	1.00 max	5xC<=Ti<=0.80; bal Fe	490-690		35	
BS 3606(78)	321S22	Corr res; Heat exch Tub	0.080 max	17.0-19.0	0.5-2.0		9.0-12.0	0.045 max	0.030 max	1.00 max	bal Fe				
BS 3606(92)	321S31	Heat exch Tub; Smls tub, Tub, Weld; 1.2-3.2mm	0.080 max	17.0-19.0	2.00 max		9.0-12.0	0.040 max	0.030 max	1.00 max	bal Fe	510-710	235	30	
BS 970/1(96)	321S31	Wrought Corr/Heat res; t<=160mm	0.08 max	17.0-19.0	2.00 max		9.00-12.0	0.045 max	0.030 max	1.00 max	5xC<=Ti<=0.80; bal Fe	510	200	35	183 HB max
BS 970/4(70)	321S12*	Obs; Corr res	0.08 max	17.0-19.0	0.50-2.00	0.7 max	9.0-12.0	0.045 max	0.03 max	0.2-1.0	Cu <=0.5; Pb <=0.15; Ti <=0.7; Ti=5C-0.7; Nb<=0.2; bal Fe				
USA															
	AISI 321	Tube	0.08 max	17.00-19.00	2.00 max		9.00-12.00	0.045 max	0.030 max	1.00 max	Ti=5xC; bal Fe				
	UNS S32100	Ti-stabilized	0.08 max	17.00-19.00	2.00 max		9.00-12.00	0.045 max	0.030 max	1.00 max	Ti>=5xC; bal Fe				
AMS 5557H(96)		Smls weld/drawn tub, OD>4.78	0.08 max	17.00-20.00	2.00 max	0.75 max	8.00-13.00	0.040 max	0.030 max	0.40-1.00	Cu <=0.75; N <=0.10; Ti <=0.70; Ti 5x(C+N); bal Fe	517-827	207	33	
AMS 5557H(96)		Smls weld/drawn tub, OD>12.70	0.08 max	17.00-20.00	2.00 max	0.75 max	8.00-13.00	0.040 max	0.030 max	0.40-1.00	Cu <=0.75; N <=0.10; Ti <=0.70; Ti 5x(C+N); bal Fe	517-724	207-35	33	
AMS 5559H(96)		Corr/Heat res; thin wall Weld tub	0.08 max	17.00-19.00	2.00 max	0.75 max	9.00-12.00	0.040 max	0.030 max	0.40-1.00	Cu <=0.75; N <=0.10; Ti <=0.70; Ti 5x(C+N); bal Fe	517-724	241	40	
AMS 5570M(92)		Corr/Heat res; Smls tub, OD<=12.70	0.08 max	17.00-19.00	2.00 max	0.75 max	9.00-13.00	0.040 max	0.030 max	0.25-1.00	Cu <=0.75; N <=0.10; Ti <=0.70; Ti 5x(C+N); bal Fe	827		30	
AMS 5570M(92)		Corr/Heat res; Smls tub, OD<=4.78	0.08 max	17.00-19.00	2.00 max	0.75 max	9.00-13.00	0.040 max	0.030 max	0.25-1.00	Cu <=0.75; N <=0.10; Ti <=0.70; Ti 5x(C+N); bal Fe	827		33	
AMS 5576H(92)		Corr/Heat res; Smls tub, OD<=12.70	0.08 max	17.00-19.00	2.00 max	0.75 max	9.00-12.00	0.040 max	0.030 max	0.25-1.00	Cu <=0.75; N <=0.10; Ti <=0.70; Ti 5x(C+N); bal Fe	827		30	
AMS 5576H(92)		Corr/Heat res; Weld tub; OD<=4.78	0.08 max	17.00-19.00	2.00 max	0.75 max	9.00-12.00	0.040 max	0.030 max	0.25-1.00	Cu <=0.75; N <=0.10; Ti <=0.70; Ti 5x(C+N); bal Fe	827		33	
AMS 5689E(95)		Wire; Corr res; UTS for straight lengths, 0.25<=D<=6.35mm	0.08 max	17.00-19.00	2.00 max	0.75 max	9.00-12.00	0.040 max	0.030 max	0.40-1.00	Cu <=0.75; N <=0.10; Ti <=0.70; Ti 5x(C+N); bal Fe	931-793			
AMS 5896(94)		Corr/Heat res; Smls or weld hydraulic tub, HT and CD										724-965	517-758	20	
ASTM A182/A182M(98)	F321	Frg/roll pipe flange valve	0.08 max	17.0 min	2.00 max		9.0-12.0	0.045 max	0.030 max	1.00 max	bal Fe	515	205	30	
ASTM A193/A193M(98)	B8T	Bolt, high-temp, Ti-stabilized, HT	0.08 max	17.0-19.0	2.00 max		9.0-12.0	0.045 max	0.030 max	1.00 max	5x(C+N)<=Ti<=0.70; bal Fe	515	205	30	223 HB
ASTM A193/A193M(98)	B8TA	Bolt, high-temp, Ti-stabilized, HT	0.08 max	17.0-19.0	2.00 max		9.0-12.0	0.045 max	0.030 max	1.00 max	5x(C+N)<=Ti<=0.70; bal Fe	515	204	30	192 HB
ASTM A194/A194M(98)	8T	Nuts, high-temp press	0.08 max	17.0-19.0	2.00 max		9.0-12.0	0.045 max	0.030 max	1.00 max	5x(C+N)<=Ti; bal Fe				126-300 HB
ASTM A194/A194M(98)	8TA	Nuts, high-temp press	0.08 max	17.0-19.0	2.00 max		9.0-12.0	0.045 max	0.030 max	1.00 max	5x(C+N)<=Ti; bal Fe				126-192 HB
ASTM A213/A213M(95)	TP321	Smls tube boiler, superheater, heat exchanger	0.06 max	17.0-20.0	2.00 max		9.00-13.00	0.040 max	0.030 max	0.75 max	Ti 5xC-0.60; bal Fe	515	205	35	

UNS numbers and US grades are provided as a means of cross referencing chemically similar alloys. Exchangability is only possible after independent examination of specifications. Tensile properties are minimum or typical as specified. UTS and YS as MPa. El as %. See Appendix for list of abbreviations used in Notes. * indicates obsolete material.

Specification	Designation	Notes	C	Cr	Mn	Mo	Ni	P	S	Si	Other	UTS	YS	El	Hard

Stainless Steel, Austenitic, 321 (Continued from previous page)

USA

Specification	Designation	Notes	C	Cr	Mn	Mo	Ni	P	S	Si	Other	UTS	YS	El	Hard
ASTM A240/A240M(98)	S32100	Ti-stabilized	0.08 max	17.00-19.00	2.00 max		9.00-12.00	0.045 max	0.030 max	1.00 max	5xC<=Ti; bal Fe	515	205	40.0	217 HB
ASTM A249/249M(96)	TP321	Weld tube; boiler, superheater, heat exch	0.08 max	17.0-20.0	2.00 max		9.00-13.00	0.04 max	0.03 max	0.75 max	Ti 5xC-0.70; bal Fe	515	205	35	90 HRB
ASTM A313/A313M(95)	321	Spring wire, t<=0.25mm	0.08 max	17.00-19.00	2.00 max		9.00-12.00	0.045 max	0.030 max	1.00 max	Ti>=5xC; bal Fe	1690-1895			
ASTM A336/A336M(98)	F321	Frg, press/high-temp	0.08 max	17.0 min	2.50 max		9.00 min	0.035 max	0.030 max	0.85 max	Ti 5xC-0.60; bal Fe	485	205	30	90 HRB
ASTM A554(94)	MT321	Weld mech tub, Rnd Ann	0.08 max	17.0-20.0	2.00 max		9.0-13.0	0.040 max	0.030 max	1.00 max	Ti>=5xC<=0.60; bal Fe	517	207	35	
ASTM A580/A580M(98)	321	Wir, Ann	0.08 max	17.0-19.00	2.00 max		9.00-12.00	0.045 max	0.030 max	1.00 max	5xC<=Ti; bal Fe	520	210	35	
ASTM A632(90)	TP321	Smls Weld tub, Ti-stabilized	0.08 max	17.0-20.0	2.00 max		9.0-13.0	0.040 max	0.030 max	0.75 max	Ti>=5xC<= 0.60; bal Fe	515	205	35	
ASTM A774/A774M(98)	TP321	Fittings; Corr service at low/mod-temp, Ti-stabilized	0.08 max	17.0-19.0	2.00 max		9.0-12.0	0.045 max	0.030 max	1.00 max	Ti>=5xC<=0.70; bal Fe	515-690	205	40.0	217 HB
ASTM A778(90)	TP321	Ti-stabilized	0.030 max	17.00-19.00	2.00 max		9.00-12.00	0.045 max	0.030 max	1.00 max	N 0; bal Fe	515	205	40	
ASTM A813/A813M(95)	TP321	Weld Pipe, Ti-stabilized	0.08 max	17.0-20.0	2.00 max		9.00-13.0	0.045 max	0.030 max	0.75 max	Ti>=5xC<=0.70, bal Fe	515	205		
ASTM A814/A814M(96)	TP321	CW Weld Pipe, Ti-stabilized	0.08 max	17.0-20.0	2.00 max		9.00-13.0	0.045 max	0.030 max	0.75 max	Ti>=5xC<=0.70, bal Fe	515	205		
ASTM A943/A943M(95)	TP321	Spray formed smls pipe	0.08 max	17.0-20.0	2.00 max		9.00-13.0	0.040 max	0.030 max	0.75 max	Ti>=5xC<=0.70; bal Fe	485-515	170-205		
ASTM A965/965M(97)	F321	Frg for press high-temp parts	0.08 max	17.0 min	2.00 max		9.0 min	0.035 max	0.030 max	0.85 max	Ti (5xC)-0.60; bal Fe	485	205	30	
MIL-T-8973(69)	321*	Obs for new design; Tube; Corr/Heat res; aerospace	0.08 max	17.0-20.0	2.00 max	0.50 max	9.0-12.0	0.040 max	0.030 max	1.00 max	Cu <=0.50; Ti:C >= 5:1; bal Fe				
SAE J405(98)	S32100	Ti-stabilized	0.08 max	17.00-19.00	2.00 max		9.00-12.00	0.045 max	0.030 max	0.75 max	N <=0.10; Ti>=5xC; bal Fe				
SAE J467(68)	321		0.04	18.5	1.00		11.0			0.40	Ti 0.40; bal Fe	586	228	58	80 HRB

Stainless Steel, Austenitic, 321H

Japan

Specification	Designation	Notes	C	Cr	Mn	Mo	Ni	P	S	Si	Other	UTS	YS	El	Hard
JIS G3214(91)	SUSF321H	Frg press ves 130-200mm diam	0.04-0.10	17.00 min	2.00 max		9.00-12.00	0.040 max	0.030 max	1.00 max	Ti 4xC-0.60; bal Fe	480	205	29	187 HB max

Mexico

Specification	Designation	Notes	C	Cr	Mn	Mo	Ni	P	S	Si	Other	UTS	YS	El	Hard
NMX-B-186-SCFI(94)	TP321H	Pipe; High-temp; <=9.5mm	0.04-0.10	17.0-20.0	2.00 max		9.00-13.0	0.040 max	0.030 max	0.75 max	Ti>=4xC, 0.60 max; bal Fe	517	207		
NMX-B-196(68)	TP321H	Smls tube for refinery	0.04-0.10	17.0-20.0	2.00 max		9.0-13.0	0.040 max	0.030 max	0.75 max	Ti>=4xC, 0.60 max; bal Fe	520	205	35	

Romania

Specification	Designation	Notes	C	Cr	Mn	Mo	Ni	P	S	Si	Other	UTS	YS	El	Hard
STAS 11523(87)	12TiNiCr180	Corr res; Heat res	0.12 max	17-19	2 max	0.2 max	9-11.5	0.045 max	0.03 max	1 max	Pb <=0.15; Ti <=0.8; 5C<=Ti<=0.8; bal Fe				

UK

Specification	Designation	Notes	C	Cr	Mn	Mo	Ni	P	S	Si	Other	UTS	YS	El	Hard
BS 1501/3(73)	321S49	Press ves; Corr res	0.04-0.09	17.0-19.0	0.5-2.0	0.15 max	9.0-12.0	0.04 max	0.03 max	0.2-1.0	Cu <=0.3; Ti <=0.7; W <=0.1; Ti=5C-0.7; bal Fe				

USA

Specification	Designation	Notes	C	Cr	Mn	Mo	Ni	P	S	Si	Other	UTS	YS	El	Hard
	UNS S32109	Ti-stabilized High-C	0.04-0.10	17.00-20.00	2.00 max		9.00-12.00	0.040 max	0.030 max	1.00 max	Ti 4xC-0.60; bal Fe				
ASTM A182/A182M(98)	F321H	Frg/roll pipe flange valve	0.04-0.10	17.0 min	2.00 max		9.0-12.0	0.045 max	0.030 max	1.00 max	bal Fe	515	205	30	
ASTM A213/A213M(95)	TP321H	Smls tube boiler, superheater, heat exchanger	0.04-0.10	17.0-20.0	2.00 max		9.00-13.00	0.040 max	0.030 max	0.75 max	Ti 4xC-0.60; bal Fe	515	205	35	
ASTM A240/A240M(98)	S32109	Ti-stabilized High-C	0.04-0.10	17.00-20.00	2.00 max		9.00-12.00	0.040 max	0.030 max	1.00 max	4C<=Ti<=0.60; bal Fe	515	205	40.0	217 HB
ASTM A249/249M(96)	TP321H	Weld tube; boiler, superheater, heat exch	0.04-0.10	17.0-20.0	2.00 max		9.00-13.00	0.04 max	0.03 max	0.75 max	Ti 4xC-0.60; bal Fe	515	205	35	90 HRB
ASTM A336/A336M(98)	F321H	Frg, press/high-temp	0.04-0.10	17.0 min	2.00 max		9.00-12.00	0.040 max	0.030 max	1.00 max	Ti 4xC-0.60; bal Fe	485	205	30	90 HRB
ASTM A813/A813M(95)	TP321H	Weld Pipe, Ti-stabilized High-C	0.04-0.10	17.0-20.0	2.00 max		9.00-13.0	0.045 max	0.030 max	0.75 max	Ti>=4xC<=0.60, bal Fe	515	205		

Specification	Designation	Notes	C	Cr	Mn	Mo	Ni	P	S	Si	Other	UTS	YS	El	Hard

Stainless Steel, Austenitic, 321H (Continued from previous page)

USA

Specification	Designation	Notes	C	Cr	Mn	Mo	Ni	P	S	Si	Other	UTS	YS	El	Hard
ASTM A814/A814M(96)	TP321H	CW Weld Pipe, Ti-stabilized High-C	0.04-0.10	17.0-20.0	2.00 max		9.00-13.0	0.045 max	0.030 max	0.75 max	Ti>=4xC<=0.60, bal Fe	515	205		
ASTM A943/A943M(95)	TP321H	Spray formed smls pipe	0.04-0.10	17.0-20.0	2.00 max		9.00-13.0	0.040 max	0.030 max	0.75 max	Ti>=4xC<=0.60; bal Fe	485-515	170-205		
ASTM A965/965M(97)	F321H	Frg for press high-temp parts	0.04-0.10	17.0 min	2.00 max		9.0-12.0	0.040 max	0.030 max	1.00 max	Ti (4xC)-0.60; bal Fe	485	205	30	
SAE J405(98)	S32109	Ti-stabilized High-C	0.04-0.10	17.00-19.00	2.00 max		9.00-12.00	0.045 max	0.030 max	0.75 max	Ti 4xC-0.60; bal Fe				

Stainless Steel, Austenitic, 330

International

Specification	Designation	Notes	C	Cr	Mn	Mo	Ni	P	S	Si	Other	UTS	YS	El	Hard
ISO 4955(94)	H17	Heat res	0.08 max	17.0-20.0	2.0 max	0.15 max	34.0-37.0	0.045 max	0.03 max	0.75-1.5	Al <=0.1; Co <=0.1; Cu <=0.3; Pb <=0.15; Ti <=0.05; V <=0.1; W <=0.1; bal Fe				

Japan

Specification	Designation	Notes	C	Cr	Mn	Mo	Ni	P	S	Si	Other	UTS	YS	El	Hard
JIS G4309	SUH330	Wir	0.15 max	14.00-17.00	2.00 max		33.00-37.00	0.040 max	0.030 max	1.50 max	bal Fe				
JIS G4311	SUH330	Heat res Bar <=180mm	0.15 max	14.00-17.00	2.00 max		33.00-37.00	0.040 max	0.030 max	1.50 max	bal Fe	560	205	40	201 HB
JIS G4312	SUH330	Heat res Plt Sh	0.15 max	14.00-17.00	2.00 max		33.00-37.00	0.040 max	0.030 max	1.50 max	bal Fe	560	205	35	201 HB max

Poland

Specification	Designation	Notes	C	Cr	Mn	Mo	Ni	P	S	Si	Other	UTS	YS	El	Hard
PNH86020	H16N36S2	Heat res	0.15 max	15-17	2 max		34-37	0.045 max	0.03 max	1.5-2	bal Fe				

USA

Specification	Designation	Notes	C	Cr	Mn	Mo	Ni	P	S	Si	Other	UTS	YS	El	Hard
	AISI 330	Tube	0.08 max	17.00-20.00	2.00 max		34.00-37.00	0.040 max	0.030 max	0.75-1.50	bal Fe				
	UNS N08330		0.08 max	17.00-20.00	2.00 max		34.0-37.0	0.03 max	0.03 max	0.75-1.50	Cu <=1.00; Pb <=0.005; Sn 0.025 max; bal Fe				
ASTM A554(94)	MT330	Weld mech tub, Rnd Ann	0.15 max	14.0-16.0	2.00 max		33.0-36.0	0.040 max	0.030 max	1.00 max	bal Fe	517	207	35	

Stainless Steel, Austenitic, 334

USA

Specification	Designation	Notes	C	Cr	Mn	Mo	Ni	P	S	Si	Other	UTS	YS	El	Hard
	UNS S33400		0.080 max	18.0-20.0	1.00 max		19.0-21.0	0.030 max	0.015 max	1.00 max	Al 0.15-0.60; Ti 0.15-0.60; bal Fe				

Stainless Steel, Austenitic, 347

Austria

Specification	Designation	Notes	C	Cr	Mn	Mo	Ni	P	S	Si	Other	UTS	YS	El	Hard
ONORM M3120	X6CrNiNb1810S	Sh Strp Plt Bar Wir Bil Tub	0.08 max	17-19	2 max		9.0-12.0	0.045 max	0.030 max	1 max	Nb>=10xC<=1.00; bal Fe				

Bulgaria

Specification	Designation	Notes	C	Cr	Mn	Mo	Ni	P	S	Si	Other	UTS	YS	El	Hard
BDS 6738(72)	0Ch18N12B	Corr res	0.08 max	17.0-19.0	2.00 max	0.30 max	10.0-13.0	0.035 max	0.025 max	0.80 max	10xC<=Nb<=1.10; bal Fe				

Canada

Specification	Designation	Notes	C	Cr	Mn	Mo	Ni	P	S	Si	Other	UTS	YS	El	Hard
CSA G110.3	347	Bar Bil	0.08 max	17-19	2 max		9-13	0.045 max	0.03 max	1 max	Nb+Ta>=10xC; bal Fe				
CSA G110.6	347	Sh Strp Plt, HR, Ann or HR/Q/A/Tmp	0.08 max	17-19	2 max		9-13	0.045 max	0.03 max	1 max	Nb+Ta>=10xC<=1.10; bal Fe				
CSA G110.9	347	Sh Strp Plt, HR, CR, Ann	0.08 max	17-19	2 max		9-13	0.045 max	0.03 max	1 max	Nb+Ta>=10xC<=1.10; bal Fe				

China

Specification	Designation	Notes	C	Cr	Mn	Mo	Ni	P	S	Si	Other	UTS	YS	El	Hard
GB 1220(92)	0Cr18Ni11Nb	Bar SA	0.08 max	17.00-19.00	2.00 max		9.00-13.00	0.035 max	0.030 max	1.00 max	Nb>=10xC; bal Fe	520	205	40	
GB 1221(92)	0Cr18Ni11Nb	Bar SA	0.08 max	17.00-19.00	2.00 max		9.00-13.00	0.035 max	0.030 max	1.00 max	Nb>=10xC; bal Fe	520	205	40	
GB 12770(91)	0Cr18Ni11Nb	Weld tube; SA	0.08 max	17.00-19.00	2.00 max		9.00-13.00	0.035 max	0.030 max	1.00 max	Nb>=10xC; bal Fe	520	210	35	
GB 12771(91)	0Cr18Ni11Nb	Weld Pipe SA	0.08 max	17.00-19.00	2.00 max		9.00-13.00	0.035 max	0.030 max	1.00 max	Nb>=10xC; bal Fe	520	210	35	
GB 13296(91)	0Cr18Ni11Nb	Smls tube SA	0.08 max	17.00-19.00	2.00 max		9.00-13.00	0.035 max	0.030 max	1.00 max	Nb>=10xC; bal Fe	520	205	35	
GB 3280(92)	0Cr18Ni11Nb	Sh Plt, CR SA	0.08 max	17.00-19.00	2.00 max		9.00-13.00	0.035 max	0.030 max	1.00 max	Nb>=10xC; bal Fe	520	205	40	

UNS numbers and US grades are provided as a means of cross referencing chemically similar alloys. Exchangability is only possible after independent examination of specifications. Tensile properties are minimum or typical as specified. UTS and YS as MPa. El as %. See Appendix for list of abbreviations used in Notes. * indicates obsolete material.

Specification	Designation	Notes	C	Cr	Mn	Mo	Ni	P	S	Si	Other	UTS	YS	El	Hard

Stainless Steel, Austenitic, 347 (Continued from previous page)

China

Specification	Designation	Notes	C	Cr	Mn	Mo	Ni	P	S	Si	Other	UTS	YS	El	Hard
GB 4237(92)	0Cr18Ni11Nb	Sh Plt, HR SA	0.08 max	17.00-19.00	2.00 max		9.00-13.00	0.035 max	0.030 max	1.00 max	Nb>=10xC; bal Fe	520	205	40	
GB 4238(92)	0Cr18Ni11Nb	Plt, HR SA	0.08 max	17.00-19.00	2.00 max		9.00-13.00	0.035 max	0.030 max	1.00 max	Nb>=10xC; bal Fe	520	205	40	
GB 4239(91)	0Cr18Ni11Nb	Sh Strp, CR SA	0.08 max	17.00-19.00	2.00 max		9.00-13.00	0.035 max	0.030 max	1.00 max	Nb>=10xC; bal Fe	520	205	40	
GB 4240(93)	0Cr18Ni11Nb	Wir Ann 0.6-1mm diam	0.08 max	17.00-19.00	2.00 max		9.00-13.00	0.035 max	0.030 max	1.00 max	Nb>=10xC; bal Fe	540-880			
GB 5310(95)	1Cr19Ni11Nb	Smls tube Pip SA	0.04-0.10	17.00-20.00	2.00 max		9.00-13.00	0.030 max	0.030 max	1.00 max	Cu 0.20; Nb+Ta>=8xC<=1.0; bal Fe	520	205	35	
GB/T 14975(94)	0Cr18Ni11Nb	Smls tube SA	0.08 max	17.00-19.00	2.00 max		9.00-13.00	0.035 max	0.030 max	1.00 max	Nb>=10xC; bal Fe	520	205	35	
GB/T 14976(92)	0Cr18Ni11Nb	Smls pipe SA	0.08 max	17.00-19.00	2.00 max		9.00-13.00	0.035 max	0.030 max	1.00 max	Nb>=10xC; bal Fe	520	205	35	

Europe

Specification	Designation	Notes	C	Cr	Mn	Mo	Ni	P	S	Si	Other	UTS	YS	El	Hard
EN 10088/2(95)	1.4550	Strp, CR Corr res; t<=6mm, Ann	0.08 max	17.00-19.00	2.00 max		9.00-12.00	0.045 max	0.015 max	1.00 max	10xC<=Nb<=1.00; bal Fe	520-720	220	40	
EN 10088/2(95)	X6CrNiNb18-10	Strp, CR Corr res; t<=6mm, Ann	0.08 max	17.00-19.00	2.00 max		9.00-12.00	0.045 max	0.015 max	1.00 max	10xC<=Nb<=1.00; bal Fe	520-720	220	40	
EN 10088/3(95)	1.4550	Bar Rod Sect, t<=160mm, Ann	0.08 max	17.00-19.00	2.00 max		9.00-12.00	0.045 max	0.015 max	1.00 max	10xC<=Nb<=1.00; bal Fe	510-740	205	40	230 HB
EN 10088/3(95)	1.4550	Bar Rod Sect; Corr res; 160<t<=250mm, Ann	0.08 max	17.00-19.00	2.00 max		9.00-12.00	0.045 max	0.015 max	1.00 max	10xC<=Nb<=1.00; bal Fe	510-740	205	30	230 HB
EN 10088/3(95)	X6CrNiNb18-10	Bar Rod Sect, t<=160mm, Ann	0.08 max	17.00-19.00	2.00 max		9.00-12.00	0.045 max	0.015 max	1.00 max	10xC<=Nb<=1.00; bal Fe	510-740	205	40	230 HB
EN 10088/3(95)	X6CrNiNb18-10	Bar Rod Sect; Corr res; 160<t<=250mm, Ann	0.08 max	17.00-19.00	2.00 max		9.00-12.00	0.045 max	0.015 max	1.00 max	10xC<=Nb<=1.00; bal Fe	510-740	205	30	230 HB

France

Specification	Designation	Notes	C	Cr	Mn	Mo	Ni	P	S	Si	Other	UTS	YS	El	Hard
AFNOR NFA35573	Z6CNNb18.10	Sh Strp Plt, CR, 5mm diam	0.08 max	17-19	2 max		9-11	0.04 max	0.03 max	1 max	Nb+Ta>=10xC<=1.0; bal Fe				
AFNOR NFA35574	Z6CNNb18.10	Bar Wir Bil	0.08 max	17-19	2 max		9-11	0.04 max	0.03 max	1 max	Nb+Ta>=10xC<=1.0; bal Fe				
AFNOR NFA36209	Z6CNNb18.10	Plt	0.08 max	17-20	2 max		9-12	0.04 max	0.03 max	1 max	Nb+Ta>=10xC<=1.0; bal Fe				
AFNOR NFA36607	Z6CNNb18.10	Frg	0.08 max	17-20	2 max	0.5 max	9-12	0.04 max	0.02 max	1 max	Cu <=1; Nb+Ta>=10xC<=1.0; bal Fe				

Germany

Specification	Designation	Notes	C	Cr	Mn	Mo	Ni	P	S	Si	Other	UTS	YS	El	Hard
DIN 17440(96)	WNr 1.4550	Plt Sh, HR Strp Bar Wir Frg	0.08 max	17.0-19.0	2.00 max		9.00-12.0	0.045 max	0.015 max	1.00 max	Nb 10xC>=1.00; bal Fe	510-740	205	30-40	
DIN 17441(97)	WNr 1.4550	CR Strp Plt Sh SHT	0.08 max	17.0-19.0	2.00 max		9.00-12.0	0.045 max	0.015 max	1.00 max	Nb 10xC>=1.00; bal Fe	510-740	205	30-40	
DIN 17455(85)	WNr 1.4550	Weld tube	0.08 max	17.0-19.0	2.00 max		9.00-12.0	0.045 max	0.015 max	1.00 max	Nb 10xC>=1.00; bal Fe	510-740	205	30-40	
DIN 17456(85)	WNr 1.4550	Smls tube	0.08 max	17.0-19.0	2.00 max		9.00-12.0	0.045 max	0.015 max	1.00 max	Nb 10xC>=1.00; bal Fe	510-740	205	30-40	
DIN 17459(92)	WNr 1.4961	Smls tube	0.04-0.10	15.0-17.0	1.50 max		12.0-14.0	0.035 max	0.015 max	0.30-0.60	Nb 10xC>=1.20; bal Fe	510-690	205	35	
DIN 17459(92)	X8CrNiNb16-13	Smls tube	0.04-0.10	15.0-17.0	1.50 max		12.0-14.0	0.035 max	0.015 max	0.30-0.60	Nb 10xC<=1.20; bal Fe	510-690	205	35	
DIN EN 10028(96)	WNr 1.4961	Flat product for Press purposes	0.04-0.10	15.0-17.0	1.50 max		12.0-14.0	0.035 max	0.015 max	0.30-0.60	Nb 10xC>=1.20; bal Fe	510-690	205	35	
DIN EN 10028(96)	X8CrNiNb16-13	Flat product for Press purposes	0.04-0.10	15.0-17.0	1.50 max		12.0-14.0	0.035 max	0.015 max	0.30-0.60	Nb 10xC<=1.20; bal Fe	510-690	205	35	
DIN EN 10088(95)	X6CrNiNb18-10	Sh Plt Strp Bar Rod Sect, SHT	0.08 max	17.0-19.0	2.00 max		9.00-12.0	0.045 max	0.015 max	1.00 max	Nb 10xC<=1.00; bal Fe	510-740	205	30-40	
TGL 39672	X5CrNiNb2010*	Obs	0.07 max	19.5-20.5	1-1.5		9.5-10.5	0.025 max	0.015 max	0.7 max	Nb 12xC<=1.00; bal Fe				

Hungary

Specification	Designation	Notes	C	Cr	Mn	Mo	Ni	P	S	Si	Other	UTS	YS	El	Hard
MSZ 4360(80)	KO34	Corr res	0.08 max	17-19	2 max	0.5 max	9-11.5	0.04 max	0.03 max	1 max	Nb=8C-1.1; bal Fe				
MSZ 4360(87)	X8CrNiNb1810	Corr res; 0-160mm; SA	0.08 max	17-19	2 max	0.5 max	9-12	0.045 max	0.03 max	1 max	Nb+Ta=10C-1.1; bal Fe	500-740	205	40L; 30T	190 HB max

Stainless Steel, Austenitic, 347 (Continued from previous page)

Specification	Designation	Notes	C	Cr	Mn	Mo	Ni	P	S	Si	Other	UTS	YS	El	Hard
India															
IS 1570/5(85)	X04Cr18Ni10Nb	Sh Plt Strp Bar Flat Band	0.08 max	17-19	2 max	0.15 max	9-12	0.045 max	0.03 max	1 max	Co <=0.1; Cu <=0.3; Pb <=0.15; W <=0.1; 10C<=Nb<=1; bal Fe	515	205	40	183 HB max
IS 1570/5(85)	X04Cr18Ni10Ti	Sh Plt Strp Bar Flat Band	0.08 max	17-19	2 max	0.15 max	9-12	0.045 max	0.03 max	1 max	Co <=0.1; Cu <=0.3; Pb <=0.15; Ti <=0.8; W <=0.1; 5C<=Ti<=0.8; bal Fe	515	205	40	183 HB max
IS 6529	04Cr18Ni10Nb40	Bil	0.08 max	17-19	2 max		9-12	0.05 max	0.03 max	1 max	10xC<=Nb<=1.00; bal Fe				
IS 6911	04Cr18Ni10Nb40	Sh Strp Plt	0.08 max	17-19	2 max		9-12	0.05 max	0.03 max	1 max	10xC<=Nb<=1.00; bal Fe				
International															
COMECON PC4-70	20	Sh Strp Plt Bar Wir Bil Tub	0.08 max	17-19	2 max		9-13	0.035 max	0.025 max	0.8 max	Nb>=10xC<=1.10; bal Fe				
ISO 683-13(74)	16*	Sh Strp Plt Bar Wir Tub, ST	0.08 max	17-19	2 max		9-12	0.045 max	0.03 max	1 max	Nb>= 10xC < 1.00; bal Fe				
Italy															
UNI 6901(71)	X6CrNiNb1811	Corr res	0.08 max	17-19	2 max		9-12	0.045 max	0.03 max	1 max	Nb+Ta=10C-1; bal Fe				
UNI 6901(71)	X8CrNiNb1811	Corr res	0.04-0.1	17-19	2 max		9-13	0.03 max	0.03 max	0.75 max	Nb+Ta=10C-1; bal Fe				
UNI 6904(71)	X6CrNiNb1811	Tube	0.08 max	17-19	2 max		9-13	0.03 max	0.03 max	0.75 max	Nb>=10xC<=1.00; bal Fe				
UNI 6904(71)	X6CrNiNb1811	Heat res	0.08 max	17-19	2 max		9-13	0.03 max	0.03 max	0.75 max	Nb+Ta=10C-1; bal Fe				
UNI 6904(71)	X8CrNiNb1811	Heat res	0.04-0.1	17-19	2 max		9-13	0.03 max	0.03 max	0.75 max	Nb+Ta8C-1; bal Fe				
UNI 7500(75)	X6CrNiNb1811	Press ves	0.08 max	17-19	2 max		9-12	0.045 max	0.03 max	1 max	Nb+Ta=10C-1; bal Fe				
UNI 7660(71)	X6CrNiNb1811KG		0.08 max	17-19	2 max		9-13	0.045 max	0.03 max	1 max	Nb>=10xC<=1.00; bal Fe				
UNI 7660(71)	X6CrNiNb1811KT		0.08 max	17-19	2 max	0.5 max	10-12	0.045 max	0.03 max	1 max	Nb>=10xC<=1.00; bal Fe				
UNI 7660(71)	X6CrNiNb1811KW		0.08 max	17-19	2 max		9-13	0.045 max	0.03 max	1 max	Nb>=10xC<=1.00; bal Fe				
UNI 8317(81)	X6CrNiNb1811	Sh Strp Plt	0.04-0.1	17-19	2 max		9-13	0.03 max	0.03 max	0.75 max	Nb>=10xC<=1.00; bal Fe				
Japan															
JIS G3214(91)	SUSF347	Frg press ves 130-200mm diam	0.08 max	17.00-20.00	2.00 max		9.00-13.00	0.040 max	0.030 max	1.00 max	Nb 10xC to 1.00; bal Fe	480	205	29	187 HB max
JIS G3446(94)	SUS347TKA	Tube, 7mm<t<8mm	0.08 max	18.00-20.00	2.00 max		8.00-11.00	0.040 max	0.030 max	1.00 max	Nb 10xC min; bal Fe	520-720	205	35	
JIS G3459(94)	SUS347TP	Pipe	0.08 max	17-19	2 max		9-13	0.04 max	0.03 max	1 max	Nb>=10xC; bal Fe	520	205	35	
JIS G3463(94)	SUS347TB	Tube	0.08 max	17.00-19.00	2.00 max		9.00-13.00	0.040 max	0.030 max	1.00 max	Nb 10xC min; bal Fe	520	205	35	90 HRB max
JIS G4303(91)	SUS347	Bar; SA; <=180mm diam	0.08 max	17.00-19.00	2.00 max		9.00-13.00	0.045 max	0.030 max	1.00 max	Nb 10xC min; bal Fe	520	205	40	187 HB
JIS G4304(91)	SUS347	Sh Plt, HR SA	0.08 max	17.00-19.00	2.00 max		9.0-13.00	0.045 max	0.030 max	1.00 max	Nb>=10xC; bal Fe	520	205	40	90 HRB max
JIS G4305(91)	SUS347	Sh Plt, CR SA	0.08 max	17.00-19.00	2.00 max		9.00-13.00	0.045 max	0.030 max	1.00 max	Nb>=10xC; bal Fe	520	205	40	187 HB max
JIS G4306	SUS347	Strp HR SA	0.08 max	17-19	2 max		9-13	0.04 max	0.03 max	1 max	Nb>=10xC; bal Fe				
JIS G4307	SUS347	Strp CR SA	0.08 max	17-19	2 max		9-13	0.04 max	0.03 max	1 max	Nb>=10xC; bal Fe				
JIS G4308(98)	SUS347	Wir rod	0.08 max	17.00-19.00	2.00 max		9.00-13.00	0.040 max	0.030 max	1.00 max	bal Fe				
JIS G4309	SUS347	Wir	0.08 max	17.00-19.00	2.00 max		9.00-13.00	0.045 max	0.030 max	1.00 max	Nb 10xC min; bal Fe				
Mexico															
DGN B-150	AE347*	Obs; Wire	0.08 max	19-21.5	1-2.5		9-11	0.03 max	0.03 max	0.25-0.6	Cu <=0.03; Nb 0.85-1.1; bal Fe				
NMX-B-171(91)	MT347	mech tub	0.08 max	17.0-20.0	2.00 max		9.0-13.0	0.040 max	0.030 max	1.00 max	(Nb+Ta)=(10xC) min, 1.00 max; bal Fe				
NMX-B-186-SCFI(94)	TP347	Smls pipe for high-temp central station service	0.08 max	17.0-20.0	2.00 max		9.00-13.0	0.040 max	0.030 max	0.75 max	Nb+Ta=10xC min, 1.00 max; bal Fe	517	207		

UNS numbers and US grades are provided as a means of cross referencing chemically similar alloys. Exchangability is only possible after independent examination of specifications. Tensile properties are minimum or typical as specified. UTS and YS as MPa. El as %. See Appendix for list of abbreviations used in Notes. * indicates obsolete material.

Stainless Steel, Austenitic, 347 (Continued from previous page)

Specification	Designation	Notes	C	Cr	Mn	Mo	Ni	P	S	Si	Other	UTS	YS	El	Hard
Mexico															
NMX-B-196(68)	TP347	Smls tube for refinery	0.08 max	17.0-20.0	2.00 max		9.0-13.0	0.040 max	0.030 max	0.75 max	Nb+Ta=10xC min, 1.0 max				
Pan America															
COPANT 513	TP347	Tube, HF, CF, 8mm diam	0.08 max	17-20	2 max		9-13	0.03 max	0.03 max	1 max	Nb+Ta>=10C<=1.00; bal Fe				
COPANT R195	TP347	Tube, HF, CF, 5.65mm diam	0.08 max	17-20	2 max		9-13	0.04 max	0.03 max	0.75 max	Nb+Ta>=10C<=1.00; bal Fe				
Poland															
PNH86020	0H18N12Nb	Corr res	0.08 max	17-19	2 max		10-13	0.045 max	0.03 max	0.8 max	Nb>=10xC<=1.10; bal Fe				
Romania															
STAS 3583(64)	7NbNC180	Corr res; Heat res	0.08 max	17-19	1-2		11-13	0.035 max	0.03 max	0.8 max	Nb <=1.2; Pb <=0.15; bal Fe				
Russia															
GOST	O8Ch18N12B	Sh Strp Plt Bar Wir Bil Tub	0.08 max	17-19	2 max	0.1 max	11-13	0.035 max	0.02 max	0.8 max	Cu <=0.3; Ti <=0.2; W <=0.2; Nb>=10xC<=1.00; bal Fe				
GOST 5632(61)	0KH18N12B	Corr res; Heat res; High-temp constr	0.08 max	17.0-19.0	1.0-2.0	0.15 max	11.0-13.0	0.035 max	0.02 max	0.8 max	Cu <=0.3; Ti <=0.05; Nb=10C-1.2; bal Fe				
GOST 5632(72)	08Ch18N12B	Corr res	0.08 max	17.0-19.0	2.0 max	0.3 max	11.0-13.0	0.035 max	0.02 max	0.8 max	Cu <=0.3; Ti <=0.2; W <=0.2; Nb=10C-1.1; bal Fe				
Spain															
UNE 36016(75)	F.3524*	see X6CrNiNb18-11	0.08 max	17.0-19.0	2.0 max	0.15 max	9.0-12.0	0.045 max	0.03 max	1.0 max	10xC<=Nb<=1.00; bal Fe				
UNE 36016(75)	X6CrNiNb18-11	Plt Bar Wir Bil Tub; Corr res	0.08 max	17.0-19.0	2.0 max	0.15 max	9.0-12.0	0.045 max	0.03 max	1.0 max	10xC<=Nb<=1.00; bal Fe				
UNE 36016(75)	X6CrNiNb18-11	Plt Bar Wir Bil Tub; Corr res	0.08 max	17.0-19.0	2.0 max	0.15 max	9.0-12.0	0.045 max	0.03 max	1.0 max	10xC<=Nb<=1.00; bal Fe				
UNE 36016/1(89)	E-347	Corr res	0.08 max	17.0-19.0	2.0 max	0.15 max	9.0-12.0	0.045 max	0.03 max	1.0 max	N <=0.1; Nb=10C-1; bal Fe				
UNE 36016/1(89)	F.3524*	Obs; E-347; Corr res	0.08 max	17.0-19.0	2.0 max	0.15 max	9.0-12.0	0.045 max	0.03 max	1.0 max	N <=0.1; 10xC<=Nb<=1.00; bal Fe				
UNE 36087(78)	F.3552*	see X7CrNiNb18-11	0.1 max	17.0-19.0	2.0 max	0.5 max	10.0-12.0	0.045 max	0.03 max	1.0 max	8xC<=Nb<=1.00; bal Fe				
UNE 36087(78)	X7CrNiNb18-11	Corr res; Sh Strp	0.1 max	17.0-19.0	2.0 max	0.5 max	10.0-12.0	0.045 max	0.03 max	1.0 max	8xC<=Nb<=1.00; bal Fe				
UNE 36087(78)	X7CrNiNb18-11	Corr res; Sh Strp	0.1 max	17.0-19.0	2.0 max	0.5 max	10.0-12.0	0.045 max	0.03 max	1.0 max	8xC<=Nb<=1.00; bal Fe				
UNE 36087/4(89)	F.3524*	see X6CrNiNb18 10	0.08 max	17-19	2 max	0.15 max	9-12	0.045 max	0.03 max	1 max	N <=0.1; 10xC<=Nb<=1.00; bal Fe				
UNE 36087/4(89)	X6CrNiNb18 10	Corr res; Sh Strp	0.08 max	17-19	2 max	0.15 max	9-12	0.045 max	0.03 max	1 max	N <=0.1; 10xC<=Nb<=1.00; bal Fe				
UK															
BS 1449/2(83)	347S31	Corr/Heat res; t<=100mm	0.08 max	17.0-19.0	2.00 max		9.00-12.0	0.045 max	0.030 max	1.00 max	10xC<=Nb<=1.00; bal Fe	510	205	35	0-200 HB
BS 1501	347S17	Plate	0.08 max	17.0-19.0	0.5-2.0		9.0-12.0	0.045 max	0.03 max	0.2-1.0	10xC<=Nb<=1.00; bal Fe				
BS 1501/3(73)	347S49	Press ves; Corr res	0.04-0.09	17.0-19.0	0.5-2.0	0.15 max	9.0-12.0	0.04 max	0.03 max	0.2-1.0	Cu <=0.03; Ti <=0.05; W <=0.1; Nb=10C-1; bal Fe				
BS 1501/3(90)	347S31	Press ves; Corr/Heat res; 0.05-100mm	0.08 max	17.0-19.0	2.0 max	0.15 max	9.0-12.0	0.045 max	0.025 max	1.0 max	Cu <=0.3; Ti <=0.05; W <=0.1; Nb=10C-1; bal Fe	510-710	205	30	
BS 1502	347S31	Bar Shp; Press ves	0.08 max	17.0-19.0	2.00 max		9.00-12.0	0.045 max	0.030 max	1.00 max	10xC<=Nb<=1.00; bal Fe				
BS 1502(82)	347S51	Bar Shp; Press ves	0.04-0.10	17.0-19.0	2.00 max		9.00-12.0	0.045 max	0.030 max	1.00 max	W <=0.1; 10xC<=Nb<=1.20; bal Fe				
BS 1503(89)	347S31	Frg press ves; t<=350mm; Quen	0.08 max	17.0-19.0	2.00 max	0.70 max	9.00-12.0	0.040 max	0.025 max	1.00 max	B <=0.005; Cu <=0.50; W <=0.1; 10xC<=Nb<=1.00; bal Fe	510-710	240	30	

Stainless Steel, Austenitic, 347 (Continued from previous page)

Specification	Designation	Notes	C	Cr	Mn	Mo	Ni	P	S	Si	Other	UTS	YS	El	Hard
UK															
BS 1506(90)	347S31	Bolt matl Pres/Corr res; t<=160mm; Hard; was En821(Nb)	0.08 max	17.0-19.0	2.00 max		9.00-12.0	0.045 max	0.030 max	1.00 max	10xC<=Nb<=1.00; bal Fe	520	205	30	
BS 1506(90)	347S31	Bolt matl Pres/Corr res; Corr res; 38<t<=44mm; SHT CD; wasEn821(Nb)	0.08 max	17.0-19.0	2.00 max		9.00-12.0	0.045 max	0.030 max	1.00 max	10xC<=Nb<=1.00; bal Fe	650	310	28	0-320 HB
BS 1506(90)	347S51	Bolt matl Pres/Corr res; t<=160mm; Hard	0.04-0.10	17.0-19.0	2.00 max		9.00-12.0	0.045 max	0.030 max	1.00 max	10xC<=Nb<=1.20; bal Fe	520	205	30	
BS 1554(90)	347S20	Wir Corr/Heat res; 6<diam<=13mm; Ann	0.08 max	17.0-21.0	2.00 max		9.00-12.0	0.045 max	0.030 max	1.00 max	W <=0.1; 10xC<=Nb<=1.00; bal Fe	0-700			
BS 1554(90)	347S20	Wir Corr/Heat res; diam<0.49mm; Ann	0.08 max	17.0-21.0	2.00 max		9.00-12.0	0.045 max	0.030 max	1.00 max	W <=0.1; 10xC<=Nb<=1.00; bal Fe	0-900			
BS 3605(73)	347S17	Corr res; Tub, Weld; Smls tub	0.08 max	17.0-19.0	0.5-2.0		9.00-12.0	0.040 max	0.030 max	0.2-1.0	10xC<=Nb<=1.00; bal Fe				
BS 3605(73)	347S18	Corr res; Tub, Weld; Smls tub	0.08 max	17.0-19.0	0.5-2.0		10.0-13.0	0.040 max	0.030 max	0.2-1.0	10xC<=Nb<=1.00; bal Fe				
BS 3606(78)	347S17	Corr res; Heat exch Tub	0.080 max	17.0-19.0	0.5-2.0		9.0-12.0	0.045 max	0.030 max	1.00 max	5xC<=Ti<=0.80; bal Fe				
BS 3606(92)	347S31	Heat exch Tub; Smls tub, Tub, Weld; 1.2-3.2mm	0.080 max	17.0-19.0	2.00 max		9.0-13.0	0.040 max	0.030 max	1.00 max	10xC<=Nb<=1.00; bal Fe	510-710	245	30	
BS 970/1(96)	347S31	Wrought Corr/Heat res; t<=160mm	0.08 max	17.0-19.0	2.00 max		9.00-12.0	0.045 max	0.030 max	1.00 max	10xC<=Nb<=1.00; bal Fe	510	205	30	183 HB max
BS 970/4(70)	347S17*	Obs; Corr res	0.08 max	17.0-19.0	0.50-2.00	0.7 max	9.0-12.0	0.045 max	0.03 max	0.2-1.0	Cu <=0.5; Pb <=0.15; Nb=10C-1; bal Fe				
USA															
	AISI 347	Tube	0.08 max	17.00-19.00	2.00 max		9.00-13.00	0.045 max	0.030 max	1.00 max	Nb>=10xC; bal Fe				
	UNS S34700	Nb stabilized	0.08 max	17.00-19.00	2.00 max		9.00-13.00	0.045 max	0.030 max	1.00 max	Nb 10xC min; bal Fe				
AMS 5512		Sh Strp Plt Bar Wir Bil Tub	0.08 max	17-19	2 max		9-13	0.045 max	0.03 max	1 max	Nb>=10xC; bal Fe				
AMS 5556G(90)		OD<=4.78; Smls or Weld tub	0.08 max	17.00-19.00	2.00 max	0.75 max	9.00-13.00	0.040 max	0.030 max	0.50-1.00	Cu <=0.75; Nb 10xC-1.10; Ta 0.05max; bal Fe	517-827	207	33	
AMS 5556G(90)		Smls Weld tub, OD>12.7	0.08 max	17.00-19.00	2.00 max	0.75 max	9.00-13.00	0.040 max	0.030 max	0.50-1.00	Cu <=0.75; Nb 10xC-1.10; Ta 0.05max; bal Fe	517-724	207	35	
AMS 5558E(96)		Corr/Heat res; thin wall Weld tub	0.08 max	17.00-19.00	2.00 max	0.75 max	9.00-12.00	0.040 max	0.030 max	0.30-1.00	Cu <=0.75; Nb 10xC-1.10; Ta 0.05max; bal Fe	517-724	241	40	
AMS 5571		Sh Strp Plt Bar Wir Bil Tub	0.08 max	17-19	2 max		9-13	0.045 max	0.03 max	1 max	Nb>=10xC; bal Fe				
AMS 5575		Sh Strp Plt Bar Wir Bil Tub	0.08 max	17-19	2 max		9-13	0.045 max	0.03 max	1 max	Nb>=10xC; bal Fe				
AMS 5654		Sh Strp Plt Bar Wir Bil Tub	0.08 max	17-19	2 max		9-13	0.045 max	0.03 max	1 max	Nb>=10xC; bal Fe				
AMS 5674		Sh Strp Plt Bar Wir Bil Tub	0.08 max	17-19	2 max		9-13	0.045 max	0.03 max	1 max	Nb>=10xC; bal Fe				
AMS 5897(94)		Corr/Heat res; Smls or weld hydraulic tub	0.08 max	17.00-20.00	2.00 max	0.75 max	9.00-13.00	0.040 max	0.030 max	1.00 max	Cu <=0.75; Nb 10xC-1.10;Ta<=0.05; bal Fe	724-965	5517-758	20	
AMS 7229		Sh Strp Plt Bar Wir Bil Tub	0.08 max	17-19	2 max		9-13	0.045 max	0.03 max	1 max	Nb>=10xC; bal Fe				
AMS 7490		Sh Strp Plt Bar Wir Bil Tub	0.08 max	17-19	2 max		9-13	0.045 max	0.03 max	1 max	Nb>=10xC; bal Fe				
ASME SA182	347	Refer to ASTM A182/A182M													
ASME SA213	347	Refer to ASTM A213/A213M(95)													
ASME SA240	347	Refer to ASTM A240(95)													
ASME SA249	347	Refer to ASTM A249/A249M(96)													
ASME SA312	347	Refer to ASTM A312/A312M(95)													

UNS numbers and US grades are provided as a means of cross referencing chemically similar alloys. Exchangability is only possible after independent examination of specifications. Tensile properties are minimum or typical as specified. UTS and YS as MPa. El as %. See Appendix for list of abbreviations used in Notes. * indicates obsolete material.

Stainless Steel, Austenitic, 347 (Continued from previous page)

Specification	Designation	Notes	C	Cr	Mn	Mo	Ni	P	S	Si	Other	UTS	YS	El	Hard
USA															
ASME SA358	347	Refer to ASTM A358/A358M(95)													
ASME SA376	347	Refer to ASTM A376/A376M(93)													
ASME SA403	347	Refer to ASTM A403/A403M(95)													
ASME SA409	347	Refer to ASTM A409/A409M(95)													
ASME SA430	347	Refer to ASTM A430/A430M(91)													
ASME SA479	347	Refer to ASTM A479/A479M(95)													
ASTM A167(96)	347*	Sh Strp Plt	0.08 max	17-19	2 max		9-13	0.045 max	0.03 max	1 max	Nb>=10xC; bal Fe				
ASTM A182	347*	Bar Frg	0.08 max	17-19	2 max		9-13	0.045 max	0.03 max	1 max	Nb>=10xC; bal Fe				
ASTM A182/A182M(98)	F347	Frg/roll pipe flange valve	0.08 max	17.0-20.0	2.00 max		9.0-13.0	0.045 max	0.030 max	1.00 max	Nb>=10xC<=1.10; bal Fe	515	205	30	
ASTM A193/A193M(98)	347*	Bolt, high-temp	0.08 max	17-19	2 max		9-13	0.045 max	0.03 max	1 max	Nb>=10xC; bal Fe				
ASTM A193/A193M(98)	B8C	Bolt, high-temp, HT	0.08 max	17.0-19.0	2.00 max		9.0-12.0	0.045 max	0.030 max	1.00 max	10xC<=Nb+Ta<=1.10; bal Fe	515	205	30	223 HB
ASTM A193/A193M(98)	B8CA	Bolt, high-temp, HT	0.08 max	17.0-19.0	2.00 max		9.0-12.0	0.045 max	0.030 max	1.00 max	10xC<=Nb+Ta<=1.10; bal Fe	515	205	30	192 HB
ASTM A194	347*	Nuts, high-temp press	0.08 max	17-19	2 max		9-13	0.045 max	0.03 max	1 max	Nb>=10xC; bal Fe				
ASTM A194/A194M(98)	8C	Nuts, high-temp press	0.08 max	17.0-19.0	2.00 max		9.0-12.0	0.045 max	0.030 max	1.00 max	10xC<=Nb+Ta; bal Fe				126-300 HB
ASTM A194/A194M(98)	8CA	Nuts, high-temp press	0.08 max	17.0-19.0	2.00 max		9.0-12.0	0.045 max	0.030 max	1.00 max	10xC<=Nb+Ta; bal Fe				126-192 HB
ASTM A213	347*	Tube	0.08 max	17-19	2 max		9-13	0.045 max	0.03 max	1 max	Nb>=10xC; bal Fe				
ASTM A213/A213M(95)	TP347	Smls tube boiler, superheater, heat exchanger	0.08 max	17.0-20.0	2.00 max		9.00-13.00	0.040 max	0.030 max	0.75 max	Ta+Nb 10xC-1.00; bal Fe	515	205	35	
ASTM A240/A240M(98)	S34700	Nb-stabilized	0.08 max	17.0-19.0	2.00 max		9.0-13.0	0.045 max	0.030 max	0.75 max	10xC<=Nb<=1.00; bal Fe	515	205	40.0	201 HB
ASTM A249/249M(96)	TP347	Weld tube; boiler, superheater, heat exch	0.08 max	17.0-20.0	2.00 max		9.00-13.00	0.04 max	0.03 max	0.75 max	Ta+Nb 10xC-1.00; bal Fe	515	205	35	90 HRB
ASTM A269	347	Smls Weld, Tube	0.08 max	17-19	2 max		9-13	0.045 max	0.03 max	1 max	Nb>=10xC; bal Fe				
ASTM A271(96)	347	Smls tube	0.08 max	17-19	2 max		9-13	0.045 max	0.03 max	1 max	Nb>=10xC; bal Fe				
ASTM A276(98)	347	Bar	0.08 max	17-19	2 max		9-13	0.045 max	0.03 max	1 max	Nb>=10xC; bal Fe				
ASTM A312/A312M(95)	347	Smls Weld, Pipe	0.08 max	17-19	2 max		9-13	0.045 max	0.03 max	1 max	Nb>=10xC; bal Fe				
ASTM A313/A313M(95)	347	Spring wire, t<=0.25mm	0.08 max	17.00-19.00	2.00 max		9.00-13.00	0.045 max	0.030 max	1.00 max	10xC<=Nb+Ta; bal Fe	1690-1895			
ASTM A314	347	Bil	0.08 max	17-19	2 max		9-13	0.045 max	0.03 max	1 max	Nb>=10xC; bal Fe				
ASTM A320	347	Bolt	0.08 max	17-19	2 max		9-13	0.045 max	0.03 max	1 max	Nb>=10xC; bal Fe				
ASTM A336/A336M(98)	F347	Frg, press/high-temp	0.08 max	17.0-19.0	2.00 max		9.00-12.00	0.040 max	0.030 max	0.85 max	Ta+Nb 10xC-1.00; bal Fe	485	205	30	90 HRB
ASTM A358/A358M(95)	347	Elect fusion weld pipe, high-temp	0.08 max	17-19	2 max		9-13	0.045 max	0.03 max	1 max	Nb>=10xC; bal Fe				
ASTM A376	347	Smls pipe	0.08 max	17-19	2 max		9-13	0.045 max	0.03 max	1 max	Nb>=10xC; bal Fe				
ASTM A403	347	Pipe	0.08 max	17-19	2 max		9-13	0.045 max	0.03 max	1 max	Nb>=10xC; bal Fe				
ASTM A409	347	Weld, Pipe	0.08 max	17-19	2 max		9-13	0.045 max	0.03 max	1 max	Nb>=10xC; bal Fe				
ASTM A430	347*	Obs 1995; see A312; Pipe	0.08 max	17-19	2 max		9-13	0.045 max	0.03 max	1 max	Nb>=10xC; bal Fe				
ASTM A473	347	Frg, Heat res SHT	0.08 max	17.00-19.00	2.00 max		9.00-13.00	0.045 max	0.030 max	1.00 max	Nb>=10xC; bal Fe	515	205	40	

Specification	Designation	Notes	C	Cr	Mn	Mo	Ni	P	S	Si	Other	UTS	YS	El	Hard

Stainless Steel, Austenitic, 347 (Continued from previous page)

USA

Specification	Designation	Notes	C	Cr	Mn	Mo	Ni	P	S	Si	Other	UTS	YS	El	Hard
ASTM A479	347	Bar	0.08 max	17-19	2 max		9-13	0.045 max	0.03 max	1 max	Nb>=10xC; bal Fe				
ASTM A493	347	Wir rod, CHd, cold frg	0.08 max	17.0-19.0	2.00 max		9.0-13.0	0.045 max	0.030 max	1.00 max	Nb>=10xC; bal Fe				
ASTM A511(96)	MT347	Smls mech tub, Ann	0.08 max	17.0-20.0	2.00 max		9.0-13.0	0.040 max	0.030 max	1.00 max	10xC<=Nb+Ta<=1.00 ; bal Fe	517	207	35	192 HB
ASTM A554(94)	MT347	Weld mech tub, Rnd Ann	0.08 max	17.0-20.0	2.00 max		9.0-13.0	0.040 max	0.030 max	1.00 max	Nb+Ta>=10xC<=1.00 ; bal Fe	517	207	35	
ASTM A580/A580M(98)	347	Wir, Ann	0.08 max	17.0-19.0	2.00 max		9.0-13.0	0.045 max	0.030 max	1.00 max	Nb+Ta>=10xC; bal Fe	520	210	35	
ASTM A632(90)	TP347	Smls Weld tub, Nb-stabilized	0.08 max	17.0-20.0	2.00 max		9.0-13.0	0.040 max	0.030 max	0.75 max	Ti>=5xC<= 0.60; bal Fe	515	205	35	
ASTM A774/A774M(98)	TP347	Fittings; Corr service at low/mod-temp, Nb-stabilized	0.08 max	17.0-19.0	2.00 max		9.0-12.0	0.045 max	0.030 max	1.00 max	Nb+Ta>=10xC<=1.10 ; bal Fe	515-690	205	40.0	202 HB
ASTM A778(90)	TP347	Nb-stabilized	0.030 max	17.00-19.00	2.00 max		9.00-13.00	0.045 max	0.030 max	1.00 max	N 0; bal Fe	515	205	40	
ASTM A813/A813M(95)	TP347	Weld Pipe, Nb-stabilized	0.08 max	17.0-20.0	2.00 max		9.00-13.0	0.045 max	0.030 max	0.75 max	Nb+Ta>=10xC<=1.00 , bal Fe	515	205		
ASTM A814/A814M(96)	TP347	CW Weld Pipe, Nb-stabilized	0.08 max	17.0-20.0	2.00 max		9.00-13.0	0.045 max	0.030 max	0.75 max	Nb+Ta>=10xC<=1.00 , bal Fe	515	205		
ASTM A943/A943M(95)	TP347	Spray formed smls pipe	0.08 max	17.0-20.0	2.00 max		9.00-13.0	0.040 max	0.030 max	0.75 max	Nb+Ta>=10xC<=1.00 ; bal Fe	515	205		
ASTM A965/965M(97)	F347	Frg for press high-temp parts	0.08 max	17.0-19.0	2.00 max		9.0-12.0	0.040 max	0.030 max	0.85 max	Nb=10xC-1.00; Ta or Nb; bal Fe	485	205	30	
FED QQ-S-763F(96)	347*	Obs; Bar Wir Shp Frg; Corr res	0.08 max	17-19	2 max		9-13	0.045 max	0.03 max	1 max	Nb>=10xC; bal Fe				
FED QQ-S-766D(93)	347*	Obs; see ASTM A240, A666, A693; Sh Strp Plt	0.08 max	17-19	2 max		9-13	0.045 max	0.03 max	1 max	Nb>=10xC; bal Fe				
MIL-P-1144D	347*	Obs; Pipe	0.08 max	17-19	2 max		9-13	0.04 max	0.03 max	1 max	Nb+Ta>=10xC<=1.00 ; bal Fe				
MIL-S-23195(A)(65)	347	Corr res; Bar frg; Rev E controlled distribution	0.08 max	17.00-19.00	2.00 max		9.00-13.00	0.035 max	0.030 max	1.00 max	Co <=0.10; Nb+Ta (10xC)-1.00; bal Fe	517	207-379	40	
MIL-S-23196	347	Controlled distribution													
MIL-S-27419(USAF)(68)	347*	Obs; Corr res ann bil	0.08 max	17-19	2 max		9-13	0.03 max	0.03 max	1 max	Cu <=0.7; Nb <=1.1; bal Fe				
MIL-S-862B	347*	Obs	0.08 max	17-19	2 max		9-13	0.045 max	0.03 max	1 max	Nb+Ta>=10xC; bal Fe				
MIL-T-6737C	347*	Obs; Tube SA	0.08 max	17-19	2 max		9-13	0.04 max	0.03 max	1 max	Cu <=0.7; Nb>=10xC; bal Fe				
MIL-T-8606C(79)	347*	Obs for new design; Tube; Corr res	0.08 max	17-19	2 max		9-13	0.04 max	0.03 max	1 max	Cu <=0.7; Nb>=10xC<=1.10; bal Fe				
MIL-T-8808B(81)	347*	Obs for new design; Tube; Corr res aircraft	0.08 max	17-19	2 max		9-13	0.04 max	0.03 max	1 max	Cu <=0.5; Nb <=1.1; bal Fe				
MIL-T-8973(69)	347*	Obs for new design; Tube; Corr/Heat res; aerospace	0.08 max	17.0-20.0	2.00 max	0.50 max	9.0-13.0	0.040 max	0.030 max	1.00 max	Cu <=0.50; Nb <=1.10; Ti <=0.75; Nb+Ta (10xC) min; bal Fe				
SAE J405(98)	S34700	Nb stabilized	0.08 max	17.00-19.00	2.00 max		9.00-13.00	0.045 max	0.030 max	0.75 max	Nb<=10xC; bal Fe				
SAE J467(68)	347		0.05	18.5	1.00		11.0			0.40	Nb 0.70; bal Fe	627	269	50	160 HB

Stainless Steel, Austenitic, 347H

Japan

Specification	Designation	Notes	C	Cr	Mn	Mo	Ni	P	S	Si	Other	UTS	YS	El	Hard
JIS G3214(91)	SUSF347H	Frg press ves 130-200mm diam	0.04-0.10	17.00-20.00	2.00 max		9.00-13.00	0.040 max	0.030 max	1.00 max	Nb 8xC to 1.00; bal Fe	480	205	29	187 HB max

Mexico

Specification	Designation	Notes	C	Cr	Mn	Mo	Ni	P	S	Si	Other	UTS	YS	El	Hard
NMX-B-186-SCFI(94)	TP347H	Pipe; High-temp	0.04-0.10	17.0-20.0	2.00 max		9.00-13.0	0.040 max	0.030 max	0.75 max	Nb+Ta=8xC min, 1.00 max; bal Fe	517	207		
NMX-B-196(68)	TP347H	Smls tube for refinery	0.04-0.10	17.0-20.0	2.00 max		9.0-13.0	0.040 max	0.030 max	0.75 max	Nb+Ta=8xC min, 1.0 max; bal Fe	520	205	35	

UK

Specification	Designation	Notes	C	Cr	Mn	Mo	Ni	P	S	Si	Other	UTS	YS	El	Hard
BS 3059/2(78)	347S59	Boiler, Superheater; Tub; Pip	0.04-0.09	17.0-19.0	0.50-2.00	0.15 max	9.0-13.0	0.040 max	0.030 max	0.20-1.00	8xC<=Nb<=1.00; bal Fe				
BS 3605(73)	347S59		0.04-0.09	17.0-19.0	0.5-2.0		11.0-14.0				10xC<=Nb<=1.00; bal Fe				

UNS numbers and US grades are provided as a means of cross referencing chemically similar alloys. Exchangability is only possible after independent examination of specifications. Tensile properties are minimum or typical as specified. UTS and YS as MPa. El as %. See Appendix for list of abbreviations used in Notes. * indicates obsolete material.

Specification	Designation	Notes	C	Cr	Mn	Mo	Ni	P	S	Si	Other	UTS	YS	El	Hard

Stainless Steel, Austenitic, 347H (Continued from previous page)

USA

Specification	Designation	Notes	C	Cr	Mn	Mo	Ni	P	S	Si	Other	UTS	YS	El	Hard
	UNS S34709	Nb stabilized, High C	0.04-0.10	17.00-20.00	2.00 max		9.00-13.00	0.040 max	0.030 max	1.00 max	Nb 8xC-1.00; bal Fe				
ASTM A182/A182M(98)	F347H	Frg/roll pipe flange valve	0.04-0.10	17.0-20.0	2.00 max		9.0-13.0	0.045 max	0.030 max	1.00 max	Nb>=8xC<=1.10; bal Fe	515	205	30	
ASTM A213/A213M(95)	TP347H	Smls tube boiler, superheater, heat exchanger	0.04-0.10	17.0-20.0	2.00 max		9.00-13.00	0.040 max	0.030 max	0.75 max	Ta+Nb 8xC-1.00; bal Fe	515	205	35	
ASTM A240/A240M(98)	S34709	Nb-stabilized High-C	0.04-0.10	17.0-20.0	2.00 max		9.0-13.0	0.040 max	0.030 max	0.75 max	8xC<=Nb<=1.00; bal Fe	515	205	40.0	201 HB
ASTM A249/249M(96)	TP347H	Weld tube; boiler, superheater, heat exch	0.04-0.10	17.0-20.0	2.00 max		9.00-13.00	0.04 max	0.03 max	0.75 max	Ta+Nb 8xC-1.00; bal Fe	515	205	35	90 HRB
ASTM A336/A336M(98)	F347H	Frg, press/high-temp	0.04-0.10	17.0-20.0	2.00 max		9.00-13.00	0.040 max	0.030 max	1.00 max	Ta+Nb 8xC-1.00; bal Fe	485	205	30	90 HRB
ASTM A813/A813M(95)	TP347H	Weld Pipe, Nb-stabilized; High-C	0.04-0.10	17.0-20.0	2.00 max		9.00-13.0	0.045 max	0.030 max	0.75 max	Nb+Ta>=8xC<= 1.0, bal Fe	515	205		
ASTM A814/A814M(96)	TP347H	CW Weld Pipe, Nb-stabilized; High-C	0.04-0.10	17.0-20.0	2.00 max		9.00-13.0	0.045 max	0.030 max	0.75 max	Nb+Ta>= 8xC<=1.0, bal Fe	515	205		
ASTM A943/A943M(95)	TP347H	Spray formed smls pipe	0.04-0.10	17.0-20.0	2.00 max		9.00-13.0	0.040 max	0.030 max	0.75 max	Nb+Ta>=8xC<=1.0; bal Fe	515	205		
ASTM A965/965M(97)	F347H	Frg for press high-temp parts	0.04-0.10	17.0-20.0	2.00 max		9.0-13.0	0.040 max	0.030 max	1.00 max	Nb+Ta (8xC)-1.00; bal Fe	450	205	30	
SAE J405(98)	S34709	Nb stabilized, High C	0.04-0.10	17.00-20.00	2.00 max		9.00-13.00	0.045 max	0.030 max	0.75 max	Nb 8xC-1.00; bal Fe				

Stainless Steel, Austenitic, 347HFG

USA

Specification	Designation	Notes	C	Cr	Mn	Mo	Ni	P	S	Si	Other	UTS	YS	El	Hard
	UNS S34710	Nb stabilized, High C	0.06-0.10	17.0-20.0	2.00 max		9.00-13.0	0.040 max	0.030 max	0.75 max	Nb 8xC-1.0; bal Fe				
ASTM A213/A213M(95)	TP347HFG	Smls tube boiler, superheater, heat exchanger	0.006-0.10	17.0-20.0	2.00 max		9.00-13.00	0.040 max	0.030 max	0.75 max	Ta+Nb 8xC-1.00; bal Fe	550	205	35	

Stainless Steel, Austenitic, 347LN

USA

Specification	Designation	Notes	C	Cr	Mn	Mo	Ni	P	S	Si	Other	UTS	YS	El	Hard
	UNS S34751	Nb stabilized Nitrogen Strengthened	0.005-0.020	17.0-20.0	2.00 max		9.0-13.0	0.040 max	0.030 max	0.75 max	N 0.06-0.10; (Nb+Ta) 0.2-0.5; (Nb+Ta)/C 15 min; bal Fe				
ASTM A213/A213M(95)	TP347LN	Smls tube boiler, superheater, heat exchanger	0.005-0.020	17.0-20.0	2.00 max		9.00-13.00	0.040 max	0.030 max	0.75 max	N 0.06-0.10; Ta+Nb 15xC-0.5; bal Fe	515	205	35	

Stainless Steel, Austenitic, 348

Bulgaria

Specification	Designation	Notes	C	Cr	Mn	Mo	Ni	P	S	Si	Other	UTS	YS	El	Hard
BDS 6738(72)	0Ch18N12B	Corr res	0.08 max	17.0-19.0	2.00 max	0.30 max	10.0-13.0	0.035 max	0.025 max	0.80 max	10xC<=Nb<=1.10; bal Fe				

Canada

Specification	Designation	Notes	C	Cr	Mn	Mo	Ni	P	S	Si	Other	UTS	YS	El	Hard
CSA G110.3	348	Bar Bil	0.08 max	17-19	2 max		9-13	0.045 max	0.03 max	1 max	Co <=0.25; Nb+Ta>=10xC; bal Fe				
CSA G110.6	348	Sh Strp Plt, HR, Ann or HR/Q/A/Tmp	0.08 max	17-19	2 max		9-13	0.045 max	0.03 max	1 max	Co <=0.2; Ta 0.1 max; Nb+Ta>=10xC<=1.10; bal Fe				
CSA G110.9	348	Sh Strp Plt, HR, CR, Ann	0.08 max	17-19	2 max		9-13	0.045 max	0.03 max	1 max	Co <=0.2; Ta 0.1 max; Nb+Ta>=10xC<=1.10; bal Fe				

Hungary

Specification	Designation	Notes	C	Cr	Mn	Mo	Ni	P	S	Si	Other	UTS	YS	El	Hard
MSZ 4360(80)	KO36Nb	Corr res	0.12 max	17-19	2 max	0.5 max	8-11	0.04 max	0.03 max	1 max	Nb=8C-1.1; bal Fe				
MSZ 4398(86)	KO36Nb	Corr res; Heat res; longitudinal weld tube; 0.8-3mm; SA	0.08 max	17-19	2 max	0.5 max	9-12	0.045 max	0.03 max	1 max	Nb+Ta=10C-1; bal Fe	510-740	205	35L	

Mexico

Specification	Designation	Notes	C	Cr	Mn	Mo	Ni	P	S	Si	Other	UTS	YS	El	Hard
DGN B-218	TP348*	Obs; Tube, CF, Q/A, Tmp, 8mm diam	0.08 max	17-20	2 max		9-13	0.04 max	0.03 max	0.75 max	Ta 0.1 max; Nb+Ta>=10xC<=1.00; bal Fe				

UNS numbers and US grades are provided as a means of cross referencing chemically similar alloys. Exchangability is only possible after independent examination of specifications. Tensile properties are minimum or typical as specified. UTS and YS as MPa. El as %. See Appendix for list of abbreviations used in Notes. * indicates obsolete material.

Stainless Steel, Austenitic, 348 (Continued from previous page)

Specification	Designation	Notes	C	Cr	Mn	Mo	Ni	P	S	Si	Other	UTS	YS	El	Hard
Mexico															
DGN B-229	TP348*	Obs; Tube	0.08 max	17-20	2 max		9-13	0.04 max	0.03 max	0.75 max	Ta 0.1 max; Nb+Ta>=10xC<=1.00; bal Fe				
NMX-B-186-SCFI(94)	TP348	Smls pipe for high-temp central station service	0.08 max	17.0-20.0	2.00 max		9.00-13.0	0.040 max	0.030 max	0.75 max	Ta 0.10; Nb+Ta=10xC min, 1.00 max; bal Fe	517	207		
Pan America															
COPANT 513	TP348	Tube, HF, CF, 8mm diam	0.08 max	17-20	2 max		9-13	0.04 max	0.03 max	1 max	Ta 0.1 max; Nb+Ta>=10xC<=1.00; bal Fe				
Poland															
PNH86020	0H18N12Nb	Corr res	0.08 max	17-19	2 max		10-13	0.045 max	0.03 max	0.8 max	Pb <=0.1; Nb>=10xC<=1.10; bal Fe				
Sweden															
SS 142338	2338	Corr res	0.08 max	17-19	2 max		9-12	0.045 max	0.03 max	1 max	Nb+1/2Ta=10C-1; bal Fe				
UK															
BS 1501/3(73)	347S67*	Press ves; Corr res	0.08 max	17.0-19.0	0.5-2.0	0.15 max	9.0-12.0	0.045 max	0.03 max	0.2-1.0	Cu <=0.3; Ti <=0.05; W <=0.1; Nb=10C-1; N=0.15-0.25; bal Fe				
BS 1501/3(90)	347S51	Press ves; Corr/Heat res; 0.05-100mm	0.04-0.1	17.0-19.0	2.0 max	0.15 max	9.0-12.0	0.045 max	0.025 max	1.0 max	Cu <=0.3; Ti <=0.05; W <=0.1; Nb=10C-1.2; bal Fe	510-710	205	30	
BS 1503(89)	347S51	Frg press ves; t<=350mm; Quen	0.04-0.10	17.0-19.0	2.00 max	0.70 max	9.00-12.0	0.040 max	0.025 max	1.00 max	B <=0.005; Cu <=0.50; W <=0.1; 10xC<=Nb<=1.20; bal Fe	510-710	240	30	
BS 3059/2(90)	347S51	Boiler, Superheater; High-temp; Smls tub; 2-11mm	0.04-0.10	17.0-19.0	2.00 max	0.15 max	9.0-13.0	0.040 max	0.030 max	1.00 max	10xC<=Nb<=1.20; bal Fe	510-710	240	35	
BS 3605/1(91)	347S31	Pip Tube press smls; t<=200mm	0.080 max	17.0-19.0	2.00 max		9.00-13.0	0.040 max	0.030 max	1.00 max	10xC<=Nb<=1.00; bal Fe	510-710		35	
BS 3605/1(91)	347S51	Pip Tube press smls; t<=200mm	0.04-0.10	17.0-19.0	2.00 max		9.00-13.0	0.040 max	0.030 max	1.00 max	10xC<=Nb<=1.00; bal Fe *	510-710		35	
USA															
	AISI 348	Tube	0.08 max	17.00-19.00	2.00 max		9.00-13.00	0.045 max	0.030 max	1.00 max	Co <=0.2; Ta 0.1 max; Nb>=10xC; bal Fe				
	UNS S34800	Nb stabilized High C Ta and Co restricted	0.08 max	17.00-19.00	2.00 max		9.00-13.00	0.045 max	0.030 max	1.00 max	Co <=0.20; Nb 10xC min; Ta 0.10 max; bal Fe				
ASME SA182	348	Refer to ASTM A182/A182M													
ASME SA213	348	Refer to ASTM A213/A213M(95)													
ASME SA240	348	Refer to ASTM A240(95)													
ASME SA249	348	Refer to ASTM A249/A249M(96)													
ASME SA312	348	Refer to ASTM A312/A312M(95)													
ASME SA358	348	Refer to ASTM A358/A358M(95)													
ASME SA376	348	Refer to ASTM A376/A376M(93)													
ASME SA403	348	Refer to ASTM A403/A403M(95)													
ASME SA409	348	Refer to ASTM A409/A409M(95)													
ASME SA479	348	Refer to ASTM A479/A479M(95)													
ASTM A167(96)	348*	Sh Strp Plt	0.08 max	17-19	2 max		9-13	0.045 max	0.03 max	1 max	Co <=0.2; Ta<=0.1; Nb>=10xC; bal Fe				
ASTM A182	348*	Bar Frg	0.08 max	17-19	2 max		9-13	0.045 max	0.03 max	1 max	Co <=0.2; Ta<=0.1; Nb>=10xC; bal Fe				
ASTM A182/A182M(98)	F348	Frg/roll pipe flange valve	0.08 max	17.0-20.0	2.00 max		9.0-13.0	0.045 max	0.030 max	1.00 max	Nb>=10xC<=1.10; bal Fe	515	205	30	

UNS numbers and US grades are provided as a means of cross referencing chemically similar alloys. Exchangability is only possible after independent examination of specifications. Tensile properties are minimum or typical as specified. UTS and YS as MPa. El as %. See Appendix for list of abbreviations used in Notes. * indicates obsolete material.

Specification	Designation	Notes	C	Cr	Mn	Mo	Ni	P	S	Si	Other	UTS	YS	El	Hard

Stainless Steel, Austenitic, 348 (Continued from previous page)

USA

Specification	Designation	Notes	C	Cr	Mn	Mo	Ni	P	S	Si	Other	UTS	YS	El	Hard
ASTM A213	348*	Tube	0.08 max	17-19	2 max		9-13	0.045 max	0.03 max	1 max	Co <=0.2; Ta<=0.1; Nb>=10xC; bal Fe				
ASTM A213/A213M(95)	TP348	Smls tube boiler, superheater, heat exchanger	0.08 max	17.0-20.0	2.00 max		9.00-13.00	0.040 max	0.030 max	0.75 max	Ta 0-0.10, Ti+Nb 10xC-1.00; bal Fe	515	205	35	
ASTM A240/A240M(98)	S34800	Nb-stabilized High-C Ta/Co restricted	0.08 max	17.0-19.0	2.00 max		9.0-13.0	0.045 max	0.030 max	0.75 max	Co <=0.20; 10xC<=Nb+Ta<=1.00; Ta<=0.10; bal Fe	515	205	40.0	201 HB
ASTM A249/249M(96)	TP348	Weld tube; boiler, superheater, heat exch	0.08 max	17.0-20.0	2.00 max		9.00-13.00	0.04 max	0.03 max	0.75 max	Ta 0-0.10, Ta+Nb 10xC-1.00; bal Fe	515	205	35	90 HRB
ASTM A269	348	Smls Weld, Tube	0.08 max	17-19	2 max		9-13	0.045 max	0.03 max	1 max	Co <=0.2; Ta<=0.1; Nb>=10xC; bal Fe				
ASTM A276(98)	348	Bar	0.08 max	17-19	2 max		9-13	0.045 max	0.03 max	1 max	Co <=0.2; Ta<=0.1; Nb>=10xC; bal Fe				
ASTM A312/A312M(95)	348	Smls Weld, Pipe	0.08 max	17-19	2 max		9-13	0.045 max	0.03 max	1 max	Co <=0.2; Ta<=0.1; Nb>=10xC; bal Fe				
ASTM A314	348	Bil	0.08 max	17-19	2 max		9-13	0.045 max	0.03 max	1 max	Co <=0.2; Ta<=0.1; Nb>=10xC; bal Fe				
ASTM A336/A336M(98)	F348	Frg, press/high-temp	0.08 max	17.0-20.0	2.00 max		9.00-13.00	0.040 max	0.030 max	1.00 max	Ta 0-0.10, Ta+Nb 10xC-1.00; bal Fe	485	205	30	90 HRB
ASTM A358/A358M(95)	348	Elect fusion weld pipe, high-temp	0.08 max	17-19	2 max		9-13	0.045 max	0.03 max	1 max	Co <=0.2; Ta<=0.1; Nb>=10xC; bal Fe				
ASTM A376	348	Smls pipe	0.08 max	17-19	2 max		9-13	0.045 max	0.03 max	1 max	Co <=0.2; Ta<=0.1; Nb>=10xC; bal Fe				
ASTM A403	348	Pipe	0.08 max	17-19	2 max		9-13	0.045 max	0.03 max	1 max	Co <=0.2; Ta<=0.1; Nb>=10xC; bal Fe				
ASTM A409	348	Weld, Pipe	0.08 max	17-19	2 max		9-13	0.045 max	0.03 max	1 max	Co <=0.2; Ta<=0.1; Nb>=10xC; bal Fe				
ASTM A479	348	Bar	0.08 max	17-19	2 max		9-13	0.045 max	0.03 max	1 max	Co <=0.2; Ta<=0.1; Nb>=10xC; bal Fe				
ASTM A580/A580M(98)	348	Wir, Ann	0.08 max	17.0-19.0	2.00 max		9.0-13.0	0.045 max	0.030 max	1.00 max	Co <=0.20; Ta<=0.10; Nb+Ta>=10xC; bal Fe	520	210	35	
ASTM A632(90)	TP348	Smls Weld tub, Nb-stabilized; High-C	0.08 max	17.0-20.0	2.00 max		9.0-13.0	0.040 max	0.030 max	0.75 max	Nb+Ta>=Cx10<=1.0; Ta=0.10 max; bal Fe	515	205	35	
ASTM A813/A813M(95)	TP348	Weld Pipe, Nb-stabilized; High-C Ta & Corr res	0.08 max	17.0-20.0	2.00 max		9.00-13.00	0.045 max	0.030 max	0.75 max	Ta<=0.10; Nb+Ta>=10xC<=1.00, bal Fe	515	205		
ASTM A814/A814M(96)	TP348	CW Weld Pipe, Nb-stabilized; High-C; Ta/Co res	0.08 max	17.0-20.0	2.00 max		9.00-13.00	0.045 max	0.030 max	0.75 max	Nb+Ta>=10xC<=1.00, Ta<=0.10, bal Fe	515	205		
ASTM A943/A943M(95)	TP348	Spray formed smls pipe	0.08 max	17.0-20.0	2.00 max		9.00-13.00	0.040 max	0.030 max	0.75 max	Nb+Ta>=10xC<=1.00; bal Fe	515	205		
ASTM A965/965M(97)	F348	Frg for press high-temp parts	0.08 max	17.0-20.0	2.00 max		9.0-13.0	0.40 max	0.30 max	1.00 max	Ta<=0.10; Nb+Ta (10xC)-1.00; bal Fe	485	205	30	
FED QQ-S-766D(93)	348*	Obs; see ASTM A240, A666, A693; Sh Strp Plt	0.08 max	17-19	2 max		9-13	0.045 max	0.03 max	1 max	Co <=0.2; Ta 0.1 max; Nb>=10xC; bal Fe				
MIL-S-23195(A)(65)	348	Corr res; Bar frg; Rev E controlled distribution	0.08 max	17.00-19.00	2.00 max		9.00-13.00	0.035 max	0.030 max	1.00 max	Co <=0.10; Nb+Ta (10xC)-1.00; Ta 0.10 max; bal Fe	483	207-379	40	
MIL-S-23196	348	Controlled distribution													
SAE J405(98)	S34800	Nb stabilized High C Ta and Co restricted	0.08 max	17.00-19.00	2.00 max		9.00-13.00	0.045 max	0.030 max	0.75 max	Co <=0.20; 10xC<=Nb+Ta<=1.00; Ta<=0.10; bal Fe				

Stainless Steel, Austenitic, 348H

USA

Specification	Designation	Notes	C	Cr	Mn	Mo	Ni	P	S	Si	Other	UTS	YS	El	Hard
	UNS S34809	Nb stabilized High C Ta and Co restricted	0.04-0.10	17.00-20.00	2.00 max		9.00-13.00	0.045 max	0.030 max	1.00 max	Co <=0.20; Nb 8xC-1.00; Ta 0.10 max; bal Fe				
ASTM A182/A182M(98)	F348H	Frg/roll pipe flange valve	0.04-0.10	17.0-20.0	2.00 max		9.0-13.0	0.045 max	0.030 max	1.00 max	Nb>=8xC<=1.10; bal Fe	515	205	30	
ASTM A213/A213M(95)	TP348H	Smls tube boiler, superheater, heat exchanger	0.04-0.10	17.0-20.0	2.00 max		9.00-13.00	0.040 max	0.030 max	0.75 max	Ta 0-0.10, Ti+Nb 8xC-1.00; bal Fe	515	205	35	

UNS numbers and US grades are provided as a means of cross referencing chemically similar alloys. Exchangability is only possible after independent examination of specifications. Tensile properties are minimum or typical as specified. UTS and YS as MPa. El as %. See Appendix for list of abbreviations used in Notes. * indicates obsolete material.

Specification	Designation	Notes	C	Cr	Mn	Mo	Ni	P	S	Si	Other	UTS	YS	El	Hard

Stainless Steel, Austenitic, 348H (Continued from previous page)

USA

Specification	Designation	Notes	C	Cr	Mn	Mo	Ni	P	S	Si	Other	UTS	YS	El	Hard
ASTM A240/A240M(98)	S34809	Nb-stabilized High-C Ta/Co restricted	0.04-0.10	17.0-20.0	2.00 max		9.0-13.0	0.045 max	0.030 max	0.75 max	Co <=0.20; 8xC<=Nb+Ta<=1.00; Ta<=0.10; bal Fe	515	205	40.0	201 HB
ASTM A249/249M(96)	TP348H	Weld tube; boiler, superheater, heat exch	0.04-0.10	17.0-20.0	2.00 max		9.00-13.00	0.04 max	0.03 max	0.75 max	Ta 0-0.10, Ta+Nb 8xC-1.00; bal Fe	515	205	35	90 HRB
ASTM A336/A336M(98)	F348H	Frg, press/high-temp	0.04-0.10	17.0-20.0	2.00 max		9.00-13.00	0.040 max	0.030 max	1.00 max	Ta 0-0.10, Ta+Nb 8xC-1.00; bal Fe	450	170	30	90 HRB
ASTM A813/A813M(95)	TP348H	Weld Pipe, Nb-stabilized; High-C Ta & Corr res	0.04-0.10	17.0-20.0	2.00 max		9.00-13.0	0.045 max	0.030 max	0.75 max	Ta<=0.10; Nb+Ta>=8xC<= 1.0, bal Fe	515	205		
ASTM A814/A814M(96)	TP348H	CW Weld Pipe, Nb-stabilized; High-C; Ta/Co res	0.04-0.10	17.0-20.0	2.00 max		9.00-13.0	0.045 max	0.030 max	0.75 max	Nb+Ta>= 8xC<=1.0, Ta<=0.10, bal Fe	515	205		
ASTM A943/A943M(95)	TP348H	Spray formed smls pipe	0.04-0.10	17.0-20.0	2.00 max		9.00-13.0	0.040 max	0.030 max	0.75 max	Nb+Ta>=8xC<=1.0; bal Fe	515	205		
ASTM A965/965M(97)	F348H	Frg for press high-temp parts	0.04-0.10	17.0-20.0	2.00 max		9.0-13.0	0.040 max	0.030 max	1.00 max	Ta<=0.1; Nb+Ta (8xC)-1.00; bal Fe	450	170	30	
SAE J405(98)	S34809	Nb stabilized High C Ta and Co restricted	0.04-0.10	17.00-20.00	2.00 max		9.00-13.00	0.045 max	0.030 max	0.75 max	Co <=0.20; 8xC<=Nb+Ta=1.00; Ta<=0.10; bal Fe				

Stainless Steel, Austenitic, 353MA

USA

Specification	Designation	Notes	C	Cr	Mn	Mo	Ni	P	S	Si	Other	UTS	YS	El	Hard
	UNS S35135	Heat res 864	0.08 max	20.0-25.0	1.00 max	4.0-4.8	30.0-38.0	0.045 max	0.015 max	0.60-1.00	Cu <=0.75; Ti 0.40-1.00; bal Fe				
ASTM A167(96)	S35315	Sh Strp Plt	0.04-0.08	24.00-26.00	2.00 max		34.00-36.00	0.040 max	0.030 max	1.20-2.00	Ce 0.03-0.08; bal Fe	650	270	40.0	217 HB

Stainless Steel, Austenitic, 384

France

Specification	Designation	Notes	C	Cr	Mn	Mo	Ni	P	S	Si	Other	UTS	YS	El	Hard
AFNOR NFA35575	26NC18.16	Wir rod, 60% Hard	0.08 max	15-17	2 max		17-19	0.04 max	0.03 max	1 max	bal Fe				
AFNOR NFA35577	Z6NC18.16	Bar Wir	0.08 max	15-17	2 max		17-19	0.04 max	0.03 max	1 max	bal Fe				

Germany

Specification	Designation	Notes	C	Cr	Mn	Mo	Ni	P	S	Si	Other	UTS	YS	El	Hard
DIN	WNr 1.4321		0.03 max	15.5-16.5	0.60-0.90		17.5-18.5	0.045 max	0.030 max	0.30-0.50	bal Fe				
DIN	X2NiCr18 16		0.03 max	15.5-16.5	0.60-0.90		17.5-18.5	0.045 max	0.030 max	0.30-0.50	bal Fe				

International

Specification	Designation	Notes	C	Cr	Mn	Mo	Ni	P	S	Si	Other	UTS	YS	El	Hard
ISO 4954(93)	X6NiCr1816E	Wir rod Bar, CH ext	0.08 max	15.0-17.0	2.00 max		17.0-19.0	0.045 max	0.030 max	1.00 max	bal Fe	600			

Japan

Specification	Designation	Notes	C	Cr	Mn	Mo	Ni	P	S	Si	Other	UTS	YS	El	Hard
JIS G4308	SUS384	Wir rod	0.08 max	15.00-17.00	2.00 max		17.00-19.00	0.045 max	0.030 max	1.00 max	bal Fe				
JIS G4315(94)	SUS384	Wir rod	0.08 max	15.00-17.00	2.00 max		17.00-19.00	0.045 max	0.030 max	1.00 max	bal Fe	460-690		20	

USA

Specification	Designation	Notes	C	Cr	Mn	Mo	Ni	P	S	Si	Other	UTS	YS	El	Hard
	AISI 384	Tube	0.08 max	15.00-17.00	2.00 max		17.00-19.00	0.045 max	0.030 max	1.00 max	bal Fe				
	UNS S38400	Low WH	0.08 max	15.00-17.00	2.00 max		17.00-19.00	0.045 max	0.030 max	1.00 max	bal Fe				
ASTM A493	384	Wir rod, CHd, cold frg	0.08 max	15.0-17.0	2.00 max		17.0-19.0	0.045 max	0.030 max	1.00 max	bal Fe	550			

Stainless Steel, Austenitic, 385

USA

Specification	Designation	Notes	C	Cr	Mn	Mo	Ni	P	S	Si	Other	UTS	YS	El	Hard
	UNS S38500*	Obs; Low WH	0.08 max	11.50-13.50	2.00 max		14.00-16.00	0.045 max	0.030 max	1.00 max	bal Fe				

Stainless Steel, Austenitic, 4565S

USA

Specification	Designation	Notes	C	Cr	Mn	Mo	Ni	P	S	Si	Other	UTS	YS	El	Hard
	UNS S34565	High-N	0.03 max	23.0-25.0	5.0-7.0	4.0-5.0	16.0-18.0	0.030 max	0.010 max	1.0 max	N 0.4-0.6; Nb <=0.1; bal Fe				
ASTM A182/A182M(98)	F49	Frg/roll pipe flange valve	0.030 max	23.0-25.0	5.0-7.0	4.0-5.0	16.0-18.0	0.030 max	0.010 max	1.00 max	N 0.4-0.6; Nb >=0.1; bal Fe	795	415	35	

UNS numbers and US grades are provided as a means of cross referencing chemically similar alloys. Exchangability is only possible after independent examination of specifications. Tensile properties are minimum or typical as specified. UTS and YS as MPa. El as %. See Appendix for list of abbreviations used in Notes. * indicates obsolete material.

Specification	Designation	Notes	C	Cr	Mn	Mo	Ni	P	S	Si	Other	UTS	YS	El	Hard

Stainless Steel, Austenitic, 4565S (Continued from previous page)

USA

Specification	Designation	Notes	C	Cr	Mn	Mo	Ni	P	S	Si	Other	UTS	YS	El	Hard
ASTM A240/A240M(98)	S34565	High-N	0.030 max	23.0-25.0	5.0-7.0	4.0-5.0	16.0-18.0	0.030 max	0.010 max	1.00 max	N 0.40-0.60; Nb<=0.10; bal Fe	795	415	35.0	241 HB
ASTM A943/A943M(95)		Spray formed smls pipe	0.03 max	23.0-25.0	5.0-7.0	4.0-5.0	16.0-18.0	0.030 max	0.010 max	1.00 max	N 0.4-0.6; Nb <=0.1; Nb+Ta<=0.1; bal Fe	795	415		
SAE J405(98)	S34565	High-N	0.030 max	23.00-25.00	5.00-7.00	4.00-5.00	16.00-18.00	0.030 max	0.010 max	1.00 max	N 0.40-0.60; Nb<=0.10; bal Fe				

Stainless Steel, Austenitic, 5642(1)

USA

Specification	Designation	Notes	C	Cr	Mn	Mo	Ni	P	S	Si	Other	UTS	YS	El	Hard
	UNS S34720	Free Machining Nb stabilized	0.08 max	17.00-19.00	2.00 max		9.00-12.00	0.040 max	0.18-0.35	1.00 max	Nb 10xC-1.10; bal Fe				

Stainless Steel, Austenitic, 5642(2)

USA

Specification	Designation	Notes	C	Cr	Mn	Mo	Ni	P	S	Si	Other	UTS	YS	El	Hard
	UNS S34723	Se-bearing Free Mach Nb stabilized	0.08 max	17.00-19.00	2.00 max		9.00-12.00	0.11-0.17	0.030 max	1.00 max	Nb 10xC-1.10; Se 0.15-0.35; bal Fe				

Stainless Steel, Austenitic, 5748

USA

Specification	Designation	Notes	C	Cr	Mn	Mo	Ni	P	S	Si	Other	UTS	YS	El	Hard
	UNS S65770	Fe Base Superalloy AFC 77	0.12-0.17	13.50-14.50	0.30 max	4.50-5.50	0.30-0.70	0.015 max	0.015 max	0.25	Co 13.0-14.0; V 0.10-0.30; bal Fe				

Stainless Steel, Austenitic, 5761

USA

Specification	Designation	Notes	C	Cr	Mn	Mo	Ni	P	S	Si	Other	UTS	YS	El	Hard
	UNS S65150	Fe Base Superalloy Pyromet X-15	0.03 max	14.50-16.00	0.10 max	2.50-3.00	0.20 max	0.015 max	0.015 max	0.10 max	Co 19.0-21.0; bal Fe				

Stainless Steel, Austenitic, 5784

USA

Specification	Designation	Notes	C	Cr	Mn	Mo	Ni	P	S	Si	Other	UTS	YS	El	Hard
	UNS S64299*	Obs; Fe Base Superalloy 29-9	0.08-0.15	27.0-31.0	1.00-2.00	0.50 max	8.50-10.50	0.040 max	0.030 max	0.75 max	Cu <=0.50; bal Fe				

Stainless Steel, Austenitic, 651

USA

Specification	Designation	Notes	C	Cr	Mn	Mo	Ni	P	S	Si	Other	UTS	YS	El	Hard
	UNS S63198	Fe Base Superalloy 19-9-DL	0.28-0.35	18.0-21.0	0.75-1.50	1.00-1.75	8.00-11.00	0.040 max	0.030 max	0.30-0.80	Cu <=0.50; Nb 0.25-0.60; Ti 0.10-0.35; W 1.00-1.75; bal Fe				
SAE J467(68)	19-9DL		0.32	18.5	1.15	1.40	9.00			0.55	Nb 0.40; Ti 0.25; W 1.35; bal Fe	814	476	56	215 HB

Stainless Steel, Austenitic, 654SMo

Europe

Specification	Designation	Notes	C	Cr	Mn	Mo	Ni	P	S	Si	Other	UTS	YS	El	Hard
EN 10088/2(95)	1.4537	Plt, HR Corr res; t<=75mm, Ann	0.020 max	24.00-26.00	2.00 max	4.70-5.70	24.00-27.00	0.030 max	0.010 max	0.70 max	Cu 1.00-2.00; N 0.17-0.25; bal Fe	600-800	290	40	
EN 10088/2(95)	1.4563	Plt, HR Corr res; t<=75mm, Ann	0.020 max	26.00-28.00	2.00 max	3.00-4.00	30.00-32.00	0.030 max	0.010 max	0.70 max	Cu 0.70-1.50; N <=0.11; bal Fe	500-700	220	40	
EN 10088/2(95)	X1CrNiMoCuN25-25-5	Plt, HR Corr res; t<=75mm, Ann	0.020 max	24.00-26.00	2.00 max	4.70-5.70	24.00-27.00	0.030 max	0.010 max	0.70 max	Cu 1.00-2.00; N 0.17-0.25; bal Fe	600-800	290	40	
EN 10088/2(95)	X1NiCrMoCu31-27-4	Plt, HR Corr res; t<=75mm, Ann	0.020 max	26.00-28.00	2.00 max	3.00-4.00	30.00-32.00	0.030 max	0.010 max	0.70 max	Cu 0.70-1.50; N <=0.11; bal Fe	500-700	220	40	
EN 10088/3(95)	1.4537	Bar Rod Sect; Corr res; 160<t<=250mm, Ann	0.020 max	24.00-26.00	2.00 max	4.70-5.70	24.00-27.00	0.030 max	0.010 max	0.70 max	Cu 1.00-2.00; N 0.17-0.25; bal Fe	600-800	300	30	250 HB
EN 10088/3(95)	1.4537	Bar Rod Sect, Corr res; t<=160mm, Ann	0.020 max	24.00-26.00	2.00 max	4.70-5.70	24.00-27.00	0.030 max	0.010 max	0.70 max	Cu 1.00-2.00; N 0.17-0.25; bal Fe	600-800	300	35	250 HB
EN 10088/3(95)	1.4563	Bar Rod Sect; Corr res; 160<t<=250mm, Ann	0.020 max	26.00-28.00	2.00 max	3.00-4.00	30.00-32.00	0.030 max	0.010 max	0.70 max	Cu 0.70-1.50; N <=0.11; bal Fe	500-750	220	30	230 HB
EN 10088/3(95)	1.4563	Bar Rod Sect, Corr res; t<=160mm, Ann	0.020 max	26.00-28.00	2.00 max	3.00-4.00	30.00-32.00	0.030 max	0.010 max	0.70 max	Cu 0.70-1.50; N <=0.11; bal Fe	500-750	220	35	230 HB
EN 10088/3(95)	X1CrNiMoCuN25-25-5	Bar Rod Sect, Corr res; t<=160mm, Ann	0.020 max	24.00-26.00	2.00 max	4.70-5.70	24.00-27.00	0.030 max	0.010 max	0.70 max	Cu 1.00-2.00; N 0.17-0.25; bal Fe	600-800	300	35	250 HB
EN 10088/3(95)	X1CrNiMoCuN25-25-5	Bar Rod Sect; Corr res; 160<t<=250mm, Ann	0.020 max	24.00-26.00	2.00 max	4.70-5.70	24.00-27.00	0.030 max	0.010 max	0.70 max	Cu 1.00-2.00; N 0.17-0.25; bal Fe	600-800	300	30	250 HB

Specification	Designation	Notes	C	Cr	Mn	Mo	Ni	P	S	Si	Other	UTS	YS	El	Hard

Stainless Steel, Austenitic, 654SMo (Continued from previous page)

Europe

Specification	Designation	Notes	C	Cr	Mn	Mo	Ni	P	S	Si	Other	UTS	YS	El	Hard
EN 10088/3(95)	X1NiCrMoCu31-27-4	Bar Rod Sect; Corr res; 160<t<=250mm, Ann	0.020 max	26.00-28.00	2.00 max	3.00-4.00	30.00-32.00	0.030 max	0.010 max	0.70 max	Cu 0.70-1.50; N <=0.11; bal Fe	500-750	220	30	230 HB
EN 10088/3(95)	X1NiCrMoCu31-27-4	Bar Rod Sect; Corr res; t<=160mm, Ann	0.020 max	26.00-28.00	2.00 max	3.00-4.00	30.00-32.00	0.030 max	0.010 max	0.70 max	Cu 0.70-1.50; N <=0.11; bal Fe	500-750	220	35	230 HB

USA

Specification	Designation	Notes	C	Cr	Mn	Mo	Ni	P	S	Si	Other	UTS	YS	El	Hard
	UNS S32654		0.020 max	24.0-25.0	2.00-4.00	7.00-8.00	21.0-23.0	0.030 max	0.005 max	0.50 max	Cu 0.30-0.60; N 0.45-0.55; bal Fe				
ASTM A240/A240M(98)	S32654		0.020 max	24.0-25.0	2.0-4.0	7.0-8.0	21.0-23.0	0.030 max	0.005 max	0.50 max	Cu 0.30-0.60; N 0.45-0.55; bal Fe	750	430	40.0	250 HB
ASTM A249/249M(96)	S32654	Weld tube; boiler, superheater, heat exch	0.020 max	24.0-25.0	2.00-4.00	7.00-8.00	21.0-23.0	0.030 max	0.005 max	0.50 max	Cu 0.30-0.60; N 0.45-0.55; bal Fe	750	430	35	100 HRB
ASTM A988(98)	S32654	HIP Flanges, Fittings, Valves/parts; Heat res	0.020 max	24.0-25.0	2.0-4.0	7.0-8.0	21.0-23.0	0.030 max	0.005 max	0.50 max	Cu 0.30-0.60; N 0.45-0.55; bal Fe	750	430	40.0	250 HB max
SAE J405(98)	S32654		0.020 max	24.00-25.00	2.00-4.00	7.00-8.00	21.0-23.0	0.030 max	0.005 max	0.50 max	Cu 0.30-0.60; N 0.45-0.55; bal Fe				

Stainless Steel, Austenitic, 662

USA

Specification	Designation	Notes	C	Cr	Mn	Mo	Ni	P	S	Si	Other	UTS	YS	El	Hard
	UNS S66220	Fe Base Superalloy Discaloy	0.08 max	12.0-15.0	1.50 max	2.50-3.50	24.0-28.0	0.040 max	0.030 max	1.00 max	Al <=0.35; B 0.0010-0.010; Cu <=0.50; Ti 1.55-2.00; bal Fe				
ASTM A638/638M(95)	662	Frg, Iron base superalloy, High-temp	0.08 max	12.00-15.00	1.50 max	2.50-3.50	24.00-28.00	0.040 max	0.030 max	1.00 max	Al <=0.35; B 0.0010-0.010; Cu <=0.50; Ti 1.55-2.00; bal Fe	860	550	15	248 HB min
ASTM A638/638M(95)	662	Bar, Iron base superalloy, High-temp	0.08 max	12.00-15.00	1.50 max	2.50-3.50	24.00-28.00	0.040 max	0.030 max	1.00 max	Al <=0.35; B 0.0010-0.010; Cu <=0.50; Ti 1.55-2.00; bal Fe	865	585	15	248 HB min
SAE J467(68)	Discaloy		0.08	13.5	0.90	2.75	26.0			0.80	Al 0.07; B 0.005; Ti 1.75; bal Fe	1000	731	19	293 HB

Stainless Steel, Austenitic, 665

Czech Republic

Specification	Designation	Notes	C	Cr	Mn	Mo	Ni	P	S	Si	Other	UTS	YS	El	Hard
CSN 417125	17125	Heat res	0.15 max	12.0-14.5	0.8 max			0.04 max	0.035 max	1.0-2.0	Al 0.6-1.2; bal Fe				

USA

Specification	Designation	Notes	C	Cr	Mn	Mo	Ni	P	S	Si	Other	UTS	YS	El	Hard
	UNS S66545	Fe Base Superalloy W545	0.08 max	12.0-15.0	1.25-2.00	1.25-2.25	24.0-28.0	0.040 max	0.030 max	0.10-0.80	Al <=0.25; B 0.01-0.07; Cu <=0.25; Ti 2.70-3.30; bal Fe				
SAE J467(68)	Stainless W 545		0.06	17.0	0.55		7.00			0.60	Al 0.20; N 0.02; Ti 0.80; bal Fe	1324	1289	13	44 HRC
SAE J467(68)	W-545		0.03	13.5	1.65	1.75	26.0			0.80	Al 0.15; B 0.02; Ti 3.00; bal Fe	1248	910	19	

Stainless Steel, Austenitic, 744X

USA

Specification	Designation	Notes	C	Cr	Mn	Mo	Ni	P	S	Si	Other	UTS	YS	El	Hard
	UNS S31100		0.05 max	25.00-27.00	1.00 max		6.00-7.00	0.030 max	0.030 max	0.60 max	Ti <=0.25; bal Fe				

Stainless Steel, Austenitic, 800H

Germany

Specification	Designation	Notes	C	Cr	Mn	Mo	Ni	P	S	Si	Other	UTS	YS	El	Hard
DIN EN 10028(96)	WNr 1.4958	Flat product for Press purposes SHT	0.03-0.08	19.0-22.0	1.50 max		30.0-32.5	0.015 max	0.010 max	0.70 max	Al 0.20-0.50; Co <=0.50; Cu <=0.50; N <=0.030; Nb <=0.10; Ti 0.20-0.50; Ni+Co 30.0-32.5; Al+Ti <=0.70; bal Fe	500-750	170	35	
DIN EN 10028(96)	X5NiCrAlTi31-20	Flat product for Press purposes SHT	0.03-0.08	19.0-22.0	1.50 max		30.0-32.5	0.015 max	0.010 max	0.70 max	Al 0.20-0.50; Co <=0.50; Cu <=0.50; N <=0.030; Nb <=0.10; Ti 0.20-0.50; Ni+Co 30.0-32.5; Al+Ti 0.70 max; bal Fe	500-750	170	35	

USA

Specification	Designation	Notes	C	Cr	Mn	Mo	Ni	P	S	Si	Other	UTS	YS	El	Hard
	UNS N08810		0.05-0.10	19.00-23.00	1.50 max		30.0-35.0	0.045 max	0.015 max	1.00 max	Al 0.15-0.60; Cu <=0.75; Ti 0.15-0.60; bal Fe				

UNS numbers and US grades are provided as a means of cross referencing chemically similar alloys. Exchangability is only possible after independent examination of specifications. Tensile properties are minimum or typical as specified. UTS and YS as MPa. El as %. See Appendix for list of abbreviations used in Notes. * indicates obsolete material.

Specification	Designation	Notes	C	Cr	Mn	Mo	Ni	P	S	Si	Other	UTS	YS	El	Hard

Stainless Steel, Austenitic, 803

UK

| BS 3076(89) | NA15 | Ni/Ni alloys; Heat res; bar | 0.10 max | 19.0-23.0 | 1.50 max | | 30.0-35.0 | | 0.015 max | 1.00 max | Al 0.15-0.60; Cu <=0.75; Ti 0.15-0.60; 30.0<=Ni+Co<=45.0; bal Fe | | | | |
| BS 3076(89) | NA15H | Ni-Fe-Cr alloy; Heat res; Bar | 0.05-0.10 | 19.0-23.0 | 1.50 max | | 30.0-35.0 | | 0.015 max | 1.00 max | Al 0.15-0.60; Cu <=0.75; Ti 0.15-0.60; 30.0<=Ni+Co<=45.0; bal Fe | | | | |

USA

| | UNS S35045 | Heat res | 0.06-0.10 | 25.0-29.0 | 1.50 max | | 32.0-37.0 | 0.045 max | 0.015 max | 1.00 max | Al 0.15-0.60; Cu <=0.75; Ti 0.15-0.60; bal Fe | | | | |

Stainless Steel, Austenitic, 864

USA

	UNS S35315	Heat res 353MA	0.04-0.08	24.0-26.0	2.0 max		34.0-36.0	0.040 max	0.030 max	1.2-2.0	N 0.12-0.18; Ce 0.03-0.08; bal Fe				
ASTM A240/A240M(98)	S35315	Heat res, 353MA	0.04-0.08	24.0-26.0	2.00 max		34.0-36.0	0.040 max	0.030 max	1.2-2.00	N 0.12-0.18; 0.03<=Ce<=0.08; bal Fe	650	270	40.0	217 HB
SAE J405(98)	S35315	Heat res 353MA	0.04-0.08	24.00-26.00	2.00 max		34.00-36.00	0.040 max	0.030 max	1.20-2.00	N 0.12-0.18; Ce 0.03-0.08; bal Fe				

Stainless Steel, Austenitic, 9

USA

| | UNS S50460 | Heat res | 0.08-0.12 | 8.0-9.5 | 0.30-0.60 | 0.85-1.05 | 0.40 max | 0.020 max | 0.010 max | 0.20-0.50 | Al <=0.04; N 0.030-0.070; Nb 0.06-0.10; V 0.18-0.25; bal Fe | | | | |

Stainless Steel, Austenitic, A167

USA

	UNS S30415	Heat res 153 MA	0.04-0.06	18.0-19.0	0.8 max	0.5 max	9.0-10.0	0.045 max	0.030 max	1.0-2.0	N 0.12-0.18; Ce 0.03-0.08; bal Fe				
ASTM A240/A240M(98)	S30415	Heat res, 153 MA	0.04-0.06	18.0-19.0	0.80 max		9.0-10.0	0.045 max	0.030 max	1.00-2.00	N 0.12-0.18; 0.03<=Ce<=0.08; bal Fe	600	290	40.0	217 HB
ASTM A249/249M(96)	S30415	Weld tube; boiler, superheater, heat exch	0.04-0.06	18.0-19.0	0.80 max		9.00-10.00	0.045 max	0.030 max	1.00-2.00	Cu 0.03-0.08; N 0.12-0.18; bal Fe	600	290	35	96 HRB
SAE J405(98)	S30415	Heat res 153 MA	0.04-0.06	18.00-19.00	0.80 max	0.5 max	9.00-10.00	0.045 max	0.030 max	1.00-2.00	N 0.12-0.18; Ce 0.03-0.08; bal Fe				

Stainless Steel, Austenitic, A182

USA

	UNS S31266	Uranus B66	0.030 max	23.0-25.0	2.0-4.0	5.0-7.0	21.0-24.0	0.035 max	0.020 max	1.00 max	Cu 0.50-3.00; N 0.35-0.60; W 1.00-3.00; bal Fe				
ASTM A240/A240M(98)	S31266	Uranus B66	0.030 max	23.0-25.0	2.0-4.0	5.0-7.0	21.0-24.0	0.035 max	0.020 max	1.00 max	Cu 1.00-3.00; N 0.35-0.60; W 1.00-2.50; bal Fe	750	420	35.0	
SAE J405(98)	S31266	Uranus B66	0.030 max	23.00-25.00	2.00-4.00	5.00-7.00	21.0-24.0	0.035 max	0.020 max	1.00 max	Cu 0.50-3.00; N 0.35-0.60; W 1.00-3.00; bal Fe				

Stainless Steel, Austenitic, A313

USA

| | UNS S20430 | Cr-Mn-Cu-Ni Grade 204CU | 0.15 max | 15.5-17.5 | 6.5-9.0 | | 1.50-3.50 | 0.060 max | 0.030 max | 1.00 max | Cu 2.00-4.00; N 0.05-0.25; bal Fe | | | | |

Stainless Steel, Austenitic, A579(51)

India

| IS 1570/7(92) | 12Cr13H | 0-160mm | 0.09-0.15 | 11.5-14 | 1 max | 0.15 max | 1 max | 0.04 max | 0.03 max | 1 max | Co <=0.1; Cu <=0.3; Pb <=0.15; W <=0.1; bal Fe | 470-670 | 265 | 20 | |
| IS 1570/7(92) | 19 | 0-160mm | 0.09-0.15 | 11.5-14 | 1 max | 0.15 max | 1 max | 0.04 max | 0.03 max | 1 max | Co <=0.1; Cu <=0.3; Pb <=0.15; W <=0.1; bal Fe | 470-670 | 265 | 20 | |

UNS numbers and US grades are provided as a means of cross referencing chemically similar alloys. Exchangability is only possible after independent examination of specifications. Tensile properties are minimum or typical as specified. UTS and YS as MPa. El as %. See Appendix for list of abbreviations used in Notes. * indicates obsolete material.

Specification	Designation	Notes	C	Cr	Mn	Mo	Ni	P	S	Si	Other	UTS	YS	El	Hard

Stainless Steel, Austenitic, A579(51) (Continued from previous page)

USA

Specification	Designation	Notes	C	Cr	Mn	Mo	Ni	P	S	Si	Other	UTS	YS	El	Hard
	UNS S41001	Superstrength	0.15 max	11.50-13.50	1.00 max	0.50 max	0.75 max	0.025 max	0.025 max	1.00 max	Al <=0.05; Cu <=0.50; Sn 0.05 max; bal Fe				
ASTM A579(96)	51	Superstrength Frg	0.15 max	11.5-13.5	1.00 max	0.50 max	0.75 max	0.025 max	0.025 max	1.00 max	Al <=0.05; bal Fe	1210	965	12	197 HB

Stainless Steel, Austenitic, A771

USA

Specification	Designation	Notes	C	Cr	Mn	Mo	Ni	P	S	Si	Other	UTS	YS	El	Hard
	UNS S38660	Reactor Grade	0.03-0.05	12.5-14.5	1.65-2.35	1.50-2.50	14.5-16.5	0.040 max	0.010 max	0.50-1.00	Al <=0.050; B <=0.0020; Co <=0.050; Cu <=0.04; N <=0.005; Nb <=0.050; Ti 0.10-0.40; V <=0.05; As<=0.030; Ta<=0.020; bal Fe				
ASTM A771/A771M(95)	A771	Smls tube; Reactor Core, CW	0.030-0.050	12.5-14.5	1.65-2.35	1.50-2.50	14.5-16.5	0.040 max	0.010 max	0.50-1.00	Al <=0.050; B <=0.0020; Co <=0.050; Cu <=0.04; N <=0.005; Nb <=0.050; Ti 0.10-0.40; V <=0.05; Ta<=0.020; As<=0.030; bal Fe	655-827	517-758	10	
ASTM A826/A826M(95)	A771	Smls duct tub; Reactor grade; CW	0.030-0.050	12.5-14.5	1.65-2.35	1.50-2.50	14.5-16.5	0.040 max	0.010 max	0.50-1.00	Al <=0.050; B <=0.0020; Co <=0.050; Cu <=0.04; N <=0.005; Nb <=0.050; Ti 0.10-0.40; V <=0.05; Ta<=0.020; As<=0.030; bal Fe	689	586	15	
ASTM A831/A831M(95)	S38660	Bar Bil Frg; reactor core, Ann	0.030-0.050	12.5-14.5	1.65-2.35	1.50-2.50	14.5-16.5	0.040 max	0.010 max	0.50-1.00	Al <=0.050; B <=0.0020; Co <=0.050; Cu <=0.04; N <=0.005; Nb <=0.050; V <=0.05; Ta 0.020, As<=0.030; bal Fe	515	205	30	

Stainless Steel, Austenitic, AL-6NX

USA

Specification	Designation	Notes	C	Cr	Mn	Mo	Ni	P	S	Si	Other	UTS	YS	El	Hard
ASTM A182/A182M(98)	F58	Frg/roll pipe flange valve	0.030 max	20.00-22.00	2.00 max	6.00-7.00	23.50-25.50	0.040 max	0.030 max	1.00 max	Cu <=0.75; N 0.18-0.25; bal Fe	655	310	30	
ASTM A194/A194M(98)	9C	Nuts, high-temp press	0.30 max	20.00-22.00	2.00 max	6.0-7.0	23.5-25.5	0.040 max	0.030 max	1.00 max	Cu <=0.75; N 0.18-0.25; bal Fe				126-300 HB
ASTM A194/A194M(98)	9CA	Nuts, high-temp press	0.30 max	20.00-22.00	2.00 max	6.0-7.0	23.5-25.5	0.040 max	0.030 max	1.00 max	Cu <=0.75; N 0.18-0.25; bal Fe				126-192 HB
ASTM A249/249M(96)	N08367	Weld tube; boiler, superheater, heat exch	0.030 max	20.00-22.00	2.00 max	6.00-7.00	23.50-25.50	0.040 max	0.030 max	1.00 max	Cu <=0.75; N 0.18-0.25; bal Fe	655-690	310	30	100 HRB
ASTM A688/A688M(96)	N08367	Weld Feedwater Heater Tub	0.030 max	20.00-22.00	2.00 max	6.00-7.00	23.50-25.50	0.040 max	0.03 max	1.00 max	Cu <=0.75; N 0.18-0.25; bal Fe	655-690	310	30	
ASTM A813/A813M(95)	N08367	Weld Pipe	0.030 max	20.00-22.00	2.00 max	6.00-7.00	23.50-25.50	0.040 max	0.030 max	1.00 max	Cu <=0.75; N 0.18-0.25; bal Fe	655-690	310		
ASTM A814/A814M(96)	N08367	CW Weld Pipe										655-690	310		
ASTM A988(98)	N08367	HIP Flanges, Fittings, Valves/parts; Heat res	0.030 max	20.0-22.0	2.00 max	6.0-7.0	23.50-25.50	0.040 max	0.030 max	1.00 max	Cu <=0.75; N 0.18-0.25; bal Fe	655	310	30.0	

Stainless Steel, Austenitic, AL-6XN

USA

Specification	Designation	Notes	C	Cr	Mn	Mo	Ni	P	S	Si	Other	UTS	YS	El	Hard
	UNS N08367		0.030 max	20.0-22.0	2.00 max	6.0-7.0	23.50-25.50	0.040 max	0.030 max	1.00 max	N 0.18-0.25; bal Fe				

Stainless Steel, Austenitic, Alloy 700Si

USA

Specification	Designation	Notes	C	Cr	Mn	Mo	Ni	P	S	Si	Other	UTS	YS	El	Hard
	UNS S70003	Fe-Ni-Cr-Si Corr res, 700Si	0.02 max	8.0-11.0	2.00 max	0.50 max	22.0-25.0	0.025 max	0.010 max	6.5-8.0	Al <=0.50; bal Fe				
ASTM A946(95)		Sh Strp Plt; Corr res; Heat res	0.02 max	8.0-11.0	2.0 max	0.50 max	22.0-25.0	0.025 max	0.010 max	6.5-8.0	bal Fe	540	240	50	95 HB

UNS numbers and US grades are provided as a means of cross referencing chemically similar alloys. Exchangability is only possible after independent examination of specifications. Tensile properties are minimum or typical as specified. UTS and YS as MPa. El as %. See Appendix for list of abbreviations used in Notes. * indicates obsolete material.

Specification	Designation	Notes	C	Cr	Mn	Mo	Ni	P	S	Si	Other	UTS	YS	El	Hard

Stainless Steel, Austenitic, Alloy 700Si (Continued from previous page)

USA

ASTM A953(96)	700Si	Smls Weld tub; Corr res, Heat res	0.020 max	8.0-11.0	2.00 max	0.50 max	22.0-25.0	0.025 max	0.010 max	6.5-8.0	bal Fe	540	240	40	220 HV max
ASTM A954(96)	700Si	Smls Weld Pipe; Corr and Heat res	0.020 max	8.0-11.0	2.00 max	0.50 max	22.0-25.0	0.025 max	0.010 max	6.5-8.0	bal Fe	540	240	40	
ASTM A968(96)		Bar Shp; Corr res; Heat res	0.02 max	8.0-11.0	2.0 max	0.50 max	22.0-25.0	0.025 max	0.010 max	6.5-8.0	bal Fe	540	240	40	95 HB

Stainless Steel, Austenitic, Alloy SX

USA

	UNS S32615		0.07 max	16.0-21.0	2.00 max	0.3-1.5	17.5-22.5	0.045 max	0.030 max	4.8-6.0	Cu 1.5-2.5; bal Fe				
ASTM A213/A213M(95)	S32615	Smls tube boiler, superheater, heat exchanger	0.07 max	16.5-19.5	2.00 max	0.3-1.5	19.0-22.0	0.045 max	0.030 max	4.8-6.0	Cu 1.5-2.5; bal Fe	550	220	25	
ASTM A240/A240M(98)	S32615		0.07 max	16.5-19.5	2.00 max	0.30-1.50	19.0-22.0	0.045 max	0.030 max	4.8-6.0	Cu 1.50-2.50; bal Fe	550	220	25	
ASTM A943/A943M(95)		Spray formed smls pipe	0.07 max	16.5-19.5	2.0 max	0.3-1.5	19.0-22.0	0.045 max	0.030 max	4.8-6.0	Cu 1.5-2.5; bal Fe	550	220		
SAE J405(98)	S32615		0.07 max	16.5-19.5	2.00 max	0.30-1.5	19.0-22.0	0.045 max	0.030 max	4.8-6.0	Cu 1.5-2.5; bal Fe				

Stainless Steel, Austenitic, B853

USA

| | UNS S31905 | with Boron | 0.05 max | 22.00-24.00 | 2.00 max | 3.00-4.00 | 17.00-19.00 | 0.03 max | 0.03 max | 1.00 max | B 0.20-0.50; N <=0.10; bal Fe | | | | |

Stainless Steel, Austenitic, ER321

UK

BS 1501/3(90)	321S51	Press ves; Corr/Heat res; 0.05-100mm	0.04-0.1	17.0-19.0	2.0 max	0.15 max	9.0-12.0	0.045 max	0.025 max	1.0 max	Cu <=0.3; Ti <=0.8; W <=0.1; Ti=5C-0.8; bal Fe	490-690	175	35	
BS 1502(82)	321S51-490	Bar Shp; Press ves	0.04-0.10	17.0-19.0	2.00 max		9.00-12.0	0.045 max	0.030 max	1.00 max	W <=0.1; 5xC<=Ti<=0.8; bal Fe				
BS 1503(89)	321S31	Frg press ves; t<=350mm; Quen	0.08 max	17.0-19.0	2.00 max	0.70 max	9.00-12.0	0.040 max	0.025 max	1.00 max	B <=0.005; Cu <=0.50; W <=0.1; 5xC<=Ti<=0.80; bal Fe	510-710	235	30	
BS 1506(90)	321S31	Bolt matl Pres/Corr res; Corr res; 38<t<=44mm; SHT CD; En821(Ti)	0.08 max	17.0-19.0	2.00 max		9.00-12.0	0.045 max	0.030 max	1.00 max	Ti <=0.8; 5xC<=Ti<=0.80; bal Fe	650	310	28	0-320 HB
BS 1506(90)	321S31	Bolt matl Pres/Corr res; t<=160mm; Hard; was En821(Ti)	0.08 max	17.0-19.0	2.00 max		9.00-12.0	0.045 max	0.030 max	1.00 max	Ti <=0.8; 5xC<=Ti<=0.80; bal Fe	520	205	35	
BS 1554(90)	321S31	Wir Corr/Heat res; diam<0.49mm; Ann	0.08 max	17.0-19.0	2.00 max		9.00-12.0	0.045 max	0.030 max	1.00 max	Ti <=0.08; W <=0.1; 5xC<=Ti<=0.80; bal Fe	0-900			
BS 1554(90)	321S31	Wir Corr/Heat res; 6<diam<=13mm; Ann	0.08 max	17.0-19.0	2.00 max		9.00-12.0	0.045 max	0.030 max	1.00 max	Ti <=0.08; W <=0.1; 5xC<=Ti<=0.80; bal Fe	0-700			
BS 3059/2(78)	321S59	Boiler, Superheater; Tub; Pip	0.04-0.09	17.0-19.0	0.50-2.00	0.15 max	9.0-13.0	0.040 max	0.030 max	0.20-1.00	4C<=Ti<=0.60; bal Fe				
BS 3605(73)	321S18	Corr res; Tub, Weld; Smls tub	0.08 max	17.0-19.0	0.5-2.0		10.0-13.0	0.040 max	0.030 max	0.2-1.0	5xC<=Ti<=0.6; bal Fe				
BS 3605(73)	321S22	Corr res; Tub, Weld; Smls tub	0.08 max	17.0-19.0	0.5-2.0		9.0-12.0	0.040 max	0.030 max	0.2-1.0	5xC<=Ti<=0.6; bal Fe				
BS 3605(73)	321S59	Corr res; Tub, Weld; Smls tub	0.04-0.09	17.0-19.0	0.5-2.0		10.0-13.0	0.040 max	0.030 max	0.2-1.0	5xC<=Ti<=0.80; bal Fe				
BS 3605/1(91)	321S31	Pip Tube press smls; t<=200mm	0.080 max	17.0-19.0	2.00 max		9.00-12.0	0.040 max	0.030 max	1.00 max	5xC<=Ti<=0.80; bal Fe	510-710		35	
BS 3605/1(91)	321S51(1010)	Pip Tube press smls; t<=200mm	0.04-0.10	17.0-19.0	2.00 max		9.00-12.0	0.040 max	0.030 max	1.00 max	5xC<=Ti<=0.80; bal Fe	510-710		35	

USA

| | UNS S32180 | Bare filler metal | 0.08 max | 18.5-20.5 | 1.00-2.50 | 0.75 max | 9.00-10.50 | 0.03 max | 0.03 max | 0.30-0.65 | Cu <=0.75; 9xC<=Ti<=1; bal Fe | | | | |

UNS numbers and US grades are provided as a means of cross referencing chemically similar alloys. Exchangability is only possible after independent examination of specifications. Tensile properties are minimum or typical as specified. UTS and YS as MPa. El as %. See Appendix for list of abbreviations used in Notes. * indicates obsolete material.

Specification	Designation	Notes	C	Cr	Mn	Mo	Ni	P	S	Si	Other	UTS	YS	El	Hard

Stainless Steel, Austenitic, EV11

India

| IS 1570/5(85) | X70Cr21Mn6Ni2N | Valve | 0.65-0.75 | 20-22 | 5.5-7 | 0.15 max | 1.4-1.9 | 0.045 max | 0.035 max | 0.45-0.85 | Co <=0.1; Cu <=0.3; Pb <=0.15; W <=0.1; bal Fe | | | | 321 HB |

International

| ISO 683-15(76) | 10* | Valve, Q, HT | 0.65-0.75 | 20-22 | 5.5-7 | 0.15 max | 1.4-1.9 | 0.05 max | 0.025-0.065 | 0.45-0.85 | Al <=0.1; Co <=0.1; Cu <=0.3; Pb <=0.15; Ti <=0.05; V <=0.1; W <=0.1; bal Fe | | | | |

USA

| | UNS S63011 | Valve, 746 | 0.65-0.75 | 20.50-22.00 | 5.50-6.90 | | 1.40-1.90 | 0.040 max | 0.035 max | 0.45-0.85 | N 0.18-0.28; bal Fe | | | | |

Stainless Steel, Austenitic, EV12

India

| IS 1570/5(85) | X55Cr21Mn8Ni2N | Valve | 0.5-0.6 | 20-22 | 7-9.5 | 0.15 max | 1.5-2.75 | 0.045 max | 0.035 max | 1 max | Co <=0.1; Cu <=0.3; Pb <=0.15; W <=0.1; bal Fe | | | | 321 HB |

USA

	UNS S63012	Valve, 21-2 N	0.50-0.60	19.25-21.50	7.00-10.00		1.50-2.75	0.050 max	0.030 max	0.25 max	N 0.20-0.40; bal Fe				
SAE J775(93)	21-2N	Engine poppet valve	0.50-0.60	19.25-21.50	7.00-10.00		1.50-2.75	0.050 max	0.030 max	0.25 max	N 0.20-0.40; bal Fe	1080	700		
SAE J775(93)	EV12*	Engine poppet valve	0.50-0.60	19.25-21.50	7.00-10.00		1.50-2.75	0.050 max	0.030 max	0.25 max	N 0.20-0.40; bal Fe	1080	700		

Stainless Steel, Austenitic, EV13

UK

BS 970/4(70)	352S52	Valve	0.48-0.58	20.0-22.0	8.00-10.00		3.25-4.50	0.040 max	0.035 max	0.45 max	N 0.38-0.50; Pb <=0.15; C+N<=0.9; bal Fe				
BS 970/4(70)	352S54	Valve	0.48-0.58	20.0-22.0	8.00-10.00		3.25-4.50	0.040 max	0.035-0.080	0.45 max	N 0.38-0.50; Pb <=0.15; C+N<=0.9; bal Fe				
BS 970/4(70)	381S34	Valve	0.15-0.25	20.0-22.0	1.50 max		10.5-12.5	0.040 max	0.030 max	0.75-1.25	N 0.15-0.30; Pb <=0.15; bal Fe				

USA

	UNS S63013	Valve, Gaman H	0.47-0.57	20.00-22.00	11.00-13.00			0.030 max	0.050 max	2.00-3.00	N 0.40-0.50; bal Fe				
SAE J775(93)	EV13*	Engine poppet valve	0.47-0.57	20.00-22.00	11.00-13.00			0.030 max	0.050 max	2.00-3.00	N 0.40-0.50; bal Fe	1080	520		
SAE J775(93)	Gaman H	Engine poppet valve	0.47-0.57	20.00-22.00	11.00-13.00			0.030 max	0.050 max	2.00-3.00	N 0.40-0.50; bal Fe	1080	520		

Stainless Steel, Austenitic, EV16

International

| ISO 683-15(92) | X33CrNiMnN238 | Valve, Q, HT | 0.28-0.38 | 22.0-24.0 | 1.5-3.5 | 0.50 max | 7.0-9.0 | 0.050 max | 0.030 max | 0.50-1.00 | N 0.25-0.35; W <=0.50; bal Fe | 1200 max | | | 360 HB max |

Spain

UNE 36022(86)	F.3210*	see X42CrNiW14-14	0.35-0.5	13-15	1.5 max	0.15 max	13-15	0.045 max	0.035 max	1-2	W 2-3; bal Fe				
UNE 36022(86)	X42CrNiW14-14	Valve	0.35-0.5	13-15	1.5 max	0.15 max	13-15	0.045 max	0.035 max	1-2	W 2-3; bal Fe				
UNE 36022(91)	F.3216*	see X33CrNiMnN23-08	0.28-0.38	22-24	1.5-3.5	0.15 max	7-9	0.03 max	0.04 max	0.5-1	N 0.25-0.4; bal Fe				
UNE 36022(91)	X33CrNiMnN23-08	Valve	0.28-0.38	22-24	1.5-3.5	0.15 max	7-9	0.03 max	0.04 max	0.5-1	N 0.25-0.4; bal Fe				

USA

	UNS S63018	Valve, 23-8N Nitronic 20	0.28-0.38	22.00-24.00	1.50-3.50	0.50 max	7.00-9.00	0.050 max	0.030 max	0.50-1.00	N 0.25-0.35; W <=0.50; bal Fe				
SAE J775(93)	23-8N	Engine poppet valve, ISO X33CrNiMnN238	0.28-0.38	22.00-24.00	1.50-3.50	0.50 max	7.00-9.00	0.050 max	0.030 max	0.50-1.00	N 0.25-0.35; W <=0.50; bal Fe	1010	580		
SAE J775(93)	EV16*	Engine poppet valve, ISOX33CrNiMnN238	0.28-0.38	22.00-24.00	1.50-3.50	0.50 max	7.00-9.00	0.050 max	0.030 max	0.50-1.00	N 0.25-0.35; W <=0.50; bal Fe	1010	580		
SAE J775(93)	ISO X33CrNiMnN238	Engine poppet valve	0.28-0.38	22.00-24.00	1.50-3.50	0.50 max	7.00-9.00	0.050 max	0.030 max	0.50-1.00	N 0.25-0.35; W <=0.50; bal Fe	1010	580		

UNS numbers and US grades are provided as a means of cross referencing chemically similar alloys. Exchangability is only possible after independent examination of specifications. Tensile properties are minimum or typical as specified. UTS and YS as MPa. El as %. See Appendix for list of abbreviations used in Notes. * indicates obsolete material.

Specification	Designation	Notes	C	Cr	Mn	Mo	Ni	P	S	Si	Other	UTS	YS	El	Hard

Stainless Steel, Austenitic, EV3

India

Specification	Designation	Notes	C	Cr	Mn	Mo	Ni	P	S	Si	Other	UTS	YS	El	Hard
IS 1570/5(85)	X20Cr2Ni12N		0.15-0.25	20-22	1.5 max	0.15 max	10.5-12.5	0.045 max	0.035 max	0.75-1.25	Co <=0.1; Cu <=0.3; Pb <=0.15; W <=0.1; bal Fe				302 HB max

USA

Specification	Designation	Notes	C	Cr	Mn	Mo	Ni	P	S	Si	Other	UTS	YS	El	Hard
	UNS S63016	Valve, 21-12	0.15-0.25	20.50-22.00	1.00-1.40		10.50-12.00	0.030 max	0.030 max	0.70 max	bal Fe				

Stainless Steel, Austenitic, EV4

USA

Specification	Designation	Notes	C	Cr	Mn	Mo	Ni	P	S	Si	Other	UTS	YS	El	Hard
	UNS S63017	Valve, 21-12N	0.15-0.25	20.00-22.00	1.00-1.50		10.50-12.50	0.045 max	0.030 max	0.70-1.25	N 0.15-0.25; bal Fe				
SAE J775(93)	21-12N	Engine poppet valve	0.15-0.25	20.00-22.00	1.00-1.50		10.50-12.50	0.045 max	0.030 max	0.70-1.25	N 0.15-0.25; bal Fe	820	430	26.2	
SAE J775(93)	EV4*	Engine poppet valve	0.15-0.25	20.00-22.00	1.00-1.50		10.50-12.50	0.045 max	0.030 max	0.70-1.25	N 0.15-0.25; bal Fe	820	430	26.2	
SAE J775(93)	JIS G4311 SUH 37	Engine poppet valve	0.15-0.25	20.50-22.50	1.00-1.60		10.00-12.00	0.040 max	0.030 max	1.00 max	N 0.15-0.30; bal Fe				

Stainless Steel, Austenitic, EV5

USA

Specification	Designation	Notes	C	Cr	Mn	Mo	Ni	P	S	Si	Other	UTS	YS	El	Hard
	UNS S63014	Valve, 10	0.30-0.45	18.0-20.0	0.80-1.30		7.75-8.25	0.030 max	0.030 max	2.75-3.25	bal Fe				

Stainless Steel, Austenitic, EV6

USA

Specification	Designation	Notes	C	Cr	Mn	Mo	Ni	P	S	Si	Other	UTS	YS	El	Hard
	UNS S63015	Valve, 10 N	0.35-0.45	18.0-20.0	0.80-1.30		7.75-8.15	0.030 max	0.030 max	2.75-3.25	N 0.15-0.25; bal Fe				

Stainless Steel, Austenitic, EV7

USA

Specification	Designation	Notes	C	Cr	Mn	Mo	Ni	P	S	Si	Other	UTS	YS	El	Hard
	UNS S63007	Valve, 21-55N	0.15-0.25	20.0-22.0	5.00-6.50		4.50-6.00	0.04 max	0.03 max	1.00 max	N 0.20-0.35; bal Fe				

Stainless Steel, Austenitic, EV8

Bulgaria

Specification	Designation	Notes	C	Cr	Mn	Mo	Ni	P	S	Si	Other	UTS	YS	El	Hard
BDS 9634	5Ch20N4AG9	Valve	0.48-0.58	20-22	8-10	0.15 max	3.25-4.5	0.035 max	0.03 max	0.45 max	Al <=0.1; Cu <=0.30; bal Fe				

China

Specification	Designation	Notes	C	Cr	Mn	Mo	Ni	P	S	Si	Other	UTS	YS	El	Hard
GB 1221(92)	5Cr21Mn9Ni4N	Bar; SA; Aged	0.48-0.58	20.00-22.00	8.00-10.00		3.25-4.50	0.040 max	0.030 max	0.35 max	N 0.35-0.50; bal Fe	885	560	8	
GB/T 12773(91)	5Cr21Mn9Ni4N	Frg SA Aged	0.48-0.58	20.00-22.00	8.00-10.00		3.25-4.50	0.040 max	0.030 max	0.35 max	N 0.35-0.50; bal Fe	950	580	8	

Czech Republic

Specification	Designation	Notes	C	Cr	Mn	Mo	Ni	P	S	Si	Other	UTS	YS	El	Hard
CSN 417465	17465	Valve	0.48-0.58	20.0-22.0	8.0-10.0	3.25-4.5		0.05 max	0.035 max	0.45 max	N 0.3-0.55; bal Fe				

International

Specification	Designation	Notes	C	Cr	Mn	Mo	Ni	P	S	Si	Other	UTS	YS	El	Hard
ISO 683-15(92)	X53CrMnNiN219	Valve, Q, HT	0.48-0.58	20.0-22.0	8.0-10.0		3.25-4.5	0.050 max	0.030 max	0.25 max	N 0.35-0.50; bal Fe	1300 max			385 HB max
ISO 683-15(92)	X53CrMnNiNb219	Valve, Q, HT	0.48-0.58	20.0-22.0	8.0-10.0		3.25-4.5	0.050 max	0.030 max	0.45 max	N 0.38-0.50; C+N>0.90;Nb+Ta 2.00-3.00; bal Fe	1300 max			385 HB max
ISO 683-15(92)	X55CrMnNiN208	Valve, Q, HT	0.50-0.60	19.5-21.5	7.0-10.0		1.5-2.75	0.050 max	0.030 max	0.25 max	N 0.20-0.40; bal Fe	1300 max			385 HB max

Italy

Specification	Designation	Notes	C	Cr	Mn	Mo	Ni	P	S	Si	Other	UTS	YS	El	Hard
UNI 3992(75)	X53CrMnNiN219	Valve	0.48-0.58	20-23	8-10		3.25-4.5	0.05 max	0.035 max	0.25 max	N 0.38-0.5; bal Fe				

Poland

Specification	Designation	Notes	C	Cr	Mn	Mo	Ni	P	S	Si	Other	UTS	YS	El	Hard
PNH86022(71)	50H21G9N4	Heat res	0.47-0.57	20-22	8-11		3.25-4.5	0.03 max	0.03 max	0.5 max	N 0.38-0.5; bal Fe				

Romania

Specification	Designation	Notes	C	Cr	Mn	Mo	Ni	P	S	Si	Other	UTS	YS	El	Hard
STAS 11311(88)	53NNiMnCr210	Valve	0.48-0.58	20-22	7-10		3.25-4.5	0.05 max	0.02-0.06	0.25 max	N 0.38-0.5; Pb <=0.15; bal Fe				

Spain

Specification	Designation	Notes	C	Cr	Mn	Mo	Ni	P	S	Si	Other	UTS	YS	El	Hard
UNE 36022(91)	F.3213*	see X53CrMnNiNb21-09	0.48-0.58	20-22	8-10	0.15 max	3.25-4.5	0.04 max	0.035 max	0.25 max	N 0.38-0.5; Nb 2-3; bal Fe				

UNS numbers and US grades are provided as a means of cross referencing chemically similar alloys. Exchangability is only possible after independent examination of specifications. Tensile properties are minimum or typical as specified. UTS and YS as MPa. El as %. See Appendix for list of abbreviations used in Notes. * indicates obsolete material.

Specification	Designation	Notes	C	Cr	Mn	Mo	Ni	P	S	Si	Other	UTS	YS	El	Hard

Stainless Steel, Austenitic, EV8 (Continued from previous page)

Spain

Specification	Designation	Notes	C	Cr	Mn	Mo	Ni	P	S	Si	Other	UTS	YS	El	Hard
UNE 36022(91)	F.3215*	see X70CrMnNiN21-06	0.65-0.75	20-22.5	5.5-7	0.15 max	1.4-1.9	0.045 max	0.035 max	0.45-0.85	N 0.18-0.28; bal Fe				
UNE 36022(91)	F.3217*	see X53CrMnNiN21-09	0.48-0.58	20-22	8-10	0.15 max	3.25-4.5	0.04 max	0.035 max	0.25 max	N 0.38-0.5; bal Fe				
UNE 36022(91)	X53CrMnNiN21-09	Valve	0.48-0.58	20-22	8-10	0.15 max	3.25-4.5	0.04 max	0.035 max	0.25 max	N 0.38-0.5; bal Fe				
UNE 36022(91)	X53CrMnNiNb21-09	Valve	0.48-0.58	20-22	8-10	0.15 max	3.25-4.5	0.04 max	0.035 max	0.25 max	N 0.38-0.5; Nb 2-3; bal Fe				
UNE 36022(91)	X70CrMnNiN21-06	Valve	0.65-0.75	20-22.5	5.5-7	0.15 max	1.4-1.9	0.045 max	0.035 max	0.45-0.85	N 0.18-0.28; bal Fe				

UK

Specification	Designation	Notes	C	Cr	Mn	Mo	Ni	P	S	Si	Other	UTS	YS	El	Hard
BS 970/4(70)	349S52	Valve	0.48-0.58	20.0-22.0	8.00-10.00		3.25-4.50	0.040 max	0.035 max	0.25 max	N 0.38-0.50; Pb <=0.15; C+N<=0.9; bal Fe				
BS 970/4(70)	349S54	Valve	0.48-0.58	20.0-22.0	8.00-10.00		3.25-4.50	0.040 max	0.035-0.080	0.25 max	N 0.38-0.50; Pb <=0.15; C+N<=0.9; bal Fe				

USA

Specification	Designation	Notes	C	Cr	Mn	Mo	Ni	P	S	Si	Other	UTS	YS	El	Hard
	UNS S63008	Valve, 21-4N	0.48-0.58	20.00-22.00	8.00-10.00		3.25-4.50	0.050 max	0.030 max	0.25 max	N 0.35-0.50; bal Fe				
SAE J775(93)	21-4N	Engine poppet valve	0.48-0.58	20.00-22.00	8.00-10.00		3.25-4.50	0.050 max	0.030 max	0.25 max	N 0.35-0.50; bal Fe	1140	740		
SAE J775(93)	EV8*	Engine poppet valve, ISO X53CrMnNiN219	0.48-0.58	20.00-22.00	8.00-10.00		3.25-4.50	0.050 max	0.030 max	0.25 max	N 0.35-0.50; bal Fe	1140	740		
SAE J775(93)	ISO X53CrMnNiN219	Engine poppet valve	0.48-0.58	20.00-22.00	8.00-10.00		3.25-4.50	0.050 max	0.030 max	0.25 max	N 0.35-0.50; bal Fe	1140	740		
SAE J775(93)	ISO X55CrMnNiN208	Engine poppet valve	0.50-0.60	19.5-21.5	7.0-1.00		1.5-2.75	0.050 max	0.030 max	0.25 max	N 0.20-0.40; bal Fe				
SAE J775(93)	JIS G4311 SUH 35	Engine poppet valve	0.48-0.58	20.00-22.00	8.00-10.00		3.25-4.50	0.040 max	0.030 max	0.35 max	N 0.35-0.50; bal Fe				

Stainless Steel, Austenitic, EV9

Bulgaria

Specification	Designation	Notes	C	Cr	Mn	Mo	Ni	P	S	Si	Other	UTS	YS	El	Hard
BDS 9634	4Ch18N9S2WG	Valve	0.4-0.5	17-20	0.8-1.4	0.15 max	8-10	0.03 max	0.03 max	2-3	Al <=0.1; Cu <=0.30; bal Fe				

India

Specification	Designation	Notes	C	Cr	Mn	Mo	Ni	P	S	Si	Other	UTS	YS	El	Hard
IS 1570/5(85)	X40Ni14Cr14W3Si2	Bar Flat Band	0.35-0.5	12-15	1 max	0.15 max	12-15	0.045 max	0.035 max	2 max	Co <=0.1; Cu <=0.3; Pb <=0.15; W 2-3; bal Fe	785	345	35	269 HB max

International

Specification	Designation	Notes	C	Cr	Mn	Mo	Ni	P	S	Si	Other	UTS	YS	El	Hard
ISO 683-15(76)	6*	Valve, Q, HT	0.4-0.5	17-20	0.8-1.5	0.15 max	8-10	0.045 max	0.03 max	2-3	Al <=0.1; Co <=0.1; Cu <=0.3; Pb <=0.15; Ti <=0.05; V <=0.1; W 0.8-1.2; bal Fe				

Italy

Specification	Designation	Notes	C	Cr	Mn	Mo	Ni	P	S	Si	Other	UTS	YS	El	Hard
UNI 3992(58)	X45CNW1909	Valve	0.4-0.5	17-20	0.8-1.15		8-10	0.035 max	0.035 max	2-3	W 0.8-1.2; bal Fe				
UNI 3992(75)	X45CrNiW189	Valve	0.4-0.5	17-20	0.8-1.5		8-10	0.035 max	0.3 max	2-3	W 0.8-1.2; bal Fe				

Spain

Specification	Designation	Notes	C	Cr	Mn	Mo	Ni	P	S	Si	Other	UTS	YS	El	Hard
UNE 36017(61)	F.321	Heat res	0.4-0.5	12-15	0.8-1.5	0.15 max	12-15	0.04 max	0.04	0.8-1.8	W 2-4; bal Fe				
UNE 36022(86)	F.3211*	see X45CrNiSiW18-09	0.4-0.5	17-19	0.8-1.5	0.15 max	8-10	0.045 max	0.03 max	2-3	W 0.8-1.2; bal Fe				
UNE 36022(86)	X45CrNiSiW18-09	Valve	0.4-0.5	17-19	0.8-1.5	0.15 max	8-10	0.045 max	0.03 max	2-3	W 0.8-1.2; bal Fe				

UK

Specification	Designation	Notes	C	Cr	Mn	Mo	Ni	P	S	Si	Other	UTS	YS	El	Hard
BS 1504(76)	330C11*	Press ves	0.35-0.55	13.0-17.0	2.0 max	1.5 max	33.0-37.0	0.04 max	0.04 max	1.5 max	Cu <=0.3; Ti <=0.05; W <=0.1; bal Fe				
BS 970/4(70)	331S40	Valve	0.35-0.50	12.0-15.0	0.50-1.00		12.0-15.0	0.040 max	0.030 max	1.00-2.00	Pb <=0.15; W 2.00-3.00; bal Fe				
BS 970/4(70)	331S42	Valve	0.37-0.47	13.0-15.0	0.50-1.00	0.40-0.70	13.0-15.0	0.040 max	0.030 max	1.00-2.00	Pb <=0.15; W 2.20-3.00; bal Fe				

USA

Specification	Designation	Notes	C	Cr	Mn	Mo	Ni	P	S	Si	Other	UTS	YS	El	Hard
	UNS S66009	Valve, TPA	0.35-0.50	12.0-15.0	1.00 max	0.20-0.50	12.0-15.0	0.045 max	0.030 max	0.30-0.80	W 1.50-3.00; if Mo not used, W 2.00-3.00; bal Fe				

Specification	Designation	Notes	C	Cr	Mn	Mo	Ni	P	S	Si	Other	UTS	YS	El	Hard
Stainless Steel, Austenitic, F10															
USA															
	UNS S33100	F-10	0.10-0.20	7.00-9.00	0.50-0.80		19.00-22.00	0.030 max	0.030 max	1.00-1.40	bal Fe				
ASTM A182/A182M(98)	F10	Frg/roll pipe flange valve	0.10-0.20	7.0-9.0	0.50-0.80		19.0-22.0	0.040 max	0.030 max	1.00-1.40	bal Fe	550	205	30	
Stainless Steel, Austenitic, F1586															
USA															
	UNS S31675	Surgical Implant	0.08 max	19.5-22.0	2.00-4.25	2.0-3.0	9.0-11.0	0.025 max	0.01 max	0.75 max	Cu <=0.25; N 0.25-0.5; Nb 0.25-0.80; bal Fe				
Stainless Steel, Austenitic, F20															
USA															
	UNS N08020		0.07 max	19.00-21.00	2.00 max	2.00-3.00	32.00-38.00	0.045 max	0.035 max	1.00 max	Cu 3.00-4.00; 8xC<=Nb<=1.00; bal Fe				
ASTM A182/A182M(98)	F20	Frg/roll pipe flange valve	0.07 max	19.0-21.0	2.00 max	2.00-3.00	32.0-38.0	0.045 max	0.035 max	1.00 max	Cu 3.0-4.0; Nb>=8xC<=1.00; bal Fe	550	240	30	
Stainless Steel, Austenitic, F44															
Europe															
EN 10088/2(95)	1.4547	Strp, CR Corr res; t<=6mm, Ann	0.020 max	19.50-20.50	1.00 max	6.00-7.00	17.50-18.50	0.030 max	0.010 max	0.70 max	Cu 0.50-1.00; N 0.18-0.25; bal Fe	650-850	320	35	
EN 10088/2(95)	X1CrNiMoCuN20-18-7	Strp, CR Corr res; t<=6mm, Ann	0.020 max	19.50-20.50	1.00 max	6.00-7.00	17.50-18.50	0.030 max	0.010 max	0.70 max	Cu 0.50-1.00; N 0.18-0.25; bal Fe	650-850	320	35	
EN 10088/3(95)	1.4547	Bar Rod Sect; Corr res; 160<t<=250mm, Ann	0.020 max	19.50-20.50	1.00 max	6.00-7.00	17.50-18.50	0.030 max	0.010 max	0.70 max	Cu 0.5-1.00; N 0.18-0.25; bal Fe	650-850	300	30	260 HB
EN 10088/3(95)	1.4547	Bar Rod Sect, Corr res; t<=160mm, Ann	0.020 max	19.50-20.50	1.00 max	6.00-7.00	17.50-18.50	0.030 max	0.010 max	0.70 max	Cu 0.50-1.00; N 0.18-0.25; bal Fe	650-850	300	35	260 HB
EN 10088/3(95)	X1CrNiMoCuN20-18-7	Bar Rod Sect, Corr res; t<=160mm, Ann	0.020 max	19.50-20.50	1.00 max	6.00-7.00	17.50-18.50	0.030 max	0.010 max	0.70 max	Cu 0.50-1.00; N 0.18-0.25; bal Fe	650-850	300	35	260 HB
EN 10088/3(95)	X1CrNiMoCuN20-18-7	Bar Rod Sect; Corr res; 160<t<=250mm, Ann	0.020 max	19.50-20.50	1.00 max	6.00-7.00	17.50-18.50	0.030 max	0.010 max	0.70 max	Cu 0.5-1.00; N 0.18-0.25; bal Fe	650-850	300	30	260 HB
Hungary															
MSZ 4360(87)	KO45ELC	Corr res; 0-160mm; SA	0.02 max	19-21	2 max	6-7	24-26	0.02 max	0.015 max	0.5 max	Cu 0.7-1; N 0.12-0.2; bal Fo	600-800	300	35L	190 HB max
MSZ 4360(87)	X2NiCrMoCuN25206	Corr res; 0-160mm; SA	0.02 max	19-21	2 max	6-7	24-26	0.02 max	0.015 max	0.5 max	Cu 0.7-1; bal Fe	600-800	300	35L	190 HB max
Sweden															
SS 142378	2378	Corr res	0.02 max	19.5-20.5	1 max	6-6.5	17.5-18.5	0.03 max		0.1 max	0.8 max	Cu 0.5-1; N 0.18-0.22; bal Fe				
USA															
	UNS S31254	254 SMO LN	0.020 max	19.50-20.50	1.00 max	6.00-6.50	17.50-18.50	0.030 max	0.010 max	0.80 max	Cu 0.50-1.00; N 0.180-0.220; bal Fe				
ASTM A182/A182M(98)	F44	Frg/roll pipe flange valve	0.020 max	19.5-20.5	1.00 max	6.0-6.5	17.5-18.5	0.040 max	0.010 max	0.80 max	Cu 0.50-1.00; N 0.18-0.22; bal Fe	650	300	35	
ASTM A193/A193M(98)	B8MLCuN	Bolt, high-temp, HT	0.020 max	19.50-20.50	1.00 max	6.0-6.5	17.5-18.5	0.030 max	0.010 max	0.80 max	Cu 0.50-1.00; N 0.18-0.22; bal Fe	550	240	30	223 HB
ASTM A193/A193M(98)	B8MLCuNA	Bolt, high-temp, HT	0.020 max	19.50-20.50	1.00 max	6.0-6.5	17.5-18.5	0.030 max	0.010 max	0.80 max	Cu 0.50-1.00; N 0.18-0.22; bal Fe	515	205	30	192 HB
ASTM A194/A194M(98)	8MLCuN	Nuts, high-temp press	0.020 max	19.5-20.5	1.00 max	6.0-6.5	17.5-18.5	0.030 max	0.010 max	0.80 max	Cu 0.50-1.00; N 0.18-0.22; bal Fe				126-300 HB
ASTM A194/A194M(98)	8MLCuNA	Nuts, high-temp press	0.020 max	19.5-20.5	1.00 max	6.0-6.5	17.5-18.5	0.030 max	0.010 max	0.80 max	Cu 0.50-1.00; N 0.18-0.22; bal Fe				126-192 HB
ASTM A240/A240M(98)	S31254	254 SMO LN	0.020 max	19.5-20.5	1.00 max	6.00-6.50	17.5-18.50	0.030 max	0.010 max	0.80 max	Cu 0.50-1.00; N 0.18-0.22; bal Fe	650	300	35.0	223 HB
ASTM A249/249M(96)	S31254	Weld tube; boiler, superheater, heat exch	0.02 max	19.5-20.5	1.00 max	6.00-6.50	17.5-18.5	0.03 max	0.01 max	0.80 max	Cu 0.50-1.00; N 0.18-0.22; bal Fe	650	300	35	96 HRB
ASTM A813/A813M(95)		Weld Pipe, 254 SMO	0.020 max	19.50-20.50	1.00 max	6.00-6.50	17.50-18.50	0.030 max	0.010 max	0.80 max	Cu 0.50-1.00; N 0.180-0.220; bal Fe	650	300		
ASTM A814/A814M(96)	S31254	CW Weld Pipe, 254 SMO	0.020 max	19.50-20.50	1.00 max	6.00-6.50	17.50-18.50	0.030 max	0.010 max	0.80 max	Cu 0.50-1.00; N 0.180-0.220; bal Fe	650	300		
ASTM A943/A943M(95)		Spray formed smls pipe	0.020 max	19.5-20.5	1.00 max	6.00-6.50	17.5-18.5	0.030 max	0.010 max	0.80 max	Cu 0.50-1.00; N 0.18-0.22; bal Fe	650	300		

UNS numbers and US grades are provided as a means of cross referencing chemically similar alloys. Exchangability is only possible after independent examination of specifications. Tensile properties are minimum or typical as specified. UTS and YS as MPa. El as %. See Appendix for list of abbreviations used in Notes. * indicates obsolete material.

Specification	Designation	Notes	C	Cr	Mn	Mo	Ni	P	S	Si	Other	UTS	YS	El	Hard

Stainless Steel, Austenitic, F44 (Continued from previous page)

USA

Specification	Designation	Notes	C	Cr	Mn	Mo	Ni	P	S	Si	Other	UTS	YS	El	Hard
ASTM A988(98)	S31254	HIP Flanges, Fittings, Valves/parts; Heat res; 254 SMO	0.020 max	19.5-20.5	1.00 max	6.0-6.5	17.5-18.5	0.030 max	0.010 max	0.80 max	Cu 0.50-1.00; N 0.18-0.22; bal Fe	650	300	35	
SAE J405(98)	S31254	254 SMO LN	0.020 max	19.50-20.50	1.00 max	6.00-6.50	17.50-18.50	0.030 max	0.010 max	0.80 max	Cu 0.50-1.00; N 0.18-0.22; bal Fe				

Stainless Steel, Austenitic, F45

Sweden

Specification	Designation	Notes	C	Cr	Mn	Mo	Ni	P	S	Si	Other	UTS	YS	El	Hard
SS 142368	2368	Corr res	0.05-0.1	20-22	0.8 max		10-12	0.04 max	0.03 max	1.4-2	N 0.14-0.2; Ce 0.03-0.08; bal Fe				

USA

Specification	Designation	Notes	C	Cr	Mn	Mo	Ni	P	S	Si	Other	UTS	YS	El	Hard
	UNS S30815	Heat res 253 MA	0.10 max	20.00-22.00	0.80 max		10.00-12.00	0.040 max	0.030 max	1.40-2.00	N 0.14-0.20; Ce 0.03-0.08; bal Fe				
ASTM A182/A182M(98)	F45	Frg/roll pipe flange valve	0.05-0.10	20.0-22.0	0.80 max		10.0-12.0	0.040 max	0.030 max	1.40-2.00	N 0.14-0.20; Ce 0.03-0.08; bal Fe	600	310	40	
ASTM A213/A213M(95)	S30815	Smls tube boiler, superheater, heat exchanger	0.05-0.10	20.0-22.0	0.80 max		10.0-12.0	0.040 max	0.030 max	1.40-2.00	N 0.14-0.20; 0.03<=Ce<=0.08; bal Fe	600	310	40	
ASTM A240/A240M(98)	S30815	Heat res, 253 MA	0.05-0.10	20.0-22.0	0.80 max		10.0-12.0	0.040 max	0.030 max	1.40-2.00	N 0.14-0.20; 0.03<=Ce<=0.08; bal Fe	600	310	40.0	217 HB
ASTM A249/249M(96)	S30815	Weld tube; boiler, superheater, heat exch	0.05-0.10	20.0-22.0	0.80 max		10.0-12.0	0.04 max	0.03 max	1.40-2.00	N 0.14-0.20; Ce 0.03-0.08; bal Fe	600	310	35	95 HRB
ASTM A813/A813M(95)		Weld Pipe, Heat res 253 MA	0.10 max	20.0-22.0	0.80 max		10.0-12.0	0.040 max	0.030 max	1.40-2.00	N 0.14-0.20; Ce 0.03-0.08; bal Fe	600	310		
ASTM A814/A814M(96)	S30815	CW Weld Pipe, Heat res 253 MA	0.10 max	20.0-22.0	0.80 max		10.0-12.0	0.040 max	0.030 max	1.40-2.00	N 0.14-0.20; Ce 0.03-0.08; bal Fe	600	310		
ASTM A943/A943M(95)		Spray formed smls pipe	0.05-0.10	20.0-22.0	0.80 max		10.0-12.0	0.040 max	0.030 max	1.4-2.00	N 0.14-0.20; Ce 0.03-0.08; bal Fe	600	310		
SAE J405(98)	S30815	Heat res 253 MA	0.05-0.10	20.00-22.00	0.80 max		10.00-12.00	0.040 max	0.030 max	1.40-2.00	N 0.14-0.20; Ce 0.03-0.08; bal Fe				

Stainless Steel, Austenitic, F46

USA

Specification	Designation	Notes	C	Cr	Mn	Mo	Ni	P	S	Si	Other	UTS	YS	El	Hard
	UNS S30600	18-15	0.018 max	17.0-18.5	2.00 max	0.20 max	14.0-15.5	0.02 max	0.02 max	3.70-4.30	Cu <=0.50; bal Fe				
ASTM A182/A182M(98)	F46	Frg/roll pipe flange valve	0.018 max	17.0-18.5	2.00 max	0.20 max	14.0-15.5	0.020 max	0.020 max	3.7-4.3	Cu <=0.50; bal Fe	540	240	40.0	
ASTM A240/A240M(98)	S30600	18-15	0.018 max	17.0-18.5	2.00 max	0.20 max	14.0-15.5	0.020 max	0.020 max	3.7-4.3	Cu <=0.50; bal Fe	540	240	40.0	
ASTM A336/A336M(98)	F46	Frg, press/high-temp	0.018 max	17.0-18.5	2.00 max	0.20 max	14.0-15.5	0.020 max	0.020 max	3.7-4.3	bal Fe	540-690	220	40	
ASTM A943/A943M(95)		Spray formed smls pipe	0.018 max	17.0-18.5	2.0 max	0.20 max	14.0-15.5	0.02 max	0.02 max	3.7-4.3	Cu <=0.50; bal Fe	540	240		
ASTM A965/965M(97)	F46	Frg for press high-temp parts	0.018 max	17.0-18.5	2.00 max	0.20 max	14.0-15.5	0.020 max	0.020 max	3.7-4.3	bal Fe	540-690	220	40	
ASTM A988(98)	S30600	HIP Flanges, Fittings, Valves/parts; Heat res; 18-15										540	240	40	
SAE J405(98)	S30600	18-15	0.018 max	17.0-18.5	2.00 max	0.20 max	14.0-15.5	0.020 max	0.020 max	3.7-4.3	Cu <=0.50; bal Fe				

Stainless Steel, Austenitic, HNV1

USA

Specification	Designation	Notes	C	Cr	Mn	Mo	Ni	P	S	Si	Other	UTS	YS	El	Hard
	UNS S64005	Valve, 2	0.50-0.60	7.50-8.50	0.20-0.60	0.60-0.90		0.020 max	0.020 max	1.25-1.75	bal Fe				

Stainless Steel, Austenitic, HNV2

USA

Specification	Designation	Notes	C	Cr	Mn	Mo	Ni	P	S	Si	Other	UTS	YS	El	Hard
	UNS S64006	Valve, F	0.35-0.50	1.50-2.50	0.20-0.60			0.030 max	0.030 max	3.50-4.50	bal Fe				

Stainless Steel, Austenitic, HNV3

Bulgaria

Specification	Designation	Notes	C	Cr	Mn	Mo	Ni	P	S	Si	Other	UTS	YS	El	Hard
BDS 9634	4Ch9S3	Valve	0.4-0.5	8-10	0.8 max	0.15 max	0.4 max	0.03 max	0.03 max	2.5-3.5	Al <=0.1; Cu <=0.30; bal Fe				

UNS numbers and US grades are provided as a means of cross referencing chemically similar alloys. Exchangability is only possible after independent examination of specifications. Tensile properties are minimum or typical as specified. UTS and YS as MPa. El as %. See Appendix for list of abbreviations used in Notes. * indicates obsolete material.

Specification	Designation	Notes	C	Cr	Mn	Mo	Ni	P	S	Si	Other	UTS	YS	El	Hard
Stainless Steel, Austenitic, HNV3 (Continued from previous page)															
China															
GB 1221(92)	4Cr9Si2	Bar Q/T	0.35-0.50	8.00-10.00	0.70 max		0.60 max	0.035 max	0.030 max	2.00-3.00	Cu <=0.30; bal Fe	885	590	19	
GB/T 12773(91)	4Cr9Si2	Bar Q/T	0.35-0.50	8.00-10.00	0.70 max		0.60 max	0.035 max	0.030 max	2.00-3.00	Cu <=0.30; bal Fe	880	590	19	
Czech Republic															
CSN 417115	17115	Heat res	0.4-0.5	8.0-10.0	0.8 max			0.04 max	0.03 max	2.8-3.5	bal Fe				
India															
IS 1570/5(85)	X45Cr9Si3	Valve	0.4-0.5	7.5-9.5	0.3-0.6	0.15 max	0.5 max	0.045 max	0.05 max	3-3.75	Co <=0.1; Cu <=0.3; Pb <=0.15; W <=0.1; bal Fe				255-293 HB
International															
ISO 683-15(76)	2*	Valve, Q, HT	0.35-0.45	9.5-11.5	0.8 max	0.7-1.3	0.4 max	0.04 max	0.03 max	1.8-3	Al <=0.1; Co <=0.1; Cu <=0.3; Pb <=0.15; Ti <=0.05; V <=0.1; W <=0.1; bal Fe				
ISO 683-15(92)	X45CrSi93	Valve, Q, HT	0.40-0.50	8.0-10.0	0.80 max		0.60 max	0.040 max	0.030 max	2.7-3.3	bal Fe				300 HB max
ISO 683-15(92)	X50CrSi82	Valve, Q, HT	0.45-0.55	7.5-9.5	0.60 max		0.60 max	0.030 max	0.030 max	1.0-2.0	bal Fe				300 HB max
ISO 683-15(92)	X85CrMoV182	Valve, Q, HT	0.80-0.90	16.5-18.5		1.5 max	2.0-2.5	0.040 max	0.030 max	1.0 max	V 0.30-0.80; bal Fe				300 HB max
Italy															
UNI 3992(58)	X43CS8	Valve	0.38-0.48	8-9.5	0.3-0.6		0.5 max	0.035 max	0.035 max	2.5-3.2	bal Fe				
UNI 3992(75)	X45CrSi8	Valve	0.4-0.5	7.5-9.5	0.8 max		0.5 max	0.035 max	0.03 max	2.8-3.5	bal Fe				
Poland															
PNH86022(71)	H9S2	Heat res	0.35-0.45	8-10	0.7 max		0.6 max	0.035 max	0.03 max	2-3	bal Fe				
Romania															
STAS 11311(88)	45SiCr90	Valve	0.4-0.5	8-10	0.8 max		0.4 max	0.04 max	0.03 max	2.7-3.3	Pb <=0.15; bal Fe				
STAS 11524(80)	45SiCr95	Valve	0.4-0.5	9-10	0.3-0.5		0.4 max	0.03 max	0.025 max	3-3.5	Pb <=0.15; bal Fe				
Spain															
UNE 36022(91)	F.3220*	see X45CrSi09-03	0.4-0.5	8-10	0.8 max	0.15 max		0.04 max	0.03 max	2.7-3.3	bal Fe				
UNE 36022(91)	F.3221*	see X40CrSiMo10-02	0.35-0.45	9-11	0.8 max	0.8-1.3		0.04 max	0.03 max	2-3	bal Fe				
UNE 36022(91)	F.3223*	see X85CrMoV18-02	0.8-0.9	16.5-18.5	1-1.5	2-2.5	0.5 max	0.03 max	0.03 max	1 max	V 0.4-0.6; bal Fe				
UNE 36022(91)	X40CrSiMo10-02	Valve	0.35-0.45	9-11	0.8 max	0.8-1.3		0.04 max	0.03 max	2-3	bal Fe				
UNE 36022(91)	X45CrSi09-03	Valve	0.4-0.5	8-10	0.8 max	0.15 max		0.04 max	0.03 max	2.7-3.3	bal Fe				
UNE 36022(91)	X85CrMoV18-02	Valve	0.8-0.9	16.5-18.5	1-1.5	2-2.5	0.5 max	0.03 max	0.03 max	1 max	V 0.4-0.6; bal Fe				
UK															
BS 970/4(70)	401S45	Valve	0.40-0.50	7.50-9.50	0.30-0.75		0.50 max	0.040 max	0.030 max	3.00-3.75	Pb <=0.15; bal Fe				
USA															
	UNS S65007	Valve, Silchrome 1	0.40-0.50	8.00-10.00	0.80 max		0.60 max	0.040 max	0.030 max	2.70-3.30	bal Fe				
SAE J775(93)	HNV3*	Engine poppet valve, Silchrome 1, ISO X45CrSi93	0.40-0.50	8.00-10.00	0.80 max			0.040 max	0.030 max	2.70-3.30	bal Fe	920	690	22	
SAE J775(93)	ISO X45CrSi93	Engine poppet valve, Silchrome 1	0.40-0.50	8.00-10.00	0.80 max			0.040 max	0.030 max	2.70-3.30	bal Fe	920	690	22	
SAE J775(93)	ISO X50CrSi82	Engine poppet valve	0.45-0.55	7.5-9.5	0.60 max		0.60 max	0.030 max	0.030 max	1.0-2.0	bal Fe				
SAE J775(93)	ISO X85CrMoV182	Engine poppet valve	0.80-0.90	16.5-18.5		1.5 max	2.0-2.5	0.040 max	0.030 max	1.0 max	V 0.30-0.60; bal Fe				
SAE J775(93)	JIS G4311 SUH 1	Engine poppet valve	0.40-0.50	7.50-9.50	0.60 max		0.60 max	0.030 max	0.030 max	3.00-3.50	bal Fe				

UNS numbers and US grades are provided as a means of cross referencing chemically similar alloys. Exchangability is only possible after independent examination of specifications. Tensile properties are minimum or typical as specified. UTS and YS as MPa. El as %. See Appendix for list of abbreviations used in Notes. * indicates obsolete material.

Specification	Designation	Notes	C	Cr	Mn	Mo	Ni	P	S	Si	Other	UTS	YS	El	Hard
Stainless Steel, Austenitic, HNV3 (Continued from previous page)															
USA															
SAE J775(93)	Sil1	Engine poppet valve, Silchrome 1, ISO X45CrSi93	0.40-0.50	8.00-10.00	0.80 max			0.040 max	0.030 max	2.70-3.30	bal Fe	920	690	22	
Stainless Steel, Austenitic, HNV5															
India															
IS 1570/5(85)	X30Cr13	.	0.26-0.35	12-14	1 max	0.15 max	1 max	0.04 max	0.03 max	1 max	Co <=0.1; Cu <=0.3; Pb <=0.15; W <=0.1; bal Fe				241 HB max
USA															
	UNS S63005	Valve, CNS	0.25-0.35	12.00-13.50	0.50 max	0.50 max	7.00-8.50	0.030 max	0.030 max	2.00-3.00	bal Fe				
Stainless Steel, Austenitic, HNV6															
China															
GB 1221(92)	8Cr20Si2Ni	Bar Q/T	0.75-0.85	19.00-20.00	0.20-0.60		1.15-1.65	0.030 max	0.030 max	1.75-2.25	Cu 0.30; bal Fe	885	685	10	
GB/T 12773(91)	8Cr20Si2Ni	Bar Q/T	0.75-0.85	19.00-20.00	0.20-0.60		1.15-1.65	0.030 max	0.030 max	1.75-2.25	Cu 0.30; bal Fe	880	680	10	
India															
IS 1570/5(85)	X80Cr20Si2Ni1		0.75-0.85	19-21	0.2-0.6	0.15 max	1.2-1.7	0.045 max	0.03	1.75-2.25	Co <=0.1; Cu <=0.3; Pb <=0.15; W <=0.1; bal Fe				255-306 HB
Italy															
UNI 3992(75)	X80CrSiNi20	Valve	0.75-0.85	19-21.	0.8 max		1-1.7	0.035 max	0.03 max	1.75-2.5	bal Fe				
Romania															
STAS 11311(88)	80SiNiCr200	Valve	0.75-0.85	19-21	1 max		1-1.75	0.03 max	0.03 max	1.75-2.5	Pb <=0.15; bal Fe				
Spain															
UNE 36022(91)	F.3222*	see X80CrSiNi20-02	0.75-0.85	19-21	0.2-0.6	0.15 max	1-1.7	0.035 max	0.035 max	1.7-2.2	bal Fe				
UNE 36022(91)	X80CrSiNi20-02	Valve	0.75-0.85	19-21	0.2-0.6	0.15 max	1-1.7	0.035 max	0.035 max	1.7-2.2	bal Fe				
UK															
BS 970/4(70)	443S65	Valve	0.75-0.85	19.0-21.0	0.30-0.75		1.20-1.70	0.040 max	0.030 max	1.75-2.25	Pb <=0.15; bal Fe				
USA															
	UNS S65006	Valve, Silchrome XB	0.75-0.90	19.00-21.00	0.80 max		1.00-1.70	0.040 max	0.040 max	1.75-2.60	bal Fe				
SAE J775(93)	HNV6*	Engine poppet valve, Silchrome XB	0.75-0.90	19.00-21.00	0.80 max		1.00-1.70	0.040 max	0.040 max	1.75-2.60	bal Fe	940	840	15.5	
SAE J775(93)	JIS G4311 SUH 4	Engine poppet valve	0.75-0.85	19.00-20.50	0.20-0.60		1.15-1.65	0.030 max	0.030 max	1.75-2.25	bal Fe				
SAE J775(93)	Sil XB	Engine poppet valve	0.75-0.90	19.00-21.00	0.80 max		1.00-1.70	0.040 max	0.040 max	1.75-2.60	bal Fe	940	840	15.5	
Stainless Steel, Austenitic, M31															
USA															
	UNS S21400	Cr-Mn Tenelon	0.12 max	17.00-18.50	14.50-16.00		0.75 max	0.045 max	0.030 max	0.30-1.00	N <=0.35; bal Fe				
ASTM A240/A240M(98)	S21400	Sht	0.12 max	17.0-18.5	14.5-16.0		1.00 max	0.045 max	0.030 max	0.30-1.00	N >=0.35; bal Fe	860	485	40.0	
ASTM A240/A240M(98)	S21400	Strp	0.12 max	17.0-18.5	14.5-16.0		1.00 max	0.045 max	0.030 max	0.30-1.00	N >=0.35; bal Fe	725	380	40.0	
ASTM A580/A580M(98)	XM-31	Wir, Ann	0.12 max	17.0-18.5	14.0-16.0		1.00 max	0.045 max	0.030 max	0.30-1.00	N <=0.35; bal Fe	690	345	40	
SAE J405(98)	S21400	Cr-Mn Tenelon	0.12 max	17.00-18.50	14.00-16.00		1.005 max	0.045 max	0.030 max	0.30-1.00	N >=0.35; bal Fe				

Specification	Designation	Notes	C	Cr	Mn	Mo	Ni	P	S	Si	Other	UTS	YS	El	Hard

Stainless Steel, Austenitic, MA 956

USA

Specification	Designation	Notes	C	Cr	Mn	Mo	Ni	P	S	Si	Other	UTS	YS	El	Hard
	UNS S67956	Fe Base Superalloy, Dispersion Strengthened MA 956	0.10 max	18.5-21.5	0.30 max		0.50 max	0.02 max	0.02 max	0.30 max	Al 3.75-5.75; Co <=0.30; Cu <=0.15; Ti 0.20-0.60; Yttria (Y2O3) 0.30-0.70% by weight; bal Fe				

Stainless Steel, Austenitic, MIL-S-46889

USA

Specification	Designation	Notes	C	Cr	Mn	Mo	Ni	P	S	Si	Other	UTS	YS	El	Hard
	UNS S30115		0.07-0.11	16.50-17.50	1.00-1.50	0.60-0.80	7.70-8.30	0.03 max	0.03 max	0.90-1.40	bal Fe				

Stainless Steel, Austenitic, N08904

Europe

Specification	Designation	Notes	C	Cr	Mn	Mo	Ni	P	S	Si	Other	UTS	YS	El	Hard
EN 10088/2(95)	1.4529	Plt, HR Corr res; t<=75mm, Ann	0.020 max	19.00-21.00	1.00 max	6.00-7.00	24.00-26.00	0.030 max	0.010 max	0.50 max	Cu 0.50-1.50; N 0.15-0.25; bal Fe	650-850	300	40	
EN 10088/2(95)	X1NiCrMoCuN25-20-7	Plt, HR Corr res; t<=75mm, Ann	0.020 max	19.00-21.00	1.00 max	6.00-7.00	24.00-26.00	0.030 max	0.010 max	0.50 max	Cu 0.50-1.50; N 0.15-0.25; bal Fe	650-850	300	40	
EN 10088/3(95)	1.4529	Bar Rod Sect; Corr res; 160<t<=250mm, Ann	0.020 max	19.00-21.00	1.00 max	6.00-7.00	24.00-26.00	0.030 max	0.010 max	0.50 max	Cu 0.50-1.50; N 0.15-0.25; bal Fe	650-850	300	35	250 HB
EN 10088/3(95)	1.4529	Bar Rod Sect; Corr res; t<=160mm, Ann	0.020 max	19.00-21.00	1.00 max	6.00-7.00	24.00-26.00	0.030 max	0.010 max	0.50 max	Cu 0.50-1.50; N 0.15-0.25; bal Fe	650-850	300	40	250 HB
EN 10088/3(95)	X1NiCrMoCuN25-20-7	Bar Rod Sect; Corr res; t<=160mm, Ann	0.020 max	19.00-21.00	1.00 max	6.00-7.00	24.00-26.00	0.030 max	0.010 max	0.50 max	Cu 0.50-1.50; N 0.15-0.25; bal Fe	650-850	300	40	250 HB
EN 10088/3(95)	X1NiCrMoCuN25-20-7	Bar Rod Sect; Corr res; 160<t<=250mm, Ann	0.020 max	19.00-21.00	1.00 max	6.00-7.00	24.00-26.00	0.030 max	0.010 max	0.50 max	Cu 0.50-1.50; N 0.15-0.25; bal Fe	650-850	300	35	250 HB

USA

Specification	Designation	Notes	C	Cr	Mn	Mo	Ni	P	S	Si	Other	UTS	YS	El	Hard
	UNS N08904		0.020 max	19.0-23.0	2.00 max	4.00-5.00	23.0-28.0	0.045 max	0.035 max	1.00 max	Cu 1.00-2.00; bal Fe				
ASTM A249/249M(96)	N08904	Weld tube; boiler, superheater, heat exch	0.020 max	19.0-23.0	2.00 max	4.0-5.0	23.0-28.0	0.045 max	0.035 max	1.00 max	Cu 1.0-2.0; N <=0.10; bal Fe	490	215	35	90 HRB

Stainless Steel, Austenitic, N08926

USA

Specification	Designation	Notes	C	Cr	Mn	Mo	Ni	P	S	Si	Other	UTS	YS	El	Hard
	UNS N08926		0.020 max	19.0-21.0	2.00 max	6.0-7.0	24.0-26.0	0.030 max	0.010 max	0.50 max	Cu 0.5-1.5; bal Fe				
ASTM A249/249M(96)	N08926	Weld tube; boiler, superheater, heat exch	0.020 max	19.00-21.00	2.00 max	6.0-7.0	24.00-26.00	0.03 max	0.01 max	0.5 max	Cu 0.5-1.5; N 0.15-0.25; bal Fe	650	295	35	100 HRB
ASTM A688/A688M(96)	N08926	Weld Feedwater Heater Tub	0.020 max	19.00-21.00	2.00 max	6.00-7.00	24.00-26.00	0.03 max	0.01 max	0.5 max	Cu 0.5-1.5; N 0.15-0.25; bal Fe	650	295	35	

Stainless Steel, Austenitic, NACE MR-01-75

USA

Specification	Designation	Notes	C	Cr	Mn	Mo	Ni	P	S	Si	Other	UTS	YS	El	Hard
	UNS S42500	Cr 15	0.08-0.20	14.0-16.0	1.0 max	0.3-0.7	1.0-2.0	0.020 max	0.010 max	1.0 max	N <=0.20; bal Fe				

Stainless Steel, Austenitic, NIC 25

UK

Specification	Designation	Notes	C	Cr	Mn	Mo	Ni	P	S	Si	Other	UTS	YS	El	Hard
BS 1501/3(90)	904S13	Press ves; Corr/Heat res; 0.05-100mm	0.03	19.0-22.0	2.00	4.00-5.00	24.0-27.0	0.040 max	0.025 max	1.00 max	Cu 1.00-2.00; W <=0.1; bal Fe	520-720	220	35	
BS 1554(90)	904S14	Wir Corr/Heat res; diam<0.49mm; Ann	0.030 max	19.5-22.0	2.00 max	4.00-5.00	24.0-27.0	0.040 max	0.030 max	1.00 max	Cu 1.00-2.00; N <=0.06; bal Fe	0-900			
BS 1554(90)	904S14	Wir Corr/Heat res; 6<diam<=13mm; Ann	0.030 max	19.5-22.0	2.00 max	4.00-5.00	24.0-27.0	0.040 max	0.030 max	1.00 max	Cu 1.00-2.00; N <=0.06; bal Fe	0-700			

USA

Specification	Designation	Notes	C	Cr	Mn	Mo	Ni	P	S	Si	Other	UTS	YS	El	Hard
	UNS S32200	Ni-Cr-Mo alloy NIC 25	0.03 max	20.0-23.0	1.0 max	2.5-3.5	23.0-27.0	0.03 max	0.005 max	0.5 max	bal Fe				

Stainless Steel, Austenitic, Nicrofer 3228

Bulgaria

Specification	Designation	Notes	C	Cr	Mn	Mo	Ni	P	S	Si	Other	UTS	YS	El	Hard
BDS 6738(72)	0Ch23N28M3D3T	Corr res	0.06 max	22.0-25.0	2.00 max	2.40-3.00	26.0-29.0	0.035 max	0.020 max	0.80 max	Cu 2.50-3.50; Ti 0.50-0.90; bal Fe				

USA

Specification	Designation	Notes	C	Cr	Mn	Mo	Ni	P	S	Si	Other	UTS	YS	El	Hard
	UNS S33228	Heat res Nicrofer 3228 NbCe, Alloy AC66	0.04-0.08	26.0-28.0	1.0 max		31.0-33.0	0.020 max	0.015 max	0.30 max	Al <=0.025; Nb 0.6-1.0; Ce 0.05-0.10; bal Fe				

UNS numbers and US grades are provided as a means of cross referencing chemically similar alloys. Exchangability is only possible after independent examination of specifications. Tensile properties are minimum or typical as specified. UTS and YS as MPa. El as %. See Appendix for list of abbreviations used in Notes. * indicates obsolete material.

Specification	Designation	Notes	C	Cr	Mn	Mo	Ni	P	S	Si	Other	UTS	YS	El	Hard

Stainless Steel, Austenitic, Nicrofer 3228 (Continued from previous page)

USA

Specification	Designation	Notes	C	Cr	Mn	Mo	Ni	P	S	Si	Other	UTS	YS	El	Hard
ASTM A167(96)	S33228	Sh Strp Plt	0.04-0.08	26.00-28.00	1.00 max		31.00-33.00	0.020 max	0.015 max	0.30 max	Nb 0.6-1.0; Ce 0.05-0.10; bal Fe	500	185	30.0	217 HB
ASTM A182/A182M(98)	F56	Frg/roll pipe flange valve	0.04-0.08	26.0-28.0	1.00 max		31.0-33.0	0.020 max	0.015 max	0.30 max	Al <=0.025; Nb 0.6-1.0; W 0.80-1.20; Ce 0.05-0.10; bal Fe	500	185	30	
ASTM A213/A213M(95)	S33228	Smls tube boiler, superheater, heat exchanger	0.04-0.08	26.0-28.0	1.0 max		31.0-33.0	0.020 max	0.015 max	0.30 max	Al <=0.025; 0.6<=Ta+Nb<=1.0, 0.05<=Ce<=0.10; bal Fe	500	185	30	
ASTM A240/A240M(98)	S33228	Heat res, Nicrofer 3228 NbCe, Alloy AC66	0.04-0.08	26.0-28.0	1.00 max		31.0-33.0	0.020 max	0.015 max	0.30 max	Al <=0.025; 0.6<=Nb<=1.0; 0.05<=Ce<=0.10; bal Fe	500	185	30.0	217 HB
ASTM A249/249M(96)	S33228	Weld tube; boiler, superheater, heat exch	0.04-0.08	26.0-28.0	1.0 max		31.0-33.0	0.020 max	0.015 max	0.30 max	Al <=0.025; Ta+Nb 0.6-1.0, Ce 0.05-0.10; bal Fe	500	185	30	90 HRB
SAE J405(98)	S33228	Heat res Nicrofer 3228 NbCe, Alloy AC66	0.04-0.08	26.0-28.0	1.0 max		31.0-33.0	0.020 max	0.015 max	0.30 max	Al <=0.025; Nb 0.6-1.0; Ce 0.05-0.10; bal Fe				

Stainless Steel, Austenitic, Nitronic 30

USA

Specification	Designation	Notes	C	Cr	Mn	Mo	Ni	P	S	Si	Other	UTS	YS	El	Hard
	UNS S20400	Cr-Ni-Mn Nitronic 30	0.03 max	15.0-17.0	7.00-9.00		1.50-3.00	0.04 min	0.03 max	1.00 max	N 0.15-0.30; bal Fe				
ASTM A240/A240M(98)	S20400	Cr-Ni-Mn Nitronic 30	0.030 max	15.0-17.0	7.0-9.0		1.50-3.00	0.040 max	0.030 max	1.00 max	N 0.15-0.30; bal Fe	655	330	35.0	241 HB
ASTM A666(96)	S20400	Sh Strp Plt Bar, Ann	0.03 max	15.00-17.00	7.00-9.00		1.50-3.00	0.040 max	0.030 max	1.00 max	N 0.15-0.30; bal Fe	655	330	35	241 HB
SAE J405(98)	S20400	Cr-Ni-Mn Nitronic 30	0.030 max	15.00-17.00	7.00-9.00		1.50-3.00	0.040 max	0.030 max	1.00 max	N 0.15-0.30; bal Fe				

Stainless Steel, Austenitic, Nitronic 60

USA

Specification	Designation	Notes	C	Cr	Mn	Mo	Ni	P	S	Si	Other	UTS	YS	El	Hard
	UNS S21800	Cr-Ni-Mn Nitronic 60	0.10 max	16.00-18.00	7.00-9.00		8.00-9.00	0.040 max	0.030 max	3.50-4.50	N 0.08-0.18; bal Fe				
ASTM A193/A193M(98)	B8S	Bolt, high-temp, HT	0.10 max	16.0-18.0	7.0-9.0		8.0-9.0	0.060 max	0.030 max	3.5-4.5	N 0.08-0.18; bal Fe	655	345	35	271 HB
ASTM A193/A193M(98)	B8SA	Bolt, high-temp, HT	0.10 max	16.0-18.0	7.0-9.0		8.0-9.0	0.060 max	0.030 max	3.5-4.5	N 0.08-0.18; bal Fe	655	345	35	271 HB
ASTM A194/A194M(98)	8S	Nuts, high-temp press	0.10 max	16.0-18.0	7.0-9.0		8.0-9.0	0.060 max	0.030 max	3.5-4.5	N 0.08-0.18; bal Fe				183-271 HB
ASTM A194/A194M(98)	8SA	Nuts, high-temp press	0.10 max	16.0-18.0	7.0-9.0		8.0-9.0	0.060 max	0.030 max	3.5-4.5	N 0.08-0.18; bal Fe				183-271 HB
ASTM A240/A240M(98)	S21800	Cr-Ni-Mn Nitronic 60	0.10 max	16.0-18.0	7.0-9.0		8.0-9.0	0.040 max	0.030 max	3.5-4.5	N 0.08-0.18; bal Fe	655	345	35.0	241 HB
ASTM A580/A580M(98)	S21800	Wir, Ann	0.10 max	16.0-18.0	7.0-9.0		8.0-9.0	0.060 max	0.030 max	3.5-4.5	N 0.08-0.18; bal Fe	655	345	35	
SAE J405(98)	S21800	Cr-Ni-Mn Nitronic 60	0.10 max	16.00-18.00	7.00-9.00		8.00-9.00	0.040 max	0.030 max	3.50-4.50	N 0.08-0.18; bal Fe				

Stainless Steel, Austenitic, No equivalents identified

Bulgaria

Specification	Designation	Notes	C	Cr	Mn	Mo	Ni	P	S	Si	Other	UTS	YS	El	Hard
BDS 6738(72)	0Ch17N20M2D2T	Corr res	0.07 max	16.5-18.5	2.00 max	2.00-2.50	19.0-21.0	0.035 max	0.025 max	1.00 max	Cu 1.80-2.20; 7xC<=Ti<=0.70; bal Fe				

Czech Republic

Specification	Designation	Notes	C	Cr	Mn	Mo	Ni	P	S	Si	Other	UTS	YS	El	Hard
CSN 417242	17242	Corr res	0.25 max	17.0-20.0	2.0 max		8.0-11.0	0.045 max	0.03 max	1.0 max	bal Fe				
CSN 417252	17252	Corr res	0.08 max	19.0-22.0	1.5 max	4.5-6.5	36.0-40.0	0.045 max	0.035 max	1.5 max	Ti <=0.6; Ti=4C-0.6; bal Fe				
CSN 417253	17253	Heat res	0.2 max	19.0-22.0	1.0 max		36.0-40.0	0.045 max	0.03 max	1.5 max	bal Fe				
CSN 417254	17254	Corr res	0.12 max	19.5-22.0	0.4-1.2		4.5-6.0	0.05 max	0.035 max	0.8 max	Ti 0.3-0.6; bal Fe				
CSN 417322	17322	Valve	0.4-0.5	12.0-15.0	0.7 max	0.2-0.4	12.0-15.0	0.04 max	0.03 max	0.8 max	W 2.0-2.75; bal Fe				
CSN 417331	17331	Corr res; Heat res	0.07-0.15	12.0-15.0	0.4-1.0	0.7-1.5	11.0-14.0	0.045 max	0.03 max	0.2-0.8	Ti <=1.0; V 0.3-0.8; W 1.0-2.0; bal Fe				

UNS numbers and US grades are provided as a means of cross referencing chemically similar alloys. Exchangability is only possible after independent examination of specifications. Tensile properties are minimum or typical as specified. UTS and YS as MPa. El as %. See Appendix for list of abbreviations used in Notes. * indicates obsolete material.

Stainless Steel, Austenitic, No equivalents identified

Specification	Designation	Notes	C	Cr	Mn	Mo	Ni	P	S	Si	Other	UTS	YS	El	Hard
Czech Republic															
CSN 417335	17335	Corr res; Heat res	0.12 max	13.5-16.5	1.0-2.0		34.0-38.0	0.045 max	0.03 max	0.8 max	Ti 1.2-1.9; W 2.7-3.7; bal Fe				
CSN 417341	17341	Heat res	0.04-0.1	16.0-18.0	2.0 max	2.0-2.8	11.5-14.0	0.045 max	0.03 max	0.8 max	bal Fe				
CSN 417345	17345	Corr res	0.15 max	16.0-19.0	2.0 max	1.5-2.5	9.0-12.0	0.045 max	0.03 max	1.5 max	bal Fe				
CSN 417481	17481	tough at subzero; high-press hydro ves	0.05-0.12	7.0-9.0	17.0-20.0		0.3 max	0.045 max	0.035 max	0.25-1.0	Ti 0.2-0.8; bal Fe				
CSN 417482	17482	High-temp const	0.05-0.12	9.5-11.5	17.0-20.0			0.045 max	0.035 max	0.25-1.0	V 0.45-0.75; bal Fe				
CSN 417483	17483	High-temp const	0.05-0.12	7.0-9.0	17.0-20.0	0.5-0.7		0.045 max	0.035 max	0.25-1.0	V 0.45-0.75; bal Fe				
CSN 417618	17618	Wear res	1.1-1.4	0.3 max	11.0-13.0			0.1 max	0.04 max	1.0 max	bal Fe				
Germany															
DIN 17459(92)	WNr 1.4958	Smls tube SHT	0.03-0.08	19.0-22.0	1.50 max		30.0-32.5	0.015 max	0.010 max	0.70 max	Al 0.20-0.50; Co <=0.50; Cu <=0.50; N <=0.030; Nb <=0.10; Ti 0.20-0.50; Ni+Co 30.0-32.5; Al+Ti <=0.70; bal Fe	500-750	170	35	
DIN 17459(92)	X5NiCrAlTi31-20	Smls tube SHT	0.03-0.08	19.0-22.0	1.50 max		30.0-32.5	0.015 max	0.010 max	0.70 max	Al 0.20-0.50; Co <=0.50; Cu <=0.50; N <=0.030; Nb <=0.10; Ti 0.20-0.50; Ni+Co 30.0-32.5; Al+Ti 0.70 max; bal Fe	500-750	170	35	
Hungary															
MSZ 4359(82)	H5Ti	Heat res; 0-60mm; SA	0.12 max	17-19	2 max		9-12	0.04 max	0.03 max	1 max	Ti <=0.8; Ti=5C-0.8; bal Fe	500	210	35L	192 HB max
MSZ 4359(82)	H6Nb	Heat res; 0-60mm; SA	0.12 max	17-19	2 max		9-12	0.04 max	0.03 max	1 max	Nb+Ta=8C-1.2; bal Fe	500	210	35L	192 HB max
MSZ 4359(82)	H7Ni	Heat res; 0-60mm; SA	0.15 max	15-17	2 max		33-37	0.04 max	0.03 max	1-2	bal Fe	550	230	30L	223 HB max
MSZ 4360(87)	KO43ELC	Corr res; 0-160mm; SA	0.02 max	16.5-18.5	2 max	0.2 max	14.0-15.5	0.02 max	0.02 max	3.7-4.3	bal Fe	540-740	240	40L	190 HB max
MSZ 4360(87)	KO44ELC	Corr res; 0-160mm; SA	0.02 max	19-21	2 max	4.5-5.5	24-26	0.02 max	0.015 max	0.5 max	Cu 1-2; bal Fe	500-700	220	35L	190 HB max
India															
IS 1570/7(92)	26	Sh Plt Frg Smls tube	0.03 max	17-19	2 max	0.15 max	9-13	0.045 max	0.03 max	1 max	bal Fe	490-690	175	30	
IS 1570/7(92)	28	Sh Plt Smls tube	0.03 max	16-18	2 max	2-3	10-14	0.045 max	0.03 max	1 max	bal Fe	490-690	185	30	
IS 1570/7(92)	3Cr17Ni12Mo3H	Sh Plt Smls tube	0.03 max	16-18	2 max	2-3	10-14	0.045 max	0.03 max	1 max	bal Fe	490-690	185	30	
IS 1570/7(92)	3Cr18Ni11H	Sh Plt Frg Smls tube	0.03 max	17-19	2 max	0.15 max	9-13	0.045 max	0.03 max	1 max	bal Fe	490-690	175	30	
International															
ISO 2604-1(75)	F46	Frg	0.03 max	17.00-19.00	2.00 max		8.00-12.00	0.045 max	0.030 max	1.00 max	bal Fe	440-640			
ISO 2604-1(75)	F47	Frg	0.02 max	17.00-19.00	2.00 max		8.00-12.00	0.045 max	0.030 max	1.00 max	bal Fe	490-690			
ISO 2604-1(75)	F49	Frg	0.02 max	17.00-19.00	2.00 max	0.50 max	8.00-12.00	0.045 max	0.030 max	1.00 max	bal Fe	490-690			
ISO 2604-1(75)	F50	Frg	0.08 max	17.00-19.00	2.00 max		9.00-13.00	0.045 max	0.030 max	1.00 max	bal Fe	490-690			
ISO 2604-1(75)	F51	Frg	0.04-0.10	17.00-19.00	2.00 max		9.00-13.00	0.045 max	0.030 max	1.00 max	bal Fe	490-690			
ISO 2604-1(75)	F52	Frg	0.08 max	17.00-19.00	2.00 max	0.50 max	9.00-13.00	0.045 max	0.030 max	1.00 max	bal Fe	490-690			
ISO 2604-1(75)	F53	Frg	0.08 max	17.00-19.00	2.00 max	0.50 max	9.00-13.00	0.045 max	0.030 max	1.00 max	bal Fe	490-690			
ISO 2604-1(75)	F54A	Frg	0.04-0.10	17.00-19.00	2.00 max		9.00-13.00	0.045 max	0.030 max	1.00 max	bal Fe	490-690			
ISO 2604-1(75)	F54B	Frg	0.04-0.10	17.00-19.00	2.00 max		9.00-13.00	0.045 max	0.030 max	1.00 max	bal Fe	490-690			

UNS numbers and US grades are provided as a means of cross referencing chemically similar alloys. Exchangability is only possible after independent examination of specifications. Tensile properties are minimum or typical as specified. UTS and YS as MPa. El as %. See Appendix for list of abbreviations used in Notes. * indicates obsolete material.

Stainless Steel, Austenitic, No equivalents identified

Specification	Designation	Notes	C	Cr	Mn	Mo	Ni	P	S	Si	Other	UTS	YS	El	Hard
International															
ISO 2604-1(75)	F55	Frg	0.08 max	17.00-19.00	2.00 max	0.50 max	9.00-13.00	0.045 max	0.030 max	1.00 max	bal Fe	490-690			
ISO 2604-1(75)	F56	Frg	0.04-0.10	15.00-17.00	2.00 max		12.00-14.00	0.045 max	0.030 max	1.00 max	bal Fe	490-690			
ISO 2604-1(75)	F64	Frg	0.04-0.09	16.00-17.50	2.00 max	2.00-3.00	10.00-14.00	0.045 max	0.030 max	1.00 max	bal Fe	490-690			
ISO 2604-1(75)	F66	Frg	0.08 max	16.50-18.50	2.00 max	2.00-3.00	11.00-14.00	0.045 max	0.030 max	1.00 max	bal Fe	490-690			
ISO 2604-5(78)	TW46	Longitudinal weld Tub	0.03 max	17.00-19.00	2.00 max		9.00-12.00	0.045 max	0.030 max	1.00 max	bal Fe	490-690	205	30	
ISO 2604-5(78)	TW47	Longitudinal weld Tub	0.07 max	17.00-19.00	2.00 max		8.00-11.00	0.045 max	0.030 max	1.00 max	Al <=0.1; Co <=0.1; Cu <=0.3; Pb <=0.15; Ti <=0.05; V <=0.1; W <=0.1; bal Fe	490-690	235	30	
ISO 2604-5(78)	TW50	Longitudinal weld Tub	0.08 max	17.00-19.00	2.00 max		9.00-12.00	0.045 max	0.030 max	1.00 max	Al <=0.1; Co <=0.1; Cu <=0.3; Pb <=0.15; Ti <=0.05; V <=0.1; W <=0.1; Nb>10xC<1.00; bal Fe	510-710	245	30	
ISO 2604-5(78)	TW53	Longitudinal weld Tub	0.08 max	17.00-19.00	2.00 max		9.00-12.00	0.045 max	0.030 max	1.00 max	Al <=0.1; Co <=0.1; Cu <=0.3; Pb <=0.15; Ti <=0.8; V <=0.1; W <=0.1; Ti>5xC<0.80; bal Fe	510-710	235	30	
ISO 2604-5(78)	TW57	Longitudinal weld Tub	0.03 max	16.00-18.50	2.00 max	2.00-2.50	11.00-14.00	0.045 max	0.030 max	1.00 max	bal Fe	490-690	215	30	
ISO 2604-5(78)	TW58	Longitudinal weld Tub	0.03 max	16.00-18.50	2.00 max	2.50-3.00	11.50-14.50	0.045 max	0.030 max	1.00 max	bal Fe	490-690	215	30	
ISO 2604-5(78)	TW60	Longitudinal weld Tub	0.07 max	16.00-18.50	2.00 max	2.00-2.50	10.50-14.00	0.045 max	0.030 max	1.00 max	Al <=0.1; Co <=0.1; Cu <=0.3; Pb <=0.15; Ti <=0.05; V <=0.1; W <=0.1; bal Fe	510-710	245	30	
ISO 2604-5(78)	TW61	Longitudinal weld Tub	0.07 max	16.00-18.50	2.00 max	2.50-3.00	11.00-14.50	0.045 max	0.030 max	1.00 max	bal Fe	510-710	245	30	
ISO 2604-5(78)	TW69	Longitudinal weld Tub	0.10 max	19.00-23.00	1.50 max		30.00-36.00	0.045 max	0.030 max	1.00 max	Al 0.15-0.60; Ti 0.15-0.60; bal Fe	480-680	235	25	
ISO 4954(93)	X6CrNiTi1810E	Wir rod Bar, CH ext	0.08 max	17.0-19.0	2.00 max		9.0-12.0	0.045 max	0.030 max	1.00 max	Ti=5xC<= 0.80; bal Fe	680			
Italy															
UNI 3992(75)	X85CrMoV193	Valve	0.75-0.9	18-20	1-1.5	2.5-4	0.5 max	0.035 max	0.03 max	0.15-0.3	V 0.4-0.6; bal Fe				
Japan															
JIS G4311	SUH31	Heat res Bar<=25mm	0.35-0.45	14.00-16.00	0.60 max		13.00-15.00	0.040 max	0.030 max	1.50-2.50	W 2.00-3.00; bal Fe	740	315	30	248 HB max
JIS G4311	SUH31	Heat res Bar 25-180mm	0.35-0.45	14.00-16.00	0.60 max		13.00-15.00	0.040 max	0.030 max	1.50-2.50	W 2.00-3.00; bal Fe	690	315	25	248 HB
JIS G4311	SUH35	Heat res Bar<=25mm	0.48-0.58	20.00-22.00	8.00-10.00		3.25-4.50	0.040 max	0.030 max	0.35 max	N 0.35-0.50; bal Fe	880	560	8	302 HB
JIS G4311	SUH36	Heat res Bar<=25mm	0.48-0.58	20.00-22.00	8.00-10.00		3.25-4.50	0.040 max	0.040-0.090	0.35 max	N 0.35-0.50; bal Fe	880	560	8	302 HB
JIS G4311	SUH37	Heat res Bar<=25mm	0.15-0.25	20.50-22.50	1.00-1.60		10.00-12.00	0.040 max	0.030 max	1.00 max	N 0.15-0.30; bal Fe	780	390	35	248 HB
JIS G4311	SUH38	Heat res Bar<=25mm	0.25-0.35	19.00-21.00	1.20 max	1.80-2.50	10.00-20.00	0.18-0.25	0.030 max	1.00 max	B 0.001-0.010; bal Fe	880	490	20	269 HB
JIS G4311	SUH661	Heat res Bar <=180mm	0.08-0.16	20.00-22.50	1.00-2.00	2.50-3.50	19.00-21.00	0.040 max	0.030 max	1.00 max	Co 18.50-21.00; N 0.10-0.20; Nb 0.75-1.25; bal Fe	690	315	35	248 HB
JIS G4311	SUH661	Heat res Bar<=75mm	0.08-0.16	20.00-22.50	1.00-2.00	2.50-3.50	19.00-21.00	0.040 max	0.030 max	1.00 max	Co 18.50-21.00; N 0.10-0.20; Nb 0.75-1.25; W 2.00-3.00; bal Fe	760	345	30	192 HB
JIS G4312	SUH661	Heat res Plt Sh	0.08-0.16	20.00-22.50	1.00-2.00	2.50-3.50	19.00-21.00	0.040 max	0.030 max	1.00 max	Co 18.50-21.00; N 0.10-0.20; Nb 0.75-1.25; W 2.00-3.00; bal Fe	690	315	35	248 HB max
Poland															
PNH86020	0H17N12M2T	Corr res	0.05 max	16-18	2 max	2-3	11-14	0.045 max	0.03 max	1 max	Ti <=0.6; Ti=5C-0.6; bal Fe				

UNS numbers and US grades are provided as a means of cross referencing chemically similar alloys. Exchangability is only possible after independent examination of specifications. Tensile properties are minimum or typical as specified. UTS and YS as MPa. El as %. See Appendix for list of abbreviations used in Notes. * indicates obsolete material.

Specification	Designation	Notes	C	Cr	Mn	Mo	Ni	P	S	Si	Other	UTS	YS	El	Hard

Stainless Steel, Austenitic, No equivalents identified

Poland

Specification	Designation	Notes	C	Cr	Mn	Mo	Ni	P	S	Si	Other	UTS	YS	El	Hard
PNH86020	0H17N16M3T	Corr res	0.08 max	16-18	2 max	3-47	14-16	0.045 max	0.03 max	0.8 max	Ti 0.3-0.6; bal Fe				
PNH86020	0H18N12Nb	Corr res	0.08 max	17-19	2 max		10-13	0.045 max	0.03 max	0.8 max	Pb <=0.1; Nb>=10xC<=1.10; bal Fe				
PNH86020	0H22N24M4TCu	Corr res	0.06 max	20-22	1.2-2	4-5	24-26	0.045 max	0.03 max	0.17-1	Cu 1.3-1.8; Ti <=0.7; Ti=5C-0.7; bal Fe				
PNH86020	0H23N28M3TCu	Corr res	0.06 max	22-25	2 max	2.5-3	26-29	0.45 max	0.03 max	0.8 max	Cu 2.5-3.5; Ti 0.5-0.9; bal Fe				
PNH86020	1H18N12T	Corr res	0.1 max	17-19	2 max		11-13	0.045 max	0.03 max	0.8 max	Ti <=0.8; Ti=5C-0.8; bal Fe				
PNH86020	2H18N9	Corr res	0.13-0.21	17-19	1-2		8-10	0.045 max	0.03 max	0.8 max	bal Fe				
PNH86022(71)	4H14N14W2M	Heat res	0.4-0.5	13-15	0.7 max	0.25-0.4	13-15	0.03	0.03	0.8 max	W 2-2.75; bal Fe				
PNH86022(71)	H18N25S2	Heat res	0.3-0.4	17-19	1.5 max		23-26	0.035 max	0.025	2-3	bal Fe				
PNH86022(71)	H23N13	Heat res	0.02 max	22-25	2 max		12-15	0.045 max	0.03 max	1 max	bal Fe				
PNH86022(71)	H26N4	Heat res	0.2 max	24-28	0.8 max		4-5	0.045 max	0.03 max	2.5 max	bal Fe				

Romania

Specification	Designation	Notes	C	Cr	Mn	Mo	Ni	P	S	Si	Other	UTS	YS	El	Hard
STAS 11311(88)	45WNiCr180	Valve	0.4-0.5	17-19	0.8-1.5		8-10	0.045 max	0.03 max	2-3	Pb <=0.15; W 0.8-1.2; bal Fe				
STAS 11311(88)	80SiNiMoWCr150	Valve	0.75-0.85	14-16	0.8 max	0.8-1.2	0.6-0.9	0.04 max	0.03 max	1.8-2.2	Pb <=0.15; W 0.8-1.2; bal Fe				
STAS 11523(87)	10AlCr180	Corr res; Heat res	0.12 max	17-19	1 max	0.2 max	0.5 max	0.04 max	0.03 max	0.7-1.4	Al 0.7-1.2; Pb <=0.15; bal Fe				
STAS 11523(87)	10AlCr240	Corr res; Heat res	0.12 max	23-26	1 max	0.2 max	0.5 max	0.04 max	0.03 max	0.7-1.4	Al 1.2-1.7; Pb <=0.15; bal Fe				
STAS 11523(87)	10AlCr70	Corr res; Heat res	0.12 max	6-8	1 max	0.2 max	0.5 max	0.04 max	0.03 max	0.5-1	Al 0.5-1; Pb <=0.15; bal Fe				
STAS 11523(87)	10TiAlCrNi320	Corr res; Heat res	0.12 max	19-23	2 max	0.2 max	30-34	0.045 max	0.03 max	1 max	Al 0.15-0.6; Pb <=0.15; Ti 0.15-0.6; bal Fe				
STAS 11523(87)	12SiCrNi360	Corr res; Heat res	0.15 max	15-17	2 max	0.2 max	34-37	0.045 max	0.03 max	1-2	Pb <=0.15; bal Fe				

Russia

Specification	Designation	Notes	C	Cr	Mn	Mo	Ni	P	S	Si	Other	UTS	YS	El	Hard
GOST	08Ch18N10T-WD	Corr res	0.08 max	17.0-19.0	1.0-2.0	0.15 max	9.0-11.0	0.015 max	0.015 max	0.8 max	Cu <=0.25; Ti <=0.6; Ti=5C-0.6; bal Fe				
GOST	08Ch18N12T-WI	Corr res	0.08 max	17.0-19.0	0.4 max	0.1 max	11.5-13.0	0.03 max	0.02 max	0.4 max	Cu <=0.2; Ti <=0.6; Ti=5C-0.6; bal Fe				
GOST	5Ch20N4AG9	Corr res	0.5-0.6	20.0-22.0	8.0-10.0	0.3 max	3.5-4.5	0.04 max	0.03 max	0.45 max	Cu <=0.3; N 0.3-0.6; Ti <=0.1; bal Fe				
GOST 11065	00Ch18N10T	Corr res	0.04 max	17.0-19.0	1.0-2.0	0.15 max	9.0-11.0	0.035 max	0.02 max	0.8 max	Cu <=0.3; Ti <=0.4; Ti=5C-0.4; bal Fe				
GOST 19277	12Ch18N10T-WD	Corr res	0.12 max	17.0-19.0	2.0 max	0.3 max	9.0-11.0	0.015 max	0.015 max	0.8 max	Cu <=0.25; Ti <=0.7; W <=0.2; Ti=5(C-0.02)-0.7; bal Fe				
GOST 19277(73)	0Ch18N10T-WD	Corr res	0.08 max	17.0-19.0	1.0-2.0	0.15 max	9.0-11.0	0.015 max	0.015 max	0.8 max	Cu <=0.3; Ti <=0.6; Ti=5C-0.6; bal Fe				
GOST 4986(79)	03Ch18N12-WI	Corr res	0.03 max	17.0-19.0	0.4 max	0.15 max	11.5-13.0	0.03 max	0.02 max	0.4 max	Cu <=0.3; Ti <=0.005; bal Fe				
GOST 5632	03Ch17N14M3	Corr res	0.03 max	16.8-18.3	1.0-2.0	2.2-2.8	13.5-15.0	0.03 max	0.02 max	0.4 max	Cu <=0.3; Ti <=0.2; W <=0.2; bal Fe				
GOST 5632	03Ch18N10T	Corr res	0.03 max	17.0-18.5	1.0-2.0	0.3 max	9.5-11.0	0.035 max	0.02 max	0.8 max	Cu <=0.3; Ti <=0.4; W <=0.2; Ti=5C-0.4; bal Fe				
GOST 5632	05Ch18N10T	Corr res	0.05 max	17.0-18.5	1.0-2.0	0.3 max	9.0-10.5	0.035 max	0.02 max	0.8 max	Cu <=0.3; Ti <=0.6; W <=0.2; Ti=5C-0.6; bal Fe				
GOST 5632	08Ch18TtsCh	Corr res	0.08 max	17.0-19.0	0.8 max	0.15 max	0.4 max	0.035 max	0.025 max	0.8 max	Cu <=0.3; Ti <=0.6; Ti=5C-0.6; Ce<=0.1; Ca<=0.05; bal Fe				
GOST 5632	12Ch18N12M3TL	Corr res	0.12 max	16.0-19.0	1.0-2.0	3.0-4.0	11.0-13.0	0.035 max	0.03 max	0.2-1.0	Cu <=0.3; Ti <=0.7; Ti=5C-0.7; bal Fe				
GOST 5632(61)	4KH14N14W2M	Corr res; Heat res; High-temp constr	0.4-0.5	13.0-15.0	0.7 max	0.25-0.4	13.0-15.0	0.035 max	0.02 max	0.8 max	Cu <=0.3; Ti <=0.1; W 2.0-2.75; bal Fe				

Specification	Designation	Notes	C	Cr	Mn	Mo	Ni	P	S	Si	Other	UTS	YS	El	Hard

Stainless Steel, Austenitic, No equivalents identified

Russia

Specification	Designation	Notes	C	Cr	Mn	Mo	Ni	P	S	Si	Other	UTS	YS	El	Hard
GOST 5632(61)	4KH9S2	Corr res; Heat res; High-temp constr	0.35-0.45	8.0-10.0	0.7 max	0.15 max	0.4 max	0.03 max	0.025 max	2.0-3.0	Cu <=0.3; Ti <=0.05; bal Fe				
GOST 5632(72)	08Ch18G8N2T	Corr res	0.08 max	17.0-19.0	7.0-9.0	0.3 max	1.8-2.8	0.035 max	0.025 max	0.8 max	Cu <=0.3; Ti 0.2-0.5; W <=0.2; bal Fe				
GOST 5632(72)	08Ch18N12T	Corr res	0.08 max	17.0-19.0	2.0 max	0.3 max	11.0-13.0	0.035 max	0.02 max	0.8 max	Cu <=0.3; Ti <=0.6; W <=0.2; Ti=5C-0.6; bal Fe				
GOST 5632(72)	08Ch18T1	Corr res; Heat res	0.08 max	17.0-19.0	0.7 max	0.3 max	0.6 max	0.035 max	0.02 max	0.8 max	Cu <=0.3; Ti 0.6-1.0; W <=0.2; bal Fe				
GOST 5632(72)	12Ch18N12T	Corr res; Heat res; High-temp constr	0.12 max	17.0-19.0	2.0 max	0.3 max	11.0-13.0	0.035 max	0.02 max	0.8 max	Cu <=0.3; Ti <=0.7; W <=0.2; Ti=5C-0.7; bal Fe				
GOST 5632(72)	12Ch18N9T	Corr res; Heat res; High-temp constr	0.12 max	17.0-19.0	2.0 max	0.3 max	8.0-9.5	0.035 max	0.02 max	0.8 max	Cu <=0.3; Ti <=0.8; W <=0.2; Ti=5C-0.8; bal Fe				
GOST 5632(72)	40Ch9S2	Heat res; High-temp constr	0.35-0.45	8.0-10.0	0.8 max	0.15 max	0.6 max	0.03 max	0.025 max	2.0-3.0	Cu <=0.3; Ti <=0.2; bal Fe				
GOST 5632(72)	45Ch14N14W2M	High-temp const	0.4-0.5	13.0-15.0	0.7 max	0.25-0.4	13.0-15.0	0.035 max	0.02 max	0.8 max	Cu <=0.3; Ti <=0.2; W 2.0-2.8; bal Fe				
GOST 5632(72)	55Ch20G9AN4	Heat res; High-temp constr	0.5-0.6	20.0-22.0	8.0-10.0	0.3 max	3.5-4.5	0.04 max	0.03 max	0.45 max	Cu <=0.3; N 0.3-0.6; Ti <=0.2; W <=0.2; bal Fe				

UK

Specification	Designation	Notes	C	Cr	Mn	Mo	Ni	P	S	Si	Other	UTS	YS	El	Hard
BS 3605(73)	304S14*	Pip Tube press smls; t<=200mm	0.03 max	17.0-19.0	0.5-2.0		10.0-13.0	0.040 max	0.030 max	0.20-1.00	bal Fe				

USA

Specification	Designation	Notes	C	Cr	Mn	Mo	Ni	P	S	Si	Other	UTS	YS	El	Hard
ASTM A213/A213M(95)	S25700	Smls tube boiler, superheater, heat exchanger	0.02 max	8.0-11.0	2.0 max	0.50 max	22.0-25.0	0.025 max	0.010 max	6.5-8.0	bal Fe	540	240	50	
ASTM A249/249M(96)	S24565	Weld tube; boiler, superheater, heat exch	0.03 max	23.0-25.0	5.0-7.0	4.0-5.00	16.0-18.0	0.030 max	0.010 max	1.00 max	Cu 1.5-2.5; N 0.4-0.6; Nb+Ta<=0.1; bal Fe	795	415	35	100 HRB
ASTM A289/A289M(97)	Class 1	Frg	0.10 max	17.5-20.0	17.5-20.0		2.00 max	0.060 max	0.015 max	0.80 max	Al <=0.04; N 0.45-0.80; Ti <=0.10; V <=0.25; bal Fe	146	135	28	
ASTM A289/A289M(97)	Class 2	Frg	0.10 max	17.5-20.0	17.5-20.0		2.00 max	0.060 max	0.015 max	0.80 max	Al <=0.04; N 0.45-0.80; Ti <=0.10; V <=0.25; bal Fe	155	145	25	
ASTM A289/A289M(97)	Class 3	Frg	0.10 max	17.5-20.0	17.5-20.0		2.00 max	0.060 max	0.015 max	0.80 max	Al <=0.04; N 0.45-0.80; Ti <=0.10; V <=0.25; bal Fe	165	160	20	
ASTM A289/A289M(97)	Class 4	Frg	0.10 max	17.5-20.0	17.5-20.0		2.00 max	0.060 max	0.015 max	0.80 max	Al <=0.04; N 0.45-0.80; Ti <=0.10; V <=0.25; bal Fe	170	165	19	
ASTM A289/A289M(97)	Class 5	Frg	0.10 max	17.5-20.0	17.5-20.0		2.00 max	0.060 max	0.015 max	0.80 max	Al <=0.04; N 0.45-0.80; Ti <=0.10; V <=0.25; bal Fe	175	170	17	
ASTM A289/A289M(97)	Class 6	Frg	0.10 max	17.5-20.0	17.5-20.0		2.00 max	0.060 max	0.015 max	0.80 max	Al <=0.04; N 0.45-0.80; Ti <=0.10; V <=0.25; bal Fe	185	180	14	
ASTM A289/A289M(97)	Class 7	Frg	0.10 max	17.5-20.0	17.5-20.0		2.00 max	0.060 max	0.015 max	0.80 max	Al <=0.04; N 0.45-0.80; Ti <=0.10; V <=0.25; bal Fe	195	190	12	
ASTM A289/A289M(97)	Class 8	Frg	0.10 max	17.5-20.0	17.5-20.0		2.00 max	0.060 max	0.015 max	0.80 max	Al <=0.04; N 0.45-0.80; Ti <=0.10; V <=0.25; bal Fe	200	195	10	
ASTM A965/965M(97)	FXM-11	Frg for press high-temp parts	0.04 max	19.0-21.5	8.0-10.0		5.5-7.5	0.060 max	0.030 max	1.00 max	N 0.15-0.40; bal Fe	620	345	40	
SAE J441(93)	SCW-12	Cut Wir shot; 0.30mm diam	0.15 max	17.00-20.00	2.00 max		8.00-10.50	0.045 max	0.030 max	1.00 max	bal Fe	2165-2370			45 HRC
SAE J441(93)	SCW-14	Cut Wir shot; 0.35mm diam	0.15 max	17.00-20.00	2.00 max		8.00-10.50	0.045 max	0.030 max	1.00 max	bal Fe	2135-2341			45 HRC
SAE J441(93)	SCW-17	Cut Wir shot; 0.45mm diam	0.15 max	17.00-20.00	2.00 max		8.00-10.50	0.045 max	0.030 max	1.00 max	bal Fe	2095-2300			45 HRC
SAE J441(93)	SCW-20	Cut Wir shot; 0.5mm diam	0.15 max	17.00-20.00	2.00 max		8.00-10.50	0.045 max	0.030 max	1.00 max	bal Fe	2068-2275			45 HRC
SAE J441(93)	SCW-23	Cut Wir shot; 0.6mm diam	0.15 max	17.00-20.00	2.00 max		8.00-10.50	0.045 max	0.030 max	1.00 max	bal Fe	2013-2220			45 HRC
SAE J441(93)	SCW-28	Cut Wir shot; 0.7mm diam	0.15 max	17.00-20.00	2.00 max		8.00-10.50	0.045 max	0.030 max	1.00 max	bal Fe	1972-2179			45 HRC

UNS numbers and US grades are provided as a means of cross referencing chemically similar alloys. Exchangability is only possible after independent examination of specifications. Tensile properties are minimum or typical as specified. UTS and YS as MPa. El as %. See Appendix for list of abbreviations used in Notes. * indicates obsolete material.

Specification	Designation	Notes	C	Cr	Mn	Mo	Ni	P	S	Si	Other	UTS	YS	El	Hard

Stainless Steel, Austenitic, No equivalents identified

USA

Specification	Designation	Notes	C	Cr	Mn	Mo	Ni	P	S	Si	Other	UTS	YS	El	Hard
SAE J441(93)	SCW-32	Cut Wir shot; 0.8mm diam	0.15 max	17.00-20.00	2.00 max		8.00-10.50	0.045 max	0.030 max	1.00 max	bal Fe	1910-2117			45 HRC
SAE J441(93)	SCW-35	Cut Wir shot; 0.9mm diam	0.15 max	17.00-20.00	2.00 max		8.00-10.50	0.045 max	0.030 max	1.00 max	bal Fe	1882-2089			45 HRC
SAE J441(93)	SCW-41	Cut Wir shot; 1.0mm diam	0.15 max	17.00-20.00	2.00 max		8.00-10.50	0.045 max	0.030 max	1.00 max	bal Fe	1855-2062			45 HRC
SAE J441(93)	SCW-47	Cut Wir shot; 1.2mm diam	0.15 max	17.00-20.00	2.00 max		8.00-10.50	0.045 max	0.030 max	1.00 max	bal Fe	1806-2013			45 HRC
SAE J441(93)	SCW-54	Cut Wir shot; 1.4mm diam	0.15 max	17.00-20.00	2.00 max		8.00-10.50	0.045 max	0.030 max	1.00 max	bal Fe	1793-1999			45 HRC
SAE J441(93)	SCW-62	Cut Wir shot; 1.6mm diam	0.15 max	17.00-20.00	2.00 max		8.00-10.50	0.045 max	0.030 max	1.00 max	bal Fe	1758-1765			45 HRC
SAE J467(68)	16-25-6		0.50	16.0	1.75	6.00	25.0				N 0.15; bal Fe	1117	986	15	326 HB
SAE J467(68)	17-14-CuMo		0.12	15.9	0.75	2.50	14.1			0.50	Cu 3.00; Nb 0.45; Ti 0.25; bal Fe	593	290	45	80 HRB
SAE J467(68)	AF71		0.30	12.5	18.0	3.00				0.30	B 0.20; N 0.20; V 0.90; bal Fe	1041	731	25	
SAE J467(68)	CRM-15D		1.00	20.0	5.00	2.00	5.00			0.50	N 0.20; Nb 2.00; W 2.00; bal Fe	793	621	1	37 HRC
SAE J467(68)	CRM-6D		1.00	20.0	5.00	1.00	5.00			0.50	Nb 1.00; W 1.00; bal Fe	759	538	2	35 HRC
SAE J467(68)	G-192		0.60	22.0	8.50					0.50	N 0.35; bal Fe	938	593	5	80 HRB
SAE J467(68)	Incoloy 805		0.12	7.50	0.60	0.50	36.0			0.50	Cu 0.10; bal Fe				
SAE J467(68)	Unitemp 212		0.08	16.0	0.05		25.0			0.15	Al 0.15; B 0.06; Nb 0.50; Ti 4.00; Zr=0.05; bal Fe	1289	896	23	39 HRC
SAE J467(68)	V-57		0.05	14.75	0.20	1.30	27.25			0.35	Al 0.20; B 0.01; Ti 3.00; V 0.30; bal Fe	1207	862	21	331 HB

Stainless Steel, Austenitic, NV8

USA

Specification	Designation	Notes	C	Cr	Mn	Mo	Ni	P	S	Si	Other	UTS	YS	El	Hard
	UNS S64007	Valve, GM-8440	0.35-0.45	1.85-2.50	0.20-0.40		0.25 max	0.030 max	0.040 max	3.60-4.20	Cu <=0.25; bal Fe				
SAE J775(93)	GM-8440	Engine poppet valve	0.35-0.45	1.85-2.50	0.20-0.40	0.10 max	0.25 max	0.030 max	0.040 max	3.60-4.20	Cu <=0.25; bal Fe	920	690		
SAE J775(93)	NV8*	Engine poppet valve	0.35-0.45	1.85-2.50	0.20-0.40	0.10 max	0.25 max	0.030 max	0.040 max	3.60-4.20	Cu <=0.25; bal Fe	920	690		

Stainless Steel, Austenitic, S20161

USA

Specification	Designation	Notes	C	Cr	Mn	Mo	Ni	P	S	Si	Other	UTS	YS	El	Hard
	UNS S20161	Cr-Mn-Ni-Si	0.15 max	15.0-18.0	4.00-6.00		4.00-6.00	0.040 max	0.040 min	3.00-4.00	N 0.08-0.20; bal Fe				
ASTM A240/A240M(98)	S20161	Cr-Mn-Ni-Si	0.15 max	15.0-18.0	4.0-6.0		4.0-6.0	0.040 max	0.040 max	3.0-4.0	N 0.08-0.20; bal Fe	860	345	40.0	255 HB
ASTM A580/A580M(98)	S20161	Wir, Ann	0.15 max	15.0-18.0	4.0-6.0		4.0-6.0	0.040 max	0.040 max	3.0-4.0	N 0.08-0.20; bal Fe	860	345	40	
SAE J405(98)	S20161	Cr-Mn-Ni-Si	0.15 max	15.00-18.00	4.00-6.00		4.00-6.00	0.040 max	0.040 max	3.00-4.00	N 0.08-0.20; bal Fe				

Stainless Steel, Austenitic, S21000

USA

Specification	Designation	Notes	C	Cr	Mn	Mo	Ni	P	S	Si	Other	UTS	YS	El	Hard
	UNS S21000	Cr-Mn-Mo-Ni-N Strengthened	0.10 max	18.0-23.0	4.00-7.00	4.00-6.00	16.0-20.0	0.03 max	0.03 max	0.60 max	Cu <=2.00; N >=0.15; bal Fe				

Stainless Steel, Austenitic, S21300

USA

Specification	Designation	Notes	C	Cr	Mn	Mo	Ni	P	S	Si	Other	UTS	YS	El	Hard
	UNS S21300	Cr-Cu-Mn-N Strengthened	0.25 max	16.0-21.0	15.0-18.0	0.50-3.00	3.00 max	0.05 max	0.05 max	1.00 max	Cu 0.50-2.00; N 0.20-0.80; bal Fe				

Stainless Steel, Austenitic, S21500

UK

Specification	Designation	Notes	C	Cr	Mn	Mo	Ni	P	S	Si	Other	UTS	YS	El	Hard
BS 3059/2(78)	1250*	Boiler, Superheater; Tub; Pip	0.06-0.15	14.0-16.0	5.50-7.00	0.80-1.20	9.0-11.0	0.040 max	0.030 max	0.20-1.00	B 0.003-0.009; Nb 0.75-1.25; V 0.15-0.40; bal Fe				
BS 3059/2(90)	215S15	Boiler, Superheater; High-temp; Smls tub; 2-11mm	0.06-0.15	14.0-16.0	5.50-7.00	0.80-1.20	9.0-11.0	0.040 max	0.030 max	0.20-1.00	B 0.003-0.009; Nb 0.75-1.25; V 0.15-0.40; bal Fe	540-740	270	35	

UNS numbers and US grades are provided as a means of cross referencing chemically similar alloys. Exchangability is only possible after independent examination of specifications. Tensile properties are minimum or typical as specified. UTS and YS as MPa. El as %. See Appendix for list of abbreviations used in Notes. * indicates obsolete material.

Specification	Designation	Notes	C	Cr	Mn	Mo	Ni	P	S	Si	Other	UTS	YS	El	Hard

Stainless Steel, Austenitic, S21500 (Continued from previous page)

UK

Specification	Designation	Notes	C	Cr	Mn	Mo	Ni	P	S	Si	Other	UTS	YS	El	Hard
BS 3605/1(91)	215S15	Pip Tube press smls; t<=200mm	0.06-0.15	14.0-16.0	5.5-7.0	0.80-1.20	9.0-11.0	0.040 max	0.030 max	0.20-1.00	B 0.003-0.009; Nb 0.75-1.25; V 0.15-0.40; bal Fe	540-740		35	

USA

Specification	Designation	Notes	C	Cr	Mn	Mo	Ni	P	S	Si	Other	UTS	YS	El	Hard
	UNS S21500	Cr-Mn-Ni-Mo Esshete 1250	0.06-0.15	14.0-16.0	5.50-7.00	0.80-1.20	9.0-11.0	0.040 max	0.030 max	0.20-1.20	B 0.003-0.009; Nb 0.75-1.25; V 0.15-0.40; bal Fe				
ASTM A213/A213M(95)	S21500	Smls tube boiler, superheater, heat exchanger	0.06-0.15	14.0-16.0	5.50-7.0	0.8-1.20	9.00-11.0	0.040 max	0.030 max	0.2-1.0	B 0.003-0.009; Nb 0.75-1.25; V 0.15-0.40; bal Fe	540	230	35	

Stainless Steel, Austenitic, S30210

USA

Specification	Designation	Notes	C	Cr	Mn	Mo	Ni	P	S	Si	Other	UTS	YS	El	Hard
	UNS S30210	Low permeability	0.26-0.33	17.0-19.0	3.00-4.00		8.0-10.0	0.18-0.33	0.035 max	1.00 max	bal Fe				

Stainless Steel, Austenitic, S30260

India

Specification	Designation	Notes	C	Cr	Mn	Mo	Ni	P	S	Si	Other	UTS	YS	El	Hard
IS 1570/7(92) •	11Cr17Ni13W3TiH	0-100mm, Smls tub Bar Frg	0.07-0.15	15.5-17.5	1 max	0.15 max	12-14.5	0.045 max	0.03 max	1 max	B <=0.006; Co <=0.1; Cu <=0.3; Pb <=0.15; Ti <=0.8; W 2.5-3.5; 4C<=Ti<=0.8; bal Fe	500-730	220	35	
IS 1570/7(92)	31	0-100mm, Smls tub Bar Frg	0.07-0.15	15.5-17.5	1 max	0.15 max	12-14.5	0.045 max	0.03 max	1 max	B <=0.006; Co <=0.1; Cu <=0.3; Pb <=0.15; Ti <=0.8; W 2.5-3.5; 4C<=Ti<=0.8; bal Fe	500-730	220	35	

USA

Specification	Designation	Notes	C	Cr	Mn	Mo	Ni	P	S	Si	Other	UTS	YS	El	Hard
	UNS S30260	Low permeability	0.15 max	16.0-18.0	1.00 max		9.5-12.0	0.20-0.40	0.040 max	1.00 max	bal Fe				

Stainless Steel, Austenitic, S30430

Australia

Specification	Designation	Notes	C	Cr	Mn	Mo	Ni	P	S	Si	Other	UTS	YS	El	Hard
AS 1449	302HQ*	Withdrawn, Wir rod	0.08 max	17-19	2 max		9-11	0.045 max	0.03 max	1 max	Cu 3-4; bal Fe				

China

Specification	Designation	Notes	C	Cr	Mn	Mo	Ni	P	S	Si	Other	UTS	YS	El	Hard
GB 1220(92)	0Cr18Ni9Cu3	Bar SA	0.08 max	17.00-19.00	2.00 max		8.50-10.50	0.035 max	0.030 max	1.00 max	Cu 3.00-4.00; bal Fe	480	177	40	
GB 4232(93)	ML0Cr18Ni9Cu3	Bar; SA; 1-3mm diam	0.08 max	17.00-19.00	2.00 max		8.50-10.50	0.035 max	0.030 max	1.00 max	Cu 3.00-4.00; bal Fe	490-640		30	

Europe

Specification	Designation	Notes	C	Cr	Mn	Mo	Ni	P	S	Si	Other	UTS	YS	El	Hard
EN 10088/3(95)	1.4567	Bar Rod Sect, Corr res; t<=160mm, Ann	0.04 max	17.00-19.00	2.00 max		8.50-10.50	0.045 max	0.030 max	1.00 max	Cu 3.00-4.00; N <=0.11; bal Fe	450-650	175	45	215 HB
EN 10088/3(95)	X3CrNiCu18-9-4	Bar Rod Sect, Corr res; t<=160mm, Ann	0.04 max	17.00-19.00	2.00 max		8.50-10.50	0.045 max	0.030 max	1.00 max	Cu 3.00-4.00; N <=0.11; bal Fe	450-650	175	45	215 HB

France

Specification	Designation	Notes	C	Cr	Mn	Mo	Ni	P	S	Si	Other	UTS	YS	El	Hard
AFNOR NFA35575	Z6CNU1810	Wir rod	0.08 max	16.5-18.5	2 max		8.5-10.5	0.04 max	0.03 max	1 max	Cu 3-4; bal Fe				

Germany

Specification	Designation	Notes	C	Cr	Mn	Mo	Ni	P	S	Si	Other	UTS	YS	El	Hard
DIN EN 10088(95)	WNr 1.4567	Bar Rod Sect	0.04 max	17.0-19.0	2.00 max		8.50-10.5	0.045 max	0.030 max	1.00 max	Cu 3.00-4.00; bal Fe				
DIN EN 10088(95)	X3CrNiCu189	Bar Rod Sect	0.04 max	17.0-19.0	2.00 max		8.50-10.5	0.045 max	0.030 max	1.00 max	Cu 3.00-4.00; N <=0.11; bal Fe				

International

Specification	Designation	Notes	C	Cr	Mn	Mo	Ni	P	S	Si	Other	UTS	YS	El	Hard
ISO 4954(93)	X3CrNiCu1893E	Wir rod Bar, CH ext	0.04 max	17.0-19.0	2.00 max		8.5-10.5	0.045 max	0.030 max	1.00 max	Cu 3.00-4.00; bal Fe	590			

Japan

Specification	Designation	Notes	C	Cr	Mn	Mo	Ni	P	S	Si	Other	UTS	YS	El	Hard
JIS G4303(91)	SUSXM 7	Bar; SA; <=180mm diam	0.08 max	17.00-19.00	2.00 max		8.50-10.50	0.045 max	0.030 max	1.00 max	Cu 3.00-4.00; bal Fe	480	175	40	187 HB
JIS G4308	SUSXM7	Wir rod	0.08 max	17.00-19.00	2.00 max		8.50-10.50	0.045 max	0.030 max	1.00 max	Cu 3.00-4.00; bal Fe				
JIS G4309	SUSXM7	Wir	0.08 max	17.00-19.00	2.00 max		8.50-10.50	0.045 max	0.030 max	1.00 max	Cu 3.00-4.00; bal Fe				
JIS G4315(94)	SUSXM7	Wir rod	0.08 max	17.00-19.00	2.00 max		8.50-10.50	0.045 max	0.030 max	1.00 max	Cu 3.00-4.00; bal Fe	440-630		30-40	

UNS numbers and US grades are provided as a means of cross referencing chemically similar alloys. Exchangability is only possible after independent examination of specifications. Tensile properties are minimum or typical as specified. UTS and YS as MPa. El as %. See Appendix for list of abbreviations used in Notes. * indicates obsolete material.

Specification	Designation	Notes	C	Cr	Mn	Mo	Ni	P	S	Si	Other	UTS	YS	El	Hard
Stainless Steel, Austenitic, S30430 (Continued from previous page)															
UK															
BS 1554(90)	394S17	Wir Corr/Heat res; diam<0.49mm; Ann	0.07 max	17.0-19.0	2.00 max		8.00-10.0	0.045 max	0.030 max	1.00 max	Cu 3.0-4.0; W <=0.1; bal Fe	0-900			
BS 1554(90)	394S17	Wir Corr/Heat res; 6<diam<=13mm; Ann	0.07 max	17.0-19.0	2.00 max		8.00-10.0	0.045 max	0.030 max	1.00 max	Cu 3.0-4.0; W <=0.1; bal Fe	0-700			
BS 3111/2(87)	394S17	Wir	0.07 max	17.0-19.0	2.00 max		8.00-10.5	0.045 max	0.030 max	1.00 max	Cu 3.00-4.00; bal Fe				
USA															
	AISI 304Cu	Tube	0.08 max	17.00-19.00	2.00 max		8.00-10.00	0.045 max	0.030 max	1.00 max	Cu 3.00-4.00; bal Fe				
	UNS S30430	Low WH 18-9-LW	0.10 max	17.00-19.00	2.00 max		8.00-10.00	0.045 max	0.030 max	1.00 max	Cu 3.00-4.00; bal Fe				
ASTM A493	XM-7	Wir rod, CHd, cold frg	0.08 max	17.0-19.0	2.00 max		8.0-10.0	0.045 max	0.030 max	1.00 max	Cu 3.0-4.0; bal Fe	605			
SAE J775(93)	302HQ	Engine poppet valve	0.10 max	17.00-19.00	2.00 max		8.00-10.00	0.045 max	0.030 max	1.00 max	Cu 3.00-4.00; bal Fe				
SAE J775(93)	EV17*	Engine poppet valve	0.10 max	17.00-19.00	2.00 max		8.00-10.00	0.045 max	0.030 max	1.00 max	Cu 3.00-4.00; bal Fe				
Stainless Steel, Austenitic, S30431															
UK															
BS 1449/2(83)	315S16	Corr/Heat res; t<=100mm	0.07 max	16.5-18.5	2.00 max	1.25-1.75	9.00-11.0	0.045 max	0.030 max	1.00 max	bal Fe	510	205	40	0-205 HB
USA															
	UNS S30431	Low WH Free Machining	0.06 max	16.0-19.0	2.00 max		9.00-11.0	0.040 max	0.14 max	1.00 max	Cu 1.30-2.40; bal Fe				
Stainless Steel, Austenitic, S30560															
USA															
	UNS S30560	Low permeability	0.10 max	17.0 min	0.2-2.5		8.5-14.0	0.035 max	0.035 max	0.2-1.5	Cu <=0.50; bal Fe				
Stainless Steel, Austenitic, S30601															
USA															
	UNS S30601	Low-C, 18-17LC	0.015 max	17.0-18.0	1.0 max		17.0-18.0	0.030 max	0.013 max	5.00-5.60	Cu <=0.35; N <=0.05; bal Fe				
ASTM A240/A240M(98)	S30601	Low-C, 18-17LC	0.015 max	17.0-18.0	0.50-0.80	0.20 max	17.0-18.0	0.030 max	0.013 max	5.0-5.6	Cu <=0.35; N <=0.05; bal Fe	540	255	30.0	
SAE J405(98)	S30601	Low-C, 18-17LC	0.015 max	17.00-18.00	0.50-0.80	0.20 max	17.00-18.00	0.030 max	0.013 max	5.00-5.60	Cu <=0.35; N <=0.050; bal Fe				
Stainless Steel, Austenitic, S30615															
Europe															
EN 10088/2(95)	1.4361	Plt, HR Corr res; t<=75mm, Ann	0.015 max	16.50-18.50	2.00 max	0.20 max	14.00-16.00	0.025 max	0.010 max	3.70-4.50	N <=0.11; bal Fe	530-730	220	40	
EN 10088/2(95)	X1CrNiSi18-15-4	Plt, HR Corr res; t<=75mm, Ann	0.015 max	16.50-18.50	2.00 max	0.20 max	14.00-16.00	0.025 max	0.010 max	3.70-4.50	N <=0.11; bal Fe	530-730	220	40	
EN 10088/3(95)	1.4361	Bar Rod Sect, Corr res; t<=160mm, Ann	0.015 max	16.50-18.50	2.00 max	0.20 max	14.00-16.00	0.025 max	0.010 max	3.70-4.50	N <=0.11; bal Fe	530-730	210	40	230 HB
EN 10088/3(95)	1.4361	Bar Rod Sect; Corr res; 160<t<=250mm, Ann	0.015 max	16.50-18.50	2.00 max	0.20 max	14.00-16.00	0.025 max	0.010 max	3.70-4.50	N <=0.11; bal Fe	530-730	210	30	230 HB
EN 10088/3(95)	X1CrNiSi18-15-4	Bar Rod Sect, Corr res; t<=160mm, Ann	0.015 max	16.50-18.50	2.00 max	0.20 max	14.00-16.00	0.025 max	0.010 max	3.70-4.50	N <=0.11; bal Fe	530-730	210	40	230 HB•
EN 10088/3(95)	X1CrNiSi18-15-4	Bar Rod Sect; Corr res; 160<t<=250mm, Ann	0.015 max	16.50-18.50	2.00 max	0.20 max	14.00-16.00	0.025 max	0.010 max	3.70-4.50	N <=0.11; bal Fe	530-730	210	30	230 HB
USA															
	UNS S30615	Heat res RA85H	0.16-0.24	17.0-19.5	2.0 max		13.5-16.0	0.03	0.03	3.2-4.0	Al 0.8-1.5; bal Fe				
ASTM A213/A213M(95)	S30615	Smls tube boiler, superheater, heat exchanger	0.16-0.24	17.0-19.5	2.00 max		13.5-16.0	0.03 max	0.03 max	3.2-4.0	Al 0.8-1.5; bal Fe	620	275	35	
ASTM A240/A240M(98)	S30615	Heat res, RA85H	0.16-0.24	17.0-19.5	2.0 max		13.5-16.0	0.030 max	0.030 max	3.2-4.0	Al 0.80-1.50; bal Fe	620	275	35.0	217 HB
ASTM A249/249M(96)	S30615	Weld tube; boiler, superheater, heat exch	0.16-0.24	17.0-19.5	2.00 max		13.5-16.0	0.03 max	0.03 max	3.2-4.0	Al 0.8-1.5; bal Fe	620	275	35	95 HRB

UNS numbers and US grades are provided as a means of cross referencing chemically similar alloys. Exchangability is only possible after independent examination of specifications. Tensile properties are minimum or typical as specified. UTS and YS as MPa. El as %. See Appendix for list of abbreviations used in Notes. * indicates obsolete material.

Specification	Designation	Notes	C	Cr	Mn	Mo	Ni	P	S	Si	Other	UTS	YS	El	Hard

Stainless Steel, Austenitic, S30615 (Continued from previous page)

USA

Specification	Designation	Notes	C	Cr	Mn	Mo	Ni	P	S	Si	Other	UTS	YS	El	Hard
SAE J405(98)	S30615	Heat res RA85H	0.16-0.24	17.0-19.5	2.0 max		13.5-16.0	0.030 max	0.030 max	3.2-4.0	Al 0.8-1.5; bal Fe				

Stainless Steel, Austenitic, S31050

USA

Specification	Designation	Notes	C	Cr	Mn	Mo	Ni	P	S	Si	Other	UTS	YS	El	Hard
	UNS S31050	Low-C, w/N	0.030 max	24.0-26.0	2.00 max	1.6-3.0	20.5-23.5	0.020 max	0.015 max	0.4 max	N 0.09-0.16; bal Fe				
ASTM A182/A182M(98)	F310MoLn	Frg/roll pipe flange valve	0.20 max	24.0-26.0	2.00 max	1.60-2.60	20.5-23.5	0.030 max	0.010 max	0.050 max	N 0.09-0.15; bal Fe	540	255	30	
ASTM A213/A213M(95)	S31050	Smls tube boiler, superheater, heat exchanger; t>6.4	0.025 max	24.0-26.0	2.00 max	1.6-2.6	20.5-23.5	0.020 max	0.015 max	0.4 max	N 0.09-0.15; bal Fe	540	255	25	
ASTM A240/A240M(98)	S31050	t<=0.25 in.	0.020 max	24.0-26.0	2.00 max	1.60-2.60	20.5-23.5	0.030 max	0.010 max	0.50 max	N 0.09-0.15; bal Fe	580	270	25	217 HB
ASTM A240/A240M(98)	S31050	t>0.25 in.	0.020 max	24.0-26.0	2.00 max	1.60-2.60	20.5-23.5	0.030 max	0.010 max	0.50 max	N 0.09-0.15; bal Fe	540	255	25	217 HB
ASTM A249/249M(96)	S31050	Weld tube; boiler, superheater, heat exch	0.025 max	24.0-26.0	2.00 max	1.6-2.6	20.5-23.5	0.020 max	0.015 max	0.4 max	N 0.09-0.15; bal Fe	540-580	255-270	25	95 HRB
ASTM A943/A943M(95)		Spray formed smls pipe	0.25 max	24.0-26.0	2.0 max	1.6-2.6	20.5-23.5	0.020 max	0.015 max	0.4 max	N 0.09-0.15; bal Fe	540-580	255-270		
SAE J405(98)	S31050	Low-C, w/N	0.030 max	24.00-26.00	2.00 max	2.00-3.00	21.00-23.00	0.030 max	0.010 max	0.50 max	N 0.10-0.16; bal Fe				

Stainless Steel, Austenitic, S31272

Spain

Specification	Designation	Notes	C	Cr	Mn	Mo	Ni	P	S	Si	Other	UTS	YS	El	Hard
UNE 36016/1(89)	E-410	Corr res	0.08-0.12	12.0-14.0	1.0 max	0.15 max	1.0 max	0.04 max	0.03 max	1.0 max	bal Fe				
UNE 36016/1(89)	F.3401*	Obs; EN10088-3(96); X12Cr13	0.08-0.12	12.0-14.0	1.0 max	0.15 max	1.0 max	0.04 max	0.03 max	1.0 max	bal Fe				

USA

Specification	Designation	Notes	C	Cr	Mn	Mo	Ni	P	S	Si	Other	UTS	YS	El	Hard
	UNS S31272		0.08-0.12	14.0-16.0	1.5-2.0	1.0-1.4	14.0-16.0	0.030 max	0.015 max	0.3-0.7	B 0.004-0.008; Ti 0.3-0.6; bal Fe				
ASTM A213/A213M(95)	S31272	Smls tube boiler, superheater, heat exchanger	0.08-0.12	14.0-16.0	1.5-2.0	1.0-1.40	14.0-16.0	0.030 max	0.015 max	0.3-0.7	B 0.004-0.008; bal Fe	450	200	35	
ASTM A943/A943M(95)		Spray formed smls pipe	0.08-0.12	14.0-16.0	1.5-2.00	1.0-14.0	14.0-16.0	0.030 max	0.015 max	0.03-0.7	Ti 0.3-0.6; bal Fe	450	200		

Stainless Steel, Austenitic, S32050

USA

Specification	Designation	Notes	C	Cr	Mn	Mo	Ni	P	S	Si	Other	UTS	YS	El	Hard
	UNS S32050	SR50A	0.030 max	22.0-24.0	1.5 max	6.0-6.8	20.0-23.0	0.035 max	0.020 max	1.0 max	Cu <=0.4; N 0.21-0.32; bal Fe				
ASTM A240/A240M(98)	S32050	SR50A	0.030 max	22.0-24.0	1.5 max	6.0-6.8	20.0-23.0	0.035 max	0.020 max	1.0 max	Cu <=0.40; N 0.21-0.32; bal Fe	675	330	40.0	250 HB
SAE J405(98)	S32050	SR50A	0.030 max	22.0-24.0	1.5 max	6.0-6.8	20.0-23.0	0.035 max	0.020 max	1.0 max	Cu <=0.4; N 0.21-0.32; bal Fe				

Stainless Steel, Austenitic, S34740

USA

Specification	Designation	Notes	C	Cr	Mn	Mo	Ni	P	S	Si	Other	UTS	YS	El	Hard
	UNS S34740*	Obs; Free Machining Nb stabilized	0.08 max	17.00-19.00	2.00 max		9.00-12.00	0.040 max	0.18-0.35	1.00 max	Nb 10xC-1.10; bal Fe				

Stainless Steel, Austenitic, S34741

USA

Specification	Designation	Notes	C	Cr	Mn	Mo	Ni	P	S	Si	Other	UTS	YS	El	Hard
	UNS S34741*	Obs; Se-bearing Free Machining Nb stabilized	0.08 max	17.00-19.00	2.00 max		9.00-12.00	0.11-0.17	0.030 max	1.00 max	Nb 10xC-1.10; Se 0.15-0.35; bal Fe				

Specification	Designation	Notes	C	Cr	Mn	Mo	Ni	P	S	Si	Other	UTS	YS	El	Hard
Stainless Steel, Austenitic, S37000															
USA															
	UNS S37000		0.030-0.050	12.5-14.5	1.65-2.35	1.50-2.50	14.5-16.5	0.040 max	0.010 max	0.50-1.00	Al <=0.050; B <=0.0020; Co <=0.050; Cu <=0.04; N <=0.005; Nb <=0.050; Ti 0.10-0.40; V <=0.05; As<=0.030; Ta<=0.020; Ti aim for 0.25; bal Fe				
Stainless Steel, Austenitic, S68000															
USA															
	UNS S68000	Fe, Cr-Ni-Mo Hardfacing Alloy Eatonite 6	1.50-2.00	26.00-30.00	1.00 max	4.00-5.00	15.00-18.00	0.025 max	0.020 max	1.10-1.50	bal Fe				
SAE J775(93)	Eatonite 6	Engine poppet valve, facing alloy	1.50-2.00	26.00-30.00	1.00 max	4.00-5.00	15.00-18.00	0.025 max	0.020 max	1.10-1.50	bal Fe				
SAE J775(93)	VF11*	Engine poppet valve, facing alloy	2.00-2.50	22.00-26.00		5.00-6.00	10.00-12.00			0.80-1.30	bal Fe				
SAE J775(93)	VMS585	Engine poppet valve, facing alloy	2.00-2.50	22.00-26.00		5.00-6.00	10.00-12.00			0.80-1.30	bal Fe				
Stainless Steel, Austenitic, SUH 11 Mod															
USA															
	UNS S64004	Valve, SUH 11M	0.47-0.55	7.50-9.50	0.60 max		0.60 max	0.030 max	0.030 max	1.00-2.00	bal Fe				
SAE J775(93)	SUH 11 SUH 11M	Engine poppet valve	0.47-0.55	7.50-9.50	0.60 max		0.60 max	0.030 max	0.030 max	1.00-2.00	bal Fe				
Stainless Steel, Austenitic, TP310HCb															
USA															
	UNS S31041	Heat res	0.04-0.10	24.0-26.0	2.00 max	0.75 max	19.0-22.0	0.045 max	0.030 max	0.75 max	Cu <=0.75; Nb+Ta 10xC min-1.10 max; bal Fe				
	UNS S31042	HR3C	0.04-0.10	24.0-26.0	2.00 max		17.0-23.0	0.030 max	0.030 max	0.75 max	N 0.15-0.35; Nb+Ta 0.20-0.60; bal Fe				
ASTM A213/A213M(95)	TP310HCb	Smls tube boiler, superheater, heat exchanger	0.04-0.10	24.00-26.00	2.00 max	0.75 max	19.00-22.00	0.045 max	0.030 max	0.75 max	Nb+Ta 10xC-1.10; bal Fe	515	205	35	
ASTM A213/A213M(95)	TP310HCbN	Smls tube boiler, superheater, heat exchanger	0.04-0.10	24.00-26.00	2.00 max		17.00-23.00	0.030 max	0.030 max	0.75 max	N 0.15-0.35; Nb+Ta 0.20-0.60; bal Fe	655	295	30	
ASTM A240/A240M(98)	S31041	Heat res	0.04-0.10	24.0-26.0	2.00 max		19.0-22.0	0.045 max	0.030 max	0.75 max	Cu <=0.7; 10xC<=Nb<=1.10; bal Fe	515	205	40.0	217 HB
ASTM A249/249M(96)	TP310HCb	Weld tube; boiler, superheater, heat exch	0.04-0.10	24.00-26.00	2.00 max	0.75 max	19.00-22.00	0.045 max	0.030 max	0.75 max	10xC<=Nb+Ta<=1.10; bal Fe	515	205	35	90 HRB
ASTM A943/A943M(95)	TP310HCb	Spray formed smls pipe	0.04-0.10	24.0-26.0	2.00 max	0.75 max	19.0-22.0	0.045 max	0.030 max	0.75 max	Nb+Ta>=10xC<=1.10; bal Fe	515	205		
SAE J405(98)	S31041	Heat res	0.04-0.10	24.00-26.00	2.00 max		19.00-22.00	0.045 max	0.030 max	0.75 max	Cu <=0.75; 10xC<=Nb<=1.10; bal Fe				
Stainless Steel, Austenitic, W-FeCrNi-1															
USA															
	UNS S30481	Thermal spray wire	0.080 max	18-20	2.0 max		8.0-11.0	0.045 max	0.045 max	1.0 max	bal Fe				
Stainless Steel, Austenitic, W-FeCrNi-2															
USA															
	UNS S30280	Thermal spray wire W-FeCrNi-2	0.12 max	17.0-19.0	1.5 max		7.0-9.0			0.50 max	bal Fe				
Stainless Steel, Austenitic, W-FeCrNi-3															
USA															
	UNS S20281	Thermal spray wire	0.15 max	17-19	7.5-10.0		4.0-6.0	0.060 max	0.03 max	1.0 max	bal Fe				

UNS numbers and US grades are provided as a means of cross referencing chemically similar alloys. Exchangability is only possible after independent examination of specifications. Tensile properties are minimum or typical as specified. UTS and YS as MPa. El as %. See Appendix for list of abbreviations used in Notes. * indicates obsolete material.

Specification	Designation	Notes	C	Cr	Mn	Mo	Ni	P	S	Si	Other	UTS	YS	El	Hard

Stainless Steel, Austenitic, W-FeCrNi-4

USA

| | UNS S20280 | Thermal spray wire | 0.06 max | 17-19 | 7.0-9.0 | | 4.0-6.0 | | | 0.80 max | bal Fe | | | | |

Stainless Steel, Austenitic, XEV-F

India

| IS 1570/5(85) | X53Cr22Mn9Ni4N | Valve | 0.48-0.58 | 20-23 | 8-10 | 0.15 max | 3.25-4.5 | 0.045 max | 0.035 max | 0.25 max | Co <=0.1; Cu <=0.3; Pb <=0.15; W <=0.1; bal Fe | | | | 321 HB |

International

| ISO 683-15(92) | X50CrMnNiNbN219 | Valve, Q, HT | 0.45-0.55 | 20.0-22.0 | 8.0-10.0 | | 3.5-5.5 | 0.050 max | 0.030 max | 0.25 max | N 0.40-0.60; W 0.80-1.50; Nb+Ta 1.80-2.50, bal Fe | 1300 max | | | 385 HB max |

Spain

| UNE 36022(91) | F.3219* | see X53CrMnNiNbW21-09 | 0.48-0.58 | 20-22 | 8-10 | 0.15 max | 3.25-4.5 | 0.04 max | 0.035 max | 0.25 max | N 0.38-0.5; Nb 2-3; W 0.8-1.2; bal Fe | | | | |
| UNE 36022(91) | X53CrMnNiNbW21-09 | Valve | 0.48-0.58 | 20-22 | 8-10 | 0.15 max | 3.25-4.5 | 0.04 max | 0.035 max | 0.25 max | N 0.38-0.5; Nb 2-3; W 0.8-1.2; bal Fe | | | | |

USA

	UNS S63019	Valve, 21-4N+Nb+W	0.45-0.55	20.00-22.00	8.00-10.00		3.50-5.50	0.050 max	0.030 max	0.45 max	N 0.40-0.60; W 0.80-1.50; Nb+Ta 1.80-2.50; bal Fe				
SAE J775(93)	21-4N+Nb+W	Engine poppet valve	0.45-0.55	20.00-22.00	8.00-10.00		3.50-5.50	0.050 max	0.030 max	0.45 max	N 0.40-0.60; W 0.80-1.50; 1.80<=Nb+Ta<=2.50; bal Fe	1190	880		
SAE J775(93)	ISO X50CrMnNiNbN219	Engine poppet valve	0.45-0.55	20.00-22.00	8.00-10.00		3.50-5.50	0.050 max	0.030 max	0.45 max	N 0.40-0.60; W 0.80-1.50; 1.80<=Nb+Ta<=2.50; bal Fe	1190	880		
SAE J775(93)	ISO X53CrMnNiNbN219	Engine poppet valve	0.48-0.58	20.0-22.0	8.0-10.0		3.25-4.5	0.050 max	0.030 max	0.45 max	N 0.38-0.50; C+N>0.90; 2.00<=Nb+Ta<=3.00; bal Fe				
SAE J775(93)	XEV-F*	Engine poppet valve, ISO X50CrMnNiNbN219	0.45-0.55	20.00-22.00	8.00-10.00		3.50-5.50	0.050 max	0.030 max	0.45 max	N 0.40-0.60; W 0.80-1.50; 1.80<=Nb+Ta<=2.50; bal Fe	1190	880		

Stainless Steel, Austenitic, XM-1

USA

	UNS S20300	Cr-Mn-Ni-Cu Free Machining 203 EZ	0.08 max	16.00-18.00	5.00-6.50	0.50 max	5.00-6.50	0.040 max	0.18-0.35	1.0 max	Cu 1.75-2.25; bal Fe				
ASTM A581/A581M(95)	XM-1	Wir rod, Ann	0.08 max	16.0-18.0	5.0-6.5		5.0-6.5	0.04 max	0.18-0.35	1.00 max	Cu 1.75-2.25; bal Fe	585-860			
ASTM A582/A582M(95)	XM-1	Bar, HF, CF, Ann	0.08 max	16.00-18.00	5.00-6.00		5.00-6.50	0.04 max	0.18-0.35	1.00 max	Cu 1.75-2.25; bal Fe				262 HB max

Stainless Steel, Austenitic, XM-10

USA

	UNS S21900	21-6-9 Nitronic 40	0.08 max	19.00-21.50	8.00-10.00		5.50-7.50	0.060 max	0.030 max	1.00 max	N 0.15-0.40; bal Fe				
AMS 5561E(98)		Corr/Heat res; Weld/drawn or Smls/drawn	0.040 max	19.00-21.50	8.00-10.00	0.75 max	5.50-7.50	0.030 max	0.030 max	1.00 max	Cu <=0.75; N 0.15-0.40; bal Fe	979-1117	827	20	
ASTM A580/A580M(98)	XM-10	Wir, Ann	0.08 max	19.0-21.5	8.0-10.0		5.5-7.5	0.060 max	0.030 max	1.00 max	N 0.15-0.40; bal Fe	620	345	45	
ASTM A813/A813M(95)	TPXM-10	Weld Pipe, 21-6-9 Nitronic 40	0.08 max	19.00-21.50	8.00-10.00		5.50-7.50	0.040 max	0.030 max	1.00 max	B <=994; N 0.15-0.40; bal Fe	620	345		
ASTM A814/A814M(96)	TPXM-10	CW Weld Pipe, 21-6-9 Nitronic 60	0.08 max	19.0-21.50	8.00-10.00		5.50-7.50	0.040 max	0.030 max	1.00 max	N 0.15-0.40; bal Fe	620	345		
ASTM A943/A943M(95)	TPXM-10	Spray formed smls pipe	0.08 max	19.0-21.5	8.00-10.00		5.50-7.50	0.040 max	0.030 max	1.00 max	N 0.15-0.40; bal Fe	620	345		

Stainless Steel, Austenitic, XM-11

Europe

| EN 10088/3(95) | 1.4578 | Bar Rod Sect, Corr res; t<=160mm, Ann | 0.04 max | 16.50-17.50 | 1.00 max | 2.00-2.50 | 10.00-11.00 | 0.045 max | 0.015 max | 1.00 max | Cu 3.00-3.50; N <=0.11; bal Fe | 450-650 | 175 | 45 | 215 HB |

UNS numbers and US grades are provided as a means of cross referencing chemically similar alloys. Exchangability is only possible after independent examination of specifications. Tensile properties are minimum or typical as specified. UTS and YS as MPa. El as %. See Appendix for list of abbreviations used in Notes. * indicates obsolete material.

Specification	Designation	Notes	C	Cr	Mn	Mo	Ni	P	S	Si	Other	UTS	YS	El	Hard

Stainless Steel, Austenitic, XM-11 (Continued from previous page)

Europe

| EN 10088/3(95) | X3CrNiCuMo17-11-3-2 | Bar Rod Sect, Corr res; t<=160mm, Ann | 0.04 max | 16.50-17.50 | 1.00 max | 2.00-2.50 | 10.00-11.00 | 0.045 max | 0.015 max | 1.00 max | Cu 3.00-3.50; N <=0.11; bal Fe | 450-650 | 175 | 45 | 215 HB |

Spain

| UNE 36087/4(89) | F.3545* | see X3CrMnNiN18 8 7 | 0.04 max | 17.0-19.0 | 6.5-8.5 | | 6-8 | 0.045 max | 0.030 max | 1.00 max | N 0.15-0.25; bal Fe | | | | |
| UNE 36087/4(89) | X3CrMnNiN18 8 7 | Corr res; Sh Strp | 0.04 max | 17.0-19.0 | 6.5-8.5 | | 6-8 | 0.045 max | 0.030 max | 1.00 max | N 0.15-0.25; bal Fe | | | | |

USA

	UNS S21904	Low-C, 21-6-9LC	0.04 max	19.00-21.50	8.00-10.00		5.50-7.50	0.060 max	0.030 max	1.00 max	N 0.15-0.40; bal Fe				
AMS 5562C(95)		Corr/Heat res; Smls tub	0.40 max	19.00-21.50	8.00-10.00	0.75 max	5.50-7.50	0.060 max	0.030 max	1.00 max	Cu <=0.75; N 0.15-0.40; bal Fe	621	345	45	
ASTM A182/A182M(98)	FXM-11	Frg/roll pipe flange valve	0.40 max	19.0-21.5	8.0-10.0		5.5-7.5	0.060 max	0.030 max	1.00 max	N 0.15-0.40; bal Fe	620	345	45	
ASTM A336/A336M(98)	FXM-11	Frg, press/high-temp	0.04 max	19.0-21.5	8.0-10.0		5.5-7.5	0.060 max	0.030 max	1.00 max	N 0.15-0.40; bal Fe	620	345	40	
ASTM A580/A580M(98)	XM-11	Wir, Ann	0.04 max	19.0-21.5	8.0-10.0		5.5-7.5	0.060 max	0.030 max	1.00 max	N 0.15-0.40; bal Fe	620	345	45	
ASTM A666(96)	XM-11	Sh Strp Plt Bar, Ann	0.04 max	19.00-21.50	8.00-10.00		5.50-7.50	0.060 max	0.030 max	0.75 max	N 0.15-0.40; bal Fe	620-690	345-415	40-45	
ASTM A813/A813M(95)	TPXM-11	Weld Pipe	0.04 max	19.00-21.50	8.00-10.00		5.50-7.50	0.040 max	0.030 max	1.00 max	B <=995; N 0.15-0.40; bal Fe	620	345		
ASTM A814/A814M(96)	TPXM-11	CW Weld Pipe	0.04 max	19.0-21.50	8.00-10.00		5.50-7.50	0.040 max	0.030 max	1.00 max	N 0.15-0.40; bal Fe	620	345		
ASTM A943/A943M(95)	TPXM-11	Spray formed smls pipe	0.04 max	19.0-21.5	8.00-10.00		5.50-7.50	0.040 max	0.030 max	1.00 max	N 0.15-0.40; bal Fe	620	345		
ASTM A988(98)	S21904	HIP Flanges, Fittings, Valves/parts; Heat res; Low-C 21-6-9 LC	0.04 max	19.0-21.5	8.0-10.0		5.5-7.5	0.045 max	0.030 max	1.00 max	N 0.15-0.40; bal Fe	620	345	45	

Stainless Steel, Austenitic, XM-14

Czech Republic

| CSN 417471 | 17471 | Corr res | 0.05-0.12 | 16.0-19.0 | 14.0-17.0 | | 1.2-2.0 | 0.045 max | 0.035 max | 0.6-1.5 | N 0.32-0.42; bal Fe | | | | |

India

| IS 1570/5(85) | X07Cr17Mn12Ni4 | Sh Plt Strp | 0.12 max | 16-18 | 10-14 | 0.15 max | 3.5-5.5 | 0.045 max | 0.03 max | 1 max | Co <=0.1; Cu <=0.3; Nb <=0.25; Pb <=0.15; W <=0.1; bal Fe | 550 | 250 | 45 | 217 HB max |

USA

| | UNS S21460 | Cr-Mn-Ni-Mo Cryogenic Tenelon | 0.12 max | 17.00-19.00 | 14.00-16.00 | | 5.00-6.00 | 0.060 max | 0.030 max | 1.00 max | N 0.35-0.50; bal Fe | | | | |
| ASTM A666(96) | XM-14 | Sh Strp Plt Bar, Ann | 0.12 max | 17.00-19.00 | 14.00-16.00 | | 5.00-6.00 | 0.060 max | 0.030 max | 0.75 max | N 0.35-0.50; bal Fe | 725 | 380 | 40 | |

Stainless Steel, Austenitic, XM-15

China

GB 1220(92)	0Cr18Ni13Si4	Bar SA	0.08 max	15.00-20.00	2.00 max		11.50-15.00	0.035 max	0.030 max	3.00-5.00	bal Fe	520	205	40	
GB 1221(92)	0Cr18Ni13Si4	Bar SA	0.08 max	15.00-20.00	2.00 max		11.50-15.00	0.035 max	0.030 max	3.00-5.00	bal Fe	520	205	40	
GB 13296(91)	0Cr18Ni13Si4	Smls tube SA	0.08 max	15.00-20.00	2.00 max		11.50-15.00	0.035 max	0.030 max	3.00-5.00	bal Fe	520	205	35	
GB 3280(92)	0Cr18Ni13Si4	Sh Plt, CR SA	0.08 max	15.00-20.00	2.00 max		11.50-15.00	0.035 max	0.030 max	3.00-5.00	bal Fe	520	205	40	
GB 4237(92)	0Cr18Ni13Si4	Sh Plt, HR SA	0.08 max	15.00-20.00	2.00 max		11.50-15.00	0.035 max	0.030 max	3.00-5.00	bal Fe	520	205	40	
GB 4238(92)	0Cr18Ni13Si4	Plt, HR SA	0.08 max	15.00-20.00	2.00 max		11.50-15.00	0.035 max	0.030 max	3.00-5.00	bal Fe	520	205	40	
GB 4239(91)	0Cr18Ni13Si4	Sh Strp, CR SA	0.08 max	15.00-20.00	2.00 max		11.50-15.00	0.035 max	0.030 max	3.00-5.00	bal Fe	520	205	40	

France

| AFNOR NFA35578 | Z15CNS20.12 | Sh Plt Bar Tub | 0.2 max | 19-21 | 2 max | | 11-13 | 0.04 max | 0.03 max | 1.5-2.5 | bal Fe | | | | |

Japan

| JIS G4303(91) | SUSXM15J1 | Bar; SA; <=180mm diam | 0.08 max | 15.00-20.00 | 2.00 max | | 11.50-15.00 | 0.045 max | 0.030 max | 3.00-5.00 | other alloying as required; bal Fe | 520 | 205 | 40 | 207 HB |

UNS numbers and US grades are provided as a means of cross referencing chemically similar alloys. Exchangability is only possible after independent examination of specifications. Tensile properties are minimum or typical as specified. UTS and YS as MPa. El as %. See Appendix for list of abbreviations used in Notes. * indicates obsolete material.

Specification	Designation	Notes	C	Cr	Mn	Mo	Ni	P	S	Si	Other	UTS	YS	El	Hard

Stainless Steel, Austenitic, XM-15 (Continued from previous page)

Japan

Specification	Designation	Notes	C	Cr	Mn	Mo	Ni	P	S	Si	Other	UTS	YS	El	Hard
JIS G4304(91)	SUSXM15J1	Plt Sh, HR	0.08 max	15.00-20.00	2.00 max		11.50-15.00	0.045 max	0.030 max	3.00-5.00	bal Fe	520	205	40	95 HRB max
JIS G4305(91)	SUSXM15J1	Plt Sh, CR	0.08 max	15.00-20.00	2.00 max		11.50-15.00	0.045 max	0.030 max	3.00-5.00	bal Fe	520	205	40	207 HB max
JIS G4306	SUSXM15J1	Strp HR	0.08 max	15-20	2 max		11.5-15	0.045 max	0.03 max	3-5	bal Fe				
JIS G4307	SUSXM15J1	Strp CR	0.08 max	15-20	2 max		11.5-15	0.045 max	0.03 max	3-5	bal Fe				
JIS G4309	SUSXM15J1	Wir	0.08 max	15.00-20.00	2.00 max		11.50-15.00	0.045 max	0.030 max	3.00-5.00	bal Fe				

Mexico

Specification	Designation	Notes	C	Cr	Mn	Mo	Ni	P	S	Si	Other	UTS	YS	El	Hard
DGN B-194	XM-15*	Obs; Tube, Q/A, 0.4-13mm diam	0.08 max	17-19	2 max		17.5-18.5	0.04 max	0.03 max	1.5-2.5	bal Fe				

USA

Specification	Designation	Notes	C	Cr	Mn	Mo	Ni	P	S	Si	Other	UTS	YS	El	Hard
	UNS S38100		0.08 max	17.00-19.00	2.00 max		17.50-18.50	0.030 max	0.030 max	1.50-2.50	bal Fe				
ASME SA213	XM-15	Refer to ASTM A213/A213M(95)													
ASME SA240	XM-15	Refer to ASTM A240(95)													
ASME SA249	XM-15	Refer to ASTM A249/A249M(96)													
ASME SA312	XM-15	Refer to ASTM A312/A312M(95)													
ASTM A167(96)	XM-15*	Sh Strp Plt	0.08 max	17-19	2 max		17.5-18.5	0.03 max	0.03 max	1.5-2.5	bal Fe				
ASTM A213/A213M(95)	XM-15	Smls tube boiler, superheater, heat exchanger	0.08 max	17.00-19.00	2.00 max		17.5-18.5	0.030 max	0.030 max	1.50-2.50	bal Fe	515	205	35	
ASTM A240/A240M(98)	S38100		0.08 max	17.0-19.0	2.00 max		17.5-18.5	0.030 max	0.030 max	1.50-2.50	bal Fe	515	205	40.0	217 HB
ASTM A249/249M(96)	TPXM-15	Weld tube; boiler, superheater, heat exch	0.08 max	17.00-19.00	2.00 max		17.5-18.5	0.03 max	0.03 max	1.50-2.50	bal Fe	515	205	35	90 HRB
ASTM A269	XM-15	Tube	0.08 max	17-19	2 max		17.5-18.5	0.03 max	0.03 max	1.5-2.5	bal Fe				
ASTM A312/A312M(95)	XM-15	Tube	0.08 max	17-19	2 max		17.5-18.5	0.03 max	0.03 max	1.5-2.5	bal Fe				
ASTM A813/A813M(95)	TPXM-15	Weld Pipe	0.08 max	17.0-19.0	2.00 max		17.50-18.50	0.030 max	0.030 max	1.50-2.50	bal Fe	515	205		
ASTM A814/A814M(96)	TPXM-15	CW Weld Pipe	0.08 max	17.0-19.0	2.00 max		17.50-18.50	0.030 max	0.030 max	1.50-2.50	bal Fe	515	205		
ASTM A943/A943M(95)	TPXM-15	Spray formed smls pipe	0.08 max	17.0-19.0	2.00 max		17.5-18.5	0.030 max	0.030 max	1.5-2.50	bal Fe	515	205		
SAE J405(98)	S38100		0.08 max	17.00-19.00	2.00 max		17.50-18.50	0.030 max	0.030 max	1.50-2.50	bal Fe				

Stainless Steel, Austenitic, XM-17

USA

Specification	Designation	Notes	C	Cr	Mn	Mo	Ni	P	S	Si	Other	UTS	YS	El	Hard
	UNS S21600	Cr-Mn-Ni-Mo 216	0.08 max	17.50-22.00	7.50-9.00	2.00-3.00	5.00-7.00	0.045 max	0.030 max	1.00 max	N 0.25-0.50; bal Fe				
ASTM A240/A240M(98)	S21600	Plt	0.08 max	17.5-22.0	7.5-9.0	2.00-3.00	5.0-7.0	0.045 max	0.030 max	0.75 max	N 0.25-0.50; bal Fe	620	345	40.0	241 HB
ASTM A240/A240M(98)	S21600	Sh Strp	0.08 max	17.5-22.0	7.5-9.0	2.00-3.00	5.0-7.0	0.045 max	0.030 max	0.75 max	N 0.25-0.50; bal Fe	690	415	40.0	241 HB
SAE J405(98)	S21600	Cr-Mn-Ni-Mo 216	0.08 max	17.50-22.00	7.50-9.00	2.00-3.00	5.00-7.00	0.045 max	0.030 max	0.75 max	N 0.25-0.50; bal Fe				

Stainless Steel, Austenitic, XM-18

USA

Specification	Designation	Notes	C	Cr	Mn	Mo	Ni	P	S	Si	Other	UTS	YS	El	Hard
	UNS S21603	Cr-Mn-Ni-Mo Low-C 216L	0.03 max	17.50-22.00	7.50-9.00	2.00-3.00	5.00-7.00	0.045 max	0.030 max	1.00 max	N 0.25-0.50; bal Fe				
ASTM A240/A240M(98)	S21603	Sh Strp	0.03 max	17.5-22.0	7.5-9.0	2.00-3.00	5.0-7.0	0.045 max	0.030 max	0.75 max	N 0.25-0.50; bal Fe	690	415	40.0	241 HB
ASTM A240/A240M(98)	S21603	Plt	0.03 max	17.5-22.0	7.5-9.0	2.00-3.00	5.0-7.0	0.045 max	0.030 max	0.75 max	N 0.25-0.50; bal Fe	620	345	40.0	241 HB
SAE J405(98)	S21603	Cr-Mn-Ni-Mo Low-C 216L	0.03 max	17.50-22.00	7.50-9.00	2.00-3.00	5.00-7.00	0.045 max	0.030 max	0.75 max	N 0.25-0.50; bal Fe				

UNS numbers and US grades are provided as a means of cross referencing chemically similar alloys. Exchangability is only possible after independent examination of specifications. Tensile properties are minimum or typical as specified. UTS and YS as MPa. El as %. See Appendix for list of abbreviations used in Notes. * indicates obsolete material.

Stainless Steel, Austenitic, XM-19

Specification	Designation	Notes	C	Cr	Mn	Mo	Ni	P	S	Si	Other	UTS	YS	El	Hard
Australia															
AS 2837(86)	209	Bar, Semi-finished product	0.06 max	20.50-23.50	4.00-6.00	1.50-3.00	11.50-13.50	0.040 max	0.030 max	1.00 max	N 0.20-0.40; Nb 0.10-0.30; V 0.10-0.30; bal Fe				
Germany															
DIN SEW 390(91)	G-X2CrNiMnMoNNb21 16 5 3	Non-magnetizable Q	0.03 max	20.0-21.5	4.00-6.00	3.00-3.50	15.0-17.0	0.025 max	0.010 max	1.00 max	N 0.20-0.35; Nb <=0.25; bal Fe	740-930	365	35	
DIN SEW 390(91)	WNr 1.3964	Non-magnetizable Q	0.03 max	20.0-21.5	4.00-6.00	3.00-3.50	15.0-17.0	0.025 max	0.010 max	1.00 max	N 0.20-0.35; Nb <=0.25; bal Fe	740-930	365	35	
USA															
	UNS S20910	Cr-Ni-Mn-Mo 22-13-5 Nitronic 50	0.06 max	20.50-23.50	4.00-6.00	1.50-3.00	11.50-13.50	0.040 max	0.030 max	1.00 max	N 0.20-0.40; Nb 0.10-0.30; V 0.10-0.30; bal Fe				
ASTM A182/A182M(98)	FXM-19	Frg/roll pipe flange valve	0.06 max	20.5-23.5	4.0-6.0	1.50-3.00	11.5-13.5	0.040 max	0.030 max	1.00 max	N 0.20-0.40; Nb 0.10-0.30; V 0.10-0.30; bal Fe	690	380	35	
ASTM A193/A193M(98)	B8R	Bolt, high-temp, HT	0.06 max	20.50-23.50	4.0-6.0	1.50-3.00	11.5-13.5	0.045 max	0.030 max	1.00 max	N 0.20-0.40; V 0.10-0.30; 0.10<=Nb+Ta<=0.30; bal Fe	690	380	35	271 HB
ASTM A193/A193M(98)	B8RA	Bolt, high-temp, HT	0.06 max	20.50-23.50	4.0-6.0	1.50-3.00	11.5-13.5	0.045 max	0.030 max	1.00 max	N 0.20-0.40; V 0.10-0.30; 0.10<=Nb+Ta<=0.30; bal Fe	690	380	35	271 HB
ASTM A194/A194M(98)	8R	Nuts, high-temp press	0.06 max	20.5-23.5	4.0-6.0	1.50-3.00	11.5-13.5	0.045 max	0.030 max	1.00 max	N 0.20-0.40; V 0.10-0.30; 0.10<=Nb+Ta<=0.30; bal Fe				183-271 HB
ASTM A194/A194M(98)	8RA	Nuts, high-temp press	0.06 max	20.5-23.5	4.0-6.0	1.50-3.00	11.5-13.5	0.045 max	0.030 max	1.00 max	N 0.20-0.40; V 0.10-0.30; 0.10<=Nb+Ta<=0.30; bal Fe				183-271 HB
ASTM A213/A213M(95)	XM-19	Smls tube boiler, superheater, heat exchanger	0.06 max	20.5-23.5	4.00-6.00	1.50-3.00	11.5-13.5	0.04 max	0.03 max	1.00 max	N 0.20-0.40; V 0.10-0.30; Ta+Nb 0.10-0.30; bal Fe	540	380	35	
ASTM A240/A240M(98)	S20910	Sh Strp	0.06 max	20.5-23.5	4.0-6.0	1.50-3.00	11.5-13.5	0.040 max	0.030 max	1.00 max	N 0.20-0.40; Nb 0.10-0.30; V 0.10-0.30; 0.10<=Nb<=0.30; bal Fe	725	415	30.0	241 HB
ASTM A240/A240M(98)	S20910	Plt	0.06 max	20.5-23.5	4.0-6.0	1.50-3.00	11.5-13.5	0.040 max	0.030 max	1.00 max	N 0.20-0.40; Nb 0.10-0.30; V 0.10-0.30; 0.10<=Nb<=0.30; bal Fe	690	380	35.0	241 HB
ASTM A249/249M(96)	TPXM-19	Weld tube; boiler, superheater, heat exch	0.06 max	20.5-23.5	4.00-6.00	1.50-3.00	11.5-13.5	0.04 max	0.03 max	1.00 max	N 0.20-0.40; V 0.10-0.30; Ta+Nb 0.10-0.30; bal Fe	690	380	35	25 HRC
ASTM A336/A336M(98)	FXM-19	Frg, press/high-temp	0.06 max	20.5-23.5	4.0-6.0	1.50-3.00	11.5-13.5	0.040 max	0.030 max	1.00 max	N 0.20-0.40; V 0.10-0.30; Ta+Nb 0.10-0.30; bal Fe	690	380	30	25 HRC
ASTM A580/A580M(98)	XM-19	Wir, Ann	0.06 max	20.5-23.5	4.0-6.0	1.50-3.00	11.5-13.5	0.040 max	0.030 max	1.00 max	N 0.20-0.40; V 0.10-0.30; bal Fe	690	380	35	25 HRC
ASTM A813/A813M(95)	TPXM-19	Weld Pipe, Cr-Ni-Mn-Mo 22-13-5 Nitronic 50	0.060 max	20.50-23.50	4.00-6.00	1.50-3.00	11.50-13.50	0.040 max	0.030 max	1.00 max	B <=997; N 0.20-0.40; V 0.10-0.30; Nb+Ta=0.10-0.30, bal Fe	690	380		
ASTM A814/A814M(96)	TPXM-19	CW Weld Pipe, Cr-Ni-Mn-Mo 22-13-5 Nitronic 50	0.060 max	20.50-23.50	4.00-6.00	1.50-3.00	11.50-13.50	0.040 max	0.030 max	1.00 max	N 0.20-0.40; V 0.10-0.30; Nb+Ta=0.10-0.30, bal Fe	690	380		
ASTM A943/A943M(95)	TPXM-19	Spray formed smls pipe	0.060 max	20.5-23.5	4.00-6.00	1.50-3.00	11.5-13.5	0.040 max	0.030 max	1.00 max	N 0.20-0.40; V 0.10-0.30; Nb+Ta= 0.10-0.30; bal Fe	690	380		
ASTM A965/965M(97)	FXM-19	Frg for press high-temp parts	0.06 max	20.5-23.5	4.0-6.0	1.50-3.00	11.5-13.5	0.040 max	0.030 max	1.00 max	N 0.20-0.40; V 0.10-0.30; Nb+Ta 0.10-0.30; bal Fe	690	380	30	
SAE J405(98)	S20910	Cr-Ni-Mn-Mo 22-13-5 Nitronic 50	0.06 max	20.50-23.50	4.00-6.00	1.50-3.00	11.50-13.50	0.040 max	0.030 max	0.75 max	N 0.20-0.40; Nb 0.10-0.30; V 0.10-0.30; bal Fe				

UNS numbers and US grades are provided as a means of cross referencing chemically similar alloys. Exchangability is only possible after independent examination of specifications. Tensile properties are minimum or typical as specified. UTS and YS as MPa. El as %. See Appendix for list of abbreviations used in Notes. * indicates obsolete material.

Specification	Designation	Notes	C	Cr	Mn	Mo	Ni	P	S	Si	Other	UTS	YS	El	Hard

Stainless Steel, Austenitic, XM-2

USA

Specification	Designation	Notes	C	Cr	Mn	Mo	Ni	P	S	Si	Other	UTS	YS	El	Hard
	UNS S30345	Free Machining 303 MA	0.15 max	17.00-19.00	2.00 max	0.40-0.60	8.00-10.00	0.05 max	0.11-0.16	1.00 max	Al 0.60-1.00; bal Fe				
ASTM A581/A581M(95)	XM-2	Wir rod, Ann	0.15 max	17.0-19.0	2.00 max	0.40-0.60	8.0-10.0	0.05 max	0.11-0.16	1.00 max	Al 0.60-1.00; bal Fe	585-860			
ASTM A582/A582M(95)	XM-2	Bar, HF, CF, Ann	0.15 max	17.00-19.00	2.00 max	0.40-0.60	8.00-10.00	0.05 max	0.11-0.16	1.00 max	Al 0.60-1.00; bal Fe				262 HB max

Stainless Steel, Austenitic, XM-21

China

Specification	Designation	Notes	C	Cr	Mn	Mo	Ni	P	S	Si	Other	UTS	YS	El	Hard
GB 1220(92)	0Cr19Ni10NbN	Bar SA	0.08 max	18.00-20.00	2.00 max		7.50-10.50	0.035 max	0.030 max	1.00 max	N 0.10-0.30; Nb <=0.15; bal Fe	685	345	35	
GB 3280(92)	0Cr19Ni10NbN	Sh Plt, CR SA	0.08 max	18.00-20.00	2.00 max		7.50-10.50	0.035 max	0.030 max	1.00 max	N 0.10-0.30; Nb <=0.15; bal Fe	685	345	35	
GB 4237(92)	0Cr19Ni10NbN	Sh Plt, HR SA	0.08 max	18.00-20.00	2.00 max		7.50-10.50	0.035 max	0.030 max	1.00 max	N 0.10-0.30; Nb <=0.15; bal Fe	685	345	35	
GB 4239(91)	0Cr19Ni10NbN	Plt, HR SA	0.08 max	18.00-20.00	2.00 max		7.50-10.50	0.035 max	0.030 max	1.00 max	N 0.10-0.30; Nb <=0.15; bal Fe	685	345	35	

USA

Specification	Designation	Notes	C	Cr	Mn	Mo	Ni	P	S	Si	Other	UTS	YS	El	Hard
	UNS S30452	High-N	0.08 max	18.00-20.00	2.00 max		8.00-10.50	0.045 max	0.030 max	1.00 max	N 0.16-0.30; bal Fe				
ASTM A240/A240M(98)	S30452	Sh Strp	0.08 max	18.0-20.0	2.00 max		8.0-10.5	0.045 max	0.030 max	0.75	N 0.16-0.30; bal Fe	620	345	30.0	241 HB
ASTM A240/A240M(98)	S30452	Plt	0.08 max	18.0-20.0	2.00 max		8.0-10.5	0.045 max	0.030 max	0.75	N 0.16-0.30; bal Fe	585	275	30.0	241 HB
SAE J405(98)	S30452	High-N	0.08 max	18.00-20.00	2.00 max		8.00-10.50	0.045 max	0.030 max	0.75 max	N 0.16-0.30; bal Fe				

Stainless Steel, Austenitic, XM-28

USA

Specification	Designation	Notes	C	Cr	Mn	Mo	Ni	P	S	Si	Other	UTS	YS	El	Hard
	UNS S24100	18-2-Mn Nitronic 32, 18-2-12	0.15 max	16.50-19.50	11.00-14.00		0.50-2.50	0.060 max	0.030 max	1.00 max	N 0.20-0.45; bal Fe				
ASTM A313/A313M(95)	XM-28	Spring wire, t<=0.23mm	0.15 max	16.50-19.00	11.00-14.00		0.50-2.50	0.060 max	0.030 max	1.00 max	bal Fe	2240-2450			
ASTM A580/A580M(98)	XM-28	Wir, Ann	0.15 max	16.5-19.0	11.0-14.0		0.5-2.5	0.040 max	0.030 max	1.00 max	N 0.20-0.45; bal Fe	690	380	30	

Stainless Steel, Austenitic, XM-29

USA

Specification	Designation	Notes	C	Cr	Mn	Mo	Ni	P	S	Si	Other	UTS	YS	El	Hard
	UNS S24000	18-3-Mn Nitronic 33, 18-3-12	0.08 max	17.00-19.00	11.50-14.50		2.50-3.75	0.060 max	0.030 max	1.00 max	N 0.20-0.40; bal Fe				
ASTM A240/A240M(98)	S24000	Plt	0.08 max	17.0-19.0	11.5-14.5		2.3-3.7	0.060 max	0.030 max	0.75 max	N 0.20-0.40; bal Fe	690	380	40.0	241 HB
ASTM A240/A240M(98)	S24000	Sh Strp	0.08 max	17.0-19.0	11.5-14.5		2.3-3.7	0.060 max	0.030 max	0.75 max	N 0.20-0.40; bal Fe	690	415	40.0	241 HB
ASTM A249/249M(96)	TPXM-29	Weld tube; boiler, superheater, heat exch	0.08 max	17.00-19.00	11.5-14.5		2.25-3.75	0.04 max	0.03 max	1.00 max	N 0.20-0.40; bal Fe	690	380	35	100 HRB
ASTM A580/A580M(98)	XM-29	Wir, Ann	0.08 max	17.0-19.0	11.5-14.5		2.3-3.7	0.060 max	0.030 max	1.00 max	N 0.20-0.40; bal Fe	690	380	30	
ASTM A688/A688M(96)	TPXM-29	Weld Feedwater Heater Tub, 18-3-Mn Nitronic 33, 18-3-12	0.060 max	17.00-19.00	11.50-14.50		2.25-3.75	0.060 max	0.03 max	1.00 max	N 0.20-0.40; bal Fe	690	380	35	
ASTM A813/A813M(95)	TPXM-29	Weld Pipe, 18-3-Mn Nitronic, 33 18-3-12	0.080 max	17.0-19.0	11.50-14.50		2.25-3.75	0.060 max	0.030 max	1.00 max	B <=998; N 0.20-0.40; bal Fe	690	380		
ASTM A814/A814M(96)	TPXM-29	CW Weld Pipe, 18-3-Mn Nitronic 33, 18-3-12	0.080 max	17.0-19.0	11.50-14.50		2.25-3.75	0.060 max	0.030 max	1.00 max	N 0.20-0.40; bal Fe	690	380		
ASTM A943/A943M(95)	TPXM-29	Spray formed smls pipe	0.080 max	17.0-19.0	11.5-14.5		2.25-3.75	0.060 max	0.030 max	1.00 max	N 0.20-0.40; bal Fe	690	380		
SAE J405(98)	S24000	18-3-Mn Nitronic 33, 18-3-12	0.08 max	17.00-19.00	11.50-14.50		2.25-3.75	0.060 max	0.030 max	0.75 max	N 0.20-0.40; bal Fe				

Stainless Steel, Austenitic, XM-3

USA

Specification	Designation	Notes	C	Cr	Mn	Mo	Ni	P	S	Si	Other	UTS	YS	El	Hard
	UNS S30360	Free Machining 303 Pb	0.15 max	17.00-19.00	2.00 max	0.75 max	8.00-10.00	0.040 max	0.12-0.30	1.00 max	Pb 0.12-0.30; bal Fe				

UNS numbers and US grades are provided as a means of cross referencing chemically similar alloys. Exchangability is only possible after independent examination of specifications. Tensile properties are minimum or typical as specified. UTS and YS as MPa. El as %. See Appendix for list of abbreviations used in Notes. * indicates obsolete material.

Specification	Designation	Notes	C	Cr	Mn	Mo	Ni	P	S	Si	Other	UTS	YS	El	Hard
Stainless Steel, Austenitic, XM-32															
India															
IS 1570/5(85)	X12Cr12	Sh Plt Strp Bar Flat Band	0.08-0.15	11.5-13.5	1 max	0.15 max	1 max	0.04 max	0.03 max	1 max	Co <=0.1; Cu <=0.3; Pb <=0.15; W <=0.1; bal Fe	450	205	20	217 HB max
IS 1570/7(92)	12Cr12Ni2Mo	0-150mm; Bar Frg Sh Plt	0.08-0.16	11-12.5	0.5-0.9	1.5-2	2-3	0.04 max	0.03 max	0.35 max	Co <=0.1; Cu <=0.3; N 0.02-0.04; Pb <=0.15; V 0.25-0.4; W <=0.1; bal Fe	930-1130	785	14	285-331 HB
IS 1570/7(92)	22	0-150mm; Bar Frg Sh Plt	0.08-0.16	11-12.5	0.5-0.9	1.5-2	2-3	0.04 max	0.03 max	0.35 max	Co <=0.1; Cu <=0.3; N 0.02-0.04; Pb <=0.15; V 0.25-0.4; W <=0.1; bal Fe	930-1130	785	14	285-331 HB
USA															
	UNS S64152	High-strength M152	0.08-0.15	11.00-12.50	0.50-0.90	1.50-2.00	2.00-3.00	0.025 max	0.025 max	0.35 max	N 0.01-0.05; V 0.25-0.40; bal Fe				
ASTM A565(97)	XM-32	Bar Frg for high-temp HT	0.08-0.15	11.00-12.50	0.50-0.90	1.50-2.00	2.00-3.00	0.025 max	0.025 max	0.35 max	N 0.01-0.05; V 0.25-0.40; bal Fe	1000	795	15	302-352 HB
Stainless Steel, Austenitic, XM-5															
USA															
	UNS S30310	Free Machining	0.15 max	17.00-19.00	2.50-4.50	0.75 max	7.00-10.00	0.20 max	0.25 min	1.00 max	bal Fe				
ASTM A581/A581M(95)	XM-5	Wir rod, Ann	0.15 max	17.0-19.0	2.5-4.5		7.0-10.0	0.20 max	0.25 min	1.00 max	bal Fe	585-860			
ASTM A582/A582M(95)	XM-5	Bar, HF, CF, Ann	0.15 max	17.00-19.00	2.50-4.50		7.00-10.00	0.20 max	0.25 min	1.00 max	bal Fe				262 HB max

UNS numbers and US grades are provided as a means of cross referencing chemically similar alloys. Exchangability is only possible after independent examination of specifications. Tensile properties are minimum or typical as specified. UTS and YS as MPa. El as %. See Appendix for list of abbreviations used in Notes. * indicates obsolete material.

Specification	Designation	Notes	C	Cr	Mn	Mo	Ni	P	S	Si	Other	UTS	YS	El	Hard

Stainless Steel, Ferritic, 25-4-4

USA

Specification	Designation	Notes	C	Cr	Mn	Mo	Ni	P	S	Si	Other	UTS	YS	El	Hard
	UNS S44635		0.025 max	24.5-26.0	1.00 max	3.50-4.50	3.50-4.50	0.040 max	0.030 max	0.75 max	N <=0.035; Nb+Ti 0.20+4(C+N)-0.80 max; bal Fe				
ASTM A240/A240M(98)	S44635		0.025 max	24.5-26.0	1.00 max	3.5-4.5	3.5-4.5	0.040 max	0.030 max	0.75 max	N <=0.035; 0.20+4(C+N)<=Ti+Nb <=0.80; bal Fe	620	515	20.0	269 HB
ASTM A803/A803M(96)	25-4-4	Weld Feedwater Heater Tub	0.025 max	24.5-26.0	1.00 max	3.5-4.5	3.5-4.5	0.040 max	0.030 max	0.75 max	N <=0.035; Ti+Nb>=0.2+4(C+N)<=0.80; bal Fe	620	515	20	270 HB max
SAE J405(98)	S44635		0.025 max	24.5-26.0	1.00 max	3.5-4.5	3.5-4.5	0.040 max	0.030 max	0.75 max	N <=0.035; 0.20+4(C+N)<=Nb+Ti <=0.80; bal Fe				

Stainless Steel, Ferritic, 26-3-3

USA

Specification	Designation	Notes	C	Cr	Mn	Mo	Ni	P	S	Si	Other	UTS	YS	El	Hard
	UNS S44660	Nb plus Ti-stabilized (SC-1)	0.030 max	25.0-28.0	1.00 max	3.00-4.00	1.00-3.50	0.040 max	0.030 max	1.00 max	N <=0.040; Nb+Ti [0.20+6(C+N)] min-1.00 max; bal Fe				
ASTM A240/A240M(98)	S44660	Nb plus Ti-stabilized (SC-1)	0.030 max	25.0-28.0	1.00 max	3.0-4.0	1.0-3.5	0.040 max	0.030 max	1.00 max	N <=0.040; 0.20+6(C+N)<=Nb+Ti <=1.00; bal Fe	585	450	18.0	241 HB
ASTM A803/A803M(96)	26-3-3	Weld Feedwater Heater Tub, Nb/titanium-stabilized (SC-1)	0.030 max	25.0-28.0	1.00 max	3.0-4.0	1.00-3.50	0.040 max	0.030 max	1.00 max	N <=0.040; Ti+Nb>=6x(C+N)>=0.20<=1.00; bal Fe	585	450	20	265 HB max
SAE J405(98)	S44660	Nb plus Ti-stabilized (SC-1)	0.030 max	25.0-28.0	1.00 max	3.00-4.00	1.0-3.5	0.040 max	0.030 max	1.00 max	N <=0.040; 0.20 and 6(C+N)<=Nb+Ti<=1.00; bal Fe				

Stainless Steel, Ferritic, 2803Mo

USA

Specification	Designation	Notes	C	Cr	Mn	Mo	Ni	P	S	Si	Other	UTS	YS	El	Hard
	UNS S32803	Ferritic Cr-Ni-Mo with Nb Cronifer	0.010 max	28.0-29.0	0.50 max	1.8-2.5	3.0-4.0	0.020 max	0.005 max	0.50 max	N <=0.025; Nb 0.15-0.50; ratio Nb/C+N=12 min, C+N = 0.030 max; bal Fe				
ASTM A176(97)	S32803	Sh Strp Plt	0.015 max	28.00-29.00	0.50 max	1.8-2.5	3.0-4.0	0.020 max	0.005 max	0.55 max	N <=0.020; Nb 0.15-0.50; Nb>=12x(C+N); bal Fe	600	500	16.0	241 HB
ASTM A240/A240M(98)	S32803	Ferr Cr-Ni-Mo with Nb Cronifer	0.010 max	28.0-29.0	0.50 max	1.8-2.5	3.0-4.0	0.020 max	0.005 max	0.50 max	N <=0.025; 0.15+12x(C+N)<=Nb <=0.50; C+N<=0.030; bal Fe	600	500	16.0	241 HB
SAE J405(98)	S32803	Ferr Cr-Ni-Mo w/Nb Cronifer	0.015 max	28.00-29.00	0.50 max	1.8-2.5	3.0-4.0	0.020 max	0.005 max	0.50 max	N <=0.025; Nb 0.15-0.50; 12(C+N)<=Nb; C+N<=0.030; bal Fe				

Stainless Steel, Ferritic, 29-4

Japan

Specification	Designation	Notes	C	Cr	Mn	Mo	Ni	P	S	Si	Other	UTS	YS	El	Hard
JIS G4303(91)	SUS447J1	Bar Ann 75mm diam	0.010 max	28.50-32.00	0.40 max	1.50-2.50	0.50 max	0.030 max	0.020 max	0.40 max	Cu <=0.20; N <=0.015; Ni+Cu<=0.50; bal Fe	450	295	20	228 HB max
JIS G4305(91)	SUS447J1		0.010 max	28.50-32.00	0.40 max	1.50-2.50		0.030 max	0.020 max	0.40 max	N <=0.015; bal Fe	450	295	22	

Mexico

Specification	Designation	Notes	C	Cr	Mn	Mo	Ni	P	S	Si	Other	UTS	YS	El	Hard
NMX-B-171(91)	29-4	mech tub	0.010 max	28.0-30.0	0.30 max	3.5-4.2	0.15 max	0.025 max	0.020 max	0.20 max	N <=0.020; bal Fe				

USA

Specification	Designation	Notes	C	Cr	Mn	Mo	Ni	P	S	Si	Other	UTS	YS	El	Hard
	UNS S44700	High-Cr+Mo	0.010 max	28.0-30.0	0.30 max	3.5-4.2	0.15 max	0.025 max	0.020 max	0.20 max	Cu <=0.15; N <=0.020; C+N 0.025 max; bal Fe				
ASTM A240/A240M(98)	S44700	High-Cr+Mo	0.010 max	28.0-30.0	0.30 max	3.5-4.2	0.15 max	0.025 max	0.020 max	0.20 max	Cu <=0.15; N <=0.020; C+N<=0.025; bal Fe	550	415	20.0	223 HB
ASTM A511(96)	29-4	Smls mech tub, Ann	0.010 max	28.0-30.0	0.30 max	3.5-4.2	0.15 max	0.025 max	0.020 max	0.20 max	Cu <=0.15; N <=0.020; bal Fe	483	379	20	207 HB
ASTM A580/A580M(98)	S44700	Wir, Ann	0.010 max	28.0-30.0	0.30 max	3.5-4.2	0.15 max	0.025 max	0.020 max	0.20 max	Cu <=0.15; N <=0.020; C+N<=0.025; bal Fe	485	380	20	

UNS numbers and US grades are provided as a means of cross referencing chemically similar alloys. Exchangability is only possible after independent examination of specifications. Tensile properties are minimum or typical as specified. UTS and YS as MPa. El as %. See Appendix for list of abbreviations used in Notes. * indicates obsolete material.

Specification	Designation	Notes	C	Cr	Mn	Mo	Ni	P	S	Si	Other	UTS	YS	El	Hard

Stainless Steel, Ferritic, 29-4 (Continued from previous page)

USA

Specification	Designation	Notes	C	Cr	Mn	Mo	Ni	P	S	Si	Other	UTS	YS	El	Hard
ASTM A803/A803M(96)	29-4	Weld Feedwater Heater Tub, High-Cr High-Mo	0.010 max	28.0-30.0	0.30 max	3.5-4.2	0.15 max	0.025 max	0.020 max	0.20 max	Cu <=0.15; N <=0.020; bal Fe	550	415	20	241 HB max
SAE J405(98)	S44700	High-Cr+Mo	0.010 max	28.0-30.0	0.30 max	3.5-4.2	0.15 max	0.025 max	0.020 max	0.20 max	Cu <=0.15; N <=0.020; (C+N)<=0.025; bal Fe				

Stainless Steel, Ferritic, 29-4-2

Mexico

Specification	Designation	Notes	C	Cr	Mn	Mo	Ni	P	S	Si	Other	UTS	YS	El	Hard
NMX-B-171(91)	29-4-2	mech tub	0.010 max	28.0-30.0	0.30 max	3.5-4.2	2.0-2.5	0.025 max	0.020 max	0.20 max	N <=0.020; (C+N) 0.025 max; bal Fe				

USA

Specification	Designation	Notes	C	Cr	Mn	Mo	Ni	P	S	Si	Other	UTS	YS	El	Hard
	UNS S44800	High-Cr	0.010 max	28.0-30.0	0.30 max	3.5-4.2	2.0-2.5	0.025 max	0.020 max	0.20 max	Cu <=0.15; N <=0.020; Nb+N 0.025 max; bal Fe				
ASTM A240/A240M(98)	S44800	High-Cr	0.010 max	28.0-30.0	0.30 max	3.5-4.2	2.0-2.5	0.025 max	0.020 max	0.20 max	Cu <=0.15; N <=0.020; Nb+N<=0.025; bal Fe	550	415	20.0	223 HB
ASTM A511(96)	29-4-2	Smls mech tub, Ann	0.010 max	28.0-30.0	0.30 max	3.5-4.2	2.0-2.5	0.025 max	0.020 max	0.20 max	Cu <=0.15; N <=0.020; bal Fe	483	379	20	207 HB
ASTM A580/A580M(98)	S44800	Wir, Ann	0.010 max	28.0-30.0	0.30 max	3.5-4.2	2.00-2.50	0.025 max	0.020 max	0.20 max	Cu <=0.15; N <=0.020; C+N<=0.025; bal Fe	485	380	20	
ASTM A803/A803M(96)	29-4-2	Weld Feedwater Heater Tub, High-Cr	0.010 max	28.0-30.0	0.30 max	3.5-4.2	2.0-2.5	0.025 max	0.020 max	0.20 max	Cu <=0.15; N <=0.020; bal Fe	550	415	20	241 HB max
SAE J405(98)	S44800	High-Cr	0.010 max	28.0-30.0	0.30 max	3.5-4.2	2.0-2.5	0.025 max	0.020 max	0.20 max	Cu <=0.15; N <=0.020; (C+N)<=0.025; bal Fe				

Stainless Steel, Ferritic, 29-4-2C

USA

Specification	Designation	Notes	C	Cr	Mn	Mo	Ni	P	S	Si	Other	UTS	YS	El	Hard
	UNS S44736	High-Cr+Mo	0.030 max	28.0-30.0	1.00 max	3.60-4.20	2.00-4.50	0.040 max	0.030 max	1.00 max	N <=0.045; Nb+Ti 0.20-1.00[6(C+N)] min; bal Fe				
ASTM A240/A240M(98)	S44736	High-Cr+Mo	0.030 max	28.0-30.0	1.00 max	3.60-4.20	2.00-4.50	0.040 max	0.030 max	1.00 max	N <=0.045; 0.20+6(C+N)<=Nb+Ti <=1.00; bal Fe				

Stainless Steel, Ferritic, 29-4C

USA

Specification	Designation	Notes	C	Cr	Mn	Mo	Ni	P	S	Si	Other	UTS	YS	El	Hard
	UNS S44735	High-Cr+Mo	0.030 max	28.0-30.0	1.00 max	3.60-4.20	1.00 max	0.040 max	0.030 max	1.00 max	N <=0.045; Nb+Ti 0.20-1.00[6(C+N)] min; bal Fe				
ASTM A240/A240M(98)	S44735	High-Cr+Mo	0.030 max	28.0-30.0	1.00 max	3.60-4.20	1.00 max	0.040 max	0.030 max	1.00 max	N <=0.045; 0.20+6(C+N)<=Nb+Ti <=1.00; bal Fe	550	415	18.0	255 HB
ASTM A803/A803M(96)	29-4C	Weld Feedwater Heater Tub, High-Cr High-Mo	0.030 max	28.0-30.0	1.00 max	3.60-4.20	1.00 max	0.040 max	0.030 max	1.00 max	N <=0.045; Ti+Nb=6(C+N)>=0.20 <=1.00; bal Fe	515	415	18	241 HB max
SAE J405(98)	S44735	High-Cr+Mo	0.030 max	28.00-30.00	1.00 max	3.60-4.20	1.00 max	0.040 max	0.030 max	1.00 max	N <=0.045; 0.20 and 6(C+N)<=Nb+Ti<=1.00; bal Fe				

Stainless Steel, Ferritic, 405

Australia

Specification	Designation	Notes	C	Cr	Mn	Mo	Ni	P	S	Si	Other	UTS	YS	El	Hard
AS 1449(94)	405	Sh Strp Plt	0.08 max	11.50-14.50	1.00 max		0.06 max	0.040 max	0.030 max	1.00 max	Al 0.1-0.3; bal Fe				

Bulgaria

Specification	Designation	Notes	C	Cr	Mn	Mo	Ni	P	S	Si	Other	UTS	YS	El	Hard
BDS 6738(72)	0Ch13Ju	Corr res	0.08 max	11.5-14.0	1.00 max	0.30 max	0.60 max	0.035 max	0.025 max	1.00 max	Al 0.10-0.30; bal Fe				
BDS 6738(72)	Ch13SJu	Corr res	0.12 max	12.0-14.0	0.80 max	0.30 max	0.60 max	0.035 max	0.025 max	1.00-1.80	Al 0.80-1.80; bal Fe				

Canada

Specification	Designation	Notes	C	Cr	Mn	Mo	Ni	P	S	Si	Other	UTS	YS	El	Hard
CSA G110.3	405	Bar Bil	0.08 max	11.5-14.5	1 max			0.04 max	0.03 max	1 max	Al 0.1-0.3; bal Fe				
CSA G110.5	405	Sh Strp Plt, HR, CR, Ann	0.08 max	11.5-14.5	1 max		0.6 max	0.04 max	0.03 max	1 max	Al 0.1-0.3; bal Fe				

UNS numbers and US grades are provided as a means of cross referencing chemically similar alloys. Exchangability is only possible after independent examination of specifications. Tensile properties are minimum or typical as specified. UTS and YS as MPa. El as %. See Appendix for list of abbreviations used in Notes. * indicates obsolete material.

Specification	Designation	Notes	C	Cr	Mn	Mo	Ni	P	S	Si	Other	UTS	YS	El	Hard

Stainless Steel, Ferritic, 405 (Continued from previous page)

Canada

| CSA G110.9 | 405 | Sh Strp Plt | 0.08 max | 11.5-14.5 | 1 max | | 0.6 max | 0.04 max | 0.03 max | 1 max | Al 0.1-0.3; bal Fe | | | | |

China

GB 1220(92)	0Cr13Al	Bar Ann	0.08 max	11.50-14.50	1.00 max		0.60 max	0.035 max	0.030 max	1.00 max	Al 0.10-0.30; bal Fe	410	177	20	
GB 1221(92)	0Cr13Al	Bar Ann	0.08 max	11.50-14.50	1.00 max		0.60 max	0.040 max	0.030 max	1.00 max	Al 0.10-0.30; bal Fe	410	177	20	
GB 3280(92)	0Cr13Al	Sh Plt, CR Ann	0.08 max	11.50-14.50	1.00 max		0.60 max	0.035 max	0.030 max	1.00 max	Al 0.10-0.30; bal Fe	410	175	20	
GB 4237(92)	0Cr13Al	Sh Plt, HR Ann	0.08 max	11.50-14.50	1.00 max		0.60 max	0.035 max	0.030 max	1.00 max	Al 0.10-0.30; bal Fe	410	177	20	
GB 4238(92)	0Cr13Al	Plt, HR Ann	0.08 max	11.50-14.50	1.00 max		0.60 max	0.035 max	0.030 max	1.00 max	Al 0.10-0.30; bal Fe	410	175	20	
GB 4239(91)	0Cr13Al	Sh Strp, CR Ann	0.08 max	11.50-14.50	1.00 max		0.60 max	0.035 max	0.030 max	1.00 max	Al 0.10-0.30; bal Fe	410	175	20	

France

| AFNOR NFA35573 | Z6CA13 | Sh Strp Plt Bar, Ann, 5-25mm | 0.08 max | 11.5-13.5 | 1 max | | | 0.04 max | 0.03 max | 1 max | Al 0.1-0.3; bal Fe | | | | |

Germany

DIN 17440(96)	WNr 1.4002	Plt Sh, HR Strp Bar	0.08 max	12.0-14.0	1.00 max			0.040 max	0.030 max	1.00 max	Al 0.10-0.30; bal Fe				
DIN 17441(97)	WNr 1.4002	CR Strp Plt Sh, Ann	0.08 max	12.0-14.0	1.00 max			0.040 max	0.030 max	1.00 max	Al 0.10-0.30; bal Fe	400-600	250	20	
DIN EN 10088(95)	WNr 1.4002	Sh Plt Strp, Ann	0.08 max	12.0-14.0	1.00 max			0.040 max	0.030 max	1.00 max	Al 0.10-0.30; bal Fe	400-600	250	20	
DIN EN 10088(95)	X6CrAl13	Sh Plt Strp, Ann	0.08 max	12.0-14.0	1.00 max			0.040 max	0.030 max	1.00 max	Al 0.10-0.30; bal Fe	400-600	250	20	
TGL 7143	X7CrA13*	Obs	0.08 max	12-14	0.8 max			0.045 max	0.03 max	0.8 max	Al 0.1-0.3; bal Fe				

Hungary

| MSZ 4359(82) | H12 | Sh Strp Plt; Heat res; 0-60mm; soft ann | 0.12 max | 12-14 | 1 max | | 0.6 max | 0.04 max | 0.03 max | 1-1.5 | Al 0.7-1.2; bal Fe | 450 | 250 | 15L | 192 HB max |

India

| IS 1570/5(85) | X04Cr12 | Sh Plt Strp Bar Flat Band | 0.08 max | 11.5-13.5 | 1 max | 0.15 max | 0.4 max | 0.04 max | 0.03 max | 1 max | Co <=0.1; Cu <=0.3; Pb <=0.15; W <=0.1; bal Fe | 415 | 205 | 22 | 183 HB max |

International

ISO 4955(94)	H3	Heat res	0.12 max	12.0-14.0	1.0 max	0.15 max	0.63-0.70	0.04 max	0.03 max	0.7-1.4	Al 0.7-1.2; Co <=0.1; Cu <=0.3; Pb <=0.15; Ti <=0.05; V <=0.1; W <=0.1; bal Fe				
ISO 683-13(74)	1*	Corr res	0.08 max	11.5-14	1 max	0.15 max	0.5 max	0.04 max	0.03 max	1 max	Al <=0.1; Co <=0.1; Cu <=0.3; Pb <=0.15; Ti <=0.05; V <=0.1; W <=0.1; bal Fe				
ISO 683-13(74)	2*	Ann	0.08 max	12-14	1 max		1 max	0.04 max	0.03 max	1 max	Al 0.1-0.3; bal Fe				

Italy

UNI 6901(71)	X10CrAl12	Heat res	0.12 max	11-13	0.6 max			0.04 max	0.03 max	1.5-2	Al 1-1.3; bal Fe				
UNI 6901(71)	X6CrAl13	Corr res	0.08 max	11.5-14	1 max		0.5 max	0.04 max	0.03 max	1 max	Al 0.1-0.3; bal Fe				
UNI 6904(71)	X10CrAl12	Heat res	0.12 max	11-13	0.6 max			0.04 max	0.03 max	1.5-2	Al 1.1-1.3; bal Fe				
UNI 6904(71)	X6CrAl13	Heat res	0.08 max	11.5-13.5	1 max		0.5 max	0.04 max	0.03 max	0.75 max	Al 0.1-0.3; bal Fe				

Japan

JIS G4303(91)	SUS405	Bar Ann 75mm diam	0.08 max	11.50-14.50	1.00 max		0.60 max	0.040 max	0.030 max	1.00 max	Al 0.10-0.30; bal Fe	410	175	20	183 HB max
JIS G4304(91)	SUS405	Sh Plt, HR Ann	0.08 max	11.50-14.50	1.00 max		0.60 max	0.040 max	0.030 max	1.00 max	Al 0.2-0.30; bal Fe	410	175	20	88 HRB max
JIS G4305(91)	SUS405	Sh Plt, CR Ann	0.08 max	11.50-14.50	1.00 max		0.60 max	0.040 max	0.030 max	1.00 max	Al 0.10-0.30; bal Fe	410	175	20	183 HB max
JIS G4306	SUS405	Strp HR Ann	0.08 max	11.5-14.5	1 max		0.6 max	0.04 max	0.03 max	1 max	Al 0.1-0.3; bal Fe				
JIS G4307	SUS405	Strp CR Ann	0.08 max	11.5-14.5	1 max		0.6 max	0.04 max	0.03 max	1 max	Al 0.1-0.3; bal Fe				

UNS numbers and US grades are provided as a means of cross referencing chemically similar alloys. Exchangability is only possible after independent examination of specifications. Tensile properties are minimum or typical as specified. UTS and YS as MPa. El as %. See Appendix for list of abbreviations used in Notes. * indicates obsolete material.

Specification	Designation	Notes	C	Cr	Mn	Mo	Ni	P	S	Si	Other	UTS	YS	El	Hard

Stainless Steel, Ferritic, 405 (Continued from previous page)

Japan

| JIS G4309 | SUS405 | Wir | 0.08 max | 11.50-14.50 | 1.00 max | | 0.60 max | 0.040 max | 0.030 max | 1.00 max | Al 0.10-0.30; bal Fe | | | | |

Mexico

| DGN B-83 | 405* | Obs; Bar, HR, CR | 0.08 max | 11.5-14.5 | 1 max | | | 0.04 max | 0.03 max | 1 max | Al 0.1-0.3; bal Fe | | | | |
| NMX-B-171(91) | MT405 | mech tub | 0.08 max | 11.5-14.5 | 1.00 max | | 0.50 max | 0.040 max | 0.030 max | 1.00 max | Al 0.10-0.30; bal Fe | | | | |

Poland

PNH86020	0H13	Corr res	0.08 max	12-14	0.08 max		0.6 max	0.04 max	0.03 max	0.8 max	bal Fe				
PNH86020	0H13J	Corr res	0.08 max	11.5-14	1 max		0.6 max	0.04 max	0.03 max	1 max	Al 0.1-0.3; bal Fe				
PNH86022(71)	H13JS	Heat res	0.12 max	12-14	0.8 max		0.5 max	0.04 max	0.03 max	1-1.3	Al 0.8-1.1; bal Fe				

Romania

| STAS 3583(87) | 7AlCr130 | Corr res; Heat res | 0.08 max | 12-14 | 1 max | 0.2 max | 0.5 max | 0.045 max | 0.03 max | 1 max | Al 0.1-0.3; Pb <=0.15; bal Fe | | | | |

Russia

| GOST 5632(72) | 10Ch13SJu | Heat resistant | 0.07-0.12 | 12.0-14.0 | 0.8 max | 0.15 max | 0.6 max | 0.03 max | 0.025 max | 1.2-2.0 | Al 1.0-1.8; Cu <=0.3; Ti <=0.2; bal Fe | | | | |

Spain

UNE 36016(75)	F.3111*	see X6CrAl13	0.08 max	11.5-14.0	1.0 max	0.15 max	0.5 max	0.04 max	0.03 max	1.0 max	Al 0.1-0.3; bal Fe				
UNE 36016(75)	X6CrAl13	Corr res	0.08 max	11.5-14.0	1.0 max	0.15 max	0.5 max	0.04 max	0.03 max	1.0 max	Al 0.1-0.3; bal Fe				
UNE 36016/1(89)	E-405	Corr res	0.08 max	12.0-14.0	1.0 max	0.15 max	1.0 max	0.04 max	0.03 max	1.0 max	Al 0.1-0.3; bal Fe				
UNE 36016/1(89)	F.3111*	Obs; E-405; Corr res	0.08 max	12.0-14.0	1.0 max	0.15 max	1.0 max	0.04 max	0.03 max	1.0 max	Al 0.1-0.3; bal Fe				
UNE 36017(85)	F.3152*	see X10CrAl13	0.12 max	12-14	1 max	0.15 max		0.04 max	0.03 max	0.7-1.4	Al 0.7-1.2; bal Fe				
UNE 36017(85)	X10CrAl13	Heat res	0.12 max	12-14	1 max	0.15 max		0.04 max	0.03 max	0.7-1.4	Al 0.7-1.2; bal Fe				

UK

| BS 1449/2(83) | 405S17 | Corr/Heat res; t<=100mm | 0.08 max | 12.0-14.0 | 1.00 max | | 1.00 max | 0.040 max | 0.030 max | 1.00 max | Al 0.10-0.30; bal Fe | 420 | 245 | 20 | 0-190 HB |
| BS 1503(89) | 405S17 | Frg press ves; t<=999mm; N/T Q/T | 0.08 max | 12.0-14.0 | 1.00 max | 0.15 max | 0.50 max | 0.040 max | 0.025 max | 0.80 max | Al 0.10-0.30; Cu <=0.30; W <=0.1; bal Fe | 420-570 | 210 | 19 | |

USA

	AISI 405	Tube	0.08 max	11.50-14.50	1.00 max			0.040 max	0.030 max	1.00 max	Al 0.10-0.30; bal Fe				
	UNS S40500	Not Hard by HT	0.08 max	11.50-14.50	1.00 max			0.040 max	0.030 max	1.00 max	Al 0.10-0.30; bal Fe				
ASME SA240	405	Refer to ASTM A240(95)													
ASME SA268	405	Refer to ASTM A268(94)													
ASME SA479	405	Refer to ASTM A479/A479M(95)													
ASTM A176(97)	405*	Sh Strp Plt	0.08 max	11.5-14.5	1 max			0.04 max	0.03 max	1 max	Al 0.1-0.3; bal Fe				
ASTM A240/A240M(98)	S40500	Not Hard by HT	0.08 max	11.50-14.50	1.00 max			0.040 max	0.030 max	1.00 max	Al 0.10-0.30; bal Fe	415	170	20.0	179 HB
ASTM A268	405	Tube	0.08 max	11.5-14.5	1 max			0.04 max	0.03 max	1 max	Al 0.1-0.3; bal Fe				
ASTM A276(98)	405	Bar	0.08 max	11.5-14.5	1 max			0.04 max	0.03 max	1 max	Al 0.1-0.3; bal Fe				
ASTM A314	405	Bil	0.08 max	11.5-14.5	1 max			0.04 max	0.03 max	1 max	Al 0.1-0.3; bal Fe				
ASTM A473	405	Frg, Heat res ann	0.08 max	11.50-14.50	1.00 max			0.040 max	0.030 max	1.00 max	Al 0.10-0.30; bal Fe	414	205	20	207 HB
ASTM A479	405	Bar	0.08 max	11.5-14.5	1 max			0.04 max	0.03 max	1 max	Al 0.1-0.3; bal Fe				
ASTM A511(96)	MT405	Smls mech tub, Ann	0.08 max	11.5-14.5	1.00 max			0.040 max	0.030 max	1.00 max	Al 0.1-0.3; bal Fe	414	207	20	207 HB

UNS numbers and US grades are provided as a means of cross referencing chemically similar alloys. Exchangability is only possible after independent examination of specifications. Tensile properties are minimum or typical as specified. UTS and YS as MPa. El as %. See Appendix for list of abbreviations used in Notes. * indicates obsolete material.

Specification	Designation	Notes	C	Cr	Mn	Mo	Ni	P	S	Si	Other	UTS	YS	El	Hard
Stainless Steel, Ferritic, 405 (Continued from previous page)															
USA															
ASTM A580/A580M(98)	405	Wir, Ann	0.08 max	11.5-14.5	1.00 max			0.040 max	0.030 max	1.00 max	Al 0.10-0.30; bal Fe	485	275	20	
DoD-F-24669/7(86)	405	Ann; Supersedes MIL-S-861	0.08 max	11.5-14.5	1.00 max			0.040 max	0.030 max	1.00 max	Al 0.10-0.30; bal Fe				250 HB max
FED QQ-S-763F(96)	405*	Obs; Bar Wir Shp Frg; Corr res	0.08 max	11.5-14.5	1 max			0.04 max	0.03 max	1 max	Al 0.1-0.3; bal Fe				
MIL-S-861	405*	Obs	0.08 max	11.5-14.5	1 max			0.04 max	0.03 max	1 max	Al 0.1-0.3; bal Fe				
MIL-S-862B	405*	Obs	0.08 max	11.5-14.5	1 max			0.04 max	0.03 max	1 max	Al 0.1-0.3; bal Fe				
SAE J405(98)	S40500	Not Hard by HT	0.08 max	11.50-14.50	1.00 max		0.60 max	0.040 max	0.030 max	1.00 max	Al 0.10-0.30; bal Fe				
Stainless Steel, Ferritic, 409															
Australia															
AS 1449(94)	409	Sh Strp Plt	0.08 max	10.50-11.75	1.00 max		0.50 max	0.045 max	0.030 max	1.00 max	Ti 6x%C min, 0.75 max; bal Fe				
International															
ISO 683-13(74)	1Ti*	Ann	0.08 max	10.5-12			1 max	0.04 max	0.03 max	1 max	Ti=6xC < 1.0; bal Fe				
Italy															
UNI 8317(81)	X6CrTi12	Corr res	0.08 max	10.5-12.5	1 max		0.5 max	0.04 max	0.03 max	1 max	Ti <=0.8; Ti=6xC-0.8; bal Fe				
Japan															
JIS G4312	SUH409L	Heat res Plt Sh	0.030 max	10.50-11.75	1.00 max		0.60 max	0.040 max	0.030 max	1.00 max	Ti 6xC to 0.75; bal Fe	360	175	25	162 HB max
JIS G4312(91)	SUH409	Sh Plt, Ann; Heat res	0.08 max	10.50-11.75	1.00 max			0.040 max	0.030 max	1.00 max	Ti 6xC min, 0.75 max; bal Fe	360	175	22	162 HB
Spain															
UNE 36016/2(89)	F.3112*	Obs; En10088-2(96); X6CrTi12	0.08 max	10.5-12.5	1.0 max	0.15 max	1.0 max	0.04 max	0.03 max	1.0 max	N <=0.035; Ti <=1.0; 6xC<=Ti<=1.80; bal Fe				
UK															
BS 1449/2(83)	409S19	Corr/Heat res; t<=100mm	0.08 max	10.5-12.5	1.00 max		1.00 max	0.040 max	0.030 max	1.00 max	bal Fe	350	200	20	0-190 HB
USA															
	AISI 409	Tube	0.08 max	10.50-11.75	1.00 max			0.045 max	0.045 max	1.00 max	Ti=6x%C min - 0.75 max; bal Fe				
	UNS S40900		0.08 max	10.50-11.75	1.00 max		0.50 max	0.045 max	0.045 max	1.00 max	Ti 6xC min, 0.75 max; bal Fe				
ASME SA268	409	Refer to ASTM A268(94)													
ASTM A176(97)	409*	Sh Strp Plt	0.08 max	10.5-11.75	1 max		0.5 max	0.045 max	0.045 max	1 max	Ti>=6x%C<=0.75; bal Fe				
ASTM A240/A240M(98)	S40900*	Replaced by S40910, S40920, S40930													
ASTM A268	409	Sh Strp Plt Bar	0.08 max	10.5-11.75	1 max		0.5 max	0.045 max	0.045 max	1 max	Ti>=6x%C<=0.75; bal Fe				
ASTM A803/A803M(96)	TP409	Weld Feedwater Heater Tub	0.08 max	10.50-11.75	1.0 max		0.50 max	0.045 max	0.45 max	1.00 max	Ti>=6xC<=0.75; bal Fe	380	205	20	207 HB max
SAE J405(98)	S40900		0.08 max	10.50-11.75	1.00 max		0.50 max	0.045 max	0.030 max	1.00 max	6xC<=Ti<=0.75; bal Fe				
Stainless Steel, Ferritic, 409Ni-Cb															
USA															
	UNS S40976	with Ni and Nb	0.030 max	10.5-11.7	1.00 max		0.75-1.00	0.040 max	0.030 max	1.00 max	N <=0.040; Nb 10x(C+N)-0.80; bal Fe				
ASTM A580/A580M(98)	S40976	Wir, Ann	0.030 max	10.5-11.7	1.00 max		0.75-1.00	0.040 max	0.030 max	1.00 max	Al 0.10-0.30; N <=0.040; 10x(C+N)<=Nb<=0.80; bal Fe	415	140	20	
Stainless Steel, Ferritic, 429															
Bulgaria															
BDS 6738(72)	0Ch13	Corr res	0.08 max	12.0-14.0	0.80 max	0.30 max	0.60 max	0.035 max	0.025 max	0.80 max	bal Fe				

UNS numbers and US grades are provided as a means of cross referencing chemically similar alloys. Exchangability is only possible after independent examination of specifications. Tensile properties are minimum or typical as specified. UTS and YS as MPa. El as %. See Appendix for list of abbreviations used in Notes. * indicates obsolete material.

Stainless Steel, Ferritic, 429 (Continued from previous page)

Specification	Designation	Notes	C	Cr	Mn	Mo	Ni	P	S	Si	Other	UTS	YS	El	Hard
Hungary															
MSZ 4360(87)	X8Cr13	Corr res; 0-25mm; soft ann	0.08 max	12-14	1 max		0.6 max	0.045 max	0.03 max	1 max	bal Fe	400-600	250	20L; 15T	185 HB max
MSZ 4360(87)	X8Cr13	Corr res; 0-25mm; Q/T	0.08 max	12-14	1 max		0.6 max	0.045 max	0.03 max	1 max	bal Fe	550-700	400	18L; 13T	
Japan															
JIS G4303(91)	SUS429*	Obs; Sh Plt, HR	0.12 max	14-16	1 max		0.60 max	0.04 max	0.03 max	1 max	bal Fe				
JIS G4304(91)	SUS429	Sh Plt, HR Ann	0.12 max	14.00-16.00	1.00 max		0.60 max	0.040 max	0.030 max	1.00 max	bal Fe	450	205	22	88 HRB max
JIS G4305(91)	SUS429	Sh Plt, CR Ann	0.12 max	14.00-16.00	1.00 max		0.60 max	0.040 max	0.030 max	1.00 max	bal Fe	450	205	22	183 HB max
JIS G4305(91)	SUS429J1		0.25-0.40	15.00-17.00	1.00 max			0.040 max	0.030 max	1.00 max	bal Fe	520	225	18	
Mexico															
DGN B-83	429*	Obs; Bar, HR, CR	0.12 max	14-16	1 max			0.04 max	0.03 max	1 max	bal Fe				
NMX-B-171(91)	MT429	mech tub	0.12 max	14.0-16.0	1.00 max		0.50 max	0.040 max	0.030 max	1.00 max	bal Fe				
Poland															
PNH86020	0H13	Corr res	0.08 max	12-14	0.08 max		0.6 max	0.04 max	0.03 max	0.8 max	bal Fe				
Spain															
UNE 36016(75)	X6Cr13	Corr res	0.08 max	11.5-14.0	1.0 max	0.15 max	0.5 max	0.04 max	0.03 max	1.0 max	bal Fe				
USA															
	AISI 429	Tube	0.12 max	14.00-16.00	1.00 max			0.040 max	0.030 max	1.00 max	bal Fe				
	UNS S42900		0.12 max	14.00-16.00	1.00 max			0.040 max	0.030 max	1.00 max	bal Fe				
ASME SA182	429	Refer to ASTM A182/A182M													
ASME SA240	429	Refer to ASTM A240(95)													
ASME SA268	429	Refer to ASTM A268(94)													
ASTM A176(97)	429*	Sh Strp Plt	0.12 max	14-16	1 max		0.75 max	0.04 max	0.03 max	1 max	bal Fe				
ASTM A182	429*	Bar Frg	0.12 max	14-16	1 max		0.75 max	0.04 max	0.03 max	1 max	bal Fe				
ASTM A182/A182M(98)	F429	Frg/roll pipe flange valve	0.12 max	14.0-16.0	1.00 max		0.50 max	0.040 max	0.030 max	0.75 max	bal Fe	415	240	20	190 HB max
ASTM A240/A240M(98)	S42900		0.12 max	14.0-16.0	1.00 max			0.040 max	0.030 max	1.00 max	bal Fe	450	205	22.0	183 HB
ASTM A268	429	Tube	0.12 max	14-16	1 max		0.75 max	0.04 max	0.03 max	1 max	bal Fe				
ASTM A276(98)	429	Bar	0.12 max	14-16	1 max		0.75 max	0.04 max	0.03 max	1 max	bal Fe				
ASTM A314	429	Bil	0.12 max	14-16	1 max		0.75 max	0.04 max	0.03 max	1 max	bal Fe				
ASTM A473	429	Frg, Heat res ann	0.12 max	14.00-16.00	1.00 max		0.75 max	0.040 max	0.030 max	1.00 max	bal Fe	450	240	23	207 HB
ASTM A493	429	Wir rod, CHd, cold frg	0.12 max	14.0-16.0	1.00 max			0.040 max	0.030 max	1.00 max	bal Fe	485			
ASTM A511(96)	MT429	Smls mech tub, Ann	0.12 max	14.0-16.0	1.00 max		0.75 max	0.040 max	0.030 max	1.00 max	bal Fe	414	241	20	190 HB
ASTM A554(94)	MT429	Weld mech tub, Rnd Ann	0.12 max	14.0-16.0	1.00 max		0.50 max	0.040 max	0.030 max	1.00 max	bal Fe	414	241	20	
ASTM A815/A815M(98)	WP429	Pipe fittings	0.12 max	14.0-16.0	1.0 max		0.50 max	0.040 max	0.030 max	0.75 max	bal Fe	415-585	240	20.0	190 HB
FED QQ-S-763F(96)	429*	Obs; Bar Wir Shp Frg; Corr res	0.12 max	14-16	1 max			0.04 max	0.03 max	1 max	bal Fe				
FED QQ-S-766D(93)	429*	Obs; see ASTM A240, A666, A693; Sh Strp Plt	0.12 max	14-16	1 max			0.04 max	0.03 max	1 max	bal Fe				
FED QQ-W-423B(85)	429*	Obs; see ASTM A313, A580; Wir Corr res	0.12 max	14-16	1 max			0.04 max	0.03 max	1 max	bal Fe				

UNS numbers and US grades are provided as a means of cross referencing chemically similar alloys. Exchangability is only possible after independent examination of specifications. Tensile properties are minimum or typical as specified. UTS and YS as MPa. El as %. See Appendix for list of abbreviations used in Notes. * indicates obsolete material.

Specification	Designation	Notes	C	Cr	Mn	Mo	Ni	P	S	Si	Other	UTS	YS	El	Hard

Stainless Steel, Ferritic, 429 (Continued from previous page)

USA

Specification	Designation	Notes	C	Cr	Mn	Mo	Ni	P	S	Si	Other	UTS	YS	El	Hard
MIL-S-862B	429*	Obs	0.12 max	14-16	1 max			0.04 max	0.03 max	1 max	bal Fe				
SAE J405(98)	S42900		0.12 max	14.00-16.00	1.00 max			0.040 max	0.030 max	1.00 max	bal Fc				

Stainless Steel, Ferritic, 430

Australia

Specification	Designation	Notes	C	Cr	Mn	Mo	Ni	P	S	Si	Other	UTS	YS	El	Hard
AS 1449(94)	430	Sh Strp Plt	0.12 max	16.00-18.00	1.00 max		0.75 max	0.040 max	0.030 max	1.00 max	bal Fe				
AS 2837(86)	430	Bar, Semi-finished product	0.08 max	16.0-18.0	1.0 max		0.50 max	0.040 max	0.030 max	1.00 max	bal Fe				

Bulgaria

Specification	Designation	Notes	C	Cr	Mn	Mo	Ni	P	S	Si	Other	UTS	YS	El	Hard
BDS 6738(72)	Ch17	Sh Strp Plt Bar	0.12 max	16.0-18.0	0.80 max	0.30 max	0.60 max	0.035 max	0.025 max	0.80 max	bal Fe				
BDS 6738(72)	Ch18SJu	Corr res	0.12 max	12.0-14.0	0.80 max	0.30 max	0.60 max	0.035 max	0.025 max	1.00-1.80	Al 0.80-1.80; bal Fe				

Canada

Specification	Designation	Notes	C	Cr	Mn	Mo	Ni	P	S	Si	Other	UTS	YS	El	Hard
CSA G110.3	430	Bar Bil	0.15 max	14-18	1 max			0.04 max	0.03 max	1 max	bal Fe				
CSA G110.5	430	Strp, CR, Ann, 1.2mm diam	0.12 max	14-18	1 max		0.75 max	0.04 max	0.03 max	1 max	bal Fe				
CSA G110.5	430	Sh, HR, Ann, Q/A, Tmp	0.12 max	14-18	1 max		0.75 max	0.04 max	0.03 max	1 max	bal Fe				
CSA G110.9	430B	Sh Strp Plt	0.12 max	16-18	1 max		0.75 max	0.04 max	0.03 max	1 max	bal Fe				

China

Specification	Designation	Notes	C	Cr	Mn	Mo	Ni	P	S	Si	Other	UTS	YS	El	Hard
GB 1220(92)	1Cr17	Bar Ann	0.12 max	16.00-18.00	1.00 max		0.60 max	0.035 max	0.030 max	0.75 max	bal Fe	450	205	22	
GB 1221(92)	1Cr17	Bar Ann	0.12 max	16.00-18.00	1.00 max		0.60 max	0.040 max	0.030 max	0.75 max	bal Fe	450	205	22	
GB 12770(91)	1Cr17	Weld tube; Ann	0.12 max	16.00-18.00	1.00 max		0.60 max	0.035 max	0.030 max	0.75 max	bal Fe	410	210	20	
GB 13296(91)	1Cr17	Smls tube Ann	0.12 max	16.00-18.00	1.00 max		0.60 max	0.035 max	0.030 max	0.75 max	bal Fe	410	245	20	
GB 3280(92)	1Cr17	Sh Plt, CR Ann	0.12 max	16.00-18.00	1.00 max		0.60 max	0.035 max	0.030 max	0.75 max	bal Fe	450	205	22	
GB 4232(93)	ML1Cr17	Wir HT 1-3mm diam, ML-rivet/bolt	0.12 max	16.00-18.00	1.00 max		0.60 max	0.035 max	0.030 max	0.75 max	bal Fe	440-460		15	
GB 4237(92)	1Cr17	Sh Plt, HR Ann	0.12 max	16.00-18.00	1.00 max		0.60 max	0.035 max	0.030 max	0.75 max	bal Fe	450	205	22	
GB 4238(92)	1Cr17	Plt, HR Ann	0.12 max	16.00-18.00	1.00 max		0.60 max	0.035 max	0.030 max	0.75 max	bal Fe	450	205	22	
GB 4239(91)	1Cr17	Sh Strp, CR Ann	0.12 max	16.00-18.00	1.00 max		0.60 max	0.035 max	0.030 max	0.75 max	bal Fe	450	205	22	
GB 4240(93)	1Cr17(-Q)	Wir HT 0.5-6mm diam, Q-TLC	0.12 max	16.00-18.00	1.00 max		0.60 max	0.035 max	0.030 max	0.75 max	bal Fe	540-790			
GB/T 14975(94)	1Cr17	Smls tube Ann	0.12 max	16.00-18.00	1.00 max		0.60 max	0.035 max	0.030 max	0.75 max	bal Fe	410	245	20	

Czech Republic

Specification	Designation	Notes	C	Cr	Mn	Mo	Ni	P	S	Si	Other	UTS	YS	El	Hard
CSN 417041	17041	Corr res	0.14 max	16.0-18.5	0.9 max		0.6 max	0.04 max	0.035 max	0.7 max	bal Fe				

France

Specification	Designation	Notes	C	Cr	Mn	Mo	Ni	P	S	Si	Other	UTS	YS	El	Hard
AFNOR NFA35586	Z8C17	Sh Strp Plt Bar Bil Tub, Ann	0.1 max	15-18	0.6 max	0.5 max	0.6 max	0.03 max	0.02 max	0.5 max	Cu <=0.5; bal Fe				

Germany

Specification	Designation	Notes	C	Cr	Mn	Mo	Ni	P	S	Si	Other	UTS	YS	El	Hard
DIN 17440(96)	WNr 1.4016	Plt Sh, HR Strp Bar Wir Frg, Ann	0.08 max	16.0-18.0	1.00 max			0.040 max	0.030 max	1.00 max	bal Fe	400-630	240	20	
DIN 17441(97)	WNr 1.4016	CR Strp Plt Sh	0.08 max	16.0-18.0	1.00 max			0.040 max	0.030 max	1.00 max	bal Fe				
DIN EN 10088(95)	WNr 1.4016	Sh Plt Strp Bar Rod, Ann	0.08 max	16.0-18.0	1.00 max			0.040 max	0.030 max	1.00 max	bal Fe	400-630	240	20	

Hungary

Specification	Designation	Notes	C	Cr	Mn	Mo	Ni	P	S	Si	Other	UTS	YS	El	Hard
MSZ 4359(82)	H13	Heat res; 0-60mm; soft ann	0.12 max	17-20	1 max		0.6 max	0.04 max	0.03 max	0.8-1.5	Al 0.7-1.2; bal Fe	500	270	15L	212 HB max
MSZ 4359(82)	H16	Sh Strp Plt Bar Wir Tub; Heat res; 0-60mm; soft ann	0.12 max	16-18	1 max		0.6 max	0.04 max	0.03 max	1 max	bal Fe	400	245	20L	212 HB max

UNS numbers and US grades are provided as a means of cross referencing chemically similar alloys. Exchangability is only possible after independent examination of specifications. Tensile properties are minimum or typical as specified. UTS and YS as MPa. El as %. See Appendix for list of abbreviations used in Notes. * indicates obsolete material.

Specification	Designation	Notes	C	Cr	Mn	Mo	Ni	P	S	Si	Other	UTS	YS	El	Hard

Stainless Steel, Ferritic, 430 (Continued from previous page)

Hungary

Specification	Designation	Notes	C	Cr	Mn	Mo	Ni	P	S	Si	Other	UTS	YS	El	Hard
MSZ 4360(87)	KO3	Corr res; 0-25mm; soft ann	0.1 max	16-18	1 max		0.6 max	0.045 max	0.03 max	1 max	bal Fe	450-600	270	20L; 18T	200 HB max
MSZ 4360(87)	X10Cr17	Corr res; 0-25mm; soft ann	0.1 max	16-18	1 max		0.6 max	0.045 max	0.03 max	1 max	bal Fe	450-600	270	20L; 18T	200 HB max
India															
IS 1570/5(85)	X07Cr17	Sh Plt Strp Bar Flat Band	0.12 max	16-18	1 max	0.15 max	0.5 max	0.04 max	0.03 max	1 max	Co <=0.1; Cu <=0.3; Pb <=0.15; W <=0.1; bal Fe	450	205	22	183 HB max
IS 6527	05Cr17	Wir rod	0.1 max	16-18	1 max		0.5 max	0.04 max	0.03 max	1 max	bal Fe				
IS 6528	05Cr17	Wir, Ann, SA	0.1 max	16-18	1 max		0.5 max	0.04 max	0.03 max	1 max	bal Fe				
IS 6529	05Cr17	Bil	0.1 max	16-18	1 max		0.5 max	0.04 max	0.03 max	1 max	bal Fe				
IS 6603	05Cr17	Bar, Ann, 5-25mm diam	0.1 max	16-18	1 max		0.5 max	0.04 max	0.03 max	1 max	bal Fe				
IS 6911	05Cr17	Sh Strp Plt, Ann, 3mm diam	0.1 max	16-18	1 max		0.5 max	0.04 max	0.03 max	1 max	bal Fe				
International															
COMECON PC4-70	7	Sh Strp Plt Bar Bil	0.12 max	16-18	0.8 max			0.035 max	0.025 max	0.8 max	bal Fe				
ISO 4955(94)	H5	Heat res	0.12 max	16.0-18.0	1.0 max	0.15 max	0.4 max	0.04 max	0.03 max	0.7-1.4	Al 0.7-1.2; Co <=0.1; Cu <=0.3; Pb <=0.15; Ti <=0.05; V <=0.1; W <=0.1; bal Fe				
ISO 683-13(74)	8*	Ann	0.08 max	16-18	1 max		1 max	0.04 max	0.03 max	1 max	bal Fe				
Italy															
UNI 6901(71)	X8Cr17	Corr res	0.1 max	16-18	1 max		0.5 max	0.04 max	0.03 max	1 max	bal Fe				
UNI 6904(71)	X8Cr17	Sh Strp Plt Bar Wir Tub, Ann; Heat res	0.1 max	16-18	1 max		0.5 max	0.04 max	0.03 max	0.75 max	bal Fe				
Japan															
JIS G4303(91)	SUS430	Bar Ann 75mm diam	0.12 max	16.00-18.00	1.00 max		0.60 max	0.040 max	0.030 max	0.75 max	bal Fe	450	205	22	183 HB max
JIS G4304(91)	SUS430	Sh Plt, HR Ann	0.12 max	16.00-18.00	1.00 max		0.60 max	0.040 max	0.030 max	0.75 max	bal Fe	450	205	22	88 HRB max
JIS G4305(91)	SUS430	Sh Plt, CR Ann	0.12 max	16.00-18.00	1.00 max		0.60 max	0.040 max	0.030 max	0.75 max	bal Fe	450	205	22	183 HB max
JIS G4305(91)	SUS430J1L		0.025 max	16.00-20.00	1.00 max			0.040 max	0.030 max	1.00 max	N <=0.025; Nb 8x(C=n) to 0.80; bal Fe	390	205	22	
JIS G4306	SUS430	Strp HR Ann	0.12 max	16-18	1 max		0.6 max	0.04 max	0.03 max	0.75 max	bal Fe				
JIS G4307	SUS430	Strp CR SA	0.12 max	16-18	1 max		0.6 max	0.04 max	0.03 max	0.75 max	bal Fe				
JIS G4308(98)	SUS430	Wir rod	0.12 max	16.00-18.00	1.00 max		0.60 max	0.040 max	0.030 max	0.75 max	bal Fe				
JIS G4309	SUS430	Wir	0.12 max	16.00-18.00	1.00 max		0.60 max	0.040 max	0.030 max	0.75 max	bal Fe				
JIS G4315	SUS430	Wir rod	0.12 max	16.00-18.00	1.00 max		0.60 max	0.040 max	0.030 max	0.75 max	bal Fe				
Mexico															
DGN B-216	TP430*	Obs; Tube	0.12 max	14-18	1 max		0.5 max	0.04 max	0.03 max	1 max	bal Fe				
DGN B-83	430*	Obs; Bar, HR, CR	0.12 max	16-18	1 max			0.04 max	0.03 max	1 max	bal Fe				
NMX-B-171(91)	MT430	CF HF Smls tub	0.12 max	16.0-18.0	1.00 max		0.50 max	0.040 max	0.030 max	1.00 max	bal Fe				
Poland															
PNH86020	2H17	Heat res	0.15 max	16-18	0.7 max		0.6 max	0.04 max	0.03 max	1.2 max	bal Fe				
PNH86020	H17	Corr res	0.1 max	16-18	0.8 max		0.6 max	0.04 max	0.03 max	0.8 max	bal Fe				
PNH86022(71)	2H17	Heat res	0.15 max	16-18	0.7 max		0.6 max	0.04 max	0.03 max	1.2 max	bal Fe				

UNS numbers and US grades are provided as a means of cross referencing chemically similar alloys. Exchangability is only possible after independent examination of specifications. Tensile properties are minimum or typical as specified. UTS and YS as MPa. El as %. See Appendix for list of abbreviations used in Notes. * indicates obsolete material.

Specification	Designation	Notes	C	Cr	Mn	Mo	Ni	P	S	Si	Other	UTS	YS	El	Hard

Stainless Steel, Ferritic, 430 (Continued from previous page)

Poland

Specification	Designation	Notes	C	Cr	Mn	Mo	Ni	P	S	Si	Other	UTS	YS	El	Hard
PNH86022(71)	H18JS	Heat res	0.12 max	17-19	0.8 max		0.5 max	0.04 max	0.03 max	0.8-1.1	Al 0.7-1.2; bal Fe				

Romania

Specification	Designation	Notes	C	Cr	Mn	Mo	Ni	P	S	Si	Other	UTS	YS	El	Hard
STAS 2583	10C170	Sh Strp Plt Bar Bil	0.12 max	16-18	0.7 max		0.6 max	0.035 max	0.03 max	0.8 max	bal Fe				
STAS 3583(87)	8Cr170	Corr res; Heat res	0.08 max	15.5-17.5	1 max	0.2 max	0.5 max	0.045 max	0.03 max	1 max	Pb <=0.15; bal Fe				

Russia

Specification	Designation	Notes	C	Cr	Mn	Mo	Ni	P	S	Si	Other	UTS	YS	El	Hard
GOST 5632	12Ch17	Sh Strp Plt Bar Wir Bil Tub, Ann	0.12 max	16-18	0.8 max		0.6 max	0.035 max	0.025 max	0.8 max	Cu <=0.3; Ti <=0.2; bal Fe				
GOST 5632(61)	KH17	Corr res; Heat res; High-temp constr	0.12 max	16.0-18.0	0.7 max	0.15 max	0.4 max	0.035 max	0.025 max	0.8 max	Cu <=0.3; Ti <=0.05; bal Fe				
GOST 5632(72)	12Ch17	Corr res; Heat res	0.12 max	16.0-18.0	0.8 max	0.15 max	0.6 max	0.035 max	0.025 max	0.8 max	Cu <=0.3; Ti <=0.2; bal Fe				
GOST 5632(72)	15Ch18SJu	Heat res	0.15 max	17.0-20.0	0.8 max	0.15 max	0.6 max	0.035 max	0.025 max	1.0-1.5	Al 0.7-1.2; Cu <=0.3; Ti <=0.2; bal Fe				

Spain

Specification	Designation	Notes	C	Cr	Mn	Mo	Ni	P	S	Si	Other	UTS	YS	El	Hard
UNE 36016(75)	F.3113*	see X8Cr17	0.1 max	16.0-18.0	1.0 max	0.15 max	0.5 max	0.04 max	0.03 max	1.0 max	bal Fe				
UNE 36016(75)	X8Cr17	Sh Strp Plt Bar Bil; Corr res	0.1 max	16.0-18.0	1.0 max	0.15 max	0.5 max	0.04 max	0.03 max	1.0 max	bal Fe				
UNE 36016/1(89)	E-430	Corr res	0.08 max	16.0-18.0	1.0 max	0.15 max	1.0 max	0.04 max	0.03 max	1.0 max	bal Fe				
UNE 36016/1(89)	F.3113*	Obs; E-430; Corr res	0.08 max	16.0-18.0	1.0 max	0.15 max	1.0 max	0.04 max	0.03 max	1.0 max	bal Fe				
UNE 36016/2(89)	F.3122*	Obs; En10088-2(96); X5CrNb17	0.07 max	16.0-18.0	1.0 max	0.15 max	1.0 max	0.04 max	0.03 max	1.0 max	N <=0.035; Nb=10C-1.2; bal Fe				
UNE 36017(85)	F.3153*	see X10CrAl18	0.12 max	17-19	1 max	0.15 max		0.04 max	0.03 max	0.7-1.4	Al 0.7-1.2; bal Fe				
UNE 36017(85)	X10CrAl18	Heat res	0.12 max	17-19	1 max	0.15 max		0.04 max	0.03 max	0.7-1.4	Al 0.7-1.2; bal Fe				

Sweden

Specification	Designation	Notes	C	Cr	Mn	Mo	Ni	P	S	Si	Other	UTS	YS	El	Hard
SS 142320	2320	Corr res	0.08 max	16-18	1 max		1 max	0.04 max	0.03 max	1 max	bal Fe				

UK

Specification	Designation	Notes	C	Cr	Mn	Mo	Ni	P	S	Si	Other	UTS	YS	El	Hard
BS 1554	430S15*	Sh Strp Plt Bar Wir Bil	0.10 max	16-18	1.00 max		0.5 max	0.040 max	0.030 max	1.00 max	bal Fe				
BS 1554(90)	430S11	Wir Corr/Heat res; diam<0.49mm; Ann	0.03 max	16.0-18.0	1.00 max		1.00 max	0.040 max	0.030 max	1.00 max	W <=0.1; bal Fe				0-650
BS 1554(90)	430S11	Wir Corr/Heat res; 6<diam<=13mm; Ann	0.03 max	16.0-18.0	1.00 max		1.00 max	0.040 max	0.030 max	1.00 max	W <=0.1; bal Fe				0-610
BS 1554(90)	430S18	Wir Corr/Heat res; diam<0.49mm; Ann	0.10 max	16.0-18.0	1.00 max		1.00 max	0.040 max	0.030 max	1.00 max	W <=0.1; bal Fe				0-650
BS 1554(90)	430S18	Wir Corr/Heat res; 6<diam<=13mm; Ann	0.10 max	16.0-18.0	1.00 max		1.00 max	0.040 max	0.030 max	1.00 max	W <=0.1; bal Fe				0-610
BS 970/1(96)	430S17	Wrought Corr/Heat res; t<=63mm	0.08 max	16.0-18.0	1.00 max		0.50 max	0.040 max	0.030 max	1.00 max	bal Fe	430	280	20	170 HB max

USA

Specification	Designation	Notes	C	Cr	Mn	Mo	Ni	P	S	Si	Other	UTS	YS	El	Hard
	AISI 430	Tube	0.12 max	16.00-18.00	1.00 max			0.040 max	0.030 max	1.00 max	bal Fe				
	UNS S43000		0.12 max	16.00-18.00	1.00 max			0.040 max	0.030 max	1.00 max	bal Fe				
AMS 5503		Bar Wir	0.12 max	16-18	1 max			0.040 max	0.03 max	1 max	bal Fe				
AMS 5627		Bar Wir	0.12 max	16-18	1 max			0.040 max	0.03 max	1 max	bal Fe				
ASME SA182	430	Refer to ASTM A182/A182M													
ASME SA240	430	Refer to ASTM A240(95)													
ASME SA268	430	Refer to ASTM A268(94)													
ASME SA479	430	Refer to ASTM A479/A479M(95)													
ASTM A176(97)	430*	Sh Strp Plt	0.12 max	16-18	1 max		0.75 max	0.04 max	0.03 max	1 max	bal Fe				

UNS numbers and US grades are provided as a means of cross referencing chemically similar alloys. Exchangability is only possible after independent examination of specifications. Tensile properties are minimum or typical as specified. UTS and YS as MPa. El as %. See Appendix for list of abbreviations used in Notes. * indicates obsolete material.

Specification	Designation	Notes	C	Cr	Mn	Mo	Ni	P	S	Si	Other	UTS	YS	El	Hard

Stainless Steel, Ferritic, 430 (Continued from previous page)

USA

Specification	Designation	Notes	C	Cr	Mn	Mo	Ni	P	S	Si	Other	UTS	YS	El	Hard
ASTM A182	430*	Bar Frg	0.12 max	16-18	1 max		0.75 max	0.04 max	0.03 max	1 max	bal Fe				
ASTM A182/A182M(98)	F430	Frg/roll pipe flange valve	0.12 max	16.0-18.0	1.00 max		0.50 max	0.040 max	0.030 max	0.75 max	bal Fe	415	240	20	190 HB max
ASTM A240/A240M(98)	S43000		0.12 max	16.0-18.0	1.00 max		0.75 max	0.040 max	0.030 max	1.00 max	bal Fe	450	205	22.0	183 HB
ASTM A268	430	Tube	0.12 max	16-18	1 max		0.75 max	0.04 max	0.03 max	1 max	bal Fe				
ASTM A276(98)	430	Bar	0.12 max	16-18	1 max		0.75 max	0.04 max	0.03 max	1 max	bal Fe				
ASTM A314	430	Bil	0.12 max	16-18	1 max		0.75 max	0.04 max	0.03 max	1 max	bal Fe				
ASTM A473	430	Frg, Heat res ann	0.12 max	16.00-18.00	1.00 max		0.75 max	0.040 max	0.030 max	1.00 max	bal Fe	485	240	20	217 HB
ASTM A479	430	Bar	0.12 max	16-18	1 max		0.75 max	0.04 max	0.03 max	1 max	bal Fe				
ASTM A493	430	Wir rod, CHd, cold frg	0.12 max	16.0-18.0	1.00 max			0.040 max	0.030 max	1.00 max	bal Fe	520			
ASTM A511(96)	MT430	Smls mech tub, Ann	0.12 max	16.0-18.0	1.00 max		0.75 max	0.040 max	0.030 max	1.00 max	bal Fe	414	241	20	190 HB
ASTM A554(94)	MT430	Weld mech tub, Rnd Ann	0.12 max	16.0-18.0	1.00 max		0.50 max	0.040 max	0.030 max	1.00 max	bal Fe	414	241	20	
ASTM A580/A580M(98)	430	Wir, Ann	0.12 max	16.0-18.0	1.00 max		0.75 max	0.040 max	0.030 max	1.00 max	bal Fe	485	275	20	
ASTM A815/A815M(98)	WP430	Pipe fittings	0.12 max	16.0-18.0	1.00 max		0.50 max	0.040 max	0.030 max	1.00 max	bal Fe	450-620	240	20.0	190 HB
FED QQ-S-763F(96)	430*	Obs; Bar Wir Shp Frg; Corr res	0.12 max	16-18	1 max			0.04 max	0.03 max	1 max	bal Fe				
FED QQ-S-766D(93)	430*	Obs; see ASTM A240, A666, A693; Sh Strp Plt	0.12 max	16-18	1 max			0.04 max	0.03 max	1 max	bal Fe				
SAE J405(98)	S43000		0.12 max	16.00-18.00	1.00 max		0.75 max	0.040 max	0.030 max	1.00 max	bal Fe				

Stainless Steel, Ferritic, 430F

Australia

Specification	Designation	Notes	C	Cr	Mn	Mo	Ni	P	S	Si	Other	UTS	YS	El	Hard
AS 2837(86)	430F	Bar, Semi-finished product	0.08 max	16.0-18.0	1.25 max		0.50 max	0.060 max	0.15-0.35	1.00 max	bal Fe				

Canada

Specification	Designation	Notes	C	Cr	Mn	Mo	Ni	P	S	Si	Other	UTS	YS	El	Hard
CSA G110.3	430F	Bar Bil	0.12 max	14-18	1.25 max	0.6 max		0.06 max	0.15-99.99	1 max	bal Fe				

China

Specification	Designation	Notes	C	Cr	Mn	Mo	Ni	P	S	Si	Other	UTS	YS	El	Hard
GB 1220(92)	Y1Cr17	Bar Ann, Y-Free cutting	0.12 max	16.00-18.00	1.25 max	0.60 max	0.60 max	0.060 max	0.15 min	1.00 max	bal Fe	450	205	22	
GB 4240(93)	Y1Cr17(-Q)	Wir HT 3-6mm diam, Q-TLC	0.12 max	16.00-18.00	1.25 max	0.60 max	0.60 max	0.060 max	0.15 min	1.00 max	bal Fe	590-880			

Europe

Specification	Designation	Notes	C	Cr	Mn	Mo	Ni	P	S	Si	Other	UTS	YS	El	Hard
EN 10088/3(95)	1.4104	Bar Rod Sect, Corr res; t<=60mm, Q/T	0.10-0.17	15.50-17.50	1.50 max	0.20-0.60		0.040 max	0.15-0.35	1.00 max	bal Fe	650-850	500	12	
EN 10088/3(95)	1.4104	Bar Rod Sect; Corr res; Ann	0.10-0.17	15.50-17.50	1.50 max	0.20-0.60		0.040 max	0.15-0.35	1.00 max	bal Fe	730 max			220 HB max
EN 10088/3(95)	X14CrMoS17	Bar Rod Sect, Corr res; t<=60mm, Q/T	0.10-0.17	15.50-17.50	1.50 max	0.20-0.60		0.040 max	0.15-0.35	1.00 max	bal Fe	650-850	500	12	
EN 10088/3(95)	X14CrMoS17	Bar Rod Sect; Corr res; Ann	0.10-0.17	15.50-17.50	1.50 max	0.20-0.60		0.040 max	0.15-0.35	1.00 max	bal Fe	730 max			220 HB max

France

Specification	Designation	Notes	C	Cr	Mn	Mo	Ni	P	S	Si	Other	UTS	YS	El	Hard
AFNOR NFA35576	Z10CF17	Bar Bil, Ann	0.12 max	16-18	1.5 max	0.2-0.6	0.5 max	0.06 max	0.15-99.99	1 max	bal Fe				

Hungary

Specification	Designation	Notes	C	Cr	Mn	Mo	Ni	P	S	Si	Other	UTS	YS	El	Hard
MSZ 4360(80)	KO18S	Corr res; 0-60mm; Q/T	0.1-0.17	16-18	1.5 max			0.04 max	0.15-0.35	1 max	bal Fe	700	450	10L	
MSZ 4360(80)	KO18S	Corr res; 0-999mm; soft ann	0.1-0.17	16-18	1.5 max			0.04 max	0.15-0.35	1 max	bal Fe				285 HB
MSZ 4360(87)	KO3S	Corr res; 0-25mm; soft ann	0.1 max	16-18	1.5 max		0.6 max	0.045 max	0.15-0.35	1 max	bal Fe	430-630	250	15L	200 HB max
MSZ 4360(87)	X10CrS17	Corr res; 0-25mm; soft ann	0.1 max	16-18	1.5 max		0.6 max	0.045 max	0.15-0.35	1 max	bal Fe	430-630	250	15L	200 HB max

UNS numbers and US grades are provided as a means of cross referencing chemically similar alloys. Exchangability is only possible after independent examination of specifications. Tensile properties are minimum or typical as specified. UTS and YS as MPa. El as %. See Appendix for list of abbreviations used in Notes. * indicates obsolete material.

Specification	Designation	Notes	C	Cr	Mn	Mo	Ni	P	S	Si	Other	UTS	YS	El	Hard

Stainless Steel, Ferritic, 430F (Continued from previous page)

International

| ISO 683-13(74) | 8a* | Ann | 0.08 max | 16-18 | 1.5 max | 0.6 max | 1 max | 0.06 max | 0.15-0.35 | 1 max | bal Fe | | | | |

Italy

| UNI 6901(71) | X10CrS17 | Corr res | 0.12 max | 16-18 | 1.5 max | 0.6 max | 0.5 max | 0.06 max | 0.15-0.35 | 1 max | bal Fe | | | | |

Japan

JIS G4303(91)	SUS430F	Bar Ann 75mm diam	0.12 max	16.00-18.00	1.25 max	0.60 max	0.60 max	0.060 max	0.15 min	1.00 max	Mo optional; bal Fe	450	205	22	183 HB max
JIS G4308(98)	SUS430F	Wir rod	0.12 max	16.00-18.00	1.25 max	0.60 max	0.60 max	0.060 max	0.15 min	1.00 max	bal Fe				
JIS G4309	SUS430F	Wir	0.12 max	16.00-18.00	1.25 max	0.60 max	0.60 max	0.060 max	0.15 min	1.00 max	bal Fe				

Mexico

| DGN B-83 | 430F* | Obs; Bar, HR, CR | 0.12 max | 16-18 | 1.25 max | 0.6 max | | 0.06 max | 0.15-99.99 | 1 max | bal Fe | | | | |

Spain

UNE 36016(75)	F.3117*	see X10CrS17	0.12 max	16.0-18.0	1.5 max	0.6 max	0.5 max	0.06 max	0.15-0.35	1.0 max	bal Fe				
UNE 36016(75)	X10CrS17	Corr res	0.12 max	16.0-18.0	1.5 max	0.6 max	0.5 max	0.06 max	0.15-0.35	1.0 max	bal Fe				
UNE 36016/1(89)	E-430S	Corr res	0.12 max	16.0-18.0	1.25 max	0.15 max		0.04 max	0.15-0.400	1.0 max	bal Fe				
UNE 36016/1(89)	F.3117*	Obs; EN10088-3(96); X8CrS17	0.12 max	16.0-18.0	1.25 max	0.15 max		0.04 max	0.15-0.400	1.0 max	bal Fe				
UNE 36016/1(89)	F.3413*	Obs; EN10088-3(96); X14CrMoS17	0.1-0.17	15.5-17.5	1.5 max	0.2-0.6	0.5 max	0.06 max	0.15-0.35	1.0 max	bal Fe				

Sweden

| SS 142383 | 2383 | Corr res | 0.1-0.17 | 16-18 | 1.5 max | 0.6 max | 0.5 max | 0.06 max | 0.15-0.35 | 1 max | bal Fe | | | | |

USA

	AISI 430F	Tube	0.12 max	16.00-18.00	1.25 max	0.60 max		0.060 max	0.15 min	1.00 max	bal Fe				
	UNS S43020	Free Machining S-bearing	0.12 max	16.00-18.00	1.25 max	0.60 max		0.060 max	0.15 min	1.00 max	bal Fe				
ASTM A314	430F	Bil	0.12 max	16-18	1.25 max			0.06 max	0.15-99.99	1 max	bal Fe				
ASTM A473	430F	Frg, Heat res ann	0.12 max	16.00-18.00	1.25 max			0.06 max	0.15 min	1.00 max	bal Fe	485	275	20	223 HB
ASTM A581/A581M(95)	430F	Wir rod, Ann	0.12 max	16.0-18.0	1.25 max			0.06 max	0.15 min	1.00 max	bal Fe	585-860			
ASTM A582/A582M(95)	430F	Bar, HF, CF, Ann	0.12 max	16.00-18.00	1.25 max			0.06 max	0.15 min	1.00 max	bal Fe				262 HB max
ASTM A895(94)	430F	Sh Strp Plt, free mach, Ann	0.12 max	16.0-18.0	1.25 max			0.06 max	0.15 min	1.00 max	bal Fe				262 HB max
MIL-S-862B	430F*	Obs	0.12 max	16-18	1.25 max			0.06 max	0.15-99.99	1 max	bal Fe				

Stainless Steel, Ferritic, 430FSe

Canada

| CSA G110.3 | 430FSe | Bar Bil | 0.12 max | 14-18 | 1.25 max | | | 0.06 max | 0.06 max | 1 max | Se 0.15; bal Fe | | | | |

International

| COMECON PC4-70 | 29 | Bar Bil | 0.15 max | 17-20 | 0.8-1.5 | | | 0.035 max | 0.025 max | 0.8 max | Se 0.7-1.2; bal Fe | | | | |

Mexico

| DGN B-83 | 430FSe* | Obs; Bar, HR, CR | 0.12 max | 16-18 | 1.25 max | | | 0.06 max | 0.06 max | 1 max | Se 0.15; bal Fe | | | | |

USA

	AISI 430FSe	Tube	0.12 max	16.00-18.00	1.25 max			0.060 max	0.060 max	1.00 max	Se>= 0.15; bal Fe				
	UNS S43023	Free Machining Se-bearing	0.12 max	16.00-18.00	1.25 max			0.060 max	0.060 max	1.00 max	Se>=0.15; bal Fe				
ASTM A314	430FSe	Bil	0.12 max	16-18	1.25 max			0.06 max	0.06 max	1 max	Se 0.15; bal Fe				
ASTM A473	430FSe	Frg, Heat res ann	0.12 max	16.00-18.00	1.25 max			0.06 max	0.06 max	1.00 max	Se>=0.15; bal Fe	485	275	20	223 HB

UNS numbers and US grades are provided as a means of cross referencing chemically similar alloys. Exchangability is only possible after independent examination of specifications. Tensile properties are minimum or typical as specified. UTS and YS as MPa. El as %. See Appendix for list of abbreviations used in Notes. * indicates obsolete material.

Specification	Designation	Notes	C	Cr	Mn	Mo	Ni	P	S	Si	Other	UTS	YS	El	Hard

Stainless Steel, Ferritic, 430FSe (Continued from previous page)

USA

Specification	Designation	Notes	C	Cr	Mn	Mo	Ni	P	S	Si	Other	UTS	YS	El	Hard
ASTM A581/A581M(95)	430FSe	Wir rod, Ann	0.12 max	16.0-18.0	1.25 max			0.06 max	0.06 max	1.00 max	Se 0.15; bal Fe	585-860			
ASTM A582/A582M(95)	430FSe	Bar, HF, CF, Ann	0.12 max	16.00-18.00	1.25 max			0.06 max	0.06 max	1.00 max	Se 0.15; bal Fe				262 HB max
ASTM A895(94)	430FSe	Sh Strp Plt, free mach, Ann	0.12 max	16.0-18.0	1.25 max			0.06 max	0.06 max	1.00 max	Se 0.15; bal Fe				262 HB max
MIL-S-862B	430FSe*	Obs	0.12 max	14-18	1.25 max			0.06 max	0.06 max	1 max	Se 0.15; bal Fe				

Stainless Steel, Ferritic, 430Ti

Bulgaria

Specification	Designation	Notes	C	Cr	Mn	Mo	Ni	P	S	Si	Other	UTS	YS	El	Hard
BDS 6738(72)	Ch17Ti	Sh Plt Bar Wir	0.10 max	16.0-18.0	0.80 max	0.30 max	0.60 max	0.035 max	0.025 max	0.80 max	5xC<=Ti<=0.80; bal Fe				

Czech Republic

Specification	Designation	Notes	C	Cr	Mn	Mo	Ni	P	S	Si	Other	UTS	YS	El	Hard
CSN 417040	17040	Corr res	0.1 max	16.0-18.5	0.9 max		0.6 max	0.04 max	0.035 max	0.7 max	Ti <=0.3; bal Fe				

France

Specification	Designation	Notes	C	Cr	Mn	Mo	Ni	P	S	Si	Other	UTS	YS	El	Hard
AFNOR NFA35578	Z8CT17	Sh Strp Plt Bar Wir	0.08 max	16-18	1 max			0.04 max	0.03 max	1 max	Ti>=7xC<=1.2; bal Fe				

Germany

Specification	Designation	Notes	C	Cr	Mn	Mo	Ni	P	S	Si	Other	UTS	YS	El	Hard
TGL 7143	X8CrTi17*	Obs	0.1 max	16-18	0.8 max			0.045 max	0.03 max	0.8 max	Ti>=7xC; bal Fe				

Hungary

Specification	Designation	Notes	C	Cr	Mn	Mo	Ni	P	S	Si	Other	UTS	YS	El	Hard
MSZ 4360(87)	KO4Ti	Corr res; 0-25mm; soft ann	0.1 max	16-18	1 max		0.6 max	0.045 max	0.03 max	1 max	Ti <=1.2; Ti=7xC-1.2	450-600	270	20L; 18T	200 HB max
MSZ 4360(87)	X10CrTi17	Corr res; 0-25mm; soft ann	0.1 max	16-18	1 max		0.6 max	0.045 max	0.03 max	1 max	Ti <=1.2; Ti=7xC-1.2	450-600	270	20L; 18T	200 HB max

International

Specification	Designation	Notes	C	Cr	Mn	Mo	Ni	P	S	Si	Other	UTS	YS	El	Hard
COMECON PC4-70	8	Sh Strp Plt Bar	0.1 max	16-18	0.8 max			0.035 max	0.025 max	0.8 max	Ti>=5xC<=0.80; bal Fe				

Italy

Specification	Designation	Notes	C	Cr	Mn	Mo	Ni	P	S	Si	Other	UTS	YS	El	Hard
UNI 8317(81)	X6CrTi17	Corr res	0.08 max	16-18	1 max		0.5 max	0.04 max	0.03 max	1 max	Ti <=0.8; Ti=5C-0.8; bal Fe				

Poland

Specification	Designation	Notes	C	Cr	Mn	Mo	Ni	P	S	Si	Other	UTS	YS	El	Hard
PNH86020	0H17T	Corr res	0.08 max	16-18	0.8 max		0.6 max	0.04 max	0.03 max	0.8 max	Ti <=0.8; Ti=5C-0.8; bal Fe				
PNH86020	0H17Ti	Sh Strp Plt Bar	0.08 max	16-18	0.8 max			0.04 max	0.03 max	0.8 max	Ti>=5xC<=0.8; bal Fe				

Romania

Specification	Designation	Notes	C	Cr	Mn	Mo	Ni	P	S	Si	Other	UTS	YS	El	Hard
STAS 3583	7TC170	Sh Plt Bar Wir	0.08 max	16-18	0.7 max		0.6 max	0.035 max	0.03 max	0.8 max	Ti>=5xC<=0.8; bal Fe				

Russia

Specification	Designation	Notes	C	Cr	Mn	Mo	Ni	P	S	Si	Other	UTS	YS	El	Hard
GOST	O8Ch17Ti	Sh Strp Plt Bar Wir	0.08 max	16-18	0.8 max		0.6 max	0.035 max	0.025 max	0.8 max	Cu <=0.3; Ti>=5xC<=1.00; bal Fe				
GOST 5632(61)	0KH17T	Corr res; Heat res; High-temp constr	0.08 max	16.0-18.0	0.7 max	0.15 max	0.4 max	0.035 max	0.025 max	0.8 max	Cu <=0.3; Ti <=0.8; Ti=5C-0.8; bal Fe				
GOST 5632(72)	08Ch17T	Corr res; Heat res	0.08 max	16.0-18.0	0.8 max	0.15 max	0.6 max	0.035 max	0.025 max	0.8 max	Cu <=0.3; Ti <=0.8; Ti=5C-0.8; bal Fe				

Spain

Specification	Designation	Notes	C	Cr	Mn	Mo	Ni	P	S	Si	Other	UTS	YS	El	Hard
UNE 36016/1(89)	E-430Ti	Corr res	0.07 max	16.0-18.0	1.0 max	0.15 max	1.0 max	0.04 max	0.03 max	1.0 max	Ti <=1.1; Ti=7xC-1.1; bal Fe				
UNE 36017(85)	F.3114*	see X8CrTi17	0.1 max	16.5-18.5	1.5 max		1 max	0.04 max	0.03 max	1.5 max	Ti 0.4-0.7; bal Fe				
UNE 36017(85)	X8CrTi17	Sh Strp Plt Bar Wir	0.1 max	16.5-18.5	1.5 max		1 max	0.04 max	0.03 max	1.5 max	Ti 0.4-0.7; bal Fe				

USA

Specification	Designation	Notes	C	Cr	Mn	Mo	Ni	P	S	Si	Other	UTS	YS	El	Hard
	UNS S43036	with Ti	0.10 max	16.00-19.50	1.00 max		0.75 max	0.040 max	0.030 max	1.00 max	Ti 5xC-0.75; bal Fe				
ASTM A268	430Ti	Tube	0.1 max	16-19.5	1 max		0.75 max	0.04 max	0.03 max	1 max	Ti>=5xC<=0.75; bal Fe				
ASTM A554(94)	MT430Ti	Weld mech tub, Rnd Ann	0.10 max	16.0-19.5	1.00 max		0.075 max	0.040 max	0.030 max	1.00 max	Ti>=5xC<=0.75; bal Fe	414	207	20	
ASTM A815/A815M(98)	WP430TI	Pipe fittings	0.10 max	16.0-19.5	1.00 max		0.75 max	0.040 max	0.030 max	1.00 max	Ti>=5xC<=0.75; bal Fe	415-585	240	20.0	190 HB

UNS numbers and US grades are provided as a means of cross referencing chemically similar alloys. Exchangability is only possible after independent examination of specifications. Tensile properties are minimum or typical as specified. UTS and YS as MPa. El as %. See Appendix for list of abbreviations used in Notes. * indicates obsolete material.

Stainless Steel, Ferritic, 434

Specification	Designation	Notes	C	Cr	Mn	Mo	Ni	P	S	Si	Other	UTS	YS	El	Hard
Canada															
CSA G110.5	434	Sh, HR, Ann, Q/A, Tmp	0.1 max	16-20	1 max	0.75-1.25	0.6 max	0.04 max	0.03 max	1 max	bal Fe				
China															
GB 1220(92)	1Cr17Mo	Bar Ann	0.12 max	16.00-18.00	1.00 max	0.75-1.25	0.60 max	0.035 max	0.030 max	1.00 max	bal Fe	450	205	22	
GB 3280(92)	1Cr17Mo	Sh Plt, CR Ann	0.12 max	16.00-18.00	1.00 max	0.75-1.25	0.60 max	0.035 max	0.030 max	1.00 max	bal Fe	450	205	22	
GB 4237(92)	1Cr17Mo	Sh Plt, HR Ann	0.12 max	16.00-18.00	1.00 max	0.75-1.25	0.60 max	0.035 max	0.030 max	1.00 max	bal Fe	450	205	22	
GB 4239(91)	1Cr17Mo	Sh Strp, CR Ann	0.12 max	16.00-18.00	1.00 max	0.75-1.25	0.60 max	0.035 max	0.030 max	1.00 max	bal Fe	450	205	22	
Czech Republic															
CSN 417353	17353	Corr res	0.1 max	16.0-18.5	2.0 max	2.5-3.0	12.0-15.0	0.045 max		0.3 max	1.0 max	Ti <=0.90; Ti>=5xC; bal Fe				
CSN 417356	17356	Corr res	0.08 max	16.0-18.0	2.0 max	3.0-4.0	13.0-16.0	0.045 max	0.03 max		1.0 max	Ti 0.3-0.90; Ti>=0.3; bal Fe				
Europe															
EN 10088/2(95)	1.4113	Strp; Corr res; CR t<=6mm, Ann	0.08 max	16.00-18.00	1.00 max	0.90-1.40		0.040 max	0.015 max	1.00 max	bal Fe	450-630	260	18	
EN 10088/2(95)	X6CrMo17-1	Strp; Corr res; CR t<=6mm, Ann	0.08 max	16.00-18.00	1.00 max	0.90-1.40		0.040 max	0.015 max	1.00 max	bal Fe	450-630	260	18	
EN 10088/3(95)	1.4113	Bar Rod Shp, Corr res; t<=100mm, Ann	0.08 max	16.00-18.00	1.00 max	0.90-1.40		0.040 max	0.030 max	1.00 max	bal Fe	440-660	280	18	200 HB max
EN 10088/3(95)	X6CrMo17-1	Bar Rod Shp, Corr res; t<=100mm, Ann	0.08 max	16.00-18.00	1.00 max	0.90-1.40		0.040 max	0.030 max	1.00 max	bal Fe	440-660	280	18	200 HB max
France															
AFNOR NFA35572	Z8CD17.01	Sh Strp Plt Bar, Ann	0.1 max	16-18	1 max	0.9-1.3	0.5 max	0.04 max	0.03 max	1 max	bal Fe				
AFNOR NFA35575	Z8CD17.01	Wir rod, 40% Stress Hardness	0.1 max	16-18	1 max	0.9-1.3	0.5 max	0.04 max	0.03 max	1 max	bal Fe				
Germany															
DIN EN 10088(95)	WNr 1.4113	Sh Strp Plt Bar Rod Ann	0.08 max	16.0-18.0	1.00 max	0.90-1.30		0.040 max	0.030 max	1.00 max	bal Fe	440-660	280	18	
DIN EN 10088(95)	X6CrMo17-1	Sh Plt Strp Bar Rod, Ann	0.08 max	16.0-18.0	1.00 max	0.90-1.30		0.040 max	0.030 max	1.00 max	bal Fe	440-660	280	18	
Hungary															
MSZ 4360(87)	KO6	Corr res; 0-25mm; soft ann	0.1 max	16-18	1 max	0.9-1.3	0.6 max	0.045 max	0.03 max	1 max	bal Fe	460-600	280	18L; 17T	205 HB max
MSZ 4360(87)	X10CrMo17	Corr res; 0-25mm; soft ann	0.1 max	16-18	1 max	0.9-1.3	0.6 max	0.045 max	0.03 max	1 max	bal Fe	460-600	280	18L; 17T	205 HB max
International															
ISO 683-13(74)	9c*	Sh Plt Bar, Ann	0.08 max	16-18	1 max	0.9-1.3	1 max	0.04 max	0.03 max	1 max	bal Fe				
Italy															
UNI 6901(71)	X8CrMo17	Sh Strp Plt Bar Wir; Corr res	0.1 max	16-18	1 max	0.9-1.3	0.5 max	0.04 max	0.03 max	1 max	bal Fe				
Japan															
JIS G4303(91)	SUS434	Bar Ann 75mm diam	0.12 max	16.00-18.00	1.00 max	0.75-1.25	0.60 max	0.040 max	0.030 max	1.00 max	bal Fe	450	205	22	183 HB max
JIS G4304(91)	SUS434	Plt-Sh, HR Ann	0.12 max	16.00-18.00	1.00 max	0.75-1.25	0.60 max	0.040 max	0.030 max	1.00 max	bal Fe	450	205	22	88 HRB max
JIS G4305(91)	SUS434	Plt Sh, CR Ann	0.12 max	16.00-18.00	1.00 max	0.75-1.25	0.60 max	0.040 max	0.030 max	1.00 max	bal Fe	450	205	22	183 HB max
JIS G4306	SUS434	Strp HR Ann	0.12 max	16-18	1 max	0.75-1.25	0.6 max	0.04 max	0.03 max	1 max	bal Fe				
JIS G4307	SUS434	Strp CR Ann	0.12 max	16-18	1 max	0.75-1.25	0.6 max	0.04 max	0.03 max	1 max	bal Fe				
JIS G4308	SUS434	Wir rod	0.12 max	16.00-18.00	1.00 max	0.75-1.25	0.060 max	0.040 max	0.030 max	1.00 max	bal Fe				
JIS G4315	SUS434	Wir rod	0.12 max	16.00-18.00	1.00 max	0.75-1.25	0.60 max	0.040 max	0.030 max	1.00 max	bal Fe				
Spain															
UNE 36016/1(89)	F.3116*	Obs; EN10088-3(96); X6CrMo171	0.08 max	16.0-18.0	1.0 max	0.9-1.3	1.0 max	0.04 max	0.03 max	1.0 max	bal Fe				

Specification	Designation	Notes	C	Cr	Mn	Mo	Ni	P	S	Si	Other	UTS	YS	El	Hard

Stainless Steel, Ferritic, 434 (Continued from previous page)

Sweden

Specification	Designation	Notes	C	Cr	Mn	Mo	Ni	P	S	Si	Other	UTS	YS	El	Hard
SS 142325	2325	Corr res	0.08 max	16-19	1 max	1.3-2	0.5 max	0.04 max	0.03 max	1 max	bal Fe				

UK

Specification	Designation	Notes	C	Cr	Mn	Mo	Ni	P	S	Si	Other	UTS	YS	El	Hard
BS 1449/2(83)	430S17	Corr/Heat res; t<=100mm	0.08 max	16-18	1.00 max		1.00 max	0.040 max	0.030 max	1.00 max	bal Fe	430	245	20	0-190 HB
BS 1449/2(83)	434S17	Corr/Heat res; t<=2.99mm	0.08 max	16-18	1.00 max	0.9-1.3	1.00 max	0.040 max	0.030 max	1.00 max	bal Fe	430	245	20	0-185 HB
BS 1449/2(83)	434S19	Sh Strp, SA, 3mm diam	0.1 max	16-18	1.00 max	0.9-1.3	0.5 max	0.04 max	0.03 max	0.8 max	bal Fe				
BS 1554(90)	434S20	Wir Corr/Heat res; diam<0.49mm; Ann	0.12 max	16.0-18.0	1.00 max	0.75-1.25		0.040 max	0.030 max	1.00 max	W <=0.1; bal Fe	0-650			
BS 1554(90)	434S20	Wir Corr/Heat res; 6<diam<=13mm; Ann	0.12 max	16.0-18.0	1.00 max	0.75-1.25		0.040 max	0.030 max	1.00 max	W <=0.1; bal Fe	0-610			

USA

Specification	Designation	Notes	C	Cr	Mn	Mo	Ni	P	S	Si	Other	UTS	YS	El	Hard
	AISI 434	Tube	0.12 max	16.00-18.00	1.00 max	0.75-1.25		0.040 max	0.030 max	1.00 max	bal Fe				
	UNS S43400		0.12 max	16.00-18.00	1.00 max	0.75-1.25		0.040 max	0.030 max	1.00 max	bal Fe				
ASTM A240/A240M(98)	S43400		0.12 max	16.0-18.0	1.00 max	0.75-1.25		0.040 max	0.030 max	1.00 max	bal Fe	450	240	22.0	
SAE J405(98)	S43400		0.12 max	16.00-18.00	1.00 max	0.75-1.25		0.040 max	0.030 max	1.00 max	bal Fe				

Stainless Steel, Ferritic, 436

Japan

Specification	Designation	Notes	C	Cr	Mn	Mo	Ni	P	S	Si	Other	UTS	YS	El	Hard
JIS G4303(91)	SUS436L*	Obs; Sh Plt, HR	0.025 max	16.00-19.00	1.00 max	0.75-1.25	0.60 max	0.040 max	0.030 max	1.00 max	N <=0.025; Nb <=0.80; Ti <=0.80; bal Fe	410	245	20	96 HRB max
JIS G4304(91)	SUS436L	Plt Sh, HR	0.025 max	16.00-19.00	1.00 max	0.75-1.25	0.60 max	0.040 max	0.030 max	1.00 max	N <=0.025; Nb <=0.80; Ti <=0.80; 8x(C+N)<=Ti+Nb+Zr<=0.80; bal Fe	410	245	20	96 HRB max
JIS G4305(91)	SUS436J1L		0.025 max	17.00-20.00	1.00 max	0.40-0.80		0.040 max	0.030 max	1.00 max	N <=0.025; 8x(C+N)<=Nb<=0.80; bal Fe	410	245	20	
JIS G4305(91)	SUS436L	Plt Sh, CR	0.025 max	16.00-19.00	1.00 max	0.75-1.25		0.040 max	0.030 max	1.00 max	N <=0.025; bal Fe	410	245	20	217 HB max
JIS G4307	SUS436L	Strp CR	0.25 max	16-19	1 max	0.75-1.25		0.04 max	0.03 max	1 max	bal Fe				

UK

Specification	Designation	Notes	C	Cr	Mn	Mo	Ni	P	S	Si	Other	UTS	YS	El	Hard
BS 1554(90)	436S20	Wir Corr/Heat res; diam<0.49mm; Ann	0.12 max	16.0-18.0	1.00 max	0.75-1.25		0.040 max	0.030 max	1.00 max	W <=0.1; 5xC<=Nb<=0.70; bal Fe	0-650			
BS 1554(90)	436S20	Wir Corr/Heat res; 6<diam<=13mm; Ann	0.12 max	16.0-18.0	1.00 max	0.75-1.25		0.040 max	0.030 max	1.00 max	W <=0.1; 5xC<=Nb<=0.70; bal Fe	0-610			

USA

Specification	Designation	Notes	C	Cr	Mn	Mo	Ni	P	S	Si	Other	UTS	YS	El	Hard
	AISI 436	Tube	0.12 max	16.00-18.00	1.00 max			0.040 max	0.030 max	1.00 max	5xC<=Nb+Ta<=0.70; bal Fe				
	UNS S43600		0.12 max	16.00-18.00	1.00 max	0.75-1.25		0.040 max	0.030 max	1.00 max	Nb+Ta 5xC-0.70				
ASTM A240/A240M(98)	S43600		0.12 max	16.0-18.0	1.00 max	0.75-1.25		0.040 max	0.030 max	1.00 max	5xC<=Nb<=0.80; bal Fe	450	240	22.0	
SAE J405(98)	S43600		0.12 max	16.00-18.00	1.00 max	0.75-1.25		0.040 max	0.030 max	1.00 max	5xC<=Nb<=0.80; bal Fe				

Stainless Steel, Ferritic, 439

France

Specification	Designation	Notes	C	Cr	Mn	Mo	Ni	P	S	Si	Other	UTS	YS	El	Hard
AFNOR NFA35573	X8CT17	Sh Strp Plt Bar Wir	0.08 max	16-18	1 max			0.04 max	0.03 max	1 max	Ti>=7xC<=1.20; bal Fe				
AFNOR NFA35578	X8CT17	Sh Strp Plt Bar Wir	0.08 max	16-18	1 max			0.04 max	0.03 max	1 max	Ti>=7xC<=1.20; bal Fe				

Germany

Specification	Designation	Notes	C	Cr	Mn	Mo	Ni	P	S	Si	Other	UTS	YS	El	Hard
DIN 17440(96)	WNr 1.4510	Plt Sh, HR Strp Ann Bar	0.05 max	16.0-18.0	1.00 max			0.040 max	0.030 max	1.00 max	Ti [4x(C+N)+0.15]<=0.80; bal Fe	450-600	270	20	

Stainless Steel, Ferritic, 439 (Continued from previous page)

Specification	Designation	Notes	C	Cr	Mn	Mo	Ni	P	S	Si	Other	UTS	YS	El	Hard
Germany															
DIN 17440(96)	X3CrTi17	CR Strp Bar Wir Frg	0.05 max	16.0-18.0	1.00 max			0.040 max	0.030 max	1.00 max	Ti [4x(C+N)+0.15]<=0.80; bal Fe	450-600	270	20	185 HB
DIN 17441(97)	WNr 1.4510	CR Strp Plt Sh Ann	0.05 max	16.0-18.0	1.00 max			0.040 max	0.030 max	1.00 max	Ti [4x(C+N)+0.15]<=0.80; bal Fe	450-600	270	20	
Hungary															
MSZ 4360(80)	KO4	Corr res; 0-60mm; Q/T	0.1 max	16-18	1 max		0.6 max	0.04 max	0.03 max	1 max	Ti <=0.8; Ti=5C-0.8; bal Fe	440	245	20L	
MSZ 4360(80)	KO4	Corr res; 0-999mm; soft ann	0.1 max	16-18	1 max		0.6 max	0.04 max	0.03 max	1 max	Ti <=0.8; Ti=5C-0.8; bal Fe				197 HB max
International															
ISO 683-13(74)	8b*	Ann	0.07 max	16-18	1 max		1 max	0.04 max	0.03 max	1 max	Ti=7x%C < 1.1; bal Fe				
Italy															
UNI 8317(81)	X6CrTi17	Corr res	0.08 max	16-18	1 max		0.5 max	0.04 max	0.03 max	1 max	Ti <=0.8; Ti=5C-0.8; bal Fe				
Japan															
JIS G4304(91)	SUS430LX	Plt Sh, HR	0.030 max	16.00-19.00	1.00 max		0.60 max	0.040 max	0.030 max	0.75	Nb <=1.00; Ti 0.10-1.00; Nb or Ti 0.-1.00; bal Fe	360	175	22	88 HRB max
JIS G4305(91)	SUS430LX	Plt Sh, CR	0.030 max	16.00-19.00	1.00 max			0.040 max	0.030 max	0.75 max	bal Fe	360	175	22	183 HB max
JIS G4306	SUS430LX	Strp HR	0.03 max	16-19	1 max			0.04 max	0.03 max	0.75	Ti or Nb 0.10-1.00; bal Fe				
JIS G4307	SUS430LX	Strp CR	0.03 max	16-19	1 max			0.04 max	0.03 max	0.75 max	Ti or Nb 0.10-1.00; bal Fe				
Poland															
PNH86020	0H17T	Corr res	0.08 max	16-18	0.8 max		0.6 max	0.04 max	0.03 max	0.8 max	Ti <=0.8; Ti=5C-0.8; bal Fe				
PNH86020	0H17Ti	Sh Strp Plt Bar	0.08 max	16-18	0.8 max			0.04 max	0.03 max	0.8 max	Ti>=5xC<=0.80; bal Fe				
Romania															
STAS 3583(87)	8TiCr170	Corr res; Heat res	0.08 max	16-18	1 max	0.2 max	0.5 max	0.045 max	0.03 max	1 max	Pb <=0.15; Ti <=1.2; 7C<=Ti<=1.2; bal Fe				
Russia															
GOST 5632	08Ch17Ti	Sh Strp Plt Bar Tub	0.08 max	16-18	0.8 max		0.6 max	0.035 max	0.025 max	0.8 max	Cu <=0.3; Ti>=5xC<=1.00; bal Fe				
Spain															
UNE 36016(75)	F.3114*	see X8CrTi17	0.1 max	16.0-18.0	1.0 max	0.15 max	0.5 max	0.04 max	0.03 max	1.0 max	Ti <=0.8; 5xC<=Ti<=0.8; bal Fe				
UNE 36016(75)	X8CrTi17	Sh Strp Plt Bar Wir; Corr res	0.1 max	16.0-18.0	1.0 max	0.15 max	0.5 max	0.04 max	0.03 max	1.0 max	Ti <=0.8; 5xC<=Ti<=0.8; bal Fe				
USA															
	AISI 439*	Obs; Tube	0.07 max	17.00-19.00	1.00 max		0.5 max	0.040 max	0.030 max	1.00 max	Al <=0.15; 12xC<=Ti<=1.10; bal Fe				
	UNS S43035		0.07 max	17.00-19.00	1.00 max		0.50 max	0.040 max	0.030 max	1.00 max	Al <=0.15; Ti 12xC-1.10; bal Fe				
ASME SA240	439	Refer to ASTM A240(95)													
ASME SA268	439	Refer to ASTM A268(94)													
ASME SA479	439	Refer to ASTM A479/A479M(95)													
ASTM A240/A240M(98)	S43035		0.07 max	17.0-19.0	1.00 max		0.50 max	0.040 max	0.030 max	1.00 max	Al <=0.15; 4(C+N)+0.20<=Ti<=1.10; bal Fe	415	205	22.0	183 HB
ASTM A268	439	Tube	0.07 max	17-19	1 max		0.5 max	0.04 max	0.03 max	1 max	Al <=0.15; Ti 12%C-1.10; bal Fe				
ASTM A479	439	Bar	0.07 max	17-19	1 max		0.5 max	0.04 max	0.03 max	1 max	Al <=0.15; Ti 12%C-1.10; bal Fe				
ASTM A803/A803M(96)	TP439	Weld Feedwater Heater Tub	0.07 max	17.00-19.00	1.00 max		0.50 max	0.040 max	0.30 max	1.00 max	Al <=0.15; N <=0.04; Ti>=0.20+4(C+N)<=1.10; bal Fe	415	205	20	207 HB max

UNS numbers and US grades are provided as a means of cross referencing chemically similar alloys. Exchangability is only possible after independent examination of specifications. Tensile properties are minimum or typical as specified. UTS and YS as MPa. El as %. See Appendix for list of abbreviations used in Notes. * indicates obsolete material.

Specification	Designation	Notes	C	Cr	Mn	Mo	Ni	P	S	Si	Other	UTS	YS	El	Hard

Stainless Steel, Ferritic, 439 (Continued from previous page)

USA

| SAE J405(98) | S43035 | | 0.07 max | 17.00-19.00 | 1.00 max | | 0.50 max | 0.040 max | 0.030 max | 1.00 max | Al <=0.15; N <=0.04; 12xC<=Ti<=1.10; bal Fe | | | | |

Stainless Steel, Ferritic, 439LT

UK

| BS 3606(92) | 439 | Heat exch Tub; t<=3.2mm | 0.04 max | 17.0-19.0 | 1.00 max | 0.15 max | 0.50 max | 0.040 max | 0.020 max | 1.00 max | Al <=0.15; Cu <=0.15; C<=N+-0.040; 15(C+N)<Ti<=0.75; bal Fe | 415-700 | 205 | 15 | |

USA

	UNS S43932	with Ti and Nb	0.020 max	17.0-19.0	1.0 max		0.50 max	0.040 max	0.030 max	1.00 max	Al <=0.15; N <=0.020; Nb 0.02-0.095; Ti 0.20-0.40; 0.20+4(C+N)<=Ti+Nb <=0.75; bal Fe				
ASTM A240/A240M(98)	S43932	with Ti and Nb	0.030 max	17.0-19.0	1.00 max		0.50 max	0.040 max	0.030 max	1.00 max	Al <=0.15; N <=0.030; Ti 0.20-0.40; 0.20+4(C+N)<=Ti+Nb <=0.75; bal Fe	415	205	22.0	183 HB
SAE J405(98)	S43932	w/Ti and Nb	0.030 max	17.0-19.0	1.0 max		0.50 max	0.040 max	0.030 max	1.00 max	Al <=0.15; N <=0.020; Ti 0.20-0.40; 0.20+4(C+N)<=Nb+Ti <=0.75; bal Fe				

Stainless Steel, Ferritic, 442

Canada

| CSA G110.5 | 442 | Sh, HR, Ann, Q/A, Tmp | 0.2 max | 18-23 | 1 max | | 0.6 max | 0.04 max | 0.03 max | 1 max | bal Fe | | | | |
| CSA G110.5 | 442 | Strp, CR, Ann | 0.2 max | 18-23 | 1 max | | 0.6 max | 0.04 max | 0.03 max | 1 max | bal Fe | | | | |

Czech Republic

| CSN 417047 | 17047 | Heat res | 0.15 max | 20.0-23.0 | 0.8 max | | 0.6 max | 0.045 max | 0.035 max | 0.8 max | Ti <=0.7; bal Fe | | | | |

UK

| BS 1449/2(83) | 442S19 | Sh Strp Plt, SA, 3mm diam | 0.1 max | 18-22 | 1.00 max | | 0.5 max | 0.04 max | 0.03 max | 0.8 max | bal Fe | | | | |

USA

	AISI 442	Tube	0.20 max	18.00-23.00	1.00 max			0.040 max	0.030 max	1.00 max	bal Fe				
	UNS S44200	Heat res	0.20 max	18.00-23.00	1.00 max			0.040 max	0.030 max	1.00 max	bal Fe				
ASTM A176(97)	442	Sh Strp Plt	0.20 max	18.00-23.00	1.00 max		0.60 max	0.040 max	0.040 max	1.00 max	bal Fe	515	275	20.0	217 HB

Stainless Steel, Ferritic, 443

Mexico

| NMX-B-171(91) | MT443 | mech tub | 0.20 max | 18.0-23.0 | 1.00 max | | 0.50 max | 0.040 max | 0.030 max | 1.00 max | bal Fe | | | | |

USA

| | UNS S44300 | High-Cr-Cu; Heat res | 0.20 max | 18.00-23.00 | 1.00 max | | 0.50 max | 0.040 max | 0.030 max | 1.00 max | Cu 0.90-1.25; bal Fe | | | | |
| ASTM A511(96) | MT443 | Smls mech tub, Ann | 0.20 max | 18.0-23.0 | 1.00 max | | 0.50 max | 0.040 max | 0.030 max | 1.00 max | Cu 0.90-1.25; bal Fe | 483 | 276 | 20 | 207 HB |

Stainless Steel, Ferritic, 444

Australia

| AS 1449(94) | 444 | Sh Strp Plt | 0.025 max | 17.5-19.5 | 1.00 max | 1.75-2.50 | 1.00 max | 0.04 max | 0.030 max | 1.00 max | N <=0.035; Nb+Ti=[0.20+4(C+N)] min, 0.80 max; bal Fe | | | | |

Europe

| EN 10088/2(95) | X2CrMoTiS18-2 | Strp; Corr res; CR t<=6mm, Ann | 0.030 max | 17.50-19.00 | 0.50 max | 2.00-2.50 | | 0.040 max | 0.15-0.35 | 1.00 max | Ti 0.30-0.80; C+N<=0.040; bal Fe | 420-640 | 300 | 20 | |

UNS numbers and US grades are provided as a means of cross referencing chemically similar alloys. Exchangability is only possible after independent examination of specifications. Tensile properties are minimum or typical as specified. UTS and YS as MPa. El as %. See Appendix for list of abbreviations used in Notes. * indicates obsolete material.

Stainless Steel, Ferritic, 444 (Continued from previous page)

Specification	Designation	Notes	C	Cr	Mn	Mo	Ni	P	S	Si	Other	UTS	YS	El	Hard
Europe															
EN 10088/3(95)	1.4523	Bar Rod Shp, Corr res; t<=100mm, Ann	0.030 max	15.00-19.00	0.50 max	2.00-2.50		0.040 max	0.15-0.35	1.00 max	Ti 0.30-0.80; C+N<=0.040; bal Fe	430-600	280	15	200 HB max
EN 10088/3(95)	X2CrMoTiS18-2	Bar Rod Shp, Corr res; t<=100mm, Ann	0.030 max	15.00-19.00	0.50 max	2.00-2.50		0.040 max	0.15-0.35	1.00 max	Ti 0.30-0.80; C+N<=0.040; bal Fe	430-600	280	15	200 HB max
Germany															
DIN EN 10088(95)	WNr 1.4523	Bar Rod Sect	0.030 max	17.5-19.0	0.50 max	2.00-2.50		0.040 max	0.15-0.35	1.00 max	Ti 0.30-0.85; C+N<=0.040; bal Fe				
International															
ISO 683-13(74)	F1*	Ann	0.025-0.05	17-19	1 max	1.75-2.5	0.6 max	0.04 max	0.03 max	1 max	N <=1; Nb+Ti=8x(C+N) min - 0.8 max; bal Fe				
Japan															
JIS G4303(91)	SUS444*	Obs; Sh Plt, HR	0.025 max	17.00-20.00	1.00 max	1.75-2.50	0.60 max	0.040 max	0.030 max	1.00 max	N <=0.025; Nb <=0.80; Ti <=0.80; bal Fe	410	245	20	96 HRB max
JIS G4304(91)	SUS444	Plt Sh, HR	0.025 max	17.00-20.00	1.00 max	1.75-2.50	0.60 max	0.040 max	0.030 max	1.00 max	N <=0.025; Nb <=0.80; Ti <=0.80; 8x(C+N)<=Ti+Nb+Zr<=0.80; bal Fe	410	245	20	96 HRB max
JIS G4305(91)	SUS444	Plt Sh, CR	0.025 max	17.00-20.00	1.00 max	1.75-2.50		0.040 max	0.030 max	1.00 max	N <=0.025; 8x(C+N)<=Ti+Nb+Zr<=0.80; bal Fe	410	245	20	217 HB max
JIS G4306	SUS444	Strp HR	0.025 max	17-20	1 max	1.75-2.5		0.04 max	0.03 max	1 max	N <=0.025; 8x(C+N)<=Ti+Nb+Zr<=0.80; bal Fe				
JIS G4307	SUS444	Strp CR	0.025 max	17-20	1 max	1.75-2.5		0.04 max	0.03 max	1 max	N <=0.025; 8x(C+N)<=Ti+Nb+Zr<=0.80; bal Fe				
Norway															
NS 14001	14115	Sh Strp Plt	0.025 max	17-19	0.5 max	2-2.5	0.5 max	0.04 max	0.02 max	1 max	N 0.025; Ti>=0+4(C+N)<=0.80; bal Fe				
Romania															
STAS 3583(87)	2TiMoCr180	Corr res; Heat res	0.025 max	17-20	1 max	1.8-2.5	0.6 max	0.04 max	0.03 max	1 max	Pb <=0.15; Ti <=0.8; 8(C+N)<=Ti<=0.8; bal Fe				
Sweden															
SS 142326	2326	Corr res	0.025 max	17-19	0.5 max	2-2.5	0.5 max	0.04 max	0.02 max	1 max	N <=0.025; Ti 0.2-0.8; Ti=0.2+4(C+N)-0.8; bal Fe				
USA															
	AISI 444*	Obs; Tube	0.025 max	17.5-19.50	1.00 max	1.75-2.5	1 max	0.040 max	0.030 max	1.00 max	Nb>=0.2+4(C+N); bal Fe				
	UNS S44400	stabilized Low-Interstitial	0.025 max	17.5-19.5	1.00 max	1.75-2.50	1.00 max	0.040 max	0.030 max	1.00 max	N <=0.025; Ti+Nb=[0.2+4(C+N)] min - 0.8 max; bal Fe				
ASTM A176(97)	S-44400*	Sh Strp Plt	0.025 max	17.5-19.5	1 max	1.75-2.5	1 max	0.04 max	0.03 max	1 max	N>=0.2+4(C%+N); bal Fe				
ASTM A213/A213M(95)	18Cr-2Mo	Smls tube boiler, superheater, heat exchanger	0.025 max	17.5-19.5	1.00 max	1.75-2.50		0.040 max	0.030 max	1.00 max	N <=0.035; Ni+Cu<=1.00, 0.20+4(C+N)<=Ti+Nb <=0.80; bal Fe	415	275	20	
ASTM A240/A240M(98)	S44400	stabilized Low-Interstitial	0.025 max	17.5-19.5	1.00 max	1.75-2.50	1.00 max	0.040 max	0.030 max	1.00 max	N <=0.035; 0.20+4(C+N)<=Ti+Nb <=0.80; bal Fe	415	275	20.0	217 HB
ASTM A268	S-44400	Tube	0.025 max	17.5-19.5	1 max	1.75-2.5	1 max	0.04 max	0.03 max	1 max	N>=0.2+4(C%+N); bal Fe				
ASTM A276(98)	S-44400	Sh Strp Plt	0.025 max	17.5-19.5	1 max	1.75-2.5	1 max	0.04 max	0.03 max	1 max	N>=0.2+4(C%+N); bal Fe				
ASTM A580/A580M(98)	S44400	Wir, Ann	0.025 max	17.5-19.5	1.00 max	1.75-2.50	1.00	0.040 max	0.030 max	1.00 max	N <=0.035; Ti+Nb: 0.20+4(C+N)<=0.80; bal Fe	485	275	20	
ASTM A803/A803M(96)	18-2	Weld Feedwater Heater Tub, stabilized low interstitial	0.025 max	17.5-19.5	1.00 max	1.75-2.50	1.00 max	0.040 max	0.030 max	1.00 max	N <=0.035; Ti+Nb>= 0.2+4(C + N)<=0.80; bal Fe	415	240	20	217 HB max
SAE J405(98)	S44400	stabilized Low-Interstitial	0.025 max	17.5-19.5	1.00 max	1.75-2.50	1.00 max	0.040 max	0.030 max	1.00 max	N <=0.025; 0.20+4(C+N)<=Nb+Ti <=0.80; bal Fe				

UNS numbers and US grades are provided as a means of cross referencing chemically similar alloys. Exchangability is only possible after independent examination of specifications. Tensile properties are minimum or typical as specified. UTS and YS as MPa. El as %. See Appendix for list of abbreviations used in Notes. * indicates obsolete material.

Specification	Designation	Notes	C	Cr	Mn	Mo	Ni	P	S	Si	Other	UTS	YS	El	Hard
Stainless Steel, Ferritic, 446															
Bulgaria															
BDS 6738(72)	Ch25T	Sh Plt Bar Bil	0.15 max	24.0-27.0	0.80 max	0.30 max	0.60 max	0.035 max	0.025 max	1.00 max	5xC<=Ti<=0.90; bal Fe				
Canada															
CSA G110.3	446	Bar Bil	0.2 max	23-27	1.5 max			0.04 max	0.03 max	1 max	N <=0.25; bal Fe				
CSA G110.5	446	Sh, HR, Ann, Q/A, Tmp	0.2 max	23-27	1.5 max			0.04 max	0.03 max	1 max	N <=0.25; bal Fe				
China															
GB 1221(92)	2Cr25N	Bar Ann	0.20 max	23.00-27.00	1.50 max		0.60 max	0.040 max	0.030 max	1.00 max	Cu <=0.30; N <=0.25; bal Fe	510	275	20	
GB 4238(92)	2Cr25N	Plt, HR Ann	0.20 max	23.00-27.00	1.50 max		0.60 max	0.040 max	0.030 max	1.00 max	Cu <=0.30; N <=0.25; bal Fe	510	275	20	
Czech Republic															
CSN 417061	17061	Heat res	0.18 max	23.0-26.0	0.8 max		0.6 max	0.045 max	0.035 max	0.8 max	Ti <=0.7; bal Fe				
France															
AFNOR NFA35586	Z10C24	Sh Plt Bar Bil, Ann	0.12 max	23-26	1 max			0.04 max	0.03 max	1.5 max	bal Fe				
Germany															
DIN	WNr 1.4083*	Obs	0.1 max	27-29	1 max			0.05 max	0.03 max	1.00 max	bal Fe				
Hungary															
MSZ 4359(82)	H14	Heat res; 0-60mm; soft ann	0.12 max	23-26	1 max		0.6 max	0.04 max	0.03 max	0.8-1.5	Al 1.2-1.7; bal Fe	520	280	10L	223 HB max
MSZ 4359(82)	H17	Sh Plt Bar Bil; Heat res; 0-60mm; soft ann	0.2 max	23-27	1.5 max		0.6 max	0.04 max	0.03 max	1 max	N 0.15-0.25; bal Fe	500	280	15L	212 HB max
India															
IS 1570/5(85)	X15Cr25N	Sh Plt Strp Bar Flat Band	0.2 max	23-27	1.5 max	0.15 max	0.4 max	0.045 max	0.03 max	1 max	Co <=0.1; Cu <=0.3; N <=0.25; Pb <=0.15; W <=0.1; bal Fe	515	275	20	217 HB max
International															
COMECON PC4-70	30	Sh Plt Bar Bil	0.15 max	24-27	0.8 max			0.035 max	0.025 max	1 max	Ti=5xC-0.90; bal Fe				
ISO 4955(94)	H6	Heat res	0.2 max	23.0-27.0	1.0 max	0.15 max	0.4 max	0.04 max	0.03 max	0.7-1.4	Al 1.2-1.7; Co <=0.1; Cu <=0.3; Pb <=0.15; Ti <=0.05; V <=0.1; W <=0.1; bal Fe				
Italy															
UNI 6901(71)	X16Cr26	Heat res	0.2 max	24-27	1.5 max			0.04 max	0.03 max	1 max	N <=0.25; bal Fe				
UNI 6904(71)	X16Cr26	Heat res	0.2 max	24-27	1.5 max		0.5 max	0.04 max	0.03 max	0.75 max	N 0.1-0.25; bal Fe				
Japan															
JIS G4309	SUH446	Wir	0.20 max	23.00-27.00	1.50 max			0.040 max	0.030 max	1.00 max	Cu <=0.30; N <=0.25; bal Fe				
JIS G4311	SUH446	Heat res Bar	0.20 max	23.00-27.00	1.50 max		0.60 max	0.040 max	0.030 max	1.00 max	Cu <=0.30; N <=0.25; bal Fe	510	275	20	201 HB
JIS G4312	SUH446	Heat res Plt Sh	0.20 max	23.00-27.00	1.50 max		0.60 max	0.040 max	0.030 max	1.00 max	N <=0.25; bal Fe	510	275	20	201 HB max
Mexico															
DGN B-83	446*	Obs; Bar, HR, CR	0.2 max	23-27	1.5 max			0.04 max	0.03 max	1 max	N <=0.25; bal Fe				
NMX-B-171(91)	MT446-1	mech tub	0.20 max	23.0-30.0	1.50 max		0.50 max	0.040 max	0.030 max	1.00 max	N <=0.25; bal Fe				
NMX-B-171(91)	MT446-2	mech tub	0.12 max	23.0-30.0	1.50 max		0.50 max	0.040 max	0.030 max	1.00 max	N <=0.25; bal Fe				
Poland															
PNH86022(71)	H24JS	Heat res	0.12 max	23-25	1 max		0.5 max	0.045 max	0.03 max	1.3-1.6	Al 1.3-1.6; bal Fe				
PNH86022(71)	H25T		0.15 max	24-27	0.8 max		0.6 max	0.045 max	0.03 max	1 max	Ti=4xC-0.80; bal Fe				
Romania															
STAS 3583(64)	12TC250	Plt Bar Bil; Corr res; Heat res	0.15 max	24-27	0.8 max		0.6 max	0.035 max	0.03 max	1 max	Pb <=0.15; Ti <=0.8; 5C<=Ti<=0.8; bal Fe				

UNS numbers and US grades are provided as a means of cross referencing chemically similar alloys. Exchangability is only possible after independent examination of specifications. Tensile properties are minimum or typical as specified. UTS and YS as MPa. El as %. See Appendix for list of abbreviations used in Notes. * indicates obsolete material.

Specification	Designation	Notes	C	Cr	Mn	Mo	Ni	P	S	Si	Other	UTS	YS	El	Hard

Stainless Steel, Ferritic, 446 (Continued from previous page)

Russia

Specification	Designation	Notes	C	Cr	Mn	Mo	Ni	P	S	Si	Other	UTS	YS	El	Hard
GOST	Ch25	Sh Plt Bar Bil	0.2 max	23-27	0.8 max		0.6 max	0.04 max	0.03 max	1 max	bal Fe				
GOST 5632	15Ch28	Sh Plt Bar Bil	0.15 max	27-30	0.8 max		0.6 max	0.035 max	0.025 max	1 max	Cu <=0.3; Ti <=0.2; bal Fe				
GOST 5632(72)	15Ch25T	Corr res; Heat res	0.15 max	24.0-27.0	0.8 max	0.15 max	0.6 max	0.035 max	0.025 max	1.0 max	Cu <=0.3; Ti <=0.9; Ti=5C-0.9; bal Fe				

Spain

Specification	Designation	Notes	C	Cr	Mn	Mo	Ni	P	S	Si	Other	UTS	YS	El	Hard
UNE 36017(85)	F.3154*	see X10CrAl24	0.12 max	23-26	1 max	0.15 max		0.04 max	0.03 max	0.7-1.4	Al 1.2-1.7; bal Fe				
UNE 36017(85)	F.3169*	see X10CrN28	0.15-0.2	26-29	1 max	0.15 max		0.04 max	0.03 max	1 max	N 0.15-0.25; bal Fe				
UNE 36017(85)	F.3308*	see X20CrNiSi25-04	0.1-0.2	24-27	2 max	0.15 max	3.5-5.5	0.045 max	0.03 max	0.18-1.15	bal Fe				
UNE 36017(85)	X10CrAl24	Heat res	0.12 max	23-26	1 max	0.15 max		0.04 max	0.03 max	0.7-1.4	Al 1.2-1.7; bal Fe				
UNE 36017(85)	X10CrN28	Heat res	0.15-0.2	26-29	1 max	0.15 max		0.04 max	0.03 max	1 max	N 0.15-0.25; bal Fe				
UNE 36017(85)	X20CrNiSi25-04	Heat res	0.1-0.2	24-27	2 max	0.15 max	3.5-5.5	0.045 max	0.03 max	0.18-1.15	bal Fe				

Sweden

Specification	Designation	Notes	C	Cr	Mn	Mo	Ni	P	S	Si	Other	UTS	YS	El	Hard
SS 142322	2322	Heat res	0.2 max	24-28	1 max			0.04 max	0.03 max	1 max	N 0.1-0.25; bal Fe				

USA

Specification	Designation	Notes	C	Cr	Mn	Mo	Ni	P	S	Si	Other	UTS	YS	El	Hard
	AISI 446	Tube	0.20 max	23.00-27.00	1.50 max			0.040 max	0.030 max	1.00 max	N <=0.25; bal Fe				
	UNS S44600	Heat res	0.20 max	23.00-27.00	1.50 max			0.040 max	0.030 max	1.00 max	N <=0.25; bal Fe				
ASME SA268	446	Refer to ASTM A268(94)													
ASTM A176(97)	446	Sh Strp Plt	0.20 max	23.00-27.00	1.50 max	0.75 max		0.040 max	0.030 max	1.00 max	N <=0.25; bal Fe	515	275	20.0	217 HB
ASTM A268	446	Tube	0.2 max	23-27	1.5 max			0.04 max	0.03 max	1 max	N <=0.25; bal Fe				
ASTM A276(98)	446	Bar	0.2 max	23-27	1.5 max			0.04 max	0.03 max	1 max	N <=0.25; bal Fe				
ASTM A314	446	Bil	0.2 max	23-27	1.5 max			0.04 max	0.03 max	1 max	N <=0.25; bal Fe				
ASTM A473	446	Frg, Heat res, ann	0.20 max	23.00-27.00	1.50 max			0.040 max	0.030 max	1.00 max	N <=0.25; bal Fe	485	275	20	223 HB
ASTM A511(96)	MT446-1	Smls mech tub, Ann	0.20 max	23.0-30.0	1.50 max	0.50 max		0.040 max	0.030 max	1.00 max	N <=0.25; bal Fe	483	276	18	207 HB
ASTM A511(96)	MT446-2	Smls mech tub, Ann	0.12 max	23.0-30.0	1.50 max	0.50 max		0.040 max	0.030 max	1.00 max	N <=0.25; bal Fe	448	276	20	207 HB
ASTM A580/A580M(98)	446	Wir, Ann	0.20 max	23.0-27.0	1.50 max			0.040 max	0.030 max	1.00 max	N <=0.25; bal Fe	485	275	20	
ASTM A815/A815M(98)	WP446	Pipe fittings	0.20 max	23.0-27.0	1.50 max	0.50 max		0.040 max	0.030 max	0.75 max	N 0.25; bal Fe	485-655	275	18.0	207 HB
FED QQ-S-763F(96)	446*	Obs; Bar Wir Shp Frg; Corr res	0.2 max	23-27	1.5 max			0.04 max	0.03 max	1 max	N <=0.25; bal Fe				
FED QQ-S-766D(93)	446*	Obs; see ASTM A240, A666, A693; Sh Strp Plt	0.2 max	23-27	1.5 max			0.04 max	0.03 max	1 max	N <=0.25; bal Fe				
MIL-S-862B	446*	Obs	0.2 max	23-27	1.5 max			0.04 max	0.03 max	1 max	N <=0.25; bal Fe				

Stainless Steel, Ferritic, A176

Spain

Specification	Designation	Notes	C	Cr	Mn	Mo	Ni	P	S	Si	Other	UTS	YS	El	Hard
UNE 36016/2(89)	F.3121*	Obs; En10088-2(96); X2CrTi12	0.03 max	10.5-12.5	1.0 max	0.15 max	1.0 max	0.04 max	0.03 max	1.0 max	N <=0.035; Ti <=1.0; Ti=6xC-1; bal Fe				

USA

Specification	Designation	Notes	C	Cr	Mn	Mo	Ni	P	S	Si	Other	UTS	YS	El	Hard
	UNS S40945	with Nb	0.030 max	10.50-11.75	1.00 max		0.50 max	0.040 max	0.030 max	1.00 max	N <=0.030; Nb 0.18-0.40; Ti 0.05-0.20; bal Fe				
	UNS S41045	with Nb	0.030 max	12.0-13.0	1.00 max		0.50 max	0.040 max	0.030 max	1.00 max	N <=0.030; Nb 9(C+N) min - 0.60 max; bal Fe				

Specification	Designation	Notes	C	Cr	Mn	Mo	Ni	P	S	Si	Other	UTS	YS	El	Hard

Stainless Steel, Ferritic, A176 (Continued from previous page)

USA

Specification	Designation	Notes	C	Cr	Mn	Mo	Ni	P	S	Si	Other	UTS	YS	El	Hard
ASTM A240/A240M(98)	S40945	with Nb	0.030 max	10.5-11.7	1.00 max		0.50 max	0.040 max	0.030 max	1.00 max	N <=0.030; Nb 0.18-0.40; Ti 0.05-0.20; 0.18<=Nb<=0.40; bal Fe	380	205	22.0	
ASTM A240/A240M(98)	S41045	with Nb	0.030 max	12.0-13.0	1.00 max		0.50 max	0.040 max	0.030 max	1.00 max	N <=0.030; 9(C+N)<=Nb<=0.60; bal Fe	380	205	22.0	
SAE J405(98)	S40945	w/Nb	0.030 max	10.50-11.75	1.00 max		0.50 max	0.040 max	0.030 max	1.00 max	N <=0.030; Nb 0.18-0.40; Ti 0.05-0.20; bal Fe				
SAE J405(98)	S41045	w/Nb	0.030 max	12.00-13.00	1.00 max		0.50 max	0.040 max	0.030 max	1.00 max	N <=0.030; (C+N)<=Nb<=0.60; bal Fe				

Stainless Steel, Ferritic, A268

Europe

Specification	Designation	Notes	C	Cr	Mn	Mo	Ni	P	S	Si	Other	UTS	YS	El	Hard
EN 10088/2(95)	1.4003	Strp; Corr res; CR t<=6mm, Ann	0.030 max	10.50-12.50	1.50 max		0.30-1.00	0.040 max	0.015 max	1.00 max	N <=0.030; bal Fe	450-650	280	20	
EN 10088/2(95)	X2CrNi12	Strp; Corr res; CR t<=6mm, Ann	0.030 max	10.50-12.50	1.50 max		0.30-1.00	0.040 max	0.015 max	1.00 max	N <=0.030; bal Fe	450-650	280	20	
EN 10088/3(95)	1.4003	Bar Rod Shp, Corr res; t<=100mm, Ann	0.030 max	10.50-12.50	1.50 max		0.30-1.00	0.040 max	0.015 max	1.00 max	N <=0.030; bal Fe	450-600	260	20	200 HB max
EN 10088/3(95)	X2CrNi12	Bar Rod Shp, Corr res; t<=100mm, Ann	0.030 max	10.50-12.50	1.50 max		0.30-1.00	0.040 max	0.015 max	1.00 max	N <=0.030; bal Fe	450-600	260	20	200 HB max

USA

Specification	Designation	Notes	C	Cr	Mn	Mo	Ni	P	S	Si	Other	UTS	YS	El	Hard
	UNS S40800		0.08 max	11.5-13.0	1.00 max		0.80 max	0.045 max	0.045 max	1.00 max	Ti 12xC-1.10; bal Fe				

Stainless Steel, Ferritic, A581

Finland

Specification	Designation	Notes	C	Cr	Mn	Mo	Ni	P	S	Si	Other	UTS	YS	El	Hard
SFS 815(86)	X2CrMoTi182	Corr res	0.025 max	17.0-19.0	2.0 max	2.0-2.5	1.0 max	0.04 max	0.02 max	1.0 max	N <=0.025; Ti 0.2-0.8; Ti=0.2+4(C+N)-0.8; bal Fe				

Sweden

Specification	Designation	Notes	C	Cr	Mn	Mo	Ni	P	S	Si	Other	UTS	YS	El	Hard
SS 142382	2382	Corr res	0.03 max	17.5-18.5	0.5 max	2-2.5	1 max	0.03 max	0.15-0.35	1 max	Ti 0.3-1; bal Fe				

USA

Specification	Designation	Notes	C	Cr	Mn	Mo	Ni	P	S	Si	Other	UTS	YS	El	Hard
	UNS S18235	Cr Low-C, Free Machining	0.025 max	17.5-18.5	0.50 max	2.00-2.50	1.00 max	0.030 max	0.15-0.35	1.00 max	N <=0.025; Ti 0.30-1.00; C+N 0.035 max; bal Fe				
ASTM A581/A581M(95)	S18235	Wir rod, Ann	0.025 max	17.5-18.5	0.50 max	2.00-2.50	1.00 max	0.030 max	0.15-0.35	1.00 max	N <=0.025; 0<=C+N<=0.035; bal Fe	415-620			
ASTM A582/A582M(95)	S18235	Bar, HF, CF, Ann	0.025 max	17.50-18.50	0.50 max	2.00-2.50	1.00	0.030 max	0.15-0.35	1.00 max	N <=0.025; Ti 0.30-1.00; C+N<=0.035; bal Fe				207 HB max

Stainless Steel, Ferritic, E4

USA

Specification	Designation	Notes	C	Cr	Mn	Mo	Ni	P	S	Si	Other	UTS	YS	El	Hard
	UNS S41050	with Ni	0.040 max	10.50-12.50	1.00 max		0.60-1.10	0.045 max	0.030 max	1.00 max	N <=0.10; bal Fe				
ASTM A240/A240M(98)	S41050	with Ni	0.040 max	10.5-12.5	1.00 max		0.60-1.10	0.045 max	0.030 max	1.00 max	N <=0.10; bal Fe	415	205	22.0	183 HB
SAE J405(98)	S41050	w/Ni	0.040 max	10.50-12.50	1.00 max		0.60-1.10	0.045 max	0.030 max	1.00 max	N <=0.10; bal Fe				

Stainless Steel, Ferritic, ER409Cb

International

Specification	Designation	Notes	C	Cr	Mn	Mo	Ni	P	S	Si	Other	UTS	YS	El	Hard
ISO 4954(93)	X6CrNb12E	Wir rod Bar, CH ext	0.08 max	10.5-12.5	1.00 max		0.50 max	0.040 max	0.030 max	1.00 max	Nb=10xC<= 1.0; bal Fe	500			

USA

Specification	Designation	Notes	C	Cr	Mn	Mo	Ni	P	S	Si	Other	UTS	YS	El	Hard
	UNS S40940	with Nb	0.08 max	10.50-11.75	1.00 max		0.50 max	0.045 max	0.040 max	1.00 max	Nb 10xC min, 0.75 max; bal Fe				
ASTM A493	409Cb	Wir rod, CHd, cold frg	0.08 max	10.5-11.75	1.00 max		0.50 max	0.045 max	0.040 max	1.00 max	10xC<=Nb<=0.75; bal Fe	485			

Specification	Designation	Notes	C	Cr	Mn	Mo	Ni	P	S	Si	Other	UTS	YS	El	Hard

Stainless Steel, Ferritic, K92930

USA

| | UNS K92930 | Ferritic | 0.06-0.14 | 10.0-12.6 | 0.70 max | 0.20-0.60 | 0.70 max | 0.030 max | 0.020 max | 0.70 max | Al <=0.040; ; B <=0.005; Cu 0.30-1.70; N 0.020-0.100; Nb 0.02-0.10; V 0.15-0.30; W 1.50-2.50; bal Fe | | | | |

Stainless Steel, Ferritic, NACE MR-01-75-92

USA

| | UNS S42400 | | 0.06 max | 12.00-14.00 | 0.50-1.00 | 0.30-0.70 | 3.50-4.50 | 0.03 max | 0.03 max | 0.30-0.60 | bal Fe | | | | |

Stainless Steel, Ferritic, No equivalents identified

Hungary

| MSZ 4359(82) | H15 | Heat res; 0-60mm; Q; Q | 0.15-0.25 | 24-27 | 2 max | | 3.5-4.5 | 0.04 max | 0.03 max | 0.8-1.5 | bal Fe | 600 | 400 | 16L | 235 HB |
| MSZ 4359(82) | H18 | Heat res; 0-60mm; soft ann | 0.08 max | 10.5-12.5 | 1 max | | 0.6 max | 0.04 max | 0.03 max | 1 max | Ti <=1; Ti=6xC-1 | 400 | 210 | 25L | 179 HB max |

International

ISO 4954(93)	X3Cr17E	Wir rod Bar, CH ext	0.04 max	16.0-18.0	1.00 max		16.0-18.0	0.040 max	0.030 max	1.00 max	bal Fe	500			
ISO 4954(93)	X6Cr17E	Wir rod Bar, CH ext	0.08 max	16.0-18.0	1.00 max		16.0-18.0	0.040 max	0.030 max	1.00 max	bal Fe	560			
ISO 4954(93)	X6CrMo171E	Wir rod Bar, CH ext	0.08 max	16.0-18.0	1.00 max	0.90-1.30	16.0-18.0	0.040 max	0.030 max	1.00 max	bal Fe	600			
ISO 4954(93)	X6CrTi12E	Wir rod Bar, CH ext	0.08 max	10.5-12.5	1.00 max		0.50 max	0.040 max	0.030 max	1.00 max	Ti=5xC<= 0.80; bal Fe	530			

Japan

| JIS G4312 | SUH21 | Heat res Plt Sh | 0.10 max | 17.00-21.00 | 1.00 max | | 0.60 max | 0.040 max | 0.030 max | 1.50 max | Al 2.00-4.00; bal Fe | 440 | 245 | 15 | 210 HB max |

Poland

| PNH86020 | 3H14 | Corr res | 0.26-0.35 | 14-17 | 0.8 max | | 0.6 max | 0.04 max | 0.03 max | 0.8 max | bal Fe | | | | |
| PNH86020 | 3H17M | Corr res | 0.33-0.43 | 15.5-17.5 | 1 max | 1-1.3 | 1 max | 0.045 max | 0.03 max | 1 max | bal Fe | | | | |

Spain

| UNE 36016/1(89) | F.3405* | Obs; E-426; Corr res | 0.42-0.5 | 12.5-14.5 | 1.0 max | 0.15 max | 1.0 max | 0.04 max | 0.03 max | 1.0 max | bal Fe | | | | |
| UNE 36016/1(89) | F.3423* | Obs; E-427; Corr res | 0.45-0.49 | 16.0-17.2 | 1.0 max | 1.0-1.3 | 1.0 max | 0.04 max | 0.03 max | 1.0 max | bal Fe | | | | |

Stainless Steel, Ferritic, S40910

USA

| | UNS S40910 | with Ti | 0.030 max | 10.5-11.7 | 1.00 max | | 0.50 max | 0.040 max | 0.010 max | 1.00 max | N <=0.030; Ti 6x(C+N) min 0.50 max; bal Fe | | | | |
| ASTM A240/A240M(98) | S40910 | with Ti | 0.030 max | 10.5-11.7 | 1.00 max | | 0.50 max | 0.040 max | 0.020 max | 1.00 max | N <=0.030; Nb <=0.17; 6x(C+N)<=Ti<=0.50; bal Fe | 380 | 170 | 20 | 179 HB |

Stainless Steel, Ferritic, S40920

USA

| | UNS S40920 | with Ti | 0.030 max | 10.5-11.7 | 1.00 max | | 0.50 max | 0.040 max | 0.010 max | 1.00 max | N <=0.030; Ti 0.15-0.50; 8x(C+N) min; bal Fe | | | | |
| ASTM A240/A240M(98) | S40920 | with Ti | 0.030 max | 10.5-11.7 | 1.00 max | | 0.50 max | 0.040 max | 0.020 max | 1.00 max | N <=0.030; Ti 0.15-0.50; 0.15+ 8x(C+N)<=Ti<=0.50; bal Fe | 380 | 170 | 20 | 179 HB |

Specification	Designation	Notes	C	Cr	Mn	Mo	Ni	P	S	Si	Other	UTS	YS	El	Hard
Stainless Steel, Ferritic, S40930															
USA															
	UNS S40930	with Ti and Nb	0.030 max	10.5-11.7	1.00 max		0.50 max	0.040 max	0.010 max	1.00 max	N <=0.030; Ti >=0.50; Ti 0.50 min; (Nb+Ti: 0.08+8x(C+N) min; 0.75 max; bal Fe				
ASTM A240/A240M(98)	S40930	with Ti and Nb	0.030 max	10.5-11.7	1.00 max		0.50 max	0.040 max	0.020 max	1.00 max	N <=0.030; Ti >=0.05; 0.08+8x(C+N)<=Nb+Ti<=0.75; bal Fe	380	170	20	179 HB
Stainless Steel, Ferritic, S40975															
Spain															
UNE 36017(85)	F.3159*	see X8CrTi18	0.1 max	16.5-18.5	1.5 max	?-0.15	1 max	0.04 max	0.03 max	1.5 max	Ti 0.4-0.7; bal Fe				
UNE 36017(85)	X8CrTi18	Heat res	0.1 max	16.5-18.5	1.5 max	?-0.15	1 max	0.04 max	0.03 max	1.5 max	Ti 0.4-0.7; bal Fe				
USA															
	UNS S40975	with Ni 409 Ni	0.030 max	10.50-11.75	1.00 max		0.50-1.00	0.040 max	0.030 max	1.00 max	N <=0.030; Ti 6x(C+N) min 0.75 max; bal Fe				
ASTM A240/A240M(98)	S40975	with Ni 409 Ni	0.030 max	10.5-11.7	1.00 max		0.50-1.00	0.040 max	0.030 max	1.00 max	N <=0.030; 6x(C+N)<=Ti<=0.75; bal Fe	415	275	20.0	197 HB
SAE J405(98)	S40975	w/Ni 409 Ni	0.030 max	10.50-11.75	1.00 max		0.50-1.00	0.040 max	0.030 max	1.00 max	N <=0.030; 6x(C+N)<=Ti<=0.75; bal Fe				
Stainless Steel, Ferritic, S40977															
USA															
	UNS S40977	with Ni	0.03 max	10.50-12.50	1.50 max		0.30-1.00	0.040 max	0.015 max	1.00 max	N <=0.030; bal Fe				
Stainless Steel, Ferritic, S41003															
USA															
	UNS S41003	Ferritic-Martensitic	0.03 max	10.5-12.5	1.50 max		1.50 max	0.040 max	0.030 max	1.00 max	N <=0.030; Ti 0.002-0.05; bal Fe				
ASTM A240/A240M(98)	S41003	Ferr-Martensitic	0.03 max	10.5-12.5	1.50 max		1.50 max	0.040 max	0.030 max	1.00 max	N <=0.030; Ti 0.002-0.05; bal Fe	455	275	18	223 HB
SAE J405(98)	S41003	Ferr-Mart	0.03 max	10.50-12.50	1.50 max		1.50 max	0.040 max	0.030 max	1.00 max	N <=0.030; Ti 0.002-0.05; bal Fe				
Stainless Steel, Ferritic, S43940															
USA															
	UNS S43940	with Ti and Nb	0.03 max	17.5-18.5	1.00 max			0.040 max	0.015 max	1.00 max	Ti 0.10-0.60; Nb=3xC+0.30 min; bal Fe				
Stainless Steel, Ferritic, S44100															
USA															
	UNS S44100	Nb-Ti stabilized	0.03 max	17.5-19.5	1.00 max		1.00 max	0.04 max	0.03 max	1.00 max	N <=0.03; Ti 0.10-0.50; Nb 0.3+(9xC) min-0.9 max; bal Fe				
Stainless Steel, Ferritic, S46800															
USA															
	UNS S46800	with Ti and Nb	0.030 max	18.00-20.00	1.00 max		0.50 max	0.040 max	0.030 max	1.00 max	N <=0.030; Nb 0.10-0.60; Ti 0.07-0.30; (Ti+Nb): [0.20+4(C+N)] min, 0.80 max; bal Fe				
ASTM A240/A240M(98)	S46800	with Ti and Nb	0.030 max	18.00-20.00	1.00 max		0.50 max	0.040 max	0.030 max	1.00 max	N <=0.030; Nb 0.10-0.60; Ti 0.07-0.30; 0.20+4(C+N)<=Ti+Nb <=0.80; bal Fe	415	205	22	
SAE J405(98)	S46800	w/Ti and Nb	0.030 max	18.00-20.00	1.00 max		0.50 max	0.040 max	0.030 max	1.00 max	N <=0.030; Nb 0.10-0.60; Ti 0.07-0.30; 0.20+4(C+N)<=Nb+Ti <=0.80; bal Fe				

UNS numbers and US grades are provided as a means of cross referencing chemically similar alloys. Exchangability is only possible after independent examination of specifications. Tensile properties are minimum or typical as specified. UTS and YS as MPa. El as %. See Appendix for list of abbreviations used in Notes. * indicates obsolete material.

Specification	Designation	Notes	C	Cr	Mn	Mo	Ni	P	S	Si	Other	UTS	YS	El	Hard

Stainless Steel, Ferritic, XM-27

China

Specification	Designation	Notes	C	Cr	Mn	Mo	Ni	P	S	Si	Other	UTS	YS	El	Hard
GB 1220(92)	00Cr27Mo	Bar Ann	0.010 max	25.00-27.00	0.40 max	0.75-1.50	0.50 max	0.030 max	0.020 max	0.40 max	N <=0.015; Ni+Cu 0.50; bal Fe	410	245	20	
GB 3280(92)	00Cr27Mo	Sh Plt, CR Ann	0.010 max	25.00-27.00	0.40 max	0.75-1.50	0.50 max	0.030 max	0.020 max	0.40 max	N <=0.015; Ni+Cu 0.50; bal Fe	410	245	20	
GB 4237(92)	00Cr27Mo	Sh Plt, HR Ann	0.010 max	25.00-27.00	0.40 max	0.75-1.50	0.50 max	0.030 max	0.020 max	0.40 max	N <=0.015; Ni+Cu 0.50; bal Fe	410	245	20	
GB 4239(91)	00Cr27Mo	Sh Strp, CR Ann	0.010 max	25.00-27.00	0.40 max	0.75-1.50	0.50 max	0.030 max	0.020 max	0.40 max	N <=0.015; Ni+Cu 0.50; bal Fe	410	245	20	

France

Specification	Designation	Notes	C	Cr	Mn	Mo	Ni	P	S	Si	Other	UTS	YS	El	Hard
AFNOR NFA35584	Z01CD26.01	Sh Strp Plt Bar	0.02 max	25-28	0.4 max	0.75-1.5		0.02 max	0.02 max	0.4 max	Ni+Cu<=0.50; bal Fe				
AFNOR NFA35584	Z61CDNb26.01	Sh Strp Plt Bar	0.02 max	25-28	1 max	0.75-1.5		0.04 max	0.03 max	1 max	Nb >=0.3; Ni+Cu<=0.50; bal Fe				

Germany

Specification	Designation	Notes	C	Cr	Mn	Mo	Ni	P	S	Si	Other	UTS	YS	El	Hard
DIN	WNr 1.4131		0.010 max	25.0-27.5	0.40 max	0.75-1.50	0.50 max	0.020 max	0.020 max	0.40 max	N <=0.015; Ni+Cu<=0.50; bal Fe				
DIN	X1CrMo26 1		0.010 max	25.0-27.5	0.40 max	0.75-1.50	0.50 max	0.020 max	0.020 max	0.40 max	N <=0.015; Ni+Cu<=0.50; bal Fe				

Japan

Specification	Designation	Notes	C	Cr	Mn	Mo	Ni	P	S	Si	Other	UTS	YS	El	Hard
JIS G4303(91)	SUSXM27	Bar Ann 75mm diam	0.010 max	25.00-27.50	0.40 max	0.75-1.50	0.50 max	0.030 max	0.020 max	0.40 max	Cu <=0.20; N <=0.015; Ni+Cu<=0.50; bal Fe	410	245	20	219 HB max
JIS G4304(91)	SUSXM27	Plt Sh, HR	0.010 max	25.00-27.50	0.40 max	0.75-1.50	0.50 max	0.030 max	0.020 max	0.40 max	Cu <=0.20; N <=0.015; Ni+Cu<=0.50; bal Fe				
JIS G4305(91)	SUSXM27	Plt Sh, CR	0.010 max	25.00-27.50	0.40 max	0.75-1.50		0.030 max	0.020 max	0.40 max	N <=0.015; bal Fe	410	245	22	192 HB max
JIS G4306	SUSXM27	Strp CR	0.01 max	25-27.5	0.4 max	0.75-1.5		0.03 max	0.02 max	0.4 max	N <=0.015; bal Fe				
JIS G4306	SUSXM27	Strp HR	0.01 max	25-27.5	0.4 max	0.75-1.5		0.03 max	0.02 max	0.4 max	N <=0.015; bal Fe				

USA

Specification	Designation	Notes	C	Cr	Mn	Mo	Ni	P	S	Si	Other	UTS	YS	El	Hard
	UNS S44625*	Obs; Low Interstital	0.01 max	25.00-27.50	0.40 max	0.75-1.50	0.50 max	0.020 max	0.020 max	0.40 max	Cu <=0.20; N <=0.015; Ni+Cu 0.50 max; bal Fe				
	UNS S44627	Low Interstital w/Nb	0.010 max	25.0-27.0	0.40 max	0.75-1.50	0.50 max	0.020 max	0.020 max	0.40 max	Cu <=0.020; N <=0.015; Nb 0.05-0.20; bal Fe				
ASME SA240	XM-27	Refer to ASTM A240(95)													
ASME SA268	XM-27	Refer to ASTM A268(94)													
ASME SA479	XM-27	Refer to ASTM A479/A479M(95)													
ASME SA731	XM-27	Refer to ASTM A731/A731M(91); Tube													
ASTM A176(97)	XM-27*	Sh Strp Plt	0.01 max	25-27	0.4 max	0.75-1.5	0.5 max	0.02 max	0.02 max	0.4 max	Cu <=0.02; N <=0.015; Nb 0.05-0.2; bal Fe				
ASTM A182/A182M(98)	FXM-27Cb	Frg/roll pipe flange valve	0.010 max	25.0-27.5	0.40 max	0.75-1.50	0.50 max	0.020 max	0.020 max	0.40 max	Cu <=0.20; N <=0.015; Nb 0.05-0.20; Ni+Cu<=0.50; bal Fe	415	240	20	190 HB max
ASTM A240/A240M(98)	S44627	Low Interstital w/Nb	0.010 max	25.0-27.0	0.40 max	0.75-1.50	0.50 max	0.020 max	0.020 max	0.40 max	Cu <=0.020; N <=0.015; Nb 0.05-0.20; Ni+Cu<=0.50; bal Fe	450	275	22.0	187 HB
ASTM A268	XM-27	Tube	0.01 max	25-27	0.4 max	0.75-1.5	0.5 max	0.02 max	0.02 max	0.4 max	Cu <=0.02; N <=0.015; Nb 0.05-0.2; bal Fe				
ASTM A276(98)	XM-27	Bar	0.01 max	25-27	0.4 max	0.75-1.5	0.5 max	0.02 max	0.02 max	0.4 max	Cu <=0.02; N <=0.015; Nb 0.05-0.2; bal Fe				
ASTM A314	XM-27	Bolt	0.01 max	25-27	0.4 max	0.75-1.5	0.5 max	0.02 max	0.02 max	0.4 max	Cu <=0.02; N <=0.015; Nb 0.05-0.2; bal Fe				
ASTM A479	XM-27	Sh Strp Plt Bar	0.01 max	25-27	0.4 max	0.75-1.5	0.5 max	0.02 max	0.02 max	0.4 max	Cu <=0.02; N <=0.015; Nb 0.05-0.2; bal Fe				

Specification	Designation	Notes	C	Cr	Mn	Mo	Ni	P	S	Si	Other	UTS	YS	El	Hard

Stainless Steel, Ferritic, XM-27 (Continued from previous page)

USA

Specification	Designation	Notes	C	Cr	Mn	Mo	Ni	P	S	Si	Other	UTS	YS	El	Hard
ASTM A731	XM-27	Tube	0.01 max	25-27	0.4 max	0.75-1.5	0.5 max	0.02 max	0.02 max	0.4 max	Cu <=0.02; N <=0.015; Nb 0.05-0.2; bal Fe				
ASTM A803/A803M(96)	TPXM-27	Weld Feedwater Heater Tub, Low interstitial w/ Cb	0.01 max	25.0-27.5	0.40 max	0.75-1.50	0.5 max	0.02 max	0.02 max	0.40 max	Cu <=0.2; N <=0.015; Nb 0.05-0.02; bal Fe	450	275	20	241 HB max
ASTM A815/A815M(98)	WP27	Pipe fittings	0.010 max	25.0-27.5	0.75 max	0.75-1.50	0.50 max	0.020 max	0.020 max	0.40 max	Cu <=0.20; N <=0.015; Nb 0.05-0.20; bal Fe	450-620	275	20.0	190 HB
SAE J405(98)	S44627	Low Interstital w/Nb	0.010 max	25.00-27.00	0.40 max	0.75-1.50	0.50 max	0.020 max	0.020 max	0.40 max	Cu <=0.20; N <=0.015; Nb 0.05-0.20; bal Fe				

Stainless Steel, Ferritic, XM-33

USA

Specification	Designation	Notes	C	Cr	Mn	Mo	Ni	P	S	Si	Other	UTS	YS	El	Hard
	UNS S44626	High-Cr+Mo 26-1	0.06 max	25.00-27.00	0.75 max	0.75-1.50	0.50 max	0.040 max	0.020 max	0.75 max	Cu <=0.20; N <=0.04; Ti 0.20-1.00; Ti 7x(C+N) min; bal Fe				
ASTM A240/A240M(98)	S44626	High-Cr+Mo 26-1	0.06 max	25.0-27.0	0.75 max	0.75-1.50	0.50 max	0.040 max	0.020 max	0.75 max	Cu <=0.20; N <=0.04; Ti 0.20-1.00; 7x(C+N)<=Ti; bal Fe	470	310	20.0	217 HB
ASTM A803/A803M(96)	TPXM-33	Weld Feedwater Heater Tub, High-Cr+Mo 26+1	0.06 max	25.0-27.0	0.75 max	0.75-1.00	0.50 max	0.040 max	0.020 max	0.75 max	Cu <=0.20; N <=0.040; Ti=7x(C+N)>=0.20<=1.00; bal Fe	470	310	20	241 HB max
ASTM A815/A815M(98)	WP33	Pipe fittings	0.06 max	25.0-27.0	0.75 max	0.75-1.50	0.50 max	0.040 max	0.020 max	0.75 max	Cu <=0.20; N <=0.040; Ti 0.20-1.00; Ti>=(7x(C+N)), bal Fe	470-640	310	20.0	241 HB
SAE J405(98)	S44626	High-Cr+Mo 26-1	0.060 max	25.00-27.00	0.75 max	0.75-1.50	0.50 max	0.040 max	0.020 max	0.75 max	Cu <=0.20; N <=0.04; Ti 0.20-1.00; 7x(C+N)<=Ti; bal Fe				

Stainless Steel, Ferritic, XM-34

Europe

Specification	Designation	Notes	C	Cr	Mn	Mo	Ni	P	S	Si	Other	UTS	YS	El	Hard
EN 10088/3(95)	1.4105	Bar Rod Shp, Corr res; t<=100mm, Ann	0.08 max	16.00-18.00	1.50 max	0.20-0.60		0.040 max	0.15-0.35	1.50 max	bal Fe	430-630	250	20	200 HB max
EN 10088/3(95)	X6CrMoS17	Bar Rod Shp, Corr res; t<=100mm, Ann	0.08 max	16.00-18.00	1.50 max	0.20-0.60		0.040 max	0.15-0.35	1.50 max	bal Fe	430-630	250	20	200 HB max

Spain

Specification	Designation	Notes	C	Cr	Mn	Mo	Ni	P	S	Si	Other	UTS	YS	El	Hard
UNE 36016/1(89)	F.3114*	Obs; E-430Ti; Corr res	0.08 max	16.0-18.0	1.5 max	0.2-0.6	1.0 max	0.06 max	0.15-0.35	1.0 max	bal Fe				

USA

Specification	Designation	Notes	C	Cr	Mn	Mo	Ni	P	S	Si	Other	UTS	YS	El	Hard
	UNS S18200	Cr-Mo Free Machining	0.08 max	17.50-19.50	1.25-2.50	1.50-2.50		0.04 max	0.15 min	1.00 max	bal Fe				
ASTM A581/A581M(95)	XM-34	Wir rod, Ann	0.08 max	17.5-19.5	2.5 max	1.50-2.50		0.04 max	0.15 min	1.00 max	bal Fe	585-860			
ASTM A582/A582M(95)	XM-34	Bar, HF, CF, Ann	0.08 max	17.50-19.50	2.50 max	1.50-2.50		0.04 max	0.15 min	1.00 max	bal Fe				285 HB max

Specification	Designation	Notes	C	Cr	Mn	Mo	Ni	P	S	Si	Other	UTS	YS	El	Hard

Stainless Steel, Martensitic, 403

Bulgaria

Specification	Designation	Notes	C	Cr	Mn	Mo	Ni	P	S	Si	Other	UTS	YS	El	Hard
BDS 6738(72)	0Ch13	Corr res	0.08 max	12.0-14.0	0.80 max	0.30 max	0.60 max	0.035 max	0.025 max	0.80 max	bal Fe				
Canada															
CSA G110.3	403	Bar Bil, Ann	0.15 max	11.5-13	1 max			0.04 max	0.03 max	0.5 max	bal Fe				
CSA G110.5	403	Sh Strp Plt	0.15 max	11.5-13	1 max		0.6 max	0.04 max	0.03 max	0.5 max	bal Fe				
China															
GB 1220(92)	1Cr12	Bar Q/T	0.15 max	11.50-13.00	1.00 max		0.60 max	0.035 max	0.030 max	0.50 max	bal Fe	590	390	25	
GB 3280(92)	1Cr12	Sh Plt, CR Ann	0.15 max	11.50-13.00	1.00 max		0.60 max	0.035 max	0.030 max	0.50 max	bal Fe	440	205	20	
GB 4237(92)	1Cr12	Sh Plt, HR Ann	0.15 max	11.50-13.00	1.00 max		0.60 max	0.035 max	0.030 max	0.50 max	bal Fe	440	205	20	
GB 4238(92)	1Cr12	Plt, HR Ann	0.15 max	11.50-13.00	1.00 max		0.60 max	0.035 max	0.030 max	0.50 max	bal Fe	440	205	20	
GB 4239(91)	1Cr12	Sh Strp, CR Ann	0.15 max	11.50-13.00	1.00 max		0.60 max	0.035 max	0.030 max	0.50 max	bal Fe	440	205	20	
Czech Republic															
CSN 417021	17021	Corr res; Heat res	0.09-0.15	12.0-14.0	0.9 max			0.04 max	0.03 max	0.7 max	bal Fe				
Europe															
EN 10088/2(95)	X6Cr17	Strp; Corr res; CR t<=6mm, Ann	0.08 max	16.00-18.00	1.00 max			0.040 max	0.015 max	1.00 max	bal Fe	450-600	260	20	
EN 10088/3(95)	1.4016	Bar Rod Shp, Corr res; t<=100mm, Ann	0.08 max	16.00-18.00	1.00 max			0.040 max	0.030 max	1.00 max	bal Fe	400-630	240	20	200 HB max
EN 10088/3(95)	X6Cr17	Bar Rod Shp, Corr res; t<=100mm, Ann	0.08 max	16.00-18.00	1.00 max			0.040 max	0.030 max	1.00 max	bal Fe	400-630	240	20	200 HB max
India															
IS 1570/7(92)	12Cr12MoH	0-75mm; Bar Frg	0.08-0.16	11.5-13	0.4-1	0.4-0.8	1 max	0.04 max	0.035 max	0.6 max	Co <=0.1; Cu <=0.3; Pb <=0.15; W <=0.1; bal Fe	680-880	490	20	192 HB
IS 1570/7(92)	12Cr12MoVH	0-150mm; Bar Frg	0.08-0.16	11.5-13	0.4-1	0.4-0.8	1 max	0.04 max	0.035 max	0.6 max	Co <=0.1; Cu <=0.3; Pb <=0.15; V 0.1-0.3; W <=0.1; bal Fe	770-930	585	15	
IS 1570/7(92)	20	0-75mm; Bar Frg	0.08-0.16	11.5-13	0.4-1	0.4-0.8	1 max	0.04 max	0.035 max	0.6 max	Co <=0.1; Cu <=0.3; Pb <=0.15; W <=0.1; bal Fe	680-880	490	20	192 HB
IS 1570/7(92)	21	0-150mm; Bar Frg	0.08-0.16	11.5-13	0.4-1	0.4-0.8	1 max	0.04 max	0.035 max	0.6 max	Co <=0.1; Cu <=0.3; Pb <=0.15; V 0.1-0.3; W <=0.1; bal Fe	770-930	585	15	
International															
ISO 683-13(74)	3*	Ann, Q/A, Tmp	0.09-0.15	11.5-13.5	1 max			0.04 max	0.03 max	1 max	bal Fe				
Japan															
JIS G4303(91)	SUS403	Bar Q/A Tmp 75mm diam	0.15 max	11.50-13.00	1.00 max		0.60 max	0.040 max	0.030 max	0.50 max	bal Fe	590	390	25	170 HB
JIS G4304(91)	SUS403	Sh, HR Ann	0.15 max	11.50-13.00	1.00 max		0.60 max	0.040 max	0.030 max	0.50 max	bal Fe	440	205	20	93 HRB max
JIS G4305(91)	SUS403	Plt Ann	0.15 max	11.50-13.00	1.00 max		0.60 max	0.040 max	0.030 max	0.50 max	bal Fe	440	205	20	201 HB max
JIS G4308	SUS403	Wir rod	0.15 max	11.50-13.00	1.00 max		0.60 max	0.040 max	0.030 max	1.00 max	Pb 0.05-0.30, bal Fe				
JIS G4309	SUS403	Wir	0.15 max	11.50-13.00	1.00 max		0.60 max	0.040 max	0.030 max	0.50 max	bal Fe				
JIS G4315	SUS403	Wir rod	0.15 max	11.50-13.00	1.00 max		0.60 max	0.040 max	0.030 max	0.50 max	bal Fe				
Mexico															
DGN B-83	403*	Obs; Bar, HR, CR	0.15 max	11.5-13	1 max			0.04 max	0.03 max	0.5 max	bal Fe				
NMX-B-171(91)	MT403	mech tub	0.15 max	11.5-13.0	1.00 max	0.60 max	0.50 max	0.040 max	0.30 max	0.50 max	bal Fe				
Russia															
GOST 5632(61)	0KH13	Corr res; Heat res; High-temp constr	0.08 max	11.0-13.0	0.6 max	0.15 max	0.4 max	0.03 max	0.025 max	0.6 max	Cu <=0.3; Ti <=0.05; bal Fe				
GOST 5632(72)	08Ch13	Corr res; High-temp constr	0.08 max	12.0-14.0	0.8 max	0.15 max	0.6 max	0.03 max	0.025 max	0.8 max	Cu <=0.3; Ti <=0.2; bal Fe				

UNS numbers and US grades are provided as a means of cross referencing chemically similar alloys. Exchangability is only possible after independent examination of specifications. Tensile properties are minimum or typical as specified. UTS and YS as MPa. El as %. See Appendix for list of abbreviations used in Notes. * indicates obsolete material.

Specification	Designation	Notes	C	Cr	Mn	Mo	Ni	P	S	Si	Other	UTS	YS	El	Hard

Stainless Steel, Martensitic, 403 (Continued from previous page)

Spain

Specification	Designation	Notes	C	Cr	Mn	Mo	Ni	P	S	Si	Other	UTS	YS	El	Hard
UNE 36016(75)	X12Cr13	Sh Strp Plt Bar Wir Frg; Corr res	0.09-0.15	11.5-14.0	1.0 max	0.15 max	1.0 max	0.04 max	0.03 max	1.0 max	bal Fe				
UNE 36016(75)	X6Cr13	Corr res	0.08 max	11.5-14.0	1.0 max	0.15 max	0.5 max	0.04 max	0.03 max	1.0 max	bal Fe				
UNE 36016/1(89)	E-403	Corr res	0.08 max	12.0-14.0	1.0 max	0.15 max	1.0 max	0.04 max	0.03 max	1.0 max	bal Fe				
UNE 36016/1(89)	F.3110*	Obs; E-403; Corr res	0.08 max	12.0-14.0	1.0 max	0.15 max	1.0 max	0.04 max	0.03 max	1.0 max	bal Fe				
UNE 36016/1(89)	F.3412*	Obs; EN10088-3(96); X12CrMoS13	0.08-0.15	12.0-14.0	1.5 max	0.2-0.6	1.0 max	0.06 max	0.15-0.35	1.0 max	bal Fe				

Sweden

Specification	Designation	Notes	C	Cr	Mn	Mo	Ni	P	S	Si	Other	UTS	YS	El	Hard
SS 142301	2301	Corr res	0.08 max	12-13.5	1 max		1 max	0.04 max	0.03 max	1 max	bal Fe				

UK

Specification	Designation	Notes	C	Cr	Mn	Mo	Ni	P	S	Si	Other	UTS	YS	El	Hard
BS 1449/2(83)	410S21	Corr/Heat res; t<=100mm	0.09 max	11.5-13.5	1.00 max		1.00 max	0.040 max	0.030 max	1.00 max	bal Fe				0-190 HB
BS 970/1(83)	403S17	Wrought Plt Corr/Heat res; t<=150mm	0.08 max	12.0-14.0	1.00 max		0.50 max	0.040 max	0.030 max	1.0 max	bal Fe	420	280	20	170 HB max

USA

Specification	Designation	Notes	C	Cr	Mn	Mo	Ni	P	S	Si	Other	UTS	YS	El	Hard
	AISI 403	Tube	0.15 max	11.50-13.00	1.00 max			0.040 max	0.030 max	0.50 max	bal Fe				
	UNS S40300	Hardenable by HT	0.15 max	11.50-13.00	1.00 max			0.040 max	0.030 max	0.50 max	bal Fe				
ASTM A176(97)	403	Sh Strp Plt	0.15 max	11.50-13.00	1.00 max		0.60 max	0.040 max	0.030 max	0.50 max	bal Fe	485	205	25.0	217 HB
ASTM A276(98)	403	Bar, HT	0.15 max	11.5-13	1 max			0.04 max	0.03 max	0.5 max	bal Fe				
ASTM A276(98)	403	Bar, Ann	0.15 max	11.5-13	1 max			0.04 max	0.03 max	0.5 max	bal Fe				
ASTM A314	403	Bil	0.15 max	11.5-13	1 max			0.04 max	0.03 max	0.5 max	bal Fe				
ASTM A473	403	Frg, Heat res ann	0.15 max	11.50-13.00	1.00 max			0.040 max	0.030 max	0.50 max	bal Fe	485	275	20	223 HB
ASTM A479	403	Bar, Ann	0.15 max	11.5-13	1 max			0.04 max	0.03 max	0.5 max	bal Fe				
ASTM A479	403	Bar, HT3	0.15 max	11.5-13	1 max			0.04 max	0.03 max	0.5 max	bal Fe				
ASTM A479	403	Bar, HT2	0.15 max	11.5-13	1 max			0.04 max	0.03 max	0.5 max	bal Fe				
ASTM A511(96)	MT403	Smls mech tub, Ann	0.15 max	11.5-13.0	1.00 max	0.60 max	0.50 max	0.040 max	0.030 max	0.5 max	bal Fe	414	207	20	207 HB
ASTM A580/A580M(98)	403	Wir, Ann	0.15 max	11.5-13.0	1.00 max			0.040 max	0.030 max	0.50 max	bal Fe	485	275	20	35 HCR
ASTM A982(98)	Grade A Class 1	Frg for turbine airfoils	0.15 max	11.5-13.0	1.0 max	0.5 max	0.75 max	0.018 max	0.015 max	0.5 max	bal Fe	690	483	20	
ASTM A982(98)	Grade A Class 2	Frg for turbine airfoils	0.15 max	11.5-13.0	1.0 max	0.5 max	0.75 max	0.018 max	0.015 max	0.5 max	bal Fe	758	552	18	
ASTM A982(98)	Grade B Class 1	Frg for turbine airfoils	0.10-0.15	11.5-13.0	0.25-0.80	0.5 max	0.75 max	0.018 max	0.015 max	0.5 max	Al <=0.025; Cu <=0.15; N <=0.08; Nb <=0.20; Ti <=0.05; W <=0.10; Sn<=0.05; bal Fe	758	621	18	
ASTM A982(98)	Grade E Class 1	Frg for turbine airfoils	0.13-0.18	11.5-13.0	0.4-0.6	0.20	0.5 max	0.030 max	0.030 max	0.5 max	Al <=0.05; Nb 0.15-0.45; bal Fe	792	517	15	
ASTM A982(98)	Grade E Class 2	Frg for turbine airfoils	0.13-0.18	11.5-13.0	0.4-0.6	0.20 max	0.5 max	0.030 max	0.030 max	0.5 max	Al <=0.05; Nb 0.15-0.45; bal Fe	758	552	18	
DoD-F-24669/7(86)	403	HT; Supersedes MIL-S-861	0.15 max	11.5-13.0	1.00 max		0.50 max	0.040 max	0.030 max	0.50 max	bal Fe	690	480	20	201-241 HB
FED QQ-S-763F(96)	403*	Obs; Bar Wir Shp Frg; Corr res	0.15 max	11.5-13	1 max			0.04 max	0.03 max	0.5 max	bal Fe				
MIL-S-861	403*	Obs	0.15 max	11.5-13	1 max			0.04 max	0.03 max	0.5 max	bal Fe				
MIL-S-862B	403*	Obs	0.15 max	11.5-13	1 max			0.04 max	0.03 max	0.5 max	bal Fe				

UNS numbers and US grades are provided as a means of cross referencing chemically similar alloys. Exchangability is only possible after independent examination of specifications. Tensile properties are minimum or typical as specified. UTS and YS as MPa. El as %. See Appendix for list of abbreviations used in Notes. * indicates obsolete material.

Specification	Designation	Notes	C	Cr	Mn	Mo	Ni	P	S	Si	Other	UTS	YS	El	Hard

Stainless Steel, Martensitic, 410

Australia

Specification	Designation	Notes	C	Cr	Mn	Mo	Ni	P	S	Si	Other	UTS	YS	El	Hard
AS 1449(94)	410	Sh Strp Plt	0.15 max	11.50-13.50	1.00 max		0.75 max	0.040 max	0.030 max	1.00 max	bal Fe				
AS 2837(86)	410	Bar, Semi-finished product	0.09-0.15	11.5-13.5	1.0 max		1.00 max	0.040 max	0.030 max	1.0 max	bal Fe				
AS G18	En56A*	Obs; see AS 1444	0.12 max	12-14	1 max		1 max	0.05 max	0.05 max	1 max	bal Fe				

Bulgaria

Specification	Designation	Notes	C	Cr	Mn	Mo	Ni	P	S	Si	Other	UTS	YS	El	Hard
BDS 6738(72)	1Ch13	Sh Strp Plt Bar Wir	0.09-0.15	12.0-14.0	0.80 max	0.30 max	0.60 max	0.035 max	0.025 max	0.80 max	bal Fe				

Canada

Specification	Designation	Notes	C	Cr	Mn	Mo	Ni	P	S	Si	Other	UTS	YS	El	Hard
CSA G110.3	410	Bar Bil	0.15 max	11.5-13.5	1 max			0.04 max	0.03 max	1 max	bal Fe				
CSA G110.5	410	Sh Strp Plt, CR, Ann, 1mm diam	0.15 max	11.5-13.5	1 max		0.75 max	0.04 max	0.03 max	1 max	bal Fe				
CSA G110.9	410	Sh Strp Plt	0.15 max	11.5-13.5	1 max		0.75 max	0.04 max	0.03 max	1 max	bal Fe				

China

Specification	Designation	Notes	C	Cr	Mn	Mo	Ni	P	S	Si	Other	UTS	YS	El	Hard
GB 1220(92)	1Cr13	Bar Q/T	0.15 max	11.50-13.50	1.00 max		0.60 max	0.035 max	0.030 max	1.00 max	bal Fe	540	345	25	
GB 1221(92)	1Cr13	Bar Q/T	0.15 max	11.50-13.50	1.00 max		0.60 max	0.035 max	0.030 max	1.00 max	Cu <=0.30; bal Fe	540	345	25	
GB 12770(91)	1Cr13	Weld tube; Ann	0.15 max	11.50-13.50	1.00 max		0.60 max	0.035 max	0.030 max	1.00 max	bal Fe	410	210	20	
GB 3280(92)	1Cr13	Sh Plt, CR Ann	0.15 max	11.50-13.50	1.00 max		0.60 max	0.035 max	0.030 max	1.00 max	bal Fe	440	205	20	
GB 4232(93)	ML1Cr13	Wir HT 1-3mm diam	0.15 max	11.50-13.50	1.00 max		0.60 max	0.035 max	0.030 max	1.00 max	bal Fe	440-460		15	
GB 4237(92)	1Cr13	Sh Plt, HR Ann	0.15 max	11.50-13.50	1.00 max		0.60 max	0.035 max	0.030 max	1.00 max	bal Fe	440	205	20	
GB 4238(92)	1Cr13	Plt, HR Ann	0.15 max	11.50-13.50	1.00 max		0.60 max	0.035 max	0.030 max	1.00 max	bal Fe	440	205	20	
GB 4239(91)	1Cr13	Sh Strp, CR Ann	0.15 max	11.50-13.50	1.00 max		0.60 max	0.035 max	0.030 max	1.00 max	bal Fe	440	205	20	
GB 4240(93)	1Cr13	Wir HT 0.5-6mm diam	0.15 max	11.50-13.50	1.00 max		0.60 max	0.035 max	0.030 max	1.00 max	bal Fe	540-790			
GB 8732(88)	1Cr13	Bar Q/T	0.15 max	11.50-13.50	1.00 max		0.60 max	0.030 max	0.030 max	1.00 max	bal Fe	540	345	25	
GB/T 14975(94)	1Cr13	Smls tube Ann	0.15 max	11.50-13.50	1.00 max		0.60 max	0.035 max	0.030 max	1.00 max	bal Fe	410	205	20	

Europe

Specification	Designation	Notes	C	Cr	Mn	Mo	Ni	P	S	Si	Other	UTS	YS	El	Hard
EN 10088/2(95)	1.4006	Sh Plt Strp, Corr res; t<=6mm Ann	0.08-0.15	11.50-13.50	1.50 max		0.75 max	0.040 max	0.015 max	1.00 max	bal Fe	600 max		20	200 HB max
EN 10088/2(95)	1.4006	Sh Plt Strp, Corr res; t<=75mm, Q/T	0.08-0.15	11.50-13.50	1.60 max		0.75 max	0.040 max	0.015 max	1.00 max	bal Fe	650-850	450	12	
EN 10088/2(95)	X12Cr13	Sh Plt Strp, Corr res; t<=75mm, Q/T	0.08-0.15	11.50-13.50	1.50 max		0.75 max	0.040 max	0.015 max	1.00 max	bal Fe	650-850	450	12	
EN 10088/2(95)	X12Cr13	Sh Plt Strp, Corr res; t<=6mm Ann	0.08-0.15	11.50-13.50	1.60 max		0.75 max	0.040 max	0.015 max	1.00 max	bal Fe	600 max		20	200 HB max

France

Specification	Designation	Notes	C	Cr	Mn	Mo	Ni	P	S	Si	Other	UTS	YS	El	Hard
AFNOR NFA35572	Z12C13	Sh Strp Plt Bar Wir rod, Q/T	0.08-0.15	11.5-13.5	1 max			0.04 max	0.03 max	1 max	bal Fe				

Germany

Specification	Designation	Notes	C	Cr	Mn	Mo	Ni	P	S	Si	Other	UTS	YS	El	Hard
DIN 17440(96)	WNr 1.4006	Plt Sh, HR Strp Bar Wir Frg, HT	0.08-0.15	11.0-13.5	1.50 max		0.75 max	0.040 max	0.030 max	1.00 max	bal Fe	650-850	450	15	
DIN 17441(97)	WNr 1.4006	CR Strp Plt Sh	0.08-0.15	11.0-13.5	1.50 max		0.75 max	0.040 max	0.030 max	1.00 max	bal Fe				
DIN EN 10088(95)	GX12Cr13	Sht Plt Strp Bar Rod HT	0.08-0.15	11.0-13.5	1.50 max		0.75 max	0.040 max	0.030 max	1.00 max	bal Fe	650-850	450	15	
DIN EN 10088(95)	WNr 1.4006	Sh Plt Strp, HT	0.08-0.15	11.0-13.5	1.50 max		0.75 max	0.040 max	0.030 max	1.00 max	bal Fe	650-850	450	15	
DIN EN 10095(95)	WNr 1.4724	Ann	0.12 max	12.0-14.0	1.00 max			0.040 max	0.030 max	0.70-1.40	Al 0.70-1.20; bal Fe	450-650	250	15	
DIN EN 10095(95)	X10CrAl13	Ann	0.12 max	12.0-14.0	1.00 max			0.040 max	0.030 max	0.70-1.40	Al 0.70-1.20; bal Fe	450-650	250	15	

Hungary

Specification	Designation	Notes	C	Cr	Mn	Mo	Ni	P	S	Si	Other	UTS	YS	El	Hard
MSZ 4360(87)	KO2	Corr res; 0-100mm; soft ann	0.09-0.15	12-14	1 max		0.6 max	0.045 max	0.03 max	1 max	bal Fe	450-650	250	20L	200 HB max

UNS numbers and US grades are provided as a means of cross referencing chemically similar alloys. Exchangability is only possible after independent examination of specifications. Tensile properties are minimum or typical as specified. UTS and YS as MPa. El as %. See Appendix for list of abbreviations used in Notes. * indicates obsolete material.

Specification	Designation	Notes	C	Cr	Mn	Mo	Ni	P	S	Si	Other	UTS	YS	El	Hard
Stainless Steel, Martensitic, 410 (Continued from previous page)															
Hungary															
MSZ 4360(87)	KO2	Corr res; 0-100mm; Q/T	0.09-0.15	12-14	1 max		0.6 max	0.045 max	0.03 max	1 max	bal Fe	600-800	410	16L	
MSZ 4360(87)	X12Cr13	Corr res; 0-100mm; Q/T	0.09-0.15	12-14	1 max		0.6 max	0.045 max	0.03 max	1 max	bal Fe	600-800	410	16L	
MSZ 4360(87)	X12Cr13	Corr res; 0-100mm; soft ann	0.09-0.15	12-14	1 max		0.6 max	0.045 max	0.03 max	1 max	bal Fe	450-650	250	20L	200 HB max
India															
IS 6527	12Cr13	Sh Strp Plt Bar Wir Bil	0.09-0.15	11.5-14	1 max		1 max	0.04 max	0.03 max	1 max	bal Fe				
IS 6528	12Cr13	Sh Strp Plt Bar Wir Bil	0.09-0.15	11.5-14	1 max		1 max	0.04 max	0.03 max	1 max	bal Fe				
IS 6529	12Cr13	Rod	0.09-0.15	11.5-14	1 max		1 max	0.04 max	0.03 max	1 max	bal Fe				
IS 6603	12Cr13	Bar, Ann, 5-100mm diam	0.09-0.15	11.5-14	1 max		1 max	0.04 max	0.03 max	1 max	bal Fe				
IS 6691	12Cr13	Rod	0.09-0.15	11.5-14	1 max		1 max	0.04 max	0.03 max	1 max	bal Fe				
International															
COMECON PC4-70 3		Sh Strp Plt Bar Wir Bil	0.09-0.15	12-14	1 max			0.035 max	0.025 max	1 max	bal Fe				
ISO 683-13(74)	3*	Ann, Q/A, Tmp	0.09-0.15	11.5-13.5	1 max			0.04 max	0.03 max	1 max	bal Fe				
Italy															
UNI 6901(71)	X12Cr13	Corr res	0.09-0.15	11.5-14	1 max		1 max	0.04 max	0.03 max	1 max	bal Fe				
UNI 6904(71)	X12Cr13	Heat res	0.15 max	11.5-13.5	1 max		0.5 max	0.04 max	0.03 max	0.75 max	bal Fe				
UNI 6942(71)	X10Cr13	Corr res	0.08-0.14	11.5-13.5	1 max		1 max	0.03 max	0.025 max	0.8 max	bal Fe				
Japan															
JIS G3214(91)	SUSF410A	Frg press ves	0.15 max	11.50-13.50	1.00 max		0.50 max	0.040 max	0.030 max	1.00 max	bal Fe	480	275	16	143-187 HB
JIS G3214(91)	SUSF410B	Frg press ves	0.15 max	11.50-13.50	1.00 max		0.50 max	0.040 max	0.030 max	1.00 max	bal Fe	590	380	16	167-229 HB
JIS G3214(91)	SUSF410C	Frg press ves	0.15 max	11.50-13.50	1.00 max		0.50 max	0.040 max	0.030 max	1.00 max	bal Fe	760	585	14	217-302 HB
JIS G3214(91)	SUSF410D	Frg press ves	0.15 max	11.50-13.50	1.00 max		0.50 max	0.040 max	0.030 max	1.00 max	bal Fe	900	760	11	262-321 HB
JIS G4303(91)	SUS410	Bar Q/A Tmp 75mm diam	0.15 max	11.50-13.50	1.00 max		0.60 max	0.040 max	0.030 max	1.00 max	bal Fe	540	345	25	159 HB
JIS G4303(91)	SUS410F2	Bar Q/A Tmp 75mm diam	0.15 max	11.50-13.50	1.00 max		0.60 max	0.040 max	0.030 max	1.00 max	Pb 0.05-0.30; bal Fe	540	345	18	159 HB
JIS G4303(91)	SUS410J1	Bar Q/A Tmp 75mm diam	0.08-0.18	11.50-14.00	1.00 max	0.30-0.60	0.60 max	0.040 max	0.030 max	0.60 max	bal Fe	690	490	20	192 HB
JIS G4303(91)	SUS410L	Bar Ann 75mm diam	0.08 max	11.00-13.50	1.00 max		0.60 max	0.040 max	0.030 max	1.00 max	bal Fe	360	195	22	183 HB max
JIS G4304(91)	SUS410	Sh Plt, HR Ann	0.15 max	11.50-13.50	1.00 max		0.60 max	0.040 max	0.030 max	1.00 max	bal Fe	440	205	20	93 HRB max
JIS G4305(91)	SUS410	Sh Plt, CR Ann	0.15 max	11.50-13.50	1.00 max		0.60 max	0.040 max	0.030 max	1.00 max	bal Fe	440	205	20	201 HB max
JIS G4305(91)	SUS410L		0.030 max	11.00-13.50	1.00 max			0.040 max	0.030 max	1.00 max	bal Fe	360	195	22	
JIS G4306	SUS410	Strp HR Ann	0.15 max	11.5-13.5	1 max		0.6 max	0.04 max	0.03 max	1 max	bal Fe				
JIS G4307	SUS410	Strp CR Ann Q/T	0.15 max	11.5-13.5	1 max		0.6 max	0.04 max	0.03 max	1 max	bal Fe				
JIS G4308(98)	SUS410	Wir rod	0.15 max	11.50-13.00	1.00 max		0.60 max	0.040 max	0.030 max	1.00 max	bal Fe				
JIS G4309	SUS410	Wir	0.15 max	11.50-13.50	1.00 max		0.60 max	0.040 max	0.030 max	1.00 max	bal Fe				
JIS G4315	SUS410	Wir rod	0.15 max	11.50-13.50	1.00 max		0.60 max	0.040 max	0.030 max	1.00 max	bal Fe				
Mexico															
DGN B-83	410*	Obs; Bar, HR, CR	0.15 max	11.5-13.5	1 max			0.04 max	0.03 max	1 max	bal Fe				
NMX-B-171(91)	MT410	mech tub	0.15 max	11.5-13.5	1.00 max		0.50 max	0.040 max	0.040 max	1.00 max	bal Fe				

UNS numbers and US grades are provided as a means of cross referencing chemically similar alloys. Exchangability is only possible after independent examination of specifications. Tensile properties are minimum or typical as specified. UTS and YS as MPa. El as %. See Appendix for list of abbreviations used in Notes. * indicates obsolete material.

Specification	Designation	Notes	C	Cr	Mn	Mo	Ni	P	S	Si	Other	UTS	YS	El	Hard

Stainless Steel, Martensitic, 410 (Continued from previous page)

Poland

Specification	Designation	Notes	C	Cr	Mn	Mo	Ni	P	S	Si	Other	UTS	YS	El	Hard
PNH86020	1H13	Corr res	0.09-0.15	12-14	0.8 max		0.6 max	0.04 max	0.03 max	0.08 max	bal Fe				
PNH86020	1H13(1Cr13)	Sh Strp Plt Bar Wir Rod	0.09-0.15	12-14	0.8 max		0.6 max	0.04 max	0.03 max	0.8 max	bal Fe				

Romania

Specification	Designation	Notes	C	Cr	Mn	Mo	Ni	P	S	Si	Other	UTS	YS	El	Hard
STAS 3583(87)	12C130	Corr res; Heat res	0.09-0.15	12-14	0.6 max		0.6 max	0.035 max	0.03 max	0.6 max	Pb <=0.15; bal Fe				

Russia

Specification	Designation	Notes	C	Cr	Mn	Mo	Ni	P	S	Si	Other	UTS	YS	El	Hard
GOST	12Ch13	Multiple Forms Ann	0.09-0.15	12-14	0.8 max		0.6 max	0.03 max	0.025 max	0.8 max	bal Fe				
GOST 5632(61)	1KH13	Heat res; Corr res; High-temp constr	0.09-0.15	12.0-14.0	0.6 max	0.15 max	0.4 max	0.03 max	0.025 max	0.6 max	Cu <=0.3; Ti <=0.05; bal Fe				
GOST 5632(72)	12Ch13	Corr res; Heat res; High-temp constr	0.09-0.15	12.0-14.0	0.8 max	0.15 max	0.6 max	0.03 max	0.025 max	0.8 max	Cu <=0.3; Ti <=0.2; bal Fe				

Spain

Specification	Designation	Notes	C	Cr	Mn	Mo	Ni	P	S	Si	Other	UTS	YS	El	Hard
UNE 36016(75)	F.3401*	see X12Cr13	0.09-0.15	11.5-14.0	1.0 max	0.15 max	1.0 max	0.04 max	0.03 max	1.0 max	bal Fe				
UNE 36016(75)	X12Cr13	Sh Strp Plt Bar Wir Frg; Corr res	0.09-0.15	11.5-14.0	1.0 max	0.15 max	1.0 max	0.04 max	0.03 max	1.0 max	bal Fe				
UNE 36016/1(89)	E-410	Corr res	0.08-0.12	12.0-14.0	1.0 max	0.15 max	1.0 max	0.04 max	0.03 max	1.0 max	bal Fe				
UNE 36016/1(89)	F.3401*	Obs; E-410; X12Cr13	0.08-0.12	12.0-14.0	1.0 max	0.15 max	1.0 max	0.04 max	0.03 max	1.0 max	bal Fe				
UNE 36017(85)	F.3162*	see X10CrSi13	0.12 max	12-14	1 max	0.15 max	1 max	0.04 max	0.03 max	1.9-2.4	bal Fe				
UNE 36017(85)	X10CrSi13	Heat res	0.12 max	12-14	1 max	0.15 max	1 max	0.04 max	0.03 max	1.9-2.4	bal Fe				

Sweden

Specification	Designation	Notes	C	Cr	Mn	Mo	Ni	P	S	Si	Other	UTS	YS	El	Hard
SS 142302	2302	Corr res	0.09-0.15	12-14	1 max		1 max	0.04 max	0.03 max	1 max	bal Fe				

UK

Specification	Designation	Notes	C	Cr	Mn	Mo	Ni	P	S	Si	Other	UTS	YS	El	Hard
BS 1449/2(83)	410S21	Corr/Heat res; t<=100mm	0.09 max	11.5-13.5	1.00 max		1.00 max	0.040 max	0.030 max	1.00 max	bal Fe				0-190 HB
BS 1503(89)	410S21	Frg press ves; t<=999mm; Q/T	0.09-0.15	11.5-13.5	1.00 max	0.30 max	1.00 max	0.040 max	0.025 max	0.80 max	Cu <=0.30; W <=0.1; bal Fe	590-740	395	17	
BS 1506(90)	410S21-690	Bolt matl Pres/Corr res; t<=100mm; Q/T	0.09-0.15	11.5-13.5	1.00 max		1.00 max	0.040 max	0.030 max	1.00 max	bal Fe	690-840	540	15	197-248 HB
BS 1506(90)	410S21-720	Bolt matl Pres/Corr res; t<=100mm; Q/T	0.09-0.15	11.5-13.5	1.00 max		1.00 max	0.040 max	0.030 max	1.00 max	bal Fe	720-870	570	15	212-262 HB
BS 1506(90)	410S21-750	Bolt matl Pres/Corr res; t<=63mm; Q/T	0.09-0.15	11.5-13.5	1.00 max		1.00 max	0.040 max	0.030 max	1.00 max	bal Fe	750-900	580	14	215-262 HB
BS 1506(90)	410S21-760	Bolt matl Pres/Corr res; t<=100mm; Q/T	0.09-0.15	11.5-13.5	1.00 max		1.00 max	0.040 max	0.030 max	1.00 max	bal Fe	760-910	585	15	217-269 HB
BS 1506(90)	410S21-770	Bolt matl Pres/Corr res; t<=29mm; Q/T	0.09-0.15	11.5-13.5	1.00 max		1.00 max	0.040 max	0.030 max	1.00 max	bal Fe	770-930	590	12	223-277 HB
BS 1554(90)	410S21	Wir Corr/Heat res; diam<0.49mm; Ann; wasEn56(A)	0.09-0.15	11.5-13.5	1.00 max		1.00 max	0.040 max	0.030 max	1.00 max	W <=0.1; bal Fe	0-780			
BS 1554(90)	410S21	Wir Corr/Heat res; 1.5<diam<=13mm; Ann; was En56(A)	0.09-0.15	11.5-13.5	1.00 max		1.00 max	0.040 max	0.030 max	1.00 max	W <=0.1; bal Fe	0-700			
BS 970/1(96)	410S21	Wrought Corr/Heat res 63<t<=150mm; Tmp	0.09-0.15	11.5-13.5	1.00 max		1.00 max	0.040 max	0.030 max	1.00 max	bal Fe	550-700	370	20	152-207 HB
BS 970/1(96)	410S21	Wrought Corr/Heat res; t<=63mm; Tmp	0.09-0.15	11.5-13.5	1.00 max		1.00 max	0.040 max	0.030 max	1.00 max	bal Fe	700-850	525	15	201-255 HB
BS DTD	161A	Bar Wir	0.12 max	12-14	1 max		1 max	0.04 max	0.03 max	1 max	bal Fe				
BS DTD	97B	Tube	0.15 max	12-14	1 max		1 max	0.04 max	0.03 max	1 max	bal Fe				

USA

Specification	Designation	Notes	C	Cr	Mn	Mo	Ni	P	S	Si	Other	UTS	YS	El	Hard
	AISI 410	Tube	0.15 max	11.50-13.50	1.00 max			0.040 max	0.030 max	1.00 max	bal Fe				
	UNS S41000	Hardenable by HT	0.15 max	11.50-13.50	1.00 max			0.040 max	0.030 max	1.00 max	bal Fe				
AMS 5591H(92)		Corr/Heat res; Smls/Weld tub, Ann	0.15 max	11.50-13.50	1.00 max	0.60 max	0.75 max	0.040 max	0.030 max	1.00 max	Al <=0.05; Cu <=0.50; N <=0.08; Sn 0.05 max; bal Fe	689		25	

UNS numbers and US grades are provided as a means of cross referencing chemically similar alloys. Exchangability is only possible after independent examination of specifications. Tensile properties are minimum or typical as specified. UTS and YS as MPa. El as %. See Appendix for list of abbreviations used in Notes. * indicates obsolete material.

Specification	Designation	Notes	C	Cr	Mn	Mo	Ni	P	S	Si	Other	UTS	YS	El	Hard

Stainless Steel, Martensitic, 410 (Continued from previous page)

USA

Specification	Designation	Notes	C	Cr	Mn	Mo	Ni	P	S	Si	Other	UTS	YS	El	Hard
AMS 7493K(96)		Rings, flashweld, martensitic and ferritic	0.15 max	11.50-13.50	.1.00 max			0.040 max	0.030 max	1.00 max	bal Fe				
ASME SA240	410	Refer to ASTM A240(95)													
ASME SA268	410	Refer to ASTM A268(94)													
ASME SA479	410	Refer to ASTM A479/A479M(95)													
ASTM A176(97)	410*	Sh Strp Plt	0.15 max	11.5-13.5	1 max		0.75 max	0.04 max	0.03 max	1 max	bal Fe				
ASTM A182/A182M(98)	F6a Class 2	Frg/roll pipe flange valve	0.15 max	11.5-13.5	1.00 max		0.50 max	0.040 max	0.030 max	1.00 max	bal Fe	585	380	18	167-229
ASTM A182/A182M(98)	F6a Class 3	Frg/roll pipe flange valve	0.15 max	11.5-13.5	1.00 max		0.50 max	0.040 max	0.030 max	1.00 max	bal Fe	760	585	15	235-302 HB
ASTM A182/A182M(98)	F6a Class 4	Frg/roll pipe flange valve	0.15 max	11.5-13.5	1.00 max		0.50 max	0.040 max	0.030 max	1.00 max	bal Fe	895	760	12	263-321
ASTM A182/A182M(98)	F6a Class1	Frg/roll pipe flange valve	0.15 max	11.5-13.5	1.00 max		0.50 max	0.040 max	0.030 max	1.00 max	bal Fe	485	275	18	143-207 HB
ASTM A193/A193M(98)	B6	Bolt, high-temp	0.15 max	11.5-13.5	1.00 max			0.040 max	0.030 max	1.00 max	bal Fe	760	585	15	
ASTM A193/A193M(98)	B6X	Bolt, high-temp	0.15 max	11.5-13.5	1.00 max			0.040 max	0.030 max	1.00 max	bal Fe	620	485	16	26 HRC
ASTM A194	410*	Nuts, high-temp press	0.15 max	11.5-13.5	1 max		0.75 max	0.04 max	0.03 max	1 max	bal Fe				
ASTM A194/A194M(98)	6	Nuts, high-temp press	0.15 max	11.5-13.5	1.00 max			0.040 max	0.030 max	1.00 max	bal Fe				228-271 HB
ASTM A240/A240M(98)	S41000	hard by HT	0.15 max	11.5-13.5	1.00 max		0.75 max	0.040 max	0.030 max	1.00 max	bal Fe	450	205	20.0	217 HB
ASTM A276(98)	410	Bar, Ann	0.15 max	11.5-13.5	1 max		0.75 max	0.04 max	0.03 max	1 max	bal Fe				
ASTM A314	410	Bil	0.15 max	11.5-13.5	1 max		0.75 max	0.04 max	0.03 max	1 max	bal Fe				
ASTM A473	410	Frg, Heat res ann	0.15 max	11.50-13.50	1.00 max		0.75 max	0.040 max	0.030 max	1.00 max	bal Fe	485	275	20	223 HB
ASTM A479	410	Bar, Ann	0.15 max	11.5-13.5	1 max		0.75 max	0.04 max	0.03 max	1 max	bal Fe				
ASTM A479	410	Bar, HT3	0.15 max	11.5-13.5	1 max		0.75 max	0.04 max	0.03 max	1 max	bal Fe				
ASTM A479	410	Bar, HT2	0.15 max	11.5-13.5	1 max		0.75 max	0.04 max	0.03 max	1 max	bal Fe				
ASTM A493	410	Wir rod, CHd, cold frg	0.15 max	11.5-13.5	1.00 max			0.040 max	0.030 max	1.00 max	bal Fe	565			
ASTM A511(96)	MT410	Smls mech tub, Ann	0.15 max	11.5-13.5	1.00 max		0.50 max	0.040 max	0.030 max	1.00 max	bal Fe	414	207	20	207 HB
ASTM A580/A580M(98)	410	Wir, Ann	0.15 max	11.5-13.5	1.00 max			0.040 max	0.030 max	1.00 max	bal Fe	485	275	20	35 HCR
ASTM A815/A815M(98)	WP410	Pipe fittings	0.15 max	11.5-13.5	1.00 max		0.50 max	0.040 max	0.030 max	1.00 max	bal Fe	485-655	205	20.0	207 HB
ASTM A837(96)	410	Frg for carburizing	0.15 max	11.50-13.50	1.00 max	0.10 max	0.50 max	0.040 max	0.030 max	1.00 max	Cu <=0.35; Si<=0.10 if VCD used; bal Fe				
ASTM A988(98)	S41000(1)	HIP Flanges, Fittings, Valves/parts; Heat res; HT hard	0.15 max	11.5-13.5	1.00 max			0.040 max	0.030 max	1.00 max	· bal Fe	485	275	18	143-187 HB
ASTM A988(98)	S41000(2)	HIP Flanges, Fittings, Valves/parts; Heat res; HT hard	0.15 max	11.5-13.5	1.00 max			0.040 max	0.030 max	1.00 max	bal Fe	585	380	18	167-229 HB
ASTM A988(98)	S41000(3)	HIP Flanges, Fittings, Valves/parts; Heat res; HT hard	0.15 max	11.5-13.5	1.00 max			0.040 max	0.030 max	1.00 max	bal Fe	760	585	15	235-302 HB
ASTM A988(98)	S41000(4)	HIP Flanges, Fittings, Valves/parts; Heat res; HT hard	0.15 max	11.5-13.5	1.00 max			0.040 max	0.030 max	1.00 max	bal Fe	895	760	12	263-321 HB
DoD-F-24669/7(86)	410	HT; Supersedes MIL-S-861	0.15 max	11.5-13.5	1.00 max			0.040 max	0.030 max	1.00 max	bal Fe	690	480	20	201-241 HB
FED QQ-S-763F(96)	410*	Obs; Bar Wir Shp Frg; Corr res	0.15 max	11.5-13.5	1 max			0.04 max	0.03 max	1 max	bal Fe				
MIL-S-861	410*	Obs	0.15 max	11.5-13.5	1 max			0.04 max	0.03 max	1 max	bal Fe				

UNS numbers and US grades are provided as a means of cross referencing chemically similar alloys. Exchangability is only possible after independent examination of specifications. Tensile properties are minimum or typical as specified. UTS and YS as MPa. El as %. See Appendix for list of abbreviations used in Notes. * indicates obsolete material.

Specification	Designation	Notes	C	Cr	Mn	Mo	Ni	P	S	Si	Other	UTS	YS	El	Hard

Stainless Steel, Martensitic, 410 (Continued from previous page)

USA

Specification	Designation	Notes	C	Cr	Mn	Mo	Ni	P	S	Si	Other	UTS	YS	El	Hard
MIL-S-862B	410*	Obs	0.15 max	11.5-13.5	1 max		·	0.04 max	0.03 max	1 max	bal Fe				
SAE J405(98)	S41000	Hardenable by HT	0.15 max	11.50-13.50	1.00 max		0.75 max	0.040 max	0.030 max	1.00 max	bal Fe				
SAE J467(68)	410		0.12	12.25	0.50	0.30	0.40			0.35	bal Fe	1082	1000	13	300 HB

Stainless Steel, Martensitic, 410S

Australia

Specification	Designation	Notes	C	Cr	Mn	Mo	Ni	P	S	Si	Other	UTS	YS	El	Hard
AS 1449(94)	410S	Sh Strp Plt	0.08 max	11.50-13.50	1.00 max		0.60 max	0.040 max	0.030 max	1.00 max	bal Fe				

Bulgaria

Specification	Designation	Notes	C	Cr	Mn	Mo	Ni	P	S	Si	Other	UTS	YS	El	Hard
BDS 6738(72)	0Ch13	Corr res	0.08 max	12.0-14.0	0.80 max	0.30 max	0.60 max	0.035 max	0.025 max	0.80 max	bal Fe				

Canada

Specification	Designation	Notes	C	Cr	Mn	Mo	Ni	P	S	Si	Other	UTS	YS	El	Hard
CSA G110.5	410S	Sh Strp, HR, Ann, Q/A, Tmp	0.08 max	11.5-13.5	1 max		0.6 max	0.04 max	0.03 max	1 max	bal Fe				
CSA G110.9	410S	Sh Strp Plt	0.08 max	11.5-13.5	1 max		0.6 max	0.04 max	0.03 max	1 max	bal Fe				

Czech Republic

Specification	Designation	Notes	C	Cr	Mn	Mo	Ni	P	S	Si	Other	UTS	YS	El	Hard
CSN 417020	17020	Corr res	0.08 max	12.0-14.0	0.9 max			0.04 max	0.035 max	0.7 max	bal Fe				

Europe

Specification	Designation	Notes	C	Cr	Mn	Mo	Ni	P	S	Si	Other	UTS	YS	El	Hard
EN 10088/2(95)	1.4000	Strp; Corr res; CR t<=6mm, Ann	0.08 max	12.00-14.00	1.00 max			0.040 max	0.015 max	1.00 max	bal Fe	400-600	240	19	
EN 10088/2(95)	X6Cr13	Strp; Corr res; CR t<=6mm, Ann	0.08 max	12.00-14.00	1.00 max			0.040 max	0.015 max	1.00 max	bal Fe	400-600	240	19	*
EN 10088/3(95)	1.4000	Bar Rod Shp, Corr res; t<=25mm, Ann	0.08 max	12.00-14.00	1.00 max			0.040 max	0.030 max	1.00 max	bal Fe	400-630	230	20	200 HB max
EN 10088/3(95)	X6Cr13	Bar Rod Shp, Corr res; t<=25mm, Ann	0.08 max	12.00-14.00	1.00 max			0.040 max	0.030 max	1.00 max	bal Fe	400-630	230	20	200 HB max

France

Specification	Designation	Notes	C	Cr	Mn	Mo	Ni	P	S	Si	Other	UTS	YS	El	Hard
AFNOR NFA35573	Z6C13	Sh Strp Plt	0.08 max	11.5-13.5	1 max			0.04 max	0.03 max	1 max	bal Fe				
AFNOR NFA35574	Z6C13	Bar	0.08 max	11.5-13.5	1 max			0.04 max	0.03 max	1 max	bal Fe				
AFNOR NFA35577	Z6C13	Bar Wir; Ann, 6mm diam	0.08 max	11.5-13.5	1 max		0.5 max	0.04 max	0.03 max	1 max	bal Fe				

Germany

Specification	Designation	Notes	C	Cr	Mn	Mo	Ni	P	S	Si	Other	UTS	YS	El	Hard
DIN	G-X7Cr13	Ann, long sample	0.08 max	13.0-15.0	1.00 max			0.045 max	0.030 max	1.00 max	bal Fe		245	20	
DIN	WNr 1.4001	Ann	0.08 max	13.0-15.0	1.00 max			0.045 max	0.030 max	1.00 max	bal Fe		245	20	
DIN 17440(96)	WNr 1.4000	HR Strp Bar Wir Frg	0.08 max	12.0-14.0	1.00 max			0.040 max	0.030 max	1.00 max	bal Fe				
DIN 17440(96)	X6Cr13	Plt Sh, HR Strp Bar Wir Frg, Ann	0.08 max	12.0-14.0	1.00 max			0.040 max	0.030 max	1.00 max	bal Fe	400-630	230	20	
DIN 17441(97)	WNr 1.4000	CR Strp Plt Sh, Ann	0.08 max	12.0-14.0	1.00 max			0.040 max	0.030 max	1.00 max	bal Fe	400-630	230	20	
DIN 17441(97)	X6Cr13	CR Strp Plt Sh	0.08 max	12.0-14.0	1.00 max			0.040 max	0.030 max	1.00 max	bal Fe				
DIN EN 10088(95)	WNr 1.4000	Sh Plt Strp Bar Rod, Ann	0.08 max	12.0-14.0	1.00 max			0.040 max	0.030 max	1.00 max	bal Fe	400-630	230	20	
TGL 7143	X7Cr13*	Obs	0.08 max	12-14	0.8 max			0.04 max	0.03 max	0.8 max	bal Fe				

Hungary

Specification	Designation	Notes	C	Cr	Mn	Mo	Ni	P	S	Si	Other	UTS	YS	El	Hard
MSZ 4360	K01	Sh Strp Plt Bar Wir Tub	0.08 max	12-14	1 max		0.6 max	0.045 max	0.03 max	1 max	bal Fe				
MSZ 4360(87)	KO1	Corr res; 0-25mm; soft ann	0.08 max	12-14	1 max		0.6 max	0.045 max	0.03 max	1 max	bal Fe	400-600	250	20L; 15T	185 HB max
MSZ 4360(87)	KO1	Corr res; 0-25mm; Q/T	0.08 max	12-14	1 max		0.6 max	0.045 max	0.03 max	1 max	bal Fe	550-700	400	18L; 13T	

India

Specification	Designation	Notes	C	Cr	Mn	Mo	Ni	P	S	Si	Other	UTS	YS	El	Hard
IS 6527	04Cr13	Wir rod	0.08 max	11.5-14.5	1 max			0.04 max	0.03 max	1 max	bal Fe				
IS 6528	04Cr13	Wir, Ann, SA	0.08 max	11.5-14.5	1 max			0.04 max	0.03 max	1 max	bal Fe				

UNS numbers and US grades are provided as a means of cross referencing chemically similar alloys. Exchangability is only possible after independent examination of specifications. Tensile properties are minimum or typical as specified. UTS and YS as MPa. El as %. See Appendix for list of abbreviations used in Notes. * indicates obsolete material.

Stainless Steel, Martensitic, 410S (Continued from previous page)

Specification	Designation	Notes	C	Cr	Mn	Mo	Ni	P	S	Si	Other	UTS	YS	El	Hard
India															
IS 6529	04Cr13	Bil	0.08 max	11.5-14.5	1 max			0.04 max	0.03 max	1 max	bal Fe				
IS 6911	04Cr13	Sh Strp Plt, Ann, 0.5-3mm diam	0.08 max	11.5-14.5	1 max			0.04 max	0.03 max	1 max	bal Fe				
International															
COMECON PC4-70	1	Sh Strp Plt Bar	0.08 max	12-14	0.8 max			0.035 max	0.025 max	0.8 max	bal Fe				
ISO 683-13(74)	1*	Corr res	0.08 max	11.5-14	1 max	0.15 max	0.5 max	0.04 max	0.03 max	1 max	Cu <=0.3; Pb <=0.15; Ti <=0.05; V <=0.1; bal Fe				
ISO 683-13(74)	1*	Corr res	0.08 max	11.5-14	1 max	0.15 max	0.5 max	0.04 max	0.03 max	1 max	Cu <=0.3; Pb <=0.15; Ti <=0.05; V <=0.1; bal Fe				
Italy															
UNI 6901(71)	X6Cr13	Corr res	0.08 max	11.5-14	1 max		0.5 max	0.04 max	0.03 max	1 max	bal Fe				
UNI 8317(81)	X6Cr13	Sh Strp Plt	0.08 max	11.5-14	1 max		0.5 max	0.04 max	0.03 max	1 max	N 0.12-0.25; bal Fe				
Japan															
JIS G4303(91)	SUS410S	Bar Q/A Tmp 75mm diam	0.08 max	11.50-13.50	1.00 max			0.040 max	0.030 max	1.00 max	bal Fe	440	205	20	88 HRB max
JIS G4304(91)	SUS410S	Plt Sh, HR	0.08 max	11.50-13.50	1.00 max			0.040 max	0.030 max	1.00 max	bal Fe	410	205	20	88 HRB max
JIS G4305(91)	SUS410S	Plt Sh, CR Ann	0.08 max	11.50-13.50	1.00 max			0.040 max	0.030 max	1.00 max	bal Fe	410	205	20	183 HB max
JIS G4306	SUS410S	Strp HR Ann	0.08 max	11.5-13.5	1 max			0.04 max	0.03 max	1 max	bal Fe				
JIS G4307	SUS410S	Strp CR	0.08 max	11.5-13.5	1 max			0.04 max	0.03 max	1 max	bal Fe				
Poland															
PNH86020	0H13	Corr res	0.08 max	12-14	0.08 max		0.6 max	0.04 max	0.03 max	0.8 max	bal Fe				
Romania															
STAS 3583(64)	7C120	Sh Strp Plt Bar Tub; Corr/Heat res	0.08 max	11-13	0.6 max		0.6 max	0.035 max	0.03 max	0.6 max	Pb <=0.15; bal Fe				
Spain															
UNE 36016(75)	F.3110*	see X6Cr13	0.08 max	11.5-14	1 max		0.5 max	0.04 max	0.03 max	1 max	bal Fe				
UNE 36016(75)	X6Cr13	Corr res	0.08 max	11.5-14	1 max		0.5 max	0.04 max	0.03 max	1 max	bal Fe				
UNE 36257(74)	AM-X12Cr13	Corr res	0.15 max	12.0-14.0	1.0 max	0.5 max	1.0 max	0.040 max	0.040 max	1.5 max	bal Fe				
UNE 36257(74)	F.8401*	see AM-X12Cr13	0.15 max	12.0-14.0	1.0 max	0.5 max	1.0 max	0.040 max	0.040 max	1.5 max	bal Fe				
UK															
BS 1449/2(83)	403S17	Corr/Heat res; t<=100mm	0.08 max	12.0-14.0	1.00 max		1.00 max	0.040 max	0.030 max	1.00 max	bal Fe	420	245	20	0-175 HB
BS 1501/3(73)	403S17*	Plate; Press ves; Corr res	0.08 max	12.0-14.0	1.0 max	0.15 max	0.5 max	0.04 max	0.03 max	0.8 max	Al <=0.1; Co <=0.1; Cu <=0.3; Pb <=0.15; Ti <=0.05; V <=0.1; W <=0.1; bal Fe				
BS 1501/3(73)	405S17*	Press ves; Corr res	0.08 max	12.0-14.0	1.0 max	0.15 max	0.5 max	0.04 max	0.03 max	0.8 max	Al 0.1-0.3; Co <=0.1; Cu <=0.3; Pb <=0.15; Ti <=0.05; V <=0.1; W <=0.1; bal Fe				
BS 1503(89)	403S17	Frg press ves t<=999mm; norm Q/T	0.08 max	12.0-14.0	1.00 max	0.30 max	0.50 max	0.040 max	0.025 max	0.80 max	Cu <=0.30; W <=0.1; bal Fe	470-620	265	18	
USA															
	AISI 410S*	Obs; Tube	0.08 max	11.50-13.50	1.00 max		0.6 max	0.040 max	0.030 max	1.00 max	bal Fe				
	UNS S41008	Hardenable by HT Low-C	0.08 max	11.5-13.5	1.00 max		0.60	0.040 max	0.030 max	1.00 max	bal Fe				
ASME SA240	410S	Refer to ASTM A240(95)													
ASTM A176(97)	410S*	Sh Strp Plt	0.08 max	11.5-13.5	1 max		0.6 max	0.04 max	0.03 max	1 max	bal Fe				

UNS numbers and US grades are provided as a means of cross referencing chemically similar alloys. Exchangability is only possible after independent examination of specifications. Tensile properties are minimum or typical as specified. UTS and YS as MPa. El as %. See Appendix for list of abbreviations used in Notes. * indicates obsolete material.

Specification	Designation	Notes	C	Cr	Mn	Mo	Ni	P	S	Si	Other	UTS	YS	El	Hard

Stainless Steel, Martensitic, 410S (Continued from previous page)

USA

Specification	Designation	Notes	C	Cr	Mn	Mo	Ni	P	S	Si	Other	UTS	YS	El	Hard
ASTM A240/A240M(98)	S41008	hard by HT Low-C	0.08 max	11.5-13.5	1.00 max		0.60 max	0.040 max	0.030 max	1.00 max	bal Fe	415	205	22.0	183 HB
ASTM A473	410S	Frg, Heat res ann	0.08 max	11.50-13.50	1.00 max		0.75 max	0.040 max	0.030 max	1.00 max	bal Fe	450	2450	22	217 HB
SAE J405(98)	S41008	Hardenable by HT Low-C	0.08 max	11.50-13.50	1.00 max		0.60 max	0.040 max	0.030 max	1.00 max	bal Fe				

Stainless Steel, Martensitic, 414

Canada

Specification	Designation	Notes	C	Cr	Mn	Mo	Ni	P	S	Si	Other	UTS	YS	El	Hard
CSA G110.3	414	Bar Bil	0.15 max	11.5-13.5	1 max		1.25-2.5	0.04 max	0.03 max	1 max	bal Fe				

Mexico

Specification	Designation	Notes	C	Cr	Mn	Mo	Ni	P	S	Si	Other	UTS	YS	El	Hard
DGN B-83	414*	Obs; Bar, HR, CR	0.15 max	11.5-13.5	1 max		1.25-2.5	0.04 max	0.03 max	1 max	bal Fe				
NMX-B-171(91)	MT414	mech tub	0.15 max	11.5-13.5	1.00 max		1.25-2.50	0.040 max	0.030 max	1.00 max	bal Fe				

USA

Specification	Designation	Notes	C	Cr	Mn	Mo	Ni	P	S	Si	Other	UTS	YS	El	Hard
	AISI 414	Tube	0.15 max	11.50-13.50	1.00 max		1.25-2.50	0.040 max	0.030 max	1.00 max	bal Fe				
	UNS S41400	Low Ni Content	0.15 max	11.50-13.5	1.00 max		1.25-2.50	0.040 max	0.030 max	1.00 max	bal Fe				
ASTM A276(98)	414	Bar, HT	0.15 max	11.5-13.5	1 max		1.25-2.5	0.04 max	0.03 max	1 max	bal Fe				
ASTM A314	414	Bil	0.15 max	11.5-13.5	1 max		1.25-2.5	0.04 max	0.03 max	1 max	bal Fe				
ASTM A473	414	Frg, Heat res ann	0.15 max	11.50-13.50	1.00 max		1.25-2.50	0.040 max	0.030 max	1.00 max	bal Fe				298 HB
ASTM A479	414	Bar	0.15 max	11.5-13.5	1 max		1.25-2.5	0.04 max	0.03 max	1 max	bal Fe				
ASTM A511(96)	MT414	Smls mech tub, Ann	0.15 max	11.5-13.5	1.00 max		1.25-2.5	0.040 max	0.030 max	1.00 max	bal Fe	689	448	15	235 HB
ASTM A580/A580M(98)	414	Wir, Ann	0.15 max	11.5-13.5	1.00 max		1.25-2.50	0.040 max	0.030 max	1.00 max	bal Fe	1035 max			42 HCR
FED QQ-S-763F(96)	414*	Obs; Bar Wir Shp Frg; Corr res	0.15 max	11.5-13.5	1 max		1.25-2.5	0.04 max	0.03 max	1 max	bal Fe				

Stainless Steel, Martensitic, 416

Australia

Specification	Designation	Notes	C	Cr	Mn	Mo	Ni	P	S	Si	Other	UTS	YS	El	Hard
AS 2837(86)	416	Bar, Semi-finished product	0.09-0.15	12.0-14.0	1.50 max	0.60 max	1.00 max	0.060 max	0.15-0.35	1.00 max	bal Fe				

Canada

Specification	Designation	Notes	C	Cr	Mn	Mo	Ni	P	S	Si	Other	UTS	YS	El	Hard
CSA G110.3	416	Bar Bil	0.15 max	12-14	1.25 max	0.6 max		0.06 max	0.15-99.99	1 max	bal Fe				

China

Specification	Designation	Notes	C	Cr	Mn	Mo	Ni	P	S	Si	Other	UTS	YS	El	Hard
GB 1220(92)	Y1Cr13	Bar Q/T	0.15 max	12.00-14.00	1.25 max	0.60 max	0.60 max	0.060 max	0.15 min	1.00 max	bal Fe	540	345	25	
GB 4240(93)	Y1Cr13(-Q)	Wir HT 3-6mm diam, Q-TLC	0.15 max	12.00-14.00	1.25 max	0.60 max	0.60 max	0.060 max	0.15 min	1.00 max	bal Fe	590-880			

Europe

Specification	Designation	Notes	C	Cr	Mn	Mo	Ni	P	S	Si	Other	UTS	YS	El	Hard
EN 10088/3(95)	1.4005	Bar Rod Sect, Corr res; t<=160mm, Q/T	0.08-0.15	12.00-14.00	1.50 max	0.60 max		0.040 max	0.15-0.35	1.00 max	bal Fe	650-850	450	12	
EN 10088/3(95)	X12CrS13	Bar Rod Sect, Corr res; t<=160mm, Q/T	0.08-0.15	12.00-14.00	1.50 max	0.60 max		0.040 max	0.15-0.35	1.00 max	bal Fe	650-850	450	12	

France

Specification	Designation	Notes	C	Cr	Mn	Mo	Ni	P	S	Si	Other	UTS	YS	El	Hard
AFNOR NFA35576	Z12CF13	Bar Wir Bil	0.08-0.15	12-14	1.5 max	0.15-0.6		0.06 max	0.15 max	1 max	bal Fe				

Germany

Specification	Designation	Notes	C	Cr	Mn	Mo	Ni	P	S	Si	Other	UTS	YS	El	Hard
DIN EN 10088(95)	WNr 1.4005	Sh Plt Strp, HT	0.08-0.15	12.0-14.0	1.50 max			0.040 max	0.15-0.25	1.00 max	bal Fe	650-850	450	12	
DIN EN 10088(95)	X12CrS13	Sh Plt Strp, HT	0.08-0.15	12.0-14.0	1.00 max			0.040 max	0.15-0.25	1.00 max	bal Fe	650-850	450	12	

International

Specification	Designation	Notes	C	Cr	Mn	Mo	Ni	P	S	Si	Other	UTS	YS	El	Hard
ISO 683-13(74)	7*	Ann, Q/A, Tmp	0.08-0.15	12-14	1.5 max	0.6 max	1 max	0.06 max	0.15-0.35	1 max	bal Fe				

Specification	Designation	Notes	C	Cr	Mn	Mo	Ni	P	S	Si	Other	UTS	YS	El	Hard

Stainless Steel, Martensitic, 416 (Continued from previous page)

Italy

Specification	Designation	Notes	C	Cr	Mn	Mo	Ni	P	S	Si	Other	UTS	YS	El	Hard
UNI 6901(71)	X12CrS13	Corr res	0.08-0.15	12-14	1.5 max	0.6 max	1 max	0.06 max	0.15-0.35	1 max	bal Fe				

Japan

Specification	Designation	Notes	C	Cr	Mn	Mo	Ni	P	S	Si	Other	UTS	YS	El	Hard
JIS G4303(91)	SUS416	Bar Q/A Tmp 75mm diam	0.15 max	12.00-14.00	1.25 max	0.60 max	0.60 max	0.060 max	0.15 min	1.00 max	bal Fe	540	345	17	159 HB
JIS G4308(98)	SUS416	Wir rod	0.15 max	12.00-14.00	1.25 max	0.60 max	0.60 max	0.060 max	0.15 min	1.00 max	bal Fe				
JIS G4309	SUS416	Wir	0.15 max	12.00-14.00	1.25 max	0.60 max	0.60 max	0.060 max	0.15 min	1.00 max	bal Fe				

Mexico

Specification	Designation	Notes	C	Cr	Mn	Mo	Ni	P	S	Si	Other	UTS	YS	El	Hard
DGN B-83	416*	Obs; Bar, HR, CR	0.15 max	12-14	1.25 max	0.6 max		0.06 max	0.15-99.99	1 max	bal Fe				

Romania

Specification	Designation	Notes	C	Cr	Mn	Mo	Ni	P	S	Si	Other	UTS	YS	El	Hard
STAS 3583(87)	10Cr130	Corr res; Heat res	0.08-0.12	12-14	1 max	0.2 max	0.5 max	0.045 max	0.03 max	1 max	Pb <=0.15; bal Fe				

Spain

Specification	Designation	Notes	C	Cr	Mn	Mo	Ni	P	S	Si	Other	UTS	YS	El	Hard
UNE 36016(75)	F.3411*	see X12CrS13	0.08-0.15	12.0-14.0	1.5 max	0.6 max	1.0 max	0.06 max	0.15-0.35	1.0 max	bal Fe				
UNE 36016(75)	X12CrS13	Bar Wir; Corr res	0.08-0.15	12.0-14.0	1.5 max	0.6 max	1.0 max	0.06 max	0.15-0.35	1.0 max	bal Fe				

Sweden

Specification	Designation	Notes	C	Cr	Mn	Mo	Ni	P	S	Si	Other	UTS	YS	El	Hard
SS 142380	2380	Corr res	0.08-0.15	12-14	1.5 max	0.6 max	1 max	0.06 max	0.15-0.35	1 max	bal Fe				

UK

Specification	Designation	Notes	C	Cr	Mn	Mo	Ni	P	S	Si	Other	UTS	YS	El	Hard
BS 1506(90)	416S29	Bolt matl Pres/Corr res; t<=63mm; Q/T	0.14-0.20	11.5-13.5	1.50 max	0.60 max	1.00 max	0.060 max	0.15-0.35	1.00 max	bal Fe				223-311 HB
BS 1554(90)	416S21	Wir Corr/Heat res; 1.5<diam<=13mm; Ann	0.09-0.15	11.5-14.0	1.50 max	0060 max	1.00 max	0.060 max	0.15-0.35	1.00 max	W <=0.1; bal Fe	0-700			
BS 1554(90)	416S21	Wir Corr/Heat res; diam<0.49mm; Ann	0.09-0.15	11.5-14.0	1.50 max	0.60 max	1.00 max	0.060 max	0.15-0.35	1.00 max	W <=0.1; bal Fe	0-780			
BS 970/1(96)	416S21	Wrought Corr/Heat res 63<t<=150mm; Tmp	0.09-0.15	11.5-13.5	1.50 max	0.60 max	1.00 max	0.060 max	0.15-0.35	1.00 max	bal Fe	550-700	370	15	152-207 HB
BS 970/1(96)	416S21	Wrought Corr/Heat res; t<=63mm; Tmp	0.09-0.15	11.5-13.5	1.50 max	0.60 max	1.00 max	0.060 max	0.15-0.35	1.00 max	bal Fe	700-850	525	11	201-255 HB
BS 970/1(96)	416S29	Wrought Corr/Heat res; t<=29mm; Tmp	0.14-0.20	11.5-13.5	1.50 max	0.60 max	1.00 max	0.060 max	0.15-0.35	1.00 max	bal Fe	775-925	585	10	223-277 HB
BS 970/1(96)	416S29	Wrought Corr/Heat res 29<t<=150mm; Tmp	0.14-0.20	11.5-13.5	1.50 max	0.60 max	1.00 max	0.060 max	0.15-0.35	1.00 max	bal Fe	700-850	525	11	201-255 HB
BS 970/1(96)	416S37	Wrought Corr/Heat res; t<=150mm; Tmp	0.20-0.28	12.0-14.0	1.50 max	0.60 max	1.00 max	0.060 max	0.15-0.35	1.00 max	bal Fe	775-925	585	10	223-277 HB

USA

Specification	Designation	Notes	C	Cr	Mn	Mo	Ni	P	S	Si	Other	UTS	YS	El	Hard
	AISI 416	Tube	0.15 max	12.00-14.00	1.25 max	0.60 max		0.060 max	0.15 min	1.00 max	bal Fe				
	UNS S41600	Free Machining	0.15 max	12.00-14	1.25 max	0.60 max		0.060 max	0.15 min	1.00 max	bal Fe				
ASTM A194	416*	Nuts, high-temp press	0.15 max	12-14	1.25 max	0.6 max		0.06 max	0.15-99.99	1 max	bal Fe				
ASTM A194/A194M(98)	6F	Nuts, high-temp press	0.15 max	12.0-14.0	1.25 max			0.060 max	0.15 min	1.00 max	bal Fe				228-271 HB
ASTM A314	416	Bil	0.15 max	12-14	1.25 max	0.6 max		0.06 max	0.15-99.99	1 max	bal Fe				
ASTM A473	416	Frg, Heat res ann	0.15 max	12.00-14.00	1.25 max	0.60 max		0.06 max	0.15 min	1.00 max	bal Fe	485	275	20	223 HB
ASTM A581/A581M(95)	416	Wir rod, Ann	0.15 max	12.0-14.0	1.25 max			0.06 max	0.15 min	1.00 max	bal Fe	585-860			
ASTM A582/A582M(95)	416	Bar, HF, CF, Ann	0.15 max	12.00-14.00	1.25 max			0.06 max	0.15 min	1.00 max	bal Fe				262 HB max
ASTM A895(94)	416	Sh Strp Plt, free mach, Ann	0.15 max	12.0-14.0	1.25 max			0.06 max	0.15 min	1.00 max	bal Fe				262 HB max
MIL-S-862B	416*	Obs	0.15 max	12-14	1.25 max	0.6 max		0.06 max	0.15-99.99	1 max	bal Fe				

Stainless Steel, Martensitic, 416Se

Canada

Specification	Designation	Notes	C	Cr	Mn	Mo	Ni	P	S	Si	Other	UTS	YS	El	Hard
CSA G110.3	416Se	Bar Bil	0.15 max	12-14	1.25 max			0.06 max	0.06 max	1 max	Se>=0.15; bal Fe				

UNS numbers and US grades are provided as a means of cross referencing chemically similar alloys. Exchangability is only possible after independent examination of specifications. Tensile properties are minimum or typical as specified. UTS and YS as MPa. El as %. See Appendix for list of abbreviations used in Notes. * indicates obsolete material.

Specification	Designation	Notes	C	Cr	Mn	Mo	Ni	P	S	Si	Other	UTS	YS	El	Hard

Stainless Steel, Martensitic, 416Se (Continued from previous page)

Mexico

Specification	Designation	Notes	C	Cr	Mn	Mo	Ni	P	S	Si	Other	UTS	YS	El	Hard
DGN B-83	416Se*	Obs; Bar, HR, CR	0.15 max	12-14	1.25 max			0.06 max	0.06 max	1 max	Se>=0.15; bal Fe				
NMX-B-171(91)	MT416SE	mech tub	0.15 max	12.0-14.0	1.25 max		0.50 max	0.060 max	0.060 max	1.00 max	bal Fe				

UK

Specification	Designation	Notes	C	Cr	Mn	Mo	Ni	P	S	Si	Other	UTS	YS	El	Hard
BS 1554(90)	416S41	Wir Corr/Heat res; diam<0.49mm; Ann	0.09-0.15	11.5-13.5	1.50 max	0.60 max	1.00 max	0.060 max	0.15-0.060	1.00 max	W <=0.1; 0.15<=Se<=0.35; bal Fe	0-780			
BS 1554(90)	416S41	Wir Corr/Heat res; 1.5<diam<=13mm; Ann	0.09-0.15	11.5-13.5	1.50 max	0.60 max	1.00 max	0.060 max	0.15-0.060	1.00 max	W <=0.1; 0.15<=Se<=0.35; bal Fe	0-700			
BS 970/1(96)	416S41	Wrought Corr/Heat res; t<=63mm; Tmp	0.09-0.15	11.5-13.5	1.50 max	0.60 max	1.00 max	0.060 max	0.060 max	1.00 max	0.15<=Se<=0.35; bal Fe	550-700	370	15	152-207 HB
BS 970/1(96)	416S41	Wrought Corr/Heat res 63<t<=150mm; Tmp	0.09-0.15	11.5-13.5	1.50 max	0.60 max	1.00 max	0.060 max	0.060 max	1.00 max	0.15<=Se<=0.35; bal Fe	700-850	525	11	201-255 HB

USA

Specification	Designation	Notes	C	Cr	Mn	Mo	Ni	P	S	Si	Other	UTS	YS	El	Hard
	AISI 416Se	Tube	0.15 max	12.00-14.00	1.25 max			0.060 max	0.060 max	1.00 max	Se>=0.15; bal Fe				
	UNS S41623	Free Machining Se-bearing	0.15 max	12.00-14	1.25 max			0.060 max	0.060 max	1.00 max	Se>=0.15; bal Fe				
ASTM A194	416Se*	Nuts, high-temp press	0.15 max	12-14	1.25 max			0.06 max	0.06 max	1 max	Se>=0.15; bal Fe				
ASTM A194/A194M(98)	6F	Nuts, high-temp press	0.15 max	12.0-14.0	1.25 max			0.060 max	0.060 max	1.00 max	Se>=0.15; bal Fe				228-271 HB
ASTM A314	416Se	Bil	0.15 max	12-14	1.25 max			0.06 max	0.06 max	1 max	Se>=0.15; bal Fe				
ASTM A473	416Se	Frg, Heat res ann	0.15 max	12.00-14.00	1.25 max			0.06 max	0.06 max	1.00 max	Se>=0.15; bal Fe	485	275	20	223 HB
ASTM A511(96)	MT416 Se	Smls mech tub, Ann	0.15 max	12.0-14.0	1.25 max		0.50 max	0.060 max	0.060 max	1.00 max	Se=0.12-0.20; bal Fe	414	241	20	230 HB
ASTM A581/A581M(95)	416Se	Wir rod, Ann	0.15 max	12.0-14.0	1.25 max			0.06 max	0.06 max	1.00 max	Se>=0.15; bal Fe	585-860			
ASTM A582/A582M(95)	416Se	Bar, HF, CF, Ann	0.15 max	12.00-14.00	1.25 max			0.06 max	0.06 max	1.00 max	Se>=0.15; bal Fe				262 HB max
ASTM A895(94)	416Se	Sh Strp Plt, free mach, Ann	0.15 max	12.0-14.0	1.25 max			0.06 max	0.06 max	1.00 max	Se>=0.15; bal Fe				262 HB max
FED QQ-S-763F(96)	416Se*	Obs; Bar Wir Shp Frg; Corr res	0.15 max	12-14	1.25 max			0.06 max	0.06 max	1 max	Se>=0.15; bal Fe				
MIL-S-862B	416Se*	Obs	0.15 max	12-14	1.25 max			0.06 max	0.06 max	1 max	Se>=0.15; bal Fe				

Stainless Steel, Martensitic, 420

Australia

Specification	Designation	Notes	C	Cr	Mn	Mo	Ni	P	S	Si	Other	UTS	YS	El	Hard
AS 1449(94)	420	Sh Strp Plt	0.30-0.40	12.00-14.00	1.00 max		0.75 max	0.040 max	0.030 max	1.00 max	bal Fe				
AS 2837(86)	420	Bar, Semi-finished product	0.20-0.28	12.0-14.0	1.00 max		1.00 max	0.040 max	0.030 max	1.00 max	bal Fe				

Bulgaria

Specification	Designation	Notes	C	Cr	Mn	Mo	Ni	P	S	Si	Other	UTS	YS	El	Hard
BDS 6738(72)	2Ch13	Sh Strp Plt Bar	0.16-0.25	12.0-14.0	0.80 max	0.30 max	0.60 max	0.035 max	0.025 max	0.80 max	bal Fe				

Canada

Specification	Designation	Notes	C	Cr	Mn	Mo	Ni	P	S	Si	Other	UTS	YS	El	Hard
CSA G110.3	420	Bar Bil	0.15 min	12-14	1 max			0.04 max	0.03 max	1 max	bal Fe				

China

Specification	Designation	Notes	C	Cr	Mn	Mo	Ni	P	S	Si	Other	UTS	YS	El	Hard
GB 1220(92)	2Cr13	Bar Q/T	0.16-0.25	12.00-14.00	1.00 max		0.60 max	0.035 max	0.030 max	1.00 max	bal Fe	635	440	20	
GB 1221(92)	2Cr13	Bar Q/T	0.16-0.25	12.00-14.00	1.00 max		0.60 max	0.035 max	0.030 max	1.00 max	bal Fe	635	440	20	
GB 3280(92)	2Cr13	Sh Plt, CR Ann	0.16-0.25	12.00-14.00	1.00 max		0.60 max	0.035 max	0.030 max	1.00 max	bal Fe	520	225	18	
GB 4237(92)	2Cr13	Sh Plt, HR Ann	0.16-0.25	12.00-14.00	1.00 max		0.60 max	0.035 max	0.030 max	1.00 max	bal Fe	520	225	18	
GB 4239(91)	2Cr13	Sh Strp, CR Ann	0.16-0.25	12.00-14.00	1.00 max		0.60 max	0.035 max	0.030 max	1.00 max	bal Fe	520	225	18	
GB 4240(93)	2Cr13(-Q)	Wir HT 3-6mm diam, Q-TLC	0.16-0.25	12.00-14.00	1.00 max		0.60 max	0.035 max	0.030 max	1.00 max	bal Fe	590-880			

UNS numbers and US grades are provided as a means of cross referencing chemically similar alloys. Exchangability is only possible after independent examination of specifications. Tensile properties are minimum or typical as specified. UTS and YS as MPa. El as %. See Appendix for list of abbreviations used in Notes. * indicates obsolete material.

Stainless Steel, Martensitic, 420 (Continued from previous page)

Specification	Designation	Notes	C	Cr	Mn	Mo	Ni	P	S	Si	Other	UTS	YS	El	Hard
China															
GB 7832(88)	2Cr13	Bar Q/T	0.16-0.24	12.00-14.00	0.60 max		0.60 max	0.030 max	0.030 max	0.60 max	Cu <=0.30; bal Fe	665	490	16	
GB/T 14975(94)	2Cr13	Smls tube Ann	0.16-0.25	12.00-14.00	1.00 max		0.60 max	0.035 max	0.030 max	1.00 max	bal Fe	470	215	19	
Czech Republic															
CSN 417022	17022	Corr res	0.16-0.25	12.0-14.0	0.8 max			0.04 max	0.03 max	0.7 max	bal Fe				
CSN 417023	17023	Corr res	0.26-0.35	12.0-14.0	0.8 max			0.04 max	0.03 max	0.7 max	bal Fe				
Europe															
EN 10088/2(95)	1.4021	Sh Plt Strp, Corr res; t<=6mm Ann	0.16-0.25	12.00-14.00	1.50 max			0.040 max	0.015 max	1.00 max	bal Fe			15	225 HB max
EN 10088/2(95)	X20Cr13	Sh Plt Strp, Corr res; t<=6mm Ann	0.16-0.25	12.00-14.00	1.50 max			0.040 max	0.015 max	1.00 max	bal Fe			15	225 HB max
France															
AFNOR NFA35572	Z20C13	Sh Strp Plt Bar Wir rod, Q/T	0.15-0.24	12-14	1 max			0.04 max	0.03 max	1 max	bal Fe				
Germany															
DIN EN 10088(95)	X30Cr13	Sh Plt Strp Bar Rod, HT	0.26-0.35	12.0-14.0	1.50 max			0.040 max	0.030 max	1.00 max	bal Fe	850-1000	650	10	
DIN 17440(96)	WNr 1.4021	Plt Sh, HR Strp Bar Wir Frg, HT	0.16-0.25	12.0-14.0	1.50 max			0.040 max	0.030 max	1.00 max	bal Fe	700-850	500	13	
DIN 17441(97)	WNr 1.4021	CR Strp Plt Sh	0.16-0.25	12.0-14.0	1.50 max			0.040 max	0.030 max	1.00 max	bal Fe				
DIN 17441(97)	WNr 1.4024	CR Strp Plt Sh	0.12-0.17	12.0-14.0	1.00 max			0.045 max	0.030 max	1.00 max	bal Fe	650-800	450	14	
DIN EN 10088(95)	WNr 1.4021	Sh Plt Strp Bar Rod, HT	0.16-0.25	12.0-14.0	1.50 max			0.040 max	0.030 max	1.00 max	bal Fe	700-850	500	13	
DIN EN 10088(95)	WNr 1.4028	Sh Plt Strp Bar Rod, HT	0.26-0.35	12.0-14.0	1.00 max			0.040 max	0.030 max	1.00 max	bal Fe	850-1000	650	10	
DIN EN 10088(95)	X20Cr13	Sh Plt Strp Bar Rod, HT	0.16-0.25	12.0-14.0	1.50 max			0.040 max	0.030 max	1.00 max	bal Fe	700-850	500	13	
Hungary															
MSZ 4360	K011	Sh Strp Plt Bar Wir	0.16-0.25	12-14	1 max			0.04 max	0.03 max	1 max	Ti <=0.2; bal Fe				
MSZ 4360(87)	KO11	Corr res; 0-100mm; soft ann	0.16-0.25	12-14	1 max		0.6 max	0.045 max	0.03 max	1 max	bal Fe	0-740			230 HB max
MSZ 4360(87)	KO11	Corr res; 0-100mm; Q/T	0.16-0.25	12-14	1 max		0.6 max	0.045 max	0.03 max	1 max	bal Fe	650-800	450	15; 11T	
MSZ 4360(87)	KO12	Corr res; 0-100mm; Q/T	0.26-0.35	12-14	1 max		0.6 max	0.045 max	0.03 max	1 max	bal Fe	800-1000	600	11L	
MSZ 4360(87)	KO12	Corr res; 0-100mm; soft ann	0.26-0.35	12-14	1 max		0.6 max	0.045 max	0.03 max	1 max	bal Fe	0-780			245 HB max
MSZ 4360(87)	X20Cr13	Corr res; 0-100mm; Q/T	0.16-0.25	12-14	1 max		0.6 max	0.045 max	0.03 max	1 max	bal Fe	650-800	450	15; 11T	
MSZ 4360(87)	X20Cr13	Corr res; 0-100mm; soft ann	0.16-0.25	12-14	1 max		0.6 max	0.045 max	0.03 max	1 max	bal Fe	0-740			230 HB max
MSZ 4360(87)	X30Cr13	Corr res; 0-100mm; soft ann	0.26-0.35	12-14	1 max		0.6 max	0.045 max	0.03 max	1 max	bal Fe	0-780			245 HB max
MSZ 4360(87)	X30Cr13	Corr res; 0-100mm; Q/T	0.26-0.35	12-14	1 max		0.6 max	0.045 max	0.03 max	1 max	bal Fe	800-1000	600	11L	
India															
IS 6529	20Cr13	Sh Strp Plt Bar Bil	0.16-0.25	12-14	1 max		1 max	0.04 max	0.03 max	1 max	bal Fe				
IS 6603	20Cr13	Bar, Ann, 5-100mm diam	0.16-0.25	12-14	1 max		1 max	0.04 max	0.03 max	1 max	bal Fe				
IS 6911	20Cr13	Sh Strp Plt, Ann	0.16-0.25	12-14	1 max		1 max	0.04 max	0.03 max	1 max	bal Fe				
International															
COMECON PC4-70	4	Sh Strp Plt Bar Wir Bil	0.15-0.26	12-14	0.8 max			0.035 max	0.025 max	0.8 max	bal Fe				
ISO 683-13(74)	4*	Ann, Q/A, Tmp 1	0.16-0.25	12-14	1 max		1 max	0.04 max	0.03 max	1 max	bal Fe				
ISO 683-13(74)	5*	Ann, Q/A, Tmp	0.26-0.35	12-14		1 max	1 max	0.04 max	0.03 max	1 max	bal Fe				

UNS numbers and US grades are provided as a means of cross referencing chemically similar alloys. Exchangability is only possible after independent examination of specifications. Tensile properties are minimum or typical as specified. UTS and YS as MPa. El as %. See Appendix for list of abbreviations used in Notes. * indicates obsolete material.

Specification	Designation	Notes	C	Cr	Mn	Mo	Ni	P	S	Si	Other	UTS	YS	El	Hard

Stainless Steel, Martensitic, 420 (Continued from previous page)

Italy

Specification	Designation	Notes	C	Cr	Mn	Mo	Ni	P	S	Si	Other	UTS	YS	El	Hard
UNI 6901(71)	X20Cr13	Corr res	0.16-0.25	12-14	1 max		1 max	0.04 max	0.03 max	1 max	bal Fe				
UNI 6901(71)	X30Cr13	Corr res	0.26-0.35	12-14	1 max		1 max	0.04 max	0.03 max	1 max	bal Fe				
UNI 6904(71)	X20Cr13	Heat res	0.16-0.25	12-14	1 max		1 max	0.04 max	0.03 max	1 max	bal Fe				

Japan

Specification	Designation	Notes	C	Cr	Mn	Mo	Ni	P	S	Si	Other	UTS	YS	El	Hard
JIS G4303(91)	SUS420J1	Bar Q/A Tmp 75mm diam	0.16-0.25	12.00-14.00	1.00 max		0.60 max	0.040 max	0.030 max	1.00 max	bal Fe	640	440	20	192 HB
JIS G4303(91)	SUS420J2	Bar Q/A Tmp 75mm diam	0.26-0.40	12.00-14.00	1.00 max		0.60 max	0.040 max	0.030 max	1.00 max	bal Fe	740	540	12	217 HB
JIS G4304(91)	SUS420J1	Plt Sh, HR Ann	0.16-0.25	12.00-14.00	1.00 max		0.60 max	0.040 max	0.030 max	1.00 max	bal Fe	520	225	18	97 HRB max
JIS G4305(91)	SUS420J1	Plt Sh, CR Ann	0.16-0.25	12.00-14.00	1.00 max		0.60 max	0.040 max	0.030 max	1.00 max	bal Fe	520	225	18	223 HB max
JIS G4305(91)	SUS420J2		0.26-0.40	12.00-14.00	1.00 max			0.040 max	0.030 max	1.00 max	bal Fe	540	225	18	
JIS G4306	SUS420J1	Strp HR Ann	0.16-0.25	12-14	1 max		0.6 max	0.04 max	0.03 max	1 max	bal Fe				
JIS G4307	SUS420J1	Strp CR	0.16-0.25	12-14	1 max		0.6 max	0.04 max	0.03 max	1 max	bal Fe				
JIS G4308	SUS420J2	Wir rod	0.26-0.40	12.00-14.00	1.00 max		0.60 max	0.040 max	0.030 max	1.00 max	bal Fe				
JIS G4308(98)	SUS420J1	Wir rod	0.16-0.25	12.00-14.00	1.00 max		0.60 max	0.040 max	0.030 max	1.00 max	bal Fe				
JIS G4309	SUS420J2	Wir	0.26-0.40	12.00-14.00	1.00 max		0.60 max	0.040 max	0.030 max	1.00 max	bal Fe				
JIS G4309(94)	SUS420J1	Wir	0.16-0.25	12.00-14.00	1.00 max		0.6 max	0.040 max	0.030 max	1.00 max	bal Fe				
JIS G4313	SUS420J2-CSP	Strp CR	0.26-0.40	12.00-14.00	1.00 max		0.60 max	0.040 max	0.030 max	1.00 max	bal Fe				

Mexico

Specification	Designation	Notes	C	Cr	Mn	Mo	Ni	P	S	Si	Other	UTS	YS	El	Hard
DGN B-83	420*	Obs; Bar, HR, CR	0.15 max	12-14	1 max			0.04 max	0.03 max	1 max	bal Fe				

Poland

Specification	Designation	Notes	C	Cr	Mn	Mo	Ni	P	S	Si	Other	UTS	YS	El	Hard
PNH86020	2H13	Corr res; Sh Strp Plt Bar Wir	0.22-0.3	1.5-1.8	0.8 max		0.6 max	0.04 max	0.03 max	0.8 max	Co 0.1; bal Fe				

Romania

Specification	Designation	Notes	C	Cr	Mn	Mo	Ni	P	S	Si	Other	UTS	YS	El	Hard
STAS 3583(87)	20Cr130	Corr res; Heat res	0.17-0.25	12-14	1 max	0.2 max	0.5 max	0.045 max	0.03 max	1 max	Pb <=0.15; bal Fe				
STAS 3583(87)	30Cr130	Corr res; Heat res	0.28-0.35	12-14	1 max	0.2 max	0.5 max	0.045 max	0.03 max	1 max	Pb <=0.15; bal Fe				

Russia

Specification	Designation	Notes	C	Cr	Mn	Mo	Ni	P	S	Si	Other	UTS	YS	El	Hard
GOST	20X13	Sh Strp Plt Bar Wir, Q/A Tmp	0.16-0.25	12-14	0.8 max			0.03 max	0.025 max	0.8 max	bal Fe				
GOST 5632(61)	2KH13	Corr res; Heat res; High-temp constr	0.16-0.24	12.0-14.0	0.6 max	0.15 max	0.4 max	0.03 max	0.025 max	0.6 max	Cu <=0.3; Ti <=0.05; bal Fe				
GOST 5632(72)	20Ch13	Corr res; High-temp constr	0.16-0.25	12.0-14.0	0.8 max	0.15 max	0.6 max	0.03 max	0.025 max	0.8 max	Cu <=0.3; Ti <=0.2; bal Fe				

Spain

Specification	Designation	Notes	C	Cr	Mn	Mo	Ni	P	S	Si	Other	UTS	YS	El	Hard
UNE 36016(75)	E-420	Sh Strp Plt Bar Wir; Corr res	0.16-0.25	12.0-14.0	1.0 max	0.15 max	1.0 max	0.04 max	0.03 max	1.0 max	bal Fe				
UNE 36016(75)	F.3402*	see X20Cr13	0.16-0.25	12.0-14.0	1.0 max	0.15 max	1.0 max	0.04 max	0.03 max	1.0 max	bal Fe				
UNE 36016(75)	F.3403*	see X30Cr13	0.26-0.35	12.0-14.0	1.0 max	0.15 max	1.0 max	0.04 max	0.03 max	1.0 max	bal Fe				
UNE 36016(75)	X20Cr13	Sh Strp Plt Bar Wir; Corr res	0.16-0.25	12.0-14.0	1.0 max	0.15 max	1.0 max	0.04 max	0.03 max	1.0 max	bal Fe				
UNE 36016(75)	X30Cr13	Corr res	0.26-0.35	12.0-14.0	1.0 max	0.15 max	1.0 max	0.04 max	0.03 max	1.0 max	bal Fe				
UNE 36016/1(89)	F.3402*	Obs; E-420; Corr res	0.17-0.23	12.0-14.0	1.0 max	0.15 max	1.0 max	0.04 max	0.03 max	1.0 max	bal Fe				
UNE 36016/1(89)	F.3415*	Obs; EN10088-3(96); X15Cr13	0.12-0.17	12.0-14.0	1.0 max	0.15 max	1.0 max	0.04 max	0.03 max	1.0 max	bal Fe				
UNE 36072/1(75)	F.5262*	see X30Cr13	0.26-0.35	12-14	1 max	0.15 max	1 max	0.03 max	0.03 max	1 max	bal Fe				

Specification	Designation	Notes	C	Cr	Mn	Mo	Ni	P	S	Si	Other	UTS	YS	El	Hard

Stainless Steel, Martensitic, 420 (Continued from previous page)

Spain

Specification	Designation	Notes	C	Cr	Mn	Mo	Ni	P	S	Si	Other	UTS	YS	El	Hard
UNE 36072/1(75)	X30Cr13	Tool; CW	0.26-0.35	12-14	1 max	0.15 max	1 max	0.03 max	0.03 max	1 max	bal Fe				

Sweden

Specification	Designation	Notes	C	Cr	Mn	Mo	Ni	P	S	Si	Other	UTS	YS	El	Hard
SS 142303	2303	Corr res	0.26-0.35	12-14	1 max		1 max	0.04 max	0.03 max	1 max	bal Fe				
SS 142304	2304	Corr res	0.26-0.35	12-14	1 max		1 max	0.04 max	0.03 max	1 max	bal Fe				

UK

Specification	Designation	Notes	C	Cr	Mn	Mo	Ni	P	S	Si	Other	UTS	YS	El	Hard
BS 1449/2(83)	420S45	Corr/Heat res; t<=100mm	0.28-0.36	12-14	1.00 max		1.00 max	0.040 max	0.030 max	1.00 max	bal Fe				0-230 HB
BS 1503(89)	420S29	Frg press ves; t<=999mm; Q/T	0.14-0.20	11.5-13.5	1.00 max	0.30 max	1.00 max	0.040 max	0.025 max	0.80 max	Cu <=0.30; W <=0.1; bal Fe	700-850	515	16	
BS 1554(90)	420S29	Wir Corr/Heat res; diam<0.49mm; Ann; wasEn56(B)	0.14-0.20	11.5-13.5	1.00 max		1.00 max	0.040 max	0.030 max *	1.00 max	W <=0.1; bal Fe	0-780			
BS 1554(90)	420S29	Wir Corr/Heat res; 1.5<diam<=13mm; Ann; was En56(B)	0.14-0.20	11.5-13.5	1.00 max		1.00 max	0.040 max	0.030 max	1.00 max	W <=0.1; bal Fe	0-700			
BS 1554(90)	420S37	Wir Corr/Heat res; 1.5<diam<=13mm; Ann; was En56(C)	0.2-0.28	12.0-14.0	1.00 max		1.00 max	0.040 max	0.030 max	1.00 max	W <=0.1; bal Fe	0-780			
BS 1554(90)	420S37	Wir Corr/Heat res; diam<0.49mm; Ann; was En56(C)	0.2-0.28	12.0-14.0	1.00 max		1.00 max	0.040 max	0.030 max	1.00 max	W <=0.1; bal Fe	0-820			
BS 1554(90)	420S45	Wir Corr/Heat res; diam<0.49mm; Ann; En56(D)	0.28-0.36	12.0-14.0	1.00 max		1.00 max	0.040 max	0.030 max	1.00 max	W <=0.1; bal Fe	0-820			
BS 1554(90)	420S45	Wir Corr/Heat res; 1.5<diam<=13mm; Ann; En56(D)	0.28-0.36	12.0-14.0	1.00 max		1.00 max	0.040 max	0.030 max	1.00 max	W <=0.1; bal Fe	0-780			
BS 2056(83)	420S45	Corr res; Spring	0.28-0.36	12.0-14.0	1.0 max	0.15 max	1.0 max	0.04 max	0.03 max	1.0 max	Al <=0.1; Co <=0.1; Cu <=0.3; Pb <=0.15; Ti <=0.05; V <=0.1; W <=0.1; bal Fe				
BS 970/1(96)	420S29	Wrought Corr/Heat res; t<=150mm; Tmp	0.14-0.20	11.5-13.5	1.00 max		1.00 max	0.040 max	0.030 max	1.00 max	bal Fe	700-850	525	15	201-255 HB
BS 970/1(96)	420S29	Wrought Corr/Heat res; t<=29mm; Tmp	0.14-0.20	11.5-13.5	1.00 max		1.00 max	0.040 max	0.030 max	1.00 max	bal Fe	775-925	585	13	223-277 HB
BS 970/1(96)	420S37	Wrought Corr/Heat res; t<=150mm; Tmp	0.20-0.28	12.0-14.0	1.00 max		1.00 max	0.040 max	0.030 max	1.00 max	bal Fe	775-925	585	13	223-277 HB

USA

Specification	Designation	Notes	C	Cr	Mn	Mo	Ni	P	S	Si	Other	UTS	YS	El	Hard
	AISI 420	Tube	0.15 min	12.00-14.00	1.25 max			0.040 max	0.030 max	1.00 max	bal Fe				
	UNS S42000	Hardenable	0.15 min	12.00-14.00	1.00 max			0.040 max	0.030 max	1.00 max	bal Fe				
ASTM A176(97)	420	Sh Strp Plt	0.15 min	12.00-14.00	1.00 max	0.50 max	0.75 max	0.040 max	0.030 max	1.00 max	bal Fe	690		15.0	217 HB
ASTM A276(98)	420	Bar	0.15 min	12-14	1 max			0.04 max	0.03 max	1 max	bal Fe				
ASTM A314	420	Bil	0.15 min	12-14	1 max			0.04 max	0.03 max	1 max	bal Fe				
ASTM A473	420	Frg, Heat res ann	0.15 min	12.00-14.00	1.00 max			0.040 max	0.030 max	1.00 max	bal Fe				223 HB
ASTM A580/A580M(98)	420	Wir, Ann	0.15 min	12.0-14.0	1.00 max			0.040 max	0.030 max	1.00 max	bal Fe	860 max			50 HCR
FED QQ-S-763F(96)	420*	Obs; Bar Wir Shp Frg; Corr res	0.15	12-14	1 max			0.04 max	0.03 max	1 max	bal Fe				
FED QQ-S-766D(93)	420*	Obs; see ASTM A240, A666, A693; Sh Strp Plt	0.15 min	12-14	1 max			0.04 max	0.03 max	1 max	bal Fe				
MIL-S-862B	420*	Obs	0.15 min	12-14	1 max			0.04 max	0.03 max	1 max	bal Fe				

Stainless Steel, Martensitic, 420F

China

Specification	Designation	Notes	C	Cr	Mn	Mo	Ni	P	S	Si	Other	UTS	YS	El	Hard
GB 1220(92)	Y3Cr13	Bar Q/T	0.26-0.40	12.00-14.00	1.25 max	0.60 max	0.60 max	0.060 max	0.15 min	1.00 max	bal Fe	540	345	25	

UNS numbers and US grades are provided as a means of cross referencing chemically similar alloys. Exchangability is only possible after independent examination of specifications. Tensile properties are minimum or typical as specified. UTS and YS as MPa. El as %. See Appendix for list of abbreviations used in Notes. * indicates obsolete material.

Specification	Designation	Notes	C	Cr	Mn	Mo	Ni	P	S	Si	Other	UTS	YS	El	Hard

Stainless Steel, Martensitic, 420F (Continued from previous page)

Europe

Specification	Designation	Notes	C	Cr	Mn	Mo	Ni	P	S	Si	Other	UTS	YS	El	Hard
EN 10088/2(95)	1.4028	Sh Plt Strp, Corr res; t<12 Ann	0.26-0.35	12.00-14.00	1.50 max			0.040 max	0.015 max	1.00 max	bal Fe	740 max		15	235 HB max
EN 10088/2(95)	1.4031	Sh Plt Strp, Corr res; t<12 Ann	0.36-0.42	12.50-14.50	1.00 max			0.040 max	0.015 max	1.00 max	bal Fe	760 max		12	240 HB max
EN 10088/2(95)	X30Cr13	Sh Plt Strp, Corr res; t<12 Ann	0.26-0.35	12.00-14.00	1.50 max			0.040 max	0.015 max	1.00 max	bal Fe	740 max		15	235 HB max
EN 10088/2(95)	X39Cr13	Sh Plt Strp, Corr res; t<12 Ann	0.36-0.42	12.50-14.50	1.00 max			0.040 max	0.015 max	1.00 max	bal Fe	760 max		12	240 HB max

Japan

Specification	Designation	Notes	C	Cr	Mn	Mo	Ni	P	S	Si	Other	UTS	YS	El	Hard
JIS G4303(91)	SUS420F	Bar Q/A Tmp 75mm diam	0.26-0.40	12.00-14.00	1.25 max	0.60 max	0.60 max	0.060 max	0.15 min	1.00 max	bal Fe	740	540	8	217 HB
JIS G4303(91)	SUS420F2	Bar Q/A Tmp 75mm diam	0.26-0.40	12.00-14.00	1.00 max		0.60 max	0.40 max	0.030 max	1.00 max	Pb 0.05-0.30; bal Fe	740	540	5	217 HB
JIS G4308	SUS420F	Wir rod	0.26-0.40	12.00-14.00	1.25 max	0.60 max	0.60 max	0.060 max	0.15 max	1.00 max	bal Fe				
JIS G4308	SUS420F2	Wir rod	0.26-0.40	12.00-14.00	1.00 max		0.60 max	0.040 max	0.030 max	1.00 max	Pb 0.05-0.30; bal Fe				
JIS G4309	SUS420F	Wir	0.26-0.40	12.00-14.00	1.25 max	0.60 max	0.60 max	0.060 max	0.15 min	1.00 max	bal Fe				
JIS G4309	SUS420F2	Wir	0.26-0.40	12.00-14.00	1.00 max		0.60 max	0.040 max	0.030 max	1.00 max	Pb 0.05-0.30; bal Fe				

Mexico

Specification	Designation	Notes	C	Cr	Mn	Mo	Ni	P	S	Si	Other	UTS	YS	El	Hard
DGN B-83	420F*	Obs; Bar, HR, CR	0.15 min	12-14	1.25 max	0.6 max		0.06 max	0.15-99.99	1 max	bal Fe				

Russia

Specification	Designation	Notes	C	Cr	Mn	Mo	Ni	P	S	Si	Other	UTS	YS	El	Hard
GOST 5632(72)	30Ch13	Corr res	0.26-0.35	12.0-14.0	0.8 max	0.15 max	0.6 max	0.03 max	0.025 max	0.8 max	Cu <=0.3; Ti <=0.2; bal Fe				

Spain

Specification	Designation	Notes	C	Cr	Mn	Mo	Ni	P	S	Si	Other	UTS	YS	El	Hard
UNE 36016(75)	F.3404*	see X40Cr13	0.36-0.45	12.5-14.5	1.0 max	0.15 max	1.0 max	0.04 max	0.03 max	1.0 max	bal Fe				
UNE 36016(75)	X40Cr13	Corr res	0.36-0.45	12.5-14.5	1.0 max	0.15 max	1.0 max	0.04 max	0.03 max	1.0 max	bal Fe				

USA

Specification	Designation	Notes	C	Cr	Mn	Mo	Ni	P	S	Si	Other	UTS	YS	El	Hard
	AISI 420F	Tube	0.15 min	12.00-14.00	1.25 max	0.60 max		0.060 max	0.15 min	1.00 max	bal Fe				
	UNS S42020	Hardenable Free Machining	0.15 min	12.00-14.00	1.25 max	0.60 max		0.060 max	0.15 min	1.00 max	bal Fe				
ASTM A582/A582M(95)	420F	Bar, HF, CF, Ann	0.15-0.40	12.00-14.00	1.25 max			0.06 max	0.15 min	1.00 max	Cu <=0.60; bal Fe				262 HB max
ASTM A582/A582M(95)	420FSe	Bar, HF, CF, Ann	0.20-0.40	12.00-14.00	1.25 max		0.50	0.06 max	0.06 max	1.00 max	Cu <=0.60; Se>=0.15; bal Fe				262 HB max
ASTM A895(94)	420F	Sh Strp Plt, free mach, Ann	0.30-0.40	12.0-14.0	1.25 max		0.50 max	0.06 max	0.15 min	1.00 max	Cu <=0.60; bal Fe				262 HB max

Stainless Steel, Martensitic, 420FSe

India

Specification	Designation	Notes	C	Cr	Mn	Mo	Ni	P	S	Si	Other	UTS	YS	El	Hard
IS 1570/5(85)	X40Cr13	Bar Flat Band	0.35-0.45	12-14	1 max	0.15 max	1 max	0.04 max	0.03 max	1 max	Co <=0.1; Cu <=0.3; Pb <=0.15; W <=0.1; bal Fe	600-750			225 HB max

USA

Specification	Designation	Notes	C	Cr	Mn	Mo	Ni	P	S	Si	Other	UTS	YS	El	Hard
	UNS S42023	Hard Cr Free Machining Se-bearing	0.30-0.40	12.0-14.0	1.25 max	0.60 max		0.06 max	0.06 max	1.00 max	Se 0.15 min; Zr or Cu 0.60 max; bal Fe				
ASTM A895(94)	420FSe	Sh Strp Plt, free mach, Ann	0.20-0.40	12.0-14.0	1.25 max		0.50 max	0.06 max	0.06 max	1.00 max	Cu <=0.60; Se 0.15; bal Fe				262 HB max

Stainless Steel, Martensitic, 422

China

Specification	Designation	Notes	C	Cr	Mn	Mo	Ni	P	S	Si	Other	UTS	YS	El	Hard
GB 1221(92)	2Cr12NiMoWV	Bar Q/T	0.20-0.25	11.00-13.00	0.50-1.00	0.75-1.25	0.50-1.00	0.035 max	0.030 max	0.50 max	Cu <=0.30; V 0.20-0.40; W 0.75-1.25; bal Fe	885	735	10	

Italy

Specification	Designation	Notes	C	Cr	Mn	Mo	Ni	P	S	Si	Other	UTS	YS	El	Hard
UNI	X22CrMoWV121	Bar	0.22 max	12 max	0.7 max	1.05 max		0.025 max	0.025 max	0.5 max	V <=0.25; W <=1.05; bal Fe				
UNI 7660(77)	X20CrMoNi1201KG	Press ves	0.23 max	11.5-12.5	0.3-1	0.7-1.2	0.3-1	0.04 max	0.03 max	0.15-0.4	V 0.2-0.35; bal Fe				

UNS numbers and US grades are provided as a means of cross referencing chemically similar alloys. Exchangability is only possible after independent examination of specifications. Tensile properties are minimum or typical as specified. UTS and YS as MPa. El as %. See Appendix for list of abbreviations used in Notes. * indicates obsolete material.

Specification	Designation	Notes	C	Cr	Mn	Mo	Ni	P	S	Si	Other	UTS	YS	El	Hard
Stainless Steel, Martensitic, 422 (Continued from previous page)															
Italy															
UNI 7660(77)	X20CrMoNi1201KW	Press ves	0.2-0.26	11.5-12.5	0.3-1	0.7-1.2	0.3-1	0.04 max	0.03 max	0.15-0.4	V 0.2-0.35; bal Fe				
Japan															
JIS G4311(91)	SUH616	Bar Q/A Tmp 75mm diam	0.20-0.25	11.00-13.00	0.5-1	0.75-1.25	0.50-1.00	0.040 max	0.030 max	0.50 max	V 0.20-0.30; W 0.75-1.25; bal Fe	880	735	25	341 HB
Poland															
PNH84024	15H12WMF	High-temp const; Bar	0.12-0.18	11-12.5	0.5-0.9	0.5-0.7	0.4-0.8	0.03 max	0.03 max	0.4 max	V 0.15-0.3; W 0.7-1.1; bal Fe				
Romania															
STAS 11523(87)	20VNiMoCr120	Corr res; Heat res	0.17-0.23	10-12.5	1.2 max	0.8-1.2	0.3-0.8	0.035 max	0.035 max	0.1-0.5	Pb <=0.15; V 0.25-0.35; bal Fe				
STAS 3583(87)	20MoCr130	Corr res	0.17-0.22	12-14	1 max	0.9-1.3	1 max	0.045 max	0.03 max	1 max	Pb <=0.15; bal Fe				
Russia															
GOST 5632	15Ch12WNMF	Bar Q/A Tmp	0.12-0.18	11-13	0.5-0.9	0.5-0.7	0.4-0.8	0.03 max	0.025 max	0.4 max	Ti <=0.2; V 0.15-0.3; W 0.7-1.1; bal Fe				
Spain															
UNE 36016/1(89)	E-422	Corr res	0.35-0.42	12.5-14.5	1.0 max	0.15 max	1.0 max	0.04 max	0.03 max	1.0 max	bal Fe				
UNE 36016/1(89)	F.3404*	Obs; E-422; Corr res	0.35-0.42	12.5-14.5	1.0 max	0.15 max	1.0 max	0.04 max	0.03 max	1.0 max	bal Fe				
USA															
	AISI 422	Tube	0.20-0.25	11.00-13.00	1.00 max	0.75-1.25	0.50-1.00	0.025 max	0.025 max	0.75 max	V 0.15-0.30; W 0.75-1.25; bal Fe				
	UNS S42200		0.2-0.25	11.00-13.50	1.00 max	0.75-1.25	0.50-1.00	0.040 max	0.030 max	0.75 max	V 0.15-0.30; W 0.75-1.25; bal Fe				
ASTM A176(97)	422	Sh Strp Plt	0.20-0.25	11.00-12.50	0.50-1.00	0.90-1.25	0.50-1.00	0.025 max	0.025 max	0.50 max	V 0.20-0.30; W 0.90-1.25; bal Fe				248 HB
ASTM A565(97)	616	Bar Frg for high-temp HT	0.20-0.25	11.00-12.50	0.50-1.00	0.90-1.25	0.50-1.00	0.025 max	0.025 max	0.50 max	V 0.20-0.30; W 0.90-1.25; bal Fe	965	760	13	302-352 HB
ASTM A982(98)	Grade D Class 1	Frg for turbine airfoils	0.20-0.25	11.0-12.5	0.5-1.0	0.9-1.25	0.5-1.0	0.020 max	0.010 max	0.20-0.50	Al <=0.025; Co <=0.20; Cu <=0.15; Nb <=0.05; Ti <=0.025; V 0.20-0.30; W 0.9-1.25; Sn=0.02; bal Fe	965	621	13	
ASTM A982(98)	Grade D Class 2	Frg for turbine airfoils	0.20-0.25	11.0-12.5	0.5-1.0	0.9-1.25	0.5-1.0	0.020 max	0.010 max	0.20-0.50	Al <=0.025; Co <=0.20; Cu <=0.15; Nb <=0.05; Ti <=0.025; V 0.20-0.30; W 0.9-1.25; Sn=0.02; bal Fe	965	690	13	
DoD-F-24669/7(86)	422	HT; Supersedes MIL-S-861	0.20-0.25	11.5-13.5	1.00 max	0.75-1.25	0.50-1.00	0.040 max	0.030 max	0.75 max	V 0.15-0.30; W 0.75-1.25; bal Fe	965	760	13	302-352 HB
MIL-S-861	422*	Obs	0.2-0.25	11.5-13.5	1 max	0.75-1.25	0.5-1	0.04 max	0.03 max	0.75 max	V 0.15-0.3; W 0.75-1.25; bal Fe				
SAE J467(68)	422	Bar	0.22	12.5	0.75	1.00	0.75	0.04 max	0.03 max	0.40	V 0.22; W 1.00; bal Fe	1027	862	18	43 HRC
SAE J775(93)	422SS	Engine poppet valve	0.20-0.25	11.00-12.50	1.00 max	0.75-1.25	0.50-1.00	0.040 max	0.030 max	0.75 max	Cu <=0.50; V 0.15-0.30; W 0.75-1.25; bal Fe	1030	860		
SAE J775(93)	HNV-8	Engine poppet valve	0.2-0.25	11.5-13.5	1 max	0.75-1.25	0.5-1	0.04 max	0.03 max	0.75 max	V 0.15-0.3; W 0.75-1.25; bal Fe				
SAE J775(93)	HNV8*	Engine poppet valve	0.20-0.25	11.00-12.50	1.00 max	0.75-1.25	0.50-1.00	0.040 max	0.030 max	0.75 max	Cu <=0.50; V 0.15-0.30; W 0.75-1.25; bal Fe	1030	860		
SAE J775(93)	JIS G4311 SUH 616	Engine poppet valve	0.20-0.25	11.00-13.00	0.50-1.00	0.75-1.25	0.50-1.00	0.040 max	0.030 max	0.50 max	V 0.20-0.30; W 0.75-1.25; bal Fe				
Stainless Steel, Martensitic, 430															
Europe															
EN 10088/2(95)	1.4016	Strp; Corr res; CR t<=6mm, Ann	0.08 max	16.00-18.00	1.00 max			0.040 max	0.015 max	1.00 max	bal Fe	450-600	260	20	

Stainless Steel, Martensitic, 431

Specification	Designation	Notes	C	Cr	Mn	Mo	Ni	P	S	Si	Other	UTS	YS	El	Hard
Australia															
AS 2837(86)	431	Bar, Semi-finished product	0.12-0.20	15.0-17.0	1.00 max		1.5-3.00	0.040 max	0.030 max	1.00 max	bal Fe				
Bulgaria															
BDS 6738(72)	1Ch17N2	Bar Bil	0.11-0.17	16.0-18.0	0.80 max	0.30 max	1.50-2.50	0.035 max	0.025 max	0.80 max	bal Fe				
BDS 6738(72)	2Ch17N2	Corr res; bar	0.17-0.25	16.0-18.0	0.80 max	0.30 max	1.50-2.50	0.035 max	0.025 max	0.80 max	bal Fe				
Canada															
CSA G110.3	431	Bar Bil	0.2 max	15-17	1 max		1.25-2.5	0.04 max	0.03 max	1 max	bal Fe				
China															
GB 1220(92)	1Cr17Ni2	Bar Q/T	0.11-0.17	16.00-18.00	0.80 max		1.50-2.50	0.035 max	0.030 max	0.80 max	bal Fe	1080		10	
GB 1221(92)	1Cr17Ni2	Bar Q/T	0.11-0.17	16.00-18.00	0.80 max		1.50-2.50	0.035 max	0.030 max	0.80 max	bal Fe	1080		10	
GB 3280(92)	1Cr17Ni2	Sh Plt, CR Ann	0.11-0.17	16.00-18.00	0.80 max		1.50-2.50	0.035 max	0.030 max	0.80 max	bal Fe	1080		10	
GB 4232(93)	ML1Cr17Ni2	Wir Ann 1-3mm diam	0.11-0.17	16.00-18.00	0.80 max		1.50-2.50	0.035 max	0.030 max	0.80 max	bal Fe	590-790		15	
GB 4240(93)	1Cr17Ni2(-R)	Wir Ann 0.05-14mm diam, R-As Ann	0.11-0.17	16.00-18.00	0.80 max		1.50-2.50	0.035 max	0.030 max	0.80 max	bal Fe	590-880			
Europe															
EN 10088/3(95)	1.4057	Bar Rod Sect; Corr res; Ann	0.12-0.22	15.00-17.00	1.50 max		1.50-2.50	0.040 max	0.030 max	1.00 max	bal Fe	950 max			295 HB max
EN 10088/3(95)	X17CrNi16-2	Bar Rod Sect; Corr res; Ann	0.12-0.22	15.00-17.00	1.50 max		1.50-2.50	0.040 max	0.030 max	1.00 max	bal Fe	950 max			295 HB max
France															
AFNOR NFA35572	Z15CN16.02	Bar, Q/A Tmp	0.1-0.2	15-17	1 max		1.5-3	0.04 max	0.03 max	1 max	bal Fe				
Germany															
DIN 17440(96)	WNr 1.4057	Bar Rod Plt Sh, HR Strp Wir Frg HT	0.12-0.22	15.0-17.0	1.50 max		1.50-2.50	0.040 max	0.030 max	1.00 max	bal Fe	800-950	600	12-14	
DIN 17440(96)	X17CrNi16-2	HT	0.12-0.22	15.0-17.0	1.50 max		1.50-2.50	0.040 max	0.030 max	1.00 max	bal Fe	800-950	600	12-14	
DIN EN 10088(95)	WNr 1.4057	Plt Sh Strp Bar Wir Frg, HT	0.12-0.22	15.0-17.0	1.50 max		1.50-2.50	0.040 max	0.030 max	1.00 max	bal Fe	800-950	600	12-14	
DIN EN 10088(95)	X17CrNi16-2	HT	0.12-0.22	15.0-17.0	1.50 max		1.50-2.50	0.040 max	0.030 max	1.00 max	bal Fe	800-950	600	12-14	
Hungary															
MSZ 4360(87)	KO16	Corr res; 0-100mm; Q/T	0.1-0.17	16-18	1 max		1.5-2.5	0.045 max	0.03 max	1 max	bal Fe	880-1080	680	11L	
MSZ 4360(87)	KO16	Corr res; 0-100mm; soft ann	0.1-0.17	16-18	1 max		1.5-2.5	0.045 max	0.03 max	1 max	bal Fe	0-950			285 HB max
India															
IS 1570/5(85)	X15Cr16Ni2	Sh Plt Strp Bar Flat Band	0.1-0.2	15-17	1 max	0.15 max	1.25-2.5	0.045 max	0.03 max	1 max	Co <=0.1; Cu <=0.3; Pb <=0.15; W <=0.1; bal Fe				285 HB max
IS 6529	15Cr16Ni2	Bil	0.1-0.2	15-18	1 max		1.5-3	0.04 max	0.03 max	1 max	bal Fe				
IS 6603	15Cr16Ni2	Ann, Bar, 5-100mm diam	0.1-0.2	15-18	1 max		1.5-3	0.04 max	0.03 max	1 max	bal Fe				
IS 6911	15Cr16Ni2	Sh Strp Plt, Ann	0.1-0.2	15-18	1 max		1.5-3	0.04 max	0.03 max	1 max	bal Fe				
International															
COMECON PC4-70	10	Bar Bil	0.17-0.25	16-18	0.8 max		1.5-2.5	0.035 max	0.025 max	0.8 max	bal Fe				
COMECON PC4-70	9	Bar Bil	0.11-0.17	16-18	0.8 max		1.5-2.5	0.035 max	0.025 max	0.8 max	bal Fe				
ISO 683-13(74)	9a*	Ann, Q/A, Tmp	0.1-0.17	15.5-17.5	1.5 max	0.6 max	1 max	0.06 max	0.15-0.35	1 max	bal Fe				
ISO 683-13(74)	9b*	Ann, Q/A, Tmp	0.14-0.23	15-17.5	1 max		1.5-2.5	0.04 max	0.03 max	1 max	bal Fe				
Italy															
UNI 6901(71)	X16CrNi16	Corr res	0.1-0.2	15-17	1 max		1.5-2.5	0.04 max	0.03 max	1 max	bal Fe				

UNS numbers and US grades are provided as a means of cross referencing chemically similar alloys. Exchangability is only possible after independent examination of specifications. Tensile properties are minimum or typical as specified. UTS and YS as MPa. El as %. See Appendix for list of abbreviations used in Notes. * indicates obsolete material.

Stainless Steel, Martensitic, 431 (Continued from previous page)

Specification	Designation	Notes	C	Cr	Mn	Mo	Ni	P	S	Si	Other	UTS	YS	El	Hard
Japan															
JIS G4303(91)	SUS431	Bar Q/A Tmp 75mm diam	0.20 max	15.00-17.00	1.00 max		1.25-2.50	0.040 max	0.030 max	1.00 max	bal Fe	780	590	15	229 HB
JIS G4308	SUS431	Wir rod	0.20 max	15.00-17.00	1.00 max		1.25-2.50	0.040 max	0.30 max	1.00 max	bal Fe				
Mexico															
DGN B-83	431*	Obs; Bar, HR, CR	0.2 max	15-17	1 max		1.25-2.5	0.04 max	0.03 max	1 max	bal Fe				
NMX-B-171(91)	MT431	mech tub	0.20 max	15.0-17.0	1.00 max		1.25-2.50	0.040 max	0.030 max	1.00 max	bal Fe				
Norway															
NS	14230	Bar Wire	0.17-0.25	16-17	1 max		1.25-2.5	0.04 max	0.03 max	1 max	bal Fe				
Poland															
PNH86020	2H17N2	Corr res	0.17-0.25	16-18	0.8 max		1.5-2.5	0.04 max	0.03 max	0.8 max	bal Fe				
PNH86020	H17N2	Corr res	0.11-0.17	16-18	0.8 max		1.5-2.5	0.04 max	0.03 max	0.8 max	bal Fe				
Romania															
STAS 3583(64)	14NC170	Bar Wir Bil; Corr res; Heat res	0.11-0.17	16-18	0.8 max		1.5-2.5	0.035 max	0.03 max	0.8 max	Pb <=0.15; bal Fe				
STAS 3583(87)	22NiCr170	Corr res	0.14-0.23	15.5-17.5	1 max	0.2 max	1.5-2.5	0.045 max	0.03 max	1 max	Pb <=0.15; bal Fe				
Russia															
GOST 5632	14Ch17N2	Bar Bil Frg	0.11-0.17	16-18	0.8 max	0.3 max	1.5-2	0.03 max	0.025 max	0.8 max	Cu <=0.3; Ti <=0.2; W <=0.2; bal Fe				
GOST 5632	20Ch17N2	Bar Bil Frg	0.17-0.25	16-18	0.8 max	0.3 max	1.5-2	0.035 max	0.025 max	0.8 max	Cu <=0.3; Ti <=0.2; W <=0.2; bal Fe				
GOST 5632(61)	1KH17N2	Heat res; Corr res; High-temp constr	0.11-0.17	16.0-18.0	0.8 max	0.15 max	1.5-2.5	0.03 max	0.025 max	0.8 max	Cu <=0.3; Ti <=0.05; bal Fe				
GOST 5632(72)	14Ch17N2	Corr res; High-temp constr	0.11-0.17	16.0-18.0	0.8 max	0.3 max	1.5-2.5	0.03 max	0.025 max	0.8 max	Cu <=0.3; Ti <=0.2; W <=0.2; bal Fe				
GOST 5632(72)	20Ch17N2	Corr res	0.17-0.25	16.0-18.0	0.8 max	0.3 max	1.5-2.5	0.035 max	0.025 max	0.8 max	Cu <=0.3; Ti <=0.2; W <=0.2; bal Fe				
Spain															
UNE 36016(75)	F.3427*	see X15CrNi16	0.1-0.2	15.0-18.0	1.0 max	0.15 max	1.5-3.0	0.04 max	0.03 max	1.0 max	bal Fe				
UNE 36016(75)	X15CrNi16	Bar Wir Bil; Corr res	0.1-0.2	15.0-18.0	1.0 max	0.15 max	1.5-3.0	0.04 max	0.03 max	1.0 max	bal Fe				
UNE 36016/1(89)	E-431	Corr res	0.14-0.23	15.5-17.5	1.0 max	0.15 max	1.5-2.5	0.04 max	0.03 max	1.0 max	bal Fe				
UNE 36016/1(89)	F.3427*	Obs; E-431; Corr res	0.14-0.23	15.5-17.5	1.0 max	0.15 max	1.5-2.5	0.04 max	0.03 max	1.0 max	bal Fe				
Sweden															
SS 142321	2321	Corr res	0.14-0.23	15.5-17.5	1 max		1.25-2.5	0.04 max	0.03 max	1 max	bal Fe				
UK															
BS 1554(90)	431S29	Wir Corr/Heat res; 6<diam<=13mm; Ann	0.12-0.20	15.0-18.0	1.00 max		2.00-3.00	0.040 max	0.030 max	1.00 max	W <=0.1; bal Fe	0-920			
BS 1554(90)	431S29	Wir Corr/Heat res; diam<1.49mm; Ann	0.12-0.20	15.0-18.0	1.00 max		2.00-3.00	0.040 max	0.030 max	1.00 max	W <=0.1; bal Fe	0-980			
BS 1554(90)	441S49	Wir Corr/Heat res; 6<diam<=13mm; Ann	0.12-0.20	15.0-18.0	1.50 max	0.60 max	2.00-3.00	0.040 max	0.030 max	1.00 max	W <=0.1; 0.15<=Se<=0.35; bal Fe	0-920			
BS 1554(90)	441S49	Wir Corr/Heat res; diam<1.49mm; Ann	0.12-0.20	15.0-18.0	1.50 max	0.60 max	2.00-3.00	0.040 max	0.030 max	1.00 max	W <=0.1; 0.15<=Se<=0.35; bal Fe	0-980			
BS 970/1(96)	431S29	Wrought Corr/Heat res; t<=150mm; Tmp	0.12-0.20	15.0-18.0	1.00 max		2.00-3.00	0.040 max	0.030 max	1.00 max	bal Fe	850-1000	680	11	248-302 HB
USA															
	AISI 431	Tube	0.20 max	15.00-17.00	1.00 max		1.25-2.50	0.040 max	0.030 max	1.00 max	bal Fe				
	UNS S43100	Ni-bearing; Hardenable	0.20 max	15.00-17.00	1.00 max		1.25-2.50	0.040 max	0.030 max	1.00 max	bal Fe				
ASTM A176(97)	431	Sh Strp Plt	0.20 max	15.00-17.00	1.00 max		1.25-2.50	0.040 max	0.030 max	1.00 max	bal Fe				285 HB

UNS numbers and US grades are provided as a means of cross referencing chemically similar alloys. Exchangability is only possible after independent examination of specifications. Tensile properties are minimum or typical as specified. UTS and YS as MPa. El as %. See Appendix for list of abbreviations used in Notes. * indicates obsolete material.

Specification	Designation	Notes	C	Cr	Mn	Mo	Ni	P	S	Si	Other	UTS	YS	El	Hard

Stainless Steel, Martensitic, 431 (Continued from previous page)

USA

Specification	Designation	Notes	C	Cr	Mn	Mo	Ni	P	S	Si	Other	UTS	YS	El	Hard
ASTM A276(98)	431	Bar	0.2 max	15-17	1 max		1.25-2.5	0.04 max	0.03 max	1 max	bal Fe				
ASTM A314	431	Bil	0.2 max	15-17	1 max		1.25-2.5	0.04 max	0.03 max	1 max	bal Fe				
ASTM A473	431	Frg, Heat res, tmp	0.20 max	15.00-17.00	1.00 max		1.25-2.50	0.040 max	0.030 max	1.00 max	bal Fe	795	620	15	321 HB
ASTM A479	431	Bar, Tmp	0.2 max	15-17	1 max		1.25-2.5	0.04 max	0.03 max	1 max	bal Fe				
ASTM A493	431	Wir rod, CHd, cold frg	0.20 max	15.0-17.0	1.00 max		1.25-2.50	0.040 max	0.030 max	1.00 max	bal Fe	760			
ASTM A511(96)	MT321	Smls mech tub, Ann	0.08 max	17.0-20.0	2.00 max		9.0-13.0	0.040 max	0.030 max	1.00 max	5xC<=Ti<=0.60; bal Fe	517	207	35	192 HB
ASTM A511(96)	MT431	Smls mech tub, Ann	0.20 max	15.0-17.0	1.00 max		1.25-2.50	0.040 max	0.030 max	1.00 max	bal Fe	724	621	20	260 HB
ASTM A579(96)	53	Superstrength Frg	0.20 max	15.0-17.0	1.00 max		1.25-2.50	0.025 max	0.025 max	1.00 max	bal Fe	1210	965	12	285 HB
ASTM A580/A580M(98)	431	Wir, Ann	0.20 max	15.0-17.0	1.00 max		1.25-2.50	0.040 max	0.030 max	1.00 max	bal Fe	965 max			40 HCR
MIL-S-18732(68)	431*	Obs for new design; Bar Wir Frg Tub	0.2 max	15-17	1 max		1.25-2.5	0.04 max	0.03 max	1 max	bal Fe				
MIL-S-862B	431*	Obs	0.2 max	15-17	1 max		1.25-2.5	0.04 max	0.03 max	1 max	bal Fe				
MIL-S-8967(68)	431*	Obs; Bar Wir Frg Tub; premium qual	0.2 max	15-17	1 max		1.25-2.5	0.04 max	0.03 max	1 max	bal Fe				

Stainless Steel, Martensitic, 440A

Canada

Specification	Designation	Notes	C	Cr	Mn	Mo	Ni	P	S	Si	Other	UTS	YS	El	Hard
CSA G110.3	440A	Bar Bil	0.6-0.75	16-18	1 max	0.75 max		0.04 max	0.03 max	1 max	bal Fe				

China

Specification	Designation	Notes	C	Cr	Mn	Mo	Ni	P	S	Si	Other	UTS	YS	El	Hard
GB 1220(92)	7Cr17	Bar Ann	0.60-0.75	16.00-18.00	1.00 max	0.75 max	0.60 max	0.035 max	0.030 max	1.00 max	bal Fe				255 max HB
GB 3280(92)	7Cr17	Sh Plt, CR Ann	0.60-0.75	16.00-18.00	1.00 max	0.75 max	0.60 max	0.035 max	0.030 max	1.00 max	bal Fe	590	245	15	
GB 4237(92)	7Cr17	Sh Plt, HR Ann	0.60-0.75	16.00-18.00	1.00 max	0.75 max	0.60 max	0.035 max	0.030 max	1.00 max	bal Fe	590	245	15	
GB 4239(91)	7Cr17	Sh Strp, CR Ann	0.60-0.75	16.00-18.00	1.00 max	0.75 max	0.60 max	0.035 max	0.030 max	1.00 max	bal Fe	590	245	15	

Europe

Specification	Designation	Notes	C	Cr	Mn	Mo	Ni	P	S	Si	Other	UTS	YS	El	Hard
EN 10088/3(95)	1.4109	Bar Rod Sect, Corr res; t<=100mm, Ann	0.65-0.75	14.00-16.00	1.00 max	0.40-0.80		0.040 max	0.030 max	0.70 max	bal Fe	900 max			280 HB max
EN 10088/3(95)	X70CrMo15	Bar Rod Sect, Corr res; t<=100mm, Ann	0.65-0.75	14.00-16.00	1.00 max	0.40-0.80		0.040 max	0.030 max	0.70 max	bal Fe	900 max			280 HB max

France

Specification	Designation	Notes	C	Cr	Mn	Mo	Ni	P	S	Si	Other	UTS	YS	El	Hard
AFNOR	Z70CD14	Bar Bil	0.6-0.75	13-15	1 max	0.5-0.6		0.04 max	0.03 max	1 max	bal Fe				

Germany

Specification	Designation	Notes	C	Cr	Mn	Mo	Ni	P	S	Si	Other	UTS	YS	El	Hard
DIN EN 10088(95)	WNr 1.4109	Bar Rod Sect	0.65-0.75	14.0-16.0	1.00 max	0.40-0.80		0.040 max	0.030 max	0.70 max	bal Fe				59 HRC
DIN EN 10088(95)	X70CrMo15		0.65-0.75	14.0-16.0	1.00 max	0.40-0.80		0.040 max	0.030 max	0.70 max	bal Fe				59 HRC

Japan

Specification	Designation	Notes	C	Cr	Mn	Mo	Ni	P	S	Si	Other	UTS	YS	El	Hard
JIS G4303(91)	SUS440A	Bar Ann Q/A Tmp	0.60-0.75	16.00-18.00	1.00 max	0.75 max	0.60 max	0.040 max	0.030 max	1.00 max	bal Fe				54 HRC
JIS G4305(91)	SUS440A	Sh Plt, CR Ann	0.60-0.75	16.00-18.00	1.00 max	0.75 max	0.60 max	0.040 max	0.030 max	1.00 max	bal Fe	590	245	15	217 HB max
JIS G4306	SUS440A	Strp HR Ann	0.6-0.75	16-18	1 max	0.75 max	0.6 max	0.04 max	0.03 max	1 max	bal Fe				
JIS G4307	SUS440A	Strp CR Ann Q/A Tmp	0.6-0.75	16-18	1 max	0.75 max	0.6 max	0.04 max	0.03 max	1 max	bal Fe				

Mexico

Specification	Designation	Notes	C	Cr	Mn	Mo	Ni	P	S	Si	Other	UTS	YS	El	Hard
DGN B-83	440A*	Obs; Bar, HR, CR	0.6-0.75	16-18	1 max	0.75 max		0.04 max	0.03 max	1 max	bal Fe				
NMX-B-171(91)	MT440A	mech tub	0.60-0.75	16.0-18.0	1.00 max	0.75 max		0.040 max	0.030 max	1.00 max	bal Fe				

UNS numbers and US grades are provided as a means of cross referencing chemically similar alloys. Exchangability is only possible after independent examination of specifications. Tensile properties are minimum or typical as specified. UTS and YS as MPa. El as %. See Appendix for list of abbreviations used in Notes. * indicates obsolete material.

Specification	Designation	Notes	C	Cr	Mn	Mo	Ni	P	S	Si	Other	UTS	YS	El	Hard

Stainless Steel, Martensitic, 440A (Continued from previous page)

Romania

Specification	Designation	Notes	C	Cr	Mn	Mo	Ni	P	S	Si	Other	UTS	YS	El	Hard
STAS 3583(87)	35MoCr165	Corr res; Heat res	0.33-0.45	15.5-17.5	1 max	0.9-1.3	1 max	0.045 max	0.03 max	1 max	Pb <=0.15; bal Fe				

Spain

Specification	Designation	Notes	C	Cr	Mn	Mo	Ni	P	S	Si	Other	UTS	YS	El	Hard
UNE 36016/2(89)	F.3424*	Obs; En10088-2(96); X55CrMoV14	0.5-0.6	13.8-15.0	1.0 max	0.45-0.6	1.0 max	0.04 max	0.03 max	1.0 max	V 0.1-0.15; bal Fe				

USA

Specification	Designation	Notes	C	Cr	Mn	Mo	Ni	P	S	Si	Other	UTS	YS	El	Hard
	AISI 440A	Tube	0.60-0.75	16.00-18.00	1.00 max	0.75 max		0.040 max	0.030 max	1.00 max	bal Fe				
	UNS S44002	Hardenable	0.6-0.75	16.00-18.00	1.00 max	0.75 max		0.040 max	0.030 max	1.00 max	bal Fe				
AMS 7445C(90)	51440A	Balls, Corr res	0.60-0.75	16.00-18.00	1.00 max	0.75 max	0.75 max	0.040 max	0.030 max	1.00 max	Cu <=0.50; bal Fe				55-64 HRC
ASTM A276(98)	440A	Bar	0.6-0.75	16-18	1 max	0.75 max		0.04 max	0.03 max	1 max	bal Fe				
ASTM A314	440A	Bil	0.6-0.75	16-18	1 max	0.75 max		0.04 max	0.03 max	1 max	bal Fe				
ASTM A473	440A	Frg, Heat res, ann	0.6-0.75	16.00-18.00	1.00 max	0.75 max		0.040 max	0.030 max	1.00 max	bal Fe				269 HB
ASTM A511(96)	MT440A	Smls mech tub, Ann	0.6-0.75	16.0-18.0	1.00 max	0.75 max		0.040 max	0.030 max	1.00 max	bal Fe	655	379	15	215 HB
ASTM A580/A580M(98)	440A	Wir, Ann	0.60-0.75	16.0-18.0	1.00 max	0.75 max		0.040 max	0.030 max	1.00 max	bal Fe	965 max			55 HCR
FED QQ-S-763F(96)	440A*	Obs; Bar Wir Shp Frg; Corr res	0.6-0.75	16-18	1 max	0.75 max		0.04 max	0.03 max	1 max	bal Fe				

Stainless Steel, Martensitic, 440B

Canada

Specification	Designation	Notes	C	Cr	Mn	Mo	Ni	P	S	Si	Other	UTS	YS	El	Hard
CSA 110.3	440B	Bar Bil	0.75-0.95	16-18	1 max	0.75 max		0.04 max	0.03 max	1 max	bal Fe				

China

Specification	Designation	Notes	C	Cr	Mn	Mo	Ni	P	S	Si	Other	UTS	YS	El	Hard
GB 1220(92)	8Cr17	Bar Ann	0.75-0.95	16.00-18.00	1.00 max	0.75 max	0.60 max	0.035 max	0.030 max	1.00 max	bal Fe				255 max HB

Europe

Specification	Designation	Notes	C	Cr	Mn	Mo	Ni	P	S	Si	Other	UTS	YS	El	Hard
EN 10088/3(95)	1.4112	Bar Rod Sect, Corr res; t<=100mm, Ann	0.85-0.95	17.00-19.00	1.00 max	0.90-1.300		0.040 max	0.030 max	1.00 max	V 0.07-0.12; bal Fe				265 HB max
EN 10088/3(95)	X90CrMoV18	Bar Rod Sect, Corr res; t<=100mm, Ann	0.85-0.95	17.00-19.00	1.00 max	0.90-1.300		0.040 max	0.030 max	1.00 max	V 0.07-0.12; bal Fe				265 HB max

France

Specification	Designation	Notes	C	Cr	Mn	Mo	Ni	P	S	Si	Other	UTS	YS	El	Hard
AFNOR NFA35579	Z85CDMV18.02	Bar Bil	0.8-0.9	16.5-18.5	1.5 max	2-2.5		0.04 max	0.03 max	1 max	V 0.3-0.6; bal Fe				

Germany

Specification	Designation	Notes	C	Cr	Mn	Mo	Ni	P	S	Si	Other	UTS	YS	El	Hard
DIN EN 10088(95)	WNr 1.4112	Bar Rod Sect	0.85-0.95	17.0-19.0	1.00 max	0.90-1.30		0.040 max	0.030 max	1.00 max	V 0.07-0.12; bal Fe				57 HRC
DIN EN 10088(95)	X90CrMoV18	Bar Rod Sect	0.85-0.95	17.0-19.0	1.00 max	0.90-1.30		0.040 max	0.030 max	1.00 max	V 0.07-0.12; bal Fe				57 HRC

India

Specification	Designation	Notes	C	Cr	Mn	Mo	Ni	P	S	Si	Other	UTS	YS	El	Hard
IS 1570/5(85)	X85Cr18Mo2V	Valve	0.8-0.9	16.5-18.5	1.5 max	2.5 max	0.4 max	0.045 max	0.035 max	1 max	Co <=0.1; Cu <=0.3; Pb <=0.15; V <=0.6; W <=0.1; bal Fe				255-306 HB

Japan

Specification	Designation	Notes	C	Cr	Mn	Mo	Ni	P	S	Si	Other	UTS	YS	El	Hard
JIS G4303(91)	SUS440B	Bar Q/A Tmp	0.75-0.95	16.00-18.00	1.00 max	0.75 max	0.60 max	0.040 max	0.030 max	1.00 max	bal Fe				56 HRC

Mexico

Specification	Designation	Notes	C	Cr	Mn	Mo	Ni	P	S	Si	Other	UTS	YS	El	Hard
DGN B-83	440B*	Obs; Bar, HR, CR	0.75-0.95	16-18	1 max	0.75 max		0.04 max	0.03 max	1 max	bal Fe				

Spain

Specification	Designation	Notes	C	Cr	Mn	Mo	Ni	P	S	Si	Other	UTS	YS	El	Hard
UNE 36022(86)	F.3223*	see X85CrMoV18-02	0.8-0.9	16-18	1-1.5	2-2.5	0.5 max	0.03 max	0.03 max	1 max	V 0.4-0.6; bal Fe				
UNE 36022(86)	X85CrMoV18-02	Bar Bil	0.8-0.9	16-18	1-1.5	2-2.5	0.5 max	0.03 max	0.03 max	1 max	V 0.4-0.6; bal Fe				

USA

Specification	Designation	Notes	C	Cr	Mn	Mo	Ni	P	S	Si	Other	UTS	YS	El	Hard
	AISI 440B	Tube	0.75-0.95	16.00-18.00	1.00 max	0.75 max		0.040 max	0.030 max	1.00 max	bal Fe				
	UNS S44003	Hardenable	0.75-0.95	16.00-18.00	1.00 max	0.75 max		0.040 max	0.030 max	1.00 max	bal Fe				

Specification	Designation	Notes	C	Cr	Mn	Mo	Ni	P	S	Si	Other	UTS	YS	El	Hard

Stainless Steel, Martensitic, 440B (Continued from previous page)

USA

Specification	Designation	Notes	C	Cr	Mn	Mo	Ni	P	S	Si	Other	UTS	YS	El	Hard
AMS 7445C(90)	51440B	Balls, Corr res	0.75-0.95	16.00-18.00	1.00 max	0.75 max	0.75 max	0.040 max	0.030 max	1.00 max	Cu <=0.50; bal Fe				55-64 HRC
ASTM A276(98)	440B	Bar	0.75-0.95	16-18	1 max	0.75 max		0.04 max	0.03 max	1 max	bal Fe				
ASTM A314	440B	Bil	0.75-0.95	16-18	1 max	0.75 max		0.04 max	0.03 max	1 max	bal Fe				
ASTM A473	440B	Frg, Heat res, ann	0.75-0.95	16.00-18.00	1.00 max	0.75 max		0.040 max	0.030 max	1.00 max	bal Fe				269 HB
ASTM A580/A580M(98)	440B	Wir, Ann	0.75-0.95	16.0-18.0	1.00 max	0.75 max		0.040 max	0.030 max	1.00 max	bal Fe	965 max			56 HCR
FED QQ-S-763F(96)	440B*	Obs; Bar Wir Shp Frg; Corr res	0.75-0.95	16-18	1 max	0.75 max		0.04 max	0.03 max	1 max	bal Fe				
MIL-S-862B	440B*	Obs	0.75-0.95	16-18	1 max	0.75 max		0.04 max	0.03 max	1 max	bal Fe				

Stainless Steel, Martensitic, 440C

Australia

Specification	Designation	Notes	C	Cr	Mn	Mo	Ni	P	S	Si	Other	UTS	YS	El	Hard
AS 2837(86)	440C	Bar, Semi-finished product	0.95-1.2	16-18	1.00 max	0.75 max	1.00 max	0.040 max	0.030 max	1.00 max	bal Fe				

Bulgaria

Specification	Designation	Notes	C	Cr	Mn	Mo	Ni	P	S	Si	Other	UTS	YS	El	Hard
BDS 6738(72)	9Ch18	Bar	0.90-1.05	17.0-19.0	0.80 max	0.30 max	0.60 max	0.035 max	0.025 max	0.80 max	bal Fe				

Canada

Specification	Designation	Notes	C	Cr	Mn	Mo	Ni	P	S	Si	Other	UTS	YS	El	Hard
CSA G110.3	440C	Bar Bil	0.95-1.2	16-18	1 max	0.75 max		0.04 max	0.03 max	1 max	bal Fe				

China

Specification	Designation	Notes	C	Cr	Mn	Mo	Ni	P	S	Si	Other	UTS	YS	El	Hard
GB 1220(92)	11Cr17	Bar Ann	0.95-1.20	16.00-18.00	1.00 max	0.75 max	0.60 max	0.035 max	0.030 max	1.00 max	bal Fe				269 max HB
GB 1220(92)	9Cr18	Bar Ann	0.90-1.00	17.00-19.00	0.80 max	0.75 max	0.60 max	0.035 max	0.030 max	0.80 max	bal Fe			1.00	255 max HB
GB 4240(92)	9Cr18(-R)	Wir Ann 0.05-14mm diam, R-As Ann	0.90-1.00	17.00-19.00	0.80 max	0.75 max	0.60 max	0.035 max	0.030 max	0.80 max	bal Fe	590-830			

Czech Republic

Specification	Designation	Notes	C	Cr	Mn	Mo	Ni	P	S	Si	Other	UTS	YS	El	Hard
CSN 417042	17042 .	Corr res	0.95-1.05	16.0-18.0	0.9 max			0.04 max	0.035 max	0.7 max	bal Fe				

Europe

Specification	Designation	Notes	C	Cr	Mn	Mo	Ni	P	S	Si	Other	UTS	YS	El	Hard
EN 10088/3(95)	1.4125	Bar Rod Sect, Corr res; t<=100mm, Ann	0.95-1.20	16.00-18.00	1.00 max	0.40-0.80		0.040 max	0.030 max	1.00 max	bal Fe				285 HB max
EN 10088/3(95)	X105CrMo17	Bar Rod Sect, Corr res; t<=100mm, Ann	0.95-1.20	16.00-18.00	1.00 max	0.40-0.80		0.040 max	0.030 max	1.00 max	bal Fe				285 HB max

France

Specification	Designation	Notes	C	Cr	Mn	Mo	Ni	P	S	Si	Other	UTS	YS	El	Hard
AFNOR NFA35575	Z100CD17	Wir rod, Q/A Tmp	0.9-1.2	16-18	1 max	0.35-0.75		0.04 max	0.03 max	1 max	bal Fe				58 HRC

Germany

Specification	Designation	Notes	C	Cr	Mn	Mo	Ni	P	S	Si	Other	UTS	YS	El	Hard
DIN EN 10088(95)	WNr 1.4125	Bar Rod Sect	0.95-1.20	16.0-18.0	1.00 max	0.40-0.80		0.040 max	0.030 max	1.00 max	bal Fe				61 HRC
DIN EN 10088(95)	X105CrMo17	Bar Rod Sect	0.95-1.20	16.0-18.0	1.00 max	0.40-0.80		0.040 max	0.030 max	1.00 max	bal Fe				61 HRC

Hungary

Specification	Designation	Notes	C	Cr	Mn	Mo	Ni	P	S	Si	Other	UTS	YS	El	Hard
MSZ 4360(87)	KO14	Corr res; 0-999mm; soft ann	0.95-1.2	16-18	1 max	0.3-0.75	0.6 max	0.045 max	0.03 max	1 max	bal Fe				285 HB max
MSZ 4360(87)	KO14	Corr res; 0-999mm; Q/T	0.95-1.2	16-18	1 max	0.3-0.75	0.6 max	0.045 max	0.03 max	1 max	bal Fe				58 HRC
MSZ 4360(87)	X105CrMo17	Corr res; 0-999mm; soft ann	0.95-1.2	16-18	1 max	0.3-0.75	0.6 max	0.045 max	0.03 max	1 max	bal Fe				285 HB max
MSZ 4360(87)	X105CrMo17	Corr res; 0-999mm; Q/T	0.95-1.2	16-18	1 max	0.3-0.75	0.6 max	0.045 max	0.03 max	1 max	bal Fe				58 HRC

India

Specification	Designation	Notes	C	Cr	Mn	Mo	Ni	P	S	Si	Other	UTS	YS	El	Hard
IS 1570/5(85)	X108Cr17Mo	Sh Plt Strp	0.95-1.2	16-18	1 max	0.75 max	0.5 max	0.045 max	0.03 max	1 max	Co <=0.1; Cu <=0.3; Pb <=0.15; W <=0.1; bal Fe				269 HB max
IS 6529	105Cr18Mo50	Bil	0.9-1.2	16-19	1 max	0.75 max	0.5 max	0.04 max	0.03 max	1 max	bal Fe				
IS 6903	105Cr18Mo50	Bar	0.9-1.2	16-19	1 max	0.75 max	0.5 max	0.04 max	0.03 max	1 max	bal Fe				

UNS numbers and US grades are provided as a means of cross referencing chemically similar alloys. Exchangability is only possible after independent examination of specifications. Tensile properties are minimum or typical as specified. UTS and YS as MPa. El as %. See Appendix for list of abbreviations used in Notes. * indicates obsolete material.

Stainless Steel, Martensitic, 440C (Continued from previous page)

Specification	Designation	Notes	C	Cr	Mn	Mo	Ni	P	S	Si	Other	UTS	YS	El	Hard
India															
IS 6911	105Cr18Mo50	Sh Strp Plt	0.9-1.2	16-19	1 max	0.75 max	0.5 max	0.04 max	0.03 max	1 max	bal Fe				
International															
COMECON PC4-70	11	Bar Bil	0.9-1.05	17-19	0.8 max			0.035 max	0.025 max	0.8 max	bal Fe				
ISO 683-17(76)	21	Ball & roller bearing, HT	0.95-1.20	16.0-18.0	1.00 max	0.35-0.75	0.50 max	0.040 max	0.030 max	1.00 max	bal Fe				
Italy															
UNI 3097(50)	X110CN17	Ball & roller bearing	1-1.2	16-18	0.3-0.5		0.5-1	0.03 max	0.03 max	0.5 max	bal Fe				
UNI 3097(75)	X105CrMo17	Ball & roller bearing	0.95-1.2	16-18	1 max	0.35-0.75	0.5 max	0.04 max	0.03 max	1 max	bal Fe				
Japan															
JIS G4303(91)	SUS440C	Bar Q/A Tmp	0.95-1.20	16.00-18.00	1.00 max	0.75 max	0.60 max	0.040 max	0.030 max	1.00 max	bal Fe				58 HRC
JIS G4303(91)	SUS440F	Bar Q/A Tmp	0.95-1.20	16.00-18.00	1.25 max	0.75 max	0.60 max	0.060 max	0.15 max	1.00 max	bal Fe				58 HRC
JIS G4308(98)	SUS440C	Wir rod	0.95-1.20	16.00-18.00	1.00 max	0.75 max	0.60 max	0.040 max	0.03 max	1.00 max	bal Fe				
JIS G4309	SUS440C	Wir	0.95-1.20	16.00-18.00	1.00 max	0.75 max	0.60 max	0.040 max	0.030 max	1.00 max	bal Fe				
Mexico															
DGN B-83	440C*	Obs; Bar, HR, CR	0.95-1.2	16-18	1 max	0.75 max		0.04 max	0.03 max	1 max	bal Fe				
Poland															
PNH86020	H18	Corr res	0.9-1.05	17-19	0.8 max		0.6 max	0.04 max	0.03 max	0.8 max	bal Fe				
Romania															
STAS 3583(87)	90Cr180	Corr res; Heat res	0.9-1	17-19	1 max	0.2 max	0.5 max	0.045 max	0.03 max	1 max	Pb <=0.15; bal Fe				
Russia															
GOST 5632	95Ch18	Bar Bil Q/A Tmp	0.9-1	17-18	0.8 max		0.6 max	0.03 max	0.025 max	0.8 max	Cu <=0.3; Ti <=0.2; bal Fe				55 HRC
GOST 5632(61)	9KH18	Corr res; Heat res; High-temp constr	0.9-1.0	17.0-19.0	0.7 max	0.15 max	0.4 max	0.03 max	0.025 max	0.8 max	Cu <=0.3; Ti <=0.05; bal Fe				
GOST 5632(72)	95Ch18	Corr res	0.9-1.0	17.0-19.0	0.8 max	0.15 max	0.6 max	0.03 max	0.025 max	0.8 max	Cu <=0.3; Ti <=0.2; bal Fe				
Spain															
UNE 36027(80)	F.3425*	see X100CrMo17	0.95-1.1	16.0-18.0	1.0 max	0.35-0.75	0.5 max	0.04 max	0.03 max	1.0 max	bal Fe				
UNE 36027(80)	X100CrMo17	Ball & roller bearing	0.95-1.1	16.0-18.0	1.0 max	0.35-0.75	0.5 max	0.04 max	0.03 max	1.0 max	bal Fe				
USA															
	AISI 440C	Tube	0.95-1.20	16.00-18.00	1.00 max	0.75 max		0.040 max	0.030 max	1.00 max	bal Fe				
	UNS S44004	Hardenable	0.95-1.20	16.00-18.00	1.00 max	0.75 max		0.040 max	0.030 max	1.00 max	bal Fe				
AMS 7445C(90)	51440C	Balls, Corr res	0.95-1.20	16.00-18.00	1.00 max	0.75 max	0.75 max	0.040 max	0.030 max	1.00 max	Cu <=0.50; bal Fe				55-64 HRC
ASTM A276(98)	440C	Bar	0.95-1.2	16-18	1 max	0.75 max		0.04 max	0.03 max	1 max	bal Fe				
ASTM A314	440C	Bil	0.95-1.2	16-18	1 max	0.75 max		0.04 max	0.03 max	1 max	bal Fe				
ASTM A473	440C	Frg, Heat res, ann	0.95-1.20	16.00-18.00	1.00 max	0.75 max		0.040 max	0.030 max	1.00 max	bal Fe				269 HB
ASTM A493	440C	Wir rod, CHd, cold frg	0.95-1.20	16.0-18.0	1.00 max	0.75 max		0.040 max	0.030 max	1.00 max	bal Fe	760			
ASTM A580/A580M(98)	440C	Wir, Ann	0.95-1.20	16.0-18.0	1.00 max	0.75 max		0.040 max	0.030 max	1.00 max	bal Fe	965 max			58 HCR
FED QQ-S-763F(96)	440C*	Obs; Bar Wir Shp Frg; Corr res	0.95-1.2	16-18	1 max	0.75 max		0.04 max	0.03 max	1 max	bal Fe				
MIL-B-913(93)	440C	Bearing, ball ring	0.95-1.20	16.00-18.00	1.00 max	0.75 max		0.040 max	0.030 max	1.00 max	bal Fe				
MIL-S-862B	440C*	Obs	0.95-1.2	16-18	1 max	0.75 max		0.04 max	0.03 max	1 max	bal Fe				
SAE J467(68)	440C		1.10	17.5	0.50	0.50				0.40	bal Fe	1965	1896	2	580 HB

UNS numbers and US grades are provided as a means of cross referencing chemically similar alloys. Exchangability is only possible after independent examination of specifications. Tensile properties are minimum or typical as specified. UTS and YS as MPa. El as %. See Appendix for list of abbreviations used in Notes. * indicates obsolete material.

Specification	Designation	Notes	C	Cr	Mn	Mo	Ni	P	S	Si	Other	UTS	YS	El	Hard

Stainless Steel, Martensitic, 440F

China

Specification	Designation	Notes	C	Cr	Mn	Mo	Ni	P	S	Si	Other	UTS	YS	El	Hard
GB 1220(92)	11Cr17	Bar Ann	0.95-1.20	16.00-18.00	1.00 max	0.75 max	0.60 max	0.035 max	0.030 max	1.00 max	bal Fe				269 max HB

Italy

| UNI 2955/3(82) | X102CrMo17KU | Alloyed tool steel | 0.95-1.1 | 16-18 | 1 max | 0.4-0.7 | | 0.07 max | 0.06 max | 1 max | bal Fe | | | | |

USA

| | UNS S44020 | Free Machining | 0.95-1.20 | 16.00-18.00 | 1.25 max | 0.40-0.60 | 0.75 max | 0.040 max | 0.10-0.35 | 1.00 max | N <=0.08; bal Fe | | | | |
| ASTM A582/A582M(95) | 440F | Bar, HF, CF, Ann | 0.95-1.20 | 16.00-18.00 | 1.25 max | | 0.50 | 0.06 max | 0.15 min | 1.00 max | Cu <=0.60; bal Fe | | | | 285 HB max |

Stainless Steel, Martensitic, 440FSe

USA

Specification	Designation	Notes	C	Cr	Mn	Mo	Ni	P	S	Si	Other	UTS	YS	El	Hard
	UNS S44023	Se-bearing Free Mach	0.95-1.20	16.00-18.00	1.25 max	0.60 max	0.75 max	0.040 max	0.030 max	1.00 max	N <=0.08; Se>=0.15; bal Fe				
ASTM A582/A582M(95)	440FSe	Bar, HF, CF, Ann	0.95-1.20	16.00-18.00	1.25 max		0.50	0.06 max	0.06 max	1.00 max	Cu <=0.60; Se>=0.15; bal Fe				285 HB max

Stainless Steel, Martensitic, 502

Romania

Specification	Designation	Notes	C	Cr	Mn	Mo	Ni	P	S	Si	Other	UTS	YS	El	Hard
STAS 8184(87)	10MoCr50	Plt Bar Bil Tub	0.15 max	4-6	0.3-0.6	0.45-0.65		0.03 max	0.03 max	0.15-0.5	Al <=0.05; Ti <=0.02; bal Fe				

Stainless Steel, Martensitic, 5612

USA

Specification	Designation	Notes	C	Cr	Mn	Mo	Ni	P	S	Si	Other	UTS	YS	El	Hard
	UNS S41041	with Nb	0.13-0.18	11.50-13.00	0.40-0.60	0.20 max	0.50 max	0.030 max	0.030 max	0.50 max	Al <=0.05; Nb 0.15-0.45; Sn 0.05 max; bal Fe				
ASTM A565(97)		Bar Frg for high-temp HT	0.13-0.18	11.50-13.00	0.40-0.60	0.20 max	0.50 max	0.030 max	0.030 max	0.50 max	Al <=0.05; Nb 0.15-0.45; bal Fe	795	515	15	277 HB max

Stainless Steel, Martensitic, 5614

USA

Specification	Designation	Notes	C	Cr	Mn	Mo	Ni	P	S	Si	Other	UTS	YS	El	Hard
	UNS S41025	with Mo	0.15 max	11.5-13.5	1.00 max	0.40-0.60	0.60 max	0.040 max	0.030 max	1.00 max	bal Fe				

Stainless Steel, Martensitic, 5617

USA

Specification	Designation	Notes	C	Cr	Mn	Mo	Ni	P	S	Si	Other	UTS	YS	El	Hard
	UNS S45503	Age Hardenable Custom 455 ELC	0.010 max	11.00-12.50	0.50 max	0.50 max	7.50-9.50	0.010 max	0.010 max	0.20 max	Cu 1.50-2.50; Ti 1.00-1.35; (Cc+Ta) 0.10-0.50; bal Fe				
ASTM A564/A564M(97)	S45503	Bar Shp, HR, CF AH	0.010 max	11.00-12.50	0.50 max	0.50 max	7.50-9.50	0.010 max	0.010 max	0.20 max	Cu 1.50-2.50; Ti 1.00-1.35; Nb+Ta 0.10-0.50; bal Fe	1410-1620	1275-1520	8-10	363-444 HB
ASTM A705/705M(95)	S45503	AH, Frg, SHT	0.010 max	11.00-12.50	0.50 max		7.50-9.50	0.010 max	0.010 max	0.20 max	Cu 1.50-2.50; Ti 1.00-1.35; Nb+Ta 0.15-0.50; bal Fe	1035	860	4	331 HB

Stainless Steel, Martensitic, 5749

USA

Specification	Designation	Notes	C	Cr	Mn	Mo	Ni	P	S	Si	Other	UTS	YS	El	Hard
	UNS S42700	Bearings	1.10-1.20	14.0-15.0	0.30-0.60	3.75-4.25	0.40 max	0.015 max	0.010 max	0.20-0.40	Cu <=0.35; V 1.10-1.30; bal Fe				

Stainless Steel, Martensitic, 5900

Spain

Specification	Designation	Notes	C	Cr	Mn	Mo	Ni	P	S	Si	Other	UTS	YS	El	Hard
UNE 36016(75)	F.3423*	see X46CrMo16	0.42-0.5	15.5-17.5	1.0 max	1.0-1.5	1.0 max	0.04 max	0.03 max	1.0 max	V <=0.2; bal Fe				
UNE 36016(75)	X46CrMo16	Corr res	0.42-0.5	15.5-17.5	1.0 max	1.0-1.5	1.0 max	0.04 max	0.03 max	1.0 max	V <=0.2; bal Fe				
UNE 36016/1(89)	F.3422*	Obs; EN10088-3(96); X45CrMoV14	0.42-0.5	13.8-15.0	1.0 max	0.45-0.6	1.0 max	0.04 max	0.03 max	1.0 max	V 0.1-0.15; bal Fe				
UNE 36027(80)	F.3406*	see X45Cr13	0.42-0.8	12.5-14.5	1.0 max	0.15 max	1.0 max	0.04 max	0.03 max	1.0 max	bal Fe				
UNE 36027(80)	X45Cr13	Ball & roller bearing	0.42-0.8	12.5-14.5	1.0 max	0.15 max	1.0 max	0.04 max	0.03 max	1.0 max	bal Fe				

UNS numbers and US grades are provided as a means of cross referencing chemically similar alloys. Exchangability is only possible after independent examination of specifications. Tensile properties are minimum or typical as specified. UTS and YS as MPa. El as %. See Appendix for list of abbreviations used in Notes. * indicates obsolete material.

Specification	Designation	Notes	C	Cr	Mn	Mo	Ni	P	S	Si	Other	UTS	YS	El	Hard
Stainless Steel, Martensitic, 5900 (Continued from previous page)															
USA															
	UNS S42800	Bearings	1.05-1.15	13.7-14.8	0.25-0.50	1.90-2.25	0.35 max	0.015 max	0.010 max	0.20-0.40	Cu <=0.35; Nb 0.25-0.35; V 0.90-1.15; bal Fe				
Stainless Steel, Martensitic, 5925															
Europe															
EN 10088/2(95)	1.4034	Sh Plt Strp, Corr res; t<12 Ann	0.43-0.50	12.50-14.50	1.00 max			0.040 max	0.015 max	1.00 max	bal Fe	780 max		12	245 HB max
EN 10088/2(95)	1.4116	Sh Plt Strp, Corr res; t<12 Ann	0.45-0.55	14.00-15.00	1.00 max	0.50-0.80		0.040 max	0.015 max	1.00 max	V 0.10-0.20; bal Fe	850 max		12	280 HB max
EN 10088/2(95)	1.4122	Sh Plt Strp, Corr res; t<=3mm, Q/T	0.33-0.45	15.50-17.50	1.50 max	0.80-1.30	1.00 max	0.040 max	0.015 max	1.00 max	bal Fe				480-580 HV
EN 10088/2(95)	1.4122	Sh Plt Strp, Corr res; t<12 Ann	0.33-0.45	15.50-17.50	1.50 max	0.80-1.30	1.00 max	0.040 max	0.015 max	1.00 max	bal Fe	900 max		12	280 HB max
EN 10088/2(95)	X39CrMo17-1	Sh Plt Strp, Corr res; t<=3mm, Q/T	0.33-0.45	15.50-17.50	1.50 max	0.80-1.30	1.00 max	0.040 max	0.015 max	1.00 max	bal Fe				480-580 HV
EN 10088/2(95)	X39CrMo17-1	Sh Plt Strp, Corr res; t<12 Ann	0.33-0.45	15.50-17.50	1.50 max	0.80-1.30	1.00 max	0.040 max	0.015 max	1.00 max	bal Fe	900 max		12	280 HB max
EN 10088/2(95)	X46Cr13	Sh Plt Strp; Corr res; Ann	0.43-0.50	12.50-14.50	1.00 max			0.040 max	0.015 max	1.00 max	bal Fe	800 max			245 HB max
EN 10088/2(95)	X50CrMoV15	Sh Plt Strp, Corr res; t<12 Ann	0.45-0.55	14.00-15.00	1.00 max	0.50-0.80		0.040 max	0.015 max	1.00 max	V 0.10-0.20; bal Fe	850 max		12	280 HB max
Spain															
UNE 36016(75)	F.3405*	see X46Cr13	0.42-0.5	12.5-14.5	1.0 max	0.15 max	1.0 max	0.04 max	0.03 max	1.0 max	bal Fe				
UNE 36016(75)	X46Cr13	Corr res	0.42-0.5	12.5-14.5	1.0 max	0.15 max	1.0 max	0.04 max	0.03 max	1.0 max	bal Fe				
USA															
	UNS S42025		0.37-0.45	15.0-16.5	0.60 max	1.50-1.90	0.30 max	0.020 max	0.005 max	0.60 max	N 0.16-0.25; V 0.20-0.40; bal Fe				
Stainless Steel, Martensitic, 5930															
USA															
	UNS S42670	Pyrowear 675	0.05-0.09	12.0-14.0	0.50-1.00	1.50-2.50	2.00-3.00	0.015 max	0.010 max	0.10-0.70	Co 4.00-7.00; V 0.40-0.80; bal Fe				
Stainless Steel, Martensitic, 5932															
USA															
	UNS S42640		0.10-0.25		1.00 max	3.00-5.00	1.75-2.75	0.020 max	0.010 max	1.00 max	Co 11.00-14.00; Nb 0.10-0.05; V 0.4-0.8; W <=0.25; bal Fe				
Stainless Steel, Martensitic, 615															
USA															
	UNS S41800	Heat res Greek Ascology	0.15-0.20	12.00-14.00	0.50 max		1.80-2.20	0.040 max	0.030 max	0.50 max	W 2.50-3.50; bal Fe				
ASTM A565(97)	615	Bar Frg for high-temp HT	0.15-0.20	12.00-14.00	0.50 max	0.50 max	1.80-2.20	0.040 max	0.030 max	0.50 max	W 2.50-3.50; bal Fe	965	760	15	302-352 HB
SAE J467(68)	Greek Ascoloy		0.15	13.0	0.40	0.15	2.00			0.30	Cu 0.15; W 3.00; bal Fe	1103	930	16	35 HRC
Stainless Steel, Martensitic, 9															
Czech Republic															
CSN 417116	17116	Heat res	0.15 max	8.0-10.0	0.3-0.6	0.7-1.1		0.03 max	0.03 max	0.25-1.0	bal Fe				
Germany															
DIN 17176(90)	WNr 1.7386	Smls tube hydrogen service	0.07-0.15	8.00-10.0	0.30-0.60	0.90-1.10		0.025 max	0.020 max	0.25-1.00	bal Fe				
DIN 17176(90)	X12CrMo9-1	Smls tube for hydrogen service	0.07-0.15	8.00-10.0	0.30-0.60	0.90-1.10		0.025 max	0.020 max	0.25-1.00	bal Fe				
Italy															
UNI 7660(77)	X12Cr9KG KW	Plt Bar Bil	0.15 max	8-10	0.3-0.6	0.9-1.1		0.03 max	0.03 max	0.5-1	bal Fe				
Romania															
STAS 11522	12CrMo90	Bar Bil Tub	0.1 max	9 max	0.45 max	1 max		0.035 max	0.03 max	0.65 max	bal Fe				

UNS numbers and US grades are provided as a means of cross referencing chemically similar alloys. Exchangability is only possible after independent examination of specifications. Tensile properties are minimum or typical as specified. UTS and YS as MPa. El as %. See Appendix for list of abbreviations used in Notes. * indicates obsolete material.

Specification	Designation	Notes	C	Cr	Mn	Mo	Ni	P	S	Si	Other	UTS	YS	El	Hard
Stainless Steel, Martensitic, 9 (Continued from previous page)															
Sweden															
SS 142203	2203		0.15 max	8-10	0.3-0.6	0.9-1.1		0.03 max	0.03 max	0.25-1	Cu <=0.25; bal Fe				
UK															
BS 1502(82)	629-590	Bar Shp; Press ves	0.08-0.15	8.00-10.0	0.30-0.60	0.90-1.10		0.030 max	0.030 max	0.25-1.00	Al <=0.020; Co <=0.1; W <=0.1; bal Fe				
BS 3059/2(90)	629-470	Boiler, Superheater; High-temp; Smls tub; 2-12.5mm	0.15 max	8.00-10.0	0.30-0.60	0.9-1.10	0.30 max	0.030 max	0.030 max	0.25-1.00	Al <=0.020; Cu <=0.25; Sn<=0.03; bal Fe	470-620	185	20	
BS 3059/2(90)	629-590	Boiler, Superheater; High-temp; Smls tub; 2-12.5mm	0.15 max	8.00-10.0	0.30-0.60	0.9-1.10	0.30 max	0.030 max	0.030 max	0.25-1.00	Al <=0.020; Cu <=0.25; Sn<=0.03; bal Fe	590-740	400	18	
USA															
	AISI 504	Tube	0.15 max	8.00-10.00	1.00 max	0.90-1.10		0.040 max	0.040 max	1.00 max	bal Fe				
	UNS S50400	Heat res	0.15 max	8.00-10.00	1.00 max	0.90-1.10		0.040 max	0.040 max	1.00 max	bal Fe				
ASTM A199	T9	Obs; see A200; tub	0.15 max	8-10	1 max	0.9-1.1		0.04 max	0.04 max	1 max	bal Fe				
ASTM A200(94)	T9	Tube	0.15 max	8-10	1 max	0.9-1.1		0.04 max	0.04 max	1 max	bal Fe				
ASTM A213	T9*	Tube	0.15 max	8-10	1 max	0.9-1.1		0.04 max	0.04 max	1 max	bal Fe				
ASTM A276(98)	T9	Bar	0.15 max	8-10	1 max	0.9-1.1		0.04 max	0.04 max	1 max	bal Fe				
ASTM A335	T9	Tube	0.15 max	8-10	1 max	0.9-1.1		0.04 max	0.04 max	1 max	bal Fe				
ASTM A473	501B	Frg, Heat res, tmp	0.15 max	8.00-10.00	1.00 max	0.90-1.10		0.040 max	0.030 max	1.00 max	bal Fe	414	205	20	201 HB
Stainless Steel, Martensitic, A276															
USA															
	UNS S42010	Hardenable Low Ni - Low Mo	0.15-0.30	13.5-15.0	1.00 max	0.40-1.00	0.25-1.00	0.040 max	0.030 max	1.00 max	bal Fe				
Stainless Steel, Martensitic, A565(619)															
Europe															
EN 10088/3(95)	1.4029	Bar Rod Sect, Corr res; t<=160mm, Q/T	0.25-0.32	12.00-13.50	1.50 max	0.60 max		0.040 max	0.15-0.25	1.00 max	bal Fe	850-100	650	9	
EN 10088/3(95)	1.4029	Bar Rod Sect, Corr res; t<=160mm, Ann	0.25-0.32	12.00-13.50	1.50 max	0.60 max		0.040 max	0.15-0.25	1.00 max	bal Fe	800 max			245 HB max
EN 10088/3(95)	X29CrS13	Bar Rod Sect, Corr res; t<=160mm, Ann	0.25-0.32	12.00-13.50	1.50 max	0.60 max		0.040 max	0.15-0.25	1.00 max	bal Fe	800 max			245 HB max
EN 10088/3(95)	X29CrS13	Bar Rod Sect, Corr res; t<=160mm, Q/T	0.25-0.32	12.00-13.50	1.50 max	0.60 max		0.040 max	0.15-0.25	1.00 max	bal Fe	850-100	650	9	
USA															
	UNS S42300	Lapelloy	0.27-0.32	11.00-12.00	0.95-1.35	2.50-3.00	0.50 max	0.025 max	0.025 max	0.50 max	V 0.20-0.30; bal Fe				
ASTM A565(97)	619	Bar Frg for high-temp HT	0.27-0.32	11.00-12.00	0.95-1.35	2.50-3.00	0.50 max	0.025 max	0.025 max	0.50 max	V 0.20-0.30; bal Fe	96	760	8	302-352 HB
SAE J467(68)	Lapelloy		0.30	12.0	1.00	2.75	0.30			0.25	V 0.25; bal Fe	1069	965	17	35 HRC
SAE J467(68)	Lapelloy C		0.22	11.5	0.80	2.75	0.20			0.25	Cu 2.00; N 0.08; bal Fe	1069	827	17	33 HRC
Stainless Steel, Martensitic, A579(52)															
USA															
	UNS S42201	Superstrength	0.20-0.25	11.0-13.5	0.50-1.00	0.75-1.25	0.75-1.25	0.025 max	0.025 max	0.20-0.60	Al <=0.05; Co <=0.25; Ti <=0.05; V 0.20-0.30; W 0.75-1.25; Sn 0.04 max; bal Fe				
ASTM A579(96)	52	Superstrength Frg	0.20-0.25	11.0-13.5	0.50-1.00	0.75-1.25	0.75-1.25	0.025 max	0.025 max	0.20-0.60	Al <=0.05; Co <=0.25; Cu <=0.50; Ti <=0.05; V 0.20-0.30; W 0.75-1.25; Sn <=0.04; bal Fe	1210-1520	965-1100	10-12	255 HB

UNS numbers and US grades are provided as a means of cross referencing chemically similar alloys. Exchangability is only possible after independent examination of specifications. Tensile properties are minimum or typical as specified. UTS and YS as MPa. El as %. See Appendix for list of abbreviations used in Notes. * indicates obsolete material.

Specification	Designation	Notes	C	Cr	Mn	Mo	Ni	P	S	Si	Other	UTS	YS	El	Hard

Stainless Steel, Martensitic, A581

USA

Specification	Designation	Notes	C	Cr	Mn	Mo	Ni	P	S	Si	Other	UTS	YS	El	Hard
	UNS S41603	Free Machining Low Hardenable	0.08 max	12.0-14.0	1.25 max			0.06 max	0.15 max	1.00 max	bal Fe				
ASTM A581/A581M(95)	S41603	Wir rod, Ann	0.08 max	12.0-17.0	1.25 max			0.06 max	0.15 min	1.00 max	bal Fe	585-860			
ASTM A582/A582M(95)	S41603	Bar, HF, CF, Ann	0.08 max	12.00-14.00	1.25 max			0.06 max	0.15 min	1.00 max	bal Fe				262 HB max

Stainless Steel, Martensitic, A756

USA

Specification	Designation	Notes	C	Cr	Mn	Mo	Ni	P	S	Si	Other	UTS	YS	El	Hard
	UNS S44025	with Mo	0.95-1.10	16.00-18.00	1.00 max	0.40-0.65	0.75 max	0.025 max	0.025 max	1.00 max	Cu <=0.50; bal Fe				
ASTM A756(94)	440C	Anti-friction bearings ann	0.95-1.10	16.00-18.00	1.00 max	0.40-0.65	0.75 max	0.025 max	0.025 max	1.00 max	Cu <=0.50; bal Fe				255 HB

Stainless Steel, Martensitic, A826

Romania

Specification	Designation	Notes	C	Cr	Mn	Mo	Ni	P	S	Si	Other	UTS	YS	El	Hard
STAS 11523(87)	20VNiWMoCr120	Corr res; Heat res	0.17-0.25	11-12.5	0.3-0.8	0.8-1.2	0.3-0.8	0.035 max	0.03 max	0.1-0.5	Pb <=0.15; V 0.25-0.35; W 0.4-0.6; bal Fe				

USA

Specification	Designation	Notes	C	Cr	Mn	Mo	Ni	P	S	Si	Other	UTS	YS	El	Hard
	UNS S42100		0.17-0.23	11.0-12.5	0.40-0.70	0.80-1.20	0.30-0.80	0.040 max	0.010 max	0.20-0.30	Al <=0.05; Nb <=0.050; V 0.25-0.35; W 0.40-0.60; bal Fe				
ASTM A771/A771M(95)	A826	Smls tube; Reactor Core	0.17-0.23	11.0-12.5	0.40-0.70	0.80-1.20	0.30-0.80	0.040 max	0.010 max	0.20-0.30	Al <=0.050; B 0; Nb <=0.050; V 0.25-0.35; W 0.40-0.60; bal Fe				
ASTM A826/A826M(95)	A826	Smls duct tub; Reactor grade; Tmp	0.17-0.23	11.0-12.5	0.40-0.70	0.80-1.20	0.30-0.80	0.040 max	0.010 max	0.20-0.30	Al <=0.050; Nb <=0.050; V 0.25-0.35; W 0.40-0.60; bal Fe	689	483	15	
ASTM A831/A831M(95)	S42100	Bar Bil Frg; reactor core, Tmp	0.17-0.23	11.0-12.5	0.40-0.70	0.80-1.20	0.30-0.80	0.040 max	0.010 max	0.20-0.30	Al <=0.050; Nb <=0.050; Ti 0.40-0.60; V 0.25-0.35; bal Fe	585	415	20	250 HB

Stainless Steel, Martensitic, A988

India

Specification	Designation	Notes	C	Cr	Mn	Mo	Ni	P	S	Si	Other	UTS	YS	El	Hard
IS 1570/5(85)	X20Cr13	0-100mm; Sh Plt Strp Bar Flat Band	0.16-0.25	12-14	1 max	0.15 max	1 max	0.04 max	0.03 max	1 max	Co <=0.1; Cu <=0.3; Pb <=0.15; W <=0.1; bal Fe	690-880	490	14	

Sweden

Specification	Designation	Notes	C	Cr	Mn	Mo	Ni	P	S	Si	Other	UTS	YS	El	Hard
SS 142317	2317	Corr res	0.18-0.24	11-12.5	0.3-0.8	0.8-1.2	0.3-0.8	0.035 max	0.035 max	0.1-0.5	V 0.25-0.35; bal Fe				

USA

Specification	Designation	Notes	C	Cr	Mn	Mo	Ni	P	S	Si	Other	UTS	YS	El	Hard
	UNS S42390	APM 2390	0.18-0.25	11.5-12.5	1.00 max	0.80-1.20	0.30-0.80	0.030 max	0.030 max	1.00 max	N 0.03-0.08; Nb 0.08-0.15; V 0.25-0.35; bal Fe				
ASTM A988(98)	S42390	HIP Flanges, Fittings, Valves/parts; Heat res; APM 2390	0.18-0.25	11.5-12.5	1.00 max	0.80-1.20	0.30-0.80	0.030 max	0.030 max	1.00 max	N 0.03-0.08; Nb 0.08-0.15; V 0.25-0.35; bal Fe	690-862	517	14.0	

Stainless Steel, Martensitic, AF913

UK

Specification	Designation	Notes	C	Cr	Mn	Mo	Ni	P	S	Si	Other	UTS	YS	El	Hard
BS 1501/3(73)	460S52*	Press ves; Corr res	0.07 max	13.2-14.7	1.0 max	1.2-2.0	5.0-6.0	0.04 max	0.03 max	0.7 max	Al <=0.1; Co <=0.1; Cu 1.2-2.0; Pb <=0.15; Ti <=0.05; V <=0.1; W <=0.1; Nb:0.2-0.7; bal Fe				

USA

Specification	Designation	Notes	C	Cr	Mn	Mo	Ni	P	S	Si	Other	UTS	YS	El	Hard
	UNS S41425	AF913	0.050 max	12.00-15.00	0.50-1.00	1.50-2.00	4.0-7.0	0.020 max	0.005 max	0.5 max	Cu <=0.30; N 0.06-0.12; bal Fe				
ASTM A565(97)		Bar Frg for high-temp HT	0.05 max	12.00-14.00	0.50-1.00	0.50-1.00	4.00-7.00	0.02 max	0.005 max	0.05 max	N 0.06-0.12; bal Fe	825	655	15	321 HB max

UNS numbers and US grades are provided as a means of cross referencing chemically similar alloys. Exchangability is only possible after independent examination of specifications. Tensile properties are minimum or typical as specified. UTS and YS as MPa. El as %. See Appendix for list of abbreviations used in Notes. * indicates obsolete material.

Specification	Designation	Notes	C	Cr	Mn	Mo	Ni	P	S	Si	Other	UTS	YS	El	Hard

Stainless Steel, Martensitic, F6b

Japan

Specification	Designation	Notes	C	Cr	Mn	Mo	Ni	P	S	Si	Other	UTS	YS	El	Hard
JIS G3214(91)	SUSF6B	Frg press ves	0.15 max	11.50-13.50	1.00 max	0.40-0.60	1.00-2.00	0.020 max	0.020 max	1.00 max	N <=0.50; bal Fe	760-930	620	15	217-285 HB

USA

Specification	Designation	Notes	C	Cr	Mn	Mo	Ni	P	S	Si	Other	UTS	YS	El	Hard
	UNS S41026	with Mo	0.15 max	11.5-13.5	1.00 max	0.40-0.60	1.0-2.0	0.02 max	0.02 max	1.00 max	Cu <=0.50; bal Fe				
ASTM A182/A182M(98)	F6b	Frg/roll pipe flange valve	0.15 max	11.5-13.5	1.00 max	0.40-0.60	1.0-2.0	0.02 max	0.02 max	1.0 max	Cu <=0.50; bal Fe	760-930	620	16	235-285 HB
ASTM A988(98)	S41026	HIP Flanges, Fittings, Valves/parts; Heat res; w/ Mo	0.15 max	11.5-13.5	1.00 max	0.40-0.60	1.00-2.00	0.02 max	0.02 max	1.0 max	Cu <=0.50; bal Fe	760-930	620	16	235-285 HB

Stainless Steel, Martensitic, F6NM

Europe

Specification	Designation	Notes	C	Cr	Mn	Mo	Ni	P	S	Si	Other	UTS	YS	El	Hard
EN 10088/2(95)	1.4313	Sh Plt Strp, Corr res; t<=75mm, Q/T	0.05 max	12.00-14.00	1.50 max	0.30-0.70	3.50-4.50	0.040 max	0.015 max	0.70 max	N >=0.020; bal Fe	900-1100	800	11	
EN 10088/2(95)	X3CrNiMo13-4	Sh Plt Strp, Corr res; t<=75mm, Q/T	0.05 max	12.00-14.00	1.50 max	0.30-0.70	3.50-4.50	0.040 max	0.015 max	0.70 max	N >=0.020; bal Fe	900-1100	800	11	

Japan

Specification	Designation	Notes	C	Cr	Mn	Mo	Ni	P	S	Si	Other	UTS	YS	El	Hard
JIS G3214(91)	SUSF6NM	Frg press ves	0.05 max	11.50-14.00	0.50-1.00	0.50-1.00	3.50-5.50	0.030 max	0.030 max	0.60 max	bal Fe	790	620	14	295 HB max

USA

Specification	Designation	Notes	C	Cr	Mn	Mo	Ni	P	S	Si	Other	UTS	YS	El	Hard
	UNS S41500	F6NM, wrought CA-6NM	0.05 max	11.5-14.0	0.50-1.00	0.50-1.00	3.50-5.50	0.030 max	0.030 max	0.60 max	bal Fe				
ASTM A182/A182M(98)	F6NM	Frg/roll pipe flange valve	0.05 max	11.5-14.0	0.5-1.0	0.5-1.0	3.5-5.5	0.030 max	0.030 max	0.60 max	bal Fe	790	620	15	295 HB max
ASTM A240/A240M(98)	S41500	F6NM, wrought CA-6NM	0.05 max	11.5-14.0	0.50-1.00	0.50-1.00	3.5-5.5	0.030 max	0.030 max	0.60 max	bal Fe	795	620	15.0	302 HB
ASTM A815/A815M(98)	S41500	Pipe fittings	0.05 max	11.5-14.0	0.50-1.00	0.50-1.00	3.5-5.5	0.030 max	0.030 max	0.60 max	W 0.50-1.00; bal Fe	760-930	620	15.0	295 HB
ASTM A988(98)	S41500	HIP Flanges, Fittings, Valves/parts; Heat res; F6NM, wrought CA-6NM	0.05 max	11.5-14.0	0.50-1.00	0.50-1.00	3.5-5.5	0.030 max	0.030 max	0.60 max	bal Fe	790	620	15	295 HB max
SAE J405(98)	S41500	F6NM, wrought CA-6NM	0.05 max	11.5-14.0	0.5-1.0	0.5-1.0	3.50-5.50	0.030 max	0.030 max	0.60 max	bal Fe				

Stainless Steel, Martensitic, NACE MR0175

USA

Specification	Designation	Notes	C	Cr	Mn	Mo	Ni	P	S	Si	Other	UTS	YS	El	Hard
	UNS S41426	SM13CRS	0.03 max	11.5-13.5	0.50 max	1.5-3.0	4.5-6.5	0.02 max	0.005 max	0.50 max	Ti 0.01-0.50; V <=0.50; bal Fe				

Stainless Steel, Martensitic, No equivalents identified

Czech Republic

Specification	Designation	Notes	C	Cr	Mn	Mo	Ni	P	S	Si	Other	UTS	YS	El	Hard
CSN 417024	17024	Corr res	0.36-0.45	12.0-14.0	0.8 max			0.04 max	0.03 max	0.7 max	bal Fe				
CSN 417027	17027	Corr res	0.15-0.25	14.0-16.0	0.9 max			0.04 max	0.035 max	0.7 max	bal Fe				
CSN 417029	17029	Corr res	0.4-0.5	14.0-16.0	0.9 max			0.04 max	0.035 max	0.7 max	bal Fe				
CSN 417030	17030	Corr res	0.54-0.63	13.0-15.0	0.9 max			0.04 max	0.035 max	0.7 max	bal Fe				
CSN 417031	17031	Corr res	0.9-1.05	13.0-15.0	1.2 max			0.04 max	0.035 max	0.4 max	bal Fe				
CSN 417113	17113	Heat res	0.12 max	6.0-7.5	0.6 max			0.04 max	0.035 max	0.8-1.3	Al 0.4-1.0; bal Fe				

Hungary

Specification	Designation	Notes	C	Cr	Mn	Mo	Ni	P	S	Si	Other	UTS	YS	El	Hard
MSZ 4360(87)	KO13	Corr res; 0-100mm; soft ann	0.36-0.45	12-14	1 max		0.6 max	0.045 max	0.03 max	1 max	bal Fe				250 HB max
MSZ 4360(87)	KO13	Corr res; 0-100mm; Q/T	0.36-0.45	12-14	1 max		0.6 max	0.045 max	0.03 max	1 max	bal Fe				50 HB

International

Specification	Designation	Notes	C	Cr	Mn	Mo	Ni	P	S	Si	Other	UTS	YS	El	Hard
ISO 4954(93)	X12Cr13E	Wir rod Bar, CH ext	0.09-0.15	11.5-13.5	1.00 max		1.0 max	0.040 max	0.030 max	1.00 max	bal Fe	600			
ISO 4954(93)	X19CrNi162E	Wir rod Bar, CH ext	0.014-0.23	15.0-17.5	1.00 max		1.5-2.5	0.040 max	0.030 max	1.00 max	bal Fe	800			

UNS numbers and US grades are provided as a means of cross referencing chemically similar alloys. Exchangability is only possible after independent examination of specifications. Tensile properties are minimum or typical as specified. UTS and YS as MPa. El as %. See Appendix for list of abbreviations used in Notes. * indicates obsolete material.

Specification	Designation	Notes	C	Cr	Mn	Mo	Ni	P	S	Si	Other	UTS	YS	El	Hard
Stainless Steel, Martensitic, No equivalents identified															
International															
ISO 683-13(74)	6*	Corr res; Martensitic	0.36-0.45	12.5-14.5	1 max	0.15 max	1 max	0.04 max	0.03 max	1 max	Al <=0.1; Co <=0.1; Cu <=0.3; Pb <=0.15; Ti <=0.05; V <=0.1; W <=0.1; bal Fe				
ISO 683-17(76)	20	Ball & roller bearing, HT	0.42-0.50	12.5-14.5	1.00 max		1.00 max	0.040 max	0.030 max	1.00 max	bal Fe				
Japan															
JIS G4311	SUH1	Heat res Bar<=75mm	0.40-0.50	7.50-9.50	0.60 max		0.60 max	0.030 max	0.030 max	3.00-3.50	Cu <=0.30; bal Fe	930	685	15	269 HB
JIS G4311	SUH11	Heat res Bar<=25mm	0.45-0.55	7.50-9.50	0.60 max		0.60 max	0.030 max	0.030 max	1.00-2.00	Cu <=0.30; bal Fe	880	685	15	262 HB
JIS G4311	SUH3	Heat res Bar 25-75mm	0.35-0.45	10.00-12.00	0.60 max	0.70-1.30	0.60 max	0.030 max	0.030 max	1.80-2.50	Cu <=0.30; bal Fe	880	635	15	262 HB
JIS G4311	SUH3	Heat res Bar<=25mm	0.35-0.45	10.00-12.00	0.60 max	0.70-1.30	0.60 max	0.030 max	0.030 max	1.80-2.50	Cu <=0.30; bal Fe	930	685	15	269 HB
JIS G4311	SUH4	Heat res Bar<=75mm	0.75-0.85	19.00-20.50	0.20-0.60		1.15-1.65	0.030 max	0.030 max	1.75-2.25	Cu <=0.30; bal Fe	880	685	10	262 HB
JIS G4311	SUH600	Heat res Bar<=75mm	0.15-0.20	10.00-13.00	0.50-1.00	0.30-0.90	0.60 max	0.040 max	0.030 max	0.50 max	Cu <=0.30; N 0.05-0.10; Nb 0.20-0.60; V 0.10-0.40; bal Fe	830	685	15	321 HB
Poland															
PNH84024	15H11MF	High-temp const	0.11-0.18	10-12	0.6 max	0.5-0.7	0.6 max	0.03 max	0.025 max	0.5 max	V 0.25-0.4; bal Fe				
PNH84024	20H12M1F	High-temp const	0.17-0.23	11-12.5	0.3-0.8	0.8-1.2	0.3-0.8	0.035 max	0.035 max	0.1-0.5	V 0.25-0.35; bal Fe				
PNH86020	1H13	Corr res	0.09-0.15	12-14	0.8 max		0.6 max	0.04 max	0.03 max	0.08 max	bal Fe				
PNH86020	1H13	Corr res	0.09-0.15	12-14	0.8 max		0.6 max	0.04 max	0.03 max	0.08 max	bal Fe				
PNH86020	2H14	Corr res	0.16-0.25	13-16	0.8 max		0.6 max	0.04 max	0.03 max	0.8 max	bal Fe				
PNH86020	4H13	Corr res	0.36-0.45	12-14	0.8 max		0.6 max	0.04 max	0.03 max	0.8 max	bal Fe				
PNH86020	4H14	Corr res	0.36-0.45	14.5-18	0.8 max		0.6 max	0.04 max	0.03 max	0.8 max	bal Fe				
PNH86022(71)	H6S2	Heat res	0.15 max	5-6.5	0.7 max		0.6 max	0.04 max	0.03 max	1.5-2	bal Fe				
Spain															
UNE 36016/1(89)	E-421	Corr res	0.28-0.35	12.0-14.0	1.0 max	0.15 max	1.0 max	0.04 max	0.03 max	1.0 max	bal Fe				
UNE 36016/1(89)	E-426	Corr res	0.42-0.5	12.5-14.5	1.0 max	0.15 max	1.0 max	0.04 max	0.03 max	1.0 max	bal Fe				
UNE 36016/1(89)	E-427	Corr res	0.45-0.49	16.0-17.2	1.0 max	1.0-1.3	1.0 max	0.04 max	0.03 max	1.0 max	bal Fe				
UNE 36016/1(89)	F.3403*	Obs; E-421; Corr res	0.28-0.35	12.0-14.0	1.0 max	0.15 max	1.0 max	0.04 max	0.03 max	1.0 max	bal Fe				
USA															
ASTM A756(94)	440C MOD	Anti-friction bearings ann	1.00-1.10	13.00-15.00	0.30-1.00	3.75-4.25	0.75 max	0.025 max	0.025 max	0.20-1.00	Cu <=0.50; bal Fe				255 HB
ASTM A831/A831M(95)	T91	Bar Bil Frg; reactor core, Tmp	0.08-0.12	8.0-9.5	0.30-0.60	0.85-1.05	0.40 max	0.020 max	0.010 max	0.20-0.50	Al <=0.04; N 0.03-0.07; Nb 0.06-0.10; V 0.18-0.25; bal Fe	690	490	10	229 HB
ASTM A982(98)	Grade C Class 1	Frg for turbine airfoils	0.10-0.17	11.25-12.75	0.65-1.05	1.5-2.0	2.25-3.25	0.020 max	0.015 max	0.10-0.35	Al <=0.025; Cu <=0.15; N 0.020-0.045; Ti <=0.05; V 0.25-0.40; W <=0.10; Sn<=0.05; bal Fe	1000	792	15	
ASTM A982(98)	Grade C Class 2	Frg for turbine airfoils	0.10-0.17	11.25-12.75	0.65-1.05	1.5-2.0	2.25-3.25	0.020 max	0.015 max	0.10-0.35	Al <=0.025; Cu <=0.15; N 0.020-0.045; Ti <=0.05; V 0.25-0.40; W <=0.10; Sn<=0.05; bal Fe	1069	827	16	
ASTM A982(98)	Grade F Class 1	Frg for turbine airfoils	0.07 max	15.0-17.5	1.0 max		3.0-5.0	0.040 max	0.030 max	1.0 max	Cu 3.0-5.0; Nb 0.15-0.45; bal Fe	931	724	16	
SAE J467(68)	14Cr-4Mo		1.05	14.5	0.50	4.00				0.30	V 0.12; bal Fe				
SAE J467(68)	422M		0.85	12.0	0.84	2.25	0.20			0.25	V 0.50; W 1.70; bal Fe				

UNS numbers and US grades are provided as a means of cross referencing chemically similar alloys. Exchangability is only possible after independent examination of specifications. Tensile properties are minimum or typical as specified. UTS and YS as MPa. El as %. See Appendix for list of abbreviations used in Notes. * indicates obsolete material.

Specification	Designation	Notes	C	Cr	Mn	Mo	Ni	P	S	Si	Other	UTS	YS	El	Hard
Stainless Steel, Martensitic, No equivalents identified															
USA															
SAE J467(68)	H-46		0.17	12.0	0.65	0.65	0.45			0.40	N 0.08; Nb 0.40; V 0.30; bal Fe	1034	883	20	302 HB
Stainless Steel, Martensitic, S41003															
Japan															
JIS G4309	SUS410F2	Wir	0.15 max	11.50-13.50	1.00 max		0.60 max	0.040 max	0.030 max	1.00 max	Pb 0.05-0.30; bal Fe				
Stainless Steel, Martensitic, S42035															
USA															
	UNS S42035	with Ti	0.08 max	13.5-15.5	1.00 max	0.20-1.20	1.00-2.50	0.045 max	0.030 max	1.00 max	Ti 0.30-0.50; bal Fe				
Stainless Steel, Martensitic, SF100															
India															
IS 1570/5(85)	X66Cr13		0.6-0.75	12-14	0.4-1	0.15 max	0.4 max	0.07 max	0.06 max	0.5 max	Co <=0.1; Cu <=0.3; Pb <=0.15; W <=0.1; bal Fe				
USA															
	UNS S42002	High-C	0.65-0.75	12.00-14.00	1.00 max	0.50 max	1.00 max	0.040 max	0.030 max	1.00 max	bal Fe				
Stainless Steel, Martensitic, W-FeCrNi-4															
Europe															
EN 10088/2(95)	1.4418	Sh Plt Strp, Corr res; t<=75mm, Q/T	0.06 max	15.00-17.00	1.50 max	0.80-1.50	4.00-6.00	0.040 max	0.015 max	0.70 max	N >=0.020; bal Fe	840-980	680	14	
EN 10088/2(95)	X4CrNiMo16-5-1	Sh Plt Strp, Corr res; t<=75mm, Q/T	0.06 max	15.00-17.00	1.50 max	0.80-1.50	4.00-6.00	0.040 max	0.015 max	0.70 max	N >=0.020; bal Fe	840-980	680	14	
Stainless Steel, Martensitic, W-FeCrNi-B															
USA															
	UNS S41683	Thermal spray wire W-FeCrNi-B	0.030 max	12-14	1.0 max		1.0 max			0.080 max	bal Fe				
Stainless Steel, Martensitic, XM-30															
USA															
	UNS S41040	with Nb	0.15 max	11.50-13.50	1.00 max			0.040 max	0.030 max	1.00 max	Nb 0.05-0.20; bal Fe				
Stainless Steel, Martensitic, XM-6															
USA															
	UNS S41610	Free Machining	0.15 max	12.00-14.00	1.50-2.50	0.60 max		0.06 max	0.15 max	1.00 max	bal Fe				
ASTM A581/A581M(95)	XM-6	Wir rod, Ann	0.15 max	12.0-14.0	1.50-2.50			0.06 max	0.15 min	1.00 max	bal Fe	585-860			
ASTM A582/A582M(95)	XM-6	Bar, HF, CF, Ann	0.15 max	12.00-14.00	1.50-2.50			0.06 max	0.15 min	1.00 max	bal Fe				262 HB max

Specification	Designation	Notes	C	Cr	Mn	Mo	Ni	P	S	Si	Other	UTS	YS	El	Hard

Stainless Steel, Precipitation-Hardening, 16-8-2-H

Mexico

| NMX-B-186-SCFI(94) | 16-8-2H | Pipe; High-temp | 0.05-0.10 | 14.5-16.5 | 2.00 max | 1.5-2.0 | 7.50-9.50 | 0.040 max | 0.030 max | 0.75 max | bal Fe | 517 | 207 | | |

USA

| | UNS S16800 | Cr-Ni-Mo 16-8-2-H | 0.05-0.10 | 14.5-16.5 | 2.00 max | 1.50-2.00 | | 0.040 max | 0.300 max | 0.75 max | N 7.50-9.50; bal Fe | | | | |

Stainless Steel, Precipitation-Hardening, 17-4 PH

Australia

| AS 2837(86) | 630 | Bar, Semi-finished product | 0.07 max | 15.5-17.5 | 1.0 max | | 3.0-5.0 | 0.040 max | 0.030 max | 1.0 max | Cu 3.0-5.0; Nb 0.15-0.45; bal Fe | | | | |

China

GB 1220(92)	0Cr17Ni4Cu4Nb	Bar; SA; Aged	0.07 max	15.50-17.50	1.00 max		3.00-5.00	0.035 max	0.030 max	1.00 max	Cu 3.00-5.00; Nb 0.15-0.45; bal Fe	1310	1180	10	
GB 1221(92)	0Cr17Ni4Cu4Nb	Bar; SA; Aged	0.07 max	15.50-17.50	1.00 max		3.00-5.00	0.035 max	0.030 max	1.00 max	Cu 3.00-5.00; Nb 0.15-0.45; bal Fe	1310	1180	10	
GB 8732(88)	0Cr16Ni4Cu4Nb	Bar Q/T	0.055 max	15.00-16.00	0.50 max		3.80-4.50	0.035 max	0.030 max	1.00 max	Al <=0.050; Cu 3.00-3.70; Nb+Ta 0.15-0.45;N 0.05; bal Fe	890-1030	755-890	16	

Europe

EN 10088/3(95)	1.4542	Bar Rod Sect, Corr res; t<=100mm, SA	0.07 max	15.00-17.00	1.50 max	0.60 max	3.00-5.00	0.040 max	0.030 max	0.70 max	Cu 3.00-5.00; 5xC<=Nb<=0.45; bal Fe	1200 max			360 HB max
EN 10088/3(95)	1.4542	Bar Rod Sect, Corr res; t<=100mm, Hard	0.07 max	15.00-17.00	1.50 max	0.60 max	3.00-5.00	0.040 max	0.030 max	0.70 max	Cu 3.00-5.00; 5xC<=Nb<=0.45; bal Fe	1070-1270	1000	10	
EN 10088/3(95)	X5CrNiCuNb16-4	Bar Rod Sect, Corr res; t<=100mm, SA	0.07 max	15.00-17.00	1.50 max	0.60 max	3.00-5.00	0.040 max	0.030 max	0.70 max	Cu 3.00-5.00; 5xC<=Nb<=0.45; bal Fe	1200 max			360 HB max
EN 10088/3(95)	X5CrNiCuNb16-4	Bar Rod Sect, Corr res; t<=100mm, Hard	0.07 max	15.00-17.00	1.50 max	0.60 max	3.00-5.00	0.040 max	0.030 max	0.70 max	Cu 3.00-5.00; 5xC<=Nb<=0.45; bal Fe	1070-1270	1000	10	

France

| AFNOR NFA35581 | Z6CNU17.04 | Sh Strp Plt Bar Frg | 0.07 max | 15.5-17.5 | 1 max | | 3-5 | 0.035 max | 0.025 max | 1 max | Cu 3-5; Nb 0.15-0.45; bal Fe | | | | |

Germany

| DIN EN 10088(95) | WNr 1.4542 | Sh Strp Plt, hard, long sample | 0.07 max | 15.0-17.0 | 1.50 max | | 3.00-5.00 | 0.040 max | 0.030 max | 0.70 max | Cu 3.00-5.00; Nb 5xC>=0.45; bal Fe | 1070-1270 | 1000 | 10 | |
| DIN EN 10088(95) | X5CrNiCuNb16-4 | Sh Plt Strp Bar Rod Sect, hard long | 0.07 max | 15.0-17.0 | 1.50 max | | 3.00-5.00 | 0.040 max | 0.030 max | 0.70 max | Cu 3.00-5.00; Nb 5xC<=0.45; bal Fe | 1070-1270 | 1000 | 10 | |

Japan

JIS G3214(91)	SUSF630	Frg press ves <200mm diam HT	0.07 max	15.00-17.50	1.00 max		3.00-5.00	0.040 max	0.030 max	1.00 max	Cu 3.00-5.00; Nb 0.15-0.45; bal Fe	930	725	12	277 HB
JIS G4303(91)	SUS630	Bar; SA; 75mm diam	0.07 max	15.00-17.50	1.00 max		3.00-5.00	0.040 max	0.030 max	1.00 max	Cu 3.00-5.00; Nb 0.15-0.45; bal Fe				38 HRC max
JIS G4305(91)	SUS630		0.07 max	15.00-17.50	1.00 max		3.00-5.00	0.040 max	0.030 max	1.00 max	Cu 3.00-5.00; Nb 0.15-0.45; bal Fe				

USA

	AISI S17400	Tube	0.07 max	15.50-17.50	1.00 max		3.00-5.00	0.040 max	0.030 max	1.00 max	Cu 3.00-5.00; 0.15<=NB=Ta<=0.45; bal Fe				
	UNS S17400	Cr-Ni-Cu 17-4 PH	0.07 max	15.0-17.5	1.00 max		3.00-5.00	0.040 max	0.030 max	1.00 max	Cu 3.00-5.00; Nb 0.15-0.45; bal Fe				
AMS 5604			0.07 max	15.5-17.5	1 max		3-5	0.040 max	0.03 max	1 max	Cu 3-5; Nb 0.15-0.45; bal Fe				
AMS 5622			0.07 max	15.5-17.5	1 max		3-5	0.040 max	0.03 max	1 max	Cu 3-5; Nb 0.15-0.45; bal Fe				
AMS 5643			0.07 max	15.5-17.5	1 max		3-5	0.040 max	0.03 max	1 max	Cu 3-5; Nb 0.15-0.45; bal Fe				
AMS 5643P(99)		Corr res; Bar Wir, Frg, Tub, Rings, Cond H1150	0.07 max	15.00-17.50	1.00 max	0.50 max	3.00-5.00	0.040 max	0.030 max	1.00 max	Cu 3.00-5.00; Nb 5xC-0.45; bal Fe	931	724	16	277-352 HB
AMS 5643P(99)		Corr res; Bar Wir, Frg, Tub, Rings, Cond H900	0.07 max	15.00-17.50	1.00 max	0.50 max	3.00-5.00	0.040 max	0.030 max	1.00 max	Cu 3.00-5.00; Nb 5xC-0.45; bal Fe	1310	1172	10	388-444 HB
ASME SA564	630	Refer to ASTM A564/A564M(92); Bar													

Specification	Designation	Notes	C	Cr	Mn	Mo	Ni	P	S	Si	Other	UTS	YS	El	Hard

Stainless Steel, Precipitation-Hardening, 17-4 PH (Continued from previous page)

USA

Specification	Designation	Notes	C	Cr	Mn	Mo	Ni	P	S	Si	Other	UTS	YS	El	Hard
ASME SA705	630	Refer to ASTM A705/A705M(93); Frg													
ASTM A564/A564M(97)	630	Bar Shp, HR, CF AH	0.07 max	15.00-17.50	1.00 max		3.00-5.00	0.040 max	0.030 max	1.00 max	Cu 3.00-5.00; Nb+Ta 0.15-0.45; bal Fe	795-1310	520-1170	10-18	255-388 HB
ASTM A693(93)	630	Sh Strp Plt, Heat res, ST	0.07 max	15.00-17.50	1.00 max		3.00-5.00	0.040 max	0.030 max	1.00 max	Cu 3.00-5.00; Nb 0.15-0.45; Nb+Ta 0.15-0.45; bal Fe	1255	1105	3	38 RHC
ASTM A705/705M(95)	630	AH, Frg, SHT	0.07 max	15.00-17.50	1.00 max		3.00-5.00	0.040 max	0.030 max	1.00 max	Cu 3.00-5.00; Nb+Ta 0.15-0.45; bal Fe				363 HB
MIL-C-24111B(77)	17-4 PH*	Obs; Bar Frg	0.07 max	15.5-17.5	1 max		3-5	0.04 max	0.03 max	1 max	Cu 3-5; Nb 0.15-0.45; bal Fe				
SAE J467(68)	17-4PH	Bar Frg	0.04	16.0	0.28		4.25	0.04 max	0.03 max	0.60	Cu 3.30; Nb 0.27; bal Fe	1400	1283	11	44 HRC

Stainless Steel, Precipitation-Hardening, 17-7 PH

China

Specification	Designation	Notes	C	Cr	Mn	Mo	Ni	P	S	Si	Other	UTS	YS	El	Hard
GB 1220(92)	0Cr17Ni7Al	Bar; SA; Aged	0.09 max	16.00-18.00	1.00 max		6.50-7.50	0.035 max	0.030 max	1.00 max	Al 0.75-1.50; Cu <=0.50; bal Fe	1140	960	5	
GB 1221(92)	0Cr17Ni7Al	Bar; SA; Aged	0.09 max	16.00-18.00	1.00 max		6.50-7.50	0.035 max	0.030 max	1.00 max	Al 0.75-1.50; Cu <=0.50; bal Fe	1140	960	5	
GB 3280(92)	0Cr17Ni7Al	Sh Plt, CR SA, Aged 3mm Thk	0.09 max	16.00-18.00	1.00 max		6.50-7.50	0.035 max	0.030 max	1.00 max	Al 0.75-1.50; Cu <=0.50; bal Fe	1140	960	3	
GB 4231(93)	0Cr17Ni7Al	Strp CR SA	0.09 max	16.00-18.00	1.00 max		6.50-7.50	0.035 max	0.030 max	1.00 max	Al 0.75-1.50; Cu <=0.50; bal Fe	1030		20	
GB 4237(92)	0Cr17Ni7Al	Sh Plt, HR SA Aged <=3mm Thk	0.09 max	16.00-18.00	1.00 max		6.50-7.50	0.035 max	0.030 max	1.00 max	Al 0.75-1.50; Cu <=0.50; bal Fe	1140	960	3	
GB 4238(92)	0Cr17Ni7Al	Plt, HR SA Aged <=3mm Thk	0.09 max	16.00-18.00	1.00 max		6.50-7.50	0.035 max	0.030 max	1.00 max	Al 0.75-1.50; Cu <=0.50; bal Fe	1140	960	3	

Europe

Specification	Designation	Notes	C	Cr	Mn	Mo	Ni	P	S	Si	Other	UTS	YS	El	Hard
EN 10088/2(95)	1.4568	Strp, Corr res; t<=6mm, CR Hard	0.09 max	16.00-18.00	1.00 max		6.50-7.80	0.040 max	0.015 max	0.70 max	Al 0.70-1.50; bal Fe	1450	1310	2	
EN 10088/2(95)	X7CrNiAl17-7	Strp, Corr res; t<=6mm, CR Hard	0.09 max	16.00-18.00	1.00 max		6.50-7.80	0.040 max	0.015 max	0.70 max	Al 0.70-1.50; bal Fe	1450	1310	2	
EN 10088/3(95)	1.4568	Bar Rod Sect; Corr res; Mart; t<=30mm, SA	0.09 max	16.00-18.00	1.00 max		6.50-7.80	0.040 max	0.015 max	0.70 max	Al 0.70-1.50; bal Fe	850 max			255 HB max
EN 10088/3(95)	X7CrNiAl17-7	Bar Rod Sect; Corr res; Mart; t<=30mm, SA	0.09 max	16.00-18.00	1.00 max		6.50-7.80	0.040 max	0.015 max	0.70 max	Al 0.70-1.50; bal Fe	850 max			255 HB max

France

Specification	Designation	Notes	C	Cr	Mn	Mo	Ni	P	S	Si	Other	UTS	YS	El	Hard
AFNOR NFA35581	Z8CNA17.07	Sh Strp Plt Bar Wir Frg	0.09 max	16-18	1 max		6.5-7.75	0.035 max	0.025 max	1 max	Al 0.75-1.5; Cu <=0.5; bal Fe				

Germany

Specification	Designation	Notes	C	Cr	Mn	Mo	Ni	P	S	Si	Other	UTS	YS	El	Hard
DIN EN 10088(95)	WNr 1.4568	Sh Plt Strp Bar Rod Sect	0.09 max	16.0-18.0	1.00 max		6.50-7.80	0.040 max	0.015 max	0.70 max	Al 0.70-1.50; bal Fe				
DIN EN 10088(95)	X7CrNiAl17-7	Sh Plt Strp Bar Rod Sect	0.09 max	16.0-18.0	1.00 max		6.50-7.80	0.040 max	0.015 max	0.70 max	Al <=1.50; bal Fe				

Japan

Specification	Designation	Notes	C	Cr	Mn	Mo	Ni	P	S	Si	Other	UTS	YS	El	Hard
JIS G4303(91)	SUS631	Bar; SA; 75mm diam	0.09 max	16.00-18.00	1.00 max		6.50-7.75	0.040 max	0.030 max	1.00 max	Al 0.75-1.50; bal Fe	1030 max	380 max	20	229 HBS/W max
JIS G4304(91)	SUS631	Sh Plt, HR SA	0.09 max	16.00-18.00	1.00 max		6.5-7.75	0.040 max	0.030 max	1.00 max	Al 0.75-1.50; bal Fe	1030 max	380 max	20	92 HRB max
JIS G4305(91)	SUS631	Sh Plt, CR SA	0.09 max	16.00-18.00	1.00 max		6.50-7.75	0.040 max	0.030 max	1.00 max	Al 0.75-1.50; bal Fe	1030 max	380 max	20	192 HB max
JIS G4306	SUS631	Strp HR SA	0.09 max	16-18	1 max		6.5-7.75	0.04 max	0.03 max	1 max	Al 0.75-1.5; bal Fe				
JIS G4307	SUS631	Strp CR SA	0.09 max	16-18	1 max		6.5-7.75	0.04 max	0.03 max	1 max	Al 0.75-1.5; bal Fe				
JIS G4307	SUS631	Strp CR Aged 3mm diam	0.09 max	16-18	1 max		6.5-7.75	0.04 max	0.03 max	1 max	Al 0.75-1.5; bal Fe				
JIS G4308(98)	SUS631J1	Wir rod	0.09 max	16.00-18.00	1.00 max		7.00-8.50	0.040 max	0.030 max	1.00 max	Al 0.75-1.50; bal Fe				
JIS G4313	SUS631-CSP	Strp CR	0.09 max	16.00-18.00	1.00 max		6.50-7.75	0.040 max	0.030 max	1.00 max	Al 0.75-1.50; bal Fe				
JIS G4313	SUS632J1-CSP	Strp CR	0.09 max	13.50-15.50	1.00 max		6.50-7.75	0.040 max	0.030 max	1.00-2.00	Cu 0.40-1.00; Ti 0.20-0.65; bal Fe				

UNS numbers and US grades are provided as a means of cross referencing chemically similar alloys. Exchangability is only possible after independent examination of specifications. Tensile properties are minimum or typical as specified. UTS and YS as MPa. El as %. See Appendix for list of abbreviations used in Notes. * indicates obsolete material.

Specification	Designation	Notes	C	Cr	Mn	Mo	Ni	P	S	Si	Other	UTS	YS	El	Hard

Stainless Steel, Precipitation-Hardening, 17-7 PH (Continued from previous page)

Russia

Specification	Designation	Notes	C	Cr	Mn	Mo	Ni	P	S	Si	Other	UTS	YS	El	Hard
GOST	O9Ch17N7Ju1	Sh Strp Plt Bar Frg	0.09 max	16.5-18	0.8 max	0.3 max	6.5-7.5	0.035 max	0.025 max	0.8 max	Al 0.7-1.1; Cu <=0.3; Ti <=0.2; W <=0.2; bal Fe				

Sweden

Specification	Designation	Notes	C	Cr	Mn	Mo	Ni	P	S	Si	Other	UTS	YS	El	Hard
SS 142388	2388	Corr res	0.09 max	16-18	1 max		6.5-7.75	0.04 max	0.03 max	1 max	Al 0.75-1.5; bal Fe				

USA

Specification	Designation	Notes	C	Cr	Mn	Mo	Ni	P	S	Si	Other	UTS	YS	El	Hard
	AISI S17700	Tube	0.09 max	16.00-18.00	1.00 max		6.50-7.75	0.040 max	0.040 max	1.00 max	Al 0.75-1.5; bal Fe				
	UNS S17700	Cr-Ni-Al-Ti 17-7 PH	0.09 max	16.00-18.00	1.00 max		6.50-7.75	0.040 max	0.040 max	1.00 max	Al 0.75-1.50; bal Fe				
AMS 5528			0.09 max	16-18	1 max		6.5-7.75	0.040 max	0.040 max	1 max	Al 0.75-1.5; bal Fe				
AMS 5529			0.09 max	16-18	1 max		6.5-7.75	0.040 max	0.040 max	1 max	Al 0.75-1.5; bal Fe				
AMS 5568			0.09 max	16-18	1 max		6.5-7.75	0.040 max	0.040 max	1 max	Al 0.75-1.5; bal Fe				
AMS 5644			0.09 max	16-18	1 max		6.5-7.75	0.040 max	0.040 max	1 max	Al 0.75-1.5; bal Fe				
AMS 5673			0.09 max	16-18	1 max		6.5-7.75	0.040 max	0.040 max	1 max	Al 0.75-1.5; bal Fe				
AMS 5678			0.09 max	16-18	1 max		6.5-7.75	0.040 max	0.040 max	1 max	Al 0.75-1.5; bal Fe				
ASME SA705	631	Refer to ASTM A705/A705M(93); Frg													
ASTM A313/A313M(95)	631	Spring wire, 0.25<t<=0.38mm	0.09 max	16.00-18.00	1.00 max		6.50-7.75	0.040 max	0.030 max	1.00 max	Al 0.75-1.50; bal Fe	2035			
ASTM A564/A564M(97)	631	Bar Shp, HR, CF AH	0.09 max	16.00-18.00	1.00 max		6.50-7.75	0.040 max	0.040 max	1.00 max	Al 0.75-1.50; bal Fe	1170-1280	965-1030	6	352-388 HB
ASTM A579(96)	62	Superstrength Frg	0.09 max	16.0-18.0	1.00 max		6.5-7.75	0.025 max	0.025 max	1.00 max	Al 0.75-1.50; B 0.003; Ca 0.06; Zr 0.02; bal Fe	1140-1240	965-1100	6	207 HB
ASTM A693(93)	631	Sh Strp Plt, Heat res, ST	0.09 max	16.00-18.00	1.00 max		6.50-7.75	0.040 max	0.030 max	1.00 max	Al 0.75-1.50; bal Fe	1035	380-450	20	92 RHB
ASTM A705/705M(95)	631	AH, Frg, SHT	0.09 max	16.00-18.00	1.00 max		6.50-7.75	0.040 max	0.030 max	1.00 max	Al 0.75-1.50; bal Fe				229 HB
MIL-S-25043	17-7 PH*	Obs; Sh Strp Bar Plt	0.09 max	16-18	1 max		6.5-7.75	0.04 max	0.04 max	1 max	Al 0.75-1.5; bal Fe				
SAE J217(94)	17-7 PH	See ASTM A313(631)													
SAE J467(68)	17-7PH	Bar Frg	0.07	17.0	0.50		7.10	0.04 max	0.04 max	0.30	Al 1.17; bal Fe	1586	1496	6	48 HRC

Stainless Steel, Precipitation-Hardening, 5603

USA

Specification	Designation	Notes	C	Cr	Mn	Mo	Ni	P	S	Si	Other	UTS	YS	El	Hard
	UNS S14800	Cr-Ni-Mo-Al PH 14-8 Mo	0.05 max	13.75-15.00	1.00 max	2.00-3.00	7.75-8.75	0.015 max	0.010 max	1.00 max	Al 0.75-1.50; bal Fe				

Stainless Steel, Precipitation-Hardening, 5774

USA

Specification	Designation	Notes	C	Cr	Mn	Mo	Ni	P	S	Si	Other	UTS	YS	El	Hard
	UNS S35080	Bare Filler Metal AM350	0.08-0.12	16.0-17.0	0.50-1.25	2.50-3.25	4.00-5.00	0.040 max	0.030 max	0.50 max	N 0.07-0.13; bal Fe				

Stainless Steel, Precipitation-Hardening, 5780

USA

Specification	Designation	Notes	C	Cr	Mn	Mo	Ni	P	S	Si	Other	UTS	YS	El	Hard
	UNS S35580	Bare Filler Metal	0.10-0.15	15.0-16.0	0.50-1.25	2.50-3.25	4.00-5.00	0.04 max	0.03 max	0.50 max	Cu <=0.50; N 0.07-0.13; bal Fe				

Stainless Steel, Precipitation-Hardening, 635

USA

Specification	Designation	Notes	C	Cr	Mn	Mo	Ni	P	S	Si	Other	UTS	YS	El	Hard
	UNS S17600	Cr-Ni-Al-Ti W	0.08 max	15.00-17.50	1.00 max		6.00-7.50	0.040 max	0.30 max	1.00 max	Al <=0.40; Ti 0.40-1.20; bal Fe				
ASTM A564/A564M(97)	635	Bar Shp, HR, CF AH	0.08 max	16.00-17.50	1.00 max		6.00-7.50	0.040 max	0.030 max	1.00 max	Al <=0.40; Ti 0.40-1.20; bal Fe	1170-1310	1035-1170	8-10	331-363 HB
ASTM A693(93)	635	Sh Strp Plt, Heat res, ST	0.08 max	16.00-17.50	1.00 max		6.00-7.50	0.040 max	0.030 max	1.00 max	Al <=0.40; Ti 0.40-1.20; bal Fe	825	515	3-5	32 RHC
ASTM A705/705M(95)	635	AH, Frg, SHT	0.08 max	16.00-17.50	1.00 max		6.00-7.50	0.040 max	0.030 max	1.00 max	Al <=0.40; Ti 0.40-1.20; bal Fe	825	515	10	302 HB

UNS numbers and US grades are provided as a means of cross referencing chemically similar alloys. Exchangability is only possible after independent examination of specifications. Tensile properties are minimum or typical as specified. UTS and YS as MPa. El as %. See Appendix for list of abbreviations used in Notes. * indicates obsolete material.

Specification	Designation	Notes	C	Cr	Mn	Mo	Ni	P	S	Si	Other	UTS	YS	El	Hard
Stainless Steel, Precipitation-Hardening, A579(61)															
USA															
	UNS S35000	AM 350	0.07-0.11	16.00-17.00	0.50-1.25	2.50-3.25	4.00-5.00	0.040 max	0.030 max	0.50 max	N 0.07-0.13; bal Fe				
AMS 5554C(94)		Corr/Heat res; Smls tub, Ann	0.08-0.12	16.00-17.00	0.50-1.25	2.50-3.25	4.00-5.00	0.040 max	0.030 max	0.50 max	N 0.07-0.13; bal Fe	1138	896	10	
ASTM A579(96)	61	Superstrength Frg	0.07 max	15.5-17.5	1.00 max		3.0-5.0	0.025 max	0.025 max	1.00 max	B 0.003; Cu 3.0-5.0; Nb 0.15-0.45; Ca 0.06; Zr 0.02; bal Fe	1140-1380	965-1240	8-12	375 HB
ASTM A693(93)	633	Sh Strp Plt, Heat res, ST	0.07-0.11	16.00-17.00	0.50-1.25	2.50-3.25	4.00-5.00	0.040 max	0.030 max	0.50 max	N 0.07-0.13; bal Fe	1380	585-620	8-12	30 RHC
SAE J467(68)	AM-350		0.10	16.5	1.00	2.75	4.25			0.40	N 0.10; bal Fe	1400	1172	13	
Stainless Steel, Precipitation-Hardening, AM355															
UK															
BS 1449/2(83)	305S19	Corr/Heat res; t<=100mm	0.10 max	17.0-19.0	2.00 max		11.0-13.0	0.045 max	0.030 max	1.00 max	bal Fe	490	185	40	0-185 HB
BS 1449/2(83)	309S24	Corr/Heat res; t<=100mm	0.15 max	22.0-25.0	2.00 max		13.0-16.0	0.045 max	0.030 max	1.00 max	bal Fe	510	205	40	0-205 HB
USA															
	UNS S35500		0.10-0.15	15.00-16.00	0.50-1.25	2.50-3.25	4.00-5.00	0.040 max	0.030 max	0.50 max	N 0.07-0.13; bal Fe				
AMS 5549F(93)*		Noncurrent; Corr/Heat res; Plt	0.10-0.15	15.00-16.00	0.50-1.25	2.50-3.25	4.00-5.00	0.040 max	0.030 max	0.50 max	N 0.07-0.13; bal Fe	1140	965	12	37-44 HRC
ASTM A564/A564M(97)	634	Bar Shp, HR, CF AH	0.10-0.15	15.00-16.00	0.50-1.25	2.50-3.25	4.00-5.00	0.040 max	0.030 max	0.50 max	N 0.07-0.13; bal Fe	1170	1070	12	341 HB
ASTM A693(93)	634	Sh Strp Plt, Heat res, ST	0.10-0.15	15.00-16.00	0.50-1.25	2.50-3.25	4.00-5.00	0.040 max	0.030 max	0.50 max	N 0.07-0.13; Nb+Ta 0.15-0.50; bal Fe				40 RHC
ASTM A705/705M(95)	634	AH, Frg, SHT	0.10-0.15	15.00-16.00	0.50-1.25	2.50-3.25	4.00-5.00	0.040 max	0.030 max	0.50 max	N 0.07-0.13; bal Fe				363 HB
SAE J467(68)	AM-355		0.15	15.5	1.00	2.75	4.25			0.40	N 0.10; bal Fe	1489	1255	19	48 HRC
Stainless Steel, Precipitation-Hardening, No equivalents identified															
Russia															
GOST 5632(72)	09Ch17N7Ju1	Corr res	0.09 max	16.5-18.0	0.8 max	0.3 max	6.5-7.5	0.035 max	0.025 max	0.8 max	Al 0.7-1.1; Cu <=0.3; Ti <=0.2; W <=0.2; bal Fe				
USA															
DoD-F-24669/5(86)	I (Cr-Ni-P)	Frg, Bar Bil, Wire, Low Magnetic; Supersedes MIL-S-17759B	0.15 max	16-18	1 max		9.5-12	0.20-0.40	0.040 max	1 max	bal Fe	830	550	18	269 HB min
DoD-F-24669/5(86)	II (Cr-Ni-Mn-P)	Frg, Bar Bil, Wire, Low Magnetic; Supersedes MIL-S-17759B	0.26-0.33	17-19	3-4		8-10	0.18-0.33	0.035 max	1 max	bal Fe	830	620	15	269 HB min
DoD-F-24669/8(91)	Grade G	Frg for Steam Turbines; Supersedes MIL-S-860B(66)	0.15 max	11.5-13.0	1.00 max	0.50 max	0.50 max	0.012 max	0.015 max	0.50 max	As<=0.020; Sb<=0.010; Sn<=0.020; bal Fe	620-758	482	18	
SAE J467(68)	14-4PH		0.03	14.1	0.35	2.38	4.25			0.75	Cu 3.25; N 0.02; Nb 0.25; bal Fe				
Stainless Steel, Precipitation-Hardening, PH 15-7 Mo															
China															
GB 1220(92)	0Cr15Ni7Mo2Al	Bar; SA; Aged	0.09 max	14.00-16.00	1.00 max	2.00-3.00	6.50-7.50	0.035 max	0.030 max	1.00 max	Al 0.75-1.50; bal Fe	1210	1100	7	
France															
AFNOR NFA35581	Z8CNDA15.07	Sh Strp Plt Bar Frg	0.09 max	14-16	1 max	2-3	6.5-7.75	0.035 max	0.025 max	1 max	Al 0.75-1.5; bal Fe				
Germany															
DIN EN 10088(95)	WNr 1.4532	Sh Strp Plt	0.10 max	14.0-16.0	1.20 max	2.00-3.00	6.50-7.80	0.040 max	0.015 max	0.70 max	Al 0.75-1.50; bal Fe				
DIN EN 10088(95)	X8CrNiMoAl15-7-2		0.10 max	14.0-16.0	1.20 max	2.00-3.00	6.50-7.80	0.040 max	0.015 max	0.70 max	Al 0.75-1.50; bal Fe				
USA															
	AISI 632*	Obs; Tube	0.09 max	14.00-16.00	1.00 max	2.00-3.00	6.5-7.75	0.040 max	0.030 max	1.00 max	Al 0.75-1.5; bal Fe				
	UNS S15700	Cr-Ni-Mo-Al PH 15-7 Mo	0.09 max	14.00-16.00	1.00 max	2.00-3.00	6.50-7.75	0.04 max	0.03 max	1.00 max	Al 0.75-1.50; bal Fe				

UNS numbers and US grades are provided as a means of cross referencing chemically similar alloys. Exchangability is only possible after independent examination of specifications. Tensile properties are minimum or typical as specified. UTS and YS as MPa. El as %. See Appendix for list of abbreviations used in Notes. * indicates obsolete material.

Specification	Designation	Notes	C	Cr	Mn	Mo	Ni	P	S	Si	Other	UTS	YS	El	Hard

Stainless Steel, Precipitation-Hardening, PH 15-7 Mo (Continued from previous page)

USA

Specification	Designation	Notes	C	Cr	Mn	Mo	Ni	P	S	Si	Other	UTS	YS	El	Hard
AMS 5520		Bar Frg	0.09 max	14-16	1 max	2-3	6.5-7.75	0.040 max	0.03 max	1 max	Al 0.75-1.5; bal Fe				
ASTM A564/A564M(97)	632	Bar Shp, HR, CF AH	0.09 max	14.00-16.00	1.00 max	2.00-3.00	6.50-7.75	0.040 max	0.030 max	1.00 max	Al 0.75-1.50; bal Fe	1240-1380	1100-1210	7-8	375-415 HB
ASTM A579(96)	63	Superstrength Frg	0.09 max	14.0-15.25	1.00 max	2.0-2.75	6.5-7.75	0.025 max	0.025 max	0.50 max	Al 0.75-1.25; B 0.003; Ca 0.06; Zr 0.02; bal Fe	1240-1550	1100-1380	5-6	241 HB
ASTM A693(93)	632	Sh Strp Plt, Heat res, ST	0.09 max	14.00-16.00	1.00 max	2.00-3.00	6.50-7.75	0.040 max	0.030 max	1.00 max	Al 0.75-1.50; bal Fe	1035	450	25	100 RHB
ASTM A705/705M(95)	632	AH, Frg, SHT	0.09 max	14.00-16.00	1.00 max	2.00-3.00	6.50-7.75	0.040 max	0.030 max	1.00 max	Al 0.75-1.50; bal Fe				269 HB
SAE J467(68)	PH 15-7 Mo	Bar Frg	0.07	15.0	0.50	2.20	7.00	0.04 max	0.03 max	0.30	Al 1.17; bal Fe	1655	1551	6	48 HRC

Stainless Steel, Precipitation-Hardening, S13800

USA

Specification	Designation	Notes	C	Cr	Mn	Mo	Ni	P	S	Si	Other	UTS	YS	El	Hard
	AISI S13800	Tube	0.05 max	12.25-13.25	0.10 max	2.00-2.50	7.50-8.50	0.01 max	0.008 max	0.10 max	Al 0.90-1.35; N <=0.010; bal Fe				
	UNS S13800	Cr-Ni-Al-Mo PH 13-8 Mo	0.05 max	12.25-13.25	0.20 max	2.00-2.50	7.50-8.50	0.01 max	0.008 max	0.10 max	Al 0.90-1.35; N <=0.01; bal Fe				
ASTM A564/A564M(97)	XM-13	Bar Shp, HR, CF AH	0.05 max	12.25-13.25	0.20 max	2.00-2.50	7.50-8.50	0.010 max	0.008 max	0.10 max	Al 0.90-1.35; N <=0.01; bal Fe	860-1520	585-1415	10-16	259-430 HB
ASTM A693(93)	XM-13	Sh Strp Plt, Heat res, ST	0.05 max	12.25-13.25	0.20 max	2.00-2.50	7.50-8.50	0.010 max	0.008 max	0.10 max	Al 0.90-1.35; N <=0.01; bal Fe				38 RHC
ASTM A705/705M(95)	XM-13	AH, Frg, SHT	0.05 max	12.25-13.25	0.20 max	2.00-2.50	7.50-8.50	0.010 max	0.008 max	0.10 max	Al 0.90-1.35; N <=0.01; bal Fe				363 HB

Stainless Steel, Precipitation-Hardening, S16600

USA

Specification	Designation	Notes	C	Cr	Mn	Mo	Ni	P	S	Si	Other	UTS	YS	El	Hard
	UNS S16600	Cr-Ni-Ti-Al Croloy 16-6 PH	0.025-0.045	15.00-16.00	0.70-0.90		7.00-8.00	0.025 max	0.025 max	0.50 max	Al 0.25-0.40; Ti 0.30-0.50; bal Fe				

Stainless Steel, Precipitation-Hardening, XM-12

Europe

Specification	Designation	Notes	C	Cr	Mn	Mo	Ni	P	S	Si	Other	UTS	YS	El	Hard
EN 10088/3(95)	1.4594	Bar Rod Sect, Corr res; t<=100mm, SA	0.07 max	13.00-15.00	1.00 max	1.20-2.00	5.00-6.00	0.040 max	0.015 max	0.70 max	Cu 1.20-2.00; Nb 0.15-0.60; bal Fe	1200 max			360 HB max
EN 10088/3(95)	1.4594	Bar Rod Sect, Corr res; t<=100mm, Hard	0.07 max	13.00-15.00	1.00 max	1.20-2.00	5.00-6.00	0.040 max	0.015 max	0.70 max	Cu 1.20-2.00; Nb 0.15-0.60; bal Fe	1070-1270	1000	10	
EN 10088/3(95)	X5CrNiCuNb14-5	Bar Rod Sect, Corr res; t<=100mm, Hard	0.07 max	13.00-15.00	1.00 max	1.20-2.00	5.00-6.00	0.040 max	0.015 max	0.70 max	Cu 1.20-2.00; Nb 0.15-0.60; bal Fe	1070-1270	1000	10	
EN 10088/3(95)	X5CrNiCuNb14-5	Bar Rod Sect, Corr res; t<=100mm, SA	0.07 max	13.00-15.00	1.00 max	1.20-2.00	5.00-6.00	0.040 max	0.015 max	0.70 max	Cu 1.20-2.00; Nb 0.15-0.60; bal Fe	1200 max			360 HB max

France

Specification	Designation	Notes	C	Cr	Mn	Mo	Ni	P	S	Si	Other	UTS	YS	El	Hard
AFNOR NFA35581	Z6CNU15.05	Sh Strp Plt Bar	0.07 max	14-15.5	1 max		3.5-5.5	0.035 max	0.025 max	1 max	Cu 2.5-4.5; Nb 0.15-0.45; bal Fe				

USA

Specification	Designation	Notes	C	Cr	Mn	Mo	Ni	P	S	Si	Other	UTS	YS	El	Hard
	AISI S15500	Tube	0.07 max	14.00-15.50	1.00 max		3.50-5.50	0.040 max	0.030 max	1.00 max	Cu 2.50-4.50; 0.15<=NB=Ta<=0.45; bal Fe				
	UNS S15500	Cr-Ni-Cu 15-5 PH	0.07 max	14.00-15.50	1.00 max		3.50-5.50	0.040 max	0.030 max	1.0 max	Cu 2.50-4.50; Nb 0.15-0.45; bal Fe				
AMS 5658			0.07 max	14-15.5	1 max		3.5-5.5	0.040 max	0.03 max	1 max	Cu 2.5-4.5; Nb 0.15-0.45; bal Fe				
AMS 5659			0.07 max	14-15.5	1 max		3.5-5.5	0.040 max	0.03 max	1 max	Cu 2.5-4.5; Nb 0.15-0.45; bal Fe				
AMS 5826			0.07 max	14-15.5	1 max		3.5-5.5	0.040 max	0.03 max	1 max	Cu 2.5-4.5; Nb 0.15-0.45; bal Fe				
AMS 5862			0.07 max	14-15.5	1 max		3.5-5.5	0.040 max	0.03 max	1 max	Cu 2.5-4.5; Nb 0.15-0.45; bal Fe				
ASME SA705	XM-12	Refer to ASTM A705/A705M(93); Frg													
ASTM A564/A564M(97)	XM-12	Bar Shp, HR, CF AH	0.07 max	14.00-15.5.0	1.00 max		3.50-5.50	0.040 max	0.030 max	1.00 max	Cu 2.50-4.50; Nb+Ta 0.15-0.45; bal Fe	795-1310	515-1170	10-18	255-388 HB
ASTM A693(93)	XM-12	Sh Strp Plt, Heat res, ST	0.07 max	14.00-15.50	1.00 max		3.50-5.50	0.040 max	0.030 max	1.00 max	Cu 2.50-4.50; Nb 0.15-0.45; Nb+Ta 0.15-0.45; bal Fe				38 RHC

Specification	Designation	Notes	C	Cr	Mn	Mo	Ni	P	S	Si	Other	UTS	YS	El	Hard

Stainless Steel, Precipitation-Hardening, XM-12 (Continued from previous page)

USA

Specification	Designation	Notes	C	Cr	Mn	Mo	Ni	P	S	Si	Other	UTS	YS	El	Hard
ASTM A705/705M(95)	XM-12	AH, Frg, SHT	0.07 max	14.00-15.50	1.00 max		3.50-5.50	0.040 max	0.030 max	1.00 max	Cu 2.50-4.50; Nb+Ta 0.15-0.45; bal Fe				363 HB

Stainless Steel, Precipitation-Hardening, XM-16

USA

Specification	Designation	Notes	C	Cr	Mn	Mo	Ni	P	S	Si	Other	UTS	YS	El	Hard
	UNS S45500	Custom 455	0.05 max	11.00-12.50	0.50 max	0.50 max	7.50-9.50	0.040 max	0.030 max	0.50 max	Cu 1.50-2.50; Nb 0.10-0.50; Ti 0.80-1.40; bal Fe				
AMS 5578D(95)		Corr/Heat res; Weld tub; t<=.51mm	0.05 max	11.00-12.50	0.50 max	0.50 max	7.50-9.50	0.015 max	0.015 max	0.50 max	Cu 1.50-2.50; Nb 0.10-0.50; Ti 0.80-1.40; bal Fe	1517	1413		42 HRC min
AMS 5672B(97)		Wire; Corr res; PH, OD 0.25-1.00mm	0.05 max	11.00-12.50	0.50 max	0.50 max	7.50-9.50	0.025 max	0.025 max	0.50 max	Cu 1.50-2.50; N <=0.015; Ti 0.80-1.40; Nb+Ta=0.10-0.50; bal Fe	2135-2345			
AMS 5672B(97)		Wire; Corr res; CD, 3.75<OD<=12.50mm	0.05 max	11.00-12.50	0.50 max	0.50 max	7.50-9.50	0.025 max	0.025 max	0.50 max	Cu 1.50-2.50; N <=0.015; Ti 0.80-1.40; Nb+Ta=0.10-0.50; bal Fe	1240			
AMS 5672B(97)		Wire; Corr res; PH, 3.75<OD<=12.50mm	0.05 max	11.00-12.50	0.50 max	0.50 max	7.50-9.50	0.025 max	0.025 max	0.50 max	Cu 1.50-2.50; N <=0.015; Ti 0.80-1.40; Nb+Ta=0.10-0.50; bal Fe	1795-2000			
AMS 5672B(97)		Wire; Corr res; CD, OD 0.25-1.00mm	0.05 max	11.00-12.50	0.50 max	0.50 max	7.50-9.50	0.025 max	0.025 max	0.50 max	Cu 1.50-2.50; N <=0.015; Ti 0.80-1.40; Nb+Ta=0.10-0.50; bal Fe	1690			
ASTM A693(93)	XM-16	Sh Strp Plt, Heat res, ST	0.05 max	11.00-12.50	0.50 max	0.50 max	7.50-9.50	0.040 max	0.030 max	0.50 max	Cu 1.50-2.50; Ti 0.80-1.40; Nb+Ta 0.15-0.50; bal Fe	1205	1105	3	36 RHC
ASTM A705/705M(95)	XM-16	AH, Frg, SHT	0.05 max	11.00-12.50	0.50 max	0.50 max	7.50-9.50	0.040 max	0.030 max	0.50 max	Cu 1.50-2.50; Ti 0.80-1.40; Nb+Ta 0.15-0.50; bal Fe	1205	1105	3	331 HB

Stainless Steel, Precipitation-Hardening, XM-25

USA

Specification	Designation	Notes	C	Cr	Mn	Mo	Ni	P	S	Si	Other	UTS	YS	El	Hard
	UNS S45000	Custom 450	0.05 max	14.00-16.00	1.00 max	0.50-1.00	5.00-7.00	0.030 max	0.030 max	1.00 max	Cu 1.25-1.75; Nb 8xC min; bal Fe				
ASTM A564/A564M(97)	XM-25	Bar Shp, HR, CF AH	0.05 max	14.00-16.00	1.00 max	0.50-1.00	5.00-7.00	0.030 max	0.030 max	1.00 max	Cu 1.25-1.75; Nb >=8xC; bal Fe	860-1240	515-1170	10-16	262-363 HB
ASTM A693(93)	XM-25	Sh Strp Plt, Heat res, ST	0.05 max	14.00-16.00	1.00 max	0.50-1.00	5.00-7.00	0.030 max	0.030 max	1.00 max	Cu 1.25-1.75; Nb>=8xC; bal Fe	1205	1035	4	33 RHC
ASTM A705/705M(95)	XM-25	AH, Frg, SHT	0.05 max	14.00-16.00	1.00 max	0.50-1.00	5.00-7.00	0.030 max	0.030 max	1.00 max	Cu 1.25-1.75; Nb>=8xC; bal Fe	1205	1035	4	311 HB

Stainless Steel, Precipitation-Hardening, XM-9

USA

Specification	Designation	Notes	C	Cr	Mn	Mo	Ni	P	S	Si	Other	UTS	YS	El	Hard
	UNS S36200	Almar 362	0.05 max	14.00-15.00	0.50 max		6.00-7.00	0.030 max	0.030 max	0.30 max	Al <=0.10; Ti 0.55-0.90; bal Fe				
ASTM A693(93)	XM-9	Sh Strp Plt, Heat res, ST	0.05 max	14.00-14.50	0.50 max		6.25-7.00	0.030 max	0.030 max	0.30 max	Al <=0.10; Ti 0.60-0.90; bal Fe	1035	860	4	38 RHC

Specification	Designation	Notes	C	Cr	Mn	Mo	Ni	P	S	Si	Other	UTS	YS	El	Hard

Stainless Steel, Unclassified, 2205

Europe

Specification	Designation	Notes	C	Cr	Mn	Mo	Ni	P	S	Si	Other	UTS	YS	El	Hard
EN 10088/2(95)	1.4462	Strp, CR Aust-Ferr Corr res; t<=6mm, Ann	0.030 max	21.00-23.00	2.00 max	2.50-3.50	4.50-6.50	0.035 max	0.015 max	1.00 max	N 0.10-0.22; bal Fe	660-950	480	20	
EN 10088/2(95)	X2CrNiMoN22-5-3	Strp, CR Aust-Ferr Corr res; t<=6mm, Ann	0.030 max	21.00-23.00	2.00 max	2.50-3.50	4.50-6.50	0.035 max	0.015 max	1.00 max	N 0.10-0.22; bal Fe	660-950	480	20	
EN 10088/3(95)	1.4462	Aust-Ferr; Corr res; t<=160mm; Ann	0.030 max	21.00-23.00	2.00 max	2.50-3.50	4.50-6.50	0.035 max	0.015 max	1.00 max	N 0.10-0.22; bal Fe	650-880	450	25	270 HB
EN 10088/3(95)	X2CrNiMoN22-5-3	Aust-Ferr; Corr res; t<=160mm; Ann	0.030 max	21.00-23.00	2.00 max	2.50-3.50	4.50-6.50	0.035 max	0.015 max	1.00 max	N 0.10-0.22; bal Fe	650-880	450	25	270 HB

Sweden

Specification	Designation	Notes	C	Cr	Mn	Mo	Ni	P	S	Si	Other	UTS	YS	El	Hard
SS 142377	2377	Corr res	0.03 max	21-23	2 max	2.5-3.5	4.5-6.5	0.03 max	0.02 max	1 max	N 0.1-0.2; bal Fe				

UK

Specification	Designation	Notes	C	Cr	Mn	Mo	Ni	P	S	Si	Other	UTS	YS	El	Hard
BS 1501/3(90)	318S13	Press ves; Corr/Heat res; 80-100mm	0.03 max	21.0-23.0	2.0 max	2.5-3.5	4.5-6.5	0.025 max	0.02 max	1.0 max	Cu <=0.3; N 0.08-0.2; Ti <=0.05; W <=0.1; bal Fe	640-840	480	25	
BS 1501/3(90)	318S13	Press ves; Corr/Heat res; 0.05-19.9mm	0.03 max	21.0-23.0	2.0 max	2.5-3.5	4.5-6.5	0.025 max	0.02 max	1.0 max	Cu <=0.3; N 0.08-0.2; Ti <=0.05; W <=0.1; bal Fe	680-880	450	25	
BS 1503(89)	318S13	Frg press ves; t<=350mm; Quen	0.030 max	21.0-23.0	2.00 max	2.50-3.50	4.50-6.50	0.025 max	0.020 max	1.00 max	Cu <=0.50; Ti <=0.10; W <=0.1; bal Fe	640-880	450	25	

USA

Specification	Designation	Notes	C	Cr	Mn	Mo	Ni	P	S	Si	Other	UTS	YS	El	Hard
	UNS S31803	Duplex Aust-Ferr	0.030 max	21.0-23.0	2.00 max	2.50-3.50	4.50-6.50	0.030 max	0.020 max	1.00 max	N 0.08-0.20; bal Fe				
	UNS S32205	Duplex Aust-Ferr, high N	0.030 max	22.0-23.0	2.00 max	3.00-3.50	4.50-6.50	0.030 max	0.020 max	1.00 max	N 0.14-0.20; bal Fe				
ASTM A182/A182M(98)	F51	Frg/roll pipe flange valve	0.030 max	21.0-23.0	2.00 max	2.5-3.5	4.5-6.5	0.030 max	0.020 max	1.00 max	N 0.08-0.20; bal Fe	620	450	25	
ASTM A240/A240M(98)	S31803	Duplex Aust-Ferr	0.030 max	21.0-23.0	2.00 max	2.50-3.50	4.50-6.50	0.030 max	0.020 max	1.00 max	N 0.08-0.20; bal Fe	620	450	25.0	293 HB
ASTM A240/A240M(98)	S32205	Duplex Aust-Ferr High-N	0.030 max	22.0-23.0	2.00 max	3.00-3.50	4.5-6.0	0.030 max	0.020 max	1.00 max	N 0.14-0.20; bal Fe	620	450	25.0	293 HB
ASTM A789/A789M(95)	2205	Smls Weld tub, Duplex Aust-Ferr	0.030 max	21.0-23.0	2.0 max	2.50-3.50	4.50-6.50	0.030 max	0.020 max	1.0 max	N 0.08-0.20; bal Fe	620	450	25	290 HB
ASTM A790/A790M(95)	2205	Smls Weld Pipe, Duplex Aust-Ferr	0.030 max	21.0-23.0	2.0 max	2.50-3.50	4.50-6.50	0.030 max	0.020 max	1.0 max	N 0.08-0.20; bal Fe	620	450	25	290 HB
ASTM A815/A815M(98)	S31803	Pipe fittings; Duplex Aust-Ferr	0.030 max	21.0-23.0	2.00 max	2.5-3.5	4.5-6.5	0.030 max	0.020 max	1.0 max	N 0.08-0.20; bal Fe	620	450	20.0	290 HB
ASTM A815/A815M(98)	S32205	Pipe fittings; Duplex Aust-Ferr	0.030 max	22.0-23.0	2.00 max	3.0-3.5	4.5-6.5	0.030 max	0.020 max	1.00 max	N 0.14-0.20; bal Fe	620	450	20.0	290 HB
ASTM A949/A949M(95)	2205	Spray formed smls pipe; Duplex Aust-Ferr	0.030 max	21.0-23.0	2.0 max	2.50-3.50	4.50-6.50	0.030 max	0.020 max	1.0 max	N 0.08-0.20; bal Fe	620	450	25	290 HB max
ASTM A988(98)	S31803	HIP Flanges, Fittings, Valves/parts; Heat res; Duplex Aust-Ferr; High-N	0.030 max	21.0-23.0	2.00 max	2.5-3.5	4.5-6.5	0.030 max	0.020 max	1.00 max	N 0.08-0.20; bal Fe	620	450	25	
ASTM A988(98)	S32205	HIP Flanges, Fittings, Valves/parts; Heat res; Duplex Aust-Ferr; High-N	0.030 max	22.0-23.0	2.00 max	3.0-3.5	4.5-6.5	0.030 max	0.020 max	1.00 max	Cu <=0.75; N 0.14-0.20; bal Fe	620	450	25.0	293 HB max
SAE J405(98)	S31803	Duplex Aust-Ferr	0.030 max	21.0-23.0	2.00 max	2.50-3.50	4.50-6.50	0.030 max	0.020 max	1.00 max	N 0.08-0.20; bal Fe				
SAE J405(98)	S32205	Duplex Aust-Ferr high N	0.030 max	22.0-23.0	2.00 max	3.00-3.50	4.50-6.50	0.030 max	0.020 max	1.00 max	N 0.14-0.20; bal Fe				

Stainless Steel, Unclassified, 2304

Europe

Specification	Designation	Notes	C	Cr	Mn	Mo	Ni	P	S	Si	Other	UTS	YS	El	Hard
EN 10088/2(95)	1.4362	Strp, CR Aust-Ferr Corr res; t<=6mm, Ann	0.030 max	22.00-24.00	2.00 max	0.10-0.60	3.50-5.50	0.035 max	0.015 max	1.00 max	Cu 0.10-0.60; N 0.05-0.20; bal Fe	600-850	420	20	
EN 10088/2(95)	X2CrNiN23-4	Strp, CR Aust-Ferr Corr res; t<=6mm, Ann	0.030 max	22.00-24.00	2.00 max	0.10-0.60	3.50-5.50	0.035 max	0.015 max	1.00 max	Cu 0.10-0.60; N 0.05-0.20; bal Fe	600-850	420	20	
EN 10088/3(95)	1.4362	Aust-Ferr; Corr res; t<=160mm; Ann	0.030 max	22.00-24.00	2.00 max	0.10-0.60	3.50-5.50	0.035 max	0.015 max	1.00 max	Cu 0.10-0.60; N 0.05-0.20; bal Fe	600-830	400	25	260 HB
EN 10088/3(95)	X2CrNiN23-4	Aust-Ferr; Corr res; t<=160mm; Ann	0.030 max	22.00-24.00	2.00 max	0.10-0.60	3.50-5.50	0.035 max	0.015 max	1.00 max	Cu 0.10-0.60; N 0.05-0.20; bal Fe	600-830	400	25	260 HB

Sweden

Specification	Designation	Notes	C	Cr	Mn	Mo	Ni	P	S	Si	Other	UTS	YS	El	Hard
SS 142327	2327	Corr res	0.03 max	22-23.5	2 max		4-5.5	0.035 max	0.02 max	1 max	N 0.05-0.15; bal Fe				

UNS numbers and US grades are provided as a means of cross referencing chemically similar alloys. Exchangability is only possible after independent examination of specifications. Tensile properties are minimum or typical as specified. UTS and YS as MPa. El as %. See Appendix for list of abbreviations used in Notes. * indicates obsolete material.

Specification	Designation	Notes	C	Cr	Mn	Mo	Ni	P	S	Si	Other	UTS	YS	El	Hard

Stainless Steel, Unclassified, 2304 (Continued from previous page)

USA

Specification	Designation	Notes	C	Cr	Mn	Mo	Ni	P	S	Si	Other	UTS	YS	El	Hard
	UNS S32304	Duplex Aust-Ferr	0.030 max	21.5-24.5	2.50 max		3.0-5.5	0.040 max	0.040 max	1.0 max	Cu 0.05-0.60; N 0.05-0.20; bal Fe				*
	UNS S32404	Duplex Aust-Ferr	0.04 max	20.5-22.5	2.0 max	2.0-3.0	5.5-8.5	0.030 max	0.010 max	1.0 max	Cu 1.0-2.0; N <=0.20; bal Fe				
ASTM A240/A240M(98)	S32304	Duplex Aust-Ferr	0.030 max	21.5-24.5	2.50 max	0.05-0.20	3.0-5.5	0.040 max	0.040 max	1.00 max	Cu 0.05-0.60; N 0.05-0.20; bal Fe	600	400	25.0	290 HB
ASTM A789/A789M(95)	2304	Smls Weld tub, Duplex Aust-Ferr	0.030 max	21.5-24.5	2.50 max	0.05-0.60	3.0-5.5	0.040 max	0.040 max	1.0 max	Cu 0.05-0.60; N 0.05-0.20; bal Fe	600-690	400-450	25	290 HB
ASTM A790/A790M(95)	2304	Smls Weld Pipe, Duplex Aust-Ferr	0.030 max	21.5-24.5	2.50 max	0.05-0.60	3.0-5.5	0.040 max	0.040 max	1.0 max	Cu 0.05-0.60; N 0.05-0.20; bal Fe	600	400	25	290 HB
ASTM A949/A949M(95)	2304	Spray formed smls pipe; Duplex Aust-Ferr	0.030 max	21.5-24.5	2.50 max	0.05-0.60	3.0-5.5	0.040 max	0.040 max	1.0 max	Cu 0.05-0.60; N 0.05-0.20; bal Fe	600	400	25	290 HB max
SAE J405(98)	S32304	Duplex Aust-Ferr	0.030 max	21.5-24.5	2.50 max	0.05-0.60	3.00-5.50	0.040 max	0.030 max	1.00 max	Cu 0.05-3.00; N 0.05-0.20; bal Fe				

Stainless Steel, Unclassified, 2507

Sweden

Specification	Designation	Notes	C	Cr	Mn	Mo	Ni	P	S	Si	Other	UTS	YS	El	Hard
SS 142328	2328	Corr res	0.03 max	22-26	1.2 max	3.5-5	6-8	0.035 max	0.02 max	0.8 max	N 0.24-0.23; bal Fe				

USA

Specification	Designation	Notes	C	Cr	Mn	Mo	Ni	P	S	Si	Other	UTS	YS	El	Hard
	UNS S32750	Duplex Aust-Ferr	0.030 max	24.0-26.0	1.20 max	3.0-5.0	6.0-8.0	0.035 max	0.020 max	0.8 max	N 0.24-0.32; bal Fe				
ASTM A182/A182M(98)	F53	Frg/roll pipe flange valve	0.030 max	24.0-26.0	1.2 max	3.0-5.0	6.0-8.0	0.035 max	0.020 max	0.8 max	Cu <=0.5; N 0.24-0.32; bal Fe	800	550	15	310 HB max
ASTM A240/A240M(98)	S32750	Duplex Aust-Ferr	0.030 max	24.0-26.0	1.20 max	3.0-5.0	6.0-8.0	0.035 max	0.020 max	0.80 max	N 0.24-0.32; bal Fe	795	550	15.0	310 HB
ASTM A789/A789M(95)	2507	Smls Weld tub, Duplex Aust-Ferr	0.030 max	24.0-26.0	1.2 max	3.0-5.0	6.0-8.0	0.035 max	0.020 max	0.8 max	Cu <=0.5; N 0.24-0.32; bal Fe	800	550	15	310 HB
ASTM A790/A790M(95)	2507	Smls Weld Pipe, Duplex Aust-Ferr	0.030 max	24.0-26.0	1.2 max	3.0-5.0	6.0-8.0	0.035 max	0.020 max	0.8 max	Cu <=0.5; N 0.24-0.32; bal Fe	800	550	15	310 HB
ASTM A815/A815M(98)	S32750	Pipe fittings; Duplex Aust-Ferr	0.030 max	24.0-26.0	1.20 max	3.0-5.0	6.0-8.0	0.035 max	0.020 max	0.8 max	Cu <=0.5; N 0.24-0.32; bal Fe	800-965	550	15	310 HB
ASTM A949/A949M(95)	2507	Spray formed smls pipe; Duplex Aust-Ferr	0.030 max	24.0-26.0	1.2 max	3.0-5.0	6.0-8.0	0.035 max	0.020 max	0.8 max	Cu <=0.5; N 0.24-0.31; bal Fe	800	550	15	310 HB max
ASTM A988(98)	S32750	HIP Flanges, Fittings, Valves/parts; Heat res; Duplex Aust-Ferr	0.030 max	24.0-26.0	1.20 max	3.0-5.0	6.0-8.0	0.035 max	0.020 max	0.80	Cu <=0.50; N 0.24-0.32; bal Fe	800	550	15	310 HB max
SAE J405(98)	S32750	Duplex Aust-Ferr	0.030 max	24.0-26.0	1.20 max	3.0-5.0	6.0-8.0	0.035 max	0.020 max	0.80 max	N 0.24-0.32; bal Fe				

Stainless Steel, Unclassified, 329

Bulgaria

Specification	Designation	Notes	C	Cr	Mn	Mo	Ni	P	S	Si	Other	UTS	YS	El	Hard
BDS 9631	Ch21N5M2L	Corr res	0.12 max	20-22	2 max	1.8-2.2	4.5-6	0.045 max	0.045 max	1.5 max	bal Fe				

China

Specification	Designation	Notes	C	Cr	Mn	Mo	Ni	P	S	Si	Other	UTS	YS	El	Hard
GB 1220(92)	0Cr26Ni5Mo2	Bar SA	0.08 max	23.00-28.00	1.50 max	1.00-3.00	3.00-6.00	0.035 max	0.030 max	1.00 max	bal Fe	590	390	18	
GB 3280(92)	0Cr26Ni5Mo2	Sh Plt, CR SA	0.08 max	23.00-28.00	1.50 max	1.00-3.00	3.00-6.00	0.035 max	0.030 max	1.00 max	bal Fe	590	390	18	
GB 4237(92)	0Cr26Ni5Mo2	Sh Plt, HR SA	0.08 max	23.00-28.00	1.50 max	1.00-3.00	3.00-6.00	0.035 max	0.030 max	1.00 max	bal Fe	590	390	18	

Europe

Specification	Designation	Notes	C	Cr	Mn	Mo	Ni	P	S	Si	Other	UTS	YS	El	Hard
EN 10088/3(95)	1.4460	Aust-Ferr; Corr res; t<=160mm; Ann	0.050 max	25.00-28.00	2.00 max	1.30-2.00	4.50-6.50	0.035 max	0.030 max	1.00 max	N 0.05-0.20; bal Fe	620-880	460	20	260 HB
EN 10088/3(95)	X3CrNiMoN27-5-2	Aust-Ferr; Corr res; * t<=160mm; Ann	0.050 max	25.00-28.00	2.00 max	1.30-2.00	4.50-6.50	0.035 max	0.030 max	1.00 max	N 0.05-0.20; bal Fe	620-880	460	20	260 HB

Finland

Specification	Designation	Notes	C	Cr	Mn	Mo	Ni	P	S	Si	Other	UTS	YS	El	Hard
SFS 391(79)	G-X12CrNiMo265	Corr res	0.12 max	23.0-27.0	2.0 max	1.3-1.8	4.5-7.0	0.045 max	0.03 max	1.5 max	bal Fe				

Germany

Specification	Designation	Notes	C	Cr	Mn	Mo	Ni	P	S	Si	Other	UTS	YS	El	Hard
DIN EN 10088(95)	WNr 1.4460	Bar Rod Sect; SHT	0.05 max	25.0-28.0	2.00 max	1.30-2.00	4.50-6.50	0.035 max	0.030 max	1.00 max	N 0.05-0.20; bal Fe	620-880	460	20	
TGL 39672	X5CrNiMo25.5*	Obs	0.07 max	25-27	0.5-1.2	1-1.5	4.5-5.5	0.025 max	0.02 max	0.5-1	bal Fe				

UNS numbers and US grades are provided as a means of cross referencing chemically similar alloys. Exchangability is only possible after independent examination of specifications. Tensile properties are minimum or typical as specified. UTS and YS as MPa. El as %. See Appendix for list of abbreviations used in Notes. * indicates obsolete material.

Specification	Designation	Notes	C	Cr	Mn	Mo	Ni	P	S	Si	Other	UTS	YS	El	Hard

Stainless Steel, Unclassified, 329 (Continued from previous page)

Japan

Specification	Designation	Notes	C	Cr	Mn	Mo	Ni	P	S	Si	Other	UTS	YS	El	Hard
JIS G4303(91)	SUS329J1	Bar <=75mm thk	0.08 max	23.00-28.00	1.50 max	1.00-3.00	3.00-6.00	0.040 max	0.030 max	1.00 max	bal Fe	590	390	18	227 HB
JIS G4303(91)	SUS329J3L	Bar <=75mm thk	0.030 max	21.00-24.00	2.00 max	2.50-3.50	4.50-6.50	0.040 max	0.030 max	1.00 max	N 0.08-0.20; bal Fe	620	450	18	302 HB
JIS G4303(91)	SUS329J4L	Bar <=75mm thk	0.030 max	24.00-26.00	1.50 max	2.50-3.50	5.50-7.50	0.040 max	0.030 max	1.00 max	N 0.08-0.30; bal Fe	620	450	18	302 HB
JIS G4305(91)	SUS329J1		0.08 max	23.00-28.00	1.50 max	1.00-3.00	3.00-6.00	0.040 max	0.030 max	1.00 max	bal Fe	590	390	18	
JIS G4305(91)	SUS329J3L		0.030 max	21.00-24.00	2.00 max	2.50-3.50	4.50-6.50	0.040 max	0.030 max	1.00 max	N 0.08-0.20; bal Fe	620	450	18	
JIS G4305(91)	SUS329J4L		0.030 max	24.00-26.00	1.50 max	2.50-3.50	5.50-7.50	0.040 max	0.030 max	1.00 max	N 0.08-0.30; bal Fe	620	450	18	
JIS G4307	SUS329J1	Strp CR SA	0.08 max	23-28	1.5 max	1-3	3-6	0.04 max	0.03 max	1 max	bal Fe				
JIS G4307	SUS329J2L	Strp CR	0.03 max	22-26	1.5 max	2.5-4	2.5-4	0.04 max	0.03 max	1 max	bal Fe				

Mexico

Specification	Designation	Notes	C	Cr	Mn	Mo	Ni	P	S	Si	Other	UTS	YS	El	Hard
DGN B-216	TP329*	Obs; Tube, As Drwn As weld, 8mm diam	0.2 max	23-28	1 max	1-2	2.5-5	0.04 max	0.03 max	0.75 max	bal Fe				

Norway

Specification	Designation	Notes	C	Cr	Mn	Mo	Ni	P	S	Si	Other	UTS	YS	El	Hard
NS 14310	14310	Bar	0.1 max	24-27	2 max	1.3-1.8	4.5-7	0.045 max	0.03 max	1 max	bal Fe				

Spain

Specification	Designation	Notes	C	Cr	Mn	Mo	Ni	P	S	Si	Other	UTS	YS	El	Hard
UNE 36016/1(89)	F.3552*	Obs; EN10088-3(96); X8CrNiMo266	0.1 max	24.0-27.0	2.0 max	1.3-1.8	4.5-7.0	0.045 max	0.03 max	1.0 max	N <=0.1; bal Fe				
UNE 36017(85)	F.3309*	see X8CrNiMo27-05	0.1 max	26-28	2 max	1.3-2	4-5	0.045 max	0.03 max	1 max	bal Fe				
UNE 36017(85)	X8CrNiMo27-05	Bar Tub; Heat res	0.1 max	26-28	2 max	1.3-2	4-5	0.045 max	0.03 max	1 max	bal Fe				

Sweden

Specification	Designation	Notes	C	Cr	Mn	Mo	Ni	P	S	Si	Other	UTS	YS	El	Hard
SS 142324	2324	Corr res	0.1 max	24-27	2 max	1.3-1.8	4.5-7	0.045 max	0.03 max	1 max	bal Fe				

USA

Specification	Designation	Notes	C	Cr	Mn	Mo	Ni	P	S	Si	Other	UTS	YS	El	Hard
	AISI 329	Tube	0.10 max	25.00-30.00	2.00 max	1.00-2.00	3.00-6.00	0.040 max	0.030 max	1.00 max	bal Fe				
	UNS S32900	Duplex Aust-Ferr	0.08 max	23.00-28.00	1.00 max	1.00-2.00	2.50-5.00	0.040 max	0.030 max	0.75 max	bal Fe				
ASME SA268	329	Refer to ASTM A268(94)													
ASTM A240/A240M(98)	S32900	Duplex Aust-Ferr	0.08 max	23.0-28.0	1.00 max	1.00-2.00	2.50-5.00	0.040 max	0.030 max	0.75 max	bal Fe	620	485	15.0	269 HB
ASTM A268	329	Tube	0.2 max	23-28	1 max	1-2	2.5-5	0.04 max	0.03 max	0.75 max	bal Fe				
ASTM A789/A789M(95)	329	Smls Weld tub, Duplex Aust-Ferr	0.08 max	23.00-28.00	1.00 max	1.00-2.00	2.50-5.00	0.040 max	0.030 max	0.75 max	bal Fe	620	485	20	271 HB
ASTM A790/A790M(95)	329	Smls Weld Pipe, Duplex Aust-Ferr	0.08 max	23.00-28.00	1.00 max	1.00-2.00	2.50-5.00	0.040 max	0.030 max	0.75 max	bal Fe	620	485	20	271 HB
ASTM A949/A949M(95)	329	Spray formed smls pipe; Duplex Aust-Ferr	0.08 max	23.00-28.00	1.00 max	1.00-2.00	2.50-5.00	0.040 max	0.030 max	0.75 max	bal Fe	620	485	20	271 HB max
SAE J405(98)	S32900	Duplex Aust-Ferr	0.08 max	23.00-28.00	1.00 max	1.00-2.00	2.50-5.00	0.040 max	0.030 max	0.75 max	bal Fe				

Stainless Steel, Unclassified, 330

UK

Specification	Designation	Notes	C	Cr	Mn	Mo	Ni	P	S	Si	Other	UTS	YS	El	Hard
BS 3076(89)	NA17	Ni/Ni alloys; Heat res; bar	0.10 max	17.0-19.0	0.80-1.50		34.5-41.0		0.030 max	1.90-2.60	Al <=0.10; Cu <=0.50; Ti <=0.20; 34.5<=Ni+Co<=41.0; bal Fe				

Stainless Steel, Unclassified, 501

Canada

Specification	Designation	Notes	C	Cr	Mn	Mo	Ni	P	S	Si	Other	UTS	YS	El	Hard
CSA 110.3	501	Bar Bil	0.1 min	4-6	1 max	0.4-0.65		0.04 max	0.03 max	1 max	bal Fe				

France

Specification	Designation	Notes	C	Cr	Mn	Mo	Ni	P	S	Si	Other	UTS	YS	El	Hard
AFNOR NFA35558	Z15CD5.05	Bar Bil	0.1-0.2	4-6	0.3-0.6	0.4-0.65		0.03 max	0.03 max	0.15-0.5	bal Fe				

Specification	Designation	Notes	C	Cr	Mn	Mo	Ni	P	S	Si	Other	UTS	YS	El	Hard

Stainless Steel, Unclassified, 501 (Continued from previous page)

Italy

Specification	Designation	Notes	C	Cr	Mn	Mo	Ni	P	S	Si	Other	UTS	YS	El	Hard
UNI 7660(77)	16CrMo205KW KG	Press ves	0.18 max	4-6	0.3-0.4	0.45-0.65		0.03 max	0.03 max	0.15-0.4	bal Fe				

Russia

| GOST 977 | 20Ch5ML | Plt Bar Bil | 0.15-0.25 | 4-6.5 | 0.4-0.6 | 0.4-0.65 | 0.4 max | 0.04 max | 0.05 max | 0.35-0.7 | Al <=0.02; Cu <=0.3; bal Fe | | | | |

UK

| BS 1503(89) | 625-50 | Frg for Press ves | 0.18 max | 4-6 | 0.3-0.8 | 0.45-0.65 | | 0.04 max | 0.04 max | 0.15-0.4 | Al <=0.02; Cu <=0.3; | | | | |

USA

Specification	Designation	Notes	C	Cr	Mn	Mo	Ni	P	S	Si	Other	UTS	YS	El	Hard
	AISI 501	Tube	0.10 min	4.00-6.00	1.00 max	0.40-0.65		0.040 max	0.030 max	1.00 max	bal Fe				
	UNS S50100	Heat res	0.10 min	4.00-6.00	1.00 max	0.40-0.65		0.040 max	0.030 max	1.00 max	bal Fe				
AMS 5502		Plt Bar Frg	0.1 min	4-6	1 max	0.4-0.65		0.040 max	0.03 max	1 max	bal Fe				
AMS 5602		Plt Bar Frg	0.1 min	4-6	1 max	0.4-0.65		0.040 max	0.03 max	1 max	bal Fe				
ASME SA387	5	Refer to ASTM A387/A387M(92); Plate													
ASTM A193/A193M(98)*	501*	Bolt, high-temp	0.1 min	4.0-6.0	1 max	0.4-0.65		0.04 max	0.03 max	1 max	bal Fe				
ASTM A193/A193M(98)	B5	Bolt, high-temp	0.1 min	4.0-6.0	1.00 max	0.40-0.65		0.040 max	0.030 max	1.00 max	bal Fe	690	550	16	
ASTM A194	501*	Nuts, high-temp press	0.1 min	4-6	1 max	0.4-0.65		0.04 max	0.03 max	1 max	bal Fe				
ASTM A194/A194M(98)	3	Nuts, high-temp press	0.1 min	4.0-6.0	1.00 max	0.40-0.65		0.040 max	0.030 max	1.00 max	bal Fe				248-352 HB
ASTM A314	501	Bolt	0.1 min	4-6	1 max	0.4-0.65		0.04 max	0.03 max	1 max	bal Fe				
ASTM A473	501	Frg, Heat res, tmp	0.10 min	4.00-6.00	1.00 max	0.40-0.65		0.040 max	0.030 max	1.00 max	bal Fe	414	205	20	201 HB

Stainless Steel, Unclassified, 502

Bulgaria

Specification	Designation	Notes	C	Cr	Mn	Mo	Ni	P	S	Si	Other	UTS	YS	El	Hard
BDS 5084	Ew-10Ch5M	Plt Bar Tub	0.12 max	4-5.5	0.4-0.7	0.4-0.6	0.3 max	0.03 max	0.03 max	0.12-0.35	bal Fe				

Canada

| CSA 110.3 | 502 | Bar Bil | 0.1 max | 4-6 | 1 max | 0.4-0.65 | | 0.04 max | 0.03 max | 1 max | bal Fe | | | | |

China

| GB 1221(92) | 1Cr5Mo | Bar Q/T | 0.15 max | 4.00-6.00 | 0.60 max | 0.45-0.60 | 0.60 max | 0.035 max | 0.030 max | 0.50 max | Cu 0.30; bal Fe | 590 | 390 | 18 | |

Czech Republic

| CSN 417102 | 17102 | Heat res | 0.15 max | 4.0-6.0 | 0.6 max | 0.45-0.65 | | 0.035 max | 0.03 max | 0.5 max | bal Fe | | | | |

France

| AFNOR NFA36206 | Z10CD5.05 | Plt, Press ves | 0.15 max | 4-6 | 0.3-0.6 | 0.45-0.65 | 0.3 max | 0.03 max | 0.03 max | 0.5 max | Cu <=0.25; V <=0.04; bal Fe | | | | |
| AFNOR NFA36602 | Z10CD5.05 | Frg, Press ves | 0.15 max | 4-6 | 0.3-0.6 | 0.45-0.65 | | 0.03 max | 0.03 max | 0.15-0.5 | Cu <=0.25; V <=0.04; bal Fe | | | | |

Germany

DIN 17176(90)	12CrMo195	Smls tube Q	0.08-0.15	4.00-6.00	0.30-0.60	0.45-0.65		0.025 max	0.020 max	0.50 max	bal Fe	590-740	390	17	
DIN 17176(90)	WNr 1.7362	Smls tube for hydrogen service	0.08-0.15	4.00-6.00	0.30-0.60	0.45-0.65		0.025 max	0.020 max	0.30-0.60	bal Fe	590-740	390	17	
DIN EN 10028(96)	WNr 1.7362	Press ves	0.08-0.15	4.00-6.00	0.30-0.60	0.45-0.65		0.025 max	0.020 max	0.30-0.60	bal Fe	590-740	390	17	
TGL 6918	12CrMo20.5*	Obs	0.08-0.15	4.5-5.5	0.3-0.6	0.45-0.55		0.035 max	0.035 max	0.3-0.5	bal Fe				

Japan

JIS G3203(88)	SFVAF5A	Frg for press ves high-temp	0.15 max	4.00-6.00	0.30-0.60	0.45-0.65		0.030 max	0.030 max	0.50 max	bal Fe	410-590	245	18	
JIS G3203(88)	SFVAF5B	Frg for press ves high-temp	0.15 max	4.00-6.00	0.30-0.60	0.45-0.65		0.030 max	0.030 max	0.50 max	bal Fe	480-660	275	18	
JIS G4109(87)	SCMV6	Plt for press ves	0.15 max	3.90-6.10	0.27-0.63	0.40-0.70		0.030 max	0.030 max	0.55 max	bal Fe	410-590	205	18	

UNS numbers and US grades are provided as a means of cross referencing chemically similar alloys. Exchangability is only possible after independent examination of specifications. Tensile properties are minimum or typical as specified. UTS and YS as MPa. El as %. See Appendix for list of abbreviations used in Notes. * indicates obsolete material.

Specification	Designation	Notes	C	Cr	Mn	Mo	Ni	P	S	Si	Other	UTS	YS	El	Hard

Stainless Steel, Unclassified, 502 (Continued from previous page)

Poland

| PNH86022(71) | H5M | Plt Bar Tub Frg | 0.15 max | 4.5-6 | 0.5 max | 0.45-0.6 | 0.5 max | 0.035 max | 0.03 max | 0.5 max | bal Fe | | | | |

UK

| BS 1503(89) | 625-520 | Frg Press ves; 0-999mm; N/T Q/T | 0.15 max | 4.0-6.0 | 0.3-0.8 | 0.45-0.65 | 0.4 max | 0.03 max | 0.025 max | 0.15-0.4 | Al <=0.02; Co <=0.1; Cu <=0.3; Pb <=0.15; Ti <=0.05; V <=0.1; W <=0.1; bal Fe | 520-670 | 365 | 18L | |
| BS 1503(89) | 625-520 | Frg for Press ves | 0.15 max | 4-6 | 0.3-0.8 | 0.45-0.65 | 0.4 max | 0.04 max | 0.04 max | 0.15-0.4 | Al <=0.02; Cu <=0.3; bal Fe | | | | |

USA

	AISI 502	Tube	0.10 max	4.00-6.00	1.00 max	0.40-0.65		0.040 max	0.030 max	1.00 max	bal Fe				
	UNS S50200	Heat res	0.10 max	4.00-6.00	1.00 max	0.40-0.65		0.040 max	0.030 max	1.00 max	bal Fe				
ASME SA387	5	Refer to ASTM A387/A387M(92); Plate													
ASTM A199	502	Obs; see A200; tub	0.1 max	4-6	1 max	0.4-0.65		0.04 max	0.03 max	1 max	bal Fe				
ASTM A200(94)	502	Tube	0.1 max	4-6	1 max	0.4-0.65		0.04 max	0.03 max	1 max	bal Fe				
ASTM A213	502*	Tube	0.1 max	4-6	1 max	0.4-0.65		0.04 max	0.03 max	1 max	bal Fe				
ASTM A314	502	Bolt	0.1 max	4-6	1 max	0.4-0.65		0.04 max	0.03 max	1 max	bal Fe				
ASTM A335	502	Plt Bar Bil Tub	0.1 max	4-6	1 max	0.4-0.65		0.04 max	0.03 max	1 max	bal Fe				
ASTM A473	502	Frg, Heat res, ann	0.10 max	4.00-6.00	1.00 max	0.40-0.65		0.040 max	0.030 max	1.00 max	bal Fe	414	205	20	201 HB
MIL-S-20146	502*	Obs; Plt Bar bil	0.1 max	4-6	1 max	0.4-0.65		0.04 max	0.03 max	1 max	bal Fe				

Stainless Steel, Unclassified, 503

USA

| | AISI 503 | Tube | 0.15 max | 6.00-8.00 | 1.00 max | 0.45-0.65 | | 0.040 max | 0.040 max | 1.00 max | bal Fe | | | | |

Stainless Steel, Unclassified, 660

China

| GB 1221(92) | 0Cr15Ni25Ti2MoAlVB | Bar; SA; Aged | 0.08 max | 13.50-16.00 | 2.00 max | 1.00-1.50 | 24.00-27.00 | 0.035 max | 0.030 max | 1.00 max | Al <=0.35; B 0.001-0.010; Ti 1.90-2.35; V 0.10-0.50; bal Fe | 900 | 590 | 15 | |
| GB 4238(92) | 0Cr15Ni25Ti2MoAlVB | Plit HR SA Aged | 0.08 max | 13.00-16.00 | 2.00 max | 1.00-1.50 | 24.00-27.00 | 0.035 max | 0.030 max | 1.00 max | Al <=0.35; B 0.001-0.010; Ti 1.90-2.35; V 0.10-0.50; bal Fe | 900 | 590 | 15 | |

France

| AFNOR NFA35577 | Z6NCTDV25.15 | Bar | 0.08 max | 13.5-16 | 2 max | 1-1.5 | 24-27 | 0.03 max | 0.015 max | 1 max | Al <=0.4; Ti 1.8-2.3; V 0.1-0.5; bal Fe | | | | |

Germany

| DIN | WNr 1.4980 | | 0.08 max | 13.5-16.0 | 2.00 max | 1.00-1.50 | 24.0-27.0 | 0.030 max | 0.030 max | 2.00 max | Al <=0.35; B 0.003-0.010; Ti 1.90-2.30; V 0.10-0.50; bal Fe | 930-1180 | 635 | 12 | |
| DIN | X5NiCrTi26-15 | | 0.08 max | 13.5-16.0 | 2.00 max | 1.00-1.50 | 24.0-27.0 | 0.030 max | 0.030 max | 2.00 max | Al <=0.35; B 0.003-0.010; Ti 1.90-2.30; V 0.10-0.50; bal Fe | 930-1180 | 635 | 12 | |

Japan

JIS G4311(91)	SUH660	Bar; SA; Aged 180mm diam	0.08 max	13.50-16.00	2.00 max	1.00-1.50	24.00-27.00	0.040 max	0.030 max	1.00 max	Al <=0.35; B 0.001-0.010; Ti 1.90-2.35; V 0.10-0.50; bal Fe	900	590	15	248 HB
JIS G4312(91)	SUH660	Sh Plt, SA Heat res	0.08 max	13.50-16.00	2.00 max	1.00-1.50	24.00-27.00	0.040 max	0.030 max	1.00 max	Al <=0.35; B 0.001-0.010; Ti 1.90-2.35; V 0.10-0.50; bal Fe	730		25	192 HB
JIS G4315	SUH660	Wir rod	0.08 max	13.50-16.00	2.00 max	1.00-1.50	24.00-27.00	0.040 max	0.030 max	1.00 max	Al <=0.35; B 0.001-0.010; bal Fe				

Sweden

| SS 142570 | 2570 | Corr res | 0.08 max | 13.5-16 | 2 max | 1-1.5 | 24-2 | 0.025 max | 0.025 max | 1 max | Al <=0.35; B 0.003-0.01; Ti 1.9-2.3; V <=0.5; bal Fe | | | | |

UNS numbers and US grades are provided as a means of cross referencing chemically similar alloys. Exchangability is only possible after independent examination of specifications. Tensile properties are minimum or typical as specified. UTS and YS as MPa. El as %. See Appendix for list of abbreviations used in Notes. * indicates obsolete material.

Specification	Designation	Notes	C	Cr	Mn	Mo	Ni	P	S	Si	Other	UTS	YS	El	Hard

Stainless Steel, Unclassified, 660 (Continued from previous page)

UK

Specification	Designation	Notes	C	Cr	Mn	Mo	Ni	P	S	Si	Other	UTS	YS	El	Hard
BS 1506(90)	286S31	Bolt matl Pres/Corr res; t<=200mm; SHT & Hard	0.08 max	13.5-16.0	2.00 max	1.00-1.50	24.0-27.0	0.045 max	0.030 max	1.00 max	Al 0.10-0.35; B 0.0030-0.010; Pb <=0.005; Ti 1.90-2.30; V 0.10-0.50; bal Fe	900	590	12	248-341 HB

USA

Specification	Designation	Notes	C	Cr	Mn	Mo	Ni	P	S	Si	Other	UTS	YS	El	Hard
	UNS S66286	Fe Base Superalloy A286	0.08 max	13.50-16.00	2.00 max	1.00-1.50	24.0-27.0	0.040 max	0.030 max	1.00 max	Al <=0.35; B 0.0010-0.010; Ti 1.90-2.35; V 0.10-0.50; bal Fe				
AMS 5528		Sh Plt Bar	0.08 max	13.5-16	2 max	1-1.5	24-27	0.040 max	0.03 max	1 max	Al <=0.35; B 0.001-0.01; Ti 1.9-2.35; V 0.1-0.50; bal Fe				
AMS 5726		Sh Plt Bar	0.08 max	13.5-16	2 max	1-1.5	24-27	0.040 max	0.03 max	1 max	Al <=0.35; B 0.001-0.01; Ti 1.9-2.35; V 0.1-0.50; bal Fe				
AMS 5731		Sh Plt Bar	0.08 max	13.5-16	2 max	1-1.5	24-27	0.040 max	0.03 max	1 max	Al <=0.35; B 0.001-0.01; Ti 1.9-2.35; V 0.1-0.50; bal Fe				
AMS 5732		Sh Plt Bar	0.08 max	13.5-16	2 max	1-1.5	24-27	0.040 max	0.03 max	1 max	Al <=0.35; B 0.001-0.01; Ti 1.9-2.35; V 0.1-0.50; bal Fe				
AMS 5734		Sh Plt Bar	0.08 max	13.5-16	2 max	1-1.5	24-27	0.040 max	0.03 max	1 max	Al <=0.35; B 0.001-0.01; Ti 1.9-2.35; V 0.1-0.50; bal Fe				
AMS 5737		Sh Plt Bar	0.08 max	13.5-16	2 max	1-1.5	24-27	0.040 max	0.03 max	1 max	Al <=0.35; B 0.001-0.01; Ti 1.9-2.35; V 0.1-0.50; bal Fe				
AMS 5804		Sh Plt Bar	0.08 max	13.5-16	2 max	1-1.5	24-27	0.040 max	0.03 max	1 max	Al <=0.35; B 0.001-0.01; Ti 1.9-2.35; V 0.1-0.50; bal Fe				
AMS 5805		Sh Plt Bar	0.08 max	13.5-16	2 max	1-1.5	24-27	0.040 max	0.03 max	1 max	Al <=0.35; B 0.001-0.01; Ti 1.9-2.35; V 0.1-0.50; bal Fe				
AMS 5853		Sh Plt Bar	0.08 max	13.5-16	2 max	1-1.5	24-27	0.040 max	0.03 max	1 max	Al <=0.35; B 0.001-0.01; Ti 1.9-2.35; V 0.1-0.50; bal Fe				
AMS 5858		Sh Plt Bar	0.08 max	13.5-16	2 max	1-1.5	24-27	0.040 max	0.03 max	1 max	Al <=0.35; B 0.001-0.01; Ti 1.9-2.35; V 0.1-0.50; bal Fe				
AMS 5895		Sh Plt Bar	0.08 max	13.5-16	2 max	1-1.5	24-27	0.040 max	0.03 max	1 max	Al <=0.35; B 0.001-0.01; Ti 1.9-2.35; V 0.1-0.50; bal Fe				
AMS 7235		Sh Plt Bar	0.08 max	13.5-16	2 max	1-1.5	24-27	0.040 max	0.03 max	1 max	Al <=0.35; B 0.001-0.01; Ti 1.9-2.35; V 0.1-0.50; bal Fe				
ASME SA638	660	Refer to ASTM A638/A638M(92); Bar Frg													
ASTM A453	660	Bar Frg	0.08 max	13.5-16	2 max	1-1.5	24-27	0.04 max	0.03 max	1 max	Al <=0.35; B 0.001-0.01; Ti 1.9-2.35; V 0.1-0.5; bal Fe				
ASTM A638/638M(95)	660	Bar Frg, Iron base superalloy, High-temp	0.08 max	13.50-16.00	2.00 max	1.00-1.50	24.00-27.00	0.040 max	0.030 max	1.00 max	Al <=0.35; B 0.0010-0.010; Ti 1.90-2.35; V 0.10-0.50; bal Fe	895	585	15	248 HB min
SAE J467(68)	A286	Bar Frg	0.05	15.0	1.40	1.30	26.0	0.04 max	0.03 max	0.40	Al 0.20; B 0.004; Ti 2.15; V 0.30; bal Fe	1000	655	24	26 HRC
SAE J775(93)	HEV-7	Engine poppet valve	0.08 max	13.5-16	2 max	1-1.5	24-27	0.04 max	0.03 max	1 max	Al <=0.35; B 0.001-0.01; Ti 1.9-2.35; V 0.1-0.5; bal Fe				

Stainless Steel, Unclassified, 7-Mo Plus

USA

Specification	Designation	Notes	C	Cr	Mn	Mo	Ni	P	S	Si	Other	UTS	YS	El	Hard
	UNS S32950	Duplex Aust-Ferr	0.03 max	26.0-29.0	2.00 max	1.00-2.50	3.50-5.20	0.035 max	0.010 max	0.60 max	N 0.15-0.35; bal Fe				
ASTM A182/A182M(98)	F52	Frg/roll pipe flange valve	0.030 max	26.0-29.0	2.00 max	1.00-2.50	3.5-5.2	0.035 max	0.010 max	0.60 max	N 0.15-0.35; bal Fe	690	485	15	
ASTM A240/A240M(98)	S32950	Duplex Aust-Ferr	0.030 max	26.0-29.0	2.00 max	1.00-2.50	3.5-5.2	0.035 max	0.010 max	0.60 max	N 0.15-0.35; bal Fe	690	485	15.0	293 HB
ASTM A789/A789M(95)	7-Mo Plus	Smls Weld tub, Duplex Aust-Ferr	0.03 max	26.00-29.00	2.00 max	1.00-2.50	3.50-5.20	0.035 max	0.010 max	0.60 max	N 0.15-0.35; bal Fe	690	480	20	290 HB

UNS numbers and US grades are provided as a means of cross referencing chemically similar alloys. Exchangability is only possible after independent examination of specifications. Tensile properties are minimum or typical as specified. UTS and YS as MPa. El as %. See Appendix for list of abbreviations used in Notes. * indicates obsolete material.

Specification	Designation	Notes	C	Cr	Mn	Mo	Ni	P	S	Si	Other	UTS	YS	El	Hard

Stainless Steel, Unclassified, 7-Mo Plus (Continued from previous page)

USA

Specification	Designation	Notes	C	Cr	Mn	Mo	Ni	P	S	Si	Other	UTS	YS	El	Hard
ASTM A790/A790M(95)	7-Mo plus	Smls Weld Pipe, Duplex Aust-Ferr	0.03 max	26.00-29.00	2.00 max	1.00-2.50	3.50-5.20	0.035 max	0.010 max	0.60 max	N 0.15-0.35; bal Fe	690	480	20	290 HB
ASTM A815/A815M(98)	S32950	Pipe fittings; Duplex Aust-Ferr	0.030 max	26.0-29.0	2.00 max	1.00-2.50	3.5-5.2	0.035 max	0.010 max	0.60 max	N 0.15-0.35; bal Fe	690	485	15.0	290 HB
ASTM A949/A949M(95)	7-Mo PLUS	Spray formed smls pipe; Duplex Aust-Ferr	0.03 max	26.00-29.00	2.00 max	1.00-2.50	3.50-5.20	0.035 max	0.010 max	0.60 max	N 0.15-0.35; bal Fe	620	485	20	290 HB max
ASTM A988(98)	S32950	HIP Flanges, Fittings, Valves/parts; Heat res; Duplex Aust-Ferr	0.030 max	26.0-29.0	2.00 max	1.00-2.50	3.5-5.2	0.035 max	0.010 max	0.60 max	N 0.15-0.35; bal Fe	690	485	15	
SAE J405(98)	S32950	Duplex Aust-Ferr	0.03 max	26.00-29.00	2.00 max	1.00-2.50	3.50-5.20	0.035 max	0.010 max	0.60 max	N 0.15-0.35; bal Fe				

Stainless Steel, Unclassified, AF918

USA

Specification	Designation	Notes	C	Cr	Mn	Mo	Ni	P	S	Si	Other	UTS	YS	El	Hard
	UNS S39277	Duplex Aust-Ferr	0.025 max	24.0-26.0		3.0-4.0	6.5-8.0	0.025 max	0.002 max	0.80 max	Cu 1.2-2.0; N 0.23-0.33; W 0.80-1.20; bal Fe				
ASTM A182/A182M(98)	F57	Frg/roll pipe flange valve	0.025 max	24.0-26.0	0.80 max	3.0-4.0	6.5-8.0	0.025 max	0.002 max	0.80 max	Cu 1.20-2.00; N 0.23-0.33; bal Fe	820	585	25	
ASTM A789/A789M(95)	AF918	Smls Weld tub, Duplex Aust-Ferr	0.025 max	24.00-26.00	0.80 max	3.00-4.00	6.50-8.00	0.025 max	0.002 max	0.80 max	Cu 1.2-2.0; N 0.23-0.33; W 0.8-1.2; bal Fe	825	620	25	290 HB
ASTM A790/A790M(95)	AF918	Smls Weld Pipe, Duplex Aust-Ferr	0.025 max	24.00-26.00	0.80 max	3.00-4.00	6.5-8.00	0.025 max	0.002 max	0.80 max	Cu 1.2-2.0; N 0.23-0.33; W 0.8-1.2; bal Fe	825	620	25	290 HB
ASTM A988(98)	S39277	HIP Flanges, Fittings, Valves/parts; Heat res; Duplex Aust-Ferr	0.025 max	24.0-26.0	0.80 max	3.0-4.0	6.5-8.0	0.025 max	0.002 max	0.80 max	Cu 1.20-2.00; N 0.23-0.33; W 0.80-1.20; bal Fe	820	585	25.0	

Stainless Steel, Unclassified, F50

Europe

Specification	Designation	Notes	C	Cr	Mn	Mo	Ni	P	S	Si	Other	UTS	YS	El	Hard
EN 10088/2(95)	1.4410	Strp, CR Aust-Ferr Corr res; t<=6mm, Ann	0.030 max	24.00-26.00	2.00 max	3.00-4.50	6.00-8.00	0.035 max	0.015 max	1.00 max	N 0.20-0.35; bal Fe	750-1000	550	15	
EN 10088/2(95)	1.4507	Strp, CR Aust-Ferr Corr res; t<=6mm, Ann	0.030 max	24.0-26.00	2.00 max	2.70-4.00	5.50-7.50	0.035 max	0.015 max	0.70 max	Cu 1.00-2.50; N 0.15-0.30; bal Fe	690-940	510	17	
EN 10088/2(95)	X2CrNiMoCuN25-6-3	Strp, CR Aust-Ferr Corr res; t<=6mm, Ann	0.030 max	24.0-26.00	2.00 max	2.70-4.00	5.50-7.50	0.035 max	0.015 max	0.70 max	Cu 1.00-2.50; N 0.15-0.30; bal Fe	690-940	510	17	
EN 10088/2(95)	X2CrNiMoN25-7-4	Strp, CR Aust-Ferr Corr res; t<=6mm, Ann	0.030 max	24.0-26.00	2.00 max	3.00-4.50	6.00-8.00	0.035 max	0.015 max	1.00 max	N 0.20-0.35; bal Fe	750-1000	550	15	
EN 10088/3(95)	1.4410	Aust-Ferr; Corr res; t<=160mm; Ann	0.030 max	24.0-26.00	2.00 max	3.00-4.50	6.00-8.00	0.035 max	0.015 max	1.00 max	N 0.20-0.35; bal Fe	730-930	530	25	290 HB
EN 10088/3(95)	1.4507	Aust-Ferr; Corr res; t<=160mm; Ann	0.030 max	24.0-26.00	2.00 max	2.70-4.00	5.50-7.50	0.035 max	0.015 max	0.70 max	Cu 1.00-2.50; N 0.15-0.30; bal Fe	700-900	500	25	270 HB
EN 10088/3(95)	X2CrNiMoCuN25-6-3	Aust-Ferr; Corr res; t<=160mm; Ann	0.030 max	24.0-26.00	2.00 max	2.70-4.00	5.50-7.50	0.035 max	0.015 max	0.70 max	Cu 1.00-2.50; N 0.15-0.30; bal Fe	700-900	500	25	270 HB
EN 10088/3(95)	X2CrNiMoN25-7-4	Aust-Ferr; Corr res; t<=160mm; Ann	0.030 max	24.0-26.00	2.00 max	3.00-4.50	6.00-8.00	0.035 max	0.015 max	1.00 max	N 0.20-0.35; bal Fe	730-930	530	25	290 HB

USA

Specification	Designation	Notes	C	Cr	Mn	Mo	Ni	P	S	Si	Other	UTS	YS	El	Hard
	UNS S31200	Ferr-Aust 44LN	0.030 max	24.0-26.0	2.00 max	1.20-2.00	5.50-6.50	0.045 max	0.030 max	1.00 max	N 0.14-0.20; bal Fe				
ASTM A182/A182M(98)	F50	Frg/roll pipe flange valve	0.030 max	24.0-26.0	2.00 max	1.2-2.0	5.5-6.5	0.045 max	0.030 max	1.00 max	N 0.14-0.20; bal Fe	690-900	450	25	
ASTM A240/A240M(98)	S31200	Ferr-Aust 44LN	0.030 max	24.0-26.0	2.00 max	1.20-2.00	5.50-6.50	0.045 max	0.030 max	1.00 max	N 0.14-0.20; bal Fe	690	450	25.0	293 HB
ASTM A789/A789M(95)	F50	Smls Weld tub, Duplex Aust-Ferr 44LN	0.030 max	24.0-26.0	2.0 max	1.20-2.00	5.50-6.50	0.045 max	0.030 max	1.0 max	N 0.14-0.20; bal Fe	690	450	25	280 HB
ASTM A790/A790M(95)	F50	Smls Weld Pipe, Ferr-aust 44LN	0.030 max	24.0-26.0	2.0 max	1.20-2.00	5.50-6.50	0.045 max	0.030 max	1.0 max	N 0.14-0.20; bal Fe	690	450	25	280 HB
ASTM A949/A949M(95)	F50	Spray formed smls pipe; Duplex Aust-Ferr 44LN	0.030 max	24.0-26.0	2.0 max	1.20-2.00	5.50-6.50	0.045 max	0.030 max	1.0 max	N 0.14-0.20; bal Fe	760	550	15	297 HB max
SAE J405(98)	S31200	Ferr-Aust 44LN	0.030 max	24.00-26.0	2.00 max	1.2-2.0	5.5-6.5	0.045 max	0.030 max	1.00 max	N 0.14-0.20; bal Fe				

Stainless Steel, Unclassified, F7

USA

Specification	Designation	Notes	C	Cr	Mn	Mo	Ni	P	S	Si	Other	UTS	YS	El	Hard
	UNS S50300	Heat res	0.15 max	6.00-8.00	1.00 max	0.45-0.65		0.040 max	0.040 max	1.00 max	bal Fe				

UNS numbers and US grades are provided as a means of cross referencing chemically similar alloys. Exchangability is only possible after independent examination of specifications. Tensile properties are minimum or typical as specified. UTS and YS as MPa. El as %. See Appendix for list of abbreviations used in Notes. * indicates obsolete material.

Specification	Designation	Notes	C	Cr	Mn	Mo	Ni	P	S	Si	Other	UTS	YS	El	Hard

Stainless Steel, Unclassified, Ferralium 255

USA

Specification	Designation	Notes	C	Cr	Mn	Mo	Ni	P	S	Si	Other	UTS	YS	El	Hard
	UNS S32550	Duplex Aust-Ferr	0.04 max	24.0-27.0	1.5 max	2.9-3.9	4.50-6.50	0.04 max	0.030 max	1.00 max	Cu 1.50-2.50; N 0.10-0.25; bal Fe				
ASTM A240/A240M(98)	S32550	Duplex Aust-Ferr	0.04 max	24.0-27.0	1.5 max	2.9-3.9	4.5-6.5	0.040 max	0.030 max	1.00 max	Cu 1.50-2.50; N 0.10-0.25; bal Fe	760	550	15.0	302 HB
ASTM A789/A789M(95)	Ferralium 25	Smls Weld tub, Duplex Aust-Ferr	0.040 max	24.0-27.0	1.5 max	2.90-3.90	4.50-6.50	0.040 max	0.030 max	1.0 max	Cu 1.5-2.5; N 0.10-0.25; bal Fe	760	550	15	297 HB
ASTM A790/A790M(95)	Ferralium 25	Smls Weld Pipe, Duplex Aust-Ferr	0.040 max	24.0-27.0	1.5 max	2.90-3.90	4.50-6.50	0.040 max	0.030 max	1.0 max	Cu 1.5-2.5; N 0.10-0.25; bal Fe	760	550	15	297 HB
ASTM A815/A815M(98)	S32550	Pipe fittings; Duplex Aust-Ferr	0.04 max	24.0-27.0	1.50 max	2.9-3.9	4.5-6.5	0.040 max	0.030 max	1.00 max	Cu 1.50-2.50; N 0.10-0.25; bal Fe	760	550	15.0	302 HB
ASTM A949/A949M(95)	Ferralium 25	Spray formed smls pipe; Duplex Aust-Ferr	0.040 max	24.0-27.0	1.5 max	2.90-3.90	4.50-6.50	0.040 max	0.030 max	1.0 max	Cu 1.5-2.5; N 0.10-0.25; bal Fe	690	450	25	280 HB max
SAE J405(98)	S32550	Duplex Aust-Ferr	0.04 max	24.0-27.0	1.5 max	2.9-3.9	4.5-6.50	0.040 max	0.030 max	1.0 max	Cu 1.5-2.5; N 0.10-0.25; bal Fe				

Stainless Steel, Unclassified, No equivalents identified

Czech Republic

Specification	Designation	Notes	C	Cr	Mn	Mo	Ni	P	S	Si	Other	UTS	YS	El	Hard
CSN 417117	17117	Corr res	0.06-0.15	8.0-10.0	0.3-0.6	0.9-1.1		0.03 max	0.03 max	0.25-1.0	V 0.09-0.2; bal Fe				
CSN 417126	17126	Heat res	0.16-0.22	11.0-12.5	0.6-1.0		0.5-1.0	0.035 max	0.035 max	0.6 max	V 0.15-0.25; W 1.6-2.2; bal Fe				
CSN 417134	17134	Corr res; Heat res	0.17-0.23	10.0-12.5	0.5-1.0	0.8-1.2	0.3-0.8	0.035 max	0.03 max	0.25-0.6	V 0.2-0.35; bal Fe				
CSN 417153	17153	Heat res	0.2 max	23.0-27.0	1.0 max		2.0 max	0.045 max	0.04 max	1.3 max	bal Fe				
CSN 417351	17351	Corr res	0.08 max	15.5-17.0	0.3-0.8		5.5-7.0	0.045 max	0.035 max	0.9 max	Al <=1.0; Ti 0.5-1.0; bal Fe				
CSN 417536	17536	Corr res	0.12 max	35.0-37.0	0.6 max			0.035 max	0.035 max	0.35 max	bal Fe				

India

Specification	Designation	Notes	C	Cr	Mn	Mo	Ni	P	S	Si	Other	UTS	YS	El	Hard
IS 1570/5(85)	X10Cr17Mn6Ni4	Bar Flat Band	0.2 max	16-18	4-8	0.15 max	3.5-5.5	0.045 max	0.03 max	1 max	Co <=0.1; Cu <=0.3; Pb <=0.15; W <=0.1; bal Fe	515	275	40	217 HB max
IS 1570/7(92)	25	Smls tube Sh Plt Bar Frg	0.04-0.1	17-20	2 max	0.15 max	8-12	0.045 max	0.03 max	0.75 max	bal Fe	490-690	195	30	
IS 1570/7(92)	29	Smls tube Bar Frg	0.04-0.1	17-20	2 max	0.15 max	9-13	0.045 max	0.03 max	0.2-0.8	Ti <=0.6; 4C<=Ti<=0.6; bal Fe	490-690	155	30	
IS 1570/7(92)	30	Smls tube Bar Frg Sh Plt	0.04-0.1	17-19	2 max	0.15 max	9-13	0.045 max	0.03 max	0.75 max	10C<=Nb+Ta<=1; bal Fe	510-710	205	30	
IS 1570/7(92)	32	0-250mm; Bar Frg Smls tube Sh Plt	0.03-0.08	13.5-16	2 max	1-1.5	24-27	0.045 max	0.03 max	1 max	Al <=0.35; B 0.003-0.01; Ti 1.9-2.3; V 0.1-0.5; Almet<=0.35; bal Fe	900-1100	600	15	248-341 HB
IS 1570/7(92)	6Ni25Cr15Ti2MoVBH	0-250mm; Bar Frg Smls tube Sh Plt	0.03-0.08	13.5-16	2 max	1-1.5	24-27	0.045 max	0.03 max	1 max	Al <=0.35; B 0.003-0.01; Ti 1.9-2.3; V 0.1-0.5; Almet<=0.35; bal Fe	900-1100	600	15	248-341 HB
IS 1570/7(92)	7Cr18Ni10H	Smls tube Sh Plt Bar Frg	0.04-0.1	17-20	2 max	0.15 max	8-12	0.045 max	0.03 max	0.75 max	bal Fe	490-690	195	30	
IS 1570/7(92)	7Cr18Ni11NbH	Smls tube Bar Frg Sh Plt	0.04-0.1	17-19	2 max	0.15 max	9-13	0.045 max	0.03 max	0.75 max	10C<=Nb+Ta<=1; bal Fe	510-710	205	30	
IS 1570/7(92)	7Cr19Ni11TiH	Smls tube Bar Frg	0.04-0.1	17-20	2 max	0.15 max	9-13	0.045 max	0.03 max	0.2-0.8	Ti <=0.6; 4C<=Ti<=0.6; bal Fe	490-690	155	30	

International

Specification	Designation	Notes	C	Cr	Mn	Mo	Ni	P	S	Si	Other	UTS	YS	El	Hard
ISO 4955(94)	H18	Heat res	0.12 max	19.0-23.0	2.0 max	0.15 max	30.0-34.0	0.045 max	0.03 max	1.0 max	Al 0.15-0.6; Co <=0.1; Cu <=0.3; Pb <=0.15; Ti 0.15-0.6; V <=0.1; W <=0.1; bal Fe				
ISO 683-15(76)	12*	Valve, Q, HT	0.08-0.16	20-22.5	1-2	2.5-3.5	19-21	0.045 max	0.03 max	1 max	Al <=0.1; Co 18.5-21.5; Cu <=0.3; N 0.1-0.2; Nb 0.75-1.25; Pb <=0.15; Ti <=0.05; V <=0.1; W 2-3; bal Fe				

Italy

Specification	Designation	Notes	C	Cr	Mn	Mo	Ni	P	S	Si	Other	UTS	YS	El	Hard
UNI 3097(75)	X45Cr13	Ball & roller bearing	0.42-0.5	12.5-14.5	1 max		1 max	0.04 max	0.03 max	1 max	bal Fe				

UNS numbers and US grades are provided as a means of cross referencing chemically similar alloys. Exchangability is only possible after independent examination of specifications. Tensile properties are minimum or typical as specified. UTS and YS as MPa. El as %. See Appendix for list of abbreviations used in Notes. * indicates obsolete material.

Specification	Designation	Notes	C	Cr	Mn	Mo	Ni	P	S	Si	Other	UTS	YS	El	Hard

Stainless Steel, Unclassified, No equivalents identified

Italy

Specification	Designation	Notes	C	Cr	Mn	Mo	Ni	P	S	Si	Other	UTS	YS	El	Hard
UNI 3992(75)	X70CrMnNiN216	Valve	0.65-0.75	20-22	5.5-7		1.4-1.9	0.04 max	0.025-0.05	0.45-0.85	N 0.18-0.28; bal Fe				
UNI 6901(71)	X14CrNi19	Corr res	0.16 max	18-20	1 max		1.5-2.5	0.04 max	0.03 max	1 max	bal Fe				
UNI 6901(71)	X40Cr14	Corr res	0.36-0.45	12.5-14.5	1 max		1 max	0.04 max	0.03 max	1 max	bal Fe				
UNI 6901(71)	X8CrNi1812	Corr res	0.1 max	17-19	2 max		11-13	0.045 max	0.03 max	1 max	bal Fe				
UNI 6901(71)	X8CrNiTi1811	Corr res	0.04-0.1	17-19	2 max		9-13	0.03 max	0.03 max	0.75 max	Ti <=0.6; Ti=4C-0.6; bal Fe				
UNI 6904(71)	X8CrNi1812	Heat res	0.1 max	17-19	2 max		11-13	0.04 max	0.03 max	0.75 max	bal Fe				
UNI 6904(71)	X8CrNiTi1811	Heat res	0.04-0.1	17-19	2 max		9-13	0.03 max	0.03 max	0.75 max	Ti <=0.6; Ti=4C-0.6; bal Fe				
UNI 7500(75)	X5CrNiMo1815	Press ves	0.06 max	17.5-19.5	2 max	3-4	13-16	0.045 max	0.03 max	1 max	bal Fe				

Japan

Specification	Designation	Notes	C	Cr	Mn	Mo	Ni	P	S	Si	Other	UTS	YS	El	Hard
JIS G3203(88)	SFVAF1	Frg for press ves high-temp	0.30 max		0.60-0.90	0.45-0.65			0.030 max	0.35 max	bal Fe	480-660	275	18	

Poland

Specification	Designation	Notes	C	Cr	Mn	Mo	Ni	P	S	Si	Other	UTS	YS	El	Hard
PNH86022(71)	H5M	Heat res	0.15 max	4.5-6	0.5 max	0.45-0.6	0.5 max	0.035 max	0.03 max	0.5 max	bal Fe				
PNH86022(71)	H5M	Heat res	0.15 max	4.5-6	0.5 max	0.45-0.6	0.5 max	0.035 max	0.03 max	0.5 max	bal Fe				

Romania

Specification	Designation	Notes	C	Cr	Mn	Mo	Ni	P	S	Si	Other	UTS	YS	El	Hard
STAS 3583(64)	40MoSC100	Corr res; Heat res	0.35-0.45	9-10.5	0.7 max	0.7-0.9	0.6 max	0.035 max	0.03 max	1.9-2.6	Pb <=0.15; bal Fe				
STAS 3583(64)	40SC90	Corr res; Heat res	0.35-0.45	8-10	0.7 max		0.6 max	0.035 max	0.03 max	2-3	Pb <=0.15; bal Fe				
STAS 3583(87)	1MoCr260	Corr res; Heat res	0.01 max	25-27.5	0.4 max	0.75-1.5	0.5 max	0.03 max	0.02 max	0.4 max	N <=0.015; Pb <=0.15; bal Fe				
STAS 3583(87)	2CuMoCrNi250	Corr res; Heat res	0.02 max	19-22	2 max	4-5	24-27	0.035 max	0.025 max	1 max	Cu 1-2; Pb <=0.15; bal Fe				
STAS 3583(87)	40Cr130	Corr res; Heat res	0.35-0.42	12.5-14.5	1 max	0.2 max	0.5 max	0.045 max	0.03 max	1 max	Pb <=0.15; bal Fe				
STAS 3583(87)	45VMoCr145	Corr res; Heat res	0.42-0.5	13.5-15	1 max	0.45-0.6	0.5 max	0.045 max	0.03 max	1 max	Al 0.1-0.15; Pb <=0.15; bal Fe				
STAS 3583(87)	90VMoCr180	Corr res; Heat res	0.85-0.95	17-19	1 max	0.9-1.3	0.5 max	0.045 max	0.03 max	1 max	Pb <=0.15; V 0.07-0.12; bal Fe				

Russia

Specification	Designation	Notes	C	Cr	Mn	Mo	Ni	P	S	Si	Other	UTS	YS	El	Hard
GOST	10Ch18N9MLS	Corr res	0.07-0.14	17.0-20.0	1.0-1.8	0.1-0.2	8.0-11.0	0.02 max	0.02 max	0.2-1.0	Cu <=0.3; Ti <=0.05; bal Fe				
GOST 5632	15Ch18N12S4TJu	Corr res	0.12-0.17	17.0-19.0	0.5-1.0	0.3 max	11.0-13.0	0.035 max	0.03 max	3.8-4.5	Al 0.13-0.35; Cu <=0.3; Ti 0.4-0.7; W <=0.2; bal Fe				
GOST 5632	36Ch18N25S2	Corr res	0.32-0.4	17.0-19.0	1.5 max	0.3 max	23.0-26.0	0.035 max	0.02 max	2.0-3.0	Cu <=0.3; Ti <=0.2; W <=0.2; bal Fe				
GOST 5632(61)	4KH10S2M	Corr res; Heat res; High-temp constr	0.35-0.45	9.0-10.5	0.7 max	0.7-0.9	0.4 max	0.03 max	0.025 max	1.9-2.6	Cu <=0.3; Ti <=0.05; bal Fe				
GOST 5632(61)	4KH13	Corr res; Heat res; High-temp constr	0.35-0.44	12.0-14.0	0.6 max	0.15 max	0.4 max	0.03 max	0.025 max	0.6 max	Cu <=0.3; Ti <=0.05; bal Fe				
GOST 5632(72)	15Ch28	Corr res; Heat res	0.15 max	27.0-30.0	0.8 max	0.15 max	0.6 max	0.035 max	0.025 max	1.0 max	Cu <=0.3; Ti <=0.2; bal Fe				
GOST 5632(72)	17Ch18N9	Corr res	0.13-0.21	17.0-19.0	2.0 max	0.3 max	8.0-10.0	0.035 max	0.02 max	0.8 max	Cu <=0.3; Ti <=0.5; W <=0.2; bal Fe				

Spain

Specification	Designation	Notes	C	Cr	Mn	Mo	Ni	P	S	Si	Other	UTS	YS	El	Hard
UNE 36017(61)	F.322*	Heat res	0.35-0.45	9.0-11.0	0.40-0.60	0.80-1.00		0.040 max	0.040 max	2.00-2.50	bal Fe				
UNE 36017(85)	F.3314*	see X10NiCrAlTi32-20	0.12 max	19.0-23.0	2.0 max		30-34	0.045 max	0.020 max	1.00 max	Al 0.15-0.6; Ti 0.15-0.60; bal Fe				
UNE 36017(85)	X10NiCrAlTi32-20	Heat res	0.12 max	19.0-23.0	2.0 max		30-34	0.045 max	0.020 max	1.00 max	Al 0.15-0.6; Ti 0.15-0.60; bal Fe				
UNE 36087(78)	F.3545*	see X9NiCr33-21*	0.21 max	19.0-23.0	2.0 max		30.0-35.0	0.045 max	0.030 max	1.00 max	Al 0.15-0.60; Ti 0.15-0.50; bal Fe				
UNE 36087(78)	X9NiCr33-21*	Heat res	0.21 max	19.0-23.0	2.0 max		30.0-35.0	0.045 max	0.030 max	1.00 max	Al 0.15-0.60; Ti 0.15-0.50; bal Fe				

UNS numbers and US grades are provided as a means of cross referencing chemically similar alloys. Exchangability is only possible after independent examination of specifications. Tensile properties are minimum or typical as specified. UTS and YS as MPa. El as %. See Appendix for list of abbreviations used in Notes. * indicates obsolete material.

Specification	Designation	Notes	C	Cr	Mn	Mo	Ni	P	S	Si	Other	UTS	YS	El	Hard

Stainless Steel, Unclassified, No equivalents identified

Sweden

Specification	Designation	Notes	C	Cr	Mn	Mo	Ni	P	S	Si	Other	UTS	YS	El	Hard
SS 142340	2340	Corr res	0.1 max	16.5-18	2 max	1.3-1.8	8-10	0.045 max	0.03 max	1 max	bal Fe				
SS 142376	2376	Corr res	0.03 max	18-19	1.2-2	2.5 3	4.3-5.2	0.03 max	0.03 max	1.4-2	N 0.05-0.1; bal Fe				
SS 142384	2384	Corr res	0.05 max	17-19	2 max	2.5-3	12.5-14	0.06 max	0.15-0.3	1 max	Cu 1-3; Ti 0.8-1.2; bal Fe				
SS 142387	2387	Corr res	0.05 max	15-17	1.5 max	0.8-1.5	4-6	0.045 max	0.03 max	1 max	bal Fe				
SS 142562	2562	Corr res	0.025 max	19-21	2 max	4-5	24-26	0.04 max	0.03 max	1 max	Cu 1.2-2; bal Fe				
SS 142564	2564	Corr res	0.06 max	19-21	2 max	4-5	24-26	0.045 max	0.03 max	1 max	Cu 3-3.5; bal Fe				
SS 142584	2584	Corr res	0.025 max	26-28	2 max	3-4	30-34	0.03 max	0.02 max	1 max	Cu 0.6-1.4; bal Fe				

USA

Specification	Designation	Notes	C	Cr	Mn	Mo	Ni	P	S	Si	Other	UTS	YS	El	Hard
ASTM A579(96)	64	Superstrength Frg	0.10-0.15	15.0-16.0	0.50-1.25	2.50-3.25	4.0-5.0	0.025 max	0.025 max	0.50 max	B 0.003; N 0.07-0.13; Ca 0.06; Zr 0.02; bal Fe	1140-1450	965-1240	10-12	321 HB
FED QQ-W-423B(85)	- - -*	Obs; Wir													

Stainless Steel, Unclassified, S31260

USA

Specification	Designation	Notes	C	Cr	Mn	Mo	Ni	P	S	Si	Other	UTS	YS	El	Hard
	UNS S31260	Duplex Aust-Ferr, DP-3	0.03 max	24.0-26.0	1.00 max	2.50-3.50	5.50-7.50	0.030 max	0.030 max	0.75 max	Cu 0.20-0.80; N 0.10-0.30; W 0.10-0.50; bal Fe				
	UNS S39226*	Obs; see S31260													
ASTM A240/A240M(98)	S31260	Duplex Aust-Ferr DP-3	0.03 max	24.0-26.0	1.00 max	2.50-3.50	5.50-7.50	0.030 max	0.030 max	0.75 max	Cu 0.14-0.80; N 0.10-0.30; W 0.10-0.50; bal Fe	690	485	20.0	290 HB
ASTM A789/A789M(95)	A240	Smls Weld tub, Duplex Aust-Ferr DP-3	0.030 max	24.0-26.0	1.00 max	2.50-3.50	5.50-7.50	0.030 max	0.030 max	0.75 max	Cu 0.20-0.80; N 0.10-0.30; W 0.10-0.50; bal Fe	690	450	25	290 HB
ASTM A790/A790M(95)	A240	Smls Weld Pipe, Duplex Aust-Ferr DP-3	0.030 max	24.0-26.0	1.00 max	2.50-3.50	5.50-7.50	0.030 max,	0.030 max	0.75 max	Cu 0.20-0.80; N 0.10-0.30; W 0.10-0.50; bal Fe	690	450	25	
ASTM A949/A949M(95)	A240	Spray formed smls pipe; Duplex Aust-Ferr DP-3	0.030 max	24.0-26.0	1.00 max	2.50-3.50	5.50-7.50	0.030 max	0.030 max	0.75 max	Cu 0.20-0.80; N 0.10-0.30; W 0.10-0.50; bal Fe	690	450	25	
SAE J405(98)	S31260	Duplex Aust-Ferr DP-3	0.03 max	24.0-26.0	1.00 max	2.50-3.50	5.50-7.50	0.030 max	0.030 max	0.75 max	Cu 0.20-0.80; N 0.10-0.30; W 0.10-0.50; bal Fe				

Stainless Steel, Unclassified, S31500

USA

Specification	Designation	Notes	C	Cr	Mn	Mo	Ni	P	S	Si	Other	UTS	YS	El	Hard
	UNS S31500	Duplex Aust-Ferr, 3RE60	0.030 max	18.0-19.0	1.20-2.00	2.50-3.00	4.25-5.25	0.030 max	0.030 max	1.40-2.00	bal Fe				
	UNS S39215*	Obs; see S31500													
ASTM A789/A789M(95)	A789	Smls Weld tub, Duplex Aust-Ferr 3RE60	0.030 max	18.0-19.0	1.20-2.00	2.50-3.00	4.25-5.25	0.030 max	0.030 max	1.40-2.00	N 0.05-0.1; bal Fe	630	440	30	290 HB
ASTM A790/A790M(95)	A789	Smls Weld Pipe, Duplex Aust-Ferr 3RE60	0.030 max	18.0-19.0	1.20-2.00	2.50-3.00	4.25-5.25	0.030 max	0.030 max	1.40-2.00	N 0.05-0.10; bal Fe	630	440	30	290 HB
ASTM A949/A949M(95)	A789	Spray formed smls pipe; Duplex Aust-Ferr 3RE60	0.030 max	18.0-19.0	1.2-2.00	2.50-3.00	4.25-5.25	0.030 max	0.030 max	1.40-2.00	N 0.05-0.10; bal Fe	635	440	30	290 HB max

Stainless Steel, Unclassified, S31803

USA

Specification	Designation	Notes	C	Cr	Mn	Mo	Ni	P	S	Si	Other	UTS	YS	El	Hard
	UNS S39205*	Obs; see S31803													

Stainless Steel, Unclassified, S32001

USA

Specification	Designation	Notes	C	Cr	Mn	Mo	Ni	P	S	Si	Other	UTS	YS	El	Hard
	UNS S32001	Duplex Aust-Ferr 19D	0.030 max	19.5-21.5	4.0-6.0	0.60 max	1.0-3.0	0.040 max	0.030 max	1.00 max	Cu <=1.00; N 0.05-0.17; bal Fe				

UNS numbers and US grades are provided as a means of cross referencing chemically similar alloys. Exchangability is only possible after independent examination of specifications. Tensile properties are minimum or typical as specified. UTS and YS as MPa. El as %. See Appendix for list of abbreviations used in Notes. * indicates obsolete material.

Specification	Designation	Notes	C	Cr	Mn	Mo	Ni	P	S	Si	Other	UTS	YS	El	Hard

Stainless Steel, Unclassified, S32001 (Continued from previous page)

USA

Specification	Designation	Notes	C	Cr	Mn	Mo	Ni	P	S	Si	Other	UTS	YS	El	Hard
ASTM A240/A240M(98)	S32001	Duplex Aust-Ferr 19D	0.030 max	19.5-21.5	4.0-6.0	0.60 max	1.0-3.0	0.040 max	0.030 max	1.00 max	Cu <=1.00; N 0.05-0.17; bal Fe	620	450	25.0	
SAE J405(98)	S32001	Duplex Aust-Ferr 19D	0.030 max	19.5-21.5	4.0-6.0	0.60 max	1.0-3.0	0.040 max	0.030 max	1.00 max	Cu <=1.00; N 0.05-0.17; bal Fe				

Stainless Steel, Unclassified, S32304

USA

Specification	Designation	Notes
	UNS S39230*	Obs; see S32304

Stainless Steel, Unclassified, S32404

USA

Specification	Designation	Notes
	UNS S39240*	Obs; see S32404

Stainless Steel, Unclassified, S32550

USA

Specification	Designation	Notes
	UNS S39253*	Obs; see S32550

Stainless Steel, Unclassified, S32750

USA

Specification	Designation	Notes
	UNS S39275*	Obs; see S32750

Stainless Steel, Unclassified, S32760

USA

Specification	Designation	Notes
	UNS S39276*	Obs; see S32760

Stainless Steel, Unclassified, S32950

USA

Specification	Designation	Notes
	UNS S39295*	Obs; see S32950

Stainless Steel, Unclassified, S39274

Europe

Specification	Designation	Notes	C	Cr	Mn	Mo	Ni	P	S	Si	Other	UTS	YS	El	Hard
EN 10088/2(95)	1.4501	Plt, HR Aust-Ferr; Corr res; t<=75mm, Ann	0.030 max	24.00-26.00	1.00 max	3.00-4.00	6.00-8.00	0.035 max	0.015 max	1.00 max	Cu 0.50-1.00; N 0.20-0.30; W 0.50-1.00; bal Fe	730-930	530	25	
EN 10088/2(95)	X2CrNiMoCuWN25-7-4	Plt, HR Aust-Ferr; Corr res; t<=75mm, Ann	0.030 max	24.00-26.00	1.00 max	3.00-4.00	6.00-8.00	0.035 max	0.015 max	1.00 max	Cu 0.50-1.00; N 0.20-0.30; W 0.50-1.00; bal Fe	730-930	530	25	
EN 10088/3(95)	1.4501	Aust-Ferr; Corr res; t<=160mm; Ann	0.030 max	24.00-26.00	1.00 max	3.00-4.00	6.00-8.00	0.035 max	0.015 max	1.00 max	Cu 0.50-1.00; N 0.20-0.30; W 0.50-1.00; bal Fe	730-930	530	25	290 HB
EN 10088/3(95)	X2CrNiMoCuWN25-7-4	Aust-Ferr; Corr res; t<=160mm; Ann	0.030 max	24.00-26.00	1.00 max	3.00-4.00	6.00-8.00	0.035 max	0.015 max	1.00 max	Cu 0.50-1.00; N 0.20-0.30; W 0.50-1.00; bal Fe	730-930	530	25	290 HB

USA

Specification	Designation	Notes	C	Cr	Mn	Mo	Ni	P	S	Si	Other	UTS	YS	El	Hard
	UNS S39274	Ferr-Aust DP3W	0.030 max	24.0-26.0	1.0 max	2.50-3.50	6.0-8.0	0.030 max	0.020 max	0.80 max	Cu 0.20-0.80; N 0.24-0.32; W 1.50-2.50; bal Fe				
ASTM A182/A182M(98)	F54	Frg/roll pipe flange valve	0.030 max	24.0-26.0	1.0 max	2.50-3.50	6.0-8.0	0.030 max	0.020 max	0.80 max	Cu 0.20-0.80; N 0.24-0.32; W 1.50-2.50; bal Fe	800	550	15	310 HB max
ASTM A789/A789M(95)	A789	Smls Weld tub, Ferr-Aust DP3W	0.030 max	24.0-26.0	1.0 max	2.50-3.50	6.0-8.0	0.030 max	0.020 max	0.80 max	Cu 0.20-0.80; N 0.24-0.32; W 1.50-2.50; bal Fe	800	550	15	310 HB
ASTM A790/A790M(95)	A789	Smls Weld Pipe, Duplex Aust-Ferr DP-3W	0.030 max	24.0-26.0	1.0 max	2.50-3.50	6.0-8.0	0.030 max	0.020 max	0.80 max	Cu 0.20-0.80; N 0.24-0.32; W 1.50-2.50; bal Fe	800	550	15	310 HB
ASTM A815/A815M(98)	S39274	Pipe fittings; Duplex Aust-Ferr	0.030 max	24.0-26.0	1.00 max	2.50-3.50	6.0-8.0	0.030 max	0.020 max	0.80 max	Cu 0.20-0.80; N 0.24-0.32; W 1.50-2.50; bal Fe	800	550	15.0	310 HB
ASTM A988(98)	S39274	HIP Flanges, Fittings, Valves/parts; Heat res; Duplex Aust-Ferr; DP3W	0.030 max	24.0-26.0	1.00 max	2.50-3.50	6.0-8.0	0.030 max	0.020 max	0.80 max	Cu 0.20-0.80; N 0.24-0.32; W 1.50-2.50; bal Fe	800	550	15	310 HB max

UNS numbers and US grades are provided as a means of cross referencing chemically similar alloys. Exchangability is only possible after independent examination of specifications. Tensile properties are minimum or typical as specified. UTS and YS as MPa. El as %. See Appendix for list of abbreviations used in Notes. * indicates obsolete material.

Specification	Designation	Notes	C	Cr	Mn	Mo	Ni	P	S	Si	Other	UTS	YS	El	Hard

Stainless Steel, Unclassified, S44500

USA

Specification	Designation	Notes	C	Cr	Mn	Mo	Ni	P	S	Si	Other	UTS	YS	El	Hard
	UNS S44500	High-Cr-Cu; Heat res; Nb stabilized	0.02 max	19.0-21.0	1.0 max		0.60 max	0.040 max	0.012 max	1.0 max	Cu 0.30-0.60; N <=0.03; Nb 10(C+N) - 0.8; bal Fe				
ASTM A240/A240M(98)	S44500	High-Cr-Cu Heat res Nb-stabilized	0.020 max	19.0-21.0	1.00 max		0.60 max	0.040 max	0.012 max	1.0 max	Cu 0.30-0.60; N <=0.03; 10(C+N)<=Nb<=0.8; bal Fe	427	205	22	
ASTM A564/A564M(97)	XM-16	Bar Shp, HR, CF AH	0.03 max	11.00-12.50	0.50 max	0.50 max	7.50-9.50	0.015 max	0.015 max	0.50 max	Cu 1.50-2.50; Ti 0.90-1.40; Nb+Ta 0.10-0.50; bal Fe	1415-1620	1275-1515	8-10	363-444 HB
SAE J405(98)	·S44500	High-Cr-Cu; Heat res; Nb stabilized	0.020 max	19.00-21.00	1.0 max		0.60 max	0.040 max	0.012 max	1.00 max	Cu 0.30-0.60; N <=0.03; 10(C+N)<=Nb<=0.80; bal Fe				

Stainless Steel, Unclassified, Safurex

USA

Specification	Designation	Notes	C	Cr	Mn	Mo	Ni	P	S	Si	Other	UTS	YS	El	Hard
	UNS S32906	Duplex Aust-Ferr	0.030 max	28.0-30.0	0.80-1.50	1.50-2.60	5.8-7.5	0.030 max	0.030 max	0.50 max	Cu <=0.80; N 0.30-0.40; bal Fe				

Stainless Steel, Unclassified, URANUS 52N+

USA

Specification	Designation	Notes	C	Cr	Mn	Mo	Ni	P	S	Si	Other	UTS	YS	El	Hard
	UNS S32520	Duplex Aust-Ferr	0.030 max	24.0-26.0	1.5 max	3.0-5.0	5.5-8.0	0.035 max	0.020 max	0.8 max	Cu 0.50-3.00; N 0.20-0.35; bal Fe				
ASTM A182/A182M(98)	F59	Frg/roll pipe flange valve	0.030 max	24.0-26.0	1.5 max	3.0-5.0	5.5-8.0	0.035 max	0.020 max	0.8 max	Cu 0.50-3.00; N 0.20-0.35; bal Fe	770	550	25	
ASTM A240/A240M(98)	S32520	Duplex Aust-Ferr	0.030 max	24.0-26.0	1.5 max	3.0-4.0	5.5-8.0	0.035 max	0.020 max	0.80	Cu 0.50-2.00; N 0.20-0.35; bal Fe	770	550	25.0	310 HB
SAE J405(98)	S32520	Duplex Aust-Ferr	0.030 max	24.0-26.0	1.50 max	3.00-5.00	5.50-8.00	0.035 max	0.020 max	0.80 max	Cu 0.50-3.00; N 0.20-0.35; bal Fe				

Stainless Steel, Unclassified, Zeron 100

USA

Specification	Designation	Notes	C	Cr	Mn	Mo	Ni	P	S	Si	Other	UTS	YS	El	Hard
	UNS S32760	Duplex Aust-Ferr	0.03 max	24.0-26.0	1.0 max	3.0-4.0	6.0-8.0	0.03 max	0.01 max	1.0 max	Cu 0.5-1.0; N 0.2-0.3; W 0.5-1.0; Cr + 3.3xMo + 16xN> 40; bal Fe				
ASTM A182/A182M(98)	F55	Frg/roll pipe flange valve	0.030 max	24.0-26.0	1.00 max	3.00-4.00	6.0-8.0	0.030 max	0.010 max	1.00 max	Cu 0.50-1.00; N 0.20-0.30; W 0.50-1.00; Cr+3.3xMo+16xN>=40; bal Fe	750-895	550	25.0	
ASTM A240/A240M(98)	S32760	Duplex Aust-Ferr	0.030 max	24.0-26.0	1.00 max	3.0-4.0	6.0-8.0	0.030 max	0.010 max	1.00 max	Cu 0.5-1.0; N 0.20-0.30; W 0.50-1.00; Cr+3.3xMo+16xN>=40; bal Fe	750	550	25.0	270 HB
ASTM A789/A789M(95)	Zeron 100	Smls Weld tub, Duplex Aust-Ferr	0.05 max	24.00-26.00	1.00 max	3.00-4.00	6.00-8.00	0.030 max	0.010 max	1.00 max	Cu 0.50-1.00; N 0.20-0.30; W 0.50-1.00; Cr+3.3xMo+16xN>=40; bal Fe	750-895	550	25	270 HB
ASTM A790/A790M(95)	Zeron 100	Smls Weld Pipe, Duplex Aust-Ferr	0.05 max	24.00-26.00	1.00 max	3.00-4.00	6.00-8.00	0.030 max	0.010 max	1.00 max	Cu 0.50-1.00; N 0.20-0.30; W 0.50-1.00; Cr+3.3xMo+16xN>=40; bal Fe	750-895	550	25	270 HB
ASTM A815/A815M(98)	S32760	Pipe fittings; Duplex Aust-Ferr	0.030 max	24.0-26.0	1.00 max	3.0-4.0	6.0-8.0	0.030 max	0.010 max	1.00 max	Cu 0.50-1.00; N 0.20-0.30; W 0.50-1.00; bal Fe	750-895	550	25.0	270 HB
ASTM A988(98)	S32760	HIP Flanges, Fittings, Valves/parts; Heat res; Duplex Aust-Ferr	0.030 max	24.0-26.0	1.00 max	3.0-4.0	6.0-8.0	0.030 max	0.010 max	1.00 max	Cu 0.50-1.00; N 0.20-0.30; W 0.50-1.00; Cr+3.3xMo+16xN>40; bal Fe	750-895	550	25.0	
SAE J405(98)	S32760	Duplex Aust-Ferr	0.030 max	24.00-26.00	1.0 max	3.0-4.0	6.0-8.0	0.03 max	0.010 max	1.00 max	Cu 0.50-1.00; N 0.2-0.3; W 0.50-1.00; Cr + 3.3xMo + 16xN> 40; bal Fe				

UNS numbers and US grades are provided as a means of cross referencing chemically similar alloys. Exchangability is only possible after independent examination of specifications. Tensile properties are minimum or typical as specified. UTS and YS as MPa. El as %. See Appendix for list of abbreviations used in Notes. * indicates obsolete material.

Specification	Designation	Notes	C	Cr	Mn	Mo	P	S	Si	V	Other	UTS	YS	El	Hard

Tool Steel, Air-Hardening Medium-Alloy Cold Work, A10

Japan

Specification	Designation	Notes	C	Cr	Mn	Mo	P	S	Si	V	Other	UTS	YS	El	Hard
JIS G3311(88)	SKS51M	CR Strp; Wood Saws	0.75-0.85	0.20-0.50	0.50 max		0.030 max	0.030 max	0.35 max		Cu <=0.25; Ni 1.30-2.00; bal Fe				200-290 HV
JIS G3311(88)	SKS5M	CR Strp; Wood Saws	0.75-0.85	0.20-0.50	0.50 max		0.030 max	0.030 max	0.35 max		Cu <=0.25; Ni 0.70-1.30; bal Fe				200-290 HV

Mexico

Specification	Designation	Notes	C	Cr	Mn	Mo	P	S	Si	V	Other	UTS	YS	El	Hard
NMX-B-082(90)	A10	Bar	1.25-1.50		1.60-2.10	1.25-1.75	0.03 max	0.03 max	1.00-1.50		Cu <=0.25; Ni 1.55-2.05; (As+Sn+Sb)<=0.040; B, Ti may be added; bal Fe				

USA

Specification	Designation	Notes	C	Cr	Mn	Mo	P	S	Si	V	Other	UTS	YS	El	Hard
	AISI A10	Ann Hard	1.25-1.50		1.60-2.10	1.25-1.75	0.030 max	0.030 max	1.00-1.50		Ni 1.55-2.05; bal Fe				235-269 HB
	UNS T30110		1.25-1.50		1.60-2.1	1.25-1.75	0.030 max	0.030 max	1.00-1.50	0.5 max	Ni 1.55-2.05; bal Fe				
ASTM A681(94)	A10	Ann	1.25-1.50		1.60-2.10	1.25-1.75	0.030 max	0.030 max	1.00-1.50		Ni 1.55-2.05; S 0.06-0.15 if spec for machining; bal Fe				269 HB
FED QQ-T-570(87)	A10*	Obs; see ASTM A681	1.25-1.5		1.6-2.1	1.25-1.75			1-1.5		Ni 1.55-2.05; bal Fe				

Tool Steel, Air-Hardening Medium-Alloy Cold Work, A11

USA

Specification	Designation	Notes	C	Cr	Mn	Mo	P	S	Si	V	Other	UTS	YS	El	Hard
	AISI A11	Ann Hard	2.40-2.50	4.75-5.50	0.35-0.60	1.10-1.50	0.030 max	0.05-0.09	0.75-1.10	9.25-10.25	W <=0.50; bal Fe				248-269 HB
	UNS T30111		2.45 min	5.25 min	0.50 min	1.30 min			0.90 min	9.75 min	bal Fe				

Tool Steel, Air-Hardening Medium-Alloy Cold Work, A2

Argentina

Specification	Designation	Notes	C	Cr	Mn	Mo	P	S	Si	V	Other	UTS	YS	El	Hard
IAS	IRAM A2	Ann	0.95-1.05	4.75-5.50	1.00 max	0.90-1.40	0.030 max	0.030 max	0.50 max	0.15-0.50	bal Fe				202-229 HB

China

Specification	Designation	Notes	C	Cr	Mn	Mo	P	S	Si	V	Other	UTS	YS	El	Hard
GB 1299(85)	Cr5Mo1V	Q/T	0.95-1.05	4.75-5.50	1.00 max	0.90-1.40	0.030 max	0.030 max	0.50 max	0.15-0.50	Cu <=0.30; Ni <=0.25; bal Fe				60 HRC

Finland

Specification	Designation	Notes	C	Cr	Mn	Mo	P	S	Si	V	Other	UTS	YS	El	Hard
SFS 908(73)	SFS908	Alloyed	0.95-1.05	5.0-5.5	0.45-0.75	1.0-1.2	0.03 max	0.02 max	0.15-0.3	0.15-0.25	bal Fe				

France

Specification	Designation	Notes	C	Cr	Mn	Mo	P	S	Si	V	Other	UTS	YS	El	Hard
AFNOR NFA35590	Z100CDV5*		0.9-1.05	4.8-5.5	0.5-0.8	0.9-1.3	0.025 max	0.025 max	0.1-0.4	0.15-0.35	bal Fe				

Germany

Specification	Designation	Notes	C	Cr	Mn	Mo	P	S	Si	V	Other	UTS	YS	El	Hard
DIN	GX100CrMoV5-1	Hard tmp to 200C	0.90-1.05	4.80-5.50	0.40-0.70	0.90-1.20	0.035 max	0.035 max	0.20-0.40	0.10-0.30	bal Fe				62 HRC
DIN	WNr 1.2363	Hard tmp to 200C	0.90-1.05	4.80-5.50	0.40-0.70	0.90-1.20	0.035 max	0.035 max	0.20-0.40	0.10-0.30	bal Fe				62 HRC

Italy

Specification	Designation	Notes	C	Cr	Mn	Mo	P	S	Si	V	Other	UTS	YS	El	Hard
UNI 2955/3(82)	X100CrMoV51KU	High-speed	0.95-1.05	4.5-5.5	0.35-0.65	0.9-1.4	0.07 max	0.06 max	0.1-0.4	0.25-0.45	bal Fe				

Japan

Specification	Designation	Notes	C	Cr	Mn	Mo	P	S	Si	V	Other	UTS	YS	El	Hard
JIS G4404(83)	SKD12		0.95-1.05	4.5-5.5	0.6-0.9	0.8-1.2			0.4	0.2-0.5	bal Fe				255 HB max

Mexico

Specification	Designation	Notes	C	Cr	Mn	Mo	P	S	Si	V	Other	UTS	YS	El	Hard
NMX-B-082(90)	A2	Bar	0.95-1.05	4.75-5.50	1.00 max	0.90-1.40	0.03 max	0.03 max	0.50 max	0.15-0.50	Cu <=0.25; Ni <=0.30; (As+Sn+Sb)<=0.040; (Cu+Ni)<=0.40; B, Ti may be added; bal Fe				

Norway

Specification	Designation	Notes	C	Cr	Mn	Mo	P	S	Si	V	Other	UTS	YS	El	Hard
NS 13860	13860	Bar Frg	0.95-1.05	5.00-5.50	0.45-0.75	1.00-1.20	0.030 max	0.020 max	0.15-0.30	0.15-0.25	bal Fe				

Pan America

Specification	Designation	Notes	C	Cr	Mn	Mo	P	S	Si	V	Other	UTS	YS	El	Hard
COPANT 337	A2	Bar, Frg	0.9-1.05	4.75-5.5	0.3-0.9	0.9-1.5	0.03	0.03	0.15-0.4	0.15-0.5	bal Fe				
COPANT 337	A2		0.9-1.05	4.75-5.5	0.3-0.9	0.9-1.5	0.03	0.03	0.15-0.4	0.15-0.3	bal Fe				

UNS numbers and US grades are provided as a means of cross referencing chemically similar alloys. Exchangability is only possible after independent examination of specifications. Tensile properties are minimum or typical as specified. UTS and YS as MPa. El as %. See Appendix for list of abbreviations used in Notes. * indicates obsolete material.

Specification	Designation	Notes	C	Cr	Mn	Mo	P	S	Si	V	Other	UTS	YS	El	Hard

Tool Steel, Air-Hardening Medium-Alloy Cold Work, A2 (Continued from previous page)

Poland

Specification	Designation	Notes	C	Cr	Mn	Mo	P	S	Si	V	Other	UTS	YS	El	Hard
PNH85023	NCLD		0.95-1.05	4.5-5.5	0.5-0.8	0.8-1.2			0.2-0.5	0.3-0.5	bal Fe				

Russia

Specification	Designation	Notes	C	Cr	Mn	Mo	P	S	Si	V	Other	UTS	YS	El	Hard
GOST 5950	Ch6WF	Alloyed	1.05-1.15	5.5-6.5	0.15-0.4	0.3 max	0.03 max	0.03 max	0.15-0.35	0.5-0.8	Cu <=0.3; Ni <=0.35; W 1.1-1.5; bal Fe				

Spain

Specification	Designation	Notes	C	Cr	Mn	Mo	P	S	Si	V	Other	UTS	YS	El	Hard
UNE	F.536		0.95-1.05	4.5-5.5	0.45-0.75	0.75-1.25	0.03 max	0.03 max	0.13-0.38	0.15-0.35	bal Fe				
UNE 36018/2(94)	F.5227*	see X100CrMoV5	0.9-1.05	4.5-5.5	0.35-0.65	0.9-1.4	0.03 max	0.02 max	0.1-0.4	0.25-0.45	bal Fe				
UNE 36018/2(94)	X100CrMoV5	Alloyed	0.9-1.05	4.5-5.5	0.35-0.65	0.9-1.4	0.03 max	0.02 max	0.1-0.4	0.25-0.45	bal Fe				
UNE 36072(75)	F.5227*	see X100CrMoV5	0.9-1.05	4.5-5.5	0.35-0.65	0.9-1.4	0.03 max	0.03 max	0.1-0.4	0.25-0.45	bal Fe				
UNE 36072(75)	X100CrMoV5		0.9-1.05	4.5-5.5	0.35-0.65	0.9-1.4	0.03 max	0.03 max	0.1-0.4	0.25-0.45	bal Fe				

Sweden

Specification	Designation	Notes	C	Cr	Mn	Mo	P	S	Si	V	Other	UTS	YS	El	Hard
SS 142260	2260		0.95-1.05	5-5.5	0.45-0.75	1-1.2	0.03 max	0.02 max	0.15-0.3	0.15-0.25	bal Fe				

UK

Specification	Designation	Notes	C	Cr	Mn	Mo	P	S	Si	V	Other	UTS	YS	El	Hard
BS 4656	BA2	Bar Rod Sh Strp Frg	0.95-1.05	4.75-5.25	0.3-0.7	0.9-1.1			0.4	0.15-0.4	bal Fe				
BS 4659	BA2		0.95-1.05	4.75-5.25	0.3-0.7	0.9-1.1			0.4	0.15-0.4	bal Fe				

USA

Specification	Designation	Notes	C	Cr	Mn	Mo	P	S	Si	V	Other	UTS	YS	El	Hard
	AISI A2	Ann Hard	0.95-1.05	4.75-5.50	1.00 max	0.90-1.40	0.030 max	0.030 max	0.15-0.50	0.15-0.50	bal Fe				201-235 HB
	UNS T30102		0.95-1.05	4.75-5.50	1.00 max	0.90-1.40	0.030 max	0.030 max	0.50 max	0.15-0.50	bal Fe				
ASTM A681(94)	A2	Ann	0.95-1.05	4.75-5.50	0.40-1.00	0.90-1.40	0.030 max	0.030 max	0.10-0.50	0.15-0.50	S 0.06-0.15 if spec for machining; bal Fe				248 HB
ASTM A685	A2		0.95-1.05	4.75-5.5	1	0.9-1.4	0.03	0.03	0.5	0.15-0.5	bal Fe				
FED QQ-T-570(87)	A2*	Obs; see ASTM A681	0.95-1.05	4.75-5.5	1	0.9-1.4			0.5	0.15-0.5	Ni 0.3; bal Fe				
SAE J437(70)	A2	Ann									bal Fe				202-229 HB
SAE J438(70)	A2		0.95-1.05	4.75-5.50	0.45-0.75	0.90-1.40			0.20-0.40	0.40 max	V optional; bal Fe				

Tool Steel, Air-Hardening Medium-Alloy Cold Work, A3

Mexico

Specification	Designation	Notes	C	Cr	Mn	Mo	P	S	Si	V	Other	UTS	YS	El	Hard
NMX-B-082(90)	A3	Bar	1.20-1.30	4.75-5.50	0.40-0.60	0.90-1.40	0.03 max	0.03 max	0.50 max	0.80-1.40	Cu <=0.25; Ni <=0.30; (As+Sn+Sb)<=0.040; (Cu+Ni)<=0.40; B, Ti may be added; bal Fe				

USA

Specification	Designation	Notes	C	Cr	Mn	Mo	P	S	Si	V	Other	UTS	YS	El	Hard
	UNS T30103*	Obs	1.20-1.30	4.75-5.50	0.40-0.60	0.90-1.40	0.030 max	0.030 max	0.50 max	0.80-1.40	bal Fe				
ASTM A681(94)	A3	Ann	1.20-1.30	4.75-5.50	0.40-0.60	0.90-1.40	0.030 max	0.030 max	0.10-0.70	0.80-1.40	S 0.06-0.15 if spec for machining; bal Fe				229 HB

Tool Steel, Air-Hardening Medium-Alloy Cold Work, A4

Germany

Specification	Designation	Notes	C	Cr	Mn	Mo	P	S	Si	V	Other	UTS	YS	El	Hard
DIN	58SiCr8	Hard tmp to 200C	0.55-0.63	0.35-0.45	0.60-0.90		0.035 max	0.035 max	1.70-2.00		bal Fe				54 HRC
DIN	90Cr3	Tmp to 200C	0.85-0.95	0.70-0.90	0.20-0.40		0.030 max	0.030 max	0.15-0.30		bal Fe				62 HRC
DIN	WNr 1.2056	Tmp to 200C	0.85-0.95	0.70-0.90	0.20-0.40		0.030 max	0.030 max	0.15-0.30		bal Fe				62 HRC
DIN	WNr 1.2103	Tmp to 200C	0.55-0.63	0.35-0.45	0.60-0.90		0.035 max	0.035 max	1.70-2.00		bal Fe				54 HRC

Specification	Designation	Notes	C	Cr	Mn	Mo	P	S	Si	V	Other	UTS	YS	El	Hard

Tool Steel, Air-Hardening Medium-Alloy Cold Work, A4 (Continued from previous page)

Mexico

Specification	Designation	Notes	C	Cr	Mn	Mo	P	S	Si	V	Other	UTS	YS	El	Hard
NMX-B-082(90)	A4	Bar	0.95-1.05	0.90-2.20	1.00-2.20	0.90-1.40	0.03 max	0.03 max	0.50 max		Cu <=0.25; Ni <=0.30; (As+Sn+Sb)<=0.040; (Cu+Ni)<=0.40; B, Ti may be added; bal Fe				

Russia

GOST 5950	ChGS	Alloyed	0.95-1.05	1.3-1.65	0.85-1.25	0.2 max	0.03 max	0.03 max	0.4-0.7	0.15 max	Cu <=0.3; Ni <=0.35; W <=0.2; bal Fe				

USA

	AISI A4	Ann Hard	0.95-1.05	0.90-2.20	1.80-2.20	0.90-1.40	0.030 max	0.030 max	0.15-0.50		bal Fe				200-241 HB
	UNS T30104		0.95-1.05	0.90-2.20	1.80-2.20	0.90-1.40	0.030 max	0.030 max	0.50 max		bal Fe				
ASTM A681(94)	A4	Ann	0.95-1.05	0.90-2.20	1.80-2.20	0.90-1.40	0.030 max	0.030 max	0.10-0.70		S 0.06-0.15 if spec for machining; bal Fe				241 HB
ASTM A685	A4		0.95-1.05	0.9-2.2	1.8-2.2	0.9-1.4	0.03	0.03	0.5		bal Fe				
FED QQ-T-570(87)	A4*	Obs; see ASTM A681	0.95-1.05	0.9-2.2	1.8-2.2	0.9-1.4			0.5		Ni 0.3; bal Fe				

Tool Steel, Air-Hardening Medium-Alloy Cold Work, A5

Mexico

Specification	Designation	Notes	C	Cr	Mn	Mo	P	S	Si	V	Other	UTS	YS	El	Hard
NMX-B-082(90)	A5	Bar	0.95-1.05	0.90-1.20	2.00-3.20	0.90-1.40	0.03 max	0.03 max	0.50 max		Cu <=0.25; Ni <=0.30; (As+Sn+Sb)<=0.040; (Cu+Ni)<=0.40; B, Ti may be added; bal Fe				

USA

	UNS T30105*	Obs	0.95-1.05	0.90-1.20	2.80-3.20	0.90-1.40	0.030 max	0.030 max	0.50 max		bal Fe				
ASTM A681(94)	A5		0.95-1.05	0.90-1.40	2.80-3.20	0.90-1.40	0.030 max	0.030 max	0.10-0.70		S 0.06-0.15 if spec for machining; bal Fe				

Tool Steel, Air-Hardening Medium-Alloy Cold Work, A6

France

Specification	Designation	Notes	C	Cr	Mn	Mo	P	S	Si	V	Other	UTS	YS	El	Hard
AFNOR NFA35590(92)	70MCD8	Soft ann	0.6-0.8	0.9-1.2	1.8-2.4	0.9-1.4	0.025 max	0.025 max	0.2-0.5		Ni <=0.4; bal Fe				248 HB max
AFNOR NFA35590(92)	70MnCrMo8	Soft ann	0.6-0.8	0.9-1.2	1.8-2.4	0.9-1.4	0.025 max	0.025 max	0.2-0.5		Ni <=0.4; bal Fe				248 HB max

Germany

DIN 17350(80)	21MnCr5	Hard tmp to 200C	0.18-0.24	1.00-1.30	1.10-1.40		0.030 max	0.030 max	0.15-0.35		bal Fe				60 HRC
DIN 17350(80)	WNr 1.2162	Hard tmp to 200C	0.18-0.24	1.00-1.30	1.10-1.40		0.030 max	0.030 max	0.15-0.35		bal Fe				60HRC

Mexico

NMX-B-082(90)	A6	Bar	0.65-0.75	0.90-1.20	1.80-2.50	0.90-1.40	0.03 max	0.03 max	0.50 max		Cu <=0.25; Ni <=0.30; (As+Sn+Sb)<=0.040; (Cu+Ni)<=0.40; B, Ti may be added; bal Fe				

Pan America

COPANT 337	A6	Bar, Frg	0.65-0.75	0.9-1.2	1.8-2.5	0.9-1.4	0.03	0.03	0.15-0.4		bal Fe				

Russia

GOST	7ChG2WM		0.68-0.76	1.5-1.8	1.8-2.3	0.5-0.8	0.03 max	0.03 max	0.2-0.4	0.1-0.25	Cu <=0.3; Ni <=0.35; W 0.5-0.9; bal Fe				

UK

BS 4659	BA6		0.65-0.75	0.85-1.15	1.8-2.1	1.2-1.6			0.4		bal Fe				

USA

	AISI A6	Ann Hard	0.65-0.75	0.90-1.20	1.80-2.50	0.90-1.40	0.030 max	0.030 max	0.15-0.50		bal Fe				217-248 HB
	UNS T30106		0.65-0.75	0.90-1.20	1.80-2.50	0.90-1.40	0.030 max	0.030 max	0.50 max		bal Fe				
ASTM A681(94)	A6	Ann	0.65-0.75	0.90-1.40	1.80-2.50	0.90-1.40	0.030 max	0.030 max	0.10-0.70		S 0.06-0.15 if spec for machining; bal Fe				248 HB

Specification	Designation	Notes	C	Cr	Mn	Mo	P	S	Si	V	Other	UTS	YS	El	Hard

Tool Steel, Air-Hardening Medium-Alloy Cold Work, A6 (Continued from previous page)

USA

Specification	Designation	Notes	C	Cr	Mn	Mo	P	S	Si	V	Other	UTS	YS	El	Hard
ASTM A685	A6		0.65-0.75	0.9-1.2	1.8-2.5	0.9-1.4	0.03	0.03	0.5		bal Fe				
FED QQ-T-570(87)	A6*	Obs; see ASTM A681	0.65-0.75	0.9-1.2	1.8-2.5	0.9-1.4			0.5		Ni 0.3; bal Fe				

Tool Steel, Air-Hardening Medium-Alloy Cold Work, A7

Mexico

Specification	Designation	Notes	C	Cr	Mn	Mo	P	S	Si	V	Other	UTS	YS	El	Hard
NMX-B-082(90)	A7	Bar	2.00-2.85	5.00-5.75	0.80 max	0.90-1.40	0.03 max	0.03 max	0.50 max	3.90-5.15	Cu <=0.25; Ni <=0.30; W 0.50-1.50; (As+Sn+Sb)<=0.040; (Cu+Ni)<=0.40; B, Ti may be added; bal Fe				

Russia

Specification	Designation	Notes	C	Cr	Mn	Mo	P	S	Si	V	Other	UTS	YS	El	Hard
GOST 5950	ChW4F	Alloyed	1.25-1.45	0.4-0.7	0.15-0.4	0.5 max	0.03 max	0.03 max	0.15-0.35	0.15-0.3	Cu <=0.3; Ni <=0.35; W 3.5-4.3; bal Fe				

USA

Specification	Designation	Notes	C	Cr	Mn	Mo	P	S	Si	V	Other	UTS	YS	El	Hard
	AISI A7	Ann Hard	2.25	5.00-5.75	0.80 max	0.90-1.40	0.030 max	0.030 max	0.15-0.50	3.90-5.15	W 0.50-1.50; bal Fe				235-269 HB
	UNS T30107		2.00-2.85	5.00-5.75	0.80 max	0.90-1.40	0.030 max	0.030 max	0.50 max	3.90-5.15	W 0.50-1.50; bal Fe				
ASTM A681(94)	A7	Ann	2.00-2.85	5.00-5.75	0.20-0.80	0.90-1.40	0.030 max	0.030 max	0.10-0.70	3.90-5.15	W 0.50-1.50; S 0.06-0.15 if spec for machining; bal Fe				269 HB
FED QQ-T-570(87)	A7*	Obs; see ASTM A681	2-2.85	5-5.75	0.8	0.9-1.4			0.5	3.9-5.15	Ni 0.3; W 0.5-1.5; bal Fe				

Tool Steel, Air-Hardening Medium-Alloy Cold Work, A8

France

Specification	Designation	Notes	C	Cr	Mn	Mo	P	S	Si	V	Other	UTS	YS	El	Hard
AFNOR NFA35590	Z38CDWV5*	Obs	0.38	5	0.3	1.25			1	0.5	W 1.25; bal Fe				

Japan

Specification	Designation	Notes	C	Cr	Mn	Mo	P	S	Si	V	Other	UTS	YS	El	Hard
JIS G4404(83)	SKD62	HR Bar Frg Ann	0.32-0.42	4.50-5.50	0.50 max	1.00-1.50	0.030 max	0.030 max	0.80-1.20	0.20-0.60	Cu <=0.25; Ni <=0.25; W 1.00-1.50; bal Fe				229 HB max

Mexico

Specification	Designation	Notes	C	Cr	Mn	Mo	P	S	Si	V	Other	UTS	YS	El	Hard
NMX-B-082(90)	A8	Bar	0.50-0.60	4.75-5.50	0.50 max	1.15-1.65	0.03 max	0.03 max	0.75-1.10		Cu <=0.25; Ni <=0.30; W 1.00-1.50; (As+Sn+Sb)<=0.040; (Cu+Ni)<=0.40; B, Ti may be added; bal Fe				

Russia

Specification	Designation	Notes	C	Cr	Mn	Mo	P	S	Si	V	Other	UTS	YS	El	Hard
GOST 5950	6Ch6W3MFS	Alloyed	0.5-0.6	5.5-6.5	0.15-0.4	0.6-0.9	0.03 max	0.03 max	0.6-0.9	0.5-0.8	Cu <=0.3; Ni <=0.35; W 2.5-3.5; bal Fe				
GOST 5950	8Ch4W2MFS2	Alloyed	0.8-0.9	4.55-5.1	0.2-0.5	0.8-1.1	0.03 max	0.03 max	1.7-2	1.1-1.4	Cu <=0.3; Ni <=0.35; W 1.8-2.3; bal Fe				

Spain

Specification	Designation	Notes	C	Cr	Mn	Mo	P	S	Si	V	Other	UTS	YS	El	Hard
UNE	F.537		0.32-0.38	4.5-5.5	0.2-0.5	1.25-1.75	0.03 max	0.03 max	0.75-1.25	0.3-0.5	W 1.25-1.75; bal Fe				

USA

Specification	Designation	Notes	C	Cr	Mn	Mo	P	S	Si	V	Other	UTS	YS	El	Hard
	AISI A8	Ann Hard	0.55	4.75-5.50	0.50 max	1.15-1.65	0.030 max	0.030 max	0.75-1.10		W 1.00-1.50; bal Fe				192-241 HB
	UNS T30108		0.50-0.60	4.75-5.50	0.50 max	1.15-1.65	0.030 max	0.030 max	0.75-1.10		W 1.00-1.50; bal Fe				
ASTM A681(94)	A8	Ann	0.50-0.60	4.75-5.50	0.20-0.50	1.15-1.65	0.030 max	0.030 max	0.75-1.10		W 1.00-1.50; S 0.06-0.15 if spec for machining; bal Fe				241 HB
FED QQ-T-570(87)	A8*	Obs; see ASTM A681	0.5-0.6	4.75-5.5	0.5	1.15-1.65			0.75-1.1		Ni 0.3; W 1-1.5; bal Fe				

Tool Steel, Air-Hardening Medium-Alloy Cold Work, A9

Brazil

Specification	Designation	Notes	C	Cr	Mn	Mo	P	S	Si	V	Other	UTS	YS	El	Hard
ABNT EA9	A9		0.5	5	0.3	1.25			0.3	1	Ni 1.5; bal Fe				

India

Specification	Designation	Notes	C	Cr	Mn	Mo	P	S	Si	V	Other	UTS	YS	El	Hard
IS 1570/6(96)	T40Ni6Cr4Mo3	CW	0.35-0.45	0.9-1.3	0.4-0.7	0.2-0.35	0.035 max	0.035 max	0.1-0.35		Ni 1.25-1.75; bal Fe				
IS 1570/6(96)	TAC15	CW	0.35-0.45	0.9-1.3	0.4-0.7	0.2-0.35	0.035 max	0.035 max	0.1-0.35		Ni 1.25-1.75; bal Fe				

Tool Steel, Air-Hardening Medium-Alloy Cold Work, A9 (Continued from previous page)

Specification	Designation	Notes	C	Cr	Mn	Mo	P	S	Si	V	Other	UTS	YS	El	Hard
Mexico															
NMX-B-082(90)	A9	Bar	0.45-0.55	4.75-5.50	0.50 max	1.30-1.80	0.03 max	0.03 max	0.95-1.15	0.80-1.40	Cu <=0.25; Ni 1.25-1.75; (As+Sn+Sb)<=0.040; B, Ti may be added; bal Fe				
Russia															
GOST 5950	8Ch6NFT	Alloyed	0.8-0.9	5-6	0.15-0.4	0.2 max	0.03 max	0.03 max	0.15-0.35	0.3-0.5	Cu <=0.3; Ni 0.9-1.3; Ti 0.05-0.15; W <=0.2; bal Fe				
USA															
	AISI A9	Ann Hard	0.45-0.55	4.75-5.50	0.50 max	1.30-1.80	0.030 max	0.030 max	0.95-1.15	0.80-1.40	Ni 1.25-1.75; bal Fe				212-248 HB
	UNS T30109		0.45-0.55	4.75-5.50	0.50 max	1.30-1.80	0.030 max	0.030 max	0.95-1.15	0.80-1.40	Ni 1.25-1.75; bal Fe				
ASTM A681(94)	A9	Ann	0.45-0.55	4.75-5.50	0.20-0.50	1.30-1.80	0.030 max	0.030 max	0.95-1.15	0.80-1.40	Ni 1.25-1.75; S 0.06-0.15 if spec for machining; bal Fe				248 HB
FED QQ-T-570(87)	A9*	Obs; see ASTM A681	0.45-0.55	4.75-5.5	0.5	1.3-1.8			0.95-1.15	0.8-1.4	Ni 1.25-1.75; bal Fe				

Tool Steel, Air-Hardening Medium-Alloy Cold Work, No equivalents identified

Specification	Designation	Notes	C	Cr	Mn	Mo	P	S	Si	V	Other	UTS	YS	El	Hard
International															
ISO 4957(80)	C105U		1.00-1.10		0.10-0.40		0.030 max	0.030 max	0.10-0.30		bal Fe				
ISO 4957(80)	C120U		1.15-1.25		0.10-0.40		0.030 max	0.030 max	0.10-0.30		bal Fe				
ISO 4957(80)	C45U		0.42-0.50		0.60-0.80		0.030 max	0.030 max	0.15-0.40		bal Fe				
ISO 4957(80)	C70U		0.65-0.75		0.10-0.40		0.030 max	0.030 max	0.10-0.30		bal Fe				
ISO 4957(80)	C80U		0.75-0.85		0.10-0.40		0.030 max	0.030 max	0.10-0.30		bal Fe				
ISO 4957(80)	C90U		0.85-0.95		0.10-0.40		0.030 max	0.030 max	0.10-0.30		bal Fe				
Poland															
PNH85023	NCLV		0.95-1.05	4.5-5.5	0.4-0.7	0.9-1.2	0.030 max	0.030 max	0.15-0.4	0.3-0.45	Ni <=0.4; bal Fe				
PNH85023	NW9		0.85-0.95	3.5-4.5	0.15-0.45		0.030 max	0.030 max	0.5 max	1.7-2.1	Ni <=0.4; W 8-10; bal Fe				

Specification	Designation	Notes	C	Cr	Mn	Mo	P	S	Si	V	Other	UTS	YS	El	Hard

Tool Steel, High-Carbon High-Chromium Cold Work, D2

Argentina

| IAS | IRAM D2 | Ann | 1.40-1.60 | 11.00-13.00 | 0.60 max | 0.70-1.20 | 0.030 max | 0.030 max | 0.60 max | 1.10 max | bal Fe | | | | 255 HB |

Bulgaria

| BDS 7938 | Ch12MF | | 0.45-0.65 | 11.0-13.0 | 0.15-0.40 | 0.40-0.60 | 0.030 max | 0.030 max | 0.15-0.35 | 0.15-0.30 | Cu <=0.30; Ni <=0.35; W <=0.15; bal Fe | | | | |

China

GB 1299	Cr12MoV*	Obs	1.45-1.7	11-12.5	0.4	0.4-0.6			0.4	0.15-0.3	Cu 0.25; Ni 0.3; bal Fe				
GB 1299(85)	Cr12Mo1V1	As Hard	1.40-1.60	11.00-13.00	0.60 max	0.70-1.20	0.030 max	0.030 max	0.60 max	1.10 max	Co <=1.00; Cu <=0.30; Ni <=0.25; bal Fe				59 HRC
YB/T 094(97)	SMCr12Mo1V1	Flat Ann	1.40-1.60	11.00-13.00	0.10-0.60	0.70-1.20	0.030 max	0.030 max	0.10-0.60	0.50-1.10	bal Fe				255 HB max

France

| AFNOR NFA35590 | Z160CDV12 | Bar Frg | 1.6 | 12 | 0.3 | 0.8 | 0.03 | 0.03 | 0.3 | 0.4 | bal Fe | | | | |

Germany

DIN	G-X165CrV12	Hard tmp to 200C	1.55-1.75	11.0-12.0	0.20-0.40		0.035 max	0.035 max	0.25-0.40	0.07-0.12	bal Fe				63 HRC
DIN	WNr 1.2201	Hard tmp to 200C	1.55-1.75	11.0-12.0	0.20-0.40		0.035 max	0.035 max	0.25-0.40	0.07-0.12	bal Fe				63 HRC
DIN	WNr 1.2609		1.55-1.75	11.0-12.0	0.20-0.40	0.50-0.70	0.035 max	0.035 max	0.25-0.40	1.10-1.30	W 0.40-0.60; bal Fe				
DIN	X165CrVMo121		1.55-1.75	11.0-12.0	0.20-0.40	0.50-0.70	0.035 max	0.035 max	0.25-0.40	1.10-1.30	W 0.40-0.60; bal Fe				
DIN 17350(80)	GX165CrMoV12	Hard tmp to 200C	1.55-1.75	11.0-12.0	0.20-0.40	0.50-0.70	0.030 max	0.030 max	0.20-0.40	0.10-0.50	W 0.40-0.60; bal Fe				61 HRC
DIN 17350(80)	WNr 1.2379	Hard tmp to 200C	1.50-1.60	11.0-12.0	0.15-0.45	0.60-0.80	0.030 max	0.030 max	0.10-0.40	0.90-1.10	bal Fe				61 HRC
DIN 17350(80)	WNr 1.2601	Hard tmp to 200C	1.55-1.75	11.0-12.0	0.20-0.40	0.50-0.70	0.030 max	0.030 max	0.20-0.40	0.10-0.50	W 0.40-0.60; bal Fe				61 HRC
DIN 17350(80)	X155CrVMo12-1	Hard tmp to 200C	1.50-1.60	11.0-12.0	0.15-0.45	0.60-0.80	0.030 max	0.030 max	0.10-0.40	0.90-1.10	bal Fe				61 HRC

Hungary

| MSZ 4352(84) | K8 | Soft ann | 1.5-1.65 | 11-13 | 0.15-0.4 | 0.6-0.8 | 0.03 max | 0.03 max | 0.1-0.4 | 0.9-1.1 | Ni <=0.35; W <=0.3; bal Fe | | | | 255 HB max |
| MSZ 4352(84) | K8 | Q/T | 1.5-1.65 | 11-13 | 0.15-0.4 | 0.6-0.8 | 0.03 max | 0.03 max | 0.1-0.4 | 0.9-1.1 | Ni <=0.35; W <=0.3; bal Fe | | | | 61 HRC |

India

| IS 3749 | T160Cr12 | Bar | 1.5-1.7 | 11-13 | 0.25-0.5 | 0.8 | 0.04 | 0.04 | 0.1-0.35 | 0.8 | bal Fe | | | | |

Italy

UNI 2955	DTC-AR X155CrVMo121KU		1.5	12	0.5	0.75			0.6	0.25	bal Fe				
UNI 2955	DTC-ARK		1.5	12	0.6	0.95			0.6	1	Co 3; bal Fe				
UNI 2955	DTC-ARW		1.65	11.5	0.4	0.6			0.4	0.3	W 0.15; bal Fe				
UNI 2955	X165CrMoW12KU		1.55-1.65	11-12	0.25-0.4	0.5-0.7			0.25-0.4	0.07-0.12	W 0.4-0.6; bal Fe				
UNI 2955/3(82)	X155CrVMo121KU		1.5-1.6	11-12.5	0.2-0.5	0.6-1	0.07 max	0.06 max	0.2-0.5	0.75-1.1	bal Fe				

Japan

| JIS G4404(83) | SKD11 | Bar HR Frg Ann | 1.40-1.60 | 11.00-13.00 | 0.60 max | 0.80-1.20 | 0.030 max | 0.030 max | 0.40 max | 0.20-0.50 | Cu <=0.25; Ni <=0.50; bal Fe | | | | 255 HB max |

Mexico

| NMX-B-082(90) | D1 | Bar | 0.90-1.10 | 11.50-12.50 | 0.20-0.40 | 0.70-0.90 | 0.03 max | 0.03 max | 0.10-0.40 | 0.30-0.60 | Cu <=0.25; Ni <=0.30; (As+Sn+Sb) <=0.040; (Cu+Ni) <=0.40; B, Ti; bal Fe | | | | |
| NMX-B-082(90) | D2 | Bar | 1.40-1.60 | 11.00-13.00 | 0.60 max | 0.70-1.20 | 0.03 max | 0.03 max | 0.60 max | 1.10 max | Co <=1.0; Cu <=0.25; Ni <=0.30; (As+Sn+Sb) <=0.040; (Cu+Ni) <=0.40; B, Ti; bal Fe | | | | |

Pan America

| COPANT 337 | D2 | Bar, Frg | 1.4-1.6 | 11-13 | 0.2-0.6 | 0.7-1.2 | 0.03 | 0.03 | 0.2-0.6 | | Co 1; bal Fe | | | | |

UNS numbers and US grades are provided as a means of cross referencing chemically similar alloys. Exchangability is only possible after independent examination of specifications. Tensile properties are minimum or typical as specified. UTS and YS as MPa. El as %. See Appendix for list of abbreviations used in Notes. * indicates obsolete material.

Specification	Designation	Notes	C	Cr	Mn	Mo	P	S	Si	V	Other	UTS	YS	El	Hard

Tool Steel, High-Carbon High-Chromium Cold Work, D2 (Continued from previous page)

Poland

Specification	Designation	Notes	C	Cr	Mn	Mo	P	S	Si	V	Other	UTS	YS	El	Hard
PNH85023	NC11LV		1.5-1.7	11-13	0.15-0.45	0.7-1	0.030 max	0.030 max	0.15-0.4	0.6-0.8	Ni <=0.4; bal Fe				

Romania

Specification	Designation	Notes	C	Cr	Mn	Mo	P	S	Si	V	Other	UTS	YS	El	Hard
STAS 3611(80)	VMoC120		1.45-1.65	11-12.5	0.15-0.45	0.4-0.6	0.03 max	0.03 max	0.15-0.35	0.15-0.3	Ni <=0.35; Pb <=0.15; bal Fe				

Russia

Specification	Designation	Notes	C	Cr	Mn	Mo	P	S	Si	V	Other	UTS	YS	El	Hard
GOST 5950	Ch12F	Alloyed	1.25-1.45	11-12.5	0.15-0.4	0.2 max	0.03 max	0.03 max	0.15-0.35	0.7-0.9	Cu <=0.3; Ni <=0.35; W <=0.2; bal Fe				
GOST 5950	Ch12MF	Alloyed	1.45-1.65	11-12.5	0.15-0.45	0.4-0.6	0.03 max	0.03 max	0.1-0.4	0.15-0.3	Cu <=0.3; Ni <=0.35; W <=0.2; bal Fe				

Spain

Specification	Designation	Notes	C	Cr	Mn	Mo	P	S	Si	V	Other	UTS	YS	El	Hard
UNE	F.520.A		1.6-1.8	12-14	0.15-0.45	0.65-0.95	0.03 max	0.03 max	0.13-0.38	0.2-0.4	bal Fe				
UNE 36018/2(94)	F.5211*	see X160CrMoV12	1.45-1.75	11-13	0.2-0.6	0.7-1	0.03 max	0.02 max	0.15-0.45	0.7-1	bal Fe				
UNE 36018/2(94)	X160CrMoV12	Alloyed	1.45-1.75	11-13	0.2-0.6	0.7-1	0.03 max	0.02 max	0.15-0.45	0.7-1	bal Fe				

Sweden

Specification	Designation	Notes	C	Cr	Mn	Mo	P	S	Si	V	Other	UTS	YS	El	Hard
SS 142310	2310	Alloyed	1.45-1.65	11-13	0.3-0.6	0.7-0.9	0.03 max	0.02 max	0.2-0.4	0.7-1	bal Fe				

UK

Specification	Designation	Notes	C	Cr	Mn	Mo	P	S	Si	V	Other	UTS	YS	El	Hard
BS 4659	(USA D2)		1.55	12		0.85				0.5	bal Fe				
BS 4659	BD2	Bar Rod Sh Strp Frg	1.4-1.6	11.5-12.5	0.6 max	0.7-1.2			0.6 max	0.25-1	bal Fe				
BS 4659	BD2A		1.6-1.9	12-13	0.6 max	0.7-0.9			0.6 max	0.25-1	bal Fe				

USA

Specification	Designation	Notes	C	Cr	Mn	Mo	P	S	Si	V	Other	UTS	YS	El	Hard
	AISI D2	Ann Hard	1.40-1.60	11.00-13.00	0.15-0.60	0.70-1.20	0.030 max	0.030 max	0.15-0.60	1.10 max	bal Fe				217-255 HB
	UNS T30402		1.40-1.60	11.00-13.00	0.60 max	0.70-1.20	0.030 max	0.030 max	0.60 max	1.10 max	Co <=1; bal Fe				
ASTM A681(94)	D2	Ann	1.40-1.60	11.00-13.00	0.10-0.60	0.70-1.20	0.030 max	0.030 max	0.10-0.60	0.50-1.10	S 0.06-0.15 for machining; bal Fe				255 HB
ASTM A685	D2		1.4-1.6	11-13	0.6	0.7-1.2	0.03	0.03	0.6	1.1	Co 1; bal Fe				
FED QQ-T-570(87)	D2*	Obs; see ASTM A681	1.4-1.6	11-13	0.6	0.7-1.2			0.6	1.1	Co 1; bal Fe				
SAE J437(70)	D2	Ann									bal Fe				207-255 HB
SAE J438(70)	D2	Air Q	1.40-1.60	11.00-13.00	0.30-0.50	0.70-1.20			0.30-0.50	0.80 max	Co <=0.60; V, Co optional; bal Fe				

Tool Steel, High-Carbon High-Chromium Cold Work, D3

Argentina

Specification	Designation	Notes	C	Cr	Mn	Mo	P	S	Si	V	Other	UTS	YS	El	Hard
IAS	IRAM D3	Ann	2.00-2.35	11.00-13.50	0.60 max		0.030 max	0.030 max	0.60 max	1.00 max	W <=1.00; bal Fe				255 HB
IAS	IRAM D6	Ann	2.00-2.20	11.50-12.50	0.20-0.40				0.70-1.00		W 0.60-0.90; bal Fe				255 HB

Bulgaria

Specification	Designation	Notes	C	Cr	Mn	Mo	P	S	Si	V	Other	UTS	YS	El	Hard
BDS 7938	Ch12		1.80-2.20	11.0-13.0	0.20-0.50		0.030 max	0.030 max	0.20-0.50	0.15 max	Cu <=0.30; Ni <=0.35; bal Fe				

Canada

Specification	Designation	Notes	C	Cr	Mn	Mo	P	S	Si	V	Other	UTS	YS	El	Hard
CSA 419437	19437		1.8-2.05	11-12.5	0.2-0.45		0.03 max	0.04 max	0.2-0.45	0.15-0.3	Ni <=0.5; W 0.6-1; bal Fe				

China

Specification	Designation	Notes	C	Cr	Mn	Mo	P	S	Si	V	Other	UTS	YS	El	Hard
GB 1299(85)	Cr12	As Hard	2.00-2.30	11.50-13.00	0.40 max		0.030 max	0.030 max	0.40 max		Cu <=0.30; Ni <=0.25; bal Fe				60 HRC

Czech Republic

Specification	Designation	Notes	C	Cr	Mn	Mo	P	S	Si	V	Other	UTS	YS	El	Hard
CSN 419436	19436	Tools for cold ext	1.8-2.05	11-12.5	0.2-0.45		0.03 max	0.035 max	0.2-0.45		Ni <=0.5; bal Fe				

Finland

Specification	Designation	Notes	C	Cr	Mn	Mo	P	S	Si	V	Other	UTS	YS	El	Hard
SFS 909(73)	SFS909	Alloyed	1.9-2.2	12.0-13.5	0.6-0.9		0.03 max	0.02 max	0.2-0.4		W 1.0-1.5; bal Fe				

France

Specification	Designation	Notes	C	Cr	Mn	Mo	P	S	Si	V	Other	UTS	YS	El	Hard
AFNOR NFA35590	Z200C12	Bar Frg	2	12	0.3		0.03	0.03	0.3		bal Fe				

Tool Steel, High-Carbon High-Chromium Cold Work, D3 (Continued from previous page)

Specification	Designation	Notes	C	Cr	Mn	Mo	P	S	Si	V	Other	UTS	YS	El	Hard
Germany															
DIN 17350(80)	WNr 1.2080	Tmp to 200C	1.90-2.20	11.0-12.0	0.15-0.45		0.030 max	0.030 max	0.10-0.40		bal Fe				62 HRC
DIN 17350(80)	WNr 1.2436	Hard tmp to 200C	2.00-2.25	11.0-12.0	0.15-0.45		0.030 max	0.030 max	0.10-0.40		W 0.60-0.80; bal Fe				62 HRC
DIN 17350(80)	X210Cr12	Tmp to 200C	1.90-2.20	11.0-12.0	0.15-0.45		0.030 max	0.030 max	0.10-0.40		bal Fe				62 HRC
DIN 17350(80)	X210CrW12	Hard tmp to 200C	2.00-2.25	11.0-12.0	0.15-0.45		0.030 max	0.030 max	0.10-0.40		W 0.60-0.80; bal Fe				62 HRC
Hungary															
MSZ 4352(72)	K1		1.9-2.2	11-13	0.15-0.4		0.03 max	0.03 max	0.15-0.35	0.15-0.3	bal Fe				
MSZ 4352(84)	K9	Q/T	1.9-2.2	11-13	0.15-0.4	0.6-0.9	0.03 max	0.03 max	0.1-0.4	0.15-0.3	Ni <=0.35; W 0.5-0.8; bal Fe				62 HRC
MSZ 4352(84)	K9	Soft ann	1.9-2.2	11-13	0.15-0.4	0.6-0.9	0.03 max	0.03 max	0.1-0.4	0.15-0.3	Ni <=0.35; W 0.5-0.8; bal Fe				255 HB max
India															
IS 1570/6(96)	TAC22	CW	2-2.3	11-13	0.25-0.5	0.8 max	0.035 max	0.035 max	0.1-0.35		Ni <=0.4; bal Fe				
IS 1570/6(96)	XT215Cr12	CW	2-2.3	11-13	0.25-0.5	0.8 max	0.035 max	0.035 max	0.1-0.35	0.8 max	Ni <=0.4; bal Fe				
Italy															
UNI 2955	DTC X205Cr12KU		2.1	13	0.5 max				0.45 max	0.2	bal Fe				
UNI 2955	DTC-W X215CrW121KU		2.15	11.5	0.4				0.45		W 0.7; bal Fe				
UNI 2955/3(82)	X205Cr12KU		1.9-2.2	11-13	0.15-0.45		0.07 max	0.06 max	0.1-0.4	0.2 max	bal Fe				
UNI 2955/3(82)	X215CrW121KU		2-2.3	11-13	0.15-0.45		0.07 max	0.06 max	0.1-0.4		W 0.8-1.1; bal Fe				
Japan															
JIS G4404(83)	SKD1	HR Frg Ann	1.80-2.40	12.00-15.00	0.60 max		0.030 max	0.030 max	0.40 max	0.30 max	Cu <=0.25; Ni <=0.50; bal Fe				269 HB max
Mexico															
NMX-B-082(90)	D3	Bar	2.00-2.35	11.00-13.50	0.60 max		0.03 max	0.03 max	0.60 max	1.00 max	Cu <=0.25; Ni <=0.30; W <=1.0; (As+Sn+Sb) <=0.040; (Cu+Ni) <=0.40; B, Ti; bal Fe				
Norway															
NS 13882	13882	Bar	1.9-2.2	12-13.5	0.6-0.9		0.03	0.02	0.2-0.4		W 1-1.5; bal Fe				
Pan America															
COPANT 337	D3	Bar, Frg	2.25	12			0.03	0.03			bal Fe				
Poland															
PNH85023	NC11	Tools for cold ext	1.8-2.1	11-13	0.15-0.45		0.030 max	0.030 max	0.15-0.4		Ni <=0.4; bal Fe				
Romania															
STAS 3611(80)	C120		1.8-2.2	11-13	0.15-0.45		0.03 max	0.025 max	0.15-0.35		Ni <=0.35; Pb <=0.15; bal Fe				
Russia															
GOST 5950	Ch12	Alloyed	2-2.2	11.5-13.5	0.15-0.45	0.2 max	0.03 max	0.03 max	0.1-0.4	0.15 max	Cu <=0.3; Ni <=0.35; W <=0.2; bal Fe				
Spain															
UNE	F.521		1.6-2	11.5-13.5	0.2-0.4		0.03 max	0.03 max	0.15-0.3		bal Fe				
UNE 36018/2(94)	F.5212*	see X210Cr12	1.9-2.2	11-13	0.15-0.45	0.15 max	0.03 max	0.02 max	0.1-0.35		bal Fe				
UNE 36018/2(94)	F.5213*	see X210CrW12	2-2.25	11-13	0.15-0.45	0.15 max	0.03 max	0.02 max	0.1-0.4	0.4 max	W 0.6-0.8; bal Fe				
UNE 36018/2(94)	X210Cr12	Alloyed	1.9-2.2	11-13	0.15-0.45	0.15 max	0.03 max	0.02 max	0.1-0.35		bal Fe				
UNE 36018/2(94)	X210CrW12	Alloyed	2-2.25	11-13	0.15-0.45	0.15 max	0.03 max	0.02 max	0.1-0.4	0.4 max	W 0.6-0.8; bal Fe				
Sweden															
SS 142314	2314	Alloyed	0.34-0.43	13-14.5	0.2-0.7		0.03 max	0.015 max	0.6-1.3	0.15-0.4	bal Fe				

Specification	Designation	Notes	C	Cr	Mn	Mo	P	S	Si	V	Other	UTS	YS	El	Hard

Tool Steel, High-Carbon High-Chromium Cold Work, D3 (Continued from previous page)

UK

Specification	Designation	Notes	C	Cr	Mn	Mo	P	S	Si	V	Other	UTS	YS	El	Hard
BS 4659	BD3	Bar Rod Sh Strp Frg	1.9-2.3	12-13	0.6 max				0.6 max	0.5 max	bal Fe				

USA

Specification	Designation	Notes	C	Cr	Mn	Mo	P	S	Si	V	Other	UTS	YS	El	Hard
	AISI D3	Ann Hard	2.00-2.35	11.00-13.50	0.15-0.60		0.030 max	0.030 max	0.15-0.60	1.00 max	W <=1.00; bal Fe				217-255 HB
	UNS T30403		2.00-2.35	11.00-13.50	0.60 max		0.030 max	0.030 max	0.60 max	1.00 max	W <=1.00; bal Fe				
ASTM A681(94)	D3	Ann	2.00-2.35	11.00-13.50	0.10-0.60		0.030 max	0.030 max	0.10-0.60	1.00 max	W <=1.00; S 0.06-0.15 for machining; bal Fe				255 HB
ASTM A685	D3		2-2.35	11-13.5	0.6		0.03	0.03	0.6	1	W 1; bal Fe				
FED QQ-T-570(87)	D3*	Obs; see ASTM A681	2-2.35	11-13.5	0.6				0.6	1	Ni 0.3; W 1; bal Fe				
SAE J437(70)	D3	Ann									bal Fe				212-255 HB
SAE J438(70)	D3	Oil Quen	2.00-2.35	11.00-13.00	0.24-0.45	0.80 max			0.25-0.45	0.80 max	W <=0.75; V, W, Mo optional; Mn may vary; bal Fe				

Tool Steel, High-Carbon High-Chromium Cold Work, D4

Czech Republic

Specification	Designation	Notes	C	Cr	Mn	Mo	P	S	Si	V	Other	UTS	YS	El	Hard
CSN 419437	19437		1.8-2.05	11-12.5	0.2-0.45		0.03 max	0.035 max	0.2-0.45	0.15-0.3	Ni <=0.5; W 0.6-1; bal Fe				

France

Specification	Designation	Notes	C	Cr	Mn	Mo	P	S	Si	V	Other	UTS	YS	El	Hard
AFNOR NFA35590	Z200CD12	Bar Frg	2	12	0.3	0.8	0.03	0.03		0.2	bal Fe				

Germany

Specification	Designation	Notes	C	Cr	Mn	Mo	P	S	Si	V	Other	UTS	YS	El	Hard
DIN	WNr 1.2884		2.00-2.30	11.5-12.5	0.20-0.40	0.30-0.50	0.035 max	0.035 max	0.20-0.40		Co 0.80-1.10; W 0.60-0.80; bal Fe				
DIN	X210CrCoW12		2.00-2.30	11.5-12.5		0.30-0.50	0.035 max	0.035 max	0.20-0.40		Co 0.80-1.10; W 0.60-0.80; bal Fe				

India

Specification	Designation	Notes	C	Cr	Mn	Mo	P	S	Si	V	Other	UTS	YS	El	Hard
IS 1570	T215Cr12		2-2.3	11-13	0.25-0.5	0.8			0.1-0.35	0.8	bal Fe				
IS 3749	T215Cr12	Bar	2-2.3	11-13	0.25-0.5	0.8	0.04	0.04	0.1-0.35	0.8	bal Fe				
IS 4367	T215Cr12	Frg	2-2.3	11-13	0.25-0.5	0.8			0.1-0.35	0.8	bal Fe				

Mexico

Specification	Designation	Notes	C	Cr	Mn	Mo	P	S	Si	V	Other	UTS	YS	El	Hard
NMX-B-082(90)	D4	Bar	2.05-2.40	11.00-13.00	0.60 max	0.70-1.20	0.03 max	0.03 max	0.60 max	1.00 max	Cu <=0.25; Ni <=0.30; (As+Sn+Sb) <=0.040; (Cu+Ni) <=0.40; V, B, Ti; bal Fe				

Poland

Specification	Designation	Notes	C	Cr	Mn	Mo	P	S	Si	V	Other	UTS	YS	El	Hard
PNH85023	NCWV		1.8-2.1	11-13	0.2-0.45				0.2-0.4	0.15-0.3	W 1-1.5; bal Fe				

Russia

Specification	Designation	Notes	C	Cr	Mn	Mo	P	S	Si	V	Other	UTS	YS	El	Hard
GOST	Ch12WM		2-2.2	11-12.5	0.15-0.4	0.6-0.9	0.03 max	0.03 max	0.2-0.4	0.15-0.3	Cu 0.3; Ni <=0.35; Ti 0.03; W 0.5-0.8; bal Fe				
GOST 5950	Ch12WMF	Alloyed	2-2.2	11-12.5	0.15-0.45	0.6-0.9	0.03 max	0.03 max	0.1-0.4	0.15-0.3	Cu <=0.3; Ni <=0.35; W 0.5-0.8; bal Fe				

Spain

Specification	Designation	Notes	C	Cr	Mn	Mo	P	S	Si	V	Other	UTS	YS	El	Hard
UNE 36018/2(94)	F.5214*	see X210CrMoV12	1.9-2.2	11-13	0.15-0.45	0.7-1	0.03 max	0.02 max	0.1-0.4	0.2-0.4	bal Fe				
UNE 36018/2(94)	X210CrMoV12	Alloyed	1.9-2.2	11-13	0.15-0.45	0.7-1	0.03 max	0.02 max	0.1-0.4	0.2-0.4	bal Fe				

Sweden

Specification	Designation	Notes	C	Cr	Mn	Mo	P	S	Si	V	Other	UTS	YS	El	Hard
SS 142312	2312	Alloyed	1.9-2.2	12-13.5	0.6-0.9		0.03 max	0.02 max	0.2-0.4		W 1-1.5; bal Fe				

UK

Specification	Designation	Notes	C	Cr	Mn	Mo	P	S	Si	V	Other	UTS	YS	El	Hard
BS 4659	(USA D4)		2.1	13	0.65	0.65			0.75		Ni 0.85; bal Fe				

USA

Specification	Designation	Notes	C	Cr	Mn	Mo	P	S	Si	V	Other	UTS	YS	El	Hard
	AISI D4	Ann Hard	2.05-2.40	11.00-13.00	0.15-0.60	0.70-1.20	0.030 max	0.030 max	0.15-0.60	1.00 max	bal Fe				217-255 HB
	UNS T30404		2.05-2.40	11.00-13.00	0.60 max	0.70-1.20	0.030 max	0.030 max	0.60 max	1.00 max	bal Fe				

UNS numbers and US grades are provided as a means of cross referencing chemically similar alloys. Exchangability is only possible after independent examination of specifications. Tensile properties are minimum or typical as specified. UTS and YS as MPa. El as %. See Appendix for list of abbreviations used in Notes. * indicates obsolete material.

Specification	Designation	Notes	C	Cr	Mn	Mo	P	S	Si	V	Other	UTS	YS	El	Hard

Tool Steel, High-Carbon High-Chromium Cold Work, D4 (Continued from previous page)

USA

Specification	Designation	Notes	C	Cr	Mn	Mo	P	S	Si	V	Other	UTS	YS	El	Hard
ASTM A681(94)	D4	Ann	2.05-2.40	11.00-13.00	0.10-0.60	0.70-1.20	0.030 max	0.030 max	0.10-0.60	0.15-1.00	S 0.06-0.15 for machining; bal Fe				255 HB
FED QQ-T-570(87)	D4*	Obs; see ASTM A681	2.05-2.4	11-13	0.6	0.7-1.2			0.6	1	Ni 0.3; bal Fe				

Tool Steel, High-Carbon High-Chromium Cold Work, D5

France

Specification	Designation	Notes	C	Cr	Mn	Mo	P	S	Si	V	Other	UTS	YS	El	Hard
AFNOR NFA35590	Z160CKDV12-03		1.5-1.75	12-14	0.15-0.45	0.7-1.1	0.025 max	0.025 max	0.1-0.4	0.15-0.3	Co 2.5-3; bal Fe				

Germany

Specification	Designation	Notes	C	Cr	Mn	Mo	P	S	Si	V	Other	UTS	YS	El	Hard
DIN	GX165CrCoMo12	Hard tmp to 200C	1.55-1.75	11.0-12.0	0.20-0.40	0.50-0.60	0.035 max	0.035 max	0.25-0.40		bal Fe				62 HRC
DIN	WNr 1.2880	Hard tmp to 200C	1.55-1.75	11.0-12.0	0.20-0.40	0.50-0.60	0.035 max	0.035 max	0.25-0.40		Co 1.20-1.40; bal Fe				62 HRC

Mexico

Specification	Designation	Notes	C	Cr	Mn	Mo	P	S	Si	V	Other	UTS	YS	El	Hard
NMX-B-082(90)	D5	Bar	1.40-1.60	11.00-13.00	0.60 max	0.70-1.20	0.03 max	0.03 max	0.60 max	1.00 max	Co 2.50-3.50; Cu <=0.25; Ni <=0.30; (As+Sn+Sb) <=0.040; (Cu+Ni) <=0.40; B, Ti; bal Fe				

USA

Specification	Designation	Notes	C	Cr	Mn	Mo	P	S	Si	V	Other	UTS	YS	El	Hard
	AISI D5	Ann Hard	1.40-1.60	11.00-13.00	0.15-0.60	0.70-1.20	0.030 max	0.030 max	0.15-0.60	1.00 max	Co 2.50-3.50; bal Fe				223-255 HB
	UNS T30405		1.40-1.60	11.00-13.00	0.60 max	0.70-1.20	0.030 max	0.030 max	0.60 max	1.00 max	Co 2.5-3.5; bal Fe				
ASTM A681(94)	D5	Ann	1.40-1.60	11.00-13.00	0.10-0.60	0.70-1.20	0.030 max	0.030 max	0.10-0.60	1.00 max	Co 2.50-3.50; S 0.06-0.15 for machining; bal Fe				255 HB
FED QQ-T-570(87)	D5*	Obs; see ASTM A681	1.4-1.6	11-13	0.6	0.7-1.2			0.6	1	Co 2.5-3.5; Ni 0.3; bal Fe				
SAE J437(70)	D5	Ann									bal Fe				207-255 HB
SAE J438(70)	D5		1.40-1.60	11.00-13.00	0.30-0.50	0.70-1.20			0.30-0.50	0.80 max	Co 2.50-3.50; V optional; bal Fe				

Tool Steel, High-Carbon High-Chromium Cold Work, D7

France

Specification	Designation	Notes	C	Cr	Mn	Mo	P	S	Si	V	Other	UTS	YS	El	Hard
AFNOR NFA35590	Z230CVD12-04	Bar Frg	2.3 max	12	0.3	1	0.03	0.03	0.3	4	bal Fe				

Germany

Specification	Designation	Notes	C	Cr	Mn	Mo	P	S	Si	V	Other	UTS	YS	El	Hard
DIN	WNr 1.2378	Hard tmp to 200C	2.15-2.30	12.0-13.0	0.25-0.40	0.80-1.00	0.035 max	0.035 max	0.15-0.30	2.00-2.30	bal Fe				61 HRC
DIN	X220CrVMo12-2	Hard tmp to 200C	2.15-2.30	12.0-13.0	0.25-0.40	0.80-1.00	0.035 max	0.035 max	0.15-0.30	2.00-2.30	bal Fe				61 HRC

Mexico

Specification	Designation	Notes	C	Cr	Mn	Mo	P	S	Si	V	Other	UTS	YS	El	Hard
NMX-B-082(90)	D7	Bar	2.15-2.50	11.50-13.50	0.60 max	0.70-1.20	0.03 max	0.03 max	0.60 max	3.00-4.40	Cu <=0.25; Ni <=0.30; (As+Sn+Sb) <=0.040; (Cu+Ni) <=0.40; B, Ti; bal Fe				

USA

Specification	Designation	Notes	C	Cr	Mn	Mo	P	S	Si	V	Other	UTS	YS	El	Hard
	AISI D7	Ann Hard	2.15-2.50	11.50-13.50	0.15-0.60	0.70-1.20	0.030 max	0.030 max	0.15-0.60	3.80-4.40	bal Fe				235-262 HB
	UNS T30407		2.15-2.50	11.50-13.50	0.60 max	0.70-1.20	0.030 max	0.030 max	0.60 max	3.80-4.40	bal Fe				
ASTM A681(94)	D7	Ann	2.15-2.50	11.50-13.50	0.10-0.60	0.70-1.20	0.030 max	0.030 max	0.10-0.60	3.80-4.40	S 0.06-0.15 for machining; bal Fe				262 HB
FED QQ-T-570(87)	D7*	Obs; see ASTM A681	2.15-2.5	11.5-13.5	0.6	0.7-1.2			0.6	3.8-4.4	Ni 0.3; bal Fe				
SAE J437(70)	D7	Ann									bal Fe				235-262HB
SAE J438(70)	D7		2.15-2.50	11.50-13.50	0.30-0.50	0.70-1.20			0.30-0.50	3.80-4.40	bal Fe				

Tool Steel, High-Carbon High-Chromium Cold Work, No equivalents identified

Poland

Specification	Designation	Notes	C	Cr	Mn	Mo	P	S	Si	V	Other	UTS	YS	El	Hard
PNH85023	NC10		1.5-1.8	11-13	0.15-0.45		0.030 max	0.030 max	0.15-0.4		Ni <=0.4; bal Fe				

UNS numbers and US grades are provided as a means of cross referencing chemically similar alloys. Exchangability is only possible after independent examination of specifications. Tensile properties are minimum or typical as specified. UTS and YS as MPa. El as %. See Appendix for list of abbreviations used in Notes. * indicates obsolete material.

Specification	Designation	Notes	C	Cr	Mn	Mo	P	S	Si	V	Other	UTS	YS	El	Hard

Tool Steel, Hot Work, CH12

India

Specification	Designation	Notes	C	Cr	Mn	Mo	P	S	Si	V	Other	UTS	YS	El	Hard
IS 1570/6(96)	TAH4		0.3-0.4	4.75-5.5	0.25-0.5	1.2-1.6	0.035 max	0.035 max	0.8-1.2	0.2-0.4	Ni <=0.4; W 1.2-1.6; bal Fe				
IS 1570/6(96)	XT35Cr5MoWiV3		0.3-0.4	4.75-5.5	0.25-0.5	1.2-1.6	0.035 max	0.035 max	0.8-1.2	0.2-0.4	Ni <=0.4; W 1.2-1.6; bal Fe				

USA

Specification	Designation	Notes	C	Cr	Mn	Mo	P	S	Si	V	Other	UTS	YS	El	Hard
	UNS T90812		0.30-0.40	4.75-5.75	0.75 max	1.25-1.75	0.030 max	0.030 max	1.50 max	0.20-0.50	Co 0.20-0.50; W 1.00-1.70; bal Fe				
ASTM A597(93)	CH-12	Cast	0.30-0.40	4.75-5.75	0.75 max	1.25-1.75	0.03 max	0.03 max	1.50 max	0.20-0.50	W 1.00-1.70; bal Fe				

Tool Steel, Hot Work, CH13

India

Specification	Designation	Notes	C	Cr	Mn	Mo	P	S	Si	V	Other	UTS	YS	El	Hard
IS 1570/6(96)	TAH3		0.3-0.4	4.75-5.5	0.25-0.5	1.2-1.6	0.035 max	0.035 max	0.8-1.2	1-1.2	Ni <=0.4; bal Fe				
IS 1570/6(96)	XT35Cr5MoV1		0.3-0.4	4.75-5.5	0.25-0.5	1.2-1.6	0.035 max	0.035 max	0.8-1.2	1-1.2	Ni <=0.4; bal Fe				

USA

Specification	Designation	Notes	C	Cr	Mn	Mo	P	S	Si	V	Other	UTS	YS	El	Hard
	UNS T90813		0.30-0.42	4.75-5.75	0.75 max	1.25-1.75	0.030 max	0.030 max	1.50 max	0.75-1.20	bal Fe				
ASTM A597(93)	CH-13	Cast	0.30-0.42	4.75-5.75	0.75 max	1.25-1.75	0.03 max	0.03 max	1.50 max	0.75-1.20	bal Fe				

Tool Steel, Hot Work, CS5

USA

Specification	Designation	Notes	C	Cr	Mn	Mo	P	S	Si	V	Other	UTS	YS	El	Hard
	UNS T91905		0.50-0.65	0.35 max	0.60-1.00	0.20-0.80	0.030 max	0.030 max	1.75-2.25	0.35 max	bal Fe				
ASTM A597(93)	CS-5	Cast	0.50-0.65	0.35 max	0.60-1.00	0.20-0.80	0.03 max	0.03 max	1.75-2.25	0.35 max	bal Fe				

Tool Steel, Hot Work, H10

Brazil

Specification	Designation	Notes	C	Cr	Mn	Mo	P	S	Si	V	Other	UTS	YS	El	Hard
ABNT	VCM		0.32	2.9	0.3	2.8			0.25	0.5	bal Fe				

China

Specification	Designation	Notes	C	Cr	Mn	Mo	P	S	Si	V	Other	UTS	YS	El	Hard
GB 1299(85)	4Cr3Mo3SiV	Q/T	0.35-0.45	3.00-3.75	0.25-0.70	2.00-3.00	0.030 max	0.030 max	0.80-1.20	0.25-0.75	Cu <=0.30; Ni <=0.25; bal Fe				62 HRC

Czech Republic

Specification	Designation	Notes	C	Cr	Mn	Mo	P	S	Si	V	Other	UTS	YS	El	Hard
CSN 419541	19541		0.27-0.37	2.8-3.4	0.2-0.5	2.7-3.5	0.03 max	0.03 max	0.3-0.7		Ni 0.5-0.8; bal Fe				

France

Specification	Designation	Notes	C	Cr	Mn	Mo	P	S	Si	V	Other	UTS	YS	El	Hard
AFNOR	30DCV28		0.3	2.8	0.3	2.8			0.3	0.5	bal Fe				
AFNOR NFA35590	32DCV28		0.28-0.35	2.6-3.3	0.2-0.5	2.5-3	0.025 max	0.025 max	0.1-0.4	0.4-0.7	Co <=0.3; W <=0.3; bal Fe				
AFNOR NFA35590(92)	32CoV12-28	Soft ann	0.28-0.35	2.6-3.3	0.2-0.5	2.5-3	0.025 max	0.025 max	0.1-0.4	0.4-0.7	Cu <=0.3; Ni <=0.4; W <=0.1; bal Fe				229 HB max
AFNOR NFA35590(92)	32CrMoV12-28	Soft ann	0.28-0.35	2.6-3.3	0.2-0.5	2.5-3	0.025 max	0.025 max	0.1-0.4	0.4-0.7	Cu <=0.3; Ni <=0.4; W <=0.1; bal Fe				229 HB max

Germany

Specification	Designation	Notes	C	Cr	Mn	Mo	P	S	Si	V	Other	UTS	YS	El	Hard
DIN 17350(80)	32CrMoV12-28	Hard tmp to 500C	0.28-0.35	2.70-3.20	0.15-0.45	2.60-3.00	0.030 max	0.030 max	0.10-0.40	0.40-0.70	bal Fe	1670			
DIN 17350(80)	WNr 1.2365	Hard tmp to 500C	0.28-0.35	2.70-3.20	0.15-0.45	2.60-3.00	0.030 max	0.030 max	0.10-0.40	0.40-0.70	bal Fe	1670			

Hungary

Specification	Designation	Notes	C	Cr	Mn	Mo	P	S	Si	V	Other	UTS	YS	El	Hard
MSZ 19731(70)	K14P	Tools in press casting mach; soft ann	0.24-0.32	2.6-3.1	0.2-0.4	2.6-3	0.03 max	0.03 max	0.2-0.4	0.45-0.55	bal Fe				230 HB max
MSZ 19731(70)	K14P	Tools in press casting mach; Quen in oil	0.24-0.32	2.6-3.1	0.2-0.4	2.6-3	0.03 max	0.03 max	0.2-0.4	0.45-0.55	bal Fe				50 HRC
MSZ 4352(84)	K14	Q/T	0.27-0.35	2.7-3.2	0.15-0.4	2.6-3	0.03 max	0.03 max	0.1-0.4	0.4-0.6	Ni <=0.35; W <=0.3; bal Fe				48 HRC
MSZ 4352(84)	K14	Q/T	0.27-0.35	2.7-3.2	0.15-0.4	2.6-3	0.03 max	0.03 max	0.1-0.4	0.4-0.6	Ni <=0.35; W <=0.3; bal Fe				57 HRC

Italy

Specification	Designation	Notes	C	Cr	Mn	Mo	P	S	Si	V	Other	UTS	YS	El	Hard
UNI 2955/4(82)	30CrMoCoV123012KU		0.26-0.34	2.6-3.4	0.3-0.6	2.6-3.4	0.07 max	0.06 max	0.1-0.4	0.7-1.1	Co 2.6-3.4; bal Fe				
UNI 2955/4(82)	30CrMoV1227KU		0.25-0.35	2.5-3.5	0.15-0.45	2.5-3	0.07 max	0.06 max	0.1-0.4	0.4-0.7	bal Fe				

UNS numbers and US grades are provided as a means of cross referencing chemically similar alloys. Exchangability is only possible after independent examination of specifications. Tensile properties are minimum or typical as specified. UTS and YS as MPa. El as %. See Appendix for list of abbreviations used in Notes. * indicates obsolete material.

Specification	Designation	Notes	C	Cr	Mn	Mo	P	S	Si	V	Other	UTS	YS	El	Hard

Tool Steel, Hot Work, H10 (Continued from previous page)

Japan

Specification	Designation	Notes	C	Cr	Mn	Mo	P	S	Si	V	Other	UTS	YS	El	Hard
JIS G4404(83)	SKD7	HR Bar Die Frg Ann	0.28-0.38	2.50-3.50	0.60 max	2.50-3.00	0.030 max	0.030 max	0.50 max	0.40-0.70	Cu <=0.25; Ni <=0.25; bal Fe				229 HB max

Mexico

| NMX-B-082(90) | H10 | Bar | 0.35-0.45 | 3.00-3.75 | 0.25-0.70 | 2.00-3.00 | 0.03 max | 0.03 max | 0.80-1.20 | 0.25-0.75 | Cu <=0.25; Ni <=0.30; (As+Sn+Sb)<=0.040; (Cu+Ni)<=0.40; B, Ti may be added; bal Fe | | | | |

Pan America

| COPANT 337 | H10 | Bar, Frg | 0.35-0.45 | 3-3.75 | 0.25-0.6 | 2-3 | 0.03 | 0.03 | 0.8-1.2 | 0.75 | bal Fe | | | | |

Poland

| PNH85021 | WLV | | 0.25-0.35 | 2.5-3.5 | 0.25-0.5 | 2.5-3 | 0.030 max | 0.030 max | 0.15-0.4 | 0.4-0.6 | Co <=0.3; Ni <=0.35; W <=0.3; bal Fe | | | | |

Romania

| STAS 3611(80) | MoVC30 | | 0.28-0.35 | 2.7-3.2 | 0.2-0.4 | 2.6-3 | 0.03 max | 0.03 max | 0.2-0.4 | 0.4-0.7 | Ni <=0.4; Pb <=0.15; bal Fe | | | | |

Russia

GOST 5950	3Ch2MNF	Alloyed	0.27-0.33	2-2.5	0.3-0.6	0.4-0.6	0.03 max	0.03 max	0.15-0.4	0.25-0.4	Cu <=0.3; Ni 1.2-1.6; W <=0.2; bal Fe				
GOST 5950	3Ch3M3F	Alloyed	0.27-0.34	2.8-3.5	0.2-0.5	2.5-3	0.03 max	0.03 max	0.1-0.4	0.4-0.6	Cu <=0.3; Ni <=0.35; W <=0.2; bal Fe				
GOST 5950	3Ch3M3F	Alloyed	0.27-0.34	2.8-3.5	0.3-0.5	2.5-3	0.03 max	0.03 max	0.2-0.4	0.4-0.6	Cu <=0.3; Ni <=0.35; W <-0.2; bal Fe				

Spain

UNE	F.520.O		0.3-0.35	2.75-3.25	0.15-0.45	2.75-3.25	0.03 max	0.03 max	0.13-0.38	0.45-0.75	bal Fe				
UNE 36072(75)	30CrMoV12	Alloyed	0.25-0.35	2.5-3.5	0.15-0.45	2.5-3	0.03 max	0.03 max	0.1-0.4	0.4-0.7	bal Fe				
UNE 36072(75)	F.5313*	see 30CrMoV12	0.25-0.35	2.5-3.5	0.15-0.45	2.5-3	0.03 max	0.03 max	0.1-0.4	0.4-0.7	bal Fe				

UK

| BS | (USA H10) | | 0.3 | 3 | 0.35 | 3 | | | 0.3 | 0.4 | bal Fe | | | | |
| BS 4659 | BH10 | Bar Rod Sh Strp Frg | 0.3-0.4 | 2.8-3.2 | 0.4 | 2.65-2.95 | | | 1.1 | 0.3-0.5 | bal Fe | | | | |

USA

	AISI H10	Ann Hard	0.35-0.45	3.00-3.75	0.25-0.70	2.00-3.00	0.030 max	0.030 max	0.80-1.20	0.25-0.75	bal Fe				192-229 HB
	UNS T20810		0.35-0.45	3.00-3.75	0.25-0.70	2.00-3.00	0.030 max	0.030 max	0.80-1.20	0.25-0.75	bal Fe				
ASTM A681(94)	H10	Ann	0.35-0.45	3.00-3.75	0.70 max	2.00-3.00	0.030 max	0.030 max	0.80-1.25	0.25-0.75	S 0.06-0.15 if spec for machining; bal Fe				229 HB
FED QQ-T-570(87)	H10*	Obs; see ASTM A681	0.35-0.45	3-3.75	0.25-0.7	2-3			0.8-1.2	0.25-0.75	Ni 0.3; bal Fe				

Tool Steel, Hot Work, H11

Argentina

Specification	Designation	Notes	C	Cr	Mn	Mo	P	S	Si	V	Other	UTS	YS	El	Hard
IAS	IRAM H11	Ann	0.33-0.43	4.75-5.50	0.20-0.50	1.10-1.60	0.030 max	0.030 max	0.80-1.20	0.30-0.60	bal Fe				192-235 HB

Bulgaria

| BDS 7938 | 4Ch5MF | | 0.33-0.43 | 4.50-5.50 | 0.40-0.60 | 1.20-1.50 | 0.030 max | 0.030 max | 0.70-1.10 | 0.30-0.50 | Cu <=0.30; Ni <=0.35; bal Fe | | | | |

Canada

| CSA 419552 | 19552 | | 0.32-0.42 | 4.5-5.5 | 0.2-0.5 | 1.1-1.6 | 0.03 | 0.03 | 0.8-1.2 | 0.35-0.6 | bal Fe | | | | |

China

| GB 1299(85) | 4Cr5MoSiV | Q/T | 0.33-0.43 | 4.75-5.50 | 0.20-0.50 | 1.10-1.60 | 0.030 max | 0.030 max | 0.80-1.20 | 0.30-0.60 | Cu <=0.30; Ni <=0.25; bal Fe | | | | 60 HRC |
| YB/T 094(97) | SM4Cr5MoSiV | Flat Ann | 0.33-0.43 | 4.75-5.50 | 0.20-0.60 | 1.10-1.60 | 0.030 max | 0.030 max | 0.80-1.25 | 0.30 | bal Fe | | | | 235 HB max |

Czech Republic

| CSN 419552 | 19552 | | 0.32-0.42 | 4.5-5.5 | 0.2-0.5 | 1.1-1.6 | 0.03 max | 0.03 max | 0.8-1.2 | 0.35-0.6 | bal Fe | | | | |

Specification	Designation	Notes	C	Cr	Mn	Mo	P	S	Si	V	Other	UTS	YS	El	Hard

Tool Steel, Hot Work, H11 (Continued from previous page)

Finland

Specification	Designation	Notes	C	Cr	Mn	Mo	P	S	Si	V	Other	UTS	YS	El	Hard
SFS 913(73)	SFS913	Alloyed	0.32-0.42	4.5-5.5	0.3-0.5	0.8-1.4	0.03 max	0.02 max	0.9-1.2	0.3-0.6	bal Fe				

France

Specification	Designation	Notes	C	Cr	Mn	Mo	P	S	Si	V	Other	UTS	YS	El	Hard
AFNOR NFA35590	FZ38CDV5		0.34-0.42	4.8-5.5	0.2-0.5	1.2-1.5	0.025 max	0.025 max	0.8-1.2	0.3-0.5	bal Fe				
AFNOR NFA35590	Z38CDV5	Bar Frg	0.38	5	0.3	1.25	0.03	0.03	1	0.5	bal Fe				
AFNOR NFA35590(92)	40CDV13	Soft ann	0.36-0.43	2.9-3.5	0.4-0.7	0.5-0.8	0.025 max	0.025 max	0.1-0.4	0.05-0.15	Ni <=0.4; bal Fe				229 HB max
AFNOR NFA35590(92)	40CrMoV13	Soft ann	0.36-0.43	2.9-3.5	0.4-0.7	0.5-0.8	0.025 max	0.025 max	0.1-0.4	0.05-0.15	Ni <=0.4; bal Fe				229 HB max
AIR 9172	40CDV20		0.38-0.43	4.75-5.25	0.2-0.4	1.2-1.4	0.02	0.01	0.8-1	0.4-0.6	bal Fe				

Germany

Specification	Designation	Notes	C	Cr	Mn	Mo	P	S	Si	V	Other	UTS	YS	El	Hard
DIN	GX38CrMoV5-3	Hard tmp to 600C	0.35-0.40	4.70-5.20	0.30-0.60	2.70-3.30	0.035 max	0.035 max	0.30-0.50	0.40-0.70	bal Fe	1860			
DIN	WNr 1.2367	Hard tmp to 600C	0.35-0.40	4.70-5.20	0.30-0.60	2.70-3.30	0.035 max	0.035 max	0.30-0.50	0.40-0.70	bal Fe	1860			
DIN	WNr 1.7783		0.38-0.43	4.75-5.25	0.20-0.40	1.20-1.40	0.015 max	0.010 max	0.80-1.00	0.40-0.60	bal Fe				
DIN	X41CrMoV5-1		0.38-0.43	4.75-5.25	0.20-0.40	1.20-1.40	0.015 max	0.010 max	0.80-1.00	0.40-0.60	bal Fe				
DIN 17350(80)	GX38CrMoV5-1	Hard tmp to 500C	0.36-0.42	4.80-5.50	0.30-0.50	1.10-1.40	0.030 max	0.030 max	0.90-1.20	0.25-0.50	bal Fe	2060			
DIN 17350(80)	WNr 1.2343	Hard tmp to 500C	0.36-0.42	4.80-5.50	0.30-0.50	1.10-1.40	0.030 max	0.030 max	0.90-1.20	0.25-0.50	bal Fe	2060			
DIN(Aviation Hdbk)	WNr 1.7784		0.38-0.43	4.75-5.25	0.20-0.40	1.20-1.40	0.015 max	0.010 max	0.80-1.00	0.40-0.60	bal Fe				

Hungary

Specification	Designation	Notes	C	Cr	Mn	Mo	P	S	Si	V	Other	UTS	YS	El	Hard
MSZ 19731(70)	K12P	Tools in press casting mach; soft ann	0.35-0.45	5-5.5	0.4-0.6	1.2-1.5	0.03 max	0.03 max	0.8-1.1	0.35-0.45	bal Fe				235 HB max
MSZ 19731(70)	K12P	Tools in press casting mach; Quen in oil	0.35-0.45	5-5.5	0.4-0.6	1.2-1.5	0.03 max	0.03 max	0.8-1.1	0.35-0.45	bal Fe				52 HRC
MSZ 4352(72)	K12		0.35-0.45	4.5-5.5	0.4-0.6	1.2-1.5	0.03 max	0.03 max	0.8-1.1	0.3-0.5	bal Fe				

India

Specification	Designation	Notes	C	Cr	Mn	Mo	P	S	Si	V	Other	UTS	YS	El	Hard
IS 3748	T35Cr5Mo1V30	Bar	0.3-0.4	4.75-5.25	0.25-0.5	1.2-1.6			0.8-1.2	0.2-0.4	bal Fe				

Italy

Specification	Designation	Notes	C	Cr	Mn	Mo	P	S	Si	V	Other	UTS	YS	El	Hard
UNI	X35CrMo05KU		0.3-0.38	4.5-5.5	0.6	1-1.5			0.7-1.2	0.3-0.5	W 1-1.6; bal Fe				
UNI 2955/4(82)	X37CrMoV51KU		0.32-0.42	4.5-5.5	0.25-0.55	1.2-1.7	0.07 max	0.06 max	0.9-1.2	0.3-0.5	bal Fe				

Japan

Specification	Designation	Notes	C	Cr	Mn	Mo	P	S	Si	V	Other	UTS	YS	El	Hard
JIS G4404(83)	SKD6	HR Bar Frg Ann	0.32-0.42	4.50-5.50	0.50 max	1.00-1.50	0.030 max	0.030 max	0.80-1.20	0.30-0.50	Cu <=0.25; Ni <=0.25; bal Fe				229 HB max

Mexico

Specification	Designation	Notes	C	Cr	Mn	Mo	P	S	Si	V	Other	UTS	YS	El	Hard
NMX-B-082(90)	H11	Bar	0.33-0.43	4.75-5.50	0.20-0.50	1.10-1.60	0.03 max	0.03 max	0.80-1.20	0.30-0.60	Cu <=0.25; Ni <=0.30; (As+Sn+Sb)<=0.040; (Cu+Ni)<=0.40; B, Ti may be added; bal Fe				

Pan America

Specification	Designation	Notes	C	Cr	Mn	Mo	P	S	Si	V	Other	UTS	YS	El	Hard
COPANT 337	H11	Bar, Frg	0.33-0.43	4.75-5.5	0.2-0.5	1.1-1.75	0.03	0.03	0.8-1.2	0.3-0.6	bal Fe				

Poland

Specification	Designation	Notes	C	Cr	Mn	Mo	P	S	Si	V	Other	UTS	YS	El	Hard
PNH85021	WCL		0.32-0.42	4.5-5.5	0.2-0.5	1.2-1.5	0.030 max	0.030 max	0.8-1.2	0.3-0.5	Ni <=0.35; W <=0.3; bal Fe				
PNH85021	WCO		0.33-0.43	4.5-5.5	0.4-0.6	1.2-1.5			0.7-1.2	0.3-0.5	Ni 0.5; bal Fe				

Romania

Specification	Designation	Notes	C	Cr	Mn	Mo	P	S	Si	V	Other	UTS	YS	El	Hard
STAS 3611(80)	MoVC50.10		0.36-0.42	4.8-5.8	0.3-0.5	0.8-1.4	0.03 max	0.03 max	0.9-1.2	0.25-0.5	Ni <=0.4; Pb <=0.15; bal Fe				

Russia

Specification	Designation	Notes	C	Cr	Mn	Mo	P	S	Si	V	Other	UTS	YS	El	Hard
GOST 5950	4Ch5MFS	Alloyed	0.32-0.4	4.5-5.5	0.2-0.5	1.2-1.5	0.03 max	0.03 max	0.9-1.2	0.3-0.5	Cu <=0.3; Ni <=0.35; W <=0.2; bal Fe				

UNS numbers and US grades are provided as a means of cross referencing chemically similar alloys. Exchangability is only possible after independent examination of specifications. Tensile properties are minimum or typical as specified. UTS and YS as MPa. El as %. See Appendix for list of abbreviations used in Notes. * indicates obsolete material.

Specification	Designation	Notes	C	Cr	Mn	Mo	P	S	Si	V	Other	UTS	YS	El	Hard

Tool Steel, Hot Work, H11 (Continued from previous page)

Russia

Specification	Designation	Notes	C	Cr	Mn	Mo	P	S	Si	V	Other	UTS	YS	El	Hard
GOST 5950	4Ch5MFS	Alloyed	0.32-0.4	4.5-5.5	0.15-0.4	1.2-1.5	0.03 max	0.03 max	0.8-1.2	0.3-0.5	Cu <=0.3; Ni <=0.35; W <=0.2; bal Fe				

Spain

Specification	Designation	Notes	C	Cr	Mn	Mo	P	S	Si	V	Other	UTS	YS	El	Hard
UNE	F.520.G		0.32-0.38	4.5-5.5	0.2-0.5	1.15-1.65	0.03 max	0.03 max	0.75-1.25	0.4-0.6	bal Fe				
UNE 36072/2(75)	F.5317*	see X37CrMoV5	0.32-0.42	4.5-5.5	0.25-0.55	1.2-1.7	0.03 max	0.03 max	0.9-1.2	0.3-0.5	bal Fe				
UNE 36072/2(75)	X37CrMoV5		0.32-0.42	4.5-5.5	0.25-0.55	1.2-1.7	0.03 max	0.03 max	0.9-1.2	0.3-0.5	bal Fe				

UK

Specification	Designation	Notes	C	Cr	Mn	Mo	P	S	Si	V	Other	UTS	YS	El	Hard
BS 4659	BH11	Strp Bar Rod Frg	0.32-0.42	4.75-5.25	0.4	1.25-1.75			0.85-1.15	0.3-0.5	bal Fe				

USA

Specification	Designation	Notes	C	Cr	Mn	Mo	P	S	Si	V	Other	UTS	YS	El	Hard
	AISI H11	Ann Hard	0.33-0.43	4.75-5.50	0.20-0.50	1.10-1.60	0.030 max	0.030 max	0.80-1.20	0.30-0.60	bal Fe				192-235 HB
	UNS T20811		0.33-0.43	4.75-5.50	0.20-0.50	1.10-1.60	0.030 max	0.030 max	0.80-1.20	0.30-0.60	bal Fe				
AMS 6437		Sh, Strp, PH	0.38-0.43	4.75-5.25	0.2-0.4	1.2-1.4			0.8-1	0.4-0.60	Cu 0.35; Ni 0.25; bal Fe				
AMS 6485		Bar, Frg	0.38-0.43	4.75-5.25	0.2-0.4	1.2-1.4			0.8-1	0.4-0.60	Cu 0.35; Ni 0.25; bal Fe				
AMS 6487		Bar, Frg	0.38-0.43	4.75-5.25	0.2-0.4	1.2-1.4			0.8-1	0.4-0.60	Cu 0.35; Ni 0.25; bal Fe				
AMS 6488		Bar, Frg	0.38-0.43	4.75-5.25	0.2-0.4	1.2-1.4			0.8-1	0.4-0.60	Cu 0.35; Ni 0.25; bal Fe				
ASTM A681(94)	H11	Ann	0.33-0.43	4.75-5.50	0.20-0.60	1.10-1.60	0.030 max	0.030 max	0.80-1.25	0.30-0.60	S 0.06-0.15 if spec for machining; bal Fe				235 HB
FED QQ-T-570(87)	H11*	Obs; see ASTM A681	0.33-0.43	4.75-5.5	0.2-0.5	1.1-1.6			0.8-1.2	0.3-0.6	Ni 0.3; bal Fe				
SAE J437(70)	H11	Ann									bal Fe				192-229 HB
SAE J438(70)	H11		0.30-0.40	4.75-5.50	0.20-0.40	1.25-1.75			0.80-1.20	0.30-0.50	bal Fe				
SAE J467(68)	H11		0.35	5.10	0.30	1.50			1.00	0.40	bal Fe	1806	1482	10	52 HRC

Tool Steel, Hot Work, H12

Canada

Specification	Designation	Notes	C	Cr	Mn	Mo	P	S	Si	V	Other	UTS	YS	El	Hard
CSA 419554	19554		0.34-0.44	4.8-5.8	0.2-0.5	1.1-1.6	0.03 max	0.03 max	0.8-1.2	0.8-1.2	bal Fe				

France

Specification	Designation	Notes	C	Cr	Mn	Mo	P	S	Si	V	Other	UTS	YS	El	Hard
AFNOR NFA35590	Z35CWDV5		0.32-0.4	4.8-5.5	0.2-0.5	1.2-1.5	0.025 max	0.025 max	0.8-1.2	0.3-0.5	W 1.1-1.6; bal Fe				

Germany

Specification	Designation	Notes	C	Cr	Mn	Mo	P	S	Si	V	Other	UTS	YS	El	Hard
DIN	GX37CrMoW51	Hard tmp to 500C	0.32-0.40	5.00-5.60	0.30-0.60	1.30-1.60	0.035 max	0.035 max	0.90-1.20	0.15-0.40	W 1.20-1.40; bal Fe	2160			
DIN	WNr 1.2606	Hard tmp to 500C	0.32-0.40	5.00-5.60	0.30-0.60	1.30-1.60	0.035 max	0.035 max	0.90-1.20	0.15-0.40	W 1.20-1.40; bal Fe	2160			

India

Specification	Designation	Notes	C	Cr	Mn	Mo	P	S	Si	V	Other	UTS	YS	El	Hard
IS 1570	T35Cr5MoW1V30		0.3-0.4	4.75-5.25	0.25-0.5	1.2-1.6			0.8-1.2	0.2-0.4	W 1.2-1.6; bal Fe				
IS 1570/6(96)	TAH2		0.3-0.4	4.75-5.5	0.25-0.5	1.2-1.6	0.035 max	0.035 max	0.8-1.2	0.2-0.4	Ni <=0.4; bal Fe				
IS 1570/6(96)	XT35Cr5Mo1V3		0.3-0.4	4.75-5.5	0.25-0.5	1.2-1.6	0.035 max	0.035 max	0.8-1.2	0.2-0.4	Ni <=0.4; bal Fe				
IS 3748	T35Cr5MoW1V30	Bar	0.3-0.4	4.75-5.25	0.25-0.5	1.2-1.6			0.8-1.2	0.2-0.4	W 1.2-1.6; bal Fe				

Italy

Specification	Designation	Notes	C	Cr	Mn	Mo	P	S	Si	V	Other	UTS	YS	El	Hard
UNI 2955	VAL105		0.52	5.15	0.4	1.4			1		W 1.2; bal Fe				
UNI 2955	X35CrMoW05KU		0.35	5.25	0.5	1.4			1	0.35	W 1.3; bal Fe				

Japan

Specification	Designation	Notes	C	Cr	Mn	Mo	P	S	Si	V	Other	UTS	YS	El	Hard
JIS G4404(83)	SKD62	HR Bar Frg Ann	0.32-0.42	4.5-5.5	0.5	1-1.5	0.03	0.03	0.8-1.2	0.2-0.6	Cu 0.25; Ni 0.25; W 1-1.5; bal Fe				229 HB max

Specification	Designation	Notes	C	Cr	Mn	Mo	P	S	Si	V	Other	UTS	YS	El	Hard

Tool Steel, Hot Work, H12 (Continued from previous page)

Mexico

Specification	Designation	Notes	C	Cr	Mn	Mo	P	S	Si	V	Other	UTS	YS	El	Hard
NMX-B-082(90)	H12	Bar	0.30-0.40	4.75-5.50	0.20-0.50	1.25-1.75	0.03 max	0.03 max	0.80-1.20	0.50 max	Cu <=0.25; Ni <=0.30; W 1.00-1.70; (As+Sn+Sb)<=0.040; (Cu+Ni)<=0.40; B, Ti may be added; bal Fe				

Pan America

| COPANT 337 | H12 | Bar, Frg | 0.3-0.4 | 4.75-5.5 | 0.2-0.5 | 1.25-1.75 | 0.03 | 0.03 | 0.2 | 0.5 | W 1-1.7; bal Fe | | | | |
| COPANT 337 | H12 | | 0.3-0.4 | 4.75-5.5 | 0.2-0.5 | 1.25-1.75 | | | 0.8-1.2 | 0.5 | W 1-1.7; bal Fe | | | | |

Romania

| STAS 3611(80) | MoWC53 | | 0.32-0.4 | 5-5.6 | 0.3-0.6 | 1.3-1.6 | 0.035 max | 0.035 max | 0.9-1.2 | 0.15-0.4 | Ni <=0.4; Pb <=0.15; W 1.2-1.4; bal Fe | | | | |

Russia

GOST 5950	4Ch4WMFS	Alloyed	0.37-0.44	3.2-4	0.2-0.5	1.2-1.5	0.03 max	0.03 max	0.6-1	0.6-0.9	Cu <=0.3; Ni <=0.6; W 0.8-1.2; bal Fe				
GOST 5950	4Ch4WMFS	Alloyed	0.37-0.44	3.2-4	0.2-0.5	1.2-1.5	0.03 max	0.03 max	0.6-1	0.6-0.9	Cu <=0.3; Ni <=0.6; W 0.8-1.2; bal Fe				
GOST 5950	4Ch5W2FS	Alloyed	0.35-0.45	4.5-5.5	0.15-0.4	0.3 max	0.03 max	0.03 max	0.8-1.2	0.6-0.9	Cu <=0.3; Ni <=0.35; W 1.6-2.2; bal Fe				
GOST 5950	4ChMFS	Alloyed	0.37-0.45	1.5-1.8	0.5-0.8	0.9-1.2	0.03 max	0.03 max	0.5-0.8	0.3-0.5	Cu <=0.3; Ni <=0.35; W <=0.2; bal Fe				
GOST 5950	4ChMNFS	Alloyed	0.35-0.42	1.25-1.55	0.15-0.4	0.65-0.85	0.03 max	0.03 max	0.7-1	0.35-0.5	; B 0.002-0.004; Cu <=0.3; Ni 1.2-1.6; W <=0.2; Zr 0.03-0.09; bal Fe				
GOST 5950	6Ch6W3MFS	Alloyed	0.5-0.6	5.5-6.5	0.15-0.4	0.6-0.9	0.03 max	0.03 max	0.6-0.9	0.5-0.8	Cu <=0.3; Ni <=0.35; W 2.5-3.2; bal Fe				
GOST 5950	6Ch6W3MFS	Alloyed	0.5-0.6	5.5-6.5	0.15-0.4	0.6-0.9	0.03 max	0.03 max	0.6-0.9	0.5-0.8	Cu <=0.3; Ni 0.35; W 2.5-3.5; bal Fe				

Spain

| UNE | F.537 | | 0.32-0.38 | 4.5-5.5 | 0.2-0.5 | 1.25-1.75 | 0.03 max | 0.03 max | 0.75-1.25 | 0.3-0.6 | W 1.25-1.75; bal Fe | | | | |

UK

| BS 4659 | (USA H12) | | 0.35 | 5 | 0.35 | 1.3 | | | 1 | 0.2 | W 1.2; bal Fe | | | | |
| BS 4659 | BH12 | Bar Rod Sh Strp Frg | 0.3-0.4 | 4.75-5.25 | 0.4 max | 1.25-1.75 | | | 0.85-1.15 | 0.5 max | W 1.25-1.75; bal Fe | | | | |

USA

	AISI H12	Ann Hard	0.30-0.40	4.75-5.50	0.20-0.50	1.25-1.75	0.030 max	0.030 max	0.80-1.20	0.50 max	W 1.00-1.70; bal Fe				192-235 HB
	UNS T20812		0.30-0.40	4.75-5.50	0.20-0.50	1.25-1.75	0.030 max	0.030 max	0.80-1.20	0.50 max	W 1.00-1.70; bal Fe				
ASTM A681(94)	H12	Ann	0.30-0.40	4.75-5.50	0.20-0.60	1.25-1.75	0.030 max	0.030 max	0.80-1.25	0.20-0.50	W 1.00-1.70; S 0.06-0.15 if spec for machining; bal Fe				235 HB
FED QQ-T-570(87)	H12*	Obs; see ASTM A681	0.3-0.4	4.75-5.5	0.2-0.5	1.25-1.75			0.8-1.2	0.5	Ni 0.3; W 1-1.7; bal Fe				
SAE J437(70)	H12	Ann									bal Fe				192-229 HB
SAE J438(70)	H12		0.30-0.40	4.75-5.50	0.20-0.40	1.25-1.75			0.80-1.20	0.10-0.50	W 1.00-1.70; bal Fe				
SAE J467(68)	H12		0.35	5.10	0.35	1.35			1.05	0.30	W 1.25; bal Fe	1413	1276	12	44 HRC

Tool Steel, Hot Work, H13

Argentina

Specification	Designation	Notes	C	Cr	Mn	Mo	P	S	Si	V	Other	UTS	YS	El	Hard
IAS	IRAM H13	Ann	0.32-0.45	4.75-5.50		1.10-1.75	0.030 max	0.030 max	0.80-1.20	0.80-1.20	bal Fe				210 HB

China

| GB 1299(85) | 4Cr5MoSiV1 | Q/T | 0.32-0.42 | 4.75-5.50 | 0.20-0.50 | 1.10-1.75 | 0.030 max | 0.030 max | 0.80-1.20 | 0.80-1.20 | Cu <=0.30; Ni <=0.25; bal Fe | | | | 60 HRC |
| YB/T 094(97) | SM4Cr5MoSiV1 | Flat Ann | 0.32-0.45 | 4.75-5.50 | 0.20-0.60 | 1.10-1.75 | 0.030 max | 0.030 max | 0.80-1.25 | 0.80-1.20 | bal Fe | | | | 235 HB max |

Czech Republic

| CSN 419554 | 19554 | | 0.34-0.44 | 4.8-5.8 | 0.2-0.5 | 1.1-1.6 | 0.03 max | 0.03 max | 0.8-1.2 | 0.8-1.2 | bal Fe | | | | |

UNS numbers and US grades are provided as a means of cross referencing chemically similar alloys. Exchangability is only possible after independent examination of specifications. Tensile properties are minimum or typical as specified. UTS and YS as MPa. El as %. See Appendix for list of abbreviations used in Notes. * indicates obsolete material.

Specification	Designation	Notes	C	Cr	Mn	Mo	P	S	Si	V	Other	UTS	YS	El	Hard

Tool Steel, Hot Work, H13 (Continued from previous page)

France

Specification	Designation	Notes	C	Cr	Mn	Mo	P	S	Si	V	Other	UTS	YS	El	Hard
AFNOR NFA35590	Z40CDV5		0.36-0.44	4.8-5.5	0.2-0.5	1.2-1.5	0.025 max	0.025 max	0.8-1.2	0.85-1.15	bal Fe				
Germany															
DIN 17350(80)	GX40CrMoV5-1	Hard tmp to 500C	0.37-0.43	4.80-5.50	0.30-0.50	1.20-1.50	0.030 max	0.030 max	0.90-1.20	0.90-1.10	bal Fe	2060			
DIN 17350(80)	WNr 1.2344	Hard tmp to 500C	0.37-0.43	4.80-5.50	0.30-0.50	1.20-1.50	0.030 max	0.030 max	0.90-1.20	0.90-1.10	bal Fe	2060			
Hungary															
MSZ 19731(70)	K13P	Tools in press casting mach; 0-999mm; air-hardened; QA	0.35-0.45	5-5.5	0.4-0.6	1.2-1.5	0.03 max	0.03 max	0.8-1.1	0.85-1.15	bal Fe				48 HRC
MSZ 19731(70)	K13P	Tools in press casting mach; Quen in oil	0.35-0.45	5-5.5	0.4-0.6	1.2-1.5	0.03 max	0.03 max	0.8-1.1	0.85-1.15	bal Fe				52 HRC
MSZ 4352(84)	K13	Q/T	0.35-0.45	4.5-5.5	0.3-0.5	1.2-1.5	0.03 max	0.03 max	0.9-1.2	0.8-1.1	Ni <=0.35; W <=0.3; bal Fe				50 HRC
MSZ 4352(84)	K13	Soft ann	0.35-0.45	4.5-5.5	0.3-0.5	1.2-1.5	0.03 max	0.03 max	0.9-1.2	0.8-1.1	Ni <=0.35; W <=0.3; bal Fe				235 HB max
MSZ 4352(84)	K13K	Soft ann	0.35-0.45	4.5-5.5	0.3-0.5	1.2-1.5	0.03 max	0.03 max	0.9-1.2	0.8-1.1	Ni <=0.35; W <=0.3; bal Fe				235 HB max
MSZ 4352(84)	K13K	Q/T	0.35-0.45	4.5-5.5	0.3-0.5	1.2-1.5	0.03 max	0.03 max	0.9-1.2	0.8-1.1	Ni <=0.35; W <=0.3; bal Fe				50 HRC
India															
IS 1570	T35Cr5MoV1		0.3-0.4	4.75-5.25	0.25-0.5	1.2-1.6			0.8-1.2	1-1.2	bal Fe				
IS 3748	T35Cr5MoV1	Bar	0.3-0.4	4.75-5.25	0.25-0.5	1.2-1.6			0.8-1.2	1-1.2	bal Fe				
IS 4367	T35Cr5MoV1	Frg	0.3-0.4	4.75-5.25	0.25-0.5	1.2-1.6			0.8-1.2	1-1.2	bal Fe				
Italy															
UNI	X35CrMoV05KU		0.3-0.38	4.5-5.5	0.6	1-1.5			0.7-1.2	0.8-1.2	bal Fe				
UNI 2955	VAL102		0.4	5	0.4	1.35			1	1	bal Fe				
UNI 2955/4(82)	X40CrMoV511KU		0.35-0.45	4.5-5.5	0.25-0.55	1.2-1.7	0.07 max	0.06 max	0.9-1.2	0.85-1.15	bal Fe				
Japan															
JIS G4404(83)	SKD61	HR Bar Frg Ann	0.32-0.42	4.50-5.50	0.50 max	1.00-1.50	0.030 max	0.030 max	0.80-1.20	0.80-1.20	Cu <=0.25; Ni <=0.25; bal Fe				229 HB max
Mexico															
NMX-B-082(90)	H13	Bar	0.32-0.45	4.75-5.50	0.20-0.50	1.10-1.75	0.03 max	0.03 max	0.80-1.20	0.80-1.20	Cu <=0.25; Ni <=0.30; (As+Sn+Sb)<=0.040; (Cu+Ni)<=0.40; B, Ti may be added; bal Fe				
Pan America															
COPANT 337	H13	Bar, Frg	0.35-0.45	4.75-5.5	0.2-0.5	1.1-1.75			0.9-1.1	0.8-1.2	bal Fe				
Poland															
PNH83161(90)	L40H5MF		0.35-0.45	4.8-5.2	0.3-0.6	1.4-1.6	0.030 max	0.030 max	0.8-1	0.5-1	Ni <=0.4; bal Fe				
PNH83166	L40H5MF		0.35-0.45	4.8-5.2	0.3-0.6	1.4-1.6	0.030 max	0.030 max	0.8-1	0.5-1	Co <=0.3; bal Fe				
PNH85021	WCLV		0.35-0.45	4.5-5.5	0.2-0.5	1.2-1.5	0.035 max	0.030 max	0.8-1.2	0.8-1.1	Co <=0.3; Ni <=0.35; W <=0.3; 3M 1.2-1.5; bal Fe				
Romania															
STAS 3611(80)	MoVC50.13		0.37-0.42	5-5.5	0.3-0.5	1.2-1.5	0.03 max	0.03 max	0.9-1.2	0.9-1.1	Ni <=0.4; Pb <=0.15; bal Fe				
Russia															
GOST 5950	4Ch5MF1S	Alloyed	0.37-0.44	4.5-5.5	0.15-0.4	1.2-1.5	0.03 max	0.03 max	0.8-1.2	0.8-1.1	Cu <=0.3; Ni <=0.35; W <=0.3; bal Fe				
GOST 5950	4Ch5MF1S	Alloyed	0.37-0.44	4.5-5.5	0.2-0.5	1.2-1.5	0.03 max	0.03 max	0.9-1.2	0.8-1.1	Cu <=0.3; Ni <=0.35; W <=0.2; bal Fe				
Spain															
UNE 36072/2(75)	F.5318*	see X40CrMoV5	0.35-0.45	4.5-5.5	0.25-0.55	1.2-1.7	0.03 max	0.03 max	0.9-1.2	0.85-1.15	bal Fe				

UNS numbers and US grades are provided as a means of cross referencing chemically similar alloys. Exchangability is only possible after independent examination of specifications. Tensile properties are minimum or typical as specified. UTS and YS as MPa. El as %. See Appendix for list of abbreviations used in Notes. * indicates obsolete material.

Specification	Designation	Notes	C	Cr	Mn	Mo	P	S	Si	V	Other	UTS	YS	El	Hard

Tool Steel, Hot Work, H13 (Continued from previous page)

Spain

| UNE 36072/2(75) | X40CrMoV5 | | 0.35-0.45 | 4.5-5.5 | 0.25-0.55 | 1.2-1.7 | 0.03 max | 0.03 max | 0.9-1.2 | 0.85-1.15 | bal Fe | | | | |

Sweden

| SS 142242 | 2242 | Alloyed | 0.35-0.42 | 5-5.5 | 0.3-0.6 | 1.2-1.6 | 0.03 max | 0.02 max | 0.8-1.2 | 0.85-1.15 | bal Fe | | | | |

UK

BS 4659	(USA H13)		0.35	5	0.5	1.3			1	0.85	bal Fe				
BS 4659	BH13	Strp Bar Rod Frg	0.32-0.42	4.75-5.25	0.4 max	1.25-1.75			0.85-1.15	0.9-1.1	bal Fe				
BS 4659	H13		0.35	5	0.35	1.3			1	0.85	bal Fe				

USA

	AISI H13	Ann Hard	0.32-0.45	4.75-5.50	0.20-0.50	1.10-1.75	0.030 max	0.030 max	0.80-1.20	0.80-1.20	bal Fe				192-229 HB
	UNS T20813		0.32-0.45	4.75-5.50	0.20-0.50	1.10-1.75	0.030 max	0.030 max	0.80-1.20	0.80-1.20	bal Fe				
AMS 6408			0.32-0.45	4.75-5.50	0.20-0.50	1.10-1.75	0.030 max	0.030 max	0.80-1.20	0.80-1.20	bal Fe				
ASTM A681(94)	H13	Ann	0.32-0.45	4.75-5.50	0.20-0.60	1.10-1.75	0.030 max	0.030 max	0.80-1.25	0.80-1.20	S 0.06-0.15 if spec for machining; Mn<=1.0 if resulf; bal Fe				235 HB
FED QQ-T-570(87)	H13*	Obs; see ASTM A681	0.32-0.45	4.75-5.5	0.2-0.5	1.1-1.75			0.8-1.2	0.8-1.2	Ni 0.3; bal Fe				
SAE J437(70)	H13	Ann									bal Fe				192-229 HB
SAE J438(70)	H13		0.30-0.40	4.75-5.50	0.20-0.40	1.25-1.75			0.80-1.20	0.80-1.20	bal Fe				
SAE J467(68)	H13		0.35	5.10	0.30	1.50			1.00	1.00	bal Fe	1482	1269	13	45 HRC

Tool Steel, Hot Work, H14

France

AFNOR NFA35590	Z40WCV5	Bar Frg	0.4	4	0.3	0.5	0.03	0.03	0.3	0.5	W 5; bal Fe				
AFNOR NFA35590(92)	40NCD16	Soft ann	0.35-0.43	1.6-2	0.3-0.6	0.3-0.5	0.025 max	0.025 max	0.1-0.4		Ni 3.7-4.2; bal Fe				227 HB max
AFNOR NFA35590(92)	40NCDV16	Soft ann	0.35-0.45	1.7-2	0.35-0.65	0.4-0.6	0.025 max	0.025 max	0.1-0.4	0.05-0.25	Ni 3.6-4.1; bal Fe				227 HB max
AFNOR NFA35590(92)	40NiCrMo16	Soft ann	0.35-0.43	1.6-2	0.3-0.6	0.3-0.5	0.025 max	0.025 max	0.1-0.4		Ni 3.7-4.2; bal Fe				227 HB max
AFNOR NFA35590(92)	40NiCrMoV16	Soft ann	0.35-0.45	1.7-2	0.35-0.65	0.4-0.6	0.025 max	0.025 max	0.1-0.4	0.05-0.25	Ni 3.6-4.1; bal Fe				227 HB max

Germany

| DIN | 30WCrV17-2 | Hard tmp to 500C | 0.25-0.35 | 2.20-2.50 | 0.20-0.40 | | 0.035 max | 0.035 max | 0.15-0.30 | 0.50-0.70 | W 4.0-4.50; bal Fe | 1690 | | | |
| DIN | WNr 1.2567 | Hard tmp to 500C | 0.25-0.35 | 2.20-2.50 | 0.20-0.40 | | 0.035 max | 0.035 max | 0.15-0.30 | 0.50-0.70 | W 4.00-4.50; bal Fe | 1690 | | | |

Hungary

| MSZ 4352 | W3 | | 0.25-0.35 | 2-2.7 | 0.3-0.5 | 0.2 max | 0.03 max | 0.03 max | 0.1-0.4 | 0.2-0.4 | Ni <=0.35; W 4-5; bal Fe | | | | |

Italy

| UNI 2955/4(82) | X30WCrV53KU | | 0.25-0.35 | 2-3 | 0.15-0.45 | | 0.07 max | 0.06 max | 0.1-0.4 | 0.4-0.7 | W 4.5-5.1; bal Fe | | | | |

Japan

| JIS G4404(83) | SKD4 | HR Bar Frg Ann | 0.25-0.35 | 2.00-3.00 | 0.60 max | | 0.030 max | 0.030 max | 0.40 max | 0.30-0.50 | Cu <=0.25; Ni <=0.25; W 5.00-6.00; bal Fe | | | | 235 HB max |

Mexico

| NMX-B-082(90) | H14 | Bar | 0.35-0.45 | 4.75-5.50 | 0.20-0.50 | | 0.03 max | 0.03 max | 0.80-1.20 | | Cu <=0.25; Ni <=0.30; W 4.00-5.25; (As+Sn+Sb)<=0.040; (Cu+Ni)<=0.40; B, Ti may be added; bal Fe | | | | |

Russia

| GOST 5950 | 4Ch2W5MF | Alloyed | 0.3-0.4 | 2.2-3 | 0.15-0.4 | 0.6-0.9 | 0.03 max | 0.03 max | 0.15-0.35 | 0.6-0.9 | Cu <=0.3; Ni <=0.35; W 4.5-5.5; bal Fe | | | | |

UNS numbers and US grades are provided as a means of cross referencing chemically similar alloys. Exchangability is only possible after independent examination of specifications. Tensile properties are minimum or typical as specified. UTS and YS as MPa. El as %. See Appendix for list of abbreviations used in Notes. * indicates obsolete material.

Specification	Designation	Notes	C	Cr	Mn	Mo	P	S	Si	V	Other	UTS	YS	El	Hard

Tool Steel, Hot Work, H14 (Continued from previous page)

USA

Specification	Designation	Notes	C	Cr	Mn	Mo	P	S	Si	V	Other	UTS	YS	El	Hard
	AISI H14	Ann Hard	0.35-0.45	4.75-5.50	0.20-0.50		0.030 max	0.030 max	0.80-1.20		W 4.00-5.25; bal Fe				207-235 HB
	UNS T20814		0.35-0.45	4.75-5.50	0.20-0.50		0.030 max	0.030 max	0.80-1.20		W 4.00-5.25; bal Fe				
ASTM A681(94)	H14	Ann	0.35-0.45	4.75-5.50	0.20-0.60		0.030 max	0.030 max	0.80-1.25		W 4.00-5.25; S 0.06-0.15 if spec for machining; bal Fe				235 HB
FED QQ-T-570(87)	H14*	Obs; see ASTM A681	0.35-0.45	4.75-5.5	0.2-0.5				0.8-1.2		Ni 0.3; W 4-5.25; bal Fe				

Tool Steel, Hot Work, H19

Germany

Specification	Designation	Notes	C	Cr	Mn	Mo	P	S	Si	V	Other	UTS	YS	El	Hard
DIN	WNr 1.2678	Hard tmp to 500C	0.40-0.50	4.00-5.00	0.30-0.50	0.40-0.60	0.025 max	0.025 max	0.30-0.50	1.80-2.10	Co 4.00-5.00; W 4.00-5.00; bal Fe	1910			
DIN	X45CoCrWV5-5-5	Hard tmp to 500C	0.40-0.50	4.00-5.00	0.30-0.50	0.40-0.60	0.025 max	0.025 max	0.30-0.50	1.80-2.10	Co 4.00-5.00; W 4.00-5.00; bal Fe	1910			

Japan

Specification	Designation	Notes	C	Cr	Mn	Mo	P	S	Si	V	Other	UTS	YS	El	Hard
JIS G4404(83)	SKD8	Bar Frg HR Ann Die	0.35-0.45	4.00-4.70	0.60 max	0.30-0.50	0.030 max	0.030 max	0.50 max	1.70-2.20	Co 3.80-4.50; Cu <=0.25; Ni <=0.25; W 3.80-4.50; bal Fe				241 HB max

Mexico

Specification	Designation	Notes	C	Cr	Mn	Mo	P	S	Si	V	Other	UTS	YS	El	Hard
NMX-B-082(90)	H19	Bar	0.32-0.45	4.00-4.75	0.20-0.50	0.30-0.55	0.03 max	0.03 max	0.20-0.50	1.75-2.20	Co 4.00-4.50; Cu <=0.25; Ni <=0.30; W 3.75-4.50; (As+Sn+Sb)<=0.040; (Cu+Ni)<=0.40; B, Ti may be added; bal Fe				

UK

Specification	Designation	Notes	C	Cr	Mn	Mo	P	S	Si	V	Other	UTS	YS	El	Hard
BS 4659	BH19 *	Bar Rod Sh Strp Frg	0.35-0.45	4-4.5	0.4	0.45			0.4	2-2.4	Co 4-4.5; W 4-4.5; bal Fe				

USA

Specification	Designation	Notes	C	Cr	Mn	Mo	P	S	Si	V	Other	UTS	YS	El	Hard
	AISI H19	Ann Hard	0.32-0.45	4.00-4.75	0.20-0.50	0.30-0.55	0.030 max	0.030 max	0.20-0.50	1.75-2.20	Co 4.00-4.50; W 3.75-4.50; bal Fe				207-241 HB
	UNS T20819		0.32-0.45	4.00-4.75	0.20-0.50	0.30-0.55	0.030 max	0.030 max	0.20-0.50	1.75-2.20	Co 4.00-4.50; W 3.75-4.50; bal Fe				
ASTM A681(94)	H19	Ann	0.32-0.45	4.00-4.75	0.20-0.50	0.30-0.55	0.030 max	0.030 max	0.15-0.50	1.75-2.20	Co 4.00-4.50; W 3.75-4.50; S 0.06-0.15 if spec for machining; bal Fe				241 HB
FED QQ-T-570(87)	H19*	Obs; see ASTM A681	0.32-0.45	4-4.75	0.2-0.5	0.3-0.55			0.2-0.5	1.75-2.2	Co 4-4.5; Ni 0.3; W 3.75-4.5; bal Fe				

Tool Steel, Hot Work, H21

Argentina

Specification	Designation	Notes	C	Cr	Mn	Mo	P	S	Si	V	Other	UTS	YS	El	Hard
IAS	IRAM H21	Ann	0.26-0.36	3.00-3.75	0.15-0.40		0.030 max	0.030 max	0.15-0.50	0.30-0.60	W 8.50-10.00; bal Fe				207-235 HB

Bulgaria

Specification	Designation	Notes	C	Cr	Mn	Mo	P	S	Si	V	Other	UTS	YS	El	Hard
BDS 7938	3Ch2B8F		0.30-0.40	2.20-2.40	0.15-0.40		0.030 max	0.030 max	0.15-0.40	0.20-0.50	Cu <=0.30; Ni <=0.35; W 7.50-9.00; bal Fe				

China

Specification	Designation	Notes	C	Cr	Mn	Mo	P	S	Si	V	Other	UTS	YS	El	Hard
GB 1299(85)	3Cr2W8V	As Hard	0.30-0.40	2.20-2.70	0.40 max		0.030 max	0.030 max	0.40 max	0.20-0.50	Cu <=0.30; Ni <=0.25; W 7.50-9.00; bal Fe				60 HRC

Czech Republic

Specification	Designation	Notes	C	Cr	Mn	Mo	P	S	Si	V	Other	UTS	YS	El	Hard
CSN 419721	19721		0.25-0.35	2.1-2.6	0.2-0.5		0.03 max	0.03 max	0.15-0.45	0.15-0.3	W 8.5-10.0; bal Fe				

France

Specification	Designation	Notes	C	Cr	Mn	Mo	P	S	Si	V	Other	UTS	YS	El	Hard
AFNOR NFA35590	Z30WCV9		0.25-0.32	2.5-3.5	0.15-0.45		0.025 max	0.025 max	0.1-0.4	0.3-0.5	W 8.5-9.5; bal Fe				

Germany

Specification	Designation	Notes	C	Cr	Mn	Mo	P	S	Si	V	Other	UTS	YS	El	Hard
DIN	WNr 1.2581	Hard tmp to 500C	0.25-0.35	2.50-2.80	0.20-0.40		0.035 max	0.035 max	0.15-0.30	0.30-0.40	W 8.00-9.00; bal Fe	1720			
DIN	X30WCrV9-3	Hard tmp to 500C	0.25-0.35	2.50-2.80	0.20-0.40		0.035 max	0.035 max	0.15-0.30	0.30-0.40	W 8.00-9.00; bal Fe	1720			

Specification	Designation	Notes	C	Cr	Mn	Mo	P	S	Si	V	Other	UTS	YS	El	Hard

Tool Steel, Hot Work, H21 (Continued from previous page)

Specification	Designation	Notes	C	Cr	Mn	Mo	P	S	Si	V	Other	UTS	YS	El	Hard
India															
IS 3748	T33W9Cr3V38	Bar	0.25-0.4	2.8-3.3	0.2-0.4		0.04	0.04	0.1-0.35	0.25-0.5	W 8-10; bal Fe				
IS 4367	T33W9Cr3V38	Frg	0.25-0.4	2.8-3.3	0.2-0.4				0.1-0.35	0.25-0.5	W 8-10; bal Fe				
Italy															
UNI	X28W09AU		0.23-0.33	2.2-3	0.5				0.4	0.2-0.4	Co 2; W 8-9.5; bal Fe				
UNI 2955/4(82)	X30WCrV93KU		0.25-0.35	2.2-3.5	0.15-0.45		0.07 max	0.06 max	0.1-0.4	0.3-0.5	W 8.1-9.5; bal Fe				
Japan															
JIS G4404(83)	SKD5	HR Bar Frg Ann	0.25-0.35	2.00-3.00	0.60 max		0.030 max	0.030 max	0.40 max	0.30-0.50	Cu <=0.25; Ni <=0.25; W 9.00-10.00; bal Fe				235 HB max
Mexico															
NMX-B-082(90)	H21	Bar	0.26-0.36	3.00-3.75	0.15-0.40		0.03 max	0.03 max	0.15-0.50	0.30-0.60	Cu <=0.25; Ni <=0.30; W 8.50-10.00; (As+Sn+Sb)<=0.040; (Cu+Ni)<=0.40; B, Ti may be added; bal Fe				
Pan America															
COPANT 337	H21	Bar, Frg	0.26-0.36	2.4-3.75	0.15-0.4	0.6	0.03	0.03	0.2-0.5	0.2-0.6	W 8.75-10; bal Fe				
Poland															
PNH85021	WWN1		0.25-0.35	2.5-3	0.25-0.5	0.02 max	0.030 max	0.030 max	0.15-0.4	0.2-0.5	Co <=0.3; Ni 1.2-1.6; W 8-10; bal Fe				
PNH85021	WWV		0.25-0.35	2.5-3	0.2-0.5	0.2 max	0.030 max	0.030 max	0.15-0.4	0.3-0.5	Co <=0.3; Ni <=1; W 8-10; bal Fe				
Romania															
STAS 3611(80)	VCW85		0.25-0.35	2.2-2.7	0.2-0.5		0.025 max	0.025 max	0.15-0.4	0.2-0.5	Ni <=0.35; Pb <=0.15; W 7.5-9; bal Fe				
Russia															
GOST 5950	3Ch2W8F	Alloyed	0.3-0.4	2.2-2.7	0.15-0.4	0.5 max	0.03 max	0.03 max	0.15-0.4	0.2-0.5	Cu <=0.3; Ni <=0.35; W 7.5-8.5; bal Fe				
GOST 5950	3Ch2W8F	Alloyed	0.3-0.4	2.2-2.7	0.15-0.4	0.5 max	0.03 max	0.03 max	0.15-0.4	0.2-0.5	Cu <=0.3; Ni <=0.35; W 7.5-8.5; bal Fe				
GOST 5950	4Ch2W2MFS	Alloyed	0.42-0.5	2-3.5	0.3-0.6	0.8-1.1	0.03 max	0.03 max	0.3-0.6	0.6-0.9	Cu <=0.3; Ni <=0.035; W 1.8-2.4; bal Fe				
GOST 5950	4Ch2W5MF	Alloyed	0.3-0.4	2-2.3	0.15-0.4	0.6-0.9	0.03 max	0.03 max	0.15-0.35	0.6-0.9	Cu <=0.3; Ni <=0.35; W 4.5-5.5; bal Fe				
Spain															
UNE	F.526		0.3-0.4	2.5-3	0.2-0.4				0.15-0.3	0.2-0.5	W 8.5-9.5; bal Fe				
UNE 36072/2(75)	F.5323*	see X30WCrV9	0.25-0.35	2.5-3.5	0.15-0.45	0.15 max	0.03 max	0.03 max	0.1-0.4	0.3-0.5	W 8.5-9.5; bal Fe				
UNE 36072/2(75)	X30WCrV9		0.25-0.35	2.5-3.5	0.15-0.45	0.15 max	0.03 max	0.03 max	0.1-0.4	0.3-0.5	W 8.5-9.5; bal Fe				
Sweden															
SS 142730	2730		0.25-0.35	2.7-3.3	0.2-0.4				0.2-0.4	0.25-0.35	W 8.5-10.5; bal Fe				
UK															
BS 4659	BH21	Bar Rod Sh Strp Frg	0.25-0.35	2.25-3.25	0.4 max	0.6 max			0.4 max	0.4 max	W 9.5-10; bal Fe				
BS 4659	H21A		0.2-0.3	2.25-3.25	0.4	0.6			0.4	0.5	Ni 2-2.5; W 8.5-10; bal Fe				
USA															
	AISI H21	Ann Hard	0.26-0.36	3.00-3.75	0.15-0.40		0.030 max	0.030 max	0.15-0.50	0.30-0.60	W 8.50-10.00; bal Fe				207-235 HB
	UNS T20821		0.26-0.36	3.00-3.75	0.15-0.40		0.030 max	0.030 max	0.15-0.50	0.30-0.60	W 8.50-10.00; bal Fe				
ASTM A681(94)	H21	Ann	0.26-0.36	3.00-3.75	0.15-0.40		0.030 max	0.030 max	0.15-0.50	0.30-0.60	W 8.50-10.00; S 0.06-0.15 if spec for machining; bal Fe				235 HB

UNS numbers and US grades are provided as a means of cross referencing chemically similar alloys. Exchangability is only possible after independent examination of specifications. Tensile properties are minimum or typical as specified. UTS and YS as MPa. El as %. See Appendix for list of abbreviations used in Notes. * indicates obsolete material.

Specification	Designation	Notes	C	Cr	Mn	Mo	P	S	Si	V	Other	UTS	YS	El	Hard

Tool Steel, Hot Work, H21 (Continued from previous page)

USA

Specification	Designation	Notes	C	Cr	Mn	Mo	P	S	Si	V	Other	UTS	YS	El	Hard
FED QQ-T-570(87)	H21*	Obs; see ASTM A681	0.26-0.36	3-3.75	0.15-0.4				0.15-0.5	0.3-0.6	Ni 0.3; W 8.5-10; bal Fe				
SAE J437(70)	H21	Ann									bal Fe				202-235 HB
SAE J438(70)	H21		0.30-0.40	3.00-3.75	0.20-0.40	1.25-1.75			0.15-0.30	0.30-0.50	W 8.75-10.00; bal Fe				

Tool Steel, Hot Work, H22

Canada

Specification	Designation	Notes	C	Cr	Mn	Mo	P	S	Si	V	Other	UTS	YS	El	Hard
CSA 419721	19721		0.25-0.35	2.1-2.6	0.2-0.5		0.03 max	0.03 max	0.15-0.45	0.15-0.3	W 8.5-10; bal Fe				

Hungary

Specification	Designation	Notes	C	Cr	Mn	Mo	P	S	Si	V	Other	UTS	YS	El	Hard
MSZ 4352	W2		0.25-0.35	2-2.7					0.2-0.4	2-2.4	W 8.5-10; bal Fe				

India

Specification	Designation	Notes	C	Cr	Mn	Mo	P	S	Si	V	Other	UTS	YS	El	Hard
IS 1570/6(96)	TAH1		0.25-0.4	2.8-3.3	0.2-0.4	0.15 max	0.035 max	0.035 max	0.1-0.35	0.25-0.5	Ni <=0.4; W 8-10; bal Fe				
IS 1570/6(96)	XT33W9Cr3V4		0.25-0.4	2.8-3.3	0.2-0.4	0.15 max	0.035 max	0.035 max	0.1-0.35	0.25-0.5	Ni <=0.4; W 8-10; bal Fe				

Japan

Specification	Designation	Notes	C	Cr	Mn	Mo	P	S	Si	V	Other	UTS	YS	El	Hard
JIS G4404(83)	SKD5	HR Bar Frg Ann	0.25-0.35	2-3	0.6				0.4	0.3-0.5	W 9-10; bal Fe				235 HB max

Mexico

Specification	Designation	Notes	C	Cr	Mn	Mo	P	S	Si	V	Other	UTS	YS	El	Hard
NMX-B-082(90)	H22	Bar	0.30-0.40	1.75-3.75	0.15-0.40		0.03 max	0.03 max	0.15-0.40	0.25-0.50	Cu <=0.25; Ni <=0.30; W 10.00-11.75; (As+Sn+Sb)<=0.040; (Cu+Ni)<=0.40; B, Ti may be added; bal Fe				

Spain

Specification	Designation	Notes	C	Cr	Mn	Mo	P	S	Si	V	Other	UTS	YS	El	Hard
UNE	F.526		0.3-0.4	2.5-3	0.2-0.4				0.15-0.3	0.2-0.5	W 8.5-9.5; bal Fe				

USA

Specification	Designation	Notes	C	Cr	Mn	Mo	P	S	Si	V	Other	UTS	YS	El	Hard
	AISI H22	Ann Hard	0.30-0.40	1.75-3.75	0.15-0.40		0.030 max	0.030 max	0.15-0.40	0.25-0.50	W 10.00-11.75; bal Fe				207-235 HB
	UNS T20822		0.30-0.40	1.75-3.75	0.15-0.40		0.030 max	0.030 max	0.15-0.40	0.25-0.50	W 10.00-11.75; bal Fe				
ASTM A681(94)	H22	Ann	0.30-0.40	1.75-3.75	0.15-0.40		0.030 max	0.030 max	0.15-0.40	0.25-0.50	W 10.00-11.75; S 0.06-0.15 if spec for machining; bal Fe				235 HB
FED QQ-T-570(87)	H22*	Obs; see ASTM A681	0.3-0.4	1.75-3.75	0.15-0.4				0.15-0.4	0.25-0.5	Ni 0.3; W 10-11.75; bal Fe				

Tool Steel, Hot Work, H23

Germany

Specification	Designation	Notes	C	Cr	Mn	Mo	P	S	Si	V	Other	UTS	YS	El	Hard
DIN	WNr 1.2625		0.30-0.35	11.5-12.5	0.20-0.40	0.40-0.60	0.025 max	0.025 max	0.15-0.30	1.00-1.10	W 11.5-12.5; bal Fe				
DIN	X33WCrVMo1212		0.30-0.35	11.5-12.5	0.20-0.40	0.40-0.60	0.025 max	0.025 max	0.15-0.30	1.00-1.10	W 11.5-12.5; bal Fe				

Mexico

Specification	Designation	Notes	C	Cr	Mn	Mo	P	S	Si	V	Other	UTS	YS	El	Hard
NMX-B-082(90)	H23	Bar	0.25-0.35	11.00-12.75	0.15-0.40		0.03 max	0.03 max	0.15-0.60	0.75-1.25	Cu <=0.25; Ni <=0.30; W 11.00-12.75; (As+Sn+Sb)<=0.040; (Cu+Ni)<=0.40; B, Ti may be added; bal Fe				

USA

Specification	Designation	Notes	C	Cr	Mn	Mo	P	S	Si	V	Other	UTS	YS	El	Hard
	AISI H23	Ann Hard	0.25-0.35	11.00-12.75	0.15-0.40		0.030 max	0.030 max	0.15-0.60	0.75-1.25	W 11.00-12.75; bal Fe				212-255 HB
	UNS T20823		0.25-0.35	11.00-12.75	0.15-0.40		0.030 max	0.030 max	0.15-0.60	0.75-1.25	W 11.00-12.75; bal Fe				
ASTM A681(94)	H23	Ann	0.25-0.35	11.00-12.75	0.15-0.40		0.030 max	0.030 max	0.15-0.60	0.75-1.25	W 11.00-12.75; S 0.06-0.15 if spec for machining; bal Fe				255 HB
FED QQ-T-570(87)	H23*	Obs; see ASTM A681	0.25-0.35	11-12.75	0.15-0.4				0.15-0.6	0.75-1.25	Ni 0.3; W 11-12.75; bal Fe				

Specification	Designation	Notes	C	Cr	Mn	Mo	P	S	Si	V	Other	UTS	YS	El	Hard
Tool Steel, Hot Work, H24															
India															
IS 1570	T5514Cr3V45		0.5-0.6	2.8-3.3	0.2-0.4				0.1-0.35	0.3-0.6	W 13-15; bal Fe				
IS 1570	T55W14Cr3V45		0.5-0.6	2.8-3.3	0.2-0.4				0.1-0.35	0.3-0.6	W 13-15; bal Fe				
Mexico															
NMX-B-082(90)	H24	Bar	0.42-0.53	2.50-3.50	0.15-0.40		0.03 max	0.03 max	0.15-0.40	0.40-0.60	Cu <=0.25; Ni <=0.30; W 14.00-16.00; (As+Sn+Sb)<=0.040; (Cu+Ni)<=0.40; B, Ti may be added; bal Fe				
Pan America															
COPANT 337	H24	Bar, Frg	0.42-0.55	2.5-3.5	0.15-0.4	1	0.03	0.03	0.15-0.4	0.4-0.6	W 14-16; bal Fe				
USA															
	AISI H24	Ann Hard	0.42-0.53	2.50-3.50	0.15-0.40		0.030 max	0.030 max	0.15-0.40	0.40-0.60	W 14.00-16.00; bal Fe				217-241 HB
	UNS T20824		0.42-0.53	2.50-3.50	0.15-0.40		0.030 max	0.03 max	0.15-0.40	0.40-0.60	W 14.00-16.00; bal Fe				
ASTM A681(94)	H24	Ann	0.42-0.53	2.50-3.50	0.15-0.40		0.030 max	0.030 max	0.15-0.40	0.40-0.60	W 14.00-16.00; S 0.06-0.15 if spec for machining; bal Fe				241 HB
FED QQ-T-570(87)	H24*	Obs; see ASTM A681	0.42-0.53	2.5-3.5	0.15-0.4				0.15-0.4	0.4-0.6	Ni 0.3; W 14-16; bal Fe				
Tool Steel, Hot Work, H25															
Mexico															
NMX-B-082(90)	H25	Bar	0.22-0.32	3.75-4.50	0.15-0.40		0.03 max	0.03 max	0.15-0.40	0.40-0.60	Cu <=0.25; Ni <=0.30; W 14.00-16.00; (As+Sn+Sb)<=0.040; (Cu+Ni)<=0.40; B, Ti may be added; bal Fe				
USA															
	UNS T20825*	Obs; H-25	0.22-0.32	3.75-4.50	0.15-0.40		0.030 max	0.030 max	0.15-0.40	0.40-0.60	W 14.00-16.00; bal Fe				
ASTM A681(94)	H25	Ann	0.22-0.32	3.75-4.50	0.15-0.40		0.030 max	0.030 max	0.15-0.40	0.40-0.60	W 14.00-16.00; S 0.06-0.15 if spec for machining; bal Fe				235 HB
Tool Steel, Hot Work, H26															
India															
IS 1570/6(96)	TAH5		0.5-0.6	2.8-3.3	0.2-0.4	0.15 max	0.035 max	0.035 max	0.1-0.35	0.3-0.6	Ni <=0.4; W 13-15; bal Fe				
IS 1570/6(96)	XT55W14Cr3V4		0.5-0.6	2.8-3.3	0.2-0.4	0.15 max	0.035 max	0.035 max	0.1-0.35	0.3-0.6	Ni <=0.4; W 13-15; bal Fe				
Mexico															
NMX-B-082(90)	H26	Bar	0.45-0.55	3.75-4.50	0.15-0.40		0.03 max	0.03 max	0.15-0.40	0.75-1.25	Cu <=0.25; Ni <=0.30; W 17.25-19.00; (As+Sn+Sb)<=0.040; (Cu+Ni)<=0.40; B, Ti may be added; bal Fe				
Pan America															
COPANT 337	H26	Bar, Frg	0.45-0.55	3.75-4.5	0.15-0.4	1	0.03	0.03	0.2-0.5	0.75-1.25	W 17.25-19; bal Fe				
UK															
BS 4659	BH26		0.5-0.6	3.75-4.5	0.4	0.6			0.4	1-1.5	Co 0.6; W 17-18; bal Fe				
USA															
	AISI H26	Ann Hard	0.45-0.55	3.75-4.50	0.15-0.40		0.030 max	0.030 max	0.15-0.40	0.75-1.25	W 17.25-19.00; bal Fe				217-241 HB
	UNS T20826		0.45-0.55	3.75-4.50	0.15-0.40		0.030 max	0.030 max	0.15-0.40	0.75-1.25	W 17.25-19.00; bal Fe				
ASTM A681(94)	H26	Ann	0.45-0.55	3.75-4.50	0.15-0.40		0.030 max	0.030 max	0.15-0.40	0.75-1.25	W 17.25-19.00; S 0.06-0.15 if spec for machining; bal Fe				241 HB

UNS numbers and US grades are provided as a means of cross referencing chemically similar alloys. Exchangability is only possible after independent examination of specifications. Tensile properties are minimum or typical as specified. UTS and YS as MPa. El as %. See Appendix for list of abbreviations used in Notes. * indicates obsolete material.

Specification	Designation	Notes	C	Cr	Mn	Mo	P	S	Si	V	Other	UTS	YS	El	Hard

Tool Steel, Hot Work, H26 (Continued from previous page)

USA

Specification	Designation	Notes	C	Cr	Mn	Mo	P	S	Si	V	Other	UTS	YS	El	Hard
FED QQ-T-570(87)	H26*	Obs; see ASTM A681	0.45-0.55	3.75-4.5	0.15-0.4				0.15-0.4	0.75-1.25	Ni 0.3; W 17.25-19; bal Fe				

Tool Steel, Hot Work, H40

Spain

Specification	Designation	Notes	C	Cr	Mn	Mo	P	S	Si	V	Other	UTS	YS	El	Hard
UNE 36011(75)	C35k	Unalloyed; Q/T	0.3-0.4		0.5-0.8	0.15 max	0.035 max	0.035 max	0.15-0.4		bal Fe				

Tool Steel, Hot Work, H41

France

Specification	Designation	Notes	C	Cr	Mn	Mo	P	S	Si	V	Other	UTS	YS	El	Hard
AFNOR NFA35590(78)	HS2-8-1		0.8-0.88	3.5-4.5	0.4 max	8-9	0.03 max	0.03 max	0.5 max	1-1.5	Co <=1; Cu <=0.3; Ni <=0.4; W 1.4-2; bal Fe				
AFNOR NFA35590(78)	Z85DCWV08-04-02-01		0.8-0.88	3.5-4.5	0.4 max	8-9	0.03 max	0.03 max	0.5 max	1-1.5	Co <=1; Cu <=0.3; Ni <=0.4; W 1.4-2; bal Fe				

Mexico

Specification	Designation	Notes	C	Cr	Mn	Mo	P	S	Si	V	Other	UTS	YS	El	Hard
NMX-B-082(90)	H41	Bar	0.60-0.75	3.50-4.00	0.15-0.40	8.20-9.20	0.03 max	0.03 max	0.20-0.45	1.00-1.30	Cu <=0.25; Ni <=0.30; W 1.40-2.10; (As+Sn+Sb)<=0.040; (Cu+Ni)<=0.40; B, Ti may be added; bal Fe				

USA

Specification	Designation	Notes	C	Cr	Mn	Mo	P	S	Si	V	Other	UTS	YS	El	Hard
	UNS T20841*	Obs	0.60-0.75	3.50-4.00	0.15-0.40	8.20-9.20	0.030 max	0.030 max	0.20-0.45	1.00-1.30	W 1.40-2.10; bal Fe				
ASTM A681(94)	H41	Ann	0.60-0.75	3.50-4.00	0.15-0.40	8.20-9.20	0.030 max	0.030 max	0.20-0.45	1.00-1.30	W 1.40-2.10; S 0.06-0.15 if spec for machining; bal Fe				235 HB

Tool Steel, Hot Work, H42

China

Specification	Designation	Notes	C	Cr	Mn	Mo	P	S	Si	V	Other	UTS	YS	El	Hard
GB 1299(85)	6W6Mo5Cr4V	As Hard	0.55-0.65	3.70-4.30	0.60 max	4.50-5.50	0.030 max	0.030 max	0.40 max	0.70-1.10	Cu <=0.30; Ni <=0.25; W 6.00-7.00; bal Fe				60 HRC

France

Specification	Designation	Notes	C	Cr	Mn	Mo	P	S	Si	V	Other	UTS	YS	El	Hard
AFNOR	Z65WDCV6.05		0.65	4	0.3	5			0.3	2	W 6; bal Fe				

Mexico

Specification	Designation	Notes	C	Cr	Mn	Mo	P	S	Si	V	Other	UTS	YS	El	Hard
NMX-B-082(90)	H42	Bar	0.55-0.70	3.75-4.50	0.15-0.40	4.50-5.50	0.03 max	0.03 max	0.20-0.45	1.75-2.20	Cu <=0.25; Ni <=0.30; W 5.50-6.75; (As+Sn+Sb)<=0.040; (Cu+Ni)<=0.40; B, Ti may be added; bal Fe				

USA

Specification	Designation	Notes	C	Cr	Mn	Mo	P	S	Si	V	Other	UTS	YS	El	Hard
	AISI H42	Ann Hard	0.55-0.70	3.75-4.50	0.15-0.40	4.50-5.50	0.030 max	0.030 max	0.20-0.45	1.75-2.20	W 5.50-6.75; bal Fe				207-235 HB
	UNS T20842		0.55-0.70	3.75-4.50	0.15-0.40	4.50-5.50	0.030 max	0.030 max	0.20-0.45	1.75-2.20	W 5.50-6.75; bal Fe				
ASTM A681(94)	H42	Ann	0.55-0.70	3.75-4.50	0.15-0.40	4.50-5.50	0.030 max	0.030 max	0.20-0.45	1.75-2.20	W 5.50-6.75; S 0.06-0.15 if spec for machining; bal Fe				235 HB
FED QQ-T-570(87)	H42*	Obs; see ASTM A681	0.55-0.7	3.75-4.5	0.15-0.4	4.5-5.5			0.2-0.45	1.75-2.2	Ni 0.3; W 5.5-6.75; bal Fe				

Tool Steel, Hot Work, H43

Mexico

Specification	Designation	Notes	C	Cr	Mn	Mo	P	S	Si	V	Other	UTS	YS	El	Hard
NMX-B-082(90)	H43	Bar	0.50-0.65	3.75-4.50	0.15-0.40	7.75-8.50	0.03 max	0.03 max	0.20-0.45	1.80-2.20	Cu <=0.25; Ni <=0.30; (As+Sn+Sb)<=0.040; (Cu+Ni)<=0.40; B, Ti may be added; bal Fe				

USA

Specification	Designation	Notes	C	Cr	Mn	Mo	P	S	Si	V	Other	UTS	YS	El	Hard
	UNS T20843*	Obs	0.50-0.65	3.75-4.50	0.15-0.40	7.75-8.50	0.030 max	0.030 max	0.20-0.45	1.80-2.20	bal Fe				
ASTM A681(94)	H43	Ann	0.50-0.65	3.75-4.50	0.15-0.40	7.75-8.50	0.030 max	0.030 max	0.20-0.45	1.80-2.20	S 0.06-0.15 if spec for machining; bal Fe				235 HB

UNS numbers and US grades are provided as a means of cross referencing chemically similar alloys. Exchangability is only possible after independent examination of specifications. Tensile properties are minimum or typical as specified. UTS and YS as MPa. El as %. See Appendix for list of abbreviations used in Notes. * indicates obsolete material.

Specification	Designation	Notes	C	Cr	Mn	Mo	P	S	Si	V	Other	UTS	YS	El	Hard
Tool Steel, Hot Work, No equivalents identified															
France															
AFNOR NFA35590(92)	40CMD8		0.35-0.45	1.75-2.15	1.35-1.65	0.15-0.3	0.025 max	0.025 max	0.15-0.4		Ni <=0.4; bal Fe				300 HB max
AFNOR NFA35590(92)	40CrMnMo8		0.35-0.45	1.75-2.15	1.35-1.65	0.15-0.3	0.025 max	0.025 max	0.15-0.4		Ni <=0.4; bal Fe				300 HB max
Hungary															
MSZ 4352(84)	NK2	Soft ann	0.5-0.6	1-1.2	0.65-0.95	0.45-0.55	0.03 max	0.03 max	0.1-0.4	0.07-0.12	Ni 1.5-1.8; W <=0.3; bal Fe				248 HB max
MSZ 4352(84)	NK2	Q/T	0.5-0.6	1-1.2	0.65-0.95	0.45-0.55	0.03 max	0.03 max	0.1-0.4	0.07-0.12	Ni 1.5-1.8; W <=0.3; bal Fe				44 HRC
India															
IS 1570/6(96)	T50NiCrMo4		0.45-0.55	0.8-1	0.5-0.8	0.3-0.4	0.035 max	0.035 max	0.1-0.35		Co <=0.1; Cu <=0.3; Ni 0.8-1; Pb <=0.15; W <=0.1; bal Fe				
IS 1570/6(96)	T55Ni6Cr3		0.5-0.6	0.5-0.8	0.5-0.8	0.15 max	0.035 max	0.035 max	0.1-0.35		Co <=0.1; Cu <=0.3; Ni 1.25-1.65; Pb <=0.15; W <=0.1; bal Fe				
IS 1570/6(96)	T55Ni6Cr3Mo3		0.5-0.6	0.5-0.8	0.5-0.8	0.25-0.35	0.035 max	0.035 max	0.1-0.35		Co <=0.1; Cu <=0.3; Ni 1.25-1.75; Pb <=0.15; W <=0.1; bal Fe				
IS 1570/6(96)	T55Ni7Cr3Mo3V1		0.5-0.6	0.6-0.8	0.65-0.95	0.25-0.35	0.035 max	0.035 max	0.1-0.35	0.07-0.12	Ni 1.5-1.8; bal Fe				
IS 1570/6(96)	T55Ni7Cr4Mo5V1		0.5-0.6	1-1.2	0.65-0.95	0.45-0.55	0.035 max	0.035 max	0.1-0.35	0.07-0.12	Ni 1.5-1.8; bal Fe				
IS 1570/6(96)	TAH6		0.5-0.6	0.6-0.8	0.65-0.95	0.25-0.35	0.035 max	0.035 max	0.1-0.35	0.07-0.12	Ni 1.5-1.8; bal Fe				
IS 1570/6(96)	TAH7		0.5-0.6	1-1.2	0.65-0.95	0.45-0.55	0.035 max	0.035 max	0.1-0.35	0.07-0.12	Ni 1.5-1.8; bal Fe				
Italy															
UNI 2955/4(82)	40NiCrMoV16KU		0.35-0.45	1.6-2	0.35-0.75	0.4-0.6	0.07 max	0.06 max	0.1-0.4	0.05-0.25	Ni 3.4-4.1; bal Fe				
UNI 2955/4(82)	56NiCrMoV7KU		0.51-0.61	1-1.2	0.65-0.95	0.45-0.55	0.07 max	0.06 max	0.1-0.4	0.07-0.12	Ni 1.5-1.8; bal Fe				
Poland															
PNH85021	WLK		0.3-0.4	2.5-3	0.25-0.5	2.5-3	0.030 max	0.030 max	0.3-0.6	0.4-0.6	Co 2.8-3.3; Ni <=0.35; W <=0.3; bal Fe				
PNH85021	WWS1		0.25-0.35	2.2-2.7	0.25-0.5	0.2 max	0.030 max	0.030 max	0.8-1.2	0.4-0.6	Co <=0.3; Ni <=0.5; W 4-5; bal Fe				

Tool Steel, Low-Alloy Special Purpose, L2

Specification	Designation	Notes	C	Cr	Mn	Ni	P	S	Si	V	Other	UTS	YS	El	Hard
Canada															
CSA 419423	19423		0.85-1	0.6-0.9	0.15-0.4	0.4 max	0.03 max	0.04 max	0.15-0.35	0.07-0.17	bal Fe				
China															
GB 1299	4SiCrV*	Obs	0.4-0.5	1.3-1.6	0.4	0.3			1.2-1.6	0.1-0.25	Cu 0.25; bal Fe				
GB 1299(85)	5CrMnMo	As Hard	0.50-0.60	0.60-0.90	1.20-1.60	0.25 max	0.030 max	0.030 max	0.25-0.60		Cu <=0.30; Mo 0.15-0.30; bal Fe				60 HRC
Czech Republic															
CSN 419423	19423		0.85-1	0.6-0.9	0.15-0.4		0.03 max	0.035 max	0.15-0.35	0.07-0.17	bal Fe				
France															
AFNOR	Y50CV4		0.5 max	1 max	0.8 max		0.03 max	0.03 max	0.3 max	0.15 max	bal Fe				
AFNOR NFA35590(92)	55CNDV4	Q/T	0.5-0.6	0.85-1.15	0.6-1	0.45-0.75	0.025 max	0.025 max	0.1-0.4	0.05-0.15	Cu <=0.3; Mo 0.3-0.5; W <=0.1; bal Fe				43 HRC
AFNOR NFA35590(92)	55CrNiMoV4	Q/T	0.5-0.6	0.85-1.15	0.6-1	0.45-0.75	0.025 max	0.025 max	0.1-0.4	0.05-0.15	Cu <=0.3; Mo 0.3-0.5; W <=0.1; bal Fe				43 HRC
Germany															
DIN	105Cr5		1.00-1.10	1.20-1.50	0.20 max		0.030 max	0.030 max	0.20-0.40		bal Fe				
DIN	105MnCr4	Hard tmp to 200C	1.00-1.10	0.70-1.00	1.00-1.20		0.035 max	0.035 max	0.15-0.30		bal Fe				62 HRC
DIN	120Cr5		1.10-1.25	1.20-1.50	0.20-0.40		0.030 max	0.030 max	0.20-0.40		bal Fe				
DIN	125CrSi5	Tmp to 200C	1.20-1.30	1.10-1.30	0.60-0.80		0.035 max	0.035 max	1.05-1.25		bal Fe				62 HRC
DIN	43MnSiMo4		0.36-0.46		0.85-1.10		0.035 max	0.035 max	0.80-1.00		Mo 0.10-0.25; bal Fe				
DIN	59CrV4	Hard tmp to 200C	0.55-0.62	0.90-1.20	0.80-1.10		0.035 max	0.035 max	0.15-0.35	0.07-0.12	bal Fe				59 HRC
DIN	61CrSiV5	Hard tmp to 200C	0.57-0.65	1.00-1.30	0.60-0.90		0.035 max	0.035 max	0.70-1.00	0.07-0.12	bal Fe				61 HRC
DIN	65MnCr4		0.60-0.68	0.60-0.80	1.00-1.20		0.035 max	0.035 max	0.30-0.50		bal Fe				
DIN	70Si7	Hard tmp to 200C	0.65-0.75		0.60-0.80		0.030 max	0.030 max	1.50-1.80		bal Fe				56 HRC
DIN	WNr 1.2059		1.10-1.25	1.20-1.50	0.20-0.40		0.030 max	0.030 max	0.20-0.40		bal Fe				
DIN	WNr 1.2060		1.00-1.10	1.20-1.50	0.20 max		0.030 max	0.030 max	0.20-0.40		bal Fe				
DIN	WNr 1.2109	Tmp to 200C	1.20-1.30	1.10-1.30	0.60-0.80		0.035 max	0.035 max	1.05-1.25		bal Fe				62 HRC
DIN	WNr 1.2125		0.60-0.68	0.60-0.80	1.00-1.20		0.035 max	0.035 max	0.30-0.50		bal Fe				
DIN	WNr 1.2127	Hard tmp to 200C	1.00-1.10	0.70-1.00	1.00-1.20		0.035 max	0.035 max	0.15-0.30		bal Fe				62 HRC
DIN	WNr 1.2242	Hard tmp to 200C	0.55-0.62	0.90-1.20	0.80-1.10		0.035 max	0.035 max	0.15-0.35	0.07-0.12	bal Fe				59 HRC
DIN	WNr 1.2243	Hard tmp to 200C	0.57-0.65	1.00-1.30	0.60-0.90		0.035 max	0.035 max	0.70-1.00	0.07-0.12	bal Fe				61 HRC
DIN	WNr 1.2382		0.36-0.46		0.85-1.10		0.035 max	0.035 max	0.80-1.00		Mo 0.10-0.25; bal Fe				
DIN	WNr 1.2823	Hard tmp to 200C	0.65-0.75		0.60-0.80		0.030 max	0.030 max	1.50-1.80		bal Fe				56 HRC
DIN 17350(80)	115CrV3	Hard tmp to 200C	1.10-1.25	0.50-0.80	0.20-0.40		0.030 max	0.030 max	0.15-0.30	0.07-0.12	bal Fe				61 HRC
DIN 17350(80)	51CrMnV4	Hard tmp to 200C	0.47-0.55	0.90-1.20	0.80-1.10		0.030 max	0.030 max	0.15-0.35	0.10-0.20	bal Fe				54 HRC
DIN 17350(80)	80CrV2	Hard tmp to 200C	0.75-0.85	0.40-0.70	0.30-0.50		0.030 max	0.030 max	0.25-0.40	0.15-0.25	bal Fe				60 HRC
DIN 17350(80)	WNr 1.2067	Tmp to 200C	0.95-1.10	1.35-1.65	0.25-0.45		0.030 max	0.030 max	0.15-0.35		bal Fe				63 HRC
DIN 17350(80)	WNr 1.2210	Hard tmp to 200C	1.10-1.25	0.50-0.80	0.20-0.40		0.030 max	0.030 max	0.15-0.30	0.07-0.12	bal Fe				61 HRC
DIN 17350(80)	WNr 1.2235	Hard tmp to 200C	0.75-0.85	0.40-0.70	0.30-0.50		0.030 max	0.030 max	0.25-0.40	0.15-0.25	bal Fe				60 HRC

UNS numbers and US grades are provided as a means of cross referencing chemically similar alloys. Exchangability is only possible after independent examination of specifications. Tensile properties are minimum or typical as specified. UTS and YS as MPa. El as %. See Appendix for list of abbreviations used in Notes. * indicates obsolete material.

Specification	Designation	Notes	C	Cr	Mn	Ni	P	S	Si	V	Other	UTS	YS	El	Hard

Tool Steel, Low-Alloy Special Purpose, L2 (Continued from previous page)

Germany

| DIN 17350(80) | WNr 1.2241 | Hard tmp to 200C | 0.47-0.55 | 0.90-1.20 | 0.80-1.10 | | 0.030 max | 0.030 max | 0.15-0.35 | 0.10-0.20 | bal Fe | | | | 54 HRC |

India

IS 1570	T50Cr1V23		0.45-0.55	0.9-1.2	0.5-0.8				0.1-0.35	0.15-0.3	bal Fe				
IS 1570	T55Cr70V15		0.5-0.6	0.6-0.8	0.6-0.8				0.1-0.35	0.1-0.2	bal Fe				
IS 1570	T80V23		0.75-0.85		0.2-0.35				0.1-0.3	0.15-0.3	bal Fe				
IS 1570	T90V23		0.85-0.95		0.2-0.35				0.1-0.3	0.15-0.3	bal Fe				

Italy

UNI 2955	51CrV3KU		0.48-0.55	0.9-1.2	0.7-1				0.1-0.4	0.1-0.2	bal Fe				
UNI 2955/3(82)	107CrV3KU		1-1.15	0.6-0.9	0.25-0.55		0.07 max	0.06 max	0.1-0.4	0.05-0.25	bal Fe				
UNI 2955/3(82)	51CrMnV4KU		0.48-0.55	0.9-1.2	0.7-1		0.07 max	0.06 max	0.1-0.4	0.1-0.2	bal Fe				
UNI 2955/4(82)	55NiCrMoV7KU	Hot work	0.5-0.6	0.65-0.9	0.65-0.95	1.5-2	0.07 max	0.06 max	0.1-0.4	0.05-0.25	Mo 0.3-0.5; bal Fe				

Japan

| JIS G4404(83) | SKT3 | HR Bar Frg Ann | 0.50-0.60 | 0.90-1.20 | 0.60-1.00 | 0.25-0.60 | 0.030 max | 0.030 max | 0.35 max | 0.20 max | Cu <=0.25; Mo 0.30-0.50; bal Fe | | | | 235 HB max |
| JIS G4410(84) | SKC11 | Rod or Bit; FF | 0.85-1.10 | 0.80-1.50 | 0.50 max | 0.20 max | 0.030 max | 0.030 max | 0.15-0.35 | 0.25 max | Cu <=0.25; Mo <=0.40; bal Fe | | | | 285-375 HB |

Mexico

NMX-B-082(90)	L1	Bar	0.90-1.10	1.20-1.60	0.10-0.40	0.30 max	0.03 max	0.03 max	0.10-0.40		Cu <=0.25; (As+Sn+Sb)<=0.040; (Cu+Ni)<=0.40; B, Ti may be added; bal Fe				
NMX-B-082(90)	L2	Bar	0.45-1.00	0.70-1.20	0.10-0.90	0.30 max	0.03 max	0.03 max	0.50 max	0.10-0.30	Cu <=0.25; Mo <=0.25; (As+Sn+Sb)<=0.040; (Cu+Ni)<=0.40; B, Ti may be added; bal Fe				
NMX-B-082(90)	L7	Bar	0.90-1.25	1.10-1.50	0.30-0.70	0.30 max	0.03 max	0.03 max	0.10-0.40		Cu <=0.25; Mo 0.30-0.50; (As+Sn+Sb)<=0.040; (Cu+Ni)<=0.40; B, Ti may be added; bal Fe				

Poland

PNH83161(90)	L75HMV		0.7-0.8	1.3-1.7	0.5-0.7				0.25-0.4	0.05-0.1	Mo 0.7-0.8; bal Fe				
PNH85021	WCV		0.45-0.55	0.4-0.7	0.4-0.7	0.35			0.15-0.35	0.1-0.25	bal Fe				
PNH85023	NCV1	Low alloy	0.75-0.85	0.4-0.7	0.3-0.6	0.4 max	0.030 max	0.030 max	0.15-0.4	0.15-0.3	bal Fe				

Romania

| STAS | VC06 | | 0.8-0.9 | 0.4-0.7 | 0.3-0.6 | 0.35 | | | 0.15-0.35 | 0.15-0.3 | bal Fe | | | | |

Russia

GOST 5950	6Ch3MFS	Alloyed	0.55-0.62	2.6-3.3	0.2-0.6	0.35 max	0.03 max	0.03 max	0.35-0.65	0.3-0.6	Cu <=0.3; Mo 0.2-0.5; W <=0.2; bal Fe				
GOST 5950	8Ch	Alloyed	0.7-0.8	0.4-0.7	0.15-0.45	0.35 max	0.03 max	0.03 max	0.1-0.4	0.15-0.3	Cu <=0.3; Mo <=0.2; W <=0.2; bal Fe				
GOST 5950	8Ch3	Alloyed	0.75-0.85	3.2-3.8	0.15-0.4	0.35 max	0.03 max	0.03 max	0.15-0.35	0.15 max	Cu <=0.3; Mo <=0.2; W <=0.2; bal Fe				
GOST 5950	8ChF	Alloyed	0.7-0.8	0.4-0.7	0.15-0.4	0.35 max	0.03 max	0.03 max	0.15-0.35	0.15-0.3	Cu <=0.3; Mo <=0.2; W <=0.2; bal Fe				
GOST 5950	9Ch	Alloyed	0.8-0.9	0.4-0.7	0.3-0.6	0.35 max	0.03 max	0.03 max	0.15-0.35	0.15-0.3	Cu <=0.3; Mo <=0.2; W <=0.2; bal Fe				
GOST 5950	9Ch1	Alloyed	0.8-0.95	1.4-1.7	0.15-0.4	0.35 max	0.03 max	0.03 max	0.25-0.45	0.15 max	Cu <=0.3; Mo <=0.2; W <=0.2; bal Fe				
GOST 5950	9ChF	Alloyed	0.8-0.9	0.4-0.7	0.3-0.6	0.35 max	0.03 max	0.03 max	0.15-0.35	0.15-0.3	Cu <=0.3; Mo <=0.2; W <=0.2; bal Fe				

Specification	Designation	Notes	C	Cr	Mn	Ni	P	S	Si	V	Other	UTS	YS	El	Hard

Tool Steel, Low-Alloy Special Purpose, L2 (Continued from previous page)

Russia

Specification	Designation	Notes	C	Cr	Mn	Ni	P	S	Si	V	Other	UTS	YS	El	Hard
GOST 5950	9ChF	Alloyed	0.8-0.9	0.4-0.7	0.3-0.6	0.35 max	0.03 max	0.03 max	0.15-0.35	0.15-0.3	Cu <=0.3; Mo <=0.15; W <=0.2; bal Fe				
GOST 5950	9ChFM	Alloyed	0.8-0.9	0.4-0.7	0.3-0.6	0.35 max	0.03 max	0.03 max	0.15-0.35	0.15-0.3	Cu <=0.3; Mo 0.15-0.25; W <=0.2; bal Fe				

Spain

Specification	Designation	Notes	C	Cr	Mn	Ni	P	S	Si	V	Other	UTS	YS	El	Hard
UNE	F.520.L		1-1.2	0.35-0.65	0.1-0.4		0.03 max	0.03 max	0.13-0.38	0.1-0.2	bal Fe				
UNE 36018/2(94)	120CrV2	Alloyed	1.1-1.29	0.4-0.7	0.15-0.35	0.25 max	0.025 max	0.02 max	0.15-0.35	0.1-0.2	Mo <=0.15; bal Fe				
UNE 36018/2(94)	F.5125*	see 120CrV2	1.1-1.29	0.4-0.7	0.15-0.35	0.25 max	0.025 max	0.02 max	0.15-0.35	0.1-0.2	Mo <=0.15; bal Fe				

USA

Specification	Designation	Notes	C	Cr	Mn	Ni	P	S	Si	V	Other	UTS	YS	El	Hard
	AISI L2	Ann Hard	0.45-1.00	0.70-1.20	0.10-0.90		0.030 max	0.030 max	0.50 max	0.10-0.30	Mo <=0.25; bal Fe				163-197 HB
	UNS T61202		0.45-1.00	0.70-1.20	0.10-0.90		0.030 max	0.03 max	0.50 max	0.10-0.30	Mo <=0.25; bal Fe				
ASTM A681(94)	L2	Ann	0.45-1.00	0.70-1.20	0.10-0.90		0.030 max	0.030 max	0.10-0.50	0.10-0.30	Mo <=0.25; S 0.06-0.15 if spec for machining; bal Fe				197 HB
FED QQ-T-570(87)	L2*	Obs; see ASTM A681	0.45-1	0.7-1.2	0.1-0.9	0.3			0.5 max	0.1-0.3	Mo 0.25; bal Fe				

Tool Steel, Low-Alloy Special Purpose, L3

Bulgaria

Specification	Designation	Notes	C	Cr	Mn	Ni	P	S	Si	V	Other	UTS	YS	El	Hard
BDS 7938	Ch		0.95-1.10	1.35-1.65	0.15-0.40	0.35 max	0.030 max	0.030 max	0.15-0.35	0.15 max	Cu <=0.30; Mo <=0.15; W <=0.15; bal Fe				

China

Specification	Designation	Notes	C	Cr	Mn	Ni	P	S	Si	V	Other	UTS	YS	El	Hard
GB 1299(85)	Cr2	As Hard	0.95-1.10	0.30-1.65	0.40 max	0.25 max	0.030 max	0.030 max	0.40 max		Cu <=0.30; bal Fe				62 HRC

France

Specification	Designation	Notes	C	Cr	Mn	Ni	P	S	Si	V	Other	UTS	YS	El	Hard
AFNOR NFA35590(92)	100Cr6	CW, Wear res; soft ann	0.95-1.1	1.35-1.6	0.2-0.4	0.4 max	0.025 max	0.025	0.1-0.35		Cu <=0.3; Mo <=0.15; bal Fe				223 HB max
AFNOR NFA35590(92)	Y100C6*	Obs see 100Cr6; CW, wear res; soft ann	0.95-1.1	1.35-1.6	0.2-0.4	0.4 max	0.025 max	0.025	0.1-0.35	0.1 max	Cu <=0.3; Mo <=0.15; W <=0.1; bal Fe				223 HB max

Hungary

Specification	Designation	Notes	C	Cr	Mn	Ni	P	S	Si	V	Other	UTS	YS	El	Hard
MSZ 4352(84)	K4	Q/T	0.95-1.1	1.3-1.65	0.15-0.4	0.35 max	0.03 max	0.03 max	0.1-0.4	0.15 max	Mo <=0.2; W <=0.3; bal Fe				62 HRC
MSZ 4352(84)	K4	Soft ann	0.95-1.1	1.3-1.65	0.15-0.4	0.35 max	0.03 max	0.03 max	0.1-0.4	0.15 max	Mo <=0.2; W <=0.3; bal Fe				229 HB max

India

Specification	Designation	Notes	C	Cr	Mn	Ni	P	S	Si	V	Other	UTS	YS	El	Hard
IS 1570/6(96)	T105Cr5Mn2	CW	0.9-1.2	1-1.6	0.4-0.8	0.4 max	0.035 max	0.035 max	0.1-0.35		Mo <=0.15; bal Fe				
IS 1570/6(96)	T110Mn4W6Cr4	CW	1-1.2	0.9-1.3	0.9-1.3	0.4 max	0.035 max	0.035 max	0.1-0.35		Mo <=0.15; W 1.25-1.75; bal Fe				
IS 1570/6(96)	TAC19	CW	1-1.2	0.9-1.3	0.9-1.3	0.4 max	0.035 max	0.035 max	0.1-0.35		Mo <=0.15; W 1.25-1.75; bal Fe				
IS 1570/6(96)	TAC7	CW	0.9-1.2	1-1.6	0.4-0.8	0.4 max	0.035 max	0.035 max	0.1-0.35		Mo <=0.15; bal Fe				
IS 1570/6(96)	THS12	High-Speed	1.05-1.2	3.5-4.5	0.4 max	0.4 max	0.03 max	0.03 max	0.5 max	1.7-2.2	Co 4.7-5.2; Mo 3.5-4.2; W 6.4-7.4; bal Fe				
IS 1570/6(96)	XT112Mo9Co3Cr4W2V1		1.05-1.2	3.5-4.5	0.4 max	0.4 max	0.03 max	0.03 max	0.5 max	0.9-1.4	Co 7.5-8.5; Mo 9-10; W 1.3-1.9; bal Fe				

Italy

Specification	Designation	Notes	C	Cr	Mn	Ni	P	S	Si	V	Other	UTS	YS	El	Hard
UNI 2955/3(82)	102Cr6KU		0.95-1.1	1.35-1.65	0.15-0.45		0.07 max	0.06 max	0.1-0.4		bal Fe				

Mexico

Specification	Designation	Notes	C	Cr	Mn	Ni	P	S	Si	V	Other	UTS	YS	El	Hard
NMX-B-082(90)	L3	Bar	0.95-1.10	1.30-1.70	0.25-0.80	0.30 max	0.03 max	0.03 max	0.50 max	0.10-0.30	Cu <=0.25; (As+Sn+Sb)<=0.040; (Cu+Ni)<=0.40; B, Ti may be added; bal Fe				

Poland

Specification	Designation	Notes	C	Cr	Mn	Ni	P	S	Si	V	Other	UTS	YS	El	Hard
PNH85023	NC4	CW	0.95-1.1	1.3-1.65	0.15-0.45	0.4 max	0.030 max	0.030 max	0.15-0.4		bal Fe				

UNS numbers and US grades are provided as a means of cross referencing chemically similar alloys. Exchangability is only possible after independent examination of specifications. Tensile properties are minimum or typical as specified. UTS and YS as MPa. El as %. See Appendix for list of abbreviations used in Notes. * indicates obsolete material.

Specification	Designation	Notes	C	Cr	Mn	Ni	P	S	Si	V	Other	UTS	YS	El	Hard

Tool Steel, Low-Alloy Special Purpose, L3 (Continued from previous page)

Romania

Specification	Designation	Notes	C	Cr	Mn	Ni	P	S	Si	V	Other	UTS	YS	El	Hard
STAS 3611(66)	C15		0.95-1.1	1.3-1.65	0.15-0.4	0.35 max	0.03 max	0.025 max	0.15-0.35		Pb <=0.15; bal Fe				

Russia

Specification	Designation	Notes	C	Cr	Mn	Ni	P	S	Si	V	Other	UTS	YS	El	Hard
GOST 5950	Ch	Alloyed	0.95-1.1	1.3-1.65	0.15-0.4	0.35 max	0.03 max	0.03 max	0.1-0.4	0.15 max	Cu <=0.3; Mo <=0.2; W <=0.2; bal Fe				

Spain

Specification	Designation	Notes	C	Cr	Mn	Ni	P	S	Si	V	Other	UTS	YS	El	Hard
UNE 36072/1(75)	100Cr6	Alloyed; CW	0.95-1.1	1.35-1.65	0.25-0.45		0.03 max	0.03 max	0.15-0.35		Mo <=0.15; bal Fe				
UNE 36072/1(75)	F.5230*	see 100Cr6	0.95-1.1	1.35-1.65	0.25-0.45		0.03 max	0.03 max	0.15-0.35		Mo <=0.15; bal Fe				

USA

Specification	Designation	Notes	C	Cr	Mn	Ni	P	S	Si	V	Other	UTS	YS	El	Hard
	UNS T61203*	Obs; Stainless	0.95-1.10	1.30-1.70	0.25-0.50		0.030 max	0.030 max	0.50 max	0.10-0.30	bal Fe				
ASTM A681(94)	L3	Ann	0.95-1.10	1.30-1.70	0.25-0.80		0.030 max	0.030 max	0.10-0.50	0.10-0.30	S 0.06-0.15 if spec for machining; bal Fe				201 HB

Tool Steel, Low-Alloy Special Purpose, L6

Argentina

Specification	Designation	Notes	C	Cr	Mn	Ni	P	S	Si	V	Other	UTS	YS	El	Hard
IAS	IRAM C2	Ann	0.50-0.60	0.80-1.10	0.60-0.70	1.60-2.10			0.15-0.40	0.10 min	Mo 0.60-0.90; bal Fe				250 HB
IAS	IRAM C4	Ann	0.55-0.65	0.65-0.95	0.60-0.70	2.55-2.85			0.15-0.40	0.15 min	Mo 0.30-0.40; bal Fe				245 HB

Brazil

Specification	Designation	Notes	C	Cr	Mn	Ni	P	S	Si	V	Other	UTS	YS	El	Hard
ABNT	VMO		0.57	1.1	0.7	1.65			0.25		Mo 0.5; Nb 0.06; bal Fe				

Bulgaria

Specification	Designation	Notes	C	Cr	Mn	Ni	P	S	Si	V	Other	UTS	YS	El	Hard
BDS 7938	5ChNM		0.50-0.60	0.50-0.80	0.50-0.80	1.40-1.80	0.030 max	0.030 max	0.15-0.35	0.15 max	Cu <=0.30; Mo 0.15-0.30; W <=0.15; bal Fe				
BDS 7938	5ChNMF		0.50-0.60	1.00-1.30	0.50-0.80	1.60-1.90	0.030 max	0.030 max	0.15-0.35	0.05-0.15	Cu <=0.30; Mo 0.50-0.70; bal Fe				

Canada

Specification	Designation	Notes	C	Cr	Mn	Ni	P	S	Si	V	Other	UTS	YS	El	Hard
CSA 419662	19662		0.5-0.6	0.5-0.9	0.5-0.9	1.5-1.9	0.03 max	0.03 max	0.3-0.6	0.1-0.25	Mo 0.15-0.3; bal Fe				
CSA 419663	19663		0.5-0.6	0.9-1.3	0.5-0.9	1.5-1.9	0.03 max	0.03 max	0.3-0.6	0.1-0.25	Mo 0.3-0.5; bal Fe				

China

Specification	Designation	Notes	C	Cr	Mn	Ni	P	S	Si	V	Other	UTS	YS	El	Hard
GB 1299(85)	5CrNiMo	As Hard	0.50-0.60	0.50-0.80	0.50-0.80	1.40-1.80	0.030 max	0.030 max	0.40 max	0.20 max	Cu <=0.30; Mo 0.15-0.30; bal Fe				60 HRC

Czech Republic

Specification	Designation	Notes	C	Cr	Mn	Ni	P	S	Si	V	Other	UTS	YS	El	Hard
CSN 419662	19662		0.5-0.6	0.5-0.9	0.5-0.9	1.5-1.9	0.03 max	0.03 max	0.3-0.6	0.1-0.25	Mo 0.15-0.3; bal Fe				
CSN 419663	19663		0.5-0.6	0.9-1.3	0.5-0.9	1.5-1.9	0.03 max	0.03 max	0.3-0.6	0.1-0.25	Mo 0.3-0.5; bal Fe				

France

Specification	Designation	Notes	C	Cr	Mn	Ni	P	S	Si	V	Other	UTS	YS	El	Hard
AFNOR NFA35590(78)	55NCDV7	Q/T	0.5-0.6	0.7-1	0.5-0.8	1.5-2	0.025 max	0.025 max	0.1-0.4	0.05-0.15	Cu <=0.3; Mo 0.3-0.5; W <=0.1; bal Fe				43 HRC

Germany

Specification	Designation	Notes	C	Cr	Mn	Ni	P	S	Si	V	Other	UTS	YS	El	Hard
DIN	45NiCr6	Hard tmp to 400C	0.40-0.50	1.20-1.50	0.50-0.80	1.50-1.80	0.035 max	0.035 max	0.15-0.35		bal Fe	1620			48 HRC
DIN	WNr 1.2710	Hard tmp to 400C	0.40-0.50	1.20-1.50	0.50-0.80	1.50-1.80	0.035 max	0.035 max	0.15-0.35		bal Fe	1620			48 HRC
DIN 17350(80)	55NiCrMoV6	Hard tmp to 400C	0.50-0.60	0.60-0.80	0.65-0.95	1.50-1.80	0.030 max	0.030 max	0.10-0.40	0.07-0.12	Mo 0.25-0.35; bal Fe	1620			47 HRC
DIN 17350(80)	56NiCrMoV7	Hard tmp to 500C	0.50-0.60	1.00-1.20	0.65-0.95	1.50-1.80	0.030 max	0.030 max	0.10-0.40	0.07-0.12	Mo 0.45-0.55; bal Fe	1570			
DIN 17350(80)	WNr 1.2713	Hard tmp to 400C	0.50-0.60	0.60-0.80	0.65-0.95	1.50-1.80	0.030 max	0.030 max	0.10-0.40	0.07-0.12	Mo 0.25-0.35; bal Fe	1620			47 HRC
DIN 17350(80)	WNr 1.2714	Hard tmp to 500C	0.50-0.60	1.00-1.20	0.65-0.95	1.50-1.80	0.030 max	0.030 max	0.10-0.40	0.07-0.12	Mo 0.45-0.55; bal Fe	1570			

Hungary

Specification	Designation	Notes	C	Cr	Mn	Ni	P	S	Si	V	Other	UTS	YS	El	Hard
MSZ 4352(84)	NK	Soft ann	0.5-0.6	0.5-0.8	0.5-0.8	1.4-1.6	0.03 max	0.03 max	0.1-0.4	0.15 max	Mo 0.15-0.3; W <=0.3; bal Fe				241 HB max
MSZ 4352(84)	NK	Q/T	0.5-0.6	0.5-0.8	0.5-0.8	1.4-1.6	0.03 max	0.03 max	0.1-0.4	0.15 max	Mo 0.15-0.3; W <=0.3; bal Fe				42 HRC

UNS numbers and US grades are provided as a means of cross referencing chemically similar alloys. Exchangability is only possible after independent examination of specifications. Tensile properties are minimum or typical as specified. UTS and YS as MPa. El as %. See Appendix for list of abbreviations used in Notes. * indicates obsolete material.

Specification	Designation	Notes	C	Cr	Mn	Ni	P	S	Si	V	Other	UTS	YS	El	Hard

Tool Steel, Low-Alloy Special Purpose, L6 (Continued from previous page)

Specification	Designation	Notes	C	Cr	Mn	Ni	P	S	Si	V	Other	UTS	YS	El	Hard
Japan															
JIS G4404(83)	SKS5	HR Ann; saws	0.75-0.85	0.20-0.50	0.50 max	0.70-1.30	0.030 max	0.030 max	0.35 max		Cu <=0.25; bal Fe				207 HB max
JIS G4404(83)	SKS51	HR Bar Frg Ann	0.75-0.85	0.20-0.50	0.50 max	1.30-2.00	0.030 max	0.030 max	0.35 max		Cu <=0.25; bal Fe				207 HB max
JIS G4404(83)	SKT4	Bar Frg HR Ann	0.50-0.60	0.70-1.00	0.60-1.00	1.30-2.00	0.030 max	0.030 max	0.35 max	0.20 max	Cu <=0.25; Mo 0.20-0.50; bal Fe				241 HB max
Mexico															
NMX-B-082(90)	L6	Bar	0.65-0.75	0.65-1.20	0.25-0.80	1.25-2.00	0.03 max	0.03 max	0.50 max		Cu <=0.25; Mo <=0.50; (As+Sn+Sb)<=0.040; B, Ti may be added; bal Fe				
Poland															
PNH83161(90)	L65HNM		0.55-0.75	0.8-1.2	0.5-0.8	0.8-1.2	0.040 max	0.040 max	0.25-0.4		Mo 0.3-0.4; bal Fe				
PNH85021	WNL		0.5-0.6	0.5-0.8	0.5-0.8	1.4-1.8	0.030 max	0.030 max	0.15-0.4		Co <=0.3; Mo 0.15-0.25; W <=0.3; bal Fe				
PNH85021	WNL1		0.5-0.6	0.5-0.8	0.5-0.8	1.4-1.8	0.030 max	0.030 max	0.15-0.4	0.05-0.12	Co <=0.3; Mo 0.15-0.25; W <=0.3; bal Fe				
Romania															
STAS 3611(80)	MoCN15		0.5-0.6	0.6-0.8	0.5-0.8	1.3-1.6	0.03 max	0.03 max	0.15-0.35		Mo 0.15-0.3; Pb <=0.15; bal Fe				
STAS 3611(80)	VMoCN17		0.52-0.62	1-1.2	0.6-0.8	1.6-1.9	0.025 max	0.03 max	0.15-0.35	0.1-0.2	Mo 0.5-0.6; Pb <=0.15; bal Fe				
Russia															
GOST	5ChNM		0.5-0.6	0.5-0.8	0.5-0.8	1.4-1.8	0.03 max	0.03 max	0.15-0.35	0.05 max	Cu <=0.3; Mo 0.15-0.3; Ti <=0.3; W <=0.2; bal Fe				
GOST 5950	5ChNM	Alloyed	0.5-0.6	0.5-0.8	0.5-0.8	1.4-1.8	0.03 max	0.03 max	0.1-0.4	0.15 max	Cu <=0.3; Mo 0.15-0.3; W <=0.2; bal Fe				
GOST 5950	5ChNM	Alloyed	0.5-0.6	0.5-0.8	0.5-0.7	1.4-1.8			0.15-0.35	0.05 max	Mo 0.15-0.3; Ti <=0.3; W <=0.2; bal Fe				
Spain															
UNE	F.520.B		0.5-0.6	0.95-1.25	0.55-0.85	1.55-1.85	0.03 max	0.03 max	0.28-0.53	0.1-0.2	Mo 0.4-0.6; bal Fe				
UNE	F.520.S		0.52-0.63	0.55-0.85	0.55-0.85	1.55-1.85	0.03 max	0.03 max	0.1-0.35	0.05-0.15	Mo 0.2-0.4; bal Fe				
UNE	F.528		0.4-0.5	0.6-0.75	0.2-0.4	1.3-1.7	0.03 max	0.03 max	0.15-0.3		Mo 0.3-0.6; bal Fe				
UNE 36072(75)	55NiCrMoV7	Alloyed	0.5-0.6	0.95-1.25	0.65-0.95	1.5-2	0.03 max	0.03 max	0.1-0.4	0.05-0.25	Mo 0.3-0.5; bal Fe				
UNE 36072(75)	F.5307*	see 55NiCrMoV7	0.5-0.6	0.95-1.25	0.65-0.95	1.5-2	0.03 max	0.03 max	0.1-0.4	0.05-0.25	Mo 0.3-0.5; bal Fe				
UNE 36072/2(75)	55NiCrMoV7		0.5-0.6	0.95-1.25	0.65-0.95	1.5-2	0.03 max	0.03 max	0.1-0.4	0.05-0.25	Mo 0.3-0.5; bal Fe				
UNE 36072/2(75)	F.5307*	see 55NiCrMoV7	0.5-0.6	0.95-1.25	0.65-0.95	1.5-2	0.03 max	0.03 max	0.1-0.4	0.05-0.25	Mo 0.3-0.5; bal Fe				
Sweden															
SS 142550	2550	Alloyed	0.52-0.6	0.9-1.1	0.3-0.5	2.8-3.2	0.03 max	0.02 max	0.2-0.4		Mo 0.25-0.35; bal Fe				
USA															
	AISI L6	Ann Hard	0.65-0.75	0.60-1.20	0.25-0.80	1.25-2.00	0.030 max	0.030 max	0.50 max		Mo <=0.50; bal Fe				183-255 HB
	UNS T61206		0.65-0.75	0.60-1.20	0.25-0.80	1.25-2.00	0.030 max	0.030 max	0.50 max		Mo <=0.50; bal Fe				
ASTM A681(94)	L6	Ann	0.65-0.75	0.60-1.20	0.25-0.80	1.25-2.00	0.030 max	0.030 max	0.10-0.50		Mo <=0.50; S 0.06-0.15 if spec for machining; bal Fe				235 HB
FED QQ-T-570(87)	L6*	Obs; see ASTM A681	0.65-0.75	0.6-1.2	0.25-0.8	1.25-2			0.5		Mo 0.5; bal Fe				
SAE J437(70)	L6	Ann									bal Fe				183-212 HB
SAE J438(70)	L6		0.65-0.75	0.65-0.85	0.55-0.85	1.25-1.75			0.20-0.40	0.25 max	Mo <=0.25; Mn may vary; V, Mo optional; bal Fe				

UNS numbers and US grades are provided as a means of cross referencing chemically similar alloys. Exchangability is only possible after independent examination of specifications. Tensile properties are minimum or typical as specified. UTS and YS as MPa. El as %. See Appendix for list of abbreviations used in Notes. * indicates obsolete material.

Specification	Designation	Notes	C	Cr	Mn	Ni	P	S	Si	V	Other	UTS	YS	El	Hard

Tool Steel, Low-Alloy Special Purpose, L7

Argentina

| IAS | IRAM L7 | Ann | 0.95-1.25 | 1.10-1.50 | 0.20-0.70 | | | | 0.10-0.40 | | Mo 0.30-0.50; bal Fe | | | | 192 HB |

USA

| SAE J437(70) | L7 | Ann | | | | | | | | | bal Fe | | | | 174-212 |
| SAE J438(70) | L7 | | 0.95-1.05 | 1.25-1.75 | 0.25-0.45 | | | | 0.20-0.40 | | Mo 0.30-0.50; bal Fe | | | | |

Tool Steel, Low-Alloy Special Purpose, No equivalents identified

Poland

| PNH85021 | WNLB | | 0.45-0.55 | 1-1.3 | 0.5-0.8 | 1.8-2.1 | 0.025 max | 0.025 max | 0.15-0.4 | 0.05-0.12 | Al 0.02-0.05; ; B 0.002-0.005; Co <=0.3; Mo 0.25-0.35; Ti 0.02-0.05; W <=0.3; bal Fe | | | | |
| PNH85021 | WNLV | | 0.5-0.6 | 1-1.3 | 0.5-0.8 | 1.8-2.1 | 0.030 max | 0.030 max | 0.15-0.4 | 0.05-0.12 | Co <=0.3; Mo 0.5-0.65; W <=0.3; bal Fe | | | | |

Russia

| GOST 5950 | 6ChS | Alloyed | 0.6-0.7 | 1-1.3 | 0.15-0.4 | 0.35 max | 0.03 max | 0.03 max | 0.6-1 | 0.15 max | Cu <=0.3; Mo <=0.2; W <=0.2; bal Fe | | | | |

Specification	Designation	Notes	C	Co	Cr	Mn	Mo	Si	V	W	Other	UTS	YS	El	Hard
Tool Steel, Molybdenum High-Speed, M1															
Brazil	•														
ABNT	EM1(M1)		0.82		3.8	0.4	8.5	0.45	1.15	1.75	bal Fe				
France															
AFNOR NFA35590	HS2-8-2		0.85 max		4 max	0.3 max	8 max	0.3 max	1.5 max	2 max	P <=0.03; S <=0.03; bal Fe				
AFNOR NFA35590(78)	HS2-9-1		0.8-0.88	1 max	3.5-4.5	0.4 max	8-9	0.5 max	1-1.5	1.4-2	Cu <=0.3; Ni <=0.4; P <=0.03; S <=0.03; bal Fe				
AFNOR NFA35590(78)	Z85DCWV08-04-02-01		0.8-0.88	1 max	3.5-4.5	0.4 max	8-9	0.5 max	1-1.5	1.4-2	Cu <=0.3; Ni <=0.4; P <=0.03; S <=0.03; bal Fe				
AFNOR NFA35590(92)	HS2-8-1	Soft ann	0.8-0.88	1 max	3.5-4.5	0.4 max	8-9	0.5 max	1 max	1.4-2	Cu <=0.3; Ni <=0.4; P <=0.03; S <=0.03; bal Fe				260 HB max
AFNOR NFA35590(92)	Z85DCWV08-04-02-01	Soft ann	0.8-0.88	1 max	3.5-4.5	0.4 max	8-9	0.5 max	1 max	1.4-2	Cu <=0.3; Ni <=0.4; P <=0.03; S <=0.03; bal Fe				260 HB max
Germany															
DIN	HS-2-9-1	As Tmp 530C	0.78-0.86		3.50-4.20	0.40 max	8.00-9.20	0.45 max	1.00-1.30	1.50-2.00	P <=0.030; S <=0.030; bal Fe				64 HRC
DIN	WNr 1.3346	As Tmp 530C	0.78-0.86		3.50-4.20	0.40 max	8.00-9.20	0.45 max	1.00-1.30	1.50-2.00	P <=0.030; S <=0.030; bal Fe				64 HRC
Hungary															
MSZ 4351(72)	R10		0.75-0.85		3.8-4.6	0.4 max	8	0.4 max	0.9-1.2	1.2-1.7	P <=0.03; S <=0.03; bal Fe				
MSZ 4352	K10		0.75-0.85		3.8-4.6	0.4	8-9	0.4	0.9-1.2	1.2-1.7	bal Fe				
Italy															
UNI	X82MoW09KU		0.75-0.9		3.5-4.5	0.5	8-9.5	0.5	0.8-1.3	1-2	bal Fe				
UNI 2955	RM1 HS 1-8-1		0.83		3.75	0.4	8.7	0.4	1.15	1.75	bal Fe				
UNI 2955/5(82)	HS1-8-1		0.77-0.87		3.5-4.5	0.4 max	8-9	0.5 max	0.9-1.4	1.4-2	P <=0.07; S <=0.06; bal Fe				
Mexico															
NMX-B-082(90)	M1	Bar	0.78-0.88		3.50-4.00	0.15-0.40	8.20-9.20	0.20-0.50	1.00-1.35	1.40-2.10	Cu <=0.25; Ni <=0.30; P <=0.03; S <=0.03; (As+Sn+Sb)<=0.040; (Cu+Ni)<=0.40; B, Ti may be added; bal Fe				
Pan America															
COPANT 337	M1	Bar, Frg	0.78-0.84		3.5-4	0.15-0.4	8.2-9.2	0.2-0.45	1-1.3	1.4-2.1	P 0.03; S 0.03; bal Fe				
Romania															
STAS 7382(88)	Rp10		0.78-0.86		3.5-4.2	0.4 max	8-9.2	0.45 max	1-1.3	1.5-2	Ni <=0.4; P <=0.03; Pb <=0.15; S <=0.03; bal Fe				
Spain															
UNE 36073(75)	2-9-2		0.95-1.05		3.5-4.5	0.4 max	8.2-9.2	0.45 max	1.5-2.2	1.5-2.1	P <=0.03; S <=0.03; bal Fe				
UNE 36073(75)	EM7		0.95-1.05		3.5-4.5	0.4 max	8.2-9.2	0.45 max	1.5-2.2	1.5-2.1	P <=0.03; S <=0.03; bal Fe				
Sweden															
SS 142715	2715		0.87-0.95		3.5-4.5	2 max	4.5-5.2	0.2-0.4	1.1-1.5	1.4-2	P <=0.03; S <=0.03; bal Fe				
UK															
BS 4659	BM1	Bar Rod Sh Strp Frg	0.75-0.85	0.6 max	3.75-4.5	0.4 max	8-9	0.4 max	1-1.25	1-2	bal Fe				
USA															
	AISI M1	Ann Hard	0.78-0.88		3.50-4.00	0.15-0.40	8.20-9.20	0.20-0.50	1.00-1.35	1.40-2.10	P <=0.03; S <=0.03; bal Fe				207-235 HB
	UNS T11301		0.78-0.88		3.50-4.00	0.15-0.40	8.20-9.20	0.20-0.50	1.00-1.35	1.40-2.10	P <=0.03; S <=0.03; bal Fe				

Specification	Designation	Notes	C	Co	Cr	Mn	Mo	Si	V	W	Other	UTS	YS	El	Hard
Tool Steel, Molybdenum High-Speed, M1 (Continued from previous page)															
USA															
ASTM A600(92)	M1	Bar Frg Plt Sh Strp, Ann	0.78-0.88		3.50-4.00	0.15-0.40	8.20-9.20	0.20-0.50	1.00-1.35	1.40-2.10	P <=0.03; S <=0.03; S 0.06-0.15 for machinability; (Ni+Cu)<=0.75; bal Fe				248 HB
FED QQ-T-590C(85)	M1*	Obs; see ASTM A600	0.78-0.88		3.5-4	0.15-0.4	8.2-9.2	0.2-0.5	1-1.35	1.4-2.1	Ni 0.3; bal Fe				
SAE J437(70)	M1	Ann									bal Fe				207-248 HB
SAE J438(70)	M1		0.75-0.85		3.75-4.50	0.20-0.40	7.75-9.25	0.20-0.40	0.90-1.30	1.15-1.85	bal Fe				
Tool Steel, Molybdenum High-Speed, M10															
Mexico															
NMX-B-082(90)	M10	Bar	0.84-0.94		3.75-4.50	0.10-0.40	7.75-8.50	0.20-0.45	1.80-2.20		Cu <=0.25; Ni <=0.30; P <=0.03; S <=0.03; (As+Sn+Sb)<=0.040; (Cu+Ni)<=0.40; B, Ti may be added; bal Fe				
Pan America															
COPANT 337	M10	Bar, Frg	0.84-0.94		3.75-4.5	0.15-0.4	7.75-8.5	0.2-0.45	1.8-2.2		P 0.03; S 0.03; bal Fe				
COPANT 337	M10 high C	Bar, Frg	0.95-1.05		3.75-4.5	0.15-0.4	7.75-8.5		1.8-2.2		P 0.03; bal Fe				
USA															
	AISI M10 high C	Ann Hard	0.95-1.05		3.75-4.50	0.10-0.40	7.75-8.50	0.20-0.45	1.80-2.20		P <=0.03; S <=0.03; bal Fe				207-255 HB
	AISI M10 reg C	Ann Hard	0.84-0.94		3.75-4.50	0.10-0.40	7.75-8.50	0.20-0.45	1.80-2.20		P <=0.03; S <=0.03; bal Fe				207-255 HB
	UNS T11310		0.84-1.05		3.75-4.50	0.10-0.40	7.75-8.50	0.20-0.45	1.80-2.20		P <=0.03; S <=0.03; bal Fe				
ASTM A600(92)	M10 high C	Bar Frg Plt Sh Strp, Ann	0.95-1.05		3.75-4.50	0.10-0.40	7.75-8.50	0.20-0.45	1.80-2.20		P <=0.03; S <=0.03; S 0.06-0.15 for machinability; (Ni+Cu)<=0.75; bal Fe				255 HB
ASTM A600(92)	M10 regular C	Bar Frg Plt Sh Strp, Ann	0.84-0.94		3.75-4.50	0.10-0.40	7.75-8.50	0.20-0.45	1.80-2.20		P <=0.03; S <=0.03; S 0.06-0.15 for machinability; (Ni+Cu)<=0.75; bal Fe				248 HB
FED QQ-T-590C(85)	M10 high C*	Obs; see ASTM A600	0.95-1.05		3.75-4.5	0.1-0.4	7.75-8.5	0.2-0.45	1.8-2.2		Ni 0.3; bal Fe				
FED QQ-T-590C(85)	M10*	Obs; see ASTM A600	0.84-0.94		3.75-4.5	0.1-0.4	7.75-8.5	0.2-0.45	1.8-2.2		Ni 0.3; bal Fe				
Tool Steel, Molybdenum High-Speed, M2															
Argentina															
IAS	IRAM M2, (High C)	Ann	0.95-1.05		3.75-4.5	0.15-0.40	4.50-5.50	0.20-0.45	1.75-2.20	5.50-6.75	P <=0.03; S <=0.03; bal Fe				260 HB
IAS	IRAM M2, (Med C)	Ann	0.86-0.94		3.75-4.5	0.15-0.40	4.50-5.50	0.20-0.45	1.75-2.20	5.50-6.75	P <=0.03; S <=0.03; bal Fe				260 HB
Brazil															
ABNT	VWM-2B high C		1		4.1	0.35	5	0.3	1.9	6.1	bal Fe				
Bulgaria															
BDS 7008(86)	R6M5		0.80-0.90	0.50 max	3.80-4.40	0.4	4.50-5.50	0.4	1.80-2.20	5.50-6.50	Ni <=0.40; P <=0.030; S <=0.030; bal Fe				
Canada															
CSA 419830	19830		0.8-0.9		3.8-4.6	0.45 max	4.5-5.5	0.45 max	1.5-2.2	5.5-7	P <=0.04; S <=0.04; bal Fe				
China															
GB 3080(82)	W6Mo5Cr4V2	Wir Q/T	0.80-0.90		3.80-4.40	0.40 max	4.50-5.50	0.40 max	1.75-2.20	5.50-6.75	Ni <=0.30; P <=0.03; S <=0.030; bal Fe				63 HRC
GB 9941(88)	W6Mo5Cr4V2	Sh Plt Ann	0.80-0.90		3.80-4.40	0.15-0.40	4.50-5.50	0.20-0.45	1.75-2.20	5.50-6.75	Ni <=0.30; P <=0.03; S <=0.030; bal Fe				255 HB max

UNS numbers and US grades are provided as a means of cross referencing chemically similar alloys. Exchangability is only possible after independent examination of specifications. Tensile properties are minimum or typical as specified. UTS and YS as MPa. El as %. See Appendix for list of abbreviations used in Notes. * indicates obsolete material.

Specification	Designation	Notes	C	Co	Cr	Mn	Mo	Si	V	W	Other	UTS	YS	El	Hard

Tool Steel, Molybdenum High-Speed, M2 (Continued from previous page)

Specification	Designation	Notes	C	Co	Cr	Mn	Mo	Si	V	W	Other	UTS	YS	El	Hard
China															
GB 9942(88)	W6Mo5Cr4V2	Frg Bar Ann	0.80-0.90		3.80-4.40	0.15-0.40	4.50-5.50	0.20-0.45	1.75-2.20	5.50-6.75	Ni <=0.30; P <=0.03; S <=0.030; bal Fe				255 HB max
GB 9943(88)	CW6Mo5Cr4V2	Bar Q/T	0.95-1.05		3.80-4.40	0.15-0.40	4.50-5.50	0.20-0.45	1.75-2.20	5.50-6.75	Ni <=0.30; P <=0.03; S <=0.030; bal Fe				65 HRC
GB 9943(88)	W6Mo5Cr4V2	Bar Q/T	0.80-0.90		3.80-4.40	0.15-0.40	4.50-5.50	0.20-0.45	1.75-2.20	5.50-6.75	Ni <=0.30; P <=0.03; S <=0.030; bal Fe				64 HRC
YB/T 084(96)	W6Mo5Cr4V2	Strp; Ann	0.80-0.90		3.80-4.40	0.15-0.40	4.50-5.50	0.20-0.45	1.75-2.20	5.50-6.75	Ni <=0.30; P <=0.03; S <=0.030; bal Fe				207-255 HB
Czech Republic															
CSN 419830	19830		0.8-0.9		3.8-4.6	0.45 max	4.5-5.5	0.45 max	1.5-2.2	5.5-7.0	P <=0.035; S <=0.035; bal Fe				
Finland															
SFS 916(73)	SFS916	Alloyed	0.82-0.9		3.5-4.5	0.2-0.4	4.5-5.5	0.15-0.4	1.7-2.1	6.0-7.0	P <=0.03; S <=0.03; bal Fe				
France															
AFNOR NFA35590(78)	HS6-5-2		0.8-0.87	1 max	3.5-4.5	0.4 max	4.6-5.3	0.5 max	1.7-2.2	5.7-6.7	Cu <=0.3; Ni <=0.4; P <=0.03; S <=0.03; bal Fe				64 HRC
AFNOR NFA35590(78)	HS6-5-2HC		0.88-0.96	1 max	3.5-4.5	0.4 max	4.6-5.3	0.5 max	1.7-2.2	5.7-6.7	Cu <=0.3; Ni <=0.4; P <=0.03; S <=0.03; bal Fe				64 HRC
AFNOR NFA35590(78)	Z85WDCV06-05-04-02		0.8-0.87	1 max	3.5-4.5	0.4 max	4.6-5.3	0.5 max	1.7-2.2	5.7-6.7	Cu <=0.3; Ni <=0.4; P <=0.03; S <=0.03; bal Fe				64 HRC
AFNOR NFA35590(78)	Z90WDCV06-05-04-02		0.88-0.96	1 max	3.5-4.5	0.4 max	4.6-5.3	0.5 max	1.7-2.2	5.7-6.7	Cu <=0.3; Ni <=0.4; P <=0.03; S <=0.03; bal Fe				64 HRC
AFNOR NFA35590(92)	HS6-5-2	Soft ann	0.8-0.87	1 max	3.5-4.5	0.4 max	4.6-5.3	0.5 max	1.7-2.2	5.7-6.7	Cu <=0.3; Ni <=0.4; P <=0.03; S <=0.03; bal Fe				260 HB max
AFNOR NFA35590(92)	HS6-5-2HC	Soft ann	0.88-0.96	1 max	3.5-4.5	0.4 max	4.6-5.3	0.5 max	1.7-2.2	5.7-6.7	Cu <=0.3; Ni <=0.4; P <=0.03; S <=0.03; bal Fe				270 HB max
AFNOR NFA35590(92)	HS6-5-4-2	Soft ann	0.67-0.73	0.5 max	4-4.5	0.4 max	4.7-5.2	0.5 max	1.8-2.1	5.5-6.4	Cu <=0.3; Ni <=0.4; P <=0.03; S <=0.03; bal Fe				250 HB max
AFNOR NFA35590(92)	Z85WDCV06-05-04-02	Soft ann	0.8-0.87	1 max	3.5-4.5	0.4 max	4.6-5.3	0.5 max	1.7-2.2	5.7-6.7	Cu <=0.3; Ni <=0.4; P <=0.03; S <=0.03; bal Fe				260 HB max
AFNOR NFA35590(92)	Z90WDCV06-05-04-02	Soft ann	0.88-0.96	1 max	3.5-4.5	0.4 max	4.6-5.3	0.5 max	1.7-2.2	5.7-6.7	Cu <=0.3; Ni <=0.4; P <=0.03; S <=0.03; bal Fe				270 HB max
AFNOR NFA35590(92)	Z90WDCV06-05-04-02	Soft ann	0.67-0.73	0.5 max	4-4.5	0.4 max	4.7-5.2	0.5 max	1.8-2.1	5.5-6.4	Cu <=0.3; Ni <=0.4; P <=0.03; S <=0.03; bal Fe				250 HB max
Germany															
DIN	S-6-5-2Si		0.84-0.92		4.80-5.40	0.40 max	4.70-5.20	0.70-1.00	1.40-1.90	5.00-6.00	P <=0.030; S <=0.030; bal Fe				
DIN	WNr 1.3345		0.84-0.92		4.80-5.40	0.40 max	4.70-5.20	0.70-1.00	1.40-1.90	5.00-6.00	P <=0.030; S <=0.030; bal Fe				
DIN	WNr 1.3353*	Obs									bal Fe				
DIN	WNr 1.3354*	Obs									bal Fe				
DIN 17350(80)	HS-6-5-2	As Tmp 540C	0.86-0.94		3.80-4.50	0.40 max	4.70-5.20	0.45 max	1.70-2.00	6.00-6.70	P <=0.030; S <=0.030; bal Fe				64 HRC
DIN 17350(80)	HS-6-5-2S	As Tmp 530C	0.86-0.94		3.80-4.50	0.40 max	4.70-5.20	0.45 max	1.70-2.00	6.00-6.70	P <=0.030; S 0.06-0.15; bal Fe				64 HRC
DIN 17350(80)	HS6-5-2CS		0.95-1.05		3.80-4.50	0.40 max	4.70-5.20	0.45 max	1.70-2.00	6.00-6.70	P <=0.030; S 0.06-0.15; bal Fe				
DIN 17350(80)	WNr 1.3340		0.95-1.05		3.80-4.50	0.40 max	4.70-5.20	0.45 max	1.70-2.00	6.00-6.70	P <=0.030; S 0.06-0.15; bal Fe				
DIN 17350(80)	WNr 1.3341	As Tmp 530C	0.86-0.94		3.80-4.50	0.40 max	4.70-5.20	0.45 max	1.70-2.00	6.00-6.70	P <=0.030; S 0.06-0.15; bal Fe				64 HRC
DIN 17350(80)	WNr 1.3343	As Tmp 540C	0.86-0.94		3.80-4.50	0.40 max	4.70-5.20	0.45 max	1.70-2.00	6.00-6.70	P <=0.030; S <=0.030; bal Fe				64 HRC
Hungary															
MSZ 4351(84)	R12	Soft ann	0.95-1.05	0.5 max	3.8-4.5	0.4 max	4.7-5.2	0.4 max	1.7-2.1	6-6.7	P <=0.03; S <=0.03; bal Fe				269 HB max

Specification	Designation	Notes	C	Co	Cr	Mn	Mo	Si	V	W	Other	UTS	YS	El	Hard

Tool Steel, Molybdenum High-Speed, M2 (Continued from previous page)

Hungary

Specification	Designation	Notes	C	Co	Cr	Mn	Mo	Si	V	W	Other	UTS	YS	El	Hard
MSZ 4351(84)	R12	Q/T	0.95-1.05	0.5 max	3.8-4.5	0.4 max	4.7-5.2	0.4 max	1.7-2.1	6-6.7	P <=0.03; S <=0.03; bal Fe				65 HRC
MSZ 4351(84)	R6	Soft ann	0.82-0.92	0.5 max	3.8-4.5	0.4 max	4.8-5.3	0.4 max	1.7-2.1	6-7	P <=0.03; S <=0.03; bal Fe				255 HB max
MSZ 4351(84)	R6	Q/T	0.82-0.92	0.5 max	3.8-4.5	0.4 max	4.8-5.3	0.4 max	1.7-2.1	6-7	P <=0.03; S <=0.03; bal Fe				64 HRC

India

Specification	Designation	Notes	C	Co	Cr	Mn	Mo	Si	V	W	Other	UTS	YS	El	Hard
IS 1570	T83MoW6Cr4V2		0.75-0.9		3.75-4.5	0.2-0.4	5.5-6.5	0.1-0.35	1.75-2	5.5-6.5	bal Fe				
IS 1570/6(96)	THS2		0.95-1.05	1 max	3.5-4.5	0.4 max	8-9	0.5 max	1.7-2.2	1.5-2.1	Ni <=0.4; P <=0.03; S <=0.03; bal Fe				
IS 1570/6(96)	XT100Mo9Cr4W2V2		0.95-1.05	1 max	3.5-4.5	0.4 max	8-9	0.5 max	1.7-2.2	1.5-2.1	Ni <=0.4; P <=0.03; S <=0.03; bal Fe				

International

Specification	Designation	Notes	C	Co	Cr	Mn	Mo	Si	V	W	Other	UTS	YS	El	Hard
ISO 683-17(76)	31	Ball & roller bearing, HT	0.78-0.86		3.80-4.50	0.40 max	4.70-5.20	0.40 max	1.70-2.00	6.00-6.70	P <=0.030; S <=0.030; bal Fe				

Italy

Specification	Designation	Notes	C	Co	Cr	Mn	Mo	Si	V	W	Other	UTS	YS	El	Hard
UNI	X82WMo0605KU		0.75-0.9		3.5-4.5	0.5	4.5-5.5	0.5	1.6-2.2	5.5-7	bal Fe				
UNI 2955	RM2 HS6-5-2		0.83		4.15	0.3	5	0.3	1.95	6.1	bal Fe				
UNI 2955/5(82)	HS6-5-2		0.82-0.92	1 max	3.5-4.5	0.4 max	4.6-5.3	0.5 max	1.7-2.2	5.7-6.7	P <=0.07; S <=0.06; bal Fe				
UNI 3097(75)	X82WMoV65	Ball & roller bearing	0.78-0.86		3.7-4.5	0.4 max	4.7-5.2	0.4 max	1.7-2	6.-6.7	P <=0.03; S <=0.03; bal Fe				

Japan

Specification	Designation	Notes	C	Co	Cr	Mn	Mo	Si	V	W	Other	UTS	YS	El	Hard
JIS G4403(83)	SKH51	Bar Frg Ann	0.80-0.90		3.80-4.50	0.40 max	4.50-5.50	0.40 max	1.60-2.20	5.50-6.70	Cu <=0.25; Ni <=0.25; P <=0.030; S <=0.030; bal Fe				255 HB max
JIS G4403(83)	SKH9*	Obs; See SKH51; Bar Frg									bal Fe				

Mexico

Specification	Designation	Notes	C	Co	Cr	Mn	Mo	Si	V	W	Other	UTS	YS	El	Hard
NMX-B-082(90)	M2	Bar	0.78-0.88		3.75-4.50	0.15-0.40	4.50-5.50	0.20-0.45	1.75-2.20	5.50-6.75	Cu <=0.25; Ni <=0.30; P <=0.03; S <=0.03; (As+Sn+Sb)<=0.040; (Cu+Ni)<=0.40; B, Ti may be added; bal Fe				

Pan America

Specification	Designation	Notes	C	Co	Cr	Mn	Mo	Si	V	W	Other	UTS	YS	El	Hard
COPANT 337	M2	Bar, Frg	0.78-0.88		3.75-4.5	0.15-0.4	4.5-5.5	0.2-0.4	1.75-2.2	5.5-6.75	P 0.03; S 0.03; bal Fe				
COPANT 337	M2 high C	Bar, Frg	0.95-1.05		3.75-4.5	0.15-0.4	4.5-5.5	0.2-0.45	1.75-2.2	5.5-6.75	P 0.03; S 0.03; bal Fe				

Poland

Specification	Designation	Notes	C	Co	Cr	Mn	Mo	Si	V	W	Other	UTS	YS	El	Hard
PNH85022	SW7M		0.82-0.92		3.8-4.8	0.4 max	4.5-5.5	0.5 max	1.7-2.2	6-7	Ni <=0.4; P <=0.030; S <=0.030; bal Fe				

Romania

Specification	Designation	Notes	C	Co	Cr	Mn	Mo	Si	V	W	Other	UTS	YS	El	Hard
STAS 7382(88)	Rp5		0.86-0.94		3.8-4.5	0.4 max	4.7-5.2	0.45 max	1.7-2	6-6.7	Ni <=0.4; P <=0.03; Pb <=0.15; S <=0.03; bal Fe				

Russia

Specification	Designation	Notes	C	Co	Cr	Mn	Mo	Si	V	W	Other	UTS	YS	El	Hard
GOST 19265	R6AM5		0.82-0.9		3.8-4.4	0.5 max	5-5.5	0.5 max	1.7-2.1	5.5-6.5	Ni <=0.4; P <=0.03; S <=0.025; bal Fe				
GOST 19265	R6AM5SHKH15		0.82-0.9	0.5 max	3.8-4.4	0.2-0.5	4.8-5.3	0.2-0.5	1.7-2.1	5.5-6.5	Cu <=0.25; N 0.05-0.1; Ni <=0.6; P <=0.03; S <=0.025; bal Fe				
GOST 19265	R6M5		0.82-0.9	0.5 max	3.8-4.4	0.2-0.5	4.8-5.3	0.2-0.5	1.7-2.1	5.5-6.5	Cu <=0.25; Ni <=0.6; P <=0.03; S <=0.025; bal Fe				
GOST 19265	R6M5		0.8-0.88	0.6	3.8-4.4	0.4	5-5.5	0.5	1.7-2.1	5.5-6.5	Ni 0.4; bal Fe				

Spain

Specification	Designation	Notes	C	Co	Cr	Mn	Mo	Si	V	W	Other	UTS	YS	El	Hard
UNE	F.550.A		0.8-0.9		3.75-4.75	0.15-0.45	4-6	0.18-0.43	1.75-2.25	5.5-7.5	P <=0.03; S <=0.03; bal Fe				
UNE 36027(80)	F.1352*	see X80WMoCrV6-5-4	0.78-0.86		3.8-4.5	0.4 max	4.7-5.2	0.4 max	1.7-2.0	6.0-6.7	P <=0.03; S <=0.03; bal Fe				

UNS numbers and US grades are provided as a means of cross referencing chemically similar alloys. Exchangability is only possible after independent examination of specifications. Tensile properties are minimum or typical as specified. UTS and YS as MPa. El as %. See Appendix for list of abbreviations used in Notes. * indicates obsolete material.

Specification	Designation	Notes	C	Co	Cr	Mn	Mo	Si	V	W	Other	UTS	YS	El	Hard

Tool Steel, Molybdenum High-Speed, M2 (Continued from previous page)

Spain

Specification	Designation	Notes	C	Co	Cr	Mn	Mo	Si	V	W	Other	UTS	YS	El	Hard
UNE 36027(80)	X80WMoCrV6-5-4	Ball & roller bearing	0.78-0.86		3.8-4.5	0.4 max	4.7-5.2	0.4 max	1.7-2.0	6.0-6.7	P <=0.03; S <=0.03; bal Fe				
UNE 36073(75)	6-5-2		0.82-0.92		3.5-4.5	0.4 max	4.6-5.3	0.45 max	1.7-2.2	5.7-6.7	P <=0.03; S <=0.03; bal Fe				
UNE 36073(75)	EM2		0.82-0.92		3.5-4.5	0.4 max	4.6-5.3	0.45 max	1.7-2.2	5.7-6.7	P <=0.03; S <=0.03; bal Fe				
UNE 36073(75)	EM2		0.82-0.92		3.5-4.5	0.4 max	4.6-5.3	0.45 max	1.7-2.2	5.7-6.7	P <=0.03; S <=0.03; bal Fe				
UNE 36073(75)	F.5603*	see 6-5-2, EM2	0.82-0.92		3.5-4.5	0.4 max	4.6-5.3	0.45 max	1.7-2.2	5.7-6.7	P <=0.03; S <=0.03; bal Fe				

Sweden

Specification	Designation	Notes	C	Co	Cr	Mn	Mo	Si	V	W	Other	UTS	YS	El	Hard
SS 142722	2722		0.82-0.9		3.5-4.5	0.2-0.4	4.5-5.5	0.3-0.5	1.7-2.1	6-7	P <=0.03; S <=0.03; bal Fe				

UK

Specification	Designation	Notes	C	Co	Cr	Mn	Mo	Si	V	W	Other	UTS	YS	El	Hard
BS 4659	(USA M2)		0.85		4		5		2	6.4	bal Fe				
BS 4659	BM2	Bar Rod Sh Strp Frg	0.8-0.9	0.6	3.75-4.5	0.4	4.75-5.5	0.4	1.75-2.05	6-6.75	bal Fe				

USA

Specification	Designation	Notes	C	Co	Cr	Mn	Mo	Si	V	W	Other	UTS	YS	El	Hard
	AISI M2 high C	Ann Hard	0.95-1.05		3.75-4.50	0.15-0.40	4.50-5.50	0.20-0.45	1.75-2.20	5.50-6.75	P <=0.03; S <=0.03; bal Fe				212-241 HB
	AISI M2 reg C	Ann Hard	0.78-0.88		3.75-4.50	0.15-0.40	4.50-5.50	0.20-0.45	1.75-2.20	5.50-6.75	P <=0.03; S <=0.03; bal Fe				212-241 HB
	UNS T11302		0.78-1.05		3.75-4.50	0.15-0.40	4.50-5.50	0.20-0.45	1.75-2.20	5.50-6.75	P <=0.03; S <=0.03; bal Fe				
ASTM A597(93)	CM-2	Cast	0.78-0.88	0.25 max	3.75-4.50	0.75 max	4.50-5.50	1.00 max	1.25-2.20	5.50-6.75	Ni <=0.25; P <=0.03; S <=0.03; bal Fe				
ASTM A600(92)	M2 high C	Bar Frg Plt Sh Strp, Ann	0.95-1.05		3.75-4.50	0.15-0.40	4.50-5.50	0.20-0.45	1.75-2.20	5.50-6.75	P <=0.03; S <=0.03; S 0.06-0.15 for machinability; (Ni+Cu)<=0.75; bal Fe				255 HB
ASTM A600(92)	M2 regular C	Bar Frg Plt Sh Strp, Ann	0.78-0.88		3.75-4.50	0.15-0.40	4.50-5.50	0.20-0.45	1.75-2.20	5.50-6.75	P <=0.03; S <=0.03; S 0.06-0.15 for machinability; (Ni+Cu)<=0.75; bal Fe				248 HB
FED QQ-T-590C(85)	M2 high C*	Obs; see ASTM A600	0.95-1.05		3.75-4.5	0.15-0.4	4.5-5.5	0.2-0.45	1.75-2.2	5.5-6.75	Ni 0.3; bal Fe				
FED QQ-T-590C(85)	M2*	Obs; see ASTM A600	0.78-0.88		3.75-4.5	0.15-0.4	4.5-5.5	0.2-0.45	1.75-2.2	5.5-6.75	Ni 0.3; bal Fe				
SAE J437(70)	M2	Ann								*	bal Fe				217-248 HB
SAE J438(70)	M2		0.78-0.88		3.75-4.50	0.20-0.40	4.50-5.50	0.20-0.40	1.60-2.20	5.50-6.75	bal Fe				
SAE J467(68)	M2		0.84		4.20	0.30	5.00	0.30	1.90	6.15	bal Fe				

Tool Steel, Molybdenum High-Speed, M3 Class 1

Argentina

Specification	Designation	Notes	C	Co	Cr	Mn	Mo	Si	V	W	Other	UTS	YS	El	Hard
IAS	IRAM M3 Class 1	Ann	1.00-1.10		3.75-4.50	0.15-0.40	4.75-6.50	0.20-0.45	2.25-2.75	5.00-6.75	P <=0.03; S <=0.03; bal Fe				223-255 HB
IAS	IRAM M35	Ann	0.78-0.84	4.75-5.25	3.75-4.50	0.15-0.40	4.50-5.50	0.20-0.45	1.80-2.20	5.50-6.75	P <=0.03; S <=0.03; bal Fe				255-270 HB

China

Specification	Designation	Notes	C	Co	Cr	Mn	Mo	Si	V	W	Other	UTS	YS	El	Hard
GB 9943(88)	W6Mo5Cr4V3	Bar Q/T	1.00-1.10		3.75-4.50	0.15-0.40	4.75-6.50	0.20-0.45	2.25-2.75	5.00-6.75	Ni <=0.30; P <=0.03; S <=0.030; bal Fe				64 HRC

Germany

Specification	Designation	Notes	C	Co	Cr	Mn	Mo	Si	V	W	Other	UTS	YS	El	Hard
DIN 17350(80)	HS-6-5-3	As Tmp 550C	1.17-1.27		3.80-4.50	0.40 max	4.70-5.20	0.45 max	2.70-3.20	6.00-6.70	P <=0.030; S <=0.030; bal Fe				64 HRC
DIN 17350(80)	HS6-5-2C	As Tmp 540C	0.95-1.05		3.80-4.50	0.40 max	4.70-5.20	0.45 max	1.70-2.00	6.00-6.70	P <=0.030; S <=0.030; bal Fe				65 HRC
DIN 17350(80)	WNr 1.3342	As Tmp 540C	0.95-1.05		3.80-4.50	0.40 max	4.70-5.20	0.45 max	1.70-2.00	6.00-6.70	P <=0.030; S <=0.030; bal Fe				65 HRC

Italy

Specification	Designation	Notes	C	Co	Cr	Mn	Mo	Si	V	W	Other	UTS	YS	El	Hard
UNI 2955	RM3 HS6-6-3		1.05		4.15	0.4	6	0.45	2.5	6.2	bal Fe				
UNI 2955/5(82)	HSC6-5-3		1-1.12		3.5-4.5	0.4 max	4.6-5.3	0.5 max	2.2-2.7	5.7-6.7	P <=0.07; S <=0.06; bal Fe				

UNS numbers and US grades are provided as a means of cross referencing chemically similar alloys. Exchangability is only possible after independent examination of specifications. Tensile properties are minimum or typical as specified. UTS and YS as MPa. El as %. See Appendix for list of abbreviations used in Notes. * indicates obsolete material.

Specification	Designation	Notes	C	Co	Cr	Mn	Mo	Si	V	W	Other	UTS	YS	El	Hard

Tool Steel, Molybdenum High-Speed, M3 Class 1 (Continued from previous page)

Japan

Specification	Designation	Notes	C	Co	Cr	Mn	Mo	Si	V	W	Other	UTS	YS	El	Hard
JIS G4403(83)	SKH52	Bar Frg Ann	1.00-1.10		3.80-4.50	0.40 max	4.80-6.20	0.40 max	2.30-2.80	5.50-6.70	Cu <=0.25; Ni <=0.25; P <=0.030; S <=0.030; bal Fe				269 HB max

Mexico

Specification	Designation	Notes	C	Co	Cr	Mn	Mo	Si	V	W	Other	UTS	YS	El	Hard
NMX-B-082(90)	M3 Class 1	Bar	1.00-1.10		3.75-4.50	0.15-0.40	4.75-6.50	0.20-0.45	2.25-2.75	5.00-6.75	Cu <=0.25; Ni <=0.30; P <=0.03; S <=0.03; (As+Sn+Sb)<=0.040; (Cu+Ni)<=0.40; B, Ti may be added; bal Fe				

Pan America

Specification	Designation	Notes	C	Co	Cr	Mn	Mo	Si	V	W	Other	UTS	YS	El	Hard
COPANT 337	M3	Bar, Frg	1-1.1		3.75-4.5	0.15-0.4	4.75-6.5	0.2-0.45	2.25-2.75	5-6.75	P 0.03; S 0.03; bal Fe				

Russia

Specification	Designation	Notes	C	Co	Cr	Mn	Mo	Si	V	W	Other	UTS	YS	El	Hard
GOST 19265	R6M5F3		0.95-1.05		3.8-4.3	0.5	5.5-6	0.5	2.3-2.7	5.7-6.7	Ni 0.4; bal Fe				
GOST 19265	R6M5F3		0.95-1.05	0.5 max	3.8-4.3	0.2-0.5	4.8-5.3	0.2-0.5	2.3-2.7	5.7-6.7	Cu <=0.25; N 0.05-0.1; Ni <=0.6; P <=0.03; S <=0.025; bal Fe				

USA

Specification	Designation	Notes	C	Co	Cr	Mn	Mo	Si	V	W	Other	UTS	YS	El	Hard
	AISI M3 Class 1	Ann Hard	1.00-1.10		3.75-4.50	0.15-0.40	4.75-6.50	0.20-0.45	2.25-2.75	5.00-6.75	P <=0.03; S <=0.03; bal Fe				223-255 HB
	UNS T11313		1.00-1.10		3.75-4.50	0.15-0.40	4.75-6.50	0.20-0.45	2.25-2.75	5.00-6.75	P <=0.03; S <=0.03; bal Fe				
ASTM A600(92)	M3 Class 1	Bar Frg Plt Sh Strp, Ann	1.00-1.10		3.75-4.50	0.15-0.40	4.75-6.50	0.20-0.45	2.25-2.75	5.00-6.75	P <=0.03; S <=0.03; S 0.06-0.15 for machinability; (Ni+Cu)<=0.75; bal Fe				255 HB
FED QQ-T-590C(85)	M3-1*	Obs; see ASTM A600	1-1.1		3.75-4.5	0.15-0.4	4.75-6.5	0.2-0.45	2.25-2.75	5-6.75	Ni 0.3; bal Fe				
SAE J437(70)	M3	Ann									bal Fe				223-255 HB
SAE J438(70)	M3		1.00-1.25		3.75-4.50	0.20-0.40	4.75-6.25	0.20-0.40	2.25-3.25	5.50-6.75	bal Fe				

Tool Steel, Molybdenum High-Speed, M3 Class 2

Argentina

Specification	Designation	Notes	C	Co	Cr	Mn	Mo	Si	V	W	Other	UTS	YS	El	Hard
IAS	IRAM M3 Class 2	Ann	1.15-1.25		3.75-4.50	0.15-0.40	4.75-6.50	0.20-0.45	2.75-3.25	5.00-6.75	P <=0.03; S <=0.03; bal Fe				223-255 HB

China

Specification	Designation	Notes	C	Co	Cr	Mn	Mo	Si	V	W	Other	UTS	YS	El	Hard
GB 9943(88)	CW6Mo5Cr4V3	Bar Q/T	1.15-1.25		3.75-4.50	0.15-0.40	4.75-6.50	0.20-0.45	2.75-3.25	5.00-6.75	Ni <=0.30; P <=0.03; S <=0.030; bal Fe				64 HRC

France

Specification	Designation	Notes	C	Co	Cr	Mn	Mo	Si	V	W	Other	UTS	YS	El	Hard
AFNOR NFA35590(78)	HS6-5-3		1.15-1.25	1 max	3.5-4.5	0.4 max	4.6-5.3	0.5 max	2.7-3.2	5.7-6.7	Cu <=0.3; Ni <=0.4; P <=0.03; S <=0.03; bal Fe				65 HRC
AFNOR NFA35590(78)	HS6-5-4	Soft ann	1.25-1.4	1 max	4-5	0.4 max	4.2-5	0.5 max	3.6-4.2	5-6	Cu <=0.3; Ni <=0.4; P <=0.03; S <=0.03; bal Fe				65 HRC
AFNOR NFA35590(78)	Z120WDCV06-05-04-03		1.15-1.25	1 max	3.5-4.5	0.4 max	4.6-5.3	0.5 max	2.7-3.2	5.7-6.7	Cu <=0.3; Ni <=0.4; P <=0.03; S <=0.03; bal Fe				65 HRC
AFNOR NFA35590(78)	Z130WDCV06-05-04-04	Soft ann	1.25-1.4	1 max	4-5	0.4 max	4.2-5	0.5 max	3.6-4.2	5-6	Cu <=0.3; Ni <=0.4; P <=0.03; S <=0.03; bal Fe				65 HRC
AFNOR NFA35590(92)	HS6-5-3	Soft ann	1.15-1.25	1 max	3.5-4.5	0.4 max	4.6-5.3	0.5 max	2.7-3.2	5.7-6.7	Cu <=0.3; Ni <=0.4; P <=0.03; S <=0.03; bal Fe				275 HB max
AFNOR NFA35590(92)	HS6-5-4	Soft ann	1.25-1.4	1 max	4-5	0.4 max	4.2-5	0.5 max	3.6-4.2	5-6	Cu <=0.3; Ni <=0.4; P <=0.03; S <=0.03; bal Fe				275 HB max
AFNOR NFA35590(92)	Z120WDCV06-05-04-03	Soft ann	1.15-1.25	1 max	3.5-4.5	0.4 max	4.6-5.3	0.5 max	2.7-3.2	5.7-6.7	Cu <=0.3; Ni <=0.4; P <=0.03; S <=0.03; bal Fe				275 HB max

Specification	Designation	Notes	C	Co	Cr	Mn	Mo	Si	V	W	Other	UTS	YS	El	Hard

Tool Steel, Molybdenum High-Speed, M3 Class 2 (Continued from previous page)

France

Specification	Designation	Notes	C	Co	Cr	Mn	Mo	Si	V	W	Other	UTS	YS	El	Hard
AFNOR NFA35590(92)	Z130WDCV06-05-04-03	Soft ann	1.25-1.4	1 max	4-5	0.4 max	4.2-5	0.5 max	3.6-4.2	5-6	Cu <=0.3; Ni <=0.4; P <=0.03; S <=0.03; bal Fe				275 HB max

Germany

Specification	Designation	Notes	C	Co	Cr	Mn	Mo	Si	V	W	Other	UTS	YS	El	Hard
DIN 17350(80)	WNr 1.3344	As Tmp 550C	1.17-1.27		3.80-4.50	0.40 max	4.70-5.20	0.45 max	2.70-3.20	6.00-6.70	P <=0.030; S <=0.030; bal Fe				64 HRC

Hungary

Specification	Designation	Notes	C	Co	Cr	Mn	Mo	Si	V	W	Other	UTS	YS	El	Hard
MSZ 4351(84)	R13	Q/T	1.17-1.27	0.5 max	3.8-4.5	0.4 max	4.7-5.2	0.4 max	2.7-3.2	6-6.7	P <=0.03; S <=0.03; bal Fe				65 HRC
MSZ 4351(84)	R13	Soft ann	1.17-1.27	0.5 max	3.8-4.5	0.4 max	4.7-5.2	0.4 max	2.7-3.2	6-6.7	P <=0.03; S <=0.03; bal Fe				300 HB max

India

Specification	Designation	Notes	C	Co	Cr	Mn	Mo	Si	V	W	Other	UTS	YS	El	Hard
IS 1570/6(96)	THS10		1.2-1.35	9.5-10.5	3.5-4.5	0.4 max	3.2-3.9	0.5 max	3-3.5	9-10	Ni <=0.4; P <=0.03; S <=0.03; bal Fe				
IS 1570/6(96)	XT127W10Co10Cr4Mo4V3		1.2-1.35	9.5-10.5	3.5-4.5	0.4 max	3.2-3.9	0.5 max	3-3.5	9-10	Ni <=0.4; P <=0.03; S <=0.03; bal Fe				

Italy

Specification	Designation	Notes	C	Co	Cr	Mn	Mo	Si	V	W	Other	UTS	YS	El	Hard
UNI 2955/5(82)	HS6-5-3		1.15-1.3	1 max	3.5-4.5	0.4 max	4.6-5.3	0.5 max	2.7-3.7	5.5-6.5	P <=0.07; S <=0.06; bal Fe				

Japan

Specification	Designation	Notes	C	Co	Cr	Mn	Mo	Si	V	W	Other	UTS	YS	El	Hard
JIS G4403(83)	SKH53	Bar Frg Ann	1.10-1.25		3.80-4.50	0.40 max	4.60-6.20	0.40 max	2.80-3.30	5.50-6.70	Cu <=0.25; Ni <=0.25; P <=0.030; S <=0.030; bal Fe				269 HB max

Mexico

Specification	Designation	Notes	C	Co	Cr	Mn	Mo	Si	V	W	Other	UTS	YS	El	Hard
NMX-B-082(90)	M3 Class 2	Bar	1.15-1.25		3.75-4.50	0.15-0.40	4.75-6.50	0.20-0.45	2.75-3.25	5.00-6.75	Cu <=0.25; Ni <=0.30; P <=0.03; S <=0.03; (As+Sn+Sb)<=0.040; (Cu+Ni)<=0.40; B, Ti may be added; bal Fe				

Pan America

Specification	Designation	Notes	C	Co	Cr	Mn	Mo	Si	V	W	Other	UTS	YS	El	Hard
COPANT 337	M3	Bar, Frg	1.15-2.25		3.75-4.5	0.15-0.4	4.75-6.5	0.2-0.45	2.75-3.25	5-6.75	P 0.03; S 0.03; bal Fe				

Spain

Specification	Designation	Notes	C	Co	Cr	Mn	Mo	Si	V	W	Other	UTS	YS	El	Hard
UNE 36073(75)	6-5-3		1.15-1.3		3.5-4.5	0.4 max	4.6-5.3	0.45 max	2.7-3.2	5.7-6.7	P <=0.03; S <=0.03; bal Fe				
UNE 36073(75)	EM3		1.15-1.3		3.5-4.5	0.4 max	4.6-5.3	0.45 max	2.7-3.2	5.7-6.7	P <=0.03; S <=0.03; bal Fe				
UNE 36073(75)	F.5605*	see EM3, 6-5-3	1.15-1.3		3.5-4.5	0.4 max	4.6-5.3	0.45 max	2.7-3.2	5.7-6.7	P <=0.03; S <=0.03; bal Fe				

Sweden

Specification	Designation	Notes	C	Co	Cr	Mn	Mo	Si	V	W	Other	UTS	YS	El	Hard
SS 142725	2725		1.23-1.33		3.5-4.5	0.5 max	4.6-5.3	0.25-0.7	2.7-3.2	5.8-6.8	P <=0.05; S <=0.17; bal Fe				
SS 142726	2726		1.23-1.33	8-9	3.5-4.5	0.5 max	4.6-5.3	0.25-0.7	2.7-3.2	5.8-6.8	P <=0.05; S <=0.17; bal Fe				

USA

Specification	Designation	Notes	C	Co	Cr	Mn	Mo	Si	V	W	Other	UTS	YS	El	Hard
	AISI M3 Class 2	Ann Hard	1.15-1.25		3.75-4.50	0.15-0.40	4.75-6.50	0.20-0.45	2.75-4.50	5.00-6.75	P <=0.03; S <=0.03; bal Fe				223-255 HB
	UNS T11323		1.15-1.25		3.75-4.50	0.15-0.40	4.75-6.50	0.20-0.45	2.75-3.25	5.00-6.75	P <=0.03; S <=0.03; bal Fe				
ASTM A600(92)	M3 Class 2	Bar Frg Plt Sh Strp, Ann	1.15-1.25		3.75-4.50	0.15-0.40	4.75-6.50	0.20-0.45	2.75-3.25	5.00-6.75	P <=0.03; S <=0.03; S 0.06-0.15 for machinability; (Ni+Cu)<=0.75; bal Fe				255 HB
FED QQ-T-590C(85)	M3-2*	Obs; see ASTM A600	1.15-1.25		3.75-4.5	0.15-0.4	4.75-6.5	0.2-0.4	2.75-3.25	5-6.75	Ni 0.3; bal Fe				
SAE J437(70)	M3	Ann									bal Fe				223-255 HB
SAE J438(70)	M3		1.00-1.25		3.75-4.50	0.20-0.40	4.75-6.25	0.20-0.40	2.25-3.25	5.50-6.75	bal Fe				

Specification	Designation	Notes	C	Co	Cr	Mn	Mo	Si	V	W	Other	UTS	YS	El	Hard

Tool Steel, Molybdenum High-Speed, M30

Mexico

Specification	Designation	Notes	C	Co	Cr	Mn	Mo	Si	V	W	Other	UTS	YS	El	Hard
NMX-B-082(90)	M30	Bar	0.75-0.85	4.50-5.50	3.50-4.25	0.15-0.40	7.75-9.00	0.20-0.45	1.00-1.40	1.30-2.30	Cu <=0.25; Ni <=0.30; P <=0.03; S <=0.03; (As+Sn+Sb)<=0.040; (Cu+Ni)<=0.40; B, Ti may be added; bal Fe				

UK

Specification	Designation	Notes	C	Co	Cr	Mn	Mo	Si	V	W	Other	UTS	YS	El	Hard
BS 4659	BM34		0.85-0.95	7.75-8.75	3.75-4.5	0.4	8-9	0.4	1.75-2.05	1.7-2.2	bal Fe				

USA

Specification	Designation	Notes	C	Co	Cr	Mn	Mo	Si	V	W	Other	UTS	YS	El	Hard
	AISI M30	Less common Ann Hard	0.80	5.00	4.00		8.00		1.25	2.00	P <=0.03; bal Fe				
	UNS T11330		0.75-0.85	4.50-5.50	3.50-4.25	0.15-0.40	7.75-9.00	0.20-0.45	1.00-1.40	1.30-2.30	P <=0.03; S <=0.03; bal Fe				
ASTM A600(92)	M30	Bar Frg Plt Sh Strp, Ann	0.75-0.85	4.50-5.50	3.50-4.25	0.15-0.40	7.75-9.00	0.20-0.45	1.00-1.40	1.30-2.30	P <=0.03; S <=0.03; S 0.06-0.15 for machinability; (Ni+Cu)<=0.75; bal Fe				269 HB
FED QQ-T-590C(85)	M30*	Obs; see ASTM A600	0.75-0.85	4.5-5.5	3.5-4.25	0.15-0.4	7.75-9	0.2-0.45	1-1.4	1.3-2.3	Ni 0.3; bal Fe				

Tool Steel, Molybdenum High-Speed, M33

Mexico

Specification	Designation	Notes	C	Co	Cr	Mn	Mo	Si	V	W	Other	UTS	YS	El	Hard
NMX-B-082(90)	M33	Bar	0.85-0.92	7.75-8.75	3.50-4.00	0.15-0.40	9.00-10.00	0.15-0.50	1.00-1.35	1.30-2.10	Cu <=0.25; Ni <=0.30; P <=0.03; S <=0.03; (As+Sn+Sb)<=0.040; (Cu+Ni)<=0.40; B, Ti may be added; bal Fe				

Spain

Specification	Designation	Notes	C	Co	Cr	Mn	Mo	Si	V	W	Other	UTS	YS	El	Hard
UNE 36073(75)	2-9-2-8		0.85-0.95	7.75-8.75	3.5-4.5	0.4 max	8-9	0.45 max	1.75-2.05	1.7-2.2	P <=0.03; S <=0.03; bal Fe				
UNE 36073(75)	EM34		0.85-0.95	7.75-8.75	3.5-4.5	0.4 max	8-9	0.45 max	1.75-2.05	1.7-2.2	P <=0.03; S <=0.03; bal Fe				

UK

Specification	Designation	Notes	C	Co	Cr	Mn	Mo	Si	V	W	Other	UTS	YS	El	Hard
BS 4659	BM34	Bar Rod Sh Strp Frg	0.85-0.95	7.75-8.75	3.75-4.5	0.4 max	8-9	0.4 max	1.75-2.05	1.7-2.2	bal Fe				

USA

Specification	Designation	Notes	C	Co	Cr	Mn	Mo	Si	V	W	Other	UTS	YS	El	Hard
	AISI M33	Ann Hard	0.85-0.92	7.75-8.75	3.50-4.00	0.15-0.40	9.00-10.00	0.15-0.50	1.00-1.35	1.30-2.10	P <=0.03; S <=0.03; bal Fe				235-269 HB
	UNS T11333		0.85-0.92	7.75-8.75	3.50-4.00	0.15-0.40	9.00-10.00	0.25-0.55	1.00-1.35	1.30-2.10	P <=0.03; S <=0.03; bal Fe				
ASTM A600(92)	M33	Bar Frg Plt Sh Strp, Ann	0.85-0.92	7.75-8.75	3.50-4.00	0.15-0.40	9.00-10.00	0.15-0.50	1.00-1.35	1.30-2.10	P <=0.03; S <=0.03; S 0.06-0.15 for machinability; (Ni+Cu)<=0.75; bal Fe				269 HB
FED QQ-T-590C(85)	M33*	Obs; see ASTM A600	0.85-0.92	7.75-8.75	3.5-4	0.15-0.4	9-10	0.15-0.5	1-1.35	1.3-2.1	Ni 0.3; bal Fe				

Tool Steel, Molybdenum High-Speed, M34

Germany

Specification	Designation	Notes	C	Co	Cr	Mn	Mo	Si	V	W	Other	UTS	YS	El	Hard
DIN	HS-2-9-2-8	As Tmp 550C	0.85-0.92	7.75-8.75	3.50-4.20	0.40 max	8.00-9.20	0.45 max	1.80-2.20	1.50-2.00	P <=0.030; S <=0.030; bal Fe				64 HRC
DIN	WNr 1.3249	As Tmp 550C	0.85-0.92	7.75-8.75	3.50-4.20	0.40 max	8.00-9.20	0.45 max	1.80-2.20	1.50-2.00	P <=0.030; S <=0.030; bal Fe				64 HRC

India

Specification	Designation	Notes	C	Co	Cr	Mn	Mo	Si	V	W	Other	UTS	YS	El	Hard
IS 1570/6(96)	THS8		0.85-0.95	4.7-5.2	3.5-4.5	0.4 max	4.6-5.3	0.5 max	1.7-2.2	5.7-6.7	Ni <=0.4; P <=0.03; S <=0.03; bal Fe				
IS 1570/6(96)	XT90W6Co5Mo5Cr4V2		0.85-0.95	4.7-5.2	3.5-4.5	0.4 max	4.6-5.3	0.5 max	1.7-2.2	5.7-6.7	Ni <=0.4; P <=0.03; S <=0.03; bal Fe				

Specification	Designation	Notes	C	Co	Cr	Mn	Mo	Si	V	W	Other	UTS	YS	El	Hard

Tool Steel, Molybdenum High-Speed, M34 (Continued from previous page)

Mexico

Specification	Designation	Notes	C	Co	Cr	Mn	Mo	Si	V	W	Other	UTS	YS	El	Hard
NMX-B-082(90)	M34	Bar	0.85-0.92	7.75-8.75	3.50-4.00	0.15-0.40	7.75-9.20	0.20-0.45	1.90-2.30	1.40-2.10	Cu <=0.25; Ni <=0.30; P <=0.03; S <=0.03; (As+Sn+Sb)<=0.040; (Cu+Ni)<=0.40; B, Ti may be added; bal Fe				

Spain

Specification	Designation	Notes	C	Co	Cr	Mn	Mo	Si	V	W	Other	UTS	YS	El	Hard
UNE 36073(75)	2-9-2-8		0.85-0.95	7.75-8.75	3.5-4.5	0.4 max	8-9	0.45 max	1.75-2.05	1.7-2.2	P <=0.03; S <=0.03; bal Fe				
UNE 36073(75)	EM34		0.85-0.95	7.75-8.75	3.5-4.5	0.4 max	8-9	0.45 max	1.75-2.05	1.7-2.2	P <=0.03; S <=0.03; bal Fe				
UNE 36073(75)	F.5611*	see 2-9-2-8, EM34	0.85-0.95	7.75-8.75	3.5-4.5	0.4 max	8-9	0.45 max	1.75-2.05	1.7-2.2	P <=0.03; S <=0.03; bal Fe				

UK

Specification	Designation	Notes	C	Co	Cr	Mn	Mo	Si	V	W	Other	UTS	YS	El	Hard
BS 4659	BM34		0.85-0.95	7.75-8.75	3.75-4.5	0.4 max	8-9	0.4 max	1.75-2.05	1.7-2.2	bal Fe				

USA

Specification	Designation	Notes	C	Co	Cr	Mn	Mo	Si	V	W	Other	UTS	YS	El	Hard
	AISI M34	Ann Hard	0.85-0.92	7.75-8.75	3.50-4.00	0.15-0.40	7.75-9.20	0.20-0.45	1.90-2.30	1.40-2.10	P <=0.03; S <=0.03; bal Fe				235-269 HB
	UNS T11334		0.85-0.92	7.75-8.75	3.50-4.00	0.15-0.40	7.75-9.20	0.20-0.45	1.9-2.30	1.40-2.10	P <=0.03; S <=0.03; bal Fe				
ASTM A600(92)	M34	Bar Frg Plt Sh Strp, Ann	0.85-0.92	7.75-8.75	3.50-4.00	0.15-0.40	7.75-9.20	0.20-0.45	1.90-2.30	1.40-2.10	P <=0.03; S <=0.03; S 0.06-0.15 for machinability; (Ni+Cu)<=0.75; bal Fe				269 HB
FED QQ-T-590C(85)	M34*	Obs; see ASTM A600	0.85-0.92	7.75-8.75	3.5-4	0.15-0.4	7.75-9.2	0.2-0.45	1.9-2.3	1.4-2.1	Ni 0.3; bal Fe				

Tool Steel, Molybdenum High-Speed, M35

France

Specification	Designation	Notes	C	Co	Cr	Mn	Mo	Si	V	W	Other	UTS	YS	El	Hard
AFNOR NFA35590(92)	HS6-5-2-5	Soft ann	0.8-0.87	4.5-5.2	3.5-4.5	0.4 max	4.6-5.3	0.5 max	1.7-2.2	5.7-6.7	Cu <=0.3; Ni <=0.4; P <=0.03; S <=0.03; bal Fe				270 HB max
AFNOR NFA35590(92)	HS6-5-2-5HC	Soft ann	0.88-0.96	4.5-5.2	3.5-4.5	0.4 max	4.6-5.3	0.5 max	1.7-2.2	5.7-6.7	Cu <=0.3; Ni <=0.4; P <=0.03; S <=0.03; bal Fe				275 HB max
AFNOR NFA35590(92)	Z85WDKCV06-05-05-04-02	Soft ann	0.8-0.87	4.5-5.2	3.5-4.5	0.4 max	4.6-5.3	0.5 max	1.7-2.2	5.7-6.7	Cu <=0.3; Ni <=0.4; P <=0.03; S <=0.03; bal Fe				270 HB max
AFNOR NFA35590(92)	Z90WDKCV06-05-05-04-02	Soft ann	0.88-0.96	4.5-5.2	3.5-4.5	0.4 max	4.6-5.3	0.5 max	1.7-2.2	5.7-6.7	Cu <=0.3; Ni <=0.4; P <=0.03; S <=0.03; bal Fe				275 HB max

Tool Steel, Molybdenum High-Speed, M36

Argentina

Specification	Designation	Notes	C	Co	Cr	Mn	Mo	Si	V	W	Other	UTS	YS	El	Hard
IAS	IRAM M36	Ann	0.80-0.90	7.75-8.75	3.75-4.50	0.15-0.40	4.50-5.50	0.20-0.45	1.75-2.25	5.50-6.50	P <=0.03; S <=0.03; bal Fe				260 HB

Finland

Specification	Designation	Notes	C	Co	Cr	Mn	Mo	Si	V	W	Other	UTS	YS	El	Hard
SFS 917(73)	SFS917	Alloyed	0.84-0.92	4.5-5.5	3.5-4.5	0.2-0.4	4.5-5.5	0.15-0.4	1.7-2.1	6.0-7.0	P <=0.03; S <=0.03; bal Fe				

France

Specification	Designation	Notes	C	Co	Cr	Mn	Mo	Si	V	W	Other	UTS	YS	El	Hard
AFNOR NFA35590	Z85WDKCV06-05-05-04		0.8-0.87	4.5-5.2	3.5-4.5	0.4 max	4.6-5.3	0.5 max	1.7-2.2	5.7-6.7	P <=0.03; S <=0.03; bal Fe				

Germany

Specification	Designation	Notes	C	Co	Cr	Mn	Mo	Si	V	W	Other	UTS	YS	El	Hard
DIN 17350(80)	HS-6-5-2-5	As Tmp 550C	0.88-0.96	4.50-5.00	3.80-4.50	0.40 max	4.70-5.20	0.45 max	1.70-2.00	6.00-6.70	P <=0.030; S <=0.030; bal Fe				64 HRC
DIN 17350(80)	WNr 1.3243	As Tmp 550C	0.88-0.96	4.50-5.00	3.80-4.50	0.40 max	4.70-5.20	0.45 max	1.70-2.00	6.00-6.70	P <=0.030; S <=0.030; bal Fe				64 HRC

Japan

Specification	Designation	Notes	C	Co	Cr	Mn	Mo	Si	V	W	Other	UTS	YS	El	Hard
JIS G4403(83)	SKH56	Bar Frg Ann	0.85-0.95	7.00-9.00	3.80-4.50	0.40 max	4.60-5.30	0.40 max	1.70-2.30	5.70-6.70	Cu <=0.25; Ni <=0.25; P <=0.030; S <=0.030; bal Fe				285 HB max

Tool Steel, Molybdenum High-Speed, M36 (Continued from previous page)

Specification	Designation	Notes	C	Co	Cr	Mn	Mo	Si	V	W	Other	UTS	YS	El	Hard
Mexico															
NMX-B-082(90)	M35	Bar	0.80-0.85	4.75-5.25	3.90-4.40	0.10-0.40	4.75-5.25	0.10-0.40	1.75-2.15	6.15-6.65	Cu <=0.25; Ni <=0.30; P <=0.03; S <=0.03; (As+Sn+Sb)<=0.040; (Cu+Ni)<=0.40; B, Ti may be added; bal Fe				
NMX-B-082(90)	M36	Bar	0.80-0.90	7.75-8.75	3.75-4.50	0.15-0.40	4.50-5.50	0.20-0.45	1.75-2.25	5.50-6.50	Cu <=0.25; Ni <=0.30; P <=0.03; S <=0.03; (As+Sn+Sb)<=0.040; (Cu+Ni)<=0.40; B, Ti may be added; bal Fe				
Pan America															
COPANT 337	M36	Bar, Frg	0.8-0.9	7.75-8.75	3.75-4.5	0.15-0.4	4.5-5.5	0.2-0.45	1.75-2.25	5.5-6.5	P 0.03; S 0.03; bal Fe				
Russia															
GOST 19265	R6M5K5		0.84-0.92	4.8-5.3	3.8-4.3	0.5 max	4.8-5.3	0.5 max	1.7-2.2	6-7	Ni <=0.4; P <=0.04; S <=0.03; bal Fe				
Spain															
UNE	F.550.C		0.8-0.9	4.25-5.25	3.75-4.75	0.15-0.45	4-6	0.18-0.43	1.75-2.25	5.25-7.25	P <=0.03; S <=0.03; bal Fe				
UNE 36073(75)	6-5-2-5		0.85-0.95	4.7-5.2	3.5-4.5	0.4 max	4.7-5.4	0.45 max	1.7-2.2	5.7-6.7	P <=0.03; S <=0.03; bal Fe				
UNE 36073(75)	EM35		0.85-0.95	4.7-5.2	3.5-4.5	0.4 max	4.7-5.4	0.45 max	1.7-2.2	5.7-6.7	P <=0.03; S <=0.03; bal Fe				
UNE 36073(75)	F.5613*	see 6-5-2-5, EM35	0.85-0.95	4.7-5.2	3.5-4.5	0.4 max	4.7-5.4	0.45 max	1.7-2.2	5.7-6.7	P <=0.03; S <=0.03; bal Fe				
Sweden															
SS 14	(USA M36)		1.1	9.5	4.3	0.3	5	0.2	2.6	6.5	bal Fe				
USA															
	AISI M36	Ann Hard	0.80-0.90	7.75-8.75	3.75-4.50	0.15-0.40	4.50-5.50	0.20-0.45	1.75-2.25	5.50-6.50	P <=0.03; S <=0.03; bal Fe				235-269 HB
	UNS T11336		0.80-0.90	7.75-8.75	3.75-4.50	0.15-0.40	4.50-5.50	0.20-0.45	1.75-2.25	5.50-6.50	P <=0.03; S <=0.03; bal Fe				
ASTM A600(92)	M36	Bar Frg Plt Sh Strp, Ann	0.80-0.90	7.75-8.75	3.75-4.50	0.15-0.40	4.50-5.50	0.20-0.45	1.75-2.25	5.50-6.50	P <=0.03; S <=0.03; S 0.06-0.15 for machinability; (Ni+Cu)<=0.75; bal Fe				269 HB
FED QQ-T-590C(85)	M36*	Obs; see ASTM A600	0.8-0.9	7.75-8.75	3.75-4.5	0.15-0.4	4.5-5.5	0.2-0.45	1.75-2.25	5.5-6.5	Ni 0.3; bal Fe				

Tool Steel, Molybdenum High-Speed, M4

Specification	Designation	Notes	C	Co	Cr	Mn	Mo	Si	V	W	Other	UTS	YS	El	Hard
Japan															
JIS G4403(83)	SKH54	Bar Frg Ann	1.25-1.40		3.80-4.50	0.40 max	4.50-5.50	0.40 max	3.90-4.50	5.30-6.50	Cu <=0.25; Ni 0.25; P <=0.030; S <=0.030; bal Fe				269 HB max
Mexico															
NMX-B-082(90)	M15	Bar	1.50-1.60	4.75-5.25	4.00-4.75	0.10-0.40	3.00-5.00	0.10-0.40	4.75-5.25	6.25-6.75	Cu <=0.25; Ni <=0.30; P <=0.03; S <=0.03; (As+Sn+Sb)<=0.040; (Cu+Ni)<=0.40; B, Ti may be added; bal Fe				
NMX-B-082(90)	M4	Bar	1.25-1.40		3.75-4.75	0.15-0.40	4.25-5.50	0.20-0.45	3.75-4.50	5.25-6.50	Cu <=0.25; Ni <=0.30; P <=0.03; S <=0.03; (As+Sn+Sb)<=0.040; (Cu+Ni)<=0.40; B, Ti may be added; bal Fe				
Pan America															
COPANT 337	M4	Bar, Frg	1.25-1.4		3.75-4.75	0.15-0.4	4.25-5.5	0.2-0.45	3.75-4.5	5.25-6.5	P 0.03; S 0.03; bal Fe				
UK															
BS 4659	BM4	Bar Rod Sh Strp Frg	1.25-1.4	0.6	3.75-4.5	0.4	4.25-5	0.4	3.75-4.25	5.75-6.5	bal Fe				

UNS numbers and US grades are provided as a means of cross referencing chemically similar alloys. Exchangability is only possible after independent examination of specifications. Tensile properties are minimum or typical as specified. UTS and YS as MPa. El as %. See Appendix for list of abbreviations used in Notes. * indicates obsolete material.

Specification	Designation	Notes	C	Co	Cr	Mn	Mo	Si	V	W	Other	UTS	YS	El	Hard

Tool Steel, Molybdenum High-Speed, M4 (Continued from previous page)

USA

Specification	Designation	Notes	C	Co	Cr	Mn	Mo	Si	V	W	Other	UTS	YS	El	Hard
	AISI M4	Ann Hard	1.25-1.40		3.75-4.75	0.15-0.40	4.25-5.50	0.20-0.45	3.75-4.50	5.25-6.50	P <=0.03; S <=0.03; bal Fe				223-255 HB
	UNS T11304		1.25-1.40		3.75-4.75	0.15-0.40	4.25-5.50	0.20-0.45	3.75-4.50	5.25-6.50	P <=0.03; S <=0.03; bal Fe				
ASTM A600(92)	M4	Bar Frg Plt Sh Strp, Ann	1.25-1.40		3.75-4.75	0.15-0.40	4.25-5.50	0.20-0.45	3.75-4.50	5.25-6.50	P <=0.03; S <=0.03; S 0.06-0.15 for machinability; (Ni+Cu)<=0.75; bal Fe				255 HB
FED QQ-T-590C(85)	M4*	Obs; see ASTM A600	1.25-1.4		3.75-4.75	0.15-0.4	4.25-5.5	0.2-0.45	3.75-4.5	5.25-6.5	Ni 0.3; bal Fe				
SAE J437(70)	M4	Ann									bal Fe				229-255 HB
SAE J438(70)	M4		1.25-1.40		4.00-4.75	0.20-0.40	4.50-5.50	0.20-0.40	3.90-4.50	5.25-6.50	bal Fe				

Tool Steel, Molybdenum High-Speed, M41

Bulgaria

Specification	Designation	Notes	C	Co	Cr	Mn	Mo	Si	V	W	Other	UTS	YS	El	Hard
BDS 7008(86)	11R3M3F2		1.02-1.12	0.50 max	3.80-4.50	0.60 max	2.50-3.00	0.50 max	2.30-2.70	2.50-3.30	Cu <=0.30; N 0.05-0.10; Nb 0.05-0.20; Ni <=0.40; P <=0.030; S <=0.030; Ti <=0.05; bal Fe				
BDS 7008(86)	11R3M3F2a		1.02-1.12	0.50 max	3.80-4.50	0.60 max	2.50-3.00	0.50 max	2.30-2.70	2.50-3.30	Cu <=0.30; Nb 0.05-0.20; Ni <=0.40; P <=0.030; S <=0.030; Ti <=0.05; bal Fe				
BDS 7008(86)	R6M5K5		0.84-0.92	4.70-5.20	3.80-4.50	0.60 max	4.80-5.30	0.50 max	1.70-2.10	5.70-6.70	Cu <=0.30; Ni <=0.40; P <=0.030; S <=0.030; Ti <=0.05; bal Fe				
BDS 7008(86)	R9K5		0.80-1.00	5.50-6.00	3.80-4.40	0.4	1.00 max	0.4	2.00-2.60	9.50-10.5	bal Fe				

China

Specification	Designation	Notes	C	Co	Cr	Mn	Mo	Si	V	W	Other	UTS	YS	El	Hard
GB 9943(88)	W7Mo4Cr4V2Co5	Bar Q/T	1.05-1.15	4.75-5.75	3.75-4.50	0.20-0.60	3.25-4.75	0.15-0.50	1.75-2.25	6.25-7.00	Ni <=0.30; P <=0.03; S <=0.030; bal Fe				66 HRC

Czech Republic

Specification	Designation	Notes	C	Co	Cr	Mn	Mo	Si	V	W	Other	UTS	YS	El	Hard
CSN 419851	19851		1.05-1.15	4.5-5.5	3.8-4.6	0.45 max	3.3-4.3	0.45 max	1.5-2.2	6.2-7.7	P <=0.035; S <=0.035; bal Fe				
CSN 419852	19852		0.8-0.9	4.3-5.2	3.8-4.6	0.45 max	4.5-5.5	0.45 max	1.5-2.2	5.5-7.0	P <=0.035; S <=0.035; bal Fe				

France

Specification	Designation	Notes	C	Co	Cr	Mn	Mo	Si	V	W	Other	UTS	YS	El	Hard
AFNOR	HS7-5-5-5		1.5	5	4	0.3	3.5	0.3	5	6.5	bal Fe				
AFNOR NFA35590	Z110WKCDV07-05-04-04		1.05-1.15	4.7-5.2	3.5-4.5	0.4 max	3.5-4.2	0.5 max	1.7-2.2	6.4-7.4	P <=0.03; S <=0.03; bal Fe				
AFNOR NFA35590(78)	HS6-5-2-5		0.8-0.87	4.5-5.2	3.5-4.5	0.4 max	4.6-5.3	0.5 max	1.7-2.2	5.7-6.7	Cu <=0.3; Ni <=0.4; P <=0.03; S <=0.03; bal Fe				64 HRC
AFNOR NFA35590(78)	HS7-4-2-5	Soft ann	1.05-1.15	4.7-5.2	3.5-4.5	0.4 max	3.5-4.2	0.5 max	1.7-2.2	6.4-7.4	Cu <=0.3; Ni <=0.4; P <=0.03; S <=0.03; bal Fe				66 HRC
AFNOR NFA35590(78)	Z110WKCDV07-05-04-04-02	Soft ann	1.05-1.15	4.7-5.2	3.5-4.5	0.4 max	3.5-4.2	0.5 max	1.7-2.2	6.4-7.4	Cu <=0.3; Ni <=0.4; P <=0.03; S <=0.03; bal Fo				66 HRC
AFNOR NFA35590(78)	Z85WDKCV06-05-05-04-02		0.8-0.87	4.5-5.2	3.5-4.5	0.4 max	4.6-5.3	0.5 max	1.7-2.2	5.7-6.7	Cu <=0.3; Ni <=0.4; P <=0.03; S <=0.03; bal Fe				64 HRC
AFNOR NFA35590(92)	HS7-4-2-5	Soft ann	1.05-1.15	4.7-5.2	3.5-4.5	0.4 max	3.5-4.2	0.5 max	1.7-2.2	6.4-7.4	Cu <=0.3; Ni <=0.4; P <=0.03; S <=0.03; bal Fe				280 HB max
AFNOR NFA35590(92)	Z110WKCDV07-05-04-04-02	Soft ann	1.05-1.15	4.7-5.2	3.5-4.5	0.4 max	3.5-4.2	0.5 max	1.7-2.2	6.4-7.4	Cu <=0.3; Ni <=0.4; P <=0.03; S <=0.03; bal Fe				280 HB max

Germany

Specification	Designation	Notes	C	Co	Cr	Mn	Mo	Si	V	W	Other	UTS	YS	El	Hard
DIN 17350(80)	HS-6-5-2-5S	As Tmp 540C	0.88-0.96	4.50-5.00	3.80-4.50	0.40 max	4.70-5.20	0.45 max	1.70-2.00	6.00-6.70	P <=0.030; S 0.06-0.15; bal Fe				64 HRC
DIN 17350(80)	HS-7-4-2-5	As Tmp 540C	1.05-1.15	4.80-5.20	3.80-4.50	0.40 max	3.60-4.00	0.45 max	1.70-1.90	6.60-7.10	P <=0.030; S <=0.030; bal Fe				64 HRC

UNS numbers and US grades are provided as a means of cross referencing chemically similar alloys. Exchangability is only possible after independent examination of specifications. Tensile properties are minimum or typical as specified. UTS and YS as MPa. El as %. See Appendix for list of abbreviations used in Notes. * indicates obsolete material.

Specification	Designation	Notes	C	Co	Cr	Mn	Mo	Si	V	W	Other	UTS	YS	El	Hard

Tool Steel, Molybdenum High-Speed, M41 (Continued from previous page)

Germany

Specification	Designation	Notes	C	Co	Cr	Mn	Mo	Si	V	W	Other	UTS	YS	El	Hard
DIN 17350(80)	WNr 1.3245	As Tmp 540C	0.88-0.96	4.50-5.00	3.80-4.50	0.40 max	4.70-5.20	0.45 max	1.70-2.00	6.00-6.70	P <=0.030; S 0.06-0.15; bal Fe				64 HRC
DIN 17350(80)	WNr 1.3246	As Tmp 540C	1.05-1.15	4.80-5.20	3.80-4.50	0.40 max	3.60-4.00	0.45 max	1.70-1.90	6.60-7.10	P <=0.030; S <=0.030; bal Fe				64 HRC

Hungary

Specification	Designation	Notes	C	Co	Cr	Mn	Mo	Si	V	W	Other	UTS	YS	El	Hard
MSZ 4351(84)	R8	Q/T	0.82-0.92	4.7-5.2	3.8-4.5	0.4 max	4.8-5.3	0.4 max	1.7-2.1	6-7	P <=0.03; S <=0.03; bal Fe				64 HRC
MSZ 4351(84)	R8	Soft ann	0.82-0.92	4.7-5.2	3.8-4.5	0.4 max	4.8-5.3	0.4 max	1.7-2.1	6-7	P <=0.03; S <=0.03; bal Fe				269 HB max
MSZ 4351(84)	R9	Q/T	1.05-1.15	4.7-5.2	3.8-4.5	0.4 max	3.8-4.3	0.4 max	2.7-3.2	6.5-7.5	P <=0.03; S <=0.03; bal Fe				66 HRC
MSZ 4351(84)	R9	Soft ann	1.05-1.15	4.7-5.2	3.8-4.5	0.4 max	3.8-4.3	0.4 max	2.7-3.2	6.5-7.5	P <=0.03; S <=0.03; bal Fe				290 HB max

Italy

Specification	Designation	Notes	C	Co	Cr	Mn	Mo	Si	V	W	Other	UTS	YS	El	Hard
UNI 2955/5(82)	HS6-5-2-5		0.85-0.95	4.7-5.2	3.5-4.5	0.4 max	4.6-5.3	0.5 max	1.7-2.2	5.7-6.7	P <=0.07; S <=0.06; bal Fe				
UNI 2955/5(82)	HS7-4-2-5		1.05-1.2	4.7-5.2	3.5-4.5	0.4 max	3.5-4.2	0.5 max	1.7-2.2	6.4-7.4	P <=0.07; S <=0.06; bal Fe				

Japan

Specification	Designation	Notes	C	Co	Cr	Mn	Mo	Si	V	W	Other	UTS	YS	El	Hard
JIS G4403(83)	SKH55	Bar Frg Ann	0.85-0.95	4.50-5.50	3.80-4.50	0.40 max	4.60-5.30	0.40 max	1.70-2.30	5.70-6.70	Cu <=0.25; Ni <=0.25; P <=0.030; S <=0.030; bal Fe				277 HB max

Mexico

Specification	Designation	Notes	C	Co	Cr	Mn	Mo	Si	V	W	Other	UTS	YS	El	Hard
NMX-B-082(90)	M41	Bar	1.05-1.15	4.75-5.75	3.75-4.50	0.20-0.60	3.25-4.25	0.15-0.50	1.75-2.25	6.25-7.00	Cu <=0.25; Ni <=0.30; P <=0.03; S <=0.03; (As+Sn+Sb)<=0.040; (Cu+Ni)<=0.40; B, Ti may be added; bal Fe				

Poland

Specification	Designation	Notes	C	Co	Cr	Mn	Mo	Si	V	W	Other	UTS	YS	El	Hard
PNH85022	SK5M		0.85-0.95	4.5-5.5	3.5-4.5	0.4 max	4.6-5.2	0.5 max	1.7-2.1	6-6.7	Ni <=0.4; P <=0.030; S <=0.030; bal Fe				

Romania

Specification	Designation	Notes	C	Co	Cr	Mn	Mo	Si	V	W	Other	UTS	YS	El	Hard
STAS	Rp1		0.9-1	5-6	3.8-4.4	0.45 max		0.3 max	0.2-0.4	2-2.6	9-10.5	Ni <=0.4; P <=0.025; S <=0.02; bal Fe				

Russia

Specification	Designation	Notes	C	Co	Cr	Mn	Mo	Si	V	W	Other	UTS	YS	El	Hard
GOST 19265	R6M5K5		0.86-0.94	4.7-5.2	3.8-4.3	0.2-0.5	4.8-5.3	0.2-0.5	1.7-2.1	5.7-6.7	Cu <=0.25; Ni <=0.6; P <=0.03; S <=0.03; bal Fe				
GOST 19265	R9K5		0.9-1.0	5-6	3.8-4.4	0.2-0.5	1 max	0.2-0.5	2.3-2.7	9-10	Cu <=0.25; Ni <=0.6; P <=0.03; S <=0.03; bal Fe				
GOST 19265	R9K5		0.9-1.0	5-6	3.8-4.4	0.5 max	1 max	0.5 max	2.2-2.6	9-10.5	Ni <=0.4; P <=0.03; S <=0.03; bal Fe				
GOST 19265	R9M4K8		1.0-1.1	7.5-8.5	3-3.6	0.2-0.5	3.8-4.3	0.2-0.5	2.3-2.7	8.5-9.5	Cu <=0.25; Ni <=0.6; P <=0.03; S <=0.03; bal Fe				
GOST 19265	R9M4K8		1.0-1.1	7.5-8.5	3-3.6	0.5 max	3.8-4.3	0.5 max	2.1-2.5	8.5-9.6	Ni <=0.4; P <=0.035; S <=0.03; bal Fe				

Spain

Specification	Designation	Notes	C	Co	Cr	Mn	Mo	Si	V	W	Other	UTS	YS	El	Hard
UNE 36073(75)	6-5-2-5		0.85-0.95	4.7-5.2	3.5-4.5	0.4 max	4.7-5.4	0.45 max	1.7-2.2	5.7-6.7	P <=0.03; S <=0.03; bal Fe				
UNE 36073(75)	7-4-2-5		1.05-1.5	4.5-5.3	3.5-4.5	0.4 max	3.5-4.2	0.45 max	1.7-2.2	6.4-7.4	P <=0.03; S <=0.03; bal Fe				
UNE 36073(75)	EM35		0.85-0.95	4.7-5.2	3.5-4.5	0.4 max	4.7-5.4	0.45 max	1.7-2.2	5.7-6.7	P <=0.03; S <=0.03; bal Fe				
UNE 36073(75)	EM41		1.05-1.5	4.5-5.3	3.5-4.5	0.4 max	3.5-4.2	0.45 max	1.7-2.2	6.4-7.4	P <=0.03; S <=0.03; bal Fe				
UNE 36073(75)	F.5615*	see EM41, 7-4-2-5	1.05-1.5	4.5-5.3	3.5-4.5	0.4 max	3.5-4.2	0.45 max	1.7-2.2	6.4-7.4	P <=0.03; S <=0.03; bal Fe				

Sweden

Specification	Designation	Notes	C	Co	Cr	Mn	Mo	Si	V	W	Other	UTS	YS	El	Hard
SS 142723	2723		0.85-0.95	4.5-5.5	3.5-4.5	0.2-0.4	4.5-5.5	0.25-0.5	1.7-2.1	6-7	P <=0.03; S <=0.03; bal Fe				
SS 142724	2724	Alloyed	0.82-0.92		3.5-4.5	0.2-0.4	2.87-3.6	0.15-0.4	1.8-2.2	6-7	P <=0.03; S <=0.03; bal Fe				

UNS numbers and US grades are provided as a means of cross referencing chemically similar alloys. Exchangability is only possible after independent examination of specifications. Tensile properties are minimum or typical as specified. UTS and YS as MPa. El as %. See Appendix for list of abbreviations used in Notes. * indicates obsolete material.

Specification	Designation	Notes	C	Co	Cr	Mn	Mo	Si	V	W	Other	UTS	YS	El	Hard

Tool Steel, Molybdenum High-Speed, M41 (Continued from previous page)

Sweden

Specification	Designation	Notes	C	Co	Cr	Mn	Mo	Si	V	W	Other	UTS	YS	El	Hard
SS 142736	2736		1.2-1.3	8.5-9.5	3.5-4.5	0.2-0.4	3.2-3.9	0.3-0.5	3-3.5	8.5-9.5	P <=0.03; S <=0.03; bal Fe				

USA

Specification	Designation	Notes	C	Co	Cr	Mn	Mo	Si	V	W	Other	UTS	YS	El	Hard
	AISI M41	Ann Hard	1.05-1.15	4.75-5.75	3.75-4.50	0.20-0.60	3.25-4.25	0.15-0.50	1.75-2.25	6.25-7.00	P <=0.03; S <=0.03; bal Fe				235-269 HB
	UNS T11341		1.05-1.15	4.75-5.75	3.75-4.50	0.20-0.60	3.25-4.25	0.15-0.50	1.75-2.25	6.25-7.00	P <=0.03; S <=0.03; bal Fe				
ASTM A600(92)	M41	Bar Frg Plt Sh Strp, Ann	1.05-1.15	4.75-5.75	3.75-4.50	0.20-0.60	3.25-4.25	0.15-0.50	1.75-2.25	6.25-7.00	P <=0.03; S <=0.03; S 0.06-0.15 for machinability; (Ni+Cu)<=0.75; bal Fe				269 HB
FED QQ-T-590C(85)	M41*	Obs; see ASTM A600	1.05-1.15	4.75-5.75	3.75-4.5	0.2-0.6	3.25-4.25	0.15-0.5	1.75-2.25	6.75-7	Ni 0.3; bal Fe				

Tool Steel, Molybdenum High-Speed, M42

Brazil

Specification	Designation	Notes	C	Co	Cr	Mn	Mo	Si	V	W	Other	UTS	YS	El	Hard
ABNT	VKM-42		1.1	8	3.75	0.3	9.5	0.3	1.15	1.5	bal Fe				

Bulgaria

Specification	Designation	Notes	C	Co	Cr	Mn	Mo	Si	V	W	Other	UTS	YS	El	Hard
BDS 7008(86)	R2M8K5F2		1.00-1.10	4.70-5.20	3.80-4.50	0.60 max	8.00-9.00	0.50 max	1.70-2.10	1.50-2.00	Cu <=0.30; N 0.05-0.10; Nb 0.10-0.30; Ni <=0.40; P <=0.030; Pb <=0.1; S <=0.030; Ti <=0.05; bal Fe				

China

Specification	Designation	Notes	C	Co	Cr	Mn	Mo	Si	V	W	Other	UTS	YS	El	Hard
GB 9943(88)	W2Mo9Cr4VCo8	Bar Q/T	1.05-1.15	7.75-8.75	3.50-4.25	0.15-0.40	9.00-10.00	0.15-0.65	0.95-1.35	1.15-1.85	Ni <=0.30; P <=0.03; S <=0.030; bal Fe				66 HRC

Germany

Specification	Designation	Notes	C	Co	Cr	Mn	Mo	Si	V	W	Other	UTS	YS	El	Hard
DIN 17350(80)	HS-2-10-1-8	As Tmp 510C	1.05-1.12	7.50-8.50	3.60-4.40	0.40 max	9.00-10.0	0.45 max	1.00-1.30	1.20-1.80	P <=0.030; S <=0.030; bal Fe				67 HRC
DIN 17350(80)	WNr 1.3247	As Tmp 510C	1.05-1.12	7.50-8.50	3.60-4.40	0.40 max	9.00-10.0	0.45 max	1.00-1.30	1.20-1.80	P <=0.030; S <=0.030; bal Fe				67 HRC

Hungary

Specification	Designation	Notes	C	Co	Cr	Mn	Mo	Si	V	W	Other	UTS	YS	El	Hard
MSZ 4351(84)	R11	Soft ann	1.05-1.15	7.5-8.5	3.8-4.5	0.4 max	8-10	0.4 max	1-1.3	1.3-1.8	P <=0.03; S <=0.03; bal Fe				285 HB max
MSZ 4351(84)	R11	Q/T	1.05-1.15	7.5-8.5	3.8-4.5	0.4 max	8-10	0.4 max	1-1.3	1.3-1.8	P <=0.03; S <=0.03; bal Fe				66 HRC

India

Specification	Designation	Notes	C	Co	Cr	Mn	Mo	Si	V	W	Other	UTS	YS	El	Hard
IS 1570/6(96)	XT112W7Co5Cr4Mo4V2		1.05-1.2	4.7-5.2	3.5-4.5	0.4 max	3.5-4.2	0.5 max	1.7-2.2	6.4-7.4	Ni <=0.4; P <=0.03; S <=0.03; bal Fe				

Italy

Specification	Designation	Notes	C	Co	Cr	Mn	Mo	Si	V	W	Other	UTS	YS	El	Hard
UNI 2955	RM42 HS2-9-1-8		1.1	8.25	3.85	0.4	9.5	0.4	1.15	1.5	bal Fe				
UNI 2955/5(82)	HS2-9-1-8		1.05-1.2	7.5-8.5	3.5-4.5	0.4 max	9-10	0.5 max	0.9-1.4	1.3-1.9	P <=0.07; S <=0.06; bal Fe				

Japan

Specification	Designation	Notes	C	Co	Cr	Mn	Mo	Si	V	W	Other	UTS	YS	El	Hard
JIS G4403(83)	SKH59	Bar Frg Ann	1.00-1.15	7.50-8.50	3.50-4.50	0.40 max	9.00-10.00	0.50 max	0.90-1.40	1.20-1.90	Cu <=0.25; Ni <=0.25; P <=0.030; S <=0.030; bal Fe				277 HB max

Mexico

Specification	Designation	Notes	C	Co	Cr	Mn	Mo	Si	V	W	Other	UTS	YS	El	Hard
NMX-B-082(90)	M42	Bar	1.05-1.15	7.75-8.75	3.50-4.25	0.15-0.40	9.00-10.00	0.15-0.65	0.95-1.35	1.15-1.85	Cu <=0.25; Ni <=0.30; P <=0.03; S <=0.03; (As+Sn+Sb)<=0.040; (Cu+Ni)<=0.40; B, Ti may be added; bal Fe				

Poland

Specification	Designation	Notes	C	Co	Cr	Mn	Mo	Si	V	W	Other	UTS	YS	El	Hard
PNH84023/08	St1E	Chain	0.07-0.12		0.3 max	0.35-0.5		0.05 max			Ni <=0.3; P <=0.045; S <=0.045; bal Fe				
PNH85022	SK8M		1.05-1.2	7.5-8.5	3.5-4.5	0.4 max	9-10	0.5 max	1-1.4	1.3-1.9	Ni <=0.4; P <=0.030; S <=0.030; bal Fe				
PNH93027	St1E	Struct; Wire Bar	0.07-0.12		0.3 max	0.35-0.5		0.05 max			Ni <=0.3; P <=0.045; S <=0.045; P+S=0.-0.09				

Specification	Designation	Notes	C	Co	Cr	Mn	Mo	Si	V	W	Other	UTS	YS	El	Hard

Tool Steel, Molybdenum High-Speed, M42 (Continued from previous page)

Spain

Specification	Designation	Notes	C	Co	Cr	Mn	Mo	Si	V	W	Other	UTS	YS	El	Hard
UNE 36073(75)	2-10-1-8		1.05-1.2	7.5-8.5	3.5-4.5	0.4 max	9-10	0.45 max	0.9-1.4	1.3-1.8	P <=0.03; S <=0.03; bal Fe				
UNE 36073(75)	EM42		1.05-1.2	7.5-8.5	3.5-4.5	0.4 max	9-10	0.45 max	0.9-1.4	1.3-1.8	P <=0.03; S <=0.03; bal Fe				
UNE 36073(75)	F.5617*	see EM42, 2-10-1-8	1.05-1.2	7.5-8.5	3.5-4.5	0.4 max	9-10	0.45 max	0.9-1.4	1.3-1.8	P <=0.03; S <=0.03; bal Fe				

Sweden

Specification	Designation	Notes	C	Co	Cr	Mn	Mo	Si	V	W	Other	UTS	YS	El	Hard
SS 142716	2716		1.05-1.2	7.5-8.5	3.5-4.5	0.4 max	9-10	0.25-0.65	0.9-1.4	1.3-1.9	P <=0.03; S <=0.03; bal Fe				

UK

Specification	Designation	Notes	C	Co	Cr	Mn	Mo	Si	V	W	Other	UTS	YS	El	Hard
BS 4659	BM42	Bar Rod Sh Strp Frg	1-1.1	7.5-8.5	3.5-4.25	0.4 max	9-10	0.4 max	1-1.3	1-2	bal Fe				

USA

Specification	Designation	Notes	C	Co	Cr	Mn	Mo	Si	V	W	Other	UTS	YS	El	Hard
	AISI M42	Ann Hard	1.05-1.15	7.75-8.75	3.50-4.25	0.15-0.40	9.00-10.00	0.15-0.65	0.95-1.35	1.15-1.85	P <=0.03; S <=0.03; bal Fe				235-269 HB
	UNS T11342		1.05-1.15	7.75-8.75	3.50-4.25	0.15-0.40	9.00-10.00	0.15-0.65	0.95-1.35	1.15-1.85	P <=0.03; S <=0.03; bal Fe				
ASTM A600(92)	M42	Bar Frg Plt Sh Strp, Ann	1.05-1.15	7.75-8.75	3.50-4.25	0.15-0.40	9.00-10.00	0.15-0.65	0.95-1.35	1.15-1.85	P <=0.03; S <=0.03; S 0.06-0.15 for machinability; (Ni+Cu)<=0.75; bal Fe				269 HB
FED QQ-T-590C(85)	M42*	Obs; see ASTM A600	1.05-1.15	7.75-8.75	3.5-4.25	0.15-0.4	9-10	0.15-0.65	0.95-1.35	1.15-1.85	Ni 0.3; bal Fe				

Tool Steel, Molybdenum High-Speed, M43

France

Specification	Designation	Notes	C	Co	Cr	Mn	Mo	Si	V	W	Other	UTS	YS	El	Hard
AFNOR NFA35590(92)	HS2-9-1-8	Soft ann	1.05-1.15	7.5-8.5	3.5-4.5	0.4 max	9-10	0.5 max	1-1.3	1.3-1.9	Cu <=0.3; Ni <=0.4; P <=0.03; S <=0.03; bal Fe				280 HB max
AFNOR NFA35590(92)	Z110DKCWV09-08-04-02-01	Soft ann	1.05-1.15	7.5-8.5	3.5-4.5	0.4 max	9-10	0.5 max	1-1.3	1.3-1.9	Cu <=0.3; Ni <=0.4; P <=0.03; S <=0.03; bal Fe				280 HB max

Mexico

Specification	Designation	Notes	C	Co	Cr	Mn	Mo	Si	V	W	Other	UTS	YS	El	Hard
NMX-B-082(90)	M43	Bar	1.15-1.25	7.75-8.75	3.50-4.25	0.20-0.40	7.50-8.50	0.15-0.65	1.50-1.75	2.25-3.00	Cu <=0.25; Ni <=0.30; P <=0.03; S <=0.03; (As+Sn+Sb)<=0.040; (Cu+Ni)<=0.40; B, Ti may be added; bal Fe				

USA

Specification	Designation	Notes	C	Co	Cr	Mn	Mo	Si	V	W	Other	UTS	YS	El	Hard
	UNS T11343*	Obs	1.15-1.25	7.75-8.75	3.50-4.25	0.20-0.40	7.50-8.50	0.15-0.65	1.50-1.75	2.25-3.00	P <=0.03; S <=0.03; bal Fe				
ASTM A600(92)	M43	Bar Frg Plt Sh Strp, Ann	1.15-1.25	7.75-8.75	3.50-4.25	0.20-0.40	7.50-8.50	0.15-0.65	1.50-1.75	2.25-3.00	P <=0.03; S <=0.03; S 0.06-0.15 for machinability; (Ni+Cu)<=0.75; bal Fe				269 HB

Tool Steel, Molybdenum High-Speed, M44

France

Specification	Designation	Notes	C	Co	Cr	Mn	Mo	Si	V	W	Other	UTS	YS	El	Hard
AFNOR NFA35590(78)	Z130KWDCV12-07-06-04-03	Q/T	1.2-1.35	11.25-12.25	3.5-4.5	0.4 max	6-6.5	0.5 max	3-3.5	6.75-7.75	Cu <=0.3; Ni <=0.4; P <=0.03; S <=0.03; bal Fe				66 HRC
AFNOR NFA35590(92)	HS10-4-3-10	Soft ann	1.2-1.35	9.5-10.5	3.5-4.5	0.4 max	3.2-3.9	0.5 max	3-3.5	9-10	Cu <=0.3; Ni <=0.4; P <=0.03; S <=0.03; bal Fe				295 HB max
AFNOR NFA35590(92)	HS7-6-3-12	Soft ann	1.2-1.35	11.25-12.25	3.5-4.5	0.4 max	6-6.5	0.5 max	3-3.5	6.75-7.75	Cu <=0.3; Ni <=0.4; P <=0.03; S <=0.03; bal Fe				295 HB max
AFNOR NFA35590(92)	Z130KWDCV12-07-06-04-03	Soft ann	1.2-1.35	11.25-12.25	3.5-4.5	0.4 max	6-6.5	0.5 max	3-3.5	6.75-7.75	Cu <=0.3; Ni <=0.4; P <=0.03; S <=0.03; bal Fe				295 HB max
AFNOR NFA35590(92)	Z130WKCDV10-10-04-04-03	Soft ann	1.2-1.35	9.5-10.5	3.5-4.5	0.4 max	3.2-3.9	0.5 max	3-3.5	9-10	Cu <=0.3; Ni <=0.4; P <=0.03; S <=0.03; bal Fe				295 HB max

Specification	Designation	Notes	C	Co	Cr	Mn	Mo	Si	V	W	Other	UTS	YS	El	Hard
Tool Steel, Molybdenum High-Speed, M44 (Continued from previous page)															
Japan															
JIS G4403(83)	SKH57	Bar Frg Ann	1.20-1.35	9.00-11.00	3.80-4.50	0.40 max	3.00-4.00	0.40 max	3.00-3.70	9.00-11.00	Cu <=0.25; Ni <=0.25; P <=0.030; S <=0.030; bal Fe				293 HB max
Mexico															
NMX-B-082(90)	M44	Bar	1.10-1.20	11.00-12.25	4.00-4.75	0.20-0.40	6.00-7.00	0.30-0.55	1.85-2.20	5.00-5.75	Cu <=0.25; Ni <=0.30; P <=0.03; S <=0.03; (As+Sn+Sb)<=0.040; (Cu+Ni)<=0.40; B, Ti may be added; bal Fe				
USA															
	UNS T11344*	Obs	1.10-1.20	11.00-12.25	4.00-4.75	0.20-0.40	6.00-7.00	0.30-0.55	1.85-2.20	5.00-5.75	P <=0.03; S <=0.03; bal Fe				
ASTM A600(92)	M44	Bar Frg Plt Sh Strp, Ann	1.10-1.20	11.00-12.25	4.00-4.75	0.20-0.40	6.00-7.00	0.30-0.55	1.85-2.20	5.00-5.75	P <=0.03; S <=0.03; S 0.06-0.15 for machinability; (Ni+Cu)<=0.75; bal Fe				285 HB
Tool Steel, Molybdenum High-Speed, M46															
Mexico															
NMX-B-082(90)	M46	Bar	1.22-1.30	7.80-8.80	3.70-4.20	0.20-0.40	8.00-8.50	0.40-0.65	3.00-3.30	1.90-2.20	Cu <=0.25; Ni <=0.30; P <=0.03; S <=0.03; (As+Sn+Sb)<=0.040; (Cu+Ni)<=0.40; B, Ti may be added; bal Fe				
USA															
	AISI M46	Ann Hard	1.22-1.30	7.80-8.80	3.70-4.20	0.20-0.40	8.00-8.50	0.40-0.65	3.00-3.30	1.90-2.20	P <=0.03; S <=0.03; bal Fe				235-269 HB
	UNS T11346		1.22-1.3	7.80-8.80	3.70-4.20	0.20-0.40	8.00-8.50	0.40-0.65	3.00-3.30	1.90-2.20	P <=0.03; S <=0.03; bal Fe				
ASTM A600(92)	M46	Bar Frg Plt Sh Strp, Ann	1.22-1.30	7.80-8.80	4.00-4.75	0.20-0.40	8.00-8.50	0.40-0.65	3.00-3.30	1.90-2.20	P <=0.03; S <=0.03; S 0.06-0.15 for machinability; (Ni+Cu)<=0.75; bal Fe				269 HB
FED QQ-T-590C(85)	M46*	Obs; see ASTM A600	1.22-1.3	7.8-8.8	3.7-4.2	0.2-0.4	8-8.5	0.4-0.65	3-3.3	1.9-2.2	Ni 0.3; bal Fe				
Tool Steel, Molybdenum High-Speed, M47															
India															
IS 1570/6(96)	THS11		1.05-1.2	7.5-8.5	3.5-4.5	0.4 max	9-10	0.5 max	0.9-1.4	1.3-1.9	Ni <=0.4; P <=0.03; S <=0.03; bal Fe				
Mexico															
NMX-B-082(90)	M47	Bar	1.05-1.15	4.75-5.25	3.50-4.00	0.15-0.40	9.25-10.00	0.20-0.45	1.15-1.35	1.30-1.80	Cu <=0.25; Ni <=0.30; P <=0.03; S <=0.03; (As+Sn+Sb)<=0.040; (Cu+Ni)<=0.40; B, Ti may be added; bal Fe				
USA															
	UNS T11347*	Obs	1.05-1.15	4.75-5.25	3.50-4.00	0.15-0.40	9.25-10.00	0.20-0.45	1.15-1.35	1.30-1.80	P <=0.03; S <=0.03; bal Fe				
ASTM A600(92)	M47	Bar Frg Plt Sh Strp, Ann	1.05-1.15	4.75-5.25	3.50-4.00	0.15-0.40	9.25-10.00	0.20-0.45	1.15-1.35	1.30-1.80	P <=0.03; S <=0.03; S 0.06-0.15 for machinability; (Ni+Cu)<=0.75; bal Fe				269 HB
Tool Steel, Molybdenum High-Speed, M48															
Finland															
SFS 918(73)	SFS918	Alloyed	1.2-1.35	9.5-11.0	3.8-4.5	0.2-0.4	3.3-3.8	0.15-0.4	3.0-3.5	9.0-10.5	P <=0.03; S <=0.03; bal Fe				
India															
IS 1570/6(96)	TAC21	CW	1.5-1.7		11-13	0.25-0.55		0.8 max	0.1-0.35	0.8 max	Ni <=0.4; P <=0.035; S <=0.035; bal Fe				

UNS numbers and US grades are provided as a means of cross referencing chemically similar alloys. Exchangability is only possible after independent examination of specifications. Tensile properties are minimum or typical as specified. UTS and YS as MPa. El as %. See Appendix for list of abbreviations used in Notes. * indicates obsolete material.

Specification	Designation	Notes	C	Co	Cr	Mn	Mo	Si	V	W	Other	UTS	YS	El	Hard
Tool Steel, Molybdenum High-Speed, M48 (Continued from previous page)															
India															
IS 1570/6(96)	XT160Cr12	CW	1.5-1.7		11-13	0.25-0.55	0.8 max	0.1-0.35	0.8 max		Ni <=0.4; P <=0.035; S <=0.035; bal Fe				
Spain															
UNE 36073(75)	10-4-3-10		1.2-1.35	9.5-10.5	3.5-4.5	0.4 max	3.2-3.9	0.45 max	3-3.5	9-10.5	P <=0.03; S <=0.03; bal Fe				
UNE 36073(75)	ET00		1.2-1.35	9.5-10.5	3.5-4.5	0.4 max	3.2-3.9	0.45 max	3-3.5	9-10.5	P <=0.03; S <=0.03; bal Fe				
UNE 36073(75)	F.5553*	see ET00, 10-4-3-10	1.2-1.35	9.5-10.5	3.5-4.5	0.4 max	3.2-3.9	0.45 max	3-3.5	9-10.5	P <=0.03; S <=0.03; bal Fe				
USA															
	AISI M48	Ann Hard	1.45-1.55	8.00-10.00	3.50-4.25	0.15-0.40	4.75-5.50	0.20-0.45	2.75-3.25	9.50-10.50	P <=0.030; S 0.05-0.09; bal Fe				285-311 HB
	UNS T11348	Powder Metallurgy M-48	1.50	9	3.75		5.25		3.1	10	bal Fe				
ASTM A600(92)	M48	Bar Frg Plt Sh Strp, Ann	1.42-1.52	8.00-10.00	3.50-4.00	0.15-0.40	4.75-5.50	0.15-0.40	2.75-3.25	9.50-10.50	P <=0.03; S <=0.07; S 0.06-0.15 for machinability; (Ni+Cu)<=0.75; bal Fe				311 HB
Tool Steel, Molybdenum High-Speed, M50															
France															
AFNOR NFA35565(84)	16NCD13		0.12-0.17		0.9-1.15	0.3-0.6	0.15-0.3	0.15-0.35	0.1 max	0.1 max	Al <=0.05; Cu <=0.35; Ni 3-3.5; P <=0.015; S <=0.008; bal Fe				
AFNOR NFA35565(84)	80DCV40*	Bearing	0.77-0.85	0.25 max	3.75-4.5	0.1-0.35	4-4.5	0.1-0.35	0.9-1.1	0.25 max	Cu <=0.2; Ni <=0.2; P <=0.015; S <=0.015; bal Fe				
AFNOR NFA35565(94)	80DCV40*	see 80MoCrV42-16	0.77-0.85	0.5 max	3.9-4.4	0.15-0.35	4-4.5	0.1-0.35	0.9-1.1	0.25 max	Cu <=0.2; Ni <=0.15; P <=0.015; S <=0.008; P<=0.015/P<=0.03; S<=0.008/S<=0.03; bal Fe				
AFNOR NFA35565(94)	80MoCrV42-16*	Ball & roller bearing	0.77-0.85	0.5 max	3.9-4.4	0.15-0.35	4-4.5	0.1-0.35	0.9-1.1	0.25 max	Cu <=0.2; Ni <=0.15; P <=0.015; S <=0.008; P<=0.015/P<=0.03; S<=0.008/S<=0.03; bal Fe				
AFNOR NFA35590	Y80DCV42.16		0.77-0.85		3.75-4.5	0.1-0.4	3.75-4.5	0.1-0.4	0.9-1.2		bal Fe				
Germany															
DIN	81MoCrV42-16	Hard tmp to 500C	0.77-0.85		3.75-4.25	0.35 max	4.00-4.50	0.25 max	0.90-1.10		P <=0.030; S <=0.030; bal Fe				61 HRC
DIN	WNr 1.2369	Hard tmp to 500C	0.77-0.85		3.75-4.25	0.35 max	4.00-4.50	0.25 max	0.90-1.10		P <=0.030; S <=0.030; bal Fe				61 HRC
DIN 17230(80)	80MoCrV42-16	Bearing	0.77-0.85	0.25 max	3.75-4.25	0.35 max	4.00-4.50	0.25 max	0.90-1.10	0.25 max	P <=0.015; S <=0.015; bal Fe				
DIN 17230(80)	WNr 1.3551	Bearing	0.77-0.85	0.25 max	3.75-4.25	0.35 max	4.00-4.50	0.25 max	0.90-1.10	0.25 max	Cu <=0.10; Ni <=0.15; P <=0.015; S <=0.015; bal Fe				
India															
IS 1570/6(96)	THS3		0.77-0.87	1 max	3.5-4.5	0.4 max	8-9	0.5 max	0.9-1.4	1.4-2	Ni <=0.4; P <=0.03; S <=0.03; bal Fe				
IS 1570/6(96)	XT82Mo8Cr4W1V1		0.77-0.87	1 max	3.5-4.5	0.4 max	8-9	0.5 max	0.9-1.4	1.4-2	Ni <=0.4; P <=0.03; S <=0.03; bal Fe				
International															
ISO 683-17(76)	30	Ball & roller bearing, HT	0.77-0.85		3.75-4.25	0.35 max	4.00-4.50	0.25 max	0.90-1.10		P <=0.025; S <=0.020; bal Fe				
Mexico															
NMX-B-082(90)	M50	Bar	0.82-0.88		4.00-4.50	0.20-0.40	3.70-4.50	0.30-0.55	1.05-1.30	0.80-1.30	Cu <=0.25; N 0.020-0.050; Ni <=0.30; P <=0.03; S <=0.03; (As+Sn+Sb)<=0.040; (Cu+Ni)<=0.40; B, Ti may be added; bal Fe				

Specification	Designation	Notes	C	Co	Cr	Mn	Mo	Si	V	W	Other	UTS	YS	El	Hard

Tool Steel, Molybdenum High-Speed, M50 (Continued from previous page)

Spain

Specification	Designation	Notes	C	Co	Cr	Mn	Mo	Si	V	W	Other	UTS	YS	El	Hard
UNE 36027(80)	80MoCrV40-16	Ball & roller bearing	0.77-0.85		3.75-4.25	0.35 max	4.0-4.5	0.25 max	0.9-1.1		P <=0.025; S <=0.02; bal Fe				
UNE 36027(80)	F.1351*	see 80MoCrV40-16	0.77-0.85		3.75-4.25	0.35 max	4.0-4.5	0.25 max	0.9-1.1		P <=0.025; S <=0.02; bal Fe				

USA

Specification	Designation	Notes	C	Co	Cr	Mn	Mo	Si	V	W	Other	UTS	YS	El	Hard
	AISI M50	Ann Hard	0.75-0.85		3.75-4.50	0.15-0.35	4.0-4.50	0.20-0.60	0.90-1.10		P <=0.03; S 0.03-0.06; bal Fe				197-235 HB
	UNS T11350	Bearing Steel	0.77-0.85	0.25 max	3.75-4.25	0.35 max	4.00-4.50	0.25 max	0.90-1.10	0.25 max	Cu <=0.1; Ni <=0.15; P <=0.015; S <=0.015; bal Fe				
AMS 6490		Bar, Frg, Tub	0.77-0.85	0.25 max	3.75-4.25	0.35 max	4-4.5	0.25 max	0.90-1.1	0.25 max	Cu 0.1; Ni <=0.15; bal Fe				
AMS 6491		Bar, Frg, Tub	0.8-0.85	0.25 max	4-4.25	0.15-0.35	4-4.5	0.25 max	0.90-1.1	0.25 max	Cu 0.1; Ni <=0.15; bal Fe				
ASTM A600(92)	M50	Bar Frg Plt Sh Strp, Ann	0.78-0.88		3.75-4.50	0.15-0.45	3.90-4.75	0.20-0.60	1.80-1.25		P <=0.03; S <=0.03; S 0.06-0.15 for machinability; (Ni+Cu)<=0.75; bal Fe				248 HB
SAE J467(68)	M10		0.87		4.00	0.20	8.25	0.30	1.90		bal Fe				
SAE J467(68)	M50		0.81		4.08	0.30	4.25	0.20	1.00		bal Fe	2834	2131	2	64 HRC

Tool Steel, Molybdenum High-Speed, M52

Mexico

Specification	Designation	Notes	C	Co	Cr	Mn	Mo	Si	V	W	Other	UTS	YS	El	Hard
NMX-B-082(90)	M52	Bar	0.80-0.90		3.60-4.20	0.20-0.40	4.20-4.80	0.20-0.60	1.75-2.20	1.00-1.50	Cu <=0.25; N 0.020-0.050; Ni <=0.30; P <=0.03; S <=0.03; (As+Sn+Sb)<=0.040; (Cu+Ni)<=0.40; B, Ti may be added; bal Fe				

Russia

Specification	Designation	Notes	C	Co	Cr	Mn	Mo	Si	V	W	Other	UTS	YS	El	Hard
GOST 19265	11R3AM3F2		1.02-1.12	0.5 max	3.8-4.3	0.2-0.5	2.5-3	0.2-0.5	2.3-2.7	2.5-3.3	Cu <=0.25; N 0.05-0.1; Nb 0.05-0.2; Ni <=0.6; P <=0.03; S <=0.03; bal Fe				

USA

Specification	Designation	Notes	C	Co	Cr	Mn	Mo	Si	V	W	Other	UTS	YS	El	Hard
	AISI M52	Ann Hard	0.85-0.95		3.75-4.25	0.15-0.35	4.15-4.75	0.20-0.60	1.75-2.10	1.05-1.45	P <=0.03; S 0.03-0.06; bal Fe				197-235 HB
	UNS T11352	Powder Metallurgy M-52	0.9		4		4		2	1.25	bal Fe				
ASTM A600(92)	M52	Bar Frg Plt Sh Strp, Ann	0.85-0.95		3.50-4.30	0.15-0.45	4.00-4.90	0.20-0.60	1.65-2.25	0.75-1.50	P <=0.03; S <=0.03; S 0.06-0.15 for machinability; (Ni+Cu)<=0.75; bal Fe				248 HB

Tool Steel, Molybdenum High-Speed, M6

Mexico

Specification	Designation	Notes	C	Co	Cr	Mn	Mo	Si	V	W	Other	UTS	YS	El	Hard
NMX-B-082(90)	M6	Bar	0.75-0.85	11.00-13.00	3.75-4.50	0.15-0.40	4.50-5.50	0.20-0.45	1.30-1.70	3.75-4.75	Cu <=0.25; Ni <=0.30; P <=0.03; S <=0.03; (As+Sn+Sb)<=0.040; (Cu+Ni)<=0.40; B, Ti may be added; bal Fe				

USA

Specification	Designation	Notes	C	Co	Cr	Mn	Mo	Si	V	W	Other	UTS	YS	El	Hard
	AISI M6	Ann Hard	0.75-0.85	11.00-13.00	3.75-4.50	0.15-0.40	4.5-5.50	0.20-0.45	1.30-1.70	3.75-4.75	P <=0.03; S <=0.03; bal Fe				248-277 HB
	UNS T11306		0.75-0.85	11.00-13.00	3.75-4.50	0.15-0.40	4.50-5.50	0.20-0.45	1.30-1.70	3.75-4.75	P <=0.03; S <=0.03; bal Fe				
ASTM A600(92)	M6	Bar Frg Plt Sh Strp, Ann	0.75-0.85	11.00-13.00	3.75-4.50	0.15-0.40	4.50-5.50	0.20-0.45	1.30-1.70	3.75-4.75	P <=0.03; S <=0.03; S 0.06-0.15 for machinability; (Ni+Cu)<=0.75; bal Fe				277 HB
FED QQ-T-590C(85) M6*		Obs; see ASTM A600	0.75-0.85	11-13	3.75-4.5	0.15-0.4	4.5-5.5	0.2-0.45	1.3-1.7	3.75-4.75	Ni 0.3; bal Fe				

Specification	Designation	Notes	C	Co	Cr	Mn	Mo	Si	V	W	Other	UTS	YS	El	Hard

Tool Steel, Molybdenum High-Speed, M61

USA

Specification	Designation	Notes	C	Co	Cr	Mn	Mo	Si	V	W	Other	UTS	YS	El	Hard
	AISI M61	Ann Hard									P <=0.03; bal Fe				
	UNS T11361	Powder Metallurgy M-61	1.6		4		6.5		5	12	bal Fe				

Tool Steel, Molybdenum High-Speed, M62

India

Specification	Designation	Notes	C	Co	Cr	Mn	Mo	Si	V	W	Other	UTS	YS	El	Hard
IS 1570/6(96)	THS5		1.15-1.3	1 max	3.5-4.5	0.4 max	4.6-5.3	0.5 max	2.7-3.2	5.7-6.7	Ni <=0.4; P <=0.03; S <=0.03; bal Fe				
IS 1570/6(96)	XT122W6Mo5Cr4V3		1.15-1.3	1 max	3.5-4.5	0.4 max	4.6-5.3	0.5 max	2.7-3.2	5.7-6.7	Ni <=0.4; P <=0.03; S <=0.03; bal Fe				

USA

Specification	Designation	Notes	C	Co	Cr	Mn	Mo	Si	V	W	Other	UTS	YS	El	Hard
	AISI M62	Ann Hard	1.25-1.35		3.50-4.25	0.15-0.40	10.00-11.00	0.15-0.40	1.80-2.20	5.75-.6.75	P <=0.030; S 0.05-0.09; bal Fe				262-285 HB
	UNS T11362	Powder Metallurgy M-62	1.3		3.75		10.5		2	6.25	bal Fe				
ASTM A600(92)	M62	Bar Frg Plt Sh Strp, Ann	1.25-1.35		3.50-4.00	0.15-0.40	10.00-11.00	0.15-0.40	1.80-2.10	5.75-6.50	P <=0.03; S <=0.07; S 0.06-0.15 for machinability; (Ni+Cu)<=0.75; bal Fe				285 HB

Tool Steel, Molybdenum High-Speed, M7

Argentina

Specification	Designation	Notes	C	Co	Cr	Mn	Mo	Si	V	W	Other	UTS	YS	El	Hard
IAS	IRAM M7	Ann	0.97-1.05		3.50-4.00	0.15-0.40	8.20-9.20	0.20-0.55	1.75-2.25	1.40-2.10	P <=0.03; S <=0.03; bal Fe				255 HB

China

Specification	Designation	Notes	C	Co	Cr	Mn	Mo	Si	V	W	Other	UTS	YS	El	Hard
GB 9943(88)	W2Mo9Cr4V2	Bar Q/T	0.97-1.05		3.50-4.00	0.15-0.40	8.20-9.20	0.20-0.55	1.75-2.25	1.40-2.10	Ni <=0.30; P <=0.03; S <=0.030; bal Fe				65 HRC

France

Specification	Designation	Notes	C	Co	Cr	Mn	Mo	Si	V	W	Other	UTS	YS	El	Hard
AFNOR NFA35590(92)	HS2-9-2	Soft ann	0.95-1.05	1 max	3.5-4.5	0.4 max	8.2-9.2	0.5 max	1.7-2.2	1.5-2.1	Cu <=0.3; Ni <=0.4; P <=0.03; S <=0.03; bal Fe				265 HB max
AFNOR NFA35590(92)	Z100DCWV09-04-02-02	Soft ann	0.95-1.05	1 max	3.5-4.5	0.4 max	8.2-9.2	0.5 max	1.7-2.2	1.5-2.1	Cu <=0.3; Ni <=0.4; P <=0.03; S <=0.03; bal Fe				265 HB max

Germany

Specification	Designation	Notes	C	Co	Cr	Mn	Mo	Si	V	W	Other	UTS	YS	El	Hard
DIN 17350(80)	HS-2-9-2	As Tmp 540C	0.97-1.07		3.50-4.20	0.40 max	8.00-9.20	0.45 max	1.80-2.20	1.50-2.00	P <=0.030; S <=0.030; bal Fe				64 HRC
DIN 17350(80)	WNr 1.3348	As Tmp 540C	0.97-1.07		3.50-4.20	0.40 max	8.00-9.20	0.45 max	1.80-2.20	1.50-2.00	P <=0.030; S <=0.030; bal Fe				64 HRC

Italy

Specification	Designation	Notes	C	Co	Cr	Mn	Mo	Si	V	W	Other	UTS	YS	El	Hard
UNI 2955	RM7 HS2-9-2		1.05		3.75	0.4	8.7	0.45	2	1.75	bal Fe				
UNI 2955/5(82)	HS2-9-2		0.95-1.05	1 max	3.5-4.5	0.4 max	8.2-9.2	0.5 max	1.7-2.2	1.5-2.1	P <=0.07; S <=0.06; bal Fe				

Japan

Specification	Designation	Notes	C	Co	Cr	Mn	Mo	Si	V	W	Other	UTS	YS	El	Hard
JIS G4403(83)	SKH58	Bar Frg Ann	0.95-1.05		3.50-4.50	0.40 max	8.20-9.20	0.50 max	1.70-2.20	1.50-2.10	Cu <=0.25; Ni <=0.25; P <=0.030; S <=0.030; bal Fe				269 HB max

Mexico

Specification	Designation	Notes	C	Co	Cr	Mn	Mo	Si	V	W	Other	UTS	YS	El	Hard
NMX-B-082(90)	M7	Bar	0.97-1.05		3.50-4.00	0.15-0.40	8.20-9.20	0.20-0.55	1.75-2.25	1.40-2.10	Cu <=0.25; Ni <=0.30; P <=0.03; S <=0.03; (As+Sn+Sb)<=0.040; (Cu+Ni)<=0.40; B, Ti may be added; bal Fe				

Pan America

Specification	Designation	Notes	C	Co	Cr	Mn	Mo	Si	V	W	Other	UTS	YS	El	Hard
COPANT 337	M7	Bar, Frg	0.98-1.05		3.5-4	0.15-0.4	8.4-9.1	0.2-0.5	1.75-2.25	1.4-2.1	P 0.03; S 0.03; bal Fe				

Spain

Specification	Designation	Notes	C	Co	Cr	Mn	Mo	Si	V	W	Other	UTS	YS	El	Hard
UNE 36073(75)	2-9-2		0.95-1.05		3.5-4.5	0.4 max	8.2-9.2	0.45 max	1.5-2.2	1.5-2.1	P <=0.03; S <=0.03; bal Fe				
UNE 36073(75)	EM7		0.95-1.05		3.5-4.5	0.4 max	8.2-9.2	0.45 max	1.5-2.2	1.5-2.1	P <=0.03; S <=0.03; bal Fe				
UNE 36073(75)	F.5607*	see EM7, 2-9-2	0.95-1.05		3.5-4.5	0.4 max	8.2-9.2	0.45 max	1.5-2.2	1.5-2.1	P <=0.03; S <=0.03; bal Fe				

Specification	Designation	Notes	C	Co	Cr	Mn	Mo	Si	V	W	Other	UTS	YS	El	Hard

Tool Steel, Molybdenum High-Speed, M7 (Continued from previous page)

Sweden

Specification	Designation	Notes	C	Co	Cr	Mn	Mo	Si	V	W	Other	UTS	YS	El	Hard
SS 1427822	2782		0.96-1.04		3.5-4.5	0.2-0.4	8.2-9.2	0.15-0.4	1.9-2.2	1.5-2	P <=0.03; S <=0.03; bal Fe				

USA

Specification	Designation	Notes	C	Co	Cr	Mn	Mo	Si	V	W	Other	UTS	YS	El	Hard
	AISI M7	Ann Hard	0.97-1.05		3.50-4.00	0.15-0.40	8.20-9.20	0.20-0.55	1.75-2.25	1.40-2.10	P <=0.03; S <=0.03; bal Fe				217-255 HB
	UNS T11307		0.97-1.05		3.50-4.00	0.15-0.40	8.20-9.20	0.20-0.55	1.75-2.25	1.40-2.10	P <=0.03; S <=0.03; bal Fe				
ASTM A600(92)	M7	Bar Frg Plt Sh Strp, Ann	0.97-1.05		3.50-4.00	0.15-0.40	8.20-9.20	0.20-0.55	1.75-2.25	1.40-2.10	P <=0.03; S <=0.03; S 0.06-0.15 for machinability; (Ni+Cu)<=0.75; bal Fe				255 HB
FED QQ-T-590C(85)	M7*	Obs; see ASTM A600	0.97-1.05		3.5-4	0.15-0.4	8.2-9.2	0.2-0.55	1.75-2.25	1.4-2.1	Ni 0.3; bal Fe				

Tool Steel, Molybdenum High-Speed, No equivalents identified

Czech Republic

Specification	Designation	Notes	C	Co	Cr	Mn	Mo	Si	V	W	Other	UTS	YS	El	Hard
CSN 419850	19850		0.9-1.0	7.3-8.7	3.8-4.6	0.45 max	4.5-5.5	0.45 max	1.7-2.4	5.5-7.0	P <=0.035; S <=0.035; bal Fe				
CSN 419861	19861		1.15-1.3	10.0-11.5	3.8-4.6	0.45 max	3.5-4.3	0.45 max	3.0-3.7		P <=0.035; S <=0.35; bal Fe				

Hungary

Specification	Designation	Notes	C	Co	Cr	Mn	Mo	Si	V	W	Other	UTS	YS	El	Hard
MSZ 4351(84)	R14	Q/T	1.20-1.35	9.5-10.5	3.5-4.5	0.4 max	3.2-3.9	0.4 max	3.0-3.5	8-10	P <=0.03; S <=0.03; bal Fe				66 HRC
MSZ 4351(84)	R14	Soft ann	1.20-1.35	9.5-10.5	3.5-4.5	0.4 max	3.2-3.9	0.4 max	3.0-3.5	8-10	P <=0.03; S <=0.03; bal Fe				300 HB max

India

Specification	Designation	Notes	C	Co	Cr	Mn	Mo	Si	V	W	Other	UTS	YS	El	Hard
IS 1570/6(96)	THS9	High-Speed	1.45-1.6	4.7-5.2	3.5-4.5	0.4 max	0.7-1	0.5 max	4.75-5.55	11.5-13	Ni <=0.4; P <=0.03; S <=0.03; bal Fe				
IS 1570/6(96)	XT152W12Co5V5Cr4Mo1	High-Speed	1.45-1.6	4.7-5.2	3.5-4.5	0.4 max	0.7-1	0.5 max	4.75-5.55	11.5-13	Ni <=0.4; P <=0.03; S <=0.03; bal Fe				

Poland

Specification	Designation	Notes	C	Co	Cr	Mn	Mo	Si	V	W	Other	UTS	YS	El	Hard
PNH85022	SK10V		1.15-1.3	9.5-10.5	3.5-4.5	0.4 max	3-3.6	0.5 max	2.7-3.2	9-11	Ni <=0.4; P <=0.030; S <=0.030; bal Fe				
PNH85022	SK5MC		1.05-1.2	4.5-5.5	3.5-4.5	0.4 max	4.5-5.5	0.5 max	1.7-2.1	6.4-7.4	Ni <=0.4; P <=0.030; S <=0.030; bal Fe				

Tool Steel, Molybdenum High-Speed, W-FeNiCr

USA

Specification	Designation	Notes	C	Co	Cr	Mn	Mo	Si	V	W	Other	UTS	YS	El	Hard
	UNS T87520	Thermal Spray Wire	0.10 max		1.0-2.0	2.5 max	1.0-3.0	0.35 max			Ni 3.7-5.0; P <=0.030; S <=0.030; bal Fe				

Specification	Designation	Notes	C	Cr	Mn	P	S	Si	V	W	Other	UTS	YS	El	Hard
Tool Steel, Oil-Hardening Cold Work, CA2															
USA															
	UNS T90102		0.95-1.05	4.75-5.50	0.75 max	0.030 max	0.030 max	1.50 max			Mo 0.90-1.40; V optional 0.20-0.50; bal Fe				
ASTM A597(93)	CA-2	Cast	0.95-1.05	4.75-5.50	0.75 max	0.03 max	0.03 max	1.50 max	0.20-0.50		Mo 0.90-1.40; V optional; bal Fe				
Tool Steel, Oil-Hardening Cold Work, CD2															
USA															
	UNS T90402		1.40-1.60	11.00-13.00	1.00 max	0.030 max	0.030 max	1.50 max			Mo 0.70-1.20; Co optional 0.70-1.00; V optional 0.40-1.00; bal Fe				
ASTM A597(93)	CD-2	Cast	1.40-1.60	11.00-13.00	1.00 max	0.03 max	0.03 max	1.50 max	0.40-1.00		Co 0.70-1.00; Mo 0.70-1.20; Co, V opt; bal Fe				
Tool Steel, Oil-Hardening Cold Work, CD5															
USA															
	UNS T90405		1.35-1.60	11.00-13.00	0.75 max	0.030 max	0.030 max	1.50 max	0.35-0.55		Co 2.50-3.50; Mo 0.70-1.20; Ni optional 0.40-0.60; bal Fe				
ASTM A597(93)	CD-5	Cast	1.35-1.60	11.00-13.00	0.75 max	0.03 max	0.03 max	1.50 max	0.35-0.55		Co 2.50-3.50; Mo 0.70-1.20; Ni 0.40-0.60; Ni optional; bal Fe				
Tool Steel, Oil-Hardening Cold Work, CO1															
USA															
	UNS T91501		0.85-1.00	0.40-1.00	1.00-1.30	0.030 max	0.030 max	1.50 max	0.30 max	0.40-0.60	bal Fe				
ASTM A597(93)	CO-1	Cast	0.85-1.00	0.40-1.00	1.00-1.30	0.03 max	0.03 max	1.50 max	0.30 max	0.40-0.60	bal Fe				
Tool Steel, Oil-Hardening Cold Work, No equivalents identified															
Poland															
PNH85023	NCMS		0.95-1.1	1.3-1.65	0.9-1.2	0.030 max	0.030 max	0.4-0.7			Ni <=0.4; bal Fe				
PNH85023	NPW		0.45-0.55	1.2-1.5	0.4-0.7	0.030 max	0.030 max	0.15-0.4	0.4-0.6	0.4-0.7	Ni 2.8-3.3; bal Fe				
PNH85023	NW1		1.1-1.25	0.3 max	0.15-0.45	0.030 max	0.030 max	0.15-0.4		1-1.5	Ni <=0.4; bal Fe				
PNH85023	NWV4		1.3-1.45	0.4-0.7	0.2-0.4	0.030 max	0.030 max	0.15-0.35	0.15-0.3	4.5-5.5	Ni <=0.4; bal Fe				
Tool Steel, Oil-Hardening Cold Work, O1															
Argentina															
IAS	IRAM O1	Ann	0.85-1.00	0.40-0.60	1.00-1.40	0.030 max	0.030 max	0.50 max	0.30 max	0.40-0.60	bal Fe				185 HB
China															
GB 1299	MnCrWV*	Obs	0.95-1.05	0.4-0.7	1-1.3			0.4	0.15-0.3	0.4-0.7	Cu 0.25; Ni 0.3; bal Fe				
GB 1299(85)	9CrWMn	As Hard	0.85-0.95	0.50-0.80	0.90-1.20	0.030 max	0.030 max	0.40 max		0.50-0.80	Cu <=0.30; Ni <=0.25; bal Fe				62 HRC
Finland															
SFS 907(73)	SFS907	Alloyed	0.85-1.0	0.4-0.6	1.1-1.3	0.03 max	0.02 max	0.2-0.4	0.05-0.15	0.4-0.6	bal Fe				
France															
AFNOR NFA35590(92)	90MnWCrV5	Q/T	0.85-1	0.35-0.65	1.05-1.35	0.025 max	0.025 max	0.1-0.4	0.05-0.2	0.4-0.7	Cu <=0.3; Mo <=0.15; Ni <=0.4; bal Fe				59 HRC
AFNOR NFA35590(92)	90MnWCrV5	Soft ann	0.85-1	0.35-0.65	1.05-1.35	0.025 max	0.025 max	0.1-0.4	0.05-0.2	0.4-0.7	Cu <=0.3; Mo <=0.15; Ni <=0.4; bal Fe				228 HB max
AFNOR NFA35590(92)	90MWCV5	Soft ann	0.85-1	0.35-0.65	1.05-1.35	0.025 max	0.025 max	0.1-0.4	0.05-0.2	0.4-0.7	bal Fe				228 HB max

UNS numbers and US grades are provided as a means of cross referencing chemically similar alloys. Exchangability is only possible after independent examination of specifications. Tensile properties are minimum or typical as specified. UTS and YS as MPa. El as %. See Appendix for list of abbreviations used in Notes. * indicates obsolete material.

Specification	Designation	Notes	C	Cr	Mn	P	S	Si	V	W	Other	UTS	YS	El	Hard

Tool Steel, Oil-Hardening Cold Work, O1 (Continued from previous page)

France

Specification	Designation	Notes	C	Cr	Mn	P	S	Si	V	W	Other	UTS	YS	El	Hard
AFNOR NFA35590(92)	90MWCV5	Q/T	0.85-1	0.35-0.65	1.05-1.35	0.025 max	0.025 max	0.1-0.4	0.05-0.2	0.4-0.7	bal Fe				59 HRC

Germany

Specification	Designation	Notes	C	Cr	Mn	P	S	Si	V	W	Other	UTS	YS	El	Hard
DIN	100MnCrW4	Hard tmp to 200C	0.90-1.05	0.50-0.70	1.00-1.20	0.035 max	0.035 max	0.15-0.35	0.05-0.15	0.50-0.70	bal Fe				62 HRC
DIN	WNr 1.2510	Hard tmp to 200C	0.90-1.05	0.50-0.70	1.00-1.20	0.035 max	0.035 max	0.15-0.35	0.05-0.15	0.50-0.70	bal Fe				62 HRC

India

Specification	Designation	Notes	C	Cr	Mn	P	S	Si	V	W	Other	UTS	YS	El	Hard
IS 1570	T90Mn2W50Cr45		0.85-0.95	0.3-0.6	1.25-1.75			0.1-0.35	0.25	0.4-0.6	bal Fe				
IS 3749	T90Mn2W50Cr45	Bar	0.85-0.95	0.3-0.6	1.25-1.75	0.04	0.04	0.1-0.35	0.25	0.4-0.6	bal Fe				
IS 4367	T90Mn2W50Cr45	Frg	0.85-0.95	0.3-0.6	1.25-1.75			0.1-0.35	0.25	0.4-0.6	bal Fe				

Italy

Specification	Designation	Notes	C	Cr	Mn	P	S	Si	V	W	Other	UTS	YS	El	Hard
UNI 2955/3(82)	95MnWCr5KU	CW	0.9-1	0.35-0.65	1.05-1.35	0.07 max	0.06 max	0.1-0.4	0.05-0.25	0.4-0.7	bal Fe				

Japan

Specification	Designation	Notes	C	Cr	Mn	P	S	Si	V	W	Other	UTS	YS	El	Hard
JIS G4404(83)	SKS21	HR Bar Frg Ann	1.00-1.10	0.20-0.50	0.50 max	0.030 max	0.030 max	0.35 max	0.10-0.25	0.50-1.00	Ni <=0.25; bal Fe				217 HB max
JIS G4404(83)	SKS3	HR Bar Frg Ann	0.90-1.00	0.50-1.00	0.90-1.20	0.030 max	0.030 max	0.35 max		0.50-1.00	Cu <=0.25; Ni <=0.25; bal Fe				217 HB max
JIS G4404(83)	SKS93	HR Bar Frg Ann	1.00-1.10	0.20-0.60	0.80-1.10	0.030 max	0.030 max	0.50 max			Ni <=0.25; bal Fe				217 HB max
JIS G4404(83)	SKS94	HR Bar Frg Ann	0.90-1.00	0.20-0.60	0.80-1.10	0.030 max	0.030 max	0.50 max			Cu <=0.25; Ni <=0.25; bal Fe				212 HB max
JIS G4404(83)	SKS95	HR Bar Frg Ann	0.80-0.90	0.20-0.60	0.80-1.10	0.030 max	0.030 max	0.50 max			Cu <=0.25; Ni <=0.25; bal Fe				212 HB max

Mexico

Specification	Designation	Notes	C	Cr	Mn	P	S	Si	V	W	Other	UTS	YS	El	Hard
NMX-B-082(90)	O1	Bar	0.85-1.00	0.40-0.60	1.00-1.40	0.03 max	0.03 max	0.50 max	0.30 max	0.40-0.60	Cu <=0.25; Ni <=0.30; (As+Sn+Sb)<=0.040; (Cu+Ni)<=0.40; B, Ti may be added; bal Fe				

Norway

Specification	Designation	Notes	C	Cr	Mn	P	S	Si	V	W	Other	UTS	YS	El	Hard
NS 13840	13840	Bar Frg	0.85-1.00	0.40-0.60	1.10-1.30	0.030 max	0.020 max	0.20-0.40	0.05-0.15	0.40-0.60	bal Fe				

Pan America

Specification	Designation	Notes	C	Cr	Mn	P	S	Si	V	W	Other	UTS	YS	El	Hard
COPANT 337	O1	Bar, Frg	0.85-1.05	0.4-0.6	1-1.4	0.03	0.03	0.15-0.4		0.4-0.6	bal Fe				

Poland

Specification	Designation	Notes	C	Cr	Mn	P	S	Si	V	W	Other	UTS	YS	El	Hard
PNH85023	NMWV		0.9-1	0.4-0.7	1-1.3	0.030 max	0.030 max	0.15-0.4	0.1-0.25	0.4-0.7	Ni <=0.4; bal Fe				

Russia

Specification	Designation	Notes	C	Cr	Mn	P	S	Si	V	W	Other	UTS	YS	El	Hard
GOST 5950	6ChWG	Alloyed	0.55-0.7	0.5-0.8	0.9-1.2	0.03 max	0.03 max	0.15-0.35	0.05 max	0.5-0.8	Cu <=0.3; Mo <=0.3; Ni <=0.35; bal Fe				
GOST 5950	6ChWG	Alloyed	0.55-0.7	0.5-0.8	0.9-1.2	0.03 max	0.03 max	0.15-0.35	0.15 max	0.5-0.8	Cu <=0.3; Mo <=0.3; Ni <=0.35; bal Fe				
GOST 5950	7ChG2WMF	Alloyed	0.68-0.76	1.5-1.8	1.8-2.3	0.03 max	0.03 max	0.2-0.4	0.1-0.25	0.5-0.9	Cu <=0.3; Mo 0.5-0.8; Ni <=0.35; bal Fe				
GOST 5950	9ChWG	Alloyed	0.85-0.95	0.5-0.8	0.9-1.2	0.03 max	0.03 max	0.15-0.35	0.15 max	0.5-0.8	Cu <=0.3; Mo <=0.3; Ni <=0.35; bal Fe				
GOST 5950	9ChWG	Alloyed	0.85-0.95	0.5-0.8	0.9-1.2			0.15-0.35	0.05	0.5-0.8	Cu 0.3; Mo 0.3; Ni 0.35; bal Fe				
GOST 5950	ChWSGF	Alloyed	0.95-1.05	0.6-1.1	0.6-0.9	0.03 max	0.03 max	0.65-1	0.05-0.15	0.5-0.8	Cu <=0.3; Mo <=0.3; Ni <=0.35; bal Fe				

Spain

Specification	Designation	Notes	C	Cr	Mn	P	S	Si	V	W	Other	UTS	YS	El	Hard
UNE	F.522.A		0.87-0.97	0.4-0.6	0.9-1.1	0.03 max	0.03 max	0.15-0.3	0.1-0.2	0.4-0.6	bal Fe				
UNE 36018/2(94)	95MnCrW5	Alloyed	0.9-1	0.35-0.65	1.05-1.35	0.025 max	0.02 max	0.1-0.4	0.05-0.25	0.4-0.7	Mo <=0.15; bal Fe				
UNE 36018/2(94)	F.5220*	see 95MnCrW5	0.9-1	0.35-0.65	1.05-1.35	0.025 max	0.02 max	0.1-0.4	0.05-0.25	0.4-0.7	Mo <=0.15; bal Fe				

Sweden

Specification	Designation	Notes	C	Cr	Mn	P	S	Si	V	W	Other	UTS	YS	El	Hard
SS 142140	2140	Alloyed	0.85-1	0.4-0.6	1.1-1.3	0.03 max	0.02 max	0.2-0.4	0.05-0.15	0.4-0.6	bal Fe				

UNS numbers and US grades are provided as a means of cross referencing chemically similar alloys. Exchangability is only possible after independent examination of specifications. Tensile properties are minimum or typical as specified. UTS and YS as MPa. El as %. See Appendix for list of abbreviations used in Notes. * indicates obsolete material.

Specification	Designation	Notes	C	Cr	Mn	P	S	Si	V	W	Other	UTS	YS	El	Hard

Tool Steel, Oil-Hardening Cold Work, O1 (Continued from previous page)

UK

Specification	Designation	Notes	C	Cr	Mn	P	S	Si	V	W	Other	UTS	YS	El	Hard
BS 4656	BO1	Bar Rod Sh Strp Frg	0.85-1	0.4-0.6	1.1-1.35			0.4 max	0.25 max	0.4-0.6	bal Fe				
BS 4659	BO1		0.85-1	0.4-0.6	1.1-1.35			0.4 max	0.25 max	0.4-0.6	bal Fe				

USA

Specification	Designation	Notes	C	Cr	Mn	P	S	Si	V	W	Other	UTS	YS	El	Hard
	AISI O1		0.85-1.00	0.40-0.60	1.00-1.40	0.030 max	0.030 max	0.50 max	0.30 max	0.40-0.60	bal Fe				
	UNS T31501		0.85-1.00	0.40-0.60	1.00-1.40	0.030 max	0.030 max	0.50 max	0.30 max	0.40-0.60	bal Fe				
ASTM A681(94)	O1	Ann	0.85-1.00	0.40-0.70	1.00-1.40	0.030 max	0.030 max	0.10-0.50	0.30 max	0.40-0.60	S 0.06-0.15 if spec for machining; bal Fe				212 HB
ASTM A685	O1		0.85-1	0.4-0.6	1-1.4	0.03	0.03	0.5	0.3	0.4-0.6	bal Fe				
FED QQ-T-570(87)	O1*	Obs; see ASTM A681	0.85-1	0.4-0.6	1-1.4			0.5	0.3	0.4-0.6	Ni 0.3; bal Fe				
SAE J437(70)	O1	Ann									bal Fe				183-212 HB
SAE J438(70)	O1		0.85-0.95	0.40-0.60	1.00-1.30			0.20-0.40	0.20 max	0.40-0.60	V optional; bal Fe				

Tool Steel, Oil-Hardening Cold Work, O2

Bulgaria

Specification	Designation	Notes	C	Cr	Mn	P	S	Si	V	W	Other	UTS	YS	El	Hard
BDS 7938	9G2F		0.80-0.90		1.70-2.10	0.030 max	0.030 max	0.15-0.40	0.10-0.20		Cu <=0.30; Ni <=0.35; bal Fe				

China

Specification	Designation	Notes	C	Cr	Mn	P	S	Si	V	W	Other	UTS	YS	El	Hard
GB 1299	9Mn2*	Obs	0.85-0.95		1.7-2			0.4	0.1-0.25		Cu 0.25; Ni 0.3; bal Fe				
GB 1299(85)	9Mn2V	As Hard	0.85-0.95	0.25 max	1.70-2.00	0.030 max	0.030 max	0.40 max	0.10-0.25		Cu <=0.30; Ni <=0.25; bal Fe				62 HRC

Czech Republic

Specification	Designation	Notes	C	Cr	Mn	P	S	Si	V	W	Other	UTS	YS	El	Hard
CSN 419312	19312		0.75-0.85	0.25 max	1.85-2.15	0.03 max	0.035 max	0.15-0.35	0.1-0.2		Ni <=0.35; bal Fe				
CSN 419313	19313		0.8-0.9	0.2-0.4	1.75-2.1	0.03 max	0.035 max	0.15-0.35	0.1-0.2		Ni <=0.35; bal Fe				
CSN 419315	19315		0.8-0.9	0.2-0.4	1.8-2.1 *	0.03 max	0.035 max	0.15-0.35	0.1-0.2	0.4-0.6	bal Fe				

France

Specification	Designation	Notes	C	Cr	Mn	P	S	Si	V	W	Other	UTS	YS	El	Hard
AFNOR NFA35590(92)	90MnV8		0.8-0.95	0.3 max	1.8-2.2	0.025 max	0.025 max	0.1-0.4	0.05-0.2	0.1 max	Cu <=0.3; Mo <=0.15; Ni <=0.4; bal Fe				
AFNOR NFA35590(92)	90MV8		0.8-0.95		1.8-2.2	0.025 max	0.025 max	0.1-0.4	0.05-0.2		Ni <=0.4; bal Fe				

Germany

Specification	Designation	Notes	C	Cr	Mn	P	S	Si	V	W	Other	UTS	YS	El	Hard
DIN 17350(80)	90MnCrV8	Hard tmp to 200C	0.85-0.95	0.20-0.50	1.90-2.10	0.030 max	0.030 max	0.10-0.40	0.05-0.15		bal Fe				60 HRC
DIN 17350(80)	WNr 1.2842	Hard tmp to 200C	0.85-0.95	0.20-0.50	1.90-2.10	0.030 max	0.030 max	0.10-0.40	0.05-0.15		bal Fe				60 HRC

Hungary

Specification	Designation	Notes	C	Cr	Mn	P	S	Si	V	W	Other	UTS	YS	El	Hard
MSZ 4352(84)	M1	Q/T	0.8-0.9	0.25 max	1.7-2.1	0.03 max	0.03 max	0.1-0.4	0.15-0.3	0.3 max	Mo <=0.2; Ni <=0.35; bal Fe				61 HRC
MSZ 4352(84)	M1	Soft ann	0.8-0.9	0.25 max	1.7-2.1	0.03 max	0.03 max	0.1-0.4	0.15-0.3	0.3 max	Mo <=0.2; Ni <=0.35; bal Fe				229 HB max

India

Specification	Designation	Notes	C	Cr	Mn	P	S	Si	V	W	Other	UTS	YS	El	Hard
IS 1570/6(96)	T90Mn6WCr2	CW	0.85-0.95	0.3-0.6	1.25-1.75	0.035 max	0.035 max	0.1-0.35	0.25 max	0.4-0.6	Mo <=0.15; Ni <=0.4; bal Fe				
IS 1570/6(96)	TAC20	CW	0.85-0.95	0.3-0.6	1.25-1.75	0.035 max	0.035 max	0.1-0.35	0.25 max	0.4-0.6	Mo <=0.15; Ni <=0.4; bal Fe				

Italy

Specification	Designation	Notes	C	Cr	Mn	P	S	Si	V	W	Other	UTS	YS	El	Hard
UNI 2955	88MnV8Ku	CW	0.8-0.95		1.8-2.2			0.4	0.1-0.2		bal Fe				
UNI 2955	SDM 90MnVCr8Ku		0.88	0.35	2			0.4	0.15		bal Fe				
UNI 2955/3(82)	90MnVCr8KU	CW	0.85-0.95	0.2-0.5	1.7-2.2	0.07 max	0.06 max	0.1-0.4	0.1-0.3		bal Fe				

Japan

Specification	Designation	Notes	C	Cr	Mn	P	S	Si	V	W	Other	UTS	YS	El	Hard
JIS G3311(88)	SKS95M	CR Strp; Spring; Cutlery	0.80-0.90	0.20-0.60	0.80-1.10	0.030 max	0.030 max	0.50 max			Cu <=0.25; Ni <=0.25; bal Fe				200-290 HV

UNS numbers and US grades are provided as a means of cross referencing chemically similar alloys. Exchangability is only possible after independent examination of specifications. Tensile properties are minimum or typical as specified. UTS and YS as MPa. El as %. See Appendix for list of abbreviations used in Notes. * indicates obsolete material.

Specification	Designation	Notes	C	Cr	Mn	P	S	Si	V	W	Other	UTS	YS	El	Hard

Tool Steel, Oil-Hardening Cold Work, O2 (Continued from previous page)

Mexico

Specification	Designation	Notes	C	Cr	Mn	P	S	Si	V	W	Other	UTS	YS	El	Hard
NMX-B-082(90)	O2	Bar	0.85-0.95	0.35 max	1.40-1.80	0.03 max	0.03 max	0.50 max	0.30 max		Cu <=0.25; Mo <=0.30; Ni <=0.30; (As+Sn+Sb)<=0.040; (Cu+Ni)<=0.40; B, Ti may be added; bal Fe				

Pan America

| COPANT 336 | O2 | Bar, Frg | 0.85-1.05 | 0.35 | 1.4-1.8 | 0.03 | 0.03 | 0.15-0.4 | | | bal Fe | | | | |

Poland

| PNH85023 | NMV | | 0.8-0.9 | 0.85-0.95 | 1.8-2.1 | 0.030 max | 0.030 max | 0.15-0.4 | 0.1-0.25 | | Ni <=0.4; bal Fe | | | | |

Romania

| STAS 3611(80) | VM18 | | 0.85-0.95 | 0.35 max | 1.7-1.9 | 0.03 max | 0.025 max | 0.35 max | 0.1-0.25 | | Ni <=0.35; Pb <=0.15; bal Fe | | | | |

Russia

| GOST 5950 | 7Ch3 | Alloyed | 0.65-0.75 | 3.2-3.8 | 0.15-0.4 | 0.03 | 0.03 | 0.15-0.35 | 0.15 max | 0.2 max | Cu <=0.3; Mo <=0.2; Ni <=0.35; bal Fe | | | | |
| GOST 5950 | 9G2F | Alloyed | 0.85-0.95 | 0.3 max | 1.7-2.2 | 0.03 max | 0.03 max | 0.1-0.4 | 0.1-0.3 | 0.2 max | Cu <=0.3; Mo <=0.2; Ni <=0.35; bal Fe | | | | |

UK

| BS 4656 | BO2 | Bar Rod Sh Strp Frg | 0.85-0.95 | | 1.5-1.8 | | | 0.4 max | 0.25 max | | bal Fe | | | | |
| BS 4659 | USA O2 | | 0.9 | 0.3 | 1.8 | | | | | 0.35 | bal Fe | | | | |

USA

	AISI O2		0.85-0.95	0.50 max	1.40-1.80	0.030 max	0.030 max	0.50 max	0.30 max		Mo <=0.30; bal Fe				
	UNS T31502		0.85-0.95	0.35 max	1.40-1.80	0.030 max	0.030 max	0.50 max	0.30 max		Mo <=0.30; bal Fe				
ASTM A681(94)	O2	Ann	0.85-0.95	0.50 max	1.40-1.80	0.030 max	0.030 max	0.50 max	0.30 max		Mo <=0.30; S 0.06-0.15 if spec for machining; bal Fe				217 HB
ASTM A685	O2	Bar	0.85-0.95	0.35	1.4-1.8	0.03	0.03	0.5	0.3		Mo 0.3; bal Fe				
FED QQ-T-570(87)	O2*	Obs; see ASTM A681	0.85-0.95	0.35	1.4-1.8			0.5	0.3		Mo 0.3; Ni 0.3; bal Fe				
SAE J437(70)	O2	Ann									bal Fe				183-212 HB
SAE J438(70)	O2		0.85-0.95	0.35 max	1.40-1.80			0.20-0.40	0.20 max		Mo <=0.30; Cr, V, Mo optional; bal Fe				

Tool Steel, Oil-Hardening Cold Work, O6

France

Specification	Designation	Notes	C	Cr	Mn	P	S	Si	V	W	Other	UTS	YS	El	Hard
AFNOR NFA35590	140SMD4	Bar Frg	1.4		1	0.03	0.03	1			Mo 0.3; bal Fe				
AFNOR NFA35590(78)	130C3	Soft ann	1.2-1.4	0.6-0.9	0.15-0.45	0.025 max	0.025 max	0.1-0.4			Ni <=0.4; bal Fe				217 HB max

India

| IS 3749 | T133 | Bar | 1.25-1.4 | | 0.2-0.35 | 0.04 | 0.04 | 0.1-0.35 | | | bal Fe | | | | |
| IS 4367 | T133 | Frg | 1.25-1.4 | | 0.2-0.35 | | | 0.1-0.3 | | | bal Fe | | | | |

Mexico

| NMX-B-082(90) | O6 | Bar | 1.25-1.55 | 0.30 max | 0.30-1.10 | 0.03 max | 0.03 max | 0.55-1.50 | | | Cu <=0.25; Mo 0.20-0.30; Ni <=0.30; (As+Sn+Sb)<=0.040; (Cu+Ni)<=0.40; B, Ti may be added; bal Fe | | | | |

Pan America

| COPANT 337 | O6 | Bar, Frg | 1.3-1.55 | 0.03 | 0.4-1 | 0.03 | 0.03 | 0.75-1.25 | | | Mo 0.2-0.3; bal Fe | | | | |

USA

| | AISI O6 | | 1.25-1.55 | 0.30 max | 0.30-1.10 | 0.030 max | 0.030 max | 0.55-1.50 | | | Mo 0.20-0.30; bal Fe | | | | |
| | UNS T31506 | | 1.25-1.55 | 0.30 max | 0.30-1.10 | 0.030 max | 0.030 max | 0.55-1.50 | | | Mo 0.20-0.30; bal Fe | | | | |

UNS numbers and US grades are provided as a means of cross referencing chemically similar alloys. Exchangability is only possible after independent examination of specifications. Tensile properties are minimum or typical as specified. UTS and YS as MPa. El as %. See Appendix for list of abbreviations used in Notes. * indicates obsolete material.

Specification	Designation	Notes	C	Cr	Mn	P	S	Si	V	W	Other	UTS	YS	El	Hard

Tool Steel, Oil-Hardening Cold Work, O6 (Continued from previous page)

USA

Specification	Designation	Notes	C	Cr	Mn	P	S	Si	V	W	Other	UTS	YS	El	Hard
ASTM A681(94)	O6	Ann	1.25-1.55	0.30 max	0.30-1.10	0.030 max	0.030 max	0.55-1.50			Mo 0.20-0.30; S 0.06-0.15 if spec for machining; bal Fe				229 HB
ASTM A685	O6	Bar	1.25-1.55	0.3	0.3-1.1	0.03	0.03	0.55-1.5			Mo 0.2-0.3; bal Fe				
FED QQ-T-570(87)	O6*	Obs; see ASTM A681	1.25-1.55	0.3	0.3-1.1						Mo 0.2-0.3; Ni 0.3; bal Fe				
SAE J437(70)	O6	Ann									bal Fe				183-212 HB
SAE J438(70)	O6		1.35-1.55		0.30-1.00					0.80-1.20	Mo 0.20-0.30; bal Fe				

Tool Steel, Oil-Hardening Cold Work, O7

Bulgaria

Specification	Designation	Notes	C	Cr	Mn	P	S	Si	V	W	Other	UTS	YS	El	Hard
BDS 7938	ChWG		0.95-1.05	0.90-1.20	0.80-1.10	0.030 max	0.030 max	0.15-0.35	0.15 max	1.20-1.60	Cu <=0.30; Ni <=0.35; bal Fe				

China

Specification	Designation	Notes	C	Cr	Mn	P	S	Si	V	W	Other	UTS	YS	El	Hard
GB 1299(85)	CrWMn	As Hard	0.90-1.05	0.90-1.20	0.80-1.10	0.030 max	0.030 max	0.40 max		1.20-1.60	Cu <=0.30; Ni <=0.25; bal Fe				62 HRC

Czech Republic

Specification	Designation	Notes	C	Cr	Mn	P	S	Si	V	W	Other	UTS	YS	El	Hard
CSN 419710	19710		1.1-1.25	0.3-0.5	0.15-0.4	0.03 max	0.035 max	0.15-0.35		0.9-1.3	Ni <=0.3; bal Fe				
CSN 419711	19711		1.1-1.15	0.15-0.35	0.15-0.4	0.03 max	0.035 max	0.15-0.35	0.15-0.3	0.9-1.3	bal Fe				
CSN 419712	19712		1.15-1.3	1.45-1.8	0.4-0.7	0.03 max	0.035 max	0.15-0.35	0.1-0.2	1.2-1.6	Ni <=0.35; bal Fe				

France

Specification	Designation	Notes	C	Cr	Mn	P	S	Si	V	W	Other	UTS	YS	El	Hard
AFNOR NFA35590(92)	105WC13*	CW, Wear res, soft ann	1-1.15	0.8-1.1	0.7-1	0.025 max	0.025 max	0.1-0.4		1-1.6	Mo <=0.15; bal Fe				228 HB max
AFNOR NFA35590(92)	105WCr5	CW, Wear res, soft ann	1-1.15	0.8-1.1	0.7-1	0.025 max	0.025 max	0.1-0.4	0.1 max	1-1.6	Cu <=0.3; Mo <=0.15; Ni <=0.4; bal Fe				228 HB max
AFNOR NFA35590(92)	130C3	Wear res; soft ann	1.2-1.4	0.6-0.9	0.15-0.45	0.025 max	0.025 max	0.1-0.4			Ni <=0.4; bal Fe				223 HB max
AFNOR NFA35590(92)	130Cr3	Wear res; soft ann	1.2-1.4	0.6-0.9	0.15-0.45	0.025 max	0.025 max	0.1-0.4	0.1 max	0.1 max	Cu <=0.3; Mo <=0.15; Ni <=0.4; bal Fe				223 HB max

Germany

Specification	Designation	Notes	C	Cr	Mn	P	S	Si	V	W	Other	UTS	YS	El	Hard
DIN	115W8	Hard tmp to 200C	1.10-1.20	0.15-0.25	0.20-0.40	0.035 max	0.035 max	0.15-0.30		1.80-2.10	bal Fe				63 HRC
DIN	120W4	Hard tmp to 200C	1.15-1.25	0.15-0.25	0.20-0.35	0.035 max	0.035 max	0.15-0.30		0.90-1.10	bal Fe				63 HRC
DIN	120WV4	Hard tmp to 200C	1.15-1.25	0.15-0.25	0.20-0.35	0.035 max	0.035 max	0.15-0.30	0.07-0.12	0.90-1.10	bal Fe				62 HRC
DIN	WNr 1.2414		1.15-1.25	0.15-0.25	0.20-0.35	0.035 max	0.035 max	0.15-0.30		0.90-1.10	bal Fe				63 HRC
DIN	WNr 1.2442	Hard tmp to 200C	1.10-1.20	0.15-0.25	0.20-0.40	0.035 max	0.035 max	0.15-0.30		1.80-2.10	bal Fe				63 HRC
DIN	WNr 1.2516	Hard tmp to 200C	1.15-1.25	0.15-0.25	0.20-0.35	0.035 max	0.035 max	0.15-0.30	0.07-0.12	0.90-1.10	bal Fe				62 HRC
DIN 17350(80)	105WCr6	Hard tmp to 200C	1.00-1.10	0.90-1.10	0.80-1.10	0.030 max	0.030 max	0.10-0.40		1.00-1.30	bal Fe				61 HRC
DIN 17350(80)	WNr 1.2419	Hard tmp to 200C	1.00-1.10	0.90-1.10	0.80-1.10	0.030 max	0.030 max	0.10-0.40		1.00-1.30	bal Fe				61 HRC

Hungary

Specification	Designation	Notes	C	Cr	Mn	P	S	Si	V	W	Other	UTS	YS	El	Hard
MSZ 4352	W9		0.95-1.1	0.9-1.2	0.8-1.1	0.03 max	0.03 max	0.15-0.4	0.15 max	1.2-1.6	Cu 0.3; Mo <=0.2; Ni <=0.35; bal Fe				

India

Specification	Designation	Notes	C	Cr	Mn	P	S	Si	V	W	Other	UTS	YS	El	Hard
IS 1570	T110W2Cr1		1-1.2	0.9-1.3	0.9-1.3			0.1-0.35		1.25-1.75	bal Fe				

Italy

Specification	Designation	Notes	C	Cr	Mn	P	S	Si	V	W	Other	UTS	YS	El	Hard
UNI 2955	100WCr		1-1.1	0.9-1.2	0.8-1.1			0.3		1.2-1.6	bal Fe				
UNI 2955	SDW 107WCr5Ku		1	1	1			0.35		1.1	bal Fe				
UNI 2955	VAL7 110W4Ku		1.15		0.35			0.4	0.15	1.05	bal Fe				
UNI 2955/3(82)	107WCr5KU		1-1.15	0.8-1.1	0.7-1	0.07 max	0.06 max	0.1-0.4		1-1.6	bal Fe				

UNS numbers and US grades are provided as a means of cross referencing chemically similar alloys. Exchangability is only possible after independent examination of specifications. Tensile properties are minimum or typical as specified. UTS and YS as MPa. El as %. See Appendix for list of abbreviations used in Notes. * indicates obsolete material.

Specification	Designation	Notes	C	Cr	Mn	P	S	Si	V	W	Other	UTS	YS	El	Hard

Tool Steel, Oil-Hardening Cold Work, O7 (Continued from previous page)

Italy

Specification	Designation	Notes	C	Cr	Mn	P	S	Si	V	W	Other	UTS	YS	El	Hard
UNI 2955/3(82)	110W4KU		0.95-1.25	0.25 max	0.35 max	0.07 max	0.06 max	0.4 max	0.2 max	0.9-1.2	bal Fe				

Japan

Specification	Designation	Notes	C	Cr	Mn	P	S	Si	V	W	Other	UTS	YS	El	Hard
JIS G3311(88)	SKS2M	CR Strp; Meat Cutting Band Saw	1.00-1.10	0.50-1.00	0.80 max	0.030 max	0.030 max	0.35 max	0.20 max	1.00-1.50	Cu <=0.25; Ni <=0.25; bal Fe				230-320 HV
JIS G3311(88)	SKS7M	CR Strp; Meat Cutting Saw	1.10-1.20	0.20-0.50	0.50 max	0.030 max	0.030 max	0.35 max			Cu <=0.25; Ni <=0.25; bal Fe				250-340 HV
JIS G4404(83)	SKS2	HR Bar Frg Ann	1.00-1.10	0.50-1.00	0.80 max	0.030 max	0.030 max	0.35 max	0.20 max	1.00-1.5	Cu <=0.25; Ni <=0.25; bal Fe				217 HB max
JIS G4404(83)	SKS31	HR Frg Ann	0.95-1.05	0.80-1.20	0.90-1.20	0.030 max	0.030 max	0.35 max		1.00-1.50	Cu <=0.25; Ni <=0.25; bal Fe				217 HB max
JIS G4404(83)	SKS7	HR Bar Frg Ann	1.10-1.20	0.20-0.50	0.50 max	0.030 max	0.030 max	0.35 max	0.20 max	2.00-2.50	Cu <=0.25; Ni <=0.25; bal Fe				217 HB max

Mexico

Specification	Designation	Notes	C	Cr	Mn	P	S	Si	V	W	Other	UTS	YS	El	Hard
NMX-B-082(90)	O7	Bar	1.10-1.30	0.35-0.85	1.00 max	0.03 max	0.03 max	0.60 max	0.40 max	1.00-2.00	Cu <=0.25; Mo <=0.30; Ni <=0.30; (As+Sn+Sb)<=0.040; (Cu+Ni)<=0.40; B, Ti may be added; bal Fe				

Pan America

Specification	Designation	Notes	C	Cr	Mn	P	S	Si	V	W	Other	UTS	YS	El	Hard
COPANT 337	O7	Bar, Frg	1.1-1.3	0.6-0.85	0.5	0.03	0.03	0.5	0.3	1.25-1.9	Mo 0.3; bal Fe				

Poland

Specification	Designation	Notes	C	Cr	Mn	P	S	Si	V	W	Other	UTS	YS	El	Hard
PNH85023	NWC		1-1.15	0.9-1.2	0.8-1.1	0.030 max	0.030 max	0.15-0.4		1.2-1.6	bal Fe				

Romania

Specification	Designation	Notes	C	Cr	Mn	P	S	Si	V	W	Other	UTS	YS	El	Hard
STAS	CW20		1-1.1	0.6-0.9	0.15-0.4	0.03 max	0.025 max	0.15-0.35		1.8-2.2	Ni <=0.35; bal Fe				

Russia

Specification	Designation	Notes	C	Cr	Mn	P	S	Si	V	W	Other	UTS	YS	El	Hard
GOST	ChWG		0.9-1.05	0.9-1.2	0.8-1.1	0.03 max	0.03 max	0.1-0.4	0.15 max	1.2-1.6	Cu <=0.3; Mo <=0.3; Ni <=0.35; bal Fe				
GOST 5950	11Ch4W2MF3S2	Alloyed	1.05-1.15	3.5-4.2	0.2-0.5	0.03 max	0.03 max	1.4-1.8	2.3-2.8	2-2.7	Cu <=0.3; Mo 0.3-0.5; Ni <=0.4; bal Fe				
GOST 5950	ChWG	Alloyed	0.9-1.05	0.9-1.2	0.8-1.1			0.15-0.35	0.05	1.2-1.6	Cu <=0.3; Mo <=0.3; Ni <=0.35; Ti 0.03; bal Fe				
GOST 5950	ChWG	Alloyed	0.9-1.05	0.9-1.2	0.8-1.1	0.03 max	0.03 max	0.1-0.4	0.15 max	1.2-1.6	Cu <=0.3; Mo <=0.3; Ni <=0.35; bal Fe				
GOST 5950	W2F	Alloyed	1.05-1.22	0.2-0.4	0.15-0.45	0.03 max	0.03 max	0.1-0.4	0.15-0.3	0.6-2	Cu <=0.3; Mo <=0.3; Ni <=0.35; bal Fe				

Spain

Specification	Designation	Notes	C	Cr	Mn	P	S	Si	V	W	Other	UTS	YS	El	Hard
UNE 36018/2(94)	105WCrV5	Alloyed	1-1.1	0.8-1.1	0.7-1	0.025 max	0.02 max	0.1-0.4	0.1-0.3	1-1.7	Mo <=0.15; bal Fe				
UNE 36018/2(94)	F.5233*	see 105WCrV5	1-1.1	0.8-1.1	0.7-1	0.025 max	0.02 max	0.1-0.4	0.1-0.3	1-1.7	Mo <=0.15; bal Fe				
UNE 36072(75)	105WCr5	Alloyed; CW	1-1.15	0.8-1.1	0.7-1	0.03 max	0.03 max	0.1-0.4		1-1.6	bal Fe				
UNE 36072(75)	F.5233*	see 105WCr5	1-1.15	0.8-1.1	0.7-1	0.03 max	0.03 max	0.1-0.4		1-1.6	bal Fe				
UNE 36072/1(75)	102WCrV5	Alloyed; CW	0.95-1.15	0.4-0.75	0.15-0.35	0.03 max	0.03 max	0.1-0.35	0.1-0.3	1-1.6	Mo <=0.15; bal Fe				
UNE 36072/1(75)	F.5237*	see 102WCrV5	0.95-1.15	0.4-0.75	0.15-0.35	0.03 max	0.03 max	0.1-0.35	0.1-0.3	1-1.6	Mo <=0.15; bal Fe				

USA

Specification	Designation	Notes	C	Cr	Mn	P	S	Si	V	W	Other	UTS	YS	El	Hard
	AISI O7		1.25-130	0.35-0.85	1.00 max	0.030 max	0.030 max	0.60 max	0.40 max	1.00-2.00	Mo <=0.30; bal Fe				
	UNS T31507		1.10-1.30	0.35-0.85	1.00 max	0.030 max	0.030 max	0.60 max	0.40 max	1.00-2.00	Mo <=0.30; bal Fe				
ASTM A681(94)	O7	Ann	1.10-1.30	0.35-0.85	0.20-1.00	0.030 max	0.030 max	0.10-0.60	0.15-0.40	1.00-2.00	Mo <=0.30; S 0.06-0.15 if spec for machining; bal Fe				241 HB
FED QQ-T-570(87)	O7*	Obs; see ASTM A681	1.1-1.3	0.35-0.85	1			0.6	0.4	1-2	Mo 0.3; Ni 0.3; bal Fe				

Specification	Designation	Notes	C	Cr	Mn	Mo	Ni	P	S	Si	Other	UTS	YS	El	Hard
Tool Steel, Mold Steel, P2															
Argentina															
IAS	IRAM C3	Ann	0.50-0.60	0.90-1.10	0.60-0.70	0.35-0.50	0.45-0.65			0.15-0.40	V >=0.20; bal Fe				240 HB
Mexico															
NMX-B-082(90)	P1	Bar	0.10 max		0.10-0.30		0.30 max	0.03 max	0.03 max	0.10-0.40	Cu <=0.25; (As+Sn+Sb)<=0.040; (Cu+Ni)<=0.40; B, Ti may be added; bal Fe				
NMX-B-082(90)	P2	Bar	0.10 max	0.75-1.25	0.10-0.40	0.15-0.40	0.10-0.50	0.03 max	0.03 max	0.10-0.40	Cu <=0.25; (As+Sn+Sb)<=0.040; B, Ti may be added; bal Fe				
USA															
	UNS T51602*	Obs	0.10 max	0.75-1.25	0.10-0.40	0.15-0.40	0.10-0.50	0.030 max	0.030 max	0.10-0.40	bal Fe				
ASTM A681(94)	P2	Ann	0.10 max	0.75-1.25	0.10-0.40	0.15-0.40	0.10-0.50	0.030 max	0.030 max	0.10-0.40	S 0.06-0.15 if spec for machining; bal Fe				100 HB
Tool Steel, Mold Steel, P20															
Argentina															
IAS	IRAM P20	Ann	0.28-0.40	1.20-2.00	0.60-1.00	0.30-0.55		0.030 max	0.030 max	0.20-0.80	bal Fe				212 HB
Brazil															
ABNT	VP-20A		0.36	1.8	0.6	0.2	1				bal Fe				
China															
GB 1299(85)	3Cr2Mo		0.28-0.40	1.40-2.00	0.60-1.00	0.30-0.55	0.25 max	0.030 max	0.030 max	0.20-0.80	Cu <=0.30; bal Fe				45 HRC
YB/T 094(97)	SM3Cr2Mo	Slab Ann	0.28-0.40	1.40-2.00	0.60-1.00	0.30-0.55		0.030 max	0.030 max	0.20-0.80	bal Fe				235 HB max
YB/T 107(97)	SM3Cr2Mo	Plt; Q/T	0.28-0.40	1.40-2.00	0.60-1.00	0.30-0.55		0.030 max	0.030 max	0.30-0.70	bal Fe	960	660	15	
France															
AFNOR NFA35590(92)	35CD8		0.3-0.4	1.5-2.2	0.5-1.5	0.4-0.6	0.4 max	0.025 max	0.025 max	0.3-0.8	bal Fe				300 HB max
AFNOR NFA35590(92)	35CMD7		0.32-0.38	1.6-2	0.8-1.2	0.35-0.6	0.4 max	0.025 max	0.025 max	0.35-0.7	Cu <=0.3; V <=0.1; W <=0.1; bal Fe				300 HB max
AFNOR NFA35590(92)	35CrMnMo7		0.32-0.38	1.6-2	0.8-1.2	0.35-0.6	0.4 max	0.025 max	0.025 max	0.35-0.7	Cu <=0.3; V <=0.1; W <=0.1; bal Fe				300 HB max
AFNOR NFA35590(92)	35CrMo8		0.3-0.4	1.5-2.2	0.5-1.5	0.4-0.6	0.4 max	0.025 max	0.025 max	0.3-0.8	bal Fe				300 HB max
AFNOR NFA35590(92)	35NCDV8		0.32-0.38	1.9-2.3	0.3-0.6	0.5-0.8	2-2.4	0.025 max	0.025 max	0.1-0.4	bal Fe				255 HB max
Germany															
DIN	35CrMo4		0.32-0.37	0.90-1.10	0.60-0.80	0.20-0.25				0.20-0.40	bal Fe				
DIN	40CrMnMo7	Hard tmp to 200C	0.35-0.45	1.80-2.10	1.30-1.60	0.15-0.25		0.035 max	0.035 max	0.20-0.40	bal Fe				50 HRC
DIN	47CrMo4		0.43-0.50	0.90-1.20	0.60-0.80	0.25-0.40		0.025 max	0.025 max	0.15-0.35	bal Fe				
DIN	WNr 1.2311	Hard tmp to 200C	0.35-0.45	1.80-2.10	1.30-1.60	0.15-0.25		0.035 max	0.035 max	0.20-0.40	bal Fe				50 HRC
DIN	WNr 1.2330		0.32-0.37	0.90-1.10	0.60-0.80	0.20-0.25				0.20-0.40	bal Fe				
DIN	WNr 1.2332		0.43-0.50	0.90-1.20	0.60-0.80	0.25-0.40		0.025 max	0.025 max	0.15-0.35	bal Fe				
DIN 1652(90)	30CrMoV9	Q/T 17-40mm	0.26-0.34	2.30-2.70	0.40-0.70	0.15-0.25		0.035 max	0.035 max	0.40 max	V 0.10-0.20; bal Fe	1200-1450	1020	9	
DIN 1652(90)	WNr 1.7707	Q/T 17-40mm	0.26-0.34	2.30-2.70	0.40-0.70	0.15-0.25		0.035 max	0.035 max	0.40 max	V 0.10-0.20; bal Fe	1200-1450	1020	9	
DIN 17204(90)	30CrMoV9	Smls tube Q/T 17-40mm	0.26-0.34	2.30-2.70	0.40-0.70	0.15-0.25		0.035 max	0.035 max	0.40 max	V 0.10-0.20; bal Fe	1200-1450	1020	9	
DIN 17204(90)	WNr 1.7707	Smls tube Q/T 17-40mm	0.26-0.34	2.30-2.70	0.40-0.70	0.15-0.25		0.035 max	0.035 max	0.40 max	V 0.10-0.20; bal Fe	1200-1450	1020	9	
DIN 17350(80)	45CrMoV7	Hard	0.42-0.47	1.70-1.90	0.85-1.00	0.25-0.30		0.030 max	0.030 max	0.20-0.30	V 0.05; bal Fe				55 HRC
DIN 17350(80)	WNr 1.2328		0.42-0.47	1.70-1.90	0.85-1.00	0.25-0.30		0.030 max	0.030 max	0.20-0.30	V 0.05; bal Fe				55 HRC

UNS numbers and US grades are provided as a means of cross referencing chemically similar alloys. Exchangability is only possible after independent examination of specifications. Tensile properties are minimum or typical as specified. UTS and YS as MPa. El as %. See Appendix for list of abbreviations used in Notes. * indicates obsolete material.

Specification	Designation	Notes	C	Cr	Mn	Mo	Ni	P	S	Si	Other	UTS	YS	El	Hard

Tool Steel, Mold Steel, P20 (Continued from previous page)

Specification	Designation	Notes	C	Cr	Mn	Mo	Ni	P	S	Si	Other	UTS	YS	El	Hard
Italy															
UNI 2955	UC12		0.4	1.9	1.5	0.2				0.4	bal Fe				
UNI 2955/3(82)	35CrMo8KU		0.3-0.4	1.5-2.2	0.5-1.5	0.4-0.6		0.07 max	0.06 max	0.3-0.8	bal Fe				
UNI 2955/4(82)	35CrMo8KU		0.3-0.4	1.5-2.2	0.5-1.5	0.4-0.6		0.07 max	0.06 max	0.3-0.8	bal Fe				
Mexico															
NMX-B-082(90)	P20	Bar	0.28-0.40	1.40-2.00	0.60-1.00	0.30-0.55	0.30 max	0.03 max	0.03 max	0.20-0.80	Cu <=0.25; (As+Sn+Sb)<=0.040; (Cu+Ni)<=0.40; B, Ti may be added; bal Fe				
Pan America															
COPANT 337	P20	Bar, Frg	0.3-0.4	1.5-1.9	0.6-0.9	0.3-0.5		0.03	0.03	0.5-0.8	V 0.3; bal Fe				
Poland															
PNH85021	WC2		0.35-0.45	1.7-2	1.1-1.5	0.35	0.35			0.15-0.35	bal Fe				
PNH85021	WCMB		0.32-0.4	2.2-2.6	1.3-1.6	0.3-0.6	0.5			0.15-0.35	bal Fe				
Romania															
STAS	MoSMC20		0.35-0.45	1.7-2.2	1.2-1.6	0.2-0.5	0.35			0.5-1	bal Fe				
Russia															
GOST	4ChS		0.35-0.45	1.3-1.6	0.15-0.4	0.15	0.35			1.2-1.6	Cu 0.3; Ti 0.03; V 0.05; W 0.2; bal Fe				
GOST	5ChGM		0.5-0.6	0.6-0.9	1.2-1.6	0.15-0.3	0.35 max	0.03 max	0.03 max	0.25-0.6	Cu <=0.3; V <=0.05; W <=0.2; bal Fe				
GOST 4543(71)	30Ch3MF	Nitriding	0.27-0.34	2.3-2.7	0.3-0.6	0.2-0.3	0.3 max	0.035 max	0.035 max	0.17-0.37	Cu <=0.3; V 0.06-0.12; bal Fe				
Spain															
UNE 36072/2(75)	35CrMo7	Alloyed; HW	0.4-0.6	1.25-2	0.5-1.5	0.4-0.6		0.03 max	0.03 max	0.3-0.8	bal Fe				
UNE 36072/2(75)	F.5303*	see 35CrMo7	0.4-0.6	1.25-2	0.5-1.5	0.4-0.6		0.03 max	0.03 max	0.3-0.8	bal Fe				
Sweden															
SS 14	(USA P20)		0.33	3	0.5	0.5				0.3	bal Fe				
UK															
BS 4659	(USA P20)		0.4	1.1	0.65	0.3					bal Fe				
USA															
	AISI P20		0.28-0.40	1.40-2.00	0.60-1.00	0.30-0.55		0.030 max	0.030 max	0.20-0.80	bal Fe				
	UNS T51620		0.28-0.40	1.40-2.00	0.60-1.00	0.30-0.55		0.030 max	0.030 max	0.20-0.80	bal Fe				
ASTM A681(94)	P20	Prehard	0.28-0.40	1.40-2.00	0.60-1.00	0.30-0.55		0.030 max	0.030 max	0.20-0.80	S 0.06-0.15 if spec for machining; bal Fe				

Tool Steel, Mold Steel, P21

Specification	Designation	Notes	C	Cr	Mn	Mo	Ni	P	S	Si	Other	UTS	YS	El	Hard
France															
AFNOR NFA35590(92)	20DN34-13		0.18-0.23	0.3 max	0.5-0.8	3.1-3.7	2.9-3.5	0.025 max	0.025 max	0.1-0.4	bal Fe				350 HB max
AFNOR NFA35590(92)	20MoNi34-13		0.18-0.23	0.3 max	0.5-0.8	3.1-3.7	2.9-3.5	0.025 max	0.025 max	0.1-0.4	bal Fe				350 HB max
AFNOR NFA35590(92)	35NiCr15		0.32-0.38	1.4-1.8	0.3-0.6	0.15 max	3.5-4	0.025 max	0.025 max	0.1-0.4	bal Fe				255 HB max
AFNOR NFA35590(92)	35NiCrMoV8		0.32-0.38	1.9-2.3	0.3-0.6	0.5-0.8	2-2.4	0.025 max	0.025 max	0.1-0.4	bal Fe				255 HB max
Mexico															
NMX-B-082(90)	P21	Bar	0.18-0.22	0.20-0.30	0.20-0.40		3.90-4.25	0.03 max	0.03 max	0.20-0.40	Al 1.05-1.25; Cu <=0.25; V 0.15-0.25; (As+Sn+Sb)<=0.040; B, Ti may be added; bal Fe				
Spain															
UNE 36072/2(75)	40NiCrMoV15		0.35-0.45	1.7-2	0.35-0.65	0.4-0.6	3.6-4.1	0.03 max	0.03 max	0.1-0.4	V 0.05-0.25; bal Fe				
UNE 36072/2(75)	F.5305*	see 40NiCrMoV15	0.35-0.45	1.7-2	0.35-0.65	0.4-0.6	3.6-4.1	0.03 max	0.03 max	0.1-0.4	V 0.05-0.25; bal Fe				

Specification	Designation	Notes	C	Cr	Mn	Mo	Ni	P	S	Si	Other	UTS	YS	El	Hard

Tool Steel, Mold Steel, P21 (Continued from previous page)

USA

	AISI P21		0.18-0.22	0.20-0.30	0.20-0.40		3.90-4.25	0.030 max	0.030 max	0.20-0.40	Al 1.05-1.25; V 0.15-0.25; bal Fe				
	UNS T51621		0.18-0.22	0.20-0.30	0.20-0.40		4.00-4.25	0.030 max	0.030 max	0.20-0.40	Al 1.05-1.25; V 0.15-0.25; bal Fe				
ASTM A681(94)	P21	Prehard	0.18-0.22	0.20-0.30	0.20-0.40		3.90-4.25	0.030 max	0.030 max	0.20-0.40	Al 1.05-1.25; V 0.15-0.25; S 0.06-0.15 if spec for machining; bal Fe				

Tool Steel, Mold Steel, P3

Mexico

| NMX-B-082(90) | P3 | Bar | 0.10 max | 0.40-0.75 | 0.20-0.60 | | 1.00-1.50 | 0.03 max | 0.03 max | 0.40 max | Cu <=0.25; (As+Sn+Sb)<=0.040; B, Ti may be added; bal Fe | | | | |

USA

| | UNS T51603* | Obs | 0.10 max | 0.40-0.75 | 0.20-0.60 | | 1.00-1.50 | 0.030 max | 0.030 max | 0.40 max | bal Fe | | | | |
| ASTM A681(94) | P3 | Ann | 0.10 max | 0.40-0.75 | 0.20-0.60 | | 1.00-1.50 | 0.030 max | 0.030 max | 0.40 max | S 0.06-0.15 if spec for machining; bal Fe | | | | 143 HB |

Tool Steel, Mold Steel, P4

Mexico

| NMX-B-082(90) | P4 | Bar | 0.12 max | 4.00-5.25 | 0.20-0.60 | 0.40-1.00 | 0.30 max | 0.03 max | 0.03 max | 0.10-0.40 | Cu <=0.25; (As+Sn+Sb)<=0.040; (Cu+Ni)<=0.40; B, Ti may be added; bal Fe | | | | |

USA

| | UNS T51604* | Obs | 0.12 max | 4.00-5.25 | 0.20-0.60 | 0.40-1.00 | | 0.030 max | 0.030 max | 0.10-0.40 | bal Fe | | | | |
| ASTM A681(94) | P4 | Ann | 0.12 max | 4.00-5.25 | 0.20-0.60 | 0.40-1.00 | | 0.030 max | 0.030 max | 0.10-0.40 | S 0.06-0.15 if spec for machining; bal Fe | | | | 131 HB |

Tool Steel, Mold Steel, P5

Mexico

| NMX-B-082(90) | P5 | Bar | 0.10 max | 2.00-2.50 | 0.20-0.60 | | 0.35 max | 0.03 max | 0.03 max | 0.40 max | Cu <=0.25; (As+Sn+Sb)<=0.040; (Cu+Ni)<=0.40; B, Ti may be added; bal Fe | | | | |

USA

| | UNS T51605* | Obs | 0.10 max | 2.00-2.50 | 0.20-0.60 | | 0.35 max | 0.030 max | 0.030 max | 0.40 max | bal Fe | | | | |
| ASTM A681(94) | P5 | Ann | 0.06-0.10 | 2.00-2.50 | 0.20-0.60 | | 0.35 max | 0.030 max | 0.030 max | 0.10-0.40 | S 0.06-0.15 if spec for machining; bal Fe | | | | 131 HB |

Tool Steel, Mold Steel, P6

France

| AFNOR | 10NC12 | | 0.1 max | 0.8 max | 0.4 max | | 3 max | 0.03 max | 0.03 max | 0.3 max | bal Fe | | | | |

Germany

DIN	14NiCr18	CH 30mm	0.10-0.17	0.90-1.30	0.40-0.70		4.25-4.75	0.035 max	0.035 max	0.15-0.35	bal Fe	1180-1370	885	7	
DIN	15NiCr14	Hard tmp to 200C	0.10-0.17	0.65-0.85	0.30-0.50		3.30-3.60	0.030 max	0.030 max	0.20-0.35	bal Fe				60 HRC
DIN	15NiCr18	Hard tmp to 200C	0.10-0.17	0.90-1.20	0.30-0.50		4.20-4.70	0.030 max	0.030 max	0.20-0.30	bal Fe				60 HRC
DIN	WNr 1.2735	Hard tmp to 200C	0.10-0.17	0.65-0.85	0.30-0.50		3.30-3.60	0.030 max	0.030 max	0.20-0.35	bal Fe				60 HRC
DIN	WNr 1.2745	Hard tmp to 200C	0.10-0.17	0.90-1.20	0.30-0.50		4.20-4.70	0.030 max	0.030 max	0.20-0.30	bal Fe				60 HRC
DIN(Military Hdbk)	WNr 1.5860	CH 30mm	0.10-0.17	0.90-1.30	0.40-0.70		4.25-4.75	0.035 max	0.035 max	0.15-0.35	bal Fe	1180-1370	885	7	

Japan

| JIS G4410(84) | SKC31 | Rod or Bit; FF | 0.12-0.25 | 1.20-1.80 | 0.60-1.20 | 0.40-0.70 | 2.80-3.20 | 0.030 max | 0.030 max | 0.15-0.35 | Cu <=0.25; bal Fe | | | | |

UNS numbers and US grades are provided as a means of cross referencing chemically similar alloys. Exchangability is only possible after independent examination of specifications. Tensile properties are minimum or typical as specified. UTS and YS as MPa. El as %. See Appendix for list of abbreviations used in Notes. * indicates obsolete material.

Specification	Designation	Notes	C	Cr	Mn	Mo	Ni	P	S	Si	Other	UTS	YS	El	Hard

Tool Steel, Mold Steel, P6 (Continued from previous page)

Mexico

Specification	Designation	Notes	C	Cr	Mn	Mo	Ni	P	S	Si	Other	UTS	YS	El	Hard
NMX-B-082(90)	P6	Bar	0.05-0.15	1.25-1.75	0.35-0.70		3.25-3.75	0.03 max	0.03 max	0.10-0.40	Cu <=0.25; (As+Sn+Sb)<=0.040; B, Ti may be added; bal Fe				

Spain

Specification	Designation	Notes	C	Cr	Mn	Mo	Ni	P	S	Si	Other	UTS	YS	El	Hard
UNE	F.153	CH	0.1-0.15	0.9-1.1	0.3-0.6	0.15 max	3.8-4.5	0.04 max	0.04 max	0.1-0.35	bal Fe				

USA

Specification	Designation	Notes	C	Cr	Mn	Mo	Ni	P	S	Si	Other	UTS	YS	El	Hard
	AISI P6		0.05-0.15	1.25-1.75	0.35-0.70		3.25-3.75	0.030 max	0.030 max	0.10-0.40	bal Fe				
	UNS T51606		0.05-0.15	1.25-1.75	0.35-0.70		3.25-3.75	0.030 max	0.030 max	0.10-0.40	bal Fe				
ASTM A681(94)	P6	Ann	0.05-0.15	1.25-1.75	0.35-0.70		3.25-3.75	0.030 max	0.030 max	0.10-0.40	S 0.06-0.15 if spec for machining; bal Fe				212 HB

Specification	Designation	Notes	C	Cr	Mn	Mo	P	S	Si	V	Other	UTS	YS	El	Hard

Tool Steel, Shock-Resisting, No equivalents identified

Poland

Specification	Designation	Notes	C	Cr	Mn	Mo	P	S	Si	V	Other	UTS	YS	El	Hard
PNH85021	WLB		0.32-0.4	2.2-2.7	1.3-1.6	0.3-0.5	0.030 max	0.030 max	0.15-0.4		; B 0.002-0.005; Co <=0.3; Ni <=0.35; W <=0.3; bal Fe				
PNH85023	NC5	CW	1.3-1.45	0.4-0.7	0.15-0.45		0.030 max	0.030 max	0.15-0.4		Ni <=0.4; bal Fe				
PNH85023	NC6	CW	1.3-1.45	1.3-1.65	0.4-0.6		0.030 max	0.030 max	0.15-0.4	0.1-0.25	Ni <=0.4; bal Fe				

Russia

Specification	Designation	Notes	C	Cr	Mn	Mo	P	S	Si	V	Other	UTS	YS	El	Hard
GOST 5950	5Ch2MNF	Alloyed	0.46-0.53	1.5-2	0.4-0.7	0.8-1.1	0.03 max	0.03 max	0.1-0.4	0.3-0.5	Cu <=0.3; Ni 1.2-1.6; W <=0.2; bal Fe				

Tool Steel, Shock-Resisting, S1

Argentina

Specification	Designation	Notes	C	Cr	Mn	Mo	P	S	Si	V	Other	UTS	YS	El	Hard
IAS	IRAM S1	Ann	0.40-0.55	1.00-1.80	0.10-0.40	0.50 max	0.030 max	0.030 max	0.15-1.20	0.15-3.00	W 1.50-3.00; bal Fe				165-212 HB

Bulgaria

Specification	Designation	Notes	C	Cr	Mn	Mo	P	S	Si	V	Other	UTS	YS	El	Hard
BDS 7938	5ChW2SF		0.40-0.50	0.90-1.20	0.20-0.40		0.030 max	0.030 max	0.80-1.10	0.15-0.30	Cu <=0.30; Ni <=0.35; W 1.70-2.00; bal Fe				
BDS 7938	6ChW2SF		0.50-0.65	0.90-1.20	0.20-0.40		0.030 max	0.030 max	0.80-1.10	0.15-0.30	Cu <=0.30; Ni <=0.35; W 1.70-2.00; bal Fe				

China

Specification	Designation	Notes	C	Cr	Mn	Mo	P	S	Si	V	Other	UTS	YS	El	Hard
GB 1299	6CrW2Si*	Obs	0.55-0.65	1-1.3	0.4				0.5-0.8		Cu 0.25; Ni 0.3; W 2-2.7; bal Fe				
GB 1299	6SiCrV*	Obs	0.4-0.5	1.3-1.6	0.4				1.2-1.6	0.1-0.25	Cu 0.25; Ni 0.3; bal Fe				
GB 1299(85)	5CrW2Si	As Hard	0.45-0.55	1.00-1.30	0.40 max		0.030 max	0.030 max	0.50-0.80		Cu <=0.30; Ni <=0.25; bal Fe				55 HRC

Czech Republic

Specification	Designation	Notes	C	Cr	Mn	Mo	P	S	Si	V	Other	UTS	YS	El	Hard
CSN 419732	19732		0.42-0.52	0.9-1.2	0.15-0.4		0.03 max	0.035 max	0.8-1.2		Ni <=0.35; W 1.7-2.2; bal Fe				
CSN 419733	19733		0.52-0.62	0.9-1.2	0.15-0.4		0.03 max	0.035 max	0.8-1.2		Ni <=0.35; W 1.7-2.2; bal Fe				
CSN 419735	19735		0.55-0.65	1.0-1.3	0.15-0.4		0.03 max	0.035 max	0.45-0.75	0.1-0.25	W 1.8-2.3; bal Fe				

Finland

Specification	Designation	Notes	C	Cr	Mn	Mo	P	S	Si	V	Other	UTS	YS	El	Hard
SFS 910(73)	SFS910	Alloyed	0.44-0.53	1.0-1.3	0.2-0.4	0.2-0.3	0.03 max	0.02 max	0.7-1.1	0.1-0.2	W 2.0-2.5; bal Fe				

France

Specification	Designation	Notes	C	Cr	Mn	Mo	P	S	Si	V	Other	UTS	YS	El	Hard
AFNOR NFA35590	55WC20		0.5-0.6	0.9-1.2	0.15-0.45		0.025 max	0.025 max	0.7-1.1		W 1.7-2.2; bal Fe				
AFNOR NFA35590(92)	45WCrV8	Soft ann	0.4-0.5	0.95-1.25	0.15-0.45	0.15 max	0.025 max	0.025 max	0.7-1.1	0.15-0.3	Cu <=0.3; Ni <=0.4; W 1.7-2.2; bal Fe				228 HB max
AFNOR NFA35590(92)	45WCV20	Soft ann	0.4-0.5	0.95-1.25	0.15-0.45	0.15 max	0.025 max	0.025 max	0.7-1.1	0.15-0.3	Cu <=0.3; Ni <=0.4; W 1.7-2.2; bal Fe				228 HB max

Germany

Specification	Designation	Notes	C	Cr	Mn	Mo	P	S	Si	V	Other	UTS	YS	El	Hard
DIN	45WCrV7	Hard tmp to 200C	0.40-0.50	0.90-1.20	0.20-0.40		0.035 max	0.035 max	0.80-1.10	0.15-0.20	W 1.80-2.10; bal Fe				56 HRC
DIN	WNr 1.2542	Hard tmp to 200C	0.40-0.50	0.90-1.20	0.20-0.40		0.035 max	0.035 max	0.80-1.10	0.15-0.20	W 1.80-2.10; bal Fe				56 HRC
DIN 17350(80)	60WCrV7	Hard tmp to 200C	0.55-0.65	0.90-1.20	0.15-0.45		0.030 max	0.030 max	0.50-0.70	0.10-0.20	W 1.80-2.10; bal Fe				58 HRC
DIN 17350(80)	WNr 1.2550	Hard tmp to 200C	0.55-0.65	0.90-1.20	0.15-0.45		0.030 max	0.030 max	0.50-0.70	0.10-0.20	W 1.80-2.10; bal Fe				58 HRC

India

Specification	Designation	Notes	C	Cr	Mn	Mo	P	S	Si	V	Other	UTS	YS	El	Hard
IS 1570	T40W2Cr1V18		0.35-0.45	1-1.5	0.2-0.4				0.5-1	0.1-0.25	W 1.75-2.25; bal Fe				
IS 1570	T50W2Cr1V18		0.45-0.55	1-1.5	0.2-0.4				0.5-1	0.1-0.25	W 1.75-2.25; bal Fe				
IS 1570/6(96)	T50Cr4V2	CW	0.45-0.55	0.9-1.2	0.5-0.8	0.15 max	0.035 max	0.035 max	0.1-0.35	0.15-0.3	Ni <=0.4; bal Fe				
IS 1570/6(96)	T50W8Cr5V2	CW	0.45-0.55	1-1.5	0.2-0.4	0.15 max	0.035 max	0.035 max	0.5-1	0.1-0.25	Ni <=0.4; W 1.75-2.25; bal Fe				
IS 1570/6(96)	TAC26	CW	0.45-0.55	0.9-1.2	0.5-0.8	0.15 max	0.035 max	0.035 max	0.1-0.35	0.15-0.3	Ni <=0.4; bal Fe				

UNS numbers and US grades are provided as a means of cross referencing chemically similar alloys. Exchangability is only possible after independent examination of specifications. Tensile properties are minimum or typical as specified. UTS and YS as MPa. El as %. See Appendix for list of abbreviations used in Notes. * indicates obsolete material.

Specification	Designation	Notes	C	Cr	Mn	Mo	P	S	Si	V	Other	UTS	YS	El	Hard

Tool Steel, Shock-Resisting, S1 (Continued from previous page)

Specification	Designation	Notes	C	Cr	Mn	Mo	P	S	Si	V	Other	UTS	YS	El	Hard
India															
IS 1570/6(96)	TAC30	CW	0.45-0.55	1-1.5	0.2-0.4	0.15 max	0.035 max	0.035 max	0.5-1	0.1-0.25	Ni <=0.4; W 1.75-2.25; bal Fe				
Italy															
UNI 2955/3(82)	45WCrV8KU	Hot work	0.4-0.5	0.9-1.2	0.15-0.45		0.07 max	0.06 max	0.8-1.1	0.1-0.3	W 1.7-2.3; bal Fe				
UNI 2955/3(82)	55WCrV8KU	CW	0.5-0.6	0.9-1.2	0.16-0.45		0.07 max	0.06 max	0.8-1.1	0.1-0.3	W 1.7-2.3; bal Fe				
Japan															
JIS G4404(83)	SKS41	HR Bar Frg Ann	0.35-0.45	1.00-1.50	0.50 max		0.030 max	0.030 max	0.35 max		Cu <=0.25; W 2.50-3.50; bal Fe				217 HB max
Mexico															
NMX-B-082(90)	S-1	Bar	0.40-0.55	1.00-1.80	0.10-0.40	0.50 max	0.03 max	0.03 max	0.15-1.20	0.15-0.30	Cu <=0.25; Ni <=0.30; W 1.50-3.00; (As+Sn+Sb)<=0.040; (Cu+Ni)<=0.40; B, Ti may be added; bal Fe				
Poland															
PNH85023	NZ2	CW	0.4-0.5	0.9-1.2	0.15-0.45		0.030 max	0.030 max	0.8-1.1	0.15-0.3	Ni <=0.4; W 1.7-2.1; bal Fe				
PNH85023	NZ3		0.5-0.6	0.9-1.2	0.15-0.45		0.030 max	0.030 max	0.8-1.1	0.15-0.3	Ni <=0.4; W 1.7-2.1; bal Fe				
Romania															
STAS 3611(80)	VSCW20		0.4-0.5	1-1.3	0.2-0.4		0.03 max	0.025 max	0.8-1.2	0.15-0.3	Ni <=0.35; Pb <=0.15; W 1.8-2.2; bal Fe				
Russia															
GOST	5ChW2S		0.45-0.55	1-1.3	0.15-0.4	0.3	0.03 max	0.03 max	0.55-0.8	0.05	Cu 0.3; Ni <=0.35; Ti 0.03; W 2-2.5; bal Fe				
GOST 5950	4Ch3WMF	Alloyed	0.4-0.48	2.8-3.5	0.3-0.6	0.4-0.6	0.03 max	0.03 max	0.6-0.9	0.6-0.9	Cu <=0.3; Ni <=0.35; W 0.6-1; bal Fe				
GOST 5950	4ChS	Alloyed	0.35-0.45	1.3-1.6	0.15-0.4	0.2 max	0.03 max	0.03 max	1.2-1.6	0.15 max	Cu <=0.3; Ni <=0.35; W <=0.2; bal Fe				
GOST 5950	4ChW2S	Alloyed	0.35-0.45	1-1.3	0.15-0.4	0.3 max	0.03 max	0.03 max	0.6-0.9	0.15 max	Cu <=0.3; Ni <=0.35; W 2-2.5; bal Fe				
GOST 5950	5Ch3W3MFS	Alloyed	0.45-0.52	2.5-3.2	0.2-0.5	0.8-1.1	0.03 max	0.03 max	0.5-0.8	1.5-1.8	Cu <=0.3; Nb 0.05-0.15; Ni <=0.35; W 3-3.6; bal Fe				
GOST 5950	5ChNW	Alloyed	0.5-0.6	0.5-0.8	0.5-0.8	0.3 max	0.03 max	0.03 max	0.15-0.35	0.15 max	Cu <=0.3; Ni 1.4-1.8; W 0.4-0.7; bal Fe				
GOST 5950	5ChNWS	Alloyed	0.5-0.6	1.3-1.6	0.3-0.6	0.3 max	0.3 max	0.03 max	0.6-0.9	0.15 max	Cu <=0.3; Ni 0.8-1.2; W 0.4-0.7; bal Fe				
GOST 5950	5ChW2SF	Alloyed	0.45-0.55	0.9-1.2	0.15-0.45	0.3 max	0.03 max	0.03 max	0.8-1.1	0.15-0.3	Cu <=0.3; Ni <=0.35; W 1.8-2.3; bal Fe				
GOST 5950	6ChW2S	Alloyed	0.55-0.65	1-1.3	0.15-0.4	0.3 max	0.03 max	0.03 max	0.5-0.8	0.15 max	Cu <=0.3; Ni <=0.35; W 2.2-2.7; bal Fe				
GOST 5950	9Ch5WF	Alloyed	0.85-1	4.5-5.5	0.15-0.4	0.3 max	0.03 max	0.03 max	0.15-0.4	0.15-0.3	Cu <=0.3; Ni <=0.35; W 0.8-1.2; bal Fe				
Spain															
UNE	F.524		0.45-0.55	0.75-1	0.25-0.35		0.03 max	0.03 max	0.9-1.1		W 1.8-2.2; bal Fe				
UNE 36018(94)	45WCrSiV8	Alloyed; CW	0.4-0.5	0.9-1.2	0.15-0.45	0.15 max	0.025 max	0.02 max	0.8-1.1	0.1-0.3	W 1.7-2.3; bal Fe				
UNE 36018(94)	F.5241*	see 45WCrSiV8	0.4-0.5	0.9-1.2	0.15-0.45	0.15 max	0.025 max	0.02 max	0.8-1.1	0.1-0.3	W 1.7-2.3; bal Fe				
UNE 36018/2(94)	60SiMoCrV8	Alloyed	0.55-0.62	0.2-0.35	0.7-1	0.3-0.5	0.025 max	0.02 max	0.7-2.2	0.1-0.3	bal Fe				
UNE 36018/2(94)	60WCrSiV8	Alloyed; CW	0.55-0.65	0.9-1.2	0.15-0.45	0.15 max	0.025 max	0.02 max	0.8-1.1	0.1-0.3	W 1.6-2.3; bal Fe				
UNE 36018/2(94)	F.5242*	see 60WCrSiV8	0.55-0.65	0.9-1.2	0.15-0.45	0.15 max	0.025 max	0.02 max	0.8-1.1	0.1-0.3	W 1.6-2.3; bal Fe				
UNE 36018/2(94)	F.5247*	see 60SiMoCrV8	0.55-0.62	0.2-0.35	0.7-1	0.3-0.5	0.025 max	0.02 max	0.7-2.2	0.1-0.3	bal Fe				
UNE 36072(75)	45WCrSi8	Alloyed; CW	0.4-0.5	0.9-1.2	0.15-0.45		0.03 max	0.03 max	0.8-1.1	0.1-0.3	W 1.7-2.3; bal Fe				
UNE 36072(75)	60WCrSi8		0.55-0.65	0.9-1.2	0.15-0.4		0.03 max	0.03 max	0.8-1.1	0.1-0.3	W 1.7-2.3; bal Fe				

Specification	Designation	Notes	C	Cr	Mn	Mo	P	S	Si	V	Other	UTS	YS	El	Hard

Tool Steel, Shock-Resisting, S1 (Continued from previous page)

Spain

Specification	Designation	Notes	C	Cr	Mn	Mo	P	S	Si	V	Other	UTS	YS	El	Hard
UNE 36072(75)	F.5241*	see 45WCrSi8	0.4-0.5	0.9-1.2	0.15-0.45		0.03 max	0.03 max	0.8-1.1	0.1-0.3	W 1.7-2.3; bal Fe				
UNE 36072(75)	F.5242*	see 60WCrSi8	0.55-0.65	0.9-1.2	0.15-0.4		0.03 max	0.03 max	0.8-1.1	0.1-0.3	W 1.7-2.3; bal Fe				

Sweden

Specification	Designation	Notes	C	Cr	Mn	Mo	P	S	Si	V	Other	UTS	YS	El	Hard
SS 142710	2710	Alloyed	0.44-0.53	1-1.3	0.2-0.4	0.2-0.3	0.03 max	0.02 max	0.7-1.1	0.1-0.2	W 2-2.5; bal Fe				

UK

Specification	Designation	Notes	C	Cr	Mn	Mo	P	S	Si	V	Other	UTS	YS	El	Hard
BS 4659	BSI	Bar Rod Sh Strp Frg	0.45-0.55	1.2-1.7	0.3-0.7				0.7-1	0.1-0.3	W 2-2.5; bal Fe				

USA

Specification	Designation	Notes	C	Cr	Mn	Mo	P	S	Si	V	Other	UTS	YS	El	Hard
	AISI S1	Oil Q/A Tmp 315 C	0.40-0.55	1.00-1.80	0.10-0.40	0.50 max	0.030 max	0.030 max	0.15-1.20	0.15-0.30	W 1.50-3.00; bal Fe				
	UNS T41901		0.40-0.55	1.00-1.80	0.10-0.40	0.50 max	0.030 max	0.030 max	0.15-1.20	0.15-0.30	W 1.50-3.00; bal Fe				
ASTM A681(94)	S1	Ann	0.40-0.55	1.00-1.80	0.10-0.40	0.50 max	0.030 max	0.030 max	0.15-1.20	0.15-0.30	W 1.50-3.00; S 0.06-0.15 if spec for machining; bal Fe				229 HB
FED QQ-T-570(87)	S-1*	Obs; see ASTM A681									bal Fe				
SAE J437(70)	S1	Ann									bal Fe				192-235 HB
SAE J438(70)	S1		0.45-0.55	1.25-1.75	0.20-0.40	0.40 max			0.25-0.45	0.15-0.30	W 1.00-3.00; Mo optional; Si may vary; bal Fe				

Tool Steel, Shock-Resisting, S2

France

Specification	Designation	Notes	C	Cr	Mn	Mo	P	S	Si	V	Other	UTS	YS	El	Hard
AFNOR NFA35571(96)	45SiCrMo6	t<1000mm; Q/T	0.42-0.5	0.5-0.75	0.5-0.8	0.2-0.3	0.025 max	0.025 max	1.3-1.7	0.1 max	Cu <=0.3; Ni <=0.4; W <=0.1; bal Fe	1550-1850	1400	6	
AFNOR NFA35590(92)	45SiCrMo6	Soft ann	0.42-0.5	0.5-0.75	0.5-0.8	0.15-0.3	0.025 max	0.025 max	1.3-1.7	0.1 max	Cu <=0.3; Ni <=0.4; W <=0.1; bal Fe				248 HB max
AFNOR NFA35590(92)	Y45SCD6	Soft ann	0.42-0.5	0.5-0.75	0.5-0.8	0.15-0.3	0.025 max	0.025 max	1.3-1.7	0.1 max	Cu <=0.3; Ni <=0.4; W <=0.1; bal Fe				248 HB max
AFNOR NFA35590(92)	Y60S7	Q/T	0.52-0.6	0.3 max	0.6-0.9	0.15 max	0.025 max	0.025 max	1.8-2.2		Ni <=0.4; bal Fe				56 HRC

India

Specification	Designation	Notes	C	Cr	Mn	Mo	P	S	Si	V	Other	UTS	YS	El	Hard
IS 1570/6(96)	50T8	CW	0.45-0.55	0.2 max	0.6-0.9	0.15 max	0.035 max	0.035 max	0.1-0.35		Cu <=0.25; Ni <=0.25; bal Fe				
IS 1570/6(96)	TC1	CW	0.45-0.55	0.2 max	0.6-0.9	0.15 max	0.035 max	0.035 max	0.1-0.35		Cu <=0.25; Ni <=0.25; bal Fe				

Mexico

Specification	Designation	Notes	C	Cr	Mn	Mo	P	S	Si	V	Other	UTS	YS	El	Hard
NMX-B-082(90)	S-2	Bar	0.40-0.55		0.30-0.50	0.30-0.60	0.03 max	0.03 max	0.90-1.20	0.50 max	Cu <=0.25; Ni <=0.30; (As+Sn+Sb)<=0.040; (Cu+Ni)<=0.40; B, Ti may be added; bal Fe				

Pan America

Specification	Designation	Notes	C	Cr	Mn	Mo	P	S	Si	V	Other	UTS	YS	El	Hard
COPANT 337	S-2	Bar, Frg	0.4-0.55		0.3-0.5	0.3-0.6	0.03	0.03	0.9-1.2	0.5	bal Fe				

Russia

Specification	Designation	Notes	C	Cr	Mn	Mo	P	S	Si	V	Other	UTS	YS	El	Hard
GOST 5950	5ChGM	Alloyed	0.5-0.6	0.6-0.9	1.2-1.6	0.15-0.3	0.03 max	0.03 max	0.25-0.6	0.15 max	Cu <=0.3; Ni <=0.35; W <=0.2; bal Fe				

Spain

Specification	Designation	Notes	C	Cr	Mn	Mo	P	S	Si	V	Other	UTS	YS	El	Hard
UNE	F.524		0.4-0.55	0.75-1	0.25-0.35		0.03 max	0.03 max	0.9-1.1		W 1.8-2.2; bal Fe				

UK

Specification	Designation	Notes	C	Cr	Mn	Mo	P	S	Si	V	Other	UTS	YS	El	Hard
BS 4659	BS2		0.45-0.65		0.3-0.5	0.3-0.6			0.9-1.2	0.1-0.3	bal Fe				
BS 970/1(83)	280M01(S)	Wrought Micro-alloyed Q/T; t<=100mm	0.30-0.55		0.60-1.50		0.035 max	0.045-0.065	0.15-0.60	0.08-0.20	Al <=0.035; 0.08<=Nb+Ti<=0.20; bal Fe	775-925	530	14	223-277 HB
BS 970/1(83)	280M01(T)	Wrought Micro-alloyed Q/T; t<=100mm	0.30-0.55		0.60-1.50		0.035 max	0.045-0.065	0.15-0.60	0.08-0.20	Al <=0.035; 0.08<=Nb+Ti<=0.20; bal Fe	925-1075	600	10	269-331 HB

UNS numbers and US grades are provided as a means of cross referencing chemically similar alloys. Exchangability is only possible after independent examination of specifications. Tensile properties are minimum or typical as specified. UTS and YS as MPa. El as %. See Appendix for list of abbreviations used in Notes. * indicates obsolete material.

Specification	Designation	Notes	C	Cr	Mn	Mo	P	S	Si	V	Other	UTS	YS	El	Hard

Tool Steel, Shock-Resisting, S2 (Continued from previous page)

USA

Specification	Designation	Notes	C	Cr	Mn	Mo	P	S	Si	V	Other	UTS	YS	El	Hard
	AISI S2		0.40-0.55		0.30-0.50	0.30-0.60	0.030 max	0.030 max	0.90-1.20	0.50 max	bal Fe				
	UNS T41902		0.40-0.55		0.30-0.50	0.30-0.60	0.030 max	0.030 max	0.90-1.20	0.50 max	bal Fe				
ASTM A681(94)	S2	Ann	0.40-0.55		0.30-0.50	0.30-0.60	0.030 max	0.030 max	0.90-1.20	0.50 max	S 0.06-0.15 if spec for machining; bal Fe				217 HB
FED QQ-T-570(87)	S-2*	Obs; see ASTM A681	0.4-0.55		0.3-0.5	0.3-0.6			0.9-1.2	0.5	Ni 0.3; bal Fe				
SAE J437(70)	S2	Ann									bal Fe				192-229 HB
SAE J438(70)	S2		0.45-0.55		0.30-0.50	0.40-0.60			0.80-1.20	0.25 max	V optional; bal Fe				

Tool Steel, Shock-Resisting, S4

Argentina

Specification	Designation	Notes	C	Cr	Mn	Mo	P	S	Si	V	Other	UTS	YS	El	Hard
IAS	IRAM S4	Ann	0.50-0.65	0.35 max	0.60-0.95		0.030 max	0.030 max	1.75-2.25	0.35 max	bal Fe				223 HB

France

Specification	Designation	Notes	C	Cr	Mn	Mo	P	S	Si	V	Other	UTS	YS	El	Hard
AFNOR NFA35590(92)	60Si8	Q/T	0.52-0.6	0.3 max	0.6-0.9	0.15 max	0.025 max	0.025 max	1.8-2.2		Ni <=0.4; bal Fe				56 HRC

India

Specification	Designation	Notes	C	Cr	Mn	Mo	P	S	Si	V	Other	UTS	YS	El	Hard
IS 1570/6(96)	55T8	CW	0.5-0.6	0.2 max	0.6-0.9	0.15 max	0.035 max	0.035 max	0.1-0.35		Cu <=0.25; Ni <=0.25; bal Fe				
IS 1570/6(96)	TC2	CW	0.5-0.6	0.2 max	0.6-0.9	0.15 max	0.035 max	0.035 max	0.1-0.35		Cu <=0.25; Ni <=0.25; bal Fe				

Mexico

Specification	Designation	Notes	C	Cr	Mn	Mo	P	S	Si	V	Other	UTS	YS	El	Hard
NMX-B-082(90)	S4	Bar	0.50-0.65	0.35 max	0.60-0.95		0.03 max	0.03 max	1.75-2.25	0.35 max	Cu <=0.25; Ni <=0.30; (As+Sn+Sb)<=0.040; (Cu+Ni)<=0.40; B, Ti may be added; bal Fe				

USA

Specification	Designation	Notes	C	Cr	Mn	Mo	P	S	Si	V	Other	UTS	YS	El	Hard
	AISI S4	Ann	0.50-0.65	0.50 max	0.60-0.95		0.030 max	0.030 max	1.75-2.25	0.35 max	bal Fe				
	UNS T41904		0.50-0.65	0.35 max	0.60-0.95		0.030 max	0.030 max	1.75-2.25	0.35 max	bal Fe				
ASTM A681(94)	S4	Ann	0.50-0.65	0.10-0.50	0.60-0.95		0.030 max	0.030 max	1.75-2.25	0.15-0.35	S 0.06-0.15 if spec for machining; bal Fe				229 HB

Tool Steel, Shock-Resisting, S5

China

Specification	Designation	Notes	C	Cr	Mn	Mo	P	S	Si	V	Other	UTS	YS	El	Hard
GB 1299	4SiCrV*	Obs	0.4-0.5	1.3-1.6	0.4				1.2-1.6	0.1-0.25	Cu 0.25; Ni 0.3; bal Fe				
GB 1299	5SiMnMoV*	Obs	0.45-0.55	0.2-0.4	0.5-0.7	0.3-0.5			1.5-1.8	0.2-0.35	Cu 0.25; Ni 0.3; bal Fe				

India

Specification	Designation	Notes	C	Cr	Mn	Mo	P	S	Si	V	Other	UTS	YS	El	Hard
IS 1570/6(96)	T55Si7	CW	0.5-0.6	0.3 max	0.8-1	0.15 max	0.035 max	0.035 max	1.5-2		Ni <=0.4; bal Fe				
IS 1570/6(96)	T55Si7Mo3	CW	0.5-0.6	0.3 max	0.8-1	0.25-0.4	0.035 max	0.035 max	1.5-2	0.12-0.2	Ni <=0.4; bal Fe				
IS 1570/6(96)	TAC27	CW	0.5-0.6	0.3 max	0.8-1	0.15 max	0.035 max	0.035 max	1.5-2		Ni <=0.4; bal Fe				
IS 1570/6(96)	TAC28	CW	0.5-0.6	0.3 max	0.8-1	0.25-0.4	0.035 max	0.035 max	1.5-2	0.12-0.2	Ni <=0.4; bal Fe				

Italy

Specification	Designation	Notes	C	Cr	Mn	Mo	P	S	Si	V	Other	UTS	YS	El	Hard
UNI 2955/3(82)	56SiMn7KU		0.52-0.6		0.6-0.9		0.07 max	0.06 max	1.5-2		bal Fe				
UNI 2955/3(82)	58SiMo8KU	CW	0.5-0.65		0.7-0.9	0.25-0.45	0.07 max	0.06 max	1.7-2.1		bal Fe				

Mexico

Specification	Designation	Notes	C	Cr	Mn	Mo	P	S	Si	V	Other	UTS	YS	El	Hard
NMX-B-082(90)	S-5	Bar	0.50-0.65	0.35 max	0.60-1.00	0.20-1.35	0.03 max	0.03 max	1.75-2.25	0.35 max	Cu <=0.25; Ni <=0.30; (As+Sn+Sb)<=0.040; (Cu+Ni)<=0.40; B, Ti may be added; bal Fe				

Specification	Designation	Notes	C	Cr	Mn	Mo	P	S	Si	V	Other	UTS	YS	El	Hard

Tool Steel, Shock-Resisting, S5 (Continued from previous page)

Pan America

Specification	Designation	Notes	C	Cr	Mn	Mo	P	S	Si	V	Other	UTS	YS	El	Hard
COPANT 337	S-5	Bar, Frg	0.5-0.6	0.35	0.6-1	0.3-0.6	0.03	0.03	1.75-2.25	0.35	bal Fe				

Russia

Specification	Designation	Notes	C	Cr	Mn	Mo	P	S	Si	V	Other	UTS	YS	El	Hard
GOST 5950	9ChS	Alloyed	0.85-0.95	0.95-1.25	0.3-0.6	0.2 max	0.03 max	0.03 max	1.2-1.6	0.15 max	Cu <=0.3; Ni <=0.35; W <=0.2; bal Fe				

Spain

Specification	Designation	Notes	C	Cr	Mn	Mo	P	S	Si	V	Other	UTS	YS	El	Hard
UNE	F.520.E		0.47-0.53		0.65-0.95	0.4-0.6	0.03 max	0.03 max	1.75-2.25		bal Fe				
UNE	F.520.F		0.9-1.1	0.9-1.2	0.15-0.45	0.17-0.27	0.03 max	0.03 max	0.13-0.38		bal Fe				
UNE 36072(75)	60SiMoCr8		0.55-0.62	0.2-0.35	0.7-1	0.3-0.5	0.03 max	0.03 max	1.7-2.2	0.1-0.3	bal Fe				
UNE 36072(75)	F.5247*	see 60SiMoCr8	0.55-0.62	0.2-0.35	0.7-1	0.3-0.5	0.03 max	0.03 max	1.7-2.2	0.1-0.3	bal Fe				

UK

Specification	Designation	Notes	C	Cr	Mn	Mo	P	S	Si	V	Other	UTS	YS	El	Hard
BS 4659	BS5		0.5-0.6		0.6-0.8	0.3-0.6			1.6-2.1	0.1-0.3	bal Fe				

USA

Specification	Designation	Notes	C	Cr	Mn	Mo	P	S	Si	V	Other	UTS	YS	El	Hard
	AISI S5	Ann	0.50-0.65	0.50 max	0.60-1.00	0.20-1.35	0.030 max	0.030 max	1.75-2.25	0.35 max	bal Fe				
	UNS T41905		0.50-0.65	0.35 max	0.60-1.00	0.20-1.35	0.030 max	0.030 max	1.75-2.25	0.35 max	bal Fe				
ASTM A681(94)	S5	Ann	0.50-0.65	0.10-0.50	0.60-1.00	0.20-1.35	0.030 max	0.030 max	1.75-2.25	0.15-0.35	S 0.06-0.15 if spec for machining; bal Fe				229 HB
FED QQ-T-570(87)	S-5*	Obs; see ASTM A681	0.5-0.65	0.35	0.6-1	0.2-1.35			1.75-2.25	0.35	Ni 0.3; bal Fe				
SAE J437(70)	S5	Ann									bal Fe				192-229 HB
SAE J438(70)	S5		0.50-0.60	0.30 max	0.60-0.90	0.30-0.50			1.80-2.20	0.25 max	Cr, V optional; bal Fe				

Tool Steel, Shock-Resisting, S6

Czech Republic

Specification	Designation	Notes	C	Cr	Mn	Mo	P	S	Si	V	Other	UTS	YS	El	Hard
CSN 419520	19520		0.35-0.45	1.7-2.2	1.2-1.6	0.2-0.4	0.03 max	0.03 max	0.5-0.9		bal Fe				

France

Specification	Designation	Notes	C	Cr	Mn	Mo	P	S	Si	V	Other	UTS	YS	El	Hard
AFNOR NFA35590(92)	45CDV6	Soft ann	0.41-0.49	1.35-1.65	0.1-0.4	0.7-1	0.025 max	0.025 max	0.1-0.4	0.15-0.35	Ni <=0.4; bal Fe				229 HB max
AFNOR NFA35590(92)	45CrMoV6	Soft ann	0.41-0.49	1.35-1.65	0.1-0.4	0.7-1	0.025 max	0.025 max	0.1-0.4	0.15-0.35	Ni <=0.4; bal Fe				229 HB max

India

Specification	Designation	Notes	C	Cr	Mn	Mo	P	S	Si	V	Other	UTS	YS	El	Hard
IS 1570/6(96)	T40W8Cr5V2	CW	0.35-0.45	1-1.5	0.2-0.4	0.15 max	0.035 max	0.035 max	0.5-1	0.1-0.25	Ni <=0.4; W 1.75-2.25; bal Fe				
IS 1570/6(96)	T45Cr5Si3	CW	0.4-0.5	1.2-1.6	0.5-0.75	0.15 max	0.035 max	0.035 max	0.8-1.1		Ni <=0.4; bal Fe				
IS 1570/6(96)	TAC24	CW	0.4-0.5	1.2-1.6	0.5-0.75	0.15 max	0.035 max	0.035 max	0.8-1.1		Ni <=0.4; bal Fe				
IS 1570/6(96)	TAC29	CW	0.35-0.45	1-1.5	0.2-0.4	0.15 max	0.035 max	0.035 max	0.5-1	0.1-0.25	Ni <=0.4; W 1.75-2.25; bal Fe				

Mexico

Specification	Designation	Notes	C	Cr	Mn	Mo	P	S	Si	V	Other	UTS	YS	El	Hard
NMX-B-082(90)	S-6	Bar	0.40-0.50	1.20-1.50	1.20-1.50	0.30-0.50	0.03 max	0.03 max	2.00-2.50	0.20-0.40	Cu <=0.25; Ni <=0.30; (As+Sn+Sb)<=0.040; (Cu+Ni)<=0.40; B, Ti may be added; bal Fe				

USA

Specification	Designation	Notes	C	Cr	Mn	Mo	P	S	Si	V	Other	UTS	YS	El	Hard
	AISI S6		0.40-0.50	1.20-1.50	1.20-1.50	0.30-0.50	0.030 max	0.030 max	2.00-2.50	0.20-0.40	bal Fe				
	UNS T41906		0.40-0.50	1.20-1.50	1.20-1.50	0.30-0.50	0.030 max	0.030 max	2.00-2.50	0.20-0.40	bal Fe				
ASTM A681(94)	S6	Ann	0.40-0.50	1.20-1.50	1.20-1.50	0.30-0.50	0.030 max	0.030 max	2.00-2.50	0.20-0.40	S 0.06-0.15 if spec for machining; bal Fe				229 HB
FED QQ-T-570(87)	S-6*	Obs; see ASTM A681	0.4-0.5	1.2-1.5	1.2-1.5	0.3-0.5			2-2.5	0.2-0.4	Ni 0.3; bal Fe				

Tool Steel, Shock-Resisting, S7

Specification	Designation	Notes	C	Cr	Mn	Mo	P	S	Si	V	Other	UTS	YS	El	Hard
China															
GB 1299	5Cr4Mo*	Obs	0.45-0.55	3.4-4	0.4	1.4-1.7			0.4		Cu 0.25; Ni 0.3; bal Fe				
Czech Republic															
CSN 419550	19550		0.5-0.6	2.9-3.6	0.6-0.9	1-1.4	0.03 max	0.035 max	0.2-0.45	0.08-0.17	bal Fe				
France															
AFNOR NFA35590(92)	50CDV13	Soft ann	0.4-0.6	3-3.5	0.55-0.85	1.3-1.7	0.025 max	0.025 max	0.2-0.5	0.15-0.35	Ni <=0.4; bal Fe				223 HB max
AFNOR NFA35590(92)	50CrMoV13	Soft ann	0.4-0.6	3-3.5	0.55-0.85	1.3-1.7	0.025 max	0.025 max	0.2-0.5	0.15-0.35	Ni <=0.4; bal Fe				223 HB max
Mexico															
NMX-B-082(90)	S-7	Bar	0.45-0.55	3.00-3.50	0.20-0.80	1.30-1.80	0.03 max	0.03 max	0.20-1.00	0.35 max	Cu <=0.25; Ni <=0.30; (As+Sn+Sb)<=0.040; (Cu+Ni)<=0.40; B, Ti may be added; bal Fe				
Pan America															
COPANT 337	S-7	Bar, Frg	0.45-0.55	2.75-3.5	0.6-0.8	1.3-1.8	0.03	0.03	0.2-0.4	0.15-0.3	bal Fe				
Russia															
GOST 5950	6Ch4M2FS	Alloyed	0.57-0.65	3.8-4.4	0.15-0.4	2-2.4	0.03 max	0.03 max	0.7-1	0.4-0.6	Cu <=0.3; Ni <=0.35; W <=0.2; bal Fe				
USA															
	AISI S7	Ann	0.45-0.55	3.00-3.50	0.20-0.80	1.30-1.80	0.030 max	0.030 max	0.20-1.00	0.35 max	bal Fe				
	UNS T41907		0.45-0.55	3.00-3.50	0.20-0.80	1.30-1.80	0.030 max	0.03 max	0.20-1.00	0.20-0.30	bal Fe				
ASTM A597(93)	CS-7	Cast	0.45-0.55	3.00-3.50	0.40-0.80	1.20-1.60	0.03 max	0.03 max	0.60-1.00		bal Fe				
ASTM A681(94)	S7	Ann	0.45-0.55	3.00-3.50	0.20-0.90	1.30-1.80	0.030 max	0.030 max	0.20-1.00	0.35 max	S 0.06-0.15 if spec for machining; bal Fe				229 HB
ASTM A685	S-7	Bar	0.45-0.55	3-3.5	0.2-0.8	1.3-1.8	0.03	0.03	0.2-1	0.2-0.3	bal Fe				
FED QQ-T-570(87)	S-7*	Obs; see ASTM A681	0.45-0.55	3-3.5	0.2-0.8	1.3-1.8			0.2-1	0.35	Ni 0.3; bal Fe				

Specification	Designation	Notes	C	Cr	Mn	P	S	Si	V	W	Other	UTS	YS	El	Hard
Tool Steel, Tungsten High-Speed, F1															
China															
GB 1299(85)	W	As Hard	1.05-1.25	0.10-0.30	0.40 max	0.030 max	0.030 max	0.40 max		0.80-1.20	Cu <=0.30; Ni <=0.25; bal Fe				62 HRC
India															
IS 1570/6(96)	T118Cr2	CW	1.1-1.25	0.3-0.6	0.4 max	0.035 max	0.035 max	0.1-0.35	0.3 max		Mo <=0.15; Ni <=0.4; bal Fe				
IS 1570/6(96)	TAC4	CW	1.1-1.25	0.3-0.6	0.4 max	0.035 max	0.035 max	0.1-0.35			Mo <=0.15; Ni <=0.4; bal Fe				
Mexico															
NMX-B-082(90)	F1	Bar	0.95-1.25		0.50 max	0.03 max	0.03 max	0.50 max		1.00-1.75	Cu <=0.25; Ni <=0.30; (As+Sn+Sb)<=0.040; (Cu+Ni)<=0.40; B, Ti may be added; bal Fe				
USA															
	UNS T60601*	Obs	0.95-1.25		0.50 max	0.030 max	0.030 max	0.50 max		1.00-1.75	bal Fe				
ASTM A681(94)	F1	Ann	0.95-1.25		0.50 max	0.030 max	0.030 max	0.10-0.50		1.00-1.75	S 0.06-0.15 if spec for machining; bal Fe				207 HB
Tool Steel, Tungsten High-Speed, F2															
Japan															
JIS G4404(83)	SKS11	High-C Special Purpose Ann	1.20-1.30	0.20-0.50	0.50 max	0.030 max	0.030 max	0.35 max	0.10-0.30	3.00-4.00	Cu <=0.25; Ni <=0.25; bal Fe				241 HB max
Mexico															
NMX-B-082(90)	F2	Bar	1.20-1.40	0.20-0.40	0.50 max	0.03 max	0.03 max	0.50 max		3.00-4.50	Cu <=0.25; Ni <=0.30; (As+Sn+Sb)<=0.040; (Cu+Ni)<=0.40; B, Ti may be added; bal Fe				
NMX-B-082(90)	F3	Bar	1.25-1.40	0.50-1.00	0.10-0.40	0.03 max	0.03 max	0.10-0.40		3.50-4.00	Cu <=0.25; Ni <=0.30; (As+Sn+Sb)<=0.040; (Cu+Ni)<=0.40; B, Ti may be added; bal Fe				
USA															
	UNS T60602*	Obs	1.20-1.40	0.20-0.40	0.50 max	0.030 max	0.030 max	0.50 max		3.00-4.50	bal Fe				
ASTM A681(94)	F2	Ann	1.20-1.40	0.20-0.40	0.10-0.50	0.030 max	0.030 max	0.10-0.50		3.00-4.50	S 0.06-0.15 if spec for machining; bal Fe				235 HB
Tool Steel, Tungsten High-Speed, No equivalents identified															
Czech Republic															
CSN 419800	19800		0.75-0.85	3.8-4.6	0.45 max	0.035 max	0.035 max	0.45 max	1.3-2.0	8.0-9.5	Mo <=0.5; bal Fe				
CSN 419802	19802		0.8-0.9	3.8-4.6	0.45 max	0.035 max	0.035 max	0.45 max	2.0-2.7	9.5-11.0	Mo <=0.5; bal Fe				
CSN 419810	19810		1.2-1.35	4.0-4.8	0.45 max	0.035 max	0.035 max	0.45 max	3.6-4.5	10.0-12.0	Mo <=0.5; bal Fe				
CSN 419820	19820		0.95-1.05	3.8-4.6	0.45 max	0.035 max	0.035 max	0.45 max	2.0-2.7	2.4-3.4	Mo 2.2-3.2; bal Fe				
CSN 419829	19829		0.98-1.08	3.8-4.6	0.45 max	0.025 max	0.035 max	0.45 max	0.5-2.2	5.5-7.0	bal Fe				
CSN 419856	19856		0.9-1.0	3.8-4.6	0.45 max	0.035 max	0.035 max	0.45 max	2.0-2.7	9.5-11.0	Co 4.5-5.5; Mo <=0.5; bal Fe				
CSN 419857	19857		0.9-1.0	3.8-4.6	0.45 max	0.035 max	0.035 max	0.45 max	2.0-2.7	9.5-11.0	Co 9.0-10.5; Mo <=0.5; bal Fe				
India															
IS 1570/6(96)	THS1	High-Speed	0.73-0.83	3.5-4.5	0.4 max	0.03 max	0.03 max	0.5 max	0.9-1.2	17.2-18.7	Mo <=0.7; Ni <=0.4; bal Fe				
IS 1570/6(96)	XT78W18Cr4V1	High-Speed	0.73-0.83	3.5-4.5	0.4 max	0.03 max	0.03 max	0.5 max	0.9-1.2	17.2-18.7	Mo <=0.7; Ni <=0.4; bal Fe				
Poland															
PNH85022	SW12		1.05-1.15	3.5-4.5	0.4 max	0.030 max	0.030 max	0.5 max	2.2-2.7	11-13	Mo <=2; Ni <=0.4; bal Fe				
PNH85022	SW2M5		0.9-1	3.5-4.5	0.4 max	0.030 max	0.3 max	0.5 max	1.1-1.4	1.5-2	Mo 4.5-5.5; Ni <=0.4; bal Fe				

UNS numbers and US grades are provided as a means of cross referencing chemically similar alloys. Exchangability is only possible after independent examination of specifications. Tensile properties are minimum or typical as specified. UTS and YS as MPa. El as %. See Appendix for list of abbreviations used in Notes. * indicates obsolete material.

Tool Steel, Tungsten High-Speed, T1

Specification	Designation	Notes	C	Cr	Mn	P	S	Si	V	W	Other	UTS	YS	El	Hard
Bulgaria															
BDS 7008(86)	R18		0.70-0.80	3.80-4.40	0.4	0.030 max	0.030 max	0.4	1.00-1.40	17.0-18.5	Mo <=1.00; Ni <=0.40; bal Fe				
Canada															
CSA 419824	19824		0.7-0.8	3.8-4.6	0.45 max	0.04 max	0.04 max	0.45 max	1-1.6	17-19	Mo <=0.5; bal Fe				
China															
GB 3080(82)	W18Cr4V	Wir Q/T	0.70-0.80	3.80-4.40	0.40 max	0.030 max	0.030 max	0.40 max	1.00-1.40	17.50-19.00	Mo <=0.30; bal Fe				63 HRC
GB 9941(88)	W18Cr4V	Sh Plt Ann	0.70-0.80	3.80-4.40	0.10-0.40	0.030 max	0.030 max	0.20-0.40	1.00-1.40	17.50-19.00	Mo <=0.30; Ni <=0.30; bal Fe				255 HB max
GB 9942(88)	W18Cr4V	Frg Bar Ann	0.70-0.80	3.80-4.40	0.10-0.40	0.030 max	0.03 max	0.20-0.40	1.00-1.40	17.50-19.00	Mo <=0.30; Ni <=0.30; bal Fe				255 HB max
GB 9943(88)	W18Cr4V	Bar Q/T	0.70-0.80	3.80-4.40	0.10-0.40	0.030 max	0.030 max	0.20-0.40	1.00-1.40	17.50-19.00	Mo <=0.30; Ni <=0.30; bal Fe				63 HRC
YB/T 084(96)	W18Cr4V	Strp; Ann	0.70-0.80	3.80-4.40	0.10-0.40	0.030 max	0.03 max	0.20-0.40	1.00-1.40	17.50-19.00	Mo <=0.30; Ni <=0.30; bal Fe				255 HB max
Czech Republic															
CSN 419824	19824		0.7-0.8	3.8-4.6	0.45 max	0.035 max	0.035 max	0.45 max	1.0-1.6	17.0-19.0	Mo <=0.5; bal Fe				
France															
AFNOR NFA35590(92)	HS18-0-1	Soft ann	0.72-0.8	3.5-4.5	0.4 max	0.03 max	0.03 max	0.5 max	1-1.3	17.2-18.7	Co <=1; Cu <=0.3; Mo <=1; Ni <=0.4; Mo<=0.15/0.5-1; bal Fe				270 HB max
AFNOR NFA35590(92)	Z80WCV18-04-01	Soft ann	0.72-0.8	3.5-4.5	0.4 max	0.03 max	0.03 max	0.5 max	1-1.3	17.2-18.7	Co <=1; Cu <=0.3; Mo <=1; Ni <=0.4; Mo<=0.15/0.5-1; bal Fe				270 HB max
Germany															
DIN	HS-18-0-1	As Tmp 550C	0.70-0.78	3.80-4.50	0.40 max	0.030 max	0.030 max	0.45 max	1.00-1.20	17.5-18.5	bal Fe				64 HRC
DIN	WNr 1.3355	As Tmp 550C	0.70-0.78	3.80-4.50	0.40 max	0.030 max	0.030 max	0.45 max	1.00-1.20	17.5-18.5	bal Fe				64 HRC
DIN 17230(80)	WNr 1.3558	Bearing	0.70-0.78	3.80-4.50	0.40 max	0.030 max	0.030 max	0.45 max	1.00-1.20	17.5-18.5	Mo <=0.60; bal Fe				
DIN 17230(80)	X75WCrV18-4-1	Bearing	0.70-0.78	3.80-4.50	0.40 max	0.030 max	0.030 max	0.45 max	1.00-1.20	17.5-18.5	Mo <=0.60; bal Fe				
Hungary															
MSZ 4351(84)	R3	Soft ann	0.73-0.83	3.8-4.5	0.4 max	0.03 max	0.03 max	0.4 max	0.9-1.2	17.0-18.5	Co <=0.5; Mo <=0.7; bal Fe				255 HB max
MSZ 4351(84)	R3	Q/T	0.73-0.83	3.8-4.5	0.4 max	0.03 max	0.03 max	0.4 max	0.9-1.2	17.0-18.5	Co <=0.5; Mo <=0.7; bal Fe				63 HRC
India															
IS 4367	T70W18Cr4V1	Frg	0.65-0.75	4-4.5	0.2-0.4			0.1-0.35	1-1.5	17.5-19	Mo 0.6; bal Fe				
International															
ISO 683-17(76)	32	Ball & roller bearing, HT	0.70-0.80	3.75-4.50	0.40 max	0.030 max	0.030 max	0.40 max	1.00-1.25	17.5-19.0	Mo <=0.60; bal Fe				
Italy															
UNI	X75W18KU		0.7-0.8	3.5-4.5	0.5 max			0.5 max	0.8-1.2	17-19	bal Fe				
UNI 2955	RC HS18-0-1		0.75	4.05	0.4			0.4	1.1	18	bal Fe				
UNI 2955/5(82)	HS18-0-1		0.73-0.83	3.5-4.5	0.4 max	0.07 max	0.06 max	0.5 max	0.9-1.2	17.2-18.7	Co <=1; Mo <=1; bal Fe				
UNI 3097(75)	X75WCrV18	Ball & roller bearing	0.7-0.8	3.75-4.5	0.4 max	0.03 max	0.03 max	0.4 max	0.9-1.2	17.5-19	Mo <=0.6; bal Fe				
Japan															
JIS G4403(83)	SKH2	Bar Frg Ann	0.73-0.83	3.80-4.50	0.40 max	0.030 max	0.030 max	0.40 max	0.80-1.20	17.00-19.00	Cu <=0.25; Ni <=0.25; bal Fe				248 HB max
Mexico															
NMX-B-082(90)	T1	Bar	0.65-0.80	3.75-4.50	0.10-0.40	0.03 max	0.03 max	0.20-0.40	0.90-1.30	17.25-18.75	Cu <=0.25; Ni <=0.30; (As+Sn+Sb)<=0.040; (Cu+Ni)<=0.40; B, Ti may be added; bal Fe				

Specification	Designation	Notes	C	Cr	Mn	P	S	Si	V	W	Other	UTS	YS	El	Hard

Tool Steel, Tungsten High-Speed, T1 (Continued from previous page)

Pan America

| COPANT 337 | T1 | Bar, Frg | 0.65-0.8 | 3.75-4.5 | 0.2-0.4 | 0.03 | 0.03 | 0.2-0.4 | 0.9-1.3 | 17.25-18.75 | bal Fe | | | | |

Poland

| PNH85022 | SW18 | Tools for cold ext; Tube bit | 0.75-0.85 | 3.5-4.5 | 0.4 max | 0.030 max | 0.030 max | 0.5 max | 1-1.4 | 17-19 | Mo <=2; Ni <=0.4; bal Fe | | | | |

Romania

| STAS 7382(88) | Rp3 | | 0.7-0.78 | 3.8-4.5 | 0.4 max | 0.03 max | 0.03 max | 0.45 max | 1-1.2 | 17.5-18.5 | Mo <=0.6; Ni <=0.4; Pb <=0.15; bal Fe | | | | |

Russia

| GOST 19265 | R18 | | 0.73-0.83 | 3.8-4.4 | 0.2-0.5 | 0.03 max | 0.03 max | 0.2-0.5 | 1-1.4 | 17-18.5 | Co <=0.5; Cu <=0.25; Mo <=1.0; Ni <=0.6; bal Fe | | | | |

Spain

UNE 36027(80)	F.1353*	see X75WCrV18-4-1	0.7-0.8	3.75-4.5	0.4 max	0.03 max	0.03 max	0.4 max	1.0-1.25	17.5-19.0	Mo <=0.6; bal Fe				
UNE 36027(80)	X75WCrV18-4-1	Ball & roller bearing	0.7-0.8	3.75-4.5	0.4 max	0.03 max	0.03 max	0.4 max	1.0-1.25	17.5-19.0	Mo <=0.6; bal Fe				
UNE 36073(75)	18-0-1	Tube bit; for cold ext	0.73-0.83	3.5-4.5	0.4 max	0.03 max	0.03 max	0.45 max	0.8-1.2	17.2-18.7	bal Fe				
UNE 36073(75)	ET1		0.73-0.83	3.5-4.5	0.4 max	0.03 max	0.03 max	0.45 max	0.8-1.2	17.2-18.7	bal Fe				
UNE 36073(75)	F.5520*	see 18-0-1, ET1	0.73-0.83	3.5-4.5	0.4 max	0.03 max	0.03 max	0.45 max	0.8-1.2	17.2-18.7	bal Fo				

Sweden

| SS 142750 | 2750 | | 0.68-0.75 | 4-5 | 0.2-0.4 | 0.07 max | 0.06 max | 0.15-0.3 | 1.1-1.3 | 17-19 | Co <=0.6; Mo <=0.5; bal Fe | | | | |

UK

| BS 4659 | BTI | Bar Rod Sh Strp Frg | 0.7-0.8 | 3.75-4.5 | 0.4 | | | 0.4 | 1-1.25 | 17.5-18.5 | Co 0.6; Mo 0.7; bal Fe | | | | |

USA

	AISI T1		0.75-0.80	3.75-4.50	0.10-0.40	0.03 max	0.03 max	0.20-0.40	0.90-1.30	117.25-18.75	bal Fe				
	UNS T12001	T-1	0.65-0.80	3.75-4.50	0.10-0.40	0.03 max	0.03 max	0.20-0.40	0.90-1.30	17.25-18.75	bal Fe				
AMS 5626		Bar, Frg	0.65-0.8	3.75-4.5	0.2-0.4			0.2-0.4	0.90-1.3	17.25-18.75	Mo 1; bal Fe				
ASTM A600(92)	T1	Bar Frg Plt Sh Strp, Ann	0.65-0.80	3.75-4.50	0.10-0.40	0.03 max	0.03 max	0.20-0.40	0.90-1.30	17.25-18.75	S 0.06-0.15 for machinability; (Ni+Cu)<=0.75; bal Fe				255 HB
FED QQ-T-590C(85)	T1*	Obs; see ASTM A600	0.65-0.8	3.75-4.5	0.1-0.4			0.2-0.4	0.9-1.3	17.25-18.25	Ni 0.3; bal Fe				
SAE J437(70)	T1	Ann									bal Fe				217-235 HB
SAE J438(70)	T1		0.65-0.75	3.75-4.50	0.20-0.40			0.20-0.40	0.90-1.30	17.25-18.75	bal Fe				

Tool Steel, Tungsten High-Speed, T15

Bulgaria

| BDS 7008(86) | R10K5F5 | | 1.45-1.55 | 4.00-4.60 | 0.4 | | | 0.4 | 4.30-5.10 | 10.0-11.5 | Co 5.50-6.00; Mo <=1.00; Ni <=0.40; bal Fe | | | | |

Canada

| CSA 419858 | 19858 | | 1.3-1.45 | 4-4.8 | 0.45 max | 0.04 max | 0.04 max | 0.45 max | 3.8-4.7 | 11-13 | Co 4.5-5.5; Mo 0.5; bal Fe | | | | |

China

| GB 9943(88) | W12Cr4V5Co5 | Bar Q/T | 1.50-1.60 | 3.75-5.00 | 0.15-0.40 | 0.03 max | 0.030 max | 0.15-0.40 | 4.50-5.25 | 11.75-13.00 | Co 4.75-5.25; Mo <=1.00; Ni <=0.30; bal Fe | | | | 65 HRC |

Czech Republic

| CSN 419858 | 19858 | | 1.3-1.45 | 4.0-4.8 | 0.45 max | 0.035 max | 0.035 max | 0.45 max | 3.8-4.7 | 11.0-13.0 | Co 4.0-4.8; Mo <=0.5; bal Fe | | | | |

France

| AFNOR | HS12-0-5-5 | | 1.5 | 4 | 0.3 | | | 0.3 | 5 | 12 | Co 5; Mo 0.5; bal Fe | | | | |

UNS numbers and US grades are provided as a means of cross referencing chemically similar alloys. Exchangability is only possible after independent examination of specifications. Tensile properties are minimum or typical as specified. UTS and YS as MPa. El as %. See Appendix for list of abbreviations used in Notes. * indicates obsolete material.

Specification	Designation	Notes	C	Cr	Mn	P	S	Si	V	W	Other	UTS	YS	El	Hard

Tool Steel, Tungsten High-Speed, T15 (Continued from previous page)

France

Specification	Designation	Notes	C	Cr	Mn	P	S	Si	V	W	Other	UTS	YS	El	Hard
AFNOR NFA35590(78)	HS12-1-5-5	High-speed, Soft ann	1.5-1.65	4-5	0.4 max	0.03 max	0.03 max	0.5 max	4.75-5.35	11.5-13	Co 4.5-5.2; Cu <=0.3; Mo 0.7-1; Ni <=0.4; bal Fe				295 HB max
AFNOR NFA35590(78)	Z160WKVC12-05-05-04	High-speed, Soft ann	1.5-1.65	4-5	0.4 max	0.03 max	0.03 max	0.5 max	4.75-5.35	11.5-13	Co 4.5-5.2; Cu <=0.3; Mo 0.7-1; Ni <=0.4; bal Fe				295 HB max
AFNOR NFA35590(92)	HS12-1-5-5	Soft ann	1.5-1.65	4-5	0.4 max	0.03 max	0.03 max	0.5 max	4.75-5.35	11.5-13	Co 4.5-5.2; Cu <=0.3; Mo 0.7-1; Ni <=0.4; bal Fe				295 HB max
AFNOR NFA35590(92)	Z160WKVC12-05-05-04	Soft ann	1.5-1.65	4-5	0.4 max	0.03 max	0.03 max	0.5 max	4.75-5.35	11.5-13	Co 4.5-5.2; Cu <=0.3; Mo 0.7-1; Ni <=0.4; bal Fe				295 HB max

Germany

Specification	Designation	Notes	C	Cr	Mn	P	S	Si	V	W	Other	UTS	YS	El	Hard
DIN 17350(80)	HS-12-1-4-5	As Tmp 560C	1.30-1.45	3.80-4.50	0.40 max	0.030 max	0.030 max	0.45 max	3.50-4.00	11.5-12.5	Co 4.50-5.00; Mo 0.70-1.00; bal Fe				65 HRC
DIN 17350(80)	WNr 1.3202	As Tmp 560C	1.30-1.45	3.80-4.50	0.40 max	0.030 max	0.030 max	0.45 max	3.50-4.00	11.5-12.5	Co 4.50-5.00; Mo 0.70-1.00; bal Fe				65 HRC

Italy

Specification	Designation	Notes	C	Cr	Mn	P	S	Si	V	W	Other	UTS	YS	El	Hard
UNI	X82WMo130505KU		1.4-1.6	4-5	0.5			0.5	4.7-5.3	12-13	Co 4.5-5.5; Mo 1-2; bal Fe				
UNI 2955	RV5K		1.55	4.5	0.4			0.4	5	12.5	Co 5; Mo 1.5; bal Fe				
UNI 2955/5(82)	HS10-4-3-10		1.2-1.35	3.5-4.5	0.4 max	0.07 max	0.06	0.5 max	3-3.5	9-10.5	Co 9.5-10.5; Mo 3.2-3.9; bal Fe				
UNI 2955/5(82)	HS12-1-5-5		1.4-1.55	3.5-4.5	0.4 max	0.07 max	0.06 max	0.5 max	4.5-5.55	11.5-13	Co 4.7-5.2; Mo 0.7 1; bal Fe				

Japan

Specification	Designation	Notes	C	Cr	Mn	P	S	Si	V	W	Other	UTS	YS	El	Hard
JIS G4403(83)	SKH10	Bar Frg Ann	1.45-1.60	3.80-4.50	0.40 max	0.030 max	0.030 max	0.40 max	4.20-5.20	11.50-13.50	Co 4.20-5.20; Cu <=0.25; Ni <=0.25; bal Fe				285 HB max

Mexico

Specification	Designation	Notes	C	Cr	Mn	P	S	Si	V	W	Other	UTS	YS	El	Hard
NMX-B-082(90)	T15	Bar	1.50-1.60	3.75-5.00	0.15-0.40	0.03 max	0.03 max	0.15-0.40	4.50-5.25	11.75-13.00	Co 4.75-5.25; Cu <=0.25; Mo <=1.00; Ni <=0.30; (As+Sn+Sb)<=0.040; (Cu+Ni)<=0.40; B, Ti may be added; bal Fe				

Pan America

Specification	Designation	Notes	C	Cr	Mn	P	S	Si	V	W	Other	UTS	YS	El	Hard
COPANT 337	T15	Bar, Frg	1.5-1.6	3.75-5	0.2-0.4	0.03	0.03	0.2-0.4	4.5-5.25	12-13	Co 4.75-5.25; Mo 1; bal Fe				
COPANT 337	T15		1.5-1.6	3.75-5	0.2-0.4			0.2-0.4	4.5-5.25	12-13	Co 4.25-5.25; Mo 1; bal Fe				

Poland

Specification	Designation	Notes	C	Cr	Mn	P	S	Si	V	W	Other	UTS	YS	El	Hard
PNH85022	SK5V		1.3-1.45	3.5-4.5	0.4 max	0.030 max	0.030 max	0.5 max	4.2-4.8	12-13.5	Co 5-6; Mo 0.7-1.2; Ni <=0.4; bal Fe				

Romania

Specification	Designation	Notes	C	Cr	Mn	P	S	Si	V	W	Other	UTS	YS	El	Hard
STAS	Rp6		1.25-1.4	3.8-4.4	0.45			0.2-0.4	3.5-4	11.5-12.5	Co 4.5-5; Mo 0.7-1.1; Ni 0.4; bal Fe				

Russia

Specification	Designation	Notes	C	Cr	Mn	P	S	Si	V	W	Other	UTS	YS	El	Hard
GOST	R10K5F5		1.45-1.55	4-6	0.4			0.5	4.3-5.1	10-11.5	Co 5-6; Mo 1; Ni 0.4; bal Fe				

Spain

Specification	Designation	Notes	C	Cr	Mn	P	S	Si	V	W	Other	UTS	YS	El	Hard
UNE 36073(75)	12-1-5-5		1.4-1.55	3.8-4.8	0.4 max	0.03 max	0.03 max	0.45 max	4.75-5.55	11.5-13	Co 4.7-5.2; Mo 0.7-1; bal Fe				
UNE 36073(75)	ET15		1.4-1.55	3.8-4.8	0.4 max	0.03 max	0.03 max	0.45 max	4.75-5.55	11.5-13	Co 4.7-5.2; Mo 0.7-1; bal Fe				
UNE 36073(75)	F.5563*	see 12-1-5-5, ET15	1.4-1.55	3.8-4.8	0.4 max	0.03 max	0.03 max	0.45 max	4.75-5.55	11.5-13	Co 4.7-5.2; Mo 0.7-1; bal Fe				

Sweden

Specification	Designation	Notes	C	Cr	Mn	P	S	Si	V	W	Other	UTS	YS	El	Hard
SS 14	(USA T15)		1.4	4.3	0.3			0.2	3.7	11.5	Co 5; bal Fe				

UK

Specification	Designation	Notes	C	Cr	Mn	P	S	Si	V	W	Other	UTS	YS	El	Hard
BS 4659	BT15	Bar Rod Sh Strp Frg	1.4-1.6	4.25-5	0.4			0.4	4.75-5.25	12-13	Co 4.5-5.5; Mo 1; bal Fe				

USA

Specification	Designation	Notes	C	Cr	Mn	P	S	Si	V	W	Other	UTS	YS	El	Hard
	AISI T15		1.50-1.60	3.75-5.00	0.15-0.40	0.03 max	0.03 max	0.15-0.40	4.50-5.25	11.75-13.00	Co 4.25-5.75; Mo <=1.00; bal Fe				

Specification	Designation	Notes	C	Cr	Mn	P	S	Si	V	W	Other	UTS	YS	El	Hard

Tool Steel, Tungsten High-Speed, T15 (Continued from previous page)

USA

Specification	Designation	Notes	C	Cr	Mn	P	S	Si	V	W	Other	Hard
	UNS T12015	T-15	1.50-1.60	3.75-5.00	0.15-0.40	0.03 max	0.03 max	0.15-0.40	4.50-5.25	11.75-13.00	Co 4.75-5.25; Mo <=1.00; bal Fe	
ASTM A600(92)	T15	Bar Frg Plt Sh Strp, Ann	1.50-1.60	3.75-5.00	0.15-0.40	0.03 max	0.03 max	0.15-0.40	4.50-5.25	11.75-13.00	Co 4.75-5.25; Mo <=1.00; S 0.06-0.15 for machinability; (Ni+Cu)<=0.75; bal Fe	277 HB
FED QQ-T-590C(85)	T15*	Obs; see ASTM A600	1.5-1.6	3.75-5	0.15-0.4			0.15-0.4	4.5-5.25	11.75-13	Co 4.75-5.25; Mo 1; Ni 0.3; bal Fe	

Tool Steel, Tungsten High-Speed, T2

India

Specification	Designation	Notes	C	Cr	Mn	P	S	Si	V	W	Other	Hard
IS 1570/6(96)	THS4		0.82-0.92	3.5-4.5	0.4 max	0.03 max	0.03 max	0.5 max	1.7-2.2	5.7-6.7	Co <=1; Mo 4.6-5.3; Ni <=0.4; bal Fe	
IS 1570/6(96)	XT87W6Mo5Cr4V2		0.82-0.92	3.5-4.5	0.4 max	0.03 max	0.03 max	0.5 max	1.7-2.2	5.7-6.7	Co <=1; Mo 4.6-5.3; Ni <=0.4; bal Fe	

Mexico

NMX-B-082(90)	T2	Bar	0.60-0.90	3.75-4.50	0.20-0.40	0.03 max	0.03 max	0.20-0.40	1.80-2.40	17.50-19.00	Cu <=0.25; Mo <=1.00; Ni <=0.30; (As+Sn+Sb)<=0.040; (Cu+Ni)<=0.40; B, Ti may be added; bal Fe	

Russia

GOST 19265	R12F3		0.95-1.05	3.8-4.3	0.2-0.5	0.03 max	0.03 max	0.2-0.5	2.5-3	12-13	Co <=0.5; Cu <=0.25; Mo <=1; Ni <=0.6; bal Fe	

USA

	UNS T12002*	Obs; T-2	0.80-0.90	3.75-4.50	0.20-0.40	0.03 max	0.03 max	0.20-0.40	1.80-2.40	17.50-19.00	Mo <=1.00; bal Fe	
ASTM A600(92)	T2	Bar Frg Plt Sh Strp, Ann	0.80-0.90	3.75-4.50	0.20-0.40	0.03 max	0.03 max	0.20-0.40	1.80-2.40	17.50-19.00	Mo <=1.00; S 0.06-0.15 for machinability; (Ni+Cu)<=0.75; bal Fe	255 HB
SAE J437(70)	T2	Ann									bal Fe	223-255 HB
SAE J438(70)	T2		0.75-0.85	3.75-4.50	0.20-0.40			0.20-0.40	1.80-2.40	17.50-19.00	Mo 0.70-1.00; bal Fe	

Tool Steel, Tungsten High-Speed, T4

Bulgaria

Specification	Designation	Notes	C	Cr	Mn	P	S	Si	V	W	Other	Hard
BDS 7008(86)	R18K5F2		0.85-0.95	3.80-4.40	0.4 max	0.030 max	0.030 max	0.4 max	1.80-2.40	17.0-18.5	Co 5.00-6.00; Mo <=1.00; Ni <=0.40; bal Fe	

China

GB 9943(88)	W18Cr4VCo5	Bar Q/T	0.70-0.80	3.75-4.50	0.10-0.40	0.03 max	0.030 max	0.20-0.40	0.80-1.20	17.50-19.00	Co 4.25-5.75; Mo 0.40-1.00; Ni <=0.30; bal Fe	63 HRC

Czech Republic

CSN 419855	19855		0.65-0.75	3.8-4.6	0.45 max	0.035 max	0.035 max	0.45 max	1.2-1.8	17.0-19.0	Co 4.2-5.2; Mo <=0.5; bal Fe	

France

AFNOR	HS18-0-1-5		0.8	4	0.3			0.3	1	18	Co 5; Mo 1; bal Fe	
AFNOR NFA35590(92)	HS18-1-1-5	Soft ann	0.77-0.85	3.5-4.5	0.4 max	0.03 max	0.03 max	0.5 max	1.1-1.6	17.2-18.7	Co 4.5-5.2; Cu <=0.3; Mo 0.7-1; Ni <=0.4; bal Fe	275 HB max
AFNOR NFA35590(92)	Z80WKCV18-05-04-01	Soft ann	0.77-0.85	3.5-4.5	0.4 max	0.03 max	0.03 max	0.5 max	1.1-1.6	17.2-18.7	Co 4.5-5.2; Cu <=0.3; Mo 0.7-1; Ni <=0.4; bal Fe	275 HB max

Germany

DIN 17350(80)	HS-18-1-2-5	As Tmp 560C	0.75-0.83	3.80-4.50	0.40 max	0.030 max	0.030 max	0.45 max	1.40-1.70	17.5-18.5	Co 4.50-5.00; Mo 0.50-0.80; bal Fe	64 HRC
DIN 17350(80)	WNr 1.3255	As Tmp 560C	0.75-0.83	3.80-4.50	0.40 max	0.030 max	0.030 max	0.45 max	1.40-1.70	17.5-18.5	Co 4.50-5.00; Mo 0.50-0.80; bal Fe	64 HRC

Hungary

MSZ 4351(72)	R2		0.74-0.84	3.8-4.6	0.4 max	0.03 max	0.03 max	0.4 max	1.2-1.5	17.5-19.0	Co 4.5-5.5; Mo 0.7-1; bal Fe	

UNS numbers and US grades are provided as a means of cross referencing chemically similar alloys. Exchangability is only possible after independent examination of specifications. Tensile properties are minimum or typical as specified. UTS and YS as MPa. El as %. See Appendix for list of abbreviations used in Notes. * indicates obsolete material.

Tool Steel, Tungsten High-Speed, T4 (Continued from previous page)

Specification	Designation	Notes	C	Cr	Mn	P	S	Si	V	W	Other	UTS	YS	El	Hard
India															
IS 1570	T75W18Co6Cr4V1Mo75		0.7-0.9	4-4.5	0.2-0.4			0.1-0.35	1-1.5	17.5-19	Co 5-6; Mo 0.5-1; bal Fe				
Italy															
UNI	X78WCo1805KU		0.7-0.85	3.5-4.5	0.5			0.5	1-1.5	17-19	Co 4.5-5.5; Mo 0.5; bal Fe				
UNI 2955	RCK5		0.8	4	0.5			0.5	1.25	18	Co 5; Mo 0.5; bal Fe				
UNI 2955/5(82)	HS18-1-1-5		0.75-0.85	3.5-4.5	0.4 max	0.07 max	0.06 max	0.5 max	1.1-1.6	17.2-18.7	Co 4.5-5.2; Mo 0.7-1; bal Fe				
Japan															
JIS G4403(83)	SKH3	Bar Frg Ann	0.73-0.83	3.80-4.50	0.40 max	0.030 max	0.030 max	0.40 max	0.80-1.20	17.00-19.00	Co 4.50-5.50; Cu <=0.25; Ni <=0.25; bal Fe				269 HB max
Poland															
PNH85022	SK5		1.05-1.15	3.5-4.5	0.4 max	0.030 max	0.030 max	0.5 max	2.1-2.6	11-13	Co 4.5-5.5; Mo <=2; Ni <=0.4; bal Fe				
Romania															
STAS 7382(88)	Rp2		0.75-0.83	3.8-4.5	0.4 max	0.03 max	0.03 max	0.45 max	1.4-1.7	17.5-18.5	Co 4.5-5; Mo 0.5-0.8; Ni <=0.4; Pb <=0.15; bal Fe				
Russia															
GOST 19265	R18K5F2		0.85-0.95	3.8-4.4	0.5 max	0.03 max	0.03 max	0.5 max	1.8-2.4	17-18.5	Co 5-6; Mo <=1; Ni <=0.4; bal Fe				
GOST 19265	R2AM9K5		1.0-1.1	3.8-4.4	0.2-0.5	0.03 max	0.03 max	0.2-0.5	1.7-2.1	1.5-2	Co 4.7-5.2; Cu <=0.25; Mo 8-9; N 0.05-0.1; Nb 0.1-0.3; Ni <=0.6; bal Fe				
Spain															
UNE 36073(75)	18-1-1-5	Tube bit	0.75-0.85	3.5-4.5	0.4 max	0.03 max	0.03 max	0.45 max	1.1-1.6	17.2-18.7	Co 4.8-5.3; Mo 0.5-0.8; bal Fe				
UNE 36073(75)	ET4	*	0.75-0.85	3.5-4.5	0.4 max	0.03 max	0.03 max	0.45 max	1.1-1.6	17.2-18.7	Co 4.8-5.3; Mo 0.5-0.8; bal Fe				
UNE 36073(75)	F.5530*	see 18-1-1-5, ET4	0.75-0.85	3.5-4.5	0.4 max	0.03 max	0.03 max	0.45 max	1.1-1.6	17.2-18.7	Co 4.8-5.3; Mo 0.5-0.8; bal Fe				
UK															
BS 4659	BT4	Bar Rod Sh Strp Frg	0.7-0.8	3.75-4.5	0.4 max			0.4 max	1-1.25	17.5-18.5	Co 4.5-5.5; Mo <=1; bal Fe				
USA															
	AISI T4		0.70-0.80	3.75-4.50	0.10-0.40	0.03 max	0.03 max	0.20-0.40	0.80-1.20	17.50-19.00	Co 4.75-5.25; Mo 0.40-1.00; bal Fe				
	UNS T12004	T-4	0.70-0.80	3.75-4.50	0.10-0.40	0.03 max	0.03 max	0.20-0.40	0.80-1.20	17.50-19.00	Co 4.25-5.75; Mo 0.40-1.00; bal Fe				
ASTM A600(92)	T4	Bar Frg Plt Sh Strp, Ann	0.70-0.80	3.75-4.50	0.10-0.40	0.03 max	0.03 max	0.20-0.40	0.80-1.20	17.50-19.00	Co 4.25-5.75; Mo 0.40-1.00; S 0.06-0.15 for machinability; (Ni+Cu)<=0.75; bal Fe				269 HB
FED QQ-T-590C(85)	T4*	Obs; see ASTM A600	0.7-0.8	3.75-4.5	0.1-0.4			0.2-0.4	0.8-1.2	17.5-19	Co 4.25-5.75; Mo 0.4-1; Ni 0.3; bal Fe				
SAE J437(70)	T4	Ann									bal Fe				229-255 HB
SAE J438(70)	T4		0.70-0.80	3.75-4.50	0.20-0.40			0.20-0.40	0.80-1.20	17.25-18.75	Co 4.25-5.75; Mo 0.70-1.00; bal Fe				

Tool Steel, Tungsten High-Speed, T5

Specification	Designation	Notes	C	Cr	Mn	P	S	Si	V	W	Other	UTS	YS	El	Hard
Argentina															
IAS	IRAM R1	Ann	1.20-1.35	3.75-5.00	0.15-0.40	0.03 max	0.03 max	0.20-0.45	3.00-3.5	9.75-11.25	Co 10.00-11.25; Mo 3.30-4.00; bal Fe				269-302 HB
IAS	IRAM T5	Ann	0.75-0.85	3.75-5.00	0.20-0.40	0.03 max	0.03 max	0.20-0.40	1.80-2.40	17.50-19.00	Co 7.00-9.50; Mo 0.50-1.25; bal Fe				235-285 HB
China															
GB 9943(88)	W18Cr4V2Co8	Bar Q/T	0.75-0.85	3.75-5.00	0.20-0.40	0.03 max	0.030 max	0.20-0.40	1.80-2.40	17.50-19.00	Co 7.00-9.50; Mo 0.50-1.25; Ni <=0.30; bal Fe				63 HRC

Tool Steel, Tungsten High-Speed, T5 (Continued from previous page)

Specification	Designation	Notes	C	Cr	Mn	P	S	Si	V	W	Other	UTS	YS	El	Hard
France															
AFNOR NFA35590(92)	HS18-0-2-9	Soft ann	0.73-0.8	3.5-4.5	0.4 max	0.03 max	0.03 max	0.5 max	1.3-1.8	17.2-18.7	Co 9-9.5; Cu <=0.3; Mo <=1; Ni <=0.4; Mo<=0.15/0.5-1; bal Fe				295 HB max
AFNOR NFA35590(92)	Z80WKCV18-10-04-02	Soft ann	0.73-0.8	3.5-4.5	0.4 max	0.03 max	0.03 max	0.5 max	1.3-1.8	17.2-18.7	Co 9-9.5; Cu <=0.3; Mo <=1; Ni <=0.4; Mo<=0.15/0.5-1; bal Fe				295 HB max
Germany															
DIN	HS-18-1-2-10	As Tmp 560C	0.72-0.80	3.80-4.50	0.40 max	0.030 max	0.030 max	0.45 max	1.40-1.70	17.5-18.5	Co 9.00-10.0; Mo 0.50-0.80; bal Fe				64 HRC
DIN	WNr 1.3265	As Tmp 560C	0.72-0.80	3.80-4.50	0.40 max	0.030 max	0.030 max	0.45 max	1.40-1.70	17.5-18.5	Co 9.00-10.0; Mo 0.50-0.80; bal Fe				64 HRC
Hungary															
MSZ 4351(72)	R1		0.74-0.84	3.8-4.6	0.4 max	0.03 max	0.03 max	0.4 max	1.2-1.5	17.5-19.0	Co 8-10; Mo 0.7-1; bal Fe				
India															
IS 1570	T75W18Co10Cr4V2Mo75		0.7-0.9	4-4.5	0.2-0.4			0.1-0.35	1.5-2	17.5-19	Co 9-10; Mo 0.5-1; bal Fe				
IS 1570/6(96)	THS6		0.75-0.85	3.5-4.5	0.4 max	0.03 max	0.03 max	0.5 max	1.3-1.8	17.2-18.7	Co 9.5-10.5; Mo <=0.7; Ni <=0.4; bal Fe				
IS 1570/6(96)	THS7		0.75-0.85	3.5-4.5	0.4 max	0.03 max	0.03 max	0.5 max	1.1-1.6	17.2-18.7	Co 4.7-5.2; Mo 0.7-1; Ni <=0.4; bal Fe				
IS 1570/6(96)	XT80W13Co10Cr4V2		0.75-0.85	3.5-4.5	0.4 max	0.03 max	0.03 max	0.5 max	1.3-1.8	17.2-18.7	Co 9.5-10.5; Mo <=0.7; Ni <=0.4; bal Fe				
IS 1570/6(96)	XT80W18Co5Cr4Mo1V1		0.75-0.85	3.5-4.5	0.4 max	0.03 max	0.03 max	0.5 max	1.1-1.6	17.2-18.7	Co 4.7-5.2; Mo 0.7-1; Ni <=0.4; bal Fe				
Italy															
UNI	X80WCo1810KU		0.75-0.85	3.5-4.5	0.5			0.5	1.5-2	17.5-19.5	Co 9-11; Mo 0.5-1; bal Fe				
UNI 2955	RCK10		0.78	4.15	0.4			0.4	1.6	18	Co 9.5; Mo 0.65; bal Fe				
UNI 2955/5(82)	HS18-0-1-10		0.75-0.85	3.5-4.5	0.4 max	0.07 max	0.06 max	0.5 max	1.3-1.9	17.2-18.7	Co 9.5-10.5; Mo <=1; bal Fe				
Japan															
JIS G4403(83)	SKH4	Bar Frg Ann	0.73-0.83	3.80-4.50	0.40 max	0.030 max	0.030 max	0.40 max	1.00-1.50	17.00-19.00	Co 9.00-11.00; Cu <=0.25; Ni <=0.25; bal Fe				285 HB max
JIS G4403(83)	SKH4A*	Obs; See SKH4; Bar Frg Ann									Cu <=0.25; bal Fe				
Mexico															
NMX-B-082(90)	T5	Bar	0.75-0.85	3.75-5.00	0.20-0.40	0.03 max	0.03 max	0.20-0.40	1.80-2.40	17.5-19.00	Co 7.00-9.50; Cu <=0.25; Mo 0.50-1.25; Ni <=0.30; (As+Sn+Sb)<=0.040; (Cu+Ni)<=0.40; B, Ti may be added; bal Fe				
Pan America															
COPANT 337	T5	Bar, Frg	0.75-0.85	3.75-5	0.2-0.4	0.03	0.03	0.2-0.4	1.8-2.4	17.5-19	Co 7-9.5; Mo 0.5-1.25; bal Fe				
Spain															
UNE 36073(75)	18-0-2-10	Tube bit	0.75-0.85	3.5-4.5	0.4 max	0.03 max	0.03 max	0.45 max	1.3-1.8	17.2-18.7	Co 9.5-10.5; Mo <=1; bal Fe				
UNE 36073(75)	ET5		0.75-0.85	3.5-4.5	0.4 max	0.03 max	0.03 max	0.45 max	1.3-1.8	17.2-18.7	Co 9.5-10.5; Mo <=1; bal Fe				
UNE 36073(75)	F.5540*	see 18-0-2-10, ET5	0.75-0.85	3.5-4.5	0.4 max	0.03 max	0.03 max	0.45 max	1.3-1.8	17.2-18.7	Co 9.5-10.5; Mo <=1; bal Fe				
Sweden															
SS 142754	2754		0.75-0.85	4-5	0.2-0.4	0.03 max	0.03 max	0.15-0.3	1.3-1.9	17.5-19.5	Co 5-6; Mo 0.9-1.6; bal Fe				
SS 142756	2756		0.7-0.85	4-5	0.2-0.4	0.03 max	0.03 max	0.15-0.3	1.3-1.9	17.5-19.5	Co 9.5-11.5; Mo 0.8-1.2; bal Fe				
UK															
BS 4659	BT5	Bar Rod Sh Strp Frg	0.75-0.85	3.75-4.5	0.4			0.4	1.75-2.05	18.5-19.5	Co 9-10; Mo 1; bal Fe				

UNS numbers and US grades are provided as a means of cross referencing chemically similar alloys. Exchangability is only possible after independent examination of specifications. Tensile properties are minimum or typical as specified. UTS and YS as MPa. El as %. See Appendix for list of abbreviations used in Notes. * indicates obsolete material.

Specification	Designation	Notes	C	Cr	Mn	P	S	Si	V	W	Other	UTS	YS	El	Hard
Tool Steel, Tungsten High-Speed, T5 (Continued from previous page)															
USA															
	AISI T5		0.75-0.85	3.75-5.00	0.20-0.40	0.03 max	0.03 max	0.20-0.40	1.80-2.40	17.50-19.00	Co 7.00-9.50; Mo 0.50-1.25; bal Fe				
	UNS T12005	T-5	0.75-0.85	3.75-5.00	0.20-0.40	0.03 max	0.03 max	0.20-0.40	1.80-2.40	17.50-19.00	Co 7.00-9.50; Mo 0.50-1.25; bal Fe				
ASTM A600(92)	T5	Bar Frg Plt Sh Strp, Ann	0.75-0.85	3.75-5.00	0.20-0.40	0.03 max	0.03 max	0.20-0.40	1.80-2.40	17.50-19.00	Co 7.00-9.50; Mo 0.50-1.25; S 0.06-0.15 for machinability; (Ni+Cu)<=0.75; bal Fe				285 HB
FED QQ-T-590C(85)	T5*	Obs; see ASTM A600	0.75-0.85	3.75-5	0.2-0.4			0.2-0.4	1.8-2.4	17.5-19	Co 7-9.5; Mo 0.5-1.25; Ni 0.3; bal Fe				
SAE J437(70)	T5	Ann									bal Fe				248-293 HB
SAE J438(70)	T5		0.75-0.85	3.75-4.50	0.20-0.40			0.20-0.40	1.80-2.40	17.50-19.00	Co 7.00-9.00; Mo 0.70-1.00; bal Fe				
Tool Steel, Tungsten High-Speed, T6															
Germany															
DIN	HS-18-1-2-15	As Tmp 560C	0.60-0.70	3.80-4.50	0.40 max	0.030 max	0.030 max	0.45 max	1.40-1.70	17.5-18.5	Co 15.0-16.0; Mo 0.50-1.00; bal Fe				64 HRC
DIN	WNr 1.3257	As Tmp 560C	0.60-0.70	3.80-4.50	0.40 max	0.030 max	0.030 max	0.45 max	1.40-1.70	17.5-18.5	Co 15.0-16.0; Mo 0.50-1.00; bal Fe				64 HRC
UK															
BS 4659	BT6		0.75-0.85	3.75-4.5	0.4			0.4	1.25-1.75	20-21	Co 11.25-12.25; Mo 1; bal Fe				
USA															
	AISI T6		0.75-0.85	4.00-4.75	0.20-0.40	0.03 max	0.03 max	0.20-0.40	1.50-2.10	18.50-21.00	Co 11.00-13.00; Mo 0.40-1.00; bal Fe				
	UNS T12006	T-6	0.75-0.85	4.00-4.75	0.20-0.40	0.03 max	0.03 max	0.20-0.40	1.50-2.10	18.50-21.00	Co 11.00-13.00; Mo 0.40-1.00; bal Fe				
ASTM A600(92)	T6	Bar Frg Plt Sh Strp, Ann	0.75-0.85	4.00-4.75	0.20-0.40	0.03 max	0.03 max	0.20-0.40	1.50-2.10	18.50-21.00	Co 11.00-13.00; Mo 0.40-1.00; S 0.06-0.15 for machinability; (Ni+Cu)<=0.75; bal Fe				302 HB
FED QQ-T-590C(85)	T6*	Obs; see ASTM A600	0.75-0.85	4-4.75	0.2-0.4			0.2-0.4	1.5-2.1	18.5-21	Co 11-13; Mo 0.4-1; Ni 0.3; bal Fe				
Tool Steel, Tungsten High-Speed, T8															
Pan America															
COPANT 337	T8	Bar, Frg	0.75-0.85	3.75-5	0.2-0.4	0.03	0.03	0.2-0.4	1.8-2.4	13.25-14.75	Co 4.25-4.75; Mo 0.4-1; bal Fe				
Poland															
PNH85022	SKC		1.05-1.15	3.8-4.8	0.5			0.5	2.1-2.6	11-13	Co 4.5-5.5; Mo 2; Ni 0.4; bal Fe				
USA															
	AISI T8		0.75-0.85	3.75-4.50	0.20-0.40	0.03 max	0.03 max	0.20-0.40	1.80-2.40	13.25-14.75	Co 4.25-5.75; Mo 0.40-1.00; bal Fe				
	UNS T12008	T-8	0.75-0.85	3.75-4.50	0.20-0.40	0.03 max	0.03 max	0.20-0.40	1.80-2.40	13.25-14.75	Co 4.25-5.75; Mo 0.40-1.00; bal Fe				
ASTM A600(92)	T8	Bar Frg Plt Sh Strp, Ann	0.75-0.85	3.75-4.50	0.20-0.40	0.03 max	0.03 max	0.20-0.40	1.80-2.40	13.25-14.75	Co 4.25-5.75; Mo 0.40-1.00; S 0.06-0.15 for machinability; (Ni+Cu)<=0.75; bal Fe				255 HB
FED QQ-T-590C(85)	T8*	Obs; see ASTM A600	0.75-0.85	3.75-4.5	0.2-0.4			0.2-0.4	1.8-2.4	13.25-14.75	Co 4.25-5.75; Mo 0.7-1; Ni 0.3; bal Fe				
SAE J437(70)	T8	Ann									bal Fe				229-255 HB
SAE J438(70)	T8		0.75-0.85	3.75-4.50	0.20-0.40			0.20-0.40	1.80-2.40	13.25-14.75	Co 4.25-5.75; Mo 0.70-1.00; bal Fe				

UNS numbers and US grades are provided as a means of cross referencing chemically similar alloys. Exchangability is only possible after independent examination of specifications. Tensile properties are minimum or typical as specified. UTS and YS as MPa. El as %. See Appendix for list of abbreviations used in Notes. * indicates obsolete material.

Specification	Designation	Notes	C	Cr	Cu	Mn	Ni	P	S	Si	Other	UTS	YS	El	Hard

Tool Steel, Water-Hardening, No equivalents identified

Argentina

Specification	Designation	Notes	C	Cr	Cu	Mn	Ni	P	S	Si	Other	UTS	YS	El	Hard
IAS	IRAM F112	Ann	1.20-1.30	0.25-0.45		0.10-0.40		0.030 max	0.030 max	0.10-0.40	V 0.15-0.30; W 1.30-1.60; bal Fe				225 HB

Poland

Specification	Designation	Notes	C	Cr	Cu	Mn	Ni	P	S	Si	Other	UTS	YS	El	Hard
PNH85020	N5		0.5-0.6	0.3 max		0.4-0.6	0.15 max	0.4 max	0.035 max	0.15 max	Mo <=0.035; bal Fe				

Tool Steel, Water-Hardening, W1

Bulgaria

Specification	Designation	Notes	C	Cr	Cu	Mn	Ni	P	S	Si	Other	UTS	YS	El	Hard
BDS 6751	U10		0.95-1.04	0.2	0.25	0.15-0.35	0.25			0.15-0.35	bal Fe				
BDS 6751	U10A		0.95-1.04	0.15	0.2	0.15-0.3	0.2			0.15-0.3	bal Fe				
BDS 6751	U11		1.05-1.14	0.2	0.25	0.15-0.35	0.25			0.15-0.35	bal Fe				
BDS 6751	U11A		1.05-1.14	0.15	0.2	0.15-0.3	0.2			0.15-0.3	bal Fe				
BDS 6751	U12		1.15-1.24	0.2	0.25	0.15-0.35	0.25			0.15-0.35	bal Fe				
BDS 6751	U12A		1.15-1.24	0.15	0.2	0.15-0.3	0.2			0.15-0.3	bal Fe				
BDS 6751	U13		1.25-1.34	0.2	0.25	0.15-0.35	0.2			0.15-0.35	bal Fe				
BDS 6751	U13A		1.25-1.34	0.15	0.2	0.15-0.3	0.2			0.15-0.3	bal Fe				
BDS 6751	U7		0.65-0.74	0.2	0.25	0.15-0.35	0.25			0.15-0.35	bal Fe				
BDS 6751	U7A		0.65-0.74	0.15		0.15-0.3	0.2			0.15-0.3	bal Fe				
BDS 6751	U8		0.75-0.84	0.2	0.25	0.15-0.35	0.25			0.15-0.35	bal Fe				
BDS 6751	U8A		0.75-0.84	0.15	0.2	0.15-0.3	0.2			0.15-0.3	bal Fe				
BDS 6751	U9		0.85-0.94	0.2	0.25	0.15-0.35	0.25			0.15-0.35	bal Fe				
BDS 6751	U9A		0.85-0.94	0.15	0.2	0.15-0.3	0.2			0.15-0.3	bal Fe				

China

Specification	Designation	Notes	C	Cr	Cu	Mn	Ni	P	S	Si	Other	UTS	YS	El	Hard
GB 1222	70*	Obs	0.62-0.75	0.25	0.25	0.5-0.8	0.25			0.17-0.37	bal Fe				
GB 1222	85*	Obs	0.82-0.9	0.25	0.25	0.5-0.8	0.25			0.17-0.37	bal Fe				
GB 1298(86)	T10	As Hard	0.95-1.04	0.25 max	0.30 max	0.40 max	0.20 max	0.035 max	0.030 max	0.35 max	bal Fe				62 HRC
GB 1298(86)	T10A	As Hard	0.95-1.04	0.25 max	0.30 max	0.40 max	0.20 max	0.030 max	0.020 max	0.35 max	bal Fe				
GB 1298(86)	T11	As Hard	1.05-1.14	0.25 max	0.30 max	0.40 max	0.20 max	0.035 max	0.030 max	0.35 max	bal Fe				62 HRC
GB 1298(86)	T11A		1.05-1.14	0.25 max	0.30 max	0.40 max	0.20 max	0.035 max	0.030 max	0.35 max	bal Fe				
GB 1298(86)	T12	As Hard	1.15-1.24	0.25 max	0.30 max	0.40 max	0.20 max	0.035 max	0.030 max	0.35 max	bal Fe				62 HRC
GB 1298(86)	T12A		1.15-1.24	0.25 max	0.30 max	0.40 max	0.20 max	0.035 max	0.030 max	0.35 max	bal Fe				
GB 1298(86)	T13	As Hard	1.25-1.35	0.25 max	0.30 max	0.40 max	0.20 max	0.035 max	0.030 max	0.35 max	bal Fe				62 HRC
GB 1298(86)	T13A		1.25-1.35	0.25 max	0.30 max	0.40 max	0.20 max	0.035 max	0.030 max	0.35 max	bal Fe				
GB 1298(86)	T7	As Hard	0.65-0.74	0.25 max	0.30 max	0.40 max	0.20 max	0.035 max	0.030 max	0.35 max	bal Fe				62 HRC
GB 1298(86)	T7A		0.65-0.74	0.25 max	0.30 max	0.40 max	0.20 max	0.030 max	0.020 max	0.35 max	bal Fe				
GB 1298(86)	T8	As Hard	0.75-0.84	0.25 max	0.30 max	0.40 max	0.20 max	0.035 max	0.030 max	0.35 max	bal Fe				62 HRC
GB 1298(86)	T8A		0.75-0.84	0.25 max	0.30 max	0.40 max	0.20 max	0.035 max	0.020 max	0.35 max	bal Fe				
GB 1298(86)	T8Mn	As Hard	0.80-0.90	0.25 max	0.30 max	0.40-0.60	0.20 max	0.035 max	0.030 max	0.35 max	bal Fe				62 HRC

UNS numbers and US grades are provided as a means of cross referencing chemically similar alloys. Exchangability is only possible after independent examination of specifications. Tensile properties are minimum or typical as specified. UTS and YS as MPa. El as %. See Appendix for list of abbreviations used in Notes. * indicates obsolete material.

Specification	Designation	Notes	C	Cr	Cu	Mn	Ni	P	S	Si	Other	UTS	YS	El	Hard
Tool Steel, Water-Hardening, W1 (Continued from previous page)															
China															
GB 1298(86)	T9	As Hard	0.85-0.94	0.25 max	0.30 max	0.40 max	0.20 max	0.035 max	0.030 max	0.35 max	bal Fe				62 HRC
GB 1298(86)	T9A		0.85-0.94	0.25 max	0.30 max	0.40 max	0.20 max	0.030 max	0.020 max	0.35 max	bal Fe				
GB 1299	MnSi*	Obs	0.95-1.05			0.6-0.9				0.65-0.95	bal Fe				
GB 5952(86)	T10	Wir Ann	0.95-1.04	0.25 max	0.30 max	0.40 max	0.20 max	0.035 max	0.030 max	0.35 max	bal Fe				197 HB max
GB 5952(86)	T11	Wir Ann	1.05-1.14	0.25 max	0.30 max	0.40 max	0.20 max	0.035 max	0.030 max	0.35 max	bal Fe				207 HB max
GB 5952(86)	T12	Wir Ann	1.15-1.24	0.25 max	0.30 max	0.40 max	0.20 max	0.035 max	0.030 max	0.35 max	bal Fe				207 HB max
GB 5952(86)	T13	Wir Ann	1.25-1.35	0.25 max	0.30 max	0.40 max	0.20 max	0.035 max	0.030 max	0.35 max	bal Fe				217 HB max
GB 5952(86)	T7	Wir Ann	0.65-0.74	0.25 max	0.30 max	0.40 max	0.20 max	0.035 max	0.030 max	0.35 max	bal Fe				187 HB max
GB 5952(86)	T8	Wir Ann	0.75-0.84	0.25 max	0.30 max	0.40 max	0.20 max	0.035 max	0.030 max	0.35 max	bal Fe				187 HB max
GB 5952(86)	T8Mn	Wir Ann	0.80-0.90	0.25 max	0.30 max	0.40-0.60	0.20 max	0.035 max	0.030 max	0.35 max	bal Fe				187 HB max
GB 5952(86)	T9	Wir Ann	0.85-0.94	0.25 max	0.30 max	0.40 max	0.20 max	0.035 max	0.030 max	0.35 max	bal Fe				192 HB max
Czech Republic															
CSN 419132	19132		0.65-0.75	0.2 max		0.25-0.45	0.25 max	0.03 max	0.035 max	0.15-0.35	bal Fe				
CSN 419133	19133		0.65-0.75	0.25 max		0.2-0.45	0.25 max	0.035 max	0.035 max	0.15-0.35	bal Fe				
CSN 419152	19152		0.75-0.9	0.2 max		0.2-0.4	0.25 max	0.03 max	0.03 max	0.15-0.35	bal Fe				
CSN 419191	19191		0.95-1.09	0.15 max		0.2-0.35	0.2 max	0.025 max	0.03 max	0.15-0.3	bal Fe				
CSN 419192	19192		0.95-1.05	0.2 max		0.2-0.4	0.2 max	0.03 max	0.035 max	0.15-0.35	bal Fe				
CSN 419221	19221		1.1-1.24	0.15 max		0.2-0.35	0.2 max	0.025 max	0.03 max	0.15-0.3	bal Fe				
CSN 419222	19222		1.05-1.2	0.2 max		0.2-0.4	0.25 max	0.03 max	0.035 max	0.15-0.35	bal Fe				
CSN 419255	19255		1.2-1.35	0.2 max		0.15-0.35	0.2 max	0.03 max	0.035 max	0.25 max	bal Fe				
CSN 419295	19295		1.2-1.35	0.2		0.15-0.35	0.2			0.25	bal Fe				
France															
AFNOR NFA35590	465*	Bar Frg	0.65			0.1-0.3		0.02 max	0.02 max	0.1-0.25	bal Fe				
AFNOR NFA35590	Y1105*	Bar Frg	1.05			0.1-0.3		0.02	0.02	0.1-0.25	bal Fe				
AFNOR NFA35590	Y1120*	Bar Frg	1.2			0.1-0.3		0.02	0.02	0.1-0.25	bal Fe				
AFNOR NFA35590	Y175*	Bar Frg	0.75			0.1-0.3		0.02	0.02	0.1-0.25	bal Fe				
AFNOR NFA35590(92)	C120E3UCr4	Quen	1.1-1.29	0.2-0.5	0.25 max	0.1-0.4	0.25 max	0.025 max	0.025 max	0.1-0.3	Mo <=0.15; V <=0.1; W <=0.1; bal Fe				64 HRC
AFNOR NFA35590(92)	C140E3U	Quen	1.3-1.5	0.2 max	0.25 max	0.1-0.4	0.25 max	0.025 max	0.025 max	0.1-0.3	Mo <=0.15; V <=0.1; W <=0.1; bal Fe				64 HRC
AFNOR NFA35590(92)	C140E3UCr4	Quen	1.3-1.5	0.2-0.5	0.25 max	0.1-0.4	0.25 max	0.025 max	0.025 max	0.1-0.3	Mo <=0.15; V <=0.1; W <=0.1; bal Fe				64 HRC
AFNOR NFA35590(92)	C48E4U	Quen	0.45-0.51	0.35 max	0.35 max	0.5-0.8	0.35 max	0.035 max	0.035 max	0.1-0.4	Mo <=0.15; V <=0.10; bal Fe				56 HRC
AFNOR NFA35590(92)	C55E4U	Quen	0.52-0.6	0.35 max	0.35 max	0.5-0.8	0.35 max	0.035 max	0.035 max	0.1-0.4	Mo <=0.15; V <=0.1; W <=0.1; bal Fe				58 HRC
AFNOR NFA35590(92)	C65E4U	Quen	0.6-0.69	0.35 max	0.35 max	0.5-0.8	0.35 max	0.035 max	0.035 max	0.1-0.4	Mo <=0.15; V <=0.1; W <=0.1; bal Fe			⁎	60 HRC
AFNOR NFA35590(92)	C70E2U	Quen	0.65-0.74	0.2 max	0.25 max	0.1-0.4	0.25 max	0.02 max	0.02 max	0.1-0.3	Mo <=0.15; V <=0.1; W <=0.1; bal Fe				61 HRC
AFNOR NFA35590(92)	C80E2U	Quen	0.75-0.84	0.2 max	0.25 max	0.1-0.4	0.25 max	0.02 max	0.02 max	0.1-0.3	Mo <=0.15; V <=0.1; W <=0.1; bal Fe				62 HRC

UNS numbers and US grades are provided as a means of cross referencing chemically similar alloys. Exchangability is only possible after independent examination of specifications. Tensile properties are minimum or typical as specified. UTS and YS as MPa. El as %. See Appendix for list of abbreviations used in Notes. * indicates obsolete material.

Specification	Designation	Notes	C	Cr	Cu	Mn	Ni	P	S	Si	Other	UTS	YS	El	Hard

Tool Steel, Water-Hardening, W1 (Continued from previous page)

France

Specification	Designation	Notes	C	Cr	Cu	Mn	Ni	P	S	Si	Other	UTS	YS	El	Hard
AFNOR NFA35590(92)	C90E2U	Quen	0.85-0.94	0.2 max	0.25 max	0.1-0.4	0.25 max	0.02 max	0.02 max	0.1-0.3	Mo <=0.15; V <=0.1; W <=0.1; bal Fe				63 HRC
AFNOR NFA35590(92)	Y170	Quen	0.65-0.74	0.2 max	0.25 max	0.1-0.4	0.25 max	0.02 max	0.02 max	0.1-0.3	Mo <=0.15; V <=0.1; W <=0.1; bal Fe				61 HRC
AFNOR NFA35590(92)	Y180	Quen	0.75-0.84	0.2 max	0.25 max	0.1-0.4	0.25 max	0.02 max	0.02 max	0.1-0.3	Mo <=0.15; V <=0.1; W <=0.1; bal Fe				62 HRC
AFNOR NFA35590(92)	Y190	Quen	0.85-0.94	0.2 max	0.25 max	0.1-0.4	0.25 max	0.02 max	0.02 max	0.1-0.3	Mo <=0.15; V <=0.1; W <=0.1; bal Fe				63 HRC
AFNOR NFA35590(92)	Y2120C	Quen	1.1-1.29	0.2-0.5	0.25 max	0.1-0.4	0.25 max	0.025 max	0.025 max	0.1-0.3	Mo <=0.15; V <=0.1; W <=0.1; bal Fe				64 HRC
AFNOR NFA35590(92)	Y2140	Quen	1.3-1.5	0.2 max	0.25 max	0.1-0.4	0.25 max	0.025 max	0.025 max	0.1-0.3	Mo <=0.15; V <=0.1; W <=0.1; bal Fe				64 HRC
AFNOR NFA35590(92)	Y2140C	Quen	1.3-1.5	0.2-0.5	0.25 max	0.1-0.4	0.25 max	0.025 max	0.025 max	0.1-0.3	Mo <=0.15; V <=0.1; W <=0.1; bal Fe				64 HRC
AFNOR NFA35590(92)	Y348	Quen	0.45-0.51	0.35 max	0.35 max	0.5-0.8	0.35 max	0.035 max	0.035 max	0.1-0.4	Mo <=0.15; V <=0.10; bal Fe				56 HRC
AFNOR NFA35590(92)	Y355	Quen	0.52-0.6	0.35 max	0.35 max	0.5-0.8	0.35 max	0.035 max	0.035 max	0.1-0.4	Mo <=0.15; V <=0.1; W <=0.1; bal Fe				58 HRC
AFNOR NFA35590(92)	Y365	Quen	0.6-0.69	0.35 max	0.35 max	0.5-0.8	0.35 max	0.035 max	0.035 max	0.1-0.4	Mo <=0.15; V <=0.1; W <=0.1; bal Fe				60 HRC
AFNOR NFA35596	Y75		0.7-0.8			0.4-0.7				0.1-0.3	bal Fe				
AFNOR NFA35596	Y90		0.85-0.95			0.4-0.7				0.1-0.3	bal Fe				

Germany

Specification	Designation	Notes	C	Cr	Cu	Mn	Ni	P	S	Si	Other	UTS	YS	El	Hard
DIN	C105W2	Ann Bar	1.00-1.10			0.10-0.35		0.030 max	0.030 max	0.10-0.30	bal Fe	640			
DIN	C125U	Ann Bar	1.20-1.35			0.10-0.35		0.030 max	0.030 max	0.10-0.30	bal Fe		710		210 HB
DIN	C135U	Ann Bar	1.30-1.45			0.10-0.35		0.030 max	0.030 max	0.10-0.30	bal Fe		780		230 HB
DIN	C55W	Ann Bar	0.50-0.58			0.30-0.50		0.030 max	0.030 max	0.15 max	bal Fe	570			
DIN	C67W	Ann Bar	0.64-0.72			0.60-0.80		0.035 max	0.035 max	0.15-0.40	bal Fe	730			
DIN	C75W	Ann Bar	0.72-0.82			0.60-0.80		0.035 max	0.035 max	0.15-0.40	bal Fe	730			
DIN	C80W2	Ann Bar	0.75-0.85			0.10-0.35		0.030 max	0.030 max	0.10-0.30	bal Fe	640			
DIN	G-31Mn4		0.28-0.33			0.80-1.00		0.035 max	0.035 max	0.40-0.60	bal Fe				
DIN	WNr 1.1554	Ann Bar	1.00-1.10			0.10-0.35		0.030 max	0.030 max	0.10-0.30	bal Fe	660			195 HB
DIN	WNr 1.1563	Ann Bar	1.20-1.35			0.10-0.35		0.030 max	0.030 max	0.10-0.30	bal Fe	710			210 HB
DIN	WNr 1.1573	Ann Bar	1.30-1.45			0.10-0.35		0.030 max	0.030 max	0.10-0.30	bal Fe	780			230 HB
DIN	WNr 1.1625	Ann Bar	0.75-0.85			0.10-0.35		0.030 max	0.030 max	0.10-0.30	bal Fe	640			
DIN	WNr 1.1645	Ann Bar	1.00-1.10			0.10-0.35		0.030 max	0.030 max	0.10-0.30	bal Fe	640			
DIN	WNr 1.1663*	Obs	1.2-1.35			0.1-0.35		0.03	0.03 max	0.1-0.3	bal Fe				
DIN	WNr 1.1673*	Obs	1.3-1.45			0.1-0.35		0.03	0.03 max	0.1-0.3	bal Fe				
DIN	WNr 1.1744	Ann Bar	0.64-0.72			0.60-0.80		0.035 max	0.035 max	0.15-0.40	bal Fe	730			
DIN	WNr 1.1750	Ann Bar	0.72-0.82			0.60-0.80		0.035 max	0.035 max	0.15-0.40	bal Fe	730			
DIN	WNr 1.1811		0.28-0.33			0.80-1.00		0.035 max	0.035 max	0.40-0.60	bal Fe				
DIN	WNr 1.1819		0.85-0.95			0.90-1.10		0.035 max	0.035 max	0.25-0.50	bal Fe				
DIN	WNr 1.1820	Ann Bar	0.50-0.58			0.30-0.50		0.030 max	0.030 max	0.15 max	bal Fe	570			
DIN 17350(80)	C105U1	Ann Bar	1.00-1.10			0.10-0.25		0.020 max	0.020 max	0.10-0.25	bal Fe	640			

Specification	Designation	Notes	C	Cr	Cu	Mn	Ni	P	S	Si	Other	UTS	YS	El	Hard
Tool Steel, Water-Hardening, W1 (Continued from previous page)															
Germany															
DIN 17350(80)	C60U	Ann Bar	0.55-0.65			0.60-0.80		0.035 max	0.035 max	0.15-0.40	bal Fe	700			
DIN 17350(80)	C80U1	Ann Bar	0.75-0.85			0.10-0.25		0.020 max	0.020 max	0.10-0.25	bal Fe	640			
DIN 17350(80)	C85U	Ann Bar	0.80-0.90			0.50-0.70		0.025 max	0.020 max	0.25-0.40	bal Fe	760			
DIN 17350(80)	WNr 1.1525	Ann Bar	0.75-0.85			0.10-0.25		0.020 max	0.020 max	0.10-0.25	bal Fe	640			
DIN 17350(80)	WNr 1.1545	Ann Bar	1.00-1.10			0.10-0.25		0.020 max	0.020 max	0.10-0.25	bal Fe	640			
DIN 17350(80)	WNr 1.1740	Ann Bar	0.55-0.65			0.60-0.80		0.035 max	0.035 max	0.15-0.40	bal Fe	700			
DIN 17350(80)	WNr 1.1830	Ann Bar	0.80-0.90			0.50-0.70		0.025 max	0.020 max	0.25-0.40	bal Fe	760			
DIN(Aviation Hdbk)	C110U	Ann Bar	1.00-1.10			0.10-0.35		0.030 max	0.030 max	0.10-0.30	bal Fe		660		195 HB
DIN(Aviation Hdbk)	WNr 1.1654*	Obs	1-1.2			0.1-0.35		0.03	0.03 max	0.1-0.3	bal Fe				
Hungary															
MSZ 4354(82)	S101	Quen	0.95-1.04	0.2 max	0.25 max	0.15-0.35	0.25 max	0.025 max	0.025 max	0.15-0.35	bal Fe				63 HRC
MSZ 4354(82)	S101	Soft ann	0.95-1.04	0.2 max	0.25 max	0.15-0.35	0.25 max	0.025 max	0.025 max	0.15-0.35	bal Fe				22 HB max
MSZ 4354(82)	S102	Soft ann	0.95-1.04	0.2 max	0.25 max	0.15-0.35	0.25 max	0.025 max	0.025 max	0.15-0.35	bal Fe				200 HB max
MSZ 4354(82)	S102	Quen	0.95-1.04	0.2 max	0.25 max	0.15-0.35	0.25 max	0.025 max	0.025 max	0.15-0.35	bal Fe				63 HRC
MSZ 4354(82)	S111	Soft ann	1.05-1.14	0.2 max	0.25 max	0.15-0.35	0.25 max	0.025 max	0.025 max	0.15-0.35	bal Fe				210 HB max
MSZ 4354(82)	S111	Quen	1.05-1.14	0.2 max	0.25 max	0.15-0.35	0.25 max	0.025 max	0.025 max	0.15-0.35	bal Fe				63 HRC
MSZ 4354(82)	S112	Quen	1.05-1.14	0.2 max	0.25 max	0.15-0.35	0.25 max	0.025 max	0.025 max	0.15-0.35	bal Fe				63 HRC
MSZ 4354(82)	S112	Soft ann	1.05-1.14	0.2 max	0.25 max	0.15-0.35	0.25 max	0.025 max	0.025 max	0.15-0.35	bal Fe				210 HB max
MSZ 4354(82)	S121	Quen	1.15-1.24	0.2 max	0.25 max	0.15-0.35	0.25 max	0.025 max	0.025 max	0.15-0.35	bal Fe				63 HRC
MSZ 4354(82)	S121	Soft ann	1.05-1.24	0.2 max	0.25 max	0.15-0.35	0.25 max	0.025 max	0.025 max	0.15-0.35	bal Fe				215 HB max
MSZ 4354(82)	S122	Soft ann	1.15-1.24	0.2 max	0.25 max	0.15-0.35	0.25 max	0.025 max	0.025 max	0.15-0.35	bal Fe				215 HB max
MSZ 4354(82)	S122	Quen	1.15-1.24	0.2 max	0.25 max	0.15-0.35	0.25 max	0.025 max	0.025 max	0.15-0.35	bal Fe				63 HRC
MSZ 4354(82)	S131	Quen	1.25-1.4	0.2 max	0.25 max	0.15-0.35	0.25 max	0.025 max	0.025 max	0.15-0.35	bal Fe				64 HRC
MSZ 4354(82)	S131	Soft ann	1.25-1.4	0.2 max	0.25 max	0.15-0.35	0.25 max	0.025 max	0.025 max	0.15-0.35	bal Fe				220 HB max
MSZ 4354(82)	S132	Soft ann	1.25-1.4	0.2 max	0.25 max	0.15-0.35	0.25 max	0.025 max	0.025 max	0.15-0.35	bal Fe				220 HB max
MSZ 4354(82)	S132	Quen	1.25-1.4	0.2 max	0.25 max	0.15-0.35	0.25 max	0.025 max	0.025 max	0.15-0.35	bal Fe				64 HRC
MSZ 4354(82)	S45	Q/T	0.42-0.5	0.2 max	0.25 max	0.6-0.8	0.25 max	0.025 max	0.025 max	0.15-0.35	bal Fe				52 HRC
MSZ 4354(82)	S45	Soft ann	0.42-0.5	0.2 max	0.25 max	0.6-0.8	0.25 max	0.025 max	0.025 max	0.15-0.35	bal Fe				190 HB max
MSZ 4354(82)	S60	Soft ann	0.57-0.65	0.2 max	0.25 max	0.6-0.8	0.25 max	0.025 max	0.025 max	0.15-0.35	bal Fe				190 HB max
MSZ 4354(82)	S60	Q/T	0.57-0.65	0.2 max	0.25 max	0.6-0.8	0.25 max	0.025 max	0.025 max	0.15-0.35	bal Fe				57 HRC
MSZ 4354(82)	S71	Soft ann	0.65-0.74	0.2 max	0.25 max	0.15-0.35	0.25 max	0.025 max	0.025 max	0.15-0.35	bal Fe				185 HB max
MSZ 4354(82)	S71	Quen	0.65-0.74	0.2 max	0.25 max	0.15-0.35	0.25 max	0.025 max	0.025 max	0.15-0.35	bal Fe				62 HRC
MSZ 4354(82)	S72	Soft ann	0.65-0.74	0.2 max	0.25 max	0.15-0.35	0.25 max	0.025 max	0.025 max	0.15-0.35	bal Fe				185 HB max
MSZ 4354(82)	S72	Quen	0.65-0.74	0.2 max	0.25 max	0.15-0.35	0.25 max	0.025 max	0.025 max	0.15-0.35	bal Fe				62 HRC

UNS numbers and US grades are provided as a means of cross referencing chemically similar alloys. Exchangability is only possible after independent examination of specifications. Tensile properties are minimum or typical as specified. UTS and YS as MPa. El as %. See Appendix for list of abbreviations used in Notes. * indicates obsolete material.

Tool Steel, Water-Hardening, W1 (Continued from previous page)

Specification	Designation	Notes	C	Cr	Cu	Mn	Ni	P	S	Si	Other	UTS	YS	El	Hard
Hungary															
MSZ 4354(82)	S81	Soft ann	0.75-0.84	0.2 max	0.25 max	0.15-0.35	0.25 max	0.025 max	0.025 max	0.15-0.35	bal Fe				190 HB max
MSZ 4354(82)	S81	Quen	0.75-0.84	0.2 max	0.25 max	0.15-0.35	0.25 max	0.025 max	0.025 max	0.15-0.35	bal Fe				62 HRC
MSZ 4354(82)	S82	Soft ann	0.75-0.84	0.2 max	0.25 max	0.15-0.35	0.25 max	0.025 max	0.025 max	0.15-0.35	bal Fe				190 HB max
MSZ 4354(82)	S82	Quen	0.75-0.84	0.2 max	0.25 max	0.15-0.35	0.25 max	0.025 max	0.025 max	0.15-0.35	bal Fe				62 HRC
MSZ 4354(82)	S91	Soft ann	0.85-0.94	0.2 max	0.25 max	0.15-0.35	0.25 max	0.025 max	0.025 max	0.15-0.35	bal Fe				195 HB max
MSZ 4354(82)	S91	Quen	0.85-0.94	0.2 max	0.25 max	0.15-0.35	0.25 max	0.025 max	0.025 max	0.15-0.35	bal Fe				62 HRC
MSZ 4354(82)	S92	Soft ann	0.85-0.94	0.2 max	0.25 max	0.15-0.35	0.25 max	0.025 max	0.025 max	0.15-0.35	bal Fe				195 HB max
MSZ 4354(82)	S92	Quen	0.85-0.94	0.2 max	0.25 max	0.15-0.35	0.25 max	0.025 max	0.025 max	0.15-0.35	bal Fe				62 HRC
India															
IS 1570	T103		0.95-1.1			0.2-0.35				0.1-0.3	bal Fe				
IS 1570	T118		1.1-1.25			0.2-0.35				0.1-0.3	bal Fe				
IS 1570	T75		0.7-0.8			0.5-0.8				0.1-0.35	bal Fe				
IS 1570	T80		0.75-0.85			0.2-0.35				0.1-0.3	bal Fe				
IS 1570	T80Mn65		0.75-0.85			0.5-0.8				0.1-0.35	bal Fe				
IS 1570	T85		0.8-0.9			0.5-0.8				0.1-0.35	bal Fe				
IS 1570	T90		0.85-0.95			0.2-0.35				0.1-0.3	bal Fe				
Italy															
UNI 2955	AB2 C70KU		0.7			0.3				0.3	bal Fe				
UNI 2955	AB3 C80KU		0.85			0.25				0.25	bal Fe				
UNI 2955	AB4S C100KU		0.98			0.4				0.35	bal Fe				
UNI 2955	L3B C120KU		1.12			0.35				0.35	bal Fe				
UNI 2955	LXB		0.95			0.3				0.3	bal Fe				
UNI 2955/2(82)	C100KU		0.95-1.09			0.35				0.3	bal Fe				
UNI 2955/2(82)	C120KU		1.1-1.29			0.35 max		0.07 max	0.06 max	0.3 max	bal Fe				
UNI 2955/2(82)	C140KU		1.3-1.5			0.35 max		0.07 max	0.06 max	0.3 max	bal Fe				
UNI 2955/2(82)	C70KU		0.65-0.74			0.35 max		0.07 max	0.06 max	0.3 max	bal Fe				
UNI 2955/2(82)	C80KU		0.75-0.84			0.35 max		0.07 max	0.06 max	0.3 max	bal Fe				
UNI 2955/2(82)	C90KU		0.85-0.94			0.35 max		0.07 max	0.06 max	0.3 max	bal Fe				
Japan															
JIS G3311(88)	SK3M	CR Strp; Hacksaw Blade	1.00-1.10	0.30 max	0.25 max	0.50 max	0.25 max	0.030 max	0.030 max	0.35 max	bal Fe				220-310 HV
JIS G3311(88)	SK5M	CR Strp; Saws	0.80-0.90	0.30 max	0.25 max	0.50 max	0.25 max	0.030 max	0.030 max	0.35 max	bal Fe				200-290 HV
JIS G3311(88)	SK6M	CR Strp; Spring; Cutlery	0.70-0.80	0.30 max	0.25 max	0.50 max	0.25 max	0.030 max	0.030 max	0.35 max	bal Fe				190-280 HV
JIS G3311(88)	SK7M	CR Strp; Spring; Cutlery	0.60-0.70	0.30 max	0.25 max	0.50 max	0.25 max	0.030 max	0.030 max	0.35 max	bal Fe				190-280 HV
JIS G4401(83)	SK1		1.30-1.50	0.30 max	0.25 max	0.50 max	0.25 max	0.030 max	0.030 max	0.35 max	bal Fe				217 HB max
JIS G4401(83)	SK2		1.10-1.30	0.30 max	0.25 max	0.50 max	0.25 max	0.030 max	0.030 max	0.35 max	bal Fe				212 HB max
JIS G4401(83)	SK3		1.00-1.10	0.30 max	0.25 max	0.50 max	0.25 max	0.030 max	0.030 max	0.35 max	bal Fe				212 HB max
JIS G4401(83)	SK4		0.90-1.00	0.30 max	0.25 max	0.50 max	0.25 max	0.030 max	0.030 max	0.35 max	bal Fe				207 HB max

UNS numbers and US grades are provided as a means of cross referencing chemically similar alloys. Exchangability is only possible after independent examination of specifications. Tensile properties are minimum or typical as specified. UTS and YS as MPa. El as %. See Appendix for list of abbreviations used in Notes. * indicates obsolete material.

Specification	Designation	Notes	C	Cr	Cu	Mn	Ni	P	S	Si	Other	UTS	YS	El	Hard

Tool Steel, Water-Hardening, W1 (Continued from previous page)

Japan

Specification	Designation	Notes	C	Cr	Cu	Mn	Ni	P	S	Si	Other	UTS	YS	El	Hard
JIS G4401(83)	SK5		0.80-0.90	0.30 max	0.25 max	0.50 max	0.25 max	0.030 max	0.030 max	0.35 max	bal Fe				207 HB max
JIS G4401(83)	SK6		0.70-0.80	0.30 max	0.25 max	0.50 max	0.25 max	0.030 max	0.030 max	0.35 max	bal Fe				201 HB max
JIS G4401(83)	SK7		0.60-0.70	0.30 max	0.25 max	0.50 max	0.25 max	0.030 max	0.030 max	0.35 max	bal Fe				201 HB max
JIS G4410(84)	SKC3	Rod; FF	0.70-0.85	0.20 max	0.25 max	0.50 max	0.25 max	0.030 max	0.15-0.35	0.15-0.35	V <=0.25; bal Fe				292-302 HB

Mexico

Specification	Designation	Notes	C	Cr	Cu	Mn	Ni	P	S	Si	Other	UTS	YS	El	Hard
NMX-B-082(90)	W1-A	Bar	1.05-1.40	0.15 max	0.25 max	0.10-0.40	0.20 max	0.03 max	0.03 max	0.10-0.40	Mo <=0.10; V <=0.10; W <=0.15; (As+Sn+Sb)<=0.040; (Cu+Ni)<=0.40; B, Ti may be added; bal Fe				
NMX-B-082(90)	W1-B	Bar	0.60-0.95	0.15 max	0.25 max	0.10-0.40	0.20 max	0.03 max	0.03 max	0.10-0.40	Mo <=0.10; V <=0.10; W <=0.15; (As+Sn+Sb)<=0.040; (Cu+Ni)<=0.40; B, Ti may be added; bal Fe				
NMX-B-082(90)	W1-M	Bar	0.95-1.05	0.15 max	0.25 max	0.10-0.40	0.20 max	0.03 max	0.03 max	0.10-0.40	Mo <=0.10; V <=0.10; W <=0.15; (As+Sn+Sb)<=0.040; (Cu+Ni)<=0.40; B, Ti may be added; bal Fe				

Pan America

Specification	Designation	Notes	C	Cr	Cu	Mn	Ni	P	S	Si	Other	UTS	YS	El	Hard
COPANT 337	W1	Bar, Frg	0.6-1.4	0.15		0.15-0.4		0.03	0.03	0.1-0.35	bal Fe				

Poland

Specification	Designation	Notes	C	Cr	Cu	Mn	Ni	P	S	Si	Other	UTS	YS	El	Hard
PNH85020	N10		0.95-1.04	0.2 max	0.25 max	0.15-0.35	0.25 max	0.030 max	0.030 max	0.15-0.35	bal Fe				
PNH85020	N10E		0.95-1.04	0.15 max	0.2 max	0.15-0.3	0.2 max	0.025 max	0.25 max	0.15-0.3	bal Fe				
PNH85020	N11		1.05-1.14	0.2 max	0.25 max	0.15-0.35	0.25 max	0.030 max	0.030 max	0.15-0.35	bal Fe				
PNH85020	N11E		1.05-1.14	0.15 max	0.2 max	0.15-0.3	0.2 max	0.025 max	0.025 max	0.15-0.3	bal Fe				
PNH85020	N12		1.15-1.24	0.2 max	0.25 max	0.15-0.35	0.25 max	0.030 max	0.030 max	0.15-0.35	bal Fe				
PNH85020	N12E		1.15-1.24	0.15 max	0.2 max	0.15-0.3	0.2 max	0.025 max	0.25 max	0.15-0.3	bal Fe				
PNH85020	N13		1.25-1.4	0.2 max	0.25 max	0.15-0.35	0.25 max	0.030 max	0.030 max	0.15-0.35	bal Fe				
PNH85020	N13E		1.25-1.34	0.15 max	0.2 max	0.15-0.3	0.2 max	0.020 max	0.020 max	0.15-0.3	bal Fe				
PNH85020	N6		0.61-0.7	0.3 max		0.3-0.5	0.4 max	0.035 max	0.035 max	0.15 max	bal Fe				
PNH85020	N7		0.65-0.74	0.2	0.25	0.15-0.35	0.25			0.15-0.35	bal Fe				
PNH85020	N7E		0.65-0.74	0.15 max	0.2 max	0.15-0.3	0.2 max	0.025 max	0.025 max	0.15-0.3	bal Fe				
PNH85020	N8	engraving	0.75-0.84	0.2 max	0.25 max	0.15-0.35	0.25 max	0.030 max	0.030 max	0.15-0.35	bal Fe				
PNH85020	N8E	engraving	0.75-0.84	0.15 max	0.2 max	0.15-0.3	0.2 max	0.025 max	0.025 max	0.15-0.3	bal Fe				
PNH85020	N9		0.85-0.94	0.2 max	0.25 max	0.15-0.35	0.25 max	0.030 max	0.030 max	0.15-0.35	bal Fe				
PNH85020	N9E		1.5-1.8	11-13		0.15-0.45	0.4 max	0.030 max	0.030 max	0.15-0.4	bal Fe				

Romania

Specification	Designation	Notes	C	Cr	Cu	Mn	Ni	P	S	Si	Other	UTS	YS	El	Hard
STAS	OSC12		1.15-1.25	0.2	0.25	0.15-0.35	0.25			0.15-0.35	bal Fe				
STAS	OSL8		0.8-0.9	0.3	0.25	0.2-0.4	0.25			0.15-0.35	bal Fe				
STAS 1699(76)	OSL10		0.95-1.05	0.3 max	0.25 max	0.2-0.4	0.25 max	0.04 max	0.04 max	0.15-0.35	Pb <=0.15; P+S<=0.07; bal Fe				
STAS 1699(76)	OSL4		0.4-0.5	0.3 max	0.25 max	0.6-0.8	0.25 max	0.04 max	0.04 max	0.15-0.35	Pb <=0.15; P+S<=0.07; bal Fe				

Specification	Designation	Notes	C	Cr	Cu	Mn	Ni	P	S	Si	Other	UTS	YS	El	Hard

Tool Steel, Water-Hardening, W1 (Continued from previous page)

Romania

Specification	Designation	Notes	C	Cr	Cu	Mn	Ni	P	S	Si	Other	UTS	YS	El	Hard
STAS 1700(80)	OSC13		1.26-1.45	0.2 max	0.25 max	0.15-0.35	0.25 max	0.03 max	0.025 max	0.15-0.35	Pb <=0.15; bal Fe				
STAS 1700(90)	OSC10		0.95-1.04	0.2 max	0.25 max	0.1-0.35	0.25 max	0.03 max	0.025 max	0.15-0.35	Pb <=0.15; bal Fe				
STAS 1700(90)	OSC11		1.05-1.14	0.2 max	0.25 max	0.1-0.35	0.25 max	0.03 max	0.025 max	0.15-0.35	Pb <=0.15; bal Fe				
STAS 1700(90)	OSC7		0.65-0.74	0.2 max	0.25 max	0.1-0.35	0.25 max	0.03 max	0.025 max	0.15-0.35	Pb <=0.15; bal Fe				
STAS 1700(90)	OSC8		0.75-0.84	0.2 max	0.25 max	0.1-0.35	0.25 max	0.03 max	0.025 max	0.15-0.35	Pb <=0.15; bal Fe				
STAS 1700(90)	OSC8M		0.8-0.9	0.2 max	0.25 max	0.35-0.8	0.25 max	0.03 max	0.025 max	0.15-0.35	Pb <=0.15; bal Fe				
STAS 1700(90)	OSC9		0.85-0.94	0.2 max	0.25 max	0.1-0.35	0.25 max	0.03 max	0.025 max	0.15-0.35	Pb <=0.15; bal Fe				

Russia

Specification	Designation	Notes	C	Cr	Cu	Mn	Ni	P	S	Si	Other	UTS	YS	El	Hard
GOST	U8G	Carbon tool	0.81-0.89	0.12	0.2	0.33-0.58	0.12			0.17-0.33	bal Fe				
GOST 1435	U10-1	Carbon tool	0.95-1.04	0.2 max	0.25 max	0.17-0.33	0.25 max	0.03 max	0.028 max	0.17-0.33	Mo <=0.15; bal Fe				
GOST 1435	U10-3	Carbon tool	0.95-1.04	0.2-0.4	0.25 max	0.17-0.33	0.25 max	0.03 max	0.028 max	0.17-0.33	Mo <=0.15; bal Fe				
GOST 1435	U10A-1	Carbon tool	0.95-1.04	0.2 max	0.25 max	0.17-0.28	0.25 max	0.025 max	0.018 max	0.17-0.33	Mo <=0.15; bal Fe				
GOST 1435	U10A-2	Carbon tool	0.95-1.04	0.12 max	0.2 max	0.17-0.28	0.12 max	0.025 max	0.018 max	0.17-0.33	Mo <=0.15; bal Fe				
GOST 1435	U10A-3	Carbon tool	0.95-1.04	0.2-0.4	0.25 max	0.17-0.28	0.25 max	0.025 max	0.018 max	0.17-0.33	Mo <=0.15; bal Fe				
GOST 1435	U11-1	Carbon tool	1.05-1.14	0.2 max	0.25 max	0.17-0.33	0.25 max	0.03 max	0.028 max	0.17-0.33	Mo <=0.15; bal Fe				
GOST 1435	U11-3	Carbon tool	1.05-1.14	0.2-0.4	0.25 max	0.17-0.33	0.25 max	0.03 max	0.028 max	0.17-0.33	Mo <=0.15; bal Fe				
GOST 1435	U11A-1	Carbon tool	1.05-1.14	0.2 max	0.25 max	0.17-0.28	0.25 max	0.025 max	0.018 max	0.17-0.33	Mo <=0.15; bal Fe				
GOST 1435	U11A-2	Carbon tool	1.05-1.14	0.12 max	0.2 max	0.17-0.28	0.12 max	0.025 max	0.018 max	0.17-0.33	Mo <=0.15; bal Fe				
GOST 1435	U11A-3	Carbon tool	1.05-1.14	0.2-0.4	0.25 max	0.17-0.28	0.25 max	0.025 max	0.018 max	0.17-0.33	Mo <=0.15; bal Fe				
GOST 1435	U12-1	Carbon tool	1.15-1.24	0.2 max	0.25 max	0.17-0.33	0.25 max	0.03 max	0.028 max	0.17-0.33	Mo <=0.15; bal Fe				
GOST 1435	U12-3	Carbon tool	1.15-1.24	0.2-0.4	0.25 max	0.17-0.33	0.25 max	0.03 max	0.028 max	0.17-0.33	Mo <=0.15; bal Fe				
GOST 1435	U12A	Carbon tool	1.16-1.23	0.12	0.2	0.17-0.28	0.2			0.17-0.33	bal Fe				
GOST 1435	U12A-1	Carbon tool	1.15-1.24	0.2 max	0.1 max	0.17-0.28	0.25 max	0.025 max	0.018 max	0.17-0.33	Mo <=0.15; bal Fe				
GOST 1435	U12A-2	Carbon tool	1.15-1.24	0.12 max	0.2 max	0.17-0.28	0.12 max	0.025 max	0.018 max	0.17-0.33	Mo <=0.15; bal Fe				
GOST 1435	U12A-3	Carbon tool	1.15-1.24	0.2-0.4	0.25 max	0.17-0.28	0.25 max	0.025 max	0.018 max	0.17-0.33	Mo <=0.15; bal Fe				
GOST 1435	U13	Carbon tool	1.26-1.34	0.12	0.25	0.17-0.33	0.25			0.17-0.33	bal Fe				
GOST 1435	U13-1	Carbon tool	1.25-1.35	0.2 max	0.25 max	0.17-0.33	0.25 max	0.03 max	0.028 max	0.17-0.33	Mo <=0.15; bal Fe				
GOST 1435	U13-3	Carbon tool	1.25-1.35	0.2-0.4	0.25 max	0.17-0.33	0.25 max	0.03 max	0.028 max	0.17-0.33	Mo <=0.15; bal Fe				
GOST 1435	U13A	Carbon tool	1.26-1.34	0.12	0.2	0.17-0.28	0.2			0.17-0.33	bal Fe				
GOST 1435	U13A-1	Carbon tool	1.25-1.35	0.2 max	0.25 max	0.17-0.28	0.25 max	0.025 max	0.018 max	0.17-0.33	Mo <=0.15; bal Fe				
GOST 1435	U13A-2	Carbon tool	1.25-1.35	0.12 max	0.2 max	0.17-0.28	0.12 max	0.025 max	0.018 max	0.17-0.33	Mo <=0.15; bal Fe				
GOST 1435	U13A-3	Carbon tool	1.25-1.35	0.2-0.4	0.25 max	0.17-0.28	0.25 max	0.025 max	0.018 max	0.17-0.33	Mo <=0.15; bal Fe				
GOST 1435	U7	Carbon tool	0.66-0.73	0.2 max	0.25 max	0.17-0.33	0.25 max	0.03 max	0.03 max	0.17-0.33	bal Fe				
GOST 1435	U7-1	Carbon tool	0.65-0.74	0.2 max	0.25 max	0.17-0.33	0.25 max	0.03 max	0.028 max	0.17-0.33	Mo <=0.15; bal Fe				

UNS numbers and US grades are provided as a means of cross referencing chemically similar alloys. Exchangability is only possible after independent examination of specifications. Tensile properties are minimum or typical as specified. UTS and YS as MPa. El as %. See Appendix for list of abbreviations used in Notes. * indicates obsolete material.

Specification	Designation	Notes	C	Cr	Cu	Mn	Ni	P	S	Si	Other	UTS	YS	El	Hard

Tool Steel, Water-Hardening, W1 (Continued from previous page)

Russia

Specification	Designation	Notes	C	Cr	Cu	Mn	Ni	P	S	Si	Other
GOST 1435	U7-3	Carbon tool	0.65-0.74	0.2-0.4	0.25 max	0.17-0.33	0.25 max	0.03 max	0.028 max	0.17-0.33	Mo <=0.15; bal Fe
GOST 1435	U7A	Carbon tool	0.66-0.73	0.2 max	0.25 max	0.17-0.33	0.2 max	0.03 max	0.02 max	0.17-0.33	bal Fe
GOST 1435	U7A-1	Carbon tool	0.65-0.74	0.2 max	0.25 max	0.17-0.28	0.25 max	0.025 max	0.018 max	0.17-0.33	Mo <=0.15; bal Fe
GOST 1435	U7A-2	Carbon tool	0.65-0.74	0.12 max	0.2 max	0.17-0.28	0.12 max	0.025 max	0.018 max	0.17-0.33	Mo <=0.15; bal Fe
GOST 1435	U7A-3	Carbon tool	0.65-0.74	0.2-0.4	0.25 max	0.17-0.28	0.25 max	0.025 max	0.018 max	0.17-0.33	Mo <=0.15; bal Fe
GOST 1435	U8	Carbon tool	0.76-0.83	0.2 max	0.25 max	0.17-0.33	0.25 max	0.03 max	0.03 max	0.17-0.33	bal Fe
GOST 1435	U8-1	Carbon tool	0.75-0.84	0.2 max	0.25 max	0.17-0.33	0.25 max	0.03 max	0.028 max	0.17-0.33	Mo <=0.15; bal Fe
GOST 1435	U8-3	Carbon tool	0.75-0.84	0.2-0.4	0.25 max	0.17-0.33	0.25 max	0.03 max	0.028 max	0.17-0.33	Mo <=0.15; bal Fe
GOST 1435	U8A	Carbon tool	0.76-0.83	0.2 max	0.2 max	0.17-0.33	0.2 max	0.03 max	0.02 max	0.17-0.33	bal Fe
GOST 1435	U8A-1	Carbon tool	0.75-0.84	0.2 max	0.25 max	0.17-0.28	0.25 max	0.025 max	0.018 max	0.17-0.33	Mo <=0.15; bal Fe
GOST 1435	U8A-2	Carbon tool	0.75-0.81	0.12 max	0.2 max	0.17-0.28	0.12 max	0.025 max	0.018 max	0.17-0.33	Mo <=0.15; bal Fe
GOST 1435	U8A-3	Carbon tool	0.75-0.81	0.2-0.4	0.25 max	0.17-0.28	0.25 max	0.025 max	0.018 max	0.17-0.33	Mo <=0.15; bal Fe
GOST 1435	U8G-1	Carbon tool	0.8-0.9	0.2 max	0.25 max	0.33-0.58	0.25 max	0.03 max	0.028 max	0.17-0.33	Mo <=0.15; bal Fe
GOST 1435	U8G-3	Carbon tool	0.8-0.9	0.2-0.4	0.25 max	0.33-0.58	0.25 max	0.03 max	0.028 max	0.17-0.33	Mo <=0.15; bal Fe
GOST 1435	U8GA	Carbon tool	0.81-0.89	0.2 max	0.2 max	0.38-0.58	0.2 max	0.03 max	0.02 max	0.17-0.33	bal Fe
GOST 1435	U8GA-1	Carbon tool	0.8-0.9	0.2 max	0.25 max	0.33-0.58	0.25 max	0.025 max	0.018 max	0.17-0.33	Mo <=0.15; bal Fe
GOST 1435	U8GA-2	Carbon tool	0.8-0.9	0.12 max	0.2 max	0.33-0.58	0.12 max	0.025 max	0.018 max	0.17-0.33	Mo <=0.15; bal Fe
GOST 1435	U8GA-3	Carbon tool	0.8-0.9	0.2-0.4	0.25 max	0.33-0.58	0.25 max	0.025 max	0.018 max	0.17-0.33	Mo <=0.15; bal Fe
GOST 1435	U9	Carbon tool	0.86-0.93	0.2 max	0.25 max	0.17-0.33	0.25 max	0.03 max	0.03 max	0.17-0.33	bal Fe
GOST 1435	U9-1	Carbon tool	0.85-0.94	0.2 max	0.25 max	0.17-0.33	0.25 max	0.03 max	0.028 max	0.17-0.33	Mo <=0.15; bal Fe
GOST 1435	U9-3	Carbon tool	0.85-0.94	0.2-0.4	0.2 max	0.17-0.33	0.12 max	0.03 max	0.028 max	0.17-0.33	Mo <=0.15; bal Fe
GOST 1435	U9A	Carbon tool	0.86-0.93	0.2 max	0.2 max	0.17-0.28	0.2 max	0.02 max	0.03 max	0.17-0.33	bal Fe
GOST 1435	U9A-1	Carbon tool	0.85-0.94	0.2 max	0.25 max	0.17-0.28	0.25 max	0.025 max	0.018 max	0.17-0.33	Mo <=0.15; bal Fe
GOST 1435	U9A-2	Carbon tool	0.85-0.94	0.12 max	0.2 max	0.17-0.28	0.12 max	0.025 max	0.018 max	0.17-0.33	Mo <=0.15; bal Fe
GOST 1435	U9A-3	Carbon tool	0.85-0.94	0.2-0.4	0.25 max	0.17-0.28	0.25 max	0.025 max	0.018 max	0.17-0.33	Mo <=0.15; bal Fe

Spain

Specification	Designation	Notes	C	Cr	Cu	Mn	Ni	P	S	Si	Other
UNE	F.520.U		0.7-0.8			0.15-0.45				0.1-0.35	bal Fe
UNE 36018/2(94)	120V2	Alloyed	1.1-1.29			0.15-0.35	0.25 max	0.025 max	0.02 max	0.15-0.35	Mo <=0.15; V 0.1-0.2; bal Fe
UNE 36018/2(94)	C102U		0.95-1.09	0.25 max		0.15-0.35	0.25 max	0.025 max	0.02 max	0.15-0.35	Mo <=0.15; bal Fe
UNE 36018/2(94)	C120U		1.1-1.29	0.25 max		0.15-0.35	0.25 max	0.025 max	0.02 max	0.15-0.35	Mo <=0.15; bal Fe
UNE 36018/2(94)	C51U	Carbon	0.45-0.54	0.25 max		0.6-0.8	0.25 max	0.025 max	0.02 max	0.15-0.35	Mo <=0.15; bal Fe
UNE 36018/2(94)	C61U	Carbon	0.55-0.64	0.25 max		0.6-0.8	0.25 max	0.025 max	0.02 max	0.15-0.35	Mo <=0.15; bal Fe
UNE 36018/2(94)	C70U	Carbon	0.65-0.74	0.25 max		0.15-0.35	0.25 max	0.025 max	0.02 max	0.15-0.35	Mo <=0.15; bal Fe
UNE 36018/2(94)	C71U	Carbon	0.65-0.74	0.25 max		0.6-0.8	0.25 max	0.025 max	0.02 max	0.15-0.35	Mo <=0.15; bal Fe

UNS numbers and US grades are provided as a means of cross referencing chemically similar alloys. Exchangability is only possible after independent examination of specifications. Tensile properties are minimum or typical as specified. UTS and YS as MPa. El as %. See Appendix for list of abbreviations used in Notes. * indicates obsolete material.

Specification	Designation	Notes	C	Cr	Cu	Mn	Ni	P	S	Si	Other	UTS	YS	El	Hard

Tool Steel, Water-Hardening, W1 (Continued from previous page)

Spain

Specification	Designation	Notes	C	Cr	Cu	Mn	Ni	P	S	Si	Other
UNE 36018/2(94)	C80U	Carbon	0.75-0.84	0.25 max		0.15-0.35	0.25 max	0.025 max	0.02 max	0.15-0.35	Mo <=0.15; bal Fe
UNE 36018/2(94)	C81U	Carbon	0.75-0.84	0.25 max		0.6-0.8	0.25 max	0.025 max	0.02 max	0.15-0.35	Mo <=0.15; bal Fe
UNE 36018/2(94)	C90U	Carbon	0.85-0.94	0.25 max		0.15-0.35	0.25 max	0.025 max	0.02 max	0.15-0.35	Mo <=0.15; bal Fe
UNE 36018/2(94)	F.5103*	see C70U	0.65-0.74	0.25 max		0.15-0.35	0.25 max	0.025 max	0.02 max	0.15-0.35	Mo <=0.15; bal Fe
UNE 36018/2(94)	F.5107*	see C80U	0.75-0.84	0.25 max		0.15-0.35	0.25 max	0.025 max	0.02 max	0.15-0.35	Mo <=0.15; bal Fe
UNE 36018/2(94)	F.5113*	see C90U	0.85-0.94	0.25 max		0.15-0.35	0.25 max	0.025 max	0.02 max	0.15-0.35	Mo <=0.15; bal Fe
UNE 36018/2(94)	F.5117*	see C102U	0.95-1.09	0.25 max		0.15-0.35	0.25 max	0.025 max	0.02 max	0.15-0.35	Mo <=0.15; bal Fe
UNE 36018/2(94)	F.5123*	see C120U	1.1-1.29	0.25 max		0.15-0.35	0.25 max	0.025 max	0.02 max	0.15-0.35	Mo <=0.15; bal Fe
UNE 36018/2(94)	F.5124*	see 120V2	1.1-1.29			0.15-0.35	0.25 max	0.025 max	0.02 max	0.15-0.35	Mo <=0.15; V 0.1-0.2; bal Fe
UNE 36018/2(94)	F.5131*	see C51U	0.45-0.54	0.25 max		0.6-0.8	0.25 max	0.025 max	0.02 max	0.15-0.35	Mo <=0.15; bal Fe
UNE 36018/2(94)	F.5132*	see C61U	0.55-0.64	0.25 max		0.6-0.8	0.25 max	0.025 max	0.02 max	0.15-0.35	Mo <=0.15; bal Fe
UNE 36018/2(94)	F.5133*	see C71U	0.65-0.74	0.25 max		0.6-0.8	0.25 max	0.025 max	0.02 max	0.15-0.35	Mo <=0.15; bal Fe
UNE 36018/2(94)	F.5137*	see C81U	0.75-0.84	0.25 max		0.6-0.8	0.25 max	0.025 max	0.02 max	0.15-0.35	Mo <=0.15; bal Fe
UNE 36071	F.510.A		0.77-0.87			0.5-0.8		0.03 max	0.03 max	0.13-0.38	bal Fe
UNE 36071	F.510.B		0.92-1.02			0.5-0.8		0.03 max	0.03 max	0.13-0.38	bal Fe
UNE 36071	F.512		0.6-0.7			0.25-0.6		0.03 max	0.03 max	0.1-0.25	bal Fe
UNE 36071	F.513		0.7-0.8			0.25-0.6		0.03 max	0.03 max		bal Fe
UNE 36071	F.514		0.8-0.9			0.25-0.6		0.03 max	0.03 max		bal Fe
UNE 36071	F.515		0.9-1			0.25-0.6		0.03 max	0.03 max		bal Fe
UNE 36071	F.516		1-1.2			0.25-0.6		0.03 max	0.03 max		bal Fe
UNE 36071	F.517		1.2-1.4			0.25-0.6		0.03 max	0.03 max		bal Fe
UNE 36071(75)	C140		1.3-1.5			0.35 max		0.03 max	0.03 max	0.35 max	Mo <=0.15; bal Fe
UNE 36071(75)	C80	Carbon	0.75-0.84			0.35 max		0.03 max	0.03 max	0.35 max	Mo <=0.15; bal Fe
UNE 36071(75)	F.5107*	see C80	0.75-0.84			0.35 max		0.03 max	0.03 max	0.35 max	Mo <=0.15; bal Fe
UNE 36071(75)	F.5127*	see C140	1.3-1.5			0.35 max		0.03 max	0.03 max	0.35 max	Mo <=0.15; bal Fe

Sweden

Specification	Designation	Notes	C	Cr	Cu	Mn	Ni	P	S	Si	Other
SS 141880	1880		0.95-1.1			0.2-0.4				0.1-0.3	bal Fe

UK

Specification	Designation	Notes	C	Cr	Cu	Mn	Ni	P	S	Si	Other
BS 4659			0.78	0.45		0.6				0.25	bal Fe
BS 4659	BW1A		0.85-0.95	0.15		0.35	0.2			0.3	Mo 0.1; bal Fe
BS 4659	BW1B		0.95-1.1	0.15		0.35	0.2			0.3	Mo 0.1; bal Fe
BS 4659	BW1C		1.1-1.3	0.15		0.35	0.2			0.3	Mo 0.1; bal Fe

USA

Specification	Designation	Notes	C	Cr	Cu	Mn	Ni	P	S	Si	Other
	AISI W1		0.70-1.50	0.15 max		0.10-0.40		0.025 max	0.025 max	0.10-0.40	Mo <=0.10; V <=0.10; W <=0.15; bal Fe
	UNS T72301		0.70-1.50	0.15 max	0.20 max	0.10-0.40	0.20 max	0.025 max	0.025 max	0.10-0.40	Mo <=0.10; V <=0.10; W <=0.15; bal Fe

UNS numbers and US grades are provided as a means of cross referencing chemically similar alloys. Exchangability is only possible after independent examination of specifications. Tensile properties are minimum or typical as specified. UTS and YS as MPa. El as %. See Appendix for list of abbreviations used in Notes. * indicates obsolete material.

Specification	Designation	Notes	C	Cr	Cu	Mn	Ni	P	S	Si	Other	UTS	YS	El	Hard

Tool Steel, Water-Hardening, W1 (Continued from previous page)

USA

Specification	Designation	Notes	C	Cr	Cu	Mn	Ni	P	S	Si	Other	UTS	YS	El	Hard
ASTM A686(92)	W1 10 Grade A	Wrought	1.00-1.10	0.15 max	0.20 max	0.10-0.40	0.20 max	0.030 max	0.030 max	0.10-0.40	Mo <=0.10; V <=0.10; W <=0.15; bal Fe				
ASTM A686(92)	W1 10 Grade C	Wrought	1.00-1.10	0.30 max	0.20 max	0.10-0.40	0.20 max	0.030 max	0.030 max	0.10-0.40	Mo <=0.10; V <=0.10; W <=0.15; bal Fe				
ASTM A686(92)	W1 11 Grade A	Wrought	1.10-1.20	0.15 max	0.20 max	0.10-0.40	0.20 max	0.030 max	0.030 max	0.10-0.40	Mo <=0.10; V <=0.10; W <=0.15; bal Fe				
ASTM A686(92)	W1 11 Grade C	Wrought	1.10-1.20	0.30 max	0.20 max	0.10-0.40	0.20 max	0.030 max	0.030 max	0.10-0.40	Mo <=0.10; V <=0.10; W <=0.15; bal Fe				
ASTM A686(92)	W1 8 Grade A	Wrought	0.80-0.90	0.15 max	0.20 max	0.10-0.40	0.20 max	0.030 max	0.030 max	0.10-0.40	Mo <=0.10; V <=0.10; W <=0.15; bal Fe				
ASTM A686(92)	W1 8 Grade C	Wrought	0.80-0.90	0.30 max	0.20 max	0.10-0.40	0.20 max	0.030 max	0.030 max	0.10-0.40	Mo <=0.10; V <=0.10; W <=0.15; bal Fe				
ASTM A686(92)	W1 9 Grade A	Wrought	0.90-1.00	0.15 max	0.20 max	0.10-0.40	0.20 max	0.030 max	0.030 max	0.10-0.40	Mo <=0.10; V <=0.10; W <=0.15; bal Fe				
ASTM A686(92)	W1 9 Grade C	Wrought	0.90-1.00	0.30 max	0.20 max	0.10-0.40	0.20 max	0.030 max	0.030 max	0.10-0.40	Mo <=0.10; V <=0.10; W <=0.15; bal Fe				
SAE J437(70)	W108	Ann									bal Fe				159-202 HB
SAE J437(70)	W109	Ann									bal Fe				159-202 HB
SAE J437(70)	W110	Ann									bal Fe				159-202 HB
SAE J437(70)	W112	Ann									bal Fe				159-202 HB
SAE J438(70)	W108	Various qual grades, std grade n, Si, Cr given	0.70-0.85	0.15 max		0.35 max				0.35 max	Mn+Si+Cr<=0.75; bal Fe				
SAE J438(70)	W109	Various qual grades, std grade n, Si, Cr given	0.85-0.95	0.15 max		0.35 max				0.35 max	Mn+Si+Cr<=0.75; bal Fe				
SAE J438(70)	W110	Various qual grades, std grade n, Si, Cr given	0.95-1.10	0.15 max		0.35 max				0.35 max	Mn+Si+Cr<=0.75; bal Fe				
SAE J438(70)	W112	Various qual grades, std grade n, Si, Cr given	1.10-1.30	0.15 max		0.35 max				0.35 max	Mn+Si+Cr<=0.75; bal Fe				

Tool Steel, Water-Hardening, W2

Argentina

Specification	Designation	Notes	C	Cr	Cu	Mn	Ni	P	S	Si	Other	UTS	YS	El	Hard
IAS	IRAM W210	Ann	0.95-1.10	0.15 max	0.20 max	0.10-0.40	0.20 max	0.030 max	0.030 max	0.10-0.40	Mo <=0.10; V 0.15-0.35; W <=0.15; bal Fe				197 HB

China

Specification	Designation	Notes	C	Cr	Cu	Mn	Ni	P	S	Si	Other	UTS	YS	El	Hard
GB 1299	V*	Obs	0.95-1.05		0.25	0.4	0.3			0.4	V 0.2-0.4; bal Fe				

Czech Republic

Specification	Designation	Notes	C	Cr	Cu	Mn	Ni	P	S	Si	Other	UTS	YS	El	Hard
CSN 419314	19314		0.9-1.0	0.45-0.7		1.0-1.3		0.03 max	0.035 max	0.2-0.4	V 0.05-0.2; W 0.45-0.7; bal Fe				
CSN 419356	19356		0.95-1.1	0.15 max		0.2-0.4	0.2 max	0.025 max	0.03 max	0.15-0.35	V 0.1-0.2; bal Fe				

France

Specification	Designation	Notes	C	Cr	Cu	Mn	Ni	P	S	Si	Other	UTS	YS	El	Hard
AFNOR NFA35565(83)	100Cr2	CW, Wear res; soft ann	0.95-1.1	0.4-0.6	0.3 max	0.2-0.4	0.4 max	0.025 max	0.025 max	0.15-0.35	Mo <=0.15; bal Fe				223 HB max
AFNOR NFA35565(83)	Y100C2*	CW, Wear res; soft ann	0.95-1.1	0.4-0.6	0.3 max	0.2-0.4	0.4 max	0.025 max	0.025 max	0.15-0.35	Mo <=0.15; V <=0.1; W <=0.1; bal Fe				223 HB max
AFNOR NFA35565(84)	100C2	Ball & roller bearing	0.95-1.1	0.4-0.6	0.35 max	0.2-0.4	0.4 max	0.03 max	0.025 max	0.15-0.35	Mo <=0.15; bal Fe				
AFNOR NFA35590	Y100C2		0.95-1.1	0.2-0.6	0.25 max	0.2-0.4		0.025 max	0.025 max	0.15-0.35	bal Fe				

UNS numbers and US grades are provided as a means of cross referencing chemically similar alloys. Exchangability is only possible after independent examination of specifications. Tensile properties are minimum or typical as specified. UTS and YS as MPa. El as %. See Appendix for list of abbreviations used in Notes. * indicates obsolete material.

Tool Steel, Water-Hardening, W2 (Continued from previous page)

Specification	Designation	Notes	C	Cr	Cu	Mn	Ni	P	S	Si	Other	UTS	YS	El	Hard
France															
AFNOR NFA35590	Y105V*		0.95-1.09	0.2 max	0.25 max	0.1-0.3	0.25 max	0.02 max	0.02 max	0.1-0.25	V 0.05-0.15; bal Fe				
AFNOR NFA35590	Y120V*		1.2			0.1-0.3				0.1-0.25	V 0.05-0.25; bal Fe				
AFNOR NFA35590	Y75V*		0.75			0.1-0.3				0.1-0.25	V 0.05-0.25; bal Fe				
AFNOR NFA35590	Y90V*		0.9			0.1-0.3				0.1-0.25	V 0.05-0.25; bal Fe				
AFNOR NFA35590(92)	100V2	CW, soft ann	0.95-1.1	0.3 max	0.3 max	0.1-0.35	0.4 max	0.025 max	0.025 max	0.1-0.3	Mo <=0.15; V 0.1-0.3; bal Fe				223 HB max
Germany															
DIN	100V1	Hard tmp to 200C	0.95-1.05			0.15-0.30		0.025 max	0.025 max	0.15-0.25	V 0.10-0.15; bal Fe				63 HRC
DIN	140CrV1	Hard tmp to 200C	1.35-1.45	0.20-0.40		0.25-0.40		0.025 max	0.025 max	0.15-0.35	V 0.10-0.15; bal Fe				62 HRC
DIN	WNr 1.2206	Hard tmp to 200C	1.35-1.45	0.20-0.40		0.25-0.40		0.025 max	0.025 max	0.15-0.35	V 0.10-0.15; bal Fe				62 HRC
DIN	WNr 1.2833	Hard tmp to 200C	0.95-1.05			0.15-0.30		0.025 max	0.025 max	0.15-0.25	V 0.10-0.15; bal Fe				63HRC
Italy															
UNI 2955	V8-C		0.8	0.55		0.5				0.4	V 0.2; bal Fe				
UNI 2955/3(82)	102V2KU		0.95-1.1			0.1-0.35		0.07 max	0.06 max	0.1-0.3	V 0.1-0.3; bal Fe				
Japan															
JIS G3311(88)	SK2M	CR Strp; Razor Blade Cutlery	1.10-1.30	0.30 max	0.25 max	0.50 max	0.25 max	0.030 max	0.030 max	0.35 max	bal Fe				220-310 HV
JIS G4404(83)	SKS43	HR Bar Frg Ann	1.00-1.10	0.20 max	0.25 max	0.30 max	0.25 max	0.030 max	0.030 max	0.25 max	V 0.10-0.25; bal Fe				217 HB max
JIS G4404(83)	SKS44	HR Bar Frg Ann	0.80-0.90	0.20 max	0.25 max	0.30 max	0.25 max	0.030 max	0.030 max	0.25 max	V 0.10-0.25; bal Fe				207 HB max
JIS G4404(83)	SKS8	HR Frg Ann	1.30-1.50	0.20-0.50	0.25 max	0.50 max	0.25 max	0.030 max	0.030 max	0.35 max	bal Fe				217 HB max
Mexico															
NMX-B-082(90)	W2-A	Bar	1.05-1.40	0.15 max	0.25 max	0.10-0.40	0.20 max	0.03 max	0.03 max	0.10-0.40	Mo <=0.10; V 0.15-0.35; W <=0.15; (As+Sn+Sb)<=0.040; (Cu+Ni)<=0.40; B, Ti may be added; bal Fe				
NMX-B-082(90)	W2-B	Bar	0.60-0.95	0.15 max	0.25 max	0.10-0.40	0.20 max	0.03 max	0.03 max	0.10-0.40	Mo <=0.10; V 0.15-0.35; W <=0.15; (As+Sn+Sb)<=0.040; (Cu+Ni)<=0.40; B, Ti may be added; bal Fe				
NMX-B-082(90)	W2-M	Bar	0.95-1.05	0.15 max	0.25 max	0.10-0.40	0.20 max	0.03 max	0.03 max	0.10-0.40	Mo <=0.10; V 0.15-0.35; W <=0.15; (As+Sn+Sb)<=0.040; (Cu+Ni)<=0.40; B, Ti may be added; bal Fe				
Pan America															
COPANT 337	W2	Bar, Frg	0.6-1.4	0.15		0.15-0.4		0.03	0.03	0.1-0.35	bal Fe				
Poland															
PNH85023	NV		0.95-1.1	0.3 max		0.15-0.4	0.4 max	0.030 max	0.030 max	0.15-0.4	V 0.15-0.3; bal Fe				
Spain															
UNE	F.520.H		0.85-0.95	0.2		1.7-2.1		0.03 max	0.03 max	0.13-0.38	V 0.1-0.35; bal Fe				
UNE	F.520.I		0.7-0.8			0.1-0.4		0.03 max	0.03 max	0.13-0.38	V 0.1-0.2; bal Fe				
UNE	F.520.K		1-1.2			0.1-0.4		0.03 max	0.03 max	0.13-0.38	V 0.1-0.2; bal Fe				
UNE	F.520.M		0.75-0.85			0.1-0.4		0.03 max	0.03 max	0.13-0.38	V 0.15-0.25; bal Fe				
UNE	F.520.N		0.9-1.1			0.1-0.4		0.03 max	0.03 max	0.13-0.38	V 0.15-0.25; bal Fe				
UNE 36018/2(94)	102V2	Alloyed	0.95-1.09			0.15-0.35	0.25 max	0.025 max	0.02 max	0.15-0.35	Mo <=0.15; V 0.1-0.2; bal Fe				

UNS numbers and US grades are provided as a means of cross referencing chemically similar alloys. Exchangability is only possible after independent examination of specifications. Tensile properties are minimum or typical as specified. UTS and YS as MPa. El as %. See Appendix for list of abbreviations used in Notes. * indicates obsolete material.

Specification	Designation	Notes	C	Cr	Cu	Mn	Ni	P	S	Si	Other	UTS	YS	El	Hard

Tool Steel, Water-Hardening, W2 (Continued from previous page)

Spain

Specification	Designation	Notes	C	Cr	Cu	Mn	Ni	P	S	Si	Other	UTS	YS	El	Hard
UNE 36018/2(94)	F.5118*	see 102V2	0.95-1.09			0.15-0.35	0.25 max	0.025 max	0.02 max	0.15-0.35	Mo <=0.15; V 0.1-0.2; bal Fe				
UNE 36071(75)	102V		0.95-1.09			0.35 max		0.03 max	0.03 max	0.35 max	V 0.1-0.35; bal Fe				
UNE 36071(75)	120V		1.1-1.29			0.35 max		0.03 max	0.03 max	0.35 max	V 0.1-0.35; bal Fe				
UNE 36071(75)	F.5118*	see 102V	0.95-1.09			0.35 max		0.03 max	0.03 max	0.35 max	V 0.1-0.35; bal Fe				
UNE 36071(75)	F.5124*	see 120V	1.1-1.29			0.35 max		0.03 max	0.03 max	0.35 max	V 0.1-0.35; bal Fe				

Sweden

Specification	Designation	Notes	C	Cr	Cu	Mn	Ni	P	S	Si	Other	UTS	YS	El	Hard
SS 14	(USA W2A)		0.75-0.85			0.2-0.4				0.1-0.3	V 0.05-0.15; bal Fe				
SS 14	(USA W2B)		1.05			0.3				0.2	V 0.13; bal Fe				
SS 14	(USA W2C)		1.05			0.3				0.2	V 0.2; bal Fe				

UK

Specification	Designation	Notes	C	Cr	Cu	Mn	Ni	P	S	Si	Other	UTS	YS	El	Hard
BS 4659	BW2		0.95-1.1	0.15 max		0.35 max	0.2 max			0.3 max	Mo <=0.1; V 0.15-0.35; bal Fe				

USA

Specification	Designation	Notes	C	Cr	Cu	Mn	Ni	P	S	Si	Other	UTS	YS	El	Hard
	AISI W2		0.70-1.50	0.15 max		0.10-0.40		0.030 max	0.030 max	0.10-0.40	Mo <=0.10; V 0.15-0.35; W <=0.15; bal Fe				
	UNS T72302		0.85-1.50	0.15 max	0.20 max	0.10-0.40	0.20 max	0.030 max	0.030 max	0.10-0.40	Mo <=0.10; V 0.15-0.35; W <=0.15; bal Fe				
ASTM A686(92)	W2 13 Grade A	Wrought	1.30-1.50	0.15 max	0.20 max	0.10-0.40	0.20 max	0.030 max	0.030 max	0.10-0.40	Mo <=0.10; V 0.15-0.35; W <=0.15; bal Fe				
ASTM A686(92)	W2 13 Grade C	Wrought	1.30-1.50	0.30 max	0.20 max	0.10-0.40	0.20 max	0.030 max	0.030 max	0.10-0.40	Mo <=0.10; V 0.15-0.35; W <=0.15; bal Fe				
ASTM A686(92)	W2 9 Grade A	Wrought	0.90-1.00	0.15 max	0.20 max	0.10-0.40	0.20 max	0.030 max	0.030 max	0.10-0.40	Mo <=0.10; V 0.15-0.35; W <=0.15; bal Fe				
ASTM A686(92)	W2 9 Grade C	Wrought	0.90-1.00	0.30 max	0.20 max	0.10-0.40	0.20 max	0.030 max	0.030 max	0.10-0.40	Mo <=0.10; V 0.15-0.35; W <=0.15; bal Fe				
FED QQ-T-580C(85)	W2*	Obs; see ASTM A686									bal Fe				
SAE J437(70)	W209	Ann									bal Fe				159-202 HB
SAE J437(70)	W210	Ann									bal Fe				159-202 HB
SAE J437(70)	W310	Ann									bal Fe				159-202 HB
SAE J438(70)	W209	Various qual grades, std grade n, Si, Cr given	0.85-0.95	0.15 max		0.35 max				0.35 max	V 0.15-0.35; Mn+Si+Cr<=0.75; bal Fe				
SAE J438(70)	W210	Various qual grades, std grade n, Si, Cr given	0.95-1.10	0.15 max		0.35 max				0.35 max	V 0.15-0.35; Mn+Si+Cr<=0.75; bal Fe				
SAE J438(70)	W310	Various qual grades, std grade n, Si, Cr given	0.95-1.10	0.15 max		0.35 max				0.35 max	V 0.35-0.50; Mn+Si+Cr<=0.75; bal Fe				

Tool Steel, Water-Hardening, W5

Bulgaria

Specification	Designation	Notes	C	Cr	Cu	Mn	Ni	P	S	Si	Other	UTS	YS	El	Hard
BDS 7938	ChF		1.10-1.25	0.50-0.80	0.30 max	0.20-0.40	0.35 max	0.030 max	0.030 max	0.15-0.30	V 0.07-0.12; bal Fe				

Canada

Specification	Designation	Notes	C	Cr	Cu	Mn	Ni	P	S	Si	Other	UTS	YS	El	Hard
CSA 419418	19418		0.75-0.85	0.45-0.65		0.3-0.5		0.03 max	0.03 max	0.2-0.4	bal Fe				

China

Specification	Designation	Notes	C	Cr	Cu	Mn	Ni	P	S	Si	Other	UTS	YS	El	Hard
GB 1299	CrO6*	Obs	1.3-1.45	0.5-0.7	0.25	0.4	0.3			0.4	bal Fe				
YB 9	GCr8*	Obs	1.05-1.15	0.4-0.7	0.3	0.2-0.4	0.25			0.15-0.35	bal Fe				

UNS numbers and US grades are provided as a means of cross referencing chemically similar alloys. Exchangability is only possible after independent examination of specifications. Tensile properties are minimum or typical as specified. UTS and YS as MPa. El as %. See Appendix for list of abbreviations used in Notes. * indicates obsolete material.

Specification	Designation	Notes	C	Cr	Cu	Mn	Ni	P	S	Si	Other	UTS	YS	El	Hard

Tool Steel, Water-Hardening, W5 (Continued from previous page)

China

Specification	Designation	Notes	C	Cr	Cu	Mn	Ni	P	S	Si	Other	UTS	YS	El	Hard
YB/T 1(81)	GCr6		1.05-1.15	0.40-0.70	0.25 max	0.20-0.40	0.027 max			0.020 max	Mo <=0.08; V 0.15-0.35; W 2.00-2.50; bal Fe				

Czech Republic

Specification	Designation	Notes	C	Cr	Cu	Mn	Ni	P	S	Si	Other	UTS	YS	El	Hard
CSN 419418	19418		0.75-0.85	0.45-0.65		0.3-0.5		0.03 max	0.03 max	0.2-0.4	bal Fe				

France

Specification	Designation	Notes	C	Cr	Cu	Mn	Ni	P	S	Si	Other	UTS	YS	El	Hard
AFNOR NFA35590	Y105C*		1.05	0.25-0.5		0.1-0.4				0.1-0.3	bal Fe				
AFNOR NFA35590(92)	C105E2U	Quen	0.95-1.09	0.2 max	0.25 max	0.1-0.4	0.25 max	0.02 max	0.02 max	0.1-0.3	Mo <=0.15; V <=0.1; W <=0.1; bal Fe				64 HRC
AFNOR NFA35590(92)	C105E2UV1		0.95-1.09	0.2 max	0.25 max	0.1-0.4	0.25 max	0.02 max	0.02 max	0.1-0.3	Mo <=0.15; V 0.05-0.20; bal Fe				
AFNOR NFA35590(92)	C120E3U	Quen	1.1-1.29	0.2 max	0.25 max	0.1-0.4	0.25 max	0.025 max	0.025 max	0.1-0.3	Mo <=0.15; V <=0.1; W <=0.1; bal Fe				64 HRC
AFNOR NFA35590(92)	Y1105	Quen	0.95-1.09	0.2 max	0.25 max	0.1-0.4	0.25 max	0.02 max	0.02 max	0.1-0.3	Mo <=0.15; V <=0.1; W <=0.1; bal Fe				64 HRC
AFNOR NFA35590(92)	Y1105V		0.95-1.09	0.2 max	0.25 max	0.1-0.4	0.25 max	0.02 max	0.02 max	0.1-0.3	Mo <=0.15; V 0.05-0.20; bal Fe				
AFNOR NFA35590(92)	Y2120	Quen	1.1-1.29	0.2 max	0.25 max	0.1-0.4	0.25 max	0.025 max	0.025 max	0.1-0.3	Mo <=0.15; V <=0.1; W <=0.1; bal Fe				64 HRC

Germany

Specification	Designation	Notes	C	Cr	Cu	Mn	Ni	P	S	Si	Other	UTS	YS	El	Hard
DIN	125Cr1	Tmp to 200C	1.20-1.30	0.30-0.40		0.25-0.40		0.030 max	0.030 max	0.15-0.30	bal Fe				63 HRC
DIN	85Cr1	Tmp to 200C	0.80-0.90	0.30-0.45		0.50-0.70		0.035 max	0.035 max	0.30-0.50	bal Fe				61 HRC
DIN	WNr 1.2002	Tmp to 200C	1.20-1.30	0.30-0.40		0.25-0.40		0.030 max	0.030 max	0.15-0.30	bal Fe				63 HRC
DIN	WNr 1.2004	Tmp to 200C	0.80-0.90	0.30-0.45		0.50-0.70		0.035 max	0.035 max	0.30-0.50	bal Fe				61 HRC

India

Specification	Designation	Notes	C	Cr	Cu	Mn	Ni	P	S	Si	Other	UTS	YS	El	Hard
IS 1570	T118Cr45		1.1-1.25	0.3-0.6		0.2-0.35				0.1-0.3	V 0.3; bal Fe				
IS 1570/6(96)	118T3	CW	1.1-1.25	0.2 max	0.25 max	0.4 max	0.25 max	0.035 max	0.035 max	0.1-0.3	Mo <=0.15; bal Fe				
IS 1570/6(96)	TC13	CW	1.1-1.25	0.2 max	0.25 max	0.4 max	0.25 max	0.035 max	0.035 max	0.1-0.3	Mo <=0.15; bal Fe				

Italy

Specification	Designation	Notes	C	Cr	Cu	Mn	Ni	P	S	Si	Other	UTS	YS	El	Hard
UNI 2955	L4B		1.25	0.35		0.3				0.3	bal Fe				
UNI 2955	L4BL		1.17	0.65		0.4				0.3	bal Fe				

Japan

Specification	Designation	Notes	C	Cr	Cu	Mn	Ni	P	S	Si	Other	UTS	YS	El	Hard
JIS G3311(88)	SK4M	CR Strp; Pen Nib String	0.90-1.00	0.30 max	0.25 max	0.50 max	0.25 max	0.030 max	0.030 max	0.35 max	bal Fe				210-300 HV

Mexico

Specification	Designation	Notes	C	Cr	Cu	Mn	Ni	P	S	Si	Other	UTS	YS	El	Hard
NMX-B-082(90)	W4	Bar	0.80-1.20	0.15-0.30	0.25 max	0.10-0.40	0.30 max	0.03 max	0.03 max	0.10-0.40	(As+Sn+Sb)<=0.040; (Cu+Ni)<=0.40; B, Ti may be added; bal Fe				
NMX-B-082(90)	W5	Bar	1.05-1.15	0.40-0.60	0.25 max	0.10-0.40	0.20 max	0.03 max	0.03 max	0.10-0.40	Mo <=0.10; V <=0.10; W <=0.15; (As+Sn+Sb)<=0.040; (Cu+Ni)<=0.40; B, Ti may be added; bal Fe				

Romania

Specification	Designation	Notes	C	Cr	Cu	Mn	Ni	P	S	Si	Other	UTS	YS	El	Hard
STAS 3611(80)	CV06		1.1-1.25	0.5-0.8		0.2-0.4	0.4 max	0.035 max	0.035 max	0.15-0.3	Pb <=0.15; V 0.07-0.12; bal Fe				

Russia

Specification	Designation	Notes	C	Cr	Cu	Mn	Ni	P	S	Si	Other	UTS	YS	El	Hard
GOST 5950	11Ch	Alloyed	1.05-1.15	0.4-0.7	0.3 max	0.4-0.7	0.35 max	0.03 max	0.03 max	0.15-0.35	Mo <=0.2; V 0.15-0.3; W <=0.2; bal Fe				
GOST 5950	11ChF	Alloyed	1.05-1.15	0.4-0.7	0.3 max	0.4-0.7	0.35 max	0.03 max	0.03 max	0.15-0.35	Mo <=0.2; V 0.15-0.3; W <=0.2; bal Fe				
GOST 5950	12Ch1	Alloyed	1.15-1.25	1.3-1.65	0.3 max	0.3-0.6	0.35 max	0.03 max	0.03 max	0.15-0.35	Mo <=0.2; V <=0.15; W <=0.2; bal Fe				
GOST 5950	13Ch	Alloyed	1.25-1.4	0.4-0.7	0.3 max	0.15-0.45	0.35 max	0.03 max	0.03 max	0.1-0.4	Mo <=0.2; V <=0.15; W <=0.2; bal Fe				

Specification	Designation	Notes	C	Cr	Cu	Mn	Ni	P	S	Si	Other	UTS	YS	El	Hard
Tool Steel, Water-Hardening, W5 (Continued from previous page)															
Spain															
UNE	F.120.K		0.95-1.15	0.35-0.65		0.15-0.45		0.04 max	0.04 max	0.13-0.38	bal Fe				
UNE	F.520.Q		1.35-1.55	0.55-0.85		0.15-0.45		0.03 max	0.03 max	0.13-0.38	bal Fe				
UNE	F.520.X		0.52-0.62	0.5-0.8		0.55-0.85		0.03 max	0.03 max	0.13-0.38	bal Fe				
UNE	F.529		0.5-0.6	0.5-1		0.6-0.9		0.03 max	0.03 max	0.2-0.5	bal Fe				
UNE	F.533		1.15-1.3	0.5-1		0.2-0.4		0.03 max	0.03 max	0.15-0.3	bal Fe				
UNE 36018/2(94)	140Cr2	Alloyed	1.3-1.5	0.4-0.7		0.15-0.35	0.25 max	0.025 max	0.02 max	0.15-0.35	Mo <=0.15; bal Fe				
UNE 36018/2(94)	F.5128*	see 140Cr2	1.3-1.5	0.4-0.7		0.15-0.35	0.25 max	0.025 max	0.02 max	0.15-0.35	Mo <=0.15; bal Fe				
UNE 36071(75)	140Cr		1.3-1.5	0.4-0.7		0.35 max		0.03 max	0.03 max	0.35 max	bal Fe				
UNE 36071(75)	F.5128*	see 140Cr	1.3-1.5	0.4-0.7		0.35 max		0.03 max	0.03 max	0.35 max	bal Fe				
USA															
	AISI W5		1.05-1.15	0.40-0.60		0.10-0.40		0.030 max	0.030 max	0.10-0.40	Mo <=0.10; V <=0.10; W <=0.15; bal Fe				
	UNS T72305		1.05-1.15	0.40-0.60	0.20 max	0.10-0.40	0.20 max	0.030 max	0.030 max	0.10-0.40	Mo <=0.10; V <=0.10; W <=0.15; bal Fe				
ASTM A686(92)	W5	Wrought	1.05-1.15	0.40-0.60	0.20 max	0.10-0.40	0.20 max	0.030 max	0.030 max	0.10-0.40	Mo <=0.10; V <=0.10; W <=0.15; bal Fe				

Specification	Designation	Notes	C	Cr	Mn	Mo	P	S	Si	V	Other	UTS	YS	El	Hard

Tool Steel, Unclassified, K23505

France

Specification	Designation	Notes	C	Cr	Mn	Mo	P	S	Si	V	Other	UTS	YS	El	Hard
AFNOR NFA35590(92)	C38E4U	Quen	0.35-0.4	0.35 max	0.5-0.8	0.15 max	0.035 max	0.035 max	0.1-0.4		Cu <=0.35; Ni <=0.35; bal Fe				50 HRC
AFNOR NFA35590(92)	C42E4U	Quen	0.4-0.45	0.35 max	0.5-0.8	0.15 max	0.035 max	0.035 max	0.1-0.4		Cu <=0.35; Ni <=0.35; bal Fe				52 HRC
AFNOR NFA35590(92)	Y338	Quen	0.35-0.4	0.35 max	0.5-0.8	0.15 max	0.035 max	0.035 max	0.1-0.4		Cu <=0.35; Ni <=0.35; bal Fe				50 HRC
AFNOR NFA35590(92)	Y342	Quen	0.4-0.45	0.35 max	0.5-0.8	0.15 max	0.035 max	0.035 max	0.1-0.4		Cu <=0.35; Ni <=0.35; bal Fe				52 HRC

India

Specification	Designation	Notes	C	Cr	Mn	Mo	P	S	Si	V	Other	UTS	YS	El	Hard
IS 1570/6(96)	T30Ni10Cr3Mo6	CW	0.25-0.35	0.5-0.8	0.4-0.7	0.4-0.7	0.035 max	0.035 max	0.1-0.35		Ni 2.25-2.75; bal Fe				
IS 1570/6(96)	T30Ni16Cr5	CW	0.25-0.45	1.1-1.4	0.4-0.7	0.15 max	0.035 max	0.035 max	0.1-0.35		Co <=0.1; Cu <=0.3; Ni 3.9-4.3; Pb <=0.15; W <=0.1; bal Fe				
IS 1570/6(96)	T40Ni10Cr3Mo6	CW	0.35-0.45	0.5-0.8	0.4-0.7	0.4-0.7	0.035 max	0.035 max	0.1-0.35		Ni 2.25-2.75; bal Fe				
IS 1570/6(96)	T40Ni14	CW	0.35-0.45	0.3 max	0.5-0.8	0.15 max	0.035 max	0.035 max	0.1-0.35		Co <=0.1; Cu <=0.3; Ni 3.2-3.6; Pb <=0.15; W <=0.1; bal Fe				
IS 1570/6(96)	TAC16	CW	0.25-0.35	0.5-0.8	0.4-0.7	0.4-0.7	0.035 max	0.035 max	0.1-0.35		Ni 2.25-2.75; bal Fe				
IS 1570/6(96)	TAC17	CW	0.35-0.45	0.5-0.8	0.4-0.7	0.4-0.7	0.035 max	0.035 max	0.1-0.35		Ni 2.25-2.75; bal Fe				

USA

Specification	Designation	Notes	C	Cr	Mn	Mo	P	S	Si	V	Other	UTS	YS	El	Hard
	UNS K23505	Non-tempering	0.32-0.39	0.55-0.75	0.35-0.50	0.50-0.70	0.025 max	0.025 max	0.50-0.70		Cu 0.50-0.70; bal Fe				

Tool Steel, Unclassified, No equivalents identified

Czech Republic

Specification	Designation	Notes	C	Cr	Mn	Mo	P	S	Si	V	Other	UTS	YS	El	Hard
CSN 419103	19103		0.5-0.6	0.25 max	0.55-0.85		0.035 max	0.035 max	0.15-0.35		Ni <=0.25; bal Fe				
CSN 419125	19125		0.6-0.7	0.3 max	0.45-0.65		0.04 max	0.04 max	0.2-0.35		bal Fe				
CSN 419340	19340		0.52-0.62	0.3 max	0.6-0.9		0.03 max	0.03 max	1.4-1.8		bal Fe				
CSN 419419	19419		0.7-0.8	0.45-0.65	0.3-0.5		0.03 max	0.03 max	0.2-0.4	0.08-0.2	bal Fe				
CSN 419420	19420		1.4-1.55	0.5-0.8	0.15-0.35		0.3 max	0.035 max	0.15-0.35		Ni <=0.3; bal Fe				
CSN 419421	19421		1.1-1.25	0.9-1.2	0.15-0.35		0.3 max	0.035 max	0.15-0.35	0.07-0.15	Ni <=0.35; bal Fe				
CSN 419422	19422		1.35-1.55	1.5-1.8	0.4-0.7		0.03 max	0.035 max	0.2-0.4	0.1-0.2	Ni <=0.35; bal Fe				
CSN 419426	19426		0.75-0.9	1.55-1.9	0.2-0.45		0.03 max	0.035 max	0.2-0.4	0.07-0.17	bal Fe				
CSN 419434	19434		0.18-0.28	12-14	0.3-0.7		0.03 max	0.03 max	0.7 max		bal Fe				
CSN 419452	19452		0.55-0.65	0.7-1	0.6-0.9		0.03 max	0.035 max	1.5-1.9		Ni <=0.35; bal Fe				
CSN 419474	19474		0.33-0.43	3.4-4.2	1.2-1.5		0.03 max	0.035 max	0.3-0.7	0.1-0.2	Ni <=0.35; bal Fe				
CSN 419486	19486		0.14-0.19	0.8-1.1	1.1-1.4		0.03 max	0.03 max	0.17-0.37		bal Fe				
CSN 419487	19487		0.17-0.22	1-1.3	1-1.3		0.03 max	0.03 max	0.17-0.37		bal Fe				
CSN 419501	19501		0.9-1.05	0.9-1.2	0.2-0.5	0.15-0.3	0.03 max	0.03 max	0.15-0.4		bal Fe				
CSN 419512	19512		0.35-0.45	2.2-2.7	0.6-0.9	0.45-0.75	0.03 max	0.03 max	0.15-0.4	0.35-0.6	bal Fe				
CSN 419561	19561		0.38-0.48	3.5-4.5	0.2-0.5	4.5-5.5	0.03 max	0.03 max	0.15-0.45	0.8-1.2	W 4.8-6.2; bal Fe				
CSN 419564	19564		0.27-0.37	2.3-2.8	0.2-0.5	0.8-1.3	0.03 max	0.03 max	0.15-0.45	0.6-0.9	W 3.2-4.2; bal Fe				
CSN 419569	19569		0.58-0.68	4.5-5.5	0.25-0.55	0.8-1.2	0.03 max	0.035 max	0.7-1.1	0.2-0.4	W <=0.6; bal Fe				

UNS numbers and US grades are provided as a means of cross referencing chemically similar alloys. Exchangability is only possible after independent examination of specifications. Tensile properties are minimum or typical as specified. UTS and YS as MPa. El as %. See Appendix for list of abbreviations used in Notes. * indicates obsolete material.

Specification	Designation	Notes	C	Cr	Mn	Mo	P	S	Si	V	Other	UTS	YS	El	Hard

Tool Steel, Unclassified, No equivalents identified

Czech Republic

Specification	Designation	Notes	C	Cr	Mn	Mo	P	S	Si	V	Other	UTS	YS	El	Hard
CSN 419572	19572		1.45-1.7	11-12.5	0.2-0.45	0.4-0.6	0.03 max	0.035 max	0.2-0.45	0.15-0.3	bal Fe				
CSN 419581	19581		2.1-2.35	4.8-5.8	0.4-0.7	0.4-0.8	0.03 max	0.035 max	0.3-0.6	4-4.8	W 1-1.4; bal Fe				
CSN 419614	19614 *		0.5-0.6	0.6-0.9	0.4-0.7		0.03 max	0.035 max	0.2-0.4	0.07-0.15	Ni 2.3-2.8; bal Fe				
CSN 419642	19642		0.3-0.4	0.7-1.1	0.4-0.7	0.2-0.4	0.03 max	0.03 max	0.15-0.4		Ni 4.4-5.2; bal Fe				
CSN 419655	19655		0.35-0.45	1.4-1.8	0.4-0.7	0.2-0.4	0.03 max	0.035 max	0.2-0.4		Ni 3.8-4.6; W 0.4-0.8; bal Fe				
CSN 419674	19674		0.24-0.34	0.5-0.9	0.2-0.45	0.3-0.6	0.03 max	0.03 max	0.3-0.6	0.15-0.3	Ni 1.1-1.6; bal Fe				
CSN 419675	19675		0.24-0.34	0.5-0.9	0.2-0.45	0.6-1	0.03 max	0.03 max	0.3-0.6	0.2-0.4	Ni 2.2-2.7; bal Fe				
CSN 419678	19678		0.24-0.34	0.5-0.9	0.2-0.45	1.1-1.7	0.03 max	0.03 max	0.15-0.4		Ni 3.9-4.6; bal Fe				
CSN 419680	19680		0.3-0.4	11.5-13.5	0.9 max		0.04 max	0.03 max	1.2-1.8	1.-1.5	Ni 11.5-13.5; W 2.2-3; bal Fe				
CSN 419714	19714		1.25-1.45	0.15-0.35	0.15-0.4		0.03 max	0.035 max	0.15-0.35	0.1-0.2	Ni <=0.35; W 4.5-5.5; bal Fe				
CSN 419720	19720		0.25-0.35	2.1-2.6	0.2-0.5		0.03 max	0.03 max	0.15-0.45	0.45-0.65	W 3.8-4.5; bal Fe				
CSN 419740	19740		0.25-0.35	1.0-1.4	0.25-0.5		0.03 max	0.03 max	0.9-1.4	0.1-0.25	W 3.3-4.0; bal Fe				

France

Specification	Designation	Notes	C	Cr	Mn	Mo	P	S	Si	V	Other	UTS	YS	El	Hard
AFNOR NFA35590(92)	Y35NC15		0.32-0.38	1.4-1.8	0.3-0.6	0.15 max	0.025 max	0.025 max	0.1-0.4		Ni 3.5-4; bal Fe				255 HB max

Hungary

Specification	Designation	Notes	C	Cr	Mn	Mo	P	S	Si	V	Other	UTS	YS	El	Hard
MSZ 4352(72)	K11	Alloyed; CW	1.45-1.65	11-13	0.15-0.4	0.4-0.6	0.03 max	0.03 max	0.15-0.35	0.15-0.3	W 0.4-0.7; bal Fe				
MSZ 4352(72)	K3	Alloyed; CW	0.8-0.9	1.6-2	0.15-0.4		0.03 max	0.03 max	0.15-0.35	0.15-0.3	bal Fe				
MSZ 4352(84)	K6	Q/T; Alloyed; CW	1.3-1.5	0.2-0.5	0.15-0.4	0.2 max	0.03 max	0.03 max	0.1-0.4	0.15 max	Ni <=0.35; W <=0.3; bal Fe				64 HRC
MSZ 4352(84)	K6	Soft ann; CW	1.3-1.5	0.2-0.5	0.15-0.4	0.2 max	0.03 max	0.03 max	0.1-0.4	0.15 max	Ni <=0.35; W <=0.3; bal Fe				229 HB max

India

Specification	Designation	Notes	C	Cr	Mn	Mo	P	S	Si	V	Other	UTS	YS	El	Hard
IS 1570/6(96)	103T3	CW	0.95-1.1	0.2 max	0.4 max	0.15 max	0.035 max	0.035 max	0.1-0.3		Cu <=0.25; Ni <=0.25; bal Fe				
IS 1570/6(96)	133T3	CW	1.25-1.4	0.2 max	0.4 max	0.15 max	0.035 max	0.035 max	0.1-0.3		Cu <=0.25; Ni <=0.25; bal Fe				
IS 1570/6(96)	60T6	CW	0.55-0.65	0.2 max	0.5-0.8	0.15 max	0.035 max	0.035 max	0.1-0.35		Cu <=0.25; Ni <=0.25; bal Fe				
IS 1570/6(96)	65T6	CW	0.6-0.7	0.2 max	0.5-0.8	0.15 max	0.035 max	0.035 max	0.1-0.35		Cu <=0.25; Ni <=0.25; bal Fe				
IS 1570/6(96)	70T3	CW	0.65-0.75	0.2 max	0.4 max	0.15 max	0.035 max	0.035 max	0.1-0.3		Cu <=0.25; Ni <=0.25; bal Fe				
IS 1570/6(96)	70T6	CW	0.65-0.75	0.2 max	0.5-0.8	0.15 max	0.035 max	0.035 max	0.1-0.35		Cu <=0.25; Ni <=0.25; bal Fe				
IS 1570/6(96)	75T6	CW	0.7-0.8	0.2 max	0.5-0.8	0.15 max	0.035 max	0.035 max	0.1-0.35		Cu <=0.25; Ni <=0.25; bal Fe				
IS 1570/6(96)	80T3	CW	0.75-0.85	0.2 max	0.4 max	0.15 max	0.035 max	0.035 max	0.1-0.3		Cu <=0.25; Ni <=0.25; bal Fe				
IS 1570/6(96)	80T6	CW	0.75-0.85	0.2 max	0.5-0.8	0.15 max	0.035 max	0.035 max	0.1-0.35		Cu <=0.25; Ni <=0.25; bal Fe				
IS 1570/6(96)	85T6	CW	0.8-0.9	0.2 max	0.5-0.8	0.15 max	0.035 max	0.035 max	0.1-0.35		Cu <=0.25; Ni <=0.25; bal Fe				
IS 1570/6(96)	90T3	CW	0.85-0.95	0.2 max	0.4 max	0.15 max	0.035 max	0.035 max	0.1-0.3		Cu <=0.25; Ni <=0.25; bal Fe				
IS 1570/6(96)	T103V2	CW	0.95-1.1	0.3 max	0.4 max	0.15 max	0.035 max	0.035 max	0.1-0.35	0.15-0.3	Ni <=0.4; bal Fe				
IS 1570/6(96)	T105Cr5	CW	0.9-1.2	1-1.6	0.2-0.4	0.15 max	0.035 max	0.035 max	0.1-0.35		Ni <=0.4; bal Fe				
IS 1570/6(96)	T135Cr2	CW	1.25-1.4	0.3-0.6	0.4 max	0.15 max	0.035 max	0.035 max	0.1-0.35	0.3 max	Ni <=0.4; bal Fe				
IS 1570/6(96)	T140W15Cr2	CW	1.3-1.5	0.3-0.7	0.25-0.5	0.15 max	0.035 max	0.035 max	0.1-0.35		Ni <=0.4; W 3.5-4.2; bal Fe				

Specification	Designation	Notes	C	Cr	Mn	Mo	P	S	Si	V	Other	UTS	YS	El	Hard

Tool Steel, Unclassified, No equivalents identified

India

Specification	Designation	Notes	C	Cr	Mn	Mo	P	S	Si	V	Other	UTS	YS	El	Hard
IS 1570/6(96)	T55Cr3	CW	0.5-0.6	0.6-0.8	0.6-0.8	0.15 max	0.035 max	0.035 max	0.1-0.35		Ni <=0.4; bal Fe				
IS 1570/6(96)	T55Cr3V2	CW	0.5-0.6	0.6-0.8	0.6-0.8	0.15 max	0.035 max	0.035 max	0.1-0.35	0.1-0.2	Ni <=0.4; bal Fe				
IS 1570/6(96)	T60Ni5	CW	0.55-0.65	0.3 max	0.5-0.8	0.15 max	0.035 max	0.035 max	0.1-0.35		Ni 1-1.5; bal Fe				
IS 1570/6(96)	T80V2	CW	0.75-0.85	0.3 max	0.4 max	0.15 max	0.035 max	0.035 max	0.1-0.35	0.15-0.3	Ni <=0.4; bal Fe				
IS 1570/6(96)	T90V2	CW	0.85-0.95	0.3 max	0.4 max	0.15 max	0.035 max	0.035 max	0.1-0.35	0.15-0.3	Ni <=0.4; bal Fe				
IS 1570/6(96)	TAC2	CW	0.85-0.95	0.3 max	0.4 max	0.15 max	0.035 max	0.035 max	0.1-0.35	0.15-0.3	Ni <=0.4; bal Fe				
IS 1570/6(96)	TAC23	CW	0.5-0.6	0.6-0.8	0.6-0.8	0.15 max	0.035 max	0.035 max	0.1-0.35		Ni <=0.4; bal Fe				
IS 1570/6(96)	TAC25	CW	0.5-0.6	0.6-0.8	0.6-0.8	0.15 max	0.035 max	0.035 max	0.1-0.35	0.1-0.2	Ni <=0.4; bal Fe				
IS 1570/6(96)	TAC3	CW	0.95-1.1	0.3 max	0.4 max	0.15 max	0.035 max	0.035 max	0.1-0.35	0.15-0.3	Ni <=0.4; bal Fe				
IS 1570/6(96)	TAC5	CW	1.25-1.4	0.3-0.6	0.4 max	0.15 max	0.035 max	0.035 max	0.1-0.35		Ni <=0.4; bal Fe				
IS 1570/6(96)	TAC6	CW	0.9-1.2	1-1.6	0.2-0.4	0.15 max	0.035 max	0.035 max	0.1-0.35		Ni <=0.4; bal Fe				
IS 1570/6(96)	TAC8	CW	1.3-1.5	0.3-0.7	0.25-0.5	0.15 max	0.035 max	0.035 max	0.1-0.35		Ni <=0.4; W 3.5-4.2; bal Fe				
IS 1570/6(96)	TAC9	CW	0.55-0.65	0.3 max	0.5-0.8	0.15 max	0.035 max	0.035 max	0.1-0.35		Ni 1.-1.5; bal Fe				
IS 1570/6(96)	TC10	CW	0.75-0.85	0.2 max	0.4 max	0.15 max	0.035 max	0.035 max	0.1-0.3		Cu <=0.25; Ni <=0.25; bal Fe				
IS 1570/6(96)	TC11	CW	0.85-0.95	0.2 max	0.4 max	0.15 max	0.035 max	0.035 max	0.1-0.3		Cu <=0.25; Ni <=0.25; bal Fe				
IS 1570/6(96)	TC12	CW	0.95-1.1	0.2 max	0.4 max	0.15 max	0.035 max	0.035 max	0.1-0.3		Cu <=0.25; Ni <=0.25; bal Fe				
IS 1570/6(96)	TC14	CW	1.25-1.4	0.2 max	0.4 max	0.15 max	0.035 max	0.035 max	0.1-0.3		Cu <=0.25; Ni <=0.25; bal Fe				
IS 1570/6(96)	TC3	CW	0.55-0.65	0.2 max	0.5-0.8	0.15 max	0.035 max	0.035 max	0.1-0.35		Cu <=0.25; Ni <=0.25; bal Fe				
IS 1570/6(96)	TC4	CW	0.6-0.7	0.2 max	0.5-0.8	0.15 max	0.035 max	0.035 max	0.1-0.35		Cu <=0.25; Ni <=0.25; bal Fe				
IS 1570/6(96)	TC5	CW	0.65-0.75	0.2 max	0.5-0.8	0.15 max	0.035 max	0.035 max	0.1-0.35		Cu <=0.25; Ni <=0.25; bal Fe				
IS 1570/6(96)	TC6	CW	0.7-0.8	0.2 max	0.5-0.8	0.15 max	0.035 max	0.035 max	0.1-0.35		Cu <=0.25; Ni <=0.25; bal Fe				
IS 1570/6(96)	TC7	CW	0.75-0.85	0.2 max	0.5-0.8	0.15 max	0.035 max	0.035 max	0.1-0.35		Cu <=0.25; Ni <=0.25; bal Fe				
IS 1570/6(96)	TC8	CW	0.8-0.9	0.2 max	0.5-0.8	0.15 max	0.035 max	0.035 max	0.1-0.35		Cu <=0.25; Ni <=0.25; bal Fe				
IS 1570/6(96)	TC9	CW	0.65-0.75	0.2 max	0.4 max	0.15 max	0.035 max	0.035 max	0.1-0.3		Cu <=0.25; Ni <=0.25; bal Fe				

Italy

Specification	Designation	Notes	C	Cr	Mn	Mo	P	S	Si	V	Other	UTS	YS	El	Hard
UNI 2955/3(82)	5CrMo16KU		0.07 max	3.5-4.5	0.05-0.25	0.4-0.6	0.07 max	0.06 max	0.05-0.25		bal Fe				
UNI 2955/3(82)	7CrNiMo8KU		0.1 max	1.8-2.1	0.2-0.5	0.1-0.3	0.07 max	0.06 max	0.1-0.4		Ni 0.4-0.6; bal Fe				
UNI 2955/3(82)	X21Cr13KU		0.16-0.25	12-14	1 max		0.07 max	0.06 max	1 max		Ni <=1; bal Fe				
UNI 2955/3(82)	X31Cr13KU		0.26-0.35	12-14	1 max		0.07 max	0.06 max	1 max		Ni <=1; bal Fe				
UNI 2955/3(82)	X38CrMo161KU		0.33-0.43	15-17	1 max	1-1.5	0.07 max	0.06 max	1 max		Ni <=1; bal Fe				
UNI 2955/3(82)	X41Cr13KU		0.36-0.45	12.5-14.5	1 max		0.07 max	0.06 max	1 max		Ni <=1; bal Fe				
UNI 2955/3(82)	X5CrMo51KU		0.07 max	4.5-5.5	0.05-0.25	0.9-1.2	0.07 max	0.06 max	0.05-0.25		bal Fe				
UNI 2955/5(82)	HS3-3-2		0.95-1.03	3.8-4.5	0.4 max	2.5-2.8	0.07 max	0.06 max	0.5 max	2.2-2.5	Co <=1; W 2.7-3; bal Fe				

UNS numbers and US grades are provided as a means of cross referencing chemically similar alloys. Exchangability is only possible after independent examination of specifications. Tensile properties are minimum or typical as specified. UTS and YS as MPa. El as %. See Appendix for list of abbreviations used in Notes. * indicates obsolete material.

Tool Steel, Unclassified, No equivalents identified

Specification	Designation	Notes	C	Cr	Mn	Mo	P	S	Si	V	Other	UTS	YS	El	Hard
Japan															
JIS G4404(83)	SKS4	HR Frg Ann	0.45-0.55	0.50-1.00	0.50 max		0.030 max	0.030 max	0.35 max		Cu <=0.25; Ni <=0.25; W 0.50-1.00; bal Fe				201 HB max
JIS G4410(84)	SKC24	Rod or Bit; FF	0.33-0.43	0.30-0.70	0.30-1.00	0.15-0.40	0.030 max	0.030 max	0.15-0.35		Cu <=0.25; Ni 2.50-3.50; bal Fe				269-352 HB
Romania															
STAS 3611(66)	CVW50		1.25-1.45	0.4-0.7	0.15-0.4		0.03 max	0.025 max	0.15-0.35	0.15-0.3	Ni <=0.35; Pb <=0.15; W 4.5-5.5; bal Fe				
STAS 3611(66)	VSCW45		0.25-0.35	1-1.3	0.2-0.5		0.03 max	0.025 max	0.6-0.9	0.15-0.3	Ni <=0.35; Pb <=0.15; W 4-5; bal Fe				
STAS 3611(80)	MCW14		0.9-1.05	0.95-1.25	0.8-1.1		0.03 max	0.025 max	0.15-0.35		Ni <=0.35; Pb <=0.15; W 1.2-1.6; bal Fe				
STAS 7382(88)	Rp9		0.95-1.03	3.8-4.5	0.4 max	2.5-2.8	0.03 max	0.03 max	0.45 max	2.2-2.5	Ni <=0.4; Pb <=0.15; W 2.7-3; bal Fe				
Spain															
UNE 36018/2(94)	F.5263*	see X40Cr13	0.36-0.45	12.5-14.5	1 max	0.15 max	0.03 max	0.02 max	1 max		Ni <=1; bal Fe				
UNE 36018/2(94)	F.5267*	see X38CrMo16	0.33-0.43	15-17	1 max	1-1.5	0.03 max	0.02 max	1 max		Ni <=1; bal Fe				
UNE 36018/2(94)	X38CrMo16	Alloyed; CW	0.33-0.43	15-17	1 max	1-1.5	0.03 max	0.02 max	1 max		Ni <=1; bal Fe				
UNE 36018/2(94)	X40Cr13	Alloyed; CW	0.36-0.45	12.5-14.5	1 max	0.15 max	0.03 max	0.02 max	1 max		Ni <=1; bal Fe				
UNE 36072/1(75)	5CrMo16	Alloyed; CW	0.07 max	3.5-4.5	0.1-0.3	0.4-0.6	0.03 max	0.03 max	0.1-0.3		bal Fe				
UNE 36072/1(75)	7CrMoNi8	Alloyed; CW	0.1 max	1.8-2.1	0.2-0.5	1.1-1.3	0.03 max	0.03 max	0.1-0.4		Ni 0.4-0.6; bal Fe				
UNE 36072/1(75)	F.5253*	see 5CrMo16	0.07 max	3.5-4.5	0.1-0.3	0.4-0.6	0.03 max	0.03 max	0.1-0.3		bal Fe				
UNE 36072/1(75)	F.5257*	see 7CrMoNi8	0.1 max	1.8-2.1	0.2-0.5	1.1-1.3	0.03 max	0.03 max	0.1-0.4		Ni 0.4-0.6; bal Fe				
UNE 36072/1(75)	F.5261*	see X20Cr13	0.16-0.25	12-14	1 max	0.15 max	0.03 max	0.03 max	1 max		Ni <=1; bal Fe				
UNE 36072/1(75)	X20Cr13	Alloyed; CW	0.16-0.25	12-14	1 max	0.15 max	0.03 max	0.03 max	1 max		Ni <=1; bal Fe				
Sweden															
SS 142727	2727	High-Speed	2.2-2.4	3.5-4.5	0.2-0.5	6.7-7.3	0.05 max	0.17 max	0.2-0.7	6.3-6.7	Co 10-11; W 6.2-6.8; bal Fe				
SS 142737	2737	High-Speed	1.35-1.45	3.7-4.4	0.2-0.4	3.4-3.8	0.03 max	0.03 max	0.25-0.5	3.2-3.6	Co 10.5-11.5; W 8.4-9.1; bal Fe				

Argentina

Instituto Argentino de Normalizacion (IRAM)
Peru 552/556
1068 Buenos Aires
Argentina

Phone: 54 11 4345 6606
Fax: 54 11 4345 3468
Email: iram4@sminter.com.ar
Internet: www.iram.com.ar

IRAM is the Argentine Standardization Institute and is authorized by the government to deal, both in national and international ambits, with all the affairs related to standardization and quality control certification. IRAM is a private society, and a member of ISO (International Organization for Standardization) and COPANT (Pan American Standards Commission).

IRAM has delegated some responsibilities for iron and steel standards to the Instituto Argentino de Siderurgia (IAS). The contact information for IAS is Carlos Maria Della Paolera 226, 1001 Buenos Aires, Argentina, Tel: 54 11 4311 6321, Fax: 54 11 4311 4016.

Argentine standards will be preceded by IRAM. Some standards in Argentina will be regional for Pan America and will use the prefix COPANT. The prefix will be followed by an alphanumeric code.

Example: IRAM W210; IRAM M7; IRAM 5121; IRAM 12 L 14; IRAM M35

Australia

Standards Australia (SAA)
P.O. Box 1055
Strathfield NSW 2135
Australia

Phone: 61(02) 9746 4700
Fax: 61 (02) 9746 8450
Email: mail@standards.com.au
Internet: www.standards.com.au

Standards Australia (SAA) was founded in 1922 and changed to its current name in 1988. It issues standards used primarily by firms doing business in Australia and the southwest Pacific area. There are currently around 6000 Australian standards, maintained by approximately 8500 voluntary experts serving on a total of 1600 technical committees, and backed up by a full-time staff of 310.

Standards Australia represents Australia on two international standardizing bodies, International Standards Organization (IS0) and International Electrotechnical Commission (IEC). Standards Australia maintains a strong relationship with Standards New Zealand and the two organizations have a formal agreement on preparing and publishing joint Standards (AS/NZS).

Example: AS 1446 X1320H; AS 1442 12L14; AS 2074 302; AS/NZS 1594 HK10B55*
*AS/NZS 1594 HK10B55 is a joint Australian/New Zealand Standard.

Austria

Osterreichisches Normungsinstitut (ON)
(Austrian Standards Institute)
Heinestrasse 38
A-1021 Wien
Austria

Phone: 43 1 213 00-627
Fax: 43 1 213 00-360
Email: infostelle@on-norm.at
Internet: www.on-norm.at

The Austrian Standards Institute (ON), founded in 1920, creates and publishes Austrian standards. The organization also recommends foreign standards for use in Austria. The ON represents Austria to the European Committee for Standardization (CEN) and the International Organization for Standardization (ISO). The Austrian standards begin with ONORM if just an Austrian standard or ONORMEN if the standard has been adopted by CEN and its members.

Example: ONORM M3124; ONORM M3108

Belgium

Institut Belge de Normalisation (IBN)
(Belgian Standardization Institute)
Avenue de la Brabanconne 29
B-1040 Bruxelles
Belgium

Phone: 32 2 738 01 11
Fax: 32 2 733 42 64
Email:
Internet:

Created in 1946, the IBN consists of approximately 5900 members, both individuals and businesses. The designations are prefixed with the letters NBN. Belgian designations are different for nonalloyed and alloyed steel. For nonalloyed steel (including carbon steel), the conventional designation usually consists of a letter, a number code, and a possible variable third part. There are three different criteria for classification: mechanical characteristics, technological characteristics, and chemical composition. For alloyed steels, the designation system varies for heavily alloyed (above 5%) and slightly alloyed steel.

Example: NBN 235-05; NBN A21-101

Brazil

Associacao Brasileira de Normas Tecnicas (ABNT)
(Brazilian Association for Technical Standards)
Avenida Treze de Maio, 13, 27o andar - Centro
20003-900 Rio de Janeiro-RJ
Brazil

Phone: 55 (021) 210 3122
Fax: 55 (021) 220 6436
          55 (021) 220 1762
Email: abnt@abnt.org.br
Internet: www.abnt.org.br

The Brazilian Association of Technical Standards (ABNT) issues national standards. These designations may begin with upper case letters NBR or ABNT. Projects are coded as Committee: Subcommittee. Working Group-Sequential number.

Example: ABNT EA9; ABNT EM1

Bulgaria

State Agency for Standardization and Metrology (BDS)
21, 6th September Str.
1000 Sofia
Bulgaria

Phone: 359 2 989 84 88
Fax: 359 2 986 17 07
Email: csm@techno-link.com
Internet:

Bulgarian standards are issued by the State Agency for Standardization and Metrology. The designation begins with the upper case letters BDS and are followed by the standard's numerical code.

Example: BDS 5785; BDS 10786; BDS 14351

Canada

CSA International (CSA)

178 Rexdale Blvd.

Etobicoke (Toronto) ON M9W 1R3

Canada

Phone: 416 747 4000

Fax: 416 747 4149

Email: certinfo@csa.ca
          sales@csa.ca

Internet: www.csa-international.org

The Canadian Standards Association has recently changed its name to CSA International. The organization is an independent, not-for-profit organization supported by more than 8,000 members. The members sit on more than 1200 committees that give input on standards for thousands of products. They have a network of offices, partners, and strategic alliances in Canada, the U.S., and around the world. Established in 1919, CSA International is involved in the field of standards development and the application of these standards through product certification, management systems registration, and information products. CSA International represents Canada to several international standards organizations including ISO.

All Canadian standards are preceded by the upper case letters CSA. The standard or designation then follows.

Examples: CSA 110.3; CSA 419824; CSA B193; CSA G110.3

China

China State Bureau of Quality and Technical Supervision (CSBTS)

4, Zhichun Road

Haidian District,
P.O. Box 8010

Beijing 100088

China

Phone: 86 10 6 203 24 24

Fax: 86 10 6 203 37 37

Email: csbts@mail.csbts.cn.net

Internet: www.csbts.cn.net

The China State Bureau of Quality and Technical Supervision is responsible for materials standards in China. The standards are all preceded by the upper case letters GB, JB, and YB. A /T following the letters means that the standard is pending. The numeric standard identification then follows with a dash and then the year the standard was approved.

Example: GB 700; GB 8492; GB/T 13796; JB/ZQ 4299; YB 9; YB/T 036.4

Czech Republic

Cesky Normalizacni Institut (CSNI)
(Czech Standard Institute)

Biskupsky dvur 5

113 47 Praha

Czech Republic

Phone: 42 (02) 21 802 111

Fax: 42 (02) 21 802 301

Email: csni@login.cz

Internet: www.csni.cz

The Czech Standards Institute (CSNI) is concerned with standardization, metrology, testing, certification, and accreditation. The CSNI is a member of the ISO, IEC, CEN and CENELEC.

Czech standards are arranged according to classes and subgroups by a six-digit reference number. All standards are preceded by CSN.

Example: CSN 422724; CSN 411264

Denmark

Dansk Standard (DS)
Danish Standardization Commission
Kollegievej 6
Charlottenlund 2920
Denmark

Phone: 45 39 96 61 01
Fax: 45 39 96 61 02
Email: dansk.standard@ds.dk
Internet: www.ds.dk

The Danish Standards Association (DS) was founded in 1926 and is involved in the standardization of all fields except telecommunications. DS is accredited to certify quality assurance systems according to ISO 9000 series.

Example: DS 11302; DS 12011; DS/EN 10025; DS/ISO 3798; DS/IEC 141-1

Europe

European Committee for Iron and Steel Standardization (ECISS)
c/o Comite Europeen de Normalisation (CEN)
(European Committee for Standardization)
36 rue de Stassart
B-1050 Brussels
Belgium

Phone: 32 2 550 08 11
Fax: 32 2 550 08 19
Email: infodesk@cenorm.be
Internet: www.cenorm.be

The European Committee for Iron and Steel Standardization (ECISS) issues standards for European steel. This effort is carried out in close cooperation with the Eruopean Committee for Standardization (CEN) which produces European Standards (EN). CEN is an association of the national standards organizations of countries of the European Union and of the European Free Trade Association. CEN's principal responsibility is to prepare and issue European Standards (EN) established on the principle of consensus and adopted by the votes of weighted majority. Adopted standards must be implemented in their entirety as national standards by each member country, regardless of the way in which the national member voted, and any conflicting national standards must be withdrawn.

The identification of European Standards in each member country begin with the reference letters of the country's national standards body, i.e. BS for BSI in the United Kingdom, DIN for DIN in Germany, NF for AFNOR in France, etc. and then is followed by the initials EN and a number of up to five digits, ie. BS EN 10025, DIN EN 10025, or NF EN 10025. An EN Standard may contain more than one document or parts due to a particular characteristic of the steel product so the numbering may also contain a suffix that further identifies the material, ie. EN 10028-1.The Standards are prepared in 3 languages: English, French, and German and may be translated by each country after adoption.

The CEN Central Secretariat in Brussels provides the focal point for the coordination and management of the organization and its processes shared between the National Members and the Associates which are industrial trade organizations.

The countries that are National Members are: Austria (ON), Belgium (IBN/BIN), Czech Republic (CSNI), Denmark (DS), Finland (SFS), France (AFNOR), Germany (DIN), Greece (ELOT), Iceland (STRI), Ireland (NSAI), Italy (UNI), Luxembourg (SEE), Netherlands (NNI), Norway (NSF), Portugal (IPQ), Spain (AENOR), Sweden (SIS), Switzerland (SNV), United Kingdom (BSI).

Affiliates are the countries from Central and Eastern Europe that are expected to join CEN as full members in the future: Albania (DSC), Bulgaria (CSM), Croatia (DZNM), Cyprus (CYS), Estonia (EVS), Hungary (MSZH), Latvia (Department of Quality Management and Structure Development), Lithuania (LST), Malta (Malta Standardization Authority), Poland (PKN), Romania (IRS), Slovakia (UNMS), Slovenia (SMIS), Turkey (TSE).

Corresponding Organizations include: Egypt, Egyptian Organization for Standardization and Quality Control (EOS); South Africa, South Africa Bureau of Standards (SABS); Ukraine, State Committee of Ukraine for Standardization, Metrology and Certification (DSTU); Yugoslavia, Federal Institution for Standardization (SZS).

Of particular interest to readers of this book may be EN 10025-1991 on classification and definition of steel grades; and EN 10027-1-1992 "Designation System for Steel--Part 1: Steel Names, Principle Symbols" and EN 10027-2-1992 "Designation System for Steel--Part 2: Numerical System"

Example: EN 10095; BS EN 10088-1:1995; DIN EN 10028-1

Finland

Finnish Standard Association (SFS)
P O Box 116
Fin-00241 Helsinki
Finland

Phone: 358 0 149 9331
Fax: 358 9 146 4925
Email: info@sfs.fi or sales@sfs.fi
Internet: www.sfs.fi

The Finnish Standards Association (SFS) is an independent, non-profit making organization co-operating with trade federations and industry, research institutes, labor market organizations, consumer organizations, and governmental and local authorities. Members of SFS include professional, commercial and industrial organizations, and the state of Finland represented by the ministries.

SFS Standards are voluntary documents drawn by technical committees of SFS. There are over 10,000 SFS Standards. SFS and its standards-writing bodies are members of the European standards organizations CEN, CENELEC and ETSI. The membership implies obligation to implement all European standards as SFS standards within six months of approval. Finnish standards and designations were preceded by the letters SFS or SFSEN if European standard has been adopted.

Example: SFS 1205; SFS 200-E; SFS M56

France

Association Francaise de Normalisation (AFNOR)
(French Association for Standardization)
Tour Europe
92049 Paris la Defense Cedex
France

Phone: 33 1 42 91 55 55
Fax: 33 1 42 91 56 56
Email: info.normes@email.afnor.fr
Internet: www.afnor.fr

The Association Francaise de Normalisation (AFNOR) is a non-profit organization founded in 1926. Of its nearly 15,600 standards, more than 1,000 relate to metallurgy and are used widely. AFNOR is the French representative to the European Committee for Standardization (CEN) and to International Standards Organization (ISO). AFNOR standards usually begin with NF and if a CEN standard has been adopted the prefix is NFEN.

Example: NFA 32055-85; NFA 35501

France

Delegation Generale pour L'Armement (AIR)
Centre de Documentation de l'Armenment
26, Boulevard Victor
00460 - Armees
France

Phone: 33 1 4552 45 24
Fax: 33 1 4552 45 74
Email:
Internet:

The French Ministry of Defense issues AIR standards. The prefix AIR in upper case letters appears with these designations.

Example: AIR 9113-A; AIR 9160-C21; AIR 9160-C83

Germany

Deutsches Institut fur Normung e.V. (DIN)
(German Standardization Institute)
Burggrafenstrasse 6-10
D-10787 Berlin
Germany

Phone: 49 30 26 01-0
Fax: 49 30 260 12 31
Email: postmaster@din.de
Internet: www.din.de

DIN standards are developed by a non-profit organization of approximately 88 standards committees with representatives from all technical circles. More than 20,000 standards have been created. Membership is voluntary and open to both German and foreign bodies.

All German standards for steel and cast iron are preceded by the upper case letters DIN and followed by a numerical or alphanumerical code. An upper case letter sometimes precedes this code. German steel designations are reported in one of two methods. One method uses a descriptive code number with chemical symbols and numbers in the designation; the second, known as the Werkstoff number, uses numbers only with a decimal point after the first digit. (The latter method was devised to be more compatible with computerization.)

DIN is a member of the European Committee for Standardization (CEN) and the International Standards Organization (ISO).

Example: DIN 17440 -- X2CrNiMo18-14-3 or DIN EN10088 -- X12CrS13 (standard and name): or EN in the case of European Standards adopted as German Standard; DIN EN 10248 -- 1.0021 (standard and Werkstoff number).

Examples: DIN 17440 -- X5CrNi18-10 or EN10088-2 -- X2CrNiMo17-12-3 (standard and name):or EN in the case of European Standards adopted as German Standard EN 10213-4 -- 1.4309 (standard and Werkstoff number).

Hungary

Magyar Szabvanyugyi Testulet (MSZT)
(Hungarian Standards Institution)
Postafiok 24
1450 Budapest 9
Hungary

Phone: 36 1 218 30 11
Fax: 36 1 218 51 25
Email: h.pongracz@helka.iif.hu
Internet: www.mszt.hu

Hungarian standards are developed by the Hungarian Standards Institution (MSZT) which was founded in 1921. MSZT is a member body of the ISO, IEC, ETSI and affiliate member in CEN and CENELEC. The basic activity of MSZT is the maintenance of the stock of national standards, the development and issue of new standards, their publication and the withdrawal of obsolete ones. A major task of MSZT is the implementation of European standards.

For irons and steels European and international standards have been implemented. Hungarian standards are preceded by the letters MSZ.

Example: MSZ 23; MSZ 2295; MSZ 120/3; MSZ ISO 2892 (ISO standard); MSZ EN (European standard adopted)

India

Bureau of Indian Standards (BIS)
Manak Bhavan
9 Bahadur Shah Zafar Marg
New Delhi 110002
India

Phone: 91 11 3230131
Fax: 91 11 323 4062
Email: bisind@del2.vsnl.net.in
Internet: wwwdel.vsnl.net.in/bis.org

Bureau of Indian Standards (BIS) is a statutory organization established under the Bureau of Indian Standards Act, 1986. The Bureau is a body corporate and comprises members representing industries, consumer organizations, scientific and research institutes, professional bodies, technical institutions, Central Ministries, State Governments and Members of Parliament. The Bureau is responsible for facilitating harmonious development of standards, product certification, quality system certification and Environmental Management Systems Certification. Indian standards begin with the prefix IS and are followed by a numerical code.

BIS is a member of International Standards Organization (ISO) and International Electrotechnical Commission (IEC).

Example: IS:3945; IS:5517

International

Council for Mutual Economic Assistance
Prospekt Kalinina, 56
Moskva SU-121205
Russia

Phone: 2909111
Fax:
Email:
Internet:

Council for Mutual Economic Assistance (COMECON) is the former international governmental organization for the coordination of economic policy among Communist nations. Founded in 1949, it adopted a formal charter in 1959 and became active in the 1960s. COMECON was dissolved in 1991 after the fall of the Communist governments of many of the European members.

Example: COMECON P

International

International Organization for Standardization (ISO)
1 rue de Varembe
Case postale 56
CH-1211 Geneve 20
Switzerland

Phone: 41 22 749 01 11
Fax: 41 22 733 34 30
Email: central@isocs.iso.ch
Internet: www.iso.ch

The International Organization for Standardization (ISO) is a worldwide federation of national standards bodies, at present comprising some 130 members, one in each country. The object of ISO is to promote the development of standardization and related activities in the world with a view to facilitating international exchange of goods and services, and to developing co-operation in the sphere of intellectual, scientific, technological and economic activity. The results of ISO technical work are published as International Standards. Among these International Standards are standards which are developed for metals and their alloys, including their designations.

Standards of particular interest may be:
-- ISO 4948/1(82) on the classification of steels into unalloyed and alloy steels based on chemical composition,
-- ISO 4948/2(82) on the classification of unalloyed and alloyed steels according to main quality classes and main property or application characteristics, and
-- ISO TR 4949 which assigns steel names based on letter symbols and includes three tables: Table 1 for unalloyed and low alloy steels characterized by mininum yield stress, Table 2 for steels characterized by application or by properties other than minimun yield stress, Table 3 for steels characterized by chemical composition.

The letters ISO alone indicate the document is a standard, ISO/R or ISO R indicates a recommendation, and ISO/TR represents a technical report.

Example: ISO 2938; ISO R683-6; ISO 683-18

Italy

Ente Nazionale Italiano di Unificazione (UNI)
(Italian National Standardization Office)
Via Battistotti Sassi 11/b
20133 Milano
Italy

Phone: 39 (02) 70 02 41
Fax: 39 (02) 70 10 61 49
Email: info@uni.unicei.it
Internet: www.unicei.it

UNI, the Italian National Standards Body, is a non-profit organization involved in the area of standardization in all industrial, commercial and tertiary sectors except for electrical and electrotechnical areas which fall within the responsibility of the CEI. It was founded in 1921, and is legally recognized both at the national and European level.

Italian standards are preceded by the upper case letters UNI and followed by an alphanumeric code. When UNI take over an international standard the upper case letter UNI is followed by ISO or EN and by an alphanumeric code.

Example: UNI 7356; UNI 2955/2; UNI ISO 9000; UNI EN 29000

Japan

Japanese Industrial Standards Committee (JISC)

c/o Secretariat: Standards Department,
Ministry of International Trade and Industry

1-3-1, Kasumigaseki, Chiyoda-ku

Tokyo 100-1921

Japan

Phone: 81 3 3501 2096
Fax: 81 3 3580 8637
Email: jisc_iso@jsa.or.jp
Internet: www.jisc.org

The JIS standards are reviewed by the Japanese Industrial Standards Committee (JISC), the Secretariat of the Standards Department which is a part of the Ministry of International Trade and Industry (MITI) and are published by the Japanese Standards Association (JSA). These organizations work closely together to publish Japanese Standards.

JISC issues standards that cover industrial or mineral products with the exception of those regulated by their own special standards organizations. The standards are divided into 17 divisions and are used both by commercial and government organizations involved in design engineering, quality assurance, research and development, construction, testing and maintenance.

JISC standards begin with the upper case letters JIS and are followed by an upper case letter which designates the standard's division. This is then followed by a space and a series of digits.

Example: JIS G 4303; JIS G 3507

Japan

Japanese Standards Association (JSA)

1-24 Akasaka 4

Minato-ku

Tokyo 107-8440

Japan

Phone: 81 3 3583 8000
Fax: 81 3 3586 2014
Email:
Internet: www.jsa.or.jp

The Japanese Standards Association (JSA) works closely with the Japanese Industrial Standards Committee to publish Japanese standards and make them available to the public.

Mexico

Direccion General de Normas (DGN)
(General Directorate of Standards)
Av. Puente de Tecamachalco No 6
Lomas de Tecamachalco, Seccion Fuentes
Naucalpan de Juarez
Mexico

Phone: 52 5 729 9300 exts. 4134 and 4157
Fax: 52 5 729 94 84
Email: cidgn@secofi.gob.mx
Internet: www.secofi.gob.mx/normas

The General Directorate of Standards (DGN) issues national standards for the country. Mexican standards begin with the upper case letters NOM (Normas Oficiales Mexicanas) or NMX (Normas Mexicanas).

A Mexican Offical Norm (NOM) is the mandatory technical regulation that contains measurement terminology, classification, characteristics, qualities, specifications, sampling and methods of test that must fulfill products and services or processes when they can constitute a risk to people or human health, animal or vegetable, general or labor environment, or, cause damages in the preservation of the natural resources. A Mexican Norm (NMX) is a voluntary technical document that is written by a national standard organization which establishes minimum specifications for goods and services. There are 22 committees that write Mexican standards as a part of SECOFI (Secretary of Commerce to Promote Industry)

Example: NOM-060-SCFI-1994; NMX-B-196-1968; NMX-B-171-1991; DGN B-297; DGN B-13

Netherlands

Nederlands Normalisatie-instituut (NNI)
(Dutch Standardization Institute)
Kalfjeslaan 2
P.O. Box 5059
2600 GB Delft
Netherlands

Phone: 31 15 2 69 03 90
Fax: 31 15 2 69 01 90
Email: info@nni.nl
Internet: www.nni.nl

The Netherlands Normalization Institute (NNI) is composed of thousands of individual firms, companies, and organizations. This association helps prepare Dutch standards and cooperates in the development of international standardization. Dutch standards are prefixed by the letters NEN and are followed by a numerical code.

Example: NEN 2733; NEN 6002-C

New Zealand

Standards New Zealand (NZS)
155 The Terrace
Private Bag 2439
Wellington
New Zealand

Phone: 64 (04) 498 5990
Fax: 64 (04) 498 5994
Email: snz@standards.co.nz
Internet: www.standards.co.nz

Standards New Zealand (SNZ) began in 1932 as the New Zealand Standards Institute. Standards New Zealand is the trading arm of the Standards Council. It is a crown entity operating under the Standards Act of 1988. SNZ is entirely self-funded, has a full-time staff of 50 and is supported by over 2000 volunteers who serve on the varied boards and committees. The Standards Council, an appointed body with representatives from all sectors of the community, oversees the development and adoption of standards and standards-related products. Standards New Zealand is involved with the development and application of national, regional and international standards, of which many are developed in partnership with Australia. SNZ is New Zealand's representative to ISO and IEC.

Example: AS/NZS 1594-1997

Norway

Norges Standardiseringsforbund (NSF)
(Norwegian Standards Association)
P.O. Box 7020
Homansbyen
N-0306 Oslo
Norway

Phone: 47 2209 9200
Fax: 47 2204 9211
Email: firmapost@standard.no
Internet: www.standard.no

The Norweigian Standards Association (NSF) is the principal organization for standardization in Norway. The standardizing bodies affiliated with NSF are as follows: The Norwegian Council for Building Standardization (NBR), the Norwegian Technology Standards Institution (NTS), the Norweigian Electrotechnical Committee (NEK), the Norwegian General Standardizing Body (NAS), the Norwegian Post and Telecommunications Authority (PT).

The Norwegian Standards Association (NSF) is the national member of ISO and CEN and the body responsible for the approval and publishing of all Norwegian standards. Norsk Standards are preceded by the upper case letters NS.

Example: NS 11301; NS 722

Pan America

Pan American Standards Commission (COPANT)
Avenida Andres Bello, Torre Fondo Comun
Piso 11
Caracas 1050
Venezuela

Phone: 58 2 5742941
Fax: 58 2 5742941
Email:
Internet:

The Pan American Standards Commission (COPANT) is comprised of national standards bodies of 18 countries (from the United States and many Latin American countries). COPANT is the Regional Standards Organization for America for the iron and steel industry and other areas such as foods and agricultural products, plastics, quality assurance, etc. For its designations the acronym COPANT in upper case letters precedes the numeric code and the year of its adoption.

Example: COPANT 514; COPANT 334; COPANT R193; COPANT 37-II

Poland

Polish Committee for Standardization (PKN)
ul. Elektoralna 2
P.O. Box 411
PL-00-950 Warszawa
Poland

Phone: 48 22 620 54 34
Fax: 48 22 624 71 22
Email: intdoc@pkn.pl
Internet: www.pkn.pl

In 1924 the Polish Committee for Standardization (PKN) was established with the task to standardize industrial products and their delivery requirements. After WWII PKN was reinstated and since that time has been reorganized several times. The most recent reorganization by the Polish Parliament in 1993 included the Standardization Law which set up the new voluntary standardization system. The Polish Committee for Standardization, Measures and Quality Control has been dissolved and divided into three independent bodies: Central Office of Measures, Polish Centre for Testing and Certification and Polish Committee for Standardization. The latter took over all obligations and duties relating to standardization.

The Committee represents Poland in international (ISO and IEC) and regional (CEN) standards organizations, participates in their works and acts abroad in questions connected with standardization. The Committee is responsible for harmonizing Polish standards with the European standards.

The Polish standards are prefixed with the upper case letters PN or PNH. The designations or standards may appear in a number of ways.

Example: PN-84024; PHN 84023/07; PN 83158

Romania

Asociatia de Standardizare Din Romania (ASRO)
Str. Mendeleev nr.21-25
70168 Bucharest 1
Romania

Phone: 40 1 211 32 96
Fax: 40 1 210 08 33
Email: irs@kappa.ro
Internet:

The standards organization in Romania is known as Asociatia de Standardizare din Romania (ASRO). The name and statute of the organization changed starting with October 31, 1998 according to the provision of a Governmental Ordinance. ASRO now enjoys the position of national standardization body in Romania, replacing the former Romanian Standards Institute, which was a governmental bugetary institution. It is a not-for-profit association, a private legal entity of public interest, recognized by the Romanian Government, and now has 200 members.

The Romanian standards are preceded by the capital letters STAS followed by a numerical code which may be followed by the year the standard was adopted.

Example: STAS 791 (92); STAS 11500/2; STAS 9277; STAS 500/2 (88)

Russia

State Committee of the Russian Federation for Standardization and Metrology
Gosstandart of Russia (GOST)
Leninsky Prospekt 9
Moskva 117049
Russian Federation

Phone: 7 095 236 40 44
Fax: 7 095 237 60 32
Email: info@gost.ru
Internet: www.gost.ru

The standardization works in Russia are based on the Law "On Standardization" of the Russian Federation promulgated in 1993 and the set of standards of the State Standardization System. These standards cover most areas of commerce, industry, agriculture and public health. The State standards and All-Russian Classifiers of Technical and Economic Information are adopted by Gosstandart of Russia; for the construction and construction materials manufacturing they are adopted by State Committee for Construction of Russia. Gosstandart of Russia is the national certification body of the Russian Federation. The standards are prefaced with the upper case letters GOST and are followed by a numerical code.

Example: GOST 977; GOST 1050; GOST 24244: GOST 10702; GOST 380

South Africa

South African Bureau of Standards (SABS)
1 Dr Lategan Rd., Groenkloof
Private Bag X191
Pretoria 0001
South Africa

Phone: 27 12 428 6925/6
Fax: 27 12 344 1568
Email: info@sabs.co.za
Internet: www.sabs.co.za

SABS is South Africia's official body for the preparation and publication of standards, and it is a statutory organization, governed by Act 29 of 1993. It renders services to government, trade, industry and the consumer by promoting quality and standardization.

The primary responsibility is development and publication of standards for products and services. They are also responsible for certification, testing, quality, training, and design.

The number of the South African standard is preceded by the letters SABS and followed by the numeric or alphanumeric material type or grade designation.

Example: SABS 719; SABS 1034; SABS 1465-2 Grade W4

Spain

Asociacion Espanola de Normalizacion y Certificacion (AENOR)
(Spanish Association for Standardization and Certification)
C Genova, 6
28004 Madrid
Spain

Phone: 34 91 432 60 00
Fax: 34 91 310 45 96
Email: info@aenor.es
Internet: www.aenor.es

The Spanish Association for Standardization and Certification (AENOR) is an independent organization of a private nature, set up to carry out Standardization and Certification activities, as a tool to improve the quality and competitiveness of products and services. AENOR is designated as a recognized body to develop Standardization and Certification (S+C) activities in Spain.

The designations begin with the letters UNE, representing the Spanish words une norm Espanola.

Example: UNE 36011; UNE 36016/1; UNE 8317; UNE 36529

Sweden

Standardiseringen i Sverige (SIS)
(Swedish Standards Institution)
S:t Eriksgatan 115
Box 6455
SE-113 82 Stockholm
Sweden

Phone: 46 8 610 30 00
Fax: 46 8 30 77 57
Email: info@sis.se
Internet: www.sis.se

SIS was founded in 1922 and is an independent non-profit association. SIS is the Swedish member of the international organizations ISO and CEN. SIS operates commercial enterprises in subsidiaries or associated companies.

SIS is the central body for standardization in Sweden and authorizes standardization bodies that are responsible for standardization within their areas of competence. Within the field of steel and iron industry there is SMS Swedish Materials and Mechanics Standards (Svensk Material & Mekanstandard).

SIS Publishing (SIS Forlag AB) is a company co-owned by SIS and the authorized standards bodies. SIS Publishing publishes, markets and sells Swedish standards.

Standards begin with the prefix SS or, if the standard was written prior to 1978, SIS. More than 95% of all new Swedish standards are based on international (global or European) and the prefix will then be SS-ISO or SS-EN.

Example: SIS 140457; SIS 141311E; SS 14; SS 141311

Switzerland

Schweizerische Normen-Vereinigung (SNV)
Mnhlebachstrasse 54
8008 Zurich
Switzerland

Phone: 41 1 254 54 54
Fax: 41 1 254 54 74
Email: info@snv.ch
Internet: www.snv.ch

The division of Schweizerische Normen-Vereinigung (SNV) that deals with the metals industry is Verein Schweizerischer Maschinen-Industrieller (VSM). Since July 1999 VSM has had a closer working relationship with ASM which is another Swiss association that also represents the Swiss mechanical and electrical engineering industries. These two organizations now share a joint address and web site under the name SWISSMEM. For information on Swiss metals standards you may also contact VSM at Kirchenweg 4, 8032 Zurich, Switzerland, Telephone 41 (01) 384 41 11, Fax 41 (01) 384 42 42, email info@swissmem.ch, and internet www.swissmem.ch.

Example: VSM 10648; VSM DST-4; VSM FH

Turkey

Turk Standardlari Enstitusu (TSE)
(Turkish Standardization Institute)
Necatibey Cad.112
06100 Bakanliklar, Ankara
Turkey

Phone: 90 312 417 83 30
Fax: 90 312 425 43 99
Email: didb@tse.org.tr
Internet: www.tse.org.tr

Founded in 1954, the Turkish Standards Institution (TSE) is a non-government state agency which was established by a private law (Republic of Turkey law 132), dedicated to the preparation and publication of Standards. TSE is also a member of the ISO and affiliate member of CEN. The prefix for Turkish Standards are the letters TS. These are followed by a code number, or, in the case of a designation, an alphanumeric code.

Example: TS 2288(97); TS 302; TS 7176

UK

British Standards Institution (BSI)
British Standards House
389 Chadwick High Road
London W4 4AL
United Kingdom

Phone: 44 (0) 181 996 9000
Fax: 44 (0) 181 996 7400
Email: info@bsi.org.uk
Internet: www.bsi.org.uk

The British Standards Institution (BSI) develops and publishes standards that are used extensively by exporters and importers. They are used both in government and industry by those who are involved in engineering, designing, production, testing and construction.

BSI represents the British industry to the ISO, IEC, CEN and CENELEC. These bodies develop the international and European standards.

The letters BS precede the standard's numerical code and may also include the alloy's designation.

Example: BS 1429(80); BS 1449/1; BS 970/1; BS S.155; BS DTD; BS 3468

USA

Aerospace Materials Specifications (AMS)
Metric Aerospace Materials Specifications (MAM)
SAE International
400 Commonwealth Drive
Warrendale PA 15096-0001
USA

Phone: 724 776 4841
Fax: 724 776 5760
Email:
Internet: www.sae.org

Aerospace Materials Specifications (AMS) or Metric Aerospace Materials Specifications (MAM), are published by SAE, International. AMS and MAN designations pertain to materials intended for aerospace applications. The specifications typically include mechanical property requirements significantly more severe than those for non aerospace applications. Processing requirements are common in AMS steels. These specifications are generally used for procurement purposes.

Example: AMS 5356, AMS 5598B, MAM 5598

USA

American Iron and Steel Institute (AISI)
1101 17th Street NW
Suite 1300
Washington DC 20036-4700
USA

Phone: 202 452 7100
Fax: 202 463 6573
Email: info@steel.org
Internet: www.steel.org

The American Iron and Steel Institute (AISI) is not a material specification writing body, although many times steels are referred to as AISI Standard steels. The steels are actually part of a designation system that refers only to the chemical composition ranges and limits of the different steels. AISI designations are reported in the same manner as the SAE steel designations, except AISI is placed in front of the code.

The most widely used system for designating carbon and alloy steels in the United States is that of the American Iron and Steel Institute (AISI) and the Society of Automotive Engineers (SAE). Although they are two separate systems, they are nearly identical and are carefully coordinated by the two groups. In this joint system, a particular designation implies the same limits and ranges of chemical composition for both an AISI steel and the corresponding SAE steel. The differences in listings occur as a result of differences in determining eligibility for listing. AISI uses production tonnage as the basis for including a steel. SAE includes a steel if it is used in significant quantity by at least two users or if it has unique engineering characteristics. The fact that a particular steel is listed by AISI or SAE implies only that it has been produced in appreciable quantity. It does not imply that other grades are unavailable, nor does it imply that any particular steel producer makes all of the listed grades.

AISI designations and standard practices are not specifications. The SAE designations are published in the annual SAE handbook under various SAE standards. These standards are comprised entirely of listings of SAE designations and the limits and ranges of chemical composition defined by these designations. Either designation contains only a portion of the information necessary to describe properly a steel product for procurement purposes.

The Steel Products Manual, consisting of individual sections covering the major steel mill products, was published for many years by the American Iron and Steel Institute (AISI), providing an authoritative source of reference material for the producers and users of steel products. AISI discontinued publication of the Manual in 1981 and in 1986, and agreement was concluded wherein the Iron & Steel Society would undertake the publication, updating and sale of the Manual. The Iron & Steel Society is well qualified to provide this publication, having served the steel industry for many years through the promotion of knowledge and the publication of literature pertaining to iron and steel, including steel production, manufacturing and research. See listing for Iron & Steel Society (ISS).

Example: Refer to the example under SAE steel designations.

USA

American National Standards Institute (ANSI)
11 West 42nd St.
New York NY 10036
USA

Phone: 212 642 4900
Fax: 212 398 0023
Email: info@ansi.org
Internet: www.ansi.org

The American National Standards Institute (ANSI) standards are used widely throughout industry. They cover a tremendous variety of items, from architectural products to consumer goods to nuclear safety standards. The Institute is the coordinator of the United States voluntary standards system and assists participants in the voluntary system to reach agreement on standards needs and priorities; arranging for competent organizations to undertake standards development work; providing fair and effective procedures for standards development; and resolving conflicts and preventing duplication of efforts.

An ANSI standard begins with the prefix ANSI. This group is followed by an alphanumeric code which begins with an upper case letter that is followed by 1 to 3 digits. These groups are then followed by additional digits that are separated by decimal points. ANSI standards can also have a standards developer's acronym in the title.

ANSI is the sole U.S. representative and dues-paying member of the two major non-treaty international standards organizations, the International Organization for Standardization (ISO), and, via the U.S. National Committee (USNC), the International Electrotechnical Commission (IEC).

USA

American Society for Testing and Materials (ASTM)
100 Barr Harbor Drive
West Conshohocken PA 19428-2959
USA

Phone: 610 832 9500
Fax: 610 832 9555
Email: service@astm.org
Internet: www.astm.org

The American Society for Testing and Materials (ASTM), founded in 1898, is a scientific and technical organization formed for the development of standards on characteristics and performance of materials, products, systems, and services. The organization issues the most widely used -- in the United States -- standard specifications for steel products, many of which are complete and generally adequate for procurement purposes. These frequently apply to specific products, which are generally oriented toward the performance of the fabricated end product.

ASTM is the world's largest source of voluntary "consensus" standards. That is, its documents represent a consensus drawn from producers, specifiers, fabricators, and users of steel mill products. In many cases, the dimensions, tolerances, mimits and restrictions in the ASTM specifications are the same as corresponding items of the standard practices in the AISI Steel products Manuals.

Many of the ASTM specifications have been adopted by the American Society of Mechanical Engineers (ASME) with little or no modification; ASME uses the prefix "S" along with the ASTM designation for these specifications. For example, ASME SA-213 and ASTM A213 are identical.

All ASTM standards begin with the prefix ASTM, followed by the actual standard code number.

Example: ASTM A311/A311M(95); ASTM A48; ASTM A372; ASTM A723 Grade 1 Class 1; ASTM A336 Grade F316

USA

American Welding Society (AWS)
550 N. W. LeJeune Road
P.O. Box 351040
Miami FL 33126
USA

Phone: 305 443 9353
          800 443 9353
Fax: 305 443 7559
Email: info@aws.org
Internet: www.aws.org

The American Welding Society (AWS) standards are used to support welding design, fabrication, testing, quality assurance and other related joining functions found in shipbuilding (design/construction), heavy construction, and a wide variety of other industries. These standards always begin with the upper case letters AWS.

Example: AWS A5.24; AWS C5.7; AWS B4.0

USA

ASME International (ASME)

Three Park Avenue

New York NY 10016-5990

USA

Phone: 800 THE ASME (US/Canada)
               95 800 843 2763 (Mexico)
               973 882 1167 (outside North America)

Fax:

Email: ASME InfoCentral

Internet: www.asme.org

Founded in 1880 as the American Society of Mechnaical Engineers, today ASME International is a nonprofit educational and technical organization serving a worldwide membership. ASME standards are used by personnel in research, testing, and design of power-producing machines such as internal combustion engines, steam and gas turbines, and jet and rocket engines. They are also used for the design and development of power-using machines such as refrigeration and air-conditioning equipment, elevators, machine tools, printing presses and, steel-rolling mills. The diversity of mechanical engineering is seen in the fact that there are 36 Technical Divisions (plus one subdivision) and 3 Institutes.

The upper case letters ASMC appear at the left of the specification followed by an alpha numeric code. When referenced by other ASME standards, standards dated prior to 1989 which carry ANSI in their designation, should be shown as the current ones (i.e. ASME followed by the alphanumeric designation given by ANSI) followed by the title and the ANSI designation in parenthesis.

Example: ASME 182; ASME SA638; ASME SA479

USA

Department of Defense

Defense Automated Printing Service (DODSSP)

700 Robbins Avenue, Bldg. 4/D

Philadelphia PA 19111-5094

USA

Phone: 215 697 2179

Fax: 215 697 1462

Email:

Internet: dodssp.daps.mil

The Department of Defense Single Stock Point (DODSSP) was created to centralize control and distribution, and provide access to extensive technical information within the collection of Military and Federal Specifications and Standards and related documents produced or adopted by the DOD. The DODSSP mission was assumed by the Defense Automated Printing Service (DAPS) Philadelphia Office, in October 1990.

Military specifications (MIL) are issued by the United States Department of Defense (DOD) to define materials, products, or services used only or predominately by military entities. Military standards provide procedures for design, manufacturing, and testing, rather than giving a particular material description.

All military specifications begin with the upper case letters MIL. The actual specification that follows begins with an upper case code letter that represents the first letter of the title for the item, followed immediately by a hyphen and then the serial number or digits.

Federal (QQ) specifications and standards are similar to military, except they are issued by the General Services Administration (GSA) and are primarily for use by federal agencies. Their use, however, is not acceptable to the United States military establishment when there are no separate MIL specifications available.

Federal specifications begin with the upper case letters QQ followed by the code numbers and letters.

Example: DoD-F-24669/1(86); MIL-B-11595E (military); FED QQ-S-698 (federal)

USA

Iron & Steel Society (ISS)
186 Thorn Hill Road
Warrendale PA 15086-7528
USA

Phone: 724 776 1535
Fax: 724 776 0430
Email: webmaster@issource.org
Internet: www.issource.org

The Iron & Steel Society was officially born on December 1, 1974, but its origin can be traced back to 1928 when the Iron and Steel Division was authorized as a part of what is now the American Institute of Mining, Metallurgical, and Petroleum Engineers, Inc (AIME). A nonprofit organization, ISS was formed with the objective to "promote the advancement of knowledge in the technical operations and processes in the iron and steel industry." To better serve the iron and steel industry ISS signed an agreement with AISI in 1986 to publish, update, and sell the Steel Products Manual. ISS is the appropriate contact organization for information on this Manual.

USA

SAE, International (SAE)
SAE World Headquarters
400 Commonwealth Dr.
Warrendale PA 15096-0001
USA

Phone: 412 776 4841
Fax: 412 776 5760
Email: info@sae.org
Internet: www.sae.org

The Society of Automotive Engineers (SAE) standards are used primarily by designers, manufacturers, and maintenance personnel in the automotive and aerospace industries. These standards are also a useful and effective series for the metals, plastics, rubber, chemical and fastener industries in their standardization efforts. Automotive SAE standards begin with the upper case letters SAE. Immediately after this prefix the letter J appears, and it is followed by a numerical code. Prefixes for other standards vary, such as AS, AIR, and ARP with a numerical code.

Example: SAE J441; SAE J775

USA

Steel Founders' Society of America (SFSA)
205 Park Avenue
Barrington IL 60010-4332
USA

Phone: 847 382 8240
Fax: 847 382 8287
Email: info@scra.org
Internet: www.sfsa.org

Founded in 1902 the Steel Founders' Society of America is a trade association representing steel foundries. SFSA members are all steel casting producers from the U.S., Canada, and Mexico. International companies may be associate members. There are four ongoing committees in the areas of Technical & Operating, Research, Marketing, and Specifications.

The Alloy Casting Institute (ACI) established a standards-designating system for cast high-alloy stainless steels. Originally a separate organization, it was absorbed by the Steel Founders' Society of America in 1970. Designations consist of an alphanumeric code.

Example: CF-8M; CF-3MA; CB-7Cu-1; CA-6NM-B

USA

Unified Numbering System (joint publication of ASTM and SAE)
SAE International
400 Commonwealth Dr.
Warrendale PA 15096-0001
USA

Phone: 724 776 4841
Fax: 724 776 5760
Email: info@sae.org
Internet: www.@sae.org

A UNS designation is not a specification because it establishes no requirements for delivery conditions such as mechanical property requirements, heat analysis tolerances, heat treatment, packaging, marking, inspection, etc. UNS designations are identifying numbers for metal and alloys for which controlling requirements are established and published by technical societies and specifying organizations such as ASTM, ASME, and AWS. There are almost 5,000 UNS designations for metals and alloys.

The UNS listing includes cross-reference specifications for many technical societies and specifying organizations, but it was not meant to be an exhaustive list. The list includes representative specifications from American Society of Mechanical Engineers (ASME), American Society of Testing and Materials (ASTM), SAE, NACE, Federal, Military, American Iron and Steel Institute (AISI), and American Welding Society (AWS).

Metals & Alloys in the Unified Numbering System is a joint publication of SAE International and ASTM.

Yugoslavia

Savezni zavod za standardizaciju (SZS)
(Yugoslavian Standardization Institute)
Kneza Milosa 20
Post Pregr. 933
11000 Beograd
Yugoslavia

Phone: 381 11 361 31 50
Fax: 381 11 361 73 41
Email: jus@szs.sv.gov.yu
Internet:

The Yugoslavian Standardization Institute (SZS) was founded in 1946 and is concerned with the adoption and application of standards, technical norms for product quality and services, and regulations covered by legislation. Yugoslavian standards begin with the prefix JUS which is followed by an alphanumeric code. The first letter of the code denotes the section under which the standard is classified. Most standards relating to metallurgy are in section C.

Example: JUSC 0.501; JUSC 3.052; JUSC.B0.500; JUSC.J3.011

Abbreviations

Abbreviation	Word or Phrase	Abbreviation	Word or Phrase	Abbreviation	Word or Phrase
*	obsolete	FN.	rimmed steel not permitted	QS.	quality steel
abr.	abrasion or abrasive	Frg.	forging(s)	qual.	quality
Abr res.	abrasion or abrasive resistant	FU.	rimmed steel	Quen.	quenched
acpt.	acceptable	gen.	general	refrg.	refrigerated or refrigeration
AH.	age hardened	Hard.	hardenable or hardened or hardness	reinf.	reinforcing
Ann.	annealed	HB.	Brinell hardness	REM.	rare earth metals
App.	applicable	HD.	hard drawn	req.	requirements
Apps.	applications	HE.	hot extruded	res.	resistant or resisting
Aust.	austenitic	Heat res.	heat resisting	Rnd.	round(s)
auto.	automotive	Hex.	hexagon(s)	SA.	solution annealed
avg.	average	HF.	hot finished or formed	sand.	sand cast
Bil.	billet	HIP.	hot isostatically pressed	SCS.	separately cast samples or specimens
Blm.	bloom	HR.	hot rolled	sect.	section
BS.	base steel	HRx.	Rockwell hardness	Sh.	sheet
CD.	cold drawn	HS.	high-strength	Shp.	shape(s)
cent.	centrifical	HSLA.	high-strength low-alloy	SHT.	solution heat treated
CEV.	carbon equivalent value may be required	HSp.	high-speed	smls.	seamless
CF.	cold finished or formed	HT.	heat treated	spec'd.	specified
CH.	case hardening or hardened	HV.	Vickers hardness	spec(s).	specification(s)
CHd.	cold heading or upsetting	HW.	hot worked or wrought	Sq.	square(s)
comm.	commercial	IH.	induction hardening	ST.	solution treated
conc.	concrete	in.	inch	ST/A.	solution treated and aged
cond.	conditioned	inv.	investment	std.	standard
const.	construction	IW.	Induction Weld	strp.	strip
corr.	corrosion	LD.	light drawn	struct.	structural
Corr res.	corrosion resistant	mach.	machine or machined	surf.	surface
CR.	cold rolled	Mart.	martensitic	surg.	surgical
CRR.	cold rerolled	matl.	material(s)	t.	thickness
CS.	commercial steel	max.	maximum	temp.	temperature
CW.	cold worked	mech.	mechanical	thk.	thickness
DDS.	deep drawing steel	med.	medium	TLC.	lightly cold work/skin pass
diam.	diameter	min.	minimum	TMCP.	thermo-mechanical control process
DS.	drawing steel	mm.	millimeter	Tmp.	tempered
El.	elongation	N/T.	normalized and tempered	Tub.	tube
elect.	electric	na.	not applicable	Ult.	ultimate tensile strength
elev.	elevated	no.	number	UTS.	tensile strength
ERW.	electric resistance weld	nom.	nominal	VCD.	vacuum carbon deoxidation
exp.	expansion	norm.	normalized	ves.	vessel
ext.	extrusion	obs.	obsolete	weld.	welded or welding
F/IH.	flame and induction hardening	Plt.	plate(s)	WH.	work hardening
Ferr.	ferritic	press.	pressure	wir.	wire
FF.	fully killed	Press ves.	pressure vessel(s)	YS.	yield strength
FH.	flame hardening	prl.	pearlitic		
flng.	flange	Q/A.	quenched and aged		
		Q/T.	quenched and tempered		

Chemical Elements

Symbol	Name	Symbol	Name	Symbol	Name
Ac.	Actinium	Ge.	Germanium	Po.	Polonium
Ag.	Silver	H.	Hydrogen	Pr.	Praseodymium
Al.	Aluminum	Ha.	Hahnium	Pt.	Platinum
Am.	Americium	He.	Helium	Pu.	Plutonium
Ar.	Argon	Hf.	Hafnium	Ra.	Radium
As.	Arsenic	Hg.	Mercury	Rb.	Rubidium
At.	Astatine	Ho.	Holmium	Re.	Rhenium
Au.	Gold	I.	Iodine	Rf.	Rutherfordium
B.	Boron	In.	Indium	Rh.	Rhodium
Ba.	Barium	Ir.	Iridium	Rn.	Radon
Be.	Beryllium	K.	Potassium	Ru.	Ruthenium
Bi.	bismuth	Kr.	Krypton	S.	Sulfur
Bk.	Berkelium	La.	Lanthanum	Sb.	Antimony
Br.	Bromine	Li.	Lithium	Sc.	Scandium
C.	Carbon	Lr (Lw).	Lawrencium	Se.	Selenium
Ca.	Calcium	Lu.	Lutetium	Si.	Silicon
Cd.	Cadmium	Md.	Mendelevium	Sm.	Samarium
Ce.	Cerium	Mg.	Magnesium	Sn.	Tin
Cf.	Californium	Mn.	Manganese	Sr.	Strontium
Cl.	Chlorine	Mo.	Molybdenum	Ta.	Tantalum
Cm.	Curium	N.	Nitrogen	Tb.	Terbium
Co.	Cobalt	Na.	Sodium	Tc.	Technetium
Cr.	Chromium	Nb (Cb).	Niobium	Te.	Tellurium
Cs.	Cesium	Nd.	Neodymium	Th.	Thorium
Cu.	Copper	Ne.	Neon	Ti.	Titanium
Dy.	Dysprosium	Ni.	Nickel	Tl.	Thallium
Er.	Erbium	No.	Nobelium	Tm.	Thulium
Es.	Einsteinium	Np.	Neptunium	U.	Uranium
Eu.	Europium	O.	Oxygen	V.	Vanadium
F.	Fluorine	Os.	Osmium	W.	Tungsten
Fe.	Iron	P.	Phosphorus	Xe.	Xenon
Fm.	Fermium	Pa.	Protactinium	Y.	Yttrium
Fr.	Francium	Pb.	Lead	Yb.	Ytterbium
Ga.	Gallium	Pd.	Palladium	Zn.	Zinc
Gd.	Gadolinium	Pm.	Promethium	Zr.	Zirconium

D-4/Designation Index

D-30/Designation Index

D-40/Designation Index

D-50/Designation Index

S-10/Specification Index

S-14/Specification Index